CLIMATE CHANGE 2001:
IMPACTS, ADAPTATION, AND VULNERABILIT

Climate Change 2001: Impacts, Adaptation, and Vulnerability is the most comprehensive and up-to-date scientific assessment of the consequences of, and adaptation responses to, climate change. The report:

- Evaluates evidence that recent observed changes in climate have already affected a variety of physical and biological systems.
- Makes a detailed study of the vulnerabilities of human populations to future climate change, including associated sea-level rise and changes in the frequency and intensity of climate extremes such as floods, droughts, heat waves and windstorms, and taking into account potential impacts on water resources, agriculture and food security, human health, coastal and other types of settlements, and economic activities.
- Assesses the potential responses of natural environments and the wildlife that inhabit them to future climate change and identifies environments at particular risk.
- Considers how adaptation to climate change might lessen adverse impacts or enhance beneficial impacts.
- Provides an overview of the vulnerabilities and adaptation possibilities by major region of the world (Africa, Asia, Australia/New Zealand, Europe, Latin America, North America, Polar Regions, and Small Island States).
- Contrasts the different vulnerabilities of the developed and developing parts of the world and explores the implications for sustainable development and equity concerns.

This latest assessment of the IPCC will again form the standard scientific reference for all those concerned with the environmental and social consequences of climate change, including students and researchers in ecology, biology, hydrology, environmental science, economics, social science, natural resource management, public health, food security, and natural hazards, and policymakers and managers in governmental, industry, and other organizations responsible for resources likely to be affected by climate change.

James J. McCarthy is Co-Chair of IPCC Working Group II, and the Alexander Agassiz Professor of Biological Oceanography and Director of the Museum of Comparative Zoology at Harvard University.

Osvaldo F. Canziani is Co-Chair of IPCC Working Group II, and Professor of Applied Meteorology of the National Universities of La Plata and Comahue, Argentina.

Neil A. Leary is the Head of the IPCC Working Group II Technical Support Unit, Washington DC, USA.

David J. Dokken is a Senior Technical Editor for the IPCC Working Group II Technical Support Unit, Washington DC, USA.

Kasey S. White is a Program Specialist for the IPCC Working Group II Technical Support Unit, Washington DC, USA.

Climate Change 2001:
Impacts, Adaptation, and Vulnerability

Edited by

James J. McCarthy
Museum of Comparative Zoology
Harvard University

Osvaldo F. Canziani
National Universities
of La Plata and Comahue

Neil A. Leary
IPCC Working Group II
Technical Support Unit

David J. Dokken
University Corporation
for Atmospheric Research

Kasey S. White
IPCC Working Group II
Technical Support Unit

Contribution of Working Group II to the Third Assessment Report
of the Intergovernmental Panel on Climate Change

Published for the Intergovernmental Panel on Climate Change

PUBLISHED BY THE PRESS SYNDICATE OF THE UNIVERSITY OF CAMBRIDGE
The Pitt Building, Trumpington Street, Cambridge, United Kingdom

CAMBRIDGE UNIVERSITY PRESS
The Edinburgh Building, Cambridge CB2 2RU, UK
40 West 20th Street, New York, NY 10011-4211, USA
10 Stamford Road, Oakleigh, VIC 3166, Australia
Ruiz de Alarcón 13, 28014 Madrid, Spain
Dock House, The Waterfront, Cape Town 8001, South Africa

http://www.cambridge.org

First published 2001

Printed in the United States of America

Typeface Times 10/12 pt. *System* QuarkXPress 4.1 [AU]

A catalog record for this book is available from the British Library

Library of Congress Cataloging-in-Publication Data available

ISBN 0 521 80768 9 hardback
ISBN 0 521 01500 6 paperback

Cover Image Credit
Earth's city lights—shown for a projection spanning Europe, Asia, and Africa—is one indicator of the scale of human influence on the world. The brightest areas of the Earth are the most urbanized, but not necessarily the most populated. Image by Robert Simmon and Craig Mayhew, Science Systems and Applications, Inc.; data courtesy of the United States Air Force Defense Meteorological Satellite Program (DMSP) and the National Oceanic and Atmospheric Administration's National Geophysical Data Center, based on research performed by Dr. Marc Imhoff of the National Aeronautics and Space Administration's Goddard Space Flight Center.

Contents

Foreword

The Intergovernmental Panel on Climate Change (IPCC) was jointly established by the World Meteorological Organization (WMO) and the United Nations Environment Programme (UNEP) in 1988. Its terms of reference includes: (i) to assess available scientific and socioeconomic information on climate change and its impacts and on the options for mitigating climate change and adapting to it, and (ii) to provide, on request, scientific/technical/socioeconomic advice to the Conference of the Parties (COP) to the United Nations Framework Convention on Climate Change (UNFCCC). From 1990, the IPCC has produced a series of Assessment Reports, Special Reports, Technical Papers, methodologies, and other products that have become standard works of reference, widely used by policymakers, scientists, and other experts.

This volume, which forms part of the Third Assessment Report (TAR), has been produced by Working Group II (WGII) of the IPCC and focuses on the environmental, social, and economic consequences of climate change and potential adaptation responses. It consists of 19 chapters covering the sensitivity, adaptive capacity, and vulnerability of natural and human systems to climate change, and the potential impacts and adaptation options at regional and global scales.

As is usual in the IPCC, success in producing this report has depended first and foremost on the knowledge, enthusiasm, and cooperation of many hundreds of experts worldwide, in many related but different disciplines. We would like to express our gratitude to all the Coordinating Lead Authors, Lead Authors, Contributing Authors, Review Editors, and Expert Reviewers. These individuals have devoted enormous time and effort to produce this report and we are extremely grateful for their commitment to the IPCC process. We would like to thank the staff of the WGII Technical Support Unit and the IPCC Secretariat for their dedication in coordinating the production of another successful IPCC report.

We are also grateful to the governments that have supported their scientists' participation in the IPCC process and that have contributed to the IPCC Trust Fund to provide for the essential participation of experts from developing countries and countries with economies in transition. We would like to express our appreciation to the governments of Australia, Japan, Malta, Morocco, Peru, Portugal, South Africa, Switzerland, and the United States that hosted drafting sessions in their countries, to the government of Switzerland that hosted the Sixth Session of Working Group II in Geneva, and to the government of the United States that funded the WGII Technical Support Unit.

We would particularly like to thank Dr. Robert Watson, Chairman of the IPCC, for his sound direction and tireless and able guidance of the IPCC, and Prof. James McCarthy and Dr. Osvaldo Canziani, the Co-Chairmen of Working Group II, for their skillful leadership of Working Group II through the production of this report.

G.O.P. Obasi

Secretary-General
World Meteorological Organization

K. Töpfer

Executive Director
United Nations Environment Programme
and
Director-General
United Nations Office in Nairobi

Preface

This volume, *Climate Change 2001: Impacts, Adaptation, and Vulnerability*, is the Working Group II (WGII) contribution to the Third Assessment Report (TAR) of the Intergovernmental Panel on Climate Change (IPCC). The companion volumes of the TAR are *Climate Change 2001: the Scientific Basis* (WGI) and *Climate Change 2001: Mitigation* (WGIII). A fourth volume of the TAR is being prepared to provide a synthesis of the findings of the three Working Groups and will focus on questions addressing particular policy issues raised in the context of the Framework Convention on Climate Change.

Since the inception of the IPCC, its Working Group II has focused on the impacts of projected climate change. The current WGII report differs somewhat in scope from earlier WGII assessments. This report examines climate change impacts, adaptation, and vulnerability across a range of systems and sectors, as was done in the Second Assessment Report (SAR, published in 1996), and includes a regional assessment, updated from the *Special Report on Regional Impacts of Climate Change* (1998). Environmental, social, and economic dimensions of these issues are assessed in this report, whereas the previous WGII report focused primarily on environmental dimensions. Efforts are made in the new assessment to address a number of issues that cut across the various sectors and systems covered by the WGII report, as well as across the three Working Groups of IPCC, such as sustainable development, equity, scientific uncertainties, costing methodologies, and decisionmaking frameworks. Mitigation of climate change, treated in previous WGII reports, is now the subject of WGIII's contribution to the TAR.

Research on climate impacts has grown considerably since the SAR, and much has been learned in the past 5 years regarding the potential risk of damage associated with projected climate change. The research has added to what we know about the vulnerabilities to climate change of a wide range of ecological systems (forests, grasslands, wetlands, rivers, lakes, and marine environments) and human systems (agriculture, water resources, coastal resources, human health, financial institutions, and human settlements).

Observational evidence of changes has accumulated in many physical and biological systems (e.g., glacial melting, shifts in geographic ranges of plant and animal species, and changes in plant and animal biology) that are highly consistent with warming observed in recent decades. These observations are adding to our knowledge of the sensitivity of affected systems to changes in climate and can help us to understand the vulnerability of systems to the greater and more rapid climate changes projected for the 21st century. A number of unique systems are increasingly recognized as especially vulnerable to climate change (e.g., glaciers, coral reefs and atolls, mangroves, boreal and tropical forests, polar and alpine ecosystems, prairie wetlands, and remnant native grasslands). In addition, climate change is expected to threaten some species with greater probability of extinction. Potential changes in the frequency, intensity, and persistence of climate extremes (e.g. heat waves, heavy precipitation, and drought) and in climate variability [e.g., El Niño Southern Oscillation (ENSO)] are emerging as key determinants of future impacts and vulnerability. The many interactions of climate change with other stresses on the environment and human populations, as well as linkages between climate change and sustainable development, are increasingly emphasized in recent research and preliminary insights from these important efforts are reflected in the report.

The value of adaptation measures to diminish the risk of damage from future climate change, and from present climate variability, was recognized in previous assessments and is confirmed and expanded upon in the new assessment. Understanding of the determinants of adaptive capacity has advanced and confirms the conclusion that developing countries, particularly the least developed countries, have lesser capacity to adapt than do developed countries. This condition contributes to relatively high vulnerability to damaging effects of climate change in these countries.

The WGII report was compiled by 183 Lead Authors between July 1998 and February 2001. In addition, 243 Contributing Authors submitted draft text and information to the Lead Author teams. Drafts of the report were circulated twice for review, first to experts and a second time to both experts and governments. Comments received from 440 reviewers were carefully analyzed and assimilated to revise the document with guidance provided by 33 Review Editors. The revised report was presented for consideration at a session of the Working Group II panel held in Geneva from 13 to 16 February 2001, in which delegates from 100 countries participated. There, the Summary for Policymakers was approved in detail and the full report accepted.

This report contains a Summary for Policymakers (SPM) and a Technical Summary (TS) in addition to the 19 chapters comprising the full report. Each paragraph of the SPM has been referenced to the supporting sections of the TS. In turn, each paragraph of the TS has been referenced to the appropriate section of the relevant chapter. The first three chapters set the stage for the report by discussing the context of climate change, methods for research and assessment, and development of scenarios. Chapters 4 through 9 assess the state of knowledge regarding climate change impacts, adaptation, and vulnerability for different natural and human systems or sectors. Chapters 10 through 17 assess vulnerabilities and key concerns of eight regions of the world: Africa, Asia, Australia/New Zealand, Europe, Latin America, North America, polar regions, and small island states. Chapter 18 presents a synthesis of adaptation challenges, options, and capacity. Chapter 19 concludes the report with a synthesis of climate change risks for unique and threatened systems, extreme climate events, uneven distribution of impacts, global aggregate impacts, and large-scale high-impact events. An electronic version of the report that can be searched for key words will be available on the web (http://www.ipcc.ch) and CD-ROM.

We wish to express our sincere appreciation to all the Coordinating Lead Authors, Lead Authors, Contributing Authors, Review Editors, and expert and government reviewers, without whose expertise, diligence, and patience and considerable investments of uncompensated time a report of this quality could never have been completed. We would also like to thank members of the Working Group II Bureau for their assistance throughout the preparation of the report.

We would particularly like to thank Neil Leary, who headed the WGII Technical Support Unit, and his staff, Dave Dokken, Kasey Shewey White, Sandy MacCracken, and Florence Ormond. Their tireless and very capable efforts to coordinate the WGII assessment ensured a final product of high scientific quality. In addition we thank Richard Moss for his invaluable contributions to the early planning phase of this work.

We would also like to thank Narasimhan Sundararaman, the Secretary of IPCC; Renate Christ, Deputy Secretary; and the staff of the IPCC Secretariat, Rudie Bourgeois, Chantal Ettori and Annie Courtin who provided logistical support for government liaison and travel of experts from the developing and transitional economy countries.

Robert T. Watson
IPCC Chairman

James J. McCarthy and Osvaldo F. Canziani
IPCC WGII Co-Chairs

SUMMARY FOR POLICYMAKERS

CLIMATE CHANGE 2001: IMPACTS, ADAPTATION, AND VULNERABILITY

A Report of Working Group II of the Intergovernmental Panel on Climate Change

This summary, approved in detail at the Sixth Session of IPCC Working Group II (Geneva, Switzerland • 13-16 February 2001), represents the formally agreed statement of the IPCC concerning the sensitivity, adaptive capacity, and vulnerability of natural and human systems to climate change, and the potential consequences of climate change.

Based on a draft prepared by:

Q.K. Ahmad, Oleg Anisimov, Nigel Arnell, Sandra Brown, Ian Burton, Max Campos, Osvaldo Canziani, Timothy Carter, Stewart J. Cohen, Paul Desanker, William Easterling, B. Blair Fitzharris, Donald Forbes, Habiba Gitay, Andrew Githeko, Patrick Gonzalez, Duane Gubler, Sujata Gupta, Andrew Haines, Hideo Harasawa, Jarle Inge Holten, Bubu Pateh Jallow, Roger Jones, Zbigniew Kundzewicz, Murari Lal, Emilio Lebre La Rovere, Neil Leary, Rik Leemans, Chunzhen Liu, Chris Magadza, Martin Manning, Luis Jose Mata, James McCarthy, Roger McLean, Anthony McMichael, Kathleen Miller, Evan Mills, M. Monirul Qader Mirza, Daniel Murdiyarso, Leonard Nurse, Camille Parmesan, Martin Parry, Jonathan Patz, Michel Petit, Olga Pilifosova, Barrie Pittock, Jeff Price, Terry Root, Cynthia Rosenzweig, Jose Sarukhan, John Schellnhuber, Stephen Schneider, Robert Scholes, Michael Scott, Graham Sem, Barry Smit, Joel Smith, Brent Sohngen, Alla Tsyban, Jean-Pascal van Ypersele, Pier Vellinga, Richard Warrick, Tom Wilbanks, Alistair Woodward, David Wratt, and many reviewers.

CONTENTS

1. Introduction

The sensitivity, adaptive capacity, and vulnerability of natural and human systems to climate change, and the potential consequences of climate change, are assessed in the report of Working Group II of the Intergovernmental Panel on Climate Change (IPCC), *Climate Change 2001: Impacts, Adaptation, and Vulnerability*.[1] This report builds upon the past assessment reports of the IPCC, reexamining key conclusions of the earlier assessments and incorporating results from more recent research.[2,3]

Observed changes in climate, their causes, and potential future changes are assessed in the report of Working Group I of the IPCC, *Climate Change 2001: The Scientific Basis*. The Working Group I report concludes, *inter alia*, that the globally averaged surface temperatures have increased by 0.6 ± 0.2°C over the 20th century; and that, for the range of scenarios developed in the IPCC *Special Report on Emission Scenarios* (SRES), the globally averaged surface air temperature is projected by models to warm 1.4 to 5.8°C by 2100 relative to 1990, and globally averaged sea level is projected by models to rise 0.09 to 0.88 m by 2100. These projections indicate that the warming would vary by region, and be accompanied by increases and decreases in precipitation. In addition, there would be changes in the variability of climate, and changes in the frequency and intensity of some extreme climate phenomena. These general features of climate change act on natural and human systems and they set the context for the Working Group II assessment. The available literature has not yet investigated climate change impacts, adaptation, and vulnerability associated with the upper end of the projected range of warming.

This Summary for Policymakers, which was approved by IPCC member governments in Geneva in February 2001, describes the current state of understanding of the impacts, adaptation, and vulnerability to climate change and their uncertainties. Further details can be found in the underlying report.[4] Section 2 of the Summary presents a number of general findings that emerge from integration of information across the full report. Each of these findings addresses a different dimension of climate change impacts, adaptation, and vulnerability, and no one dimension is paramount. Section 3 presents findings regarding individual natural and human systems, and Section 4 highlights some of the issues of concern for different regions of the world. Section 5 identifies priority research areas to further advance understanding of the potential consequences of and adaptation to climate change.

2. Emergent Findings

2.1. Recent Regional Climate Changes, particularly Temperature Increases, have Already Affected Many Physical and Biological Systems

Available observational evidence indicates that regional changes in climate, particularly increases in temperature, have already affected a diverse set of physical and biological systems in many parts of the world. Examples of observed changes include shrinkage of glaciers, thawing of permafrost, later freezing and earlier break-up of ice on rivers and lakes, lengthening of mid- to high-latitude growing seasons, poleward and altitudinal shifts of plant and animal ranges, declines of some plant and animal populations, and earlier flowering of trees, emergence of insects, and egg-laying in birds (see Figure SPM-1). Associations between changes in regional temperatures and observed changes in physical and biological systems have been documented in many aquatic, terrestrial, and marine environments. [2.1, 4.3, 4.4, 5.7, and 7.1]

The studies mentioned above and illustrated in Figure SPM-1 were drawn from a literature survey, which identified long-term studies, typically 20 years or more, of changes in biological and physical systems that could be correlated with regional changes in temperature.[5] In most cases where changes in biological and physical systems were detected, the direction of change was that expected on the basis of known mechanisms. The probability that the observed changes in the expected direction (with no reference to magnitude) could occur by chance alone is negligible. In many parts of the world, precipitation-related impacts may be important. At present, there is a lack of systematic concurrent climatic and biophysical data of sufficient length (2 or more decades) that are considered necessary for assessment of precipitation impacts.

Factors such as land-use change and pollution also act on these physical and biological systems, making it difficult to attribute changes to particular causes in some specific cases. However, taken together, the observed changes in these systems are consistent in direction and coherent across diverse localities

---

[1] *Climate change* in IPCC usage refers to any change in climate over time, whether due to natural variability or as a result of human activity. This usage differs from that in the Framework Convention on Climate Change, where *climate change* refers to a change of climate that is attributed directly or indirectly to human activity that alters the composition of the global atmosphere and that is in addition to natural climate variability observed over comparable time periods. Attribution of climate change to natural forcing and human activities has been addressed by Working Group I.

[2] The report has been written by 183 Coordinating Lead Authors and Lead Authors, and 243 Contributing Authors. It was reviewed by 440 government and expert reviewers, and 33 Review Editors oversaw the review process.

[3] Delegations from 100 IPCC member countries participated in the Sixth Session of Working Group II in Geneva on 13-16 February 2001.

[4] A more comprehensive summary of the report is provided in the Technical Summary, and relevant sections of that volume are referenced in brackets at the end of paragraphs of the Summary for Policymakers for readers who need more information.

[5] There are 44 regional studies of over 400 plants and animals, which varied in length from about 20 to 50 years, mainly from North America, Europe, and the southern polar region. There are 16 regional studies covering about 100 physical processes over most regions of the world, which varied in length from about 20 to 150 years. See Section 7.1 of the Technical Summary for more detail.

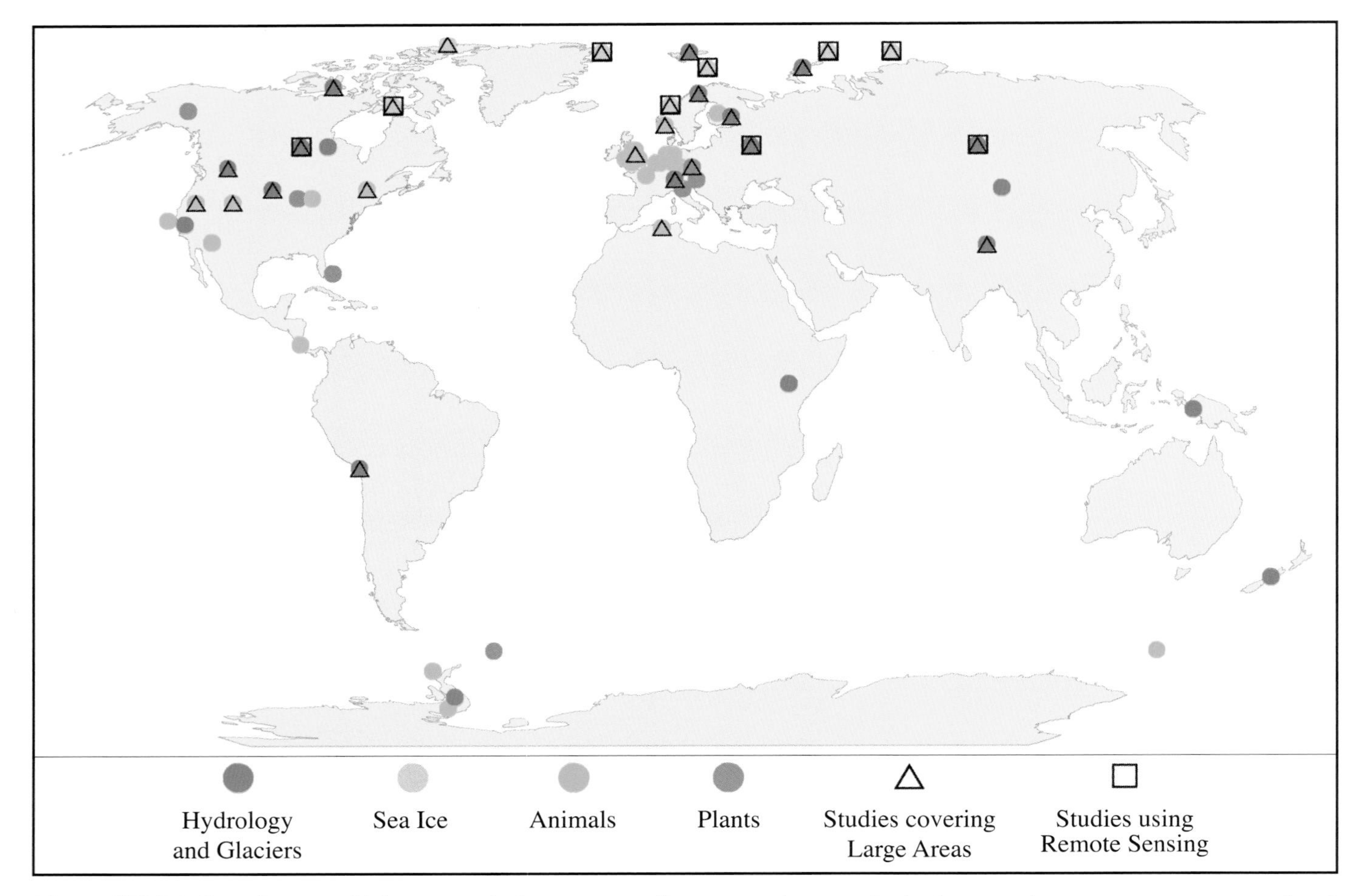

Figure SPM-1: Locations at which systematic long-term studies meet stringent criteria documenting recent temperature-related regional climate change impacts on physical and biological systems. Hydrology, glacial retreat, and sea-ice data represent decadal to century trends. Terrestrial and marine ecosystem data represent trends of at least 2 decades. Remote-sensing studies cover large areas. Data are for single or multiple impacts that are consistent with known mechanisms of physical/biological system responses to observed regional temperature-related changes. For reported impacts spanning large areas, a representative location on the map was selected.

and/or regions (see Figure SPM-1) with the expected effects of regional changes in temperature. Thus, from the collective evidence, there is *high confidence*[6] that recent regional changes in temperature have had discernible impacts on many physical and biological systems.

[6]In this Summary for Policymakers, the following words have been used where appropriate to indicate judgmental estimates of confidence (based upon the collective judgment of the authors using the observational evidence, modeling results, and theory that they have examined): *very high* (95% or greater), *high* (67-95%), *medium* (33-67%), *low* (5-33%), and *very low* (5% or less). In other instances, a qualitative scale to gauge the level of scientific understanding is used: *well established, established-but-incomplete, competing explanations*, and *speculative*. The approaches used to assess confidence levels and the level of scientific understanding, and the definitions of these terms, are presented in Section 1.4 of the Technical Summary. Each time these terms are used in the Summary for Policymakers, they are footnoted and in *italics*.

2.2. *There are Preliminary Indications that Some Human Systems have been Affected by Recent Increases in Floods and Droughts*

There is emerging evidence that some social and economic systems have been affected by the recent increasing frequency of floods and droughts in some areas. However, such systems are also affected by changes in socioeconomic factors such as demographic shifts and land-use changes. The relative impact of climatic and socioeconomic factors are generally difficult to quantify. [4.6 and 7.1]

2.3. *Natural Systems are Vulnerable to Climate Change, and Some will be Irreversibly Damaged*

Natural systems can be especially vulnerable to climate change because of limited adaptive capacity (see Box SPM-1), and some of these systems may undergo significant and irreversible damage. Natural systems at risk include glaciers, coral reefs and atolls, mangroves, boreal and tropical forests, polar and alpine ecosystems, prairie wetlands, and remnant native grasslands. While some species may increase in abundance or range, climate change will increase existing risks of extinction of some more vulnerable species and loss of biodiversity. It is *well-established*[6] that the geographical extent of the damage or loss, and the number of systems affected, will increase with the magnitude and rate of climate change (see Figure SPM-2). [4.3 and 7.2.1]

2.4. *Many Human Systems are Sensitive to Climate Change, and Some are Vulnerable*

Human systems that are sensitive to climate change include mainly water resources; agriculture (especially food security) and forestry; coastal zones and marine systems (fisheries); human settlements, energy, and industry; insurance and other financial services; and human health. The vulnerability of these systems varies with geographic location, time, and social, economic, and environmental conditions. [4.1, 4.2, 4.3, 4.4, 4.5, 4.6, and 4.7]

Projected adverse impacts based on models and other studies include:

- A general reduction in potential crop yields in most tropical and sub-tropical regions for most projected increases in temperature [4.2]
- A general reduction, with some variation, in potential crop yields in most regions in mid-latitudes for increases in annual-average temperature of more than a few °C [4.2]
- Decreased water availability for populations in many water-scarce regions, particularly in the sub-tropics [4.1]
- An increase in the number of people exposed to vector-borne (e.g., malaria) and water-borne diseases (e.g., cholera), and an increase in heat stress mortality [4.7]
- A widespread increase in the risk of flooding for many human settlements (tens of millions of inhabitants in settlements studied) from both increased heavy precipitation events and sea-level rise [4.5]
- Increased energy demand for space cooling due to higher summer temperatures. [4.5]

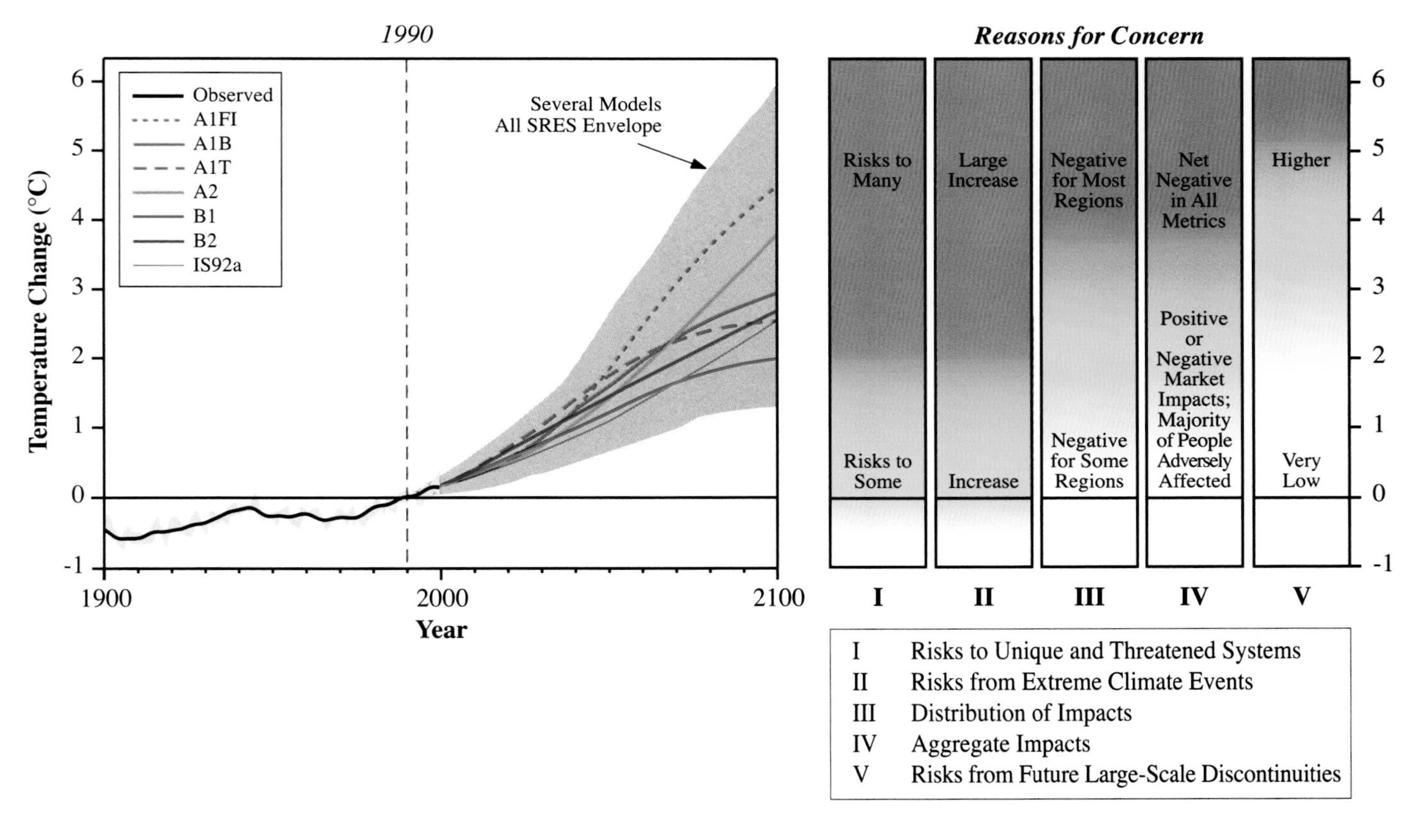

Figure SPM-2: Reasons for concern about projected climate change impacts. The risks of adverse impacts from climate change increase with the magnitude of climate change. The left part of the figure displays the observed temperature increase relative to 1990 and the range of projected temperature increase after 1990 as estimated by Working Group I of the IPCC for scenarios from the *Special Report on Emissions Scenarios*. The right panel displays conceptualizations of five reasons for concern regarding climate change risks evolving through 2100. White indicates neutral or small negative or positive impacts or risks, yellow indicates negative impacts for some systems or low risks, and red means negative impacts or risks that are more widespread and/or greater in magnitude. The assessment of impacts or risks takes into account only the magnitude of change and not the rate of change. Global mean annual temperature change is used in the figure as a proxy for the magnitude of climate change, but projected impacts will be a function of, among other factors, the magnitude and rate of global and regional changes in mean climate, climate variability and extreme climate phenomena, social and economic conditions, and adaptation.

Box SPM-1. Climate Change Sensitivity, Adaptive Capacity, and Vulnerability

Sensitivity is the degree to which a system is affected, either adversely or beneficially, by climate-related stimuli. Climate-related stimuli encompass all the elements of climate change, including mean climate characteristics, climate variability, and the frequency and magnitude of extremes. The effect may be direct (e.g., a change in crop yield in response to a change in the mean, range, or variability of temperature) or indirect (e.g., damages caused by an increase in the frequency of coastal flooding due to sea-level rise).

Adaptive capacity is the ability of a system to adjust to climate change (including climate variability and extremes) to moderate potential damages, to take advantage of opportunities, or to cope with the consequences.

Vulnerability is the degree to which a system is susceptible to, or unable to cope with, adverse effects of climate change, including climate variability and extremes. Vulnerability is a function of the character, magnitude, and rate of climate change and variation to which a system is exposed, its sensitivity, and its adaptive capacity.

Projected beneficial impacts based on models and other studies include:

- Increased potential crop yields in some regions at mid-latitudes for increases in temperature of less than a few °C [4.2]
- A potential increase in global timber supply from appropriately managed forests [4.3]
- Increased water availability for populations in some water-scarce regions—for example, in parts of southeast Asia [4.1]
- Reduced winter mortality in mid- and high-latitudes [4.7]
- Reduced energy demand for space heating due to higher winter temperatures. [4.5]

2.5. Projected Changes in Climate Extremes could have Major Consequences

The vulnerability of human societies and natural systems to climate extremes is demonstrated by the damage, hardship, and death caused by events such as droughts, floods, heat waves, avalanches, and windstorms. While there are uncertainties attached to estimates of such changes, some extreme events are projected to increase in frequency and/or severity during the 21st century due to changes in the mean and/or variability of climate, so it can be expected that the severity of their impacts will also increase in concert with global warming (see Figure SPM-2). Conversely, the frequency and magnitude of extreme low temperature events, such as cold spells, is projected to decrease in the future, with both positive and negative impacts. The impacts of future changes in climate extremes are expected to fall disproportionately on the poor. Some representative examples of impacts of these projected changes in climate variability and climate extremes are presented in Table SPM-1. [3.5, 4.6, 6, and 7.2.4]

2.6. The Potential for Large-Scale and Possibly Irreversible Impacts Poses Risks that have yet to be Reliably Quantified

Projected climate changes[7] during the 21st century have the potential to lead to future large-scale and possibly irreversible changes in Earth systems resulting in impacts at continental and global scales. These possibilities are very climate scenario-dependent and a full range of plausible scenarios has not yet been evaluated. Examples include significant slowing of the ocean circulation that transports warm water to the North Atlantic, large reductions in the Greenland and West Antarctic Ice Sheets, accelerated global warming due to carbon cycle feedbacks in the terrestrial biosphere, and releases of terrestrial carbon from permafrost regions and methane from hydrates in coastal sediments. The likelihood of many of these changes in Earth systems is not well-known, but is probably very low; however, their likelihood is expected to increase with the rate, magnitude, and duration of climate change (see Figure SPM-2). [3.5, 5.7, and 7.2.5]

If these changes in Earth systems were to occur, their impacts would be widespread and sustained. For example, significant slowing of the oceanic thermohaline circulation would impact deep-water oxygen levels and carbon uptake by oceans and marine ecosystems, and would reduce warming over parts of Europe. Disintegration of the West Antarctic Ice Sheet or melting of the Greenland Ice Sheet could raise global sea level up to 3 m each over the next 1,000 years[8], submerge many islands, and inundate extensive coastal areas. Depending on the rate of ice loss, the rate and magnitude of sea-level rise could greatly exceed the capacity of human and natural systems to adapt without substantial impacts. Releases of terrestrial carbon from permafrost regions and methane from hydrates in coastal sediments, induced by warming, would further increase greenhouse gas concentrations in the atmosphere and amplify climate change. [3.5, 5.7, and 7.2.5]

2.7. Adaptation is a Necessary Strategy at All Scales to Complement Climate Change Mitigation Efforts

Adaptation has the potential to reduce adverse impacts of climate change and to enhance beneficial impacts, but will incur costs

[7] Details of projected climate changes, illustrated in Figure SPM-2, are provided in the Working Group I Summary for Policymakers.

[8] Details of projected contributions to sea-level rise from the West Anarctic Ice Sheet and Greenland Ice Sheet are provided in the Working Group I Summary for Policymakers.

***Table SPM-1**: Examples of impacts resulting from projected changes in extreme climate events.*

Projected Changes during the 21st Century in Extreme Climate Phenomena and their Likelihood[a]	**Representative Examples of Projected Impacts**[b] *(all high confidence of occurrence in some areas*[c]*)*
Simple Extremes	
Higher maximum temperatures; more hot days and heat waves[d] over nearly all land areas (*very likely*[a])	• Increased incidence of death and serious illness in older age groups and urban poor [4.7] • Increased heat stress in livestock and wildlife [4.2 and 4.3] • Shift in tourist destinations [Table TS-4 and 5.8] • Increased risk of damage to a number of crops [4.2] • Increased electric cooling demand and reduced energy supply reliability [Table TS-4 and 4.5]
Higher (increasing) minimum temperatures; fewer cold days, frost days, and cold waves[d] over nearly all land areas (*very likely*[a])	• Decreased cold-related human morbidity and mortality [4.7] • Decreased risk of damage to a number of crops, and increased risk to others [4.2] • Extended range and activity of some pest and disease vectors [4.2 and 4.3] • Reduced heating energy demand [4.5]
More intense precipitation events (*very likely*[a] over many areas)	• Increased flood, landslide, avalanche, and mudslide damage [4.5] • Increased soil erosion [5.2.4] • Increased flood runoff could increase recharge of some floodplain aquifers [4.1] • Increased pressure on government and private flood insurance systems and disaster relief [Table TS-4 and 4.6]
Complex Extremes	
Increased summer drying over most mid-latitude continental interiors and associated risk of drought (*likely*[a])	• Decreased crop yields [4.2] • Increased damage to building foundations caused by ground shrinkage [Table TS-4] • Decreased water resource quantity and quality [4.1 and 4.5] • Increased risk of forest fire [5.4.2]
Increase in tropical cyclone peak wind intensities, mean and peak precipitation intensities (*likely*[a] over some areas)[e]	• Increased risks to human life, risk of infectious disease epidemics, and many other risks [4.7] • Increased coastal erosion and damage to coastal buildings and infrastructure [4.5 and 7.2.4] • Increased damage to coastal ecosystems such as coral reefs and mangroves [4.4]
Intensified droughts and floods associated with El Niño events in many different regions (*likely*[a]) (see also under droughts and intense precipitation events)	• Decreased agricultural and rangeland productivity in drought- and flood-prone regions [4.3] • Decreased hydro-power potential in drought-prone regions [5.1.1 and Figure TS-7]
Increased Asian summer monsoon precipitation variability (*likely*[a])	• Increased flood and drought magnitude and damages in temperate and tropical Asia [5.2.4]
Increased intensity of mid-latitude storms (little agreement between current models)[d]	• Increased risks to human life and health [4.7] • Increased property and infrastructure losses [Table TS-4] • Increased damage to coastal ecosystems [4.4]

[a] Likelihood refers to judgmental estimates of confidence used by TAR WGI: *very likely* (90-99% chance); *likely* (66-90% chance). Unless otherwise stated, information on climate phenomena is taken from the Summary for Policymakers, TAR WGI.
[b] These impacts can be lessened by appropriate response measures.
[c] High confidence refers to probabilities between 67 and 95% as described in Footnote 6.
[d] Information from TAR WGI, Technical Summary, Section F.5.
[e] Changes in regional distribution of tropical cyclones are possible but have not been established.

and will not prevent all damages. Extremes, variability, and rates of change are all key features in addressing vulnerability and adaptation to climate change, not simply changes in average climate conditions. Human and natural systems will to some degree adapt autonomously to climate change. Planned adaptation can supplement autonomous adaptation, though options and incentives are greater for adaptation of human systems than for adaptation to protect natural systems. Adaptation is a necessary strategy at all scales to complement climate change mitigation efforts. [6]

Experience with adaptation to climate variability and extremes can be drawn upon to develop appropriate strategies for adapting to anticipated climate change. Adaptation to current climate variability and extremes often produces benefits as well as forming a basis for coping with future climate change. However, experience also demonstrates that there are constraints to achieving the full measure of potential adaptation. In addition, maladaptation, such as promoting development in risk-prone locations, can occur due to decisions based on short-term considerations, neglect of known climatic variability, imperfect foresight, insufficient information, and over-reliance on insurance mechanisms. [6]

2.8 Those with the Least Resources have the Least Capacity to Adapt and are the Most Vulnerable

The ability of human systems to adapt to and cope with climate change depends on such factors as wealth, technology, education, information, skills, infrastructure, access to resources, and management capabilities. There is potential for developed and developing countries to enhance and/or acquire adaptive capabilities. Populations and communities are highly variable in their endowments with these attributes, and the developing countries, particularly the least developed countries, are generally poorest in this regard. As a result, they have lesser capacity to adapt and are more vulnerable to climate change damages, just as they are more vulnerable to other stresses. This condition is most extreme among the poorest people. [6.1; see also 5.1.7, 5.2.7, 5.3.5, 5.4.6, 5.6.1, 5.6.2, 5.7, and 5.8.1 for regional-scale information]

Benefits and costs of climate change effects have been estimated in monetary units and aggregated to national, regional, and global scales. These estimates generally exclude the effects of changes in climate variability and extremes, do not account for the effects of different rates of change, and only partially account for impacts on goods and services that are not traded in markets. These omissions are likely to result in underestimates of economic losses and overestimates of economic gains. Estimates of aggregate impacts are controversial because they treat gains for some as canceling out losses for others and because the weights that are used to aggregate across individuals are necessarily subjective. [7.2.2 and 7.2.3]

Notwithstanding the limitations expressed above, based on a few published estimates, increases in global mean temperature[9] would produce net economic losses in many developing countries for all magnitudes of warming studied (*low confidence*[6]), and losses would be greater in magnitude the higher the level of warming (*medium confidence*[6]). In contrast, an increase in global mean temperature of up to a few °C would produce a mixture of economic gains and losses in developed countries (*low confidence*[6]), with economic losses for larger temperature increases (*medium confidence*[6]). The projected distribution of economic impacts is such that it would increase the disparity in well-being between developed countries and developing countries, with disparity growing for higher projected temperature increases (*medium confidence*[6]). The more damaging impacts estimated for developing countries reflects, in part, their lesser adaptive capacity relative to developed countries. [7.2.3]

Further, when aggregated to a global scale, world gross domestic product (GDP) would change by ± a few percent for global mean temperature increases of up to a few °C (*low confidence*[6]), and increasing net losses would result for larger increases in temperature (*medium confidence*[6]) (see Figure SPM-2). More people are projected to be harmed than benefited by climate change, even for global mean temperature increases of less than a few °C (*low confidence*[6]). These results are sensitive to assumptions about changes in regional climate, level of development, adaptive capacity, rate of change, the valuation of impacts, and the methods used for aggregating monetary losses and gains, including the choice of discount rate. [7.2.2]

The effects of climate change are expected to be greatest in developing countries in terms of loss of life and relative effects on investment and the economy. For example, the relative percentage damages to GDP from climate extremes have been substantially greater in developing countries than in developed countries. [4.6]

2.9. Adaptation, Sustainable Development, and Enhancement of Equity can be Mutually Reinforcing

Many communities and regions that are vulnerable to climate change are also under pressure from forces such as population growth, resource depletion, and poverty. Policies that lessen pressures on resources, improve management of environmental risks, and increase the welfare of the poorest members of society can simultaneously advance sustainable development and equity, enhance adaptive capacity, and reduce vulnerability to climate and other stresses. Inclusion of climatic risks in the design and implementation of national and international development initiatives can promote equity and development that is more sustainable and that reduces vulnerability to climate change. [6.2]

[9] Global mean temperature change is used as an indicator of the magnitude of climate change. Scenario-dependent exposures taken into account in these studies include regionally differentiated changes in temperature, precipitation, and other climatic variables.

3. Effects on and Vulnerability of Natural and Human Systems

3.1. Hydrology and Water Resources

The effect of climate change on streamflow and groundwater recharge varies regionally and between climate scenarios, largely following projected changes in precipitation. A consistent projection across most climate change scenarios is for increases in annual mean streamflow in high latitudes and southeast Asia, and decreases in central Asia, the area around the Mediterranean, southern Africa, and Australia (*medium confidence*[6]) (see Figure SPM-3); the amount of change, however, varies between scenarios. For other areas, including mid-latitudes, there is no strong consistency in projections of streamflow, partly because of differences in projected rainfall and partly because of differences in projected evaporation, which can offset rainfall increases. The retreat of most glaciers is projected to accelerate, and many small glaciers may disappear (*high confidence*[6]). In general, the projected changes in average annual runoff are less robust than impacts based solely on temperature change because precipitation changes vary more between scenarios. At the catchment scale, the effect of a given change in climate varies with physical properties and vegetation of catchments, and may be in addition to land-cover changes. [4.1]

Approximately 1.7 billion people, one-third of the world's population, presently live in countries that are water-stressed (defined as using more than 20% of their renewable water supply, a commonly used indicator of water stress). This number is projected to increase to around 5 billion by 2025, depending on the rate of population growth. The projected climate change could further decrease the streamflow and groundwater recharge in many of these water-stressed countries—for example in central Asia, southern Africa, and countries around the Mediterranean Sea—but may increase it in some others. [4.1; see also 5.1.1, 5.2.3, 5.3.1, 5.4.1, 5.5.1, 5.6.2, and 5.8.4 for regional-scale information]

Demand for water is generally increasing due to population growth and economic development, but is falling in some countries because of increased efficiency of use. Climate change is unlikely to have a big effect on municipal and industrial water demands in general, but may substantially affect irrigation withdrawals, which depend on how increases in evaporation are offset or exaggerated by changes in precipitation. Higher temperatures, hence higher crop evaporative demand, mean that the general tendency would be towards an increase in irrigation demands. [4.1]

Flood magnitude and frequency could increase in many regions as a consequence of increased frequency of heavy precipitation events, which can increase runoff in most areas as well as groundwater recharge in some floodplains. Land-use change could exacerbate such events. Streamflow during seasonal low flow periods would decrease in many areas due to greater evaporation; changes in precipitation may exacerbate or offset the effects of increased evaporation. The projected climate change would degrade water quality through higher water temperatures and increased pollutant load from runoff and overflows of waste facilities. Quality would be degraded further where flows decrease, but increases in flows may mitigate to a certain extent some degradations in water quality by increasing dilution. Where snowfall is currently an important component of the water balance, a greater proportion of winter precipitation may fall as rain, and this can result in a more intense peak streamflow which in addition would move from spring to winter. [4.1]

The greatest vulnerabilities are likely to be in unmanaged water systems and systems that are currently stressed or poorly and unsustainably managed due to policies that discourage efficient water use and protection of water quality, inadequate watershed management, failure to manage variable water supply and demand, or lack of sound professional guidance. In unmanaged systems there are few or no structures in place to buffer the effects of hydrologic variability on water quality and supply. In unsustainably managed systems, water and land uses can add stresses that heighten vulnerability to climate change. [4.1]

Water resource management techniques, particularly those of integrated water resource management, can be applied to adapt to hydrologic effects of climate change, and to additional uncertainty, so as to lessen vulnerabilities. Currently, supply-side approaches (e.g., increasing flood defenses, building weirs, utilizing water storage areas, including natural systems, improving infrastructure for water collection and distribution) are more widely used than demand-side approaches (which alter the exposure to stress); the latter is the focus of increasing attention. However, the capacity to implement effective management responses is unevenly distributed around the world and is low in many transition and developing countries. [4.1]

3.2. Agriculture and Food Security

Based on experimental research, crop yield responses to climate change vary widely, depending upon species and cultivar; soil properties; pests, and pathogens; the direct effects of carbon dioxide (CO_2) on plants; and interactions between CO_2, air temperature, water stress, mineral nutrition, air quality, and adaptive responses. Even though increased CO_2 concentration can stimulate crop growth and yield, that benefit may not always overcome the adverse effects of excessive heat and drought (*medium confidence*[6]). These advances, along with advances in research on agricultural adaptation, have been incorporated since the Second Assessment Report (SAR) into models used to assess the effects of climate change on crop yields, food supply, farm incomes, and prices. [4.2]

Costs will be involved in coping with climate-induced yield losses and adaptation of livestock production systems. These agronomic and husbandry adaptation options could include, for example, adjustments to planting dates, fertilization rates, irrigation applications, cultivar traits, and selection of animal species. [4.2]

When autonomous agronomic adaptation is included, crop modeling assessments indicate, with *medium* to *low confidence*[6], that climate change will lead to generally positive responses at less than a few °C warming and generally negative responses for more than a few °C in mid-latitude crop yields. Similar assessments indicate that yields of some crops in tropical locations would decrease generally with even minimal increases in temperature, because such crops are near their maximum temperature tolerance and dryland/rainfed agriculture predominates. Where there is also a large decrease in rainfall, tropical crop yields would be

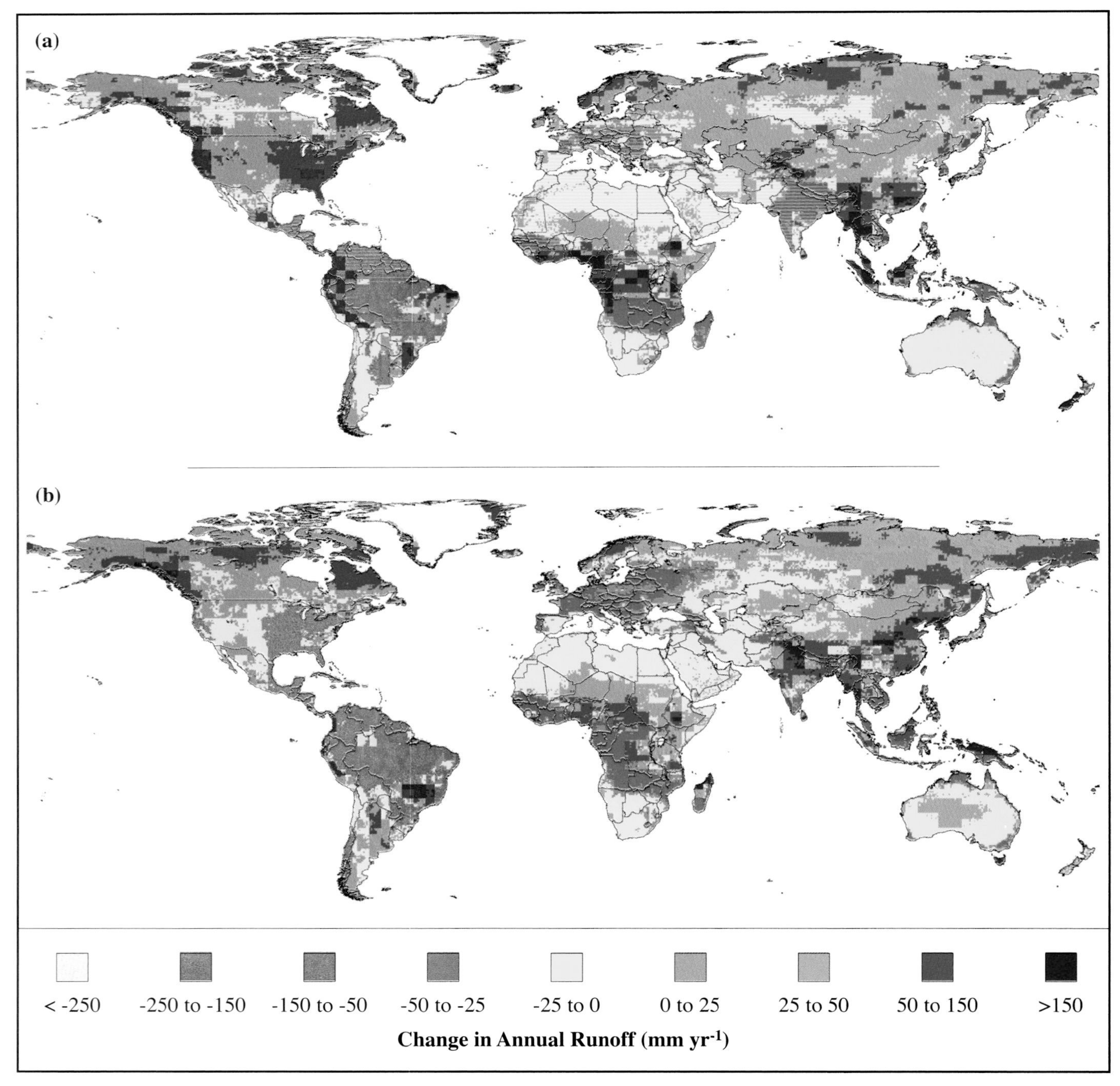

Figure SPM-3: Projected changes in average annual water runoff by 2050, relative to average runoff for 1961–1990, largely follow projected changes in precipitation. Changes in runoff are calculated with a hydrologic model using as inputs climate projections from two versions of the Hadley Centre atmosphere-ocean general circulation model (AOGCM) for a scenario of 1% per annum increase in effective carbon dioxide concentration in the atmosphere: (a) HadCM2 ensemble mean and (b) HadCM3. Projected increases in runoff in high latitudes and southeast Asia, and decreases in central Asia, the area around the Mediterranean, southern Africa, and Australia are broadly consistent across the Hadley Centre experiments, and with the precipitation projections of other AOGCM experiments. For other areas of the world, changes in precipitation and runoff are scenario- and model-dependent.

even more adversely affected. With autonomous agronomic adaptation, crop yields in the tropics tend to be less adversely affected by climate change than without adaptation, but they still tend to remain below levels estimated with current climate. [4.2]

Most global and regional economic studies not incorporating climate change indicate that the downward trend in global real commodity prices in the 20th century is likely to continue into the 21st, although confidence in these predictions decreases farther into the future. Economic modeling assessments indicate that impacts of climate change on agricultural production and prices are estimated to result in small percentage changes in global income (*low confidence*[6]), with larger increases in more developed regions and smaller increases or declines in developing regions. Improved confidence in this finding depends on further research into the sensitivity of economic modeling assessments to their base assumptions. [4.2 and Box 5-5]

Most studies indicate that global mean annual temperature increases of a few °C or greater would prompt food prices to increase due to a slowing in the expansion of global food supply relative to growth in global food demand (*established, but incomplete*[6]). At lesser amounts of warming than a few °C, economic models do not clearly distinguish the climate change signal from other sources of change based on those studies included in this assessment. Some recent aggregated studies have estimated economic impacts on vulnerable populations such as smallholder producers and poor urban consumers. These studies find that climate change would lower incomes of the vulnerable populations and increase the absolute number of people at risk of hunger, though this is uncertain and requires further research. It is established, though incompletely, that climate change, mainly through increased extremes and temporal/spatial shifts, will worsen food security in Africa. [4.2]

3.3. Terrestrial and Freshwater Ecosystems

Vegetation modeling studies continue to show the potential for significant disruption of ecosystems under climate change (*high confidence*[6]). Migration of ecosystems or biomes as discrete units is unlikely to occur; instead at a given site, species composition and dominance will change. The results of these changes will lag behind the changes in climate by years to decades to centuries (*high confidence*[6]). [4.3]

Distributions, population sizes, population density, and behavior of wildlife have been, and will continue to be, affected directly by changes in global or regional climate and indirectly through changes in vegetation. Climate change will lead to poleward movement of the boundaries of freshwater fish distributions along with loss of habitat for cold- and cool-water fishes and gain in habitat for warm-water fishes (*high confidence*[6]). Many species and populations are already at high risk, and are expected to be placed at greater risk by the synergy between climate change rendering portions of current habitat unsuitable for many species, and land-use change fragmenting habitats and raising obstacles to species migration. Without appropriate management, these pressures will cause some species currently classified as "critically endangered" to become extinct and the majority of those labeled "endangered or vulnerable" to become rarer, and thereby closer to extinction, in the 21st century (*high confidence*[6]). [4.3]

Possible adaptation methods to reduce risks to species could include: 1) establishment of refuges, parks, and reserves with corridors to allow migration of species, and 2) use of captive breeding and translocation. However, these options may have limitations due to costs. [4.3]

Terrestrial ecosystems appear to be storing increasing amounts of carbon. At the time of the SAR, this was largely attributed to increasing plant productivity because of the interaction between elevated CO_2 concentration, increasing temperatures, and soil moisture changes. Recent results confirm that productivity gains are occurring but suggest that they are smaller under field conditions than indicated by plant-pot experiments (*medium confidence*[6]). Hence, the terrestrial uptake may be due more to change in uses and management of land than to the direct effects of elevated CO_2 and climate. The degree to which terrestrial ecosystems continue to be net sinks for carbon is uncertain due to the complex interactions between the factors mentioned above (e.g., arctic terrestrial ecosystems and wetlands may act as both sources and sinks) (*medium confidence*[6]). [4.3]

Contrary to the SAR, global timber market studies that include adaptations through land and product management, even without forestry projects that increase the capture and storage of carbon, suggest that a small amount of climate change would increase global timber supply and enhance existing market trends towards rising market share in developing countries (*medium confidence*[6]). Consumers may benefit from lower timber prices while producers may gain or lose depending on regional changes in timber productivity and potential dieback effects. [4.3]

3.4. Coastal Zones and Marine Ecosystems

Large-scale impacts of climate change on oceans are expected to include increases in sea surface temperature and mean global sea level, decreases in sea-ice cover, and changes in salinity, wave conditions, and ocean circulation. The oceans are an integral and responsive component of the climate system with important physical and biogeochemical feedbacks to climate. Many marine ecosystems are sensitive to climate change. Climate trends and variability as reflected in multiyear climate-ocean regimes (e.g., Pacific Decadal Oscillation) and switches from one regime to another are now recognized to strongly affect fish abundance and population dynamics, with significant impacts on fish-dependent human societies. [4.4]

Many coastal areas will experience increased levels of flooding, accelerated erosion, loss of wetlands and mangroves, and seawater intrusion into freshwater sources as a result of climate change. The extent and severity of storm impacts, including storm-surge floods and shore erosion, will increase as a result of climate change including sea-level rise. High-latitude coasts will experience added impacts related to higher wave energy and permafrost degradation. Changes in relative sea level will vary locally due to uplift and subsidence caused by other factors. [4.4]

Impacts on highly diverse and productive coastal ecosystems such as coral reefs, atolls and reef islands, salt marshes and mangrove forests will depend upon the rate of sea-level rise relative to growth rates and sediment supply, space for and obstacles to horizontal migration, changes in the climate-ocean environment such as sea surface temperatures and storminess, and pressures from human activities in coastal zones. Episodes of coral bleaching over the past 20 years have been associated with several causes, including increased ocean temperatures. Future sea surface warming would increase stress on coral reefs and result in increased frequency of marine diseases (*high confidence*[6]). [4.4]

Assessments of adaptation strategies for coastal zones have shifted emphasis away from hard protection structures of shorelines (e.g., seawalls, groins) toward soft protection measures (e.g., beach nourishment), managed retreat, and enhanced resilience of biophysical and socioeconomic systems in coastal regions. Adaptation options for coastal and marine management are most effective when incorporated with policies in other areas, such as disaster mitigation plans and land-use plans. [4.4]

3.5. Human Health

The impacts of short-term weather events on human health have been further elucidated since the SAR, particularly in relation to periods of thermal stress, the modulation of air pollution impacts, the impacts of storms and floods, and the influences of seasonal and interannual climatic variability on infectious diseases. There has been increased understanding of the determinants of population vulnerability to adverse health impacts and the possibilities for adaptive responses. [4.7]

Many vector-, food-, and water-borne infectious diseases are known to be sensitive to changes in climatic conditions. From results of most predictive model studies, there is *medium* to *high confidence*[6] that, under climate change scenarios, there would be a net increase in the geographic range of potential transmission of malaria and dengue–two vector-borne infections each of which currently impinge on 40-50% of the world population.[10] Within their present ranges, these and many other infectious diseases would tend to increase in incidence and seasonality–although regional decreases would occur in some infectious diseases. In all cases, however, actual disease occurrence is strongly influenced by local environmental conditions, socioeconomic circumstances, and public health infrastructure. [4.7]

[10] Eight studies have modeled the effects of climate change on these diseases, five on malaria and three on dengue. Seven use a biological, or process-based approach, and one uses an empirical, statistical approach.

Projected climate change will be accompanied by an increase in heat waves, often exacerbated by increased humidity and urban air pollution, which would cause an increase in heat-related deaths and illness episodes. The evidence indicates that the impact would be greatest in urban populations, affecting particularly the elderly, sick, and those without access to air-conditioning (*high confidence*[6]). Limited evidence indicates that in some temperate countries reduced winter deaths would outnumber increased summer deaths (*medium confidence*[6]); yet, published research has been largely confined to populations in developed countries, thus precluding a generalized comparison of changes in summer and winter mortality. [3.5 and 4.7]

Extensive experience makes clear that any increase in flooding will increase the risk of drowning, diarrhoeal and respiratory diseases, and, in developing countries, hunger and malnutrition (*high confidence*[6]). If cyclones were to increase regionally, devastating impacts would often occur, particularly in densely settled populations with inadequate resources. A reduction in crop yields and food production because of climate change in some regions, particularly in the tropics, will predispose food-insecure populations to malnutrition, leading to impaired child development and decreased adult activity. Socioeconomic disruptions could occur in some regions, impairing both livelihoods and health. [3.5, 4.1, 4.2, 4.5, and 4.7]

For each anticipated adverse health impact there is a range of social, institutional, technological, and behavioral adaptation options to lessen that impact. Adaptations could, for example, encompass strengthening of the public health infrastructure, health-oriented management of the environment (including air and water quality, food safety, urban and housing design, and surface water management), and the provision of appropriate medical care facilities. Overall, the adverse health impacts of climate change will be greatest in vulnerable lower income populations, predominantly within tropical/subtropical countries. Adaptive policies would, in general, reduce these impacts. [4.7]

3.6. Human Settlements, Energy, and Industry

A growing and increasingly quantitative literature shows that human settlements are affected by climate change in one of three major ways:

1) The economic sectors that support the settlement are affected because of changes in resource productivity or changes in market demand for the goods and services produced there. [4.5]
2) Some aspects of physical infrastructure (including energy transmission and distribution systems), buildings, urban services (including transportation systems), and specific industries (such as agroindustry, tourism, and construction) may be directly affected. [4.5]
3) Populations may be directly affected through extreme weather, changes in health status, or migration. The problems are somewhat different in the largest (<1 million) and mid- to small-sized population centers. [4.5]

The most widespread direct risk to human settlements from climate change is flooding and landslides, driven by projected increases in rainfall intensity and, in coastal areas, sea-level rise. Riverine and coastal settlements are particularly at risk (*high confidence*[6]), but urban flooding could be a problem anywhere that storm drains, water supply, and waste management systems have inadequate capacity. In such areas, squatter and other informal urban settlements with high population density, poor shelter, little or no access to resources such as safe water and public health services, and low adaptive capacity are highly vulnerable. Human settlements currently experience other significant environmental problems which could be exacerbated under higher temperature/increased precipitation regimes, including water and energy resources and infrastructure, waste treatment, and transportation [4.5]

Rapid urbanization in low-lying coastal areas of both the developing and developed world is greatly increasing population densities and the value of human-made assets exposed to coastal climatic extremes such as tropical cyclones. Model-based projections of the mean annual number of people who would be flooded by coastal storm surges increase several fold (by 75 to 200 million people depending on adaptive responses) for mid-range scenarios of a 40-cm sea-level rise by the 2080s relative to scenarios with no sea-level rise. Potential damages to infrastructure in coastal areas from sea-level rise have been projected to be tens of billions US$ for individual countries—for example, Egypt, Poland, and Vietnam. [4.5]

Settlements with little economic diversification and where a high percentage of incomes derive from climate-sensitive primary resource industries (agriculture, forestry, and fisheries) are more vulnerable than more diversified settlements (*high confidence*[6]). In developed areas of the Arctic, and where the permafrost is ice-rich, special attention will be required to mitigate the detrimental impacts of thawing, such as severe damage to buildings and transport infrastructure (*very high confidence*[6]). Industrial, transportation, and commercial infrastructure is generally vulnerable to the same hazards as settlement infrastructure. Energy demand is expected to increase for space cooling and decrease for space heating, but the net effect is scenario- and location-dependent. Some energy production and distribution systems may experience adverse impacts that would reduce supplies or system reliability while other energy systems may benefit. [4.5 and 5.7]

Possible adaptation options involve the planning of settlements and their infrastructure, placement of industrial facilities, and making similar long-lived decisions in a manner to reduce the adverse effects of events that are of low (but increasing) probability and high (and perhaps rising) consequences. [4.5]

3.7. *Insurance and Other Financial Services*

The costs of ordinary and extreme weather events have increased rapidly in recent decades. Global economic losses from catastrophic events increased 10.3-fold from 3.9 billion US$ yr^{-1} in the 1950s to 40 billion US$ yr^{-1} in the 1990s (all in 1999US$, unadjusted for purchasing power parity), with approximately one-quarter of the losses occurring in developing countries. The insured portion of these losses rose from a negligible level to 9.2 billion US$ yr^{-1} during the same period. Total costs are a factor of two larger when losses from smaller, non-catastrophic weather-related events are included. As a measure of increasing insurance industry vulnerability, the ratio of global property/casual insurance premiums to weather related losses fell by a factor of three between 1985 and 1999. [4.6]

The costs of weather events have risen rapidly despite significant and increasing efforts at fortifying infrastructure and enhancing disaster preparedness. Part of the observed upward trend in disaster losses over the past 50 years is linked to socioeconomic factors, such as population growth, increased wealth, and urbanization in vulnerable areas, and part is linked to climatic factors such as the observed changes in precipitation and flooding events. Precise attribution is complex and there are differences in the balance of these two causes by region and type of event. [4.6]

Climate change and anticipated changes in weather-related events perceived to be linked to climate change would increase actuarial uncertainty in risk assessment (*high confidence*[6]). Such developments would place upward pressure on insurance premiums and/or could lead to certain risks being reclassified as uninsurable with subsequent withdrawal of coverage. Such changes would trigger increased insurance costs, slow the expansion of financial services into developing countries, reduce the availability of insurance for spreading risk, and increase the demand for government-funded compensation following natural disasters. In the event of such changes, the relative roles of public and private entities in providing insurance and risk management resources can be expected to change. [4.6]

The financial services sector as a whole is expected to be able to cope with the impacts of climate change, although the historic record demonstrates that low-probability high-impact events or multiple closely spaced events severely affect parts of the sector, especially if adaptive capacity happens to be simultaneously depleted by non-climate factors (e.g., adverse financial market conditions). The property/casualty insurance and reinsurance segments and small specialized or undiversified companies have exhibited greater sensitivity, including reduced profitability and bankruptcy triggered by weather-related events. [4.6]

Adaptation to climate change presents complex challenges, but also opportunities, to the sector. Regulatory involvement in pricing, tax treatment of reserves, and the (in)ability of firms to withdraw from at-risk markets are examples of factors that influence the resilience of the sector. Public- and private-sector actors also support adaptation by promoting disaster preparedness, loss-prevention programs, building codes, and improved land-use planning. However, in some cases, public insurance and

***Table SPM-2**: Regional adaptive capacity, vulnerability, and key concerns.*[a,b]

Region	Adaptive Capacity, Vulnerability, and Key Concerns
Africa	• Adaptive capacity of human systems in Africa is low due to lack of economic resources and technology, and vulnerability high as a result of heavy reliance on rain-fed agriculture, frequent droughts and floods, and poverty. [5.1.7] • Grain yields are projected to decrease for many scenarios, diminishing food security, particularly in small food-importing countries (*medium to high confidence*[6]). [5.1.2] • Major rivers of Africa are highly sensitive to climate variation; average runoff and water availability would decrease in Mediterranean and southern countries of Africa (*medium confidence*[6]). [5.1.1] • Extension of ranges of infectious disease vectors would adversely affect human health in Africa (*medium confidence*[6]). [5.1.4] • Desertification would be exacerbated by reductions in average annual rainfall, runoff, and soil moisture, especially in southern, North, and West Africa (*medium confidence*[6]). [5.1.6] • Increases in droughts, floods, and other extreme events would add to stresses on water resources, food security, human health, and infrastructures, and would constrain development in Africa (*high confidence*[6]). [5.1] • Significant extinctions of plant and animal species are projected and would impact rural livelihoods, tourism, and genetic resources (*medium confidence*[6]). [5.1.3] • Coastal settlements in, for example, the Gulf of Guinea, Senegal, Gambia, Egypt, and along the East–Southern African coast would be adversely impacted by sea-level rise through inundation and coastal erosion (*high confidence*[6]). [5.1.5]
Asia	• Adaptive capacity of human systems is low and vulnerability is high in the developing countries of Asia; the developed countries of Asia are more able to adapt and less vulnerable. [5.2.7] • Extreme events have increased in temperate and tropical Asia, including floods, droughts, forest fires, and tropical cyclones (*high confidence*[6]). [5.2.4] • Decreases in agricultural productivity and aquaculture due to thermal and water stress, sea-level rise, floods and droughts, and tropical cyclones would diminish food security in many countries of arid, tropical, and temperate Asia; agriculture would expand and increase in productivity in northern areas (*medium confidence*[6]). [5.2.1] • Runoff and water availability may decrease in arid and semi-arid Asia but increase in northern Asia (*medium confidence*[6]). [5.2.3] • Human health would be threatened by possible increased exposure to vector-borne infectious diseases and heat stress in parts of Asia (*medium confidence*[6]). [5.2.6] • Sea-level rise and an increase in the intensity of tropical cyclones would displace tens of millions of people in low-lying coastal areas of temperate and tropical Asia; increased intensity of rainfall would increase flood risks in temperate and tropical Asia (*high confidence*[6]). [5.2.5 and Table TS-8] • Climate change would increase energy demand, decrease tourism attraction, and influence transportation in some regions of Asia (*medium confidence*[6]). [5.2.4 and 5.2.7] • Climate change would exacerbate threats to biodiversity due to land-use and land-cover change and population pressure in Asia (*medium confidence*[6]). Sea-level rise would put ecological security at risk, including mangroves and coral reefs (*high confidence*[6]). [5.2.2] • Poleward movement of the southern boundary of the permafrost zones of Asia would result in a change of thermokarst and thermal erosion with negative impacts on social infrastructure and industries (*medium confidence*[6]). [5.2.2]

relief programs have inadvertently fostered complacency and maladaptation by inducing development in at-risk areas such as U.S. flood plains and coastal zones. [4.6]

The effects of climate change are expected to be greatest in the developing world, especially in countries reliant on primary production as a major source of income. Some countries experience impacts on their GDP as a consequence of natural disasters, with damages as high as half of GDP in one case. Equity issues and development constraints would arise if weather-related risks become uninsurable, prices increase, or availability becomes limited. Conversely, more extensive access to insurance and more widespread introduction of micro-financing schemes and development banking would increase the ability of developing countries to adapt to climate change. [4.6]

4. Vulnerability Varies across Regions

The vulnerability of human populations and natural systems to climate change differs substantially across regions and across

Table SPM-2 (continued)

Region	Adaptive Capacity, Vulnerability, and Key Concerns
Australia and New Zealand	• Adaptive capacity of human systems is generally high, but there are groups in Australia and New Zealand, such as indigenous peoples in some regions, with low capacity to adapt and consequently high vulnerability. [5.3 and 5.3.5] • The net impact on some temperate crops of climate and CO_2 changes may initially be beneficial, but this balance is expected to become negative for some areas and crops with further climate change (*medium confidence*[6]). [5.3.3] • Water is likely to be a key issue (*high confidence*[6]) due to projected drying trends over much of the region and change to a more El Niño-like average state. [5.3 and 5.3.1] • Increases in the intensity of heavy rains and tropical cyclones (*medium confidence*[6]), and region-specific changes in the frequency of tropical cyclones, would alter the risks to life, property, and ecosystems from flooding, storm surges, and wind damage. [5.3.4] • Some species with restricted climatic niches and which are unable to migrate due to fragmentation of the landscape, soil differences, or topography could become endangered or extinct (*high confidence*[6]). Australian ecosystems that are particularly vulnerable to climate change include coral reefs, arid and semi-arid habitats in southwest and inland Australia, and Australian alpine systems. Freshwater wetlands in coastal zones in both Australia and New Zealand are vulnerable, and some New Zealand ecosystems are vulnerable to accelerated invasion by weeds. [5.3.2]
Europe	• Adaptive capacity is generally high in Europe for human systems; southern Europe and the European Arctic are more vulnerable than other parts of Europe. [5.4 and 5.4.6] • Summer runoff, water availability, and soil moisture are likely to decrease in southern Europe, and would widen the difference between the north and drought-prone south; increases are likely in winter in the north and south (*high confidence*[6]). [5.4.1] • Half of alpine glaciers and large permafrost areas could disappear by end of the 21st century (*medium confidence*[6]). [5.4.1] • River flood hazard will increase across much of Europe (*medium to high confidence*[6]); in coastal areas, the risk of flooding, erosion, and wetland loss will increase substantially with implications for human settlement, industry, tourism, agriculture, and coastal natural habitats. [5.4.1 and 5.4.4] • There will be some broadly positive effects on agriculture in northern Europe (*medium confidence*[6]); productivity will decrease in southern and eastern Europe (*medium confidence*[6]). [5.4.3] • Upward and northward shift of biotic zones will take place. Loss of important habitats (wetlands, tundra, isolated habitats) would threaten some species (*high confidence*[6]). [5.4.2] • Higher temperatures and heat waves may change traditional summer tourist destinations, and less reliable snow conditions may impact adversely on winter tourism (*medium confidence*[6]). [5.4.4]
Latin America	• Adaptive capacity of human systems in Latin America is low, particularly with respect to extreme climate events, and vulnerability is high. [5.5] • Loss and retreat of glaciers would adversely impact runoff and water supply in areas where glacier melt is an important water source (*high confidence*[6]). [5.5.1] • Floods and droughts would become more frequent with floods increasing sediment loads and degrade water quality in some areas (*high confidence*[6]). [5.5] • Increases in intensity of tropical cyclones would alter the risks to life, property, and ecosystems from heavy rain, flooding, storm surges, and wind damages (*high confidence*[6]). [5.5] • Yields of important crops are projected to decrease in many locations in Latin America, even when the effects of CO_2 are taken into account; subsistence farming in some regions of Latin America could be threatened (*high confidence*[6]). [5.5.4] • The geographical distribution of vector-borne infectious diseases would expand poleward and to higher elevations, and exposures to diseases such as malaria, dengue fever, and cholera will increase (*medium confidence*[6]). [5.5.5] • Coastal human settlements, productive activities, infrastructure, and mangrove ecosystems would be negatively affected by sea-level rise (*medium confidence*[6]). [5.5.3] • The rate of biodiversity loss would increase (*high confidence*[6]). [5.5.2]

populations within regions. Regional differences in baseline climate and expected climate change give rise to different exposures to climate stimuli across regions. The natural and social systems of different regions have varied characteristics, resources, and institutions, and are subject to varied pressures that give rise to differences in sensitivity and adaptive capacity. From these differences emerge different key concerns for each of the major regions of the world. Even within regions however, impacts, adaptive capacity, and vulnerability will vary. [5]

Table SPM-2 (continued)

Region	Adaptive Capacity, Vulnerability, and Key Concerns
North America	• Adaptive capacity of human systems is generally high and vulnerability low in North America, but some communities (e.g., indigenous peoples and those dependent on climate-sensitive resources) are more vulnerable; social, economic, and demographic trends are changing vulnerabilities in subregions. [5.6 and 5.6.1] • Some crops would benefit from modest warming accompanied by increasing CO_2, but effects would vary among crops and regions (*high confidence*[6]), including declines due to drought in some areas of Canada's Prairies and the U.S. Great Plains, potential increased food production in areas of Canada north of current production areas, and increased warm-temperate mixed forest production (*medium confidence*[6]). However, benefits for crops would decline at an increasing rate and possibly become a net loss with further warming (*medium confidence*[6]). [5.6.4] • Snowmelt-dominated watersheds in western North America will experience earlier spring peak flows (*high confidence*[6]), reductions in summer flows (*medium confidence*[6]), and reduced lake levels and outflows for the Great Lakes-St. Lawrence under most scenarios (*medium confidence*[6]); adaptive responses would offset some, but not all, of the impacts on water users and on aquatic ecosystems (*medium confidence*[6]). [5.6.2] • Unique natural ecosystems such as prairie wetlands, alpine tundra, and cold-water ecosystems will be at risk and effective adaptation is unlikely (*medium confidence*[6]). [5.6.5] • Sea-level rise would result in enhanced coastal erosion, coastal flooding, loss of coastal wetlands, and increased risk from storm surges, particularly in Florida and much of the U.S. Atlantic coast (*high confidence*[6]). [5.6.1] • Weather-related insured losses and public sector disaster relief payments in North America have been increasing; insurance sector planning has not yet systematically included climate change information, so there is potential for surprise (*high confidence*[6]). [5.6.1] • Vector-borne diseases—including malaria, dengue fever, and Lyme disease—may expand their ranges in North America; exacerbated air quality and heat stress morbidity and mortality would occur (*medium confidence*[6]); socioeconomic factors and public health measures would play a large role in determining the incidence and extent of health effects. [5.6.6]
Polar	• Natural systems in polar regions are highly vulnerable to climate change and current ecosystems have low adaptive capacity; technologically developed communities are likely to adapt readily to climate change, but some indigenous communities, in which traditional lifestyles are followed, have little capacity and few options for adaptation. [5.7] • Climate change in polar regions is expected to be among the largest and most rapid of any region on the Earth, and will cause major physical, ecological, sociological, and economic impacts, especially in the Arctic, Antarctic Peninsula, and Southern Ocean (*high confidence*[6]). [5.7] • Changes in climate that have already taken place are manifested in the decrease in extent and thickness of Arctic sea ice, permafrost thawing, coastal erosion, changes in ice sheets and ice shelves, and altered distribution and abundance of species in polar regions (*high confidence*[6]). [5.7] • Some polar ecosystems may adapt through eventual replacement by migration of species and changing species composition, and possibly by eventual increases in overall productivity; ice edge systems that provide habitat for some species would be threatened (*medium confidence*[6]). [5.7] • Polar regions contain important drivers of climate change. Once triggered, they may continue for centuries, long after greenhouse gas concentrations are stabilized, and cause irreversible impacts on ice sheets, global ocean circulation, and sea-level rise (*medium confidence*[6]). [5.7]

In light of the above, all regions are likely to experience some adverse effects of climate change. Table SPM-2 presents in a highly summarized fashion some of the key concerns for the different regions. Some regions are particularly vulnerable because of their physical exposure to climate change hazards and/or their limited adaptive capacity. Most less-developed regions are especially vulnerable because a larger share of their economies are in climate-sensitive sectors and their adaptive capacity is low due to low levels of human, financial, and natural resources, as well as limited institutional and technological capability. For example, small island states and low-lying coastal areas are particularly vulnerable to increases in sea level and storms, and most of them have limited capabilities for adaptation. Climate change impacts in polar regions are expected to be large and rapid, including reduction in sea-ice extent and thickness and degradation of permafrost. Adverse changes in seasonal river flows, floods and droughts, food security, fisheries, health effects, and loss of biodiversity are among the major regional vulnerabilities and concerns of Africa, Latin America, and Asia where adaptation opportunities are generally low. Even in regions with higher adaptive capacity, such as North America and Australia and New Zealand, there are vulnerable communities, such as indigenous peoples, and the possibility of adaptation of ecosystems is very limited. In Europe, vulnerability is significantly greater in the south and in the Arctic than elsewhere in the region. [5]

Table SPM-2 (continued)

Region	Adaptive Capacity, Vulnerability, and Key Concerns
Small Island States	• Adaptive capacity of human systems is generally low in small island states, and vulnerability high; small island states are likely to be among the countries most seriously impacted by climate change. [5.8] • The projected sea-level rise of 5 mm yr^{-1} for the next 100 years would cause enhanced coastal erosion, loss of land and property, dislocation of people, increased risk from storm surges, reduced resilience of coastal ecosystems, saltwater intrusion into freshwater resources, and high resource costs to respond to and adapt to these changes (*high confidence*[6]). [5.8.2 and 5.8.5] • Islands with very limited water supplies are highly vulnerable to the impacts of climate change on the water balance (*high confidence*[6]). [5.8.4] • Coral reefs would be negatively affected by bleaching and by reduced calcification rates due to higher CO_2 levels (*medium confidence*[6]); mangrove, sea grass bed, and other coastal ecosystems and the associated biodiversity would be adversely affected by rising temperatures and accelerated sea-level rise (*medium confidence*[6]). [4.4 and 5.8.3] • Declines in coastal ecosystems would negatively impact reef fish and threaten reef fisheries, those who earn their livelihoods from reef fisheries, and those who rely on the fisheries as a significant food source (*medium confidence*[6]). [4.4 and 5.8.4] • Limited arable land and soil salinization makes agriculture of small island states, both for domestic food production and cash crop exports, highly vulnerable to climate change (*high confidence*[6]). [5.8.4] • Tourism, an important source of income and foreign exchange for many islands, would face severe disruption from climate change and sea-level rise (*high confidence*[6]). [5.8.5]

[a] Because the available studies have not employed a common set of climate scenarios and methods, and because of uncertainties regarding the sensitivities and adaptability of natural and social systems, the assessment of regional vulnerabilities is necessarily qualitative.
[b] The regions listed in Table SPM-2 are graphically depicted in Figure TS-2 of the Technical Summary.

5. Improving Assessments of Impacts, Vulnerabilities, and Adaptation

Advances have been made since previous IPCC assessments in the detection of change in biotic and physical systems, and steps have been taken to improve the understanding of adaptive capacity, vulnerability to climate extremes, and other critical impact-related issues. These advances indicate a need for initiatives to begin designing adaptation strategies and building adaptive capacities. Further research is required, however, to strengthen future assessments and to reduce uncertainties in order to assure that sufficient information is available for policymaking about responses to possible consequences of climate change, including research in and by developing countries. [8]

The following are high priorities for narrowing gaps between current knowledge and policymaking needs:

- Quantitative assessment of the sensitivity, adaptive capacity, and vulnerability of natural and human systems to climate change, with particular emphasis on changes in the range of climatic variation and the frequency and severity of extreme climate events
- Assessment of possible thresholds at which strongly discontinuous responses to projected climate change and other stimuli would be triggered
- Understanding dynamic responses of ecosystems to multiple stresses, including climate change, at global, regional, and finer scales
- Development of approaches to adaptation responses, estimation of the effectiveness and costs of adaptation options, and identification of differences in opportunities for and obstacles to adaptation in different regions, nations, and populations
- Assessment of potential impacts of the full range of projected climate changes, particularly for non-market goods and services, in multiple metrics and with consistent treatment of uncertainties, including but not limited to numbers of people affected, land area affected, numbers of species at risk, monetary value of impact, and implications in these regards of different stabilization levels and other policy scenarios
- Improving tools for integrated assessment, including risk assessment, to investigate interactions between components of natural and human systems and the consequences of different policy decisions
- Assessment of opportunities to include scientific information on impacts, vulnerability, and adaptation in decisionmaking processes, risk management, and sustainable development initiatives
- Improvement of systems and methods for long-term monitoring and understanding the consequences of climate change and other stresses on human and natural systems.

Cutting across these foci are special needs associated with strengthening international cooperation and coordination for regional assessment of impacts, vulnerability, and adaptation, including capacity-building and training for monitoring, assessment, and data gathering, especially in and for developing countries (particularly in relation to the items identified above).

TECHNICAL SUMMARY

CLIMATE CHANGE 2001: IMPACTS, ADAPTATION, AND VULNERABILITY

A Report of Working Group II of the Intergovernmental Panel on Climate Change

This summary was accepted but not approved in detail at the Sixth Session of IPCC Working Group II (Geneva, Switzerland • 13-16 February 2001). "Acceptance" of IPCC reports at a session of the Working Group or Panel signifies that the material has not been subject to line-by-line discussion and agreement, but nevertheless presents a comprehensive, objective, and balanced view of the subject matter.

Lead Authors:
K.S. White (USA), Q.K. Ahmad (Bangladesh), O. Anisimov (Russia), N. Arnell (UK), S. Brown (USA), M. Campos (Costa Rica), T. Carter (Finland), Chunzhen Liu (China), S. Cohen (Canada), P. Desanker (Malawi), D.J. Dokken (USA), W. Easterling (USA), B. Fitzharris (New Zealand), H. Gitay (Australia), A. Githeko (Kenya), S. Gupta (India), H. Harasawa (Japan), B.P. Jallow (The Gambia), Z.W. Kundzewicz (Poland), E.L. La Rovere (Brazil), M. Lal (India), N. Leary (USA), C. Magadza (Zimbabwe), L.J. Mata (Venezuela), R. McLean (Australia), A. McMichael (UK), K. Miller (USA), E. Mills (USA), M.Q. Mirza (Bangladesh), D. Murdiyarso (Indonesia), L.A. Nurse (Barbados), C. Parmesan (USA), M.L. Parry (UK), O. Pilifosova (Kazakhstan), B. Pittock (Australia), J. Price (USA), T. Root (USA), C. Rosenzweig (USA), J. Sarukhan (Mexico), H.-J. Schellnhuber (Germany), S. Schneider (USA), M.J. Scott (USA), G. Sem (Papua New Guinea), B. Smit (Canada), J.B. Smith (USA), A. Tsyban (Russia), P. Vellinga (The Netherlands), R. Warrick (New Zealand), D. Wratt (New Zealand)

Review Editors:
M. Manning (New Zealand) and C. Nobre (Brazil)

CONTENTS

1. Scope and Approach of the Assessment

1.1. *Mandate of the Assessment*

The Intergovernmental Panel on Climate Change (IPCC) was established by World Meteorological Organization and United Nations Environmental Programme (UNEP) in 1988 to assess scientific, technical, and socioeconomic information that is relevant in understanding human-induced climate change, its potential impacts, and options for mitigation and adaptation. The IPCC currently is organized into three Working Groups: Working Group I (WGI) addresses observed and projected changes in climate; Working Group II (WGII) addresses vulnerability, impacts, and adaptation related to climate change; and Working Group III (WGIII) addresses options for mitigation of climate change.

This volume—*Climate Change 2001: Impacts, Adaptation, and Vulnerability*—is the WGII contribution to the IPCC's Third Assessment Report (TAR) on scientific, technical, environmental, economic, and social issues associated with the climate system and climate change.[1] WGII's mandate for the TAR is to assess the vulnerability of ecological systems, socioeconomic sectors, and human health to climate change as well as potential impacts of climate change, positive and negative, on these systems. This assessment also examines the feasibility of adaptation to enhance the positive effects of climate change and ameliorate negative effects. This new assessment builds on previous IPCC assessments, reexamining key findings of earlier assessments and emphasizing new information and implications from more recent studies.

Box 1. Climate Change Sensitivity, Adaptability, and Vulnerability

Sensitivity is the degree to which a system is affected, either adversely or beneficially, by climate-related stimuli. Climate-related stimuli encompass all the elements of climate change, including mean climate characteristics, climate variability, and the frequency and magnitude of extremes. The effect may be direct (e.g., a change in crop yield in response to a change in the mean, range or variability of temperature) or indirect (e.g., damages caused by an increase in the frequency of coastal flooding due to sea-level rise).

Adaptive capacity is the ability of a system to adjust to climate change, including climate variability and extremes, to moderate potential damages, to take advantage of opportunities, or to cope with the consequences.

Vulnerability is the degree to which a system is susceptible to, or unable to cope with, adverse effects of climate change, including climate variability and extremes. Vulnerability is a function of the character, magnitude and rate of climate change and variation to which a system is exposed, its sensitivity, and its adaptive capacity.

1.2. *What is Potentially at Stake?*

Human activities—primarily burning of fossil fuels and changes in land cover—are modifying the concentration of atmospheric constituents or properties of the surface that absorb or scatter radiant energy. The WGI contribution to the TAR—*Climate Change 2001: The Scientific Basis*—found, "In the light of new evidence and taking into account the remaining uncertainties, most of the observed warming over the last 50 years is likely to have been due to the increase in greenhouse gas concentrations." Future changes in climate are expected to include additional warming, changes in precipitation patterns and amounts, sea-level rise, and changes in the frequency and intensity of some extreme events.

The stakes associated with projected changes in climate are high. Numerous Earth systems that sustain human societies are sensitive to climate and will be impacted by changes in climate (very high confidence). Impacts can be expected in ocean circulation; sea level; the water cycle; carbon and nutrient cycles; air quality; the productivity and structure of natural ecosystems; the productivity of agricultural, grazing, and timber lands; and the geographic distribution, behavior, abundance, and survival of plant and animal species, including vectors and hosts of human disease. Changes in these systems in response to climate change, as well as direct effects of climate change on humans, would affect human welfare, positively and negatively. Human welfare would be impacted through changes in supplies of and demands for water, food, energy, and other tangible goods that are derived from these systems; changes in opportunities for nonconsumptive uses of the environment for recreation and tourism; changes in non-use values of the environment such as cultural and preservation values; changes in incomes; changes in loss of property and lives from extreme climate phenomena; and changes in human health. Climate change impacts will affect the prospects for sustainable development in different parts of the world and may further widen existing inequalities. Impacts will vary in distribution across people, places, and times (very high confidence), raising important questions about equity.

Although the stakes are demonstrably high, the risks associated with climate change are less easily established. Risks are a function of the probability and magnitude of different types of impacts. The WGII report assesses advances in the state of

[1]*Climate change* in IPCC usage refers to any change in climate over time, whether due to natural variability or as a result of human activity. This usage differs from the definition in Article 1 of the United Nations Framework Convention on Climate Change, where *climate change* refers to a change of climate which is attributed directly or indirectly to human activity that alters the composition of the global atmosphere and which is in addition to natural climate variability observed over comparable time periods.

knowledge regarding impacts of climate stimuli to which systems may be exposed, the sensitivity of exposed systems to changes in climate stimuli, their adaptive capacity to alleviate or cope with adverse impacts or enhance beneficial ones, and their vulnerability to adverse impacts (see Box 1). Possible impacts include impacts that threaten substantial and irreversible damage to or loss of some systems within the next century; modest impacts to which systems may readily adapt; and impacts that would be beneficial for some systems.

Figure TS-1 illustrates the scope of the WGII assessment and its relation to other parts of the climate change system. Human activities that change the climate expose natural and human systems to an altered set of stresses or stimuli. Systems that are sensitive to these stimuli are affected or impacted by the changes, which can trigger autonomous, or expected, adaptations. These autonomous adaptations will reshape the residual or net impacts of climate change. Policy responses in reaction to impacts already perceived or in anticipation of potential future impacts can take the form of planned adaptations to lessen adverse effects or enhance beneficial ones. Policy responses also can take the form of actions to mitigate climate change through greenhouse gas (GHG) emission reductions and enhancement of sinks. The WGII assessment focuses on the central box of Figure TS-1—exposure, impacts, and vulnerabilities—and the adaptation policy loop.

1.3. Approach of the Assessment

The assessment process involves evaluation and synthesis of available information to advance understanding of climate change impacts, adaptation, and vulnerability. The information comes predominately from peer-reviewed published literature. Evidence also is drawn from published, non-peer-reviewed literature and unpublished sources, but only after evaluation of its quality and validity by the authors of this report.

WGII's assessment has been conducted by an international group of experts nominated by governments and scientific bodies and selected by the WGII Bureau of the IPCC for their scientific and technical expertise and to achieve broad geographical balance. These experts come from academia, governments, industry, and scientific and environmental organizations. They participate without compensation from the IPCC, donating substantial time to support the work of the IPCC.

This assessment is structured to examine climate change impacts, adaptations, and vulnerabilities of systems and regions and to provide a global synthesis of cross-system and cross-regional issues. To the extent feasible, given the available literature, climate change is examined in the context of sustainable development and equity. The first section sets the stage for the

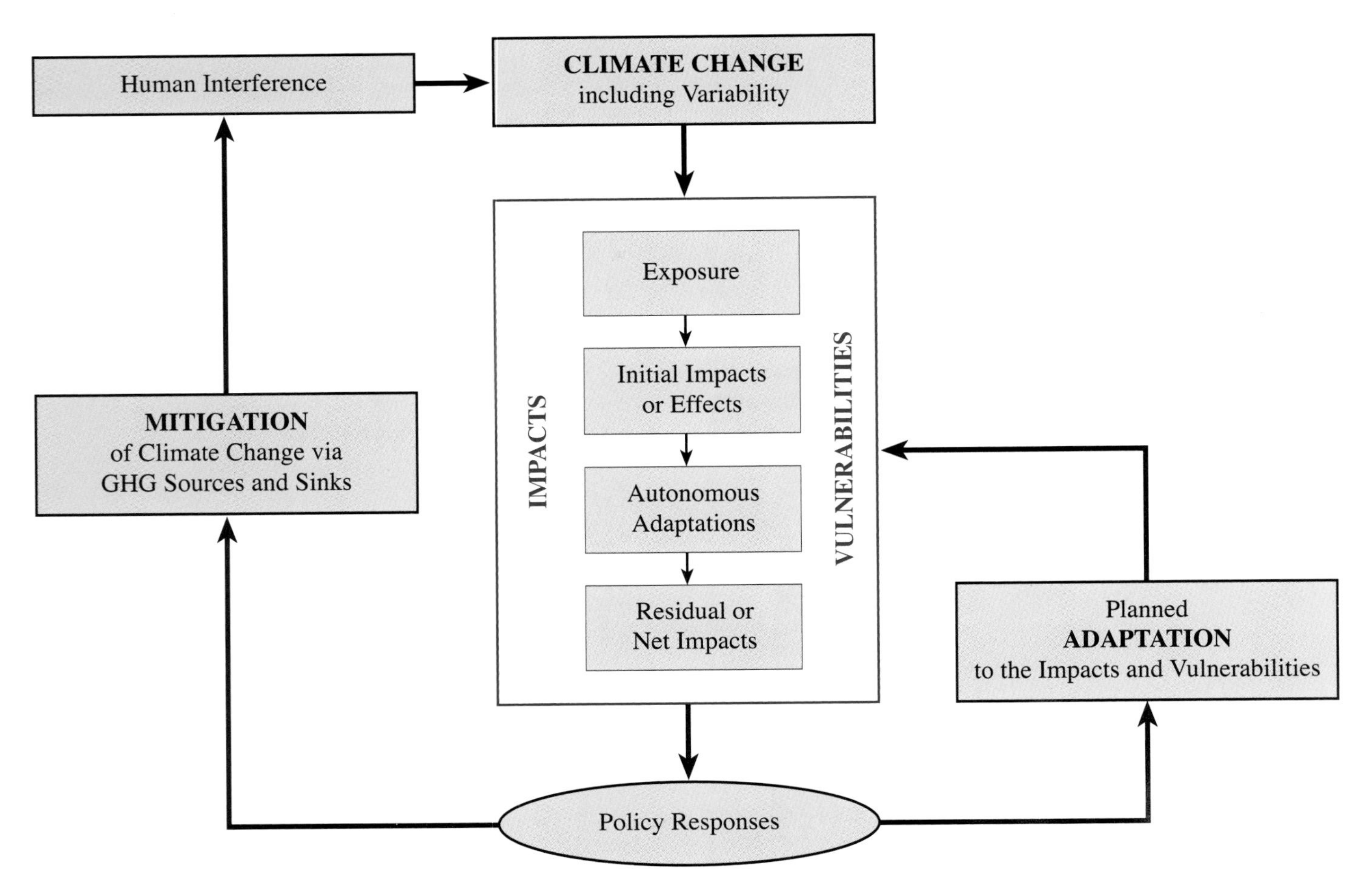

Figure TS-1: Scope of the Working Group II assessment.

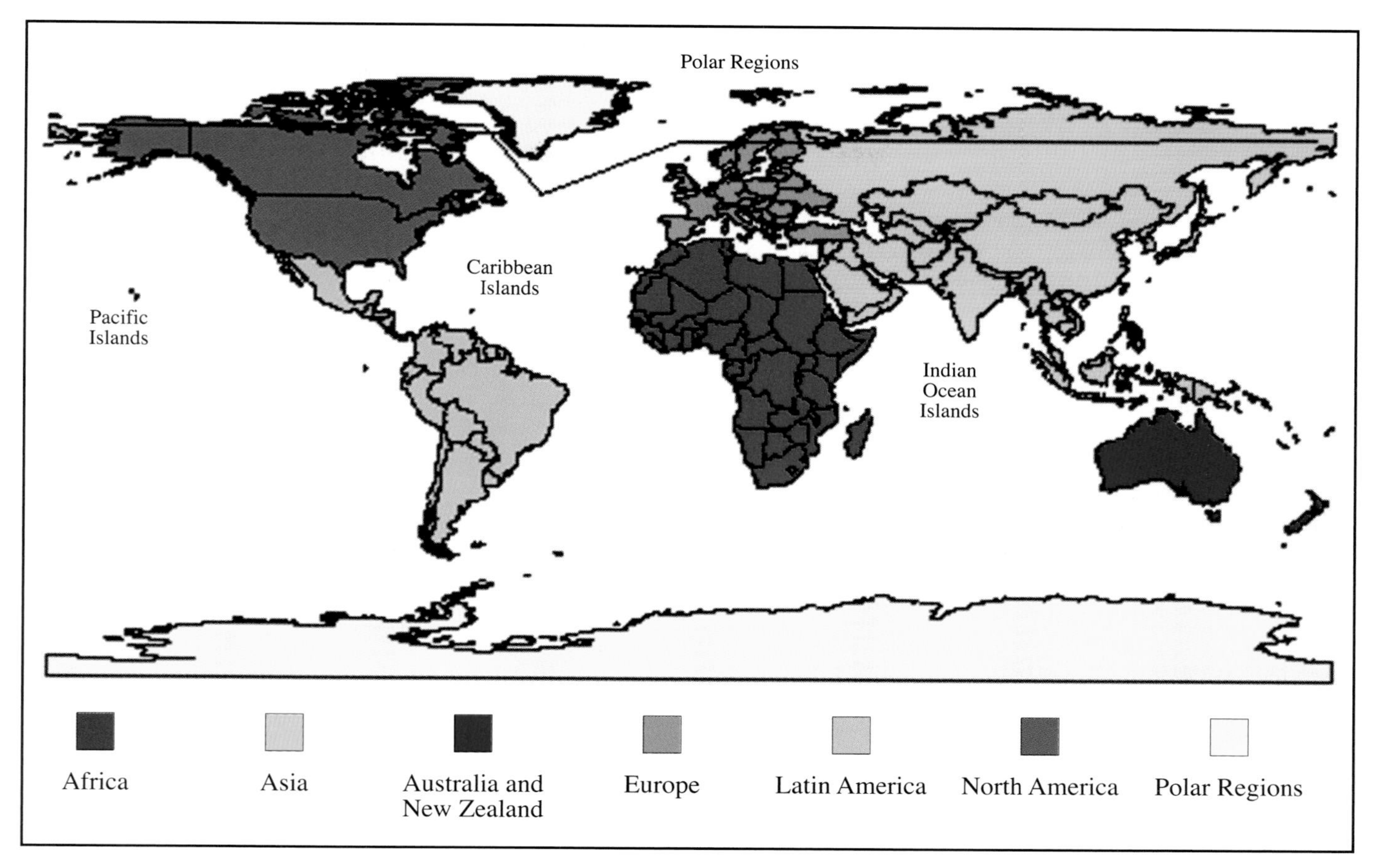

Figure TS-2: Regions for the IPCC Working Group II Third Assessment Report. Note that regions in which small island states are located include the Pacific, Indian, and Atlantic Oceans, and the Caribbean and Mediterranean Seas. The boundary between Europe and Asia runs along the eastern Ural Mountains, River Ural, and Caspian Sea. For the polar regions, the Arctic consists of the area north of the Arctic Circle, including Greenland; the Antarctic consists of the Antarctic continent, together with the Southern Ocean south of ~58°S.

assessment by discussing the context of climate change, methods and tools, and scenarios. Individual chapters assess vulnerabilities of water systems, terrestrial ecosystems (including agriculture and forestry), ocean and coastal systems, human settlements (including energy and industrial sectors), insurance and other financial services, and human health. A chapter is devoted to each of eight major regions of the world: Africa, Asia, Australia and New Zealand, Europe, Latin America, North America, polar regions, and small island states. These regions are shown in Figure TS-2. All of the regions are highly heterogeneous, and climate change impacts, adaptive capacity, and vulnerability will vary in important ways within each of the regions. The final section of the report synthesizes adaptation capacity and its potential to alleviate adverse impacts, enhance beneficial effects, and increase sustainable development and equity and reviews information that is relevant for interpretation of Article 2 of the United Nations Framework Convention on Climate Change (UNFCCC) and key provisions of international agreements to address climate change. The report also contains a Summary for Policymakers, which provides a brief synthesis of the conclusions of the report that have particular relevance to those who have responsibility for making climate change response decisions. This Technical Summary provides a more comprehensive summary of the assessment; it references sections of the underlying report in brackets at the end of the paragraphs for readers who would like more information on a particular topic. [1.1]

1.4. Treatment of Uncertainties

Since the SAR, greater emphasis has been placed on developing methods for characterizing and communicating uncertainties. Two approaches to evaluate uncertainties are applied in the WGII assessment. A quantitative approach is adopted to assess confidence levels in instances for which present understanding of relevant processes, system behavior, observations, model simulations, and estimates is sufficient to support broad agreement among authors of the report about Bayesian probabilities associated with selected findings. A more qualitative approach is used to assess and report the quality or level of scientific understanding that supports a conclusion (see Box 2). These approaches, and the rationale for them, are explained in more detail in *Third Assessment Report: Cross-Cutting Issues Guidance Papers* (http://www.gispri.or.jp), supporting material prepared by the IPCC to increase the use of consistent terms and concepts within and across the Working Group volumes of the TAR. [1.1, 2.6]

2. Methods and Tools of the Assessment

Assessment of climate change impacts, adaptations, and vulnerability draws on a wide range of physical, biological, and social science disciplines and consequently employs an enormous variety of methods and tools. Since the SAR, such methods have improved detection of climate change in biotic and physical systems and produced new substantive findings. In addition, cautious steps have been taken since the SAR to expand the "tool-box" to address more effectively the human dimensions of climate as both causes and consequences of change and to deal more directly with cross-sectoral issues concerning vulnerability, adaptation, and decisionmaking. In particular, a greater number of studies have begun to apply methods and tools for costing and valuing effects, treating uncertainties, integrating effects across sectors and regions, and applying decision analytic frameworks for evaluating adaptive capacity. Overall, these modest methodological developments are encouraging analyses that will build a more solid foundation for understanding how decisions regarding adaptation to future climate change might be taken. [2.8]

2.1. *Detecting Responses to Climate Change using Indicator Species or Systems*

Since the SAR, methods have been developed and applied to the detection of present impacts of 20th century climate change on abiotic and biotic systems. Assessment of impacts on human and natural systems that already have occurred as a result of recent climate change is an important complement to model projections of future impacts. Such detection is impeded by multiple, often inter-correlated, nonclimatic forces that concurrently affect those systems. Attempts to overcome this problem have involved the use of indicator species (e.g., butterflies, penguins, frogs, and sea anemones) to detect responses to climate change and to infer more general impacts of climate change on natural systems (e.g., in native meadows, coastal Antarctica, tropical cloud forest, and the Pacific rocky intertidal, respectively). An important component of this detection process is the search for systematic patterns of change across many studies that are consistent with expectations, based on observed or predicted changes in climate. Confidence in attribution of these observed changes to climate change increases as studies are replicated across diverse systems and geographic regions. Even though studies now number in the hundreds, some regions and systems remain underrepresented. [2.2]

To investigate possible links between observed changes in regional climate and biological or physical processes in ecosystems, the author team gathered more than 2,500 articles on climate and one of the following entities: animals, plants, glaciers, sea ice, and ice on lakes or streams. To determine if these entities have been influenced by changing climate, only studies meeting at least two of the following criteria were included:

- A trait of these entities (e.g., range boundary, melting date) shows a change over time.
- The trait is correlated with changes in local temperature.
- Local temperature changed over time.

Box 2. Confidence Levels and State of Knowledge

Quantitative Assessment of Confidence Levels

In applying the *quantitative* approach, authors of the report assign a confidence level that represents the degree of belief among the authors in the validity of a conclusion, based on their collective expert judgment of observational evidence, modeling results, and theory that they have examined. Five confidence levels are used. In the tables of the Technical Summary, symbols are substituted for words:

Very High (*****)	95% or greater
High (****)	67–95%
Medium(***)	33–67%
Low (**)	5–33%
Very Low (*)	5% or less

Qualitative Assessment of the State of Knowledge

In applying the *qualitative* approach, authors of the report evaluate the level of scientific understanding in support of a conclusion, based on the amount of supporting evidence and the level of agreement among experts about the interpretation of the evidence. Four qualitative classifications are employed:

- *Well-established*: Models incorporate known processes, observations are consistent with models, or multiple lines of evidence support the finding.
- *Established but incomplete*: Models incorporate most known processes, although some parameterizations may not be well tested; observations are somewhat consistent but incomplete; current empirical estimates are well founded, but the possibility of changes in governing processes over time is considerable; or only one or a few lines of evidence support the finding.
- *Competing explanations*: Different model representations account for different aspects of observations or evidence or incorporate different aspects of key processes, leading to competing explanations.
- *Speculative*: Conceptually plausible ideas that are not adequately represented in the literature or that contain many difficult-to-reduce uncertainties [Box 1-1]

At least two of these three criteria had to exhibit a statistically significant correlation. Only temperature was considered because it is well established in the literature how it influences the entities examined and because temperature trends are more globally homogeneous than other locally varying climatic factors, such as precipitation changes. Selected studies must also have examined at least 10 years of data; more than 90% had a time span of more than 20 years.

These stringent criteria reduced the number of studies used in the analysis to 44 animal and plant studies that cover more than 600 species. Of these species, about 90% (more than 550) show changes in traits over time. Of these 550+ species, about 80% (more than 450) show change in a direction expected given scientific understanding of known mechanisms that relate temperature to each of the species traits. The probability that more than 450 species of 550+ would show changes in the directions expected by random chance is negligible.

Sixteen studies examining glaciers, sea ice, snow cover extent/ snow melt, or ice on lakes or streams included more than 150 sites. Of these 150+ sites, 67% (100+) show changes in traits over time. Of these 100+ sites, about 99% (99+) exhibited trends in a direction expected, given scientific understanding of known mechanisms that relate temperatures to physical processes that govern change in that trait. The probability that 99+ of 100+ sites would show changes in the directions expected by chance alone is negligible. [5.2, 5.4, 19.2]

2.2. Anticipating the Effects of Future Climate Change

Since the SAR, improvements in methods and tools for studying impacts of future changes in climate have included greater emphasis on the use of process-oriented models, transient climate change scenarios, refined socioeconomic baselines, and higher resolution spatial and temporal scales. Country studies and regional assessments in every continent have tested models and tools in a variety of contexts. First-order impact models have been linked to global systems models. Adaptation has been included in many assessments, often for the first time.

Methodological gaps remain concerning scales, data, validation, and integration of adaptation and the human dimensions of climate change. Procedures for assessing regional and local vulnerability and long-term adaptation strategies require high-resolution assessments, methodologies to link scales, and dynamic modeling that uses corresponding and new data sets. Validation at different scales often is lacking. Regional integration across sectors is required to place vulnerability in the context of local and regional development. Methods and tools to assess vulnerability to extreme events have improved but are constrained by low confidence in climate change scenarios and the sensitivity of impact models to major climatic anomalies. Understanding and integrating higher order economic effects and other human dimensions of global change are required. Adaptation models and vulnerability indices to prioritize adaptation options are at early stages of development in many fields. Methods to enable stakeholder participation in assessments need improvement. [2.3]

2.3. Integrated Assessment

Integrated assessment is an interdisciplinary process that combines, interprets, and communicates knowledge from diverse scientific disciplines from the natural and social sciences to investigate and understand causal relationships within and between complicated systems. Methodological approaches employed in such assessments include computer-aided modeling, scenario analyses, simulation gaming and participatory integrated assessment, and qualitative assessments that are based on existing experience and expertise. Since the SAR, significant progress has been made in developing and applying such approaches to integrated assessment, globally and regionally.

However, progress to date, particularly with regard to integrated modeling, has focused largely on mitigation issues at the global or regional scale and only secondarily on issues of impacts, vulnerability, and adaptation. Greater emphasis on the development of methods for assessing vulnerability is required, especially at national and subnational scales where impacts of climate change are felt and responses are implemented. Methods designed to include adaptation and adaptive capacity explicitly in specific applications must be developed. [2.4]

2.4. Costing and Valuation

Methods of economic costing and valuation rely on the notion of opportunity cost of resources used, degraded, or saved. Opportunity cost depends on whether the market is competitive or monopolistic and on whether any externalities are internalized. It also depends on the rate at which the future is discounted, which can vary across countries, over time, and over generations. The impact of uncertainty also can be valued if the probabilities of different possible outcomes are known. Public and nonmarket goods and services can be valued through willingness to pay for them or willingness to accept compensation for lack of them. Impacts on different groups, societies, nations, and species must be assessed. Comparison of alternative distributions of welfare across individuals and groups within a country can be justified if they are made according to internally consistent norms. Comparisons across nations with different societal, ethical, and governmental structures cannot yet be made meaningfully.

Since the SAR, no new fundamental developments in costing and valuation methodology have taken place. Many new applications of existing methods to a widening range of climate change issues have demonstrated, however, the strengths and limitations of some of these methods. Research efforts are required to strengthen methods for multi-objective assessments. Multi-objective assessments are increasingly preferred, but the means by which their underlying metrics might more accurately reflect diverse social, political, economic, and cultural contexts must be developed. In addition, methods for integrating across these multiple metrics are still missing from the methodological repertoire. [2.5]

2.5. *Decision Analytic Frameworks*

Policymakers who are responsible for devising and implementing adaptive policies should be able to rely on results from one or more of a diverse set of decision analytical frameworks. Commonly used methods include cost-benefit and cost-effectiveness analysis, various types of decision analysis (including multi-objective studies), and participatory techniques such as policy exercises.

Very few cases in which policymakers have used decision analytical frameworks in evaluating adaptation options have been reported. Among the large number of assessments of climate change impacts reviewed in the TAR, only a small fraction include comprehensive and quantitative estimates of adaptation options and their costs, benefits, and uncertainty characteristics. This information is necessary for meaningful applications of any decision analytical method to issues of adaptation. Greater use of such methods in support of adaptation decisions is needed to establish their efficacy and to identify directions for necessary research in the context of vulnerability and adaptation to climate change. [2.7]

3. Scenarios of Future Change

3.1. *Scenarios and their Role*

A scenario is a coherent, internally consistent, and plausible description of a possible future state of the world. Scenarios are commonly required in climate change impact, adaptation, and vulnerability assessments to provide alternative views of future conditions considered likely to influence a given system or activity. A distinction is made between climate scenarios, which describe the forcing factor of focal interest to the IPCC, and nonclimatic scenarios, which provide the socioeconomic and environmental context within which climate forcing operates. Most assessments of the impacts of future climate change are based on results from impact models that rely on quantitative climate and nonclimatic scenarios as inputs. [3.1.1, Box 3-1]

3.2. *Socioeconomic, Land-Use, and Environmental Scenarios*

Nonclimatic scenarios describing future socioeconomic, land-use, and environmental changes are important for characterizing the sensitivity of systems to climate change, their vulnerability, and the capacity for adaptation. Such scenarios only recently have been widely adopted in impact assessments alongside climate scenarios.

Socioeconomic scenarios. Socioeconomic scenarios have been used more extensively for projecting GHG emissions than for assessing climate vulnerability and adaptive capacity. Most socioeconomic scenarios identify several different topics or domains, such as population or economic activity, as well as background factors such as the structure of governance, social values, and patterns of technological change. Scenarios make it possible to establish baseline socioeconomic vulnerability, pre-climate change; determine climate change impacts; and assess post-adaptation vulnerability. [3.2]

Land-use and land-cover change scenarios. Land-use change and land-cover change (LUC-LCC) involve several processes that are central to the estimation of climate change and its impacts. First, LUC-LCC influences carbon fluxes and GHG emissions, which directly alter atmospheric composition and radiative forcing properties. Second, LUC-LCC modifies land-surface characteristics and, indirectly, climatic processes. Third, land-cover modification and conversion may alter the properties of ecosystems and their vulnerability to climate change. Finally, several options and strategies for mitigating GHG emissions involve land cover and changed land-use practices. A great diversity of LUC-LCC scenarios have been constructed. Most of these scenarios do not address climate change issues explicitly, however; they focus on other issues—for example, food security and carbon cycling. Large improvements have been made since the SAR in defining current and historic land-use and land-cover patterns, as well as in estimating future scenarios. Integrated assessment models currently are the most appropriate tools for developing LUC-LCC scenarios. [3.3.1, 3.3.2]

Environmental scenarios. Environmental scenarios refer to changes in environmental factors other than climate that will occur in the future regardless of climate change. Because these factors could have important roles in modifying the impacts of future climate change, scenarios are required to portray possible future environmental conditions such as atmospheric composition [e.g., carbon dioxide (CO_2), tropospheric ozone, acidifying compounds, and ultraviolet-B (UV-B) radiation]; water availability, use, and quality; and marine pollution. Apart from the direct effects of CO_2 enrichment, changes in other environmental factors rarely have been considered alongside climate changes in past impact assessments, although their use is increasing with the emergence of integrated assessment methods. [3.4.1]

3.3. *Sea-Level Rise Scenarios*

Sea-level rise scenarios are required to evaluate a diverse range of threats to human settlements, natural ecosystems, and landscape in the coastal zone. Relative sea-level scenarios (i.e., sea-level rise with reference to movements of the local land surface) are of most interest for impact and adaptation assessments. Tide gauge and wave-height records of 50 years or more are required, along with information on severe weather and coastal processes, to establish baseline levels or trends. Recent techniques of satellite altimetry and geodetic leveling have enhanced and standardized baseline determinations of relative sea level over large areas of the globe. [3.6.2]

Although some components of future sea-level rise can be modeled regionally by using coupled ocean-atmosphere models, the most common method of obtaining scenarios is to apply global mean estimates from simple models. Changes in the occurrence of extreme events such as storm surges and wave setup, which can lead to major coastal impacts, sometimes are investigated by superimposing historically observed events onto a rising mean sea level. More recently, some studies have begun to express future sea-level rise in probabilistic terms, enabling rising levels to be evaluated in terms of the risk of exceeding a critical threshold of impact. [3.6.3, 3.6.4, 3.6.5, 3.6.6]

3.4. Climate Scenarios

Three main types of climate scenarios have been employed in impact assessments: incremental scenarios, analog scenarios, and climate model-based scenarios. Incremental scenarios are simple adjustments of the baseline climate according to anticipated future changes that can offer a valuable aid for testing system sensitivity to climate. However, because they involve arbitrary adjustments, they may not be realistic meteorologically. Analogs of a changed climate from the past record or from other regions may be difficult to identify and are seldom applied, although they sometimes can provide useful insights into impacts of climate conditions outside the present-day range. [3.5.2]

The most common scenarios use outputs from general circulation models (GCMs) and usually are constructed by adjusting a baseline climate (typically based on regional observations of climate over a reference period such as 1961–1990) by the absolute or proportional change between the simulated present and future climates. Most recent impact studies have constructed scenarios on the basis of transient GCM outputs, although some still apply earlier equilibrium results. The great majority of scenarios represent changes in mean climate; some recent scenarios, however, also have incorporated changes in variability and extreme weather events, which can lead to important impacts for some systems. Regional detail is obtained from the coarse-scale outputs of GCMs by using three main methods: simple interpolation, statistical downscaling, and high-resolution dynamical modeling. The simple method, which reproduces the GCM pattern of change, is the most widely applied in scenario development. In contrast, the statistical and modeling approaches can produce local climate changes that are different from large-scale GCM estimates. More research is needed to evaluate the value added to impact studies of such regionalization exercises. One reason for this caution is the large uncertainty of GCM projections, which requires further quantification through model intercomparisons, new model simulations, and pattern scaling methods. [3.5.2, 3.5.4, 3.5.5]

3.5. Scenarios of the 21st Century

In 2000, the IPCC completed a *Special Report on Emissions Scenarios* (SRES) to replace the earlier set of six IS92 scenarios developed for the IPCC in 1992. These newer scenarios consider the period 1990 to 2100 and include a range of socioeconomic assumptions (e.g., global population and gross domestic product). Their implications for other aspects of global change also have been calculated; some of these implications are summarized for 2050 and 2100 in Table TS-1. For example, mean ground-level ozone concentrations in July over the industrialized

***Table TS-1**: The SRES scenarios and their implications for atmospheric composition, climate, and sea level. Values of population, GDP, and per capita income ratio (a measure of regional equity) are those applied in integrated assessment models used to estimate emissions (based on Tables 3-2 and 3-9).*

Date	Global Population (billions)[a]	Global GDP (10^{12} US$ yr^{-1})[b]	Per Capita Income Ratio[c]	Ground-Level O_3 Concentration (ppm)[d]	CO_2 Concentration (ppm)[e]	Global Temperature Change (°C)[f]	Global Sea-Level Rise (cm)[g]
1990	5.3	21	16.1	—	354	0	0
2000	6.1–6.2	25–28	12.3–14.2	40	367	0.2	2
2050	8.4–11.3	59–187	2.4–8.2	~60	463–623	0.8–2.6	5–32
2100	7.0–15.1	197–550	1.4–6.3	>70	478–1099	1.4–5.8	9–88

[a] Values for 2000 show range across the six illustrative SRES emissions scenarios; values for 2050 and 2100 show range across all 40 SRES scenarios.
[b] See footnote a; gross domestic product (trillion 1990 US$ yr^{-1}).
[c] See footnote a; ratio of developed countries and economies-in-transition (Annex I) to developing countries (non-Annex I).
[d] Model estimates for industrialized continents of northern hemisphere assuming emissions for 2000, 2060, and 2100 from the A1F and A2 illustrative SRES emissions scenarios at high end of the SRES range (Chapter 4, TAR WG I).
[e] Observed 1999 value (Chapter 3, WG I TAR); values for 1990, 2050, and 2100 are from simple model runs across the range of 35 fully quantified SRES emissions scenarios and accounting for uncertainties in carbon cycle feedbacks related to climate sensitivity (data from S.C.B. Raper, Chapter 9, WG I TAR). Note that the ranges for 2050 and 2100 differ from those presented by TAR WGI (Appendix II), which were ranges across the six illustrative SRES emissions scenarios from simulations using two different carbon cycle models.
[f] Change in global mean annual temperature relative to 1990 averaged across simple climate model runs emulating results of seven AOGCMs with an average climate sensitivity of 2.8°C for the range of 35 fully quantified SRES emissions scenarios (Chapter 9, WG I TAR).
[g] Based on global mean temperature changes but also accounting for uncertainties in model parameters for land ice, permafrost, and sediment deposition (Chapter 11, WG I TAR).

continents of the northern hemisphere are projected to rise from about 40 ppb in 2000 to more than 70 ppb in 2100 under the highest illustrative SRES emissions scenarios; by comparison, the clean-air standard is below 80 ppb. Peak levels of ozone in local smog events could be many times higher. Estimates of CO_2 concentration range from 478 ppm to1099 ppm by 2100, given the range of SRES emissions and uncertainties about the carbon cycle (Table TS-1). This range of implied radiative forcing gives rise to an estimated global warming from 1990 to 2100 of 1.4–5.8°C, assuming a range of climate sensitivities. This range is higher than the 0.7–3.5°C of the SAR because of higher levels of radiative forcing in the SRES scenarios than in the IS92a-f scenarios—primarily as a result of lower sulfate aerosol emissions, especially after 2050. The equivalent range of estimates of global sea-level rise (for this range of global temperature change in combination with a range of ice melt sensitivities) to 2100 is 9–88 cm (compared to 15–95 cm in the SAR). [3.2.4.1, 3.4.4, 3.8.1, 3.8.2]

In terms of *mean changes in regional climate*, results from GCMs that have been run assuming the new SRES emissions scenarios display many similarities with previous runs. The WGI contribution to the TAR concludes that rates of warming are expected to be greater than the global average over most land areas and will be most pronounced at high latitudes in winter. As warming proceeds, northern hemisphere snow cover and sea-ice extent will be reduced. Models indicate warming below the global average in the north Atlantic and circumpolar southern ocean regions, as well as in southern and southeast Asia and southern South America in June-August. Globally, there will be increases in average water vapor and precipitation. Regionally, December-February precipitation is expected to increase over the northern extratropics, Antarctica, and tropical Africa. Models also agree on a decrease in precipitation over Central America and little change in southeast Asia. Precipitation in June-August is estimated to increase in high northern latitudes, Antarctica, and south Asia; it is expected to change little in southeast Asia and to decrease in central America, Australia, southern Africa, and the Mediterranean region.

Changes in the frequency and intensity of extreme climate events also can be expected. Based on the conclusions of the WGI report and the likelihood scale employed therein, under GHG forcing to 2100, it is very likely that daytime maximum and minimum temperatures will increase, accompanied by an increased frequency of hot days (see Table TS-2). It also is very likely that heat waves will become more frequent, and the number of cold waves and frost days (in applicable regions) will decline. Increases in high-intensity precipitation events are likely at many locations; Asian summer monsoon precipitation variability also is likely to increase. The frequency of summer drought will increase in many interior continental locations, and droughts—as well as floods—associated with El Niño events are likely to intensify. Peak wind intensity and mean and peak precipitation intensities of tropical cyclones are likely to increase. The direction of changes in the average intensity of mid-latitude storms cannot be determined with current climate models. [Table 3-10]

3.6. How can We Improve Scenarios and their Use?

Some features of scenario development and application that are now well established and tested include continued development of global and regional databases for defining baseline conditions, widespread use of incremental scenarios to explore system sensitivity prior to application of model-based scenarios, improved availability and wider application of estimates of long-term mean global changes on the basis of projections produced by specialized international organizations or the use of simple models, and a growing volume of accessible information that enables construction of regional scenarios for some aspects of global change. [3.9.1]

There also are numerous shortcomings of current scenario development, many of which are being actively investigated. These investigations include efforts to properly represent socioeconomic, land-use, and environmental changes in scenarios; to obtain scenarios at higher resolution (in time and space); and to incorporate changes in variability as well as mean conditions in scenarios. Increasing attention is required on construction of scenarios that address policy-related issues such as stabilization of GHG concentrations or adaptation, as well as improving the representation of uncertainties in projections, possibly within a risk assessment framework. [3.9.2]

4. Natural and Human Systems

Natural and human systems are expected to be exposed to climatic variations such as changes in the average, range, and variability of temperature and precipitation, as well as the frequency and severity of weather events. Systems also would be exposed to indirect effects from climate change such as sea-level rise, soil moisture changes, changes in land and water condition, changes in the frequency of fire and pest infestation, and changes in the distribution of infectious disease vectors and hosts. The sensitivity of a system to these exposures depends on system characteristics and includes the potential for adverse and beneficial effects. The potential for a system to sustain adverse impacts is moderated by adaptive capacity. The capacity to adapt human management of systems is determined by access to resources, information and technology, the skill and knowledge to use them, and the stability and effectiveness of cultural, economic, social, and governance institutions that facilitate or constrain how human systems respond.

4.1. Water Resources

There are apparent trends in streamflow volumes—increases and decreases—in many regions. However, confidence that these trends are a result of climate change is low because of factors such as the variability of hydrological behavior over time, the brevity of instrumental records, and the response of river flows to stimuli other than climate change. *In contrast, there is high confidence that observations of widespread accelerated glacier retreat and shifts in the timing of streamflow*

Table TS-2*: Examples of impacts resulting from projected changes in extreme climate events.*

Projected Changes during the 21st Century in Extreme Climate Phenomena and their Likelihood[a]	**Representative Examples of Projected Impacts**[b] *(all high confidence of occurrence in some areas*[c]*)*
Simple Extremes	
Higher maximum temperatures; more hot days and heat waves[d] over nearly all land areas (*Very Likely*[a])	• Increased incidence of death and serious illness in older age groups and urban poor • Increased heat stress in livestock and wildlife • Shift in tourist destinations • Increased risk of damage to a number of crops • Increased electric cooling demand and reduced energy supply reliability
Higher (increasing) minimum temperatures; fewer cold days, frost days, and cold waves[d] over nearly all land areas (*Very Likely*[a])	• Decreased cold-related human morbidity and mortality • Decreased risk of damage to a number of crops, and increased risk to others • Extended range and activity of some pest and disease vectors • Reduced heating energy demand
More intense precipitation events (*Very Likely*[a] over many areas)	• Increased flood, landslide, avalanche, and mudslide damage • Increased soil erosion • Increased flood runoff could increase recharge of some floodplain aquifers • Increased pressure on government and private flood insurance systems and disaster relief
Complex Extremes	
Increased summer drying over most mid-latitude continental interiors and associated risk of drought (*Likely*[a])	• Decreased crop yields • Increased damage to building foundations caused by ground shrinkage • Decreased water resource quantity and quality • Increased risk of forest fire
Increase in tropical cyclone peak wind intensities, mean and peak precipitation intensities (*Likely*[a] over some areas)[e]	• Increased risks to human life, risk of infectious disease epidemics, and many other risks • Increased coastal erosion and damage to coastal buildings and infrastructure • Increased damage to coastal ecosystems such as coral reefs and mangroves
Intensified droughts and floods associated with El Niño events in many different regions (*Likely*[a]) (see also under droughts and intense precipitation events)	• Decreased agricultural and rangeland productivity in drought- and flood-prone regions • Decreased hydro-power potential in drought-prone regions
Increased Asian summer monsoon precipitation variability (*Likely*[a])	• Increased flood and drought magnitude and damages in temperate and tropical Asia
Increased intensity of mid-latitude storms (little agreement between current models)[d]	• Increased risks to human life and health • Increased property and infrastructure losses • Increased damage to coastal ecosystems

[a] Likelihood refers to judgmental estimates of confidence used by TAR WGI: *very likely* (90–99% chance); *likely* (66–90% chance). Unless otherwise stated, information on climate phenomena is taken from the Summary for Policymakers, TAR WGI.
[b] These impacts can be lessened by appropriate response measures.
[c] Based on information from chapters in this report; high confidence refers to probabilities between 67 and 95% as described in Footnote 6 of TAR WGII, Summary for Policymakers.
[d] Information from TAR WGI, Technical Summary, Section F.5.
[e] Changes in regional distribution of tropical cyclones are possible but have not been established.

from spring toward winter in many areas are associated with observed increases in temperature. High confidence in these findings exists because these changes are driven by rising temperature and are unaffected by factors that influence streamflow volumes. Glacier retreat will continue, and many small glaciers may disappear (high confidence). The rate of retreat will depend on the rate of temperature rise. [4.3.6.1, 4.3.11]

The effect of climate change on streamflow and groundwater recharge varies regionally and among scenarios, largely following projected changes in precipitation. In some parts of the world, the direction of change is consistent between scenarios, although the magnitude is not. In other parts of the world, the direction of change is uncertain. Possible streamflow changes under two climate change scenarios are shown in Figure TS-3.

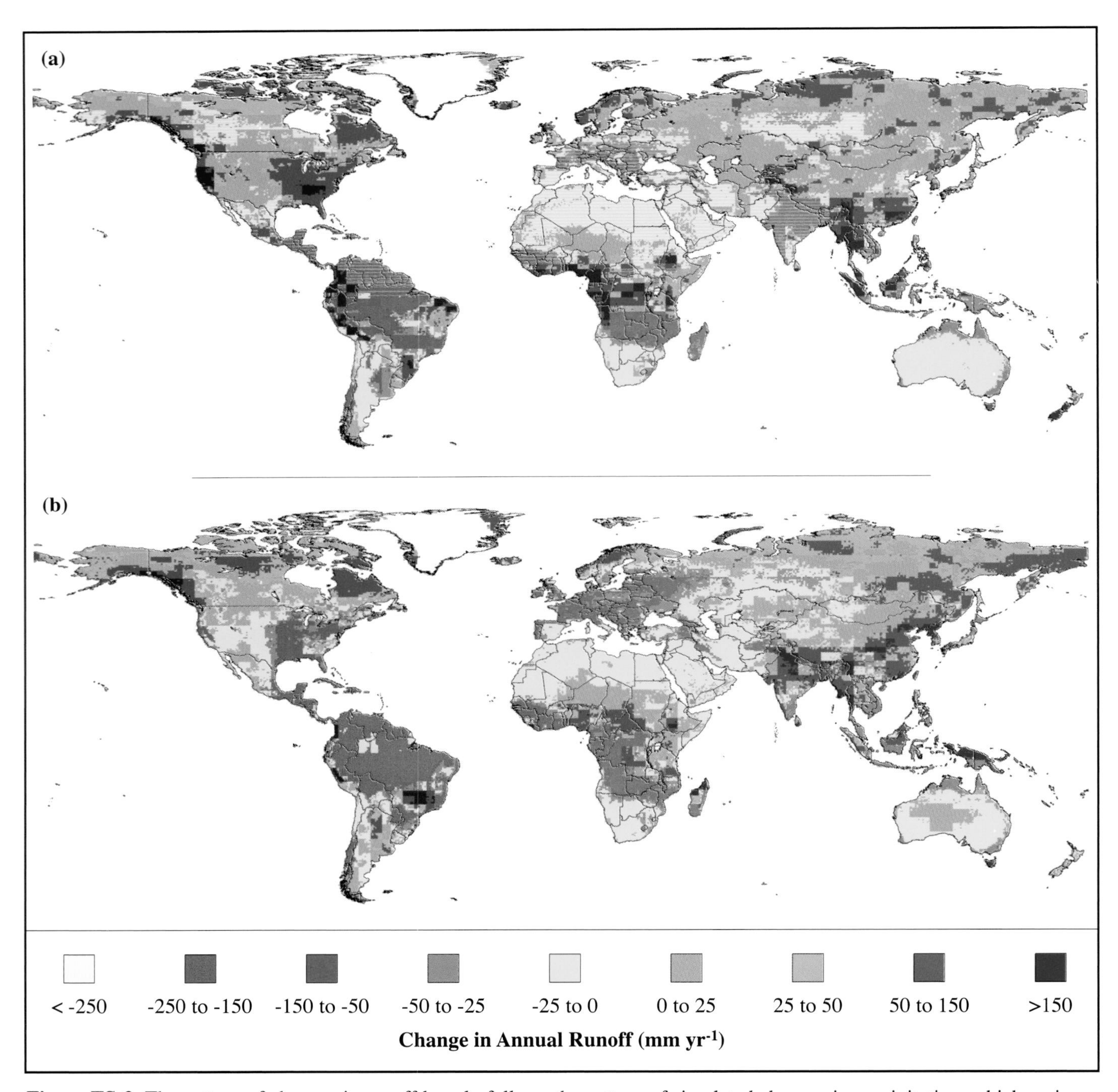

Figure TS-3: The pattern of changes in runoff largely follows the pattern of simulated changes in precipitation, which varies between climate models. The modeled increases in runoff shown in both maps [(a) HadCM2 ensemble mean and (b) HadCM3; see Section 4.3.6.2 of Chapter 4 for discussion of models and scenarios used] for high latitudes and southeast Asia, and decreases in central Asia, the area around the Mediterranean, southern Africa, and Australia are broadly consistent—in terms of direction of change—across most climate models. In other parts of the world, changes in precipitation and runoff vary between climate change scenarios.

Confidence in the projected direction and magnitude of change in streamflow and groundwater recharge is largely dependent on confidence in the projected changes in precipitation. The mapped increase in streamflow in high latitudes and southeast Asia and the decrease in streamflow in central Asia, the area around the Mediterranean, and southern Africa are broadly consistent across climate models. Changes in other areas vary between climate models. [4.3.5, 4.3.6.2]

Peak streamflow will move from spring to winter in many areas where snowfall currently is an important component of the water balance (high confidence). Higher temperatures mean that a greater proportion of winter precipitation falls as rain rather than snow and therefore is not stored on the land surface until it melts in spring. In particularly cold areas, an increase in temperature would still mean that winter precipitation falls as snow, so there would be little change in streamflow timing in these regions. The greatest changes therefore are likely to be in "marginal" zones—including central and eastern Europe and the southern Rocky Mountain chain—where a small temperature rise reduces snowfall substantially. [4.3.6.2]

Water quality generally would be degraded by higher water temperatures (high confidence). The effect of temperature on water quality would be modified by changes in flow volume, which may either exacerbate or lessen the effect of temperature, depending on the direction of change in flow volume. Other things being equal, increasing water temperature alters the rate of operation of biogeochemical processes (some degrading, some cleaning) and, most important, lowers the dissolved oxygen concentration of water. In rivers this effect may be offset to an extent by increased streamflow—which would dilute chemical concentrations further—or enhanced by lower streamflow, which would increase concentrations. In lakes, changes in mixing may offset or exaggerate the effects of increased temperature. [4.3.10]

Flood magnitude and frequency are likely to increase in most regions, and low flows are likely to decrease in many regions. The general direction of change in extreme flows and flow variability is broadly consistent among climate change scenarios, although confidence in the potential magnitude of change in any catchment is low. The general increase in flood magnitude and frequency is a consequence of a projected general increase in the frequency of heavy precipitation events, although the effect of a given change in precipitation depends on catchment characteristics. Changes in low flows are a function of changes in precipitation and evaporation. Evaporation generally is projected to increase, which may lead to lower low flows even where precipitation increases or shows little change. [4.3.8, 4.3.9]

Approximately 1.7 billion people, one-third of the world's population, presently live in countries that are water-stressed (i.e., using more than 20% of their renewable water supply—a commonly used indicator of water stress). This number is projected to increase to about 5 billion by 2025, depending on the rate of population growth. Projected climate change could further decrease streamflow and groundwater recharge in many of these water-stressed countries—for example, in central Asia, southern Africa, and countries around the Mediterranean Sea—but may increase it in some others.

Demand for water generally is increasing, as a result of population growth and economic development, but is falling in some countries. Climate change may decrease water availability in some water-stressed regions and increase it in others. Climate change is unlikely to have a large effect on municipal and industrial demands but may substantially affect irrigation withdrawals. In the municipal and industrial sectors, it is likely that nonclimatic drivers will continue to have very substantial effects on demand for water. Irrigation withdrawals, however, are more climatically determined, but whether they increase or decrease in a given area depends on the change in precipitation: Higher temperatures, hence crop evaporative demand, would mean that the general tendency would be toward an increase in irrigation demands. [4.4.2, 4.4.3, 4.5.2]

The impact of climate change on water resources depends not only on changes in the volume, timing, and quality of streamflow and recharge but also on system characteristics, changing pressures on the system, how management of the system evolves, and what adaptations to climate change are implemented. Nonclimatic changes may have a greater impact on water resources than climate change. Water resources systems are evolving continually to meet changing management challenges. Many of the increased pressures will increase vulnerability to climate change, but many management changes will reduce vulnerability. Unmanaged systems are likely to be most vulnerable to climate change. By definition, these systems have no management structures in place to buffer the effects of hydrological variability. [4.5.2]

Climate change challenges existing water resources management practices by adding uncertainty. Integrated water resources management will enhance the potential for adaptation to change. The historic basis for designing and operating infrastructure no longer holds with climate change because it cannot be assumed that the future hydrological regime will be the same as that of the past. The key challenge, therefore, is incorporating uncertainty into water resources planning and management. Integrated water resources management is an increasingly used means of reconciling different and changing water uses and demands, and it appears to offer greater flexibility than conventional water resources management. Improved ability to forecast streamflow weeks or months ahead also would significantly enhance water management and its ability to cope with a changing hydrological variability. [4.6]

Adaptive capacity (specifically, the ability to implement integrated water resources management), however, is very unevenly distributed across the world. In practice, it may be very difficult to change water management practices in a country where, for example, management institutions and market-like processes are not well developed. The challenge, therefore, is to develop ways to introduce integrated water management practices into specific institutional settings—which is necessary even in the absence of climate change to improve the effectiveness of water management. [4.6.4]

4.2. *Agriculture and Food Security*

The response of crop yields to climate change varies widely, depending on the species, cultivar, soil conditions, treatment of CO_2 direct effects, and other locational factors. It is established with medium confidence that a few degrees of projected warming will lead to general increases in temperate crop yields, with some regional variation (Table 5-4). At larger amounts of projected warming, most temperate crop yield responses become generally negative. Autonomous agronomic adaptation ameliorates temperate crop yield loss and improves gain in most cases (Figure TS-4). In the tropics, where some crops are near their maximum temperature tolerance and where dryland agriculture predominates, yields would decrease generally with even minimal changes in temperature; where there is a large decrease in rainfall, crop yields would be even more adversely affected (medium confidence). With autonomous agronomic adaptation, it is established with medium confidence that crop yields in the tropics tend to be less adversely affected by climate change than without adaptation, but they still tend to remain below baseline levels. Extreme events also will affect crop yields. Higher minimum temperatures will be beneficial to some crops, especially in temperate regions, and detrimental to other crops, especially in low latitudes (high confidence). Higher maximum temperatures will be generally detrimental to numerous crops (high confidence). [5.3.3]

Important advances in research since the SAR on the direct effects of CO_2 on crops suggest that beneficial effects may be greater under certain stressful conditions, including warmer temperatures and drought. Although these effects are well established for a few crops under experimental conditions, knowledge of them is incomplete for suboptimal conditions of actual farms. Research on agricultural adaptation to climate change also has made important advances. Inexpensive, farm-level (autonomous) agronomic adaptations such as altering of planting dates and cultivar selections have been simulated in crop models extensively. More expensive, directed adaptations—such as changing land-use allocations and developing and using irrigation infrastructure—have been examined in a small but growing number of linked crop-economic models, integrated assessment models, and econometric models.

Degradation of soil and water resources is one of the major future challenges for global agriculture. It is established with high confidence that those processes are likely to be intensified by adverse changes in temperature and precipitation. Land use and management have been shown to have a greater impact on soil conditions than the indirect effect of climate change; thus, adaptation has the potential to significantly mitigate these impacts. A critical research need is to assess whether resource degradation will significantly increase the risks faced

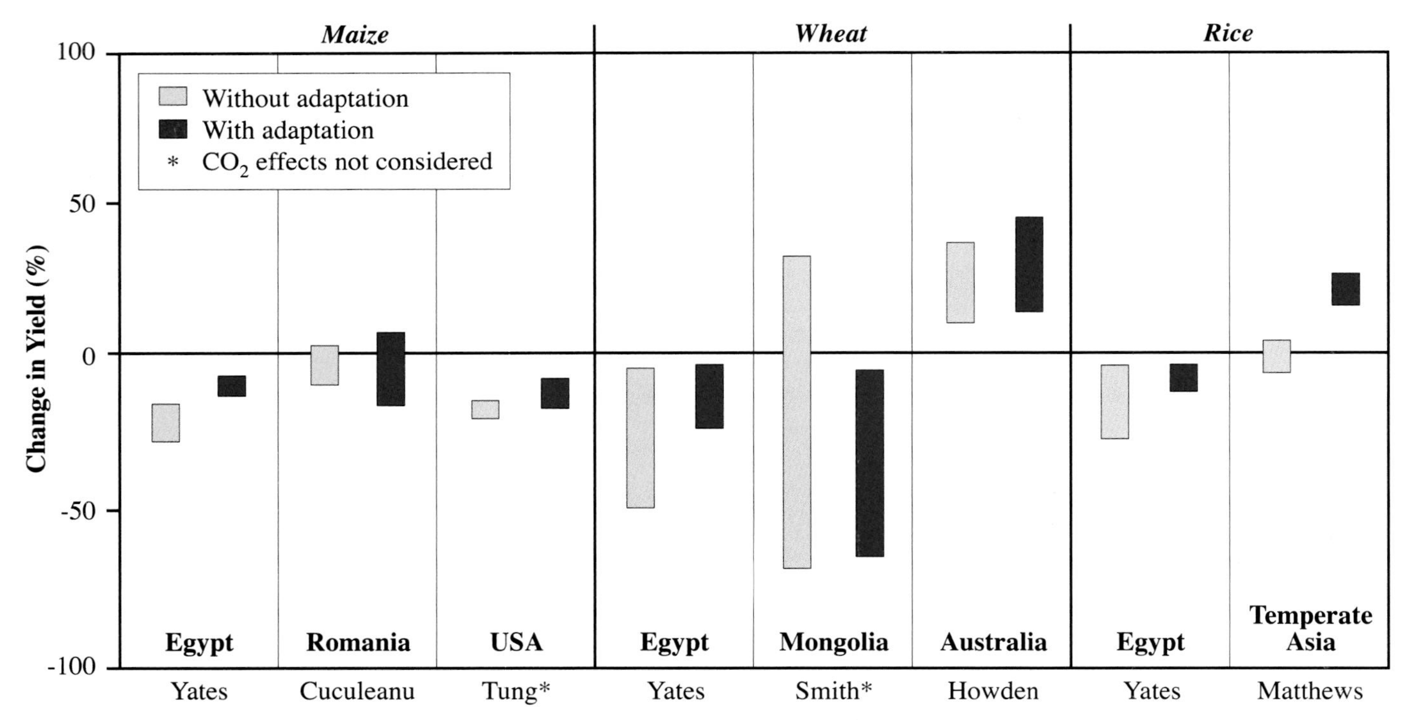

Figure TS-4: Ranges of percentage changes in crop yields (expressed in vertical extent of vertical bars only) spanning selected climate change scenarios—with and without agronomic adaptation—from paired studies listed in Table 5-4. Each pair of ranges is differentiated by geographic location and crop. Pairs of vertical bars represent the range of percentage changes with and without adaptation. Endpoints of each range represent collective high and low percentage change values derived from all climate scenarios used in the study. The horizontal extent of the bars is not meaningful. On the x-axis, the last name of the lead author is listed as it appears in Table 5-4; full source information is provided in the Chapter 5 reference list.

by vulnerable agricultural and rural populations [5.3.2, 5.3.4, 5.3.6].

In the absence of climate change, most global and regional studies project declining real prices for agricultural commodities. Confidence in these projections declines farther into the future. *The impacts of climate change on agriculture are estimated to result in small percentage changes in global income, with positive changes in more developed regions and smaller or negative changes in developing regions (low to medium confidence).* The effectiveness of adaptation (agronomic and economic) in ameliorating the impacts of climate change will vary regionally and depend a great deal on regional resource endowments, including stable and effective institutions. [5.3.1, 5.3.5]

Most studies indicate that mean annual temperature increases of 2.5°C or greater would prompt food prices to increase (low confidence) as a result of slowing in the expansion of global food capacity relative to growth in global food demand. At lesser amounts of warming than 2.5°C, global impact assessment models cannot distinguish the climate change signal from other sources of change. Some recent aggregated studies have estimated economic impacts on vulnerable populations such as smallholder producers and poor urban consumers. These studies indicate that climate change will lower the incomes of vulnerable populations and increase the absolute number of people at risk of hunger (low confidence). [5.3.5, 5.3.6]

Without autonomous adaptation, increases in extreme events are likely to increase heat stress-related livestock deaths, although winter warming may reduce neonatal deaths at temperate latitudes (established but incomplete). Strategies to adapt livestock to physiological stresses of warming are considered effective; however, adaptation research is hindered by the lack of experimentation and simulation. [5.3.3]

Confidence in specific numerical estimates of climate change impacts on production, income, and prices obtained from large, aggregated, integrated assessment models is considered to be low because there are several remaining uncertainties. The models are highly sensitive to some parameters that have been subjected to sensitivity analysis, yet sensitivity to a large number of other parameters has not been reported. Other uncertainties include the magnitude and persistence of effects of rising atmospheric CO_2 on crop yield under realistic farming conditions; potential changes in crop and animal pest losses; spatial variability in crop responses to climate change; and the effects of changes in climate variability and extreme events on crops and livestock. [Box 5-3]

4.3. Terrestrial and Freshwater Ecosystems

Ecosystems are subject to many pressures, such as land-use changes, deposition of nutrients and pollutants, harvesting, grazing by livestock, introduction of exotic species, and natural climate variability. Climate change constitutes an additional pressure that could change or endanger these systems. The impact of climate change on these systems will be influenced by land and water management adaptation and interactions with other pressures. Adaptive capacity is greater for more intensively managed lands and waters and in production of marketed goods (e.g., timber production in plantations) than in less intensively managed lands and nonmarket values of those lands and waters. [5.1, 5.2]

Populations of many species already are threatened and are expected to be placed at greater risk by the synergy between the stresses of changing climate, rendering portions of current habitat unsuitable, and land-use change that fragments habitats. Without adaptation, some species that currently are classified as "critically endangered" will become extinct, and the majority of those labeled "endangered or vulnerable" will become much rarer in the 21st century (high confidence). This may have the greatest impact on the lowest income human societies, which rely on wildlife for subsistence living. In addition, there is high confidence that loss or reduction of species would impact the services provided by wildlife through roles within an ecosystem (e.g., pollination, natural pest control), recreation (e.g., sport hunting, wildlife viewing), and cultural and religious practices of indigenous people. Possible adaptation methods to reduce risks to species could include establishment of refuges, parks, and reserves with corridors to allow migration of species, as well as use of captive breeding and translocation. However, these options may have limitations of cost. [5.4]

There are now substantial observational and experimental studies demonstrating the link between change in regional climate and biological or physical processes in ecosystems. These include a lengthening of vegetative growing season by 1.2 to 3.6 days per decade in the high northern latitudes (one factor leading to community composition changes); warming of lakes and rivers as a result of shortening duration of ice cover; upward range shifts in alpine herbs; and increased mortality and range contraction of wildlife as a result of heat stress. Others include changes in population sizes, body sizes, and migration times (see TS 2.1 and 7.1, Figure TS-11, and Table TS-16 for additional information). [5.2.1]

Vegetation distribution models since the SAR suggest that mass ecosystem or biome movement is most unlikely to occur because of different climatic tolerance of the species involved, different migration abilities, and the effects of invading species. Species composition and dominance will change, resulting in ecosystem types that may be quite different from those we see today. These changes will lag the changes in climate by years to decades to centuries (high confidence). The effects of changes in disturbances such as fire, blowdown, or pest attacks on vegetation have not been included in these studies. [5.2]

Recent modeling studies continue to show potential for significant disruption of ecosystems under climate change (high confidence). Further development of simple correlative models that were available at the time of the SAR point to areas where ecosystem disruption and the potential for ecosystem migration are high. Observational data and newer dynamic vegetation models linked to transient climate models are refining the projections. However, the precise outcomes depend on processes that are too subtle to be fully captured by current models. [5.2]

Increasing CO_2 concentration would increase net primary productivity (plant growth, litterfall, and mortality) in most systems, whereas increasing temperature may have positive or negative effects (high confidence). Experiments on tree species grown under elevated CO_2 over several years show continued and consistent stimulation of photosynthesis and little evidence of long-term loss of sensitivity to CO_2. However, changes in net ecosystem productivity (which includes plant growth, litterfall, mortality, litter decomposition, and soil carbon dynamics) and net biome productivity (which includes those effects plus the effects of fire or other disturbances) are less likely to be positive and may be generally negative. Research reported since the SAR confirms the view that the largest and earliest impacts induced by climate change are likely to occur in boreal forests, through changes in weather-related disturbance regimes and nutrient cycling. [5.6.1.1, 5.6.3.1]

Terrestrial ecosystems appear to be storing increasing amounts of carbon. At the time of the SAR, this was attributed largely to increasing plant productivity because of the interaction among elevated CO_2 concentration, increasing temperatures, and soil moisture changes. Recent results confirm that productivity gains are occurring but suggest that they are smaller under field conditions than plant-pot experiments indicate (*medium confidence*). Hence, the terrestrial uptake may be caused more by change in uses and management of land than by the direct effects of elevated CO_2 and climate. The degree to which terrestrial ecosystems continue to be net sinks for carbon is uncertain because of the complex interactions between the aforementioned factors (e.g., arctic terrestrial ecosystems and wetlands may act as sources and sinks) (*medium confidence*).

In arid or semi-arid areas (e.g., rangelands, dry forests/woodlands) where climate change is likely to decrease available soil moisture, productivity is expected to decrease. Increased CO_2 concentrations may counteract some of these losses. However, many of these areas are affected by El Niño/La Niña, other climatic extremes, and disturbances such as fire. Changes in the frequencies of these events and disturbances could lead to loss of productivity thus potential land degradation, potential loss of stored carbon, or decrease in the rate of carbon uptake (medium confidence). [5.5]

Some wetlands will be replaced by forests or heathlands, and those overlying permafrost are likely to be disrupted as a result of thawing of permafrost (high confidence). The initial net effect of warming on carbon stores in high-latitude ecosystems is likely to be negative because decomposition initially may respond more rapidly than production. In these systems, changes in albedo and energy absorption during winter are likely to act as a positive feedback to regional warming as a result of earlier melting of snow and, over decades to centuries, poleward movement of the treeline. [5.8, 5.9]

Most wetland processes are dependent on catchment-level hydrology; thus, adaptations for projected climate change may be practically impossible. Arctic and subarctic ombrotrophic bog communities on permafrost, as well as more southern depressional wetlands with small catchment areas, are likely to be most vulnerable to climate change. The increasing speed of peatland conversion and drainage in southeast Asia is likely to place these areas at a greatly increased risk of fires and affect the viability of tropical wetlands. [5.8]

Opportunities for adapting to expected changes in high-latitude and alpine ecosystems are limited because these systems will respond most strongly to globally induced changes in climate. Careful management of wildlife resources could minimize climatic impacts on indigenous peoples. Many high-latitude regions depend strongly on one or a few resources, such as timber, oil, reindeer, or wages from fighting fires. Economic diversification would reduce the impacts of large changes in the availability or economic value of particular goods and services. High levels of endemism in many alpine floras and their inability to migrate upward means that these species are very vulnerable. [5.9]

Contrary to the SAR, global timber market studies that include adaptations through land and product management suggest that climate change would increase global timber supply (medium confidence). At the regional and global scales, the extent and nature of adaptation will depend primarily on wood and non-wood product prices, the relative value of substitutes, the cost of management, and technology. On specific sites, changes in forest growth and productivity will constrain—and could limit—choices regarding adaptation strategies (high confidence). In markets, prices will mediate adaptation through land and product management. Adaptation in managed forests will include salvaging dead and dying timber, replanting new species that are better suited to the new climate, planting genetically modified species, and intensifying or decreasing management. Consumers will benefit from lower timber prices; producers may gain or lose, depending on regional changes in timber productivity and potential dieback effects. [5.6]

Climate change will lead to poleward movement of the southern and northern boundaries of fish distributions, loss of habitat for cold- and coolwater fish, and gain in habitat for warmwater fish (high confidence). As a class of ecosystems, inland waters are vulnerable to climatic change and other pressures owing to their small size and position downstream from many human activities (high confidence). The most vulnerable elements include reduction and loss of lake and river ice (very high confidence), loss of habitat for coldwater fish (very high confidence), increases in extinctions and invasions of exotics (high confidence), and potential exacerbation of existing pollution problems such as eutrophication, toxics, acid rain, and UV-B radiation (medium confidence). [5.7]

4.4. Coastal Zones and Marine Ecosystems

Global climate change will result in increases in sea-surface temperature (SST) and sea level; decreases in sea-ice cover;

and changes in salinity, wave climate, and ocean circulation. Some of these changes already are taking place. Changes in oceans are expected to have important feedback effects on global climate and on the climate of the immediate coastal area (see TAR WGI). They also would have profound impacts on the biological production of oceans, including fish production. For instance, changes in global water circulation and vertical mixing will affect the distribution of biogenic elements and the efficiency of CO_2 uptake by the ocean; changes in upwelling rates would have major impacts on coastal fish production and coastal climates. [6.3]

If warm events associated with El Niños increase in frequency, plankton biomass and fish larvae abundance would decline and adversely impact fish, marine mammals, seabirds, and ocean biodiversity (high confidence). In addition to El Niño-Southern Oscillation (ENSO) variability, the persistence of multi-year climate-ocean regimes and switches from one regime to another have been recognized since the SAR. Changes in recruitment patterns of fish populations have been linked to such switches. Fluctuations in fish abundance are increasingly regarded as biological responses to medium-term climate fluctuations in addition to overfishing and other anthropogenic factors. Similarly, survival of marine mammals and seabirds also is affected by interannual and longer term variability in several oceanographic and atmospheric properties and processes, especially in high latitudes. [6.3.4]

Growing recognition of the role of the climate-ocean system in the management of fish stocks is leading to new adaptive strategies that are based on the determination of acceptable removable percentages of fish and stock resilience. Another consequence of the recognition of climate-related changes in the distribution of marine fish populations suggests that the sustainability of many nations' fisheries will depend on adaptations that increase flexibility in bilateral and multilateral fishing agreements, coupled with international stock assessments and management plans. Creating sustainable fisheries also depends on understanding synergies between climate-related impacts on fisheries and factors such harvest pressure and habitat conditions. [6.3.4, 6.6.4]

Adaptation by expansion of marine aquaculture may partly compensate for potential reductions in ocean fish catch. Marine aquaculture production has more than doubled since 1990, and in 1997 represented approximately 30% of total commercial fish and shellfish production for human consumption. However, future aquaculture productivity may be limited by ocean stocks of herring, anchovies, and other species that are used to provide fishmeal and fish oils to feed cultured species, which may be negatively impacted by climate change. Decreases in dissolved oxygen levels associated with increased seawater temperatures and enrichment of organic matter creates conditions for the spread of diseases in wild and aquaculture fisheries, as well as outbreaks of algal blooms in coastal areas. Pollution and habitat destruction that can accompany aquaculture also may place limits on its expansion and on the survival success of wild stocks. [6.3.5]

Many coastal areas already are experiencing increased levels of sea flooding, accelerated coastal erosion, and seawater intrusion into freshwater sources; these processes will be exacerbated by climate change and sea-level rise. Sea-level rise in particular has contributed to erosion of sandy and gravel beaches and barriers; loss of coastal dunes and wetlands; and drainage problems in many low-lying, mid-latitude coastal areas. Highly diverse and productive coastal ecosystems, coastal settlements, and island states will continue to be exposed to pressures whose impacts are expected to be largely negative and potentially disastrous in some instances. [6.4]

Low-latitude tropical and subtropical coastlines, particularly in areas where there is significant human population pressure, are highly susceptible to climate change impacts. These impacts will exacerbate many present-day problems. For instance, human activities have increased land subsidence in many deltaic regions by increasing subsurface water withdrawals, draining wetland soils, and reducing or cutting off riverine sediment loads. Problems of inundation, salinization of potable groundwater, and coastal erosion will all be accelerated with global sea-level rise superimposed on local submergence. Especially at risk are large delta regions of Asia and small islands whose vulnerability was recognized more than a decade ago and continues to increase. [6.4.3, 6.5.3]

High-latitude (polar) coastlines also are susceptible to climate warming impacts, although these impacts have been less studied. Except on rock-dominated or rapidly emerging coasts, a combination of accelerated sea-level rise, more energetic wave climate with reduced sea-ice cover, and increased ground temperatures that promote thaw of permafrost and ground ice (with consequent volume loss in coastal landforms) will have severe impacts on settlements and infrastructure and will result in rapid coastal retreat. [6.4.6]

Coastal ecosystems such as coral reefs and atolls, salt marshes and mangrove forests, and submergered aquatic vegetation will be impacted by sea-level rise, warming SSTs, and any changes in storm frequency and intensity. Impacts of sea-level rise on mangroves and salt marshes will depend on the rate of rise relative to vertical accretion and space for horizontal migration, which can be limited by human development in coastal areas. Healthy coral reefs are likely to be able to keep up with sea-level rise, but this is less certain for reefs degraded by coral bleaching, UV-B radiation, pollution, and other stresses. Episodes of coral bleaching over the past 20 years have been associated with several causes, including increased ocean temperatures. Future sea-surface warming would increase stress on coral reefs and result in increased frequency of marine diseases (high confidence). Changes in ocean chemistry resulting from higher CO_2 levels may have a negative impact on coral reef development and health, which would have a detrimental effect on coastal fisheries and on social and economic uses of reef resources. [6.4.4, 6.4.5]

Few studies have examined potential changes in prevailing ocean wave heights and directions and storm waves and surges as a consequence of climate change. Such changes can be expected to have serious impacts on natural and human-modified coasts because they will be superimposed on a higher sea level than at present.

Vulnerabilities have been documented for a variety of coastal settings, initially by using a common methodology developed in the early 1990s. These and subsequent studies have confirmed the spatial and temporal variability of coastal vulnerability at national and regional levels. Within the common methodology, three coastal adaptation strategies have been identified: protect, accommodate, and retreat. Since the SAR, adaptation strategies for coastal zones have shifted in emphasis away from hard protection structures (e.g., seawalls, groins) toward soft protection measures (e.g., beach nourishment), managed retreat, and enhanced resilience of biophysical and socioeconomic systems, including the use of flood insurance to spread financial risk. [6.6.1, 6.6.2]

Integrated assessments of coastal zones and marine ecosystems and better understanding of their interaction with human development and multi-year climate variability could lead to improvements in sustainable development and management. Adaptation options for coastal and marine management are most effective when they are incorporated with policies in other areas, such as disaster mitigation plans and land-use plans.

4.5. *Human Settlements, Energy, and Industry*

Human settlements are integrators of many of the climate impacts initially felt in other sectors and differ from each other in geographic location, size, economic circumstances, and political and institutional capacity. As a consequence, it is difficult to make blanket statements concerning the importance of climate or climate change that will not have numerous exceptions. However, classifying human settlements by considering pathways by which climate may affect them, size or other obvious physical considerations, and adaptive capacities (wealth, education of the populace, technological and institutional capacity) helps to explain some of the differences in expected impacts. [7.2]

Human settlements are affected by climate in one of three major ways:

1) Economic sectors that support the settlement are affected because of changes in productive capacity (e.g., in agriculture or fisheries) or changes in market demand for goods and services produced there (including demand from people living nearby and from tourism). The importance of this impact depends in part on whether the settlement is rural—which generally means that it is dependent on one or two resource-based industries—or urban, in which case there usually (but not always) is a broader array of alternative resources. It also depends on the adaptive capacity of the settlement. [7.1]
2) Some aspects of physical infrastructure (including energy transmission and distribution systems), buildings, urban services (including transportation systems), and specific industries (such as agroindustry, tourism, and construction) may be directly affected. For example, buildings and infrastructure in deltaic areas may be affected by coastal and river flooding; urban energy demand may increase or decrease as a result of changed balances in space heating and space cooling; and coastal and mountain tourism may be affected by changed seasonal temperature and precipitation patterns and sea-level rise. Concentration of population and infrastructure in urban areas can mean higher numbers of persons and higher value of physical capital at risk, although there also are many economies of scale and proximity in ensuring well-managed infrastructure and service provision. When these factors are combined with other prevention measures, risks can be reduced considerably. However, some larger urban centers in Africa, Asia, Latin America, and the Caribbean, as well as smaller settlements (including villages and small urban centers), often have less wealth, political power, and institutional capacity to reduce risks in this way. [7.1]
3) Population may be directly affected through extreme weather, changes in health status, or migration. Extreme weather episodes may lead to changes in deaths, injuries, or illness. For example, health status may improve as a result of reduced cold stress or deteriorate as a result of increased heat stress and disease. Population movements caused by climate changes may affect the size and characteristics of settlement populations, which in turn changes the demand for urban services. The problems are somewhat different in the largest population centers (e.g., those of more than 1 million population) and mid-sized to small-sized regional centers. The former are more likely to be destinations for migrants from rural areas and smaller settlements and cross-border areas, but larger settlements generally have much greater command over national resource. Thus, smaller settlements actually may be more vulnerable. Informal settlements surrounding large and medium-size cities in the developing world remain a cause for concern because they exhibit several current health and environmental hazards that could be exacerbated by global warming and have limited command over resources. [7.1]

Table TS-3 classifies several types of climate-caused environmental changes discussed in the climate and human settlement literatures. The table features three general types of settlements, each based on the one of the three major mechanisms by which climate affects settlements. The impacts correspond to the mechanism of the effect. Thus, a given settlement may be affected positively by effects of climate change on its resource base (e.g., more agricultural production) and negatively by effects on its infrastructure (e.g., more frequent flooding of its

***Table TS-3**: Impacts of climate change on human settlements, by impact type and settlement type (impact mechanism).*[a,b]

Impact Type	Type of Settlement, Importance Rating, and Reference												Confidence[c]
	Resource-Dependent (Effects on Resources)				*Coastal-Riverine-Steeplands (Effects on Buildings and Infrastructure)*				*Urban 1+ M (Effects on Populations)*		*Urban <1 M (Effects on Populations)*		
	Urban, High Capacity	Urban, Low Capacity	Rural, High Capacity	Rural, Low Capacity	Urban, High Capacity	Urban, Low Capacity	Rural, High Capacity	Rural, Low Capacity	High Capacity	Low Capacity	High Capacity	Low Capacity	
Flooding, landslides	**L–M**	**M–H**	**L–M**	**M–H**	**L–M**	**M–H**	**M–H**	**M–H**	**M**	**M–H**	**M**	**M–H**	****
Tropical cyclone	**L–M**	**M–H**	**L–M**	**M–H**	**L–M**	**M–H**	**M**	**M–H**	**L–M**	**M**	**L**	**L–M**	***
Water quality	L–M	M	L–M	M–H	**L–M**	**M–H**	*L–M*	*M–H*	L–M	M–H	L–M	M–H	***
Sea-level rise	**L–M**	**M–H**	**L–M**	**M–H**	**M**	**M–H**	*M*	**M–H**	**L**	**L–M**	*L*	**L–M**	**** (** for resource-dependent)
Heat/cold waves	*L–M*	*M–H*	*L–M*	*M–H*	*L–M*	L–M	**L–M**	L	**L–M**	**M–H**	**L–M**	**M–H**	*** (**** for urban)
Water shortage	**L**	*L–M*	**M**	**M–H**	*L*	*L–M*	*L–M*	*M–H*	*L*	*M*	**L–M**	*M*	*** (** for urban)
Fires	*L–M*	*L–M*	**L–M**	*M–H*	*L–M*	*L–M*	*L–M*	*L–M*	**L–M**	**L–M**	*L–M*	*M*	* (*** for urban)
Hail, windstorm	**L–M**	**L–M**	**L–M**	**M–H**	*L–M*	*L–M*	*L–M*	*M*	**L–M**	**L–M**	**L–M**	**L–M**	**
Agriculture/ forestry/fisheries productivity	**L–M**	**L–M**	*L–M*	*M–H*	L	L	L	L	*L*	*L–M*	*L–M*	*M*	***
Air pollution	**L–M**	*L–M*	L	L	—	—	—	—	**L–M**	**M–H**	**L–M**	**M–H**	***
Permafrost melting	L	L	**L–M**	*L–M*	L	L	**L**	L	—	—	*L–M*	*L–M*	****
Heat islands	L	L	—	—	L	L	—	—	**M**	**L–M**	**L–M**	**L–M**	***

[a] Values in cells in the table were assigned by authors on the basis of direct evidence in the literature or inference from impacts shown in other cells. Typeface indicates source of rating: Boldface indicates direct evidence or study; italic indicates direct inference from similar impacts; regular typeface indicates logical conclusion from settlement type, but cannot be directly corroborated from a study or inferred from similar impacts.
[b] Impacts ratings: Low (L) = impacts are barely discernible or easily overcome; moderate (M) = impacts are clearly noticeable, although not disruptive, and may require significant expense or difficulty in adapting; high (H) = impacts are clearly disruptive and may not be overcome or adaptation is so costly that it is disruptive (impacts generally based on $2xCO_2$ scenarios or studies describing impact of current weather events, but have been placed in context of the IPCC transient scenarios for mid- to late 21st century). Note that "Urban 1+ M" and "Urban <1 M" refer to populations above and below 1 million, respectively.
[c] See Section 1.4 of Technical Summary for key to confidence-level rankings.

water works and overload of its electrical system). Different types of settlements may experience these effects in different relative intensities (e.g., noncoastal settlements do not directly experience impacts through sea-level rise); the impacts are ranked from overall highest to lowest importance. Most settlement effects literature is based on 2xCO_2 scenarios or studies describing the impact of current weather events (analogs) but has been placed in context of the IPCC transient scenarios. [7.1]

Climate change has the potential to create local and regional conditions that involve water deficits and surpluses, sometimes seasonally in the same geographic locations. *The most widespread serious potential impacts are flooding, landslides, mudslides, and avalanches driven by projected increases in rainfall intensity and sea-level rise.* A growing literature suggests that a very wide variety of settlements in nearly every climate zone may be affected (established but incomplete). Riverine and coastal settlements are believed to be particularly at risk, but urban flooding could be a problem anywhere storm drains, water supply, and waste management systems are not designed with enough capacity or sophistication (including conventional hardening and more advanced system design) to avoid being overwhelmed. The next most serious threats are tropical cyclones (hurricanes or typhoons), which may increase in peak intensity in a warmer world. Tropical cyclones combine the effects of heavy rainfall, high winds, and storm surge in coastal areas and can be disruptive far inland, but they are not as universal in location as floods and landslides. Tens of millions of people live in the settlements potentially flooded. For example, estimates of the mean annual number of people who would be flooded by coastal storm surges increase several-fold (by 75 million to 200 million people, depending on adaptive responses) for mid-range scenarios of a 40-cm sea-level rise by the 2080s relative to scenarios with no sea-level rise. Potential damages to infrastructure in coastal areas from sea-level rise have been estimated to be tens of billions of dollars for individual countries such as Egypt, Poland, and Vietnam. In the middle of Table TS-3 are effects such as heat or cold waves, which can be disruptive to the resource base (e.g., agriculture), human health, and demand for heating and cooling energy. Environmental impacts such as reduced air and water quality also are included. Windstorms, water shortages, and fire also are expected to be moderately important in many regions. At the lower end are effects such as permafrost melting and heat island effects—which, although important locally, may not apply to as wide a variety of settlements or hold less importance once adaptation is taken into account. [7.2, 7.3]

Global warming is expected to result in increases in energy demand for spacing cooling and in decreased energy use for space heating. Increases in heat waves add to cooling energy demand, and decreases in cold waves reduce heating energy demand. The projected net effect on annual energy consumption is scenario- and location-specific. Adapting human settlements, energy systems, and industry to climate change provides challenges for the design and operation of settlements (in some cases) during more severe weather and opportunities to take advantage (in other cases) of more benign weather. For instance, transmission systems of electric systems are known to be adversely affected by extreme events such as tropical cyclones, tornadoes, and ice storms. The existence of local capacity to limit environmental hazards or their health consequences in any settlement generally implies local capacity to adapt to climate change, unless adaptation implies particularly expensive infrastructure investment. Adaptation to warmer climate will require local tuning of settlements to a changing environment, not just warmer temperatures. Urban experts are unanimous that successful environmental adaptation cannot occur without locally based, technically and institutionally competent, and politically supported leadership that have good access to national-level resources. [7.2, 7.3, 7.4, 7.5]

Possible adaptation options involve planning of settlements and their infrastructure, placement of industrial facilities, and making similar long-lived decisions to reduce the adverse effects of events that are of low (but increasing) probability and high (and perhaps rising) consequences. Many specific conventional and advanced techniques can contribute to better environmental planning and management, including market-based tools for pollution control, demand management and waste reduction, mixed-use zoning and transport planning (with appropriate provision for pedestrians and cyclists), environmental impact assessments, capacity studies, strategic environmental plans, environmental audit procedures, and state-of-the-environment reports. Many cities have used a combination of these strategies in developing "Local Agenda 21s." Many Local Agenda 21s deal with a list of urban problems that could closely interact with climate change in the future. [7.2, 7.5]

4.6. Insurance and Other Financial Services

The financial services sector—broadly defined as private and public institutions that offer insurance and disaster relief, banking, and asset management services—is a unique indicator of potential socioeconomic impacts of climate change because it is sensitive to climate change and it integrates effects on other sectors. The sector is a key agent of adaptation (e.g., through support of building codes and, to a limited extent, land-use planning), and financial services represent risk-spreading mechanisms through which the costs of weather-related events are distributed among other sectors and throughout society. However, insurance, whether provided by public or private entities, also can encourage complacency and maladaptation by fostering development in at-risk areas such as U.S. floodplains or coastal zones. The effects of climate change on the financial services sector are likely to manifest primarily through changes in spatial distribution, frequencies, and intensities of extreme weather events (Table TS-4). [8.1, 8.2, 15.2.7]

The costs of extreme weather events have exhibited a rapid upward trend in recent decades. Yearly global economic losses from large events increased from US\$3.9 billion yr^{-1} in the 1950s to US\$40 billion yr^{-1} in the 1990s (all 1999 US\$, uncorrected for purchasing power parity). Approximately one-quarter of

***Table TS-4**: Extreme climate-related phenomena and their effects on the insurance industry: observed changes and projected changes during 21st century (after Table 3-10; see also Table 8-1).*

Changes in Extreme Climate Phenomena	Observed Changes (Likelihood)	Projected Changes (Likelihood)	Type of Event Relevant to Insurance Sector	Relevant Time Scale	Sensitive Sectors/Activities	Sensitive Insurance Branches
Temperature Extremes						
Higher maximum temperatures, more hot days and heat waves[b] over nearly all land areas	Likely[a] (mixed trends for heat waves in several regions)	Very likely[a]	Heat wave	Daily-weekly maximum	Electric reliability, human settlements	Health, life, property, business interruption
			Heat wave, droughts	Monthly-seasonal maximum	Forests (tree health), natural resources, agriculture, water resources, electricity demand and reliability, industry, health, tourism	Health, crop, business interruption
Higher (increasing) minimum temperatures, fewer cold days, frost days, and cold waves[b] over nearly all land areas	Very likely[a] (cold waves not treated by WGI)	Very likely[a]	Frost, frost heave	Daily-monthly minimum	Agriculture, energy demand, health, transport, human settlements	Health, crop, property, business interruption, vehicle
Rainfall/Precipitation Extremes						
More intense precipitation events	Likely[a] over many Northern Hemisphere mid- to high-latitude land areas	Very likely[a] over many areas	Flash flood	Hourly-daily maximum	Human settlements	Property, flood, vehicle, business interruption, life, health
			Flood, inundation, mudslide	Weekly-monthly maximum	Agriculture, forests, transport, water quality, human settlements, tourism	Property, flood, crop, marine, business interruption
Increased summer drying and associated risk of drought	Likely[a] in a few areas	Likely[a] over most mid-latitude continental interiors (lack of consistent projections in other areas)	Summer drought, land subsidence, wildfire	Monthly-seasonal minimum	Forests (tree health), natural resources, agriculture, water resources, (hydro)energy supply, human settlements	Crop, property, health

the losses occurred in developing countries. The insured portion of these losses rose from a negligible level to US$9.2 billion annually during the same period. Including events of all sizes doubles these loss totals (see Figure TS–5). The costs of weather events have risen rapidly, despite significant and increasing efforts at fortifying infrastructure and enhancing disaster preparedness. These efforts dampen to an unknown degree the observed rise in loss costs, although the literature attempting to separate natural from human driving forces has not quantified this effect. As a measure of increasing insurance industry vulnerability, the ratio of global property/casualty insurance premiums to weather-related losses—an important indicator of adaptive capacity—fell by a factor of three between 1985 and 1999. [8.3]

Part of the observed upward trend in historical disaster losses is linked to socioeconomic factors—such as population growth, increased wealth, and urbanization in vulnerable areas—and

Table TS-4 (continued)

Changes in Extreme Climate Phenomena	Observed Changes (Likelihood)	Projected Changes (Likelihood)	Type of Event Relevant to Insurance Sector	Relevant Time Scale	Sensitive Sectors/Activities	Sensitive Insurance Branches
Rainfall/Precipitation Extremes (continued)						
Increased intensity of mid-latitude storms[c]	Medium likelihood[a] of increase in Northern Hemisphere, decrease in Southern Hemisphere	Little agreement among current models	Snowstorm, ice storm, avalanche	Hourly-weekly	Forests, agriculture, energy distribution and reliability, human settlements, mortality, tourism	Property, crop, vehicle, aviation, life, business interruption
			Hailstorm	Hourly	Agriculture, property	Crop, vehicle, property, aviation
Intensified droughts and floods associated with El Niño events in many different regions (see also droughts and extreme precipitation events)	Inconclusive information	Likely[a]	Drought and floods	Various	Forests (tree health), natural resources, agriculture, water resources, (hydro)energy supply, human settlements	Property, flood, vehicle, crop, marine, business interruption, life, health
Wind Extremes						
Increased intensity of mid-latitude storms[b]	No compelling evidence for change	Little agreement among current models	Mid-latitude windstorm	Hourly-daily	Forests, electricity distribution and reliability, human settlements	Property, vehicle, aviation, marine, business interruption, life
			Tornadoes	Hourly	Forests, electricity distribution and reliability, human settlements	Property, vehicle, aviation, marine, business interruption
Increase in tropical cyclone peak wind intensities, mean and peak precipitation intensities[c]	Wind extremes not observed in the few analyses available; insufficient data for precipitation	Likely[a] over some areas	Tropical storms, including cyclones, hurricanes, and typhoons	Hourly-weekly	Forests, electricity distribution and reliability, human settlements, agriculture	Property, vehicle, aviation, marine, business interruption, life

part is linked to climatic factors such as observed changes in precipitation, flooding, and drought events. Precise attribution is complex, and there are differences in the balance of these two causes by region and by type of event. Many of the observed trends in weather-related losses are consistent with what would be expected under climate change. Notably, the growth rate in human-induced and non-weather-related losses has been far lower than that of weather-related events. [8.2.2]

Recent history has shown that weather-related losses can stress insurance companies to the point of impaired profitability, consumer price increases, withdrawal of coverage, and elevated demand for publicly funded compensation and relief. Increased uncertainty will increase the vulnerability of the insurance and government sectors and complicate adaptation and disaster relief efforts under climate change. [8.3, 15.2.7]

Table TS-4 (continued)

Changes in Extreme Climate Phenomena	Observed Changes	Projected Changes	Type of Event Relevant to Insurance Sector	Relevant Time Scale	Sensitive Sectors/Activities	Sensitive Insurance Branches
	Likelihood					
Other Extremes						
Refer to entries above for higher temperatures, increased tropical and mid-latitude storms	Refer to relevant entries above	Refer to relevant entries above	Lightning	Instant-aneous	Electricity distribution and reliability, human settlements, wildfire	Life, property, vehicle, aviation, marine, business interruption
Refer to entries above for increased tropical cyclones, Asian summer monsoon, and intensity of mid-latitude storms	Refer to relevant entries above	Refer to relevant entries above	Tidal surge (associated with onshore gales), coastal inundation	Daily	Coastal zone infrastructure, agriculture and industry, tourism	Life, marine, property, crop
Increased Asian summer monsoon precipitation variability	Not treated by WGI	Likely[a]	Flood and drought	Seasonal	Agriculture, human settlements	Crop, property, health, life

[a] Likelihood refers to judgmental estimates of confidence used by Working Group I: *very likely* (90–99% chance); *likely* (66–90% chance). Unless otherwise stated, information on climate phenomena is taken from Working Group I's Summary for Policymakers and Technical Summary. These likelihoods refer to observed and projected changes in extreme climate phenomena and likelihood shown in first three columns of table.
[b] Information from Working Group I, Technical Summary, Section F.5.
[c] Changes in regional distribution of tropical cyclones are possible but have not been established.

The financial services sector as a whole is expected to be able to cope with the impacts of future climate change, although the historic record shows that low-probability, high-impact events or multiple closely spaced events severely affect parts of the sector, especially if adaptive capacity happens to be simultaneously depleted by nonclimate factors (e.g., adverse market conditions that can deplete insurer loss reserves by eroding the value of securities and other insurer assets). There is high confidence that climate change and anticipated changes in weather-related events that are perceived to be linked to climate change would increase actuarial uncertainty in risk assessment and thus in the functioning of insurance markets. Such developments would place upward pressure on premiums and/or could cause certain risks to be reclassified as uninsurable, with subsequent withdrawal of coverage. This, in turn, would place increased pressure on government-based insurance and relief systems, which already are showing strain in many regions and are attempting to limit their exposures (e.g., by raising deductibles and/or placing caps on maximum claims payable).

Trends toward increasing firm size, diversification, and integration of insurance with other financial services, as well as improved tools to transfer risk, all potentially contribute to robustness. However, the property/casualty insurance and reinsurance segments have greater sensitivity, and individual companies already have experienced catastrophe-related bankruptcies triggered by weather events. Under some conditions and in some regions, the banking industry as a provider of loans also may be vulnerable to climate change. In many cases, however, the banking sector transfers risk back to insurers, who often purchase their debt products. [8.3, 8.4, 15.2.7]

Adaptation[2] to climate change presents complex challenges, as well as opportunities, for the financial services sector. Regulatory involvement in pricing, tax treatment of reserves, and the (in)ability of firms to withdraw from at-risk markets are examples of factors that influence the resilience of the sector. Management of climate-related risk varies by country and region. Usually it is a mixture of commercial and public arrangements and self-insurance. In the face of climate change, the relative role of each can be expected to change. Some potential response options offer co-benefits that support sustainable development and climate change mitigation objectives (e.g., energy-efficiency measures that also make buildings more resilient to natural disasters, in addition to helping the sector adapt to climate changes). [8.3.4, 8.4.2]

The effects of climate change are expected to be greatest in developing countries (especially those that rely on primary production as a major source of income) in terms of loss of life,

[2] The term "mitigation" often is used in the insurance and financial services sectors in much the same way as the term "adaptation" is used in the climate research and policy communities.

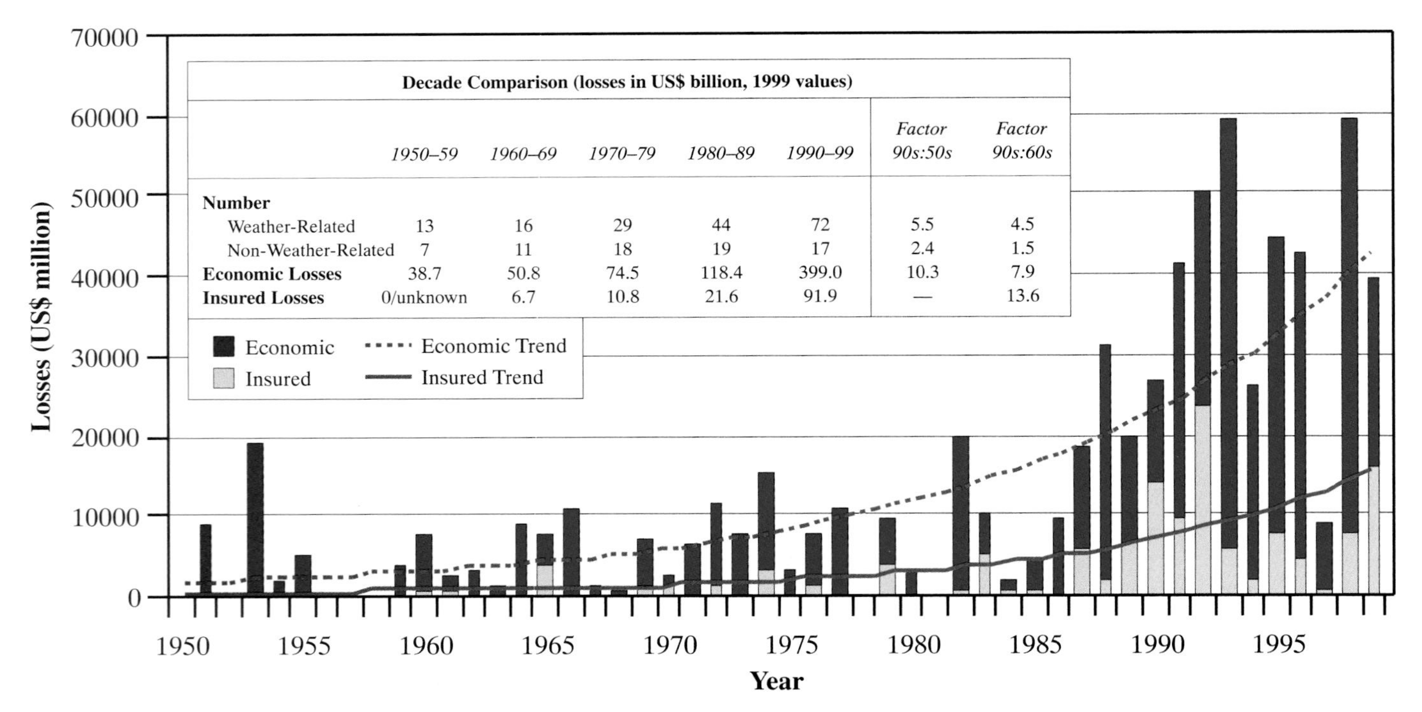

	1950–59	1960–69	1970–79	1980–89	1990–99	Factor 90s:50s	Factor 90s:60s
Number							
Weather-Related	13	16	29	44	72	5.5	4.5
Non-Weather-Related	7	11	18	19	17	2.4	1.5
Economic Losses	38.7	50.8	74.5	118.4	399.0	10.3	7.9
Insured Losses	0/unknown	6.7	10.8	21.6	91.9	—	13.6

Figure TS-5: The costs of catastrophic weather events have exhibited a rapid upward trend in recent decades. Yearly economic losses from large events increased 10.3-fold from US$4 billion yr^{-1} in the 1950s to US$40 billion yr^{-1} in the 1990s (all in 1999 US$). The insured portion of these losses rose from a negligible level to US$9.2 billion annually during the same period, and the ratio of premiums to catastrophe losses fell by two-thirds. Notably, costs are larger by a factor of 2 when losses from ordinary, noncatastrophic weather-related events are included. The numbers generally include "captive" self-insurers but not the less-formal types of self-insurance.

effects on investment, and effects on the economy. Damages from natural disasters have been as high as half of the gross domestic product (GDP) in one case. Weather disasters set back development, particularly when funds are redirected from development projects to disaster-recovery efforts. [8.5]

Equity issues and development constraints would arise if weather-related risks become uninsurable, insurance prices increase, or the availability of insurance or financing becomes limited. Thus, increased uncertainty could constrain development. Conversely, more extensive penetration of or access to insurance and disaster preparedness/recovery resources would increase the ability of developing countries to adapt to climate change. More widespread introduction of microfinancing schemes and development banking also could be an effective mechanism to help developing countries and communities adapt. [8.3]

This assessment of financial services has identified some areas of improved knowledge and has corroborated and further augmented conclusions reached in the SAR. It also has highlighted many areas where greater understanding is needed—in particular, better analysis of economic losses to determine their causation, assessment of financial resources involved in dealing with climate change damage and adaptation, evaluation of alternative methods to generate such resources, deeper investigation of the sector's vulnerability and resilience to a range of extreme weather event scenarios, and more research into how the sector (private and public elements) could innovate to meet the potential increase in demand for adaptation funding in developed and developing countries, to spread and reduce risks from climate change. [8.7]

4.7. Human Health

Global climate change will have diverse impacts on human health—some positive, most negative. Changes in the frequencies of extreme heat and cold, the frequencies of floods and droughts, and the profile of local air pollution and aeroallergens would affect population health directly. Other health impacts would result from the impacts of climate change on ecological and social systems. These impacts would include changes in infectious disease occurrence, local food production and undernutrition, and various health consequences of population displacement and economic disruption.

There is little published evidence that changes in population health status actually have occurred in response to observed trends in climate over recent decades. A recurring difficulty in identifying such impacts is that the causation of most human health disorders is multifactorial, and the "background" socioeconomic, demographic, and environmental context changes significantly over time.

Studies of the health impacts associated with interannual climate variability (particularly those related to the El Niño cycle) have provided new evidence of human health sensitivity to climate, particularly for mosquito-borne diseases. The combination of existing research-based knowledge, resultant theoretical understandings, and the output of predictive modeling leads to several conclusions about the future impacts of climate change on human population health.

If heat waves increase in frequency and intensity, the risk of death and serious illness would increase, principally in older age groups and the urban poor (high confidence). The effects of an increase in heat waves often would be exacerbated by increased humidity and urban air pollution. The greatest increases in thermal stress are forecast for mid- to high-latitude (temperate) cities, especially in populations with nonadapted architecture and limited air conditioning. Modeling of heat wave impacts in urban populations, allowing for acclimatization, suggests that a number of U.S. cities would experience, on average, several hundred extra deaths each summer. Although the impact of climate change on thermal stress-related mortality in developing country cities may be significant, there has been little research in such populations. Warmer winters and fewer cold spells will decrease cold-related mortality in many temperate countries (high confidence). Limited evidence indicates that in at least some temperate countries, reduced winter deaths would outnumber increased summer deaths (medium confidence). [9.4]

Any increases in the frequency and intensity of extreme events such as storms, floods, droughts, and cyclones would adversely impact human health through a variety of pathways. These natural hazards can cause direct loss of life and injury and can affect health indirectly through loss of shelter, population displacement, contamination of water supplies, loss of food production (leading to hunger and malnutrition), increased risk of infectious disease epidemics (including diarrhoeal and respiratory disease), and damage to infrastructure for provision of health services (very high confidence). If cyclones were to increase regionally, devastating impacts often would occur, particularly in densely settled populations with inadequate resources. Over recent years, major climate-related disasters have had major adverse effects on human health, including floods in China, Bangladesh, Europe, Venezuela, and Mozambique, as well as Hurricane Mitch, which devastated Central America. [9.5]

Climate change will decrease air quality in urban areas with air pollution problems (medium confidence). An increase in temperature (and, in some models, ultraviolet radiation) increases the formation of ground-level ozone, a pollutant with well-established adverse effects on respiratory health. Effects of climate change on other air pollutants are less well established. [9.6]

Higher temperatures, changes in precipitation, and changes in climate variability would alter the geographic ranges and seasonality of transmission of vector-borne infectious diseases—extending the range and season for some infectious diseases and contracting them for others. Vector-borne infectious diseases are transmitted by blood-feeding organisms such as mosquitoes and ticks. Such organisms depend on the complex interaction of climate and other ecological factors for survival. Currently, 40% of the world population lives in areas with malaria. In areas with limited or deteriorating public health infrastructure, increased temperatures will tend to expand the geographic range of malaria transmission to higher altitudes (high to medium confidence) and higher latitudes (medium to low confidence). Higher temperatures, in combination with conducive patterns of rainfall and surface water, will extend the transmission season in some locations (high confidence). Changes in climate, including changes in climate variability, would affect many other vector-borne infections (such as dengue, leishmansiasis, various types of mosquito-borne encephalitis, Lyme disease, and tick-borne encephalitis) at the margins of their current distributions (medium/high confidence). For some vector-borne diseases in some locations, climate change will decrease transmission via reductions in rainfall or temperatures that are too high for transmission (medium confidence). A range of mathematical models indicate, with high consistency, that climate change scenarios over the coming century would cause a small net increase in the proportion of the world's population living in regions of potential transmission of malaria and dengue (medium to high confidence). A change in climatic conditions will increase the incidence of various types of water- and food-borne infectious diseases. [9.7]

Climate change may cause changes in the marine environment that would alter risks of biotoxin poisoning from human consumption of fish and shellfish. Biotoxins associated with warmer waters, such as ciguatera in tropical waters, could extend their range to higher latitudes (low confidence). Higher SSTs also would increase the occurrence of toxic algal blooms (medium confidence), which have complex relationships with human poisoning and are ecologically and economically damaging. Changes in surface water quantity and quality will affect the incidence of diarrhoeal diseases (medium confidence). [9.8]

Changes in food supply resulting from climate change could affect the nutrition and health of the poor in some regions of the world. Studies of climate change impacts on food production indicate that, globally, impacts could be positive or negative, but the risk of reduced food yields is greatest in developing countries—where 790 million people are estimated to be undernourished at present. Populations in isolated areas with poor access to markets will be particularly vulnerable to local decreases or disruptions in food supply. Undernourishment is a fundamental cause of stunted physical and intellectual development in children, low productivity in adults, and susceptibility to infectious disease. Climate change would increase the number of undernourished people in the developing world (medium confidence), particularly in the tropics. [9.9, 5.3]

In some settings, the impacts of climate change may cause social disruption, economic decline, and population displacement that would affect human health. Health impacts associated with population displacement resulting from natural disasters

***Table TS-5**: Options for adaptation to reduce health impacts of climate change.*

Health Outcome	Legislative	Technical	Educational-Advisory	Cultural and Behavioral
Thermal stress	– Building guidelines	– Housing, public buildings, urban planning to reduce heat island effects, air conditioning	– Early warning systems	– Clothing, siesta
Extreme weather events	– Planning laws – Building guidelines – Forced migration – Economic incentives for building	– Urban planning – Storm shelters	– Early warning systems	– Use of storm shelters
Air quality	– Emission controls – Traffic restrictions	– Improved public transport, catalytic converters, smoke stacks	– Pollution warnings	– Carpooling
Vector-borne diseases		– Vector control – Vaccination, impregnated bednets – Sustainable surveillance, prevention and control programs	– Health education	– Water storage practices
Water-borne diseases	– Watershed protection laws – Water quality regulation	– Genetic/molecular screening of pathogens – Improved water treatment (e.g., filters) – Improved sanitation (e.g., latrines)	– Boil water alerts	– Washing hands and other hygiene behavior – Use of pit latrines

or environmental degradation are substantial (high confidence). [9.10]

For each anticipated adverse health impact there is a range of social, institutional, technological, and behavioral adaptation options to lessen that impact (see Table TS-5). Overall, the adverse health impacts of climate change will be greatest in vulnerable lower income populations, predominately within tropical/subtropical countries. There is a basic and general need for public health infrastructure (programs, services, surveillance systems) to be strengthened and maintained. The ability of affected communities to adapt to risks to health also depends on social, environmental, political, and economic circumstances. [9.11]

5. Regional Analysis

The vulnerability of human populations and natural systems to climate change differs substantially across regions and across populations within regions. Regional differences in baseline climate and expected climate change give rise to different exposures to climate stimuli across regions. The natural and social systems of different regions have varied characteristics, resources, and institutions and are subject to varied pressures that give rise to differences in sensitivity and adaptive capacity. From these differences emerge different key concerns for each of the major regions of the world. Even within regions, however, impacts, adaptive capacity, and vulnerability will vary. Because available studies have not employed a common set of climate scenarios and methods and because of uncertainties regarding the sensitivities and adaptability of natural and social systems, assessment of regional vulnerabilities is necessarily qualitative.

5.1. Africa

Africa is highly vulnerable to climate change. Impacts of particular concern to Africa are related to water resources, food production, human health, desertification, and coastal zones, especially in relation to extreme events. A synergy of land-use and climate change will exacerbate desertification. Selected key impacts in Africa are highlighted in Figure TS-6.

5.1.1. Water Resources

Water resources are a key area of vulnerability in Africa, affecting water supply for household use, agriculture, and industry. In shared river basins, regional cooperation protocols minimize adverse impacts and potential for conflicts. Trends in regional per capita water availability in Africa over the past half century show that water availability has diminished by 75%. Although the past 2 decades have experienced reductions in river flows, especially in sub-Saharan West Africa, the trend mainly reflects the impact of population growth—which, for most countries, quadrupled in the same period. Population growth and degradation of water quality are significant threats to water security in many parts of Africa, and the combination of continued population increases and global warming impacts is likely to accentuate water scarcity in subhumid regions of Africa.

Africa is the continent with the lowest conversion factor of precipitation to runoff, averaging 15%. Although the equatorial region and coastal areas of eastern and southern Africa are humid, the rest of the continent is dry subhumid to arid. The dominant impact of global warming will be a reduction in soil moisture in subhumid zones and a reduction in runoff. Current trends in major river basins indicate decreasing runoff of about 17% over the past decade.

Most of Africa has invested significantly in hydroelectric power facilities to underpin economic development. Reservoir storage shows marked sensitivity to variations in runoff and periods of drought. Lake storage and major dams have reached critical levels, threatening industrial activity. Model results and some reservoirs and lakes indicate that global warming will increase the frequency of such low storage as a result of flooding or drought conditions that are related to ENSO. [10.2.1]

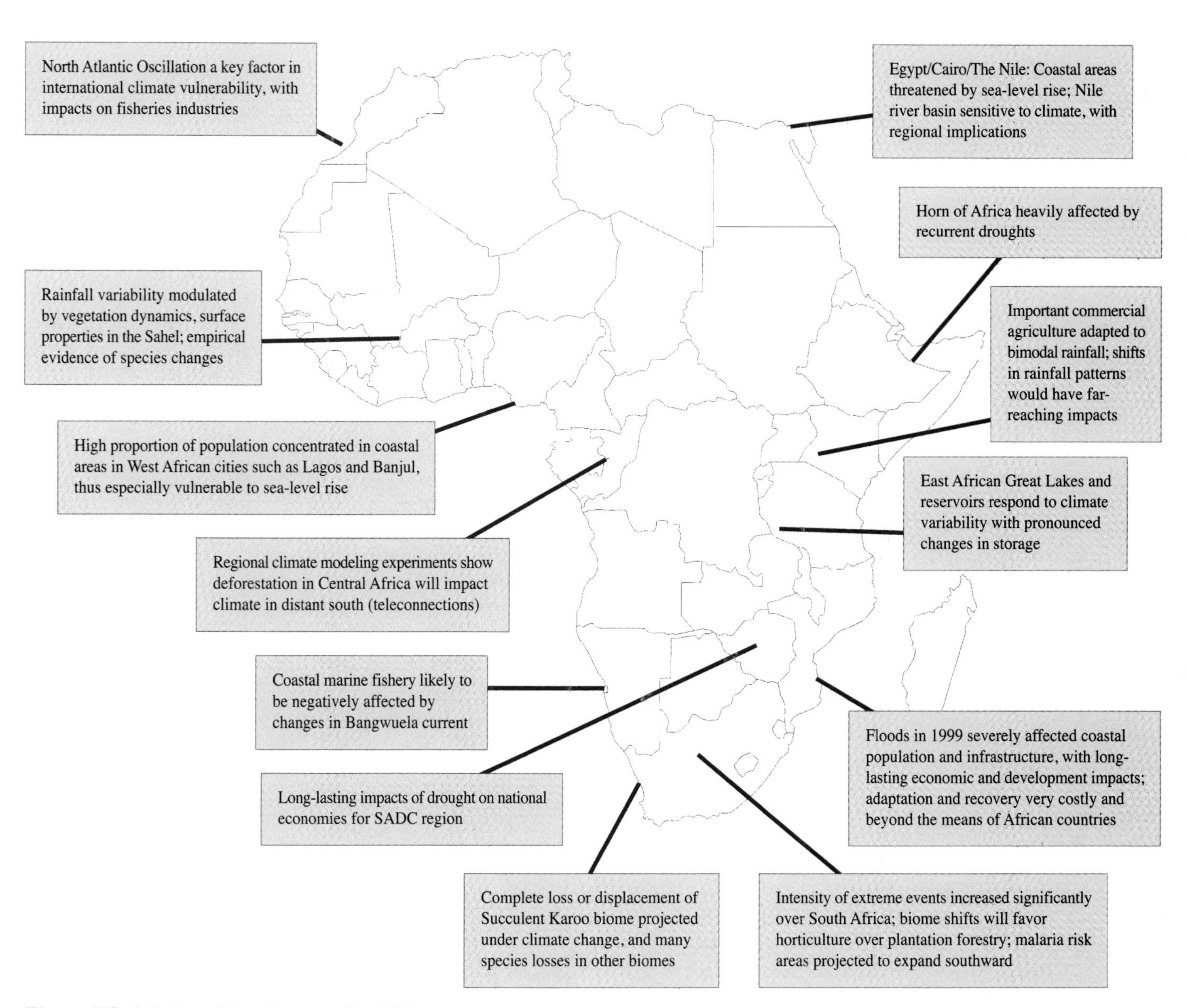

Figure TS-6: Selected key impacts for Africa.

5.1.2. Food Security

There is wide consensus that climate change will worsen food security, mainly through increased extremes and temporal/spatial shifts. The continent already experiences a major deficit in food production in many areas, and potential declines in soil moisture will be an added burden. Food-insecure countries are at a greater risk of adverse impacts of climate change. Inland and marine fisheries provide a significant contribution to protein intake in many African countries. As a result of water stress and land degradation, inland fisheries will be rendered more vulnerable to episodic drought and habitat destruction. Ocean warming is likely to impact coastal marine fisheries. [10.2.2]

5.1.3. Natural Resource Management and Biodiversity

Irreversible losses of biodiversity could be accelerated with climate change. Climate change is expected to lead to drastic shifts of biodiversity-rich biomes such as the Succulent Karoo in South Africa and many losses in species in other biomes. Changes in the frequency, intensity, and extent of vegetation fires and habitat modification from land-use change may negate natural adaptive processes and lead to extinctions. Changes in ecosystems will affect water supply, fuelwood, and other services. [10.2.3.2]

5.1.4. Human Health

Changes in temperature and rainfall will have many negative impacts on human health. Temperature increases will extend disease vector habitats. Where sanitary infrastructure is inadequate, droughts and flooding will result in increased frequency of water-borne diseases. Increased rainfall could lead to more frequent outbreaks of Rift Valley fever. Poor sanitation in urban locations and increased temperatures of coastal waters could aggravate cholera epidemics. [10.2.4.1.1, 10.2.4.4]

5.1.5. Settlements and Infrastructure

Although the basic infrastructure for development—transport, housing, and services—is inadequate in many instances, it nevertheless represents substantial investment by governments. An increase in the frequency of damaging floods, heat waves, dust storms, hurricanes, and other extreme events could degrade the integrity of such critical infrastructures at rates the economies may not be able to tolerate, leading to a serious deterioration of social, health, and economic services delivery systems. This condition will greatly compromise general human welfare. [10.2.5.3]

Sea-level rise, coastal erosion, saltwater intrusion, and flooding will have significant impacts for African communities and economies. Most of Africa's largest cities are along coasts and are highly vulnerable to extreme events, sea-level rise, and coastal erosion because of inadequate physical planning and escalating urban drift. Rapid unplanned expansion is likely to predispose large populations to infectious diseases from climate-related factors such as flooding. [10.2.5.2]

5.1.6. Desertification

Alteration of spatial and temporal patterns in temperature, rainfall, solar radiation, and winds from a changing climate will exacerbate desertification. Desertification is a critical threat to sustainable resource management in arid, semi-arid, and dry subhumid regions of Africa, undermining food and water security. [10.2.6]

5.1.7. Adaptive Capacity

Given the diversity of constraints facing many nations, the overall capacity for Africa to adapt to climate change currently is very low. National action plans that incorporate long-term changes and pursue "no regrets" strategies could increase the adaptive capacity of the region. Seasonal forecasting—for example, linking SSTs to outbreaks of major diseases—is a promising adaptive strategy that will help save lives. Current technologies and approaches, especially in agriculture and water, are unlikely to be adequate to meet projected demands, and increased climate variability will be an additional stress. It is unlikely that African countries on their own will have sufficient resources to respond effectively.

Climate change also offers some opportunities. The processes of adapting to global climate change, including technology transfer and carbon sequestration, offer new development pathways that could take advantage of Africa's resources and human potential. Regional cooperation in science, resource management, and development already are increasing, and access to international markets will diversify economies and increase food security.

This assessment of vulnerability to climate change is marked by uncertainty. The diversity of African climates, high rainfall variability, and a very sparse observational network make predictions of future climate change difficult at the subregional and local level. Underlying exposure and vulnerability to climatic changes are well established. Sensitivity to climatic variations is established but incomplete. However, uncertainty over future conditions means that there is low confidence in projected costs of climate change. This assessment can create the framework for individual states to begin to construct methodologies for estimating such costs, based on their individual circumstances.

5.2. Asia

Climate change will impose significant stress on resources throughout the Asian region. Asia has more than 60% of the world's population; natural resources already are under stress,

***Table TS-6**: Sensitivity of selected Asian regions to climate change.*

Change in Climatic Elements and Sea-Level Rise	Vulnerable Region	Primary Change	Impacts	
			Primary	Secondary
0.5–2°C (10- to 45-cm sea-level rise)	Bangladesh Sundarbans	– Inundation of about 15% (~750 km^2) – Increase in salinity	– Loss of plant species – Loss of wildlife	– Economic loss – Exacerbated insecurity and loss of employment
4°C (+10% rainfall)	Siberian permafrosts	– Reduction in continuous permafrost – Shift in southern limit of Siberian permafrost by ~100–200 km northward	– Change in rock strength – Change in bearing capacity – Change in compressibility of frozen rocks – Thermal erosion	– Effects on construction industries – Effects on mining industry – Effects on agricultural development
>3°C (>+20% rainfall)	Water resources in Kazakhstan	– Change in runoff	– Increase in winter floods – Decrease in summer flows	– Risk to life and property – Summer water stress
~2°C (-5 to 10% rainfall; 45-cm sea-level rise)	Bangladesh lowlands	– About 23–29% increase in extent of inundation	– Change in flood depth category – Change in monsoon rice cropping pattern	– Risk to life and property – Increased health problems – Reduction in rice yield

and the resilience of most sectors in Asia to climate change is poor. Many countries are socioeconomically dependent on natural resources such as water, forests, grassland and rangeland, and fisheries. The magnitude of changes in climate variables would differ significantly across Asian subregions and countries. The climate change sensitivity of a few vulnerable sectors in Asia and the impacts of these limits are presented in Table TS-6. The region's vulnerability to climate change is captured in Table TS-7 for selected categories of regions/issues.

5.2.1. Agriculture and Food Security

Food insecurity appears to be the primary concern for Asia. Crop production and aquaculture would be threatened by thermal and water stresses, sea-level rise, increased flooding, and strong winds associated with intense tropical cyclones (high confidence). In general, it is expected that areas in mid- and high latitudes will experience increases in crop yield; yields in lower latitudes generally will decrease. A longer duration of the summer season should lead to a northward shift of the agroecosystem boundary in boreal Asia and favor an overall increase in agriculture productivity (medium confidence). Climatic variability and change also will affect scheduling of the cropping season, as well as the duration of the growing period of the crop. In China, yields of several major crops are expected to decline as a result of climate change. Acute water shortages combined with thermal stress should adversely affect wheat and, more severely, rice productivity in India even under the positive effects of elevated CO_2 in the future. Crop diseases such as wheat scab, rice blast, and sheath and culm blight of rice also could become more widespread in temperate and tropical regions of Asia if the climate becomes warmer and wetter. Adaptation measures to reduce the negative effects of climatic variability may include changing the cropping calendar to take advantage of the wet period and to avoid the extreme weather events (e.g., typhoons and strong winds) during the growing season. [11.2.2.1]

Asia dominates world aquaculture, producing 80% of all farmed fish, shrimp, and shellfish. Many wild stocks are under stress as a result of overexploitation, trawling on sea-bottom habitats, coastal development, and pollution from land-based activities. Moreover, marine productivity is greatly affected by plankton shift, such as seasonal shifting of sardine in the Sea of Japan, in response to temperature changes induced during ENSO. Storm surges and cyclonic conditions also routinely lash the coastline, adding sediment loads to coastal waters. Effective conservation and sustainable management of marine and inland fisheries are needed at the regional level so that living aquatic resources can continue to meet regional and national nutritional needs. [11.2.4.4]

5.2.2. Ecosystems and Biodiversity

Climate change would exacerbate current threats to biodiversity resulting from land-use/cover change and population pressure in Asia (medium confidence). Risks to Asia's rich array of living species are climbing. As many as 1,250 of 15,000 higher plant species are threatened in India. Similar trends are evident in China, Malaysia, Myanmar, and Thailand. Many species and a

***Table TS-7**: Vulnerability of key sectors to impacts of climate change for select subregions in Asia. Key to confidence-level rankings is provided in Section 1.4 of Technical Summary.*

Regions	Food and Fiber	Biodiversity	Water Resources	Coastal Ecosystems	Human Health	Settlements
Boreal Asia	Slightly resilient ****	Highly vulnerable ***	Slightly resilient ***	Slightly resilient **	Moderately vulnerable **	Slightly or not vulnerable ***
Arid and Semi-Arid Asia						
– Central Asia	Highly vulnerable ****	Moderately vulnerable **	Highly vulnerable ****	Moderately vulnerable **	Moderately vulnerable ***	Moderately vulnerable ***
– Tibetan Plateau	Slightly or not vulnerable **	Highly vulnerable ***	Moderately vulnerable **	Not applicable	No information	No information
Temperate Asia	Highly vulnerable ****	Moderately vulnerable ***	Highly vulnerable ****	Highly vulnerable ****	Highly vulnerable ***	Highly vulnerable ****
Tropical Asia						
– South Asia	Highly vulnerable ****	Highly vulnerable ***	Highly vulnerable ****	Highly vulnerable ****	Moderately vulnerable ***	Highly vulnerable ***
– Southeast Asia	Highly vulnerable ****	Highly vulnerable ***	Highly vulnerable ****	Highly vulnerable ****	Moderately vulnerable ***	Highly vulnerable ***

large population of many other species in Asia are likely to be exterminated as a result of the synergistic effects of climate change and habitat fragmentation. In desert ecosystems, increased frequency of droughts may result in a decline in local forage around oases, causing mass mortality among local fauna and threatening their existence. With a 1-m rise in sea level, the Sundarbans (the largest mangrove ecosystems) of Bangladesh will completely disappear. [11.2.1, 11.2.1.6]

Permafrost degradation resulting from global warming would increase the vulnerability of many climate-dependent sectors affecting the economy in boreal Asia (medium confidence). Pronounced warming in high latitudes of the northern hemisphere could lead to thinning or disappearance of permafrost in locations where it now exists. Large-scale shrinkage of the permafrost region in boreal Asia is likely. Poleward movement of the southern boundary of the sporadic zone also is likely in Mongolia and northeast China. The boundary between continuous and discontinuous (intermittent or seasonal) permafrost areas on the Tibetan Plateau is likely to shift toward the center of the plateau along the eastern and western margins. [11.2.1.5]

The frequency of forest fires is expected to increase in boreal Asia (medium confidence). Warmer surface air temperatures, particularly during summer, may create favorable conditions for thunderstorms and associated lightening, which could trigger forest fires in boreal forests more often. Forest fire is expected to occur more frequently in northern parts of boreal Asia as a result of global warming. [11.2.1.3]

5.2.3. Water Resources

Freshwater availability is expected to be highly vulnerable to anticipated climate change (high confidence). Surface runoff increases during winter and summer periods would be pronounced in boreal Asia (medium confidence). Countries in which water use is more than 20% of total potential water resources available are expected to experience severe water stress during drought periods. Surface runoff is expected to decrease drastically in arid and semi-arid Asia under projected climate change scenarios. Climate change is likely to change streamflow volume, as well as the temporal distribution of streamflows throughout the year. With a 2°C increase in air temperature accompanied by a 5–10% decline in precipitation during summer, surface runoff in Kazakhstan would be substantially reduced, causing serious implications for agriculture and livestocks. Water would be a scarce commodity in many south and southeast Asian countries, particularly where reservoir facilities to store water for irrigation are minimal. Growing populations and concentration of populations in urban areas will exert increasing pressures on water availability and water quality. [11.2.3.1]

5.2.4. Extreme Weather Events

Developing countries of temperate and tropical Asia already are quite vulnerable to extreme climate events such as typhoons/cyclones, droughts, and floods. Climate change and variability would exacerbate these vulnerabilities (high confidence). Extreme weather events are known to cause adverse effects in widely separated areas of Asia. There is some evidence of increases in the intensity or frequency of some of these extreme events on regional scales throughout the 20th century. [11.1.2.2, 11.1.2.3, 11.4.1]

Increased precipitation intensity, particularly during the summer monsoon, could increase flood-prone areas in temperate and tropical Asia. There is potential for drier conditions in arid and semi-arid Asia during summer, which could lead to more severe droughts (medium confidence). Many countries in temperate and tropical Asia have experienced severe droughts and floods frequently in the 20th century. Flash floods are likely to become more frequent in many regions of temperate and tropical Asia in the future. A decrease in return period for extreme precipitation events and the possibility of more frequent floods in parts of India, Nepal, and Bangladesh is projected. [11.1.3.3, 11.2.2.2, 11.1.2.3, 11.4.1]

Conversion of forestland to cropland and pasture already is a prime force driving forest loss in tropical and temperate Asian countries. With more frequent floods and droughts, these actions will have far-reaching implications for the environment (e.g., soil erosion, loss of soil fertility, loss of genetic variability in crops, and depletion of water resources). [11.1.4.1]

Tropical cyclones and storm surges continue to take a heavy toll on life and property in India and Bangladesh. An increase in the intensity of cyclones combined with sea-level rise would result in more loss of life and property in low-lying coastal areas in cyclone-prone countries of Asia (medium confidence). The expected increase in the frequency and intensity of climatic extremes will have significant potential effects on crop growth and agricultural production, as well as major economic and environmental implications (e.g., tourism, transportation). [11.2.4.5, 11.2.6.3, 11.3]

A wide range of precautionary measures at regional and national levels, including awareness and acceptance of risk factors among regional communities, is warranted to avert or reduce the impacts of disasters associated with more extreme weather events on economic and social structures of countries in temperate and tropical Asia. [11.3.2]

5.2.5. Deltas and Coastal Zones

The large deltas and low-lying coastal areas of Asia would be inundated by sea-level rise (high confidence). Climate-related stresses in coastal areas include loss and salinization of agricultural land as a result of change in sea level and changing frequency and intensity of tropical cyclones. Estimates of potential land loss resulting from sea-level rise and risk to population displacement provided in Table TS-8 demonstrate the scale of the issue for major low-lying regions of coastal Asia. Currently, coastal erosion of muddy coastlines in Asia is not a result of sea-level rise; it is triggered largely by annual river-borne suspended sediments transported into the ocean by human activities and delta evolution. These actions could exacerbate the impacts of climate change in coastal regions of Asia. [11.2.4.2]

5.2.6. Human Health

Warmer and wetter conditions would increase the potential for higher incidence of heat-related and infectious diseases in tropical and temperate Asia (medium confidence). The rise in surface air temperature and changes in precipitation in Asia will have adverse effects on human health. Although warming would result in a reduction in wintertime deaths in temperate countries, there could be greater frequency and duration of heat stress, especially in megalopolises during summer. Global warming also will increase the incidence of respiratory and cardiovascular diseases in parts of arid and semi-arid Asia and temperate and tropical Asia. Changes in environmental temperature and precipitation could expand vector-borne diseases into temperate and arid Asia. The spread of vector-borne diseases into more northern latitudes may pose a serious

***Table TS-8**: Potential land loss and population exposed in Asian countries for selected magnitudes of sea-level rise, assuming no adaptation.*

Country	Sea-Level Rise (cm)	Potential Land Loss (km^2)	Potential Land Loss (%)	Population Exposed (million)	Population Exposed (%)
Bangladesh	45	15,668	10.9	5.5	5.0
	100	29,846	20.7	14.8	13.5
India	100	5,763	0.4	7.1	0.8
Indonesia	60	34,000	1.9	2.0	1.1
Japan	50	1,412	0.4	2.9	2.3
Malaysia	100	7,000	2.1	>0.05	>0.3
Pakistan	20	1,700	0.2	n.a.	n.a.
Vietnam	100	40,000	12.1	17.1	23.1

threat to human health. Warmer SSTs along Asian coastlines would support higher phytoplankton blooms. These blooms are habitats for infectious bacterial diseases. Waterborne diseases—including cholera and the suite of diarrheal diseases caused by organisms such as giardia, salmonella, and cryptosporidium—could become more common in many countries of south Asia in warmer climate. [11.2.5.1, 11.2.5.2, 11.2.5.4]

5.2.7. *Adaptive Capacity*

Adaptation to climate change in Asian countries depends on the affordability of adaptive measures, access to technology, and biophysical constraints such as land and water resource availability, soil characteristics, genetic diversity for crop breeding (e.g., crucial development of heat-resistant rice cultivars), and topography. Most developing countries of Asia are faced with increasing population, spread of urbanization, lack of adequate water resources, and environmental pollution, which hinder socioeconomic activities. These countries will have to individually and collectively evaluate the tradeoffs between climate change actions and nearer term needs (such as hunger, air and water pollution, energy demand). Coping strategies would have to be developed for three crucial sectors: land resources, water resources, and food productivity. Adaptation measures that are designed to anticipate the potential effects of climate change can help offset many of the negative effects. [11.3.1]

5.3. *Australia and New Zealand*

The Australia/New Zealand region spans the tropics to mid-latitudes and has varied climates and ecosystems, including deserts, rainforests, coral reefs, and alpine areas. The climate is strongly influenced by the surrounding oceans. Australia has significant vulnerability to the drying trend projected over much of the country for the next 50–100 years (Figure TS-3) because substantial agricultural areas currently are adversely affected by periodic droughts, and there already are large areas of arid and semi-arid land. New Zealand—a smaller, more mountainous country with a generally more temperate, maritime climate—may be more resilient to climate changes than Australia, although considerable vulnerability remains (medium confidence). Table TS-9 shows key vulnerabilities and adaptability to climate change impacts for Australia and New Zealand. [12.9.5]

Comprehensive cross-sectoral estimates of net climate change impact costs for various GHG emission scenarios and different societal scenarios are not yet available. Confidence remains very low in the IPCC *Special Report on Regional Impacts of Climate Change* estimate for Australia and New Zealand of -1.2 to -3.8% of GDP for an equivalent doubling of CO_2 concentrations. This estimate did not account for many of the effects and adaptations currently identified. [12.9]

Extreme events are a major source of current climate impacts, and changes in extreme events are expected to dominate the impacts of climate change. Return periods for heavy rains, floods, and sea-level surges of a given magnitude at particular locations would be modified by possible increases in intensity of tropical cyclones and heavy rain events and changes in the location-specific frequency of tropical cyclones. Scenarios of climate change that are based on recent coupled atmosphere-ocean (A-O) models suggest that large areas of mainland Australia will experience significant decreases in rainfall during the 21st century. The ENSO phenomenon leads to floods and prolonged droughts, especially in inland Australia and parts of New Zealand. The region would be sensitive to a changes towards a more El Niño-like mean state. [12.1.5]

Before stabilization of GHG concentrations, the north-south temperature gradient in mid-southern latitudes is expected to increase (medium to high confidence), strengthening the westerlies and the associated west-to-east gradient of rainfall across Tasmania and New Zealand. Following stabilization of GHG concentrations, these trends would be reversed (medium confidence). [12.1.5.1]

Climate change will add to existing stresses on achievement of sustainable land use and conservation of terrestrial and aquatic biodiversity. These stresses include invasion by exotic animal and plant species, degradation and fragmentation of natural ecosystems through agricultural and urban development, dryland salinization (Australia), removal of forest cover (Australia and New Zealand), and competition for scarce water resources. Within both countries, economically and socially disadvantaged groups of people, especially indigenous peoples, are particularly vulnerable to stresses on health and living conditions induced by climate change. Major exacerbating problems include rapid population and infrastructure growth in vulnerable coastal areas, inappropriate use of water resources, and complex institutional arrangements. [12.3.2, 12.3.3, 12.4.1, 12.4.2, 12.6.4, 12.8.5]

5.3.1. *Water Resources*

Water resources already are stressed in some areas and therefore are highly vulnerable, especially with respect to salinization (parts of Australia) and competition for water supply between agriculture, power generation, urban areas, and environmental flows (high confidence). Increased evaporation and possible decreases in rainfall in many areas would adversely affect water supply, agriculture, and the survival and reproduction of key species in parts of Australia and New Zealand (medium confidence). [12.3.1, 12.3.2, 12.4.6, 12.5.2, 12.5.3, 12.5.6]

5.3.2. *Ecosystems*

A warming of 1°C would threaten the survival of species that currently are growing near the upper limit of their temperature range, notably in marginal alpine regions and in the southwest of Western Australia. Species that are unable to migrate or relocate because of land clearing, soil differences, or topography could become endangered or extinct. Other Australian ecosystems

Table TS-9*: Main areas of vulnerability and adaptability to climate change impacts in Australia and New Zealand. Degree of confidence that tabulated impacts will occur is indicated by stars in second column (see Section 1.4 of Technical Summary for key to confidence-level rankings). Confidence levels, and assessments of vulnerability and adaptability, are based on information reviewed in Chapter 12, and assume continuation of present population and investment growth patterns.*

Sector	Impact	Vulnerability	Adaptation	Adaptability	Section
Hydrology and water supply	– Irrigation and metropolitan supply constraints, and increased salinization—****	High in some areas	– Planning, water allocation, and pricing	Medium	12.3.1, 12.3.2
	– Saltwater intrusion into some island and coastal aquifers—****	High in limited areas	– Alternative water supplies, retreat	Low	12.3.3
Terrestrial ecosystems	– Increased salinization of dryland farms and some streams (Australia)—***	High	– Changes in land-use practices	Low	12.3.3
	– Biodiversity loss notably in fragmented regions, Australian alpine areas, and southwest of WA—****	Medium to high in some areas	– Landscape management; little possible in alpine areas	Medium to low	12.4.2, 12.4.4, 12.4.8
	– Increased risk of fires—***	Medium	– Land management, fire protection	Medium	12.1.5.3, 12.5.4, 12.5.10
	– Weed invasion—***	Medium	– Landscape management	Medium	12.4.3
Aquatic ecosystems	– Salinization of some coastal freshwater wetlands—***	High	– Physical intervention	Low	12.4.7
	– River and inland wetland ecosystem changes—***	Medium	– Change water allocations	Low	12.4.5, 12.4.6
	– Eutrophication—***	Medium in inland Aus. waters	– Change water allocations, reduce nutrient inflows	Medium to low	12.3.4
Coastal ecosystems	– Coral bleaching, especially Great Barrier Reef—****	High	– Seed coral?	Low	12.4.7
	—More toxic algal blooms?—*	Unknown	—	—	12.4.7
Agriculture, grazing, and forestry	– Reduced productivity, increased stress on rural communities if droughts increase, increased forest fire risk—***	Location-dependent, worsens with time	– Management and policy changes, fire prevention, seasonal forecasts	Medium	12.5.2, 12.5.3, 12.5.4
	– Changes in global markets due to climate changes elsewhere—***, but sign uncertain	High, but sign uncertain	– Marketing, planning, niche and fuel crops, carbon trading	Medium	12.5.9
	– Increased spread of pests and diseases—****	Medium	– Exclusion, spraying	Medium	12.5.7
	– Increased CO_2 initially increases productivity but offset by climate changes later—**	Changes with time	– Change farm practices, change industry		12.5.3, 12.5.4
Horticulture	– Mixed impacts (+ and -), depends on species and location—****	Low overall	– Relocate	High	12.5.3

Table TS-9 (continued)

Sector	Impact	Vulnerability	Adaptation	Adaptability	Section
Fish	– Recruitment changes (some species)—**	Unknown net effect	– Monitoring, management	—	12.5.5
Settlements and industry	– Increased impacts of flood, storm, storm surge, sea-level rise—***	High in some places	– Zoning, disaster planning	Moderate	12.6.1, 12.6.4
Human health	– Expansion and spread of vector-borne diseases—****	High	– Quarantine, eradication, or control	Moderate to high	12.7.1, 12.7.4
	– Increased photochemical air pollution—****	Moderate (some cities)	– Emission controls	High	12.7.1

that are particularly vulnerable include coral reefs and arid and semi-arid habitats. Freshwater wetlands in coastal zones in Australia and New Zealand are vulnerable, and some New Zealand ecosystems are vulnerable to accelerated spread of weeds. [12.4.2, 12.4.3, 12.4.4, 12.4.5, 12.4.7]

5.3.3. Food Production

Agricultural activities are particularly vulnerable to regional reductions in rainfall in southwest and inland Australia (medium confidence). Drought frequency and consequent stresses on agriculture are likely to increase in parts of Australia and New Zealand as a result of higher temperatures and El Niño changes (medium confidence). Enhanced plant growth and water-use efficiency (WUE) resulting from CO_2 increases may provide initial benefits that offset any negative impacts from climate change (medium confidence), although the balance is expected to become negative with warmings in excess of 2-4°C and associated rainfall changes (medium confidence). This is illustrated in Figure TS-7 for wheat production in Australia, for a range of climate change scenarios. Reliance on exports of agricultural and forest products makes the region very sensitive to changes in production and commodity prices that are induced by changes in climate elsewhere. [12.5.2, 12.5.3, 12.5.6, 12.5.9, 12.8.7]

Australian and New Zealand fisheries are influenced by the extent and location of nutrient upwellings governed by prevailing winds and boundary currents. In addition, ENSO influences recruitment of some fish species and the incidence of toxic algal blooms. [12.5.5]

5.3.4. Settlements, Industry, and Human Health

Marked trends toward greater population and investment in exposed regions are increasing vulnerability to tropical cyclones and storm surges. Thus, projected increases in tropical cyclone intensity and possible changes in their location-specific frequency, along with sea-level rise, would have major impacts—notably, increased storm-surge heights for a given return period (medium to high confidence). Increased frequency of high-intensity rainfall would increase flood damages to settlements and infrastructure (medium confidence). [12.1.5.1, 12.1.5.3, 12.6.1, 12.6.4]

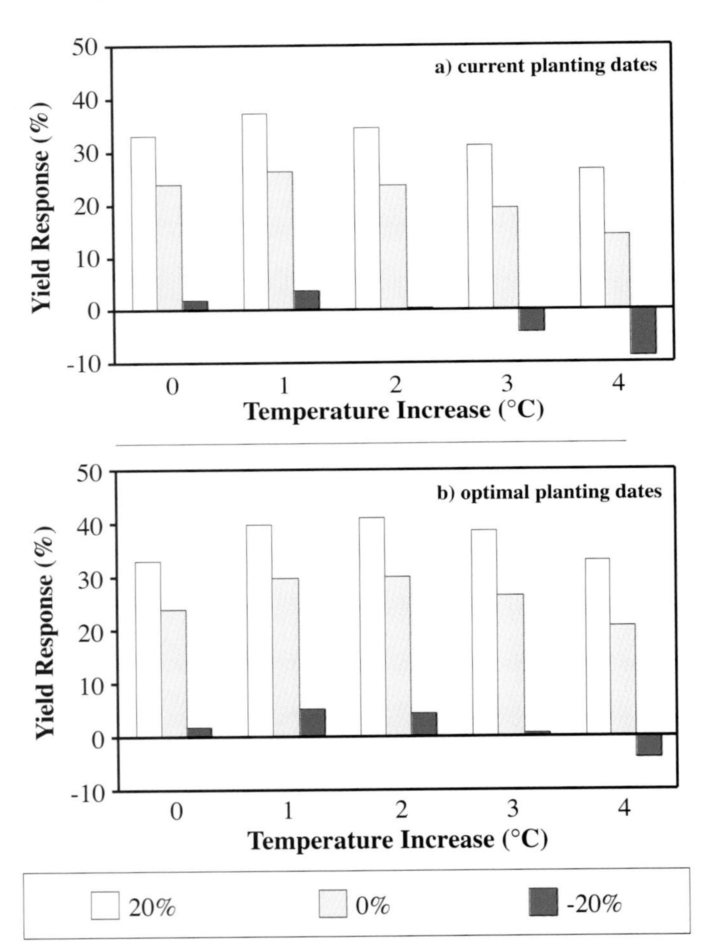

Figure TS-7: Percentage change in average annual total Australian wheat yield for CO_2 (levels of 700 ppm) and a range of changes in temperature and rainfall: a) current planting dates, and b) optimal planting dates. Yield response is shown for rainfall changes of +20% (white), 0 (light blue), and -20% (dark blue), for warmings of 0-4°C.

There is high confidence that projected climate changes will enhance the spread of some disease vectors, thereby increasing the potential for disease outbreaks such as mosquito-borne Ross River virus and Murray Valley encephalitis, despite existing biosecurity and health services. [12.7.1]

5.3.5. Key Adaptation Options

Key adaptation options include improved WUE and effective trading mechanisms for water; more appropriate land-use policies; provision of climate information and seasonal forecasts to land users to help them manage for climate variability and change; improved crop cultivars; revised engineering standards and zoning for infrastructure development; and improved biosecurity and health services. However, many natural ecosystems in Australia and New Zealand have only a limited capacity to adapt, and many managed systems will face limits on adaptation imposed by cost, acceptability, and other factors. [12.3.2, 12.3.3, 12.5.6, 12.7.4, 12.8.4, 12.8.5]

5.4. Europe

Present-day weather conditions affect natural, social, and economic systems in Europe in ways that reveal sensitivities and vulnerabilities to climate change in these systems. Climate change may aggravate such effects (very high confidence). Vulnerability to climate change in Europe differs substantially between subregions. Southern Europe and the European Arctic are more vulnerable than other parts of Europe. More-marginal and less-wealthy areas will be less able to adapt, which leads to important equity implications (very high confidence). Findings in the TAR relating to key vulnerabilities in Europe are broadly consistent with those expressed in the IPCC *Special Report on Regional Impacts of Climate Change* and the SAR, but are more specific about subregional effects and include new information concerning adaptive capacity. [13.1.1, 13.1.4, 13.4]

5.4.1. Water Resources

Water resources and their management in Europe are under pressure now, and these pressures are likely to be exacerbated by climate change (high confidence). Flood hazard is likely to increase across much of Europe—except where snowmelt peak has been reduced—and the risk of water shortage is projected to increase, particularly in southern Europe (medium to high confidence). Climate change is likely to widen water resource differences between northern and southern Europe (high confidence). Half of Europe's alpine glaciers could disappear by the end of the 21st century. [13.2.1]

5.4.2. Ecosystems

Natural ecosystems will change as a result of increasing temperature and atmospheric concentration of CO_2. Permafrost will decline; trees and shrubs will encroach into current northern tundra; and broad-leaved trees may encroach into current coniferous forest areas. Net primary productivity in ecosystems is likely to increase (also as a result of nitrogen deposition), but increases in decomposition resulting from increasing temperature may negate any additional carbon storage. Diversity in nature reserves is under threat from rapid change. Loss of important habitats (wetlands, tundra, and isolated habitats) would threaten some species (including rare/endemic species and migratory birds). Faunal shifts as a result of ecosystem changes are expected in marine, aquatic, and terrestrial ecosystems (high confidence; established but incomplete evidence). [13.2.1.4, 13.2.2.1, 13.2.2.3-5]

Soil properties will deteriorate under warmer and drier climate scenarios in southern Europe. The magnitude of this effect will vary markedly between geographic locations and may be modified by changes in precipitation (medium confidence; established but incomplete evidence). [13.2.1.2]

In mountain regions, higher temperatures will lead to an upward shift of biotic zones. There will be a redistribution of species with, in some instances, a threat of extinction (high confidence). [13.2.1.4]

Timber harvest will increase in commercial forests in northern Europe (medium confidence; established but incomplete evidence), although forest pests and disease may increase. Reductions are likely in the Mediterranean, with increased drought and fire risk (high confidence; well-established evidence). [13.2.2.1]

5.4.3. Agriculture and Food Security

Agricultural yields will increase for most crops as a result of increasing atmospheric CO_2 concentration. This increase in yields would be counteracted by the risk of water shortage in southern and eastern Europe and by shortening of the duration of growth in many grain crops because of increasing temperature. Northern Europe is likely to experience overall positive effects, whereas some agricultural production systems in southern Europe may be threatened (medium confidence; established but incomplete evidence).

Changes in fisheries and aquaculture production resulting from climate change embrace faunal shifts that affect freshwater and marine fish and shellfish biodiversity. These changes will be aggravated by unsustainable exploitation levels and environmental change (high confidence).

5.4.4. Human Settlements and Financial Services

The insurance industry faces potentially costly climate change impacts through the medium of property damage, but there is great scope for adaptive measures if initiatives are taken soon (high confidence). Transport, energy, and other industries will face changing demand and market opportunities. The concentration

***Table TS-10**: Estimates of flood exposure and incidence for Europe's coasts in 1990 and the 2080s. Estimates of flood incidence are highly sensitive to assumed protection standard and should be interpreted in indicative terms only (former Soviet Union excluded).*

Region	*1990* Exposed Population (millions)	Flood Incidence *1990* Average Number of People Experiencing Flooding (thousands yr^{-1})	Flood Incidence *2080s* Increase due to Sea-Level Rise, Assuming No Adaptation (%)
Atlantic Coast	19.0	19	50 to 9,000
Baltic Coast	1.4	1	0 to 3,000
Mediterranean Coast	4.1	3	260 to 12,0000

of industry on the coast exposes it to sea-level rise and extreme events, necessitating protection or removal (high confidence). [13.2.4]

Recreational preferences are likely to change with higher temperatures. Heat waves are likely to reduce the traditional peak summer demand at Mediterranean holiday destinations. Less-reliable snow conditions will impact adversely on winter tourism (medium confidence). [13.2.4.4]

The risk of flooding, erosion, and wetland loss in coastal areas will increase substantially, with implications for human settlement, industry, tourism, agriculture, and coastal natural habitats. Southern Europe appears to be more vulnerable to these changes, although the North Sea coast already has a high exposure to flooding (high confidence). Table TS-10 provides estimates of flood exposure and risk for Europe's coasts. [13.2.1.3]

5.4.5. *Human Health*

A range of risks is posed for human health through increased exposure to heat episodes (exacerbated by air pollution in urban areas), extension of some vector-borne diseases, and coastal and riverine flooding. Cold-related risks will be reduced (medium confidence; competing explanations). [13.2.5]

5.4.6. *Adaptive Capacity*

The adaptation potential of socioeconomic systems in Europe is relatively high because of economic conditions [high gross national product (GNP) and stable growth], a stable population (with the capacity to move within the region), and well-developed political, institutional, and technological support systems. However, the adaptation potential for natural systems generally is low (very high confidence). [13.3]

5.5. *Latin America*

There is ample evidence of climate variability at a wide range of time scales all over Latin America, from intraseasonal to long-term. In many subregions of Latin America, this variability in climate normally is associated with phenomena that already produce impacts with important socioeconomic and environmental consequences that could be exacerbated by global warming and its associated weather and climate changes.

Variations in precipitation have a strong effect on runoff and streamflow, which are simultaneously affected by melting of glaciers and snow. Precipitation variations and their sign depend on the geographical subregion under consideration. Temperature in Latin America also varies among subregions. Although these variations might depend on the origin and quality of the source data as well as on the record periods used for studies and analyses, some of these variations could be attributed to a climate change condition (low confidence). [14.1.2.1]

ENSO is responsible for a large part of the climate variability at interannual scales in Latin America (high confidence). The region is vulnerable to El Niño, with impacts varying across the continent. For example, El Niño is associated with dry conditions in northeast Brazil, northern Amazonia, the Peruvian-Bolivian Altiplano, and the Pacific coast of Central America. The most severe droughts in Mexico in recent decades have occurred during El Niño years, whereas southern Brazil and northwestern Peru have exhibited anomalously wet conditions. La Niña is associated with heavy precipitation and flooding in Colombia and drought in southern Brazil. If El Niño or La Niña were to increase, Latin America would be exposed to these conditions more often. [14.1.2]

Some subregions of Latin America frequently experience extreme events, and these extraordinary combinations of hydrological and climatic conditions historically have produced disasters in Latin America. Tropical cyclones and associated heavy rain, flooding, and landslides are very common in Central America and southern Mexico. In northwestern South America and northeastern Brazil, many of the extremes that occur are strongly related to El Niño. [14.1.2]

5.5.1. *Water Resources*

It has been well established that glaciers in Latin America have receded in the past several decades. Warming in high mountain regions could lead to disappearance of significant snow and ice surface (medium confidence), which could affect mountain sport and tourist activities. Because these areas contribute to river streamflow, this trend also would reduce water availability for irrigation, hydropower generation, and navigation. [14.2.4]

5.5.2. Ecosystems

It is well established that Latin America accounts for one of the Earth's largest concentrations of biodiversity, and the impacts of climate change can be expected to increase the risk of biodiversity loss (high confidence). Observed population declines in frogs and small mammals in Central America can be related to regional climate change. The remaining Amazonian forest is threatened by the combination of human disturbance, increases in fire frequency and scale, and decreased precipitation from evapotranspiration loss, global warming, and El Niño. Neotropical seasonally dry forest should be considered severely threatened in Mesoamerica.

Tree mortality increases under dry conditions that prevail near newly formed edges in Amazonian forests. Edges, which affect an increasingly large portion of the forest because of increased

Table TS-11*: Assessments of climate change impacts on annual crops in Latin America.*

Study[a]	Climate Scenario	Scope	Crop	Yield Impact (%)
Downing, 1992	+3°C -25% precipitation	Norte Chico, Chile	Wheat Maize Potato Grapes	decrease increase increase decrease
Baethgen, 1994	GISS, GFDL, UKMO	Uruguay	Wheat Barley	-30 -40 to -30
de Siqueira *et al.*, 1994	GISS, GFDL, UKMO	Brazil	Wheat Maize Soybeans	-50 to –15 -25 to –2 -10 to +40
Liverman and O' Brien, 1991	GFDL, GISS	Tlaltizapan, Mexico	Maize	-20 -24 -61
Liverman *et al.*, 1994	GISS, GFDL, UKMO	Mexico	Maize	-61 to –6
Sala and Paruelo, 1994	GISS, GFDL, UKMO	Argentina	Maize	-36 to -17
Baethgen and Magrin, 1995	UKMO	Argentina Uruguay (9 sites)	Wheat	-5 to -10
Conde *et al.*, 1997a	CCCM, GFDL	Mexico (7 sites)	Maize	increase-decrease
Magrin *et al.*, 1997a	GISS, UKMO, GFDL, MPI	Argentina (43 sites)	Maize Wheat Sunflower Soybean	-16 to +2 -8 to +7 -8 to +13 -22 to +21
Hofstadter *et al.*, 1997	Incremental	Uruguay	Barley Maize	-10[b] -8 to +5[c] -15[d] -13 to +10[c]

[a] See Chapter 14 reference list for complete source information.
[b] For 1°C increase.
[c] Change of -20 to +20% in precipitation.
[d] For 2°C increase.

deforestation, would be especially susceptible to the effects of reduced rainfall. In Mexico, nearly 50% of the deciduous tropical forest would be affected. Heavy rain during the 1997–1998 ENSO event generated drastic changes in dry ecosystems of northern Peru's coastal zone. Global warming would expand the area suitable for tropical forests as equilibrium vegetation types. However, the forces driving deforestation make it unlikely that tropical forests will be permitted to occupy these increased areas. Land-use change interacts with climate through positive-feedback processes that accelerate loss of humid tropical forests. [14.2.1]

5.5.3. Sea-Level Rise

Sea-level rise will affect mangrove ecosystems by eliminating their present habitats and creating new tidally inundated areas to which some mangrove species may shift. This also would affect the region's fisheries because most commercial shellfish and finfish use mangroves for nurseries and refuge. Coastal inundation that stems from sea-level rise and riverine and flatland flooding would affect water availability and agricultural land, exacerbating socioeconomic and health problems in these areas. [14.2.3]

5.5.4. Agriculture

Studies in Argentina, Brazil, Chile, Mexico, and Uruguay—based on GCMs and crop models—project decreased yields for numerous crops (e.g., maize, wheat, barley, grapes) even when the direct effects of CO_2 fertilization and implementation of moderate adaptation measures at the farm level are considered (high confidence). Predicted increases in temperature will reduce crops yields in the region by shortening the crop cycle. Over the past 40 years, the contribution of agriculture to the GDP of Latin American countries has been on the order of 10%. Agriculture remains a key sector in the regional economy because it employs 30–40% of the economically active population. It also is very important for the food security of the poorest sectors of the population. Subsistence farming could be severely threatened in some parts of Latin America, including northeastern Brazil.

It is established but incomplete that climate change would reduce silvicultural yields because lack of water often limits growth during the dry season, which is expected to become longer and more intense in many parts of Latin America. Table TS-11 summarizes studies undertaken on the region for different crops and management conditions, all under rainfed conditions; most of these results predict negative impacts, particularly for maize. [14.2.2]

5.5.5. Human Health

The scale of health impacts from climate change in Latin America would depend primarily on the size, density, location, and wealth of populations. Exposure to heat or cold waves has impacts on mortality rates in risk groups in the region (medium confidence).

Increases in temperature would affect human health in polluted cities such as Mexico City and Santiago, Chile. It is well established that ENSO causes changes in disease vector populations and in the incidence of water-borne diseases in Brazil, Peru, Bolivia, Argentina, and Venezuela. Studies in Peru and Cuba indicate that increases in temperature and precipitation would change the geographical distribution of infectious diseases such as cholera and meningitis (high confidence), although there is speculation about what the changes in patterns of diseases would be in different places. It is well established that extreme events tend to increase death and morbidity rates (injuries, infectious diseases, social problems, and damage to sanitary infrastructure), as shown in Central America with Hurricane Mitch in 1998, heavy rains in Mexico and Venezuela in 1999, and in Chile and Argentina in 2000. [14.2.5]

5.6. North America

North America will experience both positive and negative climate change impacts (high confidence). Varying impacts on ecosystems and human settlements will exacerbate subregional differences in climate-sensitive resource production and vulnerability to extreme events. Opportunities and challenges to adaptation will arise, frequently involving multiple stresses (Table TS-12). Some innovative adaptation strategies are being tested as a response to current climate-related challenges (e.g., water banks), but few cases have examined how these strategies could be implemented as regional climates continue to change. Shifting patterns in temperature, precipitation, disease vectors, and water availability will require adaptive responses—including, for example, investments in storm protection and water supply infrastructure, as well as community health services. [15.3.2, 15.4]

5.6.1. Communities and Urban Infrastructure

Potential changes in the frequency, severity, and duration of extreme events are among the most important risks associated with climate change in North America. Potential impacts of climate change on cities include fewer periods of extreme winter cold; increased frequency of extreme heat; rising sea levels and risk of storm surge; and changes in timing, frequency, and severity of flooding associated with storms and precipitation extremes. These events—particularly increased heat waves and changes in extreme events—will be accompanied by effects on health.

Communities can reduce their vulnerability to adverse impacts through investments in adaptive infrastructure, which can be expensive. Rural, poor, and indigenous communities may not be able to make such investments. Furthermore, infrastructure

***Table TS-12**: Climate change adaptation issues in North American subregions. Some unique issues for certain locations also are indicated.*

North American Subregions	Development Context	Climate Change Adaptation Options and Challenges
Most or all subregions	– Changing commodity markets – Intensive water resources development over large areas—domestic and transboundary – Lengthy entitlement/land claim/treaty agreements—domestic and transboundary – Urban expansion – Transportation expansion	– Role of water/environmental markets – Changing design and operations of water and energy systems – New technology/practices in agriculture and forestry – Protection of threatened ecosystems or adaptation to new landscapes – Increased role for summer (warm weather) tourism – Risks to water quality from extreme events – Managing community health for changing risk factors – Changing roles of public emergency assistance and private insurance
Arctic border	– Winter transport system – Indigenous lifestyles	– Design for changing permafrost and ice conditions – Role of two economies and co-management bodies
Coastal regions	– Declines in some commercial marine resources (cod, salmon) – Intensive coastal zone development	– Aquaculture, habitat protection, fleet reductions – Coastal zone planning in high demand areas
Great Lakes	– Sensitivity to lake level fluctuations	– Managing for reduction in mean levels without increased shoreline encroachment

investment decisions are based on a variety of needs beyond climate change, including population growth and aging of existing systems. [15.2.5]

5.6.2. Water Resources and Aquatic Ecosystems

Uncertain changes in precipitation lead to little agreement regarding changes in total annual runoff across North America. Modeled impacts of increased temperatures on lake evaporation lead to consistent projections of reduced lake levels and outflows for the Great Lakes-St. Lawrence system under most scenarios (medium confidence). Increased incidence of heavy precipitation events will result in greater sediment and non-point-source pollutant loadings to watercourses (medium confidence). In addition, *in regions where seasonal snowmelt is an important aspect of the annual hydrologic regime (e.g., California, Columbia River Basin), warmer temperatures are likely to result in a seasonal shift in runoff, with a larger proportion of total runoff occurring in winter, together with possible reductions in summer flows (high confidence)*. This could adversely affect the availability and quality of water for instream and out-of-stream water uses during the summer (medium confidence). Figure TS-8 shows possible impacts. [15.2.1]

Adaptive responses to such seasonal runoff changes include altered management of artificial storage capacity, increased reliance on coordinated management of groundwater and surface water supplies, and voluntary water transfers between various water users. Such actions could reduce the impacts of reduced summer flows on water users, but it may be difficult or impossible to offset adverse impacts on many aquatic ecosystems, and it may not be possible to continue to provide current levels of reliability and quality for all water users. Some regions (e.g., the western United States) are likely to see increased market transfers of available water supplies from irrigated agriculture to urban and other relatively highly valued uses. Such reallocations raise social priority questions and entail adjustment costs that will depend on the institutions in place.

5.6.3. Marine Fisheries

Climate-related variations in marine/coastal environments are now recognized as playing an important role in determining the productivity of several North American fisheries in the Pacific, North Atlantic, Bering Sea, and Gulf of Mexico regions. There are complex links between climatic variations and changes in processes that influence the productivity and spatial distribution

I. Alaska, Yukon, and Coastal British Columbia
Lightly settled/water-abundant region; potential ecological, hydropower, and flood impacts:
- Increased spring flood risks
- Glacial retreat/disappearance in south, advance in north; impacts on flows, stream ecology
- Increased stress on salmon, other fish species
- Flooding of coastal wetlands
- Changes in estuary salinity/ecology

II. Pacific Coast States (USA)
Large and rapidly growing population; water abundance decreases north to south; intensive irrigated agriculture; massive water-control infrastructure; heavy reliance on hydropower; endangered species issues; increasing competition for water:
- More winter rainfall/less snowfall—earlier seasonal peak in runoff, increased fall/winter flooding, decreased summer water supply
- Possible increases in annual runoff in Sierra Nevada and Cascades
- Possible summer salinity increase in San Francisco Bay and Sacramento/San Joaquin Delta
- Changes in lake and stream ecology—warmwater species benefitting; damage to coldwater species (e.g., trout and salmon)

III. Rocky Mountains (USA and Canada)
Lightly populated in north, rapid population growth in south; irrigated agriculture, recreation, urban expansion increasingly competing for water; headwaters area for other regions:
- Rise in snow line in winter-spring, possible increases in snowfall, earlier snowmelt, more frequent rain on snow—changes in seasonal streamflow, possible reductions in summer streamflow, reduced summer soil moisture
- Stream temperature changes affecting species composition; increased isolation of coldwater stream fish

IV. Southwest
Rapid population growth, dependence on limited groundwater and surface water supplies, water quality concerns in border region, endangered species concerns, vulnerability to flash flooding:
- Possible changes in snowpacks and runoff
- Possible declines in groundwater recharge—reduced water supplies
- Increased water temperatures—further stress on aquatic species
- Increased frequency of intense precipitation events—increased risk of flash floods

V. Sub-Arctic and Arctic
Sparse population (many dependent on natural systems); winter ice cover important feature of hydrologic cycle:
- Thinner ice cover, 1- to 3-month increase in ice-free season, increased extent of open water
- Increased lake-level variability, possible complete drying of some delta lakes
- Changes in aquatic ecology and species distribution as a result of warmer temperatures and longer growing season

VI. Midwest USA and Canadian Prairies
Agricultural heartland—mostly rainfed, with some areas relying heavily on irrigation:
- Annual streamflow decreasing/increasing; possible large declines in summer streamflow
- Increasing likelihood of severe droughts
- Possible increasing aridity in semi-arid zones
- Increases or decreases in irrigation demand and water availability—uncertain impacts on farm-sector income, groundwater levels, streamflows, and water quality

VII. Great Lakes
Heavily populated and industrialized region; variations in lake levels/flows now affect hydropower, shipping, shoreline structures:
- Possible precipitation increases coupled with reduced runoff and lake-level declines
- Reduced hydropower production; reduced channel depths for shipping
- Decreases in lake ice extent—some years w/out ice cover
- Changes in phytoplankton/zooplankton biomass, northward migration of fish species, possible extirpations of coldwater species

VIII. Northeast USA and Eastern Canada
Large, mostly urban population—generally adequate water supplies, large number of small dams, but limited total reservoir capacity; heavily populated floodplains:
- Decreased snow cover amount and duration
- Possible large reductions in streamflow
- Accelerated coastal erosion, saline intrusion into coastal aquifers
- Changes in magnitude, timing of ice freeze-up/break-up, with impacts on spring flooding
- Possible elimination of bog ecosystems
- Shifts in fish species distributions, migration patterns

IX. Southeast, Gulf, and Mid-Atlantic USA
Increasing population—especially in coastal areas, water quality/non-point source pollution problems, stress on aquatic ecosystems:
- Heavily populated coastal floodplains at risk to flooding from extreme precipitation events, hurricanes
- Possible lower base flows, larger peak flows, longer droughts
- Possible precipitation increase—possible increases or decreases in runoff/river discharge, increased flow variability
- Major expansion of northern Gulf of Mexico hypoxic zone possible—other impacts on coastal systems related to changes in precipitation/non-point source pollutant loading
- Changes in estuary systems and wetland extent, biotic processes, species distribution

Figure TS-8: Possible water resources impacts in North America.

of marine fish populations (high confidence), as well as uncertainties linked to future commercial fishing patterns. Recent experience with Pacific salmon and Atlantic cod suggests that sustainable fisheries management will require timely and accurate scientific information on environmental conditions affecting fish stocks, as well as institutional and operational flexibility to respond quickly to such information. [15.2.3.3]

5.6.4. *Agriculture*

Small to moderate climate change will not imperil food and fiber production (high confidence). There will be strong regional production effects, with some areas suffering significant loss of comparative advantage to other regions (medium confidence). Overall, this results in a small net effect. The agricultural welfare of consumers and producers would increase with modest warming. However, the benefit would decline at an increasing rate—possibly becoming a net loss—with further warming. There is potential for increased drought in the U.S. Great Plains/Canadian Prairies and opportunities for a limited northward shift in production areas in Canada.

Increased production from direct physiological effects of CO_2, and farm- and agricultural market-level adjustments (e.g., behavioral, economic, and institutional) are projected to offset losses. Economic studies that include farm- and agricultural market-level adjustments indicate that the negative effects of climate change on agriculture probably have been overestimated by studies that do not account for these adjustments (medium confidence). However, the ability of farmers to adapt their input and output choices is difficult to forecast and will depend on market and institutional signals. [15.2.3.1]

5.6.5. *Forests and Protected Areas*

Climate change is expected to increase the areal extent and productivity of forests over the next 50–100 years (medium confidence). However, climate change is likely to cause changes in the nature and extent of several "disturbance factors" (e.g., fire, insect outbreaks) (medium confidence). Extreme or long-term climate change scenarios indicate the possibility of widespread forest decline (low confidence).

There is strong evidence that climate change can lead to the loss of specific ecosystem types—such as high alpine areas and specific coastal (e.g., salt marshes) and inland (e.g., prairie "potholes") wetland types (high confidence). There is moderate potential for adaptation to prevent these losses by planning conservation programs to identify and protect particularly threatened ecosystems. Lands that are managed for timber production are likely to be less susceptible to climate change than unmanaged forests because of the potential for adaptive management. [15.2.2]

5.6.6. *Human Health*

Vector-borne diseases, including malaria and dengue fever, may expand their ranges in the United States and may develop in Canada. Tick-borne Lyme disease also may see its range expanded in Canada. However, socioeconomic factors such as public health measures will play a large role in determining the existence or extent of such infections. Diseases associated with water may increase with warming of air and water temperatures, combined with heavy runoff events from agricultural and urban surfaces. Increased frequency of convective storms could lead to more cases of thunderstorm-associated asthma. [15.2.4]

5.6.7. *Public and Private Insurance Systems*

Inflation-corrected catastrophe losses have increased eight-fold in North America over the past 3 decades (high confidence). The exposures and surpluses of private insurers (especially property insurers) and reinsurers have been growing, and weather-related profit losses and insolvencies have been observed. Insured losses in North America (59% of the global total) are increasing with affluence and as populations continue to move into vulnerable areas. Insurer vulnerability to these changes varies considerably by region.

Recent extreme events have led to several responses by insurers, including increased attention to building codes and disaster preparedness. Insurers' practices traditionally have been based primarily on historic climatic experience; only recently have they begun to use models to predict future climate-related losses, so the potential for surprise is real. Governments play a key role as insurers or providers of disaster relief, especially in cases in which the private sector deems risks to be uninsurable. [15.2.7]

5.7. *Polar Regions*

Climate change in the polar region is expected to be among the greatest of any region on Earth. Twentieth century data for the Arctic show a warming trend of as much as 5°C over extensive land areas (very high confidence), while precipitation has increased (low confidence). There are some areas of cooling in eastern Canada. The extent of sea ice has decreased by 2.9% per decade, and it has thinned over the 1978–1996 period (high confidence). There has been a statistically significant decrease in spring snow extent over Eurasia since 1915 (high confidence). The area underlain by permafrost has been reduced and has warmed (very high confidence). The layer of seasonally thawed ground above permafrost has thickened in some areas, and new areas of extensive permafrost thawing have developed. *In the Antarctic, a marked warming trend is evident in the Antarctic Peninsula, with spectacular loss of ice shelves (high confidence).* The extent of higher terrestrial vegetation on the Antarctic Peninsula is increasing (very high confidence). Elsewhere, warming is less definitive. There has been no significant change in the Antarctic sea ice since 1973, although it apparently retreated by more than 3° of latitude between the mid-1950s and the early 1970s (medium confidence). [16.1.3.2.]

The Arctic is extremely vulnerable to climate change, and major physical, ecological, and economic impacts are expected to appear rapidly. A variety of feedback mechanisms will cause an amplified response, with consequent impacts on other systems and people. There will be different species compositions on land and sea, poleward shifts in species assemblages, and severe disruptions for communities of people who lead traditional lifestyles. *In developed areas of the Arctic and where the permafrost is ice-rich, special attention will be required to mitigate the detrimental impacts of thawing, such as severe damage to buildings and transport infrastructure (very high confidence).* There also will be beneficial consequences of climatic warming, such as reduced demand for heating energy. Substantial loss of sea ice in the Arctic Ocean will be favorable for opening of Arctic sea routes and ecotourism, which may have large implications for trade and for local communities. [16.2.5.3, 16.2.7.1, 16.2.8.1, 16.2.8.2]

In the Antarctic, projected climate change will generate impacts that will be realized slowly (high confidence). Because the impacts will occur over a long period, however, they will continue long after GHG emissions have stabilized. For example, there will be slow but steady impacts on ice sheets and circulation patterns of the global ocean, which will be irreversible for many centuries into the future and will cause changes elsewhere in the world, including a rise of sea level. Further substantial loss of ice shelves is expected around the Antarctic Peninsula. Warmer temperatures and reduced sea-ice extent are likely to produce long-term changes in the physical oceanography and ecology of the Southern Ocean, with intensified biological activity and increased growth rates of fish. [16.2.3.4, 16.2.4.2]

Polar regions contain important drivers of climate change. The Southern Ocean's uptake of carbon is projected to reduce substantially as a result of complex physical and biological processes. GHG emissions from tundra caused by changes in water content, decomposition of exposed peat, and thawing of permafrost are expected to increase. Reductions in the extent of highly reflective snow and ice will magnify warming (very high confidence). Freshening of waters from increased Arctic runoff and increased rainfall, melt of Antarctic ice shelves, and reduced sea-ice formation will slow the thermohaline circulations of the North Atlantic and Southern Oceans and reduce the ventilation of deep ocean waters. [16.3.1]

Adaptation to climate change will occur in natural polar ecosystems, mainly through migration and changing mixes of species. Some species may become threatened (e.g., walrus, seals, and polar bears), whereas others may flourish (e.g., caribou and fish). Although such changes may be disruptive to many local ecological systems and particular species, the possibility remains that predicted climate change eventually may increase the overall productivity of natural systems in polar regions. [16.3.2]

For indigenous communities who follow traditional lifestyles, opportunities for adaptation to climate change are limited (very high confidence). Changes in sea ice, seasonality of snow, habitat, and diversity of food species will affect hunting and gathering practices and could threaten longstanding traditions and ways of life. Technologically developed communities are likely to adapt quite readily to climate change by adopting altered modes of transport and by increased investment to take advantage of new commercial and trade opportunities. [16.3.2]

5.8. *Small Island States*

Climate change and sea-level rise pose a serious threat to the small island states, which span the ocean regions of the Pacific, Indian, and Atlantic Oceans as well as the Caribbean and Mediterranean Seas. Characteristics of small island states that increase their vulnerability include their small physical size relative to large expanses of ocean; limited natural resources; relative isolation; extreme openness of small economies that are highly sensitive to external shocks and highly prone to natural disasters and other extreme events; rapidly growing populations with high densities; poorly developed infrastructure; and limited funds, human resources, and skills. These characteristics limit the capacity of small island states to mitigate and adapt to future climate change and sea-level rise. [17.1.2]

Many small island states already are experiencing the effects of current large interannual variations in oceanic and atmospheric conditions. As a result, the most significant and more immediate consequences for small island states are likely to be related to changes in rainfall regimes, soil moisture budgets, prevailing winds (speed and direction), short-term variations in regional and local sea levels, and patterns of wave action. These changes are manifest in past and present trends of climate and climate variability, with an upward trend in average temperature by as much as 0.1°C per decade and sea-level rise of 2 mm yr^{-1} in the tropical ocean regions in which most of the small island states are located. Analysis of observational data from various regions indicates an increase in surface air temperature that has been greater than global rates of warming, particularly in the Pacific Ocean and Caribbean Sea. Much of the variability in the rainfall record of the Pacific and Caribbean islands appears to be closely related to the onset of ENSO. However, part of the variability also may be attributable to shifts in the Intertropical and South Pacific Convergence Zone, whose influence on rainfall variability patterns must be better understood. The interpretation of current sea-level trends also is constrained by limitations of observational records, particularly from geodetic-controlled tide gauges. [17.1.3]

5.8.1. *Equity and Sustainable Development*

Although the contribution of small island states to global emissions of GHG is insignificant, projected impacts of climate change and sea-level rise on these states are likely to be

Table TS-13*: Importance of tourism for select small island states and territories.*

Country	**Number of Tourists (000s)**[a]	**Tourists as % of Population**[a]	**Tourist Receipts**[b]	
			as % of GNP	as % of Exports
Antigua and Barbuda	232	364	63	74
Bahamas	1618	586	42	76
Barbados	472	182	39	56
Cape Verde	45	11	12	37
Comoros	26	5	11	48
Cuba	1153	11	9	n/a
Cyprus	2088	281	24	49
Dominica	65	98	16	33
Dominican Republic	2211	28	14	30
Fiji	359	45	19	29
Grenada	111	116	27	61
Haiti	149	2	4	51
Jamaica	1192	46	32	40
Maldives	366	131	95	68
Malta	1111	295	23	29
Mauritius	536	46	16	27
Papua New Guinea	66	2	2	3
St. Kitts and Nevis	88	211	31	64
St. Lucia	248	165	41	67
St. Vincent	65	55	24	46
Samoa	68	31	20	49
Seychelles	130	167	35	52
Singapore	7198	209	6	4
Solomon Islands	16	4	3	4
Trinidad and Tobago	324	29	4	8
Vanuatu	49	27	19	41

[a] Data on tourist inflows and ratio to population pertain to 1997.
[b] Data for tourist receipts pertain to 1997 for the Bahamas, Cape Verde, Jamaica, the Maldives, Malta, Mauritius, Samoa, Seychelles, Singapore, and Solomon Islands; 1996 for Antigua and Barbuda, Cuba, Dominica, Dominican Republic, Fiji, Grenada, Haiti, Papua New Guinea, St. Lucia, and St. Vincent; 1995 for Barbados, Comoros, Cyprus, Trinidad and Tobago, and Vanuatu; and 1994 for St. Kitts and Nevis.

serious. The impacts will be felt for many generations because of small island states' low adaptive capacity, high sensitivity to external shocks, and high vulnerability to natural disasters. Adaptation to these changing conditions will be extremely difficult for most small island states to accomplish in a sustainable manner. [17.2.1]

5.8.2. *Coastal Zone*

Much of the coastal change currently experienced in small island states is attributed to human activities on the coast. Projected sea-level rise of 5 mm yr^{-1} over the next 100 years, superimposed on further coastal development, will have negative impacts on the coasts (high confidence). This in turn will increase the vulnerability of coastal environments by reducing natural resilience and increasing the cost of adaptation. Given that severity will vary regionally, the most serious considerations for some small island states will be whether they will have adequate potential to adapt to sea-level rise within their own national boundaries. [17.2.2.1, 17.2.3]

5.8.3. *Ecosystems and Biodiversity*

Projected future climate change and sea-level rise will affect shifts in species composition and competition. It is estimated that one of every three known threatened plants are island endemics while 23% of bird species found on islands are threatened. [17.2.5]

Coral reefs, mangroves, and seagrass beds that often rely on stable environmental conditions will be adversely affected by rising air and sea temperature and sea levels (medium confidence). Episodic warming of the sea surface has resulted in greatly stressed coral populations that are subject to widespread coral bleaching. Mangroves, which are common on low-energy, nutrient/sediment-rich coasts and embayments in the tropics, have been altered by human activities. Changes in sea levels are likely to affect landward and alongshore migration of remnants of mangrove forests that provide protection for coasts and other resources. An increase in SST would adversely affect seagrass communities, which already are under stress from land-based pollution and runoff. Changes in these systems are likely to negatively affect fishery populations that depend on them for habitat and breeding grounds. [17.2.4]

5.8.4. Water Resources, Agriculture, and Fisheries

Water resources and agriculture are of critical concern because most small island states possess limited arable land and water resources. Communities rely on rainwater from catchments and a limited freshwater lens. In addition, arable farming, especially on low islands and atolls, is concentrated at or near the coast. Changes in the height of the water table and soil salinization as a consequence of sea-level rise would be stressful for many staple crops, such as taro.

Although fishing is largely artisinal or small-scale commercial, it is an important activity on most small islands and makes a significant contribution to the protein intake of island inhabitants. Many breeding grounds and habitats for fish and shellfish—such as mangroves, coral reefs, seagrass beds, and salt ponds—will face increasing threats from likely impacts of projected climate change. Water resources, agriculture, and fisheries already are sensitive to currently observed variability in oceanic and atmospheric conditions in many small island states, and the impacts are likely to be exacerbated by future climate and sea-level change (high confidence). [17.2.6, 17.2.8.1]

5.8.5. Human Health, Settlement, Infrastructure and Tourism

Several human systems are likely to be affected by projected changes in climate and sea levels in many small island states. Human health is a major concern given that many tropical islands are experiencing high incidences of vector- and water-borne diseases that are attributable to changes in temperature and rainfall, which may be linked to the ENSO phenomenon, droughts, and floods. Climate extremes also create a huge burden on some areas of human welfare, and these burdens are likely to increase in the future. Almost all settlements, socioeconomic infrastructure, and activities such as tourism are located at or near coastal areas in small island states. Tourism provides a major source of revenue and employment for many small island states (Table TS-13). Changes in temperature and rainfall regimes, as well as loss of beaches, could devastate the economies of many small island states (high confidence). Because these areas are very vulnerable to future climate change and sea-level rise, it is important to protect and nourish beaches and sites by implementing programs that constitute wise use resources. Integrated coastal management has been identified as one approach that would be useful for many small island states for a sustainable tourism industry. [17.2.7, 17.2.9]

5.8.6. Sociocultural and Traditional Assets

Certain traditional island assets (good and services) also will be at risk from climate change and sea-level rise. These assets include subsistence and traditional technologies (skills and knowledge) and cohesive community structures that, in the past, have helped to buttress the resilience of these islands to various forms of shock. Sea-level rise and climate changes, combined with other environmental stresses, already have destroyed unique cultural and spiritual sites, traditional heritage assets, and important coastal protected areas in many Pacific island states. [17.2.10]

6. Adaptation, Sustainable Development, and Equity

Adaptation to climate change has the potential to substantially reduce many of the adverse impacts of climate change and enhance beneficial impacts, though neither without cost nor without leaving residual damage. In natural systems, adaptation is reactive, whereas in human systems it also can be anticipatory. Figure TS-9 presents types and examples of adaptation to climate change. Experience with adaptation to climate variability and extremes shows that in the private and public sectors there are constraints to achieving the potential of adaptation. The adoption and effectiveness of private, or market-driven, adaptations in sectors and regions are limited by other forces, institutional conditions, and various sources of market failure. There is little evidence to suggest that private adaptations will be employed to offset climate change damages in natural environments. In some instances, adaptation measures may have inadvertent consequences, including environmental damage. The ecological, social, and economic costs of relying on reactive, autonomous adaptation to the cumulative effects of climate change are substantial. Many of these costs can be avoided through planned, anticipatory adaptation. Designed appropriately, many adaptation strategies could provide multiple benefits in the near and longer terms. However, there are limits on their implementation and effectiveness. Enhancement of adaptive capacity reduces the vulnerability of sectors and regions to climate change, including variability and extremes, and thereby promotes sustainable development and equity. [18.2.4, 18.3.4]

Planned anticipatory adaptation has the potential to reduce vulnerability and realize opportunities associated with climate change, regardless of autonomous adaptation. Adaptation facilitated

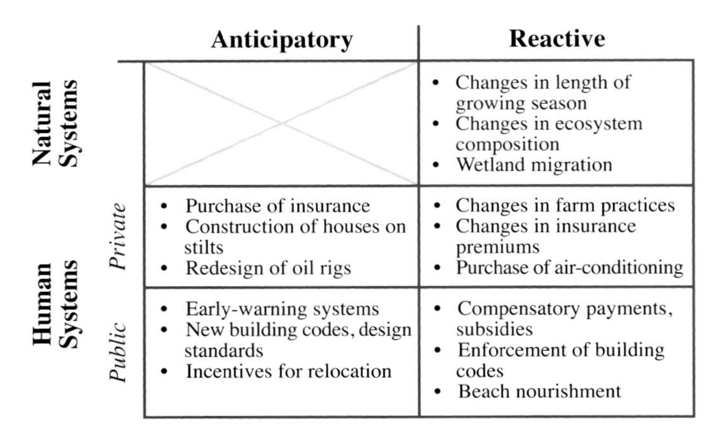

Figure TS-9: Types of adaptation to climate change, including examples.

by public agencies is an important part of societal response to climate change. Implementation of adaptation policies, programs, and measures usually will have immediate and future benefits. Adaptations to current climate and climate-related risks (e.g., recurring droughts, storms, floods, and other extremes) generally are consistent with adaptation to changing and changed climatic conditions. Adaptation measures are likely to be implemented only if they are consistent with or integrated with decisions or programs that address nonclimatic stresses. Vulnerabilities associated with climate change are rarely experienced independently of nonclimatic conditions. Impacts of climatic stimuli are felt via economic or social stresses, and adaptations to climate (by individuals, communities, and governments) are evaluated and undertaken in light of these conditions. The costs of adaptation often are marginal to other management or development costs. To be effective, climate change adaptation must consider nonclimatic stresses and be consistent with existing policy criteria, development objectives, and management structures. [18.3.5, 18.4]

The key features of climate change for vulnerability and adaptation are related to variability and extremes, not simply changed average conditions (Figure TS-10). Societies and economies have been making adaptations to climate for centuries. Most sectors, regions, and communities are reasonably adaptable to changes in average conditions, particularly if the changes are gradual. However, losses from climatic variations and extremes are substantial and, in some sectors, increasing. These losses indicate that autonomous adaptation has not been sufficient to offset damages associated with temporal variations in climatic conditions. Communities therefore are more vulnerable and less adaptable to changes in the frequency and/or magnitude of conditions other than average, especially extremes, which are inherent in climate change. The degree to which future adaptations are successful in offsetting adverse impacts of climate change will be determined by success in adapting to climate change, variability, and extremes. [18.2.2]

6.1. Adaptive Capacity

The capacity to adapt varies considerably among regions, countries, and socioeconomic groups and will vary over time. Table TS-14 summarizes adaptation measures and capacities by sector, and Table TS-15 provides this information for each region covered by the TAR. The most vulnerable regions and communities are highly exposed to hazardous climate change effects and have limited adaptive capacity. The ability to adapt and cope with climate change impacts is a function of wealth, scientific and technical knowledge, information, skills, infrastructure, institutions, and equity. Countries with limited economic resources, low levels of technology, poor information and skills, poor infrastructure, unstable or weak institutions, and inequitable empowerment and access to resources have little capacity to adapt and are highly vulnerable. Groups and regions with adaptive capacity that is limited along any of these dimensions are more vulnerable to climate change damages, just as they are more vulnerable to other stresses. [18.5, 18.7]

6.2. Development, Sustainability, and Equity

Activities required for enhancement of adaptive capacity are essentially equivalent to those promoting sustainable development. Enhancement of adaptive capacity is a necessary condition for reducing vulnerability, particularly for the most vulnerable regions, nations, and socioeconomic groups. Many sectors and regions that are vulnerable to climate change also are under pressure from forces such as population growth and resource depletion. Climate adaptation and sustainability goals can be jointly advanced by changes in policies that lessen pressure on resources, improve management of environmental risks, and enhance adaptive capacity. Climate adaptation and equity goals can be jointly pursued through initiatives that promote the welfare of the poorest members of society—for example, by improving food security, facilitating access to safe water and health care, and providing shelter and access to other resources. Development decisions, activities, and programs play important roles in modifying the adaptive capacity of communities and regions, yet they tend not to take into account risks associated with climate variability and change. Inclusion of climatic risks

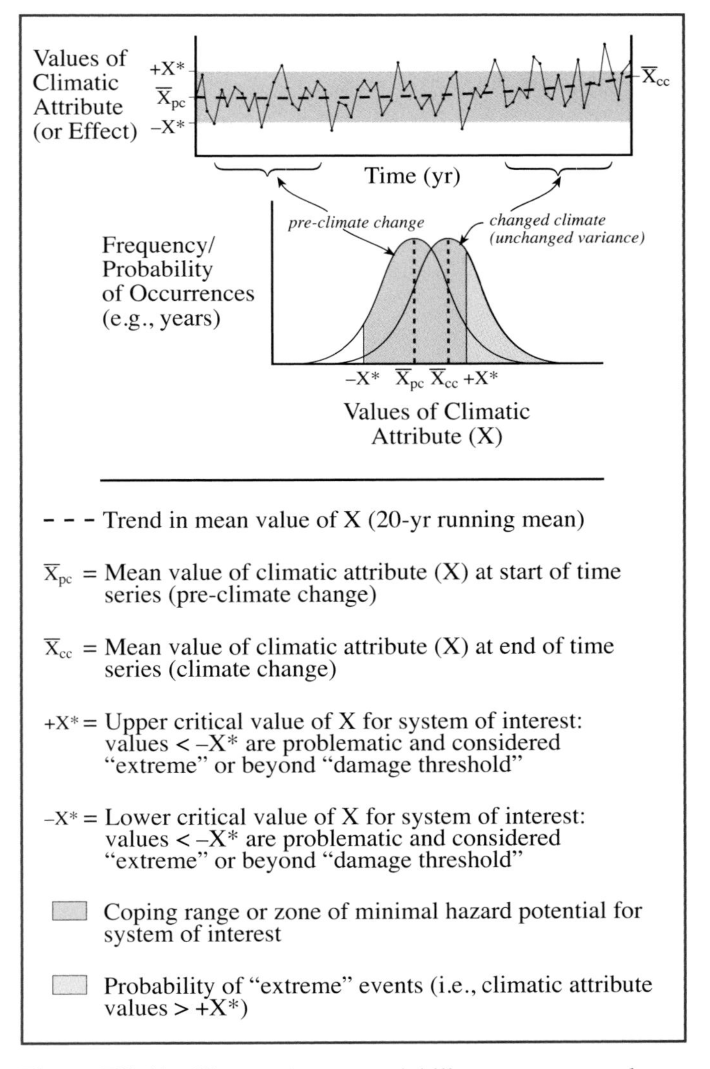

Figure TS-10: Climate change, variability, extremes, and coping range.

***Table TS-14**: Adaptation and adaptive capacity in sectors (key findings from Chapters 4 through 9).*

Sector	Key Findings
Water Resources	– Water managers have experience with adapting to change. Many techniques exist to assess and implement adaptive options. However, the pervasiveness of climate change may preclude some traditional adaptive strategies, and available adaptations often are not used. – Adaptation can involve management on the supply side (e.g., altering infrastructure or institutional arrangements) and on the demand side (changing demand or risk reduction). Numerous no-regret policies exist, which will generate net social benefits regardless of climate change. – Climate change is just one of numerous pressures facing water managers. Nowhere are water management decisions taken solely to cope with climate change, although it is increasingly considered for future resource management. Some vulnerabilities are outside the conventional responsibility of water managers. – Estimates of the economic costs of climate change impacts on water resources depend strongly on assumptions made about adaptation. Economically optimum adaptation may be prevented by constraints associated with uncertainty, institutions, and equity. – Extreme events often are catalysts for change in water management, by exposing vulnerabilities and raising awareness of climate risks. Climate change modifies indicators of extremes and variability, complicating adaptation decisions. – Ability to adapt is affected by institutional capacity, wealth, management philosophy, planning time scale, organizational and legal framework, technology, and population mobility. – Water managers need research and management tools aimed at adapting to uncertainty and change, rather than improving climate scenarios.
Ecosystems and Their Services	– Adaptation to loss of some ecosystem services may be possible, especially in managed ecosystems. However, adaptation to losses in wild ecosystems and biodiversity may be difficult or impossible. – There is considerable capacity for adaptation in agriculture, including crop changes and resource substitutions, but adaptation to evolving climate change and interannual variability is uncertain. – Adaptations in agriculture are possible, but they will not happen without considerable transition costs and equilibrium (or residual) costs. – Greater adverse impacts are expected in areas where resource endowments are poorest and the ability of farmers to adapt is most limited. – In many countries where rangelands are important, lack of infrastructure and investment in resource management limit options for adaptation. – Commercial forestry is adaptable, reflecting a history of long-term management decisions under uncertainty. Adaptations are expected in land-use management (species-selection silviculture) and product management (processing-marketing). – Adaptation in developed countries will fare better, while developing countries and countries in transition, especially in the tropics and subtropics, will fare worse.
Coastal Zones	– Without adaptations, the consequences of global warming and sea-level rise would be disastrous. – Coastal adaptation entails more than just selecting one of the technical options to respond to sea-level rise (strategies can aim to protect, accommodate, or retreat). It is a complex and iterative process rather than a simple choice. – Adaptation options are more acceptable and effective when they are incorporated into coastal zone management, disaster mitigation programs, land-use planning, and sustainable development strategies. – Adaptation choices will be conditioned by existing policies and development objectives, requiring researchers and policymakers to work toward a commonly acceptable framework for adaptation. – The adaptive capacity of coastal systems to perturbations is related to coastal resilience, which has morphological, ecological, and socioeconomic components. Enhancing resilience—including the technical, institutional, economic, and cultural capability to cope with impacts—is a particularly appropriate adaptive strategy given future uncertainties and the desire to maintain development opportunities. – Coastal communities and marine-based economic sectors with low exposure or high adaptive capacity will be least affected. Communities with lower economic resources, poorer infrastructure, less-developed communications and transportation systems, and weak social support systems have less access to adaptation options and are more vulnerable.

Table TS-14 (continued)

Sector	Key Findings
Human Settlements, Energy, and Industry	– Larger and more costly impacts of climate change occur through changed probabilities of extreme weather events that overwhelm the design resiliency of human systems. – Many adaptation options are available to reduce the vulnerability of settlements. However, urban managers, especially in developing countries, have so little capacity to deal with current problems (housing, sanitation, water, and power) that dealing with climate change risks is beyond their means. – Lack of financial resources, weak institutions, and inadequate or inappropriate planning are major barriers to adaptation in human settlements. – Successful environmental adaptation cannot occur without locally based, technically competent, and politically supported leadership. – Uncertainty with respect to capacity and the will to respond hinder assessment of adaptation and vulnerability.
Insurance and Other Financial Services	– Adaptation in financial and insurance services in the short term is likely to be to changing frequencies and intensities of extreme weather events. – Increasing risk could lead to a greater volume of traditional business and development of new financial risk management products, but increased variability of loss events would heighten actuarial uncertainty. – Financial services firms have adaptability to external shocks, but there is little evidence that climate change is being incorporated into investment decisions. – The adaptive capacity of the financial sector is influenced by regulatory involvement, the ability of firms to withdraw from at-risk markets, and fiscal policy regarding catastrophe reserves. – Adaptation will involve changes in the roles of private and public insurance. Changes in the timing, intensity, frequency, and/or spatial distribution of climate-related losses will generate increased demand on already overburdened government insurance and disaster assistance programs. – Developing countries seeking to adapt in a timely manner face particular difficulties, including limited availability of capital, poor access to technology, and absence of government programs. – Insurers' adaptations include raising prices, non-renewal of policies, cessation of new policies, limiting maximum claims, and raising deductibles—actions that can seriously affect investment in developing countries. – Developed countries generally have greater adaptive capacity, including technology and economic means to bear costs.
Human Health	– Adaptation involves changes in society, institutions, technology, or behavior to reduce potential negative impacts or increase positive ones. There are numerous adaptation options, which may occur at the population, community, or personal levels. – The most important and cost-effective adaptation measure is to rebuild public health infrastructure—which, in much of the world, has declined in recent years. Many diseases and health problems that may be exacerbated by climate change can be effectively prevented with adequate financial and human public health resources, including training, surveillance and emergency response, and prevention and control programs. – Adaptation effectiveness will depend on timing. "Primary" prevention aims to reduce risks before cases occur, whereas "secondary" interventions are designed to prevent further cases. – Determinants of adaptive capacity to climate-related threats include level of material resources, effectiveness of governance and civil institutions, quality of public health infrastructure, and preexisting burden of disease. – Capacity to adapt also will depend on research to understand associations between climate, weather, extreme events, and vector-borne diseases.

in the design and implementation of development initiatives is necessary to reduce vulnerability and enhance sustainability. [18.6.1]

7. Global Issues and Synthesis

7.1. Detection of Climate Change Impacts

Observational evidence indicates that climate changes in the 20th century already have affected a diverse set of physical and biological systems. Examples of observed changes with linkages to climate include shrinkage of glaciers; thawing of permafrost; shifts in ice freeze and break-up dates on rivers and lakes; increases in rainfall and rainfall intensity in most mid- and high latitudes of the Northern Hemisphere; lengthening of growing seasons; and earlier flowering dates of trees, emergence of insects, and egg-laying in birds. Statistically significant associations between changes in regional climate and observed changes in physical and biological systems have been documented in freshwater, terrestrial, and marine environments on all continents. [19.2]

***Table TS-15**: Adaptation and capacity in regions (key findings from Chapters 10 through 17).*

Sector	Key Findings
Africa	– Adaptive measures would enhance flexibility and have net benefits in water resources (irrigation and water reuse, aquifer and groundwater management, desalinization), agriculture (crop changes, technology, irrigation, husbandry), and forestry (regeneration of local species, energy-efficient cook stoves, sustainable community management). – Without adaptation, climate change will reduce the wildlife reserve network significantly by altering ecosystems and causing species' emigrations and extinctions. This represents an important ecological and economic vulnerability in Africa. – A risk-sharing approach between countries will strengthen adaptation strategies, including disaster management, risk communication, emergency evacuation, and cooperative water resource management. – Most countries in Africa are particularly vulnerable to climate change because of limited adaptive capacity as a result of widespread poverty, recurrent droughts, inequitable land distribution, and dependence on rainfed agriculture. – Enhancement of adaptive capacity requires local empowerment in decisionmaking and incorporation of climate adaptation within broader sustainable development strategies.
Asia	– Priority areas for adaptation are land and water resources, food productivity, and disaster preparedness and planning, particularly for poorer, resource-dependent countries. – Adaptations already are required to deal with vulnerabilities associated with climate variability, in human health, coastal settlements, infrastructure, and food security. Resilience of most sectors in Asia to climate change is very poor. Expansion of irrigation will be difficult and costly in many countries. – For many developing countries in Asia, climate change is only one of a host of problems to deal with, including nearer term needs such as hunger, water supply and pollution, and energy. Resources available for adaptation to climate are limited. Adaptation responses are closely linked to development activities, which should be considered in evaluating adaptation options. – Early signs of climate change already have been observed and may become more prominent over 1 or 2 decades. If this time is not used to design and implement adaptations, it may be too late to avoid upheavals. Long-term adaptation requires anticipatory actions. – A wide range of precautionary measures are available at the regional and national level to reduce economic and social impacts of disasters. These strategies include awareness-building and expansion of the insurance industry. – Development of effective adaptation strategies requires local involvement, inclusion of community perceptions, and recognition of multiple stresses on sustainable management of resources. – Adaptive capacities vary between countries, depending on social structure, culture, economic capacity, and level of environmental disruptions. Limiting factors include poor resource and infrastructure bases, poverty and disparities in income, weak institutions, and limited technology. – The challenge in Asia lies in identifying opportunities to facilitate sustainable development with strategies that make climate-sensitive sectors resilient to climate variability. – Adaptation strategies would benefit from taking a more systems-oriented approach, emphasizing multiple interactive stresses, with less dependence on climate scenarios.
Australia and New Zealand	– Adaptations are needed to manage risks from climatic variability and extremes. Pastoral economies and communities have considerable adaptability but are vulnerable to any increase in the frequency or duration of droughts. – Adaptation options include water management, land-use practices and policies, engineering standards for infrastructure, and health services. – Adaptations will be viable only if they are compatible with the broader ecological and socioeconomic environment, have net social and economic benefits, and are taken up by stakeholders. – Adaptation responses may be constrained by conflicting short- and long-term planning horizons. – Poorer communities, including many indigenous settlements, are particularly vulnerable to climate-related hazards and stresses on health because they often are in exposed areas and have less adequate housing, health care, and other resources for adaptation.

Table TS-15 (continued)

Sector	Key Findings
Europe	– Adaptation potential in socioeconomic systems is relatively high because strong economic conditions, stable population (with capacity to migrate), and well-developed political, institutional, and technological support systems. – The response of human activities and the natural environment to current weather perturbations provides a guide to critical sensitivities under future climate change. – Adaptation in forests requires long-term planning; it is unlikely that adaptation measures will be put in place in a timely manner. – Farm-level analyses show that if adaptation is fully implemented large reductions in adverse impacts are possible. – Adaptation for natural systems generally is low. – More marginal and less wealthy areas will be less able to adapt; thus, without appropriate policies of response, climate change may lead to greater inequities.
Latin America	– Adaptation measures have potential to reduce climate-related losses in agriculture and forestry. – There are opportunities for adapting to water shortages and flooding through water resource management. – Adaptation measures in the fishery sector include changing species captured and increasing prices to reduce losses.
North America	– Strain on social and economic systems from rapid climate and sea-level changes will increase the need for explicit adaptation strategies. In some cases, adaptation may yield net benefits, especially if climate change is slow. – Stakeholders in most sectors believe that technology is available to adapt, although at some social and economic cost. – Adaptation is expected to be more successful in agriculture and forestry. However, adaptations for water, health, food, energy, and cities are likely to require substantial institutional and infrastructure changes. – In the water sector, adaptations to seasonal runoff changes include storage, conjunctive supply management, and transfer. It may not be possible to continue current high levels of reliability of water supply, especially with transfers to high-valued uses. Adaptive measures such as "water markets" may lead to concerns about accessibility and conflicts over allocation priorities. – Adaptations such as levees and dams often are successful in managing most variations in weather but can increase vulnerability to the most extreme events. – There is moderate potential for adaptation through conservation programs that protect particularly threatened ecosystems, such as high alpines and wetlands. It may be difficult or impossible to offset adverse impacts on aquatic systems.

The presence of multiple causal factors (e.g., land-use change, pollution) makes attribution of many observed impacts to regional climate change a complex challenge. Nevertheless, studies of systems subjected to significant regional climate change—and with known sensitivities to that change—find changes that are consistent with well-established relationships between climate and physical or biological processes (e.g., shifts in the energy balance of glaciers, shifts in the ranges of animals and plants when temperatures exceed physiological thresholds) in about 80% of biological cases and about 99% of physical cases. Table TS-16 shows ~450 changes in processes or species that have been associated with regional temperature changes. Figure TS-11 illustrates locations at which studies have documented regional temperature change impacts. These consistencies enhance our confidence in the associations between changes in regional climate and observed changes in physical and biological systems. Based on observed changes, there is high confidence that 20th century climate changes have had a discernible impact on many physical and biological systems. Changes in biota and physical systems observed in the 20th century indicate that these systems are sensitive to climatic changes that are small relative to changes that have been projected for the 21st century. High sensitivity of biological systems to long-term climatic change also is demonstrated by paleorecords. [19.2.2.]

Signals of regional climate change impacts are expected to be clearer in physical and biotic systems than in social and economic systems, which are simultaneously undergoing many complex non-climate-related stresses, such as population growth and urbanization. Preliminary indications suggest that some social and economic systems have been affected in part by 20th century regional climate changes (e.g., increased damages by floods and droughts in some locations, with apparent increases in insurance impacts). Coincident or alternative explanations for such observed regional impacts result in only low to medium

Table TS-15 (continued)

Sector	Key Findings
Polar Regions	– Adaptation will occur in natural polar ecosystems through migration and changing mixes of species. Species such as walrus, seals, and polar bears will be threatened; while others, such as fish, may flourish. – Potential for adaptation is limited in indigenous communities that follow traditional lifestyles. – Technologically developed communities are likely to adapt quite readily, although the high capital investment required may result in costs in maintaining lifestyles. – Adaptation depends on technological advances, institutional arrangements, availability of financing, and information exchange.
Small Island States	– The need for adaptation has become increasingly urgent, even if swift implementation of global agreements to reduce future emissions occurs. – Most adaptation will be carried out by people and communities that inhabit island countries; support from governments is essential for implementing adaptive measures. – Progress will require integration of appropriate risk-reduction strategies with other sectoral policy initiatives in areas such as sustainable development planning, disaster prevention and management, integrated coastal zone management, and health care planning. – Strategies for adaptation to sea-level rise are retreat, accommodate, and protect. Measures such as retreat to higher ground, raising of the land, and use of building set-backs appear to have little practical utility, especially when hindered by limited physical size. – Measures for reducing the severity of health threats include health education programs, health care facilities, sewerage and solid waste management, and disaster preparedness plans. – Islanders have developed some capacity to adapt by application of traditional knowledge, locally appropriate technology, and customary practice. Overall adaptive capacity is low, however, because of the physical size of nations, limited access to capital and technology, shortage of human resource skills, lack of tenure security, overcrowding, and limited access to resources for construction. – Many small islands require external financial, technical, and other assistance to adapt. Adaptive capacity may be enhanced by regional cooperation and pooling of limited resources.

confidence about determining whether climate change is affecting these systems. [19.2.2.4]

7.2. *Five Reasons for Concern*

Some of the current knowledge about climate change impacts, vulnerability, and adaptation is synthesized here along five reasons for concern: unique and threatened systems, global aggregate impacts, distribution of impacts, extreme weather events, and large-scale singular events. Consideration of these reasons for concern contribute to understanding of vulnerabilities and potential benefits associated with human-induced climate change that can aid deliberations by policymakers of what could constitute dangerous interference with the climate system in the context of Article 2 of the UNFCCC. No single dimension is paramount.

Figure TS-12 presents qualitative findings about climate change impacts related to the reasons for concern. At a small increase in global mean temperature,[3] some of the reasons for concern show the potential for negative impacts, whereas others show little adverse impact or risk. At higher temperature increases, all lines of evidence show a potential for adverse impacts, with impacts in each reason for concern becoming more negative at increasing temperatures. There is high confidence in this general relationship between impacts and temperature change, but confidence generally is low in estimates of temperature change thresholds at which different categories of impacts would happen. [19.8]

7.2.1. *Unique and Threatened Systems*

Small increases in global average temperature may cause significant and irreversible damage to some systems and species, including possible local, regional, or global loss. Some plant and animal species, natural systems, and human settlements are highly sensitive to climate and are likely to be adversely affected by climate changes associated with scenarios of <1°C mean global warming. Adverse impacts to species and systems

[3]Intervals of global mean temperature increase of 0–2, 2–3, and >3°C relative to 1990 are labeled small, moderate, and large, respectively. The relatively large range for the "small" designation results because the literature does adequately address a warming of 1–2°C. These magnitudes of change in global mean temperature should be taken as an approximate indicator of when impacts might occur; they are not intended to define absolute thresholds or to describe all relevant aspects of climate-change impacts, such as rate of change in climate and changes in precipitation, extreme climate events, or lagged (latent) effects such as rising sea levels.

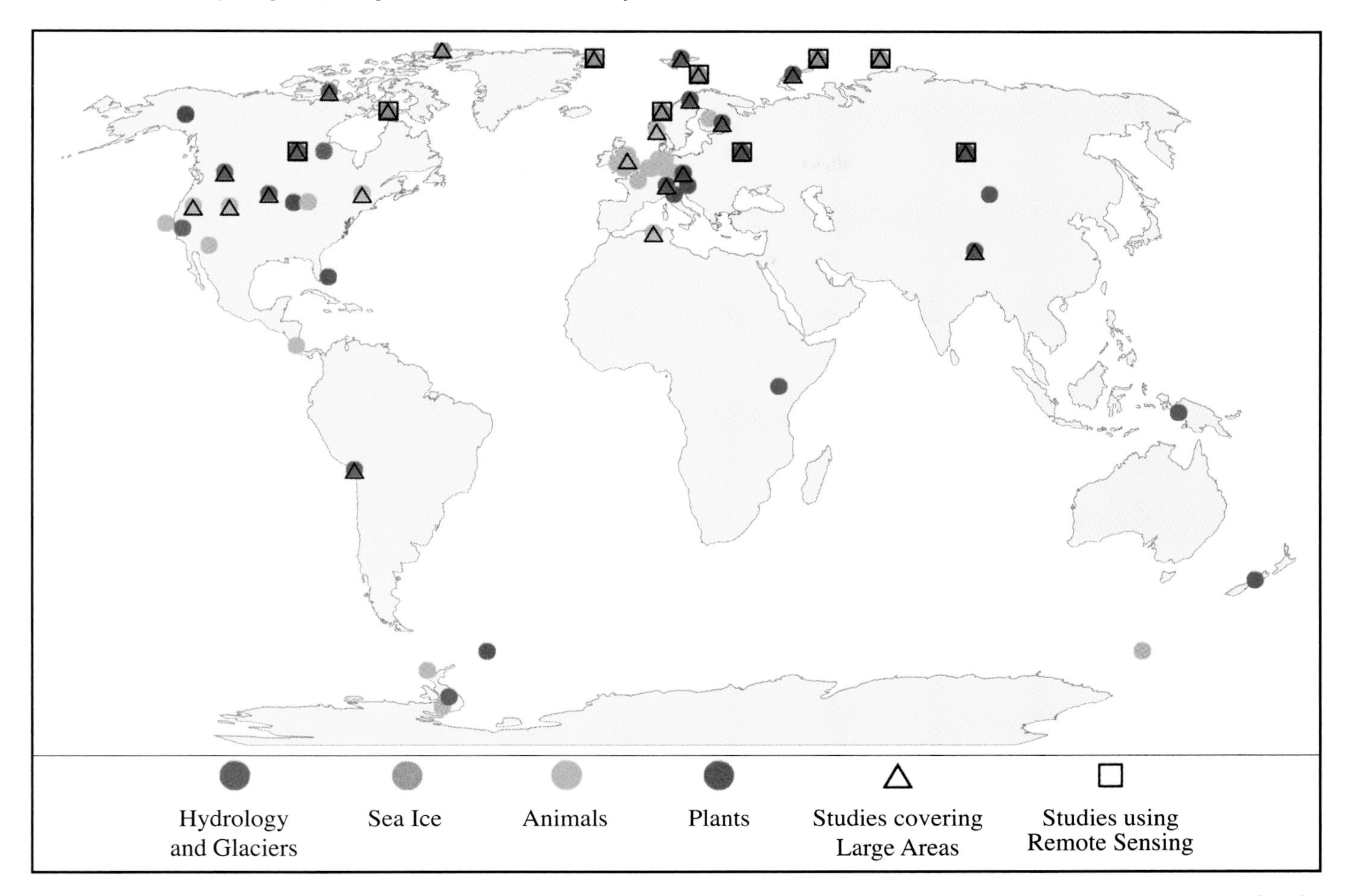

Figure TS-11: Locations at which systematic long-term studies meet stringent criteria documenting recent temperature-related regional climate change impacts on physical and biological systems. Hydrology, glacial retreat, and sea-ice data represent decadal to century trends. Terrestrial and marine ecosystem data represent trends of at least 2 decades. Remote-sensing studies cover large areas. Data are for single or multiple impacts that are consistent with known mechanisms of physical/biological system responses to observed regional temperature-related changes. For reported impacts spanning large areas, a representative location on the map was selected.

would become more numerous and more serious for climatic changes that would accompany a global mean warming of 1–2°C and are highly likely to become even more numerous and serious at higher temperatures. The greater the rate and magnitude of temperature and other climatic changes, the greater the likelihood that critical thresholds of systems would be surpassed. Many of these threatened systems are at risk from climate change because they face nonclimate pressures such as those related to human land use, land-use change, and pollution. [19.3]

Species that may be threatened with local or global extinction by changes in climate that may accompany a small mean global temperature increase include critically endangered species generally, species with small ranges and low population densities, species with restricted habitat requirements, and species for which suitable habitat is patchy in distribution, particularly if under pressure from human land-use and land-cover change. Examples of species that may be threatened by small changes include forest birds in Tanzania, the Resplendent Quetzal in Central America, the mountain gorilla in Africa, amphibians that are endemic to cloud forests of the neotropics, the spectacled bear of the Andes, the Bengal tiger and other species that are endemic to the Sundarban wetlands, and rainfall-sensitive plant species that are endemic to the Cape Floral Kingdom of South Africa. Natural systems that may be threatened include coral reefs, mangroves, and other coastal wetlands; montane ecosystems that are restricted to the upper 200–300 m of mountainous areas; prairie wetlands; remnant native grasslands; coldwater and some coolwater fish habitat; ecosystems overlying permafrost; and ice edge ecosystems that provide habitat for polar bears and penguins. Human settlements that may be placed at serious risk by changes in climate and sea level that may be associated with medium to large mean warming include some settlements of low-lying coastal areas and islands, floodplains, and hillsides—particularly those of low socioeconomic status such as squatter and other informal settlements. Other potentially threatened settlements include traditional peoples that are highly dependent on natural resources that are sensitive to climate change. [19.3]

***Table TS-16**: Processes and species found in studies to be associated with regional temperature change.[a]*

Region	Glaciers, Snow Cover/ Melt, Lake/ Stream Ice[b]		Vegetation		Invertebrates		Amphibians and Reptiles		Birds		Mammals	
Africa	1	0	—	—	—	—	—	—	—	—	—	—
Antarctica	3	2	2	0	—	—	—	—	2	0	—	—
Asia	14	0	—	—	—	—	—	—	—	—	—	—
Australia	1	0	—	—	—	—	—	—	—	—	—	—
Europe	29	4	13	1	46	1	7	0	258	92	7	0
North America	36	4	32	11	—	—	—	—	17	4	3	0
Latin America	3	0	—	—	—	—	22	0	15	0	—	—
Total	87	10	47	12	46	1	29	0	292	96	10	0

[a] The columns represent the number of species and processes in each region that were found in each particular study to be associated with regional temperature change. For inclusion in the table, each study needed to show that the species or process was changing over time and that the regional temperature was changing over time; most studies also found a significant association between how the temperature and species or processes were changing The first number indicates the number of species or processes changing in the manner predicted with global warming. The second number is the number of species or processes changing in a manner opposite to that predicted with a warming planet. Empty cells indicate that no studies were found for this region and category.
[b] Sea ice not included.

7.2.2. Aggregate Impacts

With a small temperature increase, aggregate market-sector impacts could amount to plus or minus a few percent of world GDP (medium confidence); aggregate nonmarket impacts could be negative (low confidence). The small net impacts are mainly the result of the fact that developed economies, many of which could have positive impacts, contribute the majority of global production. Applying more weight to impacts in poorer countries to reflect equity concerns, however, can result in net aggregate impacts that are negative even at medium warming. It also is possible that a majority of people will be negatively affected by climate change scenarios in this range, even if the net aggregate monetary impact is positive. With medium to higher temperature increases, benefits tend to decrease and damages increase, so the net change in global economic welfare becomes negative—and increasingly negative with greater warming (medium confidence). Some sectors, such as coastal and water resources, could have negative impacts in developed and developing countries. Other sectors, such as agriculture and human health, could have net positive impacts in some countries and net negative impacts in other countries. [19.5]

Results are sensitive to assumptions about changes in regional climate, levels of development, adaptive capacity, rates of change, valuation of impacts, and methods used for aggregating losses and gains, including the choice of discount rate. In addition, these studies do not consider potentially important factors such as changes in extreme events, advantageous and complementary responses to the threat of non-climate-driven extreme events, rapid change in regional climate (e.g., resulting from changes in ocean circulation), compounding effects of multiple stresses, or conflicting or complementary reaction to those stresses. Because these factors have yet to be accounted for in estimates of aggregate impacts and estimates do not include all possible categories of impacts, particularly nonmarket impacts, estimates of aggregate economic welfare impacts of climate change are considered to be incomplete. Given the uncertainties about aggregate estimates, the possibility of negative effects at a small increase in temperature cannot be excluded. [19.5]

7.2.3. Distribution of Impacts

Developing countries tend to be more vulnerable to climate change than developed countries (high confidence). Developing countries are expected to suffer more adverse impacts than developed countries (medium confidence). A small temperature increase would have net negative impacts on market sectors in many developing countries (medium confidence) and net positive impacts on market sectors in many developed countries (medium confidence). The different results are attributable partly to differences in exposures and sensitivities (e.g., present temperatures are below optimal in mid- and high latitudes for many crops but at or above optimal in low latitudes) and partly to lesser adaptive

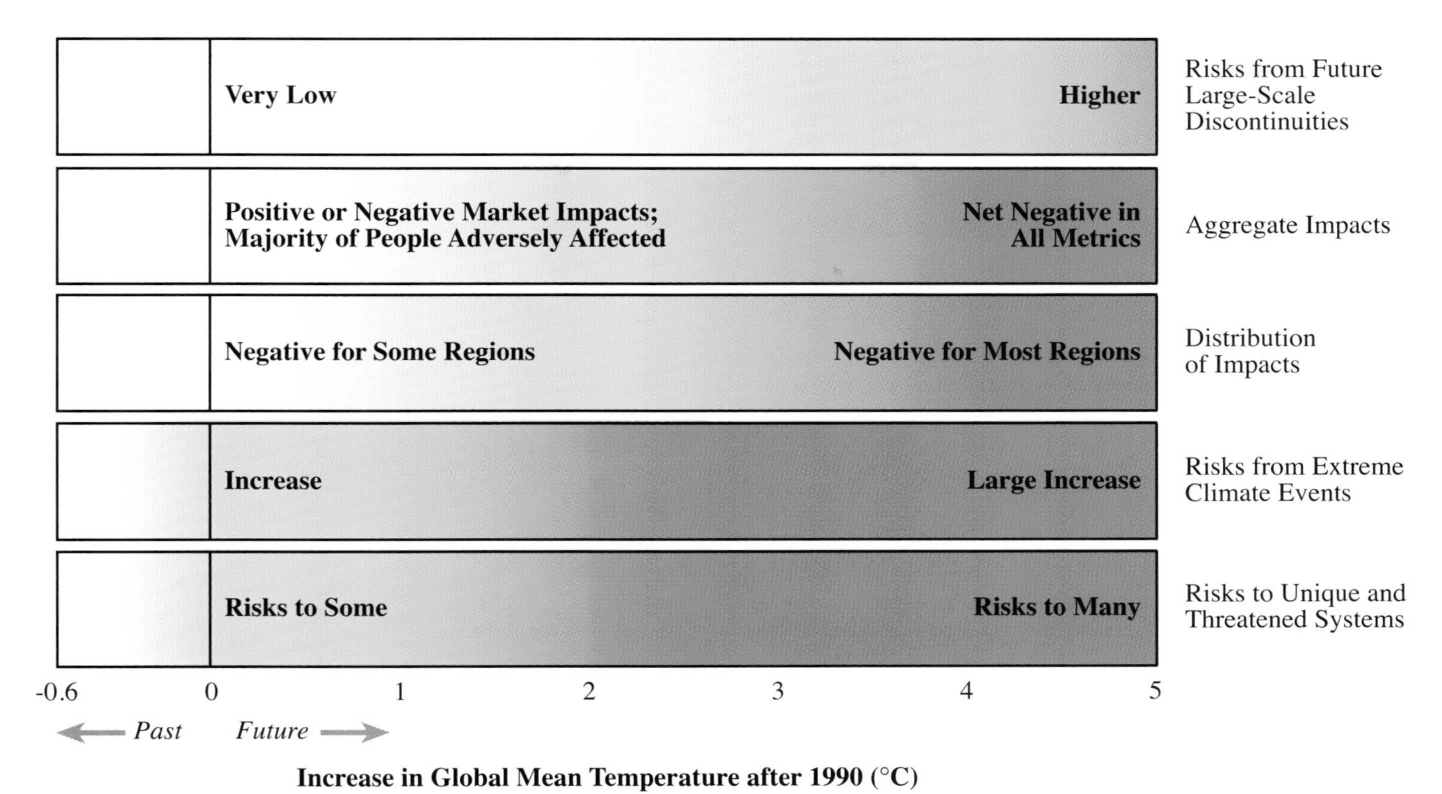

Figure TS-12: Impacts of or risks from climate change, by reason for concern. Each row corresponds to a reason for concern, and shades correspond to severity of impact or risk. White means no or virtually neutral impact or risk, yellow means somewhat negative impacts or low risks, and red means more negative impacts or higher risks. Global-averaged temperatures in the 20th century increased by 0.6°C and led to some impacts. Impacts are plotted against increases in global mean temperature after 1990. This figure addresses only how impacts or risks change as thresholds of increase in global mean temperature are crossed, not how impacts or risks change at different rates of change in climate. These temperatures should be taken as approximate indications of impacts, not as absolute thresholds.

capacity in developing countries relative to developed countries. At a medium temperature increase, net positive impacts would start to turn negative and negative impacts would be exacerbated (high confidence). The results of these studies do not fully take into account nonmarket impacts of climate change such as impacts on natural systems, which may be sensitive to small amounts of warming. Particularly vulnerable regions include deltaic regions, low-lying small island states, and many arid regions where droughts and water availability are problematic even without climate change. Within regions or countries, impacts are expected to fall most heavily, in relative terms, on impoverished persons. The poorest members of society can be inferred to be most vulnerable to climate change because of their lack of resources with which to cope and adapt to impacts, but few studies have explicitly examined the distribution of impacts on the poor relative to other segments of society. [19.4]

Impacts on unmanaged systems are likely to increase in severity with time, but impacts on managed systems could increase or decrease through the 21st century. The distribution of impacts over the 21st century is influenced by several factors. As GHG concentrations increase, the magnitude of exposure to change in climate stimuli also would increase. Nonclimate pressures on natural and social systems, which increase the vulnerability of systems, also may grow through time as a result of population growth and increased demands for land, water, public infrastructure, and other resources. Increased population, incomes, and wealth also mean that more people and human-made resources potentially would be exposed to climate change, which would tend to increase market-sector damages in absolute dollar terms; this has been the case historically. Counteracting these tendencies are factors such as increased wealth and technology and improved institutions, which can raise adaptive capacity and reduce vulnerability to climate change. [8, 19.4]

Whether impacts and vulnerability increase or decrease with time is likely to depend in part on the rates of climate change and development and may differ for managed and unmanaged systems. The more rapid the rate of climate change, the greater would be future exposure to potentially adverse changes and the greater the potential for exceeding system thresholds. The more rapid the rate of development, the more resources would be exposed to climate change in the future—but so too would the adaptive capacity of future societies. The benefits of increased adaptive capacity are likely to be greater for intensively managed systems than for systems that presently are unmanaged or lightly managed. For this reason, and because of the possibility that nonclimate pressures on natural systems may increase in the future, the vulnerability of natural systems is expected to increase with time (medium confidence). [19.4.2, 19.4.3]

Future development paths, sustainable or otherwise, will shape future vulnerability to climate change, and climate change impacts may affect prospects for sustainable development in different parts of the world. Climate change is one of many stresses that confront human and natural systems. The severity of many of these stresses will be determined in part by the development paths followed by human societies; paths that generate lesser stresses are expected to lessen the vulnerability of human and natural systems to climate change. Development also can influence future vulnerability by enhancing adaptive capacity through accumulation of wealth, technology, information, skills, and appropriate infrastructure; development of effective institutions; and advancement of equity. Climate change impacts could affect prospects for sustainable development by changing the capacity to produce food and fiber, the supply and quality of water, and human health and by diverting financial and human resources to adaptation. [18]

7.2.4. Extreme Weather Events

Many climatic impacts are related to extreme weather events, and the same will hold for the impacts of climate change. The large damage potential of extreme events arises from their severity, suddenness, and unpredictability, which makes them difficult to adapt to. Development patterns can increase vulnerability to extreme events. For example, large development along coastal regions increases exposure to storm surges and tropical cyclones, increasing vulnerability.

The frequency and magnitude of many extreme climate events increase even with a small temperature increase and will become greater at higher temperatures (high confidence). Extreme events include, for example, floods, soil moisture deficits, tropical cyclones, storms, high temperatures, and fires. The impacts of extreme events often are large locally and could strongly affect specific sectors and regions. Increases in extreme events can cause critical design or natural thresholds to be exceeded, beyond which the magnitude of impacts increases rapidly (high confidence). Multiple nonextreme consecutive events also can be problematic because they can lessen adaptive capacity by depleting reserves of insurance and reinsurance companies. [8, 19.6.3.1]

An increase in the frequency and magnitude of extreme events would have adverse effects throughout sectors and regions. Agriculture and water resources may be particularly vulnerable to changes in hydrological and temperature extremes. Coastal infrastructure and ecosystems may be adversely affected by changes in the occurrence of tropical cyclones and storm surges. Heat-related mortality is likely to increase with higher temperatures; cold-related mortality is likely to decrease. Floods may lead to the spread of water-related and vector-borne diseases, particularly in developing countries. Many of the monetary damages from extreme events will have repercussions on a broad scale of financial institutions, from insurers and reinsurers to investors, banks, and disaster relief funds. Changes in the statistics of extreme events have implications for the design criteria of engineering applications (e.g., levee banks, bridges, building design, and zoning), which are based on estimates of return periods, and for assessment of the economic performance and viability of particular enterprises that are affected by weather. [19.6.3.1]

7.2.5. Large-Scale Singular Events

Human-induced climate change has the potential to trigger large-scale changes in Earth systems that could have severe consequences at regional or global scales. The probabilities of triggering such events are poorly understood but should not be ignored, given the severity of their consequences. Events of this type that might be triggered include complete or partial shutdown of the North Atlantic and Antarctic Deep Water formation, disintegration of the West Antarctic and Greenland Ice Sheets, and major perturbations of biosphere-regulated carbon dynamics. Determining the timing and probability of occurrence of large-scale discontinuities is difficult because these events are triggered by complex interactions between components of the climate system. The actual discontinuous impact could lag the trigger by decades to centuries. These triggers are sensitive to the magnitude and rate of climate change. Large temperature increases have the potential to lead to large-scale discontinuities in the climate system (medium confidence).

These discontinuities could cause severe impacts on the regional and even global scale, but indepth impact analyses are still lacking. Several climate model simulations show complete shutdown of the North Atlantic thermohaline circulation with high warming. Although complete shutdown may take several centuries to occur, regional shutdown of convection and significant weakening of the thermohaline circulation may take place within the next century. If this were to occur, it could lead to a rapid regional climate change in the North Atlantic region, with major societal and ecosystem impacts. Collapse of the West Antarctic Ice Sheet would lead to a global sea-level rise of several meters, which may be very difficult to adapt to. Although the disintegration might take many hundreds of years, this process could be triggered irreversibly in the next century. The relative magnitude of feedback processes involved in cycling of carbon through the oceans and the terrestrial biosphere is shown to be distorted by increasing temperatures. Saturation and decline of the net sink effect of the terrestrial biosphere—which is projected to occur over the next century—in step with similar processes, could lead to dominance of positive feedbacks over negative ones and strong amplification of the warming trend. [19.6.3.2]

8. Information Needs

Although progress has been made, considerable gaps in knowledge remain regarding exposure, sensitivity, adaptability, and vulnerability of physical, ecological, and social systems to climate change. Advances in these areas are priorities for advancing understanding of potential consequences of climate change for human society and the natural world, as well as to support analyses of possible responses.

Exposure. Advances in methods for projecting exposures to climate stimuli and other nonclimate stresses at finer spatial scales are needed to improve understanding of potential consequences of climate change, including regional differences, and stimuli to which systems may need to adapt. Work in this area should draw on results from research on system sensitivity, adaptability, and vulnerability to identify the types of climate stimuli and nonclimate stresses that affect systems most. This research is particularly needed in developing countries, many of which lack historical data, adequate monitoring systems, and research and development capabilities. Developing local capacity in environmental assessment and management will increase investment effectiveness. Methods of investigating possible changes in the frequency and intensity of extreme climate events, climate variability, and large-scale, abrupt changes in the Earth system such as slowing or shutdown of thermohaline circulation of oceans are priorities. Work also is needed to advance understanding of how social and economic factors influence the exposures of different populations.

Sensitivity. Sensitivity to climate stimuli is still poorly quantified for many natural and human systems. Responses of systems to climate change are expected to include strong nonlinearities, discontinuous or abrupt responses, time-varying responses, and complex interactions with other systems. However, quantification of the curvature, thresholds, and interactions of system responses is poorly developed for many systems. Work is needed to develop and improve process-based, dynamic models of natural, social, and economic systems; to estimate model parameters of system responses to climate variables; and to validate model simulation results. This work should include use of observational evidence, paleo-observations where applicable, and long-term monitoring of systems and forces acting on them. Continued efforts to detect impacts of observed climate change is a priority for further investigation that can provide empirical information for understanding of system sensitivity to climate change.

Adaptability. Progress has been made in the investigation of adaptive measures and adaptive capacity. However, work is needed to better understand the applicability of adaptation experiences with climate variability to climate change, to use this information to develop empirically based estimates of the effectiveness and costs of adaptation, and to develop predictive models of adaptive behavior that take into account decision making under uncertainty. Work also is needed to better understand the determinants of adaptive capacity and to use this information to advance understanding of differences in adaptive capacity across regions, nations, and socioeconomic groups, as well as how capacity may change through time. Advances in these areas are expected to be useful for identifying successful strategies for enhancing adaptation capacity in ways that can be complementary to climate change mitigation, sustainable development, and equity goals.

Vulnerability. Assessments of vulnerability to climate change are largely qualitative and address the sources and character of vulnerability. Further work is needed to integrate information about exposures, sensitivity, and adaptability to provide more detailed and quantitative information about the potential impacts of climate change and the relative degree of vulnerability of different regions, nations, and socioeconomic groups. Advances will require development and refinement of multiple measures or indices of vulnerability such as the number or percentage of persons, species, systems or land area negatively or positively affected; changes in productivity of systems; the monetary value of economic welfare change in absolute and relative terms; and measures of distributional inequities.

Uncertainty. Large gaps remain in refining and applying methods for treating uncertainties, particularly with respect to providing scientific information for decisionmaking. Improvements are required in ways of expressing the likelihood, confidence, and range of uncertainty for estimates of outcomes, as well as how such estimates fit into broader ranges of uncertainty. Methods for providing "traceable accounts" of how any aggregated estimate is made from disaggregated information must be refined. More effort is needed to translate judgments into probability distributions in integrated assessment models.

1

Overview of Impacts, Adaptation, and Vulnerability to Climate Change

STEPHEN SCHNEIDER (USA) AND JOSE SARUKHAN (MEXICO)

Lead Authors:

J. Adejuwon (Nigeria), C. Azar (Sweden),W. Baethgen (Uruguay), C. Hope (UK), R. Moss (USA), N. Leary (USA), R. Richels (USA), J.-P. van Ypersele (Belgium)

Contributing Authors:

K. Kuntz-Duriseti (USA), R.N. Jones (Australia)

Review Editors:

J. Bruce (Canada) and B. Walker (Australia)

CONTENTS

1.1. Overview of the Assessment

The world community faces many risks from climate change. Clearly, it is important to understand the nature of those risks, where natural and human systems are likely to be most vulnerable, and what may be achieved by adaptive responses. To understand better the potential impacts and associated dangers of global climate change, Working Group II of the Intergovernmental Panel on Climate Change (IPCC) offers this Third Assessment Report (TAR) on the state of knowledge concerning the sensitivity, adaptability, and vulnerability of physical, ecological, and social systems to climate change. Building on the Second Assessment Report (SAR), this new report reexamines key findings of the earlier assessment and emphasizes new information and implications on the basis of more recent studies.

Human activities—primarily burning of fossil fuels and changes in land cover—are modifying the concentration of atmospheric constituents or properties of the Earth's surface that absorb or scatter radiant energy. In particular, increases in the concentrations of greenhouse gases (GHGs) and aerosols are strongly implicated as contributors to climatic changes observed during the 20th century and are expected to contribute to further changes in climate in the 21st century and beyond. These changes in atmospheric composition are likely to alter temperatures, precipitation patterns, sea level, extreme events, and other aspects of climate on which the natural environment and human systems depend.

One of several primary issues this report has been organized to address is a key question before the United Nations Framework Convention on Climate Change (UNFCCC): What are the potential impacts for societies and ecosystems of different atmospheric concentrations of GHGs and aerosols that absorb and scatter sunlight (United Nations, 1992)? Answering this question is a necessary step in assessing what constitutes "dangerous anthropogenic interference in the climate system." This report does not make any judgments about what level of concentrations is "dangerous" because that is not a question of science *per se* but a value judgment about relative risks and tradeoffs. The task is to make the evidence about relative risks as clear as possible. This report therefore describes what is known about the distribution of impacts; how, why, and to what extent they differ from region to region or place to place; and how this relates to the distribution of vulnerability and capacity to adapt. However, it critically assesses the literature to help inform policymakers about effects associated with different concentration levels, so they may judge what levels of risk are acceptable. Assessment of what constitutes dangerous interference in the climate systems will require analysis of the interactions of climate change and social and economic conditions, which are inextricably linked. Understanding the role of socioeconomic factors, particularly adaptive responses and capacity, is critical.

Part of the justification for a TAR at this time is the abundance of new evidence that has come to the attention of the expert community since publication of the SAR. The evidence is drawn predominantly from published, peer-reviewed scientific literature. Evidence also is drawn from published, non-peer-reviewed literature and unpublished sources such as industry journals; reports of government agencies, research institutions, and other organizations; proceedings of workshops; working papers; and unpublished data sets. The quality and validity of information from non-peer-reviewed and unpublished sources have been assessed by authors of this report prior to inclusion of information from these sources in the report. The procedures for the use of information from non-peer-reviewed and unpublished sources are described in IPCC (1999a) and discussed in Skodvin (2000).

Although this report builds on previous assessments, including the SAR and the IPCC's *Special Report on Regional Impacts of Climate Change* (IPCC, 1998), the TAR departs from them in important respects. In comparison to previous assessments, greater attention is given to climate change adaptation; multiple pressures on systems; links between climate change, sustainable development, and equity; and characterization of the state-of-the-science and confidence levels associated with key conclusions of the assessment (see Box 1-1). This overview chapter does not attempt to provide a comprehensive summary of the principal findings of the TAR, but it helps to illustrate basic concepts by selectively reporting on a few key conclusions, as well as providing a more comprehensive road map to the materials presented later in the report:

- Part I sets the stage for assessment of impacts, adaptation, and vulnerability by discussing the context of climate change, methods for research on impacts, and development of scenarios. These are important additions to what was emphasized in the SAR. Consideration of the context of change (in this chapter) draws attention to the relationship of climate change and sustainable development, including interactions of climate variability and change with other environmental changes and evolving demographic, social, and economic conditions that affect driving forces of change and resources available for adaptation. Assessment of methods (Chapter 2) and approaches for developing and applying scenarios (Chapter 3) cover scientific and technical aspects of research on impacts, providing a review of the science underlying the topics covered in other chapters of the report.
- Part II assesses recent advances in experimental work, observations, and modeling that contribute to the current state of knowledge of baseline trends, vulnerabilities, and adaptation options in six sectors or resource areas. This section of the report integrates material that had been covered in 18 chapters in the SAR, focusing to a greater extent on cross-sectoral issues. For example, Chapter 5, Ecosystems and their Goods and Services, integrates what had been separate chapters on forests, rangelands, deserts, mountain regions, wetlands, agriculture, food security, and other systems; it also adds assessment of climate change impacts on

wildlife—an issue not covered in the SAR. Integrating these issues into a single chapter provides an opportunity for improved assessment of interactions across these systems, the effects of change on landscapes, and the distribution of land use and cover. Information on the impacts of natural variability and the potential for nonlinear interactions also is included in the chapters in Part II.

- Chapters in Part III build on key findings of the IPCC's *Special Report on Regional Impacts of Climate Change*. Each chapter and subchapter in this part of the report explores what has been learned regarding the context of change, sensitivity, adaptation, and vulnerability of key sectors. A chapter is devoted to each of eight regions of the world: Africa, Asia, Australia and New Zealand, Europe, Latin America, North America, polar regions, and small island states (see Figure 1-1). These regions are chosen to correspond to continents or—in the case of polar regions and small island states—to bring together in one chapter areas that share important attributes related to climate change vulnerability. This regionalization of the world is convenient for organizing the report, but it must be recognized that there is a high degree of heterogeneity within each of these regions in terms of climate, ecosystems, culture, and social, economic, and political systems. Consequently, climate change impacts, adaptive capacity, and vulnerability vary markedly within each of these regions. These chapters provide an opportunity to place vulnerability and adaptation in the context of multiple stresses and regional resources for adaptation. This is an extremely important development because it calls attention to the issues that regional and local decisionmakers in the private and public sectors will be facing in each of the regions.

Areas of important new findings include detection of impacts of observed climatic changes on environmental systems, transient scenarios, vulnerability to changes in climate variability, and vulnerability to strongly nonlinear, complex, and discontinuous responses to climate change. Another distinction from previous assessments is the recognition in the TAR that the many complexities of analysis logically lead to a focus on ranges of outcomes and characterizations, using subjective probabilities of events, rather than primary emphasis on "best guesses," point estimates, single "optimum," or aggregate conclusions.

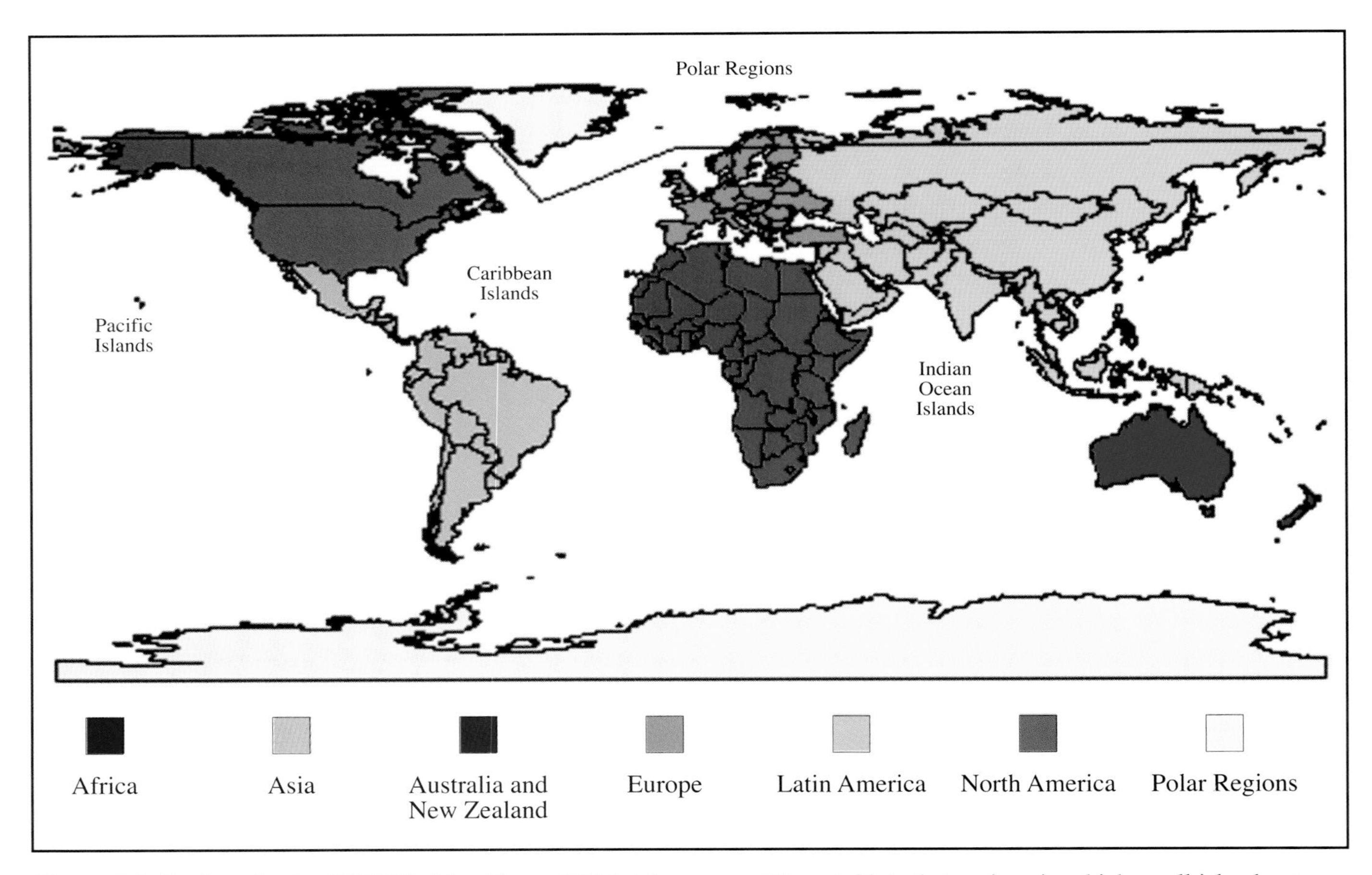

Figure 1-1: Regions for the IPCC Working Group II Third Assessment Report. Note that regions in which small island states are located include the Pacific, Indian, and Atlantic Oceans, and the Caribbean and Mediterranean Seas. The boundary between Europe and Asia runs along the eastern Ural Mountains, River Ural, and Caspian Sea. For the polar regions, the Arctic consists of the area north of the Arctic Circle, including Greenland; the Antarctic consists of the Antarctic continent, together with the Southern Ocean south of ~58°S.

Box 1-1 Uncertainties and Confidence Scale

The many conclusions presented in this report are subject to varying degrees of uncertainty. The degree of uncertainty attached to conclusions in this report are assessed and reported in two different ways. One is to assess and report a confidence level for a conclusion, using a Bayesian probability framework. (Bayesian assessments of probability distributions would lead to the following interpretation of probability statements: The probability of an event is the degree of belief that exists among lead authors and reviewers that the event will occur, given observations, modeling results, and theory currently available.) The second is to assess and report the quality or level of scientific understanding that supports a conclusion.

The 5-point confidence scale below is used to assign confidence levels to selected conclusions. The confidence levels are stated as Bayesian probabilities, meaning that they represent the degree of belief among the authors of the report in the validity of a conclusion, based on their collective expert judgment of all observational evidence, modeling results, and theory currently available to them.

5-Point Quantitative Scale for Confidence Levels

95% or greater	Very High Confidence
67–95%	High Confidence
33–67%	Medium Confidence
5–33%	Low Confidence
5% or less	Very Low Confidence

For some conclusions, the 5-point quantitative scale is not appropriate as a characterization of associated uncertainty. In these instances, authors qualitatively evaluate the level of scientific understanding in support of a conclusion, based on the amount of supporting evidence and the level of agreement among experts about the interpretation of the evidence. The matrix below has been used to characterize the level of scientific understanding.

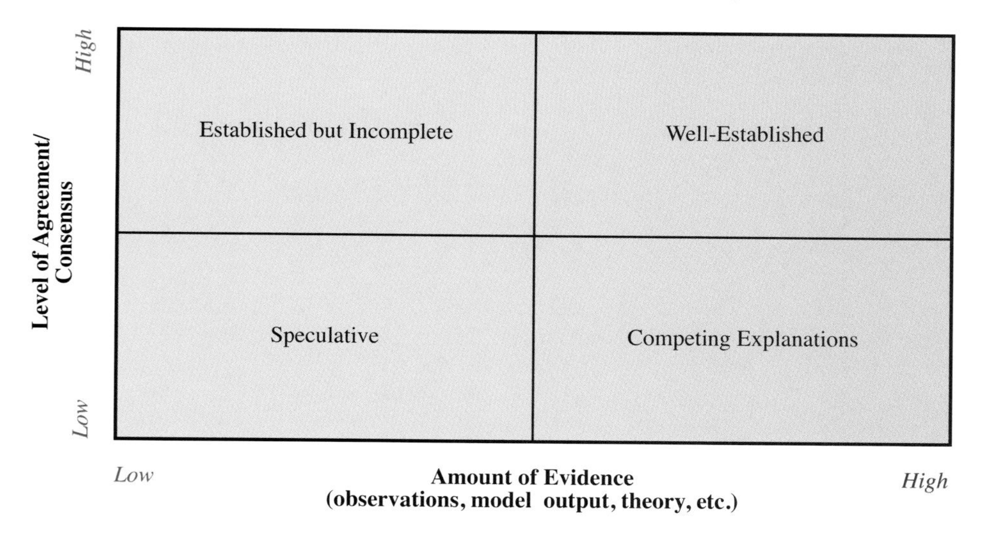

Key to Qualitative "State of Knowledge" Descriptors

- *Well-Established*: Models incorporate known processes; observations are consistent with models; or multiple lines of evidence support the finding.
- *Established but Incomplete:* Models incorporate most known processes, although some parameterizations may not be well tested; observations are somewhat consistent but incomplete; current empirical estimates are well founded, but the possibility of changes in governing processes over time is considerable; or only one or a few lines of evidence support the finding.
- *Competing Explanations*: Different model representations account for different aspects of observations or evidence or incorporate different aspects of key processes, leading to competing explanations.
- *Speculative*: Conceptually plausible ideas that haven't received much attention in the literature or that are laced with difficult to reduce uncertainties.

Box 1-2. Cross-Cutting Issues Guidance Papers

Four cross-cutting guidance papers (Pachauri *et al.*, 2000) were available to all Lead Authors of all three IPCC working groups. Many of the concepts in these papers were previously unfamiliar to a large number of the Lead Authors. Significant efforts were made to incorporate the uncertainties guidance and at least consider the guidance in other papers by each working group. Future assessments will be increasingly able to benefit from the suggestions and frameworks in these four cross-cutting guidance papers.

Development, Sustainability, and Equity (DSE) (Munasinghe, 2000)

DSE is relevant to Working Group II with respect to three questions: How do development paths and equity conditions influence vulnerability to climate change and the capacity to adapt to or cope with climate change? How might climate change impacts and adaptation responses affect prospects for attaining sustainable development and equity goals? What types of policies are capable of reducing climate change vulnerability and promoting sustainable development and equity objectives?

DSE is a response to principles contained in UNFCCC Article 3.4 (to promote sustainable development). Article 3.2 takes into account the special needs and circumstances of developing countries. Article 4 deals, for example, with the responsibility of developed nations, competing priorities for developing nations, and "common, but differentiated responsibilities," and Article 2 says to avoid "dangerous" interference with the climate system (United Nations, 1992). DSE is closely tied to sustainable development with respect to three underlying dimensions: economic, social, and environmental.

Development has been characterized as "qualitative improvement" (Ishida, 1998), including economic growth and social dimensions. Sustainability of a system refers to its durability or its capacity to withstand and recover from disturbances (WCED, 1987)—in other words, its resilience. Equity refers to procedural as well as distributional issues. Procedural issues relate to how decisions are made (e.g., internal equity and governance structures within nations could have significant effects on adaptive capacity). Distributional equity, on the other hand, relates to how the costs of impacts, mitigation, and adaptation are shared. Equity considerations are important in addressing global climate change for several reasons, including moral and ethical concerns; facilitating cooperation because equitable decisions carry greater legitimacy; the social dimension of sustainable development; and the UNFCCC itself, which considers equity as one of its basis principles (in Article 3.1).

Climate change could undermine social welfare, equity, and the sustainability of future development. In particular, it is generally believed that developing countries and disadvantaged groups within all countries are more vulnerable to the impacts of climate change (e.g., Chapter 18) as a result of limited resources and low adaptive capacity.

Uncertainty (Moss and Schneider, 2000)

Anticipating the imperfect nature of available information, UNFCCC Article 3.3 provides guidance to the effect that "where there are threats of serious or irreversible damage, lack of full scientific certainty should not be used as a reason for postponing measures to anticipate, prevent, or minimize the causes of climate change and mitigate its adverse effects" (United Nations, 1992). The uncertainties guidance paper develops a unified approach for assessing, characterizing, and reporting uncertainties in the TAR. The most important contribution of the uncertainties guidance paper is the construction of confidence schemes and qualitative terms to describe the state of science, which are reproduced in Box 1-1. The goal is to promote consistency in evaluating the judgments of scientific experts, to facilitate communication of these judgments to nonspecialists, and to provide peer-reviewed guidance for policymakers. Thus, a great deal of importance is attached to assessing the scientific merit of information in the literature and "explicitly distinguishing and communicating which findings are well understood, which are somewhat understood, and which are speculative" (Moss and Schneider, 1997). Section 2.6 discusses in more detail the differences between the well-calibrated ranges in the literature and the much larger, full range of uncertainty as well as the cascade of uncertainty that occurs when ranges in climate scenarios are cascaded with uncertainties in each successive step of assessment. In Chapters 4–19, selected findings are assigned levels of confidence, using either the scale for assessing confidence level quantitatively or the matrix for assessing the state of knowledge qualitatively.

Especially in the regional chapters, uncertainty about future climate is the dominant cause of uncertainty about the character and magnitude of impacts. In such cases, confidence estimates are evaluated conditionally on a specific climate change scenarios to avoid "cascades" (see Figure 2-2) in which confidence in the occurrence of an event does not include compounded uncertainties in each factor that contributes to the final outcome. Instead, the assessment evaluates each step in the cascade separately—what is called a "traceable account"—and is particularly appropriate for any aggregate conclusions.

Finally, the TAR has circulated cross-cutting "guidance papers" to all three working groups to try to achieve more consistency in dealing with four areas: development, sustainability, and equity; uncertainties; costing methodologies; and decision analytic frameworks (see Box 1-2).

The IPCC's charge to Working Group II for the TAR implies that consideration of the impacts of climate change in the SAR is insufficient *per se* as a basis for decisionmaking. In general, the SAR was able to address the implications of climate change only for single economic sectors or environmental components. With this in mind, Chapters 4–19 consider the various decision analysis framework tools to improve upon the responses to the impacts of climate change provided in the SAR. In the current exercise, not only are possible implications of climate change for the various economic sectors or environmental components assessed, options to alleviate identified impacts are investigated. In addition, direct and indirect costs of adaptation options are explored, and an extensive assessment of direct and indirect benefits is provided. Where monetary values can be assigned, CBA is employed to determine the optimal value of adaptation measures, including sensitivity analysis to critical parameters, and CEA is adopted to identify the least-cost solution to targeted mitigation objectives.

The concluding section of this report (Chapters 18 and 19) examines global issues and offers a synthesis. A significant addition to previous assessments is Chapter 18, which is devoted

Box 1-2. Cross-Cutting Issues Guidance Papers (continued)

Costing Methodologies (Markandya and Halsnaes, 2000)

This guidance paper discusses many issues related to assessing the impacts of climate change, including identifying the costs of mitigation and adaptation and benefits from avoiding climate damages (market versus nonmarket values; bottom-up versus top-down; willingness-to-pay versus willingness-to-accept-payment); recognizing value-laden assumptions associated with the dollar metric; addressing problems with aggregating costs and benefits that confound differentiated costs and benefits across groups; and determining how to account for ancillary (or co-)benefits and costs. Many of these issues are discussed more fully in the text of this chapter; see also Chapter 2. As noted in the uncertainties guidance paper, a "traceable account" of all aggregations is essential to maintain transparency of any conclusions.

"Cost" refers to any adverse consequence that humans would be willing to expend resources to avoid. "Benefit," on the other hand, is defined as any advantageous consequence that humans would be willing to expend resources to secure. With respect to adaptation, costs are attached primarily to measures employed to alleviate the impacts of climate change. Real costs, however, are computed as the difference between total expenses on adaptation measures and the value of benefits generated in the process of alleviating impacts. Costs also are attached to damages that would be sustained in the event of no response or inadequate response.

Section 2.5 provides details about methods of costing and valuation used in the TAR. Elements of costing and valuation described include opportunity costs and discounting. Costing of market impacts, nonmarket impacts, and uncertainty are discussed and their applications in decision analyses, scenario analyses, and integrated assessments outlined. In Chapters 4–19, cost-benefit analysis is used to present and evaluate information about impacts and their consequences on ecosystems, socioeconomic sectors, and human health, convenience, and comfort. Where information is available, costs and benefits are quantified and given in monetary units. Often the magnitude of costs and benefits is linked explicitly to specific climate scenarios. As much as possible, the sensitivity of costs and benefits to different magnitudes and rates of climate change and the potential for nonlinear costs or benefits are assessed. For example, the sensitivity of costs and benefits to socioeconomic conditions and differences between developed and developing countries are assessed—with obvious implications for DSE. The contributions of costing and valuation to uncertainty levels in the major conclusions also are described.

Decision Analysis Frameworks (Toth, 2000)

Details on methods and tools of decision analysis frameworks that are applicable in the context of adaptation are provided in Section 2.7, which describes major decision analysis frameworks—including decision analysis (DA), cost-benefit analysis (CBA), cost-effectiveness analysis (CEA), and the policy exercise approach (PE). DA integrates utility theory, probability, and mathematical optimization in a procedure designed to select the best pathway. CBA involves valuing all costs and benefits of a proposed project over time on the basis of WTP and specifying a decision criterion to accept or turn down the project. CEA takes a predetermined objective and seeks ways to accomplish it as inexpensively as possible. PE synthesizes and assesses knowledge in several relevant fields of science for policy purposes in light of complex practical management problems. Application of these decision analysis frameworks cuts across the assignments of Working Groups II and III.

to assessment of opportunities for and barriers to adaptation. This chapter considers determinants of adaptive capacity; lessons from adaptation to present-day climate variability and extremes; the potential effectiveness of adaptation measures; and global-, national-, and local-scale options for strengthening adaptive capacity, especially for vulnerable populations, countries, or zones. Chapter 19, a synthesis, also is new to the TAR. It draws on the analyses of other chapters, synthesizing information that is important for interpretation of Article 2 of the UNFCCC and key provisions of international agreements to address climate change. Potential global impacts of different stabilization levels of atmospheric concentrations of GHGs are assessed. Chapter 19 assesses vulnerability within the framework of sustainable development and equity, acknowledging common but differentiated responsibilities.

The issue of what constitutes sustainable development was advanced in 1987 by the World Commission on Environment and Development (the so-called Brundtland Commission; WCED, 1987). The commission defined *sustainable development* as "development that meets the needs of the present without compromising the ability of future generations to meet their own needs" and notes that "even the narrow notion of physical sustainability implies a concern for social equity between generations, a concern that must logically be extended to equity within each generation." The goal of sustainable development is a stable human environmental system in which available resources are sufficient to meet the needs of society in perpetuity. Questions have been asked about whether "needs," as conceived in the Brundtland Commission report, should be limited to environment-dependent basic necessities of food, clothing, shelter, and health or should include more qualitative aspects such as comfort, convenience, or other "quality of life" measures. There is no consensus in the literature regarding what constitutes the limits of "needs" in this context.

Because available studies have not employed a common set of climate scenarios and methods and because of uncertainties regarding the sensitivities and adaptability of natural and human-dominated systems, assessment of regional vulnerabilities is necessarily qualitative. Whenever possible, quantitative estimates of the impacts of climate change are cited in this report. Field or experimental data often provide quantitative underpinnings for specific circumstances, but rarely are complex systems sufficiently described by the limited number of cases in which a large quantity of "hard data" is available. Thus, most quantitative estimates are still dependent on the specific assumptions employed regarding future changes in climate, as well as the particular methods and models applied in the analyses. On the other hand, issues in which there is a great deal of relevant field or lab data are likely to carry higher confidence in any such quantitative estimate. To interpret these estimates, it is important to bear in mind uncertainties regarding the character, magnitude, and rates of future climate change that will affect society's degree of exposure. Of comparable importance are uncertainties associated with future states of the human condition—for example, the extent and quality of economic development throughout the world and the evolution of traditions and institutions in societies—that will affect profoundly the capacity for coping and adaptation, hence level of vulnerability. These uncertainties impose limitations on the ability of the research community to project the impacts of climate change, particularly at regional and smaller scales.

This introductory chapter is organized to address a series of questions: What is potentially at stake as a result of changes in climate (Section 1.2)? How has society responded to the risks and potential opportunities (Section 1.3)? How are impacts, adaptation, and vulnerability assessed in the report (Section 1.4)? How do the complexities of analysis affect the assessment (Section 1.5)? How can this assessment be used to address policy-relevant questions (Section 1.6)?

1.2. What is Potentially at Stake?

The stakes surrounding anthropogenic climate change can be very high in terms of the vulnerabilities of some sectors and regions and in terms of the distributional consequences of actions taken to deal with these possibilities. The context is that humankind already is challenged today to provide the opportunity for this and future generations to achieve a more sustainable and equitable standard of living. Billions of people today live without adequate nourishment, access to clean water, modern energy services, and other basic human needs (see, e.g., UNDP, 1999). Providing for the increasing well-being of humans, especially the poor, in the context of sustainable and equitable development is one of the great challenges of the 21st century. Unabated climate change is likely to make meeting this challenge significantly more difficult. On the other hand, it also is argued (e.g., Grossman and Krueger, 1995) that increasing economic growth may lead to reductions in population growth and environmental degradation. Throughout the past century, however, per capita carbon dioxide (CO_2) emissions from combustion of fossil fuels have been driven primarily by growth in gross domestic product (GDP) per capita (although the growth rate in CO_2 emissions generally has not been as fast as the growth in GDP, owing to improvements in energy and carbon intensities of industrial economies (e.g., Hoffert *et al.*, 1998).

The impact (I) of a given population on the environment can be decomposed into the product of three factors: population size (P), affluence per capita (A), and unit impact per unit of affluence, which is related to technologies used (T) (Ehrlich and Holdren, 1971; Ehrlich and Ehrlich, 1990). Rising per capita consumption and a growing world population have resulted in unprecedented human resource use, which is altering global systems, including climate (Bartiaux and van Ypersele, 1993; Yang and Schneider, 1998). According to all of the scenarios considered in the IPCC's *Special Report on Emissions Scenarios* (SRES) (IPCC, 2000), the human population will continue to grow until at least 2050, reaching a population that is 60–100% larger than it was in 1990. The SRES scenarios describe futures that generally are more affluent than today; many of the SRES scenarios assume a narrowing of income differences (in relative but not absolute terms) among world regions. This implies that the third factor in the "I=PAT" identity, the unit impact per unit of affluence, will have a very important role in assessment of the global impact of human activities. Increasing population and affluence, if not accompanied by significant decreases in unit impact per unit of affluence, will make the challenge of promoting sustainable development even more difficult—particularly in developing countries, where most of the increase in population is projected to take place.

We have reached the point that the cumulative interaction of several factors related to human activities (e.g., land-use changes and emissions of GHGs, ozone-depleting substances, and local air pollutants) increases the risk of causing or aggravating potentially irreversible events, such as loss of species, forests, human settlements, glaciers, or heritage sites near coastlines and, in the long term, altered oceanic circulation regimes.

Although some regions may experience beneficial effects of climate change (e.g., increasing agricultural productivity at high latitudes), previous IPCC assessments have concluded that net negative climate impacts are more likely in most parts of the world (assessment of potential positive and negative impacts is one of the main purposes of this report; see Sections 2.5.6 and 2.6.4 and subsequent sections in this chapter for a discussion of uses of and problems with net monetary aggregation of impacts, and see Chapter 19 for a synthesis). These impacts will affect human welfare directly and indirectly, in many cases undercutting efforts to promote sustainable development that, in turn, serve as driving forces of environmental change.

Moreover, the time scales of change vary tremendously. For environmental systems, these time scales range from decades (for restoration of slightly disturbed ecosystems) to many centuries (for equilibration of the climate system and sea level), even with a stable level of atmospheric GHG concentrations. These environmental time scales imply that human activities in the short term will set in motion a chain of events with long-term consequences for the climate that cannot be reversed quickly, if at all.

For most human institutions, the time scales range from years for very short electoral cycles that determine the tenure of a government to a half-century or more for the useful lifetimes of buildings and major infrastructure such as irrigation projects, transportation networks, or energy supply systems. It may take a generation or more to effect significant changes to institutions. Because of these time scales, some decisions taken during the next few decades may limit the range of possible options in the future with respect to emissions reduction and adaptation, whereas other decisions may expand this range of options. During this period, many more insights into the effects and impacts of climatic changes will emerge. However, it is well established that uncertainties will remain, and efforts to manage risks in the face of considerable uncertainty will be a characteristic of climatic change assessments for decades more.

Working Group II's contribution to the TAR focuses principally on the time horizon reaching from the present to the year 2100—which reflects the preponderance of studies on this time period in the literature and the high degree of uncertainty about the state of socioenvironmental systems beyond the 21st century. By 2100, most projections of human-induced climate change fall into ranges of about 1.4 to almost 5.8°C increase in annual global mean surface temperature (see Figure 5d in the TAR WGI Summary for Policymakers) compared to 1990 (although estimates that are outliers to both ends of even this large range can be found in the literature; Morgan and Keith, 1995) and about 10- to 90-cm rise in mean sea level (Figure 5e, TAR WGI Summary for Policymakers). By the time of doubling of CO_2-equivalent concentration, the global mean precipitation is projected to be about 1–5% higher than in 1990. These global numbers hide complex spatial patterns of changes (especially for temperature and precipitation), which are summarized in Chapter 3. In some regions, temperature increases are projected to be three times the global mean. In addition, high confidence is attached to "projected changes in extreme weather and climatic events" (e.g., see Table 1 in the TAR WGI Summary for Policymakers). Such changes, particularly at the higher ends of the ranges given, represent significant deviations from the climatic conditions of recent centuries. As noted above, warming of the climate and sea-level rise would continue for centuries beyond 2100, even if atmospheric concentrations of GHGs stabilize during the 21st century. For perspective, it should be noted that since the early Miocene (about 24 million years ago), atmospheric CO_2 concentrations appear to have remained below 500 ppmv (Pearson and Palmer, 2000). If human emissions of GHG until 2100 remain at or—as in many scenarios in the literature—increase well beyond current levels, CO_2 concentrations will be significantly above this value. It can therefore be remarked that climate changes in the 22nd century could exceed any experienced in more than 1 million years (see, e.g., Crowley, 1990; Crowley and North, 1991). Indeed, these authors estimate that global temperature was never significantly warmer than the present during the past 2 million years and that one would need to return to the early Pliocene (3–5 million years ago) or even the Miocene (5–25 million years ago) to find a climate that is warmer than today by more than 2°C. The potential impacts of these very large projected changes cannot be disregarded, even though it is difficult to imagine what human societies would look like in the 22nd century (see, e.g., Cline, 1992). However—reflecting the scarcity of studies of climatic impacts beyond 2100, despite their potential relevance to Article 2 of the UNFCCC—these impacts are not a major focus in the TAR (although Chapter 19 does focus on the possibilities of abrupt, nonlinear, and/or irreversible climatic changes in the centuries ahead).

1.2.1. Climatic Change Represents Opportunities and Risks for Human Development

Climate change is likely to present opportunities for some sectors and regions. For example, agriculture could expand into regions where it currently is limited by low temperatures, if adequate soils are present (see Chapter 5). Thinning of Arctic sea ice might allow surface navigation in areas that previously were accessible only to submarines and icebreakers (see Chapter 16). The increase in winter temperature could decrease heating demand or mortality from cold spells (see Chapter 9). However, climate change also is likely to have numerous negative effects on human development and well-being. This is documented in Chapers 4–17 and reflected in the first sentence of the UNFCCC, which states that "changes in the Earth's climate and its adverse effects are a common concern of humankind" (United Nations, 1992). The very existence of the UNFCCC demonstrates that the international community exhibits great concern for the risks that climatic change represents for human development and well-being, despite the potential opportunities it offers. Those risks are classified in Article 2 of the UNFCCC, which describes the Convention's ultimate objective (preventing "dangerous anthropogenic interference with the climate system"). That article mentions the need "to allow ecosystems to adapt naturally to climate change, to ensure that food production is not threatened, and to enable economic development to proceed in a sustainable manner." Because a key function of the IPCC's assessment reports is to help decisionmakers determine what constitutes "dangerous anthropogenic interference with the climate system" (as evidenced by approval at a 1999 IPCC Panel meeting of a series of Policy-Related Questions directed to the Synthesis Report authors; IPCC, 1999b), the discussion that follows is framed according to the aforementioned three categories of impacts.

1.2.1.1. Allow Ecosystems to Adapt Naturally to Climate Change

The speed and magnitude of climate change affect the success of species, population, and community adaptation. The rate of climatic warming may exceed the rate of shifts in certain species ranges; these species could be seriously affected or even disappear because they are unable to adapt (Chapter 5). Some plant and animal species (such as endangered species generally and species adapted to narrow niches for which habitat is discontinuous and barriers impede or block migration) and natural systems (such as coral reefs, mangroves, and other coastal wetlands; prairie wetlands; remnant native grasslands; montane ecosystems near ridges and mountaintops; and ecosystems overlying permafrost) could be adversely affected by regional climatic variations that correspond to a less than 1°C mean global warming by 2100. With mean warming of 1–2°C by 2100, some regional changes would be significant enough so that adverse impacts to some of these highly sensitive species and systems would become more severe and increase the risk of irreversible damage or loss, and additional species and systems would begin to be adversely impacted. Warming beyond 2°C would further compound the risks (note discussions and citations in Chapters 5 and 19).

1.2.1.2. Ensure that Food Production is not Threatened

Human production factors notwithstanding, food production is influenced mostly by the availability of water and nutrients, as well as by temperature. Increases in temperature could open new areas to cultivation, but they also could increase the risk of heat or drought stress in other areas. Livestock (e.g., cattle, swine, and poultry) are all susceptible to heat stress and drought (Gates, 1993). The effects of climatic changes—even smooth trends—will not be uniform in space or time. For smoothly evolving climatic scenarios, recent literature (see Chapter 5) tends to project that high latitudes may experience increases in productivity for global warming up to a 1°C increase, depending on crop type, growing season, changes in temperature regimes, and seasonality of precipitation. In the tropics and subtropics—where some crops already are near their maximum temperature tolerance and where dryland, non-irrigated agriculture predominates—the literature suggests that yields will tend to decrease with even nominal amounts of climate change (IPCC, 1998; Chapter 5). Moreover, the adaptive capacity of less-developed countries in the tropics is limited by financial and technological constraints that are not equally applicable to more temperate, developed countries. This would increase the disparity in food production between developed and developing countries. For global warming greater than 2.5°C, Chapter 5 reports that most studies agree that world food prices—a key indicator of overall agricultural vulnerability—would increase. Much of the literature suggests that productivity increases in middle to high latitudes will diminish, and yield decreases in the tropics and subtropics are expected to be more severe (Chapters 5 and 19). These projections are likely to be underestimates, and our confidence in them cannot be high because they are based on scenarios in which significant changes in extreme events such as droughts and floods are not fully considered or for which rapid nonlinear climatic changes have not been assumed (Section 2.3.4 notes that vulnerability to extreme events generally is higher than vulnerability to changing mean conditions).

Water availability can be regarded as another component of food security. Water quantity and distribution depends to a large extent on rainfall and evaporation, which are both affected in a changing climate. Typically, estimated patterns of changes for 2100 under SRES scenarios include rainfall increases in high latitudes and some equatorial regions and decreases in many mid-latitude, subtropical, and semi-arid regions—which would increase water stress (the ratio between water usage and renewable flow) in the latter regions and decrease it in the former. As noted in Chapter 4, negative trends in water availability have the potential to induce conflict between different users (e.g., Kennedy *et al*., 1998). For perspective, it should be remembered that the capability of current water supply systems and their ability to respond to changes in water demand determine to a large extent the severity of possible climate change impacts on water supply. Currently, 1.3 billion people do not have access to adequate supplies of safe water, and 2 billion people do not have access to adequate sanitation (Gleick, 1998; UNDP, 1999). In addition to changes in average water supply, climate extremes such as droughts and floods often are projected to become a larger problem in many temperate and humid regions (IPCC, 1998; Table SPM-1, WGII TAR Summary for Policymakers).

1.2.1.3. Enable Economic Development to Proceed in a Sustainable Manner

Sustainable development, as noted earlier, implies "meeting the needs of the present without compromising the ability of future generations to meet their own needs" (WCED, 1987). Besides food and water, essential needs include a space to live, good health, respite from extreme events, peace and basic freedom, energy and natural resources that allow development, and so forth. Each of these factors could be affected by climatic change. For example, if there were no significant adaptive responses, a 1-m sea-level rise would decrease the area of Bangladesh by 17.5% or that of the Majuro Atoll in the Marshall Islands by 80%. Human health impacts of global climatic change include changes in the geographic range and seasonality of various infectious diseases (with positive and negative impacts), increases in mortality and morbidity associated with heat waves, and effects on malnutrition and starvation in some regions as a result of redistribution of food and water resources (Chapter 9). The possibility of improved conditions in other regions remains, but, as noted in Chapter 5, the literature tends to project that positive effects in agriculture would be concentrated in high latitudes and negative effects in lower latitudes—precisely where problems of hunger already exist. The frequency and severity of extreme events such as heat waves, high rainfall intensity events, summer droughts, tropical cyclones, windstorms, storm surges, and possibly El Niño-like conditions are likely to increase in a warmer world (Table SPM-1, WGII TAR Summary for Policymakers), which would have a range of adverse impacts and would affect the conditions of development. Migration of populations affected by extreme events or average changes in the distribution of resources might increase the risks of political instabilities and conflicts (e.g., Myers, 1993; Kennedy *et al*., 1998; Rahman, 1999). For each of these potential impacts, the relative vulnerability of different regions to adverse impacts of climatic change is largely determined by their access to resources, information, and technology and by the stability and effectiveness of their institutions. This means that possibilities to promote sustainable development will be affected more negatively by climatic change in developing countries and among less-privileged populations. Thus, climatic change could make satisfying the essential needs of these populations more difficult, in the short term and in the long term. In that sense, climatic change is likely to increase world and country-scale inequity, within the present generation and between present and future generations, particularly in developing countries. Given this potential vulnerability, steps to strengthen adaptive and mitigative (see TAR WGIII Chapter 1) capacity and to lessen nonclimatic stressors could well enhance sustainable development.

1.2.2. Human-Environment Systems: Implications for Development, Equity, and Sustainability

The TAR attempts to place the issue of climate change more centrally within the evolving socioeconomic context. This context is critical to evaluation of the vulnerability of sectors or regions to climatic changes and thus must be borne in mind by anyone who attempts such assessments, as well as policymakers who will need to consider the wide range of implications of technological or organizational choices on the resilience of natural and social systems to climatic changes.

Development of social institutions and technological innovations over the past 10,000 years (the era of civilization after the glacial age when ice largely disappeared) has led to rapid advancement in material well-being but also, very importantly, population growth and resource pressures (e.g., Cohen, 1995; Meyer, 1996). This development process has accelerated and become much larger in scale in recent decades. Globally, growth in annual per capita income has been estimated to have risen from about 0.6% in the 19th-century period of industrial expansion to more than 2% yr^{-1} in the post-World War II era of high technological innovation and global economic cooperation (Cooper, 2000). Some analysts attribute this boom to the combination of stabilization of national economies by governmental management and liberalization of trade allowed by international organizations. Indeed, in this vision—which has been labeled the "cornucopian world view" (Ehrlich and Ehrlich, 1996)—a competitive system that fosters and rewards innovation has led to a prolonged period of development and growth that will increasingly embrace currently less-developed nations as well. Furthermore, according to Kates (1996), it is possible to achieve a world without famine, with little seasonal or chronic undernutrition and virtually no nutrient deficiencies or nutrition-related illnesses. In the area of energy and natural resources, according to von Weizsäcker *et al.* (1998), it is possible to increase resource productivity by a factor of four: The world would then enjoy twice the wealth that is currently available, simultaneously halving the stress placed on our natural environment.

However, this level of development continues to be an elusive goal for a large fraction of the world's population. There is a noticeable disparity between the levels of development that have been achieved in various societies. These differences are obscured by globally averaged income growth data such as those reported by Cooper (2000). For example, gaps between the rich and the poor are widening between developed and developing countries and within tropical African, Asian, and Latin American countries (UNDP, 1999). Although there have been notable successes, many countries in these regions have experienced increases in economic instability, social insecurity, environmental degradation, and endemic poverty. Despite spectacular gains in the means of development—such as advances in science, technology, and medicine during the past century—development planning at national and global levels has not always alleviated poverty and inequity (see Box 1-2; Munasinghe, 2000).

Global food security clearly has improved in recent years as the focus of famine has shifted from large, heavily populated countries to sparsely populated and small nations, but the number of people at risk of hunger still is very high, even in parts of heavily populated nations. Chen and Kates (1996) estimate that the population at risk of outright starvation could be as high as 35 million; FAO (1999) estimates that about 800 million people in developing countries and 34 million people in developed countries suffer from undernourishment. Achieving global food security is complicated by growth in human population and political instability that disrupts food delivery systems. Projections vary widely, ranging from stabilization of population at near-present levels sometime in the 21st century to a greater than three-fold increase by the end of the century (e.g., Fischer *et al*., 1996; Lutz, 1996; IPCC, 2000). At current population growth rates, world food production must double within the next 40 years to feed this population; such a doubling of food production may require expansion of agricultural land into forests and areas that presently are considered marginal for agriculture (but not necessarily marginal in other respects). On the other hand, some authors (e.g., Waggoner *et al*., 1996) have argued it is possible to increase the dietary standard of all humans up to a doubling of current populations and at the same time to "spare land for nature" (Ausubel, 1996). This expansive vision follows from the belief that resources can be made available for the extension of current intensive agricultural practices to currently low-technology regions. The extent to which such practices can be extended is debated, on social and environmental grounds, by those who hold an opposing worldview, which has been called "limits to growth" (e.g., Ehrlich *et al*., 1995). Moreover, even if such agricultural intensification were to occur, there is no guarantee that extensive land use for economic development activities other than growing food would not simultaneously occur. Thus, the hope of "sparing land for nature" via intensification likewise is a controversial vision.

The need to improve productivity per unit area has led to more intensive methods in developing countries—which, together with low or negative economic support for agricultural products, often has driven smallholders off their land and led to emigration to urban centers (WCED, 1987). The influx of poor, unskilled, and often unemployable people has led to explosive and difficult-to-manage growth in these centers (O'Meara, 1999). This sets the stage for the gestation of a new set of environmental problems, including substandard housing, squatter settlements, solid waste buildups, unsatisfactory sewage disposal, urban floods, and urban water pollution, as well as the characteristic problems of large cities such as crime and social insecurity.

In opposition to the aforementioned expansive visions, others (e.g., Daily, 1997) express concern that services provided by ecosystems to society may be undermined by a combination of unsustainable population growth, destruction of natural habitats, and pollution of air, soils, and waters. Three decades ago, debate raged about whether indefinite economic expansion would be limited by environmental and other resource constraints. Meadows *et al*. (1972) postulated in a controversial work that environmental protection and economic growth are not compatible; there are "limits to growth." For those holding this worldview, current development patterns will not allow continued improvement of the human condition for much longer; instead, such development will ensure continuing degradation of natural assets such as biodiversity (e.g., Pimm, 1991). Thus, it is feared that the environment may be losing part of its capacity to support life and therefore may be imposing another set of constraints on the development process—disturbances to air, waters, soils, and species distributions brought about by human activities—that will require responses to reduce additional risks. Several sharp critiques appeared (e.g., Cole *et al*., 1973), noting that the "limits" paradigm ignored enhanced productivity brought about by innovation and that although limits eventually might become a problem, increased knowledge generated by economic expansion could create substitutes for resources that were being used nonrenewably, and much less energy and materials would be needed to produce economic growth as technology blossomed (e.g., Grossman and Krueger, 1995; but see Myers and Simon, 1994). Moreover, it has been argued that enhanced wealth and knowledge also can reduce vulnerability to environmental stresses such as climatic change.

Subsequently, a modified view that considered both the "cornucopian" and "limits" paradigms emerged: the strategy of sustainable development. It is designed to promote conservation of resources and protection of the environment while sustaining a healthy society whose needs are securely provided. In response to requests from governments participating in the IPCC process, the TAR is attentive to the concept of sustainable development.

Technology and organization clearly have reduced the vulnerability of humans in some countries to a variety of hazards. In the context of the IPCC process, this would include, for example, flood control engineering projects that have reduced lives lost in catastrophic flooding. However, pioneering analyses in the natural hazards literature (e.g., Burton *et al*., 1993) note that large-scale dependence of massive populations on the functioning of giant engineering projects or social institutions often has simply transformed our risks from the predevelopment state of high-frequency, low-amplitude risk (many localized threats to small numbers of people in each instance) to the present state of low-frequency, high-amplitude vulnerability (where a rare levee failure or the simultaneous occurrence of drought in several major exporting granaries poses the risk of infrequent but very catastrophic losses). Moreover, the consequences of these risks are unlikely to be equitably distributed within and across income groups and nations, which requires assessment of the distributional implications of developmental risks and benefits (e.g., Box 1-2). In many developing regions, population pressures and poverty have led to occupation of hazardous lands (e.g., steep slopes, valley bottoms) and has greatly increased vulnerability to climate extremes. Of course, many factors other than those mentioned above can contribute to vulnerability (e.g., Etkin, 1999).

In addition to this huge list of challenges, potential threats to the global environment are connected to the development process. The TAR identifies scientific and policy linkages among key global environmental issues, one of which is climate change. Other global environmental issues include loss of biological diversity, stratospheric ozone depletion, marine environment and resource degradation, and persistent organic pollutants (Watson *et al.*, 1998). Other contemporary issues are evident in many places across the globe—though each instance is not global in scale (e.g., Turner *et al.*, 1990)—such as freshwater degradation, desertification, land degradation, deforestation, and unsustainable use of forest resources. None of these threats implies that the *net* effects of human developments are necessarily negative, only that embedded in many development activities are a host of negative aspects that many analysts and policymakers believe must be considered in development planning. Strategies to modify the amount and/or kinds of development activities to account for these threats are considered more comprehensively in the report of Working Group III. The TAR also focuses on linkages between climate change on one hand and local and regional environmental issues—for example, urban air pollution and regional acid deposition—on the other. (Strategies to deal with these issues that also help with adaptive or mitigative capacity for climate change often are called co-benefits.) Among the new areas of emphasis in the TAR are linkages between global environmental issues and the challenges of meeting key human needs such as adequate food, clean water, clean air, and adequate and affordable energy services.

1.3. How has Society Responded?

1.3.1. International Responses

A primary response to concerns about climate change has been *international* action to address the issue, particularly through the UNFCCC. International action to date has focused mainly on mitigation, although adaptation is mentioned in UNFCCC Article 4.1 (e) and in funding by the Global Environment Facility (GEF) of adaptation studies (e.g., the Caribbean Planning for Adaptation to Climate Change program). Multinational action is required because no single country or small group of countries can reduce emissions sufficiently to stop GHG concentrations from continuing to grow and because wherever emissions originate, they affect climate globally. Because the extent and urgency of action required to mitigate emissions depends on our vulnerability, a key question is the degree to which human society and the natural environment are vulnerable to the potential effects of climate change.

At the first meeting of the Conference of Parties to the UNFCCC in 1995, governments reviewed the adequacy of existing international commitments to achieve this goal and decided that additional commitments were required. They established the Ad Hoc Group on the Berlin Mandate (AGBM, 1995) to identify appropriate actions for the period beyond 2000, including strengthening of commitments through adoption of a protocol or another legal instrument. The AGBM process culminated in adoption of the Kyoto Protocol in December 1997 (United Nations, 1997). In the Kyoto Protocol, industrialized countries (Annex I Parties to the UNFCCC) agreed to reduce their overall emissions of six GHGs by an average of 5% below 1990 levels between 2008 and 2012. The Protocol also allowed the Parties to account for the removal of GHGs by sinks resulting from direct, human-induced land-use change and forestry activities, emissions trading, "joint implementation" (JI) between developed countries, and a "clean development mechanism" (CDM) to encourage joint emissions reduction projects between developed and developing countries and a commitment to provide assistance in meeting the costs of adaptation for countries deemed most vulnerable to the adverse effects of climate change using the proceeds of the CDM (Article 12). To date, the Kyoto Protocol has not entered into force. The UNFCCC also established national reporting requirements for all Parties regarding their emissions and their potential vulnerabilities/adaptation options. These reporting obligations are being fulfilled through preparation of National Communications to the UNFCCC.

The foundation for any policy to address the climate change problem is information on GHG emissions, the climate system and how it may change, likely impacts on human activities and the environment, and the costs and co-benefits (e.g., protecting primary forests not only retains stored carbon in the trees but also confers the "co-benefit" of biodiversity protection; Kremen *et al.*, 2000) of taking steps to reduce GHG emissions or to change land use. To provide the best available scientific information for policymakers and the public, governments established the IPCC to periodically assess and summarize the state of knowledge in the literature related to climate change. The IPCC completed comprehensive assessments in 1990 and 1995 of the effects of human activities on the climate system, potential consequences of climate for natural and human systems, and the effectiveness and costs of response options (IPCC, 1990, 1996a,b,c). In addition, the IPCC has prepared numerous special reports, technical papers, and methodologies on topics ranging from radiative forcing of climate to technologies, policies, and measures for emissions mitigation. As knowledge has progressed, IPCC assessments have added a regional focus by assessing regional climate modeling and regional sensitivities and adaptive capacity.

Other international bodies also are taking up the challenge of climate change. These organizations include the World Bank, the United Nations Environment Programme (UNEP), the UN Development Programme (UNDP), and the GEF, as well as a variety of regional institutions. Although a primary audience for this report is those who are involved in negotiating and implementing the UNFCCC and, to some extent, other international agreements on global environmental problems, the report also contains information that is useful to other international institutions. The report has been designed to be useful in assessing potential projects and opportunities for investment that will be robust to potential negative effects and to emerging opportunities from climate change.

1.3.2. National and Local Governmental Responses

Governments have initiated a spectrum of responses, ranging from international assessments of climate science, impacts, and abatement strategies (the United States, for example) to implementation of a legally binding mitigation policy (Sweden, for example, has implemented a domestic carbon tax on direct fuel use and on fuel use in the transportation sector; see also OECD, 1999). Governments also have produced country studies, vulnerability assessments of sea-level rise, and national communications; carried out GHG reductions in other countries; and created research opportunities and fora for exchanges of ideas and data. Such management of climate-related research and educational activities has accelerated in the wake of climatic assessment that suggested a discernible human influence on climate (IPCC, 1996a). Similarly, many countries have implemented policies for reasons unrelated to climate change that nevertheless have led to reductions of GHG emissions (e.g., the ethanol program in Brazil, support to renewable energy and energy efficiency in a large number of countries). With regard to adaptation, the first National Communications to UNFCCC from most countries contained analyses of vulnerability and adaptation options.

At the local level, dozens of cities—mainly in industrialized countries—have adopted GHG emission reduction targets and have taken measures to implement them, mostly in the energy and transport sector. In many cases, these policies have been defined by coupling climate protection objectives with other, more local objectives: co-benefits such as reducing air pollution, traffic congestion, or waste production. Some measures, such as water conservation, are adaptive (more resilient to drought) and reduce emissions (less energy for pumping). The use of "social" policy instruments such as public awareness campaigns, information, and technical assistance is commonplace (OECD, 1999). With regard to adaptation, for example, the Federation of Canadian Municipalities is promoting adaptation as well as mitigation measures.

Many countries have developed national climate strategies that are based on a diverse range of policy instruments such as economic instruments, regulation, research and development, and public awareness and information. Energy efficiency, fuel switching, public transportation, and renewable energies typically are promoted. The government sector itself is an increasingly common target for GHG mitigation, and "greening" of government purchasing policies has started to take place in some countries (OECD, 1999).

Overall, these policies and measures to date have had limited effect on emissions, probably because of their lack of integration in a truly global, long-term framework, as well as continued economic growth around the world (AGBM, 1995).

1.3.3. Organizational Responses

Numerous private businesses have developed plans to facilitate trading of permits for carbon emissions or have set up schemes to help manage CDM transactions if the Kyoto Protocol is ratified or some other instrument of carbon policy is put in place by some nations. Moreover, large multinational corporations such as Shell International and BP Amoco have declared that they will voluntarily observe elements of the Kyoto Protocol (van der Veer, 1999; Browne, 2000).

International scientific organizations have responded to the prospect of climate changes for more than 2 decades, from the second objective of the Global Atmospheric Research Program (GARP) to a series of World Climate Conferences sponsored by the World Meteorological Organization and UNEP. The International Council of Scientific Unions and dozens of national scientific societies have responded by creating journals to publish the results of climatic assessments, organizing many meetings and symposia to further our understanding of climate-related scientific issues, and creating committees to help steer research in promising directions.

Similarly, environmental nongovernmental organizations (NGOs) around the world have initiated climate campaigns with the aim of convincing citizens and governments to strengthen the Kyoto Protocol. Meanwhile, direct advertisements have appeared in the media—primarily sponsored by organizations that are attempting to influence public opinion to oppose the Kyoto Protocol.

1.3.4. Adaptive Responses

Natural and human systems have adapted to spatial differences in climate. There also are examples of adaptation (with varying degrees of success) to temporal variations—notably, deviations from annual average conditions. Many social and economic systems—including agriculture, forestry, settlements, industry, transportation, human health, and water resource management—have evolved to accommodate some deviations from "normal" conditions, but rarely the extremes.

Adaptations come in a huge variety of forms. Autonomous adaptations invariably take place in reactive response (after initial impacts are manifest) to climatic stimuli as a matter of course, without directed intervention by a public agency. The extent to which society can rely on autonomous, private, or market adaptation to reduce the costs of climate change impacts to an acceptable or nondangerous level is an issue of great interest. There is little evidence to date that efficient and effective adaptations to climate change risks will be undertaken autonomously (see Chapter 18).

Planned adaptations can be reactive or anticipatory (undertaken before impacts are apparent). Potential adaptations include sharing losses, modifying threats, preventing or decreasing effects, changing use, and changing location. There are many lists of adaptation measures, initiatives, or strategies that have potential to moderate impacts, if they were implemented. Such lists indicate the range of strategies and measures that represent possible adaptations to climate change risks in particular sectors and regions. Only in a few cases have such lists of potential adaptations considered who might undertake them, under what conditions might they be implemented, and how effective might they be.

Knowledge of processes by which individuals or communities actually adapt to changes in conditions over time comes largely from analog and other empirical analyses. These studies indicate that autonomous adaptations tend to be incremental and ad hoc, take multiple forms, occur in response to multiple stimuli (usually involving a particular catalyst, rarely climate alone), and are constrained by economic, social, technological, institutional, and political conditions.

Although an impressive variety of adaptation initiatives have been undertaken across sectors and regions, responses are not universally or equally available. Adaptation options generally occur in socioeconomic sectors and systems in which turnover of capital investment and operating costs is shorter, and less often where long-term investment is required. Examples include purchase of more efficient irrigation equipment by individual farmers in anticipation of increased evapotranspiration in a warmer climate, design of bridges or dams to account for an expected increase in sea level or extremes of drought and flood, purchase of insurance, abandonment of insurance coverage to people living in high-risk areas such as coastlines, and creation of migration corridors for species expected to be forced to migrate with climate change.

Often more than one adaptation option is available. People rarely seem to choose the best responses—those among available options that would most effectively reduce losses—often because of an established preference for, or aversion to, certain options. In some cases, there is limited knowledge of risks or alternative adaptation strategies. In other cases, adoption of adaptive measures is constrained by other priorities, limited resources, or economic or institutional barriers. Recurrent vulnerabilities, in many cases with increasing damages, illustrate less than perfect adaptation of systems to climatic variations and risks. Chapter 18 describes some evidence that the costs of adaptations to climate conditions are growing.

Current adaptation strategies with clear applications to climate change in agriculture include moisture-conserving practices, hybrid selection, and crop substitution. In the water resources sector, current management practices often represent useful adaptive strategies for climate change. Some analysts go further to point out that certain adaptations to climate change not only address current hazards but may be additionally beneficial for other reasons. Such evaluations are further complicated by the existence of secondary impacts, related to the adaptation itself. For example, water development projects (adaptations to water supply risks) can have significant effects on local transmission of parasitic diseases. Improved water supply in some rural areas of Asia has resulted in a dramatic increase in Aedes mosquito breeding sites and, consequently, outbreaks of dengue (Section 18.4.4).

1.4. How are Impacts, Adaptation, and Vulnerability Assessed in this Report?

An important new category of issues on climatic impacts pertains to methodological advances; Chapter 2 provides more details. These advances include methods for assessing impacts and vulnerability, methods for detecting biotic response to climate change in natural ecological systems by using indicator species, and detection and attribution of observed changes in environmental systems to climatic changes, as distinct from other possible causal factors (Chapter 5). Other new methods relate to costing and valuation, decision analytic techniques and frameworks, uncertainties assessment, and consistent characterization of levels of confidence that could be attached to observations or conclusions (see Box 1-2 and Chapter 2). All of these developments extend methodological considerations beyond those typically employed in the SAR. Development and application of regional-scale scenarios to climate change impacts, adaptations, and vulnerability (as described in Chapter 3) represent a new emphasis of a technique that was limited in the SAR to the science of climate change. The directive in the IPCC's charge to Working Group II to emphasize regional issues is reflected in the eight regional chapters (Chapters 10–17). This directive calls for utilization of new advances in knowledge, including integrated methods, to assess the most cost-effective approaches to adapt to changes in climate at the regional level. This section elaborates briefly on several issues that are related to assessment of impacts, vulnerability, and adaptation and illustrates several of the foregoing methodological considerations.

1.4.1. Sensitivity, Adaptability, and Vulnerability

This report assesses recent advances in our understanding of the vulnerability of major sectors, systems, and regions to climate change. Consistent with common usage and definitions in the SAR, *vulnerability* is defined as the extent to which a natural or social system is susceptible to sustaining damage from climate change. Vulnerability is a function of the *sensitivity* of a system to changes in climate (the degree to which a system will respond to a given change in climate, including beneficial and harmful effects), *adaptive capacity* (the degree to which adjustments in practices, processes, or structures can moderate or offset the potential for damage or take advantage of opportunities created by a given change in climate), and the degree of *exposure* of the system to climatic hazards (Figure 1-2). Under this framework, a highly vulnerable system would be a system that is very sensitive to modest changes in climate, where the sensitivity includes the potential for substantial harmful effects, and for which the ability to adapt is severely constrained. *Resilience* is the flip side of vulnerability—a resilient system or population is not sensitive to climate variability and change and has the capacity to adapt.

Adaptation is recognized as a crucial response because even if current agreements to limit emissions are implemented, they will not stabilize atmospheric concentrations of GHG emissions and climate (Wigley, 1998). Hence, adaptation is considered here, along with mitigation—the more widely considered response to climate change—as a key component of an integrated and balanced response to climate variability and change (MacIver, 1998). Adaptations, which can be autonomous or policy-driven, are adjustments in practices, processes, or structures to take

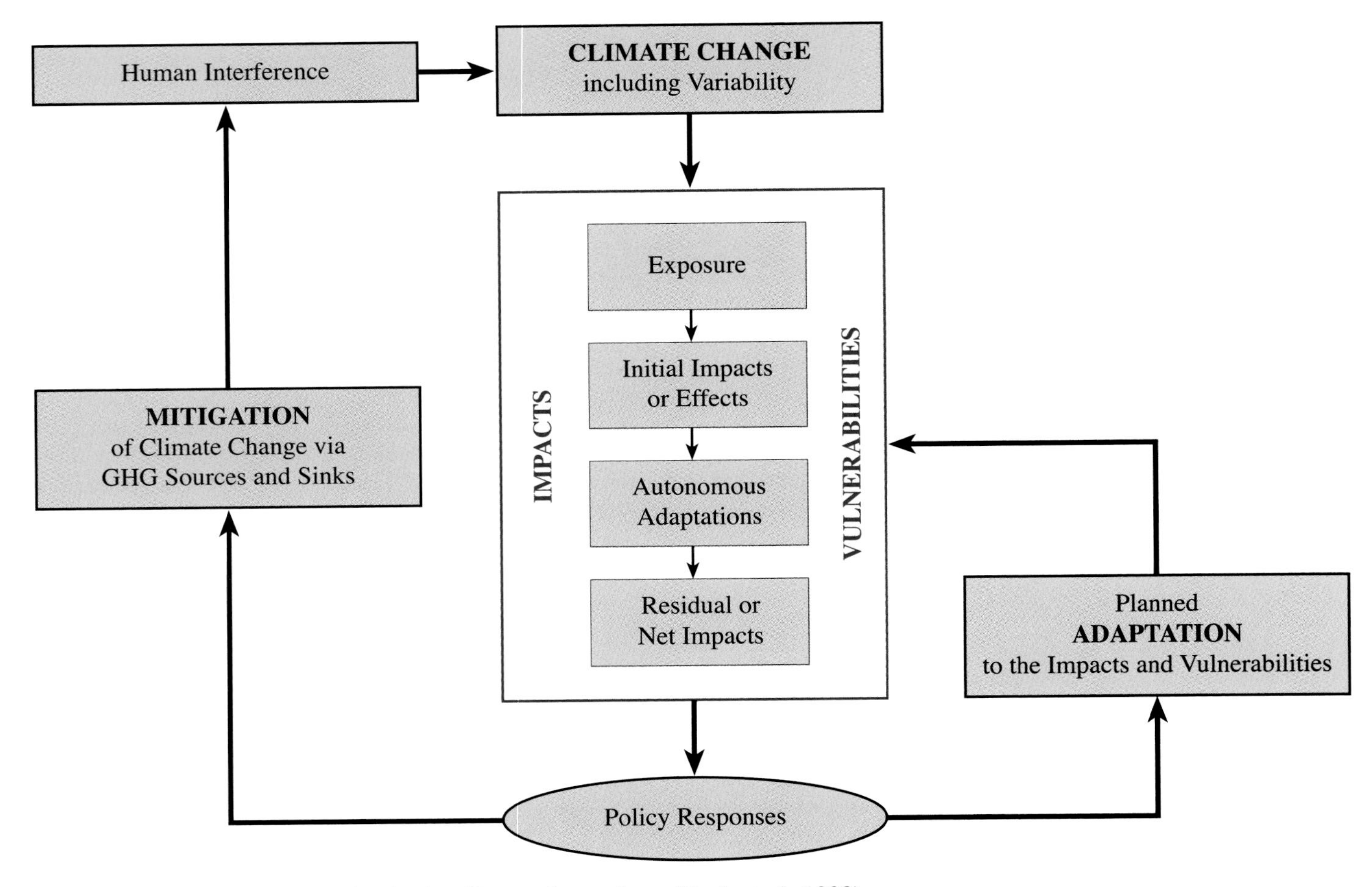

Figure 1-2: Places of adapatation in the climate change issue (Smit *et al.*, 1999).

account of changing climate conditions. Impacts, however, sometimes are difficult to identify, let alone quantify, in part because of the nonlinear nature of climate change itself. Impacts can be subtle but nonetheless significant, and their consequences can differ for different members of the same community—as when some individuals or groups perceive an opportunity with change and others experience a loss, thereby changing community dynamics and complicating decisions about how to adapt and the apportionment of costs of adaptation. Negative impacts often are observed as chance occurrences (surprises) beyond critical values (thresholds) of accustomed weather parameters. They can be conceived as risks or the "probability of occurrence of a damaging event," such as flood, drought, strong winds, heat wave, subfreezing temperatures, or forest or bush fire. "The extent to which natural ecosystems, global food supplies and sustainable development are in danger depends partly on the nature of climate change and partly on the ability of the impacted systems to adapt" to these events (Smit *et al.*, 1999).

The capacity of a sector or region to adapt to climatic changes depends on several factors (see Figure 1-2 and Chapter 18). The literature emphasizes that studies that neglect adaptive potential are likely to overestimate the costs of climatic impacts (e.g., Reilly *et al.*, 1996). However, more recent literature also notes that maladaptations are possible—particularly when information about future climatic and other conditions is much less than perfect—as a response to an incorrect perception of such changes, often driven by a masking of slowly evolving trends by large natural variability or extreme events (West and Dowlatabadi, 1999; Schneider *et al.*, 2000a; West *et al.*, 2001). Maladaptations can increase the costs of impacts relative to those when adaptive agents have perfect foresight or when adaptive responses are absent. On the other hand, appropriate adaptations can reduce negative impacts or take advantage of new opportunities presented by changing climate conditions. The SAR assessed technical options for adaptation but did not evaluate the feasibility of these options for different regions and circumstances because little information was available in the literature. The *Special Report on Regional Impacts of Climate Change* focuses to a greater extent on the regional dimensions of adaptation; because the report is based largely on the SAR, as well as early and preliminary results from country studies and national communications to the UNFCCC, however, many questions about the capacities required to implement theoretically promising adaptation options remain. Hence, in this report, greater attention has been focused on societal determinants of adaptive capacity and vulnerability. To the extent possible, the report seeks to examine information in the literature on the interaction of these factors to develop options for bolstering adaptive capacity.

Previous IPCC assessments conclude that most systems are sensitive to the magnitude and the rate of climate change. Sensitive systems include, for example, aspects of food and fiber production, water resources, ecosystems of all types, coastal systems, human settlements, and human health. This sensitivity includes adverse effects in many regions, as well as potentially beneficial effects in some regions and sectors (high confidence). Social systems generally are more resilient than natural systems because of the potential for deliberate adaptation (high confidence). However, confidence in most specific aggregate estimates of impacts remains low because of uncertainties and complexities of analysis (as detailed further in succeeding sections).

For systems that already are exposed to increasing resource demands, unsustainable management, and pollution, exposure to climate change is an important additional pressure. Systems that are exposed to multiple pressures (synergistic effects) usually are more vulnerable to climate change than systems that are not (high confidence).

1.4.2. Detection and Attribution of Impacts to Climate Change

Many observed changes in ecosystems, animal (e.g., butterfly and bird patterns) and plant (e.g., timing of flowering) species, and physical systems (e.g., glaciers or river runoff) have been associated with observed changes in climate (not necessarily anthropogenic changes) in recent decades (high confidence). Moreover, as described in Chapters 5 and 19, such observed changes often are in the directions expected as a response to climate stimuli, based on understandings expressed in the literature about biophysical processes that govern responses to climate (e.g., Root and Schneider, 2001; Root *et al*., 2001). This consistency has led the authors of such studies to conclude that surface temperature trends of recent decades are likely to be discernible at regional scales through observed changes in biological and physical entities for several systems (varying confidence, depending on which specific system is considered; see, e.g., Chapters 5 and 19). From these observed responses to the relatively small climate changes observed to date (as compared to changes projected for the next century), it is concluded that many environmental systems can be highly sensitive to climate change. However, determination of a potential causal relationship between the response of a physical or biological system to observed recent climatic changes does not imply that regional climate changes were a direct result of anthropogenic global climatic trends, although the latter are likely to have had significant influence on many regional trends. Working Group II does not focus on evaluating the likelihood that regional observed climatic variations are caused by anthropogenic climate changes; detection and attribution assessment of climatic changes is primarily a Working Group I activity. As noted, however, Working Group II does address attribution of observed changes to environmental systems to observed climate changes, even if the connection to possible anthropogenic climate changes is not specifically addressed here.

1.4.3. Key Determinants of Impacts

1.4.3.1. Magnitude of Change

Early studies concentrated on impacts caused by changes in global mean temperature. Often these studies were carried out at a few elevated temperatures—typically, 2°C and 4°C (corresponding to the bulk of the range of IPCC Working Group I SAR equilibrium temperature rise expected for a doubling of CO_2 concentration above pre-industrial levels). Global mean temperature still is a significant variable, serving as a modulus of change against which to compare climate sensitivities and impacts. In addition to mean quantities, however, other characteristics of climate measures, such as climate variability or runs of unusually warm weather, have become important variables for analysis (e.g., Mearns *et al*., 1984; Colombo *et al*., 1999), as has specification of changes in regional temperatures, sea level, and precipitation. These expanded measures of climatic change are routinely included in recent impact studies (e.g., IPCC, 1998). Less often considered are changes in extreme events (but see Table SPM-1, WGII TAR Summary for Policymakers), despite their potential importance.

1.4.3.2. Rate of Change

It is essentially undisputed that a sustained 2°C temperature change occurring in a decade would have a more profound impact than one occurring over a century. The effect of rates of change on impacts is still under active investigation (see Chapter 19). Early results have suggested that rates of change exceeding the ability of ecosystems to migrate would be particularly damaging (see Chapter 5). Adaptation of coastal dwellers to rapid climatic changes or a high background "noise level" of natural variability would be more difficult relative to slowly occurring changes or smoothly varying climates (e.g., West *et al*., 2001). Finally, as noted by IPCC Working Group I (1996a, p. 7), "nonlinear systems, when rapidly forced, are particularly subject to unexpected behavior." In other words, the adaptability of various decision agents would be reduced if any change is unexpected; thus, a rapid rate of change is more likely to generate "surprises" that inhibit effective adaptation by natural and managed systems. Table 1-1 describes several extreme events that can substantially influence the vulnerability of sectors or regions to climatic changes (see also Table SPM-1, WGII TAR Summary for Policymakers).

Economists also have suggested that the transient stage of moving from one equilibrium climate to another could cause the greatest economic impacts, even if the static impacts of the new equilibrium climate were small (Nordhaus and Boyer, 2000).

1.4.3.3. Transient Scenarios

Climate sensitivity—the globally averaged response of the surface temperature to a fixed doubling of CO_2—is based on static or equilibrium calculations in which the climatic model

Table 1-1: Typology of climate extremes.

Type	Description	Examples of Events	Typical Method of Characterization[a]
Simple extremes	Individual local weather variables exceeding critical level on a continuous scale	Heavy rainfall, high/low temperature, high wind speed	Frequency/return period, sequence and/or duration of variable exceeding a critical level
Complex extremes	Severe weather associated with particular climatic phenomena, often requiring a critical combination of variables	Tropical cyclones, droughts, ice storms, ENSO-related events	Frequency/return period, magnitude, duration of variable(s) exceeding a critical level, severity of impacts
Unique or singular phenomena	A plausible future climatic state with potentially extreme large-scale or global outcomes	Collapse of major ice sheets, cessation of thermohaline circulation, major circulation changes	Probability of occurrence and magnitude of impact

[a] Stakeholders also can be engaged to define extreme circumstances via thresholds that mark a critical level of impact for the purposes of risk assessment. Such critical levels often are locally specific, so they may differ between regions.

is allowed to reach a steady state after the CO_2 increase is applied. The real Earth, on the other hand, is being forced by a time-evolving forcing of GHGs and other global change forcings; this, combined with the time-evolving response of the climate system to any forcing, means that the amount of global climatic warming, as well as the time-evolving patterns of climatic changes, are likely to be different during the transient phase of climatic change than in equilibrium. Recent studies of climate change impacts have made use of transient or time-dependent scenarios of climate change that are derived from fully coupled, ocean-atmosphere general circulation models (AOGCMs). These studies indicate that many systems would be notably affected (see Chapter 19)—some adversely and some beneficially—by changes in climate within the next 2 to 3 decades (high confidence). Farther into the 21st century, as radiative forcing on the climate builds, the magnitude of adverse impacts would increase, the number and scale of many beneficial effects would decrease (Chapter 19), and the probability that adverse impacts would predominate would increase (high confidence). Transient scenarios are just entering the climate impacts literature, which unfortunately tends to lag the climate effects literature by several years; thus, much of the impacts literature still is based on equilibrium climate change scenarios. To the extent possible, Working Group II has assessed literature that uses transient scenarios. It is important to use transient scenarios as much as possible because the climate effects literature suggests that static calculations (typically, CO_2 held fixed at double pre-industrial concentrations) do not produce the same time-evolving regional patterns of climatic changes as do transients—and because, of course, the actual Earth is undergoing a transient response to anthropogenic forcings.

1.4.3.4. Climate Variability and Extreme Events

Most studies of climate change impacts have focused on changes in mean climate conditions. However, global climate change is likely to bring changes in climate variability and extreme events as well. This is relevant here because decisionmakers often consider hedging strategies to be prepared for the possibility of low-probability but high-consequence events—a risk management framework. Features of projected changes in extreme weather and climate events in the 21st century include more frequent heat waves, less frequent cold spells (barring so-called singular events), greater intensity of heavy rainfall events, more frequent midcontinental summer drought, greater intensity of tropical cyclones, and more intense El Niño-Southern Oscillation (ENSO) events (Table SPM-1, WGII TAR Summary for Policymakers).

A small number of studies have investigated the potential impacts of hypothesized changes in climate variability and/or extreme events. Results of these studies, coupled with observations of impacts from historical events (e.g., Chapter 8), suggest that changes in climate variability and extremes are likely to be at least as important as changes in mean climate conditions in determining climate change impacts and vulnerability (high confidence). The literature suggests that omission of changes in extreme events and/or climate variability will yield underestimates of climate change impacts and vulnerability. In its assessment of potential vulnerabilities and adaptation options, Working Group II has focused on the interactions of natural climate variability and anthropogenic change and the potential for "win-win" adaptation options that would increase resilience to both phenomena.

1.4.3.5. Thresholds

In many environmental fields, there are thought to be thresholds below which only minor effects occur. Critical levels in acid rain are one example (Brodin and Kuylenstierna, 1992). These kinds of thresholds also are possible in climate change and are incorporated into some models as "tolerable" levels that must be exceeded before significant impacts occur (Hope *et al.*, 1993).

However, in climate change, thresholds have been proposed that are much more complicated. Below the threshold, there may be some impacts, but they will be smoothly varying with the change in climate. Some positive effects might even be observed in some regions or sectors for a small global warming, giving the impression that there is little impact. Above the threshold, however, potentially damaging events may occur. For example, most models show (by 2100) a weakening of thermohaline circulation that transports warmer water to the North Atlantic (see TAR WGI Summary for Policymakers) but only very low confidence that there will be full collapse of the thermohaline circulation by 2100—although some rapid greenhouse buildup scenarios suggest that emissions during the 21st century could trigger a collapse in the following century (e.g., Rahmstorf, 1999; Schneider and Thompson, 2000). Likewise, only very low confidence is given to the prospect of substantial collapse by 2100 of the West Antarctic Ice Sheet (TAR WGI Summary for Policymakers). Other examples of potential threshold phenomena can be found in the literature for regional situations. For example, Wang and Eltahir (2000) demonstrate that rainfall in the Sahel region of Africa can have several equilibrium values, depending on the level of disturbance to vegetation cover. For vegetation removal of less than a threshold value, the system recovers within a few years. For vegetation removal above a threshold, however, there is a new steady-state rainfall regime that is much reduced from "normal" conditions. These thresholds may be, as characterized previously, a result of rapid transient forcing of the climate system, in terms of altered radiative properties of the atmosphere or characteristics of the land surface. Although such threshold events remain somewhat speculative, their impacts clearly would be more severe than smoothly varying (and thus more adaptable) events. Some thresholds in impacts, however, are much less speculative, such as hospital admissions for heat conditions above a threshold temperature—and these threshold temperatures vary regionally as there is some acclimatization to heat stress (Chapter 9)—or species living near mountaintops that would be forced out of existence, even by smooth climatic warming, because they reached the threshold of having no place to move up into (e.g., Still *et al.*, 1999).

Sometimes the expression "threshold" is used as an approximation when the response actually is more likely to be smooth but strongly nonlinear. The release of methane from gas hydrates trapped in deep sediments and the health impacts of thermal stress would be examples of this category. Working Group II assesses potential thresholds for ecological and human systems.

1.4.3.6. Surprises

By definition, it is difficult to give examples of the surprises that might be created under a changed climate. Such surprises, however, can make even the most careful calculation of impacts extremely inaccurate, as noted previously. Surprises have been classified by many authors in many contexts (see Schneider *et al.*, 1998, for a review of the literature and many citations). In particular, low-probability events—or those whose probability is difficult to assess—often are labeled rhetorically as "surprises," even though the event has been classified or identified as known. Strictly speaking, such events are more accurately called "imaginable surprises;" true surprises are wholly unexpected events. Another useful category is "imaginable conditions for surprise" (Schneider *et al.*, 1998), where the specific event in question is unexpected but a set of conditions that increases the likelihood of surprises can be assessed; increasing the rate of forcing of the climatic system is one example, as noted in Section 1.4.3.2.

1.4.3.7. Nonlinear, Complex, and Discontinuous Responses

Investigations into climate change and its potential consequences have begun to highlight the importance of strongly nonlinear, complex, and discontinuous responses. These types of responses, called singularities, can occur at all temporal and spatial scales of systems influenced by climate change (high confidence can be given to the likelihood that some such singularities will occur, but low confidence usually is assigned to any specific example of a possible abrupt event; see Chapter 19). Strongly nonlinear responses are characterized by thresholds—which, if exceeded by a stimulus, result in substantially greater sensitivity to further stimulus or dramatic change, explosive growth, or collapse. Complex responses involve interactions of many intricate elements that yield outcomes that are not easily predicted. Examples of these types of responses include coral bleaching, collapse of fish stocks, disease outbreaks, changes in fire and other disturbance regimes in vegetation systems, crop failure, malnutrition and hunger, and collapse of pastoral communities. Advances in our understanding of these types of responses are largely qualitative, but they are important in understanding the character of dangers posed by climate change. Omission of potential nonlinear and complex responses from climate change impact assessments is expected (well-established, but incomplete) to yield underestimates of impacts (see Chapters 5 and 19). Because of the magnitude of their potential consequences, large-scale discontinuous responses warrant careful consideration in evaluations of climate change dangers. Working Group II points to the potential for such occurrences and their potential consequences for human and natural systems, but it is unable to provide detailed assessments of potential effects, given the paucity of information in the literature.

1.4.4. Synergies and Tradeoffs

1.4.4.1. Synergies and Tradeoffs between Climate Change and Other Environmental Issues

Climate change is only one issue among many. The early stages of economic development typically lead to an increase in many pollutants, and actions taken to reduce one can have ancillary benefits caused by simultaneous reduction of others. Assessments that neglect these synergies can seriously underestimate the justification for cutbacks. On the other hand, impacts from climate change can depend on the levels of other pollutants. For example, forests weakened by acid rain are likely to be more vulnerable to changes in rainfall brought on by climate change or warming, and lake acidification can have a synergy with ultraviolet radiation penetration into the water (e.g., Schindler *et al.*, 1996). While maintaining its primary focus on decadal to centennial-scale climate change, Working Group II has examined linkages among climate change and other environmental issues, including climate variability, loss of biodiversity, deforestation, and desertification.

1.4.4.2. Synergies and Tradeoffs between Adaptation and Mitigation

It is often argued in the literature that there is a tradeoff between adaptation and mitigation in that resources committed to one are not available for the other. This is debatable in practice because the people who bear emission reduction costs or benefits often are different from those who pay for and benefit from adaptation measures. Arguments are given on both sides of this issue. On one hand, in a straight comparison, several factors point to the wisdom of initially committing resources to adaptation. Insofar as no level of mitigation will completely prevent some climate change, some adaptation will be necessary. The benefits from adaptation are received in the country that incurs the costs, so there is no "free-rider" problem; climate change from GHG emissions that already have occurred means that adaptation will be required even if quite stringent mitigation also is agreed on; many adaptation options, such as switching agricultural crops and strengthening seawalls, are relatively cheap options for some (but not all—e.g., for small island states), and there may be ancillary benefits of the adaptation action even if climatic change effects turn out to be small (e.g., "no regrets" policies such as improving the efficiency of irrigation equipment).

On the other hand, it has been argued that climatic changes today still are relatively small, thus there is little need for adaptation, although there is considerable need for mitigation to avoid more severe future damages. By this logic, it is more prudent to invest the bulk of the resources for climate policy in mitigation, rather than adaptation.

It is reasonable to assume that many adaptation options will be pursued. This means that the baseline against which mitigation options should be assessed is one with adaptation also occurring. If the adaptations were effective in reducing the costs of climatic impacts, this can significantly reduce the benefits that otherwise would have been attributable to mitigation. On the other hand, as Section 1.4.1 notes, lack of perfect foresight about future climatic or other relevant social trends can lead to maladaptations. This situation would then argue for more emphasis on mitigation because maladaptations in the future would increase the costs of climatic impacts thus justify stronger abatement efforts. Furthermore, it has been argued that early steps toward mitigation can lower long-term costs of carbon abatement by reducing the rate at which the energy-intensive capital stock has to be turned over, by inducing research and development, and/or by enhancing learning by doing (Grubb *et al.*, 1994; Azar, 1998; Goulder and Schneider, 1999). Others have argued that delayed abatement is more cost-effective because the bulk of the climate damages are likely to occur in the future, whereas the costs of immediate abatement occur in the nearer term; thus, discounting reduces the present value of the benefits of avoided climate damage versus less discounted abatement costs (e.g., Wigley *et al.*, 1996). Working Group III explores these issues in more depth, but in the context of the Working Group II mandate it must be recognized that many factors that still contain considerable uncertainty enter the debate about tradeoffs between timing and magnitudes of adaptation and mitigation efforts.

1.4.5. Integrated Assessment

Given the multi-sectoral, multi-regional, multidisciplinary, and multi-institutional nature of the integration of climatic change assessments of effects, impacts, and policy options, methods to perform "end-to-end" analyses have been developed and often are labeled "integrated assessments" (see, e.g., Weyant *et al.*, 1996; Morgan and Dowlatabadi, 1996, and references therein). Integrated assessment models (IAMs) have been developed to provide the logical consequences of a variety of explicit assumptions that undergird any formal assessment technique. IAMs seek to combine knowledge from several disciplines that is relevant to climate change in mathematical representations of the determinants of GHG emissions, responses of the climate system and feedbacks to emissions, effects on socioeconomic activities and ecosystems, and potential policies and responses (Parson and Fisher-Vanden, 1997). To date, IAMs have relied primarily on highly aggregated representations that directly link monetized measures of projected impacts to mean climate variables—principally, annual global mean temperature. Over time, these sorts of estimates have been extended by introducing variation between regions, by separating market and nonmarket damages, or by introducing other climate variables such as precipitation (Parson and Fisher-Vanden, 1997). A few IAMs adopt a process-based, geographically explicit approach to modeling, thus have more detailed representation of impacts, often including changes in physical units (e.g., crop yields) as measures of impact. These models do not translate impacts into a common metric, such as money. This makes comparing the level of impacts depicted in the two different modeling approaches very difficult (Tol and Fankhauser, 1998).

IAMs have evolved from a variety of disciplinary tools that often were developed for purposes other than assessments of climatic changes. IAMs have been classified into a hierarchy of five levels (Schneider, 1997). This classification scheme does not imply that each successive level of modeling along the hierarchy (see Section 2.3.8) incorporates all of the elements at lower levels or that incorporation of additional levels of comprehensiveness or complexity provides more fidelity in the model's simulation skills; that depends on the validity of the underlying assumptions and the accuracy of methods used to formally solve the equations that represent those assumptions. Finally, difficulties are encountered in aggregating costs or benefits across the many categories of impacts or opportunities, and a traceable account of any aggregations must be paramount to maintain transparency of any analytic methods such as IAMs (see Sections 1.5.6 and 2.6.4).

Despite these complexities, IAMs are a principal tool for studying systematic sets of interactions that are believed to be important in explaining systems behavior or simulating the consequences of various policies on the magnitude and distribution of risks and benefits of climatic changes or policies to enhance adaptation or encourage mitigation. The goal of IAMs has been to provide insights about the possible interactions of many factors in a complex socionatural system, rather than "answers" to specific scientific or policy questions.

1.5. How do the Complexities of Analysis Affect the Assessment?

The threat posed by climate change must be considered in the context of efforts by countries around the world to achieve sustainable development (see Section 1.1). Improved analysis of impacts of and adaptation to climate change is important for the development of appropriate policy measures. However, the chain of events from human behaviors that give rise to disturbances to the climatic system; to atmospheric changes; to impacts on humans, societies, other species, ecosystems, and their adaptive responses is very complex (as noted in Chapters 2 and 19). Uncertainty is a common feature in the discussion of complexity, and it is compounded by the complex interactions of many subsystems that constitute the socionatural system, each of which has its own inherent uncertainties (see Box 1-2). This section summarizes some of the complexities that make it difficult to provide very many highly confident projections about climatic impacts—assessments that are directly relevant to the oft-asked policy question: "What should we do about climate change?" (see Chapter 2 and references therein for more complete treatment).

1.5.1. Regional Climate Uncertainties

At the regional level, there is a wide range of projected changes in temperature and precipitation simulated from a doubling of CO_2 concentrations because of large model-to-model differences. Annex B of the *Special Report on Regional Impacts of Climate Change* (IPCC, 1998) provides the following conclusion regarding the confidence that can be placed in regional climate projections:

> "Analysis of surface air temperature and precipitation results from regional climate change experiments carried out with AOGCMs indicates that the biases in present-day simulations of regional climate change and the inter-model variability in the simulated regional changes are still too large to yield a high level of confidence in simulated change scenarios. The limited number of experiments available with statistical downscaling techniques and nested regional models has shown that complex topographical features, large lake systems, and narrow land masses not resolved at the resolution of current GCMs significantly affect the simulated regional and local change scenarios, both for precipitation and (to a lesser extent) temperature (IPCC, 1996a). This adds a further degree of uncertainty in the use of GCM-produced scenarios for impact assessments. In addition, most climate change experiments have not accounted for human-induced landscape changes and only recently has the effect of aerosols been vigorously investigated. Both these factors can further affect projections of regional climate change."

The wide range of projected changes in temperature and precipitation would affect the degree of exposure of systems and populations to climatic stimuli and hence their vulnerability to climate change. This range suggests that high confidence will not often be assigned to any regional impact assessments that are based on GCM results. Difficulty in obtaining many highly confident outcomes is why the term "climate scenarios" has been adopted in most impact assessments. Such scenarios should be regarded as internally consistent patterns of plausible future climates, not predictions carrying assessed probabilities (see Section 2.6 and Chapter 3). Decisionmakers need to be aware of the large range of plausible climate projections when they formulate strategies to cope with the risks of climate change. However, in the absence of some explicit estimation of the likelihood of various scenarios by those who produce them, users of the many decision frameworks in the literature (see Box 1-2 and Section 2.7) often have to impute likelihood to various scenarios to apply many of these methods.

The review chapters in this report summarize impact studies that are based on a range of climate scenarios, when available. As noted earlier, transient scenarios are particularly valuable because the Earth currently is undergoing a transient response to global change disturbances. Great care is required in interpreting and comparing results from research or assessments that use different climate scenarios, particularly when some conclusions follow from static scenarios and others from transient scenarios. Unfortunately, such mixed use of scenarios is still a problem in the literature and in assessments of it.

1.5.2. Socioeconomic Uncertainties

An often overlooked source of uncertainty in assessments of impacts and vulnerability is the wide difference in assumptions (often not even stated) in the initial conditions and trends of environmental systems and socioeconomic conditions. These assumptions include information on population and related variables (e.g., population density), economic trends (e.g., income levels, sectoral composition of GDP, or levels of trade), other social indicators (e.g., education levels, private- and public-sector institutions), culture, land cover and use, and availability and use of other resources such as water. They are important not only for determining the forces driving global changes but also for understanding the general capabilities available to societies for adaptation. Projections of these factors for time periods such as the middle of the 21st century are at least as uncertain as projections of future climate; hence, it is probably most advisable to use such information as scenarios of change, or conditioning assumptions (IPCC, 1998). Moreover, culture exerts important influences on socioeconomic processes, problemsolving methods, and the like. The formation of coalitions, social movements, and educational programs directed toward changing institutional norms that might influence people's behavior concerning climatic change is culturally determined, like other complex social and psychological processes. Cultural processes and economic behavior, for example, can be modeled to capture some of the complexity of the social processes, structures, and cognitive behavior involving culture (e.g., Rotmans and van Asselt, 1996; Koizumi and Lundstedt, 1998). Thus, it is simply impossible to predict with high confidence how societies and economies will develop in the future—hence the extent to which they will have the capacity needed for adaptation. The use of scenarios to assess driving forces and adaptive capacity is one way to explicitly acknowledge these kinds of structural uncertainties (e.g., IPCC, 2000). Socioeconomic scenarios, as already noted, are not predictions of future states of the world but consistent and plausible sets of assumptions about issues such as population growth, economic development, values, and institutions.

Although the emphasis on adaptation to reduce vulnerability and take advantage of emerging opportunities is increasing in impact assessment, many uncertainties remain regarding the effectiveness of different options, the relationship between adaptation to short-term climate fluctuations and long-term climate change, and constraints and opportunities that will be imposed by factors such as existing institutional structures, economic and financial limitations, and cultural resistance (IPCC, 1998).

1.5.3. Risk and Uncertainty

Uncertainties are pervasive throughout climate change impact assessment. For some sectors, such as agriculture, uncertainty is large enough to prevent a highly confident assessment of even the sign of the impacts. Until a few years ago, uncertainties in assessments were so great that few researchers were willing to carry their analysis through to numerical estimates of monetary impacts. Even today, as the applicability of subjective probabilities is becoming more accepted, impact estimates with explicit confidence intervals are the exception rather than the rule (a few exceptions are Peck and Teisberg, 1992; Hope *et al.*, 1993; Nordhaus, 1994a; Manne and Richels, 1995; Morgan and Dowlatabadi, 1996; Titus and Narayanan, 1996; Roughgarden and Schneider, 1999). Figure 2-2 (Moss and Schneider, 2000) graphically depicts how uncertainties in emissions scenarios feed into uncertainties in carbon cycle response, climate sensitivity, regional climate responses, and ranges of impacts in an "explosion" or "cascade" of widening uncertainty bounds. However, despite this daunting expansion of uncertainty, methods to classify and formally treat such uncertainties via subjective probability distributions are available in the literature (see Box 1-2 and Section 2.6) and can help to clarify which subcomponents of the overall human-environment system are most critical to integrated assessments of the costs and benefits of climatic changes or climate policies.

1.5.4. Low-Probability Catastrophic Events

Efforts to deal with low-probability, potentially catastrophic events in integrated assessments of climate change are not well-represented in the literature. One possibility would be to treat these risks like any hazard and use methods from risk analysis: The value of the risk is the probability of occurrence multiplied by the consequences of the event. For rare and catastrophic possibilities, there is very little frequency data; thus, probabilities assessed are based largely on subjective methods (e.g., Nordhaus, 1994b; Roughgarden and Schneider, 1999). Equally important, under these conditions the expected cost estimate would be very sensitive to the analyst's (subjective) assumptions about the costs of catastrophic events. Subjective probabilities can vary widely from analyst to analyst under such conditions. This partly explains why most analysts have been reluctant to include low-probability but potentially catastrophic events in integrated assessments (for a recent exception, see Mastrandrea and Schneider, 2001). However, absence of analysis does not necessarily imply absence of risk, and many risk management decisions in the private and public sectors are based on strategic hedging against low-probability but highly costly possibilities, such as insurance and deterrence (see Chapter 8). However, the expected cost approach would imply a risk neutrality—an uncomfortable position for those holding risk averse values in the face of possibilities such as collapse of the "conveyor belt" circulation in the North Atlantic Ocean (e.g., Broecker, 1997; Rahmstorf, 1999; Chapter 19) or melting of the West Antarctic Ice Sheet (e.g., Oppenheimer, 1998). Risk-averse individuals often worry about the possibility that a forecast for a high-consequence event is either accurate or an underestimate—the "type 2 error." Such individuals have argued that a better way to treat the possibility of catastrophe is to ensure that all possible efforts are taken to avoid it—the "precautionary principle" (see, e.g., Wiener, 1995). However, spending valuable, limited resources to hedge against possible catastrophic outcomes with a low probability of occurring is infeasible in practice; scarce resources could have been used more productively elsewhere, including dealing with more probable climatic threats. People who are concerned about "squandering" resources on what they perceive to be unlikely threats or even an erroneous forecast—the "type 1 error"—often are engaged in contentious debates with those more concerned with type 2 errors—a situation that is well-known in risk management disciplines. Thus, it is difficult to apply the precautionary principle unambiguously to justify a hedging strategy against a potential catastrophic climatic event without also applying it to the possibility of negative outcomes from the hedging strategy itself, then weighing the relative risks of type 1 versus type 2 errors (Wiener, 1995).

1.5.5. Valuation Methods—Monetary Measures or Multiple Numeraires

Although much progress on valuation techniques is being made, as noted in Box 1-2, uncertainties are still large, and many impact estimates are "highly speculative" (Nordhaus and Boyer, 2000). Impacts can be divided into market and nonmarket impacts.

Market impacts occur in sectors or activities such as agriculture, forestry, provision of water, insurance against extreme events, transportation, tourism, and activities that use low-lying coastal land. Where these activities produce marketed goods, a monetary estimate of impacts (in units of dollars per °C, for example) sometimes can be made with fairly straightforward techniques, at least under present-day conditions; this has been the most common approach in impact studies to date (e.g., Mendelsohn *et al*., 2000). Market prices, adjusted to correct for market distortions (e.g., externalities), are the appropriate measure for unit impacts. Although the techniques are well established, the numbers obtained still are approximate as a result of all the uncertainties that surround impact assessments. Working out how the impacts will unfold in the distant future is much less straightforward. Impacts could increase as the intensity and scale of the activity increases (e.g., loss of coastal property) or decrease as more modern and robust systems replace existing ones (e.g., new crop strains are introduced with more climatic adaptability). Also, as noted in Box 1-2, impacts expressed in economic terms embed the values people attribute to the impacts across several numeraires, as well as the values of future generations (see Section 2.5.6 for further elaborations).

For example, the use of highly aggregated decision analysis frameworks (see Box 1-2 and Chapter 2) can be controversial because aggregation of positive and negative costs of even a limited number of market category sectors involves the arithmetic sum of many subelements that contain large uncertainties and are related to different regions. Furthermore, important market costs could be incurred by political instability (e.g., Kennedy *et al*., 1998), migration of displaced persons (e.g., Myers, 1993), diminished capacity of damaged ecosystems to provide accustomed services (e.g., Daily, 1997), or loss of heritage sites from sea-level rise (e.g., Schneider *et al*., 2000b). Moreover, losses in nonmonetary categories (i.e., other numeraires such as biodiversity lost, lives lost, quality of life degraded, or inequity generated—all per °C) are very controversial (e.g., Goulder and Kennedy, 1997, discuss attempts to estimate the intrinsic value of species). Any aggregation over such numeraires into a common metric—usually the dollar—cannot be accomplished transparently unless a variety of assumptions are explicitly given for the valuation of each of these numeraires before aggregation hides the underlying assumptions of how valuation was accomplished.

1.5.6. Damage Aggregation and Distributional Effects

Aggregation of various damages into a single estimate sometimes is appropriate to provide policymakers with information about the magnitude of damages that can be expected on a global scale. However, as noted in Box 1-2, Section 1.5.5, and Section 2.6.4, there also is the risk that such aggregation conceals rather than highlights some of the critical issues and value-laden assumptions that are at stake.

As a hypothetical but concrete example, assume that climatic change would cause destruction of lives, ecosystems, and property in Bangladesh, corresponding to a loss of 80% of its GDP. This loss to Bangladesh would amount to roughly 0.1% of global GDP. If the global economy grows at 2% yr^{-1}, this assumed impact on Bangladesh would correspond to a delay in global income growth of less than 3 weeks. It is debatable whether adding, say, the possible benefits for temperate agriculture to the losses of lives resulting from sea-level rise in Bangladesh helps to assess the severity of climate change impacts because the "winner" does not compensate the "loser" (i.e., benefits for temperate agriculture offer little relief to those who have been affected by sea-level rise in other regions). Authors in the literature have expressed concern about trading the costs of emission reduction in some countries (e.g., more efficient end-use energy technologies) with large-scale losses of lives and human health in others (e.g., Munasinghe, 2000). Still, this is implicitly done in most conventional cost-benefit analyses of climate change available in the literature. As noted above and in Section 2.6.4, this points to the necessity of using appropriately disaggregated cost and benefit data to make the analysis more transparent. Possible ways of incorporating equity concerns include use of distributional weights in cost-benefit analysis (e.g., Azar and Sterner, 1996; Fankhauser *et al*., 1997; Azar, 1999).

Owing to the complexities of valuation and aggregation analyses described above and in the preceding subsection, the TAR authors are cautious about the applicability of single "optimal" answers. Instead, they attempt to examine ranges of outcomes calculated under a variety of assumptions available in the literature, for which alternative valuation methods can be applied to different categories across various numeraires.

1.5.7. Discounting

Comparing impact, mitigation, and adaptation costs that occur at different points in time requires them to be discounted. There is longstanding debate about the appropriate rate of discount to use (e.g., Arrow *et al*., 1996; Portney and Weyant, 1999). Uncertainty regarding the discount rate relates not to calculation of its effects, which is mathematically precise, but to a value judgment about the appropriateness of the present generation valuing various services for future generations (see Section 2.3.1 for elaboration).

Two different approaches to discounting are presented in the SAR (Arrow *et al*., 1996). The *descriptive* approach focuses on intertemporal cost-efficiency, and the discount rate is based on observed market interest rates. The *prescriptive* approach emphasizes that normative issues are involved in valuing the future. One important problem for both approaches is the fact that we cannot observe future market interest rates or know the level of income that will prevail in the future (at least for time horizons involved in the climate change debate). Most analysts have resolved this dilemma by using constant discount rates over the entire horizon, despite the fact that they are likely to change. Others have suggested or used non-fixed discount rates that apply strong short-term discounting but entail little further discounting for the very long-term future (e.g., Azar and Sterner, 1996; Heal, 1997). That would cause events a decade or two hence to be significantly discounted but would not cause events a century hence to be reduced in value by powers of 10, as is the occurrence with conventional exponential (compound interest) discounting. Because the largest costs from climate change usually are believed to occur many decades in the future, conventional discounting renders the present value of such future damages very small, whereas non-fixed discount rates (e.g., "hyperbolic discounting") would cause present generations to take serious notice of very large potential damages, even a century hence. Because both the value of the discount rate and the choice of discounting approach involve value judgments about the ethics of intergenerational transfers, it is important for all assessments to be clear about what discounting formulations have been used and the sensitivity of the conclusions to alternative formulations.

1.5.8. "Safe Emission Levels," Cost-Effectiveness Analysis, and the Timing of Emission Abatement

Several issues raised in this section are discussed primarily in the report of Working Group III. However, because this chapter is intended to provide a context for impact, adaptation, and vulnerability issues, this section briefly reviews several emissions abatement complexities that have a bearing on the adaptation/mitigation tradeoff issues (see Section 1.4.4.2 and Chapter 2). Because estimates of the monetary costs of impacts span a wide range of values given the many uncertainties and often are value laden, some analysts have argued that climate change targets should be based on physical or social, rather than economic, indicators—for example, past fluctuations in temperature or expected climate-related deaths or some general reference to sustainability or the precautionary principle (see Section 1.5.4). This precautionary approach is used in European negotiations on emissions of acidifying substances and is acknowledged in Article 3, paragraph 3, of the UNFCCC, which states as a principle that "The Parties should take precautionary measures to anticipate, prevent or minimize the causes of climate change and mitigate its adverse effects. Where there are threats of serious or irreversible damage, lack of full scientific certainty should not be used as a reason for postponing such measures...." Such threshold levels (see Section 1.4.3.5) also have been used as upper ceilings on the amount of warming considered "tolerable" in the academic sphere (see Alcamo and Kreileman, 1996; Azar and Rodhe, 1997) and the political sphere (for instance, the European Union has adopted a maximum of 2°C temperature change above pre-industrial levels or a maximum of 550 ppm CO_2 concentration target). Implicit in this approach is the assumption of the possibility of very nonlinear damage functions. One drawback with this approach is that necessary tradeoffs between climate damage avoidance and the opportunity costs of resources used to mitigate that climate change often are not made explicit.

Even if the precautionary approach were taken, cost-efficiency analysis would be used to identify the lowest cost of meeting the predefined target. Several studies have made an argument that "where" and "when" flexibility in emissions reductions can greatly reduce its costs (Wigley *et al.*, 1996). Ha-Duong *et al.* (1997) and Goulder and Schneider (1999) show that preexisting market failures in the energy sector could reduce the costs of immediate climate policies substantially or that neglect of inducing technological changes by delaying incentives associated with immediate climate policies could reverse the conclusions that delayed abatement is more cost-effective. Unfortunately, there is very little literature on how climate policies might induce technological change (see WGIII TAR). Another reason for the controversy in the literature about abatement timing is a misreading of Wigley *et al.* (1996) that they do not endorse efforts over the next 30 years to make abatement cheaper in the future. Azar (1998) argues, however, that if stabilization targets would be at or below 450 ppm CO_2, early abatement (not just efforts to make future abatement cheaper) would be cost-efficient, even in the Wigley *et al.* (1996) model.

Furthermore, the problem of valuing impacts in monetary terms cannot be avoided entirely even under the cost-efficiency approach. Different trajectories toward the stabilization target have different impacts and costs associated with them. How does delaying mitigation affect the impacts, including distributive consequences? The answer to this question is unclear, partly because of large remaining uncertainties about the extent to which rapid forcing of the climate system could trigger threshold events (e.g., Tol, 1995). Moreover, the difference in impacts between early and delayed mitigation responses appears to be sensitive to assumptions about sulfate aerosol cooling and whether small transient temperature differences can have significant effects.

1.5.9. Validation

Validation of models and assessments that deal with projections over many decades is a serious issue. Often it is not helpful in the context of sustainable development to suggest postponing policy responses until model predictions can be directly compared against reality because that would require experiencing the consequences without amelioration. Instead, models and assessments are subjected to varying levels of quality control, intercomparison with standard assumptions, comparisons with experiments, and extensive peer review. Some authors have argued (e.g., Oreskes *et al.*, 1994) that it is impossible in principle to "validate" models for future events when the processes that determine the model projections contain structural uncertainties (see Boxes 1-1-and 2-1). Although the impossibility of direct before-the-fact validation is strictly true, this does not mean that models cannot be rigorously tested. Several stages are involved. First, how well known are the data used to construct model parameters? Second, have the individual processes been tested against lab experiments, field data, or other more comprehensive models? Third, has the overall simulation skill of the model been tested against known events? Fourth, has the model been tested for sensitivity to known shocks (e.g., an oil price hike in an economic model or a paleoclimatic abrupt change in a climate model)? For example, crop yield models are tested against actual yield variation data (Chapter 5), and sea-level increase models are tested for their ability to reproduce observed changes in the 20th century. The ability of a model to reproduce past conditions is a necessary, but not necessarily sufficient, condition for a highly confident forecast of future conditions, unless the underlying processes that gave rise to the phenomena observed in the past will be fully operative in the future and the model captures the influence of such phenomena. Finally, has the comparison between model and data been done at commensurate scales, so that small-scale data are first aggregated to the scale of the lowest resolved element of the model before attempting evaluation (e.g., Root and Schneider, 1995)? When such validation protocols are performed and a model performs "well," subjective confidence that assessment teams can assign to various projections based on such models increases considerably (see Section 2.6), even if "definitive proof" of a specific forecast before the fact is impossible in principle.

All of these considerations demonstrate how the complexities of analysis have led Working Group II TAR authors to emphasize risk management approaches to climate change and policy assessment, rather than just an optimizing framework (e.g., see Section 2.7). These complexities of analysis are not problematic only for the assessment of impacts, vulnerabilities, and adaptability; they also carry forward to questions of tradeoffs between investments in adaptation and mitigation strategies and make a connection between the purviews of Working Groups II and III.

1.6. How can this Assessment be Used to Address Policy-Relevant Questions? A Users' Guide

1.6.1. United Nations Framework Convention on Climate Change

An important audience for this report is the UNFCCC Conference of Parties and Subsidiary Bodies, through which implementation of the provisions of the Convention (United Nations, 1992) and associated protocols will be negotiated. The major issue is contained in Article 2 of the UNFCCC and relates to identifying the level for stabilization of GHG concentrations. As stated in that Article, the level for stabilization is set in terms of impacts of climate change. Hence, the focus of this report is on identifying impacts potentially associated with different rates and levels of climate change. It is important to reiterate that readers will not find any magnitude or rate of climate change defined as "dangerous" by this report. As noted earlier, this is because such a designation is necessarily political for two important reasons. First, the impacts associated with any given concentration target or emissions trajectory will be unevenly distributed across countries, ecosystems, and socioeconomic sectors. Thus, some sectors or regions may receive some benefit from a particular pattern of climate change, whereas others will be harmed. It is not the role of the scientific community to determine whether a particular pattern of impacts constitutes "dangerous" interference; that is a political judgment to be negotiated among participating governments and institutions. Second, there are scientific uncertainties associated with climate change scenarios and our knowledge of impacts that may result. Thus, it is not possible to state in absolute terms that particular impacts will be associated with a given concentration target or stabilization pathway. Instead, information about impacts will be conditional and is best considered in a risk management framework—that is, different stabilization targets or pathways pose different risks to food production, ecosystems, and economic development, and such risks are likely to vary by region and over time. There is no way to determine scientifically what level of risk is acceptable under the UNFCCC. This, too, will be a matter for negotiation by governments. However, information on the state of the science presented in IPCC assessments is widely believed to help put such decisionmaking exercises on a firmer factual basis (see discussion of guidelines for practitioners from an international social science assessment of human choice and climate change in Rayner and Malone, 1998).

The TAR focuses on the vulnerability of different systems and regions to various rates and magnitudes of climate change. Assessment of vulnerability and adaptation is relevant not only to identifying impacts associated with different targets but also to identifying "developing country Parties that are particularly vulnerable to the adverse effects of climate change" (Article 12; United Nations, 1997); these countries are to be compensated from the proceeds of the CDM to help meet the costs of adaptation.

1.6.2. Links to Biodiversity Loss, Desertification, Deforestation and Unsustainable Use of Forests, Stratospheric Ozone Depletion, and Other Global Environmental Issues

Climate change is not an isolated issue; it is intimately connected to other recognized natural hazards and global environmental problems. Separate international conventions and processes exist to address these issues; in several cases, these include successful scientific assessment mechanisms. This report contains information of relevance to these bodies and processes, although it is not the intention of the report to supercede or contradict information developed in those assessments. The purpose of incorporating information of relevance to these issues is to highlight scientific and policy links among them, so that unnecessary tradeoffs can be avoided and potential multiple benefits can be realized (e.g., Orlando and Smeardon, 1999; Kremen *et al.*, 2000). For example, several international conventions and agreements call for sustainable management and use of land and water resources, with varying goals (such as enhancing GHG sinks and reservoirs, protecting biological diversity, safeguarding aquatic ecosystems, managing forests to meet human needs, and halting desertification). To the extent that these objectives are potentially affected by climate change, and to the extent that options to adapt to changing climate conditions can be structured to help attain additional environmental or socioeconomic objectives associated with these other agreements (i.e., co-benefits), this is highlighted in the relevant sections of the TAR.

1.6.3. Resource Planners, Managers in National and Regional Institutions, and Actors in Specialized International Agencies

Although the primary audiences of this report are involved in negotiating and implementing the UNFCCC (United Nations, 1992) and, to some extent, other international agreements on global environmental problems, the report also contains information that is useful to resource managers in national governments; regional institutions such as regional development or lending agencies; and specialized international agencies such as the World Bank, UNEP, UNDP, or the GEF. In the chapters that focus on sectors or systems of climate change (e.g., Chapters 4–9, which cover advances in our understanding of impacts and adaptation options in water resources, agriculture, health, ecosystems, and so forth), planners and managers in national ministries or regional planning authorities will find information on how their mandates—such as encouraging agriculture, providing freshwater, protecting endangered species, or increasing energy production—could be affected by climate change. To the extent provided in the literature, these chapters also include detailed technical and cost information on adaptation options and factors that will influence their implementation. In chapters that focus on regional analyses, managers and planners at regional and international agencies will find information on baselines and trends (climate, socioeconomic, and other environmental); each chapter also highlights particular vulnerabilities and opportunities for adaptation that may occur in each region. It is hoped that this information will be useful in assessing potential projects and opportunities for investment, so that these can be structured to be more robust to potential negative effects of climate change or to take advantage of emerging opportunities. In addition, this report will be useful in the education of the media and the general public about climate, the environment, and development issues.

References

AGBM, 1995: *Report of the Ad Hoc Group on the Berlin Mandate on the Work of Its First Session Held at Geneva From 21 to 25 August 1995*. Available online at http://www.unfccc.int/resource/docs/1995/agbm/02.htm.

Alcamo, J. and E. Kreileman, 1996: Emission scenarios and global climate protection. *Global Environmental Change*, **6**, 305–334.

Arrow, K.J., W. Cline, K.G. Mäler, M. Munasinghe, R. Squitieri, and J. Stiglitz, 1996: Intertemporal equity, discounting and economic efficiency. In: *Climate Change 1995: Economic and Social Dimensions of Climate Change, Second Assessment of the Intergovernmental Panel on Climate Change* [Bruce, J.P., H. Lee, and E.F. Haites (eds.)]. Cambridge University Press, Cambridge, United Kingdom and New York, NY, USA, pp. 135–144.

Ausubel, J.H., 1996: Can technology spare the earth? *American Scientist*, **84(2),** 166–178.

Azar, C., 1999: Weight factors in cost benefit analysis of climate change. *Environmental and Resource Economics*, **13,** 249–268.

Azar, C., 1998: The timing of CO_2 emissions reduction—the debate revisited. *International Journal of Environment and Pollution*, **10,** 508–521.

Azar, C. and H. Rodhe, 1997: Targets for stabilization of atmosperic CO_2. *Science*, **276,** 1818–1819.

Azar, C. and T. Sterner, 1996: Discounting and distributional considerations in the context of global warming. *Ecological Economics*, **19,** 169–185.

Bartiaux, F. and J.P. van Ypersele, 1993: The role of population growth in global warming. In: *Proceedings of the International Population Conference* (ISBN 2-87108-030-5). International Union for the Scientific Study of Population, Liège, Belgium, **4,** 33–54.

Brodin, Y.-W. and J.C.I. Kuylenstierna, 1992: Acidification and critical loads in Nordic countries: a background. *Ambio*, **21**, 332–338.

Broecker, W.S., 1997: Thermohaline circulation, the Achilles heel of our climate system: will man-made CO_2 upset the current balance? *Science*, **278,** 1582–1588.

Browne, J., 2000: Rethinking corporate responsibility. *Reflections: The SoL Journal*, **1(4),** 48–53.

Burton, I., R.W. Kates, and G.F. White, 1993: *The Environment as Hazard*. Guilford Press, New York, NY, USA, 2nd. ed., 290 pp.

Chen, R.S. and R.W. Kates, 1996: Trends in agriculture and food security. In: *Climate Change and World Food Security* [Downing, T.E. (ed.)]. Springer-Verlag, Heidelberg, Germany, pp. 23–52.

Cline, W.R., 1992: *The Economics of Global Warming*. Institute of International Economics, Washington, DC, USA, 399 pp.

Cohen, J., 1995: *How Many People Can the Earth Support?* W.W. Norton, New York, NY, USA, 532 pp.

Cole, H.S.D., C. Freeman, M. Jahoda, and K.L.R. Pavitt, 1973: *Models of Doom: A Critique of the Limits to Growth*. Universe Books, New York, NY, USA, 244 pp.

Colombo, A.F., D. Etkin, and B.W. Karney, 1999: Climate variability and the frequency of extreme temperature events for nine sites across Canada: implications for power usage. *Journal of Climate*, **12,** 2490–2502.

Cooper, R.N., 2000: *Prospects for the World Economy*. Working Paper #26, UCLA Center for International Relations, Los Angeles, CA, USA, 27 pp.

Crowley, T.J., 1990: Are there any satisfactory geologic analogs for a future greenhouse warming? *Climate*, **3,** 1282–1292.

Crowley, T.J. and G.R. North, 1991: *Paleoclimatology*. Oxford University Press, Oxford, United Kingdom, 339 pp.

Daily, G.C., 1997: *Nature's Services: Societal Dependence on Natural Ecosystems*. Island Press, Washington, DC, USA, 392 pp.

Ehrlich, P.R. and A.H. Ehrlich, 1996: *Betrayal of Science and Reason*. Island Press, Washington, DC, USA, 335 pp.

Ehrlich, P.R. and A.H. Ehrlich, 1990: *The Population Explosion*. Simon and Schuster, New York, NY, USA, 320 pp.

Ehrlich, P.R. and J.P. Holdren, 1971: Impact of Population Growth. *Science*, **171,** 1212–1217.

Ehrlich, P.R., A.H. Ehrlich, and G. Daily, 1995: *The Stork and the Plow*. Putnam, New York, NY, USA, 364 pp.

Etkin, D.A., 1999: Risk transference and related trends: driving forces towards more mega-disasters. *Environmental Hazards*, **1,** 69–75.

Fankhauser, S., R.S.J. Tol and D.W. Pearce, 1997: The aggregation of climate change damages: a welfare theoretic approach. *Environmental and Resource Economics*, **10**, 249–266.

FAO, 1999: *The State of Food Insecurity in the World 1999*. Food and Agriculture Organization of the United Nations, Rome, Italy. Available online at http://www.fao.org/FOCUS/E/SOFI/home-e.htm.

Fischer, G., K. Frohberg, M.L. Parry, and C. Rosenzweig, 1996: Impacts of potential climate change on global and regional food production and vulnerability. In: *Climate Change and World Food Security* [Downing, T.E. (ed.)]. Springer-Verlag, Heidelberg, Germany, pp. 115–160.

Gates, D., 1993: *Climate Change and its Biological Consequences*. Sinauer Associates, Sunderland, MA, USA, 280 pp.

Gleick, P.H., 1998: *The World's Water 1998-1999: The Biennial Report on Freshwater Resources*. Island Press, Washington, DC, USA, 307 pp. Available online at http://www.worldwater.org.

Goulder, L.H. and D. Kennedy, 1997: Valuing ecosystems: philosophical bases and empirical methods. In: *Nature's Services: Societal Dependence on Natural Ecosystems* [Daily, G.C. (ed.)]. Island Press, Washington, DC, USA, pp. 23–48.

Goulder, L.H. and S.H. Schneider, 1999: Induced technological change and the attractiveness of CO_2 emissions abatement policies. *Resource and Energy Economics*, **21,** 211–253.

Grossman, G.M. and A.B. Krueger, 1995: Economic growth and the environment. *Quarterly Journal of Economics*, **110,** 353–377.

Grubb, M., M. Ha-Duong, and T. Chapuis, 1994: Optimizing climate change abatement responses: on inertia and induced technology development. In: *Integrative Assessment of Mitigation, Impacts, and Adaptation to Climate Change* [Nakicenovic, N., W.D. Nordhaus, R. Richels, and F.L. Toth (eds.)]. International Institute for Applied Systems Analysis, Laxenburg, Austria, pp. 513–534.

Ha-Duong, M., M. Grubb., and J.-C. Hourcade, 1997: Influence of socioeconomic inertia and uncertainty on optimal CO_2 emission abatement. *Nature*, **390,** 270–273.

Heal, G., 1997: Discounting and climate change, an editorial essay. *Climatic Change*, **37(2),** 335–343.

Hoffert, M.I., K. Caldeira, A.K. Jain, L.D.D. Harvey, E.F. Haites, S.D. Potter, M.E. Schlesinger, S.H. Schneider, R.G. Watts, T.M.L. Wigley, and D.J. Wuebbles, 1998: Energy implications of future stabilization of atmospheric CO_2 content. *Nature*, **395,** 881–884.

Hope, C.W., J. Anderson, and P. Wenman, 1993: Policy analysis of the greenhouse effect—an application of the PAGE model. *Energy Policy*, **15,** 328–338.

IPCC, 2000: *Emissions Scenarios. Special Report of the Intergovernmental Panel on Climate Change* [Nakicenovic, N. and R. Swart (eds.)]. Cambridge University Press, Cambridge, United Kingdom and New York, NY, USA, 599 pp.

IPCC, 1999a: *Procedures for the Preparation, Review, Acceptance, Adoption, Approval and Publication of IPCC Reports. Appendix A to the Principles Governing IPCC Work*. Available online at http://www.ipcc.ch/about/procd.htm.

IPCC, 1999b: *Scientific, Technical and Socioeconomic Questions Selected by the Panel*. Available online at http://www.ipcc.ch/activity/tarquestion.html.

IPCC, 1998: *The Regional Impacts of Climate Change. An Assessment of Vulnerability. A Special Report of IPCC Working Group II* [Watson, R.T., M.C. Zinyowera, and R.H. Moss (eds.)]. Cambridge University Press, Cambridge, United Kingdom and New York, NY, USA, 517 pp.

IPCC, 1996a: *Climate Change 1995: The Science of Climate Change. Contribution of Working Group I to the Second Assessment Report of the Intergovernmental Panel on Climate Change* [Houghton, J.T., L.G. Meira Filho, B.A. Callander, N. Harris, A. Kattenberg, and K. Maskell (eds.)]. Cambridge University Press, Cambridge, United Kingdom and New York, NY, USA, 572 pp.

IPCC, 1996b: *Climate Change 1995: Impacts, Adaptations and Mitigation of Climate Change: Scientific-Technical Analyses. Contribution of Working Group II to the Second Assessment Report of the Intergovernmental Panel on Climate Change* [Watson, R.T., M.C. Zinyowera, and R.H. Moss (eds.)]. Cambridge University Press, Cambridge, United Kingdom and New York, NY, USA, 878 pp.

IPCC, 1996c: *Climate Change 1995: Economic and Social Dimensions of Climate Change. Contribution of Working Group III to the Second Assessment Report of the Intergovernmental Panel on Climate Change* [Bruce, J.P., H. Lee, and E.F. Haites (eds.)]. Cambridge University Press, Cambridge, United Kingdom and New York, NY, USA, 448 pp.

IPCC, 1990: *Climate Change: The IPCC Scientific Assessment* [Houghton, J.T., G.J. Jenkins, and J.J. Ephraims (eds.)]. Cambridge University Press, Cambridge, United Kingdom and New York, NY, USA, 365 pp.

Ishida, Y., 1998: Context and problems of sustainable development. In: *"Society" in 2050: Choice for Global Sustainability*. Global Industrial and Social Progress Research Institute, Asahi Glass Foundation, Japan, pp. 5–14.

Kates, R.W., 1996: Ending hunger: current status and future prospects. *Consequences*, **2(2),** 3–12. Available online at http://www.gcrio.org/cgi-bin/showcase?/Consequences/IntroCon.html.

Kennedy, D., D. Holloway, E. Weinthal, W. Falcon, P. Ehrlich, R. Naylor, M. May, S.H. Schneider, S. Fetter, and J. Choi, 1998: *Environmental Quality and Regional Conflict*. Carnegie Commission on Preventing Deadly Conflict, Washington, DC, USA, 72 pp.

Koizumi, T. and S.B. Lundstedt, 1998: Mind, culture and economy: the morphology of socio-economic systems. *Cybernetica*, **XLI(2/3/4),** 1121–151.

Kremen, C., J.O. Niles, M.G. Dalton, G.C. Daily, P.R. Ehrlich, J.P. Fay, D. Grewal, and R.P. Guillery, 2000: Economic incentives for rain forest conservation across scales. *Science*, **288,** 1828–1832.

Lutz, W. (ed.), 1996: *The Future Population of the World: What Can We Assume Today?* Earthscan, London, United Kingdom, 500 pp.

MacIver, D.C., 1998: *Adaptation to Climate Variability and Change*. IPCC Workshop Summary, San Jose, Costa Rica, 29 March - 1 April 1998. Atmospheric Environment Service, Environment Canada, Downsview, Ontario, Canada, 55 pp.

Manne, A. and R. Richels, 1995: The greenhouse debate: economic efficiency, burden sharing and hedging strategies. *The Energy Journal*, **16(4),** 1–37.

Markandya, A. and K. Halsnaes, 2000: Costing methodologies. In: *Guidance Papers on the Cross Cutting Issues of the Third Assessment Report of the IPCC* [Pachauri, R., T. Taniguchi, and K. Tanaka (eds.)]. Intergovernmental Panel on Climate Change, Geneva, Switzerland, pp. 15–31. Available online at http://www.gispri.or.jp.

Mastrandrea, M. and S.H. Schneider, 2001: Integrated assessment of abrupt climatic changes. *Climate Policy* (in press).

Meadows, D.H., D.L. Meadows, J. Randers, and W.W. Behrens, 1972: *The Limits to Growth: A Report for the Club of Rome's Project on the Predicament of Mankind*. Universe Books, New York, NY, USA, 205 pp.

Mearns, L.O., R.W. Katz, and S.H. Schneider, 1984: Extreme high temperature events: changes in their probabilities and changes in mean temperature. *Journal of Climate and Applied Meterology*, **23,** 1601–1613.

Mendelsohn, R., N.G. Androva, W. Morrison, and M.E. Schlesinger, 2000: Country-specific market impacts of climate change. *Climatic Change*, **45(3–4),** 553–569.

Meyer, W.B., 1996: *Human Impact on the Earth*. Cambridge University Press, Cambridge, United Kingdom, 253 pp.

Morgan, G. and H. Dowlatabadi, 1996: Learning from integrated assessment of climate change, *Climatic Change*, **34(3–4),** 337–368.

Morgan, M.G. and D.W. Keith, 1995: Subjective judgments by climate experts. *Environmental Science and Technology*, **29,** 468A–477A.

Moss, R.H. and S.H. Schneider, 2000: Uncertainties in the IPCC TAR: recommendations to lead authors for more consistent assessment and reporting. In: *Guidance Papers on the Cross Cutting Issues of the Third Assessment Report of the IPCC* [Pachauri, R., K. Tanaka, and T. Taniguchi (eds.)]. Intergovernmental Panel on Climate Change, Geneva, Switzerland, pp. 33–51. Available online at http://www.gispri.or.jp.

Moss, R.H. and S.H. Schneider, 1997: Characterizing and communicating scientific uncertainty: building on the IPCC second assessment. In: *Elements of Change* [Hassol, S.J. and J. Katzenberger (eds.)]. Aspen Global Change Institute, Aspen, CO, USA, pp. 90–135.

Munasinghe, M., 2000: Development, equity and sustainability (DES) in the context of climate change. In: *Guidance Papers on the Cross Cutting Issues of the Third Assessment Report of the IPCC* [Pachauri, R., K. Tanaka, and T. Taniguchi (eds.)]. Intergovernmental Panel on Climate Change, Geneva, Switzerland, pp. 69–110. Available online at http://www.gispri.or.jp.

Myers, N., 1993: *Ultimate Security—The Environmental Basis of Political Stability*. W.W. Norton, New York, NY, USA, 308 pp.

Myers, N. and J. Simon, 1994: *Scarcity or Abundance? A Debate on the Environment*. W.W. Norton, New York, NY, USA, 254 pp.

Nordhaus, W.D. and J. Boyer, 2000: *Warming the World: Economic Models of Global Warming*. MIT Press, Boston, MA, USA, 232 pp.

Nordhaus, W.D., 1994a: *Managing the Global Commons*. MIT Press, Cambridge, MA, USA, 213 pp.

Nordhaus, W.D., 1994b: Expert opinion on climatic change. *American Scientist*, **82,** 45–52.

OECD, 1999: *National Climate Policies and the Kyoto Protocol*. Organisation for Economic Co-operation and Development, Paris, France, 87 pp.

O'Meara, M., 1999: Exploring a new vision for cities. In: *State of the World* [Brown, L.R., C. Flavin, H.F. French, and L. Starke (eds.)]. Worldwatch Institute, Earthscan, London, United Kingdom, pp. 133–150.

Oppenheimer, M., 1998: Global warming and the stability of the west Antarctic ice sheet. *Nature,* **393,** 325–332.

Oreskes, N.K., K. Belitz, and K. Shrader-Frechette, 1994: Verification, validation and confirmation of numerical models in the earth sciences. *Science,* **263,** 641–646.

Orlando, B.M. and L. Smeardon (eds.), 1999: *Report of the Eleventh Global Biodiversity Forum: Exploring Synergy Between the UN Framework Convention on Climate Change and the Convention on Biological Diversity.* IUCN—The World Conservation Union, Gland, Switzerland and Cambridge, United Kingdom, 46 pp. Available online at http://www.gbf.ch.

Pachauri, R., T. Taniguchi, and K. Tanaka (eds.), 2000: *Guidance Papers on the Cross Cutting Issues of the Third Assessment Report of the IPCC.* Geneva, Switzerland, 138 pp. Available online at http://www.gispri.or.jp.

Parson, E.A. and K. Fisher-Vanden, 1997: Integrated assessment models of global climate change. In: *Annual Review of Energy and the Environment,* **22,** 589–628.

Pearson, P.N. and M.R. Palmer, 2000: Atmospheric carbon dioxide concentrations over the past 60 million years. *Nature,* **406,** 695–699.

Peck, S.C. and T.J. Teisberg, 1992: CETA: a model of carbon emissions trajectory assessment. *The Energy Journal,* **13(1),** 55–77.

Pimm, S., 1991: *The Balance of Nature? Ecological Issues in the Conservation of Species and Communities.* University of Chicago Press, Chicago, IL, USA, 434 pp.

Portney, P.R. and J.P. Weyant (eds.), 1999: *Discounting and Intergenerational Equity.* Resources for the Future, Washington, DC, USA, 186 pp.

Rahman, A., 1999: Climate change and violent conflicts. In: *Ecology, Politics, and Violent Conflicts* [Suliman, M. (ed.)]. Zed Books, London, United Kingdom, pp. 181–210.

Rahmstorf, S., 1999: Shifting seas in the greenhouse? *Nature,* **399,** 523–524.

Rayner, S. and E.L. Malone, 1998: Ten suggestions for policymakers: guidelines from an international social science assessment of human choice and climate change. In: *Human Choice and Climate Change* [Malone, E.L. and S. Rayner (eds.)]. Batelle Press, Columbus, OH, USA, **4,** 109–138.

Reilly, J., W. Baethgen, F.E. Chege, S.C. van de Geijn, L. Erda, A. Iglesias, G. Kenny, D. Patterson, J. Rogasik, R. Rötter, C. Rosenzweig, W. Sombroek, J. Westbrook, D. Bachelet, M. Brklacich, U. Dämmgen, M. Howden, R.J.V. Joyce, P.D. Lingren, D. Schimmelpfennig, U. Singh, O. Sirotenko, and E. Wheaton, 1996: Agriculture in a changing climate: impacts and adaptation. In: *Climate Change 1995: Impacts, Adaptations and Mitigation of Climate Change: Scientific-Technical Analyses. Contribution of Working Group II to the Second Assessment Report of the Intergovernmental Panel on Climate Change* [Watson, R.T., M.C. Zinyowera, and R.H. Moss (eds.)]. Cambridge University Press, Cambridge, United Kingdom and New York, NY, USA, pp. 427–467.

Root, T.L. and S.H. Schneider, 2001: Climate change: overview and implications for wildlife. In: *Wildlife Responses to Climate Change: North American Case Studies* [Schneider, S.H. and T.L. Root (eds.)]. Island Press, Washington, DC, USA, (in press).

Root, T.L. and S.H. Schneider, 1995: Ecology and climate: research strategies and implications. *Science,* **269,** 331–341.

Root, T.L., J.T. Price, K.R. Hall, C. Rosenzweig, and S.H. Schneider, 2001: The impact of climatic change on animals and plants: a meta-analysis. *Nature* (submitted).

Rotmans, J. and M. van Asselt, 1996: Integrated assessment: a growing child on its way to maturity—an editorial. *Climatic Change,* **34(3–4),** 327–336.

Roughgarden, T. and S.H. Schneider, 1999: Climate change policy: quantifying uncertainties for damages and optimal carbon taxes. *Energy Policy,* **27(7),** 415–429.

Schindler, D.W., P.J. Curtis, B.R. Parker, and M.P. Stainton, 1996: Consequences of climate warming and lake acidification for UVB penetration in North American boreal lakes. *Nature,* **379,** 705–708.

Schneider, S.H., 1997: Integrated assessment modeling of global climate change: transparent rational tool for policy making or opaque screen hiding value-laden assumptions? *Environmental Modeling and Assessment,* **2(4),** 229–248.

Schneider, S.H., W.E. Easterling, and L.O. Mearns, 2000a: Adaptation: sensitivity to natural variability, agent assumptions and dynamic climate changes. *Climatic Change,* **45(1),** 203–221.

Schneider, S.H., K. Kuntz-Duriseti, and C. Azar, 2000b: Costing nonlinearities, surprises and irreversible events. *Pacific and Asian Journal of Energy,* **10(1),** 81–106.

Schneider, S.H. and S.L. Thompson, 2000: A simple climate model for use in economic studies of global change. In: *New Directions in the Economics and Integrated Assessment of Global Climate Change* [DeCanio, S.J., R.B. Howarth, A.H. Sanstad, S.H. Schneider, and S.L. Thompson (eds.)]. Pew Center on Global Climate Change, Arlington, VA, USA, pp. 47–63.

Schneider, S.H., B.L. Turner, and H. Morehouse Garriga, 1998: Imaginable surprise in global change science. *Journal of Risk Research,* **1(2),** 165–185.

Skodvin, T., 2000: Revised rules of procedure for the IPCC process. *Climatic Change,* **46,** 409–415.

Smit, B., I. Burton, and J.T.K. Richard, 1999: The science of adaptation: a framework for assessment. *Mitigation and Adaptation Strategies for Global Change,* **4(3-4),** 199–213.

Still, C.J., P.N. Foster, and S.H. Schneider, 1999: Simulating the effects of climate change on tropical montane cloud forests. *Nature,* **398,** 608–610.

Titus, J. and V. Narayanan, 1996: The risk of sea level rise: a delphic Monte Carlo analysis in which twenty researchers specify subjective probability distributions for model coefficients within their respective areas of expertise. *Climatic Change,* **33(2),** 151–212.

Tol, R.S.J., 1995: The damage costs of climate change: toward more comprehensive calculations. *Environmental and Resource Economics,* **5,** 353–374.

Tol, R.S.J. and S. Fankhauser, 1998: On the representation of impact in integrated assessment models of climate change. *Environmental Monitoring and Assessment,* **3,** 63–74.

Toth, F., 2000: Decision analysis frameworks in TAR. In: *Guidance Papers on the Cross Cutting Issues of the Third Assessment Report of the IPCC* [Pachauri, R., K. Tanaka, and T. Taniguchi (eds.)]. Intergovernmental Panel on Climate Change, Geneva, Switzerland, pp. 53–68. Available online at http://www.gispri.or.jp.

Turner, B.L. II, R.E. Kasperson, W.B. Meyer, K.M. Dow, D. Golding, J.X. Kasperson, R.C. Mitchell, and S.J. Ratick, 1990: Two types of global environmental change: definitional and spatial-scale issues in their human dimensions. *Global Environmental Change,* **1(1),** 14–22.

UNDP, 1999: *Human Development Report 1999.* United Nations Development Programme, Oxford University Press, Oxford, United Kingdom, 262 pp. Available online at http://www.undp.org/hdro/99.htm.

United Nations, 1997: *Kyoto Protocol to the United Nations Framework Convention on Climate Change.* Kyoto, Japan, 60 pp. Available online at http://www.unfccc.int/text/resource/docs/cop3/07a01.pdf.

United Nations, 1992: *United Nations Framework Convention on Climate Change.* Rio de Janeiro, Brazil, 33 pp. Available online at http://www.unfccc.int/text/resource/docs/convkp/conveng.pdf.

van der Veer, J., 1999: Profits and principles, the experiences of an industry leader. In: *Proceedings of the Greenport '99 Conference, April 1999.* World Business Council for Sustainable Development, Geneva, Switzerland. Available online at Available online at http://www.wbcsd.ch/ Speech/s73.htm.

von Weizsäcker, E., A.B. Lovins, and L.H. Lovins, 1998: *Factor Four—Doubling Wealth, Halving Resource Use. Report to the Club of Rome.* Earthscan, London, United Kingdom, 322 pp.

Waggoner, P.E., J.H. Ausubel, and I.K. Wernick, 1996: Lightening the tread of population on the land: American examples. *Population and Development Review,* **22(3),** 531–546.

Wang, G. and E.A.B. Eltahir, 2000: Role of vegetation dynamics in enhancing the low-frequency varibility of the Sahel rainfall. *Water Resources Research,* **36 (4),** 1013–1021.

Watson, R.T, J.A. Dixon, S.P. Hamburg, A.C. Janetos, and R.H. Moss, 1998: *Protecting Our Planet, Securing Our Future.* United Nations Environment Programme, U.S. National Aeronautics and Space Administration, and The World Bank, Washington, DC, USA, 95 pp.

WCED, 1987: *Our Common Future.* The World Commission on Environment and Development, Oxford University Press, Oxford, United Kingdom, 400 pp.

Wiener, J.B., 1995: Protecting the global environment. In: *Risk versus Risk: Tradeoffs in Protecting Health and the Environment* [Graham, J.D. and J.B. Wiener (eds.)]. Harvard University Press, Cambridge, MA, USA, pp. 193–225.

West, J.J. and H. Dowlatabadi, 1999: Assessing economic impacts of sea level rise. In: *Climate Change and Risk* [Downing, T.E., A.A. Olsthoorn, and R. Tol (eds.)]. Routledge, London, United Kingdom, pp. 205–220.

West, J.J., M.J. Small, and H. Dowlatabadi, 2001: Storms, investor decisions and the economic impacts of sea level rise. *Climatic Change*, **48(2-3)**, 317–342.

Weyant, J., O. Davidson, H. Dowlatabadi, J. Edmonds, M. Grubb, E.A. Parson, R. Richels, J. Rotmans, P.R. Shukla, and R.S.J. Tol, 1996: Integrated assessment of climate change: an overview and comparison of approaches and results. In: *Climate Change 1995: Economic and Social Dimensions of Climate Change. Contribution of Working Group III to the Second Assessment Report of the Intergovernmental Panel on Climate Change* [Bruce, J.P., E.F. Haites, and H. Lee (eds.)]. Cambridge University Press, Cambridge, United Kingdom and New York, NY, USA, pp. 367–396.

Wigley, T.M.L., 1998: The Kyoto protocol: CO_2, CH_4 and climate implications. *Geophysical Research Letters*, **25(13),** 2285–2288.

Wigley, T.M.L., J. Edmonds, and R. Richels, 1996: Economic and environmental choices in the stabilization of atmospheric CO_2 concentrations. *Nature*, **379(6582),** 240–243.

Yang, C. and S.H. Schneider, 1998: Global carbon dioxide emissions scenarios: sensitivity to social and technological factors in three regions. *Mitigation and Adaptation Strategies for Global Change*, **2,** 373–404.

2

Methods and Tools

Q.K. AHMAD (BANGLADESH) AND RICHARD A. WARRICK (NEW ZEALAND)

Lead Authors:
T.E. Downing (UK), S. Nishioka (Japan), K.S. Parikh (India), C. Parmesan (USA), S.H. Schneider (USA), F. Toth (Germany), G. Yohe (USA)

Contributing Authors:
A.U. Ahmed (Bangladesh), P. Ayton (UK), B.B. Fitzharris (New Zealand), J.E. Hay (New Zealand), R.N. Jones (Australia), G. Morgan (USA), R. Moss (USA), W. North (USA), G. Petschel-Held (Germany), R. Richels (USA)

Review Editors:
I. Burton (Canada) and R. Kates (USA)

CONTENTS

EXECUTIVE SUMMARY

The purpose of this chapter is to address several overarching methodological issues that transcend individual sectoral and regional concerns. In so doing, this chapter focuses on five related questions: How can current effects of climate change be detected? How can future effects of climate change be anticipated, estimated, and integrated? How can impacts and adaptations be valued and costed? How can uncertainties be expressed and characterized? What frameworks are available for decisionmaking? In addressing these questions, each section of Chapter 2 seeks to identify methodological developments since the Second Assessment Report (SAR) and to identify gaps and needs for further development of methods and tools.

Detection of Response to Climate Change by Using Indicator Species or Systems

Assessment of the impacts on human and natural systems that already have occurred as a result of recent climate change is an important complement to model projections of future impacts. Such detection is impeded by multiple, often inter-correlated, nonclimatic forces that are concurrently affecting those systems. Attempts to overcome this problem have involved the use of indicator species to detect responses to climate change and to infer more general impacts of climate change on natural systems. An important component of the detection process is the search for systematic patterns of change across many studies that are consistent with expectations, based on observed or predicted changes in climate. Confidence in attribution of these observed changes to climate change increases as studies are replicated across diverse systems and geographic regions.

Since the SAR, approaches to analysing and synthesizing existing data sets from abiotic and biotic systems have been developed and applied to detection of present impacts of 20th-century climate change. Even though studies now number in the hundreds, some regions and systems are underrepresented. However, there is a substantial amount of existing data that could fill these gaps. Organized efforts are needed to identify, analyze, and synthesize those data sets.

Anticipating the Effects of Climate Change

A wide range of methods and tools are now used and available for studies of local, regional, and global impacts. Since the SAR, improvements have included greater emphasis on the use of process-oriented models and transient climate change scenarios, refined socioeconomic baselines, and higher resolution assessments. Country studies and regional assessments in every continent have tested models and tools in a variety of contexts. First-order impact models have been linked to global systems models. Adaptation has been included in many assessments, often for the first time.

Methodological gaps remain concerning scales, data, validation, and integration. Procedures for assessing regional and local vulnerability and long-term adaptation strategies require high-resolution assessments, methodologies to link scales, and dynamic modeling that uses corresponding and new data sets. Validation at different scales often is lacking. Regional integration across sectors is required to place vulnerability in the context of local and regional development. Methods and tools to assess vulnerability to extreme events have improved but are constrained by low confidence in climate change scenarios and the sensitivity of impact models to major climatic anomalies. Understanding and integrating higher order economic effects and other human dimensions of global change are required. Adaptation models and vulnerability indices to prioritize adaptation options are at early stages of development in many fields. Methods to enable stakeholder participation in assessments need improvement.

Integrated Assessment

Integrated assessment is an interdisciplinary process that combines, interprets, and communicates knowledge from diverse scientific disciplines in an effort to investigate and understand causal relationships within and between complicated systems. Methodological approaches employed in such assessments include computer-aided modeling, scenario analyses, simulation gaming and participatory integrated assessment, and qualitative assessments that are based on existing experience and expertise.

Since the SAR, significant progress has been made in developing and applying these approaches to integrated assessment, globally and regionally. However, the emphasis in such integrated assessments, particularly in integrated modeling, has been on mitigation; few existing studies have focused on adaptation and/or determinants of adaptive capacity. Methods designed to include adaptation and adaptive capacity explicitly in specific applications need to be developed.

Costing and Valuation

Methods of economic costing and valuation rely on the notion of opportunity cost of resources used, degraded, or saved. Opportunity cost depends on whether the market is competitive or monopolistic and whether any externalities are present. It also depends on the rate at which future costs are discounted, which can vary across countries, over time, and over generations. The impact of uncertainty also can be valued if the probabilities of different possible outcomes are known. Public and nonmarket goods and services can be valued through willingness to pay for them or willingness to accept compensation for lack of them. Impacts on different groups, societies, nations, and species need to be assessed. Comparison of alternative distributions of welfare across individuals and groups within a country can be justified if they are made according to internally consistent norms. Comparisons across nations with different societal, ethical, and governmental structures cannot yet be made meaningfully.

No new fundamental developments in costing and valuation methodology have taken place since the SAR. Many new applications of existing methods to a widening range of climate change issues, however, have demonstrated the strengths and limitations of some of these methods. For example, many contingent valuation studies have raised questions about the reliability of such evaluations. Similarly, more attention is now paid to the limitations of methods that underlie efforts to reduce all impacts to one monetary value and/or to compare welfare across countries and cultures. Multi-objective assessments are preferred, but means by which their underlying metrics might more accurately reflect diverse social, political, economic, and cultural contexts need to be developed. In addition, methods for integrating across these multiple metrics are still missing from the methodological repertoire.

Treatment of Uncertainties

The Earth's linked climate and social-natural systems are very complex; thus, there are many unresolved uncertainties in nearly all aspects of the assessment of climatic impacts, vulnerabilities, and adaptation. Subjective judgments are inevitable in most estimates of such complex systems. Since the SAR, more consistent treatment of uncertainties and assessment of biases in judgments have been attempted. Progress also has been made in developing methods for expressing confidence levels for estimates, outcomes, and conclusions, based on more consistent quantitative scales or consistently defined sets of terms to describe the state of the science. Notable attempts to provide "traceable accounts" of how disaggregated information has been incorporated into aggregated estimates have been made, but more work is needed. Greater attention to eliminating inconsistent use of confidence terms or including a full range of uncertainty for key results is still needed in future assessments. Whereas significant progress on issues of uncertainty has been achieved in the context of impacts and vulnerability, a major challenge now lies in addressing uncertainties associated with adaptability.

Decision Analytic Frameworks

Policymakers who are responsible for devising and implementing adaptive policies should be able to rely on results from one or more of a diverse set of decision analytical frameworks. Commonly used methods include cost-benefit and -effectiveness analyses, various types of decision analysis (including multi-objective studies), and participatory techniques such as policy exercises, but there are many other possible approaches. Among the large number of assessments of climate change impacts reviewed in this volume, only a small fraction include comprehensive and quantitative estimates of adaptation options and their costs, benefits, and uncertainty characteristics. This information is necessary for meaningful applications of any decision analytical method. Very few cases in which decision analytic frameworks have been used in evaluating adaptation options have been reported. Greater use of methods in support of adaptation decisions is needed to establish their efficacy and identify directions for necessary research in the context of vulnerability and adaptation to climate change.

2.1. Introduction

In assessing impacts, vulnerability, and adaptation to climate change, a large array of methods and tools pertain to specific sectors, scales of analysis, and environmental and socioeconomic contexts. In this chapter, the term *methods* refers to the overall process of assessment, including tool selection and application; the term *tools* refers to the formulated means of assessment. It is not the intent of this chapter to comprehensively canvas this full array of methods and tools; clearly, such appraisal falls more properly within the purview of the individual chapters in this volume. The purpose of this chapter is to address several overarching methodological questions that transcend individual sectoral and regional concerns. In so doing, this chapter focuses on five related questions:

- *How can the current effects of climate change be detected?* Is climate change already having a discernible effect? One of the key methodological problems is how to unequivocally identify a climate change signal in indicators of change in biotic and abiotic systems. This problem is exemplified in Section 2.2 by focusing on biological indicators and methodological advances that have been made since the Second Assessment Report (SAR).
- *How can the future effects of climate change be anticipated, estimated, and integrated*? Since the SAR, an explosion of climate change vulnerability and adaptation studies has occurred around the world, stimulated in large part by the United Nations Framework Convention on Climate Change (UNFCCC) and its national reporting requirements, as well as the availability of international donor support to non-Annex I countries. Section 2.3 reflects on methodological developments and needs for such vulnerability and adaptation studies, and Section 2.4 focuses on methods for regional and cross-sectoral integration.
- *How can impacts and adaptations be valued and costed*? Ultimately, decisions to avoid or reduce the adverse effects of climate change (or enhance the benefits) require some means of appraisal (monetary or otherwise) of projected impacts and alternative adaptation options. Section 2.5 reviews various methods for valuing and costing, including issues of nonmarket effects, equity, integration, and uncertainty.
- *How can uncertainties be expressed and characterized*? From the science of climate change to assessments of its impacts, uncertainties compound, resulting in a "cascade of uncertainty" that perplexes decisionmaking. Section 2.6 canvasses the problems of, and methods for, incorporating uncertainty into policy-relevant assessments.
- *What frameworks are available for decisionmaking*? Once adaptations have been valued, the choice of adaptation requires methods of weighing and balancing options. Section 2.7 summarizes the main decision analytic frameworks (DAFs) that can be used in this context.

In addressing these questions, the following sections seek to furnish a brief description of the state of methods at hand, methodological developments that have occurred since the SAR, and needs and directions for applications and methods development for the future. Section 2.8 contains concluding remarks.

2.2. Detection of Response to Climate Change by Using Indicator Species or Systems

Climate change may cause responses in many human and natural systems, influencing human health (disease outbreaks, heat/cold stress), agriculture (yield, pest outbreaks, crop timing), physical systems (glaciers, icepack, streamflow), and biological systems (distributions/abundances of species, timing of events). In intensely human-managed systems, the direct effects of climate change may be either buffered or so completely confounded with other factors that they become impossible to detect. Conversely, in systems with little human manipulation, the effects of climate change are most transparent. Systems for which we have a good process-based understanding of the effects of climate and weather events, and have had minimal human intervention, may act as indicators for the more general effects of climate change in systems and sectors where they are less readily studied.

2.2.1. Detection in Natural Systems

2.2.1.1. Predicted Physical Responses to Climatic Warming Trends

The cryosphere is very sensitive to climate change because of its proximity to melting. Consequently, the size, extent, and position of margins of various elements of the cryosphere (sea ice, river and lake ice, snow cover, glaciers, ice cores, permafrost) are frequently used to indicate past climates and can serve as indicators of current climate change (Bradley and Jones, 1992; Fitzharris, 1996; Everett and Fitzharris, 1998). In particular, former glacier extent is indicative of past glacials and the Little Ice Age. At high latitudes and high altitudes, ice cores have provided high-resolution annual (and, in some cases, seasonal) records of past precipitation, temperatures, and atmospheric composition. These records stretch back for many hundreds of years, well before the instrumental period, so they have proven to be very valuable in documenting past climates. Borehole measurements provide data on permafrost warming. Later freeze-up and earlier breakup of river and lake ice is measurable at high latitudes.

Interpretation of climate change resulting from changes in the cryosphere is seldom simple. For example, in the case of glaciers, glacier dynamics and extent are influenced by numerous factors other than climate. Different response times are observed for the same climate forcing, so some glaciers can be in retreat while others are advancing. Changes in glacier size can be caused by changes in temperature or in precipitation—or even a nonlinear combination of both. Similarly, changes in sea ice can be a result of changes in ice dynamics (winds, currents) as much as thermodynamics (temperature). Thus, attribution of the exact nature of climate change from changes in the cryosphere is quite complicated.

With many measures of the cryosphere, there frequently is large interannual variability. This makes determination of possible anthropogenic climate trends difficult to distinguish from the natural noise of the data. Another problem is that high-resolution records usually are not available, except from polar or high-altitude ice cores. Changes in the extent of sea ice and seasonal snow are best observed with satellites, but such records are relatively short (from about 1970), so long-term climate change is difficult to distinguish from short-term, natural variability. The first records of cryospheric extent and changes often come from documentary sources such as old diaries, logbooks of ships, company records, and chronicles (Bradley and Jones, 1992). Although these sources are fraught with difficulty of interpretation, they clearly demonstrate climate changes such as the medieval warm period and the Little Ice Age (see Section 5.7, Chapter 16, and Section 19.2).

2.2.1.2. Predicted Biological Responses to Climatic Warming Trends

All organisms are influenced by climate and weather events. Physiological and ecological thresholds shape species distributions (i.e., where species can survive and reproduce) and the timing of their life cycles (i.e., periods of growth, reproduction, and dormancy) (Uvarov, 1931; MacArthur, 1972; Precht *et al.*, 1973; Weiser, 1973; Brown *et al.*, 1996; Hoffman and Parsons, 1997; Saether, 1997) In the face of a local environmental change, such as a systematic change in the climate, wild species have three possible responses:

- Change geographical distribution to track environmental changes
- Remain in the same place but change to match the new environment, through either a plastic or genetic response [a plastic response is a reversible change within an individual, such as a shift in phenology (timing of growth, budburst, breeding, etc.); a genetic response is an evolutionary change within a population over several generations, such as an increase in the proportion of heat-tolerant individuals]
- Extinction.

In many individual studies, careful experimental design or direct tests of other possible driving factors make attribution of response to climate change possible with medium to high confidence. These studies address three questions (see Sections 5.4 and 19.2):

- Are changes observed in natural systems during the 20th century in accord with predictions from known effects of climate and from bioclimatic theory?
- Given that species, communities, and ecosystems are responding to a complex function of factors, do statistical analyses identify climatic components that statistically explain most of the observed change?
- If so, how can these results guide future predictive models of biotic response to climate change?

Studies of responses to past large-scale climatic changes during the Pleistocene ice ages and the early Holocene provide a good basis for predicting biotic responses to current climate change. Overwhelmingly, the most common response was for a species to track the climatic change such that it maintained, more or less, a species-specific climatic envelope in which it lived or bred. Typically, a species' range or migratory destination shifted several hundreds of kilometers with each 1°C change in mean annual temperature, moving poleward and upward in altitude during warming trends (Barnosky, 1986; Woodward, 1987; Goodfriend and Mitterer, 1988; Davis and Zabinski, 1992; Graham, 1992; Baroni and Orombelli, 1994; Coope, 1995; Ashworth, 1996; Brandon-Jones, 1996). Extinctions of entire species, as well as observable evolutionary shifts, were rare. Phenological shifts may have occurred but cannot be detected with Pleistocene data.

For very mobile or migratory animals—such as many birds, large mammals, pelagic fish, and some insects—shifts of species range occur when individuals move or migration destinations change. Thus, these movements actually track yearly climatic fluctuations. In contrast, most wild species, especially plants, are sedentary, living their lives in a single spot because they have limited mobility or because they lack behavioral mechanisms that would cause them to disperse from their site of birth. Rather than occurring by individual movements, range changes in sedentary species operate by the much slower process of population extinctions and colonizations. Intertidal organisms represent a mix of these two extremes: Adults frequently are completely sedentary, but many species have free-floating planktonic larvae. The dispersal of this early life-history phase is heavily governed by ocean currents. As a result, changes in distribution are driven by a combination of changes in strength and pathways of currents as well as general changes in sea temperature.

2.2.1.3. Bioclimatic Models

A variety of modeling techniques have been used to determine the strength of association between suites of biotic and abiotic variables and species distributions. These associations can then be used to predict responses to environmental change, including climatic change. Bioclimatic models encompass a wide range of complexity. The simplest model is described as a "climate envelope." It is designed to describe static associations between a species' distribution and a single set of climatic variables (Grinnell, 1924, 1928). Modern statistical analyses and improved computer power have facilitated determination of complex suites of climatic and nonclimatic variables that correlate with the range boundaries for a given species (e.g., software such as BIOCLIM and GARP) (Stockwell and Noble, 1991). These models incorporate biological realism, such as local adaptation and differences in the nature of range limitations at different edges. Modern biogeographic models have demonstrated a high level of predictive power in cross-validation tests (Peterson and Cohoon, 1999).

2.2.1.4. Strengths and Limitations of Data

In assessing the strengths of studies as indicators of response to climate change, it is helpful to consider where they lie along axes of time, space, and replication (numbers of populations, numbers of species, etc.). To assess changes in species distributions, data over large geographic areas are important, especially for areas that represent the boundaries of a species' range or migratory destination. To assess trends through time, frequent (yearly is ideal) observations over many decades are most informative. And to assess the generality of the result, good replication is necessary, with many populations/census sites per species to indicate distributional changes within species or many species per community to indicate community shifts.

2.2.1.5. Data and Response Types

Ranges of migratory or mobile species can be very sensitive to climate when individuals show an immediate response in their migratory destinations. As with climatic data itself, one then needs long time series to distinguish year-to-year variation (noise) from long-term trends. Distributions of sedentary species have an inherent lag time stemming from limited dispersal abilities. Neither the numbers of populations nor the geographic location of the range limit may fluctuate strongly between adjacent years, and detectable shifts in species ranges may take decades or even centuries. In such cases, data often are not continuous through time, although data for a single year can be taken as representative of the state of the species during the surrounding multi-year period.

In addition to these shifts in species distributions, a suite of more subtle "plastic" responses allow organisms to adjust seasonally to natural variations of climate. Phenological changes—that is, shifts in the timing of events—can be assessed. These events include dates of budburst, flowering, seed set, fruit ripening, hibernation, breeding, and migration (Yoshino and Ono, 1996; Bradley *et al.*, 1999; Menzel and Fabian, 1999). Changes in phenologies can be detected in a wide variety of organisms, but this requires studies conducted over several years, in which weekly or daily observations should be made before and during the target event (e.g., flowering). Remote-sensing data have the advantage that they can be analyzed for such effects years after the events, but they are limited to very general, community-wide questions such as dates when the ground begins to turn "green" from spring growth. They indicate trends only for the past 30 years because satellites with suitable detection equipment have been in place only since the early 1970s (Myneni *et al.*, 1997). There are very long-term records (i.e., centuries) in a few unusual cases (Lauscher, 1978; Hameed, 1994; Sparks and Carey, 1995), but most monitoring data also are in the realm of the past 30 years (see Sections 5.4 and 19.2).

A different type of rapid response—probably nongenetic—is exemplified by changes in body size of small mammals and lizards (Sullivan and Best, 1997; Smith *et al.*, 1998). Body size becomes smaller with general warming and larger with either cooling or increased variability of climate. This source of information has been studied with reference to historical climate (Morgan *et al.*, 1995; Hadly, 1997; Badgley, 1998); it has been unexplored with respect to current trends and should be given greater attention.

Attribution of an observed biological trend to effects of climate change rests on several grounds (Easterling *et al.*, 2000; Parmesan *et al.*, 2000), namely:

- Known fundamental mechanistic links between thermal/precipitation tolerances and species in the studies
- A large body of theory that links known regional climate changes to observed biotic changes
- Direct observations of climate effects in some studies.

2.2.2. Interpretation of Causation from Correlative Data

2.2.2.1. Lines of Evidence

Attribution of observed changes in natural systems to the effects of climate change is analogous to attribution of anthropogenic greenhouse gases (GHGs) as causal factors of recent climate trends. Within the climate realm, the lines of evidence are as follows:

1) Knowledge of fundamental processes of atmospheric forcing by different gases and radiative features
2) Geological evidence that shows changes in particular atmospheric gases associated with changes in global climate
3) General circulation models (GCMs) that accurately "predict" climatic trends of the 20th century, based on fundamental principles of atmospheric forcing
4) Analyses of global mean temperature and precipitation records that indicate large variances within and among station data as a result of genuine climate variance, as well as errors and biases resulting from instrument change, location changes, or local urbanization. There are large differences in the length of records because stations have been added over the century. Total record length may vary widely. This necessitates large-scale analyses that average the effects over many hundreds or thousands of stations so that the true climate signal can emerge.

Analogs in the biological realm are as follows:

1) Knowledge of fundamental responses of organisms to climate and extreme events. This knowledge is based

on experimental work in the laboratory on physiological thresholds and metabolic costs of different thermal/water regimes, as well as experimental work in the field on ecological thresholds and fitness costs of different temperature/water treatments. In addition to these controlled, manipulated experiments, there are onsite observations of individuals and populations before and after particular weather events (e.g., documentation of population evolution of body size in birds caused by a single winter storm or a single extreme drought, or population extinctions of butterflies caused by a single midseason freeze or a single extreme drought year). The biological community generally accepts the assertion that climate is a major influence on the abundances and distributions of species.

2) Geological evidence that shows changes in global mean temperature associated with changes in the distributions of species. Species' ranges typically shifted toward the poles by about 400–2,000 km between glacial and interglacial periods (change of 4°C).
3) Ecological and biogeographic theory and models that accurately "predict" current distributions of species, based on fundamental principles of climatic tolerances.
4) Analyses of biological records starting from the 1700s, when the first researchers began to systematically record the timing of biological events and the locations of species. There are some variances within and among the historical records for any given species or locality as a result of genuine variance of the biological trait as well as small errors resulting from changes in the recorder, methods of recording, local urbanization, and other landscape changes. There are large differences in the length of records because interest in taking such records gradually has increased over the centuries. Total record length may vary from 300 to <10 years. This necessitates large-scale syntheses that assess the effects over many hundreds of species or studies so that any true global climate signal can emerge.

2.2.2.2. *Complex Systems and Responses*

Interpretation of changes in marine organisms is difficult because of the strong influence of oceanic currents on dispersal and local temperatures. As the links between ocean currents and atmospheric conditions become better understood, linking changes in marine biota to climate change will become easier.

Tree rings provide long series of yearly data spanning centuries, and data are easily replicated across taxa and geographic regions. The width of an annual ring indicates growth for that year, but growth is affected by disease, herbivory, acidification, nitrification, and atmospheric conditions [carbon dioxide (CO_2), ozone (O_3), ultraviolet (UV) radiation], as well as by yearly climate (Bartholomay *et al.*, 1997; Jacoby and D'Arrigo, 1997; Briffa *et al.*, 1998). Correlative studies can be conducted to assess the relative impacts of different climatic variables on tree rings by focusing on the 20th century, for which independent climate data exist. If the correlations are strong, one can then attempt to reconstruct past climates (prior to the existence of climate stations) from the derived relationship. One cannot distinguish the primary cause of changes in ring width in any single case (Vogel *et al.*, 1996; Brooks *et al.*, 1998). However, because excellent geographic replication is possible, these complex causal factors can be statistically reduced to those with very large-scale effects; general global climatic conditions are one of the few factors that could simultaneously affect very distant organisms (Feng and Epstein, 1996; Tessier *et al.*, 1997; Briffa *et al.*,1998).

Finally, an evolutionary response (a genetic change in a population/species) is possible (Berthold and Helbig, 1992; Rodríguez-Trelles *et al.*, 1996, 1998a). Modern molecular techniques make it possible to sequence DNA from small samples taken from museum specimens, which could then be compared to the DNA of current populations. Unfortunately, for most species, scientists do not yet know which genes are associated with climatic adaptations, so this method cannot provide useful data for more than a handful of species that have been intensively studied genetically (Rodríguez-Trelles *et al.*, 1996; Rodríguez-Trelles and Rodriguez, 1998).

2.2.2.3. *Methodological Considerations*

Studies that relate observed changes in natural biota to climatic changes are necessarily correlational. It is not possible to address this question through a standard experimental approach, so direct cause-and-effect relationships cannot be established. However, the level of uncertainty can be reduced until it is highly unlikely that any force other than climate change could be the cause of the observed biotic changes. Studies can reduce uncertainty in three ways:

- Maximize statistical power
- Design to control for major confounding factors
- For confounding factors that remain, directly analyze whether they could explain the biotic changes and, if so, quantify the strength of that relationship.

Statistical power is gained by using:

- Large sample sizes (numbers of populations/numbers of species)
- Data gathered over a large region
- Studies conducted over multiple regions
- Studies conducted on multiple taxa (different families, orders, phyla, etc.)
- Selecting populations or species without *a priori* knowledge of changes to minimize sample bias
- Data gathered over a long time period such that bi-directional responses to opposite climatic trends may be detected.

Confounding factors can be addressed in a correlational study. Biologists know that many nonclimatic anthropogenic forces affect population dynamics, community stability, and species distributions. These forces fall largely under the main headings of land-use change, hydrological changes, pollution, and invasive species. The term "land-use change" comprises a suite of human interventions that eliminate or degrade natural habitats, leading to loss of species that are dependent on those habitats. Habitat loss can be overt destruction, as occurs with urbanization, conversion to agriculture, or clear-cut logging. Habitat degradation is more subtle; it usually results from changes in land management, such as changes in grazing intensity/timing, changes in fire intensity/frequency, or changes in forestry practices (coppicing, logging methods, reforestation strategies), as well as irrigation dams and associated flood control. Loss of habitat by either means not only causes extinctions at that site but endangers surrounding good habitat patches by increased fragmentation. As good habitat patches become smaller and more isolated from other good patches, the populations on those patches are more likely to become permanently extinct.

The main airborne pollutants that are likely to affect distributions and compositions of natural biotic systems are sulfates, which lead to acid rain; nitrates, which fertilize the soil; and CO_2, which affects basic plant physiology (particularly the carbon/nitrogen ratio). Urban areas, in addition to having locally high amounts of sulfates and nitrates, are artificial sources of heat. Aquatic and coastal marine systems suffer from runoff of fertilizers and pesticides from agricultural areas and improperly treated sewage, as well as fragmentation from dams and degradation resulting from trawling, dredging, and silting.

These confounding factors cannot be completely eliminated, but their influences can be minimized by (Parmesan, 1996, 2001; Parmesan *et al.*, 1999):

- Conducting studies away from large urban or agricultural areas
- Conducting studies in large natural areas (e.g., northern Canada, Alaska, areas in Australia)
- Choosing individual sites in preserved areas (national parks/preserves, field stations)
- Eliminating from consideration extreme habitat specialists or species known to be very sensitive to slight human modifications of the landscape.

If a particular confounding factor cannot be greatly minimized, it should be measured and analyzed alongside climatic variables to assess their relative effects.

Ideal target species, communities, or systems in which to look for biotic responses to climate change meet the following criteria (DeGroot *et al.*, 1995; Parmesan, 2001):

- Basic research has led to a process-based understanding of underlying mechanisms by which climate affects the organism or community. This knowledge may come from experimental laboratory or field studies of behavior and physiology or from correlational studies between field observations climatic data.
- The target is relatively insensitive to other anthropogenic influences, so the effects of possible confounding factors are minimized.
- Short (decadal) or no lag time is expected between climate change and response [e.g., tree distributional responses may have a lag time of centuries, so they may not be ideal for looking for distributional changes over recent decades (Lavoie and Payette, 1996)].
- There are good historical records, either from being a model system in basic research or by having a history of amateur collecting.
- Current data are available (from monitoring schemes, long-term research) or are easy to gather.

Use of indicator species or communities is crucial for defining the level of climate change that is important to natural systems and for giving baseline data on impacts. However, caution is advisable when extending these results to predictive scenarios because the indicators often are chosen specifically to pinpoint simple responses to climate change. Thus, these studies purposefully minimize known complexities of multiple interacting factors, such as:

- The direct effects of CO_2 on plants may vary with temperature.
- The outcome of competitive interactions between species is different under different thermal regimes (Davis *et al.*, 1998), and, conversely, competitive environment can affect sensitivity and response to particular climatic variables (Cescatti and Piutti, 1998).
- Different species have different lag times for response, which inevitably will cause the breakup of traditional communities (Davis and Zabinski, 1992; Overpeck *et al.*, 1992; Root and Schneider, 1995).
- The ability of wild plant and animal life to respond to climate change through movement is likely to be hindered by human-driven habitat fragmentation; those with lowest dispersal will be most affected (Hanski, 1999).

2.2.3. *Detection in Managed Systems*

2.2.3.1. Human Health

Because many wild organisms serve as vectors for human diseases, and these diseases are very well documented historically (with records going back hundreds of years), one might think of using the distribution and intensity of disease occurrence as an indicator of shifts in wild vector distributions or altered dynamics of pathogen transmission. Many disease vectors are known to be strongly influenced by climate (e.g., the anopheline mosquitoes that carry malaria).

The problem with using disease records is that the presence of the vector is necessary but not sufficient to cause disease transmission. Socioeconomic factors—such as sanitation systems, vaccination programs, nutritional conditions, and so forth—largely determine whether the presence of the disease in wild vectors actually leads to outbreaks of disease in nearby human populations. In fact, transmission and virulence of disease are themselves directly affected by climate. Thus, although disease is a potentially important component of climate change impacts, it is not a useful indicator of the direct effects of climate change (see Chapter 9 and Section 19.2).

2.2.3.2. *Agriculture*

Crop plants, like plants in general, are more strongly affected by the direct effects of increased atmospheric CO_2 than are animals. Increased CO_2 alters the physical structures and the carbon/nitrogen balance in plants—which in turn alters the plant's growth rate, yield, susceptibility to pest attack, and susceptibility to water stress. These effects interact with the effects of climate change itself in complex ways. In addition, the effects of climate change are buffered in agricultural systems as farming methods are altered to adjust to current climate conditions (e.g., irrigation practices, crop varieties used) (see Sections 5.3 and 19.2). A few selected attributes and systems may be possible indicators of climate change effects. Possible traits are leafing dates of grapevines in orchard with old stock, and planting dates of yearly crops in areas that have not changed seed variety over a given length of time.

2.2.4. *Advances since the SAR and Future Needs*

Since the SAR, methods have been developed and applied to the detection of present impacts of 20th-century climate change on abiotic and biotic systems. Assessment of impacts on human and natural systems that already have occurred as a result of recent climate change is an important complement to model projections of future impacts. How can such effects be detected? Such detection is impeded by multiple, often intercorrelated, nonclimatic forces that concurrently affect those systems. Attempts to overcome this problem have involved the use of indicator species to detect responses to climate change and infer more general impacts of climate change on natural systems. An important component of this detection process is the search for systematic patterns of change across many studies that are consistent with expectations on the basis of observed or predicted changes in climate. Confidence in attribution of these observed changes to climate change increases as studies are replicated across diverse systems and geographic regions. Even though studies now number in the hundreds, some regions and systems are underrepresented. However, there is a substantial amount of existing data that could fill these gaps. Organized efforts are needed to identify, analyze, and synthesize those data sets.

2.3. Anticipated Effects of Climate Change

2.3.1. *Background*

This section outlines recent developments of methods and tools that are used to anticipate the effects of climate change—the broad approaches to climate change impact and vulnerability assessment. It considers future research and development needs, particularly to facilitate more informed policy decisions.

Based on interviews with experts and reviews of impacts methodologies, seven questions frame recent progress and needs for impacts methods and tools:

1) What are the appropriate scales of analysis for impact assessments?
2) What should be the baselines for comparison?
3) How should integrated scenarios of climatic and socioeconomic change be used?
4) What are the prospects for assessing the impacts of climatic extremes and variability?
5) How can transient effects be included in methods and tools?
6) What is the recent progress in methods for assessing adaptive capacity?
7) How can vulnerability be related to policies for reducing GHG emissions?

Conclusions are provided in the following subsections; the succeeding subsections provide further insight.

2.3.2. *What are the Appropriate Scales of Analysis for Impact Assessments?*

Climate change impact assessments must begin with decisions about the scope and scale of the assessment: What are the main policy issues? What and who are exposed to climate change impacts? What is the appropriate scale—time frame, geographical extent, and resolution? Considerable progress has been achieved since the SAR in raising such framing questions at the outset of an assessment cycle, often in conjunction with representative stakeholders (see Carter *et al.*, 1994; Downing *et al.*, 2000).

Methods for identifying policy issues include checklists and inventories, document analysis, surveys and interviews, and simulations. The process of determining the scope of assessment should be iterative. The project design should specify what and who is exposed to climate change impacts—economic sectors, firms, or individuals. Evaluation of adaptation strategies should be cognizant of actors involved in making decisions or suffering consequences.

The choice of temporal scales, regional extent, and resolution should be related to the focus of the assessment. Often, more than one scale is required, under methods such as strategic scale cycling (Root and Schneider, 1995) or multi-level modeling (e.g., Easterling *et al.*, 1998). Linkage to global assessments may be necessary to understand the policy and economic context (e.g., Darwin *et al.*, 1995).

The most common set of methods and tools remains various forms of dynamic simulation modeling, such as crop-climate models or global vegetation dynamic models. A major improvement in impact modeling has been applicaton of process-oriented models, often with geographically explicit representations, instead of models that are based on correlations of climatic limits. Data for running and validating models is a recurrent issue. Intermodel comparisons have been undertaken in some areas (e.g., Mearns *et al.*, 1999), but much remains to be done.

Climate change is likely to have multiple impacts across sectors and synergistic effects with other socioeconomic and environmental stresses, such as desertification, water scarcity, and economic restructuring. Most studies (especially as reported in the SAR) have focused on single-sector impacts. Relatively few studies have attempted to integrate regionally or even identified segments of the population that are most at risk from climate change.

Vulnerability assessment may be one way of integrating the various stresses on populations and regions arising from climate change (see Briguglio, 1995; Clark *et al.,* 1998; Huq *et al.,* 1999; Kaly *et al.,* 1999; Mimura *et al.*, 2000; Downing *et al.*, 2001). There are some areas in which formal methods for vulnerability assessment have been well developed (e.g., famine monitoring and food security, human health) and applied to climate change. However, methods and tools for evaluating vulnerability are in formative stages of development.

Further development of methods and tools for vulnerability assessment appears warranted, especially for the human dimensions of vulnerability, integration of biophysical and socioeconomic impacts, and comparison of regional vulnerability. Conceptual models and applications of the evolution of vulnerability on the time scale of climate change are required. Formal methods of choosing indicators and combining them into meaningful composite indices must be tested. Combining qualitative insight and quantitative information is difficult but essential to full assessments. Finally, improved methods and tools should facilitate comparison of vulnerability profiles between at-risk regions and populations and highlight potential reductions in vulnerability, through policy measures or the beneficial effects of climate change.

2.3.3. *What should be the Baseline for Comparison?*

Climate change impacts generally are agreed to be the difference between conditions with and without climate change. However, there is controversy among researchers about how to set the baseline for estimating impacts (or evaluating adaptation).

Most studies apply scenarios of future climate change but estimate impacts on the basis of *current* environmental and socioeconomic baselines. Although this approach is expedient and provides information about the sensitivity of current systems, it skirts the issue of evolving sensitivity to climatic variations (Parry and Carter, 1998). Even without climate change, the environment and societal baselines will change because of ongoing socioeconomic development and, with climate change, because of system responses and autonomous adaptation (e.g., as described for Bangladesh—Warrick and Ahmad, 1996). Strictly speaking, the effects of climate change should be evaluated by taking the moving baseline into account (further discussion on socioeconomic, climate, and sea-level rise scenarios appears in Chapter 3).

Given the uncertainty of the future and the complexity of the various driving forces affecting any given exposure unit, a wide range of different assumptions about future baselines is plausible. The emission scenarios in the *Special Report on Emissions Scenarios* (SRES) reflect this perspective and are based on multiple projections of "alternative futures" (see Chapter 3). Framing local concerns for adaptation to changing risks may require exploratory scenarios, extending the coarse driving forces inherent in the SRES suite. For example, coping with water shortages in Bangladesh is sensitive to scenarios of regional collaboration with India and Nepal (e.g., Huq *et al.*, 1999). For vulnerability and adaptation assessment, there is little apparent consistency regarding elements or procedures for development of these future baselines, including who is exposed, how to select sensitive sectors, and the drivers of social and institutional change at the scale of stakeholders exposed to climate impacts.

2.3.4. *How should Integrated Scenarios of Climatic and Socioeconomic Change be Used?*

As a result of time lags in the impact assessment research cycle, impact assessment studies included in this Third Assessment Report (TAR) do not necessarily employ the set of Intergovernmental Panel on Climate Change (IPCC) reference scenarios outlined in Chapter 3. This time lag is unavoidable because it takes almost half a year to define emissions of GHG after setting socioeconomic scenarios. Following that, it usually takes several months to produce local climate change data used in impact assessment studies. Thus, most of the impact studies reported in the TAR are based on the set of IS92 emission scenarios developed for the IPCC in 1992 and included in the SAR (e.g., Parry and Livermore, 1999).

To assist researchers, the IPCC took the initiative to create the IPCC Data Distribution Centre (<http://ipcc-ddc.cru.uea.ac.uk/>) and posted the SRES scenarios on the Consortium for International Earth Science Information Network (CIESEN) Web site (<http://sres.ciesin.org/index.html>). The IPCC is responsible for distributing consistent scenarios, including socioeconomic trends and regional climate change data. Consistent use of common scenarios provides a consistent reference for comparing and interpreting the results of different studies.

Vulnerability assessments can be conducted on temporal and spatial scales where the effects of climate change could feed back to GHG emissions and climatic changes. In such cases, there may be reason to ensure that scenarios of climate change that are based on GHG emissions and scenarios of changing social, economic, and technological conditions are consistent. This is essential for global assessments and integrated assessment (see Section 2.4). It may not be as critical for studies of local adaptation where there is little feedback between mitigation and adaptation, particularly over a typical planning horizon of several decades. Downscaling the global reference scenarios to local socioeconomic and political conditions remains a significant methodological challenge.

2.3.5. What are the Prospects for Assessing the Impacts of Climatic Extremes and Variability?

Discrete climatic events cause substantial damage. Heavy losses of human life, property damage, and other environmental damages were recorded during the El Niño-Southern Oscillation (ENSO) event of 1997–1998. Details are reported in the regional chapters on Africa, Asia, and the Americas, and Chapter 8 assesses the damages from a financial services perspective. For many policymakers and stakeholders, the impacts of climatic extremes and variability are a major concern (Downing *et al.*, 1999b). The uneven impacts of climatic hazards raises humanitarian concerns for development and equity.

An increase in variability and frequency of extreme events could have greater impacts than changes in climate means (e.g., Katz and Brown, 1992; Mearns, 1995; Semenov and Porter, 1995; Wang and Erda, 1996). Extreme events are a major source of climate impacts under the present climate, and changes in extreme events are expected to dominate impacts under a changing climate (see Section 12.1).

Methodological issues concerning extreme events in the context of climate change include developing climate scenarios, estimating impacts, evaluating responses, and looking at large-scale effects.

2.3.5.1. Developing Scenarios of Changes in Variability and Extreme Events

Working Group I discusses methodologies for estimating changes in variability from the results of GCMs (see Sections 9.3.2 and 13.4.2). Despite certain shortcomings, GCMs can provide estimates of trends in climatic variability (TAR WGI Section 9.1.5). Using extreme events from historical data as analogs also is useful.

The frequency of extreme events is likely to change as mean values shift, even without changes in variability. Chapter 3 reviews potential changes in different climatic elements (see Table 3-10).

From the instrumental record, some regional changes in extremes have been identified, although it is difficult to say whether they are related to GHG-induced climate change. For example, there has been a recent increase in heavy and extreme precipitation in the mid to high-latitude countries of the northern hemisphere, and in several regions of east Asia a decrease in the frequency of temperature extremes together with heavy and extreme precipitation have been observed (see TAR WGI Chapter 2).

2.3.5.2. Estimating First-Order Impacts

Many models that validate well for present climate conditions may not respond realistically to future climatic conditions and subsequent changes in extreme events. For some sectoral impacts, however, methods for evaluating a system's response to changes in variability change are improving. One example is estimation of changes in flooding by using 10-year return periods given by transient GCMs and applied to a watershed model (Takahashi *et al.*, 1998).

2.3.5.3. Analyzing Institutional and Stakeholder Responses

Because of nonlinear relationships, an increase in variability can result in a substantial increase in the frequency of extreme impacts. If a climate element exceeds an acceptable risk threshold (e.g., when the design risk threshold for water storage is exceeded and water shortage is experienced with higher frequency), vulnerability will become "unacceptable." One issue in adaptation is the level at which to set acceptable risks in the future. Stakeholder-determined thresholds are an emerging area of research in Australia (see Section 12.1), and methods to evaluate stakeholder and institutional learning in response to changing climatic hazards are being developed (see Bakker *et al.*, 1999; see also <http://www.eci.ox.ac.uk>). Decision analytical techniques are described below (Section 2.7). An alternative is an inverse approach that focuses on sensitivity to present risks, characterization of the kinds of changes in hazards that would have large effects, and evaluation of response capacities (Downing *et al.*, 1999a).

Research on discrete climatic events is an area that also needs further research. Present GCM resolutions have not achieved the ability to estimate the intensity, route, and frequency of discrete events such as hurricanes (or tropical cyclones) (TAR WGI Sections 9.3.6 and 13.4.2.2). Though there are some indications from GCMs that ENSO-like conditions will become more persistent with global warming (Timmermann *et al.*, 1999), it is still difficult to incorporate these estimates into vulnerability assessments (TAR WGI Sections 9.3.6.3 and 13.4.2.1).

Empirical/analog methods are suitable for assessment of discrete events. Such methods were applied for detailed analyses of damages incurred by ENSO in 1997–1998, as well as the 1998 cyclones in Bangladesh. This method is applied to "if-then" (i.e., if climate change occurs, then such and such impacts may be induced) simulations. For example, analogs from the 1930s Dust Bowl period detailing water shortages and reductions in agriculture yields have been used to simulate the impacts of climate change in the U.S. corn belt (Rosenberg, 1993).

2.3.5.4. Large-Scale Effects

Because unique or singular events, referred to as fiasco scenarios (see Section 19.5.3.3)—such as changes in the thermohaline circulation (Broecker, 1997) and potential destabilization of the West Antarctic Ice Sheet (Oppenheimer, 1998)—have not been proven implausible, there is a need for further studies of potential catastrophic events and unacceptable impacts. However, limited knowledge of such large-scale impacts poses a challenge; to date, systematic vulnerability assessments have not been carried out.

2.3.6. How can Transient Effects be Included in Methods and Tools?

Transient climate scenarios are now widely used in impact assessment—an improvement on earlier use of equilibrium climate scenarios (even if scaled to temporal projections) (see Chapter 3). A corresponding shift from static to dynamic, process-oriented impact models is apparent (as shown in the sectoral chapters).

Many models applied for predicting climate change effects on the behavior of an exposure unit are derived from equilibrium models. These include many basin watershed models, crop models for potential agricultural productivity, and potential vegetation models. With such models, the change effected on the unit at a fixed point in time is estimated, ignoring potentially relevant processes of change.

Systems often consist of elements with different time responses to climate change. This means that the present equilibrium will not be maintained in the next point of time. The velocity of change is a key factor in deciding this transient pattern.

Some terrestrial biosphere research illustrates that the world's biomes will not shift as homogeneous entities in response to changing climate and land use (see Section 5.2.1). Competition between individuals and species, modified disturbance regimes (e.g., fires, windstorms), and migration of species all lead to significant time lags in biospheric responses. Furthermore, if mortality from increased disturbance occurs faster than regrowth of other vegetation, there will be a net release of carbon to the atmosphere, which will change climate forcing. Responses may be a function of spatial scale as well. Dynamic global vegetation models illustrate the shift to transient, scalable impact models (e.g., Woodward *et al.*, 1995).

To ensure a temporally sensitive assessment, impact models should include the different time responses of the system. For example, impacts of malaria depend on human tolerance to repeated infection (Martens *et al.*, 1999). Alternatively, the value of climate change damages could be related to the rate of change rather than solely to the magnitude of climate change (Tol and Fankhauser, 1996). Understanding of the temporal interactions between climate change, impacts, and responses in a truly transient methodology is still a major methodological challenge.

2.3.7. What Recent Progress has been Made in Assessing Adaptive Capacity?

In recent years, assessment of adaptive capacity has emerged as a critical focus of attention, for two reasons: the realization that the Kyoto Protocol is inadequate to prevent substantial changes in climate, and the rising expectation that social and natural systems can cope with climate change, at least within limits, and that adaptation is a viable option to reduce GHG emissions.

Although there are numerous examples of model calculations for adaptive shifts in flora, far less attention has been paid to assessing the adaptability of the system as a whole (e.g., White *et al.*, 1999). In contrast with other phenomena, such as changes in the water cycle, changes in natural ecosystems are related to a long-term process of adaptation and extinction. As noted above, transient climate change scenarios have become a mainstream research procedure.

The recent literature concerned with the impacts of climate change on the managed environment [e.g., on agriculture (see Section 5.3) and coastal zones (see Section 6.7)] generally considers adaptive strategies (e.g., Rosenzweig and Parry, 1994). Water management, for example, has a long history of evaluation of strategies for adapting to climate change and variation (Frederick *et al.*, 1997). However, adaptation often is approached narrowly in terms of technological options. Adaptation *processes*—including the environmental, behavioral, economic, institutional, and cultural factors that serve as barriers or incentives to adaptation over time—often are not considered.

Five methodological directions could enhance future work on adaptation (see Chapter 18). First, methods for increasing understanding of the relationship between adaptation, individual decisionmaking, and local conditions are required. For example, adaptation by farmers could avoid more than half of the potential impacts of climate change on agriculture (e.g., Darwin *et al.*, 1995; El-Shaer *et al.*, 1997). The mix of appropriate measures depends, however, on the local context of soils, climates, economic infrastructure, and other resources (Rosenzweig and Tubiello, 1997) and how they are perceived by farmers. Assessments of adaptation could address these issues of site-scale characteristics and local knowledge, perhaps through participatory methods (e.g., Cohen, 1997, 1998) or interviews and expert opinion (as in the UK Climate Impacts Programme—Mackenzie-Hedger *et al.*, 2000; see <http://www.ukcip.org>).

Second, interactions across scale are likely to be significant for adaptation. In the agricultural sector, for example, adaptive strategies are influenced by multi-scale factors—at the farm, national, and global levels—and their integration into decisionmaking. Methods and tools for examining these multi-scale interactions and their implications for adaptation are required, such as multi-level modeling (Easterling *et al.*, 1998), integrated assessment (see Section 2.4), and agent-based simulation (Downing *et al.*, 2000).

Third, specific measures (such as changing planting dates and cultivars) and longer term adaptation strategies and processes (such as monitoring and research) need to be addressed. Many studies focus on the former; assessing the latter is a major methodological challenge.

Fourth, comparative frameworks are required for assessing the priority of adaptation strategies across populations, regions, and sectors, in addition to evaluating specific measures. Fankhauser (1998) devised a list of adaptation policy options, discussed conceptual issues of economic evaluation, and illustrated typical cost/benefit calculation methods. Section 2.5 considers the use of economic evaluation methods, but nonmonetary frameworks are alternatives (see Huq *et al.*, 1999, for a case example). Issues of equity and valuation on indirect benefits and costs are salient.

Fifth, adaptation to extremes and variability already are important areas of assessment (see above) but need to be more explicitly tied to longer term climate change.

Sixth, stakeholder evaluation of adaptation strategies and measures is required—for example, using decision analytical tools, as noted in Section 2.7. Indicators of vulnerability could be used to monitor the effectiveness of adaptative strategies and measures (see Downing *et al.*, 2001).

2.3.8. *How can Vulnerability Assessments be Related to Policies for Reducing GHG Emissions?*

One approach to mitigation policy is to evaluate targets for reducing GHG emissions on the basis of reductions in vulnerability, rather than GHG concentrations or similar indirect measures of dangerous climate change. By applying existing methods for impact analysis, it is possible to invert the assessment procedure and start with defined sets or windows of impacts that are judged to be tolerable for humankind. This procedure results in emission corridors that embrace all future GHG emissions that are compatible with changes defined to be tolerable—the "safe landing" approach (WBGU, 1995; Alcamo and Kreileman, 1996; Petschel-Held *et al.*, 1999). This approach can be extended to include economic, social, or equity aspects—that is, to define tolerable windows for climate-related facets of these sectors and obtain emission corridors that simultaneously satisfy all possible windows (Toth *et al.*, 1997).

Such approaches require that climate impacts should be differentiated between smooth changes and thresholds that mark abrupt shifts in the system's functioning. In the latter case, the definition of tolerable windows appears to be quite obvious: Damaging, abrupt shifts should be avoided. In the case of smooth changes, specification of tolerable windows is more difficult, confounded in part by uncertainty about adaptation.

Normative decisions on tolerable windows must be a consultative process involving scientists in close cooperation with stakeholders, decisionmakers, nongovernmental organizations (NGOs), and others. There are various designs for this participatory process, such as policy exercises (Toth, 1986, 1988a,b) or what is known as the Delft Process (van Daalen *et al.*, 1998). Nevertheless, specification of windows remains somewhat arbitrary and preferably is used as an assumption in an "if-then" analysis rather than as an ultimate specification.

Methodological challenges include development and validation of reduced-form models, devising robust damage functions, identifying thresholds in adaptive systems, and concerns for equity in relating the distribution of impacts to systemic vulnerability.

2.4. Integrated Assessment

Policymakers require a coherent synthesis of all aspects of climate change. Researchers have spent the past decade developing integrated assessment methods to meet these needs of policymakers. An overview of the framework, including examination of impact and vulnerability, is in the SAR (Weyant *et al.*, 1996). In addition, Rotmans and Dowlatabadi (1998) have concentrated on the broader social science components of integrated assessment; as a result, they came closer to presenting a view within which impacts and adaptation might be most fully investigated with and without relying on models. They assert, "Integrated assessment is an interdisciplinary process of combining, interpreting, and communicating knowledge from diverse scientific disciplines in such a way that the whole set of cause-effect interactions of a problem can be evaluated." Current integrated assessment efforts generally adopt one or more of four distinct methodological approaches:

- Computer-aided modeling in which interrelationships and feedbacks are mathematically represented, sometimes with uncertainties incorporated explicitly (see Chapter 19)
- Scenario analyses that work within representations of how the future might unfold [the MINK study, based on a climate analog of the dust bowl climate of the 1930s, is a classic example (see Rosenberg, 1993, for details)]
- Simulation gaming and participatory integrated assessment, including policy (see Parson and Ward, 1998, for a careful review)
- Qualitative assessments that are based on limited and heterogeneous data and built from existing experience and expertise. Cebon *et al.* (1998) contains a collection of papers that offer similar qualitative coverage; their insights can serve as the basis for a long-run research agenda that looks for regions and sectors in which uncertain futures most significantly cloud our view of where and when impacts might be most severe.

Schneider (1997) has developed a taxonomy of integrated assessments that creates an historically rooted taxonomy of modeling approaches. It begins with "premethodogical assessments" that worked with deterministic climate change, with direct causal links and without feedbacks. It ends with "fifth-generation" assessments that try to include changing values explicitly. In between are three other stages of development, differentiated in large measure by the degree to which they integrate disaggregated climate impacts, subjective human responses, and endogenous policy and institutional evolution.

Methodological bias is an issue in interpreting the results of integrated assessments, as it is in every research endeavor. Schneider (1997) also warns that models composed of many submodules adopted from a wide range of disciplines are particularly vulnerable to misinterpretation and misrepresentation. He underscores the need for validation protocols and explorations of predictability limits. At the very least, integrated assessments must record their underlying value-laden assumptions as transparently as possible. Including decisionmakers and other citizens early in the development of an assessment project can play an essential role in analytical processes designed to produce quality science and facilitate appropriate incorporation of their results into downstream decisions.

In the past decade, several research teams have been working on the development of such frameworks (see Tol and Fankhauser, 1998, for a compendium of current approaches). Known as integrated assessment models (IAMs), these frameworks have been used to evaluate a variety of issues related to climate policy. Although the current generation of IAMs vary greatly, in scope and in level of detail, they all attempt to incorporate key human and natural processes required for climate change policy analysis. More specifically, a full-scale IAM includes submodels for simulating:

- Activities that give rise to GHG emissions
- The carbon cycle and other processes that determine atmospheric GHG concentrations
- Climate system responses to changes in atmospheric GHG concentrations
- Environmental and economic system responses to changes in key climate-related variables.

Although IAMs provide an alternative approach to impact assessment, it is important to note that there is no competition between such integrated approaches and the more detailed sectoral and country case studies discussed in preceding sections. Each approach has its strengths and weaknesses and its comparative advantage in answering certain types of questions. In addition, there are considerable synergies between the two types of studies. Integrated approaches depend on more disaggregated efforts for specification and estimation of aggregate functions and, as such, can be only as good as the disaggregated efforts. Reduced-form integrated approaches make it relatively easy to change assumptions on the "causal chain." That is, one can identify critical assumptions upon which a policy analysis might turn.

In conducting such sensitivity analyses, one can identify where the value of information is highest and where additional research may have the highest payoff from a policy perspective. This can provide some useful guidance to the impacts community about where to direct their efforts to resolve uncertainty. At the same time, integrated models become more useful as uncertainty is narrowed (through the contributions of partial impact assessments); hence, the reduced-form representations become more realistic.

2.4.1. *Integrated Assessment Analyses*

There are many different approaches within the family of integrated models. This diversity is important for a balanced understanding of the issues because different types of models can shed light on different aspects of the same problem. For example, many analyses start with a particular emissions baseline and examine the economic and ecological implications of meeting a given emissions target (e.g., Alcamo, 1994; Edmonds *et al.*, 1997; Morita *et al.*, 1997b; Murty *et al.*, 1997; Yohe, *et al.*, 1998; Jacoby and Wing, 1999; Nordhaus and Boyer, 1999; Tol, 1999a,b; Yohe and Jacobsen, 1999). In such analyses, impacts are first assessed under a so-called "business-as-usual" or reference-case scenario. The analysis is then repeated with a constraint on the future. The change in impacts represents the climate-related benefits of the policy.

Other approaches select a different starting point. For example, Wigley *et al.* (1996) begin with atmospheric CO_2 concentrations and explore a range of stabilization targets. For each target they employ "inverse methods" to determine the implications for global CO_2 emissions. Recognizing that a particular concentration target can be achieved through a variety of emission pathways and that impacts may be path-dependent, they identify the implications of the choice of emissions pathway on temperature change and sea-level rise.

Two other approaches—"tolerable windows" (Toth *et al.*, 1997; Yohe, 1997; Petschel-Held *et al.*, 1999; Yohe and Toth, 2000) and "safe corridors" (Alcamo *et al.*, 1998)—also utilize inverse methods but begin further down the causal chain. Here the focus is on the range of emissions that would keep emission reduction costs and climate change impacts within "acceptable" limits. Working with policymakers, the analysts identify the set of impacts for consideration. Bounds are specified, and the cost of achieving the objective is calculated. If mitigation costs are deemed too costly, policymakers have the opportunity to relax the binding constraint. In this way, the team is able to move iteratively toward an acceptable solution.

Integrated assessment analyses also can be distinguished by their approach to optimization. For example, the focus of the UNFCCC is cost-effectiveness analysis. Article 2 states that the ultimate goal is "stabilization of greenhouse gas concentrations in the atmosphere at a level that would prevent dangerous anthropogenic interference with the climate system." Mitigation cost is more of a consideration in how the target is to be achieved. The Convention states that policies and measures to deal with climate change should be cost-effective to ensure global benefits at the lowest possible cost. Several analysts, beginning with Nordhaus (1991), have identified the least-cost path for achieving a particular concentration target.

Despite the goal of the UNFCCC, several integrated assessment frameworks are designed for benefit-cost analyses. These models identify the emissions pathway that minimizes the sum of mitigation costs and climate change damages. Such policy optimization models have been developed by Nordhaus (1991, 1992, 1994b), Peck and Teisberg (1992, 1994, 1995), Chattopadhyay and Parikh (1993), Parikh and Gokarn (1993), Maddison (1995), Manne *et al.* (1995), Manne and Richels (1995), Nordhaus and Yang (1996), Yohe (1996), Edmonds *et al.* (1997), Tol (1997, 1999c,d), and Nordhaus and Boyer (1999).

2.4.2 *The State of the Art*

Treatment of impacts in these models also varies greatly. Generally, however, impacts are one of the weakest parts of IAMs. To a large extent this is a reflection of the state of the art of the underlying research, but it also reflects the high complexity of the task at hand (see Tol and Fankhauser, 1998, for a survey). Despite the growing number of country-level case studies, our knowledge about climate change and climate change impacts at the regional level remains limited. A coherent global picture, based on a uniform set of assumptions, has yet to emerge. The basis of most global impact assessments remain studies undertaken in developed countries (often the United States), which are then extrapolated to other regions. Such extrapolation is difficult and will be successful only if regional circumstances are carefully taken into account, including differences in geography, level of development, value systems, and adaptive capacity. Not all analyses are equally careful in undertaking this task, and not all models rely on the latest available information in calibrating their damage functions.

The actual functional relationships applied in many integrated models remain simple and often ad hoc. This reflects our still poor understanding of how impacts change over time and as a function of climate parameters. Impacts usually are a linear or exponential function of absolute temperature, calibrated around static "snapshot" estimates (such as $2xCO_2$) without distinguishing the different dynamics that may govern impacts in different sectors. Developing a better understanding of these relationships is one of the most important challenges for integrated model development.

Baseline trends—such as economic development, population growth, technological progress, changes in values, natural climate fluctuations, and increased stress on natural ecosystems—have strong repercussions for climate change vulnerability (e.g., Mendelsohn and Neumann, 1999). They must be better understood and their effect incorporated in the models. Unfortunately, these trends are inherently difficult, if not impossible, to predict over the longer term. This generic problem will not go away, but it can be overcome, at least partly, through broad scenario and sensitivity analysis.

Another key challenge is taking adaptation into account. Adaptation can significantly reduce people's vulnerability to climate change, as shown in Chapter 18. However, adaptation can take many different forms and is correspondingly difficult to model (see Section 19.4). To date there are no IAMs available that can adequately represent or guide the full range of adaptation decisions.

2.5. Methods for Costing and Valuation

Since the SAR, costing and valuation methods have been used increasingly to quantify the cost of potential impacts; these costs include the costs of adaptations required specifically to respond to climate change and climate variability, as well as the costs of residual damages. The bulk of this section focuses on the foundation of economic costs, but there are metrics other than the economic paradigm; these are reviewed briefly in Section 2.3.6.

Researchers have adapted fundamental costing and valuation techniques drawn from the economic paradigm to handle the complexities of increasingly intricate applications. Market mechanisms provide important ways with which we can aggregate across a diversity of individual valuations, but they are tied to historical distributions of resources. Other mechanisms have been exercised, and this section begins by displaying their conceptual foundations within the economic context from which they have all evolved. It proceeds by suggesting how relaxing each underlying economic assumption has been a conceptual challenge. Many of the responses to these challenges, however, are now part of the general economic paradigm.

2.5.1. *Elements of Costing and Valuation Methods*

2.5.1.1. *Opportunity Cost and the Foundations of Valuation Methods*

Opportunity cost is the fundamental building block of modern economic analysis. The true economic cost of one unit of some good X reflects the cost of opportunities foregone by devoting resources to its production. This cost measures the economic value of outputs, goods, and services that would have been possible to produce elsewhere with the resources used to produce the last unit of good X. The social opportunity cost of employing a resource for which there is no alternative economic use is thus zero, even if its price is positive, and opportunity cost will be different under conditions of full employment than under circumstances involving large quantities of visible or invisible unemployment. Moreover, opportunity cost applies only to small "marginal" changes from equilibrium in systems for which there are multiple equilibria. Likewise, the marginal benefit from consuming good X is the value of the last unit purchased, measured in terms of a real price that reflects the welfare that would have been enjoyed if the requisite expenditure had been devoted to consuming another good (or goods).

These concepts may appear circular, but that is an artifact of the circular nature of economic systems. Suppliers of some economic goods are consumers of others. The opportunity cost of a good to the producer and the marginal benefit to the consumer are equal when all of the following conditions are obtained:

- All markets are perfectly competitive.
- Markets are comprehensively established in the sense that all current and future property rights are assigned.
- Marketed goods are exclusive (ownership is singular and well defined) and transferable (goods can be bought, sold, or given away).
- The underlying social and legal systems guarantee that property rights are (reasonably) secure.
- There are no transaction costs involved in creating and/or maintaining any current or future market.
- There is perfect and complete information about all current and future markets.

Under these conditions, the marginal opportunity cost of any good with multiple uses or multiple demanders is equal to its marginal benefit. Marginal (opportunity) cost and marginal benefit then match the accounting price that can be read from the market, and economic efficiency is assured in the sense that nobody can be made better off without hurting somebody.

It is not difficult, of course, to think of circumstances in which one or more of these conditions do not hold (and this is not news to the economics profession). Much of modern economics has been devoted to exploring how to measure and compare costs and benefits when these conditions break down. For researchers interested in impacts, however, theoretical results are less important than some practical insight into what to do.

Theory instructs, for example, that producers who have some monopoly power in imperfectly competitive markets would restrict output compared to the quantity that would prevail in a competitive market. Consequently, marginal opportunity cost would fall short of marginal benefit even if all of the other assumptions held, and the market price would overestimate marginal cost by an amount that is related to the price elasticity of demand.

Markets can fail if production or consumption produces a positive or negative externality (i.e., if either provides extra benefit or imposes extra cost on some other actor in the economy). Externalities occur, for example, when a producer who pollutes the air or water or contributes to GHG-induced warming does not pay the cost that this pollution imposes on others. Theory tells us that the private opportunity cost that might be reflected in the market price of even a competitive market would then underestimate or overestimate the true social (opportunity) cost, depending on whether the externality were positive or negative. By how much? There is the rub.

Goods whose consumption is not exclusive tend to be provided publicly. But how much should be provided? Theory reports that public goods should be provided up to the point at which the sum of marginal benefits across all consumers equals the marginal opportunity cost of provision. How much should be charged for "consuming" such a good? That price could fall to zero if the good is truly nonexclusive. In these cases, people have an incentive not to reveal their true preferences, so there is a tendency for such goods to be underprovided.

Transaction costs can drive a measurable wedge between marginal opportunity cost and marginal benefit, sometimes to the point at which markets fail completely. Can opportunity cost or marginal benefit be measured when markets do not exist? If not, what then? The growing field of nonmarket valuation might then apply (see Section 2.3.3).

Uncertainty causes problems as well. Theory speaks of risk premiums and offers models of how people make decisions under uncertainty. Information can reduce risk and uncertainty, so it has value. Uncertainty can even cause markets to fail. The key is to keep track of who knows what and when they know it. It also is essential to understand why people and institutions find some information credible and other information incredible. These are questions whose answers confound cost accounting and valuation exercises.

The passage of time and the prevalence of asymmetric information raise issues of completeness and comprehensiveness. All of the markets that are necessary to sustain efficiency probably do not exist, particularly if future property rights and future participants are not reflected in the current workings of existing markets.

Finally, economic efficiency says little about equity. Indeed, the second theorem of welfare economics indicates that the aforementioned conditions (plus a few technicalities) are sufficient to guarantee that a market-based equilibrium derived from any initial distribution of economic resources will be efficient in the sense that nobody can be made better off without making somebody else worse off (see Varian, 1992, Section 17.7). This does not mean that the market equilibrium would be equitable. Nor does it mean that economics has nothing to say about equity. It also does not mean that there is no cost associated with inequity (even in economic terms). It does mean, however, that care must be taken to keep track of the distribution of resources and to highlight the possible tradeoff between equity and efficiency.

2.5.1.2. Specifying the Baseline

Each of the foregoing assumptions represents a qualitative dimension along which the baseline of an impact assessment must be defined. A researcher who wants to estimate the costs or benefits of changing conditions must define as fully as possible the socioeconomic, political, institutional, and cultural environments within which the change will be felt. A "first-best" analysis assumes that everything works efficiently in response to changing conditions in the context of all of the right information; results of first-best analyses reflect benchmarks of "best-news" scenarios. Second-best analyses assume that distortions caused by the failure of some or all of these assumptions to hold will diminish the efficiency of the first-best world; they can produce dramatically different answers to cost and valuation questions. Indeed, baselines that are constructed to reflect the global externalities of climate change by definition reflect second-best circumstances.

It may be reasonable to assume that distortions will persist as change occurs over the short run. Making the same assumption over the long run could be a mistake, however. Will information not improve over time? If distortions are costly, they may persist over the long term if the beneficiaries have sufficient power to preserve their advantage. There is no right way to do second-best analysis; it is simply incumbent on the researcher to report precisely what assumptions define the baseline.

2.5.1.3. *Discounting the Future*

The discount rate allows costs and values occurring at different times to be compared by converting future economic values into their equivalent present values. Formally, the present value of some cost C_t that will come due in *t* years is

$$C_t / (1+d)^t ,$$

where d is the appropriate discount rate. The discount rate is non-negative because resources invested today in physical and human capital usually can be transformed into more resources later on. The IPCC and others have focused an enormous amount of attention on the discount rate (for detailed discussions see Arrow *et al.*, 1996; Portnoy and Weyant, 1999; Chapter 1 of this volume). Toth (2000b) provides a review of this and other more recent literature, with particular emphasis on the implications of discounting to issues of intergenerational equity.

2.5.2. **Market Impacts**

Cost and valuation exercises work best when competitive markets exist. Even when markets are distorted, they provide some useful information. This section offers brief insights into how the elements described can be applied in these situations.

2.5.2.1. *Deadweight Loss*

Deadweight loss is a measure of the value of aggregate economic welfare that is lost when marginal social opportunity cost does not equal marginal social benefit. Aggregate economic welfare can be regarded as the sum of the total benefit derived from consuming a specific quantity of a specific good, net of the total opportunity cost of its production. Aggregate welfare is maximized in a competitive market. Deadweight loss therefore can be computed as the difference between economic welfare generated in a distorted market and economic welfare attained at the social optimum of a competitive market. More specifically, it is estimated as the area under a demand curve that reflects marginal social benefits and above a supply curve that reflects marginal social cost between the observed or anticipated outcome and the social optimum—the outcome that would equate marginal social costs and benefits. Moreover, changes in deadweight loss can be deduced by computing the appropriate areas even if the social optimum cannot be identified. In either case, deadweight loss simply is the sum of a change in private benefits, differences between social and private benefits, a change in private costs, and differences between social and private costs.

2.5.2.2. *Preexisting Distortions*

Market-based exercises that evaluate the costs and benefits of change must carefully account for preexisting distortions in markets. In the presence of one distortion, in fact, creation of another might actually improve welfare. Changes may or may not work to reduce preexisting distortions, so they actually can produce benefits that would be missed entirely if analyses were confined to competitive conditions. Goulder and Schneider (1999), for example, have noted that preexisting subsidies to conventional energy industries reduce the costs of climate policies but that preexisting subsidies to alternative energy industries would increase costs. Moreover, they point out that the opportunity costs of research and development (R&D) could be reduced or even reversed if there were an ample supply of R&D providers rather than a scarcity.

2.5.3. **Nonmarket Impacts**

Many impacts involve changes in the direct and/or indirect flows of valued services to society. These services can offer a wide range of valuable attributes, but they frequently go unpriced in the economic sense. Markets simply do not exist for some attributes and some services; contemplating markets for some others (e.g., health services) has been questioned even given extensive competiton for services and products. For others, markets that do exist fall short of being comprehensive or complete in the presence of externalities of production or consumption. In either case (and others), researchers have recognized the need to develop alternative means with which to assess value. More precisely, they have tried to extend the scope of the economic paradigm so that implicit and explicit tradeoffs between development and conservation of unpriced resources can be explored within the structures of standard decision analytic tools such as cost-benefit analysis, cost-effectiveness analysis, and so on. Parikh and Parikh (1997, 1998) provide a primer on valuation with case studies.

To be more specific, economists have built a theory of choice on the basis of the notions of consumer sovereignty and rationality. Economists assume, therefore, that individuals are able to value changes in nonmarket goods and services as easily as they can value changes in marketed goods and services. The only difference between the two cases is that markets provide the researcher with some indirect data with which to assess individuals' values of marketed products. Nevertheless, individuals should be able to tell researchers what they would be willing to pay for changes in nonmarket conditions or willing to accept as compensation for those changes. In fact, willingness to accept (WTA) payment for foregoing a good and willingness to pay (WTP) for a good are the two general yardsticks against which values are judged.

It should be noted that WTA and WTP are seldom the same for most nonmarket goods or services. In fact, WTA and WTP can give wildly different estimates of the value of these services if there are no perfect substitutes (i.e., if it is impossible to fully compensate individuals unit by unit for their loss). When such a substitute does not exist, WTA > WTP. By how much? Cummings *et al.* (1986) report that it is not uncommon for estimated WTA to be more than 10 times larger than estimated WTP. These differences might be derivative of the method of estimation, but they also reflect the fact that WTA and WTP are two different concepts that need not match.

It also should be noted that WTA and WTP have analogs in the market context. Compensated variation (CV) is the extra income that individuals would require to accept an increase in the price of some marketed good; CV is the analog of WTA. Equivalent variation (EV) is the income that individuals would be willing to forego to see the price of some marketed good fall; EV is the analog of WTP. These measures sometimes are used in market-based analysis. It should be no surprise that EV < CV unless the good in question has a perfect substitute.

2.5.3.1. *Direct Methods of Valuation*

Valuation methods usually are divided into two distinct approaches. Direct methods try to judge individuals' value for nonmarketed goods by asking them directly. Contingent valuation methods (CVMs), for example, ask people for their maximum WTP to effect a positive change in their environments or their minimum WTA to endure a negative change. Davis (1963) authored the first paper to report CVM results for environmental goods. Comprehensive accounts of these methods appear in Mitchell and Carson (1989), Hanley and Spash (1993), and Bateman and Willis (1995). This is a controversial method, and current environmental and resource literature continues to contain paper after paper confronting or uncovering problems of consistency, bias, truth-revelation, embedding, and the like. Hanley *et al.* (1997) offer a quick overview of these discussions and a thorough bibliography.

2.5.3.2. *Indirect Methods of Valuation*

Indirect methods of valuation try to judge individuals' value for nonmarketed goods by observing their behavior in related markets. Hedonic pricing methods, for example, assume that a person buys goods for their various attributes. Thus, for example, a house has attributes such as floor area; number of bathrooms; the view it provides; access to schools, hospitals, entertainment, and jobs; and air quality . By estimating the demand for houses with different sets of attributes, we can estimate how much people value air quality. One can thus estimate "pseudo-demand curves" for nonmarketed goods such as air quality. Travel costs are another area in which valuation estimates of the multiple criteria on which utility depends can be finessed out of observable behavior. The hedonic method was first proposed by Lancaster (1966) and Rosen (1974). Tiwari and Parikh (1997) have estimated such a hedonic demand function for housing in Bombay. Mendelsohn *et al.* (2000) brought the hedonic approach to the fore in the global change impacts arena. Braden and Kolstad (1991) and Hanley and Spash (1993) offer thorough reviews of both approaches. Is there a scientific consensus on the state of the science for these methods? Not really. There is, instead, a growing literature that warns of caveats in their application and interpretation (e.g., health services) and/or improves their ability to cope with these caveats. Smith (2000) provides a careful overview of this literature and an assessment of progress over the past 25 years.

2.5.4. ***The Cost of Uncertainty***

This section reviews the primary methods for incorporating uncertainty into analyses of climate impacts. Here we look at how to judge the cost associated with uncertainty. Cost and valuation depend, in general, on the entire distribution of the range of outcomes.

2.5.4.1. *Insurance and the Cost of Uncertainty*

Risk-averse individuals who face uncertainty try to buy insurance to protect themselves from the associated risk (e.g., different incomes next year or over the distant future, depending on the state of nature that actually occurs). How much? Assuming the availability of "actuarially fair" coverage (i.e., coverage available from an insurance provider for which the expected cost of claims over a specified period of time equals the expected income from selling coverage), individuals try to insure themselves fully so that the uncertainty would be eliminated. How? By purchasing an amount of insurance that is equal to the difference between the expected monetary value of all possible outcomes and the certainty-equivalent outcome that insurance would guarantee—the income for which utility equals the *expected utility of all possible outcomes*.

For a risk-averse person, the certainty-equivalent income is less than the expected income, so the difference can be regarded as WTP to avoid risk. In a real sense, therefore, willingly paid insurance premiums represent a measure of the cost of uncertainty. Therefore, they can represent society's WTP for the assurance that nondiversifiable uncertainty would disappear (if that were possible). Thus, this is a precise, utility-based measure of economic cost. The cost of uncertainty would be zero if the objective utility function were risk-neutral; indeed, the WTP to avoid risk is positive only if the marginal utility of economic activity declines as income increases. Moreover, different agents could approach the same uncertain circumstance with different subjective views of the relative likelihoods of each outcome and/or different utility functions. The amount of insurance that they would be willing to purchase would be different in either case. Application of this approach to society therefore must be interpreted as the result of contemplating risk from the perspective of a representative individual. Yohe *et al.* (2000), for example, apply these structures to and offer interpretations for the distributional international impact of Kyoto-style climate policy.

2.5.4.2. *The Value of Information*

A straightforward method of judging the value of information in an uncertain environment has been developed and applied (see Manne and Richels, 1992, for an early and careful description). The idea is simply to compute the expected cost of uncertainty with and without the information and compare the outcomes. For example, it might be that improved information about the range of uncertainty might change the mean and the variance of associated costs. If the researcher were interested only in the resulting change in costs, however, the value of information would simply be the difference between expected cost with and without the new information, and only the mean would matter. If the same researcher wanted to represent the value of information in terms of welfare that displays some degree of risk aversion so that variance also plays a role, however, a comparison of insurance-based estimates of the WTP to avoid uncertainty would be more appropriate.

2.5.4.3. *Uncertainty and Discounting*

Uncertainty about costs and/or values that are incurred or enjoyed over time can be handled in two ways. One method calculates the present value across the full range of possibilities; means and distributions of present values are the result. The second method, reported in Arrow *et al.* (1996), converts outcomes at each point in time into their certainty equivalents and then applies discounting techniques. This approach raises the possibility of including risk aversion into the calculation according to the foregoing definition.

The story is quite different when uncertainty surrounds selection of the discount rate itself. It may not be appropriate, in these sorts of cases, to use a certainty-equivalent discount rate (or an average over the range of possible rates). Weitzman (1998) has noted, in particular, that the "lowest possible" discount rate should be used for discounting the far-distant future. The reason, quite simply, is that the expected value of present value over a range of discount rates is not equal to the present value calculated with an average rate. Moreover, the difference between the two is exaggerated in the distant future. Present values computed with low rates, in fact, can dominate those computed with high rates by orders of magnitude when the future is extended; thus, their contribution to the expected value must be recognized explicitly in the selection of a discount rate.

2.5.5. *Equity and Distribution*

Assessments of the impacts of alternative climate change scenarios require assessments of their impacts on different groups, societies, nations, and even species. Indeed, this report reveals that many sectors and/or regions are at greater risk to climate change than others. This section addresses this need.

2.5.5.1. *Interpersonal Comparisons*

First principles of economic theory offer two approaches for comparing situations in which different people are affected differently. In the first—the utilitarian approach attributed to Bentham (1822) and expanded by Mills (1861)—a situation in which the sum of all individual utilities is larger is preferred. Because Bentham's view of utility reflected "pleasure" and "pain," this approach embraces the "greatest happiness principle." Many objections have been raised against it, however, primarily because the whole notion of interpersonal comparisons of utility is problematic. Indeed, Arrow (1951 and 1963) objected strenuously in arguing that "interpersonal comparisons in the measurement of utilities has *no meaning* and, in fact, there is no meaning relevant to welfare comparisons in the measurability of individual utility." For example, it is impossible to compare the pleasure that a person receives from listening to a concert with what another gets from watching a dance. Second, maximizing the sum total of utility, if it were possible, would require that the marginal utilities of all individuals be equal. But this would say nothing about the level of utility for each individual. They could be quite different, so the utilitarian rule is insensitive to distributional issues except in the special case in which all individuals have identical utility functions.

These difficulties led to the development of a second approach—the welfarist approach, in which a social welfare function of individual utilities is postulated. Utilitarianism is thus a special case in which the social welfare function is simply the sum of individual utilities. There are other options, of course. The Gandhian principle, for example, can support a function that judges every possible action on the basis of its impact on the poorest of the poor.

It also is possible to compare two situations without defining an explicit social welfare function and without making interpersonal comparisons of individual utilities. The Pareto principle offers one method, by which one judges any situation better than another if at least one person is better off and no one else is worse off. A partial social ordering with which unambiguous comparisons can be made in some (but not all) cases can be constructed from the Pareto principle if cardinal utilities can be added across individuals, if society accepts the principle of anonymity (i.e., only the distribution matters, not which particular person is in a particular place), and if there is an aversion to regressive transfers (i.e., transfers from the poor to the rich). To see how, consider two situations, X and Y. Assume that there are n individuals ordered from poorest to richest. Let them have incomes (or utilities) $\{X_1, ..., X_n\}$ and $\{Y_1, ..., Y_n\}$ in X and Y, respectively. X can be deemed preferable to Y if $X_1 \geq Y_1$, $[X_1 + X_2] \geq [Y_1 + Y_2]$, and so on through $[X_1 +...+ X_n] \geq [Y_1 +...+ Y_n]$, with at least one strict inequality holding. Note that showing that X is not preferred to Y is not sufficient to show that Y is preferred to X.

Rothschild and Stiglitz (1973) took these notions further by showing three alternative but equivalent ways of comparing distributions X and Y. They concluded that X would be preferred to Y if all of the following obtain:

- The Lorenz curve for X were inside the Lorenz curve for Y.
- All those who valued equality preferred X to Y.
- Y could be obtained from X by transfers from the poor to the rich.

Note, in passing, that Lorenz curves simply plot the percentage of income received by various percentiles of populations when they are ordered from least to greatest. Rothschild and Stiglitz (1973) also point out, however, that these measures apply only to a one-good economy. This requirement is equivalent to assuming that income is desired by all individuals and there are no externalities; the implications of more than one good are "substantial."

None of these measures speaks to estimating the cost of inequity when comparisons can be made. But just as insurance can be used as a utility-based measure of the cost of uncertainty, similarly constructed estimates that are based on social welfare functions that display aversion to inequality can be constructed. Insurance premiums computed in these cases simply represent a measure of what society would willingly pay to eliminate inequality. Such an approach assumes the possibility of defining an international social welfare function. Let us now look at the difficulties involved in defining it.

2.5.5.2. *Comparisons Across Nations*

Comparisons of interpersonal well-being across nations have been the focus of increasing attention over the past few years (see, e.g., Tol, 1999a,b), but it is clear that these comparisons involve more than one element. The conventional approach to making such comparisons is to use purchasing power parity (PPP) to adjust the calculation of gross domestic product (GDP). The technique is flawed, however, in many ways. First, GDP is now widely recognized to be a poor indicator of well-being (e.g., UNDP, 1990). This recognition has inspired many attempts to create other measures, such as the physical quality of life index (PQLI) (Morris, 1979) and various versions of the human development index (HDI) by the United Nations Development Programme (UNDP). However, many researchers, including Srinivasan (1994), have criticized the HDI for theoretical inadequacies. Nevertheless, the major point that GDP misses too much continues to be emphasized exclusively. Calculations of the Index of Sustainable Economic Welfare (ISEW) by Daly and Cobb (1994) have shown, for example, that the ISEW for the United States has fallen since 1970 even though GDP has grown substantially.

In addition, real-world comparisons must account for many commodities, services, and attributes. This causes enormous index number problems in computing conversion factors such as the PPP. Indeed, one country's income can be higher or lower than another depending on which country is used as the base for the PPP index.

Third, different societies, cultures, and nations have different social structures, mores, and public institutions. The public goods, services, and safety net provisions of each are different. Moreover, activity outside the marketplace can differ substantially. With industrial development, for example, the clan seems to change to a joint family structure, then to a nuclear family, and perhaps to temporary nuclear families in postindustrial societies. More to the point, the nature of social and human capital and the scope of the marketplace are very different from place to place, depending on the stage of development. And if welfare involves having, being, doing, relating, and caring, a more complex measure of welfare is required to accommodate the multiple stresses of climate change.

Fourth, Sen (1985) suggests that equality in persons' "capabilities" that are determined by income and access to public goods, services, social capital, and institutions should be a global objective. Each of these determinants clearly varies from nation to nation.

Fifth, the principle of "anonymity" that is used in welfare comparisons is highly suspect. Deliberations of climate impacts and climate policy clearly should keep track of who is affected and where (within and across countries) they live.

2.5.5.3. *Ensuring Equity*

All of the complications outlined in Section 2.5.5.2 lead to a sad conclusion: Economics may be able to highlight a large menu of distributional issues that must be examined, but it has trouble providing broad answers to measuring and accounting for inequity, particularly across nations. Recourse to ethical principles clearly is in order.

2.5.6. *Alternative Metrics for Measuring Costs*

Application and extension of the economic paradigm certainly focuses attention on cost measures that are denominated in currency, but practitioners have been criticized on the grounds that these measures inadequately recognize nonmarket costs. Schneider *et al.* (2000), for example, have listed five numeraires or metrics with which the costs of climate change might be captured. Their list includes monetary losses, loss of life, changes in quality of life (including a need to migrate, conflict over resources, cultural diversity, loss of cultural heritage sites, etc.), species or biodiversity loss, and distributional equity.

Chapter 19 recognizes the content of these diverse numeraires in exploring magnitudes and/or rates of climate change that might be dangerous according to three lines of evidence: threatened systems, distributions of impacts, and aggregate impacts. The implications of the fourth line of evidence, large-scale discontinuous events, are then traced along these three dimensions.

When all is said and done, however, costs denominated in one numeraire must be weighed, at least subjectively, with costs denominated in another—and there are no objective quantitative methods with which to do so. A survey conducted by Nordhaus (1994a), however, offered some insight into 15 researchers' subjective views of the relative importance of several different measures along three different "what if" scenarios. Table 2-1 displays some of the results in terms of anticipated cost denominated in lost world GDP, the likelihood of high-consequence impacts, the distribution of costs across the global population, and the proportion of costs that would be captured by national income accounts. The survey results shows wide disagreement across the first three metrics; this disagreement generally can be explained in terms of a dichotomy of views between mainstream economists and natural scientists. Nonetheless, Nordhaus (1994a) reports that a majority of respondents held the view that a high proportion of costs would be captured in national accounts. It would seem, therefore, that natural scientists think that mainstream economists not only underestimate the severity of nonmarket impacts but also that the implications of those impacts into the monetized economy do not follow.

Multi-attribute approaches also could be applied in climate impact analysis. They have not yet found their way into the literature, however, except to the degree to which they are captured in indirect methods outlined above. Chapter 1 also notes that cultural theory can serve as a valuation framework.

2.6. Characterizing Uncertainty and "Levels of Confidence" in Climate Assessment

Uncertainty—or, more generally, debate about the level of certainty required to reach a "definitive" conclusion—is a perennial issue in science. Difficulties in explaining uncertainty have become increasingly salient as society seeks policy advice to deal with global environmental change. How can science be useful when evidence is incomplete or ambiguous, the subjective judgments of experts in the scientific and popular literature differ, and policymakers seek guidance and justification for courses of action that could cause—or prevent—significant environmental and societal changes? How can scientists improve

***Table 2-1**: Subjective expert opinion on climate change (Nordhaus, 1994a).*

Cost Metric	Scenario A[a]	Scenario B[b]	Scenario C[c]
a) Loss in gross world product[d]			
– Mean	1.9	4.1	5.5
– Median	3.6	6.7	10.4
– High	21.0	35.0	62.0
– Low	0.0	0.0	0.8
b) Probability of high-consequence event[e]			
– Mean	0.5	3.0	5.0
– Median	4.8	12.1	17.5
– High	30.0	75.0	95.0
– Low	0.0	0.2	0.3
c) Top to bottom ratio of impacts[f]			
– Mean	4.2		
– Median	3.5		
– High	10.0		
– Low	1.0		
d) Percentage of total in national accounts			
– Mean	62.4	66.6	65.6
– Median	62.5	70.0	80.0

[a] Scenario A postulated 3°C warming by 2090.
[b] Scenario B postulated scenario A continuing to produce 6°C warming by 2175.
[c] Scenario C postulated 6°C warming by 2090.
[d] Percentage of global world product lost as a result of climate change.
[e] Likelihood of a high-consequence event (a loss of 25% of gross world product, comparable to the Great Depression).
[f] Proportion of loss felt by the poorest quintile of income distribution relative to the loss felt by the richest quintile; a value of 1 signifies an equal distribution of burden.

their characterization of uncertainties so that areas of slight disagreement do not become equated with paradigmatic disputes, and how can individual subjective judgments be aggregated into group positions? In short, how can the full spectrum of the scientific content of public policy debates be fairly and openly assessed?

The term "uncertainty" implies anything from confidence just short of certainty to informed guesses or speculation. Lack of information obviously results in uncertainty; often, however, disagreement about what is known or even knowable is a source of uncertainty. Some categories of uncertainty are amenable to quantification, whereas other kinds cannot be expressed sensibly in terms of probabilities (see Schneider *et al.*, 1998, for a survey of literature on characterizations of uncertainty). Uncertainties arise from factors such as lack of knowledge of basic scientific relationships, linguistic imprecision, statistical variation, measurement error, variability, approximation, and subjective judgment (see Box 2-1). These problems are compounded by the global scale of climate change, but local scales of impacts, long time lags between forcing and response, low-frequency variability with characteristic times that are greater than the length of most instrumental records, and the impossibility of before-the-fact experimental controls also come into play. Moreover, it is important to recognize that even good data and thoughtful analysis may be insufficient to dispel some aspects of uncertainty associated with the different standards of evidence (Morgan, 1998; Casman *et al.*, 1999).

This section considers methods to address such questions: first by briefly examining treatments of uncertainties in past IPCC assessments, next by reviewing recommendations from a guidance paper on uncertainties (Moss and Schneider, 2000) prepared for the TAR, and third by briefly assessing the state of the science concerning the debate over the quality of human judgments (subjective confidence) when empirical evidence is insufficient to form clear "objective" statements of the likelihood that certain events will occur.

Box 2-1. Examples of Sources of Uncertainty

Problems with Data

1) Missing components or errors in the data
2) "Noise" in data associated with biased or incomplete observations
3) Random sampling error and biases (nonrepresentativeness) in a sample

Problems with Models

4) Known processes but unknown functional relationships or errors in structure of model
5) Known structure but unknown or erroneous values of some important parameters
6) Known historical data and model structure but reasons to believe parameters or model structure will change over time
7) Uncertainty regarding predictability (e.g., chaotic or stochastic behavior) of system or effect
8) Uncertainties introduced by approximation techniques used to solve a set of equations that characterize the model

Other Sources of Uncertainty

9) Ambiguously defined concepts and terminology
10) Inappropriate spatial/temporal units
11) Inappropriateness of/lack of confidence in underlying assumptions
12) Uncertainty resulting from projections of human behavior (e.g., future consumption patterns or technological change), as distinct from uncertainty resulting from "natural" sources (e.g., climate sensitivity, chaos)

2.6.1. *Treatments of Uncertainties in Previous IPCC Assessments*

The IPCC function is to assess the state of our understanding and to judge the confidence with which we can make projections of climate change and its impacts. These tentative projections will aid policymakers in deciding on actions to mitigate or adapt to anthropogenic climate change, which will need to be re-assessed on a regular basis. It is recognized that many remaining uncertainties need to be reduced in each of (many) disciplines, which is why IPCC projections and scenarios are often expressed with upper and lower limits. These ranges are based on the collective judgment of the IPCC authors and the reviewers of each chapter, but it may be appropriate in the future to draw on formal methods from the discipline of decision analysis to achieve more consistency in setting criteria for high and low range limits (McBean *et al.*, 1996; see Raiffa, 1968, for an introduction to decision analysis).

Although the SAR on impacts, adaptation, and mitigation (IPCC, 1996b) explicitly links potentially serious climate change with mitigation and adaptation assessment in its Technical Summary, the body of the report is restricted mostly to describing sensitivity and vulnerability assessments (see also Carter *et al.*, 1994). Although this methodology is appropriate for testing sensitivity and vulnerability of systems, it is poorly suited for planning or policy purposes. IAMs available to SAR authors (e.g., Weyant *et al.*, 1996) generate outcomes that are plausible but typically contain no information on the likelihood of outcomes or much information on confidence in estimates of outcomes, how each result fits into broader ranges of uncertainty, or what the ranges of uncertainty may be for each outcome (see Chapter 1 and Section 2.4 for further discussions of integrated assessment issues). However, several studies since the SAR do use probability distributions (e.g., Morgan and Dowlatabadi, 1996, and citations in Schneider, 1997).

IPCC Working Group I (WGI) in its contribution to the SAR (IPCC, 1996a) uses two different methods or techniques to estimate climate change: scenarios and projections. A scenario is a description of a plausible future without estimation of its likelihood (e.g. the individual IS92a-f emission scenarios or climate scenarios generated by GCMs in which a single emission path is used). Scenarios may contain several sources of uncertainty but generally do not acknowledge them explicitly.

Careful reading of the SAR WGI Technical Summary (IPCC, 1996a) reveals that the term projection is used in two senses:

1) A single trajectory over time produced from one or more scenarios (e.g., projected global temperature using the IS92a emissions scenario with a climate sensitivity of 2.5°C)
2) A range of projections expressed at a particular time in the future, incorporating one or more sources of uncertainty (e.g., projected global warming of 0.8–3.5°C by 2100, based on IS92a-f emission scenarios and a climate sensitivity of 1.5–4.5°C at $2xCO_2$).

Projections are used instead of *predictions* to emphasize that they do not represent attempts to forecast the most likely evolution of climate in the future, only *possible* evolutions (IPCC, 1996a, Section F.1). In the SAR, *projection* and *scenario* are used to describe possible future states, with projections used mainly in terms of climate change and sea-level rise. This usage defines climate projection as a *single* trajectory of a subset of scenarios. When used as input into impact assessments, the same climate projections commonly are referred to as climate scenarios.

Projected *ranges* are constructed from two or more scenarios in which one or more sources of uncertainty may be acknowledged. Examples include projections of atmospheric CO_2 derived from the IS92a-f emission scenarios (IPCC, 1996a), global temperature ranges (IPCC, 1996a), and regional temperature

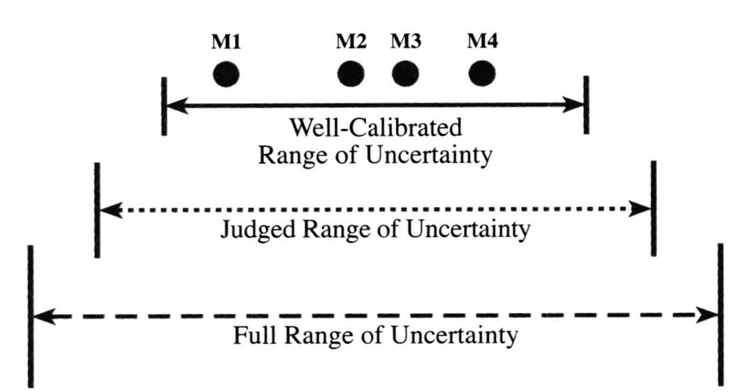

Figure 2-1: Schematic depiction of the relationship between "well-calibrated" scenarios, the wider range of "judged" uncertainty that might be elicited through decision analytic techniques, and the "full" range of uncertainty, which is drawn wider to represent overconfidence in human judgments. M1 to M4 represent scenarios produced by four models (e.g., globally averaged temperature increases from an equilibrium response to doubled CO_2 concentrations). This lies within a "full" range of uncertainty that is not fully identified, much less directly quantified by existing theoretical or empirical evidence (modified from Jones, 2000).

ranges (CSIRO, 1996). A range of projections will always be more likely to encompass what actually will transpire than a single scenario. Although projected ranges are more likely to occur than single scenarios, they are not full-fledged forecasts because they incorporate only part of the total uncertainty space. The relationship between scenarios and projected ranges as treated in the SAR is shown schematically in Figure 2-1.

A projected range is a quantifiable range of uncertainty situated within a population of possible futures that cannot be fully identified (termed "knowable" and "unknowable" uncertainties by Morgan and Henrion, 1990). The limits of this total range of uncertainty are unknown but may be estimated subjectively (e.g., Morgan and Keith, 1995). Given the finding in the cognitive psychology literature that experts define subjective probability distributions too narrowly because of overconfidence (see Section 2.6.5.3), the inner range represents the "well-calibrated" range of uncertainty. Thus, the wider range of uncertainty represents a "judged" range of uncertainty, based on expert judgments—which may not encompass the full range of uncertainty given the possibility of cognitive biases such as overconfidence. Although the general point remains that there is always a much wider uncertainty range than the envelope developed by sets of existing model runs, it also is true that there is no distinct line between "knowable" and "unknowable" uncertainties; instead, it is a continuum. The actual situation depends on how well our knowledge (and lack thereof) has been integrated into assessment models. Moreover, new information—particularly empirical data, if judged reliable and comprehensive—eventually may narrow the range of uncertainty to well inside the well-calibrated range by falsifying certain outlier values.

If the full range of uncertainty in Figure 2-1 were known, the probability of a particular outcome could be expressed as a forecast (provided we can state the probability). Although there are significant sources of uncertainty that cannot yet be quantified, decision analytic elicitation procedures (Section 2.4) can estimate the full range of uncertainties and conditional probabilities (see Section 2.5.5 for an assessment of the state of the science concerning human judgment). Conditional probabilities may be calculated within a projected range even though the probability of the range itself remains unknown.

Moss and Schneider (1997) document several cases in which the SAR authors in each of the three Working Groups use ranges to describe uncertain outcomes but unfortunately had no consistent criteria for assigning probabilities to range limits or for identifying outlier outcomes (those occurring beyond the well-calibrated range limits). Moreover, there was no consistent use of terms to characterize levels of confidence in particular outcomes or common methods for aggregating many individual judgments into a single collective assessment. Recognition of this shortcoming of the SAR led to preparation of a guidance paper on uncertainties (Moss and Schneider, 2000) for use by all three TAR Working Groups and which has been widely reviewed and debated.

Attempts to achieve more consistency in establishing end points of ranges and outlier values (or the distribution of subjective probabilities within and beyond the range) have not received much attention. Despite the difficulty of assigning a distribution of probabilities for uncertain outcomes or processes, the scientific complexity of the climate change issue and the need for information that is useful for policy formulation requires researchers and policymakers to work together toward improved management of uncertainties. A common basis for characterizing sources of uncertainties is one step; Box 2-1 represents an attempt in the uncertainties guidance paper (Moss and Schneider, 2000) to provide such a common basis.

In this situation, the research community must bear in mind that users of IPCC reports often assume for themselves what they think the authors believe to be the distribution of probabilities when the authors do not specify it themselves. The decision analytic literature (e.g., Morgan and Henrion, 1990) often suggests that it is preferable for scientists debating the specifics of a topic to provide their best estimates of probability distributions and possible outliers, based on their assessment of the literature, than to have users make their own guesses. This information, along with an appraisal of the limitations of the models, would make the ranges more meaningful to other scientists, the policy community, and the public.

2.6.2. "Objective" and "Subjective" Probabilities are not Always Explicitly Distinguished

Some scientists have expressed concern that scientific investigation requires a long sequence of observational records, replicable trials, or model runs (e.g., Monte Carlo simulations) so the results can be specified by a formal statistical characterization of the frequency and frequency distribution of outcomes being assessed. In statistical terms, "objective" science means attempting to verify any hypothesis through a series of experiments and recording the frequency with which that particular outcome occurs. The idea of a limitless set of identical and independent trials that is "objectively out there" is a heuristic device that we use to help us rigorously quantify uncertainty by using frequentist statistics. Although there may be a large number of trials in some cases, however, this is not the same as a "limitless" number, and these trials rarely are truly identical or independent.

Most interesting complex systems cannot possibly be put to every conceivable test to find the frequency of occurrence of some socially or environmentally salient event. The popular philosophical view of "objective science" as a series of "falsifications" breaks down when it confronts systems that cannot be fully tested. For example, because climate change forecasts are not empirically determinable (except by "performing the experiment" on the real Earth—Schneider, 1997), scientists must rely on "surrogate" experiments, such as computer simulations of the Earth undergoing volcanic eruptions or paleoclimatic changes. As a result of these surrogate experiments and many additional tests of the reliability of subcomponents of such models, scientists attain confidence to varying degrees about the likelihood of various outcomes (e.g., they might assign with high confidence a low probability to the occurrence of extreme climate outcomes such as a "runaway greenhouse effect"). These confidence levels are not frequentist statistics but "subjective probabilities" that represent degrees of belief that are based on a combination of objective and subjective subcomponents of the total system. Because subjective characterization of the likelihood of many potentially important climatic events—especially those that might be characterized by some people as "dangerous"—is unavoidable, "Bayesian" or "subjective" characterization of probability will be more appropriate.

Bayesian assessments of probability distributions would lead to the following interpretation of probability statements: The probability of an event is the degree of belief that exists among lead authors and reviewers that the event will occur, given the observations, modeling results, and theory currently available, all of which contribute to estimation of a "prior" probability for the occurrence of an outcome. As new data or theories become available, revised estimates of the subjective probability of the occurrence of that event—so-called "posterior probability"—can be made, perhaps via the formalism of Bayes theorem (see, e.g., Edwards, 1992, for a philosophical basis for Bayesian methods; for applications of Bayesian methods, see, e.g., Howard *et al.*, 1972; Anderson, 1998; Tol and de Vos, 1998; Malakoff, 1999).

2.6.3. Making Estimates

2.6.3.1. Identifying Extreme Values, Ranges, and Thresholds

It is worth noting that by failing to provide an estimate of the full range of outcomes (i.e., not specifying outliers that include rapid nonlinear events), authors of previous assessments were not conveying to potential users a representation of the full range of uncertainty associated with the estimate. This has important implications with regard to the extent to which the report accurately conveyed to policymakers potential benefits or risks that may exist, even if at a low or unknown probability (see Figure 2-1). If it were necessary to truncate the range, it should have been clearly explained what the provided range includes and/or excludes. Furthermore, the authors might have specified how likely it is that the answer could lie outside the truncated distribution.

Pittock and Jones (1999) recommend construction of thresholds that can be linked to projected ranges of climate change. Such thresholds can account for biophysical and/or socioeconomic criteria in the initial stages of an assessment but must be expressed in climatic terms (e.g., above a certain temperature, rainfall frequency, water balance, or combination of several factors). Further analysis compares these thresholds with projected regional climate change. Similar approaches are contained in concepts of tolerable climate change (see Hulme and Brown, 1998).

2.6.3.2. Valuation Issues

Any comprehensive attempt to evaluate the societal value of climate change should include, in addition to the usual monetary value of items or services traded in markets, measures of valued items or services that are not easily marketed. Schneider *et al*. (2000) refer to this costing problem in vulnerability analysis as "The Five Numeraires": monetary loss, loss of human life, reductions in quality of life (including forced migration, conflicts over environmentally dependent resources, loss of cultural diversity, loss of cultural heritage sites, etc.), loss of species/biodiversity, and increasing inequity in the distribution of material well-being. There is little agreement on how to place a monetary value on the nonmarket impacts of climate change, yet such valuation is essential to several analytic techniques to assess the efficiency or cost-effectiveness of alternative climate policy proposals (see Section 2.5.6).

One such technique for valuation is to survey expert opinion on subjective assessment of probability distributions of climate damage estimates (see Nordhaus, 1994a; Morgan and Keith, 1995; Titus and Narayanan, 1996, for examples of decision analytic elicitations of climate effects and impacts; see Morgan and Dowlatabadi, 1996, for examples of how such elicited subjective probability distributions can be incorporated into IAMs that examine "optimal" policies). An alternative valuation framework is to use "cultural theory" (Douglas and Wildavsky, 1982) to identify different value perspectives in designing policy strategies (see van Asselt and Rotmans, 1995, for an application to population growth). With this technique, subjective judgment about uncertainties may be described from the viewpoints of different cultural perspectives. Preferred policy options depend on the perspective adopted. Real policy choice, of course, depends on the logic and consistency of formulating a basis for policy choices (i.e., the role of decision analysis tools) and on the values of decisionmakers at all levels. More formal and explicit incorporation of uncertainties into decision analysis is the emphasis here.

2.6.4. Aggregation and the Cascade of Uncertainty

A single aggregated damage function or a "best guess" climate sensitivity estimate is a very restricted representation of the wide range of beliefs available in the literature or among lead authors about climate sensitivity or climate damages. If a causal chain includes several different processes, the aggregate distribution might have very different characteristics than the various distributions that constitute the links of the chain of causality (see Jones, 2000). Thus, poorly managed projected ranges in impact assessment may inadvertently propagate uncertainty. The process whereby uncertainty accumulates throughout the process of climate change prediction and impact assessment has been described as a "cascade of uncertainty" (Schneider, 1983) or the "uncertainty explosion" (Henderson-Sellers, 1993). The cascade of uncertainty implied by coupling the separate probability distributions for emissions and biogeochemical cycle calculations to arrive at concentrations needed to calculate radiative forcing, climate sensitivity, climate impacts, and valuation of such impacts into climate damage functions has yet to be produced in the literature (see Schneider, 1997, Table 2). When the upper and lower limits of projected ranges of uncertainty are applied to impact models, the range of possible impacts commonly becomes too large for practical application of adaptation options (Pittock and Jones, 1999). This technique is less explicitly applied in assessments where two or more scenarios (e.g., M1 to M4 in Figure 2-1) are used and the results expressed as a range of outcomes. If an assessment is continued through to economic and social outcomes, even larger ranges of uncertainty can be accumulated (see Figure 2-2).

Because of the lack of consistent guidance on the treatment of uncertainties, diversity of subject areas, methods, and stage of development of the many fields of research to be assessed in the SAR, it was not possible to agree on a single set of terms to describe the confidence that should be associated with the many outcomes and/or processes that had been assessed. Thus, the uncertainties guidance paper (see also Box 1-1) suggests that the TAR authors agree on two alternative sets of terms from which writing teams can select (see Figures 2 and 3 in Moss and Schneider, 2000). As noted in the decision analysis literature (e.g., Morgan and Henrion, 1990), *it is important to attach a quantitative range to each verbal characterization* to assure that different users of the same language mean the same degree of confidence.

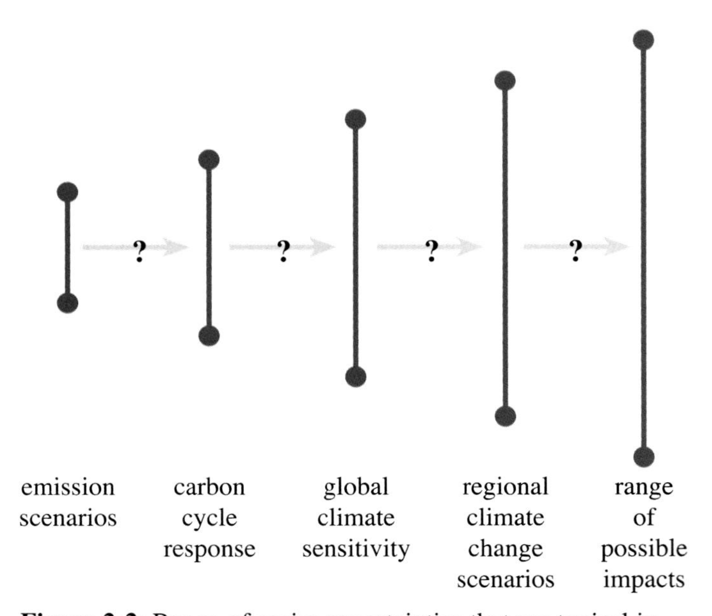

Figure 2-2: Range of major uncertainties that are typical in impact assessments, showing the "uncertainty explosion" as these ranges are multiplied to encompass a comprehensive range of future consequences, including physical, economic, social, and political impacts and policy responses (modified after Jones, 2000, and "cascading pyramid of uncertainties" in Schneider, 1983).

2.6.5. *Debate over the Quality of Human Judgment*

2.6.5.1. *Deficiencies in Human Judgment*

At some level, human judgment is an unavoidable element of all human decisions. The question, then, naturally arises: How good is human judgment? Psychological studies of human judgment provide evidence for shortcomings and systematic biases in human decisionmaking. Furthermore, not only do people—including experts—suffer various forms of myopia; they also often are oblivious of the fact. Indeed, statistical linear models summarizing the relationship between a set of predictor variables and a predicted outcome often (repeatedly) perform better than intuitive expert judgments (or subjective expert opinions). Burgeoning empirical evidence suggests that humans, including experts, can be inept at making judgments, particularly under conditions of high uncertainty.

Since the early 1970s, psychologists repeatedly have demonstrated human judgmental error and linked these errors to the operational nature of mental processes. The idea, spelled out in Kahneman *et al.* (1982), is that, because of limited mental processing capacity, humans rely on strategies of simplification, or mental heuristics, to reduce the complexity of judgment tasks. Although this strategy facilitates decisionmaking, these procedures are vulnerable to systematic error and bias.

2.6.5.2. *Violation of Probability Laws*

In a classic series of publications, Tversky and Kahneman (1974, 1983) and Kahneman and Tversky (1979, 1996) claim that human judgment under uncertainty violates normative rules of probability theory. For example, Tversky and Kahneman (1983) invoke the "judgment by a representativeness" heuristic to explain evidence for the *conjunction fallacy,* whereby a conjunction of events is judged to be more likely than one of its constituents. This is a violation of a perfectly simple principle of probability logic: If A includes B, the probability of B cannot exceed A. Nevertheless, respondents consistently give a higher likelihood to the possibility of a subset or joint event than to the whole set, thereby violating the conjunction rule. Typically, respondents judge likelihood by representativeness (or stereotypes) and thus fail to integrate statistically relevant factors.

However, Gigerenzer (1994, 1996) argues that people are naturally adapted to reasoning with probabilities in the form of frequencies and that the conjunction fallacy "disappears" if reasoning is in the form of frequencies. Several studies report that violations of the conjunction rule are rare if respondents are asked to consider the relative frequency of events rather than the probability of a single event.

Kahneman and Tversky (1996) disagree and argue that the frequency format provides respondents with a powerful cue to the relation of inclusion between sets that are explicitly compared or evaluated in immediate succession. When the structure of the conjunction is made more apparent, respondents who appreciate the constraint supplied by the rule will be less likely to violate it.

Kahneman and Lovallo (1993) argue that people have a strong tendency to regard problems as unique although they would be viewed more advantageously as instances of a broader class. People pay particular attention to the distinguishing features of a particular case and reject analogies to other instances of the same general type as crudely superficial and unappealing. Consequently, they fall prey to fallacies of planning by anchoring their estimates on present values or extrapolations of current trends. Despite differing causal theories, both approaches find evidence for poor judgment under uncertainty or, alternatively, evidence that people are better off not attempting to assess probabilities for single events.

Nonetheless, public understanding of likelihood seems to be improved by adoption of frequentist formats. Several studies have shown that experts have great difficulty reasoning with subjective probabilities for unique or single events. However, respondents apparently are much more successful when the same problems are presented with frequencies rather than probabilities. Although experts have difficulties with the probability version—most give wrong answers—most undergraduates readily provide the correct answer to similar problems constructed with frequencies.

Psychological research suggests that measures of risk that are communicated in terms of frequencies rather than probabilities will be more readily understood and rationally responded to, although IAMs need to translate these frequencies into probability distributions (e.g., Morgan and Dowlatabadi, 1996) to portray the wide range of outcomes that currently reflect estimates in the literature and by most IPCC authors.

2.6.5.3. *Overconfidence*

Overconfidence is another cognitive illusion that has been reported to plague experts' judgments. In the 1970s and 1980s, a considerable amount of evidence was amassed for the view that people suffer from an overconfidence bias. The common finding is that respondents are correct less often than their confidence assessments imply.

However, "ecological" theorists (*cf.* McClelland and Bolger, 1994) claim that overconfidence is an artifact of artificial experimental tasks and nonrepresentative sampling of stimulus materials. Gigerenzer *et al.* (1991) and Juslin (1994) claim that individuals are well adapted to their environments and do not make biased judgments. Overconfidence is observed because the typical general knowledge quiz used in most experiments contains a disproportionate number of misleading items. These authors have found that when knowledge items are randomly sampled, the overconfidence phenomenon disappears. Juslin *et al.* (2000) report a meta-analysis comparing 35 studies in which items were randomly selected from a defined domain with 95 studies in which items were selected by experimenters.

Although overconfidence was evident for selected items, it was close to zero for randomly sampled items—which suggests that overconfidence is not simply a ubiquitous cognitive bias. This analysis suggests that the appearance of overconfidence may be an illusion created by research, not a cognitive failure by respondents.

Furthermore, in cases of judgments of repeated events (weather forecasters, horse race bookmakers, tournament bridge players), experts make well-calibrated forecasts. In these cases, respondents might be identifying relative frequencies for sets of similar events rather than judging the likelihood of individual events. If we compare studies of the calibration of probability assessments concerning individual events (e.g., Wright and Ayton, 1992) with those in which subjective assessments have been made for repetitive predictions of events (Murphy and Winkler, 1984), we observe that relatively poor calibration has been observed in the former, whereas relatively good calibration has been observed in the latter.

It might be concluded that a frequentist rather than a Bayesian approach should be adopted when attempting to elicit judgment. However, there are occasions when there will be events for which no obvious reference class exists and one will be unable to assess likelihood by adopting the frequentist approach. This particularly applies to novel situations for which there is no actuarial history. One might well be able to account for the (no doubt varying) subjective probabilities offered by a sample of people by identifying mental heuristics. However, note that, without a reference class, we have no means of evaluating the validity of any judgments that might be offered. Consequently, any probability given to a unique event remains somewhat ambiguous.

2.6.6. *Building Experience with Subjective Methods In a "Science for Policy" Assessment*

Although one might be tempted to infer from the foregoing arguments that judgments of likelihood should be considered only with caution, for some decision analytic frameworks that often appear in the climate policy literature (.g., cost-benefit analysis and IAMs), there often are few viable alternatives. However, as noted in the decision analysis frameworks guidance paper (Toth, 2000a; see also Section 2.4), several alternative decisional analytic methods are less dependent on subjective probability distributions; virtually all frameworks do require subjective judgments, however. Although physical properties such as weight, length, and illumination have objective methods for their measurement, there are no objective means for assessing in advance the probability of such things as the value future societies will put on now-endangered species or the circulation collapse of the North Atlantic Ocean from anticipated anthropogenic emissions. Even a highly developed understanding of probability theory would be of little avail because no empirical data set exists, and the underlying science is not fully understood. Some authors have argued that under these circumstances, for any practical application one ought to abandon any attempt to produce quantitative forecasts and instead use more qualitative techniques such as scenario planning (e.g., Schoemaker, 1991; van der Heijden, 1998) or argumentation (Fox, 1994). On the other hand, others—though noting the cognitive difficulties with estimation of unique events—have argued that quantitative estimations are essential in environmental policy analyses that use formal and explicit methods (e.g., Morgan and Henrion, 1990).

Given its potential utility in applied and conservation ecology, it seems surprising that Bayesian analysis is relatively uncommon. However, logical and theoretical virtue is not sufficient to encourage its use by managers and scientists. The spread of a new idea or practice is an example of cultural evolution (in this case, within the scientific community). It is best understood as a social and psychological phenomenon (Anderson, 1998).

Helping to achieve such penetration of awareness of uncertainty analyses will be a multi-step process that includes "1) consistent methods for producing verbal summaries from quantitative data, 2) translation of single-event probabilities into frequencies with careful definition of reference classes, 3) attention to different cognitive interpretations of probability concepts, and 4) conventions for graphic displays" (Anderson, 1998). The latter also is advocated in the uncertainties guidance paper (Moss and Schneider, 2000), and an example is provided in Chapter 7 (Figure 7-2).

Although all arguments in the literature agree that it is essential to represent uncertainties in climatic assessments, analysts disagree about the preferred approach. Some simply believe that until empirical information becomes available, quantitative estimates of uncertain outcomes should be avoided because "science" is based on empirical testing, not subjective judgments. It certainly is true that "science" itself strives for "objective" empirical information to test, or "falsify," theory and models (caveats in Section 2.5.2 about frequentism as a heuristic notwithstanding). At the same time, "science for policy" must be recognized as a different enterprise than "science" itself. Science for policy (e.g., Ravetz, 1986) involves being responsive to policymakers' needs for expert judgment at a particular time, given information currently available, even if those judgments involve a considerable degree of subjectivity. The methods outlined above and in Moss and Schneider (2000) are designed to make such subjectivity more consistently expressed (linked to quantitative distributions when possible, as needed in most decision analytic frameworks) across the TAR and more explicitly stated so that well-established and highly subjective judgments are less likely to get confounded in media accounts or policy debates. The key point is that *authors should explicitly state their approach in each case*. Transparency is the key to accessible assessments.

2.7. Decision Analytic Methods and Frameworks

This section presents basic principles of decision analytic frameworks that have been or could be used in assessing adaptation decisions in sectors and regions. Thus, it provides a common base for decision analysis-related discussions in sectoral and regional chapters of this report.

2.7.1. *Decision Analysis to Support Adaptive Decisions—Introduction to Frameworks and Principles*

Decisionmakers who are responsible for climate-sensitive economic sectors (e.g., forestry, agriculture, health care, water supply) or environmental assets (e.g., nature reserves) face questions related to undertaking adaptation measures on the basis of what impacts might be expected if global GHG emissions continue unabated or as a result of globally agreed mitigation action at different levels of control. The starting point for adaptation decisions is to explore the possible range of impacts to which one would need to adapt. This is a complex task in itself because it involves understanding possible regional patterns of climate change, the evolution of key socioeconomic and biophysical components of the sector or region under consideration, and the dynamics of the impacts of changing climatic conditions on the evolving social system.

Adaptation decisions in private sectors operating under free-market conditions will be made largely as part of a business-as-usual approach and will rely on analytical frameworks that are compatible with the management culture. The emerging literature on adaptation describes this as autonomous adaptation. The flexibility of private-sector actors and thus the range of options they are able to consider in adapting to any external impact (not only climatic ones) can be severely constrained by market distortions or by a lack of resources to implement any transformation. Under such circumstances, the potential for autonomous adaptation is limited. Planned adaptation will be required, and the importance of public policy is larger.

A standard example for autonomous adaptation is the farmer who switches from one cultivar of a given crop to another or from one crop to another in response to perceived changes in weather patterns, simultaneously considering changes in relative prices of input factors and agricultural commodities, the evolving technological and agronomic conditions behind them, and other factors affecting his profits. However, if prices are distorted by a quota system or state subsidy, decisions are excessively dominated by these considerations, which could lead to maladaptation. Similarly, subsistence farmers in less-developed countries are not profit maximizers, and they may not possess the resources required to make even minor shifts in response to changes in their external conditions. Under these conditions, the only possibility for them might be to give up their livelihood altogether.

With a view to the already significant atmospheric load of GHGs since the industrial revolution, as well as the huge inertia and long delays characterizing the atmosphere-ocean-biosphere system, adaptation to anthropogenic climate change appears inevitable over the coming decades. It is worth looking at some of the key differences in applications of DAFs that deal with climate change mitigation (reducing GHG emissions) and adaptation (managing and counterbalancing the impacts of climate change).

The most crucial difference between mitigation- and adaptation-oriented DAF applications is to whom the benefits of action accrue. Except for "no-regret" options, benefits of mitigation will become a globally shared public good. Adaptation actions will predominantly benefit agents who adapt, in the case of private actors, or gains will be shared by the community in the case of local/regional public goods and services, such as flood protection.

The second important difference between mitigation and adaptation decisions is related to the timing of policy options. If climate protection were needed, as many scientists and policymakers maintain, policies and technologies that help reduce GHG emissions at the lowest possible social costs would be required immediately. On the adaptation side, in contrast, the bulk of more significant impacts of climate change may be felt 30, 50, or 100 years from now. This leaves a longer time period (compared to mitigation action) to steer the development of climate-sensitive sectors so that their climate vulnerability will be lower and, more important, to develop technologies that will help reduce remaining negative impacts by the time they really happen. Nevertheless, forethought and action might well be required in sectors with long-lived infrastructure and large social inertia (e.g., changing institutions such as misallocated property rights) to foster adaptation to future climate.

In terms of public policies, the foregoing analysis implies urgency on the mitigation side to formulate and put in place appropriate measures; by contrast, in general there is more time on the adaptation side to sort out potential impacts, adaptation needs, institutional and technological options involved in various adaptation measures, and public policies to develop and deliver them. At the same time, however, in many countries around the world that suffer from current climatic variability and extremes, there is an urgent need for appropriate adaptation policies and programs to be designed and implemented now to lessen adverse impacts; such actions also will help to build adaptive capacity for future climate changes.

Several other differences between mitigation and adaptation decisions must be considered in framing DAFs appropriately. Mitigation decisions are to be crafted globally, and their implementation may entail a global spread to reduce costs, whereas adaptation decisions are more limited to nations or subnational regions. The region in this context is intentionally defined loosely and given a broad interpretation. In considering climate change impacts and adaptation policies, a region typically would be a sociogeographical unit under the jurisdiction of a legally recognized policy entity within a country. However, a region in this sense also could correspond to a whole country, especially if it is relatively small and geographically homogeneous. Moreover, regional climate impact and adaptation studies also are conceivable (and in some cases have been undertaken) at the level of a supranational region, provided it has a recognized policymaking entity (e.g., the European Union).

Many market-sector impacts can be relieved at least partially by a combination of regional adjustments and interregional trade (especially in the agriculture and livestock sectors), but regions remain the prime focus of adaptation policies even in these cases. In terms of the public policy agenda, mitigation decisions have to be made today in the context of current short-term economic problems, social challenges, and policy debates. Adaptation and adaptation-related analyses will have to be developed in the context of long-term socioeconomic and technological development, with a special view to economic and technological trends in climate-sensitive economic sectors and environmental systems. This makes adaptation-oriented DAF applications easier because options to factor them in are much broader, but it also makes them more difficult because the future is difficult to predict and there is a clear need for policies that will be successful across a broad range of plausible futures (these policies commonly are called robust). The information base for adaptation decisions will improve over time, whereas DAFs to support near-term mitigation decisions must cope with current knowledge plagued with enormous uncertainties.

2.7.2. Major DAFs and their Use in Adaptation Studies

A broad range of DAFs could be used in principle; to date, however, only a few have been used in practice to provide substantial information to policymakers who are responsible for adaptation decisions at various levels. This subsection lists DAFs that appear to be most relevant for analyzing adaptation decisions. Many DAFs overlap in practice, and clear classification of practical applications sometimes is difficult. The IPCC Guidance Paper on DAFs (Toth, 2000a) provides a more comprehensive, yet incomplete, catalog.

Just as in analyzing decision options for overall climate policy (i.e., at what level should concentrations of GHGs be stabilized, considering the costs and benefits involved?) or for mitigation decisions (timing, location, ways and means of emission reductions), the proper mode to conduct analyses to support adaptation decisions also is sequential decisionmaking under uncertainty and considering future learning. The principal task is to identify adaptation strategies that will take regions or sectors to the best possible position for revising those strategies at later dates in light of new information about expected patterns of regional climate change, socioeconomic development, and changes in climate-sensitive sectors. Consequently, applications of all DAFs in adaptation studies should be formulated in the sequential decisionmaking mode.

The complexities involved in climate change decisionmaking and selecting appropriate tools to support it stem from the interconnectedness of the various realms of decisionmaking. Analysts provide advice for setting the global climate policy target at the global scale; these targets become external constraints when adaptation strategies are sought at the regional scale that are socially just, environmentally sustainable, and compatible with regional development objectives.

DAFs that are applicable in adaptation assessments can be distinguished according to whether they rely solely on "desk studies" (involving or not involving formal models) or entail participation of clients, stakeholder groups, or others. Model-based DAFs tend to focus primarily on structuring the problem, apply convenient simplifications, and find efficient solutions to the problem. Participatory DAFs, in contrast, can better accommodate diverse views on climate change impacts and often conflicting interests and options to restrain them. Insights from both kinds of studies are crucial for policymakers to craft effective and broadly acceptable policies.

2.7.2.1. Decision Analysis

Decision analysis (DA) is the product of integrating utility theory, probability, and mathematical optimization (see French, 1990; Morgan and Henrion, 1990; Chechile and Carlisle, 1991; Keeney and Raiffa, 1993; Kleindorfer *et al.*, 1993; Marshall and Oliver, 1995; Clemen, 1996). The process starts with problem identification and preparation of a possibly comprehensive list of decision options. Structural analysis would organize options into a decision tree, carefully distinguishing decision nodes (points at which the outcome is chosen by the decisionmaker) and chance nodes (points at which the outcome results from stochastic external events). Next, uncertainty analysis would assign subjective probabilities to chance nodes, and utility analysis would stipulate cardinal utilities (in terms of absolute values) for outcomes. Finally, optimization produces the best outcome according to a selected criterion, typically maximizing expected utility or any other that best reflects the risk attitude of the decisionmaker.

Advanced DA provides various extensions of the foregoing conceptual framework and supports a huge diversity of applications. In the literature, some features (sequential decisionmaking, hedging), specific versions (multi-criteria analysis), distinctive applications [risk assessment (RA)], or basic components (multi-attribute utility theory) of DA sometimes are emphasized and taken as separate DAFs, although they all are rooted in the same theoretical framework. As indicated, sequential decisionmaking is an indispensable mode of analysis of climate change in any DAF. It refers to framing of the analysis rather than a distinctive DAF. DA can be performed with a single criterion or with multiple criteria; multi-attribute utility theory provides the conceptual underpinnings for the latter. Finally, DA adapted to managing technological, social, or environmental hazards constitutes part of RA, in which a range of other methods also is available. RA involves estimation of the nature and size of risks. Its objective is to identify quantitative measures of hazards in terms of magnitude and probability. RA methods are diverse; the choice depends on the disciplinary focus and the nature of the hazard to be assessed, but all methods rely on extrapolation (see Kates and Kasperson, 1983). See Chapter 12 for applications of RA in climate impact assessment.

DA is a promising DAF for use in adaptation assessments. Problem formulation in DA allows for consideration of a broad range of uncertain outcomes, different probability distributions assigned to them, and a variety of possible adaptation actions. Structuring the DA model in an intertemporal fashion is helpful for identifying robust adaptation strategies that prove to be effective under a broad range of possible futures and retain a sufficient degree of freedom for course correction.

2.7.2.2. Cost-Benefit Analysis

Cost-benefit analysis (CBA) involves valuing all costs and benefits of a proposed project over time on the basis of willingness to pay (or willingness to accept compensation) on the part of project beneficiaries (affected people) and specifying a decision criterion to accept or turn down the project (see Ray, 1984; Morgenstern, 1997). This criterion usually is the compensation principle, implying that those who benefit from the project should be able to compensate the losers. The applicability of CBA as a DAF for climate policy has been a fiercely debated issue. Although the debate continues about the extent to which traditional CBA can provide useful information for global-level decisionmaking, there is more agreement on its usefulness in adaptation decisions at the national and regional scales.

In practical applications, all costs (C) and benefits (B) are defined as follows:

$$B = \sum_{t=0}^{n} \frac{B_t}{(1 + i)^t} \quad \text{and} \quad C = \sum_{t=0}^{n} \frac{C_t}{(1 + i)^t}$$

where i is the social discount rate, n is the project life, and t denotes the year. One can use different cost-benefit criteria for ranking projects or choosing the best among them: the cost-benefit ratio, $CBR = B/C > 1$; the net present value, $NPV = B - C > 0$; and the internal rate of return, $IRR > i$, where IRR is the discount rate to make $B = C$. When we evaluate a single project, these criteria lead to the same conclusion. In choosing the most desirable alternative, however, these criteria indicate different orders of desirability.

A CBA in the adaptation context takes potential regional climate change scenarios and their impacts as its starting point. The next step is to establish costs of alternative adaptive measures as a function of their scales of application—the marginal cost curve. A related task is to estimate how much damage can be averted by increasing the adaptation effort—WTP (marginal benefit curve). The decision principle suggests undertaking adaptive measures as long as marginal averted damages (benefits) exceed marginal costs. This rule of thumb is easier to apply in sectoral adaptation decisions, in which costs and benefits can be derived from market prices. Difficulties arise in nonmarket sectors in which the valuation behind the marginal cost and benefit curves often is debated. Difficulties multiply, in a regional context, when costs and benefits must be aggregated across many sectors.

A frequent critique of CBA and its applicability in adaptation studies is that the underlying measurements are incomplete (especially in regional studies, which do not cover all important aspects), inaccurate (even the costs and benefits of adaptive actions included in the analysis are impossible to measure precisely), and debated (related to the two preceding points; the inclusion and exact valuation of many costs and benefits involve inherently subjective value judgments). These criticisms are largely valid. However, it is still better to get at least the measurable components right and complement them with a combination of judgments on hard-to-measure items and sensitivity tests to assess their implications than to abandon the whole method simply because it does not get everything perfect. Nevertheless, it is important that users of these tools and their results fully understand the limitations and confidence attached to them. Duke and Kammen (1999) argue that accounting for dynamic feedback between the demand response and price reductions from production experience can be used to account for deadweight loss and other market dynamics that determine the benefit-cost ratio of economic and policy measures to expand the market for clean energy technologies. These results further support a broader role for market transformation programs to commercialize new environmentally attractive technologies. The same dynamic feedback processes also are relevant for CBA applications to adaptation decisions. For example, consider changing precipitation patterns that would increase the frequency of high-water conditions. Take flood-related damages as the function of flood return periods: Annual flooding may cause the least damages, whereas a 5-year return flood will cause somewhat more, a 20-year return flood even more, and so on. Adaptation costs increase along the same axis because it takes higher dikes and larger flood protection reservoirs to control a 50-year return flood than a 5-year return flood. The level at which a given society will decide to protect itself against floods depends on local economic conditions and geographical and technological endowments. A CBA suggests that it should be in the neighborhood of where marginal costs of additional flood protection would be equal to WTP for additional flood protection.

2.7.2.3. Cost-Effectiveness Analysis

Cost-effectiveness analysis (CEA) takes a predetermined objective (often an outcome negotiated by key stakeholder groups in a society) and seeks ways to accomplish it as inexpensively as possible. The thorny issues of compensations and actual transfers boil down to less complex but still contentious issues of burden sharing.

CBA will always be controversial because of the intricacies of valuing benefits of many public policies, especially intangible benefits of environmental policies, properly. CEA takes the desired level of a public good as externally given (a vertical marginal benefit curve) and minimizes costs across a range of possible actions. Like other target-based approaches, CEA often turns into an implicit CBA, especially if even the minimum costs turn out to be too high and beyond the ability to pay of the society. In this case, the target is iteratively revised until an acceptable solution is found.

Consider the foregoing example of changing precipitation pattern induced by climate change and resulting high-water conditions. In many countries, legally binding criteria exist regarding the level of flood protection (e.g., protection against a 50- or 100-year return flood). CEA would take these or other socially agreed flood protection targets and seek the mix of dams, reservoirs, and other river basin management options that would minimize the costs of achieving the specified target.

2.7.2.4. *Policy Exercise Approach*

The policy exercise (PE) approach involves a flexibly structured process that is designed as an interface between academics and policymakers. Its function is to synthesize and assess knowledge accumulated in several relevant fields of science for policy purposes in light of complex practical management problems. At the heart of the process are scenario writing ("future histories," emphasizing nonconventional, surprise-rich, but still plausible futures) and scenario analyses via interactive formulation and testing of alternative policies that respond to challenges in the scenario. These scenario-based activities take place in an organizational setting that reflects the institutional features of the issues addressed. Throughout the exercise, a wide variety of hard (mathematical and computer models) and soft methods are used (Brewer, 1986; Toth, 1988a,b; Parson, 1997).

The product of a PE is not necessarily new scientific knowledge or a series of explicit policy recommendations but a new, better structured view of the problem in the minds of participants. The exercise also produces statements concerning priorities for research to fill gaps of knowledge, institutional changes that are needed to cope more effectively with the problems, technological initiatives that are necessary, and monitoring and early warning systems that could ease some of the problems in the future. In recent years we have witnessed increasing use of the PE approach to address climate change at the national scale (see Klabbers *et al.,* 1995, 1996) and at the global level.

2.7.3. *Relevance and Use of DAFs in Sectoral Adaptation Decisions—Selected Examples*

Working Group II has reviewed a huge volume of climate impact assessment studies conducted to date. Most of these studies investigate possible implications of climate change for a single economic sector or environmental component. An increasing, yet still small, fraction of these studies lists options to alleviate impacts, but few take even the next step of exploring direct and indirect costs of those adaptation options. Even fewer studies provide comprehensive assessments of direct and indirect benefits.

Although these studies qualitatively indicate that many policy options proposed as adaptation measures to reduce negative impacts of climate change would be justified even in the absence of climate change (dubbed "no regret" measures on the impacts adaptation side), to date very few have been developed to the point at which comprehensive and quantitative assessment of adaptation options would be possible. Nevertheless, they are a prerequisite for establishing appropriate applications of the more quantitative DAFs reviewed in Section 2.7.2. The main reason is that, despite uncertainties of regional climate change patterns and resulting impacts, some information is generated about possible biophysical impacts. However, little is known about future socioeconomic sensitivity and even less about future adaptive capacity. Resolving this would require fairly detailed regional development scenarios to provide the broader context for sectoral assessments. All these factors together make rigorous applications of quantitative DAFs difficult.

A simple ranking of climate impact and adaptation studies according to how far they get in using DA tools would start with those that are preoccupied almost exclusively with impacts and casually mention some obvious adaptation options. The next category would be studies that attempt to produce a comprehensive list of possible adaptive measures. More advanced studies would explore positive and, if they exist, negative effects of listed options and try to establish at least a qualitative ranking. By assigning monetary values to those comprehensive effects, CBA could help determine the optimal level of adaptation measures; CEA would select the least-cost solution to provide a predetermined level of adaptation objective.

Perhaps the most crucial area of public policy in climate change adaptation is water resource management. A set of papers arranged by Frederick *et al.* (1997a) looks at different aspects of climate change and water resources planning. Their general conclusion is that DAFs adopted in public policy procedures of water management are largely "appropriate for planning and project evaluation under the prospect of climate change, but new applications and extensions of some criteria may be warranted" (Frederick *et al.*, 1997b). The authors mention nonstationarity, interest rates, and multiple objectives as issues on which progress is required to support better assessments of climate change adaptation decisions.

Water is an important factor to consider in most other sectoral impact and adaptation assessments, even if their primary focus is on a single sector. With a view to the complexity of interactions among sectoral impacts on one hand and adaptation measures on the other, integrated regional assessments increasingly are considered to be indispensable to understand climate-related risks.

2.7.4. *Relevance and Use of DAFs in Regional Adaptation Decisions—Selected Examples*

Sectoral adaptation decisions must be considered in the broader regional context in which evolution in related sectors and their responses to changing climatic conditions represent additional factors to consider in planning a given sector's own adaptation strategy. This process is likely to involve a broad mix of private and public stakeholders and their interactions. From the perspectives of regional planners and policymakers who are responsible for the overall socioeconomic development of a specific region, the challenge is to create conditions under which relevant sectoral actors can formulate their own adaptive strategies efficiently and install public policies that will help adaptation in sectors that provide public services and manage public resources.

Designing and implementing regional climate change studies that incorporate full-fledged DAFs to support the development of regional climate adaptation policies has proven to be an insurmountable challenge to date. This is understandable, in view of the difficulties involved, and indicates a crucial research area for the future.

Most statements on regional adaptation policies in the literature stem from limited but logical extensions of sectoral climate impact assessment studies. Once possible biophysical changes and their direct or indirect socioeconomic consequences are established, impact assessors mention a few options that could mitigate those impacts or moderate their consequences. Seldom are these lists comprehensive, and they scarcely entail even direct cost estimates, let alone assessments of indirect costs and ancillary benefits involved in the specified adaptation options.

The study by Ringius *et al.* (1996) on climate change vulnerability and adaptation in Africa is a good example. Focusing on impacts on agriculture and water, the authors develop a typology of adaptive responses and discuss their effectiveness from the perspectives of different stakeholders. Although the study is extremely useful in pointing out that a convenient and crucial starting point for decisions on adapting to expected climate change in Africa is to reduce present vulnerability and enhance the capacity to respond to any environmental and economic perturbations (not just climate and weather), no attempt has been made to evaluate the costs and benefits of different options or to rank them in terms of their effectiveness.

An early policy-oriented impact assessment study adopted the PE approach to synthesize results of sectoral studies in a DAF in selected countries in southeast Asia (Toth, 1992a,b). The project included data collection, modeling, completion of first-order impact assessments, analysis of socioeconomic impacts on the impact assessment side, development of background scenarios, and pre-interviews with "policy" participants as preparations for PE workshops. The results of these workshops indicate that the PE approach might be a useful tool in structuring the numerous uncertain facets that are related to developing robust regional adaptation policies.

A partially integrated regional cost-benefit assessment has been prepared for the entire coastal area of Poland (Zeidler, 1997). Scenarios of sea-level rise have been combined with different assumptions about socioeconomic development in the potentially affected coastal region to explore mainly direct and relatively easy-to-estimate costs and benefits of three specifically defined adaptation strategies: retreat (no adaptation), limited protection, and full protection. Because of its numerous merits and despite its limitations, this study has demonstrated the feasibility of using CBA to formulate climate change adaptation problems in a simple DAF and the potential usefulness of its results to policymakers.

2.7.5. Contribution of DAFs in Adaptation to Integrated Climate Change Decisions on Balancing Mitigation and Adaptation

Information generated in applying DAFs in sectoral and regional climate impact assessment studies is oriented primarily toward decisionmakers who have the mandate to initiate and implement public policies to reduce future adverse impacts of climate change. Just the attempt to integrate adaptation options into selected DAFs would force analysts to think comprehensively and achieve internal consistency, to consider broader factors beyond the influence of sectoral or regional stakeholders. Even though a comprehensive CBA or DA remains difficult to develop, the overall quality of the impact assessment improves.

A second, equally important use of these results is to help define GHG mitigation objectives. National and regional positions at global negotiations on long-term climate stabilization targets (with respect to anthropogenic forcing) apparently are influenced by perceived risks involved in climate change as well as net damage remaining even after plausible and affordable adaptation options have been considered.

Admittedly, it is a difficult task to formulate impact/adaptation studies properly in any DAF. This explains the modest progress in the field since the SAR. Regional and sectoral chapters in this volume review a small number of DAF applications, whereas there was hardly any application on which to report in the SAR.

2.8. Conclusion

In the decade prior to the SAR, the preponderance of studies employed methods and tools largely for the purpose of ascertaining the biophysical impacts of climate change, usually on a sectoral basis. Thus, the methods included models and other means for examining the impacts of climate change on water resources, agriculture, natural ecosystems, or coasts. Such methods have improved with regard to detection of climate change in biotic and physical systems and produced new substantive findings. In addition, since the SAR, cautious steps have been taken to expand the "toolbox" to address more effectively the human dimensions of climate as cause and consequence of change and to deal more directly with cross-sectoral issues concerning vulnerability, adaptation, and decisionmaking. In particular, more studies have begun to apply methods and tools for costing and valuing effects, treating uncertainties, integrating effects across sectors and regions, and applying DAFs to evaluate adaptive capacity. Overall, these modest methodological developments are encouraging analyses that will build a more solid foundation for understanding how decisions regarding adaptation to future climate change might be taken.

References

Alcamo, J. (ed.), 1994: *IMAGE 2.0—Integrated Modeling of Global Climate Change*. J. Kluwer Academic Publishers, Dordrecht, The Netherlands, 328 pp.

Alcamo, J., R. Leemans, and G.J.J. Kreileman, 1998: *Global Change Scenarios of the 21st Century. Results from the Image 2.1 Model*. Pergamon & Elseviers Science, London, United Kingdom.

Alcamo, J. and E. Kreileman, 1996: Emission scenarios and global climate protection. *Global Environmental Change*, **6,** 305–334.

Anderson, J.L., 1998: Embracing uncertainty: the interface of Bayesian statistics and cognitive psychology. *Conservation Ecology*, **2(1),** 2. Available online at http://www.consecol.org/vol2/iss1/art2.

Arnell, N., B. Bates, H. Lang, J.J. Magnuson, and P. Mulholland, 1996: Hydrology and freshwater ecology. In: *Climate Change 1995: Impacts, Adaptations and Mitigation of Climate Change: Scientific-Technical Analyses. Contribution of Working Group II to the Second Assessment Report of the Intergovernmental Panel on Climate Change* [Watson, R.T., M.C. Zinyowera, and R.H. Moss (eds.)]. Cambridge University Press, Cambridge, United Kingdom and New York, NY, USA, pp. 325–364.

Arrow, K.J., 1951: *Social Choice and Individual Values*. John Wiley and Sons, Chichester, United Kingdom, 124 pp.

Arrow, K.J., 1963: *Social Choice and Individual Values*. John Wiley and Sons, London, Chichester, United Kingdom, 2nd. ed., 124 pp.

Arrow, K.J., W.R. Cline, K.G. Mäler, M. Munasinghe, R. Squitieri, and J.E. Stiglitz, 1996: Discounting. In: *Climatic Change 1995: Economic and Social Dimensions of Climate Change, Second Assessment of the Intergovernmental Panel on Climate Change* [Bruce, J.P., H. Lee, and E.F. Haites (eds.)]. Cambridge University Press, Cambridge, United Kingdom and New York, NY, USA, pp. 129–144.

Ashworth, A.C., 1996: The response of arctic Carabidae (*Coleoptera*) to climate change based on the fossil record of the Quaternary period. *Annales Zoologici Fennici*, **33,** 125–131.

Atkinson, A.B., 1970: On the measurement of inequality. *Journal of Economic Theory*, **2,** 244–263.

Badgley, C., 1998: Vertebrate indicators of paleoclimate. *Journal of Vertebrate Paleontology*, **18(3),** 25A.

Bakker, K., T. Downing, A. Garrido, C. Giansante, E. Iglesias, L. del Moral, G. Pedregal, P. Riesco, and the SIRCH Team, 1999: *A Framework for Institutional Analysis*. Environmental Change Unit, Oxford University, Oxford, United Kingdom, 52 pp.

Barnosky, A.D., 1986: "Big game" extinction caused by late Pleistocene climatic change: Irish elk (*Megaloceros giganteus*) in Ireland. *Quaternary Research*, **25,** 128–135.

Baroni, C. and G. Orombelli, 1994: Abandoned penguin rookeries as Holocene paleoclimatic indicators in Antarctica. *Geology*, **22,** 23–26.

Bartholomay, G.A., R.T. Eckert, and K.T. Smith, 1997: Reductions in tree-ring widths of white pine following ozone exposure at Acadia National Park, Maine, U.S.A. *Canadian Journal of Forest Research*, **27(3),** 361–368.

Bateman, I. and K. Willis (eds.), 1995: *Valuing Environmental Preferences: Theory and Practice of the Contingent Valuation Method*. Oxford University Press, Oxford, United Kingdom, 645 pp.

Berthold, P. and A.J. Helbig, 1992: The genetics of bird migration: stimulus, timing and direction. *Ibis*, **1,** 35–40.

Braden, J. and C. Kolstad, 1991: *Measuring the Demand for Environmental Quality*. Elsevier Press, Amsterdam, The Netherlands, 370 pp.

Bradley, R.S. and P.D. Jones (eds.), 1992: *Climate since A.D. 1500*. Routledge, London, United Kingdom and New York, NY, USA, 706 pp.

Bradley, N.L., A.C. Leopold, J. Ross, and H. Wellington, 1999: Phenological changes reflect climate change in Wisconsin. *Proceedings of the National Academy of Sciences*, **96,** 9701–9704.

Brandon-Jones, D., 1996: The Asian Colobinae (*Mammalia: Cercopithecidae*) as indicators of Quaternary climate change. *Biological Journal of the Linnean Society*, **59,** 327–350.

Brewer, G.D., 1986: Methods for synthesis: policy exercises. In: *Sustainable Development of the Biosphere* [Clark, W.C. and R.E. Munn (eds.)]. Cambridge University Press, Cambridge, United Kingdom and New York, NY, USA, pp. 455–473.

Briffa, K.R., F.H. Schweingruber, P.D. Jones, T.J. Osborn, I.C. Harris, S.G. Shiyatov, E.A. Vaganov, and H. Grudd, 1998: Trees tell of past climates: but are they speaking less clearly today? *Philosophical Transactions of the Royal Society of London B Biological Sciences,* **353(365),** 65–73.

Briguglio, L., 1995: Small island states and their economic vulnerabilities. *World Development,* **23,** 1615–1632.

Broecker, W.S., 1997: Thermohaline circulation, the Achilles heel of our climate system: will man-made CO_2 upset the current balance? *Science,* **278,** 1582–1588.

Brooks, J.R., L.B. Flanagan, and J.R. Ehleringer, 1998: Responses of boreal conifers to climate fluctuations: indications from tree-ring widths and carbon isotope analyses. *Canadian Journal of Forest Research*, **28(4),** 524–533.

Brown, J.H., G.C. Stevens, and D.M. Kaufman, 1996: The geographic range: size, shape, boundaries and internal structure. *Annual Review of Ecological Systems,* **27,** 597–623.

Carter, T.R., M.L. Parry, H. Harasawa, and S. Nishioka, 1994: *IPCC Technical Guidelines for Assessing Climate Change Impacts and Adaptations*. University College London, Centre for Global Environmental Research, and National Institute for Environmental Studies, Tsukuba, Japan, 59 pp.

Casman, E.A., M.G. Morgan, and H. Dowlatabadi, 1999: Mixed levels of uncertainty in complex policy models. *Risk Analysis,* **19(1),** 33–42.

Cebon, P., U. Dahinden, H.C. Davies, D.M. Imboden, and C.C. Jaeger (eds.), 1998: *Views from the Alps. Towards Regional Assessments of Climate Change*. MIT Press, Cambridge, MA, USA, 536 pp.

Cescatti, A. and E. Piutti, 1998: Silvicultural alternatives, competition regime and sensitivity to climate in a European beech forest. *Forest Ecology and Management*, **102(2–3),** 213–223.

Chattopadhyay, D. and J.K. Parikh, 1993: CO_2 emissions reduction from power system in India. *Natural Resources Forum*, **17(4),** 251–261.

Chechile, A. and S. Carlisle (eds.), 1991: *Environmental Decisionmaking: A Multidisciplinary Perspective*. Van Nostrand Reinhold, New York, NY, USA, 296 pp.

Clark, G.E., S.C. Moser, S.J. Ratick, K. Dow, W.B. Meyer, S. Emani, W. Jin, J.X. Kasperson, R.E. Kasperson, and H.E. Schwarz, 1998: Assessing the vulnerability of coastal communities to extreme storms: the case of Revere, MA, USA. *Mitigation and Adaptation Strategies for Global Change*, **3,** 59–82.

Clemen, R.T., 1996: *Making Hard Decisions: An Introduction to Decision Analysis*. Duxbury Press, Belmont, CA, USA, 664 pp.

Cohen, S., 1998: Scientist-stakeholder collaboration in integration assessment of climate change: lessons from a case study of northwest Canada. *Environmental Modeling and Assessment,* **2,** 281–293.

Cohen, S., 1997: *Mackenzie Basin Impact Study—Final Report*. Environment Canada and the University of British Columbia, Vancouver, BC, Canada, 372 pp.

Coope, G.R., 1995: Insect faunas in ice age environments: why so little extinction? In: *Extinction Rates* [Lawton, J.H. and R.M. May (eds.)]. Oxford University Press, Oxford, United Kingdom, pp. 55–74.

CSIRO, 1996: *Climate Change Scenarios for the Australian Region*. Climate Impact Group, CSIRO Division of Atmospheric Research, Melbourne, Australia, 8 pp.

CSIRO, 1992: *Climate Change Scenarios for the Australian Region*. Climate Impact Group, CSIRO Division of Atmospheric Research, Melbourne, Australia, 6 pp.

Cummings, R., D. Brookshire, and W. Schultze, 1986: *Valuing Environmental Goods: An Assessment of the Contingent Valuation Method*. Rowman & Littlefield, Lanham, MD, USA, 462 pp.

Daly, H.E. and J.B. Cobb, 1994: *For the Common Good: Redirecting the Economy toward Community, the Environment, and a Sustainable Future*. Beacon Press, Boston, MA, USA, 534 pp.

Darwin, R., M. Marinos, J. Lewandrowski, and A. Raneses, 1995: *World Agriculture and Climate Change—Economic Adaptations*. Department of Energy, Economic Research Service Report 703, U.S. Department of Agriculture, Washington, DC, USA, 86 pp.

Davis, A.J., L.S. Jenkinson, J.H. Lawton, B. Shorrocks, and S. Wood, 1998: Making mistakes in predicting responses to climate change. *Nature,* **391,** 783.

Davis, M.B., 1988: *Ecological Systems and Dynamics in Toward an Understanding of Global Change*. National Academy Press, Washington, DC, USA, pp. 69–106.

Davis, M.B. and C. Zabinski, 1992: Changes in geographical range resulting from greenhouse warming: effects on biodiversity in forests. In: *Global Warming and Biological Diversity* [Peters, R.L. and T.E. Lovejoy (eds.)]. Yale University Press, New Haven, CT, USA, pp. 297–307.

Davis, R., 1963: Recreational planning as an economic problem. *Natural Resources Journal*, **3,** 239–249.

DeGroot, R.S., P. Ketner, and A.H. Ovaa, 1995: Selection and use of bio-indicators to assess the possible effects of climate change in Europe. *Journal of Biogeography*, **22,** 935–943.

Douglas, M. and A. Wildavsky, 1982*: Risk and Culture: An Essay on the Selection of Technical and Environmental Danger.* University of California Press, Berkeley, CA, USA, 221 pp.

Downing, T.E., (ed.), 1996: *Climate Change and World Food Security*. Springer-Verlag, Heidelberg, Germany, 662 pp.

Downing, T.E., R. Butterfield, S. Cohen, S. Huq, R. Moss, A. Rahman, Y. Sokona, and L. Stephen, 2001: *Climate Change Vulnerability: Toward a Framework for Understanding Adaptability to Climate Change Impacts*. United Nations Environment Programme, Nairobi, Kenya, and Environmental Change Institute, Oxford University, Oxford, United Kingdom.

Downing, T.E., S. Moss, and C. Pahl-Wostl, 2000: Understanding climate policy using participatory agent-based social simulation. In: *Proceedings of the Second International Workshop on Multi-Agent Based Simulation. Lecture Notes in Artificial Intelligence* 1979 [Davidson, P. and S. Moss (eds.)]. pp. 198–215.

Downing, T.E., M.J. Gawith, A.A. Olsthorn, R.S.J. Tol, and P. Vellinga, 1999a: Introduction. In: *Climate, Change and Risk* [Downing, T.E., A.A. Olsthoorn, and R.S.J. Tol (eds.)]. Routledge, London, United Kingdom, pp. 1–18.

Downing, T.E., A.A. Olsthoorn, and R.S.J. Tol (eds.), 1999b: *Climate, Change and Risk*. Routledge, London, United Kingdom, 407 pp.

Duke, R.D. and D.M. Kammen, 1999: The economics of energy market transformation initiatives. *The Energy Journal*, **20(4),** 15–64.

Easterling, D.R., G.A. Meehl, C. Parmesan, S. Chagnon, T. Karl, and L. Mearns, 2000: Climate extremes: observations, modeling, and impacts. *Science*, **289,** 2068–2074.

Easterling, W.E., C. Polsky, D. Goodin, M. Mayfield, W.A. Muraco, and B. Yarnal, 1998: Changing places, changing emissions: the cross-scale reliability of greenhouse gas emissions inventories in the U.S. *Local Environment*, **3(3),** 247–262.

Edmonds, J.A., S.H. Kim, C.N. MacCracken, R.D. Sands, and M.A. Wise, 1997: *Return to 1990: The Cost of Mitigating United States Carbon Emissions in the Post-2000 Period*. PNNL-11819, Pacific Northwest National Laboratory, Washington, DC, USA, 109 pp.

Edwards, A.W.F., 1992: *Likelihood*. The Johns Hopkins University Press, Baltimore, MD, USA, expanded edition, 296 pp. (Originally published in 1972 by Cambridge University Press, Cambridge, United Kingdom.)

El-Shaer, H.M., C. Rosenzweig, A. Iglesias, M.H. Eid, and D. Hillel, 1997: Impact of climate change on possible scenarios for Egyptian agriculture in the future. *Mitigation and Adaptation Strategies for Global Change*, **1(3),** 223–250.

Everett, J.T. and B.B. Fitzharris, 1998: The Arctic and the Antarctic. In: *The Regional Impacts of Climate Change: An Assessment of Vulnerability. A Special Report of IPCC Working Group II* [Watson, R.T., M.C. Zinyowera, and R.H. Moss (eds.)]. Cambridge University Press, Cambridge, United Kingdom and New York, NY, USA, pp. 85–103.

Fankhauser, S., 1998: *The Cost of Adapting Climate Change*. Working Paper 16, Global Environment Facility, Washington, DC, USA.

Feng, X. and S. Epstein, 1996: Climatic trends from isotopic records of tree rings: the past 100–200 years. *Climatic Change*, **33(4),** 551–562.

Fitzharris, B.B., 1996: The cryosphere: changes and their impacts. In*: Climate Change 1995: Impacts, Adaptations, and Mitigation of Climate Change: Scientific-Technical Analyses. Contribution of Working Group II of the Second Assessment Report for the Intergovernmental Panel on Climate Change* [Watson, R.T., M.C. Zinyowera, and R.H. Moss (eds)]. Cambridge University Press, Cambridge, United Kingdom and New York, NY, USA, pp. 241–266.

Fox, J., 1994: On the necessity of probability: Reasons to believe and grounds for doubt. In: *Subjective Probability* [Wright, G. and P. Ayton (eds.)]. John Wiley and Sons, Chichester, United Kingdom, pp. 75–104.

Frederick, K.D., D.C. Major, and E.Z. Stakhiv (eds.), 1997a: Climate change and water resources planning criteria. *Climatic Change (Special Issue)*, **37,** 1–313.

Frederick, K.D., D.C. Major, and E.Z. Stakhiv, 1997b: Introduction. *Climatic Change (Special Issue)*, **37,** 1–5.

French, S. 1990: *Reading in Decision Analysis*. Chapman and Hall, London, United Kingdom, 240 pp.

Gigerenzer, G., 1996: On narrow norms and vague heuristics: a rebuttal to Kahneman and Tversky. *Psychological Review*, **103,** 592–596.

Gigerenzer, G., 1994: Why the distinction between single event probabilities and frequencies is important for psychology and vice-versa. In: *Subjective Probability* [Wright, G. and P. Ayton (eds.)]. John Wiley and Sons, Chichester, United Kingdom, pp. 129–161.

Gigerenzer, G., U. Hoffrage, and H. Kleinbölting, 1991: Probabilistic mental models: A Brunswikian theory of confidence. *Psychological Review*, **98,** 506–528.

Goodfriend, G.A. and R.M. Mitterer, 1988: Late quaternary land snails from the north coast of Jamaica: local extinctions and climatic change. *Palaeogeography Palaeoclimatology Palaeoecology*, **63,** 293–312.

Goulder, L. and S.H. Schneider, 1999: Induced technological change and the attractiveness of CO_2 abatement policies. *Resource and Energy Economics*, **21,** 211–253.

Graham, R.W., 1992: Late Pleistocene faunal changes as a guide to understanding effects of greenhouse warming on the mammalian fauna of North America. In: *Global Warming and Biological Diversity* [Peters, R.L. and T.E. Lovejoy (eds.)]. Yale University Press, New Haven, CT, USA, pp. 76–87.

Grinnell, J., 1928: Presence and absence of animals. *University of California Chronicle*, **30,** 429.

Grinnell, J., 1924: Geography and evolution. *Ecology*, **5,** 225.

Grubb, M., M. Ha-Duong, and T. Chapuis, 1994: Optimizing climate change abatement responses: on inertia and induced technology development. In: *Integrative Assessment of Mitigation, Impacts, and Adaptation to Climate Change*. [Nakicenovic, N., W.D. Nordhaus, R. Richels, and F.L. Toth (eds.)]. International Institute for Applied Systems Analysis, Laxenburg, Austria, pp. 513–534.

Hadly, E.A., 1997: Evolutionary and ecological response of pocket gophers (*Thomomys talpoides*) to late-Holocene climatic changes. *Biological Journal of the Linnean Society,* **60,** 277–296.

Hameed, S., 1994: Variation of spring climate in lower-middle Yangtse River Valley and its relation with solar-cycle length. *Geophysical Research Letters*, **21,** 2693–2696.

Hanley, N.J., J.F. Shogren, and B. White, 1997: *Environmental Economics in Theory and Practice*. Oxford University Press, Oxford, United Kingdom, 464 pp.

Hanley, N.J. and C. Spash, 1993: *Cost-Benefit Analysis and the Environment*. Edward Elgar, Aldershot, The Netherlands, 385 pp.

Hanski, I., 1999: *Metapopulation Ecology.* Oxford University Press, Oxford, United Kingdom, 313 pp.

Henderson-Sellers, A., 1993: An Antipodean climate of uncertainty. *Climatic Change,* **25,** 203–224.

Hoffman, A.A. and P.A. Parsons, 1997: *Extreme Environmental Change and Evolution*. Cambridge University Press, Cambridge, United Kingdom and New York, NY, USA, 259 pp.

Howard, R.A., J.E. Matheson, and D.W. North, 1972: The decision to seed hurricanes. *Science,* **176,** 1191–1202.

Hulme, M. and O. Brown, 1998: Portraying climate scenario uncertainties in relation to tolerable regional climate change. *Climate Research,* **10,** 1–14.

Huq, S., Z. Karim, M. Asaduzaman, and F. Mahtab (eds.), 1999: *Vulnerability and Adaptation to Climate Change in Bangladesh*. J. Kluwer Academic Publishers, Dordrecht, The Netherlands, 164 pp.

IPCC, 1996a: *Climate Change 1995: The Science of Climate Change. Contribution of Working Group I to the Second Assessment Report of the Intergovernmental Panel on Climate Change* [Houghton, J.T., L.G. Meira Filho, B.A. Callander, N. Harris, A. Kattenberg, and K. Maskell (eds.)]. Cambridge University Press, Cambridge, United Kingdom and New York, NY, USA, 572 pp.

IPCC, 1996b: *Climate Change 1995: Impacts, Adaptations and Mitigation of Climate Change: Scientific-Technical Analyses. Contribution of Working Group II to the Second Assessment Report of the Intergovernmental Panel on Climate Change* [Watson, R.T., M.C. Zinyowera, and R.H. Moss (eds.)]. Cambridge University Press, Cambridge, United Kingdom and New York, NY, USA, 879 pp.

Jacoby, G.C. and R.D. D'Arrigo, 1997: Tree rings, carbon dioxide, and climatic change. *Proceedings of the National Academy of Sciences*, **94(16),** 8350–8353.

Jacoby, H.D. and I.S. Wing, 1999: Adjustment time, capital malleability and policy cost. In: *The Energy Journal, Special Issue–The Costs of the Kyoto Protocol: A Multi-Model Evaluation*, **3,** 79–92.

Jones, R.N., 2000: Managing uncertainty in climate change projections: Issues for impact assessment. *Climatic Change*, **45(3–4),** (in press).

Juslin, P., 1994: The overconfidence phenomenon as a consequence of informal experimenter-guided selection of almanac items. *Organizational Behavior and Human Decision Processes*, **57,** 226–246.

Juslin, P., A. Winman, and H. Olsson, 2000: Naive empiricism and dogmatism in confidence research: a critical examination of the hard-easy effect. *Psychological Review*, **107(2),** 22–28.

Kahneman, D. and D. Lovallo, 1993: Timid choices and bold forecasts: a cognitive perspective on risk taking. *Management Science*, **39,** 17–31.

Kahneman, D. and A. Tversky, 1996: On the reality of cognitive illusions: A reply to Gigerenzer's critique. *Psychological Review*, **103,** 582–591.

Kahneman, D. and A. Tversky, 1979: Intuitive prediction: biases and corrective procedures. *Management Science*, **12,** 313–327.

Kahneman, D., P. Slovic, and A. Tversky (eds.), 1982: *Judgement under Uncertainty: Heuristics and Biases*. Cambridge University Press, Cambridge, United Kingdom and New York, NY, USA, 555 pp.

Kaly, U., L. Briguglio, H. McLeod, S. Schmall, C. Pratt, and R. Pal, 1999: *Environmental Vulnerability Index (EVI) to Summarise National Environmental Vulnerability Profiles*. SOPAC Technical Report 275, South Pacific Applied Geoscience Commission, Suva, Fiji, 73 pp.

Kates, R.W. and J.X. Kasperson, 1983: Comparative risk analysis of technological hazards (a review). *Proceedings of National Academy of Sciences*, **80,** 7027–7038.

Katz, R.W. and B. Brown, 1992: Extreme events in a changing climate: variability is more important than average. *Climate Change*, **21,** 289–302.

Keeney, R.L. and H. Raiffa, 1993: *Decisions with Multiple Objectives*. Cambridge University Press, Cambridge, United Kingdom, 569 pp.

Klabbers, J.H.G., R.J. Swart, A.P. Van Ulden, and P. Vellinga, 1995: Climate policy: management of organized complexity through gaming. In: *Simulation and Gaming across Disciplines and Cultures: ISAGA at a Watershed* [Crookall, D. and K. Arai (eds.)]. Sage, Thousand Oaks, CA, USA, pp. 122–133.

Klabbers, J.H.G., C. Bernabo, M. Hisschemsller, and B. Moomaw, 1996: Climate change policy development: enhancing the science/policy dialogue. In: *Simulation Now! Learning through Experience: The Challenge of Change* [Watts, F. and A. Garcia (eds.)]. Carbonell, Diputacio de Valencia, Valencia, Spain, pp. 285–297.

Kleindorfer, P.R., H.C. Kunreuther, and P.J.H. Shoemaker, 1993: *Decision Sciences: An Integrative Perspective*. Cambridge University Press, Cambridge, United Kingdom, 480 pp.

Lancaster, K., 1966: A new approach to consumer theory. *Journal of Political Economy*, **74,** 132–157.

Lauscher, F., 1978: Neue Analysen ältester und neuerer phänologischer Reihen. *Arch. Meteorol. Geophysik Bioklimatol.*, **28,** 373–385.

Lavoie, C. and S. Payette, 1996: The long-term stability of the boreal forest limit in subarctic Quebec. *Ecology*, **77(4),** 1226–1233.

MacArthur, R.M., 1972: *Geographical Ecology.* Harper & Row, New York, NY, USA, 269 pp.

Mackenzie-Hedger, M., I. Brown, R. Connell, T. Downing, and M. Gawith (eds.), 2000: *Climate Change: Assessing the Impacts—Identifying Responses. The First Three Years of the UK Climate Impacts Programme*. UKCIP Technical Report, United Kingdom Climate Impacts Programme and Department of the Environment, Transport, and the Regions, Oxford, United Kingdom.

Maddison, D.J., 1995: A cost-benefit analysis of slowing climate change. *Energy Policy*, **23(4/5),** 337–346.

Malakoff, D., 1999: Bayes offers a "new" way to make sense of numbers. *Science*, **286,** 1460–1464.

Manne, A.S. and R. Richels, 1995: The greenhouse debate: economic efficiency, burden sharing and hedging strategies. *The Energy Journal*, **16(4),** 1–37.

Manne, A.S. and R. Richels, 1992: *Buying Greenhouse Insurance: The Economic Costs of CO_2 Emissions Limits*. MIT Press, Cambridge, MA, USA, 314 pp.

Manne, A.S., R.O. Mendelsohn, and R.G. Richels, 1995: MERGE—a model for evaluating regional and global effects of GHG reduction policies. *Energy Policy*, **23(1),** 17–34.

Marshall, K.T. and R.M. Oliver, 1995: *Decisionmaking and Forecasting*. McGraw-Hill, New York, NY, USA, 407 pp.

Martens, P., R.S. Kovats, S. Nijhof, P. De Vries, M.T.J. Livermore, D.J. Bradley, J. Cox, and A.J. McMichael, 1999: Climate change and future populations at risk of malaria. *Global Environmental Change*, **9,** S89–S107.

McBean, G.A., P.S. Liss, and S.H. Schneider, 1996: Advancing our understanding. In: *Climate Change 1995: The Science of Climate Change. Contribution of Working Group I to the Second Assessment Report of the Intergovernmental Panel on Climate Change* [Houghton, J.T., L.G. Meira Filho, B.A. Callander, N. Harris, A. Kattenberg, and K. Maskell (eds.)]. Cambridge University Press, Cambridge, United Kingdom and New York, NY, USA, pp. 517–530.

McClelland, A.G.R. and F. Bolger, 1994: The calibration of subjective probabilities: theories and models 1980–1994. In: *Subjective Probability* [Wright, G. and P. Ayton (eds.)]. John Wiley and Sons, New York, NY, USA, pp. 453–482.

Mearns, L.O., 1995: Research issues in determining the effects of changing climatic variability on crop yields. *Climate Change and Agriculture: Analysis of Potential International Impacts*. American Society of Agronomy Special Publication No. 58, Madison, WI, USA, pp. 123–146.

Mearns, L.O., T. Mavromatis, E. Tsvetsinskaya, C. Hays, and W. Easterling, 1999: Comparative responses of EPIC and CERES crop models to high and low spatial resolution climate change scenarios. *Journal of Geophysical Research—Atmospheres*, **104(D6),** 6623–6646.

Mendelsohn, R., A. Andronova, W. Morrison, and M. Schlesinger, 2000: Country-specific market impacts of climate change. *Climatic Change*, **45,** 553–569.

Mendelsohn, R. and J. Neumann (eds.), 1999: *The Impacts of Climate Change on the U.S. Economy*. Cambridge University Press, Cambridge, United Kingdom and New York, NY, USA, 331 pp.

Menzel, A. and P. Fabian, 1999: Growing season extended in Europe. *Nature*, **397,** 659.

Mimura, N., K. Satoh, and S. Machida, 2000: Asian and Pacific vulnerability assessment—an approach to integrated regional assessment. In: *Proceedings of the Thai-Japanese Geological Meeting. The Comprehensive Assessments on Impact of Sea-Level Rise*. pp. 123–128.

Mitchell, R. and R. Carson, 1989: *Using Surveys to Value Public Goods: The Contingent Valuation Method*. Resources for the Future, Washington, DC, USA, 268 pp.

Morgan, G. and H. Dowlatabadi, 1996: Learning from integrated assessment of climate change, *Climatic Change*, **34(3–4),** 337–368.

Morgan, M.E., C. Badgley, G.F. Gunnell, P.D. Gingerich, J.W. Kappelman, M.C. Maas, 1995: Comparative paleoecology of Paleogene and Neogene mammalian faunas: body-size structure. *Palaeogeography Palaeoclimatology Palaeoecology*, **115(1–4),** 287–317.

Morgan, M.G., 1998: Uncertainty analysis in risk assessment. *Human and Ecological Risk Assessment*, **4(1),** 25–39.

Morgan, M.G. and M. Henrion, 1990: *Uncertainty: A Guide to Dealing with Uncertainty in Quantitative Risk and Policy Analysis*. Cambridge University Press, Cambridge, United Kingdom and New York, NY, USA, 332 pp.

Morgan, M.G. and D.W. Keith, 1995: Subjective judgments by climate experts. *Environmental Science and Technology*, **29,** 468A–476A.

Morgenstern, R.D., 1997: *Economic Analysis at EPA: Accessing Regulatory Impact*. Resources for the Future, Washington, DC, USA, 478 pp.

Morita, T.S., P.R. Shukla, and O.K. Cameron, 1997b: Epistemological gaps between integrated assessment models and developing countries. *Proceedings of the IPCC Asia-Pacific Workshop on Integrated Assessment Models*. Center for Global Environmental Research, National Institute for Environmental Studies, Tsukuba, Japan, pp. 125–138.

Morris, M.D., 1979: *Measuring the Condition of the World's Poor: The Physical Quality of Life Index*. Pergamon Press, New York, NY, USA, 176 pp.

Moss, R.H. and S.H. Schneider, 2000: Uncertainties in the IPCC TAR: recommendations to lead authors for more consistent assessment and reporting. In: *Guidance Papers on the Cross Cutting Issues of the Third Assessment Report of the IPCC* [Pachauri, R., T. Taniguchi, and K. Tanaka (eds.)]. World Meteorological Organization, Geneva, Switzerland, pp. 33–51.

Moss, R.H. and S.H. Schneider, 1997: Characterizing and communicating scientific uncertainty: building on the IPCC second assessment. In: *Elements of Change* [Hassol, S.J. and J. Katzenberger (eds.)]. Aspen Global Change Institute, Aspen, CO, USA, pp. 90–135.

Murphy, A.H. and R.L. Winkler, 1984: Probability forecasting in meteorology. *Journal of the American Statistical Association*, **79,** 489–500.

Murty, N.S., M. Panda, and J. Parikh, 1997: Economic development, poverty reduction, and carbon emissions in India. *Energy Economics*, **19,** 327–354.

Myneni, R.B., G. Asrar, C.D. Keeling, R.R. Nemani, and C.J. Tucker, 1997: Increased plant growth in the northern high latitudes from 1981 to 1991. *Nature*, **386,** 698–702.

Nordhaus, W.D., 1994a: Expert opinion on climatic change. *American Scientist*, **82,** 45–52.

Nordhaus, W.D., 1994b: *Managing the Global Commons: The Economics of Climate Change*. MIT Press, Cambridge, MA, USA, 239 pp.

Nordhaus, W.D., 1992: An optimal transition path for controlling greenhouse gases. *Science*, **258,** 1315–1319.

Nordhaus, W.D., 1991: To slow or not to slow: the economics of the greenhouse effect. *Economic Journal*, **101,** 920–937.

Nordhaus, W.D. and J. Boyer, 1999: Requiem for Kyoto: An economic analysis of the Kyoto Protocol. *The Energy Journal, Special Issue—The Costs of the Kyoto Protocol: A Multi-Model Evaluation*, **20,** pp. 93–130.

Nordhaus, W.D. and Z. Yang, 1996: RICE: a regional dynamic general equilibrium model of optimal climate-change policy. *American Economic Review*, **86(4),** 741–765.

Oppenheimer, M., 1998: Global warming and the stability of the west Antarctic ice sheet. *Nature*, **393,** 325–332.

Overpeck, J.T., R.S. Webb, and T. Webb III, 1992: Mapping eastern North American vegetation change over the past 18,000 years: no analogs and the future. *Geology*, **20,** 1071–1074.

Parikh, J. and S. Gokarn, 1993: Climate change and India's energy policy options: new perspectives on sectoral CO_2 emissions and incremental costs. *Global Environmental Change*, **3,** 3.

Parikh, J.K. and K.S. Parikh, 1998: *Accounting and Valuation of Environment: Vol. II, Case Studies from the ESCAP Region*. Economic and Social Commission for Asia and the Pacific Region, United Nations, New York, NY, USA.

Parikh, J.K. and K.S. Parikh, 1997: *Accounting and Valuation of Environment: Vol. I, A Primer for Developing Countries*. Economic and Social Commission for Asia and the Pacific Region, United Nations, New York, NY, USA.

Parmesan, C., 2001: Butterflies as bio-indicators for climate change impacts. In: *Evolution and Ecology Taking Flight: Butterflies as Model Systems* [Boggs, C.L., P.R. Ehrlich, and W.B. Watt (eds.)]. University of Chicago Press, Chicago, IL, USA, (in press).

Parmesan, C., 1996: Climate and species' range. *Nature*, **382,** 765–766.

Parmesan, C., T.L. Root, and M. Willig, 2000: Impacts of extreme weather and climate on terrestrial biota. *Bulletin of the American Meteorological Society*, **81,** 443–450.

Parmesan, C., N. Ryrholm, C. Stefanescu, J. Hill, C. Thomas, H. Descimon, B. Huntley, L. Kaila, J. Kullberg, T. Tammaru, W.J. Tennent, J.A. Thomas, and M. Warren, 1999: Poleward shifts of butterfly species ranges associated with regional warming. *Nature*, **399(6736),** 579–583.

Parry, M. and T. Carter, 1998: *Climate Impact and Adaptation Assessment*. Earthscan Publications, London, United Kingdom, 166 pp.

Parry, M. and M. Livermore, 1999: A new assessment of the global effect of climate change. *Global Environmental Change, Supplementary Issue*. Pergamon Press, New York, NY, USA, **9,** S1–S107.

Parson, E.A., 1997: Informing global environmental policy making: A plea for new of assessment and synthesis. *Environmental Modeling and Assessment*, **2,** 267–279.

Parson, E.A. and H. Ward, 1998: Games and simulations. In: *Human Choice and Climate Change, Vol. 3* [Rayner, S. and E. Malone (eds.)]. Battelle Press, Columbus, OH, USA, pp. 105–140.

Peck, S.C. and T.J. Teisberg, 1995: Optimal CO_2 control policy with stochastic losses from temperature rise. *Climatic Change*, **31,** 19–34.

Peck, S.C. and T.J. Teisberg, 1994: Optimal carbon emissions trajectories when damages depend on the rate or level of global warming. *Climatic Change*, **28,** 289–314.

Peck, S.C. and T.J. Teisberg, 1992: CETA: a model of carbon emissions trajectory assessment. *The Energy Journal,* **13 (1),** 55–77.

Peterson, A.T. and K.P. Cohoon, 1999: Sensitivity of distributional prediction algorithms to geographic data completeness. *Ecological Modelling*, **117,** 159–164.

Petschel-Held, G., H.J. Schellnhuber, T. Bruckner, F. Toth, and K. Hasselmann, 1999: The tolerable windows approach: theoretical and methodological foundations. *Climatic Change*, **41(3–4),** 303–331.

Pittock, A.B. and R.N. Jones, 1999: Adaptation to what, and why? *Environmental Monitoring and Assessment*, **61,** 9–35.

Portnoy, P. and J. Weyant (eds.), 1999: *Discounting and Intergenerational Equity*. Resources for the Future, Washington, DC, USA, 209 pp.

Precht, H., J. Christophersen, H. Hensel, and W. Larcher, 1973: *Temperature and Life*. Springer-Verlag, New York, NY, USA, 779 pp.

Raiffa, H., 1968: *Decision Analysis: Introductory Lectures on Choices under Uncertainty*. Addison Wesley, Reading, MA, USA, 309 pp.

Ravetz, J.R., 1986: Usable knowledge, usable ignorance: incomplete science with policy implications. In: *Sustainable Development of the Biosphere* [Clark, W. and R.E. Munn (eds.)]. Cambridge University Press, Cambridge, United Kingdom and New York, NY, USA, pp. 415–432.

Ray, A. 1984: *Issues in Cost-Benefit Analysis: Issues and Methodologies*. Johns Hopkins University Press, Baltimore, MD, USA, 176 pp.

Ringius, L., T.E. Downing, M. Hulme, D. Waughray, and R. Selrod, 1996: *Climate Change in Africa—Issues and Challenges in Agriculture and Water for Sustainable Development*. Report 1996-08, CICERO, Oslo, Norway, 179 pp.

Rodríguez-Trelles, F. and M.A. Rodríguez, 1998: Rapid micro-evolution and loss of chromosomal diversity in *Drosophila* in response to climate warming. *Evolutionary Ecology,* **12,** 829–838.

Rodríguez-Trelles, F., M.A. Rodríguez, and S. M. Scheiner, 1998: Tracking the genetic effects of global warming: Drosophila and other model systems. *Conservation Ecology*, **2(2)**. Available online at http://www.consecol.org/Journal/vol2/iss2/art2/.

Rodríguez-Trelles, F., G. Alvarez, and C. Zapata, 1996: Time-series analysis of seasonal changes of the O inversion polymorphism of Drosophila subobscura. *Genetics*, **142,** 179–187.

Root, T.L. and S.H. Schneider, 1995: Ecology and climate: research strategies and implications. *Science*, **269,** 331–341.

Rosen, S., 1974: Hedonistic prices and implicit markets: product differentiation in pure competition. *Journal of Political Economy*, **82,** 34–55.

Rosenberg, N.J. (ed.), 1993: Towards an integrated impact assessment of climate change: the MINK study. *Climatic Change (Special Issue)*, **24,** 1–173.

Rosenzweig, C. and M. Parry, 1994: Potential impact of climate change in world food supply. *Nature*, **367,** 133–138.

Rosenzweig, C. and F.N. Tubiello, 1997: Impacts of global climate change on Mediterranean agriculture: current methodologies and future direction—an introductory essay. *Mitigation and Adaptation Strategies for Global Change*, **1(3),** 219–232.

Rothschild, M. and J.E. Stiglitz, 1973: Some further results in the measurement of inequality. *Journal of Economic Theory*, **6,** 188–204.

Rotmans, J. and H. Dowlatabadi, 1998: Integrated assessment modeling. In: *Human Choice and Climate Change, Vol. 3* [Rayner, S. and E. Malone (eds.)]. Battelle Press, Columbus, OH, USA, pp. 291–378.

Saether, B.-E., 1997: Environmental stochasticity and population dynamics of large herbivores: a search for mechanisms. *Trends in Ecology and Evolution*, **12,** 143–149.

Schneider, S.H., 1997: Integrated assessment modeling of global climate change: transparent rational tool for policymaking or opaque screen hiding value-laden assumptions? *Environmental Modeling and Assessment*, **2(4),** 229–248.

Schneider, S.H., 1983: CO_2, climate and society: a brief overview. In: *Social Science Research and Climate Change: In Interdisciplinary Appraisal* [Chen, R.S., E. Boulding, and S.H. Schneider (eds.)]. D. Reidel, Boston, MA, USA, pp. 9–15.

Schneider, S.H., K. Kuntz-Duriseti, and C. Azar, 2000: Costing non-linearities, surprises, and irreversible events. *Pacific and Asian Journal of Energy,* **10(1),** 81–106.

Schneider, S.H., B.L. Turner, and H. Morehouse Garriga, 1998: Imaginable surprise in global change science. *Journal of Risk Research,* **1(2),** 165–185.

Schoemaker, P.J.H., 1991: When and how to use scenario planning: a heuristic approach with illustration. *Journal of Forecasting,* **10,** 549–564.

Semenov, M.A. and J.R. Porter, 1995: Climatic variability and modeling of crop yields. *Agricultural and Forest Meteorology,* **73,** 265–283.

Sen, A., 1985: *Commodities and Capabilities.* Professor Dr. P. Hennipman Lectures. North Holland, Amsterdam, The Netherlands.

Smith, F.A., H. Browning, and U.L. Shepherd, 1998: The influence of climate change on the body mass of woodrats Neotoma in an arid region of New Mexico, USA. *Ecography,* **21(2),** 140–148.

Smith, V.K., 2000: JEEM and non-market valuation. *Journal of Environmental Economics and Management,* **39,** 351–374.

Sparks, T.H. and P.D. Carey, 1995: The responses of species to climate over two centuries: an analysis of the Marsham phenological record, 1736–1947. *Journal of Ecology,* **83,** 321–329.

Srinivasan, T.N., 1994: Human development: a new paradigm or reinvention of the wheel? *American Economic Review,* **84(2),** 238–243.

Stockwell, D.R.B. and I.R. Noble, 1991: Induction of sets of rules from animal distribution data: a robust and informative method of data analysis. *Mathematics and Computers in Simulation,* **32,** 249–254.

Sullivan, R.M. and T.L. Best, 1997: Best effects of environment of phenotypic variation and sexual dimorphism in Dipodomys simiulans (Rodentia: heteromyidae). *Journal of Mammalogy,* **78,** 798–810.

Takahashi, K., Y. Matsuoka, and H. Harasawa,, 1998: Impacts of climate change on water resources, crop production, and natural ecosystems in the Asia and Pacific region. *Journal of Global Environment Engineering,* **4,** 91–103.

Tessier, L., F. Guibal, and F.H. Schweingruber, 1997: Research strategies in dendroecology and dendroclimatology in mountain environments. *Climatic Change,* **36(3–4),** 499–517.

Timmermann, A., J. Oberhuber, A. Bacher, M. Esch, M. Latif, and E. Roeckner, 1999: Increased El Niño frequency in a climate model forced by future greenhouse warming. *Nature,* **398,** 694–696.

Titus, J. and V. Narayanan, 1996: The risk of sea level rise: a delphic Monte Carlo analysis in which twenty researchers specify subjective probability distributions for model coefficients within their respective areas of expertise. *Climatic Change,* **33(2),** 151–212.

Tiwari, P. and J. Parikh, 1997: Demand for housing in the Bombay metropolitan region. *Journal of Policy Modelling,* **19(3),** 295–321.

Tol, R.S.J., 1999a: *New Estimates of the Damage Costs of Climate Change. Part I: Benchmark Estimates.* Working Paper D99-01, Institute for Environmental Studies, Vrije Universiteit, Amsterdam, The Netherlands, 32 pp.

Tol, R.S.J., 1999b: *New Estimates of the Damage Costs of Climate Change. Part II: Dynamic Estimates.* Working Paper D99-02, Institute for Environmental Studies, Vrije Universiteit, Amsterdam, The Netherlands, 31 pp.

Tol, R.S.J., 1999c: The marginal costs of greenhouse gas emissions. *The Energy Journal,* **20(1),** 61–81.

Tol, R.S.J., 1999d: Time discounting and optimal control of climate change—an application of FUND. *Climatic Change,* **41(3–4),** 351–362.

Tol, R.S.J., 1997: On the optimal control of carbon dioxide emissions: an application of FUND. *Environmental Modeling and Assessment,* **2,** 151–163.

Tol, R.S.J. and S. Fankhauser, 1998: On the representation of impact in integrated assesment models of climate change. *Environmental Modeling and Assessment,* **3,** 63–74.

Tol, R.S.J. and A.F. de Vos, 1998: A Bayesian statistical analysis of the enhanced greenhouse effect. *Climatic Change,* **38,** 87–112.

Toth, F., 2000a: Decision analysis frameworks in TAR: a guidance paper for IPCC. In: *Guidance Papers on the Cross Cutting Issues of the Third Assessment Report of the IPCC* [Pachauri, R., T. Taniguchi, and K. Tanaka (eds.)]. Intergovernmental Panel on Climate Change, Geneva, Switzerland, pp. 53–68.

Toth, F.L., 2000b: Intergenerational equity and discounting. *Integrated Assessment,* **2,** 127–136.

Toth, F.L., 1992a: Policy implications. In: *The Potential Socioeconomic Effects of Climate Change in South-East Asia.* [Parry, M.L., M. Blantran de Rozari, A.L. Chong, and S. Panich (eds).]. United Nations Environment Programme, Nairobi, Kenya, pp. 109–121.

Toth, F.L., 1992b: Policy responses to climate change in Southeast Asia. In: *The Regions and Global Warming: Impacts and Response Strategies* [Schmandt, J. and J. Clarkson (eds).]. Oxford University Press, New York, NY, USA. pp. 304–322.

Toth, F.L., 1988a: Policy exercises: objectives and design elements. *Simulation & Games,* **19(3),** 235–255.

Toth, F.L., 1988b: Policy exercises: procedures and implementation. *Simulation & Games,* **19(3),** 256–276.

Toth, F.L., 1986: *Practicing the Future: Implementing the "Policy Exercise" Concept.* Working Paper 86-23, International Institute for Applied Systems Analysis, Laxenburg, Austria, 31 pp.

Toth, F.L., T. Bruckner, H.-M. Füssel, M. Leimbach, G. Petschel-Held, and H.-J. Schellnhuber, 1997: The tolerable windows approach to integrated assessments. In: *Climate Change and Integrated Assessment Models: Bridging the Gaps—Proceedings of the IPCC Asia Pacific Workshop on Integrated Assessment Models, United Nations University, Tokyo, Japan, March 10–12, 1997* [Cameron, O.K., K. Fukuwatari, and T. Morita (eds.)]. National Institute for Environmental Studies, Tsukuba, Japan, pp. 403–430.

Tversky, A. and D. Kahneman, 1983: Extensional versus intuitive reasoning: the conjunction fallacy in probability judgment. *Psychological Review,* **90,** 293–315.

Tversky, A. and D. Kahneman, 1974: Judgment under uncertainty: heuristics and biases. *Science,* **185,** 1124–1131.

UNDP, 1990: Human Development Report. United Nations Development Programme, Oxford University Press, New York, NY, USA.

UNEP, 1996: *UNEP Handbook on Methods for Climate Change Impact Assessment and Adaptation Strategies (draft).* Atmosphere Unit of United Nations Environment Programme, Nairobi, Kenya.

Uvarov, B.P., 1931: Insects and climate. *Royal Entomology Society of London,* **79,** 174–186.

van Asselt, M. and J. Rotmans, 1995: *Uncertainty in Integrated Assessment Modelling: A Cultural Perspective-based Approach.* GLOBO Report Series No.9, RIVM, Bilthoven, The Netherlands.

van Daalen, E., W. Thissen, and M. Berk, 1998: The Delft process: experiences with a dialogue between policymakers and global modellers. In: *Global Change Scenarios for the 21st Century. Results from the IMAGE2.1 Model* [Alcamo, J., E.Kreileman, and R. Leemans (eds.)]. Pergamon Press, Oxford, United Kingdom, pp. 267–286.

van der Heijden, K. 1998: Scenario planning: scaffolding disorganised ideas about the future. In: *Forecasting with Judgment* [Wright, G. and P. Goodwin (eds.)]. John Wiley and Sons, Chichester, United Kingdom, pp. 549–572.

Varian, H., 1992: *Microeconomic Analysis.* W.W. Norton, New York, NY, USA, 3rd. ed., 506 pp.

Vogel, R.B., H. Egger, and F.H. Schweingruber, 1996: Interpretation of extreme tree ring values in Switzerland based on records of climate between 1525 and 1800 A.D. *Vierteljahrsschrift der Naturforschenden Gesellschaft in Zuerich,* **141(2),** 65–76.

Wang, J. and L. Erda, 1996: The impacts of potential climate change and climate variability on simulated maize production in China. *Water, Air, and Soil Pollution,* **92,** 75–85.

WBGU, 1995: *Scenarios for the Derivation of Global CO_2 Reduction Targets and Implementation Strategies.* German Advisory Council on Global Change, Bremerhaven, Germany.

Weiser, W. (ed.), 1973: *Effects of Temperature on Ectothermic Organisms.* Springer-Verlag, New York, NY, USA, 298 pp.

Weitzman, M.L., 1998: Why the far distant future should be discounted at its lowest possible rate. *Journal of Environmental Economics and Management,* **36,** 201–208.

Weyant, J., O. Davidson, H. Dowlatabadi, J. Edmonds, M. Grubb, E.A. Parson, R. Richels, J. Rotmans, P.R. Shukla, and R.S.J. Tol, 1996: Integrated assessment of climate change: an overview and comparison of approaches and results. In: *Climate Change 1995: Economic and Social Dimensions of Climate Change. Contribution of Working Group III to the Second Assessment Report of the Intergovernmental Panel on Climate Change* [Bruce, J.P., H. Lee, and E.F. Haites (eds.)]. Cambridge University Press, Cambridge, United Kingdom and New York, NY, USA, pp. 367–396.

White, A., M.G.R. Cannell, and A.D. Fried, 1999: Climate change impacts on ecosystems and the terrestrial carbon sink: a new assessment. *Global Environmental Change, Supplementary Issue*, **9,** S21–S30.

Wigley, T., R. Richels, and J.A. Edmonds, 1996: Economic and environmental choices in the stabilization of atmospheric CO_2 concentrations. *Nature*, **379(6582),** 240–243.

Woodward, F.I., 1987: *Climate and Plant Distribution*. Cambridge University Press, Cambridge, United Kingdom and New York, NY, USA, 174 pp.

Woodward, F.I., T.M. Smith, and W.R. Emanuel, 1995: A global primary productivity and phytogeography model. *Global Biogeochemical Cycles*, **9,** 471–490.

Wright, G. and P. Ayton, 1992: Judgmental forecasting in the immediate and medium term. *Organizational Behavior and Human Decision Processes*, **51,** 344–363.

Yohe, G., 1997: Uncertainty, short term hedging and the tolerable window approach. *Global Environmental Change*, **7,** 303–315.

Yohe, G., 1996: Exercises in hedging against the extreme consequences of global change and the expected value of information. *Global Environmental Change*, **6,** 87–101.

Yohe, G. and M. Jacobsen, 1999: Meeting concentration targets in the post-Kyoto world: does Kyoto further a least cost strategy? *Mitigation and Adaptation Strategies for Global Change*, **4,** 1–23.

Yohe, G. and F. Toth, 2000: Adaptation and the guardrail approach to tolerable climate change. *Climatic Change*, **45,** 103–128.

Yohe, G., D. Montgomery, and P. Bernstein, 2000: Equity and the Kyoto Protocol: measuring the distributional effects of alternative emissions trading regimes. *Global Environmental Change*, **10,** 121–132.

Yohe, G., T. Malinowski, and M. Yohe, 1998: Fixing global carbon emissions: choosing the best target year. *Energy Policy*, **26,** 219–231.

Yoshino, M. and H.P. Ono, 1996: Variations in the plant phenology affected by global warming. In: *Climate Change and Plants in East Asia* [Omasa, K., H. Kai, Z. Taoda, and M. Uchijima (eds.)]. Springer-Verlag, Dordrecht, The Netherlands, 93–107.

Zeidler, R.B., 1997: Climate change vulnerability and response strategies for the coastal zone of Poland. *Climatic Change*, **36,** 151–173.

3

Developing and Applying Scenarios

TIMOTHY R. CARTER (FINLAND) AND EMILIO L. LA ROVERE (BRAZIL)

Lead Authors:
R.N. Jones (Australia), R. Leemans (The Netherlands), L.O. Mearns (USA), N. Nakicenovic (Austria), A.B. Pittock (Australia), S.M. Semenov (Russian Federation), J. Skea (UK)

Contributing Authors:
S. Gromov (Russian Federation), A.J. Jordan (UK), S.R. Khan (Pakistan), A. Koukhta (Russian Federation), I. Lorenzoni (UK), M. Posch (The Netherlands), A.V. Tsyban (Russian Federation), A. Velichko (Russian Federation), N. Zeng (USA)

Review Editors:
Shreekant Gupta (India) and M. Hulme (UK)

CONTENTS

EXECUTIVE SUMMARY

What are Scenarios and What is Their Role?

A scenario is a coherent, internally consistent, and plausible description of a possible future state of the world. Scenarios commonly are required in climate change impact, adaptation, and vulnerability assessments to provide alternative views of future conditions considered likely to influence a given system or activity. A distinction is made between climate scenarios—which describe the forcing factor of focal interest to the Intergovernmental Panel on Climate Change (IPCC)—and nonclimatic scenarios, which provide socioeconomic and environmental "context" within which climate forcing operates. Most assessments of the impacts of future climate change are based on results from impact models that rely on quantitative climate and nonclimatic scenarios as inputs.

Types of Scenarios

Socioeconomic scenarios can serve multiple roles within the assessment of climate impacts, adaptation, and vulnerability. Until recently, they have been used much more extensively to project greenhouse gas (GHG) emissions than to assess climate vulnerability and adaptive capacity. Most socioeconomic scenarios identify several different topics or domains, such as population or economic activity, as well as background factors such as the structure of governance, social values, and patterns of technological change. Scenarios make it possible to establish baseline socioeconomic vulnerability, pre-climate change; determine climate change impacts; and assess post-adaptation vulnerability.

Land-use and land-cover scenarios should be a major component of scenarios for climate change impact and adaptation assessments. A great diversity of land-use and land-cover change scenarios have been constructed. However, most of these scenarios do not address climate change issues explicitly; they focus instead on other issues—for example, food security and carbon cycling. Large improvements have been made since the Second Assessment Report (SAR) in defining current and historic land-use and land-cover patterns, as well as in estimating future scenarios. Integrated assessment models currently are the most appropriate tools for developing land-use and land-cover change scenarios.

Environmental scenarios embrace changes in environmental factors other than climate that will occur in the future regardless of climate change. Because these changes could have an important role in modifying the impacts of future climate change, scenarios are required to portray possible future environmental conditions, such as atmospheric composition [e.g., carbon dioxide (CO_2), tropospheric ozone (O_3), acidifying compounds, and ultraviolet (UV)-B radiation]; water availability, use, and quality; and marine pollution. Apart from the direct effects of CO_2 enrichment, changes in other environmental factors rarely have been considered alongside climate changes in past impact assessments, although their use is increasing with the emergence of integrated assessment methods.

Climate scenarios of three main types have been employed in impact assessments: incremental scenarios, analog scenarios, and climate model-based scenarios. Of these, the most common use outputs from general circulation models (GCMs) and usually are constructed by adjusting a baseline climate (typically based on regional observations of climate over a reference period such as 1961–1990) by the absolute or proportional change between the simulated present and future climates. Most recent impact studies have constructed scenarios on the basis of transient GCM outputs, although some still apply earlier equilibrium results. Regional detail is obtained from the coarse-scale outputs of GCMs by using three main methods: simple interpolation, statistical downscaling, and high-resolution dynamic modeling. The simple method, which reproduces the GCM pattern of change, is the most widely applied in scenario development. In contrast, the statistical and modeling approaches can produce local climate changes that are different from the large-scale GCM estimates. More research is needed to evaluate the value added to impact studies of such regionalization exercises. One reason for this caution is the large uncertainty of GCM projections, which requires further quantification through model intercomparisons, new model simulations, and pattern-scaling methods. Such research could facilitate future evaluation of impacts in a risk assessment framework.

Sea-level rise scenarios are required to evaluate a diverse range of threats to human settlements, natural ecosystems, and landscape in coastal zones. Relative sea-level scenarios (i.e., sea-level rise with reference to movements of the local land surface) are of the most interest for impact and adaptation assessments. Tide gauge and wave height records of 50 years or more are required, along with information on severe weather and coastal processes, to establish baseline levels or trends. Although some components of future sea-level rise can be modeled regionally, using coupled ocean-atmosphere models, the most common method of obtaining scenarios is to apply global mean estimates from simple models. Changes in the occurrence of extreme events such as storm surges and wave set-up, which can lead to major coastal impacts, sometimes are investigated by superimposing historically observed events onto rising mean sea level. More recently, some studies have begun to express future sea-level rise in probabilistic terms, enabling rising levels to be evaluated in terms of the risk that they will exceed a critical threshold of impact.

How Useful have Scenarios Been in Past Impact and Adaptation Assessments?

Study of past assessments has highlighted problems of compatibility in the development and application of scenarios. These problems include difficulties in obtaining credible and compatible projections over long time horizons across different scientific disciplines, inconsistencies in scenarios adopted and their methods of application between different impact assessments, and time lags between reporting of recent climate science and the use of this science in developing scenarios for impact assessment. Furthermore, the use of nonclimatic scenarios at the regional level, alongside more conventional climate scenarios, is only a relatively recent introduction to impact assessment, and methods of scenario development (especially of socioeconomic scenarios) still are at a rudimentary level.

Scenarios of the 21st Century

The IPCC recently completed the *Special Report on Emissions Scenarios* (SRES) to replace the earlier set of six IS92 scenarios developed for the IPCC in 1992. These new scenarios consider the period 1990–2100 and include a range of socioeconomic assumptions [e.g., a global population by 2100 of 7.0–15.1 billion; average gross domestic product (GDP) of $197–550 trillion (1990 US$)]. Their implications for other aspects of global change also have been calculated. For example, mean ground-level O_3 concentrations in July over the industrialized continents of the northern hemisphere are projected to rise from about 40 ppb in 2000 to more than 70 ppb in 2100 under the highest illustrative SRES emissions scenarios. Estimates of CO_2 concentration range from 478 to 1099 ppm by 2100, accounting for the range of SRES emissions and uncertainties about the carbon cycle. This range of implied radiative forcing gives rise to an estimated global warming from 1990 to 2100 of 1.4–5.8°C, assuming a range of climate sensitivities. This range is somewhat higher than the 0.7–3.5°C of the SAR because of higher levels of radiative forcing in the SRES scenarios than in the IS92a-f scenarios, primarily because of lower sulfate aerosol emissions, especially after 2050. The equivalent range of estimates of global sea-level rise (for this range of global temperature change in combination with a range of ice-melt sensitivities) to 2100 is 9–88 cm (compared to 15–95 cm in the SAR).

In terms of *mean changes of climate*, results from GCMs that have been run to date—assuming the new SRES emissions scenarios—display many similarities with previous runs. Rates of warming are expected to be greater than the global average over most land areas and most pronounced at high latitudes in winter. As warming proceeds, northern hemisphere snow cover and sea-ice extent will be reduced. Models indicate warming below the global average in the North Atlantic and circumpolar southern ocean regions, as well as in southern and southeast Asia and southern South America in June–August. Globally, there will be increases in average water vapor and precipitation. Regionally, December–February precipitation is expected to increase over the northern extratropics and Antarctica and over tropical Africa. Models also agree on a decrease in precipitation over Central America and little change in southeast Asia. Precipitation in June–August is expected to increase in high northern latitudes, Antarctica, and south Asia; to change little in southeast Asia; and to decrease in Central America, Australia, southern Africa, and the Mediterranean region.

Changes in the frequency and intensity of extreme climate events also can be expected. Based on the conclusions of the Working Group I report and the likelihood scale employed there, under GHG forcing to 2100, it is very likely that daytime maximum and minimum temperatures will increase, accompanied by an increased frequency of hot days. It also is very likely that heat waves will become more frequent; the number of cold waves and frost days (in applicable regions) will decline. Increases in high-intensity precipitation events are likely at many locations, and Asian summer monsoon precipitation variability also is likely to increase. The frequency of summer drought will increase in many interior continental locations, and it is likely that droughts, as well as floods, associated with El Niño events will intensify. The peak wind intensity and mean and peak precipitation intensities of tropical cyclones are likely to increase. The direction of changes in the average intensity of mid-latitude storms cannot be determined with current climate models.

How can We Improve Scenarios and Their Use?

Methods of scenario construction and application are evolving rapidly, but numerous deficiencies still must be addressed:

- Representing and integrating future nonclimatic (socioeconomic, environmental, and land-use) changes in scenarios for impact assessment
- Treatment of scenario uncertainties
- The requirement for scenario information at higher spatial and temporal resolution
- Representing changes in variability and the frequency of extreme events in scenarios
- Reducing time lags between climate modeling and climate scenario development
- Increasing attention on the construction of policy-relevant scenarios (e.g., stabilization of atmospheric CO_2 concentration)
- Recognizing linkages between scenarios for studies of mitigation, impacts, and adaptation
- Improving guidance material and training in construction and application of scenarios.

3.1. Definitions and Role of Scenarios

3.1.1. *Introduction*

This chapter examines the development and application of scenarios required for assessment of climate change impacts, adaptation, and vulnerability. Scenarios are one of the main tools for assessment of future developments in complex systems that often are inherently unpredictable, are insufficiently understood, and have high scientific uncertainties. The central goals of the chapter are to set out the different approaches to scenario use, to evaluate the strengths and weaknesses of these approaches, and to highlight key issues relating to scenario application that should be considered in conducting future assessments.

Recognizing the central role of scenarios in impact and adaptation studies, scenarios are treated separately for the first time by Working Group II.[1] This chapter builds on Chapter 13 of the WGI contribution to the Third Assessment Report (TAR), which describes construction of climate scenarios, by embracing scenarios that portray future developments of any factor (climatic or otherwise) that might have a bearing on climate change vulnerability, impacts, and adaptive capacity. A distinction is drawn between climate scenarios, which describe the forcing factor of key interest in this report, and nonclimatic scenarios (e.g., of projected socioeconomic, technological, land-use, and other environmental changes), which provide the "context"—a description of a future world on which the climate operates. Many early impact assessments tended to focus on climate forcing without properly considering the context, even though this might have an important or even dominant role in determining future vulnerability to climate.

In addition to serving studies of impacts, scenarios are vital aids in evaluating options for mitigating future emissions of greenhouse gases (GHGs) and aerosols, which are known to affect global climate. For instance, projections of future socioeconomic and technological developments are as essential for obtaining scenarios of future emissions as they are for evaluating future vulnerability to climate (see TAR WGIII Chapter 2). Thus, although the focus of this chapter is on the development and use of scenarios in impact and adaptation assessment, reference to scenarios that have been developed for purposes of addressing mitigation is important and unavoidable.

There is a varied lexicon for describing future worlds under a changing climate; alternative terms often reflect differing disciplinary origins. Therefore, for the sake of consistency in this chapter, working definitions of several terms are presented in Box 3-1.

[1]Hereafter the Working Group I, II, and III contributions to the Third Assessment Report (TAR) are referred to as WGI, WGII, and WGIII, respectively.

Box 3-1. Definitions

Projection. The term "projection" is used in two senses in this chapter. In general usage, a projection can be regarded as any description of the future and the pathway leading to it. However, a more specific interpretation was attached to the term "climate projection" throughout the Second Assessment Report (SAR) to refer to model-derived estimates of future climate.

Forecast/Prediction. When a *projection* is branded "most likely," it becomes a forecast or prediction. A forecast is often obtained by using deterministic models—possibly a set of such models—outputs of which can enable some level of confidence to be attached to projections.

Scenario. A scenario is a coherent, internally consistent, and plausible description of a possible future state of the world (IPCC, 1994). It is not a *forecast*; each scenario is one alternative image of how the future can unfold. A *projection* may serve as the raw material for a scenario, but scenarios often require additional information (e.g., about *baseline* conditions). A set of scenarios often is adopted to reflect, as well as possible, the range of uncertainty in projections. Indeed, it has been argued that if probabilities can be assigned to such a range (while acknowledging that significant unquantifiable uncertainties outside the range remain), a new descriptor is required that is intermediate between *scenario* and *forecast* (Jones, 2000). Other terms that have been used as synonyms for scenario are "characterization" (*cf.* Section 3.8), "storyline" (*cf.* Section 3.2), and "construction."

Baseline/Reference. The baseline (or reference) is any datum against which change is measured. It might be a "current baseline," in which case it represents observable, present-day conditions. It also might be a "future baseline," which is a projected future set of conditions, excluding the driving factor of interest. Alternative interpretations of reference conditions can give rise to multiple baselines.

3.1.2. *Function of Scenarios in Impact and Adaptation Assessment*

Selection and application of baseline and scenario data occupy central roles in most standard methodological frameworks for conducting climate change impact and adaptation assessment (e.g., WCC, 1993, 1994; IPCC, 1994; Smith *et al.*, 1996; Feenstra *et al.*, 1998; see Section 2.1). Many assessments treat scenarios *exogenously*, as an input, specifying key future socioeconomic and environmental baselines of importance for an exposure unit,[2] possibly with some aspects of adaptation potential also considered. Other assessments—especially those that use integrated assessment models (IAMs)—generate projections (e.g., of emissions, concentrations, climate, sea level) *endogenously* as outcomes, requiring only prior specification of the key driving variables (e.g., economic development, population). Outputs from such assessments might be applied themselves as scenarios for downstream analysis. Moreover, in IAMs, some of the original driving variables may be modified through modeled feedbacks.

Scenarios are widely used in climate change-related assessments. For some uses, scenarios are qualitative constructions that are intended to challenge people to think about a range of alternative futures that might go beyond conventional expectations or "business as usual" (BAU). Some of the socioeconomic and technological assumptions underlying GHG emissions scenarios are of this type (see TAR WGIII Chapter 2). For other uses, scenarios may be mainly quantitative, derived by running models on the basis of a range of different input assumptions. Most assessments of the impacts of future climate change are based on results from impact models that rely on quantitative climate and nonclimatic scenarios as inputs. Some scenario exercises blend the two approaches. However, not all impact assessments require a scenario component; in some cases, it may be sufficient that system sensitivities are explored without making any assumptions about the future.

3.1.3. Approaches to Scenario Development and Application

The approaches employed to construct scenarios vary according to the purpose of an assessment. For instance, scenarios may be required for:

- Illustrating climate change (e.g., by depicting the future climate expected in a given region in terms of the present-day climate currently experienced in a familiar neighboring region)
- Communicating potential consequences of climate change (e.g., by specifying a future changed climate to estimate potential future shifts in natural vegetation and identifying species at risk of local extinction)
- Strategic planning (e.g., by quantifying possible future sea-level and climate changes to design effective coastal or river flood defenses)
- Guiding emissions control policy (e.g., by specifying alternative socioeconomic and technological options for achieving some prespecified GHG concentrations)
- Methodological purposes (e.g., by describing altered conditions, using a new scenario development technique, or to evaluate the performance of impact models).

A broad distinction can be drawn between exploratory scenarios, which project anticipated futures, and normative scenarios, which project prescribed futures. In practice, however, many scenarios embrace aspects of both approaches.

3.1.3.1. *Exploratory Scenarios*

Exploratory (or descriptive) scenarios describe how the future might unfold, according to known processes of change or as extrapolations of past trends. They are sometimes described as BAU scenarios; often they involve no major interventions or paridigm shifts in the organization or functioning of a system but merely respect established constraints on future development (e.g., finite resources, limits on consumption). However, the term "business-as-usual" may be misleading because exploratory scenarios also can describe futures that bifurcate at some point (an example might be uptake or rejection of a new technology) or that make some assumptions about regulation and/or adaptation of a system. The simplest model is a direct extrapolation of past trends (e.g., projection of future agricultural crop productivity often is based on extrapolation of recorded increases in productivity; Mela and Suvanto, 1987; Alexandratos, 1995). Most climate scenarios considered in this report can be regarded as exploratory: They are future climates that might occur in the absence of explicit policies of GHG reduction.

3.1.3.2. *Normative Scenarios*

Normative (or prescriptive) scenarios describe a prespecified future, presenting "a picture of the world achievable (or avoidable) only through certain actions. The scenario itself becomes an argument for taking those actions" (Ogilvy, 1992). Normative scenarios span a wide spectrum, according to their degree of prescriptiveness. At one end of the spectrum are scenarios that are constrained in only one or a few dimensions. For example, scenarios that lead to a substantial degree of climate change sometimes are used as a reference for assessing the "worst case" as far as impacts are concerned (e.g., scenarios that explore extreme events and tails of frequency distributions).

At the other extreme of the spectrum are comprehensive, multidimensional normative scenarios that are constructed to meet the constraints of a prescribed target world. Examples are scenarios that constrain emissions within bounds ("safe emissions corridors") that avoid inducing a critical climate change, defined according to a subjectively selected impact criterion (Alcamo and Kreileman, 1996). Most of the emissions stabilization scenarios explored by the IPCC in recent assessments (IPCC, 1996a; Schimel *et al.*, 1997a) are founded on similar premises.

3.1.4. What Changes are Being Considered?

The types of scenarios examined in this chapter are depicted schematically in Figure 3-1; they include scenarios of:

- *Socioeconomic factors* (Section 3.2), which are the major underlying anthropogenic cause of environmental

[2] An exposure unit is an activity, group, region, or resource that is exposed to significant climatic variations (IPCC, 1994).

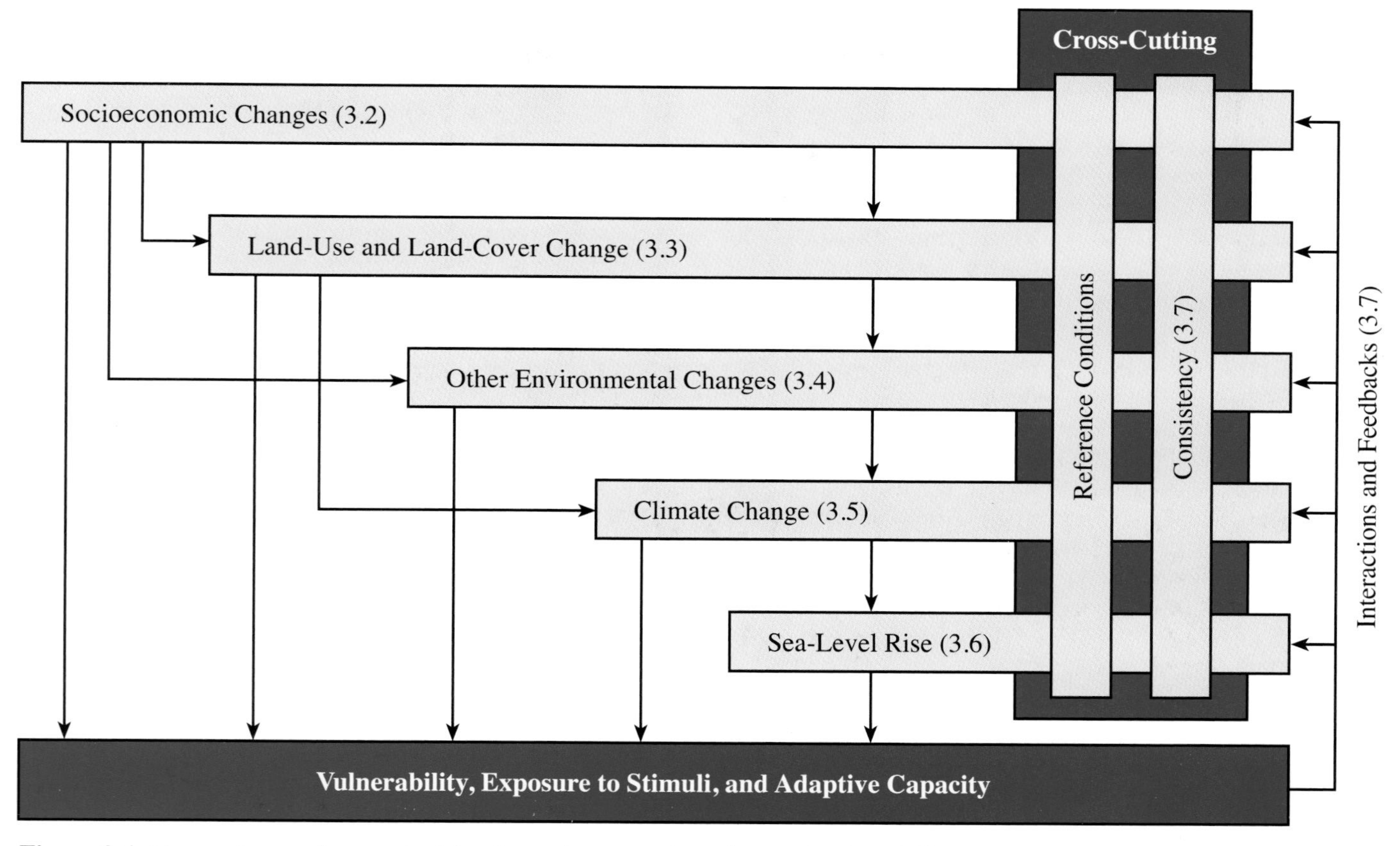

Figure 3-1: Types of scenarios required in climate impact, adaptation, and vulnerability assessment and their interactions. Numbers in parentheses refer to sections of this chapter.

change and have a direct role in conditioning the vulnerability of societies and ecosystems to climatic variations and their capacity to adapt to future changes.

- *Land use and land cover* (Section 3.3), which currently are undergoing rapid change as a result of human activities. Climate change itself may induce land-use and land-cover changes, with probable feedbacks to the climate system. Furthermore, future land cover may be influenced by efforts to sequester carbon and offset GHG emissions into the atmosphere.
- *Other environmental factors* (Section 3.4), which is a catch-all for a range of nonclimatic changes in the natural environment (e.g., CO_2 concentration, air pollution, stratospheric ozone depletion, and freshwater availability) that are projected to occur in the future and could substantially modify the vulnerability of a system or activity to impacts from climate change.
- *Climate* (Section 3.5), which is the focus of the IPCC and underpins most impact assessments reported in this volume.
- *Sea level* (Section 3.6), which generally is expected to rise relative to the land (with some regional exceptions) as a result of global warming—posing a threat to some low-lying coasts and islands.

Issues that are common to all scenarios concerning scenario consistency and the interactions and feedbacks between scenarios are treated in Section 3.7. Characterizations of future climate and related conditions during the 21st century, based on the new IPCC emissions scenarios, are introduced in Section 3.8, and the chapter closes with a brief examination of key gaps in knowledge and emerging new methods of scenario development.

3.2. Socioeconomic Scenarios

3.2.1. Purpose

The main purposes of socioeconomic scenarios in the assessment of climate impacts, adaptation, and vulnerability are:

- To characterize demographic, socioeconomic, and technological driving forces underlying anthropogenic GHG emissions that cause climate change
- To characterize the sensitivity, adaptive capacity, and vulnerability of social and economic systems in relation to climate change.

This section focuses on the second use. However, in integrated global assessments, scenarios underpinning these two applications should be consistent with one another. Many key parameters, such as population and economic growth, are common to both types of exercise. More flexibility with regard to consistency may be appropriate at local and regional scales. Regional trends may be diverse, and developments in a specific region may diverge from those at the global level.

The use of socioeconomic scenarios in assessing vulnerability to climate change is less well developed than their use in exploring GHG emissions. The IPCC *Technical Guidelines for Assessing Climate Change Impacts and Adaptations* (IPCC, 1994) recommend the use of socioeconomic scenarios, with and without climate change, to assess impacts and adaptive responses. At that time, few studies had reached that ideal. As new frameworks for characterizing vulnerability develop (Downing *et al.*, 1999), impact studies can begin to use more consistent, global scenario approaches.

Socioeconomic scenarios in impact assessment have tended to focus on quantitative characterization of key parameters and to ignore the qualitative "storyline" elements of a fully developed scenario approach. If the implications of climate change impacts and adaptation on sustainable development are to be assessed (Munasinghe, 2000), much more sophisticated descriptions of vulnerable impact units will be required, along with better understanding of institutional and economic coping capacity. Section 3.2.4 provides examples of emerging work of this kind.

Socioeconomic scenarios in general have been developed to aid decisionmaking under conditions of great complexity and uncertainty in which it is not possible to assign levels of probability to any particular state of the world at a future point in time. Therefore, it usually is not appropriate to make a statement of confidence concerning a specific socioeconomic scenario (Moss and Schneider, 2000). However, this does not mean that all scenarios are equally likely. Some, used to test sensitivities, may be at the limits of the range of plausibility. More robust statements may be possible about the level of confidence in specific quantitative indicators, such as population or GDP, associated with given scenarios.

3.2.2. Representing Baseline Conditions

The socioeconomic baseline describes the present or future state of all nonenvironmental factors that influence an exposure unit. The factors may be geographical (land use or communications), technological (pollution control, water regulation), managerial (forest rotation, fertilizer use), legislative (water-use quotas, air quality standards), economic (income levels, commodity prices), social (population, diet), or political (levels and styles of decisionmaking). The IPCC has published a set of baseline statistics for 195 countries that are representative of the early to mid-1990s (IPCC, 1998). The data were collected from a variety of sources, such as the World Bank, the United Nations Environment Programme (UNEP), and the Food and Agriculture Organization (FAO) (see Table 3-1). These are only selected, summary data; individual impact studies are likely to require information on other factors or at a much higher spatial resolution.

Climate change impact assessment requires sound understanding of current socioeconomic vulnerabilities. These vulnerabilities have implications for deliberate adaptations that "involve conscious actions to mitigate or exploit the effects of climate change" (Adger, 1999). Many of those who are exposed will be vulnerable to a range of other stresses, *irrespective* of climate change (e.g., high population growth, rapid urbanization, environmental degradation, ambient air pollution, social inequality, infrastructure degradation, and health hazards). In time, stresses associated with the development process may reinforce those generated by climate change. For instance, sea-level rise causes saltwater intrusion, which can be aggravated by diverting freshwater outflows to satisfy the needs of agriculture, energy, and human consumption.

3.2.3. Constructing Socioeconomic Scenarios

Socioeconomic scenarios can be constructed in the same variety of ways and for the same variety of purposes as global change scenarios in general. In practice, a variety of approaches may be combined in a single exercise. The UNEP country studies program has developed detailed guidance on construction and use of socioeconomic scenarios (Tol, 1998). This guidance emphasizes the importance of avoiding simple extrapolation—especially for developing countries, which may be undergoing demographic or economic transition; the role of formal modeling in filling in, but not defining, scenarios; and the role of expert judgment in blending disparate elements into coherent and plausible scenarios.

Most socioeconomic scenarios cover several different topics or domains, such as population or economic activity. Table 3-1 shows the range of issues covered in recent scenario or scenario-based exercises.

3.2.3.1. Basic Drivers

Population and economic activity are characterized in quantitative terms in most scenario exercises. The degree of disaggregation according to world region, country, or sector varies from one study to another. Coverage of other socioeconomic domains also can vary markedly among different activities.

3.2.3.2. Underlying Socioeconomic Drivers

Some scenarios incorporate explicit assumptions about underlying socioeconomic drivers of change such as social values and governance institutions. These scenarios usually are generated through synthetic or expert judgment-led approaches, expressed in qualitative terms. Social values can affect the willingness of societies to preserve ecosystems or protect biodiversity. Institutional and governance factors affect the capacity of a society to organize and direct the resources needed to reduce climate vulnerability (Adger, 1999). Qualitative factors such as institutional effectiveness and social values are key determinants of the effectiveness of coping strategies for adapting to climate change (see Chapter 18). They

***Table 3-1**: Dimensions and attributes of socioeconomic scenarios reported in some recent climate change impact and adaptation assessments.*

Scenarios	**IPCC Base**[a]	**SRES**[b]	**Pakistan**[c]	**UKCIP**[d]	**ACACIA**[e]	**USNACC**[f]
Time frame/horizon	*Early 1990s*	*1990-2100*	*2020/2050*	*2020s/2050s*	*2020s/2050s/2080s*	*2050/2100*
Focus	*Impacts*	*Emissions*	*Impacts*	*Impacts*	*Impacts*	*Both*
Scenario attributes[g]						
– Economic growth	✔	✔	✔	✔	✔	✔
– Population	✔	✔	✔	✔		✔
– Land use	✔	✔	✔	✔		✔
– Energy	✔	✔	✔	✔		
– Agriculture/ food production	✔		✔	✔		
– Technological change		✔		✔	✔	✔
– Water	✔			✔		
– Level of governance		✔		✔	✔	
– Social values		✔		✔	✔	
– Contextual data						✔
– Institutional change						✔
– Biodiversity	✔			✔		
– Coastal zone management				✔		
– Settlement patterns				✔		
– Political organization					✔	
– Social policy		✔			✔	
– Environmental policy		✔			✔	
– Regional development		✔			✔	
– Literacy			✔			
– Health care			✔			

[a] IPCC Baseline Statistics (IPCC, 1998).
[b] IPCC Special Report on Emissions Scenarios (Nakicenovic *et al.*, 2000).
[c] UNEP Pakistan Country Study (Government of Pakistan, 1998).
[d] United Kingdom Climate Impacts Programme (Berkhout *et al.*, 1999).
[e] A Concerted Action Towards A Comprehensive Climate Impacts and Adaptations Assessment for the European Union (Parry, 2000).
[f] U.S. National Assessment of the Potential Consequences of Climate Variability and Change national-scenarios; additional scenarios were developed for individual regions and sectors (<http://www.nacc.usgcrp.gov/>).
[g] Categories, some of which overlap, used by authors of the scenarios.

determine adaptive capacity and hence the vulnerability of socioeconomic systems. They are critical in any assessment of the implications of climate change for development, equity, and sustainability (Munasinghe, 2000).

3.2.3.3. Technological Change

Technology critically affects the capacity to adapt to climate change; it confers opportunities and risks. For example, genetic modification of crops and other developments in agricultural technology could enhance that sector's ability to adapt to different climatic conditions. However, excessive reliance on one particular strain of plant might increase vulnerability.

Technological change must be characterized in quantitative and qualitative terms. It may be very difficult to identify specific features of a technology that could affect vulnerability to climate change. Expert judgments are needed about the direction in which change takes place, public acceptability of different options, and the rate of adoption in the marketplace. Quantitative assumptions will be needed about the rate of improvement of a technology, including its cost, overall efficiency in using resources to meet the need for a given service, and possible impacts at various scales.

3.2.3.4. Time Horizons

The appropriate time horizon for socioeconomic scenarios depends on the use to which they are put. Climate modelers often use scenarios that look forward 100 years or more. Socioeconomic scenarios with similar time horizons may be needed to drive models of climate change, climate impacts, and land-use change. However, policymakers also may wish to use socioeconomic scenarios as decision tools in framing current policies for climate change adaptation. In this context, time horizons on the order of 20 years may be more appropriate, reflecting the immediate needs of decisionmakers.

Short-term socioeconomic scenarios can still be very uncertain. "Surprises" such as economic slumps or booms, wars, or famines frequently occur in social and economic systems. Over the course of 50–100 years, even the most basic scenario drivers, such as population and aggregate economic activity, are highly uncertain, and their future development can be projected with any credibility only by using alternative scenarios. Moreover, technologies will have been replaced at least once, and those in use 100 years hence could have unimagined effects on climate sensitivity and vulnerability. Politically led developments in local, regional, and international systems of governance also will unfold along unpredictable paths.

3.2.3.5. *Spatial Resolution*

Global emissions scenarios form the framework for predicting climate change and variability impacts at the national level. To assess vulnerability and adaptation potential, national scenarios must account for biophysical and socioeconomic impacts. The potential for autonomous adaptations must be understood, reflecting the ability of nature and society to cope with climate change and climate variability. Many of the impacts of climate change on the coping ability of human systems are likely to be location-specific. Impact assessors therefore should make use of local/regional scenarios, where appropriate, and be wary of generalizing experiences from one location to another. Matching of regional scenarios may be difficult, however—for example, if data on population and land use are available at different levels of resolution.

3.2.3.6. *Sectoral Scenarios*

As illustrated in Table 3-1, scenario exercises often make specific assumptions about individual sectors. These sectors usually are chosen because they are considered particularly sensitive to climate change (e.g., water, agriculture/food) or because they are important sources or sinks for GHGs (e.g., energy, forestry). Detailed quantitative assumptions often are made about levels of future economic activity or the price of key commodities, which will influence adaptation strategies.

Formal modeling work generally is used to improve the detail, coherence, and internal consistency of socioeconomic variables that are susceptible to quantification. Expert judgment or stakeholder consultations may be used to build consensus around the characterization of more subjective and less quantifiable variables that relate to values and institutions. Stakeholder engagement also can provide a wealth of local expertise about specific impacts and vulnerabilities.

3.2.4. *Use of Socioeconomic Scenarios*

This section presents a set of case studies that illustrates a range of specific approaches to the construction of socioeconomic scenarios that are relevant to climate impact assessment.

3.2.4.1. *IPCC Special Report on Emissions Scenarios*

The IPCC's *Special Report on Emissions Scenarios* (SRES) (Nakicenovic *et al.*, 2000) was prepared to improve on the earlier set of six IS92 scenarios developed in 1992 (Leggett *et al.*, 1992). The SRES describes 40 scenarios in all—based on an extensive literature assessment, six alternative modeling approaches, and an "open process" that solicited worldwide participation and feedback. The scenarios (which are described in more detail in Section 3.8) cover the main demographic, economic, technological, and land-use driving forces of future emissions. They include emissions of all relevant GHGs plus sulfur dioxide (SO_2), carbon monoxide (CO), nitrogen oxides (NO_x), and nonmethane volatile organic hydrocarbons (VOCs). The IPCC specified that the scenarios should not include future policies that explicitly address climate change. However, they necessarily encompass other policies that may indirectly influence GHG sources and sinks. The scenarios suggest that assumptions about technology, rather than population and economic development, may be the most important driving force of future emissions.

The SRES emissions scenarios serve several purposes. First, they provide baselines of socioeconomic, technological, and land-use change, in combination with emissions trajectories, for the assessment of mitigation policies and measures (see TAR WGIII Chapter 2). Second, they can be used to drive the assessment of climate change (see TAR WGI and Section 3.8). Third, they provide a global socioeconomic framework for regional-scale assessment of impacts and adaptation [e.g., see the United Kingdom Climate Impacts Programme (UKCIP) and European ACACIA examples, below].

3.2.4.2. *UNEP Pakistan Country Study*

The Pakistan Environment Ministry has produced a suite of three socioeconomic scenarios to inform national climate impact and adaptation planning (Government of Pakistan, 1998). The scenarios focus on two reference years—2020 and 2050—and include a combination of quantitative and qualitative indicators. Quantitative scenarios are presented for population, economic growth, agricultural production, energy demand, and industrial output. Variations in future rates of literacy, health care, import tariffs, forest cover, and infrastructure are expressed in qualitative terms. The 2020 scenario is the more detailed of the two; it is a composite of existing national projections and scenarios produced for "nonclimate" policymaking. The 2020–2050 scenarios were developed for the sole purpose of informing climate impact assessments and are much less detailed (Tol, 1998).

3.2.4.3. UKCIP "Nonclimate Scenarios" for Climate Impact Assessment

A set of "nonclimate scenarios" has been developed to provide a common framework for assessing climate impacts and adaptation under the stakeholder-led UKCIP (Berkhout *et al.*, 1999). The scenarios were based on a broader "Environmental Futures" exercise (UK National Foresight Programme, 1999)—which, in turn, had drawn on emerging SRES work.

Four scenarios for the 2020s and 2050s were defined by two factors affecting the capacity and willingness of society to adapt to climate change: the extent to which social values reflect environmental concern and the effectiveness of governance institutions. Development of the scenarios involved extensive consultations with stakeholders (Lorenzoni *et al.*, 2000). As a result, detailed scenario characterization was confined to the 2020s. For each scenario, key national indicators were developed. These indicators included population and GDP, as well as more specific variables relating to land-use change, patterns of agricultural activity, water demand, and investment in coastal defense. In addition, climate vulnerability was assessed qualitatively in several "impact domains," including agriculture, water, biodiversity, coastal zone management, and infrastructure and the built environment.

The framework scenarios were found to be a useful starting point for subsequent studies. However, the scenarios needed to be articulated in more detail to be useful at the regional or sectoral level. More quantification generally was required. This exercise underlined the need for scenarios to be tailored for end users, while maintaining broad consistency about key indicators such as population and GDP.

3.2.4.4. ACACIA Scenarios for Europe

ACACIA (A Concerted Action Towards A Comprehensive Climate Impacts and Adaptations Assessment for the European Union) assessed climate impacts and potential adaptation in Europe to the 2080s (Parry, 2000). ACACIA elaborated four scenarios on the basis of a combination of the UKCIP and SRES approaches (Jordan *et al.*, 2000; see also Chapter 13). This analysis concluded that certain systems will thrive under some scenarios and will be inherently more vulnerable in others, *independent of* climate change. Adaptive strategies are likely to differ across the four scenarios. In addition, the manner in which society values different parts of the human and physical environment is markedly different under the different scenarios, with clear implications for adaptation policies.

3.2.4.5. U.S. National Assessment

The approach to socioeconomic scenarios adopted by the U.S. National Assessment of Climate Impacts was determined by the nature of the assessment process, with a national synthesis linking separate analyses in nine U.S. regions and five sectors (National Assessment Synthesis Team, 1998). Recognizing that the sensitivity of particular regions or sectors may depend on highly specific socioeconomic characteristics, the assessment adopted a two-part approach to scenario development. First, to allow national aggregation, high, medium, and low scenarios were specified for variables such as population and GDP to be used by all subnational analyses (NPA Data Services, 1999). Second, teams were asked to identify a small number of additional socioeconomic variables that would have the strongest and most direct influence on their particular region or sector. They developed and documented their own assumptions for these variables, following a consistent template developed by the National Assessment Synthesis Team (NAST) (Parson, 1999). High and low values then could be assumed for each key impact variable, without having to specify what combination of demographic, market, ecosystem, and technological factors caused it to take a particular value. Teams were advised to construct a small set of high- and low-impact scenarios on the basis of different combinations of assumptions about key impact variables. Instead of an idealized approach to scenario development, which would have attempted to specify all factors consistently across different sectors and regions, the more pragmatic and pluralistic approach adopted in the U.S. National Assessment allowed regional and sectoral specificities to be reflected.

3.3. Land-Use and Land-Cover Change Scenarios

3.3.1. Purpose

The land cover of the Earth has a central role in many important biophysical and socioeconomic processes of global environmental change. Contemporary land cover is changed mostly by human use; therefore, understanding of land-use change is essential in understanding land-cover change (Turner *et al.*, 1995). Land use is defined through its purpose and is characterized by management practices such as logging, ranching, and cropping. Land cover is the actual manifestation of land use (i.e., forest, grassland, cropland) (IPCC, 2000). Land-use change and land-cover change (LUC-LCC) involve several processes that are central to the estimation of climate change and its impacts (Turner *et al.*, 1995). First, LUC-LCC influences carbon fluxes and GHG emissions (Houghton, 1995; Braswell *et al.*, 1997). This directly alters atmospheric composition and radiative forcing properties. Second, LUC-LCC changes land-surface characteristics and, indirectly, climatic processes (Bonan, 1997; Claussen, 1997). Third, LUC-LCC is an important factor in determining the vulnerability of ecosystems and landscapes to environmental change (Peters and Lovejoy, 1992). LCC, for example—through nitrogen addition, drainage and irrigation, and deforestation (Skole and Tucker, 1993; Vitousek *et al.*, 1997)—may alter the properties and possible responses of ecosystems. Finally, several options and strategies for mitigating GHG emissions involve land cover and changed land-use practices (IPCC, 1996b).

The central role of LUC-LCC highlights the importance of its inclusion in scenario development for assessing global change impacts. To date this has not been done satisfactorily in most assessments (Leemans *et al.*, 1996a). For instance, in earlier emission scenarios (e.g., Leggett *et al.*, 1992), constant emission factors were applied to define land use-related methane (CH_4) and nitrous oxide (N_2O) emissions. Furthermore, linear extrapolations of observed deforestation rates were assumed, along with an averaged carbon content in deforested areas. The SRES scenarios (Nakicenovic *et al.*, 2000) have improved on the underlying LUC-LCC assumptions, considerably enhancing scenario consistency. Unfortunately, these SRES scenarios provide highly aggregate regional LUC-LCC information, which is difficult to use in impact assessments. A comprehensive treatment of the other roles of LUC-LCC in the climate system is still deficient. To highlight these shortcomings, this section reviews studies and approaches in which LUC-LCC information is applied to develop scenarios for both impact and mitigation assessment.

3.3.2. Methods of Scenario Development

3.3.2.1. Baseline Data

The SAR evaluated land-use and land-cover data sets and concluded that they often were of dubious quality (Leemans *et al.*, 1996a). Since the SAR, many statistical data sources have been upgraded and their internal consistency improved (e.g., FAO, 1999), although large regional differences in quality and coverage remain. In addition, the high-resolution global database, DISCover, has become available (Loveland and Belward, 1997). This database is derived from satellite data and consists of useful land-cover classes. Furthermore, attempts also have been made to develop historical land-use and land-cover databases (Ramankutty and Foley, 1999; Klein Goldewijk, 2001). These databases use proxy sources—such as historic maps, population-density estimates, and infrastructure—to approximate land-cover patterns. All of these improvements to the information base are important for initializing and validating the models used in scenario development for global change assessments.

3.3.2.2. Regional and Sector-Specific Approaches

A large variety of LUC-LCC scenarios have been constructed. Many of them focus on local and regional issues; only a few are global in scope. Most LUC-LCC scenarios, however, are developed not to assess GHG emissions, carbon fluxes, and climate change and impacts but to evaluate the environmental consequences of different agrosystems (e.g., Koruba *et al.*, 1996), agricultural policies (e.g., Moxey *et al.*, 1995), and food security (e.g., Penning de Vries *et al.*, 1997) or to project future agricultural production, trade, and food availability (e.g., Alexandratos, 1995; Rosegrant *et al.*, 1995). Moreover, changes in land-cover patterns are poorly defined in these studies. At best they specify aggregated amounts of arable land and pastures.

One of the more comprehensive attempts to define the consequences of agricultural policies on landscapes was the "Ground for Choices" study (Van Latesteijn, 1995). This study aimed to evaluate the consequences of increasing agricultural productivity and the Common Agricultural Policy in Europe and analyzed the possibilities for sustainable management of resources. It concluded that the total amount of agricultural land and employment would continue to decline—the direction of this trend apparently little influenced by agricultural policy. Many different possibilities for improving agricultural production were identified, leaving room for development of effective measures to preserve biodiversity, for example. This study included many of the desired physical, ecological, socioeconomic, and regional characteristics required for comprehensive LUC-LCC scenario development but did not consider environmental change.

Different LUC-LCC scenario studies apply very different methods. Most of them are based on scenarios from regression or process-based models. In the global agricultural land-use study of Alexandratos (1995), such models are combined with expert judgment, whereby regional and disciplinary experts reviewed all model-based scenarios. If these scenarios were deemed inconsistent with known trends or likely developments, they were modified until a satisfactory solution emerged for all regions. This approach led to a single consensus scenario of likely agricultural trends to 2010. Such a short time horizon is appropriate for expert panels; available evidence suggests that expert reviews of longer term scenarios tend to be conservative, underestimating emerging developments (Rabbinge and van Oijen, 1997).

3.3.2.3. Integrated Assessment Models

Most scenarios applied in climate change impact assessments fail to account satisfactorily for LUC-LCC. By incorporating land-use activities and land-cover characteristics, it becomes feasible to obtain comprehensive estimates of carbon fluxes and other GHG emissions, the role of terrestrial dynamics in the climate system, and ecosystem vulnerability and mitigation potential. Currently, the only tools for delivering this are IAMs (Weyant *et al.*, 1996; Parson and Fisher-Vanden, 1997; Rotmans and Dowlatabadi, 1998; see also Section 1.4.6), but only a few successfully incorporate LUC-LCC, including Integrated Climate Assessment Model (ICAM—Brown and Rosenberg, 1999), Asian-Pacific Integrated Model (AIM—Matsuoka *et al.*, 1995), Integrated Model for the Assessment of the Greenhouse Effect (IMAGE—Alcamo *et al.*, 1998b), and Tool to Assess Regional and Global Environmental and Health Targets for Sustainability (TARGETS—Rotmans and de Vries, 1997). These models simulate interactions between global change and LUC-LCC at grid resolution (IMAGE, AIM) or by regions (ICAM, TARGETS). All of these models, however, remain too coarse for detailed regional applications.

LUC-LCC components of IAMs generally are ecosystem and crop models, which are linked to economic models that specify changes in supply and demand of different land-use products for different socioeconomic trends. The objectives of each model differ, which has led to diverse approaches, each characterizing a specific application.

ICAM, for example, uses an agricultural sector model, which integrates environmental conditions, different crops, agricultural practices, and their interactions (Brown and Rosenberg, 1999). This model is implemented for a set of typical farms. Productivity improvements and management are explicitly simulated. Productivity levels are extrapolated toward larger regions to parameterize the production functions of the economic module. The model as a whole is linked to climate change scenarios by means of a simple emissions and climate module. A major advantage of ICAM is that adaptive capacity is included explicitly. Furthermore, new crops, such as biomass energy, can be added easily. Land use-related emissions do not result from the simulations. ICAM is used most effectively to assess impacts but is less well suited for the development of comprehensive spatially explicit LUC-LCC scenarios.

IMAGE uses a generic land-evaluation approach (Leemans and van den Born, 1994), which determines the distribution and productivity of different crops on a 0.5° grid. Achievable yields are a fraction of potential yields, set through scenario-dependent regional "management" factors. Changing regional demands for land-use products are reconciled with achievable yields, inducing changes in land-cover patterns. Agricultural expansion or intensification lead to deforestation or afforestation. IMAGE simulates diverse LUC-LCC patterns, which define fluxes of GHGs and some land-climate interactions. Changing crop/vegetation distributions and productivity indicate impacts. Emerging land-use activities (Leemans *et al.*, 1996a,b) and carbon sequestration activities defined in the Kyoto Protocol, which alter land-cover patterns, are included explicitly. This makes the model very suitable for LUC-LCC scenario development but less so for impact and vulnerability assessment because IMAGE does not explicitly address adaptive capacity.

3.3.3. Types of Land-Use and Land-Cover Change Scenarios

3.3.3.1. Driving Forces of Change

In early studies, the consequences of LUC often were portrayed in terms of the CO_2 emissions from tropical deforestation. Early carbon cycle models used prescribed deforestation rates and emission factors to project future emissions. During the past decade, a more comprehensive view has emerged, embracing the diversity of driving forces and regional heterogeneity (Turner *et al.*, 1995). Currently, most driving forces of available LUC-LCC scenarios are derived from population, income, and agricultural productivity assumptions. The first two factors commonly are assumed to be exogenous variables (i.e., scenario assumptions), whereas productivity levels are determined dynamically. This simplification does not yet characterize all diverse local driving forces, but it can be an effective approximation at coarser levels (Turner *et al.*, 1995).

3.3.3.2. Processes of LUC-LCC

The central role of LUC-LCC in determining climate change and its impacts has not fully been explored in the development of scenarios. Only limited aspects are considered. Most scenarios emphasize arable agriculture and neglect pastoralism, forestry, and other land uses. Only a few IAMs have begun to include more aspects of land use. Most scenarios discriminate between urban and rural population, each characterized by its specific needs and land uses. Demand for agricultural products generally is a function of income and regional preferences. With increasing wealth, there could be a shift from grain-based diets toward more affluent meat-based diets. Such shifts strongly alter land use (Leemans, 1999). Similar functional relations are assumed to determine the demand for nonfood products. Potential productivity is determined by climatic, atmospheric CO_2, and soil conditions. Losses resulting from improper management, limited water and nutrient availability, pests and diseases, and pollutants decrease potential productivity (Penning de Vries *et al.*, 1997). Most models assume constant soil conditions. In reality, many land uses lead to land degradation that alters soil conditions, affecting yields and changing land use (Barrow, 1991). Agricultural management, including measures for yield enhancement and protection, defines actual productivity. Unfortunately, management is demonstrably difficult to represent in scenarios.

Most attempts to simulate LUC-LCC patterns combine productivity calculations and demand for land-use products. In this step, large methodological difficulties emerge. To satisfy increased demand, agricultural land uses in some regions intensify (i.e., increase productivity), whereas in others they expand in area. These processes are driven by different local, regional, and global factors. Therefore, subsequent LCC patterns and their spatial and temporal dynamics cannot be determined readily. For example, deforestation is caused by timber extraction in Asia but by conversion to pasture in Latin America. Moreover, land-cover conversions rarely are permanent. Shifting cultivation is a common practice in some regions, but in many other regions agricultural land also has been abandoned in the past (Foster *et al.*, 1998) or is abandoned regularly (Skole and Tucker, 1993). These complex LUC-LCC dynamics make the development of comprehensive scenarios a challenging task.

The outcome of LUC-LCC scenarios is land-cover change. For example, the IMAGE scenarios (Alcamo *et al.*, 1998b) illustrate some of the complexities in land-cover dynamics. Deforestation continues globally until 2050, after which the global forested area increases again in all regions except Africa and Asia. Pastures expand more rapidly than arable land, with large regional differences. One of the important assumptions in these scenarios is that biomass will become an important energy source. This requires additional cultivated land.

3.3.3.3. Adaptation

Adaptation is considered in many scenarios that are used to estimate future agricultural productivity. Several studies (Rosenberg, 1993;

Rosenzweig and Parry, 1994; Brown and Rosenberg, 1999; Mendelsohn and Neumann, 1999) assume changes in crop selection and management and conclude that climate change impacts decrease when available measures are implemented. Reilly *et al.* (1996) conclude that the agricultural sector is not very vulnerable because of its adaptive capability. However, Risbey *et al.* (1999) warn that this capability is overestimated because it assumes rapid diffusion of information and technologies.

In contrast, most impact studies on natural ecosystems draw attention to the assumed fact that LCC will increase the vulnerability of natural systems (Peters and Lovejoy, 1992; Huntley *et al.*, 1997). For example, Sala *et al.* (2000) use scenarios of LUC-LCC, climate, and other factors to assess future threats to biodiversity in different biomes. They explicitly address a biome's adaptive capacity and find that the dominant factors that determine biodiversity decline will be climate change in polar biomes and land use in tropical biomes. The biodiversity of other biomes is affected by a combination of factors, each influencing vulnerability in a different way.

3.3.4. *Application and Interpretation of Scenarios and their Uncertainties*

LUC-LCC scenarios are all sensitive to underlying assumptions of future changes in, for example, agricultural productivity and demand. This can lead to large differences in scenario conclusions. For example, the FAO scenario (Alexandratos, 1995) demonstrates that land as a resource is not a limiting factor, whereas the IMAGE scenarios (Alcamo *et al.*, 1996) show that in Asia and Africa, land rapidly becomes limited over the same time period. In the IMAGE scenarios, relatively rapid transitions toward more affluent diets lead to rapid expansion of (extensive) grazing systems. In contrast, the FAO study does not specify the additional requirement for pastureland. The main difference in assumptions is that animal productivity becomes increasingly dependent on cereals (FAO) compared to pastures (IMAGE). This illustrates how varying important assumptions may lead to discrepancies and inconsistencies between scenario conclusions. In interpreting LUC-LCC scenarios, their scope, underlying assumptions, and limitations should be carefully and critically evaluated before resulting land-cover patterns are declared suitable for use in other studies. A better perspective on how to interpret LUC-LCC both as a driving force and as a means for adaptation to climate change is strongly required. One of the central questions is, "How can we better manage land and land use to reduce vulnerability to climate change and to meet our adaptation and mitigation needs?" Answering this question requires further development of comprehensive LUC-LCC scenarios.

3.4. Environmental Scenarios

3.4.1. *Purpose*

Observations during the 20th century have demonstrated clearly the multifaceted nature of anthropogenic environmental changes. Therefore, it is reasonable to expect that changes in climate anticipated for the future will occur in combination with other changes in the environment. Some of these changes will occur independently of climate change (e.g., groundwater depletion, acidification); others are a cause of climate change (e.g., changing atmospheric CO_2 concentration); and still others are a direct consequence of climate change (e.g., sea-level rise). All of these could have a role in modifying the impacts of future climate change. Hence, realistic scenarios of nonclimatic environmental factors are required to facilitate analysis of these combined effects and quantify them in impact assessments.

This section introduces environmental changes that are of importance at scales from subcontinental to global and describes how scenarios commonly are constructed to represent them. Requirements for environmental scenarios are highly application- and region-specific. For example, scenarios of CO_2 concentration may be important in considering future vegetation productivity under a changing climate but are unlikely to be required for assessment of human health impacts. Most of the scenarios treated here relate to atmospheric composition: CO_2, SO_2, sulfur and nitrogen deposition, tropospheric O_3, and surface UV-B radiation. Scenarios of water resources and marine pollution also are examined. Changes in the terrestrial environment are addressed in Section 3.3, and changes in sea level are addressed in Section 3.6.

3.4.2. *CO_2 Scenarios*

3.4.2.1. *Reference Conditions*

Aside from its dominant role as a greenhouse gas, atmospheric CO_2 also has an important direct effect on many organisms, stimulating photosynthetic productivity and affecting water-use efficiency in many terrestrial plants. In 1999, the concentration of CO_2 in the surface layer of the atmosphere (denoted as $[CO_2]$) was about 367 ppm (see Table 3-2), compared with a concentration of approximately 280 ppm in preindustrial times (see TAR WGI Chapter 3). CO_2 is well mixed in the atmosphere, and, although concentrations vary somewhat by region and season (related to seasonal uptake by vegetation), projections of global mean annual concentrations usually suffice for most impact applications. Reference levels of $[CO_2]$ between 300 and 360 ppm have been widely adopted in CO_2-enrichment experiments (Cure and Acock, 1986; Poorter, 1993; see Table 3-2) and in model-based impact studies. $[CO_2]$ has increased rapidly during the 20th century, and plant growth response could be significant for responsive plants, although the evidence for this from long-term observations of plants is unclear because of the confounding effects of other factors such as nitrogen deposition and soil fertility changes (Kirschbaum *et al.*, 1996).

3.4.2.2. *Development and Application of $[CO_2]$ Scenarios*

Projections of $[CO_2]$ are obtained in two stages: first, the rate of emissions from different sources is evaluated; second, concentrations are evaluated from projected emissions and sequestration of carbon. Because CO_2 is a major greenhouse gas, CO_2 emissions have been projected in successive IPCC scenarios (Scenarios A–D—Shine *et al.*, 1990; IS92 scenarios—Leggett *et al.*, 1992; SRES scenarios—Nakicenovic *et al.*, 2000). To obtain scenarios of future [CO_2] from those of emissions, global models of the carbon cycle are required (e.g., Schimel *et al.*, 1995). Some estimates of [CO_2] for the SRES emissions scenarios are given in Table 3-2.

In recent years, there has been growing interest in emissions scenarios that lead to [CO_2] stabilization (see Section 3.8.4). Typically, levels of [CO_2] stabilized between 350 and 1000 ppm have been examined; these levels usually are achieved during the 22nd or 23rd century, except under the most stringent emissions targets (Schimel *et al.*, 1997a). Work to develop storylines for a set of stabilization scenarios is reported in Chapter 2 of WGIII. Whatever scenarios emerge, it is likely to be some time before a set of derivative CO_2-stabilization impact and adaptation assessments are completed, although a few exploratory studies already have been conducted (UK-DETR, 1999).

Experimental CO_2-enrichment studies conventionally compare responses of an organism for a control concentration representing current [CO_2] with responses for a fixed concentration assumed for the future. In early studies this was most commonly a doubling (Cure and Acock, 1986), to coincide with equilibrium climate model experiments (see Section 3.5). However, more recent transient treatment of future changes, along with the many uncertainties surrounding estimates of future [CO_2] and future climate, present an infinite number of plausible combinations of future conditions. For example, Table 3-2 illustrates the range of [CO_2] projected for 2050 and 2100 under the SRES emissions

***Table 3-2**: Some illustrative estimates of reference and future levels of atmospheric constituents that typically are applied in model-based and experimental impact studies. Global values are presented, where available. European values also are shown to illustrate regional variations at the scale of many impact studies.*

Scenario	**[CO_2][a] (ppm)**	**[SO_2][b] ($\mu g\ m^{-3}$)**	**S-Deposition[c] ($meq\ m^{-2}\ a^{-1}$)**	**N-Deposition[c] ($meq\ m^{-2}\ a^{-1}$)**	**Ground-Level [O_3][d] (ppb)**
Reference/Control					
– Global/hemispheric	367	0.1–10	26	32	40
– Europe	—	5–100+	12–165 (572)	11–135 (288)	28–50 (72)
– Experiments	290–360	0–10	—	—	10–25
Future					
– Experiments	490–1350	50–1000	—	—	10–200
2010/2015					
– Global/hemispheric	388–395	—	26	36	—
– Europe	—	—	7–63 (225)	5–95 (163)	—
2050/2060					
– Global/hemispheric	463–623	—	—	—	~60
– Europe	—	—	8–80 (280)	5–83 (205)	—
2100					
– Global/hemispheric	478–1099	—	—	—	>70
– Europe	—	—	6–49 (276)	4–60 (161)	—

[a] **Carbon dioxide concentration**. *Reference*: Observed 1999 value (Chapter 3, WG I TAR). *Experiments*: Typical ranges used in enrichment experiments on agricultural crops. Some controls used ambient levels; most experiments for future conditions used levels between 600 and 1000 ppm (Strain and Cure, 1985; Wheeler *et al.*, 1996). *Future*: Values for 2010, 2050, and 2100 are for the range of emissions from 35 SRES scenarios, using a simple model (data from S.C.B. Raper, Chapter 9, TAR WGI); note that these ranges differ from those presented by TAR WGI (see Footnote c of Table 3-9 for an explanation).

[b] **Sulphur dioxide concentration**. *Reference*: Global values are background levels (Rovinsky and Yegerov, 1986; Ryaboshapko *et al.*, 1998); European values are annual means at sites in western Europe during the early 1980s (Saunders, 1985). *Experiments*: Typical purified or ambient (control) and elevated (future) concentrations for assessing long-term SO_2 effects on plants (Kropff, 1989).

[c] **Deposition of sulphur/nitrogen compounds**. *Reference*: Global values are mean deposition over land areas in 1992, based on the STOCHEM model (Collins *et al.*, 1997; Bouwman and van Vuuren, 1999); European values are based on EMEP model results (EMEP, 1998) and show 5th and 95th percentiles of grid box (150 km) values for 1990 emissions, assuming 10-year average meteorology (maximum in parentheses). *Future*: Global values for 2015 are from the STOCHEM model, assuming current reduction policies; European values are based on EMEP results for 2010, assuming a "current legislation" scenario under the Convention on Long-Range Transboundary Air Pollution (UN/ECE, 1998) and, for 2050 and 2100, assuming a modification of the preliminary SRES B1marker emissions scenario (B1-SR scenario—Mayerhofer *et al.*, 2000).

[d] **Ground-level ozone concentration**. *Reference*: Global/hemispheric values are model estimates for industrialized continents of the northern hemisphere, assuming 2000 emissions (Chapter 4, TAR WGI); European values are based on EMEP model results (Simpson *et al.*, 1997) and show 5th and 95th percentiles of mean monthly grid box (150 km) ground-level values for May-July during 1992–1996 (maximum in parentheses). *Experiments*: Typical range of purified or seasonal background values (control) and daily or subdaily concentrations (future) for assessing O_3 effects on agricultural crops (Unsworth and Hogsett, 1996; Krupa and Jäger, 1996). *Future*: Model estimates for 2060 and 2100 assuming the A1FI and A2 illustrative SRES emissions scenarios (Chapter 4, TAR WGI).

scenarios, using simple models. To cover these possibilities, although doubled [CO_2] experiments are still common, alternative concentrations also are investigated (Olesen, 1999)—often in combination with a range of climatic conditions, by using devices such as temperature gradient tunnels (Wheeler *et al.*, 1996).

3.4.3. Scenarios of Acidifying Compounds

3.4.3.1. Reference Conditions

Sulfur dioxide and nitrogen compounds are among the major air pollutants emitted by industrial and domestic sources. SO_2 is further oxidized to sulfate, which exists in the atmosphere mainly as aerosols. The main anthropogenic components of emissions of nitrogen compounds to the atmosphere are NO_x and ammonia (NH_3). Increased atmospheric SO_2 concentrations from anthropogenic sources are known to have negative effects on tree growth and crop yield (Kropff, 1989; Semenov *et al.*, 1998) and are described below. Concentrations of nitrogen compounds are not considered because scenarios seldom are required for impact studies. However, wet and dry deposition of sulfur and nitrogen from the atmosphere onto the Earth's surface can lead to acidification, with detrimental effects on soils, surface waters, building materials, and ecosystems (Grennfelt *et al.*, 1996). Nitrogen deposition may serve simultaneously as a plant fertilizer, positively influencing carbon gain in forests (Reich *et al.*, 1990; Woodward, 1992; Petterson *et al.*, 1993). Thus, deposition scenarios also are important.

Current global background concentrations of SO_2 are monitored at stations belonging to the Background Atmospheric Pollution Monitoring Network (BAPMoN), established by the World Meteorological Organization (WMO) and UNEP, as well as in regional networks. Annual mean SO_2 concentrations ([SO_2]) over land areas are estimated to be approximately 0.1–10 mg m^{-3} (Rovinsky and Yegorov, 1986; Ryaboshapko *et al.*, 1998). However, they can be much higher locally (Table 3-2). For example, annual average values of more than 80 mg m^{-3} were measured at some sites in Czechoslovakia in the 1970s (Materna, 1981). Model results have shown that [SO_2] averaged over the vegetative season reached 35 mg m^{-3} in some regions of Europe during 1987–1993 (Semenov *et al.*, 1998, 1999). In recent years reductions of SO_2 and NO_2 emissions have been recorded in many regions, accompanied by large-scale decreases in concentrations, especially evident in remote areas (Whelpdale and Kaiser, 1997). Typical rates of regional total (dry + wet) deposition of sulfur and nitrogen compounds, based on model simulations, are shown in Table 3-2.

Reference concentrations of SO_2 adopted in impact assessments vary according to the objective of the study. For example, in some field experiments an enhanced [SO_2] treatment is compared to a control case at ambient background concentrations. The latter concentrations can vary from year to year, depending on ambient weather and air quality conditions (Kropff, 1989). Alternatively, other experiments at locations close to pollution sources have used air purification systems to attain preindustrial levels of [SO_2] in closed chambers, comparing plant responses to those under (locally high) ambient concentrations.

3.4.3.2. Development and Application of Sulfur and Nitrogen Scenarios

Several models have been developed to project atmospheric concentrations and deposition of sulfur and (in some cases) nitrogen compounds. At the regional scale these models include: for Europe, RAINS (Alcamo *et al.*, 1990; Schöpp *et al.*, 1999) and ASAM (ApSimon *et al.*, 1994), both of which use output from mechanistic models developed by the Co-operative Programme for Monitoring and Evaluation of the Long-Range Transmission of Air Pollutants in Europe (EMEP); for Asia, RAINS-Asia (Foell *et al.*, 1995); and for North America and Asia, ATMOS (Arndt *et al.*, 1997). There also are global models: GRANTOUR (Penner *et al.*, 1994), MOGUNTIA (Langner and Rodhe, 1991), ECHAM (Feichter *et al.*, 1996), and STOCHEM (Collins *et al.*, 1997).

There have been few studies of the joint impacts of acidifying compounds and climate change. Some of these studies are multifactorial model simulations of plant response (e.g., Semenov *et al.*, 1998). There also have been some modeling studies based on the IS92a emissions scenario (Posch *et al.*, 1996; Fischer and Rosenzweig, 1996), under which a substantial increase in annual sulfur deposition is projected to occur by 2050, with commensurate suppression of modeled GHG warming in some regions. However, this scenario is now thought to overestimate future emissions of sulfur (Grübler, 1998), as reflected in the new SRES scenarios (see Section 3.8.1). Not all of the models used in developing the SRES scenarios provide information on nitrogen emissions, but those that do can be used to produce consistent scenarios of [NO_x], [SO_2], sulfur and nitrogen deposition, and climate change for impact studies (Mayerhofer *et al.*, 2000; Stevenson *et al.*, 2000; see Table 3-2).

3.4.4. Scenarios of Tropospheric Ozone

3.4.4.1. Reference Conditions

Tropospheric ozone forms part of the natural shield that protects living organisms from harmful UV-B rays. In the lowest portion of the atmosphere, however, excess accumulations of ozone can be toxic for a wide range of plant species (Fuhrer, 1996; Semenov *et al.*, 1998, 1999).Ozone is produced by a chain of chemical and photochemical reactions involving, in particular, NO, NO_2, and VOCs (Finlayson-Pitts and Pitts, 1986; Derwent *et al.*, 1991; Alexandrov *et al.*, 1992; Simpson, 1992, 1995a; Peters *et al.*, 1995). These chemical precursors of ozone can be human-derived (e.g., energy production, transport) or natural (e.g., biogenic emissions, forest fires). Surface ozone concentrations are highly variable in space and time (Table 3-2); the highest values typically are over industrial regions and large cities.

Global background concentrations of ground-level ozone (annual means) are about 20–25 ppb (Semenov *et al.*, 1999). Background concentrations have increased in Europe during the 20th century from 10–15 to 30 ppb (Grennfelt, 1996). In the northern hemisphere as a whole, trends in concentrations since 1970 show large regional differences: increases in Europe and Japan, decreases in Canada, and only small changes in the United States (Lelieveld and Thompson, 1998). In an effort to reverse the upward trends still recorded in many regions, a comprehensive protocol to abate acidification, eutrophication, and ground-level ozone was signed in 1999, setting emissions ceilings for sulfur, NO_x, NH_3, and VOCs for most of the United Nations Economic Commission for Europe (UN/ECE) region.

3.4.4.2. *Development and Application of Tropospheric Ozone Scenarios*

Results from the first intercomparison of model-based estimates of global tropospheric ozone concentration assuming the new SRES emissions scenarios (see Section 3.8.1) are reported in TAR WGI Chapter 4. Estimates of mean ground-level O_3 concentrations during July over the industrialized continents of the northern hemisphere under the SRES A2 and A1FI scenarios are presented in Table 3-2. These scenarios produce concentrations at the high end of the SRES range, with values in excess of 70 ppb for 2100 emissions (TAR WGI Chapter 4). Local smog events could enhance these background levels substantially, posing severe problems in achieving the accepted clean-air standard of <80 ppb in most populated areas.

Regional projections of ozone concentration also are made routinely, assuming various emissions reduction scenarios (e.g., SEPA, 1993; Simpson, 1995b; Simpson *et al.*, 1995). These projections sometimes are expressed in impact terms—for example, using AOT40 (the integrated excess of O_3 concentration above a threshold of 40 ppb during the vegetative period), based on studies of decline in tree growth and crop yield (Fuhrer, 1996; Semenov *et al.*, 1999).

There are few examples of impact studies that have evaluated the joint effects of ozone and climate change. Some experiments have reported on plant response to ozone and CO_2 concentration (Barnes *et al.*, 1995; Ojanperä *et al.*, 1998), and several model-based studies have been conducted (Sirotenko *et al.*, 1995; Martin, 1997; Semenov *et al.*, 1997, 1998, 1999).

3.4.5. UV-B Radiation Scenarios

3.4.5.1. *Reference Conditions*

Anthropogenic emissions of chlorofluorocarbons (freons) and some other substances into the atmosphere are known to deplete the stratospheric ozone layer (Albritton and Kuijpers, 1999). This layer absorbs ultraviolet solar radiation within a wavelength range of 280–320 nm (UV-B), and its depletion leads to an increase in ground-level flux of UV-B radiation (Herman *et al.*, 1996; Jackman *et al.*, 1996; McPeters *et al.*, 1996; Madronich *et al.*, 1998; McKenzie *et al.*, 1999). Enhanced UV-B suppresses the immune system and may cause skin cancer in humans and eye damage in humans and other animal species (Diffey, 1992; de Gruijl, 1997; Longstreth *et al.*, 1998). It can affect terrestrial and marine ecosystems (IASC, 1995; Zerefos and Alkiviadis, 1997; Caldwell *et al.*, 1998; Hader *et al.*, 1998; Krupa *et al.*, 1998) and biogeochemical cycles (Zepp *et al.*, 1998) and may reduce the service life of natural and synthetic polymer materials (Andrady *et al.*, 1998). It also interacts with other atmospheric constituents, including GHGs, influencing radiative forcing of the climate (see TAR WGI Chapters 4, 6, and 7).

Analyses of ozone data and depletion processes since the early 1970s have shown that the total ozone column has declined in northern hemisphere mid-latitudes by about 6% in winter/spring and 3% in summer/autumn, and in southern hemisphere mid-latitudes by about 5% on a year-round basis. Spring depletion has been greatest in the polar regions: about 50% in the Antarctic and 15% in the Arctic (Albritton and Kuijpers, 1999). These five values are estimated to have been accompanied by increases in surface UV-B radiation of 7, 4, 6, 130, and 22%, respectively, assuming other influences such as clouds to be constant. Following a linear increase during the 1980s, the 1990s springtime ozone depletion in Antarctica has continued at about the same level each year. In contrast, a series of cold, protracted winters in the Arctic have promoted large depletions of ozone levels during the 1990s (Albritton and Kuijpers, 1999).

3.4.5.2. *Development and Application of UV-B Scenarios*

Scenarios of the future thickness of the ozone column under given emissions of ozone-depleting gases can be determined with atmospheric chemistry models (Alexandrov *et al.*, 1992; Brasseur *et al.*, 1998), sometimes in combination with expert judgment. Processes that affect surface UV-B flux also have been investigated via models (Alexandrov *et al.*, 1992; Matthijsen *et al.*, 1998). Furthermore, several simulations have been conducted with coupled atmospheric chemistry and climate models, to investigate the relationship between GHG-induced climate change and ozone depletion for different scenarios of halogenated compounds (Austin *et al.*, 1992; Shindell *et al.*, 1998). It is known that potential stratospheric cooling resulting from climate change may increase the likelihood of formation of polar stratospheric clouds, which enhance the catalytic destruction of ozone. Conversely, ozone depletion itself contributes to cooling of the upper troposphere and lower stratosphere (see TAR WGI Chapter 7).

Serious international efforts aimed at arresting anthropogenic emissions of ozone-depleting gases already have been undertaken—namely, the Vienna Convention for the Protection of the Ozone Layer (1985) and the Montreal Protocol on Substances that Deplete the Ozone Layer (1990) and its Amendments. The abundance of ozone-depleting gases in the atmosphere peaked in the late 1990s and now is expected to decline as a result of these measures (Montzka *et al*., 1996), recovering to pre-1980 levels around 2050 (Albritton and Kuijpers, 1999). Without these measures, ozone depletion by 2050 was projected to exceed 50% in northern mid-latitudes and 70% in southern mid-latitudes—about 10 times larger than today. UV-B radiation was projected to double and quadruple in northern and southern mid-latitudes, respectively (Albritton and Kuijpers, 1999).

There have been numerous experimental artificial exposure studies of the effects of UV-B radiation on plants (Runeckles and Krupa, 1994). There also have been a few investigations of the joint effects of enhanced UV-B and other environmental changes, including climate (Unsworth and Hogsett, 1996; Gwynne-Jones *et al*., 1997; Sullivan, 1997). A study of the impacts of UV-B on skin cancer incidence in The Netherlands and Australia to 2050, using integrated models, is reported by Martens (1998), who employed scenarios of future ozone depletion based on the IS92a emissions scenario and two scenarios assuming compliance with the London and Copenhagen Amendments to the Montreal Protocol.

3.4.6. Water Resource Scenarios

3.4.6.1. Reference Conditions

Water is a resource of fundamental importance for basic human survival, for ecosystems, and for many key economic activities, including agriculture, power generation, and various industries. The quantity and quality of water must be considered in assessing present-day and future resources. In many parts of the world, water already is a scarce resource, and this situation seems certain to worsen as demand increases and water quality deteriorates, even in the absence of climate change. Abundance of the resource at a given location can be quantified by water availability, which is a function of local supply, inflow, consumption, and population. The quality of water resources can be described by a range of indicators, including organic/fecal pollution, nutrients, heavy metals, pesticides, suspended sediments, total dissolved salts, dissolved oxygen, and pH.

Several recent global analyses of water resources have been published (Raskin *et al*., 1997; Gleick, 1998; Shiklomanov, 1998; Alcamo *et al*., 2000). Some estimates are shown in Table 3-3. For regional and local impact studies, reference conditions can be more difficult to specify because of large temporal variability in the levels of lakes, rivers, and groundwater and human interventions (e.g., flow regulation and impoundment, land-use changes, water abstraction, effluent return, and river diversions; Arnell *et al*., 1996).

Industrial wastes, urban sewage discharge, application of chemicals in agriculture, atmospheric deposition of pollutants, and salinization negatively affect the quality of surface and groundwaters. Problems are especially acute in newly industrialized countries (UNEP/GEMS, 1995). Fecal pollution of freshwater basins as a result of untreated sewage seriously threatens human health in some regions. Overall, 26% of the population (more than 1 billion people) in developing countries still do not have access to safe drinking water, and 66% do not have adequate environmental sanitation facilities—contributing

***Table 3-3**: Estimates of global and regional water intensity and water withdrawals in 1995 and scenarios for 2025.*

Aggregate World Regions	Water Intensity (m^3 cap^{-1} yr^{-1})[a] 1995[b]	2025 BAU[b,c]	TEC[b,d]	VAL[b,e]	CDS[f]	Total Water Withdrawals (km^3) 1995[b]	2025 BAU[b,c]	TEC[b,d]	VAL[b,e]	CDS[f]
Africa	5678	2804	2859	2974	2858	167	226	228	204	240
Asia	3884	2791	2846	3014	2778	1913	2285	2050	1499	2709
Central America	6643	4429	4507	4895	4734	126	171	140	112	145
CIS[g]	17049	16777	17124	17801	14777	274	304	226	186	480
Europe	4051	3908	3922	4119	3765	375	359	256	201	415
North America	17625	14186	14186	15533	14821	533	515	323	245	668
Oceania	64632	46455	46455	51260	42914	27	27	28	20	32
South America	30084	21146	21576	23374	21176	157	208	162	128	211
World	7305	5167	5258	5563	5150	3572	4095	3413	2595	4899

[a] Calculated by using estimates of water availability from UN Comprehensive Assessment of the Freshwater Resources of the World (Shiklomanov, 1998) and population from footnoted source.
[b] World Commission on Water for the 21st Century (Alcamo *et al*., 2000).
[c] Business-as-usual scenario (domestic water intensity increases, then stabilizes with increasing incomes, some increase in water-use efficiency).
[d] Technology, Economics, and Private Sector scenario (relative to BAU: similar population and income level; domestic water-use intensity one-third lower; higher water-use efficiency in industrialized countries).
[e] Values and Lifestyles scenario (relative to BAU: lower population and higher income; domestic water-use intensity two-thirds lower; much higher water-use efficiency in all countries).
[f] Conventional Development scenario (Raskin *et al*., 1997—population slightly higher than in BAU scenario; per capita water use falls in developed world and rises in developing world).
[g] Commonwealth of Independent States.

to almost 15,000 deaths each day from water-related diseases, nearly two-thirds of which are diarrheal (WHO, 1995; Gleick, 1998; see Chapter 9).

3.4.6.2. Development and Application of Water Resource Scenarios

Water resource scenarios have been developed at different time and space scales. For example, projections to 2025 on the basis of national water resource monitoring data have been reported by Shiklomanov (1998). Model-based projections of water use and availability to 2025 at the river basin scale have been made by Alcamo *et al.* (2000), assuming a BAU scenario and two alternative, normative scenarios that focus on water conservation. Some results of these scenario exercises are shown in Table 3-3. Among the most developed scenarios of water quality are model-based scenarios of acidification of freshwaters in Europe (e.g., NIVA, 1998). More general normative scenarios describing rural and urban access to safe drinking water by 2025 and 2050 are presented by Raskin *et al.* (1998). Scenarios of water availability have been applied in several climate change impact studies. Most of these are in the water resources sector and are reported in Chapter 4. However, they are increasingly being applied in multi-sectoral and integrated assessments (e.g., Strzepek *et al.*, 1995).

3.4.7. Scenarios of Marine Pollution

3.4.7.1. Reference Conditions

Marine pollution is the major large-scale environmental factor that has influenced the state of the world oceans in recent decades. Nutrients, oxygen-demanding wastes, toxic chemicals (such as heavy metals, chlorinated hydrocarbons, potential endocrine-disrupting chemicals, and environmental estrogens), pathogens, sediments (silt), petroleum hydrocarbons, and litter are among the most important contaminants leading to degradation of marine ecosystems (Izrael and Tsyban, 1989; GESAMP, 1990; Tsyban, 1997). The following ranges of concentrations of heavy metals are characteristic of open ocean waters: mercury (0.3–7 ng l^{-1}), cadmium (10–200 ng l^{-1}), and lead (5–50 ng l^{-1}); levels of chlorinated hydrocarbons are a few ng l^{-1}. Chemical contaminants and litter are found everywhere in the open ocean, from the poles to the tropics and from beaches to abyssal depths. Nonetheless, the open ocean still remains fairly clean relative to coastal zones, where water pollution and the variability of contaminant concentrations are much higher (often by one to two orders of magnitude; specific values depend on the pattern of discharge and local conditions).

3.4.7.2. Development and Application of Marine Pollution Scenarios

Data characterizing the state of the marine environment have been obtained through national as well as international monitoring programs in recent decades, and analysis of tendencies may serve as an initial basis for developing environmental scenarios. At present, expert judgment appears to be the most promising method of scenario development because modeling methods are insufficiently developed to facilitate prediction.

In qualitative terms, trends in marine pollution during the 21st century could include enhanced eutrophication in many regions, enhancement of exotic algal blooms, expanded distribution and increased concentration of estrogens, invasion of nonindigenous organisms, microbiological contamination, accumulation of pathogens in marine ecosystems and seafood, and increases of chemical toxicants (Izrael and Tsyban, 1989; Goldberg, 1995).

3.5. Climate Scenarios

3.5.1. Purpose

The purpose of this section is to provide a summary of major methodological issues in the science of climate scenario development and to relate these developments to applications of scenarios in this report. We distinguish between a climate scenario, which refers to a plausible future climate, and a climate change scenario, which implies the *difference* between some plausible future climate and the present-day climate, though the terms are used interchangeably in the scientific literature. This brief overview is distilled largely from material presented in TAR WGI Chapter 13. See that chapter, as well as TAR WGI Chapters 8, 9, and 10, for more complete coverage of this subject.

3.5.2. Methods

Methods of climate scenario development largely have been ignored in earlier IPCC assessments, although some aspects of scenario development have been alluded to (e.g., palaeoclimatic analogs in Folland *et al.*, 1990; downscaling methods in Kattenburg *et al.*, 1996). Table 3-4 provides an overview of the main methods, which also are discussed in TAR WGI Chapter 13. Thus, we present only a very brief summary of three major methods. A fourth method, expert judgement, that also has been used in developing climate scenarios (NDU, 1978; Morgan and Keith, 1995), is discussed further in Section 3.5.5.

3.5.2.1. Incremental Scenarios for Sensitivity Studies

In this approach, particular climatic (or related) elements are changed by realistic but arbitrary amounts. They are commonly applied to study the sensitivity of an exposure unit to a wide range of variations in climate and to construct impact response surfaces over multivariate climate space. Most studies have adopted incremental scenarios of constant changes throughout the year (e.g., Terjung *et al.*, 1984; Rosenzweig *et al.*, 1996), but some have introduced seasonal and spatial variations in the changes (e.g., Rosenthal *et al.*, 1995); others have examined

Table 3-4: The role of various types of climate scenarios and an evaluation of their advantages and disadvantages according to the five criteria described in the text. Note that in some applications, a combination of methods may be used—for example, regional modeling and a weather generator (WGI TAR Chapter 13, Table 13.1).

Scenario Type or Tool	Description/Use	Advantages[a]	Disadvantages[a]
Incremental	• Testing system sensitivity • Identifying key climate thresholds	• Easy to design and apply (5) • Allows impact response surfaces to be created (3)	• Potential for creating unrealistic scenarios (1,2) • Not directly related to GHG forcing (1)
Analog			
Palaeoclimatic	• Characterizing warmer periods in past	• Physically plausible changed climate that really did occur in the past of a magnitude similar to that predicted for ~2100 (2)	• Variables may be poorly resolved in space and time (3,5) • Not related to GHG forcing (1)
Instrumental	• Exploring vulnerabilities and some adaptive capacities	• Physically realistic changes (2) • Can contain a rich mixture of well-resolved, internally consistent, variables (3) • Data readily available (5)	• Not necessarily related to GHG forcing (1) • Magnitude of climate change usually quite small (1) • No appropriate analogs may be available (5)
Spatial	• Extrapolating climate/ecosystem relationships • Pedagogic	• May contain a rich mixture of well-resolved variables (3)	• Not related to GHG forcing (1,4) • Often physically implausible (2) • No appropriate analogs may be available (5)
Climate Model-Based			
Direct AOGCM outputs	• Starting point for most climate scenarios • Large-scale response to anthropogenic forcing	• Information derived from the most comprehensive, physically based models (1,2) • Long integrations (1) • Data readily available (5) • Many variables (potentially) available (3)	• Spatial information poorly resolved (3) • Daily characteristics may be unrealistic except for very large regions (3) • Computationally expensive to derive multiple scenarios (4,5) • Large control run biases may be a concern for use in certain regions (2)
High-resolution/ stretched grid (AGCM)	• Providing high-resolution information at global/continental scales	• Provides highly resolved information (3) • Information derived from physically based models (2) • Many variables available (3) • Globally consistent and allows for feedbacks (1,2)	• Computationally expensive to derive multiple scenarios (4,5) • Problems in maintaining viable parameterizations across scales (1,2) • High resolution dependent on SSTs and sea ice margins from driving model (AOGCM) (2) • Dependent on (usually biased) inputs from driving AOGCM (2)
Regional models	• Providing high spatial/temporal resolution information	• Provides very highly resolved information (spatial and temporal) (3) • Information derived from physically based models (2) • Many variables available (3) • Better representation of some weather extremes than in GCMs (2,4)	• Computationally expensive, thus few multiple scenarios (4,5) • Lack of two-way nesting may raise concern regarding completeness (2) • Dependent on (usually biased) inputs from driving AOGCM (2)

Table 3-4 (continued)

Scenario Type or Tool	Description/Use	Advantages[a]	Disadvantages[a]
Climate Model-Based (cont.)			
Statistical downscaling	• Providing point/ high spatial resolution information	• Can generate information on high-resolution grids or nonuniform regions (3) • Potential, for some techniques, to address a diverse range of variables (3) • Variables are (probably) internally consistent (2) • Computationally (relatively) inexpensive (5) • Suitable for locations with limited computational resources (5) • Rapid application to multiple GCMs (4)	• Assumes constancy of empirical relationships in the future (1,2) • Demands access to daily observational surface and/or upper air data that span range of variability (5) • Not many variables produced for some techniques (3,5) • Dependent on (usually biased) inputs from driving AOGCM (2)
Climate scenario generators	• Integrated assessments • Exploring uncertainties • Pedagogic	• May allow for sequential quantification of uncertainty (4) • Provides "integrated" scenarios (1) • Multiple scenarios easy to derive (4)	• Usually rely on linear pattern-scaling methods (1) • Poor representation of temporal variability (3) • Low spatial resolution (3)
Weather Generators	• Generating baseline climate time series • Altering higher order moments of climate • Statistical downscaling	• Generates long sequences of daily or subdaily climate (2,3) • Variables usually are internally consistent (2) • Can incorporate altered frequency/intensity of ENSO events (3)	• Poor representation of low-frequency climate variability (2,4) • Limited representation of extremes (2,3,4) • Requires access to long observational weather series (5) • In absence of conditioning, assumes constant statistical characteristics (1,2)
Expert Judgment	• Exploring probability and risk • Integrating current thinking on changes in climate	• May allow for "consensus" (4) • Has potential to integrate very broad range of relevant information (1,3,4) • Uncertainties can be readily represented (4)	• Subjectivity may introduce bias (2) • Representative survey of experts may be difficult to implement (5)

[a]Numbers in parentheses within the Advantages and Disadvantages columns indicate that they are relevant to the criteria described. The five criteria follow: 1) *Consistency* at regional level with global projections; 2) *physical plausibility and realism*, such that changes in different climatic variables are mutually consistent and credible and spatial and temporal patterns of change are realistic; 3) *appropriateness* of information for impact assessments (i.e., resolution, time horizon, variables); 4) *representativeness* of potential range of future regional climate change; and 5) *accessibility* for use in impact assessments.

arbitrary changes in interannual, within-month, and diurnal variability, as well as changes in the mean (e.g., Williams *et al.*, 1988; Mearns *et al.*, 1992, 1996; Semenov and Porter, 1995). Some of these studies are discussed in Chapter 5.

3.5.2.2. *Analog Approaches*

Temporal and spatial analogs also have been used in constructing climate scenarios. Temporal analogs make use of climatic information from the past as an analog of possible future climate (Pittock, 1993). They are of two types: palaeoclimatic analogs and instrumentally based analogs.

Palaeoclimatic analogs: Palaeoclimatic analogs are based on reconstructions of past climate from fossil evidence, such as plant or animal remains and sedimentary deposits. Two periods have received particular attention: the mid-Holocene (~5–6 ky BP[3]), when northern hemisphere temperatures are estimated to

[3]ky BP = 1,000 years before present.

have been about 1°C warmer than today, and the Last (Eemian) Interglacial (~120–130 ky BP), when temperatures were about 2°C warmer.

The major disadvantages of this method are the causal differences between past changes in climate and posited future changes (Crowley, 1990; Mitchell, 1990) and the large uncertainties about the quality of palaeoclimatic reconstructions (Covey, 1995; Kneshgi and Lapenis, 1996; Borzenkova, 1998). However, these scenarios continue to be used occasionally in impact assessments (Anisimov and Nelson, 1996; Budyko and Menzhulin, 1996) and are useful for providing insights about system vulnerability to climate change.

Instrumentally based analogs: Periods of observed regional or global-scale warmth during the historical period also have been used as an analog of a GHG-induced warmer world. Scenarios are constructed by estimating the difference between the regional climate during the warm period and that of the long-term average or that of a similarly selected cold period (Lough *et al.*, 1983; Rosenberg *et al.,* 1993). Major objections to the use of these analogs include the relatively minor changes in climate involved (although small changes could be adequate for examining near-term climate change) and, again, differences between the causes of historical fluctuations and those of posited larger future climate changes (Glantz, 1988; Pittock, 1989).

Spatial analogs: These are regions that today have a climate analogous to that anticipated in the study region in future. For example, Bergthórsson *et al.* (1988) used temperatures in northern Britain as a spatial analog for the potential future temperatures over Iceland. The approach is severely restricted, however, by the frequent lack of correspondence between other important features (climatic and nonclimatic) of the two regions. Nevertheless, spatial analogs are still adopted in a few studies—for example, to assess potential effects of climate change on human health (see Chapter 9).

3.5.2.3. Use of Climate Model Outputs

The most common method of developing climate scenarios for quantitative impact assessments is to use results from GCM experiments. Most estimates of impacts described in this report rely on this type of scenario. GCMs are three-dimensional mathematical models that represent physical and dynamical processes that are responsible for climate. All models are first run for a control simulation that is representative of the present-day or preindustrial times. They have been used to conduct two types of "experiment" for estimating future climate: equilibrium and transient-response experiments. In the former, the equilibrium response (new stable state) of the global climate following an instantaneous increase (e.g., doubling) of atmospheric CO_2 concentration or its radiative equivalent, including all GHGs, is evaluated (Schlesinger and Mitchell, 1987; Mitchell *et al.*, 1990). Transient experiments are conducted with coupled atmosphere-ocean models (AOGCMs), which link, dynamically, detailed models of the ocean with those of the atmosphere. AOGCMs are able to simulate time lags between a given change in atmospheric composition and the response of climate (see TAR WGI Chapter 8). Most recent evaluations of impacts, as reflected in this report, are based on scenarios formed from results of transient experiments as opposed to equilibrium experiments.

3.5.3. Baseline Climatologies

3.5.3.1. Baseline Period

Any climate scenario must adopt a reference baseline period from which to calculate changes in climate. This baseline data set serves to characterize the sensitivity of the exposure unit to present-day climate and usually serves as the base on which data sets that represent climate change are constructed. Among the possible criteria for selecting the baseline period (IPCC, 1994), it should be representative of the present-day or recent average climate in the study region and of a sufficient duration to encompass a range of climatic variations, including several significant weather anomalies (e.g., severe droughts or cool seasons).

A popular climatological baseline period is a 30-year "normal" period, as defined by the WMO. The current WMO normal period is 1961–1990, which provides a standard reference for many impact studies. Note, however, that in some regions, observations during this time period may exhibit anthropogenic climate changes relative to earlier periods.

3.5.3.2. Sources and Characteristics of Data

Sources of baseline data include a wide variety of observed data, reanalysis data (a combination of observed and model-simulated data), control runs of GCM simulations, and time series generated by stochastic weather generators. Different impact assessments require different types and resolutions of baseline climatological data. These can range from globally gridded baseline data sets at a monthly time scale to single-site data at a daily or hourly time scale. The variables most often required are temperature and precipitation, but incident solar radiation, relative humidity, windspeed, and even more exotic variables sometimes may be needed.

Two important issues in the development of baseline data sets are their spatial and temporal resolution and uncertainties related to their accuracy (New, 1999) (see TAR WGI Section 13.3.2 for further details). Evaluation of the differences between baseline data sets recently has become an important step in scenario development because these differences can have an important bearing on the results obtained in an impact assessment (Arnell, 1995; Pan *et al.*, 1996).

3.5.4. Construction of Scenarios

Techniques for constructing climate scenarios (i.e., scenario information that is directly usable in impact studies) have evolved very slowly during the past 2 decades. However, in the past few years several new developments in climate modeling and scenario development have expanded the array of techniques for scenario formation. The following subsections discuss some of these issues and present some background illustrative material.

3.5.4.1. Choosing Variables of Interest

In principle, GCM-based scenarios can be constructed for a wide range of variables at time resolutions down to subdaily time steps. In practice, however, not all data are available at the desired temporal and spatial resolutions. Most scenarios are conventionally based on changes in monthly mean climate, although with greater quantities of model output now being saved operationally, daily output and information on certain types of extreme events (e.g., mid-latitude cyclone intensities) can be accessed readily. However, consideration must be given to whether model output regarding a particular phenomenon is deemed "meaningful." For example, although information on changes in the frequency and intensity of El Niño-Southern Oscillation (ENSO) events may be desirable from an impacts point of view, analyses of possible future changes in this oscillation still are very preliminary (see TAR WGI Chapter 9).

3.5.4.2. Selecting GCM Outputs

Many equilibrium and transient climate change experiments have been performed with GCMs (Kattenberg *et al.*, 1996; TAR WGI Chapter 9). Several research centers now serve as repositories of GCM information (see, e.g., Hulme *et al.*, 1995; CSIRO, 1997). The IPCC Data Distribution Centre (IPCC-DDC, 1999) complements these existing sources. Table 3-5 lists GCM

Table 3-5*: Catalog of GCM experiments used to develop scenarios applied by impact studies referenced in this report. Columns show the acronym of the modeling center; the common model acronym found in the impacts literature; a code for the model experiment; reference number for the experiment from Chapter 8, WGI TAR; main reference sources; type of experiment (EQ = equilibrium; TRS = transient with simple ocean; TRC = transient cold start with dynamic ocean; TRW = transient warm start with dynamic ocean); increase in CO_2-equivalent concentration; effective climate sensitivity [equilibrium warming at CO_2-doubling from AOGCM experiments (see Chapter 9, WG I TAR); in some cases this differs from climate sensitivities cited elsewhere derived from atmosphere-only GCMs]; and availability from IPCC Data Distribution Centre.*

Center	Model	Expt	WG I	Reference	Type	Forcing	ΔT_{2xCO2} (°C)	DDC
CCCma	CCC	a	—	McFarlane *et al.* (1992)	EQ	2 x CO_2	3.5	—
	CGCM1	b	6	Boer *et al.* (2000)	TRW	1% a^{-1}	3.6	✔
CCSR/NIES	CCSR-98	c	5	Emori *et al.* (1999)	TRW	1% a^{-1}	3.5	✔
CSIRO	CSIRO	d	—	Watterson *et al.* (1997)	EQ	2 x CO_2	4.3	—
	CSIRO-Mk2	e	10	Gordon and O'Farrell (1997)	TRW	1% a^{-1}	3.7	✔
DKRZ	ECHAM1	f	13	Cubasch *et al.* (1992)	TRC	IPCC90A	2.6	—
	ECHAM3	g	14	Cubasch *et al.* (1996)	TRW	IPCC90A	2.2	✔
	ECHAM4	h	15	Roeckner *et al.* (1996)	TRW	IPCC90A	2.6	✔
GFDL	GFDL	i	—	Wetherald and Manabe (1986)	EQ	2 x CO_2	4.0	—
	GFDLTR	j	—	Manabe *et al.* (1991)	TRC	1% a^{-1}	4.0	—
	GFDL-R15	k	16	Haywood *et al.* (1997)	TRW	1% a^{-1}	4.2	✔
GISS	GISS	l	—	Hansen *et al.* (1983)	EQ	2 x CO_2	4.2	—
	GISSTR	m	—	Hansen *et al.* (1988)	TRS	1.5% a^{-1}	4.2	—
NCAR	NCAR	n	—	Washington and Meehl (1984)	EQ	2 x CO_2	4.0	—
	NCAR1	o	28	Washington and Meehl (1996)	TRW	1% a^{-1}	4.6	✔
OSU	OSU	p	—	Schlesinger and Zhao (1989)	EQ	2 x CO_2	2.8	—
UKMO	UKMO	q	—	Wilson and Mitchell (1987)	EQ	2 x CO_2	5.2	—
	UKHI	r	—	Haarsma *et al.* (1993)	EQ	2 x CO_2	3.5	—
	UKTR	s	—	Murphy (1995)	TRC	1% a^{-1}	2.7	—
	HadCM2	t	22	Mitchell and Johns (1997)	TRW	1% a^{-1}	2.5	✔
	HadCM3	u	23	Gordon *et al.* (2000)	TRW	1% a^{-1}	3.0	✔

experiments that have been used to develop scenarios for impacts studies evaluated in this report.

Four criteria for selecting GCM outputs from such a large sample of experiments are suggested by Smith and Hulme (1998):

1) *Vintage*: Recent model simulations are likely (though by no means certain) to be more reliable than those of an earlier vintage since they are based on recent knowledge and incorporate more processes and feedbacks.
2) *Resolution*: In general, increased spatial resolution of models has led to better representation of climate.
3) *Validation*: Selection of GCMs that simulate the present-day climate most faithfully is preferred, on the premise that these GCMs are more likely (though not guaranteed) to yield a reliable representation of future climate.
4) *Representativeness of results*: Alternative GCMs can display large differences in estimates of regional climate change, especially for variables such as precipitation. One option is to choose models that show a range of changes in a key variable in the study region.

3.5.4.3. Constructing Change Fields

Because climate model results generally are not sufficiently accurate (in terms of absolute values) at regional scales to be used directly (Mearns *et al.*, 1997), mean differences between the control (or current climate) run and the future climate run usually are calculated and then combined with some baseline observed climate data set (IPCC, 1994). Conventionally, differences (future climate minus control) are used for temperature variables, and ratios (future climate/control) are used for other variables such as precipitation, solar radiation, relative humidity, and windspeed. Most impact applications consider one or more fixed time horizon(s) in the future (e.g., the 2020s, the 2050s, and the 2080s have been chosen as 30-year time windows for storing change fields in the IPCC-DDC). Some other applications may require time-dependent information on changes, such as vegetation succession models that simulate transient changes in plant composition (e.g., VEMAP members, 1995).

3.5.4.4. Spatial Scale of Scenarios

One of the major problems in applying GCM projections to regional impact assessments is the coarse spatial scale of the gridded estimates—on the order of hundreds of kilometers—in relation to many of the exposure units being studied (often at one or two orders of magnitude finer resolution). Concern about this issue is raised in Chapters 4 and 5. Several solutions have been adopted to obtain finer resolution information.

3.5.4.4.1. Simple methods

Conventionally, regional "detail" in climate scenarios has been incorporated by appending changes in climate from the nearest coarse-scale GCM grid box to the study area (observation point or region) (e.g., Rosenzweig and Parry, 1994) or by interpolating from GCM grid box resolution to a higher resolution grid or point location (Leemans and van den Born, 1994; Harrison and Butterfield, 1996).

Three major methods have been developed to produce higher resolution climate scenarios at the sub-GCM grid scale: regional climate modeling (Giorgi and Mearns, 1991, 1999; McGregor, 1997), statistical downscaling (von Storch *et al.*, 1993; Rummukainen, 1997; Wilby and Wigley, 1997), and variable- and high-resolution GCM experiments (Fox-Rabinovitz *et al.*, 1997). All three methods are presented in Table 3-4 and discussed in detail in TAR WGI Chapter 10, but we briefly review here the first two, since they have been most commonly applied to impact assessments. Both methods are dependent on large-scale circulation variables from GCMs. Large-scale circulation refers to the general behavior of the atmosphere at large (i.e., continental) scales.

3.5.4.4.2. Regional climate modeling

The basic strategy with regional models is to rely on the GCM to reproduce the large-scale circulation of the atmosphere and to use the regional model, run at a higher resolution, to simulate sub-GCM scale regional distributions of climate. In numerous experiments with regional models driven by control and doubled CO_2 output from GCMs for regions throughout the world, the spatial pattern of changed climate—particularly changes in precipitation—simulated by the regional model departs from the more general pattern over the same region simulated by the GCM (TAR WGI Chapter 10).

3.5.4.4.3. Statistical methods

Statistical methods are much less computationally demanding than dynamic methods; they offer an opportunity to produce ensembles of high-resolution climate scenarios (for reviews, see von Storch, 1995; Wilby and Wigley, 1997). However, these techniques rely on the (questionable) assumption that observed statistical relationships will continue to be valid under future radiative forcing—that is, they are time-invariant (Wilby, 1997).

Although regional modeling and statistical techniques have been available for at least a decade—their developers claiming use in impact assessments as one of their important applications—it is only recently that they have actually provided scenarios for impact assessments (Mearns *et al.*, 1998, 1999, 2001; Sælthun *et al.*, 1998; Hay *et al.*, 1999; Brown *et al.*, 2000; Whetton *et al.*, 2001). Mearns *et al.* (1999, 2001) demonstrate that a high-resolution scenario results in agricultural impacts that differ from those produced with a coarser resolution GCM scenario (discussed in Chapter 5). Hay *et al.* (1999) found differences in runoff calculations, based on a GCM-scenario and a statistically downscaled scenario.

3.5.4.5. Temporal Resolution (Mean versus Variability)

For the most part, climate changes calculated from climate model experiments have been mean monthly changes in relevant variables. Techniques for generating changes in variability emerged in the 1990s (Mearns *et al.*, 1992, 1996, 1997; Wilks, 1992; Semenov and Barrow, 1997). The most common technique involves manipulation of the parameters of stochastic weather generators to simulate changes in variability on daily to interannual time scales (e.g., Bates *et al.*, 1994, 1996). Several studies have found important differences in the estimated impacts of climate change when effects of variance change were included (Mearns *et al.*, 1997; Semenov and Barrow, 1997). Combined changes in mean and variability also are evident in a broad suite of statistical downscaling methods (Katz and Parlange, 1996; Wilby *et al.*, 1998). Other types of variance change still are difficult to incorporate, such as possible changes in the frequency and intensity of El Niño events (Trenberth and Hoar, 1997). However, where ENSO signals are strong, weather generators can be conditioned on ENSO phases, enabling scenarios of changed ENSO frequency to be generated stochastically (e.g., Woolhiser *et al.*, 1993). However, climate models still are not capable of clearly indicating how ENSO events might change in the future (TAR WGI Chapter 9).

3.5.4.6. Incorporation of Extremes in Scenarios

Whereas changes in both the mean and higher order statistical moments (e.g., variance) of time series of climate variables affect the frequency of extremes based on these variables (e.g., extreme high daily or monthly temperatures; drought and flood episodes), other types of extremes are based on complex atmospheric phenomena (e.g., hurricanes). Given the importance of the more complex extremes—such as hurricanes, tornadoes, and storm surges (see Table 1-1)—it would be desirable to incorporate changes in the frequency of such phenomena into scenarios. Unfortunately, very little work has been performed on how to accomplish this, and there is only limited information on how the frequency, intensity, and spatial characteristics of such phenomena might change in the future (see Section 3.8.5).

An example of an attempt to incorporate such changes into impact assessments is the study of McInnes *et al.* (2000), who developed an empirical/dynamical model that gives return period versus height for tropical cyclone-related storm surges for a location on the north Australian coast. The model can accept changes in tropical cyclone characteristics that may occur as a result of climate change, such as changes in cyclone intensity. Other methods for incorporating such changes into quantitative climate scenarios remain to be developed; further advances in this area of research can be expected over the next few years.

3.5.4.7. Surprises: Low-Probability, High-Impact Events

Several types of rapid, nonlinear response of the climate system to anthropogenic forcing, sometimes referred to as "surprises," have been suggested. These include reorganization of the thermohaline circulation, rapid deglaciation, and fast changes to the carbon cycle (e.g., Stocker and Schmittner, 1997). For instance, it has been suggested that a sudden collapse of the thermohaline circulation in the North Atlantic—an event that has not been simulated by any AOGCM (TAR WGI Chapter 9) but cannot be ruled out on theoretical grounds (TAR WGI Chapter 7)—could cause major disruptions in regional climate over northwest Europe. Such a possibility has been used to create synthetic arbitrary climate scenarios to investigate possible extreme impacts (Alcamo *et al.*, 1994; Klein Tank and Können, 1997).

3.5.5. Uncertainties of Climate Scenarios

The concept of uncertainty is implicit in the philosophy of climate scenario development, and characterization and quantification of uncertainty has become one of the most vigorous and dynamic branches of climate scenario research. Some important sources of uncertainty are detailed in TAR WGI Chapter 13, of which three major sources are:

1) Uncertainties in future GHG and aerosol emissions. The IS92 and SRES emissions scenarios described in Section 3.8 exemplify these uncertainties; each scenario implies different atmospheric compositions and hence different radiative forcing.
2) Uncertainties in global climate sensitivity,[4] mainly as a result of differences in the way physical processes and feedbacks are simulated in different models. This means that some GCMs simulate greater mean global warming per unit of radiative forcing than others.
3) Uncertainties in regional climate changes, which are apparent from differences in regional estimates of climate change by different GCMs for the same mean global warming.

Many early impact studies employed a climate scenario derived from a single GCM. However, it was recognized early on that different GCMs yield different regional climate responses, even when they are perturbed with identical forcing (e.g., Smith and Tirpak, 1989). Therefore, various approaches have been used to capture this range of responses in impact studies. These approaches include using all available GCM results (e.g., Santer, 1985; Yohe *et al.*, 1999); using a selected subset of GCM experiments, in some cases based on the performance of the GCMs at simulating the current climate (e.g., Robock *et al.*, 1993; Risbey and Stone, 1996; Smith *et al.*, 1996); using results from different GCMs that have been "pattern-scaled" in conjunction with simple climate models to represent different types of uncertainty (e.g., Barrow *et al.*, 2000; see also Section 3.8.3); or using the mean or median GCM response (e.g., Rotmans

---

[4]Climate sensitivity is the long-term (equilibrium) change in global mean surface temperature following a doubling of atmospheric equivalent CO_2 concentration.

et al., 1994). The effect is to generate a range of future impacts. Much of the quantitative, scenario-based, impacts literature assessed in IPCC (1990) and IPCC (1996b) reported these kinds of analyses. More recently, impact studies have begun to consider the impacts of anthropogenic climate change alongside the effects of natural multi-decadal climate variability (Hulme *et al.*, 1999a). This creates a distribution of impact indicator values for the present day to compare with the range of future impacts under alternative climate scenarios.

There have been a few preliminary attempts to derive frequency distributions of future climate by using expert judgment (Morgan and Keith, 1995; Hulme and Carter, 1999) or by projecting the statistical fit of modeled versus observed 20th-century climate onto modeled future changes (Allen *et al.*, 2000). This information may be useful for impact assessment because it offers an opportunity to express impacts in terms of risk—for example, the risk of exceeding a given threshold impact (Jones, 2000; Pittock, 1999).

3.6. Sea-Level Rise Scenarios

3.6.1. Purpose

Sea-level rise scenarios are constructed to assess climate change impacts and adaptations in the coastal zone. Variations in sea level are measured in two ways. Eustatic sea level represents the level of the ocean independent of land movements. Relative sea level is measured relative to the local land surface (Klein and Nicholls, 1998), so it consists of two components: eustatic sea-level change and local land movements. Climate modelers largely concentrate on estimating eustatic sea-level change, whereas impact researchers focus on relative sea-level change.

3.6.2. Baseline Conditions

Based on historical tide gauge records and allowing for land movements, eustatic sea level has risen at an estimated rate of 1.0–2.0 mm yr^{-1} during the past century (TAR WGI Chapter 11). This rate of sea-level rise is consistent with recent satellite altimeter data (Nerem *et al.*, 1997), which directly measures eustatic variations in sea level. Tide gauge records are the main source of information on relative sea level; records are archived by the Permanent Service for Mean Sea Level (PSMSL) (Spencer and Woodworth, 1993). These records exhibit variations in interannual and multi-decadal variability (e.g., Delcroix, 1998; Bell *et al.*, 1999; Nerem, 1999). The land surface forming the coastline at any point may be subsiding, static, or rising. Subsidence can be caused by tectonic movements, isostatic subsidence, compaction of sediments, or extraction of groundwater, oil, and/or gas. Uplift, as a result of postglacial isostatic rebound or tectonic processes, reduces or reverses relative sea-level rise. To allow for these influences, Douglas (1997) recommends that tide gauge records be at least 50 years in length before they are used to establish long-term trends or a nonstationary baseline.

Most studies of vulnerability to sea-level rise use the mean sea level at a reference date. For instance, studies employing the IPCC Common Methodology (WCC 1993, 1994) use the level in 1990 (Nicholls, 1995; Bijlsma, 1996). For more comprehensive assessments of coastal vulnerability, however, baseline time series of sea-level variability are required. These reflect tidal variations and the influences of water temperature, wind, air pressure, surface waves, and Rossby and Kelvin waves in combination with the effects of extreme weather events. Baseline information for coastal processes also may be necessary where the coastline is accreting, eroding, or changing in form as a result of previous environmental changes. Where an earlier climate or sea-level shift can be related directly to a response in coastal or adjacent marine processes, this may serve as a historical or palaeo-analog for assessment of future changes.

3.6.3. Global Average Sea-Level Rise

The major components of average global sea-level rise scenarios are thermal expansion, glaciers and small ice caps, the Greenland and Antarctic ice sheets, and surface and groundwater storage (Warrick *et al.*, 1996; TAR WGI Chapter 11). These phenomena usually are modeled separately. Using GCM output, the thermal component of sea-level rise has been estimated by Bryan (1996), Sokolov *et al.* (1998), and Jackett *et al.* (2000). Contributions from glaciers and ice sheets usually are estimated via mass-balance methods that use coupled atmosphere-ocean and atmosphere-ice relationships. Such studies include: for glaciers and the Greenland ice sheet, Gregory and Oerlemans (1998); for Greenland only, Van de Wal and Oerlemans (1997) and Smith (1998); for the Antarctic ice sheet, Smith *et al.* (1998); and for Greenland and Antarctica, Ohmura *et al.* (1996) and Thompson and Pollard (1997).

Simple models that integrate these separate components through their relationship with climate, such as the upwelling diffusion-energy balance model of Wigley and Raper (1992, 1993, 1995) used in Warrick *et al.* (1996), can be used to project a range of total sea-level rise. De Wolde *et al.* (1997) used a two-dimensional model to project a smaller range than in Warrick *et al.* (1996); the major differences were related to different model assumptions. Sokolov and Stone (1998) used a two-dimensional model to achieve a larger range. Some new estimates are presented in Section 3.8.2.

3.6.4. Regional Sea-Level Rise

Regional sea-level rise scenarios require estimates of regional sea-level rise integrated with estimates of local land movements. Currently there are too few model simulations to provide a range of regional changes in sea level, restricting most scenarios to using global mean values (de Wolde, 1999). An exception is Walsh *et al.* (1998), who produced scaled scenarios of regional sea-level rise for the Gold Coast of eastern Australia on the basis of a suite of runs from a single GCM. Because relative sea-level rise scenarios are needed for coastal impact studies, local land movements also must be estimated. This requires long-term tide gauge records with associated ground- or satellite-based geodetic leveling. Geophysical models of isostatic effects, incorporating the continuing response of the Earth to ice-loading during the last glaciation, also provide estimates of long-term regional land movements (Peltier, 1998; Zwartz *et al.*, 1999).

3.6.5. Scenarios Incorporating Variability

Most impacts on the coast and near coastal marine environments will result from extreme events affecting sea level, such as storm surges and wave set-up. The magnitude of extreme events at any particular time is influenced by tidal movements, storm severity, decadal-scale variability, and regional mean sea level. These phenomena are additive. Because it is impossible to provide projections of all of these phenomena with any confidence, many assessments of coastal impacts simply add projections of global average sea level to baseline records of short-term variability (e.g., Ali, 1996; McDonald and O'Connor, 1996; McInnes and Hubbert, 1996; Lorenzo and Teixiera, 1997). Moreover, several coastal processes also are stochastic, and locally specific scenarios may have to be constructed for these (e.g., Bray and Hooke, 1997).

3.6.6. Application of Scenarios

3.6.6.1. Simple Scenarios

Simple scenarios are based on one or several estimates of sea-level rise consistent with IPCC-projected ranges of global sea-level rise for a particular date. Usually a mid-range or upper estimate is chosen. The application of a eustatic scenario, where a relative scenario is required, discounts the impact of regional sea-level change and local land movements, although it is possible to add the latter explicitly where estimates exist (Gambolati *et al.*, 1999). Assessments that use simple scenarios usually test whether a coastal region is sensitive and/or vulnerable to a plausible upper limit of climate change (e.g., Zeidler, 1996; El Raey *et al.*, 1997; Olivo, 1997).

3.6.6.2. Projected Ranges

A range of global sea-level rise can be applied, bounded by its upper and lower extremes, for a particular date (e.g., Ali, 1996; Nicholls *et al.*, 1999). This will project a likely range of impacts but without any reference to the likelihood of that range or specific changes within that range (Section 2.5). The major disadvantage of this technique is the large range of uncertainty that is produced, making it difficult for policymakers and planners to decide on a concrete response.

3.6.6.3. Risk and Integrated Assessment

Risk assessment aims to produce meaningful outcomes under conditions of high uncertainty. For sea-level rise, two approaches to risk assessment have been reported. The first approach is to construct a probability distribution for a single outcome. For example, Titus and Narayanan (1996) conclude that a sea-level rise of 10–65 cm by 2100 has an 80% probability of occurring; the 99th percentile was associated with a 104-cm rise. The second approach is to calculate the probability of exceedance above a given threshold identified as a hazard. Pittock and Jones (2000) suggest the use of critical thresholds, which link an unacceptable level of harm with a key climatic or climate-related variable. For coastal impacts, the critical threshold is then linked to a projected range of sea-level scenarios, through key climatic and marine variables, and its risk of exceedance calculated (Jones *et al.*, 1999).

IAMs attempt to represent the interaction of human activities with socioeconomic and biophysical systems on a global scale (see Section 3.3.2.3). In the TARGETS model (Rotmans and de Vries, 1997), various human activities that affect a succession of phenomena are simulated to produce scenarios of sea-level rise, which then lead to calculations of people and capital at risk in low-lying coastal regions (Hoekstra, 1997). The IMAGE 2 integrated model applies baseline scenarios of global environmental change (Alcamo *et al.*, 1996) to project several global outcomes, one of which is sea-level rise. Yohe and Schlesinger (1998) used a model of global economic activity to produce emissions profiles, which they then used to calculate temperature and sea-level changes and integrated with an economic damages model for the U.S. coastline. The scenarios of sea-level rise were probabilistically weighted from a sample of 280 to calculate the 10th and 90th percentiles and the median estimate, producing several ranges similar in magnitude to that of Titus and Narayanan (1996).

3.7. Representing Interactions in Scenarios and Ensuring Consistency

3.7.1. Introduction

There is great diversity in the scenarios adopted in impact assessments. This diversity is valuable in providing alternative views of the future, although it can hamper attempts to summarize and interpret likely impacts by introducing inconsistencies within or between studies. Moreover, there are certain key dependencies in climate change science that have resulted in time lags and inconsistencies in the application of scientific results between different research areas. This has been reflected in the IPCC process (see Table 3-6). Thus, although TAR WGI reviews recent projections of future climate, these results are not yet available to the impacts community to prepare and publish their analyses, on which the TAR WGII assessment is based. Instead, most impact studies have relied on earlier, more rudimentary climate projections. Similarly, the simplified assumptions used in climate model simulations about changes in radiative forcing of the climate from changing GHG and aerosol concentrations represent only a limited subset of plausible atmospheric conditions under a range of emissions scenarios reviewed by TAR WGIII.

***Table 3-6**: Approximate chronology of IPCC process in relation to GCM simulations, their adoption in impact studies, and the development of IPCC emissions scenarios. Abbreviations follow: AGCM = atmospheric GCM with simple ocean; AOGCM = coupled atmosphere-ocean GCM; GHG = greenhouse gas; IS92 = IPCC emissions scenarios published in 1992 (Leggett et al., 1992); SRES = Special Report on Emissions Scenarios (Nakicenovic et al., 2000).*

Date	IPCC Process	Working Group I GCM Simulations	Working Group II GCM-Based Scenarios used in Impact Studies	Working Group III Emissions Scenarios
1988–1990	First Assessment Report (FAR), 1990	Equilibrium high-resolution AGCM	Equilibrium low–resolution 2 x CO_2	Scenarios A-D (A = Business-as-Usual)
1991–1992	FAR Supplement, 1992	Transient AOGCM cold start GHG-only (Scenario A emissions)	Equilibrium low-resolution 2 x CO_2	IS92a-f
1993–1996	Second Assessment Report (SAR), 1996	Transient AOGCM warm-start GHG + aerosol (0.5 or 1% per year emissions)	Equilibrium low/high-resolution; transient cold-start	IS92a-f (modified)
1997–1998	Regional Impacts Special Report, 1998	Transient AOGCM ensemble/ multi-century control	Equilibrium low/high-resolution; transient cold-start/warm-start	IS92a-f (modified)
1999–2001	Third Assessment Report (TAR), 2001	Transient AOGCM CO_2-stabilization; SRES-forced	Transient warm-start; multi-century control and ensembles	SRES; stabilization

Creation of comprehensive scenarios that encompass the full complexity of global change processes and their interactions (including feedbacks and synergies) represents a formidable scientific challenge. This section addresses some components of this complexity. First it treats generally accepted biogeochemical processes; second, it addresses emerging climate-system processes; and third, it reviews rarely considered interactions between anthropogenic and natural driving forces. Finally, the importance of comprehensiveness and compatibility in scenario development is discussed.

3.7.2. Representing Processes and Interactions in Scenarios

3.7.2.1. Generally Considered Interactions

Emissions of greenhouse gases have increased their atmospheric concentrations, which alter the radiative properties of the atmosphere and can change the climate (see TAR WGI Chapters 3–8). Determination of atmospheric concentrations from emissions is not straightforward; it involves the use of models that represent biogeochemical cycles and chemical processes in the atmosphere (Harvey *et al*., 1997; TAR WGI Chapters 3–5). Several atmosphere-ocean interactions are considered in defining the future transient response of the climate system (Sarmiento *et al*., 1998; TAR WGI Chapter 8). For the purposes of scenario development, CO_2 occupies a special role, as a greenhouse gas (IPCC, 1996a) and by directly affecting carbon fluxes through CO_2 fertilization and enhanced water-use efficiency (see Section 3.4.2). These direct responses are well known from experimentation (Kirschbaum *et al*., 1996). Biospheric carbon storage is further strongly influenced by climate, land use, and the transient response of vegetation. All of these interactions define the final CO_2 concentrations in the atmosphere and subsequent levels of climate change (see Table 3-7).

The early simple climate models that were used in the IPCC's First and Second Assessment Reports all emphasized the importance of CO_2 fertilization but few other biogeochemical interactions (Harvey *et al*., 1997). Inclusion of more realistic responses of the carbon cycle in climate scenarios still is an evolving research area (Walker *et al*., 1999), but most interactions now are adequately represented.

3.7.2.2. Less Considered Interactions

Interactions between land, vegetation, and the atmosphere have been studied extensively in deforestation and desertification model experiments (Charney *et al*., 1977; Bonan *et al*., 1992; Zhang *et al*., 1996; Hahmann and Dickinson, 1997). Changes in surface characteristics such as snow/ice and surface albedo and surface roughness length modify energy, water, and gas fluxes and affect atmospheric dynamics. These interactions occur at various scales (Hayden, 1998), but although their importance is well appreciated (Eltahir and Gong, 1996; Manzi and Planton, 1996; Lean and Rowntree, 1997; Zeng, 1998) they still generally are ignored in scenario development.

Climate modeling studies (e.g., Henderson-Sellers *et al.*, 1995; Thompson and Pollard, 1995; Sellers *et al.*, 1996) suggest an additional warming of about 0.5°C after deforestation on top of the radiative effects of GHG, but these effects are not necessarily additive on regional scales. Betts *et al.* (1997) concur that vegetation feedbacks can be significant for climate on regional scales. More recent studies, however, tend to predict smaller changes, partly as a result of the inclusion of more interactions such as the cloud radiative feedback. Field experiments show large changes in surface hydrology and micrometeorological conditions at deforested sites (Gash *et al.*, 1996). On the other hand, observations have not provided direct evidence of changes in overall climate in the Amazon basin (Chu *et al.*, 1994) or in Sahel surface albedo (Nicholson *et al.*, 1998), but the available data series are too short to be conclusive.

Palaeoclimatic reconstructions, using empirical data and model results, provide better opportunities to study vegetation-atmosphere interactions. Climate models that incorporate dynamic vegetation responses simulate larger vegetation shifts for changed past climates than expected by the orbitally forced climate effect alone. For example, an additional 200–300 km poleward displacement of forests simulated for 6,000 ky BP in North America was triggered by changes in surface albedo (Kutzbach *et al.*, 1996; Texier *et al.*, 1997; Ganopolski *et al.*, 1998). However, these shifts are not observed in all model experiments (e.g., Broström *et al.*, 1998). Other modeling results suggest that oceans also play a prominent role (Hewitt and Mitchell, 1998). Thus, vegetation-ocean-climate interaction seems to be important in defining regional climate change responses.

Most vegetation models used in scenario development are equilibrium models (i.e., for a given climate they predict a fixed vegetation distribution). The latest dynamic vegetation models attempt to include plant physiology, biogeochemistry, and land surface hydrology (e.g., Goudriaan *et al.*, 1999), and some explicitly treat vegetation structure and succession. Foley *et al.* (1998) coupled one such model to a GCM and found that the most climatically sensitive zones were the desert/grassland and forest/tundra ecotones. These zones also tend to be exposed to large disturbances and natural climate variability (Schimel *et al.*, 1997b). In another model experiment, Zeng and Neelin (1999) found that interannual and inter-decadal climate variability helps to keep the African savannah region from getting either too dry or too wet, through nonlinear vegetation-atmosphere interactions. Few of these models contain simulations of disturbances, such as fire regimes (Crutzen and Goldammer, 1993; Kasischke and Stocks, 2000), which rapidly alter vegetation patterns and influence vegetation responses. Unfortunately, hardly any of these insights are included routinely in scenario development.

3.7.2.3. *Rarely Considered Interactions*

Most scenarios emphasize systemic interactions within nonhuman components of the climate system. These interactions are relatively well studied. The response of society to changes in the climate system is much less well studied. Land-use and land-cover change is an exception, but its treatment in climate scenarios still is far from ideal (see Section 3.3). The difficulty of including such interactions in scenario development is that many are not precisely specified and act indirectly. For example, warmer climates would change heating and cooling requirements of buildings. Such effects frequently are listed as impacts but are not factored in as adjustments to energy use and thus emission levels. Another example is population migration, which can be treated as an impact of environmental or socioeconomic change while also serving as a scenario of demographic change affecting future regional vulnerability (Döös, 1997).

A model that accounts for such societal interactions with the climate system is TARGETS (Rotmans and de Vries, 1997), which evolved from the WORLD model of Meadows *et al.*

***Table 3-7**: Illustration of importance of some different feedback processes. Values are for the year 2100, obtained from a baseline scenario implemented in the IMAGE-2 integrated assessment model (adapted from Alcamo et al., 1998a). The no-feedbacks case excludes CO_2 fertilization and accelerated ice melt and includes an intermediate adaptation level of vegetation.*

Simulation	$[CO_2]$ (ppm)	Net Ecosystem Productivity (Pg a^{-1})[a]	Temperature Change (°C)	Sea-Level Rise (cm)	Vegetation Shift (%)[b]
All feedbacks	737	6.5	2.8	43	41
No CO_2 fertilization	928	0.1	3.6	52	39
Vegetation adapts immediately	724	7.0	3.1	45	40
No adaptation of vegetation	762	5.3	3.2	46	41
No land-use change	690	6.9	2.9	41	39
No feedbacks	937	0.0	3.5	29	45
No land-use change/no feedbacks	889	0.2	3.4	28	45
Range	690–937	0.0–7.0	2.8–3.6	28–52	39–45

[a] 1 Pg a^{-1} = 10^{15} grams per year.
[b] Percentage of vegetated area for which climate change induces a change of vegetation class.

(1992). TARGETS is a highly aggregated model (only two regions and few resource classes) with simple relationships between population, economic development, and resource use; between environmental conditions and population/health/wealth; and between emissions, concentrations, climate, and impacts. The model generates globally averaged emissions and climate change scenarios. The strength of the model is that different interactions—including controversial ones, such as the effects of climate change on food availability and health and their interactions with population—can be explored easily, but its use for developing scenarios for impact assessment is limited.

3.7.3. Tools Capable of Addressing Interactions

Most climate scenarios for impact assessments were developed by using outputs from AOGCMs (Hulme and Brown, 1998). Often these results are scaled toward the desired emission levels with simple climate models (Hulme *et al.*, 1995; Harvey *et al.*, 1997; see Section 3.8.3). In this simple approach, most interactions are neglected.

The only models that can be used to develop more consistent scenarios that incorporate most of the important interactions are IAMs (see Section 3.3.2.3). IAMs have been developed with different levels of complexity, from extremely simple to highly complex (Harvey *et al.*, 1997). Different interactions are included, although no single model provides a fully comprehensive treatment. The models are most commonly used for emission scenario development and mitigation policy assessment (Schimel *et al.*, 1997a; Alcamo *et al.*, 1998a; Pepper *et al.*, 1998). All simulate a causal chain (e.g., human activities, emissions, climate change, sea-level rise, and other impacts). Emissions, climate change, impact, and mitigation scenarios derived from these models have been published (Schimel *et al.*, 1997a; Leemans *et al.*, 1998; Pepper *et al.*, 1998) and collected in several databases (Alcamo *et al.*, 1995; Nakicenovic *et al.*, 2000). Unfortunately, it is not always clear which interactions are explicitly included in individual IAMs. This reduces the comparability of individual IAM-derived scenarios and thus their utility.

Depending on assumed interactions during scenario development, a wide range of estimates of climate change and its impacts is possible (see Table 3-7). However, within this range certain responses are more likely than others. To define appropriate and realistic levels of interactions, expert judgment and sensitivity experiments with models could be very valuable (van der Sluijs, 1997). Innovative, objective, and systematic approaches have to be developed to evaluate underlying scenario assumptions and to validate the scenario results. This is still an immature area of scenario development.

3.7.4. Problems of Compatibility between Scenarios

One difficulty faced by authors in attempting to summarize and synthesize the results of impact studies for previous IPCC assessments (i.e., IPCC, 1996b, 1998) has been a lack of consistency in projections. Different climate projections have been adopted in different studies, in different regions (or within the same region), and in different sectors. Moreover, even where the same climate projections are assumed, they might not be applied in the same way in different impact studies. Finally, some studies also are inconsistent in their methods of projecting changes in climate alongside concurrent changes in related socioeconomic and environmental conditions.

For example, GHG concentrations often are transformed into CO_2-equivalent concentrations to determine radiative forcing levels and climate change. The GCM community often presents climate change simulations as "doubled CO_2" anomalies.

Box 3-2. The Global Impact of Climate Change on Five Sectors (Parry and Livermore, 1999)

In this assessment, the prospective effects of unmitigated climate change during the 21st century are estimated at a global scale in five sectoral studies (see Table 3-8). Each study has different scenario requirements, though some are common to several studies. For example, the ecosystems study estimates potential biomass on the basis of scenarios of climate, CO_2 concentration, and nitrogen deposition, but it ignores future land-cover and land-use changes that would be expected regardless of climate change. In contrast, the study on food security examines the effects on crop productivity of the same scenarios of climate (though for fewer variables) and CO_2 concentration; it too ignores likely land-cover and land-use changes and does not consider effects of nitrogen deposition, although it adopts a range of socioeconomic and technological scenarios to evaluate the number of persons at risk from hunger.

Notably, across all of the studies the scenarios adopted are designed to be mutually consistent. For instance, the population and GDP scenarios are those adopted in constructing the IS92a emissions scenario (Leggett *et al.*, 1992). An approximation of the IS92a emissions scenario is used to force the HadCM2 and HadCM3 GCMs that were employed to construct the climate and sea-level scenarios (Hulme *et al.*, 1999b). Other scenarios are chosen to be broadly consistent with these assumptions. The scenarios are required as inputs to global impact models, and results from these are described elsewhere in this report. Finally, it also should be noted that although these studies are compatible and consistent, they are not integrated across sectors. For example, climate-induced changes in water resources for irrigation are not accounted for in estimates of future food security.

Depending on the scenario, however, 5–40% of the forcing is caused by non-CO_2 GHGs (30% in 1990). The doubled-CO_2 scenarios often are interpreted as CO_2 only (e.g., Cramer *et al.*, 1997); others add an explicit distinction between CO_2 and non-CO_2 gases (e.g., Downing *et al.*, 1999). In determining the impacts of direct CO_2 effects and climate change, this can easily lead to inconsistencies. Similar discrepancies exist for other types of interactions.

Finally, it is a significant challenge to integrate climate or sea-level rise scenarios, with a time horizon of decades to hundreds of years, with nonclimatic scenarios of social, economic, and technological systems that can change rapidly over a time scale of years. For instance, it is difficult to devise credible socioeconomic scenarios that extend beyond the lifetime of current infrastructure and institutions. Moreover, social/economic actors who need to be involved in the scenario development process (e.g., business, governments) often find long time horizons difficult to contemplate. Box 3-2 illustrates a recent example of an attempt to harmonize climate change, sea level, atmospheric composition, and socioeconomic scenarios in a multi-sectoral global impact assessment.

3.8. Scenarios of the 21st Century

This section summarizes recent developments that are likely to affect the construction of scenarios over the coming few years. One of these developments is construction of the new SRES emission scenarios. Some features of these scenarios and their implications for atmospheric composition, global climate, and sea level are described below. In addition, a brief review of possible regional climate changes during the 21st century is presented, followed by discussions of stabilization scenarios and changes in climate variability and extreme events—key issues in constructing scenarios for policy-relevant impact and adaptation assessments.

3.8.1. SRES Storylines and Emissions Scenarios

Development of the SRES scenarios (Nakicenovic *et al.*, 2000) is outlined in Section 3.2.4.1. The 40 scenarios, 35 of which are fully quantified, are based on four different narrative storylines and associated scenario families. Each storyline describes a different world evolving through the 21st century, and each

***Table 3-8**: Summary of scenarios adopted in an assessment of global impacts on five sectors (Parry and Livermore, 1999).*

Scenario Type (up to 2100)	Ecosystems[a]	Water Resources[b]	Food Security[c]	Coastal Flooding[d]	Malaria Risk[e]
Socioeconomic/technological					
– Population	—	✔	✔	✔	✔
– GDP	—	—	✔	✔	—
– GDP per capita	—	—	✔	✔	—
– Water use	—	✔	—	—	—
– Trade liberalization	—	—	✔	—	—
– Yield technology	—	—	✔	—	—
– Flood protection	—	—	—	✔	—
Land-cover/land-use change	—	—	—	✔	—
Environmental					
– CO_2 concentration	✔	—	✔	—	—
– Nitrogen deposition	✔	—	—	—	—
Climate					
– Temperature	✔	✔	✔	—	✔
– Precipitation	✔	✔	✔	—	✔
– Humidity	✔	✔	—	—	—
– Cloud cover/radiation	✔	✔	—	—	—
– Windspeed	—	✔	—	—	—
– Diurnal temperature range	✔	—	—	—	—
Sea level	—	—	—	✔	—

[a] White *et al.* (1999) and see Chapter 5.
[b] Arnell (1999) and see Chapters 4 and 19.
[c] Parry *et al.* (1999) and see Chapters 5 and 19.
[d] Nicholls *et al.* (1999) and see Chapters 6, 7, and 19.
[e] Martens *et al.* (1999) and see Chapters 8 and 18.

may lead to quite different GHG emissions trajectories. Four of the scenarios are designated as "markers," each characterizing one of four "scenario families"; two additional scenarios illustrate alternative energy developments in one of the families. The storylines and scenario families are as follows:

- *A1:* A future world of very rapid economic growth, global population that peaks mid-century and declines thereafter, and rapid introduction of new and more efficient technologies. Major underlying themes are economic and cultural convergence and capacity-building, with a substantial reduction in regional differences in per capita income. The A1 scenario family develops into three groups that describe alternative directions of technological change in the energy system: fossil-intensive (A1FI), nonfossil energy sources (A1T), and a balance across all sources (A1B).
- *A2:* A differentiated world. The underlying theme is self-reliance and preservation of local identities. Fertility patterns across regions converge very slowly, resulting in continuously increasing population. Economic development is primarily regionally orientated, and per capita economic growth and technological change are more fragmented and slower than other storylines.
- *B1:* A convergent world with rapid change in economic structures toward a service and information economy, reductions in material intensity, and introduction of clean technologies. The emphasis is on global solutions to economic, social, and environmental sustainability, including improving equity, but without additional climate change policies.
- *B2:* A world in which the emphasis is on local solutions to economic, social, and environmental sustainability. This is a world with continuously increasing global population at a lower rate than in scenario A2, intermediate levels of economic development, and less rapid and more diverse technological change than in the A1 and B1 storylines. Although this scenario also is orientated toward environmental protection and social equity, it focuses on the local and regional levels.

Measures of global population, economic development (expressed in annual GDP), and equity (per capita income ratio) for 2050 and 2100 that are implied under the SRES scenarios are shown in Table 3-9, alongside the IS92a scenario and estimates for the present day. Attempts are underway to "downscale" aspects of these global scenarios for use in regional impact assessment (e.g., Lorenzoni *et al.*, 2000).

3.8.2. Implications of SRES Scenarios for Atmospheric Composition and Global Climate

Estimates of atmospheric composition resulting from the SRES emissions scenarios are presented in TAR WGI Chapters 3–5. Information on CO_2 and ground-level O_3 concentrations is given in Tables 3-2 and 3-9. More detailed regional estimates of pollutant concentrations and deposition of acidifying compounds based on these scenarios also are beginning to emerge (e.g., Mayerhofer *et al.*, 2000; see Section 3.4).

To interpret the possible range of global temperature and sea-level response to the SRES scenarios, estimates have been made with simple models for all 35 of the quantified SRES scenarios (Table 3-9; see also TAR WGI Chapters 9 and 11). Estimates of global warming from 1990 to 2100 give a range of 1.4–5.8°C—somewhat higher than the 0.7–3.5°C of the SAR. The main reason for this increase is that the levels of radiative forcing in the SRES scenarios are higher than in the IS92a-f scenarios, primarily because of lower sulfate aerosol emissions, especially after 2050. The temperature response also is calculated differently; rather than using the conventional idealized, equilibrium climate sensitivity range of 1.5–4.5°C (IPCC, 1996a), the simple model is tuned to the effective climate sensitivities of a sample of individual AOGCMs (see TAR WGI Chapter 9 for details). Sea-level rise between 1990 and 2100 is estimated to be 9–88 cm, which also accounts for uncertainties in ice-melt parameters (see TAR WGI Chapter 11).

3.8.3. Implications of SRES Scenarios for Regional Mean Climate

3.8.3.1. Regional Information from AOGCMs

Estimates of regional climate change to 2100 based on AOGCM experiments are described in TAR WGI Chapters 9 and 10. The results of nine AOGCMs run with the A2 and B2 SRES scenarios[5] display many similarities with previous runs that assume IS92a-type emissions, although there also are some regional differences (see below). Overall, rates of warming are expected to be greater than the global average over most land areas and most pronounced at high latitudes in winter. As warming proceeds, northern hemisphere snow cover and sea-ice extent will be reduced. Models indicate warming below the global average in the North Atlantic and circumpolar southern ocean regions, as well as in southern and southeast Asia and southern South America in June–August. Globally there will be increases in average water vapor and precipitation. Regionally, December–February precipitation is expected to increase over the northern extratropics and Antarctica and over tropical Africa. Models also agree on a decrease in precipitation over Central America and little change in southeast Asia. Precipitation in June–August is projected to increase in high northern latitudes, Antarctica, and south Asia; change little in southeast Asia; and decrease in Central America, Australia, southern Africa, and the Mediterranean region.

The main differences between the SRES-based and IS92-based runs concern greater disagreement in the SRES runs on the magnitude of warming in some tropical and southern hemisphere regions and differing intermodel agreement on the

[5]Preliminary marker emissions scenarios released in 1998 for use in climate modeling (Nakicenovic *et al.*, 2000).

magnitude of precipitation change in a few regions, possibly as a result of aerosol effects. However, there are no cases in which the SRES and IS92a results indicate precipitation changes of opposite direction (see TAR WGI Chapter 10).

3.8.3.2. *Regional Climate Characterizations*

Only a limited number of AOGCM results based on the SRES emissions scenarios have been released and analyzed to date (i.e., results for the A2 and B2 scenarios), and none were available for the impact studies assessed in this report. In the meantime, alternative approaches have been used to gain an impression of possible regional changes in climate across a wider range of emissions scenarios. One method uses results from existing AOGCM simulations and scales the pattern of modeled regional climate change up or down according to the range of global temperature changes estimated by simple climate models for different emissions scenarios or assumptions about climate sensitivity (Santer *et al*., 1990; Mitchell *et al*., 1999;

Table 3-9*: Some aspects of the SRES emissions scenarios and their implications for CO_2 concentration, global temperature and sea-level rise by 2050 and 2100 compared to the IS92a emissions scenario (Leggett et al., 1992). Data in columns 2–4 are taken from Nakicenovic et al. (2000). Calculations in columns 6–7 are relative to 1990. ΔT is change in mean annual temperature averaged across simple climate model runs emulating results of seven AOGCMs with average climate sensitivity of 2.8°C (Chapter 9, TAR WGI). CO_2 concentrations were estimated by using the same model runs (data from S.C.B. Raper, Chapter 9, TAR WGI). Sea-level rise estimates are based on temperature changes (Chapter 11, TAR WGI). SRES-min and SRES-max are minimum and maximum estimates across all 40 SRES scenarios (35 fully quantified scenarios for CO_2, ΔT, and sea level). High and low estimates of CO_2 concentration and temperature change account for uncertainties in climate sensitivity (across the range 1.7–4.2°C). Sea-level rise range also accounts for uncertainties in model parameters for land ice, permafrost, and sediment deposition. Note that scenario values are mutually consistent along all rows except for SRES-min and SRES-max.*

Emissions Scenario	Global Population (billions)	Global GDP[a] (10^{12} US\$ a^{-1})	Per Capita Income Ratio[b]	CO_2 Concentration[c] (ppm)	Global ΔT (°C)	Global Sea-Level Rise (cm)
1990	5.3	21	16.1	354	0	0
2000	6.1–6.2	25–28[d]	12.3–14.2[d]	367[e]	0.2	2
2050						
– SRESA1FI	8.7	164	2.8	573	1.9	17
– SRESA1B	8.7	181	2.8	536	1.6	17
– SRESA1T	8.7	187	2.8	502	1.7	18
– SRESA2	11.3	82	6.6	536	1.4	16
– SRESB1	8.7	136	3.6	491	1.2	15
– SRESB2	9.3	110	4.0	478	1.4	16
– IS92a	10.0	92	9.6	512	1.0	—
– SRES-min	8.4	59	2.4	463	0.8	5
– SRES-max	11.3	187	8.2	623	2.6	32
2100						
– SRESA1FI	7.1	525	1.5	976	4.5	49
– SRESA1B	7.1	529	1.6	711	2.9	39
– SRESA1T	7.1	550	1.6	569	2.5	37
– SRESA2	15.1	243	4.2	857	3.8	42
– SRESB1	7.0	328	1.8	538	2.0	31
– SRESB2	10.4	235	3.0	615	2.7	36
– IS92a	11.3	243	4.8	721	2.4	–
– SRES-min	7.0	197	1.4	478	1.4	9
– SRES-max	15.1	550	6.3	1099	5.8	88

[a] Gross domestic product (1990 US\$ trillion yr^{-1}).
[b] Ratio of developed countries and economies in transition (Annex I) to developing countries (Non-Annex I).
[c] Modeled values are not the same as those presented by TAR WGI, Appendix II, which are based on simulations using two different carbon cycle models for the six illustrative SRES emissions scenarios. Both models produce very similar results to the model applied here for a mid-range climate sensitivity; discrepancies in the high and low estimates are attributable to differences in the modeled climate-carbon cycle feedback.
[d] Modeled range across the six illustrative SRES scenarios.
[e] Observed 1999 value (Chapter 3, WG I TAR).

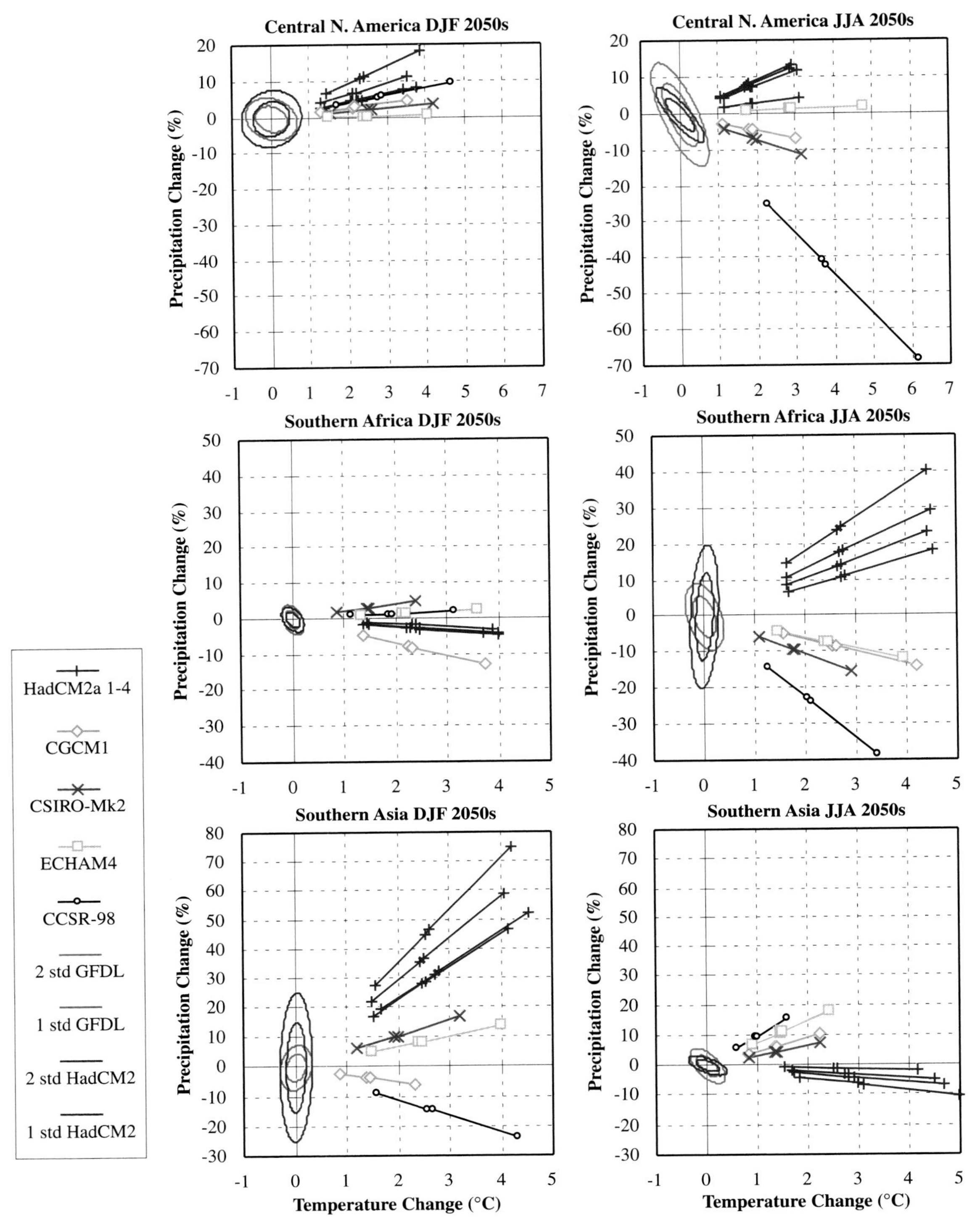

Figure 3-2: Scaled outputs of mean December-February (left) and June-August (right) temperature and precipitation change by the 2050s relative to 1961–1990 over land grid boxes representing Central North America (top), Southern Africa (middle), and Southern Asia (bottom) from eight simulations with five AOGCMs (experiments b, c, e, h, and a four-member ensemble from t; see Table 3-5). Simulations assume forcing by greenhouse gases but not aerosols, and are standardized according to the climate sensitivity of each AOGCM. Lines connect four points for each simulation, all in the same order from the origin: B1-low, B2-mid, A1-mid, A2-high. Each point represents the standardized regional changes in climate from the AOGCM, linearly scaled according to the global warming estimated with a simple climate model for one of four preliminary SRES marker emissions scenarios (B1, B2, A1, and A2) and a value of the climate sensitivity (low = 1.5°C; mid = 2.5°C, and high = 4.5°C). Also plotted are ±1 and ±2 standard deviation ellipses from the 1400-year HadCM2 and 1000-year GFDL unforced simulations, which are used to indicate natural multi-decadal variability and are orientated according to the correlation between modeled 30-year mean temperature and precipitation. Results from two other AOGCMs did not extend to the 2050s (Carter *et al.*, 2000).

see also detailed discussion in TAR WGI Chapter 13). This "pattern-scaling" method has been employed by Carter *et al.* (2000), using results from simulations with seven AOGCMs, all assuming a radiative forcing approximating the IS92a emissions scenario (for GHGs but excluding aerosols) scaled across a range of global temperature changes estimated by using a simple climate model for the four preliminary marker SRES emissions scenarios.

Regional-scale summary graphs of scaled temperature and precipitation changes were constructed for 32 world regions, at subcontinental scale, chosen to represent the regions being assessed by Working Group II (Carter *et al.*, 2000). Examples of individual plots are shown in Figure 3-2. Changes are plotted alongside estimates of "natural" multi-decadal variability of temperature and precipitation, extracted from two multi-century unforced AOGCM simulations. The graphs thus provide a quick assessment of the likely uncertainty range *and* significance of each AOGCM projection; they also show the extent to which different AOGCMs agree or disagree with regard to regional response to a given magnitude of global warming. Although a preliminary comparison of these results with SRES AOGCM runs (which also include aerosol forcing) suggests broad agreement on regional temperature and precipitation changes, more rigorous comparison remains to be carried out, offering a useful test of the pattern-scaling method.

3.8.4. *Stabilization Scenarios*

The SRES scenarios assume no climate policy intervention, but nations already are engaged in negotiations to reduce emissions of GHGs. Targets for stabilization of GHG concentrations in the atmosphere are being investigated by scientists and policymakers. TAR WGIII Chapter 2 reviews more than 120 mitigation scenarios, most of which aim to stabilize emissions of CO_2 at some target level. Simple climate models, as well as some AOGCMs, have been used to estimate the climate and sea-level response to stabilization (see Harvey *et al.*, 1997; TAR WGI Chapters 9 and 11). Relative to most reference emissions scenarios (e.g., the SRES scenarios), stabilization scenarios reduce global warming, especially beyond 2100. However, even for the lowest stabilization targets considered (450 ppm), based on long simulations by AOGCMs, the climate system and oceans may continue to respond for many centuries after stabilization of atmospheric concentrations of GHGs. Furthermore, because of regional variations in the time lag of response, regional patterns of climate change might be quite different from the unmitigated case (Whetton *et al.*, 1998).

3.8.5. *Scenarios of Changes in Climate Variability and Extreme Events*

It is demonstrated throughout this report that changes in climatic variability and extremes often play a dominant role in climate change impacts. Moreover, the magnitude and frequency of extreme events can change rapidly with only relatively small changes in climatic averages (see Section 3.5.4.6). However, climate modelers have more confidence in estimates of changes in averages than in changes in variability and extremes (see TAR WGI Chapters 8–10 and 13). Thus, impact assessors need to look carefully at the extent to which changes in variability and extremes are covered implicitly by changes in averages; when this is not the case, they must incorporate possible changes in these phenomena into scenarios. Table 3-10 summarizes projected changes in several types of extreme climate events and their likelihood taken from TAR WGI Technical Summary (see Table 1-1 for a typology of extremes). Table 3-10 also provides representative examples, drawn from different sectors and regions, of impacts that would be expected with high confidence, conditional on the occurrence of a given change in climate extremes. All of this information is reported in other chapters in this report.

3.9. State of the Science and Future Needs for Scenario Development

This chapter outlines the current practice of scenario development for climate impact, vulnerability, and adaptation assessment. Methods of scenario construction and application are evolving rapidly, so it is useful to identify which aspects are well developed and which aspects still are deficient.

3.9.1. *Well-Developed Features*

Some features of scenario development and application are well established and tested:

- Extensive monitoring efforts and continued development of global and regional databases has improved the quality and consistency of baseline observational data required for some scenario exercises, even in some data-sparse regions.
- Many impact studies apply incremental scenarios to explore the sensitivity of an exposure unit to a range of climate futures; studies seldom rely exclusively on a single, model-based scenario.
- Estimates of long-term mean global changes are available and widely applied for a limited number of variables (e.g., population, economic development, CO_2 concentration, global mean temperature), based on projections produced by specialized international organizations or the use of simple models.
- A growing volume of information now is available to enable scientists to construct regional scenarios of many features of global change, even though uncertainties in most projections remain high. A notable example is the IPCC-DDC, which was established in 1998 to facilitate the timely distribution of a consistent set of up-to-date projections of changes in climate and related environmental and socioeconomic factors for use in climate impact and adaptation assessment. Some of the studies reported in this volume use scenarios derived from information held in the DDC (see, e.g., Table 3-5).

***Table 3-10**: Examples of impacts resulting from projected changes in extreme climate events.*

Projected Changes during the 21st Century in Extreme Climate Phenomena and their Likelihood[a]	Representative Examples of Projected Impacts[b] *(all high confidence of occurrence in some areas[c])*
Simple Extremes	
Higher maximum temperatures; more hot days and heat waves[d] over nearly all land areas (*Very Likely*[a])	• Increased incidence of death and serious illness in older age groups and urban poor • Increased heat stress in livestock and wildlife • Shift in tourist destinations • Increased risk of damage to a number of crops • Increased electric cooling demand and reduced energy supply reliability
Higher (increasing) minimum temperatures; fewer cold days, frost days, and cold waves[d] over nearly all land areas (*Very Likely*[a])	• Decreased cold-related human morbidity and mortality • Decreased risk of damage to a number of crops, and increased risk to others • Extended range and activity of some pest and disease vectors • Reduced heating energy demand
More intense precipitation events (*Very Likely*[a] over many areas)	• Increased flood, landslide, avalanche, and mudslide damage • Increased soil erosion • Increased flood runoff could increase recharge of some floodplain aquifers • Increased pressure on government and private flood insurance systems and disaster relief
Complex Extremes	
Increased summer drying over most mid-latitude continental interiors and associated risk of drought (*Likely*[a])	• Decreased crop yields • Increased damage to building foundations caused by ground shrinkage • Decreased water resource quantity and quality • Increased risk of forest fire
Increase in tropical cyclone peak wind intensities, mean and peak precipitation intensities (*Likely*[a] over some areas)[e]	• Increased risks to human life, risk of infectious disease epidemics, and many other risks • Increased coastal erosion and damage to coastal buildings and infrastructure • Increased damage to coastal ecosystems such as coral reefs and mangroves
Intensified droughts and floods associated with El Niño events in many different regions (*Likely*[a]) (see also under droughts and intense precipitation events)	• Decreased agricultural and rangeland productivity in drought- and flood-prone regions • Decreased hydro-power potential in drought-prone regions
Increased Asian summer monsoon precipitation variability (*Likely*[a])	• Increase in flood and drought magnitude and damages in temperate and tropical Asia
Increased intensity of mid-latitude storms (little agreement between current models)[d]	• Increased risks to human life and health • Increased property and infrastructure losses • Increased damage to coastal ecosystems

[a] Likelihood refers to judgmental estimates of confidence used by TAR WGI: *very likely* (90-99% chance); *likely* (66-90% chance). Unless otherwise stated, information on climate phenomena is taken from the Summary for Policymakers, TAR WGI.

[b] These impacts can be lessened by appropriate response measures.

[c] Based on information from chapters in this report; high confidence refers to probabilities between 67 and 95% as described in Footnote 6 of TAR WGII, Summary for Policymakers.

[d] Information from TAR WGI, Technical Summary, Section F.5.

[e] Changes in regional distribution of tropical cyclones are possible but have not been established.

3.9.2. Deficiencies in Knowledge and Future Needs

There are many shortcomings of current scenario development, but there also are promising new methods that may address these problems and require further attention. These include:

- Future socioeconomic, environmental, and land-use changes have not been represented satisfactorily in many recent impact studies and need to be integrated into the process of scenario development.
- Many impact studies fail to consider adequately uncertainties embedded in the scenarios they adopt. New techniques are emerging to explore the role of scenarios, conditional probabilities, and conditional forecasts in providing policy-relevant advice in impact assessments in an environment of high uncertainty.
- There is a mismatch between the time and space scales at which scenario information commonly is provided and the resolution at which it is required for impact assessments. Methods of obtaining higher resolution scenarios of global change from broad-scale projections are being actively developed and refined. However, in some regions of the world the coverage and availability of baseline global change data are still poor, which has hampered efforts at scenario development.
- Most global change scenarios consider long-term and broad-scale changes in mean conditions. Scenarios of changes in variability and the frequency of extreme events (climatic or nonclimatic) seldom are constructed because it is difficult to simulate such events and because they are complicated to formulate and explain. More research is required into methods of representing variability change in scenarios.
- Scenarios for impact studies lag new developments in climate modeling. There is a need to reduce this time lag to deliver up-to-date scenarios for impact assessment (e.g., constructing regional climate and sea-level scenarios by using outputs from AOGCM simulations that are based on SRES emissions scenarios).
- Few comprehensive scenarios have been developed to date for examining the consequences of stabilizing GHG concentrations at different concentrations, in line with Article 2 of the United Nations Framework Convention on Climate Change (UNFCCC).
- Climate change mitigation conventionally has been treated separately from impacts and adaptation, except in some studies that use IAMs. However, these two methods of responding to climate change are inextricably linked, and this linkage should be reflected in scenarios. Efforts to develop the SRES scenarios with well-elaborated narratives and improved appreciation of important interactions in the climate system seem likely to generate greater consistency among scenarios.
- Few scenarios directly address adaptation, but existing scenario methods could be refined to do so (e.g., by combining scenarios of climate change with decision support and similar systems being used to foster adaptation under current climate variability).
- Improved guidance material and training is required in the construction of integrated global change scenarios (see, e.g., IPCC-TGCIA, 1999; Hulme *et al.*, 2000), especially concerning the development of nonclimatic scenarios.

References

Adger, W.N., 1999: Social vulnerability to climate change and extremes in coastal Vietnam. *World Development*, **27(2),** 249–269.

Albritton, D. and L. Kuijpers (eds.), 1999: *Synthesis of the Reports of the Scientific, Environmental Effects, and Technology and Economic Assessment Panels of the Montreal Protocol. A Decade of Assessments for Decision Makers Regarding the Protection of the Ozone Layer: 1988–1999.* United Nations Environment Programme, Ozone Secretariat, Nairobi, Kenya, 161 pp.

Alcamo, J. and E. Kreileman, 1996: Emission scenarios and global climate protection. *Global Environmental Change*, **6,** 305–334.

Alcamo, J., R.W. Shaw, and L. Hordijk, (eds.), 1990: *The RAINS Model of Acidification.* Kluwer Academic Publishers, Dordrecht, The Netherlands, 402 pp.

Alcamo, J., G.J. van den Born, A.F. Bouwman, B.J. de Haan, K. Klein Goldewijk, O. Klepper, J. Krabec, R. Leemans, J.G.J. Olivier, A.M.C. Toet, H.J.M. de Vries, and H.J. van der Woerd, 1994: Modeling the global society-biosphere-climate system: part 2: computed scenarios. *Water, Air, and Soil Pollution*, **76,** 37–78.

Alcamo, J., A. Bouwman, J. Edmonds, A. Grübler, T. Morita, and A. Sugandhy, 1995: An evaluation of the IPCC IS92 emission scenarios. In: *Climate Change 1994: Radiative Forcing of Climate Change and an Evaluation of the IPCC IS92 Emissions Scenarios* [Houghton, J.T., L.G. Meira Filho, J. Bruce, H. Lee, B.A. Callander, E. Haites, N. Harris, and K. Maskell (eds.)]. Cambridge University Press, Cambridge, United Kingdom and New York, NY, USA, pp. 247–304.

Alcamo, J., G.J.J. Kreileman, J.C. Bollen, G.J. van den Born, R. Gerlagh, M.S. Krol, A.M.C. Toet, and H.J.M. de Vries, 1996: Baseline scenarios of global environmental change. *Global Environmental Change*, **6,** 261–303.

Alcamo, J., E. Kreileman, M. Krol, R. Leemans, J.C. Bollen, M. Schaeffer, J. van Minnen, A.M.C. Toet, and H.J.M. de Vries, 1998a: Global modelling of environmental change: an overview of IMAGE 2.1. In: *Global Change Scenarios of the 21st Century. Results from the IMAGE 2.1 Model* [Alcamo, J., R. Leemans, and E. Kreileman (eds.)]. Pergamon, London, UK, pp. 3–94.

Alcamo, J., R. Leemans, and E. Kreileman (eds.), 1998b: *Global Change Scenarios of the 21st Century. Results from the IMAGE 2.1 Model.* Pergamon, London, United Kingdom, 296 pp.

Alcamo, J., T. Henrichs, and T. Rösch, 2000: *World Water in 2025—Global Modeling and Scenario Analysis for the World Commission on Water for the 21st Century.* Report A0002, Center for Environmental Systems Research, University of Kassel, Germany, 48 pp.

Alexandratos, N. (ed.), 1995: *Agriculture: Towards 2010: An FAO Study.* John Wiley and Sons, Chichester, United Kingdom, 488 pp.

Alexandrov, E.L., Yu.A. Izrael, I.L. Karol, and A.H. Khrgian, 1992: *The Ozone Shield of the Earth and Its Changes.* St. Petersburg, Gidrometeoizdat, Russia, 288 pp. (in Russian).

Ali, A., 1996: Vulnerability of Bangladesh to climate change and sea level rise through tropical cyclones and storm surges. *Water, Air, and Soil Pollution*, **92,** 171–179.

Allen, M.R., P.A. Stott, J.F.B. Mitchell, R. Schnur, and T.L. Delworth, 2000: Quantifying the uncertainty in forecasts of anthropogenic climate change. *Nature*, **407,** 617–620.

Andrady, A.L., S.H. Hamid, X. Hu, and A. Torikai, 1998: Effects of increased solar ultraviolet radiation on materials. *Journal of Photochemistry and Photobiology B: Biology*, **46,** 96–103.

Anisimov, O.A. and F.E. Nelson, 1996: Permafrost and global warming: strategies of adaptation. In: *Adapting to Climate Change: An International Perspective* [Smith, J.B., N. Bhatti, G. Menzhulin, R. Benioff, M. Campos, B. Jallow, F. Rijsberman M.I. Budyko, and R.K. Dixon (eds.)]. Springer-Verlag, New York, NY, USA, pp. 440–449.

ApSimon, H.M., R.F. Warren, and J.J.N. Wilson, 1994: The abatement strategies assessment model—ASAM: applications to reductions of sulphur dioxide emissions across Europe. *Atmospheric Environment*, **28(4)**, 649–663.

Arndt, R.L., G.R. Carmichael, D.G. Streets, and D.G. Bhatti, 1997: Sulfur dioxide emissions and sectorial contributions to sulfur deposition in Asia. *Atmospheric Environment*, **31(10)**, 1553–1572.

Arnell, N.W., 1995: Grid mapping of river discharge. *Journal of Hydrology*, **167**, 39–56.

Arnell, N.W., 1999: Climate change and global water resources. *Global Environmental Change*, **9**, S31–S49.

Arnell, N., B. Bates, H. Lang, J.J. Magnuson, and P. Mulholland, 1996: Hydrology and freshwater ecology. In: *Climate Change 1995: Impacts, Adaptations, and Mitigation of Climate Change: Scientific-Technical Analyses. Contribution of Working Group II to the Second Assessment Report of the Intergovernmental Panel on Climate Change* [Watson, R.T., M.C. Zinyowera, and R.H. Moss (eds.)]. Cambridge University Press, Cambridge, United Kingdom and New York, NY, USA, pp. 325–363.

Austin, J., N. Butchart, and K.P. Shine, 1992: Possibility of an Arctic ozone hole in a doubled-CO_2 climate. *Nature*, **360,** 221–225.

Barnes, J.D., J.H. Ollerenshaw, and C.P. Whitfield, 1995: Effects of elevated CO_2 and/or O_3 on growth, development and physiology of wheat (*Triticum aestivum* L.). *Global Change Biology*, **1,** 101–114.

Barrow, C.J., 1991: *Land Degradation*. Cambridge University Press, Cambridge, United Kingdom and New York, NY, USA, 295 pp.

Barrow, E.M., M. Hulme, M.A. Semenov, and R.J. Brooks, 2000: Climate change scenarios. In: *Climate Change, Climate Variability and Agriculture in Europe: An Integrated Assessment* [Downing, T.E., P.E. Harrison, R.E. Butterfield, and K.G. Lonsdale (eds.)]. Research Report No. 21, Environmental Change Institute, University of Oxford, Oxford, United Kingdom, pp. 11–27.

Bates, B.C., S.P. Charles, N.R. Sumner, and P.M. Fleming, 1994: Climate change and its hydrological implications for South Australia. *Transactions of the Royal Society of South Australia*, **118,** 35–43.

Bates, B.C., A.J. Jakeman, S.P. Charles, N.R. Summer, and P.M. Fleming, 1996: Impacts of climate change on Australia's surface water resources. In: *Greenhouse: Coping with Climate Change* [Bouma, W.J., G.I. Pearman, and M.R. Manning (eds.)]. Commonwealth Scientific and Industrial Research Organisation Publishing, Collingwood, Victoria, Australia, pp. 248–262.

Bell, R.G., D.G. Goring, and W.P. de Lange, 1999: Rising and extreme sea levels around New Zealand. *Proceedings of the IPENZ Congress 1999*. The Institution of Professional Engineers New Zealand, Wellington, New Zealand, pp. 121–131.

Bergthórsson, P., H. Björnsson, O. Dyrmundsson, B. Gudmundsson, A. Helgadóttir, and J.V. Jónmundsson, 1988: The effects of climatic variations on agriculture in Iceland. In: *The Impact of Climatic Variations on Agriculture. Volume 1. Assessments in Cool Temperate and Cold Regions* [Parry, M.L., T.R. Carter, and N.T. Konijn (eds.)]. Kluwer Academic Publishers, Dordrecht, The Netherlands, pp. 381–509.

Berkhout, F., J. Hertin, I. Lorenzoni, A. Jordan, K. Turner, T. O'Riordan, D. Cobb, L. Ledoux, R. Tinch, M. Hulme, J. Palutikof, and J. Skea, 1999: *Non-Climate Futures Study: Socio-Economic Futures Scenarios for Climate Impact Assessment*. Report produced for the United Kingdom Department of Environment, Transport and the Regions, Science and Technology Policy Research (SPRU), University of Sussex, Brighton, United Kingdom, 81 pp.

Betts, R.A., P.M. Cox, S.E. Lee, and F.I. Woodward, 1997: Contrasting physiological and structural vegetation feedbacks in climate change simulations. *Nature*, **387,** 796–799.

Bijlsma, L., 1996: Coastal zones and small islands. In: *Climate Change 1995: Impacts, Adaptations, and Mitigation of Climate Change: Scientific-Technical Analyses. Contribution of Working Group II to the Second Assessment Report of the Intergovernmental Panel on Climate Change* [Watson, R.T., M.C. Zinyowera, and R.H. Moss (eds.)]. Cambridge University Press, Cambridge, United Kingdom and New York, NY, USA, pp. 289–324.

Boer, G.J., G. Flato, and D. Ramsden, 2000: A transient climate change simulation with greenhouse gas and aerosol forcing: projected climate to the twenty-first century. *Climate Dynamics*, **16,** 427–450.

Bonan, G.B., 1997: Effects of land use on the climate of the United States. *Climatic Change*, **37,** 449–486.

Bonan, G.B., D. Pollard, and S.L. Thompson, 1992: Effects of boreal forest vegetation on global climate. *Nature*, **359,** 716–718.

Borzenkova, I.I., 1998: The accuracy of paleoclimatic information. *Meteorology and Hydrology*, **9,** 51–61 (in Russian).

Bouwman, A.F. and D.P. van Vuuren, 1999: *Global Assessment of Acidification and Eutrophication of Natural Ecosystems*. United Nations Environment Programme, Division of Environmental Information, Assessment, and Early Warning (UNEP/DEIA&EW), Nairobi, Kenya, and National Institute of Public Health and the Environment (RIVM), Bilthoven, The Netherlands, 52 pp.

Brasseur, G.P., D.A. Hauglustaine, S. Walters, P.J. Rasch, J.-F. Müller, C. Granier, and X.X. Tie, 1998: MOZART, a global chemical transport model for ozone and related chemical tracers, 1, model description. *Journal of Geophysical Research*, **103,** 28265–28289.

Braswell, B.H., D.S. Schimel, E. Linder, and B. Moore II, 1997: The response of global terrestrial ecosystems to interannual temperature variability. *Science*, **278,** 870–872.

Bray, M.J. and J.M. Hooke, 1997: Prediction of soft-cliff retreat with accelerating sea-level rise. *Journal of Coastal Research*, **13,** 453–467.

Broström, A., M. Coe, S.P. Harrison, R. Gallimore, J.E. Kutzbach, J. Foley, I.C. Prentice, and P. Behling, 1998: Land surface feedbacks and palaeomonsoons in northern Africa. *Geophysical Research Letters*, **25,** 3615–3618.

Brown, R.A. and N.J. Rosenberg, 1999: Climate change impacts on the potential productivity of corn and winter wheat in their primary United States growing regions. *Climatic Change*, **41,** 73–107.

Brown, R.A., N.J. Rosenberg, W.E. Easterling, C. Hays, and L.O. Mearns, 2000: Potential production and environmental effects of switchgrass and traditional crops under current and greenhouse-altered climate in the MINK region of the central United States. *Ecology and Agriculture Environment*, **78,** 31–47.

Bryan, K., 1996: The steric component of sea level rise associated with enhanced greenhouse warming: a model study. *Climate Dynamics*, **12,** 545–555.

Budyko, M.I. and G.V. Menzhulin, 1996: Climate change impacts on agriculture and global food production: options for adaptive strategies. In: *Adapting to Climate Change: An International Perspective* [Smith, J.B., N. Bhatti, G. Menzhulin, R. Benioff, M. Campos, B. Jallow, F. Rijsberman M.I. Budyko, and R.K. Dixon (eds.)]. Springer-Verlag, New York, NY, USA, pp. 188–203.

Caldwell, M.M., L.O. Bjorn, J.F. Bornman, S.D. Flint, G. Kulandaivelu, A.H. Teramura, and M. Tevini, 1998: Effects of increased solar ultraviolet radiation on terrestrial ecosystems. *Journal of Photochemistry and Photobiology B: Biology*, **46,** 40–52.

Carter, T.R., M. Hulme, J.F. Crossley, S. Malyshev, M.G. New, M.E. Schlesinger, and H. Tuomenvirta, 2000: *Climate Change in the 21st Century - Interim Characterizations based on the New IPCC Emissions Scenarios*. The Finnish Environment 433, Finnish Environment Institute, Helsinki, 148 pp.

Charney, J.G., W.J. Quirk, S.H. Chow, and J. Kornfield, 1977: A comparative study of the effects of albedo change on drought in semi-arid regions. *Journal of the Atmospheric Sciences*, **34,** 1366–1385.

Chu, P.-S., Z.-P. Yu, and S. Hastenrath, 1994: Detecting climate change concurrent with deforestation in the Amazon Basin: which way has it gone? *Bulletin of the American Meteorological Society*, **75,** 579–583.

Claussen, M., 1997: Modeling bio-geophysical feedback in the African and Indian monsoon region. *Climate Dynamics*, **13,** 247–257.

Collins, W.J, D.S. Stevenson, C.E. Johnson, and R.G. Derwent, 1997: Tropospheric ozone in a global-scale three-dimensional Lagrangian model and its response to NO_X emission controls. *Journal of Atmospheric Chemistry*, **26,** 223–274.

Covey, C., 1995: Using paleoclimates to predict future climate: how far can analogy go? *Climatic Change*, **29,** 403–407.

Cramer, W., D.W. Kicklighter, A. Bondeau, B. Moore, G. Churkina, A. Ruimy, and A. Schloss, 1997: *Comparing Global Models of Terrestrial Net Primary Productivity (NPP): Overview and Key Results*. PIK Report No. 30, Potsdam Institute for Climate Impact Research (PIK), Potsdam, The Netherlands, 38 pp.

Crowley, T.J., 1990: Are there any satisfactory geologic analogs for a future greenhouse warming? *Journal of Climate*, **3,** 1282–1292.

Crutzen, P.J. and J.G. Goldammer (eds.), 1993: *Fire in the Environment: The Ecological, Atmospheric, and Climatic Importance of Vegetation Fires.* John Wiley and Sons, Chichester, United Kingdom, 400 pp.

CSIRO, 1997: *OzClim—A Climate Scenario Generator and Impacts Package for Australia.* Commonwealth Scientific and Industrial Research Organisation, Division of Atmospheric Research, Climate Impact Group, Aspendale, Australia, 4 pp. Available online at http://www.dar.csiro.au/publications/ozclim.htm.

Cubasch, U., K. Hasselmann, H. Höck, E. Maier-Reimer, U. Mikolajewicz, B.D. Santer, and R. Sausen, 1992: Time-dependent greenhouse warming computations with a coupled ocean-atmosphere model. *Climate Dynamics*, **7,** 55–69.

Cubasch, U., G.C. Hegerl, and J. Waszkewitz, 1996: Prediction, detection and regional assessment of anthropogenic climate change. *Geophysica*, **32,** 77–96.

Cure, J.D. and B. Acock, 1986: Crop responses to carbon dioxide doubling: a literature survey. *Agricultural and Forest Meteorology*, **38,** 127–145.

de Gruijl, F.R., 1997: Health effects from solar UV radiation. *Radiation Protection Dosimetry*, **72,** 177–196.

Delcroix, T., 1998: Observed surface oceanic and atmospheric variability in the tropical Pacific at seasonal and ENSO timescales: a tentative overview. *Journal of Geophysical Research*, **103,** 18611–18633.

Derwent, R.G., P. Grennfelt, and Ø. Hov, 1991: Photochemical oxidants in the atmosphere. Status Report Prepared for the Nordic Council of Ministers. *Nord*, **7,** Nordic Council of Ministers, Copenhagen, 72 pp.

de Wolde, J.R., 1999: Uncertainties in sea-level scenarios. In: *Representing Uncertainty in Climate Change Scenarios and Impact Studies.* Proceedings of the ECLAT-2 Helsinki Workshop, 14–16 April, 1999 [Carter, T.R., M. Hulme, and D. Viner (eds.)]. Climatic Research Unit, Norwich, United Kingdom, pp. 71–73.

de Wolde, J.R., P. Huybrechts, J. Oerlemans, and R.S.W. VandeWal, 1997: Projections of global mean sea level rise calculated with a 2D energy-balance model and dynamic ice sheet models. *Tellus Series A*, **49,** 486–502.

Diffey, B.L., 1992: Stratospheric ozone depletion and the risk of non-melanoma skin cancer in a British population. *Physics in Medicine and Biology*, **37,** 2267–2279.

Döös, B.R., 1997: Can large-scale environmental migrations be predicted? *Global Environmental Change*, **7,** 41–61.

Douglas, B.C., 1997: Global sea rise: a redetermination. *Surveys in Geophysics*, **18,** 279–292.

Downing, T.E., R.E. Butterfield., S. Cohen, S. Huq, R. Moss, A. Rahman, Y. Sokona, and L. Stephen, 2000: *Climate Change Vulnerability: Toward a Framework for Understanding Adaptability to Climate Change Impacts.* UN Environment Programme, Nairobi, Kenya, and Environmental Change Institute, Oxford, United Kingdom.

El Raey, M., Y. Fouda, and S. Nasr, 1997: GIS assessment of the vulnerability of the Rosetta area, Egypt to impacts of sea rise. *Environmental Monitoring and Assessment*, **47,** 59–77.

Eltahir, E.A.B. and C.L. Gong, 1996: Dynamics of wet and dry years in West Africa. *Journal of Climate*, **9,** 1030–1042.

EMEP, 1998: Transboundary acidifying air pollution in Europe. In: *Research Report 66, EMEP/MSC-W Status Report 1998, Parts 1 and 2.* Cooperative Programme for Monitoring and Evaluation of the Long-Range Transmission of Air Pollutants in Europe (EMEP), Meteorological Synthesizing Centre–West (MSC-W), Norwegian Meteorological Institute, Blindern, Norway.

Emori, S., T. Nozawa, A. Abe-Ouchi, A. Numaguti, M. Kimoto, and T. Nakajima, 1999: Coupled ocean-atmosphere model experiments of future climate change with an explicit representation of sulfate aerosol scattering. *Journal of the Meteorological Society of Japan*, **77,** 1299–1307.

FAO, 1999: *FAOSTAT Statistics Online Database.* Food and Agriculture Organization, Rome, Italy. Available online at http://apps.fao.org.

Feenstra, J., I. Burton, J.B. Smith, and R.S.J. Tol (eds.), 1998: *Handbook on Methods of Climate Change Impacts Assessment and Adaptation Strategies.* United Nations Environment Programme, Nairobi, Kenya, and Institute for Environmental Studies, Amsterdam, The Netherlands, 448 pp.

Feichter, J., E. Kjellstrom, H. Rodhe, F. Dentener, J. Lelieveld, and G.-J. Roelofs, 1996: Simulation of the tropospheric sulfur cycle in a global climate model. *Atmospheric Environment*, **30(11),** 1693–1707.

Finlayson-Pitts, B.J. and J.N. Pitts, 1986: *Atmospheric Chemistry Fundamentals and Experimental Techniques.* John Wiley and Sons, New York, NY, USA, 1098 pp.

Fischer, G. and C. Rosenzweig, 1996: The impacts of climate change, carbon dioxide, and sulfur deposition on agricultural supply and trade: an integrated impact assessment. In: *Climate Change: Integrating Science, Economics and Policy* [Nakicenovic, N., W.D. Nordhaus, R. Richels, and F.L. Toth (eds.)]. IIASA Collaborative Paper CP-96–01, International Institute of Applied Systems Analysis, Laxenburg, Austria, pp. 83–110.

Foell, W.K., C. Green, M. Amann, S. Bhattacharya, G. Carmichael, M. Chadwick, S. Cinderby, T. Haugland, J.-P. Hettelingh, L. Hordijk, J. Kuylenstierna, J. Shah, R. Shrestha, D. Streets, and D. Zhao, 1995: Energy use, emissions, and air pollution reduction strategies in Asia. *Water, Air, and Soil Pollution*, **85(4),** 2277–2282.

Foley, J.A., S. Levis, I.C. Prentice, D. Pollard, and S.L. Thompson, 1998: Coupling dynamic models of climate and vegetation. *Global Change Biology*, **4,** 561–579.

Folland, C.K., T.R. Karl, and K.Ya. Vinnikov, 1990: Observed climate variations and change. In: *Climate Change: The IPCC Scientific Assessment* [Houghton, J.T., G.J. Jenkins, and J.J. Ephraums (eds.)]. Cambridge University Press, Cambridge, United Kingdom and New York, NY, USA, pp. 195–238.

Foster, D.R., G. Motzkin, and B. Slater, 1998: Land-use history as long-term broad-scale disturbance: regional forest dynamics in Central New England. *Ecosystems*, **1,** 96–119.

Fox-Rabinovitz, M.S., G. Stenchikov, and L.L.Takacs, 1997: A finite-difference GCM dynamical core with a variable resolution stretched grid. *Monthly Weather Review*, **125,** 2943–2961.

Fuhrer, J., 1996: The critical levels for effects of ozone on crops, and the transfer to mapping. In: *Critical Levels for Ozone in Europe: Testing and Finalizing the Concepts. UN-ECE Workshop Report* [Kärenlampi, L. and L. Skärby (eds.)]. University of Kuopio, Finland, pp. 27–43.

Gambolati, G., P. Teatini, L. Tomasi, and M. Gonella, 1999: Coastline regression of the Romagna region, Italy, due to natural and anthropogenic land subsidence and sea level rise. *Water Resources Research*, **35,** 163–184.

Ganopolski, A., C. Kubatzki, M. Claussen, V. Brovkin, and V. Petoukhov, 1998: The influence of vegetation-atmosphere-ocean interaction on climate during the mid-Holocene. *Science*, **280,** 1916–1919.

Gash, J.H.C., C.A. Nobre, J.M. Roberts, and R.L. Victoria (eds.), 1996: *Amazonian Deforestation and Climate.* John Wiley and Sons, New York, NY, USA, 611 pp.

GESAMP, 1990: *The State of the Marine Environment.* GESAMP Report No. 39, IMO/FAO/Unesco/WMO/IAEA/UNU/UNEP, Joint Group of Experts on the Scientific Aspects of Marine Pollution (GESAMP), United Nations Environment Programme, Nairobi, Kenya, 111 pp.

Giorgi, F. and L.O. Mearns, 1991: Approaches to the simulation of regional climate change: a review. *Reviews of Geophysics*, **29,** 191–216.

Giorgi, F. and L.O. Mearns, 1999: Regional climate modeling revisited: an introduction to the special issue. *Journal of Geophysical Research (Special Issue on New Developments and Applications with the NCAR Regional Climate Model [RegCM])*, **104(D6),** 6335–6352.

Glantz, M. (ed.), 1988: *Societal Responses to Regional Climatic Change: Forecasting by Analogy.* Westview Press, Boulder, Colorado, USA, 428 pp.

Gleick, P.H., 1998: *The World's Water 1998–1999: The Biennial Report on Freshwater Resources.* Island Press, Washington, DC, USA, 319 pp.

Goldberg, E.D., 1995: Emerging problems in the coastal zone for the twenty-first century. *Marine Pollution Bulletin*, **31,** 152–158.

Gordon, C., C. Cooper, C.A. Senior, H.T. Banks, J.M. Gregory, T.C. Johns, J.F.B. Mitchell, and R.A. Wood, 2000: The simulation of SST, sea ice extents and ocean heat transports in a version of the Hadley Centre coupled model without flux adjustments. *Climate Dynamics*, **16,** 147–168.

Gordon, H.B. and S.P. O'Farrell, 1997: Transient climate change in the CSIRO coupled model with dynamic sea ice. *Monthly Weather Review*, **125,** 875–907.

Government of Pakistan, 1998: *Study on Climate Impact and Adaptation Strategies for Pakistan.* Ministry of Environment, Local Government and Rural Development, Islamabad, Pakistan.

Goudriaan, J., H.H. Shugart, H. Bugmann, W. Cramer, A. Bondeau, R.H. Gardner, L.A. Hunt, W.K. Lauwenroth, J.J. Landsberg, S. Linder, I.R. Noble, W.J. Parton, L.F. Pitelka, M. Stafford Smith, R.W. Sutherst, C. Valentin, and F.I. Woodward, 1999: Use of models in global change studies. In: *The Terrestrial Biosphere and Global Change: Implications for Natural and Managed Ecosystems* [Walker, B., W. Steffen, J. Canadell, and J. Ingram (eds)]. Cambridge University Press, Cambridge, United Kingdom and New York, NY, USA, pp. 106–140.

Gregory, J.M. and J. Oerlemans, 1998: Simulated future sea-level rise due to glacier melt based on regionally and seasonally resolved temperature changes. *Nature*, **391**, 474–476.

Grennfelt, P., 1996: The second NO_x protocol: how to link science to policy? In: *Critical Levels for Ozone in Europe: Testing and Finalizing the Concepts. UN-ECE Workshop Report* [Kärenlampi, L. and L. Skärby (eds.)]. University of Kuopio, Finland, pp. 108–109.

Grennfelt, P., H. Rodhe, E. Thörnelöf, and J. Wisniewski (eds.), 1996: Acid reign '95? Proceedings of the 5th international conference on acidic deposition, Göteborg, Sweden, 26–30 June 1995. *Water, Air, and Soil Pollution*, **85(1–4)**, 2730 pp.

Grübler, A., 1998: A review of global and regional sulfur emission scenarios. *Journal of Mitigation and Adaptation Strategies for Global Change*, **3**, 383–418.

Gwynne-Jones, D., J.A. Lee, and T.V. Callaghan, 1997: Effects of enhanced UV-B radiation and elevated carbon dioxide concentrations on a sub-arctic forest heath ecosystem. *Plant Ecology*, **128**, 242–249.

Haarsma, R.J., J.F.B. Mitchell, and C.A. Senior, 1993: Tropical disturbances in a GCM. *Climate Dynamics*, **8**, 247–257.

Hader, D.-P., H.D. Kumar, R.C. Smith, and R.C. Worrest, 1998: Effects on aquatic ecosystems. *Journal of Photochemistry and Photobiology B: Biology*, **46**, 53–68.

Hahmann, A.N. and R.E. Dickinson, 1997: RCCM2–BATS model over tropical South America: applications to tropical deforestation. *Journal of Climate*, **10**, 1944–1964.

Hansen, J., G. Russell, D. Rind, P. Stone, A. Lacis, S. Lebedeff, R. Ruedy, and L. Travis, 1983: Efficient three-dimensional global models for climate studies: models I and II. *Monthly Weather Review*, **111**, 609–662.

Hansen, J., I. Fung, A. Lacis, D. Rind, S. Lebedeff, R. Ruedy, and G. Russell, 1988: Global climate changes as forecast by Goddard Institute for Space Studies three-dimensional model. *Journal of Geophysical Research*, **93(D8)**, 9341–9364.

Harrison, P.A. and R.E. Butterfield, 1996: Effects of climate change on Europe-wide winter wheat and sunflower productivity. *Climate Research*, **7**, 225–241.

Harvey, L.D.D., J. Gregory, M. Hoffert, A. Jain, M. Lal, R. Leemans, S.C.B. Raper, T.M.L. Wigley, and J.R. de Wolde, 1997: An introduction to simple climate models used in the IPCC Second Assessment Report. *IPCC Technical Paper II* [Houghton, J.T., L.G. Meira Filho, D.J. Griggs, and K. Maskell (eds.)]. Intergovernmental Panel on Climate Change, Geneva, Switzerland, 50 pp.

Hay, E.L., R.B. Wilby, and G.H. Leavesy, 1999: A comparison of delta change and downscaled GCM scenarios: implications for climate change scenarios in three mountainous basins in the United States. In: *Proceedings of the AWRA Specialty Conference on Potential Consequences of Climatic Variability and Change to Water Resources of the United States, May, 1999, Atlanta, GA* [Adams, D.B. (ed.)]. American Water Resources Association, Middleburg, VA, USA, 424 pp.

Hayden, B.P., 1998: Ecosystem feedbacks on climate at the landscape scale. *Philosophical Transactions of the Royal Society of London, Series B*, **353**, 5–18.

Haywood, J.M., R.J. Stouffer, R.T. Wetherald, S. Manabe, and V. Ramaswamy, 1997: Transient response of a coupled model to estimated changes in greenhouse gas and sulfate concentrations. *Geophysical Research Letters*, **24**, 1335–1338.

Henderson-Sellers, A., K. McGuffie, and C. Gross, 1995: Sensitivity of global climate model simulations to increased stomatal resistance and CO_2 increases. *Journal of Climate*, **8**, 1738–1756.

Herman, J.R., P.K. Bhartia, J. Ziemke, Z. Ahmad, and D. Larko, 1996: UV-B radiation increases (1979–1992) from decreases in total ozone. *Geophysical Research Letters*, **23**, 2117–2120.

Hewitt, C.D. and J.F.B. Mitchell, 1998: A fully coupled GCM simulation of the climate of the mid-Holocene. *Geophysical Research Letters*, **25**, 361–364.

Hoekstra, A.Y., 1997: The water submodel: AQUA. In: *Perspectives on Global Change: The TARGETS Approach* [Rotmans, J. and B. de Vries (eds.)]. Cambridge University Press, Cambridge, United Kingdom and New York, NY, USA, pp. 109–134.

Houghton, R.A., 1995: Land use change and the carbon cycle. *Global Change Biology*, **1**, 275–287.

Hulme, M. and O. Brown, 1998: Portraying climate scenario uncertainties in relation to tolerable regional climate change. *Climate Research*, **10**, 1–14.

Hulme, M. and T.R. Carter, 1999: Representing uncertainty in climate change scenarios and impact studies. In: *Representing Uncertainty in Climate Change Scenarios and Impact Studies. Proceedings of the ECLAT-2 Helsinki Workshop, 14–16 April, 1999* [Carter, T.R., M. Hulme, and D. Viner (eds.)]. Climatic Research Unit, University of East Anglia, Norwich, United Kingdom, pp. 11–37.

Hulme, M., T. Jiang, and T. Wigley, 1995: *SCENGEN: A Climate Change SCENario GENerator. Software User Manual, Version 1.0*. Climatic Research Unit, University of East Anglia, Norwich, United Kingdom, 38 pp.

Hulme, M., E.M. Barrow, N. Arnell, P.A. Harrison, T.E. Downing, and T.C. Johns, 1999a: Relative impacts of human-induced climate change and natural climate variability. *Nature*, **397**, 688–691.

Hulme, M., J. Mitchell, W. Ingram, J. Lowe, T. Johns, M. New, and D. Viner, 1999b: Climate change scenarios for global impacts studies. *Global Environmental Change*, **9**, S3–S19.

Hulme, M., T.M.L. Wigley, E.M. Barrow, S.C.B. Raper, A. Centella, S. Smith, and A.C. Chipanshi, 2000: *Using a Climate Scenario Generator for Vulnerability and Adaptation Assessments: MAGICC and SCENGEN, Version 2.4 Workbook*. Climatic Research Unit, East Anglia University, Norwich, United Kingdom, 52 pp.

Huntley, B., W.P. Cramer, A.V. Morgan, I.C. Prentice, and J.R.M. Allen (eds.), 1997: *Past and Future Rapid Environmental Changes: The Spatial and Evolutionary Responses of Terrestrial Biota*. Springer-Verlag, Berlin, Germany, 523 pp.

IASC, 1995: *Effects of Increased Ultraviolet Radiation in the Arctic*. IASC Report 2, International Arctic Science Committee, Oslo, Norway, 56 pp.

IPCC, 1990: *Climate Change: The IPCC Impacts Assessment* [Tegart, W.J.McG., G.W. Sheldon, and D.C. Griffiths (eds.)]. Australian Government Publishing Service, Canberra, Australia, 210 pp.

IPCC, 1994: *IPCC Technical Guidelines for Assessing Climate Change Impacts and Adaptations. Part of the IPCC Special Report to the First Session of the Conference of the Parties to the UN Framework Convention on Climate Change, Working Group II, Intergovernmental Panel on Climate Change* [Carter, T.R., M.L. Parry, H. Harasawa, and S. Nishioka (eds.)]. University College London, United Kingdom and Center for Global Environmental Research, National Institute for Environmental Studies, Tsukuba, Japan, 59 pp.

IPCC, 1996a: *Climate Change 1995. The Science of Climate Change. Contribution of Working Group I to the Second Assessment Report of the Intergovernmental Panel on Climate Change* [Houghton, J.T., L.G. Meira Filho, B.A. Callander, N. Harris, A. Kattenberg, and K. Maskell (eds.)]. Cambridge University Press, Cambridge, United Kingdom and New York, NY, USA, 572 pp.

IPCC, 1996b: *Climate Change 1995: Impacts, Adaptations, and Mitigation of Climate Change: Scientific-Technical Analyses. Contribution of Working Group II to the Second Assessment Report of the Intergovernmental Panel on Climate Change* [Watson, R.T., M.C. Zinyowera, and R.H. Moss (eds.)]. Cambridge University Press, Cambridge, United Kingdom and New York, NY, USA, 880 pp.

IPCC, 1998: *The Regional Impacts of Climate Change: An Assessment of Vulnerability. Special Report of IPCC Working Group II* [Watson, R.T., M.C. Zinyowera, and R.H. Moss (eds.)]. Cambridge University Press, Cambridge, United Kingdom and New York, NY, USA, 517 pp.

IPCC, 2000: *Land Use, Land-Use Change, and Forestry. Special Report of the Intergovernmental Panel on Climate Change* [Watson, R.T., I.R. Noble, B. Bolin, N.H. Ravindranath, D.J. Verardo, and D.J. Dokken (eds.)]. Cambridge University Press, Cambridge, United Kingdom and New York, NY, USA, 377 pp.

IPCC-DDC, 1999: *The IPCC Data Distribution Centre: Providing Climate Change and Related Scenarios for Impacts Assessments*, CD-ROM Version 1.0, Climatic Research Unit, University of East Anglia, Norwich, United Kingdom. Available online at http://ipcc-ddc.cru.uea.ac.uk.

IPCC-TGCIA, 1999: *Guidelines on the Use of Scenario Data for Climate Impact and Adaptation Assessment. Version 1* [Carter, T.R., M. Hulme, and M. Lal (eds.)]. Intergovernmental Panel on Climate Change, Task Group on Scenarios for Climate Impact Assessment, 69 pp. Available online at http://ipcc-ddc.cru.uea.ac.uk/cru_data/support/guidelines.html.

Izrael, Yu.A. and A.V. Tsyban, 1989: *Anthropogenic Ecology of the Ocean*. Leningrad, Gidrometeoizdat, Russia, 528 pp. (in Russian).

Jackett, D.R., T.J. McDougall, M.H. England, and A.C. Hirst, 2000: Thermal expansion in ocean and coupled general circulation models. *Journal of Climate*, **13,** 1384–1405.

Jackman, C.H., E.L. Flemming, S. Chandra, D.B. Considine, and J.E. Rosenfeld, 1996: Past, present and future modeled ozone trends with comparisons to observed trends. *Journal of Geophysical Research*, **101,** 28753–28767.

Jones, R.N., 2000: Managing uncertainty in climate change projections—issues for impact assessment. *Climatic Change*, **45,** 403–419.

Jones, R.N., K.J. Hennessy, and D.J. Abbs, 1999: *Climate Change Analysis Relevant to Jabiluka, Attachment C—Assessment of the Jabiluka Project*. Report of the Supervising Scientist to the World Heritage Committee, Environment Australia, Canberra, Australia, 24 pp.

Jordan, A., T. O'Riordan, K. Turner, and I. Lorenzoni, 2000: Europe in the new millennium. In: *Assessment of Potential Effects and Adaptation for Climate Change in Europe: The Europe ACACIA Project* [Parry, M.L. (ed.)]. Jackson Environment Institute, University of East Anglia, Norwich, United Kingdom, pp. 35–45.

Kasischke, E.S. and B.J. Stocks (eds.), 2000: *Fire, Climate Change and Carbon Cycling in the Boreal Forest*. Springer-Verlag, New York, NY, USA, 461 pp.

Kattenberg, A., F. Giorgi, H. Grassl, G.A. Meehl, J.F.B. Mitchell, R.J. Stouffer, T. Tokioka, A.J. Weaver, and T.M.L. Wigley, 1996: Climate models—projections of future climate. In: *Climate Change 1995: The Science of Climate Change. Contribution of Working Group I to the Second Assessment Report of the Intergovernmental Panel on Climate Change* [Houghton, J.T., L.G. Meira Filho, B.A. Callander, N. Harris, A. Kattenberg, and K. Maskell (eds.)]. Cambridge University Press, Cambridge, United Kingdom and New York, NY, USA, pp. 285–357.

Katz, R.W. and M.B. Parlange, 1996: Mixtures of stochastic processes: application to statistical downscaling. *Climate Research*, **7,** 185–193.

Kirschbaum, M.U.F., P. Bullock, J.R. Evans, K. Goulding, P.G. Jarvis, I.R. Noble, M. Rounsevell, and T.D. Sharkey, 1996: Ecophysiological, ecological, and soil processes in terrestrial ecosystems: a primer on general concepts and relationships. In: *Climate Change 1995: Impacts, Adaptations and Mitigation of Climate Change: Scientific-Technical Analyses. Contribution of Working Group II to the Second Assessment Report of the Intergovernmental Panel on Climate Change* [Watson, R.T., M.C. Zinyowera, and R.H. Moss (eds.)]. Cambridge University Press, Cambridge, United Kingdom and New York, NY, USA, pp. 57–74.

Klein, R.J.T. and R.J. Nicholls, 1998: Coastal zones. In: *UNEP Handbook on Methods for Climate Change Impact Assessment and Adaptation Studies* [Burton, I., J.F. Feenstra, J.B. Smith, and R.S.J. Tol (eds.)]. Version 2.0, United Nations Environment Programme and Institute for Environmental Studies, Vrije Universiteit, Amsterdam, The Netherlands, chapter 7, pp. 1–36. Available online at http://www.vu.nl/english/o_o/instituten/IVM/research/climatechange/Handbook.htm.

Klein Goldewijk, C.G.M., 2001: Estimating global land use change over the past 300 years: the HYDE 2.0 database. *Global Biogeochemical Cycles*, **14,** (in press).

Klein Tank, A.M.G. and G.P. Können, 1997: Simple temperature scenario for a Gulf Stream-induced climate change. *Climatic Change*, **37,** 505–512.

Kneshgi, H.S. and A.G. Lapenis, 1996: Estimating the accuracy of Russian paleotemperature reconstruction. *Palaeogeography, Palaeoclimatology, Palaeoecology*, **121,** 221–237.

Koruba, V., M.A. Jabbar, and J.A. Akinwumi, 1996: Crop-livestock competition in the West African derived savannah: application of a multi-objective programming model. *Agricultural Systems*, **52,** 439–453.

Kropff, M., 1989: *Quantification of SO_2 Effects on Physiological Processes, Plant Growth and Crop Production*. Diss. Wageningen Agricultural University, CIP-Gegevens Koninklijke Bibliotheek, The Hague, The Netherlands, 201 pp.

Krupa, S.V. and H.-J. Jäger, 1996: Adverse effects of elevated levels of ultraviolet (UV)-B radiation and ozone (O_3) on crop growth and productivity. In: *Global Climate Change and Agricultural Production: Direct and Indirect Effects of Changing Hydrological, Pedological and Plant Physiological Processes* [Bazzaz, F. and W. Sombroek (eds.)]. John Wiley, Chichester, 141–169.

Krupa, S.V., R.N. Kickert, and H.-J. Jäger, 1998: *Elevated Ultraviolet (UV)-B Radiation and Agriculture*. Springer-Verlag, Berlin and Heidelberg, Germany, 296 pp.

Kutzbach, J., G. Bonan, J. Foley, and S.P. Harrison, 1996: Vegetation and soil feedbacks on the response of the African monsoon to orbital forcing in the early to middle Holocene. *Nature*, **384,** 623–626.

Langner, J. and H. Rodhe, 1991: A global three-dimensional model of the tropospheric sulfur cycle. *Journal of Atmospheric Chemistry*, **13,** 225–263.

Lean, J. and P.R. Rowntree, 1997: Understanding the sensitivity of a GCM simulation of Amazonian deforestation to the specification of vegetation and soil characteristics. *Journal of Climate*, **10,** 1216–1235.

Leemans, R., 1999: Modelling for species and habitats: new opportunities for problem solving. *The Science of the Total Environment*, **240,** 51–73.

Leemans, R. and G.J. van den Born, 1994: Determining the potential global distribution of natural vegetation, crops and agricultural productivity. *Water, Air, and Soil Pollution*, **76,** 133–161.

Leemans, R., S. Agrawala, J.A. Edmonds, M.C. MacCracken, R.M. Moss, and P.S. Ramakrishnan, 1996a: Mitigation: cross-sectoral and other issues. In: *Climate Change 1995: Impacts, Adaptations and Mitigation of Climate Change: Scientific-Technical Analyses. Contribution of Working Group II to the Second Assessment Report of the Intergovernmental Panel on Climate Change* [Watson, R.T., M.C. Zinyowera, and R.H. Moss (eds.)]. Cambridge University Press, Cambridge, United Kingdom and New York, NY, USA, pp. 799–797.

Leemans, R., W. Cramer, and J.G. van Minnen, 1996b: Prediction of global biome distribution using bioclimatic equilibrium models. In: *Effects of Global Change on Coniferous Forests and Grasslands* [Melillo, J.M. and A. Breymeyer (eds.)]. John Wiley and Sons, New York, NY, USA, pp. 413–450.

Leemans, R., E. Kreileman, G. Zuidema, J. Alcamo, M. Berk, G.J. van den Born, M. den Elzen, R. Hootsmans, M. Janssen, M. Schaeffer, A.M.C. Toet, and H.J.M. de Vries, 1998: *The IMAGE User Support System: Global Change Scenarios from IMAGE 2.1*. RIVM Publication 4815006, CD-ROM, National Institute of Public Health and the Environment, Bilthoven, The Netherlands.

Leggett, J., W.J. Pepper, and R.J. Swart, 1992: Emissions Scenarios for IPCC: an update. In: *Climate Change 1992. The Supplementary Report to the IPCC Scientific Assessment* [Houghton, J.T., B.A. Callander, and S.K. Varney (eds.)]. Cambridge University Press, Cambridge, United Kingdom and New York, NY, USA, pp. 69–95.

Lelieveld, J. and A.M. Thompson, 1998: Tropospheric ozone and related processes. Scientific assessment of ozone depletion. In: *Global Ozone Research and Monitoring Project* [Albritton, D.L., P.J. Aucamp, G. Megie, and R.T. Watson (eds.)]. Report No. 44, World Meteorological Organization, Geneva, Switzerland.

Longstreth, J., F.R. de Gruijl, M.L. Kripke, S. Abseck, F. Arnold, H.I. Slaper, G. Velders, Y. Takizawa, and J.C. van der Leun, 1998: Health risks. *Journal of Photochemistry and Photobiology B: Biology*, **46,** 20–39.

Lorenzo, E. and L. Teixiera, 1997: Sensitivity of storm waves in Montevideo (Uruguay) to a hypothetical climate change. *Climate Research*, **9,** 81–85.

Lorenzoni, I., A. Jordan., M. Hulme., R.K. Turner, and T. O'Riordan, 2000: A co-evolutionary approach to climate change impact assessment: part I: integrating socio-economic and climate change scenarios. *Global Environmental Change*, **10,** 57–68.

Lough, J.M., T.M.L. Wigley, and J.P. Palutikof, 1983: Climate and climate impact scenarios for Europe in a warmer world. *Journal of Climatology and Applied Meteorology*, **22,** 1673–1684.

Loveland, T.R. and A.S. Belward, 1997: The IGBP-DIS global 1 km land cover data set, DISCover: first results. *International Journal of Remote Sensing*, **18,** 3291–3295.

Madronich, S., R.L. McKenzie, L.O. Bjorn, and M.M. Caldwell, 1998: Changes in biologically active ultraviolet radiation reaching the Earth's surface. *Journal of Photochemistry and Photobiology B: Biology*, **46,** 5–19.

Manabe, S., R.J. Stouffer, M.J. Spelman, and K. Bryan, 1991: Transient responses of a coupled ocean-atmosphere model to gradual changes of atmospheric CO_2: part I: annual mean response. *Journal of Climate*, **4,** 785–818.

Manzi, A.O. and S. Planton, 1996: A simulation of Amazonian deforestation using a GCM calibrated with ABRACOS and ARME data. In: *Amazonian Deforestation and Climate* [Gash, J.H.C., C.A. Nobre, J.M. Roberts, and R.L. Victoria (eds.)]. John Wiley and Sons, New York, NY, USA, pp. 505–529.

Martens, P., 1998: *Health and Climate Change: Modelling the Impacts of Global Warming and Ozone Depletion*. Earthscan Publications, London, United Kingdom, 176 pp.

Martens, P., R.S. Kovats, S. Nijhof, P. de Vries, M.T.J. Livermore, D.J. Bradley, J. Cox, and A.J. McMichael, 1999: Climate change and future populations at risk of malaria. *Global Environmental Change*, **9,** S89–S107.

Martin, M.J., 1997: *Models of the Interactive Effects of Rising Ozone, Carbon Dioxide and Temperature on Canopy Carbon Dioxide Exchange and Isoprene Emission*. Diss. University of Essex, Colchester, United Kingdom, 165 pp.

Materna, J., 1981: Concentration of sulfur dioxide in the air and sulfur content in Norway spruce needles (*Picea abies* Karst.). *Communicationes Instituti Forestalis Cechosloveniae*, **12,** 137–146.

Matsuoka, Y., M. Kainuma, and T. Morita, 1995: Scenario analysis of global warming using the Asian Pacific Integrated Model (AIM). *Energy Policy*, **23,** 357–371.

Matthijsen, J., K. Suhre, R. Rosset, F. Eisele, R. Mauldin, and D. Tanner, 1998: Photodissociation and UV radiative transfer in a cloudy atmosphere: modeling and measurements. *Journal of Geophysical Research*, **103,** 16665–16676.

Mayerhofer, P., J. Alcamo, J.G. van Minnen, M. Posch, R. Guardans, B.S. Gimeno, T. van Harmelen, and J. Bakker, 2000: *Regional Air Pollution and Climate Change in Europe: an Integrated Analysis (AIR-CLIM)*. Progress Report 2, Center for Environmental Systems Research, University of Kassel, Germany, 59 pp. Available online at http://www.usf.uni-kassel.de/service/bibliothek.htm.

McDonald, N.J. and B.A. O'Connor, 1996: Changes in wave impact on the Flemish coast due to increased mean sea level. *Journal of Marine Systems*, **7,** 133–144.

McFarlane, N.A., G.J. Boer, J.-P. Blanchet, and M. Lazare, 1992: The Canadian Climate Centre second-generation general circulation model and its equilibrium climate. *Journal of Climate*, **5,** 1013–1044.

McGregor, J.J., 1997: Regional climate modeling. *Meteorological Atmospheric Physics*, **63,** 105–117.

McInnes, K.L. and G.D. Hubbert, 1996: *Extreme Events and the Impact of Climate Change on Victoria's Coastline*. Report No. 488, Victorian Environment Protection Authority, Melbourne, Australia, 69 pp.

McInnes, K.L., K.J.E. Walsh, and A.B. Pittock, 2000: *Impact of Sea Level Rise and Storm Surges on Coastal Resorts. CSIRO Tourism Research: Final Report*. Division of Atmospheric Research, Commonwealth Scientific, Industrial and Research Organisation, Aspendale, Australia, 13 pp.

McKenzie, R.L., B. Connor, and G. Bodeker, 1999: Increased summertime UV radiation in New Zealand in response to ozone loss. *Science*, **285,** 1709–1711.

McPeters, R.D., S.M. Hollandsworth, L.E.Flynn, J.R.Herman, and C.J.Seftor, 1996: Long-term ozone trends derived from the 16-year combined Nimbus 7/Meteor 3 TOMS Version 7 record. *Geophysical Research Letters*, **23,** 3699–3702.

Meadows, D.H., D.L. Meadows, and J. Randers, 1992: *Beyond the Limits: Global Collapse of a Sustainable Future*. Earthscan Publications Ltd, London, United Kingdom, 300 pp.

Mearns, L.O., C. Rosenzweig, and R. Goldberg, 1992: Effect of changes in interannual climatic variability on CERES-Wheat yields: sensitivity and $2xCO_2$ general circulation model studies. *Agricultural and Forest Meteorology*, **62,** 159–189.

Mearns, L.O., C. Rosenzweig, and R. Goldberg, 1996: The effect of changes in daily and interannual climatic variability on CERES-Wheat: a sensitivity study. *Climatic Change*, **32,** 257–292.

Mearns, L.O., C. Rosenzweig, and R. Goldberg, 1997: Mean and variance change in climate scenarios: methods, agricultural applications, and measures of uncertainty. *Climatic Change*, **35,** 367–396.

Mearns, L.O., W. Easterling, and C. Hays, 1998: The effect of spatial scale of climate change scenarios on the determination of impacts: an example of agricultural impacts on the Great Plains. In: *Proceedings of the International Workshop on Regional Modeling of the General Monsoon System in Asia, Beijing, October 20–23, 1998*. START Regional Committee for Temperate East Asia Report No. 4, Global Change System for Analysis, Research, and Training (START) Regional Center for Temperate East Asia, Beijing, China, pp. 70–73.

Mearns, L.O., T. Mavromatis, E. Tsvetsinskaya, C. Hays, and W. Easterling, 1999: Comparative responses of EPIC and CERES crop models to high and low resolution climate change scenario. *Journal of Geophysical Research*, **104(D6),** 6623–6646.

Mearns, L.O., W. Easterling, and C. Hays, 2001: Comparison of agricultural impacts of climate change calculated from high and low resolution climate model scenarios: part I: the uncertainty of spatial scale. *Climatic Change*, (in press).

Mela, T. and T. Suvanto, 1987: *Peltokasvien Satoennuste Vuoteen 2000 (Field Crop Yield Forecast to the Year 2000)*. Publication No. 14, Department of Crop Husbandry, University of Helsinki, Helsinki, Finland, 201 pp. (in Finnish).

Mendelsohn, R. and J. Neumann, 1999: *The Impact of Climate Change on the United States Economy*. Cambridge University Press, Cambridge, United Kingdom and New York, NY, USA, 344 pp.

Mitchell, J.F.B., 1990: Greenhouse warming: is the mid-Holocene a good analogue? *Journal of Climate*, **3,** 1177–1192.

Mitchell, J.F.B. and T.C. Johns, 1997: On the modification of global warming by sulphate aerosols. *Journal of Climate*, **10,** 245–267.

Mitchell, J.F.B., S. Manabe, V. Meleshko, and T. Tokioka, 1990: Equilibrium climate change—and its implications for the future. In: *Climate Change: The IPCC Scientific Assessment* [Houghton, J.T., G.J. Jenkins, and J.J. Ephraums (eds.)]. Cambridge University Press, Cambridge, United Kingdom and New York, NY, USA, pp. 131–164.

Mitchell, J.F.B., T.C. Johns, M. Eagles, W.J. Ingram, and R.A. Davis, 1999: Towards the construction of climate change scenarios. *Climatic Change*, **41,** 547–581.

Montzka, S.A., J.H. Butler, R.C. Myers, T.M.Thompson, T.H. Swanson, A.D. Clarke, L.T Lock, and J.W. Elkins, 1996: Decline in the tropospheric abundance of halogens from halocarbons: implications for stratospheric ozone depletion. *Science*, **272,** 1318–1322.

Morgan, M.G. and D. Keith, 1995: Subjective judgements by climate experts. *Environmental Science and Technology*, **29,** 468–476.

Moss, R.H. and S.H. Schneider, 2000: Uncertainties in the IPCC TAR: recommendations to lead authors for more consistent assessment and reporting. In: *Guidance Papers on the Cross Cutting Issues of the Third Assessment Report of the IPCC* [Pachauri, R., T. Taniguchi, and K. Tanaka (eds.)]. Intergovernmental Panel on Climate Change, Geneva, Switzerland, pp. 33–51.

Moxey, A.P., B. White, and J.R. O'Callaghan, 1995: CAP reform: an application of the NELUP Economic model. *Journal of Environmental Planning and Management*, **38,** 117–123.

Munasinghe, M., 2000: Development, equity and sustainability (DES) in the context of climate change. In: *Guidance Papers on the Cross Cutting Issues of the Third Assessment Report of the IPCC* [Pachauri, R., T. Taniguchi, and K. Tanaka (eds.)]. Intergovernmental Panel on Climate Change, Geneva, Switzerland, pp. 69–90.

Murphy, J.M., 1995: Transient response of the Hadley Centre coupled ocean-atmosphere model to increasing carbon dioxide: part I: control climate and flux correction. *Journal of Climate*, **8,** 36–56.

Nakicenovic, N., J. Alcamo, G. Davis, B. de Vries, J. Fenhann, S. Gaffin, K. Gregory, A. Grübler, T.Y. Jung, T. Kram, E.L. La Rovere, L. Michaelis, S. Mori, T. Morita, W. Pepper, H. Pitcher, L. Price, K. Raihi, A. Roehrl, H.-H. Rogner, A. Sankovski, M. Schlesinger, P. Shukla, S. Smith, R. Swart, S. van Rooijen, N. Victor, and Z. Dadi, 2000: *Emissions Scenarios. A Special Report of Working Group III of the Intergovernmental Panel on Climate Change*. Cambridge University Press, Cambridge, United Kingdom and New York, NY, USA, 599 pp.

National Assessment Synthesis Team, 1998: *US National Assessment: Socio-Economic Scenarios Guidance Document*. United States Global Change Research Program, National Assessment Coordination Office, Washington, D.C., USA. Available online at http://www.nacc.usgcrp.gov/meetings/socio-econ.html.

NDU, 1978: *Climate Change to the Year 2000*. National Defense University, Washington, DC, USA.

Nerem, R.S., 1999: Measuring very low frequency sea level variations using satellite altimeter data. *Global and Planetary Change*, **20,** 157–171.

Nerem, R.S., B.J. Haines, H. Hendricks, J.F. Minster, G.T. Mitchum, and W.B. White, 1997: Improved determination of global mean sea level variations using TOPEX/POSEIDON altimeter data. *Geophysical Research Letters*, **24,** 1331–1334.

New, M., 1999: Uncertainty in representing the observed climate. In: *Representing Uncertainty in Climate Change Scenarios and Impact Studies. Proceedings of the ECLAT-2 Helsinki Workshop, 14–16 April, 1999* [Carter, T.R., M. Hulme, and D. Viner (eds.)]. Climatic Research Unit, University of East Anglia, Norwich, United Kingdom, pp. 59–66.

NIVA, 1998: *Critical Loads and their Exceedances for ICP-Waters Sites*. Programme Centre of the Norwegian Institute for Water Research (NIVA), Oslo, Norway, 35 pp.

Nicholls, R.J., 1995: Synthesis of vulnerability analysis studies. In: *Preparing to Meet the Coastal Challenges of the 21st Century, Vol. 1. Proceedings of the World Coast Conference, November 1–5, 1993, Noordwijk, The Netherlands*. CZM-Centre Publication No. 4, Ministry of Transport, Public Works and Water Management, The Hague, The Netherlands, pp. 181–216.

Nicholls, R.J., F.M.J. Hoozemans, and M. Marchand, 1999: Increasing flood risk and wetland losses due to global sea-level rise: regional and global analyses. *Global Environmental Change*, **9,** S69–S87.

Nicholson, S.E., C.J. Tucker, and M.B. Ba, 1998: Desertification, drought, and surface vegetation: an example from the West African Sahel. *Bulletin of the American Meteorological Society*, **79,** 815–829.

NPA Data Services, 1999: *Analytic documentation of three alternate socio-economic projections, 1997–2050*. NPA Data Services, Inc., Washington, DC, USA.

Ogilvy, J., 1992: Future studies and the human sciences: the case for normative scenarios. *Futures Research Quarterly*, **8(2),** 5–65.

Ohmura, A., M. Wild, and L. Bengtsson, 1996: A possible change in the mass balance of Greenland and Antarctic Ice Sheets in the coming century. *Journal of Climate*, **9,** 2124–2135.

Ojanperä, K., E. Pätsikkä, and T. Yläranta, 1998: Effects of low ozone exposure of spring wheat in open-top chambers on net CO_2—uptake, rubisco, leaf senescence and grain filling. *New Phytologist*, **138,** 451–460.

Olesen, J.E., 1999: Uncertainty in impact studies: agroecosystems. In: *Representing Uncertainty in Climate Change Scenarios and Impact Studies. Proceedings of the ECLAT-2 Helsinki Workshop, 14–16 April, 1999* [Carter, T.R., M. Hulme, and D. Viner (eds.)]. Climatic Research Unit, University of East Anglia, Norwich, United Kingdom, pp. 78–82.

Olivo, M.D., 1997: Assessment of the vulnerability of Venezuela to sea-level rise. *Climate Research*, **9,** 57–65.

Pan, Y.A., A.D. McGuire, D.W. Kicklighter, and J.D. Melillo, 1996: The importance of climate and soils for estimates of net primary production: a sensitivity analysis with the terrestrial ecosystem model. *Global Change Biology*, **2,** 5–23.

Parry, M.L. (ed.), 2000: *Assessment of Potential Effects and Adaptation for Climate Change in Europe: The Europe ACACIA Project*. Jackson Environment Institute, University of East Anglia, Norwich, United Kingdom, 320 pp.

Parry, M. and M. Livermore (eds.), 1999: A new assessment of the global effects of climate change. *Global Environmental Change*, **9,** S1–S107.

Parry, M., C. Rosenzweig, A. Iglesias, G. Fischer, and M. Livermore, 1999: Climate change and world food security: a new assessment. *Global Environmental Change*, **9,** S51–S67.

Parson, E.A., 1999: Assessment methods II: socio-economic scenarios. *Acclimations,* **4,** 7. Newsletter of the U.S. National Assessment of Climate Variability and Change available online at http://www.nacc.usgcrp.gov/newsletter/1999.02/SocEc.htm.

Parson, E.A. and K. Fisher-Vanden, 1997: Integrated assessment models of global climate change. *Annual Review of Energy and the Environment*, **22,** 589–628.

Peltier, W.R., 1998: Postglacial variations in the level of the sea: implications for climate dynamics and solid-earth geophysics. *Reviews of Geophysics*, **36,** 603–689.

Penner, J.E., C.A. Atherton, and T.E. Graedel, 1994: Global emissions and models of photochemically active compounds. In: *Global Atmospheric-Biospheric Chemistry* [Prinn, R.G., (ed.)]. Plenum Press, New York, NY, USA, pp. 223–248.

Penning de Vries, F.W.T., R. Rabbinge, and J.J.R. Groot, 1997: Potential and attainable food production and food security in different regions. *Philosophical Transactions of the Royal Society of London, Series B*, **352,** 917–928.

Pepper, W., W. Barbour, A. Sankovski, and B. Braatz, 1998: No-policy greenhouse gas emission scenarios: revisiting IPCC 1992. *Environmental Science and Policy*, **1,** 289–312.

Peters, R.L. and T.E. Lovejoy (eds.), 1992: *Global Warming and Biological Diversity*. Yale University Press, New Haven, CT, USA, 386 pp.

Peters, L.K., C.M. Berkowitz, G.R. Carmichael, R.C. Easter, G. Fairweather, S.J. Ghan, G.M. Hales, L.R. Laung, W.R. Pennell, F.A. Potra, and R.D. Saylor, 1995: The current state and future direction of eulerian modelings in simulation the tropospheric chemistry and transport of trace species: a review. *Atmospheric Environment*, **29,** 189–222.

Petterson, R., A.J.S. McDonald, and I. Stadenberg, 1993: Response of small birch plants (*Betula pendula* Roth.) to elevated CO_2 and nitrogen supply. *Plant, Cell and Environment*, **16,** 1115–1121.

Pittock, A.B., 1989: Book review: societal responses to regional climatic change [Glantz, M. (ed.)]. *Bulletin of the American Meteorological Society*, **70,** 1150–1152.

Pittock, A.B., 1993: Climate scenario development. In: *Modelling Change in Environmental Systems* [Jakeman, A.J., M.B. Beck, and M.J. McAleer (eds.)]. John Wiley and Sons, New York, NY, USA, pp. 481–503.

Pittock, A.B., 1999: Climate change: question of significance. *Nature*, **397,** 657.

Pittock, A.B. and R.N. Jones, 2000: Adaptation to what and why? *Environmental Monitoring and Assessment*, **61,** 9–35.

Poorter, H., 1993: Interspecific variation in the growth response of plants to an elevated ambient CO_2 concentration. *Vegetatio*, **104/105,** 77–97.

Posch, M., J.-P. Hettelingh, J. Alcamo, and M. Krol, 1996: Integrated scenarios of acidification and climate change in Asia and Europe. *Global Environmental Change*, **6,** 375–394.

Rabbinge, R. and M. Van Oijen, 1997: Scenario studies for future agriculture and crop protection. *European Journal of Plant Pathology*, **103,** 197–201.

Ramankutty, N. and J.A. Foley, 1999: Estimating historical changes in global land cover: croplands from 1700 to 1992. *Global Biogeochemical Cycles*, **13,** 997–1027.

Raskin, P., P., Gleick, P., Kirshen, G., Pontius, and K. Strzepek, 1997: *Water Futures: Assessment of Long-Range Patterns and Problems. Background Report for the Comprehensive Assessment of the Freshwater Resources of the World*. Stockholm Environment Institute, Stockholm, Sweden, 78 pp.

Raskin, P., G. Gallopin, P. Gutman, A. Hammond, and R. Swart, 1998: *Bending the Curve: Toward Global Sustainability*. Global Scenario Group, Stockholm Environment Institute, Stockholm, Sweden, 90 pp. (plus appendices).

Reich, P.B., D.S. Ellsworth, B.D. Kleoppel, J.H. Fowner, and G.T. Grower, 1990: Vertical variation in canopy structure and CO_2 exchange of oak-maple forests: influence of ozone, nitrogen, and other factors on simulated canopy carbon gain. *Tree Physiology*, **7,** 329–345.

Reilly, J., W. Baethgen, F.E. Chege, S. van de Geijn, L. Erda, A. Iglesias, G. Kenny, D. Patterson, J. Rogasik, R. Rötter, W. Sombroek, J. Westbrook, D. Bachelet, M. Brklacich, U. Dämmgen, M. Howden, R.J.V. Joyce, P.D. Lingren, D. Schimmelpfennig, U. Singh, O. Sirotenko, and E. Wheaton, 1996: Agriculture in a changing climate: impacts and adaptation. In: *Climate Change 1995: Impacts, Adaptations and Mitigation of Climate Change: Scientific-Technical Analyses. Contribution of Working Group II to the Second Assessment Report of the Intergovernmental Panel on Climate Change* [Watson, R.T., M.C. Zinyowera, and R.H. Moss (eds.)]. Cambridge University Press, Cambridge, United Kingdom and New York, NY, USA, pp. 427–467.

Risbey, J.S. and P.H. Stone, 1996: A case study of the adequacy of GCM simulations for input to regional climate change assessments. *Journal of Climate*, **9,** 1441–1467.

Risbey, J., M. Kandlikar, and H. Dowlatabadi, 1999: Scale, context, and decision making in agricultural adaptation to climate variability and change. *Mitigation and Adaptation Strategies for Global Change*, **4,** 137–165.

Robock, A., R.P. Turco, M.A. Harwell, T.P. Ackerman, R. Andressen, H-S. Chang, and M.V.K. Sivakumar, 1993: Use of general circulation model output in the creation of climate change scenarios for impact analysis. *Climatic Change*, **23,** 293–335.

Roeckner, E., K. Arpe, L. Bengtsson, M. Christoph, M. Claussen, L. Dümenil, M. Esch, M. Giorgetta, U. Schlese, and U. Schulzweida, 1996: *The Atmospheric General Circulation Model ECHAM-4: Model Description and Simulation of Present-Day Climate*. Report No. 218, Max-Planck Institute for Meteorology, Hamburg, Germany, 90 pp.

Rosegrant, M.W., M. Agcaoili-Saombilla, and N.D. Perez, 1995: *Global Food Projections to 2020: Implications for Investment, Food, Agriculture and the Environment*. 2020 Vision Discussion Paper 5, International Food Policy Research Institute (IFPRI), Washington DC, USA, 54 pp.

Rosenberg, N.J. (ed.), 1993: Towards an integrated assessment of climate change: the MINK Study. *Climatic Change (Special Issue)*, **24,** 1–173.

Rosenberg, N.J., P.R. Crosson, K.D. Frederick, W.E. Easterling, M.S. McKenney, M.D. Bowes, R.A. Sedjo, J. Darmstadter, L.A. Katz, and K.M. Lemon, 1993: The MINK methodology: background and baseline: paper 1. *Climatic Change*, **24,** 7–22.

Rosenthal, D.H., H.K. Gruenspecht, and E.A. Moran, 1995: Effects of global warming on energy use for space heating and cooling in the United States. *Energy Journal*, **16,** 41–54.

Rosenzweig, C. and M.L. Parry, 1994: Potential impact of climate change on world food supply. *Nature*, **367,** 133–138.

Rosenzweig, C., J. Phillips, R. Goldberg, J. Carroll, and T. Hodges, 1996: Potential impacts of climate change on citrus and potato production in the U.S. *Agricultural Systems*, **52,** 455–479.

Rotmans, J. and B. de Vries (eds.), 1997: *Perspectives on Global Change: The TARGETS Approach*. Cambridge University Press, Cambridge, United Kingdom and New York, NY, USA, 479 pp.

Rotmans, J. and H. Dowlatabadi, 1998: Integrated assessment modelling. In: *Human Choice and Climate Change. Volume 3: The Tools for Policy Analysis* [Rayner, S. and E.L. Malone (eds.)]. Batelle Press, Columbus, Ohio, USA, pp. 291–377.

Rotmans, J., M. Hulme, and T.E. Downing, 1994: Climate change implications for Europe: an application of the ESCAPE model. *Global Environmental Change*, **4,** 97–124.

Rovinsky, F.Ya. and V.I. Yegorov, 1986: *Ozone and Oxides of Sulphur and Nitrogen in the Lower Atmosphere*. Leningrad, Gidrometeoizdat, Russia, 184 pp. (in Russian).

Rummukainen, M., 1997: *Methods for Statistical Downscaling of GCM Simulations*. Report No. 80, Swedish Meteorological and Hydrological Institute, Norrköping, Sweden, 29 pp.

Runeckles, V.C. and S.V. Krupa, 1994: The impact of UV-B radiation and ozone on terrestrial vegetation. *Environmental Pollution*, **83,** 191–213.

Ryaboshapko, A.G., L. Gallardo, E. Klellstrom, S. Gromov, S. Paramonov, O. Afinogenova, and H. Rodhe, 1998: Balance of oxidized sulfur and nitrogen over the former Soviet Union territory. *Atmospheric Environment*, **32(4),** 647–658.

Sala, O.E., F.S. Chapin III, J.J. Armesto, E. Berlow, J. Bloomfield, R. Dirzo, E. Huber-Sanwald, L.F. Huenneke, R. Jackson, A. Kinzig, R. Leemans, D. Lodge, H.A. Mooney, M. Oesterheld, L. Poff, M.T. Sykes, B.H. Walker, M. Walker, and D. Wall, 2000: Global biodiversity scenarios for the year 2100. *Science*, **287,** 1770–1774.

Sælthun, N.R., P. Aittoniemi, S. Bergström, K. Einarsson, T. Jóhannesson, G. Lindström, P.-E. Ohlsson, T. Thomsen, B. Vehviläinen, and K.O. Aamodt, 1998: Climate change impacts on runoff and hydropower in the Nordic countries. In: *TemaNord*, **552.** Nordic Council of Ministers, Copenhagen, Denmark, 170 pp.

Santer, B., 1985: The use of general circulation models in climate impact analysis—a preliminary study of the impacts of a CO_2-induced climatic change on West European agriculture. *Climatic Change*, **7,** 71–93.

Santer, B.D., T.M.L. Wigley, M.E. Schlesinger, and J.F.B. Mitchell, 1990: *Developing Climate Scenarios from Equilibrium GCM Results*. Report No. 47, Max-Planck-Institut-für-Meteorologie, Hamburg, Germany, 29 pp.

Sarmiento, J.L., T.M.C. Hughes, R.J. Stouffer, and S. Manabe, 1998: Simulated response of the ocean carbon cycle to anthropogenic climate warming. *Nature*, **393,** 245–249.

Saunders, P.W.J., 1985: Regulations and research on SO_2 and its effects on plants in the European Communities. In: *Sulfur Dioxide and Vegetation: Physiology, Ecology and Policy Issues* [Winner, W.E., H.A. Mooney, and R.A. Goldstein (eds.)]. Stanford University Press, Stanford, CA, USA, pp. 37–55.

Schimel, D., I. Enting, M. Heimann, T. Wigley, D. Raynaud, D. Alves, and U. Siegenthaler, 1995: CO_2 and the carbon cycle. In: *Climate Change 1994: Radiative Forcing of Climate Change and an Evaluation of the IPCC IS92 Emission Scenarios* [Houghton, J.T., L.G. Meira Filho, J. Bruce, H. Lee, B.A. Callander, E. Haites, N. Harris, and K. Maskell (eds.)]. Cambridge University Press, Cambridge, United Kingdom and New York, NY, USA, pp. 35–71.

Schimel, D., M. Grubb, F. Joos, R. Kaufmann, R. Moss, W. Ogana, R. Richels, T. Wigley, R. Cannon, J. Edmonds, E. Haites, D. Harvey, A. Jain, R. Leemans, K. Miller, R. Parkin, E. Sulzman, R. Tol, J. de Wolde, and M. Bruno, 1997a: Stabilization of atmospheric greenhouse gases: physical, biological, and socio-economic implications. *IPCC Technical Paper III* [Houghton, J.T., L.G. Meira Filho, D.J. Griggs, and K. Maskell (eds)]. Intergovernmental Panel on Climate Change, Geneva, Switzerland, 51 pp.

Schimel, D.S., VEMAP members, and B.H. Braswell, 1997b. Continental scale variability in ecosystem processes: models, data, and the role of disturbance. *Ecological Monographs*, **67,** 251–271.

Schlesinger, M.E. and J.F.B. Mitchell, 1987: Climate model simulations of the equilibrium climatic response to increased carbon dioxide. *Reviews of Geophysics*, **25,** 760–798.

Schlesinger, M.E. and Z.C. Zhao, 1989: Seasonal climate changes induced by doubled CO_2 as simulated by the OSU atmospheric GCM mixed layer ocean model. *Journal of Climate*, **2,** 459–495.

Schöpp, W., M. Amann, J. Cofala, C. Heyes, and Z. Klimont, 1999: Integrated assessment of European air pollution emission control strategies. *Environmental Modelling and Software*, **14,** 1–9.

Sellers, P.J., L. Bounoua, G.J. Collatz, D.A. Randall, D.A. Dazlich, S.O. Los, J.A. Berry, I. Fung, C.J. Tucker, C.B. Field, and T.G. Jensen, 1996: Comparison of radiative and physiological effects of doubled atmospheric CO_2 on climate. *Science*, **271,** 1402–1406.

Semenov, M.A. and E.M. Barrow, 1997: Use of a stochastic weather generator in the development of climate change scenarios. *Climatic Change*, **35,** 397–414.

Semenov, M.A. and J.R. Porter, 1995: Climatic variability and the modelling of crop yields. *Agricultural and Forest Meteorology*, **73,** 265–283.

Semenov, S.M., B.A. Koukhta, and A.A. Rudkova, 1997: Assessment of the tropospheric ozone effect on higher plants. *Russian Meteorology and Hydrology*, **12,** 36–40.

Semenov, S.M., I.M. Kounina, and B.A. Koukhta, 1998: An ecological analysis of anthropogenic changes in ground-level concentrations of O_3, SO_2, and CO_2 in Europe. *Doklady Biological Sciences*, **361,** 344–347.

Semenov, S.M., I.M. Kounina, and B.A. Koukhta, 1999: *Tropospheric Ozone and Plant Growth in Europe*. Publishing Center, Meteorology and Hydrology, Moscow, Russia, 88 pp. (in Russian with English summary).

SEPA, 1993: *Ground-Level Ozone and Other Photochemical Oxidants in the Environment*. Swedish Environmental Protection Agency, Stockholm, Sweden, 68 pp.

Shiklomanov, I.A., 1998: *Assessment of Water Resources and Water Availability in the World. Background Report for the Comprehensive Assessment of the Freshwater Resources of the World*. Stockholm Environment Institute, Stockholm, Sweden, 88 pp.

Shindell, D.T., D. Rind, and P. Lonergan, 1998: Climate change and the middle atmosphere: part IV: ozone response to doubled CO_2. *Journal of Climate*, **11,** 895–918.

Shine, K.P., R.G. Derwent, D.J. Wuebbles, and J-J. Morcrette, 1990: Radiative forcing of climate. In: *Climate Change: The IPCC Scientific Assessment* [Houghton, J.T., G.J. Jenkins, and J.J. Ephraums (eds.)]. Cambridge University Press, Cambridge, United Kingdom and New York, NY, USA, pp. 41–68.

Simpson, D., 1992: Long-period modelling of photochemical oxidants in Europe: calculations for July 1985. *Atmospheric Environment*, **26A,** 1609–1634.

Simpson, D., 1995a: Hydrocarbon reactivity and ozone formation in Europe. *Journal of Atmospheric Chemistry*, **20,** 163–177.

Simpson, D., 1995b: Biogenic emissions in Europe 2: implications for ozone control strategies. *Journal of Geophysical Research*, **100(D11),** 22891–22906.

Simpson, D., A. Guenther, C.N. Hewitt, and R. Steinbrecher, 1995: Biogenic emissions in Europe 1; estimates and uncertainties. *Journal of Geophysical Research*, **100(D11),** 22875–22890.

Simpson, D., K. Olendrzynski, A. Semb, E. Støren, and S. Unger, 1997: *Photochemical Oxidant Modelling in Europe: Multi-Annual Modelling and Source-Receptor Relationships*. EMEP/MSC-W Report 3/97, Norwegian Meteorological Institute, Oslo, Norway.

Sirotenko, O.D., E.V. Abashina, and V.N. Pavlova, 1995: Sensitivity of agriculture of Russia to changes in climate, chemical composition of the atmosphere and soil fertility. *Meteorology and Hydrology*, **4,** 107–114 (in Russian).

Skole, D. and C. Tucker, 1993: Tropical deforestation and habitat fragmentation in the Amazon: satellite data from 1978 to 1988. *Science*, **260,** 1905–1910.

Spencer, N.E., and P.L. Woodworth, 1993: *Data Holdings of the Permanent Service for Mean Sea Level*. Permanent Service for Mean Sea Level, Bidston, Birkenhead, United Kingdom, 81 pp.

Smith, I.N., 1998: Estimating mass balance components of the Greenland ice sheet from a long-term GCM simulation. *Global and Planetary Change*, **20,** 19–32.

Smith, J.B. and M. Hulme, 1998: Climate change scenarios. In: *UNEP Handbook on Methods for Climate Change Impact Assessment and Adaptation Studies* [Burton, I., J.F. Feenstra, J.B. Smith, and R.S.J. Tol (eds.)]. Version 2.0, United Nations Environment Programme and Institute for Environmental Studies, Vrije Universiteit, Amsterdam, pp. 3-1 to 3-40. Available online at http://www.vu.nl/english/o_o/instituten/IVM/research/climatechange/Handbook.htm

Smith, J.B. and D.A. Tirpak (eds.), 1989: *The Potential Effects of Global Climate Change on the United States*. Report to Congress, United States Environmental Protection Agency, EPA-230–05–89–050, Washington, DC, USA, 409 pp.

Smith, J.B., S. Huq, S. Lenhart, L.J. Mata, I. Nemesová, and S. Toure (eds.), 1996: *Vulnerability and Adaptation to Climate Change. Interim Results from the U.S. Country Studies Program*. Kluwer Academic Publishers, Dordrecht, The Netherlands, 366 pp.

Smith, I.N., W.F. Budd, and P. Reid, 1998: Model estimates of Antarctic accumulation rates and their relationship to temperature changes. *Annals of Glaciology*, **27,** 246–250.

Sokolov, A.P. and P.H. Stone, 1998: A flexible climate model for use in integrated assessments, *Climate Dynamics*, **14,** 291–303.

Sokolov, A., C. Wang, G. Holian, P. Stone, and R. Prinn, 1998: Uncertainty in the oceanic heat and carbon uptake and its impact on climate projections. *Geophysical Research Letters*, **25,** 3603–3606.

Spencer, N.E. and P.L. Woodworth, 1993: *Data Holdings of the Permanent Service for Mean Sea Level (November 1993)*. Permanent Service for Mean Sea Level, Bidston, Birkenhead, United Kingdom, 81 pp.

Stevenson, D.S., C.E. Johnson, W.J. Collins, R.G. Derwent, and J.M. Edwards, 2000: Future tropospheric ozone radiative forcing and methane turnover—the impact of climate change. *Geophysical Research Letters*, **27,** 2073–2076.

Stocker, T.F. and A. Schmittner, 1997: Influence of CO_2 emission rates on the stability of the thermohaline circulation. *Nature*, **388,** 862–865.

Strain, B.R. and J.D. Cure (eds.), 1985: *Direct Effects of Increasing Carbon Dioxide on Vegetation*. DOE/ER-0238, U.S. Department of Energy, Office of Energy Research, Washington, DC, USA, 286 pp.

Strzepek, K.M., S.C. Onyeji, M. Saleh, and D. Yates, 1995: An assessment of integrated climate change impact on Egypt. In: *As Climate Changes: International Impacts and Implications* [Strzepek, K.M. and J. Smith (eds.)]. Cambridge University Press, Cambridge, United Kingdom and New York, NY, USA, pp. 180–200.

Sullivan, J.H., 1997: Effects of increasing UV-B radiation and atmospheric CO_2 on photosynthesis and growth: implications for terrestrial ecosystems. *Plant Ecology*, **128,** 195–206.

Terjung, W.H., D.M. Liverman, J.T. Hayes, P.A. O'Rourke, and P.E. Todhunter, 1984: Climatic change and water requirements for grain corn in the North American Great Plains. *Climatic Change*, **6,** 193–220.

Texier, D., N. de Noblet, S.P. Harrison, A. Haxeltine, D. Jolly, S. Joussaume, F. Laarif, I.C. Prentice, and P. Tarasov, 1997: Quantifying the role of biosphere-atmosphere feedbacks in climate change: a coupled model simulation for 6000 yr BP and comparison with paleodata for northern Eurasia and northern Africa. *Climate Dynamics*, **13,** 865–881.

Thompson, S.L. and D. Pollard, 1995: A global climate model (GENESIS) with a land-surface-transfer scheme (LSX): part II: CO_2 sensitivity. *Journal of Climate*, **8,** 1104–1121.

Thompson, S.L. and D. Pollard, 1997: Greenland and Antarctic mass balances for present and doubled CO_2 from the GENESIS version 2 global climate model. *Journal of Climate*, **10,** 871–900.

Titus, J. and V. Narayanan, 1996: The risk of sea level rise: a delphic Monte Carlo analysis in which twenty researchers specify subjective probability distributions for model coefficients within their respective areas of expertise. *Climatic Change,* **33(2),** 151–212.

Tol, R., 1998: Socio-economic scenarios. In: *UNEP Handbook on Methods for Climate Change Impact Assessment and Adaptation Studies* [Burton, I., J.F. Feenstra, J.B. Smith, and R.S.J. Tol (eds.)]. Version 2.0, United Nations Environment Programme and Institute for Environmental Studies, Vrije Universiteit, Amsterdam, The Netherlands, Chapter 2, pp. 1–19. Available online at http://www.vu.nl/english/o_o/instituten/IVM/research/climatechange/Handbook.htm.

Trenberth, K.E. and T.J. Hoar, 1997: El Niño and climate change. *Geophysical Research Letters*, **24,** 3057–3060.

Tsyban, A.V., 1997: Ecological problems associated with human activities affecting the World Ocean. In: *Proceedings of the Annual Meeting of the Oceanographic Society of Japan*. Oceanographic Society of Japan, Tsukuba, Japan, pp. 58–68.

Turner, B.L., D.L. Skole, S. Sanderson, G. Fischer, L. Fresco, and R. Leemans, 1995: *Land-Use and Land-Cover Change: Science/Research Plan*. IGBP Report No. 35 and HDP Report No. 7, International Geosphere-Biosphere Programme and Human Dimensions of Global Environmental Change Programme, Stockholm, Sweden, 132 pp.

UK-DETR, 1999: *Climate Change and its Impacts: Stabilisation of CO_2 in the Atmosphere*. United Kingdom Department of the Environment, Transport and the Regions, The Met Office, Bracknell, UK, 28 pp.

UK National Foresight Programme, 1999: *Environmental Futures*. Office of Science and Technology, Department of Trade and Industry, London, United Kingdom, 24 pp.

UN/ECE, 1998: *Integrated Assessment Modelling*. Document EB.AIR/1998/1, UN Economic Commission for Europe, Geneva, Switzerland, 27 pp.

UNEP/GEMS, 1995: *Water Quality of World River Basins* [Fraser, A.S., M. Meybeck, and E.D. Ongley (eds.)]. UNEP Environment Library No. 14, United Nations Environment Programme, Nairobi, Kenya, 40 pp.

Unsworth, M.H. and W.E. Hogsett, 1996: Combined effects of changing CO_2, temperature, UV-B radiation and O_3 on crop growth. In: *Global Climate Change and Agricultural Production: Direct and Indirect Effects of Changing Hydrological, Pedological and Plant Physiological Processes* [Bazzaz, F. and W. Sombroek (eds.)]. John Wiley and Sons, Chichester, United Kingdom, pp. 171–197.

Van de Wal, R.S.W. and J. Oerlemans, 1997: Modelling the short-term response of the Greenland ice-sheet to global warming. *Climate Dynamics*, **13,** 733–744.

van der Sluijs, J.P., 1997: *Anchoring Amid Uncertainty. On the Management of Uncertainties in Risk Assessment of Anthropogenic Climate Change*. Diss. University of Utrecht, The Netherlands, 260 pp.

Van Latesteijn, H.C., 1995: Assessment of future options for land use in the European Community. *Ecological Engineering*, **4,** 211–222.

VEMAP members, 1995: Vegetation/ecosystem modeling and analysis project: comparing biogeography and biogeochemistry models in a continental-scale study of terrestrial ecosystem responses to climate change and CO_2 doubling. *Global Biogeochemical Cycles*, **9,** 407–437.

Vitousek, P.M., J.D. Aber, R.W. Howarth, G.E. Likens, P.A. Matson, D.W. Schindler, W.H. Schlesinger, and D.G. Tilman, 1997: Human alteration of the global nitrogen cycle: sources and consequences. *Ecological Applications*, **7,** 737–750.

von Storch, H., 1995: Inconsistencies at the interface of climate impact studies and global climate research. *Meteorologische Zeitschrift*, N.F., **4,** 72–80.

von Storch, H., E. Zorita, and U. Cubasch, 1993: Downscaling of global climate change estimates to regional scales: an application to Iberian rainfall in wintertime. *Journal of Climate*, **6,** 1161–1171.

Walker, B., W. Steffen J. Canadell, and J. Ingram (eds.), 1999: *The Terrestrial Biosphere and Global Change: Implications for Natural and Managed Ecosystems*. Cambridge University Press, Cambridge, United Kingdom and New York, NY, USA, 439 pp.

Walsh, K.J.E., D.R. Jackett, T.J. McDougall, and A.B. Pittock, 1998: *Global Warming and Sea Level Rise on the Gold Coast*. Report Prepared for the Gold Coast City Council, CSIRO Atmospheric Research, Mordialloc, Australia, 34 pp.

Warrick, R.A., C. Le Provost, M.F. Meier, J. Oerlemans, and P.L. Woodworth, 1996: Changes in sea level. In: *Climate Change 1995: The Science of Climate Change. Contribution of Working Group I to the Second Assessment Report of the Intergovernmental Panel on Climate Change* [Houghton, J.T., L.G. Meira Filho, B.A. Callander, N. Harris, A. Kattenberg, and K. Maskell (eds.)]. Cambridge University Press, Cambridge, United Kingdom and New York, NY, USA, 359–405.

Washington, W.M. and G.A. Meehl, 1984: Seasonal cycle experiment on the climate sensitivity due to a doubling of CO_2 with an atmospheric general circulation model coupled to a simple mixed layer ocean model. *Journal of Geophysical Research*, **89,** 9475–9503.

Washington, W.M. and G.A. Meehl, 1996: High-latitude climate change in a global coupled ocean-atmosphere-sea ice model with increased atmospheric CO_2. *Journal of Geophysical Research*, **101,** 12795–12801.

Watterson, I.G., S.P. O'Farrell, and M.R. Dix, 1997: Energy transport in climates simulated by a GCM which includes dynamic sea-ice. *Journal of Geophysical Research*, **102(D10),** 11027–11037.

WCC'93, 1994: *Preparing to Meet the Coastal Challenges of the 21st Century. Report of the World Coast Conference, Noordwijk, 1–5 November 1993*. Ministry of Transport, Public Works and Water Management, The Hague, The Netherlands, 49 pp. (plus appendices).

Wetherald, R.T. and S. Manabe, 1986: An investigation of cloud cover change in response to thermal forcing. *Climatic Change*, **8,** 5–23.

Weyant, J., O. Davidson, H. Dowlabathi, J. Edmonds, M. Grubb, E.A. Parson, R. Richels, J. Rotmans, P.R. Shukla, R.S.J. Tol, W. Cline, and S. Fankhauser, 1996: Integrated assessment of climate change: an overview and comparison of approaches and results. In: *Climate Change 1995: Economic and Social Dimensions of Climate Change* [Bruce, J.P., H. Lee, and E.F. Haites (eds.)]. Cambridge University Press, Cambridge, United Kingdom and New York, NY, USA, pp. 367–396.

Wheeler, T.R., T.D. Hong, R.H. Ellis, G.R. Batts, J.I.L. Morison, and P. Hadley, 1996: The duration and rate of grain growth, and harvest index, of wheat (*Triticum aestivum*) in response to temperature and CO_2. *Journal of Experimental Botany*, **47,** 623–630.

Whelpdale, D.M. and M.S. Kaiser (eds.), 1997: *Global Acid Deposition Assessment*. Global Atmosphere Watch No. 106, WMO-TD 777, World Meteorological Organization, Geneva, Switzerland, 241 pp.

Whetton, P.H., J.J. Katzfey, K. Nguyen, J.L. McGregor, C.M. Page, T.I. Eliot, and K.J. Hennessy, 1998: *Fine Resolution Climate Change Scenarios for New South Wales. Part 2: Climatic Variability*. Commonwealth Scientific and Industrial Research Organisation, Division of Atmospheric Research, Climate Impact Group, Aspendale, Victoria, Australia, 51 pp.

Whetton, P.H., J.J. Katzfey, K.J. Hennesey, X. Wu, J.L. McGregor, and K. Nguyen, 2001: Developing scenarios of climate change for southeastern Australia: an example using regional climate model output. *Climate Research*, (in press).

White, A., M.G.R. Cannell, and A.D. Friend, 1999: Climate change impacts on ecosystems and the terrestrial carbon sink: a new assessment. *Global Environmental Change*, **9,** S21–S30.

WHO, 1995: *Community Water Supply and Sanitation: Needs, Challenges and Health Objectives*. Report A48/INF.DOC./2 of the Director-General, World Health Organization, World Health Assembly, Geneva, Switzerland.

Wigley, T.M.L. and S.C.B. Raper, 1992: Implications for climate and sea level of revised IPCC emissions scenarios. *Nature*, **357,** 293–300.

Wigley, T.M.L. and S.C.B. Raper, 1993: Future changes in global mean temperature and thermal-expansion-related sea level rise. In: *Climate and Sea Level Change: Observations, Predictions and Implications* [Warrick, R.A., E.M. Barrow, and T.M.L. Wigley (eds.)]. Cambridge University Press, Cambridge, United Kingdom and New York, NY, USA, pp. 111–133.

Wigley, T.M.L. and S.C.B Raper, 1995: An heuristic model for sea level rise due to the melting of small glaciers. *Geophysical Research Letters*, **22,** 2749–2752.

Wilby, R.L., 1997: Non-stationarity in daily precipitation series: implications for GCM down-scaling using atmospheric circulation indices. *International Journal of Climatology*, **17,** 439–454.

Wilby, R.L. and T.M.L. Wigley, 1997: Downscaling general circulation model output: a review of methods and limitations. *Progress in Physical Geography*, **21,** 530–548.

Wilby, R.L., T.M.L. Wigley, D. Conway, P.D. Jones, B.C. Hewitson, J. Main, and D.S. Wilks, 1998: Statistical downscaling of general circulation model output: a comparison of methods. *Water Resources Research*, **34,** 2995–3008.

Wilks, D.S., 1992: Adapting stochastic weather generation algorithms for climate change studies. *Climatic Change*, **22,** 67–84.

Williams, G.D.V., R.A. Fautley, K.H. Jones, R.B. Stewart, and E.E. Wheaton, 1988: Estimating effects of climatic change on agriculture in Saskatchewan, Canada. In: *The Impact of Climatic Variations on Agriculture. Volume 1. Assessments in Cool Temperate and Cold Regions* [Parry, M.L., T.R. Carter, and N.T. Konijn (eds.)]. Kluwer Academic Publishers, Dordrecht, The Netherlands, pp. 219–379.

Wilson, C.A. and J.F.B. Mitchell, 1987: Simulated CO_2 induced climate change over western Europe. *Climatic Change*, **10,** 11–42.

Woodward, F.I., 1992: Predicting plant response to global environmental change. *New Phytologist*, **122,** 230–251.

Woolhiser, D.A., T.O. Keefer, and K.T. Redmond, 1993: Southern oscillation effects on daily precipitation in the southwestern United States. *Water Resources Research*, **29,** 1287–1295.

Yohe, G.W. and M.E. Schlesinger, 1998: Sea-level change: the expected economic cost of protection or abandonment in the United States. *Climatic Change*, **38,** 337–472.

Yohe, G., M. Jacobsen, and T. Gapotchenko, 1999: Spanning "not-implausible" futures to assess relative vulnerability to climate change and climate variability. *Global Environmental Change*, **9,** 233–249.

Zeidler, R.B., 1996: Climate change vulnerability and response strategies for the coastal zone of Poland. *Climatic Change*, **36,** 151–173.

Zeng, N., 1998: Understanding climate sensitivity to tropical deforestation in a mechanistic model. *Journal of Climate*, **11,** 1969–1975.

Zeng, N. and J.D. Neelin, 1999: A land-atmosphere interaction theory for the tropical deforestation problem. *Journal of Climate*, **12,** 857–887.

Zepp, R.G., T.V. Callaghan, and D.J. Erickson, 1998: Effects of enhanced solar ultraviolet radiation on biogeochemical cycles. *Journal of Photochemistry and Photobiology B: Biology*, **46,** 69–82.

Zerefos, C.S. and F.B. Alkiviadis, 1997: *Solar Ultraviolet Radiation: Modelling, Measurements, and Effects*. NATO Asi Series I, Global Environmental Change, Springer-Verlag, Berlin and Heidelberg, Germany, **52,** 336 pp.

Zhang, H., A. Henderson-Sellers, and K. McGuffie, 1996: Impacts of tropical deforestation: part I: process analysis of local climatic change. *Journal of Climate*, **9,** 1497–1517.

Zwartz, D., P. Tregoning, K. Lambeck, P. Johnston, and J. Stone, 1999: Estimates of present-day glacial rebound in the Lambert Glacier region, Antarctica. *Geophysical Research Letters*, **26,** 1461–1464.

4

Hydrology and Water Resources

NIGEL ARNELL (UK) AND CHUNZHEN LIU (CHINA)

Lead Authors:
R. Compagnucci (Argentina), L. da Cunha (Portugal), K. Hanaki (Japan), C. Howe (USA), G. Mailu (Kenya), I. Shiklomanov (Russia), E. Stakhiv (USA)

Contributing Author:
P. Döll (Germany)

Review Editors:
A. Becker (Germany) and Jianyun Zhang (China)

CONTENTS

EXECUTIVE SUMMARY

- There are apparent trends in streamflow volume—both increases and decreases—in many regions. These trends cannot all be definitively attributed to changes in regional temperature or precipitation. However, widespread accelerated glacier retreat and shifts in streamflow timing in many areas from spring to winter are more likely to be associated with climate change.
- The effect of climate change on streamflow and groundwater recharge varies regionally and between scenarios, largely following projected changes in precipitation. In some parts of the world, the direction of change is consistent between scenarios, although the magnitude is not. In other parts of the world, the direction of change is uncertain.
- Peak streamflow is likely to move from spring to winter in many areas where snowfall currently is an important component of the water balance.
- Glacier retreat is likely to continue, and many small glaciers may disappear.
- Water quality is likely generally to be degraded by higher water temperature, but this may be offset regionally by increased flows. Lower flows will enhance degradation of water quality.
- Flood magnitude and frequency are likely to increase in most regions, and low flows are likely to decrease in many regions.
- Demand for water generally is increasing as a result of population growth and economic development, but it is falling in some countries. Climate change is unlikely to have a large effect on municipal and industrial demands but may substantially affect irrigation withdrawals.
- The impact of climate change on water resources depends not only on changes in the volume, timing, and quality of streamflow and recharge but also on system characteristics, changing pressures on the system, how the management of the system evolves, and what adaptations to climate change are implemented. Nonclimatic changes may have a greater impact on water resources than climate change.
- Unmanaged systems are likely to be most vulnerable to climate change.
- Climate change challenges existing water resources management practices by adding additional uncertainty. Integrated water resources management will enhance the potential for adaptation to change.
- Adaptive capacity (specifically, the ability to implement integrated water resources management), however, is distributed very unevenly across the world.

4.1. Introduction and Scope

This chapter assesses our understanding of the implications of climate change for the hydrological cycle, water resources, and their management. Since the beginnings of concern over the possible consequences of global warming, it has been widely recognized that changes in the cycling of water between land, sea, and air could have very significant impacts across many sectors of the economy, society, and the environment. The characteristics of many terrestrial ecosystems, for example, are heavily influenced by water availability and, in the case of instream ecosystems and wetlands, by the quantity and quality of water in rivers and aquifers. Water is fundamental to human life and many activities—most obviously agriculture but also industry, power generation, transportation, and waste management—and the availability of clean water often is a constraint on economic development. Consequently, there have been a great many studies into the potential effects of climate change on hydrology (focusing on cycling of water) and water resources (focusing on human and environmental use of water). The majority of these studies have concentrated on possible changes in the water balance; they have looked, for example, at changes in streamflow through the year. A smaller number of studies have looked at the impacts of these changes for water resources—such as the reliability of a water supply reservoir or the risk of flooding—and even fewer explicitly have considered possible adaptation strategies. This chapter summarizes key findings of research that has been conducted and published, but it concentrates on assessing opportunities and constraints on adaptation to climate change within the water sector. This assessment is based not only on the few studies that have looked explicitly at climate change but also on considerable experience within different parts of the water sector in adapting to changing circumstances in general.

This chapter first summarizes the state of knowledge of climate change impacts on hydrology and water resources (Section 4.2), before assessing effects on the hydrological cycle and water balance on the land (Section 4.3). Section 4.4 examines potential changes in water use resulting from climate change, and Section 4.5 assesses published work on the impacts of climate change for some water resource management systems. Section 4.6 explores the potential for adaptation within the water sector. The final two sections (Sections 4.7 and 4.8) consider several integrative issues as well as science and information requirements. The implications of climate change on freshwater ecosystems are reviewed in Chapter 5, although it is important to emphasize here that water management is increasingly concerned with reconciling human and environmental demands on the water resource. The hydrological system also affects climate, of course. This is covered in the Working Group I contribution to the Third Assessment Report (TAR); the present chapter concentrates on the impact of climate on hydrology and water resources.

At the outset, it is important to emphasize that climate change is just one of many pressures facing the hydrological system and water resources. Changing land-use and land-management practices (such as the use of agrochemicals) are altering the hydrological system, often leading to deterioration in the resource baseline. Changing demands generally are increasing pressures on available resources, although per capita demand is falling in some countries. The objectives and procedures of water management are changing too: In many countries, there is an increasing move toward "sustainable" water management and increasing concern for the needs of the water environment. For example, the Dublin Statement, agreed at the International Conference on Water and the Environment in 1992, urges sustainable use of water resources, aimed at ensuring that neither the quantity nor the quality of available resources are degraded. Key water resources stresses now and over the next few decades (Falkenmark, 1999) relate to access to safe drinking water, water for growing food, overexploitation of water resources and consequent environmental degradation, and deterioriation in water quality. The magnitude and significance of these stresses varies between countries. The late 1990s saw the development of several global initiatives to tackle water-related problems: The UN Commission on Sustainable Development published the "Comprehensive Assessment of the Freshwater Resources of the World" (WMO, 1997), and the World Water Council asked the World Commission for Water to produce a vision for a "water-secure world" (Cosgrove and Rijbersman, 2000). A series of periodical reports on global water issues was initiated (Gleick, 1998). The impacts of climate change, and adaptation to climate change, must be considered in the context of these other pressures and changes in the water sector.

4.2 State of Knowledge of Climate Change Impacts on Hydrology and Water Resources: Progress since the Second Assessment Report

4.2.1 Introduction

Over the past decade—and increasingly since the publication of the Second Assessment Report (SAR) (Arnell *et al.*, 1996; Kaczmarek, 1996)—there have been many studies into climate change effects on hydrology and water resources (see the online bibliography described by Chalecki and Gleick, 1999), some coordinated into national programs of research (as in the U.S. National Assessment) and some undertaken on behalf of water management agencies. There are still many gaps and unknowns, however. The bulk of this chapter assesses current understanding of the impacts of climate change on water resources and implications for adaptation. This section highlights significant developments in three key areas since the SAR: methodological advances, increasing recognition of the effect of climate variability, and early attempts at adaptation to climate change.

4.2.2 Estimating the Impacts of Climate Change

The impacts of climate change on hydrology usually are estimated by defining scenarios for changes in climatic inputs to a hydrological model from the output of general circulation models

(GCMs). The three key developments here are constructing scenarios that are suitable for hydrological impact assessments, developing and using realistic hydrological models, and understanding better the linkages and feedbacks between climate and hydrological systems.

The heart of the scenario "problem" lies in the scale mismatch between global climate models (data generally provided on a monthly time step at a spatial resolution of several tens of thousands of square kilometers) and catchment hydrological models (which require data on at least daily scales and at a resolution of perhaps a few square kilometers). A variety of "downscaling" techniques have been developed (Wilby and Wigley, 1997) and used in hydrological studies. These techniques range from simple interpolation of climate model output (as used in the U.S. National Assessment; Felzer and Heard, 1999), through the use of empirical/statistical relationships between catchment and regional climate (e.g., Crane and Hewitson, 1998; Wilby *et al.*, 1998, 1999), to the use of nested regional climate models (e.g., Christensen and Christensen, 1998); all, however, depend on the quality of simulation of the driving global model, and the relative costs and benefits of each approach have yet to be ascertained. Studies also have looked at techniques for generating stochastically climate data at the catchment scale (Wilby *et al.*, 1998, 1999). In principle, it is possible to explore the effects of changing temporal patterns with stochastic climate data, but in practice the credibility of such assessments will be strongly influenced by the ability of the stochastic model to simulate present temporal patterns realistically.

Considerable effort has been expended on developing improved hydrological models for estimating the effects of climate change. Improved models have been developed to simulate water quantity and quality, with a focus on realistic representation of the physical processes involved. These models often have been developed to be of general applicability, with no locally calibrated parameters, and are increasingly using remotely sensed data as input. Although different hydrological models can give different values of streamflow for a given input (as shown, for example, by Boorman and Sefton, 1997; Arnell, 1999a), the greatest uncertainties in the effects of climate on streamflow arise from uncertainties in climate change scenarios, as long as a conceptually sound hydrological model is used. In estimating impacts on groundwater recharge, water quality, or flooding, however, translation of climate into response is less well understood, and additional uncertainty is introduced. In this area, there have been some reductions in uncertainty since the SAR as models have been improved and more studies conducted (see Sections 4.3.8 and 4.3.10). The actual impacts on water resources—such as water supply, power generation, navigation, and so forth—depend not only on the estimated hydrological change but also on changes in demand for the resource and assumed responses of water resources managers. Since the SAR, there have been a few studies that have summarized potential response strategies and assessed how water managers might respond in practice (see Section 4.6).

There also have been considerable advances since the SAR in the understanding of relationships between hydrological processes at the land surface and processes within the atmosphere above. These advances have come about largely through major field measurement and modeling projects in different geographical environments [including the First ISLSCP Field Experiment (FIFE), LAMBADA, HAPEX-Sahel, and NOPEX; see www.gewex.com], coordinated research programs (such as those through the International Geosphere-Biosphere Programme (IGBP; see www.igbp.se) and large-scale coupled hydrology-climate modeling projects [including GEWEX Continental-Scale International Project (GCIP), Baltic Sea Experiment (BALTEX), and GEWEX Asian Monsoon Experiment (GAME); see www.gewex.com/projects.html]. The ultimate aim of such studies often is to lead to improved assessments of the hydrological effects of climate change through the use of coupled climate-hydrology models; thus far, however, the benefits to impact assessments have been indirect, through improvements to the parameterizations of climate models. A few studies have used coupled climate-hydrology models to forecast streamflow (e.g., Miller and Kim, 1996), and some have begun to use them to estimate effects of changing climate on streamflow (e.g., Miller and Kim, 2000).

4.2.3. Increased Awareness of the Effect of Climatic Variability on Hydrology and Water Resources

Since the SAR, many studies have explored linkages between recognizable patterns of climatic variability—particularly El Niño and the North Atlantic Oscillation—and hydrological behavior, in an attempt to explain variations in hydrological characteristics over time. These studies in North America (McCabe, 1996; Piechota *et al.*, 1997; Vogel *et al.*, 1997; Olsen *et al.*, 1999), South America (Marengo, 1995; Compagnucci and Vargas, 1998), Australasia (Chiew *et al.*, 1998), Europe (e.g., Shorthouse and Arnell, 1997), and southern Africa (Shulze, 1997) have emphasized variability not just from year to year but also from decade to decade, although patterns of variability vary considerably from region to region. Most studies focus on the past few decades with recorded hydrological data, but an increasing number of studies have reconstructed considerably longer records from various proxy data sources (e.g., Isdale *et al.*, 1998; Cleaveland, 2000). Such research is extremely valuable because it helps in interpretation of observed hydrological changes over time (particularly attribution of change to global warming), provides a context for assessment of future change, and opens up possibilities for seasonal flow prediction (e.g., Piechota *et al.*, 1998) hence more efficient adaptation to climatic variability. It also emphasizes that the hydrological "baseline" cannot be assumed to be constant, even in the absence of climate change.

4.2.4. Adaptation to Climate Change in the Water Sector

Water management is based on minimization of risk and adaptation to changing circumstances (usually taking the form of altered demands). A wide range of adaptation techniques has been developed and applied in the water sector over decades. One widely used classification distinguishes between increasing capacity (e.g., building reservoirs or structural flood defenses), changing operating rules for existing structures and systems, managing demand, and changing institutional practices. The first two often are termed "supply-side" strategies, whereas the latter two are "demand-side." Over the past few years, there has been a considerable increase in interest in demand-side techniques. International agencies such as the World Bank (World Bank, 1993) and initiatives such as the Global Water Partnership are promoting new ways of managing and pricing water resources to manage resources more effectively (Kindler, 2000).

This work is going on largely independently of climate change, but changes in water management practices will have a very significant impact on how climate change affects the water sector. Water managers in some countries are beginning to consider climate change explicitly, although the methodologies for doing so are not yet well defined and vary between and within countries depending on the institutional arrangements for long-term water resources planning. In the UK, for example, water supply companies were required by regulators in 1997 to "consider" climate change in estimating their future resource, hence investment, projections (Subak, 2000). In the United States, the American Water Works Association urged water agencies to explore the vulnerability of their systems to plausible climate changes (AWWA, 1997).

Clearly, however, the ability of water management agencies to alter management practices in general or to incorporate climate change varies considerably between countries. This issue is discussed further in Section 4.6.

4.3. Effects on the Hydrological Cycle

4.3.1. Introduction

This section summarizes the potential effects of climate change on the components of the water balance and their variability over time.

4.3.2. Precipitation

Precipitation is the main driver of variability in the water balance over space and time, and changes in precipitation have very important implications for hydrology and water resources. Hydrological variability over time in a catchment is influenced by variations in precipitation over daily, seasonal, annual, and decadal time scales. Flood frequency is affected by changes in the year-to-year variability in precipitation and by changes in short-term rainfall properties (such as storm rainfall intensity). The frequency of low or drought flows is affected primarily by changes in the seasonal distribution of precipitation, year-to-year variability, and the occurrence of prolonged droughts.

TAR WGI Section 2.5 summarizes studies into trends in precipitation. There are different trends in different parts of the world, with a general increase in Northern Hemisphere mid- and high latitudes (particularly in autumn and winter) and a decrease in the tropics and subtropics in both hemispheres. There is evidence that the frequency of extreme rainfall has increased in the United States (Karl and Knight, 1998) and in the UK (Osborn *et al.*, 2000); in both countries, a greater proportion of precipitation is falling in large events than in earlier decades.

Current climate models simulate a climate change-induced increase in annual precipitation in high and mid-latitudes and most equatorial regions but a general decrease in the subtropics (Carter and Hulme, 1999), although across large parts of the world the changes associated with global warming are small compared to those resulting from natural multi-decadal variability, even by the 2080s. Changes in seasonal precipitation are even more spatially variable and depend on changes in the climatology of a region. In general, the largest percentage precipitation changes *over land* are found in high latitudes, some equatorial regions, and southeast Asia, although there are large differences between climate models.

Until recently, very few projections of possible changes in year-to-year variability as simulated by climate models have been published, reflecting both the (until recently) short model runs available and the recognition that climate models do not necessarily reproduce observed patterns of climatic variability. Recent developments, however, include the increasing ability of some global climate models to reproduce features such as El Niño (e.g., Meehl and Washington, 1996) and open up the possibility that it may be feasible to estimate changes in year-to-year variability. Recent scenarios for the UK, derived from HadCM2 experiments, indicate an increase in the relative variability of seasonal and annual rainfall totals resulting from global warming (Hulme and Jenkins, 1998).

Potential changes in intense rainfall frequency are difficult to infer from global climate models, largely because of coarse spatial resolution. However, there are indications (e.g., Hennessy *et al.*, 1997; McGuffie *et al.*, 1999) that the frequency of heavy rainfall events generally is likely to increase with global warming. Confidence in this assertion depends on the confidence with which global climate models are held. More generally, uncertainty in GCM projections of precipitation largely determines the uncertainty in estimated impacts on hydrological systems and water resources.

Increasing temperatures mean that a smaller proportion of precipitation may fall as snow. In areas where snowfall currently is marginal, snow may cease to occur—with consequent, very significant, implications (discussed below) for hydrological regimes. This projection is considerably less uncertain than possible changes in the magnitude of precipitation.

4.3.3. Evaporation

Evaporation from the land surface includes evaporation from open water, soil, shallow groundwater, and water stored on vegetation, along with transpiration through plants. The rate of evaporation from the land surface is driven essentially by meteorological controls, mediated by the characteristics of vegetation and soils, and constrained by the amount of water available. Climate change has the potential to affect all of these factors—in a combined way that is not yet clearly understood—with different components of evaporation affected differently.

The primary meteorological controls on evaporation from a well-watered surface (often known as potential evaporation) are the amount of energy available (characterized by net radiation), the moisture content of the air (humidity—a function of water vapor content and air temperature), and the rate of movement of air across the surface (a function of windspeed). Increasing temperature generally results in an increase in potential evaporation, largely because the water-holding capacity of air is increased. Changes in other meteorological controls may exaggerate or offset the rise in temperature, and it is possible that increased water vapor content and lower net radiation could lead to lower evaporative demands. The relative importance of different meteorological controls, however, varies geographically. In dry regions, for example, potential evaporation is driven by energy and is not constrained by atmospheric moisture contents, so changes in humidity are relatively unimportant. In humid regions, however, atmospheric moisture content is a major limitation to evaporation, so changes in humidity have a very large effect on the rate of evaporation.

Several studies have assessed the effect of changes in meteorological controls on evaporation (e.g., Chattopadhyary and Hulme, 1997), using models of the evaporation process, and the effect of climate change has been shown to depend on baseline climate (and the relative importance of the different controls) and the amount of change. Chattopadhyary and Hulme (1997) calculated increases in potential evaporation across India from GCM simulations of climate; they found that projected increases in potential evaporation were related largely to increases in the vapor pressure deficit resulting from higher temperature. It is important to emphasize, however, that different evaporation calculation equations give different estimates of absolute evaporation rates and sensitivity to change. Therefore, it can be very difficult to compare results from different studies. Equations that do not consider explicitly *all* meteorological controls may give very misleading estimates of change.

Vegetation cover, type, and properties play a very important role in evaporation. Interception of precipitation is very much influenced by vegetation type (as indexed by the canopy storage capacity), and different vegetation types have different rates of transpiration. Moreover, different vegetation types produce different amounts of turblence above the canopy; the greater the turbulence, the greater the evaporation. A change in catchment vegetation—directly or indirectly as a result of climate change—therefore may affect the catchment water balance (there is a huge hydrological literature on the effects of changing catchment vegetation). Several studies have assessed changes in biome type under climate change (e.g., Friend *et al.*, 1997), but the hydrological effects of such changes—and, indeed, changes in agricultural land use—have not yet been explored.

Although transpiration from plants through their stomata is driven by energy, atmospheric moisture, and turbulence, plants exert a degree of control over transpiration, particularly when water is limiting. Stomatal conductance in many plants falls as the vapor pressure deficit close to the leaf increases, temperature rises, or less water becomes available to the roots—and transpiration therefore falls. Superimposed on this short-term variation in stomatal conductance is the effect of atmospheric carbon dioxide (CO_2) concentrations. Increased CO_2 concentrations reduce stomatal conductance in C_3 plants (which include virtually all woody plants and temperate grasses and crops), although experimental studies show that the effects vary considerably between species and depend on nutrient and water status. Plant water-use efficiency (WUE, or water use per unit of biomass) therefore may increase substantially (Morison, 1987), implying a reduction in transpiration. However, higher CO_2 concentrations also may be associated with increased plant growth, compensating for increased WUE, and plants also may acclimatize to higher CO_2 concentrations. There have been considerably fewer studies into total plant water use than into stomatal conductance, and most empirical evidence to date is at the plant scale; it is difficult to generalize to the catchment or regional scale (Field *et al.*, 1995; Gifford *et al.*, 1996; Amthor, 1999). Free-air CO_2 enrichment (FACE) experiments, however, have allowed extrapolation at least to the 20-m plot scale. Experiments with cotton, for example (Hunsaker *et al.*, 1994), showed no detectable change in water use per unit land area when CO_2 concentrations were increased to 550 ppmv; the 40% increase in biomass offset increased WUE. Experiments with wheat, however, indicated that increased growth did not offset increased WUE, and evaporation declined by approximately 7% (although still less than implied by the change in stomatal conductance; Kimball *et al.*, 1999). Some model studies (e.g., Field *et al.*, 1995, for forest; Bunce *et al.*, 1997, for alfalfa and grass; Cao and Woodward, 1998, at the global scale) suggest that the net direct effect of increased CO_2 concentrations at the catchment scale will be small (Korner, 1996), but others (e.g., Pollard and Thompson, 1995; Dickinson *et al.*, 1997; Sellers *et al.*, 1997; Raupach, 1998, as discussed by Kimball *et al.*, 1999) indicate that stomata have more control on regional evaporation. There clearly is a large degree of uncertainty over the effects of CO_2 enrichment on catchment-scale evaporation, but it is apparent that reductions in stomatal conductance do not necessarily translate into reductions in catchment-scale evaporation.

The *actual* rate of evaporation is constrained by water availability. A reduction in summer soil water, for example, could lead to a reduction in the rate of evaporation from a catchment despite an increase in evaporative demands. Arnell (1996) estimated for a sample of UK catchments that the rate of actual evaporation would increase by a smaller percentage than the atmospheric demand for evaporation, with the greatest difference in the "driest" catchments, where water limitations are greatest.

4.3.4. Soil Moisture

The amount of water stored in the soil is fundamentally important to agriculture and is an influence on the rate of actual evaporation, groundwater recharge, and generation of runoff. Soil moisture contents are directly simulated by global climate models, albeit over a very coarse spatial resolution, and outputs from these models give an indication of possible directions of change. Gregory *et al.* (1997), for example, show with the HadCM2 climate model that a rise in greenhouse gas (GHG) concentrations is associated with reduced soil moisture in Northern Hemisphere mid-latitude summers. This was the result of higher winter and spring evaporation, caused by higher temperatures and reduced snow cover, and lower rainfall inputs during summer.

The local effects of climate change on soil moisture, however, will vary not only with the degree of climate change but also with soil characteristics. The water-holding capacity of soil will affect possible changes in soil moisture deficits; the lower the capacity, the greater the sensitivity to climate change. Climate change also may affect soil characteristics, perhaps through changes in waterlogging or cracking, which in turn may affect soil moisture storage properties. Infiltration capacity and water-holding capacity of many soils are influenced by the frequency and intensity of freezing. Boix-Fayos *et al.* (1998), for example, show that infiltration and water-holding capacity of soils on limestone are greater with increased frost activity and infer that increased temperatures could lead to increased surface or shallow runoff. Komescu *et al.* (1998) assess the implications of climate change for soil moisture availability in southeast Turkey, finding substantial reductions in availability during summer.

4.3.5. Groundwater Recharge and Resources

Groundwater is the major source of water across much of the world, particularly in rural areas in arid and semi-arid regions, but there has been very little research on the potential effects of climate change. This section therefore can be regarded as presenting a series of hypotheses.

Aquifers generally are replenished by effective rainfall, rivers, and lakes. This water may reach the aquifer rapidly, through macro-pores or fissures, or more slowly by infiltrating through soils and permeable rocks overlying the aquifer. A change in the amount of effective rainfall will alter recharge, but so will a change in the duration of the recharge season. Increased winter rainfall—as projected under most scenarios for mid-latitudes—generally is likely to result in increased groundwater recharge. However, higher evaporation may mean that soil deficits persist for longer and commence earlier, offsetting an increase in total effective rainfall. Various types of aquifer will be recharged differently. The main types are unconfined and confined aquifers. An unconfined aquifer is recharged directly by local rainfall, rivers, and lakes, and the rate of recharge will be influenced by the permeability of overlying rocks and soils. Some examples of the effect of climate change on recharge into unconfined aquifers have been described in France, Kenya, Tanzania, Texas, New York, and Caribbean islands. Bouraoui *et al.* (1999) simulated substantial reductions in groundwater recharge near Grenoble, France, almost entirely as a result of increases in evaporation during the recharge season. Macro-pore and fissure recharge is most common in porous and aggregated forest soils and less common in poorly structured soils. It also occurs where the underlying geology is highly fractured or is characterized by numerous sinkholes. Such recharge can be very important in some semi-arid areas (e.g., the Wajir region of Kenya; Mailu, 1993). In principle, "rapid" recharge can occur whenever it rains, so where recharge is dominated by this process it will be affected more by changes in rainfall amount than by the seasonal cycle of soil moisture variability. Sandstrom (1995) modeled recharge to an aquifer in central Tanzania and showed that a 15% reduction in rainfall—with no change in temperature—resulted in a 40–50% reduction in recharge; he infers that small changes in rainfall could lead to large changes in recharge and hence groundwater resources. Loaiciga *et al.* (1998) explored the effect of a range of climate change scenarios on groundwater levels in the Edwards Balcones Fault Zone aquifer in Texas, a heavily exploited aquifer largely fed by streamflow seepage. They show that, under six of the seven GCM-based scenarios used, groundwater levels and springflows would reduce substantially as a result of lower streamflow. However, they use 2xCO_2 scenarios that represent changes in temperature that are considerably greater than those projected even by the 2080s under current scenarios (Carter and Hulme, 1999), so the study considerably overstates the effect of climate change in the next few decades.

Shallow unconfined aquifers along floodplains, which are most common in semi-arid and arid environments, are recharged by seasonal streamflows and can be depleted directly by evaporation. Changes in recharge therefore will be determined by changes in the duration of flow of these streams—which may locally increase or decrease—and the permeability of the overlying beds, but increased evaporative demands would tend to lead to lower groundwater storage. In semi-arid areas of Kenya, flood aquifers have been improved by construction of subsurface weirs across the river valleys, forming subsurface dams from which water is tapped by shallow wells. The thick layer of sands substantially reduces the impact of evaporation. The wells have become perennial water supply sources even during the prolonged droughts (Mailu, 1988, 1992).

Sea-level rise will cause saline intrusion into coastal aquifers, with the amount of intrusion depending on local groundwater gradients. Shallow coastal aquifers are at greatest risk (on Long Island, New York, for example). Groundwater in low-lying islands therefore is very sensitive to change. In the atolls of the Pacific Ocean, water supply is sensitive to precipitation patterns and changes in storm tracks (Salinger *et al.*, 1995). A reduction in precipitation coupled with sea-level rise would not only cause a diminution of the harvestable volume of water; it also would reduce the size of the narrow freshwater lense (Amadore *et al*, 1996). For many small island states, such as some Caribbean islands, seawater intrusion into freshwater aquifers has been observed as a result of overpumping of aquifers. Any sea-level rise would worsen the situation.

It will be noted from the foregoing that unconfined aquifers are sensitive to local climate change, abstraction, and seawater intrusion. However, quantification of recharge is complicated by the characteristics of the aquifers themselves as well as overlying rocks and soils.

A confined aquifer, on the other hand, is characterized by an overlying bed that is impermeable, and local rainfall does not influence the aquifer. It is normally recharged from lakes, rivers, and rainfall that may occur at distances ranging from a few kilometers to thousands of kilometers. Recharge rates also vary from a few days to decades. The Bahariya Oasis and other groundwater aquifers in the Egyptian Desert, for example, are recharged at the Nubian Sandstone outcrops in Sudan; such aquifers may not be seriously affected by seasonal or interannual rainfall or temperature of the local area.

Attempts have been made to calculate the rate of recharge by using carbon-14 isotopes and other modeling techniques. This has been possible for aquifers that are recharged from short distances and after short durations. However, recharge that takes place from long distances and after decades or centuries has been problematic to calculate with accuracy, making estimation of the impacts of climate change difficult. The medium through which recharge takes place often is poorly known and very heterogeneous, again challenging recharge modeling. In general, there is a need to intensify research on modeling techniques, aquifer characteristics, recharge rates, and seawater intrusion, as well as monitoring of groundwater abstractions. This research will provide a sound basis for assessment of the impacts of climate change and sea-level rise on recharge and groundwater resources.

4.3.6. River Flows

By far the greatest number of hydrological studies into the effects of climate change have concentrated on potential changes on streamflow and runoff. The distinction between "streamflow" and "runoff" can be vague, but in general terms streamflow is water within a river channel, usually expressed as a rate of flow past a point—typically in $m^3 s^{-1}$—whereas runoff is the amount of precipitation that does not evaporate, usually expressed as an equivalent depth of water across the area of the catchment. A simple link between the two is that runoff can be regarded as streamflow divided by catchment area, although in dry areas this does not necessarily hold because runoff generated in one part of the catchment may infiltrate before reaching a channel and becoming streamflow. Over short durations, the amount of water leaving a catchment outlet usually is expressed as streamflow; over durations of a month or more, it usually is expressed as runoff. In some countries, "runoff" implies surface runoff only (or, more precisely, rapid response to an input of precipitation) and does not include the contribution of discharge from groundwater to flow, but this is a narrow definition of the term.

This section first considers recent trends in streamflow/runoff and then summarizes research into the potential effects of future climate change.

4.3.6.1. *Trends in Observed Streamflow*

Since the SAR, there have been many notable hydrological events—including floods and droughts—and therefore many studies into possible trends in hydrological data. Table 4-1 summarizes some of these studies and their main results.

In general, the patterns found are consistent with those identified for precipitation: Runoff tends to increase where precipitation has increased and decrease where it has fallen over the past few years. Flows have increased in recent years in many parts of the United States, for example, with the greatest increases in low flows (Lins and Slack, 1999). Variations in flow from year to year have been found to be much more strongly related to precipitation changes than to temperature changes (e.g., Krasovskaia, 1995; Risbey and Entekhabi, 1996). There are some more subtle patterns, however. In large parts of eastern Europe, European Russia, central Canada (Westmacott and Burn, 1997), and California (Dettinger and Cayan, 1995), a major—and unprecedented—shift in streamflow from spring to winter has been associated not only with a change in precipitation totals but more particularly with a rise in temperature: Precipitation has fallen as rain, rather than snow, and therefore has reached rivers more rapidly than before. In cold regions, such as northern Siberia and northern Canada, a recent increase in temperature has had little effect on flow timing because precipitation continues to fall as snow (Shiklomanov, 1994; Shiklomanov *et al.*, 2000).

However, it is very difficult to identify trends in hydrological data, for several reasons. Records tend to be short, and many data sets come from catchments with a long history of human intervention. Variability over time in hydrological behavior is very high, particularly in drier environments, and detection of any signal is difficult. Variability arising from low-frequency climatic rhythms is increasingly recognized (Section 4.2), and researchers looking for trends need to correct for these patterns. Finally, land-use and other changes are continuing in many catchments, with effects that may outweigh any climatic trends. Changnon and Demissie (1996), for example, show that human-induced changes mask the effects of climatic variability in a sample of midwest U.S. catchments. Even if a trend is identified, it may be difficult to attribute it to global warming because of other changes that are continuing in a catchment. A widespread lack of data, particularly from many developing

***Table 4-1**: Recent studies into trends in river flows.*

Study Area	Data Set	Key Conclusions	Reference(s)
Global	– 161 gauges in 108 major world rivers, data to 1990	– Reducing trend in Sahel region but weak increasing trend in western Europe and North America; increasing relative variability from year to year in several arid and semi-arid regions	– Yoshino (1999)
Russia			
– European Russia and western Siberia	– 80 major basins, records from 60 to 110 years	– Increase in winter, summer, and autumn runoff since mid-1970s; decrease in spring flows	– Georgiyevsky *et al.* (1995, 1996, 1997); Shiklomanov and Georgiyevsky (2001)
– European former Soviet Union	– 196 small basins, records up to 60 years	– Increase in winter, summer, and autumn runoff since mid-1970s; decrease in spring flows	– Georgiyevsky *et al.* (1996)
Baltic Region			
– Scandinavia		– Increase in winter, summer and autumn runoff since mid-1970s; decrease in spring flows	– Bergstrom and Carlsson (1993)
– Baltic states		– Increase in winter, summer and autumn runoff since mid-1970s; decrease in spring flows	– Tarend (1998)
Cold Regions			
– Yenesei, Siberia	– Major river basin	– Little change in runoff or timing	– Shiklomanov (1994)
– Mackenzie, Canada	– Major river basin	– Little change in runoff or timing	– Shiklomanov *et al.* (2000)
North America			
– United States	– 206 catchments	– 26 catchments with significant trends: half increasing and half decreasing	– Lins and Slack (1999)
– California	– Major river basins	– Increasing concentration of streamflow in winter as a result of reduction in snow	– Dettinger and Cayan (1995); Gleick and Chalecki (1999)
– Mississippi basin	– Flood flows in major basins	– Large and significant increases in flood magnitudes at many gauges	– Olsen *et al.* (1999)
– West-central Canada	– Churchill-Nelson river basin	– Snowmelt peaks earlier; decreasing runoff in south of region, increase in north	– Westmacott and Burn (1997)
South America			
– Colombia	– Major river basins	– Decrease since 1970s	– Marengo (1995)
– Northwest Amazon	– Major river basins	– Increase since 1970s	– Marengo *et al.* (1998)
– SE South America	– Major river basins	– Increase since 1960s	– Genta *et al.* (1998)
– Andes	– Major river basins	– Increase north of 40°S, decrease to the south	– Waylen *et al.* (2000)
Europe			
– UK	– Flood flows in many basins	– No clear statistical trend	– Robson *et al.* (1998)
Africa			
– Sahelian region	– Major river basins	– Decrease since 1970s	– Sircoulon (1990)
Asia			
– Xinjiang region, China	– Major river basins	– Spring runoff increase since 1980 from glacier melt	– Ye *et al.* (1999)
Australasia			
– Australia	– Major basins	– Decrease since mid-1970s	– Thomas and Bates (1997)

countries, and consistent data analysis makes it impossible to obtain a representative picture of recent patterns and trends in hydrological behavior. Monitoring stations are continuing to be closed in many countries. Reconstructions of long records, stretching back centuries, are needed to understand the characteristics of natural decadal-scale variability in streamflow.

4.3.6.2. *Effects of Climate Change on River Flows*

By far the majority of studies into the effects of climate change on river flows have used GCMs to define changes in climate that are applied to observed climate input data to create perturbed data series. These perturbed data are then fed through a hydrological model and the resulting changes in river flows assessed. Since the SAR, there have been several global-scale assessments and a large number of catchment-scale studies. Confidence in these results is largely a function of confidence in climate change scenarios *at the catchment scale*, although Boorman and Sefton (1997) show that the use of a physically unrealistic hydrological model could lead to misleading results.

Arnell (1999b) used a macro-scale hydrological model to simulate streamflow across the world at a spatial resolution of 0.5°x0.5°,

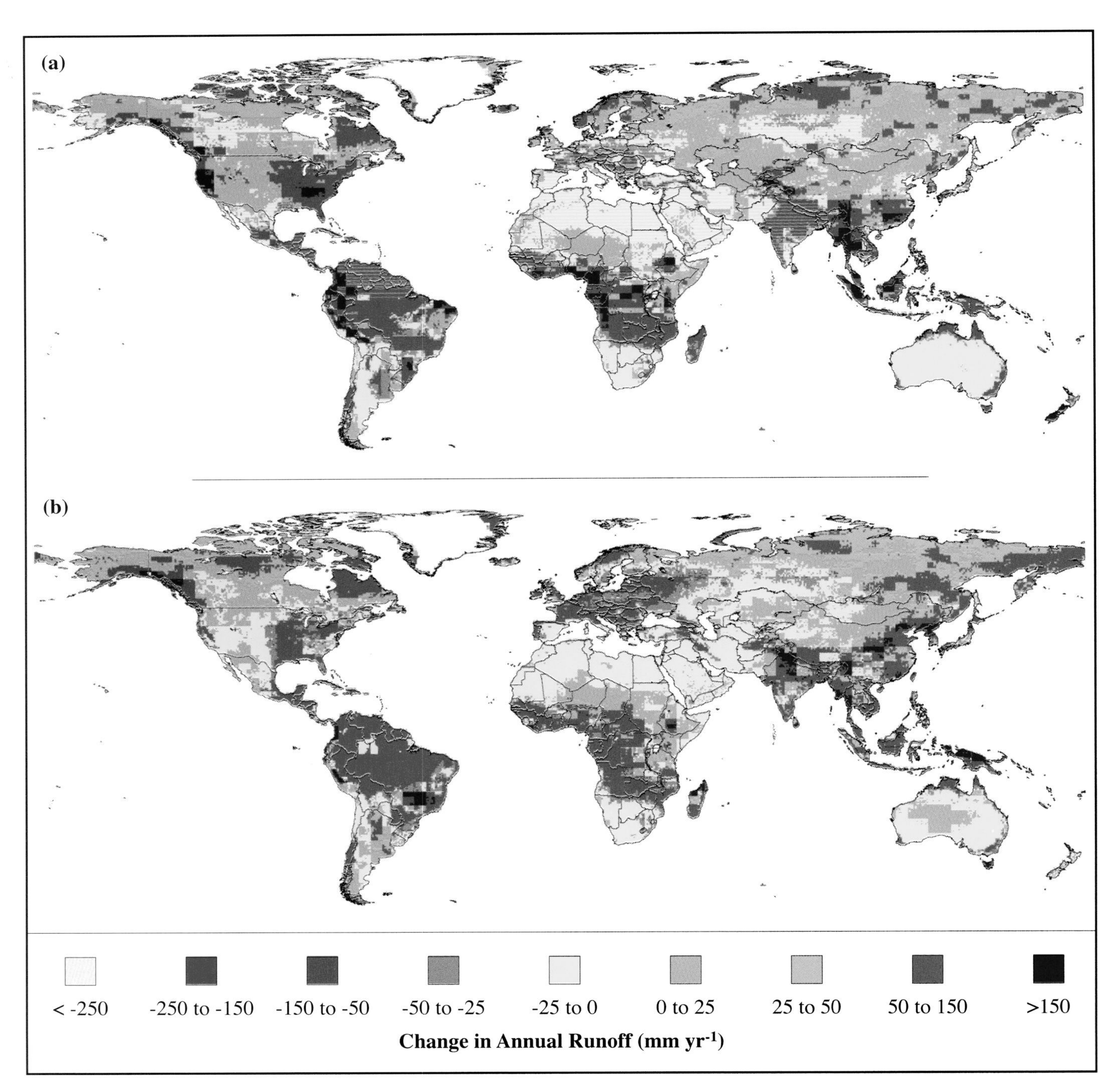

Figure 4-1: Change in average annual runoff by 2050 under HadCM2 ensemble mean (a) and HadCM3 (b) (Arnell, 1999b).

under the 1961–1990 baseline climate and under several scenarios derived from HadCM2 and HadCM3 experiments. Figure 4-1 shows the absolute change in annual runoff by the 2050s under the HadCM2 and HadCM3 scenarios: Both have an increase in effective CO_2 concentrations of 1% yr^{-1}. The patterns of change are broadly similar to the change in annual precipitation—increases in high latitudes and many equatorial regions but decreases in mid-latitudes and some subtropical regions—but the general increase in evaporation means that some areas that see an increase in precipitation will experience a reduction in runoff. Alcamo *et al.* (1997) also simulated the effects of different climate change scenarios on global river flows, showing broadly similar patterns to those in Figure 4-1.

Rather than assess each individual study, this section simply tabulates catchment-scale studies published since the SAR and draws some general conclusions. As in the SAR, the use of different scenarios hinders quantitative spatial comparisons. Table 4-2 summarizes the studies published since the SAR, by continent. All of the studies used a hydrological model to estimate the effects of climate scenarios, and all used scenarios based on GCM output. The table does not include sensitivity studies (showing the effects of, for example, increasing precipitation by 10%) or explore the hydrological implications of past climates. Although such studies provide extremely valuable insights into the sensitivity of hydrological systems to changes in climate, they are not assessments of the potential effects of future global warming.

It is clear from Table 4-2 that there are clear spatial variations in the numbers and types of studies undertaken to date; relatively few studies have been published in Africa, Latin America, and southeast Asia. A general conclusion, consistent across many studies, is that the effects of a given climate change scenario vary with catchment physical and land-cover properties and that small headwater streams may be particularly sensitive to change—as shown in northwestern Ontario, for example, by Schindler *et al.* (1996).

4.3.6.2.1. Cold and cool temperate climates

These areas are characterized by precipitation during winter falling as snow and include mountainous and low-lying regions. A major proportion of annual streamflow is formed by snow melting in spring. These areas include large parts of North America, northern and eastern Europe, most of Russia, northern China, and much of central Asia. The most important climate change effect in these regions is a change in the timing of streamflow through the year. A smaller proportion of precipitation during winter falls as snow, so there is proportionately more runoff in winter and, as there is less snow to melt, less runoff during spring. Increased temperatures, in effect, reduce the size of the natural reservoir storing water during winter. In very cold climates (such as in Siberia and northern Russia), there is little change in the timing of streamflow because winter precipitation continues to fall as snow with higher temperatures.

Table 4-2*: Catchment-scale studies since the Second Assessment Report addressing the effect of climate change on hydrological regimes.*

Region/Scope	Reference(s)
Africa	
– Ethiopia	Hailemariam (1999)
– Nile Basin	Conway and Hulme (1996); Strzepek *et al.* (1996)
– South Africa	Schulze (1997)
– Southern Africa	Hulme (1996)
Asia	
– China	Ying and Zhang (1996); Ying *et al.* (1997); Liu (1998); Shen and Liang (1998); Kang *et al.* (1999)
– Himalaya	Mirza and Dixit (1996); Singh and Kumar (1997); Singh (1998)
– Japan	Hanaki *et al.* (1998)
– Philippines	Jose *et al.* (1996); Jose and Cruz (1999)
– Yemen	Alderwish and Al-Eryani (1999)
Australasia	
– Australia	Bates *et al.* (1996); Schreider *et al.* (1996); Viney and Sivapalan (1996)
– New Zealand	Fowler (1999)
Europe	
– Albania	Bruci and Bicaj (1998)
– Austria	Behr (1998)
– Belgium	Gellens and Roulin (1998); Gellens *et al.* (1998)
– Continent	Arnell (1999a)
– Czech Republic	Hladny *et al.* (1996); Dvorak *et al.* (1997); Buchtele *et al.* (1998)
– Danube basin	Starosolszky and Gauzer (1998)
– Estonia	Jaagus (1998); Jarvet (1998); Roosare (1998)
– Finland	Lepisto and Kivinen (1996); Vehviläinen and Huttunen (1997)
– France	Mandelkern *et al* (1998)
– Germany	Daamen *et al.* (1998)
– Greece	Panagoulia and Dimou (1996)
– Hungary	Mika *et al.* (1997)
– Latvia	Butina *et al.* (1998); Jansons and Butina (1998)
– Nordic region	Saelthun *et al.* (1998)
– Poland	Kaczmarek et al. (1996, 1997)
– Rhine basin	Grabs (1997)
– Romania	Stanescu *et al.* (1998)
– Russia	Georgiyevsky *et al.*, (1995, 1996, 1997); Kuchment (1998); Shiklomanov (1998)
– Slovakia	Hlaveova and Eunderlik (1998); Petrovic (1998)

Table 4-2 (continued)

Region/Scope	Reference
Europe (continued)	
– Spain	Avila *et al.* (1996); Ayala-Carcedo (1996)
– Sweden	Xu (1998); Bergstrom *et al.* (2001)
– Switzerland	Seidel *et al.* (1998)
– UK	Arnell (1996); Holt and Jones (1996); Arnell and Reynard (1996, 2000); Sefton and Boorman (1997); Roberts (1998); Pilling and Jones (1999)
Latin America	
– Continent	Yates (1997); Braga and Molion (1999)
– Panama	Espinosa *et al.* (1997)
North America	
– USA	Bobba *et al.* (1997); Hanratty and Stefan (1998); Chao and Wood (1999); Hamlet and Lettenmaier (1999); Lettenmaier *et al.* (1999); Leung and Wigmosta (1999); Miller *et al.* (1999); Najjar (1999); Wolock and McCabe (1999); Miller and Kim (2000); Stonefelt *et al.* (2000)
– Mexico	Mendoza *et al.* (1997)

The largest effects are in the most "marginal" snow-dominated regime areas.

The effects of climate change on the magnitude of annual runoff and flows through the year are much less consistent than the effect on streamflow timing because they depend not on the temperature increase but on the change in precipitation. In general, precipitation increases in high-latitude areas under most scenarios, but in lower latitudes precipitation may decrease. Kazcmarek *et al.* (1997), for example, show a decrease in annual runoff in Poland under a Geophysical Fluid Dynamics Laboratory (GFDL)-based scenario (by around 20% by the 2050s) but an increase under a Goddard Institute for Space Studies (GISS) scenario (by as much as 20%); in both cases, the season of maximum flow shifts from spring to winter.

Similar patterns are found for rivers in mountainous regions or draining from mountains. The Rhine and Danube, for example, would both see a reduction in spring flows and an increase in winter runoff (Grabs, 1997; Starosolszky and Gauzer, 1998), as would rivers draining east and west from the Rocky Mountains in North America.

4.3.6.2.2. *Mild temperate climates*

Hydrological regimes in these regions are dominated by the seasonal cycles of rainfall and evaporation; snowfall and snowmelt are not important. Here, climate change tends to affect the magnitude of flows in different seasons—by an amount that depends on the change in rainfall—and may lead to an exaggerated seasonal cycle, but it generally does not affect the timing of flows through the year. In the UK, for example, most scenarios result in an increase in winter runoff and, particularly in the south, a decrease in summer runoff (Arnell and Reynard, 1996); similar patterns are found across most of western Europe under most scenarios (Arnell, 1999a). Low flows tend to occur during summer, and changes in low-flow frequency are closely related to changes in the balance between summer rainfall and summer evaporation. Across most mid-latitude temperate regions, summer rainfall would decline with global warming, leading to a reduction in low flows.

The detailed effect of a given change in climate, however, depends to a large extent on the geological characteristics of the catchment. Studies in the UK (Arnell and Reynard, 1996) and Belgium (Gellens and Roulin, 1998) have indicated that in catchments with considerable groundwater, changes in summer flows are largely a function of the change not in summer rainfall but in rainfall during the winter recharge season.

4.3.6.2.3. *Arid and semi-arid regions*

River flows in arid and semi-arid regions are very sensitive to changes in rainfall: A given percentage change in rainfall can produce a considerably larger percentage change in runoff. There have been relatively few studies in such regions since the SAR, but work has been done in southern Africa (Schulze, 1997), Australia (Bates *et al*., 1996), northern China (Ying *and* Huang, 1996), and southern Russia (Georgiyevsky *et al*., 1996; Shiklomanov, 1998).

4.3.6.2.4. *Humid tropical regions*

Runoff regimes in these regions are very much influenced by the timing and duration of the rainy season or seasons. Climate change therefore may affect river flows not only through a change in the magnitude of rainfall but also through possible changes in the onset or duration of rainy seasons (such as those caused by monsoon).

4.3.7. Lakes

Lakes are particularly vulnerable to changes in climate parameters. Variations in air temperature, precipitation, and other meteorological components directly cause changes in evaporation, water balance, lake level, ice events, hydrochemical and hydrobiological regimes, and the entire lake ecosystem. Under some climatic conditions, lakes may disappear entirely. There are many different types of lakes, classified according to lake formation and origin, the amount of water exchange, hydrochemistry, and so forth.

An important distinction is drawn between closed (endorheic) lakes, with no outflow, and exorheic lakes, which are drained by outflowing rivers. Endorheic lakes are very dependent on the balance of inflows and evaporation and are very sensitive to change in either (whether driven by climate change, climatic variability, or human interventions). This also means that they are very important indicators of climate change and can provide records of past hydroclimatic variability over a large area (e.g., Kilkus, 1998; Obolkin and Potemkin, 1998). Small endorheic lakes are most vulnerable to a change in climate; there are indications that even relatively small changes in inputs can produce large fluctuations in water level (and salinity) in small closed lakes in western North America (Laird *et al.*, 1996).

The largest endorheic lakes in the world are the Caspian and Aral Seas, Lake Balkash, Lake Chad, Lake Titicaca, and the Great Salt Lake. Some of the largest east African lakes, including Lakes Tanganyika and Malawi, also can be regarded as practically endorheic. Changes in inflows to such lakes can have very substantial effects: The Aral Sea, for example, has been significantly reduced by increased abstractions of irrigation water upstream, the Great Salt Lake in the United States has increased in size in recent years as a result of increased precipitation in its catchment, and Qinghai Lake in China has shrunk following a fall in catchment precipitation. Many endorheic lake systems include significant internal thresholds, beyond which change may be very different. Lake Balkash, for example, currently consists of a saline part and a fresh part, connected by a narrow strait. Several rivers discharge into the fresh part, preventing salinization of the entire lake. A reduction in freshwater inflows, however, would change the lake regime and possibly lead to salinization of the freshwater part; this would effectively destroy the major source of water for a large area.

Exorheic lakes also may be sensitive to changes in the amount of inflow and the volume of evaporation. Evidence from Lake Victoria (east Africa), for example, indicates that lake levels may be increased for several years following a short-duration increase in precipitation and inflows. There also may be significant thresholds involving rapid shifts from open to closed lake conditions. Progressive southward expansion of Lake Winnipeg under postglacial isostatic tilting was suppressed by a warm dry climate in the mid-Holocene, when the north basin of the lake became closed (endorheic) and the south basin was dry (Lewis *et al.*, 1998). A trend of progressively moister climates within the past 5,000 years caused a return from closed to open (overflowing) lake conditions in the north basin and rapid flooding of the south basin about 1,000 years later. Other examples include Lake Manitoba, which was dry during the warm mid-Holocene (Teller and Last, 1982). Computations of sustainable lake area under equilibrium water balance (after Bengtsson and Malm, 1997) indicate that a return to dry conditions comparable to the mid-Holocene climate could cause this 24,400-km^2 lake draining a vast area from the Rocky Mountains east almost to Lake Superior to become endorheic again (Lewis *et al.*, 1998).

Climate change also is likely to have an effect on lake water quality, through changes in water temperature and the extent and duration of ice cover. These effects are considered in Section 4.3.10.

4.3.8. Changes in Flood Frequency

Although a change in flood risk is frequently cited as one of the potential effects of climate change, relatively few studies since the early 1990s (e.g., Nash and Gleick, 1993; Jeton *et al.*, 1996) have looked explicitly at possible changes in high flows. This largely reflects difficulties in defining credible scenarios for change in the large rainfall (or snowmelt) events that trigger flooding. Global climate models currently cannot simulate with accuracy short-duration, high-intensity, localized heavy rainfall, and a change in mean monthly rainfall may not be representative of a change in short-duration rainfall.

A few studies, however, have tried to estimate possible changes in flood frequencies, largely by assuming that changes in monthly rainfall also apply to "flood-producing" rainfall. In addition, some have looked at the possible additional effects of changes in rainfall intensity. Reynard *et al.* (1998), for example, estimated the change in the magnitude of different return period floods in the Thames and Severn catchments, assuming first that all rainfall amounts change by the same proportion and then that only "heavy" rainfall increases. Table 4-3 summarizes the changes in flood magnitudes in the Thames and Severn by

***Table 4-3**: Percentage change in magnitude of peak floods in Severn and Thames catchments by the 2050s (Reynard et al., 1998).*

	Return Period				
Catchment	2-Year	5-Year	10-Year	20-Year	50-Year
Thames					
– GGx-x[a]	10	12	13	14	15
– GGx-s[b]	12	13	14	15	16
Severn					
– GGx-x[a]	13	15	16	17	20
– GGx-s[b]	15	17	18	19	21

[a] GGx-x = HadCM2 ensemble mean scenario with proportional change in rainfall.
[b] GGx-s = HadCM2 ensemble mean scenario with change in storm rainfall only.

the 2050s: Flood risk increases because winter rainfall increases, and in these relatively large catchments it is the total volume of rainfall over several days, not the peak intensity of rainfall, that is important. Schreider *et al.* (1996) in Australia assessed change in flood risk by assuming that all rainfall amounts change by the same proportion. They found an increase in flood magnitudes under their wettest scenarios—even though annual runoff totals did not increase—but a decline in flood frequency under their driest scenarios.

Panagoulia and Dimou (1997) examined possible changes in flood frequency in the Acheloos basin in central Greece. Floods in this catchment derive from snowmelt, and an increase in winter precipitation—as indicated under the scenarios used—results in more frequent flood events of longer duration. The frequency and duration of small floods was most affected. Saelthun *et al.* (1998) explored the effect of fixed increases in temperature and precipitation in 25 catchments in the Nordic region. They show that higher temperatures and higher precipitation increases flood magnitudes in parts of the region where floods tended to be generated from heavy rainfall in autumn but decrease flood magnitudes where floods are generated by spring snowmelt. In some cases, the peak flood season shifts from spring to autumn. This conclusion also is likely to apply in other environments where snow and rain floods both occur.

Mirza *et al.*(1998) investigated the effects of changes in precipitation resulting from global warming on future flooding in Bangladesh. Standardized precipitation change scenarios from four GCMs were used for the analysis. The most extreme scenario showed that for a 2°C rise in global mean temperature, the average flood discharge for the Ganges, Brahmaputra, and Meghna could be as much as 15, 6, and 19% higher, respectively.

4.3.9. Changes in Hydrological Drought Frequency

Droughts are considerably more difficult to define in quantitative terms than floods. Droughts may be expressed in terms of rainfall deficits, soil moisture deficits, lack of flow in a river, low groundwater levels, or low reservoir levels; different definitions are used in different sectors. A "hydrological" drought occurs when river or groundwater levels are low, and a "water resources" drought occurs when low river, groundwater, or reservoir levels impact water use. Low river flows in summer may not necessarily create a water resources drought, for example, if reservoirs are full after winter; conversely, a short-lived summer flood may not end a water resources drought caused by a prolonged lack of reservoir inflows. Water resources droughts therefore depend not only on the climatic and hydrological inputs but critically on the characteristics of the water resource system and how droughts are managed. This section focuses on hydrological drought, particularly on low river flows. Different studies have used different indices of low river flows, including the magnitude of minimum flows, the frequency at which flows fall below some threshold, the duration of flow below a threshold, and the cumulative difference between actual flows and some defined threshold.

At the global scale, Arnell (1999b) explored the change in the minimum annual total runoff with a return period of 10 years under several scenarios, based on HadCM2 and HadCM3 GCMs. He shows that the pattern of this measure of "low flow" (which is relatively crude) changes in a similar way to average annual runoff (as shown in Figure 4-1) but that the percentage changes tend to be larger. Arnell (1999a) mapped a different index of low flow across Europe—the average summed difference between streamflow and the flow exceeded 95% of the time, while flows are below this threshold—under four scenarios. The results suggest a reduction in the magnitude of low flows under most scenarios across much of western Europe, as a result of lower flows during summer, but an amelioration of low flows in the east because of increased winter flows. In these regions, however, the season of lowest flows tends to shift from the current winter low-flow season toward summer.

Döll *et al.* (1999) also modeled global runoff at a spatial resolution of 0.5°x0.5°, not only for average climatic conditions but also for typical dry years. The annual runoff exceeded in 9 years out of 10 (the 10-year return period "drought" runoff) was derived for each of more than 1,000 river basins covering the whole globe. Then the impact of climate change on these runoff values was computed by scaling observed temperature and precipitation in the 1-in-10 dry years with climate scenarios of two different GCMs (Chapter 3), ECHAM4/OPYC3 and GFDL-R15. Climate variability was assumed to remain constant. For the same GHG emission scenario, IS92a, the two GCMs compute quite different temperature and more so precipitation changes. With the GFDL scenario, runoff in 2025 and 2075 is simulated to be higher in most river basins than with the ECHAM scenario. The 1-in-10 dry year runoff is computed to decrease between the present time (1961–1990 climate) and 2075 by more than 10% on 19% (ECHAM) or 13% (GFDL) of the global land area (Table 4-4) and to increase by more than 50% on 22% (ECHAM) or 49% (GFDL) of the global land area. These results underline the high sensitivity of computed future runoff changes to GCM calculations.

Table 4-4*: Computed change of 1-in-10 dry year runoff under emission scenario IS92a between the present time (1961–90) and 2075: Influence of climate scenarios computed by two GCMs (Döll et al., 1999).*

Change in Runoff between Present and 2075 (%, decrease negative)	Fraction of Global Land Area, where Runoff will have Changed (%), using Climate Scenarios of	
	MPI	GFDL
Increase by more than 200%	8.4	14.4
+50 to +200	13.4	34.9
+10 to + 50	39.5	24.0
-10 to +10	19.9	14.0
-50 to -10	12.1	10.1
Decrease by more than 50%	6.7	2.5

There have been several other studies into changes in low flow indicators at the catchment scale. Gellens and Roulin (1998), for example, simulated changes in low flows in several Belgian catchments under a range of GCM-based scenarios. They show how the same scenario could produce rather different changes in different catchments, depending largely on the catchment geological conditions. Catchments with large amounts of groundwater storage tend to have higher summer flows under the climate change scenarios considered because additional winter rainfall tends to lead to greater groundwater recharge (the extra rainfall offsets the shorter recharge season). Low flows in catchments with little storage tend to be reduced because these catchments do not feel the benefits of increased winter recharge. Arnell and Reynard (1996) found similar results in the UK. The effect of climate change on low flow magnitudes and frequency therefore can be considered to be very significantly affected by catchment geology (and, indeed, storage capacity in general). Dvorak *et al.* (1997) also showed how changes in low flow measures tend to be proportionately greater than changes in annual, seasonal, or monthly flows.

4.3.10. Water Quality

Water in rivers, aquifers, and lakes naturally contains many dissolved materials, depending on atmospheric inputs, geological conditions, and climate. These materials define the water's chemical characteristics. Its biological characteristics are defined by the flora and fauna within the water body, and temperature, sediment load, and color are important physical characteristics. Water "quality" is a function of chemical, physical, and biological characteristics but is a value-laden term because it implies quality in relation to some standard. Different uses of water have different standards. Pollution can be broadly defined as deterioration of some aspect of the chemical, physical, or biological characteristics of water (its "quality") to such an extent that it impacts some use of that water or ecosystems within the water. Major water pollutants include organic material, which causes oxygen deficiency in water bodies; nutrients, which cause excessive growth of algae in lakes and coastal areas—known as eutrophication (leading to algal blooms, which may be toxic and consume large amounts of oxygen when decaying); and toxic heavy metals and organic compounds. The severity of water pollution is governed by the intensity of pollutants and the assimilation capacity of receiving water bodies—which depends on the physical, chemical, and biological characteristics of streamflow—but not all pollutants can be degraded, however.

Chemical river water quality is a function of the chemical load applied to the river, water temperature, and the volume of flow. The load is determined by catchment geological and land-use characteristics, as well as by human activities in the catchment: Agriculture, industry, and public water use also may result in the input of "polluting" substances. Agricultural inputs are most likely to be affected by climate change because a changing climate might alter agricultural practices. A changing climate also may alter chemical processes in the soil, including chemical weathering (White and Blum, 1995). Avila *et al.* (1996) simulated a substantial increase in base cation weathering rates in Spain when temperature and precipitation increased (although if precipitation were reduced, the effects of the higher temperature were offset). This, in turn, resulted in an increase in concentrations of base cations such as calcium, sodium, and potassium and an increase in streamwater alkalinity. Warmer, drier conditions, for example, promote mineralization of organic nitrogen (Murdoch *et al.*, 2000) and thus increase the potential supply to the river or groundwater. Load also is influenced by the processes by which water reaches the river channel. Nitrates, for example, frequently are flushed into rivers in intense storms following prolonged dry periods.

River water temperature depends not only on atmospheric temperature but also on wind and solar radiation (Orlob *et al.*, 1996). River water temperature will increase by a slightly lesser amount than air temperature (Pilgrim *et al.*, 1998), with the smallest increases in catchments with large contributions from groundwater. Biological and chemical processes in river water are dependent on water temperature: Higher temperatures alone would lead to increases in concentrations of some chemical species but decreases in others. Dissolved oxygen concentrations are lower in warmer water, and higher temperatures also would encourage the growth of algal blooms, which consume oxygen on decomposition.

Streamwater quality, however, also will be affected by streamflow volumes, affecting both concentrations and total loads. Carmichael *et al.* (1996), for example, show how higher temperatures and lower summer flows could combine in the Nitra River, Slovakia, to produce substantial reductions in dissolved oxygen concentrations. Research in Finland (Frisk *et al.*, 1997; Kallio *et al.*, 1997) indicates that changes in stream water quality, in terms of eutrophication and nutrient transport, are very dependent on changes in streamflow. For a given level of inputs, a reduction in streamflow might lead to increases in peak concentrations of certain chemical compounds. Cruise *et al.* (1999) simulated increased concentrations of nitrate in the southeast United States, for example, but the total amount transported from a catchment might decrease. Hanratty and Stefan (1998) simulated reductions in nitrate and phosphate loads in a small Minnesota catchment, largely as a result of reductions in runoff. Alexander *et al.* (1996) suggest that nutrient loadings to receiving coastal zones would vary primarily with streamflow volume. Increased streamflow draining toward the Atlantic coast of the United States under many scenarios, for example, would lead to increased nutrient loadings. An increased frequency of heavy rainfall would adversely affect water quality by increasing pollutant loads flushed into rivers and possibly by causing overflows of sewers and waste storage facilities. Polluting material also may be washed into rivers and lakes following inundation of waste sites and other facilities located on floodplains.

Water temperature in lakes responds to climate change in more complicated ways because thermal stratification is formed in summer, as well as in colder regions in winter. Meyer *et al.* (1999) evaluated the effect of climate change on thermal stratification by simulation for hypothetical lakes. They show that lakes in subtropic zones (about latitude 30 to 45°) and in subpolar zones (latitude 65 to 80°) are subject to greater relative changes in thermal stratification patterns than mid-latitude or equatorial lakes and that deep lakes are more sensitive than shallow lakes in the subtropic zones. Hostetler and Small (1999) simulated potential impacts on hypothetical shallow and deep lakes across North America, showing widespread increases in lake water temperature slightly below the increase in air temperature in the scenarios used. The greatest increases were in lakes that were simulated to experience substantial reductions in the duration of ice cover; the boundary of ice-free conditions shifted northward by 10° of latitude or more (1,000 km). Fang and Stefan (1997) show by simulation that winter stratification in cold regions would be weakened and the anoxic zone would disappear. Observations during droughts in the boreal region of northwestern Ontario show that lower inflows and higher temperatures produce a deepening of the thermocline (Schindler *et al.*, 1996).

The consequences of these direct changes to water quality of polluted water bodies may be profound, as summarized by Varis and Somlyody (1996) for lakes. Increases in temperature would deteriorate water quality in most polluted water bodies by increasing oxygen-consuming biological activities and decreasing the saturation concentration of dissolved oxygen. Hassan *et al.* (1998a,b) employed a downscaled climate model combined with GCM output to predict future stratification for Suwa Lake, Japan, on a daily basis, as well as for the prolonged summer stratification period. They predict increased growth of phytoplankton and reduced dissolved oxygen concentrations at different depths in the lake. Analysis of past observations in Lake Biwa in Japan (Fushimi, 1999) suggests that dissolved oxygen concentrations also tend to reduce when air (and lake water) temperature is higher.

Water quality in many rivers, lakes, and aquifers, however, is heavily dependent on direct and indirect human activities. Land-use and agricultural practices have a very significant effect on water quality, as do management actions to control point and nonpoint source pollution and treat wastewaters discharged into the environment. In such water bodies, future water quality will be very dependent on future human activities, including water management policies, and the direct effect of climate change may be very small in relative terms (Hanratty and Stefan, 1998). Considerable effort is being expended in developed and developing countries to improve water quality (Sections 4.5 and 4.6), and these efforts will have very significant implications for the impact of climate change on water quality.

Confidence in estimates of change in water quality is determined partly by climate change scenarios (and their effects on streamflow), but additional uncertainty is added by current lack of detailed understanding of some of the process interactions involved.

4.3.11. Glaciers and Small Ice Caps

Valley glaciers and small ice caps represent storages of water over long time scales. Many rivers are supported by glacier melt, which maintains flows through the summer season. The state of a glacier is characterized by the relationship between the rate of accumulation of ice (from winter snowfall) and the rate of ablation or melt. Most, but not all, valley glaciers and small ice caps have been in general retreat since the end of the Little Ice Age, between 100 and 300 years ago—for example, in Switzerland (Greene *et al.*, 1999), Alaska (Rabus and Echelmeyer, 1998), the Canadian Rockies (Schindler, 2001), east Africa (Kaser and Noggler, 1991), South America (Ames and Hastenrath, 1996; see also Chapter 14), the arid region of northwest China (Liu *et al.*, 1999), and tropical areas as a whole (Kaser, 1999). Temperature appears to be the primary control (Greene *et al.*, 1999), and rates of retreat generally are accelerating (Haeberli *et al.*, 1999). The World Glacier Monitoring Service (see http://www.geo.unizh.ch/wgms) monitors glacier mass balances and publishes annual reports on glacier fluctuations.

The effect of future climate change on valley glaciers and small ice caps depends on the extent to which higher temperatures are offset by increased winter accumulation. At the global scale, Gregory and Oerlemans (1998) simulate a general decline in valley glacier mass (and consequent rise in sea level), indicating that the effects of higher temperatures generally are more significant than those of additional winter accumulation. Model studies of individual glaciers have shown general retreat with global warming. Wallinga and van de Wal (1998) and Haerberli and Beniston (1998), for example, both simulated retreat in Alpine glaciers with higher temperatures and changes in winter accumulation. Davidovich and Ananicheva's (1996) simulation results show retreat of Alaskan glaciers but also a substantial increase in mass exchange (and therefore rate of movement) as a result of increased winter accumulation.

Oerlemans *et al.* (1998) simulated the mass balance of 12 valley glaciers and small ice sheets distributed across the world. They found that most scenarios result in retreat (again showing that temperature changes are more important than precipitation changes) but showed that it was very difficult to generalize results because the rate of change depends very much on glacier hypsometry (i.e., variation in altitude across the glacier). Their simulations also show that, in the absence of a change in precipitation, a rise in temperature of 0.4°C per decade would virtually eliminate all of their study glaciers by 2100, but a rise of 0.1°C per decade would "only" lead to a reduction in glacier volume of 10–20%.

Tropical glaciers are particularly exposed to global warming. Kaser *et al.* (1996) show that the equilibrium line altitude (ELA)—the line separating the accumulation zone from the ablation zone—of a tropical glacier is relatively more sensitive to changes in air temperature than that of a mid-latitude glacier. This is because of the lack of seasonality in tropical temperatures and the fact that ablation is significant year-round. To illustrate, a 1°C rise in temperature during half of the year only will have a direct impact on total ablation, annual mass balance, and ELA of a tropical glacier. In the case of a mid-latitude glacier, this increase may occur during winter when temperatures may be well below freezing over much (if not all) of the glacier. As a result, there may be no significant change in ablation or position of the ELA, even though the annual temperature will have increased.

Glacier retreat has implications for downstream river flows. In rivers fed by glaciers, summer flows are supported by glacier melt (with the glacier contribution depending on the size of the glacier relative to basin area, as well as the rate of annual melt). If the glacier is in equilibrium, the amount of precipitation stored in winter is matched by melt during summer. However, as the glacier melts as a result of global warming, flows would be expected to increase during summer—as water is released from long-term storage—which may compensate for a reduction in precipitation. As the glacier gets smaller and the volume of melt reduces, summer flows will no longer be supported and will decline to below present levels. The duration of the period of increased flows will depend on glacier size and the rate at which the glacier melts; the smaller the glacier, the shorter lived the increase in flows and the sooner the onset of the reduction in summer flows.

4.3.12. River Channel Form and Stability

Patterns of river channel erosion and sedimentation are determined largely by variations in streamflow over time—in particular, the frequency of floods. There is considerable literature on past changes in streamflow—caused by human influences or natural climatic variability—and associated river channel changes (Rumsby and Mackin, 1994) but very little on possible future channel changes. This largely reflects a lack of numerical models to simulate erosion and sedimentation processes; assessments of possible future channel changes that have been made have been inferred from past changes. In northern England, for example, Rumsby and Mackin (1994) show that periods with large numbers of large floods are characterized by channel incision, whereas periods with few floods were characterized by lateral reworking and sediment transfer. Increased flooding in the future therefore could be associated with increased channel erosion.

The density of the drainage network reflects the signature of climate on topography. Moglen *et al.* (1998) show that drainage density is sensitive to climate change but also that the direction of change in density depends not only on climate change but also on the current climate regime.

Hanratty and Stefan (1998) simulated streamflow and sediment yield in a small catchment in Minnesota. The scenario they used produced a reduction in sediment yield, largely as a result of reduced soil erosion, but their confidence in the model results was low. In fact, the lack of physically based models of river channel form and sediment transport means that the confidence in estimates of the effect of climate change on river channels is low in general.

4.3.13. Climate Change and Climatic Variability

Even in the absence of a human-induced climate change, hydrological behavior will vary not only from year to year but also from decade to decade (see Section 4.2). Hulme *et al.* (1999) simulated streamflow across Europe under four climate change scenarios for the 2050s (based on four different simulations from the HadCM2 climate model) and seven scenarios representing different 30-year climates extracted from a long run of the HadCM2 model with no GHG forcing. They show that natural multi-decadal (30-year) variability in average annual runoff is high across most of Europe and that this natural variability in runoff in mid-latitude Europe is greater than the simulated signal of climate change. In northern and southern Europe, the magnitude of climate change by the 2050s is greater than the magnitude of natural variability. However, the spatial patterns of climate change and climatic variability are very different, with a much more coherent (usually north-south) pattern in the climate change signal. Nevertheless, the results indicate that, for individual catchments in certain areas, the magnitude of climate change effects on some indicators of streamflow may be smaller than natural climatic variability for several decades, whereas in other areas, the climate change signal will be larger than past experience.

4.4. Effects on Water Withdrawals

4.4.1. Introduction

The consequences of climate change for water resources depend not only on possible changes in the resource base—as indicated in Section 4.3—but also on changes in the *demand*, both human and environmental, for that resource. This section assesses the potential effects of climate change on water withdrawals and use, placing these effects in the context of the many nonclimatic influences that are driving demand.

It must be noted that "demand" in its economic sense means willingness to pay for a particular service or commodity and is a function of many variables—particularly price, income (for households), output (for industries or agriculture), family composition, education levels, and so forth. The usefulness of the demand function is found in the ability to predict the effects of changes in causal variables and in measurement of the demanding party's "willingness to pay" as a measure of gross benefits to the demanding party of various quantities. This willingness to pay is measured as the area under the demand function in the price-quantity plane. The quantities *actually* purchased (the quantities of water withdrawn or used) over time are the result of the interaction of factors affecting demand as defined above and conditions of supply (or availability). Thus, for example, the fact that the quantity purchased over time increases could be the result of falling costs of supply (a shift in the supply curve) rather than an increase in demand (shift in the demand curve). In this section, the term "demand" often is used as a synonym for "requirements;" this reflects usage of the term in large parts of the water sector.

Demands can be classified along two dimensions: instream or offstream, and consumptive or nonconsumptive. Instream demands use water within the river channel (or lake) and do not involve withdrawal. Examples include ecosystem uses, navigation, hydropower generation, recreation, and use of the water course for waste assimilation. Offstream demands extract water from the river channel, lake, or aquifer. They include domestic, industrial, and agricultural demands, as well as extractions for industrial and power station cooling. These demands can be consumptive or nonconsumptive. Consumptive demands "use" the water so it cannot be entirely returned to the river; nonconsumptive demands return the water to the river, although it may be returned to a different catchment or at a different quality. The primary consumptive demands are for irrigation and some types of industrial cooling (where the water is evaporated to the atmosphere rather than returned to the river).

4.4.2. *World Water Use*

Figure 4-2 shows estimated total water withdrawals, by sector, from 1900 to 1998 (Shiklomanov, 1998; Shiklomanov *et al.*, 2000). Agricultural use—primarily for irrigation—is by far the largest proportion, accounting in 1995 for 67% of all withdrawals and 79% of all water consumed. Municipal, or domestic, use represents only about 9% of withdrawals. There are large differences, of course, between continents, with the greatest absolute volume of irrgation withdrawals in Asia.

Over the past few years there have been many projections of future water withdrawals; virtually all have overestimated the actual rate of increase (Shiklomanov, 1998). Figure 4-2 also shows projected total global water withdrawals estimates made for the UN Comprehensive Assessment of the Freshwater Resources of the World (Raskin *et al.*, 1997). The central projection represents a "Conventional Development Scenario" (CDS), with "best-guess" estimates of future population growth, economic development, and water-use intensity. The upper and lower lines represent high and low cases, where the assumed rates of growth are altered. Under the core CDS, global water withdrawals would increase by about 35% over 1995 values by 2025, with low and high estimates of 23 and 49%, respectively. The greatest rates of growth are projected to be in Africa and the Middle East, with the lowest growth in developed economies. These projections are very dependent not only on the assumed rate of population growth but also on the different assumed rates of water usage. They do not take potential climate change into account.

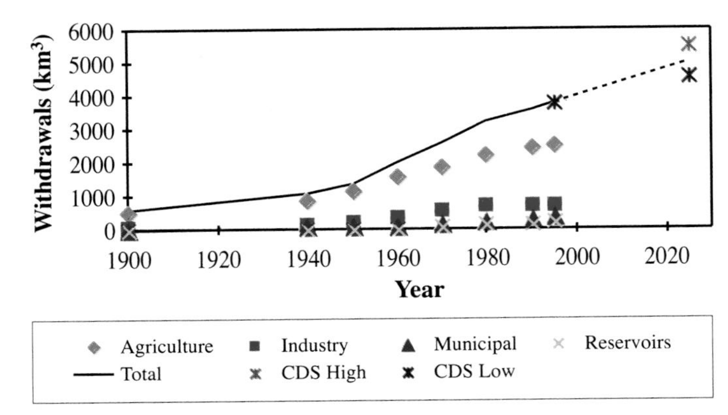

Figure 4-2: Global water withdrawals, 1900–1995, with projected future total withdrawals to 2025 [data from Shiklomanov *et al.*, 2000 (after Raskin *et al.*, 1997)].

The amount of municipal water withdrawals is dependent on the number of urban dwellers, the level of development (related to the availability of a piped distribution network and sewage systems, amongst other things), income levels, and price where actually applied to water. Per capita withdrawals in many developed countries are falling, and this may more than offset an increasing population: Total municipal withdrawals in the United States, for example, are falling largely as a result of increasing prices, conservation education provided by water utilities, and increasing use of water-efficient appliances. However, municipal withdrawals in developing countries can be expected to increase with living standards; under the CDS, per capita withdrawals in these countries are projected to rise toward those in more economically developed countries. Rapid urbanization in developing countries is expected to lead to very substantial increases in total municipal water withdrawals.

The industrial sector currently accounts for approximately 20% of current withdrawals. This water is used primarily either in processing (food processing or heavy industry) or for cooling; the relative proportions vary between countries. Future industrial water use is expected to rise substantially as industrial development continues, but the increase probably will be less than the increase in industrial production as water is used with greater efficiency (using less per unit of production, for example, or relocating power generation plants to coastal areas to use seawater for cooling). Under the CDS, total industrial water use increases; this increase is concentrated largely in Asia and Latin America (Raskin *et al.*, 1997).

The amount of water used for agriculture is dependent primarily on the level of irrigation development, the pricing of water, and the reliability of supply. Future irrigation uses are a function of the rate of expansion of irrigated land, irrigation efficiency, and pricing practices. Efficiency is projected to improve—partly as a result of changes in the cost of water and partly as a result of technological developments—but this may be more than offset by increases in the area under irrigation. The amount of increase, however, is very uncertain; it depends on, among other things, assumed rates of population growth (increasing demand) and assumed changes in world agricultural markets. Previous estimates of future water use have tended to overestimate the rate of increase because the rate of expansion of irrigated land has been overestimated (Shiklomanov, 1998). The expanded use of pricing to reflect water scarcity is being strongly promoted by international organizations and will affect the quantities used. However, many regions that historically have been dependent on rainfall are using *supplemental irrigation*—a factor that will increase irrigation use.

Estimates of future water withdrawals are notoriously uncertain, largely reflecting uncertainties in the future rate of population and economic growth. There is an analogy here with GHG emissions scenarios. Also important, however, are possible changes in the way water is priced. Much agricultural water, for example, currently is heavily subsidized, and a shift toward a more "economic" price for water is likely to have a very significant effect on use. The World Bank and other economic development agencies are pushing for major reforms in the way water is priced and sold, as well as the use of water markets as an efficient way of reallocating existing supplies and motivating efficiency.

4.4.3. *Sensitivity of Demand to Climate Change*

Climate change is another potential influence on the demand for water. Municipal demand is related to climate to a certain extent. Shiklomanov (1998) notes different rates of use in different climate zones, although in making comparisons between cities it is difficult to account for variation in nonclimatic controls. The sensitivity of municipal demand to climate change is likely to be very dependent on the uses to which the water is put. The most sensitive areas are increased personal washing and—more importantly in some cultures—increased use of water in the garden and particularly on the lawn. Studies in the UK (Herrington, 1996) suggest that a rise in temperature of about 1.1°C by 2025 would lead to an increase in average per capita domestic demand of approximately 5%—in addition to nonclimatic trends—but would result in a larger percentage increase in peak demands (demands for garden watering may be highly concentrated). Boland (1997) estimated the effects of climate change on municipal demand in Washington, D.C., under a range of different water conservation policies. Table 4-5 summarizes percentage change in summer water use under the range of scenarios considered. Boland (1997) concludes that the effect of climate change is "small" relative to economic development and the effect of different water conservation policies.

***Table 4-5**: Percentage change in average summer water use from 1990 by 2030: Washington, D.C. (Boland, 1997).*[a]

	Policy 1[b]	Policy 2[c]	Policy 3[d]
No change in climate	+100	+61	+45
Additional Change over Baseline Climate			
GISS A scenario	+8	+8	+8
GISS B scenario	-13	-13	-13
GFDL scenario	+15	+15	+15
Max Planck scenario	+17	+16	+17
Hadley Centre scenario	+19	+19	+19

[a] See original paper for scenario definitions.
[b] Policy 1 = 1990 measures.
[c] Policy 2 = Policy 1 plus increased recycling, public education, and altered plumbing regulations.
[d] Policy 3 = Policy 2 plus 50% real increase in water tariffs.

Industrial use for processing purposes is insensitive to climate change; it is conditioned by technologies and modes of use. Demands for cooling water, however, may be affected by climate change. Increased water temperatures will reduce the efficiency of cooling, perhaps necessitating increased abstraction (or, of course, changes in cooling technologies to make them more efficient).

Agricultural demand, particularly for irrigation water, is considerably more sensitive to climate change. There are two potential effects. First, a change in field-level climate may alter the need for and timing of irrigation: Increased dryness may lead to increased demands, but demands could be reduced if soil moisture content rises at critical times of the year. Döll and Siebert (1999) applied a global irrigation water-use model with a spatial resolution of 0.5°x0.5° to assess the impact of climate change on net irrigation requirements per unit irrigated area, with a climate change scenario based on the ECHAM4 GCM. Figure 4-3 shows the relative change of net irrigation requirements between the present time (1961–1990) and 2025 in all areas equipped for irrigation in 1995. Under this scenario—and similarly under the corresponding HadCM3 scenario—net irrigation requirements per unit irrigated area generally would decrease across much of the Middle East and northern Africa as a result of increased precipitation, whereas most irrigated areas in India would require more water. The extra irrigation requirements per unit area in most parts of China would be small; the HadCM3 scenario leads to a greater increase in northern China. Other climate models would give different indications of regional changes in irrigation requirements. On the global scale, increases and decreases in net irrigation requirements largely cancel, and there is less difference between different climate models; under two scenarios considered by Döll and Siebert (2001), global net irrigation requirements would increase, relative to the situation without climate change, by 3.5–5% by 2025 and 6–8% by 2075. Actual changes in withdrawals would be dependent on changes in the efficiency of irrigation water use.

The second potential effect of climate change on irrigation demand is through increasing atmospheric CO_2 concentrations (Chapter 5). Higher CO_2 concentrations lower plant stomatal conductance, hence increase WUE; but as indicated in Section 4.3.3, this may be offset to a large extent by increased plant growth.

Hatch *et al.*(1999) assessed irrigation water requirements in Georgia, USA, using a climate change scenario derived from HadCM2. This scenario produced increased rainfall in most seasons, which, together with a shorter growing season and the assumed effect of CO_2 enrichment, resulted in a *decrease* in irrigation demand, ranging from just 1% by 2030 for soybean

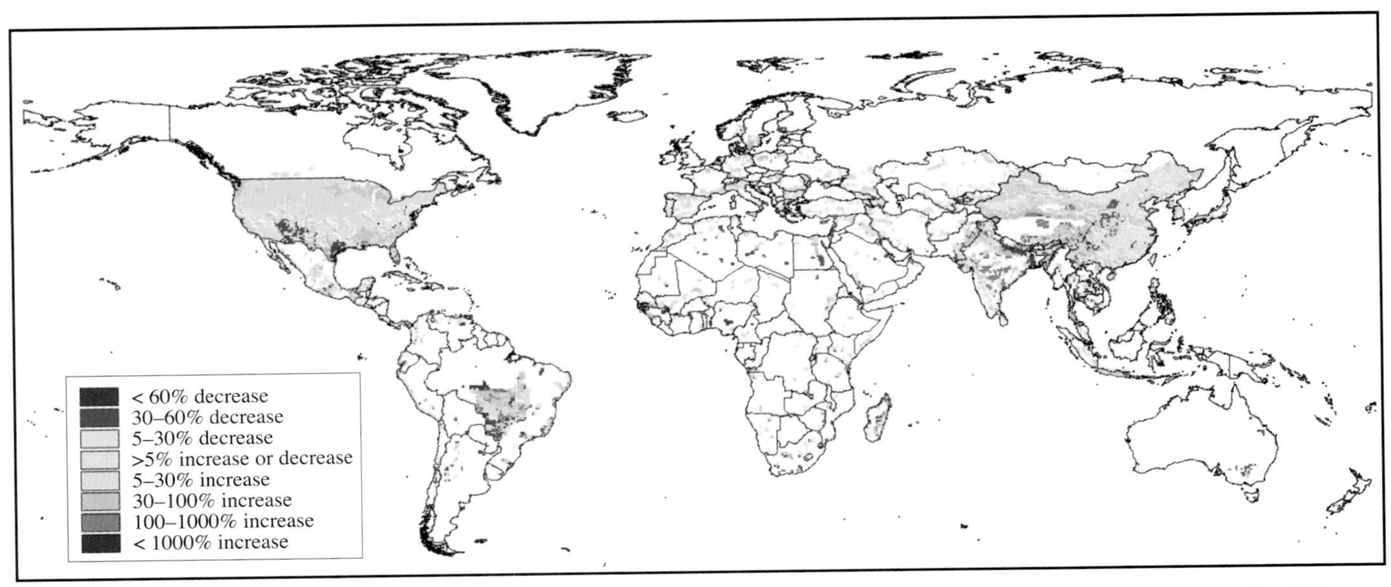

Figure 4-3: Relative change of annual net irrigation requirement between present time (1961–1990) and 2025 as a result of climate change (MPI climate scenario; areas equipped for irrigation in 1995 shown) (Döll and Siebert, 2001).

to as much as 20% by 2030 for corn. Along the Gulf Coast of the United States, however, the same scenario implies an increase in irrigation demands (Ritschard *et al.*, 1999). Strzepek *et al.* (1999) also simulated decreases in irrigation requirements across the U.S. cornbelt under two of three scenarios (with the decrease depending on assumed irrigation use efficiency) but an increase under the third scenario. These three studies together indicate considerable uncertainty in estimated future irrigation withdrawals.

4.5. Impacts on Water Resources and Hazards

4.5.1. Introduction

The preceding sections have assessed the potential effect of climate change on river flows, groundwater recharge and other biophysical components of the water resource base, and demands for that resource. The consequences, or impacts, of such changes on risk or resource reliability depend not only on the biophysical changes in streamflow, recharge, sea-level rise, and water quality but also on the characteristics of the water management system. This section considers what possible changes in hydrology and demand will mean for water supply, flood risk, power generation, navigation, pollution control, recreation, habitats, and ecosystems services *in the absence of planned adaptation to climate change*. In practice, of course, the *actual* impacts of climate change will be rather different because water managers will make incremental or autonomous adaptations to change—albeit on the basis of imperfect knowledge—and the impact of change will be a function of adaptation costs and residual impacts. However, very few studies have incorporated deliberate adaptation strategies (Alexandrov, 1998, is one), and studies that do not consider adaptation provide a base case for assessing the magnitude of the climate change "problem." More significant, some studies have not accounted for nonclimatic changes in the way water resources are managed or systems are operated and have applied the future climate to the present management system. This is unrealistic, but the extent of adaptation by many water managers is uncertain. It is important to assess the effect of climate change by, say, the 2050s in the context of the water management system that would exist by then in the absence of climate change—considering, for example, changes in demand or legislative requirements.

The sensitivity of a water resource system to climate change is a function of several physical features and, importantly, societal characteristics. Physical features that are associated with maximum sensitivity include:

- A current hydrological and climatic regime that is marginal for agriculture and livestock
- Highly seasonal hydrology as a result of either seasonal precipitation or dependence on snowmelt
- High rates of sedimentation of reservoir storage
- Topography and land-use patterns that promote soil erosion and flash flooding conditions
- Lack of variety in climatic conditions across the territory of the national state, leading to inability to relocate activities in response to climate change.

Societal characteristics that maximize susceptibility to climate change include:

- Poverty and low income levels, which prevent long-term planning and provisioning at the household level
- Lack of water control infrastructures
- Lack of maintenance and deterioration of existing infrastructure
- Lack of human capital skills for system planning and management

- Lack of appropriate, empowered institutions
- Absence of appropriate land-use planning
- High population densities and other factors that inhibit population mobility
- Increasing demand for water because of rapid population growth
- Conservative attitudes toward risk [unwillingness to live with some risks as a tradeoff against more goods and services (risk aversion)]
- Lack of formal links among the various parties involved in water management.

This section first considers the global-scale implications of climate change on broad measures of water resources then assesses in more detail potential impacts on defined systems.

4.5.2. Impacts of Climate Change on Water Resources: A Global Perspective

There are several indicators of water resource stress, including the amount of water available per person and the ratio of volume of water withdrawn to volume of water potentially available. When withdrawals are greater than 20% of total renewable resources, water stress often is a limiting factor on development (Falkenmark and Lindh, 1976); withdrawals of 40% or more represents high stress. Similarly, water stress may be a problem if a country or region has less than 1,700 m^3 yr^{-1} of water per capita (Falkenmark and Lindh, 1976). Simple numerical indices, however, give only partial indications of water resources pressures in a country or region because the consequences of "water stress" depend on how the water is managed.

At the global scale, assessments of water stress usually are made by country because that is the unit at which water-use data generally are available. In 1990, approximately one-third of the world's population lived in countries using more than 20% of their water resources, and by 2025 about 60% of a larger total would be living in such stressed countries, in the absence of climate change (WMO, 1997), largely because population growth. Arnell (1999b, 2000) estimates the effect of a number of climate change scenarios on national water resource availability and compares this with estimated future demands for water (increasing following the CDS outlined in Section 4.4). Table 4-6 shows the numbers of people living in countries using more than 20% of their water resources in 2025 and 2050 and in which the amount of resources decreases by more than 10% as a result of climate change. There is considerable variability between scenarios, essentially reflecting how resources change in populous countries, but by the 2020s the table indicates that about 0.5 billion people could see increased water resources stress as a result of climate change. Significant geographic variations are hidden in Table 4-6. Under most of the scenarios considered, climate change increases stresses in many countries in southern and western Africa and the Middle East, whereas it ameliorates stresses in parts of Asia. Alcamo *et al.* (1997) found broadly similar results.

Figure 4-4 shows water resources per capita in 1990 and 2050 for a set of countries, as listed in Table 14-3 of the WGII contribution to the SAR, showing resources per capita in 2050 without climate change (long line) and under eight climate change scenarios (short lines) (Arnell, 2000). There are some differences with the earlier table because of the use of updated data sets, but similar conclusions can be drawn. Climate change tends to have a small effect relative to population growth, and the range of magnitudes of effect between scenarios also is little changed; the effects are still uncertain. For most of the example countries, climate change may result in either an increase or a decrease, although for some the climate change signal is more consistent (reductions in South Africa, Cyprus, and Turkey, for example, and increases in China). Note that these figures represent national averages, and different parts of each country may be differently affected.

Table 4-7 gives an indication of the potential effect of stabilizing GHG concentrations on the total number of people living in water-stressed countries adversely affected by climate change (Arnell *et al.*, 2001). The results are conditional on the climate model used and the stabilization scenario, but this study—using just the HadCM2 climate model—suggests that by the 2050s the "weaker" stabilization target has little effect on the total number of impacted people, and although the "stronger" target reduces the impact of climate change, it does not eliminate it. The changes by the 2020s are very much affected by climatic variability between the various GCM runs.

4.5.3. Catchment and System Case Studies

Although there have been many assessments of the effect of climate change on river flows and (to a much lesser extent)

***Table 4-6**: Number of people living in water-stressed countries that are adversely affected by climate change, under a "business-as-usual" emissions scenario (IS92a) (Arnell, 2000).*

	Total Population (millions)	Population in Water-Stressed Countries[a] (millions)	Number of People (millions) in Water-Stressed Countries with Increase in Water Scarcity							
			HadCM2	HadCM3	ECHAM4	CGCM1	CSIRO	CCSR	GFDL	NCAR
2025	8055	5022	338–623	545	488	494	746	784	403	428
2050	9505	5915	2209–3195	1454	662	814	1291	1439	—	—

[a] Water-stressed countries use more than 20% of their available resources.

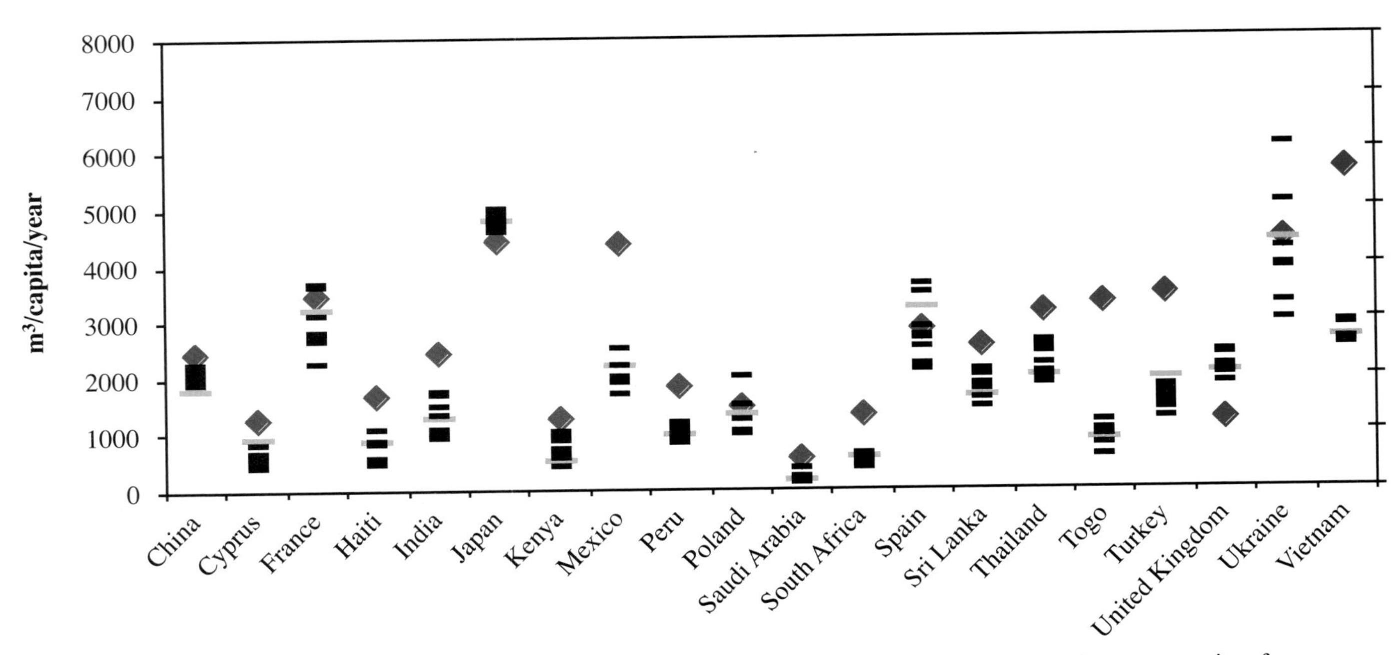

Figure 4-4: National water resources per capita (m³ yr⁻¹), in 1990 and 2050 under several climate change scenarios, for some countries (Arnell, 2000). Blue diamonds represent 1990; long pink bars 2050 with no climate change; and short black bars 2050 under different climate change scenarios.

groundwater recharge, there have been few published quantitative studies into impacts on real water resource systems. Table 4-8 lists studies published in particular aspects of the water sector. Rather than summarize a large number of studies—which use different scenarios and different methodologies—this section gives a description of a few particularly comprehensive studies.

Perhaps the most comprehensive set of studies conducted to date was published by Lettenmaier *et al.* (1999). This study looked at potential climate change impacts on six dimensions of water resource (power generation, municipal water supply, recreation, flood defense, navigation, and environmental flow regulation) in six major U.S. basins, using realistic models of the water system operation and scenarios for possible future nonclimatic changes in demands and objectives (not all the six dimensions were assessed in each basin). Table 4-9 summarizes the results, showing changes in a range of indicators (which varied between basins) by 2050 under three climate change scenarios and a varying number of operational and demand scenarios. The results clearly show considerable variability between scenarios and catchments; they also show that climate change might produce water resources benefits. The results also indicate that, in most sectors and basins, the impacts of different demand and operational assumptions by 2050 are greater than, or of similar magnitude to, the potential impacts of climate change.

Kaczmarek *et al.* (1996) assessed the impact of climate change on the water supply system in the Warta River basin in Poland, looking at two climate change scenarios in the context of increasing demand for water (particularly from irrigation). In the absence of climate change, they show that there would be supply problems in part of the system by 2050, simply because of the increase in demand. Under one of the scenarios, inflows to supply reservoirs would increase sufficiently to prevent supply problems; under the other scenario, the risk of shortage would increase substantially (the probability of an annual deficit of 10% would increase from 4 to ~25%, for example). Kaczmarek *et al.* (1996) also looked at the feasibility of one adaptation option—transferring water from one reservoir to another—and showed how it could lessen the likelihood of shortage.

***Table 4-7**: Effect of stabilization of CO_2 concentrations on numbers of people living in water-stressed countries adversely affected by climate change (Arnell et al., 2001). Climate change under each emissions scenario is simulated with HadCM2 general circulation model; other climate models could give different indications of the effect of stabilization.*

	Total Population (millions)	Population in Water-Stressed Countries (millions)	Number of People (millions) in Water-Stressed Countries with Increase in Water Scarcity		
			IS92a	S750	S550
2025	8055	5022	338–623	242	175
2050	9505	5915	2209–3195	2108	1705

Table 4-8: Studies of impact of climate change on water resources and hazards (published since the SAR).

Impact	Catchment/Region	Reference
Public water supply		
– Water supply systems	Four U.S. basins	Lettenmaier *et al.* (1999)
– Supply reservoirs	Czech Republic	Dvorak *et al.* (1997)
– Supply reservoir	Philippines	Jose *et al.* (1996)
– Supply reservoir	Yangtze basin, China	Shen and Liang (1998)
– Supply reservoirs	UK, Iran	Adeloye *et al.* (1998)
– Supply system	Poland	Kaczmarek *et al.* (1996)
– Groundwater supply	Texas, USA	Loaiciga *et al.* (1998)
– Water supply system	New York City, USA	Blake *et al.* (2000)
– Supply system	Rhine River	Grabs (1997)
Irrigation		
– Impacts on crop yields	New York, Indiana, and Oklahoma, USA	Tung and Haith (1998)
– Impacts on crop yields	Southern European Russia	Georgiyevsky *et al.* (1996)
– Rice irrigation	Senegal River basin, West Africa	Venema *et al.* (1997)
– Irrigated agriculture	Columbia River, USA/Canada	Cohen *et al.* (2000)
– Irrigated cropland	Georgia, USA	Hatch *et al.* (1999)
– Irrigated cropland	USA	Strzepek *et al.* (1999)
Power generation		
– Hydropower (reservoir)	Southeast USA	Robinson (1997)
– Hydropower (reservoir)	Nordic region	Saelthun *et al.* (1998)
– Hydropower (lake)	Great Lakes	Chao and Wood (1999)
– Hydropower (lake)	Four U.S. basins	Lettenmaier *et al.* (1999)
– Hydropower (reservoir)	Columbia River, USA/Canada	Cohen *et al.* (2000)
– Hydropower (reservoir)	Tana River, Kenya	Mutua (1998)
Navigation		
– River navigation	Rhine River	Grabs (1997)
– River navigation	Three U.S. basins	Lettenmaier *et al.* (1999)
– River navigation	Columbia River, USA/Canada	Cohen *et al.* (2000)
– Lake navigation	Great Lakes	Chao and Wood (1999)
Flood risk		
– Riverine flood risk	Rhine basin	Grabs (1997)
– Riverine flood risk	Eastern Australia	Minnery and Smith (1996)
– Riverine flood risk	Columbia River, USA/Canada	Cohen *et al.* (2000)
– Riverine flood risk	Five U.S. basins	Lettenmaier *et al.* (1999)

The River Rhine is a very important transport route within Europe. Grabs (1997) considered the effect of two climate change scenarios on navigation opportunities, having translated climate into streamflow by using a catchment water balance model. Table 4-10 summarizes the results: Under one of the scenarios, there would be little obvious effect on navigation opportunities, but under the other movement could be curtailed, particularly by the middle of the 21st century.

The vast majority of the impact assessments in Table 4-8 describe the effects of climate change on the reliability of an existing system. Very few explore the costs of these impacts, primarily because of difficulties in deciding the basis for calculation. Are the costs of climate change equal to the cost of continuing to provide the current standard of service? Are the costs of services foregone (in terms of extra flood damages or reduced use of water), or are they incurred in providing services at a new economically-optimum level? In other words, estimates of the cost of climate change must consider explicitly the measures used to adapt to that change, and the economic costs of climate change will depend on the adaptation strategies adopted. Carmichael *et al.* (1996) present one of the few studies that has tried to cost the implications of climate change. They investigated the treatment costs necessary to maintain a given water quality standard (expressed in terms of dissolved oxygen content) in a river in Slovakia and calculated the least costly treatment under the present hydrological regime and under one scenario for the 2020s. They showed that costs would be little

Table 4-8 (continued)

Impact	Catchment/Region	Reference
Pollution control and water quality management		
– Wastewater treatment	Slovakia	Carmichael *et al.* (1996)
Low flows and instream needs		
– Fishery impacts	Columbia River, USA/Canada	Cohen *et al.* (2000)
– Environmental low flows	Midwest USA	Eheart *et al.* (1999)
General overview		
– Review	UK	Arnell (1998)
– Review	USA	Gleick (2000)
– Review	Canada	Bruce *et al.* (2000)
– Review	Central Great Plains, USA	Ojima *et al.* (1999)
– Review	Mexico	Mendoza *et al.* (1997)
– Review	Estonia	Jarvet (1998)
– Review	South Asia	Mirza (1999)
– Review	Kenya	Awuor (1998)

different if the aim were to meet a 4 ppm dissolved oxygen target under average summer conditions but would rise by a factor of about 14 (at current prices) if the aim were to meet the same target under low-flow conditions, even taking a least-cost approach.

Aggregated estimates of the cost of impacts of climate change on water resources have been prepared for Spain, the UK, and the United States. Ayala-Carcedo and Iglesias-Lopez (2000) estimate that the reduction in water supplies under one scenario would cost nearly US$17 billion (2000 values) between 2000 and 2060, or about US$280 million yr^{-1} in terms of increased expenditure to maintain supplies and lost agricultural production. A study in the UK estimated the costs of climate change for water supply and flood protection (ERM, 2000). Table 4-11 shows the costs (converted to US$) involved in making up shortfalls of 5, 10, and 20% in the supply or demand across Britain, under several different types of approaches (see Section 4.6.2). The study assumes that the same change in water availability occurred across all of Britain—which probably overstates the costs because many parts of Britain are projected to have increased runoff—and estimated costs on the basis of standardized costs per unit of water. The study does not consider the feasibility of each of the potential adaptations. The cost of demand management measures increases substantially for large reductions in demand because more expensive technologies are needed. Note that a 5% reduction in demand represents just more than half the water of a 5% increase in supply; reducing domestic demand by 20% has a similar effect to increasing supply by 10%. The ERM study assumes that annual riverine flood damages would increase, because of increased flooding, by about US$80–170 million yr^{-1} over the next 30 years (compared to a current figure of about US$450 million), and the average annual cost of building structural works to prevent this extra flooding would be about US$40 million.

There have been two sets of estimates of the aggregate cost of climate change for water resources in the United States, using different approaches. Hurd *et al.* (1999) examined four river basins under nine climate change scenarios (defining fixed changes in temperature and precipitation) and extrapolated to the United States as a whole. Their study uses detailed economic and hydrological modeling and suggests that the largest costs would arise through maintaining water quality at 1995 standards—US$5.68 billion yr^{-1} (1994 US$) by 2060 with a temperature increase of 2.5°C and a 7% increase in precipitation—and through lost hydropower production (US$2.75 billion yr^{-1} by 2060, under the same scenario). Costs of maintaining public water supplies would be small, and although loss of irrigation water would impact agricultural users, changed cropping and irrigation patterns would mean that the economic losses to agriculture would be less than US$0.94 billion yr^{-1} by 2060. However, this study extrapolates from the four study catchments to the entire United States by assuming that the same climate change would apply across the whole country.

Frederick and Schwarz (1999) take a different approach, looking at 18 major water resource regions and 99 assessment subregions, with two climate change scenarios for the 2030s based on climate model simulations. Water scarcity indices were developed for each assessment subregion, comparing scarcities under "desired streamflow conditions" and "critical streamflow conditions" on the demand side with "mean streamflows" and "dry-condition streamflows" on the supply side. These indices played a key role in determining the costs of meeting various streamflow targets. A supply-demand balance in each region is achieved through supply- and demand-side measures, each of which has an assumed unit cost. Three strategies were defined for each region: "environmental," focusing on protecting the environment; "efficient," maintaining supplies to users; and "institutional," placing limits on

***Table 4-9**: Impact of climate change scenarios by 2050 on various water resource indices in six U.S. basins (Lettenmaier et al., 1999).*[a]

	Savannah	Apalachicola-Chattanooga-Flint (ACF)	Missouri	Columbia	Tacoma	Boston Water Supply
Energy production						
– GFDL	+26%	+5%	-8%	-5%		
– Hadley Centre	+3%	-10%	-13%	-4%		
– MPI	-5%	-1%	-33%	-12%		
– Demand/operational	-3%	-5 to +3%	not calculated	-10 to -15%		
Municipal and industrial supply						
– GFDL			-9%	-5%	0	0
– Hadley Centre			-5%	-1%	0	0
– MPI			-15%	-14%	0	-5%
– Demand/Operational			not calculated	0	-15%	-3 to -40%
Flood risk						
– GFDL	+70%	+32%	+4%		+40%	
– Hadley Centre	+50%	-3%	+10%		+40%	
– MPI	+16%	-10%	+12%		+35%	
Navigation						
– GFDL		+3%	-8%	-2%		
– Hadley Centre		-2%	-10%	-5%		
– MPI		-1%	-10%	+5%		
– Demand/operational		-2 to -17%	not calculated	-4%		
Instream flow requirements						
– GFDL	0			-4%	-1%	
– Hadley Centre	-1%			-5%	-6%	
– MPI	-3%			-10%	-8%	
– Demand/operational	-1 to -3%			0 to +12%	-8 to -9%	
Recreation						
– GFDL	+7%	+22%	0	-10%		
– Hadley Centre	+5%	-5%	0	-8%		
– MPI	+3%	+4%	0	-12%		
– Demand/operational	+18%	-25 to +28%	not calculated	-9 to +3%		

[a] See original paper for detailed summary of scenarios used.

changes in environmental indicators and the area of irrigation. The total national cost of climate change was determined under each strategy by aggregating least-cost measures in each subregion. Table 4-12 summarizes the estimated national costs under the three strategies and two scenarios. The costs are considerably greater under the drier CGCM1 scenario than under the wetter HadCM2 scenario (which, in fact, implies a benefit), and they vary with management strategy. Costs under the drier scenario are considerably higher than those estimated by Hurd *et al.* (1999), reflecting partly the different approaches used and partly the spatial variability in the effect of climate change considered by Frederick and Schwarz (1999).

4.5.4. *Impacts of Climate Change on Water Resources: An Overview*

This section explores the global-scale implications of climate change for water resources stress and summarizes a few studies into climate change impacts on several real-world water management systems. However, few published studies consider impacts in quantitative terms on real-world systems; most published studies infer changes in water resources from changes in streamflow.

It is very difficult to draw quantitative conclusions about the impacts of climate change, for several reasons. Different studies

***Table 4-10**: Effect of climate change on navigation opportunities on River Rhine (Grabs, 1997).*

	Average Annual Number of Days when Large Boat-Trains can Move (flows between 2000 and 5500 m^3 sec^{-1}) UKHI	CCC
1990	168	
2020	164	170
2050	156	170
2100	148	166

have used different methodologies and different scenarios, but, most important, different systems respond very differently to climate change. It is possible, however, to make some qualitative generalizations:

- In systems with large reservoir capacity, changes in resource reliability may be proportionately smaller than changes in river flows.
- The potential impacts of climate change must be considered in the context of other changes that affect water management. Few studies have explicitly compared climate change with other pressures, but in many environments it is likely that over a time horizon of less than 20 years, climate change impacts will be very small relative to other pressures. This will depend on the system.
- The implications of climate change are likely to be greatest in systems that currently are highly stressed.

By far the majority of studies of the impact of climate change on water resources have concentrated on human aspects of the water environment. Only a very few (e.g., Eheart *et al.*, 1999; Meyer *et al.*, 1999) have considered impacts on the aquatic environment. Some of these studies are considered in Chapter 5, but it must be remembered that water resources systems in many parts of the world increasingly are being managed to maintain instream and wetland ecosystems. This either increases effective water demand or decreases water availability.

Confidence in estimated quantitative impacts of climate change on water resources generally is low, reflecting initial confidence in climate change scenarios and low confidence in estimates of future pressures on water resources (as a result of factors such as changes in demand or legislative requirements). However, techniques for estimating the impacts of a given scenario are now well established.

4.6. Adaptation Options and Management Implications

4.6.1. Introduction

The preceding sections have assessed the possible effects of climate change on the water resource base and on the demand for water, as well as the potential impacts on water users. Most published studies have looked at impacts in the absence of planned adaptation to climate change, and the few studies that have tried to cost impacts have had to make assumptions about adaptation. This section assesses opportunities in the water sector for adapting to climate change and explores any constraints which may exist.

Water management has always adapted to change (especially following extreme events or in response to increased demand), and climate change is just one of the pressures facing water managers. Other pressures include increasing demands for water resources or protection against hazard, changing water management objectives (which recently have included increasing

***Table 4-11**: Estimated national average annual costs (US$ million) of impacts of climate change on water resources and riverine flooding, UK, over next 30 years (ERM, 2000).*

	5% reduction in supply by 2030	10% reduction in supply by 2030	20% reduction in supply by 2030
Volume of water (Ml day^{-1})	757	1514	3028
Supply-side			
– Reservoirs	3.3–25	6–50	12–100
– Conjunctive use schemes	140–1200	280–2430	570–4900
– Bulk transfers	0.5–90	1–175	2–360
– Desalination	4–12	10–24	19–48
	5% reduction in municipal demand by 2030	**10% reduction in municipal demand by 2030**	**20% reduction in municipal demand by 2030**
Volume of water (Ml day^{-1})	420	835	1670
Demand management measures	0.5	1	9

***Table 4-12**: National average annual cost of maintaining water supply-demand balance in the USA (Frederick and Schwarz, 1999). Values in 1994 US$ billion.*

Management Strategy	HadCM2	CGCM1
"Efficient"	-4.7	105
"Environmental"	-4.7	251
"Institutional"	not calculated	171

recognition of the importance of meeting environmental needs as well as those of offstream demands), changing water management technologies, and altered legislative environments.

It is important to distinguish between development of adaptive options for meeting changing demands and resources and assessment of the abilities of a given water management agency (interpreted broadly) actually to adapt to climate change. Over the years, a wide range of adaptive techniques has been developed, largely in response to the need to meet increased demands. Broad distinctions can be drawn among "supply-side" adaptive techniques (changing structures, operating rules, and institutional arrangements) and "demand-side" techniques (which change the demand for water or protection against risk and include institutional changes as well). Examples of supply-side adaptations include increasing flood defenses, building weirs and locks to manage water levels for navigation, and modifying or extending infrastructure to collect and distribute water to consumers. Demand-side techniques include water demand management (such as encouraging water-efficient irrigation and water pricing initiatives), changing water allocations (Miller *et al.*, 1997), and nonstructural flood management measures (such as land-use controls). Distinctions also can be drawn between anticipatory and reactive actions. The former are taken in advance of some change, the latter in response to a change. Reactive actions include short-term operational adaptations, such as temporary exploitation of new sources, and longer term measures. A major flood or drought, for example, often triggers a change in water management. However, although many adaptive options do exist, knowledge of these options and the expertise of officials to execute them may be limited in some situations.

The optimum extent of adaptation can be characterized in terms of the benefits and costs of adaptation. The extremes of adaptation are "no adaptation" and "adaptation sufficient to eliminate all effects" (which usually is not physically possible). The optimum level of adaptation minimizes the combined costs of adaptation and residual negative effects, with the most cost-effective steps taken first.

Water managers long have had access to many techniques for assessing options and implementing adaptive strategies. However, the techniques used have changed over time and vary between countries, and they are very much influenced by institutional arrangements in place in a country. Factors that affect adaptive capacity in a country include institutional capacity, wealth, management philosophy (particularly management attitudes toward supply-side versus demand-side strategies, as well as "sustainable" management), planning time scale, and organizational arrangements (adaptation will be harder, for example, when there are many different "managers" involved or where water managers do not have sound professional guidance).

This section looks first at water management options, then at management techniques. It contends that water managers generally are aware of technical and institutional options—although for many reasons may not have access to all of them—and that climate change challenges management techniques for assessing and selecting options, rather than the technical and institutional options themselves.

4.6.2. Water Management Options

Table 4-13 summarizes some supply- and demand-side adaptive options, by water-use sector. Each option has a set of economic, environmental, and political advantages and disadvantages.

Most of these strategies are being adopted or considered in many countries in the face of increasing demands for water resources or protection against risk. In the UK, for example, water supply companies currently are pursuing the "twin track" of demand management and supply management in response to potential increases in demand for water (although there is a conflict between different parts of the water management system over the relative speeds with which the two tracks should be followed). These management strategies also are potentially feasible in the face of climate change. Nowhere, however, are water management actions being taken explicitly and solely to cope with climate change, although in an increasing number of countries climate change is being considered in assessing future resource management. In the UK, for example, climate change is one of the factors that must be considered by water supply companies in assessing their future resource requirements—although companies are highly unlikely to have new resources justified at present on climate change alone.

The continuing debate in water management (Easter *et al.*, 1998) is between the practicalities and costs of supply-side versus demand-side options, and this debate is being pursued indepedently of climate change. The tide is moving toward the use of demand-side options because they are regarded as being more environmentally sustainable, cost-effective, and flexible (Frederick, 1986; World Bank, 1993; Young *et al.*, 1994; Anderson and Hill, 1997). "Smart" combinations of supply-side and demand-side approaches are needed, although in many cases new supply-side infrastructure may be necessary. This is particularly the case in developing countries, where the challenge often is not to curb demand but to meet minimum human health-driven standards.

There do appear, however, to be numerous "no regret" policies that warrant immediate attention. In this context, a "no regret" policy is one that would generate net social benefits regardless

Table 4-13: *Supply-side and demand-side adaptive options: some examples.*

Supply-Side		Demand-Side	
Option	*Comments*	*Option*	*Comments*
Municipal water supply			
– Increase reservoir capacity	– Expensive; potential environmantal impact	– Incentives to use less (e.g., through pricing)	– Possibly limited opportunity; needs institutional framework
– Extract more from rivers or groundwater	– Potential environmental impact	– Legally enforceable water use standards (e.g., for appliances)	– Potential political impact; usually cost-inefficient
– Alter system operating rules	– Possibly limited opportunity	– Increase use of grey water	– Potentially expensive
– Inter-basin transfer	– Expensive; potential environmental impact	– Reduce leakage	– Potentially expensive to reduce to very low levels, especially in old systems
– Desalination	– Expensive (high energy use)		
– Seasonal forecasting	– Increasingly feasible	– Development of non-water-based sanitation systems	– Possibly too technically advanced for wide application
Irrigation			
– Increase irrigation source capacity	– Expensive; potential environmental impact	– Increase irrigation-use efficiency	– By technology or through increasing prices
		– Increase drought-toleration	– Genetic engineering is controversial
		– Change crop patterns	– Move to crops that need less or no irrigation
Industrial and power station cooling			
– Increase source capacity	– Expensive	– Increase water-use efficiency and water recycling	– Possibly expensive to upgrade
– Use of low-grade water	– Increasingly used		
Hydropower generation			
– Increase reservoir capacity	– Expensive; potential environmental impact	– Increase efficiency of turbines; encourage energy efficiency	– Possibly expensive to upgrade
– Seasonal forecasting	– May not be feasible		
Navigation			
– Build weirs and locks	– Expensive; potential environmental impact	– Alter ship size and frequency	– Smaller ships (more trips, thus increased costs and emissions)
– Increased dredging	– Potential environmental impact		
Pollution control			
– Enhance treatment works	– Potentially expensive	– Reduce volume of effluents to treat (e.g., by charging discharges)	
		– Catchment management to reduce polluting runoff	– Requires management of diffuse sources of pollution
Flood management			
– Increase flood protection (levees, reservoirs)	– Expensive; potential environmental impact	– Improve flood warning and dissemination	– Technical limitations in flash-flood areas, and unknown effectiveness
– Catchment source control to reduce peak discharges	– Most effective for small floods	– Curb floodplain development	– Potential major political problems

of whether there was climate change. Examples include elimination of subsidies to agriculture and floodplain occupancy and explicit recognition of environmental values in project design and evaluation. The effect of successful demand-side policies is to reduce the need for supply augmentation, although they may not prevent such needs entirely if changes are large. Such policy changes represent the minimum package of "anticipatory policy changes" in response to climate change.

4.6.3. Implications of Climate Change for Water Management Policy

Climate change exaggerates current pressures in water management—adding to the debate on sound management strategies—and adds a new component. This new component relates to uncertainty in climate change: How can water management efficiently adapt to climate change, given that the magnitude (or possibly even the direction) of change is not known? Conventionally, water resource managers assume that the future resource base will be the same as that of the past and therefore that estimates of indices such as average reservoir yield or probable maximum flood that are based on past data will apply in the future. There are two issues: assessing alternatives in the face of uncertainty and making decisions on the basis of this assessment.

Techniques for assessing alternatives include scenario analysis and risk analysis. Scenario analysis is central to climate change impact assessment, but it is not widely used in water resource assessment (although there are some very important exceptions, such as at the federal level in the United States). Scenario analysis, as in climate change impact assessment, tends to involve simulation of the effects of different scenarios, although in water resources assessment these tend to be different demand and operational scenarios rather than different climate scenarios. Stakhiv (1998) argues that if water managers already adopt a scenario-based approach, as at the federal level in the United States (Lins and Stakhiv, 1998), climate change therefore does not cause any additional *conceptual* challenges to water management: Climate change can be regarded simply as an extra type of scenario. However, the uncertain nature of climate change and the potential for nonlinearities in impact mean not only that the range of scenarios conventionally considered may be too narrow but also that a larger number of scenarios must be evaluated. In practice, scenario-based approaches are used in few water management agencies, and adoption of scenario analysis would challenge conventional water management practices in many countries.

Risk analysis involves assessment of the risk of certain thresholds being crossed under different possible futures (Major, 1998). It generally involves stochastic simulation of hydrological data to develop a sampling distribution of possible futures. In principle, climate change can be incorporated into risk analysis by changing the underlying population from which data are generated according to climate change scenarios. Matalas (1997) discusses the role of stochastic simulation in the context of climate change and argues that given the wide range in futures that often is simulated by assuming a stationary climate, the operational assumption of stationarity may remain appropriate in the face of climate change in some regions. However, it is possible that climate change could generate futures outside those produced under stationarity, and it cannot be assumed that climate change can be ignored in all circumstances.

The second main issue is that of decisionmaking under uncertainty. This issue was widely investigated during the 1960s and 1970s, largely in the context of uncertainties about demands or the precise distribution of floods and droughts over the short and medium terms. Climate change has revived interest in decisionmaking under uncertainty, and several analyses of different techniques have been published (e.g., Fisher and Rubio, 1997; Frederick, 1997; Hobbs, 1997; Hobbs *et al.*, 1997; Luo and Caselton, 1997; Chao *et al.*, 1999). There still is considerable debate. Hobbs (1997), for example, concludes that Bayesian approaches involving allocation of probabilities to specific outcomes are more suitable than Dempster-Shafer reasoning (which requires the analyst to assign probabilities to ranges—perhaps overlapping—of outcomes), but Luo and Caselton (1997) conclude the reverse. Particularly significant is the issue of assigning probabilities to alternative possible futures. Hobbs *et al.* (1997) note unease among water planners in assigning subjective probabilities to different futures.

Planners of water resource and flood protection schemes conventionally cope with uncertainty by adding a safety factor to design estimates. This safety factor usually is defined arbitrarily.

***Table 4-14**: Headroom "score" characterizing effect of climate change on resource zone yield: an approach used in UK (UKWIR, 1998).*

Range in Resource Zone Yield between Four Defined Scenarios[a]	Case 1: Two Scenarios Above and Two Below Mean	Case 2: Three Scenarios Below and One Above Mean	Case 3: Three Scenarios Above and One Below Mean
<15%	2	3	1
15–25%	4	6	2
25–35%	6	9	3
>35%	8	10	4

[a]As percentage of "best estimate" of yield.

As part of a review of water resource design practices in the UK, a more formal approach to calculation of this safety factor, or "headroom," has been developed (UKWIR, 1998). This procedure identifies eight sources of supply-side uncertainty and three sources of demand-side uncertainty, each of which is given a score. The total score is summed and converted into a percentage value for the headroom allowance (with a maximum of 20%). Climate change is included as one of the supply-side uncertainties; its score depends on the range of estimates of supply-yield under four defined climate change scenarios (Table 4-14). Although this approach has many arbitrary elements, it does represent a systematic approach to the treatment of climate change uncertainties in water resources assessment.

Different aspects of the water sector have different planning horizons and infrastructure lifetimes. The parts of the water sector with long horizons and lifetimes need to take a different approach to climate change than parts with shorter lead times; one assessment and decision methodology will not be suitable for all managers.

4.6.4. Factors Affecting Adaptive Capacity

From the beginning of human attempts to shape the water environment to human benefit, water management has dealt with the variability of the native supply of water and the variability of demands for the use of water (Stakhiv, 1998). Great strides have been made in dealing with even extreme water regimes—particulary droughts—through interventions on the supply and demand sides (e.g., Stern and Easterling, 1999). Drought management planning is playing an increasing role in many water management agencies, lowering their susceptibility to drought impacts. Thus, in some ways the prospects of a change in the resource base—perhaps characterized by lower mean supplies and higher variability—represent only a sharpening of traditional challenges to water management. There are three important differences, however. First, future climate change is highly uncertain at spatial and temporal scales that are relevant to water management: All we know is that the future may not necessarily be like the recent past. Second, as noted above, the potential pervasiveness of these changes across large regions presents challenges that preclude some traditional steps of adaptation and requires innovative approaches that go beyond experience to date. Third, climate-induced effects may be nonlinear, carrying potential for surprises beyond those incorporated in traditional water management.

The ability to adapt to climate variability and climate change is affected by a range of institutional, technological, and cultural features at the international, national, regional, and local levels, in addition to specific dimensions of the change being experienced. Among the most important features are the following:

1) The capacity of water-related institutions, consisting of water agencies' authority to act, skilled personnel, the capability and authority to consider a wide range of alternatives (including but not limited to supply-side and demand-side interventions) in adapting to changed conditions, the capability and authority to use multi-objective planning and evaluation procedures in the assessment of policy alternatives, procedures for conflict resolution, and incentives to undertake serious *ex post* analysis of policies and projects to learn what has really worked (OECD, 1985). For example, O'Connor *et al.* (1999) found in the Susquehanna River Basin, USA, that experienced full-time water managers are more likely to consider future scenarios in their planning than part-time managers.
2) The legal framework for water administration that always constrains, for better and for worse, the options that are open to water management. Naturally, laws change as needs change, but the changes are slow and greatly lag changing needs. In many countries, the legal framework for water management is moving toward increasing environmental protection (e.g., the European Union's habitats directive). Such a direction poses further constraints on options to address climate change, but if the move reflects an increasing concern with *sustainable* water management (however defined), opportunities for considering adaptation to climate change are increased.
3) The wealth of nations in terms of natural resources and ecosystems, human-made capital (especially in the form of water control systems), and human capital (including trained personnel) that determines what nations can "afford to commit" to adaptation. This should include the ability and willingness to transfer wealth among population groups and regions within a country and among nations. This is the major constraint on adaptation to climate change in poorer countries.
4) The state of technology and the framework for the dissemination (or monopolization) of technology, especially in the fields of bioengineering of drought- and salt-resistant varieties of plants and techniques for the desalination of seawater.
5) Mobility of human populations to change residential and work locations in response to severe climate events or climate change. This is a major factor in coastal and island areas. Mobility is severely hampered by population pressures, especially in tropical island settings.
6) The *speed* of climate change is crucial in determining the capabilities of societies to adapt and change water management practices. Speed of change and the *cumulative extent* of change affect the impacts on society in nonlinear fashions (Howe *et al.*, 1990; National Research Council, 1992).
7) The complexity of management arrangements also may be a factor in response. In principle, the fewer agencies involved in water management, the easier it will be to implement an adaptation strategy (although the structure within the agencies will be very important). If there are many stakeholders to involve—perhaps with conflicting requirements, management goals, and perceptions and each with some management control over part of the water system—it may be more difficult

to adapt to changing circumstances. There is evidence that in some mature infrastructure systems, there may be substantial oportunities for increasing the resilience of water resource systems through institutional changes as well (Hansler and Major, 1999).

8) The ability of water managers to assess current resources and project future resources. This requires continuing collection of data and the ability to use scenarios with hydrological models to estimate possible future conditions.

Whether adaptation takes place or not may be heavily influenced by the occurrence of extreme events. Such events often are catalysts for change in management and may serve two roles. First, they may expose failings in the current water management system. Second, they may raise the perception among decision makers of the possibility of climate change—even if they cannot be attributed directly to climate change.

Recent experience with extreme events (e.g., the Chinese floods of 1998, the Rhine floods of 1996 and 1997, the eastern European floods of 1997 and 1998, and the Mozambique floods of 2000) shows that many societies are extremely exposed to loss and damage during extreme events, especially floods. At first, it may appear that this implies that existing adaptive techniques, as widely used by water managers, are not working as expected to minimize risk and loss (some loss will always be inevitable because no flood protection scheme can provide complete protection): Adaptation is not working. However, there is extensive evidence that social vulnerability to extreme events is serious and increasing (Munasinghe and Clark, 1995; Hewitt, 1997; Tobin and Montz, 1997; Haughton, 1998; La Red, 1999; Mileti, 1999) and that this exposure to hazards has been significantly increased by public and private development with insufficient regard for known hazards (Hewitt, 1997; Marsden, 1997; Pulwarty and Riebsame, 1997). In the United States there was more damage from hurricanes between 1990 and 1995 than there was between 1970 and 1990, after adjustments for inflation (Pielke, 1997), even though both periods had low hurricane frequency (Landsea *et al.*, 1996). Changnon *et al.* (1997) analyzed the dramatic increase in dollar losses of insured property in the United States, which reached US$840 billion in the 1990–1994 period, and conclude that changes in weather and climate were not primary causes. Detailed meteorological analyses came to the same conclusion for flooding losses (Changnon, 1998; Karl and Knight, 1998).

Thus, societies' failure to adapt to extreme events *in the broadest sense* (i.e., by "allowing" risk-prone development) appears to have been largely responsible for increased damages, and that failure has not improved with time (Changnon and Changnon, 1998; Pielke and Landsea, 1998; Kunkel *et al.* 1999). It also appears that political decisions may have produced maladaptive results (Wiener, 1996; Hewitt, 1997; Mileti, 1999). In the United States, insurance has been a leading instrument for hazard awareness and post-event recovery. After 30 years of promotion, education, and subsidized premiums, only 20% of residents in floodplains were insured by the late 1990s (LeCompte and Gahagan, 1998; Pasterick, 1998). These failures to take advantage of insurance suggest that even wealthy societies adapt poorly to foreseeable hazards.

The residual damages of hazard events also are inequitably distributed across populations. This was shown clearly by studies of Hurricane Andrew in Florida (Peacock *et al.*, 1997), leading the director of the Pan American Health Organization to state that "those who lost the most had the least to lose" (PAHO, 1999). Hurricane Mitch devastated Central America in 1998, exhibiting the extreme vulnerability of that region (La Red, 1999; UNICEF, 1999). Among the responsible factors were lack of land-use planning, deforestation, and inappropriate consumption and production systems (Hewitt, 1997; Mileti, 1999; PAHO, 1999).

Thus, available evidence concerning the effectiveness of adaptation to meteorological and geologic hazards indicates poor levels of individual and social adaptation to hazards. This failing extends well beyond the water management sector as conventionally defined and can be argued to reflect weaknesses in development control, planning guidance, public education, and fiscal incentives. The foregoing examples indicate that having the ability to adapt to change is not the same as actually adapting to change: The tools often are not used, for a variety of reasons.

4.6.5. Adaptation to Climate Change in the Water Sector: an Overview

Water managers are accustomed to adapting to changing circumstances, many of which can be regarded as analogs of future climate change, and a wide range of adaptive options has been developed. Supply-side options are more familiar to most water managers, but demand-side options increasingly are being implemented. Water management is evolving continually, and this evolution will affect the impact of climate change in practice. For reasons noted above, climate change is likely to challenge existing water management practices, especially in countries with less experience in incorporating uncertainty into water planning. The generic issue is incorporation of climate change into the types of uncertainty traditionally treated in water planning.

Integrated water resources management (IWRM) (Bogardi and Nachtnebel, 1994; Kindler, 2000) increasingly is regarded as the most effective way to manage water resources in a changing environment with competing demands. IWRM essentially involves three major components: explicit consideration of *all* potential supply-side and demand-side actions, inclusion of all stakeholders in the decision process, and continual monitoring and review of the water resources situation. IWRM is an effective approach in the absence of climate change, and there already are many good reasons for it to be implemented. Adopting integrated water resources management will go a long way toward increasing the ability of water managers to adapt to climate change.

There are three final points to make:

1) "Upstream" adaptation may have implications for "downstream" uses. In other words, the impact of climate change on one user may be very much determined by the actions of other users in response to climate change. This emphasizes the need for basin-scale management.
2) The emphasis in this section has been on managed water systems. In many countries, particularly in rural parts of the developing world, water supply is "managed" at the household level, utilizing local water sources. There is a need to look at the implications of climate change in circumstances of this type in which investment in substantial infrastructure is unlikely.
3) Adaptation to climate change to reduce vulnerability in the water sector should involve far more than just water managers. Increasing social vulnerability to water stress (in terms of drought and flood) in many parts of the world reflects a wide range of pressures, many of which are outside the responsibility of water managers. Reducing vulnerability to climate change-induced flood and drought will require decisions about issues such as development and planning control, fiscal incentives (such as subsidized insurance or government disaster relief) to occupy (and continue to occupy after loss) hazard-prone land, and wealth enhancement.

4.7. Integration: Water and Other Sectors

4.7.1. The Nonclimate Context

The impact of climate change in the water sector is a function of biophysical changes in water quantity and composition, the use to which the water is put, and the way in which those uses are managed. The implications of climate change for water resources therefore must be considered in the context of the many other pressures on water resources and their management. These pressures—and management responses to them—are evolving rapidly, and the water management system (legal, infrastructural, and institutional) in the future may be very different in many countries from that at present. Considerable efforts are underway in many international agencies and organizations (e.g., Global Water Partnership, World Bank) to improve the way water is used and managed; these actions will have very significant consequences not only for economies, access to safe water, and the environment but also for the impacts of climate change. *Adaptation to climate change in the water sector must be considered in the context of these other changes*—and, of course, climate change must be considered as a factor in the development of improved management techniques.

4.7.2. Water and Other Related Sectors

Water is a fundamental component of many economic activities. The impact of climate change on the quality and quantity of water therefore will be felt by such economic activities in one way or another. Examples of such linkages are given in the following subsections.

4.7.2.1. Ecosystems (TAR Chapter 5)

Changes in hydrological characteristics will lead to changes in aquatic and wetland ecosystems (as reviewed in Chapter 5)—as, indeed, may some of the actions taken by water managers to adapt to climate change. In practice, much water management increasingly focuses on ensuring that human use of water does not adversely impact the water environment, and maintaining and enhancing environmental quality is regarded as a legitimate management goal. Environmental demands, of course, will alter as climate changes.

Agriculture also will be affected by water availability, and actions taken by farmers in response to climate change may impact the water environment. For example, climate change may increase demands for irrigation from the agricultural sector, and if these extra needs are withdrawn from rivers or aquifers, there will be an effect on hydrological and ecological regimes: The "direct" effect of climate change on hydrological regimes and ecosystems may be enhanced. On the other hand, a lack of water resulting from climate change might mean that increased irrigation demands cannot be met, and changes in the water sector therefore are impacting directly on agricultural response to climate change. In addition, changes in agricultural land use resulting directly or indirectly from climate change may affect catchment water balance and water quality. These effects may be more substantial than the direct effects of climate change on hydrology.

4.7.2.2. Coastal and Marine Zones (TAR Chapter 6)

The ecology and morphology of river deltas reflect a balance between coastal and upstream processes. Changes in freshwater flow regimes will impact deltas, although the effects probably will be smaller than those of sea-level rise. Estuary characteristics also are affected by inflows from upstream, and the relative effects of sea-level rise and changes in river flows may be similar. Saline intrusion along estuaries, associated with higher sea levels and perhaps exacerbated by lower river flows, could threaten low-lying freshwater intakes, although adaptive options (relocation) are easy to implement. Saline intrusion into coastal aquifers also is a possibility, creating severe adaptation challenges in some settings—particularly low-lying islands such as atolls. Finally, rivers bring large quantities of nutrients and other materials to the coastal zone, and these fluxes are likely to be affected by changes in streamflow volumes in particular.

4.7.2.3. Settlements (TAR Chapter 7)

Provision of water to cities—especially the mega-cities emerging in some parts of the developing world—may become increasingly problematic, with consequent effects on city growth and access to safe water. Altered river flows also may affect the ability of settlements to dispose of waste safely. Urban storm drainage is potentially very sensitive to changes in short-duration rainfall and is both expensive to install and difficult to upgrade. Finally, changes in flood flows imply changes in urban flood risk; indications are that the risk generally will increase.

The most vulnerable parts of the mega-cities are the informal settlements that do not have planned water distribution and sanitation systems. Rural populations also are exposed to climate change, and it is possible that their sensitivity to change may be greater: The urban population enjoys planned water supply systems that can adapt to changes of climate change better than unplanned systems in rural areas.

4.7.2.4. Financial Services (TAR Chapter 8)

The main linkage with the finance sector is through insurance and public disaster relief. Insurance against flood losses is available in some countries, and major flood events in these countries could challenge—at least temporarily—local and perhaps international insurers.

4.7.2.5. Health (TAR Chapter 9)

Changes in hydrological regimes have the potential to alter health risks. Most important are potential changes in access to safe drinking water, but that is likely to be more affected by factors other than climate change (such as provision of water distribution systems and improved sanitation). Water-borne diseases and water-related insect vector diseases are more sensitive to changes in hydrological patterns (e.g., Patz *et al*., 1998; Checkley *et al*., 2000). Floods have associated health problems, and climate change also has the potential to alter contamination of water supplies (through changes in flow pathways that lead to increased leaching of pollutants and through reduced flows that lead to increased concentrations) and contamination of shellfish and fish.

4.7.3. Water and Conflict

A change in water availability has the potential to induce conflict between different users (Biswas, 1994; Dellapena, 1999). These users may be in the same area—cities versus farmers, for example—or they may be in different parts of the river basin. Much has been written about the potential for international conflict (hot or cold) over water resources (e.g., Gleick, 1998); where there are disputes, the threat of climate change is likely to exacerbate, rather than ameliorate, matters because of uncertainty about the amount of future resources that it engenders. One major implication of climate change for agreements between competing users (within a region or upstream versus downstream) is that allocating rights in absolute terms may lead to further disputes in years to come when the total absolute amount of water available may be different.

4.8. Science and Information Needs

4.8.1. Introduction

In the water sector, it is important to distinguish between the needs of those who wish to estimate the potential magnitude of climate change impacts on hydrology and water resources—to meet IPCC concerns, for example—and the more pragmatic needs of water managers who need to consider how best to adapt to climate change. The two sets of requirements are linked, but there are some important differences in emphasis.

4.8.2. Estimating Future Impacts of Climate Change

Some climate change analysts are essentially concerned with estimating what would actually happen under different climate futures: What are the impacts, for example, of continued growth of emissions of GHGs at 1% yr^{-1}, and what would be the impact of stabilizing CO_2 concentrations at, say, 550 ppmv by 2150? How do changes in variability affect the water environment? These impacts—and their costs—then could be compared with the impacts, costs, and benefits of mitigation. Such studies, in principle, could allow identification of "dangerous" levels of climate change. There also are important science questions concerning the processes by which climate change might impact the water environment. For example, how might flow pathways through soils change?

Such research questions need developments in the following areas:

- *Creation of credible climate change scenarios*. This involves improvements to GCMs so that they simulate present climate and its multi-decadal variability even better, development of conceptually sound downscaling techniques (in the absence of high-resolution global climate models), and characterization of potential changes in variability at time scales from daily to decadal. These requirements are common to all impact sectors and (with the exception of downscaling) are central to improving the understanding of climate change in the most general sense.
- *Characterization of natural climatic and hydrological variability*. Potential future climate changes resulting from increasing concentrations of GHGs need to be placed in context by appreciation of "natural" climatic and hydrological variability. Much needs to be learned about linkages between different components of the climate system in different parts of the world, which requires joint use of observational data (including remotely sensed data), palaeoclimatic data, and model simulations. Palaeoclimatic and palaeohydrological reconstructions can provide very useful information

on the variability in "natural" hydrological systems, as well as insights into nonlinear relationships between climate forcing and hydrological response.

- *Improved hydrological models*. Particularly important is development and application of process-based models of hydrological processes that include realistic representations of processes that generate streamflow and recharge and determine water quality. Key issues include development of models that do not need catchment calibration (but may require remotely sensed inputs) to assess the effects of climate change in parts of the world with limited hydrological data and development of coupled climate-hydrology models (which also are important for the improvement of climate model performance and for seasonal forecasting). The international collaborative research efforts summarized in Section 4.2.2 are extremely important.
- *Characterization of uncertainty*. How important are the different sources of uncertainty—in emissions, global climate response, and regional climate change—for estimated effects of climate change? Is downscaling cost-effective, given the wide range of changes in climate that might result from different emissions scenarios, for example? What can ensemble climate model experiments contribute? There has been little systematic analysis to date of the relative importance of different sources of uncertainty.
- *Impacts on real-world water systems*. Section 4.5 notes that there have been relatively few published studies on the impacts of climate change on real-world water resources systems, and inferences about impacts generally have been made from estimates of changes in streamflow alone. This may give a very misleading impression of the actual impacts of change because the characteristics of the water management system are a very important buffer between hydrological effect and impact on users and the environment. Therefore, more studies into real-world systems are needed.
- *Effects of adaptation*. Most impact studies have ignored adaptation by water managers, and in opposition it often is asserted that water managers will be able to adapt. However, how will managers make adaptation decisions in practice on the basis of incomplete information, and what would be the effects of inefficient adaptation on the impacts of climate change?

4.8.3. Adapting to Climate Change

Water managers are beginning to consider adapting to climate change. Some—but not all—water management plans and infrastructure have long lead times and long design lives. Improved understanding of the "science" of climate change impacts in the water sector is important but is not in itself enough to enable efficient adaptation. This is because it will *never* be feasible to base decisions on just one future climate scenario, particularly for time horizons greater than a decade. This is partly a result of incomplete knowledge but largely because of inherent uncertainty in future emissions of GHGs. Therefore, water managers always will be dealing with a range of scenarios, and research aimed at enabling efficient adaptation consequently must focus largely on appropriate analytical and management tools to cope with uncertainty and change rather than on "improving" climate change science and scenarios *per se*. In some aspects of water management—particularly associated with water quality—scientific research into processes is fundamental to allowing efficient adaptation.

Efficient adaptation to climate change in the water sector requires effort in five main areas:

- *Data for monitoring*. Adaptive water management requires reliable data on which to make decisions, calibrate models, and develop projections for the future. These data should cover not just hydrological characteristics but also indicators of water use.
- *Understanding patterns of variability*. An understanding of patterns of variability—in particular, the stability of a "baseline" climate—is important for medium-term water management. It is increasingly recognized that even in the absence of climate change, the recent past may not be a reliable guide to the hydrological resource base of the near future.
- *Analytical tools*. Effective water management requires numerous tools to assess options and the future. These tools include scenario analysis and risk analysis, which are used in some parts of water management but currently are by no means widespread.
- *Decision tools*. Scenario and risk analysis provides information on possible futures and their consequences. They must be supplemented with tools such as Bayesian and other decisionmaking tools to make decisions on the basis of the information provided. Again, techniques for decisionmaking under uncertainty are not widely used in water management at present, and some of the approaches being used are not very sophisticated.
- *Management techniques*. These are the techniques that are actually implemented to meet management objectives. The broad spectrum of techniques (such as building a reservoir or managing demand) is well known, but there is a need for research into specific aspects of many demand-side approaches in particular, as well as into opportunities for seasonal flow forecasting and innovative water supply and treatment technologies (such as desalination). It also is necessary to undertake research to determine how to enhance the range of techniques considered by water managers.

Note that the above efforts are needed to improve water management even in the absence of climate change, and there is an overarching need to improve the exchange of information between hydrological science and water managers.

Water managers have long been accustomed to dealing with change, although until recently this has been primarily change resulting from changes in demand and altered legislative or statutory requirements. Climate change does not in itself stimulate development of new adaptive strategies, but it encourages a more adaptive, incremental, risk-based approach to water management. More precisely, it provides further encouragement for a trend that already is gathering pace.

References

Adeloye, A.J., N.R. Nawaz, and M. Montaseri, 1998: Sensitivity of storage-yield of multiple reservoirs to climate change induced variations in reservoir surface evaporation and rainfall fluxes. In: *Proceedings of the Second International Conference on Climate and Water, Espoo, Finland, August 1998*. Helsinki University of Technology, Helsinki, Finland, pp. 1525–1535.

Alcamo, J., P. Döll, F. Kaspar, and S. Siebert, 1997: *Global Change and Global Scenarios of Water Use and Availability: An Application of Water GAP1.0*. University of Kassel, Kassel, Germany, 47 pp. (plus appendices).

Alderwish, A. and M. Al-Eryani, 1999: An approach for assessing the vulnerability of the water resources of Yemen to climate change. *Climate Research*, **21,** 85–89.

Alexander, R.B., P.S. Murdoch, and R.A. Smith, 1996: Streamflow-induced variations in nitrate flux in tributaries to the Atlantic coastal zone. *Biogeochemistry*, **33(3),** 149–177.

Alexandrov, V., 1998: A strategy evaluation of irrigation management of maize crop under climate change in Bulgaria. In: *Proceedings of the Second International Conference on Climate and Water, Espoo, Finland, August 1998*. Helsinki University of Technology, Helsinki, Finland, pp. 1545–1555.

Ames, A. and S. Hastenrath, 1996: Diagnosing the imbalance of Glacier Santa Rosa, Cordillera Raura, Peru. *Journal of Glaciology*, **42,** 212–218.

Amthor, J.S., 1999: Increasing atmospheric CO_2 concentration, water use, and water stress: scaling up from the plant to the landscape. In: *Carbon Dioxide and Environmental Stress* [Luo, Y. and H.A. Mooney (eds.)]. Academic Press, San Diego, CA, USA, pp. 33–59.

Anderson, T.L. and P.J. Hill, 1997: *Water Marketing: The Next Generation*. Roman and Littlefield, Publishers, Inc., Lanham, MA, USA, 216 pp.

Amadore, L., W.C. Bolhofer, R.V. Cruz, R.B. Feir, C.A. Freysinger, S. Guill, K.F. Jalal, A. Iglesias, A. Jose, S. Leatherman, S. Lenhart, S. Mukherjee, J.B. Smith, and J. Wisniewski, 1996. Climate change vulnerability and adaptation in Asia and the Pacific: workshop summary. *Water, Air, and Soil Pollution*, **92,** 1–12.

Arnell, N.W., 1996: *Global Warming, River Flows and Water Resources*. John Wiley and Sons, Chichester, United Kingdom, 226 pp.

Arnell, N.W., 1998: Climate change and water resources in Britain. *Climatic Change*, **39,** 83–110.

Arnell, N.W., 1999a: The effect of climate change on hydrological regimes in Europe: a continental perspective. *Global Environmental Change*, **9,** 5–23.

Arnell, N.W., 1999b: Climate change and global water resources. *Global Environmental Change*, **9,** S31–S49.

Arnell, N.W., 2000: *Impact of climate change on global water resources: Volume 2, unmitigated emissions*. Report to Department of the Environment, Transport and the Regions, University of Southampton, Southampton, United Kingdom, 53 pp.

Arnell, N.W. and N.S. Reynard, 2000: Climate change and UK hydrology. In: *The Hydrology of the UK: A Study of Change* [Acreman, M.C. (ed.)]. Routledge, London, United Kingdom, pp. 3–29.

Arnell, N.W. and N.S. Reynard, 1996: The effects of climate change due to global warming on river flows in Great Britain. *Journal of Hydrology*, **183,** 397–424.

Arnell, N.W., B.C. Bates, H. Lang, J.J. Magnuson, and P. Mulholland, 1996: Hydrology and freshwater ecology. In: *Climate Change 1995: Impacts, Adaptations, and Mitigation of Climate Change: Scientific-Technical Analyses. Contribution of Working Group II to the Second Assessment Report of the Intergovernmental Panel on Climate Change* [Watson, R.T., M.C. Zinyowera, and R.H. Moss (eds.)]. Cambridge University Press, Cambridge, United Kingdom and New York, NY, USA, pp. 325–363.

Arnell, N.W., M.G.R. Cannell, M. Hulme, J.F.B. Mitchell, R.S. Kovats, R.J. Nicholls, M.L. Parry, M.T.J. Livermore, and A. White, 2001: The consequences of CO_2 stabilization for the impacts of climate change. *Climatic Change*, (in press).

Avila, A., C. Neal, and J. Terradas, 1996: Climate change implications for streamflow and streamwater chemistry in a Mediterranean catchment. *Journal of Hydrology*, **177,** 99–116.

Awuor, V.O., 1998: Kenyan water resources in a changing climate: impacts and vulnerability. In: *Proceedings of the Second International Conference on Climate and Water, Espoo, Finland, August 1998*. Helsinki University of Technology, Helsinki, Finland, pp. 1598–1609.

AWWA, 1997: Climate change and water resources. *Journal of the American Water Works Association*, **89,** 107–110.

Ayala-Carcedo, F.J., 1996: Reduction of water resources in Spain due to climate change. *Tecnoambiente*, **64,** 43–48 (in Spanish).

Ayala-Carcedo, F.J. and A. Iglesias-López, 2000: Impacts of climate change on water resources, design and planning in Spain Peninsula. *El Campo de las Ciencias y las Artes*, **137,** 201–222 (in Spanish).

Bates, B.C., A.J. Jakeman, S.P. Charles, N.R. Sumner, and P.M. Fleming, 1996: Impact of climate change on Australia's surface water resources. In: *Greenhouse: Coping with Climate Change* [Bouma, W.J., M.R. Manning, and G.I. Pearman (eds.)]. Commonwealth Scientific and Industrial Research Organization, Collingwood, Victoria, Australia, pp. 248–262.

Behr, O., 1998: Possible climate impacts on the water resources of the Danube river Basin. In: *Proceedings of the Second International Conference on Climate and Water, Espoo, Finland, August 1998*. Helsinki University of Technology, Helsinki, Finland, pp. 829–838.

Bengtsson, L. and J. Malm, 1997: Using rainfall-runoff modeling to interpret lake level. *Journal of Paleolimnology*, **18,** 235–248.

Bergstrom, S. and B. Carlsson, 1993: *Hydrology of the Baltic Basin*. Swedish Meteorological and Hydrological Institute Reports. *Hydrology*, **7,** 21.

Bergstrom, S., B. Carlsson, M. Gardelin, G. Lindstrom, A. Pettersson, and M. Rummukainen, 2001: Climate change impacts on runoff in Sweden: assessment by global climate models, dynamic downscaling and hydrological model. *Climate Research*, **16,** 101–112.

Biswas, A.K. (ed.), 1994: *International Waters of the Middle East: from Euphrates-Tigris to Nile*. Oxford University Press, Oxford, United Kingdom.

Blake, R., R. Khanbilvardi, and C. Rosenzweig, 2000: Climate change impacts on New York City's water supply system. *Journal of the American Water Resources Association*, 36, 279–292.

Bobba, A.G., V.P. Singh, D.S. Jeffries, and L. Bengtsson, 1997: Application of a watershed runoff model to North East Pond River, Newfoundland, study water balance and hydrological characteristics owing to atmospheric change. *Hydrological Processes*, **11,** 1573–1593.

Bogardi, J.J. and H.-P. Nachtnebel (eds.), 1994: *Multicriteria Decision Analysis in Water Resources Management*. International Hydrological Programme, UNESCO, Paris, France.

Boix-Fayos, C., A. Calvo-Cases, A.C. Imeson, M.D. Soriano Soto, and I.R. Tiemessen, 1998: Spatial and short-term temporal variations in runoff, soil aggregation and other soil properties along a Mediterranean climatological gradient. *Catena*, **33,** 123–138.

Boland, J.J., 1997: Assessing urban water use and the role of water conservation measures under climate uncertainty. *Climatic Change*, **37,** 157–176.

Boorman, D.B. and C.E. Sefton, 1997: Recognizing the uncertainty in the quantification of the effects of climate change on hydrological response. *Climatic Change*, **35,** 415–434.

Bouraoui, F., G. Vachaud, L.Z.X. Li, H. LeTreut, and T. Chen, 1999: Evaluation of the impact of climate changes on water storage and groundwater recharge at the watershed scale. *Climate Dynamics*, **15,** 153–161.

Braga, B.P.F. and L.C.B. Molion, 1999: Assessment of the impacts of climate variability and change on the hydrology of South America. In: *Impacts of Climate Change and Climate Variability on Hydrological Regimes* [van Dam, J.C. (ed.)]. UNESCO, International Hydrology Series, Cambridge University Press, Cambridge, United Kingdom and New York, NY, USA, pp. 21–35.

Bruce, J., I. Burton, H. Martin, B. Mills, and L. Mortsch, 2000: *Water Sector: Vulnerability and Adaptation to Climate Change, Final Report*. GCSI and the Meteorological Service of Canada, Toronto, ON, Canada, 144 pp.

Bruci, D.E. and M. Bicaj, 1998: Implication of the expected climate change on the water resources in Albania. In: *Proceedings of the Second International Conference on Climate and Water, Espoo, Finland, August 1998*. Helsinki University of Technology, Helsinki, Finland, pp. 1471–1477.

Buchtele, J., M. Buchtelova, and M. Fortova, 1998: Possible runoff changes simulated using climate scenarios UKHI, UKTR and XCCC in the Czech part of the Elbe River Basin. In: *Proceedings of the Second International Conference on Climate and Water, Espoo, Finland, August 1998*. Helsinki University of Technology, Helsinki, Finland, pp. 208–214.

Bunce, J.A., K.B. Wilson, and T.N. Carlson, 1997: The effect of doubled CO_2 on water use by alfalfa and orchard grass: simulating evapotranspiration using canopy conductance measurements. *Global Change Biology*, **3,** 81–87.

Butina, M., G. Melnikova, and I. Stikute, 1998: Potential impact of climate change on the hydrological regime in Latvia. In: *Proceedings of the Second International Conference on Climate and Water, Espoo, Finland, August 1998*. Helsinki University of Technology, Helsinki, Finland, pp. 1610–1617.

Cao, M.K. and F.I. Woodward, 1998: Dynamic responses of terrestrial ecosystem carbon cycling to global climate change. *Nature*, **393,** 249–252.

Carmichael, J.J., K.M. Strzepek, and B. Minarik, 1996: Impacts of climate change and seasonal variability on economic treatment costs: a case study of the Nitra River Basin, Slovakia. *International Journal of Water Resources Development*, **12,** 209–227.

Carter, T.R., M. Hulme, J.F. Crossley, S. Malyshev, M.G. New, M.E. Schlesinger, and H. Tuomenvirta, 2000: *Climate Change in the 21st Century - Interim Characterizations based on the New IPCC Emissions Scenarios*. The Finnish Environment 433, Finnish Environment Institute, Helsinki, 148 pp.

Chalecki, E.L. and P.H. Gleick, 1999: A framework of ordered climate effects on water resources: a comprehensive bibliography. *Journal of the American Water Resources Association*, **35,** 1657–1665.

Changnon, S.A., 1998: Comments on "secular trends of precipitation amount, frequency, and intensity in the United States" by Karl and Knight. *Bulletin of the American Meteorological Society*, **79,** 2550–2552.

Changnon, S.A. and D. Changnon, 1998: Climatological relevance of major USA weather losses during 1991–1994. *International Journal of Climatology*, **18,** 37–48.

Changnon, S.A. and M. Demissie, 1996: Detection of changes in streamflow and floods resulting from climate fluctuations and land use-drainage changes. *Climatic Change*, **32,** 411–421.

Changnon, S.A., D. Changnon, E.R. Fosse, D.C. Hoganson, R.J. Roth, and J.M. Totsch, 1997: Effects of recent weather extremes on the insurance industry; major implications for the atmospheric sciences. *Bulletin of the American Meteorological Society*, **78,** 425–435.

Chao, P. and A.W. Wood, 1999: *Water management implications of global warming: 7. The Great Lakes-St. Lawrence River Basin*. Institute for Water Resources, U.S. Army Corps of Engineers, Alexandria, VA, USA, 102 pp.

Chao, P.T., B.F. Hobbs, B.N. Venkatesh, 1999: How climate uncertainty should be included in Great Lakes management: modelling workshop results. *Journal of the American Water Resources Association*, **35,** 1485–1497.

Chattopadhyary, N. and M. Hulme, 1997: Evaporation and potential evapotranspiration in India under conditions of recent and future climate change. *Agricultural and Forest Meteorology*, **87,** 55–73.

Checkley, W., L.D. Epstein, R.H. Gilman, D. Figueroa, R.I. Cama, J.A. Patz, and R.E. Black, 2000: Effect of El Niño and ambient temperature on hospital admissions for diarrhoeal diseases in Peruvian children. *Lancet*, **355,** 442–450.

Chiew, F.H.S., T.C. Piechota, J.A. Dracup, and T.A. McMahon, 1998: El Niño Southern Oscillation and Australian rainfall, streamflow and drought—links and potential for forecasting. *Journal of Hydrology*, **204,** 138–149.

Christensen, O. and J. Christensen, 1998: Climate simulations with the HIRHAM limited area regional climate model over Scandinavia. In: *Proceedings of the Second International Conference on Climate and Water, Espoo, Finland, August 1998*. Helsinki University of Technology, Helsinki, Finland, pp. 10–19.

Cleaveland, M.K., 2000: A 963–year reconstruction of summer (JJA) streamflow in the White River, Arkansas, USA, from tree rings. *The Holocene*, **10,** 33–41.

Cohen, S.J., K.A. Miller, A.F. Hamlet, and W. Avis, 2000: Climate change and resource management in the Columbia River Basin. *Water International*, **25,** 253–272.

Compagnucci, R.H. and W.M. Vargas, 1998: Interannual variability of the Cuyo River's streamflow in the Argentinian Andean mountains and ENSO events. *International Journal of Climatology*, **18,** 1593–1609.

Conway, D. and M. Hulme, 1996: The impacts of climate variability and future climate change in the Nile Basin on water resources in Egypt. *Water Resources Development*, **12,** 277–296.

Cosgrove, W.J. and F.R. Rijbersman, 2000: *World Water Vision: Making Water Everybody's Business*. World Water Council, Earthscan Publications Ltd., London, United Kingdom, 108 pp.

Crane, R.G. and B.C. Hewitson, 1998: Doubled CO_2 precipitation changes for the Susquehanna River Basin: downscaling from the GENESIS general circulation model. *International Journal of Climatology*, **18,** 65–66.

Cruise, J.F., A.S. Limaye, and N. Al Abed, 1999: Assessment of impacts of climate change on water quality in the southeastern United States. *Journal of the American Water Resources Association*, **35,** 1539–1550.

Daamen, K., P. Krahe, and K. Wilke, 1998: Impacts of possible climate change on the discharge in German low mountain range catchments. In: *Proceedings of the Second International Conference on Climate and Water, Espoo, Finland, August 1998*. Helsinki University of Technology, Helsinki, Finland, pp. 805–819.

Davidovich, N.V. and M.D. Ananicheva, 1996: Prediction of possible changes in glacio-hydrological characteristics under global warming: Southeastern Alaska, USA. *Journal of Glaciology*, **42,** 407–412.

Dellapenna, J.W., 1999: Adapting the law of water management to global climate change and other hydropolitical stresses. *Journal of the American Water Resources Association*, **35,** 1301–1326.

Dettinger, M.D. and D.R. Cayan, 1995: Large-scale forcing of recent trends toward early smowmelt runoff in California. *Journal of Climate*, **8,** 606–623.

Dickinson, R.E., A.N. Hahmann, and Q. Shao, 1997: Commentary on Mecca sensitivity studies. In: *Assessing Climate Change* [Howe, W. and A. Henderson-Sellers (eds.)]. Gordon and Breach Science Publishers, Sydney, Australia, pp. 195–206.

Döll, P. and S. Siebert, 2001: *Global Modeling of Irrigation Water Requirements*. University of Kassel, Kassel, Germany.

Döll, P., F. Kaspar, and J. Alcamo, 1999: Computation of global water availability and water use at the scale of large drainage Basins. *Mathematische Geologie*, **4,** 111–118.

Dvorak, V., J. Hladny, and L. Kasparek, 1997: Climate change hydrology and water resources impact and adaptation for selected river Basins in the Czech Republic. *Climatic Change*, **36,** 93–106.

Easter, K.W., M. Rosegrant, and A. Dinar (eds.), 1998: *Markets for Water:Potential and Performance*. Kluwer Academic Publishers, Dordrecht, Germany.

Eheart, J.W., A.J. Wildermuth, and E.E. Herricks, 1999: The effects of climate change and irrigation on criterion low streamflows used for determining total maximum daily loads. *Journal of the American Water Resources Association*, **35,** 1365–1372.

ERM, 2000: *Potential UK Adaptation Strategies for Climate Change*. UK Department of the Environment, Transport and the Regions, London, United Kingdom, 66 pp.

Espinosa, D., A. Mendez, I Madrid, and R. Rivera, 1997: Assessment of climate change impacts on the water resources of Panama: the case of the La Villa, Chiriqui and Chagres River Basins. *Climate Research*, **9,** 131–137.

Falkenmark, M. and G. Lindh, 1976: *Water for a Starving World*. Westview Press, Boulder, CO, USA.

Falkenmark, M., 1999: Forward to the future: a conceptual framework for water dependence. *Ambio*, **28,** 356–361.

Fang, X. and H.G. Stefan, 1997: Simulated climate change effects on dissolved oxygen characteristics in ice-covered lakes. *Ecological Modelling*, **103(2–3),** 209–229.

Felzer, B. and P. Heard, 1999: Precipitation differences amongst GCMs used for the U.S. national assessment. *Journal of the American Water Resources Association*, **35,** 1327–1339.

Field, C.B., R.B. Jackson, and H.A. Mooney, 1995: Stomatal responses to increased CO_2—implications from the plant to the global scale. *Plant Cell and Environment*, **18,** 1214–1225.

Fisher, A.C. and S.J. Rubio, 1997: Adjusting to climate change: implications of increased variability and asymetric adjustment costs for investment in water reserves. *Journal of Environmental Economics and Management*, **34,** 207–227.

Fowler, A., 1999: Potential climate change impacts on water resources in the Auckland Region (New Zealand). *Climate Research*, **11,** 221–245.

Frederick, K.D., 1997: Adapting to climate impacts on the supply and demand for water. *Climatic Change*, **37,** 141–156.

Frederick, K.D. (ed.), 1986: *Scarce Water and Institutional Change*. Resources for the Future, Inc., Washington, DC, USA.

Frederick, K.D. and G.E. Schwarz, 1999: Socioeconomic impacts of climate change on U.S. water supplies. *Journal of the American Water Resources Association*, **35,** 1563–1583.

Friend, A.D., A.K. Stevens, R.G. Knox, and M.G.R. Cannell, 1997: A process-based, terrestrial biosphere model of ecosystem dynamics (HYBRID v3.0). *Ecological Modelling*, **95,** 249–287.

Frisk, T., Ä. Bilaletdin, K. Kallio, M. Saura, 1997. Modelling the effects of climatic change on lake eutrophication. *Boreal Environment Research*, **2,** 53–67.

Fushimi, H. 1999: Water resources and environmental problems of Lake Biwa, Japan. In: *Limnology: Textbook for the Ninth IHP Training Course in 1999: International Hydrological Programme* [Terai, H. (ed.)]. Institute for Hydrospheric-Atmospheric Sciences, Nagoya University and UNESCO, Nagoya, Japan, Chapter 9.

Gellens, D. and E. Roulin, 1998: Streamflow response of Belgian catchments to IPCC climate change scenarios. *Journal of Hydrology*, **210,** 242–258.

Gellens, D., E. Roulin, and F. Gellens-Meulenberghs, 1998: Impact of climate change on the water balance in the river Meuse Basin (Belgium). In: *Proceedings of the Second International Conference on Climate and Water, Espoo, Finland, August 1998*. Helsinki University of Technology, Helsinki, Finland, pp. 820–828.

Genta, J.L., G. Perez-Iribarren, and C.R. Mechoso, 1998. A recent increasing trend in the streamflow of rivers in southeastern South America. *Journal of Climate*, **11,** 2858–2862.

Georgiyevsky, V.Yu., S.A. Zhuravin, and A.V. Ezhov, 1995: Assessment of trends in hydrometeorological situation on the Great Russian Plain under the effect of climate variations. In: *Proceedings of American Geophysical Union, 15th Annual Hydrology Days*, pp. 47–58.

Georgiyevsky, V.Yu., A.V. Yezhov, A.L. Shalygin, A.I. Shiklomanov, and I.A. Shiklomanov, 1996: Evaluation of possible climate change impact on hydrological regime and water resources of the former USSR rivers. *Russian Meteorology and Hydrology*, **11,** 89–99.

Georgiyevsky, V.Yu., A.V. Yezhov, and A.L. Shalygin, 1997: An assessment of changing river runoff due to man's impact and global climate warming. In: *River Runoff Calculations, Report at the International Symposium*. UNESCO, pp. 75–81.

Gifford, R.M., D.J. Barrett, J.L. Lutze, and A.B. Samarakoon, 1996: Agriculture and global change: scaling direct carbon dioxide impacts and feedbacks through time. In: *Global Change and Terrestrial Ecosystems* [Walker, B.R. and W. Steffen (eds.)]. Cambridge University Press, Cambridge, United Kingdom and New York, NY, USA, pp. 229–259.

Gleick, P.H., 1998: *The World's Water. 1998/99*. Island Press, Washington DC, USA, 307 pp.

Gleick, P.H., 2000: *Water: Potential Consequences of Climate Variability and Change for the Water Resources of the United States*. Report of the Water Sector Assessment Team of the National Assessment of the Potential Consequences of Climate Variability and Change. Pacific Institute for Studies on Development, Economics, and Security, Oakland, CA, USA, 151 pp.

Gleick, P.H. and E.L. Chalecki, 1999: The impacts of climatic changes for water resources of the Colorado and Sacramento-San Jaoquin River Basins. *Journal of the American Water Resources Association*, **35,** 1429–1441.

Grabs, W. (ed.) 1997: *Impact of climate change on hydrological regimes and water resources management in the Rhine Basin*. International Commission for the Hydrology of the Rhine Basin (CHR), Report I-16, Ledystad, The Netherlands, 172 pp.

Greene, A.M., W.S. Broecker, and D. Rind, 1999: Swiss glacier recession since the Little Ice Age: reconciliation with climate records. *Geophysical Research Letters*, **26,** 1909–1912.

Gregory, J.M. and J. Oerlemans, 1998: Simulated future sea level rise due to glacier melt based on regionally and seasonally resolved temperature changes. *Nature*, **391,** 474–476.

Gregory, J.M., J.F.B. Mitchell, and A.J. Brady, 1997: Summer drought in northern midlatitudes in a time-dependent CO_2 climate experiment. *Journal of Climate*, **10,** 662–686.

Haerberli, W. and M. Beniston, 1998: Climate change and its impacts on glaciers and permafrost in the Alps. *Ambio*, **27,** 258–265.

Haeberli, W., R. Frauenfelder, M. Hoelzle, and M. Maisch, 1999: On rates and acceleration trends of global glacier mass changes. *Geografisker Annaler Series A – Physical Geography*, **81A,** 585–591.

Hailemariam, K., 1999: Impact of climate change on the water resources of Awash River Basin, Ethiopia. *Climate Research*, **6,** 91–96.

Hamlet, A.F. and D.P. Lettenmaier, 1999: Effects of climate change on hydrology and water resources in the Columbia River Basin. *Journal of the American Water Resources Association*, **35,** 1597–1623.

Hanaki, K., K. Takara, T. Hanazato, H. Hirakuchi, and H. Kayanne, 1998: Impacts on hydrology/water resources and water environment. In: *Global Warming—The Potential Impact on Japan* [Nishioka, S. and H. Harasawa (eds.)]. Springer-Verlag, Tokyo, Japan, pp. 131–163.

Hanratty, M.P. and H.G. Stefan, 1998: Simulating climate change effects in a Minnesota agricultural watershed. *Journal of Environmental Quality*, **27,** 1524–1532.

Hansler, G. and D.C. Major, 1999: Climate change and the water supply systems of New York City and the Delaware Basin: planning and action considerations for water managers. In: *Proceedings of the Specialty Conference on Potential Consequences of Climate Variability and Change to Water Resources of the United States* [Briane Adams, D. (ed.)]. American Water Resources Association, Herndon, VA, USA, pp. 327–330.

Hassan, H., K. Hanaki, and T. Matsuo, 1998a: A modelling approach to simulate the impact of climate change on lake water quality: phytoplankton growth rate assessment. *Water Science and Technology*, **37(2),** 177–185.

Hassan, H., T. Aramaki, K. Hanaki, T. Matsuo, and R. Wilby, 1998b: Lake stratification and temperature profiles simulated using downscaled GCM output. *Water Science and Technology*, **38(11),** 217–226.

Hatch, U., S. Jagtap, J. Jones, and M. Lamb, 1999: Potential effects of climate change on agricultural water use in the Southeast U.S. *Journal of the American Water Resources Association*, **35,** 1551–1561.

Haughton, G., 1998: Private profits—public drought: the creation of a crisis in water management for West Yorkshire. *Transactions of the Institute of British Geographers*, **NS 23,** 410–435.

Hennessy, R.J., J.M. Gregory, and J.F.B. Mitchell, 1997: Changes in daily precipitation under enhanced greenhouse conditions. *Climate Dynamics*, **13,** 667–680.

Herrington, P., 1996: *Climate Change and the Demand for Water*. Her Majesty's Stationery Office, London, United Kingdom, 164 pp.

Hewitt, K., 1997: *Regions of Risk: A Geographical Introduction to Disasters*. Addison-Wesley Longman, Essex, United Kingdom.

Hladny, J., J. Buchtele, M. Doubkova, V. Dvorak, L. Kasparek, O. Novicky, E. Prensilova, 1996: *Impact of a Potential Climate Change on Hydrology and Water Resources: Country Study for the Czech Republic*. Publication No. 20, National Climate Program of the Czech Republic, Prague, Czech Republic, 137 pp.

Hlaveova, K. and J. Eunderlik, 1998: Impact of climate change on the seasonal distribution of runoff in mountainous Basins in Slovakia. *Hydrology, Water Resources and Ecology in Headwaters*, **248,** 39–46.

Hobbs, B.F., 1997: Bayesian methods for analysing climate change and water resource uncertainties. *Journal of Environmental Management*, **49,** 53–72.

Hobbs, B.F., P.T. Chao, and B.N. Venkatesh, 1997: Using decision analysis to include climate change in water resources decision making. *Climatic Change*, **37,** 177–202.

Holt, C.P. and J.A.A. Jones, 1996: Equilibrium and transient global warming scenarios: implications for water resources in England and Wales. *Water Resources Bulletin*, **32,** 711–721.

Hostetler, S.W. and E.E. Small, 1999: Response of North American freshwater lakes to simulated future climates. *Journal of the American Water Resources Association*, **35,** 1625–1637.

Howe, C.W., J.K. Lazo, and K.R. Weber, 1990: The economic impacts of agriculture-to-urban water transfers on the area of origin: a case study of the Arkansas River Valley in Colorado. *American Journal of Agricultural Economics*, **72,** 1200–1204.

Hulme, M. (ed.), 1996: *Climate Change and Southern Africa*. Climatic Research Unit, University of East Anglia, Norwich, United Kingdom, 104 pp.

Hulme, M. and G. Jenkins, 1998: *Climate Change Scenarios for the United Kingdom: Scientific Report*. UKCIP Technical Report No. 1. Climatic Research Unit, University of East Anglia, Norwich, United Kingdom, 80 pp.

Hulme, M., E.M. Barrow, N.W. Arnell, P.A. Harrison, T.C. Johns, and T.E. Downing, 1999: Relative impacts of human-induced climate change and natural climate variability. *Nature*, **397,** 688–691.

Hunsaker, D.J., G.R. Hendrey, B.A. Kimball, K.F. Lewin, J.R. Mauney, and J. Nagy, 1994: Cotton evapotranspiration under field conditions with CO_2 enrichment and variable soil moisture regimes. *Agricultural and Forest Meteorology*, **70,** 247–258.

Hurd, B., N. Leary, R. Jones, and J. Smith, 1999: Relative regional vulnerability of water resources to climate change. *Journal of the American Water Resources Association*, **35,** 1399–1409.

Isdale, P.J., B.J. Stewart, K.S. Tickle, and J.M. Lough, 1998. Palaeohydrological variation in a tropical river catchment: a reconstruction using fluorescent bands in corals of the Great Barrier Reef, Australia. *The Holocene*, **8,** 1–8.

Jaagus, J., 1998: Modelled changes in river runoff using various climate scenarios for Estonia. In: *Proceedings of the Second International Conference on Climate and Water, Espoo, Finland, August 1998*. Helsinki University of Technology, Helsinki, Finland, pp. 94–103.

Jarvet, A., 1998: Estimation of possible climate change impact on water management in Estonia. In: *Proceedings of the Second International Conference on Climate and Water, Espoo, Finland, August 1998*. Helsinki University of Technology, Helsinki, Finland, pp. 1449–1458.

Jansons, V. and M. Butina, 1998: Potential impacts of climate change on nutrient loads from small catchments. In: *Proceedings of the Second International Conference on Climate and Water, Espoo, Finland, August 1998*. Helsinki University of Technology, Helsinki, Finland, pp. 932–939.

Jeton, A.E., M.D. Dettinger, and J. LaRue Smith, 1996: Potential effects of climate change on streamflow, eastern and western slopes of the Sierra Nevada, California and Nevada. *U.S. Geological Survey, Water Resources Investigations Report*, **95-4260,** 44 pp.

Jose, A.M. and N.A. Cruz, 1999: Climate change impacts and responses in the Philippines: water resources. *Climate Research*, **12,** 77–84.

Jose, A.M., L.M. Sosa, and N.A. Cruz, 1996: Vulnerability assessment of Angat water reservoir to climate change. *Water, Air, and Soil Pollution*, **92,** 191–201.

Kaczmarek, Z., 1996: Water resources management. In: *Climate Change 1995: Impacts, Adaptations, and Mitigation of Climate Change: Scientific-Technical Analyses. Contribution of Working Group II to the Second Assessment Report of the Intergovernmental Panel on Climate Change* [Watson, R.T., R.H. Moss, and M.C. Zinyowera (eds.)]. Cambridge University Press, Cambridge, United Kingdom and New York, NY, USA, pp. 469–486.

Kaczmarek, Z., D. Jurak, and J.J. Napiórkowski, 1997: *Impact of Climate Change on Water Resources in Poland*. Public Institute of Geophysics, Polish Academy of Sciences, Warsaw, Poland, **295,** 51 pp.

Kaczmarek, Z., J. Napiórkowski, and K. Strzepek, 1996: Climate change impacts on the water supply system in the Warta River catchment, Poland. *International Journal of Water Resources Development*, **12,** 165–180.

Kallio, K., S. Rekolainen, P. Ekholm, K. Granlund, Y. Laine, H. Johnsson, and M. Hoffman, 1997: Impacts of climatic change on agricultural nutrient losses in Finland. *Boreal Environment Research*, **2,** 33–52.

Kang, E., G. Cheng, Y. Lan, and H. Jin, 1999: A model for simulating the response of runoff from the mountainous watersheds of northwest China to climatic changes. *Science in China*, **42.**

Karl, T.R. and R.W. Knight, 1998: Secular trends of precipitation amount, frequency and intensity in the United States. *Bulletin of the American Meteorological Society*, **79,** 231–241.

Kaser, G., 1999: A review of the modern fluctuations of tropical glaciers. *Global and Planetary Change*, **22,** 93–103.

Kaser, G., S. Hastenrath, and A. Ames, 1996: Mass balance profiles on tropical glaciers. *Zeitschrift für Gletscherkunde und Glazialgeologie*, **32,** 75–81.

Kaser, G. and B. Noggler, 1991: Observations on Speke Glacier, Ruwenzori Range, Uganda. *Journal of Glaciology*, **37,** 313–318.

Kilkus, K., 1998: Lakes of temperate regions as climate change indicators. In: *Proceedings of the Second International Conference on Climate and Water, Espoo, Finland, August 1998*. Helsinki University of Technology, Helsinki, Finland, pp. 588–596.

Kimball, B.A., R.L. LaMorte, P.J. Pinter Jr., G.W. Wall, D.J. Hunsaker, F.J. Adamsen, S.W. Leavitt, T.L. Thompson, A.D. Matthias, and T.J. Brooks, 1999: Free-air CO_2 enrichment and soil nitrogen effects on energy balance and evapotranspiration of wheat. *Water Resources Research*, **35,** 1179–1190.

Kindler, J., 2000: Integrated water resources management: the meanders. *Water International*, **25,** 312–319.

Komescu, A.U., A. Erkan, and S. Oz, 1998: Possible impacts of climate change on soil moisture availability in the Southeast Anatolia Development Project Region (GAP): an analysis from an agricultural drought perspective. *Climatic Change*, **40,** 519–545.

Korner, C., 1996: The response of complex multispecies systems to elevated CO_2. In: *Global Change and Terrestrial Ecosystems* [Walker, B.R. and W. Steffen (eds.)]. Cambridge University Press, Cambridge, United Kingdom, pp. 20–42.

Krasovskaia, I., 1995: Quantification of the stability of river flow regimes. *Hydrological Sciences Journal*, **40,** 587–598.

Kuchment, L.S., 1998: The estimation of the sensitivity of the river runoff to climate change for different physiogeographic regions of Russia and an experience of prediction of possible runoff change. In: *Proceedings of the Second International Conference on Climate and Water, Espoo, Finland, August 1998*. Helsinki University of Technology, Helsinki, Finland, pp. 171–177.

Kunkel, K.E., R.A. Pielke Jr., and S.A. Changnon, 1999: Temporal fluctuations in weather and climate extremes that cause economic and human health impacts: a review. *Bulletin of the American Meteorological Society*, **80,** 1077–1098.

La Red, 1999: *La Red de Estudios Sociedades en Prevencion de Desastres*. International Non-Governmental Organization, Lima, Peru. Available online at http:www.lared.org.pe.

Laird, K.R., S.C. Fritz, E.C. Grimm, and P.G. Mueller, 1996: Century-scale paleoclimatic reconstruction from Moon Lake, a closed-Basin lake in the northern Great Plains. *Limnology and Oceanography*, **41,** 890–902.

Landsea, C.W., N. Nicholls, W.M. Gray, and L.A. Avila, 1996: Downward trend in the frequency of intense Atlantic hurricanes during the past five decades. *Geophysical Research Letters*, **23,** 1697–1700.

LeCompte, E. and K. Gahagan, 1998: Hurricane insurance protection in Florida. In: *Paying the Price* [Kunreuther, H. and R.J. Roth (eds.)]. The Joseph Henry Press, Washington, DC, USA.

Lepisto, A. and Y. Kivinen, 1997: Effects of climate change on hydrological patterns of a forested catchment: a physically-based modelling approach. *Boreal Environment Research*, **2,** 19–31.

Lettenmaier, D.P., A.W. Wood, R.N. Palmer, E.F. Wood, and E.Z. Stakhiv, 1999: Water resources implications of global warming: a U.S. regional perspective. *Climatic Change*, **43,** 537–579.

Leung, L.R. and M.S. Wigmosta, 1999: Potential climate change impacts on mountain watersheds in the Pacific Northwest. *Journal of the American Water Resources Association*, **35,** 1463–1471.

Lewis, C.F.M., D.L. Forbes, E. Nielsen, L.M. Thorliefson, A.M. Telka, R.E. Vance, and B.J. Todd, 1998: Mid-Holocene Lake Winnipeg: where was it? In: *Proceedings of GSA 1998 Annual Meeting, Toronto, October 1998*. Geological Society of America, Boulder, USA. Abstracts 30, p. A168.

Lins, H.F. and J.R. Slack, 1999: Streamflow trends in the United States. *Geophysical Research Letters*, **26,** 227–230.

Lins, H. and E.Z. Stakhiv, 1998: Managing the nation's water in a changing climate. *Journal of the American Water Resources Association*, **34,** 1255–1264.

Liu, C., 1998: The potential impact of climate change on hydrology and water resources in China. In: *Proceedings of the Second International Conference on Climate and Water, Espoo, Finland, August 1998*. Helsinki University of Technology, Helsinki, Finland, pp. 1420–1434.

Liu, C., E. Kang, and S. Liu, 1999: Study on glacier variation and its runoff responses in the arid region on Northwest China. *Science in China*, **42.**

Loaiciga, L.C., D.R. Maidment, and J.B. Valdes, 1998: *Climate change impacts on the water resources of the Edwards Balcones Fault Zone aquifer, Texas*. ASCE/USEPA Cooperative Agreement CR824540-01-0, American Society of Civil Engineers, Reston VA, USA, 72 pp. (plus figures).

Luo, W.B. and B. Caselton, 1997: Using Dempster-Shafer theory to represent climate change uncertainties. *Journal of Environmental Management*, **49,** 73–93.

Mailu, G.M., 1988: Groundwater potential in Mandera District. *Kenya Engineer*, 25–27.

Mailu, G.M., 1992: Impact of rock catchments on water resources of Kitui District. In: *Proceedings of the Second National Conference on Rainwater Catchment Systems in Kenya*. Nairobi, Kenya, pp. 220–230.

Mailu, G.M., 1993: The climatic impact on water resources in Wajir District, Kenya. In: *Proceedings of the First Iinternational Conference of the African Meteorological Society*. Nairobi, Kenya, pp. 925–931.

Major, D.C., 1998: Climate change and water resources: the role of risk management methods. *Water Resources Update*, **112,** 47–50.

Mandelkern, S., S. Parey, and N. Tauveron, 1998: Hydrologic consequences of greenhouse effect, a case study of the river Doubs. In: *Proceedings of the Second International Conference on Climate and Water, Espoo, Finland, August 1998*. Helsinki University of Technology, Helsinki, Finland, pp. 1505–1514.

Marengo, J.A., 1995: Variations and change in South American streamflow. *Climatic Change*, **31,** 99–117.

Marengo, J.A., J. Tomasella, and C.R. Uvo, 1998: Trends in streamflow and rainfall in tropical South America: Amazonia, eastern Brazil and northwestern Peru. *Journal of Geophysical Research—Atmospheres*, **103,** 1775–1783.

Marsden, T.K., 1997: Reshaping environments: agriculture and water interactions and the creation of vulnerability. *Transactions of the Institute of British Geographers*, **22,** 321–337.

Matalas, N.C., 1997: Stochastic hydrology in the context of climate change. *Climatic Change*, **37,** 89–101.

McCabe, G.J., 1996: Effects of winter atmospheric circulation on temporal and spatial variability in annual streamflow in the western United States. *Hydrological Sciences Journal*, **41,** 873–888.

McGuffie, K., A. Henderson-Sellers, N. Holbrook, Z. Kothavala, O. Balachova, and J. Hoekstra, 1999: Assessing simulations of daily temperature and precipitation variability with global climate models for present and enhanced greenhouse climates. *International Journal of Climatology*, **19,** 1–26.

Meehl, G.A. and W.M. Washington, 1996: El Niño-like climate change in a model with increased atmospheric CO_2 concentrations. *Nature*, **382,** 56–60.

Meyer, J.L., M.J. Sale, P.J. Mulholland, and N.L. Poff, 1999: Impacts of climate change on aquatic ecosystems functioning and health. *Journal of the American Water Resources Association*, **35,** 1373–1386.

Mendoza, V.M., E.E. Villanueva, and J. Adem, 1997: Vulnerability of basins and watersheds in Mexico to global climate change. *Climate Research*, **9,** 139–145.

Mileti, D., 1999: *Disasters by Design: A Reassessment of Natural Hazards in the United States*. The Joseph Henry Press, Washington, DC, USA, 250 pp.

Miller, K.A., S.L. Rhodes, and L.J. MacDonnell, 1997: Water allocation in a changing climate: institutions and adaptation. *Climatic Change*, **35,** 157–177.

Miller, N.L. and J. Kim, 1996: Numerical prediction of precipitation and streamflow over the Russian River watershed during the January 1995 California storms. *Bulletin of the American Meteorological Society*, **77,** 101–105.

Miller, N.L. and J. Kim, 2000: Climate change sensitivity analysis for two California watersheds. *Journal of the American Water Resources Association*, **36,** 657–661.

Miller, N.L., J. Kim, R.K. Hartman, and J. Farrara, 1999: Downscaled climate and streamflow study of the southwestern United States. *Journal of the American Water Resources Association*, **35,** 1525–1538.

Minnery, J.R. and F.I. Smith, 1996: Climate change, flooding and urban infrastructure. In: *Greenhouse: Coping with Climate Change* [Bouma, W.J., M. Manning, and G.I. Pearman (eds.)]. Commonwealth Scientific and Industrial Research Organization Publishing, Melbourne, Australia, pp. 235–247.

Mirza, M.Q., 1999: Climate change and water resources in South Asia. *Asia Pacific Journal of Environment and Development*, **7,** 17–29.

Mirza, M.Q. and A. Dixit, 1996: Climate change and water management in the GBM Basins. *Water Nepal*, **5,** 71–100.

Mirza, M.Q., R.A. Warrick, N.J. Ericksen, and G.J. Kenny, 1998: Trends and persistence in precipitation in the Ganges, Brahmaputra and Meghna Basins in South Asia. *Hydrological Sciences Journal*, **43,** 845–858.

Moglen, G.E., E.A.B. Eltahir, and R.L. Bras, 1998: On the sensitivity of drainage density to climate change. *Water Resources Research*, **34,** 855–862.

Morison, J.I.L., 1987: Intercellular CO_2 concentration and stomatal response to CO_2. In: *Stomatal Function* [Zeiger, E. and G.D. Farquhar (eds.)]. Stanford University Press, Stanford, CA, USA, pp. 229–251.

Munasinghe, M. and C. Clarke (eds.), 1995: *Disaster Prevention for Sustainable Development: Economic and Policy Issues; A Report from the Yokohama World Conference on Natural Disaster Reduction, May 1994*. World Bank, Washington, DC, USA.

Murdoch, P.S., J.S. Baron, and T.L. Miller, 2000: Potential effects of climate change on surface water quality in North America. *Journal of the American Water Resources Association*, **36,** 347–366.

Mutua, F.M., 1998: Sensitivity of the hydrologic cycle in Tana River Basin to climate change. In: *Vulnerability and adaptation to potential impacts of climate change in Kenya* [Omenda, T.O., J.G. Kariuki, and P.N. Mbuthi (eds.)]. pp. 22–30.

Najjar, R.G., 1999: The water balance of the Susquehanna River Basin and its response to climate change. *Journal of Hydrology*, **219,** 7–19.

Nash, L.L. and P.H. Gleick, 1993: *The Colorado River Basin and Climatic Change: The Sensitivity of Streamflow and Water Supply to Variations in Temperature and Precipitation*. EPA230-R-93-009, U.S. Environmental Protection Agency, Washington, DC, USA, 121 pp.

National Research Council, 1992: *Water Transfers in the West: Efficiency, Equity and the Environment*. National Academy Press, Washington, DC, USA.

Obolkin, V. and V. Potemkin, 1998: The impact large lakes on climate: in past and present. In: *Proceedings of the Second International Conference on Climate and Water, Espoo, Finland, August 1998*. Helsinki University of Technology, Helsinki, Finland, pp. 1217– 1221.

O'Connor, R.E., B. Yarnal, R. Neff, R. Bond, N. Wiefek, C. Reenoik, R. Shudak, C.L. Jocoy, P. Pascate, and C.G. Knight, 1999: Weather and climate extremes, climate change and planning: views of community water system managers in Pennsylvania's Susquehanna River Basin. *Journal of the American Water Resources Association*, **35,** 1411–1419.

Oerlemans, J., B. Anderson, A. Hubbard, P. Huybrechts, T. Johannesson, W.H. Krap, M. Schmeits, A.P. Stroeven, R.S.W. van der Wal, J. Wallinga, and Z. Zuo, 1998: Modelling the response of glaciers to climate warming. *Climate Dynamics*, **14,** 267–274.

Ojima, D., L. Garcia, E. Eigaali, K. Miller, T.G.F. Kittel, and J. Lackelt, 1999: Potential climate change impacts on water resources in the Great Plains. *Journal of the American Water Resources Association*, **35,** 1443–1454.

Olsen, J.R., J.R. Stedinger, N.C. Matalas, and E.Z. Stakhiv, 1999: Climate variability and flood frequency estimation for the Upper Mississippi and Lower Missouri Rivers. *Journal of the American Water Resources Association*, **35,** 1509–1523.

OECD, 1985: *Gestion des Projets D'Amenagement des Eaux*. Organisation de Cooperation et de Developpment Economiques, Paris, France.

Orlob, G.T., G.K. Meyer, L. Somlyody, D. Jurak, and K. Szesztay, 1996: Impact of climate change on water quality. In: *Water Resources Management in the Face of Climatic/Hydrologic Uncertainties* [Kaczmarek, Z., K. Strzepek, and L. Somlyody (eds.)]. Kluwer Academic Publishers, Dordrecht, The Netherlands, pp. 70–105.

Osborn, T.J., M. Hulme, P.D. Jones, and T.A. Basnet, 2000: Observed trends in the daily intensity of United Kingdom precipitation. *International Journal of Climatology*, **20,** 347–364.

PAHO, 1999: *Conclusions and Recommendations: Meeting on Evaluation of Preparedness and Response to Hurricanes Georges and Mitch, 16–19 February, 1999: Santo Domingo, Dominican Republic*. Pan-American Health Organization, Washington, DC. Available online at http://www.paho.org/english/ped/pedhome.htm.

Panagoulia, D. and G. Dimou, 1996: Sensitivities of groundwater-streamflow interaction to global climate change. *Hydrological Sciences Journal*, **41,** 781–796.

Panagoulia, D. and G. Dimou, 1997: Sensitivity of flood events to global climate change. *Journal of Hydrology*, **191,** 208–222.

Pasterick, E.T., 1998: The national flood insurance program. In: *Paying the Price* [Kunreuther, H. and R.J. Roth (eds.)]. The Joseph Henry Press, Washington, DC.

Patz, J.A., K. Strzepek, and L. Lele, 1998: Predicting key malaria transmission factors, biting and entomological inoculation rates, using modelled soil moisture in Kenya. *Tropical Medicine and International Health*, **3,** 818–827.

Peacock, W., B. Morrow, and H. Gladwin (eds.), 1997: *Hurricane Andrew: Ethnicity, Gender and the Sociology of Disasters*. Routledge, London, United Kingdom.

Petrovic, P., 1998: Possible climate impacts on the water resources of the Danube River Basin, case study: subBasin of the Nitra River. In: *Proceedings of the Second International Conference on Climate and Water, Espoo, Finland, August 1998*. Helsinki University of Technology, Helsinki, Finland, pp. 981–990.

Piechota, T.C., J.A. Dracup, and R.G. Fovell, 1997: Western U.S. streamflow and atmospheric circulation patterns during El Niño-Southern Oscillation. *Journal of Hydrology*, **201,** 249–271.

Piechota, T.C., J.A. Dracup, F.H.S. Chiew, and T.A. McMahon, 1998: Seasonal streamflow forecasting in eastern Australia and the El Niño-Southern Oscillation. *Water Resources Research*, **34,** 3035–3044.

Pielke, R.A. Jr., 1997: Asking the right questions: atmospheric research and societal needs. *Bulletin of the American Meteorological Society*, **78,** 255–264.

Pielke, R.A. Jr. and C.W. Landsea, 1998 Normalized hurricane damages in the United States: 1925–95. *Weather and Forecasting*, **13,** 621–631.

Pilgrim, J.M., X. Fang, and H.G. Stefan, 1998: Stream temperature correlations with air temperatures in Minnesota: implications for climate warming. *Journal of the American Water Resources Association*, **34,** 1109–1121.

Pilling, C. and J.A.A. Jones, 1999: High resolution climate change scenarios: implications for British runoff. *Hydrological Processes*, **13,** 2877–2895.

Pollard, D. and S.L. Thompson, 1995: Use of a land-surface transfer scheme (LSX) in a global climate model: the response to doubling stomatal conductance. *Global and Planetary Change*, **10,** 129–161.

Pulwarty, R.S. and W.E. Riebsame, 1997: The political ecology of vulnerability to hurricane-related hazards. In: *Hurricanes: Climate and Socioeconomic Impacts* [Diaz, H. and R. Pulwarty (eds.)]. Springer-Verlag, New York, NY, USA.

Rabus, B.T. and K.A. Echelmeyer, 1998: The mass balance of McCall Glacier, Brooks Range, Alaska, USA: its regional relevance and implications for climate change in the Arctic. *Journal of Glaciology*, **44,** 333–351.

Raskin, P., P. Gleick, P. Kirshen, G. Pontius, and K. Strzepek, 1997: *Water Futures: Assessment of Long-Range Patterns and Problems. Background Report for the Comprehensive Assessment for the Freshwater Resources of the World*. Stockholm Environment Institute, Stockholm, Sweden, 78 pp.

Raupach, M.R., 1998: Influences of local feedbacks on land-air exchanges of energy and carbon. *Global Change Biology*, **4,** 477–494.

Reynard, N.S., C. Prudhomme, and S.M. Crooks, 1998: The potential impacts of climate change on the flood characteristics of a large catchment in the UK. In: *Proceedings of the Second International Conference on Climate and Water, Espoo, Finland, August 1998*. Helsinki University of Technology, Helsinki, Finland, pp. 320–332.

Risby, J.S. and D. Entekhabi, 1996: Observed Sacremento Basin streamflow response to precipitation and temperature changes and its relevance to climate impact studies. *Journal of Hydrology*, **184,** 209–223.

Ritschard, R.L., J.F. Cruise, and L.U. Hatch, 1999: Spatial and temporal analysis of agricultural water requirements in the Gulf Coast of the United States. *Journal of the American Water Resources Association*, **35,** 1585–1596.

Roberts, G., 1998: The effects of possible future climate change on evaporation losses from four contrasting UK water catchment areas. *Hydrological Processes*, **12,** 727–739.

Robinson, P.J., 1997: Climate change and hydropower generation. *International Journal of Climatology*, **17,** 983–996.

Robson, A.J., T.K. Jones, D.W. Reed, and A.C. Bayliss, 1998: A study of national trend and variation in UK floods. *International Journal of Climatology*, **18,** 165–182.

Roosare, J., 1998: Local-scale spatial interpration of climate change impact on river runoff in Estonia. In: *Proceedings of the Second International Conference on Climate and Water, Espoo, Finland, August 1998*. Helsinki University of Technology, Helsinki, Finland, pp. 86–93.

Rumsby, B.T. and M.G. Mackin, 1994: Channel and floodplain response to recent abrupt climate change—the Tyne Basin, Northern England. *Earth Surface Processes and Landforms*, **19,** 499–515.

Saelthun, N.R., P. Aittoniemi, S. Bergstrom, K. Einarsson, T. Johannesson, G. Lindstrom, P.-O. Ohlsson, T. Thomsen, B. Vehriläinen, and K.O. Aamodt, 1998: Climate change impacts on runoff and hydropower in the Nordic countries. *TemaNord*, **552,** 170 pp.

Salinger, M.J., R. Basher, B. Fitzharris, J. Hay, P.D. Jones, I.P. Macveigh, and I. Schmideley-Lelu, 1995: Climate trends in the south-west Pacific. *International Journal of Climatology*, **15,** 285–302.

Sandstrom, K., 1995: Modeling the effects of rainfall variability on groundwater recharge in semi-arid Tanzania. *Nordic Hydrology*, **26,** 313–330.

Schindler, D.W., 2001: The cumulative effect of climate warming and other human stresses on Canadian freshwaters in the new millennium. *Canadian Journal of Fisheries and Aquatic Science*, **58,** 1–12.

Schindler, D.W., S.E. Bayley, B.R. Barker, K.G. Beaty, D.R. Cruikshank, E.J. Fee, E.U. Schindler, and M.P. Stainton, 1996: The effects of climatic warming on the properties of boreal lakes and streams at the Experimental Lakes Area, northwestern Ontario. *Limnology and Oceanography*, **41,** 1004–1017.

Schreider, S.Y., A.J. Jakeman, A.B. Pittock, and P.H. Whetton, 1996: Estimation of the possible climate change impacts on water availability, extreme flow events and soil moisture in the Goulburn and Ovens Basins, Victoria. *Climatic Change*, **34,** 513–546.

Schulze, R.E., 1997: Impacts of global climate change in a hydrologically vulnerable region: challenges to South African hydrologists. *Progress in Physical Geography*, **21,** 113–136.

Sefton, C.E.M. and D.B. Boorman, 1997: A regional investigation into climate change impacts on UK streamflows. *Journal of Hydrology*, **195,** 26–44.

Seidel, K., C. Ehrler, and J. Martinec, 1998: Effects of climate change on water resources and runoff in an Alpine Basin. *Hydrological Processes*, **12,** 1659–1669.

Sellers, P.J., R.E. Dickinson, D.A. Randall, A.K. Betts, F.G. Hall, J.A. Berry, G.J. Collatz, A.S. Denning, H.A. Mooney, C.A. Nobre, N. Sato, C.B. Field, and A. Henderson-Sellers, 1997: Modeling the exchanges of energy, water, and carbon between continents and the atmosphere. *Science*, **275,** 502–509.

Shen, D. and R. Liang, 1998: Global warming effects on Hanjaing river hydrological regimes and water resources. In: *Proceedings of the Second International Conference on Climate and Water, Espoo, Finland, August 1998*. Helsinki University of Technology, Helsinki, Finland, pp. 769–777.

Shiklomanov, A.I., 1994: The influence of anthropogenic changes in global climate on the Yenisey River Runoff. *Russian Meteorology and Hydrology*, **2,** 84–93.

Shiklomanov, I.A., 1998: *Asssessment of water resources and water availability in the world. Background Report for the Comprehensive Assessment of the Freshwater Resources of the World*. Stockholm Environment Institute, Stockholm, Sweden, 88 pp.

Shiklomanov, I.A. and V.Yu. Georgiyevsky, 2001: Anthropogenic global climate change and water resources. In: *World Water Resources at the Beginning of the 21st Century*. UNESCO, Paris, France, (in press).

Shiklomanov, I.A., A.I. Shiklomanov, R.B. Lammers, B.J. Peterson, and C. Vorosmarty, 2000: The dynamics of river water inflow to the Arctic Ocean. In: *The Freshwater Budget of the Arctic Ocean*. Kluwer Academic Publishers, Dordrecht, The Netherlands, (in press).

Shorthouse, C. and N.W. Arnell, 1997: Spatial and temporal variability in European river flows and the North Atlantic Oscillation. *FRIEND'97: International Association of Hydrological Science Publications*, **246,** 77–85.

Singh, P., 1998: Effect of global warming on the streamflow of high-altitude Spiti River. In: *Ecohydrology of High Mountain Areas* [Chalise, S.R. A. Herrman, N.R. Khanal, H. Lang, L. Molnar, and A.P. Pokhrel (eds.)]. International Centre for Integrated Mountain Development (ICIMOD), Kathmandu, Nepal, pp. 103–104.

Singh, P. and N.Kumar, 1997: Impact assessment of climate change on the hydrological response of a snow and glacier melt runoff-dominated Himalayan river. *Journal of Hydrology*, **193,** 316–350.

Sircoulon, J., 1990: *Impact Possible des Changements Climatiques à Venir sur les Ressources en eau des Régions Arides et Semi-Arides. Comportement des Cours d'eau Tropicaux, des Rivières et des Lacs en Zone Sahèlienne.* World Clmate Action Programme 12, WMO Technical Document 380, World Meteorological Organisation, Geneva, Switzerland, (in French).

Stakhiv, E.Z., 1998: Policy implications of climate change impacts on water resources management. *Water Policy*, **1,** 159–175.

Stanescu, V., C. Corbus, V. Ungureanu, and M. Simota, 1998: Quantification of the hydrological regime modification in the case of climate changes. In: *Proceedings of the Second International Conference on Climate and Water, Espoo, Finland, August 1998.* Helsinki University of Technology, Helsinki, Finland, pp. 198–207.

Starosolszky, O. and B. Gauzer, 1998: Effect of precipitation and temperature changes on the flow regime of the Danube river. In: *Proceedings of the Second International Conference on Climate and Water, Espoo, Finland, August 1998.* Helsinki University of Technology, Helsinki, Finland, pp. 839–848.

Stern, P.C. and E.W. Easterling, 1999: *Making Climate Forecasts Matter.* National Academy Press, Washington, DC.

Stonefelt, M.D., T.A. Fontaine, and R.H. Hotchkiss, 2000: Impacts of climate change on water yield in the Upper Wind Basin. *Journal of the American Water Resources Association*, **36,** 321–336.

Strzepek, K.M., D.N. Yates, and D.E. El Quosy, 1996: Vulnerability assessment of water resources in Egypt to climatic change in the Nile Basin. *Climate Research*, **6,** 89–95.

Strzepek, K.M., D.C. Major, C. Rosenzweig, A. Iglesias, D.N. Yates, A. Holt, and D. Hillel, 1999: New methods of modelling water availability for agriculture under climate change: the U.S. Cornbelt. *Journal of the American Water Resources Association*, **35,** 1639–1655.

Subak, S., 2000: Climate change adaptation in the U.K. water industry: managers' perceptions of past variability and future scenarios. *Water Resources Management*, **14,** 137–156.

Tarend, D.D., 1998: Changing flow regimes in the Baltic States. In: *Proceedings of the Second International Conference on Climate and Water, Espoo, Finland, August 1998.* Helsinki University of Technology, Helsinki, Finland.

Teller, J. and W. Last, 1982: Pedogenic zones in post-glacial sediments of Lake Manitoba, Canada. *Earth Surface Processes and Landforms*, **7,** 367–379.

Thomas, I.F. and B.C. Bates, 1997: Responses to the variability and increasing uncertainty of climate in Australia. *The Third IHP/IAHS G. Covach Colloquium: Risk, Reliability, Uncertainty and Robustness of Water Resources Systems, 19–21 September, 1996.* UNESCO, Paris, France.

Tobin, G.A. and B. Montz, 1997: *Natural Hazards: Explanation and Integration.* Guilford Press, New York, NY, USA.

Tung, C.P. and D.A. Haith, 1998: Climate change, irrigation and crop response. *Journal of American Water Resources Association*, **34,** 1071–1085.

UKWIR, 1998: *A Practical Method for Converting Uncertainty into Headroom.* UK Water Industry Research Ltd. Report 98/WR/13/1, UK Water Industry Research Ltd., London, United Kingdom, 132 pp.

UNICEF, 1999: Available online at http://www.unicefusa.org/alert/emergency/mitch.

Varis, O. and L. Somlyody, 1996: Potential impact of climate change on lake and reservoir water quality. In: *Water Resources Management in the Face of Climatic/Hydrologic Uncertainties* [Kaczmarek, Z., K. Strzepek, and L. Somlyody (eds.)]. Kluwer Academic Publishers, Dordrecht, The Netherlands, pp. 46–69.

Vehviläinen, B. and M. Huttunen, 1997: Climate change and water resources in Finland. *Boreal Environment Research*, **2,** 3–18.

Venema, H.D., E.J. Schiller, K. Adamowski, and J.M. Thizy, 1997: A water resources planning response to climate change in the Senegal River Basin. *Journal of Environmental Management*, **49,** 125–155.

Viney, N.R. and M. Sivapalan, 1996: The hydrological response of catchments to simulated changes in climate. *Ecological Modelling*, **86,** 189–193.

Vogel, R.M., C.J. Bell, and N.M. Fennessey, 1997: Climate, streamflow and water supply in the northeastern United States. *Journal of Hydrology*, **198,** 42–68.

Wallinga, J. and R.S.W. van de Wal, 1998: Sensitivity of Rhonegletscher, Switzerland, to climate change: experiments with a one-dimensional flowline model. *Journal of Glaciology*, **44,** 383–393.

Waylen, P., R.H. Compagnucci, and M. Caffera, 2001: Inter-annual and inter-decadal variability in stream flow from the Argentine Andes. *Physical Geography*, (in press).

Westmacott, J.R. and D.H. Burn, 1997: Climate change effects on the hydrologic regime within the Churchill-Nelson River Basin. *Journal of Hydrology*, **202,** 263–279.

White, A.F. and A.E. Blum, 1995: Effects of climate on chemical weathering in watersheds. *Geochimica et Cosmochimica Acta*, **59,** 1729–1747.

Wiener, J.D., 1996: Research opportunities in search of federal flood policy. *Policy Sciences*, **29,** 321–344.

Wilby, R. and T.M.L. Wigley, 1997: Downscaling general circulation model output: a review of methods and limitations. *Progress in Physical Geography*, **21,** 530–548.

Wilby, R., T.M.L. Wigley, D. Conway, P.D. Jones, B.C. Hewitson, J. Main, and D.S. Wilks, 1998: Statistical downscaling of general circulation model output: a comparison of methods. *Water Resources Research*, **34,** 2995–3008.

Wilby, R.L., L.E. Hay, and G.H. Leavesley, 1999: A comparison of downscaled and raw GCM output: implications for climate change scenarios in the San Juan River Basin, Colorado. *Journal of Hydrology*, **225,** 67–91.

Wolock, D.M. and G.J. McCabe, 1999: Esitmates of runoff using water balance and atmospheric general circulation models. *Journal of the American Water Resources Association*, **35,** 1341–1350.

World Bank, 1993: *Water Resources Management.* World Bank Policy Paper, World Bank, Washington, DC, USA, 140 pp.

WMO, 1997: *Comprehensive Assessment of the Freshwater Resources of the World.* World Meteorological Organisation, Geneva, Switzerland, 34 pp.

Xu, C., 1998: Hydrological responses to climate change in central Sweden. In: *Proceedings of the Second International Conference on Climate and Water, Espoo, Finland, August 1998.* Helsinki University of Technology, Helsinki, Finland, pp. 188–197.

Yates, D.N., 1997: Climate change impacts on the hydrologic response of South America: an annual, continental scale assessment. *Climate Research*, **9,** 147–155.

Ye, B., Y. Ding, E. Kang, G. Li, and T. Han, 1999: Response of the snowmelt and glacier runoff to the climate warming up in the last 40 years in the Xinjiang Autonomous Region, China. *Science in China*, **42.**

Ying, A. and G. Zhang, 1996: Response of water resources to climate change in Liaohe Basin. *Advances in Water Science*, **7,** 67–72.

Yoshino, F., 1999: Studies on the characteristics of variation and spatial correlation of the long-term annual runoff in the world rivers. *Journal of the Japanese Society for Hydrology and Water Resources*, **12,** 109–120.

Young, G.J., J.C.I. Dooge, and J.C. Rodda, 1994: *Global Water Resources Issues.* Cambridge University Press, Cambridge, United Kingdom and New York, NY, USA, 194 pp.

5

Ecosystems and Their Goods and Services

HABIBA GITAY (AUSTRALIA), SANDRA BROWN (USA),
WILLIAM EASTERLING (USA), AND BUBU JALLOW (THE GAMBIA)

Lead Authors:
J. Antle (USA), M. Apps (Canada), R. Beamish (Canada), T. Chapin (USA), W. Cramer (Germany), J. Frangi (Argentina), J. Laine (Finland), Lin Erda (China), J. Magnuson (USA), I. Noble (Australia), J. Price (USA), T. Prowse (Canada), T. Root (USA), E. Schulze (Germany), O. Sirotenko (Russia), B. Sohngen (USA), J. Soussana (France)

Contributing Authors:
H. Bugmann (Switzerland), C. Egorov (Russia), M. Finlayson (Australia), R. Fleming (Canada), W. Fraser (USA), L. Hahn (USA), K. Hall (USA), M. Howden (Australia), M. Hutchins (USA), J. Ingram (UK), Ju Hui (China), G. Masters (UK), P. Megonigal (USA), J. Morgan (USA), N. Myers (UK), R. Neilson (USA), S. Page (UK), C. Parmesan (USA), J. Rieley (UK), N. Roulet (Canada), G. Takle (USA), J. van Minnen (The Netherlands), D. Williams (Canada), T. Williamson (Canada), K. Wilson (USA)

Review Editors:
A. Fischlin (Switzerland) and S. Diaz (Argentina)

CONTENTS

EXECUTIVE SUMMARY

Ecosystems are subject to many pressures (e.g., land-use change, resource demands, population changes); their extent and pattern of distribution is changing, and landscapes are becoming more fragmented. Climate change constitutes an additional pressure that could change or endanger ecosystems and the many goods and services they provide.

There now is a substantial core of observational and experimental studies demonstrating the link between climate and biological or physical processes in ecosystems (e.g., shifting range boundaries, flowering time or migration times, ice break-up on streams and rivers), most evident in high latitudes. Recent modeling studies continue to show the potential for significant disruption of ecosystems under climate change. Further development of simple correlative models that were available at the time of the Second Assessment Report (SAR) point to areas where ecosystem disruption and the potential for ecosystem migration are high. Observational data and newer dynamic vegetation models linked to transient climate models are refining the projections. However, the precise outcomes depend on processes that are too subtle to be fully captured by current models.

At the time of the SAR, the interaction between elevated carbon dioxide (CO_2), increasing temperatures, and soil moisture changes suggested a possible increase in plant productivity through increased water-use efficiency (WUE). Recent results suggest that the gains might be small under field conditions and could be further reduced by human management activities. Many ecosystems are sensitive to the frequency of El Niño-Southern Oscillation (ENSO) and other extreme events that result in changes in productivity and disturbance regimes (e.g., fires, pest and disease outbreak).

Agriculture

Most global and regional economic studies—with and without climate change—indicate that the downward trend in real commodity prices in the 20th century is likely to continue into the 21st century, although confidence in these predictions decreases farther into the future (see Section 5.3.1).

Impacts

- Experiments have shown that relative enhancement of productivity caused by elevated CO_2 usually is greater when temperature rises but may be less for crop yields at above-optimal temperatures (established but incomplete). Although the beneficial effects of elevated CO_2 on the yield of crops are well established for the experimental conditions tested, this knowledge is incomplete for numerous tropical crop species and for crops grown under suboptimal conditions (low nutrients, weeds, pests and diseases). In experimental work, grain and forage quality declines with CO_2 enrichment and higher temperatures (high confidence) (see Sections 5.3.3, 5.4.3, and 5.5.3).
- Experimental evidence suggests that relative enhancement of productivity caused by elevated CO_2 usually is greater under drought conditions than in wet soil. Nevertheless, a climate change-induced reduction in summer soil moisture (see Table 3-10)—which may occur even in some cases of increased summer precipitation—would have detrimental effects on some of the major crops, especially in drought-prone regions (medium confidence).
- Soil properties and processes—including organic matter decomposition, leaching, and soil water regimes—will be influenced by temperature increase (high confidence). Soil erosion and degradation are likely to aggravate the detrimental effects of a rise in air temperature on crop yields. Climate change may increase erosion in some regions, through heavy rainfall and through increased windspeed (competing explanations) (see Section 5.3.3).
- Model simulations of wheat growth indicate that greater variation in temperature (change in frequency of extremes) under a changing climate reduces average grain yield. Moreover, recent research emphasizes the importance of understanding how variability interacts with changes in climate means in determining yields (established but incomplete) (see Section 5.3.4).
- Crop modeling studies that compare equilibrium scenarios with transient scenarios of climate change report significant yield differences. The few studies that include comparable transient and equilibrium climate change scenarios generally report greater yield loss with equilibrium climate change than with the equivalent transient climate change. Even these few studies are plagued with problems of inconsistency in methodologies, which make comparisons speculative at this time (see Section 5.3.4).

Adaptation and Vulnerability

- Prospects for adaptation of plant material to increased air temperature through traditional breeding and genetic modification appear promising (established but incomplete). More research on possible adaptation of crop species to elevated CO_2 is needed before more certain results can be presented (see Section 5.3.3).

- Simulations without adaptation suggest more consistent yield losses from climate change in tropical latitudes than temperate latitudes. Agronomic adaptation abates extreme yield losses at all latitudes, but yields tend to remain beneath baseline levels after adaptation more consistently in the tropics than in temperate latitudes (moderate confidence) (see Section 5.3.4).
- The ability of livestock producers to adapt their herds to the physiological stress of climate change is not known conclusively, in part because of a general lack of experimentation and simulations of livestock adaptation to climate change (see Section 5.3.3).
- Crop and livestock farmers who have sufficient access to capital and technologies are expected to adapt their farming systems to climate change (medium to low confidence) (see Section 5.3.4). Substantial shifts in their mix of crops and livestock production may be necessary, however, and considerable costs could be involved in this process—*inter alia*, in learning and gaining experience with different crops or if irrigation becomes necessary. In some cases, a lack of water resulting from climate change might mean that increased irrigation demands cannot be met (see Section 4.7.2). Although this conclusion is speculative because of lack of research, it is intuitive that the costs of adaptation should depend critically on the rate of climate change.
- Impacts of climate change on agriculture after adaptation are estimated to result in small percentage changes in global income; these changes tend to be positive for a moderate global warming, especially when the effects of CO_2 fertilization are taken into account (low confidence) (see Section 5.3.5).
- The effectiveness of adaptation in ameliorating the economic impacts of climate change across regions will depend critically on regional resource endowments. It appears that developed countries will fare better in adapting to climate change; developing countries and countries in transition, especially in the tropics and subtropics, will fare worse. This finding has particularly significant implications for the distribution of impacts within developing countries, as well as between more- and less-developed countries. These findings provide evidence to support the hypothesis advanced in the SAR that climate change is likely to have its greatest adverse impacts on areas where resource endowments are poorest and the ability of farmers to respond and adapt is most limited (medium confidence) (see Section 5.3.5).
- Degradation of soil and water resources is one of the major future challenges for global agriculture (see Section 5.3.2). These processes are likely to be intensified by adverse changes in temperature and precipitation. Land use and management have been shown to have a greater impact on soil conditions than the direct effects of climate change; thus, adaptation has the potential to significantly mitigate these impacts (see Section 5.3.4). A critical research need is to assess whether resource degradation will significantly increase the risks faced by vulnerable agricultural and rural populations (see Section 5.3.6).
- It is concluded with low confidence that a global temperature rise of greater than 2.5°C will result in rising commodity prices. Similarly, a global temperature rise of greater than 2.5°C increases by 80 million the absolute number of people at risk of hunger. It should be noted, however, that these hunger estimates are based on the assumption that food prices will rise with climate change, which is highly uncertain (see Section 5.3.6).

Wildlife

Recent estimates indicate that 25% of the world's mammals and 12% of birds are at significant risk of global extinction. Climate change is only one of a long list of pressures on wildlife. Other pressures include exploitation of animals, pollution and other biochemical poisonings, extreme climatic events, wildlife diseases, collisions with towers and other structures, anthropogenic barriers to dispersal, and war and other civil conflicts. Alone or in combination, these pressures will greatly increase species' vulnerabilities to rarity and extinction (high confidence). Habitat conversion and degradation affect nearly 89% of all threatened birds and 83% of all threatened mammals. About one-fifth of threatened mammals in Australia and the Americas and the world's birds are affected by introduced species.

Impacts, Adaptations, and Vulnerabilities

- Many animals already may be responding to local climatic changes. The types of changes already observed include poleward and elevational movement of ranges, changes in animal abundance, changes in body size, and shifts in the timing of events, such as earlier breeding in spring. Possible climatically associated shifts in animal ranges and densities have been noted on many continents and within each major taxonomic group of animals (see Table 5-3).
- Laboratory and field studies have demonstrated that climate plays a strong role in limiting species' ranges (high confidence). Even though only a small fraction of all species have been monitored long enough to detect significant trends, changes exhibited over the past few decades in the bulk of these species are consistent with local warming and expected physiological responses (medium confidence). However, possible *specific* changes in wildlife from climate change can be projected only with low confidence for most species because of many possible contributing factors, such as habitat destruction and introduction of exotic species. Some species clearly are responding to global change (see Section 5.4.3), and many more changes probably have gone undetected. Researchers are in the process of coupling these discernible changes with various biological theories regarding climate and species spatial and temporal patterns; through this process, we expect that reliable *general* projections can be and in fact are being made.
- Protecting threatened and endangered species requires measures that, in general, reverse the trend toward rarity. Without management, rapid climate change—in conjunction

with other pressures—is likely to cause many species that currently are classified as critically endangered to become extinct and several labeled endangered or vulnerable to become much rarer, and thereby closer to extinction, in the 21st century (high confidence).

- Concern over species becoming rare or extinct is warranted because of the goods and services provided by ecosystems and the species themselves. Most of the goods and services provided by wildlife (e.g., pollination, natural pest control) are derived from their roles within systems. Other valuable services are provided by species contributing to ecosystem stability or to ecosystem health and productivity. The recreational value (e.g., sport hunting, wildlife viewing) of species is large in market and nonmarket terms. Species loss also could impact the cultural and religious practices of indigenous peoples around the world. Losses of species can lead to changes in the structure and function of affected ecosystems and loss of revenue and aesthetics. Understanding the role each species plays in ecosystem services is necessary to understand the risks and possible surprises associated with species loss. Without this information, the probability of surprises associated with species loss is high (medium confidence).
- Humans may need to adapt not only in terms of wildlife conservation but also to replace lost ecological services normally provided by wildlife. It may be necessary to develop adaptations to losses to natural pest control, pollination, and seed dispersal. Although replacing providers of these services sometimes may be possible, the alternatives may be costly. Finding replacements for other services, such as contributions to nutrient cycling and ecosystem stability/biodiversity, are much harder to imagine. In many cases, such as the values of wildlife associated with subsistence hunting and cultural and religious ceremonies, any attempt at replacement may represent a net loss. In many countries, climate change impacts, such as reductions in wildlife populations, may have the greatest impact on the lowest income groups—those with the least ability to adapt if hunting opportunities decline.

Rangelands

Most rangelands in the world have been affected by human activity, and many are degraded in some way. Desertification tends to be associated with land degradation in rangelands; however, desertification combines many land degradation processes and can be exacerbated by climate change. Many of the rangelands of the world are affected by ENSO events and are sensitive to the frequency of these events, resulting in changes in productivity of these systems.

Impacts, Adaptations, and Vulnerabilities

- Based on observations and modeling studies, the effects of elevated CO_2 and climate change could result in increased plant productivity and thus an increase in soil carbon sequestration in many rangelands. However, some of the gains in productivity would be offset by increases in temperatures and by human management activities (medium confidence).
- Modeling studies and observations suggest that plant production, species distribution, disturbance regimes (e.g., frequencies of fires, insect/pest outbreaks), grassland boundaries, and nonintensive animal production would be affected by potential changes in climate and land use. The impacts of climate change are likely to be minor compared to those of land degradation (high confidence).
- Irrigation in semi-arid climates is a major cause of secondary salinization. Elevated CO_2 may reduce the impacts of secondary salinization, although experimental work shows that any increase in temperature may negate these benefits and may even exacerbate problems of secondary salinity (medium confidence).
- In many parts of the world that are dominated by rangelands, lack of infrastructure and investment in resource management limits available options for adaptation and makes these areas more sensitive and vulnerable to the impacts of climate change (high confidence). Some adaptation options (e.g., integrated land management) could be implemented irrespective of technology and infrastructure. Other adaptation options could be implemented through active involvement of communities in the management of rangelands.

Forests and Woodlands

Loss in forest cover appears to have slowed in recent years relative to 1980–1995. However, fragmentation, nonsustainable logging of mature forests, degradation, and development of infrastructure—all leading to losses of biomass—have occurred over significant areas in developing and developed countries (high confidence). Pressure from disturbances such as fires appears to be increasing around the world. Fire suppression in temperate managed and unmanaged forests with access to infrastructure and human capital has been largely successful, but regions with comparatively less infrastructure have been more susceptible to natural and human-caused fires. Deforestation will continue to be the dominant factor influencing land-use change in tropical regions. Timber harvests near roads and mills in tropical regions are likely to continue to fragment and damage natural forests (high confidence).

Non-wood forest products (NWFP) such as edible mushrooms, nuts, fruits, palm hearts, herbs, spices, gums, aromatic plants, game, fodder, rattan, medicinal and cosmetic products, resins, and the like make important contributions to household income, food security, national economies, and the environmental objectives of conservation of biodiversity.

Impacts, Adaptations, and Vulnerabilities

- Forest response to climate change and other pressures will alter future carbon storage in forests, but the global extent and direction of change is unknown.

- Recent experimental evidence suggests that the net balance between net primary productivity (NPP—usually assumed to increase with warming, but challenged by recent studies), heterotrophic respiration (often assumed to increase with warming, but also challenged by recent studies), and disturbance releases (often ignored, but shown to be important in boreal estimates of net biome productivity) is no longer as clear as stated in the SAR.
- Research reported since the SAR confirms the view that the largest and earliest impacts induced by climate change are likely to occur in boreal forests, where changes in weather-related disturbance regimes and nutrient cycling are primary controls on productivity (high confidence). The effect of these changes on NPP and carbon storage is uncertain.
- Since the SAR, free-air CO_2 enrichment (FACE) experiments suggest that tree growth rates may increase, litterfall and fine root increment may increase, and total NPP may increase, but these effects are expected to saturate because forest stands tend toward maximum carrying capacity, and plants may become acclimated to increased CO_2 levels (medium confidence).
- Questions of saturation of the CO_2 response can be addressed through longer term experiments on tree species grown under elevated CO_2 in open-top chambers under field conditions over several growing seasons. Results from these experiments show continued and consistent stimulation of photosynthesis and little evidence of long-term loss of sensitivity to CO_2; the relative effect on aboveground dry mass was highly variable and greater than indicated by seedling studies, and the annual increase in wood mass per unit of leaf area increased (high confidence).
- Contrary to the SAR, global timber market studies that include adaptation suggest that climate change will increase global timber supply and enhance existing market trends toward rising market share in developing countries. Consumers are likely to benefit from lower timber prices; producers may gain or lose, depending on regional changes in timber productivity and potential dieback effects. Studies that do not consider global market forces, timber prices, or adaptation predict that supply in boreal regions could decline (medium confidence).
- Industrial timber harvests are predicted to increase by 1–2% yr^{-1}. The area of industrial timber plantations is likely to continue to expand, and management in second-growth forests in temperate regions is likely to continue to intensify, taking pressure off natural forests for harvests (high confidence).
- At the regional and global scale, the extent and nature of adaptation will depend primarily on wood and non-wood product prices, the relative value of substitutes, the cost of management, and technology. On specific sites, changes in forest growth and productivity will constrain—and could limit—choices of adaptation strategies (high confidence). In markets, prices will mediate adaptation through land and product management. Adaptation in managed forests will include salvaging dead and dying timber and replanting with new species that are better suited to the new climate.

Lakes and Rivers

Capture, culture, and recreational fisheries are reported to land about 23 Mt yr^{-1} of biomass, but the actual numbers probably are twice that. Fish species in freshwater total about 11,800. High levels of endemism are common for many different groups of freshwater organisms. In addition to climatic changes, lakes and rivers are impacted by pressures such as land and water use, pollution, capture and culture fisheries, water extraction, and hydrologic engineering structures such as dams, dykes, and channelization (well established). These pressures interact with pressures from climate change and vary in their impact even at local levels mediated through changes in hydrology (medium confidence).

Impacts, Adaptations, and Vulnerabilities

- A 150-year trend in 26 lakes and rivers in the northern hemisphere averaging 9 days later freeze and 10 days earlier ice breakup has resulted from a 1.8°C increase in air temperature (very high confidence).
- Empirical and simulation studies show that elevated water temperature increases summer anoxia in deep waters of stratified lakes (high confidence).
- There has been an observed poleward movement of southern and northern boundaries of fish distributions (high confidence).
- There is a loss of habitat for cold- and coolwater fishes and gain in habitat for warmwater fishes (high confidence).
- There are complex relations between warmer temperature, more episodic rainfall, and poleward movement of warm-water zooplanktivorous fishes, which may exacerbate eutrophication of lakes and rivers (medium confidence).
- Human activities to manage water flow may exacerbate the impact on lakes and rivers (medium confidence); for example, the increase in hydrologic engineering structures is likely to result in fewer free-flowing streams (medium confidence). Attempts to manage the poleward movement of fauna and flora, especially in lakes whose isolation is likely to slow down species from moving poleward, is likely to be contentious and produce frequent surprises and unexpected dynamics of freshwater communities (medium confidence).
- New opportunities may be provided for aquaculture of warmer water species in more poleward locations (medium confidence).
- As a class of ecosystems, inland waters are vulnerable to climatic change and other pressures, owing to their small size and position downstream from many human activities (high confidence). The most vulnerable elements include reduction and loss of lake and river ice (very high confidence), loss of habitat for coldwater fish (very high confidence), increases in extinctions and invasions of exotics (high confidence), and potential exacerbation of existing pollution problems such as eutrophication, toxics, acid rain, and ultraviolet-B (UV-B) radiation (medium confidence).

Inland Wetlands

Wetlands play an important role in maintaining biological diversity by providing a habitat for many plant and animal species, some of which are endemic or endangered. They also have significant scientific value that goes beyond their plant and animal communities. Peat-accumulating wetlands are important for global change because of the large carbon store accumulated over the millennia and the risk that this store would be released to the atmosphere in conditions modified by global change. Besides being carbon sinks, wetlands are sources of methane to the atmosphere.

We can state with high confidence that the pressures of climate change on habitat, biodiversity, and carbon sink on most wetland types will be largely indirect, operating through changes in water level. On peatlands underlain by permafrost, however, direct temperature impacts can be more important. Food and fiber production are ecosystem services that simultaneously form a pressure (land-use change) on other services (habitat, biodiversity, carbon sink).

Impacts, Adaptations, and Vulnerabilities

- A warmer, drier climate is not necessarily likely to lead to a large loss of stored peat for all peatland types (e.g., boreal peatlands), but there are competing explanations because feedback mechanisms between climate and peatland hydrology and the autogenic nature of peatland development are poorly understood.
- Peatlands underlain by permafrost are likely to become net carbon sources rather than sinks, mainly because of melting of permafrost and lowering of the water table (high confidence). Work in tussock and wet tundra has shown that these systems already may be net annual sources of 0.19 Gt C.
- Extensive seasonally inundated freshwater swamps, which are major biodiversity foci, could be displaced if predicted sea-level rises of 10–30 cm by 2030 occur (high confidence).
- In the southeast Asian region, droughts in recent years related to ENSO events have lowered local water tables and thus increased the severity of fires in tropical peatlands (high confidence).
- The effects of water-level drawdown after drainage for forestry indicate that the shift in species composition from wetland species to forest species only slightly affects plant species richness of individual sites; in regions dominated by forests, however, there would be a clear reduction in regional diversity as landscapes become homogenized after water-level drawdown (medium confidence).
- Most wetland processes are dependent on catchment-level hydrology. Thus, adaptations to projected climate change may be practically impossible. For degrading key habitats, small-scale restoration may be possible if sufficient water is available.
- Arctic and subarctic ombrotrophic bog communities on permafrost, together with more southern depressional wetlands with small catchment areas, are likely to be most vulnerable to climate change. The increasing speed of peatland conversion and drainage in southeast Asia is likely to place these areas at a greatly increased risk of fires and affect the viability of tropical wetlands.

Arctic and Alpine Ecosystems

Observations show that high-latitude warming has occurred in the Arctic region since the 1960s. Precipitation and surface evaporation have increased at high latitudes. Permafrost temperatures have warmed in western North America by 2–4°C from 1940 and in Siberia by 0.6–0.7°C from 1970 to 1990, whereas permafrost cooled in northeastern Canada. These patterns roughly parallel recent trends in air temperature. Carbon flux measurements in Alaska suggest that the recent warming trend may have converted tundra from a net carbon sink to a source of as much as 0.7 Gt C yr^{-1}. During recent decades, the peak-to-trough amplitude in the seasonal cycle of atmospheric CO_2 concentrations has increased, and the phase has advanced at arctic and subarctic CO_2 observation stations north of 55°N. This change in carbon dynamics in the atmosphere probably reflects some combination of increased uptake during the first half of the growing season—which could explain the observed increase in biomass of some shrubs, increased winter efflux, and increased seasonality of carbon exchange associated with disturbance. This "inverse" approach generally has concluded that mid-northern latitudes were a net carbon sink during the 1980s and early 1990s. At high northern latitudes, these models give a wider range of estimates; some analyses point to a net and others to a sink.

Climatic changes observed in alpine areas generally have paralleled climatic patterns in surrounding regions, with the most pronounced warming at high latitudes, in the Alps, and in Asia and the least pronounced changes in tropical alpine regions. Precipitation generally has increased, with the most pronounced changes in winter, leading to increased snow depth. Regional trends in climate have led to shrinkage of alpine and subpolar glaciers equivalent to 0.25 ± 0.1 mm yr^{-1} of sea-level change—or 16% of the sea-level rise in the past 100 years. Net mass reduction of the alpine glaciers has been most pronounced since 1980, when regional warming was greatest. Climatic warming observed in the Alps has been associated with upward movement of some plant taxa of 1–4 m per decade on mountaintops and the loss of some taxa that formerly were restricted to high elevations. In general, direct human impacts on alpine vegetation from grazing, tourism, and nitrogen deposition are so strong that climatic effects on the goods and services provided by alpine ecosystems are difficult to detect.

Impacts, Adaptations, and Vulnerabilities

- Projected climatic warming of 4–10°C by the end of the 21st century is likely to cause substantial increases in decomposition, nutrient release, and primary production (high confidence). Many of these changes in productivity may be threshold effects; they also may be mediated by changes in species composition and therefore are likely to lag changes in climate by years, decades, or centuries (medium confidence).

- Important changes in diversity in the Arctic include changes in the abundance of caribou, waterfowl, and other subsistence resources (medium confidence).
- Tundra has a three- to six-fold higher winter albedo than boreal forest, but summer albedo and energy partitioning differ more strongly among ecosystems within either tundra or boreal forest than between these two biomes (high confidence). Changes in albedo and energy absorption during winter are likely to act as a positive feedback to regional warming.
- Disturbances such as fires are likely to change with changes in regional climate and affect biophysical properties (high confidence).
- In alpine areas, warming is likely to create a shortened snowmelt season, with rapid water release creating floods and later growing-season droughts, affecting productivity in these areas (medium confidence).
- Changes in goods and services in alpine ecosystems are likely to be dominated by changes in land use associated with grazing, recreation, and other direct impacts because of their proximity to population centers (high confidence). Many of the alpine zones with greatest biodiversity, such as the Caucasus and Himalayas, are areas where human population pressures may lead to most pronounced land-use change. Overgrazing, trampling, and nutrient/pollution deposition may tend to destabilize vegetation, leading to erosion and loss of soils that are the long-term basis of the productive capacity of alpine ecosystems.
- Opportunities for adapting to expected changes in Arctic and alpine ecosystems are limited because these systems will respond most strongly to globally induced changes in climate (medium confidence). Opportunities for mitigation will include protection of peatlands, yedoma sediments, and other carbon-rich areas from large-scale hydrological change, land use, and pollutant levels. Careful management of wildlife resources could minimize climatic impacts on indigenous peoples. Many Arctic regions depend strongly on one or a few resources, such as timber, oil, reindeer, or wages from fighting fires. Economic diversification would reduce the impacts of large changes in the availability or economic value of particular goods and services.
- The goods and services provided by many Arctic regions depend on the physical integrity of permafrost and therefore are vulnerable to climate change (high confidence). The large carbon stocks in these regions are vulnerable to loss to the atmosphere as CO_2 or methane.
- The high levels of endemism in many alpine floras and their inability to migrate upward means that these species are most vulnerable (medium confidence).

5.1. Introduction and Scope

Ecosystems provide many products and services that are crucial to human survival (Daily, 1997; UNEP, 1998; WRI, 2000). Ecosystems affect biogeochemical and physical feedbacks to the biosphere and atmosphere, hence are important for the functioning of the Earth's systems. Ecosystems form a landscape and are connected in many ways, often by streams, rivers, and wildlife. Thus, landscape fragmentation, along with other human activities, affects ecosystems' ability to meet human needs and will continue to do so for the future, possibly at a faster rate (UNEP, 1998). Changes in global climate and atmospheric composition are likely to have an impact on most of these goods and services, with significant impacts on socioeconomic systems (Winnett, 1998).

This chapter assesses the impacts of climate change on ecosystem goods and services from sectors such as agriculture, forests, and wetlands. Inland aquatic systems are covered from an ecosystem perspective in this chapter; hydrology and water as a physical resource are covered in Chapter 4. Marine and coastal systems are considered in Chapter 6. When ecosystems are highly managed, as in agriculture and forestry, or their goods and services are traded in markets, the social and economic consequences of climate change that naturally arise are assessed explicitly.

Biomass production, biogeochemical cycling, soil and water relationships, and animal-plant interactions (including biodiversity) are considered to be some of the major functions of ecosystems. Within these functions, various products (goods) and services can be identified, including food, fiber, fuel and energy, fodder, medicines, clean water, clean air, flood/storm control, pollination, seed dispersal, pest and disease control, soil regeneration, biodiversity, and recreation/amenity (UNEP, 1998; WRI, 2000). Society places values on these goods and services, directly or indirectly (Table 5-1). Ecosystems provide many of these goods and services simultaneously. For example, agricultural systems provide much of our food, fiber, and fuel needs and at the same time influence biogeochemical cycling, soil and water quality, and biodiversity. Many services from ecosystems lie outside market systems, making it difficult to price them (Bawa and Gadgil, 1997; Goulder and Kennedy, 1997; National Research Council, 1999). However, these nonmarket values are likely to be larger (as much as 1,000-fold; WRI, 2000) than the value of services provided by markets in total and at many specific sites (Costanza *et al.*, 1997). Although several studies estimating different values for nonmarket services from ecosystems exist (see Table 5-2), they can be applied only with low to medium confidence (Goulder and Kennedy, 1997). Valuation of ecosystem services is complex because many goods and services occur simultaneously. Thus, it is not sufficient to consider, for example, the timber value of

***Table 5-1**: Ecosystems function with links to good/services and possible societal value (modified from Ewel et al., 1998).*

Function	Goods/Service	Value[a]
Production	– Food – Fiber (timber and non-wood products) – Fuel – Fodder	Direct
Biogeochemical cycling	– Nutrient cycling (especially N and P absorption/deposition) – Carbon sinks	Mostly indirect, although future values have to be considered
Soil and water conservation	– Flood and storm control – Erosion control – Clean water – Clean air – Water for irrigation – Organic matter or sediment export – Pollution control – Biodiversity	Mostly indirect, although future values have to be considered
Animal-plant interactions	– Pollination – Animal migration – Biodiversity	Mostly indirect, future, bequest, and existence values have to be considered
Carrier	– Lansdcape connectivity – Animal migration – Biodiversity – Aesthetic/spiritual/cultural service	Mostly indirect and existence, but bequest may have to be considered

[a] Value definitions are from Pearce and Moran (1994); see Table 5-2.

***Table 5-2**: Examples of goods and services with possible uses and values (adapted from Pearce and Moran, 1994).*

Value	Examples of Goods and Services
Direct use	Food, fiber, fuel, fodder, water supply, recreation, non-wood forest products
Indirect use	Biodiversity, biogeochemical cycles, tourism, flood and storm control, clean water supply, pollution control
Option	Future discoveries (i.e., pharmacological and biotechnological), future recreation
Bequest	Intergenerational and sustainable development
Existence	Mostly conservation, aesthetic, spiritual

the forest; we also must consider the soil/water protection that the trees provide, the habitat for pollinators, or the bequest value of the forest (WRI, 2000).

The Earth is being subjected to many human-induced and natural changes, often referred to as global change. These changes include pressures from increased demand for resources driven by economic growth, increased human population, land-use and land-cover change, the accelerated rate of anthropogenic nitrogen production and other air pollutants, and urbanization and industrialization; resulting fossil fuel emissions contribute to a discernible impact on global climate (Naiman *et al.*, 1995a; Vitousek *et al.*, 1997a; IPCC, 1998; UNEP, 1998; Walker *et al.*, 1999). For ecosystems, the impacts of climate change include changes in atmospheric composition and disturbance regimes, such as frequencies of fires, storms, floods, and drought. The impacts of other pressures often lead to increased demand for access to land, water, and wildlife resources. The result is a change in the state of the Earth's land surface, the services humans receive, and the landscapes where humans live at regional and global scales. Governance and equity issues are important (UNEP, 1998) in overcoming some of these; these are covered in detail in TAR WGIII Chapter 1.

Understanding the current status of ecosystems, pressures on them, and their responses is important for assessing the impact of the additional pressure of climate change. The State-Pressure-Response model of the Organisation for Economic Cooperation and Development (OECD) has been used as a framework in structuring many subsequent sections of this chapter (see Figure 5-1). Using this model, the "state" refers to assessments of current status and recent trends of each sector and "pressures" include direct and indirect human pressures. The concept of "response" has been modified to include automatic responses of ecosystems (referred to as automatic adaptation in Chapter 18) to the impacts of climate change (including that of natural climatic variability). Deliberate "adaptation" options to climate change to overcome some of these impacts-responses could alleviate some of the pressures. Systems that are not able to adapt are likely to become vulnerable (see Chapter 18); thus, vulnerability also is assessed.

In many sections, four goods and services have been emphasized because of the availability of literature and space constraints: food, fiber, carbon storage, and biodiversity. Many other topics—for example, related to water cycling and hydrology—could have been considered. Some of these are covered in other chapters of this report (e.g., Chapter 4 and as key regional concerns in Chapters 10–17).

Some impacts of global change apply across many sectors and are considered in Section 5.2. Many of the impacts of climate change are projected by using models, and Section 5.2 includes a critique of several of these models. Some studies in the literature already have identified regional impacts of post-industrial climate change on some ecosystems. These are discussed in the most appropriate sections under "impacts-response" but are dealt with mainly in Sections 5.2 and 5.4.

5.2. Effects of Global Change on the Terrestrial Biosphere

Terrestrial ecosystems consist of plants, animals, and soil biota and their environment. The distribution of biota within and across ecosystems is constrained by the physical and chemical conditions of the atmosphere, the availability of nutrients and/or pollutants, and disturbances from natural origin (fire, wind-throw, etc.) or human land use. Global change affects all of these factors, but through widely differing pathways and at different scales. Global climate change, for example, affects local weather and climate in ways that are strongly dependent on location. Increased atmospheric CO_2 concentration, on the other hand, is geographically more uniform.

At the global scale, the sum of all ecosystems (including the marine biosphere) exerts a significant role on the balance of carbon and water in the atmosphere. Feedbacks exist between climate-driven changes in biospheric functioning, such as enhanced or reduced primary productivity, and the amount of greenhouse gas (GHG)-related radiative forcing. In addition, changes in land surface characteristics affect the atmosphere by altering the radiation balance at the reflective surface, creating further feedbacks. These biospheric feedbacks are discussed in TAR WGI Chapter 3.

In trying to understand the effects of global change on the biosphere as a whole, scientists often focus on higher level entities such as ecosystems or biomes (the collection of ecosystems within a particular climatic zone with similar structure but differing species—e.g., the temperate forest biome). However, all factors of change act on individual organisms that are part of a complex web of interactions within and between species and their ecosystem and within landscapes that contain

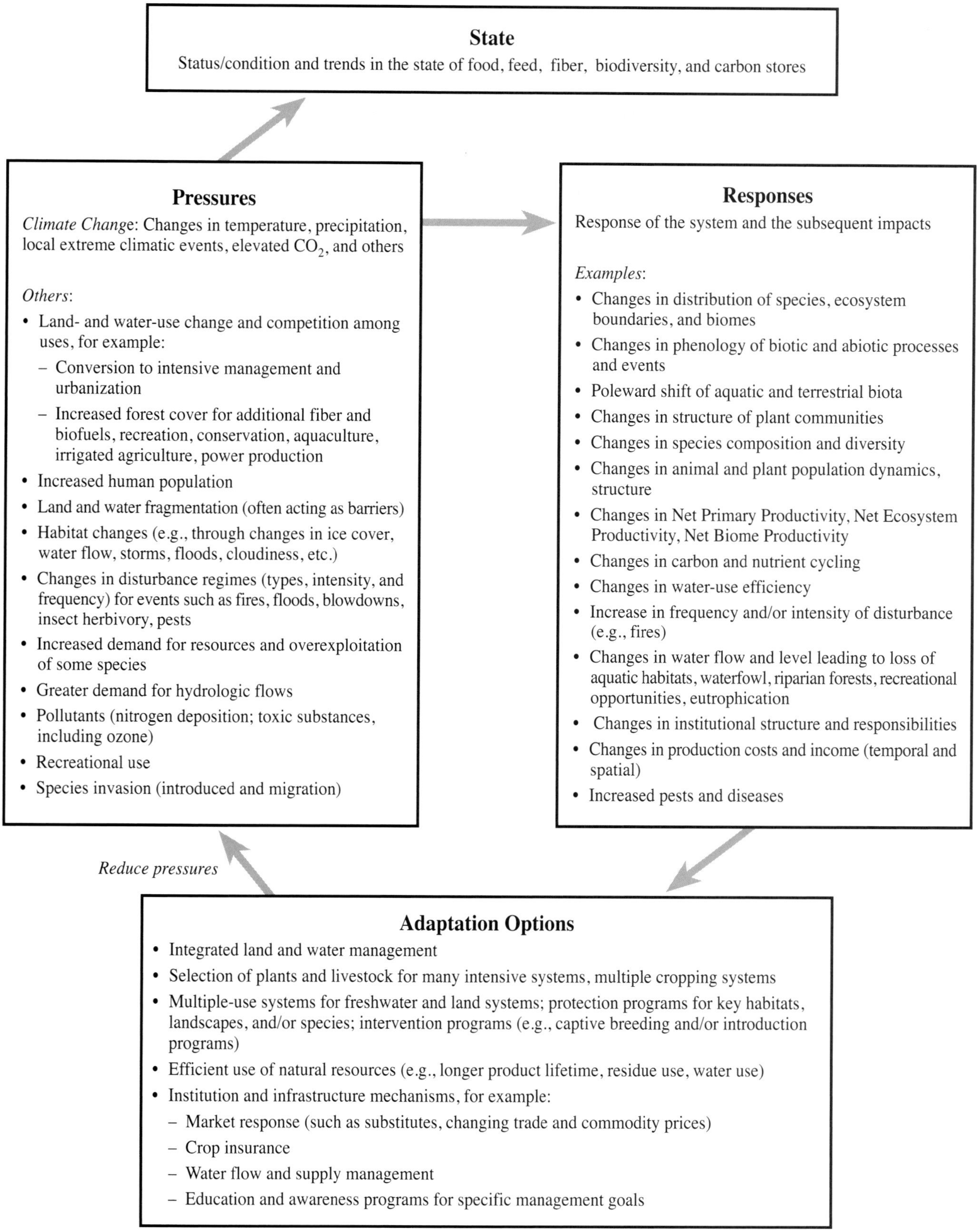

State

Status/condition and trends in the state of food, feed, fiber, biodiversity, and carbon stores

Pressures

Climate Change: Changes in temperature, precipitation, local extreme climatic events, elevated CO_2, and others

Others:

- Land- and water-use change and competition among uses, for example:
  - Conversion to intensive management and urbanization
  - Increased forest cover for additional fiber and biofuels, recreation, conservation, aquaculture, irrigated agriculture, power production
- Increased human population
- Land and water fragmentation (often acting as barriers)
- Habitat changes (e.g., through changes in ice cover, water flow, storms, floods, cloudiness, etc.)
- Changes in disturbance regimes (types, intensity, and frequency) for events such as fires, floods, blowdowns, insect herbivory, pests
- Increased demand for resources and overexploitation of some species
- Greater demand for hydrologic flows
- Pollutants (nitrogen deposition; toxic substances, including ozone)
- Recreational use
- Species invasion (introduced and migration)

Responses

Response of the system and the subsequent impacts

Examples:

- Changes in distribution of species, ecosystem boundaries, and biomes
- Changes in phenology of biotic and abiotic processes and events
- Poleward shift of aquatic and terrestrial biota
- Changes in structure of plant communities
- Changes in species composition and diversity
- Changes in animal and plant population dynamics, structure
- Changes in Net Primary Productivity, Net Ecosystem Productivity, Net Biome Productivity
- Changes in carbon and nutrient cycling
- Changes in water-use efficiency
- Increase in frequency and/or intensity of disturbance (e.g., fires)
- Changes in water flow and level leading to loss of aquatic habitats, waterfowl, riparian forests, recreational opportunities, eutrophication
- Changes in institutional structure and responsibilities
- Changes in production costs and income (temporal and spatial)
- Increased pests and diseases

Reduce pressures

Adaptation Options

- Integrated land and water management
- Selection of plants and livestock for many intensive systems, multiple cropping systems
- Multiple-use systems for freshwater and land systems; protection programs for key habitats, landscapes, and/or species; intervention programs (e.g., captive breeding and/or introduction programs)
- Efficient use of natural resources (e.g., longer product lifetime, residue use, water use)
- Institution and infrastructure mechanisms, for example:
  - Market response (such as substitutes, changing trade and commodity prices)
  - Crop insurance
  - Water flow and supply management
  - Education and awareness programs for specific management goals

Figure 5-1: Generalized diagram of the state of specific goods and services that ecosystems provide; how these goods and services are affected by the multiple pressures of climate change and human activities; and how the system responds (autonomous adaptation as in Chapter 19), thus affecting the provision of goods and services. Adaptation options reduce the impacts and thus change the vulnerability of the system.

mosaics of different ecosystems. It is not feasible to model the impacts of global change at global, or even regional or landscape, scales at an individual-by-individual level. Thus, most models of global change have dealt with impacts at the ecosystem or biome level. This is in contrast with limitations to observational and experimental studies in which only a selection of individuals and species of a few ecosystems can be included.

5.2.1. *Observational Studies*

There is now a substantial number of observational and experimental studies that demonstrate the link between climate and biological or physical processes in ecosystems. The authors of this chapter assembled a database of more than 2,500 studies that address climate and either a physical process (e.g., melting of ice on lakes) or a biological factor (e.g., spring arrival time) of an animal or plant. Most of these studies address experiments that are valuable primarily in helping to understand the biological mechanisms prompting the responses of plants and animals to climate but are not helpful in detecting patterns of change. Many of these studies were conducted over a period of shorter than 10 years. Because at least 10 years of data are needed to show a possible trend, this narrowed the number of studies to approximately 500. Because temperature is the variable that can most reliably be predicted with increasing GHGs, only studies that addressed temperature as the climatic variable were examined, leaving approximately 250 studies. All 250 studies then were examined to determine if they met at least two of the following criteria:

- The authors found a statistically significant correlation between temperature and a species trait (e.g., egg-laying date, location of range boundaries) or physical process.
- The authors found a statistically significant change in the species trait over time.
- The authors found a statistically significant change in temperature over time.

Table 5-3*: The number of species and processes in each region that were found in each particular study to be significantly associated with regional temperature change. For inclusion in the table, each study had to meet two of the following three criteria: species or processes changing over time; regional temperature changing over time; and significant association between how the temperature and species or processes were changing. The first number indicates the number of species or processes changing in the manner predicted with global warming. The second number is the number of species or processes changing in a manner opposite to that predicted with a warming planet.* ***When considering those species that have shown a change, 80% are changing in the manner expected with global warming, while 20% are changing in the opposite direction****. Note that about 61 of all species examined did not show a statistically significant change. References for each cell are located below the table and collated by row number and column number (e.g., references for European birds are under* ***E,5****—row E, column 5). "—" indicates that no studies were found for this region and category.*

Region	Column 1: Lake and Stream Ice		Column 2: Vegetation		Column 3: Invertebrates		Column 4: Amphibians and Reptiles		Column 5: Birds		Column 6: Mammals	
Row A: Africa	—	—	—	—	—	—	—	—	—	—	—	—
Row B: Antarctica	—	—	2	0	—	—	—	—	2	0	—	—
Row C: Asia	3	0	—	—	—	—	—	—	—	—	—	—
Row D: Australia	—	—	—	—	—	—	—	—	—	—	—	—
Row E: Europe	8	0	13	1	46	1	7	0	258	92	7	0
Row F: North America	18	0	32	11	—	—	—	—	17	4	3	0
Row G: Latin America	—	—	—	—	—	—	22	0	15	0	—	—
Total	29	0	47	12	46	1	29	0	292	96	10	0

Notes: **B,2**. Smith (1994); **B,5**. Fraser *et al*. (1992), Cunningham and Moors (1994), Smith *et al*. (1999); **C,1**. Magnuson *et al*. (2000); **E,1**. Magnuson *et al*. (2000); **E,2**. Grabherr *et al*. (1994), Ross *et al*. (1994), Hasenauer *et al*. (1999), Menzel and Fabian (1999); **E,3**. Fleming and Tatchell (1995), Zhou *et al*. (1995), de Jong and Brakefield (1998), Rodriguez-Trelles and Rodriguez (1998), Visser *et al*. (1998), Parmesan *et al*. (1999), Parmesan *et al*. (2000); **E,4**. Beebee (1995), Reading and Clarke (1995), Reading (1998), Sparks (1999); **E,5**. Jarvinen (1989, 1994), Gatter (1992), Bezzel and Jetz (1995), Mason (1995), Winkel and Hudde (1996, 1997), Crick *et al*. (1997), Ludwichowski (1997), Forchhammer *et al*. (1998), McCleery and Parrins (1998), Prop *et al*. (1998), Visser *et al*. (1998), Bergmann (1999), Crick and Sparks (1999), Slater (1999), Sparks (1999), Thomas and Lennon (1999); **E,6**. Post and Stenseth (1999); **F,1**. Magnuson *et al*. (2000); **F,2**. Barber *et al*. (2000), Bradley *et al*. (1999); **F,5**. Bradley *et al*. (1999), Brown *et al*. (1999), Dunn and Winkler (1999); **F,6**. Post and Stenseth (1999); **G,4**. Pounds *et al*. (1999); **G,5**. Pounds *et al*. (1999).

In some cases, the criteria were met by two companion papers rather than a single paper. These criteria narrowed the qualifying studies to 60; seven were companion papers. Among the 60 studies, 16 look at physical processes, 10 examine vegetation changes, eight look at invertebrates, six investigate amphibians and reptiles, 26 examine birds, and one addresses mammals. Some of these studies investigate multiple taxa (e.g., bird and insect) in the same paper. A total of 39 physical processes, 117 plants, 65 insects, 63 amphibians and reptiles, 209 birds, and 10 mammal species were examined in the 43 studies (summarized in Table 5-3). Approximately 39% of these species showed no change. Changes in the other 61% included earlier ice-off and later freeze dates in inland lakes and streams, earlier breeding times, shifting to higher elevations or latitudes, and changes in densities, development, morphologies, and genetics.

Several lines of evidence indicate lengthening of the vegetative growing season by 1.2–3.6 days per decade in the Northern Hemisphere, particularly at higher latitudes where temperature rise also has been greatest. This lengthening involved earlier onset of spring and later onset of fall. Summer photosynthetic activity [based on Normalized Differential Vegetation Index (NDVI) estimates from satellite data] increased from 1981 to 1991 (Myneni *et al.*, 1997), concurrent with an advance (by 7 days) and an increase in amplitude of the annual CO_2 cycle since the 1960s, most intensely during the 1980s (Keeling *et al.*, 1996). Phenological/climate models for Finland indicate an overall increase in growing season length since 1900 (Carter, 1998). These physical measures are in accord with observations on organisms. In controlled, mixed-species gardens across Europe, a lengthening of the growing season by 10.8 days occurred from 1959 to 1993 (Menzel and Fabian, 1999). Likewise, a study of 36 species in the central United States documented advances in flowering dates by an average of 7.3 days from 1936 to 1998 (Bradley *et al.*, 1999).

Responses to increased atmospheric CO_2 have been detected in increased stomatal densities in the leaves of temperate woodland plants (Beerling and Kelly, 1997). Recent changes (over 9- to 30-year periods) in community composition have occurred at protected sites in the lower United States and Alaska, concurrent with local warming trends (Chapin *et al.*, 1995; Brown *et al.*, 1997a; Alward *et al.*, 1999). Results of warming experiments coupled with previous knowledge of species' habitat requirements implicate climate as one factor in these community reorganizations, but additional effects of multiple pressures have led to complex responses that were not always predicted by bioclimatic theory (Schneider and Root, 1996).

Multiple studies of treelines at high latitudes in the northern hemisphere have shown 20th century poleward shifts, often measured as increased growth at northern boundaries and decreased growth at southern boundaries. Interpretation of these trends is not straightforward because most change occurred during the early 20th century warming and the trends have been less pronounced or absent in recent warm decades (Kullman, 1986, 1990; Hamburg and Cogbill, 1988; Innes, 1991; Lescop-Sinclair and Payette, 1995; Jacoby and D'Arrigo, 1995; Briffa *et al.*, 1998). These authors have hypothesized that the general lack of response to recent warming is a result of increases in water stress, severity of insect attack, and UV radiation and trends toward earlier snowmelt or to sunlight becoming a limiting growth factor. In addition, some localities that showed warming and increased growth in the early 20th century have shown cooling and stable growth since the1970s (Kullman, 1991, 1993). In contrast, simple predictions of range shifts have been fulfilled in alpine herbs, which have moved to higher altitudes concurrent with warming in Switzerland (Grabherr *et al.*, 1994), and loss of low-elevation pine forests in Florida as sea-level rise has caused toxic levels of salination near coastal areas (Ross *et al.*, 1994).

5.2.2. *Current Models of Ecosystem Change*

A large literature is developing on modeling the response of ecosystems to climate and global changes. Most of these models simulate changes in a small patch of land. These models are reviewed as appropriate in other sections of this chapter and elsewhere in this report. The focus here is on modeling changes in ecosystem composition, structure, and function at global or regional scales. There are several reasons for developing such models. One is to estimate carbon fluxes and their contribution to the global carbon cycle. This involves making estimates of NPP, net ecosystem productivity (NEP), and net biome productivity (NBP) (see Box 5-1). Another is to develop models of feedbacks

Box 5-1. Plant Productivity: Terms and Definitions

Plants are responsible for the vast majority of carbon uptake by terrestrial ecosystems. Most of this carbon is returned to the atmosphere via a series of processes, including respiration, consumption (followed by animal and microbial respiration), combustion (i.e., fires), and chemical oxidation. Gross primary productivity (GPP) is the total uptake through photosynthesis, whereas net primary productivity (NPP) is the rate of accumulation of carbon after losses from plant respiration and other metabolic processes in maintaining the plant's living systems are taken into account. Consumption of plant material by animals, fungi, and bacteria (heterotrophic respiration) returns carbon to the atmosphere, and the rate of accumulation of carbon over a whole ecosystem and over a whole season (or other period of time) is called net ecosystem production (NEP). In a given ecosystem, NEP is positive in most years and carbon accumulates, even if only slowly. However, major disturbances such as fires or extreme events that cause the death of many components of the biota release greater than usual amounts of carbon. The average accumulation of carbon over large areas and/or long time periods is called net biome productivity (NBP) (see also Box 3-1 in TAR WGI).

between the atmosphere and the land surface (Van Minnen *et al*., 1995; Foley *et al*., 1998). Neither of these applications are covered in detail in this chapter (see TAR WGI Chapter 3). Instead, we concentrate on their application in forecasting the impacts of climate change on biodiversity and the provision of other ecosystem goods and services.

5.2.2.1. *Two Paradigms Describing Ecosystem Responses to Climate Change*

There are two paradigms about the way ecosystems (thus biomes) will respond to global change. The ecosystem movement paradigm assumes that ecosystems will migrate relatively intact to new locations that are closer analogs to their current climate and environment. This paradigm clearly is a gross simplification of what will actually happen, but it has the advantage that the well-demonstrated relationship between ecosystem range and existing climate can be used to project new ecosystem distributions under changed climate scenarios.

Basic ecological knowledge suggests that the ecosystem movement paradigm is most unlikely to occur in reality because of different climatic tolerance of species involved, including intra-species genetic variability (Crawford, 1993); different longevities, including clonal regeneration (e.g., survival over 2,000 years by *Carex curvula*; Steinger *et al*., 1996); different migration abilities (Pitelka and Plant Migration Workshop Group, 1997); and the effects of invading species (Dukes and Mooney, 1999). It is an idealized working paradigm that is useful for screening scenarios of climate change for potential significant effects.

The alternative paradigm, ecosystem modification, assumes that as climate and other environmental factors change there will be *in situ* changes in species composition and dominance. These changes will occur as some species decline in abundance or become locally extinct (Jackson and Weng, 1999) and others increase in abundance. The longevity of individuals, the age structure of existing populations, and the arrival of invading species will moderate these changes. The outcome will be ecosystem types that may be quite different from those we see today. Paleoecological data indicate that ecosystem types broadly similar to those seen today did exist in the past (Pregitzer *et al*., 2000), but that there also occurred combinations of dominant species that are not observed today (Davis, 1981; Jablonski and Sepkoski, 1996; Ammann *et al*., 2000; Prieto, 2000).

Numerous paleoecological studies provide evidence of important species within an ecosystem responding differently to climate change. For example, the postglacial migration pattern for Sierra lodgepole pine (*Pinus contorta* ssp. *murrayana*) was largely elevational, with little migrational lag, whereas the more widely distributed Rocky Mountain subspecies (*P. contorta* var. *latifolia*) migrated both latitudinally and elevationally (Anderson, 1996). Colinvaux *et al*. (1997) have interpreted pollen data to show that during glacial cooling, Andean vegetation did not move upslope and downslope as belts but that plant associations were reorganized as temperature-sensitive species found different centers of distribution with changing temperature. Similarly, heat-intolerant plants have moved in and out of the Amazonian rainforests during periods of cooling and warming (Colinvaux *et al*., 2000). Colinvaux *et al*. (2000) also argue that expulsion of heat-intolerant species from the lowland forests in this postglacial warming already is complete and that the forest property of maintaining its own microhabitat will allow high species richness to survive more global warming, provided large enough tracts of forest are conserved. This conclusion is at odds with some modeling studies (e.g., White *et al*., 1999).

The problem with the ecological modification paradigm is that it is very difficult to use in practical forecasting of possible trends. Thus, most global and regional studies assessing the potential impacts of climate change have had to use the ecosystem paradigm, as illustrated in Box 5-2. They also tend to be limited to projecting changes in vegetation distributions, with the implicit assumption that animal populations will track the vegetation components of an ecosystem. However, observational and experimental studies show many cases in which animals are responding to climate and environmental change well before any significant changes in the vegetation (see Section 5.4).

5.2.2.2. *Climate or Environment Envelope Models*

Climate envelope models have been used since at least the 1980s (e.g., Box, 1981; Busby, 1988). Usually they are based on describing the climate or environment encompassing the current distribution of a species or ecosystem (the environmental envelope or climatic envelope), then mapping the location of this same envelope under a climate change scenario. Sometimes, other forms of bioclimatic correlations or categorization systems are used, but they are based on the same assumption of a close correlation between climate and ecosystems associated with it. These models are useful as a first screening device to point to potential significant changes in ecosystem composition resulting from climate change (e.g., Huntley, 1995, and Huntley *et al*., 1995 for European plants and birds; Somaratne and Dhanapala, 1996, for Sri Lankan plants; Brereton *et al*., 1995, for Australia mammals; Gignac *et al*., 1998, for Canadian peatlands; Eeley *et al*., 1999, for South African woody plants; Iverson and Prasad, 1998, Iverson *et al*., 1999, for eastern U.S. trees).

A basic problem with these models is that every species has a "potential niche"—that is, a location in which it could survive and reproduce under the climatic conditions (Kirschbaum *et al*., 1996). What is observed and used as the basis for the model, however, is the "realized niche," which is a more limited area in which the species is found given the effects of competitors, predators, and diseases. If climate change results in species remaining within their potential niche and the competitors and predators change, the species may be able to survive *in situ*. In some models, some locations are predicted to have climates that are not encompassed by any of the climate envelopes of any of the vegetation types (Lenihan and Neilson, 1995). Used cautiously, however, the approach can point to

Box 5-2. Illustration of Use and Limitation of Ecosystem Movement Models

The study by Malcolm and Markham (2000) is a good example of modeling that uses the ecosystem movement paradigm, but it also demonstrates the inherent weaknesses of the approach.

The study uses two models of existing ecosystem distributions (MAPSS and BIOME3; Neilson *et al.*, 1998) and compares predicted distributions at present CO_2 levels with the equilibrium climate associated with doubled CO_2 as projected by several general circulation models (GCMs). It avoids the naive assertion that the latter climate constitutes a forecast of the future distribution of ecosystems; instead, it uses the two predictions to calculate the necessary rate of migration (m yr^{-1}) for species in the ecosystems to migrate to the new locations within 100 years (other time frames also are explored in a sensitivity analysis). It then maps these required rates to show areas where unusually high rates may be required in the future if a "climatically appropriate" ecosystem is to be established (referred to as "migration-stressed" locations).

The study predicted that about 20% of the Earth's surface will require migration rates ≥1 km yr^{-1}, which is equivalent to the highest rates observed in the geological past. The effects of natural barriers (e.g., lakes) or barriers resulting from land-cover modification by humans are globally small but can be regionally significant. Their approach also gives an indication of which regions of the globe may be most likely to be migration-stressed by climate change. It shows that much of the Earth's surface will be "stressed" in at least one of the 14 combinations of vegetation and climate models used. For some regions in the northern boreal zones of Eurasia and North America, most of the models predict such stress.

The study then goes on to deal with locations where the models predict that under climates applying a doubled CO_2 scenario, current ecosystems will fall outside their climatic range (referred to as "climate-stressed" locations). One must be careful not to attribute specific impacts or changes to climate-stressed locations. Biomes (or, more correctly, species constituting ecosystems of the biome) may be able to tolerate the new climatic conditions (i.e., new conditions fall within the potential niche) and thus may be relatively little changed. This same proviso should be applied to migration-stressed locations (i.e., existing vegetation may continue to occupy the site; thus, the migratory restriction does not come into play for decades to centuries).

The authors move on to equate climate-stressed locations with habitat loss and conclude that 36% of the land area will be affected. For aforementioned reasons, this must be regarded as an upper bound. The authors then attempt to estimate the reduction in habitat patch size by counting pixels affected by climate-stress in contiguous blocks of the same biome type, then applying a simple species area relationship to estimate species loss (see McDonald and Brown, 1992). Little reliance should be placed on these estimates given the foregoing provisos and the caveats listed by the authors themselves.

priority areas for further study. For example, Lassiter *et al.* (2000) analyzed data for 200 woody plant species in the eastern United States and found the usual result: a northward shift in predicted range. Excluding water limitations (i.e., allowing increased WUE under elevated CO_2 to compensate for a drier climate), however, most species survive in their current locations. This implies the need to better understand water-use and water-availability relationships under climate change. If species are able to persist, there may be little spread into new regions and thus little vegetation change for many decades.

In a study of 80 species of trees in the United States, using 100,000 plots to describe current distribution and importance, nearly half of the species assessed showed the potential for ecological optima to shift at least 100 km to the north (Iverson *et al.*, 1999). Whether these species will be able to achieve these potential distributions will depend on their migration rates through fragmented landscapes. When Iverson *et al.* (1999) incorporated a migration model, they found severely limited migration in regions of high forest fragmentation, particularly when the species is low in abundance near the range boundary.

Kirilenko and Solomon (1998) also have developed a bioclimate correlative model that incorporates a migration component. They demonstrate the importance of incorporating migration by showing that in simulations in which tree migration is delayed, the estimated global terrestrial carbon stock decreases by 7–34 Gt C, in contrast to an increase in carbon stock projected under nonlimiting migration (Solomon and Kirilenko, 1997).

Climate envelopes continue to be a useful tool in identifying potential changes in species distributions but are limited by the foregoing problems. Emphasis is shifting to more mechanistically based dynamic models of vegetation change that are linked directly to transient outputs from GCMs.

5.2.2.3. *Toward Dynamic Global and Regional Ecosystem Models*

Since the SAR, there have been significant changes in vegetation models and GCMs. A major modeling comparison project—Vegetation/Ecosystem Modeling and Analysis Project (VEMAP)—compared six equilibrium models of vegetation distribution and biogeochemistry (VEMAP Members, 1995). In the IPCC *Special Report on the Regional Impacts of Climate Change*, Neilson *et al.* (1998) compared two leading vegetation models (MAPSS and BIOME3) run against transient outputs from recent GCMs.

In MAPPS and BIOME3, potential vegetation is simulated by first calculating carbon flux (plant growth) on the basis of climate and hydrology information derived from the GCM. The models calculate a leaf area index that represents the capacity of the site to support plant canopy. Physiologically based rules are then used to classify the site into a vegetation type (e.g., forest, shrub, or grassland; evergreen or deciduous), leading to allocation to one of 45 vegetation types in MAPSS or 18 in BIOME3. These vegetation models are equilibrium models in that they calculate which vegetation type might be most suited to the climate and do not consider how the existing vegetation type might change to that new type. Details of the model outputs and maps are presented in IPCC (1998) and Neilson *et al.* (1998).

Neilson *et al.* (1998) compared the more recent combination of vegetation and climate models with those being used in the SAR. They found that the newer, transient GCMs produce a cooler climate than earlier versions, mostly because lag effects are simulated and the temperature does not increase to its equilibrium value within the model runs. This means that changes in predicted vegetation tend to be less than those reported in the SAR. Nevertheless, there are significant poleward shifts in cold-limited vegetation types. Vegetation types that are limited by water availability showed more complex changes, depending on the balance between precipitation change, hydrological balance, and physiological adjustment under higher CO_2.

The newer modeling combinations continue to predict significant potential changes in the distribution of most ecosystem types and increases in the area of tropical and temperate forests, but it cannot be determined whether this potential will be observed until transient (or dynamic) global vegetation models are developed. These models simulate the change in abundance of important species or "functional groups" of species on a year-by-year (or similar) basis in response to the output of the GCM (Cramer *et al.*, 1999).

Such models are being developed and used for assessments of overall carbon storage potential of the land biosphere (Cramer *et al.*, 2001), but at this stage it is too early to place much reliance on the outputs for specific biomes or ecosystems. The results show the sensitivity of ecosystems to the treatment of water use and especially the balance between changes in water availability resulting from climate change (often decreased availability in a warmer climate) and response to higher CO_2 concentrations in the atmosphere (often increased water-use efficiency). This means that model output can vary significantly, depending on the GCM used, because these models have tended to produce different interannual variability in precipitation and thus water availability. Other challenges are to simulate loss of vegetation from disturbances such as fire, blowdown, or pest attacks and migration of species or groups of species to new locations.

Van Minnen *et al.* (2000) have shown that by modifying the IMAGE2 model to include unlimited migration, limited migration and no migration result in significantly different patterns of vegetation change, especially in high-latitude regions.

5.2.3. *Impacts on Biodiversity*

Biodiversity is assessed quantitatively at different levels—notably at the genetic level (i.e., the richness of genetically different types within the total population), the species level (i.e., the richness of species in an area), and the landscape level (i.e., the richness of ecosystem types within a given area). Overall, biodiversity is forecast to decrease in the future as a result of a multitude of pressures, particularly increased land-use intensity and associated destruction of natural or semi-natural habitats (Heywood and Watson, 1996). The most significant processes are habitat loss and fragmentation (or reconnection, in the case of freshwater bodies); introduction of exotic species (invasives); and direct effects on reproduction, dominance, and survival through chemical and mechanical treatments. In a few cases, there might be an increase in local biodiversity, but this usually is a result of species introductions, and the longer term consequences of these changes are hard to foresee.

These pressures on biodiversity are occurring independent of climate change, so the critical question is: How much might climate change enhance or inhibit these losses in biodiversity? There is little evidence to suggest that processes associated with climate change will slow species losses. Palaeoecology data suggest that the global biota should produce an average of three new species per year, with large variation about that mean between geological eras (Sepkoski, 1998). Pulses of speciation sometimes appear to be associated with climate change, although moderate oscillations of climate do not necessarily promote speciation despite forcing changes in species' geographical ranges.

Dukes and Mooney (1999) conclude that increases in nitrogen deposition and atmospheric CO_2 concentration favor groups of species that share certain physiological or life history traits that are common among invasive species, allowing them to capitalize on global change. Vitousek *et al.* (1997b) are confident that the doubling of nitrogen input into the terrestrial nitrogen cycle as a result of human activities is leading to accelerated losses of biological diversity among plants adapted to efficient use of nitrogen and animals and microorganisms that depend on them. In a risk assessment of Switzerland alpine flora, Kienast *et al.* (1998) conclude that species diversity could increase or at least remain unchanged, depending on the precise climate change scenario used.

5.2.3.1. Global Models of Biodiversity Change

Several general principles describe global biodiversity patterns in relation to climate, evolutionary history, isolation, and so forth. These principles continue to be the subject of considerable ecological theory and testing; the Global Biodiversity Assessment

(Heywood and Watson, 1996) and the Encyclopaedia of Biodiversity (Levin, 2000) contain detailed reviews.

Kleidon and Mooney (2000) have developed a process-based model that simulates the response of randomly chosen parameter combinations ("species") to climate processes. They demonstrate that the model mimics the current distribution of biodiversity under current climate and that modeled "species" can be grouped into categories that closely match currently recognized biomes. Sala *et al.* (2000) used expert assessment and a qualitative model to assess biodiversity scenarios for 2100. They conclude that Mediterranean climate and grassland ecosystems are likely to experience the greatest proportional change in biodiversity because of the substantial influence of all drivers of biodiversity change. Northern temperate ecosystems are estimated to experience the least biodiversity change because major land-use change already has occurred.

Modeling to date demonstrates that the global distribution of biodiversity is fundamentally constrained by climate (see Box 5-2). Future development along these lines (e.g., adding competitive relations and migration processes) could provide useful insights into the effect of climate change on biodiversity and the effects of biodiversity on fluxes of carbon and water on a global scale.

5.2.4. *Challenges*

There has been considerable progress since the SAR on our understanding of effects of global change on the biosphere. Observational and experimental studies of the effects of climate change on biological and physical processes have increased significantly, providing greater insights into the nature of the relationships. Greater biological realism has been incorporated into models of small patches of vegetation (point models), and more realistic biological representations have been incorporated into regional and global change models. The main improvement has been development of dynamic representations of biological processes that respond directly to climate. Nevertheless, several major challenges remain before fully effective models of the interaction between climate and biophysical processes will be available.

5.2.4.1. *Landscape Processes*

Most vegetation models still treat patches of vegetation as a matrix of discrete units, with little interaction between each unit. However, modeling studies (Noble and Gitay, 1996; Rupp *et al.,* 2000) have shown that significant errors in predicting vegetation changes can occur if spatial interactions of landscape elements are treated inadequately. For example, the spread of fires is partly determined by the paths of previous fires and subsequent vegetation regrowth. Thus, the fire regime and vegetation dynamics generated by a point model and a landscape model with the same ignition frequencies can be very different. There has been considerable progress in modeling of spatial patterns of disturbances within landscapes (Bradstock *et al.,* 1998; He and Mladenoff, 1999; Keane *et al.,* 1999), but it is not possible to simulate global or regional vegetation change at the landscape scale. Thus, the challenge is to find rules for incorporating landscape phenomena into models with much coarser resolution.

Another challenge is to develop realistic models of plant migration. On the basis of paleoecological, modeling, and observational data, Pitelka and Plant Migration Workshop Group (1997) conclude that dispersal would not be a significant problem for most species in adapting to climate change, provided that the matrix of suitable habitats was not too fragmented. However, in habitats fragmented by human activities that are common over much of the Earth's land surface, opportunities for migration will be limited and restricted to only a portion of the species pool (Björkman, 1999).

5.2.4.2. *Will Organisms have to Migrate?*

Section 5.2.2.2 raises the question of whether organisms will need to migrate under climate change or whether many species will be able to survive *in situ* under new climatic conditions (Woodward and Beerling, 1997). Some species do occupy sites that are on the limits of their physiological tolerance, and if climate change takes local climate beyond that threshold, clearly they will not be able to persist at that site (see Section 5.6 for examples). However, there is mounting paleoecological evidence of vegetation types persisting through significant climate changes.

Lavoie and Payette (1996) conclude that the stability of the black spruce (*Picea mariana* [Mill.] BSP.) forest boundary during warm and cold periods of the late Holocene [warm approximately 2,000 years before present (BP), medieval times, and this century; cold approximately 3,000 years BP, 1,300 years BP, and the Little Ice Age] demonstrates that mechanisms that allow forest boundaries to advance or retreat are not easily triggered by climatic change. The black spruce old growth forest persisted for more than 1,500 years through many variations in climate probably because of the buffering effect of the trees on the local environment (Arseneault and Payette, 1997). Only a fire in 1568 AD broke this forest influence on microclimate and local growth conditions and caused the forest vegetation to shift to krummholz. Such studies emphasize the importance of effectively incorporating sufficient biophysical detail to capture climate-ameliorating effects and realistic disturbance regimes.

5.2.4.3. *Human Land-Use Issues*

No projection of the future state of the earth's ecosystems can be made without taking into account past, present, and future human land-use patterns. Human land-use will endanger some ecosystems, enhance the survival of others, and greatly affect the ability of organisms to adapt to climate change via migration.

Leemans (1999) used IMAGE2 to forecast possible global shifts in vegetation types and land-use change. Using a mid-level scenario of human responses to global change, he forecasts that nondomesticated land (his proxy for biodiversity) will decrease from 71 to 62% of the land area between 2000 and 2025, then remain approximately stable until the end of the 21st century. Losses in Africa and parts of Asia from 2000 to 2025 may be as much as 20–30% of remaining nondomesticated land.

5.2.4.4. *Testing Models of Ecosystem Response*

It is clear that ecosystem change models still contain assumptions that are not fully tested, and most models inevitably work better for the geographical regions and time periods for which they were constructed (Hurtt *et al.*, 1998). This derives from a mixture of fitting parameters to available data and deliberate and subconscious bias in selecting processes to include in a model. However, testing of ecosystem response models is gradually improving. Bugmann and Solomon (1995) tested the behavior of a model developed for European ecosystems by running it for a comparable North American site. The results were broadly in agreement, with useful indicators about where the model could be improved.

Another test is to compare different approaches to modeling for a particular purpose, as Yates *et al.* (2000) did for correlative models such as the Holdridge climatic correlation model and more mechanistic models of vegetation distribution. The strengths, weaknesses, and appropriate areas of application can be determined. Another approach is to conduct sensitivity studies in which the sensitivity of model outputs to changes in the input data and assumptions is evaluated. Hallgren and Pitman (2000) have carried out a sensitivity analysis of BIOME3 (Haxeltine and Prentice, 1996); they conclude that parameters that affect photosynthesis, water use, and NPP change the competitive interactions between specific plant groups (e.g., C_3 versus C_4 plants). Numerous studies have found that the outputs of ecosystem response models are sensitive to the precise treatment of water availability and water use by vegetation (Gao and Yu, 1998; Churkina *et al.*, 1999; Hallgren and Pitman, 2000; Lassiter *et al.*, 2000). This points to areas in which further development is needed.

Beerling *et al.* (1997) used a climate change experiment (CLIMEX) that exposed an entire catchment of boreal vegetation to elevated CO_2 and temperature for 3 years to test their Sheffield dynamic global vegetation model. There generally was a good match between observations and predictions, but longer runs of such experiments are needed to test such models thoroughly. As paleoecological data sets improve, there is increasing opportunity to test models against these data. However, this is a multiple test because the validity of paleoclimate models themselves also is under test. For example, Kohfeld and Harrison (2000) tested the ability of GCMs to describe changes in data collected for the environments of the last glacial maximum (21,000 years BP) and/or mid-Holocene (6,000 years BP). They conclude that better land-surface (including vegetation) response models are needed to capture the detail observed in the paleoecological data.

In summary, recent studies show that the potential for significant disruption of ecosystems under climate change remains. Further development of simple correlative models that were available at the time of the SAR points to areas where ecosystem disruption and the potential for ecosystem migration are high. Observational data and newer dynamic vegetation models linked to transient climate models are refining the projections. However, the precise outcomes depend on processes that are too subtle to be fully captured by current models.

5.3. Agriculture

Conclusions regarding the consequences of climate change for the agriculture sector in the SAR (Reilly *et al.*, 1996) provide an important benchmark for this section. The focus in this section is on basic mechanisms and processes that regulate the sensitivity of agriculture to climate change, relying mostly on research results since the SAR. Specifically, we ask how the conclusions of the SAR have stood the test of new research. Research advances since the SAR have brought several new issues to light—for example, understanding the adaptation of agriculture to climate change.

The discussion in this section is guided by the State-Pressure-Impact-Response-Adaptation model (see Figure 5-1). The pace of social, economic, and technological change in the agriculture sector will steadily transform the setting in which climate change is likely to interact with sensitive features of the food system. The current *state* of the sector and important trends that would transform it provide a baseline against which to examine the potential consequences of climate change (Section 5.3.1). Multiple *pressures* are being exerted on the agriculture sector, including the need to meet rising demand for food and fiber, resource degradation, and a variety of environmental changes (Section 5.3.2). Agricultural impacts, response, and adaptation are discussed concurrently because they are inseparable parts of the calculus of the vulnerability of agricultural systems to climate change. Hence, we consider the response and adaptive potential of agriculture in each of the succeeding sections. Agriculture is likely to *respond* initially to climate change through a series of automatic mechanisms. Some of these mechanisms are biological; others are routine adjustments by farmers and markets. Note that we equate response with automatic adaptation, as discussed in Chapter 18.

Climate change will *impact* agriculture by causing damage and gain at scales ranging from individual plants or animals to global trade networks. At the plant or field scale, climate change is likely to interact with rising CO_2 concentrations and other environmental changes to affect crop and animal physiology (Section 5.3.3). Impacts and adaptation (agronomic and economic) are likely to extend to the farm and surrounding regional scales (Section 5.3.4). Important new work also models agricultural impacts and adaptation in a global economy (Section 5.3.5).

Finally, the vulnerabilities of the agriculture sector, which persist after taking account of adaptation, are assessed (Section 5.3.6).

5.3.1. State of the Global Agricultural Sector

As Reilly *et al.* (1996) argue in the SAR, one of the foremost goals for global agriculture in coming decades will be expansion of the global capacity of food and fiber in step with expansion of global demand. Agriculture in the 20th century accomplished the remarkable achievement of increasing food supply at a faster rate than growth in demand, despite rapidly growing populations and per capita incomes. Key summary indicators of the balance between global demand and supply are world prices for food and feed grains. Johnson (1999) and Antle *et al.* (1999a) show that during the second half of the 20th century, real (inflation-adjusted) prices of wheat and feed corn have declined at an average annual rate of 1–3%. Climate change aside, several recent studies (World Bank, 1993; Alexandratos, 1995; Rosegrant *et al.*, 1995; Antle *et al.*, 1999a; Johnson, 1999) anticipate that aggregate food production is likely to keep pace with demand, so that real food prices will be stable or slowly declining during the first 2 decades of the 21st century.

According to the U.S. Department of Agriculture (1999), food security[1] has improved globally, leading to a decline in the total number of people without access to adequate food. The declining real price of food grains has greatly improved the food security of the majority of the world's poor, who spend a large share of their incomes on these staples. The global number, however, masks variation in food security among regions, countries, and social groups that are vulnerable because of low incomes or a lack of access to food (FAO, 1999a). In lower income countries, political instability and inadequate physical and financial resources are the root causes of the food security problem (see Section 5.3.6). In higher income, developing countries, food insecurity stems from unequal distribution of food that results from wide disparities in purchasing power.

Agricultural production and trade policies also affect global food availability and food security. There is a widespread tendency for high-income countries to maintain policies that effectively subsidize agricultural production, whereas low-income countries generally have policies that tax or discourage agricultural production (Schiff and Valdez, 1996). Many low-income countries also pursue policies that promote food self-sufficiency. Although all of these policies tend to reduce the efficiency of agricultural resource utilization in low- and high-income countries, they have not changed long-run trends in global supply and demand (Antle, 1996a).

Relatively few studies have attempted to predict likely paths for food demand and supply beyond 2020. There are reasons for optimism that growth in food supply is likely to continue apace with demand beyond 2020. For example, population growth rates are projected to decline into the 21st century (Bos *et al.*, 1994; Lutz *et al.*, 1996; United Nations, 1996), and multiple lines of evidence suggest that agricultural productivity potential is likely to continue to increase. Rosegrant and Ringler (1997) project that current and future expected yields will remain below theoretical maximums for the foreseeable future, implying opportunities for further productivity growth.

Other analysts are less optimistic about long-term world food prospects. For example, there is evidence that the Asian rice monoculture may be reaching productivity limits because of adverse impacts on soils and water (Pingali, 1994). Tweeten (1998) argues that extrapolation of the downward trend in real food prices observed in the latter half of the 20th century could be erroneous because the supply of the best arable land is being exhausted and rates of productivity growth are declining. At the same time, demand is likely to continue to grow at reasonably high rates well into the 21st century. Other studies indicate concerns about declining rates of investment in agricultural productivity and their impacts on world food production in some major producing and consuming areas (Hayami and Otsuka, 1994; Rozelle and Huang, 1999). Ruttan (1996) indicates that despite advances in biotechnology, most yield improvements during the first decades of the 21st century are likely to continue to come from conventional plant and animal breeding techniques. These concerns about future productivity growth, if correct, mean that simple extrapolation of yield for impact assessment (e.g., Alexandratos, 1995) may be overoptimistic. The implication is that confidence in predictions of the world food demand and supply balance and price trends beyond the early part of the 21st century is low.

5.3.2. Pressures on Agriculture Sector

5.3.2.1. Degradation of Natural Resources

Degradation of natural resources—taken here as soils, forests, marine fisheries, air, and water—diminishes agricultural production capacity (Pinstrup-Andersen and Pandya-Lorch, 1998). Soil degradation emerges as one of the major challenges for global agriculture. It is induced via erosion, chemical depletion, water saturation, and solute accumulation. In the post-World War II period, approximately 23% of the world's agricultural land, permanent pastures, forests, and woodland were degraded as defined by the United Nations Environment Programme (UNEP) (Oldeman *et al.*, 1991). Various estimates put the annual loss of land at 5–10 Mha yr^{-1} (Scherr and Yadav, 1997). Although irrigated land accounts for only 16% of the world's cropland, it produces 40% of the world's food. There are signs of a slowing in the rate of expansion of irrigation: 10–15% of irrigated land is degraded to some extent by waterlogging and salinization (Alexandratos, 1995). Degradation of natural resources is likely to hinder increases in agricultural productivity and could dim optimistic assessments of the prospects of satisfying growing world food demand at acceptable environmental cost.

[1]Food security often is defined as "access by all peoples at all times to enough food for an active, healthy life" (Chen and Kates, 1996).

5.3.2.2. *Other Global Change Factors*

Regional scenarios of seasonal temperature and precipitation change for 32 world regions analyzed in Chapter 3 show the current variability of climate and the range of changes predicted by GCMs for 30-year time periods centered on 2025, 2055, and 2085. This background information is essential to interpret the potential impacts of climate change on crops and livestock production. Equally important background information is provided by agroclimatic indices. Agroclimatic indices are useful in conveying climate variability and change in terms that are meaningful to agriculture. They give a first approximation of the potential effects of climate change on agricultural production and should continue to be used (Sirotenko *et al.*, 1995; Sirotenko and Abashina, 1998; Menzhulin, 1998).

Several other climate-related global environmental changes are likely to affect the agriculture sector in coming years. Reilly *et al.* (1996) reviewed the exposure of crops to tropospheric ozone (O_3). Progress in sorting out interactions between O_3, CO_2, and climate variability is reviewed below.

Climate change is likely to interact with other global changes, including population growth and migration, economic growth, urbanization, and changes in land use and resource degradation. Döös and Shaw (1999) use an accounting system to estimate the sensitivity of agricultural production to various aspects of global change, including loss of cropland from soil degradation and urbanization. Imhoff *et al.* (1997) use remote-sensing techniques and soils data to show that urbanization in the United States has occurred primarily on high-quality agricultural lands.

5.3.3. *Response of Crops and Livestock and Impacts on Food and Fiber*

5.3.3.1. *Interaction between Rising CO_2 Concentrations and Climate Change*

Advances in knowledge of CO_2 effects on crop and forage plants establish convincingly, although incompletely, that it is no longer useful to examine the impacts of climate change absent their interactions with rising atmospheric CO_2 (see Boxes 5-3 and 5-4). Crop and forage plants are likely to be forced to deal with the combined effects of climate change and rising atmospheric CO_2 concentrations. In this section, emphasis is placed on understanding basic interactions between plant productivity, climate change, and rising CO_2 concentrations. The direct effects of climate change on livestock also are considered.

5.3.3.1.1. *Interactive effects of temperature increase and atmospheric CO_2 concentration*

Because temperature increase enhances photorespiration in C_3 species (Long, 1991), the positive effects of CO_2 enrichment on photosynthetic productivity usually are greater when temperature rises (Bowes *et al.*, 1996; Casella *et al.*, 1996). A rise in mean global nighttime temperatures (Horton, 1995)

Box 5-3. Impacts of Climate Change and Elevated CO_2 on Grain and Forage Quality from Experimentation

The importance of climate change impacts on grain and forage quality emerges from new research. For rice, the amylose content of the grain—a major determinant of cooking quality—is increased under elevated CO_2 (Conroy *et al.*, 1994). Cooked rice grain from plants grown in high-CO_2 environments would be firmer than that from today's plants. However, concentrations of iron and zinc, which are important for human nutrition, would be lower (Seneweera and Conroy, 1997). Moreover, the protein content of the grain decreases under combined increases of temperature and CO_2 (Ziska *et al.*, 1997).

With wheat, elevated CO_2 reduces the protein content of grain and flour by 9–13% (Hocking and Meyer, 1991; Conroy *et al.*, 1994; Rogers *et al.*, 1996a). Grain grown at high CO_2 produces poorer dough of lower extensibility and decreased loaf volume (Blumentahl *et al.*, 1996), but the physiochemical properties of wheat starch during grain fill are not significantly modified (Tester *et al.*, 1995). Increases in daily average temperatures above 30°C, even applied for periods of up to 3 days, tend to decrease dough strength (Randall and Moss, 1990). Hence, for breadmaking, the quality of flour produced from wheat grain developed at high temperatures and in elevated CO_2 degrades.

With high-quality grass species for ruminants, elevated CO_2 and temperature increase have only minor impacts on digestibility and fiber composition of cut material (Akin *et al.*, 1995; Soussana *et al.*, 1997). The large increase in water-soluble carbohydrates in elevated CO_2 (Casella and Soussana, 1997) could lead to faster digestion in the rumen, whereas declines in nitrogen concentration occurring mainly with C_3 species (Owensby *et al.*, 1994; Soussana *et al.*, 1996; Read *et al.*, 1997) reduce the protein value of the forage. The protein-to-energy ratio has been shown to be more critical in tropical climates than in temperate countries (Leng, 1990). Livestock that graze low protein-containing rangeland forage therefore may be more detrimentally affected by increased C:N ratios than energy-limited livestock that graze protein-rich pastures (Gregory *et al.*, 1999). Basically, lowering of the protein-to-energy ratio in forage could reduce the availability of microbial protein to ruminants for growth and production, leading to more inefficient utilization of the feed base and more waste, including emissions of methane.

Box 5-4. Elevated CO_2 Impacts on Crop Productivity: Recent Estimates with Field-Grown Crops under FACE Experimentation

The short-term responses to elevated CO_2 of plants grown in artificial conditions are notoriously difficult to extrapolate to crops in the field (Körner, 1995a). Moreover, with field-grown plants, enclosures tend to modify the plant's environment (Kimball *et al*., 1997). However, even the most realistic free-air CO_2 enrichment (FACE) experiments undertaken to date create a modified area (Kimball *et al*., 1993), analogous to a single irrigated field in a dry environment, and impose an abrupt change in CO_2 concentration. A cotton crop exposed to FACE increased biomass and harvestable yield by 37 and 48%, respectively, in elevated (550 ppm) CO_2. This effect was attributed to increased early leaf area, more profuse flowering, and a longer period of fruit retention (Mauney *et al*., 1994). At 550 ppm CO_2, spring wheat increased grain yields by 8–10% under well-watered conditions (Pinter *et al*., 1996). More recent studies with optimal nitrogen and irrigation increased final grain yield by 15 and 16% for two growing seasons at elevated CO_2 concentration (550 ppm), compared with control treatments (Pinter *et al*., 1996). If these latter results are linearly extrapolated to the possible effect of a doubling (700 ppm) of the current atmospheric CO_2 concentration, yields under ideal conditions would be 28% greater—in agreement with previous statements by Reilly *et al*. (1996). In grass-clover mixtures, the proportion of legume increased significantly under elevated CO_2 (Hebeisen *et al*.,1997)—a conclusion also reached by several experimental studies with temperate and fertile managed grasslands (Newton *et al*., 1996; Soussana and Hartwig, 1996; Stewart and Potvin, 1996).

could enhance carbon losses from crops by stimulating shoot dark respiration (Amthor, 1997). Despite possible short-term effects of elevated CO_2 on dark respiration (Amthor, 1997; Drake *et al*., 1997), the long-term ratio of shoot dark respiration to photosynthesis is approximately constant with respect to air temperature and CO_2 concentration (Gifford, 1995; Casella and Soussana, 1997). With moderate temperatures, long-term doubling of current ambient CO_2 under field-like conditions leads to a 30% enhancement in the seed yield of rice, despite a 5–10% decline in the number of days to heading (Horie *et al*., 2000). The grain yield of CO_2-enriched rice shows about a 10% decline for each 1°C rise above 26°C. This decline is caused by a shortening of growth duration and increased spikelet sterility. Similar scenarios have been reported for soybean and wheat (Mitchell *et al*., 1993; Bowes *et al*., 1996). With rice, the effects of elevated CO_2 on yield may even become negative at extremely high temperatures (above 36.5 °C) during flowering (Horie *et al*., 2000). However, in some cropping systems with growth in the cooler months, increased rates of phenological development with warm temperatures and/or earlier planting dates may tend to move the grain fill period earlier into the year during the cooler months, offsetting at least part of the deleterious effects of higher temperatures (Howden *et al*., 1999a).

5.3.3.1.2. Interactive effects of water availability and atmospheric CO_2 concentration

Although stomatal conductance is decreased under elevated CO_2, the ratio of intercellular to ambient CO_2 concentration usually is not modified, and stomata do not appear to limit photosynthesis more in elevated CO_2 compared to ambient CO_2 (Drake *et al*., 1997). Elevated-CO_2 effects on crop evapotranspiration per unit land area (E) have been small with cotton (Dugas *et al*., 1994; Hunsaker *et al*., 1994; Kimball *et al*., 1994) and spring wheat (Kimball *et al*.,1995, 1999) crops supplied with ample nitrogen fertilizer. With rice, under field-like conditions, CO_2 enrichment reduced seasonal total E by 15% at 26°C but increased E by 20% at 29.5°C (Horie *et al*., 2000). A larger decline (-22%) in the daily E of a C_4-dominated tallgrass prairie was reported by Ham *et al*. (1995), and a strong reduction in water use per plant also was observed for maize (Samarakoon and Gifford, 1996), a C_4 plant. The consequences of these direct effects of elevated CO_2 concentrations on E are still unclear at the catchment scale (see Section 4.3.3).

Relative enhancement of growth owing to CO_2 enrichment might be greater under drought conditions than in wet soil because photosynthesis would be operating in a more CO_2-sensitive region of the CO_2 response curve (André and Du Cloux, 1993; Samarakoon and Gifford, 1995). In the absence of water deficit, C_4 photosynthesis is believed to be CO_2 saturated at present atmospheric CO_2 concentration (Bowes, 1993; see also Kirschbaum *et al*., 1996). However, as a result of stomatal closure, it can become CO_2-limited under drought. Some of the literature examples in which C_4 crop species, such as maize, have responded to elevated CO_2 may have involved (possibly unrecognized) minor water deficits (Samarakoon and Gifford, 1996). Therefore, CO_2-induced growth enhancement in C_4 species (e.g., Poorter, 1993) may be caused primarily by improved water relations and WUE (Samarakoon and Gifford, 1996) and secondarily by direct photosynthetic enhancement and altered source-sink relationships (Ruget *et al*., 1996; Meinzer and Zhu, 1998). With rice, at the optimal temperature for growth, a doubling of CO_2 increases crop WUE by about 50%. However, this increase in WUE declines sharply as temperature increases beyond the optimum (Horie *et al*., 2000). Although increased productivity from increased WUE is the major response to elevated CO_2 in a C_3 or C_4 crop that is exposed frequently to water stress (Idso and Idso, 1994; Ham *et al*., 1995; Drake *et al*., 1997), changes in climatic factors (temperature, rainfall) may interact with elevated CO_2 to alter soil water status, which in turn will influence hydrology and nutrient relations. Therefore, to realistically project impacts on crop yields and regional evaporation (see Chapter 4), more research is needed on the interactions of elevated CO_2, high temperature, and precipitation.

5.3.3.1.3. Interactive effects of atmospheric chemistry and CO_2 concentration

An exposure-response model that linearly relates a change in gas exposure over a time period to log-scale change in biomass increment of a plant during the same period suggests that a decline in recent yields of grain crops caused by an increase in surface ozone concentrations may have reached 20% in some parts of Europe (Semenov *et al.*, 1997, 1998, 1999). Recent research has shown that multiple changes in atmospheric chemistry can lead to compensating or synergistic effects on some crops. Heagle *et al.* (1999) used field studies to examine the impact of higher O_3 levels on cotton growth under higher CO_2 conditions. They found that higher CO_2 compensates for growth suppression resulting from elevated O_3 levels. With wheat, elevated CO_2 fully protects against the detrimental effects of O_3 on biomass but not yield (McKee *et al.*, 1997). Similar results have been reported with soybean (Fiscus *et al.*, 1997) and tomato (Reinert *et al.*, 1997). Meyer *et al.* (1997) measured responses of spring wheat to different levels of ozone in chambers at different growth stages and found that photosynthesis and carbohydrate accumulations were strongly affected during anthesis, especially after a period of heat stress.

Mark and Tevini (1997) observed combined effects of UV-B, temperature, and CO_2 in growth chambers in seedlings of sunflower and maize. They found that a 4°C rise in daily maximum temperature (from 28 to 32°C), with or without higher CO_2, compensated for losses from enhanced UV-B. Teramura *et al.* (1990) report that yield increases with elevated CO_2 are suppressed by UV-B more in cereals than in soybean; rice also loses its CO_2-enhanced WUE. However, Unsworth and Hogsett (1996) assert that many research studies from the preceding decade used unrealistic UV-B exposures, and they conclude that UV-B does not pose a threat to crops alone or in combination with other stressors.

5.3.3.2. Interactive Effects of CO_2 Concentrations, Climate Change, Soils, and Biotic Factors

5.3.3.2.1. Interactive effects of CO_2 concentrations with soils

There is not yet any clear consensus regarding the magnitude and sign of interactions between elevated CO_2 and nutrient availability for crop growth. Reviews of available data indicate that, on average, plants grown at high nutrient supply respond more strongly to elevated CO_2 than nutrient-stressed plants (Poorter, 1993, 1998). Nevertheless, the current rise in atmospheric CO_2 concentration may help plants cope with soil nutritional deficiencies (Idso and Idso, 1994) and especially with low nitrogen availability (Lloyd and Farquhar, 1996; Drake *et al.*, 1997). Several authors emphasize that a strong increase in biomass production under elevated CO_2 cannot be sustained in low fertilizer input systems without an appropriate increase in nutrients assimilation (Comins and McMurtrie, 1993; Gifford, 1994; Schimel, 1998). When other nutrients are not strongly limiting, a decline in nitrogen availability could be prevented by an increase in biological N_2 fixation under elevated CO_2 (Gifford, 1992, 1994). In fertile grasslands, legumes benefit more from elevated CO_2 than nonfixing species, resulting in significant increases in symbiotic N_2 fixation (Soussana and Hartwig, 1996; Zanetti *et al.*, 1996).

Plants grown under elevated CO_2 generally increase the allocation of photosynthates to roots (Rogers *et al.*, 1996b; Murray, 1997), which increases the capacity and/or activity of belowground carbon sinks (Rogers *et al.*, 1994; Canadell *et al.*, 1996; Körner, 1996), enhancing root turnover (Pregitzer *et al.*, 1995; Loiseau and Soussana, 1999b), rhizodeposition (Cardon, 1996), and mycorrhizal development (Dhillion *et al.*, 1996) in some but not all systems. Some measurements also have shown an increase in soil N cycling (Hungate *et al.*, 1997a), in response to short-term enrichment in CO_2, although other studies have shown either no detectable change (Prior *et al.*, 1997) or even a reduction in soil N mineralization (Loiseau and Soussana, 2001). The relationships between C and N turnover in soils after exposure to elevated CO_2 therefore are not fully understood, and it is still a matter of debate whether the availability of soil nitrogen for crop plants is reduced after a step increase in atmospheric CO_2 concentration.

Soil organic carbon (SOC) stocks result from the balance between inputs and decomposition of soil organic matter (SOM). Residues of cotton (Torbert, *et al.*, 1995), soybean, and sorghum (Henning *et al.*, 1996) display increased C:N ratios from growth under elevated CO_2, which may reduce their rate of decomposition in the soil and lead to an increment in ecosystem carbon stocks, similar to that observed in fertile grasslands (Casella and Soussana, 1997; Loiseau and Soussana, 1999a). However, some studies (Newton *et al.*, 1996; Ross *et al.*, 1996, Hungate *et al.*, 1997b) suggest higher carbon turnover rather than a substantial net increase in soil carbon under elevated CO_2. Predicted increased air and soil temperatures can be expected to increase the mineralization rate of SOM fractions that are not physically or chemically protected. The degree of protection of SOM varies with several soil-specific factors, including structure, texture, clay mineralogy, and base cation status. This may lead in the long term to negative effects on structural stability, water-holding capacity, and the availability of certain nutrients in the soil (see Reilly *et al.*, 1996). Organic matter decomposition tends to be more responsive than NPP to temperature, especially at low temperatures (Kirschbaum, 2000). Within this range, any warming would stimulate organic matter decomposition (carbon loss) more than NPP (carbon gain); the net response would be a loss of soil carbon. Mineralization rates also are influenced by soil water content. For example, lower soil moisture in Mediterranean regions (see Chapter 3) would compensate temperature increase effects on carbon and nitrogen mineralization (Leiros *et al.*,1999).

As a result of these interactions with soil processes, experiments that impose sudden changes in temperature or CO_2 and last only a few years are unlikely to predict the magnitude of long-term responses in crop productivity, soil nutrients (Thornley and Cannell, 1997), and carbon sequestration (Luo and Reynolds,

1999). This may imply—in agreement with Walker *et al.* (1999)—that the actual impact of elevated CO_2 on crop yields in farmers' fields could be less than in earlier estimates that did not take into account limitations of nutrient availability and plant-soil interactions.

5.3.3.2.2. Interactions between effects of climate change and soil degradation

Land management will continue to be the principal determinant of SOM content and susceptibility to erosion during the next few decades, but changes in vegetation cover resulting from short-term changes in weather and near-term changes in climate are likely to affect SOM dynamics and erosion, especially in semi-arid regions (Valentin, 1996; Gregory *et al.*, 1999).

The severity, frequency, and extent of erosion are likely to be altered by changes (see Table 3-10) in rainfall amount and intensity and changes in wind (Gregory *et al.*, 1999). Models demonstrate that rill erosion is directly related to the amount of precipitation but that wind erosion increases sharply above a threshold windspeed. In the U.S. corn belt, a 20% increase in mean windspeed greatly increases the frequency with which the threshold is exceeded and thus the frequency of erosion events (Gregory *et al.*, 1999). Thus, the frequency and intensity of storms would have substantial effects on the amount of erosion expected from water and wind (Gregory *et al.*, 1999). Different conclusions might be reached for different regions. Thus, before predictions can be made, it is important to evaluate models for erosion and SOM dynamics (Smith *et al.*, 1997). By reducing the water-holding capacity and organic matter contents of soils, erosion tends to increase the magnitude of nutrient and water stress. Hence, in drought-prone and low-nutrient environments such as marginal croplands, soil erosion is likely (high confidence) to aggravate the detrimental effects of a rise in air temperature on crop yields.

5.3.3.2.3. Interactions with weeds, pests, and diseases

Modest progress has been made in understanding of pest (weeds, insects, pathogens) response to climate change since the SAR. Oerke *et al.* (1995) estimate preharvest losses to pests in major food and cash crops to be 42% of global potential production. Rosenzweig *et al.* (2000) suggest that ranges of several important crop pests in the United States have expanded since the 1970s, including soybean cyst nematode and corn gray leaf blight; these expansions are consistent with enabling climate trends, although there are competing explanations. Promising work linking generic pest damage mechanisms with crop models is reported by Teng *et al.* (1996). For example, Luo *et al.* (1995) linked the BLASTSIM and CERES-RICE models to simulate the effects of climate change on rice leaf blast epidemics. They found that elevated temperature increases maximum blast severity and epidemics in cool subtropical zones; it inhibits blast development in warm humid subtropics. Such model linkages have been used to examine climate change impacts on weed-crop competition (e.g., for rice-weed interactions see Graf *et al.*, 1990) and insect pests (Venette and Hutchison, 1999; Sutherst *et al.*, 2000). Any direct yield gain caused by increased CO_2 could be partly offset by losses caused by phytophagous insects, pathogens, and weeds. Fifteen studies of crop plants showed consistent decreases in tissue nitrogen in high CO_2 treatments; the decreases were as much as 30%. This reduction in tissue quality resulted in increased feeding damage by pest species by as much as 80% (Lincoln *et al.*, 1984, 1986; Osbrink *et al.*, 1987; Coviella and Trumble, 1999). Conversly, seeds and their herbivores appear unaffected (Akey *et al.*, 1988). In general, leaf chewers (e.g., lepidoptera) tend to perform poorly (Osbrink *et al.*, 1987; Akey and Kimball, 1989; Tripp *et al.*, 1992; Boutaleb Joutei *et al.*, 2000), whereas suckers (e.g., aphids) tend to show large population increases (Heagle *et al.*, 1994; Awmack *et al.*, 1997a; Bezemer and Jones, 1998)—indicating that pest outbreaks may be less severe for some species but worse for others under high CO_2. It is important to consider these biotic constraints in studies on crop yield under climate change. Nearly all previous climate change studies excluded pests (Coakley *et al.*, 1999).

5.3.3.3. Impacts on Livestock

Recent research supports the major conclusions of Reilly *et al.* (1996) on animal husbandry. Farm animals are affected by climate directly and indirectly. Direct effects involve heat exchanges between the animal and its environment that are linked to air temperature, humidity, windspeed, and thermal radiation. These linkages influence animal performance (e.g., growth, milk and wool production, reproduction), health, and well-being. Indirect effects include climatic influences on quantity and quality of feedstuffs such as pastures, forages, and grain and the severity and distribution of livestock diseases and parasites. When the magnitudes (intensity and duration) of adverse environmental conditions exceed threshold limits with little or no opportunity for relief (recovery), animal functions can become impaired by the resulting stress, at least in the short term (Hahn and Becker, 1984; Hahn and Morrow-Tesch, 1993; Hahn, 1999). Genetic variation, life stage, and nutritional status also influence the level of vulnerability to potential environmental stresses. These relationships form the basis for developing biological response functions that can be used to estimate performance penalties associated with direct climate factors (Hahn, 1976, 1981, 1995). Earlier work (Hahn *et al.*, 1992; Klinedinst *et al.*, 1993) used such response functions with the Goddard Institute for Space Studies (GISS), Geophysical Fluid Dynamics Laboratory (GFDL), and United Kingdom Meteorological Office (UKMO) scenarios and found substantial reductions in dairy cow performance with climate change. For example, milk production of moderate- to high-producing shaded dairy cows in hot/hot-humid southern regions of the United States might decline an additional 5–14% beyond expected summer reductions. Conception rates of dairy cows were reduced by as much as 36% during the summer season in the southeastern United States. Short-term extreme events (e.g., summer heat waves, winter storms) can result in the death of vulnerable animals (Balling, 1982; Hahn and Mader, 1997; Hahn, 1999), which can have substantial financial impacts on livestock producers.

5.3.3.4. Response of Plant Crops and Possible Adaptation Options

Very little work has investigated prospects for natural adaptation of crop species to climate change, and the results of the few studies that do have been inconclusive. However, there appears to be a wide range of resistance to high-temperature stress within and among crop species. For example, moderately large genetic variation in the tolerance to high-temperature induced spikelet sterility has been reported among and between indica- and japonica-type rice genotypes (Matsui *et al.,* 1997). Some rice cultivars have the ability to flower early in the morning, thereby potentially avoiding the damaging effects of higher temperatures later in the day (Imaki *et al*., 1987).

Prospects for managed genetic modification appear to be more optimistic than for natural adaptation. Intraspecific variation in seed yield of soybean in response to elevated CO_2 was observed by Ziska *et al*. (1998). Differences in carbon partitioning among soybean cultivars may influence reproductive capacity and fecundity as atmospheric CO_2 increases, with subsequent consequences for future agricultural breeding strategies (Ziska *et al.,* 1998). However, no significant intraspecific variability in responses to elevated CO_2 was detected in studies with wheat and temperate forage species (Lüscher and Nösberger, 1997; Batts *et al.,* 1998). To promote adaptation to an environment of high CO_2 and high temperature, plant breeders have suggested selection of cultivars that exhibit heat tolerance during reproductive development, high harvest index, small leaves, and low leaf area per unit ground (to reduce heat load) (Hall and Allen, 1993). However, prospects to improve adaptation of crop species to elevated CO_2 remain very uncertain, and more research in this direction is required.

5.3.4. Impacts and Adaptation at Farm to Subnational Regional Scales

5.3.4.1. Modeling Crop Yield Impact

The number of studies that model the yield impacts of climate change (with and without CO_2 direct effects and with and without adaptation) across individual sites in regions has continued to grow since the SAR. Of particular note is the expansion of studies that explicitly model the effects of change in climate variability and means simultaneously versus change in climate means only (Southworth *et al*., 1999), use transient climate change scenarios, and report modeling of agronomic and socioeconomic adaptation. A selection of major global and regional model-based studies reported since the SAR is summarized in Table 5-4.

Table 5-4 yields are reported as ranges of percentage change over the climate change scenarios, modeling sites, and crop as noted. Thirteen of the yield ranges—without adaptation—are from studies of tropical crops. Of the 13 ranges, 10 encompass changes that are exclusively lower than current yields. In three ranges, a portion of the range is approximately no different from current yields or slightly above. In the tropics, most crops are at or near theoretical temperature optimums, and any additional warming is deleterious to yields. Thirty ranges of percentage changes in temperate crop yields also appear in Table 5-4. Of these 30 ranges, six encompass changes that are exclusively higher than current yields. In another seven, half or more of the changes were more than current yields. In yet another seven, less than half of the changes extended above current yields. The remaining 10 ranges encompassed changes that were exclusively less than current yields. Hence, in two-thirds of the cases, temperate crop yields benefited at least some of the time from climate change.

New work on climate change scenarios (Mitchell *et al*., 2000) generated with stabilized radiative forcing at 550 and 750 ppm equivalent-CO_2 and unstabilized radiative forcing (i.e., unmitigated emissions) in the HadCM2 model simulated major cereal yield response globally in 2080 (Arnell *et al*., 2001). The pattern of yield changes with unstabilized forcing duplicates the pattern described above: Generally positive changes at mid- and high latitudes overshadowed by reductions in yields at low latitudes. Stablization at 550 ppm ameliorates yield reductions everywhere, although substantial reductions persist in many low-latitude countries. Stabilization at 750 ppm produces a pattern of yield response that is intermediate relative to the 550 ppm and unstabilized forcing scenarios, with anomalous yield increases in mid-latitudes relative to 550 ppm as a result of interactions between atmospheric CO_2, temperature, and moisture. More studies are needed before confidence levels can be assigned to understanding of the agricultural consequences of stabilization, although this work is an important step.

In all agricultural regions, the effects of natural climate variability are likely to interact with human-induced climate change to determine the magnitude of impacts on agricultural production. Some analyses postulate an increase in weather variability (Mearns *et al*., 1992, 1995; Rosenzweig *et al*., 2000); simulations of wheat growth indicate that greater interannual variation of temperature reduces average grain yield (Semenov and Porter, 1995). Hulme *et al*. (1999) simulated natural climate variability in a multi-century control climate for comparison with changed variability in a set of transient climate change simulations. Wheat yields were simulated with control and climate change scenarios. For some regions, the impacts of climate change on wheat yields were undetectable relative to the yield impacts of the natural variability of the control climate (see Table 5-4). Greater efforts to take account of the "noise" of natural climate variability are indicated (Semenov *et al*., 1996).

The importance of diurnal climate variability has emerged since the SAR (Reilly *et al*., 1996). Cold temperatures presently limit the yield of rice in all temperate rice-growing regions. Jacobs and Pearson (1999) provide new field results on irreversible

***Table 5-4**: Recent agricultural studies: a) studies with explicit global economics and/or global yields; b) studies of yield and production in developed regions, nations, and subnational regions; and c) studies of yield and production in economies-in-transition and developing regions, nations, and subnational regions.*

Study	Scope	Crops	Climate Scenario[a]	Yield Impact w/o Adaptation[b]	Yield Impact w/ Adaptation[b]	Socioeconomic Impact	Comments
a) Studies with Explicit Global Economics and/or Global Yields							
Parry *et al*. (1999)	Global	Wheat, rice, maize, soybeans	Transient scenarios: 4 HadCM2 ensemble scenarios, 1 HadCM3 (both assume IS92a forcing)		All cereals by 2080s[c]: NA (-10 to +3%); LA (-10 to +10%); WE (0 to +3%); EE (-10 to +3%); AS (-10 to +5%); AF (-10 to +3%)	By the 2080s: global cereal production (-4 to -2%), cereal prices (+13 to +45%), number of people at risk of hunger (+36 to +50%)	Farm-level adaptations (changes in plant date, varieties, irrigation, fertilizer); economic adjustments (increased investment, reallocation of resources, more land in production); no feedback between economic adjustments and yields; CO_2 direct effects included
Darwin *et al*. (1995)	Global	13 commodities	UKMO, GISS			Agriculture prices [wheat (-10 to -3%), other grains (-6 to -4%)]; global GDP (+0.3 to +0.4%)	Adaptation through market-induced land-use change; CO_2 effect not included
Darwin (1999)	Global	Same as Darwin *et al*. (1995)	OSU, GFDL, GISS, UKMO			Qualitative impacts: world (positive for temperature change <2°C, negative for temperature change >2°C), regional (positive for high latitudes, negative for tropics)	Same as Darwin *et al*. (1995)
Darwin and Kennedy (2000)	Global	Same as Darwin *et al*. (1995)	CO_2 effect on yields only, no climate change	Yield changes with full CO_2 effect: wheat (7%), rice (19%), soybeans (34%), other crops (25%)		Previous studies' estimates of economic value of CO_2 fertilization effect overstated by 61–166%	Scenarios run for CO_2 effect on yields ranging from very low to full effect
Adams *et al*. (1998)	USA	Various	+2.5°C, +7% ppt.; +5°C, +0% ppt.			Agricultural price changes (-19 to +15%), GDP (+0 to +0.8%)	Includes direct effects of CO_2

effects and retardation (but recoverable) impacts of cold temperatures on various physiological processes in rice. On the other hand, rice spikelet sterility above 35°C at flowering (usually during daytime) puts rice at risk from increased daily maximum temperatures (Horie *et al*., 1996). In light of recent observed rises in temperatures that are larger for daily minima

Table 5-4 (continued)

Study	Scope	Crops	Climate Scenario[a]	Yield Impact w/o Adaptation[b]	Yield Impact w/ Adaptation[b]	Socioeconomic Impact	Comments
a) Studies with Explicit Global Economics and/or Global Yields (continued)							
Yates and Strzepek (1998)	Egypt	Wheat, rice, maize, soybean, fruit	GFDL and UKMO $2xCO_2$ equilibrium scenarios, GISS-A transient scenario at $2xCO_2$	Yield changes: wheat (-51 to -5%), rice (-27 to -5%), maize (-30 to -17%), soybean (-21 to -1%), fruit (-21 to -3%)	Yield changes: wheat (-25 to -3%), rice (-13 to -3%), maize (-15 to -8%), soybean (-10 to 0%), fruit (-10 to -2%)	Change in selected economic indicators[d]: consumer-producer surplus (-3 to +6%), calories per day (-1 to +5%), trade balance (-15 to +36%)	Includes direct effects of CO_2; adaptations (shift in plant date, increased fertilizer, new varieties)
Rosenzweig and Iglesias (1998)	global (same sites as Parry *et al.*, 1999)	Wheat, rice, soybean, maize	Sensitivity analysis (+2, +4°C)	+2°C[e] [+8% (maize) to +16% (soybean)]; +4°C[e] [-8% (rice) to -2% (wheat)]	Adaptation more successful at high and mid-latitudes than at low latitudes		Includes direct effects of CO_2; transient yield response highly nonlinear
			GISS-A transient scenario, GISS $2xCO_2$ scenario	Wheat[f] [2050 (-18 to +25%), $2xCO_2$ (-32 to +27%)]; maize[f] [2050 (-26 to +13%), $2xCO_2$ (-35 to +23%)]; soybean[f] [2050 (+23 to +24%), $2xCO_2$ (+13 to +17%)]			

than daily maxima (Easterling *et al.*, 1997), Dhakhwa *et al.* (1997) and Dhakhwa and Campbell (1998) conclude that, compared to equal day-night warming, differential warming leads to less water loss through evapotranspiration and better WUE. This is likely to lead to enhanced photosynthesis, crop growth, and yield—although at a possible loss of nutritional quality (Murray, 1997). Possible reduction of frost incidence is not normally considered in these studies. On the negative side, higher nighttime temperatures could extend the overwintering range for some insect pests and broaden the range of other temperature-sensitive pathogens.

Substantial progress has been made in development of transient (time-evolving) scenarios of climate change for use in agricultural impact assessment. An important question arises regarding whether 2 years with exactly the same climate, one produced by a transient scenario and the other by an equilibrium scenario, would give different production system responses. Many crop models contain cumulative functions that retain environmental information over several years (e.g., water balance, soil nutrients). This factor alone could account for substantial yield response differences between transient and equilibrium climate change scenarios. Only a few studies deliberately have compared simulated yields with transient and equilibrium climate change scenarios. Using the UKHIV equilibrium scenario with increased interannual variability at Rothamsted, Semenov *et al.* (1996) simulate a loss of wheat yield relative to current with two crop models and no change with a third. With the UKTR transient scenario, all three models show yield increases relative to current climate. The U.S. Country Studies Program (Smith *et al.*, 1996a) used the Clouds and Earth's Radiant Energy System (CERES) model to simulate larger average increases in winter wheat across Kazakhstan with the GFDL transient climate change scenario (for the 10th decade) (+21% winter wheat yield) than the GFDL equilibrium scenario (+17% winter wheat yield). Spring wheat yields decreased with both scenarios; again, however, yields simulated with the transient climate change were not as adversely affected as those simulated with the equilibrium climate change. Rosenzweig and Iglesias (1998) also found that wheat, maize, and soybean yields are less adversely affected by transient climate change than equilibrium climate change. Lack of consistency

Table 5-4 (continued)

Study	Scope	Crops	Climate Scenario[a]	Yield Impact w/o Adaptation[b]	Yield Impact w/ Adaptation[b]	Socioeconomic Impact	Comments
a) Studies with Explicit Global Economics and/or Global Yields (continued)							
Winters *et al.* (1999)	Africa, Asia, Latin America	Maize, rice, wheat, coarse grains, soybean, "cash crops"	GISS, GFDL, UKMO		*Africa* [maize (-29 to -23%), rice (0%), wheat (-20 to -15%), coarse grains (-30 to -25%), soybean (-2 to +10%), cash crops (-10 to -4%)]; *Asia* [maize (-34 to -20%), rice (-12 to -3%), wheat (-54 to -8%), coarse grains (-34 to -22%), soybean (-9 to +10%), cash crops (-13 to +2%)]; *Latin America* [maize (-26 to -18%), rice (-26 to -9%), wheat (-34 to -24%), coarse grains (-27 to -19%), soybean (-8 to +12%), cash crops (-20 to -5%)]	*Africa* [total agricultural production (-13 to -9%), GDP per capita (-10 to -7%), agricultural prices (-9 to +56%)]; *Asia* [total agricultural production (-6 to 0%), GDP per capita (-3 to 0%), agricultural prices (-17 to +48%)]; *Latin America* [total agricultural production (-15 to -6%), GDP per capita (-6 to -2%), agricultural prices (-8 to +46%)]	Yield impacts based on Rosenzweig and Parry (1994) values for "level 1" (farm-level) adaptations and CO_2 direct effects; yield impacts are weighted (by production) average of country-level yield changes; values for total agricultural production and per capita GDP include both yield and price impacts; range for agricultural prices is across food and cash crops, and GCMs

in application of transient climate change scenarios to impact modeling between studies results in competing explanations about differences in impact estimates between the two types of climate change scenarios.

5.3.4.2. Historical Analogs of Adaptation

The agriculture sector historically has shown enormous capacity to adjust to social and environmental stimuli that are analogous to climate stimuli. Historical analogs of the adaptability of agriculture to climate change include experience with historical climate fluctuations, deliberate translocation of crops across different agroclimatic zones, rapid substitution of new crops for old ones, and resource substitutions induced by scarcity (Easterling, 1996). In the Argentine Pampas, the proportion of land allocated to crops has increased markedly at the expense of grazing land during historic humid periods, and vice versa during dry periods (Viglizzo *et al*., 1997). Historical expansion of hard red winter wheat across thermal and moisture gradients of the U.S. Great Plains provides an example of crop translocation (Rosenberg, 1982). At present, the northern boundary of winter wheat in China is just south of the Great Wall and the north edge of China, where large temperature increases are expected under climate change (Lin, 1997). Winter wheat planting has shifted from Dalian (38°54'N) to Shenyang (41°46'N) in Liaoning province. The shift was aided by introduction of freeze-resistant winter wheat varieties from high-latitude countries such as Russia, the United States, and Canada into Liaoning province (Hou, 1994, 1995; Chen and Libai, 1997). Rapid introduction of canola in Canadian agriculture in the 1950s and 1960s shows how rapidly farmers can modify their production systems to accommodate a new crop (National Research Council, 1991). Adaptation to declining groundwater tables by substituting dryland for irrigated crops in regions of the U.S. Great Plains is an example of substitutions to deal with water

Table 5-4 (continued)

Study	Scope	Crops	Climate Scenario[a]	Yield Impact w/o Adaptation[b]	Yield Impact w/ Adaptation[b]	Socioeconomic Impact	Comments
b) Studies of Yield and Production in Developed Regions, Nations, and Subnational Regions							
Hulme *et al*. (1999)	Europe	Wheat	HadCM2—moderate (1% yr^{-1}) and low forcing (0.5% yr^{-1}) simulations for 2050	+9 to +39%[g] (note that climate change impacts are indistinguishable from climate variability for 4 of 10 countries)			Includes direct effects of CO_2
Antle *et al*. (1999b), Paustian *et al*. (1999)	Montana, USA	Winter wheat, spring wheat, barley	CCC	Climate change only (-50 to -70 %); CO_2 fertilization only (+17 to +55%); climate change + CO_2 (-30 to +30%)		*With adaptation* [mean returns (-11 to +6%), variability of returns (+7 to +25%)]; *without adaptation* [mean returns (-8 to -31%), variability of returns (+25 to +83%)]	Scenarios include climate change plus CO_2 fertilization; yield impacts from Century model; adaptation modeled as change in crop rotation and management
Barrow and Semenov (1995)	1 site in UK; 1 site in Spain	Wheat	Sensitivity analysis (+2,+4°C); downscaled UKMO high-resolution transient run (UKTR)	UK site only [+2°C (-7%), +4°C (-10%)]; both sites [+3°C (-14 to -5%), UKTR (-5 to +1%)]			Direct effects of CO_2 not considered
Dhakhwa *et al*. (1997)	North Carolina, USA (1 site)	Maize	GFDL, UKMO with equal and unequal day/night warming	-28 to -2%			Includes direct effects of CO_2
Tung and Haith (1998)	New York, Indiana, and Oklahoma (1 site each)	Corn	GFDL	-24 to -15%[h]	-19 to -9%[h]		Direct effects of CO_2 not considered; water supply also modeled; adaptations (change in variety, plant date, irrigation amount); assumes management practices currently optimal
Howden *et al*. (1999a)	Australia	Wheat	CSIRO 1996	9 to 37%	13 to 46%	Gross margins (28 to 95%)	Assumes prices unchanged

Table 5-4 (continued)

Study	Scope	Crops	Climate Scenario[a]	Yield Impact w/o Adaptation[b]	Yield Impact w/ Adaptation[b]	Socioeconomic Impact	Comments
b) Studies of Yield and Production in Developed Regions, Nations, and Subnational Regions (continued)							
Brown and Rosenberg (1999)	USA corn and wheat regions	Corn, wheat	Three GCM-based 2xCO_2 scenarios distributed over time[i] (GISS, UKTR, BMRC)	Corn[j] [+1°C (-6 to +7%), +3°C (-17 to +4%), +5°C (-34 to -3%)]; wheat[j] [+1°C (-8 to +47%), +3°C (-20 to +37%), +5°C (-70 to -11%)]		Change in production: Corn[j] [+1°C (-10 to +10%), +3°C (-20 to +5%), +5°C (-35 to -5%)]; wheat[j] [+1°C (-10 to +55%), +3°C (-25 to +45%), +5°C (-75 to -8%)]	CO_2 level corresponds to temperature change (365–750 ppm); dryland cropping only; planting date and growing season length allowed to vary in response to climate
c) Studies of Yield and Production in Economies-in-Transition and Developing Regions, Nations, and Sub-National Regions							
Alexandrov (1999)	Bulgaria (2 sites)	Winter wheat, maize	GISS, GFDL R-30, CCC, OSU, UK89, HCGG, and HCGS equilibrium scenarios;	Maize (-35 to -1%), wheat (+8 to +20%)	Maize (-24 to -10%)[k]	Net return with adaptation: maize (-29 to -12%)[k]	Includes CO_2 direct effects; adaptation (change in planting date)
			GFDL-T transient scenario at 2060s	Maize (-22%), wheat (+14%)	Maize (-21%)[l]	Maize (-26%)[l]	
Cuculeanu *et al*. (1999)	Romania (5 sites)	Winter wheat, maize	CCC, GISS	Wheat (+15 to +21%), dry maize (+43 to +84%), irrigated maize (-12 to +4%)	Irrigated maize (-18 to +8%)[m]		Includes CO_2 direct effects; adaptation (new cultivars and changes in plant date, crop density, fertilizer amount)
Matthews *et al.* (1997)	Asia	Rice	Sensitivity analysis (+1, +2, +4°C);	+1°C (-7 to +26%)[n], +4°C (-31 to -7%)[n]			Includes CO_2 direct effects; adaptation (single to double cropping system, planting date shift, change in variety)
			GFDL, GISS, UKMO	-8 to +5%[o]	+14 to +27%[o] (with change in variety)[o]	Change in production: China with change in cropping system (+37 to +44%), region with change in variety (+13 to +25%)[o]	

Table 5-4 (continued)

Study	Scope	Crops	Climate Scenario[a]	Yield Impact w/o Adaptation[b]	Yield Impact w/ Adaptation[b]	Socioeconomic Impact	Comments
c) Studies of Yield and Production in Economies-in-Transition and Developing Regions, Nations, and Subnational Regions (cont.)							
Smith *et al.* (1996a)	The Gambia	Maize, millet-early, millet-late, groundnuts	CCC, GFDL, GISS	-26 to -15% -44 to -29% -21 to -14% +40 to +52%			CO_2 direct effects considered in all cases but Mongolia; adaptation in Mongolia consists of earlier seeding
	Zim-babwe	Maize	CCC, GFDL	-14 to -12%			
	Kazakh-stan	Spring wheat, winter wheat	CCC, GFDL[p], incremental scenarios	-70 to -25% -35 to +17%			
	Mon-golia	Spring wheat	GFDL, GISS	-74 to +32%	-67 to -5%[q]		
	Czech Republic	Winter wheat	Incremental scenarios	-3 to +16%			
Singh and Mayaar (1998)	Trinidad	Sugar cane	4 synthetic scenarios, 1 GCM-based (CCC equilibrium)	-42 to -18%			Direct effects of CO_2 not considered
Magrin *et al.* (1997)	Argen-tina (pampas region)	Wheat, maize, soybean	GISS scenario for 2050	Wheat (-15 to +15%), maize (-30 to -5%), soybean (+10 to +70%)		Change in production: wheat (+4%), maize (-16%), soybean (+21%)	Includes direct effects of CO_2 (550 ppm)
Amien *et al.* (1996)	Indo-nesia	Rice	GISS transient	2050 (-14 to -9%)[r]			Includes direct effects of CO_2
Saseen-dran *et al.* (2000)	India (5 sites)	Rice	Synthetic (+1.5°C, +2 mm day^{-1} precipitation)	-15 to -3%[s]			Includes direct effects of CO_2 (460 ppm)
Lal *et al.* (1999)	India	Soybean	Sensitivity analysis (+2,+4°C; ±20, ±40% precipitation)	-22 to +18%			Includes direct effects of CO_2
Buan *et al.* (1996)	Philip-pines (6 sites)	Rice, corn	CCC, GFDL, GISS, UKMO	Rice (-13 to +9%), corn (-14 to -8%)			Includes direct effects of CO_2

Table 5-4 (continued)

Study	Scope	Crops	Climate Scenario[a]	Yield Impact w/o Adaptation[b]	Yield Impact w/ Adaptation[b]	Socioeconomic Impact	Comments
c) Studies of Yield and Production in Economies-in-Transition and Developing Regions, Nations, and Subnational Regions (cont.)							
Karim *et al.* (1996)	Bangla-desh	Rice, wheat	CCC, GFDL	Rice (-17 to -10%), wheat (-61 to -20%)			CO_2 direct effects not considered
Jinghua and Erda (1996)	China	Maize	GFDL, UKMO, MPI	-19 to +5%[t]		Change in production: -6 to -3%	CO_2 direct effects not considered

[a] GCMs are $2xCO_2$, unless otherwise noted.
[b] Unless otherwise noted, range is across GCM scenarios and site values are averaged; for sensitivity analyses, yield range is across sites.
[c] Range across countries and GCM scenarios; NA = North America, LA = Latin America, WE = Western Europe, EE = Eastern Europe, AS = Asia, AF = Africa.
[d] Range across three GCM scenarios, two levels of adaptation, and two baseline scenarios for 2060 (optimistic and pessimistic).
[e] World yields (weighted by production); range across crops.
[f] Range across countries in study.
[g] Range across countries and GCM scenarios.
[h] Range across sites.
[i] Temporal distribution accomplished using SCENGEN scenario generator.
[j] Change in global mean temperature (used as surrogate for time); regional temperature change may be higher or lower.
[k] One site only; range across three $2xCO_2$ equilibrium GCM scenarios (two sites and seven GCMs considered in *without adaptation* values).
[l] One site (two sites considered in *without adaptation* values).
[m] Range across two climate scenarios and degree of adjustment (e.g., different planting date options).
[n] Range across two crop models and CO_2 levels (1, 1.5, and $2xCO_2$).
[o] Range across crop model and GCM scenarios.
[p] Transient scenario.
[q] Adaptation not applied when climate change increases yields in the no-adaptation case.
[r] Average of "normal" and El Niño years; range across two sites.
[s] Range across sites.
[t] Range across sites and GCM scenarios.

becoming a scarce production resource (Glantz and Ausubel, 1988). None of these examples, however, deals specifically with an evolving climate change and all are historic—which limits confidence in extending their conclusions to future climate change.

5.3.4.3. Agronomic Adaptation of Yields

Increasing numbers of studies have investigated the effectiveness of agronomic adaptation strategies (e.g., adjustments in planting dates, fertilization rates, irrigation applications, cultivar traits) in coping with climate-induced yield losses and gains since the SAR (see Table 5-4). Considerable costs could be involved in this process, however—for example, in learning about and gaining experience with different crops or if irrigation becomes necessary. In some cases, a lack of water resulting from climate change might mean that increased irrigation demands cannot be met (see Section 4.7.2).

Methodologically, there has been little progress since the SAR in modeling agronomic adaptations. On one hand, the adaptation strategies being modeled are limited to a small subset of a much larger universe of possibilities, which may underestimate adaptive capacity. On the other hand, the adaptations tend to be implemented unrealistically, as though farmers are perfectly clairvoyant about evolving climate changes, which may inflate their effectiveness (Schneider *et al.*, 2000). Some studies find agronomic adaptation to be most effective in mid-latitude developed regions and least effective in low-latitude developing regions (Rosenzweig and Iglesias, 1998; Parry *et al.*, 1999). This finding clearly is supported across the studies summarized in Table 5-4, although the number of studies that include adaptation is not large. A small number of studies in Table 5-4 compare yield changes with and without agronomic adaptation. Percentage changes in yields across a range of climate change scenarios for those studies are shown in Figure 5-2. Each pair of vertical bars represents the range of percentage changes by crop, with and without adaptation, for each study. Clearly, adaptation ameliorates yield loss (and enhances yield gains) in most instances. The median adapted yields (mid-point of the vertical bars) shift upward relative to the median unadapted yields in six of the eight studies. Two studies do not show such an upward shift (Mongolia, Romania) because of peculiarities in the modeling (see figure caption). Adaptation ameliorates the worst yield losses in seven of the eight studies.

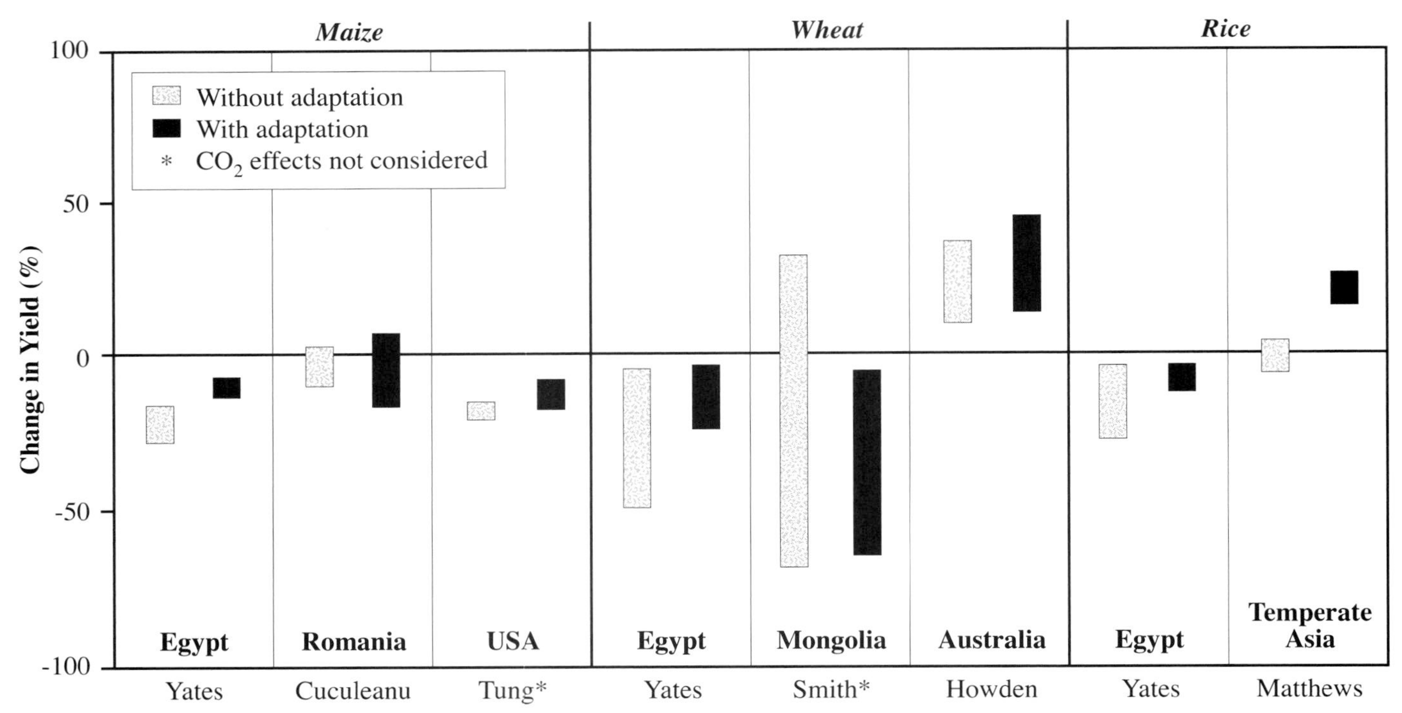

Figure 5-2: Ranges of percentage changes in crop yields (expressed in vertical extent of the vertical bars only) spanning selected climate change scenarios—with and without agronomic adaptation—from paired studies in Table 5-4. Each pair of ranges is differentiated by geographic location and crop. Pairs of vertical bars represent the range of percentage changes with and without adaptation. Endpoints of each range represent collective high and low percentage change values derived from all climate scenarios used in the study. Horizontal extent of the bars is not meaningful. Note that on the x-axis the last name of the lead author is listed as it appears on Table 5-4. See Table 5-4 for details on climate scenarios used and types of adaptation strategies modeled in each study. In the case of Mongolia, adaptation was not modeled when climate change resulted in positive yield change. In Romania, earlier planting of irrigated maize results in slightly lower yields than with current planting dates.

It is important to note, however, that differences in modeling methodology and aggregation of results often lead to conflicting conclusions in specific regions. For example, in two studies that used the same GCM scenarios, Matthews *et al.* (1997) simulate large increases, whereas Winters *et al.* (1999) simulate large decreases in rice yield with adaptation across several countries in Asia (see Table 5-4c). Hence, confidence in these simulations is low.

Important work has investigated the geographic distribution of crop potential under climate change. Carter and Saarikko (1996) used crop models to demonstrate a poleward shift in Finnish potential cereal cultivation by 100–150 km for each 1°C increase in mean annual temperature. Reyenga *et al.* (1999) indicate that climate change accompanied by a doubling of CO_2 is likely to enable expansion of existing Finnish wheat-growing areas into dry margins, considerably extending potential cropping areas. This expansion is moderated by dry conditions but not effectively enhanced by wetter conditions.

5.3.4.4. *Adaptation of Livestock*

Significant costs can be incurred to counter the effects of climate change on animal production; moreover, the impact of a warmer climate in terms of costs is not linear: Larger changes in climate can increase costs exponentially (Hahn and Morgan, 1999). Possible benefits of climate change during cooler seasons are not well documented, but the benefits are likely to be less than the consequential negative hot weather impacts (Hahn *et al.*, 1992). The ability of livestock managers to cope with climate is demonstrated daily in their coping with normally varying conditions. A variety of management adaptations are available for livestock production systems. For example, Hahn and Mader (1997) outline a series of proactive management countermeasures that can be taken during heat waves (e.g., shades and/or sprinklers) to reduce excessive heat loads. Historical success in coping with climate variability suggests that livestock producers are likely to adjust to climate change successfully. Johnson (1965) provides examples from advances in genetics and breeding as related to the environment. These capabilities should allow adaptation to changing, less favorable circumstances associated with projected rates of climate change. However, coping can entail significant dislocation costs for certain producers. For individual producers, uncertainties associated with potential climate change imply additional risks related to how and when to adapt current production practices (Lewandrowski and Schimmelpfennig, 1999). Confidence in the foregoing projections of the ability of livestock producers to adapt their herds to the physiological stresses of climate change is difficult to judge.

The general lack of simulations of livestock adaptation to climate change is problematic. This absence of a well-developed livestock counterpart to crop modeling of adaptation assessments suggests a major methodological weakness. Hence, we give only low to moderate confidence in projections of successful livestock adaptability.

5.3.4.5. Adaptation Effects on Farm and Regional Land Use and Income

Reilly *et al.* (1996) review several studies that analyzed the effects of climate scenarios at the farm scale. One important implication of these studies is that changes in production activities and management adaptations could play an important role in mitigating the impacts of climate change.

Survey-based studies on farm-level decisionmaking in response to climate change have propagated since the SAR. Case studies in Canada suggest several farm-level adaptive strategies to which farmers may resort (Brklacich *et al.*, 1997; Smithers and Smit, 1997; Chiotti, 1998). In southern Ontario, Smit *et al.* (1997) found that farmers' choice of cultivar to plant in the current year is highly conditioned by the climate experienced the year before. There have been few quantitative, farm-level economic studies of response and adaptation to climate change since the SAR. In the U.S. midwest, Doering *et al.* (1997) used a crop-livestock linear programming model linked to the Century biogeochemistry model to investigate the impacts of climate change on 11 representative farms in the region under a doubled-CO_2 scenario. The analysis considered these farms under static technology and adapted technology; thus, the analysis could be regarded as representing farm-level economic responses to climate change as well as farm-level combined technological and economic adaptation. It showed that climate change may cause substantial shifts in the mix of crops grown in the upper midwest, with much less land planted to a corn-soybean rotation and more land devoted to wheat than now. Earlier planting of corn increased returns; hence, a more frost-resistant corn variety was found to be important to farm-level adaptation. Using a similar methodology, Parry *et al.* (1996) predict optimal agricultural land use in response to climate change in England and Wales.

Antle *et al.* (1999b) used an econometric-process simulation model of the dryland grain production system in Montana linked to the Century model (as reported in Paustian *et al.*, 1999) to assess the economic impacts of climate change in that region. Unlike other farm-level studies, this analysis is based on a statistically representative sample of commercial grain farms in the region, not on a small number of representative farms, so the results can be interpreted as representing the entire population of farms in the study area. Moreover, instead of using linear programming models, this analysis combines site-specific data and econometric production models in a stochastic simulation model, allowing representation of physical and economic heterogeneity in the region. Simulations were conducted for baseline and doubled CO_2 (Canadian Climate model) with observed production technology, with and without land-use adaptation and with and without CO_2 fertilization.

Box 5-5. Extending Uncertainty in Crop Models to Uncertainty in Economic Analysis

Several factors contribute uncertainty to modeling of impacts of climate change on agricultural systems (Parry *et al.*, 1999). Crop modeling studies invariably highlight the need to develop confidence that the outputs are not "model-dependent." There is uncertainty because of the fact that yield estimates obtained in climate impact assessments vary from one impact model to another. In some cases, simulations across models may exhibit good agreement. Comparison of rice models showed that their predictions for potential production were quite close to observed values (Peng *et al.*, 1995). The Erosion Productivity Impact Calculator (EPIC), a generalized crop model, predicts observed yields most closely during years with extreme warmth, lending confidence to its ability to predict yields under climate change conditions (Easterling *et al.*, 1996). In other situations, agreement is not as good. In comparisons between wheat simulation models (Goudriaan *et al.*, 1994; Wolf *et al.*, 1996), grain yield predictions were markedly different between models. Such findings have stimulated work to compare the performance of different models and to analyze the underlying reasons for differences (Gregory *et al.*, 1999; Mearns *et al.*, 1999).

Spatial resolution of crop models is another important source of uncertainty in crop models. Crop models simulate processes that regulate growth and development at fine scales (a few kilometers), whereas climate change scenarios that drive crop models typically are produced by climate models operating at coarse scales (1,000 km or more) (Barrow and Semenov, 1995; Easterling *et al.*, 1998). Studies that use statistical downscaling techniques and nested limited area numerical models to increase the resolution of GCM scenarios of climate change have shown large simulated yield discrepancies between coarse-resolution (GCM) and fine-resolution (downscaled) climate change scenarios. Mearns *et al.* (1999) demonstrate for a site in Iowa that yields can change algebraic sign depending on the resolution of the climate change scenario: Maize yield decreases (-11%) with low-resolution climate change (CSIRO model) and increases (2%) with high-resolution climate change (RegCM limited area model nested within CSIRO model). Causes of variance in model output must be identified.

Box 5-5. Extending Uncertainty in Crop Models to Uncertainty in Economic Analysis (continued)

Some analysts use statistical approaches to relate historical climate to yields as a basis for projecting yield response to future climate change. Spatial analog models use reduced-form econometric models that are based on historical data and estimate the relationship between economic variables such as asset values or value of production and climate variables. These models then can be used to simulate the effects of climate change on economic outcomes (Mendelsohn *et al.*, 1994). A strength of the spatial analog approach is that it embeds the complex responses of agricultural decisionmakers to spatial variations in climate. A disadvantage of some spatial analog models is that they do not represent market equilibrium. They can be used, however, to simulate price scenarios or be linked to an equilibrium model (e.g., see Adams *et al.*, 1999). Another key limitation is that these models do not represent productivity explicitly, so they cannot be linked to process models to incorporate the important effects of CO_2 fertilization on productivity and economic behavior. Kaufmann and Snell (1997) relate underlying determinants of yields to historical climate, to exploit the advantages of statistical analysis in capturing adaptation while utilizing the robustness of process-based crop modeling.

Integrated assessments of agricultural impacts of climate change (Table 5-4) often link outputs from climate models with crop process models and economic models that together predict changes in land use, crop choice, production, prices, and impacts on economic welfare of producers and consumers. The studies listed in Table 5-4 provide some indication of sensitivity to alternative climate models by comparing results from several different models or model variants. These comparisons show that model results are highly sensitive to alternative climate model inputs, reflecting the wide range of yield impacts that have been estimated with crop models. For example, Adams *et al.* (1999) report that net U.S. welfare changes from climate change range from –$16 billion to +$117 billion, depending on the climate scenario used. This surely illustrates Schneider and Moss' (Chapter 2) cascading uncertainty. By changing the assumptions about demand for U.S. exports, Adams *et al.* (1999) also find that the impacts of climate change on U.S. agriculture are highly sensitive to assumptions about the demand for U.S. exports, with impacts ranging from –$1.3 billion to +$123 billion. Many other assumptions are described in these studies (and, given the large size and complexity of these models, many other assumptions are not discussed explicitly) that are embedded in these models and that have not been subjected to sensitivity analysis. The limited sensitivity analyses that have been performed suggest that uncertainties in economic models alone are large and further imply that the economic impacts of climate change on agriculture, such as those presented in the SAR (Reilly *et al.*, 1996) and in Table 5-4, are given low confidence.

In view of the significant potential impacts of CO_2 on crop yields and the many interactions with climate and soil factors (Section 5.3.3), another significant question is how this effect is incorporated into integrated assessments of agricultural impacts. Some of the studies cited in Table 5-4 do not incorporate CO_2 fertilization effects; others use estimates of yield changes that are based on experiments or models that do incorporate these effects. Another question is whether experimental or simulated estimates of the CO_2 fertilization effect on yields is a realistic estimate of on-farm yield changes. Darwin and Kennedy (2000) found that estimates of the economic benefits of CO_2 fertilization are sensitive to the magnitude of yield change.

Taken together, it is clear that the results of existing agricultural impact studies must be assigned a low degree of confidence. How uncertain these results are has not been quantified. The implication is that econometric and integrated assessment modelers need to undertake sensitivity analysis to determine key assumptions and parameters and focus quantitative uncertainty analysis on those dimensions of the models.

With climate change, CO_2 fertilization, and adaptation, mean returns change by –11 to + 6% relative to the base climate and variability in returns increases by +7 to +25%, whereas without adaptation mean returns change by –8 to –31% and variability increases by +25 to 83%. These findings provide support for the hypothesis that ability to adapt plays a critical role not only in mean impacts but also in the spatial variability of impacts. They provide empirical support to the hypothesis advanced in the SAR that climate change is likely to have its greatest adverse impacts on areas where resource endowments are poorest and the ability of farmers to respond and adapt is most limited.

5.3.4.6. Environmental and Natural Resource Consequences of Responses and Adaptation to Climate Change

None of the economic studies in Reilly *et al.* (1996) analyze the environmental consequences of adaptation to climate change, such as increased demands on land and water resources. Adams *et al.* (1998) observe that this shortcoming remains true of most recent studies. However, Darwin *et al.* (1995) do include an analysis of impacts of climate change on land and water resources in a global model with the world subdivided into eight regions. They argue that competition from crop production could aggravate direct climate-induced losses of forests in moist tropical regions. IMAGE 2.0 simulations with future scenarios of limited CO_2 emissions show that increased deforestation increases agricultural capacity because of a smaller CO_2 fertilizaton effect on crops than if emissions continue on the current trajectory (Leemans, 1997).

Lewandrowski and Schimmelpfinnig (1999) draw implications for impacts on land and water resources, wild species, and natural ecosystems from the literature. They suggest that increased demand for irrigation predicted by these studies is likely to increase the opportunity cost of water and possibly reduce water availability for wildlife and natural ecosystems. However, it is difficult to go beyond such generalities with these aggregate models because most environmental impacts of agriculture are site-specific. Strzepek *et al.* (1999) show that some scenarios of climate change may reduce irrigation system reliability in the lower Missouri River in the U.S. corn belt, which may induce instream environmental stress. In many developing countries, current irrigation efficiencies are very low by developed-country standards. Irrigation effeciency in the Philippines in 1990 was 18%, compared to the global average of 43% (Asian Development Bank, 1998). Currently, 3,480–5,000 liters of water are used to produce 1.0 kg rough rice (equivalent to 640 g milled rice) in the Philippines (Baradas, 1999) and some neighboring countries. At those irrigation efficiencies, increased irrigation demand caused by climate change would strain irrigation supplies. Hence, one adaptation strategy is to increase irrigation efficiency.

Agriculture is a source and a sink of GHGs; hence, climate-induced agricultural land-use change is likely to impact soil carbon stocks in agricultural soils (Paustian *et al.,* 1996; Lal *et al.*, 1998; IPCC, 2000). Antle *et al.* (1999b) and Paustian *et al.* (1999) link a field-scale econometric-process simulation model to the Century ecosystem model to assess the impacts of climate change (from the Canadian Climate Centre model) on soil carbon in the dryland grain production system of the U.S. northern Great Plains. In a related set of studies, Antle *et al.* (2000) and Paustian *et al.* (2000) link a regional economic agricultural land-use model to the Century ecosystem model to assess the impacts of climate change on soil carbon in central U.S. cropland. These studies demonstrate that adaptive changes in land use and management are likely to have greater impacts on soil carbon than the direct effects of climate. Thus, adverse effects of climate change on soil carbon tend to be offset by the adaptive changes in land use that would be made by farmers in response to climate change. The degree of this offset depends on the magnitude of CO_2 fertilization effects on crop yields.

5.3.4.7. Note on Costs of Adaptation

Adaptation is unlikely to come without cost. In a literature survey, Tol *et al.* (1998) conclude that adaptation costs (as opposed to net costs of damages) are not reported in most impact studies, especially in agriculture. Yet transition costs (e.g., to retrain farmers in new practices) and equilibrium costs (e.g., to develop additional irrigation or apply more fertilizer) may be considerable. The absence of a benefit-cost calculus for agricultural adaptation is a key deficiency. Existing studies also fail to account for the process of long-term, endogenous adaptation of technology in ways that are consistent with the extensive economic literature on that subject (Antle, 1996b). This process also will involve significant costs. An extensive body of economic research has studied the benefits and costs of agricultural research and has shown that institutions that are responsible for agricultural research adapt agricultural technology across space and time in response to relative resource scarcity (Hayami and Ruttan, 1985). Quiggin and Horowitz (1999) argue that changes in fixed capital for on-farm and off-farm infrastructure may be the most significant cost associated with adaptation to climate change.

5.3.5. *Modeling Impacts and Adaptation in a Global Economy*

Relatively few studies cited in the SAR linked estimates of yield responses to climate change with regional or global economic models to estimate production and welfare impacts. New studies that make such links provide important information on climate change-induced impacts on agriculture and on global and regional well-being (summaries are in Darwin *et al.,* 1995; Adams *et al.,* 1998; Lewandrowski and Schimmelpfennig, 1999). Recent contributions incorporate more crops and livestock, utilize Geographic Information System (GIS)-based land-use data, and link structural and spatial analogs (Table 5-4a). As noted in Section 5.3.2, the price of agricultural commodities is a useful statistic to summarize the net impacts of climate change on the regional or global supply/demand balance and on food security. Table 5-4a shows that the global model used by Darwin *et al.* (1995) and the U.S. model developed by Adams *et al.* (1998) predict that, with the rate of average warming expected by IPCC scenarios over the next century, agricultural production and prices are likely to continue to follow the downward path observed in the 20th century (see Section 5.3.2). As a result, impacts on aggregate welfare are a small percentage of GDP and tend to be positive, especially when the effects of CO_2 fertilization are incorporated. The only study that predicts real price increases with only modest amounts of climate change is Parry *et al.* (1999).

An important limitation of studies summarized in Reilly *et al.* (1996) is their focus on high-income regions of the world. Antle (1996b), Reilly *et al.* (1996), and Smith *et al.* (1996a) have suggested that impacts of climate change may be larger and more adverse in poorer parts of the world, where farmers and consumers are less able to adapt. The first study to address this question quantitatively in an aggregate regional analysis is Winters *et al.* (1999; they studied the impacts of climate change on Africa, Asia, and Latin America by using a computable general equilibrium model. Their analysis used larger price increases than those predicted by more recent studies in Table 5-4a, so this analysis should show more adverse impacts than an analysis based on prices from the more recent global studies.

The results summarized in Table 5-4a focus on the most vulnerable groups in poor countries—poor farmers and urban poor consumers. The results show that impacts on the incomes of these vulnerable groups would tend to be negative and in the range of 0 to –10%; in contrast, the impacts on consumer and producer groups predicted for the United States by Adams *et al.* (1998) ranged from –0.1 to +1%. Darwin (1999) reports results disaggregated by region and concludes that developing regions are likely to have welfare effects that are less positive or more negative than more-developed regions.

5.3.6. Vulnerability of the Agricultural Sector

A population, region, or sector is vulnerable to climate change when serious deficits or unused opportunities remain after taking account of adaptation. Assessment of agricultural vulnerability to climate change calls attention to populations, regions, and sectors that may lose the means to satisfy basic needs (food security, progress toward development, a healthy environment) or fail to seize opportunities to improve social welfare. Particularly when such cases result in part from market failure, issues of consequential equity arise that signal policy interventions. Although no single measure of agricultural vulnerability exists, several indices together provide a sketch of vulnerability, including crop yields, crop prices, production, income, number of people at risk of hunger, rates of erosion, and irrigation demand.

5.3.6.1. How Much Warming can Global Agriculture Absorb Before Prices Rise?

Is there an amount of climate change to which the global food production system can adapt with little harm but beyond which it is likely to impose serious hardship? An answer can be sketched only with very low confidence at this time because of the combination of uncertainties noted above. As noted in Section 5.3.2, prices are the best indicator of the balance between global food supply and demand. They determine the access of a majority of the world's population to an adequate diet. Two of three global studies reviewed here project that real agricultural output prices will decline with a mean global temperature increase of as much as 2.5°C, especially if accompanied by modest increase in precipitation (Darwin *et al.*, 1995; Adams *et al.*, 1998). Another study (Parry *et al.*, 1999) projects that output prices will rise with or without climate change, and even a global mean temperature increase of ~1°C (projected by 2020) causes prices to rise relative to the case with no climate change. When studies from the SAR are included with these more recent ones, there is general agreement that a mean global temperature rise of more than 2.5°C could increase prices (Reilly *et al.*, 1996; Adams *et al.*, 1998; Parry *et al.*, 1999), with one exception (Darwin *et al.*, 1995). Thus, with very low confidence, it is concluded from these studies that a global temperature rise of greater than 2.5° C is likely to exceed the capacity of the global food production system to adapt without price increases. However, results are too mixed to support a defensible conclusion regarding the vulnerability of the global balance of agricultural supply and demand to smaller amounts of warming than 2.5°C.

5.3.6.2. Vulnerable Regions and Populations

Although one may be reasonably optimistic about the prospects of adapting the global agricultural production system to the early stages of warming, the distribution of vulnerability among regions and people is likely to be uneven. As pointed out in Section 5.3.3, in the tropics—where some crops are near their maximum temperature tolerance and where dryland, nonirrigated agriculture predominates—yields are likely to decrease with even small amounts of climate change. The livelihoods of subsistence farmers and pastoral people—who make up a large proportion of rural populations in some regions, particularly in the tropics, and who are weakly coupled to markets—also could be negatively affected. In regions where there is a likelihood of decreased rainfall, agriculture could be substantially affected regardless of latitude. However, regional economic analysis (see Table 5-4) indicates that aggregate impacts on incomes even in the most vulnerable populations may not be large.

Clearly, in addition to the foregoing generality on productivity, other features of agricultural vulnerability are likely to vary widely among people, regions, nations, and continents (see Chapters 10–17). As noted in several places in this section and elsewhere (Downing *et al.*, 1996a), the poor—especially those living in marginal environments—will be most vulnerable to climate-induced food insecurity. Parry *et al.* (1999) assessed the consequences of climate change for the number of people at risk of hunger as defined by the Food and Agriculture Organization (FAO, 1988) (see Table 5-4 for details of the study). By the 2080s, the additional number of people at risk of hunger as a result of climate change is estimated to be about 80 million. However, some regions (particularly in the arid and subhumid tropics) may be affected more. Africa is projected to experience marked reductions in yield, decreases in production, and increases in the risk of hunger as a result of climate change. The continent can expect to have 55–65 million extra people at risk of hunger by the 2080s under the HadCM2 climate scenarios. Under the HadCM3 climate scenario, the effect is even more severe, producing an estimated additional 70+ million people at risk of hunger in Africa. It should be noted, however, that these hunger estimates are based on the assumption that food prices will rise with climate change, which (as noted above) is highly uncertain as far as 80 years into the future.

Who are these extra people at risk of hunger likely to be? Downing *et al.* (1996b) suggest the following classes: rural smallholder producers, pastoralists, rural wage laborers, urban poor, and refugees and displaced people. In addition, they point to particular kinds of individuals: rural women, malnourished children, handicapped and infirm people, and the elderly.

5.4. Wildlife in Ecosystems

The overall mobility of wildlife and the fact that they are physiologically constrained by temperature and moisture make them effective indicators of climatic changes (Root, 1988a; Parmesan *et al.*, 2000). Evidence is presented below that shows that many different taxa from around the world already are exhibiting recognizable changes, such as poleward and elevational range shifts and changes in the timing of events such as breeding (see Table 5-3). Many factors (e.g., habitat conversion, pollution) pressure animals (see Figure 5-1). The information that follows indicates that the changes from such pressures could result in patterns that differ from those created by rapid climate change, which are created by the physiological constraints of organisms in response to climatic variables. For example, the pattern many species are showing of general poleward movement is not likely to be created by habitat conversion because such conversion generally does not occur less frequently along the poleward sides of many species ranges (e.g., along the northern boundaries in the northern hemisphere) than along those nearer the equator. Consequently, the balance of evidence suggests that, for animals that are exhibiting significant large-scale patterns of changes, the most consistent explanation is recent climatic change. Thus, like the proverbial "canaries in the coal mine," wildlife seem to be providing an important early indicator of how ecosystems might respond to the discernible human impact on climate that is contributing to its change (Santer *et al.*, 1996).

Much of the early work on the effects of climate change on ecosystems focused on vegetation. Animals (nondomestic animals or wildlife—these terms are used interchangeably) are important members of ecosystems and are affected by weather and climate (Andrewartha and Birch, 1954). Consequently, concerns about the impacts that rapid climatic change may have on wildlife and the risks these changes may impose on ecosystem services are assessed and summarized in this section.

5.4.1. State of Wildlife

5.4.1.1. Current Status of Endangered/Extinct Animals

Recent estimates indicate that 25% (~1,125 species) of the world's mammals and 12% (~1,150 species) of birds are at a significant risk of global extinction (Stattersfield *et al.*, 1998; UNEP, 2000). One indicator of the magnitude of this problem is the speed at which species at risk are being identified. For example, the number of birds considered at risk has increased by almost 400 since 1994, and current population sizes and trends suggest an additional 600–900 soon could be added to these lists (IUCN, 1994; UNEP, 2000). The number of animals threatened with extinction varies by region (see Table 5-5). Global patterns of total diversity are reflected in the number of species at risk in each region, in that areas with more total species are likely to have more at risk. The number of threatened invertebrates in Table 5-5 is unrealistically low. The extinction rate of invertebrates in tropical forests alone has been estimated at 27,000 yr^{-1}, largely because of habitat conversion (Wilson, 1992).

5.4.1.2. Species Status from Secure to Extinction: Ranking Risks

Extinction often is caused by a combination of pressures acting over time (Wilson, 1992). Three traits of species populations that contribute to endangerment status are range size, distribution of suitable habitat within the range, and population size. Species that are most at risk often have small ranges, inhabit a unique type of habitat or one found in isolated areas (patchy in distributions), and/or typically occur at low population densities (Rabinowitz, 1981; Rabinowitz *et al.*, 1986). Using these criteria, signs that a species may be at risk include shrinking range, decreased availability of habitat within the range, and local or widespread population declines.

Species with restricted habitat requirements typically are most vulnerable to extinction (Pimm *et al.*, 1995), including many

***Table 5-5**: State of some of the world's vertebrate wildlife. For each region, the table lists the number of critically endangered, endangered, and vulnerable species, separated by slashes (UNEP, 2000).*

Region	Totals	Amphibians	Reptiles	Birds	Mammals
Africa	102 / 109 / 350	0 / 4 / 13	2 / 12 / 34	37 / 30 / 140	63 / 63 / 163
Asia and the Pacific	148 / 300 / 739	6 / 18 / 23	13 / 24 / 67	60 / 95 / 366	69 / 163 / 283
Europe and Central Asia	23 / 43 / 117	2 / 2 / 8	8 / 11 / 10	6 / 7 / 40	7 / 23 / 59
Western Asia	7 / 11 / 35	0 / 0 / 0	2 / 4 / 2	2 / 0 / 20	3 / 7 / 13
Latin America[a] and the Caribbean	120 / 205 / 394	7 / 3/ 17	21 / 20/ 35	59 / 102 / 192	33 / 80 / 150
North America[a]	38 / 85 / 117	2 / 8 /17	3 / 12 / 20	19 / 26 / 39	14 / 39 / 41

[a] UNEP data place Mexico in North America because of similarity of biomes.

endemic species that could be lost with loss of their habitat (Wyman, 1991; Bibby *et al.*, 1992; Stattersfield *et al.*, 1998). For example, the Sundarban, the only remaining habitat of Bengal tigers (*Panthera tigris tigris*) in Bangladesh, is projected to decrease considerably in size as a result of rising sea levels; Milliman *et al.* (1989) estimate a loss of 18% of the land by 2050 and as much as 34% by 2100. For tigers and the many other species that inhabit these forested wetland habitats, migration to higher ground probably would be blocked by human habitation of adjacent lands (Seidensticker, 1987; ADB, 1994). Many mountainous areas also have endemic species with narrow habitat requirements (Dexter *et al.*, 1995; Stattersfield *et al.*, 1998). With warming, habitats may be able to move up in elevation if the mountain is high enough. If not, the habitat could be lost (Still *et al.*, 1999). Some montane species that are susceptible to this change include forest birds in Tanzania (Seddon *et al.*, 1999), Resplendent Quetzal (*Pharomachrus mocinno*) in Central America (Hamilton, 1995), mountain gorilla (*Gorilla gorilla beringei*) in Africa, and spectacled bear (*Tremarctos ornatus*) in the Andes (Hamilton, 1995, and references therein). Protecting species that currently are vulnerable, endangered, or critically endangered (see Table 5-5) requires measures that, in general, reverse the trend toward rarity. Without management, there is high confidence that rapid climate change, in conjunction with other pressures, probably will cause many species that currently are classified as critically endangered to become extinct and several of those that are labeled endangered or vulnerable to become much rarer, and thereby closer to extinction, in the 21st century (Rabinowitz, 1981).

5.4.1.3. *Wildlife Ties to Goods and Services*

Concern that species will become rare or extinct is warranted because of the goods and services provided by intact ecosystems and the species themselves. Most of the goods and services provided by wildlife (e.g., pollination, natural pest control; see Table 5-1) derive from their roles within systems. Other valuable services are provided by species that contribute to ecosystem stability or to ecosystem health and productivity. The recreational value (e.g., sport hunting, wildlife viewing) of species is large in market and nonmarket terms. Losses of species can lead to changes in the structure and function of the affected ecosystems, as well as loss of revenue and aesthetics (National Research Council, 1999).

5.4.2. ***Pressures on Wildlife***

Rapid climate change is only one of a long list of pressures on wildlife. Alone or in combination, these pressures may greatly increase species' vulnerabilities to rarity and extinction. Pressures such as land use and land-use change, introduction of exotic species, pollution/poisoning, and extreme climatic events and recent rapid climate change are of major concern and are discussed below. Other pressures include wildlife diseases, human persecution (e.g., overharvest, harassment), collisions with towers and other structures, collisions with cars and other forms of transportation, electrocutions, anthropogenic barriers to dispersal, war and other civil conflict, and wildlife trade (see Price *et al.*, 2000, for more information and references).

Conversion of natural and semi-natural habitats currently is quite extensive. Although climate change is starting to have observable effects on wildlife and the predicted future impact of climate change is expected to be large, the immediate principal current threat to the world's wildlife is habitat conversion (Vitousek *et al.*, 1997a). Roughly 80% of forests that originally covered the Earth have been cleared or degraded, and logging, mining, or other large-scale developments threaten 39% of what remains (UNEP, 2000). Roughly 65% of Asia's wildlife habitat has been converted to other uses, as well as nearly 75% of Australian rainforests (UNEP, 2000). Habitat conversion and degradation affect nearly 89% of all threatened birds and 83% of all threatened mammals (IUCN, 2000). Nearly 75% of all threatened bird species are found in forests (Stattersfield *et al.*, 1998), and tropical forests are the most species-rich terrestrial habitats; an estimated 90% of the world's species occur in moist tropical forests, which cover only 8% of the land area (UNEP, 2000).

Introduction of exotic species, intentionally and accidentally, has had deleterious effects on wildlife populations. Overall, 18% of threatened mammals in Australia and the Americas and 20% of the world's birds are challenged by introduced species (UNEP, 2000). Most extinctions on islands can be tied to introduction of species that prey on native species or destroy critical habitats (Stattersfield, *et al.*, 1998).

Pollution and other biochemical poisonings have direct and indirect effects on wildlife. Chemical contaminants have been detected in the tissues of species from around the globe, in regions such as Brazilian Pantanal wetlands (Alho Cleber and Vieira Luiz, 1997) and remote arctic habitats (UNEP, 2000). The most obvious effects of chemicals are direct poisoning events. These frequently are side effects of pesticide applications and can lead to losses of thousands of individuals (Biber and Salathe, 1991). In addition to affecting wildlife directly, contaminants can indirectly affect them by modifying their habitats.

Stochastic and extreme climatic events can cause deaths of large numbers of individuals and contribute significantly to determining species composition in ecosystems (Parmesan *et al.*, 2000). Hurricanes can lead to direct mortality, and their aftermath may cause declines because of loss of resources required for foraging and breeding (Wiley and Wunderle, 1994). Many of these extreme climatic events are cyclical in nature, such as sea surface temperature (SST) changes that are associated with the El Niño-Southern Oscillation (ENSO). Sea temperature increases associated with ENSO events have been implicated in reproductive failure in seabirds (Wingfield *et al.*, 1999), reduced survival and reduced size in iguanas (Wikelski and Thom, 2000), and major shifts in island food webs (Stapp *et al.*, 1999). Extreme drought in Africa was thought to have contributed to declines in the populations of many Palearctic migratory birds that wintered in the savanna and steppe zones of the Sahel (Biber and Salathe, 1991).

Climate and climate change are strong drivers of biotic systems. The distribution and survival of most species are moderated by climate (Root, 1988a,b,c; Martin, 1998; Duellman,1999). Although species have responded to climatic changes throughout evolutionary time (Harris, 1993), the primary concern today is the projected rapid rate of change. High species richness appears to be related to stable conditions; abrupt impoverishment of species has occurred during times of rapid change (Tambussi *et al.*, 1993).

Synergistic effects are likely to be quite damaging to animals. As habitat becomes more fragmented, barriers to dispersal or expansions of species' ranges could occur. This could force individuals to remain in inhospitable areas, decreasing the range and population size of species and ultimately leading to extinction (Rabinowitz, 1981). Fragmentation also may facilitate movement of invasive species into an area, leading to potential population declines through predation, competition, or transmission of disease (e.g., May and Norton, 1996). Increasing urbanization also could lead to increasing exposure to contaminants, which may make species less fit to survive changes in environmental conditions or weaken their immune systems (Pounds and Crump, 1987; Berger *et al.,* 1998). Human responses to climate change also may contribute to synergistic effects; for example, if new pest outbreaks are countered with increased pesticide use, nontarget species might have to endure climate- and contaminant-linked stressors.

5.4.3. Responses of Wildlife and Impacts on Goods and Services

Findings indicate that many animals already may be responding to local climatic changes. Types of changes already observed include poleward and elevational movement of ranges, changes in animal abundance, changes in body size, and shifts in the timing of events such as breeding to earlier in the spring. These responses have been identified by a group of studies from around the world in a variety of different species (see Table 5-3). Far more information is available than can be summarized here. More detail on these changes is available in Hughes (2000) and Price *et al.* (2000).

5.4.3.1. Changes Exhibited by Animals

Results from most studies that use large-scale data sets provide circumstantial (e.g., correlational) evidence about the association between changes in climate-related environmental factors and animal numbers or activities. Circumstantial evidence, though insufficient by itself, is highly suggestive when multiple studies examining a myriad of different species on all continents find similar results. Combined with smaller scale studies, experimental studies, and modeling studies that examine mechanistic connections between animals and climate change, the weight of evidence becomes even stronger. Such is the case for wildlife already exhibiting change related to climate forcings (see Table 5-3). The information given in the following subsections is a sampling of the types of studies that have examined the potential impacts of climate change on animals. The studies were selected for taxonomic and geographic inclusiveness and are not inclusive of the breadth of range of published studies. Information on more studies can be found in Table 5-3 and in Price *et al.* (2000).

5.4.3.1.1. Shifts in animal ranges and abundances

Ranges and abundances of prehistoric animals are known to have changed significantly over time (Goodfriend and Mitterer, 1988; Baroni and Orombelli, 1994; Coope, 1995). Currently, many species are undergoing range changes because of habitat conversion, land degradation (e.g., grazing, changes in fire regime), climate change, or a combination of factors. Possible climatically associated shifts in animal ranges and densities have been noted on three continents (Antarctica, Europe, and North America) and within each major taxonomic group of animals (see Table 5-3).

Invertebrates: Insect dispersal to favorable areas to make effective use of microclimatic differences is a common response to changing climate (e.g., Fielding *et al.*, 1999). The ranges of butterflies in Europe and North America have been found to shift poleward and upward in elevation as temperatures have increased (Pollard, 1979; Parmesan, 1996; Ellis *et al.,* 1997; Parmesan *et al.,* 1999). Warming and changed rainfall patterns also may alter host plant-insect relations, through community or physiological responses (e.g., host plant food quality) (Masters *et al.*, 1998).

Amphibians and Reptiles: Amphibians may be especially susceptible to climatic change because they have moist, permeable skin and eggs and often use more than one habitat type and food type in their lifetimes (Lips, 1998). Many amphibious species appear to be declining, although the exact causes (e.g., climate change, fungus, UV radiation, or other stresses) are difficult to determine (Laurance, 1996; Berger *et al.,* 1998; Houlahan *et al.,* 2000). Disappearance of the golden toad (*Bufo periglenes*) and the harlequin frog (*Atelopus varius*) from Costa Rica's Monteverde Cloud Forest Reserve appear to be linked to extremely dry weather associated with the 1986–1987 ENSO event (Pounds and Crump, 1994). Correlation between warming, reduced frequency of dry-season mist, and the timing of population crashes of four other frog species and two lizard species from the same cloud forest also has been found (Pounds *et al.*, 1999).

Birds: Bird ranges reportedly have moved poleward in Antarctica (Emslie *et al.,* 1998), North America (Price, 2000), Europe (Prop *et al.,* 1998), and Australia (Severnty, 1977). For example, the spring range of Barnacle Geese (*Branta leucopsis*) has moved north along the Norwegian coast, correlated with a significant increase in the number of April and May days with temperatures above 6°C (Prop *et al.*, 1998). The elevational range of some Costa Rican tropical cloud forest birds also apparently are shifting (Pounds *et al.*, 1999).

Mammals: Changes in mammal abundance can occur through changes in food resources caused by climate-linked changes or changes in exposure to disease vectors. For example, the Australian quokka (*Setonix brachyurus*) differs in susceptibility to Salmonella infections depending on climatic environmental conditions (Hart *et al.*, 1985).

5.4.3.1.2. Changes in timing (phenology)

Invertebrates: Warmer conditions during autumn and spring adversely affect the phenology of some cold-hardy species. Experimental work on spittlebugs (*Philaenus spumarius*) found that they hatched earlier in winter-warmed (3°C above ambient) grassland plots (Masters *et al.*, 1998).

Amphibians: Two frog species, at their northern range limit in the UK, spawned 2–3 weeks earlier in 1994 than in 1978 (Beebee, 1995). These changes were correlated with temperature, which also showed increasing trends over the same period.

Birds: Changes in phenology, or links between phenology and climate, have been noted for earlier breeding of some birds in Europe, North America, and Latin America (see Table 5-3). Changes in migration also have been noted, with earlier arrival dates of spring migrants in the United States (Ball, 1983; Bradley *et al.*, 1999), later autumn departure dates (Bezzel and Jetz, 1995), and changes in migratory patterns in Europe (Gatter, 1992).

5.4.3.1.3. Changes in morphology, physiology, and behavior

Amphibians and Reptiles: Correlations between temperature and calling rates have been found in Egyptian frogs (Akef Mamdouh and Schneider, 1995). Indian tree frogs show differences in behaviors that depend on their level of hydration (Lillywhite *et al.*, 1998). Painted turtles grew larger in warmer years, and during warm sets of years turtles reached sexual maturity faster (Frazer *et al.*, 1993). Physiological effects of temperature, primarily sex determination, also can occur while reptiles are still within their eggs (Gutzke and Crews, 1988).

Birds: Spring and summer temperatures have been linked to variations in the size of eggs of the Pied Flycatcher (*Ficedula hypoleuca*). Early summer mean temperatures explain ~34% of the annual variation in egg size between the years 1975 and 1994 (see Figures 5-3 to 5-5; Jarvinen, 1996).

Mammals: Body size is correlated with many life-history traits, including reproduction, diet, and size of home ranges. North American wood rat (*Neotoma spp.*) body weight has shown a significant decline that is inversely correlated with a significant increase in temperature over the past 8 years (Smith *et al.*, 1998). Juvenile red deer (*Cervus elaphus*) in Scotland grew faster in warm springs, leading to increases in adult body size (Albon and Clutton-Brock, 1988).

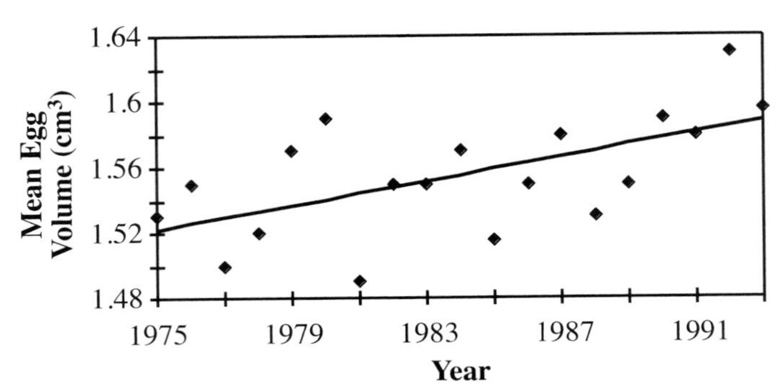

Figure 5-3: Change in egg volume of Pied Flycatcher birds over time. Regression line is y = 1.24 + 0.004 x (P<0.01).

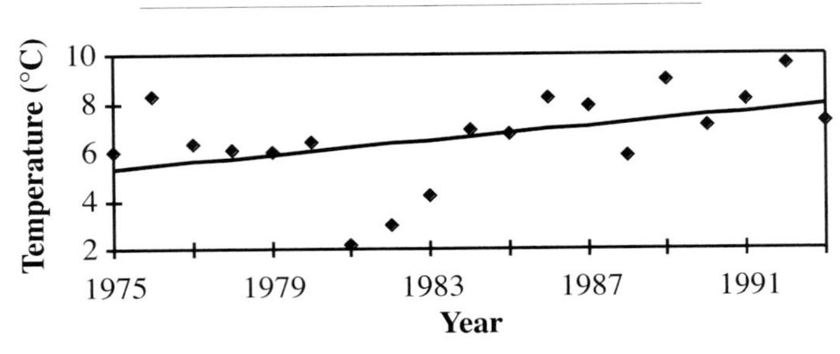

Figure 5-4: Change in the mean air temperature during the egg laying time period of the Pied Flycatcher. Regression line is y = -5.5 + 0.14 x (P<0.7).

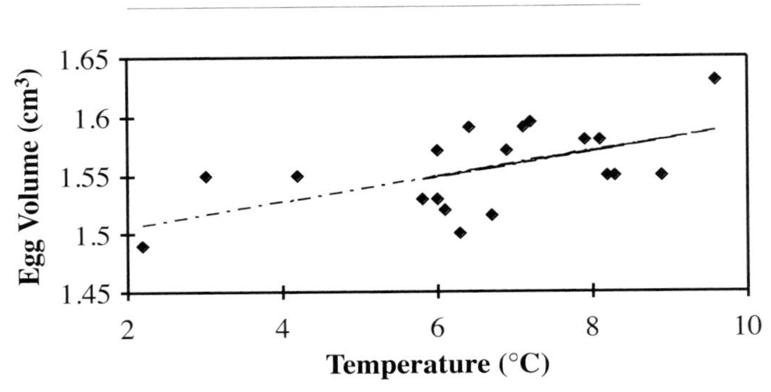

Figure 5-5: Correlation between egg volume of the Pied Flycatcher over the years and mean air temperature (regression is significant at P<0.01).

5.4.3.1.4. Changes in community composition, biotic interactions, and behavior

Differential responses by species could cause existing animal communities to undergo restructuring. This occurred in prehistoric plant communities because no analogous communities exist today (Davis, 1990). Similarly, Graham and Grimm (1990) found no present-day mammal communities that are analogous to some prehistoric animal communities.

Invertebrates: Potato aphids grown on plants kept in elevated CO_2 (700 ppm) showed a reduced response to alarm pheromones in comparison to those grown on plants in ambient CO_2 (350 ppm). Aphids were more likely to remain on leaves, possibly making them more susceptible to predators (Awmack *et al.*, 1997b).

Amphibians: Temperature and dissolved oxygen concentrations can alter the behavior of amphibian larvae, and changes in thermal environments can alter the outcome of predator-prey interactions (Manjarrez, 1996; Moore and Townsend, 1998).

Birds: In the UK, climate change may be causing a mismatch in the timing of breeding of Great Tits (*Parus major*) with other species in their communities (Visser *et al.*, 1998). The phenology of plants and some animals in the study area has advanced over the past 23 years, whereas breeding timing of Great Tits has not changed. This decoupling could lead to birds hatching when food supplies may be in low abundance (Visser *et al.*, 1998).

Mammals: Post *et al.* (1999) document a positive correlation between gray wolf (*Canis lupus*) winter pack size and snow depth on Isle Royale (USA). Greater pack size leads to three times more moose kills than in years with less snow. Fewer moose mean less browsing and thus better growth of understory balsam fir (*Abies balsamea*).

Box 5-6. Penguins as Indicators of Climate Warming in Western Antarctic

Midwinter surface air temperatures in the Western Antarctic Peninsula (WAP) region have increased by 4–5°C over the past 50 years (Smith *et al.*, 1996b). Studies confirm that the spatial and temporal patterns of winter sea-ice development in the WAP have changed during this time in response to rapid warming (Fraser *et al.*, 1992; de la Mare, 1997; Jacobs and Comiso, 1997; Loeb *et al.*, 1997). Chinstrap (*Pygoscelis antarctica*) and Adélie (*P. adeliae*) penguin populations also have changed during the past 25 years.

Although these two species are ecologically very similar, with diets and breeding ranges that overlap in the WAP (Volkman *et al.*, 1980), their winter habitat preferences are radically different. Adélies are obligate inhabitants of the pack ice, whereas Chinstraps are ice-intolerant, preferring to remain in close association with open water (Fraser *et al.*, 1992; Ainley *et al.*, 1994). The quality and availability of winter habitat is an essential determinant of survival and therefore a key factor regulating seabird populations (Birkhead and Furness, 1984). Adélie penguins have decreased by 22% whereas Chinstrap penguins have increased by more than 400% over the past 25 years (Fraser and Patterson, 1997; Smith *et al.*, 1999). This pattern supports the hypothesis that the increasing availability of open water as a result of warmer winters is favoring the survival of Chinstraps over the ice-dependent Adélies (see Fraser *et al.*, 1992).

5.4.3.2. *Model Projections of Wildlife Responses*

Invertebrates: Because changes in the distribution of disease vectors, such as mosquitoes, and crop and forest pests can have major human health and economic impacts, most modeling of changes in insect range and abundance have focused on pest species (Sutherst *et al.*, 1995). In southern Africa, models project changes in the ranges of tsetse flies, ticks, and mosquitoes (Rogers, 1996). Potential range changes of other disease vectors are discussed in Chapter 9.

High proportions of boreal forest insect pests overwinter as eggs. The strong link between patterns of minimum temperature and the location of outbreaks are best explained by the fact that eggs are killed when temperatures dip below a species' tolerance threshold (Sullivan, 1965; Austara, 1971; Virtanen *et al.*, 1996). Modeling work that is based on these observations and projected climate data suggests that increasing nighttime winter temperatures may increase the frequency of these pest species in northern areas (Tenow, 1996; Virtanen *et al.*, 1996, 1998), although warmer summer temperatures may reduce the intensity of outbreaks (Niemelä, 1980; Neuvonen *et al.*, 1999; Virtanen and Neuvonen, 1999; but see Ayres, 1993).

Amphibians and Reptiles: In studies on altitudinal (Pettus and Angleton, 1967; Licht, 1975; Bizer, 1978; Berven, 1982a,b) and latitudinal (Collins, 1979) gradients, a general pattern of faster metamorphosis at smaller sizes occurs at high elevations and northern latitudes. Changes in these life history attributes may affect species' abundances as a result of susceptibility to predators or environmental extremes or changes in reproductive output (Calef, 1973; Travis, 1981). Species that inhabit high-altitude areas may be at particular risk from climate change because as temperatures increase, their habitats may disappear (Hamilton, 1995; Pounds *et al.*, 1999). In Australia, frog distributions are strongly correlated with patterns of annual rainfall, implying that frogs in these areas may be able to expand their ranges if precipitation increases (Tyler, 1994). Reptile ranges often correlate with temperature (Nix, 1986; Owen and Dixon, 1989; Yom-Tov and Werner, 1996), suggesting that ranges may shift with temperature change. Desert tortoises (*Testudo graeca graeca*) in southern Morocco already have shifted their ranges in response to drier conditions possibly resulting from land-use changes (Bayley and Highfield, 1996).

Birds: In the prairie pothole region of the United States and Canada, a significant correlation between wetlands, duck numbers, and the Palmer Drought Severity Index has been found (Sorenson *et al.*, 1998). Projections of warming and drying for this region suggest that the number of wetlands and, correspondingly, the number of breeding ducks could be reduced.

Similar losses of wetlands have been projected for Africa (Magadza, 1996) and Australia (Hassal and Associates, 1998).

Mammals: Population reductions in mammals in African arid lands are possible if the incidence of drought increases (IPCC, 1998). In Australia, declines in several mammal species may occur if droughts increase in frequency or intensity (Caughley *et al.*, 1985; Roberston, 1986; Gordon *et al.*, 1988). Mountains (see Section 5.4.3.1), patchy habitats, and oceans can be barriers to range shifts. The Arctic Ocean is an obstacle to 25 species of Canadian mammals, with the collared lemming (*Dicrostonyx groenlandicus*) possibly losing at least 60% of its available habitat to climate change (Kerr and Packer, 1998).

The nutritional quality of some plant species has been found to decrease with increased CO_2 availability (Bazzaz, 1996). This could mean that herbivores might have to eat more. Most studies have dealt with insect or domestic mammals, but similar results are likely to hold for wild herbivorous mammals (Baker *et al.*, 1993; Bolortsetseg and Tuvaansuren, 1996). Relative changes in major plant lifeforms also can affect species distributions and populations densities; for example, in high-latitude rangelands, shrub abundance may increase and forb abundance may decrease (Chapin *et al.*, 1995), possibly leading to limitations on food supplies available to migrating caribou (White and Trudell, 1980).

5.4.3.3. Impacts on Goods and Services with Market Values

5.4.3.3.1. Control of pest species and disease vectors

Wildlife plays a role in natural and agricultural ecosystems through reduction of injurious insect pests and disease vectors. In some forested ecosystems, birds, together with parasitoids and other insect predators, are effective predators of insect pests (Crawford and Jennings, 1989; see Price *et al.*, 2000). Insects, ants, and spiders also play key roles in reducing populations of pest insects. For example, when ants were removed from maize plots in Nicaragua, two pest species increased in abundance, and damage to the maize increased (Perfecto, 1991). The annual economic value of wildlife control of insect pests has been estimated to be hundreds of billions of U.S. dollars (de Groot, 1992; Pimentel *et al.,* 1992, 1993; Pimentel, 1997). This includes destruction of pests that are injurious to crops (estimated at US$90 billion year in the United States alone), as well as those affecting forests.

Climate change could impact many of these systems by decoupling predators from their prey (Root and Schneider, 1995). Studies in North America project reductions in the extent of distribution size of some of the species that feed on pests in forest, grassland, and agricultural ecosystems (see Price, 1995; Price *et al.*, 2000). This could lead to an increasing need to use pesticides, with accompanying health risks (human and wildlife) and economic costs (Kirk *et al.*, 1996; Colburn *et al.*, 1997; Herremans, 1998).

5.4.3.3.2. Pollinators

Pollination is crucial to the reproduction of many plants, thus to maintenance of functioning ecosystems and biodiversity. Many factors influence the occurrence and density of insects (e.g., habitat conversion, excessive pesticide use). The principal mode of pollination of many plant species is by insects. Worldwide, an estimated 400 crop species are pollinated by bees and more than 30 other animal genera; possible crop loss in some species would be more than 90% in the absence of bees (Southwick, 1992; Buchmann and Nabhan, 1996). More than 100,000 different animals around the world pollinate 250,000 types of plants (Buchmann and Nabhan, 1996). The estimated annual value of wildlife pollination to commercial crops and pasture grasses is tens of billions of U.S. dollars (de Groot, 1992; Pimentel *et al.*, 1992, 1995). If pollination of noncommercial plants were added, the figure would be significantly higher.

5.4.3.3.3. Seed dispersal

Seeds of some plants are dispersed by one or a few animal species. In Costa Rican tropical deciduous forests, as many as 60% of plants have their seeds dispersed by birds; the numbers are 75% in subtropical evergreen forests and 80% in montane evergreen forests (Bawa, 1995). On Samoa, the vast majority of seed dispersal in the dry season is mediated by flying foxes (Cox *et al.*, 1991). Plants commonly disperse via seeds passing through the digestive tracts of animals or with animals that cache seeds (Robinson and Handel, 1993; Lanner, 1996). Consequently, if the ranges of the appropriate animals become disjunct from even part of the ranges of specific plants, dispersal of the plants may suffer (see Price *et al.*, 2000, for other examples).

5.4.3.3.4. Decomposers and soil maintainers

Invertebrates and microorganisms living on or below the soil surface provide needed goods and services to human societies (e.g., mixing and aeration of soils, decomposition of materials and human waste) (Daily *et al.*, 1997). These processes contribute to creation of fertile topsoil from organic matter and mineral components of the soil. Some of these organisms (e.g., ants) are susceptible to climatic changes, especially droughts (Folgarait, 1998). On a global scale, the estimated cost of replacing waste elimination services performed by these organisms is hundreds of billions of U.S. dollars annually; the estimated cost of replacing topsoil production services is tens of billions of U.S. dollars annually (de Groot, 1992; Crosson *et al.,* 1995; Pimentel *et al.,* 1995; Daily *et al.,* 1997).

5.4.3.3.5. Protection of endangered species

Another way of assessing the value of wildlife is by examining how much is spent on its conservation. In fiscal year 1995, approximately US$330 million was spent on the conservation of threatened and endangered species in the United States and its territories (U.S. Fish and Wildlife Service, 1995). As pressures facing wildlife increase, the number of species that require conservation attention also will most likely increase.

5.4.3.3.6. Subsistence hunting

People in many parts of the world depend on wildlife for their daily nutritional needs. This is most pronounced in less-developed areas. For example, the Cree along James Bay in Canada harvest approximately 800,000 kg of animal food annually. The per capita replacement value of this harvest was estimated to be CDN$6,000 in 1986 dollars (Scott, 1987). The abundance of caribou (*Rangifer tarandus groenlandicus*) available for harvest by indigenous peoples could decrease as a result of increased temperatures, snowfall, and potential shifts in the timing of precipitation (Brotton and Wall, 1997; Ferguson, 1999). Adverse impacts also have been projected for other subsistence species, including marine birds, seals, polar bears (Stirling, 1997), tundra birds (Jefferies *et al.*, 1992), and other tundra-grazing ungulates (Jeffries, *et al.*, 1992; Gunn, 1995).

Wildlife species also are a significant source of food and medicine for people in many temperate and tropical countries, such as Botswana and Nigeria (McNeely *et al.*, 1990) and Australia (Bomford and Caughley, 1996). Among the Boran (Kenya), birds are used for medicines (ostrich oils) and their feathers for cleaning wounds (Isack, 1987). In many countries, climate change impacts such as reductions in wildlife populations may have the greatest impact on the lowest-income groups—those with the least ability to adapt if hunting opportunities decline (Arntzen and Ringrose, 1996).

5.4.3.3.7. Recreational use of animals and ecotourism

In many African countries, ecotourism to view wildlife is a major contributor to gross national product (GNP). Worldwide, ecotourism is estimated to provide US$500 billion to 1 trillion annually to the global economy (Munasinghe and McNeely, 1994). Changes in climate could reduce the populations of some of the species people are willing to pay to see (Mills *et al.*, 1995; Allen-Diaz, 1996).

5.4.3.4. Impacts on Goods and Services with Nonmarket Valuations

Though difficult to measure, nonmarket values must be taken into account in discussing the state of wildlife. These values can be described in terms of the cultural, religious, scientific, and aesthetic importance of wildlife species (Bawa and Gadgil, 1997; Goulder and Kennedy, 1997; National Research Council, 1999). At times, monetary value can be assigned to nonmarket aspects, such as a tourist's willingness to pay to see wildlife in natural habitats (e.g., Edwards, 1991). Although this is difficult, monetary values might be able to be assigned to some of the wildlife services described in this section. The fact that a monetary value is difficult to assign should not diminish the importance of the service; for many, no substitute exists for the services provided by wildlife.

Tools used to assess nonmarket monetary values vary, but a few approaches have become common. One attempts to determine "existence value": people's willingness to pay to know that an animal simply continues to exist, even if they never see it (National Research Council, 1999). Similarly, an "option value" is based on a person's desire to be able to potentially interact with the animal some time in the future, and a "bequest value" is being assured that future generations will be able to use or view an animal (National Research Council, 1999). Listed below are some areas where it is particularly difficult to estimate a monetary value for animal-related services.

5.4.3.4.1. Cultural and religious importance of animals

Many indigenous peoples use wildlife as integral parts of their cultural and religious ceremonies. For example, birds are strongly integrated into Pueblo Indian (United States) communities. Birds are regarded as messengers to the gods and a connection to the spirit realm. Among Zuni Indians (United States), prayer sticks with feathers from 72 different species of birds are used as offerings to the spirit realm (Tyler, 1991). In Boran (Kenya) ceremonies, selection of tribal leaders involves rituals that require Ostrich (*Struthio camelus*) feathers. Birds also are used for tribal cosmology, meteorology, religion, and cultural ceremonies (Isack, 1987). Wildlife plays similar roles in cultures elsewhere in the world. Thus, shifts in the timing or the ranges of wildlife species could impact the cultural and religious lives of some indigenous peoples.

5.4.3.4.2. Wildlife and scientific information

Researchers have been studying spider silk to learn how this strong and elastic material could be manufactured (Xu and Lewis, 1990). Similarly, scientists have examined the structure of the small scales on butterfly wings to understand how they reflect light and dissipate heat. These studies may help engineers design better processes for manufacturing and designing computer chips (Miaoulis and Heilman, 1998). Researchers also have studied the regenerative properties of cells within rhinoceros horns and have identified potent antimicrobial chemicals from frogs and toads (Cruciani *et al.*, 1992; Moore *et al.*, 1992; Boskey, 1998).

5.4.3.4.3. Ecosystem function and biodiversity maintenance

Other valuable services are provided by species that contribute to ecosystem health and productivity. Reductions in or losses of species can lead to reduced local biodiversity and changes in the structure and function of affected ecosystems (National Research Council, 1999). The most well-known example of this kind of effect comes from marine systems, where the presence or absence of a starfish species has been found to greatly influence the species composition of intertidal habitats (Paine, 1974). Species in terrestrial systems also can have a strong influence on the biodiversity of their ecosystems; in many cases these effects are related to their functions as pollinators or seed dispersers (see Price *et al.*, 2000).

5.4.4. *Adaptation by Humans for Changes in Goods and Services*

The ability of wildlife to adapt naturally to climate change is largely a function of available genetic diversity and the rate of change. This section deals not with natural adaptation but with human adaptation to changes in wildlife populations, in terms of conserving wildlife and replacing some of the goods and services that wildlife provides.

5.4.4.1. *Parks and Reserves*

One typical method to adapt to declines in wildlife populations has been establishment of refuges, parks, and reserves. Placement of reserves, however, rarely has taken into account potential rapid climate change, even though the problems of such change and reserve placement were discussed in the mid-1980s (Peters and Darling, 1985). Managers of current reserves and parks need to be encouraged to consider rapid climate change in developing future management plans (Solomon, 1994; Halpin, 1997). Developing a series of bioindicators to monitor the potential impacts of rapid climate change on parks and reserves may be possible (de Groot and Ketner, 1994).

In the United Kingdom, the Institute of Terrestrial Ecology has estimated that 10% of all designated areas (i.e., nature reserves) could be lost (e.g., to habitat degradation) within 30–40 years and that species distribution in 50% of designated areas could change significantly over the same period (UK DETR, 1999). In light of these changes, there is a need for a robust nature conservation system that can accommodate climate change.

In part, the disparity between siting reserves where wildlife species currently are versus where they may be in the future may stem from uncertainties in the rate and amount of projected climate change. If a species' range shifts out of a reserve created for its survival, the current reserve placement could even be considered maladaptive. However, if reserves are not created and species are lost to other pressures, the potential effects of climate change on species distributions are moot (see Box 5-7).

Box 5-7. Biodiversity Hotspots and Climate Change

Biodiversity hotspots are areas with exceptional concentrations of endemic species facing extraordinary threats of habitat destruction. Twenty-five hotspots contain the sole remaining habitats for 133,149 (44%) vascular land plants and 9,732 (36%) terrestrial vertebrates (Myers *et al.*, 2000). The number of invertebrates is not known, but in light of the many interactions (e.g., pollination) between plants and invertebrates, especially insects, the concentrations of these animals probably parallel those of plants. Nine of these hot spots occur on islands, making them particularly vulnerable to sea-level rise and limiting or preventing the opportunity for many terrestrial animals to modify their range (Myers *et al.*, 2000). Because these 25 hotspots constitute only 1.4% of the Earth's land surface (Myers *et al.*, 2000), they provide an opportunity for planners to respond to the biodiversity crisis by giving priority status to conserving them.

Unfortunately, synergistic effects of climate change and requirements to conserve areas for species survival may complicate the situation. With warming temperatures, many species are expected to move poleward (Parmesan *et al.*, 1999) or upward in altitude (Pounds *et al.*, 1999). This implies that the locations of hotspot reserves may need to allow for such movement, which may require a somewhat larger region to be conserved. Even with these efforts, some species may not be conserved because they presently are as far poleward (e.g., fynbos region at the southern tip of South Africa) or as high in altitude (e.g., cloud forests in Costa Rica) as they can be.

5.4.4.2. *Captive Breeding and Translocations*

Another way in which humans have dealt with endangered wildlife populations has been through captive breeding and translocations. These techniques have been put forward in the past as methods to deal with future population pressures caused by climate change (Peters, 1992). However, although captive breeding and translocation are effective tools for conservation of some species, they may be appropriate for only a handful of species (see Box 5-8).

5.4.4.3. *Replacing Lost Ecosystem Services*

Humans may need to adapt not only in terms of wildlife conservation but also to replace lost ecological services normally provided by wildlife. It may be necessary to develop adaptations to losses of natural pest control, pollination, and seed dispersal. Although replacing providers of these three services sometimes may be possible, the alternatives may be costly (Buchmann and Nabhan, 1996). Finding a replacement for other services, such as contributions to nutrient cycling and ecosystem stability/biodiversity, are much harder to imagine. In many cases, such as the values of wildlife associated with subsistence hunting and cultural and religious ceremonies, any attempt at replacement may represent a net loss.

In many agricultural/silvicultural systems, pesticides are used to prevent losses to pests (insects, pathogens, some vertebrates). In the past 50 years, pesticide use worldwide has increased more than 25-fold (Worldwatch Institute, 1999). The estimated cost of pesticide use in the United States in the mid-1990s was US$11.9 billion (equivalent to 4.5% of total U.S. farm production expenditures), and worldwide use was US$30.6 billion (Aspelin and Grube, 1999). Given that these values are for systems that still had some natural pest control, changes in wildlife distributions might necessitate changes in economic expenditures for pesticides.

However, pesticides often kill more than the target species, possibly eliminating natural predators that keep pest populations low. For example, increased pesticide use in Indonesia between 1980 and 1985 led to the destruction of the natural enemies of the brown planthopper. Subsequent increases in planthopper numbers caused reductions in rice yields estimated to cost US$1.5 billion (FAO figures cited in Pimentel *et al.*, 1992).

Adaptation to loss of natural pollinators may be possible in some cases. Farmers sometimes lease bee colonies to pollinate their crops. Although this may be an option for the ~15% of

Box 5-8. Limitations of Captive Breeding and Translocation to Conserve Biological Diversity Threatened by Climate Change

In some cases, threatened populations of sensitive species could be augmented or reestablished through captive breeding for reintroduction, especially if the degree of climate change proves to be small or moderate. In combination with habitat restoration, such efforts may be successful in preventing the extinction of small numbers of key selected taxa. Similarly, translocation of wildlife between areas within their natural range also might mitigate the effects of small to moderate climate change. This strategy has been applied successfully to augment or restore depleted populations of various species (Boyer and Brown, 1988).

Captive breeding for reintroduction and translocation are likely to be less successful if climate change is more dramatic. Such change could result in large-scale modifications of environmental conditions, including loss or significant alteration of existing habitat over some or all of a species' range (Myers *et al.*, 2000). Captive breeding and translocation therefore should not be perceived as panaceas for the loss of biological diversity that might accompany dramatic climate change, especially given the current state of the environment. Populations of many species already are perilously small; further loss of habitat and stress associated with severe climate change may push many taxa to extinction.

One limitation to captive breeding is the lack of space available to hold wildlife for breeding purposes. Zoos and offsite breeding facilities can be expected to accommodate no more than a small fraction of the number of species that might be threatened. Recent studies have indicated that no more than 16 snake species and 141 bird species could be accommodated and sustained in accredited North American zoos and aquariums in long-term management programs (Quinn and Quinn, 1993; Sheppard, 1995).

Captive breeding programs are expensive, and locating funding to support large numbers of programs could be difficult (Hutchins *et al.*, 1996). For example, it costs US$22,000 to raise a single golden lion tamarin in the United States and reintroduce it to its native Brazil (Kleiman *et al.*, 1991). Part of the cost associated with such programs includes the extensive scientific studies that must be conducted for the program to be successful. Reintroduction is technologically difficult and unlikely to be successful in the absence of knowledge about the species' basic biology and behavior (Hutchins *et al.*, 1996). Rearing and release strategies must be tested experimentally, and released animals must be monitored to assess the efficacy of various methods (Beck *et al.*, 1994). In the case of black-footed ferrets (*Mustela nigripes*) and golden lion tamarins, it took more than a decade to develop the knowledge base required for success.

If wildlife translocation involves moving species outside their natural ranges, other problems may ensue. Exotic species can have devastating effects on host ecosystems, including extinction of native fauna (McKnight, 1993). The unpredictable consequences of species introductions means that translocation is severely limited in its ability to conserve species that are threatened by climate change.

Finally, reintroduction and translocation programs cannot be successful if there is no appropriate habitat left for captive-bred or translocated animals to be released into (Hutchins *et al.*, 1996). Not all of the habitat components that are necessary for a species to survive can be translocated. Entire suites of plant and invertebrate species may be critical elements for a species to succeed in a new environment, but no techniques exist for translocating intact biological communities. Although captive breeding and translocation have potential value for well-studied animals, these strategies appear to be impractical for the vast number of species threatened by rapid climate change.

crops fertilized by domestic honeybees, it may not be an option for crops typically fertilized by wild pollinators or for the 250,000 types of wild plants that are pollinated by 100,000 different invertebrate species (Buchmann and Nabhan, 1996).

5.4.5. *Equity Issues*

People in many parts of the world are dependent on wildlife for all or part of their daily nutritional needs. A typical adaptive response to this situation would be to replace all or part of this food with store-bought products. This might be feasible in areas near developed societies but could become increasingly difficult in more remote communities. However, there is more to subsistence hunting than the capture of food. Subsistence hunting plays a major role in the culture of Cree communities in northern Canada. "The killing, preparation, sharing, and consumption of game is central to the seasonal renewal of social relations in Cree villages, and of a relationship to the land which is both secular and sacred in importance" (Scott, 1987). Even if compensations or substitutions for subsistence uses could be made, there still would be equity issues stemming from the loss of culture associated with this way of life.

Many of the aforementioned potential adaptations are more applicable to developed countries than in developing countries. For example, the use of leased honeybees is not applicable to crops fertilized by flying foxes or other wild animals. The same can be said of many forms of seed dispersal. Increased use of pesticides may require more capital than is available to small farmers in some developing countries (Pimentel *et al.*, 1992). Adaptations that may be practical for developed countries simply may not be equitable for developing countries.

5.4.6. *Vulnerabilities, Sensitivities, Uncertainties*

In trying to understand and predict potential impacts of climate change on wildlife species, some species and geographic areas are found to be at greater risk than others. Species with small populations, restricted ranges, and specific habitat requirements often are most vulnerable (see Section 5.4.1.2).

Migratory species may be especially vulnerable because they require separate breeding, wintering, and migration habitats. In many cases, one or more of these habitats could be at risk because of climate change and other habitat loss. For example, a large portion of the eastern population of the monarch butterfly (*Danaus plexippus*) winters in a small region of warm-temperate dry forest in Mexico. With climate change, this area is projected to contain trees that are more typical of a subtropical dry forest—probably unsuitable for wintering monarchs (Villers-Ruíz and Trejo-Vázquez, 1998). The relative vulnerability of shorebird migration sites in the United States varies, depending on local geomorphologic and anthropogenic factors, and these factors could exacerbate the effects of sea-level rise. For example, southern San Francisco Bay could lose most of its intertidal feeding habitat with a 2°C average temperature rise (medium confidence) (Galbraith *et al.*, 2001).

One key region of concern is the Arctic and Antarctic, where the temperature increase is projected to be large and changes to habitat availability and accessibility (e.g., freezing and thawing of sea ice and tundra) are expected. Such changes may hamper migration, reproduction, and survival of many species, including birds, polar bears (*Ursus maritimus*), caribou, and musk-oxen (Jefferies *et al.*, 1992; Stirling and Derocher, 1993; Gunn and Skogland, 1997; Stirling, 1997).

Many biological uncertainties exist in the understanding of ecosystem processes. Nevertheless, the balance of evidence suggests that projecting impacts of climatic change on a variety of wildlife species is possible (medium confidence). Laboratory and field studies have demonstrated that climate plays a strong role in limiting species' ranges (high confidence). Only a small fraction of all species have been monitored long enough to detect significant trends. Most monitored species that show significant trends have exhibited changes over the past few decades that are consistent with local warming and expected physiological responses (high confidence). However, potential *specific* changes in wildlife resulting from climate change can be projected only with low confidence for most species because of many possible contributing factors, such as habitat destruction and exotic invasive species. Some species clearly are responding to global change (see Section 5.4.3), and many more changes probably have gone undetected. Researchers are in the process of coupling these discernible changes with various biological theories regarding climate and species spatial and temporal patterns; through this process, we expect that reliable *general* projections can be and in fact are being made (high confidence).

Scientists also need to develop a better understanding of how all of the components of ecosystems work together. The role each species plays in ecosystem services, in wild and managed systems, is necessary to understand risks and possible surprises associated with species loss. Without this information, the probability of surprises associated with species loss is high (medium confidence).

5.5. Rangelands (Grasslands, Savannas, and Deserts)

Rangelands here are taken to include deserts (cold, hot, and tundra), grasslands (unimproved), scrub, chaparral, and savannas (after, e.g., Allen-Diaz, 1996). This section does not consider improved grasslands or croplands in detail because they are covered in Section 5.3. It does partly cover tundra because it is an important grazing system; that ecosystem is discussed in more detail in Section 5.9.

Ecosystems within rangelands are characterized by low-stature vegetation because of temperature and moisture restrictions; they are found on every continent (Allen-Diaz, 1996). They are adapted to great variations in temperature and rainfall on an annual and interannual basis, but they generally are confined to areas that have about one unit of precipitation to every 16 units of evapotranspiration (the ratio that is characteristic of drylands; Noble and Gitay, 1996). They often are referred to as "pulse systems" (Noble and Gitay, 1996).

In many countries, human activities in rangelands have evolved in response to variable and often unpredictable climate. Human practices include pastoralism; subsistence farming (Allen-Diaz, 1996); and, more recently, commercial ranching (Canziani and Diaz, 1998). Rangelands also are important for many national economies in terms of foreign cash (e.g., through tourism). They are important stores of biodiversity, including ancestors of many of the cereals (World Bank, 1995), and have high levels of endemism (Barnard *et al.*, 1998). Rangelands are used primarily for grazing and hence livestock production (Squires and Sidahmed, 1997). Thus, they are important for food (mostly livestock, but also wild fruits), fuelwood, wood poles for construction, and feed (Campbell *et al.*, 1997). Other key services are biodiversity, water cycle, and carbon stores. Some of these products and services can be given economic valuation; however, only a small component of the total economic value is represented by products that can be given a market value (Campbell *et al.,* 1997), which suggests that nonmarket values are quite important for some rangelands.

Rangelands are adapted to grazing and other disturbances, such as fire, flood, and insect herbivory (Allen-Diaz, 1996). Vegetation tends to be sparse and thus is not considered worth mechanical harvesting; however, the sparse grass/herbaceous cover is efficiently harvested by grazers (Williams, 1986). In many cases, episodic fires (Bock *et al.*, 1995) are important for providing new and lush growth for grazers, and fire sometimes is used to manage grass-woody shrub balance (Noble *et al.*, 1996), which is important for livestock and meat and wool production (see, e.g., Chapter 12). Human management can be critical to the status of these systems with or without climate change. Previous IPCC reports concluded that fluctuating rainfall and temperatures along with increased human activity (especially in more tropical systems) has led to land degradation and eventually desertification in many areas (Bullock and Le Houérou, 1996; Gitay and Noble, 1996; Canziani and Diaz, 1998).

In this section, food/fiber, biodiversity, and carbon stores are examined in detail (see Table 5-1). Water as a resource is considered elsewhere in the report, at the global level (Chapter 4) and in many of the regional chapters (Chapters 10-12).

5.5.1. Current Status of Key Goods and Services

Estimates for rangeland cover vary between 31 and 51% of the land surface of the Earth (Allen-Diaz, 1996; WRI, 2000); the upper estimate includes tundra grasslands. Rangelands support human populations at low densities (Batchelor *et al.*, 1994) on almost every populated continent. The latest figures suggest a population of at least 938 million people—or about 17% of the world's population (WRI, 2000).

The World Resources Institute report (WRI, 2000) assesses the food/fiber production and biodiversity of rangelands as "fair" (on a scale of excellent, good, fair, poor, and bad) over the past 20–30 years, but the underlying biological ability of the rangelands to continue to support that productivity and biodiversity is declining, suggesting that productivity and the biodiversity assessment may not hold in the future. Livestock production in rangelands is estimated to be 65 Mt in 1998 (WRI, 2000), with global meat production estimated at 225 Mt. Nineteen percent of the world's centers for plant diversity are found in rangelands. For carbon stores, the WRI assessment of rangelands is "good"; again, however, the underlying ability is judged to be declining (WRI, 2000). IPCC (2000) estimates carbon stores in rangelands as 84 Gt C in vegetation (of a global total of 466 Gt), 750 Gt C in soils (of a global total of 2,011 Gt). WRI (2000) gives a range of 405–806 Gt for total carbon stores. Both estimates suggest that rangelands are important carbon stores.

Many of the people in rangelands rely on fuelwood for their daily cooking and heating needs. Estimated fuelwood use as a percentage of total energy use in 1993 for countries dominated by rangelands is about 60–90% (e.g., in Afghanistan, Mozambique, Swaziland); the world average is 6%, and Africa's average is 35% (WRI, 1998).

5.5.2. Major Pressures on Key Goods and Services

Climate change is a pressure on key goods and services in a system that is responsive to climatic fluctuations. It is possible that climate change would lead to increased frequency of extremes of climatic events (such as drought and floods) driven through change in the frequency of ENSO events. Apart from climate change, other direct and indirect pressures from human activities can be important in the delivery of services from rangelands. These pressures include land-use change, which often leads to fragmentation (Allen-Diaz, 1996; Gitay and Noble, 1998; WRI, 2000); changes in the densities of livestock (WRI, 2000); competition for land and water; and altered fire regimes (Russell-Smith *et al.*, 1997). These pressures would have subsequent impacts that cannot be disaggregated from impacts of climate change.

5.5.3. Responses of Rangelands and Impacts on Goods and Services

It has been suggested that for rangelands the possible effect of climate change may be trivial compared with the past and present impacts of human activities, including livestock grazing (Le Houérou, 1996). This may not be entirely true for all rangelands, but it does suggest that it would be difficult to separate the impacts of climate change from the impacts of many of the other pressures that are acting and will continue to act on the system. Subsequent sections assess the direct impacts of climate change; impacts from other aspects of global change, especially the pressures listed above, also are assessed (see

Box 5-9). Major impacts of climate change on key goods and services are mediated through changes in NPP; changes in plant community composition, structure, and forage quality (e.g., through changes in C_3 and C_4 plants); and changes in plant herbivory and phenology.

5.5.3.1. *Impacts of Changes in ENSO and Related Events*

Many of the world's rangelands are affected by ENSO events. There have been several ENSO events in the 1980s and 1990s that followed each other closely (Polis *et al.*, 1997). Thus, rangelands have been subjected to prolonged drought conditions with little recovery time. Simulation results have shown that ENSO events are likely to intensify (Chapter 2) under a doubled CO_2 scenario, with the result that dry areas within rangelands are likely to become drier and mesic areas will become wetter during ENSO events (Noble and Gitay, 1996). Thus, some rangelands are likely to experience more extremes of events, with subsequent changes in vegetation and water availability. These changes often are tracked by insect herbivory (Polis *et al.*, 1997), leading to additional impacts.

5.5.3.2. *Impacts of Increases in CO_2 and Climate Change on Plant Productivity, Species Composition, Decomposition, and Carbon Stores*

The effect of elevated CO_2 concentrations on decomposition, plant productivity, and carbon storage could be as large as the

Box 5-9. Impacts of Some Pressures on Rangelands

Land-Use Change

Major factors in land-use change are conversion of rangelands to croplands (Allen-Diaz, 1996; WRI, 2000) and increased human settlements, especially urbanization (Gitay and Noble, 1998), which lead to fragmentation (WRI, 2000). There have been large-scale changes in land use: For example, in the South Platte Basin in the United States, 40% of the land cover has been converted from rangelands to croplands. This can alter carbon stores, sometimes leading to soil carbon loss of as much as 50% (Allen-Diaz, 1996), but it also can lead to increased plant productivity through irrigated grain production (Baron *et al.*, 1998). Baron *et al.* (1998) conclude that subsequent impacts on the biogeochemical cycles of the basin and on land-atmosphere interactions can affect many rangelands. In many rangelands, native species that occur at low density are used for fuelwood. In some cases, fuelwood collection can lead to decreases in woody vegetation cover and possible land degradation.

Livestock Production

Some rangelands have high densities (>100 km^{-2}) of livestock, with livestock being moved to take advantage of the periodic growth, especially after rain and/or fires (WRI, 2000). Modeling studies show that increased grazing pressure (i.e., overgrazing) would cause grass and herbaceous productivity to fall below a certain threshold, resulting in increased and rapid rates of land degradation especially under drier and/or hotter climate conditions (Abel, 1997).

Competition for Land and Water

There is increased demand for water for direct human consumption and for irrigation (WRI, 2000). Food production obviously is positively affected by increased water use (see Section 5.3 and Chapter 4), but this water use is an added pressure on many rangeland ecosystems. Nomadic pastoralism, which was common in many rangelands until recently, allowed pastoralists to cope with the variable climate of the rangelands they inhabited. Land-use changes (to permanent agriculture, urban areas, conservation, and game reserves that have included loss of sources of permanent water) have led to overall loss of land available. Together with increased human population, this has led to competition for land and changes from pastoral communities to more market-orientated and cash-based economies (Allen-Diaz, 1996).

Altered Fire Regimes

The SAR did an extensive review of fires and rangelands. The projected increase in variability in climate led to the conclusion that the frequency and severity of fires will increase in rangelands (Allen-Diaz, 1996), provided that drought and grazing do not lead to a reduction in vegetation biomass. There is historical evidence that fire frequency has changed (increased and decreased, depending on vegetation biomass) in recent decades (see, e.g., Russell-Smith *et al.*, 1997), leading to changes in vegetation composition. Fires also are started by humans; for example, in Africa 25–80% of rangelands are burned every year, often to induce new plant growth (WRI, 2000). This has implications for short-term productivity but possibly long-term land degradation.

impact of climate change alone (Ojima *et al.*, 1993; Parton *et al.*, 1994; Hall *et al.*, 1995).

5.5.3.2.1. Plant productivity

At the global level, plant production is projected to increase in grasslands when climate change and elevated CO_2 are combined (Parton *et al.*, 1994) but could be affected by potential changes in disturbance regimes and by land-use practices in some rangelands (Parton *et al.*, 1994, 1995). As CO_2 concentrations increase, transpiration per unit leaf area is expected to decrease and WUE will increase (Morgan *et al.*, 1994a; Read *et al.*, 1997; Wand *et al.*, 1999). Thus, increased CO_2 could lead to enhanced productivity (e.g., Hunt *et al.*, 1996; Owensby *et al.*, 1996), especially under low soil moisture conditions that are characteristic of rangelands (Wand *et al.*, 1999). However, Bolortsetseg and Tuvaansuren (1996)—using 30-year climatic data as a baseline with projected climate change scenarios that incorporate elevated CO_2—report a decrease in productivity in deserts but an increase in colder areas; Gao and Yu (1998)—using a regional model with elevated CO_2, a 20% increase in precipitation, and a 4°C increase in temperature—project that NPP of most steppes in China will decrease by 15–20% and NPP of woodland and shrublands and desert grasslands will increase by 20–115%. Productivity also is likely to increase because elevated CO_2 could result in enhanced nitrogen uptake (Jones and Jongen, 1996; Coughenour and Chen, 1997). However, the response varies between species: Dominant species show a less enhanced response than rarer species, suggesting a possible change in the composition and structure of the vegetation in rangelands (Jones and Jongen 1996; Berntson *et al.*, 1998). Slower depletion of soil water under increased CO_2 concentrations should favor plants that otherwise might do poorly under water stress (Polley *et al.*, 1997), thus altering the species mix.

5.5.3.2.2. Carbon stores

Many rangelands tend to have large belowground stores (Tate and Ross, 1997). Studies suggest a strong interaction between CO_2 and temperature on soil carbon fluxes, possibly mediated through NPP. Soil carbon losses of about 2 Gt over 50 years under combined climate change and elevated CO_2 have been projected. These losses compare with carbon losses of about 4 Gt from climate change alone (Hall *et al.*, 1995; Parton *et al.*, 1995). Hall *et al.* (1995) project that tropical savannas will be soil carbon sinks under elevated CO_2 and climate change. For temperate grasslands, Thornley and Cannell (1997) suggest an annual carbon sink of 0.5–1.5 t ha^{-1} yr^{-1} under a scenario of elevated CO_2 and a temperature increase of 5°C; however, grasslands are likely to be a net source under that temperature increase alone. Based on some experimental studies and supported by observations around CO_2-venting spring areas in temperate grasslands of New Zealand, future carbon storage could be favored in soils of moderate nutrient status, moderate to high clay content, and low to moderately high soil moisture status (Tate and Ross, 1997). In contrast, Cook *et al.* (1999) show that long-term CO_2 enrichment by a natural CO_2-venting spring in a subarctic grassland is likely to result in slower soil carbon accumulation compared to the ambient CO_2 atmosphere. Soil carbon is greatly affected by management practices, and appropriate management in many rangeland systems could result in carbon sequestration (see below).

Productivity of rangelands also can be affected by changes (which could be a result of climate change) in soil fauna and flora, as well as microbial activity (Yeates *et al.*, 1997; Zaller and Arnone, 1997; Kandeler *et al.*, 1998). These studies, which have not looked at arid systems within rangelands, suggest that elevated CO_2 is likely to cause an increase in soil microbial activity and increased soil CO_2 production. NPP has been reported to be enhanced by elevated CO_2 in nodulated and mycorhizae-infected species (Allen-Diaz, 1996), although specific infection rates often are unchanged.

5.5.3.2.3. Changes in species composition, C_3 and C_4 plants

C_3 and C_4 plants react differently to elevated CO_2 and climatic factors. In paleostudies in the northern Chihuahan Desert, Buck and Monger (1999) observe a major shift from C_4 grasses to C_3 desert shrub-dominated vegetation about 7,000–9,000 years BP, associated with a possible increase in aridity and a subsequent increase in C_4 grasses, with an increase in moisture about 4,000 years BP. Photosynthesis in C_3 plants is expected to respond more strongly to CO_2 enrichment than in C_4 plants (Mayeux *et al.*, 1991). If this is the case, it is likely to lead to an increase in the geographic distribution of C_3 plants (many of which are woody plants) at the expense of C_4 grasses (Noble and Gitay, 1996; Ehleringer *et al.*, 1997; Polley *et al.*, 1997). However, the impacts are not that simple. In pot experiments, elevated CO_2 is reported to improve water relations and enhance productivity in the C_4 shortgrass steppe grass *Bouteloua gracilis* (Morgan *et al.*, 1994b). In modeling and experimental studies, NPP of C_3 and C_4 grasses increase under elevated CO_2 for a range of temperatures and precipitation (Chen *et al.*, 1996; Hunt *et al.*, 1996; Owensby *et al.*, 1999) but could result in relatively small changes in their geographical distributions (Howden *et al.*, 1999b). There are additional interactions with soil characteristics and climatic factors. Epstein *et al.* (1998)—using existing distribution of grasslands in the U.S. Great Plains combined with modeling studies but not including the effect of elevated CO_2—found that C_3 plants are more productive in cooler, drier conditions and do particularly well in soils with high clay. There was a differentiation of the response of C_4 plants into two height classes: shortgrasses and tallgrasses. Productivity of C_4 shortgrasses increased with increasing mean annual temperature but decreased with mean annual precipitation and sand content, whereas C_4 tallgrass productivity tended to increase with mean annual precipitation and sand content and was highest at intermediate values of mean annual precipitation. Coffin and Lauenroth (1996) used gap models linked to a soil water model and found that C_4 grasses increase in dominance as a result of increases in temperature in all months. The coolest sites that currently are dominated by C_3 grasses are predicted to shift to dominance by C_4 grasses, whereas sites that currently are dominated by C_4 grasses have an increase in importance of this group. When the number of frost days is decreased, subtropical C_4 grasses may invade the more palatable C_3 grasses, leading to a decrease in forage quality. The rate and duration of this change is likely to be affected by human activity; high grazing pressure may mean more establishment sites for C_4 grasses (Panario and Bidegain, 1997).

5.5.3.2.4. Woody species, grass, and weeds mix

There have been some suggestions that woody weed encroachment into herbaceous parts of rangeland (up to three-fold in some cases) may be a result of changes in regional climate (Brown *et al.*, 1997b) and may be facilitated by the increase in atmospheric CO_2 concentration since industrialization (Polley *et al.*,1996). However, the more plausible explanation for the present observed increase in woody weeds in many rangelands is that it is a result of land-use change (especially increased grazing and changes in fire regime) as well as land degradation (Bond *et al.*, 1994; Archer *et al.*, 1995; Brown *et al.*, 1997b; Gill and Burke 1999). A modeling study of semi-arid woodlands in Australia under elevated CO_2 and climate change suggests that future potential burning opportunities may increase as a result of the CO_2 response of the grass layer, which allows adequate fuel to build up more frequently, provided stocking rates are not increased commensurately (Howden *et al.*, 1999c). Thus, regional climate change and elevated CO_2, as concluded in the SAR, may change the balance from more herbaceous species (grasses and herbs) to more woody species (mainly shrubs), subsequently affecting productivity, decomposition, and fire frequency of the system, as well as forage quality (Allen-Diaz, 1996; Noble and Gitay, 1996) and regional carbon cycling (Gill and Burke, 1999).

5.3.3.2.5. Changes in phenology

Increased variability of rainfall and temperature is likely to affect the phenology of plant and animal species (Gitay and Noble, 1998). It also would affect animal abundance and feeding behavior (see Section 5.4) and would be critical for some pests and diseases and availability of forage for livestock and other mammals (Watt *et al.*, 1995). Species composition changes also could occur. Brown and Carter (1998) manipulated the climate of small areas of temperate grasslands and found that weeds become established in gaps created through increased occurrence of spring or summer drought. Warmer winters induce a late spring drought—which, when combined with an imposed summer drought—leads to significant reduction in cover of nonweeds especially and an increase in weedy species. However, if summer moisture is increased, grasses become dominant. Thus, if climate change results in changes in summer and winter soil moisture and land use creates gaps, this is likely to affect species composition in some grasslands. Fuller and Prince (1996)—using satellite data and NDVI to detect vegetation change—found that there is an early greening of the Miombo woodlands. These woodlands are sensitive to the arrival of spring rains, thus might undergo changes if there is a shift in rainfall patterns.

5.5.3.3. Biogeographical Shifts and Land Degradation

Section 5.2 summarizes the main outputs of recent model outputs. Previous IPCC reports have documented boundary shifts in rangeland vegetation with adjoining vegetation as a response to past climate changes (see, e.g., Gitay and Noble, 1998). The SAR concluded that model outputs suggest significant changes in the boundaries of grasslands and deserts, with expansion of warm grass/shrub types and a decrease in tundra systems (Allen-Diaz, 1996). More recent studies by Allen and Breshears (1998) do not support this rapid change, except perhaps in cases of high drought-induced mortality. Human activities are likely to affect the final changes, especially in tropical and subtropical areas (Allen-Diaz, 1996; Villers-Rúiz and Trejo-Vázquez, 1998), and will negate any gains resulting from amelioration of climate (Gitay and Noble, 1998).

Palaeoecological evidence shows shifts in rangeland vegetation—for example, between the last glacial maximum (LGM) and the Holocene in the Sahara-Gobi desert belt (Lioubimtseva *et al.*, 1998). These shifts have resulted in changes in the carbon stores. Dry and cool conditions during the LGM (about 20,000 years BP) resulted in the spread of arid and semi-arid ecosystems at northern and southern margins of the desert belt. The southern limit of the Sahara migrated southward at least 400 km relative to its present position and almost 1,000 km southward compared to the mid-Holocene (about 9,000 years BP). The northern margin of the temperate deserts and dry steppes of central Asia shifted northward by about 300 km over Kazakhstan, southern Siberia, and Mongolia. During the last world deglaciation, the Sahara-Gobi desert belt was a sink for approximately 200 Gt of atmospheric carbon, but since the mid-Holocene it has been a source.

Modeling studies by Ni (2000), using the BIOME3 model (see Section 5.2) with present climate for the Tibetan Plateau and elevated CO_2 (550 ppm), suggest that there is likely to be a large reduction in temperate deserts, alpine steppe, desert, and ice/polar desert; a large increase in temperate shrubland/meadow and temperate steppe; and a general poleward shift of vegetation zones. Villers-Rúiz and Trejo-Vázquez (1998) found that projected impacts of climate change (using two climate change scenarios generated by the Canadian Climate Center) would result in expansion of some rangeland systems into moist forest areas in Mexico. However, under a GFDL scenario there would be an increase in the distribution of tropical humid and wet forests into rangelands. Zimov *et al.* (1995) suggest that shifts in vegetation boundaries in the rangelands could be influenced by mammalian grazing (in this case, the study area did not have livestock); thus, model projections should include the impact of mammal populations for realistic projections of future vegetation changes (see also Section 5.4).

Desertification tends to be associated with land degradation in rangelands (Pickup, 1998). However, desertification aggregates many land degradation processes and can be exacerbated by climate change. The main processes involved in land degradation include physical, chemical, and biological degradation (Lal *et al.*, 1989, 1999; see also Section 5.3.3) and result in soil compaction, destruction of soil surfaces (which often are kept intact by the presence cyanobacteria, lichens, and mosses), and salinization (Dregne, 1995; Pickup, 1998). These processes are sensitive to livestock trampling. The end result can be decreased water penetration, increased surface runoff, and exposure of soils to wind and water erosion, thereby changing the water and nutrient cycles (Belnap, 1995). In some cold deserts, changes to the soil surface and thus water penetration are quite critical because low vegetation cover, low surface rooting plants, few nitrogen fixers, and low soil temperatures result in slow recovery in nutrient cycles (Belnap, 1995). Under many circumstances, recovery of the soil surface in many arid and semi-arid parts of the world is estimated to take several hundred years, with nitrogen-fixing capability (mostly from the soil crust) taking at least 50 years—making them generally vulnerable to further desertification and soil degradation.

Land degradation is a nonlinear process with thresholds that can be triggered by climatic factors and human activities (Puigdefabregas, 1998). This has implications for the sensitivities and vulnerability of these systems. Villers-Rúiz and Trejo-Vázquez (1998) suggest that the impacts of climate change are likely to be minor compared to land degradation in parts of rangelands in Mexico where conversion of forests into grasslands is mainly for cattle ranching. Lavee *et al.* (1998) suggest on the basis of field and experimental work that if climate becomes more arid in Mediterranean to arid areas of Israel (through decreases in annual precipitation and the frequency and intensity of precipitation, together with increasing temperature), productivity would decrease—leading to decreased organic matter, soil permeability,and rates of infiltration and thus land degradation. The rate of change of these variables along the climatic transect is nonlinear. A step-like threshold exists at the semi-arid area, which sharply separates the Mediterranean climate and arid systems. This means that only a relatively small climatic change would be needed to shift the borders between these two systems. Because many regions of Mediterranean climate lie adjacent to semi-arid areas, the former are threatened by desertification in the event of climate change.

Irrigation in semi-arid areas is a major cause of secondary (or dryland) salinity. Elevated CO_2 increases WUE, so it may reduce irrigation needs, whereas warming may enhance water demand and exacerbate problems of secondary salinity (Yeo, 1999). Increased subsoil drainage under elevated CO_2 may increase the risk of secondary salinization and areas potentially affected (Howden *et al.*, 1999d).

5.5.3.4. Changes in Biodiversity

This subject is reviewed extensively in the SAR and the *Special Report on Regional Impacts of Climate Change*. More recent studies mostly suggest that recently observed changes in rangelands, especially animals (see Section 5.4), are occurring as a result of land degradation. For example, in Australia, land degradation from pastoral activity as well as the impacts of introduced feral mammals (goats, foxes, rabbits) has resulted in extinctions of small- and medium-sized mammals (Pickup, 1998). There have been past changes in plant biodiversity as a result of changes in rangeland vegetation. For example, Prieto (2000), using paleorecords from eastern plains of Argentina, found that there was a replacement of dry steppe by humid grasslands during the late glacial-Holocene transition. The plant composition indicates a frequently disturbed habitat with a lot of weedy species that did not exist in the records in previous ages.

5.5.4. Adaptation Options

Human societies in rangelands would have to adapt to changes in climate, especially temperature and water availability. The SAR concludes that the lack of infrastructure and investment in resource management in many countries dominated by rangelands makes some adaptation options problematic (Allen-Diaz, 1996) but also makes these areas more sensitive to impacts of climate change (Gitay and Noble, 1998). Nevertheless, some adaptation options are available for many of the rangelands.

Specific examples of the interaction between climate change and management decisions may be highlighted better at the regional level. For example, in Australia, Pickup (1998) found that substantial shifts in rainfall have occurred over the past 100 years. If climate change results in further shifts in rainfall patterns, the major impacts are likely to be related to increased climate variability. Pastoral management decisions in these rangelands tend to be taken over the short term; wetter periods generate unrealistic expectations about land use and high stocking rates, which drier periods are unable to support. This has and would lead to land degradation.

5.5.4.1. Landscape Management

Rangelands consist of a mosaic of various ecosystem types (WRI, 2000) with soil and water processes as well as associated nutrient cycles that operate at the landscape or regional scale (Coughenour and Ellis, 1993). Human use of rangelands often affects landscape processes (e.g., water flow, soil erosion) and changes in processes such as productivity, decomposition, and fire. Thus, possible future adaptation options might have to be sought at the landscape level (Aronson *et al.*, 1998) and over long time frames (Allen-Diaz, 1996). Because many rangelands are in semi-arid and arid parts of the world, actions to reduce destruction of the soil crust (which are important for soil stabilization and nitrogen fixation) and thus land degradation are extremely important. These actions could include adjustment in the time and intensity of grazing (Belnap and Gillette, 1998). Restoration of degraded soils has vast potential to sequester carbon in soil and aboveground biomass (Lal *et al.*, 1999), although restoration could be costly (Puigdefabregas, 1998).

5.5.4.2. *Selection of Plants and Livestock*

Selection of plants (e.g., legume-based systems) and animal species and better stock management are likely to be the most positive management options (Chapman *et al.*, 1996; Gitay and Noble, 1998). Rotational cropping and decreased use of marginal lands might be necessary in rangeland management. This might mean more intensive land management in some areas, leading to more reliable food supplies and perhaps reduction in methane production from livestock (decrease in methane production would be caused by improved forage quality; Allen-Diaz, 1996). Potential stocking rates could be determined via satellite imagery (Oesterheld *et al.*, 1998). However, the decision on stocking rates might still be made on a social basis, especially given the values associated with livestock in many pastoral rangeland communities (Turner, 1993).

5.5.4.3. *Multiple Cropping System and Agroforestry*

As human population in rangelands increases and land use changes, some traditional practices are becoming less appropriate. Some of the options for sustainable agriculture could include efficient small-scale or garden irrigation, more effective rain-fed farming, changing cropping patterns, intercropping, or using crops with lower water demand (Lal, 1989; Batchelor *et al.*, 1994; Dixon *et al.*, 1994a; Dabbagh and Abdelrahman, 1998). Conservation-effective tillage is an option that could help to achieve improved productivity (Benites and Ofori, 1993). Agroforesry, using potential fuelwood species, also is an obvious option to alleviate land degradation as well as to meet some social needs. Management (e.g., appropriate coppicing in Uganda) or other practices (e.g., collecting only dead or fallen wood; Benjaminsen, 1993) are considered essential for maintenance of fuelwood species. Where woody weeds are increasing, they could be used for domestic fuel supply; otherwise, management options (e.g., use of fire along with regulation of grazing—Archer, 1995; Brown and Archer, 1999) might have to be implemented to control woody weeds.

5.5.4.4. *Role of Community Participation and Public Policy*

Community participation in decisionmaking and management, along with public policy, can be a favorable and critical issue in implementing some adaptation options. This could result in better management of rangelands (Pringle, 1995; Allen-Diaz, 1996; Thwaites *et al.*, 1998), thereby probably meeting conservation objectives (Pringle, 1995). Decisions to be made might include:

- Determinations about appropriate stocking rates, which might require discussion and negotiations among stakeholders, especially because stocking rates might be more social than technically oriented (Abel, 1997)
- Choosing some agroforestry practices that fulfil local needs, especially because many communities rely on fuelwood (Benjaminsen, 1993)
- Diversification, since some communities could get or have already gotten involved in tourism as a way of highlighting some of the unique flora, fauna, and landscape features, thereby conserving the systems, reducing some of the impacts, and obtaining cash (Hofstede, 1995).

In dealing with options for reducing the consequences of land degradation in the future, public policy may have a crucial role (Hess and Holechek, 1995), especially because decisions at the landscape level (which are likely to include many different land tenures) are going to be increasingly important. Policies could be developed to address multiple pressures and, over the long term, to encompass sustainable land management and could include investments by governments to improve rangeland status (Morton *et al.*, 1995; Pickup, 1998).

5.5.5. *Vulnerabilities and Sensitivity to Climate Change*

For the future of rangelands, it is important to reduce the vulnerability of these systems to climate change. This is likely to be achieved by considering social and economic factors that determine land use by human populations (Allen-Diaz, 1996). Soil stability and thus maintenance of water and nutrient cycles are essential in reducing the risk of desertification. Any changes in these processes could make rangelands particularly vulnerable to climate change. Land degradation is a nonlinear process with thresholds that make these systems sensitive and vulnerable (Puigdefabregas, 1998). Prevention of land degradation might be a cheaper option than restoration, which can be costly (Puigdefabregas, 1998). Some studies suggest that changes in rainfall pattern may make some vegetation types within rangelands more vulnerable (e.g., Miombo woodlands—Fuller and Prince, 1996) if growing periods could not shift or if these growing periods coincide with insect outbreak.

5.6. Forests and Woodlands

Forests and woodlands[2] provide many goods and services that society values, including food, marketable timber and non-wood products (fuel, fiber, construction material), medicines, biodiversity, regulation of biogeochemical cycles, soil and water conservation (e.g., erosion prevention), carbon reservoirs, recreation, research opportunities, and spiritual and cultural values. Forests play a key role in the functioning of the biosphere—for example, through carbon and water cycles (the latter is discussed in Chapter 4)—and hence indirectly affect the provision of many other goods and services (Woodwell and MacKenzie, 1995). Changes in global climate and atmospheric composition are likely to have an impact on most of these goods and services, with significant impacts on socioeconomic systems (Winnett, 1998).

[2]In this section, "forest" includes both forest land and woodlands, unless otherwise specified.

Since the SAR, many studies have dealt with changes in the structure, composition, and spatial patterns of forests (e.g., VEMAP Members, 1995; Smith *et al*., 1996a; Bugmann, 1997; Shriner *et al*., 1998). The biogeochemical literature has focused on the carbon cycle (e.g., Apps and Price, 1996; Fan *et al*., 1998; Steffen *et al*., 1998; Tian *et al*., 1998; IPCC, 2000; Schimel *et al*., 2000). There is an expectation that directed land-management practices can either increase or retain carbon stocks in forests, thereby helping to mitigate increases in atmospheric CO_2 levels; this is discussed elsewhere (IPCC, 2000; see also TAR WGIII Chapter 4 and TAR WGI Chapter 3).

The influence of climate change on forests and associated goods and services is difficult to separate from the influence of other global change pressures such as atmospheric changes, land use, and land-use change resulting from human activities. The State-Pressure-Impacts approach outlined in Section 5.1 is used here as a framework to examine interrelated responses to global change and expected changes in supply of services from forests and woodlands (see Table 5-1). In this section, the focus is on some of the important pressures, impacts, and responses for three goods and services provided by forests and woodland ecosystems: carbon, timber, and non-wood goods and services. The impact on biodiversity in forests is covered in Sections 5.2 and 5.4. The state of the sector and the pressures acting on it, as well as possible responses, impacts, and adaptation opportunities, will differ among the regions of the world; here the focus is on global commonality.

5.6.1. Current Status and Trends

This section presents an overview of the current status and trends for forests in general; specific regional trends are presented in some of the regional chapters (Chapters 10–17).

The world's forests cover approximately 3,500 Mha (FAO, 1997a), or about 30% of the total land area (excluding Greenland and Antarctica). About 57% of the world's forests, mostly tropical, are located in developing countries. About 60% are located in seven countries (in order): the Russian Federation, Brazil, Canada, the United States, China, Indonesia, and the Democratic Republic of Congo.

In 1995, plantation forests were estimated to cover 81 Mha (2.3% of total estimated cover) in developing countries and about 80–100 Mha in developed countries (FAO, 1997a). They play an important role, particularly in the production of industrial roundwood and fuelwood, restoration of degraded lands, and provision of non-wood products (FAO, 1997a). Some countries obtain 50–95% of their industrial roundwood production from plantations that cover 1–17% of their total forest area (Sedjo, 1999). In many developing countries, plantations often occur as community woodlots, farm forests, and agroforestry operations.

Between 1980 and 1995, the area of the world's forests decreased by about 180 Mha (5% loss of total forest area in 15 years) as a result of human activities (FAO, 1997a). About 200 Mha were converted to agriculture (subsistence agriculture, cash crops, and ranching), but this loss was partially offset by about 40 Mha increase in plantations. In developed countries, forests increased over the same period by about 20 Mha through afforestation and natural regeneration on land no longer in use by agriculture, despite losses of forests to urbanization and infrastructure development. Loss of native forest in developing countries (tropical and nontropical) appears to have slowed during 1990–1995, with an overall loss of about 65 Mha (FAO, 1997a). However, other changes such as fragmentation, nonsustainable logging of mature forests, degradation, and development of infrastructure—all leading to losses of biomass—have occurred over large areas. Of about 92 Mha of tropical closed forest that underwent a change in cover class during 1980–1990, 10% became fragmented forest and 20% was converted to open forest or extended forest fallow (FAO, 1996). There are no global estimates of forest degradation, but data from specific areas give an indication of the extent of degradation. Logging practices damage and degrade more than 1 Mha yr^{-1} of forest in the Brazilian Amazon; surface fires (e.g., those in 1998) may burn large areas of standing forest in these regions (Cochrane *et al*., 1999; Nepstad *et al*., 1999; see also Chapter 14). These authors conclude that present estimates of annual deforestation for Brazilian Amazonia capture less than half of the forest area that is impoverished each year—and even less during years of severe drought. In the boreal zone, there has been continuing encroachment by agriculture and development of infrastructure (roads, survey lines, wellheads, etc.) that open access to primary forests. Preliminary estimates of these effects for Canada, for example, indicate a net loss of 54,000–81,000 ha yr^{-1} of forest over the period 1990–1998 as a result of various activities (Robinson *et al*., 1999).

5.6.1.1. Carbon Pools and Flux

Carbon pools in the world's forests are estimated to be 348 Gt C in vegetation and 478 Gt C in soil (to 1 m) (updated since Dixon *et al.*, 1994b, by Brown, 1998). Bolin and Sukumar (2000) based their numbers on Dixon *et al.* (1994b). The largest vegetation and soil carbon pools are in tropical forests (60 and 45% of the total, respectively) because of their large extent and relatively high carbon densities. Carbon stocks in forests vary, depending on the type of forest in relation to climate, soil, management, frequency of disturbances, and level of human-caused degradation.

Based on traditional carbon inventories, terrestrial ecosystems were shown to be carbon sources during the 1980s in the SAR (Brown *et al*., 1996); a high- and mid-latitude forest sink was exceeded by the source from low-latitude forests. However, recent work—using atmospheric measurements and modeling—suggests that terrestrial ecosystems appear to be net sinks for atmospheric carbon, even when losses from land-use change are taken into account (Bolin and Sukumar, 2000). For the 1980s, a net storage increase of 0.2 ± 1.0 Gt C yr^{-1} by terrestrial ecosystems (largely in forests) was estimated as the difference between a net emission of 1.7 ± 0.8 Gt C yr^{-1} from land-use changes (primarily in the tropics) and global terrestrial uptake of 1.9 ± 1.3 Gt C yr^{-1} (Bolin and Sukumar, 2000). In the 1990s, estimated net emissions from land-use change decreased slightly to 1.6 ± 0.8 Gt C yr^{-1}, and global terrestrial uptake increased to 2.3 ± 1.3 Gt C yr^{-1}, resulting in a net terrestrial uptake of 0.7 ±1.0 Gt C yr^{-1} (Bolin and Sukumar, 2000). This terrestrial net sink of carbon arises as the net effect of land-use practices (agricultural abandonment and regrowth, deforestation, and degradation); the indirect effects of human activities (e.g., atmospheric CO_2 fertilization and nutrient deposition); and the effects of changing climate, climatic variation, and disturbances. The relative importance of these different processes is known to vary strongly from region to region.

Regional source and sink relationships also have been inferred by techniques of inverse modeling of observed atmospheric CO_2 gradients and circulation patterns (Ciais *et al.*, 1995; Fan *et al.*, 1998). These estimates are relatively imprecise and are difficult to relate to those that are based on forest inventory data. For example, the study by Fan *et al.* (1998) suggests that 1.4 Gt C yr^{-1} was taken up by terrestrial biota in North America in the 1980s. However, mechanistic models and measurements based on forest inventories do not agree with the magnitude or spatial distribution of this carbon sink (Holland and Brown, 1999; Potter and Klooster, 1999).

Temperate forests are considered to be net carbon sinks at present, with estimates of 1.4–2.0 t C ha^{-1} yr^{-1} (Nabuurs *et al.*, 1997; Brown and Schroeder, 1999; Schulze *et al.*, 1999). These findings are consistent with recent estimates of carbon in woody biomass, based on statistics for 55 temperate and boreal countries; these statistics indicate a general increase in forest biomass from the 1980s to the 1990s (UN-ECE/FAO, 2000c). Changes in forest management (reduction of harvest levels, increased regeneration effort, and administrative set-asides), as well as changes in the environment (N and CO_2 fertililization), appear to have contributed to this trend, but the relative contribution of different factors varies among forest regions and countries (Kauppi *et al.*, 1992; Houghton *et al.*, 1998, 1999; Brown and Schroeder, 1999; Liski *et al.*, 1999; Nadelhoffer *et al.*, 1999).

In boreal forests, carbon budgets vary strongly among different forest types (Apps *et al.*, 1993; Bonan, 1993, Shvidenko and Nilsson, 1994). Although some boreal forest regions currently appear to be net carbon sources (Shepashenko *et al.*, 1998; Kurz and Apps, 1999), others appear to be net sinks, varying between 0.5 and 2.5 t C ha^{-1} yr^{-1} (Shvidenko and Nilsson, 1994; Jarvis *et al.*, 1997). The difference between annual increment and net fellings reported to the FAO (reported in UN-ECE/FAO, 2000c) does not account for changes in the frequency and severity of disturbances that have a large influence on source and sink relationships in boreal forests (Kasischke *et al.*, 1995; Kurz and Apps, 1999; Bhatti, 2001). For example, detailed analyses of forest inventory data together with observed changes in disturbance over time indicate that Canadian forest ecosystems changed from a modest sink (0.075 Gt C yr^{-1}) from 1920–1970 to a small net source of 0.050 Gt C yr^{-1} in 1994 (Kurz and Apps, 1999). Similarly, in Russia between 1983 and 1992, managed forest ecosystems in the European portion, where disturbances were relatively controlled, were a sink of 0.051 Gt C yr^{-1} but a net source of 0.081–0.123 Gt C yr^{-1} in the less intensively managed Siberian forests of the east (Shepashenko *et al.*, 1998). Factors that were not included in these analyses that may offset losses of biotic carbon from disturbed forests include nitrogen deposition and CO_2 fertilization (Chen *et al.*, 2000; Schimel *et al.*, 2000), but experimental verification of these influences is not yet possible from inventory data.

In the tropics, forests are still reported to be a net carbon source as a result of land-use change. Although some studies suggest net carbon uptake in some tropical forests (Grace *et al.*, 1995; Phillips *et al.*, 1998), losses associated with high rates of forest clearing and degradation exceed such gains. The magnitude of the net carbon source from the tropics has been reported to be about 0.1 Gt C yr^{-1} lower in the first half of the 1990s than in the 1980s (Houghton and Hackler, 1999; Houghton *et al.*, 2000; Houghton, 2001), mostly because of reduced rates of deforestation in the 1990s (FAO, 1996, 1997a).

In summary, carbon stored in forest ecosystems appears to be increasing, and at an increasing rate with about 0.2 Gt C yr^{-1} being stored in the 1980s and 0.7 Gt C yr^{-1} in the 1990s. Most storage occurs in temperate forests, with a small net source from tropical forests and boreal forests varying depending on the disturbance regime they experience.

5.6.1.2. *Timber and Non-Wood Products*

Forests contribute to GDP in three main ways: industrial wood products, fuelwood, and the economic impacts of recreation and non-wood products (e.g., mushrooms). Regional estimates generally are available only for the first (industrial) component, which captures the direct value of harvesting timber and the value added by manufacturing. At a global level, forestry contributes approximately 2% to GDP (FAO, 1997a)—6% in Africa, 3% in South America, 2% in North and Central America, and 1% in Europe. In developing countries, forestry contributes 4% to GDP; in developed countries the contribution is 1%. Total industrial timber production in 1997 was 1.5 billion m^3, with more than 60% coming from developed countries (FAO, 1997a).

Although income and population growth influence demand for industrial timber, recycling and technological change (e.g., use of wood chips for manufactured products) can affect the quantity harvested from forests. Total industrial wood harvests have remained relatively constant over the past 20 years, even as global population and incomes have increased (FAO, 1998). Global per capita consumption of wood (including fuelwood and roundwood) is about 0.6 m^3 yr^{-1}; this level of consumption has been relatively stable over the past 40 years (Solberg *et al.*, 1996). Global fuelwood production in 1996 is estimated to be 1.9 billion m^3, with 90% of this production occurring in developing countries (FAO, 1997a). In 1994, annual per capita fuelwood consumption in developing countries was 0.39 m^3, versus 0.16 m^3 in developed countries (FAO, 1997a). It is estimated that 2 billion people rely on wood and charcoal for fuel (mostly derived from forests), and ensuring an adequate and sustainable supply will continue to be an important pressure on forests.

Non-wood forest products (NWFP)—such as edible mushrooms, nuts, fruits, palm hearts, herbs, spices, gums, aromatic plants, game, fodder, rattan, medicinal and cosmetic products, resins, and the like—make important contributions to household income, food security, national economies, and environmental objectives of conservation of biodiversity (FAO, 1997a). It is estimated that about 80% of the population of the developing world depends on NWFP to meet some of their health and nutritional needs. Several million households worldwide depend heavily on these products for subsistence consumption and income.

5.6.2. *Pressures on Forests and Woodlands*

Forests have many pressures acting on them that result in changes to their structure and composition, as well as their function (see Figure 5-1). These structural changes, in turn, alter the function of forests in the physical climate system (Sellers *et al.*, 1990; Apps, 1993).

5.6.2.1. *Climate Variability and Climate Change*

Changes in climatic conditions affect all productivity indicators of forests (NPP, NEP, and NBP; see Box 5-1) and their ability to supply goods and services to human economies. The effects on forested area and forest productivity, however, vary from location to location, with gains in some regions and losses in others. Furthermore, the impacts vary among different measures of ecosystem productivity. For example, in boreal and alpine forests—where short growing seasons and heat sums are limiting factors to growth—NPP of many forest stands may increase with increasing temperature (Bugmann, 1997; but see Barber *et al.*, 2000), whereas NEP decreases as a result of increased decomposition (Schimel *et al.*, 1994; Valentini *et al.*, 2000; but see Giardina and Ryan, 2000). If higher temperatures lead to summer drought, even NPP may decrease as a result of lowered photosynthetic rates associated with reduced stomatal conductance (Sellers *et al.*, 1997), exacerbating the decrease in NEP from decomposition. If drier conditions also result in increased fires, biomass and soil carbon losses may result in negative NBP (Wirth *et al.*, 1999; Apps *et al.*, 2000).

Projected changes in forest area, structure, NPP, and NEP as a result of climate change vary by forest type and biome (Neilson *et al.*, 1998). Climate change also is likely to include changes in seasonality (Myneni *et al.*, 1997), timing of freeze-thaw patterns (Goulden *et al.*, 1998), the length of the growing season, nutrient feedbacks (Tian *et al.*, 1998), disturbance regimes (Kurz *et al.*, 1995), and diurnal temperature patterns (Clark and Clark, 1999). Changes in intra-annual variation that fall outside the historical norm for a particular region also may have catastrophic effects—for example, through local climatic extremes or through late and early frost (Repo *et al.*, 1996; Ogren *et al.*, 1997; Colombo, 1998). These factors are likely to influence the distributional range of some tree species (Macdonald *et al.*, 1998; Rehfeldt *et al.*, 1999a). Changes in precipitation may not have immediate effects on mature and old-growth forests, which have well-established root systems, but are likely to have pronounced effects on regeneration success for some species following disturbance, such as harvest or fire (Hogg and Schwarz, 1997; Price *et al.*, 1999a,b).

5.6.2.2 *Changes in Disturbance Regimes*

At the landscape scale, changes in the disturbance regime introduce instabilities in forest age-class distributions (Bhatti *et al.*, 2000) and eventually in the distribution of plant species. If changes in disturbances are caused or accompanied by changes in environmental conditions, the responses of the forest ecosystem can be very complex. Changes in disturbance regimes, spatially and over time, can be exacerbated or mitigated by human activities (e.g., by fire ignitions and fire suppression).

5.6.2.2.1. *Pressures from fires*

Large areas of mixed savanna-woodlands in dry tropical zones of Africa, South America, Australia, and large areas of tropical humid forests burn every year (WRI, 2000). These fires are part of the natural seasonal cycle of growth, decay, and combustion and are ignited by lightning strikes. However, humans have long played a significant role in modifying fire regimes by changing the season and frequency of burning and consequently vegetation composition and structure (Goldammer and Price, 1998). During the 1990s in tropical humid areas, major fires have occurred in the Brazilian Amazon, Mexico, and Indonesia (Kalimantan and Sumatra) and were particularly severe in 1997–1998 during an El Niño episode (FAO, 1997a; Nepstad *et al.*, 1999). Fire also is a serious threat to native forests in many parts of tropical and nontropical developing countries. China, for example, suffered large losses in a single fire event in 1987, with 1 Mha (Anon., 1987) to 1.3 Mha (Cahoon *et al.*, 1994) burned. Fire prevention and suppression capability depends on available infrastructure such as imagery, roads, machinery, and human capital. In general, developed countries are better able to manage fire in regions with roads; developing countries may lack one or all of the factors.

The Indonesian fires of 1997–1998 were associated with a significant, but not unique, drought over much of the region. Estimates of the area burned vary from 96,000 ha to more than 8 Mha (Harwell, 2000). Most of this area was not forest but scrub, grassland, and agricultural lands. Almost no fires occurred deep within undisturbed primary forest; most were associated with land-clearing for new settlements or plantations or with logging operations. The most persistent fires were seven clusters of fires along the edges of degraded peat-swamp forests in southern Sumatera and Kalimantan (Legg and Laumonier, 1999). The extent and persistence of these fires, and similar fires in Brazil, show the importance of interaction between climate and human actions in determining the structure and composition of tropical forests, land-use patterns, and carbon emissions.

In the boreal zone of Canada, there has been a marked increase in fire since about 1970, after a 5-decade decrease (Kurz *et al.*, 1995). The area of boreal forest burned annually in western North America has doubled in the past 20 years (0.28% in the 1970s to 0.57% in the 1990s), in parallel with the observed warming trend in the region (Kasischke *et al.*, 1999), despite much improved detection and suppression technology. Similar trends have been noted for Eurasian forests (Shvidenko and Nilsson, 1994, 1997; Kasischke *et al.*, 1999). Whether these changes are the result of human-induced climate change or are a result of natural climatic variability is not certain (Clark *et al.*, 1996; Flannigan *et al.*, 1998). Changes in the disturbance regime over periods of decades result in changes in forest age-class distribution to younger versus old forests (Kurz and Apps, 1999). These changes will reduce the landscape-averaged biomass stock and dead organic matter pools, including soils (Bhatti *et al.*, 2000). Hence, changes in NBP occur on scales of years to decades.

Fire frequency is expected to increase with human-induced climate change, especially where precipitation remains the same or is reduced (Stocks *et al.*, 1998). A general but moderate increase in precipitation, together with increased productivity, also could favor generation of more flammable fine fuels. Miranda (1994) suggests an increase in risk, severity, and frequency of forest fires in Europe. Stocks *et al.* (1998) used four GCMs and found similar predictions of an earlier start to the fire season and significant increases in the area experiencing high to extreme fire danger in Canada and Russia. Some regions may experience little change or even decreases in fire frequency, where precipitation increases or temperature does not rise (as in eastern Canada, where regional cooling has led to decreased fire frequency—Flannigan *et al.*, 1998). In most regions, there is likely to be an increased risk of forest fires, resulting in a change in vegetation structure that in turn exacerbates this risk (Cramer and Steffen, 1997).

During the past decade, forest fires in developed countries generally have become smaller, with the exception of the former Soviet Union (FAO, 1997a), Canada (Kurz *et al.*, 1995; Kurz and Apps, 1999), and the United States (Sampson and DeCoster, 1998). Where observed, the slight decline in forest areas burned per year is believed to be in part a result of improved prevention, detection, and control of fires. However, many such protected forests have increasing fuel loads and an abundance of dead and dying trees that eventually will make them more susceptible to catastrophic fires (e.g., Sampson and DeCoster, 1998). Several authors suggest that climate change is likely to increase the number of days with severe burning conditions, prolong the fire season, and increase lightning activity, all of which lead to probable increases in fire frequency and areas burned (Price and Rind, 1994; Goldammer and Price, 1998; Stocks *et al.*, 1998).

5.6.2.2.2. Pressures from diseases and insect herbivory

Insect outbreaks can be extremely important disturbance factors (Hall and Moody, 1994); during outbreaks, trees often are killed over vast areas (Hardy *et al.*, 1986; Candau *et al.*, 1998). Under climate change, damage patterns caused by insects may change considerably, particularly those of insects whose temporal and spatial distributions strongly depend on climatic factors. The ecological interactions are complex, however (see Box 5-10).

In temperate and tropical regions, insect and disease outbreaks are reported mostly for plantation forests; relatively less is known about native forests (FAO, 1997a). In the boreal zone, insect-induced mortality was a significant part of the changing disturbance regime for Canada in the period 1920–1995 (Kurz *et al.*, 1995). Insect mortality accounted for the loss of approximately 76 Mha in that period, with a near tripling of the average annual rate after 1970 (Kurz and Apps, 1999). Similar trends have been observed for Russian forests, where recent annual insect damage and disease mortality affecting as much as 4 Mha was reported by Shvidenko *et al.* (1995). In Siberian and Canadian forests, insect damage is estimated to be of the same magnitude as fire loss (Krankina *et al.*, 1994; Fleming and Volney, 1995; Kurz *et al.*, 1995; Shvidenko *et al.*, 1995; Shvidenko, 2000). Changes in drought conditions appear to play an important role in insect outbreaks (Volney, 1988; Sheingauz, 1989; Isaev, 1997).

There is likely to be an increase in declines and dieback syndromes (Manion, 1991) caused by changes in disease patterns, involving a variety of diseases. For example, in temperate and boreal regions, there may be increased incidence of canker diseases in poplars and other tree species. Some canker diseases increase in severity with decreased bark moisture content brought on by drought (Bloomberg, 1962). As another example, *Armillaria* root disease is found throughout the world and causes significant damage on all forested continents (Kile *et al.*, 1991), through mortality and growth loss. This disease—one of the largest threats to regeneration in the productive forest of the Pacific Northwest of North America—has surfaced as a result of past management practices (Filip, 1977) but may be exacerbated by changing climate. Under present conditions, *Armillaria* root disease causes losses of 2–3 million m^3 yr^{-1} in the forests of Canada's Pacific Northwest (Morrison and Mallett, 1996). More recently, Mallet and Volney (1999) report a 43% decrease in annual volume increment and a 23% loss in annual height increment in lodgepole pine caused by *Armillaria* root disease. The incidence of *Armillaria* root disease can be expected to change under warmer or drier conditions. Significant damage has been observed in forests that have undergone drought stress (Wargo and Harrington, 1991). Morevoer, in regions such as the Pacific Northwest where the mean annual temperature presently is below the optimum

Box 5-10. Complex Interactions: North America's Southern Boreal Forests, Pests, Birds, and Climate Change

The eastern spruce budworm (*Choristoneura fumiferana*) is estimated to defoliate approximately 2.3 Mha in the United States (Haack and Byler, 1993) and affects 51 million m^3 of timber in Canada annually (Fleming and Volney, 1995). Although the budworm usually is present at low densities, budworm densities can reach 22 million larvae ha^{-1} during periodic outbreaks (Crawford and Jennings, 1989). Outbreaks can extend over 72 Mha and last for 5–15 years, killing most trees in mature stands of balsam fir (Crawford and Jennings, 1989; Fleming and Volney, 1995).

Weather is thought to play a role in determining the budworm's range. Outbreaks frequently follow droughts (Mattson and Haack, 1987) or start after hot, dry summers (Fleming and Volney, 1995). Drought stresses host trees and changes plant microhabitats (Mattson and Haack, 1987). Moreover, the number of spruce budworm eggs laid at 25°C is 50% greater than the number laid at 15°C (Jardine, 1994). In some areas, drought and higher temperatures also shift the timing of reproduction in budworms so that they may no longer be affected by some of their natural parasitoid predators (Mattson and Haack, 1987). Weather, at least in central Canada, also may play a role in stopping some outbreaks if late spring frosts kill the tree's new growth on which the larvae feed.

Control of some populations of eastern spruce budworm may be strongly aided by bird predators, especially some of the wood warblers (Crawford and Jennings, 1989; but see Royama, 1984). Birds can consume as much as 84% of budworm larvae and pupae when budworm populations are low (approximately 100,000 ha^{-1}), but once larvae populations exceed 1,000,000 ha^{-1}, bird predation is unable to substantially effect budworm populations. This predatory action of birds works in concert with those of other predators, mostly insects.

The spruce budworm's northern range may shift northward with increasing temperatures—which, if accompanied by increased drought frequency, could lead to outbreaks of increasing frequency and severity that lead to dramatic ecological changes (Fleming and Volney, 1995). Increasing temperatures also might reduce the frequency of late spring frosts in southern boreal forests, perhaps increasing outbreak duration in some of those areas.

A changing climate also might decouple some budworm populations from those of their parasitoid and avian predators (Mattson and Haack, 1987; Price, 2000). Distributions of many of the warblers that feed on spruce budworms could shift poleward, perhaps becoming extirpated from latitudes below 50°N (Price, 2000). Replacing biological control mechanisms with chemical control mechanisms (e.g., pesticides) ultimately may yield a different set of problems; there are economic and social issues relating to large-scale pesticide application (see Section 5.4.4.3).

(25°C) for *Armillaria* growth, a warmer climate is likely to result in increased root disease and rate of spread.

5.6.2.2.3. Pressures from other disturbances

Additional disturbances are associated with extreme weather events such as hurricanes, tornadoes, unexpected drought or heavy rainfall, flooding, and icestorms that lead to extensive mortality and ecosystem change (e.g., Lugo *et al.*, 1990; Walker and Waide, 1991). Such events generally are highly localized and take place in a relatively short period of time but have long-term economic impacts (Haight *et al.*, 1995) and effects on ecosystems (Pontailler *et al.*, 1997). There is some evidence of recent increases in damage from such extreme events (Berz, 1999; see also Chapters 8 and 9).

5.6.2.3. Changing Demand for Forest Goods and Services

Future demand for industrial wood products depends on income growth, population growth, technological change, growth in human capital, changes in tastes and preferences, and institutional and political change (Solberg *et al.*, 1996). Changes in other markets also can influence demand for wood products. For example, increases in the price of substitutes, such as steel and concrete building materials, would increase the demand for industrial timber. In light of these driving factors, recent timber market assessments have predicted that industrial harvests will increase by 1–2% yr^{-1} (Solberg *et al.*, 1996; Brooks, 1997; FAO, 1997b; Sohngen *et al.*, 1999). These results contrast with those in the SAR (Solomon *et al.*, 1996), which concluded that global demand for industrial fiber would exceed global supply in the next century.

There is some debate about which forests are likely to be harvested in the future. Some authors contend that most supply will come from new industrial plantations, secondary growth forests, and enhanced management, rather than from native forests (FAO, 1997b; Sohngen *et al.*, 1999). The proportion of global timber from subtropical plantations (presently 10%) may increase to 40% by 2050 (Sohngen *et al.*, 1999). Non-native species, such as eucalypts and pines, are favored in these regions because the costs of management and harvesting are low compared to those in temperate and boreal forests (Sedjo and Lyon, 1990; Sedjo, 1999). Recent estimates use global timber market models that incorporate management responses to prices across a wide range of forests. Under most price scenarios, subtropical plantations with 5- to 20-year rotations are a financially attractive alternative (Sedjo, 1999). However, higher demand still may put pressures on native forests even if plantation establishment and forest management responds to price increases (Solberg *et al.*, 1996; Brooks, 1997).

Policies that raise prices for substitute products, such as non-wood building materials made from steel or plastics, may increase timber demand, increase non-native plantation establishment, and cause additional pressures on native forests. Substituting non-wood products for wood products could increase carbon emissions as well (Marland and Schlamadinger, 1995; Schlamadinger *et al.,* 1997).

Increased reliance on plantations may have positive and negative ancillary consequences. For example, most of the plantations established for industrial purposes involve nonindigenous species, and the environmental effects of these plantations are not fully evident. However, most plantations have been established on former agricultural lands, which begins the process of restoring forests (Lugo *et al.*, 1993). Furthermore, plantations may reduce harvest pressures on natural forests. Despite increased reliance on plantations, however, industrial harvests in native forests along accessible roadways are likely to continue (Johns *et al.*, 1996).

The relationship between income and fuelwood demand is nonlinear. As incomes rise and infrastructure grows, households substitute alternative fuel sources (e.g., natural gas, fuel oil). Brooks *et al.* (1996), for example, suggest that fuelwood harvests will increase by 17% by 2050 under a low-GDP-growth scenario, but only by 4% under a high-growth scenario. Fuelwood harvest depends on the extent of substitution by alternative methods of heating and cooking. Currently, use of wood for energy on a large scale does not appear to be cost-effective relative to other energy sources, but if future energy prices rise, the demand for wood as a source of energy could rise.

5.6.2.4. Land-Use Change

Land-use change, including deforestation, is still considered to be a large pressure on global forests (Alcamo *et al.,* 1998). Several factors contribute to deforestation in tropical regions, including income and population growth, road-building policies, and other government policies (Southgate, 1998). Conversion to agriculture is impractical in boreal systems because of low productivity and high access costs, but some boreal forests continue to be converted to second-growth, managed forests. In temperate regions, conversion of agricultural lands back to forest has increased with agricultural productivity and falling prices (Kuusela, 1992; UN-ECE/FAO, 2000a). There is some evidence that these trends will continue, although at a lower rate (UN-ECE/FAO, 2000a). Factors affecting these trends include urbanization, agricultural yields and prices, timber prices, access and conversion costs, and subsidy programs (such as those that promote afforestation for environmental reasons, including mitigation of climate change impacts).

5.6.2.5. Other Pressures

Fragmentation of forest landscapes as a result of climate change, land-use practices, and disturbance is expected to take place in advance of larger scale biome shifts (Fahrig and Merriam, 1994; Shriner *et al.*, 1998). Fragmentation can change biodiversity and resiliency (Sala *et al.*, 2000). Fragmentation of the landscape can occur as a result of disturbance (natural or anthropogenic) or more gradually from successional responses to environmental changes (NBIOME SSC, 1992).

Pressures from air pollution and air quality: There is evidence of decline in forest condition as a result of air pollution, especially in areas adjacent to industrial areas and large cities (e.g., deposition of heavy metals, sulfur, nitrogen, and ozone) (Nilsson *et al.*, 1998). Of major concern is increased nitrogen deposition caused by industrial processes and agriculture (Vitousek *et al.*, 1997b). Nitrogen deposition is higher in northern Europe than elsewhere (Vitousek *et al.*, 1997b). Low-level increases in nitrogen deposition associated with air pollution have been implicated in productivity increases over large regions (Schindler and Bayley, 1993; Vitousek *et al.*, 1997b). Temperate and boreal forests, which historically have been nitrogen-limited, appear to be most affected (Townsend and Rastetter, 1996; Vitousek *et al.*, 1997b). In other areas that become nitrogen-saturated, other nutrients are leached from the soil, resulting in forest dieback (Vitousek *et al.*, 1997b)—counteracting, or even overwhelming, any growth-enhancing effects of CO_2 enrichment.

Tropospheric ozone has been shown to impact the structure and productivity of forest ecosystems throughout industrialized countries (Chameides *et al.*, 1994; Weber *et al.*, 1994; Grulke *et al.*, 1998) and is likely to increase in extent with further industrial development and agriculture management (Chameides *et al.*, 1994). It has been suggested that the impact of ozone damage is reduced, but not eliminated, by increasing CO_2 (Tingey *et al.*, 2001).

In developed countries, the major impacts of air pollution on forest services are likely to be on recreation and non-wood products. Air pollution has been shown not to have significant impacts on industrial timber markets in the United States (Haynes and Kaiser, 1990), although European timber market studies suggest potentially larger local effects (Nilsson *et al.*, 1992). Although increasing industrialization in developing countries that have less restrictive air pollution requirements could have effects on local industrial or fuelwood markets, these increases are not expected to have major effects on global timber supply.

5.6.3. Responses by Forests and Woodlands and Impacts on their Goods and Services

Assessment of responses and impacts must distinguish between transient and equilibrium situations. Because forests are composed of long-lived organisms, responses to climate change and resulting impacts may take a long time to propagate through the system. To the extent that global change (climate and land-use change) proceeds faster than the life cycle of many late-succession trees, transient responses will predominate. Forest structure today is the result of activities and events that occurred many years (>100) ago; hence, responses and parameters measured today are not in equilibrium with present conditions. Many state variables, such as carbon pools, are expected to change with a time delay. Furthermore, forest responses to climate change and resulting impacts may extend longer than the change in climate.

5.6.3.1. Responses and Impacts: Carbon Storage

There are numerous ways in which forests respond to climate change and other pressures. These responses affect their ability to store carbon. Responses include changes in species distribution, NPP, NEP, and NBP; pests and disease outbreaks; and elevated CO_2, as well as changes in climate variability and weather extremes.

5.6.3.1.1. Forest and species distribution

Models that predict changes in species distribution suggest reduced forest carbon storage as climate changes (King and Neilson, 1992; Smith and Shugart, 1993; Kirilenko and Solomon, 1998; Woodward *et al.*, 1998), although the change in forest carbon stocks depends on species migration rates (Solomon and Kirilenko, 1997; see also Section 5.2). Where seed availability and dispersal are impeded (e.g., by fragmented landscapes), achieved/realized productivity may remain below the potential for some time (resulting in carbon losses) unless aided by human intervention (Iverson and Prasad, 1998; Sohngen *et al.*, 1998; Iverson *et al.*, 1999). Pitelka and Plant Migration Workshop Group (1997) point out, however, that increases in weed species that take advantage of human mobility may be an adverse effect.

Changes in forest distribution as a result of climate change are likey to occur subtly and nonlinearly (Davis and Botkin, 1985; Prentice *et al.*, 1993; Neilson and Marks, 1994; Tchbekova *et al.*, 1994; Bugmann *et al.*, 1996; Neilson and Drapek, 1998). Prediction of changes in species distribution is complicated by the current lack of precise predictions of environmental changes themselves (especially precipitation) and responses of species to these changes. Price *et al.* (1999b) show that responses to precipitation are greatly dependent on assumptions made about species parameters and the temporal pattern of rainfall. The most rapid changes are expected where they are accelerated by changes in natural and anthropogenic disturbance patterns (Overpeck *et al.*, 1991; Kurz *et al.*, 1995; Flannigan and Bergeron, 1998).

At the stand level, climate-induced changes in competitive relationships are likely to lead to dieback and replacement of maladapted species, causing changes in stand population and productivity (e.g., Cumming and Burton, 1996; Rehfeldt *et al.*, 1999b). In addition, increases in locally extreme events and disturbances (fires, insects, diseases and other pathogens) developing over different time scales (Wein *et al.*, 1989; Campbell and McAndrews, 1993; Campbell and Flannigan, 2000) also lead to regionally specific increases in mortality and dieback. Resulting changes in age-class distributions and productivity are likely to have short- and long-term impacts on carbon stocks (Kurz *et al.*, 1995; Fleming, 1996; Hogg, 1997, 1999; Fleming and Candau, 1998).

5.6.3.1.2. NPP and NEP

Some modeled responses of forest ecosystems to climate change suggest that forests could increase carbon storage during climate change (Xiao *et al.*, 1997; Prinn *et al.*, 1999). However, recent work has provided new experimental evidence on the response of vegetation uptake (NPP) and ecosystem losses (affecting both NEP and NBP) to observed changes in climate over the past century. The net balance between NPP (usually assumed to increase with warming, though this assumption is challenged by data offered by Barber *et al.*, 2000), heterotrophic respiration (often assumed to increase with warming, though this is challenged, for example, by Giardina and Ryan, 2000, and Liski *et al.*, 1999—at least for mineral soil components) and disturbance releases (often ignored but shown—for example, by Kurz and Apps, 1999—to be important in boreal estimates of NBP) is no longer as clear as the SAR asserts. Present research continues to improve scaling of localized responses (at the stand level, where increases and decreases in NPP and NEP are observed) to the global scale.

Research reported since the SAR confirms the view that the largest and earliest impacts induced by climate change are likely to occur in boreal forests, where changes in weather-related disturbance regimes and nutrient cycling are primary controls on productivity (Kasischke *et al.*, 1995; Kurz *et al.*, 1995; Yarie, 1999). The impacts are exacerbated by characteristic ecosystem time constants (rotation length, mean residence time of SOM, etc.) that are long compared to other forest ecosystems. In boreal forests, recent warming has been shown to change seasonal thaw patterns (Goulden *et al.*, 1998; Osterkamp and Romanovsky, 1999); increase growing season length (Keeling *et al.*, 1996; Myneni *et al.*, 1997); and, if accompanied by summer drought, reduce NPP (Sellers *et al.*, 1997; Barber *et al.*, 2000). These trends are expected to continue on average (see Table 3-10), although short-term modulations will occur (e.g., in association with ENSO events) (Black *et al.*, 2000).

The SAR (Kirschbaum *et al.*, 1996) concludes that in lowland humid tropics, temperatures already are close to optimum ranges for year-round growth. Hence, an increase in temperature as a result of climate change is likely to have a marginal effect on forest processes. However, research in a lowland tropical forest in Costa Rica has shown that the annual growth in six major species of this forest (with markedly different growth rates and life histories), over a period of 13 years, was highly negatively correlated with annual mean minimum (nighttime) temperatures (Clark and Clark, 1999). Although annual tree growth varied among the six species, there was a highly significant interannual coherence of growth among species.

5.6.3.1.3. Insect herbivory, pests, and diseases

Some evidence suggests that insect populations already are responding to climate change (Fleming and Tatchell, 1995). In general, current forecasts of the response of forest insects and other pathogens to climate change are based on historical relationships between outbreak patterns and climate. These forecasts suggest more frequent or longer outbreaks (Thomson and Shrimpton, 1984; Thomson *et al.*, 1984; Mattson and Haack, 1987; Volney and McCullough, 1994; Carroll *et al.*, 1995; Cerezke and Volney, 1995; Brasier, 1996; Roland *et al.*, 1998). Outbreaks also may involve range shifts northward, poleward, or to higher elevations (Williams and Liebhold, 1997). All of these responses will tend to reduce forest productivity and carbon stocks, although the quantitative extent of these changes is hard to predict (see Box 5-10).

5.6.3.1.4. Elevated CO_2

At the time of the SAR, no experiments on intact forest ecosystems exposed to elevated CO_2 had been performed. Since then, several FACE experiments have been implemented and are beginning to show interesting results. In a 13-year-old loblolly pine plantation (North Carolina), CO_2 levels have been maintained at 200 ppm above ambient. After 2 years, the growth rate of the dominant trees increased by about 26% relative to trees under ambient conditions (DeLucia *et al.*, 1999). Litterfall and fine root increment also increased under the CO_2-enriched conditions. Total NPP increased by 25%. The study concludes, however, that stimulation is expected to saturate not only because each forest stand tends toward its maximum carrying capacity (limited by nutrient capital) but also because plants may become acclimated to increased CO_2 levels.

Research on CO_2 fertilization, however, has taken place only over a short fraction of the forest ecosystem's life cycle. Questions of saturation of response can be addressed through longer term experiments on tree species grown under elevated CO_2 in open-top chambers under field conditions over several growing seasons (Norby *et al.*, 1999). A review of such experiments by Norby *et al.* (1999) found that the evidence shows continued and consistent stimulation of photosynthesis, with little evidence of long-term loss of sensitivity to CO_2; the relative effect on aboveground dry mass was highly variable but greater than indicated by seedling studies, and the annual increase in wood mass per unit of leaf area increased. Norby *et al.* (1999) also found that leaf nitrogen concentrations were lower in CO_2-enriched trees, but not as low as seedling studies indicated, and the leaf litter C/N ratio did not increase. In the majority of CO_2 chamber experiments, the decrease in the percentage of nitrogen in litter was matched by an increase in the percentage of lignin. Moore *et al.* (1999) have suggested, however, that lower litter quality caused by CO_2 fertilization may have offset the expected temperature-induced increase in decomposition. A longer term perspective still is needed because long-term trends cannot be extrapolated directly from relatively short-term experiments on individual trees (Idso, 1999). Field experiments on elevated CO_2 provide inconclusive evidence at this time to predict overall changes in carbon storage in forests.

5.6.3.1.5. Climate-induced changes in variability and weather extremes

The net rate of carbon storage in forests varies as a result of interannual variability in rainfall, temperature, and disturbance regimes. For example, transient simulations with the Terrestrial Ecosystem Model (TEM) suggest that forests of the Amazon basin during hot, dry El Niño years were a net source of carbon (of as much as 0.2 Gt C yr^{-1}), whereas in other years they were a net sink (as much as 0.7 Gt C yr^{-1}) (Tian *et al.*, 1998). Notably, source and sink strength varied across the basin, indicating regional variation. Similarly for the United States, the net rate of carbon storage varied from a net source of about 0.1 Gt C yr^{-1} to a net sink of 0.2 Gt C yr^{-1} (Schimel *et al.*, 2000). The rate of carbon accumulation in undisturbed forests of the Amazon basin reported by Tian *et al.* (1998) was approximately equal to the annual source from deforestation in the same area. For the United States, Houghton *et al.* (1999)—using inventory data—estimate a sink of 0.35 Gt yr^{-1} during the 1980s as a result of forest management and regrowth on abandoned agricultural lands, whereas Schimel *et al.* (2000) indicate that 0.08 Gt yr^{-1} stored from 1980 to 1993 could have been a result of carbon fertilization and climate effects. Thus, for the United States, the rate of carbon accumulation in forest regrowth on abandoned agricultural and harvested managed forest lands appears to be as large as or larger than the direct effects of CO_2 and climate (Schimel *et al.*, 2000). The SAR (Kirschbaum *et al.*, 1996) suggests that modeled trends described above are unlikely to continue under projected climate change and high elevated CO_2 concentrations. However, climatic variability (including ENSO events) is likely to increase under a changed climate (see Table 3-10), which may increase interannual variation in regional carbon uptake—as demonstrated by Tian *et al.* (1998) and Schimel *et al.* (2000).

5.6.3.2. Responses and Impacts: Timber and Non-Wood Goods and Services

5.6.3.2.1. Response to locally extreme events

In addition to fire and insect predation, other episodic losses may become increasingly important in response to locally extreme weather events. For example, in Europe, wind-throw damage has appeared to increase steadily from negligible values prior to about 1950; wind-throws exceeding 20 million m^3 have occurred 10 times since then (UN-ECE/FAO, 2000b). Losses in 1990 and 1999 were estimated at 120 million and 193 million m^3, respectively—the latter equivalent to 2 years of harvest and the result of just three storms over a period of 3 days (UN-ECE/FAO, 2000b). The unusual 1998 icestorms in eastern North America caused heavy damage to infrastructure as well as large areas of forest, the extent of which is still being assessed (Irland, 1998).

Financial returns to forest landowners decline if these episodic events (including fire and insect predation) increase (Haight *et al.*, 1995), although the impact on global timber supply of such episodic losses is unlikely to be significant. Local effects on timber and non-wood goods and services may be significant, although timber loss may be ameliorated through salvage logging. Moreover, the location of the loss is particularly important for non-wood products and services. For example, forest fires in recreational areas are known to have an impact (Englin *et al.*, 1996).

5.6.3.2.2. *Industrial timber*

Recent industrial timber studies link equilibrium ecological models to economic models to measure market impacts to potential climate change and include adaptation (Joyce *et al.*, 1995; Perez-Garcia *et al.*, 1997; Sohngen and Mendelsohn, 1998; McCarl *et al.*, 2000; Sohngen *et al.*, 2000). Conclusions from regional studies are similar to those in the SAR (Solomon *et al.*, 1996) for temperate regions, suggesting that wood supply in these regions will not be reduced by climate change. Studies in the United States that consider only changes in forest growth find small negative or positive impacts on timber supply (Joyce *et al.*, 1995; McCarl *et al.*, 2000). Studies that consider growth effects and species redistribution effects (dieback of existing forests, followed by redistribution), as well as alternative economic scenarios, find that welfare economic impacts in U.S. timber markets could change by –1 to +11% (Sohngen and Mendelsohn, 1998). Generally, consumers are predicted to benefit from increased supply and lower prices. Producers in some regions of the United States may lose because of lower prices and dieback, although productivity gains could offset lower prices (Sohngen, 2001).

In contrast to the SAR (Solomon *et al.*, 1996), more recent global market studies suggest that climate change is likely to increase global timber supply and enhance existing market trends toward rising market share in developing countries (Perez-Garcia *et al.*, 1997; Sohngen *et al.*, 2000). One study that uses TEM (Melillo *et al.*, 1993) to predict growth changes finds that global timber growth rises, global timber supply increases, prices fall, and consumers and mill owners benefit (Perez-Garcia *et al.*, 1997). Landowners in regions where increased timber growth does not offset lower prices perceive losses. A study using the BIOME3 model (Haxeltine and Prentice, 1996) suggests that producers in temperate and boreal forests could be susceptible to economic losses from short-term dieback effects and lower prices, although long-term (>50 years) supply from these regions is predicted to increase (Sohngen *et al.*, 2000). Alternatively, studies that do not consider global market forces, timber prices, or adaptation predict that supply in boreal regions is likely to decline (Solomon and Leemans, 1997).

5.6.3.2.3. *Recreation and non-wood forest products*

There is considerably less published literature available to assess the effects of climate change on nonmarket services from forests (Wall, 1998), including recreation and non-wood forest products. Climate change is likely to have direct effects on forest-based recreation. For example, lengthening of the summer season may increase forest recreation (Wall, 1998). Changes in the mean and variance in daily temperature and precipitation during peak seasons will affect specific activities differently, however. In the United States, some studies suggest that higher temperatures may negatively affect camping, hiking, and skiing but positively affect fishing (Loomis and Crespi, 1999; Mendelsohn and Markowski, 1999; Joyce *et al.*, 2000).

Climate change also will have indirect effects on forest recreation. For example, changes in the structure and function of natural forests that are used for recreation could alter visitation patterns by causing users to substitute alternative sites. Alternatively, some recreational industries may have large adaptation costs—for example, snow-making costs in skiing areas may increase (Irland *et al.*, 2001). In addition, institutional factors are likely to play a strong role in mediating the response of recreation because much of it occurs on public, natural forests where adaptation may be less economically feasible. Many forest-based activities, such as hunting, rely on management decisions by agencies that will have to adapt to climate change as well (Brotton and Wall, 1997).

5.6.4. *Adaptation Options and Vulnerability of Goods and Services*

5.6.4.1. *Adaptation in Timber and Non-Wood Goods and Services*

In markets, prices mediate adaptation through land and product management. At the regional and global scales, the extent and nature of adaptation will depend primarily on prices, the relative value of substitutes, the cost of management, and technology (Joyce *et al.*, 1995; Binkley *et al.*, 1997; Perez-Garcia *et al.*, 1997; Skog and Nicholson, 1998; Sohngen and Mendelsohn, 1998). On specific sites, changes in forest growth and productivity will constrain, and could limit, choices of adaptation strategies (Lindner, 1999, 2000).

Forest management has a history of long-term decisions under uncertain future market and biological conditions (e.g., prices, pest infestations, or forest fires). Most adaptation in land management will occur in managed forests; it will include salvaging dead and dying timber, replanting new species that are better suited to the new climate, planting genetically modified species, intensifying or decreasing management, and other responses (e.g., Binkley, 1988; Joyce *et al.*, 1995; Perez-Garcia *et al.*, 1997; Sohngen and Mendelsohn, 1998; Lindner, 1999). Climate change is not likely to cause humans to convert highly productive agricultural land to forests (McCarl *et al.*, 2000), even with afforestation incentives (e.g., the Kyoto Protocol) (Alig *et al.*, 1997). Adaptation estimates are sensitive to assumed rates of salvage and choices made in regenerating species during climate change (Sohngen and Mendelsohn, 1998; Lindner, 1999). In product management, adaptation includes substituting species in the production process for solid wood and pulpwood products, shifting harvests from one region to another, and developing new technologies and products, such as wood products manufactured with adhesives (McCarl *et al.*, 2000; Irland *et al.*, 2001). In some cases, producers may need to adapt to changes in wood quality (Gindl *et al.*, 2000).

In addition to adaptation through traditional management, agroforestry, small woodlot management, and windrows (or shelterbelts) could provide numerous adaptation options for maintaining tree cover and fuelwood supplies in developing countries. Afforestation in agroforestry projects designed to mitigate climate change may provide important initial steps towards adaptation (Sampson *et al.*, 2000). Although agroforestry is not expected to play a large role in global industrial timber supplies, it may have important regional implications.

Recreation users can adapt by substituting recreational sites as forests respond to climate change, but this will have impacts on recreational industries. Substitution will be easier in regions where transportation networks are well established. Adaptation will be more difficult in regions where recreation depends on particular forest structures (e.g., old-growth forests) that are negatively affected by climate change and for which there are no close substitutes (Wall, 1998).

5.6.4.2. *Vulnerability Associated with NPP, NEP, and NBP*

Given their inertia, forested systems may exhibit low vulnerability and low climate sensitivity, unless drought and disturbance are driving factors (Peterken and Mountford, 1996). Extant forests may persist and appear to exhibit low vulnerability and low climate sensitivity (see Section 5.2), but they may be climate sensitive in ways that are not immediately apparent. Thus, their vulnerability may occur as a reduction in quality (degradation) even where the forest persists as an entity. Increases in disturbances, however, may lead to rapid structural changes of forests, with replacement by weedy species (Overpeck *et al.*, 1990). For example, an increase in area burned by fire or destroyed by invading insects and disease could rapidly undergo changes in species composition, successional dynamics, rates of nutrient cycling, and many other aspects of forest ecosystems—with impacts on goods and services provided by these forests. From this perspective, in the context of all of the direct and indirect impacts of climatic change and their interactions, the potential vulnerability of forests is high.

The main processes that determine the carbon balance of forests—photosynthesis, plant and heterotrophic respiration, and disturbance releases—are regulated by different environmental factors. Carbon assimilation is a function mainly of available light, temperature, nutrients, and CO_2, and this function shows a saturation characteristic. Respiration is a function of temperature; it increases exponentially with temperature. Thus, close correlation between GPP and NPP does not exist (allocation plays an important role). Similarly, there is no strict correlation between NPP and NEP (high NEP is possible at high and low rates of assimilation and respiration). Disturbances such as harvest or fire export carbon from forests, bypassing respiration; thus, NBP is not expected to correlate with NEP and NPP. Photosynthesis, plant and heterotrophic respiration, and disturbance are vulnerable in different ways and are quite sensitive to climate change and other global change forces.

5.6.4.3. *Vulnerability of Unmanaged Systems*

Increases in disturbances such as insect infestations and fires can lead to rapid structural and functional changes in forests (species composition, successional dynamics, rates of nutrient cycles, etc.), with replacement by weedy species. The effects of these vulnerabilities in unmanaged systems on goods and services vary. They are not likely to have large effects on market products given that unmanaged forests constitute an increasingly small portion of timber harvests. They could have large impacts on local provision of timber products, NWFP, and fuelwood. Local services from forests could be highly vulnerable, particularly if the services are tied to specific forest functions that change (e.g., hiking in old-growth forests).

5.6.4.4. *Vulnerability in Managed Systems*

Managed systems are vulnerable to direct impacts on NPP (related to volume production) and species composition (related to timber quality), as well as market forces. The global supply of market products exhibits low vulnerability because timber markets have high capacity to adapt to change. Temperate and boreal producers are vulnerable to dieback effects and lower prices caused by potential global increases in timber growth. Managed forests in subtropical regions have low vulnerability given their high growth rates and short rotation periods, which provide multiple opportunities to adapt to changes.

5.7. Lakes and Streams

Lake and stream ecosystems include many familiar places: large and small lakes, permanent and temporary ponds, and streams—from tiny, often temporary rivulets at headwaters to powerful floodplain rivers discharging from our continents.

Freshwaters (lakes and rivers), which are so valuable for the sustainability of life as we know it, constitute only 0.0091% of the Earth's surface waters by volume. This is a great deal less than 0.5% for groundwater and 97.3% for oceans (Cole, 1994). Lakes and rivers are used intensively for recreation and are aesthetically valued. These values include fishing, hunting, swimming, boating, skating, and simply enjoying the view.

This section focusses on physical and biological processes and how they affect goods and services from lakes and streams that are regulated by these processes. The emphasis is on food, carbon, and biodiversity. Hydrology and water supply are covered in Chapter 4; oceans and coastal systems are covered in Chapter 6. Hydrological goods and services are covered largely in Chapter 4, but are considered here as an interactive influence of climate change on inland waters. Wetlands are covered in Section 5.8.

5.7.1. *Status of Goods and Services*

Products from lakes and streams include fisheries (fish, amphibians, crustaceans, and mollusks) and aquaculture; services include biodiversity, recreation, aesthetics, and biogeochemical cycling. A detailed list of services from freshwater ecosystems includes many items that are undervalued in economic terms and often are unrecognized and unappreciated—for example, their role in the carbon cycle.

Reported catches in freshwater and inland fisheries were 7.7 Mt of biomass in 1996 (FAO, 1999b) and 15.1 Mt from aquaculture. China accounts for 80% of aquaculture production. Actual landings in capture fisheries are believed to be two to three times larger owing to nonreporting (FAO, 1999c). About 100 fish species are reported in world catches—primarily cyprinids, cichlids, snakeheads, catfish, and barbs. Asia and China report the largest catches; Africa is second; North America ranks relatively low. River and large reservoir fisheries are important.

Recreational fish catches are included in the foregoing world catches and were reported to be 0.48 Mt in 1990 (FAO, 1992). This is likely to be an underestimate because only 30 (of 200) countries reported recreational catches. Recreational fishing is increasing in developed and developing countries. The number of anglers is estimated at 21.3 million in 22 European countries, 29.7 million in the United States, and 4.2 million in Canada (U.S. Fish and Wildlife Service, 1996; Department of Fisheries and Oceans, Canada, 1998). Total expenditures for recreational angling are in the billions of dollars—for example, $38 billion in the United States in 1996 (U.S. Fish and Wildlife Service, 1996) and $4.9 billion in Canada in 1995.

Freshwaters are known for high biodiversity and endemism owing to their island-like nature, which leads to speciation and reduces invasions of competitors and predators. Of about 28,000 fish species known on Earth, 41% are freshwater species and 58% are marine, of which 1% spend part of their lives in freshwater (Moyle and Cech, 1996). Individual east African Rift Valley lakes contain species flocks of almost 250 cichlid species. In Lake Baikal in Russia, 35% of the plants and 65% of the animals are endemic (Burgis and Morris, 1987). At the global level, biodiversity in many lakes has been decreasing in recent decades, with many species becoming extinct (Naiman *et al.*, 1995b). The trend is likely to continue, with many species that now are listed as endangered or threatened becoming extinct (IUCN, 1996). The causes for these extinctions are likely to be related to the many pressures listed below.

Inland waters play a major role in biogeochemical cycling of elements and compounds such as carbon, sulfur, nitrogen, phosphorous, silica, calcium, and toxic substances. The general roles are storage, transformation, and transport (Stumm and Morgan, 1996). Storage is important because sediments and associated minerals accumulate in the bottom sediments of lakes, reservoirs, and floodplains. Transformation includes organic waste purification and detoxification of various human created compounds such as insecticides. Water movement redistributes these spatially.

Of special interest here is the role of freshwaters in carbon storage and CO_2 and methane (CH_4) release. Organic carbon from primary production in lakes and adjacent riparian lands accumulates in sediments; estimates are that 319 Mt yr^{-1} are buried in the 3.03 million km^2 of small and large lakes, reservoirs, and inland seas worldwide (USGS, 1999; see also Stallard, 1998). This estimate excludes amounts for peatlands (96 Mt yr^{-1}). It is three times greater than estimates for the ocean in absolute terms (97 Mt yr^{-1}) even given the relatively small area of inland waters. Lakes also are a source of GHGs. Lakes become supersaturated with dissolved CO_2, and net gas exchange is from the lake to the atmosphere (Cole *et al.*, 1994). For example, during summer in Lake Pääjärv, Finland, the amount of carbon from respiration in the water column was greater than that produced by phytoplankton and sedimentation combined (Kankaala *et al.*, 1996). Pulses of CO_2 from the water column and CH_4 from sediments are released to the atmosphere during spring and fall mixing of the water column of dimictic lakes (Kratz *et al.*, 1987; Riera *et al.*, 1999; Kortelainen *et al.*, 2001). Methane releases, especially from the littoral zone, can be significant in lakes and reservoirs (Fearnside, 1995, 1997; Alm *et al.*, 1997a; Hyvönen *et al.*, 1998).

Hydroelectric power plants generally are assumed to emit less CO_2 than fossil fuel plants. However, a hydroelectric reservoir may contribute more to GHGs over 100 years of operation than a fossil fuel plant that produces an equivalent amount of electricity (Fearnside, 1997). Emissions are likely to be high in the first few decades and then decrease. This is exemplified by a Brazilian hydroelectric reservoir that is simulated to release a large quantity of CO_2 during the first 10 years after filling (*ca.* 5–27 Mt CO_2 gas yr^{-1}) but relatively low amounts from years 30 to 100. Releases of CH_4 were simulated to be high for at least 100 years (about 0.05–0.1 Mt CH_4 yr^{-1}, with actual estimates for 1990 of 0.09 Mt) (Fearnside, 1995, 1997).

5.7.2. *Pressures on Goods and Services*

People expect drinkable, swimmable, boatable, and fishable freshwaters. However, rivers and lakes have many pressures. These pressures include land and water use for urbanization, agriculture, and aquaculture; hydrologic engineering structures such as dams, dykes, channelization, and construction of drainage and connecting canals; water extractions for industry, drinking water, irrigation, and power production; water pollution with toxics, excess nutrients, and suspended sediments; capture fisheries; and invasion of exotics. UV-B is an interactive pressure with climate change owing to the change in water clarity influenced by drought. Under these conditions, UV-B penetrates farther and causes more damage in clear waters than in murky waters.

Human demand for water in many areas will increase more rapidly with climate change and population increases and leave fewer waters unmodified by water projects (see Chapter 4). Water projects interact with many aspects of climate change as related to natural resource and environmental management. Additional dams will increase the difficulty of managing migratory fish populations in streams. Sedimentation that occurs above dams will reduce downstream transport of sand, sediments, and toxic substances. Lakes and reservoirs with increased water withdrawals will reduce the suitability of the littoral zone for fish spawning and nursery areas. Diversions of water by canals, ships, or pipes will transport exotics into new watersheds and confound biodiversity and exotics issues.

5.7.3 *Responses of Lakes and Streams and Impacts on their Goods and Services*

Many responses of lakes and streams to climate change were documented in the SAR by Arnell *et al.* (1996), and more have been added by Cushing (1997) and Domoto *et al.* (2000). Responses include warming of waters; reductions in ice cover; reduction in dissolved oxygen in deep waters; changes in the interaction between waters and their watersheds; changes in biogeochemical cycling; greater frequencies of extreme events, including flood and drought; changes in growth, reproduction, and distribution of organisms; and poleward movement of climate zones for organisms. Only a few are mentioned in the following subsections.

5.7.3.1. *Physical Conditions*

5.7.3.1.1. *Ice cover*

To date, the only limnological properties measured or simulated at a global scale are lake and river ice phenologies. Ice-cover durations for inland waters have decreased over the entire northern hemisphere (Magnuson *et al.*, 2000) (see Figure 5-6). Change from 1846 to 1995 averaged 8.7 days later freeze and 9.8 days earlier break-up; these changes correspond to a 1.8°C increase in air temperature. Lake Suwa in Japan, Tornionoki River in Finland, and Angara River in Siberia have longer records; ice phenologies have been changing in the direction of warming since about the early 1700s, but at slower rates than during 1846–1995. Interannual variability in freeze dates, thaw dates, and ice duration are increasing (Kratz *et al.*, 2001). Interannual variability was greater from 1971 to 1990 than from 1951 to 1970 in 184 lakes in the northern hemisphere.

For the Baltic Sea, which contains freshwater communities near shore and to the north, simulation until 2050 reduced the extent of ice cover from 38% of the area to 10%, and simulation until 2100 reduced ice cover to zero (Haapala and Leppäranta, 1997). For U.S. lakes, ice-on date, ice-off date, ice duration, ice thickness, and the continuity of ice cover all responded to a $2xCO_2$ change (Fang and Stefan, 1998). Mean duration at 209 locations distributed evenly across the country declined by 45 days (60%), and ice thickness by 21 cm (62%). Simulated

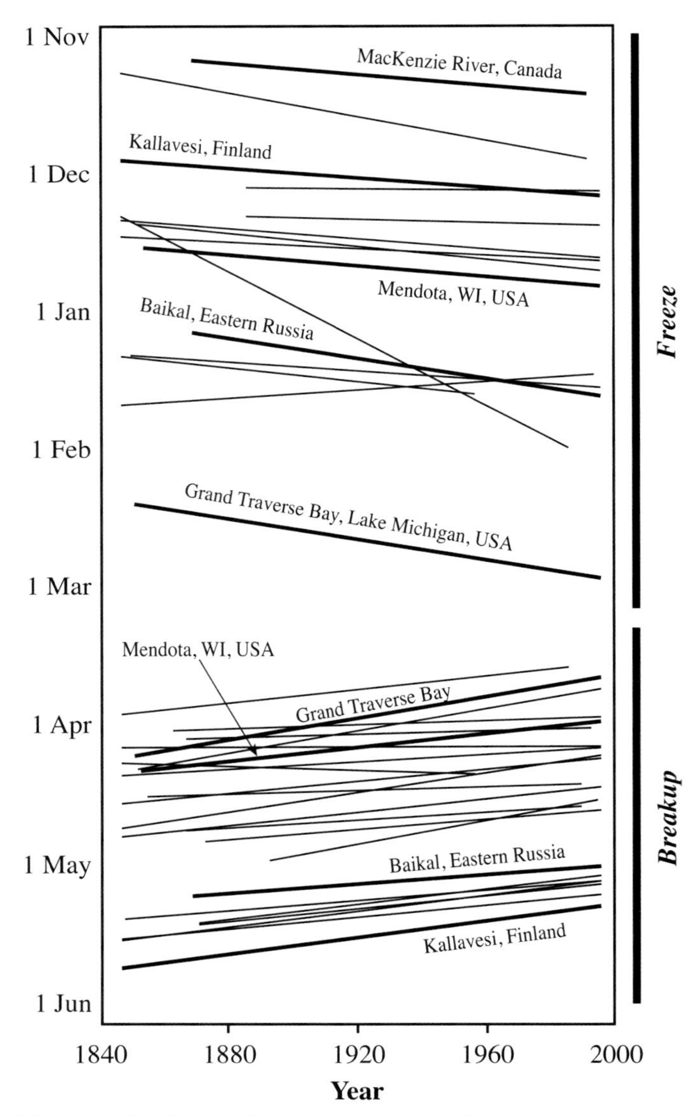

Figure 5-6: Historical trends in freeze and breakup dates of lakes and rivers in the Northern Hemisphere; 37 of the 39 trend slopes are in the direction of warming (modified from Magnuson *et al.*, 2000).

profiles of deepwater oxygen concentrations and water temperatures also changed dramatically in lakes across the United States (Fang and Stefan, 1997).

Lake ice influences biogeochemical cycling, including gas exchange with the atmosphere, fish habitat availability (through changes in pH and dissolved oxygen), biodiversity, and seasonal succession (Arnell *et al*., 1996; Cushing, 1997). Decreased ice cover should reduce winterkill of fish.

Ice break-up date influences the productivity of diatoms beneath the ice of Lake Baikal (Granin *et al.,* 2001). Seasonal succession of phytoplankton in small lakes is altered partly because of variation in the incoculum of algae beneath the ice in different years (Adrian and Hintze, 2001). The mechanism for enhanced diatom productivity in Lake Baikal is a combination of nutrient enrichment immediately beneath the ice owing to the extrusion of salts as the water freezes and vertical mixing that results from the greater density of the water with extruded salts immediately beneath the ice. Reductions in ice thickness and increases in snow would be expected to reduce diatom production.

River ice plays an important biological role by regulating flow aeration and oxygen concentrations under the ice (Prowse, 1994; Chambers *et al*., 1997); dissolved oxygen can approach critical levels for river biota (e.g., Power *et al*., 1993). Decreases in the duration of the river-ice season or increases in the size and frequency of open-water sections where re-aeration can occur will reduce anoxia. Warmer winters favor the formation of mid-winter break-ups produced by rapid snowmelt runoff, particularly those initiated by rain-on-snow events. Such events significantly impact benthic invertebrate and fish populations where late-season break-ups are the norm (Cunjak *et al*., 1998).

Ecological impacts will be influenced most by changes in break-up timing and intensity. Physical disturbances from break-up scouring and flooding influence nutrient and organic matter dynamics, spring water chemistry, and the abundance and diversity of river biota (Scrimgeour *et al*., 1994; Cunjak *et al*., 1998). Ice-induced flooding supplies the flux of sediment, nutrients, and water that are essential to the health of freshwater delta ecosystems (e.g., Lesack *et al*., 1991; Prowse and Conly, 1998). Even the mesoscale climate of delta ecosystems depends on the timing and severity of break-up flooding (Prowse and Gridley, 1993).

Climate-induced change in ice cover or the timing and severity of its break-up will affect the movement and deposition of sediment and associated contaminants. Stable cover leads to deposition of sediment particles that would remain in suspension without ice cover. Environmental contaminants with an affinity for adsorption to fine-grained sediments are more likely to be deposited during ice-covered, low-flow season than during open-water season. The break-up plume contains the winter-long accumulation of contaminated sediment (Milburn and Prowse, 1996). Sediment deposition zones often are where biological productivity is greatest, such as on river deltas (Milburn and Prowse, 1998). Because growth rates of aquatic organisms in cold regions are low and because higher trophic level animals are longer lived, potential exists for greater life-long accumulation of contaminants in these aquatic ecosystems than in more temperate regions.

River ice is a key agent of geomorphologic change; it is responsible for creating numerous erosional and depositional features within river channels and on channel floodplains (e.g., Prowse and Gridley, 1993; Prowse, 1994). During break-up, these processes cause channel enlargement and removal and succession of riparian vegetation. Climatic conditions that alter the severity of such events will greatly influence river morphology and riparian vegetation.

5.7.3.1.2. Direct use of water

Intensification in direct uses of water will negatively influence "natural" waters and could seriously reduce services provided by these ecosystems. Some values and services of lakes and rivers would be degraded by human responses to their greater demands for water for direct human use through dam construction, dyke and levee construction, water diversions, and wetland drainage (Postel and Carpenter, 1997). Such activities can alter water temperatures, recreation, pollution dilution, hydropower, transportation, and the timing and quantity of river flows; lower water levels; destroy the hydrologic connection between the river and the river floodplain; reduce natural flood control, nutrient and sediment transport, and delta replenishment; eliminate key components of aquatic environments; and block fish migration. At risk are aquatic habitat, biodiversity habitat, sport and commercial fisheries, waterfowl, natural water filtration, natural floodplain fertility, natural flood control, and maintenance of deltas and their economies (Ewel, 1997).

Water-level increases as well as decreases can have negative influences on lakes and streams. A recent example is from Lake Baikal, where the combined influences of a dam on the outlet and increased precipitation has increased water level by 1.5 m and decreased biodiversity and fishery production (Izrael *et al*., 1997; ICRF, 1998).

5.7.3.2. Fisheries and Biodiversity

Potential effects of climate change on fisheries are documented in the SAR by Everett *et al*. (1996); freshwater fisheries in small rivers and lakes are considered to be more sensitive to changes in temperature and precipitation than those in large lakes and rivers. Changes in temperature and precipitation resulting from climate change are likely to have direct impacts on freshwater fisheries through changes in abundance, distribution, and species composition. Because current exploitation rates tend to be high or excessive, any impact that concentrates fish will increase their catchability and further stress the population.

Assessing the potential impacts of climate change on aquaculture is uncertain, in part because the aquaculture industry is mobile and in a period of rapid expansion. There is no question that large abundances relative to traditional wild harvests can be produced in small areas and that in many cases these sites can be moved to more favorable locations. Changes in groundwater may be especially significant for aquaculture. In tropical areas, crustaceans are cultured in ponds (Thia-Eng and Paw, 1989); fish frequently are cultured in cages. Commonly cultured species such as carp and tilapia may grow faster at elevated temperatures, but more food is required and there is an increased risk of disease. Temperature is a key factor affecting growth, but other factors relating to water quality and food availability can be important. Modeling studies suggest that for every 1°C average increase, the rate of growth of channel catfish would increase by about 7%, and the most favorable areas for culture would move 240 km northward in North America (McCauley and Beitinger, 1992; see Figure 5-7, right panel). The southern boundary also would move northward as surface water temperatures increase, perhaps exceeding lethal ranges on occasion. Growth would increase from about 13 to 30°C and then fall off rapidly as the upper lethal limit of about 35°C is reached (McCauley and Beitinger, 1992). Thus, aquaculture for traditional species at a specific location may have to switch to warmer water species.

Warmer conditions are more suitable for warmer loving flora and fauna and less suitable for cold-loving flora and fauna. Warmer temperatures, however, lead to higher metabolic rates, and if productivity of prey species does not increase, reductions in growth would occur at warmer temperatures (Arnell *et al.*, 1996; Magnuson *et al.*, 1997; Rouse *et al.*, 1997). Rates of natural dispersal across land barriers of less mobile species poleward or to higher altitudes are not likely to keep up with rates of change in freshwater habitats. Species most affected would include fish and mollusks; in contrast, almost all aquatic insects have an aerial life history stage, thus are less likely to be restricted. Some streams have a limited extent and would facilitate only limited poleward or altitudinal dispersal. This could be especially problematic as impoundments restricting movements of organisms increase in number. Coldwater species and many coolwater species would be expected to be extirpated or go extinct in reaches where temperatures are at the warmer limits of a species range. Many lakes do not have surface water connections to adjacent waters, especially in headwater regions where interlake movement would be limited without human transport of organisms across watershed boundaries.

Exotics will become a more serious problem for lake and stream ecosystems with warming. In the northern hemisphere, for example, range extensions occur along the northern boundaries of species ranges and extinctions occur along the southern boundaries, in natural waters and in aquaculture operations (Arnell *et al.*, 1996; Magnuson *et al.*, 1997). In addition, loss of habitat for biodiversity will result from warmer, drier conditions interacting with increases in impoundment construction.

Distributions of fish are simulated to move poleward across North America and northern Europe. In northern Europe, Lehtonen (1996) forecasts a shrinking range for 11 coldwater species and an expanding range for 16 cool- and warmwater

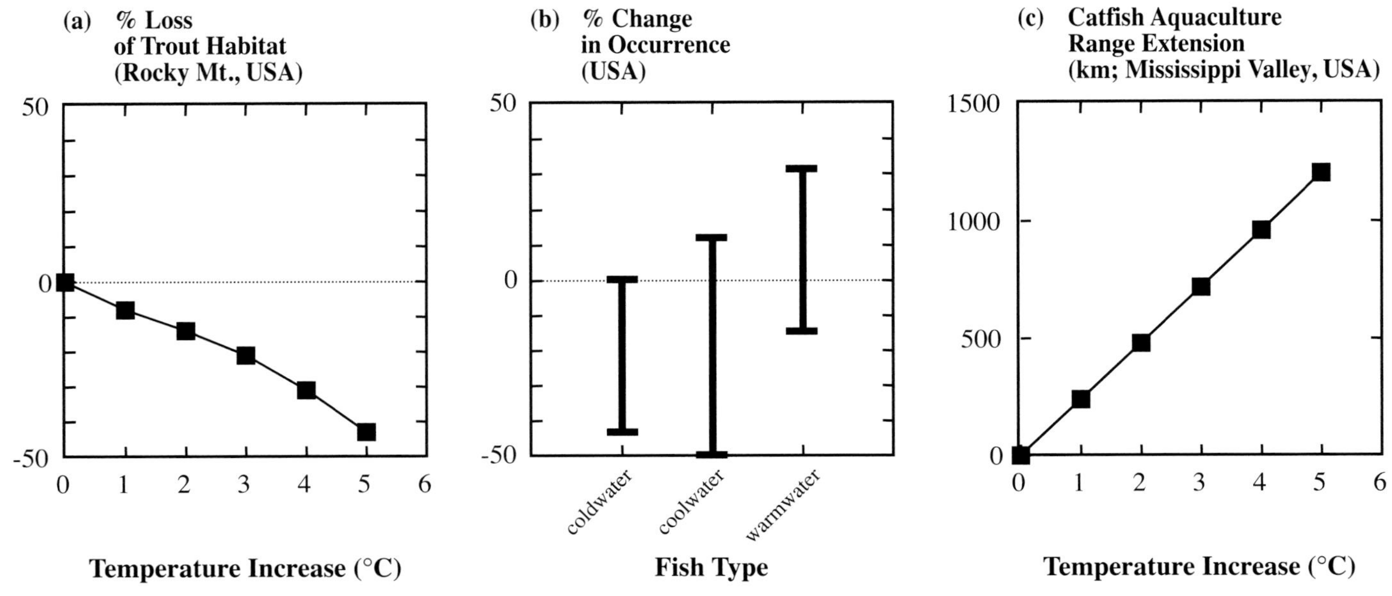

Figure 5-7: Simulated changes in thermal habitat for fish in the continental United States. Left panel is modified from Keleher and Rahel (1996). Center panel is modified from Mohseni and Stefan (2000) and Fang *et al.*(1998). Right panel is modified from McCauley and Beitinger (1992). Simulation is for a $2xCO_2$ climate from the Canadian Climate Model and represents air temperature increases of 3–6.5°C in different parts of the United States. Coldwater fish include trout and salmon; coolwater fish include yellow perch, walleye, northern pike, and white sucker; and warmwater fish include sunfish (black basses, bluegill, pumpkinseed) and common carp.

fish. In simulations for Finland by Lappalainen and Lehtonen (1995), lake whitefish lost habitats progressively toward the north, whereas brown trout did better, at least in the north. Coolwater fish were forecast to spread northward through the country. In simulation studies, boundaries of individual warmwater species ranges in were projected to move northward by 400–500 km in Ontario, Canada (Minns and Moore, 1995), and southern boundaries of coldwater fishes were projected to move 500–600 km northward in the southeastern United States (estimated from Fang *et al.,* 1998). In simulations based on an elevated CO_2 scenario using the Canadian Climate Centre model, warmwater fish were projected to benefit in shallow eutrophic and mesotrophic lakes around the United States, owing to reduction in winterkill, but habitable lakes and streams for coolwater fish and especially coldwater fish were projected to decline, owing to summer kill (Fang *et al.,* 1998; Mohseni and Stefan, 2000). Habitat changes from various studies over large regions (see Figure 5-7, center panel) projected 0–43% reductions for coldwater species, 50% reductions to 12% increases for coolwater species, and 14% reductions to 31% increases for warmwater fish. Changes differ among ecosystem types and areas, depending on latitude and altitude.

A dramatic picture of the regional decline in trout habitat in the Rocky Mountain region of the western United States is provided by Keleher and Rahel (1996) (see Figure 5-7, left panel). Even a 1°C increase in mean July air temperatures is simulated to decrease the length of streams inhabitable by salmonid fish by 8%; a 2°C increase causes a reduction of 14%, a 3°C increase causes a 21% decline, a 4°C increase causes a 31% reduction, and a 5°C increase causes a 43% reduction. There also is likely to be a increased fragmentation of inhabitable areas for the North Platte River Drainage in Wyoming (Rahel *et al.*, 1996).

5.7.3.3. Biogeochemical Cycling and Pollution

Understanding of interactions of climate change with other human-caused pressures on lakes and streams is still in its infancy. No simulation models have been developed to assess the combined effects of these pressures on these systems.

5.7.3.3.1. Eutrophication

Nutrient cycling would be altered by climate change in ways that could exacerbate existing water quality problems such as eutrophication (see Figure 5-8). Eutrophication of lakes results when nutrient inputs from catchments and recycling from bottom sediments are large. The result is excessive production of algae; blue-green algae reduce water quality for recreation and drinking. Deep coldwater habitats become anoxic, owing to greater rates of decomposition of sinking organic mater (Horne and Goldman, 1994).

Interaction between climate change and eutrophication is complex, and projections are somewhat contradictory (see Figure 5-8) because climate-influenced processes have interacting and often opposing effects (Magnuson *et al.*, 1997; Schindler, 1997). Consider, for example, phosphate release from anoxic sediments. In a warmer climate, the longer period of summer stratification would increase the likelihood that anoxia develops below the thermocline (Stefan and Fang, 1993); this would increase the solubility of phosphates in sediment and increase nutrient recycling. At the same time, warmer climates would reduce the duration of ice cover in lakes, which would reduce winter anoxia and decrease sediment phosphate release in winter. This is further complicated by water column stability.

In Lake Mendota, Wisconsin (Lathrop *et al.*, 1999), it is not surprising that one-third of observed year-to-year variation in summer water clarity is associated with variability in runoff (see Figure 5-8); more nutrient input leads to higher populations of phytoplankton, which reduce water clarity. Runoff would be influenced by differences in precipitation and the frequency of extreme rainfall events during autumn, winter, and spring. Precipitation trends differ around the world, and there is evidence for increased frequency of extreme rainfall events that may occur in different seasons of the year (see Chapter 4). Wetter climates or climates with more extreme rainfall events would increase export of nutrients and sediment to lakes and streams; dryer climates or those with more even rainfall would reduce export to lakes and streams. Extreme rainfall events would export more if they occurred at seasons when the earth was bare in agricultural watersheds.

In Lake Mendota, the other two-thirds of the variation in water clarity also is related to climate (see Figure 5-8). One-third is through climatic influences on vertical mixing, where warmer summers lead to greater water column stability, less recycling from deep water, and greater water clarity. The remaining one-third is related to the abundance of herbivorous zooplankton that eat phytoplankton. A warming climate would allow invasion of new species of fish that forage on the herbivorous zooplankton; in Lake Mendota, it probably would be the gizzard shad that lives in reservoirs south of Wisconsin (see Dettmers and Stein, 1996).

Several other complexities would lead to different results, depending on the change in precipitation patterns. For example, in a dryer climate, increases in water residence time (Schindler *et al.*, 1996a) would increase the importance of nutrient recycling within lakes, and storage of nutrients in the sediments of lakes would be reduced (Hauer *et al*., 1997). This is complicated by changes in light penetration that occur if dissolved organic carbon (DOC) inputs are reduced (Magnuson *et al.*, 1997; Schindler, 1997). DOC can reduce light penetration, so a reduction in DOC input would increase light availability and could lead to increases in primary production in deeper water. This influence would depend on whether climate becomes dryer or wetter.

Several empirical and simulation studies support the idea of increased eutrophication with climate change. Results of these studies contradict the expectations of reductions in nutrient loading from catchments in drier climates and greater stability of the water column in warmer climates.

Empirical relations (Regier *et al.*, 1990; Lin and Regier, 1995) suggest that annual primary production by phytoplankton, zooplankton biomass, and sustained yields of fisheries all increase with temperature. Simulations for Lake Erie (Blumberg and Di Toro, 1990) and smaller lakes (Stefan and Fang, 1993) indicate that climate change leads to more eutrophic conditions with respect to loss of oxygen beneath the thermocline in summer. Ogutu-Ohwayo *et al.* (1997) suggest that recent changes in the regional climate of Lake Victoria in Africa may have reduced physical mixing and contributed to increases in deepwater anoxia and thus nutrient recycling contributing to eutrophication.

5.7.3.3.2. Acidification

Acidification of streams and recovery of acidified lakes would be altered by climate changes (Yan *et al.*, 1996; Magnuson *et al.*, 1997; Dillon *et al.*, 2001). In addition to direct atmospheric deposition of acids, sulfates deposited in the catchment are transported to streams during storm events as pulses of acidity. Lakes would receive less buffering materials in dryer climates and more in wetter climates. Lakes high in the landscape that

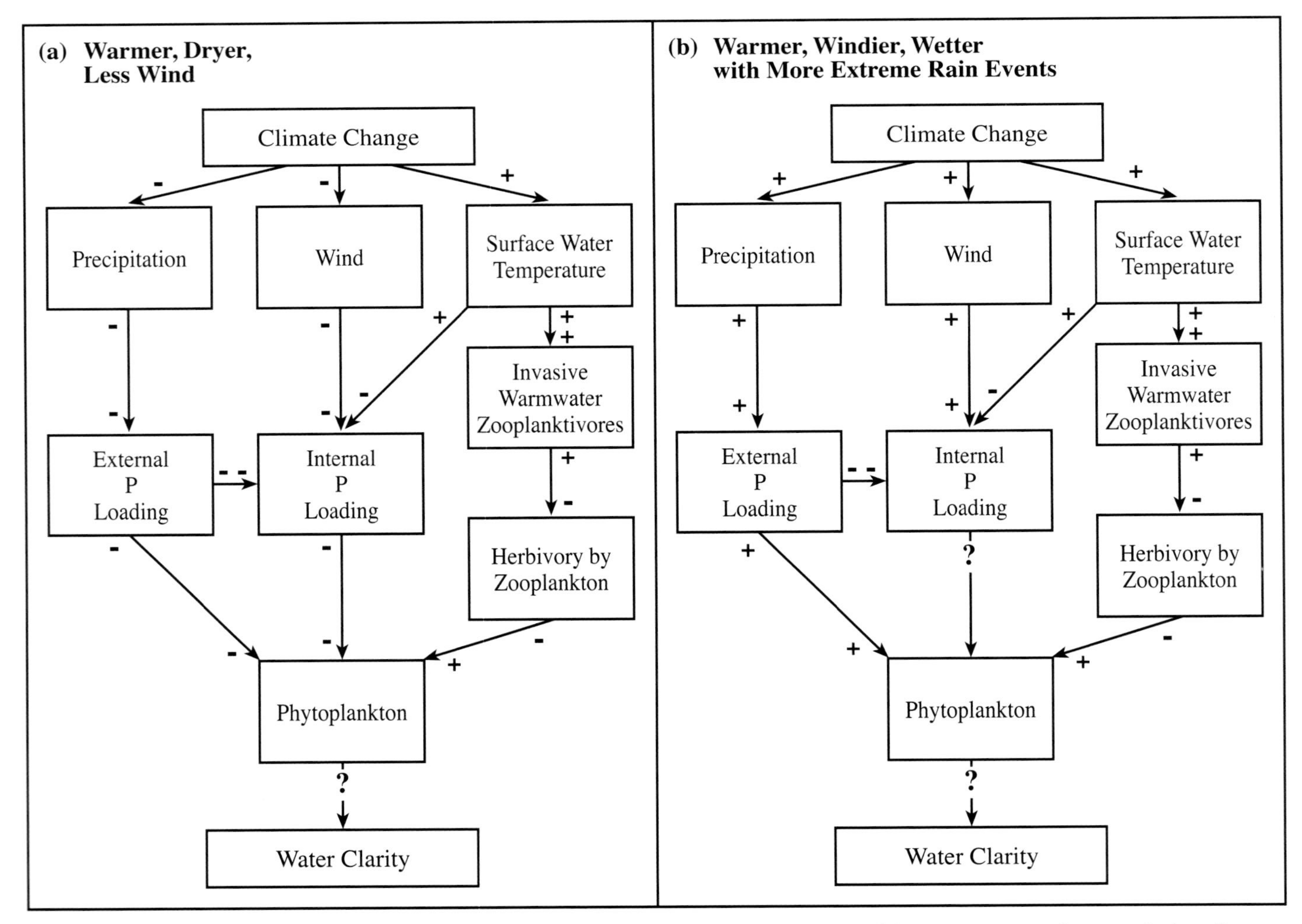

Figure 5-8: Diagram of complex interactions between climate change, watershed and lake processes, and water clarity of a eutrophic lake [modified from Lathrop (1998) and information in Lathrop *et al.* (1999)]. The left panel diagrams a warmer, dryer climate with less wind; the right panel diagrams a warmer, windier, and wetter climate with more extreme rain events. In both cases, the altered climate would be expected to change the water quality of the lake, but the complexity of relations leads to uncertain future water clarities. A "**+**" means an increase and a "**-**" means a decrease in the condition or process; a "**?**" means conflicting expectations. Greater blooms of phytoplankton lead to lower water clarity, and reduced blooms lead to greater clarity. Blooms depend on external and internal loading of phosphorus (P). Dryer climates lead to less external loading, whereas wetter climates or more episodic rains lead to more external loading. Warmer climates lead to warmer surface waters and increased vertical stability in the water column, thus less mixing and internal loading. Warmer waters also allow potential invasion by warmer loving, zooplaktivorous fish that can reduce zooplankton species that in turn reduce algal populations. Windier climates result in increased vertical mixing, thus greater internal loading.

receive less groundwater in dry years would be more vulnerable to acidification (Webster *et al.*, 1996).

5.7.3.3.3. *Toxics*

Climate change would interact with biogeochemical transport and transformation of toxics such as mercury, zinc, and pesticides (see brief reviews in Magnuson *et al.*, 1997; Schindler, 1997; and related physiological understanding in Wood and McDonald, 1997). The expected influences are poorly known as well as complex and variable, depending on the toxin, the organism, and the climate scenario. Enough is known to say that toxic stresses are not independent of climate change.

5.7.3.4. *Other*

5.7.3.4.1. *Recreation*

Winter recreational opportunities would decline with warmer climates. Declines in safe ice conditions would reduce all ice-related activities such as ice fishing, ice skating, ice boating, and snowmobiling on lakes and rivers. Recreational uses of lakes that become more eutrophic are likely to be degraded by lower water clarity and increased blooms of noxious blue-green algae. Nonmarket values for water-based recreation are in direct conflict with greater direct uses of that water and a warmer and drier climate. These values ranged from $3–65 per thousand m^3 of water for fishing, rafting, and river recreation in general in Colorado (Postel and Carpenter, 1997).

UV-B radiation can be harmful to freshwater organisms (Bothwell *et al.*, 1994; Williamson and Zagarese, 1994; Williamson *et al.*, 1996). Absorption of UV-B is lower in clearwater lakes. Reduction of colored DOC entering lakes during drier conditions results in greater transmission and thus greater harm to organisms (Schindler *et al.*, 1996a; Yan *et al.*, 1996; Schindler, 1997; Williamson *et al.*, 1996).

5.7.3.4.2. *Heterogeneity in response*

Responses of lakes and streams to climate change are spatially heterogeneous. Much of the local heterogeneity depends on the type of water body being considered: lake or stream, large or small, shallow or deep, eutrophic or oligotrophic, and so forth. Spatial heterogeneity also results from spatial differences in climate drivers themselves with latitude, altitude, and distance from the coast or a large lake. Inappropriate responses with large expenditures of local resources will result unless the spatial heterogeneity of local responses is understood and can be predicted.

Two ideas that help explain local heterogeneity in responses of lakes and streams to climate change from a landscape perspective are the stream continuum (Vannote *et al.*, 1980; Minshall *et al.*, 1985) and the position of a lake in the landscape (Kratz *et al.*, 1991; Kratz *et al.*, 1997; Magnuson and Kratz, 2000; Riera *et al.*, 2000). Both concepts are geomorphic legacies resulting from the location of the water body in the hydrologic flow field. Headwater streams will be more shielded from warming relative to lowland streams because cool groundwater sources are more important; in forest catchments, they are more shaded from radiation, and in mountain catchments they are at higher, cooler altitudes (Hauer *et al.*, 1997). For lakes, changes in chemical inputs with changes in rainfall will be greater in upland lakes than in lowland lakes. Upland lakes are supplied more by dilute precipitation than by solute-rich groundwater or overland flow. The chemistry of headwater lakes is extremely responsive to climate-driven changes in groundwater inputs because they tend to have smaller volumes and shorter residence times (Krabbenhoft and Webster, 1995).

Differences in the extent of connected wetlands or other sources of DOC contribute to local heterogeneity in responses. Export of DOC to lakes and streams increases in wetter times and decreases during drought. This, in turn, changes light penetration and vertical distribution of solar heating and, in lakes, the depth of thermocline and the relative magnitude of cold and warm thermal habitat for fish (Schindler *et al.*, 1996b). Penetration of UV-B, thus the damaging effects of that radiation (Schindler, 1997), also will differ among lakes. Lakes without large sources of DOC that leach in during wetter periods will respond less to climate changes in these respects.

Shorter term, more stochastic patterns in catchments alter the behavior of lake ecosystems. Consider a drought that increases the likelihood of forest fires in the watersheds of boreal lakes in Ontario (Schindler, 1997). Burned areas typically are patchy and may or may not include the catchment of a given lake. If they do include the lake's watershed, there is an initial increase in the input of solutes to streams and the lake. The lake also is under greater influence of wind mixing with trees gone, which would deepen the thermocline and again alter the relative magnitude of cold and warm thermal habitat for fish.

Predicted responses of threatened anadromous Pacific salmonid stocks in the Columbia basin of the northwestern United States are varied. Climate-related factors that are important to successful reproduction include temperature, the river hydrograph (peak and annual flow), and sedimentation (Neitzel *et al.*, 1991). Based on expected changes in these streams and 60 stocks across the basin, impacts on 23% were judged to be negative, 37% positive, and 40% neutral.

5.7.4. *Adaptation Options*

Human responses to climate change could further exacerbate the negative impact on aquatic ecological systems. For example, human responses to a warmer climate and the variety of precipitation scenarios, as well as human population increases, are likely to place greater demands on freshwaters to meet water needs for drinking, industry, and irrigation, along with an increase in water management projects (see Chapter 4).

Potential results would include fewer free-flowing streams and greater fluctuations in water level. These changes would cause a loss of ecosystem services and products from "natural/ unmodified" lakes and streams. Conflicts between developers and those wishing to reduce development pressure on lakes and streams probably would intensify as freshwater becomes either more scarce or more abundant.

One adaptation to climate change would entail poleward transportation of fish, mollusks, and other less vagil organisms across watershed boundaries to cooler waters (Magnuson *et al.*, 1997). Historical introductions and range expansions of species have resulted in extinction or extirpation of preexisting fauna owing to ecological interactions, especially predation by the invader. Well-known examples include the sea lamprey, alewife, smelt, and zebra mussel to the Laurentian Great Lakes in North America (Mills *et al.*, 1994); the peacock bass in Panama (Zaret and Paine, 1973); and the Nile perch into Lake Victoria (Kaufman, 1992). Other examples being documented include the decline or extirpation of cisco and yellow perch in small Wisconsin lakes in the United States with the arrival of rainbow smelt (Hrabik *et al.,* 1998) and the loss of cyprinids (minnows) and the decline of lake trout with the introduction of black bass and rockbass into Ontario lakes in Canada (Casselman, 2000; Jackson, 2000; Vander Zanden, 2000). The history of invasions exemplifies the homogenization of continental faunas (Rahel, 2000), losses in native organisms, and large changes in ecosystem processes and structure—all with economic consequences.

Fisheries (capture, aquaculture, and recreational) and associated introductions would exacerbate biodiversity and exotic problems in a warming climate. A substantial number of inland fisheries are associated with introduced species and stock enhancement. Tilapia, carp, grass carp, and rainbow trout are commonly introduced or stocked. An estimated 38% of the recreational catch is of non-native fishes. The practice of introducing species apparently has resulted in 1,354 introductions of 237 species into 140 countries from pre-1900 to the mid-1980s (FAO, 1988). Despite recognition of the importance of wild native species and increasing appreciation of the importance of biodiversity, impacts of climate change are likely to be managed by introducing better adapted species, if past management practices prevail. In the United States, an average of 38% of the recreational fishery is for non-native species, 75% of which now reproduce naturally (Horak, 1995). As climate change impacts affect the abundance of native (and non-native) species, it is probable that the past precedent of introducing non-native species would be applied to maintaining catches and recreational fishing stability.

Moving species poleward to adapt to changing climate zones for lakes and streams is fraught with scientific uncertainties and human conflict. The exotic can become a superabundant pest species, with negative effects on native organisms—including extirpation and extinction. The invasion ecology of organisms is not a predictive science; many surprises would be expected. The case has been made that managing with exotics increases the instability of the fish community and fish management problems and includes many unexpected consequences (Magnuson, 1976). Introduction of warmer water fauna on top of regional fauna that are having increasing problems from warming climates is likely to be a controversial adaptation.

Climate changes may provide development opportunities for aquaculture if traditional wild fisheries are less stable and markets favor the stability of the aquaculture product. Aquaculture represents a mobile technology that potentially can move with changing climate to the best conditions for culture (see Figure 5-7).

5.7.5. *Vulnerability*

Specific vulnerable elements include reduction and loss of lake and river ice, loss of habitat for coldwater fish, increases in extinctions and invasions of exotics, and potential exacerbation of existing pollution problems such as eutrophication.

Adaptation to climate change may induce other negative effects related to secondary pressures from new hydrologic engineering structures, poleward transport by humans of fauna and flora adapted to warmer lakes and streams, and interactions resulting from increased stocking and relocation of recreational and aquacultural endeavors.

Inland waters are affected hydrologically, physically, chemically, and biologically by climate change (Arnell *et al.*, 1996, Cushing, 1997). One reason for their vulnerability is that lakes, rivers, and wetlands integrate and reflect human and natural events in their watersheds and airsheds (Naiman *et al.*, 1995b). Interactions with changes to their watersheds, riparian shorelines, and human use of water combine to make lake and stream ecosystems vulnerable. Potential changes in quantity and quality of water reduce the ability of these waters to provide goods and services.

5.8. Inland Wetlands

Wetlands are defined here as any area of land where the water table is at or near the surface for some defined period of time, leading to unique physiochemical and biological processes and conditions that are characteristic of shallow flooded systems (Mitch and Gosselink, 1993; Oquist *et al.*, 1996). However, it should be noted that many other definitions also are in use, and care is required in comparing wetland areas in different regions of the world (Lugo *et al.*, 1990; Finlayson and van der Valk, 1995; Finlayson and Davidson, 1999).

Wetlands described in this section include those dominated by forested and nonforested vegetation; those on highly organic soils such as peatlands (partially decomposed plant material); and those on mineral soils, as is often typical of riverine systems. Peatlands are peat-accumulating wetlands and usually are divided into bogs and fens, both of which may be forested.

Peatlands are found at all latitudes, from the Arctic to the tropics. The growing vegetation layer of bogs is totally dependent on atmospheric inputs for their water and solute supply; consequently, surface peat layers are acidic and poor in nutrients. Fens receive additional water from overland flow and/or groundwater and typically have near-neutral pH and higher nutrient levels.

Wetlands provide many services and goods, many of which are indirect values (e.g., recreation, education, biodiversity, wild food) and are difficult to quantify. Key ecosystem services and goods considered here include biodiversity, carbon sinks, food, and fiber (peat and wood).

5.8.1. State of Wetland Services

5.8.1.1. Habitat and Biodiversity

Global estimates for the area of wetlands vary according to the definition of wetland used. Spiers (1999) reports global estimates for natural freshwater wetlands of 5.7 million km^2. The estimate by Matthews and Fung (1987) is 5.3 million km^2, which represents approximately 4% of the Earth's land surface, although a more recent estimate by Lappalainen (1996) is somewhat larger, at 6.4 million km^2. This compares well with Finlayson and Davidson's (1999) estimate of about 7 million km^2, including 1.3 million km^2 of rice paddy.

Maltby and Proctor (1996) estimate that peatlands cover about 4 million km^2 (±4%), constituting about 75% of wetlands. More than 90% of peatlands are in temperate, boreal, and subarctic regions. The total area of tropical peatlands is estimated to be 0.37–0.46 million km^2 (i.e., approximately 10 % of the global resource), but the full extent is uncertain (Immirzi *et al.*, 1992).

Many wetlands have irregular wetting and drying cycles, driven by climate. To date, little attention has been given to the impact of climate change on these less regular cycles of wetlands in semi-arid and arid regions (Sahagian and Melack, 1998). Changes in the area of these wetlands can be immense but could be monitored by using area-based parameters—for example, functional parameters and wetland extent expressed in terms of ha-days (Sahagian and Melack, 1998).

Species that form wetland plant communities are adapted to varying degrees to life in a flooded environment. These phenomena show large spatial variability, and different species show varying degrees of susceptibility to them, so it is not surprising that wetland vegetation exhibits such a high degree of variation in species composition (Crawford, 1983). Peatland plant communities have been observed to change over long periods of time, reflecting the peat accumulation process and leading to gradually drier conditions. This inherent changeability of wetland communities results largely from their occurrence in environments where a single extremely variable habitat factor—water supply—is predominant (Tallis, 1983). Consequently, land use and climate change impacts on these ecosystems can be expected to be mediated through changes in the hydrological regime.

Primary production in wetland communities is highly variable (Bradbury and Grace, 1983; Lugo *et al.*, 1988). Generally, wetland communities—which are dominated by trees, sedges, and grasses—have higher production rates than those characterized by shrubs and mosses. Organic matter produced in many wetlands is accumulated partially (2–16%—Päivänen and Vasander, 1994) as peat. A necessary antecedent condition for peat formation and accumulation is an excess of water stored on the mineral soil or sediment surface. This arises in humid climatic regions where precipitation exceeds evaporation (Ivanova, 1981; Clymo, 1984) or in more arid climatic regions where lateral inputs of water via surface runoff and/or groundwater seepage are sufficient to exceed evaporative demand (Glaser *et al.*, 1996).

Tropical peatlands play an important role in maintenance of biological diversity by providing a habitat for many tree, mammal, bird, fish, and reptile species (Prentice and Parish, 1992; Rieley and Ahmad-Shah, 1996; Page *et al.*, 1997); some of these species may be endemic or endangered. In common with other peatlands, tropical systems have a significant scientific value that goes beyond their plant and animal communities.

Within the peat also lies a repository of paleoenvironmental and paleogeochemical information that is extremely important in understanding past climatic conditions. These paleorecords are used to estimate rates of peat formation or degradation, former vegetation, climatic conditions, and depositional environments (Morley, 1981; Cecil *et al.*, 1993; Moore and Shearer, 1997).

5.8.1.2. Carbon Sink

The importance of peat-accumulating wetlands to global change is via the large carbon store accumulated over millennia—and the risk that this store would be released to the atmosphere in conditions modified by global change (e.g., fires). The carbon store in boreal and subarctic peatlands alone has been estimated at 455 Gt with an annual sink of slightly less than 0.1 Gt (Gorham, 1991). Tropical peatlands also are a considerable store (total of 70 Gt), containing as much as 5,000 t C ha^{-1}, compared with an average of 1,200 t C ha^{-1} for peatlands globally (Immirzi *et al.*, 1992; Diemont *et al.*, 1997). Estimates of the annual sink in tropical peatlands vary from 0.01 Gt (Sorensen, 1993) and 0.06 Gt (Franzen, 1994) to 0.09 Gt (Immirzi *et al.*, 1992)—emphasizing the lack of reliable data.

Optimal conditions for carbon sequestration appear to be in areas with mean annual temperatures between 4 and 10°C (Clymo *et al.*, 1998), which prevail in much of the southern boreal and temperate zones. The present carbon accumulation rate for boreal and subarctic bogs and fens is estimated as 0.21 t C ha^{-1} yr^{-1} (Clymo *et al.*, 1998). Rapid carbon accumulation rates also have been estimated for some tropical peatlands (Neuzil, 1997); retrospective values of Indonesian peatlands range from 0.61 to 1.45 t C ha^{-1} yr^{-1} (Neuzil, 1997).

Another important long-term (>100 years) sink for carbon in forested wetlands is wood biomass. Based on growth data (Shepard *et al.*, 1998) and conversion factors (Turner *et al.*, 1995) for bottomland hardwood forests, southern U.S. swamps, for example, sequester 0.011 Gt C yr^{-1}.

A small but significant proportion of organic matter in wetland soils is transformed into methane in the metabolism of methanogenic bacteria. Methane production is a characteristic feature in all wetland soils; the rate is governed largely by substrate availability and temperature (Shannon and White, 1994; Mikkelä *et al.*, 1995; Schimel, 1995; Bergman *et al.*, 1998; Komulainen *et al.*, 1998). Part of the methane produced in anoxic soil is oxidized by methanotrophic bacteria in aerobic surface layers.

The ratio of methane production to consumption determines the magnitude of the flux from the soil to the atmosphere; this rate is governed largely by the depth of the aerobic layer (Roulet *et al.*, 1993; Shannon and White, 1994). The role of vascular plants in providing continuous substrate flux for methanogenesis and as a transport pathway to the atmosphere has been stressed (Whiting and Chanton, 1993; Schimel, 1995; Frenzel and Rudolph, 1998). Slow fermentation of organic matter in growing peat layers of bogs has been cited as the factor that theoretically limits the final volume a bog may reach during its development (Clymo, 1984).

5.8.1.3. *Food and Fiber Production*

Wild berries growing on peatlands are an important natural resource in many regions of the boreal zone (Reier, 1982; Yudina *et al.*, 1986; Salo, 1996). In Finnish peatlands alone, the annual biological yield of wild berries may exceed 150 million kg, of which approximately 10% is picked, with a value of US$13.5 million (Salo, 1996). In North America, cranberry (*Vaccinium macrocarpon*) is commercially cultivated in peatlands (Johnson, 1985). The use of peatlands for agriculture has a long history; presently, 10,000 km^2 are under this land use (Immirzi *et al.*, 1992).

The importance of wetlands in North America as waterfowl habitat has long been recognized (Mitch and Gosselink, 1993). The commercial value of these wetlands is not direct; it comes through the added value of the activities of hunters in local economies. Although there are few estimates of this service, much of the wetland conservation effort in North America has focused on conservation, enhancement, and creation of habitat.

Products from tropical forested wetlands include rattans, resins, latex, fungi, fruit, honey, and medicinal plants, sale of which provides revenue for local communities. Exploitation of fish from swamp forests and associated waterways also can supply a modest income and is an important source of protein for local human populations (Immirzi *et al.*, 1996; Lee and Chai, 1996).

Forested wetlands are valuable for wood production, mostly as a result of wetland modification. For example, in Finland, these wetlands produce about 18 million m^3 of timber annually—nearly 25% of the total annual increment (Tomppo, 1998). Forested wetlands in the southern United States produce 39 million m^3 of timber, of which 33 million m^3 are removed annually (Shepard *et al.*, 1998).

Direct harvest of forest resources from tropical peatlands yields several important products, ranging from timber and bark to non-timber products (Immirzi *et al.*, 1996; Lee and Chai, 1996). Southeast Asia's peat swamps yield some of the most valuable tropical timbers—in particular, ramin (*Gonystylus bancanus*) (Ibrahim, 1996).

Peat has been used as a domestic energy source in northwestern and central Europe for centuries (Feehan and O'Donovan, 1996). The present volume of industrial peat harvesting is estimated at 71 million m^3, most of it in Finland and Ireland (Asplund, 1996). The peatland area occupied by peat harvesting is rather small in comparison to the area of land uses such as agriculture and forestry; for example, in Finland 24 million m^3 of energy peat are harvested from an area of 530 km^2. Employment aspects may be important in peat harvesting because most peat sites are located in remote areas where few industrial jobs are available. Peat harvesting may offer an estimated 550 permanent/seasonal jobs for each 1 million m^3 of produced peat (Nyrönen, 1996). Canada has a small but prosperous peat harvesting industry for horticulture and medical uses (Rubec *et al.*, 1988; Keys, 1992). Between 1986 and 1990, Canadian shipments were 662,000–812,000 t yr^{-1} (Keys, 1992).

5.8.2. **Pressures on Wetland Services**

Pressures on wetlands are likely to be mediated through changes in hydrology, direct and indirect effects of changes in temperature, and land-use change. There would be interactions among these pressures and subsequent impacts on services and good from these ecosystems.

5.8.2.1. *Changes in Hydrology*

Climate change will affect the hydrology of individual wetland ecosystems mostly through changes in precipitation and temperature regimes. Because the hydrology of the surface layer of bogs is dependent on atmospheric inputs (Ingram, 1983), changes in the ratio of precipitation to evapotranspiration may be expected to be the main factor in ecosystem change. However, work on the large peatland complexes of the former Glacial Lake Aggazzi region indicate that the hydrology of bogs cannot be considered in isolation or independent of local and regional groundwater flow systems (Siegel and Glaser, 1987; Branfireun and Roulet, 1998) and that groundwater flow reversals, even in ombrotrophic peatlands, can have an impact on their water storage and biogeochemistry (Siegel *et al.*, 1995;

Devito *et al.*, 1996). From the perspective of assessment of climate variability and change of peatlands, these systems need to be viewed in the broader context of their hydrogeological setting.

Fen, marsh, and floodplain wetlands receive additional water influx from the surrounding basin, including underground sources—sometimes from a considerable distance. This means that climate change impacts are partially mediated through changes in the whole basin area or even further afield where groundwater reserves do not correspond with surface basins. These changes also may affect the geochemistry of wetlands. Recharge of local and regional groundwater systems, the position of the wetland relative to the local topography, and the gradient of larger regional groundwater systems are critical factors in determining the variability and stability of moisture storage in wetlands in climatic zones where precipitation does not greatly exceed evaporation (Winter and Woo, 1990). Changes in recharge external to the wetland may be as important to the fate of the wetland under changing climatic conditions as the change in direct precipitation or evaporation on the wetland itself (Woo *et al.*, 1993).

5.8.2.2. *Changes in Temperature*

Temperature is an important factor controlling many ecological and physical functions of wetlands. Primary productivity and microbial activity are both controlled to a certain extent by temperature conditions. Temperature also affects evapotranspiration rates and has an impact on the water regime. Because higher temperatures and drying of the surface soil usually occur together and interactively affect the ecosystem processes, it is not always possible to separate their impacts.

Although Gorham (1991) suggests that the effects of temperature will be overshadowed by the impacts of water-level drawdown on northern peatlands, the impacts of temperature increases on wetlands on permafrost may be drastic (Gorham, 1994a). A fairly small increase in temperature might initiate large-scale melting of permafrost, with thermokarst erosion and changed hydrological regimes as a consequence (Billings, 1987). Work by Vitt *et al.* (1994) and Halsey *et al.* (1997) has demonstrated clearly the dynamic association between the distribution of peatlands, peatland types, and the presence or absence of permafrost in North America. This association is strong enough that it has been used as a proxy method for inferring climatic variability during the Holocene (Halsey *et al.*, 1995). This might imply shifts from black spruce/*Sphagnum*/lichen communities on permafrost to wetter fen communities, with subsequent changes in carbon cycling.

5.8.2.3. *Land-Use Change*

Land-use change may create multiple pressures on wetland habitats. Area estimates of the scale of direct development of tropical peatlands vary and provide only an imprecise picture of the current situation (Immirzi *et al.*, 1992; Maltby and Immirzi, 1996). In southeast Asia, agriculture and forestry are the major peatland land uses. Toward the end of the 1980s, it was estimated that in Indonesia alone 3.7 Mha (18% of the total peat swamp forest) had undergone some form of development (Silvius *et al.*, 1987).

Cultivation of tropical peatlands involves measures that radically change the hydrological regime and consequently influence vegetation and soil processes. Forests are cleared and effective drainage installed. In many places in southeast Asia, cultivation of horticultural and estate crops has met with mixed success, and some previously converted peatlands have been abandoned, although peat-forming vegetation has failed to reestablish (Immirzi *et al.*, 1992). Reasons for failure include poor water management and persistent infertility of the soil (Rijksen *et al.*, 1997).

The total area of tropical peatland drained or otherwise altered during forestry management is not known. Silvius *et al.* (1987) suggest that as much as 0.11 million km^2 of peatlands in Indonesia (i.e., as much as 50% of the total resource) are possibly being exploited for forestry purposes. In Malaysia, most of the remaining peat swamp forest outside limited conservation areas has been logged (Immirzi *et al.*, 1992). Sustainable-yield forestry is likely to be the most appropriate form of land use for peat swamp forest, but such methods applicable to peat swamps have yet to be developed, let alone implemented (Immirzi *et al.*, 1992).

Use of peatlands for forestry usually brings about smaller changes in the ecosystem. In floodplain swamps and peatlands of the more continental areas of North America, often only tree stands are managed (Dahl and Zoltai, 1997), but in northwestern Europe and the southeastern United States, forestry use includes artificial drainage (Richardson and McCarthy, 1994; Päivänen, 1997). Some 0.15 million km^2 have been drained for forestry, mostly in Scandinavia and Russia (Päivänen, 1997). In these cases, much of the original vegetation (Laine *et al.*, 1995) and functions (Aust and Lea, 1991; Minkkinen *et al.*, 1999) are preserved during forestry management. About 70% of the expansive peatlands in North Carolina (6,000 km^2) have been entirely or partially degraded through draining, ditching, or clearing (Richardson and Gibbons, 1993).

Peat harvesting for energy or horticultural use has the most drastic impact on the ecosystem; vegetation is removed with the topsoil prior to harvesting, and most of the accumulated peat gradually is extracted (Nyrönen, 1996).

5.8.3. ***Impacts on Wetland Services***

Generally, climatic warming is expected to start a drying trend in wetland ecosystems. According to Gorham (1991), this largely indirect influence of climate change, leading to alteration in the water level, would be the main agent in ecosystem change and would overshadow the impacts of rising temperature and longer growing seasons in boreal and subarctic peatlands. Monsoonal areas are more likely to be affected by more intense rain events over shorter rainy seasons, exacerbating flooding and erosion in catchments and the wetlands themselves. Similarly, longer dry seasons could alter fire regimes and loss of organic matter to the atmosphere (Hogenbirk and Wein, 1991).

5.8.3.1. Habitat and Biodiversity

Climate change may be expected to have clear impacts on wetland ecosystems, but there are only a few studies available for assessing this. There are some laboratory studies concerned with the responses of individual plant species or groups of species (Jauhiainen *et al.*, 1997, 1998a,b; van der Heijden *et al.*, 1998). Based on these studies, however, it is difficult to predict the responses of plant communities formed by species with somewhat varying environmental requirements.

The response of wetland plant communities to drought has received some attention in temperate freshwater wetlands (Greening and Gerritsen, 1987; Streng *et al.*, 1989). Stratigraphical studies in peatlands have shown hydroseral succession whereby swamp and fen communities gradually develop into bog communities (Tallis, 1983). These changes are largely autogenic, connected to growth of wetland communities and caused by past climatic variability or artificial drainage. An alternative approach to observe the vegetation-environmental change succession has been to use space as a time substitute by mapping different plant communities onto climatic and hydrological surfaces (Gignac *et al.*, 1991). This approach has shown tight coupling between various peatland plants, climate, hydrology, and resultant chemistry—and even for trace gas exchange (Bubier, 1995)—and has been used to infer certain aspects of peatland development through macrofossil analysis (Gorham and Janssens, 1992; Kuhry *et al.*, 1993).

Much is known about how vegetation changes as a result of water-level drawdown following drainage for forestry in northwestern Europe. Drying of surface soil initiates a secondary succession whereby original wetland species gradually are replaced by species that are typical of forests and heathlands (Laine and Vanha-Majamaa, 1992; Vasander *et al.*, 1993, 1997; Laine *et al.*, 1995). Plants living on wet surfaces are the first to disappear, whereas hummock-dwelling species may benefit from drying of surface soil. In nutrient-poor peatlands, bog dwarf-shrubs dominate after water-level drawdown; at more nutrient-rich sites, species composition develops toward upland forest vegetation (Laine and Vanha-Majamaa, 1992; Minkkinen *et al.*, 1999).

The effect of sea-level rise on wetlands has been addressed in several assessments. In northern Australia, extensive seasonally inundated freshwater swamps and floodplains are major biodiversity foci (Finlayson *et al.*, 1988). They extend for approximately 100 km or more along many rivers but could be all but displaced if predicted sea-level rises of 10–30 cm by 2030 occur and are associated with changes in rainfall in the catchment and tidal/storm surges (Bayliss *et al.*, 1997; Eliot *et al.*, 1999). Expected changes have been demonstrated by using information collected from the World Heritage-listed Kakadu National Park, but the scenario of massive displacement of these freshwater wetlands can be extended further afield given similarities in low relief, monsoonal rainfall, and geomorphic processes (Finlayson and Woodroffe, 1996; Eliot *et al.*, 1999). In fact, the potential outcome of such change can be seen in the nearby Mary River system, where saline intrusion, presumably caused by other anthropogenic events, already has destroyed 17,000 ha of freshwater woodland and sedge/grassland (Woodroffe and Mulrennan, 1993; Jonauskas, 1996).

Mechanisms by which environmental factors and biotic interactions control wetland biodiversity are not well understood (Gorham, 1994b). The effects of water-level drawdown after drainage for forestry indicate that the shift in species composition from bog and fen species to forest species only slightly affects plant species richness of individual sites (Laine *et al.*, 1995). In regions dominated by forests, there would be clear reduction in regional diversity as landscapes become homogenized after drainage (Vasander *et al.*, 1997).

The inherent changeability of wetland communities, resulting from spatial and temporal variability in water supply (Tallis, 1983), may be the key factor in the response of wetland communities to climate change. Because there may be differences between species in adaptation potential, community structures would change, and there would be profound effects on the nature of the affected wetlands, as discussed by Gorham (1994a). The response of wetland plant communities to changing environment may have fundamental effects on the species diversity of these ecosystems.

Because of spatial and temporal variability in ecosystem processes, development of systems models for wetlands is becoming an important assessment tool. A fully coupled peatland-climate model has not yet been developed, but there have been some significant advances in modeling various components of the peatland and/or wetland biogeochemical system (Harris and Frolking, 1992; Roulet *et al.*, 1992; Christensen and Cox, 1995; Christensen *et al.*, 1996; Walter *et al.*, 1996; Granberg, 1998), and several process-level models are now used at the global scale (Cao *et al.*, 1996; Potter *et al.*, 1996; Potter, 1997).

Elevated CO_2 levels will increase photosynthetic rates in some types of vegetation (e.g., C_3 trees and emergent macrophytes) (Bazzaz *et al.*, 1990; Idso and Kimball, 1993; Drake *et al.*, 1996; Megonigal and Schlesinger, 1997; see also Section 5.6.3.1). Responses of nonvascular vegetation, such as sphagna, have been less clear (Jauhiainen *et al.*, 1994, 1997, 1998a; Jauhiainen and Silvola, 1999). A study in Alaskan tussock tundra found that photosynthetic rates in *Eriophorum vaginatum* quickly adjusted downward such that rates with elevated and ambient CO_2 were similar after 1 year (Tissue and Oechel, 1987). However, a sustained increase in net ecosystem carbon sequestration was observed when elevated CO_2 treatments were combined with a 4°C increase in temperature (Oechel *et al.*, 1994).

Many C_3 plants respond to elevated CO_2 with a decrease in stomatal conductance (Curtis, 1996), which could reduce transpiration rates. Because transpiration is an important pathway for water loss from many ecosystems (Schlesinger, 1997), including wetlands (Richardson and McCarthy, 1994), reductions in transpiration rate could affect the position of the aerobic-anaerobic interface in wetland soils (Megonigal and Schlesinger, 1997).

5.8.3.2. Carbon Sink

It has been suggested that water-level drawdown and increased temperature will decrease carbon sequestration in subarctic and boreal peatlands, especially in more southern latitudes (Gorham, 1991). However, conclusions are hampered by the diversity of possible responses.

Impacts of climate change on wetland carbon sink can be measured directly by using the eddy covariance method or chamber techniques (Crill *et al.*, 1988; Fowler *et al.*, 1995; Alm *et al.*, 1997b; Aurela *et al.*, 1998) and modeling the balances (Frolking *et al.*, 1998). In recent years, several high quality, continuous, snow-free season time series of net ecosystem exchange have been obtained for wetlands and peatlands in the subarctic zone (Burton *et al.*, 1996; Friborg *et al.*, 1997; Schreader *et al.*, 1998), the boreal zone (Shurpali *et al.*, 1995; Jarvis *et al.*, 1997; Kelly *et al.*, 1997; Lafleur *et al.*, 1997; Roulet *et al.*, 1997; Suyker *et al.*, 1997; Goulden *et al.*, 1998), and the temperate zone (Happell and Chanton, 1993; Pulliam, 1993). Extrapolations of temporally detailed enclosure measurements along with estimates of the export of DOC have been undertaken to derive peatland carbon balances (Waddington and Roulet, 1996; Carroll and Crill, 1997). These studies point out the difficulty of assessing the sink/source strength of a peatland within a reasonable level of certainty because of errors introduced in the scaling process (Waddington and Roulet, 2000).

For northern peatlands, much is known about the effects of water-level drawdown on the carbon balance, based on studies carried out in peatlands drained for forestry (Glenn *et al.*, 1993; Roulet *et al.*, 1993; Laine *et al.*, 1995; Martikainen *et al.*, 1995; Minkkinen and Laine, 1998). These studies may be used cautiously to represent the climate change impact because the effect of drainage on ecosystem structure and functioning is similar to that predicted after drying caused by climate change in northern latitudes (Laine *et al.*, 1996). The change observed in vegetation structure after water-level drawdown directs biomass production to the shrub and tree layer; in most cases, primary production and biomass increased (Laiho and Finér, 1996; Laiho and Laine, 1997; Sharitz and Gresham, 1998). Simultaneously, litter production increased (Laiho and Finér, 1996; Laiho and Laine, 1997; Finér and Laine, 1998), and the litter was more resistant to decomposition (Meentemeyer, 1984; Berg *et al.*, 1993; Couteaux *et al.*, 1998). There is some evidence that part of the carbon from decomposing litter is stored in the peat profile down to 0.5-m depth (Domisch *et al.*, 1998). These alterations to vegetation production and litter flow into soil have been observed to keep the net carbon accumulation rate into the soil of boreal bogs in most cases at the level prior to water-level drawdown, sometimes exceeding this level (Minkkinen and Laine, 1998). However, increased duration and shortened return periods of extreme droughts may have detrimental effects on the peat carbon balance, as indicated by the results of Alm *et al.* (1997b).

Water-level drawdown will cause a decrease in CH_4 emissions as substrate flux to anoxic layers is decreased and consumption of CH_4 in the thicker aerobic layer is enhanced (Glenn *et al.*, 1993; Roulet *et al.*, 1993; Martikainen *et al.*, 1995; Roulet and Moore, 1995). It has been suggested that reduced CH_4 emissions after water-level drawdown, together with an increase in tree biomass and a fairly small change in carbon sequestration into peat, may even decrease the greenhouse effect of these ecosystems (Laine *et al.*, 1996).

Therefore, it is not immediately clear that a warmer, drier climate necessarily will lead to a large loss of stored peat for all peatland types. The feedback between climate and peatland hydrology and the autogenic nature of peatland development is poorly understood. Clymo (1984) first developed this idea, and it has recently been used to show how surface topography on peatlands is preserved (Belyea and Clymo, 1999). Hilbert *et al.* (1998) have expanded on the work of Clymo (1984) and developed a model of peatland growth that explicitly incorporates hydrology and feedbacks between moisture storage and peatland production and decomposition. Their studies suggest that some peatland types (e.g., most bogs) will adjust relatively quickly to perturbations in moisture storage.

Future rates of carbon sequestration in swamps on mineral soils will depend largely on the response of trees to changes in hydrology, temperature, and elevated CO_2 concentrations. The aboveground productivity of temperate zone swamp forests is strongly regulated by the extent of soil saturation and flooding, and the long-term effect of changes in hydrology on growth will depend on the position of forests along the current hydrological gradient (Megonigal *et al.*, 1997). Drier conditions may increase NPP on extensively flooded sites and decrease it on dry and intermediate sites.

The combination of high soil carbon density and rapid warming in peatlands underlain by permafrost have raised concerns that northern peatlands may become net carbon sources rather than sinks (Lal *et al.*, 2000). Indeed, working in tussock and wet tundra, Oechel *et al.* (1993) estimate that these systems are now net sources of 0.19 Gt C, caused mainly by melting of permafrost and lowering of the water table. Botch *et al.* (1995) report that peatlands of the former Soviet Union are net sources of 0.07 Gt C yr^{-1}. Other parts of the boreal zone may have become enhanced sinks as a result of recent warming (Myneni *et al.*, 1997), and continued warming could change the equation to favor net carbon storage—as suggested by warming experiments in arctic ecosystems (Hobbie, 1996).

Higher temperatures may affect carbon cycling of other wetlands as well. Increased photosynthetic activity of deep-rooted wetland plants, such as sedges, may enhance substrate availability for methanogenesis—which, together with higher temperatures, might lead to higher CH_4 emissions (Valentine *et al.*, 1994; Bergman *et al.*, 1998; Segers, 1998) where water level would remain near the soil surface.

Land uses such as agriculture and forestry always change carbon fluxes in ecosystems. High carbon losses have been reported for agricultural crop production on drained wetland in Europe and North America, as much as 10–20 t C ha^{-1} yr^{-1} (Armentano and Menges, 1986). Studies have shown that agriculture on peat soils may contribute significantly to nitrous oxide (N_2O) emission (Nykänen *et al.*, 1995a). Kasimir-Klemedtsson *et al.*, (1997) conclude that agricultural practices on organic soils lead to a net increase in emissions of GHGs because of large fluxes of CO_2 and N_2O, over decreases in emissions of CH_4.

As agricultural management fundamentally alters the processes of wetlands and gradually leads to decreases in wetland area (Okruszko, 1996). Large areas of wetlands have been lost in Russia, Europe, and North America by complete drainage and conversion to other land uses. It has been estimated that 53% of the original 89 Mha of wetlands in the coterminous United States were lost by the 1980s (Shepard *et al.*, 1998), much of it to agricultural conversion. Development for agriculture also can have offsite effects—for example, reduced water quality that impacts fisheries (Notohadiprawiro, 1998). Arable agriculture always transforms wetlands into sources of GHGs to the atmosphere (Armentano and Menges, 1986; Okruszko, 1996), with the exception of CH_4.

Consequences of the development of tropical peatlands include lowering of the water table, which promotes peat oxidation and decomposition. Peat loss and subsidence can occur at very fast rates—as much as 0.9 cm per month (Dradjad *et al.*, 1986). Eventually, shrinkage and oxidation may lead to loss of the entire peat profile and exposure of underlying nutrient-poor substrates or potential acid sulphate soils (Maltby *et al.*, 1996; Rieley *et al.*, 1996). In subcoastal situations, this may be followed by marine inundation.

When wetland use for forestry involves only management of existing tree stands, the impacts on functions and processes may be small, and the ecosystem may remain within the wetland concept (Aust and Lea, 1991; Minkkinen *et al.*, 1999). However, if artificial drainage is included, decay of organic matter is enhanced, with consequent increases in CO_2 emissions from peat (Glenn *et al.*, 1993; Silvola *et al.*, 1996). The results of the carbon balance change reported are highly variable, depending on methods used and climatological differences. Losses of peat carbon have been reported by Braekke and Finer (1991) and Sakovets and Germanova (1992), whereas increased post-drainage carbon stores have been reported by Anderson *et al.* (1992) and Vompersky *et al.* (1992). Based on a large cross-sectional data set from forest drainage areas in Finland, it was shown that carbon accumulation in peat soil increased in southern parts of the country (annual mean temperature 3–4.5°C) but decreased in northern Finland (with mean temperature of 0–1°C). The carbon accumulation increase was clearest for nutrient-poor bog sites (Minkkinen and Laine, 1998). Cannell and Dewar (1995) have concluded that drainage and planting of conifers on organic soils produces little long-term change in soil carbon stores because enhanced organic matter oxidation is compensated by increased litter production of the tree stand.

Water-level drawdown after drainage decreases CH_4 emissions from peatland (Glenn *et al.*, 1993; Roulet *et al.*, 1993; Martikainen *et al.*, 1995; Roulet and Moore, 1995). Increased consumption in the surface soil may even form a small CH_4 sink in some cases (Glenn *et al.*, 1993; Roulet *et al.*, 1993; Fowler *et al.*, 1995; Martikainen *et al.*, 1995; Roulet and Moore, 1995; Komulainen *et al.*, 1998). The effect of drainage on N_2O emissions has been reported to be fairly small and restricted to fen sites (Martikainen *et al.*, 1993).

Forest harvesting in tropical swamp forests can result in changes to the quality and quantity of organic matter inputs from vegetation, and—as the work of Brady (1997) has shown—if tree root mats decline, net accumulation of peat also may decline. Where selective logging is combined with artificial drainage, decomposition and subsidence of peat may proceed at rates of 3.5–6.0 cm yr^{-1} (Brady, 1997). In contrast with peatlands of the temperate and boreal zone, there has been poor success with establishing forestry plantations on tropical peatlands.

Peat harvesting totally changes the structure and functioning of the original ecosystem by removing the vegetation and finally most of the accumulated peat deposit. This has a fundamental impact on the GHG balances of harvesting sites: CH_4 emissions almost stop (Nykänen *et al.*, 1995b), but the whole accumulated carbon store forms a CO_2 source to the atmosphere during harvesting and combustion (Rodhe and Svensson, 1995), even if peat combustion may replace imported energy in countries with no other major domestic energy sources.

Recent findings have shown that restoration of cut-away peatlands after harvesting soon initiates colonization of peatland plants (Smart *et al.*, 1989; Tuittila and Komulainen, 1995; Campeau and Rochefort, 1996; Wheeler, 1996; Wind-Mulder *et al.*, 1996; Ferland and Rochefort, 1997; LaRose *et al.*, 1997; Price *et al.*, 1998) and may restart carbon accumulation (Tuittila *et al.*, 1999).

Elevated CO_2 is likely to stimulate CH_4 emissions in a wide variety of wetland ecosystems, including freshwater marshes (Megonigal and Schlesinger, 1997) and rice paddies (Ziska *et al.*, 1998). Studies in northern peatlands have been equivocal: One study reports a maximum increase of 250% (Hutchin *et al.*, 1995) and another no increase (Saarnio *et al.*, 1998). Because CH_4 is a more powerful GHG than CO_2, wetlands amplify the greenhouse effect of elevated CO_2 by converting a portion of this gas to CH_4.

Increased nitrogen deposition may alter the species composition of wetland communities (Aerts *et al.*, 1992) and their production, leading to higher production and net accumulation rates in peatlands where production is nitrogen limited (Aerts *et al.*, 1995). There is some indication that nitrogen inputs may affect trace gas emissions from peat soils by increasing emissions of N_2O and sometimes decreasing those of CO_2 and CH_4 (Aerts, 1997; Aerts and Ludwig, 1997; Aerts and Toet, 1997; Regina *et al.*, 1998).

5.8.3.3. *Food and Fiber*

Services that involve artificial drainage might even benefit from climatic warming and additionally lowered water levels. Moya *et al.* (1998) report that rice biomass and seed yield is increased by CO_2 concentrations of 200 and 300 ppm above ambient, but these increases are diminished or reversed when air temperature is elevated by 4°C. In the northern latitudes of Scandinavia, forest production in drained peatlands clearly is favored by higher temperatures (Keltikangas *et al.*, 1986) and lower water-table levels.

5.8.4. *Adaptation Options for Wetlands*

Most wetland processes are dependent on catchment-level hydrology, which is being changed by land-use changes at fairly large scales. Thus, it may be very difficult if not impossible to adapt to the consequences of projeted climate change. For key habitats, small-scale restoration may be possible if sufficient water is available. In cases where wetlands are used for arable agriculture, the impact on the carbon balance could be controlled by the choice of cropping method, including alternative crops and depth of drainage.

5.8.5. *Vulnerability of Functions and Key Services of Wetlands*

The types of inland wetlands that are most vulnerable to global change (i.e., experience the largest changes) are difficult to ascertain. As concluded in the SAR, arctic and subarctic ombrotrophic bog communities on permafrost would change drastically after thawing of the frost layer and might be considered prime candidates in the vulnerability assessment, together with more southern depressional wetlands with small catchment areas. The increasing speed of peatland conversion and drainage in southeast Asia will place these areas at a greatly increased risk of fire. This will be one of the principal factors in determining the viability of tropical systems.

Global change impacts on wetlands would cause changes in many of the ecosystem services of wetlands. Especially vulnerable are functions that depend on a high degree of water availability. Services that involve artificial drainage might even benefit from climatic warming and additionally lowered water levels. For instance, in northern latitudes of Scandinavia, forest production in drained peatlands clearly is favored by higher temperatures (Keltikangas *et al.*, 1986).

5.9. Arctic and Alpine Ecosystems

Arctic and alpine ecosystems are characterized by low human population densities. However, they provide important goods and services locally and globally. At the local scale, they are the resource base of many indigenous cultures and provide recreation, food, and fiber to people in adjacent regions. At the global level, arctic and boreal regions play an important role in the world's climate system. They contain 40% of the world's reactive soil carbon (McGuire *et al.*, 1995); influence global heat transport through their impact on regional water and energy exchange with the atmosphere; and determine freshwater input to the Arctic Ocean, which influences bottom-water formation and thermohaline circulation of the oceans. Alpine regions are important sources of freshwater and hydropower for surrounding lowlands. Changes in these goods and services would have socioeconomic impacts throughout the world.

Chapter 16 presents information on the effects of climate on the physical environment of polar regions. The impacts of climate change on high-mountain systems and arctic tundra are extensively reviewed in the SAR by Beniston and Fox (1996) and Allen-Diaz (1996), respectively. This section emphasizes the effects of climatic change particularly on ecosystem production, carbon stores, biodiversity, and water flow—which, in turn, can affect the productivity especially of alpine ecosystems.

5.9.1. *State and Trends of Goods and Services*

5.9.1.1. *Arctic Ecosystems*

Arctic and alpine tundra each occupy 4 million km^2 (Körner, 1999). Approximately 25% of the tundra is ice-covered, 25% shrub-dominated, and the rest dominated by herbaceous plants. Soil carbon stocks in boreal and tundra peatlands are large (see Sections 5.6 and 5.8).

High-latitude warming that has occurred in the Arctic region since the 1960s (Chapman and Walsh, 1993; Serreze *et al.*, 2000) is consistent with simulations of climate models that predict increased greenhouse forcing (Kattenberg *et al.*, 1996). Climatic change has been regionally variable, with cooling in northeastern North America and warming in northwestern North America and northern Siberia. The warming results from a change in the frequency of circulation modes rather than gradual warming (Palmer, 1999). Precipitation (P) and surface evaporation (E) have increased at high latitudes, with no significant temporal trend in the balance between the two (P-E) (Serreze *et al.*, 2000).

Permafrost underlies 20–25% of the northern hemisphere land area (Brown *et al.*, 1997a). Ice that forms during periods of cold climate frequently constitutes a high proportion (20–30%)

of the volume of these frozen soils (Brown *et al.*, 1997a). Consequently, melting of permafrost can lead to surface collapse of soils forming thermokarst, an irregular topography of mounds, pits, troughs, and depressions that may or may not be filled with water, depending on topography. Permafrost temperatures have warmed in western North America by 2–4°C from 1940 to 1980 (Lachenbruch and Marshall, 1986) and in Siberia by 0.6–0.7°C from 1970 to 1990 (Pavlov, 1994), whereas permafrost cooled in northeastern Canada (Wang and Allard, 1995)—patterns that roughly parallel recent trends in air temperature. However, the magnitude of warming and patterns of interannual variation (roughly 10-year oscillations) in permafrost temperatures are not readily explained as a simple response to regional warming (Osterkamp *et al.*, 1994). Changes in permafrost regime probably reflect undocumented changes in the thickness or thermal conductance of snow or vegetation, in addition to changes in air temperature (Osterkamp and Romanovsky, 1999). Thermokarst features are developing actively in the zone of discontinuous permafrost (Osterkamp and Romanovsky, 1999), particularly in association with fire and human disturbance, but there are no long-term records from which to detect trends in the regional frequency of thermokarst.

In contrast to the long-term trend in tundra carbon accumulation during the Holocene, flux measurements in Alaska suggest that recent warming may have converted tundra from a net carbon sink to a source of as much as 0.7 Gt C yr^{-1} (Oechel *et al.*, 1993; Oechel and Vourlitis, 1994). The direction and magnitude of the response of carbon exchange to warming may be regionally variable, depending on climate, topography, and disturbance regime (Zimov *et al.*, 1999, McGuire *et al.*, 2000). During recent decades, peak-to-trough amplitude in the seasonal cycle of atmospheric CO_2 concentrations has increased, and the phase has advanced at arctic and subarctic CO_2 observation stations north of 55°N (Keeling *et al.*, 1996). This change in carbon dynamics in the atmosphere probably reflects some combination of increased uptake during the first half of the growing season (Randerson *et al.*, 1999), increased winter efflux (Chapin *et al.*, 1996), and increased seasonality of carbon exchange associated with disturbance (Zimov *et al.*, 1999). This "inverse" approach generally has concluded that mid-northern latitudes were a net carbon sink during the 1980s and early 1990s (Tans *et al.*, 1990; Ciais *et al.*, 1995; Fan *et al.*, 1998; Bousquet *et al.*, 1999; Rayner *et al.*, 1999). At high northern latitudes, these models give a wider range of estimates, with some analyses pointing to a net source (Ciais *et al.*, 1995; Fan *et al.*, 1998) and others to a sink (Bousquet *et al.*, 1999; Rayner *et al.*, 1999).

High-latitude wetlands and lakes account for 5–10% of global CH_4 fluxes to the atmosphere (Reeburgh and Whalen, 1992). These fluxes increase dramatically with thermokarst (Zimov *et al.*, 1997), acting as a potentially important positive feedback to global warming.

Satellite imagery suggests an increase in NDVI (a measure of "greenness") from 1981 to 1991 (Myneni *et al.*, 1997), although interpretation is complicated by changes in sensor calibration (Fung, 1997). If these changes in satellite imagery are an accurate reflection of vegetation activity, changes in NDVI could help explain the increase in seasonal amplitude in atmospheric CO_2 observed at high northern latitudes (Keeling *et al.*, 1996; Randerson *et al.*, 1999). This also is consistent with the increased biomass of shrubs in the arctic tundra (Chapin *et al.*, 1995).

5.9.1.2. *Alpine Ecosystems*

Climatic changes observed in alpine areas generally have paralleled climatic patterns in surrounding regions, with the most pronounced warming at high latitudes, in the Alps, and in Asia and the least pronounced changes in tropical alpine regions (Diaz and Bradley, 1997). Precipitation generally has increased, with the most pronounced changes in winter, leading to increased snow depth (Beniston, 1997).

Regional trends in climate have led to shrinkage of alpine and subpolar glaciers, equivalent to a 0.25 ± 0.1 mm yr^{-1} of sea-level change or 16% of the sea-level rise in the past 100 years (Dyurgerov and Meier, 1997). This trend has been regionally variable, with Asia contributing 45% of this sea-level rise and arctic islands an additional 18%. The net mass reduction of the alpine glaciers has been most pronounced since 1980, when regional warming was greatest.

Climatic warming observed in the Alps has been associated with upward movement of some plant taxa of 1–4 m per decade on mountaintops and loss of some taxa that formerly were restricted to high elevations (Braun-Blaunquet, 1956; Grabherr *et al.*, 1994). In general, direct human impacts on alpine vegetation from grazing, tourism, and nitrogen deposition are so strong that climatic effects on goods and services provided by alpine ecosystems are difficult to detect (Körner, 1999). Soil carbon stocks per unit area in alpine ecosystems are only one-third as great as those in the arctic because greater topographic relief promotes greater drainage and decomposition than in the arctic (Körner, 1995b).

5.9.2. *Responses of Arctic and Alpine Ecosystems and Impacts on their Goods and Services*

5.9.2.1. *Impacts Resulting from Changes in Climate on Arctic Ecosystems*

Changes in climate are likely to be the greatest cause of changes in goods and services in the arctic (Walker *et al.*, 2001). Projected climatic warming of 4–10°C by the end of the century probably would cause substantial increases in decomposition, nutrient release, and primary production. The net effect on the carbon balance will depend primarily on soil moisture (McKane *et al.*, 1997; McGuire *et al.*, 2000), which cannot be projected with confidence. In general, surface soils on slopes are expected to become drier as thaw depth increases. Lowlands may experience substantial thermokarst, impoundment of water, and reduced aeration.

Plant production frequently is limited in the arctic by excessive moisture and slow turnover of nutrients in soils, and warming and drying of soils is likely to enhance decomposition, nutrient mineralization, and productivity. Many of these changes in productivity may be mediated by changes in species composition and therefore are likely to lag changes in climate by years to decades (Chapin *et al.*, 1995; Shaver *et al.*, 2000). Threshold changes in productivity associated with poleward movement of the treeline is likely to experience time lags of decades to centuries because of limitations in dispersal and establishment of trees (Starfield and Chapin, 1996; Chapin and Starfield, 1997).

The net effect of warming on carbon stores in high-latitude ecosystems depends on changes in the balance between production and decomposition. Decomposition initially may respond more rapidly than production, causing trends toward net carbon efflux (Shaver *et al.*, 1992; Smith and Shugart 1993).

Warming-induced thermokarst is likely to increase CH_4 flux to the atmosphere in lowlands, particularly peatlands of northern Canada and western Siberia (Gorham, 1991; Roulet and Ash, 1992; see also Section 5.8) and the loess-dominated "yedoma" sediments of central and eastern Siberia (Zimov *et al.*, 1997). Fires and other disturbances are likely to affect the thermokarsts; however, the role of these disturbances—which can be mediated with changes in regional climate in inducing thermokarst—are poorly understood.

Changes in community composition associated with warming are likely to alter feedbacks to climate. Tundra has a three- to six-fold higher winter albedo than boreal forest, but summer albedo and energy partitioning differ more strongly among ecosystems within tundra or boreal forest than between these two biomes (Betts and Ball, 1997; Eugster *et al.*, 2000). If regional surface warming continues, changes in albedo and energy absorption during winter are likely to act as positive feedbacks to regional warming as a result of earlier melting of snow and, over the long term, poleward movement of the treeline. Surface drying and a change in dominance from mosses to vascular plants also would enhance sensible heat flux and regional warming in tundra (Lynch *et al.*, 1999; Chapin *et al.*, 2000).

Poleward migration of taxa from boreal forest to the Arctic tundra will depend not only on warming climate but also on dispersal rates, colonization rates, and species interactions and therefore may exhibit substantial time lags. The arctic historically has experienced fewer invasions of weeds and other exotic taxa than other regions (Billings, 1973). Some of the most important changes in diversity in the arctic may be changes in the abundance of caribou, waterfowl, and other subsistence resources (see also Section 5.4). Changes in community composition and productivity may be particularly pronounced in the high arctic, where much of the surface currently is unvegetated and is prone to establishment and expansion of additional vegetation (Wookey *et al.*, 1993; Callaghan and Jonasson, 1995).

5.9.2.2. *Impacts of Land-Use Change*

Extraction of oil and mineral resources is likely to be the greatest direct human disturbance in the arctic. Although the spatial extent of these disturbances is small, their impacts can be far-reaching because of road and pipeline systems associated with them (Walker *et al.*, 1987). Roads in particular open previously inaccessible areas to new development, either directly related to tourism and hunting or to support facilities for resource extraction.

Changes in goods and services in alpine ecosystems are likely to be dominated by changes in land use associated with grazing, recreation, and other direct impacts, as a result of their proximity to population centers (Körner, 1999; Walker *et al.*, 2001). Many of the alpine zones with greatest biodiversity, such as the Caucasus and Himalayas, are areas where human population pressures may lead to most pronounced land-use change in alpine zones (Akakhanjanz and Breckle, 1995). Direct impacts from human activities are likely to be most pronounced in lower elevational zones that are most accessible to people (Körner, 1999). Improved road access to alpine areas often increases human use for recreation, mining, and grazing and increased forestry pressure at lower elevations (Miller *et al.*, 1996). Overgrazing and trampling by people and animals may tend to destabilize vegetation, leading to erosion and loss of soils that are the long-term basis of the productive capacity of alpine ecosystems.

Alpine areas that are downwind of human population or industrial centers experience substantial rates of nitrogen deposition and acid rain (Körner, 1999). Continued nitrogen deposition at high altitudes, along with changes in land use that lead to soil erosion, can threaten provision of clean water to surrounding regions. Nitrogen deposition occurs primarily during the winter and is transmitted directly to streams during snowmelt, so it readily enters water supplies (Körner, 1999). Many places that use water from alpine areas depend on slow release of the water by melting of snowfields in spring and summer. Warming is likely to create a shortened snowmelt season, with rapid water release creating floods and, later, growing-season droughts. These changes in seasonality, combined with increased harvest in montane forests, would amplify floods.

5.9.3. *Adaptation Options*

Opportunities for adapting to expected changes in high-latitude ecosystems are limited because these systems will respond most strongly to globally induced changes in climate rather than to regionally controlled and regulated factors. The effects of climate and disturbance regime on regional productivity and carbon balance may be difficult to alter at the local or regional scale. The most important opportunities for mitigation will be protection of peatlands, yedoma sediments, and other carbon-rich areas from large-scale hydrological change. It is unlikely that expected changes in fire or thermokarst regime can be altered except in areas of high population densities. Adaptation options for forests that occur in these areas are presented in Section 5.6.

The most important opportunities for adaptation in arctic ecosystems may exist for culturally important resources such as reindeer, caribou, waterfowl, and specific plant taxa (see Section 5.4). Careful management of these resources could minimize climatic impacts on indigenous peoples. Another adaptation option is diversification. Many high-latitude regions depend strongly on one or a few resources—such as timber, oil, reindeer, or wages from fighting fires. Economic diversification would reduce the impacts of large changes in the availability or economic value of particular goods and services.

In alpine regions, major opportunities for adaptation relate to protection from large-scale changes in land use or pollutant levels. Because these causes of change are local to regional, wise management could minimize changes.

5.9.4. *Vulnerability*

Goods and services provided by many arctic regions depend on the physical integrity of permafrost and therefore are vulnerable to warming-induced thermokarst. The large carbon stocks in these regions are vulnerable to loss to the atmosphere as CO_2 or CH_4, which could act as positive feedbacks to global warming. A few key taxa of animals—such as reindeer, caribou, and waterfowl—are critical cultural resources. Often these taxa are concentrated in particular habitats, such as riparian areas or high-Arctic oases. Ecological changes that modify the population dynamics of these species or modify the impact on their habitats could have important social consequences.

High levels of endemism in many alpine flora and their inability to migrate upward mean that these species are most vulnerable. In addition, alpine soils are vulnerable to losses from erosion, which would radically reduce the goods and services that these regions could provide.

5.10. Research Needs across Ecosystems

Most of the foregoing sections mention the need for further research in many areas. Results are needed for future assessments of the impacts of climate change on goods and services provided by ecosystems, adaptation options, or identification of vulnerable systems or regions. It also is clear that most of the research to date has focused on impacts and some on adaptation, with little effort on the vulnerability of systems or regions. Key areas where research is needed to improve future assessments and decrease uncertainties in existing knowledge are as follows (order does not imply priority):

- Long-term monitoring of agricultural processes for detection of impacts of climate change and climate variability and development of possible adaptations to these impacts
- Coupling of pest and crop/animal process modeling to estimate climate-induced pest impacts on crops and livestock
- Pest and disease interaction with climate change and the impact of these interactions on many ecosystems
- Development of coupled biophysical-economic modeling of farm-level decisionmaking under climate change
- Improved understanding of agricultural vulnerability and adaptive capacity
- Improved understanding of interactive effects of climate change and management on carbon source-sink relations in many ecosystems
- Determination of mechanisms that create the often complex changes in wildlife populations, through a systems-type approach to problemsolving such as strategic cyclical scaling (SCS)[3]
- Development of better understanding of the link between biodiversity and ecosystems functions (role each species plays in ecosystem goods and services is necessary to understand the risks and possible surprises associated with species loss)
- The impact of land-use change on biodiversity and the consequence of these impacts on good and services from ecosystems
- The impact of the added stress of climate change on many ecosystems that are under pressure from human activities
- Exploration of management options (including individual species range) that could be used to adapt to the impacts of global change (allowing ecosystems to continue to provide essential goods and services in a changed and rapidly changing world)
- Exploration of the role of integrated management as a means for providing better management options for many ecosystems
- Long-term experiments on intact natural ecosystems across their native range to study the interactive effect of climate change, elevated CO_2, nutrient/pollution deposition, soil moisture, and WUE of plants
- The interaction of elevated CO_2, increased temperature, and changes in soil moisture with nitrogen deposition and land-use change in influencing the goods and services provided by ecosystems
- Valuation of nonmarket goods and services, such as recreation and NWFP
- Management of exotics and achieving desired community structures (research and alternative models for such activities needed in terrestrial and aquatic systems)

[3]SCS has three basic steps: detection, attribution, and validation (Root and Schneider, 1995). Many types of ecological data would improve the ability to conduct studies based on the SCS approach. First, additional monitoring systems are needed that would allow long-term data to be collected over a broad scale. Such data would provide the means to detect patterns of change more efficiently. Second, specific small-scale studies examining possible cause of a shifting pattern are needed. After several such studies are completed, these data could be used to validate the cause or mechanism.

- Development of a predictive approach for identifying and dealing with spatial heterogeneity, since responses to climate change differ greatly in adjacent areas (increasing the risk of inappropriate actions with large expenditures of local resources)
- Development of systems models for peatlands, including analysis and modeling of the dynamic interactions between climate change impacts, vegetation development, and carbon exchange between wetland ecosystems and the atmosphere at several spatial scales (from stands to regions to the globe)
- The role of fire and other disturbances on many ecosystems and their role in trace-gas budgets
- Time lags by which productivity, decomposition, and disturbance in ecosystems respond to climatic change
- Assessment of the vulnerability of unique biological resources that are culturally important to many indigenous peoples.

Acknowledgments

Thanks to Mark O'Brien, Kimberley Reade Hall, Alison Saunders, Margo Davies, Rukshika Parera, Sophie Maupin, Mary McKenney Easterling, the Center for Integrated Regional Assessment, and many more for their assistance. Thanks to all our families for their tolerance and understanding while the authors worked long hours as "volunteers" to complete this chapter. Thanks to the numerous reviewers for being so thorough in their reading and their constructive comments.

References

Abel, N., 1997: Mis-measurement of the productivity and sustainability of African communal rangelands—a case study and some principles from Botswana. *Ecological Economics*, **23(2),** 113–133.

Adams, R.M., B.H. Hurd, S. Lenhart, and N. Leary, 1998: Effects of global climate change on agriculture: an interpretive review. *Climate Research*, **11,** 19–30.

Adams, R.M., B.A. McCarl, K. Seerson, C. Rosenzweig, K.J. Bryant, B.L. Dixon, R. Conner, R.E. Evenson, and D. Ojima, 1999: The economic effects of climate change on U.S. agriculture. In: *Impacts of Climate Change on the U.S. Economy* [Mendelsohn, R. and J. Neumann (eds.)]. Cambridge University Press, Cambridge, United Kingdom and New York, NY, USA, pp. 55-74.

ADB, 1998: *The Bank's Policy on Water*. Working paper, Asian Development Bank, Manila, Philippines.

ADB, 1994: *Climate Change in Asia: Bangladesh Country Report, Volume 2: ADB Regional Study on Global Environmental Issues*. Asian Development Bank, Manila, Philippines.

Adrian, R. and T. Hintze, 2001: Effects of winter air temperature on the ice phenology of the Muggelsee (Berlin, Germany). *Verhandlungen der Internationalen Vereinigung fuer Limnologie*, (in press).

Aerts, R., 1997: Atmospheric nitrogen deposition affects potential denitrification and N_2O emission from peat soils in the Netherlands. *Soil Biology and Biochemistry*, **29,** 1153–1156.

Aerts, R. and F. Ludwig, 1997: Water-table changes and nutritional status affect trace gas emissions from laboratory columns of peatland soils. *Soil Biology and Biochemistry*, **29,** 1691–1698.

Aerts, R. and S. Toet, 1997: Nutritional controls of carbon dioxide and methane emission from Carex-dominated peat soils. *Soil Biology and Biochemistry*, **29,** 1683–1690.

Aerts, R., B. Wallén, and N. Malmer, 1992: Growth-limiting nutrients in Sphagnum-dominated bogs subject to low and high atmospheric nitrogen supply. *Journal of Ecology*, **80,** 131–140.

Aerts, R., R. van Logtestijn, M. van Staalduinen, and S. Toet, 1995: Nitrogen supply effects on productivity and potential leaf litter decay of Carex species from peatlands differing in nutrient limitation. *Oecologia*, **104,** 447–453.

Agakhanjanz, O. and S.W. Breckle, 1995: Origin and evolution of the mountain flora in middle Asia and neighboring mountain regions. In: *Arctic and Alpine Biodiversity: Patterns, Causes and Ecosystem Consequences* [Chapin, F.S. III and C. Körner (eds.)]. Springer-Verlag, Berlin, Germany, pp. 63–80.

Ainley, D.G., C.R. Ribic, and W.R. Fraser, 1994: Ecological structure among migrant and resident seabirds of the Scotia-Weddell confluence region. *Journal of Animal Ecology*, **63,** 347–364.

Akef Mamdouh, S.A. and H. Schneider, 1995: Calling behavior and mating call pattern in the mascarene frog, Ptychadena mascareniensis (Amphibia, Anura, Ranidae) in Egypt. *Journal of African Zoology*, **109,** 225–229.

Akey, D.H. and B.A. Kimball, 1989: Growth and development of the beet armyworm on cotton grown in enriched carbon dioxide atmospheres. *Southwest Entomology*, **14,** 255–260.

Akey, D.H., B.A. Kimball, and J.R. Mauny, 1988: Growth and development of the pink bollworm, Pectinophora gossypiella (Lepidoptera: Gelechiidae), on bolls of cotton grown in enriched carbon dioxide atmospheres. *Environmental Entomology*, **17,** 452–455.

Akin, D.E., B.A. Kimball, W.R. Windham, P.J. Pinter Jr., G.W. Wall, R.L. Garcia, R.L. La Morte, and W.H. Morrison III, 1995: Effect of free-air CO_2 enrichment (FACE) on forage quality. *Animal Feed Science and Technology*, **53,** 29–43.

Albon, S.D. and T.H. Clutton-Brock, 1988: Climate and population dynamics of red deer in Scotland. In: *Ecological Change in the Uplands* [Usher, M.B. and D.B.A. Thompson (eds.)]. Blackwell Scientific Publications, Oxford, United Kingdom, pp. 93–107.

Alcamo, J., R. Leemans, and E. Kreileman (eds.), 1998: *Global Change Scenarios of the 21st Century: Results from the IMAGE 2.1 Model*. Pergamon Press, Elsevior Science, Oxford, United Kingdom, 296 pp.

Alexandratos, N. (ed.), 1995: *World Agriculture: Towards 2010. An FAO Study*. John Wiley and Sons, Chichester, United Kingdom, 488pp.

Alexandrov, V., 1999: Vulnerability and adaptation of agronomic systems in Bulgaria. *Climate Research*, **12,** 161–173.

Alho Cleber, J.R. and M. Vieira Luiz, 1997: Fish and wildlife resources in the pantanal wetlands of Brazil and potential disturbances from the release of environmental contaminants. *Environmental Toxicology and Chemistry*, **16,** 71–74.

Alig, R., D. Adams, B. McCarl, J. Callaway, and S. Winnett, 1997: Assessing effects of mitigation strategies for global climate change with an intertemporal model of the U.S. forest and agriculture sectors. *Critical Reviews in Environmental Science and Technology*, **27,** S97–S111.

Allen, C.D. and D.D. Breshears, 1998: Drought-induced shift of a forest-woodland ecotone: rapid landscape response to climate variation. *Proceedings of the National Academy of Sciences of the United States of America*, 95, 14839–14842.

Allen-Diaz, B., 1996: Rangelands in a changing climate; impacts, adaptations, and mitigation. In: *Climate Change 1995: Impacts, Adaptions, and Mitigation of Climate Change: Scientific-Technical Analyses. Contribution of Working Group II to the Second Assessment Report for the Intergovernmental Panel on Climate Change* [Watson, R.T., M.C. Zinyowera, and R.H. Moss (eds.)]. Cambridge University Press, Cambridge, United Kingdom and New York, NY, USA, pp. 131–158.

Alm, J.J.T. Huttunen, H. Nykänen, J. Silvola, and P.J. Martikainen, 1997a: Greenhouse gas emissions from small eutrophic lakes in Finland. In: *Abstracts of Invited and Volunteer Papers for XIII International Symposium on Environmental Biogeography, Matter and Energy Fluxes in the Anthropogenic Environment, Sept. 21–26, 1997, Monopoli (Bari), Italy* [Senesi, N. and T.M. Miano (eds.)].

Alm, J., A. Talanov, S. Saarnio, J. Silvola, E. Ikkonen, H. Aaltonen, H. Nykänen, and P.J. Martikainen, 1997b: Reconstruction of the carbon balance for microsites in a boreal oligotrophic pine fen, Finland. *Oecologia*, **110,** 423–431.

Alward, R.D., J.K. Detling, and D.G. Milchunas, 1999: Grassland vegetation changes and nocturnal global warming. *Science*, **283(5399)**, 229–231.

Amien, I., P. Redjekiningrum, A. Pramudia, and E. Susanti, 1996: Effect of interannual climate variability and climate change on rice yield in Java, Indonesia. *Water, Air, and Soil Pollution*, **92**, 29–39.

Ammann, B., H.J.B. Birks, S.J. Brooks, U. Eicher, B. von Grafenstein, U. Hofmann, W. Lemdahl, G. Schwander, J. Tobolski, and K.L. Wick, 2000: Quantification of biotic responses to rapid climatic changes around the Younger Dryas—a synthesis. *Palaeogeography, Palaeoclimatology, Palaeoecology*, **159**, 313–347.

Amthor, J.S., 1997: Plant respiratory responses to elevated carbon dioxide partial pressure. *Advances in Carbon Dioxide Research*, **61**, 35–77.

Anderson, A.R., D.G. Pyatt, J.M. Sayers, S.R. Blackhall, and H.D. Robinson, 1992: Volume and mass budgets of blanket peat in the north of Scotland. *Suo*, **43**, 195–198.

Anderson, R.S., 1996: Postglacial biogeography of Sierra lodgepole pine (Pinus contorta var murrayana) in California. *Ecoscience*, **3**, 343–351.

André, M. and H. Du Cloux, 1993: Interaction of CO_2 enrichment and water limitations on photosynthesis and water efficiency in wheat. *Plant Physiology and Biochemistry*, **31**, 103–112.

Andrewartha, H.G. and L.C. Birch, 1954: *The Distribution and Abundance of Animals*. University of Chicago Press, Chicago, IL, USA, 782 pp.

Anon., 1987: *Chinese Forestry Yearbook, 1987*. Beijing, China.

Antle, J.M., 1996a: Why is World agriculture still in disarray? In: *Economics of Agriculture, Vol 2* [Antle, J.M. and D.A. Summer (eds)]. Chicago, University of Chicago Press, pp. 400-419.

Antle, J.M., 1996b: Methodological issues in assessing potential impacts of climate change on agriculture. *Agricultural and Forest Meteorology*, **80**, 67–85.

Antle, J.M., S.M. Capalbo, and J. Hewitt, 2000: Testing hypotheses in integrated impact assessments: climate variability and economic adaptation in great plains agriculture. National Institute for Global Environmental Change, University of California, Davis, CA, USA, Available online at http://nigec.ucdavis.edu.

Antle, J.M., D. Hayes, S. Mohanty, P. Vavra, and V.H. Smith, 1999a: Long-term supply and demand trends: whither the real price of Wheat? In: *The Economics of World Wheat Markets* [Antle, J.M. and V.H. Smith (eds.)]. CAB International, Montana State University, USA, pp. 39–76.

Antle, J.M., S.M. Capalbo, and J. Hewitt, 1999b: Testing hypotheses in integrated impact assessments: climate variability and economic adaptation in great plains agriculture. National Institute for Global Environmental Change, Nebraska Earth Science Education Network, University of Nebraska-Lincoln, Lincoln, NE, USA, pp. T5-4 and 21. Available online at http://nesen.unl.edu/nigec/facts/projects/Antle98.html.

Apps, M.J., 1993: NBIOME: a biome-level study of biospheric response and feedback to potential climate changes. *World Research Review*, **5(1)**, 41–65.

Apps, M.J. and D.T. Price, 1996: *Forest Ecosystems, Forest Management, and the Global Carbon Cycle*. NATO ASI Series, Subseries 1: Global Environmental Change, Springer-Verlag, Heidelberg, Germany, Vol. 40, 452 pp.

Apps, M.J., J.S. Bhatti, D. Halliwell, H. Jiang, and C. Peng, 2000: Simulated carbon dynamics in the boreal forest of central Canada under uniform and random disturbance regimes. In: *Global Climate Change and Cold Regions Ecosystems* [Lal, R., J.M. Kimble, and B.A. Stewart (eds.)]. Advances in Soil Science, Lewis Publishers, Boca Raton, FL, USA, pp. 107–121.

Apps, M.J., W.A. Kurz, R.J. Luxmoore, L.O. Nilsson, R.A. Sedjo, R. Schmidt, L.G. Simpson, and T.S. Vinson, 1993: Boreal forests and tundra. *Water, Air, and Soil Pollution*, **70(1–4)**, 39–53.

Archer, S., 1995: Herbivore mediation of grass-woody plant interactions (Harry Stobbs Memorial Lecture, 1993). *Tropical Grasslands*, **29(4)**, 218–235.

Archer, S., D.S. Schimel, and E.A. Holland, 1995: Mechanisms of shrubland expansion—land use, climate or CO_2. *Climatic Change*, **29(1)**, 91–99.

Armentano, T.V. and E.S. Menges, 1986: Patterns of change in the carbon balance of organic soil -wetlands of the temperate zone. *Journal of Ecology*, **74**, 755–774.

Arnell, N., B. Bates, H. Lang, J. Magnuson, P. Mulholland, S. Fisher, C. Liu, D. McKnight, O. Starosolszky, and M. Taylor, 1996: Hydrology and freshwater ecology. In: *Climate Change 1995: Impacts, Adaptions, and Mitigation of Climate Change: Scientific-Technical Analyses. Contribution of Working Group II to the Second Assessment Report for the Intergovernmental Panel on Climate Change* [Watson, R.T., M.C. Zinyowera, and R.H. Moss (eds.)]. Cambridge University Press, Cambridge, United Kingdom and New York, NY, USA, pp. 327–363.

Arnell, N.W., M.G.R. Cannell, M. Hulme, J.F.B. Mitchell, R.S. Kovats, R.J. Nicholls, M.L. Parry, M.T.J. Livermore, and A. White, 2001: The consequences of CO_2 stabilisation for the impacts of climate change. *Climate Change* (in press).

Arntzen, J. and S. Ringrose, 1996: Changes in rangelands. In: *Climate Change and Southern Africa: An Exploration of Some Potential Impacts and Implications in the SADC Region* [Hulme, M. (ed.)]. Climatic Research Unit, University of East Anglia, Norwich, United Kingdom and World Wildlife Fund International, Gland, Switzerland, pp. 62–71.

Aronson, J., E. Lefloch, J.F. David, S. Dhillion, M. Abrams, J.L. Guillerm, and A. Grossmann, 1998: Restoration ecology studies at Cazarils (southern France): biodiversity and ecosystem trajectories in a mediterranean landscape. *Landscape and Urban Planning*, **41(3–4)**, 273–283.

Arseneault, D. and S. Payette, 1997: Reconstruction of millennial forest dynamics from tree remains in a subarctic tree line peatland. *Ecology*, **78**, 1873–1883.

Aspelin, A.L. and A.H. Grube, 1999: *Pesticide Industry Sales and Usage: 1996 and 1997 Market Estimates*. Environmental Protection Agency, Washington, DC, USA, 40 pp.

Asplund, D., 1996: Energy use of peat. In: *Global Peat Resources* [Lappalainen, E. (ed.)]. International Peat Society, Jyväskylä, Finland, pp. 319–325.

Aurela, M., J.-P. Tuovinen, and T. Laurila, 1998: Carbon dioxide exchange in a sub-Arctic peatland ecosystem in northern Europe measured by the eddy covariance technique. *Journal of Geophysical Research*, **103**, 11289–11301.

Aust, W.M. and R. Lea, 1991: Soil temperature and organic matter in a disturbed forested wetland. *Soil Science Society of America Journal*, **55**, 1741–1746.

Austarå, Ö., 1971: Cold hardiness in eggs of Neodiprion sertifer (Geoffroy) (Hym., Diprionidae) under natural conditions. *Norsk Ent. Tidsskr*, **18**, 45–48.

Awmack, C.S., R. Harrington, and S.R. Leather, 1997a: Host plant effects on the performance of the aphid Aulacorthum solani (Kalt.) (Homoptera: Aphidae) at ambient and elevated CO_2. *Global Change Biology*, **3**, 545–549.

Awmack, C.S., C.M. Woodcock, and R. Harrington, 1997b: Climate change may increase vulnerability of aphids to natural enemies. *Ecological Entomology*, **22**, 366–368.

Ayres, M.P., 1993: Plant defense, herbivory, and climate change. In: *Biotic Interactions and Global Change* [Kareiva, P.M., J.G. Kingsolver, and R.B. Huey (eds.)]. Sinauer Associates Inc., Sunderland, MA, USA, pp. 75–94.

Baker, B.B., J.D. Hanson, R.M. Bourdon, and J.B. Eckert, 1993: The potential effects of climate change on ecosystem processes and cattle production on U.S. rangelands. *Climatic Change*, **25**, 97–117.

Ball, T., 1983: The migration of geese as an indicator of climate change in the southern Hudson Bay region between 1715 and 1851. *Climatic Change*, **5**, 85–93.

Balling, R.C. Jr., 1982: Weight gain and mortality in feedlot cattle as influenced by weather conditions: refinement and verification of statistical models. In: *Center for Agricultural Meteorology and Climatology Report 82–1*. University of Nebraska-Lincoln, Lincoln, NE, USA, 52 pp.

Baradas, M.W., 1999: Using PVC pipes to double the irrigated area in the Philippines. In: *Proceedings, Philippine Society of Agricultural Engineers' National Convention, General Santos City, Philippines, April 26–29, 1999*.

Barber, V.A., G.P. Juday, and B.P. Finney, 2000: Reduced growth of Alaskan white spruce in the twentieth century from temperature-induced drought stress. *Nature*, **405**, 668–673.

Barnard, P., C.J. Brown, A.M. Jarvis, A. Robertson, and L. Vanrooyen, 1998: Extending the namibian protected area network to safeguard hotspots of endemism and diversity. *Biodiversity and Conservation*, **7(4)**, 531–547.

Baron, J.S., M.D. Hartman, T.G.F. Kittel, L.E. Band, D.S. Ojima, and R.B. Lammers, 1998: Effects of land cover, water redistribution, and temperature on ecosystem processes in the South Platte Basin. *Ecological Applications*, **8(4),** 1037–1051.

Baroni, C. and G. Orombelli, 1994: Abandoned penguin rookeries as Holocene paleoclimatic indicators in Antarctica. *Geology*, **22,** 23–26.

Barrow, E.M. and M.A. Semenov, 1995: Climate change scenarios with high spatial and temporal resolution for agricultural application. *Forestry*, **68,** 349–360.

Batchelor, C.H., C.J. Lovell, and A.J. Semple, 1994: Garden irrigation for improving agricultural sustainability in dryland areas. *Land Use Policy*, **11(4),** 286–293.

Batts, G.R., R.H. Ellis, J.I.L. Morison, and P. Hadley, 1998: Canopy development and tillering of field grown crops of two contrasting cultivars of winter wheat (Triticum aestivum) in response to CO_2 and temperature. *Annals of Applied Biology*, **133,** 101–109.

Bawa, K.S., 1995: Pollination, seed dispersal, and diversification of angiosperms. *Trends in Ecology and Evolution*, **10,** 311–312.

Bawa, K.S. and M. Gadgil, 1997: Ecosystem services in subsistence economies and conservation of biodiversity. In: *Nature's Services: Societal Dependence on Natural Ecosystems* [Daily, G. (ed.)]. Island Press, Washington, DC, USA, pp. 295–310.

Bayley, J.R. and A.C. Highfield, 1996: Observations on ecological changes threatening a population of Testudo graeca graeca in the Souss Valley, Southern Morocco. *Chelonian Conservation and Biology*, **2,** 36–42.

Bayliss, B.L., K.G. Brennan, I. Eliot, C.M. Finlayson, R.N. Hall, T. House, R.W.J. Pidgeon, D. Walden, and P. Waterman, 1997: *Vulnerability Assessment of Predicted Climate Change and Sea Level Rise in the Alligator Rivers Region, Northern Territory, Australia*. Report 123, Supervising Scientist, Canberra, Australia, 134.

Bazzaz, A., J.S. Coleman, and S.R. Morse, 1990: Growth responses of seven major co-occuring tree species of the northeastern United States to elevated CO_2. *Canadian Journal of Forest Research*, **20,** 1479–1484.

Bazzaz, F.A., 1996: Succession and global change: will there be a shift toward more early successional systems? In: *Plants in Changing Environments: Linking Physiological, Population, and Community Ecology* [Bazaaz, F.A. (ed)]. Cambridge University Press, Cambridge, United Kingdom and New York, NY, USA, pp. 264–279.

Beck, B.B., L.G. Rapaport, M.R. Stanley Price, and A.C. Wilson, 1994: Reintroduction of captive-born animals. In: *Creative Conservation: Interactive Management of Wild and Captive Animals* [Olney, P.J.S., G.M. Mace, and A.T.C. Feistner (eds.)]. Chapman and Hall, London, United Kingdom, pp. 265–286.

Beebee, T.J.C., 1995: Amphibian breeding and climate. *Nature*, **374,** 219–220.

Beerling, D.J., F.I. Woodward, M. Lomas, and A.J. Jenkins, 1997: Testing the responses of a dynamic global vegetation model to environmental change—a comparison of observations and predictions. *Global Ecology and Biogeography Letters*, **6,** 439–450.

Beerling, D.J. and C.K. Kelly, 1997: Stomatal density repsonses of temperate woodland plants over the past seven decades of CO_2 increase: a comparison of Salisbury (1927) with contemporary data. *American Journal of Botany*, **84,** 1572–1583.

Belnap, J., 1995: Surface disturbances—their role in accelerating desertification. *Environmental Monitoring and Assessment*, **37(1–3),** 39–57.

Belnap, J. and D.A. Gillette, 1998: Vulnerability of desert biological soil crusts to wind erosion—the influences of crust development, soil texture, and disturbance. *Journal of Arid Environments*, **39(2),** 133–142.

Belyea, L.R. and R.S. Clymo, 1999: Do hollows control the rate of peat bog growth? In: *Patterned Mires: Origin and Development, Flora and Fauna* [Meade, R., V. Standen, and J.H. Tallis (eds.)]. British Ecological Society Mires Research Group, London, United Kingdom, pp. 1–15.

Beniston, M., 1997: Variation of snow depth and duration in the Swiss Alps over the last 50 years: links to changes in large-scale climatic forcings. *Climatic Change*, **36,** 281–300.

Beniston, M. and D.G. Fox, 1996: Impacts of climate change on mountain regions. In: *Climate Change 1995: Impacts, Adaptions, and Mitigation of Climate Change: Scientific-Technical Analyses. Contribution of Working Group II to the Second Assessment Report* [Watson, R.T., M.C. Zinyowera, and R.H. Moss (eds.)]. Cambridge University Press, Cambridge, United Kingdom and New York, NY, USA, pp. 191–214.

Benites, J.R. and C.S. Ofori, 1993: Crop production through conservation-effective tillage in the tropics. *Soil and Tillage Research*, **27(1–4),** 9–33.

Benjaminsen, T.A., 1993: Fuelwood and desertification—sahel orthodoxies(1) discussed on the basis of field data from the gourma region in mali. *Geoforum*, **24(4),** 397–409.

Berg, B., M.P. Berg, P. Bottner, E. Box, A. Breymeyer, R. Calvo de Anta, M.M. Couteaux, A. Escudero, A. Gallardo, W. Kratz, M. Madeira, E. Mälkönen, C. McGlaugherty, V. Meentemeyer, F. Munoz, P. Piussi, J. Remacle, and A. Virzo de Santo, 1993: Litter mass loss rates in pine forests of Europe and eastern United States: some relationships with climate and litter quality. *Biogeochemistry*, **20,** 127–159.

Berger, L., R. Speare, P. Daszak, D.E. Green, A.A. Cunningham, C.L. Goggin, R. Slocombe, M.A. Ragan, A.D. Hyatt, K.R. McDonald, H.B. Hines, K.R. Lips, G. Marantelli, and H. Parkes, 1998: Chytridiomycosis causes amphibian mortality associated with population declines in the rain forests of Australia and Central America. *Proceedings of the National Academy of Sciences of the United States of America*, **95,** 9031–9036.

Bergman, I., B.H. Svensson, and M. Nilsson, 1998: Regulation of methane production in a Swedish acid mire by pH, temperature and substrate. *Soil Biology and Biochemistry*, **30,** 729–741.

Berntson, G.M., N. Rajakaruna, and F.A. Bazzaz, 1998: Growth and nitrogen uptake in an experimental community of annuals exposed to elevated atmospheric CO_2. *Global Change Biology*, **4(6),** 607–626.

Berven, K.A., 1982a: The genetic basis of altitudinal variation in the wood frog Rana sylvatica, I: an experimental analysis of life history traits. *Evolution*, **36,** 962–983.

Berven, K.A., 1982b: The genetic basis of altitudinal variation in the wood frog Rana sylvatica, II: an experimental analysis of larval development. *Oecologia*, **52,** 360–369.

Berz, G.A, 1999: Catastrophes and climate change: concerns and possible countermeasures of the insurance industry. *Mitigation and Adaptation Stategies for Global Change*, **4,** 283–293.

Betts, A.K. and J.H. Ball, 1997: Albedo over the boreal forest. *Journal of Geophysical Research—Atmospheres*, **102,** 28901–28909.

Bezemer, T.M. and T.H. Jones, 1998: Plant-insect herbivore interactions in elevated atmospheric CO_2: quantitative analyses and guild effects. *Oikos*, **82,** 212-22.

Bezzel, E. and W. Jetz, 1995: Delay of the autumn migratory period in the Blackcap (Sylvia atricappila) 1966–1993: a reaction to global warming? *Journal Für Ornithologie*, **136,** 83–87.

Bhatti, J.S., M.J. Apps, and H. Jiang, 2000: Examining the carbon stocks of boreal forest ecosystems at stand and regional scales. In: *Assessment Methods for Soil C Pools* [Lal, R., J.M. Kimble, R.F. Follett, and B.A. Stewart (eds.)]. Advances in Soil Science Series, CRC Press, Boca Raton, FL, USA, pp. 513–531.

Bibby, C.J., N.J. Collar, M.J. Crosby, M.F. Heath, C. Imboden, T.H. Johnson, A.J. Long, A.J. Stattersfield, and S.J. Thurgood, 1992: *Putting Biodiversity on the Map: Priority Areas for Global Conservation*. International Council for Bird Preservation, Cambridge, United Kingdom, 90 pp.

Biber, J.P. and T.T. Salathe, 1991: Threats to migratory birds. In: *Conserving Migratory Birds. ICBP Technical Publication No. 12* [Salathe, T. (ed.)]. International Council for Bird Preservation, Cambridge, United Kingdom, pp. 17–35.

Billings, W.D., 1987: Carbon balance of Alaskan tundra and taiga ecosystems: past, present and future. *Quaternary Science Reviews*, **6,** 165–177.

Billings, W.D., 1973: Arctic and alpine vegetation: similarities, differences, and susceptibility to disturbance. *BioScience*, **23,** 697–704.

Binkley, C.S., M.J. Apps, R.K. Dixon, P.E. Kauppi, and L.O. Nilsson, 1997: Sequestering carbon in natural forests. *Critical Reviews in Environmental Science and Technology*, **27,** S23–S45.

Binkley, C.S, 1988: A case study of the effects of CO_2-induced climatic warming on forest growth and the forest sector, B: economic effects on the world's forest sector. In: *The Impact of Climatic Variations on Agriculture* [Parry, M.L., T.R. Carter, and N.T. Konijn (eds.)]. J. Kluwer Academic Publishers, Dordrecht, The Netherlands.

Birkhead, T.R. and R.W. Furness, 1984: Regulation of seabird populations. In: *Behavioural Ecology, Ecological Consequences of Adaptive Behaviour* [Sibly, R.M. and R.H. Smith (eds.)]. Blackwell Scientific Publications, London, United Kingdom, pp. 145–67.

Bizer, J.R., 1978: Growth rates and size at metamorphosis of high elevation populations of Ambystoma tigrinum. *Oecologia*, **34,** 175–184.

Björkman, L., 1999: The establishment of *Fagus sylvatica* at the stand-scale in southern Sweden. *Holocene*, **9,** 237–245.

Black, T.A., W.J. Chen, A.G. Barr, M.A. Arain, Z. Chen, Z. Nesic, E.H. Hogg, H.H. Neumann, and P.C. Yang, 2000: Increased carbon sequestration by a boreal deciduous forest in years with a warm spring. *Geophysical Research Letters*, **27,** 1271–1274.

Bloomberg, W.J., 1962: Cytospora canker of poplars: factors influencing the development of the disease. *Canadian Journal of Botany*, **40,** 1271–1280.

Blumberg, A.F. and D.M. Di Toro, 1990: Effects of climate warming on dissolved oxygen concentrations in Lake Erie. *Transactions of the American Fisheries Society,* **119(2),** 210–223.

Blumentahl, C., H.M. Rawson, E. McKenzie, P.W. Gras, E.W.R. Barlow, and C.W. Wrigley, 1996: Changes in wheat grain quality due to doubling the level of atmospheric CO_2. *Cereal Chemistry*, **73,** 762–766.

Bock, C.E., J.H. Bock, M.C. Grant, and T.R. Seastedt, 1995: Effects of fire on abundance of eragrostis intermedia in a semi-arid grassland in southeastern Arizona. *Journal of Vegetation Science*, **6,** 325–328.

Bolin, B. and R. Sukumar, 2000: Global perspective. In: *Land Use, Land-Use Change, and Forestry* [Watson, R.T., I.R. Noble, B. Bolin, N.H. Ravindranath, D.J. Verardo, and D.J. Dokken (eds.)]. Special Report of the Intergovernmental Panel on Climate Change, Cambridge University Press, Cambridge, United Kingdom and New York, NY, USA, pp. 23–51.

Bolortsetseg, B. and G. Tuvaansuren, 1996: The potential impacts of climate change on pasture and cattle production in Mongolia. *Water, Air, and Soil Pollution*, **92,** 95–105.

Bomford, M. and J. Caughley (eds.), 1996: *Sustainable Use of Wildlife by Aboriginal Peoples and Torres Strait Islanders*. Australian Government Publishing Service, Canberra, Australia, 216 pp.

Bonan, G.B., 1993: Physiological controls of the carbon balance of boreal ecosystems. *Canadian Journal of Forest Research,* **23,** 1453-1471.

Bond, W.J., W.D. Stock, and M.T. Hoffman, 1994: Has the karoo spread? A test for desertification using carbon isotopes from soils. *South African Journal of Science*, **90(7),** 391–397.

Bos, E., M. Vu, E. Massiah, and R. Bulatao, 1994: *World Population Projections, 1994–95*. World Bank, Johns Hopkins University Press, Baltimore, MD, USA, 532 pp.

Boskey, A.L., 1998: Will biomimetics provide new answers for old problems of calcified tissues? *Calcified Tissue International*, **63,** 179–182.

Botch, M.S., K.I. Kobak, T.S. Vinson, and T.P. Kolchugina, 1995: Carbon pools and accumulation in peatlands of the former Soviet Union. *Global Biogeochemical Cycles*, **9,** 37–46.

Bothwell, M.L., D.M.J. Sherbot, and C.M. Pollack, 1994: Ecosystem response to solar ultraviolet-B radiation: influence of trophic-level interactions. *Science*, **265,** 97–100.

Bousquet, P., P. Ciais, P. Peylin, M. Ramonet, and P. Monfray, 1999: Inverse modeling of annual atmospheric CO_2 sources and sinks, part 1: method and control inversion. *Journal of Geophysical Research*, **104,** 26161–26178.

Boutaleb Joutei, A., J. Roy, G. Van Impe, and P. Lebrun, 2000: Effect of elevated CO_2 on the demography of a leaf-sucking mite feeding on bean. *Oecologia*, **23(1),** 75-81.

Bowes, G., 1993: Facing the inevitable: plants and increasing atmospheric CO_2. *Annual Review of Plant Physiology*, **44,** 309–332.

Bowes, G., J.C.V. Vu, M.W. Hussain, A.H. Pennanen, and L.H. Allen Jr., 1996: An overview of how rubisco and carbohydrate metabolism may be regulated at elevated atmospheric (CO_2) and temperature. *Agricultural and Food Science in Finland*, **5,** 261–270.

Box, E.O., 1981: *Macroclimate and Plant Forms: An Introduction to Predictive Modeling in Phytogeography*. Dr. W. Junk Publishers, The Hague, The Netherlands, 272 pp.

Boyer, D.A. and R.D. Brown, 1988: A survey of translocation of mammals in the United States. In: *Translocation of Wild Animals* [Nielson, L. and R.D. Brown (eds.)]. Wisconsin Humane Society, Milwaukee, WI, and Caesar Kleberg Wildlife Research Institute, Kingsville, TX, USA, pp. 1–11.

Bradbury, I.K. and J. Grace, 1983: Primary production in wetlands. In: *Ecosystems of the World, 4A, Mires: Swamp, Bog, Fen and Moor. General Studies* [Gore, A.J.P. (ed.)]. Elsevier Scientific Publishing Company, Amsterdam, The Netherlands; Oxford, United Kingdom; and New York, NY, USA, pp. 285–310.

Bradley, N.L., A.C. Leopold, J. Ross, and W. Huffaker, 1999: Phenological changes reflect climate change in Wisconsin. *Proceedings of the National Academy of Sciences of the United States of America*, **96,** 9701–9704.

Bradstock, R., A.M. Bedward, B.J. Kenny, and J. Scott 1998: Spatially-explicit simulation of the effect of prescribed burning on fire regimes and plant extinctions in shrublands typical of south-eastern Australia. *Biological Conservation*, **86,** 83–95.

Brady, M.A., 1997: Effects of vegetation changes on organic matter dynamics in three coastal peat deposits in Sumatra, Indonesia. In: *Biodiversity and Sustainability of Tropical Peatlands* [Rieley, J.O. and S.E. Page (eds.)]. Samara Publishing, Cardigan, Wales, United Kingdom, pp. 113–134.

Braekke, F.H. and L. Finer, 1991: Fertilization effects on surface peat of pine bogs. *Scandinavian Journal of Forest Research*, **6,** 433–449.

Branfireun, B.A. and N.T. Roulet, 1998: The baseflow and stormflow hydrology of a Precambrian Shield headwater peatland. *Hydrological Processes*, **12,** 57–72.

Brasier, C.M., 1996: Phytophthora cinnamomi and oak decline in southern Europe: environmental constraints including climate change. *Annales des Sciences Forestieres*, **53(2–3),** 347–358.

Braun-Blaunquet, J., 1956: Ein Jahrhundert Florenwandel am Piz Linard (3414 m). *Bulleti Jardin Botanique de l'Etat Bruxelles*, **26,** 221–232.

Brereton, R., S. Bennett, and I. Mansergh, 1995: Enhanced greenhouse climate change and its potential effect on selected fauna of south-eastern Australia—a trend analysis. *Biological Conservation*, **72,** 339–354.

Briffa, K.R., F.H. Schweingruber, P.D. Jones, T.J. Osborn, S.G. Shiyatov, and E.A. Vaganov, 1998: Reduced sensitivity of recent tree-growth to temperature at high northern latitudes. *Nature*, **391,** 678–682.

Brklacich, M., D. McNaab, C. Bryant, and J. Dumnski, 1997: Adaptability of agricultural systems to global climate change: a Renfrew County, Ontario, Canada pilot study. In: *Agricultural Restrecturing and Sustainability: A Geographical Perspective* [Ilbery, B. and T. Rickard (eds.).] CAB International, Wallingford, United Kingdom.

Brooks, D., 1997: The outlook for demand and supply of wood: implications for policy and sustainable management. *Commonwealth Forestry Review*, **76(1),** 31–36.

Brooks, D.J., H. Pajuoja, T.J. Peck, B. Solberg, and P.A. Wardle, 1996: Long term trends and prospects in world supply and demand for wood. In: *Long Term Trends and Prospects for World Supply and Demand for Wood and Implications for Sustainable Management* [Solberg, B. (ed.)]. Research Report 6, European Forest Institute and Norwegian Forest Research Insitute, Joensuu, Finland, pp. 75–106.

Brotton, J. and G. Wall, 1997: Climate change and the Bathurst caribou herd in the Northwest Territories, Canada. *Climatic Change*, **35,** 35–52.

Brown, J., O.J.J. Ferrians, J.A. Heginbottom, and E.S. Melnikov, 1997a: *International Permafrost Association Circum Arctic Map of Permafrost and Ground Ice Conditions*. U.S. Geological Survey, Circum Pacific Map Series, Map CP 45, Scale 1:10,000,000, Washington, DC, USA.

Brown, J.H., T.J. Valone, and C.G. Curtin, 1997b: Reorganization of an arid ecosystem in response to recent climate change. *Proceedings of the National Academy of Sciences of the United States of America*, **94,** 9729–9733.

Brown, J.L., S.-H. Li, and N. Bhagabati, 1999: Long-term trend toward earlier breeding in an American bird: a response to global warming? *Proceedings of the National Academy of Sciences of the United States of America*, **96,** 5565–5569.

Brown, J.R. and S. Archer, 1999: Shrub invasion of grassland: recruitment is continuous and not regulated by herbaceous biomass or density. *Ecology*, **80,** 2385–2396.

Brown, J.R. and J. Carter, 1998: Spatial and temporal patterns of exotic shrub invasion in an Australian tropical grassland. *Landscape Ecology*, **13(2),** 93–102.

Brown, R.A. and N.J. Rosenberg, 1999: Climate change impacts on the potential productivity of corn and winter wheat in their primary United States growing regions. *Climatic Change*, **41,** 73–107.

Brown, S., 1998: Present and future role of forests in global climate change. In: *Ecology Today: An Anthology of Contemporary Ecological Research* [Goapl, B., P.S. Pathak, and K.G. Saxena (eds.)]. International Scientific Publications, New Delhi, India, pp. 59–74.

Brown, S. and P.E. Schroeder, 1999: Spatial patterns of aboveground production and mortality of woody biomass for eastern U.S. forests. *Ecological Applications*, **9**, 968–980.

Brown, S., J. Sathaye, M. Cannell, and P. Kauppi, 1996: Management of forests for mitigation of greenhouse gas emissions. In: *Climate Change 1995: Impacts, Adaptions, and Mitigation of Climate Change: Scientific-Technical Analyses. Contribution of Working Group II to the Second Assessment Report* [Watson, R.T., M.C. Zinyowera, and R.H. Moss (eds.)]. Cambridge University Press, Cambridge, United Kingdom and New York, NY, USA, 878pp.

Buan, R.D., A.R. Maglinao, P.P. Evangelista, and B.G. Pajuelas, 1996: Vulnerability of rice and corn to climate change in The Philippines. *Water, Air, and Soil Pollution*, **92**, 41–51.

Bubier, J.L., 1995: The relationship of vegetation to methane emission and hydrochemical gradients in northern peatlands. *Journal of Ecology*, **83**, 403–420.

Buchmann, S.L. and G.P. Nabhan, 1996: *The Forgotten Pollinators*. Island Press, Washington, DC, USA, 320pp.

Buck, B.J. and H.C. Monger, 1999: Stable isotopes and soil-geomorphology as indicators of Holocene climate change, northern Chihuahuan Desert. *Journal of Arid Environments*, **43**, 357–373.

Bugmann, H.K.M., 1997: Sensitivity of forests in the European Alps to future climatic change. *Climate Research*, **8**, 35–44.

Bugmann, H.K.M. and A.M. Solomon, 1995: The use of a European forest model in North America: a study of ecosystem response to climate gradients. *Journal of Biogeography*, **2(2–3)**, 477–484.

Bugmann, H.K.M., H.A. Fischlin, and F. Kienast, 1996: Model convergence and state variable update in forest gap models. *Ecological Modelling*, **89**, 197–208.

Bullock, P. and H. Le Houérou, 1996: Land degredation and desertification. In: *Climate Change 1995: Impacts, Adaptions, and Mitigation of Climate Change: Scientific-Technical Analyses. Contribution of Working Group II to the Second Assessment Report* [Watson, R.T., M.C. Zinyowera, and R.H. Moss (eds.)]. Cambridge University Press, Cambridge, United Kingdom and New York, NY, USA, pp. 171–190.

Burgis, M.J. and P. Morris, 1987: *The Natural History of Lakes*. Cambridge University Press, Cambridge, United Kingdom and New York, NY, USA, 218 pp.

Burton, K.L., W.R. Rouse, and L.D. Boudreau, 1996: Factors affecting the summer carbon dioxide budget of sub-Arctic wetland tundra. *Climate Research*, **6**, 203–213.

Busby, J.R., 1988: Potential impacts of climate change on Australia's flora and fauna. In: *Greenhouse: Planning for Climate Change* [Pearman, G.I. (ed.)]. Commonwealth Scientific and Industrial Research Organisation, Melbourne, FL, USA, pp. 387–398.

Cahoon, D.R., B.J. Stocks, J.S. Levine, W.R. Cofer, and J.M. Pierson, 1994: Satellite analysis of the severe 1987 forest fires in northern China and southeastern Siberia. *Journal of Geophysical Research*, **99(D9)**, 18627–18638.

Calef, G.W., 1973: Natural mortality of tadpoles in a population of Rana aurora. *Ecology*, **54**, 741–758.

Callaghan, T.V. and S. Jonasson, 1995: Implications for changes in Arctic plant biodiversity from environmental manipulation experiments. In: *Arctic and Alpine Biodiversity: Patterns, Causes, and Ecosystem Consequences* [Chapin, F.S. III and C. Körner (eds.)]. Springer-Verlag, Berlin, Germany, pp. 151–166.

Campbell, B.M., M. Luckert, and I. Scoones, 1997: Local-level valuation of savanna resources: a case study from zimbabwe. *Economic Botany*, **51(1)**, 59–77.

Campbell, I.D. and J.H. McAndrews, 1993: Forest disequilibrium caused by rapid Little Ice Age Cooling. *Nature*, **366**, 336–338.

Campbell, I.D. and M.D. Flannigan, 2000: Long-term perspectives on fire-climate-vegetation relationships in the North American boreal forest. In: *Boreal Forests* [Stocks, B.J. and E. Kasischke (eds.)]. Springer-Verlag, New York, NY, USA, pp. 1–29.

Campeau, S. and L. Rochefort, 1996: Sphagnum regeneration on bare peat surfaces: field and greenhouse experiments. *Journal of Applied Ecology*, **33**, 599–608.

Canadell, J.G., L.F. Pitelka, and J.S.I. Ingram, 1996: The effects of elevated CO_2 on plant-soil carbon below-ground: a summary and synthesis. *Plant and Soil*, **187**, 391–400.

Cannell, M.G.R. and R.C. Dewar, 1995: The carbon sink provided by plantation forests and their products in Britain. *Forestry*, **68**, 35–48.

Candau, J.-N., R.A. Fleming, and A.A. Hopkin, 1998: Spatio-temporal patterns of large-scale defoliation caused by the spruce budworm in Ontario since 1941. *Canadian Journal of Forest Research*, **28**, 1–9.

Canziani, O.F. and S. Diaz, 1998: Latin America. In: *The Regional Impacts of Climate Change: An Assessment of Vulnerability. Special Report of IPCC Working Group II* [Watson, R.T., M.C. Zinyowera, and R.H. Moss (eds.)]. Cambridge University Press, Cambridge, United Kingdom and New York, NY, USA, pp. 187–230.

Cao, M., S. Marshal, and K. Gregson, 1996: Global carbon exchange and methane emissions from natural wetlands: application of a process-based model. *Journal of Geophysical Research*, **101**, 14399–14414.

Cardon, Z.G., 1996: Influence of rhizodeposition under elevated CO_2 on plant nutrition and soil organic matter. *Plant and Soil*, **187**, 277–288.

Carroll, A.L., J. Hudak, J.P. Meades, J.M. Power, T. Gillis, P.J. McNamee, C.H.R. Wedeles, and G.D. Sutherland, 1995: EHLDSS—a decision support system for management of the eastern hemlock looper. In: *Proceedings of Decision Support 2001* [Power, J.M. and T.C. Daniel (eds.)]. American Society for Photogrammetry and Remote Sensing, Bethesda, MD, USA, pp. 807–824.

Carroll, P. and P.M. Crill, 1997: Carbon balance of a temperate poor fen. *Global Biogeochemical Cycles*, **11**, 349–356.

Carter, T., 1998: Changes in the thermal growing season in Nordic countries during the past century and prospects for the future. *Agricultural and Food Science in Finland*, **7**, 161–179.

Carter, T.R. and R.A. Saarikko, 1996: Estimating regional crop potential in Finland under a changing climate. *Agricultural and Forest Meteorology*, **79**, 301–313.

Casella, E. and J.F. Soussana, 1997: Long-term effects of CO_2 enrichment and temperature increase on the carbon balance of a temperate grass sward. *Journal of Experimental Botany*, **48**, 1309–1321.

Casella, E., J.F. Soussana, and P. Loiseau, 1996: Long term effects of CO_2 enrichment and temperature increase on a temperate grass sward, I: productivity and water use. *Plant and Soil*, **182**, 83–99.

Casselman, J., 2000: Quantitative fish community changes after establishment of exotic species—impact of bass introductions. In: *Abstracts, American Fisheries Society 130th Annual Meeting, August 20–24, 2000, St. Louis, Missouri, USA*.

Caughley, G., G.C. Grigg, and L. Smith, 1985: The effect of drought on kangaroo populations. *Journal of Wildlife Management*, **49**, 679–685.

Cecil, C.B., F.T. Dulong, J.C. Cobb, and X.X. Supardi, 1993: Allogenic and autogenic controls on sedimentation in the Central Sumatra Basin as an analogue for Pennsylvanian coal-bearing strata in the Appalachian Basin. In: *Modern and Ancient Coal-Forming Environments* [Cobb, J.C. and C.B. Cecil (eds.)]. Special Paper 286, Geological Society of America, Boulder, CO, USA, pp. 3–22.

Cerezke, H.F. and W.J.A. Volney, 1995: Forest insect pests in the northwest region. In: *Forest Insect Pests in Canada* [Armstrong, J.A. and W.G.H. Ives (eds.)]. Canadian Forest Service, Ottawa, ON, Canada, pp. 59–72.

Chambers, P.A., G.J. Scrimgeour, and A. Pietroniro, 1997: Winter oxygen conditions in ice-covered rivers: the impact of pulp mill and municipal effluents. *Canadian Journal of Fisheries and Aquatic Sciences*, **54**, 2796–2806.

Chameides, W.L., P.S. Kasibhatla, J. Yienger, and H. Levy II, 1994: Growth of continental-scale metro-agro-plexes, regional ozone pollution, and world food production. *Science*, **264**, 74–77.

Chapin, F.S. III and A.M. Starfield, 1997: Time lags and novel ecosystems in response to transient climatic change in Arctic Alaska. *Climatic Change*, **35**, 449–461.

Chapin, F.S. III, G.R. Shaver, A.E. Giblin, K.G. Nadelhoffer, and J.A. Laundre, 1995: Responses of Arctic tundra to experimental and observed changes in climate. *Ecology*, **76**, 694–711.

Chapin, F.S. III, W. Eugster, J.P. McFadden, A.H. Lynch, and D.A. Walker, 2000: Summer differences among Arctic ecosystems in regional climate forcing. *Journal of Climate*, **13,** 2002–2010.

Chapin, F.S. III, S.A. Zimov, G.R. Shaver, and S.E. Hobbie, 1996: CO_2 fluctuation at high latitudes. *Nature*, **383,** 585–586.

Chapman, A.L., J.D. Sturtz, A.L. Cogle, W.S. Mollah, and R.J. Bateman, 1996: Farming systems in the Australian semi-arid tropics: a recent history. *Australian Journal of Experimental Agriculture*, **36,** 915–928.

Chapman, W.L. and J.E. Walsh, 1993: Recent variations of sea ice and air temperature in high latitudes. *Bulletin of the American Meteorological Society*, **74,** 33–47.

Chen, H. and H. Libai, 1997: Investigating about varieties filtering in winter wheat northward shifting. *Journal of Chenyang Agricultural University*, **28,** 175–179.

Chen, D.X., H.W. Hunt, and J.A. Morgan, 1996: Responses of a C_3 and C_4 perennial grass to CO_2 enrichment and climate change: comparison between model predictions and experimental data. *Ecological Modelling*, **87,** 11–27.

Chen, J.M., W.J. Chen, J. Liu, J. Cihlar, and S. Gray, 2001: Annual carbon balance of Canada's forest during 1895–1996. *Global Biogeochemical Cycles, (*in press).

Chen, R.S. and R.W. Kates, 1996: Towards a food secure world: prospect and trends. In: *Climate Change and World Food Security* [Downing T.E. (ed)]. Springer-Verlag, Heidelberg, Germany, pp. 23–51.

Chiotti, Q., 1998: An assessment of the regional impacts and opportunities from climate change in Canada. *Canadian Geographer*, **42,** 380–393.

Christensen, T.R. and P. Cox, 1995: Response of methane emission from Arctic tundra to climatic change: results from a model simulation. *Tellus*, **47B,** 301–309.

Christensen, T.R., I.C. Prentice, J. Kaplan, A. Haxeltine, and S. Sitch, 1996: Methane flux from northern wetlands and tundra: an ecosystem source modelling approach. *Tellus*, **48B,** 652–661.

Churkina, G., S.W. Running and A.L. Schloss, 1999: Comparing global models of terrestrial net primary productivity (NPP): the importance of water availability. *Global Change Biology*, **5(Suppl. 1),** 46–55.

Ciais, P., P.P. Tans, M. Trolier, J.W.C. White, and R.J. Francey, 1995: A large northern hemisphere terrestrial CO_2 sink indicated by the $^{13}C/^{12}C$ ratio of atmospheric CO_2. *Nature*, **269,** 1098–1102.

Clark, J.S., P.D. Royall, and C. Churmbley, 1996: The role of fire during cliamte change in an eastern deciduous forest at Devils's Bathtub, New York. *Ecology*, **77,** 2148–2166.

Clark, D.A. and D.B. Clark, 1999: Assessing the growth of tropical rain forest trees: issues for forest modeling and management. *Ecological Applications*, **9,** 981–997.

Clymo, R.S., 1984: The limits to peat bog growth. *Transactions of the Royal Society of London B*, **303,** 605–654.

Clymo, R.S., J. Turunen, and K. Tolonen, 1998: Carbon accumulation in peatland. *Oikos*, **81,** 368–388.

Coakley, S.M., H. Scherm, and S. Chakraborty, 1999: Climate change and plant disease management. *Annual Review of Phytopathology*, **37,** 399–426.

Cochrane, M.A., A. Alencar, M.D. Schulze, C.M. Souza, D.C. Nepstad, P. Lefebvre, and E.A. Davidson, 1999: Positive feedbacks in the fire dynamic of closed canopy tropical forests. *Science*, **284,** 1832–1835.

Coffin, D.P. and W.K. Lauenroth, 1996: Transient responses of North American grasslands to changes in climate. *Climatic Change*, **34,** 269–278.

Colborn, T., D. Dumanoski, J.P. Myers and A. Gore Jr. 1997: *Our Stolen Future*. Penguin Books, New York, NY, USA, 336 pp.

Cole, G.A., 1994: *Textbook of Limnology*. Waveland Press Inc., Prospect Heights, IL, USA, 4th. ed., 412 pp.

Cole, J.J., N.F. Caraco, G.W. Kling, and T.K. Kratz, 1994: Carbon dioxide supersaturation in the surface waters of lakes. *Science*, **265,** 1568–1570.

Colinvaux, P.A., P.E. De Oliveira, and M.B. Bush, 2000: Amazonian and neotropical plant communities on glacial time-scales: the failure of the aridity and refuge hypotheses. *Quaternary Science Reviews*, **19,** 141–169.

Colinvaux, P.A., M.B. Bush, M. Steinitzkannan, and M.C. Miller, 1997: Glacial and postglacial pollen records from the Ecuadorian Andes and Amazon. *Quaternary Research*, **48,** 69–78.

Collins, J.P., 1979: Intrapopulation variation in the body size at metamorphosis of and timing of metamorphosis in the bullfrog, Rana catesbeiana. *Ecology*, **60,** 738–749.

Colombo, S.J., 1998: Climatic warming and its effect on bud burst and risk of frost damage to white spruce in Canada. *Forestry Chronicle*, **74,** 567–577.

Comins, H.N. and R.E. McMurtrie, 1993: Long term response of nutrient limited forests to CO_2 enrichment: equilibrium behaviour of plant-soil models. *Ecological Applications*, **3,** 666–681.

Conroy, J.P., S. Seneweera, A.S. Basra, G. Rogers, and B. Nissenwooller, 1994: Influence of rising atmospheric CO_2 concentrations and temperature on growth, yield and grain quality of cereal crops. *Australian Journal of Plant Physiology*, **21,** 741–758.

Cook, A.C., W.C. Oechel, J.S. Amthor, S.E. Trumbore, and J.R. Southon, 1999: Effects of long-term elevated CO_2 on ecosystem carbon accumulation. *EOS*, **80(17),** S128.

Coope, G.R., 1995: The effects of quaternary climate change on insect populations: lessons from the past. In: *Insects in a Changing Environment* [Harrington, R. and N.E. Stork (eds.)]. Academic Press, San Diego, CA, USA, pp. 30–39.

Costanza, R., R. d'Arge, R. de Groot, S. Farber, M. Grasso, B. Hannon, K. Limburg, S. Naeem, R. V. O'Neill, J. Paruelo, R. G. Raskin, P. Sutton, and M. van den Belt, 1997: The value of the world's ecosystem services and natural capital. *Nature*, **387,** 253-60.

Coughenour, M.B. and D.X. Chen, 1997: Assessment of grassland ecosystem responses to atmospheric change using plant-soil process models. *Ecological Applications*, **7,** 802–827.

Coughenour, M.B. and J.E. Ellis, 1993: Landscape and climatic control of woody vegetation in a dry tropical ecosystem—Turkana District, Kenya. *Journal of Biogeography*, **20(4),** 383–398.

Couteaux, M.M., K.B. McTiernan, B. Berg, D. Szuberla, P. Dardenne, and P. Bottner, 1998: Chemical composition and carbon mineralisation potential of Scots pine needles at different stages of decomposition. *Soil Biology and Biochemistry*, **30,** 583–595.

Coviella, C.E. and J.T. Trumble, 1999: Elevated atmospheric CO_2 will alter the impact of Bacillus thuringiensis. In: *Proceedings of the Ecological Society of America Poster Session, Snowbird, Utah, 2-4th August 2000*. Ecological Society of America, USA, 428pp.

Cox, P.A., T. Elmquist, E.D. Pierson, and W.D. Rainey, 1991: Flying foxes as strong interactors in South Pacific Island ecosystems: a conservation hypothesis. *Conservation Biology*, **5,** 448–454.

Cramer, W. and W. Steffen. 1997: Forecast changes in the global environment: What they mean in terms of ecosystem responses on different time-scales. Pages 415-426 in B. Huntley, W. Cramer, A. V. Morgan, H. C. Prentice, and J. R. M. Allen, editors. Past and future rapid environmental changes: The spatial and evolutionary responses of terrestrial biota. Springer-Verlag, Berlin.

Cramer, W., H.H. Shugart, I.R. Noble, F.I. Woodward, H. Bugmann, A. Bondeau, J.A. Foley, R.H. Gardner, W.K. Laurenroth, L.F Pitelka, and R.W. Sutherst, 1999: Ecosystem composition and structure. In: *The Terrestrial Biosphere and Global Change: Implications for Natural and Managed Ecosystems* [Walker, B., W. Steffen, J. Canadell, and J. Ingram (eds.)]. Cambridge University Press, Cambridge, United Kingdom and New York, NY, USA, pp. 190–228.

Cramer, W., A. Bondeau, F.I. Woodward, I.C. Prentice, R.A. Betts, V. Brovkin, P.M. Cox, V. Fisher, J. Foley, A.D. Friend, C. Kucharik, M.R. Lomas, N. Ramankutty, S. Sitch, B. Smith, A. White, and C. Young-Molling, 2001: Global response of terrestrial ecosystem structure and function to CO_2 and climate change: results from six dynamic global vegetation models. *Global Change Biology*, (in press).

Crawford, H.S. and D.T. Jennings, 1989: Predation by birds on spruce budworm Choristoneura fumiferana: functional, numerical and total responses. *Ecology*, **70,** 152–163.

Crawford, R.M.M., 1983: Root survival in flooded soils. In: *Ecosystems of the World, 4A, Mires: Swamp, Bog, Fen and Moor. General Studies* [Gore, A.J.P. (ed.)]. Elsevier Scientific Publishing Company, Amsterdam, The Netherlands; Oxford, United Kingdom; and New York, NY, USA, pp. 257–283.

Crick, H.Q.P. and T.H. Sparks, 1999: Climate change related to egg-laying trends. *Nature*, **399,** 423–424.

Crick, H.Q., C. Dudley, D.E. Glue, and D.L. Thomson, 1997: UK birds are laying eggs earlier. *Nature*, **388**, 526.

Crill, P.M., K.B. Bartlett, R.C. Harriss, E. Gorham, E.S. Verry, D.I. Sebacher, R. Madzar, and W. Sanner, 1988: Methane fluxes from Minnesota peatlands. *Global Biogeochemical Cycles*, **2,** 371–384.

Crosson, P., D. Pimentel, and C. Harvey, 1995: Soil erosion estimates and costs. *Science*, **269,** 461–465.

Cruciani, R.A., J.L. Barker, S.R. Durell, G. Raghunathan, H.R. Guy, M. Zasloff, and E.F. Stanley, 1992: Magainin 2: a natural antibiotic from frog skin forms ion channels in lipid bilayer membranes. *European Journal of Pharmacology-Molecular Pharmacology Section*, **8,** 287–296.

Cuculeanu, V., A. Marica, and C. Simota, 1999: Climate change impact on agricultural crops and adaptation options in Romania. *Climate Research*, **12,** 153–160.

Cumming, S.G. and P.J. Burton, 1996: Phenology-mediated effects of climatic change on some simulated British Columbia forests. *Climatic Change*, **34(2),** 213–222.

Cunjak, R.A., T.D. Prowse, and D.L. Parrish, 1998: Atlantic salmon in winter: "the season of parr discontent." *Canadian Journal of Fisheries and Aquatic Sciences*, **55(1),** 161–180.

Cunningham, D.M. and P.J. Moors, 1994: The decline of Rockhopper penguins (Eudyptes chrysocome) at Campbell Island, Southern Ocean and the influence of rising sea temperatures. *Emu*, **94,** 27–36.

Curtis, P.S., 1996: A meta-analysis of leaf gas exchange and nitrogen in trees grown under elevated carbon dioxide. *Plant, Cell and Environment*, **19,** 127–137.

Cushing, C.E. (ed.), 1997: *Freshwater Ecosystems and Climate Change in North America: A Regional Assessment*. John Wiley and Sons, New York, NY, USA, 262 pp.

Dabbagh, A.E. and W.A. Abdelrahman, 1997: Management of groundwater resources under various irrigation water use scenarios in Saudi Arabia. *Arabian Journal for Science and Engineering*, **22(1C),** 47–64.

Dahl, T.E. and S.C. Zoltai, 1997: Forested northern wetlands of North America. In: *Northern Forested Wetlands: Ecology and Management* [Trettin, C.C., M.F. Jurgensen, D.F. Grigal, M.R. Gale, and J.K. Jeglum (eds.)]. Lewis Publishers, Boca Raton, FL, USA; New York, NY, USA; London, United Kingdom; and Tokyo, Japan, pp. 3–17.

Daily, G. (ed.), 1997: *Nature's Services: Societal Dependence on Natural Ecosystems*. Island Press, Washington, DC, USA, 392 pp.

Daily, G.C., P.A. Matson, and P.M. Vitousek, 1997: Ecosystem services supplied by soils. In: *Nature's Services: Societal Dependence on Natural Ecosystems* [Daily, G. (ed.)]. Island Press, Washington, DC, USA, pp. 113–132.

Darwin, R., 1999: A farmer's view of the Ricardian approach to measuring agricultural effects of climatic change. *Climatic Change*, **41,** 371–411.

Darwin, R. and D. Kennedy, 2000: Economic effects of CO_2 fertilization of crops: transforming changes in yield into changes in supply. *Environmental Modeling and Assessment*. **5 (3),** 157-168.

Darwin, R.F., M. Tsigas, J. Lewandrowski, and A. Raneses, 1995: *World Agriculture and Climate Change: Economic Adaptations*. Agricultural Economic Report Number 703, U.S. Department of Agriculture, Economic Research Service, Washington, DC, USA, 86 pp.

Davis, M.B., 1990: Climate change and the survival of forest species. In: *The Earth in Transition: Patterns and Processes of Biotic Impoverishment* [Woodwell, G.M. (ed.)]. Cambridge University Press, Cambridge, United Kingdom and New York, NY, USA, 544 pp.

Davis, M.B., 1981: Quaternary history and the stability of forests. In: *Forest Succession: Concepts and Application* [West, D.C., H.H. Shugart, and D.B. Botkin (eds.)]. Springer-Verlag, New York, NY, USA, pp. 132–153.

Davis, M.B. and D.B. Botkin, 1985: Sensitivity of cool-temperate forests and their fossil pollen record to rapid temperature change. *Quaternary Research*, **23,** 327–340.

de Groot, R.S., 1992: *Functions of Nature: Evaluation of Nature in Environmental Planning, Management, and Decision Making*. Wolters-Noordhoff, Groningen, The Netherlands.

de Groot, R.S. and P. Ketner, 1994: Senstivity of NW European species and ecosystems to climate change and some implications for nature conservation and management. In: *Impacts of Climate Change on Ecosystems and Species: Implications for Protected Areas* [Pernetta, J., R. Leemans, D. Elder, and S. Humphrey (eds.)]. International Union for Conservation of Nature and Natural Resources. Gland, Switzerland, pp. 29–53.

de la Mare, W.K., 1997: Abrupt mid-twentieth-century decline in Antarctic sea-ice extent from whaling records. *Nature*, **389,** 57–60.

De Jong, P.W. and P.M. Brakefield, 1998: Climate and change in clines for melanism in the two-spot ladybird, Adalia bipunctata (Coleoptera: Coccinellidae). *Proceedings of the Royal Society of London B*, **265,** 39–43.

DeLucia, E.H., J.G. Hamilton, S.L. Naidu, R.B. Thomas, J.A. Andrews, A. Finzi, M. Lavine, R. Matamala, J.E. Mohan, G.R. Hendrey, and W.H. Schlesinger, 1999: Net primary production of a forest ecosystem with experimental CO_2 enrichment. *Science*, **284,** 1177–1179.

Department of Fisheries and Oceans, 1998: *1995 Survey Highlights: Survey of Recreational Fishing in Canada*. Department of Fisheries and Oceans, Canada.

Dettmers, J.M. and R.A. Stein, 1996: Quantifying linkages among gizzard shad, zooplankton, and phytoplankton in reservoirs. *Transactions of the American Fisheries Society*, **125,** 27–41.

Devito, K.J., A.R. Hill, and N.T. Roulet, 1996: Groundwater-surface water interactions in headwater forested wetlands of the Canadian Shield. *Journal of Hydrology*, **181,** 127–147.

Dexter, E.M., A.D. Chapman, and J.R. Busby, 1995: *The Impact of Global Warming on the Distribution of Threatened Vertebrates*. Environment Australia, Department of Environment and Heritage, Canberra, Australia. Available online at http://www.environment.gov.au/life/end_vuln/animals/climate/climate_change/ccch5.html

Dhakhwa, G.B. and C.L. Campbell, 1998: Potential effects of differential day-night warming in global climate change on crop production. *Climatic Change*, **40,** 647–667.

Dhakhwa, G.B., C.L. Campbell, S.K. LeDuc, and E.J. Cooter, 1997: Maize growth: assessing the effects of global warming and CO_2 fertilization with crop models. *Agricultural Forest Meteorology*, **87,** 251–270.

Dhillion, S.S., J. Roy, and M. Abrams, 1996: Assessing the impact of elevated CO_2 on soil microbial activity in a Mediterranean model ecosystem. *Plant and Soil*, **187,** 333–342.

Diaz, H.F. and R. Bradley, 1997: Temperature variations during the last century at high elevation sites. *Climatic Change*, **36,** 253–279.

Diemont, W.H., G.J. Nabuurs, J.O. Rieley, and H.D. Rijksen, 1997: Climate change and management of tropical peatlands as a carbon reservoir. In: *Biodiversity and Sustainability of Tropical Peatlands* [Rieley, J.O. and S.E. Page (eds.)]. Samara Publishing, Cardigan, Wales, United Kingdom, pp. 363–368.

Dillon, P.J., L.A. Mollot, and M. Futter, 2001: The effect of El Niño-related drought on the recovery of acidified lakes. *Environmental Monitoring and Assessment*, (in press).

Dixon, R.K., J.K. Winjum, K.J. Andrasko, J.J. Lee, and P.E. Schroeder, 1994a: Integrated land-use systems: assessment of promising agroforest and alternative land-use practices to enhance carbon conservation and sequestration. *Climatic Change*, **27(1),** 71–92.

Dixon, R.K., S. Brown, R.A. Houghton, A.M. Solomon, M.C. Trexler, and J. Wisniewski, 1994b: Carbon pools and flux of global forest ecosystems. *Science*, **263,** 185–190.

Doering, O.C., M. Habeck, J. Lowenberg-DeBoer, J.C. Randolph, J.J. Johnston, B.Z. Littlefield, M.A. Mazzocco, and R. Pfeifer, 1997: Mitigation strategies and unforseen consequences: a systematic assessment of the adaptation of upper midwest agriculture to future climate change. *World Resources Review*, **9,** 447–459.

Domisch, T., L. Finér, M. Karsisto, R. Laiho, and J. Laine, 1998: Relocation of carbon from decaying litter in drained peat soils. *Soil Biology and Biochemistry*, **30,** 1529–1536.

Domoto, A., K. Iwatsuki, T. Kawamichi, and J. McNeely (eds.), 2000: *A Threat to Life: The Impact of Climate Change on Japan's Biodiversity*. Tsukiji-Shokan Publishing Co. Ltd., Japan and International Union for Conservation of Nature and Natural Resources. Gland, Switzerland and Cambridge, United Kingdom.

Döös, B.R. and R. Shaw, 1999: Can we predict the future food production? A sensitivity analysis. *Global Environmental Change*, **9,** 261–283.

Downing, T.E., M.J. Watts, and H.G. Bhole, 1996a: Climate change and food insecurity: towards a sociology and geography of vulnerability. In: *Climate Change and World Food Security* [Downing, T.E. (ed.)]. North Atlantic Treaty Organisation. Scientific Affairs Division, Springer-Verlag, University of Oxford, United Kingdom, pp. 182–206.

Downing, T.E., A.A. Olsthoorn, and R.S.J. Tol (eds.), 1996b: *Climate Change and Extreme Events: Altered Risk, Socio-Economic Impacts and Policy Responses*. Environmental Change Unit, Oxford, United Kingdom and Institute for Environmental Studies, Amsterdam, The Netherlands, 309 pp.

Dradjad, M., S. Soekodarmodjo, M.S. Hidyat, and M. Nitisapto, 1986: Subsidence of peat soils in the tidal swamp lands of Barambai, South Kalimantan. In: *Proceedings of the Symposium on Lowland Development in Indonesia*. Research Papers, International Institute for Land Reclamation and Improvement, Wageningen, The Netherlands.

Drake, B.G., M.A. Gonzales-Meler, and S.P. Long, 1997: More efficient plants: a consequence of rising atmospheric CO_2? *Annual Review of Plant Physiology and Plant Molecular Biology*, **48,** 607–637.

Drake, B.G., G. Peresta, E. Beugeling, and R. Matamala, 1996: Long-term elevated CO_2 exposure in a Chesapeake Bay wetland: ecosystem gas exchange, primary production, and tissue nitrogen. In: *Carbon Dioxide and Terrestrial Ecosystems* [Koch, G.W. and H.A. Mooney (eds.)]. Academic Press, San Diego, CA, USA, pp. 197–214.

Dregne, H.E., 1995: Desertification control: a framework for action. *Environmental Monitoring and Assessment*, **37(1–3),** 111–122.

Duellman, W.E., 1999: Global distribution of amphibians: patterns, conservation, and future challenges. In: *Patterns of Distribution of Amphibians: A Global Perspective* [Duellman, W.E. (ed.)]. Johns Hopkins University Press, Baltimore, MD, USA, pp. 1–30.

Dugas, W.A., M.L. Heuer, D. Hunsaker, B.A. Kimball, K.F. Lewin, J. Nagy, and M. Johnson, 1994: Sap flow measurements of transpiration from cotton grown under ambient and enriched CO_2 concentrations. *Agricultural and Forest Meteorology*, **70,** 231–245.

Dukes, J.S. and H.A. Mooney, 1999: Does global change increase the success of biological invaders? *Trends in Ecology and Evolution*, **14,** 135–139.

Dunn, P.O. and D.W. Winkler, 1999: Climate change has affected the breeding date of tree swallow throughout North America. *Proceedings of the Royal Society of London B*, **266,** 2487–2490.

Dyurgerov, M.B. and M.F. Meier, 1997: Year-to-year fluctuation of global mass balance of small glaciers and their contributions to sea level changes. *Arctic and Alpine Research*, **29,** 392–402.

Easterling, D.R., B. Horton, P.D. Jones, T.C. Peterson, T.R. Karl, D.E. Parker, M.J. Salinger, V. Razuvayev, N. Plummer, P. Jamason, and C.K. Folland, 1997: Maximum and minimum trend for the globe. *Science*, **277,** 364–367.

Easterling, W.E. III, 1996: Adapting North American agriculture to climate change in review. *Agricultural and Forest Meteorology*, **80,** 1–53.

Easterling, W.E., A. Weiss, C.J. Hays, and L.O. Mearns, 1998: Spatial scales of information for simulating wheat and maize productivity: the case of the U.S. great plains. *Agricultural and Forest Meteorology*, **90,** 51–63.

Easterling, W.E., X. Chen, C.J. Hays, J.R. Brandle, and H. Zhang, 1996: Improving the validation of model-simulated crop yield response to climate change: an application to the EPIC model. *Climate Research*, **6,** 263–273.

Edwards, S.F., 1991: The demand for Galapagos vacations: estimation and application to wilderness preservation. *Coastal Management*, **19,** 155–199.

Eeley, H.A.C., M.J. Lawes, and S.E. Piper, 1999: The influence of climate change on the distribution of indigenous forest in KwaZulu-Natal, South Africa. *Journal of Biogeography*, **26,** 595–617.

Ehleringer, J.R., T.E. Cerling, and B.R. Helliker, 1997: C_4 photosynthesis, atmospheric CO_2 and climate (review). *Oecologia*, **112(3),** 285–299.

Eliot, I., P. Waterman, and C.M. Finlayson, 1999: Predicted climate change, sea level rise and wetland management in the Australian wet-dry tropics. *Wetlands Ecology and Management*.

Ellis, W.N., J.H. Donner, and J.H. Kuchlein, 1997: Recent shifts in phenology of microlepidoptera, related to climatic change (Lepidoptera). *Entomologische Berichten (Amsterdam)*, **57,** 66–72.

Emslie, S.D., W. Fraser, R.C. Smith, and W. Walker, 1998: Abandoned penguin colonies and environmental change in the Palmer Station area, Anvers Island, Anatarctic Peninsula. *Antarctic Science*, **10,** 257–268.

Englin, J., P. Boxall, K. Chakraborty, and D. Watson, 1996: Valuing the impacts of forest fires on backcountry forest recreation. *Forest Science* **42,** 450–455.

Epstein, H.E., W.K. Lauenroth, I.C. Burke, and D.P. Coffin, 1998: Regional productivities of plant species in the great plains of the United States. *Plant Ecology*, **134,** 173–195.

Eugster, W., W.R. Rouse, R.A. Pielke, J.P. McFadden, D.D. Baldocchi, T.G.F. Kittel, F.S. Chapin III, G.E. Liston, P.L. Vidale, E. Vaganov, and S. Chambers, 2000: Land-atmosphere energy exchange in Arctic tundra and boreal forest: available data and feedbacks to climate. *Global Change Biology*, **6(1),** 84–115.

Everett, J.T., A. Krovnin, D. Lluch-Belda, E. Okemwa, H.A. Regier, and J.-P. Troadec, 1996: Fisheries. In: *Climate Change 1995: Impacts, Adaptions, and Mitigation of Climate Change: Scientific-Technical Analyses. Contribution of Working Group II to the Second Assessment Report for the Intergovernmental Panel on Climate Change* [Watson, R.T., M.C. Zinyowera, and R.H. Moss (eds.)]. Cambridge University Press, Cambridge, United Kingdom and New York, NY, USA, pp. 511–537.

Ewel, K.C., 1997: Water quality improvement by wetlands. In: *Natures Services, Societal Dependence on Natural Ecosystems* [Daily, G.C. (ed.)]. Island Press, Washington, DC, USA, pp. 329–344.

Ewel, K., R. Twilley, and J. Ong, 1998: Different kinds of mangrove forests different kinds of goods and services. *Global Ecology and Biogeography Letters*, **7(1),** 83–94.

Fahrig, L. and G. Merriam, 1994: Conservation of fragmented populations. *Conservation Biology*, **8,** 50–59.

Fan, S., M. Gloor, J. Mahlman, S. Pacala, J. Sarmiento, T. Takahashi, and P. Tans, 1998: A large terrestrial carbon sink in North America implied by atmospheric and oceanic carbon dioxide data and models. *Science*, **282,** 442–446.

Fang, X. and H.G. Stefan, 1998: Projections of climate change effects on ice covers of small lakes in the contiguous U.S. *Cold Regions Science and Technology*, **27,** 119–140.

Fang, X. and H.G. Stefan, 1997: Simulated climate change effects on dissolved oxygen characteristics in ice-covered lakes. *Ecological Modelling*, **103,** 209–229.

Fang, X., S.R. Alam, and H.G. Stefan, 1998: *Continental-Scale Projections of Potential Climate Change Effects on Small Lakes in the Contiguous U.S., Vol. 3: Effects of Climate Conditions on Fish Habitat*. Project 421, University of Minnesota, St. Anthony Falls Laboratory, Minneapolis, MN, USA, 37 pp.

FAO, 1999a: *The State of Food Security in the World, 1999*. United Nations Food and Agriculture Organization, Rome, Italy, 32 pp.

FAO, 1999b: *The State of the World Fisheries and Aquaculture, 1998*. United Nations Food and Agriculture Organization, Rome, Italy, 125 pp.

FAO, 1999c: *Review of the State of the World Fisheries Resources: Inland Fisheries*. FAO Fisheries Circular No. 942, United Nations Food and Agriculture Organization, Rome, Italy, 59 pp.

FAO, 1998: *FAO Yearbook of Forestry Products, 1996*. United Nations Food and Agriculture Organization, Rome, Italy.

FAO, 1997a: *State of the World's Forests, 1997*. United Nations Food and Agriculture Organization, Rome, Italy.

FAO, 1997b: *FAO Provisional Outlook for Global Forest Products Consumption, Production, and Trade to 2010*. Food and Agricultural Organization of the United Nations, Rome, Italy, pp. 54–57.

FAO, 1996: *Forest Resources Assessment 1990: Survey of Tropical Forest Cover and Study of Change Processes*. FAO Forestry Paper 130, United Nations Food and Agriculture Organization, Rome, Italy.

FAO, 1992: *Co-ordinating Working Party on Atlantic Fishery Statistics: Recreational Fisheries*. CWP-15/10, Food and Agricultural Organization of the United Nations, Rome, Italy, 6 pp.

FAO, 1988: *International Introductions of Inland Aquatic Species*. Technical Paper 294, United Nations Food and Agriculture Organization, Rome, Italy, 318 pp.

Fearnside, P.M., 1997: Greenhouse-gas emissions from Amazonian hydro-electric reservoirs: the example of Brazil's Tucurui Dam as compared to fossil fuel alternatives. *Environmental Conservation*, **24(1),** 64–75.

Fearnside, P.M., 1995: Hydrolectric dams in the Brazilian Amazon as sources of 'greenhouse' gases. *Environmental Conservation*, **22,** 7–19.

Feehan, J. and G. O'Donovan, 1996: *The Bogs of Ireland. An Introduction to the Natural, Cultural and Industrial Heritage of Irish Peatlands*. The Environmental Institute, University College Dublin, Dublin, Ireland.

Ferguson, M.A.D., 1999: Arctic tundra caribou and climatic change: questions of temporal and spatial scales. *Geoscience Canada*, **23,** 245–252.

Ferland, C. and L. Rochefort, 1997: Restoration techniques for Sphagnum-dominated peatlands. *Canadian Journal of Botany*, **75,** 1110–1118.

Fielding, C.A., J.B. Whittaker, J.E.L. Butterfield, and J.C. Coulson, 1999: Predicting responses to climate change: the effect of altitude and latitude on the phenology of the Spittlebug Neophilaenus lineatus. *Functional Ecology*, **13,** 65–73.

Filip, G.M., 1977: An Armillaria epiphytotic on the Winema National Forest, Oregon. *Plant Disease Reporter,* **61,** 708–711.

Finér, L. and J. Laine, 1998: Fine root dynamics at drained peatland sites of different fertility in southern Finland. *Plant and Soil*, **201,** 27–36.

Finlayson, C.M. and N.C. Davidson, 1999: Global review of wetland resources and priorities for wetland inventory: project description and methodology. In: *Global Review of Wetland Resources and Priorities for Wetland Inventory* [Finlayson, C.M. and A.G. Spiers (eds.)]. Supervising Scientist Report No. 144, Supervising Scientist, Canberra, Australia, pp. 15–64.

Finlayson, C.M. and A.G. van der Valk, 1995: Wetland classification and inventory: a summary. In: *Classification and Inventory of the World's Wetlands. Advances in Vegetation Science 16* [Finlayson, C.M. and A.G. van der Valk (eds.)]. J. Kluwer Academic Publishers, Dordrecht, The Netherlands, pp. 185–192.

Finlayson, C.M. and C.D. Woodroffe, 1996: Wetland vegetation. In: *Landscape and Vegetation Ecology of the Kakadu Region, Northern Australia. Geobotany 23* [Finlayson, C.M. and I. von Eortzen (eds.)]. J. Kluwer Academic Publishers, Dordrecht, The Netherlands, pp. 81–112.

Finlayson, C.M., B.J. Bailey, W.J. Freeland, and M. Fleming, 1988: Wetlands of the Northern Territory. In: *The Conservation of Australian Wetlands* [McComb, A.J. and P.S. Lake (eds.)]. Surrey Beatty and Sons, Sydney, Australia, pp. 103–116.

Fiscus, E.L., C.D. Reid, J.E. Miller, and A.S. Heagle, 1997: Elevated CO_2 reduces O_3 flux and O_3-induced yield losses in soybeans: possible implications for elevated CO_2 studies. *Journal of Experimental Botany*, **48,** 307–313.

Flannigan, M.D. and Y. Bergeron, 1998: Possible role of disturbance in shaping the northern distribution of Pinus resinosa. *Journal of Vegetation Science*, **9,** 477–482.

Flannigan, M.D., Y. Bergeron, O. Engelmar, and B.M. Wotton, 1998: Future wildfire in circimboreal forests in relation to global warming. *Journal of Vegetation Science*, **9,** 469–476.

Fleming, R.A., 1996: A mechanistic perspective of possible influences of climate change on defoliating insects in North America's boreal forests. *Silva Fennica*, **30,** 281–294.

Fleming, R.A. and J.-N. Candau, 1998: Influences of climate change on some ecological processes of an insect outbreak system in Canada's boreal forests and the implications for biodiversity. *Environmental Monitoring and Assessment*, **49,** 235–249.

Fleming, R.A. and G.M. Tatchell, 1995: Shifts in the flight periods of British aphids: a response to climate warming? In: *Insects in a Changing Environment* [Harrington, R. and N.E. Stork (eds.)]. Academic Press, San Diego, CA, USA, pp. 505–508.

Fleming, R.A. and W.J.A. Volney, 1995: Effects of climate change on insect defoliator population processes in Canada's boreal forest: some plausible scenarios. *Water, Air, and Soil Pollution*, **82,** 445–454.

Foley, J.A., S. Levis, I.C. Prentice, D. Pollard, and S.L. Thompson, 1998: Coupling dynamic models of climate and vegetation. *Global Change Biology*, **4,** 561–579.

Folgarait, P.J., 1998: Ant biodiversity and its relationship to ecosystem functioning: a review. *Biodiversity and Conservation*, **7,** 1221–1244.

Forchhammer, M.C., E. Post, and N.C. Stenseth, 1998: Breeding phenology and climate. *Nature*, **391,** 29–30.

Fowler, D., K.J. Hargreaves, J.A. Macdonald, and B. Gardiner, 1995: Methane and CO_2 exchange over peatland and the effects of afforestation. *Forestry*, **68,** 327–334.

Franzen, L.G., 1994: Are wetlands the key to the ice-age cycle enigma? *Ambio*, **23,** 300–308.

Fraser, W.R. and D.L. Patterson, 1997: Human disturbance and long-term changes in Adélie Penguin populations: a natural experiment at Palmer Station, Antarctica. In: *Antarctic Communities: Species, Structure, Survival* [Battaglia, B., J. Valencia, and D. Walton (eds.)]. Cambridge University Press, Cambridge, United Kingdom and New York, NY, USA, pp. 445–452.

Fraser, W.R., W.Z. Trivelpiece, D.C. Ainley, and S.G. Trivelpiece, 1992: Increases in Antarctic penguin populations: reduced competition with whales or a loss of sea ice due to environmental warming? *Polar Biology*, **11,** 525–531.

Frazer, N.B., J.L. Greene, and J.W. Gibbons, 1993: Temporal variation in growth rate and age at maturity of male painted turtles, Chrysemys picta. *American Midland Naturalist*, **130,** 314–324.

Frenzel, P. and J. Rudolph, 1998: Methane emission from a wetland plant: the role of CH_4 oxidation in Eriophorum. *Plant and Soil*, **202,** 27–32.

Friborg, T., T.R. Christensen, and H. Sogaard, 1997: Rapid response of greenhouse gas emission to early spring thaw in a sub-Arctic mire as shown by micrometeorological techniques. *Geophysical Research Letters*, **24,** 3061–3064.

Frolking, S.E., J.L. Bubier, T.R. Moore, T. Ball, L.M. Bellisario, A. Bhardwaj, P. Carroll, P.M. Crill, P.M. Lafleur, J.H. McCaughey, N.T. Roulet, A.E. Suyker, S.B. Verma, J.M. Waddington, and G.J. Whiting, 1998: Relationship between ecosystem productivity and photosynthetically active radiation for northern peatlands. *Global Biogeochemical Cycles*, **12,** 115–126.

Fuller, D.O. and S.D. Prince, 1996: Rainfall and foliar dynamics in tropical southern Africa: potential impacts of global climatic change on savanna vegetation. *Climatic Change*, **33,** 69–96.

Fung, I., 1997: A greener north. *Nature*, **386,** 659–660.

Galbraith, H., R. Jones, R. Park, S. Herrod-Julius, J. Clough, G. Page, and B. Harrington, 2001: Potential impacts of sea level rise due to global climate change on migratory shorebird populations at coastal sites in the United States. *Climate Change*, (in press).

Gao, Q. and M. Yu, 1998: A model of regional vegetation dynamics and its application to the study of northeast China transect (NECT) responses to global change. *Global Biogeochemical Cycles*, **12,** 329–344.

Gatter, W., 1992: Timing and patterns of visible autumn migration: can effects of global warming be detected? *Journal Für Ornithologie*, **133,** 427–436.

Giardina, C.P. and M.G. Ryan, 2000: Evidence that decomposition rates of organic carbon in mineral soil do not vary with temperature. *Nature*, **404,** 858–861.

Gifford, R.M., 1995: Whole plant respiration and photosynthesis of wheat under increased CO_2 concentration and temperature long-term vs. short-term distinctions for modelling. *Global Change Biology*, **1,** 385–396.

Gifford, R.M., 1994: The global carbon cycle: a viewpoint on the missing sink. *Australian Journal Plant Physiology*, **21,** 1–15.

Gifford, R.M., 1992: Interaction of carbon dioxide with growth-limiting environmental factors in vegetation productivity: implications for the global carbon cycle. *Advances in Bioclimatology*, **1,** 25–58.

Gignac, L.D., B.J. Nicholson, and S.E. Bayley, 1998: The utilization of bryophytes in bioclimatic modeling: predicted northward migration of peatlands in the MackenzieRiver Basin, Canada, as a result of global warming. *Bryologist*, **101,** 572–587.

Gignac, L.D., D.H. Vitt, S.C. Zoltai, and S.E. Bayley, 1991: Bryophyte response surface along climatic, chemical, and physical gradients in peatlands of western Canada. *Nova Hedwigia*, **53,** 27–71.

Gill, R.A. and I.C. Burke, 1999: Ecosystem consequences of plant life form changes at threesites in the semiarid United States. *Oecologia,* **121(4),** 551–563.

Gindl, W., M. Grabner and R. Wimmer, 2000: The influence of temperature on latewood lignin content in treeline Norway spruce compared with maximum density ring width. *Trees- Structure and Function*, **14,** 409–414.

Gitay, H. and I.R. Noble, 1998: Middle East and arid Asia. In: *The Regional Impacts of Climate Change: An Assessment of Vulnerability. Special Report of IPCC Working Group II* [Watson, R.T., M.C. Zinyowera, and R.H. Moss (eds.)]. Intergovernmental Panel on Climate Change, Cambridge University Press, Cambridge, United Kingdom and New York, NY, USA, pp. 231–252.

Glantz, M.H. and J.H. Ausubel, 1988: Impact assessment by analogy: comparing the Ogallala aquifer depletion and CO_2-induced climate change. In: *Social Response to Regional Climate Change* [Glantz, M.H. (ed.)]. A Westview Special Study, Westview Press, Boulder, CO, USA, pp. 113–142.

Glaser, P.H., P.C. Jansen, and D.I. Siegel, 1996: The response of vegetation to chemical and hydrologic gradients in the Lost River peatland, northern Minnesota. *Journal of Ecology*, **78,** 1021–1048.

Glenn, S., A. Heyes, and T.R. Moore, 1993: Carbon dioxide and methane emissions from drained peatland soils, southern Quebec. *Global Biogeochemical Cycles*, **7,** 247–258.

Goldammer, J.G. and C. Price, 1998: Potential impacts of climate change on fire regimes in the tropics based on MAGICC and a GISS GCM-derived lightning model. *Climatic Change*, **39,** 273–296.

Goodfriend, G.A. and R.M. Mitterer, 1988: Late quaternary land snails from the north coast of Jamaica: local extinctions and climatic change. *Palaeogeography, Palaeoclimatology, Palaeoecology*, **63,** 293–312.

Gordon, G., A.S. Brown, and T. Pulsford, 1988: A koala (Phascolartos cinereus Goldfuss) population crash during drought and heatwave conditions in south-western Queensland. *Australian Journal of Ecology*, **13,** 451–461.

Gorham, E., 1991: Northern peatlands: role in the carbon cycle and probable responses to climatic warming. *Ecological Applications*, **1,** 182–195.

Gorham, E., 1994a: The future of research in Canadian peatlands: a brief survey with particular reference to global change. *Wetlands*, **14,** 206–215.

Gorham, E., 1994b: The biogeochemistry of northern peatlands and its possible responses to global warming. In: *Biotic Feedbacks in the Global Climatic System: Will the Warming Feed the Warming?* [Woodwell, G.M. and F.T. Mackenzie (eds.)]. Oxford University Press, Oxford, United Kingdom and New York, NY, USA, pp. 169–187.

Gorham, E. and J.A. Janssens, 1992: The paleorecord of geochemistry and hydrology in northern peatlands and its relation to global change. *Suo*, **43,** 117–126.

Goudriaan, J., S.C. van de Geijn, and J.S.I. Ingram, 1994: *GCTE Focus 3 Wheat Modelling and Experimental Data Comparison*. Workshop Report, Global Change and Terrestrial Ecosystems. Focus 3 Office, Oxford, United Kingdom, 16 pp.

Goulden, M.L., S.C. Wofsy, J.W. Harden, S.E. Trumbore, P.M. Crill, S.T. Gower, T. Fries, B.C. Daube, S.M. Fan , D.J. Sutton, A. Bazza and J.W. Munger, 1998: Sensitivity of boreal forest carbon balance to soil thaw. *Science*, **279,** 214–216.

Goulder, L.H. and D. Kennedy, 1997: Valuing ecosystem services: philosophical bases and empirical methods. In: *Nature's Services: Societal Dependence on Natural Ecosystems* [Daily, G. (ed.)]. Island Press, Washington, DC, USA, pp. 23–48.

Grabherr, G., M. Gottfried, and H. Pauli, 1994: Climate effects on mountain plants. *Nature*, **369,** 448.

Grace, J., J. Lloyd, J. McIntyre, A.C. Miranda, P. Meir, H. Miranda, C. Nobre, J.B. Moncrieff, J. Massheder, Y. Malhi, I.R. Wright, and J. Gash, 1995: Carbon dioxide uptake by an undisturbed tropical rain forest in south-west Amazonia, 1992–1993. *Science*, **270,** 778–780.

Graf, B., A.P. Gutierrez, O. Rakotobe, P. Zahner, and Y. Delucchi, 1990: A simulation model for the dynamics of rice growth and development, II: the competition with weeds for nitrogen and light. *Agricultural Systems*, **32,** 367–392.

Graham, R.W. and E.C. Grimm, 1990: Effects of global climate change on the patterns of terrestrial biological communities. *Trends in Ecology and Evolution*, **5,** 289–292.

Granberg, G., 1998: *Environmental Controls of Methane Emission from Boreal Mires*. Swedish University of Agricultural Sciences, No. 146, Umeå, Sweden.

Granin, N.G., D.H. Jewson, A.A. Zhdanov, L.A. Levin, A.I. Averin, R.Y. Gnatovsky, L.A. Tcekhanovsky, and N.P. Minko, 2001: Physical processes and mixing of algal cells under the ice of Lake Baikal. *Verhandlungen der Internationalen Vereinigung fuer Limnologie*, (in press).

Greening, H.S. and J. Gerritsen, 1987: Changes in macrophyte community structure following drought in the Okefenokee Swamp, Georgia, USA. *Aquatic Botany*, **28,** 113–128.

Gregory, P., J. Ingram, B. Campbell, J. Goudriaan, T. Hunt, J. Landsberg, S. Linder, M. Stafford-Smith, B. Sutherst, and C. Valentin, 1999: Managed production systems. In: *The Terrestrial Biosphere and Global Change. Implications for Natural and Managed Ecosystems. Synthesis Volume* [Walker, B., W. Steffen, J. Canadell, and J. Ingram (eds.)]. International Geosphere-Biosphere Program Book Series 4, Cambridge, United Kingdom, pp. 229–270.

Grulke, NE., C.P. Andersen, M.E. Fen, and P.R. Miller, 1998: Ozone exposure and nitrogen deposition lowers root biomass of ponderosa pine in the San Bernardino Mountains, California. *Environmental Pollution*, **103,** 63–73.

Gunn, A., 1995: Responses of Arctic ungulates to climate change. In: *Human Ecology and Climate Change—People and Resources in the Far North* [Peterson, D.L. and D.R. Johnson (eds.)]. Taylor and Francis, Washington, DC, USA, pp. 89–104.

Gunn, A. and T. Skogland, 1997: Responses of caribou and reindeer to global warming. *Ecological Studies*, **124,** 189–200.

Gutzke, W.H.N. and D. Crews, 1988: Embryonic temperature determines adult sexuality in a reptile. *Nature*, **332,** 832–834.

Haack, R.A. and J.W. Byler, 1993: Insects and pathogens - regulators of forest ecosystems. *Journal of Forestry*, **91 (9),** 32–37.

Haapala, J. and M. Leppäranta, 1997: The Baltic Sea ice season in changing climate. *Boreal Environment Research*, **2(1),** 93–108.

Hahn, G.L., 1999: Dynamic responses of cattle to thermal heat loads. *Journal of Animal Science*, **77(2),** 10–20.

Hahn, G.L., 1995: Environmental influences on feed intake and performance, health and well-being of livestock. *Japanese Journal of Livestock Management*, **30,** 113–127.

Hahn, G.L., 1981: Housing and management to reduce climatic impacts on livestock. *Journal of Animal Science*, **52,** 175–186.

Hahn, G.L., 1976: Rational environmental planning for efficient livestock production. *Biometeorology*, **6(II),** 106–114.

Hahn, G.L. and B.A. Becker, 1984: Assessing livestock stress. *Agricultural Engineering*, **65,** 15–17.

Hahn, G.L. and T.L. Mader, 1997: Heat waves in relation to thermoregulation, feeding behavior, and mortality of feedlot cattle. In: *Proceedings of the 5th International Livestock Environment Symposium, Minneapolis, MN, USA*. pp. 563–571.

Hahn, G.L. and J.A. Morgan, 1999: Potential consequences of climate change on ruminant livestock production. In: *Proceedings of Workshop on Global Change Impacts in the Great Plains, February 25, 1999, Omaha, NE, USA*.

Hahn, G.L. and J. Morrow-Tesch, 1993: Improving livestock care and well-being. *Agricultural Engineering*, **74,** 14–17.

Hahn, G.L., P.L. Klinedinst, and D.A. Wilhite, 1992: *Climate Change Impacts on Livestock Production and Management*. American Society of Agricultural Engineers, St. Joseph, MI, USA, 16 pp.

Haight, R.G., W.D. Smith, and T.J. Straka, 1995: Hurricanes and the economics of Loblolly pine plantations. *Forest Science*, **41,** 675–688.

Hall, A.E. and L.H. Allen Jr., 1993: Designing cultivars for the climatic conditions of the next century. In: *International Crop Science I* [Buxton, D.R., R. Shibles, R.A. Forsberg, B.L. Blad, K.H. Asay, G.M. Paulsen, and R.F. Wilson (eds.)]. Crop Science Society of America, Madison, WI, USA, pp. 291–297.

Hall, D.O., D.S. Ojima, W.J. Parton, and J.M.O. Scurlock, 1995: Response of temperate and tropical grasslands to CO_2 and climate change. *Journal of Biogeography*, **22,** 537–547.

Hall, J.P. and B.H. Moody, 1994: *Forest Depletions Caused by Insects and Diseases in Canada, 1982–1987*. Information Report ST-X-8, Canadian Forest Service, Science and Sustainable Development Directorate, Ottawa, ON, Canada, 14 pp.

Hallgren, W.S. and A.J. Pitman, 2000: The uncertainty in simulations by a Global Biome Model (BIOME3) to alternative parameter values. *Global Change Biology*, **6,** 483–495.

Halpin, P.N., 1997: Global climate change and natural-area protection: management responses and research directions. *Ecological Applications*, **7,** 828–843.

Halsey, L., D.H. Vitt, and S.C. Zoltai, 1997: Climatic and physiographic controls on wetland type and distribution in Manitoba, Canada. *Wetlands*, **17,** 243–262.

Halsey, L., D.H. Vitt, and S.C. Zoltai, 1995: Disequilibrium response of permafrost in boreal continental western Canada to climate change. *Climatic Change*, **30,** 57–73.

Ham, J.M., C.E. Owensby, P.I. Coyne, and D.J. Bremer, 1995: Fluxes of CO_2 and water vapor from a prairie ecosystem exposed to ambient and elevated atmospheric CO_2. *Agricultural and Forest Meteorology*, **77,** 73–93.

Hamburg, S.P. and C.V. Cogbill, 1988: Historical decline of red spruce population and climatic warming. *Nature*, **331,** 428–431.

Hamilton, L.S., 1995: Mountain cloud forest conservation and research: a synopsis. *Mountain Research and Development*, **15,** 259–266.

Happell, J.D and J.P. Chanton, 1993: Carbon remineralization in a north Florida swamp forest - effects of water level on the pathways and rates of soil organic matter decomposition. *Global Biogeochemical Cycle,.* **7(3),** 475-490.

Hardy, Y., M. Mainville, and D.M. Schmitt, 1986: *An Atlas of Spruce Budworm Defoliation in Eastern North America, 1938–1980*. Miscellaneous Publication No. 1449, U.S. Department of Agriculture, Forest Service, Cooperative State Research Service, Washington, DC, USA, 51 pp.

Harris, A.H., 1993: Wisconsinan pre-pleniglacial biotic change in southeastern New Mexico. *Quaternary Research (Orlando)*, **40,** 127–133.

Harris, W.F. and S.E. Frolking, 1992: The sensitivity of methane emissions from northern freshwater wetlands to global warming. In: *Climate Change and Freshwater Ecosystems* [Firth, P. and S. Fisher (eds.)]. Springer-Verlag, New York, NY, USA, pp. 48–67.

Hart, R.P., S.D. Bradshaw, and J.B. Iveson, 1985: Salmonella infections in a marsupial, the Quokka (Setonix brachyurus), in relation to seasonal changes in condition and environmental stress. *Applied and Environmental Microbiology*, **49,** 1276–1281.

Harwell, E.E., 2000: Remote sensibilities: discourses of technology and the making of Indonesia's natural disaster. *Development and Change*, **31(1),** 307–340.

Hasenauer, H., R.R. Nemani, K. Schadauer, and S.W. Running, 1999: Forest growth response to changing climate between 1961 and 1990 in Austria. *Forest Ecology and Management*, **122,** 209–219.

Hassal and Associates, 1998 *Climate Change Scenarios and Managing the Scarce Water Resources of the Macquarie River*. Report for the Department of the Environment, Sports, and Territories, Under the Climate Impacts and Adaptation Grants Program, Canberra, Australia, 113 PP.

Hauer, F.R., J.S. Baron, D.H. Campbell, K.D. Fausch, S.W. Hostetler, G.H. Leavesley, P.R. Leavitt, D.M. McKnight, and J.A. Stanford, 1997: Assesment of climate change and freshwater ecosystems of the Rocky Mountains, USA, and Canada. *Hydrological Processes*, **11,** 85–106.

Haxeltine, A. and I.C. Prentice, 1996: BIOME3: An equilibirium terrestrial biosphere model based on ecophysiological constraints, resource availability, and competition among plant functional types. *Global Biogeochemical Cycles*, **10,** 693–709.

Hayami, Y. and K. Otsuka, 1994: Beyond the green revolution: agricultural development strategy into the new Century. In: *Agricultural Technology: Policy Issues for the International Community* [Anderson, J.R. and U.K. Wallingford (eds.)]. CAB International and World Bank, Washington, DC, USA, pp. 15–42.

Hayami, Y. and V.W. Ruttan, 1985: *Agricultural Development: An International Perspective*. The John Hopkins University Press, Baltimore, MD, USA, 506 pp.

Haynes, R.W. and H.F. Kaiser, 1990: Forests: methods for valuing acidic deposition and air pollution effects. In: *Acidic Deposition: State of Science and Technology, Volume IV: Control Technologies, Future Emissions, and Effects Valuation* [Irving, P.M. (ed.)]. U.S. Government Printing Office, Washington, DC, USA.

He, H.S. and D.J. Mladenoff, 1999: Spatially explicit and stochastic simulation of forest-landscape fire disturbance and succession. *Ecology*, **80,** 81–99.

Heagle, A.S., R.L. Brandenburg, J.C. Burns, and J.E. Miller, 1994: Ozone and carbon dioxide effects on spider mites in white clover and peanut. *Journal of Environmental Quality*, **23,** 1168–1176.

Heagle, A.S., J.E. Miller, F.L. Booker, and W.A. Pursley, 1999: Ozone stress, carbon dioxide enrichment, and nitrogen fertility interactions in cotton. *Crop Science*, **39,** 731–741.

Hebeisen, T., A. Lüscher, S. Zanetti, B.U. Fischer, U.A. Hartwig, M. Frehner, G.R. Hendrey, H. Blum, and J. Nösberger, 1997: Growth response of Trifolium repens L. and Lolium perenne L. as monocultures and bi-species mixture to free air CO_2 enrichment and management. *Global Change Biology*, **3,** 149–160.

Henning, F.P., C.W. Wood, H.H. Rogers, G.B. Runion, and S.A. Prior, 1996: Composition and decomposition of soybean and sorghum tissues grown under elevated atmospheric carbon dioxide. *Journal of Environmental Quality*, **25,** 822–827.

Herremans, M., 1998: Conservation status of birds in Botswana in relation to land use. *Biological Conservation*, **86,** 139–160.

Hess, K. and J.L. Holechek, 1995: Policy roots of land degradation in the arid region of the United States—an overview. *Environmental Monitoring and Assessment*, **37(1–3),** 123–141.

Heywood, V.H. and R.T. Watson, 1996: *Global Biodiversity Assessment*. Cambridge University Press, Cambridge, United Kingdom and New York, NY, USA, 1152 pp.

Hilbert, D.W., N.T. Roulet, and T.R. Moore, 1998: Modelling and analysis of peatlands as dynamical systems. *Journal of Ecology*, **88,** 230–242.

Hobbie, J.E., 1996: Temperature and plant species control over litter decomposition in Alaskan tundra. *Ecological Monographs*, **66,** 503–522.

Hocking, P.J. and C.P. Meyer, 1991: Carbon dioxide enrichment decreases critical nitrate and nitrogen concentrations in wheat. *Journal of Plant Nutrition*, **14,** 571–584.

Hofstede, R.G.M., 1995: Effects of livestock farming and recommendations for management and conservation of paramo grasslands (Colombia). *Land Degradation and Rehabilitation*, **6,** 133–147.

Hogenbirk, J.C. and R.W. Wein, 1991: Fire and drought experiments in a northern wetland: a climate change analogue. *Canadian Journal of Forest Research*, **21,** 1689–1693.

Hogg, E.H, 1999: Simulation of interannual responses of trembling aspen stands to climatic variation and insect defoliation in western Canada. *Ecological Modelling*, **114,** 175–193.

Hogg, E.H., 1997: Temporal scaling of moisture and the forest-grassland boundary in western Canada. *Agricultural and Forest Meteorology*, **84,** 115–122.

Hogg, E.H. and A.G. Schwarz, 1997: Regeneration of planted conifers across climatic moisture gradients on the Canadian prairies: implications for distribution and climate change. *Journal of Biogeography*, **24,** 527–534.

Holland, E.A. and S. Brown, 1999: North American carbon sink. *Science*, **283,** 1815a.

Horak, D., 1995: Native and non-native fish species used in state fisheries management programs in the United States. *American Fisheries Society Symposium*, **15,** 61–67.

Horie, T., J.T. Baker, H. Nakagawa, and T. Matsui, 2000: Crop ecosystem responses to climatic change: rice. In: *Climate Change and Global Crop Productivity* [Reddy, K.R and H.F. Hodges (eds)]. CAB International, Wallingford, United Kingdom, pp. 81–106.

Horie, T., T. Matsui, H. Nakagawa, and K. Omasa, 1996: Effects of elevated CO_2 and global climate change on rice yield in Japan. In: *Climate Change and Plants in East Asia* [Omasa, K., K. Kai, H. Taoda, Z. Uchijima, and M. Yoshino (eds.)]. Springer-Verlag, Tokyo, Japan, 215 pp.

Horne, A.J. and C.R. Goldman, 1994: *Limnology*. McGraw-Hill, New York, NY, USA, 2nd. ed., 576 pp.

Horton, B., 1995: Geographical distribution of changes in maximum and minimum temperatures. *Atmospheric Research*, **37,** 101–117.

Hou, L., 1995: About the feasibility and practice of winter wheat northward shifting in Liaoning province. *Foreign Agronomy-Wheat Crop*, **3,** 42–44.

Hou, L., 1994: The initial appraise of winter wheat over wintering in Shenyang region. *Agricultural Science of Liaoning*, **5,** 36–40.

Houghton, R.A., 2001: A new estimate of global sources and sinks of carbon from land-use change. *AGU Weekly (EOS Supplement)*, (in press).

Houghton, R.A. and J.L. Hackler, 1999: Emissions of carbon from forestry and land-use change in tropical Asia. *Global Change Biology*, **5,** 481–492.

Houghton, R.A., E.A. Davidson, and G.M. Woodwell, 1998: Missing sinks, feedbacks and understanding the role of terrestrial ecosystems in the global carbon balance. *Global Biogeochemistry Cycles*, **12,** 25–34.

Houghton, R.A., J.L. Hackler, and K.T. Lawrence, 1999: The U.S. carbon budget: contributions from land-use change. *Science*, **285**, 574–578.

Houghton, R.A., D.L. Skole, C.A. Nobre, J.L. Hackler, K.T. Lawrence, and W.H. Chomentowski, 2000: Annual fluxes of carbon from deforestation and regrowth in the Brazilian Amazon. *Nature*, **403**, 301–304.

Houlahan, J.E., C.S. Findlay, B.R. Schmidt, A.H. Meyer, and S.L. Kuzmin, 2000: Quantitative evidence for global amphibian population declines. *Nature*, **752**, 752–755.

Howden, S.M., P.J. Reyenga, and H. Meinke, 1999a: *Global Change Impacts on Australian Wheat Cropping*. Working Paper Series 99/04, Commonwealth Scientific and Industrial Research Organisation. Resource Futures Program, Australia, 122 pp.

Howden, S.M., G.M. McKeon, J.O. Carter, and A. Beswick, 1999b: Potential global change impacts on C_3-C_4 grass distribution in eastern Australian rangelands. In: *People and Rangelands: Building the Future. Proceedings of the VI International Rangeland Congress, Townsville, Australia, July 19-23 1999* [Eldridge, D. and D. Freudenberger (eds.)]. The VI International Rangelands Congress, Aithkenval, Australia, pp. 41–43.

Howden, S.M., J.L. Moore, G.M. McKeon, P.J. Reyenga, J.O. Carter, and J.C. Scanlan, 1999c: Dynamics of mulga woodlands in south-west Queensland: global change impacts and adaptation. In: *Modsim'99 International Congress on Modelling and Simulation Proceedings, December 6–9, Hamilton, New Zealand*. pp. 637–642.

Howden, S.M., G.M. McKeon, L. Walker, J.O. Carter, J.P. Conroy, K.A. Day, W.B. Hall, A.J. Ash, and O. Ghannoum, 1999d: Global change impacts on native pastures in south-east Queensland, Australia. *Environmental Modelling and Software*, **14**, 307–316.

Hrabik, T.R., J.J. Magnuson, and A.S. Mclain, 1998: Predicting the effects of rainbow smelr on native fishes: evidence from long-term research on two lakes. *Canadian Journal of Fisheries and Aquatic Science*, **55**, 1364–1371.

Hughes, L., 2000: Biological consequences of global warming: is the change already apparent? *Trends in Ecology and Evolution*, **15**, 56–61.

Hulme, M., E.M. Barrow, N.W. Arnell, P.A. Harrison, T.C. Johns, and T.E. Downing, 1999: Relative impacts of human-induced climate change and natural climate variability. *Nature*, **397**, 688–691.

Hungate, B.A., F.S. Chapin, E.A. Zhong, E.A. Holland, and C.B. Field, 1997a: Stimulation of grassland nitrogen cycling under carbon dioxide enrichment. *Oecologia*, **109**, 149–153.

Hungate, B.A., E.A. Holland, R.B. Jackson, F.S. Chapin, H.A. Mooney, and C.B. Field, 1997b: The fate of carbon in grasslands under carbon dioxide enrichment. *Nature*, **388**, 576–579.

Hunsaker, D.J., G.R. Hendrey, B.A. Kimball, K.F. Lewin, J.R. Mauney, and J. Nagy, 1994: Cotton evapotranspiration under field conditions with CO_2 enrichment and variable soil moisture regimes. *Agricultural and Forest Meteorology*, **70**, 247–258.

Hunt, H.W., E.T. Elliott, J.K. Detling, J.A. Morgan, and D.-X. Chen, 1996: Responses of a C_3 and C_4 perennial grass to elevated CO_2 and climate change. *Global Change Biology*, **2**, 35–47.

Huntley, B., 1995: Plant species response to climate change: implications for the conservation of European birds. *Ibis*, **137(1)**, 127–138.

Huntley, B., P.M. Berry, W. Cramer, and A.P. McDonald, 1995: Modelling present and potential future ranges of some European higher plants using climate response surfaces. *Journal of Biogeography*, **22**, 967–1001.

Hurtt, G.C., P.R. Moorcroft, S.W. Pacala, and S.A. Levin, 1998: Terrestrial models and global change - Challenges for the future. *Global Change Biology*, **4(5)**, 581-590.

Hutchin, P.R., M.C. Press, J.A. Lee, and T.W. Ashenden, 1995: Elevated concentrations of CO_2 may double methane emissions from mires. *Global Change Biology*, **1**, 125–128.

Hutchins, M., R. Wiese, and K. Willis, 1996: Why we need captive breeding. In: *1996 AZA Annual Conference Proceedings*. American Zoo and Aquarium Association, Bethesda, MD, USA, pp. 77–86.

Hyvönen, T., A. Ojala, P. Kankaala, and P.J. Martikainen, 1998: Methane release from stands of water horestail (Equisetum fluviatile) in a boreal lake. *Freshwater Biology*, **40**, 275–284.

Ibrahim, S., 1996: Forest management systems in peat swamp forest: a Malaysian perspective. In: *Tropical Lowland Peatlands of Southeast Asia* [Maltby, E., C.P. Immirzi, and R.J. Safford (eds.)]. IUCN, Gland, Switzerland, pp. 175–180.

ICRF, 1998: *Second National Communication*. Roshydromet, Interagency Commission of the Russian Federation on Climate Change Problems, Moscow, Russia, 67 pp.

Idso, K.E. and S.B. Idso, 1994: Plant responses to atmospheric CO_2 enrichment in the face of environmental constraints: a review of the past ten year's research. *Agricultural Forest Meteorology*, **69**, 153–203.

Idso, S.B., 1999: The long-term response of trees to atmospheric CO_2 enrichment. *Global Change Biology*, **5**, 593–595.

Idso, S.B. and B.A. Kimball, 1993: Tree growth in carbon dioxide enriched air and its implications for global carbon cycling and maximum levels of atmospheric CO_2. *Global Biogeochemical Cycles*, **7**, 537–555.

Imaki, T., S. Tokunaga, and S. Obara, 1987: High temperature-induced spikelet sterility of rice in relation to flowering time. *Japanese Journal of Crop Science*, **56**, 209–210 (in Japanese).

Imhoff, M.L., E. Levine, M.V. Privalsky, V. Brown, W.T. Lawrence, C.D. Elvidge, and T. Paul, 1997: Using nighttime DMSP/OLS images of city lights to estimate the impact of urban land use on soil resources in the United States. *Remote Sensing of Environment*, **59**, 105–117.

Immirzi, C.P., E. Maltby, and P. Vijarnsorn, 1996: Development problems and perspectives in the peat swamps of Southern Thailand: results of a village survey. In: *Tropical Lowland Peatlands of Southeast Asia* [Maltby, E., C.P. Immirzi, and R.J. Safford (eds.)]. IUCN, Gland, Switzerland, pp. 199–246.

Immirzi, C.P., E. Maltby, and R.S. Clymo, 1992: *The Global Status of Peatlands and Their Role in Carbon Cycling*. A report for the Friends of the Earth by the Wetland Ecosystems Research Group, Department of Geology, University of Exeter, London, United Kingdom.

Ingram, H.A.P., 1983: Hydrology. In: *Ecosystems of the World 4B Mires: Swamp, Bog, Fen and Moor. Regional Studies* [Gore, A.J.P. (ed.)]. Elsevier, Amsterdam, The Netherlands, pp. 67–158.

Innes, J.L. 1991: High-altitude and high-latitude tree growth in relation to past, present and future global climate change. *The Holocene*, **12,** 168–173.

IPCC, 2000: *Land Use, Land-Use Change, and Forestry. A Special Report of the IPCC* [Watson, R.T., I.R. Noble, B. Bolin, N.H. Ravindranath, D.J. Verardo, and D.J. Dokken (eds.)]. Cambridge University Press, Cambridge, United Kingdom and New York, NY, USA, 377 pp.

IPCC, 1998: *The Regional Impacts of Climate Change: An Assessment of Vulnerability. Special Report of IPCC Working Group II* [Watson, R.T., M.C. Zinyowera, and R.H. Moss (eds.)]. Intergovernmental Panel on Climate Change, Cambridge University Press, Cambridge, United Kingdom and New York, NY, USA, 517 pp.

IPCC, 1990: *Climate Change: The IPCC Scientific Assessment, Working Group I* [Houghton, J.T., G.J. Jenkins, and J.J. Ephraums (eds.)]. Cambridge University Press, Cambridge, United Kingdom and New York, NY, USA, 364 pp.

Irland, L., 1998: Ice storm 1998 and the forests of the Northeast. *Journal of Forestry*, **96**, 22–30.

Irland, L., D. Adams, R. Alig, C.J. Betz, C. Chen, M. Muchins, B.A. McCarl, K. Skog, and B. Sohngen, 2001: Assessing socioeconomic impacts of climate change on U.S. forests, wood product markets, and forest recreation. *BioScience*, (in press).

Isack, H.A., 1987: The cultural and economic importance of birds among the Boran people of northern Kenya. In: *The Value of Birds. ICBP Technical Publication No. 6* [Diamond, A.W. and F.L. Filion (eds.)]. International Council for Bird Preservation, Cambridge, United Kingdom and New York, NY, USA, pp. 89–98.

Isaev, A.S. (ed.), 1997: *Program of Extraordinary Activities on Biolgical Fighting with Dangerous Insects in Forests of the Krasnoyarsk Kray*. Project Report, Russian Federal Forest Service, Moscow, Russia, 154 pp. (in Russian).

IUCN, 2000: *2000 IUCN Red List of Threatened Animals*. IUCN, Gland, Switzerland.

IUCN, 1996: The 1996 IUCN Red List of Threatened Animals. IUCN,Gland, Switzerland.

IUCN, 1994: *1994 IUCN Red List of Threatened Animals*. IUCN, Gland, Switzerland.

Ivanova, K.E., 1981: *Water Movement in Mirelands*. Academic Press, London, United Kingdom, 276 pp.

Iverson, L.R. and A.M. Prasad, 1998: Predicting abundance of 80 tree species following climate change in the eastern United States. *Ecological Monographs*, **68,** 465–485.

Iverson, L.R., A. Prasad, and M.W. Schwartz, 1999: Modeling potential future individual tree-species distributions in the eastern United States under a climate change scenario: a case study with Pinus virginiana. *Ecological Modelling*, **115,** 77–93.

Izrael, Y., Y. Anokhin, and A.D. Eliseev, 1997: Adaptation of water management to climate change. In: *Global Changes of Environment and Climate: Collection of Selected Scientific Papers* [Laverov, N.P. (ed.)]. The Federal Research Program of Russia, Russian Academy of Sciences, Moscow, Russia, pp. 373–392.

Jablonski, D. and J.J. Sepkoski, 1996: Paleobiology, community ecology, and scales of ecological pattern. *Ecology*, **77,** 1367–1378.

Jackson, S.T. and C.Y. Weng, 1999: Late quaternary extinction of a tree species in eastern North America. *Proceedings of the National Academy of Sciences of the United States of America*, **96,** 13847–13852.

Jackson, D., 2000: Now that we have bass everywhere, what are they doing? In: *Abstracts of the American Fisheries Society 130th Annual Meeting, August 20–24, 2000, St. Louis, Missouri, USA*.

Jacobs, B.C. and C.J. Pearson, 1999: Growth, development and yield of rice in response to cold temperature. *Journal of Agronomy and Crop Science*, **182,** 79–88.

Jacobs, S.S. and J.C. Comiso, 1997: Climate variability in the Amundsenand Bellingshausen Seas. *Journal of Climate*, **10,** 697–709.

Jacoby, G.C. and R.D. D'Arrigo, 1995: Tree ring width and density evidence of climatic and potential forest change in Alaska. *Global Biogeochemical Cycles*, **9,** 227–234.

Jardine, K., 1994: Finger on the carbon pulse. *The Ecologist*, **24,** 220–224.

Jarvinen, A., 1996: Correlation between egg size and clutch size in the Pied Flycatcher Ficedula hypoleuca in cold and warm summers. *Ibis*, **138,** 620–623.

Jarvinen, A., 1994: Global warming and egg size of birds. *Ecography*, **17(1),** 108–110.

Jarvinen, A., 1989: Patterns and causes of long-term variation in reproductive traits of the Pied Flycatcher Ficedula hypoleuca in Finnish Lapland. *Ornis Fennica*, **66,** 24–31.

Jarvis, P.G., J.M. Massheder, S.E. Hale, J.B. Moncrief, M. Rayment, and S.L. Scott, 1997: Seasonal variation of carbon dioxide, water vapour, and energy exchanges of a boreal black spruce forest. *Journal of Geophysical Research*, **102(D4),** 28953–28966.

Jauhiainen, J. and J. Silvola, 1999: Photosynthesis of *Sphagnum fuscum* at long-term raised CO_2 concentrations. *Annales Botanici Fennici*, **36(1),** 11-19.

Jauhiainen, J., J. Silvola, and H. Vasander, 1998a: The effects of increased nitrogen deposition and CO_2 on Sphagnum angustifolium and S. warnstorfii. *Annales Botanici Fennici*, **35(4),** 247–256.

Jauhiainen, J., H. Vasander, and J. Silvola, 1998b: Nutrient concentration in Sphagna at increased N-deposition rates and raised atmospheric CO_2 concentrations. *Plant Ecology*, **138,** 149–160.

Jauhiainen, J., J. Silvola, K. Tolonen, and H. Vasander, 1997: Response of Sphagnum fuscum to water levels and CO_2 concentration. *Journal of Bryology*, **19,** 391–400.

Jauhiainen, J., H. Vasander, and J. Silvola, 1994: Response of Sphagnum fuscum to N deposition and increased CO_2. *Journal of Bryology*, **18,** 83–95.

Jefferies, R.L., J. Svoboda, G. Henry, M. Raillard, and R. Ruess, 1992: Tundra grazing systems and climatic change. In: *Arctic Ecosystems in a Changing Climate: An Ecophysiological Perspective* [Chapin, R.S., R.L. Jeffries, J.F. Reynolds, G.R. Shaver, J. Svoboda, and E. Chu (eds.)]. Academic Press, San Diego, USA, pp. 391–412.

Jinghua, W. and L. Erda, 1996: The impacts of potential climate change and climate variability on simulated maize production in China. *Water, Air, and Soil Pollution*, **92,** 75–85.

Johns, J., P. Barreto, and C. Uhl, 1996: Logging damage during planned and unplanned logging operations in the eastern Amazon. *Forest Ecology and Management*, **89,** 59–77.

Johnson, C.W., 1985: *Bogs of the Northeast*. University Press of New England, Hanover and London, NH, USA,269 pp.

Johnson, D.G., 1999: Food security and world trade prospects. *American Journal of Agricultural Economics*, **80,** 941–947.

Johnson, H.D., 1965: Response of animals to heat. *Meteorological Monographs*, **6,** 109–122.

Jonauskas, P., 1996: Making multiple land use work. In: *Proceedings of the Wetlands Workshop, December 1994, Darwin* [Jonauskas, P. (ed.)]. Department of Lands, Planning and the Environment, Darwin, Australia, 120 pp.

Jones, M.B. and M. Jongen, 1996: Sensitivity of temperate grassland species to elevated atmospheric CO_2 and the interaction with temperature and water stress. *Agricultural and Food Science in Finland*, **5,** 271–283.

Joyce, L., J. Aber, S. McNulty, V. Dale, A. Hansen, L. Irland, R. Neilson, and K. Skog, 2000: Potential consequences of climate variability and change for the forests of the United States. In: *Climate Change Impacts on the United States: Foundation Report*. National Assessment Synthesis Team, Cambridge University Press, Cambridge, United Kingdom, and New York, NY, USA, (in press).

Joyce, L.A., J.R. Mills, L.S. Heath, A.D. McGuire, R.W. Haynes, and R.A. Birdsey, 1995: Forest sector impacts from changes in forest productivity under climate change. *Journal of Biogeography*, **22,** 703–713.

Kandeler, E., D. Tscherko, R.D. Bardgett, P.J. Hobbs, C. Kampichler, and T.H. Jones, 1998: The response of soil microorganisms and roots to elevated CO_2 and temperature in a terrestrial model ecosystem. *Plant and Soil*, **202,** 251–262.

Kankaala, P., L. Arvola, T. Tulonen, and A. Ojala, 1996: Carbon budget for the pelagic food web of the euphotic zone in a boreal lake (Lake Paajarvi). *Canadian Journal of Fisheries and Aquatic Sciences*, **53,** 1663–1674.

Karim, Z., S.G. Hussain, and M. Ahmed, 1996: Assessing impacts of climatic variations on foodgrain production in Bangladesh. *Water, Air, and Soil Pollution*, **92,** 53–62.

Kasimir-Klemedtsson, A., L. Klemedtsson, K. Berglund, P. Martikainen, J. Silvola, and O. Ocenema, 1997: Greenhouse gas emissions from farmed organic soils: a review. *Soil Use and Management*, **13,** 245–250.

Kasischke, E.S., K. Bergen, R. Fennimore, F. Sotelo, G. Stephens, A. Janetos, and H.H. Shugart, 1999: Satellite imagery gives clear picture of Russian's boreal forest fires. *Transactions of the American Geophysical Union*, **80,** 141–147.

Kasischke, E.S., N.L. Christensen, and B.J. Stocks, 1995: Fire, global warming and the carbon balance of boreal forests. *Ecological Applications*, **5,** 437–451.

Kattenberg, A., F. Giorgi, H. Grassl, G.A. Meehl, J.F.B. Mitchell, R.J. Stouffer, T. Kokioka, A.J. Weaver, and T.M.L. Wigley, 1996: Climate models: projections of future climate. In: *Climate Change 1995: The Science of Climate Change. Contribution of Working Group I to the Second Assessment Report of the Intergovernmental Panel on Climate Change* [Houghton, J.T., L.G. Meira Filho, B.A. Callander, N. Harris, A. Kattenberg, and K. Maskell (eds.)]. Cambridge University Press, Cambridge, United Kingdom and New York, NY, USA, pp. 285–357.

Kaufman, L.S., 1992: Catastrophic change in species-rich freshwater ecosystems, the lessons of Lake Victoria. *Bioscience*, **42,** 846–852.

Kaufmann, R.K. and S.E. Snell, 1997: A biophysical model of corn yield: integrating climatic and social determinants. *American Journal of Agricultural Economics*, **79,** 178–190.

Kauppi, P., K. Mielikainen, and K. Kuusela, 1992: Biomass and carbon budget of European forests, 1971–1990. *Science*, **256,** 70–74.

Keane, R.E., P. Morgan, and J.D. White 1999: Temporal patterns of ecosystem processes on simulated landscapes in Glacier National Park, Montana, USA. *Landscape Ecology*, **14,** 311–329.

Keeling, C.D., J.F.S. Chin, and T.P. Whorf, 1996: Increased activity of northern vegetation inferred from atmospheric CO_2 measurements. *Nature*, **382,** 146–149.

Keleher, C.J. and F.J. Rahel, 1996: Thermal limits to salmonid distributions in the Rocky Mountain Region and the potential habitat loss due to global warming: a geographic information system (GIS) approach. *Transactions of the American Fisheries Society*, **125,** 1–13.

Kelly, C.A., J.W.M. Rudd, R.A. Bodaly, N.T. Roulet, V.L. St. Louis, A. Heyes, T.R. Moore, R. Aravena, B. Dyck, R. Harris, S. Schiff, B. Warner, and G. Edwards, 1997: Increases in fluxes of greenhouse gases and methyl mercury following flooding of an experimental reservoir. *Environmental Science and Technology*, **31,** 1334–1344.

Keltikangas, M., J. Laine, P. Puttonen, and K. Seppälä, 1986: Vuosina 1930–1978 metsäojitetut suot: ojitusalueiden inventoinnin tuloksia. (Summary: Peatlands drained for forestry during 1930–1978: results from field surveys of drained areas). *Acta Forestalia Fennica*, **193**, 1–94.

Kerr, J. and L. Packer, 1998: The impact of climate change on mammal diversity in Canada. *Environmental Monitoring and Assessment*, **49**, 263–270.

Keys, D., 1992: *Canadian Peat Harvesting and the Environment*. North American Wetlands Conservation Council, Ottawa, ON, Canada.

Kienast, F., O. Wildi, and B. Brzeziecki, 1998: Potential impacts of climate change on species richness in mountain forests: an ecological risk assessment. *Biological Conservation*, **83**, 291–305.

Kile, G.A., G.I. McDonald, and J.W. Byler, 1991: Ecology and disease in natural forests. In: *Agricultural Handbook 691*. U.S. Dept. of Agriculture, Washington, DC, USA, pp. 102–121.

Kimball, B.A., F.S. Mauney, F.S. Nakayama, and S.B. Idso, 1993: Effects of elevated CO_2 and climate variables on plants. *Journal of Soil Water Conservation*, **48**, 9–14.

Kimball, B.A., P.J. Pinter Jr., G.W. Wall, R.L. Garcia, R.L. La Morte, P.M.C. Jak, K.F. Arnoud Frumau, and H.F. Vugts, 1997: *Comparison of Responses of Vegetation to Elevated Carbon Dioxide in Free Air and Open-Top Chamber Facilities: Advances in Carbon Dioxide Research*. ASA Special Publication, USA, 61 pp.

Kimball, B.A., P.J. Pinter Jr., R.L. Garcia, R.L. LaMorte, G.W. Wall, D.J. Hunsaker, G. Wechsung, F. Wechsung, and T. Karschall, 1995: Productivity and water use of wheat under free-air CO_2 enrichment. *Global Change Biology*, **1**, 429–442.

Kimball, B.A., R.L. LaMorte, R.S. Seay, J. Pinter, P.J. Pinter, R.R. Rokey, D.J. Hunsaker, W.A. Dugas, M.L. Heuer, and J.R. Mauney, 1994: Effects of free-air CO_2 enrichment on energy balance and evapotranspiration of cotton. *Agricultural and Forest Meteorology*, **70**, 259–278.

Kimball, B.A., R.L. LaMorte, J. Pinter, P.J.Pinter, G.W. Wall, D.J. Hunsaker, F.J. Adamsen, S.W. Leavitt, T.L. Thompson, A.D. Matthias, and T.J. Brooks, 1999: Free-air CO_2 enrichment and soil nitrogen effects on energy balance and evapotranspiration of wheat. *Water Resources Research*, **35**, 1179–1190.

King, G.A. and R.P. Neilson, 1992: The transient response of vegetation to climate change: a potential source of CO_2 to the atmosphere. *Water, Air, and Soil Pollution*, **64(1–2)**, 365–383.

Kirilenko, A.P. and A.M. Solomon, 1998: Modeling dynamic vegetation response to rapid climate change using bioclimatic classification. *Climatic Change*, **38**, 15–49.

Kirk, D.A., M.D. Evenden, and P. Mineau, 1996: Past and current attempts to evaluate the role of birds as predators of insect pests in temperate agriculture. In: *Current Ornithology* [Nolan, V. Jr. and E. Ketterson (eds.)]. Plenum Press, New York, NY, USA, pp. 175–269.

Kirschbaum, M.U.F., 2000: Will changes in soil organic matter act as a positive or negative feedback on global warming? *Biogeochemistry*, **48**, 21–51.

Kirschbaum, M.U.F., A. Fischlin, M.G.R. Cannell, R.V.O. Cruz, W. Cramer, A. Alvarez, M.P. Austin, H.K.M. Bugmann, T.H. Booth, N.W.S. Chipompha, W.M. Cisela, D. Eamus, J.G. Goldammer, A. Henderson-Sellers, B. Huntley, J.L. Innes, M.R. Kaufmann, N. Kräuchi, G.A. Kile, A.O. Kokorin, C. Körner, J. Landsberg, S. Linder, R. Leemans, R.J. Luxmoore, A. Markham, R.E. McMurtrie, R.P. Neilson, R.J. Norby, J.A. Odera, I.C. Prentice, L.F. Pitelka, E.B. Rastetter, A.M. Solomon, R. Stewart, J. van Minnen, M. Weber, and D. Xu, 1996: Climate change impacts on forests. In: *Climate Change 1995: Impacts, Adaptations and Mitigation of Climate Change: Scientific-Technical Analyses. Contribution of Working Group II to the Second Assessment Report of the Intergovernmental Panel on Climate Change* [Watson, R.T., M.C. Zinyowera, and R.H. Moss (eds.)]. Cambridge University Press, Cambridge, United Kingdom and New York, NY, USA, pp. 95–129.

Kleidon, A. and H.A. Mooney, 2000: A global distribution of biodiversity inferred from climatic constraints: results from a process-based modelling study. *Global Change Biology*, **6**, 507–523.

Kleiman, D.G., B.B. Beck, J.M. Dietz, and L.A. Dietz, 1991: Costs of a reintroduction and criteria for success: accounting and accountability in the golden lion tamarin conservation program. *Symposium of the Zoological Society of London*, **62**, 125–142.

Klinedinst, P.L., D.A. Wilhite, G.L. Hahn, and K.G. Hubbard, 1993: The potential effects of climate change on summer season dairy cattle milk production and reproduction. *Climatic Change*, **23**, 21–36.

Kohfeld, K.E. and S.P. Harrison, 2000: How well can we simulate past climates? Evaluating the models using global palaeoenvironmental datasets. *Quaternary Science Reviews*, **19**, 321-346.

Komulainen, V.-M., H. Nykänen, P.J. Martikainen, and J. Laine, 1998: Short-term effect of restoration on vegetation succession and methane emissions from peatlands drained for forestry in southern Finland. *Canadian Journal of Forest Research*, **28**, 402–411.

Körner, C., 1999: *Alpine Plant Life*. Springer-Verlag, Berlin, Germany, 330 pp.

Körner, C., 1996: The response of complex multispecies systems to elevated CO_2. In: *Global Change and Terrestrial Ecosystems* [Walker, B. and W. Steffen (eds.)]. IGBP book series, Cambridge University Press, Cambridge, United Kingdom and New York, NY, USA, pp. 20–42.

Körner, C., 1995a: Towards a better experimental basis for upscaling plant responses to elevated CO_2 and climate warming. *Plant Cell and Environment*, **18**, 1101–1110.

Körner, C., 1995b: Alpine plant diversity: a global survey and functional interpretations. In: *Arctic and Alpine Biodiversity: Patterns, Causes, and Ecosystem Consequences* [Chapin, F.S. III and C. Körner (eds.)]. Springer-Verlag, Berlin, Germany, pp. 45–62.

Kortelainen, P., J. Huttunen, T. Vaisanen, T. Mattsson, P. Karjalainen, and P. Martikainen, 2001: CH_4, CO_2, and N_2O supersaturation in 12 Finnish lakes before and after ice-melt. *Verhandlungen der Internationalen Vereinigung fuer Limnologie*, (in press).

Krabbenhoft, D.P. and K.E. Webster, 1995: Transient hydrogeological controls in the chemistry of a seepage lake. *Water Resources Journal*, **31(9)**, 2295–2305.

Krankina, O.N., R.K. Dixon, A.Z. Shvidenko, A.V. Selikhovkin, 1994: Forest dieback in Russia: causes, distribution and implications for the future. *World Resource Review*, **6(4)**, 524–534.

Kratz, T.K., R.B. Cook, C.J. Bowser, and P.L. Brezonik, 1987: Winter and spring pH depressions in northern Wisconsin lakes caused by increases in pCO_2. *Canadian Journal of Fisheries and Aquatic Sciences*, **44(5)**, 1082–1088.

Kratz, T.K., B.P. Hayden, B.J. Benson, and W.Y.B. Chang, 2001: Patterns in the interannual variability of lake freeze and thaw dates. *Verhandlungen der Internationalen Vereinigung fuer Limnologie*, (in press).

Kratz, T.K., B.J. Benson, E.R. Blood, G.L. Cunningham, and R.A. Dahlgren, 1991: The influence of landscape position on temporal variability in four North American ecosystems. *The American Naturalist*, **138(2)**, 355–378.

Kratz, T.K., K.E. Webster, C.J. Bowser, J.J. Magnuson, and B.J. Benson, 1997: The influence of landscape position on lakes in northern Wisconsin. *Freshwater Biology*, **37**, 209–217.

Kuhry, P., B.J. Nicholson, L.D. Gignac, D.H. Vitt, and S.E. Bayley, 1993: Development of Sphagnum dominated peatlands in boreal continental Canada. *Canadian Journal of Botany*, **71**, 10–22.

Kullman, L., 1993: Tree limit dynamics of Betula pubescens spp. tortuosa in relation to climate variability: evidence from central Sweden. *Journal of Vegetation Science*, **4**, 765–772.

Kullman, L., 1991: Pattern and process of present tree-limits in the Tärna region, southern Swedish Lapland. *Fennia*, **169**, 25–38.

Kullman, L., 1990: Dynamics of altitudinal tree-limits in Sweden: a review. *Norsk Geografisk Tidsskrift*, **44**, 103–116.

Kullman, L., 1986: Recent tree-limit history of Picea abies in the southern Swedish Scandes. *Canadian Journal of Forest Research*, **16**, 761–771.

Kurz, W.A. and M.J. Apps, 1999: A 70-year retrospective analysis of carbon fluxes in the Canadian forest sector. *Ecological Applications*, **9(2)**, 526–547.

Kurz, W.A., M.J. Apps, B.J. Stocks, and W.J.A. Volney, 1995: Global climatic change: disturbance regimes and biospheric feedbacks of temperate and boreal forests. In: *Biospheric Feedbacks in the Global Climate System: Will the Warming Feed the Warming?* [Woodwell, G.F. and F. McKenzie (eds.)]. Oxford University Press, New York, NY, USA, pp. 119–133.

Kuusela, K., 1992: *The Dynamics of Boreal Forests*. Finnish National Fund for Research and Development, SITRA, Helsinki, Finland, 172 pp.

Lachenbruch, A.H. and B.V. Marshall, 1986: Climate change: geothermal evidence from permafrost in the Alaskan Arctic. *Science*, **34**, 689–696.

Lafleur, P.M., J.H. McCaughey, D.W. Joiner, P.A. Bartlett, and D.E. Jelinski, 1997: Seasonal trends in energy, water and carbon dioxide fluxes at a northern boreal wetland. *Journal of Geophysical Research*, **102(D24),** 29009–29020.

Laiho, R. and L. Finér, 1996: Changes in root biomass after water-level drawdown on pine mires in Southern Finland. *Scandinavian Journal of Forest Research*, **11,** 251–260.

Laiho, R. and J. Laine, 1997: Tree stand biomass and carbon content in an age sequence of drained pine mires in southern Finland. *Forest Ecology and Management*, **93,** 161–169.

Laine, J. and I. Vanha-Majamaa, 1992: Vegetation ecology along a trophic gradient on drained pine mires in southern Finland. *Annales Botanici Fennici*, **29,** 213–233.

Laine, J., H. Vasander, and T. Laiho, 1995: Long-term effects of water level drawdown on the vegetation of drained pine mires in southern Finland. *Journal of Applied Ecology*, **32,** 785–802.

Laine, J., J. Silvola, K. Tolonen, J. Alm, H. Nykänen, H. Vasander, T. Sallantaus, I. Savolainen, J., Sinisalo, and P.J. Martikainen, 1996: Effect of water-level drawdown in northern peatlands on the global climatic warming. *Ambio*, **25,** 179–184.

Lal, R, 1989: Conservation tillage for sustainable agriculture. *Advanced Agronomy*, **42,** 85–197.

Lal, R., G.F. Hall, and F.P. Miller, 1989: Soil degradation, I: basic processes. *Land Degradation and Rehabilitation*, **1,** 51–69.

Lal, R., H.M. Hassan, and J. Dumanski, 1999: Desertification control to sequester C and mitigate the greenhouse effect. In: *Carbon Sequestration in Soils: Science, Monitoring and Beyond* [Rosenberg, R.J., R.C. Izaurralde, and E.L. Malone (eds.)]. Battelle Press, Columbus, OH, USA, pp. 83–107.

Lal, R., L. M. Kimble, R. F. Follett, and C. V. Cole, 1998: *The Potential of U.S. Crop land for carbon sequestration and greenhouse effect mitigation*. USDA-NRCS, Washington DC. Ann Arbor Press, Chelsea, Michigan, 128 pp.

Lal, R., J.M. Kimble, and B.A. Stewart (eds.), 2000: *Global Climate Change and Cold Regions Ecosystems*. CRC/Lewis Publishers, Boca Raton, FL, USA, 280 pp.

Lanner, R.M., 1996: *Made for Each Other: A Symbiosis of Birds and Pines*. Oxford University Press, Oxford, United Kingdom, 160 pp.

Lappalainen, E., 1996: General review on world peatland and peat resources. In: *Global Peat Resources* [Lappalainen, E. (ed.)]. International Peat Society, Jyväskylä, Finland, pp. 53–56.

Lappalainen, J. and H. Lehtonen, 1995: Year-class strength of pikeperch (Stizostedion lucioperca L.) in relation to environmental factors in a shallow Baltic Bay. *Annales Zoologici Fennici*, **32,** 411–419.

LaRose, S., J. Price, and L. Rochefort, 1997: Rewetting of a cutover peatland: hydrologic assessment. *Wetlands*, **17,** 416–423.

Lassiter, R.R., E.O. Box, R.G. Wiegert, J.M. Johnston, J. Bergengren, and L.A. Suarez, 2000: Vulnerability of ecosystems of the mid-Atlantic region, USA, to climate change. *Environmental Toxicology and Chemistry*, **19,** 1153–1160.

Lathrop, R.C., S.R. Carpenter, and D.M. Robertson, 1999: Summer water clarity responses to phosphorus, Daphnia grazing, and internal mixing in Lake Mendota. *Limnology and Oceanography*, **44(1),** 137–146.

Lathrop, R.C., 1998: *Water Clarity Responses to Phosphorus and Daphnia in Lake Mendota*. Diss. University of Wisconsin-Madison, Madison, WI, USA, 140 pp.

Laurance, W.F., 1996: Catastrophic declines of Australian rainforest frogs: is unusual weather reponsible? *Biological Conservation*, **77,** 203–212.

Lavee, H., A.C. Imeson, and P. Sarah, 1998: The impact of climate change on geomorphology and desertification along a Mediterranean-arid transect. *Land Degradation and Development*, **9(5),** 407–422.

Lavoie, C. and S. Payette, 1996: The long-term stability of the boreal forest limit in subarctic Quebec. *Ecology*, **77,** 1226–1233.

Le Houérou, H.N., 1996: Climate change, drought and desertification (review). *Journal of Arid Environments*, **34(2),** 133–185.

Lee, H.S. and F. Chai, 1996: Production functions of peat swamp forests in Sarawak. In: *Tropical Lowland Peatlands of Southeast Asia* [Maltby, E., C.P. Immirzi, and R.J. Safford (eds.)]. IUCN, Gland, Switzerland, pp. 129–136.

Leemans, R., 1999: Modelling for species and habitats: new opportunities for problem solving. *The Science of the Total Environment*, **240,** 51–73.

Leemans, R., 1997: Effects of global change on agricultural land use: scaling up from physiological processes to ecosystem dynamics. In: *Ecology in Agriculture* [Jackson, L. (ed.)]. Academic Press, San Diego, CA, USA, pp. 415–452.

Legg, C.A. and Y. Laumonier, 1999: Fires in Indonesia, 1997: a remote sensing perspective. *Ambio*, **28(6),** 479–485.

Lehtonen, H., 1996: Potential effects of global warming on northern European freshwater fish and fisheries. *Fisheries Management and Ecology*, **3(1),** 59–71.

Leiros, M.C., C. Trasar-Cepeda, S. Seoane, and F. Gil-Sotres, 1999: Dependence of mineralization of soil organic matter on temperature and moisture. *Soil Biology and Biochemistry*, **31,** 327–335.

Leng, R.A., 1990: Factors affecting the utilisation of 'poor quality' forage by ruminants particularly under tropical conditions. *Nutrition Research Reviews*, **3,** 277–303.

Lenihan, J.M. and R.P. Neilson, 1995: Canadian vegetation sensitivity to projected climatic change at three organizational levels. *Climatic Change*, **30,** 27–56.

Lesack, L.F.W., R.E. Hecky, and P. Marsh, 1991: The influence of frequency and duration of flooding on the nutrient chemistry of Mackenzie Delta lakes. In: *Mackenzie Delta, Environmental Interactions and Implications of Development* [Dixon, J. (ed.)]. National Hydrology Research Institute, NHRI Symposium No. 4, Environment Canada, Saskatoon, SK, Canada, pp. 19–36.

Lescop-Sinclair, K. and S. Payette, 1995: Recent advance of the arctic treeline along the eastern coast of Hudson Bay. *Journal of Ecology*, **83,** 929–936.

Levin, S.A. (ed.), 2000: *Encyclopedia of Biodiversity*. Academic Press, New York, NY, USA, 4700 pp.

Lewandrowski, J. and D. Schimmelpfennig, 1999: Economic implications of climate change for U.S. agriculture: assessing recent evidence. *Land Economics*, **75,** 39–57.

Licht, L.E., 1975: Comparative life history features of the western spotted frog, Rana pretiosa, from low- and high-elevation populations. *Canadian Journal of Zoology*, **53,** 1254–1257.

Lillywhite, H.B., A.K. Mittal, T.K. Garg, and I. Das, 1998: Basking behavior, sweating and thermal ecology of the Indian tree frog Polypedates maculatus. *Journal of Herpetology*, **32,** 169–175.

Lin, E., 1997: *Modeling Chinese Agriculture Impact Under Globe Climate Change*. Chinese Agricultural Science and Technology Press, Beijing, China, pp. 61–68.

Lin, P. and H.A. Regier, 1995: Use of Arrhenius models to describe temperature dependence of organismal rates in fish. *Canadian Special Publication of Fisheries and Aquatic Sciences*, **121,** 211–225.

Lincoln, D.E., D. Couvet, and N. Sionit, 1986: Response of an insect herbivore to host plants grown in carbon dioxide enriched atmosphere. *Oecologia*, **69,** 556–560.

Lincoln, D.E., N. Sionit, and B.R. Strain, 1984: Growth and feeding response of Psuedoplusia includens (Lepidoptera: Noctuidae) to host plants grown in controlled carbon dioxide atmoshperes. *Environmental Entomology*, **13,** 1527–1530.

Lindner, M., 2000: Developing adaptive forest management strategies to cope with climate change. *Tree Physiology*, **20(5–6),** 299–307.

Lindner, M., 1999: Forest management strategies in the context of potential climate change. *Forstwissenschafliches Centralblatt*, **118(1),** 1–13.

Lioubimtseva, E., B. Simon, H.M. Faure, L. Fauredenard, and J.M. Adams, 1998: Impacts of climatic change on carbon storage in the Sahara-Gobi Desert Belt since the last glacial maximum. *Global and Planetary Change*, **17,** 95–105.

Lips, K.R., 1998: Decline of a tropical montane amphibian fauna. *Conservation Biology*, **12,** 106–117.

Liski, J., H. llvesnieme, A. Makela, and C.J. Westman, 1999: CO_2 emissions from soil in response to climatic warming are overestimated: the decomposition of old soil organic matter is tolerant of temperature. *Ambio*, **28(2),** 171–174.

Lloyd, J. and G.D. Farquhar, 1996: The CO_2 dependence of photosynthesis, plant-growth responses to elevated atmospheric CO_2 concentrations and their interaction with soil nutrient status, I: general principles and forest ecosystems. *Functional Ecology*, **10,** 4–32.

Loeb, V., V. Siegel, O. Holm-Hansen, R. Hewitt, W. Fraser, W. Trivelpiece, and S. Trivelpiece, 1997: Effects of sea-ice extent and krill or salp dominance on the Antarctic food web. *Nature*, **387,** 897–900.

Loiseau, P. and J.-F. Soussana, 2001: Effect of CO_2 temperature and N fertilization on nitrogen fluxes in a temperate grassland ecosystem. *Global Change Biology*, (in press).

Loiseau, P. and J.-F. Soussana, 1999a: Effects of elevated CO_2 temperature and N fertilizer on the accumulation of below-ground carbon in a temperate grassland ecosystem. *Plant and Soil*, **212,** 123–134.

Loiseau, P. and J.-F. Soussana, 1999b: Effect of elevated CO_2 temperature and N fertilizer on the turnover of below-ground carbon in a temperate grassland ecosystem. *Plant and Soil*, **212,** 233–247.

Long, S.P., 1991: Modification of the response of photosynthetic productivity to rising temperature by atmospheric CO_2 concentrations: has its importance been underestimated? *Plant, Cell and Environment*, **14,** 729–739.

Loomis, J. and J. Crespi, 1999: Estimated effect of climate-change on selected outdoor recreation activities in the United States. In: *The Impact of Climate Change on the United States Economy* [Mendelsohn, R. and J.E. Neumann (eds.)]. Cambridge University Press, Cambridge, United Kingdom, and New York, NY, USA.

Ludwichowski, I., 1997: Long-term changes of wing-length, body mass and breeding parameters in first-time breeding females of goldeneyes (Bucephala clangula clangula) in northern Germany. *Vogelwarte*, **39,** 103–116.

Lugo, A.E., M.M. Brinson, and S. Brown (eds.), 1990: *Forested Wetlands, Vol. 15: Ecosystems of the World*. Elsevier Scientific Publishing Co., Amsterdam, The Netherlands, 527 pp.

Lugo, A.E., S. Brown, and M.M. Brinson, 1988: Forested wetlands in freshwater and salt-water environments. *Limnology and Oceanography*, **33,** 894–909.

Lugo, A.E., J.A. Parotta, and S. Brown, 1993: Loss in species caused by tropical deforestation and their recovery through management. *Ambio*, **22,** 106–109.

Luo, Y. and J.F. Reynolds, 1999: Validity of extrapolating field CO_2 experiments to predict carbon sequestration in natural ecosystems. *Ecology*, **80,** 1568–1583.

Luo, Y., D.O. TeBeest, P.S. Teng, and N.G. Fabellar, 1995: The effect of global temperature change on rice leaf blast epidemics: a simulation study in three agroecological zones. *Agriculture, Ecosystema and Environment*, **68,** 187–196.

Lüscher, A. and J. Nösberger, 1997: Interspecific and intraspecific variability in the response of grasses and legumes to free air CO_2 enrichment. *Acta Oecologica*, **18,** 269–275.

Lutz, W., W. Sanderson, S. Scherbov, and A. Goujon, 1996: World population scenarios for the twenty-first century. In: *The Future Population of the World* [Lutz, W. (ed.)]. International Institute for Applied Systems Analysis, Laxenburg, Austria, pp. 361–396.

Lynch, A.H., G.B. Bonan, F.S. Chapin III, and W. Wu, 1999: The impact of tundra ecosystems on the surface energy budget and climate of Alaska. *Journal of Geophysical Research—Atmospheres*, **104,** 6647–6660.

Macdonald, G.M., J.M. Szeicz, J. Claricoates, and K.A. Dale, 1998: Response of the central canadian treeline to recent climatic changes. *Annals of the Association of American Geographers*, **88(2),** 183–208.

Magadza, C.H.D., 1996: Climate change: some likely multiple impacts in southern Africa. In: *Climate Change and World Food Security* [Downing, T.E. (ed.)]. Springer-Verlag, Heidelberg, Germany, pp. 449–483.

Magnuson, J.J., 1976: Managing with exotics: a game of chance. *Transactions of the American Fisheries Society*, **105,** 1–9.

Magnuson, J.J. and T.K. Kratz, 2000: Lakes in the landscape: approaches to regional limnology. *Verhandlungen der Internationalen Vereinigung fuer Limnologie*. **27,** 1–14.

Magnuson, J.J., K.E. Webster, R.A. Assel, C.J. Bowser, P.J. Dillon, J.G. Eaton, H.E. Evans, E.J. Fee, R.I. Hall, L.R. Mortsch, D.W. Schindler, and F.H. Quinn, 1997: Potential effects of climate changes on aquatic systems: Laurentian Great Lakes and precambrian shield region. *Hydrological Processes*, **11,** 825–871.

Magnuson, J.J., D.M. Robertson, B.J. Benson, R.H. Wynne, D.M. Livingstone, T. Arai, R.A. Assel, R.G. Barry, V. Card, E. Kuusisto, N.G. Granin, T.D. Prowse, K.M. Stewart, and V.S. Vuglinski, 2000: Historical trends in lake and river cover in the Northern Hemisphere. *Science*, **289,** 1743–1746.

Magrin, G.O., M.I. Travasso, R.A. Díaz, and R.O. Rodríguez, 1997: Vulnerability of the agricultural systems of Argentina to climate change. *Climate Research*, **9,** 31–36.

Malcolm, J.R. and A. Markham, 2000: *Global Warming and Terrestrial Biodiversity Decline*. World Wildlife Fund, Gland, Switzerland, 34 pp.

Maltby, E. and C.P. Immirzi, 1996: The sustainable utilisation of tropical peatlands. In: *Tropical Lowland Peatlands of Southeast Asia* [Maltby, E., C.P. Immirzi, and R.J. Safford (eds.)]. IUCN, Gland, Switzerland, pp. 1–14.

Maltby, E. and M.C.F. Proctor, 1996: Peatlands: their nature and role in the biosphere. In: *Global Peat Resources* [Lappalainen, E. (ed.)]. International Peat Society, Jyväskylä, Finland, pp. 11–19.

Maltby, E., P. Burbridge, and A. Fraser, 1996: Peat and acid sulphate soils: a case study from Vietnam. In: *Tropical Lowland Peatlands of Southeast Asia* [Maltby, E., C.P. Immirzi, and R.J. Safford (eds.)]. IUCN, Gland, Switzerland, pp. 187–198.

Mallett, K.I. and W.J.A. Volney, 1999: The effect of Armillaria root disease on lodgpole pine tree growth. *Canadian Journal of Forest Research*, **29,** 252–259.

Manion, P.D., 1991: *Tree Disease Concepts*. Prentice-Hall, Englewood Cliffs, NJ, USA, 399 pp.

Manjarrez, J., 1996: Temperature limited activity in the garter snake *Thamnophis melanogaster* (Colubridae). *Ethology*, **102,** 146–56.

Mark, U., and M. Tevini, 1997: Effects of solar ultraviolet-B radiation, temperature and CO_2 on growth and physiology of sunflower and maize seedlings. *Plant Ecology*, **128,** 224–234.

Marland, G. and B. Schlamadinger, 1995: Biomass fuels and forest-management strategies: how do we calculate the greenhouse-gas emissions benefits? *Energy*, **20(11),** 1131–1140.

Martikainen, P.J., H. Nykänen, J. Alm, and J. Silvola, 1995: Change in fluxes of carbon dioxide, methane and nitrous oxide due to forest drainage of mire sites of different trophy. *Plant and Soil*, **168–169,** 571–577.

Martikainen, P.J., H. Nykänen, P.M. Crill, and J. Silvola, 1993: Effect of a lowered water table on nitrous oxide fluxes from northern peatlands. *Nature*, **366,** 51–53.

Martin, T.E., 1998: Are microhabitat preferences of coexisting species under selection and adaptive? *Ecology*, **79,** 656–670.

Mason, C.F., 1995: Long-term trends in the arrival dates of spring migrants. *Bird Study*, **42,** 182–189.

Masters, G.J., V.K. Brown, I.P. Clarke, and J.B. Whittaker, 1998: Direct and indirect effects of climate change on insect herbivores: Auchenorrhyncha (Homoptera). *Ecological Entomology*, **23,** 45–52.

Matsui, T., O.S. Namuco, L.H. Ziska, and T. Horie, 1997: Effects of high temperature and CO_2 concentration on spikelet sterility in indica rice. *Field Crops Research*, **5,** 213–219.

Matthews, E. and I. Fung, 1987: Methane emission from natural wetlands: global distribution, area, and environmental characteristics of sources. *Global Biogeochemical Cycles*, **1,** 61–86.

Matthews, R.B., M.J. Kropff, and D. Bachelet, 1997: Simulating the impact of climate change on rice production in Asia and evaluating options for adaptation. *Agricultural Systems*, **54,** 399–425.

Mattson, W.J. and R.A. Haack, 1987: The role of drought in outbreaks of plant-eating insects. *BioScience*, **37,** 110–118.

Mauney, J.R., B.A. Kimball, P.J. Pinter, R.L. Lamorte, K.F. Lewin, J. Nagy, and G.R. Hendrey, 1994: Growth and yield of cotton in response to a free-air carbon dioxide enrichment (FACE) environment. *Agricultural and Forest Meteorology*, **70,** 49–67.

May, S.A. and T.W. Norton, 1996: Influence of fragmentation disturbance on the potential impact of feral predators on native fauna in Australian forest ecosystems. *Wildlife Research*, **23,** 387–400.

Mayeux, H.S., H.B. Johnson, and H.W. Polley, 1991: Global change and vegetation dynamics. In: *Noxious Range Weeds* [James, L.F., J.O. Evans, M.H. Ralphs, and R.D. Child (eds.)]. Westview Press, Boulder, CO, USA, pp. 62–74.

McCarl, B.A., D.M. Burton, D.M. Adams, R.J. Alig, and C.C. Chen, 2000: Forestry effects of global climate change. *Climate Research*, **15(3),** 195–205.

McCauley, R. and T. Beitinger, 1992: Predicted effects of climate warming on the commercial culture of the channel catfish, Ictalurus punctatus. *Geojournal*, **28(1),** 29–37.

McCleery, R.H. and C.M. Perrins, 1998: Temperature and egg-laying trends. *Nature*, **391,** 30–31.

McDonald, K.A. and J.H. Brown, 1992: Using montane mammals to model extinctions due to global change. *Conservation Biology*, **6,** 409–415.

McGuire, A.D., J.W. Melillo, D.W. Kicklighter, and L.A. Joyce, 1995: Equilibrium responses of soil carbon to climate change: empirical and process-based estimates. *Journal of Biogeography*, **22,** 785–796.

McGuire, A.D., J.S. Clein, J.M. Melillo, D.W. Kicklighter, R.A. Meier, C.J. Vorosmarty, and M.C. Serreze, 2000: Modeling carbon responses of tundra ecosystems to historical and projected climate: sensitivity of pan-Arctic carbon storage to temporal and spatial variation in climate. *Global Change Biology*, **6(Suppl 1),** 141-159.

McKane, R.B., E.B. Rastetter, G.R. Shaver, K.J. Nadelhoffer, A.E. Giblin, J.A. Laundre, and F.S. Chapin III, 1997: Climatic effects on tundra carbon storage inferred from experimental data and a model. *Ecology*, **78,** 1170–1187.

McKee, I.F., J.F. Bullimore, and S.P. Long, 1997: Will elevated CO_2 concentrations protect the yield of wheat from O_3 damage? *Plant, Cell and Environment*, **20,** 77–84.

McKnight, B., 1993: *Biological Pollution: The Control and Impact of Invasive Exotic Species*. Indiana Academy of Sciences, Indianapolis, IN, USA.

McNeely, J., K. Miller, W. Reid, and T. Werner, 1990: *Conserving the World's Biological Diversity*. World Resources Institute, IUCN, Conservation International, World Wildlife Fund, Gland, Switzerland, and World Bank, Washington, DC, USA, 193 pp.

Mearns, L.O., F. Giorgi, L. McDaniel, and C. Shields, 1995: Analysis of climate variability and diurnal temperature in a nested regional climate model: comparison with observations and doubled CO_2 results. *Climate Dynamics*, **11,** 193–209.

Mearns, L.O., C. Rosenzweig, and R. Goldberg, 1992: Effects of changes in interannual climatic variability on CERES-wheat yields: sensitivity and $2xCO_2$ general circulation model studies. *Agricultural Forest Meteorology*, **62,** 159–189.

Mearns, L.O., T. Mavromatis, E. Tsvetsinskaya, C. Hays, and W. Easterling, 1999: Comparative response of EPIC and CERES Crop Models to high and low resolution climate change scenarios. *Journal of Geophysical Research*, **104,** 6623–6646.

Meentemeyer, V., 1984: The geography of organic decomposition rates. *Annals of the Association of American Geographers*, **74,** 551–560.

Megonigal, J.P. and W.H. Schlesinger, 1997: Enhanced CH_4 emissions from a wetland soil exposed to elevated CO_2. *Biogeochemistry*, **37,** 77–88.

Megonigal, J.P., W.H. Conner, S. Kroeger, and R.R. Sharitz, 1997: Aboveground production in southeastern floodplain forests: a test of the subsidy-stress hypothesis. *Ecology*, **78,** 370–384.

Meinzer, F.C. and J. Zhu, 1998: Nitrogen stress reduces the efficiency of the C_4-CO_2 concentrating system, and therefore quantum yield, in Saccharum (sugarcane) species. *Journal of Experimental Botany*, **49,** 1227–1234.

Melillo, J.M., A.D. McGuire, D.W. Kicklighter, B. Moore III, C.J. Vorosmarty, and A.L. Schloss, 1993: Global climate change and terrestrial net primary production. *Nature*, **363,** 234–240.

Mendelsohn, R. and M. Markowski, 1999: The impact of climate change on outdoor recreation. In: *The Impact of Climate Change on the United States Economy* [Mendelsohn, R. and J.E. Neumann (eds.)]. Cambridge University Press, Cambridge, United Kingdom and New York, NY, USA.

Mendelsohn, R., W. Nordhaus, and D. Shaw, 1994: Impact of global warming on agriculture: a Ricardian analysis. *American Economic Review*, **84,** 753–771.

Menzel, A. and P. Fabian, 1999: Growing season extended in Europe. *Nature*, **397,** 659.

Menzhulin, G.V., 1998: Impact of global warming of climate on agriculture in Russia. In: *Influence of Global Change in the Environment and Climate on Functioning of Economy of Russia*. Moscow, Russia, pp. 48–73 (in Russian).

Meyer, U., B. Kollner, J. Willenbrink, and G.H.M. Krause, 1997: Physiological changes on agricultural crops induced by different ambient ozone exposure regimes, I: effects on photsynthesis and assimilate allocation in spring wheat. *New Phytologist*, **136,** 645–652.

Miaoulis, I.N. and B.D. Heilman, 1998: Butterfly thin films serve as solar collectors. *Annals of the Entomological Society of America*, **91,** 122–127.

Mikkelä, C., I. Sundh, B.H. Svensson, and M. Nilsson, 1995: Diurnal variation in methane emission in relation to the water table, soil temperature, climate and vegetation cover in a Swedish acid mire. *Biogeochemistry*, **28,** 93–114.

Milburn, D. and T.D. Prowse, 1998: An assessment of a northern delta as a hydrologic sink for sediment-bound contaminants. *Nordic Hydrology*, **29(4/5),** 64–71.

Milburn, D. and T.D. Prowse, 1996: The effect of river-ice break-up on suspended sediment and select trace-element fluxes. *Nordic Hydrology*, **27(1/2),** 69–84.

Miller, J.R., L.A. Joyce, and R.M. King, 1996: Forest roads and landscape structure in the southern Rocky Mountains. *Landscape Ecology*, **11,** 115–127.

Milliman, J.D., J.M. Broadus, and F. Gable, 1989: Environmental and economic implications of rising sea level and subsiding deltas: the Nile and Bengal examples. *Ambio*, **18,** 340–345.

Mills, E.L., J.H. Leach, J.T. Carlton, and C.L. Secor, 1994: Exotic species and the integrity of the Great Lakes: lessons from the past. *Bioscience*, **44(10),** 666–676.

Mills, M.G.L., H.C. Biggs, and I.J. Whyte, 1995: The relationship between rainfall, lion predation and population trends in African herbivores. *Wildlife Research*, **22,** 75–88.

Minkkinen, K. and J. Laine, 1998: Long-term effect of forest drainage on the peat carbon stores of pine mires in Finland. *Canadian Journal of Forest Research*, **28,** 1267–1275.

Minkkinen, K., H. Vasander, S. Jauhiainen, M. Karsisto, and J. Laine, 1999: Post-drainage changes in vegetation composition and carbon balance in Lakkasuo mire, Central Finland. *Plant and Soil*, **207,** 107–120.

Minns, C.K. and J.E. Moore, 1995: Factors limiting the distributions of Ontario's freshwater fishes: the role of climate and other variables, and the potential impacts of climate change. In: *Climate Change and Northern Fish Populations* [Beamish, R.J. (ed.)]. Canadian Special Publication, Fish Aquatic Sciences, Canada, pp. 137–160.

Minshall, G.W., K.W. Cummins, and R.C. Peterson, 1985: Developments in stream ecosystem theory. *Canadian Journal of Fisheries and Aquatic Science*, **42,** 1045–1055.

Miranda, A.I., 1994: Forest fire emissions in Portugal: a contribution to global warming? *Environmenal Pollution*, **83,** 121–123.

Mitch, W.J. and J.G. Gosselink, 1993: *Wetlands*. Van Nostrand Reinhold, New York, NY, USA, 2nd. ed., 920 pp.

Mitchell, J.F.B., T.C. Johns, W.J. Ingram, and J.A. Lowe, 2000: The effect of stabilising atmospheric carbon dioxide concentrations on global and regional climate change. *Geophysical Research Letters*, **27,** 2977–2980.

Mitchell, R.A.C., V.J. Mitchell, S.P. Driscoll J. Franklin, and D.W. Lawlor, 1993: Effects of increased CO_2 concentration and temperature on growth and yield of winter wheat at 2 levels of nitrogen application. *Plant Cell Environment*, **16,** 521–529.

Mohseni, O. and H.G. Stefan, 2000: *Projections of Fish Survival in U.S. Streams After Global Warming*. Project Report 441, St. Anthony Falls Laboratory, University of Minnesota, Minneapolis, MN, USA, 100 pp.

Moore, K.S., C.L. Bevins, N. Tomassini, K.M. Huttner, K. Sadler, J.E. Moreira, J. Reynolds, and M. Zasloff, 1992: A novel peptide-producing cell in Xenopus multinucleated gastric mucosal cell strikingly similar to the granular gland of the skin. *Journal of Histochemistry and Cytochemistry*, **40,** 367–378.

Moore, M.K. and V.R. Townsend, 1998: The interaction of temperature, dissolved oxygen and predation pressure in an aquatic predator-prey system. *Oikos*, **81,** 329–336.

Moore, T.A. and Shearer, J.C. 1997: Evidence for aerobic degradation of Palangka Raya peat and implications for its sustainability. In: *Biodiversity and Sustainability of Tropical Peatlands* [Rieley, J.O. and S.E. Page (eds.)]. Samara Publishing, Cardigan, Wales, United Kingdom, pp. 157–168.

Moore, T.R., T.A. Trofymow, B. Taylor, C. Prescott, C. Camire, L. Duschene, J. Fyles, L. Kozak, M. Kranabetter, I. Morrison, M. Siltanen, S. Smith, B. Titus, S. Visser, R. Wein, and S. Zoltai, 1999: Rates of litter decomposition in Canadian forests. *Global Change Biology*, **5,** 75–82.

Morgan, J.A., H.W. Hunt, C.A. Monz, and D.R. LeCain, 1994a: Consequences of growth at two carbon dioxide concentrations and temperatures for leaf gas exchange of Pascopyrum smithii (C_3) and Bouteloua gracilis (C_4). *Plant, Cell and Environment*, **17,** 1023–1033.

Morgan, J.A., W.G. Knight, L.M. Dudley, and H.W. Hunt, 1994b: Enhanced root system C-sink activity, water relations and aspects of nutrient acquisition in mycotrophic Bouteloua gracilis subjected to CO_2 enrichment. *Plant and Soil*, **165,** 139–146.

Morley, R.J., 1981: Development and vegetation dynamics of a lowland ombrogenous peat swamp in Kalimantan Tengah. *Journal of Biogeography*, **8,** 383–404.

Morrison, D.J. and K.I. Mallett, 1996: Silvicultural management of Armillaria root disease in western Canadian forests. *Canadian Journal of Plant Pathology*, **18,** 194–199.

Morton, S.R., D.M.S. Smith, M.H. Friedel, G.F. Griffin, and G. Pickup, 1995: The stewardship of arid Australia: ecology and landscape management. *Journal of Environmental Management*, **43(3),** 195-217.

Moya, T.B., L.H. Ziska, O.S. Namuco, and D. Olszyk, 1998: Growth dynamics and genotypic variation in tropical, field-grown paddy rice (Oryza sativa L.) in response to increasing carbon dioxide and temperature. *Global Change Biology*, **4,** 645–656.

Moyle, P.B. and J.J. Cech Jr., 1996: *Fishes: An Introduction to Ichthyology*. Prentice-Hall, Upper Saddle River, NJ, USA, 590 pp.

Munasinghe, M. and J. McNeely (eds.), 1994: *Protected Area Economics and Policy: Linking Conservation and Sustainable Development*. World Conservation Union, IUCN, Gland, Switzerland and World Bank, Washington, DC, USA, 364 pp.

Murray, D.R., 1997: *Carbon Dioxide and Plant Response*. Research Studies Press Ltd., John Wiley and Sons, New York, NY, USA, 275 pp.

Myers, N., R.A. Mittermeier, C.G. Mittermeier, G.A.B. da Fonseca, and J. Kent, 2000: Biodiversity hotspots for conservation priorities. *Nature*, **403,** 853–858.

Myneni, R.B., C.D. Keeling, C.J. Tucker, G. Asrar, and R.R. Nemani, 1997: Increased plant growth in the northern high latitudes from 1981–1991. *Nature*, **386,** 698–702.

Nabuurs, G.J., R. Päivinen, R. Sikkema, and G.M. Mohren, 1997: The role of European forests in the global carbon cycle: a review. *Biomass and Bioenergy*, **13,** 345–358.

Nadelhoffer, K.J., B.A. Emmett, P. Gundersen, C.J. Koopmans, P. Schleppi, A. Tietema and R.F. Wright, 1999: Nitrogen deposition makes a minor contribution to carbon sequestration in temperate forests. *Nature*, **398,** 145–148.

Naiman, R.J., J.J. Magnuson, D.M. McKnight, J.A. Stanford, and J.R. Karr, 1995a: Freshwater ecosystems and their management: a national initiative. *Science*, **270,** 584–585.

Naiman, R.J., J.J. Magnuson, D.M. McKnight, and J.A. Stanford, 1995b: *The Freshwater Imperative: A Research Agenda*. Island Press, Washington, DC, USA, 157 pp.

National Research Council, 1999: *Perspectives on Biodiversity: Valuing Its Role in an Everchanging World*. National Academy Press, Washington, DC, USA, 168 pp.

National Research Council, 1991: *Policy Implications of Greenhouse Warming*. Report of the Committee on Science, Engineering and Public Policy, National Research Council/National Academy of Sciences (NRC/NAS), National Academy Press, Washington, DC, USA, 127 pp.

NBIOME SSC, 1992: *Northern Biosphere Observation and Modelling Experiment*. Incidental Report Series No. IR93-1, NBIOME Science Steering Committee, Canadian Global Change Program, Science Plan, Royal Society of Canada, Ottawa, ON, Canada, 61 pp.

Neilson, R.P. and R.J. Drapek, 1998: Potentially complex biosphere responses to transient global warming. *Global Change Biology*, **4,** 505–521.

Neilson, R.P. and D. Marks, 1994: A global perspective of regional vegetation and hydrologic sensitivities from climatic change. *Journal of Vegetation Science*, **5(5),** 715–730.

Neilson, R.P., I.C. Prentice, B. Smith, T. Kittel, and D. Viner, 1998: Simulated changes in vegetation distribution under global warming. In: *The Regional Impacts of Climate Change: An Assessment of Vulnerability. Special Report of IPCC Working Group II* [Watson, R.T., M.C. Zinyowera, and R.H. Moss (eds.)]. Cambridge University Press, Cambridge, United Kingdom and New York, NY, USA, pp. 441–446.

Neitzel, D.A., M.J. Scott, S.A. Shankle, and J.C. Chatters, 1991: The effect of climate change on stream environments: the Salmonid resource of the Columbia River Basin. *The Northwest Environmental Journal*, **7,** 271–293.

Nepstad, D.C., A. Verissimo, A. Alancar, C. Nobre, E. Lima, P.A. Lefebve, P. Schlesinger, C. Potter, P. Moutinho, E. Mendoza, M. Cochrane, and V. Brooke, 1999: Large-scale impoverishment of Amazonian forests by logging and fire. *Nature*, **348,** 505–508.

Neuvonen, S., P. Niemelä, and T. Virtanen, 1999: Climatic change and insect outbreaks in boreal forests: the role of winter temperatures. *Ecological Bulletins*, **47**.

Neuzil, S.G., 1997: Onset and rate of peat and carbon accumulation in four domed ombrogenous peat deposits, Indonesia. In: *Biodiversity and Sustainability of Tropical Peatlands* [Rieley, J.O. and S.E. Page (eds.)]. Samara Publishing, Cardigan, Wales, United Kingdom, pp. 55–72.

Newton, P.C.D., H. Clark, C.C. Bell, and E.M. Glasgow, 1996: Interaction of soil moisture and elevated CO_2 on above-ground growth rate, root length density and gas exchange of turves from temperate pasture. *Journal of Experimental Botany*, **47,** 771–779.

Ni, J., 2000: A simulation of biomes on the Tibetan Plateau and their responses to global climate change. *Mountain Research and Development*, **20(1),** 80–89.

Niemelä, P., 1980: Dependence of Oporinia autumnata (Lep., Geometridae) outbreaks on summer temperature. *Reports From the Kevo Sub-Arctic Research Station*, **16,** 27–30.

Nilsson, S., O. Sallnäs, and P. Duinker, 1992: *Future Forest Resources of Western and Eastern Europe*. International Institute for Applied Systems Analysis, Vienna, Austria.

Nilsson, S., K. Blauberg, E. Samarskaia, and V. Kharuk, 1998: Pollution stress of Siberian forests. In: Air Pollution in the Ural Mountains [Linkov, I. and R. Wilson (eds.)]. J. Kluwer Academic Publishers, Dordrecht, The Netherlands, pp. 31–54.

Nix, H., 1986: A biogeographic analysis of Australian elapid snakes. In: *Atlas of Elapid Snakes of Australia* [Longmore, R. (ed.)]. Australian Government Pulic Service, Canberra, Australia, pp. 4–15.

Noble, I.R. and H. Gitay, 1996: Functional classifications for predicting the dynamics of landscapes. *Journal of Vegetation Science*, **7,** 329–336.

Noble, I.R., M. Barson, R. Dumsday, M. Friedel, R. Hacker, N. McKenzie, G. Smith, M. Young, M. Maliel, and C. Zammit, 1996: Land resources. In: *Australia: State of the Environment 1996* CSIRO Publishing, Melbourne, Australia, pp. 6.1–6.55.

Norby, R.J., S.D. Wullschleger, C.A. Gunderson, D.W. Johnson, and R. Ceulemans, 1999: Tree response to rising CO_2 in field experiments: implications for the future forest. *Plant Cell and Environment*, **22,** 683–714.

Notohadiprawiro, T., 1998: Conflict between problem-solving and optimising approach to land resources development policies: the case of Central Kalimantan wetlands. In: *The Spirit of Peatlands: 30 Years of the International Peat Society* [Sopo, R. (ed.)]. International Peat Society, Jyväskylä, Finland, pp. 14–24.

Nykänen, H., J. Alm, K. Lång, J. Silvola, and P.J. Martikainen, 1995a: Emissions of CH_4, N2O and CO_2 from a virgin fen and a fen drained for grassland in Finland. *Journal of Biogeography*, **22,** 351–357.

Nykänen, H., J. Silvola, J. Alm, and P.J. Martikainen, 1995b: Fluxes of greenhouse gases CH_4, CO_2 and N_2O from some peat harvesting areas. In: *Proceedings: Peat Industry and Environment, Pärnu, Estonia, 12–15 September 1995*. Ministry of Environment, Environment Information Centre, Tallinn, Estonia, pp. 41–43.

Nyrönen, T., 1996: Peat production. In: *Global Peat Resources* [Lappalainen, E. (ed.)]. International Peat Society, Jyväskylä, Finland, pp. 315–318.

Oechel, W.C. and G.L. Vourlitis, 1994: The effects of climate change on Arctic tundra ecosystems. *Trends in Ecology and Evolution*, **9,** 324–329.

Oechel, W.C., S.J. Hastings, G. Vourlitis, M. Jenkins, G. Riechers, and N. Grulke, 1993: Recent change of Arctic tundra ecosystems from a net carbon dioxide sink to a source. *Nature*, **361,** 520–523.

Oechel, W.C., S. Cowles, N. Grulke, S.J. Hastings, B. Lawrence, T. Prudhomme, G. Riechers, B. Strain, D.T. Tissue, and G. Vourlitis, 1994: Transient nature of CO_2 fertilization in Arctic tundra. *Nature*, **371,** 500–502.

Oerke, E.C., H.W. Dehne, F. Schohnbeck, and A. Weber, 1995: *Crop Production and Crop Protection: Estimated Losses in Major Food and Cash Crops*. Elesvier, Amsterdam, The Netherlands, 808 pp.

Oesterheld, M., C.M. Dibella, and H. Kerdiles, 1998: Relation between NOAA-AVHRR satellite data and stocking rate of rangelands. *Ecological Applications*, **8(1),** 207–212.

Ogren, E., T. Nilsson, and L.G. Sunblad, 1997: Relationship between respiratory depletion of sugars and loss of cold hardiness in coniferous seedlings over-wintering at raised temperatures: indications of different sensitivities of spruce and pine. *Plant Cell and Environment*, **20(2),** 247–253.

Ogutu-Ohwayo, R., R.E. Hecky, A.S. Cohen, and L. Kaufman, 1997: Human impacts on the African Great Lakes. *Environmental Biology of Fishes*, **50,** 117–131.

Ojima, D.S., W.J. Parton, D.S. Schimel, J.M.O. Scurlock, and T.G.F. Kittel, 1993: Modeling the effects of climatic and CO_2 changes on grassland storage of soil-C. *Water, Air, and Soil Pollution*, **70,** 643–657.

Okruszko, H., 1996: Agricultural use of peatlands. In: *Global Peat Resources* [Lappalainen, E. (ed.)]. International Peat Society, Jyväskylä, Finland, pp. 303–309.

Oldeman, R.L., T.A. Hakkeling, and W.G. Sombroek, 1991: *World Map of the Status of Human-Induced Soil Degradation, 2nd. Rev. Ed*. International Soil Reference and Information Centre, Wageningen, The Netherlands.

Oquist, M.G., B.H. Svensson, P. Groffman, M. Taylor, K.B. Bartlett, M. Boko, J. Brouwer, O.F. Canziani, C.B. Craft, J. Laine, D. Larson, P.J. Martikainen, E. Matthews, W. Mullie, S. Page, C.J. Richardson, J. Rieley, N. Roulet, J. Silvola, and Y. Zhang, 1996: Assessment of impacts and adaptation options: non-tidal wetlands. In: *Climate Change 1995: Impacts, Adaptions, and Mitigation of Climate Change: Scientific-Technical Analyses. Contribution of Working Group II to the Second Assessment Report* [Watson, R.T., M.C. Zinyowera, and R.H. Moss (eds.)]. Cambridge University Press, Cambridge, United Kingdom and New York, NY, USA, pp. 214–239.

Osbrink, W.L.A., J.T. Trumble, and R.E. Wagner, 1987: Host suitability of Phaseolus lunata for Trichoplusia ni (Lepidoptera: Noctuidae) in controlled carbon dioxide atmospheres. *Environmental Entomology*, **16,** 639–644.

Osterkamp, T.E. and V.E. Romanovsky, 1999: Evidence for warming and thawing of discontinuous permafrost in Alaska. *Permafrost and Periglacial Processes*, **10,** 17–37.

Osterkamp, T.E., T. Zhang, and V.E. Romanovsky, 1994: Thermal regime of permafrost in Alaska and predicted global warming. *Journal of Cold Regions Engineering*, **4,** 38–42.

Overpeck, J.T., P.J. Bartlein, and T. Webb III, 1991: Potential magnitude of future vegetation change in eastern North America: comparisons with the past. *Science*, **254,** 692–695.

Overpeck, J.T., D. Rind, and R. Goldberg, 1990: Climate-induced changes in forest disturbance and vegetation. *Nature*, **343,** 51–53.

Owen, J.G., and J.R. Dixon, 1989: An ecogeographic analysis of the herpetofauna of Texas (USA). *Southwestern Naturalist*, **34,** 165–180.

Owensby, C.E., L.M. Auen, and P.I. Coyne, 1994: Biomass production in a nitrogen-fertilised tallgrass prairie ecosystem exposed to ambient and elevated levels of CO_2. *Plant and Soil*, **165,** 105–113.

Owensby, C.E., J.M. Ham, A.K. Knapp, and L.M. Auen, 1999: Biomass production and species composition change in a tallgrass prairie ecosystem after long-term exposure to elevated atmospheric CO_2, *Global Change Biology*, **5,** 497–506.

Owensby, C.E., J.M. Ham, A. Knapp, C.W. Rice, P.I. Coyne, and L.M. Auen, 1996: Ecosystem-level responses of tallgrass prairie to elevated CO_2. In: *Carbon Dioxide and Terrestrial Ecosystems* [Körner, C. and F.A. Bazzaz (eds.)]. Academic Press, San Diego, CA, USA, pp. 147–162.

Page, S.E., J.O. Rieley, K. Doody, S. Hodgson, S. Husson, P. Jenkins, H. Morrough-Bernard, S. Otway, and S. Wilshaw, 1997: Biodiversity of tropical peat swamp forest: a case study of animal diversity in the Sungai Sebangau catchment of Central Kalimantan, Indonesia. In: *Biodiversity and Sustainability of Tropical Peatlands* [Rieley, J.O. and S.E. Page (eds.)]. Samara Publishing, Cardigan, Wales, United Kingdom, pp. 231–242.

Paine, R.T., 1974: Intertidal community structure: experimental studies on the relationship between a dominant competitor and its principal competitor. *Oecologia*, **15,** 93–120.

Päivänen, J., 1997: Forested mires as a renewable resource: toward a sustainable forestry practice. In: *Northern Forested Wetlands: Ecology and Management* [Trettin, C.C., M.F. Jurgensen, D.F. Grigal, M.R. Gale, and J.K. Jeglum (eds.)]. CRC Lewis Publishers, Boca Raton, FL and New York, NY, USA; London, United Kingdom; and Tokyo, Japan, pp. 27–44.

Päivänen, J. and H. Vasander, 1994: Carbon balance in mire ecosystems. *World Resource Review*, **6,** 102–111.

Palmer, T.N., 1999: A nonlinear dynamical perspective on climate prediction. *Journal of Climate*, **12,** 575–591.

Panario, D. and M. Bidegain, 1997: Climate change effects on grasslands in Uruguay. *Climate Research*, **9(1–2),** 37–40.

Parmesan, C., 1996: Climate and species' range. *Nature*, **382,** 765–766.

Parmesan, C., T.L. Root, and M.R. Willig, 2000: Impacts of extreme weather and climate on terrestrial biota. *Bulletin of the American Meteorological Society*, **81,** 443–450.

Parmesan, C., N. Ryrholm, C. Stefanescu, J.K. Hill, C.D. Thomas, H. Descimon, B. Huntley, L. Kaila, J. Kullberg, T. Tammaru, W.J. Tennent, J.A. Thomas, and M. Warren, 1999: Poleward shifts in geographical ranges of butterfly species associated with regional warming. *Nature*, **399,** 579–583.

Parry, M., C. Fischer, M. Livermore, C. Rosenzweig, and A. Iglesias, 1999: Climate change and world food security: a new assessment. *Global Environmental Change*, **9,** S51–S67.

Parry, M.L., J.E. Hossell, P.J. Jones, T. Rehman, R.B. Tranter, J.S. Marsh, C. Rosenzweig, G. Fischer, I.G. Carson, and R.G.H. Bunce, 1996: Integrating global and regional analyses of the effects of climate change: a case study of land use in England and Wales. *Climatic Change*, **32,** 185–198.

Parton, W.J., D.S. Ojima, and D.S. Schimel, 1994: Environmental change in grasslands: assessment using models (review). *Climatic Change*, **28,** 111–141.

Parton, W.J., J.M.O. Scurlock, D.S. Ojima, D.S. Schimel, D.O. Hall, and SCOPEGRAM Group Members, 1995: Impact of climate change on grassland production and soil carbon worldwide. *Global Change Biology*, **1,** 13–22.

Paustian, K., E.T. Elliott, and L. Hahn, 1999: Agroecosystem boundaries and C dynamics with global change in the central United States. In: *Great Plains Region Annual Progress Reports, Reporting Period: 1 July 1998 to 30 June 1999*. National Institute for Global Environmental Change, U.S. Department of Energy, University of California, Davis, CA, USA, Available online at http://nigec.ucdavis.edu/publications/annual99/greatplains/GPPaustian0.html.

Paustian, K., E. T. Elliott, and L. Hahn, 2000: *Agroecosystem Boundaries and C Dynamics with Global Change in the Central United States*." FY 1999/2000 Progress Report, National Institute for Global Environmental Change. Available online at http://nigec.ucdavis.edu.

Paustian, K., E.T. Elliott, G.A. Peterson, and K. Killian, 1996: Modelling climate, CO_2 and management impacts on soil carbon in semi-arid agroecosystems. *Plant and Soil*, **187,** 351–365.

Pavlov, A.V., 1994: Current changes of climate and permafrost in the Arctic and sub-Arctic of Russia. *Permafrost and Periglacial Processes*, **5,** 101–110.

Pearce, D. and D. Moran, 1994: *The Economic Value of Biodiversity*. Earthscan, London, United Kingdom, 172 pp.

Peng, S., K.T. Ingram, H.-U. Neue, and L.H. Ziska (eds.), 1995: *Climate Change and Rice*. International Rice Research Institute. Los Baños, The Phillipines, 374 pp.

Perez-Garcia, J., L.A. Joyce, C.S. Binkley, and A.D. McGuire, 1997: Economic impacts of climate change on the global forest sector: an integrated ecological/economic assessment. *Critical Reviews in Environmental Science and Technology*, **27,** S123–S138.

Perfecto, I., 1991: Ants (Hymenoptera, Formicidae) as natural control agents of pests in irrigated maize in Nicaragua. *Journal of Economic Entomology*, **84,** 65–70.

Peterken, G.F. and E.P. Mountford, 1996: Effects of drought on beech in Lady Park Wood, an unmanaged mixed deciduous forest woodland. *Forestry*, **69,** 125–136.

Peters, R.L., 1992: Conservation of biological diversity in the face of climate change. In: *Global Warming and Biological Diversity* [Peters, R.L. and T.E. Lovejoy (eds.)]. Yale University Press, New Haven, CT, USA, pp. 15–30.

Peters, R.L. and J.D. Darling, 1985: The greenhouse effect and nature reserves. *Bioscience*, **35(11)**, 707–717.

Pettus, D. and G.M. Angleton, 1967: Comparative reproductive biology of montane and piedmont chorus frogs (Pseudacris triseriata, Hylidae). *Evolution*, **21,** 500–507.

Phillips, O.L., Y. Malhi, N. Higuchi, W.F. Laurence, P.V. Nunez, R.M. Vasquez, S.G. Laurence, L.V. Ferreira, M. Stern, S. Brown, and J. Grace, 1998: Changes in the carbon balance of tropical forests: evidence from long-term plots. *Science*, **282,** 439–442.

Pickup, G., 1998: Desertification and climate change: the Australian perspective. *Climate Research*, **11,** 51–63.

Pimentel, D., 1997: *Techniques for Reducing Pesticide Use Economic and Environmental Benefits*. John Wiley and Sons, Chichester, United Kingdom.

Pimentel, D., C. Harvey, and P. Resosudarmo, 1995: Environmental and economic costs of soil erosion and conservation benefits. *Science*, **267,** 1117–1123.

Pimentel, D., L. McLaughlin, and A. Zepp, 1993: Environmental and economic effects of reducing pesticide use in agriculture. *Agriculture, Ecosystems and Environment*, **46,** 273–288.

Pimentel, D., H. Acquay, M. Biltonen, P. Rice, M. Silva, J. Nelson, V. Lipner, S. Giordano, A. Horowitz, and M. D'Amore, 1992: Environmental and economic costs of pesticide use. *BioScience*, **42,** 750–760.

Pimm, S.L., G.J. Russell, and J.L. Gittleman, 1995: The future of biodiversity. *Science*, **269,** 347–350.

Pingali, P., 1994: Technological prospects for reversing the declining trend in Asia's rice productivity. In: *Agricultural Technology: Policy Issues for the International Community* [Anderson, J.R. (ed.)]. CAB International and World Bank, Wallingford, United Kingdom, pp. 384–401.

Pinstrup-Andersen, P. and R. Pandya-Lorch, 1998: Food security and sustainable use of natural resources: a 2020 vision. *Ecological Economics*, **26,** 1–10.

Pinter, P.J. Jr., B.A. Kimball, R.L. Garcia, G.W. Wall, D.J. Hunsaker, and R.L. LaMorte, 1996: Free-air CO_2 enrichment: responses of cotton and wheat crops. In: *Carbon Dioxide and Terrestrial Ecosystems* [Koch, G.W. and H.A. Mooney (eds.)]. Academic Press, San Diego, CA, USA, pp. 215–249.

Pitelka, L.F. and Plant Migration Workshop Group, 1997: Plant migration and climate change. *American Scientist*, **85,** 464–473.

Polis, G.A., S.D. Hurd, C.T. Jackson, and F.S. Pinero, 1997: El Niño effects on the dynamics and control of an island ecosystem in the Gulf of California. *Ecology*, **78(6),** 1884–1897.

Pollard, E., 1979: Population ecology and change in range of the white admiral butterfly Ladoga camilla L. in England. *Ecological Entomology*, **4,** 61–74.

Polley, H.W., H.B. Johnson, H.S. Mayeux, and C.R. Tischler, 1996: Are some of the recent changes in grassland communities a response to rising CO_2 concentrations? In: *Carbon Dioxide, Populations, and Communities* [Körner, C. and F.A. Bazzaz (eds.)]. Academic Press, London, United Kingdom, pp. 177–195.

Polley, H.W., H.S. Mayeux, H.B. Johnson, and C.R. Tischler, 1997: Viewpoint: atmospheric CO_2, soil water, and shrub/grass ratios on rangelands. *Journal of Range Management*, **50,** 278–284.

Pontailler, J.-V., A. Faille, and G. Lemée, 1997: Storms drive successional dynamics in natural forests: a case study in Fontainebleau Forest (France). *Forest Ecology and Management*, **98,** 1–15.

Poorter, H., 1993: Interspecific variation in the growth response of plants to an elevated ambient CO_2 concentration. *Vegetatio*, **104,** 77-97.

Poorter, H., 1998: Do slow-growing species and nutrient stressed plants respond relatively strongly to elevated CO_2? *Global Change Biology*, **4,** 639–697.

Post, E. and N.C. Stenseth, 1999: Climatic variability, plant phenology, and northern ungulates. *Ecology*, **80,** 1322–1339.

Post, E., R.O. Peterson, N.C. Stenseth, and B.E. McLaren, 1999: Ecosystem consequences of wolf behavioural response to climate. *Nature*, **401,** 905–907.

Postel, S. and S. Carpenter, 1997: Freshwater ecosystem services. In: *Natures Services, Societal Dependence on Natural Ecosystems* [Daily, G.C. (ed.)]. Island Press, Washington, DC, USA, pp. 195–214.

Potter, C.S., 1997: An ecosystem siumulation model for methane production and emission from wetlands. *Global Biogeochemical Cycles*, **11,** 495–506.

Potter, C.S., E.A. Davidson, and L.V. Verchot, 1996: Estimation of global biogeochemical controls and seasonality in soil methane consumption. *Chemosphere*, **32,** 2219–2246.

Potter, S.C. and S.A. Klooster, 1999: North American carbon sink. *Science*, **283,** 1815a.

Pounds, J.A. and M.L. Crump, 1994: Amphibian decline and climate disturbance: the case of the golden toad and the harlequin frog. *Conservation Biology*, **8,** 72–85.

Pounds, J.A. and M.L. Crump, 1987: Harlequin frogs along a tropical montane stream: aggregation and the risk of predation by frog-eating flies. *Biotropica*, **19,** 306–309.

Pounds, J.A., M.P.L. Fogden, and J.H. Campbell, 1999: Biological response to climate change on a tropical mountain. *Nature*, **398,** 611–615.

Power, G., R. Cunjak, J. Flannagan, and C. Katopodis, 1993: Biological effects of river ice. In: *Environmental Aspects of River Ice* [Prowse, T.D. and N.C. Gridley (eds.)]. NHRC Science Report No. 3, National Hydrology Research Institute, Environment Canada, Saskatoon, SK, Canada, pp. 97–119.

Pregitzer, K.S., D.R. Zak, P.S. Curtis, M.E. Kubiske, J.A. Teeri, and C.S. Vogel, 1995: Atmospheric CO_2, soil nitrogen and turnover of fine roots. *New Phytologist*, **129,** 579–585.

Pregitzer, K.S., D.D. Reed, T.J. Bornhorst, D.R. Foster, G.D. Mroz, J.S. Mclachlan, P.E. Laks, D.D. Stokke, P.E. Martin, and S.E. Brown, 2000: A buried spruce forest provides evidence at the stand and landscape scale for the effects of environment on vegetation at the Pleistocene/Holocene boundary. *Journal of Ecology*, **88,** 45–53.

Prentice, I.C. and D. Parish, 1992: Conservation of peat swamp forest: a forgotten ecosystem. In: *Proceedings of the International Conference on Tropical Biodiversity*. Malaysian Naturalist Society, Kuala Lumpur, Malaysia.

Prentice, I.C., M.T. Sykes, M. Lautenschlager, S.P. Harrison, O. Denissenko and P.J. Bartlein, 1993: A simulation model for the transient effects of climate change on forest landscape. *Ecological Modelling*, **65,** 51–70.

Price, C. and D. Rind, 1994: Possible implications of global climate change on global lightning distributions and frequencies. *Journal of Geophysical Research*, **99,** 108–123.

Price, D.T., D.H. Halliwell, M.J. Apps, and C.H. Peng, 1999a: Adapting a patch model to simulate the sensitivity of central-Canadian boreal ecosystems to climate variability. *Journal of Biogeography*, **26,** 1101–1113.

Price, D.T., C. Peng, M.J. Apps, and D.H. Halliwell, 1999b: Simulating effects of climate change on boreal ecosystem carbon pools in central Canada. *Journal of Biogeography*, **26,** 1237–1248.

Price, J., 2000: Climate change, birds and ecosystems—why should we care? In: *Proceedings of the International Health Conference, Sacramento, CA, USA, August 1999*.

Price, J., 1995: *Potential Impacts of Global Climate Change on the Summer Distributions of Some North American Grassland Birds*. Wayne State University, Detroit, MI, USA.

Price, J., L. Rochefort, and F. Quinty, 1998: Energy and moisture considerations on cutover peatlands: surface microtopography, mulch cover and Sphagnum regeneration. *Ecological Engineering*, **10,** 293–312.

Price, J.T., T.L. Root, K.R. Hall, G. Masters, L. Curran, W. Fraser, M. Hutchins, and N. Myers, 2000: *Climate Change, Wildlife and Ecosystems*. Supplemental information prepared for IPCC Working Group II, Intergovernmental Panel on Climate Change. Available online at http://www.usgcrp.gov/ipcc/html/ecosystem.pdf.

Prieto, A.R., 2000: Vegetational history of the late glacial-Holocene transition in the grasslands of eastern Argentina. *Palaeogeography, Palaeoclimatology, Palaeoecology*, **157,** 167–188.

Pringle, H.J.R., 1995: Pastoralism, nature conservation and ecological sustainability in western Australia's southern shrubland rangelands. *International Journal of Sustainable Development and World Ecology*, **2(1),** 26-44.

Prinn, R., H. Jacoby, A. Sokolov, C. Wang, X. Xiao, Z. Yang, R. Eckhaus, P. Stone, D. Ellerman, J. Melillo, J. Fitzmaurice, D. Kicklighter, G. Holian, and Y. Liu, 1999: Integrated global system model for climate policy assessment: feedbacks and sensitivity studies. *Climatic Change*, **41,** 469–546.

Prior, S.A., H.A. Torbert, G.B. Runion, H.H. Rogers, C.W. Wood, B.A. Kimball, R.L. LaMorte, P.J. Pinter, and G.W. Wall, 1997: Free-air carbon dioxide enrichment of wheat: soil carbon and nitrogen dynamics. *Journal of Environmental Quality*, **26,** 1161–1166.

Prop, J., J.M. Black, P. Shimmings, and M. Owen, 1998: The spring range of barnacle geese Branta leucopsis in relation to changes in land management and climate. *Biological Conservation*, **86,** 339–346.

Prowse, T.D., 1994: Environmental significance of ice to streamflow in cold regions. *Freshwater Biology*, **32,** 241–259.

Prowse, T.D. and M. Conly, 1998: Impacts of climatic variability and flow regulation on ice-jam flooding of a northern delta. *Hydrological Processes*, **12(10–11),** 1589–1610.

Prowse, T.D. and N.C. Gridley (eds.), 1993: *Environmental Aspects of River Ice*. NHRI Science Report No. 5, National Hydrology Research Institute, Environment Canada, Saskatoon, SK, Canada, 155 pp.

Puigdefabregas, J., 1998: Ecological impacts of global change on drylands and their implications for desertification. *Land Degradation and Development*, **9(5),** 393–406.

Pulliam, W.M., 1993: Carbon dioxide and methane exports from a southeastern floodplain swamp. *Ecological Monographs*, **63,** 29–53.

Quiggin, J. and J.K. Horowitz, 1999: The impact of global warming on agriculture: a Ricardian analysis: comment. *American Economic Review*, **89,** 1044–1045.

Quinn, H. and H. Quinn, 1993: Estimated number of snake species that can be managed by species survival plans in North America. *Zoo Biology*, **12,** 243–255.

Rabinowitz, D., 1981: Seven forms of rarity. In: *Biological Aspects of Rare Plant Conservation* [Synge, H. (ed.)]. John Wiley and Sons, Chichester, United Kingdom, pp. 205–217.

Rabinowitz, D., S. Cairns, and T. Dillon, 1986: Seven forms of rarity and their frequency in the flora of the British Isles. In: *Conservation Biology: The Science of Scarcity and Diversity* [Soulé, M. (ed.)]. Sinauer, Sunderland, MA, USA, pp. 182–204.

Rahel, F.J., 2000: Homogenation of fish faunas across the United States. *Science*, **288,** 854–856.

Rahel, F.J., C.J. Keleher, and J.L. Anderson, 1996: Potential habitat loss and population fragmentation for cold water fish in the North Platte River drainage of the Rocky Mountains: response to climate warming. *Limnology and Oceanography*, **41,** 1116–1123.

Randall, P.J. and H.J. Moss, 1990: Some effects of temperature regimes during grain filling on wheat quality. *Australian Journal of Agricultural Research*, **41,** 603–617.

Randerson, J.T., C.B. Field, I.Y. Fung, and P.P. Tans, 1999: Increases in early season net ecosystem uptake explain changes in the seasonal cycle of atmospheric CO_2 at high northern latitudes. *Geophysical Research Letters*, **26,** 2765–2768.

Rayner, P.J., I.G. Enting, R.J. Francey, and R. Langenfelds, 1999: Reconstructing the recent carbon cycle from atmospheric CO_2, $D^{13}C$, and O_2/N_2 observations. *Tellus*, **51B,** 213–232.

Read, J.J., J.A. Morgan, N.J. Chatterton, and P.A. Harrison, 1997 : Gas exchange and carbohydrate and nitrogen concentrations in leaves of Pascopyrum smithii (C_3) and Bouteloua gracilis (C_4) at different carbon dioxide concentrations and temperatures. *Annals of Botany*, **79,** 197–206.

Reading, C.J., 1998: The effect of winter temperatures on the timing of breeding activity in the common toad Bufo bufo. *Oecologia*, **117,** 469–475.

Reading, C.J. and R.T. Clarke, 1995: The effects of density, rainfall and environmental temperature on body condition and fecundity in the common toad, Bufo bufo. *Oecologia*, **102,** 453–459.

Reeburgh, W.S. and S.C. Whalen, 1992: High latitude ecosystems as CH_4 sources. *Ecological Bulletin*, **42,** 62–70.

Regier, H.A., J.A. Holmes, and D. Pauly, 1990: Influence of temperature changes on aquatic ecosystems: an interpretation of empirical data. *Transactions American Fisheries Society*, **119,** 374–389.

Regina, K., H. Nykänen, M. Maljanen, and J. Silvola, 1998: Emissions of N_2O and NO and net nitrogen mineralization in a boreal forested peatland treated with different nitrogen compounds. *Canadian Journal of Forest Research*, **28,** 132–140.

Rehfeldt, G.E., C.C. Ying, D.L. Spittlehouse, and D.A. Hamilton, 1999a: Genetic responses to climate in Pinus contorta: niche breadth, climate change, and reforestation. *Ecological Monographs*, **69,** 375–407.

Rehfeldt, G.E., N.M. Tchebakova, and L.K. Barnhardt, 1999b: Efficacy of climate transfer functions: introduction of Eurasian populations of Larix into Alberta. *Canadian Journal of Forest Research*, **29,** 1660–1668.

Reier, Ü., 1982: *Murakad*. Valgus, Tallinn, Estonia.

Reilly, J., 1996: Agriculture in a changing climate: impacts and adaptation. In: *Climate Change 1995: Impacts, Adaptations, and Mitigation of Climate Change: Scientific-Technical Analyses. Contribution of Working Group II to the Second Assessment Report of the Intergovernmental Panel on Climate Change* [Watson, R.T., M.C. Zinyowera, and R.H. Moss (eds.)]. Cambridge University Press, Cambridge, United Kingdom and New York, NY, USA, pp. 429–467.

Reinert, R.A., G. Eason, and J. Barton, 1997: Growth and fruiting of tomato as influenced by elevated carbon dioxide and ozone. *New Phytology*, **137,** 411–420.

Repo, T., H. Hanninen, and S. Kellomaki, 1996: The effects of long-term elevation of air temperature and co2 on the frost hardiness of Scots Pine. *Plant, Cell and Environment*, **19(2),** 209–216.

Reyenga, P.J., S.M. Howden, H. Meinke, and W.B. Hall, 1999: Global change impacts on wheat production along an environmental gradient in South Australia. In: *Modsim '99 International Congress on Modelling and Simulation Proceedings, December 6–9, 1999, Hamilton, New Zealand*. pp. 753–758.

Richardson, C.J. and E.J. McCarthy, 1994: Effect of land development and forest management on hydrologic response in southeastern coastal wetlands: a review. *Wetlands*, **14,** 56–71.

Richardson, C.J. and J.W. Gibbons, 1993: Pocosins, Carolina bays, and mountain bogs. In: *Biodiversity of the Southeastern United States: Lowland Terrestrial Communities* [Martin, W.H., S.G. Boyce, and A.C. Echternacht (eds.)]. John Wiley and Sons, Chichester, United Kingdom; New York, NY, USA; Brisbane, Australia; Tokyo, Japan; and Singapore, pp. 257–310.

Rieley, J.O. and A.A. Ahmad-Shah, 1996: The vegetation of tropical peat swamp forests. In: *Tropical Lowland Peatlands of Southeast Asia* [Maltby, E., C.P. Immirzi, and R.J. Safford (eds.)]. IUCN, Gland, Switzerland, pp. 55–74.

Rieley, J.O., A.A. Ahmad-Shah, and M.A. Brady, 1996: The extent and nature of tropical peat swamps. In: *Tropical Lowland Peatlands of Southeast Asia* [Maltby, E., C.P. Immirzi, and R.J. Safford (eds.)]. IUCN, Gland, Switzerland, pp. 17–54.

Riera, J.L., J.E. Schindler, and T.K. Kratz, 1999: Seasonal dynamics of carbon dioxide and methane in two clear-water lakes and two bog lakes in northern Wisconsin, USA. *Canadian Journal of Fisheries and Aquatic Sciences*, **56(2),** 265–274.

Riera, J.L., J.J. Magnuson, T.K. Kratz, and K.E. Webster, 2000: A geomorphic template for the analysis of lake districts applied to Northern Highland Lake District, Wisconsin, USA. *Freshwater Biology*, **43,** 301–318.

Rijksen, H.D., W.H. Diemont, and M. Griffith, 1997: The Singkil Swamp: the kidneys of the Leuser ecosystem in Aceh, Sumatra, Indonesia. In: *Biodiversity and Sustainability of Tropical Peatlands* [Rieley, J.O. and S.E. Page (eds.)]. Samara Publishing, Cardigan, Wales, United Kingdom, pp. 355–362.

Roberston, G., 1986: The mortality of kangaroos in drought. *Australian Wildlife and Restoration*, **13,** 349–354.

Robinson, D.C.E., W.A. Kurz, and C. Pinkham, 1999: *Estimating the Carbon Losses from Deforestation in Canada*. ESSA Technologies Ltd., National Climate Change Secretariat, Ottawa, ON, Canada, 81 pp. Available online at http://www.nccp.ca/html/index.htm.

Robinson, G.R. and S.N. Handel, 1993: Forest restoration on a closed landfill: rapid addition of new species by bird dispersal. *Conservation Biology*, **7,** 271–278.

Rodhe, H. and B.H. Svensson, 1995: Impact on the greenhouse effect of peat mining and combustion. *Ambio*, **24,** 221–225.

Rodriguez-Trelles, F. and M.A. Rodriguez, 1998: Rapid micro-evolution and loss of chromosomal diversity in Drosophila in response to climate warming. *Evolutionary Ecology*, **12,** 829–838.

Rogers, D., 1996: Changes in disease vectors. In: *Climate Change and Southern Africa: An Exploration of Some Potential Impacts and Implications in the SADC Region* [Hulme, M. (ed.)]. Climatic Research Unit, University of East Anglia, Norwich, United Kingdom, and World Wildlife Fund International, Gland, Switzerland.

Rogers, G., P.J. Milham, M. Gillings, and J.P. Conroy, 1996a: Sink strength may be the key to growth and nitrogen responses in N-deficient wheat at elevated CO_2. *Australian Journal of Plant Physiology*, **23,** 253–264.

Rogers, G., P.J. Milham, M.C. Thibaud, and J.P. Conroy, 1996b: Interactions between rising CO_2 concentration and nitrogen supply in cotton, I: growth and leaf nitrogen concentration. In *Australian journal of plant physiology*, **23,** 119–125.

Rogers, H.H., G.B. Runion, and S.V. Krupa, 1994: Plant responses to atmospheric CO_2 enrichment with emphasis on roots and the rhizosphere. *Environmental Pollution*, **83,** 155–189.

Roland, J., B.G. Mackey, and B. Cooke, 1998: Effects of climate and forest structure on duration of forest tent caterpillar outbreaks across central Ontario, Canada. *Canadian Entomologist*, **130,** 703–714.

Root, T.L., 1988c: Energy constraints on avian distributions and abundances. *Ecology*, **69,** 330–339.

Root, T.L., 1988a: Environmental factors associated with avian distributional boundaries. *Journal of Biogeography*, **15,** 489–505.

Root, T.L., 1988b: *Atlas of Wintering North American Birds: An Analysis of Christmas Bird Count Data*. University of Chicago Press, Chicago, IL, USA, 312 pp.

Root, T.L. and S.H. Schneider, 1995: Ecology and climate: research strategies and implications. *Science*, **269,** 334–341.

Rosegrant, M.W. and C. Ringler, 1997: World food markets into the 21st century: environmental and resource constraints and policies. *Australian Journal of Agricultural and Resource Economics*, **41,** 401–428.

Rosegrant, M.W., M. Agcaoili-Sombilla, and N.D. Perez, 1995: *Global Food Projections to 2020: Implications for Investment. 2020 Vision for Food, Agriculture, and the Environment*. Discussion Paper No. 5, International Food Policy Research Institute, Washington, DC, USA.

Rosenberg, N.J., 1982: The increasing CO_2 concentration in the atmosphere and its implication on agricultural productivity, part II: effects through CO_2-induced climatic change. *Climatic Change*, **4,** 239–254.

Rosenzweig, C. and A. Iglesias, 1998: The use of crop models for international climate change impact assessment. In: *Understanding Options for Agriculture Production* [Tsuji, G.Y., G. Hoogrnboom, and P.K. Thorton (eds.)]. J. Kluwer Academic Publishers, Dordrecht, The Netherlands, pp. 267–292.

Rosenzweig, C., A. Iglesias, X. Yang, P. Epstein, and E. Chivian, 2000: *Climate Change and U.S. Agriculture: The Impacts of Warming and Extreme Weather Events on Productivity, Plant Diseases, and Pests*. Center for Health and the Global Environment, Harvard Medical School, Boston, MA, USA, 46 pp.

Ross, D.J., S. Saggar, K.R. Tate, C.W. Feltham, and P.C.D. Newton, 1996: Elevated CO_2 effects on carbon and nitrogen cycling in grass/clover turves of a Psammaquent soil. *Plant and Soil*, **182,** 185–198.

Ross, M.S., J.J. O'Brien, L. Da Silveira, and L. Sternberg, 1994: Sea-level rise and the reduction in pine forests in the Florida keys. *Ecological Applications*, **4,** 144–156.

Roulet, N.T. and R. Ash, 1992: Low boreal wetlands as a source of atmospheric methane. *Journal of Geophysical Research*, **97,** 3739–3749.

Roulet, N.T. and T.R. Moore, 1995: The effect of forestry drainage practices on the emission of methane from northen peatlands. *Canadian Journal of Forest Research*, **25,** 491–499.

Roulet, N.T., R. Ash, W. Quinton, and T.R. Moore, 1993: Methane flux from drained northern peatland: effect of persistent water table lowering on flux. *Global Biogeochemical Cycles*, **7,** 749–769.

Roulet, N.T., T.R. Moore, J.L. Bubier, and P. Lafleur, 1992: Northern fens: methane flux and climatic change. *Tellus*, **44B,** 100–105.

Roulet, N.T., P.M. Crill, N.T. Comer, A. Dove, and R.A. Boubonniere, 1997: CO_2 and CH_4 flux between a boreal beaver pond and the atmosphere. *Journal of Geophysical Research*, **102,** 29313–29319.

Rouse, W.R., M.S.V. Douglas, R.E. Hecky, A.E. Hershey, G.W. Kling, L. Lesack, P. Marsh, M.M. McDonald, B.J. Nicholson, N.T. Roulet, and J.P. Smol, 1997: Effects of climate change on the freshwaters of Arctic and Subarctic North America. In: *Freshwater Ecosystems and Climate Change in North America* [Cushing, C.E. (ed.)]. John Wiley and Sons, New York, NY, USA, pp. 55–84.

Royama, T., 1984: Population dynamics of the spruce budworm Choristoneura fumiferana. *Ecological Monographs*, **54,** 429–462.

Rozelle, S. and D.D. Huang, 1999: Wheat in China: supply, demand and trade in the 21st century. In: *The Economics of World Wheat Markets* [Antle, J.M. and V.H. Smith (eds.)]. CAB International, Wallingford, United Kingdom.

Rubec, C.A., P. Lynch-Stewart, G.M. Wickware, and I. Kessel-Taylor, 1988: Wetland utilization in Canada. In: *Wetlands of Canada* [Group, N.W.W. (ed.)]. Polyscience Publications and Environment Canada, Montreal, QB, Canada, Vol. 24, pp. 381–412.

Ruget, F., O. Bethenod, and L. Combe, 1996: Repercussions of increased atmospheric CO_2 on maize morphogenesis and growth for various temperature and radiation levels. *Maydica*, **41,** 181–191.

Rupp, T.S., A.M. Starfield, and F.S. Chapin III, 2000: A frame-based spatially explicit model of subarctic vegetation response to climatic change: comparison with a point model. *Landscape Ecology*, **15,** 383–400.

Russell-Smith, J., D. Lucas, M. Gapindi, B. Gunbunuka, N. Kapirigi, G. Namingum, K. Lucas, P. Giuliani, and G. Chaloupka, 1997: Aboriginal resource utilization and fire management practice in western arnhem land, monsoonal northern Australia: notes for prehistory, lessons for the future (review). *Human Ecology*, **25(2),** 159–195.

Ruttan, V.W., 1996: Constraints on sustainable growth in agricultural production: into the twenty-first century. In: *The Economics of Agriculture, Volume 2: Papers in Honor of D. Gale Johnson* [Antle, J.M. and D.A. Sumner (eds.)]. The University of Chicago Press, Chicago, IL, USA, pp. 204–221.

Saarnio, S., J. Alm, P.J. Martikainen, and J. Silvola, 1998: Effects of raised CO_2 on potential CH_4 production and oxidation in, and CH_4 emission from, a boreal mire. *Journal of Ecology*, **86,** 261–268.

Sahagian, D. and J. Melack, 1998: Global wetland distribution and functional characterization: trace gases and the hydrologic cycle. In: *Report from the Joint GAIM, BAHC, IGBP-DIS, IGAC, and LUCC Workshop, Santa Barbara, CA, USA, May 1996*. IGBP Report 46, Stockholm, Sweden.

Sakovets, V.V. and N.I. Germanova, 1992: Changes in the carbon balance of forested mires in Karelia due to drainage. *Suo*, **43,** 249–252.

Sala, O.E., F.S. Chapin III, J.J. Armesto, E. Berlow, J. Bloomfield, R. Dirzo, E. Huber-Sanwald, L.F. Huenneke, R.B. Jackson, A. Kinzig, R. Leemans, D.M. Lodge, H.A. Mooney, M. Oesterheld, N.L. Poff, M.T. Sykes, B.H. Walker, M. Walker, and D.H. Wall, 2000: Biodiversity: global biodiversity scenarios for the year 2100. *Science*, **287,** 1770–1774.

Salo, K., 1996: Peatland berries: a valuable nourishing resource. In: *Peatlands in Finland* [Vasander, H. (ed.)]. Finnish Peatland Society, Helsinki, Finland, pp. 39–44.

Samarakoon, A. and R.M. Gifford, 1996: Elevated CO_2 effects on water use and growth of maize in wet and drying soil. *Australian Journal of Plant Physiology*, **23,** 53–62.

Samarakoon, A.B. and R.M. Gifford, 1995: Soil water content under plants at high CO_2 concentration and interactions with the direct CO_2 effects: a species comparison. *Journal of Biogeography*, **22,** 193–202.

Sampson, R.N. and L.A. DeCoster, 1998: *Forest Health in the United States*. American Forests, Washington, DC, USA, 76 pp.

Sampson, R.N., R.J. Scholes, C. Cerri, L. Erda, D.O. Hall, M. Handa, P. Hill, M. Howden, H. Janzen, J. Kimble, R. Lal, G. Marland, K. Minami, K. Paustian, P. Read, P.A. Sanchez, C. Scoppa, B. Solberg, M.A. Trossero, S. Trumbore, O. Van Cleemput, A. Whitmore, and D. Xu, 2000: Additional human-induced activities. In: *Land Use, Land-Use Change, and Forestry* [Watson, R.T., I.R. Noble, B. Bolin, N.H. Ravindranath, D.J. Verardo, and D.J. Dokken (eds.)]. Intergovernmental Panel on Climate Change, Cambridge University Press, Cambridge, United Kingdom and New York, NY, USA, 181-281.

Santer, B.D., T.M.L. Wigley, T.P. Barnett, and E. Anyamba, 1996: Detection of climate change and attribution of causes. In: *Climate Change 1995: The Science of Climate Change. Contribution of Working Group I to the Second Assessment Report of the Intergovernmental Panel on Climate Change* [Houghton, J.T., L.G. Meira Filho, B.A. Callander, N. Harris, A. Kattenberg, and K. Maskell (eds.)]. Cambridge University Press, Cambridge, United Kingdom and New York, NY, USA, pp. 407–33.

Saseendran, S.A., K.K. Singh, L.S. Rathore, S.V. Singh, and S.K. Sinha, 2000: Effect of climate change on rice production in the tropical humid climate of Kerala, India. *Climatic Change*, **44,** 495–514.

Scherr, S.J. and S. Yadav, 1997: *Land Degradation in the Developing World: Issues and Policy Options for 2020*. No. 44, International Food Policy Research Institute (IFPRI), Washington, DC, USA, 2 pp.

Schiff, M. and A. Valdez, 1996: Agricultural incentives and growth in developing countries: a cross-country perspective. In: *The Economics of Agriculture, Vol. 2: Papers in Honor of D. Gale Johnson* [Antle, J.M. and D.A. Sumner (eds.)]. University of Chicago Press, Chicago, IL, USA, pp. 386–399.

Schimel, D.S., 1998: The carbon equation. *Nature*, **393,** 208–209.

Schimel, D.S., B.H. Braswell, E.A. Holland, R. Mckeown, D.S. Ojima, T.H. Painter, W.J. Parton, and A.R. Townsend, 1994: Climatic, edaphic and biotic controls over storage and turnover of carbon in soils. *Global Biogeochemical Cycles*, **8,** 279–293.

Schimel, D.S., J.M. Melillo, H.Q. Tian, A.D. McGuire, D. Kicklighter, T. Kittel, N. Rosenbloom, S. Running, P. Thornton, D. Ojima, W. Parton, R. Kelly, M. Sykes, R. Neilson, and B. Rizzo, 2000: Contrbution of increasing CO_2 and climate to carbon storage by ecosystems in the United States. *Science*, **287,** 2004–2006.

Schimel, J.P., 1995: Plant transport and methane production as controls on methane flux from Arctic wet meadow tundra. *Biogeochemistry*, **28,** 183–200.

Schindler, D.W., 1997: Widespread effects of climatic warming on freshwater ecosystems in North America. *Hydrological Processes*, **11,** 825–871.

Schindler, D.W. and S.E. Bayley, 1993: The biosphere as an increasing sink for atmospheric carbon: estimates from increased nitrogen deposition. *Global Biogeochemical Cycles*, **7(4),** 717–733.

Schindler, D.W., P.J. Curtis, B.R. Parker, and M.P. Stainton, 1996a: Consequences of climate warming and lake acidification for UV-B penetration in North American boreal lakes. *Nature*, **379**, 705–708.

Schindler, D.W., S.E. Bayley, B.R. Parker, K.G. Beaty, D.R. Cruikshank, E.J. Fee, E.U. Schindler, and M.P. Stainton, 1996b: The effects of climatic warming on the properties of boreal lakes and streams at the experimental lakes area, northwestern Ontario. *Limnology and Oceanography*, **41,** 1004–1017.

Schlamadinger, B., M. Apps, F. Bohlin, L. Gustavsson, G. Jungmeier, G. Marland, K. Pingoud, and I. Savolainen, 1997: Towards a standard methodology for greenhouse gas balances of bioenergy systems in comparison with fossil energy systems. *Biomass and Bioenergy*, **13,** 359–375.

Schlesinger, W.H., 1997: *Biogeochemistry: An Analysis of Global Change*. Academic Press, San Diego, CA, USA, 588 pp.

Schneider, S.H. and T.L. Root, 1996: Ecological implications of climate change will include surprises. *Biodiversity and Conservation*, **5,** 1109–1119.

Schneider, S.H., W.E. Easterling, and L.O. Mearns, 2000: Adaptation: sensitivity to natural variability, agent assumptions and dynamic climate changes. *Climatic Change*, **45,** 203–221.

Schreader, C.P., W.R. Rouse, T.J. Giffis, L.D. Boudreau, and P.D. Blanken, 1998: Carbon dioxide fluxes in a northern fen during a hot, dry summer. *Global Biogeochemical Cycles*, **12,** 729–740.

Schulze, E.D., J. Lloyd, F.M. Kelliher, C. Wirth, C. Rebmann, B. Lühker, M. Mund, A. Knohl, I.M. Milyukova, W. Schulze, W. Ziegler, A.B. Varlagin, A.F. Sogachev, R. Valentini, S. Dore, S. Grigoriev, O. Kolle, M.I. Panfyorov, N. Tchebakova, and N.N. Vygodskaya, 1999: Productivity of forests in the Eurosiberian boreal region and their potential to act as a carbon sink: a synthesis. *Global Change Biology*, **5,** 703–722.

Scott, C.H., 1987: The socio-economic significance of waterfowl among Canada's aboriginal Cree: native use and local management. In: *The Value of Birds. ICBP Technical Publication No. 6* [Diamond, A.W. and F.L. Filion (eds.)]. International Council for Bird Preservation, Cambridge, United Kingdom, pp. 49–62.

Scrimgeour, G.J., T.D. Prowse, J.M. Culp, and P.A. Chambers, 1994: Ecological effects of river ice break-up: a review and perspective. *Freshwater Biology*, **32,** 261–275.

Seddon, N., J.M.M. Ekstrom, D.R. Capper, I.S. Isherwood, R. Muna, R.G. Pople, E. Tarimo, and J. Timothy, 1999: The importance of the Nilo and Nguu North Forest Reserves for the conservation of montane forest birds in Tanzania. *Biological Conservation*, **87,** 59–62.

Sedjo, R. and K. Lyon, 1990: *Long Term Adequacy of World Timber Supply*. Resources for the Future, Washington, DC, USA, 230 pp.

Sedjo, R.A., 1999: The potential of high-yield plantation forestry for meeting timber needs. *New Forests*, **17(1–3),** 339–359.

Segers, R., 1998: Methane production and methane consumption: a review of processes underlying wetland methane fluxes. *Biogeochemistry*, **41,** 23–51.

Seidensticker, J., 1987: Managing tigers in the Sundarbans: experience and opportunity. In: *Tigers of the World: The Biology, Biopolitics, Management, and Conservation of an Endangered Species* [Tilson, R.L. and U.S. Seal (eds.)]. Noyes Publications, Park Ridge, NJ, USA, pp. 416–426.

Sellers, P.J., J.C. Cihlar, M.J. Apps, B. Goodison, D. Leckie, F. Hall, E. LeDrew, P. Matson and S. Running,, 1990: Charting the boreal forest's role in global change. *EOS*, **22(4),** 33–34.

Sellers, P.J., F.G. Hall, R.D. Kelly, A. Black, D. Baldocchi, J. Berry, M. Ryan, K.J. Ranson, P.M. Crill, D.P. Lettenmaier, H. Margolis, J. Cihlar, J. Newcomer, D. Fitzjarrald, P.G. Jarvis, S.T. Gower, D. Halliwell, D. Williams, B. Goodison, D.E. Wickland, and F.E. Guertin, 1997: BOREAS in 1997: experiment overview, scientific results, and future directions. *Journal of Gephysical Research*, **102(D24),** 731–728 and 769.

Semenov, M.A. and J.R. Porter, 1995: Climatic variability and the modelling of crop yields. *Agricultural Forest Meteorology*, **38,** 127–145.

Semenov, M.A., J. Wolf, L.G. Evans, H. Eckersten, and A. Eglesias, 1996: Comparison of wheat simulation models under climate change, 2: application of climate change scenarios. *Climate Research*, **7,** 271–281.

Semenov, S.M., I.M. Kounina, and B.A. Koukhta, 1999: Tropospheric ozone and plant growth in Europe. *Meteorology and Hydrology*, 208 pp. (in Russian).

Semenov, S.M., I.M. Kounina, and B.A. Koukhta, 1998: An ecological analysis of anthropogenic changes in ground-level concentrations of O_3, SO_2, and CO_2 in Europe. *Doklady Biological Sciences*, **361,** 344–347.

Semenov, S.M., B.A. Kukhta, and A.A. Rudkova, 1997: Assessment of the tropospheric ozone effect on higher plants. *Meteorology and Hydrology*, **12,** 36–40.

Seneweera, S.P. and J.P. Conroy, 1997: Growth, grain yield and quality of rice (Oryza sativa L.) in response to elevated CO_2 and phosphorus nutrition. *Soil Science and Plant Nutrition*, **43,** 1131–1136.

Sepkoski, J.J., 1998: Rates of speciation in the fossil record. *Transactions of the Royal Society of London B*, **353,** 315–326.

Serreze, M.C., J.E. Walsh, F.S. Chapin III, T. Osterkamp, M. Dyurgerov, V. Romanovsky, W.C. Oechel, J. Morison, T. Zhang, and R.G. Barry, 2000: Observational evidence of recent change in the northern high-latitude environment. *Climatic Change*, **46,** 159–207.

Severnty, D.L., 1977: The use of data on the distribution of birds to monitor climatic changes. *Emu*, **77,** 162–166.

Shannon, R.D. and J.R. White, 1994: A three-year study of controls of methane emissions from two Michigan peatlands. *Biogeochemistry*, **27,** 35–60.

Sharitz, R.R. and C.A. Gresham, 1998: Pocosins and Carolina bays. In: *Southern Forested Wetlands: Ecology and Management* [Messina, M.G. and W.H. Conner (eds.)]. Lewis Publishers, Boca Raton, FL and New York, NY, USA; London, United Kingdom; and Tokyo, Japan, pp. 343–377.

Shaver, G.R., W.D. Billings, F.S. Chapin III, A.E. Giblin, K.J. Nadelhoffer, W.C. Oechel, and E.B. Rastetter, 1992: Global change and the carbon balance of Arctic ecosystems. *BioScience*, **61,** 415–435.

Shaver, G.R., J. Canadell, F.S. Chapin III, J. Gurevitch, J. Harte, G. Henry, P. Ineson, S. Jonasson, J. Melillo, L. Pitelka, and L. Rustad, 2000: Global warming and terrestrial ecosystems: a conceptual framework for analysis. *BioScience*, **50,** 871–882.

Sheingauz, A.S., 1989: *Forest Resources of the Far East Economic Region: State, Utilization, Reproduction*. Far Eastern Forestry Research Institute, Khabarovsk, Russia, 42 pp. (in Russian).

Shepard, J.P., S.J. Brady, N.D. Cost, and C.G. Storrs, 1998: Classification and inventory. In: *Southern Forested Wetlands: Ecology and Management* [Messina, M.G. and W.H. Conner (eds.)]. Lewis Publishers, Boca Raton, FL and New York, NY, USA; London, United Kingdom; and Tokyo, Japan, pp. 3–28.

Sheparshenko, D., A. Shvidenko, and S. Nilsson, 1998: Phytomass (live biomass) and carbon of Siberian Forests. *Biomass and Bioenergy*, **14(1),** 21–31.

Sheppard, C., 1995: Propagation of endangered birds in U.S. institutions: how much space is there? *Zoo Biology*, **14,** 197–210.

Shriner, D.S., R.B. Street, R. Ball, D.K. D'Amours, D. Kaiser, A. Maarouf, M.P. Mortsch, R. Neilson, J.A. Patz, J.D. Scheraga, J.G. Titus, H. Vaughan, M. Weltz, R. Adams, R. Alig, J. Andrey, M.J. Apps, M. Brklacich, D. Brooks, A.W. Diamon, A. Grambsch, D. Goodrich, L. Joyce, M.R. Kidwell, G. Koshida, J. Legg, J. Malcolm, D.L. Martell, R.J. Norby, H.W. Polley, W.M. Post, M.J. Sale, M. Scott, B. Sohngen, B.J. Stocks, W. Van Winkle, and S. Wullshleger, 1998: North America. In: *The Regional Impacts of Climate Change: An Assessment of Vulnerability. Special Report of IPCC Working Group II* [Watson, R.T., M.C. Zinyowera, and R.H. Moss (eds.)]. Cambridge University Press, Cambridge, United Kingdom and New York, NY, USA, pp. 253–330.

Shurpali, N.J., S.B. Verma, J. Kim, and T.J. Arkebauer, 1995: Carbon dioxide exchange in a peatland ecosystem. *Journal of Geophysical Research*, **100,** 14319–14326.

Shvidenko, A., 2000: Global significance of disturbances in boreal forests. In: *Disturbances in Boreal Forest Ecosystems: Human Impacts and Natural Resources* [Conrad, S.G. (ed.)]. USDA - Forest Service, North Central Research Station, St Paul, USA, General Technical ReportNC-209, 17–29.

Shvidenko, A. and S. Nilsson, 1997: Are the Russian forests disappearing? *Unasylva,* **48,** 57–64.

Shvidenko, A. and S. Nilsson, 1994: What do we really know about the Siberian forests? *Ambio*, **23(7),** 396–404.

Shvidenko, A., S. Nilsson, V. Rojkov, 1995: *Possibilities for Increased Carbon Sequestration Through Improved Protection of Russian Forests*. IIASA WP-95-86, International Institute for Applied Systems Analysis, Laxenburg, Austria, 27 pp.

Siegel, D.I. and P.H. Glaser, 1987: Groundwater flow in a bog-fen complex, Lost River Peatland, Northern Minnesota. *Journal of Ecology*, **75,** 743–754.

Siegel, D.I., A.S. Reeve, P.H. Glaser, and E.A. Romanowicz, 1995: Climate-driven flushing of pore water in peatlands. *Nature*, **374,** 531–533.

Silvius, M.J., A.P.J.M. Steeman, E.T. Berczy, E. Djuharsa, and A.W. Tanfik, 1987: *The Indonesian Wetland Inventory*. PHPA AWB/INTERWADER, EDWIN, Bogor, Indonesia.

Silvola, J., J. Alm, U. Ahlholm, H. Nykänen, and P.J. Martikainen, 1996: CO_2 fluxes from peat in boreal mires under varying temperature and moisture conditions. *Journal of Ecology*, **84,** 219–228.

Singh, B. and M. El Maayar, 1998: Potential impacts of greenhouse gas climate change scenarios on sugar cane yields in Trinidad. *Tropical Agriculture*, **75,** 348–353.

Sirotenko, O.D. and E.V. Abashina, 1998: Agroclimatic resources and physical-geographic zones of Russia under global warming. *Meteorology and Hydrology*, **3,** 92–103 (in Russian).

Sirotenko, O.D., E.V. Abashina, and V.N. Pavlova, 1995: Sensitivity of Russian agriculture to changes in climate, chemical composition of the atmospere, and soil fertility. *Meteorology and Hydrology*, **4,** 107–114 (in Russian).

Skog, K.E. and G.A. Nicholson, 1998: Carbon cycling through wood products: the role of wood and paper products in carbon sequestration. *Forest Products Journal*, **48(7-8),** 75-83.

Slater, F.M., 1999: First-egg date fluctuations for the Pied Flycatcher Ficedula hypoleuca in the woodlands of mid-Wales in the twentieth century. *Ibis*, **141,** 497–499.

Smart, P.J., B.D. Wheeler, and A.J. Willis, 1989: Revegetation of peat excavations in a derelict raised bog. *New Phytologist*, **111,** 733–748.

Smit, B., R. Blain, and K. Philip, 1997: Corn hybrid selection and climatic variability: gambling with nature? *The Canadian Geographer*, **41,** 429–438.

Smith, F.A., H. Browning, and U.L. Shepherd, 1998: The influence of climate change on the body mass of woodrats Neotoma in an arid region of New Mexico. *Ecography*, **21,** 140–148.

Smith, J.B., S. Huq, S. Lenhart, L.J. Mata, I. Nemesova, and S. Toure, 1996a: *Vulnerability and Adaptation to Climate Change: Interim Results from the U.S. Country Studies Program*. J. Kluwer Academic Publishers, Dordrecht, The Netherlands and Boston, MA, USA, 366 pp.

Smith, R.C., S.E. Stammerjohn, and K.S. Baker, 1996b: Surface air temperature variations in the western Anatarctic Peninsula region. In: *Foundations for Ecological Research West of the Antarctic Peninsula* [Ross, R.M., E.E. Hofmannm, and L.B. Quetin (eds.)]. No. 70, Antarctic Research Series, American Geophysical Union, Washington, DC, USA, pp. 105–121.

Smith, P., J.U. Smith, D.S. Powlson, J.R.M. Arah, O.G. Chertov, K. Coleman, U. Franko, S. Frolking, D.S. Jenkinson, L.S. Jensen, R.H. Kelyy, H. Kelin-Gunnewick, A.S. Komarov, C. Li, J.A.E. Molina, T. Mueller, W.S., Parton, J.H.M. Thornley, and A.P. Whitmore, 1997: A comparison of the performance of nine soil organic matter models using datasets from seven long-term experiments. *Geoderma*, **81(1-2)**, 153-225.

Smith, R.C., D.G. Ainley, K. Baker, E. Domack, S.D. Emslie, W.R. Fraser, J. Kennet, A. Leventer, E. Mosley-Thompson, S.E. Stammerjohn, and M. Vernet, 1999: Marine ecosystem sensitivity to historical climate change in the Antarctic Peninsula. *BioScience*, **49,** 393–404.

Smith, R.I.L., 1994: Vascular plants as bioindicators of regional warming in Antarctic. *Oecologia*, **99,** 322–328.

Smith, T.M. and H.H. Shugart, 1993: The transient response of terrestrial carbon storage to a perturbed climate. *Nature*, **361(6412),** 523–526.

Smithers, J. and B. Smit, 1997: Human adaptation to climatic variability and change. *Global Environmental Change and Policy Dimensions*, **7,** 129–146.

Sohngen, B., 2001: Timber: ecological-economic analysis. In: *Global Warming and the American Economy: A Regional Assessment of Climate Change* [Mendelsohn, R. (ed.)]. Edward Elgar Publishing, United Kingdom.

Sohngen, B. and R. Mendelsohn, 1998: Valuing the market impact of large scale ecological change: the effect of climate change on U.S. timber. *American Economic Review*, **88(4),** 689–710.

Sohngen, B., R. Mendelsohn, and R. Neilson, 1998: Predicting CO_2 emissions from forests during climate change: a comparison of human and natural response models. *Ambio*, **27(7),** 509–513.

Sohngen, B., R. Mendelsohn, and R. Sedjo, 2000: *Measuring Climate Change Impacts with a Global Timber Model*. Working paper, Department of Agricultural, Environmental, and Development Economics, Ohio State University, Columbus, OH, USA.

Sohngen, B., R. Mendelsohn, and R. Sedjo, 1999: Forest management, conservation, and global timber markets. *American Journal of Agricultural Economics*, **81(1),** 1–13.

Solberg, B., D. Brooks, H. Pajuoja, T.J. Peck, and P.A. Wardle, 1996: Long-term trends and prospects in world supply and demand for wood and impolications for sustainable forest management: a synthesis. In: *Long-Term Trends and Prospects in World Supply and Demand for Wood and Implications for Sustainable Forest Management* [Solberg, B. (ed.)]. European Forest Institute Research Report 6, Joensuu, Finland.

Solomon, A.M., 1994: Management and planning of terrestrial parks and reserves during climate change. In: *Impacts of Climate Change on Ecosystems and Species: Implications for Protected Areas* [Pernetta, J., R. Leemans, D. Elder, and S. Humphrey (eds.)]. IUCN, Gland, Switzerland, pp. 1–12.

Solomon, A.M. and A.P. Kirilenko, 1997: Climate change and terrestrial biomass: what if trees do not migrate? *Global Ecology and Biogeography Letters*, **6,** 139–148.

Solomon, A.M. and R. Leemans, 1997: Boreal forest carbon stocks and wood supply: past, present, and future responses to changing climate, agriculture, and species availability. *Agricultural and Forest Meteorology*, **84,** 137–151.

Solomon, A.M., N.H. Ravindranath, R.B. Stewart, M. Weber, S. Nilsson, P.N. Duinker, P.M. Fearnside, P.J. Hall, R. Ismail, L.A. Joyce, S. Kojima, W.R. Makundi, D.F.W. Pollard, A. Shvidenko, W. Skinner, B.J. Stocks, R. Sukumar, and X. Deying, 1996: Wood production under changing climate and land use. In: *Climate Change 1995: Impacts, Adaptions, and Mitigation of Climate Change: Scientific-Technical Analyses. Contribution of Working Group II to the Second Assessment Report for the Intergovernmental Panel on Climate Change* [Watson, R.T., M.C. Zinyowera, and R.H. Moss (eds.)]. Cambridge University Press, Cambridge, United Kingdom and New York, NY, USA, 487-510.

Somaratne, S. and A.H. Dhanapala, 1996: Potential impact of global climate change on forest distribution in Sri Lanka. *Water, Air, and Soil Pollution*, **92,** 129–135.

Sorensen, K.W., 1993: Indonesian peat swamp forests and their role as a carbon sink. *Chemosphere*, **27,** 1065–1082.

Sorenson, L., R. Goldberg, T.L. Root, and M.G. Anderson, 1998: Potential effects of global warming on waterfowl populations breeding in the northern Great Plains. *Climatic Change*, **40,** 343–369.

Soussana, J.F. and U. Hartwig, 1996: The effects of elevated CO_2 on symbiotic N_2 fixation: a link between the C and N cycles in grassland ecosystems. *Plant and Soil*, **187,** 321–332.

Soussana, J.F., E. Casella, and P. Loiseau, 1996: Long-term effects of CO_2 enrichment and temperature increase on a temperate grass sward, II: plant nitrogen budgets and root fraction. *Plant and Soil*, **182,** 101–114.

Soussana, J.F., J.M. Besle, I. Chabaux, and P. Loiseau, 1997: Long-term effects of CO_2 enrichment and temperature on forage quality in a temperate grassland. In: *Proceeding of the XVIII International Grassland Congress, Saskatoon, Canada, 8–17 June* [Buchanan-Smith, J.G., L.D. Bailey, and P. McCaughey (eds.)]. pp. 23–24.

Southgate, D., 1998: *Tropical Forest Conservation: An Economic Assessment of the Alternatives in Latin America*. Oxford University Press, Oxford, United Kingdom, 175 pp.

Southwick, E. E., 1992: Estimating the economic value of honey bees (Hymenoptera:Apidae) as agricultural pollinators in the United States. *Journal of Economic Entomology* , **85**, 621-33.

Southworth, J., J.C. Randolph, M. Habeck, O. Doering, R. Pfeifer, D. Rao, and J. Johnson, 1999: Consequences of future climate change and changing climate variability on maize yields in the midwestern United States. *Agriculture, Ecosystems and the Environment,* **82(1-3),** 139-158.

Sparks, T.H., 1999: Phenology and the changing pattern of bird migration in Britain. *International Journal of Biometeorology*, **42,** 134–138.

Spiers, A.G., 1999: Review of international/continental wetland resources. In: *Global Review of Wetland Resources and Priorities for Wetland Inventory* [Finlayson, C.M. and A.G. Spiers (eds.)]. Supervising Scientist Report, Canberra, Australia.

Squires, V.R. and A. Sidahmed, 1997: Livestock management in dryland pastoral systems: prospects and problems. *Annals of Arid Zone*, **36(2),** 79–96.

Stallard, R.F., 1998: Terrestrial sedimentation and the carbon cycle: coupling weathering and erosion to carbon burial. *Global Biogeochemical Cycles*, **12,** 231–257.

Stapp, P., G.A. Polis, and F. Sánchez Piñero, 1999: Stable isotopes reveal strong marine and El Niño effects on island food webs. *Nature*, **401,** 467–469.

Starfield, A.M. and F.S. Chapin III, 1996: Model of transient changes in Arctic and boreal vegetation in response to climate and land use change. *Ecological Applications*, **6,** 842–864.

Stattersfield, A.J., M.J. Crosby, A.J. Long, and D.C. Wege, 1998: *Endemic Bird Areas of the World: Priorities for Biodiversity Conservation. BirdLife Conservation Series No. 7*. BirdLife International, Cambridge, United Kingdom.

Stefan, H.G. and X. Fang, 1993: Model simulations of dissolved oxygen characteristics of Minnesota lakes: past and future. *Environmental Management*, **18,** 73–92.

Steffen, W., I. Noble, J. Canadell, M. Apps, E.D. Schulze, P.G. Jarvis, D. Baldocchi, P. Ciais, W. Cramer, J. Ehleringer, G. Farquhar, C.B. Field, A. Ghazi, R. Gifford, M. Heimann, R. Houghton, P. Kabat, C. Körner, E. Lambin, S. Linder, H.A. Mooney, D. Murdiyarso, W.M. Post, I.C. Prentice, M.R. Raupach, D.S. Schimel, A. Shvidenko and R. Valentini, 1998: The terrestrial carbon cycle: implications for the Kyoto Protocol. *Science*, **280,** 1393–1394.

Steinger, T., C. Körner, and B. Schmid, 1996: Long-term persistence in a changing climate: DNA analysis suggests very old ages of clones of alpine Carex curvula. *Oecologia*, **105,** 94–99.

Stewart, J. and C. Potvin, 1996: Effects of elevated CO_2 on an artificial grassland community: competition, invasion and neighbourhood growth. *Functional Ecology*, **10,** 157–166.

Still, C.J., P.N. Foster, and S.H. Schneider, 1999: Simulating the effects of climate change on tropical montane cloud forests. *Nature*, **398,** 608–610.

Stirling, I., 1997: The importance of polynas, ice edges, and leads to marine mammals and birds. *Journal of Marine Systems*, **10,** 9–21.

Stirling, I., and A. E. Derocher, 1993: Possible impacts of climatic warming on polar bears. *Arctic*, **46**, 240-245.

Stocks, B.J., M.A. Fosberg, T.J. Lynham, L. Mearns, B.M. Wotton, Q. Yang, J.Z.Jin, K. Lawrence, G.R. Hartley, J.A. Mason and D.W. McKenney, 1998: Climate change and forest fire potential in Russian and Canadian boreal forests. *Climatic Change*, **38,** 1–13.

Streng, D.R., J.S. Glitzenstein, and P.A. Harcombe, 1989: Woody seedling dynamics in an east Texas floodplain forest. *Ecological Monographs*, **59(2),** 177-204.

Strzepek, K.M., C.D. Major, C. Rosenzweig, A. Iglesias, D.Yates, A. Holt, and D. Hillel, 1999: New methods of modeling water availability for agriculture under climate change: the U.S. corn belt. *Journal of American Water Resources Association*, **35(6),** 1639–1655.

Stumm, W. and J.J. Morgan, 1996: *Aquatic Chemistry*. John Wiley and Sons, New York, NY, 3rd. ed., 1022 pp.

Sullivan, C.R., 1965: Laboratory and field investigations on the ability of eggs of the European pine sawfly, Neodiprion sertifer (Geoffroy) to withstand low winter temperatures. *Canadian Entomologist*, 978–993.

Sutherst, R.W., B.S. Collyer, and T. Yonow, 2000: The vulnerability of Australian horticulture to the Queensland fruit fly, Bactrocera (Dacus) tryoni, under climate change. *Australian Journal of Agricultural Research*, **51,** 467–480.

Sutherst, R.W., G.F. Maywald, and D.B. Skarratt, 1995: Predicting insect distributions in a changed climate. In: *Insects in a Changing Environment* [Harrington, R. and N.E. Stork (eds.)]. Academic Press, San Diego, CA, USA, pp. 60–93.

Suyker, A.E., S.B. Verma, and T.J. Arkebauer, 1997: Season-long measurements of carbon dioxide exchange in a boreal fen. *Journal of Geophysical Research*, **102(D24),** 29021–29028.

Tallis, J.H., 1983: Changes in wetland communities. In: *Ecosystems of the World 4A. Mires: Swamp, Bog, Fen and Moor. General Studies* [Gore, A.J.P. (ed.)]. Elsevier Scientific Publishing Company, Amsterdam, The Netherlands; Oxford, United Kingdom; and New York, NY, USA, pp. 311–347.

Tambussi, C.P., J.I. Noriega, and E.P. Tonni, 1993: Late Cenozoic birds of Buenos Aires Province (Argentina): an attempt to document quantitative faunal changes. *Palaeogeography, Palaeoclimatology, Palaeoecology*, **101,** 117–129.

Tans, P.P., I.Y. Fung, and T. Takahashi, 1990: Observational constraints on the global CO_2 budget. *Science*, **247,** 1431–1438.

Tate, K.R. and D.J. Ross, 1997: Elevated CO_2 and moisture effects on soil carbon storage and cycling in temperate grasslands. *Global Change Biology*, **3,** 225–235.

Tchbakova, N.M., R.A. Monserud and D.I. Nazimova, 1994: A Siberian vegetation model based on climatic parameters. *Canadian Journal of Forest Research*, **24,** 1597–1607.

Teng, P.S., K.L. Heong, M.J. Kropff, F.W. Nutter, and R.W. Sutherst, 1996: Linked pest-crop models under global change. In: *Global Change in Terrestrial Ecosystems* [Walker, B. and W. Steffen (eds.)]. Cambridge University Press, Cambridge, United Kingdom and New York, NY, USA, pp. 291–316.

Tenow, O., 1996: Hazards to a mountain birch forest—Abisko in perspective. *Ecological Bulletins*, **45,** 104–114.

Teramura, A.H., J.H. Sullivan, and L.H. Ziska, 1990: Interactions of ultraviolet B radiation and CO_2 on productivity and photosynthetic characteristics in wheat, rice, and soybean. *Plant Physiology*, **94,** 470–475.

Tester, R.F., W.R. Morrison, R.H. Ellis, J.R. Pigott, G.R. Batts, T.R. Wheeler, J.I.L. Morison, P. Hadley, and D.A. Ledward, 1995: Effects of elevated growth temperature and carbon dioxide levels on some physiochemical properties of wheat starch. *Journal of Cereal Science*, **22,** 63–71.

Thia-Eng, C. and J.N. Paw, 1989: The impact of global climate change on the aquaculture of tropical species. *INFOFISH International*, **6,** 44–47.

Thomas, C.D. and J.J. Lennon, 1999: Birds extend their ranges northwards. *Nature*, **399,** 213

Thomson, A.J., and D.M. Shrimpton, 1984: Weather associated with the start of mountain pine beetle outbreaks. *Canadian Journal of Forest Research*, **14,** 255–258.

Thomson, A.J., R.F. Shepherd, J.W.E. Harris, and R.H. Silversides, 1984: Relating weather to outbreaks of western spruce budworm, Choristoneura occidentalis (Lepidoptera: Tortricidae), in British Columbia. *Canadian Entomologist*, **116,** 375–381.

Thornley, J.H.M. and M.G.R. Cannell, 1997: Temperate grassland responses to climate change—an analysis using the Hurley Pasture Model. *Annals of Botany*, **80,** 205–221.

Thwaites, R., T. Delacy, L.Y. Hong, and L.X. Hua, 1998: Property rights, social change, and grassland degradation in Xilongol Biosphere Reserve, Inner Mongolia, China. *Society and Natural Resources*, **11(4),** 319–338.

Tian, H., J.M. Melillo, D.W. Kicklighter, D.A. McGuire, J.V.K. Helfrich, B. Moore III, and C.J. Vorosmarty, 1998: Effect of interannual climate variability on carbon storage in Amazonian ecosystems. *Nature*, **396,** 664–667.

Tingey, D.T., J. Lawrence, J.A. Weber, J. Greene, W.E. Hogsett, S, Brown, and E.H. Lee, 2001: Elevated CO_2 and temperature alter the response of Pinus ponderosa to ozone: a simulation analysis. *Ecological Applications*, (in press).

Tissue, D.T. and W.C. Oechel, 1987: Response of Eriophorum vaginatum to elevated CO_2 and temperature in the Alaskan tussock tundra. *Ecology*, **68,** 401–410.

Tol, R.S.J., S. Frankhauser, and J.B. Smith, 1998: The scope for adaptation to climate change: what can we learn from the impact literature? *Global Environmental Change*, **8,** 109–123.

Tomppo, E., 1998: Finnish forest resources in 1989–1994 and their changes from 1951. In: *Environment Change and the Vitality of Forests* [Mälkönen, E. (ed.)]. Research Papers 691, Finnish Forest Research Institute, pp. 9–24 (in Finnish).

Torbert, H.A., S.A. Prior, and H.H. Rogers, 1995: Elevated atmospheric carbon dioxide effects on cotton plant residue decomposition. *Journal of the American Soil Science Society*, **59,** 1321–1328.

Townsend, A.R. and E.B. Rastetter, 1996: Nutrient constrains on carbon storage in forested ecosystems. In: *Forest Ecosystems, Forest Management and the Global Carbon Cycle* [Apps, M.J. and D.T. Price (eds.)]. Springer-Verlag, Berlin and Heidelberg, Germany, pp. 35–45.

Travis, J., 1981: Control of larval growth variation in a population of Psuedacris triseriata. *Evolution*, **35,** 423–432.

Tripp, K.E., W.K. Kroen, M.M. Peet, and D.H. Willits, 1992: Fewer whiteflies found on CO_2-enriched greenhouse tomatoes with high C:N ratios. *Horticultural Science*, **27,** 1079–1080.

Tuittila, E.-S. and V.-M. Komulainen, 1995: Vegetation and CO_2 balance in an abandoned harvested peatland in Aitoneva, southern Finland. *Suo*, **46,** 69–80.

Tuittila, E.-S., V.-M. Komulainen, H. Vasander, and J. Laine, 1999: Restored cut-away peatland as a sink for atmospheric CO_2. *Oecologia*, **120,** 563–574.

Tung, C.-P. and D.A. Haith, 1998: Climate change, irrigation, and crop response. *Journal of the American Water Resources Association*, **34(5),** 1071–1085.

Turner, M., 1993: Overstocking the range: a critical analysis of the environmental science of Sahelian Pastoralism (review). *Economic Geography*, **69(4),** 402–421.

Turner, D.P., G.J. Koerper, M.E. Harmon, and J.J. Lee, 1995: A carbon budget for forests of the conterminous United States. *Ecological Applications*, **5,** 421–436.

Tweeten, L., 1998: Dodging a Malthusian bullet in the 21st century. *Agribusiness*, **14,** 15–32.

Tyler, H.A., 1991: *Pueblo Birds and Myths*. Northland Publishing, Flagstaff, AZ, USA.

Tyler, M. J., 1994: Climatic change and its implications for the amphibian fauna. *Transactions of the Royal Society of South Australia*, **118,** 53-57.

UK-DETR, 1999: *Climate Change and its Impacts: Stabilisation of CO_2 in the Atmosphere*. United Kingdom Department of the Environment, Transport and the Regions, The Met Office, Bracknell, UK, 28 pp.

UN-ECE/FAO, 2000a: *Forest Resources of Europe, CIS, North America, Australia, Japan and New Zealand*. Contribution to the Global Forest Resources Assessment 2000, United Nations Economic Commission for Europe and Food and Agriculture Organization, Geneva, Switzerland.

UN-ECE/FAO, 2000b: *Forest Products Annual Market Review 1999–2000*. ECE/TIM/BULL 53(3), Vol. LIII, Timber Bulletin, United Nations Economic Commission for Europe and Food and Agriculture Organization, pp. 23–37.

UNEP, 2000: *Global Environment Outlook 2000*. United Nations Environment Program, Nairobi, Kenya, 398 pp.

UNEP, 1998: *Protecting Our Planet, Securing Our Future* [Watson, R.T., J.A. Dixon, S.P. Hanburg, A.C. Janetos, and R.H. Moss (eds.)]. United Nations Environment Program, NASA, and World Bank, Washington, DC, USA, 95 pp.

United Nations, 1996: *Indicators of Sustainable Development: Framework and Methodologies*. United Nations Publications, New York, NY, USA, 428 pp.

Unsworth, M.H. and W.E. Hogsett, 1996: Combined effects of changing CO_2, temperature, UV-B radiation, and O_3 on crop growth. In: *Global Climate Change and Agricultural Production* [Bazzaz, F. and W. Sombroek (eds.)]. Food and Agriculture Organization, John Wiley and Sons, West Sussex, 345 pp.

U.S. Department of Agriculture, 1999: *Food Security Assessment, Situation and Outlook Series*. GFA-11, December, Economic Research Service, U.S. Department of Agriculture, Washington, DC, USA.

U.S. Fish and Wildlife Service, 1996: National Survey of Fishing, Hunting, and Wildlife-Associated Recreation. U.S. Department of the Interior, U.S. Fish and Wildlife Service and U.S. Department of Commerce, Bureau of the Census, Washington, DC, USA.

U.S. Fish and Wildlife Service, 1995: *Federal and State Endangered Species Expenditures—Fiscal Year 1995*. U.S. Department of the Interior, U.S. Fish and Wildlife Service, Washington, DC, USA.

USGS, 1999: *Magnitude and Significance of Carbon Burial in Lakes, Reservoirs, and Northern Peatlands*. USGS Fact Sheet FS-058-99, U.S. Geological Survey, U.S. Department of the Interior, Washington, DC, USA, 2 pp.

Valentin, C., 1996: Soil erosion under global change. In: *Global Change and Terrestrial Ecosystems* [Walker, B.H. and W.L. Steffen (eds.)]. IGBP Book Series, No. 2, Cambridge University Press, Cambridge, United Kingdom, pp. 317–338.

Valentine, D.W., E.A. Holland, and J.P. Schimel, 1994: Ecosystem and physiological controls over methane production in northern wetlands. *Journal of Geophysical Research*, **99,** 1563–1571.

Valentini, R., G. Matteucci, A.J. Dolman, E-D. Schulze, C. Rebmann, E.J. Moors, A. Granier, P. Gross, N.O. Jensen, K. Pilegaard, A. Lindroth, A. Grelle, C. Bernhofer, T. Grunwald, M. Aubinet, R. Ceulemans, A.S. Kowalski, T. Vesala, U. .Rannik, P. Berbigier, D. Loustau, J. Guomundsson, H. Thorgeirsson, A. Ibrom and K. Morgenstern, 2000: Respiration as the main determinant of carbon balance in European forests. *Nature*, **404,** 861–864.

Vander Zanden, M., 2000: Consequences of smallmouth bass introductions in Ontario lakes. In: *Abstracts, American Fisheries Society, 130th Annual Meeting, August 20–24, 2000, St. Louis, Missouri, USA*.

van der Heijden, E., J. Jauhiainen, J. Matero, M. Eekhof, and E. Mitchell, 1998: Effects of elevated CO_2 and nitrogen deposition on Sphagnum species. In: *Responses of Plant Metabolism to Air Pollution and Global Change* [De Kok, L.J. and I. Stulen (eds.)]. Backhuys Publishers, Leiden, The Netherlands, pp. 475–478.

Van Minnen, J.G., R. Leemans, and F. Ihle, 2000: Defining the importance of including transient ecosystem responses to simulate C-cycle dynamics in a global change model. *Global Change Biology*, **6,** 595–611.

Van Minnen, J.G., K.K. Goldewijk, and R. Leemans, 1995: The importance of feedback processes and vegetation transition in the terrestrial carbon cycle. *Journal of Biogeography*, **22,** 805–814.

Vannote, R.L., G.W. Minshall, K.W. Cummins, J.R. Sedell, and C.E. Cushing, 1980: The river continuum concept. *Canadian Jounal of Fisheries and Aquatic Sciences*, **37,** 130–137.

Vasander, H., J. Kuusipalo, and T. Lindholm, 1993: Vegetation changes after drainage and fertilization in pine mires. *Suo*, **44,** 1–9.

Vasander, H., R. Laiho, and J. Laine, 1997: Changes in species diversity in peatlands drained for forestry. In: *Northern Forested Wetlands: Ecology and Management* [Trettin, C.C., M. Jurgensen, D. Grigal, M. Gale, and J. Jeglum (eds.)]. Lewis Publishers/CRC Press, Boca Raton, FL and New York, NY, USA; London, United Kingdom; and Tokyo, Japan, pp. 113–123.

VEMAP Members, 1995: Vegetation/ecosystem modelling and analysis project: comparing biogeography and biogeochemistry models in a continental-scale study of terrestrial ecosystem responses to climate change and CO_2 doubling. *Global Biogeochemical Cycles*, **9,** 407–437.

Venette, R. C, and W.D. Hutchison, 1999: Assessing the risk of establishment by Pink Bollworm (*Lepidoptera gelechiidae*) in the Southeastern United States. *Environmental Entomology*, **28,** 445–455.

Viglizzo, E.F., Z. Roberto, F. Lertora, G. Lopez, and J. Bernardos, 1997: Climate and land use change in field-crop ecosystems of Argentina. *Agriculture, Ecosystems and the Environment*, **66,** 61–70.

Villers-Ruíz, L. and I. Trejo-Vázquez, 1998: Climate change on Mexican forests and natural protected areas. *Global Environmental Change: Human and Policy Dimensions*, **8,** 141–157.

Virtanen, T., and S. Neuvonen, 1999: Performance of moth larvae on birch in relation to altitude, climate, host quality and parasitoids. *Oecologia*, **120,** 92–101.

Virtanen, T., S. Neuvonen, and A. Nikula, 1998: Modelling topoclimatic patterns of egg mortality of Epirrita autumnata (Lep., Geometridae) with Geographical Information System: predictions for current climate and warmer climate scenarios. *Journal of Applied Ecology*, **35,** 311–322.

Virtanen, T., S. Neuvonen, A. Nikula, M. Varama, and P. Niemelä, 1996: Climate change and the risks of Neodiprion sertifer outbreaks on Scots Pine. *Silva Fennica*, **30,** 169–177.

Visser, M.E., A.J. Vannoordwijk, J.M. Tinbergen, and C.M. Lessells, 1998: Warmer springs lead to mistimed reproduction in Great Tits (Parus major). *Proceedings of the Royal Society of London B*, **265,** 1867–1870.

Vitousek, P.M., H.A. Mooney, J. Lubchenco, and J.M. Melillo, 1997a: Human domination of Earth's ecosystems. *Science*, **277,** 494–499.

Vitousek, P.M., J.D. Aber, R.W. Howarth, G.E. Likens, P.A. Matson, D.W. Schindler, W.H. Schlesinger, and D.G. Tilman, 1997b: Human alteration of the global nitrogen cycle: sources and consequences. *Ecological Applications*, **7,** 737–750.

Vitt, D.H., L.A. Halsey, and S.C. Zoltai, 1994: The bog landform of continental western Canada in relation to climate and permafrost features. *Arctic and Alpine Research*, **26,** 1–13.

Volkman, N.J., P. Presler, and W.Z. Trivelpiece, 1980: Diests of pygoscelid penguins at King George Island, Antarctica. *Condor*, **82,** 373–378.

Volney, W.J.A., 1988: Analysis of historic jack pine budworm outbreaks in the Prairie provinces of Canada. *Canadian Journal of Forest Research*, **18,** 1152–1158.

Volney, W.J.A. and D.G. McCullough, 1994: Jack pine budworm population behaviour in northwestern Wisconsin. *Canadian Journal of Forest Research*, **24,** 502–510.

Vompersky, S.E., M.V. Smagina, A.I. Ivanov, and T.V. Glukhova, 1992: The effect of forest drainage on the balance of organic matter in forest mires. In: *Peatland Ecosystems and Man: An Impact Assessment* [Bragg, O.M., P.D. Hulme, H.A.P. Ingram, and R.A. Robertson (eds.)]. British Ecological Society, International Peat Society, Department of Biological Sciences, University of Dundee, Dundee, United Kingdom, pp. 17–22.

Waddington, J.M. and N.T. Roulet, 2000: Carbon balance of a boreal patterned peatland. *Global Change Biology*, **6,** 87–98.

Waddington, J.M. and N.T. Roulet, 1996: Atmosphere-wetland carbon exchange: scale dependency of CO_2 and CH_4 exchange on the developmental topography of the peatland. *Global Biogeochemical Cycles*, **10,** 233–245.

Walker, B.H., W.L. Steffen, and J. Langridge, 1999: Interactive and integrated effects of global change on terrestrial ecosystems. In: *The Terrestrial Biosphere and Global Change: Implications for Natural and Managed Ecosystems. Synthesis Volume* [Walker, B., W. Steffen, J. Canadell, and J. Ingram (eds.)]. International Geosphere-Biosphere Program Book Series 4, Cambridge, United Kingdom, pp. 329–375.

Walker, D.A., P.J. Webber, E.F. Binnian, K.R. Everett, N.D. Lederer, E.A. Nordstrand, and M.D. Walker, 1987: Cumulative impacts of oil fields on northern Alaskan landscapes. *Science*, **238,** 757–761.

Walker, L.R. and R. Waide, 1991: Special issue: ecosystems, plant, and animal responses to hurricanes in the Caribbean. *Biotropica*, **24.**

Walker, M.D., W.A. Gould, and F.S. Chapin III, 2001: Scenarios of biodiversity change in Arctic and alpine tundra. In: Scenarios of Future Biodiversity [Chapin, F.S. III, O. Sala, and E. Huber-Sannwald (eds.)]. Springer-Verlag, New York, NY, USA, (in press).

Wall, G., 1998: Implications of global climate change for tourism and recreation in wetland areas. *Climatic Change*, **40,** 371–389.

Walter, B.P., M. Heimann, R.D. Shannon, and J.R. White, 1996: *A Process-Based Model to Derive Methane Emissions from Natural Wetlands*. Max-Planck-Institut für Meteorologie, Hamburg, Germany.

Wand, S.J.E., G.F. Midgley, M.H. Jones, and P.S. Curtis, 1999: Responses of wild C_4 and C_3 grass (Poaceae) species to elevated atmospheric CO_2 concentration: a meta-analytic test of current theories and perceptions. *Global Change Biology*, **5,** 723–741.

Wang, B. and M. Allard, 1995: Recent climatic trend and thermal response of permafrost at Salluit, Northern Quebec, Canada. *Permafrost and Periglacial Processes*, **6,** 221–234.

Wargo, P.M. and T.C. Harrington, 1991: Host stress and susceptibility. In: *Armillaria Root Disease* [Shaw, C.G. III, and G.A. Kile (eds.)]. Agriculture Handbook No. 691, U.S. Department of Agriculture, Forest Service, pp. 88–101.

Watt, A.D., J.B. Whittaker, M. Docherty, G. Brooks, E. Lindsay, and D.T. Salt, 1995: The impact of elevated CO_2 on insect herbivores. In: *Insects in a Changing Environment* [Harrington, R. and N.E. Stork (eds.)]. Academic Press, San Diego, USA, pp. 198–251.

Weber, J.M., D.T. Tingey, and C.P. Andersen, 1994: Plant response to air pollution. In: Plant-Environment Interactions [Wilkinson, R.E. (ed.)]. Marcel Dekker, New York, NY, USA, pp. 357–387.

Webster, K.E., T.K. Kratz, C.J. Bowser, and J.J. Magnuson, 1996: The influence of landscape position on lake chemical responses to drought in northern Wisconsin. *Limnology and Oceanography*, **41(5),** 977–984.

Wein, R.W., M.A. El-Bayoumi, and J. Da Silva, 1989: Simulated predictions of forest dynamics in Fundy National Park Canada. *Forest Ecology and Management*, **28,** 47–60.

Wheeler, B.D., 1996: Conservation of peatlands. In: *Global Peat Resources* [Lappalainen, E. (ed.)]. International Peat Society, Jyväskylä, Finland, pp. 285–301.

White, A.M., G.R. Cannell, and A.D. Friend, 1999: Climate change impacts on ecosystems and the terrestrial carbon sink: a new assessment. *Global Environmental Change: Human and Policy Dimensions*, **9,** S21–S30.

White, R.G. and J. Trudell, 1980: Habitat preference and forage consumption by reindeer and caribou near Atkasook, Alaska. *Arctic and Alpine Research*, **12,** 511–529.

Whiting, G.J. and J.P. Chanton, 1993: Primary production control of methane emission from wetlands. *Nature*, **364,** 794–795.

Wikelski, M. and C. Thom, 2000: Marine iguanas shrink to survive El Niño. *Nature*, **403,** 37.

Wiley, J.W. and J.M. Wunderle Jr., 1994: The effects of hurricanes on birds, with special reference to Caribbean islands. *Bird Conservation International*, **3,** 319–349.

Williams, D.W. and A.M. Liebhold, 1997: Latitudinal shifts in spruce budworm (Lepidoptera=: Tortricidae) outbreaks and spruce-fir forest distributions with climate change. *Acta Phytopathologica et Entomologica Hungarica*, **32,** 205–215.

Williams, G.D.V., 1986: Land use and agroecosystem management in semi-arid conditions. In: *Land Use and Agroecosytem Management Under Severe Climatic Conditions*. Techincal Note No. 184, WMO No. 633, World Meteorological Organization, Geneva, Switzerland, pp. 70–90.

Williamson, C.E. and H.E. Zagarese, 1994: The impact of UV-B radiation on pelagic freshwater ecosystems. *Archiv fuer Hydrolbiologie und Limnologie*, **43,** IX-XI.

Williamson, C.E., R.S. Stemberger, D.P. Morris, T.M. Frost, and S.G. Paulsen, 1996: Ultraviolet radiation in North American lakes: attenuation estimates from DOC measurements and implications for plankton communities. *Limnology and Oceanography*, **41(5),** 1024–1034.

Wilson, E.O., 1992: *The Diversity of Life*. W.W. Norton Company, New York, NY, USA, 424 pp.

Wind-Mulder, H.L., L. Rochefort, and D.H. Vitt, 1996: Water and peat chemistry comparisons of natural and post-harvested peatlands across Canada and their relevance to peat restoration. *Ecological Engineering*, **7,** 161–181.

Wingfield, J.C., G. Ramos-Fernandez, A. Nunez-de la Mora, and H. Drummond, 1999: The effects of an El Niño-Southern Oscillation event on reproduction in male and female blue-footed boobies, Sula nebouxii. *General and Comparative Endocrinology*, **114,** 163–172.

Winkel, W. and H. Hudde, 1997: Long-term trends in reproductive traits of tits (Parus major, P. caeruleus) and pied flycatchers (Ficedula hypoleuca). *Journal of Avian Biology*, **28,** 187–190.

Winkel, W. and H. Hudde, 1996: Long-term changes of breeding parameters of Nuthatches Sitta europaea in two study areas of northern Germany. *Journal Für Ornithologie*, **137,** 193–202.

Winnett, S.M., 1998: Potential effects of climate change on U.S. forests: a review. *Climate Research*, **11,** 39–49.

Winter, T.C. and M.K. Woo, 1990: Hydrology of lakes and wetlands. In: *Surface Water Hydrology* [Wolman, M.G. and H.C. Riggs (eds.)]. Geological Society of America, Boulder, CO, USA, Vol. O-1, pp. 159–188.

Winters, P., R. Murgai, A. de Janvry, E. Sadoulet, and G. Frisvold, 1999: Climate change and agriculture: effects on developing countries. In: *Global Environmental Change and Agriculture* [Frisvold, G. and B. Kuhn (eds.)]. Edward Elgar Publishers, Cheltenham, United Kingdom.

Wirth, C., E.D. Schulze, W. Schulze, K. von Stünzner-Karbe, W. Ziegler, I.M. Miljukova, A. Sogachev, A.B. Varlagin, M. Panvyorov, S. Grigoriev, W. Kusnetzova, M. Siry, G. Hardes, R. Zimmermann, and N.N. Vygodskaya, 1999: Above-ground biomass and structure of pristine Siberian Scots pine forests as controlled by competition and fire. *Oecologia*, **121,** 66–80.

Wolf, J., A. Iglesias, L.G. Evans, M.A. Semenov, and H. Eckersten, 1996: Comparison of wheat simulation models under climate change, I: model calibration and sensitivity analyses. *Climate Research*, **7,** 253–270.

Woo, M.K., R.D. Rowsell, and R.G. Clark, 1993: *Hydrological Classification of Canadian Prairie Wetlands and Prediction of Wetland Inundation in Response to Climatic Variability*. Canadian Wlidlife Service, Environment Canada, Ottawa, ON, Canada.

Wood, C.M. and D.G. McDonald (eds.), 1997: *Global Warming: Implications for Freshwater and Marine Fish*. Seminar Series 61, Society for Experimental Biology, Cambridge University Press, Cambridge, United Kingdom, 425 pp.

Woodroffe, C.D. and M.E. Mulrennan, 1993: *Geomorphology of the Lower Mary River Plains, Northern Territory*. Australian National University, North Australian Research Unit and the Conservation Commission of the Northern Territory, Darwin, Australia, 152 pp.

Woodward, F.I. and D.J. Beerling, 1997: The dynamics of vegetation change: health warnings for equilibrium dodo models. *Global Ecology and Biogeography Letters*, **6,** 413–418.

Woodward, F.I., M.R. Lomas, and R.A. Betts, 1998: Vegetation-climate feedbacks in a greenhouse world. *Transactions of the Royal Society of London—B*, **353,** 20–39.

Woodwell, G.M. and F.T. Mackenzie (eds.), 1995: *Biotic Feedbacks in the Global Climatic System: Will the Warming Speed the Warming?* Oxford University Press, New York, NY, USA, 416 pp.

Wookey, P.A., A.N. Parsons, J.M. Welker, J.A. Potter, T.V. Callaghan, J.A. Lee, and M.C. Press, 1993: Comparative responses of phenology and reproductive development to simulated environmental change in sub-Arctic and high Arctic plants. *Oikos*, **67,** 490–502.

World Bank, 1995: *World Bank Country Study: Tajikistan*. World Bank, Washington, DC, USA, 240 pp.

World Bank, 1993: *Water Resources Management*. World Bank Policy Study, Washington, DC, USA, 140 pp.

Worldwatch Institute, 1999: *State of the World 1999*. W.W. Norton Company, New York, NY, USA.

WRI, 2000: *World Resources 2000–2001. People and Ecosystems: The Fraying Web of Life*. World Resources Institute, Washington, DC, USA, 400 pp.

WRI, 1998: *World Resources Report 1998–99. Environmenal Change and Human Health*. World Resources Institute, Oxford University Press, Oxford, United Kingdom, 384 pp.

Wyman, R.L., 1991: Multiple threats to wildlife: climate change, acid precipitation, and habitat fragmentation. In: *Global Climate Change and Life on Earth* [Wyman, R.L. (ed.)]. Chapman and Hall, New York, NY, USA, pp. 134–155.

Xiao, X., D.W. Kicklighter, J.M. Melillo, A.D. McGuire, P.H. Stone, and A.P. Sokolov, 1997: Linking a global terrestrial biogeochemical model and a 2-dimensional climate model: implications for the global carbon budget. *Tellus*, **49B,** 18–37.

Xu, M. and R.V. Lewis, 1990: Structure of a protein superfiber: spider dragline silk. *Proceedings of the National Academy of Sciences of the United States of America*, **87,** 7120–7124.

Yan, N.D., W. Keller, N.M. Scully, D.R.S. Lean, and P.J. Dillon, 1996: Increased UV-B penetration in a lake owing to drought-induced acidification. *Nature*, **381,** 141–143.

Yarie, J., 1999: Nitrogen productivity of Alaskan tree species at an individual tree and landscape level. *Ecology*, **78,** 2351–2358.

Yates, D.N. and K.M. Strzepek, 1998: An assessment of integrated climate change impacts on the agricultural economy of Egypt. *Climatic Change*, **38,** 261–287.

Yates, D.N., T.G.F. Kittel, and R.F. Cannon, 2000: Comparing the correlative Holdridge Model to mechanistic biogeographical models for assessing vegetation distribution response to climatic change. *Climatic Change*, **44,** 59–87.

Yeates, G.W., K.R. Tate, and P.C.D. Newton, 1997: Response of the fauna of a grassland soil to doubling of atmospheric carbon dioxide concentration. *Biology and Fertility of Soils*, **25,** 307–315.

Yeo, A., 1999: Predicting the interaction between the effects of salinity and climate change on crop plants (review). *Scientia Horticulturae*, **78(1–4),** 159–174.

Yom-Tov, Y. and Y.L. Werner, 1996: Environmental correlates of geographical distribution of terrestrial vertebrates in Israel. *Israel Journal of Zoology*, **42,** 307–315.

Yudina, V.F., E.M. Vahrameyeva, P.N. Tokarev, and T.A. Maksimova, 1986: *Klyukva b Karelii*. Kareliya, Petrozavodsk, Russia (in Russian).

Zaller, J.G. and J.A. Arnone, 1997: Activity of surface-casting earthworms in a calcareous grassland under elevated atomospheric CO_2. *Oecologia*, **111,** 249–254.

Zanetti, S., U.A. Hartwig, A. Luscher, T. Hebeisen, M. Frehner, B.U. Fischer, G.R. Hendrey, H. Blum, and J. Nösberger, 1996 : Stimulation of symbiotic N_2 fixation in Trifolium repens L. under elevated atmospheric pCO_2 in a grassland ecosystem. *Plant Physiology*, **112,** 575–583.

Zaret, T.M. and R.T. Paine, 1973: Species introductions in a tropical lake: a newly introduced piscivore can produce population changes in a wide range of trophic levels. *Science*, **182,** 449–455.

Zhou, X., R. Harrington, I.P. Woiwod, J.N. Perry, J.S. Bale, and S.J. Clark, 1995: Effects of temperature on aphid phenology. *Global Change Biology*, **1,** 303–313.

Zimov, S.A., V.I. Chuprynin, A.P. Oreshko, F.S. Chapin, J.F. Reynolds, and M.C. Chapin, 1995: Steppe-tundra transition: a herbivore-driven biome shift at the end of the Pleistocene (review). *American Naturalist*, **146,** 765–794.

Zimov, S.A., S.P. Davidov, G.M. Zimova, A.I. Davidova, F.S. Chapin III, and M.C. Chapin, 1999: Contribution of disturbance to high-latitude amplification of atmospheric CO_2. *Science*, **284,** 1973–1976.

Zimov, S.A., Y.V. Voropaev, I.P. Semiletov, S.P. Davidov, S.F. Prosiannikov, F.S. Chapin III, M.C. Chapin, S. Trumbore, and S. Tyler, 1997: North Siberian lakes: a methane source fueled by Pleistocene carbon. *Science*, **277,** 800–802.

Ziska, L.H., J.A. Bunce, and F. Caufield, 1998: Intraspecific variation in seed yield of soybean (Glycine max) in response to increased atmospheric carbon dioxide. *Australian Journal of Plant Physiology*, **25,** 801–807.

Ziska, L.H., O. Namuco, T. Moya, and J. Quilang, 1997: Growth and yield response of field-grown tropical rice to increasing carbon dioxide and air temperature. *Agronomy Journal*, **89,** 45–53.

6

Coastal Zones and Marine Ecosystems

R.F. MCLEAN (AUSTRALIA) AND ALLA TSYBAN (RUSSIAN FEDERATION)

Lead Authors:
V. Burkett (USA), J.O. Codignotto (Argentina), D.L. Forbes (Canada), N. Mimura (Japan), R.J. Beamish (Canada), V. Ittekkot (Germany)

Review Editors:
L. Bijlsma (The Netherlands) and I. Sanchez-Arevalo (Spain)

CONTENTS

EXECUTIVE SUMMARY

Global climate change will affect the physical, biological, and biogeochemical characteristics of the oceans and coasts, modifying their ecological structure, their functions, and the goods and services they provide. Large-scale impacts of global warming on the oceans will include:

- Increases in sea level and sea-surface temperature
- Decreases in sea-ice cover
- Changes in salinity, alkalinity, wave climate, and ocean circulation.

Feedbacks to the climate system will occur through changes in ocean mixing, deep water production, and coastal upwelling. Collectively, these changes will have profound impacts on the status, sustainability, productivity, and biodiversity of the coastal zone and marine ecosystems.

Scientists recently have recognized the persistence of multi-year climate-ocean regimes and shifts from one regime to another. Changes in recruitment patterns of fish populations and the spatial distribution of fish stocks have been linked to climate-ocean system variations such as the El Niño-Southern Oscillation (ENSO) and decadal-scale oscillations. Fluctuations in fish abundance increasingly are regarded as a biological response to medium-term climate-ocean variations, and not just as a result of overfishing and other anthropogenic factors. Of course, such factors can exacerbate natural fluctuations and damage fish stocks. Global warming will confound the impact of natural variation and fishing activity and make management more complex.

Growing recognition of the role of understanding the climate-ocean system in the management of fish stocks also is leading to new adaptive strategies that are based on the determination of stock resilience and acceptable removable percentages of fish. We need to know more about these interactions. Climate-ocean–related changes in the distribution of fish populations suggest that the sustainability of the fishing industries of many countries will depend on increasing flexibility in bilateral and multilateral fishing agreements, coupled with international stock assessments and management plans.

Marine mammals and seabirds are large consumers of fish and have been shown to be sensitive to inter-annual and longer term variability in oceanographic and atmospheric parameters. Several marine mammal and bird species, including polar bears and some seabirds, may be threatened by long-term climate change.

Marine aquaculture production has more than doubled since 1990 and is expected to continue its upward trend. However, aquaculture may be limited if key fish species used in feed production are negatively impacted by climate change. Increases in seawater temperature may directly impact aquaculture; such increases already have been associated with increases in diseases and algal blooms.

The adaptive capacities of marine and coastal ecosystems varies among species, sectors, and geographical regions. In the broader oceans, marine organisms will be relatively free to move to new geographical areas; organisms in enclosed seas and coastal zones are more constrained by the physical features of the shore, making natural adaptation more difficult.

Coastal zones are among the world's most diverse and productive environments. With global warming and sea-level rise, many coastal systems will experience:

- Increased levels of inundation and storm flooding
- Accelerated coastal erosion
- Seawater intrusion into fresh groundwater
- Encroachment of tidal waters into estuaries and river systems
- Elevated sea-surface and ground temperatures.

Tropical and subtropical coastlines, particularly in areas that are already under stress from human activities, are highly susceptible to global warming impacts. Particularly at risk are the large delta regions—especially in Asia, where vulnerability was recognized more than a decade ago and continues to increase. Mid-latitude temperate coasts often comprise coastal plains and barriers and soft sedimentary cliffs and bluffs that have been the subject of historical and model studies, virtually all of which confirm the high vulnerability of these coasts. High-latitude coastlines also are susceptible, although the impacts in these areas have been less studied. A combination of accelerated sea-level rise, increased melting of ground ice, decreased sea-ice cover, and associated more energetic wave conditions will have severe impacts on coastal landforms, settlements, and infrastructure.

Coastal areas also include complex ecosystems such as coral reefs, mangrove forests, and salt marshes. In such environments, the impact of accelerated sea-level rise will depend on vertical accretion rates and space for horizontal migration, which may be limited by the presence of infrastructure. Many mangrove forests are under stress from excessive exploitation, and salt marshes are under stress from reclamation. Many coral reefs already are degraded. In such situations, ecosystem resilience will be greatly reduced through human impacts as well as rising sea levels, increasing sea temperatures, and other climate-ocean–related changes, including prevailing wave activity and storm waves and surges.

Progress in evaluating the potential effects of climate change and sea-level rise on socioeconomic systems has not been as substantial as that relating to biogeophysical impacts. With reference to coastal zones, socioeconomic impacts have been considered in several ways, including:

- As a component of vulnerability assessment of natural systems
- With an emphasis on market-oriented or nonmarket-oriented approaches
- With a focus on costs for infrastructure and adaptation options.

Three coastal adaptation strategies have been identified previously: protect, accommodate, and retreat. In the past few years, structural shore-protection measures have been reevaluated, and there has been greater interest in managing coastal retreat. Enhancement of biophysical and socioeconomic resilience in coastal regions is increasingly regarded as a desirable adaptive strategy but appears not to be feasible in many of the world's coastal zones. Additional insights can be gained by understanding adaptation to natural variability.

Although some countries and coastal communities have the adaptive capacity to minimize the impacts of climate change, others have fewer options; the consequences may be severe for them. Geographic and economic variability leads to inequity in the vulnerability of coastal communities and potentially in intergenerational access to food, water, and other resources. Techniques for the integration of biophysical and socioeconomic impact assessment and adaptation are developing slowly, while human population growth in many coastal regions is increasing socioeconomic vulnerability and decreasing the resilience of coastal ecosystems.

Integrated assessment and management of open marine and coastal ecosystems and a better understanding of their interaction with human development will be important components of successful adaptation to climate change. Also important will be integration of traditional practices into assessments of vulnerability and adaptation.

6.1. Introduction and Scope

The oceans cover 70% of the Earth's surface and play a vital role in the global environment. They regulate the Earth's climate and modulate global biogeochemical cycles. They are of significant socioeconomic value as suppliers of resources and products worth trillions of dollars each year (IPCC, 1998). Oceans function as areas of recreation and tourism, as a medium for transportation, as a repository of genetic and biological information, and as sinks for wastes. These functions are shared by the coastal margins of the oceans.

Approximately 20% of the world's human population live within 30 km of the sea, and nearly double that number live within the nearest 100 km of the coast (Cohen *et al.*, 1997; Gommes *et al.*, 1998). Nicholls and Mimura (1998) have estimated that 600 million people will occupy coastal floodplain land below the 1,000-year flood level by 2100.

Any changes associated with global warming should be considered against the background of natural variations, such as long-term variations caused by solar and tectonic factors, as well as short- and mid-term changes related to atmospheric and oceanic conditions. Climate change will affect the physical, biological, and biogeochemical characteristics of the oceans and coasts at different time and space scales, modifying their ecological structure and functions. These changes, in turn, will exert significant feedback on the climate system. The world's oceans already are under stress as a result of a combination of factors—such as increased population pressure in coastal areas, habitat destruction, and increased pollution from the atmosphere, from land-based sources, and from river inputs of nutrients and other contaminants (Izrael and Tsyban, 1983). These factors, along with increased UV-B radiation resulting from stratospheric ozone depletion, are expected to impair the resilience of some marine ecosystems to climate change.

This chapter reviews the potential impacts of climate change on the coastal zone, marine ecosystems, and marine fisheries. It provides an assessment of the latest scientific information on impacts and adaptation strategies that can be used to anticipate and reduce these impacts. Emphasis is placed on scientific work completed since 1995. This chapter builds on earlier IPCC reports but differs in several significant ways. First, the content of the present chapter was covered in three separate chapters in the Second Assessment Report (SAR: Chapter 8, Oceans; Chapter 9, Coastal Zones and Small Islands; Chapter 16, Fisheries) and in two chapters in the First Assessment Report (FAR: World Oceans and Coastal Zones, Working Group II; Coastal Zone Management, Working Group III). Second, in the *Special Report on Regional Impacts of Climate Change* (IPCC, 1998), impacts on the oceans and coasts were considered in each of the regional chapters; those on the Small Island States and the Arctic and Antarctic were of particular relevance to the present chapter. Third, whereas previous IPCC reports highlighted sea-level rise, vulnerability assessment, biogeophysical effects, and single-sector impacts, this chapter covers several other topics—including a range of methodologies; climate-change parameters; physical, biological, and socioeconomic sensitivities; and adaptation mechanisms. Additional and regionally specific coastal and marine details are included in the regional chapters of this Third Assessment Report (TAR).

6.2. State of Knowledge

In the past decade there has been considerable improvement in our knowledge of the impacts of climate change on coastal zones and marine ecosystems. This improvement has been far from uniform, either thematically or in regional coverage, and there are still substantial gaps in our understanding.

The First and Second Assessment Reports on ocean systems (Tsyban *et al.*, 1990; Ittekkot *et al.*, 1996) conclude that global warming will affect the oceans through changes in sea-surface temperature (SST), sea level, ice cover, ocean circulation, and wave climate. A review of the global ocean thermohaline circulation system—for which the term "ocean conveyor belt" has been coined (Broecker, 1994, 1997)—emphasizes the role of the global ocean as a climate regulator. Ittekkot *et al.* (1996) notes that the oceans also function as a major heat sink and form the largest reservoir of the two most important greenhouse gases [water vapor and carbon dioxide (CO_2)], as well as sustaining global biogeochemical cycles.

Climate-change impacts on the ocean system that were projected with confidence by Ittekkot *et al.* (1996) include SST-induced shifts in the geographic distribution of marine biota and changes in biodiversity, particularly in high latitudes; future improvement of navigation conditions in presently ice-infested waters; and sea-level changes resulting from thermal expansion and changes in terrestrial ice volume. Regional variations caused by dynamic processes in the atmosphere and ocean also were identified with some confidence. Less confident predictions include changes in the efficiency of carbon uptake through circulation and mixing effects on nutrient availability and primary productivity; changes in ocean uptake and storage capacity for greenhouse gases; and potential instability in the climate system caused by freshwater influx to the oceans and resultant weakening of the thermohaline circulation.

The SAR includes a comprehensive review of climate-change impacts on fisheries (Everett *et al.,* 1996). Principal impacts are believed to be compounded by overcapacity of fishing fleets, overfishing, and deterioration of aquatic habitats. The authors also note that the impacts of natural climate variability on the dynamics of fish stocks is being considered as an important component of stock management, although the nature and magnitude of that variability are not clear.

In the present report, we identify new information about the impacts of climate change that has accumulated since the SAR. We include assessments of impacts on fisheries, marine mammals, sea birds, aquaculture, and marine diseases. We show that more recent information identifies natural multi-year climate-ocean trends as an essential consideration in fisheries management and stewardship of marine ecosystems. We point out that separating the impacts of natural climate variability and regime shifts from those associated with long-term climate change will be important, although distinguishing between the two will be a difficult task.

The potential impacts of sea-level rise on coastal systems have been emphasized in recent years. Much less attention has been given to the effects of increases in air and sea-surface temperatures; and changes in wave climate, storminess, and tidal regimes. There are at least two reasons for this lack of attention. First, low-lying coastal areas such as deltas, coastal plains, and atoll islands are regarded as particularly vulnerable to small shifts in sea level. Second, global sea-level rise is regarded as one of the more certain outcomes of global warming and already is taking place. Over the past 100 years, global sea level has risen by an average of 1–2 mm yr^{-1}, and scientists anticipate that this rate will accelerate during the next few decades and into the 22nd century.

The FAR (Tsyban *et al.*, 1990) regards sea-level rise as the most important aspect of climate change at the coast and identifies seven key impacts:

1) Lowland inundation and wetland displacement
2) Shoreline erosion
3) More severe storm-surge flooding
4) Saltwater intrusion into estuaries and freshwater aquifers
5) Altered tidal range in rivers and bays
6) Changes in sedimentation patterns
7) Decreased light penetration to benthic organisms.

In the SAR, Bijlsma *et al.* (1996) acknowledge the importance of these impacts and further conclude, with high confidence, that natural coastal systems will respond dynamically to sea-level rise; responses will vary according to local conditions and climate; and salt marshes and mangroves may survive where vertical accretion equals sea-level rise—but built infrastructure limits the potential for landward migration of coastal habitats. That report also provides summaries of national and global vulnerability assessments, focusing on numbers of people, land areas, and assets at risk. Several coastal adaptation strategies are identified. The importance of resilience in coastal systems is hinted at and subsequently has become an important consideration in vulnerability analysis of sectors and geographical regions.

6.3 Marine Ecosystems

6.3.1. *Habitat*

The oceans have significant adaptive capacity to store heat and are the largest reservoir of water vapor and CO_2, although the storage capacity for CO_2 in the Southern Ocean recently has been questioned (Caldeira and Duffy, 2000). In the oceans, climate change will induce temperature changes and associated adjustments in ocean circulation, ice coverage, and sea level. Changes in the frequency of extreme events also may be expected. These changes, in turn, will affect marine ecosystem structure and functioning, with feedback to global biogeochemical cycles and the climate system.

Recent investigations have shown that there has been a general warming of a large part of the world ocean during the past 50 years (Levitus *et al.*, 2000). Analysis of historical SST data by Cane *et al.* (1997) shows an overall increase associated with land-based global temperature trends. Regional differences exist such that over the past century a cooling was observed in the eastern equatorial Pacific, combined with a strengthening of the zonal SST gradient.

Global mean sea-level has risen by about 0.1–0.2 mm yr^{-1} over the past 3,000 years and by 1–2 mm yr^{-1} since 1900, with a central value of 1.5 mm yr^{-1}. TAR WGI Chapter 11 projects that for the full range of the six illustrative scenarios in the IPCC's *Special Report on Emissions Scenarios,* sea level will rise by 0.09–0.88 m between 1990 and 2100. This range is similar to the total range of projections given in the SAR of 0.13–0.94 m. Higher mean sea level will increase the frequency of existing extreme levels associated with storm waves and surges.

The El Niño-Southern Oscillation is a natural part of the Earth's climate. A major issue is whether the intensity or frequency of ENSO events might change as a result of global warming. Timmermann *et al.* (1999) suggested an increased frequency of El Niño-like conditions under future greenhouse warming and stronger "cold events" in the tropical Pacific Ocean. Cooling has been observed in the eastern equatorial Pacific, not reproduced in most GCMs, and has been explained by an increase in upwelling from the strengthening of trade winds because of a uniform warming of the atmosphere (Cane *et al.*, 1997). If temperature differences between the tropics and polar regions are reduced, however, a weakening of the atmospheric circulation patterns that cause upwelling could be expected.

In recent years there have been several studies of global ocean wind and wave climates (e.g., Young, 1999), but analyses of changes over the past few decades have been limited to a few regions. In the past 30 years there has been an increase in wave height over the whole of the North Atlantic, although scientists are not certain that global change is the cause of this phenomenon (Guley and Hasse, 1999). Similarly, analyses of wave buoy data along the entire west coast of North America demonstrate that the heights of storm-generated waves have increased significantly during the past 3 decades (Komar *et al.*, 2000). On the U.S. east coast, analyses have shown that there has been no discernible long-term trend in the number and intensity of coastal storms during the past century, although there has been considerable interdecadal variation (Zhang *et al.*, 2000). The sensitivity of storm waves to a hypothetical sea-level rise and increase in wind strength recently has been modeled for Uruguay (Lorenzo and Teixeira, 1997).

Projected changes in tropical cyclone frequency and intensity remain inconclusive, although some studies have suggested that the maximum intensity of tropical cyclones may rise by 10–20% (Henderson-Sellars *et al.*, 1998; Knutson *et al.*, 1998). Walsh and Pittock (1998) and Walsh and Katzfey (2000) also suggest that once formed, tropical cyclone-like vortices might travel to higher latitudes and persist for longer as a result of increased SST.

Increased precipitation intensity in extreme events is suggested by climate models under doubled CO_2 for Europe (Jones *et al.*, 1997) and the United States (Mearns *et al.*, 1995), and there is firm evidence that moisture in the atmosphere is increasing over China, the Caribbean region, and the western Pacific (Trenberth, 1999). Heavy rainfall increased during the 20th century in the United States (Karl and Knight, 1998), and there is evidence for increased precipitation rates in Japan and Australia (Iwashima and Yamamoto, 1993; Suppiah and Hennessy, 1998). Changes in the probability of heavy precipitation also are regarded as important indicators of climate change.

Globally, oceanic thermohaline circulation plays an important role in controlling the distribution of heat and greenhouse gases. This circulation is driven by differences in seawater temperature and salinity. There is some evidence that the global thermohaline circulation will weaken as a result of climate change, although views on this issue are still evolving.

Sea ice covers about 11% of the ocean, depending on the season. It affects albedo, salinity, and ocean-atmosphere thermal exchange. The latter determines the intensity of convection in the ocean and, consequently, the mean time scale of deep-ocean processes affecting CO_2 uptake and storage. Projected changes in climate should produce large reductions in the extent, thickness, and duration of sea ice. Major areas that are now ice-bound throughout the year are likely to have lengthy periods during which waters are open and navigable. Observations in the northern hemisphere already have shown a significant decrease in spring and summer sea-ice extent by about 10–15% since the 1950s. It also has been suggested that the decline in ice volume is underestimated because of significant thinning of sea ice in the Arctic (Rothrock *et al.*, 1999). Evidence from whaling records implies a decline in Antarctic ice extent by as much as 25% between the mid-1950s and the early 1970s (de la Mare, 1997).

The foregoing physical responses in the ocean-climate system have implications for habitat and ecology in the oceans and coastal seas. Projected climate changes have the potential to become a major factor affecting marine living resources over the next few decades. The degree of the impact is likely to vary within a wide range, depending on the species and community characteristics and the regional specific conditions. Smith *et al.* (1999) review the sensitivity of marine ecosystems to climate change.

6.3.2. Biological Processes

Marine biota have an important role in shaping climate. Marine biological processes sequester CO_2 and remove carbon from surface waters to the ocean interior through the settling of organic particles and as ocean currents transport dissolved organic matter. This process, which is called the biological pump, reduces the total carbon content of the surface layers and increases it at depth. This process may be partially offset by biocalcification in reefs and organisms in the open ocean, which increases surface layer CO_2 by reducing bicarbonate alkalinity.

Projected global warming through the 21st century is likely to have an appreciable effect on biological processes and biodiversity in the ocean. A rise in temperature will result in acceleration of biodegradation and dispersal of global organic pollutants (petroleum and chlorinated hydrocarbons, for example). This process would promote their removal from the photic zone of the ocean, as has been demonstrated by Tsyban (1999a) for the Bering and Chukchi Seas.

Physiochemical and biological processes regulate uptake and storage of CO_2 by oceans. The Arctic Ocean, for instance, is an important CO_2 source in winter and sink in summer (Tsyban, 1999b). Climate change is expected to affect the processes that control the biogeochemical cycling of elements. Uptake and storage of CO_2 by the ocean via the biological pump therefore may change. Any changes that do occur are expected to feed back into the carbon cycle (Ittekkot *et al.*, 1996).

Photosynthesis, the major process by which marine biota sequester CO_2, is thought to be controlled by the availability of nutrients and trace elements such as iron (de Baar *et al.*, 1995; Behrenfeld *et al.*, 1996; Coale *et al.*, 1996; Falkowsky *et al.*, 1998). Changes in freshwater runoff resulting from climate warming could affect the inputs of nutrients and iron to the ocean, thereby affecting CO_2 sequestration. Impacts are likely to be greatest in semi-enclosed seas and bays.

Climate change can cause shifts in the structure of biological communities in the upper ocean—for example, between coccoliths and diatoms. In the Ross Sea, diatoms (primarily *Nitzshia subcurvata*) dominate in highly stratified waters, whereas *Phaeocystis antarctica* dominate when waters are more deeply mixed (Arrigo *et al.*, 1999). Such shifts alter the downward fluxes of organic carbon and consequently the efficiency of the biological pump.

6.3.3. Marine Carbon Dioxide Uptake

The oceans are estimated to have taken up approximately 30% (with great uncertainties) of CO_2 emissions arising from fossil-fuel use and tropical deforestation between 1980 and 1989, thereby slowing down the rate of greenhouse global warming (Ittekkot *et al.*, 1996). An important process in the oceans is burial of organic carbon in marine sediments, which removes atmospheric CO_2 for prolonged time periods. Studies of the Southern Ocean by Caldeira and Duffy (2000) have shown high fluxes of anthropogenic CO_2 but very low storage. Model results imply that if global climate change reduces the density of surface waters in the Southern Ocean, isopycnal surfaces that now outcrop may become isolated from the atmosphere, which would tend to diminish Southern Ocean carbon uptake.

Using models of the effects of global warming on ocean circulation patterns, Sarmiento and Le Quéré (1996) analyzed the potential for changes in oceanic CO_2 uptake. They found that a weakening of the thermohaline circulation could reduce the ocean's ability to absorb CO_2 whereby, under a doubled-CO_2 scenario, oceanic uptake of CO_2 dropped by 30% (exclusive of biological effects) over a 350-year period. In simulations with biological effects under the same CO_2 conditions and time frame, they found that the oceanic uptake was reduced only by 14%. Confirmation that a collapse of global thermohaline circulation could greatly reduce the uptake of CO_2 by the ocean has been reported by Joos *et al.* (1999).

Sarmiento *et al.* (1998) modeled carbon sequestration in the ocean with increasing CO_2 levels and changing climate from 1765 to 2065. They found substantial changes in the marine carbon cycle, especially in the Southern Ocean, as a result of freshwater inputs and increased stratification, which in turn reduces the downward flux of carbon and the loss of heat to the atmosphere

6.3.4. Marine Fish

Climatic factors affect the biotic and abiotic elements that influence the numbers and distribution of fish species. Among the abiotic factors are water temperature, salinity, nutrients, sea level, current conditions, and amount of sea ice—all of which are likely to be affected by climate change. Biotic factors include food availability and the presence and species composition of competitors and predators. Clearly the relationship between climatic factors and the fish-carrying capacity of the marine environment is complicated, although water temperature can be used as a basis for forecasting the abundance and distribution of many species (Lehodey *et al.*, 1997). Water temperature also can have a direct effect on spawning and survival of larvae and juveniles as well as on fish growth, by acting on physiological processes. Sea temperature also affects the biological production rate thus food availability in the ocean, which is a powerful regulator of fish abundance and distribution.

The question of large-scale, long-term fluctuations in the abundance of marine organisms, primarily those of considerable commercial importance, recently has gained attention. Research has shown that variations (with cycles of 10–60 years or more) in the biomass volume of marine organisms depend on sea temperature and climate (Ware, 1995). Examples include periodic fluctuations in the climate and hydrographic regime of the Barents Sea, which have been reflected in variations in commercial production over the past 100 years. Similarly, in the northwest Atlantic Ocean results of fishing for cod during a period of 300 years (1600–1900) showed a clear correlation between water temperature and catch, which also involved changes in the population structure of cod over cycles of 50–60 years. Shorter term variations in North Sea cod have been related to a combination of overfishing and warming over the past 10 years (O'Brien *et al.*, 2000).

From 1987 to 1996, the world catch of all marine fishes averaged 74.5 Mt. From 1987 to 1993, catches were relatively stable, ranging between 71.6 and 75.9 Mt. There was a small increase over the period 1994–1996, ranging from 77.1 to 78.6 Mt (FAO, 1998). The 10 species with the largest landings represented 37.4% of the catch in 1996 and the next 10 species an additional 10.9%. Fluctuations in abundance of species representing the 10 largest landings often have been considered to result from overfishing and occasionally from a combination of ocean environment changes and fishing effects. However, there is increasing evidence to suggest that the impacts of climate variations are also having an important effect (O'Brien *et al.*, 2000).

The collapse of the Peruvian anchovy *(Engraulis ringens)* fishery from the mid-1970s to the mid-1980s was widely accepted as an example of overfishing and poor management. The increases in recent years, to catches slightly smaller than the large catches prior to the collapse, provide an example of the important impact of natural fluctuations and the difficulty of sorting out the impacts of fishing and climate-ocean-induced changes. Caddy and Rodhouse (1998) reported an increase in world cephalopod landings as world marine fish catches stabilized. These increases were believed to be related to reduced predation from overfishing of groundfish stocks, although warmer oceans were considered an important factor. These examples emphasize the importance of considering the ecosystem impacts of climate variations, as well as changes for individual species.

McGowan *et al.* (1998) show that there are large-scale biological responses in the ocean to climate variations. Off California, the climate-ocean regime shift in 1976–1977 (Ebbesmeyer *et al.*, 1991) resulted in a reduced rate of supply of nutrients to a shallower mixing layer, decreasing productivity and zooplankton and causing reductions in kelp and sea birds. Although there is no question that fishing has impacts on the dynamics of fish populations, the recent evidence of climate-related impacts is beginning to confound past interpretations of fishing effects. McGowan *et al.* (1998) point out that the success of future fish stock assessments would depend, to a large extent, on the ability to predict the impacts of climate change on the dynamics of marine ecosystems. The assumption that marine ecosystems are stable is no longer acceptable, which raises questions about the definition of sustained yield (O'Brien *et al.*, 2000).

Weather impacts and seasonal rhythms have long been recognized by the global fishing industry, but decadal-scale regime changes have been acknowledged only recently as a factor in fish and ocean ecosystem dynamics. The concept of distinct states in climate-ocean environments, which after periods of persistence switch abruptly to other states, have been called regimes and regime shifts, respectively. More formally, regimes (Steele, 1996) can be defined as multi-year periods of linked recruitment patterns in fish populations or as a stable mean in physical

***Table 6-1**: Largest marine fisheries in 1996 (FAO, 1998).*

Species	Landings (t)	% of Total
Peruvian anchovy (*Engraulis ringens*)	8,864,000	11.3
Walleye pollock (*Theragra chalcogramma*)	4,533,000	5.8
Chilean Jack mackerel (*Trachurus murphyi*)	4,379,000	5.8
Atlantic herring (*Clupea harengus*)	2,331,000	3.0
Chub mackerel (*Scomber japonicus*)	2,168,000	2.8
Capelin (*Mallotus villosus*)	1,527,000	1.9
South American pilchard (*Sardinops sagax*)	1,494,000	1.9
Skipjack tuna (*Katsuwonus pelamis*)	1,480,000	1.9
Atlantic cod (*Gadus morhua*)	1,329,000	1.7
Largehead hairtail (*Trichiarus lepturus*)	1,275,000	1.6

data. A regime shift is a change in the mean of a data series. The existence of decadal-scale regimes in the environment has been documented (e.g., Gargett, 1997; Gu and Philander, 1997; Mantua *et al.*, 1997). States even longer than the decadal-scale may exist (Ware, 1995; Marsh *et al.*, 1999). Adkinson *et al.* (1996) and Beamish *et al.* (1997) have documented a large-scale response in fish populations to regimes and regime shifts for Pacific salmon.

Among the most important groups of marine fishes are herrings *(Clupea sp.)*, sardines and pilchards *(Sardinops sp.)*, and anchovies *(Engraulis sp.)* (see Table 6-1). These fish tend to be short-lived species that mature at an early age. Large fluctuations in abundance have been associated with changes in the climate-ocean environment, although it has not been possible to discover the mechanisms that link climate-ocean changes to recruitment (Cole and McGlade, 1998). One of the most convincing relationships of large-scale, synchronous responses in major fisheries resulting from changes in climate-ocean states exists for sardine *(Sardinops sp.)*. The decadal variability in the Japanese sardine catch was synchronous with decadal-scale variability in the ocean and climate of the North Pacific; these phenomena also were synchronous with the fluctuations of sardine catches off Chile and California (Kawasaki, 1991; Hiyama *et al.*, 1995) and with trends in Pacific salmon catches (Beamish *et al.*, 1999) (see Box 6-1).

Fluctuations in the abundance and distribution of herring (*Clupea harengus*) and sardine (*Sadinella pilchardus*) in the North and Baltic Seas have been linked to variations of the North Atlantic Oscillation and the resulting strength and pattern of southwesterly winds (Alheit and Hagen, 1997).

Most fishing regime changes can be related directly to sea-temperature changes, but changes in other physical attributes also can have an impact. For instance, a decrease in wind stress off Tasmania that reduced large zooplankton production affected the density of Jack mackerel (*Trachurus declivis*), which eliminated the possibility of a commercially viable mackerel fishery (Harris *et al.*, 1992).

The aforementioned climate-related fluctuations in the Japanese sardine occurred at the same time as shifts in the migratory patterns of the northern bluefin tuna *(Thunnus thynnus*), with a higher proportion of bluefin remaining in the western Pacific when sardine abundance was high (Polovina, 1996). The migratory pattern of albacore tuna (*T. alalunga*) also was altered by decadal-scale climate changes (Kimura *et al.*, 1997).

Nearly 70% of the world's annual tuna harvest comes from the Pacific Ocean. In 1996, Skipjack tuna *(Katsuwonus pelamis)* was the eighth-largest marine fishery in the world (see Table 6-1). Catches of skipjack are highest in the western equatorial Pacific warm pool; Lehodey *et al.* (1997) have shown that major spatial shifts in the skipjack population can be linked to large zonal displacements of the warm pool (see Box 6-2).

Welch *et al.* (1998) propose that continued warming of the North Pacific Ocean would compress the distributions of Sockeye salmon (*Oncorhynchus nerka*), essentially squeezing them out of the North Pacific and into the Bering Sea. Some modeling of future impacts of greenhouse gas increases, however, has shown an intensification of the Aleutian Low, which has been associated with mid-ocean cooling (Deser *et al.*, 1996) and increased Pacific salmon production. The warmer surface waters could reduce growth if bioenergetic costs are higher and less food is available as a consequence of ocean habitat changes. The potential impact of climate change on Pacific salmon is expected to occur in freshwater and ocean situations (Hinch *et al.*, 1995). This fact is important because production of more juveniles in hatcheries would not mitigate changes in the ocean carrying capacity for Pacific salmon. The most effective strategy to manage the impacts of climate change on Pacific salmon may be to ensure that wild salmon are preserved and protected, rather than to produce more salmon through artificial enhancement. It is possible that the variety of life history types and genetic traits of the wild stocks are inherent biological solutions to changing freshwater and marine habitats (Bisbal and McConnaha, 1998).

The potential deepening of the Aleutian Low and increase in the amplitude of the Pacific Decadal Oscillation would result in major changes in marine ecosystems (Mantua *et al.*, 1997).

Box 6-1. Regimes and Regime Shifts: Salmon and Sardine Catch

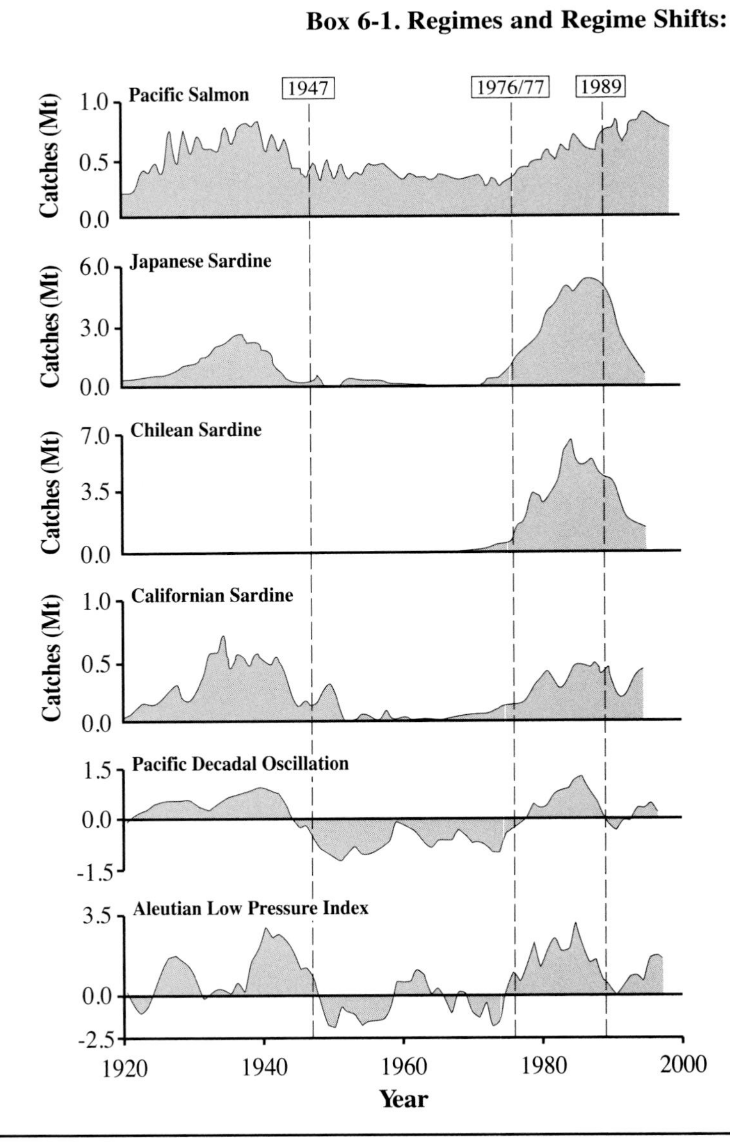

As the figure to the left shows, catches (Mt) of Pacific salmon (*Oncorhynchus sp.*) and sardines (*Sardinops sp.*) fluctuate synchronously with large-scale climate-ocean changes in the North Pacific, as indicated by the Aleutian Low Pressure Index and SST as expressed in the Pacific Decadal Oscillation both of which are shown as a 5-year running average (data from Beamish *et al.*, 1999, and Mantua *et al.*, 1997). Persistent states or regimes are separated by regime shifts, which occurred about 1947, 1976–1977, and 1989. These shifts are shown by the vertical dashed lines. The well-documented 1976–1977 regime shift has been associated with changing abundance trends of other marine organisms such as plankton. It also has been related to an abrupt change in the southernmost extent of sea ice in the Bering Sea, in turn affecting the distribution of Walleye pollock (*Theragra chalcogramma*) and Arctic cod (*Boreogadus saida*) (Wyllie-Echeverria and Wooster, 1998).

The concepts of regimes, regime shifts, and natural trends in the abundance of animals indicates that it is important to assess the potential impact of global warming on decadal-scale processes as well as on species specific responses.

As ecosystems change, there may be impacts on the distribution and survival of fishes. Any changes in natural mortality would be associated with increased predation and other factors such as disease. Improved growth in the early life stages would improve survival, whereas decreased growth could facilitate increased mortality.

Sea temperature is an important regulator of fish behaviors. Wood and McDonald (1997) provide examples of how climate change could induce temperature responses in fish, but there are several areas where less certainty exists. The effect that global climate change will have on trends in the Aleutian Low Pressure system in the Pacific Ocean is an example. Although there are clear linkages between the intensity and position of the low and production trends of many of the commercially important fish species (Kawasaki *et al.*, 1991; Polovina *et al.*, 1995; Gargett, 1997; Mantua *et al.*, 1997; Francis *et al.*, 1998), a reduction in equator-to-pole temperature gradients would probably weaken winds and consequently reduce open-ocean upwelling. Important changes in species distributions in surface waters could result.

There is now a cautious acceptance that climate change will have major positive and negative impacts on the abundance and distribution of marine fish. Thus, the impacts of fishing and climate change will affect the dynamics of fish and shellfish such as abalone in Mexico (Shepherd *et al.*, 1998). Fishing impacts may be particularly harmful if natural declines in productivity occur without corresponding reductions in exploitation rates. Changes in fish distributions and the development of aquaculture may reduce the value of some species, however—as it has for wild Pacific salmon—and these changes may reduce fishing pressures in some areas.

Key to understanding the direction of change for world fisheries is the ability to incorporate decadal-scale variability into general circulation models (GCMs). Although progress has occurred, it still is not possible to assess regional responses to shifts in climate

Box 6-2. Tuna Migration and Climate Variability

Skipjack tuna (*Katsuwonus pelamis)* dominate the world's catch of tuna. The habitat supporting the densest concentrations of skipjack is the western equatorial Pacific warm pool, with SST of 29°C and warmer. Panel (a) clearly shows the association of skipjack tuna catch (shaded and cross-hatched areas indicate January–June catch of 200,000+ t) and mean SST [data from Lehodey *et al.* (1997)]. The figure also shows that the location of the warm pool is linked to ENSO and that it changes during El Niño and La Niña events. For instance, the catch area in the first half of 1989 (La Niña period), which is shown by cross-hatch, was centered around Palau and the Federated States of Micronesia; in the first half of 1992 (El Niño period), the center of abundance had shifted to the east, to the Marshall Islands and Kiribati (shown by shading).

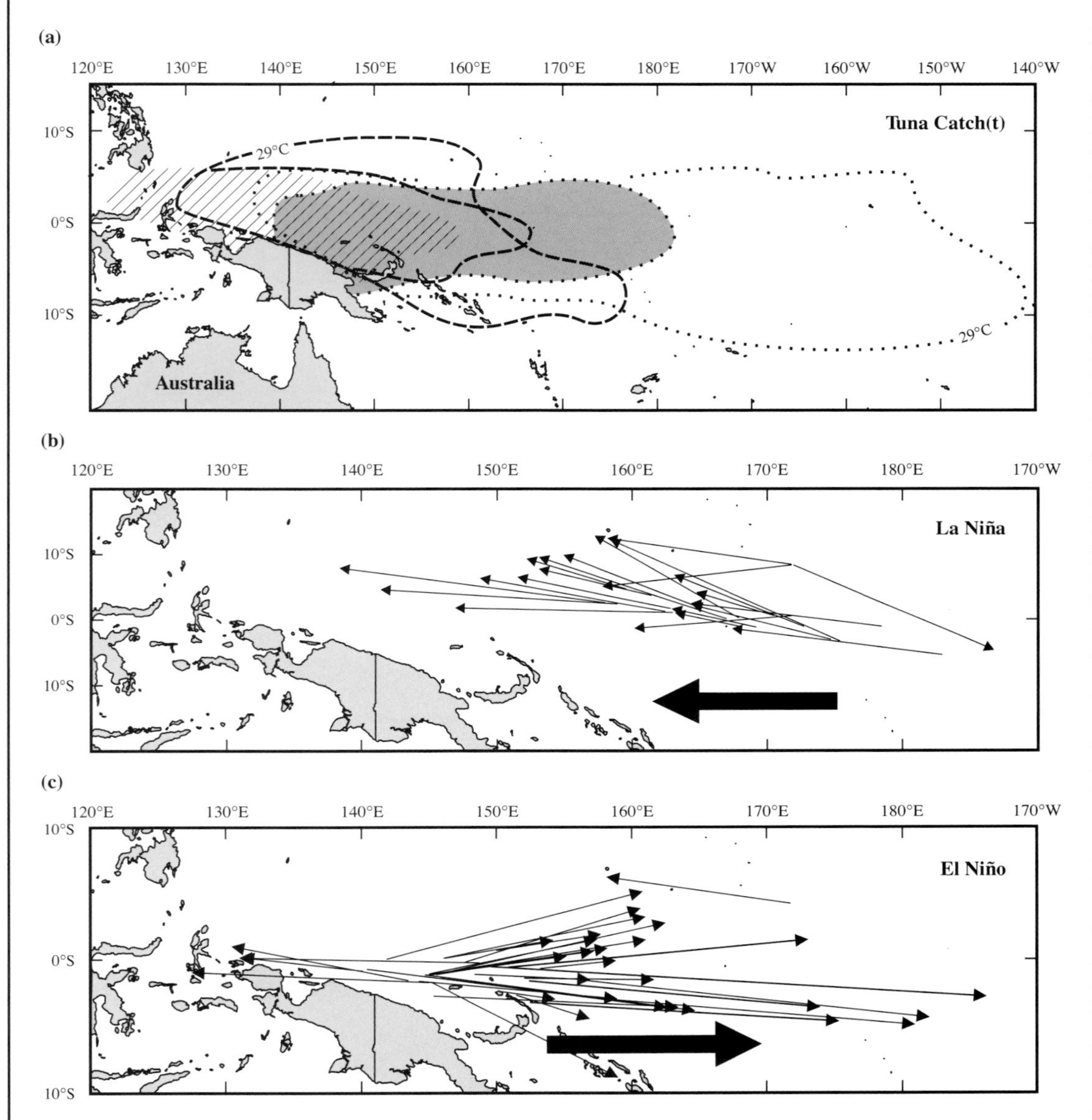

Panels (b) and (c) indicate the scale of tuna migration during a La Niña period and an El Niño period, respectively. These figures were compiled from records of a large-scale skipjack tagging program carried out by the Secretariat of the Pacific Community (SPC). The illustrations are from Lehodey *et al.* (1997, 1998).

The close association of skipjack tuna catch and ENSO is evidence that climate variability profoundly affects the distribution pattern of tuna and resulting fishing opportunities. Scientists do not know how projected climate changes will affect the size and location of the warm pool in the western and central Pacific, but if more El Niño-like conditions occur an easterly shift in the center of tuna abundance may become more persistent.

trends, and it is unknown if a general warming will increase or decrease the frequency and intensity of decadal-scale changes in regions where national fisheries occur. Recent studies have not produced evidence to change the conclusion from the SAR (Everett *et al.,* 1996) that future saltwater fisheries production is likely to be about the same as at present, though changes in distribution could affect who catches a particular stock. However, if aquaculture becomes the major source of fish flesh and management of fisheries becomes more precautionary, the exploitation rate of wild marine fish may decrease in some areas.

6.3.5. Aquaculture

Marine aquaculture production more than doubled from 4.96 Mt in 1990 to 11.14 Mt in 1997. Similar trends were exhibited in freshwater aquaculture, with increases from 8.17 Mt in 1990 to 17.13 Mt in 1997, while yields from marine and freshwater fisheries remained relatively constant. The net result was that aquaculture production represented approximately 30% of total fish and shellfish production for human consumption in 1997. Aquaculture production is expected to continue its upward trend in the foreseeable future, although in many areas (such as in Thailand) there is a boom and bust pattern to aquaculture.

About 30% (29.5 Mt) of the world fish catch is used for nonhuman consumption, including the production of fishmeal and fish oils that are employed in agriculture, in aquaculture, and for industrial purposes. Fishmeal and fish oils are key diet components for aquaculture production; depending on the species being cultured, they may constitute more than 50% of the feed. Climate change could have dramatic impacts on fish production, which would affect the supply of fishmeal and fish oils. Unless alternative sources of protein are found, future aquaculture production could be limited by the supply of fishmeal or fish oils if stocks of species used in the production of fishmeal are negatively affected by climate change and live-fish production. The precise impacts on future aquaculture production also will depend in part on other competing uses for fishmeal and fish oils. Usage of other sources of protein or developments in synthetic oils for industrial applications could reduce demands on fishmeal and fish oils, thereby reducing potential impacts on aquaculture.

Climate change is expected to have physical and ecosystem impacts in the freshwater and marine environments in which aquaculture is situated. Water and air temperatures in mid- to high latitudes are expected to rise, with a consequent lengthening of the growing season for cultured fish and shellfish. These changes could have beneficial impacts with respect to growth rate and feed conversion efficiency (Lehtonen, 1996). However, increased water temperatures and other associated physical changes, such as shifts in dissolved oxygen levels, have been linked to increases in the intensity and frequency of disease outbreaks and may result in more frequent algal blooms in coastal areas (Kent and Poppe, 1998). Any increases in the intensity and frequency of extreme climatic events such as storms, floods, and droughts will negatively impact aquaculture production and may result in significant infrastructure damage. Sea-level rise can be expected to have a negative effect on pond walls and defenses.

Elevated temperatures of coastal waters also could lead to increased production of aquaculture species by expanding their range. These species could be cultivated in higher latitudes as well as in existing aquafarms as a result of a longer warm season during which water temperature will be near optimal. A decrease in sea-ice cover could widen the geographical boundaries, allowing cultivation of commercially valuable species in areas hitherto not suitable for such developments.

6.3.6. Ocean Ranching

Ocean ranching is used to increase the production of several fish species. The primary difference between ocean ranching and aquaculture is that in ocean ranching the fish are cultured for only a portion of their life cycle and then released prior to maturity. The cultured fish are then captured as "common property" in a variety of fisheries. These cultured or "enhanced" fish interact with the wild fish in the ecosystem and compete for the finite food and habitat resources available (often referred to as "carrying capacity"). Climate change will alter carrying capacity, but the impacts associated with ocean ranching or stock enhancement activities also need to be considered when examining overall changes to the ecosystem.

Although the precise impacts are species-dependent, the addition of billions of enhanced fish into the marine ecosystem may have significant consequences from genetic and ecological perspectives. In the Pacific, for example, large numbers of salmon are released from hatcheries in Russia, Japan, Canada, and the United States (Mahnken *et al.*, 1998). Beamish *et al.* (1997) estimate that 83% of the catch of chum salmon by all countries is from hatchery production. If climate change increases SST and reduces winter wind mixing in the upper layers of the ocean, the feeding areas for salmon may be less productive because of increased surface layer stability. Reductions in overall production and catches of salmon—and likely other species—would result (Gargett, 1997).

6.3.7. Marine Mammals and Seabirds

Marine mammals and seabirds are sensitive indicators of changes in ocean environments. Springer (1998) concluded that synchrony in extreme fluctuations of abundance of marine birds and mammals across the North Pacific and Western Arctic were a response to physical changes, including climate warming. The linkages with climate change were compelling enough for Springer to suggest that fluctuations in marine bird and mammal populations in the North Pacific are entirely related to climate variations and change.

The climate variations beginning in the 1990s and associated with El Niño conditions (Trenberth and Hoar, 1996), in combination with overfishing, have been linked to behavioral changes in killer whales. These changes drastically reduced sea otter abundance along the Aleutian Islands, which in turn changed the ecology of the kelp forests (Estes *et al.*, 1998). The changes in prey resulting from persistent changes in climate appear to be one of the important impacts of a changing climate on the marine mammals that feed from the top of the food chain.

Climate change also may have an effect on access to prey among marine mammals. For instance, extended ice-free seasons in the Arctic could prolong the fasting of polar bears (*Ursus maritimus*), with possible implications for the seal population (Stirling *et al.*, 1999). Reduced ice cover and access to seals would limit hunting success by polar bears and foxes, with resulting reductions in bear and fox populations. This dynamic could have negative effects on the lifestyle, food, and health standards of some indigenous peoples (Hansell *et al.*, 1998). Because global climate change is likely to have profound impacts on sea-ice extent and duration, it is in this habitat where the initial impacts on marine mammals may be first evident. Reductions in sea ice have been predicted to alter the seasonal distributions, geographic ranges, migration patterns, nutritional status, reproductive success, and ultimately the abundance of Arctic marine mammals (Tynan and DeMaster, 1997). Studies recognizing multi-year to decadal variability in marine biotic systems include Mullin (1998) on zooplankton over 5 decades in the eastern Pacific (with connections to El Niño), and Tunberg and Nelson (1998) on soft bottom macrobenthic communities in the northeast Atlantic (with connections to the North Atlantic Oscillation). Sagarin *et al.* (1999) have argued that changes in the distribution of intertidal macroinvertebrates on rocky shores in California over the past 60 years have been caused by climate change.

Seabirds are an integral part of marine ecosystems, where they may consume vast amounts of fish. It has been estimated that seabirds consume 600,000 t yr^{-1} of food in the North Atlantic (Hunt and Furness, 1996). Modeling studies have shown that in several marine ecosystems, seabirds eat 20–30% of the annual pelagic fish production. The dependence on some species of fish, particularly during breeding, and their large abundance make seabirds a good indicator of ecosystem change. Where changes in breeding success or mortality occur, however, distinguishing the climate impact from fishing impacts can be difficult (Duffy and Schneider, 1994). Very few decadal-scale studies of seabirds are available to assess the impacts of long-term variations in climate, however.

In general, seabirds have evolved to adapt to weather patterns (Butler *et al.*, 1997). The ability of a species to alter its migration strategy appears to be important to survival in a changing climate. Food resources appear to be critical to general survival, especially for young seabirds. Dolman and Sutherland (1994) proposed that feeding rate affects the ability of individuals to survive winter. The change in marine ecosystem described by Roemmich and McGowan (1995) was associated with a mortality resulting in a 40% decline in seabird abundance within the California current system from 1987 to 1994 (Veit *et al.*, 1996). The decline was largely related to a dramatic (90%) decline of sooty shearwaters *(Puffinus girseus),* but the response in the ecosystem was not characterized only by declines. There was a northward movement of some species, and in offshore waters the abundance of the most common species, Leach's storm petrels *(Oceanodroma leucorhoa)*, increased over the same period. The authors were careful to note that the changes in abundance they described could not be related directly to population dynamics because of complex migratory patterns and the size of the habitat.

Such changes are evidence of the sensitivity of seabirds to climate-ocean changes and that survival and distribution impacts will occur as climates shift. The anomalous cold surface waters that occurred in the northwest Atlantic in the early 1990s changed the fish species composition in the surface waters on the Newfoundland shelf. These changes were readily detected in the diets of northern gannet *(Sula bassana)*. The sensitivity of the distribution patterns of the pelagic prey of fish-feeding and plankton-feeding seabirds imply to Montevecchi and Myers (1997) that small changes in the ocean environment resulting from climate changes could affect seabird reproductive success. Changes in fish-feeding seabird abundance in the eastern Bering Sea are related to the abundance of juvenile pollock (Springer, 1992). It has been argued that long lifespans and genetic variation within populations enable seabirds to survive adverse short-term environmental events, as evidenced by the response to El Niño and La Niña events in the tropical Pacific (Ribic *et al.*, 1997). However, small populations tied to restricted habitat, such as the Galapagos Penguin (*Spheniscus mendiculus*), may be threatened by long-term climate warming (Boersma, 1998).

6.3.8. Diseases and Toxicity

Changes in precipitation, pH, water temperature, wind, dissolved CO_2, and salinity can affect water quality in estuarine and marine waters. Some marine disease organisms and algal species are strongly influenced by one or more of these factors (Anderson *et al.*, 1998). In the past few decades there has been an increase in reports of diseases affecting closely monitored marine organisms, such as coral and seagrasses, particularly in the Caribbean and temperate oceans. The worldwide increase in coral bleaching in 1997–1998 was coincident with high water temperatures associated with El Niño, but Harvell *et al.* (1999) suggest that the demise of some corals might have been accelerated by opportunistic infections affecting the temperature-stressed reef systems. Talge *et al.* (1995) report a new disease in reef-dwelling foraminifera, with implications for coastal sedimentation.

ENSO cycles and increased water temperatures have been correlated with Dermo disease (caused by the protozoan parasite *Perkinsus marinus*) and MSX (multinucleated spore unknown) disease in oysters along the U.S. Atlantic and Gulf coasts. In addition to affecting marine hosts, several viruses, protozoa, and bacteria affected by climatic factors can affect people, by direct contact or by seafood consumption. Many of the reported cases of water-borne diseases involve gastrointestinal illnesses; some can be fatal in infants, elderly people, and people with weakened immune systems (ASM, 1998).

The bacterium *Vibrio vulnificus*, which is found in oysters and is potentially lethal to humans with immune-system deficiencies, becomes more abundant as water temperature increases (Lipp and Rose, 1997). The incidence and severity of cholera (*Vibrio cholerae*) epidemics associated with marine plankton also has been linked with prolonged elevated water temperature. Annual epidemics of cholera in Bangladesh have been correlated with increased SST and sea-surface height (Harvell *et al.*, 1999).

6.4. Coastal Systems

6.4.1. *General Considerations*

Coastal environments occupy one of the most dynamic interfaces on Earth, at the boundary between land and sea, and they support some of the most diverse and productive habitats. These habitats include natural ecosystems, in addition to important managed ecosystems, economic sectors, and major urban centers. The existence of many coastal ecosystems is dependent on the land-sea connection or arises directly from it (e.g., deltas and estuaries). Coastal ecosystems can encompass a wide range of environmental conditions over short distances, particularly of salinity (from fresh to hypersaline) and energy (from sheltered wetlands to energetic wave-washed shorelines). At a much coarser geographical scale, there is a spectrum of climate types—from tropical to polar—with concomitant broad-scale differences in biogeophysical processes and features. Coastal environments, settlements, and infrastructure are exposed to land-sourced and marine hazards such as storms (including tropical cyclones), associated waves and storm surges, tsunamis, river flooding, shoreline erosion, and influx of biohazards such as algal blooms and pollutants. All of these factors need to be recognized in assessing climate-change impacts in the coastal zone.

A summary of potential impacts appears in Box 6-3. Note, however, that owing to the great diversity of coastal environments; regional and local differences in projected relative sea level and climate changes; and differences in the resilience and adaptive capacity of ecosystems, sectors, and countries, the impacts summarized here will be highly variable in time and space and will not necessarily be negative in all situations.

Box 6-3. Potential Impacts of Climate Change and Sea-Level Rise on Coastal Systems

Biophysical impacts can include the following:

- Increased coastal erosion
- Inhibition of primary production processes
- More extensive coastal inundation
- Higher storm-surge flooding
- Landward intrusion of seawater in estuaries and aquifers
- Changes in surface water quality and groundwater characteristics
- Changes in the distribution of pathogenic microorganisms
- Higher SSTs
- Reduced sea-ice cover.

Related socioeconomic impacts can include the following:

- Increased loss of property and coastal habitats
- Increased flood risk and potential loss of life
- Damage to coastal protection works and other infrastructure
- Increased disease risk
- Loss of renewable and subsistence resources
- Loss of tourism, recreation, and transportation functions
- Loss of nonmonetary cultural resources and values
- Impacts on agriculture and aquaculture through decline in soil and water quality.

Some natural features of the shore zone provide significant coastal protection, including coral reefs (the most extensive, massive, and effective coastal protection structures in the world); sand and gravel beaches, which function as wave energy sinks; and barrier beaches, which act as natural breakwaters. Coastal dunes form natural buffers and sand repositories, from which sand may be extracted during storms without major shoreline retreat; coastal vegetation often absorbs wind or wave energy, retarding shoreline erosion. Even the value of salt marsh as a sea defense (King and Lester, 1995) and mangroves as a sediment trap (Solomon and Forbes, 1999) have been recognized. These functions of natural coastal systems contribute to resilience, as discussed in Section 6.6.2.

Bijlsma *et al*. (1996) and the various regional reports in IPCC (1998) identify the areas of greatest sensitivity to accelerated sea-level rise. These areas comprised low-elevation coral atolls and reef islands, as well as low-lying deltaic, coastal plain, and barrier coasts, including sandy beaches, coastal wetlands, estuaries, and lagoons. To this list can be added coarse gravel beaches and barriers, especially if sediment-starved; cliffed coasts in unlithified deposits, particularly where the proportion of sand and gravel is limited; and ice-rich cliffed coasts in high latitudes. Bold and rock-dominated coasts are relatively less vulnerable but often include coastal reentrants with beaches, estuaries, or deltas, which may represent areas of localized vulnerability. On such coasts, wave runup and overtopping can be a factor that threatens infrastructure situated well above mean sea level (Forbes, 1996).

It is important to recognize that vulnerable coastal types in many parts of the world already are experiencing relative sea-level rise, from a combination of subsidence and the global component of sea-level rise identified to date. Submergence rates of 2.5 mm yr^{-1} or more are not uncommon, and higher rates apply locally, such as in parts of China (Ren, 1994), the United States (Dean, 1990), Canada (Shaw *et al*., 1998a), and Argentina (Codignotto, 1997). Although this sea-level rise implies enhanced vulnerability, it also provides a basis for assessing coastal response to various rates of relative sea-level rise, where similar coastal types, boundary conditions, and system properties can be identified. Numerous studies along the U.S. Atlantic coast, where relative sea level is rising at rates of 2–4 mm yr^{-1}, have demonstrated common patterns of barrier beach retreat by washover and ephemeral inlet processes (Leatherman *et al*., 2000). More rapid retreat is recorded in delta-margin settings characterized by rapid subsidence (e.g., Stone and McBride, 1998).

In addition to submergence, seawater intrusion into freshwater aquifers in deltaic and nondeltaic areas is an increasing problem with rising sea level (Moore, 1999). This intrusion has been documented in diverse environments such as the arid Israeli coast, the humid Thailand coast, the Chinese Yangtze Delta, the Vietnamese Mekong Delta, and low-lying atolls (e.g., Melloul and Goldberg, 1997; Chen and Stanley, 1998; Singh and Gupta, 1999).

Although some low, sediment-starved, gravel barrier beaches show rapid retreat under rising sea level, this process is highly nonlinear and in some cases is more closely related to storm event frequency and severity (Forbes *et al.*, 1997a). The response of coasts to storm-related sea-level variations around the North Sea has not been determined, although past increases in the winter means of high water levels of the order of 1–2 mm yr^{-1} have taken place (Langenberg *et al.*, 1999).

6.4.2. Beaches, Barriers, and Cliff Coasts

Sandy coasts shaped and maintained primarily by wave and tidal processes occupy about 20% of the global coastline (Bird, 1993). A smaller proportion consists of gravel and cobble-boulder beaches and related landforms, occurring in tectonically active and high-relief regions and in mid- to high-latitude areas of former glaciation. Coral rubble beaches and islands are common in low-latitude reefal areas. Any analysis of climate-change impacts on the coastal zone should include beaches and barriers of sand and/or gravel as well as coastal cliffs and bluffs.

Over the past 100 years or so, about 70% of the world's sandy shorelines have been retreating, about 20–30% have been stable, and less than 10% have been advancing. Bird (1993) argues that with global warming and sea-level rise there will be tendencies for currently eroding shorelines to erode further, stable shorelines to begin to erode, and accreting shorelines to wane or stabilize. Local changes in coastal conditions and particularly in sediment supply may modify these tendencies, although Nicholls (1998) has indicated that accelerated sea-level rise in coming decades makes general erosion of sandy shores more likely.

Previous discussions of shoreline response to climate change have considered the well-known simple relations between sea-level rise and shoreline retreat of Bruun (1962). This two-dimensional model assumes maintenance of an equilibrium nearshore profile in the cross-shore direction as sea level rises. Some papers have supported this approach for long-term shoreline adjustment (Mimura and Nobuoka, 1996; Leatherman *et al.*, 2000); others have suggested various refinements (Komar, 1998a). Although the model's basic assumptions are rarely satisfied in the real world (Bruun, 1988; Eitner, 1996; Trenhaile, 1997), its heuristic appeal and simplicity have led to extensive use in coastal vulnerability assessments for estimating shoreline retreat under rising sea levels, with varying degrees of qualification (Richmond *et al.*, 1997; Lanfredi *et al.*, 1998; Stewart *et al.*, 1998). Erroneous results can be expected in many situations, particularly where equilibrium profile development is inhibited, such as by the presence of reefs or rock outcrops in the nearshore (Riggs *et al.*, 1995). Moreover, Kaplin and Selivanov (1995) have argued that the applicability of the Bruun Rule, based on an equilibrium approach, will diminish under possible future acceleration of sea-level rise.

Few models of shoreline response incorporate large-scale impacts of sea-level rise coupled to changes in sediment availability. Efforts to address this shortcoming have been pioneered by Cowell and Thom (1994) for sandy barrier-dune complexes and Forbes *et al.* (1995) for gravel barriers. Although these parametric models incorporate sediment supply as well as sea-level change, they are still in the early stages of development and are useful primarily to indicate general patterns of response. A multifaceted approach is needed to incorporate other factors such as longshore and cross-shore variability in shore-zone morphology, sediment supply, texture and composition, nonlinear shore-zone response to storms and storm sequences (Forbes *et al.*, 1995), tectonic history of the site, and the presence or absence of biotic protection such as mangroves or other strand vegetation.

Impact assessment, adaptation actions, and other management decisions must consider all of these factors within a coastal systems context. Temporal variation in storminess and wind climate can produce significant coastal adjustments (Forbes *et al.*, 1997a). Another important component of analysis involves historical trends of shoreline change, including variability caused by storms or other anomalous events (Douglas *et al.*, 1998; Gorman *et al.*, 1998). This analysis can provide essential baseline data to enable comparisons in the future, albeit prior to anticipated climate-change impacts.

Field studies and numerical simulation of long-term gravel barrier evolution in formerly glaciated bays of eastern Canada (Forbes *et al.*, 1995) have revealed how sediment supply from coastal cliffs may be positively correlated with the rate of relative sea-level rise. In this case, rising relative sea level favors barrier progradation, but the system switches to erosional retreat when the rate of sea-level rise diminishes, cliff erosion ceases, and no new sediment is supplied to the beach. Along the South American coast, El Niño events are linked to higher-than-average precipitation causing increased sediment discharge to the Peruvian coast, leading to the formation of gravel beach-ridge sequences at several sites (Sandweiss *et al.*, 1998).

In assessing coastal response to sea-level rise, the relevant sedimentary system may be defined in terms of large-scale coastal cells, bounded by headlands or equivalent transitions—typically one to several tens of kilometers in length and up to hundreds of kilometers in some places (Wijnberg and Terwindt, 1995). Within such cells, coastal orientation in relation to dominant storm wind and wave approach direction can be very important (Héquette *et al.*, 1995; Short *et al.*, 2000), and sediment redistribution may lead to varying rates and/or directions of shoreline migration between zones of sediment erosion and deposition.

Changes in wave or storm patterns may occur under climate change (Schubert *et al.*, 1998). In the North Atlantic, a multidecadal trend of increased wave height is observed, but the cause is poorly understood and the impacts are unclear. Changing atmospheric forcing also has been suggested as a process contributing to increases in mean water level along the North Sea coast, independent of eustatic and isostatic contributions to relative sea level. Changes in large-scale ocean-atmospheric circulation and climate regimes such as ENSO and the Pacific Decadal Oscillation have implications for coastal beach and barrier stability (see Box 6-4).

Erosion of unlithified cliffs is promoted by rising sea levels but may be constrained or enhanced by geotechnical properties and other antecedent conditions (Shaw *et al.*, 1998a; Wilcock *et al.*, 1998). Bray and Hooke (1997) review the possible effects of sea-level rise on soft-rock cliffs over a 50- to 100-year planning scale. They evaluate different methods of analyzing historical recession rates and provide simple predictive models to estimate cliff sensitivity to sea-level rise in southern England. Historical observations of cliff erosion under an accelerating sea level suggest, however, that the results of such methods must be interpreted carefully.

If El Niño-like conditions become more prevalent (Timmermann *et al.*, 1999), increases in the rate of cliff erosion may occur along the Pacific coasts of North and South America (Kaminsky *et al.*, 1998; Komar, 1998a,b). For example, El Niño events raise sea level along the California coast and are marked by the presence of larger than average, and more damaging, waves and increased precipitation. These conditions and the changed direction of wave attack combine to increase sea-cliff erosion on the central California coast, particularly on southerly or southwesterly facing cliffs. An increase in El Niño-like conditions with global warming would very likely increase sea-cliff erosion along this section of coast and endanger infrastructure and property (Storlazzi and Griggs, 2000).

6.4.3. Deltaic Coasts

In addition to increasing erosion, many of the world's low-lying coastal regions will be exposed to potential inundation. Deltas that are deteriorating as a result of sediment starvation, subsidence, and other stresses are particularly susceptible to accelerated inundation, shoreline recession, wetland deterioration, and interior land loss (Biljsma *et al.*, 1996; Day *et al.*, 1997).

River deltas are among the most valuable, heavily populated, and vulnerable coastal systems in the world. Deltas develop where rivers deposit more sediment at the shore than can be carried away by waves. Deltas are particularly at risk from climate change—partly because of natural processes and partly because of human-induced stresses. Deltaic deposits naturally dewater and compact as a result of sedimentary loading. When compaction is combined with isostatic loading or other tectonic effects, rates of subsidence can reach 20 mm yr^{-1} (Alam, 1996). Human activities such as draining for agricultural development; levee building to prevent flooding; and channelization, damming, and diking of rivers to impede sediment transfers have made deltas more vulnerable to sea-level rise. Examples of sediment starvation include the Rhone and Ebro deltas (Jimenez and Sanchez-Arcilla, 1997) and polder projects in the Ganges-Brahmaputra (Jelgersma, 1996). Sediment transport by the Nile, Indus, and Ebro Rivers has been reduced by 95% and in the Mississippi by half in the past 200 years, mostly since 1950 (Day *et al.*, 1997). Further stress has been caused by subsurface fluid withdrawals and draining of wetland soils. In the Bangkok area of the Chao Phraya delta, groundwater extraction during 1960–1994 increased average relative sea-level rise by 17 mm yr^{-1} (Sabhasri and Suwarnarat, 1996). Similar severe land subsidence has been experienced in the Old Huange and Changjiang deltas of China (Chen, 1998; Chen and Stanley, 1998). In the latter case, groundwater removal was curtailed, leading to a reduction in subsidence rates.

Where local rates of subsidence and relative sea-level rise are not balanced by sediment accumulation, flooding and marine processes will dominate. Indeed, Sanchez-Arcilla and Jimenez (1997) suggest that in the case of largely regulated deltas, the main impacts of climate change will be marine-related because impacts related to catchment areas will be severely damped by river regulation and management policies. In such cases, significant land loss on the outer delta can result from wave erosion; prominent examples include the Nile (Stanley and Warne, 1998), Mackenzie (Shaw *et al.*, 1998b), and Ganges (Umitsu, 1997). In South America, large portions of the Amazon, Orinoco, and Paraná/Plata deltas will be affected if sea-level rise accelerates as projected (Canziani *et al.*, 1998). If vertical accretion rates resulting from sediment delivery and *in situ* organic matter production do not keep pace with sea-level rise, waterlogging of wetland soils will lead to death of emergent vegetation, a rapid loss of elevation because of decomposition of the belowground root mass, and, ultimately, submergence and erosion of the substrate (Cahoon and Lynch, 1997).

In some situations, saltwater intrusion into freshwater aquifers also is a potentially major problem, as demonstrated by a three-scenario climate change and sea-level rise model study of the Nile delta (Sherif and Singh, 1999). In other places, saltwater intrusion is already taking place (Mulrenna and Woodroffe, 1998). In the Yangtze delta, one consequence of saltwater incursion will be that during dry seasons shortages of freshwater for agriculture are likely to be more pronounced and agricultural yields seriously reduced particularly around Shanghai (Chen and Zong, 1999).

6.4.4. Coastal Wetlands

An estimate by Nicholls *et al.* (1999) suggests that by the 2080s, sea-level rise could cause the loss of as much as 22% of the world's coastal wetlands. Although there would be significant regional variations (Michener *et al.*, 1997), such losses would reinforce other adverse trends of wetland loss resulting primarily from direct human action—estimated by DETR (1999) to be

Box 6-4. Changes in Wave Climate, Storm Waves, and Surges

Over the long term, beaches and coastal barriers are adjusted in plan shape, profile morphology, and geographical position to factors such as sediment type and availability; wave climate, including prevailing wave energy and direction; and episodic storm waves and storm surge events. Few studies have been made of potential changes in prevailing ocean wave heights and directions as a consequence of climate change and sea-level rise, even though such changes can be expected. Similarly, changes in the magnitude of storm waves and surges with a higher sea level can be expected to reach to higher elevations on land than at present, as well as to extend further inland. Changes cannot be expected to be uniform, however, and impacts will vary locally and regionally.

The following case studies illustrate these points and highlight variations in prevailing storm wave/surge trends, differences in attribution and in the nature of past and potential coastal erosion and accretion impacts.

Changes in wind generated ocean waves in North Atlantic Ocean
Over the past 30 years, visual estimates from merchant ships and instrumental records suggest that significant wave height increases of 0.1–0.3 m have occurred over the whole of the North Atlantic except the west and central subtropics. The coastal response to this change in wave climate has not been documented.
Sources: WASA Group (1998); Guley and Hasse (1999).

Changes in extreme storm surges off Western Europe: The recent record
Storm surge activity in the Irish Sea and North Sea during the 1960s and 1970s reached levels unprecedented since the 1900s. These levels were followed by a sharp decline in the 1980s, taking the number of surges back to the levels of decades before the 1960s. Changes in pressure conditions could be a manifestation of shifts in storm tracks. Changes are part of natural variability on decadal time scales rather than long-term climate change resulting from anthropogenic influences.
Source: Holt (1999).

Changes in waves and storm activity off Western Europe with climate change
A high-resolution climate change experiment mimicking global warming resulted in a weak increase in storm activity and extreme wave heights in the Bay of Biscay and the North Sea; waves and storm action decreased slightly along the Norwegian coast. A weak increase in storm surges in the North Sea can be expected.
Source: WASA Group (1998).

Beach rotation and the Southern Oscillation in eastern Australia
Beach profiles measured at monthly intervals along Narrabeen Beach (Sydney) from 1976 to 1999 suggest a cyclic beach oscillation, with two cycles over the 23-year period; the profiles also suggest that the beach is rotating in a cyclic pattern around a central point. These changes have been related to ENSO. When the Southern Oscillation Index (SOI) is positive, there is a greater prevalence of east to northerly waves; these waves help build out the southern beach. When the SOI is negative, southerly waves dominate the wave climate, leading to a northerly shift of sand, thus feeding beach accretion in the north while the south end of the beach is eroded. If more El Niño-like conditions prevail in the future, a net change in shoreline position can be expected.
Source: Short et al. (2000).

Venice and the northern Adriatic coast: reduction in storm surges as a result of recent climate change?
Coastal flooding and damaging storm surges generated by the *bora* and other easterly winds affect the northern Adriatic Sea. Analysis of wind records from Trieste (1957–1996) show a decline in frequency of such winds. This decline may be caused in part by interdecadal variability, though their persistence suggests that it may be a consequence of recent global warming and less frequent drifts of polar cold air toward middle latitudes.
Source: Pirazzoli and Tomasin (1999).

Sensitivity of storm waves in Montevideo (Uruguay) to future climate change
Outputs from a simple storm wave generation model that uses real-time wind data for the 1980s have been compared with simulations representing a 10% higher wind strength and a 1-m sea-level rise. Under this scenario, storm waves would increase in height; their angle of incidence would remain unchanged.
Source: Lorenzo and Teixeira (1997).

about 40% of 1990 values by the 2080s. Stabilization scenarios developed by DETR (1999) show a large reduction in wetland losses—to 6–7%, compared with unmitigated emissions (13%). Two main types of tidal wetland—mangrove forest and salt marsh—are considered here, although serious impacts on other coastal vegetation types, including subtidal seagrasses, can be expected (Short and Neckles, 1999).

Mangrove forests often are associated with tropical and subtropical deltas, but they also occur in low- to mid-latitude lagoon and estuary margins, fringing shorelines from Bermuda in the north to northern New Zealand in the south. Mangroves have important ecological and socioeconomic functions, particularly in relation to seafood production, as a source of wood products, as nutrient sinks, and for shoreline protection (Rönnbäck, 1999). Moreover, different kinds of mangrove provide different goods and services (Ewel *et al.*, 1998). The function and conservation status of mangroves has been considered in special issues of two journals, introduced by Field *et al.* (1998) and Saenger (1998).

Many mangrove forests are being exploited and some are being destroyed, reducing resilience to accommodate future sea-level rise. In Thailand, 50% of mangrove has been lost in the past 35 years (Aksornkoae, 1993); yet with greatly increased sediment supply to the coastal zone in some places, mangrove colonization has expanded seaward in suitable habitats (Panapitukkul *et al.*, 1998). As for other shore types, this example emphasizes the importance of sediment flux in determining mangrove response to sea-level rise. Ellison and Stoddart (1991), Ellison (1993), and Parkinson *et al.* (1994) suggest that mangrove accretion in low- and high-island settings with low sediment supply may not be able to keep up with future sea-level rise, whereas Snedaker *et al.* (1994) suggest that low-island mangroves may be able to accommodate much higher rates of sea-level rise. This ability may depend on stand composition and status (e.g., Ewel *et al.*, 1998; Farnsworth, 1998) and other factors, such as tidal range and sediment supply (Woodroffe, 1995, 1999; Miyagi *et al.*, 1999). In some protected coastal settings, inundation of low-lying coastal land may promote progressive expansion of mangroves with sea-level rise (Richmond *et al.*, 1997). In contrast, Alleng (1998) predicts the complete collapse of a mangrove wetland in Jamaica under rapid sea-level rise.

The response of tidal marshes to sea-level rise is similarly affected by organic and inorganic sediment supply and the nature of the backshore environment (Brinson *et al.*, 1995; Nuttle *et al.*, 1997). In general, tidal marsh accretion tracks sea-level rise and fluctuations in the rate of sea-level rise (e.g., van de Plassche *et al.*, 1998, 1999; Varekamp and Thomas, 1998). Marsh accretion also reflects marsh growth effects (Varekamp *et al.*, 1999). Bricker-Urso *et al.* (1989) estimate a maximum sustainable accretion rate of 16 mm yr^{-1} in salt marshes of Rhode Island (assuming that vertical accretion rates are controlled mainly by *in situ* production of organic matter); this rate is an order of magnitude higher than rates reported by others. Orson *et al.* (1998) also emphasizes the effects of variability between marsh species types.

Temporal and spatial variability in rates of relative sea-level rise also is important. Stumpf and Haines (1998) report rates of >10 mm yr^{-1} in the Gulf of Mexico over several years, where the long-term mean rate of relative sea-level rise is 2 mm yr^{-1} or less. Forbes *et al.* (1997b) report multi-year fluctuations in sea-level rise at Halifax, Nova Scotia, of as much as 10 mm yr^{-1} and occasionally higher, superimposed on a long-term mean of 3.6 mm yr^{-1}. Thus, short-term fluctuations in sea-level rise may approach the maximum limit of accretion, although the drowning of marsh surfaces is unlikely to be a major concern. Higher local rates of sea-level change have been recorded over the past 100 years or so in a few places, one of which is the Caspian Sea. Here, the response of riparian vegetation to the sea-level fall in the early part of the 20th century was a rapid seaward progression of vegetation. This progression ceased with the rise in Caspian sea level averaging 120 mm yr^{-1} from 1978 to 1996, but it did not result in a similar rapid regression of vegetation. Instead, the vegetation consolidated its position, which has been partly explained by the wide flooding tolerance of the major emergent plant species, with floating vegetation increasing in extent with more favorable (higher water level) conditions (Baldina *et al.*, 1999).

In some areas, the current rate of marsh elevation gain is insufficient to offset relative sea-level rise. For instance, model results from a wetland elevation model designed to predict the effect of an increasing rate of sea-level rise on wetland sustainability in Venice Lagoon revealed that for a 0.48-m rise in the next 100 years, only one site could maintain its elevation relative to sea level; for a 0.15-m rise, seven sites remained stable (Day *et al.*, 1999).

Maintenance of productive marsh area also depends on horizontal controls discussed by Nuttle *et al.* (1997) and Cahoon *et al.* (1998). For example, in settings with sufficient sediment influx, the wetland may expand toward the estuary, while also expanding landward if the backshore slope is sufficiently low and not backed by fixed infrastructure (Brinson *et al.*, 1995). If sediment supply is low, however, marsh front erosion may occur (Dionne, 1986).

Although determining the threshold for such erosion is difficult, this erosion is regarded as a negative impact on many wetlands, particularly those constrained by artificial structures on the landward side. Nicholls and Branson (1998) use the term "coastal squeeze" to describe the progressive loss and inundation of coastal habitats and natural features located between coastal defenses and rising sea levels. They believe that intertidal habitats will continue to disappear progressively, with adverse consequences for coastal biological productivity, biodiversity, and amenity value. Where sediment influx is insufficient to sustain progradation, there is potential for significant loss of coastal wetlands. After considering two "what if" climate change scenarios, Mortsch (1998) found that key wetlands around the Great Lakes of Canada-USA are at risk—particularly those that are impeded from adapting to the new water-level conditions by artificial structures or geomorphic conditions.

6.4.5. Tropical Reef Coasts

Coral reefs occur in a variety of fringing, barrier, and atoll settings throughout the tropical and subtropical world. Coral reefs constitute important and productive sources of biodiversity; they harbor more than 25% of all known marine fish (Bryant *et al.*, 1998), as well as a total species diversity containing more phyla than rainforests (Sale, 1999). Reefs also represent a significant source of food for many coastal communities (Wilkinson *et al.*, 1999). Coral reefs serve important functions as atoll island foundations, coastal protection structures, and sources of beach sand; they have economic value for tourism (which is increasingly important for many national economies) and support emerging opportunities in biotechnology. Moberg and Folke (1999) have published a comprehensive list of goods and ecological services provided by coral reef ecosystems.

The total areal extent of living coral reefs has been estimated at about 255,000 km^2 (Spalding and Grenfell, 1997). As much as 58% (rising locally to >80% in southeast Asia) are considered at risk from human activities, such as industrial development and pollution, tourism and urbanization, agricultural runoff, sewage pollution, increased sedimentation, overfishing, coral mining, and land reclamation (Bryant *et al.*, 1998), as well as predation and disease (e.g., Antonius, 1995; Richardson *et al.*, 1998). In the past these local factors, together with episodic natural events such as storms, were regarded as the primary cause of degradation of coral reefs. Now, Brown (1997), Hoegh-Guldberg (1999), and Wilkinson (1999), for instance, invoke global factors, including global climate change, as a cause of coral reef degradation.

Previous IPCC assessments have concluded that the threat of sea-level rise to coral reefs (as opposed to reef islands) is minor (Bijlsma *et al.*, 1996; Nurse *et al.*, 1998). This conclusion is based on projected rates of global sea-level rise from Warrick *et al.* (1996) on the order of 2–9 mm yr^{-1} over the next 100 years. Reef accretion at these rates has not been widely documented, largely because most reefs have been growing horizontally under stable or falling sea levels in recent years (Wilkinson and Buddemeier, 1994). Schlager (1999) reports an approximate upper limit of vertical reef growth during the Holocene of 10 mm yr^{-1}, suggesting that healthy reef flats are able to keep pace with projected sea-level rise. The situation is less clear for the large numbers of degraded reefs in densely populated regions of south and southeast Asia, eastern Africa, and the Caribbean (Bryant *et al.*, 1998), as well as those close to population centers in the Pacific (Zann, 1994).

Positive trends of SST have been recorded in much of the tropical ocean over the past several decades, and SST is projected to rise by 1–2°C by 2100. Many coral reefs occur at or close to temperature tolerance thresholds (Goreau, 1992; Hanaki *et al.*, 1998), and Brown (1997) has argued that steadily rising SST will create progressively more hostile conditions for many reefs. This effect, along with decreased $CaCO_3$ saturation state (as CO_2 levels rise), represent two of the most serious threats to reefs in the 21st century (Hoegh-Guldberg, 1999; Kleypas *et al.*, 1999).

Several authors regard an increase in coral bleaching as a likely result of global warming. However, Kushmaro *et al.* (1998) cite references that indicate it is not yet possible to determine conclusively that bleaching episodes and the consequent damage to reefs are caused by global climate change. Corals bleach (i.e., pale in color) because of physiological shock in response to abrupt changes in temperature, salinity, and turbidity. This paling represents a loss of symbiotic algae, which make essential contributions to coral nutrition and clarification. Bleaching often may be temporary, with corals regaining color once stressful environmental conditions ameliorate. Brown *et al.* (2000) indicate that some corals in the Indian Ocean, Pacific Ocean, and Caribbean Sea are known to bleach on an annual basis in response to seasonal variations in temperature and irradiance. Major bleaching events can occur when SSTs exceed seasonal maximums by >1°C (Brown *et al.*, 1996). Mortality for small excursions of temperature is variable and, in some cases, apparently depth-related (Phongsuwan, 1998); surviving coral has reduced growth and reproductive capacity. More extensive mortality accompanies temperature anomalies of 3°C or more over several months (Brown and Suharsono, 1990). Hoegh-Guldberg (1999) found that major episodes of coral bleaching over the past 20 years were associated with major El Niño events, when seasonal maximum temperatures were exceeded by at least 1°C.

Corals weakened by other stresses may be more susceptible to bleaching (Glynn, 1996; Brown, 1997), although Goreau (1992) found in Jamaica that anthropogenically stressed areas had lower bleaching frequencies. More frequent and extensive bleaching decreases live coral cover, leading to reduced species diversity (Goreau, 1992; Edinger *et al.*, 1998) and greater susceptibility to other threats (e.g., pathogens and emergent diseases as addressed by Kushmaro *et al.*, 1996, 1998; Aronson *et al.*, 2000). In the short term, this bleaching may set back reef communities to early successional stages characterized by noncalcifying benthos such as algae, soft corals, and sponges (Done, 1999). Reefs affected by coral bleaching may become dominated by physically resilient hemispherical corals because branching corals are more susceptible to elevated SST, leading to a decrease in coral and habitat diversity (Brown and Suharsono, 1990). Differential susceptibility to bleaching among coral taxa has been reported during the large-scale event in 1998 on the Great Barrier Reef (Marshall and Baird, 2000).

The 1998 bleaching event was unprecedented in severity over large areas of the world, especially the Indian Ocean. This event is interpreted by Wilkinson *et al.* (1999) as ENSO-related and could provide a valuable indicator of the potential effects of global climate change. However, the 1998 intense warming in the western Indian Ocean has been associated with shifts in the Indian dipole rather than ENSO.

Attempts to predict bleaching have met with variable success. Winter *et al.* (1998) compared a 30-year record of SST at La Parguera, Puerto Rico, with coral bleaching events at the same location but could not forecast coral bleaching frequency from the temperature record. On the other hand, analyses of recent sea temperature anomalies, based on satellite data, have been used to predict the mass coral bleaching extent during 1997–1998 (Hoegh-Guldberg, 1999).

Recently it has been suggested that a doubling of CO_2 levels could reduce reef calcification, but this effect is very difficult to predict (Gattuso *et al.*, 1999). Kleypas *et al.* (1999) argue that such effects could be noticed by 2100 because of the decreased availability of $CaCO_3$ to corals. In combination with potentially more frequent bleaching episodes, reduced calcification could impede a reef's ability to grow vertically in pace with sea-level rise.

The implications for reef-bound coasts in terms of sediment supply, shore protection, and living resources may be complex, either positive or negative, and are difficult to predict at a global scale. However, there have been suggestions that fishing yields will be reduced as reef viability decreases, leading to reduced yields of protein for dependent human populations, and that the effects of reducing the productivity of reef ecosystems on birds and marine mammals are expected to be substantial (Hoegh-Guldberg, 1999).

6.4.6. *High-Latitude Coasts*

High-latitude coasts are highly susceptible to a combination of climate change impacts in addition to sea-level rise, particularly where developed in ice-bonded but otherwise unlithified sediments. In this context, atmospheric warming affects ground-surface temperatures and thaw, as well as SST and sea ice.

Ground temperatures determine the presence of perennially frozen ground (permafrost), which often contains large volumes of excess ice that may occur in the form of massive ice. The seasonal cycle of ground and nearshore seawater temperatures determines the depth of the seasonally active thaw layer in high-latitude beaches and the nearshore, with implications for limiting beach scour during storms (Nairn *et al.*, 1998). Deepening of the active layer (Vyalov *et al.*, 1998) also can lead to melting of near-surface massive ice and may trigger additional coastal slope failure (Dallimore *et al.*, 1996; Shaw *et al.*, 1998b).

Rapid coastal retreat already is common along ice-rich coasts of the Beaufort Sea in northwestern Canada (e.g., Dallimore *et al.*, 1996), the United States, and the Russian Arctic (e.g., Are, 1998). Where communities are located in ice-rich terrain along the shore, warmer temperatures combined with increased shoreline erosion can have a very severe impact. For example, at Tuktoyaktuk—the main community, port, and offshore supply base on the Canadian Beaufort sea coast—many structures are located over massive ice along eroding shore (Wolfe *et al.*, 1998).

Coastal recession rates along the Arctic coast also are controlled by wave energy during the short open-water season. An early study based on historical records of shoreline recession, combined with hindcast waves derived from measured wind and observed ice distributions, showed that coastal recession rates at several sites are correlated with open-water fetch, storminess, and wave energy. Using estimates of less extensive ice distribution under a doubled-CO_2 atmosphere, Solomon *et al.* (1994) were able to demonstrate an increase in coastal erosion rates to a mean value comparable to the maximum observed rates under present climate conditions. In the Canadian Arctic Archipelago region, where many fine-grained (mud and sand) shorelines and deltas now experience almost zero wave energy (e.g., Forbes and Taylor, 1994), any increase in open water will lead to rapid reworking and potentially substantial shoreline retreat.

Sea ice may erode the seabed in the nearshore zone, but it also may supply shoreface sediments to the nearshore and beach (Reimnitz *et al.*, 1990; Héquette *et al.*, 1995). Thinner ice or later freeze-up resulting from climate warming may lead to changes in nearshore ice dynamics and associated sediment transport. Warmer temperatures and associated changes in winter sea-ice distribution at mid-latitudes are expected to have a negative impact on coastal stability. Rapidly eroding sandy coasts in the southern Gulf of St. Lawrence are partially protected in winter by development of an icefoot and nearshore ice complex. The strongest storms and most persistent onshore winds occur in winter, partially overlapping the ice season. Severe erosion in recent years has been linked to warmer winters with late freeze-up—an anticipated outcome of greenhouse warming (Forbes *et al.*, 1997b).

6.5. Socioeconomic Impacts of Climate Change

In the past decade, some progress has been made in evaluating potential socioeconomic impacts of climate change and sea-level rise on coastal and marine systems. This progress, however, has not been as substantial as that relating to biogeophysical impacts; nor has it been especially comprehensive (Turner *et al.*, 1995, 1996). To date, emphasis has been in three areas. First, research has focused on the coastal zone itself (we are not aware of any studies of the socioeconomic impact of climate change on open-ocean marine ecosystems). Second, there has been an emphasis on the potential socioeconomic impact of sea-level rise but little on any other climate change variables. Third, emphasis has been on economic effects, not on impacts on social and cultural systems. These emphases are evident in the following review, in which we consider socioeconomic impacts initially as a component of the methodology for vulnerability assessment and then through economic cost-benefit analyses of coastal zones in general and infrastructure developments in particular. In these cases, "benefits" derive from the inclusion of adaptation options—primarily shore protection—into the analyses to derive some net cost. Finally, we consider attempts that have been made to "value" natural systems, as well as the potential social and cultural impacts of climate change.

6.5.1. *Socioeconomic Impacts as Part of Vulnerability Assessment*

In the SAR, Bijlsma *et al.* (1996) reviewed several country case studies that had applied the IPCC Common Methodology for assessing the vulnerability of coastal areas to sea-level rise. These case studies offered important insights into potential impacts and possible responses. Many of the assessments emphasized the severe nature of existing coastal problems such as beach erosion, inundation, and pollution, as well as the effects of climate change acting on coastal systems that already are under stress.

The Common Methodology defined *vulnerability* as a country's degree of capability to cope with the consequences of climate change and accelerated sea-level rise. The methodology of seven consecutive analytical steps allowed for identification of coastal populations and resources at risk and the costs and feasibility of possible responses to adverse impacts. The SAR also identified the strengths and weaknesses of the Common Methodology. More recently, Klein and Nicholls (1999) have evaluated the IPCC's approach and results, concluding that the Common Methodology has contributed to understanding the consequences of sea-level rise and encouraged long-term thinking about coastal zones. They went on to develop a new conceptual framework for coastal vulnerability assessment that identifies the main components of the natural system and the socioeconomic system, as well as the linkages between them and climate change and other change variables. This framework is outlined in Box 6-5.

In the SAR, data on socioeconomic impacts were derived from country and global vulnerability assessment studies. Figures for several countries were given relating to the population affected, capital value at loss, and adaptation/protection costs (Bijlsma *et al.*, 1996, Table 9-3). For the coastal zone, the authors concluded that:

- There will be negative impacts on several sectors, including tourism, freshwater quality and supply, fisheries and aquaculture, agriculture, human settlements, financial services, and human health.
- The number of people potentially affected by storm-surge flooding is expected to double (or triple) in the next century, ignoring potential adaptation and population growth.
- Protection of low-lying island states and nations with large deltaic areas is likely to be very costly.
- Adaptation to sea-level rise and climate change will involve important tradeoffs, which may include environmental, economic, social, and cultural values.

Since the SAR, there have been several summaries of the socioeconomic results of the vulnerability assessment studies, presenting data at local, regional, and global levels. Examples include case studies of Poland and Estonia (Kont *et al.*, 1997; Zeider, 1997), the Philippines (Perez *et al.*, 1999), Bangladesh (Ali, 1999), Egypt (El-Raey *et al.*, 1999), and The Gambia and Abidjan (Jallow *et al.*, 1999), as well as the regional analyses and global synthesis of Nicholls and Mimura (1998). The initial global vulnerability assessment has been revised on the basis of scenarios for global sea-level rise derived from the Hadley Centre's HadCM2 ensemble simulations and HadCM3 simulations for GHG-only forcing (Nicholls *et al.*, 1999). This assessment indicated that by the 2080s, the potential number of people flooded by storm surge in a typical year will be more than five times higher than today (using a sea-level rise of 0.38 m from 1990 to 2080) and that between 13 million and 88 million people could be affected even if evolving protection is included. Broadly similar results are given in the study undertaken by DETR (1999). However, they note that the flood impacts of sea-level rise are reduced by emissions scenarios that lead to stabilization of CO_2. By the 2080s, the annual number of people flooded is estimated to be 34 million under the 750-ppm scenario and 19 million under the 550-ppm scenario.

Klein and Nicholls (1999) have categorized the potential socioeconomic impacts of sea-level rise as follows

- Direct loss of economic, ecological, cultural, and subsistence values through loss of land, infrastructure, and coastal habitats
- Increased flood risk of people, land, and infrastructure and the aforementioned values
- Other impacts related to changes in water management, salinity, and biological activities.

They also developed a methodology that has not yet been applied in any case studies.

6.5.2. *Economic Costs of Sea-Level Rise*

In the early 1990s, several studies examined the cost of sea-level rise in the United States, based on uniform national response strategies of holding back the sea (Titus *et al.*, 1992) or not holding back the sea (Yohe, 1990). A series of nationwide studies in other countries (see Bijlsma *et al.*, 1996) followed much the same approach. Turner *et al.* (1995) attempted to assess the cost of sea-level rise in general. Yohe *et al.* (1996) analyzed the economic cost of sea-level rise for developed property in the United States. They conclude that their cumulative figure of US$20.4 billion (1990 US$) cumulative is much lower than earlier estimates because those estimates had not considered offsetting factors, such as the cost-reducing potential of natural, market-based, adaptation measures or the efficiency of discrete decisions to protect (or not to protect) small tracts of property on the basis of individual economic merit. In most analyses, those "offsetting factors," or adaptations, are shore protection works. The expected economic cost of protection or abandonment in the United States also has been assessed (Yohe and Schlesinger, 1998).

Saizar (1997) assesses the potential impacts of a 0.5-m sea-level rise on the coast of Montevideo, Uruguay. Given no adaptive response, the cost of such a rise is estimated to be US$23 million, with a shoreline recession of 56 m and loss of only 6.8 ha of land. Olivo (1997) determined the potential economic impacts of a sea-level rise of 0.5 m on the coast of Venezuela. At six study sites, he identified land and infrastructure at risk—such

Box 6-5. Conceptual Framework for Coastal Vulnerability Assessment

In the scheme illustrated below, analysis of coastal vulnerability starts with a notion of the natural system's *susceptibility* to the biogeophysical effects of sea-level rise and its capacity to cope with these effects (*resilience* and *resistance*). Susceptibility reflects the coastal system's potential to be affected by sea-level rise; resilience and resistance determine the system's robustness or ability to continue functioning in the face of possible disturbance. Together, these factors determine the *natural vulnerability* of the coastal zone.

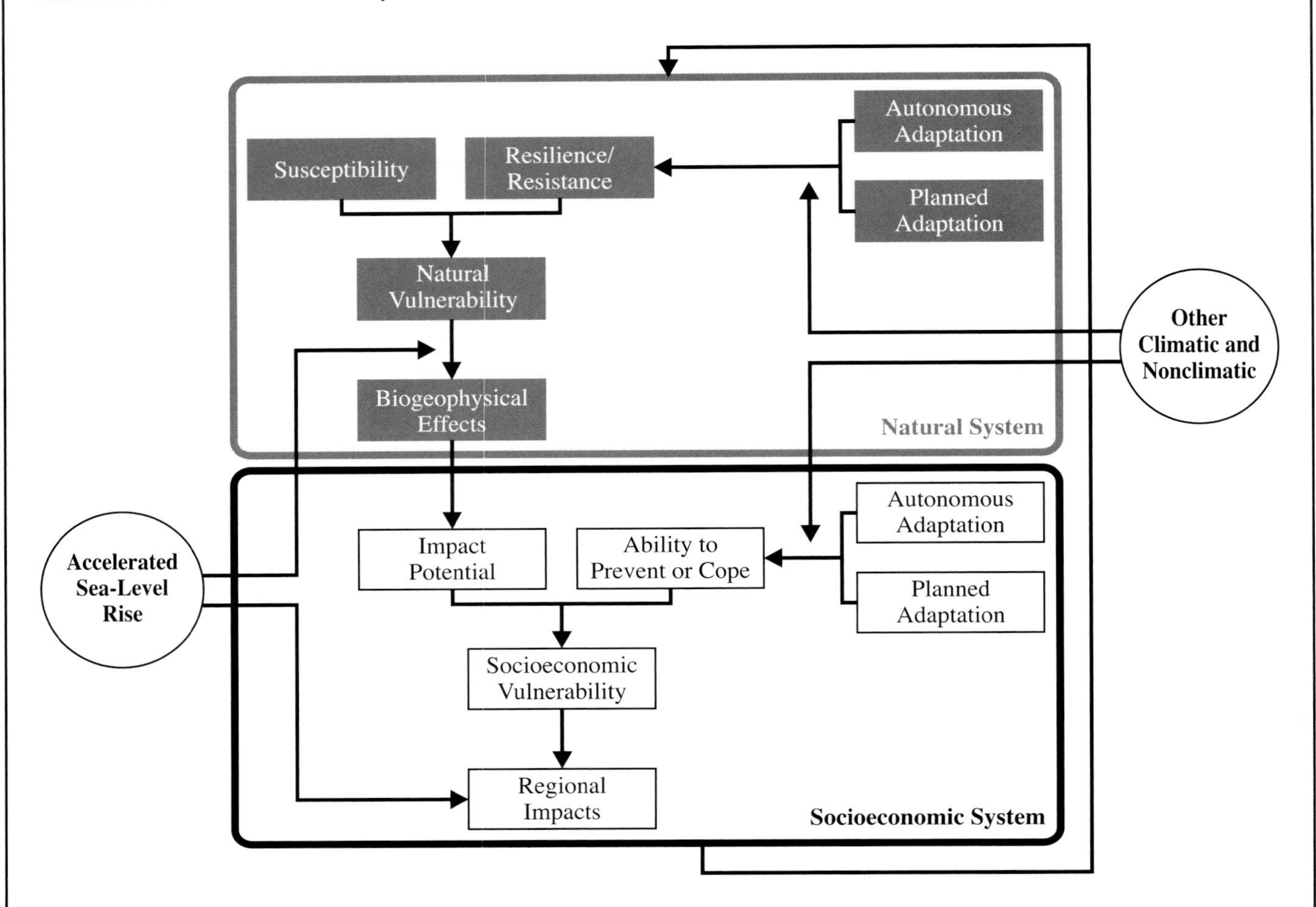

Resilience and resistance are functions of the natural system's capacity for *autonomous adaptation*, which represents the coastal system's natural adaptive response. Resilience and resistance often are affected by human activities, which need not only be negative: *Planned adaptation* can reduce natural vulnerability by enhancing the system's resilience and resistance, thereby adding to the effectiveness of autonomous adaptation.

The biogeophysical effects of sea-level rise impose a range of potential socioeconomic impacts. This *impact potential* is the socioeconomic equivalent of susceptibility; it is dependent on human influences. S*ocioeconomic vulnerability* is determined by the impact potential and society's technical, institutional, economic, and cultural ability to prevent or cope with these impacts. As with the natural system's resilience and resistance, the potential for *autonomous adaptation* and *planned adaptation* determines this ability to prevent or cope.

Dynamic interaction takes place between natural and socioeconomic systems. Instead of being considered as two independent systems, they are increasingly regarded as developing in a co-evolutionary way, as shown by the feedback loop from the socioeconomic system to the natural system.

Sources: Klein and Nicholls (1999); Mimura and Harasawa (2000).

as oil infrastructure, urban areas, and tourist infrastructure—and, after evaluating four scenarios, concludes that Venezuela cannot afford the costs of sea-level rise, either in terms of land and infrastructure lost under a no-protection policy or in terms of the costs involved in any of three protection policies. In Poland, Zeider (1997) estimates the total cost of land-at-loss at US$30 billion; the cost of full protection for the 2,200 km of coast would be US$6 billion.

Several different methods have been used to estimate economic costs. Yohe and Neumann (1997) focus attention on the cost-benefit procedures applied by coastal planners to evaluate shoreline protection projects in relation to sea-level rise. They develop three alternative adaptive responses to the inundation threat from climate-induced sea-level rise: cost-benefit with adaptation foresight (CBWAF), cost-benefit absent adaptation (CBAA), and protection guaranteed (PG). On economic grounds, CBWAF became the preferred option; CBAA conforms most closely with the routine application of existing procedures. "Adaptation foresight" in Yohe and Neumann's cost-benefit procedures assumes that erosion resulting from sea-level rise is a gradual process and that storm impacts do not change as sea level rises. West and Dowlatabadi (1999) suggest, however, that although a rise in sea level may be gradual and predictable, the effects of storms on coastal shorelines and structures are often stochastic and uncertain, in part because of sea-level rise effects.

Sea-level rise can increase the damage caused by storms because mean water level (the base level for storm effects) is higher, waves can attack higher on the shore profile, and coastal erosion often is accelerated, bringing structures nearer the shoreline and potentially removing protection offered by dunes and other protective features. The Heinz Center (2000) estimates that roughly 1,500 homes in the United States will be lost to coastal erosion each year for several decades, at a cost to property owners of US$530 million yr^{-1}. Most of the losses over the next 60 years will be in low-lying areas that also are subject to flooding, although some damage will be along eroding bluffs or cliffs. West *et al.* (2001) estimate that the increase in storm damage because of sea-level rise increases the direct damages of sea-level rise from erosion of the shoreline by 5%, but storm damage could be as much as 20% of other sea-level rise damages. They also developed a method for evaluating the effects of investor decisions to repair storm damage on the net economic impacts of rising sea level in the United States. Neumann *et al.* (2000) have estimated that a 0.5-m sea-level rise by 2100 could cause cumulative impacts to U.S. coastal property of US$20 billion to US$150 billion and that more extensive damage could result if climate change increases storm frequency or intensity.

Morisugi *et al.* (1995) attempt to evaluate the household damage cost in Japan from increased storm surges and the potential benefit generated by countermeasures, using a microeconomic approach. In this approach, household utility is expressed as a function of disaster occurrence probability, income, and other variables, and the change of utility level from sea-level rise and/or countermeasures is translated into monetary terms. For a 0.5-m sea-level rise, the damage cost without countermeasures in Japan would be about US$3.4 billion yr^{-1}, based on the comparison of the utility level between no sea-level rise and a 0.5-m sea-level rise without countermeasures. When the utility level is calculated for the case with countermeasures, the damage cost is reduced to about –US$1.3 billion yr^{-1}, which means that a benefit is created. Therefore, the benefit created by countermeasures for a 0.5-m sea-level rise, which is defined as the decrease of damage cost, would be about US$4.7 billion yr^{-1} at the national level. Because the annual expense for countermeasures is estimated to be about US$1.9 billion yr^{-1} (Mimura *et al.*, 1998), the countermeasures are still beneficial after expenses are considered. This and other examples given here suggest that more robust assessments of the economic impacts of sea-level rise are possible and that they can improve the quality of adaptation strategies.

6.5.3. *Impacts on Coastal Infrastructure*

A large portion of the human population now lives in coastal areas, and the rate of population growth in these areas is higher than average (Cohen *et al.*, 1997; Gommes *et al.*, 1998). Many large cities are located near the coast (e.g., Tokyo, Shanghai, Jakarta, Bombay, New York), and Nicholls and Mimura (1998) have argued that the future of the subsiding megacities in Asia, particularly those on deltas, is among the most challenging issues relating to sea-level rise. People in developed coastal areas rely heavily on infrastructure to obtain economic, social, and cultural benefits from the sea and to ensure their safety against natural hazards such as high waves, storm surges, and tsunamis. Their well-being is supported by systems of infrastructure that include transportation facilities, energy supply systems, disaster prevention facilities, and resorts in coastal areas. Significant impacts of climate change and sea-level rise on these facilities would have serious consequences (see Chapter 7). This analysis applies not only in highly developed nations but in many developing economies and small island states (Nunn and Mimura, 1997). The vulnerability of waste facilities, septic systems, water quality and supply, and roads is a particular concern in many places (Solomon and Forbes, 1999).

Mimura *et al.* (1998) summarize Japanese studies on the impacts on infrastructure. Several studies suggested that disaster prevention facilities such as coastal dikes, water gates, and drainage systems and coastal protection structures such as seawalls, breakwaters, and groins will become less functional because of sea-level rise and may lose their stability. A common concern relates to the bearing capacity of the soil foundation for structures. For instance, the increased water table resulting from sea-level rise decreases the bearing capacity of the soil foundation and increases the possibility of liquefaction, which results in higher instability of coastal infrastructures to earthquakes (Shaw *et al.*, 1998a). In the United Kingdom, sea defenses and shore protection works around 4,300 km of coast cost approximately US$500 million yr^{-1} to maintain at present—a figure that Turner *et al.* (1998) suggest will continue to rise in the future.

Port facilities are another type of infrastructure that will be affected by climate change and sea-level rise. Higher sea level probably will decrease the effectiveness of breakwaters against wave forces, and wharves may have to be raised to avoid inundation. When such effects are anticipated, countermeasures can be implemented to maintain function and stability. Therefore, the real impacts will occur as an additional expenditure to reinforce the infrastructure. The total expenditure to keep the present level of functions and stability for about 1,000 Japanese ports is estimated to be US$110 billion for a 1-m sea-level rise (Mimura *et al.*, 1998).

6.5.4. Socioeconomic Impacts and Natural Systems

Turner *et al.* (1995) assert that relationships between the physical impacts of climate change and socioeconomic implications in the coastal zone have not been fully encompassed in recent work. This statement remains valid to date. Some attempts have been made to express the value of coastal features that are normally regarded as nonmarket goods (Costanza *et al.*, 1997; Alexander *et al.*, 1998). Several assessments of mangrove and reef ecosystems have highlighted their economic value on the basis of ecosystem goods and services, as well as natural capital value (e.g., Moberg and Folke, 1999).

Estimates of the monetary value of wetlands and information about attitudes toward wetland conservation can be used in policy decisions (e.g., Söderqvist, 2000). "Use" and "non-use" values may be determined (Stein *et al.*, 2000). For example, Rönnbäck (1999) suggests that mangrove systems alone account for US$800–16,000 ha^{-1} in seafood production. Streever *et al.* (1998) sought public attitudes and values for wetland conservation in New South Wales, Australia, and found that a conservative estimate of the aggregate value of these wetlands, based on willingness-to-pay criteria, was US$30 million yr^{-1} for the next 5 years. They also refer to an earlier study on the value of marketable fish in mangrove habitats of Moreton Bay, Queensland, estimated at more than US$6,000 ha^{-1} yr^{-1} (Marton, 1990). Stein *et al.* (2000) have developed a framework for crediting and debiting wetland values that they suggest provides an ecologically effective and economically efficient means to fulfill compensatory mitigation requirements for impacts to aquatic resources.

Climate change impacts on natural systems can have profound effects on socioeconomic systems (Harvey *et al.*, 1999). One example cited by Wilkinson *et al.* (1999) is the 1998 coral bleaching event in the Indian Ocean. This event was unprecedented in severity; mortality rates reached as high as 90% in many shallow reefs, such as in the Maldives and the Seychelles. Such severe impacts are expected to have long-term socioeconomic consequences as a result of changed fish species mix and decreased fish stocks and negative effects on tourism as a result of degraded reefs. Degradation of reefs also will lead to diminished natural protection of coastal infrastructure against high waves and storm surges on low-lying atolls. Wilkinson *et al.* (1999) estimate the costs of the 1998 bleaching event to be between US$706 million (optimistic) and US$8,190 million (pessimistic) over the next 20 years. The Maldives and the Seychelles are identified as particularly affected, because of their heavy reliance on tourism and fishery.

Some economic impacts of marine diseases and harmful algal blooms influenced by climate variations have been evaluated since the SAR. An outbreak in 1997 of the toxic dinoflagellate *Pfiesteria piscida*, which has been associated with increased nutrients and SSTs, caused large fish kills on the U.S. Atlantic coast that resulted in public avoidance and economic losses estimated at US$60 million (CHGE, 1999). A persistent brown tide bloom in the Peconic Estuary system of New York blocked light and depleted oxygen in the water column, severely affecting seagrass beds and reducing the value of the Peconic Bay scallop fishery by approximately 80% (CHGE, 1999). Harmful algal blooms associated with increased SST and the influx of nutrients into an estuary can result in economic harm through shellfish closures, impacts on tourism, reduction of estuarine primary productivity, deterioration of fishery habitat (e.g., seagrass beds), and mortality of fish and shellfish.

6.5.5. Social and Cultural Impacts

In some coastal societies, the significance of cultural values is equal to or even greater than that of economic values. Thus, some methodologies have been developed that include traditional social characteristics, traditional knowledge, subsistence economy, close ties of people to customary land tenure, and the fact that these factors are intrinsic components of the coastal zone (e.g., Kay and Hay, 1993). As such, they must be taken into account in certain contexts, including many South Pacific island countries (e.g., Yamada *et al.*, 1995; Solomon and Forbes, 1999) and indigenous communities in northern high latitudes (e.g., Peters, 1999), among other examples.

Patterns of human development and social organization in a community are important determinants of the vulnerability of people and social institutions to sea-level rise and other coastal hazards. This observation does not mean that all people in a community share equal vulnerability; pre-event social factors determine how certain categories of people will be affected (Heinz Center, 1999). Poverty is directly correlated with the incidence of disease outbreaks (CHGE, 1999) and the vulnerability of coastal residents to coastal hazards (Heinz Center, 1999).

Examples of inequitable vulnerability to coastal hazards include population shifts in Pacific island nations such as Tonga and Kiribati. In Tonga, people moving from outer islands to the main island of Tongatapu were forced to settle in low-lying areas, including the old dumping site, where they were proportionally more vulnerable to flooding and disease (Fifita *et al.*, 1992). Storm-surge flooding in Bangladesh has caused very high mortality in the coastal population (e.g., at least 225,000 in November 1970 and 138,000 in April 1991), with the highest mortality among the old and weak (Burton *et al.*, 1993). Land that is subject to flooding—at least 15% of the

Bangladesh land area—is disproportionately occupied by people living a marginal existence with few options or resources for adaptation.

El-Raey *et al.* (1997, 1999) studied the effects of a 0.5-m sea-level rise on the Nile delta. In addition to economic costs from loss of agricultural land and date palms, they identify social and cultural impacts. Under a no-protection policy, El-Raey *et al.* (1997) predict that the population would suffer from the loss of residential shelter (32% of urban areas flooded) and employment (33.7% of jobs lost). In addition, a substantial number of monuments and historic sites would be lost (52%).

6.6. Adaptation

6.6.1. Evolution of Coastal Adaptation Options

In the SAR, Biljsma *et al.* (1996) identified three possible coastal response options:

- *Protect,* which aims to protect the land from the sea so that existing land uses can continue, by constructing hard structures (e.g., seawalls) as well as using soft measures (e.g., beach nourishment)
- *Accommodate,* which implies that people continue to occupy the land but make some adjustments (e.g., elevating buildings on piles, growing flood- or salt-tolerant crops)
- *Retreat,* which involves no attempt to protect the land from the sea; in an extreme case, the coastal area is abandoned.

An evaluation of such strategies was regarded as a crucial component of the vulnerability assessment Common Methodology. Klein and Nicholls (1999) argue, however, that as far as adaptation is concerned, that methodology has been less effective in assessing the wide range of technical, institutional, economic, and cultural elements in different localities. Indeed, they indicated that there has been concern that the methodology emphasizes a protection-oriented response rather than consideration of the full range of adaptation options.

Klein *et al.* (2000) develop a methodology that seeks to address some of these comments. They argue that successful coastal adaptation embraces more than just selecting one of the technical options to respond to sea-level rise; it is a more complex and iterative process, with a series of policy cycles. Four steps can be distinguished in the process of coastal adaptation:

1) Information collection and awareness raising
2) Planning and design
3) Implementation
4) Monitoring and evaluation.

In reality, however, adaptive responses often are undertaken reactively rather in a step-wise, planned, and anticipatory fashion.

The process of coastal adaptation can be conceptualized by showing that climate change and/or climate variability, together with other stresses on the coastal environment, produce actual and potential impacts. These impacts trigger efforts of mitigation, to remove the cause of the impacts, or adaptation to modify the impacts. Bijlsma *et al.* (1996) noted that climate-related changes represent potential additional stresses on systems that already are under pressure. Climate change generally will exacerbate existing problems such as coastal flooding, erosion, saltwater intrusion, and degradation of ecosystems. At the same time, nonclimate stresses can be an important cause of increasing coastal vulnerability to climate change and variability. Given such interactive effects, adaptation options to be most effective should be incorporated with policies in other areas, such as disaster mitigation plans, land-use plans, and watershed resource plans. In other words, adaptation options are best addressed when they are incorporated in integrated coastal management and sustainable development plans.

Policy criteria and coastal development objectives condition the process of adaptation. Other critical influences include values, awareness, and factors such as historical legacies, institutions, and laws. There is growing recognition of the need for researchers, policymakers, residents, and other key stakeholders to work together to establish a framework for adaptation that is integrated within current coastal management processes and practices and takes a broader view of the subject. Collaborative efforts of this kind can support a process of shared learning and joint problem solving, thereby enabling better understanding, anticipation of, and response to climate change. Cash and Moser (2000) identify some of the deficiencies in integrating science and policy. They suggest the following guidelines for meeting the challenge: Use "boundary organizations" that can link researchers and decisionmakers at various scales, capitalize on particular scale-specific capabilities, and develop adaptive assessment and management strategies through long-term iterative processes of integrated assessment and management.

6.6.2. Resilience and Vulnerability

In the context of climate change and coastal management, vulnerability is now a familiar concept. On the other hand, the concept of coastal resilience is less well known but has become much more important in recent years (Box 6-5). Coastal resilience has ecological, morphological, and socioeconomic components, each of which represents another aspect of the coastal system's adaptive capacity to external disturbances. We have identified several natural features that contribute to resilience of the shore-zone by providing ecological buffers, including coral reefs, salt marsh, and mangrove forest and morphological protection in the form of sand and gravel beaches, barriers, and coastal dunes.

Socioeconomic resilience is the capability of a society to prevent or cope with the impacts of climate change and sea-level rise, including technical, institutional, economic, and cultural ability (as indicated in Box 6-5). Enhancing this resilience is equivalent to reducing the risk of the impacts on society. This resilience can be strengthened mainly by decreasing the probability of occurrence of hazard (managed retreat or protection); avoiding or reducing its potential effects (accommodation or protection), and facilitating recovery from the damages when impacts occur. Among these options, managed retreat has gained some prominence in the past 2 decades (see Box 6-6); Clark (1998) has argued that flood insurance is an appropriate management strategy to enhance coastal resilience in the UK.

Technological capacity is a component of social and economic resilience, although adaptation strategies may involve more than engineering measures. Technological options can be implemented efficiently only in an appropriate economic, institutional, legal, and sociocultural context. A list of technologies that could be effective for adaptation appears in Klein *et al.* (2000). Indigenous (traditional) technologies should be considered as an option to increase resilience and, to be effective, must fit in with traditional social structures (Veitayaki, 1998; Nunn *et al.*, 1999).

Enhancing coastal resilience in these ways increasingly is regarded as an appropriate way to prepare for uncertain future changes, while maintaining opportunities for coastal development (although some tradeoffs are involved, and the political discourse is challenging). In short, enhancing resilience is a potentially powerful adaptive measure.

6.6.3. Adaptation in the Coastal Zone

The purpose of adaptation is to reduce the net cost of climate change and sea-level rise, whether those costs apply to an economic sector, an ecosystem, or a country. A simple schematic of the objective of adaptation appears in Figure 6-1.

Adaptation within natural systems has been considered a possibility only recently; it results in part from considerations of coastal resilience. An example is provided by coral reefs. Applying the two types of adaptation discussed in Box 6-5, Pittock (1999) suggests that "autonomous adaptation" is what reefs would do by themselves, whereas "planned adaptation" involves conscious human interference to assist in the persistence of some desirable characteristics of the coral reef system. The first type of adaptation may involve more rapid growth of coral, changes in species composition, or evolution of particular species in response to changed temperatures or other conditions. Planned adaptation might involve "seeding" of particular reefs with species adapted to higher temperatures or attempts to limit increased sediment, pollutant, or freshwater flow onto reefs. For reef communities that presently are under stress and are likely to be particularly vulnerable to climate change, the design of managed (or planned) adaptation should involve an evaluation of the extent of autonomous adaptation that can be expected given the current and probable future status of the reef system.

Box 6-6. Adaptation through Managed Retreat

Managed retreat generally is designed to avoid hazards and prevent ecosystems from being squeezed between development and the advancing sea. The most common mechanisms for managed retreat are *setbacks* that require new development to be a minimum distance from the shore, *density restrictions* that limit development, and *rolling easement* policies that allow development on the condition that it be removed to enable wetlands to migrate landward (Titus, 1998). These strategies may all become elements of an integrated coastal management policy. Setback could be considered a managed retreat strategy, particularly in cases in which the setback line is shifted inland as the shoreline recedes. Other measures of managed retreat can include conditional phased-out development, withdrawal of government subsidies, and denial of flood insurance.

Examples of managed retreat and related measures as adaptation to sea-level rise include the following:

- ***Canada:*** New Brunswick completed remapping of the entire coast of the province to delineate the landward limit of coastal features. Setback for new development is defined from this limit. Some other provinces have adopted a variety of setback policies, based on estimates of future coastal retreat.
- ***Barbados:*** A national statute establishes a minimum building setback along sandy coasts of 30 m from mean high-water mark; along coastal cliffs the setback is 10 m from the undercut portion of the cliff.
- ***Aruba and Antigua:*** Setback established at 50 m inland from high-water mark.
- ***Sri Lanka:*** Setback areas and *no-build zones* identified in Coastal Zone Management Plan. Minimum setbacks of 60 m from line of mean sea level are regarded as good planning practice.
- ***United Kingdom:*** House of Commons in 1998 endorsed the concept of *managed realignment* as the preferred long-term strategy for coastal defense in some areas.
- ***United States:*** The states of Maine, Massachusetts, Rhode Island, and South Carolina have implemented various forms of *rolling easement* policies to ensure that wetlands and beaches can migrate inland as sea level rises.
- ***Australia:*** Several states have coastal setback and minimum elevation policies, including those to accommodate potential sea-level rise and storm surge. In South Australia, setbacks take into account the100-year erosional trend plus the effect of a 0.3-m sea-level rise to 2050. Building sites should be above storm-surge flood level for the 100-year return interval.

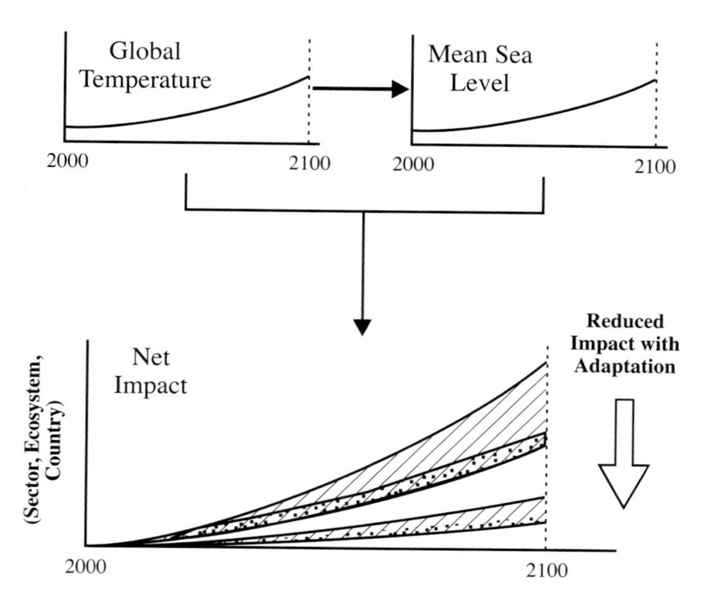

Figure 6-1: The role of adaptation in reducing potential impacts in the coastal zone from global temperature increase and sea-level rise to the year 2100. The bottom panel is a schematic that shows the increasing cost or loss to an economic sector, ecosytem, or country. The area shown by cross-hatch indicates the range of possible impacts and how net impact can be reduced with adaptation. Stipple within the cross-hatched areas indicates the importance of sector, ecosystem, or country resilience as a component of net impact.

Some adaptation measures handle uncertainty better than others. For example, beach nourishment can be implemented as relative sea level rises and therefore is more flexible than a dike or seawall; expansion of the latter may require removal or addition of structures. Any move from "hard" (e.g., seawall) to "soft" (e.g., beach nourishment) shore protection measures must be accompanied, however, by a much better understanding of coastal processes that prevail in the area (Leafe *et al.*, 1998). Rolling easements are more robust than setbacks (Titus, 1998) but may be impractical for market or cultural reasons. Maddrell (1996) found that over time scales of 35 and 100 years, managed coastal retreat is the most cost-effective adaptation option in reducing flood risks and protection costs for nuclear power facilities on the shingle foreland at Dungeness, UK. Flood insurance can discourage a flexible response if rates are kept artificially low or fixed at the time of initial construction, as they are in the United States (Crowell *et al.*, 1999).

Reevaluations of the efficacy of hard shore protection schemes as a long-term response to climate change and sea-level rise are increasingly being undertaken. Chao and Hobbs (1997) have considered the role of decision analysis of shore protection under climate change uncertainty; Pope (1997) has suggested several ways of responding to coastal erosion and flooding that have relevance in the context of climate change. Documented changes in tidal characteristics as a result of the construction of sea dikes and seawalls also have implications for shore protection in the face of rising sea level. Several alternatives to seawalls have been suggested as adaptation measures to reduce coastal erosion and saltwater intrusion from rising sea levels in Shanghai, including improving drainage quality and channel capacity, increasing pumping facilities to reduce the water table, constructing a barrier across the mouth of the river, and developing new crops that are tolerant of a higher groundwater table (Chen and Zong, 1999).

It also should be noted, however, that Doornkamp (1998) has argued that in some situations past management decisions about human activities in the coastal zone (including flood defenses, occupance of flood-prone lands, extraction of groundwater and natural gas) have had an impact on relative land and sea levels and have done more to increase the risk of coastal flooding than damage that can be assigned to global warming to date.

6.6.4. Adaptation in Marine Ecosystems

Adaptation of the fishing industry to climate change is closely connected with investigations of the consequences of the effect of climatic anomalies and climate change scenarios. Because the effects of changes in climate factors will have different consequences for various species, development of special measures aimed at adaptation of the fish industry is regional in character and falls into the category of important socioeconomic problems.

Possible adverse effects of climate change can be aggravated by an inadequate utilization of fish reserves. For example, if a fish stock decreases as a result of the combined effect of climate change and overfishing, and the commercial catch remains high, species abundance may decrease dramatically and the commercial catch may become unprofitable. In such circumstances, some measures may need to be taken to protect fish reserves, such as the precautionary measures suggested by O'Brien *et al.* (2000) to give the North Sea cod fishery a chance to rebuild. Several sustainability indicators of marine capture species are discussed in Garcia and Staples (2000). Aquaculture also can be regarded as an adaptation; though to be an ecologically sustainable industry, it must emphasize an integrated approach to management (Carvalho and Clarke, 1998). Another adaptation is fish stock enhancement through ocean ranching (see Section 6.3.6).

Fish reserves rank among the most important economic resources in many countries. Approximately 95% of the world catch falls within the 200-mile economic zones of maritime states. Environmental impacts in those zones as a consequence of climate change could affect the catch volume and national economies. It should be noted that gains and losses at different levels of social organization can occur not only as a consequence of climate change but also as a result of human society's responses to this change. In some regions, for instance, special measures may be taken to promote adaptation and to reduce the negative consequences of climate change—which adds another dimension to fish management.

Adaptation measures that are relevant to the fishing industry may include the following:

- Establishment of national and international fishery management institutions that will be able to manage expected changes
- Expansion of aquaculture as a way of meeting increasing demand for seafoods of an increasing world population
- Support for innovative research and integrated management of fisheries within coastal and open marine ecosystems
- Improvement and development of an integrated monitoring system in the most productive areas, aimed at obtaining systematic information on hydrophysical, hydrochemical, and hydrobiological processes
- Organization of data banks on the results of integrated ecological monitoring to identify anthropogenic changes, including climate change, and predict fish productivity
- Modification and improvement of the technology of the fishing industry and management of the fish trade as required to adapt to climate change
- Organization of marine biosphere reserves and protected areas for the habitat of marine mammals
- Use of emerging predictive information related to natural climate variability (e.g., ENSO) to support fishery management and planning.

Adoption of some adaptation options to the potential impacts of climate change is not a panacea, however. Fish often are transboundary resources in that they may cross international and state boundaries in their oceanic migrations. In the case of Pacific salmon, for instance, problems have arisen in the agreement between the United States and Canada that are attributable in part to the effects of large-scale climate fluctuations (see Box 6-1). Miller (2000) suggests that the Pacific salmon case demonstrates that it may not be a simple matter for the fishing industry or governments to respond effectively to climate change. She concludes that adaptation is difficult when a resource is exploited by multiple competing users who possess incomplete information about the resource. If their incentives to cooperate are disrupted by the impacts of climate variation, dysfunctional breakdown in management rather than efficient adaptation may occur (Miller, 2000).

6.7. Synthesis and Integration

This chapter is concerned with two closely related but geographically different environments. The oceans—which cover more than 70% of the Earth's surface—are open, expansive, and spatially continuous. By contrast, coastal zones are long, narrow, and discontinuous. As a result, climate change impacts on marine ecosystems may be accommodated more readily in the open ocean (e.g., by migration) than in coastal regions, where mobility is restricted, there are more environmental constraints, and human impacts may be more severe.

The potential biological and physical impacts of climate change and sea-level rise vary considerably between the oceans and coastal regions. The least vulnerable coastal and marine ecosystems have low exposure or high resilience to the impacts. Similarly, coastal communities and marine-based economic sectors that have low exposure or high adaptive capacity will be least affected. Countries, communities, and individuals in the higher range of economic well-being have access to technology, insurance, construction capital, transportation, communication, social support systems, and other assets that enhance their adaptive capacity. Those that do not have access have limited adaptive capacity. Unequal access to adaptation options, therefore unequal vulnerability, are attributable largely to different socioeconomic conditions. Poor adaptation or "maladaptation" also may lead to increased impacts and vulnerability in the future, with implications for intergenerational equity. These concepts are summarized in Figure 6-2.

Whereas the estimated costs of sea-level rise and other climate-related impacts in developed countries typically are limited to

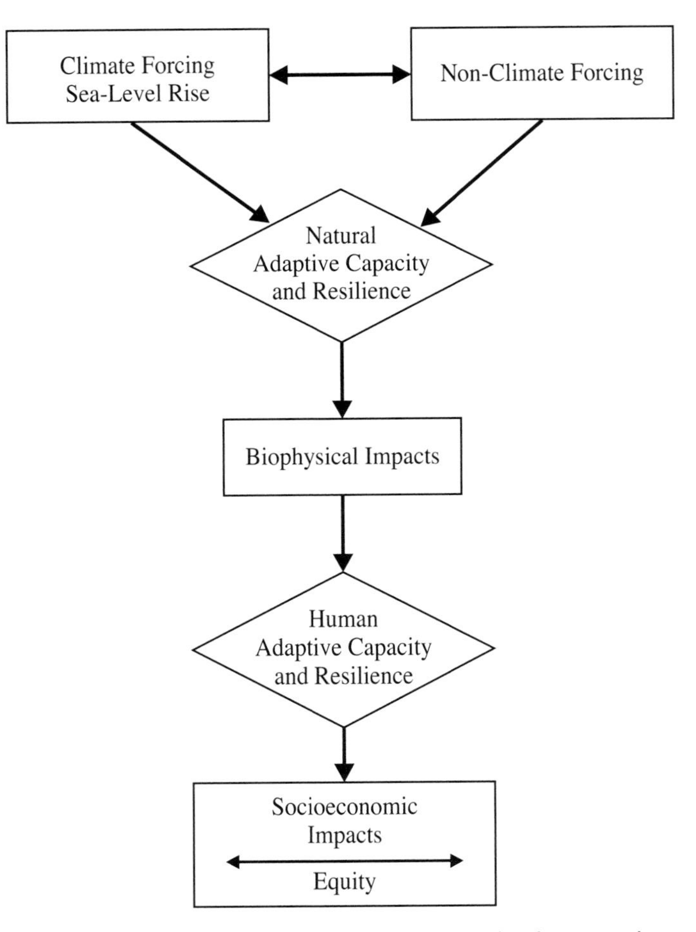

Figure 6-2: The role of natural and human adaptive capacity and resilience on the socioeconomic impacts of climate change following climate forcing, sea-level rise, and nonclimate forcing. The equity arrow in the bottom box indicates that impacts will not be uniform and that there will be wide inequalities, depending on socioeconomic conditions.

property losses, the reported outcomes of coastal floods in developing countries often include disease and loss of life. Vulnerability assessments in the developing world often do not consider the costs of business interruptions and failures, social disruption and dislocation, health care, evacuation, or relocation. A full accounting of the economic costs associated with lost, diminished, or disrupted lives would require estimates of the economic productivity losses they represent. Because this level of cost accounting is rare in developing countries, inequities may be significantly underestimated.

Turner *et al.* (1996) note that coastal zones are under increasing stress because of an interrelated set of planning failures, including information, economic market, and policy intervention failures. Moreover, moves toward integrated coastal management are urgently required to guide the co-evolution of natural and human systems. Acknowledging that forecasts of sea-level rise have been scaled down, they note that (1) much uncertainty remains over, for example, combined storm-surge and other events; and (2) within the socioeconomic analyses of the problem, resource valuations have been only partial at best and have failed to incorporate sensitivity analysis in terms of the discount rates utilized. They suggest that these factors would indicate an underestimation of potential damage costs and conclude that a precautionary approach is justified, based on the need to act ahead of adequate information acquisition and use economically efficient resource pricing and proactive coastal planning.

More recently, Turner *et al.* (1998) have aimed to elicit the main forces influencing the development of coastal areas and the means available to assess present use and manage future exploitation of the coastal zone. Their way of analyzing coastal change and resource management is through the pressure-state-impact-responses (P-S-I-R) conceptual framework. They analyze a variety of pressures and trends (including climate change, population changes, port development, marine aggregate extraction, and pollution). In the P-S-I-R framework, all of these factors are examined in the context of sustainable use of coastal resources and on the basis of an interdisciplinary ecological economics approach.

Several changes that might be expected to accelerate with global warming in the future already are detected in some regions and systems. Examples include global sea-level rise, increases in SST, and regional decreases in sea-ice cover. Impacts associated with these changes have included shoreline erosion, wetland loss, seawater intrusion into freshwater lenses, and some impacts on coral reefs. These impacts provide contemporary analogs for potential impacts in the future, recognizing that future changes and impacts may be of greater magnitude than those experienced so far and that changes and impacts may become more geographically widespread, including expansion into new areas that have not experienced such conditions previously. The contemporary environment gives us some insights into potential ecosystem impacts, some idea of costs, and some experience with potentially useful adaptation strategies.

We have found very few studies that indicate benefits of climate change and sea-level rise in coastal and marine systems. Recent studies, however, point to possible economic benefits from adaptation measures. Such benefits are likely to be restricted, particularly in the areas most at risk—including a large number of developing countries. Furthermore, the extent of impacts in those regions and the range of potentially effective adaptation measures remain poorly defined. Although there is growing acceptance of the need for integrated management strategies, progress has been slow in implementing these concepts in many jurisdictions. Part of the reason is limited development of understanding and tools for integrated assessment and management needs, involving various levels and aspects of integration, each of which may be difficult to implement. For instance, integration between the different disciplines involved in coastal and marine impact and adaptation analysis has been identified as a key issue by Capobianco *et al.* (1999). This and other integration needs are summarized in Box 6-7.

Some progress has been made since the SAR in developing and refining methodologies for assessing impacts of sea-level rise. Environments under particular threat include deltas, low coastal plains, coastal barriers, heavily utilized seas, tropical reefs and mangroves, and high-latitude coasts where impacts from warming may occur sooner or more rapidly.

Some topic areas rarely have been addressed, however. For instance, only a few case studies attempt to integrate potential impacts of sea-level rise and increased precipitation and runoff in coastal watersheds in assessing coastal vulnerability. Techniques for similar integration between biophysical and socioeconomic impacts are developing slowly, while human development and population growth in many regions have increased socioeconomic vulnerability and decreased the resilience of coastal ecosystems. Few studies provide details or any quantitative measures. We believe that integrated assessment and management of open marine and coastal ecosystems and a better understanding of their interaction with human development could lead to improvements in the quality of sustainable development strategies.

Global climate change will affect the biogeophysical characteristics of the oceans and coasts, modifying their ecological structure and affecting their ability to sustain coastal residents and communities. Impacts in the coastal zone will reflect local geological, ecological, and socioeconomic conditions within a broader regional or global context. Shorelines are inherently dynamic, responding to short- and long-term variability and trends in sea level, wave energy, sediment supply, and other forcing. Coastal communities—particularly on low-lying deltas, atolls, and reef islands—face threats of inundation, increased flooding, and saltwater intrusion, with impacts on health and safety, water supply, artisanal fisheries, agriculture, aquaculture, property, transportation links, and other infrastructure. In some coastal areas, particularly in developed nations, a shift in emphasis toward managed retreat appears to have gained momentum. Enhancement of biophysical and socioeconomic resilience in coastal regions increasingly is regarded as a cost-effective and

Box 6-7 Integration for Assessment and Management of Marine and Coastal Systems

Integration of marine, terrestrial, and coastal processes and a better understanding of their interactions with human development could lead to substantial improvements in the quality of adaptation strategies. Integration must take place in several areas, including the following:

- ***Subject/topic-area integration*** (e.g., climate-change related stresses plus non-climate stresses; biophysical and socioeconomic susceptibility, resilience, vulnerability, impacts)
- ***Geographical/spatial integration*** (e.g., linkages between terrestrial, coastal, and oceanic systems and feedbacks; global, regional, local scales)
- ***Methodological integration*** (e.g., integrating physical, social, and economic models)
- ***Integrated implications*** (e.g., for sustainable development, intergenerational equity and ethics)
- ***Integration of science, impacts, and policy.***

Estuaries as an example of the need for integration
Changes in salinity, temperature, sea level, tides, and freshwater inflows to estuaries are considered likely consequences of climate change on estuarine systems. Estuaries are among the world's most-stressed ecosystems because of their close proximity to areas of population growth and development. Understanding of regional differences in the physical drivers that will cause changes in estuarine ecosystems and their ecological functions is limited. Uncertainty also exists regarding changes to dissolved carbon, nutrient delivery and pollutant loading, and their interactions. For example, intensive forestry and agriculture that may be implemented as some regions adapt to climatic change could increase the transport of nutrients such as nitrogen and phosphorus to estuaries. Linkage of hydrological models for surface waters with ocean-atmosphere models is needed to integrate marine and terrestrial ecosystem change. Estuaries illustrate the need for vertical integration among the foregoing subject areas and issues, spatial scales, and methodological approaches, with implications for habitation and use of coastal environments and ecosystems.

desirable adaptive strategy. Growing recognition of the role of the climate-ocean system in the management of fish stocks is leading to new adaptive strategies that are based on determination of accepable removable percentages in relation to climate change and stock resilience.

Vulnerability to climate change and sea-level rise has been documented for a variety of coastal settings via common methodologies developed in the early 1990s; these assessments have confirmed the spatial and temporal variability of coastal vulnerability at national and regional levels. New conceptual frameworks include biophysical and socioeconomic impacts and highlight adaptation and resilience as components of vulnerability. Recent advances include models to evaluate economic costs and benefits that incorporate market and nonmarket values. Sustainable approaches to integrated coastal management can now include new financial accounting approaches that include ecological services and traditional cultural values. Nevertheless, adaptive choices will be conditioned by policy criteria and development objectives, requiring researchers and policymakers to work toward a commonly acceptable framework for adaptation.

References

Adkinson, M.D., R.M. Peterman, M.F. Lapointe, D.M. Gillis, and J. Korman, 1996: Alternative models of climatic effects on sockeye salmon (*Oncorhynchus nerka*) productivity in Bristol Bay, Alaska, and the Fraser River, British Columbia. *Fisheries Oceanography*, **5,** 137–152.

Aksornkoae, S., 1993: *Ecology and Management of Mangroves*. International Union for Conservation of Nature and Natural Resources, Bangkok, Thailand, 176 pp.

Alam, M., 1996: Subsidence of the Ganges-Brahmaputra delta of Bangladesh and associated drainage, sedimentation and salinity problems. In: *Sea-Level Rise and Coastal Subsidence: Causes, Consequences and Strategies* [Milliman, J.D. and B.U. Haq (eds.)]. J. Kluwer Academic Publishers, Dordrecht, The Netherlands, pp. 169–192.

Alheit, J. and E. Hagen, 1997: Long-term climate forcing of European herring and sardine populations. *Fisheries Oceanography*, **6,** 130–139.

Alleng, G.P., 1998: Historical development of the Port Royal mangrove wetland, Jamaica. *Journal of Coastal Research*, **14,** 951–959.

Alexander, A.M., J.A. Last, M. Margolis, and R.C. d'Arge, 1998: A method for valuing global ecosystem services. *Ecological Economics*, **27,** 161–170.

Ali, A., 1999: Climate change impacts and adaptation assessment in Bangladesh. In: *National Assessment Results of Climate Change: Impacts and Responses* [Mimura N. (ed.)]. *Climate Research (Special Issue)*, **6,** 109–116.

Anderson, D.M., A.D. Cembella, and G.M. Hallegraeff (eds.), 1998: Physiological ecology of harmful algal blooms. In: *Proceedings of NATO Advanced Study Institute, Bermuda, 1996*. Springer-Verlag, Berlin, Germany, 662 pp.

Antonius, A., 1995: Pathologic syndromes on reef corals: a review. In: *Coral Reefs in the Past, the Present and the Future* [Geister, J. and B. Lathuillere (eds.)]. *Publications du Service géologique de Luxembourg*, **29,** 161–169.

Are, F.E., 1998: The contribution of shore thermoabrasion to the Laptev Sea sediment balance. In: *Permafrost: Proceedings of the Seventh International Conference, Yellowknife, NWT, June 23-27, 1998* [Lewkowicz, A.G. and M. Allard (eds.)]. Centre d'études nordiques, Université Laval, Québec, Canada, pp. 25–30.

Aronson, R.B, W.F. Precht, I.G. Macintyre, and T. Murdoch, 2000: Coral bleach-out in Belize. *Nature*, **405,** 36.

Arrigo, K.R., D.H. Robinson, R.B. Dunbar, G.R. Di Tullo, M. van Woert, and M.P. Lizotte, 1999: Phytoplankton community structure and the drawdown of nutrients and CO_2 in the Southern Ocean. *Science*, **283,** 365–367.

ASM, 1998: *Microbial Pollutants in Our Nation's Water: Environmental and Public Health Issues*. American Society for Microbiology, Office of Public Affairs, Washington, DC, USA, 16 pp.

Baldina, E.A., J. de Leeuw, A.K. Gorbunov, I.A. Labutina, A.F. Zhivogliad, and J.F. Kooistra, 1999: Vegetation change in the Astrkhanskiy Biosphere Reserve (Lower Volta Delta, Russia) in relation to Caspian Sea. *Environmental Conservation,* **26,** 169–178.

Beamish, R.J., C. Mahnken, and C.M. Neville, 1997: Hatchery and wild production of Pacific salmon in relation to large-scale, natural shifts in the productivity of the marine environment. *International Council for the Exploration of the Sea Journal of Marine Science*, **54,** 1200–1215.

Beamish, R.J., D.J. Noakes, G.A. McFarlane, L. Klyashtorin, V.V. Ivanov, and V. Kurashov, 1999: The regime concept and natural trends in the production of Pacific salmon. *Canadian Journal of Fisheries and Aquatic Sciences,* **56,** 516–526.

Behrenfeld, M.J., A.J. Bale, Z.S. Kolber, J. Aiken, and P.G. Falkowsky, 1996: Confirmation of iron limitation of phytoplankton photosynthesis in the equatorial Pacific Ocean. *Nature*, **383,** 508–516.

Bijlsma, L., C.N. Ehler, R.J.T. Klein, S.M. Kulshrestha, R.F. McLean, N. Mimura, R.J. Nicholls, L.A. Nurse, H. Pérez Nieto, E.Z. Stakhiv, R.K. Turner, and R.A. Warrick, 1996: Coastal zones and small islands. In: *Climate Change 1995: Impacts, Adaptations, and Mitigation of Climate Change: Scientific-Technical Analyses. Contribution of Working Group II to the Second Assessment Report of the Intergovernmental Panel on Climate Change* [Watson, R.T., M.C. Zinyowera, and R.H. Moss (eds.)]. Cambridge University Press, Cambridge, United Kingdom and New York, NY, USA, pp. 289–324.

Bird, E.C.F., 1993: *Submerging Coasts: The Effects of a Rising Sea Level on Coastal Environments*. John Wiley and Sons, Chichester, United Kingdom, 184 pp.

Bisbal, G.A. and W.E. McConnaha, 1998: Consideration of ocean conditions in the management of salmon. *Canadian Journal of Fisheries and Aquatic Sciences,* **55,** 2178–2186.

Boersma, P.D., 1998: Populations trends of the Galapagos Penguin—Impacts of El Niño and La Niña. *Condor*, **100(2),** 245–253.

Bray, M.J. and J.M. Hooke, 1997: Prediction of soft-cliff retreat with accelerating sea level rise. *Journal of Coastal Research*, **13(2),** 453–467.

Bricker-Urso, S.S., W. Nixon, J.K. Cochran, D.J. Hirschberg, and C. Hunt, 1989: Accretion rates and sediment accumulation in Rhode Island salt marshes. *Estuaries*, **12,** 300–317.

Brinson, M.M., R.R. Christian, and L.K. Blum, 1995: Multiple states in the sea-level induced transition from terrestrial forest to estuary. *Estuaries*, **18,** 648–659.

Broecker, W.S., 1994: An unstable superconveyor. *Nature*, **367,** 414–415.

Broecker, W.S., 1997: Thermohaline circulation, the achilles heel of our climate system: will man-made CO_2 upset the current balance? *Science*, **278,** 1582–1588.

Brown, B.E., 1997: Coral bleaching: causes and consequences. *Coral Reefs*, **16,** 129–138.

Brown, B.E., R.P. Dunne, and H. Chansang, 1996: Coral bleaching relative to elevated sea surface temperature in the Andaman Sea (Indian Ocean) over the last 50 years. *Coral Reefs*, **15,** 151–152.

Brown, B.E. and Suharsono, 1990: Damage and recovery of coral reefs affected by El Niño-related seawater warming in the Thousand Islands, Indonesia. *Coral Reefs*, **8,** 163–170.

Brown, B.E., R.P. Dunne, M.S. Goodson, and A.E. Douglas, 2000: Bleaching patterns in reef corals. *Nature*, **404,** 142–143.

Bruun, P., 1962: Sea level rise as a cause of shore erosion. *Journal of Waterways and Harbors Division—ASCE*, **88,** 117–130.

Bruun, P., 1988: The Bruun Rule of erosion by sea-level rise: a discussion of large-scale two and three-dimensional useages. *Journal of Coastal Research*, **4,** 627–648.

Bryant, D., L. Burke, J.W. McManus, and M. Spalding, 1998: *Reefs at Risk: A Map-Based Indicator of Threats to the World's Coral Reefs*. World Resources Institute, Washington, DC, USA, 56 pp. Available online at http://www.wri.org/wri/indictrs/reefrisk.htm.

Burton, I., R.W. Kates, and G.F. White, 1993: *The Environment as Hazard*. The Guilford Press, New York, USA, 2nd. ed., 90 pp.

Butler, R., W.T.D. Wiliams, N. Warnock, and M.A. Bishop, 1997: Wind assistance: a requirement for migration of shorebirds? *The Auk*, **114(3),** 456–466.

Caddy, J.F. and P.G. Rodhouse, 1998: Cephalopod and groundfish landings: evidence for ecological change in global fisheries? *Reviews in Fish Biology and Fisheries*, **8,** 431–444.

Cahoon, D.R. and J.C. Lynch, 1997: Vertical accretion and shallow subsidence in a mangrove forest of southwestern Florida, USA. *Mangroves and Salt Marshes*, **1,** 173–186.

Cahoon, D.R., J.W. Day Jr., D.J. Reed, and R.S. Young, 1998: Global climate change and sea-level rise: estimating the potential for submergence of coastal wetlands. In: *Vulnerability of Coastal Wetlands in the Southeastern United States: Climate Change Research Results* [Guntenspergen, G.R. and B.A. Varin (eds.)]. U.S. Geological Survey, Biological Resources Division, Biological Science Report, Washington, DC, USA, pp. 21–35.

Caldeira, K. and P.B. Duffy, 2000: The role of the Southern Ocean in uptake and storage of anthropogenic carbon dioxide. *Science*, **287,** 620–622.

Cane, M.A., A.C. Clement, A. Kaplan, Y. Kushir, R. Pozdayakov, S. Seager, and E. Zebiak, 1997: Twentieth century sea surface temperature trends. *Science*, **275,** 957–960.

Canziani, O.F., S. Diaz, E. Calvo, M. Campos, R. Carcavallo, C.C. Cerri, C. Gay-García, L.J. Mata, and A. Saizar, 1998: Latin America. In: *The Regional Impacts of Climate Change: An Assessment of Vulnerability. Special Report of IPCC Working Group II* [Watson, R.T., M.C. Zinyowera, and R.H. Moss (eds.)]. Intergovernmental Panel on Climate Change, Cambridge University Press, Cambridge, United Kingdom and New York, NY, USA, pp. 187–230.

Capobianco, M., H.J, DeVriend, R.J. Nicholls, and M.J.F. Stive, 1999: Coastal area impact and vulnerability assessment: the point of view of a morphodynamic modeler. *Journal of Coastal Research,* **15,** 701–716.

Carvalho, P. and B. Clarke, 1998: Ecological sustainability of the South Australian coastal aquaculture management policies. *Coastal Management,* **26,** 281–290.

Cash, D.W. and S.C. Moser, 2000: Linking global and local scales: designing dynamic assessment and management processes. *Global Environmental Change,* **10(2),** 109–120.

Chao, P.T. and B.F. Hobbs, 1997: Decision analysis of shoreline protection under climate change uncertainty. *Water Resources Research*, **33(4),** 817–829.

Chen, X., 1998: Changjian (Yangtze) River Delta, China. *Journal of Coastal Research*, **14,** 838–857.

Chen, X. and D.J. Stanley, 1998: Sea-level rise on eastern China's Yangtze Delta. *Journal of Coastal Research*, **14,** 360–366.

Chen, X. and Y. Zong, 1999: Major impacts of sea-level rise on agriculture in the Yangtze Delta area around Shanghai. *Applied Geography,* **19,** 69–84.

CHGE, 1999: *Extreme Weather Events: The Health and Consequences of the 1997/98 El Niño and La Niña* [Epstein, P.R. (ed.)]. Center for Health and the Global Environment, Harvard Medical School, Boston, MA, USA, 41 pp.

Clark, M.J., 1998: Flood insurance as a management strategy for United Kingdom coastal resilience. *Geographical Journal*, **164,** 333–343.

Coale, K.H., K.S. Johnson, S. Fitzwater, R. Gordon, S. Tanner, F. Chavez, L. Ferioli, P. Nightingale, D. Cooper, W. Cochlan, M. Landry, J. Constatinou, G. Rollwagen, A. Trasvina, and R. Kudela, 1996: A massive phytoplankton bloom induced by an ecosystem scale iron fertilization experiment in the equatorial Pacific Ocean. *Nature*, **383,** 495–501.

Codignotto, J.O., 1997: Geomorfologia y dinámica costera (Geomorphology and coastal dynamics). In: *El Mar Argentino y Sus Recursos Pesqueros* [Boschi, E.E. (ed.)]. Instituto Nacional de Investigacion y Desarrollo Pesquero, Mar del Plata, Argentina, **1,** 89–105, (in Spanish).

Cohen, J.E., C. Small, A. Mellinger, J. Gallup, and J. Sachs, 1997: Estimates of coastal populations. *Science*, **278,** 1211–1212.

Cole, J. and J. McGlade, 1998: Clupeoid population variability, the environment and satellite imagery in coastal upwelling systems. *Reviews in Fish Biology and Fisheries*, **8,** 445–471.

Costanza, R., R. d'Arge, R. de Groot, S. Farber, M. Grasso, B. Hannon, K. Linburg, S. Naeem, R.V. O'Neill, J. Paruelo, R.G. Raskin, P. Sutton, and M. van den Belt, 1997: The value of the world's ecosystem services and natural capital. *Nature*, **387,** 253–260.

Cowell, P.J. and B.G. Thom, 1994: Morphodynamics of coastal evolution. In: *Coastal Evolution: Late Quaternary Shoreline Morphodynamics* [Carter, R.W.G. and C.D. Woodroffe (eds.)]. Cambridge University Press, Cambridge, United Kingdom and New York, NY, USA, pp. 33–86.

Crowell, M., H. Leiken, and M.K. Buckley, 1999: Evaluation of coastal erosion hazards study: an overview. *Journal of Coastal Research (Special Issue)*, **28,** 2–9.

Dallimore, S.R., S. Wolfe, and S.M. Solomon, 1996: Influence of ground ice and permafrost on coastal evolution, Richards Island, Beaufort Sea coast, NWT. *Canadian Journal of Earth Sciences*, **33,** 664–675.

Day, J.W., J.F. Martin, L. Cordoch, and P.H. Templet, 1997: System functioning as a basis for sustainable management of deltaic ecosystems. *Coastal Management* **25,** 115–153.

Day, J.W., J. Rybvzyk, F. Scarton, A. Rismondo, D. Are, and G. Cecconi, 1999: Soil accretionary dynamics, sea-level rise and the survival of wetlands in Venice Lagoon: a field and modelling approach. *Estuarine and Coastal Shelf Science,* **49,** 607–628.

de Baar, H.J.W., J.T.M. de Jong, D.C.E. Bakker, B.M. Löscher, C. Veth, U. Bathmann, and V. Semtacek, 1995: Importance of iron for plankton blooms and carbon dioxide drawdown in the Southern Ocean. *Nature*, **373,** 412–415.

de la Mare, W.K., 1997: Abrupt mid-twentieth-century decline in Antarctic sea ice extent from whaling records. *Nature*, **389,** 57–60.

Dean, R.G., 1990: Beach response to sea level change. In: *The Sea* [le Méhauté, B. and D.M. Hanes (eds.)]. John Wiley and Sons, New York, USA, **9,** 869–887.

Deser, C., M.A. Alexander, and M.S. Timlin, 1996: Upper-ocean thermal variations in the North Pacific during 1970–1991. *Journal of Climate*, **9(8),** 1840–1855.

DETR, 1999: *Climate Change and its Impacts: Stabilisation of CO_2 in the Atmosphere*. United Kingdom Department of the Environment, Transport and the Regions, The Met Office, Bracknell, UK, 28 pp.

Dionne, J.-C., 1986: Érosion récente des marais intertidaux de l'estuaire du Saint-Laurent (Recent erosion of intertidal marshes in the St. Lawrence estuary). *Géographie Physique et Quaternaire*, **40,** 307–323 (in French).

Done, T.J., 1999: Coral community adaptability to environmental change at scales of regions, reefs and reef zones. *American Zoologist,* **39,** 66–79.

Dolman, P.M. and W.J. Sutherland, 1994: The response of bird populations to habitat loss. *Ibis*, **137,** S38–S46.

Doornkamp, J.C., 1998: Coastal flooding, global warming and environmental management. *Journal of Environmental Management*, **52(4),** 327–333.

Douglas, B.C., M. Crowell, and S.P. Leatherman, 1998: Considerations for shoreline position prediction. *Journal of Coastal Research*, **14,** 1025–1033.

Duffy, D.C. and D.C. Schneider, 1994: Seabird-fishery interactions: a manager's guide In: *Seabirds on Islands: Threats, Case Studies and Action Plans* [Nettleship, D.N., J. Burger, and M. Gochfeld (eds.)]. BirdLife Conservation, Series 1, BirdLife International, London, United Kingdom, pp. 26–38.

Ebbesmeyer, C.C., D.R. Cayan, D.R. McLain, F.H. Nichols, D.H. Peterson, and K.T. Redmond, 1991: 1976 step in the Pacific climate: forty environmental changes between 1968–1975 and 1977–1984. In: *Proceedings of the Seventh Annual Pacific Climate (PACLIM) Workshop, April 1990* [Betancourt, J.L. and V.L. Tharp (eds.)]. California Department of Water Resources, Interagency Ecological Studies Program, Technical Report No. 26, Asilomar, CA, USA, pp. 115–126.

Edinger, E.N., J. Jompa, G.V. Limmon, W. Widjatmoro, and M.J. Risk, 1998: Reef degradation and coral biodiversity in Indonesia: effects of land-based pollution, destructive fishing practices and changes over time. *Marine Pollution Bulletin*, **36,** 617–630.

Eitner, V., 1996: Geomorphological response of the East Frisian barrier islands to sea-level rise: an investigation of past and future evolution. *Geomorphology,* **15,** 57–65.

Ellison, J.C., 1993: Mangrove retreat with rising sea level, Bermuda. *Estuarine, Coastal and Shelf Science*, **37,** 75–87.

Ellison, J.C. and D.R. Stoddart, 1991: Mangrove ecosystem collapse during predicted sea-level rise: Holocene analogues and implications. *Journal of Coastal Research*, **7,** 151–165.

El-Raey, M., K.Dewidar, and M. El Hattab, 1999: Adaptation to the impacts of sea level rise in Egypt. *Climate Research (Special Issue)*, **6,** 117–128.

El-Raey, M., Y. Fouda, and S. Nasr, 1997: GIS assessment of the vulnerability of the Rosetta area, Egypt to impacts of sea rise. *Environmental Monitoring and Assessment*, **47(1),** 59–77.

El-Raey, M., O. Frihy, S.M. Nasr, and K.H. Dewidar, 1999: Vulnerability assessment of sea-level rise over Port Said governate, Egypt. *Environmental Monitoring and Assessment,* **56,** 113–128.

Estes, J.A., M.T. Tinkewr, T.M. Williams, and D.F. Doak, 1998: Killer whale predation on sea otters linking oceanic and nearshore ecosystems. *Science*, **282,** 473–475.

Everett, J.T., A. Krovnin, D. Lluch-Belda, E. Okemwa, H.A. Regier, and J.-P. Troadec, 1996: Fisheries. In: *Climate Change 1995: Impacts, Adaptations, and Mitigation of Climate Change: Scientific-Technical Analyses. Contribution of Working Group II to the Second Assessment Report of the Intergovernmental Panel on Climate Change* [Watson, R.T., M.C. Zinyowera, and R.H. Moss (eds.)]. Cambridge University Press, Cambridge, United Kingdom and New York, NY, USA, pp. 511–537.

Ewel, K.C., R.R. Twilley, and J.E. Ong, 1998: Different kinds of mangrove forests provide different goods and services. *Global Ecology and Biogeography Letters*, **7,** 83–94.

Falkowsky, P.G., R.T. Barber, V. Semtacek, 1998: Biogeochemical controls and feedbacks on ocean primary production. *Science*, **281,** 200–206.

FAO, 1998: *Yearbook of Fishery Statistics 1996 Capture Production*. Food and Agriculture Organization of the United Nations, Rome, Italy.

Farnsworth, E.J., 1998: Issues of spatial, taxonomic and temporal scale in delineating links between mangrove diverstiy and ecosystem function. *Global Ecology and Biogeography Letters*, **7,** 15–25.

Field, C.D., J.G. Osborn, L.L. Hoffman, J.F. Polsenberg, D.A. Ackerly, J.A. Berry, O. Bjorkman, A. Held, P.A. Matson, and H.A. Mooney, 1998: Mangrove biodiversity and ecosystem function. *Global Ecology and Biogeography Letters,* **7(1),** 3–14.

Fifita, N.P., N. Mimura, and N. Hori, 1992: Assessment of the vulnerability of the Kingdom of Tonga to sea-level rise. In: *Global Climate Change and the Rising Challenge of the Sea. Proceedings of the International Workshop held on Margarita Island, Venezuela, March 9-13, 1992*. National Oceanic and Atmospheric Administration, Silver Spring, MD, USA, pp. 119–139.

Forbes, D.L., 1996: *Coastal geology and hazards of Niue*. South Pacific Applied Geoscience Commission, Technical Report 233, Suva, Fiji, 100 pp. (plus appendices and map).

Forbes, D.L. and R.B. Taylor, 1994: Ice in the shore zone and the geomorphology of cold coasts. *Progress in Physical Geography*, **18,** 59–89.

Forbes, D.L., J.D. Orford, R.W.G. Carter, J. Shaw, and S.C. Jennings, 1995: Morphodynamic evolution, self-organization, and instability of coarse-clastic barriers on paraglacial coasts. *Marine Geology*, **126,** 63–85.

Forbes, D.L., J.D. Orford, R.B. Taylor, and J. Shaw, 1997a: Interdecadal variation in shoreline recession on the Atlantic coast of Nova Scotia. In: *Proceedings of the Canadian Coastal Conference '97, Guelph, Ontario*. Canadian Coastal Science and Engineering Association, Ottawa, ON, Canada, pp. 360–374.

Forbes, D.L., J. Shaw, and R.B. Taylor, 1997b: Climate change impacts in the coastal zone of Atlantic Canada. In: *Climate Variability and Climate Change in Atlantic Canada* [Abraham, J., T. Canavan, and R. Shaw, (eds.)]. Canada Country Study: Climate Impacts and Adaptation, Environment Canada, Ottawa, ON, Canada, **6,** 51–66.

Francis, R.C., S.R. Hare, A.B. Hollowed, and W.S. Wooster, 1998: Effects of interdecadal climate variability on the oceanic ecosystems of the North East Pacific. *Fisheries Oceanography,* **7(1),** 1–21.

Garcia, S.M. and D.J. Staples, 2000: Sustainability indicators in marine capture species: introduction to the special issue. *Marine and Freshwater Research,* **51,** 381–384.

Gargett, A.E., 1997: The optimal stability window: a mechanism underlying decadal fluctuations in North Pacific salmon stocks. *Fisheries Oceanography*, **6,** 109–117.

Gattuso, J.-P., D. Allemand, and M. Frankignoulle, 1999: Photosynthesis and calcification at cellular, organismal and community level in coral reefs: a review on interactions and control by carbonate chemistry. *American Zoologist,* **39,** 160–183.

Glynn, P.W., 1996: Coral reef bleaching: facts, hypotheses and implications. *Global Change Biology*, **2,** 495–509.

Gommes, R., J. du Guerny, F. Nachtergaele, and R. Brinkman, 1998: *Potential Impacts of Sea-Level Rise on Populations and Agriculture*. Food and Agriculture Organization of the United Nations, SD (Sustainable Development) Dimensions/Special. Available online at http://www.fao.org/WAICENT/FAOINFO/SUSTDEV/Eldirect/Elre0045.htm.

Goreau, T.J., 1992: Bleaching and reef community change in Jamaica: 1951–1991. *American Zoologist*, **32,** 683–695.

Gorman, L., A. Morang, and R. Larson, 1998: Monitoring the coastal environment; part IV: mapping, shoreline changes, and bathymetric analysis. *Journal of Coastal Research*, **14,** 61–92.

Gu, D. and S.G.H. Philander, 1997: Interdecadal climatic fluctuations that depend on exchanges between the tropics and extratropics. *Science*, **275,** 805–807.

Guley, S.K. and L. Hasse, 1999: Changes in wind waves in the North Atlantic over the last 30 years. *International Journal of Climatology*, **19,** 1091–1117.

Hanaki, K., K. Takara, T. Hanazato, H. Hirakuchi, and H. Kayanne, 1998: Impacts on hydrology/water resources and water environment. In: *Global Warming: The Potential Impact on Japan* [Nishioka, S. and H. Harasawa (eds.)]. Springer-Verlag, Tokyo, Japan, pp. 131–163.

Hansell, R.I.C., J.R. Malcolm, H. Welch, R.L. Jefferies, and P.A. Scott, 1998: Atmospheric change and biodiversity in the Arctic. *Environmental Monitoring and Assessment*, **49,** 303–325.

Harris, G.P., F.B. Griffiths, and L. Clementson, 1992: Climate and fisheries off Tasmania: interactions of physics, food chains and fish. *South African Journal of Marine Scoience*, **12,** 585–597.

Harvell, C.D., K. Kim, J.M. Burkholder, R.R. Colwell, P.R. Epstein, J. Grimes, E.E. Hofman, E. Lipp, A.D.M.E. Osterhaus, R.M. Overstreet, J.W. Porter, G.W. Smith, and G. Vasta, 1999: Emerging marine diseases: climate links and anthropogenic factors. *Science*, **285,** 1505–1510.

Harvey, N., B. Clouston, and P. Carvalho, 1999: Improving coastal vulnerability assessment methodologies for integrated coastal zone management. *Australian Geographical Studies,* **37(1),** 50–69.

Heinz Center, 1999: *The Hidden Costs of Coastal Hazards: Implications for Risk Assessment and Mitigation*. Island Press, Washington, DC, USA, 220 pp.

Heinz Center, 2000: *Evaluation of Erosion Hazard*. Report prepared for Federal Emergency Management Agency (FEMA). Heinz Center, Washington, DC, USA, 205 pp. Also available online at http://www.heinzctr.org/publications/erosion/erosnrpt.pdf.

Henderson-Sellars, A., H. Zhang, G. Berz, K. Emanuel, W. Gray, C. Landsea, G. Holland, J. Lighthill, S-L. Shieh, P. Webster, and K. McGuffie, 1998: Tropical cyclones and global climate change: a post-IPCC assessment. *Bulletin of the American Meteorological Society*, **79,** 19–38.

Héquette, A., M. Desrosiers, and D.L. Forbes, 1995: The role of shoreline configuration and coastal geomorphology in nearshore sediment transport under storm combined flows, Canadian Beaufort Sea. In: *Proceedings of the International Conference on Coastal Change*. Workshop Report 105, Intergovernmental Oceanographic Commission, Paris, France, pp. 563–570.

Hinch, S., G.M.C. Healey, R.E. Diewert, K.A. Thomson, R. Hourston, M.A. Henderson, and F. Juanes, 1995: Potential effects of climate change on marine growth and survival of Fraser River sockeye salmon. *Canadian Journal of Fisheries and Aquatic Sciences*, **52,** 2651–2659.

Hiyama, Y., H. Nishida, and T. Goto, 1995: Interannual fluctuations in recruitment of growth of the sardine, *Sardinops melanostictus*, in the Sea of Japan and adjacent waters. *Research in Population Biology,* **37(2),** 177–183.

Hoegh-Guldberg, O., 1999: Climate change, coral bleaching and the future of the world's coral reefs. *Marine and Freshwater Research,* **50,** 839–866.

Holt, T., 1999: A classification of ambient climatic conditions during extreme surge events off Western Europe. *International Journal of Climatology,* **19,** 725–744.

Hunt, G.L. and R.W. Furness, 1996: *Seabirds/Fish Interactions, with Particular Reference to Seabirds in the North Sea*. International Council for the Exploration of the Sea (ICES) Cooperative Research Report No. 216, 87 pp.

IPCC, 1998: *The Regional Impacts of Climate Change: An Assessment of Vulnerability. Special Report of IPCC Working Group II* [Watson, R.T., M.C. Zinyowera, and R.H. Moss (eds.)]. Intergovernmental Panel on Climate Change, Cambridge University Press, Cambridge, United Kingdom and New York, NY, USA, 517 pp.

Ittekkot, V., S. Jilan, E. Miles, E. Desa, B.N. Desai, J.T. Everett, J.J. Magnuson, A. Tsyban, and S. Zuta, 1996: Oceans. In: *Climate Change 1995: Impacts, Adaptations, and Mitigation of Climate Change: Scientific-Technical Analyses. Contribution of Working Group II to the Second Assessment Report of the Intergovernmental Panel on Climate Change* [Watson, R.T., M.C. Zinyowera, and R.H. Moss (eds.)]. Cambridge University Press, Cambridge, United Kingdom and New York, NY, USA, pp. 267–288.

Iwashima, T. and R. Yamamoto, 1993: A statistical analysis of the extreme events: Long-term trend of heavy daily precipitation. *Journal of the Meteorological Society of Japan*, **71,** 637–640.

Izrael, Yu.A. and A.V. Tsyban, 1983: *Anthropogenic Ecology of the Ocean*. Gidrometeoizdat, Leningrad, Russian Federation, 528 pp. (in Russian).

Jallow, B.P., S. Tours, M.M.K. Barrow, and A.A. Mthieu, 1999: Coastal zone of The Gambia and the Abidjan region in Côte d'Ivoir: sea level rise vulnerability, response strategies, and adaptation options. In: *National Assessment Results of Climate Change: Impacts and Responses* [Mimura, N. (ed.)]. *Climate Research (Special Issue)*, **6,** 137–143.

Jelgersma, S., 1996: Land subsidence in coastal lowlands. In *Sea-Level Rise and Coastal Subsidence* [Milliman, J.D. and B.U. Haq (eds.)]. J. Kluwer Academic Publishers, Dordrecht, The Netherlands, pp. 47–62.

Jimenez, J.A. and A. Sanchez-Arcilla, 1997: Physical impacts of climatic change on deltaic coastal systems (II): driving terms. *Climatic Change*, **35,** 95–118.

Jones, R.G., J.M. Murphy, M. Noguer, and S.B. Keen, 1997: Simulation of climate change over Europe using a nested regional-climate model: comparison of driving and regional model responses to a doubling of carbon dioxide. *Quarterly Journal of the Royal Meteorological Society,* **123,** 265–292.

Joos, F., G.K. Plattner, T.F. Stocker, O. Marchal, and A. Schmittner, 1999: Global warming and marine carbon cycle feedbacks on future atmospheric CO_2. *Science*, **284,** 464–467.

Kaminsky, G.M., P. Ruggiero, and G. Gelfenbaum, 1998: Monitoring coastal change in southwest Washington and northwest Oregon during the 1997/98 El Niño. *Shore & Beach*, **66(3),** 42–51.

Kaplin, P.A. and A.O. Selivanov, 1995: Recent coastal evolution of the Caspian Sea as a natural model for coastal responses to the possible acceleration of global sea-level rise. *Marine Geology,* **124,** 161–175.

Karl, T.R. and R.W. Knight, 1998: Secular trends of precipitation amount, frequency and intensity in the United States. *Bulletin of the American Meteorological Society,* **79,** 231–241.

Kawasaki, T., 1991: The relation of the change in pelagic fish populations and their environment. In: *Proceedings of the International Symposium* [Kawasaki, T. (ed.)]. Pergamon Press, Tokyo, Japan, pp. 47–60 (in Japanese).

Kawasaki, T., S. Tanaka, Y. Toba, and A. Taniguchi (eds), 1991*: Long-Term Variability of Pelagic Fish Populations and Their Environment*. Pergamon Press, Oxford, United Kingdom, 402 pp.

Kay, R. and J. Hay, 1993: A decision support approach to coastal vulnerability and resilience assessment: a tool for integrated coastal management. In: *Vulnerability Assessment to Sea-Level Rise and Coastal Zone Management. Proceedings of the IPCC Eastern Hemisphere Workshop, Tsukuba, Japan, 3–6 August 1993* [McLean, R.F. and N. Mimura (eds.)]. Department of Environment, Sport and Territories, Canberra, Australia, pp. 213–125.

Kent, M.L. and T.T. Poppe, 1998: *Diseases of Seawater Netpen-Reared Salmonid Fishes*. Fisheries and Oceans Canada, Pacific Biological Station, Nanaimo, BC, Canada, 137 pp.

Kimura, S., N. Munenori, and T. Sugimoto, 1997: Migration of albacore, *Thunnus alalunga*, in the North Pacific Ocean in relation to large oceanic phenomena. *Fisheries Oceanography*, **6(2),** 51–57.

King, S.E. and J.M. Lester, 1995: The value of salt-marsh as a sea defence. *Marine Pollution Bulletin*, **30(3),** 180–189.

Klein, R.J.T., E.N. Buckley, R.J. Nicholls, S. Ragoonaden, J. Aston, M. Capobianco, M.Mizutani, and P.D. Nunn, 2000: Coastal adaptation. In: *Methodologies and Technological Issues in Technology Transfer. A Special Report of IPCC Working Group III* [Metz, B., O.R. Davidson, J.W. Martens, S.N. van Rooijen, and L.Van Wie McGrory (eds.)]. Cambridge University Press, Cambridge, United Kingdom and New York, USA, pp.349–372.

Klein, R.J.T. and R.J. Nicholls, 1999: Assessment of coastal vulnerability to climate change. *Ambio,* **28(2),** 182–187.

Kleypas, J.A., R.W. Buddemeier, D. Archer, J. Gattuso, C. Landon, and B.N. Opdyke, 1999: Geochemical consequences of increased carbon dioxide in coral reefs. *Science,* **284,** 118–120.

Knutson, T.R., R.E. Tuleya, and Y. Kurihara, 1998: Simulated increase of hurricane intensities in a CO_2-warmed climate. *Science,* **279,** 1018–1020.

Komar, P.D., 1998a: *Beach Processes and Sedimentation.* Prentice Hall, Upper Saddle River, NJ, USA, 2nd. ed., 544 pp.

Komar, P.D., 1998b: The 1997–98 El Niño and erosion on the Oregon coast. *Shore & Beach,* **66(3),** 33–41.

Komar, P.D., J. Allan, G.M. Dias-Mendez, J.J. Marra, and P. Ruggiero, 2000: El Niño and La Niña: erosion processes and impacts. In: *Abstracts of the 27th International Conference on Coastal Engineering, July 16–22, 2000, Sydney, Australia.* Vol. 1, Paper 2. Australian Institution of Engineers, Australia, Barton, ACT, Australia

Kont, A., U. Ratas, and E. Puurmann, 1997: Sea-level rise impact of coastal areas of Estonia. *Climatic Change,* **36,** 175–184.

Kushmaro, A., Y. Loya, M. Fine, and E. Rosenberg, 1996: Bacterial infection and coral bleaching. *Nature,* **380,** 396.

Kushmaro, A., E. Rosenberg, M. Fine, Y. Ben Haim, and Y. Loya, 1998: Effect of temperature on bleaching of the coral *Oculina patagonica* by *Vibrio* AK-1. *Marine Ecology Progress Series,* **171,** 131–137.

Lanfredi, N.W., J.L. Pousa, and E.E.D. d'Onofrio, 1998: Sea-level rise and related potential hazards on the Argentine Coast. *Journal of Coastal Research,* **14(1),** 47–60.

Langenberg, H., A. Pfizenmayer, H. von Storch, and J. Sündermann, 1999: Storm-related sea level variations along the North Sea coast: natural variability and anthropogenic change. *Continental Shelf Research,* **19,** 821–842.

Leafe, R., J. Pethick, and I. Townsend, 1998: Realizing the benefits of shoreline management. *Geographical Journal,* **164,** 282–290.

Leatherman, S.P., K. Zhang, and B.C. Douglas, 2000: Sea level rise shown to drive coastal erosion. *Eos (Transactions American Geophysical Union),* **81(6),** 55–57.

Lehodey, P., J.M. Andre, M. Bertignac, J. Hampton, A. Stoens, C. Menkes, L. Memery, and N. Grima, 1998: Predicting skipjack tuna forage distributions in the equatorial Pacific using a coupled dynamical bio-geochemical model. *Fisheries Oceanography,* **7,** 317–325.

Lehodey, P., M. Bertignac, J. Hampton, A. Lewis, and J. Picaut, 1997: El Niño-Southern Oscillation and tuna in the western Pacific. *Nature,* **389,** 715–717.

Lehtonen, H., 1996: Potential effects of global warming on northern European freshwater fish and fisheries. *Fisheries Management and Ecology,* **3,** 59–71.

Levitus, S., J.I. Antonov, T.P. Boyer, and C. Stephens, 2000: Warming of the world ocean. *Science,* **287,** 2225–2229.

Lipp, E.K. and J.B. Rose, 1997: The role of seafood in foodborne diseases in the United States of America. *Reviews of Science and Technology,* **16(2),** 620–640.

Lorenzo, E. and L. Teixeira, 1997: Sensitivity of storm waves in Montevideo (Uruguay) to a hypothetical climate change. *Climate Research,* **9,** 81–85.

Maddrell, R.J., 1996: Managed coastal retreat, reducing flood risks and protection costs, Dungeness Nuclear Power Station, UK. *Coastal Engineering,* **28,** 1–15.

Mahnken, C., G. Ruggerone, W. Waknitz, and T. Flagg, 1998: A historical perspective on salmonid production from Pacific rim hatcheries. In: *Assessment and Stock Status of Pacific Rim Salmonid Stocks* [Welch, D.W., D.M. Eggers, K. Wakabayashi, and V.I. Karpenko (eds.)]. Bulletin Number 1, North Pacific Anadromous Fish Commission, Vancouver, BC, Canada, pp. 38–53.

Mantua, N.J., S.R. Hare, Y. Zhang, J.M. Wallace, and R.C. Francis, 1997: A Pacific interdecadal climate oscillation with impacts on salmon production. *Bulletin American Meteorological Society,* **78,** 1069–1079.

Marsh, R., B. Petrie, C.R. Weidman, R.R. Dickson, J.W. Loder, C.G. Hannah, K. Frank, and K. Drinkwater, 1999: The 1882 tilefish kill—a cold event in shelf waters off the north-eastern United States? *Fisheries Oceanography,* **8(1),** 39–49.

Marshall, P.A. and A.H. Baird, 2000: Bleaching of corals on the Great Barrier Reef: differential susceptibilities among taxa. *Coral Reefs,* **19,** 155–163.

Marton, R.M., 1990: Community structure, density and standing crop of fishes in a subtropical Australian mangrove area. *Marine Biology,* **105,** 385–394.

McGowan, J.A., D.R. Cayan, L.M. Dorman, 1998: Climate-ocean variability and ecosystem response in North Pacific. *Science,* **281,** 201–217.

Mearns, L.O., F. Giorgio, L. McDaniel, and C. Shields, 1995: Analysis of daily variability of precipitation in a nested regional climate model: comparison with observations and doubled CO_2 results. *Global Planetary Change,* **10,** 55–78.

Melloul, A.J. and L.C. Goldberg, 1997: Monitoring of seawater intrusion in coastal aquifers: basis and local concerns. *Journal of Environmental Management,* **51,** 73–86.

Michener, W.K., E.R. Blood, K.L. Bildstein, M.M. Brinson, and L.R. Gardner, 1997: Climate change, hurricanes and tropical storms and rising sea level in coastal wetlands (review). *Ecological Adaptations,* **7,** 770–801.

Miller, K.A., 2000: Pacific salmon fisheries: climate, information and adaptation in a conflict-ridden context. *Climatic Change,* **45,** 37–61.

Mimura, N. and H. Harasawa (eds.), 2000: *Data Book of Sea-Level Rise 2000.* Center for Global Environmental Research, National Institute for Environmental Studies, Tsukuba, Japan, 128 pp.

Mimura, N. and H. Nobuoka, 1996: Verification of the Bruun Rule for the estimation of shoreline retreat caused by sea-level rise. In: *Proceedings of Coastal Dynamics '95, Gdansk, Poland, September 4-8, 1995.* American Society of Civil Engineers, Reston, VA, USA, pp. 607–616.

Mimura, N., J. Tsutsui, T. Ichinose, H. Kato, and K. Sakaki, 1998: Impacts on infrastructure and socio-economic system. In: *Global Warming: The Potential Impact on Japan* [Nishioka, S. and H. Harasawa (eds.)]. Springer-Verlag, Tokyo, Japan, pp. 165–201.

Miyagi, T., C. Tanavud, P. Pramojanee, K. Fujimoto, and Y. Mochida, 1999: Mangrove habitat dynamics and sea-level change—a scenario and GIS mapping of the changing process of the delta and estuary type mangrove habitat in southwestern Thailand. *Tropics,* **8,** 179–196.

Moberg, F. and C. Folke, 1999: Ecological goods and services of coral reef ecosystems. *Ecological Economics,* **29,** 215–233.

Montevecchi, W.A. and R.A. Myers, 1997: Centurial and decadal oceanographic influences on changes in northern gannet populations and diets in the north-west Atlantic: implications for climate change. *International Council for the Exploration of the Sea Journal of Marine Science,* **54,** 608–614.

Moore, W.S., 1999: The subterranean estuary: a reaction zone of ground water and sea water. *Marine Chemistry,* **65,** 111–125.

Morisugi, H., E. Ohno, K. Hoshi, A. Takagi, and Y. Takahashi, 1995: Definition and measurement of a household's damage cost caused by an increase in storm surge frequency due to sea level rise. *Journal of Global Environment Engineering,* **1,** 127–136.

Mortsch, L.D., 1998; Assessing the impact of climate change on the Great Lakes shoreline wetlands. *Climatic Change,* **40,** 391–416.

Mullin, M.M., 1998: Interannual and interdecadal variation in California Current zooplankton: *Calanus* in the late 1950s and early 1990s. *Global Change Biology,* **4,** 115–119.

Mulrenna, M.E. and C.D. Woodroffe, 1998: Saltwater intrusion into the coastal plains of the Lower Mary River, Northern Territory, Australia. *Journal of Environmental Management,* **54,** 169–188.

Nairn, R.B., S. Solomon, N. Kobayashi, and J. Virdrine, 1998: Development and testing of a thermal-mechanical numerical model for predicting Arctic shore erosion processes. In: *Permafrost: Proceedings of the Seventh International Conference, Yellowknife, NWT, June 23-27, 1998* [Lewkowicz, A.G. and M. Allard (eds.)]. Centre d'études nordiques, Université Laval, Québec, Canada, pp. 789–795.

Neumann, J.E., G. Yohe, R. Nicholls, and M.Manion, 2000: *Sea-Level Rise and Global Climate Change: A Review of Impacts to U.S. Coasts.* Pew Center on Global Climate Change, Arlington, VA, USA, 32 pp.

Nicholls, R.J., 1998: Assessing erosion of sandy beaches due to sea-level rise. In: *Geohazard in Engineering Geology* [Maund, J.G. and M. Eddleston (eds.)]. Engineering Geology Special Publications, Geological Society, London, United Kingdom, **15,** 71–76.

Nicholls, R.J. and J. Branson, 1998: Coastal resilience and planning for an uncertain future: an introduction. *Geographical Journal,* **164(3),** 255–258.

Nicholls, R.J. and N. Mimura, 1998: Regional issues raised by sea-level rise and their policy implications. *Climate Research*, **11(1)**, 5–18.

Nicholls, R.J., F.M.J. Hoozemans, and M. Marchand, 1999: Increasing flood risk and wetland losses due to sea-level rise: regional and global analyses. *Global Environmental Change*, **9,** S69–S87.

Nunn, P.D. and N. Mimura, 1997: Vulnerability of South Pacific island nations to sea level rise. *Journal of Coastal Research (Special Issue)*, **24,** 133–151.

Nunn, P.D., J. Veitayaki, V. Ram-Bidesi, and A. Venisea, 1999: Coastal issues for oceanic islands: implications for human futures, *Natural Resources Forum*, **23,** 195–207.

Nurse, L.A., R.F. McLean, and A.G. Suarez, 1998: Small island states. In: *The Regional Impacts of Climate Change: An Assessment of Vulnerability. Special Report of IPCC Working Group II* [Watson, R.T., M.C. Zinyowera, and R.H. Moss (eds.)]. Intergovernmental Panel on Climate Change, Cambridge University Press, Cambridge, United Kingdom and New York, NY, USA, pp. 331–354.

Nuttle, W., M. Brinson, and D. Cahoon, 1997: Processes that maintain coastal wetlands in spite of rising sea level. *Eos (Transactions American Geophysical Union)*, **78(25),** 257–261.

O'Brien, C.M., C.J. Fox, B. Planque, and J. Casey, 2000: Climate variability and North Sea cod. *Nature*, **404,** 142.

Olivo, M.D., 1997: Assessment of the vulnerability of Venezuela to sea-level rise. *Climate Research*, **9(1–2),** 57–65.

Orson, R.A., R.S. Warren, and W.A. Niering, 1998: Interpreting sea level rise and rates of vertical marsh accretion in a southern New England tidal salt marsh. *Estuarine, Coastal and Shelf Science*, **47,** 419–429.

Panapitukkul, N., C.M. Duarte, U. Thampanya, P. Kheowvonngsri, N. Srichai, O. Geertz-Hansen, J. Terrados, and S. Boromthanarath, 1998: Mangrove colonization: mangrove progression over the growing Pak Phanang (SE Thailand) mud flat. *Estuarine, Coastal and Shelf Science*, **47,** 51–61.

Parkinson, R.W., R.D. DeLaune, and J.C. White, 1994: Holocene sea-level rise and the fate of mangrove forests within the wider Caribbean region. *Journal of Coastal Research*, **10,** 1077–1086.

Perez, R.T., L.A. Amadore, and R.B. Feir, 1999: Climate change impacts and responses in the Philippines coastal sector. *Climate Research (Special Issue)*, **6,** 97–107.

Peters, E.J., 1999: Native people and the environmental regime in the James Bay and Northern Quebec Agreement. *Arctic*, **52(4),** 395–410.

Phongsuwan, N., 1998: Extensive coral mortality as a result of bleaching in the Andaman Sea in 1995. *Coral Reefs*, **17,** 70.

Pirazzoli, P. and A. Tomasin, 1999: Recent abatement of easterly winds in the northern Adriatic. *International Journal of Climatology*, **19,** 1205–1219.

Pittock, A.B., 1999: Coral reefs and environmental change: adaptation to what? *American Zoologist*, **39(1),** 110–129.

Polovina, J.J., 1996: Decadal variation in the trans-Pacific migration of northern bluefin tuna (*Thunnus thynnus*) coherent with climate-induced change in prey abundance. *Fisheries Oceanography*, **5(2),** 114–119.

Polovina, J.J., G.T. Mitchum, and G.T. Evans, 1995: Decadal and basin-scale variation in mixed layer depth and the impact on biological production in the central and north Pacific, 1960–1988. *Deep Sea Research*, **42,** 1701–1716.

Pope, J., 1997: Responding to coastal erosion and flooding damages. *Journal of Coastal Research*, **13(3),** 704–710.

Reimnitz, E., P.W. Barnes, and J.R. Harper, 1990: A review of beach nourishment from ice transport of shoreface materials, Beaufort Sea, Alaska. *Journal of Coastal Research*, **6,** 439–470.

Ren, M., 1994: Relative sea-level rise in China and its socioeconomic implications. *Marine Geodesy*, **17,** 37–44.

Ribic, C.A., D.G. Ainley, and L.B. Spear, 1997: Scale-related seabird-environmental relationships in Pacific equatorial waters, with reference to El Niño-Southern Oscillation events. *Marine Ecology Progress Series*, **156,** 183–203.

Richardson, L.L., W.M. Goldberg, K.G. Kuta, R.B. Aronson, G.W. Smith, J.C. Halas, J.S. Feingold, and S.L. Miller, 1998: Florida's mystery coral-killer identified. *Nature*, **392,** 557–558.

Richmond, B.M., B. Mieremet, and T.E. Reiss, 1997: Yap Islands natural coastal systems and vulnerability to potential accelerated sea-level rise. *Journal of Coastal Research (Special Issue)*, **24,** 153–173.

Riggs, S.R., W.J. Cleary, and S.W. Snyder, 1995: Influence of inherited geologic framework on barrier shoreface morphology and dynamics. *Marine Geology*, **126,** 213–234.

Roemmich, D. and J. McGowan, 1995: Climate warming and the decline of zooplantkon in the California current. *Science*, **267,** 1324–1326.

Rönnbäck, P., 1999: The ecological basis for economic value of seafood production supported by mangrove ecosystems. *Ecological Economics*, **29,** 235–252.

Rothrock, D.A, Y. Yu, and G.A. Maykut, 1999: Thinning of the Arctic sea-ice cover. *Geophysical Research Letters*, **26,** 3469–3472.

Sabhasri, S. and K. Suwarnarat, 1996: Impact of sea level rise on flood control in Bangkok and vicinity. In: *Sea-Level Rise and Coastal Subsidence: Causes, Consequences and Strategies* [Milliman, J.D. and B.U. Haq (eds.)]. J. Kluwer Academic Publishers, Dordrecht, The Netherlands, pp. 343–355.

Saenger, P., 1998: Mangrove vegetation: an evolutionary perspective. *Marine and Freshwater Research*, **49(4),** 277–286.

Sagarin, R.D., J.P. Barry, S.E. Gilman, and C.H. Baxter, 1999: Climate-related change in an intertidal community over short and long time scales. *Ecological Monographs*, **69(4),** 465–490.

Saizar, A., 1997: Assessment of a potential sea-level rise on the coast of Montevideo, Uruguay. *Climate Research*, **9(1–2),** 73–79.

Sale, P.F., 1999: Coral reefs: recruitment in space and time. *Nature*, **397,** 25–27.

Sanchez-Arcilla, A. and J.A. Jimenez, 1997: Physical impacts of climatic change on deltaic coastal systems: 1: an approach. *Climatic Change*, **35,** 71–93.

Sandweiss, D.H., K.A. Maasch, D.F. Belknap, J.B. Richardson III, and H.B. Rollins, 1998: Discussion of Lisa E. Wells, 1996: The Santa Beach Ridge Complex. *Journal of Coastal Research*, **14,** 367–373.

Sarmiento, J.L. and C. Le Quéré, 1996: Oceanic carbon dioxide uptake in a model of century sale global warming. *Science*, **274,** 1346–1350.

Sarmiento, J.L., T.M.C. Hughes, R.J. Stouffler, and S. Manabe, 1998: Simulated response of the ocean carbon cycle to anthropogenic climate warming. *Nature*, **393,** 245–249.

Schlager, W., 1999: Scaling of sedimentation rates and drowning of reefs and carbonate platforms. *Geology*, **27,** 183–186.

Schubert, M., R. Blender, K. Fraedrich, F. Lunkeit, and J. Pertwitz, 1998: North Atlantic cyclones in CO_2-induced warm climate simulations: frequency, intensity and tracks. *Climate Dynamics*, **14,** 827–837.

Shaw, J., R.B. Taylor, S. Solomon, H.A. Christian, and D.L. Forbes, 1998a: Potential impacts of global sea-level rise on Canadian coasts. *The Canadian Geographer*, **42,** 365–379.

Shaw, J., R.B. Taylor, D.L. Forbes, M.-H. Ruz, and S.M. Solomon, 1998b: *Sensitivity of the Coasts of Canada to Sea-Level Rise.* Bulletin 505, Geological Survey of Canada, Ottawa, ON, Canada, 79 pp. (plus map).

Shepherd, S.A., J.R. Turrubiates-Morales, and K. Hall, 1998: Decline of the abalone fishery at La Natividad, Mexico: overfishing or climate change? *Journal of Shellfish Research*, **17(3),** 839–846.

Sherif, M.M. and V.P. Singh, 1999: Effect of climate change on sea water intrusion in coastal aquifers. *Hydrological Processes*, **13,** 1277–1287.

Short, A.D., A. Trembanis, and I. Turner, 2000: Beach oscillation, rotation and the Southern Oscillation: Narrabeen Beach, Australia. *Abstracts of the 27th International Conference on Coastal Engineering, July 16-22, 2000, Sydney, Australia.* Vol. 1, Paper 4. Australian Institution of Engineers, Australia, Barton, ACT, Australia.

Short, F.T. and H.A. Neckles, 1999: The effects of global climate change on seagrasses. *Aquatic Botany*, **63,** 169–196.

Singh, V.S. and C.P. Gupta, 1999: Groundwater in a coral island. *Environmental Geology*, **37,** 72–77.

Smith, R.C., D. Ainley, K. Baker, E. Domack, S. Emslie, B. Fraser, J. Kennett, A. Leventer, E. Mosley-Thompson, S. Stammerjohn, and M. Vernet, 1999: Marine ecosystem sensitivity to climate change. *Bioscience*, **49,** 393–404.

Snedaker, S.C., J.F. Meeder, R.S. Ross, and R.G. Ford, 1994: Discussion of Ellison and Stoddart, 1991. *Journal of Coastal Research*, **10,** 497–498.

Söderqvist, T., W.J.Mitsch, and R.K.Turner, 2000: Valuation of wetlands in a landscape and institutional perspective. *Ecological Economics*, **35,** 1–6.

Solomon, S.M. and D.L. Forbes, 1999: Coastal hazards and associated management issues on South Pacific islands. *Ocean & Coastal Management*, **42,** 523–554.

Solomon, S.M., D.L. Forbes, and B. Kierstead, 1994: *Coastal Impacts of Climate Change: Beaufort Sea Erosion Study*. Canadian Climate Centre, Report 94-2, Environment Canada, Downsview, Ontario, Canada, 34 pp.

Spalding, M.D. and A.M. Grenfell, 1997: New estimates of global and regional coral reef areas. *Coral Reefs*, **16,** 225–230.

Springer, A.M., 1998: Is it all climate change? Why marine bird and mammal populations fluctuate in the North Pacific. In: *Biotic Impacts of Extratropical Climate Variability in the Pacific* [Holloway, G., P. Muller, and D. Henderson (eds.)]. National Oceanic and Atmospheric Administration and the University of Hawaii, USA, pp. 109–120.

Springer, A.M., 1992: Walleye pollock in the North Pacific: how much difference do they really make? *Fisheries Oceanography*, **1,** 80–96.

Stanley, D.J. and A.G. Warne, 1998: Nile Delta in its destruction phase. *Journal of Coastal Research*, **14,** 794–825.

Steele, J.H., 1996: Regime shifts in fisheries management. *Fisheries Research*, **25,** 19–23.

Stein, E.D., F.Tabatabai, and R.F. Ambrose, 2000: Wetland mitigation banking: a framework for crediting and debiting. *Environmental Management*, **26(3),** 233–250.

Stewart, R.W., B.D. Bornhold, H. Dragert, and R.E. Thomson, 1998: Sea level change. In: *The Sea* [Brink, K.H. and A.R. Robinson (eds.)]. John Wiley and Sons, New York, NY, USA, **10,** 191–211.

Stirling, I., N.J. Lunn, and JJ. Iacozza, 1999: Long-term trends in the population ecology of polar bears in western Hudson Bay in relation to climatic change. *Arctic*, **52(3),** 294–306.

Stone, G.W. and R.A. McBride, 1998: Louisiana barrier islands and their importance in wetland protection: forecasting shoreline change and subsequent response of wave climate. *Journal of Coastal Research*, **14,** 900–915.

Storlazzi, C.D. and G.B. Griggs, 2000: Influence of El Niño-Southern Oscillation (ENSO) events on the evolution of central California's shoreline. *Geological Society of America Bulletin*, **112(2),** 236–249.

Streever. W.J., M. Callaghan-Perry, A. Searles, T. Stevens, and P. Svoboda, 1998: Public attitudes and values for wetland conservation in New South Wales, Australia. *Journal of Environmental Management*, **54,** 1–14.

Stumpf, R.P. and J.W. Haines, 1998: Variations in tidal level in the Gulf of Mexico and implications for tidal wetlands. *Estuarine, Coastal and Shelf Science*, **46,** 165–173.

Suppiah, R. and K.J. Hennessy, 1998: Trends in total rainfall, heavy rain events, and number of dry days in Australia, 1910–1990. *International Journal of Climatology*, **18(10),** 1141–1164.

Talge, H.K., E.M. Cockey, and R.G. Muller, 1995: A new disease in reef dwelling foraminifera: implications for coastal sedimentation. *Journal of Foraminiferal Research*, **25,** 280–286.

Timmermann, A., J. Oberhuber, A. Bacher, M. Esch, M. Latif, and E. Roeckner, 1999: Increased El Niño frequency in a climate model forced by future greenhouse warming. *Nature*, **398,** 694–696.

Titus, J.G., 1998: Rising seas, coastal erosion, and the takings clause: how to save wetlands and beaches without hurting property owners. *Maryland Law Review*, **57,** 1279–1399.

Titus, J., R. Park, S. Leatherman, J. Weggle, M. Greene, S. Brown, C. Gaunt, M. Trehan, and G. Yohe, 1992: Greenhouse effect and sea level rise: the cost of holding back the sea. *Coastal Management*, **19,** 219–233.

Trenberth, K.E., 1999: Conceptual framework for changes of extreme of the hydrological cycle with climate change. *Climatic Change*, **42,** 327–339.

Trenberth, K. and T. Hoar, 1996: The 1990–1995 El Niño-Southern Oscillation event: largest on record. *Geophysical Research Letters*, **23,** 57–60.

Trenhaile, A.S., 1997: *Coastal Dynamics and Landforms*. Clarendon Press, Oxford, United Kingdom, 366 pp.

Tsyban, A.V., 1999a: The BERPAC Project: development and overview of ecological investigations in the Bering and Chukchi Seas. In: *Dynamics of the Bering Sea* [Loughlin, T.R. and O. Kiyotaka (eds.)]. North Pacific Marine Science Organization, University of Alaska Press, Fairbanks, AK, USA, pp. 713–729.

Tsyban, A.V., 1999b: Ecological investigations in the Russian Arctic seas: results and perspectives. *Aquatic Conservation: Marine and Freshwater Ecosystems*, **9,** 503–508.

Tsyban, A.V., J.T. Everett, and J.G. Titus, 1990: World oceans and coastal zones. In: *Climate Change: The IPCC Impacts Assessment. Contribution of Working Group II to the First Assessment Report of the Intergovernmental Panel on Climate Change* [Tegart, W.J.McG., G.W. Sheldon, and D.C. Griffiths (eds.)]. Australian Government Publishing Service, Canberra, Australia, chapter 6, pp. 1–28.

Tunberg, B.G. and W.G. Nelson, 1998: Do climatic oscillations influence cyclical patterns of soft bottom macrobenthic communities on the Swedish west coast? *Marine Ecology Progress Series*, **170,** 85–94.

Turner, R.K., P. Doktor, and W.N. Adger, 1995: Assessing the economic costs of sea level rise. *Environment and Planning A*, **27,** 1777–1796.

Turner, R.K., S. Subak, and W.N. Adger, 1996: Pressures, trends, and impacts in coastal zones: interactions between socioeconomic and natural systems. *Environmental Management*, **20(2),** 159–173.

Turner, R.K., I. Lorenzoni, N. Beaumont, I.J. Bateman, I.H. Langford, and A.L. McDonald, 1998: Coastal management for sustainable development: analysing environmental and socio-economic changes on the UK coast. *Geographical Journal*, **164(3),** 269–281.

Tynan, C.T. and D.P. DeMaster, 1997: Observations and predictions of Arctic climate change: potential effects on marine mammals. *Arctic*, **50(4),** 308–322.

Umitsu, M., 1997: Landforms and floods in the Ganges Delta and coastal lowland of Bangladesh. *Marine Geodesy*, **20,** 77–87.

van de Plassche, O., K. van der Borg, and A.F.M. de Jong, 1999: Sea level-climate correlation during the past 1400 years: reply. *Geology*, **27,** 190.

van de Plassche, O., K. van der Borg, and A.F.M. de Jong, 1998: Sea level-climate correlation during the past 1400 years. *Geology*, **26,** 319–322.

Varekamp, J.C. and E. Thomas, 1998: Sea level rise and climate change over the last 1000 years. *Eos (Transactions American Geophysical Union)*, **79,** 69–75.

Varekamp, J.C., E. Thomas, and W.G. Thomson, 1999: Sea level-climate correlation during the past 1400 years: comment. *Geology*, **27,** 189–190.

Veit, R R., P. Pyle, and J.A. McGowan, 1996: Ocean warming and long-term change in pelagic bird abundance within the California current system. *Marine Ecology Progress Series*, **139,** 11–18.

Veitayaki, J., 1998: Traditional and community-based marine resources management system in Fiji: an evolving integrated process. *Coastal Management*, **26,** 47–60.

Vyalov, S.S., A.S. Gerasimov, and S.M. Fotiev, 1998: Influence of global warming on the state and geotechnical properties of permafrost. In: *Permafrost: Proceedings of the Seventh International Conference, Yellowknife, NWT, June 23-27,1998* [Lewkowicz, A.G. and M. Allard (eds.)]. Centre d'études nordiques, Université Laval, Québec, Canada, pp. 1097–1102.

Walsh, K.J.E. and J.J. Katzfey, 2000: The impact of climate change on the poleward movement of tropical cyclone-like vortices in a regional climate model. *Journal of Climate*, **13,** 1116–1132.

Walsh, K.J.E and A.B. Pittock, 1998: Potential changes in tropical storms, hurricanes, and extreme rainfall events as a result of climate change. *Climatic Change*, **39,** 199–213.

Ware, D.M., 1995: A century and a half of change in the climate of the North East Pacific. *Fisheries Oceanography*, **4,** 267–277.

Warrick, R.A., C. Le Provost, M.F. Meier, J. Oerlemans, and P.L. Woodworth, 1996: Changes in sea level. In: *Climate Change 1995: The Science of Climate Change. Contribution of Working Group I to the Second Assessment Report of the Intergovernmental Panel on Climate Change* [Houghton, J.T., L.G. Meira Filho, B.A. Callander, N. Harris, A. Kattenberg, and K. Maskell (eds.)]. Cambridge University Press, Cambridge, United Kingdom and New York, NY, USA, pp. 359–405.

WASA, 1998: Changing waves and storms in the Northeast Atlantic. *Bulletin of the American Meteorological Society*, Waves and Storms in the North Atlantic Group, **79(5),** 741–760.

Webster, P.J., A.M. Moore, J.P. Loschnigg, and R.R. Leben, 1999: Coupled ocean-temperature dynamics in the Indian Ocean during 1997–1998. *Nature*, **401,** 356–360.

Welch, D.W., Y. Ishida, and K. Nagasawa, 1998: Thermal limits and ocean migrations of sockeye salmon (*Oncorhynchus nerka*): long-term consequences of global warming. *Canadian Journal of Fish and Aquatic Science,* **55(1),** 937–948.

West, J.J. and H. Dowlatabadi, 1999: On assessing the economic impacts of sea-level rise on developed coasts. In: *Climate Change and Risk* [Downing, T.E., A.J. Olsthoorn. and R.S.J. Tols (eds.)]. Routledge, London, United Kingdom, pp. 205–220.

West, J.J., M.J.Small, and H. Dowlatabadi, 2001: Storms, investor decisions, and the economic impacts of sea-level rise. *Climatic Change,* **48,** 317–342.

Wijnberg, K.M. and J.H.J. Terwindt, 1995: Extracting decadal morphological behaviour from high-resolution, long-term bathymetric surveys along the Holland coast using eigenfunction analysis. *Marine Geology,* **126,** 301–330.

Wilcock, P.R., D.S. Miller, R.H. Shea, and R.T. Kerkin, 1998: Frequency of effective wave activity and the recession of coastal bluffs: Calvert Cliffs, Maryland. *Journal of Coastal Research,* **14,** 256–268.

Wilkinson, C.R., 1999: Global and local threats to coral reef functioning and existence: review and predictions. *Marine and Freshwater Research,* **50,** 867–878.

Wilkinson, C.R. and R.W. Buddemeier, 1994: *Global Climate Change and Coral Reefs: Implications for People and Reefs. Report of the UNEP-IOC-ASPEI-IUCN Global Task Team on the Implications of Climate Change on Coral Reefs.* International Union for Conservation of Nature and Natural Resources (IUCN), Gland, Switzerland, 124 pp.

Wilkinson, C.R., O. Linden, H. Cesar, G. Hodgson, J. Rubens, and A. Strong, 1999: Ecological and socioeconomic impacts of 1998 coral mortality in the Indian Ocean: an ENSO impact and a warning of future change? *Ambio,* **28(2),** 190–196.

Winter, A., R.S. Appeldoorn, A. Bruckner, E.H. Williams Jr., and C. Goenaga, 1998: Sea surface temperatures and coral reef bleaching off La Paguera, Puerto Rico (northeastern Caribbean Sea). *Coral Reefs,* **17,** 377–382.

Wolfe, S.A., S.R. Dallimore, and S.M. Solomon, 1998: Coastal permafrost investigations along a rapidly eroding shoreline, Tuktoyaktuk, NWT. In: *Permafrost: Proceedings of the Seventh International Conference, Yellowknife, NWT, June 23-27, 1998* [Lewkowicz, A.G. and M. Allard (eds.)]. Centre d'études nordiques, Université Laval, Québec, Canada, pp. 1125–1131.

Wood, C.M. and D.G. McDonald (eds.), 1997: *Global Warming: Implications for Freshwater and Marine Fish.* Cambridge University Press, Cambridge, United Kingdom and New York, NY, USA, 425 pp.

Woodroffe, C.D., 1995: Response of tide-dominated mangrove shorelines in northern Australia to anticipated sea-level rise. *Earth Surface Processes and Landforms,* **20,** 65–86.

Woodroffe, C.D., 1999: Response of mangrove shorelines to sea-level change. *Tropics,* **8,** 159–177.

Wyllie-Echeverria, T. and W.S. Wooster, 1998: Year-to-year variations in Bering Sea ice cover and some consequences for fish distributions. *Fisheries Oceanography,* **7(2),** 159–170.

Yamada, K., P.D. Nunn, N. Mimura, S. Machida, and M. Yamamoto, 1995: Methodology for the assessment of vulnerability of South Pacific island countries to sea-level rise and climate change. *Journal of Global Environment Engineering,* **1,** 101–125.

Yohe, G.W., 1990: The cost of not holding back the sea: toward a national sample of economic vulnerability. *Coastal Management,* **18,** 403–431.

Yohe, G.W. and J. Neumann, 1997: Planning for sea-level rise and shore protection under climate uncertainty. *Climatic Change,* **37(1),** 243–270.

Yohe, G.W. and M.E. Schlesinger, 1998: Sea level change: the expected economic cost of protection or abandonment in the United States. *Climatic Change,* **38,** 447–472.

Yohe, G.W., J. Neumann, P. Marshall, and H. Ameden, 1996: The economic cost of greenhouse-induced sea-level rise for developed property in the United States. *Climatic Change,* **32,** 387–410.

Young, I., 1999: Seasonal variability of the global ocean wind and wave climate. *International Journal of Climatology,* **19,** 931–950.

Zann, L., 1994: The status of coral reefs in south western Pacific islands. *Marine Pollution Bulletin,* **29,** 53–55.

Zeider, R.B., 1997: Climate change vulnerability and response strategies for the coastal zone of Poland. *Climatic Change,* **36,** 151–173.

Zhang, K., B.C. Douglas, and S.P. Leatherman, 2000: Twentieth-century storm activity along the U.S. east coast. *Journal of Climate,* **13,** 1748–1761.

7

Human Settlements, Energy, and Industry

MICHAEL SCOTT (USA) AND SUJATA GUPTA (INDIA)

Lead Authors:
E. Jáuregui (Mexico), J. Nwafor (Nigeria), D. Satterthwaite (UK), Y.A.D.S. Wanasinghe (Sri Lanka), T. Wilbanks (USA), M. Yoshino (Japan)

Contributing Author:
U. Kelkar (India)

Review Editors:
L. Mortsch (Canada) and J. Skea (UK)

CONTENTS

EXECUTIVE SUMMARY

Climate change will affect human settlements against a very dynamic background of other environmental and socioeconomic factors. Human settlements are expected to be among the sectors that could be most easily adapted to climate change, given appropriate planning and foresight and appropriate technical, institutional, and political capacity. This chapter covers the same general topic areas as Chapters 11 and 12 of the Second Assessment Report (SAR); however, this chapter analyzes a wider variety of settlement types, provides a specific assessment of uncertainty and confidence in findings and adaptive capacity in human settlements, and places much of the discussion in a context of development, sustainability, and equity (DSE) (see Munasinghe, 2000). Energy and industry are treated as part of settlements. Figure 7-1 characterizes the conclusions of this chapter on two dimensions: scientific support (evidence available in the literature to support the finding) and consensus in that literature. Both scales are described more fully in Box 1-1 of Chapter 1 and in Moss and Schneider (2000).

Major Effects on Human Settlements

Infrastructure would have increased vulnerability to urban flooding and landslides (established but incomplete). Detailed modeling of rainfall event frequency and intensity in the context of global warming has been linked to increased intensity and frequency of urban flooding, with considerable damage to infrastructure (see Chapters 4, 8, and 10–15). Although not definitive for any part of the world, the model-based analysis is plausible and demonstrates that flooding could be an increased threat for riverine settlements under climate change. More predictable is loss of snow pack in many regions, combined with more winter flooding. Landslides are a current threat in many hilly areas and could be more so with more intense rainfall events.

Tropical cyclones would be more destructive under climate change (established but incomplete). Close behind floods and landslides are tropical cyclones (hurricanes or typhoons), which could have higher peak intensity in a warmer world with warmer oceans (see Chapters 3 and 8, and TAR WGI Chapter 10). Tropical cyclones combine the effects of heavy rainfall, high winds, storm surge, and sea-level rise in coastal areas and can be disruptive far inland but are not as universally distributed as floods and landslides.

Water supplies for human settlements would be vulnerable to increased warming, dryness, and flooding (established but

Level of Agreement/Consensus	Amount of Evidence: Low	Amount of Evidence: High
High	**Established but Incomplete** • Increased vulnerability of infrastructure to urban flooding and landslides • Tropical cyclones more destructive • Fire danger to urban/wildland fringe infrastructure increased • Sea-level rise increases cost/vulnerability of resource-based industry • Water supplies more vulnerable	**Well-Established** • Sea-level rise increases cost/vulnerability of coastal infrastructure • Energy demand sensitive; parts of energy supply vulnerable • Local capacity critical to successful adaptation • Infrastructure in permafrost regions vulnerable
Low	**Speculative** • Fire damage to key resources increased • More hail and windstorm damage	**Competing Explanations** • Agroindustry and artisanal fisheries vulnerable • Heat waves more serious for human health, resources • Nonclimate effects more important than climate • Heat island effects increase summer energy demand, reduce winter energy demand • Increased air and water quality problems

Low — **Amount of Evidence (observations, model output, theory)** — High

Figure 7-1: Human settlements impacts, categorized by state of scientific knowledge.

incomplete). There is reasonable consensus among experts that settlements in regions of the world that already are water-deficient (e.g., much of north Africa, the Middle East, southwest Asia, portions of western North America, and some Pacific islands) would face still higher demands for water with a warmer climate, with no obvious low-cost ways in which to obtain increased physical supplies. Observations on current water supply balances tend to back up this conclusion (see Chapters 4 and 10–13). However, theory and model output, though consistent with this view, are too weak quantitatively to offer much support, especially for urban areas. Repeated flooding also could create water quality problems in other areas.

Fire danger in settlements could increase with climate change (speculative for resource-dependent settlements; established but incomplete for infrastructure). Examples include forested and wildland/urban fringes in boreal regions (e.g., Canada, Alaska, Russia) and in Mediterranean climates in both hemispheres (e.g., California, southern Spain and France, and Australia) that could be affected (see Chapters 11, 12, 13, and 15). Although general circulation model (GCM)-projected summer climate in many regions looks similar to the hot, dry "fire weather" in many warm years of recent memory and economic activity in forests sometimes is restricted to reduce fire danger, impacts on the resource base have not been demonstrated, research has not shown what future fuel loadings would be, and it is unclear whether future economic activity and settlement infrastructure would be more vulnerable to fire.

Hail and windstorm could cause more damage to settlements (speculative). Although there is potential for more (and more severe) extreme weather episodes in a warmer atmosphere, modeling and data have not demonstrated a higher incidence of storms or of more severe storms (see Chapters 3, 8, 12, and 15).

Agroindustry and artisanal fisheries are sensitive to and in many cases vulnerable to climate change (well-established overall; competing explanations in specific regions). This conclusion dates back to the First Assessment Report (FAR). Additional studies and analysis conducted in the past 10 years have modified the details of the conclusion but have not overturned it. As described in Chapter 5, agriculture itself is sensitive to climate change. In some cases, yields may be reduced by as much as several tens of percent as a result of hotter weather, greater evaporation, and lower precipitation in mid-continental growing regions in particular. However, other regions may benefit, with higher yields possible. Impacts on agricultural processors and suppliers would tend to follow the impacts on agriculture itself. Changes in ocean conditions from El Niño episodes have demonstrated that changes such as ocean warming have substantial impacts on the locations and types of species available for fisheries, especially artisanal fisheries, but other regions could benefit (see Chapters 5, 6, and 10–17).

Heat waves would have more serious effects on human health and productivity (competing explanations). The impact of heat waves is most severe on the weakest parts of the populations (old, chronically ill, very young) that are not acclimated, but effects on future overall death rates are less clear (see Chapters 9, 11,13, 14, and 15). Because anthropogenic warming is projected to be greater at night than during the day, it would deprive sufferers of nighttime relief. Projections for several temperate climates show increased risk of severe heat waves (Chapter 3). As the weather becomes very warm, economic productivity of unprotected and outdoor populations declines.

Sea-level rise increases the cost/vulnerability of infrastructure and coastal resource-based industry (well-established for infrastructure; established but incomplete for resources). Although the amount of sea-level rise to be expected as a result of global warming by any given date and in any given location is uncertain, some studies are beginning to discuss likely ranges and probability distributions (e.g., Titus and Narayanan, 1995). The sensitivity of human infrastructure in coastal zones to given levels of sea-level rise is backed by theory, model results, and data on current rates of increase. In addition, several industries—such as tourism and recreation (the principal industry in many island economies)—are dependent on coastal resources (see Chapters 6, 8, and 10–17). Effective types of adaptive responses also are known in some circumstances, but vulnerability with adaptation is difficult to assess because the capacity and will to respond are uncertain or in doubt in many instances.

Energy demand in some locations is sensitive, and parts of the supply system are vulnerable (well-established). Modeling, theory, data, and expert opinion all say that warming of 1–5°C would considerably reduce the amount of energy that would be needed to heat buildings at mid- and high latitudes and altitudes, whereas cooling energy use would increase (see Chapters 10–15 and 17). The net overall impact on energy use would depend on local circumstances. If temperature increases take place primarily at night and during winter months, heating demand would be smaller and the increase in demand for energy for cooling and irrigation would be somewhat smaller than otherwise. Future climate is expected to include more intense rainfall events (which would require more conservative water storage strategies to prevent flood damage), greater probability of water deficits (less hydroelectric production), and less precipitation falling as snow (less water available during warm months) (see Chapter 4). All three factors point to less (or, at least, less flexible) hydroelectric capacity at current powerhouses. Reduced flows in rivers and higher temperatures reduce the capabilities of thermal electric generation, and high temperatures may reduce transmission capabilities as well.

There will be increased air and water pollution impacts (competing explanations). Climate change could contribute to water pollution problems in human settlements through drought or flooding, although not by simple increases in flow (which offers more dilution for pollutants). If droughts and floods become more frequent (see Chapter 3), so would instances of poor water quality (see Chapters 4, 10–15, and 17). Air pollution could be exacerbated if climate change alters the stability of air sheds and permits greater buildup of atmospheric pollutants (see Chapters 10–15). However, the outcomes remain largely theoretical, unsupported by data or modeling.

Infrastructure in permafrost regions is vulnerable to warming (well-established). Data from circumpolar regions and model results suggest that permafrost areas would see some melting of permafrost. Permafrost melting is a threat to infrastructure in these regions because of increased landslides and loss of foundation stability for structures, as well as increased damage from freeze-thaw cycles, among other impacts. In addition, melting permafrost is thought to be a source of methane (CH_4) and carbon dioxide (CO_2) gases (see Chapters 15 and 16).

Heat island effects could increase heat stress, increase summer energy demand, and reduce winter energy demand (competing explanations). As discussed in Chapters 3 and 9, heat waves may increase in frequency and severity in a warmer world, leading directly to increases in mortality among sensitive populations that are not acclimated. Heat island effects exacerbate the oppressive effects of heat waves by increasing temperatures experienced in the summer by up to several °C; at the same time, increased demand for air conditioning increases the demand for electricity and the severity of the heat island itself through thermal electric production. Winter energy use for heating would be reduced by the same phenomenon (see Chapters 11, 13, 14, and 15). Effects in specific regions are far less clear.

Other Observations

Local capacity is critical to successful adaptation (well-established). Adaptation means local tuning of settlements to a changing environment, not just warmer temperatures. Urban experts are unanimous that successful environmental adaptation cannot occur without locally based, technically and institutionally competent, and politically supported leadership. Local adaptive capacity generally is strongly correlated with the wealth, human capital, and institutional strength of the settlement. In addition, capacity depends in part on the settlement's access to national resources. Attempts to impose environmental solutions on settlements from the international or national level frequently have been maladapted to local circumstances. The most effective sustainable solutions are strongly supported and often developed locally, with technical assistance and institutional support from higher level bodies (see Chapters 10, 11, 14, 17, and 18).

Nonclimate effects are likely to be more important than climate change (competing explanations). The effects of climate change would occur against a background of other socioeconomic and environmental change that is itself very uncertain and complex (see Chapter 3). Model results, the current rate of environmental change, and economic theory all suggest that climate would be a relatively small additional uncertainty for most human settlements. Climate change in isolation also is unlikely to be as important a factor for DSE effects as other aspects of development, such as economic and technological change. In combination with other stresses from other processes such as population growth, however, climate change is likely to exacerbate total stresses in a multi-stress context. Particularly important could be effects of climate change on equity because relatively advantaged parts of global and local societies are likely to have better coping capacities than less advantaged parts.

Managing growth to ensure that it is sustainable and equitably distributed currently is a greater problem for most countries than the impacts of climate change. However, some experts are not in agreement on this point for the future, pointing out that the economic models do not show climate feedback to the economy and that climate effects are so uncertain that they could well dominate in some regions, especially by the end of the 21st century.

7.1. Introduction and Purpose

Humans live in a wide variety of settlements, ranging from hunter-gatherer camps and villages of a handful of families to modern megacities and metropolitan regions of tens of millions of inhabitants. Settlement economic and social structure—and the components of infrastructure that support settlements: energy, water supply, transportation, drains, waste disposal, and so forth—have varying degrees of vulnerability to climate change and generally are evolving far more quickly than the natural environment. Settlements can be affected directly through changes in human health and infrastructure and indirectly through impacts on the environment, natural resources, and local industries such as tourism or agriculture. Furthermore, these effects on human settlements theoretically could lead to tertiary impacts such as altered land use, redistribution of population and activities to other regions, and altered trade patterns among regions, resulting in still further changes in natural resources and other activities. Tertiary effects, however, are largely speculative at the current state of knowledge. Some of these tertiary effects could be either positive or negative at the regional level.

7.1.1. Overview of the SAR

This chapter builds on Chapters 11 and 12 of the IPCC Second Assessment Report (IPCC, 1996), and on the findings in the *Special Report on Regional Impacts of Climate Change* (RICC) (IPCC, 1998). The SAR identifies the most vulnerable types of communities, many examples of which are documented in RICC. The SAR states that the most vulnerable communities are not only poorer coastal and agrarian communities in arid areas identified in the First Assessment Report in 1990; they also include a great variety of settlements, most of them informal or illegal and with a predominance of low-income residents, built on hazardous sites such as wetlands or steep hillsides in or around many urban areas in the developing world.

The SAR and RICC also identify two categories of climate-sensitive industries. Sectors with activities that are sensitive to climate include construction, transportation operations and infrastructure, energy transportation and transmission, offshore oil and gas, thermal power generation, water availability for industry, pollution control, coastal-sited industry, and tourism and recreation. Sectors in which economic activity is dependent on climate-sensitive resources are agroindustry, biomass, and other renewable energy.

The SAR notes that infrastructure typically is designed to tolerate a reasonable level of variability within the climate regime that existed when it was designed and built. However, climate change could affect both average conditions and the probability of extreme events.

This Third Assessment Report (TAR) confirms most of these conclusions. However, the analyses in the SAR and RICC are concerned mostly with identifying and documenting potential effects. The TAR assesses their relative importance and the certainty/confidence of the conclusions reached.

Although literature published since the SAR was issued has not changed the catalog of potential impacts, much more has been learned about the quantitative details of many of the effects, which are being studied more systematically than was true 5–10 years ago. The results are becoming somewhat more quantitative, and it is becoming possible to assign confidence ratings to many of the effects for the first time. More also is known concerning adaptation options. It is now possible to describe many of the options more quantitatively and in the context of development, sustainability, and equity (see Munasinghe, 2000). Energy, industry, and infrastructure are treated as part of settlements in the TAR.

7.1.2. Overview of Types of Effects

Human settlements integrate many climate impacts initially felt in other sectors and differ from each other in geographic location, size, economic circumstances, and technical, political, institutional, and social capacities. Climate affects human settlements by one of three major pathways, which provides an organizational structure for the settlements effects discussion in this chapter:

1) Changes in productive capacity (e.g., in agriculture or fisheries) or changes in market demand for goods and services produced in settlements (including demand from those living nearby and from tourism). The importance of this impact depends on the range of economic alternatives. Rural settlements generally depend on one or two resources, whereas urban settlements usually (but not always) have a broader array of alternative resources. Impacts also depend on the adaptive capacity of the settlement, which in turn depends on socioeconomic factors such as the wealth, human capital, and institutional capability of the settlement.
2) Physical infrastructure or services may be directly affected (e.g., by flooding). Concentration of population and infrastructure in urban areas can mean higher numbers of persons and value of physical capital at risk, although there also are many economies of scale and proximity that help to assure well-managed infrastructure and provision of services such as fire protection and may help reduce risk. Smaller settlements (including villages and small urban centers) and many larger urban centers in Africa and much of Asia, Latin America, and the Caribbean often have less wealth, political power, and institutional capacity to reduce risks in this way.
3) Populations may be directly affected through extreme weather, changes in health status, or migration. Extreme weather episodes may lead to changes in deaths, injuries, or illness. Health status may improve as a result of less cold stress, for example, or deteriorate as a result of more heat stress and disease.

The discussion of impacts on human settlements, energy, and industry that follows begins with a discussion of nonclimate trends that affect settlements. The discussion then assesses potential impacts of climate change on three general types of settlements: resource-dependent settlements; riverine, coastal, and steeplands settlements; and urban settlements. This discussion is followed by a discussion of impacts on the energy sector and industries that may be particularly affected by climate change and an assessment of potential impacts on infrastructure. The chapter next discusses management and adaptation issues and integration of impacts across sectors, and it closes with a review of science and information needs.

7.2. State of Knowledge Regarding Climate Change Impacts on Human Populations

The TAR differs from the two previous assessments in that the literature has begun to quantify several of the climate-related risks to human settlements that previously were identified only in qualitative terms. Additional attention and research has been devoted to adaptation mechanisms that provide resistance to climate-related impacts and ability to recover from them. Several economic and social trends that are specific to development and change in human settlements will interact with the effects of climate change in the future and may exacerbate or mitigate the effects of climate change alone.

7.2.1. Nonclimate Trends Affecting Vulnerability to Climate

Population growth: Except for parts of Europe and the Russian Federation, most regions are expected to experience population growth. Although *Special Report on Emission Scenarios* (SRES) marker scenarios in Chapter 3 do not span the entire realm of possibilities and have not been assigned probabilities, they do show that under plausible conditions, future regional population growth rates will range from modest (Europe and North America, where projected rates are just above or below replacement) to 3% or more (portions of Latin America and especially Africa).

Urbanization (proportion of population living in urban areas) is expected to continue, especially in the developing world. Close to half of the world's population now lives in urban areas, and the likely trend toward a more urban world means that the impacts of climate change on human settlements, if they occur, increasingly will affect urban populations. The most rapid urban growth rates are occurring in the developing world, where urban populations are estimated to be growing at 2.7% yr^{-1}, compared to 0.5% yr^{-1} in more developed regions (UN, 2000). There also is a growing concentration of population in cities with more than 1 million inhabitants. The number of such cities worldwide grew from 80 in 1950 to more than 300 by 1990 and is expected to exceed 500 by 2010 (UNCHS, 1996; UN, 2000). Most cities with more than 1 million inhabitants are now in the developing world, although—as in more developed regions—they are heavily concentrated in its largest economies (UNCHS, 1996).

Cities also are reaching unprecedented sizes. However, the future world may be less dominated by "megacities" (cities of more than 10 million population) than previously predicted. Megacities are likely to be smaller than previously predicted and still contain a small proportion of the world's population—less than 4% in 1990, the last date for which there is census data for most nations (UNCHS, 1996; UN, 2000). Most of the world's urban population live in the 40,000–50,000 urban centers with fewer than 1 million inhabitants (UNCHS, 1996). In 1990, cities with more than 1 million inhabitants had just more than one-third of the world's urban population and just more than one-seventh of its total population (UN, 2000). Urban population increases were particularly sharp in the second half of the 20th century in some regions where urbanization had been held down by policy, such as China (Institute of Land Development and Regional Economy, State Planning Committee, 1998). Trends toward urbanization mean that the impacts of climate change on human settlements in most countries, if they occur, increasingly will affect urban populations, not rural or traditional settlements.

Poverty is becoming increasingly urbanized, as a growing proportion of the population suffering from absolute poverty lives in urban areas. In more developed regions and in much of Latin America (e.g., 36% in Latin America—ECLAC, 2000), poverty is concentrated in urban areas. In other regions, the number of rural poor still exceeds the number of urban poor, although the proportion of absolute poor living in urban areas is growing. In addition, the scale and depth of urban poverty frequently is underestimated, in part because official income-based poverty lines are set too low in relation to the cost of living (or the income needed to avoid deprivation) in most urban centers and in part because no provision is made to include housing conditions, access to services, assets, and aspects of social exclusion within most government poverty definitions (Satterthwaite 1997). Where it occurs, urban poverty reduces the capacity of urban populations to take action to adapt to climate change; poverty also may exacerbate many of its effects.

Market systems and privatization increasingly are being used to provide new infrastructure and maintain older systems (World Bank, 1994), giving government a smaller direct role in providing infrastructure for energy, environmental residuals, communications, and other key urban services. Governments that are trying to adapt settlements to climate change increasingly may have to work indirectly through markets and regulation of private providers to adapt buildings and infrastructure to climate change.

Energy systems are changing in some places, helping to determine which mechanisms are salient in human settlements impacts (Schipper and Meyers, 1992; Hall *et al.*, 1993; World Energy Council, 1993a):

- Use of biomass fuels for cooking and space heating in many developing countries remains significant, which has added to deforestation and environmental destruction

in some places but not others (Leach and Mearns, 1989; Tiffen and Mortimore, 1992). Biomass growth may be stimulated by warming, if precipitation remains adequate, but may fall otherwise.
- The increase in natural gas use in Europe and North America (and nuclear power in France) over the past 2 decades has held down the rate of use of coal and oil and has reduced coal use by 20% in western Europe. Accelerated coal use is expected in developing Asia (EIA, 1998). Much of the increase is related to increasing electricity demand, which would be compounded by climate warming.
- An increasing market share for electricity is occurring in new homes in all regions. Between 1995 and 2020, the world's annual consumption of electricity is projected to rise from 12 trillion to 23 trillion kWh. The greatest increases are expected in developing Asia and in Central and South America (EIA, 1998). Climate warming in these regions would increase the demand for space cooling, which is primarily fueled by electricity, at the same time that rapid electrification already is stretching capacity.
- Air conditioning in the commercial sector already accounts for a greater proportion of final energy demand than in the residential sector in developed countries. Commercial sector energy use also is increasing as a percentage of the total in developing countries. Some of this is a result of computerization of commerce.

Transportation activity and associated energy consumption are growing very rapidly in nearly every region. Except for economies in transition, the amount of goods traveling by road increased between 1990 and 1996. The increases were 50% or more, and total paved roadways worldwide rose from 39 to 46% of the total (World Bank, 1999). In all Organisation for Economic Cooperation and Development (OECD) countries, car ownership continues to rise steadily, but much of the growth in vehicle ownership is expected in developing countries and transition economies—especially in east Asia and the Pacific, and especially in urban areas (World Resources Institute, 1996). This trend contributes to local air pollution (which can be exacerbated by warm weather episodes) and to greenhouse gas (GHG) emissions.

A poleward intensification of agricultural, forestry, and mining activities is occurring, resulting in increased population and intensified settlement patterns in Canada's mid-north, for example, and even in arctic areas. Climate change could profoundly affect settlements in these regions, if climate change is greater toward the poles (Cohen, 1997). For example, some arctic and subarctic activities such as mining depend on snow roads, which would have to be replaced with more conventional transport.

Impact of urban wealth: Many of the worst city-level problems—such as sanitation and water supply—have been addressed in high-income cities such as those in Europe and North America, but not in many developing world cities (WHO, 1992; Hardoy *et al.*, 2000; McGranahan and Satterthwaite, 2000). A wealthy city can more easily afford the public finance and administration required to regulate more perceptible forms of pollution than a poor one. However, although the ambient environment of high-income cities may be more benign in terms of health impacts of pollution, these cities exert a far greater toll on the regional and global environment (UNCHS, 1996).

7.2.2 Sensitivity and Vulnerability of Human Settlements to Direct and Indirect Impacts of Climate Change

This chapter highlights some of the key processes through which climate impacts could occur; individual regional chapters categorize settlements based on size, location, or complete coverage of the population.

As a result of research that has been done on settlements since the SAR and RICC, as well as additional interpretation of older research, it is becoming clearer where many of the key vulnerabilities of human settlements, energy, and industry occur, although it is still very difficult to provide more than qualitative guidance. Table 7-1 provides an overview of these vulnerabilities for the years between approximately 2050 and 2080; much of the available literature concentrates on the effects of climate change of a magnitude roughly corresponding to that time period. The table divides human settlements into general size categories and economic function in a hierarchy of settlements. The table emphasizes the most salient effects that appear to be characteristic of certain types of settlements and mechanisms that might make the settlements more or less sensitive to climate change.

Implications of climate change for development of settlements, energy, and industry are highly location-specific. For instance, as shown in Table 7-1, climate change is more likely to have important impacts on the development of settlements in resource-dependent regions or coastal or riverine locations. Most of the concerns are about possible negative impacts on development (e.g., on the comparative advantage of a settlement for economic growth compared with other locations), although impacts on some areas are likely to be positive. Impacts on sustainability depend very largely on how climate change interacts with other processes related to multiple stresses and opportunities—such as economic, demographic, and technological change—except in low-lying areas that may be subject to sea-level rise or polar regions whose physical conditions will be more directly affected by global warming. Equity effects are of considerable concern because the ability to cope with negative impacts or to take advantage of positive impacts is likely to be greater among advantaged groups than among disadvantaged groups, within regions and between regions. As a result, climate change has the potential to enlarge equity-related gaps in human settlements and systems.

In general, country studies that have been completed since the SAR was published have provided more specific regional details concerning sensitivities and vulnerabilities to climate

***Table 7-1**: Impacts of climate change on human settlements, by impact type and settlement type (impact mechanism). Typeface indicates source of rating: Bold indicates direct evidence or study; italics indicates direct inference from similar impacts; and plain text indicates logical conclusion from settlement type, but cannot be directly corroborated from a study or inferred from similar impacts. Impacts generally are based on $2xCO_2$ scenarios or studies describing the impact of current weather events (analogs) but have been placed in context of the IPCC transient scenarios for the mid- to late 21st century.*

	Type of Settlement, Importance Rating, and Reference												
	Resource-Dependent (Effects on Resources)				*Coastal-Riverine-Steeplands (Effects on Buildings and Infrastructure)*				*Urban 1+ M (Effects on Populations)*		*Urban <1 M (Effects on Populations)*		
Impact Type	Urban, High Capacity	Urban, Low Capacity	Rural, High Capacity	Rural, Low Capacity	Urban, High Capacity	Urban, Low Capacity	Rural, High Capacity	Rural, Low Capacity	High Capacity	Low Capacity	High Capacity	Low Capacity	**Confidence**
Flooding, landslides	**L–M[1]**	**M–H[2]**	**L–M[1]**	**M–H[2]**	**L–M[1]**	**M–H[2]**	**M–H[1]**	**M–H[2]**	**M[1]**	**M–H[2]**	**M[1]**	**M–H[2]**	H
Tropical cyclone	**L–M[3]**	**M–H[4]**	**L–M[3]**	**M–H[4]**	**L–M[3]**	**M–H[4]**	**M[3]**	**M–H[4]**	**L–M[3]**	**M[4]**	**L[3]**	**L–M[4]**	M
Water quality	L–M	M	L–M	M–H	**L–M[5]**	**M–H[6]**	*L–M*	*M–H*	L–M	M–H	L–M	M–H	M
Sea-level rise	**L–M[7]**	**M–H[6]**	**L–M[7]**	**M–H[6]**	**M[8]**	**M–H[9]**	*M*	**M–H[6]**	**L[8]**	**L–M[6]**	*L*	**L–M[6]**	H (L for resource-dependent)
Heat/cold waves	*L–M*	*M–H*	*L–M*	*M–H*	*L–M[10]*	L–M	**L–M[10]**	L	**L–M[10]**	**M–H[11]**	**L–M[10]**	**M–H[11]**	M (H for urban)
Water shortage	**L[12]**	*L–M*	**M[12]**	**M–H[13]**	*L*	*L–M*	*L–M*	*M–H*	*L*	*M*	**L–M[12]**	*M*	M (L for urban)
Fires	*L–M*	*L–M*	**L–M[14]**	*M–H*	*L–M*	*L–M*	*L–M*	*L–M*	**L–M[15]**	**L–M[16]**	*L–M*	*M*	VL (M for urban)
Hail, windstorm	**L–M[17]**	**L–M[18]**	**L–M[17]**	**M–H[18]**	*L–M*	*L–M*	*L–M*	*M*	**L–M[17]**	**L–M[18]**	**L–M[17]**	**L–M[18]**	L
Agriculture/ forestry/fisheries productivity	**L–M[19]**	**L–M[20]**	*L–M*	*M–H*	L	L	L	L	*L*	*L–M*	*L–M*	*M*	L
Air pollution	**L–M[21]**	*L–M*	L	L	—	—	—	—	**L–M[10]**	**M–H[22]**	**L–M[10]**	**M–H[22]**	M
Permafrost melting	L	L	**L–M[23]**	*L–M*	L	L	**L[23]**	L	—	—	*L–M*	*L–M*	H
Heat islands	L	L	—	—	L	L	—	—	**M[24]**	**L–M[24]**	**L–M[25]**	**L–M[25]**	M

change (e.g., IPCC, 1998; see Chapters 10–17). Because of variability in settlements across the world, it is virtually impossible to create rankings of impacts that do not contain numerous exceptions. However, the impact ratings in Table 7-1 provide a framework that can be adapted to local circumstances. Table 7-1 shows the author team's judgments, based on the available literature, about the vulnerability of different types of settlement to various aspects of climate change. The horizontal axis differentiates vulnerability according to type of settlement, capacity to adapt, and the mechanism through which the settlement is affected by climate change. For example, the resource base of settlements that are economically dependent on activities such as agriculture, forestry, fishing, hunting and gathering, or tourism may be affected; housing and infrastructure may be affected in coastal areas, riverine floodplains, islands that are sensitive to flooding, steeplands that are sensitive to landslides, and urban/wildland boundaries that are sensitive to fires; and the health and productivity of urban populations may be affected directly through air pollution, heat waves, and heat island effects. The vertical axis identifies 12 different types of climate change impact in descending order of global importance. Vulnerabilities are rated as low, medium, or high magnitude as described in Box 7-1. The information in Table 7-1 generally is presented as a range, reflecting the diversity of settlements within each broad class. The final column shows the level of confidence that the author team assigns to each type of climate impact. Table 7-1 depicts vulnerabilities for the years between approximately 2050 and 2080. Much of the available human settlements literature is silent on the timing of impacts; the choice of the years 2050–2080 in Table 7-1 is based on the size of the impacts or amount of climate change addressed in the literature reviewed by the author team. Table 7-1 takes into account the number and type of settlements affected worldwide and the likely strength of these effects by mid-to-late 21st century, as well as the financial, technical, and institutional capacity of settlements to respond. Figure 7-2 provides confidence scores for the impacts on individual scales described more fully in Box 7-1 (see also Moss and Schneider, 2000).

The negative impacts in Table 7-1 generally would be less negative or even positive in some regions before 2050 but greater than shown and becoming more negative in more regions after 2100. The table is not intended to show that only specific types of settlements would be harmed (or helped) in certain ways by certain changes; it is intended to show that settlements of certain types probably are likely to be affected by certain impact mechanisms and are likely to be particularly vulnerable to certain types of climate changes or conditions.

Many of the effects in Table 7-1 are quite likely for some communities in some places; other effects are extremely uncertain, controversial, or inapplicable. Key articles that underlie the ratings are provided as footnotes to Table 7-1.

Confidence in the main conclusions of this chapter in Table 7-1 is rated in Figure 7-2 from very high (5) to very low (1) in four dimensions: support from theory, support from model results, support from data or trends in the existing environment, and the degree of consensus in expert opinion. Although these ratings reflect the subjective judgments of the chapter's authors concerning the weight that can be given to each element that increases confidence in the findings, the figure is useful in depicting the dimensions of the underlying literature that are particularly strong or weak in support of the chapter's conclusions. Confidence levels vary widely:

- Results that are very high on all dimensions, as in the expected vulnerability of at least some coastal settlements to sea-level rise
- Results that are very strong on most dimensions, such as local capacity being very important in practice for successful adaptation to environmental problems (even though the theory has not really been applied to climate change)
- Results that are very high on one or two dimensions, such as human health effects, where theory and model results are strongly supportive of the conclusion, whereas data are weaker or ambiguous and experts are somewhat divided
- Results for which there is some evidence, but most of it is only modestly supportive; one or two modestly

≺ Table 7-1 Notes

1. Changnon (1996b), Yohe *et al.* (1996), Evans and Clague (1997), FEMA (1997), Smith *et al.* (1999); **2.** Choudhury (1998), Rosquillas (1998), Magaña (1999); **3.** Landsea *et al.* (1996), Pielke (1996), Pielke and Landsea (1998); **4.** Yohe *et al.* (1996), Hurricane Mitch cost Honduras 80% of its GDP and Nicaragua 49% (FAO, 1999), Swiss Re (1999); **5.** in general, wealthier areas substitute new locations from which to draw water (WG2 SAR Section 10.5.4; Changnon and Glantz, 1996; Arnell, 1998); **6.** Meehl (1996), Nicholls and Hoozemans (1996), Nicholls and Mimura (1998); **7.** Mimura *et al.* (1998); **8.** FEMA (1991), Scott (1996), Rosenzweig and Solecki (2000); **9.** Ren (1994), Nicholls *et al.* (1999), see also Chapters 6 and 11; **10.** Phelps (1996), Chestnut *et al.* (1998), Duncan *et al.* (1999), Kerry *et al.* (1999); **11.** despite acclimatization, Indian cities have lost dozens to hundreds of people to heat-related deaths in recent years—more than 1,300 in 1998 (De and Mukhopadhyay, 1998); **12.** Wheaton and Arthur (1989), Rosenberg (1993), Lettenmaier *et al.* (1998), Gleick (2000); **13.** Meehl (1996), Scott (1996), Lewis *et al.* (1998); **14.** Hirsch (1999); **15.** the 1991 Oakland Hills and the 1994 Sydney fires are examples of losses sustained at urban interface in developed countries [in Oakland, a wildfire destroyed approximately 600 ha and more than 2,700 structures in the hills surrounding East Bay, took 25 lives, and caused more than US$1.68 billion in damages (see <www.firewise.org>, sponsored by the U.S. Forest Service); in the Sydney area, 800,000 ha burned, more than 200 houses—mostly in urban areas—were destroyed, and two firefighters and two civilians were killed (see Australian National University's FIRENET Web site)]; **16.** EEPSEA (2000), Wheeler (2000); **17.** Andrey and Mills (1999), Dorland *et al.* (1999), Changnon (2000); **18.** for example, on the Indian subcontinent on 26 April 1989, a single severe storm—locally known as "nor'westers" or kal'boishakhi—and a tornado north of Dhaka killed 1,300 and injured 12,000; **19.** Rosenberg (1993); **20.** Meltzoff *et al.* (1997); **21.** Scott (1996); **22.** WRI (1999); **23.** Cohen (1997), Andrey and Mills (1999); **24.** Quattrochi (1996), Chestnut *et al.* (1998); **25.** Jáuregui (1997), Chestnut *et al.* (1998), Lam (1999).

Box 7-1. Development of Scales for Assessing Potential Vulnerability of Human Settlements to Effects of Climate Change and Confidence in the Certainty of Impacts

Climate affects the stability of resources that support human systems. One way to assess the potential impact of climate change on human systems is by using a qualitative scale that expresses the vulnerability of settlements to various kinds of climate effects (e.g., floods) in terms of how potentially disruptive these climate effects are expected to be for various types of human settlements (based on differences in their economic base, location, size, and adaptability). The definitions in the rating system below are derived from standard environmental impact assessment language and are intended to apply to local climate impacts. However, the scale may be used nationally if the nation is small and homogeneous or if most of the population lives in settlements of a certain type.

Magnitude Ratings (Size of Impacts)

- *Low:* Impacts of changed climate are not distinguishable from normal background variability in weather impacts or there is little noticeable effect.
- *Moderate:* Resources or sectors are affected noticeably, even substantially, but the effect is not destabilizing and recovery is rapid.
- *High:* Impacts are large and sometimes catastrophic. Resources or settlements are destabilized, with little hope for near-term recovery.

A semi-quantitative approach is used with a 5-point confidence scale to indicate the certainty of the effects of climate change. The author team subjectively rated confidence on the basis of the literature in four dimensions: consensus among experts (*consensus*), the extent to which underlying theory and data is developed (*theory*), the quality of model results (*model results*), and the consistency of observational evidence (*observations*). The scores were used to create a four-sided polygon, as shown in Figure 7-2 on the facing page. All four dimensions were weighted equally to determine the area of the polygon and an overall confidence score.

$$\textit{Polygon Area} = 0.5 \bullet (\textit{Theory} \bullet \textit{Observations} + \textit{Observations} \bullet \textit{Model Results} + \textit{Model Results} \bullet \textit{Consensus} + \textit{Consensus} \bullet \textit{Theory})$$

The overall confidence score assigned was based on the area of the polygon. For example, to rate a 4 for "high confidence," the polygon had to have an area between 16 and 25—the area of a polygon with ratings of greater than 4 but less than 5 on all four dimensions.

Confidence Rating (Certainty of Impacts)

1) *Very Low:* Impacts are extremely difficult to predict (confidence < 5%) (*Polygon Area* = 0–8).
2) *Low:* Impacts are regularly much greater or less than the median value (confidence < 33%) (*Polygon Area* = 8–18).
3) *Medium:* Impacts are regularly greater or less than the median value (confidence ≥ 33%) (*Polygon Area* = 18–32).
4) *High:* There is noticeable variation in the size of impacts (confidence ≥ 67%) (*Polygon Area* = 32–50).
5) *Very High:* There is little variation in impact among scenarios, within a settlement type (confidence = 95%) (*Polygon Area* = 50).

strong elements lead to the conclusion, but confidence is weak.

7.2.2.1. Resource-Dependent Settlements

What makes resource-dependent settlements unique is the extent to which they are dependent on and tuned to the natural resources of the region and the extent to which they are vulnerable to changes in these natural resources. Resource dependency emphasizes the impact of climate change on the economic livelihood of inhabitants. An extreme form of resource-dependent settlements are settlements of traditional peoples, including hunter-gatherer communities, subsistence agricultural settlements, artisanal fishing communities, and the like. The issues are somewhat different in different locations, as indicated below.

7.2.2.1.2. Arid and semi-arid regions

Human societies have been developing by adaptation to the arid environment in desert regions for centuries. An example is the oasis water supply system called *karez, qanats,* or *foggaras*. For adaptation or mitigation, windbreaks are a powerful method to reduce the effects of the strong winds, dust storms, and sand dune movements (Du *et al.*, 1996; Maki *et al.*, 1997a,b). In addition, trees in windbreaks provide materials for

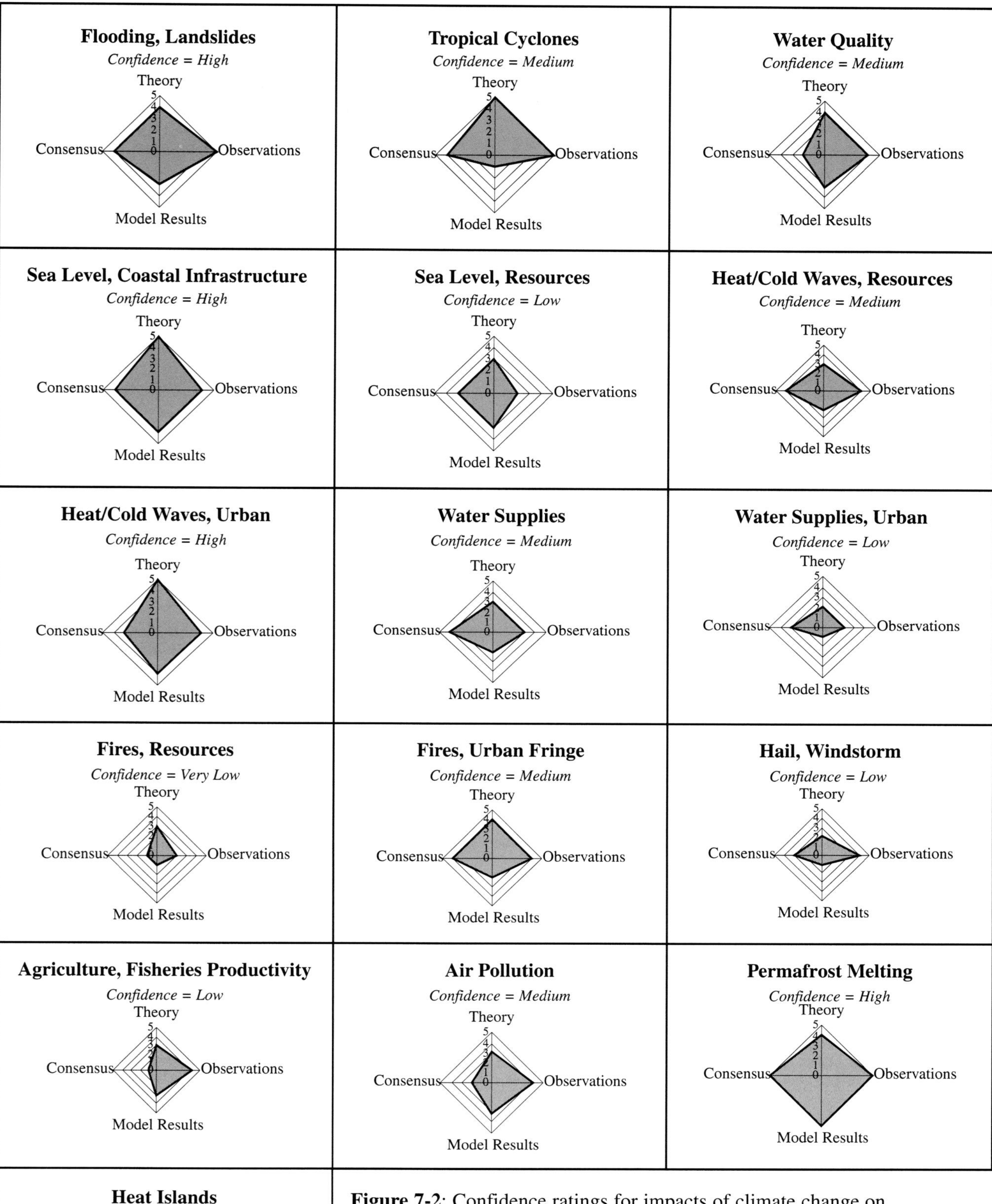

Figure 7-2: Confidence ratings for impacts of climate change on human settlements.

construction and firewood for people, keep a better hygienic environment in the settlements, and improve conditions for crops and livestock.

Social vulnerabilities in arid regions include low income among residents as a result of underdevelopment of industry in the oases, difficulty in telecommunication and transportation between oases, and growing human pressure on the limited land base, as illustrated by the scarcity of suitable irrigable land in oases in the Taklimakan Desert, northwest China (Yoshino, 1997c, 1998). On the northern rim of the Taklimakan desert, only 4–4.5% of the land is available for irrigation, and only about 2% on the south rim is available. Although these limit values vary from desert to desert because of the amount of available water, scarcity of suitable land is a vulnerable feature of settlements in all arid regions. In northern Africa, southwest Asia, and parts of the Middle East, there is a belt of entire countries that currently suffer from inadequate water supplies (see Chapter 4). Increasing temperatures and reduction in rainfall would further limit these settlements.

Climate change could reduce water availability in the semi-arid savanna ecosystem of tropical Africa, affecting farmers, herders, and tourist-industry workers, which in turn will impact human settlements. Conflict already occurs between herdsmen and farmers in this region, and the SAR discussed the impacts on major cities when people leave the land (see Chapter 10).

Small settlements in arid/semi-arid regions occasionally confront a higher risk of damage by flooding than their counterparts in more humid environments. This is usually because of the longer return periods or rarity of extreme rainfall. However, in a warmer world the frequency of these intense storms in semi-arid and arid regions may increase (Smith, 1996; Smith and Handmer, 1996; Smith *et al*., 1999). During extreme rainfall events, houses, roads, irrigation systems, and other constructions are destroyed and human settlements in oases are isolated because telecommunication and traffic connections are broken. Because settlements have been adapted to dry conditions, the total numbers of deaths and the total amount of damages caused by drought and other impacts of dry conditions often are smaller than those caused by heavy rainfall.

7.2.2.1.2. Polar and subpolar settlements

There are two major types of polar and subpolar settlements: traditional indigenous communities that are based on hunting and gathering activities such as whaling, caribou hunting, and seal hunting and modern "outpost" settlements such as military, mining, and oil camps. The more traditional settlements are economically vulnerable to changes in the regional ecology that might occur as a result of climate change (changes in sea ice, migration routes, or abundance of game species). Although military and mining operations generally would not be concerned with game species, impacts of warming on permafrost areas, sea lanes, and flying weather could significantly improve or reduce the efficiency of resource extraction (Cohen, 1997). Infrastructure in both types of settlements is vulnerable to permafrost melting, desiccation during warming, landslides, and flooding (see Chapter 16).

7.2.2.1.3. Forest settlements

Traditional settlements of indigenous peoples still exist in (mostly tropical) portions of Africa, Asia and nearby islands, and Latin America. Although these societies engage in subsistence agriculture and some cash activities such as guiding tourists, much of their economy is based on subsistence hunting and gathering. Already under threat from growth in farming, mining, and commercial forestry activities (which may themselves be affected by climate change), under climate change the traditional forest communities would face the additional challenge of changed ecology, which could change the availability of key species and adversely affect the sustainability of these communities.

7.2.2.1.4. Agricultural and fisheries settlements and related industry

Agroindustry is considered to be among the more adaptable industries affected by climate change. Brown and Rosenberg (1996) and Brown *et al*. (1998) were able to show, for example, that growing switchgrass (*Panicum virgatum*) for biomass is an effective adaptation in the warmer, drier climates expected in central North America under climate change. Chapter 5 also describes agriculture as adaptable and discusses adaptive responses. However, the capacity to adapt varies significantly among regions and among groups of farmers within regions. Adaptability and sustainability of agriculture and agroindustry depend not only on resources such as favorable climate, water, soils, and amount of land but also on the adaptability of whole market chains, which in turn are influenced by wealth levels, technical and financial sophistication, and institutional flexibility and strength (Tweeten, 1995; Nellis, 1996).

Changes in ocean conditions from El Niño episodes have demonstrated that changes such as ocean warming have substantial impacts on the locations and types of marine species available (UNEP, 1989; Meltzoff *et al*., 1997; Suman, 2001). Fishery-based settlements would be strongly affected as fishermen follow high-valued stocks to other locations, make do with lower valued stocks, or even abandon fishing altogether (see Chapters 6, 10, 14, 15, and 17). Other communities may benefit if high-valued stocks become more accessible.

What often is not appreciated about resource-dependent settlements is the complexity of the economic and social relationships among participants. Changes in the technology of farming, transportation, and communications have integrated farming communities into national and world markets as never before and increasingly have made farming a part-time activity in some parts of the developed world (e.g., Sakamoto, 1996; Sasaki, 1996) and the developing world. The complexity of interactions and the increasing degree of integration offer additional means of adaptation to climate change at the regional level, but these factors also mean that agriculture and agroindustry are being challenged from several directions at once, possibly complicating the impact of climate change alone (see, for example, Chapter 10).

To illustrate some of the complexities involved, in Japan agrarian structures have changed drastically during the past 30 years (Kumagai, 1996). Small farms have evolved through cooperative groups facing environmental requirements, double-cropping, cheaper imports, and high national economic growth (Sakamoto, 1996) into a regional system of farm households, hamlets, traditional villages, villages, and towns (Yoshino, 1997a,b), with significant part-time labor (Sasaki, 1996), and economics of large-scale production in farming, reductions in investment risks costs, diversification within farming, diversification between farming and other activities, and increases in nonfarm income.

Japanese rice-growing illustrates some constraints on adaptation and community sustainability. Inefficient agriculture (ironically) ties the younger generation to the rural community, where they play important roles in making the community active and viable. Provision of modern facilities and infrastructure does not lead to sustainable communities; more important, one must consider residents' pride or attachment to the community. Realization of a sustainable community requires investments in time, talent, and money for the future of the community (Tabayashi, 1996). Autonomous adaptation here, as elsewhere, might involve numerous social and economic dimensions that involve the entire settlement. It is unlikely to be simple or straightforward and may not result in sustaining every settlement.

7.2.2.1.5. Tourism and recreation

The SAR noted that tourism—a major and growing industry in many regions—will be affected by changes in precipitation patterns, severely affecting income-generating activities (IPCC, 1998). The outcome in any particular area depends in part on whether the tourist activity is summer- or winter-oriented and, for the latter, the elevation of the area and the impact of climate on alternative activities and destinations. For example, in spring 1997, when conditions in alternative destinations in the Alps were poor, the number of skiers in the High Atlas in Morrocco increased (Parish and Funnell, 1999). Scotland has been predicted to have less snow cover at its lower elevation ski areas with global warming but may have drier and warmer summers for hill-walking and other summer activities (Harrison *et al*., 1999) (see regional chapters for other examples).

The impacts of sea-level rise on coastal tourism are compounded by the fact that tourist facility development planning and execution in many cases has been inadequate even for current conditions, leading to environmental problems such as water shortages (Wong, 1998). Furthermore, tourism businesses, which usually are location-specific, have a lower potential than tourists themselves (who have a wide variety of options) to adapt to climate change (Wall, 1998). Small island states may find themselves especially vulnerable to changes in the tourism economy because of their often high economic dependence on tourism, concentration of assets and infrastructure in the coastal zone, and often poor resident population (see Chapter 17).

Whetton *et al*. (1996) quantified the effects of climate change on snow cover in the Australian Alps, which illustrates the problems of snow-based recreation activities. Under the best-case scenario for 2030, simulated average snow-cover duration and the frequency of more than 60 days cover annually decline at all sites considered. However, this effect is not very marked at higher sites (above 1,700 m). For the worst-case scenario, at higher sites, simulated average snow cover roughly halves by 2030 and approaches zero by 2070. At lower sites, near-zero average values are simulated by 2030.

7.2.2.2. Riverine, Coastal, and Steeplands Settlements—Impacts on Infrastructure

Riverine and coastal settlements are notable largely for the potential that flooding and especially sea-level rise can have on them; steeplands in many regions are expected to become more vulnerable to landslides. The mechanisms of these effects depend on the settlement being located in harm's way. River floods can arise from intense local rainfall events or rapid snowmelt, for which long-term probabilities are difficult to forecast (see Chapter 4; Georgakakos *et al*., 1998). Rapid snowmelt from rain-on-snow events or warm periods in the middle of winter is a potential threat in a warmer world in some heavily settled, snow-fed river systems such as the Rhine in Europe (see Chapter 13), whereas steeplands may suffer more landslides and snow avalanches (e.g., Evans and Clague, 1997). The mechanisms of adaptation are similar: defend against flooding or landslides with increasingly expensive protection structures; retreat from the floodplain and unstable areas to safe ground; or accommodate flooding and landslides in structure design, land-use planning, and evacuation plans (see also Chapters 9, 10, 11, 12, 13, 15, and 17).

The most widespread serious potential impact of climate change on human settlements is believed to be flooding. A growing literature suggests that a very wide variety of settlements in nearly every climate zone may be affected, although specific evidence is still very limited. Riverine and coastal settlements are believed to be particularly at risk, but urban flooding could be a problem anywhere storm drains, water supply, and waste management systems are not designed with enough system capacity or sophistication to avoid being overwhelmed. Urbanization itself explains much of the increase in runoff relative to precipitation in settled areas (Changnon and Demissie, 1996) and contributes to flood-prone situations.

In coastal regions (especially on river deltas and small islands), sea-level rise will be the most fundamental challenge of global warming that human settlements face. Some additional national and regional analyses of coastal vulnerability to sea-level rise have been published since the SAR. Many of these studies are summarized in Chapters 6 and 8 because of the importance of the issue to these sectors. In general, estimates of potential damages continue to increase because of increased movement of people and property into the coastal zone, even though the expected degree of sea-level rise has decreased. Worldwide, depending on the degree of adaptive response, the number of people at risk from annual flooding as a result of a 40-cm sea-level rise and population increase in the coastal zone is expected to increase from today's level of 10 million to 22–29 million by the 2020s, 50–80 million by 2050s, and 88–241 million by the 2080s (Nicholls *et al.*, 1999). Without sea-level rise, the numbers were projected at 22–23 million in the 2020s, 27–32 million in the 2050s, and 13–36 million in the 2080s. The 40-cm sea-level rise is consistent with the middle of the range currently being projected for 2100 by Working Group I. In 2050, more than 70% (90% by the 2080s) of people in settlements that potentially would be flooded by sea-level rise are likely to be located in a few regions: west Africa, east Africa, the southern Mediterranean, south Asia, and southeast Asia. In terms of relative increase, however, some of the biggest impacts are in the small island states (Nicholls *et al.*, 1999).

Although a 1-m sea-level rise is not considered likely before 2100, it often is used to calibrate many damage estimates. For example, a macroscopic analysis of coastal vulnerability, areas, population, and amount of assets at risk from sea-level rise and storm surges for Japan shows about 861 km^2 of land currently is below the high-water level, with 2 million people and 54 trillion Japanese yen in assets. A 1-m sea-level rise would expand the area at risk 2.7 times, to 2,339 km^2, and increase population and assets at risk to 4.1 million and 109 trillion Japanese yen, respectively (Mimura *et al.*, 1998). El-Raey (1997) identifies potential impacts on Egypt: 2 million persons and 214,000 jobs affected, US$35 billion in land value, property, and tourism income lost for a 50-cm sea-level rise. For a 100-cm rise, Zeidler (1997) identifies potential land losses of US$30 billion, plus US$18 billion at risk of flooding and as much as US$6 billion of "full protection" in Poland. (See also Nicholls *et al.,* 1999, for potential damages to settlements on the world's coasts; Adger, 1999, specifically for damages to human settlements in Vietnam; Weerakkody, 1997, for Sri Lanka; and Liu, 1997, for China.)

If sea-level changes occur slowly, economically rational decisions could be made to protect only property that is worth more than its protection costs. With foresight, settlements can be planned to avoid much of the potential cost of protection, given that between 50 and 100 years are expected to pass before a 1-m sea-level rise would be expected. Yohe and Neumann (1997) offer a method by which this planning might be applied. This method can reduce the costs of protection by more than an order of magnitude. Yohe *et al.* (1996) estimate discounted (at 3% yr^{-1}) cumulative U.S. national protection costs plus property abandonment costs for a 1-m sea-level rise by the end of the 21st century at US$5–6 billion, as opposed to previous estimates of $73–111 billion (Smith and Tirpak, 1989).

Sea-level rise exacerbates beach erosion, changes sedimentation patterns, increases river floors in estaurine zones, and inundates wetlands and tidal flats. Groundwater salinization also is a serious problem in coastal zones and many small islands, where recharge does not keep up with usage even under current conditions (Liu, 1997). Higher sea levels are projected to further reduce the size of coastal fresh aquifers and exacerbate the problem. Directly, it causes drinking water quality problems for people in settlements; indirectly, it may limit agriculture in coastal zones (e.g., Dakar, Senegal—Timmerman and White, 1997). Groundwater pumping makes matters worse in many coastal zones by contributing to serious land subsidence. Examples of human settlements affected range from modern European cities (Venice) to large coastal settlements in developing countries (Alexandria, Tianjin, Jakarta, and Bangkok) (Timmerman and White, 1997; Nicholls *et al.*, 1999).

Vulnerabilities of settlements in coastal regions to higher sea levels are compounded by severe wind damage and storm surge caused by tropical cyclones and extra-tropical cyclones. The possibility of increasingly frequent (in some regions) or more intense tropical cyclones also cannot be rejected (see Chapter 3). Even if cyclones do not increase in intensity or frequency, with sea-level rise they would be expected to be an increasingly severe problem in low-lying coastal regions (e.g., for settlements along the North Sea coast in northwest Europe, the Seychelles, parts of Micronesia, the Gulf Coast of the United States and Mexico, the Nile Delta, the Gulf of Guinea, and the Bay of Bengal—specific vulnerable regions are identified in the FAR, SAR, RICC, and regional chapters of this report). Infrastructure hardening costs can be high if a decision were made to protect everything: The costs of protecting port facilities and coastal structures, raising wharves and quays, and reconstructing water gates and pumping stations for a 1-m sea-level rise in 39 prefectures in Japan has been estimated at 22 trillion Japanese yen (US$194 billion), or about 7% of annual GDP (Mimura and Harasawa, 2000). In cases in which neither retreat nor defense is feasible, flooding can be accommodated through infrastructure that is designed to reduce damage and evacuation planning to reduce loss of life. Bangladesh, for example, provides hardened storm shelters (Choudhury, 1998).

When extreme weather disasters happen in these regions as a result of tropical or extra-tropical cyclones, the total costs of damages become very large and, where insured, often cause serious problems for insurance carriers (see Chapter 8). The United States' direct annual insured and uninsured costs for tropical cyclones (hurricanes), adjusting for inflation, averaged $1.6 billion (1995 US$) from 1950 to 1989 and $6.2 billion from 1989 to 1995 (Hebert *et al.*, 1996). Estimates of worldwide annual direct costs have been placed at $10–15 billion annually (Pielke, 1997). Increased losses in the United States and elsewhere have occurred during a period in which the number and intensity of tropical cyclones actually was declining (Landsea *et al.*, 1996; Pielke and Landsea, 1998). Thus, any future climate change-induced increase in tropical cyclone frequency or intensity remains a matter of great concern to insurers and coastal facilities planners. A single US$50 billion storm is not considered unlikely (Pielke and Landsea, 1998).

7.2.2.3. Urban Settlements

Urban settlements feature many of the same impacts of climate change as other settlements—such as air and water pollution, flooding, or consequences of increasingly viable disease vectors. These impacts may take unusual or extremely costly forms in urban areas—for example, flooding that results not from river flooding but from overwhelmed urban storm drains and sewers during extreme rainfall events (which may become more common in the future). Urban settlements also experience the consequences of accommodating migrant populations, the unique aspects of urban heat islands (which affect human health and energy demand), and some of the more severe aspects of air and water pollution. To some extent, the effects of climate change anywhere in the world are integrated through world markets, social and political changes, and migration. Many of these social and economic effects appear in the world's cities, including some of the world's largest (Rosenzweig and Solecki, 2000).

7.2.2.3.1. Migration

The SAR discussed at some length the potential impact of population movement in response to environmental impact and attractions of large urban areas. Human populations show significant tendencies to adapt to interannual variability of climate via migration, although migration may be the last of a complex set of coping strategies (Meze-Hausken, 2000). For example, decreases in crop and rice yields as a result of a prolonged dry season under ENSO conditions in Indonesia causes farmers to leave the villages to work in the surrounding cities. Population subsequently recovers (Yoshino, 1996b; Yoshino *et al.*, 1997). In some cases, immigration is more permanent and does not involve large urban areas. For example, after three successive typhoons hit Tau Island in American Samoa in 1987, 1990, and 1991, about one-third of the population abandoned their homes and moved to Pago Pago on Tutuila Island, putting more population pressure on the limited economic opportunities and services of that island (Meehl, 1996).

A school of thought based on observations of several ethnic conflicts in the developing world suggests that environmental degradation, loss of access to resources, and resulting human migration (including "environmental refugees") in some circumstances could become a source of political and even military conflict (see Chapters 10 and 11). The result is possible, but the many intervening and contributory causes of intergroup and intragroup conflicts allow only low confidence in predictions of increases in such conflicts as a result of climate change, even where environmental resources are scarce and threatened (e.g., Wolf, 1998).

7.2.2.3.2. Human health

The impact of climate change on human heath in settlements is a complicated mechanism that involves the interaction of physical attributes of settlements and precursors for direct effects of heat stress, vector-borne diseases such as malaria, and enteric diseases such as cholera (see Chapter 9). These impacts are common to all types of settlements, including traditional ones. The 1997–98 El Niño event provided a way to derive and test process models for the impacts of climate change. In Latin America, for example, outbreaks of malaria and Dengue fever appeared to be related to anomalously high nighttime minimum temperatures (Epstein *et al.*, 1998). Settlements provide disease vectors and organisms with habitat in the form of standing water, garbage dumps, and space sheltered from the elements. Flooding can flush organisms into settlements' clean water supplies, causing disease outbreaks. Heavy rainfall in normally dry areas leads to rapid increases in rodent populations, in turn leading to increases in rodent-borne diseases such as hantavirus (Glass *et al.*, 2000). Cholera-harboring marine plankton blooms also can be triggered by riverine flooding, which provides extra nutrients to the coastal environment (Colwell, 1996). Extremely dry conditions reduce the quantities and quality of water available for sanitary and drinking purposes, which also can trigger cholera and diarrheal outbreaks. Historically, this has been a problem on various Pacific islands during drought. Poverty, crowding, and poor sanitation in settlements add to these problems, as reported in the SAR.

7.2.2.3.3. Heat islands and human health

Warming of urban air increases in intensity and area as cities grow (Oke, 1982). This growing "heat island" tends to aggravate the risk of more frequent heat waves, as well as their impacts. Research indicates that variability in summer nighttime minimum temperature (temperatures above 32ºC at night)—combined with lack of acclimatization, high humidity, and poorly ventilated and insulated housing stock—may be the most important factor in urban heat deaths (Chestnut *et al.*, 1998). Elderly people, the very young, people in ill health, and poor people are most likely to be affected (see Chapter 9 for these and other health effects). Because climate change is expected to raise nighttime minimum temperatures more than daytime highs, urban heat islands would be a significant health concern in the largest human settlements.

Conversely, during the rainy season (except in a few cities, such as Cairo, where practically no rain occurs) the heat island enhances the intensity and frequency of rain showers (Changnon, 1992; Jáuregui and Romales, 1996), leading to higher risk of street flooding or mudslides where the urban poor live. Moreover, warmer and drier climates may aggravate air pollution seasonally because of wind erosion of bare soil areas in cities with semi-arid or arid climates (e.g., Mexico City, Beijing, Delhi, and cities located in sub-Saharan Africa). Blowing dust and high summer temperatures are likely to increase the incidence of heat stroke, respiratory illness, and transmission of disease by deposition of airborne bacteria in lungs and on food.

In warmer and drier climates, local minimum temperatures tend to be higher, which tends to attenuate the intensity (and depth) of temperature inversions formed by nocturnal radiation cooling and reduce the risk of poor air quality. However—and especially in large cities located in valleys (e.g., Mexico City, Santiago, Beijing, Delhi)—this attenuation effect could be compensated by a higher rate of radiation cooling to an air layer with less moisture content, aggravating the air pollution situation. Elevated subsidence inversions such as those on the descending branch of semi-permanent anticyclones and limiting vertical dispersion of pollutants in cities such as São Paulo, Los Angeles, or Tijuana are less likely to significantly change their thermal structure in a warmer world.

7.2.2.3.4. Water pollution

Despite massive investment in water treatment in the developed world and increasingly in the developing world over the past century, many settlements throughout the world (especially in rural areas) still are without adequate water treatment (UNCHS, 1999). In the case of drought, reduced water availability could force people to use polluted water sources in settlements at the same time that reduced flow rates reduce the rate of dilution of water contaminants. In the opposite case, flooding frequently damages water treatment works and floods wells, pit latrines and septic tanks, and agricultural and waste disposal areas and sometimes simply overwhelms treatment systems, contaminating water supplies.

7.2.2.3.5. Air pollution

Air pollution is a serious problem in many cities of the world, even under the current climate. The following issues emerge from a review of developing country cities that are members of the 69 urban agglomerations with population of more than 3 million in 1990 (UNEP, 1992):

- Population trends have not yet stabilized, which means cities will continue to extend their urban area (and urban heat island).
- Motor vehicles now constitute the main source of pollutants in most cities of the industrialized world. Although the rate of growth in the number of vehicles in some settlements in the developing world has been very rapid (Simon, 1996), cities in developing countries (with exceptions) exhibit greater variety in air pollution sources. This depends on the level of motorization and the level, density, and type of industry present. Cities in Latin America, for example, tend to have high vehicle densities and high vehicle-to-total pollution loads. The major sources of air pollution in Delhi are 2.8 million vehicles, thermal power plants, industries, and domestic

Box 7-2. Air Pollution Problems of Large Cities in the Developing World—Update on the Case of Mexico City

The SAR notes the complex air pollution problems of Mexico City and expresses concern that increases in air stability episodes under climate change could exacerbate an already difficult situation. Although these problems remain severe, as noted below, Mexico City also shows that adaptation (mitigation of air pollution) is possible and effective.

Mexico City is located in an elevated inland valley (approximately 2,250 m above sea level) in central Mexico. The climate is subhumid tropical, tempered by the altitude. Industrial activity contributes 20% of the city's air pollution; 3 million motor vehicles that consume 17.3 million liters of gasoline and 5 million liters of diesel daily (in 1997) generate 75% of the pollution (Office of the Environment Annual Report, 1997). Typical anticyclonic weather prevailing during the dry season contributes to frequent thermal inversions that prevent dispersion of pollutants. Moreover, abundant insolation prevailing during the dry warm period (March to May) favors activation of precursors [mainly oxides of nitrogen (NO_x) and hydrocarbons (HC)] to produce high levels of ozone. This also is the season for dust storms and blowing dust (Jáuregui, 1989). Although transport may generate 75% of the airborne pollutants by weight, estimates made in 1989 regarding transport's contribution to air pollution in terms of toxicity suggested its contribution was 42.4%. A considerable contribution also comes from vegetation and topsoil (12% of all airborne pollutants by weight in 1994) (see Connolly, 1999).

The Mexico City Ministry of the Environment subsequently found that ozone—one of the most serious threats to health—was still above acceptable levels on 300 days in 1999 but that some progress had been achieved. During the worst days from 1990 to 1992, pollutants hit emergency levels on as many as 177 days annually. Emergency levels occurred on 5 days in 1999. This achievement was considered to be a result of anti-pollution efforts by the local government. For example, in the early 1990s, lead was removed from gasoline sold in the Valley of Mexico; laws restricted the use of cars without catalytic converters to 4 weekdays; and inspections of factories to reduce pollution doubled in 1999, to 152. The average ozone reading fell from 197.6 to 144, where the air quality norm is 100 points of ozone—equivalent to exposure to 0.11 ppm for 1 hour (Mexico City Ministry of the Environment, 2000). This progress in improving air quality suggests that even huge developing world cities can begin to reduce pollution.

fuel combustion. The problem has been compounded by unplanned development, inadequate public transport, poor road conditions, lack of traffic management, inadequate vehicle/engine maintenance, use of old vehicles, and poor fuel quality. Many developing countries' vehicle fleets tend to be older and poorly maintained (and not easily replaced because of the low incomes of their owners)—a factor that will increase the significance of motor vehicles as a pollution source (UNEP, 1992).

- Cities with seasonally warm, calm air and sunny weather with high traffic densities tend to be especially prone to the net formation of ozone and other photochemical oxidants, although it is not yet clear whether such conditions will be more or less prevalent under climate change. Volatile organic carbons (VOCs) from biogenic and anthropogenic sources such as automobiles increase at high temperatures, and thermal decomposition of peroxyacetyl nitrate (PAN) also increases (Samson *et al.*, 1989).
- Atmospheric and air-shed modeling exercises suggest that higher temperatures and more stable air episodes under global warming of 4°C could lead to 1–20% increases in peak urban ozone concentrations in the United States (Morris *et al.*, 1989; Penner *et al.*, 1989) and increased violation of clean air standards (Morris *et al.*, 1995). See Chapter 9 for the health effects of elevated ozone levels.

Box 7-2 shows that adaptive measures can be effective in reducing many of the precursors to adverse air quality under current climate. These measures also would help in the context of unfavorable atmospheric conditions as a result of climate change.

7.3. Energy, Transportation, and Other Climate-Sensitive Industry

7.3.1. Energy Supply and Demand

The SAR notes that climate change would impact energy supply and demand. Subsequent studies confirm this sensitivity. Hydropower generation is the energy source that is most likely to be impacted because it is sensitive to the amount, timing, and geographical pattern of precipitation as well as temperature (rain or snow, timing of melting) (see Chapter 4). Where reduced streamflows occur, they are expected to negatively impact hydropower production; greater streamflows, if they are timed correctly, might help hydroelectric production. In some regions, change of streamflow timing from spring to winter may increase hydropotential more in the winter than it reduces it in the spring and summer, but there is a question of whether the electric system can take advantage of the increase in winter flows and whether storage would be adequate. Hydroelectric projects generally are designed for a specific river flow regime, including a margin of safety. Projected climate changes are expected to change flow regimes—perhaps outside these safety margins in some instances (see Chapter 4). Although it is not yet possible to provide reliable forecasts of shifts in flow regimes for world river systems as a consequences of climate change, what is known suggests more intense rainfall events (which would require more conservative water storage strategies to prevent flood damage), greater probability of drought (less hydroelectric production), and less precipitation falling as snow (less water available during warm months). All three factors point to less (or, at least, less flexible) hydroelectric capacity at current powerhouses. Reduced flows in rivers and higher temperatures reduce the capabilities of thermal electric generation (Herrington *et al.*, 1997); high temperatures also reduce transmission capabilities.

Some advanced energy technologies also may be affected. For example, the United States and Japan are trying to learn how to exploit the potential of methane hydrates. If global warming leads to warmer oceans or warms areas that currently are permafrost regions, these compounds are likely to become less stable, making it more problematic to attempt to recover methane from them (Kripowicz, 1998).

Increased cloudiness can reduce solar energy production. Wind energy production would be reduced if wind speeds increase above or fall below the acceptable operating range of the technology. Changes in growing conditions could affect the production of biomass, as well as prospects for carbon sequestration in soils and forest resources. Climate change could worsen current trends in depletion of biomass energy stocks in Africa, which is expected to become drier (see Chapters 3 and 10). The impact on biomass elsewhere is less clear; it may include enhancement of growth because of higher rainfall in Africa as well.

The portion of total energy supply from renewable energy sources varies among countries, developed and developing. In the United States in 1998, renewable sources provided roughly 7% of gross energy consumption—about half of that as hydroelectric energy (EIA, 1999a). In other countries, developed and developing, the percentages vary. For example, biomass accounts for 5% of north African, 15% of south African, and 86% of sub-Saharan (minus South Africa) energy consumption; in Cote d'Ivoire, the Democratic Republic of Congo, Ethiopia, Mozambique, and Zambia, the vast majority of on-grid electricity generation comes from hydropower (EIA, 1999b). Hydroelectricity represents the primary source of electricity in Canada and most South and Central American countries, with the highest reliance in Paraguay and Brazil (99 and 87% of generating capacity, respectively) (EIA, 1999c). Although renewable energy sources may be adaptable to new climate, larger percentages of renewables (especially hydroelectricity) in a country's energy supply might make the country relatively more sensitive to climate (see Chapters 10, 11, and 12). However, fossil fuel extraction may be adversely affected by increased wind and wave action, heavy precipitation, shoreline erosion, and permafrost melting in regions where this applies (see Chapter 16). In addition, thermal power plants can be adversely affected by loss of cooling water as a result of low flows (see Chapter 12).

If a warmer climate is characterized by more extreme weather events such as windstorms, ice storms, floods, tornadoes, and hail, the transmission systems of electric utilities may experience a higher rate of failure, with attendant costs (see, however, Chapter 3 and TAR WGI Chapter 10) These failures can be extremely costly, as illustrated by the great eastern Canada ice storm of January 1998, which toppled hundreds of transmission towers and downed 120,000 km of power lines—in some cases for a month to 6 weeks—and cost CDN$3 billion in economic damage (only half of which was insured) (Kerry *et al.*, 1999). A 5-week power failure in the central business district of Auckland, New Zealand, occurred in February–March 1998 when four high-voltage transmission cables failed (Ministry of Commerce of New Zealand, 1998). Hot weather contributed to high demand and less-than-optimal operating conditions of these cables as a result of high soil temperature and dryness, although it was not ruled the direct cause. Transmission and distribution systems can be hardened to respond to greater risk, but only at substantial cost.

The SAR notes that on the demand side, space-cooling demand would increase and space-heating demand would decrease. Electrical system expansion (generation, transmission, and distribution) may be required to meet greater summer peaks. In warmer areas, it is expected that the demand for electricity will certainly increase, as may the demand for energy overall. Urbanization, rising incomes, and warmer climates could combine to increase energy used for space cooling—already a major concern in tropical and subtropical cities, most of which are in developing countries (e.g., as much as 60% of total electricity use in the commercial sector in Hong Kong, 60% of all electric energy in Riyadh—see Al-Rabghi *et al.*, 1999; Lam, 1999, 2000). Besides having a major impact on the energy sector, air conditioning would tend to enhance heat island effects because of the energy used. At the same time, research has found that air conditioning, where available and affordable, is a statistically significant factor in reducing the chances of hot-weather-related mortality (Chestnut *et al.*, 1998).

The heat island phenomenon may have a positive impact in cities with seasonally cool to cold winters. For example, during the 20th century, the long-term impact of the urban heat island has been to reduce potential energy demand for space heating by as much as 50% in the central quarters of megacities such as Tokyo and Mexico City (Jáuregui, 1998). Urban warming and increased demand for cooling is expected, even though urban aerosol production (e.g., from power plants) does have a cooling effect (*Science News*, 1992; Jáuregui and Luyando, 1999).

Additional studies that have been published since the SAR continue to show that whether *net* energy consumption will increase or decrease as a result of climate change depends very much on location—in particular, whether energy consumption includes larger heating loads or cooling loads. The north-south orientation of Japan provides some insight into this question. Ichinose (1996) (quoted in Mimura *et al.*, 1998) has shown for Japan that reduction in heating would be about 30% in Sapporo on the northern island of Hokkaido, whereas it would be only 10% in Tokyo on the central island of Honshu. On the other hand, electricity consumption for cooling would increase hardly at all on Honshu and several percent in Naha on the southern island of Okinawa. The direction of net change also is sensitive to the future market penetration of air conditioning and to energy prices. In one Japanese study (Hattori *et al.*, 1991, summarized in Mimura *et al.*, 1998) the sensitivity of peak electric power demand to air temperature was shown to have increased 2.3 times during the 15 years between 1975 and 1990, largely as a result of the increase in the market penetration and unit size of air conditioners. Amano (1996, summarized in Mimura *et al.*, 1998) points out that a decline in energy prices contributed to the increase in the sensitivity of electricity consumption to climate.

Belzer *et al.* (1996) is among the few studies since the SAR that has estimated the effect of climate change on energy demand by the commercial sector. The study projects the change in demand at the national level for the United States in 2030. Accounting for changes in the building stock, a 4°C increase in average annual temperature, holding other loads constant, leads to an estimated 0–5% reduction in total energy consumption by the commercial sector in the year 2030 (note that 4°C was then considered possible at mid-latitudes if worldwide temperatures increased 2.5°C; now it is probably at the upper end of potential increases).

The SAR notes that energy used for irrigation would increase. Peart *et al.* (1995) studied the effects of climate change on energy efficiency in agriculture (including irrigation) in the southeastern United States. Results indicate that climate change would cause an increase in energy inputs required to produce a given amount of maize, soybeans, and peanuts.

Only a handful of studies since the SAR have looked at the effects of climate change on overall energy demand. Mendelsohn and Schlesinger (1999) estimated climate response functions and economic welfare for the entire energy sector in the United States, based on the cross-sectional study of household and firm energy expenditures in Mendelsohn and Neumann (1998). Two approaches were used: laboratory experiments coupled with process-based simulation models, and cross-sectional studies to substitute for impacts of climate change over time. Economic welfare associated with energy was found to have a quadratic relationship with temperature, with a maximum at 10°C. Although the experimental method succeeded in isolating the effect of climate from other variables, it failed to fully incorporate adaptive responses. The cross-sectional studies found that annual energy expenditures were minimized with an annual temperature of 12.8°C in the commercial energy sector and with 11.7°C in the residential energy sector. The cross-sectional approach, of course, does not allow for the transient response of the climate system or the actual dynamics of the energy sector in response to climate. It substitutes static history for a dynamic future and cannot deal with irreversibilities, higher moments of climatic changes such as alterations to diurnal or seasonal cycles, synergic responses (see Section 7.6), or extreme events. (e.g., Schneider, 1997).

7.3.2. *Transportation*

Changnon (1996a) studied the effects of potential shifts in summer precipitation on transportation in Chicago, using data for 1977–79 and assuming continued use of current modes of transport. The study suggests that a future climate with more summer rainy days, somewhat higher rain rates, and more rainstorms would increase total vehicular accidents and total injuries in vehicular accidents, reduce travel on public transportation systems, and cause more aircraft accidents and delays. A drier climate probably would experience fewer moderate to heavy rain events, but results show that rain events during drier conditions produce a greater frequency of accidents and injuries per event than during wetter conditions. If high-heat events became more common with warmer climate, they also could become a problem. They have been known to soften asphalt roads, "explode" or buckle concrete roads, warp railroad rails, close airports because of lack of "lift" in extremely hot air, and increase mechanical failures in automobiles and trucks. On the other hand, there might be fewer mechanical failures resulting from extreme cold (Adams, 1997). Floods are costly to transportation systems, as they are to other infrastructure. Although the effect of climate change on flying weather is not clear, transportation by air is known to be sensitive to adverse weather conditions; major systemwide effects sometimes follow from flight cancellations, rerouting, or rescheduling. For example, one diverted flight can cause anywhere from 2 to 50 flight delays, and one canceled flight can result in 15–20 flight delays. The cost of a diverted flight can be as much as US$150,000, and a cancellation can cost close to US$40,000. The corresponding direct annual costs to 16 U.S. airlines are US$47 million and US$222 million, respectively (Qualley, 1997). Several additional examples of impacts on transportation are cited in Chapter 13.

7.3.3. *Construction*

Flooding and other extreme weather events that damage buildings and infrastructure could cost the world's economies billions of dollars under climate change simply to replace the damage—a cost that could divert funds from other needed investment (see Chapter 8). However, Mimura *et al.* (1998) note that cost increases for disaster rehabilitation and countermeasures against natural calamities could expand the market for the construction industry. Although no direct studies have been done, it is likely that a greater incidence of summer heat waves would reduce the productivity of this sector, but a lower incidence of cold waves and snowy conditions would increase the amount of year-round construction that could be accomplished in climates that currently have long, cold winters. Changes in design requirements for infrastructure, leading to additional requirements for construction, are discussed in the SAR.

7.3.4. *Manufacturing*

Manufacturing industries that are not directly dependent on natural resources generally would not be affected by climate unless key infrastructure is destroyed by flood or landslides, or unless shipments of inputs and outputs are affected (e.g., by snow blocking roads, airports, and train tracks; flooding or low flow that make river transportation untenable; or low water supplies that make process cooling and environmental activities more difficult). However, manufacturers are influenced by climate change in two other ways. First, they would be affected through the impact of government policies pertaining to climate change, such as carbon taxes (thereby increasing the cost of inputs). Second, they could be affected through consumer behavior that in turn is affected by climatic variations. For example, less cold-weather clothing and more warm-weather clothing might be ordered. Manufacture that depends on climate-sensitive natural resources would be affected by impacts on those resources. For example, food processing activity would follow the success of agriculture. Very little is known concerning the effects of warming on industry, and most information is highly speculative.

7.3.5. *Financial Services and Insurance*

Climate change increases risks for the insurance sector, but the effect on profitability is not likely to be severe because insurance companies are capable of shifting changed risks to the insured, provided that they are "properly and timely informed" on the consequences of climate change (Tol, 1998). For example, during the great storms in the early 1990s, the insurance sector reacted to increased risk and large losses by restricting coverage and raising premiums. Tucker (1997) also shows that increased climatic variability necessitates higher insurance premiums to account for the higher probability of damages. However, insurance companies still can be destabilized by large losses in a major weather-related catastrophe in a region where actuarial tables and estimated risks do not adequately reflect true weather risk (including greater variability), and companies therefore may not have made adequate provision for losses. See Chapter 8 for a description of impacts on financial services.

7.3.6. *Estimating and Valuing Effects*

Valuation of climate impacts remains difficult on three grounds. The first is uncertainty associated with determining physical changes and responses to these changes. The second is economic valuations of these changes that vary across regions. Fankhauser *et al.* (1998) show that damage cost estimates are sensitive to assumptions made on the basis of valuation (willingness to pay versus willingness to accept), accountability for impacts, differentiation of per unit values, and aggregation of damage costs over diverse regions. A third problem can be expressed as follows: "Which metric?" Five popular metrics are used: market costs, lives lost, species lost, changes in the distribution of costs/benefits, changes in quality of life (loss of heritage sites, environmental refugees, etc). Schneider *et al.* (2000) conclude that when aggregation exercises are undertaken, disaggregation of all estimated effects into each of five numeraires is needed first, followed by a traceable account of any aggregation so others holding different weighting schemes for each numeraire can re-aggregate. This is done rarely, if ever.

The Workshop on the Social and Economic Impacts of Weather at the National Center for Atmospheric Research, 2–4 April 1997, in Boulder, CO, estimated that property losses from extreme weather of all types currently costs the United States about \$15 billion yr^{-1} (\$6.2 billion related to hurricanes), as well as about 1,500 deaths (about half resulting from cold events); the worst flood and hurricane years yield about \$30–40 billion in property losses.

Smith (1996) standardized estimates of climate change damages for the United States for a 2.5°C warming, a 50-cm sea-level rise, 1990 income and population, and a 4% real rate of return on investments. Total damage estimates are slightly less than 1% of United States gross national product (GNP) in 1990. Within individual sectors such as agriculture and electricity, however, standardized damages differ by more than an order of magnitude. This level of uncertainty appears to apply among experts as well. For example, Nordhaus (1994) surveyed experts, and their damage estimates ranged over more than an order of magnitude.

Yohe *et al.* (1996) calculated the cost of a 50-cm sea-level rise trajectory for developed property along the U.S. coastline. Transient costs in 2065 were estimated to be approximately \$70 million (undiscounted and measured in constant 1990 US\$). These costs are nearly an order of magnitude lower than estimates published prior to 1995 (e.g., Fankhauser, 1995). This is because Yohe *et al.* (1996) incorporated the cost-reducing potential of market-based adaptation in anticipation of the threat of sea-level rise. In addition, they assumed efficient discrete decisions to protect or abandon small tracts of property, based on their economic merit. Some work since suggests that maladaptation may cause the costs of sea-level rise to be somewhat higher (West and Dowlatabadi, 1998).

7.3.7. *Tools/Methods/Approaches/Models Used in Developing New Knowledge, including Assumptions, Sensitivities, and Scenarios Used in Models*

Current impact assessment methods focus on comparing current conditions to a single alternative steady state—that associated with doubling of GHGs. Mendelsohn and Schlesinger (1999) attempt to estimate climate response functions for market sectors in the United States that reflect how damages change as climate changes through a range of values. Impacts are generated by using national climate values, rather than global values, and the timing of climate change is included in the modeling of capital-intensive sectors such as coastal resources and timber, which cannot adjust quickly. Empirical estimates of climate response functions are based on laboratory experiments coupled with process-based simulation models and cross-sectional studies (Mendelsohn and Neumann, 1998). Both methods indicate that agriculture, forestry, and energy have a bell-shaped relationship to temperature. Similarly, an increase in precipitation is likely to be beneficial to some agriculture, forestry, and water sectors, although this effect is reversed at sufficiently high levels. However, this work captures neither the transient response of the climate system nor the actual dynamics of the energy sector in response to climate (e.g., Schneider, 1997).

7.4. Infrastructure

7.4.1. *Water Supply and Demand*

Increases in average atmospheric temperature accelerate the rate of evaporation and demand for cooling water in human settlements, thereby increasing overall water demand, while simultaneously either increasing or decreasing water supplies (depending on whether precipitation increases or decreases and whether additional supply, if any, can be captured or simply runs off and is lost). Shimizu (1993, quoted in Mimura *et al.*, 1998) showed that daily water demand in Nagoya, Japan, would increase by 10% as the highest daily temperature rose from 25 to 30°C. Boland (1997) looked at several climate transient forecasts for the Washington, DC, metropolitan area for the year 2030 and estimated increases in summertime use of 13 to 19% and annual use of –8 to +11% relative to a future increase from 1990 without climate change of approximately 100%. In China, using four GCMs for the year 2030, water deficiency under normal (50%) and extreme dry (95%) hydrological conditions for various basins was predicted as -1.6×10^8 to 1.43×10^9 m^3 in the Beijing-Tianjin-Tangshan area and -1.1×10^8 to 121.2×10^8 m^3 in the Yellow River Basin (China Country Study Team, 1999).

Estimates of effects on water supply mostly have dealt with linking atmospheric and hydrologic models in an attempt to produce more plausible forecasts with statistical variability. Although they do not directly forecast water supplies to human settlements, Kwadijk and Rotmans (1995) do show an increase in variability of supply in the Rhine River with climate change, but little change in estimated annual flow. The study is unusual in that it attempts to directly link impacts to mitigation policy. A more conventional study that still links water supply (directly for a municipal water system) and climate scenarios is Wood *et al.* (1997).

7.4.2. *Buildings, Transportation, and Other Infrastructure*

Additional research since the SAR seems to have added to concerns about increased intensity of rainfall and urban flooding (e.g., for the highly urbanized northeast United States—Rosensenzweig and Solecki, 2000). Increases in intensity are projected by Fowler and Hennessy (1995) and Hennessy *et al.*

(1997). Smith *et al.* (1999) have performed a series of four case studies that combine climate and hydrological modeling to directly model flood frequency and magnitude and economic losses under enhanced greenhouse rainfall intensities. The study concludes that there would be little change in forecast flood damage by the year 2030 but that there would be substantial increase in flood risk (shortened average return interval) and flood damage (as a result of building inundation and failure) by the year 2070. The estimation technique was an improvement over earlier studies that assumed that changes in intensities of rainfall events would be associated with changes in flood frequency (e.g., Minnery and Smith, 1996).

Generally speaking, climate change will change the level and type of climatic effects that need to be covered by infrastructure design codes. This could affect infrastructure durability and energy usage. Potential changes in humidity and climate may change distribution in factors such as termites, for example—with potential degradation of structures and more serious impacts from given extreme events. There also could be adverse effects of storms, heat, and humidity on walls and insulation, though perhaps less winter damage.

7.4.3. Estimating and Valuing Infrastructure Effects

There still are relatively few reports that estimate the impact of global warming on the value of economic losses resulting from effects on infrastructure. However, Smith *et al.* (1999) and Penning-Powswell *et al.* (1996) provide estimates that show that increases in damages from urban riverine flooding probably would be substantial. See also the discussion on sea-level rise in Section 7.2.2.2.

7.5. Management and Adaptation of Human Settlements

Social and natural sustainability are important for sustainable development of human settlements (Yoshino, 1994). Coping with flooding and drought; getting potable water, breathable air, and a stable environment; and so forth have been prime concerns of urban planners, engineers, governments, and citizenry for thousands of years (Priscoli, 1998). Climate change simply adds to the challenge. Some of the adaptations probably would take place autonomously, but some adaptations may be much improved by taking climate into account explicitly (Wood *et al.*, 1997).

7.5.1. Adaptation

Questions such as “adapt to what?”, “who or what adapts?”, and “how does adaptation occur?” (Smit *et al.*, 1998) are still difficult to answer in a strict sense. Management, adaptation, and vulnerabilities have been discussed for settlements in coastal (Fukuma, 1999/2000), arid, agrarian (Douguédroit, 1997; Douguédroit *et al.*, 1997; Le Treut 1997), and urban regions (Maunder, 1995). To be successful, adaptations must be consistent with economic development, they must be environmentally and socially sustainable over time, and they must be equitable (that is, not have significantly deleterious effects on disadvantaged groups) (Munasinhge, 2000).

7.5.2. Adaptation to What and Why?

In most cases, human settlements have designed into them the ability to withstand most of the consequences of some environmental variability. In most regions, climate change would change the probability of certain weather conditions. The only effect for which average change would be important is sea-level rise, under which there could be increased risk of inundation of coastal settlements from average (higher) sea levels. Human settlements for the most part would have to adapt to more or less *frequent* or intense rain conditions or more or less *frequent* mild winters and hot summers, although individual days’ weather may be well within the range of current weather variability and thus not require exceptionally costly adaptation measures. The larger, more costly impacts of climate change on human settlements would occur through increased (or decreased) probability of extreme weather events that overwhelm the designed resiliency of human systems.

Much of the management of urban centers as well as the governance structures that direct and oversee them are related to reducing environmental hazards, including those posed by extreme weather events and other natural hazards. Most regulations and management practices related to buildings, land use, waste management, and transportation have important environmental aspects. So too do most public and private investments in infrastructure. A significant part of health care and emergency services exists to limit the health impacts of environmental hazards. Local capacity to limit environmental hazards or their health consequences in any settlement generally implies local capacity to adapt to climate change, unless adaptation implies particularly expensive infrastructure investment.

An increasing number of urban centers are developing more comprehensive plans to manage the environmental implications of urban development. Many techniques can contribute to better environmental planning and management including: market-based tools for pollution control, demand management and waste reduction, mixed-use zoning and transport planning (with appropriate provision for pedestrians and cyclists), environmental impact assessments, capacity studies, strategic environmental plans, environmental audit procedures, and state-of-the-environment reports (Haughton, 1999). Many cities have used a combination of these techniques in developing “Local Agenda 21s.” Many Local Agenda 21s deal with a list of urban problems that could closely interact with climate change in the future. Examples of these problems include (WRI, 1996; Velasquez, 1998):

- Transport and road infrastructure systems that are inappropriate to the settlement’s topography (could be

damaged by landslides or flooding with climate change)
- Dwellings that are located in high-risk locations for floods, landslides, air and water pollution, or disease (vulnerable to flood or landslides; disease vectors more likely)
- Industrial contamination of rivers, lakes, wetlands, or coastal zones (vulnerable to flooding)
- Degradation of landscape (interacts with climate change to produce flash floods or desertification)
- Shortage of green spaces and public recreation areas (enhanced heat island effects)
- Lack of education, training, or effective institutional cooperation in environmental management (lack of adaptive capacity).

7.5.3. *Sustainable Cities Activities*

The following generic lessons from Curitiba, Brazil—which come from the context of "sustainable cities" under *existing* conditions—may be applicable to future adaptation responses (Rabinovitch, 1998):

- Top priority should be given to public transportation rather than to automobiles and other light-duty vehicles, and to pedestrians and cyclists rather than to motorized vehicles. This reduces air pollution and some other forms of pollution. It was noted that some alternative fuels such as hydrogen are particularly attractive for reducing local air quality problems, as well as mitigating GHG emissions.
- There can be an action plan for each set of urban problems, but solutions within a city are connected, not isolated.
- Action plans must be participatory, with partnerships involving all responsible parties [government, private sector, nongovernmental organizations (NGOs), individuals].
- Creativity can substitute for financial resources (labor-intensive and creative ideas can substitute for capital).
- Even during rapid demographic growth, physical expansion can be guided by integrated road planning, investment in public transportation, and enforcement of appropriate land-use legislation.
- Technological solutions and standards for everything from public transit to recycling should be chosen on the basis of affordability (cost-effectiveness, combined with sensitivity to total cost).
- Public information and awareness are essential.

The most effective pathways for adaptation that result in sustainable development are likely to arise out of an informed evolution of existing institutions. Several authors emphasize the importance of the support and will of local public officials in developing successful environmental solutions (e.g., Gilbert *et al.*, 1996: Foronda, 1998). Others emphasize the need in traditional societies to build from and integrate modern techniques into traditional management practices and kinship and community networks, to effectively collect and disseminate data needed for assessing impacts, to open public participation processes for formulating policy, and to provide a process for strengthening financial, legal, institutional, and technical elements (Huang, 1997).

7.5.4. *Adaptation Options*

Adaptation to climate changes involves planning of settlements and their infrastructure, placement of industrial facilities, and making similar long-lived decisions to reduce the adverse effects of events that are of low (but increasing) probability and high (and perhaps rising) consequences. The adaptation response consists of planning to reduce the sensitivity of key assets, designing resilience and flexibility into the public and private infrastructure on appropriate time scales, and managing settlements and institutions in a climate-resilient manner. The following discussion of example options and strategies is divided into Planning and Design, Management, and Institutional Frameworks (see Table 7-2 for a summary):

- *Planning and Design*
  - Take advantage of rapidly increasing populations in many regions. Growing populations and cities provide economic advantages, not just costs (Satterthwaite, 1998). In the case of infrastructure, growing populations provide an opportunity for new construction that can be designed for increased resilience and flexibility with respect to climate change. With good planning and building practices in new construction, considerable energy and environmental cost can be saved at relatively small incremental construction costs compared with later retrofit or protection (Rabinovitch, 1998). It can be less costly to design and build flood works "oversized" in the beginning than to rebuild them later to add capacity (this is not a foregone conclusion and depends on local uncertainties; see Wood *et al.*, 1997).
  - Take advantage of replacement schedules. Many short-lived assets such as consumer goods, motor vehicles, and space heating/cooling systems will be replaced several times in the course of a few decades, offering considerable opportunities for adaptation. Even medium-life assets such as industrial plants, oil and gas pipelines, and conventional power stations are likely to be completely replaced over such a time scale, though there will be less opportunity for adaptation through upgrades and relocation.
  - Community design tools such as floodplain and hillside building practices and landscape design (zoning in developed countries; perhaps land-use planning in developing countries) can be improved to limit damage. Reducing heat islands (through judicious use of vegetation and light-colored surfaces, reducing motor transportation, and taking advantage of solar resources) also should be included in the package of possibilities.

Table 7-2*: Planning and design, management, and institutional frameworks actions for human settlements, by type of settlement.*

	Resource-Dependent Settlements	Coastal, Riverine, and Steeplands Settlements	Urban Settlements
Planning and Design	– Increase economic diversification – Oasis development – Windbreaks – Develop irrigation and water supply – Rural planning – Redevelop tourism and recreation industry – Take advantage of replacement schedules for buildings and infrastructure	– Zoning in developed countries; perhaps land-use planning in developing countries – Better building codes to limit impact of extreme events, reduce resource use – Soft and hard measures to reduce risk of floods: • Reconstruction of harbor facilities and infrastructure • Flood barriers • Managed retreat (acquisition of properties; fiscal and financial incentives) • Hazard mapping • Tsunami damage-prevention facilities – Take advantage of rapidly increasing populations for sizing infrastructure – Take advantage of replacement schedules for buildings and infrastructure – Use community design tools such as floodplain and hillside building practices, public transportation – Improve sanitation, water supply, electric power distribution systems – Employ design practices to prevent fire damage (development densities and/or lot sizes, setbacks, etc.)	– Site designs and building materials and technologies that moderate temperature extremes indoors – Improve infrastructure and services, including water, sanitation, storm and surface water drainage, and solid waste collection and disposal – For higher temperatures: • Building and planning regulations and incentives that encourage building measures to limit development of "heat islands" – Take advantage of rapidly increasing populations for sizing infrastructure
Management	– Employ countermeasures for desertification – Increase environmental education – Improve landscape management – Develop agricultural and fisheries cooperatives to reduce risk – Preserve and maintain environmental quality – Institute emergency preparedness and improve neighborhood response systems	– Provide warning systems and evacuation plans; salvage; emergency services; insurance and flood relief – Better implement/enforce existing building codes – Employ special measures to promote adaptation and disaster preparedness in sites or cities at high risk from such events – Institute market-like mechanisms and more efficient management of water supplies (e.g., fix leaks) – Institute neighborhood water wholesaling and improve delivery	– Institute neighborhood water wholesaling and improve delivery systems – Improve sanitation and waste disposal – Create and enforce pollution controls for solid, liquid, and gaseous wastes – Efficiently operate public transportation systems – Institute emergency preparedness and improve neighborhood response systems – Improve health education

Table 7-2 (continued)

	Resource-Dependent Settlements	Coastal, Riverine, and Steeplands Settlements	Urban Settlements
Management (continued)		– Institute emergency preparedness and improve neighborhood response systems – Improve health education	
Institutional Frameworks	– Build institutional capacity in environmental management – Create partnerships between all responsible parties (government, private sector, NGOs, individuals) – Regularize property rights for informal settlements and other measures to allow low-income groups to buy, rent, or build good quality housing on safe sites – Improve technology of farm machinery, herbicides, computers, etc.	– Build institutional capacity in environmental management – Create partnerships between all responsible parties (government, private sector, NGOs, individuals) – Regularize property rights for informal settlements and other measures to allow low-income groups to buy, rent, or build good quality housing on safe sites	– Build institutional capacity in environmental management – Create partnerships between all responsible parties (government, private sector, NGOs, individuals) – Regularize property rights for informal settlements and other measures to allow low-income groups to buy, rent, or build good quality housing on safe sites

– Improved sanitation, water supply, and electric power distribution systems can be planned in an integrated manner, with sensitivity to the location of air sheds and water sheds and efficient utilization of plant (e.g., properly price water to reflect scarcity and reduce leakage in water supply systems) (Wood *et al.*, 1997; Lettenmaier *et al.*, 1998; Tindleni, 1998). In some cases, this requires "hardening" the system. In other cases, it requires flexible approaches to plans for cleanup and movement of waste facilities in the flood zone throughout the watershed and, for example, sanitation (Bartone, 1998). There are more opportunities to do this in developing countries, which are still acquiring their basic urban infrastructure. According to the United Nations Center for Human Settlements (UNCHS) Global Urban Indicators Program Web site in 1999, which developed city data collection systems in 237 cities in 110 countries, about 38.5% of African urban households were connected to water systems, 15.4% to sewerage, 46.9% to electricity, and 14.1% to telephones; in Asia the corresponding connection rates were 52.4, 33.2, 82.5, and 26.5%, respectively, and in the industrialized countries the percentages were 99.4, 95.8, 99.2, and 78.2% (UNCHS, 1999).

– Measures can be taken by governments to anticipate floods (Smith and Handmer, 1984), including "hard" engineering measures such as dams, levees, diversions, channels, and retarding basins and "soft" or "nonstructural" methods such as acquisition of properties, fiscal and financial incentives, regulations such as zoning and building regulations, information and education campaigns, forecasts/warning systems/evacuation plans, salvage, emergency services, insurance, and flood relief. "Designing with nature" is an effective strategy for curbing flood losses even with current climate (Rabinovitch, 1998). In some cases, this may require cleanup and movement of waste facilities in the flood zone throughout the watershed, an issue also discussed in the SAR. FEMA (1997) has concluded that the process of establishing and implementing state and community comprehensive development and land-use plans provides significant opportunities to mitigate damages caused by natural hazards. Losses from floods in the upper Mississippi valley in 1996 (after some of these actions had been taken) were two orders of magnitude lower than in 1993 for similar-sized floods. Albergel and Dacosta (1996) propose analyzing runoff patterns from non-normal storm events as a basis for more sustainable water resource management in the specific instance of Senegal to deal with extreme events more effectively. Colombo, Sri Lanka, has an ambitious plan to redesign and rebuild its flood management system, incorporating improved maintenance of the existing system, development of flood retention areas, cost-effective defined flood safety margins, movement of people out of flood-prone lands immediately adjacent to canals, and even a plan to move industries and public organizations out of Colombo to hold constant the amount of unbuilt lands in the face of increased urbanization (Gooneratne, 1998).

- Put key infrastructure in less vulnerable areas and (as appropriate) harden against fire. Urban/wildlands interface fires at locations such as Oakland/East Bay Hills in California and Sydney, Australia, have yielded examples of design practices to prevent such occurrences, including limiting development densities and/or requiring large lot sizes; setting buildings back from flood, landslide, and fault hazard zones; requiring adequate minimum paved street widths and grades; requiring second access points; restricting the lengths of cul-de-sacs; developing adequate water supply, flows, and redundant storage; and using open space easements for fire breaks, equipment staging, and evacuation areas (Topping, 1996). Wildlands fires also may affect settlements at a distance. In 1997, health and haze effects of wildfires alone cost Indonesia US$1 billion (EEPSEA, 1999; Wheeler, 2000). Total costs were about US$4.5 billion.
- Diversifying economic activity could be an important precautionary response that would facilitate successful adaptation to climate change for resource-dependent settlements.
- Use suitable design techniques to reduce cooling demand in many buildings. For example, the U.S. Department of Energy's Building America Program uses a systems engineering approach and works with the home building industry to produce quality homes that use 30–50% less energy without increasing building costs, that reduce construction time and waste by as much as 50%, and that improve builder productivity. Eliminating inefficiencies is not strictly an adaptation option, but these improvements also would have the benefit of reducing the impact of warming on energy use. Adaptive measures that can effectively reduce energy use relating to space cooling include better design of building envelopes (which can cut cooling energy consumption by more than one-third in current subtropical conditions or 5–30% under current European conditions—see Chan and Chow, 1998; Balarus *et al.*, 2000); high-albedo roofs (field tests in Florida show savings of 2–40%—Akbari *et al.*, 1999); and development of urban trees and other greenery (Avissar, 1996; Spronkensmith and Oke, 1998; Upmais *et al.*, 1998), reducing summertime urban cooling loads by 3–5% (Sailor, 1998). See also Angioletti (1996) for criteria for sustainable building practices. Low-income energy assistance and weatherization can play a role in assuring equity (Miller *et al.*, 2000).

- *Management*
  - To stretch water supplies further, especially in high-income parts of developed countries, institute market-like mechanisms and more efficient management of water supplies. Experiments with market systems in the United States have yielded water surpluses during drought in California, but such schemes must consider distribution and equity effects when water consumption is very low already (Hardoy *et al.*, 2000), as well as external environmental effects on third parties (Frederick, 1997). In some urban water systems, a significant percentage of water put into the system is lost through leakage (e.g., 23%—or about eight times the expected increase in demand resulting from climate change—in the case of the southern part of Britain). Urban water systems may lose as much as 40–60% of their water in distribution (Cairncross, 1990; Daniere, 1996; Rahman *et al.*, 1997). Simply fixing leaks in cases such as this would go a long way toward making sure that supplies are available (Arnell, 1998). Boland (1997) found that mandating and widely promoting conservation measures in the Washington, DC, area (such as a 50% increase in tariff level, industrial/commercial water reuse and recycling, and a moderate conservation-oriented plumbing code) could cut the increase in baseline water use in the year 2030 from 100% to about 45%. This would more than offset any increase in use resulting from climate warming. In developing-world cities' poor neighborhoods, where water use already may be below levels that are conducive to good health, a mixed response involving improving pipes, instituting neighborhood water wholesaling, and improving water vendors and kiosks may be effective and affordable, if coordinated with organized liquid waste disposal (Hardoy *et al.*, 2000).
  - Provide flood control and other forms of property assurance (not necessarily insurance). On the insurance side, Germany and France have taken very different approaches, with Germans adopting private insurance and minimal flood assistance and the French adopting compulsory extended coverage under the "cat 'nat" system. Neither system has proved to be able "to provide sufficient coverage at reasonable cost (Gardette, 1997). The United States has been able to provide flood insurance through a national system—at the cost of not controlling losses. Much more activity is now going into loss prevention in the United States (FEMA, 2000). The FEMA Hazard Mitigation Grant Program (HMGP) provides grants to states and local governments to reduce loss of life and property from natural disasters and to enable mitigation measures to be implemented during the immediate recovery from a disaster (FEMA, 2000). Some authors argue that loss prevention could be anticipatory (Miller *et al.*, 2000).
  - Institute emergency preparedness and improve neighborhood response systems and mutual assistance. This strategy is featured in many of

the local initiatives cataloged by the International Council for Local Environmental Initiatives (ICLEI). Increasing access to cooling stations and city heat emergency response plans for urban heat island effects are an example. Plans for extreme events have improved considerably in the past 50 years, especially in developed countries, and can be very effective against some of the anticipated consequences of climate change. For example, preparation of disaster prevention organization in Japan, including forecast techniques and information systems between weather bureau/meteorological offices and individuals through local government/offices, has resulted in a dramatic decrease in the number of deaths and houses destroyed, especially for middle-sized typhoons (Class III) in the past 4 decades. This trend suggests that disaster prevention systems such as river improvement, establishment of information systems, improvement of typhoon forecasting, and so forth were most effective for middle-sized typhoons (Fukuma, 1996; Yoshino, 1996a). In most developing countries, local governments are particularly weak and ineffective at environmental management and have little capacity to integrate disaster preparedness into their current tasks and responsibilities. For example, no significant change has been found in deaths per violent hurricane in Bangladesh (Fukuma, 1999/2000), whereas there has been a significant decline in the United States (Pielke, 1997). However, some of the differences may be partly a result of a reduction in violent hurricanes making landfall in populated areas. U.S. property damages increased while the incidence of hurricanes fell (see Pielke and Landsea, 1998). Analysis of 15 years of simulation and observations show that typhoon frequency has decreased off the Philippine Islands and that the location of their generation shifts toward the east during El Niño years (Matsuura *et al.*, 1999). If similar conditions in typhoon regions also occur under global warming, more effective prediction and tracking of tropical typhoons in the 21st century could reduce typhoon disasters.

– Improve health education. Because the consequences of flood and drought on water quantity and quality are fairly straightforward to predict, concerted public awareness campaigns can significantly reduce adverse heath consequences on populations from water-borne enteric diseases. For example, in the Marshall Islands in 1997–1998, a public awareness campaign to boil all water supplies was triggered by public health officials' concerns over dwindling water availability. The number of hospitalizations for diarrheal disease was lower than normal (Lewis *et al.*, 1998).

- *Institutional Frameworks*
  – An institutional framework can be put in place that is more "friendly" for adaptation strategies. Some key features of such a strategy include regularizing property rights for informal settlements and other measures to allow low-income groups to buy, rent, or build good-quality housing on safe sites. Much of the poor-quality settlement infrastructure in the developing world in particular is traceable to the questionable legal status of housing in these settlements (e.g., Jaglin, 1994; Acho-Chi, 1998; Perlman, 1998). Thus, the occupants of such housing units can be reluctant to spend much on quality construction and may have little legal standing to demand municipal government services such as piped water, sewerage, or waste collection. Perhaps more important, even if they are financially able to do so, governments may be unable or reluctant to extend services to households and communities lacking legal standing, especially if such extension of service thereby "validates" the settlement.
  – Build institutional capacity in environmental management. Capacity to adapt to climate change, as with other environmental problems, will be realized only if the necessary information is available, enterprises and organizations have the institutional and financial capacity to manage change, and there is an appropriate framework within which to operate. In this respect, autonomous adaptation cannot necessarily be relied on. Governments may have a role in terms of disseminating information (Miranda and Hordijk, 1998) and in any case should not stand in the way through indifference, hostility, or inefficient or corrupt management (Foronda, 1998). Building efficient environmental institutions requires coherent policy, planning, mechanisms for implementation, procedures for monitoring and corrective action, and a means for review. Mechanisms include use of market-based regulatory instruments to limit contributory pollution and to organize land use; development of appropriate roles for central and local governments; involvment of communities and civil societies in adaptive strategies (including traditional environmental knowledge and informal regulation as appropriate); and expansion of the scope for international cooperation (World Bank, 1999).

7.5.5. Barriers and Opportunities for Adaptation

Most urban authorities in developing regions have very little investment capacity despite rapid growth in their populations and the need for infrastructure. Problems arise from inadequate and inappropriate planning for settlements. Yet the need for planning becomes even more pressing in light of increased social, economic, and environmental impacts of urbanization; growing consumption levels; and renewed concern for sustainable development since the adoption of Agenda 21 (UNCHS, 1996). Environmental management tends to be more difficult in very large cities (WRI, 1996). Although there are commitments from developed countries in Article 4.4 of the United Nations Framework Convention on Climate Change (UNFCCC) to assist particularly vulnerable countries with adaptation, the financial resources needed to provide services to tens of millions of people are daunting.

Increasingly, settlements are exchanging ideas concerning methods and experiences for community design and management to improve sustainability and livability. For example, ICLEI is an association of local governments dedicated to prevention and solution of local, regional, and global environmental problems through local action. More than 300 cities, towns, counties, and their associations from around the world are members of the Council (ICLEI, 1995).

Environmentally sound land-use planning is central to achievement of healthy, productive, and socially accountable human settlements within societies whose draw on natural resources and ecosystems is sustainable. The challenge is not only how to direct and contain urban growth but also how to mobilize human, financial, and technical resources to ensure that social, economic, and environmental needs are addressed adequately (UNCHS, 1996).

7.6. Integration

There are multiple pressures on human settlements that interact with climate change. The discussion in this chapter shows that these other effects are more important in the short run; climate is a *potential* player in the long run. For example, urban population in the least-developed countries currently is growing at about 5% yr^{-1}, compared with 0.3% yr^{-1} in highly industrialized countries.[1] Providing for this rapidly urbanizing population will be much higher on most countries' agendas than longer term issues with climate change.

Other environmental problems will tend to interact with climate change, adversely affecting human settlements. For example, 25–90% of domestic energy supply in the developing world is met by biomass resources, especially in small urban centers (Barnes *et al*., 1998). In some countries, 11–20% of all deforestation may be attributable to charcoal production (Ribot, 1993), much of it to meet urban needs. If biomass growth is slowed via climate change effects, the impacts on biomass may be compounded.

Deforestation and cultivation of marginal lands can compound the effects of extreme events. For example, floods resulting from Hurricane Mitch, though not caused by climate change, illustrate the fact that poor watershed management can contribute to flooding and landslides—which, in turn, causes loss of life and destroys infrastructure and the means of livelihood. Mitch cost Honduras 80% and Nicaragua 49% of one year's GDP (FAO, 1999). Poor watershed management and technical failures has contributed to loss of life in landslides in Sri Lanka, Peru, Brazil, several European countries, and the United States (Katupotha, 1994)

Urban water resources already are in extremely short supply in 19 Middle Eastern and African countries (IPCC, 1998) and in cities in many parts of the world, where as many as 60% of poorer residents may not have access to reliable water supplies (Foronda, 1998). Poor urban water management may be responsible for losses through leakage of 20–50% in cities in the developing world and even in some cities in the industrialized world (WRI, 1996). If climate change makes water more scarce—by increasing demand (even in regions that currently are not particularly short of water, such as Great Britain—see Arnell, 1998) or by reducing supply (reduced surface runoff, exacerbation of water quality problems as a result of warmer temperatures and reduced flows, or salinization of coastal aquifers resulting from sea-level rise)—water supply problems would be exacerbated.

Liquid waste disposal is a significant problem in urban areas as diverse as Chimbote, Peru (Foronda, 1998); Buenos Aires, Argentina (Pirez, 1998); Cotonou, Benin (Dedehouanou, 1998); and Chicago, USA (Changnon and Glantz, 1996). There are two ways in which climate could interact with this problem: reductions in supplies of water with which wastes are diluted and the impact of more severe flooding episodes that overtop sewer systems and treatment plants (Walsh and Pittock, 1998). Where most inhabitants rely on pit latrines and wells, flooding spreads excreta from pit latrines everywhere and contaminates wells (Boko, 1991, 1993, 1994). Land-use changes associated with urbanization also have reduced the absorptive capacity of many river basins, increasing the ratio of runoff to precipitation and making flooding more likely (Changnon and Demissie, 1996).

Large urban areas, especially in the developed world, depend on extended "linkage systems" for their viability (Timmerman and White, 1997; Rosenzweig and Solecki, 2000). They depend on imports from the local area, region, nation, and even the world for everything from raw material and food, to product and waste exports and communications. These linkage systems often are vulnerable to severe storms, floods, and other severe weather events. Management, redundancy, and robustness of these interlocking systems is a top priority for developed world settlements especially, but increasingly so for the developing world (Timmerman and White, 1997).

7.6.1. Key Vulnerabilities

Key climate-related sensitivities in urban areas of the world include water supply and the effects of extreme events (primarily

[1] See <www.sustainabledevelopment.org>, which provides data on growth for 237 world cities by development level, based on the UNCHS (Habitat) Global Urban Indicators database.

flooding) on infrastructure in river floodplains and coastal zones. These areas should be considered sensitive to climate change. To the extent that sensitive settlements coincide with conditions of poverty and lack of technical infrastructure, these settlements also will be particularly vulnerable to climate change.

7.6.2. *Potential for Nonlinear Interactions and Synergistic Effects*

Because of their role as centers for administration and commerce, urban areas integrate all of the environmental effects that visit a society and to some extent buffer their human occupants from natural environmental fluctuations. However, these urban areas still may be affected by several stresses that interact with each other in a nonlinear fashion (Rosenzweig and Solecki, 2000; Wilbanks and Wilkinson, 2001). Whatever cash economy there is in a country tends to reside in its biggest cities, and trade routes also focus on these areas. These are two very important coping mechanisms. Thus, for example, climate-related food shortages are more likely to be experienced in urban areas as an increase in migrants from the countryside or loss of business in agriculture-related business rather than as famine *per se*.

Once populations are housed in urban settlements, there are other potential interactions among climate effects that lead to nonlinear impacts on them. Flooding events that are beyond the designed capacity of settlement infrastructure are a case in point, especially in cases in which systems already may be degraded. Urban flooding can overwhelm sewage treatment systems, thereby increasing the risk of disease at the same time that water treatment systems are compromised, health services are disrupted, disease vector species are driven into close contact with people, and people are exposed to the elements because of lost housing. Outbreaks of epidemics are always a risk under such circumstances, whereas the breakdown of any single one of the affected systems might be merely inconvenient. Although all of these effects and mechanisms are well understood, however, it is not possible to predict impacts *quantitatively* at this time.

Table 7-3 shows the primary synergistic effects between climate-related factors that may affect human settlements and the primary types of settlements or industries affected. Each cell identifies a synergistic effect between the climate impact featured in the row and another effect shown in the column. For example, climate-related impacts of flooding, landslides, and fire are compounded when they occur in settlements that also might be crowded by migration. Likewise, flooding in particular exacerbates water pollution and human health impacts and probably would compound problems in obtaining drinking water and transportation. In addition, the agricultural base and energy supplies could be affected in regions that already are water-deficient.

Air and water pollution effects of climate change would be worse if the health of human populations already is compromised (e.g., asthma attacks may be more severe or prolonged in a weakened population). Charlot-Valdieu *et al.* (1999) argue for reducing some stresses in a multiple-stress context to handle other stresses in a more sustainable manner.

Access to energy, clean water, sanitation, and selected other resources is essential to maintain human settlements and the health of the populations within. Flooding, landslides, or fire resulting from extreme weather could compound the shortage of resources by destroying critical infrastructure (floods in Honduras in 1998 and Mozambique in 2000 are examples of the phenomenon under current climate) and, in the case of water, polluting the sources. Similarly, water pollution reduces the effective water supply by making some sources unusable.

7.7. Science and Information Needs

Our ability to answer questions about climate change, vulnerability, and adaptation on the basis of research evidence is very limited

***Table 7-3**: Matrix of synergistic effects, by type of effect and settlement and industry type.*[a,b]

	Synergistic Impact Mechanism, Settlement Type or Industry				
Primary Impact Mechanism	Migration	Flooding, Landslides, Fire	Air and Water Pollution	Human Health	Energy, Water, Other Resources
Migration	—	U,RCS	U,RCS	U,RCS	U,E,A
Flooding , Landslides, Fire	U,RCS,TR	—	U,RD,RCS	U,RD,RCS	U,RD,RCS,TR
Air and Water Pollution	U,RCS	U,RD,RCS,RE	—	U,RD,RCS,RE,A	RE
Human Health	U	U,RD,RCS	U,RC,RE	—	U
Energy, Water, Other Resources	U,RD,RCS,A	U,RD,RCS	U,RD,RCS	—	—

[a] Settlement types: RD = resource-dependent, RCS = riverine, coastal, steeplands, U = urban.
[b] Industry types: A = agroindustry, E = energy, TR = transportation, RE = recreation.

for human settlements, energy, and industry. Energy has been regarded mainly as an issue for Working Group III, related more to causes of climate change than to impacts. Industry generally has been considered relatively insensitive to most primary climate change impacts, although some sectors (e.g., agroindustry) are dependent on supply streams that could be vulnerable to climate change impacts. Impacts of climate change on human settlements are hard to forecast, at least partly because the ability to project climate change at an urban or smaller scale has been so limited. As a result, more research is needed on impacts and adaptations in human settlements. Several activities also have been developed by governments in the area of "sustainable communities," which are designed primarily to reduce the impact of human settlements on the environment. Many of the actions recommended also reduce the vulnerability of settlements to global warming. Some areas of information required to support these programs have been identified by organizations such as the United Nations' International Decade for Natural Disaster Reduction (IDNDR Secretariat), the ICLEI, and the U.S. President's Council on Sustainable Development. Others were identified during preparation of the FAR, SAR, RICC, and this report.

The highest priority needs for research on impacts, adaptation, and vulnerabilities in human settlements are as follows:

- A much larger number and variety of bottom-up empirical case studies of climate change impacts and possible responses in settlements in the developing and industrialized world
- More reliable climate change scenarios at the scale of urban and even smaller areas
- Improved understanding of how climate change interacts with integrated multiple-stress contexts in human settlements, including possible ramifications of global urbanization
- Improved understanding of adaptation pathways, their costs and benefits, and what can reasonably be expected from them, especially in resource-constrained developing regions (includes autonomous and planned adaptation, as well as traditional and local adaptation; for example, analysis of water demand lags analysis of energy use in most countries)
- Improved understanding of the effects of climate on human migration and the effects of migration on source and destination settlements
- Improved understanding of critical climate change vulnerabilities in settlements, including conceivable low-probability, high-impact effects of climate change (need for continuing research and capacity-building efforts to improve preparedness and strengthen early warning and other mitigation aspects; establishment of a tropical cyclone landfall program is regarded as a logical vehicle for carrying research and development initiatives into the 21st century)
- Better understanding of the particular vulnerabilities of livelihoods and settlements of low-income and marginalized groups
- Improved understanding of the implications of climate variability and change for the well-being of human settlements as they relate to other sectors, other places, and the broader sustainable development process
- Improved understanding of the cascading of climate change through primary, secondary, and tertiary impacts within human settlements (a conclusion of the RICC, but not yet addressed effectively)
- Improved analytical capability to incorporate uncertainty, ambiguity, and indeterminacy in assessments of impacts and response strategies, at least partly by strengthening the science base for integrating quantitative and qualitative analysis, including undertakings such as scenario development and stakeholder participation.

Other key challenges to be faced include development of essential scientific and technical capacity in vulnerable regions, establishment and maintenance of adequate meteorological and hydrological monitoring networks, and improvement of seasonal and interannual prediction.

References

Acho-Chi, 1998: Human interference and environmental instability: addressing the environmental consequences of rapid urban growth in Bamenda, Cameroon. *Environment and Urbanization*, **10(2)**, 161–174.

Adams, C.R., 1997: *Impacts of Temperature Extremes: Workshop on the Social and Economic Impacts of Weather. 2–4 April 1997, Boulder, CO.* Economic and Social Impacts Group, National Center for Atmospheric Research, Boulder, CO, USA.

Adger, W.N., 1999: Social vulnerability to climate change and extremes in coastal Vietnam. *World Development*, **27(2),** 249–269.

Akbari, H., S. Konopacki, and M. Pomerantz, 1999: Cooling energy savings potential of reflective roofs for residential and commercial buildings in the United States. *Energy*, **24,** 391–407.

Albergel, J. and H. Dacosta, 1996: Les Ecoulements Non Perennes sur les Petits Bassins du Senegal. In: *Hydrologie Tropicale: Geoscience et Outil pour le Developpement* [Chevallier, P. and B. Pouyard (eds.)]. Publication 238, International Association of Hydrologic Sciences/Association Internationale des Sciences Hydrologiques, Wellngford, UK, 436 pp.

Al-Rabghi, O.M., M.H. Al-Beirutty, and K.A. Fathalah, 1999: Estimation and measurement of electric energy consumption due to air conditioning cooling load. *Energy Conversion and Management*, **40,** 1527–1542.

Amano, A., 1996: Energy price change and energy intensity. *Nihon Keizai Kenkyu*, **30(7),** 184–1990 (in Japanese.)

Angioletti, R., 1996: Vingt-quatre criteres pour concevoir et construire un batiment dans un logique de developpement durable. *Cahiers du Centre Scientifique et Technique du Batiment*, **366,** 29 (in French).

Arnell, N.W., 1998: Climate change and water resources in Britain. *Climatic Change*, **39,** 83–110.

Avissar, R., 1998: Potential effects of vegetation on the urban thermal environment. *Atmospheric Environment*, **30(3),** 437–448.

Balarus, C.A., K. Droutsa, A.A. Argiriou, and D.N. Asimakopoulos, 2000: Potential for energy conservation in apartment buildings. *Energy and Buildings*, **31,** 143–154.

Barnes, D.F., J. Dowd, L. Qian, K. Krutilla, and W. Hyde, 1998: *Urban Energy Transitions, Poverty, and the Environment: Understanding the Role of the Urban Household Energy in Developing Countries.* ESMAP, The World Bank, Washington, DC.

Bartone, C., 1998: Urban environmental management strategies and action plans in São Paulo and Kumasi. In: *Environmental Strategies for Sustainable Development in Urban Areas: Lessons from Africa and Latin America* [Fernandes, E. (ed.)]. Ashgate, Aldershot, United Kingdom, pp. 83–97.

Belzer, D.B., M.J. Scott, and R.D. Sands, 1996: Climate change impacts on U.S. commercial building energy consumption: an analysis using sample survey data. *Energy Sources*, **18(2),** 177–201.

Boko, M., 1994: Housing and water pollution in urban areas of south Benin (West Africa): a case study of lack of environmental education. In: *Integrated Land and Water Management: Challenges and New Opportunities. Proceedings of the Fourth Stockholm Water Symposium (9–13 August, 1994).* Publication No. 4, Stockholm International Water Institute, Stockholm, Sweden, pp. 153–162.

Boko, M., 1993: Geomorphology, rainfall pattern, human settlements and water pollution in the rural areas of the coastal plain of Benin (West Africa). In: *Proceedings of the Third International Water Symposium, Stockholm, Sweden.* Stockholm International Water Institute, Stockholm, Sweden, pp. 409–416.

Boko, M., 1991: Valuation and forecast of climatic risks in hydroagricultural settlements management: methodological aspects; international seminar on climatic fluctuations and water management, Cairo, 1989. *Water Resources Development*, **7(1),** 60–63.

Boland, J.J., 1997: Assessing urban water use and the role of water conservation measures under climate uncertainty. *Climatic Change*, **37,** 157–167.

Brown, R.A. and N.J. Rosenberg, 1996: The potential for biomass energy production in the Missouri-Iowa-Nebraska-Kansas (MINK) region. In: *Proceedings of the 1995 Society of American Foresters Convention, Octover 28–November 1, 1995, Portland, ME.* Society of American Foresters, Bethesda, MD, USA, pp. 204–209.

Brown, R.A., N.J. Rosenberg,, W.E. Easterling III, and C. Hays, 1998: *Potential Production of Switchgrass and Traditional Crops under Current and Greenhouse-Altered Climate in the "MINK" Region of the Central United States. Report to Texas A&M Under Subcontract 22023.* PNWD-2432, Pacific Northwest National Laboratory, Richland, WA, USA, 71 pp.

Cairncross, S., 1990: Water supply and the urban poor. In: The Poor Die Young: Housing and Health in Third World Cities [Cairncross, S., J. Hardoy, and D. Satterthwaite(eds.)]. Earthscan Publications, London, United Kingdom, pp. 109–126.

Chan, K.T and W.K. Chow, 1998: Energy impact of commercial -building envelopes in the sub-tropical climate. *Applied Energy*, **60,** 21–39.

Changnon, S.A., 2000: Impacts of hail in the United States. In: *Storms, Vol. II* [Pielke Jr., R.A. and R.A. Pielke Sr. (eds.)]. Routledge, London, United Kingdom, , pp. 163–191.

Changnon, S.A., 1996a: Effects of summer precipitation on urban transportation. *Climatic Change*, **32(4),** 481–494.

Changnon, S.A. (ed.), 1996b: *The Great Flood of 1993: Causes, Impacts, Responses.* Westview Press, Boulder, CO, USA, 321 pp.

Changnon, S.A., 1992: Inadvertent weather modification in urban areas: lessons for global climate change. *Bulletin of the American Meteorological Association*, **73,** 6619–6627.

Changnon, S.A. and M. Demissie, 1996: Detection of changes in streamflow and floods resulting from climate fluctuations and land use-drainage changes. *Climatic Change*, **32,** 411–421.

Changnon, S.A. and M.H. Glantz, 1996: The Great Lakes diversion at Chicago and its implications for climate change. *Climatic Change*, **32,** 199–214.

Changnon, S., R.A. Pielke Jr., D. Changnon, R.T. Sylves, and R.Pulwarty, 2000: Human factors explain the increased losses from weather and climate extremes. *Bulletin of the American Meteorological Society*, **81,** 437–442.

Charlot-Valdieu, C. and P. Outrequin, 1999: La ville et le developpement durable. *Cahiers du CSTB*, **3,** 35 pp.

Chestnut, L.G., W.S. Breffle, J.B. Smith, and L.S. Kalkstein, 1998: Analysis of differences in hot-weather-related mortality across 44 U.S. metropolitan areas. *Environmental Science and Policy*, **59,** 59–78.

China Country Study Team, 1999: *China Climate Change Country Study.* Research Team of China Climate Change Country Study, Bejing, China.

Choudhury, S.H.M., 1998: *Report on the Bangladesh Flood 1998: Chronology, Damages, Reponse.* Management Information and Monitoring (MIM) Division, Disaster Management Bureau, Government of Bangladesh, Dhaka, Bangladesh.

Cohen, S.J. (ed.), 1997: *Mackenzie Basin Impact Study Final Report.* Environment Canada, Downsview, ON, Canada, 372 pp.

Colwell, R.R., 1996: Global climate and infectious disease: the cholera paradigm. *Science*, **274,** 2025–2031.

Connolly, P., 1999: Mexico City: our common future? *Environment and Urbanization*, **11(1),** 53–78.

Daniere, A., 1996: Growth, inequality, and poverty in southeast Asia: the case of Thailand. *Third World Planning Review*, **18(4),** 373–395.

De, U.S. and R.K. Mukhopadhyay, 1998: Severe heat wave over the Indian subcontinent in 1998: perspective of global climate. *Current Science*, **75(12),** 1308–1311.

Devas, N. and C. Rakodi, 1993: The urban challenge. In: *Managing Fast-Growing Cities* [Devas, N. and C. Rakodi (eds.)]. Longman Group, Essex, United Kingdom, and John Wiley & Sons, Inc., New York, NY, USA, 387 pp.

Dockery, D.W., C.A. Pope, X. Xu, J.D. Spangler, J.H. Ware, M.E. Fay, B.G. Ferris, and F.E. Speizer, 1993: An association between air pollution and mortality in six U.S. cities. *New England Journal of Medicine*, **329(4),** 1753–1759.

Dorland, C., R.S.J. Tol, and J.P. Palutikof, 1999: Vulnerability of the Netherlands and northwest Europe to storm damage under climate change: a model approach based on storm damage in the Netherlands. *Climatic Change*, **43,** 513–535.

Douguédroit, A., 1997: On relationships between climate variability and change and societies. In: *Climate and Societies: A Climatological Perspective* [Yoshino, M., M. Domrös, A. Douguédroit, J. Paszynski, and L.C. Nkemdim (eds.)]. Kluwer Academic Publishers, Dortrecht, The Netherlands, pp. 21–41.

Douguédroit, A., J.-P. Marchand, M.F. de Saintignon, and A. Vidal, 1997: Impacts of climate variability on human activities. In: *Climate and Societies: A Climatological Perspective* [Yoshino, M., M. Domrös, A. Douguédroit, J. Paszynski, and L.C. Nkemdim (eds.)]. Kluwer Academic Publishers, Dortrecht, The Netherlands, pp. 119–150.

Du, M., M. Yoshino, Y. Fujita, S. Arizono, T. Maki, and J. Lei, 1996: Climate change and agricultural activities in the Taklimakan Desert, China, in recent years. *Journal of Arid Land Studies*, **5(2),** 173–183.

Duncan, K., T. Guidotti, W. Chung,, K. Naidoo, G. Gibson, L. Kalkstein, S. Sheridan, D. Walter-Toews, S. MacEachern, and J. Last, 1999: *Health Sector. Canada Country Study: Climate Impacts and Adaptation, Volume VII* [Koshida, G. and W. Avis (eds.)]. Environment Canada, Downsview, ON, Canada, pp. 501–590.

ECLAC, 2000: The equity gap: a second look. In: *Proceedings of the Second Regional Conference in Follow-up to the World Summit for Social Development, Santiago, Chile, 15–17 May, 2000* [Deampo, J.A. and R. Franco (eds.)]. United Nations Economic Commission for Latin America and the Caribbean, Santiago, Chile, 303 pp. (in Spanish and English).

ECLAC, 1994: *Social Panorama of Latin America, 1994.* United Nations Economic Commission for Latin America and the Caribbean, Santiago, Chile.

EEPSEA, 2000: *The Indonesian Fires and Haze of 1997: The Economic Toll and the World Wide Fund for Nature.* Economy and Environment Program for SE Asia, World Wildlife Fund, International Development Research Centre, Ottawa, Canada, 9 pp. Available via e-mail at info@idrc.ca.

EIA, 1999a: *Country Analysis Briefs: North America.* U.S. Department of Energy, Energy Information Administration, Washington, DC, USA. Available online at http://www.eia.doe.gov/emeu/cabs/special.html.

EIA, 1999b: Energy in Africa. In: *Country Analysis Briefs: Special Topics.* U.S. Department of Energy, Energy Information Administration, Washington, DC, USA, 44 pp. Available online at http://www.eia.doe.gov/emeu/cabs/special.html.

EIA, 1999c: Energy in the Americas. In: *Country Analysis Briefs.* U.S. Department of Energy, Energy Information Administration, Washington, DC, USA, 50 pp. Available online at http://www.eia.doe.gov/emeu/cabs/special.html.

EIA, 1998: *International Energy Outlook.* DOE/EIA-0484(98), U.S. Department of Energy, Energy Information Administration, Washington, DC, USA, 217 pp.

El-Raey, M., 1997: Vulnerability assessment of the coastal zone of the Nile Delta of Egypt, to the impacts of sea level rise. *Ocean and Coastal Management*, **37(1),** 29–40.

Epstein, P.R., H.F. Diaz, S. Eleas, G. Grabherr, N.E. Graham, W.J.M. Martens, E. Mosely-Thompson, and J. Suskind, 1998: Biological and physical signs of of climate change focus on mosquito-borne disease. *Bulletin of the American Meteorological Society*, **78,** 409–417.

Evans, S.G. and J.G. Clague, 1997: The impact of climate change on catastrophic geomorphic processes in the mountains of British Columbia, Yukon, and Alberta. In: *Responding to Global Climate Change in British Columbia and Yukon. Canada Country Study: Climate Impacts and Adaptation, Volume I* [Taylor, E. and B. Taylor (eds.)]. Environment Canada, Toronto, ON, Canada, 363 pp.

Fankhauser, S., R.S.J. Tol, and D.W. Pearce, 1998: Extensions and alternative to climate change impact valuation: on the critique of IPCC Working Group III's impact estimates. *Environment and Development Economics*, **3(1),** 59–81.

FAO, 1999: *The State of Food Insecurity in the World, 1999*. Food and Agriculture Organization, Rome, Italy, 35 pp.

FEMA, 2000: *Mitigation Programs and Activities: Implementing Regulations at U.S. Code of Federal Regulations 44 CFR 206.430*. Federal Emergency Management Agency, Washington, DC, USA. Available online at http://www.fema.gov/mit/program.htm.

FEMA, 1997: *Report on the Cost and Benefits of Natural Disaster Mitigation*. Federal Emergency Management Agency, Washington, DC, USA, 57 pp.

Foronda, M.E., 1998: Chimbote's Local Agenda 21: Initiatives to Support Its Development and Implementation. *Environment and Urbanization*, **10(2),** 129–147.

Fowler, A.M. and K.J. Hennessy, 1995: Potential impacts of global warming on the frequency and magnitude of heavy precipitation. *Natural Hazards*, **11,** 283–303.

Frederick, K.D., 1997: Adapting to climate impacts on the supply and demand for water. *Climatic Change*, **37,** 141–156.

Fukuma, Y., 1999/2000: Comparison among countries of countermeasures against tropical cyclone disasters and evaluation methods. *Global Environment Research*, **3(1),** 59–67.

Fukuma, Y., 1996: Effectiveness of prevention measures against typhoons in Wakayama Prefecture based on calibrated wind speed data. *Journal of Meteorological Research*, **48(3),** 103–110 (in Japanese with English abstract).

Gardette, J.M., 1997: Insurability of river floods? Legal comparison between France and Germany. *Zeitschrift für die Gesamit Versicherungswissenschaf*, **86,** 211–232.

Gavidia, J., 1994: Housing and land in large cities of Latin America. In: *Enhancing the Management of Metropolitan Living Environments in Latin America*. United Nations Centre for Regional Development, Nagoya, Japan, pp. 19–27.

Georgakakos, A.P., H. Yao, M.G. Mullusky, and K.P. Georgakakos, 1998: Impacts of climate variability on the operational forecast and management of the upper Des Moines River basin. *Water Resources Research*, **34(4),** 799–821.

Gilbert, R., D. Stevenson, H. Girardet, and R. Stren, 1996: *Pour des Villes Durables, le Rôle des Autorités Locales dans L'environnement Urbain (For Sustainable Towns, the Role of Local Authorities in the Urban Environment)*. La Fédération mondiale des Cités unies, Paris, France, 154 pp. (in French).

Glass, G.E., J.E. Cheek, J.A. Patz, T.M. Shields, T.J. Doyle, D.A. Thoroughman, D.K. Hunt, R.E. Enscore, K.L. Gage, C. Irland, C.J. Peters, and R. Bryan, 2000: Using remotely sensed data to identify areas at risk for hantavirus pulmonary syndrome. *Emerging Infectious Disease*, **6(3),** 238–247.

Gleick, P.H., 2000: *Water: The Potential Consequences of Climate Variability and Change for the Water Resources of the United States*. Report of the Water Sector Assessment Team of the National Assessment of the Potential Consequences of Climate Variability and Change. Pacific Institute for Studies in Development, Environment, and Security, San Francisco, CA, USA, 151 pp.

Gooneratne, N.V., 1998: Why does Colombo City get flooded? In: *Volume I: Proceedings of the Workshop on the Role of R&D Institutions in Natural Disaster Management, Colombo, Sri Lanka, 10–11 September, 1998*. pp. 6-1 to 6-7.

Gugler, J., 1988: Overurbanization reconsidered. In: *Urbanization of the Third World* [Gugler, J. (ed.)]. Oxford University Press, Oxford, United Kingdom, pp. 74–92.

Hardoy, J.E., D. Mitlin, and D. Satterthwaite, 2000: *Environmental Problems in an Urbanizing World: Local Solutions for City Problems in Africa, Asia, and Latin America*. Earthscan Publications, London, United Kingdom, 442 pp.

Harrison, S.J., C. Sheppard, and S.J. Winterbottom, 1999: The potential effects of climate change on the Scottish tourist industry. *Tourism Management*, **20,** 203–211.

Hattori, T.O. Kadota, and N. Watanabe, 1991: Analysis and prospect of tight power demand in summer season. *Energy Keizai*, **17(4),** 16–22 (in Japanese).

Haughton, G., 1999: Information and participation within environmental management. *Environment and Urbanization*, **11(2),** 51–62.

Hennessy, K.J., J.M. Gregory, and J.F.B. Mitchell, 1997: Changes in daily precipitation under enhanced greenhouse conditions. *Climate Dynamics*, **13,** 667–680.

Herrington, R., B. Johnson, and F. Hunter, 1997: *Responding to Global Climate Change in the Prairies. Canada Country Study: Climate Impacts and Adaptation, Volume III* [Taylor, E. and B. Taylor (eds.)]. Environment Canada, Downsview, ON, Canada, 270 pp.

Hirsch, K., 1999: *Canada's Wildland Urban Interface: Challenges and Solutions* [Atlack, I. (ed.)]. Bombadier Aerospace, Montreal, Canada, pp. 2–5.

Huang, J.C.K., 1997: Climate change and integrated coastal management: a challenge for small island states. *Ocean and Coastal Management*, **37(1),** 95–107.

Ichinose, T., 1996: *Analyses on Energy Consumption in Urban Area and Urban Thermal Environment Based on Very Precise Digital Land Use Data Set*. Diss. University of Tokyo, Tokyo, Japan, 247 pp. (in Japanese).

ICLEI, 1995: *Limiting Automobile Use Through Integrated Transportation Demand Management: Republic of Singapore, Case Study No. 38*. International Council for Local Environmental Initiatives, Toronto, ON, Canada, 35 pp.

Institute of Land Development and Regional Economy, 1998: *Report on Population, Resources and Environment of China*. Institute of Land Development and Regional Economy, State Planning Committee, Chinese Environmental Science Press, Beijing, China, 142 pp. (in Chinese).

IPCC, 1998: *The Regional Impacts of Climate Change: An Assessment of Vulnerability. Special Report of IPCC Working Group II* [Watson, R.T., M.C. Zinyowera, and R.H. Moss (eds.)]. Intergovernmental Panel on Climate Change, Cambridge University Press, Cambridge, United Kingdom and New York, NY, USA, 517 pp.

IPCC, 1996: *Climate Change 1995: Impacts, Adaptations, and Mitigation of Climate Change: Scientific-Technical Analyses. Contribution of Working Group II to the Second Assessment Report of the Intergovernmental Panel on Climate Change* [Watson, R.T., M.C. Zinyowera, and R.H. Moss (eds.)]. Cambridge University Press, Cambridge, United Kingdom and New York, NY, USA, 880 pp.

Jaglin, S., 1994: Why mobilize town dwellers? Joint management in Ouagadougou (1983–1990). *Environment and Urbanization*, **6(2),** 113–114.

Jáuregui, E., 1998: Long-term effects of urbanization on the thermal climate of two megacities and impact on space energy demand. In: *Book of Abstracts International Symposium on Human Biometrorology, Fuji-Yoshida, Yamanashi, Japan, 31 August–2 September* [Shibata, M., M. Iriki, K. Kanosue, and Y. Inaba (eds.)]. Yamashi Institute of Environmental Science, Yamashi, Japan, p. 10.

Jáuregui, E., 1997: Climates of tropical and subtropical cities. In: *Climate and Societies: A Climatological Perspective* [Yoshino, M., M. Domrös, A. Douguédroit, J. Paszynski, and L.C. Nkemdim (eds.)]. Kluwer Academic Publishers, Dortrecht, The Netherlands, pp. 361–373.

Jáuregui, E. and E. Luyando, 1999: Global radiation attenuation by air pollution and its effects on the therrnal climate in Mexico City region. *International Journal of Climatology*, **19,** 683–694.

Katupotha, J., 1994: Landslides in Sri Lanka in the 21st century. In: *Proceedings of the National Symposium on Landslides in Sri Lanka, Colombo, Sri Lanka, March 1994*. National Building Research Organization, Ministry of Housing, Construction, and Urban Development, Colombo, Sri Lanka, pp. 161–168.

Kerry, M., G. Kelk, D. Etkin, I. Burton, and S. Kalhok, 1999: Glazed over: Canada copes with the ice storm of 1998. In: *A Report from the Adaptation Learning Experiment* [Kerry, M., I. Burton, S. Kalhok, and G. Kelk (eds.)]. Environment Canada, Toronto, ON, Canada, pp. 51–58.

Kripowicz, R.S., 1998: *Statement of Robert S. Kripowicz, Acting Assistant Secretary for Fossil Energy, U.S. Department of Energy, Before the Subcommittee on Energy and Environment Committee on Science, U.S. House of Representatives, September 15, 1998 [S. 1418, The Methane Hydrate Research and Development Act]*. Congressional Record, Washington, DC, USA, 6 pp.

Kumagai, H., 1996: Concept of "sustainable and regional agriculture" in Japan. In: *Geographical Perspectives on Sustainable Rural Systems* [Sasaki, H., T. Morimoto, I. Saito, and A. Tabayashi (eds.)]. Kaisei Publications, Tokyo, Japan, pp. 36–43.

Lam, J.C., 2000: Energy analysis of commercial buildings in subtropical climates. *Building and Environment*, **35,** 19–26.

Lam, J.C., 1999: Climatic influences on the energy performance of air-conditioned buildings. *Energy Conversion and Management*,**40,** 39–49.

Landsea, C.W., N. Nicholls, W.M. Gray, and L.A. Avila, 1996: Downward trends in the frequency of intense Atlantic hurricanes during the past five decades. *Geophysical Research Letters*, **23,** 1697–1700.

Leach, G. and R. Mearns, 1989: *Beyond the Woodfuel Crisis—People, Land and Trees in Africa*. Earthscan Publications, London, United Kingdom, 309 pp.

Le Treut, H., 1997: Climate of the future: an evaluation of the current uncertainties. In: *Climate and Societies: A Climatological Perspective* [Yoshino, M., M. Domrös, A. Douguédroit, J. Paszynski, and L.C. Nkemdim (eds.)]. Kluwer Academic Publishers, Dortrecht, The Netherlands, pp. 99–117.

Lettenmaier, D.P., A.E. Keizur, R.N. Palmer, A.W. Wood, and S.M. Fisher, 1998: *Water Management Implications of Global Warming: 2. The Boston Water Supply System*. Institute for Water Resources, U.S. Army Corps of Engineers, Fort Belvoir, VA, USA.

Lewis, N., M. Hamnett, U. Prasad, L. Tran, and A. Hilton, 1998: Climate, ENSO, and health in the Pacific: research in progress. *Pacific Health Dialog*, **5(1),** 187–190.

Liu, S.K., 1997: Using coastal models to estimate effects of sea level rise. *Ocean and Coastal Management*, **37(1),** 85–94.

Magaña, V. (ed.), 1999: *The Impacts of El Niño in Mexico*. Kluwer Academic Publishers, Dordrecht The Netherlands, p. 207.(in Spanish)

Maki, T., M. Du, and B. Pan, 1997a: Desertification of agricultural land, arid climate, crop growth and prevention of sand movement in Xinjiang of Northwest China. *Journal of Arid Land Studies*, **7(2),** 273–276.

Maki, T., M. Du, R. Sameshima, and B. Pan, 1997b: Sand dune movement in Xinjiang of Northwest China and prevention of desertification by wind break acilities in arid lands. *Journal of Agricultural Meteorology*, **52(5),** 633–636.

Matsuura, T., M. Yumoto, S. Iizuka, and R. Kawamura, 1999: Typhoon and ENSO simulation using a high-revolution coupled GCM. *Geophysical Research Letters*, **26(12),** 1755–1758.

McGranahan, G. and D. Satterthwaite, 2000: Environmental health or sustainability? Reconciling the brown and green agendas in urban development. In: *Sustainable Cities in Developing Countries* [Pugh, C. (ed.)]. Earthscan Publications, London United Kingdom, pp. 73–90.

Meehl, G.A., 1996: Vulnerability of freshwater resources to climate change in the tropical Pacific region. *Water, Air, and Soil Pollution*, **92,** 203–213.

Mendelsohn, R. and J. Neumann (eds.), 1998: *The Economic Impact of Climate Change on the United States Economy*. Cambridge University Press, Cambridge, United Kingdom and New York, NY, USA, 320 pp.

Mendelsohn, R. and M.E. Schlesinger, 1999: Climate response functions. *Ambio*, **28(4),** 362–366.

Meltzoff, S.K., K. Broad, and C.M. Farias, 1997: ENSO and Chilean fisheries: a case study in social complexity. *The ENSO Signal*, **6,** 4–8. Available online at at http://www.ogp.noaa.gov/library/ensosig6.htm.

Mexico City Ministry of the Environment, 2000: *Bimonthly Report on Mexico City's Air Quality*. Office of the Prevention and Control of Air Pollution, Mexico City Ministry of the Environment, Mexico City, Mexico.

Meze-Hausken, E., 2000: Migration caused by climate change: how vulnerable are people in dryland areas? A case study of northern Ethiopia. *Mitigation and Adaptation Strategies of Global Change*, **5(4),** 379–406.

Miller, A., S. Gautam, and G.H. Wolff, 2000: *What's Fair? Consumers and Climate Change*. Redefining Progress. San Francisco, CA, USA, 67 pp.

Miller, K.A., 2000: Managing supply variability: the use of water banks in the western United States. In: *Drought* [Wilhite, D.A. (ed.)]. Routledge, London, United Kingdom, Vol. II, pp. 70–86.

Mimura, N. and H. Harasawa (eds.), 2000: *Data Book of Sea-Level Rise 2000*. Center for Global Environmental Research, National Institute for Environmental Studies, Tsukuba, Japan, 128 pp.

Mimura, N., T. Ichinose, H. Kato, J. Tsutsui, and K. Sakaki, 1998: Impacts on infrastructure and socio-economic system. In: *Global Warming: The Potential Impact on Japan* [Nishioka, S. and H. Harasawa (eds.)]. Springer-Verlag, Tokyo, Japan, pp. 165–201.

Ministry of Commerce of New Zealand, 1998: *Inquiry into the Auckland Power Supply Failure*. Ministry of Commerce of New Zealand, Wellington, New Zealand, 193 pp.

Minnery, J.R. and D.I. Smith, 1996: Climatic change, flooding, and urban infrastructure. In: *Greenhouse: Coping with Climate Change* [Bouma, W.J., M.R. Manning, and G.I. Pearman (eds.)]. Commonwealth Scientific and Industrial Research Organisation Publishing, Melbourne, Australia, pp. 235–247.

Miranda, L. and M. Hordijk, 1998: Let us build cities for life: the national campaign of local agenda 21s in Peru. *Environment and Urbanization*, **10(2),** 69–102.

Morris, R.E., P.D. Guthrie, and C.A. Knopes, 1995: Photochemical analysis under global warming conditions. In: *Proceedings of the 88th Air and Waste Management Association Annual Meeting, June 18–23, 1995, San Antonio, Texas*. Air and Waste Management Association, Pittsburgh, PA, USA.

Morris, R.E., M.W. Gery, M.-K. Liu, G.E. Moore, C. Daly, and S.M. Greenfield, 1989: Sensitivity of a regional oxidant model to variations in climate parameters. In: *The Potential Effects of Global Climate Change on the United States, Appendix F: Air Quality* [Smith, J.B. and D.A. Tirpak (eds.)]. EPA-230-05-89-056, Environmental Protection Agency, Washington, DC, USA.

Moss, R., and S.H. Schneider, 2000: Uncertainties in the IPCC TAR: recommendations to lead authors for a more consistent assessment and reporting. In: *Guidance Papers on the Cross Cutting Iissues of the Third Assessment Report of the IPCC* [Pachauri, R., T. Taniguchi, and K. Tanaka (eds.)]. Intergovernmental Panel on Climate Change, Geneva, Switzerland, pp. 33–51.

Munasinghe, M., 2000: Development, equity and sustainability (DES) and climate change. In: *Guidance Papers on the Cross Cutting Issues of the Third Assessment Report of the IPCC* [Pachauri, R., T. Taniguchi, and K. Tanaka (eds.)]. Intergovernmental Panel on Climate Change, Geneva, Switzerland, pp. 69–110.

National Research Council, 1995: *Mexico City's Water Supply: Improving the Outlook for Sustainability*. Academía de la Investigación Científica, Academía Nacional de Ingeniería, National Academy Press, Washington, DC, USA, 256 pp.

National Research Council, 1993: *Managing Wastewater in Coastal Urban Areas*. Committee on Wastewater Management for Coastal Urban Areas, National Academy Press, Washington, DC, USA, 496 pp.

Nellis, M.D., 1996: The sustainability of agricultural systems: geographic perspectives. In: *Geographical Perspectives on Sustainable Rural Systems* [Sasaki, H., I. Saito, A. Tabayashi, and T. Morimoto (eds.)]. Kaisei Publications, Tokyo, Japan, pp. 7–13.

Nicholls, R.J. and M.J. Hoozemans, 1996: The Mediterranean: vulnerability to coastal implications of climate change. Ocean and Coastal Management, **31,** 105–132.

Nicholls, R.J. and N. Mimura, 1998: Regional issues raised by sea-level rise and their policy implications. *Climate Research*, **11(1),** 5–18.

Nicholls, R.J., F.M.J. Hoozemans, and M. Marchand. 1999: Increasing flood risk and wetland losses due to global sea-level rise: regional and global analyses. *Global Environmental Change*, **9,** S69–S87.

Parish, R. and D.C. Funnell, 1999: Climate change in mountain regions: some possible consequences in the Moroccan High Atlas. *Climatic Change*, **9(1),** 15–58.

Peart, R.M., R.B. Curry, C. Rosenzweig, J.W. Jones, K.J. Boote, and L.H. Allen Jr., 1995: Energy and irrigation in south eastern U.S. agriculture under climate change. *Journal of Biogeography*, **22,** 635–642.

Penner, J.E., P.S. Connell, D.J. Wuebbles, and C.C. Covey, 1989: Climate change and its interaction with air chemistry: perspectives and research needs. In: *The Potential Effects of Global Climate Change on the United States, Appendix F: Air Quality* [Smith, J.B. and D.A. Tirpak (eds.)]. EPA-230-05-89-056, Environmental Protection Agency, Washington, DC, USA.

Penning-Powswell, E.G., J. Handmer, and S. Tapsell, 1996: Extreme events and climate change: floods. In: *Climatic Change and Extreme Events* [Downing, T.E., A.A. Olsthoorn, and R.S.J. Tol (eds.)]. Vrije Universiteit, Amsterdam, The Netherlands, pp. 97–127.

Perlman, J., 1998: Towards sustainable megacities in Latin America and Africa. In: *Environmental Strategies for Sustainable Development in Urban Areas: Lessons from Africa and Latin America* [Fernandes, E. (ed.)]. Ashgate, Aldershot, United Kingdom, pp. 109–135.

Phelps, P., 1996: *Conference on Human Health and Climate Change: Summary of the Proceeding*. National Academy Press, Washington, DC, 64 pp.

Piel, C., I. Perez, and T. Maytraud, 1998: Trois exemples d'espaces temporairement inondables in milieu urbain dense: une application du developpement durable. *Nouvelles Technologies en Assainissement Pluvial*, **4** (in French).

Pielke, R.A. Jr., 1997: Trends in hurricane impacts in the United States. In: *Report of Workshop on the Social and Economic Impacts of Weather, 2-4 April, 1997, Boulder, CO*. Environmental and Societal Impacts Group, National Center for Atmospheric Research, Boulder, CO, USA. Available online at http://www.esig.ucar.edu/socasp/weather1/index.html.

Pielke, R.A. Jr., 1996: *Midwest Flood of 1993: Weather, Climate, and Societal Impacts*. Environmental and Societal Impacts Group, National Center for Atmospheric Research, 159 pp.

Pielke, R.A. Jr. and C.W. Landsea, 1998: Normalized hurricane damages in the United States: 1925–95. *Weather and Forecasting*, **13,** 621–631.

Pope, C.A. III, M.J. Thun, M.M. Namboordiri, D.W. Dockery, J.S. Evans, F.E. Speizer, and C.W. Heath,, 1995: Particulate air pollution as a predictor of mortality in a prospective study of U.S. adults. *American Journal of Respiratory and Critical Care Medicine*, **151(3),** 669–674.

Priscoli, J.D., 1998: Water and civilization: using history to reframe water policy debates and to build a new ecological realism. *Water Policy*, **1,** 623–636.

Qualley, W., 1997: *Impact of Weather on and Use of Weather Information by Commercial Airline Operations. Workshop on the Social and Economic Impacts of Weather, 2–4 April, 1997, Boulder, CO*. National Center for Atmospheric Research, Boulder, CO, USA.

Quattrochi, D.A., 1996: Cities as urban ecosystems. In: *Proceedings of the Pecora 13 Symposium, 20-22 August 1996, Sioux Falls, South Dakota*. APRS, Bethesda, MD, USA.

Rabinovitch, J., 1998: Global, rgional, and local perspectives towards sustainable urban and rural development. In: *Environmental Strategies for Sustainable Development in Urban Areas: Lessons from Africa and Latin America* [Fernandes, E. (ed.)]. Ashgate, Aldershot, United Kingdom, pp. 16–44.

Rahman, A., H.K. Lee, and M.A. Khan, 1997: Domestic water contamination in rapidly growing megacities of asia: case of Karachi, Pakistan. *Environmental Monitoring and Assessment*, **44(1–3),** 339–360.

Ren, M., 1994: Relative sea-level rise in China and its socioeconomic implications. *Marine Geodesy*, **17,** 37–44.

Ribot, J.C., 1993: Forestry policy and charcoal production in Senegal. *Energy Policy*, **21(5),** 559–585.

Rosenberg, N.J. (ed.), 1993: *Towards an Integrated Impact Assessment of Climate Change: The MINK Study*. Kluwer Academic Publishers, Dordrecht, The Netherlands, 173 pp.

Rosenzweig, C. and W.D. Solecki (eds.), 2000: *Climate Change and a Global City: The Metropolitan East Coast Regional Assessment*. Columbia Earth Insitute, New York, NY, USA, 4 pp.

Rosquillas, A.H., 1998: Effects of El Niño in Tijuana, Mexico. *Prevención*, **21,** 32–36.

Sailor, D.J., 1998: Simulations of annual degree day impacts of urban vegetative augmentation. *Atmospheric Environment*, **32(1),** 43–52.

Sakamoto, H., 1996: Some experiences of sustainable agriculture in Japan. In: *Geographical Perspectives on Sustainable Rural Systems* [Sasaki, M., A. Tabayashi, and T. Morimoto (eds.)]. Kaisei Publications, Tokyo, Japan, pp. 203–209.

Samson, P.J., S.J. Augustine, and S. Sillman, 1989: Linkages between global climate warming and ambient air quality. In: *Global Change Linkages: Acid Rain, Air Quality, and Stratospheric Ozone* [White, J.C. (ed.)]. Elsevier, New York, NY, USA, 262 pp.

Sasaki, H., 1996: Struggle against urbanization of rice cultivation in Niigata Plain, Japan. In: *Geographical Perspectives on Sustainable Rural Systems* [Sasaki, M., A. Tabayashi, and T. Morimoto (eds.)]. Kaisei Publications, Tokyo, Japan, pp. 216–223.

Satterthwaite, D., 1997: Urban poverty: reconsidering its scale and nature. *IDS Bulletin*, **28(2),** 9–23.

Schneider, S.H., 1997: Integrated assessment modeling of global climate change: transparent rational tool for policy making or opaque screen hiding value-laden assumptions? *Environmental Modeling and Assessment*, **2(4),** 229–248.

Schneider, S.H., K. Kuntz-Duriseti, and C. Azar, 2000: Costing non-linearities, surprises, and irreversible events. *Pacific and Asian Journal of Energy*, **10(1),** 81–106.

Shimizu, Y., 1993: *Assessment of Global Warming on Water Demand and Water Resources Management. Proceedings of Earth Environmental Symposium*. Japanese Society of Civil Engineers, Tokyo, Japan, pp. 246–253 (in Japanese with English summary).

Silas, J., 1992: Government-community partnerships in Kampung improvement programme in Surabaya. *Environment and Urbanization*, **4(2),** 35–36.

Simon, D., 1996: *Transport and Development in the Third World*. Routledge, New York, NY, USA, 194 pp.

Smit, B. and J. Smithers, 1993: Sustainable agriculture: interpretations, analyses and prospects. *Canadian Journal of Regional Science*, **16(3),** 499–524.

Smit, B., I. Burton, and R.J.T. Klein, 1998: *The Science of Adaptation: A Framework for Assessment. Working Paper Prepared for IPCC Workshop on Adaptation to Climatic Variability and Change, Costa Rica, 29 March - 1 April 1998*. Intergovernmental Panel on Climate Change, Washington, DC, 15 pp.

Smith, D.I. and J.W. Handmer, 1996: Urban flooding in Australia: policy development and implementation. *Disasters*, **8(2),** 105–117.

Smith, D.I., S. Schreider, A.J. Jakeman, A. Zerger, B.C. Bates, and S.P. Charles, 1999: *Urban Flooding: Greenhouse-Induced Impacts, Methodology and Case Studies*. Resource and Environmental Studies No. 17, Centre for Resource and Environmental Studies, The Australian National University, Canberra, Australia, 68 pp.

Smith, J.B., 1996: Standardized estimates of climate change damages for the United States. *Climatic Change*, **32(3),** 313–326.

Smith, J.B. and D.A. Tirpak (eds.), 1989: *The Potential Effects of Global Climate Change on the United States*. Report to Congress, United States Environmental Protection Agency, EPA-230–05–89–050, Washington, DC, USA, 409 pp.

Spronken-Smith, R. and T. Oke, 1998: The thermal regime of urban parks in two cities with different summer climates. *International Journal of Remote Sensing*, **19,** 2085–2104.

Suman, D., 2001: Las reacciones del sector pesquero Chileno al fenómeno El Niño, 1997–98 (Reactions of the Chilean fishery sector to the El Niño phenomenon, 1997–98). In: *"El Niño" en América Latina, sus Impactos Biológicos y Sociales*. CONSYTEC, Lima, Peru, (in press).

Tabayashi, A., 1996: Sustainability of rice-growing communities in Central Japan. In: *Geographical Perspectives on Sustainable Rural Systems* [Sasaki, M., A. Tabayashi, and T. Morimoto (eds.)]. Kaisei Publications, Tokyo, Japan, pp. 224–240.

Tiffen, M. and M. Mortimore, 1992: Environment, population growth and productivity in Kenya: a case study of Machakos District. *Development Policy Review*, **10,** 359–387.

Timmerman, P. and R. White, 1997: Megahydropolis: coastal cities in the context of global environmental change. *Global Environmental Change*, **7(3),** 205–234.

Tindleni, V., 1998: The role of NGOs and CBOs in a sustainable development strategy for metropolitan Cape Town, South Africa. In: *Environmental Strategies for Sustainable Development in Urban Areas: Lessons from Africa and Latin America* [Fernandes, E. (ed.)]. Ashgate, Aldershot, United Kingdom, 45–61.

Titus, J.G. and V. Narayanan, 1995: *The Probability of Sea Level Rise*. EPA 230-R95-008, U.S. Environmental Protection Agency, Washington, DC, USA, 186 pp.

Tol, R.S.J., 1998: Climate change and insurance: a critical appraisal. *Energy Policy*, **26(3)**, 257–262.

Topping, K., 1996: Mitigation from the ground up. *Disaster Research*,**197**, 1.

Tucker, M., 1997: Climate change and the insurance industry: the cost of increased risk and the impetus for action. *Ecological Economics*, **22(2)**, 85–96.

Tweeten, L., 1995: The structure of agriculture: implications for soil and water conservation. *Journal of Soil and Water Conservation*, **50(4)**, 347–351.

United Nations, 2000: *World Urbanization Prospects: The 1999 Revision*. ESA/P/WP.161, Population Division, Department of Economic and Social Affairs, United Nations, New York, NY, USA.

UNCHS, 1999: *Global Urban Indicators Database*. United Nations Centre for Human Settlements (HABITAT), Oxford University Press, Oxford, United Kingdom and New York, NY, USA. Available online at http://www.urbanobservatory.org/indicators/database.

UNCHS, 1996: *An Urbanizing World: Global Report on Human Settlements 1996*. United Nations Centre for Human Settlements (HABITAT), Oxford University Press, Oxford, United Kingdom and New York, NY, USA, 559 pp.

UNEP, 1992: *Urban Air Pollution in Megacities of the World*. United Nations Environment Programme and the World Health Organization, Blackwell Publishers, Oxford, United Kingdom, 230 pp.

UNEP, 1989: *Implications of Climate Changes in the Wider Caribbean Region. CEP Technical Report No. 3*. United Nations Environment Programme, Caribbean Environment Programme, Kingston, Jamaica, 17 pp.

Upmanis, H., I. Eliasson, and S. Lindqvist, 1998: The influence of green areas on nocturnal temperatures in a high latitude city. *International Journal of Climatology*, **18,** 681–700.

Wall, G., 1998: Implications of global climate change for tourism and recreation in wetland areas. *Climatic Change*, **40(2)**, 371–389.

Walsh, K. and A.B. Pittock, 1998: Potential changes in tropical storms, hurricanes, and extreme rainfall events as a result of climate change. *Climatic Change*, **39,** 199–213.

Weerakkody, U., 1997: Potential impact of accelerated sea-level rise on beaches of Sri Lanka. *Journal of Coastal Research (Special Issue)*, **24,** 225–242.

West, J.J. and H. Dowlatabadi, 1999: On assessing the economic impacts of sea-level rise on developed coasts. In: *Climate Change and Risk* [Downing, T.E., A.J. Olsthoorn, and R.S.J. Tol (eds.)]. Routledge, London, United Kingdom, pp. 205–220.

Wheeler, C., 2000: Counting the cost of the 1997 haze. In: *Reports: Science from the Developing World*. International Development Research Centre, Ottawa, ON, Canada, 4 pp. Available online at http://www.idrc.ca/reports/index.cfm.

Whetton, P.H., R. Galloway, and M.R. Haylock, 1996: Climate change and snow cover duration in the Australian Alps. *Climatic Change*, **32(4)**, 447–479.

WHO, 1992: *Urban Health Problems. Report of the Panel on Urbanization, WHO Commission on Health and the Environment*. World Health Organization, Geneva, Switzerland, 160 pp.

Wolf, A.T., 1998: Conflict and cooperation along international waterways. *Water Policy*, **1(2)**, 251–265.

Wong, J.C.Y., 1998: *Content Coverage and Time-Space Pattern of Environmental News: Taiwan Area as Example, 1986–1995*. Department of Geography, National Taiwan Normal University, Taipei, China, pp. 1–31 (in Chinese with English abstract).

Wong, P.P., 1998: Coastal tourism development in Southeast Asia: relevance and lessons for coastal zone management. *Ocean and Coastal Management*, **38,** 89–109.

Wood, A.W., D.P. Lettenmaier, and R.N. Palmer, 1997: Assessing climate change implications for water resources planning. *Climatic Change*, **37,** 203–228.

World Bank, 1999: *World Development Report 1998/99: Knowledge for Development*. The World Bank, Washington, DC, USA.

World Bank, 1994: *World Development Report 1994: Infrastructure for Development*. The World Bank, Washington, DC, USA.

WRI, 1999: Urban air: health effects of particulates, SO_2, and O_3. In: *China's Health and Environment: Air Pollution and Health Effects*. World Resources Institute, Washington, DC, USA.

WRI, 1996: *World Resources 1996–97*. World Resources Institute, Oxford University Press, New York, NY, USA.

Yohe, G.W. and J. Neumann, 1997: Planning for sea-level rise and shore protection under climate uncertainty. *Climatic Change*, **37(1)**, 243–270.

Yohe, G.W., H. Ameden, P. Marshall, and J. Neumann, 1996: The economic cost of greenhouse-induced sea-level rise for developed property in the United States. *Climatic Change*, **32,** 387–410.

Yoshino, M., 1998: Nature and human life in the Desert of Taklimakan. *Journal of Arid Land Studies*, **8(2)**, 85–94 (in Japanese with English abstract).

Yoshino, M., 1997a: Human activities and environmental change: a climatologist's view. In: *Climate and Societies: A Climatological Perspective* [Yoshino, M., M. Domrös, A. Douguédroit, J. Paszynski, and L.C. Nkemdim (eds.)]. Kluwer Academic Publishers, Dortrecht, The Netherlands, pp. 3–17.

Yoshino, M., 1997b: Agricultural land use and local climate. In: *Climate and Societies: A Climatological Perspective* [Yoshino, M., M. Domrös, A. Douguédroit, J. Paszynski, and L.C. Nkemdim (eds.)]. Kluwer Academic Publishers, Dortrecht, The Netherlands, pp. 381–400.

Yoshino, M., 1997c: *Desertification of China*. Taimeido, Tokyo, Japan, 300 pp. (in Japanese).

Yoshino, M., 1996a: Decadal change of number of deaths and disasters caused by typhoons in Japan. *INCEDE Report*, **1,** 113–121.

Yoshino, M., 1996b: Impacts of ENSO on agricultural production, meteorological disasters and phenology in Asia: some examples. In: *Proceedings of the 28th International Geographical Congress, IGU, 5–7 August, 1996, The Hague, The Netherlands*. pp. 97–110.

Yoshino, M., 1994: Urban systems in a large scale climate context. In: *Report of the Technical Conference on Tropical Urban Climates, 28 March–2 April, 1993, Dhaka, Bangladesh*. WCA SP-30, WMO/TD-No.647, 83–94.

Yoshino, M., K. Urushibara-Yoshino, and W. Suratman, 1997: ENSO and its impacts on agricultural production and population: an Indonesian example. In: *Climate and Life in the Asia-Pacific* [Sirinanda, K.U. (ed.)]. University of Brunei, Dar Es Salaam, Brunei, pp. 251–267.

Zeidler, R., 1997: Climate change vulnerability and response strategies for the coastal zone of Poland. *Climatic Change*, **36,** 151–173.

8

Insurance and Other Financial Services

PIER VELLINGA (THE NETHERLANDS) AND EVAN MILLS (USA)

Lead Authors:
G. Berz (Germany), L. Bouwer (The Netherlands), S. Huq (Bangladesh), L.A. Kozak (USA), J. Palutikof (UK), B. Schanzenbächer (Switzerland), G. Soler (Argentina)

Contributing Authors:
C. Benson (UK), J. Bruce (Canada), G. Frerks (The Netherlands), P. Huyck (USA), P. Kovacs (Canada), A. Olsthoom (The Netherlands), A. Peara (USA), S. Shida (Japan)

Review Editor:
A. Dlugolecki (UK)

CONTENTS

EXECUTIVE SUMMARY

The financial services sector—defined as private and public institutions that offer insurance, banking, and asset management services—is a unique qualitative indicator of the potential socioeconomic impacts of climate change because the sector is sensitive to climate change and offers an integrator of effects on other sectors. This assessment highlights insurance and other components of the financial services sector because they represent a risk-spreading mechanism through which the costs of weather-related events are distributed among other sectors and throughout society. The effects of natural and human-induced climate change on the financial services sector are likely to become manifest primarily through changes in the spatial distribution, frequencies, and intensities of ordinary and catastrophic weather events. There is high confidence that climate change and anticipated changes in weather-related events that are perceived to be linked to climate change would increase actuarial uncertainty in risk assessment and thus in the functioning of insurance markets.

The costs of ordinary and catastrophic weather events have exhibited a rapid upward trend in recent decades. Yearly global economic losses[1] from catastrophic events increased from US$4 billion in the 1950s to US$40 billion yr^{-1} in the 1990s (all 1999 US$). Including events of all sizes increases these totals by approximately two-fold. The insured portion of these losses rose from a negligible level to US$9.2 billion annually during the same period, with a significantly higher insured fraction in industrialized countries. As a measure of increasing insurance industry vulnerability, the ratio of global property/casualty insurance premiums to weather-related losses—an important indicator of adaptive capacity—fell by a factor of three between 1985 and 1999. Chapter 15 discusses insurance issues for North America in depth.

The costs of weather events have risen rapidly despite significant and increasing efforts at fortifying infrastructure and enhancing disaster preparedness. These efforts dampen the observed rise in loss costs to an unknown degree, although the literature attempting to separate natural from human driving forces has not quantified this effect. Demographic and socioeconomic trends are increasing society's exposure to weather-related losses. Part of the observed upward trend in historical disaster losses is linked to socioeconomic factors such as population growth, increased wealth, and urbanization in vulnerable areas, and part is linked to climatic factors such as observed changes in precipitation, flooding, and drought events (e.g., see Section 8.2.2 and Chapter 10). Precise attribution is complex, and there are differences in the balance of these two causes by region and by type of event. Notably, the growth rate in the damage cost of non-weather-related and anthropogenic losses was one-third that of weather-related events for the period 1960–1999 (Munich Re, 2000). Many of the observed upward trends in weather-related losses are consistent with what would be expected under human-induced climate change.

Recent history has shown that weather-related losses can stress insurance companies to the point of bankruptcies, elevated consumer prices, withdrawal of insurance coverage, and elevated demand for publicly funded compensation and relief. Increased uncertainty regarding the frequency, intensity, and/or spatial distribution of weather-related losses will increase the vulnerability of the insurance and government sectors and complicate adaptation efforts.

The financial services sector as a whole is expected to be able to cope with the impacts of future climate change, although low-probability, high-impact events or multiple closely spaced events could severely affect parts of the sector. Trends toward increasing firm size, greater diversification, greater integration of insurance with other financial services, and improved tools to transfer risk all potentially contribute to this robustness. However, the property/casualty insurance and reinsurance

[1] Total economic losses are dominated by direct damages (insured and uninsured)—defined as damage to fixed assets (including property or crops), capital, and inventories of finished and semi-finished goods or raw materials and finished products—that occur simultaneously or as a direct consequence of the natural phenomenon causing a disaster. Economic loss data also can include indirect or other secondary damages such as business interruptions, personal loss (e.g., injuries and death), or temporary relocation expenses for displaced households and businesses, as well as the effect on flow of goods that will not be produced and services that will not be provided. More loosely related damages such as impacts on national gross domestic product (GDP) are not included. Insured losses are a subset of economic losses. The data presented here are based on a diversity of sources compiled by the Geosciences Group at Munich Re for the period 1950–1999, and are unadjusted for purchasing power parity. The particular costs included can vary somewhat among countries and over time. In some cases, country definitions of losses set minimum thresholds for inclusion; thus, the totals presented here are underestimates of actual losses. For example, because of the minimum cost threshold of US$5 million until 1996 and US$25 million thereafter in the United States, no winter storms were included in the statistics for the 46-year period 1949–1974, and few were included thereafter (Kunkel *et al.*, 1999). Although large in aggregate, highly diffuse losses resulting from structural damages from land subsidence (e.g., approaching as much as US$1 billion yr^{-1} during periods of low rainfall in the UK; see Figure 8-3) also would rarely be captured in these statistics.

segments have greater sensitivity, and small, specialized, or undiversified companies even run the risk of bankruptcy. The banking industry as a provider of loans may be vulnerable to climate change under some conditions and in some regions. However, in many cases the banking sector transfers its risk back to the insurers who often purchase debt products.

Adaptation to climate change presents complex challenges, but it also presents opportunities to the sector. [It is worth noting that the term "mitigation" often is used in the insurance and financial services sectors in much the same way that the term "adaptation" is used in the climate research and policy communities.] Regulatory involvement in pricing, tax treatment of reserves, and the (in)ability of firms to withdraw from at-risk markets are examples of factors that influence the resilience of the sector. Management of climate-related risk varies by country and region. Usually it is a mixture of commercial and public arrangements and self-insurance. In the face of climate change, the relative role of each can be expected to change. Some potential response options offer co-benefits (e.g., stemming from climate change mitigation opportunities), in addition to helping the sector adapt to climate changes.

The effects of climate change—in terms of loss of life, effects on investment, and effects on the economy—are expected to be greatest in developing countries. Several countries experience impacts on their GDP as a consequence of natural disasters; damages have been as high as half of GDP in one case. Weather disasters set back development, particularly when funds are redirected from development projects to recovery projects.

Equity issues and development constraints would arise if weather-related risks become uninsurable, prices increase, or availability becomes limited. Increased uncertainty could constrain the availability of insurance and investment funds and thus development. Conversely, more-extensive penetration of or access to insurance would increase the ability of developing countries to adapt to climate change. More widespread introduction of microfinancing schemes and development banking also could be an effective mechanism in helping developing countries and communities adapt.

The need for financial resources for adaptation in developing countries is addressed in the United Nations Framework Convention on Climate Change (UNFCCC) and the Kyoto Protocol. However, development of financing arrangements and analysis of the role of the financial services sector in developed and developing countries still is a relatively unexplored area.

This assessment of financial services identifies some areas of improved knowledge and has corroborated and further augmented conclusions reached in the Intergovernmental Panel on Climate Change's Second Assessment Report (Dlugolecki *et al.*, 1996). It also highlights many areas in which greater understanding is needed—in particular, improved knowledge of future patterns of extreme weather; better analysis of economic losses to determine their causation; exploration of financial resources involved in dealing with climate change damage and adaptation; evaluation of alternative methods to generate such resources; deeper investigation of the sector's vulnerability and resilience to a range of extreme weather event scenarios; and more research into how the sector (private and public elements) could innovate to meet the potential increase in demand for adaptation funding in developed and developing countries, both to spread and to reduce risks from climate change.

8.1. Introduction

Our definition of the financial services sector includes private and public institutions that offer insurance, disaster preparedness/recovery, banking, and asset management services. Analysis of the financial services sector provides a unique opportunity to quantify the potential socioeconomic impacts of climate change and offers a barometer of effects on other sectors (including the government sector). The Intergovernmental Panel on Climate Change (IPCC) Third Assessment Report (TAR) highlights insurance and other components of the financial services sector because they represent a risk-spreading mechanism through which the costs of weather-related events are distributed among other sectors and throughout society. The sector also is among the world's largest and is captured less effectively in other parts of the TAR. The financial services sector also stands to play a central part in adaptation and mitigation activities and is a major source of global and regional data on the costs of weather-related events (Mills, 1996; Changnon *et al.*, 2000; Kunreuther, 2000).

This chapter is about the impact of climate change on the financial services sector, as well as the way this sector can adapt and help society to adapt to climate change. Still, little can be said about the total financial cost of adapting to climate change. Short-term effects are likely to be felt most through changing frequencies and intensities of ordinary and catastrophic weather events.

The Second Assessment Report (SAR) chapter on financial services concluded that "within financial services the property insurance industry is most likely to be directly affected by climate change, since it is already vulnerable to variability in extreme weather events" (Dlugolecki *et al.*, 1996). Experience and analyses over the past 5 years has confirmed the trend of growing weather-related damage costs since the 1950s (see Section 8.2).

The vulnerability of and challenges for the insurance sector, private and public, are addressed in Section 8.3. Section 8.4 discusses the implications for other financial services, such as corporate, retail, and investment banking. There is evidence that the banking and insurance industries have become more aware of opportunities and threats with regard to climate change since the SAR. However, little information is available on climate change impact and adaptation implications for the banking sector.

Climate change impacts are expected to be greatest in the developing world. There is only limited penetration of or access to insurance in these regions. This situation makes these regions more vulnerable and will impair their ability to adapt. Over the past few years, several multilateral organizations and banks have taken initiatives to develop new financial schemes for coping with natural disasters in developing countries (see Section 8.5).

Issues regarding funding for adaptation are addressed in Section 8.6. Although knowledge about the financial services sector, private and public, generally has increased since the SAR, major questions remain. Research could help explore the potential roles of the sector in helping society respond to the challenge of climate change (see Section 8.7).

8.2. Climate Change and Extreme Events that are Relevant to the Financial Services Sector

8.2.1. Present-Day Conditions

Present-day impacts of weather events on financial services are caused mainly by extreme events. Differences in vulnerability exist, caused by geographical location, population distribution, and national wealth. In developing countries, there may be very high mortality from extreme weather but relatively small costs to the financial sector because of low insurance penetration. In developed nations, the loss of life may be much less but may have enormous—even catastrophic—costs to the insurance industry (see Section 8.3.1). Swiss Re (2000b) has compiled lists of the 40 worst catastrophes between 1970 and 1999 in terms of insurance losses and fatalities. These lists show that:

- Of the 40 worst insured losses since 1970, only six were not weather related.
- Nineteen of the weather-related catastrophes affected the United States.
- Twenty-eight were related to windstorm (tropical and temperate latitudes).

In contrast, of the 40 worst events in terms of fatalities, only 16 were weather related, of which 13 occurred in Asia. A list of natural disasters causing billion-dollar losses drawn up by Munich Re (2000; see Table 8-3) shows that, of 30 such disasters, 15 affected the United States and seven affected Europe. Eighteen were related to windstorm. With the exception of earthquakes, all were weather related.

In recent decades, economic and insured losses related to weather extremes have increased rapidly (see Figure 8-1). An important part of this trend is related to socioeconomic factors; another part may be explained by climatic factors. Where trends in climate variables do occur, there are two possible principle causes:

- Variability in the natural modes of variability of the global climate system—for example, the Southern Oscillation, with its two characteristic modes of El Niño and La Niña. In the 1980s and 1990s, El Niño events occurred more frequently and lasted longer. The longest El Niño of the 20th century persisted from 1991 to 1995 and was rapidly succeeded by the most intense El Niño of the 20th century, in 1997–1998 (WMO, 1999).
- Anthropogenic global warming, which may be expected to lead to changes in all attributes of the climate system. Most obviously, we would expect it to lead to an increased frequency of high-temperature

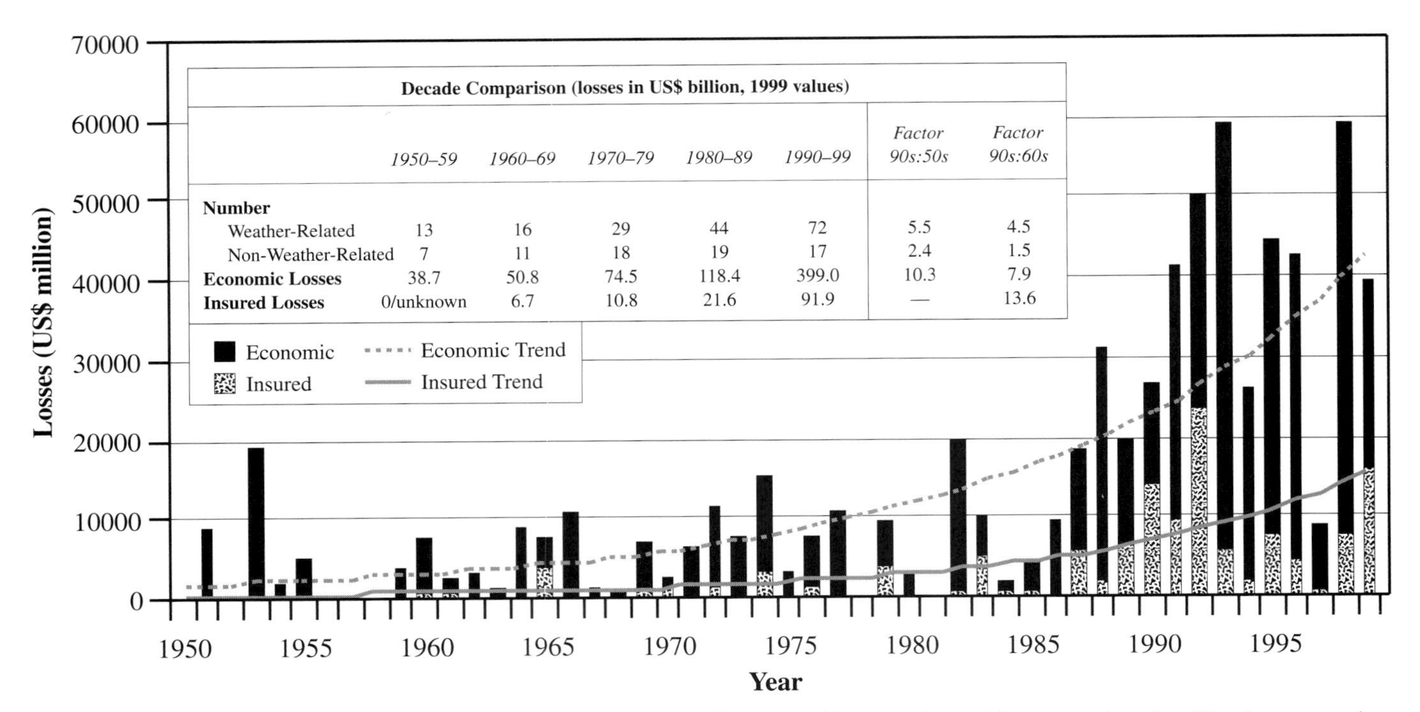

Decade Comparison (losses in US$ billion, 1999 values)

	1950–59	*1960–69*	*1970–79*	*1980–89*	*1990–99*	*Factor 90s:50s*	*Factor 90s:60s*
Number							
Weather-Related	13	16	29	44	72	5.5	4.5
Non-Weather-Related	7	11	18	19	17	2.4	1.5
Economic Losses	38.7	50.8	74.5	118.4	399.0	10.3	7.9
Insured Losses	0/unknown	6.7	10.8	21.6	91.9	—	13.6

Figure 8-1: The costs of catastrophic weather events have exhibited a rapid upward trend in recent decades. Yearly economic losses from large events increased 10.3-fold from US$4 billion in the 1950s to US$40 billion per year in the 1990s (all in 1999 US$). The insured portion of these losses rose from a negligible level to US$9.2 billion annually during the same period, and the ratio of premiums to catastrophe losses fell by two-thirds. Notably, costs are larger by a factor of 2 when losses from ordinary, noncatastrophic weather-related events are included (e.g., as shown in Figure 8-6). The numbers generally include "captive" self-insurers but not the less-formal types of self-insurance (Munich Re, 2000).

extremes and a reduction in days with very low temperatures. There is evidence that the latter trend already is occurring (Easterling *et al*., 2000b).

Whatever the cause, it is important to note that a relatively small change in the mean of a climate variable can lead to a large change in the occurrence of extremes. Meehl *et al.* (2000a) explore the implications for extremes of changes in the mean and/or variance; they show clearly that the relationship between a change in the mean and a change in the occurrence of extremes is nonlinear, as illustrated in Figure 8-2.

8.2.2. Attribution Analyses of Loss Trends

Weather-related events of all magnitudes resulted in US$707 billion in insured and uninsured economic losses between 1985 and 1999 (Munich Re, 2000). A longer term comparison of large catastrophic events over the past 50 years reveals that economic losses (adjusted for inflation) increased by a factor of 10.3 (Figure 8-1). Over this same period, population grew by a factor of 2.4.

One of the vexing dilemmas in analyzing such historical data is disentangling causal factors related to human-induced climatic change, natural variability, and those having to do with human activity that could accelerate or dampen measured impacts. Numerous human factors are in operation that contribute to the upward trends in real economic losses, including population growth, rising standard of living, urbanization and industrialization in high-risk regions, vulnerability of modern societies and technologies, environmental degradation, penetration of insurance, and changing societal attitudes toward compensation (the latter two factors may lead to an increase in losses reported). Data on the numbers of events also show an increase in many cases. The number of disasters (defined as annual requests from states for federal disaster declarations) has roughly doubled in the United States since the early 1980s (Anderson, 2000). It is relevant to note here that such requests involve considerations of significant social effects (Kunkel *et al*., 1999); as a consequence, it is an indirect and subjective proxy for the frequency of events.

Growth trends in non-climate-related losses have been relatively constant over the past 3 decades. Losses from human-induced catastrophes have remained relatively constant (Swiss Re, 1999a). Earthquake losses have increased, but more slowly than weather-related losses (Bruce *et al*., 1999). The number of disasters causing more than 1% GDP damage to affected countries has increased two to three times as rapidly for weather-related disasters as for earthquakes in the period 1963–1992 (United Nations, 1994).

Insurers have pointed out that local environmental factors such as soil degradation, loss of biodiversity, lack of drinkable water, pollution, deforestation, forest degradation, and land-use changes can amplify the impacts of weather-related catastrophes

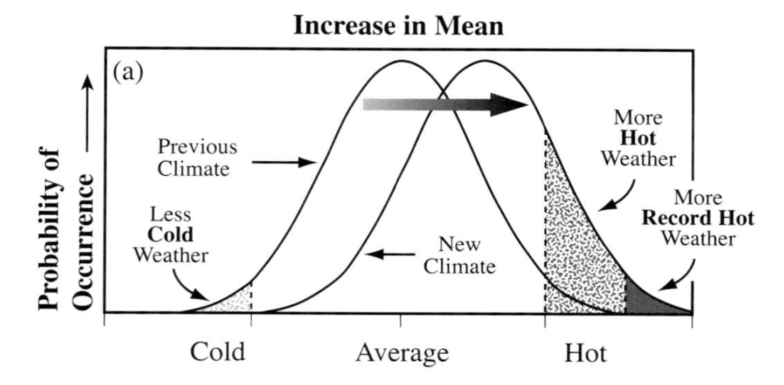

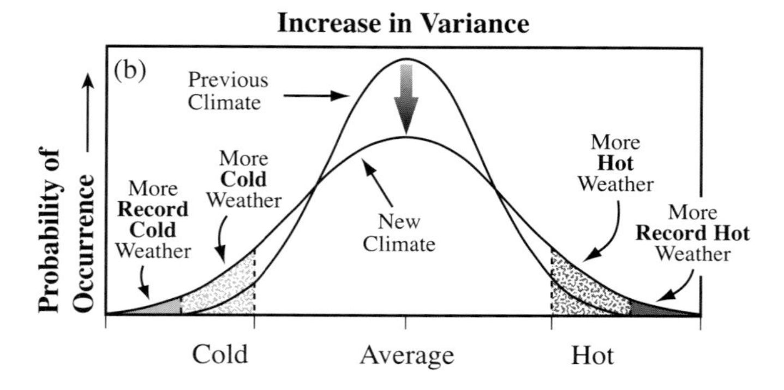

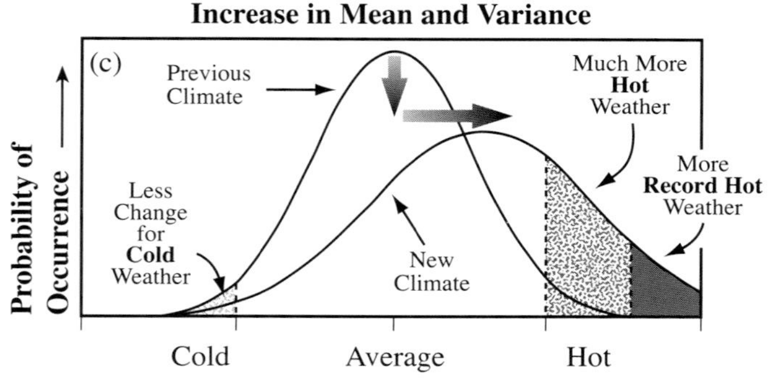

Figure 8-2: Schematic showing effect on extreme temperatures when (a) mean temperature increases, (b) variance increases, and (c) when both mean and variance increase for a normal distribution of temperature (TAR WGI, Figure 2.32).

(Zeng and Kelly, 1997). As an illustration, the extent of flood losses from Hurricane Mitch was attributed in part to deforestation in Central America.

Attempts to analyze the underlying causes of trends in natural disasters also must allow for the effects of human activities that offset growth factors (Kunkel *et al.,* 1999). A considerable leveling off or reduction in loss of life during U.S. disasters is one indicator that mitigation has been effective (Easterling *et al.,* 2000a). Loss-reduction efforts—typically unaccounted for in analyses we have seen—include considerable efforts to avert or reduce natural disaster impacts (e.g., coastal protection structures along coastlines; cloud seeding to deflect hailstorms; improved building codes; tightened land-use zoning; enhanced fire-suppression capacity; improved weather forecasts and early-warning systems; and improved disaster preparedness, response, and recovery). Within the insurance arena, increasing deductibles (the initial tier of loss costs paid for by the insured) and withdrawal of coverage from particularly high-risk areas have reduced observed losses. The literature has not attempted to quantify the contribution of these activities.

The relative contributions of human and climatic factors to the changing patterns of losses varies, depending on place and type of event (see Table 8-1; also see Easterling *et al.,* 2000a,b for a review). U.S. studies have found that demography largely explains increases in losses for hurricanes, wind, hail, and tornado events, whereas winter storm damage has mixed causation (Pielke and Landsea, 1998; Changnon, 1999; Changnon and Changnon, 1999; Kunkel *et al.*, 1999). In addition, decadal-scale trends have been discerned for tropical cyclones. There is good evidence that the intensity and frequency of precipitation and flood-related extreme events in the United States is increasing (Zeng and Kelly, 1997; Karl and Knight, 1998; Pielke and Downton, 2000). This trend also has been found for precipitation in many other parts of the world (see Chapter 3). In a study of hailstorms in France, Dessens (1995) used insurance loss information as a proxy for storm occurrence and found a statistically significant upward trend between 1946 and 1990.

In one global analysis, Munich Re (1999b) estimates that economic losses from large natural disasters increased two-fold between the 1970s and 1990s, after correcting for inflation, insurance penetration and pricing effects, and increases in the material standard of living. A similar result was reported for UK buildings in the SAR (Dlugolecki *et al*., 1996).

Based on the findings of TAR WGI, the information summarized in Table 8-1, and the analysis presented above, we conclude that some part of the upward trend in the cost of weather-related disasters illustrated in Figure 8-1 is linked to socioeconomic factors (increased wealth, shifts of population to the coasts, etc.) and some part is linked to climatic factors such as observed changes in precipitation and drought events. There are regional differences in the balance of these two causes.

8.2.3. Climate Events that are Relevant to the Insurance and Other Financial Services Sectors

Most weather extremes have relevance for the financial sector, as shown in Table 8-1. Column 6 summarizes the impacts of extremes on the main sectors of activity considered by TAR WGII. The ways in which these impacts affect the insurance industry are shown in Column 7.

Hot Temperature Extremes. Hot summers are likely to become more common as a result of global warming. The nonlinear effect of global warming on extreme events (see Figure 8-2) can be clearly illustrated by the example of temperature. Hulme (1997) estimates for the UK that the change in mean annual temperature in 2035 relative to the 1961–1990 mean will be approximately 1°C. Yet as a result, temperature conditions similar to those in the exceptional (1-in-300 years) summer of 1995 should occur once every 10 years on average between

***Table 8-1**: Extreme climate-related phenomena and their effects on the insurance industry: observed changes and projected changes during the 21st century [after Table 3-10; Munich Re, 1999b (p. 106)].*

Changes in Extreme Climate Phenomena	Observed Changes (Likelihood)	Projected Changes (Likelihood)	Type of Event Relevant to Insurance Sector	Time Scale	Sensitive Sectors/Activities	Sensitive Insurance Branches[b]
Temperature Extremes						
Higher maximum temperatures, more hot days and heat waves[c] over nearly all land areas	Likely[a] (mixed trends for heatwaves in several regions)	Very likely[a]	Heat wave	Daily-weekly maximum	Electric reliability, human settlements	Health, life, property, business interruption
			Heat wave, droughts	Monthly-seasonal maximum	Forests (tree health), natural resources, agriculture, water resources, electricity demand and reliability, industry, health, tourism	Health, crop, business interruption
Higher (increasing) minimum temperatures, fewer cold days, frost days, and cold waves[c] over nearly all land areas	Very likely[a] (cold waves not treated by WGI)	Very likely[a]	Frost, frost heave	Daily-monthly minimum	Agriculture, energy demand, health, transport, human settlements	Health, crop, property, business interruption, vehicle
Rainfall/Precipitation Extremes						
More intense precipitation events	Likely[a] over many Northern Hemisphere mid- to high-latitude land areas	Very likely[a] over many areas	Flash flood	Hourly-daily maximum	Human settlements	Property, flood, vehicle, business interruption, life, health
			Flood, inundation, mudslide	Weekly-monthly maximum	Agriculture, forests, transport, water quality, human settlements, tourism	Property, flood, crop, marine, business interruption
Increased summer drying and associated risk of drought	Likely[a] in a few areas	Likely[a] over most mid-latitude continental interiors (lack of consistent projections in other areas)	Summer drought, land subsidence, wildfire	Monthly-seasonal minimum	Forests (tree health), natural resources, agriculture, water resources, (hydro) energy supply, human settlements	Crop, property, health

2021 and 2050. Insurance claims could rise because of land subsidence, business interruption, and crop failure. Although heat waves have been shown to lead to an increase in daily mortality and morbidity (see Section 9.4.1)—an impact that may be compounded by poor air quality—the effect is likely to be too small to noticeably affect the financial services sector.

Cold Temperature Extremes. As a result of global warming, cold extremes of winter weather are likely to become rarer. In temperate latitudes, this development generally would be beneficial for business activities in, for example, the construction and transport sectors, with concomitant reductions in claims for business interruption. Although cold conditions should become rarer, a more active hydrological cycle might lead to more episodes of heavy snowfall, provided that temperatures remain below freezing. Regional shifts in the occurrence of phenomena such as ice storms may be expected. Ice storms occur when precipitation falls as rain but freezes on contact

Table 8-1 (continued)

Changes in Extreme Climate Phenomena	Observed Changes (Likelihood)	Projected Changes (Likelihood)	Type of Event Relevant to Insurance Sector	Time Scale	Sensitive Sectors/Activities	Sensitive Insurance Branches[b]
Rainfall/Precipitation Extremes (continued)						
Increased intensity of mid-latitude storms[c]	Medium likelihood[a] of increase in Northern Hemisphere, decrease in Southern Hemisphere	Little agreement among current models	Snowstorm, ice storm, avalanche	Hourly-weekly	Forests, agriculture, energy distribution and reliability, human settlements, mortality, tourism	Property, crop, vehicle, aviation, life, business interruption
			Hailstorm	Hourly	Agriculture, property	Crop, vehicle, property, aviation
Intensified droughts and floods associated with El Niño events in many different regions (see also droughts and extreme precipitation events)	Inconclusive information	Likely[a]	Drought and floods	Various	Forests (tree health), natural resources, agriculture, water resources, (hydro) energy supply, human settlements	Property, flood, vehicle, crop, marine, business interruption, life, health
Wind Extremes						
Increased intensity of mid-latitude storms[c]	No compelling evidence for change	Little agreement among current models	Mid-latitude windstorm	Hourly-daily	Forests, electricity distribution and reliability, human settlements	Property, vehicle, aviation, marine, business interruption, life
			Tornadoes	Hourly	Forests, electricity distribution and reliability, human settlements	Property, vehicle, aviation, marine, business interruption
Increase in tropical cyclone peak wind intensities, mean and peak precipitation intensities[d]	Wind extremes not observed in the few analyses available; insufficient data for precipitation	Likely[a] over some areas	Tropical storms, including cyclones, hurricanes, and typhoons	Hourly-weekly	Forests, electricity distribution and reliability, human settlements, agriculture	Property, vehicle, aviation, marine, business interruption, life

with a solid surface. Air temperatures close to freezing are ideal for ice storm occurrence. Thus, in colder regions where the weather currently is well below freezing in the winter, ice storms may become more common as a result of global warming, although they could become less frequent in areas where they occur at present (Francis and Hengeveld, 1998). An ice storm that occurred 7-10 January 1998, in the northeastern United States and eastern Canada, led to insured damage estimated at US$1.2 billion (Lecomte *et al.*, 1998).

Heavy Rainfall and Flooding. TAR WGI Chapter 9 indicates that "many models" now project that conditions in the tropical Pacific may become more El Niño-like, with associated changes in precipitation patterns (Meehl *et al.*, 2000b). This would lead to more frequent patterns of El Niño-like floods and drought conditions in areas where teleconnections to the El Niño-Southern Oscillation (ENSO) exist. Observational studies assessed in TAR WGI Chapter 2 suggest that there has been a widespread increase in heavy and extreme precipitation

Table 8-1 (continued)

Changes in Extreme Climate Phenomena	Observed Changes (Likelihood)	Projected Changes (Likelihood)	Type of Event Relevant to Insurance Sector	Time Scale	Sensitive Sectors/Activities	Sensitive Insurance Branches[b]
Other Extremes						
Refer to entries above for higher temperatures, increased tropical and mid-latitude storms	Refer to relevant entries above	Refer to relevant entries above	Lightning	Instant-aneous	Electricity distribution and reliability, human settlements, wildfire	Life, property, vehicle, aviation, marine, business interruption
Refer to entries above for increased tropical cyclones, Asian summer monsoon, and intensity of mid-latitude storms	Refer to relevant entries above	Refer to relevant entries above	Tidal surge (associated with onshore gales), coastal inundation	Daily	Coastal zone infrastructure, agriculture and industry, tourism	Life, marine, property, crop
Increased Asian summer monsoon precipitation variability	Not treated by WGI	Likely[a]	Flood and drought	Seasonal	Agriculture, human settlements	Crop, property, health, life

[a] Likelihood refers to judgmental estimates of confidence used by Working Group I: *very likely* (90–99% chance); *likely* (66–90% chance). Unless otherwise stated, information on climate phenomena is taken from Working Group I's Summary for Policymakers and Technical Summary. These likelihoods refer to observed and projected changes in extreme climate phenomena and likelihood shown in first three columns of this table.
[b] All findings in this column are high confidence, as described in Section 1.4 of the Technical Summary.
[c] Information from Working Group I, Technical Summary, Section F.5.
[d] Changes in regional distribution of tropical cyclones are possible but have not been established.

events in regions where total precipitation has increased (i.e., the middle and high latitudes of the Northern Hemisphere). Flooding is responsible for 40% of total economic losses and 10% of weather-related insurance losses globally.

Tropical hurricanes can lead to landslides. Hurricane Mitch probably is the most well-known event in recent years. This system, the strongest ever October tropical storm in the Atlantic Basin, stalled over Central America and produced more than 600 mm of rainfall in 48 hours. Resulting landslides and mudslides led to an estimated 9,000 deaths and insured losses of US$513 million (Swiss Re, 2000b). In disasters of this magnitude, preparedness and planning can make a huge difference in loss of life and the amount of damage sustained.

Large river basin floods develop over huge areas following weeks of unusually high rainfall. In July and August 1997, flooding in central Europe caused 54 fatalities in Poland and required the evacuation of 162,000 people (Kundzewicz *et al.*, 1999). The value of the economic losses throughout central Europe amounted to approximately US$5 billion, with insured losses of US$940 million. The intensity of such flood events is driven not only by climatology but also by human management of the watershed.

Low Rainfall—Drought, Land Subsidence, and Wildfire. Drought is important for the financial sector through impacts on commercial agriculture, building foundations, and wildfire occurrence. Figure 8-3 shows the cost of subsidence claims to the industry from 1975 to 1997 in England and Wales. There is a clear relationship with rainfall (with some lag effects). Similar effects are seen in France (Radevsky, 1999). Where insurance is used as the mechanism to finance repairs to building

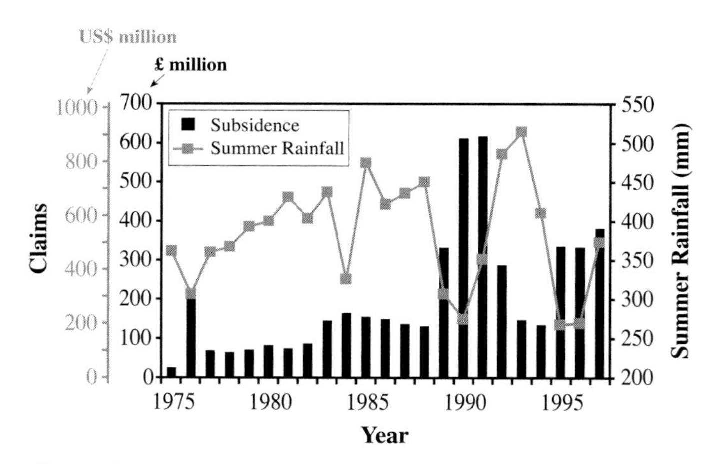

Figure 8-3: Summer rainfall and subsidence claims in the UK: 1975–1997. Rainfall data are for England and Wales, April to September (from Climatic Research Unit, University of East Anglia, UK). Subsidence claim costs are in original-year values (from Association of British Insurers).

foundations, as in the UK and France, costs for domestic properties can be higher than where the damage is not insured, as in Australia. Adaptive responses such as stronger foundations in new buildings and repairs to older housing capital should reduce the problem.

The worst drought of recent decades has occurred (indeed, it persists) in the Sahelian region of West Africa, where since 1968 rainfall has been below the long-term average in almost every year (Nicholson *et al.*, 2000). The strength and persistence of this deficit is unparalleled in recent times. Despite the drought's severity, it has had minimal impact on the commercial financial sector because of the low penetration of insurance in the region. However, the drought's role in the development of the region has been significant.

Wildfire is an increasingly important insurance issue, as illustrated by the US$140 million economic losses sustained in the Los Alamos fire of 2000 (Hofmann, 2000b). Outdoor fire occurrence is likely to increase in a future warmer climate, particularly along the increasingly popular urban-rural fringe (Swiss Re, 1992; Torn *et al.*, 1998). Whereas in Europe most wildfires are of human origin (either deliberate or accidental), lightning (see below) is still the leading cause of forest fires in the western United States and Alaska (the regions of North America with the greatest number of wildfires).

Even if rainfall amounts are unchanged by global warming, higher temperatures will increase the level of risk associated with these hazards because of increased water loss through evaporation and transpiration.

Lightning Strikes. Model experiments are not able to tell us anything directly about changes in lightning occurrence as a result of global warming. Any increase in convective activity should lead to more frequent electrical storms and lightning discharges, and it seems likely that global warming will have such an effect in the tropics (Lal *et al.*, 1998) and in extratropical latitudes (White and Etkin, 1997). Reeve and Toumi (1999) suggest that a 1°C increase in average wet-bulb temperature can be accompanied in mid-latitudes by a 40% increase in lightning. Of relevance to insurers, lightning is a cause of fires and damage to electrical equipment, with associated business interruption claims (Mills *et al.*, 2001).

Tropical and Extratropical Windstorm. Experiments with climate models to date have not produced a consensus regarding the likely future occurrence of tropical and extratropical wind storms. Both have a very large capacity to cause damage. Hurricane Andrew, for example, occurred in 1992 in the Atlantic Basin and made landfall over the United States, causing US$21 billion (1999 US$) in insured damage. Hurricane Floyd, which caused US$2.2 billion in insured losses in 1999, required the evacuation of 2 million people and imposed huge stress on infrastructure, resources, and ultimately health. The most damaging extratropical windstorm was Daria in 1990, which caused US$6.8 billion in insured losses in northwestern Europe. In December 1999, windstorms Martin and Lothar tracked south of the normal route, affecting France, northern Spain, and central Europe. Together they caused 140 fatalities and US$8.4 billion in insured damage.

Sea-Level Rise. Increases in sea level pose a major potential risk to coastal zones (TAR WGI Chapter 6), especially if they are associated with an increase in storminess. The mid-range increase in sea level by the year 2100 as a result of anthropogenic climate change is 49 cm, taking into account atmospheric aerosol concentrations, with estimates ranging from 26 to 72 cm (TAR WGI Chapter 11). The main risk to the financial sector is in the effect that this change in mean sea level may imply for the occurrence of tidal surges, which already cause enormous damage and loss of life, especially in the developing world (see Box 8-4). One of Europe's greatest natural disasters in terms of loss of life was the 1953 storm surge in the North Sea, which led to almost 2,000 fatalities in The Netherlands and the UK.

8.3. Private and Public Insurance

This section examines the sensitivity, vulnerability, and adaptability of private- and public-sector insurance to climate change. Activities within these segments are significantly interrelated, and the role of each varies widely from country to country and over time (Van Schoubroeck, 1997; Ryland, 2000). Government programs exist primarily to correct market failures in the private sector, when insurance cannot be provided at a reasonable rate, or when insufficient capacity exists to pay claims (Mittler, 1992). In addition, the nature of events anticipated under climate change (e.g., increased flooding) draws into question their very insurability by private companies (Denenberg, 1964; Mittler, 1992; White and Etkin, 1997; Hausmann, 1998; Kunreuther, 1998; Nuttall, 1998).

Insurers are sensitive to a diversity of potential climate changes (Ross, 2000). Understanding and adapting to weather-related losses are high priorities in the insurance industry. Loss growth has resulted in the absence of commercial insurance for the most vulnerable risks, such as flood or crop damage in many countries. Changes in weather-related events associated with global climate change would increase the sector's vulnerability (Vellinga and Tol, 1993; Changnon *et al.*, 2000; TAR WGI Chapters 9 and 10). Recent history has shown that weather-related losses can stress insurance firms to the point of elevated prices, withdrawal of coverage, and insolvency (bankruptcy).

The private insurance sector is highly heterogeneous, and the penetration of insurance varies dramatically across regions and within countries, as does the exposure and vulnerability of human populations and property to natural disaster events. Analyses that are meaningful to local policymakers, governments, and economies must adopt a variety of perspectives: regional, state, municipality, company, and the growing number who are self-insured.

Based on observations over the past decade, the property/casualty (P/C) segment is more vulnerable to weather-related events than the life/health segment (Table 8-2). The P/C segment is extremely diverse. The single most vulnerable branch appears to be property insurance, including business interruption (Bowers, 1998). Other lines, such as personal automobile insurance, have more limited exposure.

Of 8,820 loss events analyzed worldwide by Munich Re between 1985 and 1999, 85% were weather related, as were 75% of the economic losses and 87% of the insured losses (Munich Re, 1999b, 2000). The weather-related share of total losses is as high as 100% in Africa and 98% in Europe. Global weather-related insurance losses from large events[2] have escalated from a negligible level in the 1950s to an average of US\$9.2 billion yr^{-1} in the 1990s (Figure 8-1)—13.6-fold for the 1960–1999 period for which detailed data are available. Insurance losses have grown significantly faster than total economic losses and insurance reserves and assets (i.e., adaptive capacity). Since the 1950s, the decadal number of catastrophic weather-related events experienced by the insurance sector has grown 5.5-fold.

***Table 8-2**: Distribution of the global insurance market, including life/health and property/casualty, by region (Swiss Re, 1999b). Note that weightings between property/casualty and life/health vary considerably among countries. Swiss Re (1999b) provides detailed information by country. In some cases (e.g., Japan), life insurance premiums include annuities, which eventually are reimbursed to the insured.*

Total Business	Premiums in 1998 (US$M)	Share of World Market in 1998 (%)	Premiums as % of GDP in 1998	Premiums per capita in 1998 (US$)	Property/Casualty Premiums as % of Total
America	817,858	38.0	7.7	1,021	54
– North America	779,593	36.2	9.0	2,592	53
– Latin America	38,265	1.8	2.0	77	72
Europe	699,474	32.5	6.9	614	42
– Western Eurpe	684,848	31.8	7.3	1,466	42
– Central/Eastern Europe	14,626	0.7	2.1	23	75
Asia	571,272	26.5	7.8	36	23
– Japan	453,093	21.0	11.7	3,584	20
– South and East Asia	107,430	5.0	3.8	34	31
– Middle East	10,749	0.5	1.7	42	67
Africa	28,792	1.3	4.8	36	25
Oceania	37,872	1.8	9.4	1,378	41
World	2,155,269	100.0	7.4	271	41
– Industrialized countries[a]	1,955,406	90.7	8.8	2,132	41
– Emerging markets[b]	199,863	9.3	3.0	37	43
OECD[c]	2,016,084	93.5	8.5	1,805	41
G7[d]	1,725,007	80.0	8.9	2,498	41
EU[e]	672,939	31.2	7.4	1,651	40
NAFTA[f]	785,901	36.5	8.3	1,960	53
ASEAN[g]	11,711	0.5	2.6	26	42

[a] North America, Western Europe, Japan, Oceania.
[b] Latin America and Caribean, Central and Western Europe, South and East Asia, Middle East, Africa.
[c] 29 members.
[d] USA, Canada, UK, Germany, France, Italy, Japan.
[e] 15 members.
[f] USA, Canada, Mexico.
[g] Singapore, Malaysia, Thailand, Indonesia, The Philippines, Vietnam; the three remaining members—Brunei, Laos, and Myanmar—are not included.

These trends would be exacerbated by increased vulnerability resulting from development of high-hazard zones and increasingly sensitive infrastructure (Swiss Re, 1998a; Hooke, 2000; see Chapter 4).

Insurers have differing views on climate change (Mills *et al.*, 2001). Although several insurers have devoted significant attention to the issue (especially in Europe and Asia), the vast majority have given it little visible consideration. Some have taken definitive precautionary positions in stating that there is a material threat (Swiss Re, 1994; UNEP, 1995, 1996; Jakobi, 1996; Nutter, 1996; Zeng and Kelly, 1997; Berz, 1999; Bruce *et al.*, 1999; Munich Re, 1999b; Storebrand, 2000), whereas others have taken a different view (Mooney, 1998; Unnewehr, 1999). Some have elected to focus on disaster preparedness; others have adopted a "wait-and-see" stance.

8.3.1. Major Market Segments: Property/Casualty and Life/Health

The world insurance market enjoyed revenues of US$2.155 trillion in 1998 (7.4% of global GDP) (Table 8-2). Although insurance penetration is relatively low in developing countries and economies in transition, their insurance market growth rate averages approximately twice that in industrialized countries. Expenditures on insurance in developing countries typically represent between 0.5 and 4% of GDP, compared to 5–15% percent in developed countries (Swiss Re, 1999c). With 36% of total global insurance premiums, North America is the largest regional market (see Chapter 15), closely followed by Western Europe at 32%. Reinsurance is particularly focused on high-value loss situations, in developing countries, or for smaller primary insurers. Reinsurers typically collect US$100 billion in premiums globally each year from primary insurers from whom they assume various (mostly property) risks.

The P/C insurance segment represented 41% of global industry premiums collected in 1998. As shown in Figure 8-4, the segment as a whole exhibits sensitivity to major natural disaster events, as evidenced by the reductions in U.S. insurer profitability during 1992 (Hurricane Andrew and Iniki) and 1994 (Northridge earthquake). A list of the most costly events is presented in Table 8-3. Over the past 15 years, the global ratio of P/C premium income to natural catastrophe losses has decreased from 351:1 to 122:1—almost a three-fold rise in "exposure" (Figure 8-5; see Figure 15-6 for North America).

Climate- and weather-related risks faced by life/health insurers include injuries or death resulting from extreme weather episodes, water- or vector-borne diseases, degraded urban air quality, pressure on the quality and adequacy of food and water supplies, and increased vulnerability to power failures (see Chapters 4, 5, 9, 15; TAR WGIII Chapter 8; World Bank, 1997a; Epstein, 1999). In some areas, climate changes may yield health benefits, but negative health impacts are expected to outweigh positive ones if no actions are taken to adjust (Chapter 9). Such impacts will not be significant for the global financial sector in the near term, because life/heath insurance penetration currently is low in developing countries; the burden will fall largely on the informal and government sectors.

Owing to structural changes underway in the industry, the financial distinction between life and P/C insurers is blurring somewhat as a result of consolidation and mergers. Life insurers also are major holders of real estate and providers of mortgage lending; thus, they participate as property owners in weather-related property risks and may additionally assume property risk as investors in catastrophe bonds or other weather derivatives.

8.3.2. Risk Sharing between the Private and Public Sectors

The private insurance industry is part of a larger community that bears the costs of weather-related events (Ryland, 2000). The nature and cost of weather-related losses vary considerably around the globe, as does the portion of the loss that is privately insured. Private insurance pays a higher proportion of benefits for storm-related losses than for any other weather-related event, although flood insurance has a particularly low rate of coverage (Figure 8-6).

Insurers bear only 20% of the total economic costs of weather-related events globally. The ratio is far lower in developing countries (e.g., 7% in Africa and 4% in Asia for the year 1998) (Munich Re, 1999b). Even in countries where insurance penetration is high, insurance can account for less than half of the weather-related payouts—for example, 27% in Europe, 30% in the United States, 34% in Australia (Munich Re, 1999b), and 20% in Canada (EPC, 2000). In a review of four major wildfire and flood catastrophes in Australia, Leigh *et al.* (1998a,b) found that private-sector insurers bore 9–39% of the total economic losses; a comparable amount was provided by local and federal governments. Other entities assuming such costs include federal disaster relief providers, local governments, and uninsured property owners (Pielke and

---

[2]Economic losses are defined in footnote 1. The definition of "large" weather-related events is those in which the response capacity is overtaxed and interregional or international assistance becomes necessary, often in cases where thousands of people are killed, hundreds of thousands are left homeless, or the economic loss is substantial (Munich Re, 2000). Thus, events that are small but frequent tend to be excluded from these statistics. For example, land subsidence losses from two droughts during the 1990s in France resulted in losses of US$2.5 billion, and even more in the UK, but these losses are largely absent from the "large" event data series. A similar case involves frequent but relatively small winterstorm events in northern latitudes and their losses. Figure 8-6 includes a fuller range of events, which tend to result in an adjusted loss level that is approximately twice that indicated by data on "large" events alone. "Large" events represent only 1% of the total number of events globally.

Landsea, 1998)—as in the case of Hurricane Andrew, in which only half of the losses were insured (Pielke, 1997).

One important risk-assuming group, the corporate self-insurance market, is growing rapidly. In the United States, such premiums are approaching the level of the traditional commercial insurance market (roughly US$134 billion) (Best's Review, 1998; Bowers, 1999).

Where insurers will not or are directly or indirectly regulated not to accept specific catastrophe risks, governments in many countries—including Belgium, France, Japan, The Netherlands, New Zealand, Norway, Spain, and the United States—may adopt the role of insurer or reinsurer or of regulator in establishing risk-pooling mechanisms (III, 2000b). Programs in France, Japan, and New Zealand explicitly define the governments' role as paying for "uninsurable damages" (CCR, 1999; Gastel, 1999). In some countries (e.g., Canada, Finland, France, Norway, the United States) this is the case for drought or other agricultural risks, and in others (e.g., Japan) this is limited to earthquake risks. Such schemes can grow rapidly, as illustrated by the jump in the numbers of policies under the Florida Windstorm Underwriting Association from 62,000 to 417,000 between 1992 and 1997 (Anderson, 2000).

Government's role in providing resources for disaster preparedness and recovery and insurance products related to natural disasters also is a key moderating factor in insurers' involvement in such risks. It can be a two-edged sword: It provides a platform for private industry to participate, but it

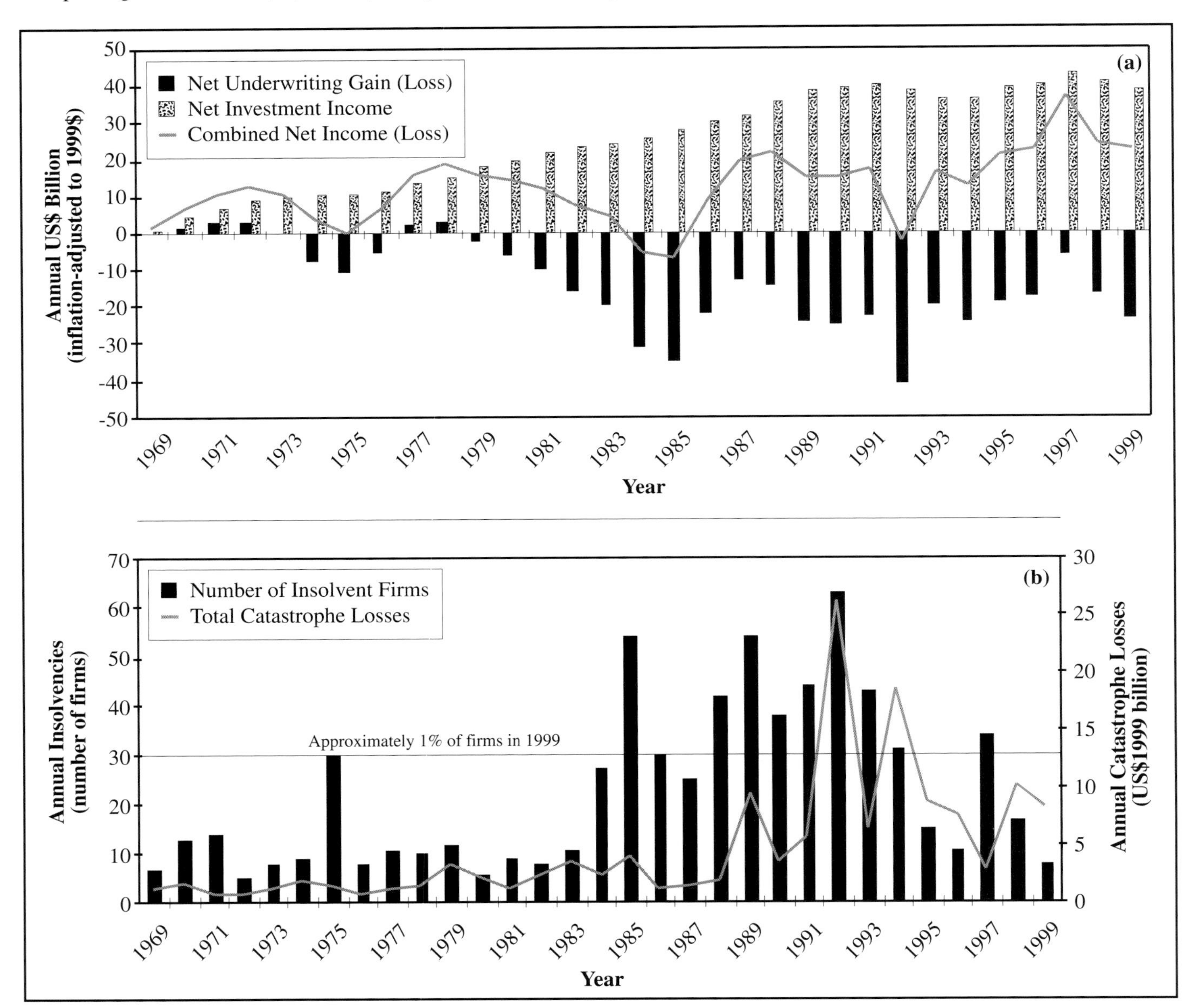

Figure 8-4: Trends in U.S. natural disasters, insurance industry profitability, and solvency, 1969–1999: (a) Sensitivity of property/casualty insurance sector net financial results to investment income and underwriting gain/loss. Upper bars indicate investment income; lower bars indicate net result of core business (premium revenues vs. claims paid). Curve is the net result. (b) Annual number of insolvencies and natural disaster losses (Mills *et al.*, 2001).

***Table 8-3**: Billion-dollar and larger insurance losses, 1970–1999, as of December 2000 (Munich Re, 2000). Figures are adjusted for inflation (1999 values).*

Year	Event	Area	Insured Losses (US$M)	Economic losses (US$M)	Ratio of Insured/ Economic Losses
1992	Hurricane Andrew	USA	20,800	36,600	0.57
1994	Northridge earthquake	USA	17,600	50,600	0.35
1991	Typhoon Mireille	Japan	6,900	12,700	0.54
1990	Winterstorm Daria	Europe	6,800	9,100	0.75
1989	Hurricane Hugo	Caribbean, USA	6,300	12,700	0.50
1999	Winterstorm Lothar	Europe	5,900	11,100	0.53
1987	Winterstorm	Western Europe	4,700	5,600	0.84
1998	Hurricane Georges	Caribbean, USA	3,500	10,300	0.34
1995	Earthquake	Japan	3,400	112,100	0.03
1999	Typhoon Bart	Japan	3,400	5,000	0.60
1990	Winterstorm Vivian	Europe	2,800	4,400	0.64
1999	Winterstorm Martin	Europe	2,500	4,100	0.61
1995	Hurricane Opal	USA	2,400	3,400	0.71
1999	Hurricane Floyd	USA	2,200	4,500	0.49
1983	Hurricane Alicia	USA	2,200	3,500	0.63
1991	Oakland fire	USA	2,200	2,600	0.85
1993	Blizzard	USA	2,000	5,800	0.34
1992	Hurricane Iniki	Hawaii	2,000	3,700	0.54
1999	Winterstorm Anatol	Europe	2,000	2,300	0.87
1996	Hurricane Fran	USA	1,800	5,700	0.32
1990	Winterstorm Wiebke	Europe	1,800	3,000	0.60
1990	Winterstorm Herta	Europe	1,800	2,600	0.69
1995	Hurricane Luis	Caribbean	1,700	2,800	0.61
1999	Tornadoes	USA	1,485	2,000	0.74
1998	Hailstorm, tempest	USA	1,400	1,900	0.74
1995	Hailstorm	USA	1,300	2,300	0.57
1993	Floods	USA	1,200	18,600	0.06
1998	Ice storm	Canada, USA	1,200	2,600	0.46
1999	Hailstorm	Australia	1,100	1,500	0.67
1998	Floods	China	1,050	30,900	0.03

also can drive consumers away from commercial market solutions (Klein, 1997; Pullen, 1999a). The absolute value of government payments for natural disasters is poorly documented, and the statistical record is fragmented. The United States made disaster-related payments of US$119 billion (1993 US$) over the 1977–1993 period, equivalent to an average of US$7 billion yr^{-1} (Anderson, 2000). The Japanese government has devoted 5–9% of its national budget to disaster preparedness and recovery in recent decades (Sudo *et al*., 2000).

Flood insurance merits special mention, given the magnitude of risks and losses, the difficulty of establishing fair and actuarially based rates, and the connection between flood and climate change (see Chapter 4; Aldred, 2000). Recent analyses in the United States found that 25% of homes and other structures within 150 m of the coastline will fall victim to the effects of erosion within 60 years (Heinz Center, 2000). Sea-level rise will impact flood insurance through inundation and erosion resulting from storm surge (see Chapter 6). Countries differ widely with regard to their approach to defining and financing flood risks via private-sector (re)insurance versus public mechanisms (Van Schoubroeck, 1997; Gaschen *et al*., 1998; Hausmann, 1998). Hybrid public-private systems and government-only systems also can be found (e.g., in the United States), as can systems with no formal flood insurance whatsoever.

A central question is whether changes in natural disaster-related losses will generate increased reliance on already overburdened government-provided insurance mechanisms and disaster assistance. Governments already are showing decreased willingness to assume new weather-related liabilities, and tensions concerning risk-sharing between local and federal government bodies also are evident (Fletcher, 2000).

8.3.3. Insurers' Vulnerability and Capacity to Absorb Losses

A central component of vulnerability for public and private insurers alike is actuarial uncertainty in the dimensions, location, or timing of extreme weather events. This is particularly true

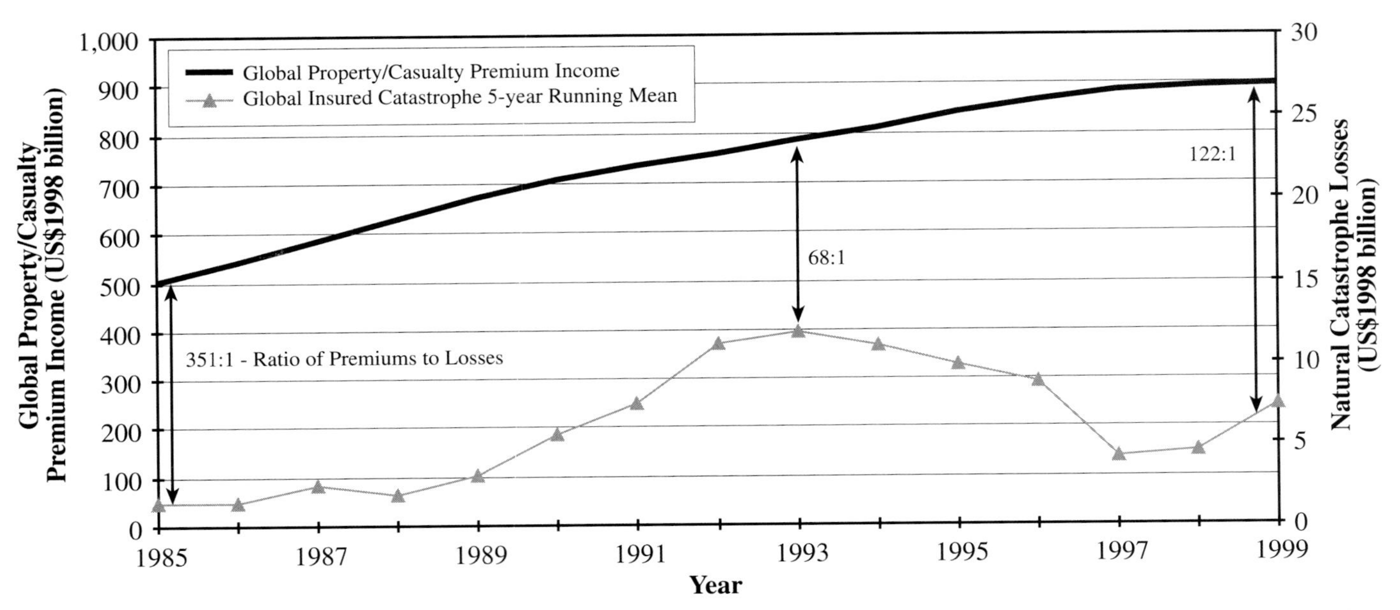

Figure 8-5: Global insured natural catastrophe losses (right-hand scale) vs. property/casualty premium income (left-hand scale), using a 5-year running mean. Global losses are from Munich Re (2000) and premiums from Swiss Re (1999b and earlier years). Note that these data include only major weather-related losses (approximately half of total weather-related losses). Premiums include considerable revenues (and associated reserves and surplus, not usable to pay catastrophe losses) from non-weather-related business segments and from self-insurers. The numbers generally include "captive" self-insurers but not the less-formal types of self-insurance. Exposure—measured as the ratio of premiums to losses—increased by a factor of 2.9 between the endpoints and by 5.2 in the worst single year (1993) within this time interval.

for insurance where the rate of damage rises faster than the driving weather phenomenon. Examples include the relationships between peak wind speeds and structural damages (Dlugolecki *et al.*, 1996), average temperature changes and lightning strokes (Price and Rind, 1994; Dinnes, 1999; Reeve and Toumi, 1999), extreme temperature events and electric power reductions or crop damages (Colombo *et al.*, 1999) and heat stress mortality (see Chapter 9), and precipitation and flooding (White and Etkin, 1997).

Changes in the spatial distribution of natural disasters pose special risks and challenges for the insurance sector. Localities to which risks shift will tend to be relatively inexperienced and unprepared to handle such risks, potentially resulting in a net societal increase in losses. A given insurer's vulnerability often extends internationally. For example, U.S. insurers collected nearly 15% of their premiums overseas in 1997, and the ratio has been growing (III, 1999). Reinsurers have a particularly high degree of international exposure.

8.3.3.1. Quantifying Vulnerability and Adaptive Capacity

For insurers, vulnerability can be viewed broadly in terms of the sector's capacity to pay for extreme events, together with the temporal sequence of such events. The key to vulnerability is the probable maximum loss (PML), which is the best estimate of the cost that is likely to emanate from an event with a specified probability of occurrence. In recent times, PMLs often have been revised upward significantly. The European winter storms Lothar and Martin of 1999 (US$8.4 billion insured losses) caught European insurers and reinsurers offguard, presenting losses that substantially exceeded prevailing expectations. These storms constituted the most serious natural disaster ever covered by insurance in France, with about 3 million claims (FFSA, 2000). One recent estimate for the United States was a combined PML of US$155 billion for 1-in-100-year (i.e., 1% yr^{-1} likelihood) for all types of natural disasters nationally (see Figure 15-8).

Unnewehr (1999) segmented the market and estimated that 17% of 1997 U.S. P/C insurance premiums were associated with "significant" exposure to weather-related loss. The paper did not explore other measures or sources of vulnerability and exposure, such as total insured property values at risk (US$4 trillion in insured property in the Gulf and Atlantic coastal counties of the United States) (Hooke, 2000), or the extent of insolvency risk. These results are not transferable to other regions, where insurance systems and natural hazards can be very unlike those in the United States (see Figure 8-6).

The particular role of weather in vehicle-related losses is not well studied. Vehicle insurance represents 48% of U.S. P/C premiums; it is ranked in the aforementioned study as having "minor" weather sensitivity. Of total vehicle-related accidents, 16% of those in the United States are caused by adverse weather conditions (NHTSA, 1999); 33% of those in Canada are weather related (White and Etkin, 1997). Physical damage to vehicles during U.S. natural catastrophes between 1996 and mid-2000 represented an additional $3.4 billion (10%) of total insured property losses, ranging as high as 55% for individual events (PCS, 2000; Mills *et al.*, 2001).

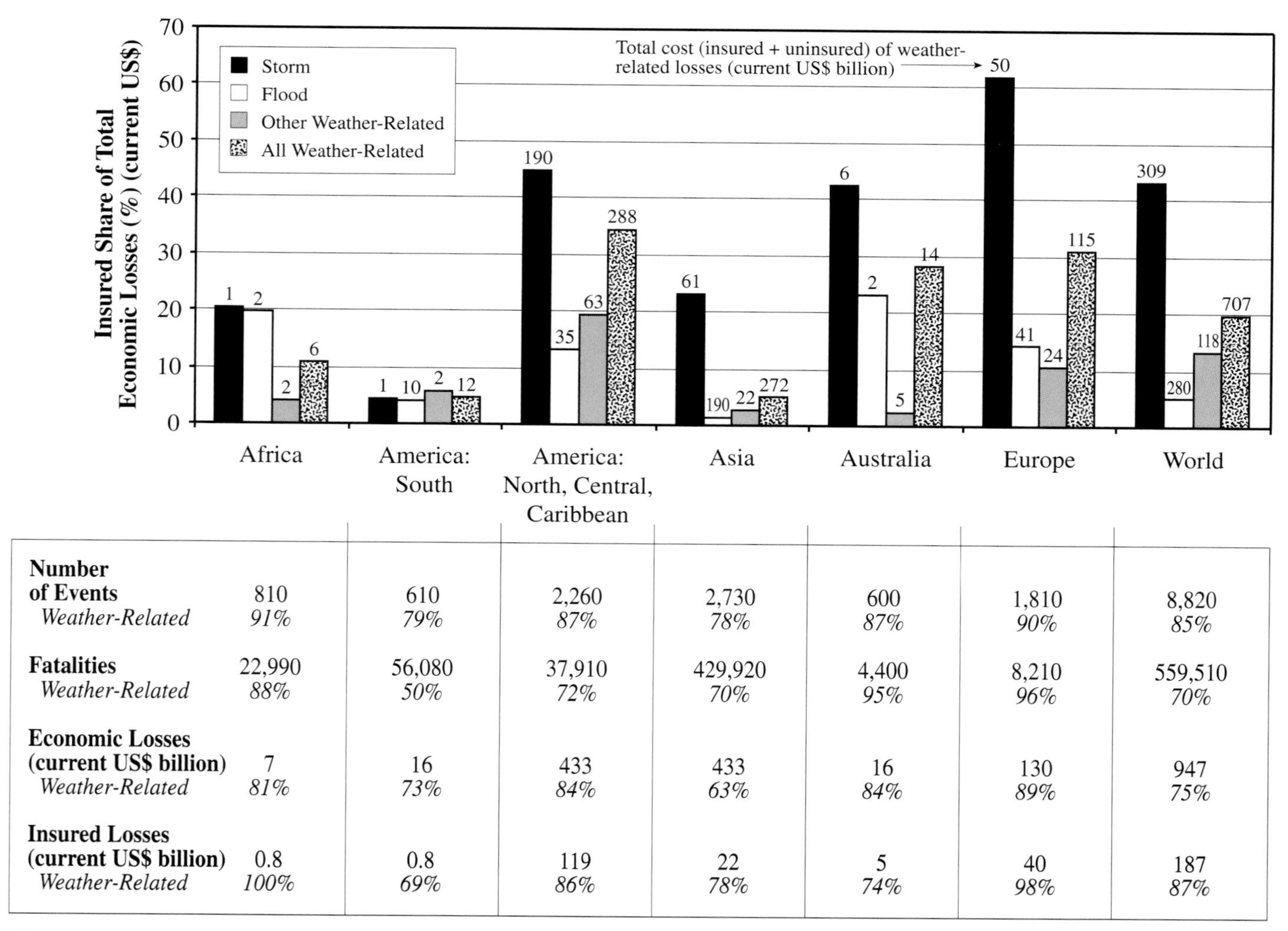

	Africa	America: South	America: North, Central, Caribbean	Asia	Australia	Europe	World
Number of Events	810	610	2,260	2,730	600	1,810	8,820
Weather-Related	*91%*	*79%*	*87%*	*78%*	*87%*	*90%*	*85%*
Fatalities	22,990	56,080	37,910	429,920	4,400	8,210	559,510
Weather-Related	*88%*	*50%*	*72%*	*70%*	*95%*	*96%*	*70%*
Economic Losses (current US$ billion)	7	16	433	433	16	130	947
Weather-Related	*81%*	*73%*	*84%*	*63%*	*84%*	*89%*	*75%*
Insured Losses (current US$ billion)	0.8	0.8	119	22	5	40	187
Weather-Related	*100%*	*69%*	*86%*	*78%*	*74%*	*98%*	*87%*

Figure 8-6: Regional insurance coverage for weather- and non-weather-related natural disasters, 1985–1999. The role of insurance in paying weather-related losses varies by event type and region, generally dominated by windstorm (Munich Re, 1999b). "Other" includes weather-related events such as wildfire, landslides, land subsidence, avalanches, extreme temperature events, droughts, lightning, frost, and ice/snow damages (Munich Re, 2000). The numbers generally include "captive" self-insurers but not the less-formal types of self-insurance. Total costs are higher than those summarized in Figure 8-1 because of the restriction of Figure 8-1 losses to those from large catastrophic events. Rounding errors may appear in data labels.

Although aggregate industry assessments are useful, analyses of vulnerability clearly must take into account the complexity and specialized structure of the insurance sector (GAO, 2000a). Although an *aggregate* U.S. insurance surplus of US$200–350 billion often is cited (Doherty, 1997; GAO, 2000a), roughly 80% of this surplus is required for non-weather-sensitive branches (e.g., workers' compensation), assuming proportionality with premium-based risk figures quoted by Unnewehr (1999). In addition, insurers are independent and have radically different mixes of risks, so individual firms may become insolvent long before losses approach the industry's aggregate capacity (Doherty, 1997; Klein, 1997). Single-state PML events at the 1% likelihood level would result in economic stress ranging from 5 to 60% of insurers by market share (Pullen, 1999b). Moreover, catastrophes can disrupt insurance markets and harm insurance companies and consumers even in cases in which all claims are paid (GAO, 2000a; Ryland, 2000).

Reinsurance adequacy is another issue in vulnerability assessment. Swiss Re (1997) concluded that the availability of reinsurance coverage for natural disasters in 14 major markets was insufficient and that following a major event, primary insurers' (the customers of reinsurers) equity base (surplus) would come under considerable strain. For PML windstorm events in Australia, Japan, and the United States, the impact on aggregate surplus would be reductions of 24, 41, and 11%, respectively (Swiss Re, 1997). Solvency analyses typically give only "partial credit" to primary insurers for reinsurance (e.g., 50% in the European Union) because of the uncertain viability of reinsurance contracts or reinsurers themselves following catastrophic losses (Doherty *et al.*, 1992; Swiss Re, 2000a).

Aside from issues of solvency, past extreme weather events clearly have measurable short- to medium-term impacts on insurance and reinsurance profitability—even at a national scale (Figure 8-4a) and on the availability of insurance following the event (Davidson 1996; Pullen, 1999b). Catastrophe losses during 1999 and 2000 contributed to marked short-term depressions in earnings and stock prices for several large insurers and reinsurers (Edgecliffe-Johnson, 1999; Carpenter, 2000; Lonkevich, 2000). This development can restrict insurers' ability to raise new capital for expansion or even to continue the operations of highly exposed branches.

The overarching insurance business environment also is a key factor in determining vulnerability. Cyclical pressures or incidental broad-based stresses on the industry—such as major tobacco litigation (Bradford, 2000; Clow, 2000; Hofmann, 2000a), the crisis in environmental liability insurance (U.S. Superfund, asbestos, and lead paint claims), the Asian financial crisis, or increased competition from Internet sales (Ceniceros, 2000)—could place considerable demands on surplus (Mooney, 1999; GAO, 2000a; Swiss Re, 1998b, 2000a). Developments in financial markets can influence the level and availability of insurance surplus (Cummins *et al.*, 1999; GAO, 2000b; Swiss Re, 2000a). More than three-quarters of the growth in the U.S. insurance industry's surplus since 1995 resulted from capital gains (GAO, 2000a).

On one hand, the trend toward convergence between banking and insurance potentially increases diversification and robustness. On the other hand, it exposes one sector to risks faced in the other, and, in some cases, geographical diversification of a company's insurance business has moved it into the path of increased disaster losses (Berry, 2000; Greenwald, 2000; Howard, 2000b; Lonkevich, 2000). Weather-related vulnerability could increase if insurers participate in emerging capital market alternatives for risk financing (Marcon, 1999). In general, such convergence is more likely for the life insurance segment.

8.3.3.2. Natural Catastrophes and Insurer Solvency

Historical weather-related insolvencies illustrate the vulnerability of large and small insurers to the types of natural disasters that potentially are associated with climate change (Stipp, 1997; Swiss Re, 2000a; Mills *et al.*, 2001). Nearly 650 U.S. insurers became insolvent between 1969 and 1998 (Figure 8-4b) (Matthews *et al.*, 1999).

Of the 36 of 426 specifically attributed insolvencies occurring primarily as a result of natural catastrophes, more than half occurred between 1989 and 1993—the period of Hurricanes Hugo, Iniki, and Andrew—despite increased insurer capacity (Davidson, 1997; Doherty, 1997; Matthews *et al.*, 1999; Swiss Re, 2000b). Given the multi-factorial nature of most insolvencies, weather-related losses were no doubt a contributing factor in other cases as well. Although small or geographically specialized firms are most vulnerable, insolvencies of larger and more regionally diversified companies have occurred in the European Union (Swiss Re, 2000a) and in the United States (see Section 15.2.7). As a result of Hurricane Andrew, the largest U.S. home insurer, State Farm Fire and Casualty, was brought to the brink of insolvency by a US$4 billion loss and had to be rescued by its parent company (Stipp, 1997). The second largest U.S. home insurer, Allstate, paid out US$1.9 billion (which was US$500 million more than it had made in profits from its Florida operations from all types of insurance, including investment income, over the 53 years it had been in business) and also had to be rescued by its parent company (III, 2000a).

Little analysis of historic insolvencies in relation to natural catastrophes has been conducted outside the United States. Data for property/casualty firms in France, Germany, the UK, and the United States show that annual "baseline" insolvencies range from 0 to 0.5% of all firms to more than 2% in years with larger natural disasters (Swiss Re, 2000a). Natural disasters contributed to the severe difficulties experienced by the London market, including Lloyd's of London's near insolvency.

Future-oriented analyses of insurer solvency also have been conducted; these analyses show PMLs of US$45–100 billion, which—despite several recognized conservatisms in the analyses—would result in serious levels of insolvency in the industry (ISO, 1996; Cummins *et al.*, 1999; Kelly and Zeng, 1999). As many as 45% of insurers in the United States (representing 62% of the market share) could be placed in this position (GAO, 2000a,b). These findings are comparable to an earlier study showing that the rise in PMLs is stretching insurers' adaptive capacity (AIRAC, 1986).

Although much attention is focused on catastrophic loss events, "small" loss events are responsible for 50% of total economic and insured losses from weather-related events globally (Munich Re, 2000). If such events are closely spaced, they also can generate insolvencies or deplete surplus (Swiss Re, 1997; Ryland, 2000). Hybrid events involving multiple sources of insurance losses are of particular concern (White and Etkin, 1997; Francis and Hengeveld, 1998). This concern is exemplified in the case of ENSO events. A series of small events could be worse for insurers, in fact, than a single large event because individual losses per event often are capped (Stipp, 1997). Very little has been published on this subject since the mid-1980s (AIRAC, 1986).

8.3.3.3. Vulnerability of Reinsurers

Reinsurance provides a significant and essential form of risk-spreading capacity for primary insurers. For natural catastrophes, this risk-spreading normally takes the form of an "excess" contract; primary insurers retain the first tier of losses up to a "trigger point" above which they purchase reinsurance, which operates up to a specified "exit point" or upper limit. After the catastrophes of the past 2 decades, reinsurers are leaving more of the risks with primary insurers, by increasing trigger points and lowering exit points (Stipp, 1997).

Many of the vulnerabilities experienced by primary insurers also apply to reinsurers. Several reinsurers became insolvent or were absorbed by larger firms during the crisis period of

1989–1993 (ISO, 1999; Mooney, 2000). The unexpectedly costly European windstorms of 1999 caused further problems (Andrews, 2000). For example, an already weakened Australian reinsurer covering these storms became insolvent despite total assets of US$2.3 billion (Howard, 2000a). According to the Insurance Information Institute (III, 2000a), the world's catastrophe reinsurance industry "...lacks the capacity to insure mega-losses in excess of US$50 billion." Government reinsurance systems also have shown signs of stress—as evidenced in France, where reserves fell by 50% during the 1990s and reinsurance rates rose sharply (CCR, 1999).

8.3.3.4. Regulatory Uncertainties

An additional source of vulnerability arises from regulatory uncertainties, such as the degree of flexibility afforded in withdrawing from markets and risks and in raising insurance prices (Davidson, 1996; Insurance Regulator, 1998; III, 2000a; Ryland, 2000). In some jurisdictions, regulators have restricted policy cancellations and nonrenewals following natural disaster losses such as Hurricane Andrew (ISO, 1994a,b; Lecomte and Gahagan, 1998). Recent requests from Florida insurers to double rates to protect insurers from hurricane risks also have been resisted by regulators (III, 2000b). On the other hand, under some conditions regulators can force insurers to withdraw from markets or otherwise change their business practices so they maintain minimum solvency requirements (GAO, 2000a). Pre-event accumulation and taxation of reserves also is an important issue, and policies vary by country (Eley, 1996; Davidson, 1997).

8.3.3.5. Vulnerability of Local, State, and Federal Governments as Providers of Insurance and Relief Assistance

Under climate change, sustained increases in the frequency and/or intensity of extreme weather events could stress the government sector itself as a provider of insurance, a provider of domestic and international disaster preparedness/recovery services, and an entity that itself manages property and undertakes weather-sensitive activities (e.g., ranging from mail delivery to operation of military facilities near coastlines or waterways). Increasingly, governments seek to cap or reduce existing exposures (ISO 1994b, 1999; Gastel, 1999; Pullen 1999b; FEMA, 2000; III, 2000b). Governments in developing countries participate especially deeply in weather-related risks, given the low level of private insurance availability and often a higher level of government-owned infrastructure.

Disaster relief provided by the U.S. government has totaled $30 billion since 1953 (Changnon and Easterling 2000). Nearly half of these losses have occurred since 1990, and inflation-corrected payments rose six-fold between the late 1960s and the early 1990s (Easterling et *al.*, 2000a). These costs do not include temporary housing, unemployment insurance, and small business loans also provided by government.

Governments are particularly sensitive to changes in flood- and crop-related losses because they often are the primary or sole providers of such insurance, and climate changes are expected to exacerbate these losses (see Chapters 4 and 5; Rosenzweig *et al.*, 2000). U.S. government-insured crop/hail losses grew 11-fold between the 1950s and the 1990s (Easterling *et al.*, 2000a). In Japan, the majority of international relief—7–8 billion yen in 1990—is related to floods (Sudo *et al.*, 2000). Solvency is a material issue for government programs, as exemplified by the $810 million deficit in the U.S. flood insurance program in the mid-1990s (Anderson, 2000). U.S. crop and flood insurance programs have never been profitable (GAO, 2000a; Heinz Center, 2000). The French catastrophe reinsurance fund (Caisse Centrale de Réassurance) had become depleted as of the late 1990s and could no longer deal with a major catastrophe from accumulated surplus (CCR, 1999).

8.3.4. *Adaptation*

Insurance losses are paid out of premiums and from surplus (net assets). The ability to generate premiums and rebuild surpluses cannot be increased quickly in response to changes in the incidence of losses. In a developing country context, where insurance markets are nascent, this problem is particularly acute.

Insurers have many tools for reducing their financial vulnerability to losses (Mooney, 1998; Berz, 1999; Bruce *et al.*, 1999; Unnewehr, 1999; III, 2000b). These tools include raising prices, nonrenewal of existing policies, cessation of writing new policies, limiting maximum losses claimable, paying for the depreciated value of damaged property instead of new-replacement value, or raising deductibles. The additional strategies of improved pricing and better claims-handling were reviewed in some detail in the SAR (Dlugolecki *et al.*, 1996). Many adaptation strategies in use or under discussion make good sense for insurers irrespective of potential changes in the climate resulting from human activities (Sarewitz *et al.*, 2000) (see Box 8-1).

Insurance prices exhibit sensitivity to disaster events (Paragon Reinsurance Risk Management Services, as cited in Klein, 1997; Edgecliffe-Johnson, 1999). Reinsurance prices rose by approximately 250% following Hurricane Andrew (see Section 15.2.7). Following the upsurge in catastrophe losses in 1999, the trend once again is toward upward pressure on prices (Mooney, 2000).

Following the period of (upward) price adjustments in response to a major natural disaster, however, insurers often enter or re-enter a battered market that offers substantial nonactuarially based discounts, resulting in inadequate prices for all players in the market (Matthews *et al.*, 1999). Similar behavior has been observed among reinsurers (Stipp, 1997). Insurers also may reduce risk management efforts and incentives in the face of competitive pressures on prices. Competitive pressures can cause some insurers to assume greater risk to offer more attractive

Box 8-1. Co-Benefits that Are Relevant for the Insurance and Other Financial Services Sectors

Co-benefits are discussed elsewhere in the Third Assessment Report (TAR WGIII Chapters 3 and 8). Several adaptation mechanisms that are relevant to public and private disaster risk management possess important co-benefits, but these mechanisms are rarely accounted for in cost-effectiveness analyses. Though they normally are associated with mitigation (e.g., emissions reductions or enhanced carbon sinks), some also stand to enhance adaptive capacity or otherwise benefit insurers and other parties in the financial services sector (Sarewitz *et al.*, 2000). Further research on this topic is merited.

- *Energy End-Use Efficiency.* Various co-benefits pertaining to energy-efficient technologies have been documented (Mills and Rosenfeld, 1996; Vine *et al.*, 1999, 2000; Changnon and Easterling, 2000; Zwirner, 2000; TAR WGIII Chapter 5). Improved insulation and equipment efficiency can reduce the vulnerability of structures to extreme temperature episodes and contribute to reduced greenhouse gas emissions. Other examples include linkages between public transit and reduced speed limits and improved highway safety (Unnewehr, 1999; TAR WGIII Chapter 9); energy-efficient ultraviolet water disinfection to conserve fuelwood and reduce deforestation (Gadgil *et al.*, 1997); and emission reductions resulting in improved air quality and reduced respiratory disease (see Chapter 9).
- *Renewable Energy and Distributed Energy Systems.* Certain renewable and distributed energy supply technologies have attributes that are relevant to risk management and disaster recovery (Mills, 1996, 1999; Mills and Knoepfel, 1997). For example, low-power/energy-efficient technologies can reduce business interruption risks by extending the reliability and operating range of backup power systems (Stauffer, 1995; Kats, 1998; Lecomte and Gahagan, 1998; Vine *et al.*, 1999; Deering and Thornton, 2000). Substitution of biofuels for fossil fuels can yield improved air quality and reduced flood risk (IPCC, 2000; TAR WGIII Chapter 9).
- *Sustainable Forestry, Agriculture, and Wetlands Management.* Enhancing organic soil content benefits crop insurance as well as contributing to improved water quality and food security. Sustainable forestry practices yield benefits of watershed management and flood/mudflow control, which are necessary foundations for establishing a modern economy (Scott, 1996; IFRC-RCS, 1999b; IPCC, 2000; Hamilton, 2000; see also Chapter 5). Wetlands restoration helps to protect against flooding and coastal erosion, although methane release from wetlands also must be considered (IPCC, 2000).
- *"Green" Financial Products.* Initiatives such as innovative financing of energy-efficiency improvements, insurance products that promote better environmental management, or insurance for adaptation/mitigation projects under the U.N. Framework Convention on Climate Change (UNFCCC) can simultaneously support adaptation and mitigation objectives (Hugenschmidt and Janssen, 1999; Mills, 1999; UNEP, 1999; Zwirner, 2000). However, considerable business risk and liability may be associated with UNFCCC projects if measurement and verification are poor or issues of buyer/seller liability are not addressed by insurers in the drafting of insurance contracts.

prices and products to consumers, through acquisitions of weakened companies and destabilizing growth rates (Matthews *et al.*, 1999).

Favorable underwriting or investment experience may generate surpluses, but many legislatures do not permit insurers explicitly to fund pre-event catastrophe reserves to account for anticipated changes in climate and weather. Alternatively, insurers may try to raise more capital or reduce dividends paid to shareholders, but such actions will not be acceptable to financial markets if the risk-to-reward ratio is not competitive with that of other companies or sectors. The trend toward consolidation within the insurance sector is sometimes regarded as a factor that reduces insurer vulnerability to catastrophic losses.

8.3.4.1. Adaptation Mechanisms: Risk-Spreading

Public and private insurance is inherently a risk-spreading mechanism. Insurers also can spread risks through reinsurance, depending on its availability and price. Losses associated with uninsurable risks, or unpaid claims in the event of insurer insolvencies, often are partly spread to the community through disaster relief or guaranty ("solvency") funds. State-managed guaranty funds—to which insurers must contribute—are used for specified catastrophe losses in France, Germany, Japan, The Netherlands, the UK, and the United States (III, 2000a; Swiss Re, 2000a). Of the 25 largest U.S. P/C insolvencies (amounting to US$5 billion in claims), 29% of the losses were recoverable through guaranty funds; national capacity was only US$3.4 billion as of 1998 (NCIGF, 1999). In the United States, the property insurance residual markets known as Fair Access to Insurance Requirements (FAIR Plans), Beach or Windstorm Plans, and joint underwriting associations (JUAs) represented insured property value (exposure) of US$24 billion in 1970 and US$285 billion in 1998 (III, 1999; Gastel, 2000).

Although risks also can be spread between public and private insurers, governments have elected to cap their exposures by formally limiting government-paid losses for weather-related events in the United States (GAO, 1994; Pullen, 1999b; III, 2000b) and earthquake losses in Japan (Gastel, 1999).

Governments also are trying to reduce their insurance and disaster recovery spending (ISO, 1994b, 1999; FEMA, 2000).

Nonconventional "alternative risk transfer" (ART) mechanisms have begun to emerge and are regarded by some banks and insurers as playing a role in the continued viability of insurance (see Section 8.4). On the other hand, some insurers, consumers, and members of the financial community question the efficacy and attractiveness of these new risk-spreading mechanisms (Tol, 1998; Peara, 1999; Swiss Re, 1999b; Bantwal and Kunreuther, 2000; Freeman, 2000; GAO, 2000a; Jamison, 2000; Nutter, 2000).

"Moral hazard"—a pervasive issue in the industry—results when, by virtue of adaptation efforts or the very availability of insurance (or reinsurance or government aid), the insured feels less compelled to prevent losses (White and Etkin, 1997; Ryland, 2000). Government programs have been faulted for unintentionally encouraging such maladaptation and risky behavior (Anderson, 2000; Changnon and Easterling, 2000). For example, it is estimated that one-quarter of the development over the past 20 years in at-risk areas along the U.S. coastline is a result of the presence of the National Flood Insurance Program (Heinz Center, 2000). Moral hazard also has been ascribed to primary insurers or reinsurers who rely excessively on state-maintained guaranty funds (Kunreuther and Roth, 1998; Swiss Re, 2000a).

8.3.4.2. Adaptation Mechanisms: Risk Reduction

Although risk-spreading is largely an economic and distributional process, risk reduction focuses more on technology, environmental management, land-use planning, engineered disaster preparedness/recovery, and predictive modeling. Hooke (2000) provides a good overview of the challenges facing risk-reduction initiatives. The UN's International Decade for Natural Disaster Reduction (IDNDR) is a leading example of international cooperation in this area.

The insurance industry is an important participant in partnership with other public and private entities (Ryland, 2000). Examples include the use of geographic information systems to better understand and pinpoint risks, land-use planning, flood control programs, early warning systems, sustainable forest management, coastal defense, and wind-resistant construction techniques supported by building codes (Bourrelier *et al.*, 2000; Davenport, 2000; Hamilton, 2000; Hooke, 2000; Sudo *et al.*, 2000). However, the scale of effort has been much smaller than that anticipated for global climatic changes, and loss prevention generally has focused on fortifying the individual against perils, rather than reducing the peril itself (Kunreuther and Roth, 1998) and on post-disaster actions (Ryland, 2000).

Any discussion of vulnerability, impacts, and adaptation also should include insurance brokers, agents, risk managers, and trade associations. In 1998, there were more than 750,000 such workers in the United States alone (III, 1999).

A key but often untapped opportunity is to rebuild damaged structures in a more disaster-resistant fashion following loss events, as in the U.S. National Flood Insurance Program. Pervasive problems with building code enforcement and compliance have emerged following natural disasters. For example, 70% of the losses from Hurricane Alicia were traced to lax code enforcement (III, 2000a). Building industry stakeholders often resist new codes. Reinvigorating businesses and other forms of economic activity also is central to disaster recovery (Carrido, 2000).

Energy systems can have important implications for economic and insured losses through the vulnerability (reliability and/or physical damage) of energy generation, transmission, and distribution technologies (Epps, 1997; Keener, 1997; Deering and Thornton, 2000). Hydroelectric power resources, for example, are weather sensitive (see Chapter 15). Climate change may confound the actuarial basis for weather-related insurance provided to energy producers and for utility interruption insurance provided to energy users. Energy-related business interruption (via lightning damages, interrupted operations, inventory spoilage, event cancellation, disrupted tourism, etc.) is a significant weather-related exposure faced by the insurance sector (as evidenced by the extended power failure faced by Auckland, New Zealand, following a major heat wave in 1998). The North American ice storm of 1998 offers another dramatic example of the role of power disruption in disaster-related insurance losses (Lecomte and Gahagan, 1998; Table 15-5). Improved appraisal of the physical vulnerability of existing energy systems and of new technologies deployed for emission-reduction projects (e.g., as part of Clean Development Mechanism or Joint Implementation) would help to reduce vulnerability to extreme weather events and other losses (World Bank, 1999; Zwirner, 2000). The aftermath of Hurricane Andrew illustrated the complex nature of losses caused by natural disasters. About 20% of insured economic losses were related to business interruption (40% in the case of Hurricane Hugo) (Mills, 1996).

Effective risk reduction requires foresight. The insurance sector participates in a limited way in weather- and climate-related research and modeling (Kelly and Zeng, 1999). The Risk Prediction Initiative and the World Institute for Disaster Risk Management are two examples of insurer-funded research centers. Insurers' catastrophe models are not presently used in association with climate-prediction tools such as general circulation models (Peara and Mills, 1999). Their predictive power is poorly validated (Pielke, 1998; Pielke *et al.*, 1999) and often exhibits significant unexplained model-to-model variation (Matthews *et al.*, 1999; GAO, 2000a). Insurance regulators in the United States have resisted efforts to include them in ratemaking proceedings (III, 2000b). Thus, the insurance community may stand to benefit from analytical collaboration with the natural sciences community (Nutter 1996; Changnon *et al.*, 1997; Zeng, 2000; Mills *et al.*, 2001). Formal solvency analyses conducted by insurance regulators also could benefit from more explicit treatment of future climate scenarios.

Box 8-2. Equity Issues that are Relevant for the Insurance and Other Financial Services Sectors

Equity is a material issue facing the financial services sector systems, within and among countries. For example, inequities can be created when the premiums paid by insureds become severely decoupled from the risks they face. On the other hand, strictly equalized insurance payments can result in a problem known as "adverse selection," wherein only those with higher-than-average risk will actively purchase insurance, causing the system to become ineffective.

The burden of natural disasters tends to fall disproportionately on economically disadvantaged people, especially in developing countries (Hooke, 2000; Kreimer and Arnold, 2000). However, access to the benefits of insurance is correlated with income level. Lower income consumers in poor and wealthy countries alike have difficulty affording insurance or financing even at current rates (Miller *et al.*, 2000) and often live and work in more vulnerable locations. Immigrant cultural groups, as well as aboriginal peoples, may have less access to pre-disaster information and be more vulnerable to natural disasters themselves (Solis *et al.*, 1997).

As an illustration of price-related stresses, projected increases in coastal erosion in the United States would require a doubling of current insurance rates—probably requiring cross-subsidies among insureds (Heinz Center, 2000).

In developing nations, the availability of insurance and financing has considerably lower penetration than in wealthy nations. At the global scale, one form of inequity arises in which a greater share of the costs of extreme weather events are borne by governments and consumers in the "south" than in the "north." Rising uncertainties could reduce the availability of insurance in some areas and impede the expansion of adaptive capacity offered by insurance markets in developing countries. Governments' ability to compensate by providing more insurance and disaster relief would be similarly strained.

Although much progress has been made in risk-reduction technology *per se*, attention is increasingly focused on problems of implementation. Key issues identified by IDNDR include public awareness of risks, training of practitioners, commitment by public officials, and justification and financing of risk-reduction strategies (Hamilton, 2000; Hooke, 2000)—all areas where the financial services sector can play a part.

8.4. Impacts and the Role of the Banking Industry

In the private sector, the insurance and banking industries play leading roles as investors, although they focus on different aspects of this business. The role of banks is to cover the credit part (by providing loans), whereas insurance companies act as investors on the capital markets, as well as in the property/casualty branches; they also insure projects financed by banks. This section focuses on the banking industry.

8.4.1. Climate Change Impacts

Environmental issues such as climate change may have substantive impacts on the global economy. From a financial point of view, such problems are regarded as environmentally induced economic risks (Figge, 1998). In general, the size of the players, their diversification, and increasingly sophisticated techniques of risk reduction make it unlikely that banks and asset managers will perceive climate change as presenting any material threat to their economic viability.

On the positive side, banks could provide services and develop financing techniques that accommodate and facilitate adaptation to weather extremes (e.g., private insurance, catastrophe bonds, weather-related trading). Assessment of expected benefits of an investment decision—whether it is a direct investment, through financing of an infrastructure project, or an indirect investment that involves investing in shares—is core to financial institutions. Economic assessment of an investment is based on three different factors: expected revenues, operating costs, and risks. Climate change can have an impact on all three aspects but is probably more important for the risk side of an investment decision (Figge, 1998; Mag, 1990).

Lending and Climate Change

Most private and corporate loans are secured by property. If a region becomes more exposed to climate-related natural disasters such as floods or windstorms, the prices for property could go down—which could result in a loss of confidence in the local economy and may even trigger a credit crunch (Grabbe, 1998; Heinz Center, 2000). As an indirect effect, other types of business such as management of private assets and granting of private loans that are not backed by property also will be affected (Bender, 1991; Thompson, 1996).

In terms of the impacts of climate change on the banking industry, there is no clear scientific evidence on how this sector will be affected. One view is that the banking sector is likely to be largely unaffected by climate change because the sector increasingly transfers loans directly to the capital market through asset-backed securities and similar instruments. The major commercial banks are large and diversified and are getting more so as the industry concentrates in the face of global competition. They prefer not to keep any substantial portion of the loans they make on a long-term basis. Instead, driven by capital constraints and return requirements, they actively syndicate and/or securitize their loan commitments (i.e., sell down the loans or shift the loan exposure to other banks and institutions). Even the portion they retain is increasingly likely to be held for a shorter period of time (a maturity under 1 year is better from a capital requirement standpoint) than other institutional lenders such as insurance companies and pension funds. The question still remains: At the end of the day, who will bear the risk of climate change on investments? It is particularly the insurance and asset management sectors that invest in asset-backed securities. So the insurance industry may even get hit twice, first through direct losses in property-related claims but also through impacts on their investments (Salt, 2000). Detailed information on what this increased vulnerability means for insurers and asset managers must be further explored.

On the other hand, it is obvious that banks could be affected indirectly as climate change affects their customers' operations, consumption, and financial circumstances. Any investment activity could be affected if property insurers withdraw coverage or drastically increase premiums, as happened in Florida and the Caribbean. Sectors that are likely to be affected by a drastic change in the local climate are agriculture and tourism. Warm winters in Europe already have negatively impacted the performance of skiing resorts in the Alps and have led some banks to review their credit applications in view of possible impacts of climate change (Credit Suisse Group, 1999).

8.4.2. Adaptation Issues

To date, the literature does not explicitly address financial services firms outside the insurance sector. There is emerging evidence that some investors or businesses as a group are modifying their risk perception to incorporate the potential for climate change. Partly this is driven by pension funds that are filing shareholder resolutions against polluting companies or banks that finance such practices (Behn, 2000). Similarly, there also is emerging evidence that financial services firms are including consideration of potential climate change as a risk factor in evaluating investments or developing new products (World Bank, 1999; Jeucken and Bouma, 2000). However, history has shown that the ability of banks and asset management firms to respond and adapt to external shocks is strongly tied to the ability of those institutions to diversify risk, both for themselves and for their customers. Over the past 25 years or so, financial services firms have changed significantly in response to a variety of circumstances, including macroeconomic disturbances of local and global proportions, advances in communications and information technologies, and changing regulatory regimes (Kaufman, 1992; Downing *et al.*, 1999). Several types of tools can be identified for managing risk: improved information and research, diversification, building up reserves, and new product development.

The Role of New Product Development

Over the years, banks as a whole have demonstrated their ability to continuously develop new products and services to respond to changes in their own business environment as well as the changing needs of their customers (Folkerts-Landau and Mathieson, 1988; Haraf and Kushmeider, 1988; Jeucken and Bouma, 2000). The ability of those firms to respond and adapt to any impacts of potential climate change will be determined largely by their ability to identify any changes in their customers' views of asset risk and to develop new products to hedge and diversify that risk. Again, the literature does not discuss explicitly which specific existing products might be useful in responding to changes in risk stemming from potential climate change or what types of new products might be developed to respond to such potential changes in risk. However, the industry continues to apply basic concepts—including options, swaps, and futures contracts—in new and different ways to create new products that provide investors and businesses with useful tools for reducing well-known and understood risks (Mills, 1999, Vine *et al.*, 1999). These products can range from environmentally and socially screened investment funds to very sophisticated derivatives that hedge against weather risks.

In the past few years, such weather derivatives have seen rapidly growing use to hedge the risks of businesses whose sales and revenues are strongly affected by the weather. Securitization is becoming more and more widely used as a means of spreading risk and obtaining resources for investment banking with a secure flow of income in the future. Financial institutions other than insurance companies have been developing and offering such instruments in the form of catastrophe bonds, for example (see Box 8-3).

In summary, the banking industry is more likely to see climate change and the possible response more as an opportunity than as a threat. In the new global competition, banks and asset managers are likely to be less concerned about any possible threat posed to their existing portfolios by weather extremes induced by climate change and more preoccupied with adjusting to a rapidly changing and increasingly competitive global market in which failure to adjust leads rapidly to loss of market share and net revenue and a decline in share price and shareholder value. They have little incentive to try to change the rules, but they are highly motivated to respond once changes are imminent or implemented.

8.4.3. The Role of UNEP Financial Services Initiatives in the Climate Change Debate

The United Nations Environment Programme (UNEP) has brokered statements of environmental commitment by banks and insurance companies that have been endorsed by many of the major players in these industries. These statements have

Box 8-3. Capital Market Alternatives for Risk Financing

Alternative catastrophe risk financing mechanisms through weather derivatives have begun to emerge and are regarded by some observers as playing a role in the continued viability of the insurance sector. Other authors suggest that these instruments will continue to be a niche product because of inability to come up with adequate pricing for these mechanisms as offered through the capital markets. Such products also raise the awareness and visibility of natural disasters and climate change issues within the financial markets (Swiss Re, 1996; Credit Suisse Group, 1998; Lester, 1999; Mahoney, 1999; Punter, 1999).

Contingent Capital Securities. The two types of capital contingency securities available to investors are contingent surplus notes and catastrophe equity puts. Investors in these securities become—at the insurer's option—creditors of or equity investors in the insurer. The exercised "notes" and "puts" are shown as surplus on an insurer's balances sheet and thus increase assets without an offsetting increase in the liability portion of the balance sheet. The insurer can draw from surplus to pay unreserved catastrophe losses and have the funds (surplus) necessary to take on new exposures.

Catastrophe Risk Securities. Two forms of "cat risk securities" are available that transfer underwriting risk to investors: catastrophe bonds and catastrophe insurance options. Primary insurers and reinsurers can make use of these securities. Both benefit insurers by making monies available to offset catastrophe losses. In contrast to contingent capital securities, these instruments do not bolster an insurer's surplus; they provide funds for the payment of losses. They are reflected as both an asset and as a liability on the insurer's financial statements.

These approaches are relatively new, and their efficacy and robustness must be evaluated (see Tol, 1998; Peara, 1999; Swiss Re, 1999b; Bantwal and Kunreuther, 2000; GAO, 2000a; Nutter, 2000; Jamison, 2000; Mills *et al.*, 2001). Among the questions to address include:

- In a more competitive environment, would insurers and reinsurers be inclined to participate in or encourage (subsidize) risk-reduction measures?
- Do derivatives signal a potential means by which self-insurers can expand their capacity, thereby providing greater competition for primary insurers and reinsurers?
- Will the occurrence(s) of catastrophic weather-related events turn away investors after an event?
- Do existing catastrophe and climate modeling techniques yield information necessary to adequately evaluate financial risks and thus the prices of these derivatives?

Of 11 major trends in investing, catastrophe bonds were rated by members of the International Securities Market Association as least likely to have significant impacts on securities markets in the future (Freeman, 2000).

Despite doubts about these new instruments, banks and insurance companies consider this a growing business. In 1999, the cumulative volume of weather-related bonds/derivatives reached US$3 billion. It can be assumed that an increasing number of such instruments will be available to hedge against climate risks. This, in turn, will allow banks to get the "insurance" coverage they need for their lending activities (Nicholls, 2000).

now been signed by almost 300 banks and insurance companies from all parts of the world (most from Europe and Asia) (UNEP, 2000). By signing the statement, companies undertake to make every effort to incorporate environmental considerations into their internal and external processes (UNEP, 1995; Schanzenbächer, 1997). Measures implemented by signatories range from reduction of energy consumption of buildings under their management to incorporation of environmental issues in credit business and risk management considerations. One might think that financial services is a clean industry with very little direct impact on climate change, but insurance companies in particular own huge physical assets (e.g., ~500 million ft^2 of building space in the United States alone, which corresponds to an energy bill of US$750 million a year) (Mills and Knoepfel, 1997).

8.5. Special Issues in Developing Countries

8.5.1. *Statistics on Disasters*

Although the vast majority of weather-related insurance losses occur in wealthy countries, most of the human suffering occurs in poor countries (Figure 8-6). Whereas 45% of the natural disaster losses between 1985 and 1999 took place in wealthy countries (those with per capita income of more than US$9,360), these countries represent 57% of the US$984 billion in total economic losses and 92% of the US$178 billion in insured losses (Munich Re, 1999b). In contrast, 25% of the economic losses and 65% of the 587,000 deaths took place in the poorest countries (those with per capita income below US$760).

Other literature sources, using slightly different definitions and different time periods, conclude that about 90% of deaths from natural disasters from 1973 to 1997 occurred in Africa and Asia (IFRC-RCS, 1999a). Figures from the World Disasters Report 1999 (IFRC-RCS, 1999a) indicate that, in the period from 1973 to 1997, on average nearly 85,000 persons were killed each year by natural disasters; the number of otherwise affected (impoverished, homeless, injured) was more than 140 million annually. The record of disasters (see Figure 8-6) is a further illustration of the geographic distribution of weather-related disasters.

As indicated in Chapter 3, climate change comes with changing frequencies and intensities of extreme weather events. The most vulnerable regions and communities are those that are both highly exposed to hazardous climate change effects and have limited adaptive capacity. Countries with limited economic resources, low levels of technology, poor information and skills, poor infrastructure, unstable or weak institutions, and inequitable empowerment and access to resources have little capacity to adapt and are highly vulnerable (see Chapter 18). The regional chapters in this volume (Chapters 10–17) indicate that developing countries, because of their limited or nonexistent financial buffers, are particularly vulnerable to the effects of climate change. Human-induced climate change is expected to result in a further upward trend of disaster losses.

Developing countries—especially those that are reliant on primary production as a major source of income—are particularly vulnerable because these countries and their communities hardly have any financial buffer and there is very little penetration of insurance (see also Figure 8-6). The conditions facing private insurance markets and government disaster relief differ considerably in developing countries. The penetration of private insurance is extremely low in most cases, although it is growing quickly. The degree of preparedness also is low. The government sector is far less able to operate as a surrogate insurer, even in areas such as crop and flood insurance where governments traditionally are essential in the developed world. In developing countries, the economic and social impacts of catastrophic weather events can pose a material impediment to development. Increased frequency or intensity of such events as a result of climate change could render these markets less attractive than they are at present for private insurers, in turn compounding the adverse impact on development. Thus, developing countries tend to have greater vulnerability and less adaptive capacity than developed countries.

8.5.2. Disaster Relief

Because of the lack of insurance, disaster relief is the major input for disaster recovery in many developing countries. After a disaster, the first relief usually is provided by the national government in the form of assistance by the military, the police, and other government services. Often, governments also act as the insurer for uninsured damages in these cases. When the capacity of local disaster relief institutions is exceeded, countries tend to call for help from international institutions. In the period 1992–2000, a yearly average of US$330 million was transferred from country to country for disaster aid (United Nations, 2000).

The institutional setting of international disaster relief is complicated. Presently, 16 UN agencies have a mandate that allows them to work in emergency situations. The UN Office for the Coordination of Humanitarian Affairs (UN-OCHA) is supposed to coordinate efforts in disaster relief. The International Committee of the Red Cross and the International Federation of Red Cross and Red Crescent Societies (IFRC-RCS) have a basis in international law. Médecins sans Frontières (MSF) and OXFAM are examples of internationally operating nongovernmental organizations (NGOs), of which there are hundreds. In addition, all types of local NGOs may be involved in the relief work, along with the national government and local authorities. In a typical disaster situation, one has to cope with a multitude of different agencies (Frerks *et al.*, 1999). Donor governments and agencies as well as international organizations provide the funds; substantial amounts may be raised directly from the public at large.

The large amount of relief amount for cyclones in 1998 was largely a result of Hurricane Mitch, which struck Central America in that year. In the same year, Bangladesh and China were struck by very large flood disasters. The foregoing numbers show that on an annual basis, international relief is in hundreds of millions of dollars. This is a small number compared with total global damage from natural disasters (tens of billions of dollars).

8.5.3. Natural Disasters and Development

Disasters may have a significant impact on the national economy of the country concerned. Some countries lose annually up to 1% or more of their annual GDP as a consequence of recurring natural disaster; in individual cases, damages have been as high as 50% of GDP. The typical Chinese loss experience in bad years is in the range of 5–7% of annual GDP. In 1974, losses in Honduras from Hurricane Fifi were equivalent to 50% of the country's 1973 GDP (Hooke, 2000). Setbacks in development may have been up to 1 decade or more. The majority of these damages usually are covered by the affected population itself or from other domestic sources (United Nations, 1994). In some cases, the relation between GDP and disasters is ambiguous because post-disaster investments may increase GDP. Long-term problems arise when the return period of a disaster is the same order of magnitude or smaller than the time needed for reconstruction. In such cases, the economy of a country or a specific region is likely to spiral downward (Downing *et al.*, 1999). In fact, GDP is a very limited way of describing the impact of weather-related disasters. For example, the UN has defined a disaster as *large* when the ability of the region to cope with the effects of the disaster on its own is exceeded.

Urban and rural infrastructure loss in the developing world as a result of natural disasters has impacted the activity of the world's international lending institutions. The World Bank has estimated that it has loaned US$14 billion to developing countries in the past 20 years for damages from natural disasters. This amount is nearly 2.5 times the amount loaned by the Bank for relief from civil disturbance worldwide (Kreimer *et al.*, 1998). The Asian Development Bank (ADB) has estimated that between 1988 and 1998, 5.6% of ADB loans were for disaster rehabilitation. In 1992, nearly 20% of ADB loans were for rehabilitative assistance to natural disaster recovery (Arriens and Benson, 1999). The World Bank has estimated that during the past decade in Mexico, as much as 35% of its lending earmarked for infrastructure has been diverted to pay for the costs of (Mexican) natural catastrophes (Freeman, 1999). In recent years, the World Bank has recognized the importance of disaster prevention and mitigation for development and poverty reduction (Kreimer and Arnold, 2000).

8.5.4. Vulnerability and Financial Adaptation in Developing Countries

Spreading the risks of catastrophes presents special difficulties in developing countries, particularly rural areas. In general, there is very limited use of commercial insurance because of long histories of economic instability, fluctuating and prohibitive insurance costs as related to agricultural prices, lack of enforcement of building codes and land-use regulations, subjective evaluation of risk by consumers ("it won't happen to me"), and non-monitored economies. In some developing countries, government-organized crop and disaster insurance exists on paper, but with large debt loads and weak economies, many such programs are inactive. In many cases, governments are unable to respond to public expectations.

A World Bank/UN Development Programme (UNDP) workshop reports that disaster mitigation is evolving from the phases of relief and contingency planning, technical preparedness, and structural solutions to a phase in which there is a greater emphasis on reducing social and economic vulnerabilities and investing in long-term mitigation activities. However, formal sector mechanisms may completely bypass the poorest households. Therefore, the need to develop informal and flexible financial instruments such as microfinance for disaster mitigation has become extremely important (World Bank, 2000).

Although targeted microfinance programs have been able to meet the financial needs of individual households, the same attributes of microfinance also could be applied to deal with natural disaster reduction. There is a potential for microfinance to provide explicit and implicit insurance to households (World Bank, 2000). However, limitations of microfinance as a risk-reduction mechanism arise from issues of moral hazard (see Section 8.3.4), inadequate monitoring of credit programs after large spatial shocks, and reduction in informal insurance arrangements provided by social networks. There also is the possibility of governments committing much less to relief programs in the wake of a disaster if affected communities are served by microfinance institutions. Small microfinance programs without access to reinsurance may collapse in a natural disaster. For nationwide disasters, even the largest microfinance programs may require international arrangements (World Bank, 2000).

Several microfinance organizations in Bangladesh have been seriously affected by the floods in 1998 in terms of maintaining savings mobilization, credit repayment, and cash availability. Larger microfinance organizations with greater capitalization and preparedness cope better with disasters than small microfinance organizations, many of which get completely wiped out. There is now a recognition of the need for providing a financial cushion for the unexpected. It could be provided through a Central Reserve Fund/Emergency Fund and bigger microfinance organizations such as Grameen Bank setting aside part of their funds to meet the contingencies of natural disasters (World Bank, 2000).

8.6. Issues that are Related to Funding for Adaptation

Although some discussion of adaptation appears in virtually all chapters of the WGII report, Chapter 18 addresses core concepts and considerations. It states that key factors affecting the adaptive capacity of a region or country include economic resources, technology, information and skills, infrastructure, and institutions; it notes that there is considerable variability among countries with regard to their ability to adapt to climate change. However, Chapter 18 does not address specifics of how adaptation is likely to be funded in developed and developing countries or how the need to fund adaptation will affect the financial services sector.

Chapter 18 also states that adaptive actions are most likely to be implemented when they are components of or changes to existing resource management or development programs (see World Bank, 1999). Klein (1998) has suggested that investments for adaptation should essentially be incremental to projects justified for other reasons, where a project has value even if climate change were not to occur. When the impact of climate change is uncertain, this is probably the best way to proceed. The financial sector would then play its traditional role, and financial institutional arrangements would be the same as for present-day investments in infrastructure. However, when projections of climate are more certain and/or when the assets exposed are of high value or when many people are at risk, dedicated additional climate change adaptation investments may be required and special funding arrangements may be developed. In these situations, new climate change-related financing schemes may emerge whereby funds are generated through fees on emissions of greenhouse gases or fees on trading of greenhouse gases. The Clean Development Mechanism developed in the framework of the Kyoto Protocol is an example of the development of such a financing arrangement. Such arrangements could generate new roles for the financial sector.

An example of early cost estimates of measures meant to protect people and properties against the risk of increasing rainfall,

Box 8-4. Case Study: Bangladesh Flooding 1998

Bangladesh witnessed 35 cyclones from 1960 to 1991 (Haider *et al.*, 1991) and seven major floods from 1974 to 1998 (Matin, 1998). The flood of 1998 is considered to be one of the worst natural disasters experienced by the country in the 20th century. It occurred from July 12 to September 14, a duration of 65 days (Choudhury, 1998). The flood affected about 100,000 km^2 (68% of the country's geographical area). The numbers of affected families and population were more than 5,700,000 and 30,900,000, respectively (Choudhury, 1998). The flood caused 918 fatalities and disease among 242,500 people. Approximately 1.3 Mha of standing crops were fully or partially damaged. Total economic losses amounted to US$3.3 billion (8% of GDP, 1998 value), according to a study by Choudhury *et al.* (1999). The study also shows that there is a wide discrepancy between its estimates and estimates by other agencies, which is mainly a result of coverage error.

Generally, victims have to depend on their own resources to rehabilitate themselves. During the emergency period, however, the government and NGOs mobilized considerable financial resources to provide relief in the form of food, clothing, and building materials.

To reduce the damage from natural catastrophes, planned activities by the government and NGOs (national and international) include construction of an adequate number of cyclone shelters, embankments, and other shelters in coastal areas, especially in the offshore islands.

With regard to insurance against such calamities, there is not much available except for the large industries and the commercial sector. Flood victims were paid US$27.7 million as compensation by the insurance companies, of which about 70% went to large industrial units. There was virtually no insurance coverage for losses in the agricultural sector. Losses incurred by shrimp farms and water transports, however, received sizeable compensation by the insurance systems, according to government sources (Choudhury *et al.*, 1999).

increasing river run-off, and sea-level rise, using an integral approach, is the white paper of The Netherlands National Committee on Water Management. This white paper projects the additional cost of water management in The Netherlands, taking into account climate change and long-term land subsidence and land-use changes. The paper concludes that a budget of approximately US$2.5 billion is required for investments until 2015 and an additional US$8 billion for the 2015–2050 period to maintain adequate safety levels for people and property (Netherlands National Committee on Water Management, 2000).

Developing countries seeking to adapt in a timely manner face major needs, including availability of capital and access to technology. Given the present state of knowledge, many actions for adaptation are likely to be integrated with and incremental to projects that already are occurring for other reasons. The World Bank (1999) states that "there is no case to be made for 'stand-alone' projects on adaptation to climate change." It also has noted that projects for adaptation should be designed as incremental to projects that are justified for economic development purposes. However, providing financing for projects in developing countries is a complex matter. Even for projects for which the risks and expected returns are commensurate with the requirements of the financial markets, matching investors that have available funds with projects seeking funding is by no means easy. Most simplistically, this process involves linking investors with projects via appropriate sets of institutional and financial intermediaries. The ability to do this successfully depends, in part, on the level of development of financial markets and the financial services sector in the country where the project will be implemented (World Bank, 1997b).

However, returns that can be expected from many prospective projects are not sufficient for investors to assume the risks that they believe are inherent in any individual project. This complicates the process further. If such projects are going to be funded, some creative modification must be made to bring each project's risk/return profile in line with the requirements of the financial markets. Unfortunately, there is no straightforward, standardized means for identifying and implementing needed changes. The process is guided in part by the principle that risks should be assumed by the party best able to manage them (IFC, 1996).

This need for financial resources for adaptation in developing countries is addressed in the UNFCCC (or "Convention") and the Kyoto Protocol. The Convention explicitly states that:

- All Parties have responsibilities to make and implement plans for adapting to any human-induced climate change.
- The developed countries shall assist developing countries in meeting the costs of adapting to any adverse effects of such climate change.[3]

Both accords also address this notion more generally in identifying potential actions to aid developing countries, including provision

[3]See Articles 4.1(b), 4.1(e), 4.1(f), 4.3, 4.4, and 4.5 of the Convention.

of "environmentally sound" technology.[4] In addition, the Protocol indicates that a portion of the proceeds from Clean Development Mechanism (CDM) projects is to be used to meet the needs of "particularly vulnerable" Parties for Adaptation[5] (UNFCCC, 1992, 1997). Taken together, provisions in these two accords provide new sources of public sector funding for developing countries to implement adaptation measures.

The Global Environment Facility (GEF), as the main focus of financial commitments under the Convention thus far, has been the institutional mechanism for this funding. GEF projects provide financial models for promoting technology diffusion in developing countries, with some projects designed to mobilize private-sector financing (UNFCCC, 1999). However, GEF activity generally has not addressed the adaptation elements of the Convention. This lack of activity is driven by internal requirements that GEF projects have global benefits, as well as directives that such funding should cover only planning activities that are associated with adaptation (Yamin, 1998). Caribbean Planning for Adaptation to Climate Change is one example of a GEF project that is addressing adaptation. This US$6.3 million project is focusing largely on planning and capacity-building needs for addressing adaptation in the Caribbean (GEF, 1998).

However, there are still many issues to be addressed in connection with both the Convention and the Protocol (Werksman, 1998; Yamin, 1998). Differing interpretations of various provisions of the accords remain.

For example, detailed provisions of the CDM have yet to be worked out, including those related to adaptation funding. One key issue is the size of the "set-aside" from CDM projects that is dedicated to funding adaptation. If this set-aside is too large, it will make otherwise viable mitigation projects uneconomic and serve as a disincentive to undertake projects. This would be counterproductive to the creation of a viable source of funding for adaptation. There also have been no decisions on how these "set-aside" funds would be allocated to adaptation projects. They could be used to fully fund projects or leveraged to simply supplement other sources of funding. Any resulting allocation will be driven by more technical and financial elements of the merits of alternative projects, as well as political considerations of equity and fairness. As a result, it may be some time before any of these provisions can produce a viable source of funding for adaptation. An overview and analysis of the literature on climate change policies and equity appears in Banuri *et al.* (1996). Linnerooth-Bayer and Amendola (2000) propose that subsidized risk transfer can be an efficient and equitable way for industrialized countries to assume partial responsibility for increasing disaster losses in developing countries. Review of the literature indicates that understanding of adaptation and the financial resources involved is still in its early stages. As knowledge grows, the potential role(s) for the financial sector will become clearer.

[4] See Articles 4.8 and 4.9 of the Convention and Article 3.14 of the Protocol.

[5] See Article 12 of the Protocol.

8.7. Future Challenges and Research Needs

This assessment reviews our improved knowledge since the SAR. However, it also identifies many areas in which greater understanding is still needed and suggests several challenges for the research community. These challenges can be summarized as follows.

Improve the transfer of knowledge from the scientific community studying climate change and weather forecasting to the financial services community. There clearly is a need for better understanding of how extreme weather events that are important to financial service firms could be affected by climate change. New knowledge should be communicated to the financial community and society for practical use. There is a specific need to:

- Develop ways for the insurance sector to blend information from the scientific community's climate models, as they evolve, with its own loss estimation models
- Improve daily, seasonal, and annual forecasting of extreme weather, and adapt it for use in disaster prevention.

Advance the understanding of the relative global and regional vulnerability and adaptability of insurance and other financial services to climate change. The trend in losses from extreme weather events has raised questions about the insurance sector's vulnerability to climate change in some respects, although as a whole the industry could be quite resilient. Even less is known about the relative vulnerability or resiliency of other segments of the financial services sector. A more definitive assessment of the industry's strengths and weaknesses in the face of climate change is necessary, with specific needs to:

- Continue analysis to disaggregate climate change, socioeconomic, and any other non-climate drivers of observed trends in historic economic (insured and uninsured) losses
- Explore specific aspects of the industry's vulnerability and resiliency, including maximum probable insurance losses; insurer surplus available for paying claims; insolvency risk in local insurance markets; and ability to raise rates, reduce coverage, or otherwise decrease losses by shifting risk to others
- Assess how climate change could affect the actual and perceived risk of existing loan and investment portfolios
- Understand if or how investors are changing their perceptions of investment risk in light of potential climate change and explore actions that investors are taking in light of any changes in the perception of risk.

Explore the role of the financial services sector in dealing with risks to society from climate change. As intermediaries and risk experts, the financial services sector could play a positive role in efforts to deal with the risks of climate change. The sector also could play an important role in identifying potential synergies and conflicts regarding funding for adaptation and mitigation measures. Work is needed to:

- Assess and develop financial instruments that can spread and hedge against the risks of climate change for developing and developed countries
- Quantify the need for financial resources for investment in adaptation
- Determine how the availability of funds for adaptation could be affected by the use of funds for mitigation activities and vice versa
- Identify any synergies between options for adaptation and for mitigation.

Explore the range of possible financing arrangements to cover the cost of adapting to climate change:

- Investigate potential financial resources needed over the next few decades to cover the cost of damage from climate change and adaptation to climate change
- Evaluate alternative methods of covering such costs
- Develop innovative finance schemes for issues of risk and security regarding long-term investments
- Investigate the potential role of and effects on private and public financial services providers.

Acknowledgments

The author team thanks Patrick Bidan, Hessel Dooper, Emily Horninge, Richard Radevsky, and Caroline van Schoubroeck for their participation in the workshop on natural disasters and financial services in Amsterdam on 10 March 2000, and their help in identifying topics and providing useful literature. Els Hunfeld is thanked for her many helping hands and dedication during the writing process. Grace Koshida and Angelika Wirtz also are thanked for their assistance. Finally, we thank the several dozen reviewers and experts from whom we received many useful comments and constructive suggestions.

References

AIRAC, 1986: *Catastrophic Losses: How the Insurance System Would Handle Two $7 Billion Hurricanes*. All-Industry Research Advisory Council (now the Insurance Research Council), Malvern, PA, USA, 73 pp.

Aldred, C., 2000: Huge flood risk predicted for English coasts. *Business Insurance*, **April 10,** 19.

Anderson, D.R., 2000: Catastrophe insurance and compensation: remembering basic principles. *CPCU Journal*, **53(2)** 76–89.

Andrews, C., 2000: Eurostorms hit results. *Reinsurance*, **March,** 10.

Arriens, W.T.L. and C. Benson, 1999: Post disaster rehabilitation: the experience of the Asian Development Bank. In: *Proceedings of the IDNDR-ESCAP Regional Meeting for Asia: Risk Reduction and Society in the 21st Century, Bangkok, 23–26 February 1999*. Asian Development Bank, Manila, Phillipines, 17 pp.

Bantwal, V.J. and H.C. Kunreuther, 2000: A cat bond premium puzzle? *Journal of Psychology and Financial Markets*, **1(1),** 76–91.

Banuri, T., K. Göran-Mäler, M. Grubb, H.K. Jacobson, and F. Yamin, 1996: Equity and social considerations. In: *Economic and Social Dimensions of Climate Change. Contribution of Working Group III to the Second Assessment Report of the Intergovernmental Panel on Climate Change* [Bruce, J.P., H. Lee, and E.F. Haites (eds.)]. Cambridge University Press, Cambridge, United Kingdom and New York, NY, USA, pp. 79–124.

Behn, S., 2000: *Environmental, Human Rights Activists Target Banks. Agence France Presse*, **6 April**. Available online at http://www.irn.org/programs/threeg/000406.discover.html.

Bender, S.O., 1991: Managing natural hazards. In: *Managing Natural Disasters and the Environment* [Kreimer, A. and M. Munasinghe (eds.)]. Proceedings of a colloquium sponsored by the World Bank, Washington DC, USA, pp. 182–185.

Berry, A., 2000: Future shock? An industry forecast. *Risk Management*, **April,** 25–32.

Berz, G., 1999: Catastrophes and climate change: concerns and possible countermeasures of the insurance industry. *Mitigation and Adaptation Strategies for Global Change*, **4(3/4),** 283–293.

Best's Review, 1998: Slow growth—by the numbers. *Best's Review—Property/Casualty*, **July,** 29–45.

Bourrelier, P.H.B, G. Deneufbourg, and B. de Vanssay, 2000: IDNDR objectives: French technical sociological contributions. *Natural Hazards Review*, **1(1),** 18–26.

Bowers, B., 1998: We interrupt this disaster... *Best's Review—Property/Casualty Edition*, **August,** 34–38.

Bowers, B., 1999: The new face of the alternative market. *Best's Review—Property/Casualty Edition*, **February,** 28–35.

Bradford, M., 2000: Tobacco firms set strategy for appeal. *Business Insurance*, **July 24,** 1.

Bruce, J.P., I. Burton, and M. Egener, 1999: *Disaster Mitigation and Preparedness in a Changing Climate*. Research Paper No. 2, Institute for Catastrophic Loss Reduction, Ottawa, ON, Canada, 36 pp. Available online at http://www.epc-pcc.gc.ca/research/down/DisMit_e.pdf.

Carpenter, D., 2000: Allstate blames storms for losses. *Associated Press*, April 20.

Carrido, M.L., 2000: An international disaster recovery business alliance. *Natural Hazards Review*, **(1)1,** 50–55.

CCR, 1999: *Natural Disasters in France*. Caisse Centrale de Réassurance, Paris, France, 22 pp. (in French and English).

Ceniceros, R., 2000: Internet privacy liability growing. *Business Insurance*, **May 22,** 1.

Changnon, S.A., 1999: Factors affecting temporal fluctuations in damaging storm activity in the United States based on insurance loss data. *Meteorological Applications*, **6,** 1–10.

Changnon, S.A. and D. Changnon, 1999: Record-high losses for weather disasters in the United States during the 1990s: how excessive and why? *Natural Hazards*, **18,** 287–300.

Changnon, S.A. and D.R. Easterling, 2000: U.S. policies pertaining to weather and climate extremes. *Science*, **289,** 2053–2054.

Changnon, S.A., R.A. Pielke, D. Changnon, R.T. Sylves, and R. Pulwarthy, 2000: Human factors explain the increased losses from weather and climate extremes. *Bulletin of the American Meteorological Society*, **(81)3,** 437–442. Available online at http://ams.allenpress.com.

Changnon, S.A., D. Changnon, E.R. Fosse, D.C. Hoganson, R.J. Roth Sr., and J.M. Totsch, 1997: Effects of recent weather extremes on the insurance industry: major implications for the atmospheric sciences. *Bulletin of the American Meteorological Society*, **78(3),** 425–431. Available online at http://ams.allenpress.com.

Choudhury, S.H.M., 1998: *Report on Bangladesh Flood 1998. Chronology, Damages and Response*. Management Information and Monitoring (MIM) Division, Disaster Management Bureau, Government of Bangladesh, Dhaka, Bangladesh, 3 pp.

Choudhury, O.H., K.M. Nabiul Islam, and D. Bhattacharya, 1999: *The Losses of 1998 Flood and its Impact on National Economy*. Bangladesh Institute for Development Studies (BIDS), Dhaka, Bangladesh, 18 pp.

Clow, R., 2000: Big tobacco, smoked in Florida, suing insurers. *New York Post*, **July 24,** 34. Available online at http://www.nypost.com/07242000/business/33725.htm.

Colombo, A.F., D. Etkin, and B.W. Karney, 1999: Climate variability and the frequency of extreme temperature events for nine sites across Canada: implications for power usage. *Journal of the American Meteorological Society*, **12,** 2490–2502.

Credit Suisse Group, 1998: *Corporate Environmental Report 1997/98*. Credit Suisse Group, Zurich, Switzerland, 43 pp.

Credit Suisse Group, 1999: *Umweltrelevante Brancheninformationen Tourismus*. Credit Suisse Group, Zurich, Switzerland, 2 pp. (in German).

Cummins, J.D., N.A. Doherty, and A. Lo, 1999: *Can Insurers Pay for the Big One? Measuring the Capacity of the Insurance Market to Respond to Catastrophic Losses*. Wharton School, University of Pennsylvania, Philadelphia, PA, USA, 46 pp. Available online at http://fic.wharton.upenn.edu/fic/wfic/papers/98/pcat01.html.

Davenport, A.G., 2000: The decade for natural disaster reduction in Canada. *Natural Hazards Review*, **1(1),** 27–36.

Davidson, R.J., 1996: Tax-deductible, pre-event catastrophe reserves. *Journal of Insurance Regulation*, **15(2),** 195–190.

Davidson, R., 1997: The disaster double threat. *Contingencies: Journal of the American Academy of Actuaries*, **September/October,** 14–19.

Deering, A. and J.P. Thornton, 2000: Solar solutions for natural disasters. *Risk Management*, **February,** 28–33.

Denenberg, H.S., 1964: *Risk and Insurance*. Prentice-Hall, Englewood Cliffs, NJ, USA, 615 pp.

Dessens, J., 1995: Severe convective weather in the context of a global night-time warming. *Geophysical Research Letters*, **22(10),** 1241–1244.

Dinnes, D., 1999: *Blitzgefahrdung in Deutschland*. Diss. University of Munich, Meteorological Institute, 135 pp. (in German).

Dlugolecki, A.F., K.M. Clark, F. Knecht, D. McCauley, J.P. Palutikof, and W. Yambi, 1996: Financial Services. In: *Climate Change 1995: Impacts, Adaptation, and Mitigation of Climate Change: Scientific-Technical Analyses. Contribution of Working Group II to the Second Assessment Report of the Intergovernmental Panel on Climate Change* [Watson, R.T., M.C. Zinyowera, and R.H. Moss (eds.)]. Cambridge University Press, Cambridge, United Kingdom and New York, NY, USA, pp. 539–560.

Doherty, N., 1997: Insurance markets and climate change. *The Geneva Papers on Risk and Insurance*, **22(83),** 223–237.

Doherty, N., L. Lipowski Posey, and A.E. Kleffner, 1992: *Insurance Surplus: Its Function, Its Accumulation and Its Depletion*. National Committee on Property Insurance, Boston, MA, USA, 71 pp.

Downing, T.E., A.J. Olsthoorn, and R.S.J. Tol, 1999: *Climate, Change and Risk*. Routledge, London, United Kingdom, 407 pp.

Easterling, D.R., J.L. Evans, P.Y. Groisman, T.R. Karl, K.E. Kunkel, and P. Ambenje, 2000a: Observed variability and trends in extreme climate events, a brief review. *Bulletin of the American Meteorological Society*, **81(3),** 417–425. Available online at http://ams.allenpress.com.

Easterling, D.R., G.A. Meehl, C. Parmesan, S. Changnon, T.R. Karl, and L.O. Mearns, 2000b: Climate extremes, observations, modelling and impacts. *Science*, **289,** 2068–2074.

Edgecliffe-Johnson, A., 1999: U.S. insurers are struck by hurricane. *Financial Times*, **September 25,** 25.

Eley, D., 1996: Creating catastrophe reserves: a balancing act. *Journal of Insurance Regulation*, **15(2),** 191.

EPC, 2000: *EPC Disaster Database, Version 2.0*. Emergency Preparedness Canada, Ottawa, Ontario Canada. Available online at http://www.epc-pcc.gc.ca/research/epcdatab.html.

Epps, D., 1997: Weather impacts on energy activities in the U.S. Gulf Coast. In: *The Social and Economic Impacts of Weather* [Pielke, R.A. Jr. (ed.)]. Proceedings of a workshop at the University Corporation for Atmospheric Research, Boulder, CO, USA, 3 pp. Available online at http://www.esig.ucar.edu/socasp/weather1/epps.html.

Epstein, P. (ed.), 1999: *Extreme Weather Events: The Health and Economic Consequences of the 1997/98 El Niño and La Niña*. Center for Health and the Global Environment, Harvard Medical School, Boston, MA, USA, 42 pp. Available online at http://chge2.med.harvard.edu/enso/disease.html.

FEMA, 2000: *Disaster Assistance; Insurance Requirements for the Public Assistance Program. Advanced Notice of Proposed Rulemaking*. 44 CFR Part 206, RIN 3067-AC90, Federal Emergency Management Agency, Washington, DC, USA, 5 pp. Available online at http://www.fema.gov/library/3AC90anpr.pdf.

FFSA, 2000: *French Insurance in 1999*. Fédération Française des Sociétés d'Assurances, Paris, France, 98 pp. (in French). Available online at http://www.ffsa.fr/pdf/rapport/99.pdf.

Figge, F., 1998: *Systematisation of Economic Risks Through Global Environmental Problems—A Threat to Financial Markets?* WWZ/Sarasin and CIE Studie No. 56, Basel, Switzerland, 26 pp. (in German). Available online at http://www.sustainablevalue.com/Risk/Risk.htm.

Fletcher, L., 2000: FEMA's Witt urges public sector to insure, manage property risks. *Business Insurance*, **June 19,** 3.

Folkerts-Landau, D.F.I. and D.J. Mathieson, 1988: Innovation, institutional changes, and regulatory response in international financial markets. In: *Restructuring Banking and Financial Services in America* [Haraf, W.S. and R.-M. Kushmeider (eds.)]. American Enterprise Institute for Public Policy Research, Washington, DC, USA, pp. 392–423.

Francis, D. and H. Hengeveld, 1998: *Extreme Weather and Climate Change*. Climate Change Digest Series CCD 98-01, Environment Canada, Toronto, ON, Canada, 31 pp. Available online at http://www.msc-smc.ec.gc.ca/saib/climate/ccsci_e.cfm.

Freeman, A., 2000: *The Risk Revolution and Its Impacts on Securities Markets*. International Securities Market Association, Zurich, Switzerland, 82 pp.

Freeman, P.K., 1999: *Infrastructure, Natural Disasters and Poverty*. Paper presented for a Consultative Group for Global Disaster Reduction, Paris, France, June 1–2, 1999, 7 pp.

Frerks, G., D. Hilhorst, and A. Moreyra, 1999: *Natural Disasters: Framework for Analysis and Action*. Report for MSF-Holland. Wageningen University, Disaster Studies, Wageningen, The Netherlands, 68 pp.

Gadgil, A., A. Drescher, D. Greene, P. Miller, C. Motau, and F. Stevens, 1997: Field-testing UV disinfection of drinking water. In: *Proceedings of the 23rd WEDC Conference: Water and Sanitation for All, September 1–5 1997, Durban, South Africa* [Pickford, J., P. Barker, B. Elson, M. Ince, P. Larcher, D. Miles, J. Parr, B. Reed, K. Sansom, D. Saywell, M. Smith, and I. Smout (eds.)]. Water Engineering Development Center, University of Loughborough, United Kingdom, pp. 153–156. Available online at http://www.lboro.ac.uk/departments/cv/wedc/papers/23/groupd/gadgil.pdf.

GAO, 2000a: *Insurers' Ability to Pay Catastrophe Claims*. Report B-284252, U.S. General Accounting Office, Washington, DC, USA, 32 pp. Available online at http://www.gao.gov/corresp/gg00057r.pdf.

GAO, 2000b: *Surplus Would Help P/C Insurers Survive Major Catastrophe*. U.S. General Accounting Office, Washington, DC, USA, 20 pp.

GAO, 1994: *Federal Disaster Insurance: Goals Are Good, But Insurance Programs Would Expose the Federal Government to Large Potential Losses*. U.S. General Accounting Office, Washington, DC, USA, 33 pp.

Gaschen, S., P. Hausmann, I. Menzinger, and W. Schaad, 1998: *Floods—An Insurable Risk? A Market Survey*. Swiss Reinsurance Company, Zurich, Switzerland, 48 pp. Available online at http://www.swissre.com/e/publications/publications/natural1/floods.html.

Gastel, R., 2000: Insolvencies/guaranty funds. *Insurance Issues Update*. Insurance Information Institute, New York, NY, USA, 22 pp.

Gastel, R., 1999: Catastrophes: insurance issues. *Insurance Issues Update (September)*. Insurance Information Institute, New York, NY, USA, 16 pp.

GEF, 1998: *Project Performance Report 1998*. Global Environment Facility, Washington, DC, USA, 89 pp. Available online at http://www.gefweb.org/M&E/ppr/1998PPR-E.htm.

Grabbe, J.O., 1998: The credit crunch. *Laissez Faire City Times*, **2(33).** Available online at http://www.zolatimes.com/v2.33/CreditCr.html.

Greenwald, J., 2000: P/C trade unlikely to attract banks. *Business Insurance*, **March 27,** 10.

Haider, R., A.A. Rahman, and S. Huq (eds.), 1991: *Cyclone '91. An Environmental and Perceptional Study*. Bangladesh Centre for Advanced Studies (BCAS), Dhaka, Bangladesh, 91 pp.

Hamilton, R.M., 2000: Science and technology for natural disaster reduction. *Natural Hazards Review*, **1(1),** 56–60.

Haraf, W.S. and R.-M. Kushmeider, 1988: Redefining Financial Markets. In: *Restructuring Banking and Financial Services in America* [Haraf, W.S. and R.-M. Kushmeider (eds.)]. American Enterprise Institute for Public Policy Research, Washington, DC, USA, pp. 1–33.

Hausmann, P., 1998: *Floods: An Insurable Risk?* Swiss Reinsurance Company, Zurich, Switzerland, 45 pp. Available online at http://www.swissre.com/e/publications/publications/natural1/floods.html.

Heinz Center, 2000: *Evaluation of Erosion Hazards*. Federal Emergency Management Agency, Washington, DC, USA, 252 pp. Available online at http://www.fema.gov/nwz00/erosion.htm.

Hofmann, M.A., 2000a: Sparking a fire: award may fuel tort reform. *Business Insurance,* **July 24,** 1.

Hofmann, M.A., 2000b: Risk managers urged to make wildfires a hot topic. *Business Insurance*, **August 14,** 3.

Hooke, W.H., 2000: U.S. participation in international decade for natural disaster reduction. *Natural Hazards Review*, **1(1),** 2–9.

Howard, L., 2000a: European storms ruin reinsurance Australia. *National Underwriter*, **104(10),** 2.

Howard, L., 2000b: Could P-C move into financial services backfire? *National Underwriter*, **104(16),** 2.

Hugenschmidt, H. and J. Janssen, 1999: Kyoto Protocol: new market opportunities or new risks? *Swiss Derivatives Review*, **11,** 22–23.

Hulme, M., 1997: Climate. In: *Economic Impacts of the Hot Summer and Unusually Warm Year of 1995* [Palutikof, J.P., M.D. Agnew, and S. Subak (eds.)]. Department of the Environment, University of East Anglia, Norwich, United Kingdom, pp. 5–14.

IFC, 1996: *Financing Private Infrastructure*. International Finance Corporation, World Bank, Washington, DC, 128 pp.

IFRC-RCS, 1999a: *World Disasters Report*. International Federation of Red Cross and Red Crescent Societies, Geneva, Switzerland, 198 pp.

IFRC-RCS, 1999b: *International Red Cross Predicts More Global "Super Disasters."* American Red Cross Washington, D.C., USA. Available online at http://www.redcross.org/news/inthnews/99/6%2D25%2D99.html.

III, 2000a: *Catastrophes Background*. Insurance Information Institute, New York, NY, USA, 5 pp. Available online at http://www.iii.org/inside.pl5?media=issues=/media/issues/catastrophe_background.html.

III, 2000b: *Catastrophes: Insurance Issues (February)*. Insurance Information Institute, New York, NY, USA, 6 pp. Available online at http://www.iii.org/inside.pl5?media=issues=/media/issues/catastrophes.html.

III, 1999: *The Fact Book, 2000: Property/Casualty Insurance Facts*. Insurance Information Institute, New York, NY, USA, 156 pp.

Insurance Regulator, 1998: Arbitration panel overturns Nelson's rate hike rejection. *Insurance Regulator*, **9(11),** 1.

IPCC, 2000: *Land Use, Land Use Change, and Forestry* [Watson, R.T., I.R. Noble, B. Bolin, N.H. Ravindranath, D.J Verardo, and D.J. Dokken (eds.)]. Cambridge University Press, Cambridge, United Kingdom and New York, NY, USA, 377 pp.

ISO, 1999: *Financing Catastrophe Risk: Capital Market Solutions*. Insurance Services Office, New York, NY, USA, 62 pp. Available online at http://www.iso.com/docs/stud013.htm.

ISO, 1996: *Managing Catastrophe Risk*. Insurance Services Office, New York, NY, USA, 42 pp. Available online at http://www.iso.com/docs/stud001.htm.

ISO, 1994a: *The Impact of Catastrophes on Property Insurance*. Insurance Services Office, New York, NY, USA, 50 pp. Available online at http://www.iso.com/dos.stud006.htm.

ISO, 1994b: *Catastrophes: Insurance Issues Surrounding the Northridge Earthquake and Other Natural Disasters*. Insurance Services Office, New York, NY, USA, 42 pp. Available online at http://www.iso.com/docs/stud005.htm.

Jakobi, W., 1996: Taking care of the future. *Reinsurance Market Report*, **335(6),** 282.

Jamison, K.S., 2000: Consumers leery about financial integration. *National Underwriter*, **104(34),** 10.

Jeucken, M.H.A. and J.J. Bouma, 2000: The changing environment of banks. *Greener Management International*, **27,** 21–35.

Karl, T.R. and R.W. Knight, 1998: Secular trends of precipitation amount, frequency, and intensity in the United States. *Bulletin of the American Meteorological Society*, **79(2),** 231–241. Available online at http://ams.allenpress.com.

Kats, D., 1998: Solar technology ripe for loss control. *National Underwriter*, **102(32),** 3.

Kaufman, G.G., 1992: Banking and financial intermediary markets in the United States: where from, where to? In: *Regulating International Financial Markets: Issues and Policies* [Edwards, F.R. and H.T. Patrick (eds.)]. J. Kluwer Academic Publishers, Dordrecht, The Netherlands, pp. 85–103.

Keener, R.N., 1997: The estimated impact of weather on daily electric utility operations. In: *Social and Economic Impacts of Weather* [Pielke, R. Jr. (ed.)]. Proceedings of a workshop at the University Corporation for Atmospheric Research, Boulder, CO, USA. Available online at http://www.esig.ucar.edu/socasp/weather1/keener.html.

Kelly, J.P. and L. Zeng, 1999: The temporal and spatial variability of economic losses due to intense hurricanes—a comparative analysis. *Presented at the 79th Annual Meeting of the American Meteorological Society, Dallas, Texas, 10-15 January 1999*. Arkwright Mutual Insurance Company, Waltham, MA, USA, 22 pp. Available online at http://www.atmos.washington.edu/~lixin/KZ1999.pdf.

Klein, R.W., 1997: *Catastrophe Risk Problems*. Written testimony presented to the house subcommittee on housing and community opportunity, June 24, 1997. Available online at http://www.house.gov/banking/62497kle.htm.

Klein, R.J.T., 1998: Towards better understanding, assessment, and funding of climate adaptation. *Change,* **44,** 15–19. Available online at http://www.nop.nl/gb/nieuwsbrief/change44/n44p8.htm.

Kreimer, A. and M. Arnold, 2000: World Bank's role in reducing impacts of natural disasters. *Natural Hazards Review*, **1(1),** 37–42.

Kreimer, A., J. Eriksson, R. Muscat, and M. Arnold, 1998: *The World Bank's Experience with Post-Conflict Reconstruction*. World Bank, Washington, DC, USA, 120 pp.

Kundzewicz, Z.W., K. Szamalek, and P. Kowalczak, 1999: The great flood of 1997 in Poland. *Hydrological Sciences Journal*, **44(6),** 855–870.

Kunkel, K.E., R.A. Pielke Jr., and S.A. Changnon, 1999: Temporal fluctuations in weather and climate extremes that cause economic and human health impacts: a review. *Bulletin of the American Meteorological Society*, **80(6),** 1077–1098. Available online at http://ams.allenpress.com.

Kunreuther, H., 2000: Insurance as a cornerstone for public-private sector partnerships. *Natural Hazards Review*, **1(2),** 126–136.

Kunreuther, H., 1998: Insurability conditions and the supply of coverage. In: *Paying the Price: The Status and Role of Insurance Against Natural Disasters in the United States* [Kunreuther, H. and R. Roth. (eds.)]. Joseph Henry Press, Washington, DC, USA, pp. 17–50.

Kunreuther, H. and R. Roth, 1998: *Paying the Price: The Status and Role of Insurance Against Natural Disasters in the United States* [Kunreuther, H. and R. Roth. (eds.)]. Joseph Henry Press, Washington, DC, USA, 320 pp.

Lal, M., P.H. Whetton, B. Chakraborty, and A.B. Pittock, 1998: The greenhouse gas-induced climate change over the Indian subcontinent as projected by general circulation model experiments. *Terrestrial, Atmospheric and Oceanic Sciences*, **9,** 673–690.

Lecomte, E., A.W. Pang, and J.W. Russell, 1998: *Ice Storm '98*. Institute for Catastrophic Loss Reduction (Canada) and Institute for Business and Home Safety, USA, 39 pp. (in French and English).

Lecomte, E. and K. Gahagan, 1998: Hurricane insurance protection in Florida. In: *Paying the Price: The Status and Role of Insurance Against Natural Disasters in the United States* [Kunreuther, H. and R. Roth. (eds.)]. Joseph Henry Press, Washington, DC, USA, pp. 97–124.

Leigh, R., R. Taplin, and G. Walker, 1998a: Insurance and climate change: the implications for Australia with respect to natural hazards. *Australian Journal of Environmental Management*, **5,** 81–96.

Leigh, R., E. Cripps, R. Taplin, and G. Walker, 1998b: *Adaptation of the Insurance Industry to Climate Change and Consequent Implications. Final Report for Dept. of Environment, Sport and Territories, Government of Australia, Canberra*. Climatic Impacts Centre, Macquarie University, Sydney, Australia, 180 pp.

Lester, R., 1999: The World Bank and natural catastrophe funding. In: *The Changing Risk Landscape: Implications for Insurance Risk Management* [Britton, N.R. (ed.)]. Proceedings of a Conference sponsored by Aon Group Australia Ltd., Sydney, Australia, pp. 173–194. Available online at http://www.aonre.com.au/pdf/World%20Bank-Lester.pdf.

Linnerooth-Bayer, J. and A. Amendola, 2000: Global change, natural disasters and loss-sharing: issues of efficiency and equity. *The Geneva Papers on Risk and Insurance: Issues and Practice*, **(25)2,** 203–219.

Lonkevich, D., 2000: Cat losses hammer Bermuda in fourth quarter. *National Underwriter*, **104(2),** 3.

Mag, W., 1990: Risiko und unsicherheit. In: *Handwörterbuch der Wirtschaftswissenschaften*. Bd. 6, Stuttgart, Germany, pp. 72–74 (in German).

Mahoney, D.L., 1999: Meeting the challenges of the changing landscape of risk: implications for insurance. In: *The Changing Risk Landscape: Implications for Insurance Risk Management* [Britton, N.R. (ed.)]. Proceedings of a Conference sponsored by Aon Group Australia Ltd., Sydney, Australia, pp. 195–217. Available online at http://www.aonre.com.au/pdf/Implications-Mahoney.pdf.

Marcon, F., 1999: Shoring up cat risk: insurers can use capital, reinsurance and catastrophe options to reduce the cost of financing their catastrophe risks. *Best's Review—Property/Casualty Edition*, **February,** 60–68.

Matin, M.A., 1998: Some lessons to be learnt from 1998 flood. *The Daily Star*, **September 1,** Dhaka, Bangladesh.

Matthews, P.B., M.P. Sheffield, J.E. Andre, J.H. Lafayette, J.M. Roethen, and E. Dobkin, 1999: Insolvency: will historic trends return? *Best's Review—Property/Casualty Edition*, **March,** 59. Available online at http://www.bestreview.com/pc/1999–03/trends.html.

Meehl, G.A., T. Karl, D.R. Easterling, S. Changnon, R. Pielke Jr., D. Changnon, J. Evans, P.Y. Groisman, T.R. Knutson, K.E. Kunkel, L.O. Mearns, C. Parmesan, R. Pulwarty, T. Root, R.T. Sylves, P. Whetton, and F. Zwiers, 2000a: An introduction to trends in extreme weather and climate events: observations, socioeconomic impacts, terrestrial ecological impacts and model projections. *Bulletin of the American Meteorological Society*, **81(3),** 413–416. Available online at http://ams.allenpress.com.

Meehl, G.A., F. Zwiers, J. Evans, T. Knutson, L. Mearns, and P. Whetton, 2000b: Trends in extreme weather and climate events: issues related to modeling extremes in projections of future climate change. *Bulletin of the American Meteorological Society*, **81(3),** 427–436. Available online at http://ams.allenpress.com.

Miller, A, G. Sethi, and G.H. Wolff. 2000: *What's Fair? Consumers and Climate Change*. Redefining Progress, San Francisco, CA, USA, 67 pp. http://www.rprogress.org/pubs/pdf/wf_consumers.pdf.

Mills, E., 1999: The insurance and risk management industries: new players in the delivery of energy-efficient products and services. In: *Proceedings of the ECEEE 1999 Summer Study, European Council for an Energy-Efficient Economy, May 31–June 4 1999, Mandelieu, France, and for the United Nations Environment Programme's 4th International Conference of the Insurance Industry Initiative, Natural Capital at Risk: Sharing Practical Experiences from the Insurance and Investment Industries, July 10–11 1999, Oslo, Norway*. European Council for an Energy-Efficient Economy, Copenhagen, Denmark, pp. 2–12.

Mills, E., 1996: Energy efficiency: no-regrets climate change insurance for the insurance industry. *Journal of the Society of Insurance Research*, **9(3),** 21–58. Available online at http://eande.lbl.gov/CBS/Insurance/ClimateInsurance.html.

Mills, E. and I. Knoepfel, 1997: Energy-efficiency options for insurance loss-prevention. In: *Proceedings of the 1997 ECEEE Summer Study, European Council for an Energy-Efficient Economy, Copenhagen, Denmark*. Lawrence Berkeley National Laboratory Report No. 40426, Center for Building Science, Environmental Energy Technologies Division, Lawrence Berkeley National Laboratory, Berkeley, California, USA, 23 pp. Available online at http://eande.lbl.gov/CBS/PUBS/no-regrets.html.

Mills, E. and A. Rosenfeld, 1996: Consumer non-energy benefits as a motivation for making energy-efficiency improvements. *Energy—The International Journal*, **21(7/8),** 707–720.

Mills, E., E. Lecomte, and A. Peara., 2001: U.S. Insurance Industry Perspectives on Climate Change. Lawrence Berkeley National Laboratory Technical Report No. LBNL-45185, Berkeley, California, USA.

Mittler, E., 1992: *A Fiscal Responsibility Analysis of a National Earthquake Insurance Program*. National Committee on Property Insurance, Boston, MA, USA, 54 pp.

Mooney, S., 1998: Insurers should be the experts, not activists, in climate change debate. *National Underwriter—Property & Casualty/Risk & Benefits Management*, **102(23),** 43–44.

Mooney, S., 1999: Should we worry about loss reserve drop? *National Underwriter*, **103(13),** 15.

Mooney, S., 2000: Reinsurance not driving the cycle this time. *National Underwriter*, **104(9),** 19.

Munich Re, 2000: *Topics—Annual Review of Natural Disasters 1999* (supplementary data and analyses provided by Munich Reinsurance Group/Geoscience Research Group, MRNatCatSERVICE). Munich Reinsurance Group, Munich, Germany, 46 pp.

Munich Re, 1999a: *Topics—Annual Review of Natural Catastrophes 1998*. Munich Reinsurance Group, Geoscience Research Group, Munich, Germany, 20 pp.

Munich Re, 1999b: *Topics 2000—Natural Catastrophes, The Current Position* (published statistics updated by Munich Re to reflect adjustments for 1999 year-end loss accounting). Munich Reinsurance Group, Geoscience Research Group, Munich, Germany, 126 pp.

NCIGF, 1999: *1998 Assessment Information by Guaranty Fund*. National Conference of Insurance Guaranty Funds, Indianapolis, IN, USA, 41 pp. Available online at http://www.ncigf.org/assesshist/ashistory.htm.

Netherlands National Committee on Water Management, 2000: *Water Management for the 21st Century. Advice of the Water Management Commission*. Department of Traffic and Water-Management, The Hague, The Netherlands, 126 pp. (in Dutch).

NHTSA, 1999: *Traffic Safety Facts 1998*. National Highway Traffic Safety Administration, U.S. Department of Transportation, 226 pp. Available online at http://www.nhtsa.dot.gov/people/ncsa/tsf-1998.pdf.

Nicholls, M., 2000: The bankers are coming. *Environmental Finance*, **September,** 14–15. Available online at http://environmental-finance.webserver.org/envfin/featsept.htm.

Nicholson, S.E., B. Some, and B. Kone, 2000: An analysis of recent rainfall conditions in West Africa, including the rainy seasons of the 1997 El Niño and the 1998 La Niña years. *Journal of Climate*, **13,** 2628–2640.

Nuttall, N., 1998: Climate disaster map pinpoints "no-go" areas for insurers. *Times of London*, **November 9,** 12. Available online at http://www.times-archive.co.uk/news/pages/tim/1998/11/09/timfgname02002.html.

Nutter, F.W., 2000: U.S. reinsurers deserve level playing field. *National Underwriter*, **104(30),** 20.

Nutter, F.W., 1996: Insurance and the natural sciences: partners in the public interest. *Journal of Society of Insurance Research*, **9(6),** 15–19.

PCS, 2000: *Catastrophe Loss Database*. Property Claim Services, Insurance Service Office, Rahwy, NJ, USA, 27 pp.

Peara, A.T., 1999: Climate change: risks and opportunities for U.S. property/casualty, life and health insurance companies. Diss. University of Oregon, Eugene, OR, USA, 167 pp.

Peara, A. and E. Mills, 1999: Climate for change: an actual perspective on global warming and its potential impact on insurers. *Contingencies: Journal of the American Academy of Actuaries* (Lawrence Berkeley National Laboratory Report No. 42580), **10(6),** 16–23.

Pielke, R.A. Jr., 1998: Catastrophe models: boon or bane? *Weather Zine*, **14,** 1–2. Available online at http://www.esig.ucar.edu/socasp/zine/14.html.

Pielke, R.A. Jr., 1997: Trends in hurricane impacts in the United States. In: *Proceedings of Workshop on the Social and Economic Impacts of Weather* [Pielke, R. Jr. (ed.)]. University Corporation for Atmospheric Research, Boulder, CO, USA, 4 pp. Available online at http://www.esig.ucar.edu/socasp/weather1/pielke.html.

Pielke, R.A. Jr. and M.W. Downton, 2000: Precipitation and damaging floods: trends in the United States, 1932–1997. *Journal of Climate*, **13(20),** 3625–3637. Available online at http://www.esig.ucar.edu/trends/index.html.

Pielke, R.A. Jr. and C.W. Landsea, 1998: Normalized hurricane damages in the United States, 1925–1995. *Weather and Forecasting*, **13(3),** 621–631.

Pielke, R.A. Jr., C.W. Landsea, R. Musulin, and M.W. Downton, 1999: A methodology for the evaluation of catastrophe models. *Journal of Insurance Regulation*, **18,** 177–194.

Price, C. and D. Rind, 1994: The impact of a $2xCO_2$ climate on lightning-caused fires. *Journal of Climate*, **7(10),** 1484–1494.

Pullen, R., 1999a: Catastrophe bailout proposed. *Best's Review—Property/Casualty Edition*, **September,** 12.

Pullen, R., 1999b: Saving up for a Stormy Day. *Best's Review—Property/Casualty Edition*, **April,** 14.

Punter, A., 1999: *Alternative Risk Financing: Changing the Face of Insurance*. Jim Bannister Developments Ltd., in association with Aon Group and Zurich International, London, United Kingdom, 2nd. ed., 92 pp.

Radevsky, R., 1999: *Subsidence, a Global Perspective*. Association of British Insurers, London, United Kingdom, 44 pp. Summary available online at http://www.abi.org.uk/Books/Report1.pdf.

Reeve, N. and R. Toumi, 1999: Lightning activity as an indicator of climate change. *Quarterly Journal of the Royal Meteorological Society*, **124,** 893–903.

Rosenzweig, C., A. Iglesias, X.B. Yang, P.R. Epstein, and E. Chivian, 2000: *Climate Change and U.S. Agriculture: The Impacts of Warming and Extreme Weather Events on Productivity, Plant Diseases, and Pests.* Center for Health and the Global Environment, Harvard Medical School, Boston, MA, USA, 47 pp. Available online at http://www.med.harvard.edu/chge/reports/climate_change_us_ag.pdf.

Ross, A., 2000: *Reflections on the Future: Climate Change and its Impacts on the Insurance Industry.* Paper No. 8, Institute for Catastrophic Loss Reduction (ICLR), Toronto, ON, Canada, 9 pp.

Ryland, H.G., 2000: A piece of the puzzle: insurance industry perspective on mitigation. *Natural Hazards Review*, **1(1),** 43–49.

Salt, J., 2000: Why the insurers should wake up. *Environmental Finance*, **September,** 24–25.

Sarewitz, D., R.A. Pielke Jr., and R. Byerly (eds.), 2000: *Prediction: Science, Decision Making and the Future of Nature.* Island Press, Washington, DC, USA, 400 pp.

Schanzenbächer, B., 1997: Impacts of Environmental Issues on the Financial Services Sector. In: *Proceedings of the Washington SkyShip Conference on Banks and Environment, Manila, Philippines, 22 November 1997.* Asian Institute of Technology (AIT), Manila, Phillipines, pp. 11–16.

Scott, M.J., 1996: Human settlements in a changing climate: impacts and adaptation. In: *Climate Change 1995: Impacts, Adaptations, and Mitigation of Climate Change: Scientific-Technical Analysis. Contribution of Working Group II to the Second Assessment Report of the Intergovernnmental Panel on Climate Change* [Watson, R.T., M.C. Zinyowera, and R.H. Moss (eds.)]. Cambridge University Press, Cambridge, United Kingdom and New York, NY, USA, pp. 401–426.

Solis, G., H.C. Hightower, and J. Kawaguchi, 1997: *Guidelines on Cultural Diversity and Disaster Management.* Prepared by The Disaster Preparedness Resources Centre, University of British Columbia, for Emergency Preparedness Canada, within the Canadian Framework for the International Decade for Natural Disaster Reduction, Ottawa, ON, Canada, 26 pp. Available online at http://www.epc-pcc.gc.ca/research/scie_tech/guid_cult.html.

Stauffer, R.F., 1995: *Nature's Power on Demand: Renewable Energy Systems as Emergency Power Sources.* U.S. Department of Energy, Washington, DC, USA, 19 pp. Available online at http://www.sustainable.doe.gov/freshstart/articles/enrgsyst.htm.

Stipp, D., 1997: A new way to bet on disasters. *Fortune*, **136(5),** 124–132. Available online at http://www.fortune.com/fortune/1997/970908/hur.html.

Storebrand, 2000: *Environmental Status 1998-2000.* Storebrand, Oslo, Norway, 24 pp. Available online at http://www.storebrand.com/storebrand/com/publications.nsf/81d62b95df443a23c1256839006736e3/48fb9a407f4b712cc12569bd002b0318/$FILE/Environmetal_report.pdf.

Sudo, K., H. Kameda, and Y. Ogawa, 2000: Recent history of Japan's disaster mitigation and the impact of IDNDR. *Natural Hazards Review*, **1(1),** 10–17.

Swiss Re, 2000a: *Solvency of Non-Life Insurers: Balancing Security and Profitability Expectations.* Sigma Report 1, Swiss Reinsurance Company, Zurich, Switzerland, 36 pp. Available online at http://www.swissre.com/e/publications/publications/sigma1/sigma010300.html.

Swiss Re, 2000b: *Natural catastrophes and manmade disasters in 1999.* Sigma Report 2, Swiss Reinsurance Company, Zurich, Switzerland, 32 pp. Available online at http://www.swissre.com/e/publications/publications/sigma1/sigma060300.html.

Swiss Re, 1999a: *Natural Catastrophes and Man-Made Disasters 1998.* Sigma Report 1, Swiss Reinsurance Company, Zurich, Switzerland, 36 pp. Available online at http://www.swissre.com/e/publications/publications/sigma1/sigma9901.html.

Swiss Re, 1999b: *Alternative Risk Transfer for Corporations: A Passing Fashion or Risk Management for the 21st Century?* Sigma Report 2, Swiss Reinsurance Company, Zurich, Switzerland, 39 pp. Available online at http://www.swissre.com/e/publications/publications/sigma1/sigma9902.html.

Swiss Re, 1999c: *World Insurance in 1998.* Sigma Report 7, Swiss Reinsurance Company, Zurich, Switzerland, 28 pp. Available online at http://www.swissre.com/e/publications/publications/sigma1/sigma9907.html.

Swiss Re, 1998a: *Climate Research Does Not Remove Uncertainty: Coping with the Risks of Climate Change.* Swiss Reinsurance Company, Zurich, Switzerland, 8 pp. Available online at http://www.swissre.com/e/publications/publications/flyers1/archive6/climate.html.

Swiss Re, 1998b: *The Global Reinsurance Market in the Midst of Consolidation.* Sigma Report 9, Swiss Reinsurance Company, Zurich, Switzerland, 35 pp. Available online at http://www.swissre.com/e/publications/publications/sigma1/sigma_09.html.

Swiss Re, 1997: *Too Little Reinsurance of Natural Disasters in Many Markets.* Sigma Report 7, Swiss Reinsurance Company, Zurich, Switzerland, 20 pp. Available online at http://www.swissre.com/e/publications/publications/sigma1/sigma_071.Paras.0014.File.pdf.

Swiss Re, 1996: *Rethinking Risk Financing.* Swiss Reinsurance Company, Zurich, Switzerland, 26 pp. Available online at http://www.swissre.com/e/publications/publications/integrated1/rethinking.html.

Swiss Re, 1994: *Climate Change—Element of Risk.* Swiss Reinsurance Company, Zurich, Switzerland, 8 pp.

Swiss Re, 1992: *Fire of the Future.* Swiss Reinsurance Company, Zurich, Switzerland, 28 pp.

Thompson, H., 1996: The financial sector. In: *Review of the Potential Effects of Climate Change in the UK—Second Report of the Climate Change Impacts Review Group.* Department of the Environment, Her Majesty's Stationery Office, London, United Kingdom, pp. 179–187.

Tol, R.S.J., 1998: Climate change and insurance: a critical appraisal. *Energy Policy*, **26(3),** 257–262.

Torn, M., E. Mills, and J. Fried, 1998: Will climate change spark more wildfire damages? *Contingencies: Journal of the American Academy of Actuaries*, **11(4),** 34–43. Available online at http://eande.lbl.gov/CBS/EMills/wild.html.

UNEP, 2000: *List of Signatories to the UNEP Financial Services Statements.* United Nations Environment Program, Geneva, Switzerland, 1 p. Available online at http://www.unep.ch/etu/finserv/insura/Signatories-by-Association.htm.

UNEP, 1999: *The Kyoto Protocol and Beyond: Potential Implications for the Insurance Industry.* United Nations Environment Program Insurance Industry Initiative for the Environment, Geneva, Switzerland, 42 pp. Available online at http://www.unep.ch/etu/finserv/insura/KYOTO-HPT-FINAL1.html.

UNEP, 1996: Insurance Industry Position Paper on Climate Change. United Nations Environment Program, Environmental and Trade, Geneva, Switzerland, 2 pp. Available online at http://www.unep.ch/etu/finserv/insura/position.htm.

UNEP, 1995: *Statement of Environmental Commitment by the Insurance Industry.* United Nations Environment Program, Environment and Trade, Geneva, Switzerland, 2 pp. Available online at http://www.unep.ch/etu/finserv/insura/statemen.htm.

UNFCCC, 1999: *Development and Transfer of Technologies, Submissions from Parties: Part 1.* United Nations Framework Convention on Climate Change, Subsidiary Body for Scientific and Technological Advice (SBSTA), Bonn, Germany, 119 pp. Available online at http://www.unfccc.int/resource/docs/1999/sbsta/misc05.pdf.

UNFCCC, 1997: *Kyoto Protocol.* United Nations Framework Convention on Climate Change, Bonn, Germany, 23 pp. Available online at http://www.unfccc.de/resource/docs/cop3/l07a01.pdf.

UNFCCC, 1992: *United Nations Framework Convention on Climate Change.* United Nations, New York, NY, USA, 32 pp. Available online at http://www.unfccc.int/resource/docs/convkp/conveng.pdf.

United Nations, 2000: *Reliefweb.* United Nations, Geneva, Switzerland. Available online at http://www.reliefweb.int/fts/index.html.

United Nations, 1994: *Disasters Around the World—A Global and Regional View.* IDNDR Information Paper No. 4, United Nations World Conference on Natural Disaster Reduction, 23-27 May 1994, Yokohama, Japan, 87 pp.

Unnewehr, D., 1999: *Property-Casualty Insurance and the Climate Change Debate: A Risk Assessment.* American Insurance Association, Washington, DC, USA, 9 pp. Available online at http://www.aiadc.org/media/april/041999ppr.htm.

Van Schoubroeck, C., 1997: Legislation and practice concerning natural disasters and insurance in a number of European countries. *The Geneva Papers on Risk and Insurance*, **22(8),** 238–267.

Vellinga, P. and R.S.J. Tol, 1993: Climate change: extreme events and society's response. *Journal of Reinsurance*, **1(2),** 59–72.

Vine, E., E. Mills, and A. Chen, 2000: Energy-Efficient and Renewable Energy Options for Risk Management & Insurance Loss Reduction. *Energy*, **25,** 131–147.

Vine, E., E. Mills, and A. Chen, 1999: Tapping into energy: new technologies and procedures that use energy more efficiently or supply renewable energy offer a largely untapped path to achieving risk management objectives. *Best's Review—Property/Casualty Edition*, **May,** 83–85. Available online at http://eetd.lbl.gov/cbs/insurance/lbnl-41432.html.

Werksman, J., 1998: Compliance issues under the Kyoto Protocol's clean development mechanism. In: *The Clean Development Mechanism. Draft Working Papers, October.* WRI/FIELD/CSDA. FIELD (Foundation for International Environmental Law and Development), London, United Kingdom, pp. 32–39. Available online at http://www.field.org.uk/papers/pdf/ciuk.pdf.

White, R. and D. Etkin, 1997: Climate change, extreme events and the Canadian insurance industry. *Natural Hazards*, **16,** 135–163.

WMO, 1999: *WMO Statement on the Status of the Global Climate in 1998.* WMO-No. 896, World Meteorological Organization, Geneva, Switzerland, 12 pp.

World Bank, 2000: *A Colloquium on Microfinance: Disaster Reduction for the Poor*. World Bank and United Nations Development Programme, 2 February 2000, Washington, DC, USA, 7 pp.

World Bank, 1999: *Come Hell or High Water—Integrating Climate Change Vulnerability and Adaptation into Bank Work*. World Bank, Washington, DC, USA, 60 pp.

World Bank, 1997a: *Clear Water, Blue Skies*. ISBN 0-8213-4044-1, World Bank, Washington, DC, USA, 122 pp.

World Bank, 1997b: *Private Capital flows to Developing Countries*. World Bank, Washington, DC, USA, 300 pp. Summary available online at http://www.worldbank.org/html/extpb/pcf.htm.

Yamin, F., 1998: Adaptation and the clean development mechanism In: *The Clean Development Mechanism. Draft Working Papers, October.* WRI/FIELD/CSDA. FIELD (Foundation for International Environmental Law and Development), London, United Kingdom, pp. 41–52. Available online at http://www.field.org.uk/papers/pdf/cdma.pdf.

Zeng, L., 2000: Weather derivatives and weather insurance: concept, application, and analysis. *Bulletin of the American Meteorological Society*, **81(9),** 2075–2082. Available online at http://ams.allenpress.com.

Zeng, L. and P.J. Kelly, 1997: *A Preliminary Investigation of the Trend of Flood Events in the United States*. Paper presented at the National Association of Real Estate Investment Managers Senior Officer Property Insurance Forum, Boston, March 18–19, 1997, Arkwright Mutual Insurance Company, Norwood, MA, USA, 6 pp. Available online at http://www.atmos.washington.edu/~lixin/ZK1997a.pdf.

Zwirner, O., 2000: Impact of greenhouse-gas mitigation on the insurance industry. In: *Sectoral Economic Costs and Benefits of GHG Mitigation* [Bernstein, L. and J. Pan (eds.)]. Proceedings of an IPCC Expert Meeting, Eisenach, Germany, 14–15 February 2000, pp. 260–269.

9

Human Health

ANTHONY MCMICHAEL (UK) AND ANDREW GITHEKO (KENYA)

Lead Authors:
R. Akhtar (India), R. Carcavallo (Argentina), D. Gubler (USA), A. Haines (UK), R.S. Kovats (UK), P. Martens (The Netherlands), J. Patz (USA), A. Sasaki (Japan)

Contributing Authors:
K.L. Ebi (USA), D. Focks (USA), L. Kalkstein (USA), E. Lindgren (Sweden), S. Lindsay (UK), R. Sturrock (UK)

Review Editors:
U. Confalonieri (Brazil) and A. Woodward (New Zealand)

CONTENTS

EXECUTIVE SUMMARY

Global climate change will have a wide range of health impacts. Overall, negative health impacts are anticipated to outweigh positive health impacts. Some health impacts would result from changes in the frequencies and intensities of extremes of heat and cold and of floods and droughts. Other health impacts would result from the impacts of climate change on ecological and social systems and would include changes in infectious disease occurrence, local food production and nutritional adequacy, and concentrations of local air pollutants and aeroallergens, as well as various health consequences of population displacement and economic disruption.

There is little published evidence that changes in population health status actually have occurred as yet in response to observed trends in climate over recent decades. A recurring difficulty in identifying such impacts is that the causation of most human health disorders is multifactorial and the "background" socioeconomic, demographic, and environmental context varies constantly. A further difficulty is foreseeing all of the likely types of future health effects, especially because for many of the anticipated future health impacts it may be inappropriate to extrapolate existing risk-function estimates to climatic-environmental conditions not previously encountered. Estimation of future health impacts also must take account of differences in vulnerability between populations and within populations over time.

Research since the Second Assessment Report (SAR) mainly has described the effect of climate variability, particularly daily and seasonal extremes, on health outcomes. Studies of health impacts associated with the El Niño-Southern Oscillation (ENSO) have identified interannual climate-health relationships for some epidemic diseases. The upward trend in worldwide numbers of people adversely affected by weather disasters has been characterized by peak impacts during El Niño events. Meanwhile, there has been an expanded effort to develop, test, and apply mathematical models for predicting various health outcomes in relation to climate scenarios. This mix of epidemiological studies and predictive modeling leads to the following conclusions.

An increase in the frequency or intensity of heat waves will increase the risk of mortality and morbidity, principally in older age groups and the urban poor (high confidence). The greatest increases in thermal stress are forecast for higher latitude (temperate) cities, especially in populations that have limited resources, such as access to air conditioning. The pattern of acclimatization to future climate regimes is difficult to estimate. Recent modeling of heat wave impacts in U.S. urban populations, allowing for acclimatization, suggests that several U.S. cities would experience, on average, several hundred extra deaths per summer. Poor urban populations in developing countries may be particularly vulnerable to the impacts of increased heat waves, but no equivalent predictions are available. Warmer winters and fewer cold spells, because of climate change, will decrease cold-related mortality in many temperate countries (high confidence). The reduction in winter deaths will vary between populations. Limited evidence indicates that, in at least some temperate countries, reduced winter deaths would outnumber increased summer deaths.

Any regional increases in climate extremes (storms, floods, cyclones, etc.) associated with climate change would cause physical damage, population displacement, and adverse effects on food production, freshwater availability and quality, and would increase the risks of infectious disease epidemics, particularly in developing countries (very high confidence/ well-established). Over recent years, several major climate-related disasters have had major adverse effects on human health—including floods in China, Mozambique, Bangladesh, and Europe; famine in Sudan; and Hurricane Mitch, which devastated Central America. Although these events cannot be confidently attributed to climate change, they indicate the susceptibility of vulnerable populations to the adverse effects of such events.

Climate change will cause some deterioration in air quality in many large urban areas, assuming that current emission levels continue (medium to high confidence). Increases in exposure to ozone and other air pollutants (e.g., radon, forest fire particulates) could increase known morbidity and mortality effects.

Vector-borne diseases are maintained in complex transmission cycles involving blood-feeding arthropod vectors (and usually reservoir hosts) that depend on specific ecological conditions for survival. These diseases are sensitive to climatic conditions, although response patterns vary between diseases. In areas with limited or deteriorating public health infrastructure, and where temperatures now or in the future are permissive of disease transmission, an increase in temperatures (along with adequate rainfall) will cause certain vector-borne diseases (including malaria, dengue, and leishmaniasis) to extend to higher altitudes (medium to high confidence) and higher latitudes (medium to low confidence). Higher temperatures, in combination with conducive patterns of rainfall and surface water, will prolong transmission seasons in some endemic locations (medium to high confidence). In other locations, climate change will decrease transmission via reductions in rainfall or temperatures that are too high for transmission (low to medium confidence). In all such situations, the actual health impacts of changes in potential infectious disease transmission will be strongly determined by the effectiveness of the public health system.

Mathematical models indicate that climate change scenarios over the coming century would modestly increase the proportion of world population living in regions of potential transmission of malaria and dengue (medium to high confidence). These models are limited by their reliance on climate factors, without reference to modulating influences of environmental, ecological, demographic, or socioeconomic factors. Although the most recent of several biologically based model studies suggests that the increase in population living in regions of potential malaria transmission would be on the order of an extra 260–320 million people in 2080 (against a baseline expectation of about 8 billion), a recent statistically based modeling study, which incorporated conservative assumptions, estimated that there would be no net change in actual transmission of malaria by 2080, assuming a business-as-usual climate scenario and adaptation. In the latter study, regional increases and decreases would approximately cancel out.

Changes in climate, including changes in climate variability, would affect many other vector-borne infections (such as various types of mosquito-borne encephalitis, Lyme disease, and tick-borne encephalitis) at the margins of current distributions (medium to high confidence). For some diseases—such as malaria in the Sahel, Western equine encephalitis in North America, and tick-borne encephalitis in Europe—a net decrease may occur. Changes in surface water quantity and quality will affect the incidence of diarrheal diseases (medium confidence). Ocean warming will facilitate transmission of cholera in coastal areas (low confidence; speculative).

Fish and shellfish poisoning is closely associated with marine ecology. There is some evidence that sea-surface warming associated with El Niño increases the risk to humans of ciguatera poisoning and the occurrence of toxic (and ecologically harmful) algal blooms. Climate change will increase the incidence of ciguatera poisoning and shellfish poisoning (low confidence).

Climate change represents an additional pressure on the world's food supply system and is expected to increase yields at higher latitudes and lead to decreases at lower latitudes. These regional differences in climate impacts on agricultural yield are likely to grow stronger over time, with net beneficial effects on yields and production in the developed world and net negative effects in the developing world. This would increase the number of undernourished people in the developing world (medium confidence).

In some settings, the impacts of climate change may cause social disruption, economic decline, and displacement of populations. The ability of affected communities to adapt to such disruptive events will depend on the social, political, and economic situation of the country and its population. The health impacts associated with such social-economic dislocation and population displacement are substantial [high confidence; well-established].

For each anticipated adverse health impact there is a range of social, institutional, technological, and behavioral adaptation options to lessen that impact. There is a basic and general need for public health infrastructure (programs, services, surveillance systems) to be strengthened and maintained. It also is crucial for nonhealth policy sectors to appreciate how the social and physical conditions of living affect population health.

Our scientific capacity to model the various potential health outcomes of climate change is limited. Nevertheless, it is clear that for many health outcomes—especially for those that result indirectly from a sequence of environmental and social impacts—precise and localized projections cannot yet be made. In the meantime, a precautionary approach requires that policy development proceed on the basis of available—though often limited and qualitative—evidence of how climate change will affect patterns of human population health. Furthermore, high priority should be assigned to improving the public health infrastructure and developing and implementing effective adaptation measures.

9.1. Introduction and Scope

This chapter assesses how climatic changes and associated environmental and social changes are likely to affect human population health. Such an assessment necessarily takes account of the multivariate and interactive ecological framework within which population health and disease are determined. This *ecological* perspective recognizes that the foundations of long-term good health lie in the continued stability and functioning of the biosphere's natural systems—often referred to as "life-support systems."

Deliberate modification of these ecological and physical systems by human societies throughout history has conferred many social, economic, and public health benefits. However, it also has often created new risks to health, such as via mobilization of infectious agents, depletion of freshwater supplies, and reduced productivity of agroecosystems (Hunter *et al.*, 1993; Gubler, 1996). Consider, for example, the chain of consequences from clearance of tropical forests. In the first instance, it typically leads to a warmer and drier local climate. The consequent drying of soil and loss of its organic structure predisposes the area to increased water runoff during heavy rainfall. This, in turn, can endanger human health via flooding, water contamination, impaired crop yields, and altered patterns of vector-borne infectious diseases. Meanwhile, forest clearance also contributes to the atmospheric buildup of carbon dioxide (CO_2) and hence to climate change and its health impacts.

Today, as the scale of human impact on the environment increases, a range of population health impacts can be expected from these large-scale changes in the Earth's life-support systems (Watson *et al.*, 1998). That is the complex context within which actual and potential health impacts of global climate change must be assessed.

9.1.1. Summary of IPCC Second Assessment Report (1996): Potential Health Impacts of Climate Change

The IPCC Second Assessment Report (McMichael *et al.*, 1996a) relied on the relatively limited scientific literature that had emerged during the late 1980s and early 1990s. Most published studies were on health impacts associated with climate variability (e.g., El Niño) and extreme events (natural disasters and heat waves). Predictive modeling of future health impacts was in an early developmental stage.

The SAR noted the many inherent uncertainties in forecasting the potential health impacts of climate change. This included recognition that various other changes in social, economic, demographic, technological, and health care circumstances would unfold over coming decades and that these developments would "condition" the impact of climatic and environmental changes on human health. However, such accompanying changes can be foreseen neither in detail nor far into the future.

The overall assessment was that the likely health impacts would be predominantly adverse. Reflecting the published literature, most of the specific assessments were nonquantitative and relied on expert judgment. They drew on reasoned extrapolations from knowledge of health hazards posed by extreme weather events, increases in temperature-dependent air pollution, summertime increases in certain types of food poisoning, and the spectrum of public health consequences associated with economic disruption and physical displacement of populations. It was noted that the projected effects of climate change on agricultural, animal, and fishery productivity could increase the prevalence of malnutrition and hunger in food-insecure regions experiencing productivity downturns.

For two of the anticipated health impacts, the published literature available by 1995 allowed a more quantitative approach. The relevant conclusions were as follows:

- An increase in the frequency or severity of heat waves would cause a short-term increase in (predominantly cardiorespiratory) deaths and illness. In some very large cities (e.g., Atlanta, Shanghai) by about 2050, this would result in up to several thousand extra heat-related deaths annually. This heat-related mortality increase would be offset by fewer cold-related deaths in milder winters, albeit to an extent that was not yet adequately estimated and likely to vary between populations.
- Climate-induced changes in the geographic distribution and biological behavior of vector organisms of vector-borne infectious diseases (e.g., malaria-transmitting mosquitoes) and infective parasites would alter—usually increase—the potential transmission of such diseases. For example, simulations with global/regional mathematical models indicated that, in the absence of demographic shifts, the proportion of the world's population living within the potential malaria transmission zone would increase from ~45% in the 1990s to ~60% by 2050. Some localized decreases in malaria transmissibility also may occur in response to climate change.

9.1.2. Population Health and its Significance as an Outcome of Climate Change

This is the last of the sector-impact chapters in this volume. This is appropriate because human population health is influenced by an extensive "upstream" range of environmental and social conditions. Indeed, over time, the level of health in a population reflects the quality of social and natural environments, material standards of living, and the robustness of the public health and health service infrastructure. Therefore, population health is an important integrating index of the effects of climate change on ecosystems, biological processes, physical environmental media, and the social-economic environment.

Two other points are important. First, the causation of most human diseases is complex and multifactorial. Second, there is great heterogeneity in the types of disease: acute and chronic;

infectious and noninfectious; physical injury and mental health disorders. These two considerations explain some of the difficulties in fully understanding and quantifying the influences of climate on human health.

Profiles of health and disease vary greatly between regions and countries and over time. Currently, noncommunicable diseases (including mental health disorders) predominate in developed countries, with cardiovascular diseases and cancer accounting for more than half of all deaths. In poorer countries, infectious diseases (especially in childhood) remain important, even as noncommunicable diseases increase in urbanizing populations that are exposed to changes in lifestyle and environmental and occupational exposures. Globally, infectious diseases remain a major cause of human morbidity and are responsible for approximately one-third of all deaths (WHO, 1999a). Many of these water-, food-, and vector-borne infectious diseases are sensitive to climate.

9.2. Research into the Relationship between Climate Change and Health: Caveats and Challenges

9.2.1. New Knowledge about Climate Change Impacts on Health

Since the SAR, much of the additional research on health impacts has examined natural climate variability in relation to interannual variations in infectious diseases—particularly vector-borne diseases—and the relationship between daily weather and mortality in various urban populations. Predictive modeling of the impact of climate scenarios on vector-borne disease transmissibility has undergone further development. Meanwhile, however, data sets that allow study of the effects of the health impacts of observed longer term trends in climate remain sparse.

9.2.2. Characteristics and Methodological Difficulties

The research task of assessing the actual and potential health impacts of climate change has several distinctive characteristics and poses four major challenges to scientists:

1) Anticipated anthropogenic climate change will be a gradual and long-term process. This projected change in mean climate conditions is likely to be accompanied by regional changes in the frequency of extreme events. Changes in particular health outcomes already may be occurring or soon may begin to occur, in response to recent and ongoing changes in world climate. Identification of such health effects will require carefully planned epidemiological studies.
2) In epidemiological studies (in which associations are observed with or without knowledge of likely causal mechanisms), there often are difficulties in estimating the role of climate *per se* as a cause of change in health status. Changes in climate typically are accompanied by various other environmental changes. Because most diseases have multiple contributory causes, it often is difficult to attribute causation between climatic factors and other coexistent factors. For example, in a particular place, clearing of forest for agriculture and extension of irrigation may coincide with a rise in regional temperature. Because all three factors could affect mosquito abundance, it is difficult to apportion between them the causation of any observed subsequent increase in mosquito-borne infection. This difficulty is well recognized by epidemiologists as the "confounding" of effects.
3) It is equally important to recognize that certain factors can modify the vulnerability of a particular population to the health impacts of climate change or variability. This type of effect-modification (or "interaction") can be induced by endogenous characteristics of the population (such as nutritional or immune status) or contextual circumstances that influence the "sensitivity" of the population's response to the climate change (such as unplanned urbanization, crowding, or access to air conditioning during heat waves). Deliberate social, technological, or behavioral adaptations to reduce the health impacts of climate change are an important category of effect-modifying factor.
4) Simulation of scenario-based health risks with predictive models entails three challenges. These challenges relate to validity, uncertainty, and contextual realism:
   - Valid representation of the main environmental and biological relationships and the interacting ecological and social processes that influence the impact of those relationships on health is difficult. A balance must be attained between complexity and simplicity.
   - There are various sources of (largely unavoidable) uncertainty. There is uncertainty attached to the input scenarios of climate change (and of associated social, demographic, and economic trends). Subsequently, there are three main types of uncertainties in the modeling process itself: "normal" statistical variation (reflecting stochastic processes of the real world); uncertainty about the correct or appropriate values of key parameters in the model; and incomplete knowledge about the structural relationships represented in the model.
   - Climate change is not the sole global environmental change that affects human health. Various large-scale environmental changes now impinge on human population health simultaneously, and often interactively (Watson *et al.*, 1998). An obvious example is vector-borne infectious diseases, which are affected by climatic conditions, population movement, forest clearance and land-use patterns, freshwater surface configurations, human population density, and the population density of insectivorous predators (Gubler, 1998b). In accordance with point 2 above, each change in health outcome must be appropriately apportioned between climate and other influences.

9.3. Sensitivity, Vulnerability, and Adaptation

There are uncertainties regarding the sensitivity (i.e., rate of change of the outcome variable per unit change in the input/exposure variable) of many health outcomes to climate or climate-induced environmental changes. Relatively little quantitative research, with estimation of exposure-response relationships, has been done for outcomes other than death rates associated with thermal stress and changes in the transmission potential of several vector-borne infectious diseases. There has been increased effort to map the current distribution of vectors and diseases such as malaria by using climate and other environmental data (including satellite data).

Continuation of recent climatic trends soon may result in some shifts in the geographic range and seasonality of diseases such as malaria and dengue. In reality, however, such shifts also would depend on local topographical and ecological circumstances, other determinants of local population vulnerability, and the existence and level of adaptive public health defenses. There has been some recent debate in the scientific literature about whether there is any evidence of such shifts yet (Epstein *et al.*, 1997; Mouchet *et al.*, 1998; Reiter, 1998a,b). It is not yet clear what criteria are most appropriate for assessment of climatic influences on such changes in infectious disease patterns. A balance is needed between formal, statistically based analysis of changes within a particular local setting and a more synthesizing assessment of the consistency of patterns across diverse settings and across different systems—physical, biotic, social, and public health. As with climate change itself, there is an inherent difficulty in detecting small climate-induced shifts in population health outcomes and in attributing the shift to a change in climate.

Population vulnerability is a function of the extent to which a health outcome in that particular environmental-demographic setting is sensitive to climate change and the capacity of the population to adapt to new climate conditions. Determinants of population vulnerability to climate-related threats to health include level of material resources, effectiveness of governance and civil institutions, quality of public health infrastructure, access to relevant local information on extreme weather threats, and preexisting burden of disease (Woodward *et al.*, 1998). Thus, vulnerability is determined by individual, community, and geographical factors:

- *Individual factors include:*
  - Disease status (people with preexisting cardiovascular disease, for example, may be more vulnerable to direct effects such as heat waves)
  - Socioeconomic factors (in general, the poor are more vulnerable)
  - Demographic factors (the elderly are more vulnerable to heat waves, for example, and infants are more vulnerable to diarrheal diseases).
- *Community factors may include:*
  - Integrity of water and sanitation systems and their capacity to resist extreme events
  - Local food supplies and distribution systems
  - Access to information, including early warnings of extreme climate events
  - Local disease vector distribution and control programs.
- *Geographical factors may include:*
  - The influence of El Niño cycle or the occurrence of extreme weather events that are more common in some parts of the world
  - Low-lying coastal populations more vulnerable to the effects of sea-level rise
  - Populations bordering current distributions of vector-borne disease particularly vulnerable to changes in distribution
  - Rural residents often with less access to adequate health care, and urban residents more vulnerable to air pollution and heat island effects
  - Environmentally degraded and deforested areas more vulnerable to extreme weather events.

Understanding a population's capacity to adapt to new climate conditions is crucial to realistic assessment of the potential health impacts of climate change (Smithers and Smit, 1997). This issue is addressed more fully in Section 9.11.

9.4. Thermal Stress (Heat Waves, Cold Spells)

9.4.1. *Heat Waves*

Global climate change is likely to be accompanied by an increase in the frequency and intensity of heat waves, as well as warmer summers and milder winters (see Table 3-10). The impact of extreme summer heat on human health may be exacerbated by increases in humidity (Gaffen and Ross, 1998; Gawith *et al.*, 1999).

Daily numbers of deaths increase during very hot weather in temperate regions (Kunst *et al.*, 1993; Ando, 1998a,b). For example, in 1995, a heat wave in Chicago caused 514 heat-related deaths (12 per 100,000 population) (Whitman *et al.*, 1997), and a heat wave in London caused a 15% increase in all-cause mortality (Rooney *et al.*, 1998). Excess mortality during heat waves is greatest in the elderly and people with preexisting illness (Sartor *et al.*, 1995; Semenza *et al.*, 1996; Kilbourne, 1997; Ando *et al.*, 1998a,b). Much of this excess mortality from heat waves is related to cardiovascular, cerebrovascular, and respiratory disease. The mortality impact of a heat wave is uncertain in terms of the amount of life lost; a proportion of deaths occur in susceptible persons who were likely to have died in the near future. Nevertheless, there is a high level of certainty that an increase in the frequency and intensity of heat waves would increase the numbers of additional deaths from hot weather. Heat waves also are associated with nonfatal impacts such as heat stroke and heat exhaustion (Faunt *et al.*, 1995; Semenza *et al.*, 1999).

Heat waves have a much bigger health impact in cities than in surrounding suburban and rural areas (Kilbourne, 1997; Rooney *et al.*, 1998). Urban areas typically experience higher—and nocturnally sustained—temperatures because of the "heat island" effect (Oke, 1987; Quattrochi *et al.*, 2000). Air pollution also is typically higher in urban areas, and elevated pollution levels often accompany heat waves (Piver *et al.*, 1999) (see also Section 9.6.1.2 and Chapter 8).

The threshold temperature for increases in heat-related mortality depends on the local climate and is higher in warmer locations. A study based on data from several European regions suggests that regions with hotter summers do not have significantly different annual heat-related mortality compared to cold regions (Keatinge *et al.*, 2000). However, in the United States, cities with colder climates are more sensitive to hot weather (Chestnut *et al.*, 1998). Populations will acclimatize to warmer climates via a range of behavioral, physiological, and technological adaptations. Acclimatization will reduce the impacts of future increases in heat waves, but it is not known to what extent. Initial physiological acclimatization to hot environments can occur over a few days, but complete acclimatization may take several years (Zeisberger *et al.*, 1994).

Weather-health studies have used a variety of derived indices—for example, the air mass-based synoptic approach (Kalkstein and Tan, 1995) and perceived temperature (Jendritzky *et al.*, 2000). Kalkstein and Greene (1997) estimated future excess mortality under climate change in U.S. cities. Excess summer mortality attributable to climate change, assuming acclimatization, was estimated to be 500–1,000 for New York and 100–250 for Detroit by 2050, for example. Because this is an isolated study, based on a particular method of treating meteorological conditions, the chapter team assigned a medium level of certainty to this result.

The impact of climate change on mortality from thermal stress in developing country cities may be significant. Populations in developing countries (e.g., in Mexico City, New Delhi, Jakarta) may be especially vulnerable because they lack the resources to adapt to heat waves. However, most of the published research refers to urban populations in developed countries; there has been relatively little research in other populations.

9.4.2. Decreased Mortality Resulting from Milder Winters

In many temperate countries, there is clear seasonal variation in mortality (Sakamoto-Momiyama, 1977; Khaw, 1995; Laake and Sverre, 1996); death rates during the winter season are 10–25% higher than those in the summer. Several studies indicate that decreases in winter mortality may be greater than increases in summer mortality under climate change (Langford and Bentham, 1995; Martens, 1997; Guest *et al.*, 1999). One study estimates a decrease in annual cold-related deaths of 20,000 in the UK by the 2050s (a reduction of 25%) (Donaldson *et al.*, 2001). However, one study estimates that increases in heat-related deaths will be greater than decreases in cold-related death in the United States by a factor of three (Kalkstein and Greene, 1997).

Annual outbreaks of winter diseases such as influenza, which have a large effect on winter mortality rates, are not strongly associated with monthly winter temperatures (Langford and Bentham, 1995). Social and behavioral adaptations to cold play an important role in preventing winter deaths in high-latitude countries (Donaldson *et al.*, 1998). Sensitivity to cold weather (i.e., the percentage increase in mortality per 1°C change) is greater in warmer regions (e.g., Athens, southern United States) than in colder regions (e.g., south Finland, northern United States) (Eurowinter Group, 1997). One possible reason for this difference may be failure to wear suitable winter clothing. In North America, an increase in mortality is associated with snowfall and blizzards (Glass and Zack, 1979; Spitalnic *et al.*, 1996; Gorjanc *et al.*, 1999) and severe ice storms (Munich Re, 1999).

The extent of winter-associated mortality that is directly attributable to stressful weather therefore is difficult to determine and currently is being debated in the literature. Limited evidence indicates that, in at least some temperate countries, reduced winter deaths would outnumber increased summer deaths. The net impact on mortality rates will vary between populations. The implications of climate change for nonfatal outcomes is not clear because there is very little literature relating cold weather to health outcomes.

9.5. Extreme Events and Weather Disasters

Major impacts of climate change on human health are likely to occur via changes in the magnitude and frequency of extreme events (see Table 3-10), which trigger a natural disaster or emergency. In developed countries, emergency preparedness has decreased the total number of tropical cyclone-related deaths (see Section 7.2.2). However, in developed countries, studies indicate an increasing trend in the number and impacts (deaths, injuries, economic losses) of all types of natural disasters (IFRC, 1998; Munich Re, 1999). Some of the interannual variability in rates of persons affected by disasters may be associated with El Niño (Bouma *et al.*, 1997a). The average annual number of people killed by natural disasters between 1972 and 1996 was about 123,000. By far the largest number of people affected (i.e., in need of shelter or medical care) are in Asia, and one study reveals that Africa suffers 60% of all disaster-related deaths (Loretti and Tegegn, 1996).

Populations in developing countries are much more affected by extreme events. Relative to low socioeconomic conditions, the impact of weather-related disasters in poor countries may be 20–30 times larger than in industrialized countries. For example, floods and drought associated with the El Niño event of 1982–1983 led to losses of about 10% in gross national product (GNP) in countries such as Bolivia, Chile, Ecuador, and Peru (50% of their annual public revenue) (Jovel, 1989).

Disasters occur when climate hazards and population vulnerability converge. Factors that affect vulnerability to disasters are shown in Figure 9-1. The increase in population vulnerability

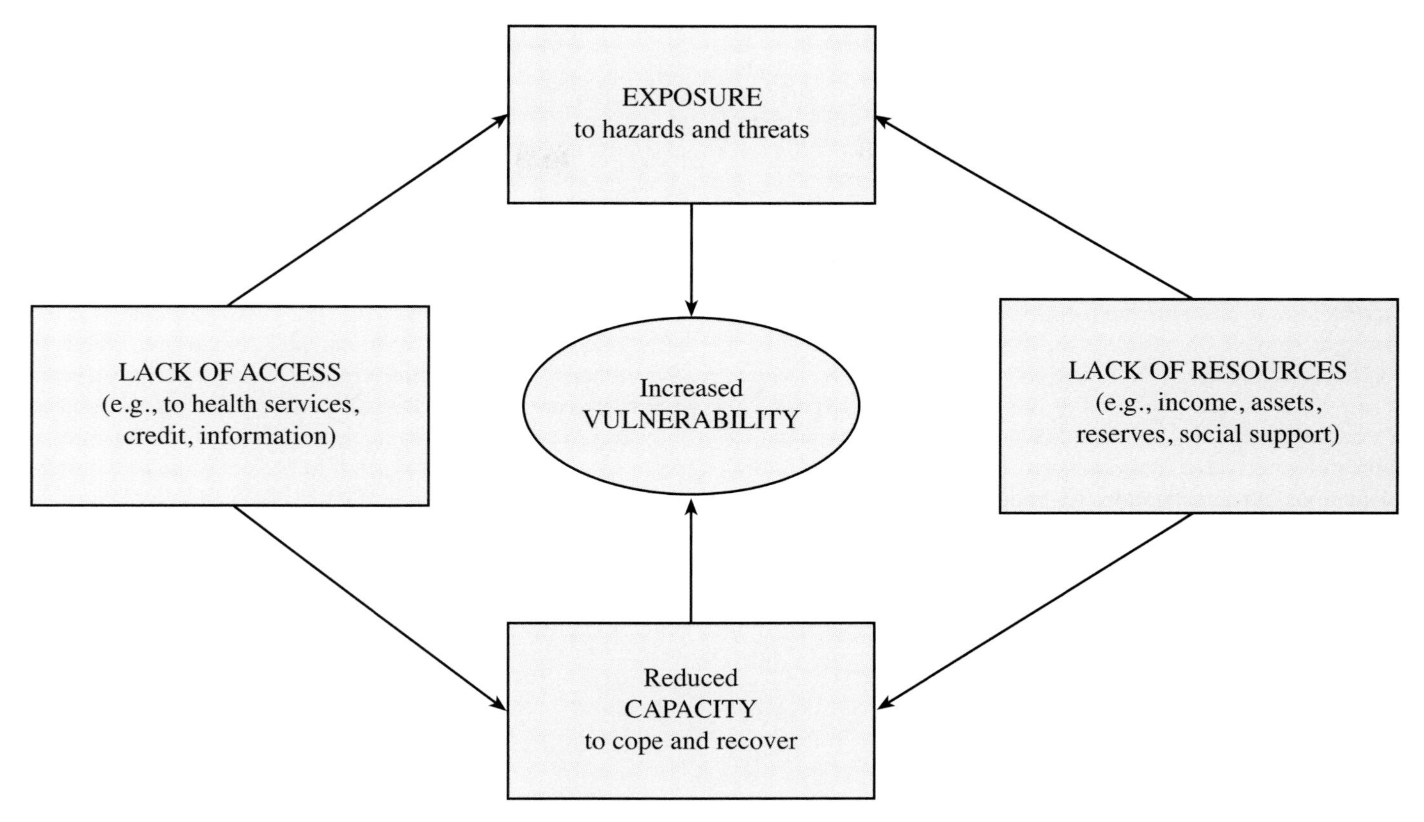

Figure 9-1: Diagrammatic illustration of vulnerability to disasters (McMichael *et al.*, 1996b).

to extreme weather is primarily caused by the combination of population growth, poverty, and environmental degradation (Alexander, 1993). Concentration of people and property in high-risk areas (e.g., floodplains and coastal zones) also has increased. Degradation of the local environment also may contribute to vulnerability (see Chapter 7).

The health impacts of natural disasters include (Noji, 1997):

- Physical injury
- Decreases in nutritional status, especially in children
- Increases in respiratory and diarrheal diseases resulting from crowding of survivors, often with limited shelter and access to potable water
- Impacts on mental health, which in some cases may be long-lasting
- Increased risk of water-related diseases as a result of disruption of water supply or sewage systems
- Release and dissemination of dangerous chemicals from storage sites and waste disposal sites into floodwaters.

Extreme weather events cause death and injury directly. However, substantial indirect health impacts also occur because of damage to the local infrastructure and population displacement (see also Section 9.10). Following disasters, fatalities and injuries can occur as residents return to clean up damage and debris (Philen *et al.*, 1992). Bereavement, property loss, and social disruption may increase the risk of depression and mental health problems (WHO, 1992). For example, cases of post-traumatic stress disorder were reported in the United States up to 2 years after Hurricane Andrew (Norris *et al.*, 1999).

9.5.1. Floods

Floods are associated with particular dangers to human populations (Menne *et al.*, 1999). Climate change may increase the risk of river and coastal flooding (see Chapters 4 and 6). The health impacts of floods may be divided into the immediate, medium, and long terms. Immediate effects are largely death and injuries caused by drowning and being swept against hard objects. Medium-term effects include increases in communicable diseases such as those caused by ingestion of contaminated water (e.g., cholera, hepatitis A), contact with contaminated water (e.g., leptospirosis—see Section 9.7.9.1), or respiratory diseases resulting from overcrowding in shelters. A study in populations displaced by catastrophic floods in Bangladesh in 1988 found that diarrhea was the most common illness, followed by respiratory infection. Watery diarrhea was the most common cause of death for all age groups under 45 (Siddique *et al.*, 1991). In rural Bangladesh and Khartoum, Sudan, the proportion of severely malnourished children increased after flooding (Woodruff *et al.*, 1990; Choudhury and Bhuiya, 1993). Also, in the aftermath of flooding, molds and fungi may grow on interior surfaces, providing a potent stimulus to allergic persons (American Academy of Pediatrics, 1998).

In China, floods experienced over the past few years have been particularly severe. In 1996, official national statistics showed

200 million people affected by flooding: There were more than 3,000 deaths, and 363,800 people were injured; 3.7 million houses were destroyed, and 18 million houses were damaged. Direct economic loses exceeded US$12 billion (IFRC 1997). In 1998, official national statistics showed 200 million people affected by flooding, more than 3,000 deaths, and 4 million houses damaged; direct economic losses exceeded US$20 billion (National Climate Centre of China, 1998). Nevertheless, the vulnerability of the Chinese population has been reduced by a combination of better preparedness, including sophisticated warning systems, and relief efforts. In the longer term, reforestation may reduce the risk of flooding in these regions.

In developed countries, physical and disease risks from flooding are greatly reduced by a well-maintained flood control and sanitation infrastructure and public health measures, such as monitoring and surveillance activities to detect and control outbreaks of infectious disease. However, the experience of the central European floods of 1997, when more than 100 people died, showed that even in industrialized countries floods can have a major impact on health and welfare. In Poland, 6,000 km^2 were flooded, and 160,000 people were evacuated from their homes. The cost of the damage was estimated at US$3 billion [2.7% of 1996 gross domestic product (GDP)]. In the Czech Republic, 50,000 people were evacuated and damage was estimated at US$1.8 billion (3.7% of GDP) (IFRC, 1998). There was an increase in cases of leptospirosis in the Czech Republic (Kriz *et al.*, 1998). Floods also have an important impact on mental health in the affected community (WHO, 1992; Menne *et al.*, 1999). Increases in suicide, alcoholism, and psychological and behavioral disorders, particularly among children, were reported following floods in Poland in 1997 (IFRC 1998).

9.5.2. *Storms and Tropical Cyclones*

Impoverished and high-density populations in low-lying and environmentally degraded areas are particularly vulnerable to tropical cyclones (also called hurricanes and typhoons). Many of the most serious impacts of tropical cyclones in the 20th century have occurred in Bangladesh because of the combination of meteorological and topographical conditions, along with the inherent vulnerability of this low-income, poorly resourced population. Tropical cyclones also can cause landslides and flooding. Most deaths are caused by drowning in the storm surge (Alexander, 1993; Noji, 1997). The impacts of cyclones in Japan and other developed countries have been decreasing in recent years because of improved early warning systems. However, the experience of Hurricane Mitch demonstrated the destructive power of an extreme event on a densely populated and poorly resourced region (PAHO, 1999).

9.5.3. *Droughts*

The health impacts of drought on populations occur primarily via impacts on food production. Famine often occurs when a preexisting situation of malnutrition worsens. The health consequences of drought include diseases resulting from malnutrition (McMichael *et al.*, 1996b). In times of shortage, water is used for cooking rather than hygiene. In particular, this increases the risk of diarrheal diseases (as a result of fecal contamination) and water-washed diseases (e.g., trachoma, scabies). Outbreaks of malaria can occur during droughts as a result of changes in vector breeding sites (Bouma and van der Kaay, 1996). Malnutrition also increases susceptibility to infection.

In addition to adverse environmental conditions, political, environmental, or economic crises can trigger a collapse in food marketing systems. These factors may have a cumulative or synergistic effect. For example, a breakdown in the reserve food supply system resulting from the sale of grain or livestock reserves might be exacerbated by conflict and breakdown in law and order. The major food emergency in Sudan during 1998 illustrates the interrelationship between climatic triggers of famine and conflict. Land mines made portions of major roads in southern Sudan impassable and contributed to poor access for relief supplies. By July 1998, the World Food Programme's air cargo capacity had increased to more than 10,000 t to overcome the transport difficulties. These air cargoes were supplemented by barge convoys and road repair projects (WFP, 1999). Vulnerability to drought and food shortages can be greatly reduced through the use of seasonal forecasts as part of an early warning system (see Section 9.11.1).

9.6. Air Pollution

9.6.1. *Gases, Fine Particulates*

Weather conditions influence air pollution via pollutant (or pollutant precursor) transport and/or formation. Weather conditions also can influence biogenic (e.g., pollen production) and anthropogenic (e.g., as a result of increased energy demand) air pollutant emissions. Exposure to air pollutants can have many serious health effects, especially following severe pollution episodes. Studies that are relevant to climate change and air pollution can be divided into two categories: those that estimate the combined impact of weather and air pollutants on health outcomes and those that estimate future air pollution levels. Climate change may increase the concentration of ground-level ozone, but the magnitude of the effect is uncertain (Patz *et al.*, 2000). For other pollutants, the effects of climate change and/or weather are less well studied.

Current air pollution problems are greatest in developing country cities. For example, nearly 40,000 people die prematurely every year in India because of outdoor air pollution (World Bank, 1997). Air quality also is one of the main concerns for environmental health in developed countries (Bertollini *et al.*, 1996; COMEAP, 1998).

Radon is an inert radioactive gas. The rate at which it is emitted from the ground is sensitive to temperature (United Nations, 1982). High indoor exposures are associated with an increased risk of lung cancer (IARC, 1988). There is some evidence from modeling experiments that climate warming may increase radon concentrations in the lower atmosphere (Cuculeanu and Iorgulescu, 1994).

9.6.1.1. *Effects of Air Pollution, Season, and Weather on Health*

The six standard air pollutants that have been extensively studied in urban populations are sulfur dioxide (SO_2), ozone (O_3), nitrogen dioxide (NO_2), carbon monoxide (CO), lead, and particulates. The impact of some air pollutants on health is more evident during the summer or during high temperatures (Bates and Sizto, 1987; Bates *et al.*, 1990; Castellsague *et al.*, 1995; Bobak and Roberts, 1997; Katsouyanni *et al.*, 1997; Spix *et al.*, 1998; de Diego Damia *et al.*, 1999; Hajat *et al.*, 1999). For example, the relationship between SO_2 and total and cardiovascular mortality in Valencia (Ballester *et al.*, 1996) and Barcelona, Spain (Sunyer *et al.*, 1996), and Rome, Italy (Michelozzi *et al.*, 1998), was found to be stronger during hot periods than during winter. However, Moolgavkar *et al.* (1995) conclude that, in Philadelphia, SO_2 had the strongest health effects in spring, autumn, and winter. Increases in daily mortality and morbidity (indicated by hospital admissions) are associated with high ozone levels on hot days in many cities (e.g., Moolgavkar *et al.*, 1995; Sunyer *et al.*, 1996; Touloumi *et al.*, 1997).

High temperatures also have acute effects on mortality (see Section 9.4.1). Some studies have found evidence of an interaction between the effects of ozone and the effects of higher temperatures (e.g., Katsouyanni *et al.*, 1993; Sartor *et al.*, 1995). Other studies addressing the combined effects of weather and particulate air pollution have not found evidence of such an interaction (e.g., Samet *et al.*, 1998). Correlations between climate and site-specific air quality variables must be further evaluated and, in some instances, need to include temperature, pollution, and interaction terms in regression models.

Climate change is expected to increase the risk of forest and rangeland fires (see Section 5.6.2.2.1). Haze-type air pollution therefore is a potential impact of climate change on health. Majors fires in 1997 in southeast Asia and the Americas were associated with increases in respiratory and eye symptoms (Brauer, 1999; WHO, 1999b). In Malaysia, a two- to three-fold increase in outpatient visits for respiratory disease and a 14% decrease in lung function in school children were reported. In Alta Floresta, Brazil, there was a 20-fold increase in outpatient visits for respiratory disease. In 1998, fires in Florida were linked to significant increases in emergency department visits for asthma (91%), bronchitis (132%), and chest pain (37%) (CDC, 1999). However, a study of 1994 bushfires in western Sydney showed no increase in asthma admissions to emergency departments (Smith *et al.*, 1996).

9.6.1.2. *Future Changes in Air Quality*

Weather has a major influence on the dispersal and ambient concentrations of air pollutants. Large high-pressure systems often create an inversion of the normal temperature profile, trapping pollutants in the shallow boundary layer at the Earth's surface. It is difficult to predict the impact of climate change on local urban climatology and, therefore, on average local air pollution concentrations. However, any increase in anticyclonic conditions in summer would tend to increase air pollution concentrations in cities (Hulme and Jenkins, 1998).

Box 9-1. Stratospheric Ozone Depletion and Exposure to Ultraviolet Radiation

Stratospheric ozone destruction is an essentially separate process from greenhouse gas (GHG) accumulation in the lower atmosphere. However, not only are several of the anthropogenic GHGs [e.g., chlorofluorocarbons (CFCs) and N_2O] also ozone-depleting gases but tropospheric warming apparently induces stratospheric cooling, which exacerbates ozone destruction (Shindell *et al.*, 1998; Kirk-Davidoff *et al.*, 1999). Stratospheric ozone shields the Earth's surface from incoming solar ultraviolet radiation (UVR), which has harmful effects on human health. Long-term decreases in summertime ozone over New Zealand have been associated with significant increases in ground-level UVR, particularly in the DNA-damaging waveband (McKenzie *et al.*, 1999). In a warmer world, patterns of personal exposure to solar radiation (e.g., sunbathing in temperate climates) also are likely to change.

Many epidemiological studies have implicated solar radiation as a cause of skin cancer (melanoma and other types) in fair-skinned humans (IARC, 1992; WHO, 1994). The most recent assessment by UNEP (1998) projects significant increases in skin cancer incidence as a result of stratospheric ozone depletion. High-intensity UVR also damages the eye's outer tissue, causing "snowblindness"—the ocular equivalent of sunburn. Chronic exposure to UVR is linked to conditions such as pterygium (WHO, 1994). The role of UV-B in cataract formation is complex. Some cataract subtypes appear to be associated with UVR exposure, whereas others do not. In humans and experimental animals, UVR can cause local and whole-body immunosuppression (UNEP, 1998). Cellular immunity has been shown to be affected by ambient doses of UVR (Garssen *et al.*, 1998). Concern exists that UVR-induced immunosuppression could influence patterns of infectious disease. Nevertheless, no direct evidence exists for such effects in humans, and uncertainties remain about the underlying biological processes.

Formation and destruction of ozone is accelerated by increases in temperature and ultraviolet radiation. Existing air quality models have been used to examine the effect of climate change on ozone concentrations (e.g., Morris *et al.*, 1989; Penner *et al.*, 1989; Morris *et al.*, 1995; Sillman and Samson, 1995). The models indicate that decreases in stratospheric ozone and elevated temperature increase ground-level ozone concentration. An increase in occurrence of hot days could increase biogenic and anthropogenic emissions of volatile organic compounds (e.g., from increased evaporative emissions from fuel-injected automobiles) (Sillman and Samson, 1995). These studies of the impact of climate change on air quality must be considered indicative but by no means definitive. Important local weather factors may not be adequately represented in these models.

9.6.2. *Aeroallergens (e.g., Pollen)*

Daily, seasonal, and interannual variation in the abundance of many aeroallergens, particularly pollen, is associated with meteorological factors (Emberlin, 1994, 1997; Spieksma *et al.*, 1995; Celenza *et al.*, 1996). The start of the grass pollen season can vary between years by several weeks according to the weather in the spring and early summer. Pollen abundance, however, is more strongly associated with land-use change and farming practices than with weather (Emberlin, 1994). Pollen counts from birch trees (the main cause of seasonal allergies in northern Europe) have been shown to increase with increasing seasonal temperatures (Emberlin, 1997; Ahlholm *et al.*, 1998). In a study of Japanese cedar pollen, there also was a significant increase in total pollen count in years in which summer temperatures had risen (Takahashi *et al.*, 1996). However, the relationship between meteorological variables and specific pollen counts can vary from year to year (Glassheim *et al.*, 1995). Climate change may affect the length of the allergy season. In addition, the effect of higher ambient levels of CO_2 may affect pollen production. Experimental research has shown that a doubling in CO_2 levels, from about 300 to 600 ppm, induces an approximately four-fold increase in the production of ragweed pollen (Ziska and Caulfield, 2000a,b).

High pollen levels have been associated with acute asthma epidemics, often in combination with thunderstorms (Hajat *et al.*, 1997; Newson *et al.*, 1998). Studies show that the effects of weather and aeroallergens on asthma symptoms are small (Epton *et al.*, 1997). Other assessments have found no evidence that the effects of air pollutants and airborne pollens interact to exacerbate asthma (Guntzel *et al.*, 1996; Stieb *et al.*, 1996; Anderson *et al.*, 1998; Hajat *et al.*, 1999). Airborne pollen allergen can exist in subpollen sizes; therefore, specific pollen/asthma relationships may not be the best approach to assessing the risk (Beggs, 1998). One study in Mexico suggests that altitude may affect the development of asthma (Vargas *et al.*, 1999). Sources of indoor allergens that are climate-sensitive include the house dust mite, molds, and cockroaches (Beggs and Curson, 1995). Because the causation of initiation and exacerbation of asthma is complex, it is not clear how climate change would affect this disease. Further research into general allergies (including seasonal and geographic distribution) is required.

9.7. Infectious Diseases

The ecology and transmission dynamics of infectious diseases are complex and, in at least some respects, unique for each disease within each locality. Some infectious diseases spread directly from person to person; others depend on transmission via an intermediate "vector" organism (e.g., mosquito, flea, tick), and some also may infect other species (especially mammals and birds).

The "zoonotic" infectious diseases cycle naturally in animal populations. Transmission to humans occurs when humans encroach on the cycle or when there is environmental disruption, including ecological and meteorological factors. Various rodent-borne diseases, for example, are dependent on environmental conditions and food availability that determine rodent population size and behavior. An explosion in the mouse population following extreme rainfall from the 1991–1992 El Niño event is believed to have contributed to the first recorded outbreak of hantavirus pulmonary syndrome in the United States (Engelthaler *et al.*, 1999; Glass *et al.*, 2000).

Many important infectious diseases, especially in tropical countries, are transmitted by vector organisms that do not regulate their internal temperatures and therefore are sensitive to external temperature and humidity (see Table 9-1). Climate change may alter the distribution of vector species—increasing or decreasing the ranges, depending on whether conditions are favorable or unfavorable for their breeding places (e.g., vegetation, host, or water availability). Temperature also can influence the reproduction and maturation rate of the infective agent within the vector organism, as well as the survival rate of the vector organism, thereby further influencing disease transmission.

Changes in climate that will affect potential transmission of infectious diseases include temperature, humidity, altered rainfall, and sea-level rise. It is an essential but complex task to determine how these factors will affect the risk of vector- and rodent-borne diseases. Factors that are responsible for determining the incidence and geographical distribution of vector-borne diseases are complex and involve many demographic and societal—as well as climatic—factors (Gubler, 1998b). An increase in vector abundance or distribution does not automatically cause an increase in disease incidence, and an increase in incidence does not result in an equal increase in mortality (Chan *et al.*, 1999). Transmission requires that the reservoir host, a competent arthropod vector, and the pathogen be present in an area at the same time and in adequate numbers to maintain transmission. Transmission of human diseases is dependent on many complex and interacting factors, including human population density, housing type and location, availability of screens and air conditioning on habitations, human behavior, availability of reliable piped water, sewage and waste management systems, land use and irrigation systems, availability and efficiency of vector control programs, and general environmental hygiene. If all of these factors are favorable for transmission, several meteorological factors may influence the intensity of transmission (e.g., temperature, relative humidity, and precipitation patterns). All of the foregoing factors influence the transmission dynamics of a disease and play a role in determining whether endemic or epidemic transmission occurs.

The resurgence of infectious diseases in the past few decades, including vector-borne diseases, has resulted primarily from demographic and societal factors—for example, population growth, urbanization, changes in land use and agricultural practices, deforestation, international travel, commerce, human and animal movement, microbial adaptation and change, and breakdown in public health infrastructure (Lederberg *et al.*, 1992; Gubler, 1989, 1998a). To date, there is little evidence that climate change has played a significant role in the recent resurgence of infectious diseases.

The following subsections describe diseases that have been identified as most sensitive to changes in climate. The majority of these assessments rely on expert judgment. Where models have been developed to assess the impact of climate change, these also are discussed.

9.7.1. Malaria

Malaria is one of the world's most serious and complex public health problems. The disease is caused by four distinct species of plasmodium parasite, transmitted between individuals by Anopheline mosquitoes. Each year, it causes an estimated 400–500 million cases and more than 1 million deaths, mostly

***Table 9-1**: Main vector-borne diseases: populations at risk and burden of disease (WHO data).*

Disease	Vector	Population at Risk	Number of People Currently Infected or New Cases per Year	Disability-Adjusted Life Years Lost[a]	Present Distribution
Malaria	Mosquito	2400 million (40% world population)	272,925,000	39,300,000	Tropics/subtropics
Schistosomiasis	Water Snail	500–600 million	120 million	1,700,000	Tropics/subtropics
Lymphatic filariasis	Mosquito	1,000 million	120 million	4,700,000	Tropics/subtropics
African trypanosomiasis (sleeping sickness)	Tsetse Fly	55 million	300,000–500,000 cases yr^{-1}	1,200,000	Tropical Africa
Leishmaniasis	Sandfly	350 million	1.5–2 million new cases yr^{-1}	1,700,000	Asia/Africa/ southern Europe/ Americas
Onchocerciasis (river blindness)	Black Fly	120 million	18 million	1,100,000	Africa/Latin America/ Yemen
American trypanosomiasis (Chagas' disease)	Triatomine Bug	100 million	16–18 million	600,000	Central and South America
Dengue	Mosquito	3,000 million	Tens of millions cases yr^{-1}	1,800,000[b]	All tropical countries
Yellow fever	Mosquito	468 million in Africa	200,000 cases yr^{-1}	Not available	Tropical South America and Africa
Japanese encephalitis	Mosquito	300 million	50,000 cases yr^{-1}	500,000	Asia

[a] Disability-Adjusted Life Year (DALY) = a measurement of population health deficit that combines chronic illness or disability and premature death (see Murray, 1994; Murray and Lopez, 1996). Numbers are rounded to nearest 100,000.
[b] Data from Gubler and Metzer (1999).

***Table 9-2**: Effect of climate factors on vector- and rodent-borne disease transmission.*

Climate Factor	Vector	Pathogen	Vertebrate Host and Rodents
Increased temperature	– Decreased survival, e.g., *Culex. tarsalis* (Reeves *et al*., 1994) – Change in susceptibility to some pathogens (Grimstad and Haramis, 1984; Reisen, 1995); seasonal effects (Hardy *et al*., 1990) – Increased population growth (Reisen, 1995) – Increased feeding rate to combat dehydration, therefore increased vector–human contact – Expanded distribution seasonally and spatially	– Increased rate of extrinsic incubation in vector (Kramer *et al*., 1983; Watts *et al*., 1987) – Extended transmission season (Reisen *et al*., 1993, 1995) – Expanded distribution (Hess *et al*., 1963)	– Warmer winters favor rodent survival
Decreases in precipitation	– Increase in container-breeding mosquitoes because of increased water storage – Increased abundance for vectors that breed in dried-up river beds (Wijesunder, 1988) – Prolonged droughts could reduce or eliminate snail populations	– No effect	– Decreased food availability can reduce populations – Rodents may be more likely to move into housing areas, increasing human contact
Increases in precipitation	– Increased rain increases quality and quantity of larval habitat and vector population size – Excess rain can eliminate habitat by flooding – Increased humidity increases vector survival – Persistent flooding may increase potential snail habitats downstream	– Little evidence of direct effects – Some data on humidity effect on malarial parasite development in *Anopheline* mosquito host	– Increased food availability and population size (Mills *et al*., 1999)
Increase in precipitation extremes	– Heavy rainfall events can synchronize vector host-seeking and virus transmission (Day and Curtis, 1989) – Heavy rainfall can wash away breeding sites	– No effect	– Risk of contamination of flood waters/runoff with pathogens from rodents or their excrement (e.g., *Leptospira* from rat urine)
Sea-level rise	– Coastal flooding affects vector abundance for mosquitoes that breed in brackish water (e.g., *An. subpictus* and *An. sundaicus* malaria vectors in Asia)	– No effect	– No effect

in children (WHO, 1998a). Malaria is undergoing a global resurgence because of a variety of factors, including complacency and policy changes that led to reduced funding for malaria control programs in the 1970s and 1980s, the emergence of insecticide and drug resistance, human population growth and movement, land-use change, and deteriorating public health infrastructure (Lindsay and Birley, 1996). Variation in malaria transmission also is associated with changes in temperature, rainfall, and humidity as well as the level of immunity (Lindsay and Birley, 1996). All of these factors can interact to affect adult mosquito densities and the development of the parasite within the mosquito (see Table 9-2).

Very high temperatures are lethal to the mosquito and the parasite. In areas where mean annual temperature is close to the physiological tolerance limit of the parasite, a small temperature increase would be lethal to the parasite, and malaria transmission would therefore decrease. However, at low temperatures, a small increase in temperature can greatly increase the risk of malaria transmission (Bradley, 1993; Lindsay and Birley, 1996).

Micro- and macroenvironmental changes can affect malaria transmission. For example, deforestation may elevate local temperatures (Hamilton, 1989). Changes in types of housing may change indoor temperatures where some vectors spend most of the time resting (Garnham, 1945). In Africa, deforestation, vegetation clearance, and irrigation can all provide the open sunlit pools that are preferred by important malaria vectors and thus increase transmission (Chandler and Highton, 1975;

Walsh *et al.*, 1993; Githeko *et al.*, 1996; Lindsay and Birley, 1996).

Malaria currently is present in 101 countries and territories (WHO, 1998a). An estimated 40% (i.e., 2.4 billion people) of the total world population currently lives in areas with malaria. In many malaria-free countries with a developed public health infrastructure, the risk of sustained malaria transmission after reintroduction is low in the near term. Other areas may become at risk as a result of climate change if, for example, malaria control programs have broken down or if transmission currently is limited mainly by temperature. Environmental conditions already are so favorable for malaria transmission in tropical African countries that climate change is unlikely to affect overall mortality and morbidity rates in endemic lowland regions (MARA, 1998). Furthermore, reductions in rainfall around the Sahel may decrease transmission in this region of Africa (Mouchet *et al.*, 1996; Martens *et al.*, 1999). Future climate change may increase transmission in some highland regions, such as in East Africa (Lindsay and Martens, 1998, Mouchet *et al.*, 1998; Cox *et al.*, 1999; see Box 9-2). Studies that map malaria in Africa indicate that, at the broad scale, distribution of the disease is determined by climate, except at the southern limit (MARA, 1998). Malaria transmission currently is well within the climatic limits of its distribution in mid- to high-latitude developed countries because of effective control measures and other environmental changes. However, in South America the southern limits of malaria distribution may be affected by climate change. The southern geographical distribution limit of a major malaria vector in South America (*An. darlingi*) coincides with the April mean isotherm of 20°C. If temperature and rainfall increase in Argentina, *An. darlingi* may extend its distribution in southern Argentina, whereas if rainfall decreases, conditions may become unfavorable for *An. darlingi* (Carcavallo and Curto de Casas, 1996).

Malaria was successfully eradicated from Australia, Europe, and the United States in the 1950s and 1960s, but the vectors were not eliminated (Bruce-Chwatt and de Zulueta, 1980; Zucker, 1996). In regions where the vectors persist in sufficient abundance, there is a risk of locally transmitted malaria. This small risk of very localized outbreaks may increase under climate change. Conditions currently exist for malaria transmission in those countries during the summer months, but few nonimported cases have been reported (Holvoet *et al.*, 1983; Zucker, 1996; Baldari *et al.*, 1998; Walker, 1998). Malaria could become established again under the prolonged pressures of climatic and other environmental-demographic changes if a strong public health infrastructure is not maintained. A particular concern is the reintroduction of malaria in countries of the former Soviet Union with economies in transition, where public health infrastructure has diminished (e.g., Azerbaijan, Russia).

9.7.1.1 Modeling the Impact of Climate Change on Malaria

Classical epidemiological models of infectious disease use the basic reproduction rate, R_0. This measure is defined as the

Box 9-2. Have Recent Increases in Highland Malaria been Caused by Climate Warming?

"Highland malaria" usually is defined as malaria that occurs around its altitudinal limit, exhibiting an unstable fluctuating pattern. There has been considerable debate about the causes of the resurgence of malaria in the African highlands. Early in the 20th century, malaria epidemics occurred at elevations of 1,500–2,500 m in Africa, South America, and New Guinea (Mouchet *et al.*, 1998; Reiter, 1998a). Highland malaria in Africa was effectively controlled in the 1950s and 1960s, mainly through the use of DDT and improved medical care. Important changes that have contributed to the subsequent resurgence include changes in land use, decreasing resources for malaria control and treatment, and population growth and movement (Lindsay and Martens, 1998; Malakooti *et al.*, 1998; Mouchet *et al.*, 1998; Reiter, 1998a). There are insufficient historical data on malaria distribution and activity to determine the role of warming, if any, in the recent resurgence of malaria in the highlands of Kenya, Uganda, Tanzania, and Ethiopia (Cox *et al.*, 1999).

That malaria is sensitive to temperature in some highland regions is illustrated by the effect of El Niño. Increases in malaria have been attributed to observed El Niño-associated warming in highland regions in Rwanda (Loevinsohn, 1994) and Pakistan (Bouma *et al.*, 1996). However, increases in rainfall (sometimes associated with El Niño) also trigger highland epidemics (e.g., Uganda—Lindblade *et al.*, 1999). Lindsay *et al.* (2000) found a reduction in malaria infection in Tanzania associated with El Niño when heavy rainfall may have flushed out Anopheline mosquitoes from their breeding sites.

Most increases in malaria transmission entail single epidemics or a sequence of epidemics that occur over a 1- to 2-year period. Although many epidemics are triggered by transient increases in temperature and/or rainfall, the short time scale of events and the difficulty of linking different epidemics in different parts of the world make it difficult to say if long-term climate change is a factor. Furthermore, there has been little work that identifies where malaria transmission currently is limited by temperature and therefore where highland populations are at risk of malaria as a result of climate change. To determine the role of climate in the increase in highland malaria, a comprehensive research effort is required, together with implementation of a sustainable disease surveillance system that combines trend analyses across multiple sites to account for substantial local factors.

number of new cases of a disease that will arise from one current case when introduced into a nonimmune host population during a single transmission cycle (Anderson and May, 1992). The basic reproduction rate—or a related concept, "vectorial capacity"—can provide a relative index of the impact of different climate scenarios on the transmissibility of vector-borne diseases such as malaria. Vectorial capacity, however, is determined by complex interactions of many host, vector, pathogen, and environmental factors. Some of the variables are sensitive to temperature, including mosquito density, feeding frequency, mosquito survival, and the extrinsic incubation period (EIP) of the parasite (plasmodium) in the mosquito (Martens *et al.*, 1999). The EIP is especially important, and, within the lower temperature range, it is very temperature-sensitive.

Biological (or process-based) models have been used to estimate the potential transmission of malaria. This is a measure of the extent to which the natural world (the global environment-climate complex) would allow the transmission of malaria if there were no other human-imposed constraints on transmission. However, in some areas where human-imposed constraints have occurred as a result of economic growth, or were put in place purposely, malaria transmission has been successfully controlled, regardless of suitable local temperatures. There has been considerable evolution of models since the SAR (Martens *et al.*, 1995, 1997, 1999; Martin and Lefebvre, 1995). One model (Martens *et al.*, 1999) includes vector-specific information regarding the temperature-transmission relationship and mosquito distribution limits. Recent studies using that revised model applied to the HadCM2 climate scenarios project a global increase of 260–320 million people in 2080 living in the potential transmission zone (against a baseline expectation of about 8 billion—that is, a 2–4% increase in the number of people at risk) (Martens *et al.*, 1999; McMichael *et al.*, 2000a). This projection, by design, does not take into account the fact that much of this additional population at risk is in middle- or high-income countries where human-imposed constraints on transmission are greatest and where potential transmission therefore is unlikely to become actual transmission. The model also projects regional increases and a few decreases in the seasonal duration of transmission in current and prospective areas of malaria transmission. Constraining of GHG emissions to achieve CO_2 stabilization within the range 550–750 ppm would reduce those projected increases by about one-third (Arnell *et al.*, 2001).

On a global scale, all biological models show net increases in the potential transmission zone of malaria and changes in seasonal transmission under various climate scenarios (Martens *et al.*, 1995, 1999; Martin and Lefebvre, 1995). Some local decreases in malaria transmission also are predicted to occur where declines in rainfall would limit mosquito survival. The outputs of these malaria models are very sensitive to assumptions about the minimum rainfall or humidity levels needed for malaria transmission.

Another global modeling study (Rogers and Randolph, 2000) used a statistical-empirical approach, in contrast to the aforementioned biological models. The outcome variable in this model is the number of people living in an *actual* transmission zone, as opposed to a *potential* transmission zone (as estimated by biological models). Using an IS92a (unmitigated) climate scenario, this study estimated no significant net change by 2080 in the portion of the world's population living in actual malaria transmission zones; modeled malaria transmission increased in some areas and decreased in others. This study made the assumption that the actual geographic distribution of malaria in today's world is a satisfactory approximation of its historical distribution prior to modern public health interventions. This assumption is likely to have biased the estimation of the underlying multivariate relationship between climatic variables and malaria occurrence because the sensitive climate-malaria relationship in the lower temperature range in temperate zones (especially Europe and the southern United States) would have been excluded from the empirically derived equation. Hence, the use of that derived equation to predict malaria risk in 2080 would have been relatively inert to marginal climatic changes at the fringes of the current geographic distribution.

Another type of modeling addresses changes in the distribution of mosquito vector species only. The CLIMEX model estimates changes in global and national (Australia) distribution of malaria vectors under a range of climate scenarios, based on the vectors' temperature and moisture requirements (Bryan *et al.*, 1996; Sutherst, 1998). The distribution of *Anopheles gambiae* complex is projected to undergo a net increase in distribution in southern Africa under three climate change scenarios (Hulme, 1996). However, these models do not address complex ecological interactions, such as competition between species.

None of these models have been adequately validated at global or regional levels. Modeling to date has not satisfactorily addressed regional vulnerability to malaria or changes in risk in highland regions (Lindsay and Martens, 1998). This is principally because it is difficult to obtain sufficiently detailed geographic distribution maps of mosquitoes and malaria occurrence over time. An important criticism of biological models is that undue emphasis is placed on temperature changes, without consideration to other ecological complexities—including those influenced by rainfall, humidity, and host exposure—that influence transmission dynamics. Furthermore, the equations within a global model may be inappropriate for particular local conditions, and there is a need for cross-validation of large-scale and small-scale studies (Root and Schneider, 1995). Some attempts to apply these integrated modeling techniques to smaller scale regional settings have attempted to take account of local/regional conditions (Lindsay and Martens, 1998). None of the modeling to date has incorporated the modulating effect of public health strategies and other social adaptive responses to current or future malaria risk (Sutherst *et al.*, 1998). Nevertheless, it remains a legitimate and important question to estimate, under scenarios of climate change, change in the extent to which the natural world (the global environment-climate complex) would allow transmission of malaria if there were no other human-imposed constraints on transmission.

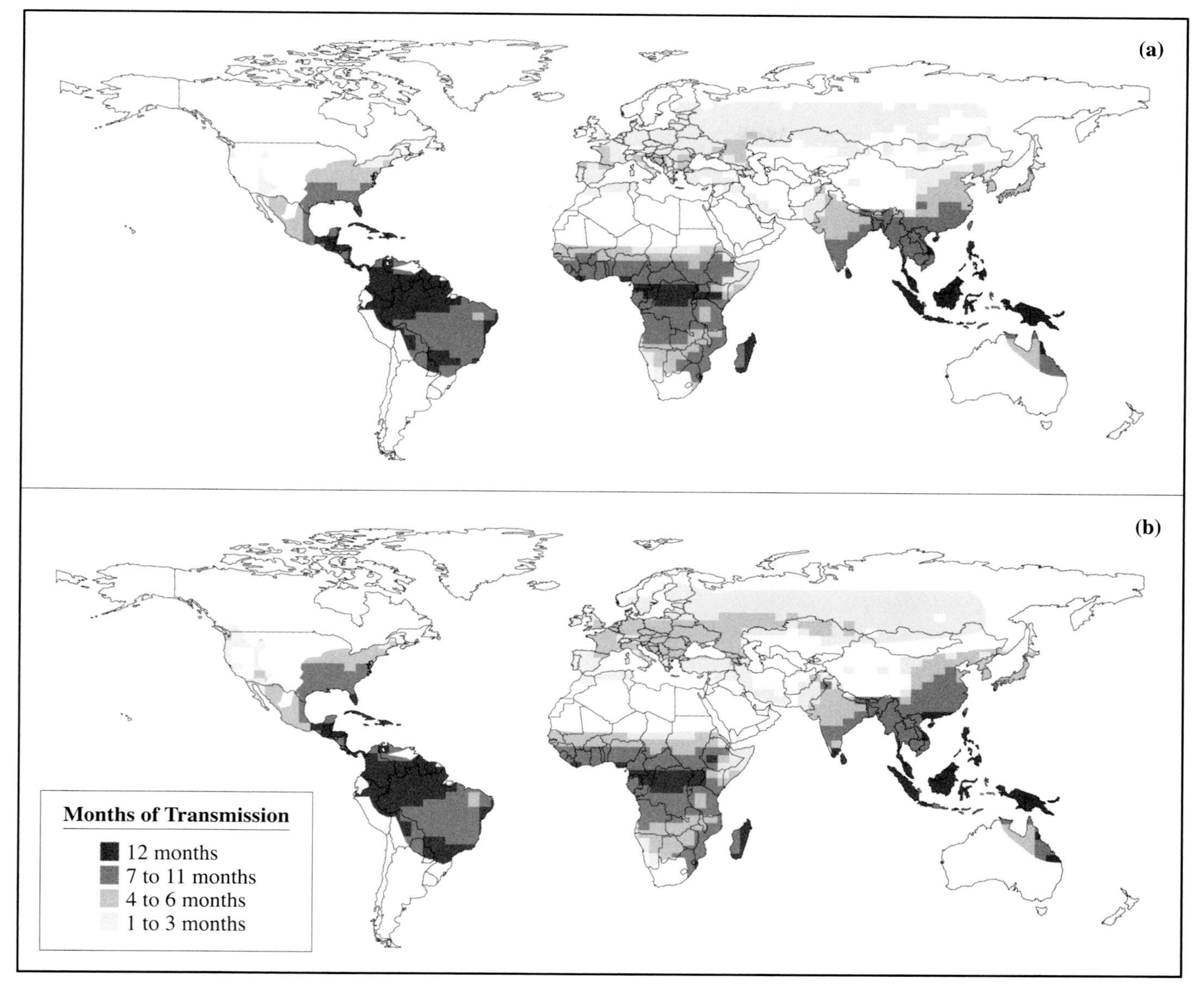

Figure 9-2: Potential impact of climate change on seasonal transmission of falciparum malaria. Output from MIASMA v2.0 malaria model (Martens *et al.*, 1999) indicates the number of months per year when climate conditions are suitable for falciparum transmission and where there is competent mosquito vector: (a) months of potential transmission under current climate (1961–1990); (b) months of potential transmission under a GHG-only climate scenario (HadCM2 ensemble mean) in the 2080s. Future changes in mosquito distributions are not modeled. This model does not take into account control or eradication activities that have significantly limited the distribution of malaria.

9.7.2. Dengue

Dengue is a disease that is caused by four closely related viruses that are maintained in a human-*Aedes aegypti*-human cycle in most urban centers of the tropics (Gubler, 1997). The geographic distribution of the dengue viruses and mosquito vectors *(Aedes aegypti* and *A albopictus*) has expanded to the point that dengue has become a major tropical urban health problem (Gubler, 1997, 1998b). Dengue is primarily an urban disease; more than half of the world's population lives in areas of risk (Gubler, 1997, 1998b). In tropical areas of the world, dengue transmission occurs year-round but has a seasonal peak in most countries during months with high rainfall and humidity. Major factors causing epidemics include population growth, rapid urbanization, lack of effective mosquito control, and movement of new dengue virus strains and serotypes between countries (Gubler, 1997, 1998b).

The global resurgence of dengue in recent years has resulted in increased imported dengue and cases of local transmission in the United States and Australia. As with malaria, the number of cases is small and sporadic (Gubler, 1989, 1997, 1998b). By contrast, Mexican states bordering the United States have had repeated large epidemics of dengue (Gubler, 1989, 1998b; Reiter, 1997; Rawlins *et al.*, 1998). The difference in vulnerability may be caused by differences in living standards and human behavior, which in the United States decrease the probability that vector mosquitoes will feed on humans. It is unlikely that climate change will affect these factors and cause increased epidemic dengue activity in temperate zone developed countries.

9.7.2.1 *Modeling the Impact of Climate Change on Dengue*

To date, all published studies regarding evaluations of the possible impact of global climate change on dengue transmission have involved modification of the standard equation for vectorial capacity (VC) (Jetten and Focks, 1997; Martens *et al.*, 1997; Patz *et al.*, 1998a). Temperature affects the rate of mosquito larval development, adult survival, vector size, and gonotrophic cycle, as well as the EIP of the virus in the vector (Focks *et al.*, 1993a,b, 1995).

Modeling studies (Jetten and Focks, 1997; Martens *et al.*, 1997; Patz *et al.*, 1998a) suggest that a warming projection of 2°C by 2100 will result in a net increase in the potential latitudinal and altitudinal range of dengue and an increase in duration of the transmission season in temperate locations. However, they also ignore the complex epidemiological and ecological factors that influence transmission of dynamics of dengue. Changes in potential transmission in areas that currently are endemic for dengue are projected to be limited. As with malaria, models indicate that the areas of largest change of potential transmission intensity as a result of temperature rise are places where mosquitoes already occur but where development of the virus is limited by temperature during part of the year. However, these models do not incorporate demographic, societal, and public health factors that have been responsible for eliminating dengue from temperate areas. Transmission intensity in tropical endemic countries is limited primarily by herd immunity, not temperature; therefore, projected temperature increases are not likely to affect transmission significantly. Moreover, in subtropical developed areas, where transmission is limited primarily by demographic and societal factors, it is unlikely that the anticipated temperature rise would affect endemicity (Gubler, 1998b).

9.7.3. ***Other Mosquito-Borne Viruses***

Mosquitoes transmit many viruses, more than 100 of which are known to infect humans—causing illness ranging from acute viral syndrome to severe and sometimes fatal encephalitis and hemorrhagic fever. The natural transmission cycles of these viruses are complex and usually involve birds or rodents as well as several mosquito species; each region of the world has its own unique viruses (Gubler and Roehrig, 1998). These viruses have become important global emergent/resurgent public health problems in recent years, causing widespread epidemics (Gubler 1996, 1998a).

Yellow fever—a virus that occurs naturally in the rain forests of Africa and South America in an enzootic cycle involving lower primates and mosquitoes—also can cause major urban epidemics in a cycle involving *Aedes aegypti* that is identical to dengue (Monath, 1988). As such, it has similar weather and climate sensitivity to dengue. Yellow fever was effectively controlled in the 1950s and 1960s through vaccination (Africa) and mosquito control (the Americas). With reinvasion of most large American tropical urban centers by *Aedes aegypti* in the past 30 years (Gubler, 1989), the region is at its highest risk for urban epidemics in 50 years (Gubler, 1998c). Once urban epidemics of yellow fever begin to occur in tropical America, it is expected that this virus will move very quickly via modern transportation to Asia and the Pacific, where it has never occurred (Gubler, 1998c).

Several mosquito-borne viruses cause encephalitis, including eastern equine encephalitis (EEE), western equine encephalitis (WEE), St. Louis encephalitis (SLE), La Crosse encephalitis (LAC), and Venezuelan equine encephalitis (VEE) in the Americas; Japanese encephalitis (JE) in Asia; Murray Valley encephalitis (MVE) and Kuniin (KUN) in Australia; and West Nile (WN) and Rift Valley fever (RVF) viruses in Africa (Gubler and Roehrig 1998). WN virus also occurs in west and central Asia, the Middle East, and Europe and recently was introduced into the United States, where it caused a major epidemic in New York City (Asnis *et al.*, 2000; Komar, 2000). All of these viruses have birds (EEE, WEE, SLE, JE, MVE, KUN, WN) or rodents (LAC, VEE) as natural reservoir hosts. The natural host for RVF is not known, but large ungulates act as amplifying hosts.

Epidemics of these diseases occur when their natural ecology is disturbed in some way (Gubler and Roehrig, 1998). This could include environmental changes such as meteorological changes or forest clearing, changes in population densities and structure of the mosquito or vertebrate host, or genetic changes in the viruses. All of these diseases are very climate-sensitive, but it is difficult to know how climate change will influence their distribution and incidence because of the complexities of their transmission cycles in nature. For example, in the United States, WEE and SLE could expand their geographic distribution northward, and WEE could disappear from most of the country (Reeves *et al.*, 1994). Climate change also may have an effect on endemic/enzootic arboviruses in Australia (Russell, 1998; Tong *et al.*, 1998; Bi *et al.*, 2000). Thus, there probably would be positive and negative impacts, depending on the disease.

Floods may cause an immediate decrease in mosquito populations because of loss of breeding sites. However, disease risk may rise as floodwaters recede and vector populations increase, but only if the virus is present (Nasci and Moore, 1998). This underscores the need to have effective surveillance systems and prevention strategies in place to monitor disease and control vector activity, as well as the need for more research on the transmission dynamics of vector-borne diseases.

9.7.4. ***Leishmaniasis***

There are two principal clinical types of leishmaniasis—visceral and cutaneous—which is caused by a range of species of *Leishmania* parasites. The parasites are transmitted by sandflies,

Table 9-3: Temperature thresholds of pathogens and vectors. T_{min} is minimum temperature required for disease transmission. T_{max} for the pathogen is upper threshold beyond which temperatures are lethal. T_{max} for vectors are not provided. Temperatures are in degrees Celsius. Note that temperatures assume optimum humidity; vector survival decreases rapidly as dryness increases. There is considerable variation in these thresholds within and between species (Purnell, 1966; Pfluger, 1980; Curto de Casas and Carcavallo, 1984; Molineaux, 1988; Rueda et al., 1990).

Disease	Pathogen	T_{min}	T_{max}	Vector	T_{min} for Vector
Malaria	*Plasmodium falciparum*	16–19	33–39	*Anopheles*	8–10 (biological activity)
Malaria	*Plasmodium vivax*	14.5–15	33–39	*Anopheles*	8–10 (biological activity)
Chagas' disease	*Trypanosoma cruzi*	18	38	*Triatomine* bugs	2–6 (survival) 20 (biological activity)
Schistosomiasis	Cercaria	14.2	>37	Snails (*Bulinus* and others)	5 (biological activity) 25±2 (optimum range)
Dengue fever	Dengue virus	11.9	not known	*Aedes*	6–10
Lyme disease	*Borrelia burdorferi*	Not yet determined		*Ixodes* ticks	5–8

of which the two most important genera are *Phlebotomus* in Europe and Asia and *Lutzomyia* in the Americas. In central Asia and Europe, leishmaniasis has become an important co-infection with human imunodeficiency virus (HIV) (Alvar *et al.*, 1997; WHO/UNAIDS, 1998). Sandflies are very sensitive to temperature, and increases in temperature also may increase daily mortality rates. *Phlebotominae* are sensitive to sudden temperature changes and prefer regions with small differences between maximum and minimum temperatures. Thomson *et al.* (1999) mapped *P. orientalis* in Sudan and found that the geographic distribution was best explained by mean annual maximum daily temperature and soil type. One study on leishmaniasis in Italy indicates that climate change may expand the range of one vector (*P. perniciosus*) but decrease the range of another (*P. perfiliewi*) (Kuhn, 1997). A 3°C increase in temperature could increase the geographic and seasonal distribution of *P. papatasi* in southwest Asia, provided other ecological requirements are met (Cross and Hyams, 1996; Cross *et al.*, 1996).

The southern limit of leishmaniasis and vectors in South America is the extreme north of Argentina (Curto de Cassas and Carcavallo, 1995; Marcondes *et al.*, 1997). There have been no systematic studies of the relationship between climate parameters and vectors or human cases in the Americas. Climate change could affect the geographical distribution of these vector species in Brazil, Paraguay, Bolivia, and Argentina (Carcavallo and Curto de Casas, 1996).

9.7.5. Schistosomiasis

Schistosomiasis, which is caused by five species of the trematode (flat worm) *Schistosoma*, requires water snails as an intermediate host. Worldwide prevalence has risen since the 1950s largely as a result of expansion of irrigation systems in hot climates where viable snail populations can survive and the parasite can find human parasite carriers (Hunter *et al.*, 1993). All three genera of snail hosts (*Bulinus*, *Biomphalaria,* and *Oncomelania*) can tolerate a wide temperature range. At low temperatures, snails are effectively dormant and fecundity is virtually zero, but survival is good. At high temperatures, births (egg production) increase, but so does mortality (Table 9-3). However, snails are mobile and can move to avoid extreme temperatures within their habitats; water can act as an efficient insulator (Hairston, 1973; Gillett, 1974; Schiff *et al.*, 1979). The precise conditions within water bodies that determine transmission depend on a host of environmental factors, including local geology and topography, the general hydrology of the region, the presence or absence of aquatic vegetation, and local agricultural usage (Appleton and Stiles, 1976; Appleton, 1977). In east Africa, colonies of *Biomphalaria* and *Bulinus spp*. persist at altitudes of 2,000 m or more, but transmission—if it occurs at all—is restricted to brief warm seasons. Climate change might allow schistosomiasis transmission to extend its range to higher altitudes. Conversely, increasing temperatures at sea level could decrease transmission unless the snails move to cooler refuges.

Water shortages resulting from climate change could create greater need for irrigation, particularly in arid regions. If irrigation systems expand to meet this need, host snail populations may increase (Schorr *et al.*, 1984), leading to greater risk of human infection with the parasite. However, this impact could be reduced by constructing irrigation systems that are not conducive to snail breeding.

9.7.6. Chagas' Disease

The geographical distribution of American trypanosomiasis (Chagas' disease) is limited to the Americas, ranging from the southern United States to southern Argentina and Chile (Carcavallo *et al.*, 1998, 1999). Chagas' disease is transmitted by triatomine bugs (see Table 9-3). Temperature affects the major components of VC (reviewed by Zeledón and Rabinovich, 1981; Carcavallo, 1999). If temperatures exceed 30°C and humidity does not increase sufficiently, the bugs increase their feeding rate to avoid dehydration. If indoor temperatures rise, vector species in the domestic environment may develop shorter life cycles and higher population densities (Carcavallo and Curto de Casas, 1996). High temperatures also accelerate development of the pathogen, *Trypanosoma cruzi*, in the vector (Asin and Catalá, 1995). Many vector species are domesticated. Lazzari *et al*. (1998) found that in the majority of structures, differences between inside and outside temperature were small, although differences in humidity were significant. Triatomine dispersal also is sensitive to temperature (Schofield *et al.*, 1992). Population density of domestic vectors also is significantly affected by human activities to control or eradicate the disease (e.g., replastering of walls, insecticide spraying). The southern limits of *Triatoma infestans* and Chagas' disease distributions recently have been moved significantly inside their climatically suitable limits by large-scale control campaigns (Schofield and Dias, 1999).

9.7.7. Plague

Plague is a bacterial disease that is transmitted by the bite of infected fleas (*Xenopsylla cheopis*), by inhaling infective bacteria, and, less often, by direct contact with infected animals (Gage, 1998). Plague exists focally in all regions except Europe. Notable plague outbreaks have occurred in several Asian, African, and South American countries in the past 10 years (John, 1996; WHO, 1997; Gage, 1998; PAHO, 1998). It is unclear whether climate change may affect the distribution and incidence of plague. There does appear to be a correlation between rainfall patterns and rodent populations (Parmenter *et al.*, 1999; see also Section 9.7.9). Prospective field research studies must be conducted to confirm this.

9.7.8. Tick-Borne Diseases

Tick-borne diseases—in particular, Lyme disease, Rocky Mountain spotted fever, ehrlichiosis, and tick-borne encephalitis (TBE)—are the most common vector-borne diseases in temperate zones in the northern hemisphere. Ticks are ectoparasites; their geographical distribution depends on the distribution of suitable host species—usually mammals or birds (Glass *et al.*, 1994; Wilson, 1998). Species that transmit these diseases have complex life cycles that require 3 years and three different hosts species—one for each stage of the cycle (larvae, nymph, and adult). Climate directly and indirectly influences the tick vector, its habitat, host and reservoir animals, time between blood meals, and pathogen transmission. Bioclimatic threshold temperatures set limits for tick distribution and are of importance for the magnitude of disease occurrence (Table 9-3). Temperatures must be sufficiently high for completion of the tick's life cycle. Humidity must be sufficient to prevent tick eggs from drying out. Temperatures above the optimum range reduce the survival rate of ticks. In temperate countries, tick vectors are active in the spring, summer, and early autumn months.

Over the past 2 decades, marked increases have been reported in the abundance of ticks and the incidence of tick-borne disease in North America and Europe. In North America, these changes have been attributed to an increase in awareness of tick-borne diseases and increased abundance of wild tick hosts (principally deer), as reforestation has expanded areas of suitable habitat (Dennis, 1998). There is some evidence that the northern limit of distribution of the tick vector (*Ixodes ricinus*) and tick density increased in Sweden between the early 1980s and 1994, concurrent with an increased frequency of milder winters (Talleklint and Jaenson, 1998; Lindgren *et al.*, 2000). In New York state, *Ixodes scapularis* has expanded its geographic distribution northward and westward in the past 10 years. The reasons for this expansion are unknown.

9.7.8.1. Lyme Disease

Lyme disease is caused by infection with the spirochete *Borrelia burgdorferi*. It is transmitted by ticks of the *Ixodes ricinus* complex (Dennis, 1998). Lyme disease has a global distribution in temperate countries of North America, Europe, and Asia. The transmission cycle of Lyme disease involves a range of mammalian and avian species, as well as tick species—all of which are affected by local ecology. Under climate change, a shift toward milder winter temperatures may enable expansion of the range of Lyme disease into higher latitudes and altitudes, but only if all of the vertebrate host species required by the tick vector also are able to expand their distribution. A combination of milder winters and extended spring and autumn seasons would be expected to prolong seasons for tick activity and enhance endemicity, but this would not be expected to change disease activity because humans usually are infected by the nymphal stage, which feeds at a specific time during the second year of the cycle.

9.7.8.2. Tick-Borne Encephalitis

Tick-borne encephalitis (TBE) is caused by two closely related but biologically distinct viruses (Gubler and Roehrig, 1998). The eastern subtype is transmitted by *Ixodes persulcatus* and causes Russian spring-summer encephalitis. It occurs from China to eastern Europe and is highly focal in its distribution. The western subtype is transmitted by *Ixodes ricinus* and causes central European encephalitis, a milder form of the disease. It occurs within discrete foci from Scandinavia in the north to Croatia in the south, with only occasional cases further south. A related virus, Powassan, occurs in Canada and the United States and is transmitted by *Ixodes scapularis*. Humans usually become infected when they are exposed to ticks in habitats where the viruses are maintained. The viruses also may be transmitted directly through ingestion of raw goat milk.

It is possible that warming would extend the transmission season for TBE in Europe. The aforementioned study showed a northward extension of the tick population in Sweden in association with warmer winters, accompanied by an increase in the annual number of cases of tick-borne encephalitis reported within Sweden. Most transmission to humans is by the nymphal ticks, each of which feeds for a few days during spring-summer before dropping to the ground and molting to adult ticks, which feed primarily on deer and other large mammals. All tick stages have well-defined seasons of feeding activity, which vary geographically and may be prolonged in regions with mild winters.

Unlike Lyme disease, sustainable transmission of TBE requires a high level of coincident feeding of larval and nymphal ticks. This seasonal synchrony depends on a particular seasonal profile of land surface temperature—specifically, a rapid rate of cooling in the autumn (Randolph *et al.*, 2000). Synchrony may be disrupted by climate change as patterns of overwinter development by ticks are changed. A statistical model, based on the current distribution of TBE, indicates significant net contraction in the geographic distribution of TBE under mid-range climate scenarios by the 2050s (Randolph and Rogers, 2000). The model indicates that although disease foci spread to higher latitudes and altitudes, current foci in central Europe largely disappear as a result of disruption of the tick seasonal dynamic by climate change. Thus, one model suggests that it is unlikely that warming would increase the incidence or net geographic distribution of TBE in Europe.

9.7.9. Rodent-Borne Diseases

Rodent-borne diseases are zoonoses that are transmitted directly to humans by contact with rodent urine, feces, or other body fluids (Mills and Childs, 1998; Peters, 1998). Rodents are principle hosts for arthropod vectors such as fleas (see Section 9.7.7) and ticks (see Section 9.7.8). Environmental factors that affect rodent population dynamics include unusually high rainfall, drought, and successful introduction of exotic plant species. Rodent-borne pathogens are affected indirectly by ecological determinants of food sources that affect rodent population size (Williams *et al.*, 1997; Engelthaler *et al.*, 1999).

9.7.9.1. *Leptospirosis*

Leptospirosis is an acute febrile disease caused by the bacteria *Leptospira*. It probably is the most widespread zoonotic disease in the world and is particularly common in the tropics (PAHO, 1998). Infection is caused by exposure to water, damp soil, or vegetation contaminated with the urine of infected wild and domestic animals (e.g., rodents and dogs) (Thiermann, 1980). Outbreaks often occur after heavy rainfall and during floods (Kriz *et al.*, 1998; Trevejo *et al.*, 1998). Therefore, any increase in flooding associated with climate change may affect the incidence of this disease.

9.7.9.2. *Hantaviruses*

Several hantaviruses are capable of causing severe, often fatal, illness in humans (PAHO, 1998). Each has a specific geographic distribution that is determined by that of the primary rodent host (Schmaljohn and Hjelle, 1997). Humans are infected by aerosol exposure to infectious excreta or occasionally by bites. The better known of these diseases are hemorrhagic fever with renal syndrome, caused by Hantaan virus, in China and Korea and hantavirus pulmonary syndrome in the Americas, caused by several viruses that are specific to their rodent host (Schmaljohn and Hjelle, 1997). Outbreaks of disease may be associated with weather that promotes rapid increases in rodent populations, which may vary greatly between seasons and from year to year (Glass *et al.*, 2000). Many hantavirus infections occur in persons of lower socioeconomic status, where poorer housing and agricultural activities favor closer contact between humans and rodents (Schmaljohn and Hjelle, 1997). Arenaviruses (Lassa, Junin, Machupo, etc.), which are ecologically similar to hantaviruses, may respond similarly (Mills and Childs, 1998).

9.7.10. Water-Related Infectious Diseases

There are complex relationships between human health and problems of water quality, availability, sanitation, and hygiene. Predicting the potential impacts of climate change on water-related diseases therefore is difficult because access to a clean safe water supply is determined primarily by socioeconomic factors. Extreme weather—floods or droughts—can increase the risk of disease via contamination of water resources, poor hygiene, or other mechanisms. Currently, the World Health Organization (WHO) estimates that more than 1 billion people worldwide are without access to safe drinking water and that every year as many as 4 million die prematurely because they do not have access to safe drinking water and sanitation. Increases in water stress are projected under climate change in certain countries (see Chapter 4), but it is difficult to translate such indicators directly into the attributable risk for water-related diseases. Water scarcity may necessitate use of poorer quality sources of freshwater, such as rivers, which often are contaminated. Decreases in water supplies could reduce the water available for drinking and washing and lower the efficiency of local sewerage systems, leading to increased concentration of pathogenic organisms in raw water supplies.

Excessive precipitation can transport terrestrial microbiological agents into drinking-water sources. For example, some outbreaks of cryptosporidiosis, giardia, and other infections have been triggered by heavy rainfall events in the UK and United States (Lisle and Rose, 1995; Atherholt *et al.*, 1998; Rose *et al.*, 2000; Curriero *et al.*, 2001). Significant correlation between the cumulative monthly distribution of cholera cases and the monthly distribution of precipitation has been observed in Guam (Borroto and Haddock, 1998). In many countries, handling of sewage is not separate from the drainage system for stormwaters. It is important that water resource management can adapt to changes in the frequency of precipitation extremes to minimize the risk of microbiological contamination of the public water supply.

Cholera is a water- and food-borne disease and has a complex mode of transmission. In tropical areas, cases are reported year-round. In temperate areas, cases are reported mainly in the warmest season. The seventh cholera pandemic currently is spreading across Asia, Africa, and South America. A new serogroup (*V. cholerae* O139) appeared in 1992 and is responsible for large epidemics in Asia. During the 1997–1998 El Niño, excessive flooding caused cholera epidemics in Djibouti, Somalia, Kenya, Tanzania, and Mozambique (WHO, 1998b). Birmingham *et al.* (1997) found a significant association between bathing and drinking water from Lake Tanganyika and the risk of infection with cholera. Warming in the African Great Lakes may cause conditions that increase the risk of cholera transmission in the surrounding countries (WHO, 1998b). See Section 9.8 for a discussion of cholera in coastal waters.

9.7.11. Other Infectious Diseases

Water- and food-borne diseases tend to show marked seasonality, with peaks in early spring or summer. Higher temperatures favor microorganism proliferation and often are associated with an increase in gastrointestinal infections. Above-average temperatures in Peru during the 1997–1998 El Niño were associated with a doubling in the number of children admitted to the hospital with diarrhea (Checkley *et al.*, 2000). Higher temperatures also can trigger spore maturation (e.g., *Cyclospora cayetanensis*—Ortega *et al.*, 1993; Smith *et al.*, 1997). In Peru, the incidence of cyclosporosis peaks in the summer months (Madico *et al.*, 1997). Because climate change is expected to entail warmer springs and summers, additional cases of food-borne disease may occur, if current trends continue (Bentham and Langford, 1995). In most developed countries, food-borne disease incidence is increasing as a result of changes in behavior, consumption patterns, and commerce.

Major epidemics of meningococcal infection usually occur every 5–10 years within the African "meningitis belt;" they usually start in the middle of the dry season and end a few months later with the onset of the rains (Greenwood *et al.*, 1984). Between February and April 1996, the disease affected thousands of people in parts of northern Nigeria, many of whom died (Angyo and Okpeh, 1997). The epidemic spread from the original meningitis belt to Kenya, Uganda, Rwanda, Zambia, and Tanzania (Hart and Cuevas, 1997). One of the environmental factors that predispose to infection and epidemics is low humidity (Tikhoumirov *et al.*, 1997). However, a climate-meningitis association was not clear in parts of the Gulf of Guinea (Besancenot, 1997). The fact that this disease has been limited to semi-arid areas of Africa suggests that its transmission could be affected by warming and reduced precipitation.

Warm and humid conditions can promote fungal skin infections such as sporotrichosis (Conti Diaz, 1989). Decreases in humidity can lead to increased dispersion of particulate fungal spores, thereby increasing the risk of pneumonia caused by coccidioidomycosis (Durry *et al.*, 1997; Schneider *et al.*, 1997).

9.8. Coastal Water Issues

Pathogens often are found in coastal waters; transmission occurs though shellfish consumption or bathing. Coastal waters in developed and developing countries frequently are contaminated with untreated sewage. Higher temperatures encourage microorganism proliferation. The presence of *Vibrio spp*. (some of which are pathogens that cause diarrhea) has been associated with higher sea-surface temperature (SST) (Lipp and Rose, 1997). *Vibrio vulnificus* is a naturally occurring estuarine bacterium that may be more often transmitted to humans under conditions of higher SST (Patz *et al.*, 2000).

Acute poisoning can occur following consumption of fish and shellfish contaminated with biotoxins (WHO, 1984). Phytoplankton organisms respond rapidly to changes in environmental conditions and therefore are sensitive biological indicators of the combined influences of climate change and environmental change (Harvell *et al.*, 1999). Algal blooms are associated with several environmental factors, including sunlight, pH, ocean currents, winds, SSTs, and runoff (which affects nutrient levels) (Epstein *et al.*, 1993; NRC, 1999). Algal blooms can be harmful to fish and other aquatic life, often causing severe economic damage, and are reported to have increased globally in the past several decades (Hallegraeff, 1993; Sournia, 1995), although some of the observed increase is attributed to changes in monitoring, effluent, and land use.

There is no straightforward relationship between the presence of an algal bloom and an outbreak of poisoning. Human poisoning can occur in the absence of a bloom. Two main types of biotoxin poisoning are associated with temperate climates and colder coastal waters: paralytic shellfish poisoning and diarrheic shellfish poisoning. If water temperatures rise as a result of climate change, shifts in the distribution of these diseases could follow. Biotoxins associated with warmer waters, such as ciguatera in tropical waters, could extend their range to higher latitudes (Tester, 1994). An association has been found between ciguatera (fish poisoning) and SST in some Pacific islands (Hales *et al.*, 1999a).

Recent evidence suggests that species of copepod zooplankton provide a marine reservoir for the cholera pathogen and facilitate its long-term persistence in certain regions, such as the estuaries of the Ganges and Bramaputra in Bangladesh (Colwell, 1996). The seasonality of cholera epidemics may be linked to the seasonality of plankton (algal blooms) and the marine food chain. Studies using remote-sensing data have shown a correlation between cholera cases and SST in the Bay of Bengal (Lobitz *et al.*, 2000). Interannual variability in cholera incidence in Bangladesh also is linked to ENSO and regional temperature anomalies (Pascual *et al.*, 2000). Epidemiological evidence further suggests a widespread environmental cause of the 1991 epidemic in Peru, rather than point-source contamination (Seas *et al.*, 2001). There is some evidence for a link between warmer sea surfaces and cholera risk in the Bay of Bengal, but it is not possible to extrapolate such findings to cholera incidence inland or in other regions. The potential impact of long-term climate warming on cholera incidence or risk of epidemics remains uncertain.

Climate-related ecological changes may enhance primary and secondary transmission of cholera in developing countries, particularly among populations settled in low-lying coastal areas in the tropics. However, the causal link between sea temperature, plankton blooms, and human disease requires further elucidation and confirmation.

9.9. Food Yields and Nutrition

Background climate and annual weather patterns are key factors in agricultural productivity, despite technological advances such as improved crop varieties and irrigation systems. As temperature, rainfall, and soil moisture change, plant physiology is affected; so too is the much less predictable risk of a change in patterns of plant pests and pathogens. There are many social, economic, and environmental influences on agricultural, horticultural, and livestock productivity. Climate change represents an additional pressure on the world food supply system. That system, which has yielded an overall increase in per capita food supplies over the past 4 decades, has shown signs of faltering over the past decade. There is ongoing scientific debate about the relative importance of economic, technical, and ecological influences on current food yields (Waterlow *et al.*, 1998; Dyson, 1999). Optimists point to falling food prices; pessimists point to falling soil fertility.

Modeling studies (reviewed in Chapter 5) indicate that, under climate change, yields of cereal grains (the world's dominant food commodity) would increase at high and mid-latitudes but decrease at lower latitudes. Furthermore, this disparity would become more pronounced as time progresses. The world's food system may be able to accommodate such regional variations at the global level, with production levels, prices, and the risk of hunger relatively unaffected by the additional stress of climate change. To minimize possible adverse consequences, a dual development program is desirable. Adaptation should be undertaken via continued development of crop breeding and management programs for heat and drought conditions. These will be immediately useful in improving productivity in marginal environments today. Mitigation strategies should be implemented to try to reduce further enhanced global warming. However, recent work suggests that the main benefits of mitigation will not accrue until late in the 21st century (Parry *et al.*, 1998).

The United Nations Food and Agriculture Organization (FAO) estimates that in the late 1990s, 790 million people in developing countries did not have enough to eat (FAO, 1999). The FAO report on food insecurity has identified population groups, countries, and regions that are vulnerable. For example, nearly half the population in countries of central, southern, and east Africa are undernourished. Environmental factors, including natural factors and those that are a consequence of human activities, can limit agricultural potential. These factors include extremely dry or cold climates, poor soil, erratic rainfall, steep slopes, and severe land degradation. The FAO report further states that undernutrition and malnutrition prevail in regions where environmental, economic, and other factors expose populations to a high risk of impoverishment and food insecurity.

Undernutrition is a fundamental cause of stunted physical and intellectual development in children, low productivity in adults, and susceptibility to infectious disease in everyone. Decreases in food production and increases in food prices associated with climate change would increase the number of undernourished people. Conversely, if food production increases and food prices decrease, the number of undernourished people would fall, but populations in isolated areas with poor access to markets still may be vulnerable to locally important decreases or disruptions in food supply.

9.10. Demographic and Economic Disruption

Health impacts associated with population displacement fall under two general categories: health impacts resulting from the new ecological environment and health impacts resulting from the living environment in refugee camps (Prothero, 1994). Even displacement from longer term cumulative environmental deterioration is associated with such health impacts. Cumulative changes that may cause population displacement include land degradation, salinity, deforestation, waterlogging, desertification, and water scarcity. When pastoralists in west Africa were forced to move because of reduced pasture and water, they were faced with new ecological conditions. They experienced psychological stress and were more at risk of infectious diseases (Stock, 1976; Prothero, 1994). Climate change may affect human security via changes in water supplies and/or agricultural productivity (Lonergan, 1998, 1999). An increase in the magnitude and frequency of extreme events also would be disruptive to political stability.

Immediate environmental catastrophes can force sudden displacement of a population. In these cases, adverse health impacts usually result from living in refugee camps in overcrowded, poor accommodations with inadequate food, water supplies, sanitation, and waste disposal (Shears *et al.*, 1985; Noji, 1997). These conditions predispose people to parasitic and communicable diseases such as malaria and cholera, respiratory infections, intestinal disorders, malnutrition, and psychological stress (Prothero, 1994).

The potential impacts of sea-level rise on the health and well-being of coastal populations are an important consideration (Klein and Nicholls, 1999). Estimates of the potential number of people at risk from sea-level rise are addressed elsewhere in TAR WGI and this volume. For example, a 0.5-m rise in sea level along the Nile delta would flood 32% of urban areas, resulting in a significant loss of shelter and forced migration (El-Raey *et al.*, 1999; see Chapter 6). In some locations, sea-level rise could disrupt stormwater drainage and sewage disposal and result in salinization of freshwater supplies. It can affect health indirectly by reducing food production—for example, by reducing rice production in low-lying coastal rice paddies. Sea-level rise also could affect the distribution of vector-borne diseases—for example, some of the coastal wetlands of the United States may be flooded, thereby destroying the habitat of the EEE virus. Populations with limited economic, technical, and social resources have increased vulnerability to various infectious, psychological, and other adverse health consequences.

9.11. Adaptation Options

Adaptation measures can be used effectively to greatly reduce many of the potential health impacts of climate change (Gubler, 1998d; McMichael and Kovats, 2000; WHO, 2000). The most important, cost-effective, and urgently needed measure is to rebuild public health infrastructure. In very many countries of the world, this infrastructure has declined in recent years. Many diseases and public health problems that otherwise may be exacerbated by climate change could be prevented substantially or completely with adequate financial and public health resources. These resources would encompass public health training programs, research to develop and implement more effective surveillance and emergency response systems, and sustainable prevention and control programs.

Understanding vulnerability to changes in ranges or rates of diseases is the first step in addressing adaptive capacity. Adaptation involves the ability to change behavior or health infrastructure to reduce these potential negative impacts or increase potential positive impacts of climate change. Interventions early in the causal chain of disease are preferred (e.g., "primary" prevention to remove or reduce risks before any human cases occur). To the extent that this is not always feasible (or the risk factors unknown), "secondary prevention" or surveillance for early warning to prevent any further cases also is important.

Adaptation is a function of several societal systems, including access to financial resources (for individuals and populations), technical knowledge, public health infrastructure, and the capacity of the health care system. Note that there is much similarity in the determinants of adaptive capacity and those of vulnerability (see Section 9.3). Adaptation can occur via two routes: autonomous adaptation, which is the natural or spontaneous response to climate change by affected individuals, and purposeful adaptation, which is composed of planned responses to projected climate change—typically by governmental or other institutional organizations (MacIver and Klein, 1999). Purposeful adaptation also can occur via deliberate modification of personal, family, and community lifestyles, particularly in response to public education programs. Anticipatory adaptations are planned responses that take place in advance of climate change.

Adaptation to the impacts of climate change may occur at the population, community, or personal level (see Table 9-4). The capacity to adapt to potential changes in the climate will depend on many factors, including improving the current level of public health infrastructure; ensuring active surveillance for important diseases; and continuing research to further our understanding of associations between weather, extreme events, and vector-borne diseases. In addition, continuing research into medical advances required for disease prevention, control, and treatment—such as vaccines, methods to deal with drug-resistant strains of infectious agent, and mosquito control—is needed. More generally, research is needed to identify adaptation needs, evaluate adaptation measures, assess their environmental and health implications, and set priorities for adaptation strategies. The following subsections outline adaptive measures that have been developed for two areas of climate change impacts on health.

9.11.1. Extreme Events and Natural Disasters

Major impacts on human health may occur via changes in the magnitude and frequency of extreme events (see Table 3-10 and TAR WGI Chapter 9). Following Hurricanes George and Mitch, a range of policies to reduce the impacts of such extreme events has been identified (PAHO, 1999):

- Undertaking vulnerability studies of existing water supply and sanitation systems and ensuring that new systems are built to reduce vulnerability
- Developing improved training programs and information systems for national programs and international cooperation on emergency management
- Developing and testing early warning systems that should be coordinated by a single national agency and involve vulnerable communities providing and evaluating mental health care, particularly for those who may be particularly vulnerable to the adverse psychosocial effects of disasters (e.g., children, the elderly, and the bereaved).

Adaptation strategies to reduce heat-related mortality in vulnerable cities around the world include weather-based early warning systems (WMO, 1997; Ortiz *et al.*, 1998). A different system must be developed for each city, based on that city's specific meteorology. Specific weather/health thresholds are determined and used to call health warnings or advisories. Many systems are based on synoptic methodology; specific "offensive" air masses are identified and forecasts are developed to determine if they will intrude into a city within the next 60 hours. Two systems are under construction for Rome, Italy, and Shanghai (WMO, 1997).

Institutional and cultural barriers to the use of seasonal forecast information remain. Decisionmakers should be educated or encouraged to use scientific information that may lead to reduction in losses from natural disasters (Pfaff *et al.*, 1999).

Table 9-4: Options for adaptation to reduce health impacts of climate change.

Health Outcome	Legislative	Technical	Educational-Advisory	Cultural and Behavioral
Thermal stress	– Building guidelines	– Housing, public buildings, urban planning to reduce heat island effects, air conditioning	– Early warning systems	– Clothing, siesta
Extreme weather events	– Planning laws – Building guidelines – Forced migration – Economic incentives for building	– Urban planning – Storm shelters	– Early warning systems	– Use of storm shelters
Air quality	– Emission controls – Traffic restrictions	– Improved public transport, catalytic converters, smokestacks	– Pollution warning	– Carpooling
Vector-borne diseases		– Vector control – Vaccination, impregnated bednets – Sustainable surveillance, prevention and control programs	– Health education	– Water storage practices
Water-borne diseases	– Watershed protection laws – Water quality regulation	– Genetic/molecular screening of pathogens – Improved water treatment (e.g., filters) – Improved sanitation (e.g., latrines)	– Boil water alerts	– Washing hands and other hygiene behavior – Use of pit latrines

9.11.2. Malaria Epidemics

Malaria prevention illustrates approaches to adaptation that also apply to other vector-borne disease threats. To reduce the increased risks of malaria, human populations must take adaptive measures to diminish the impacts. Although malaria epidemics can be triggered by changes in meteorological or socioeconomic conditions, many health services fail to monitor these variables because indicators of risk for epidemic-prone areas have not been determined (Najera *et al.*, 1998). Malaria surveillance and epidemic preparedness may benefit from recently developed tools that predict the seasonality and risks of epidemics by using satellite or ground-based meteorological data (e.g., Hay *et al.*, 1998; Patz *et al.*, 1998b). New approaches to mapping the distribution of malaria vectors over large areas may facilitate species-specific vector control activities. It has been shown in western Kenya that the risk of malaria transmission in the highlands can be predicted with a simple rainfall- and temperature-dependent predictive model (Githeko *et al.*, 2000).

Epidemics are focal in nature and often may be controlled by limited application of safe and effective residual insecticides. Parasite resistance to antimalarials is a threat to malaria control programs; therefore, it is essential that drug sensitivity is reviewed regularly. At the personal level, insecticide-protected fabrics (e.g., bednets) have been shown to be effective against infective mosquito bites (Legeler, 1998).

9.12. Secondary Health Benefits of Mitigation Policies

Actions taken to reduce GHG emissions are very likely to benefit population health (Wang and Smith, 1999; WHO, 1999c; OECD, 2000; see also TAR WGIII Chapter 9). Fossil fuel combustion releases local hazardous air pollutants (especially particulates, ozone, nitrogen oxides, and sulfur dioxide) and GHGs. Hence, policies to reduce GHG emissions via reductions in vehicle exhausts or an increase in the efficiency of indoor household cookstoves would yield great benefits to health (see also TAR WGIII Section 9.2.8.4). Controlling road traffic also would benefit health through reductions in road traffic accidents—a leading cause of death worldwide (Murray and Lopez, 1996).

The benefits to health from mitigation are highly dependent on the technologies and sectors involved. A study by Wang and Smith (1999) indicates that a significant number of premature

Box 9-3. Understanding El Niño Can Help Adaptation to Climate Change: Seasonal Climate Forecasting

There is evidence of an association between El Niño and epidemics of vector-borne diseases such as malaria and dengue in some areas where El Niño affects the climate (Kovats *et al.*, 1999). Malaria transmission in unstable areas is particularly sensitive to changes in climate conditions, such as warming or heavy rainfall (Akhtar and McMichael, 1996; Gupta, 1996; Najera *et al.*, 1998). In Venezuela and Colombia, malaria morbidity and mortality increases in the year following the onset of El Niño (Bouma and Dye, 1997; Bouma *et al.*, 1997b; Poveda *et al.*, 2000). ENSO also has been shown to affect dengue transmission in some Pacific islands (Hales *et al.*, 1999b), though not in Thailand (Hay *et al.*, 2000). However, in many of the studies that have found a relationship between El Niño and disease, the specific climate drivers or mechanisms have not been determined. There also are other climate oscillations that are less well studied. Furthermore, there are other important explanations of cyclic epidemics, such as changes in herd immunity (Hay *et al.*, 2000).

The ENSO phenomenon provides opportunities for early warning of extreme weather, which could improve epidemic preparedness in the future. Seasonal forecasting methods and information have the potential to be used to far greater effect by the health sector (IRI, 1999; Kovats *et al.,* 1999). In addition to these direct applications, attention to the impacts of interannual climate variability associated with the ENSO phenomenon would help countries develop the necessary capacity and preparedness to address longer term impacts associated with global climate change (Hales *et al.*, 2000). On the other hand, there are limitations to using ENSO interannual climate variability to assess potential impacts of long-term climate change.

deaths can be prevented via reductions in particulate emissions in the household sector (i.e., domestic fuel use) in China. The Working Group on Public Health and Fossil Fuel Combustion (1997) estimates that a worldwide reduction in outdoor exposure to particulate matter (PM10), under a Kyoto-level (but global) emissions mitigation scenario, would avert 700,000 premature deaths annually by 2020 compared to a business-as-usual scenario. This figure, however, can be regarded only as indicative, given the broad assumptions and many uncertainties that underlay the estimation. Large numbers of people lack access to clean energy. Renewable energy sources—particularly solar and wind—could help provide this much needed energy while minimizing GHG emissions and maximizing health gain (Haines and Kammen, 2000).

9.13 Research and Information Needs, including Monitoring

Research on the health impacts of global climate change should be conducted within an international network of scientists. Climatic-environmental changes will vary by geographic location, and local populations vary in their vulnerability to such changes. Therefore, the patterns of health gains and losses will be very context-dependent. This type of research requires maximum exchange of information and cross-fertilization of ideas and techniques among scientists, agencies, and institutes. In particular, forecasting the likely health outcomes of exposure to future climate-environmental scenarios requires development of predictive models that can integrate across disparate systems. This will require an interdisciplinary approach. There is an urgent need to focus research efforts more sharply. Particular tasks include:

- Epidemiological studies of ongoing climatic variability and trends in relation to health
- Development of mathematical models to forecast likely health outcomes in relation to projected climatic/environmental changes, accounting for concurrent social and economic circumstances and their projected changes
- Development of monitoring methods and systems to detect early evidence of health-related changes and further inform epidemiological and predictive modeling studies.

Monitoring of the potential impacts of climate change on health is important for several reasons (Campbell-Lendrum *et al.*, 2000; Kovats and Martens, 2000):

- Early detection of the health impacts of climate change
- Improved analysis of relationships between climate and health
- Validation of predictive models
- Increased understanding of vulnerability
- Assessment of effectiveness of adaptation strategies.

Epidemiological data are necessary to inform policymakers about the magnitude of actual or potential impacts of climate change. Most current infection surveillance systems have been designed to detect particular causes, such as food-borne disease, and individual risk factors, such as overseas travel. Monitoring of climate change requires a more comprehensive approach to infection etiology, examining the possible influence of climate on the environmental sources of pathogens and on human behavior (WHO-ECEH, 1998a,b). Another challenge for climate study is the size of data sets required. Although trends in any one country will be a starting point, improved coordination of infection data across regions will be needed. Epidemiological data also would help to determine the requirements for and the effectiveness of preventive actions.

Bioindicators of health risk also need to be developed, to detect early or unanticipated health impacts of climate change and stratospheric ozone depletion. For example, mapping and monitoring of vector species could be strengthened to detect early changes in their distribution associated with climate change (Campbell-Lendrum *et al.*, 2000). The effect of extreme weather events such as heat waves and floods need to be included in enhanced surveillance for assessment of future impacts.

Populations vary in their vulnerability to health impacts and in the resources available for adaptive responses (McMichael *et al.*, 2000b). These differences in vulnerability, between and within populations, reflect a wide range of demographic, cultural, political, socioeconomic, and technological circumstances. In the future, national impact assessments should describe and identify means by which the vulnerability of populations and subgroups could be reduced and select priorities for monitoring.

9.14. Cross-Cutting Issues

9.14.1 Costing the Health Impacts of Climate Change

Costing the health impacts of climate change is complex and controversial. It is complex because of the great heterogeneity of the health impacts, which include death, infectious disease, nutritional deprivation, and post-traumatic stress disorders. It is controversial because of difficulties in assigning money values to a diverse range of health deficits, doing so across varied cultures and economies, and taking account of the full "stream" of health impacts into the future (with appropriate time-discounting). During the 1990s, an attempt was made to develop a more standardized approach to measurement of the population health deficit by combining chronic illness or disability and premature death, via weighting procedures, into an integrated index—the Disability-Adjusted Life Year (Murray, 1994; Murray and Lopez, 1996).

To date, however, there is negligible scientific literature on the population burden of disease attributable to current or future climate change. There is no such literature on the DALY-based impact. Hence, there is no basis for making overall estimates of the direct costs to society of the health impacts of climate change. Nevertheless, some approximate estimations have been published of the impacts on national economies of major infectious disease outbreaks, such as might occur more often under conditions of climate change. For example, the outbreak of plague-like disease in Surat, northwest India, in 1994 cost an estimated US$3 billion in lost revenues to India alone (John, 1996; WHO, 1997). The cost of the 1994 Dengue Haemorrhagic Fever (DHF) epidemic in Thailand was estimated to be US$19–51 million (Sornmani, *et al.*, 1995). The cost of the 1994 epidemic of dengue/DHF in Puerto Rico was estimated to be US$12 million for direct hospitalization costs alone (Rodriguez, 1997; Meltzer *et al.*, 1998).

9.14.2 Development, Sustainability, and Equity

The ideas of development, sustainability, and equity inform much of the content of this chapter. It has been noted repeatedly that health impacts will tend to occur unevenly in the world and that the impacts in poorer populations, especially in the least-developed countries, often will be augmented by the heightened vulnerability of those populations. That is one of several reasons why—in today's world in which the gap between rich and poor is widening (UNDP 1999), in association with the nonredistributive character of market-dominated global economics (McMichael and Beaglehole, 2000)—new ways of redressing the imbalance in wealth and knowledge should be found.

The chapter also notes that development on a broad front—social, economic, technological, and provision of public health services and capacities—is crucial to a population's adaptive capacity to lessen the impacts of climate change.

Indeed, the health of a population is a key indicator of "sustainability." The capacity of the global population to achieve and maintain good health is an index of how well the natural and social environments are being managed. Wealthy local populations can afford to subsidize their health maintenance, drawing on resources imported from elsewhere. At a global level, however, health indicators provide a more valid indication of the extent to which the "carrying capacity" of the biosphere is being maintained.

9.15. Conclusions

The prospect of global climate change affecting patterns of human health poses a central challenge to scientists and policymakers. For scientists, the causation of most of the health outcomes considered in this chapter—from respiratory and cardiovascular disease to various types of infectious diseases—is complex: Various social, technological, demographic, behavioral, and environmental factors influence the risk of occurrence of these diseases. For that reason, it will remain difficult in the near future to identify any early impacts of the current climate trends on health. This complex causation of human disease also means that predictive modeling of future climatic impacts should take realistic account of the coexistent and modulating effects of nonclimate factors.

Over the past 5 years, we have acquired better understanding of direct temperature effects on health (heat and cold), temperature effects on air pollutant production, the seasonality of certain infectious diseases, and the public health consequences (and situational modifiers) of extreme weather events. Predictive modeling of how scenarios of future climate change would affect the patterns and impacts of vector-borne diseases has evolved, as has modeling of impacts on regional agricultural yields and the geography of world hunger.

Policymakers should appreciate that although our scientific capacity to foresee and model these various health outcomes of climate change continues to evolve, it is not possible to make precise and localized projections for many health outcomes—especially those that result indirectly from a sequence of impacts. In the meantime, a precautionary approach requires that policy development proceed on the basis of the available—though often limited and qualitative—evidence of how climate change will affect patterns of human population health. Furthermore, high priority should be assigned to improving the public health infrastructure and developing and implementing effective adaptation measures.

References

Ahlholm, J.U., M.L. Helander, and J. Savolainen, 1998: Genetic and environmental factors affecting the allergenicity of birch (*Betula pubescens ssp. czerepanovii* [Orl.] Hamet-Ahti) pollen. *Clinical and Experimental Allergy,* **28,** 1384–1388.

Akhtar, R. and A.J. McMichael, 1996: Rainfall and malaria outbreaks in western Rajasthan. *Lancet,* **348,** 1457–1458.

Alexander, D., 1993: *Natural Disasters*. University College London Press, London, United Kingdom, 632 pp.

Alvar, J., C. Canavate, B. Gutierrez-Solar, M. Jimenez, F. Laguna, R. Lopez-Velez, R. Molina, and J. Moreno, 1997: *Leishmania* and human immunodeficiency virus co-infection: the first 10 years. *Clinical Microbiology Review,* **10,** 298–319.

American Academy of Pediatrics, 1998: Committee on environmental health: toxic effects of indoor molds. *Pediatrics,* **101,** 712–714.

Anderson, R.M. and R.M. May, 1992: *Infectious Diseases of Humans, Dynamics and Control*. Oxford University Press, Oxford, United Kingdom, 757 pp.

Anderson, H.R., A. Ponce de Leon, J.M. Bland, J.S. Bower, J. Emberlin, and D.P. Strachan, 1998: Air pollution, pollens, and daily admission for asthma in London, 1987–92. *Thorax,* **53,** 842–8.

Ando, M., I. Uchiyama, and M. Ono, 1998a: Impacts on human health. In: *Global Warming: The Potential Impact on Japan*. [Nishioka, S. and H. Harasawa (eds.)]. Springer-Verlag, Tokyo, Japan, pp. 203–213.

Ando, M., I.N. Kobayashi, I. Kawahara, S. Asanuma, and C.K. Liang, 1998b: Impacts of heat stress on hyperthermic disorders and heat stroke. *Global Environmental Research,* **2,** 111–120.

Angyo, I.A. and E.S. Okpeh, 1997: Clinical predictors of epidemic outcome in meningococcal infection in Jos, Nigeria. *East African Medical Journal,* **74,** 423–426.

Appleton, C.C., 1977: The influence of temperature on the life-cycle and distribution of Biomphalaria pfeifferi and Bulinus sp. *International Journal for Parasitology,* **7,** 335–345.

Appleton, C.C. and G. Stiles, 1976: Geology and geomorphology in relation to the distribution of snail intermediate hosts in South Africa. *Annals of Tropical Medicine and Parasitology,* **70,** 189–198.

Arnell, N.W., M.G.R. Cannell, M. Hulme, J.F.B. Mitchell, R.S. Kovats, R.J. Nicholls, M.L. Parry, M.T.J. Livermore, and A. White, 2001: The consequences of CO_2 stabilisation for the impacts of climate change. *Climatic Change* (in press).

Asin, S. and S.S. Catalá, 1995: Development of *Trypanosoma cruzi* in *Triatoma infestans*: influence of temperature and blood consumption. *Journal of Parasitology,* **81,** 1–7.

Asnis, D.S., R. Conetta, A.A. Teixeira, G. Walman, and B.A. Sampson, 2000: The West Nile virus outbreak of 1999 in New York: the Flushing hopital experience. *Clinical Infectious Diseases,* **30,** 413–418.

Atherholt, T.B., M.W. LeChevallier, W.D. Norton, and J.S. Rosen, 1998: Effect of rainfall on giardia and cryptosporidium. *Journal of American Water Works Association,* **90,** 66–80.

Baldari, M., A. Tamburro, G. Sabatinelli, R. Romi, C. Severini, G. Cuccagna, G. Fiorilli, M.P. Allegri, C. Buriani, and M. Toti, 1998: Malaria in Maremma, Italy. *Lancet,* **351,** 1246–1247.

Ballester, F., D. Corella, S. Pérez Hoyos, M. Sáez and A. Hervás, 1996: Air pollution and mortality in Valencia, Spain: a study using the APHEA methodology. *Journal of Epidemiology and Community Medicine,* **50,** 527–533.

Bates, D.V. and R. Sizto, 1987: Air pollution and hospital admissions in Southern Ontario: the acid summer haze effect. *Environmental Research,* **43,** 317–331.

Bates, D.V., M. Baker-Anderson, and R. Sizto, 1990: Asthma attack periodicity: a study of hospital emergency visits in Vancouver. *Environmental Research,* **51,** 51–70.

Beggs, P.J., 1998: Pollen and pollen antigen as triggers of asthma: what to measure? *Atmospheric Environment,* **32,** 1777–1783.

Beggs, P.J. and P.H. Curson, 1995: An integrated environmental asthma model. *Archives of Environmental Health,* **50,** 87–94.

Bentham, G. and I.H. Langford, 1995: Climate change and the incidence of food poisoning in England and Wales. *International Journal of Biometeorology,* **39,** 81–86.

Bertollini, R., C. Dora, and M. Kryzanowski, 1996: *Environment and Health 1: Overview and Main European Issues*. WHO European Centre for Environment and Health, Rome/European Environment Agency, Copenhagen, Denmark, 56 pp.

Besancenot, J.P., 1997: Tropical climate pathology. *Medicine Tropicale (Mars),* **57(4),** 431–435.

Bi, P., S.L. Tong, K. Donald, K. Parton, and J. Hobbs, 2000: Southern Oscillation Index and the transmission of Barmah Forest virus infection. *Journal of Epidemiology and Community Health,* **54,** 69–70.

Birmingham, M.E., L.A. Lee, N. Ndayimirije, S. Nkurikiye, B.S. Hersh, J.G. Wells, and M.S. Deming, 1997: Epidemic cholera in Burundi: patterns of transmission in the Great Rift Valley Lake region. *Lancet,* **349,** 981–5. (Published erratum appears in *Lancet,* **349(9067),** 1776.)

Bobak, M. and A. Roberts, 1997: Heterogeneity of air pollution effects is related to average temperature. *British Medical Journal,* **315,** 1161–1162.

Borroto, R.R. and R.L. Haddock, 1998: Seasonal pattern of cholera in Guam and survival of *Vibrio cholerae* in aquatic environments. *Journal of Environment, Disease and Health Care Planning,* **3,** 1–9.

Bouma, M.J. and C. Dye, 1997: Cycles of malaria associated with El Niño in Venezuela. *Journal of the American Medical Association,* **278,** 1772–1774.

Bouma, M.J. and H.J. van der Kaay, 1996: The El Niño Southern Oscillation and the historic malaria epidemics in the Indian Subcontinent and Sri Lanka: an early warning system for future epidemics. *Tropical Medicine and International Health,* **1,** 86–96.

Bouma, M.J., C. Dye, and H.J. Van Der Kaay, 1996: Falciparum malaria and climate change in the North Frontier Province of Pakistan. *American Journal of Tropical Medicine and Hygiene,* **55,** 131–137.

Bouma, M.J., S. Kovats, S.A. Goubet, J. Cox, and A. Haines, 1997a: Global assessment of El Niño's disaster burden. *Lancet,* **350,** 1435–1438.

Bouma, M.J., G. Poveda, W. Rojas, D. Chavasse, M. Quinones, and J.A. Patz, 1997b: Predicting high-risk years for malaria in Colombia using parameters of El Niño Southern Oscillation. *Tropical Medicine and International Health,* **2,** 1122–1127.

Bradley, D.J., 1993: Human tropical diseases in a changing environment. In: *Environmental Change and Human Health* [Lake, J., K. Ackrill, and G. Bock (eds.)]. Discussion 162–70, Ciba Foundation Symposium, CIBA Foundation, London, United Kingdom, pp. 146–162.

Brauer, M., 1999: Health impacts of air pollution from vegetation fires. In: *Health Guidelines for Vegetation Fire Events: Background Papers* [Goh, K.-T., D. Schwela, J. Goldammer, and O. Simpson (eds.)]. Institute of Environmental Epidemiology, Singapore/WHO.

Bruce-Chwatt, L.J. and J. de Zulueta, 1980: *The Rise and Fall of Malaria in Europe: A Historico-Epidemiological Study*. Oxford University Press, Oxford, United Kingdom, 240 pp.

Bryan, J.H., D.H. Foley, and R.W. Sutherst, 1996: Malaria transmission and climate change in Australia. *Medical Journal of Australia,* **164,** 345–347.

Campbell-Lendrum, D., P. Wilkinson, K. Kuhn, R.S. Kovats, A. Haines, B. Menne, and T. Parr, 2001: Monitoring the health impacts of global climate change. In: *Health Impacts of Global Environmental Change: Concepts and Methods* [Martens, P. and A.J. McMichael (eds.)]. Cambridge University Press, Cambridge, United Kingdom and New York, NY, USA, (in press).

Carcavallo, R.U., 1999: Climate factors related to Chagas disease transmission. *Memorias do Instituto Oswaldo Cruz,* **94(I),** 367–369.

Carcavallo, R.U. and S.I. Curto de Casas, 1996: Some health impacts of global warming in South America: vector-borne diseases. *Journal of Epidemiology,* **6(4),** S153–S157.

Carcavallo, R.U., I. Galindez, J. Jurberg, and H. Lent (eds.), 1999: *Atlas of Chagas Disease Vectors in the Americas, Volume 3*. Editora Fiocruz, Rio de Janeiro, Brazil, 733 pp.

Carcavallo, R.U., I. Galindez, J. Jurberg, and H. Lent (eds.), 1998: *Atlas of Chagas Disease Vectors in the Americas, Volume 2*. Editora Fiocruz, Rio de Janeiro, Brazil, 733 pp.

Castellsague, J., J. Sunyer, M. Saez, and J.M. Anto, 1995: Short-term association between air pollution and emergency room visits for asthma in Barcelona. *Thorax*, **50,** 1051–1056.

Celenza, A., J. Fothergill, E. Kupek, and R.J. Shaw, 1996: Thunderstorm associated asthma: a detailed analysis of environmental factors. *British Medical Journal*, **312,** 604–607.

CDC, 1999: Surveillance of morbidity during wildfires: Central Florida, 1999. Centers for Disease Control and Prevention (CDC), *Morbidity and Mortality Weekly Report*, **281,** 789–90.

Chan, N.Y., K.L. Ebi, F. Smith, T.F. Wilson, and A.E. Smith, 1999: An integrated assessment framework for climate change and infectious diseases. *Environmental Health Perspectives*, **107,** 329–337.

Chandler, J.A. and R.B. Highton, 1975: The succession of mosquitoes species (Diptera: Culicidae) in rice fields in western Kenya. *Bulletin of Entomological Research*, **65,** 295–302.

Checkley W, L.D. Epstein, R.H. Gilman, D. Figueroa, R.I. Cama, J.A. Patz, and R.E. Black, 2000: Effects of El Niño and ambient temperature on hospital admissions for diarrhoeal diseases in Peruvian children. *Lancet*, **355,** 442–450.

Chestnut, L.G., W.S. Breffle, J.B. Smith, and L.S. Kalkstein, 1998: Analysis of differences in hot-weather-related mortality across 44 U.S. metropolitan areas. *Environmental Science and Policy*, **1,** 59–70.

Choudhury, A.Y. and A. Bhuiya, 1993: Effects of biosocial variables on changes in nutritional status of rural Bangladeshi children pre- and post-monsoon flooding. *Journal of Biosocial Science,* **25,** 351–357.

Colwell, R.R., 1996: Global climate and infectious disease: the cholera paradigm. *Science*, **274,** 2025–2031.

COMEAP, 1998: *Quantification of the Effects of Air Pollution on Health in the United Kingdom*. Great Britain Committee on the Medical Effects of Air Pollutants, Her Majesty's Stationery Office (HMSO), London, United Kingdom, 78 pp.

Conti Diaz, I.A., 1989: Epidemiology of sporotrichosis in Latin America. *Mycopathologia,* **108,** 113–6.

Cox, J., M. Craig, D. le Sueur, and B. Sharp, 1999: *Mapping Malaria Risk in the Highlands of Africa*. Mapping Malaria Risk in Africa/Highland Malaria Project (MARA/HIMAL) Technical Report, MARA/Durban, London School of Hygiene and Tropical Medicine, London, 96 pp.

Cross, E.R. and K.C. Hyams, 1996: The potential effect of global warming on the geographic and seasonal distribution of *Phlebotomus papatasi* in Southwest Asia. *Environmental Health Perspectives,* **104,** 724–727.

Cross, E.R., W.W. Newcomb, and C.J. Tucker, 1996: Use of weather data and remote sensing to predict the geographic and seasonal distribution of *Phlebotomus papatasi* in Southwest Asia. *American Journal of Tropical Medicine and Hygiene,* **54,** 530–536.

Cuculeanu, V. and D. Iorgulescu, 1994: Climate change impact on the radon activity in the atmosphere. *Romanian Journal of Meteorology*, **1,** 55–58.

Curriero, F.C., J.A. Patz, J.B. Rose, and S. Lele, 2001: Analysis of the association between extreme precipitation and waterborne disease outbreaks in the United States, 1948-1994. *American Journal of Public Health*, (in press).

Curto de Casas, S.I. and R.U. Carcavallo, 1984: Limites del triatomismo en la Argentina. I: Patagonia (The limits of triatominae in Argentina. I: Patagonia). *Chagas*, **1,** 35–40.

Curto de Casas, S.I. and R.U. Carcavallo, 1995: Global distribution of American pathogenic complexes. In: *The Health of Nations* [Iyun, B.F., Y. Verhasselt, and J.A. Hellen (eds.)]. Averbury, Aldershot, United Kingdom, pp. 21–32.

Day, J.F. and G.A. Curtis, 1989: Influence of rainfall on *Culex nigripalpus* (Diptera: Culicidae) blood-feeding behavior in Indian River County, Florida. *Annals of the Entomological Society of America,* **82,** 32–37.

de Diego Damia, A., M. Leon Fabregas, M. Perpina Tordera, and L. Compte Torrero, 1999: Effects of air pollution and weather conditions on asthma exacerbation. *Respiration*, **66,** 52–58.

Dennis, D.T., 1998: Epidemiology, ecology, and prevention of Lyme disease. In: *Lyme Disease* [Rahn, D.W. and J. Evans (eds.)]. American College of Physicians, Philadelphia, PA, USA, pp. 7–34.

Donaldson, G.C., R.S. Kovats, W.R. Keatinge, and A.J. McMichael, 2001: Heat- and cold-related mortality and morbidity and climate change. In: *Health Effects of Climate Change in the UK*. Department of Health, London, United Kingdom.

Donaldson, G.C., V.E. Tchernjavskii, S.P. Ermakov, K. Bucher, and W.R. Keatinge, 1998: Winter mortality and cold stress in Yekaterinberg, Russia: interview survey. *British Medical Journal,* **316,** 514–518.

Durry, E., D. Pappagianis, S.B. Werner, L. Hutwagner, R.K. Sun, M. Maurer, M.M. McNeil, and R.W. Pinner, 1997: Coccidioidomycosis in Tulare County, California, 1991: reemergence of an endemic disease. *Journal of Medical and Veterinary Mycology*, **35,** 321–326.

Dyson, T., 1999: Prospects for feeding the world. *British Medical Journal*, **319,** 988–990.

El-Raey, M., K. Dewidar, and M. El-Hattab, 1999: Adaptation to the impacts of sea-level rise in Egypt. *Mitigation and Adaptation Strategies for Global Change*, **4(3–4),** 343–361.

Emberlin, J., 1997: The trend to earlier birch pollen seasons in the UK: a biotic response to changes in weather conditions. *Grana,* **36,** 29–33.

Emberlin, J, 1994: The effects of patterns in climate and pollen abundance on allergy. *Allergy,* **94,** 15–20.

Engelthaler, D.M., D.G. Mosley, J.E. Cheek, C.E. Levy, K.K. Komatsu, P. Ettestad, T. Davis, D.T. Tanda, L. Miller, J.W. Frampton, R. Porter, and R.T. Bryan, 1999: Climatic and environmental patterns associated with hantavirus pulmonary syndrome, Four Corners Region, United States. *Emerging Infectious Diseases*, **5,** 87–94.

Epstein, P.R., T.E. Ford, and R.R. Colwell, 1993: Marine ecosystems. *Lancet*, **342,** 1216–1219.

Epstein, P.R., H.F. Diaz, S.A. Elias, G. Grabherr, N.E. Graham, W.J.M. Martens, E. Mosley-Thompson, and J. Susskind, 1997: Biological and physical signs of climate change: focus on mosquito-borne diseases. *Bulletin of the American Meteorological Society,* **78,** 409–417.

Epton, M.J., I.R. Martin, P. Graham, P.E. Healy, H. Smith, R. Balasubramaniam, I.C. Harvey, D.W. Fountain, J. Hedley, and G.I. Town, 1997: Climate and aeroallergen levels in asthma: a 12 month prospective study. *Thorax,* **52,** 528–534.

Eurowinter Group, 1997: Cold exposure and winter mortality from ischaemic heart disease, cerebrovascular disease, respiratory disease, and all causes in warm and cold regions of Europe. *Lancet,* **349,** 1341–1346.

FAO, 1999: *The State of Food Insecurity in the World 1999*. Food and Agriculture Organization of the United Nations, Rome, Italy, 32 pp.

Faunt, J.D., T.J. Wilkinson, P. Aplin, P. Henschke. M. Webb, and R.K. Penhal, 1995: The effete in the heat: heat-related hospital presentations during a ten-day heatwave. *Australian and New Zealand Journal of Medicine*, **25,** 117–121.

Focks, D.A., E. Daniels, D.G. Haile, and J.E. Keesling, 1995: A simulation model of epidemiology of urban dengue: literature analysis, model development, preliminary validation, and samples of simulation results. *American Journal of Tropical Medicine and Hygiene,* **53,** 489–506.

Focks, D.A., E. Daniels, D.G. Haile, and G.A. Mount, 1993a: Dynamic life table model for *Aedes aegypti* (L.) (Diptera: Culicidae): analysis of the literature and model development. *Journal of Medical Entomology,* **30,** 1003–1017.

Focks, D.A., E. Daniels, D.G. Haile, and G.A. Mount, 1993b: Dynamic life table model for *Aedes aegypti* (L.) (Diptera: Culicidae): simulation results and validation. *Journal of Medical Entomology,* **30,** 1018–1028.

Gaffen, D.J. and R.J. Ross, 1998: Increased summertime heat stress in the U.S. *Nature*, **396,** 529–530.

Gage, K.L., 1998: Plague. In: *Topley and Wilson's Microbiology and Microbial Infections, 9th edition, Volume 3: Bacterial Infections* [Hausler, W.J. Jr. and M. Sussman (eds.)]. Arnold, London, United Kingdom, pp. 886–903.

Garnham, P.C.C., 1945: Malaria epidemics at exceptionally high altitudes in Kenya. *British Medical Journal*, **ii,** 45–47.

Garssen, J., M. Norval, A. el Ghorr, N.K. Gibbs, C.D. Jones, D. Cerimele, C. De Simone, S. Caffiere, F. Dall'Acqua, F.R. De Gruijl, Y. Sontag, and H. Van Loveren, 1998. Estimation of the effect of increasing UVB exposure on the human immune system and related resistance to infectious diseases and tumours. *Journal of Photochemistry and Photobiology B,* **42,** 167–179.

Gawith, M.J., T.E. Downing, and T.S. Karacostas, 1999: Heatwaves in a changing climate. In: *Climate, Change and Risk* [Downing, T.E., A.J. Olsthoorn, and R.S.J. Tol (eds.)]. Routledge, London, United Kingdom, pp. 279–307.

Gillett, J.D., 1974: Direct and indirect influences of temperature on the transmission of parasites from insects to man. In: *The Effects of Meteorological Factors upon Parasites* [Taylor, A.E.R. and R. Muller (eds.)]. Blackwell Scientific Publications, Oxford, United Kingdom, pp. 79–95.

Githeko, A.K., M.W. Service, C.M. Mbogo, and F.K. Atieli, 1996: Resting behavior, ecology and genetics of malaria vectors in a large-scale agricultural areas of western Kenya. *Parassitologia*, **38,** 481–490.

Githeko, A.K., S.W. Lindsay, U.E. Confalonieri, and J.A. Patz, 2000: Climate change and vector-borne diseases: a regional analysis. World Health Organization, *Bulletin of WHO*, **78,** 1136–1147.

Glass, R.T. and M.M.J. Zack, 1979: Increase in deaths from ischaemic heart disease after blizzards. *Lancet*, **1,** 485–487.

Glass, G.E., F.P. Amerasinghe, J.M. Morgan, and T.W. Scott, 1994: Predicting *Ixodes scapularis* abundance on white-tailed deer using geographic information systems. *American Journal of Tropical Medicine and Hygiene,* **51,** 538–544.

Glass, G., J. Cheek, J.A. Patz, T.M. Shields, T.J. Doyle, D.A. Thoroughman, D.K. Hunt, R.E. Ensore, K.L. Gage, C. Ireland, C.J. Peters, and R. Bryan, 2000: Predicting high risk areas for hantavirus pulmonary syndrome with remotely sensed data: the Four Corners outbreak, 1993. *Emerging Infectious Diseases,* **6,** 238–247.

Glassheim, J.W., R.A. Ledoux, T.R. Vaughan, M.A. Damiano, D.L. Goodman, H.S. Nelson, and R.W. Weber, 1995: Analysis of meteorologic variables and seasonal aeroallergen pollen counts in Denver, Colorado. *Annals of Allergy and Asthma Immunology,* **75,** 149–156.

Gorjanc, M.L., W.D. Flanders, J. Vanderslice, J. Hersh, and J. Malilay, 1999: Effects of temperature and snowfall on mortality in Pennsylvania. *American Journal of Epidemiology*, **149,** 1152–1160.

Greenwood, B.M., I.S. Blakebrough, A.K. Bradley, S. Wali, and H.C. Whittle, 1984: Meningococcal disease and season in sub-Saharan Africa. *Lancet,* **1,** 1339–1342.

Grimstad, P.R. and L.D. Haramis, 1984: *Aedes triseriatus* (Diptera: Culicidae) and La Crosse virus, III: enhanced oral transmission by nutrition-deprived mosquitoes. *Journal of Medical Entomology,* **30,** 249–256.

Gubler, D.J., 1998a: Resurgent vector borne diseases as a global health problem. *Emerging Infectious Diseases,* **4,** 442–450.

Gubler, D.J., 1998b: Dengue and dengue hemorrhagic fever. *Clinical Microbiology Reviews*, **11,** 480–496.

Gubler, D.J., 1998c: Yellow fever. In: *Textbook of Pediatric Infectious Diseases* [Feigin. R.D. and J.D. Cherry (eds.)]. W.B. Saunders Co., Philadelphia, PA, USA, pp. 1981–1984.

Gubler, D.J., 1998d: Climate change: implications for human health. *Health and Environment Digest,* **12,** 54–55.

Gubler, D.J., 1997: Dengue and dengue hemorrhagic fever: its history and resurgence as a global public health problem. In: *Dengue and Dengue Hemorrhagic Fever* [Gubler, D.J. and G. Kuno (eds.)]. CAB International, New York, NY, USA, pp. 1–22.

Gubler, D.J., 1996: The global resurgence of arboviral diseases. *Transactions of the Royal Society of Tropical Medicine and Hygiene*, **90,** 449–451.

Gubler, D.J., 1989: *Aedes aegypti* and *Aedes aegypti*-borne disease control in the 1990s: top down and bottom up. *American Journal of Tropical Medicine and Hygiene*, **40,** 571–578.

Gubler, D.J. and M. Meltzer, 1999: The impact of dengue/dengue hemorrhagic fever in the developing world. In: *Advances in Virus Research, Vol. 53* [Maramorosch, K., F.A. Murphy, and A.J. Shatkin (eds.)]. Academic Press, San Diego, CA, USA, pp. 35–70.

Gubler, D.J. and J.T. Roehrig, 1998: Arboviruses (Togaviridae and Flaviviridae). In: *Topley and Wilson's Microbiology and Microbial Infections, Vol. 1.* [Balows, A. and M. Sussman (eds.)]. Arnold Publishing, London, United Kingdom, pp. 579–600.

Guest, C.S., K. Willson, A. Woodward, K. Hennessy, L.S. Kalkstein, C. Skinner, and A.J. McMichael, 1999: Climate and mortality in Australia: retrospective study, 1979–1990, and predicted impacts in five major cities in 2030. *Climate Research*, **13,** 1–15.

Guntzel, O., U. Bollag, and U. Helfenstein, 1996: Asthma and exacerbation of chronic bronchitis: sentinel and environmental data in a time series analysis. *Zentralblatt fur Hygiene und Umweltmedizin,* **198,** 383–393.

Gupta, R., 1996: Correlation of rainfall with upsurge of malaria in Rajasthan. *Journal of the Association of Physicians of India*, **44,** 385–389.

Haines, A. and D. Kammen, 2000: Sustainable energy and health. *Global Change and Human Health*, **1,** 2–11.

Hairston, N.G., 1973: The dynamics of transmission. In: *Epidemiology and Control of Schistosomiasis* [Ansari, N. (ed.)]. Karger, Basel, Switzerland, pp. 250–336.

Hajat, S., S.A. Goubet, and A. Haines, 1997: Thunderstorm-associated asthma: the effect on GP consultations. *British Journal of General Practitioners*, **47,** 639–641.

Hajat, S., A. Haines, S.A. Goubet, R.W. Atkinson, and H.R. Anderson, 1999: Association of air pollution with daily GP consultations for asthma and other lower respiratory conditions in London. *Thorax*, **54,** 597–605.

Hales, S., R.S. Kovats, and A. Woodward, 2000: What El Niño can tell us about human health and climate change. *Global Change and Human Health*, **1,** 66–77.

Hales, S., P. Weinstein, and A. Woodward, 1999a: Ciguatera fish poisoning, El Niño and Pacific sea surface temperatures. *Ecosystem Health*, **5,** 20–25.

Hales, S., P. Weinstein, and A. Woodward, 1999b: El Niño and the dynamics of vector-borne disease transmission. *Environmental Health Perspectives*, **107,** 99–102.

Hallegraeff, G.M., 1993: A review of harmful algal blooms and their apparent increase. *Phycologica*, **32,** 79–99.

Hamilton, A.C., 1989: The climate of the East Usambaras. In: *Forest Conservation in the East Usambaras* [Hamilton, A.C. and R. Benste-Smith (eds.)]. International Union of the Conservation of Nature, Gland, Switzerland, pp. 97–102.

Hardy, J.L., R.P. Meyer, S.B. Presser, and M.M. Milby, 1990: Temporal variations in the susceptibility of a semi-isolated population of *Culex tarsalis* to peroral infection with western equine encephalomyelitis and St. Louis encephalitis viruses. *American Journal of Tropical Medicine and Hygiene,* **42,** 500–511.

Hart, C.A. and L.E. Cuevas, 1997: Meningococcal disease in Africa. *Annals of Tropical Medicine and Parasitology*, **91,** 777–785.

Harvell, C.D., K. Kim, J.M Burkholder, R.R. Colwell, P.R. Epstein, J. Grimes, E.E. Hofmann, E. Lipp, A.D.M.E. Osterhaus, R. Overstreet, J.W. Porter, G.W. Smith, and G. Vasta, 1999: Diseases in the ocean: emerging pathogens, climate links, and anthropogenic factors. *Science*, **285,** 1505–1510.

Hay, S.I., R.W. Snow, and D.J. Rogers, 1998: Predicting malaria seasons in Kenya using multi-temporal meteorological satellite sensor data. *Transactions of the Royal Society of Tropical Medicine and Hygiene,* **92,** 12–20.

Hay, S.I., M. Myers, D.S. Burke, D.W. Vaughn, T. Endy, N. Ananda, G.D. Shanks, R.W. Snow, and D.J. Rogers, 2000: Etiology of interepidemic periods of mosquito-borne disease. *Proceedings of the National Accademy of Sciences,* **97,** 9335–9339.

HEED, 1998: *Marine Ecosystems: Emerging Diseases as Indicators of Change*. Year of the Ocean Special Report: Health of the Oceans from Labrador to Venezuela. Health and Ecological Dimensions (HEED) of Global Change Program. **Hess,** A.D., C.E. Cherubin, and L.C. LaMotte, 1963: Relation of temperature to activity of western and St. Louis encephalitis viruses. *American Journal of Tropical Medicine and Hygiene,* **12,** 657–667.

Holvoet, G., P. Michielsen, and J. Vandepitte, 1983: Autochthonous falciparum malaria in Belgium. *Annales de la Societe Belge de Medicine Tropicale*, **63,** 111–117.

Hulme, M., 1996: *Climate Change and Southern Africa: An Exploration of Some Potential Impacts and Implications in the SADC Region*. Climatic Research Unit, University of East Anglia, Norwich, United Kingdom, 104 pp.

Hulme, M. and G.J. Jenkins, 1998: *Climate Change Scenarios for the UK: scientific report*. UKCIP Technical Report No. 1, Climatic Research Unit, University of East Anglia, Norwich, United Kingdom, 80 pp.

Hunter, J.M.L., L. Rey, K. Chu, E.O. Adekolu-John, and K.E. Mott, 1993: *Parasitic diseases in water resources development: the need for intersectoral negotiation*. World Health Organization, Geneva, Switzerland, 152 pp.

IARC, 1992: *Solar and Ultraviolet Radiation: IARC Monograph on the Evaluation of Carcinogenic Risks to Humans, Vol. 55*. International Agency for Research on Cancer, Lyon, France, 316 pp.

IARC, 1988: *Radon: IARC Monograph on the Evaluation of Carcinogenic Risks to Humans, Vol. 43*. International Agency for Research on Cancer, Lyon, France, 300 pp.

IFRC, 1998: *World Disaster Report 1998*. International Federation of Red Cross and Red Crescent Societies, Oxford University Press, Oxford, United Kingdom and New York, NY, USA, 198 pp.

IFRC, 1997: *World Disaster Report 1997*. International Federation of Red Cross and Red Crescent Societies, Oxford University Press, Oxford, United Kingdom and New York, NY, USA, 173 pp.

IRI, 1999: *Climate Prediction and Disease/Health in Africa: Results from a Regional Training Course*. IRI-CR-99/2, International Research Institute for Climate Prediction, Columbia University, New York, NY, USA, 73 pp.

Jendritzky, G., K. Bucher, G. Laschewski, and H. Walther, 2000: Atmospheric heat exhange of the human being, bioclimate assessments, mortality and heat stress. *International Journal of Circumpolar Health*, **59,** 222–227.

Jetten, T.H. and D.A. Focks, 1997: Potential changes in the distribution of dengue transmission under climate warming. *American Journal of Tropical Medicine and Hygiene,* **57,** 285–287.

John, T.J., 1996: Emerging and re-emerging bacterial pathogens in India. *Indian Journal of Medical Research,* **103,** 4–18.

Jovel, J.R., 1989: *Natural disasters and their economic and social impact*. Economic Commissions, Latin America and the Caribbean (CEPAL Review No. 38), Santiago, Chile.

Kalkstein, L.S. and J.S. Greene, 1997: An evaluation of climate/mortality relationships in large U.S. cities and the possible impacts of a climate change. *Environmental Health Perspectives*, **105,** 84–93.

Kalkstein, L.S and G. Tan, 1995: Human health. In: *As Climate Changes: International Impacts and Implications* [Strzepek, K.M. and J.B. Smith (eds.)]. Cambridge University Press, Cambridge, United Kingdom and New York, NY, USA, pp.124–145.

Katsouyanni, K., A. Pantazopoulou., G. Touloumi, I. Tselepidaki, K. Moustris, D. Asimakopoulos, G. Poulopoulou, and D. Trichopoulos, 1993: Evidence for interaction between air pollution and high temperature in the causation of excess mortality. *Archives of Environmental Health,* **48,** 235–242.

Katsouyanni, K., G. Touloumi, C. Spix, J. Schwartz, F. Balducci, S. Medina, G. Rossi, B. Wojtyniak, J. Sunyer, L. Bacharova, J.P. Schouten, A. Ponka, and H.R. Anderson, 1997: Short-term effects of ambient sulphur dioxide and particulate matter on mortality in 12 European cities: results from time series data from the APHEA project; air pollution and health: a European approach. *British Medical Journal*, **314,** 1658–1663.

Keatinge, W.R., G.C. Donaldson, E. Cordioli, M. Martinelli, A.E. Kunst, J.P. Mackenbach, S. Nayha, and I. Vuori, 2000: Heat related mortality in warm and cold regions of Europe: observational study. *British Medical Journal,* **81,** 795–800.

Khaw, K.T., 1995: Temperature and cardiovascular mortality. *Lancet*, **345,** 337–338.

Kilbourne, E.M., 1997: Heatwaves. In: *The Public Health Consequences of Disasters* [Noji, E. (ed.)]. Oxford University Press, Oxford, United Kingdom and New York, NY, USA, pp. 51–61.

Kirk-Davidoff, D.B., E.J. Hintsa, J.G. Anderson, and D.W. Keith, 1999: The effect of climate change on ozone depletion through changes in stratospheric water vapour. *Nature*, **402,** 399–401.

Klein, R.J. and R.J. Nicholls, 1999: Assessment of coastal vulnerability to climate change. *Ambio*, **28,** 182–187.

Komar, N., 2000: West Nile viral encephalitis. *Revue Scientific et Technicale, Office Internationale des Epizootics*, **19,** 166–176.

Kovats, R.S. and Martens, P., 2000: Human health. In: *Assessment of Potential Effects and Adaptations for Climate Change in Europe: The Europe ACACIA Project* [Parry, M.L. (ed.)]. Jackson Environment Institute, University of East Anglia, Norwich, United Kingdom, pp. 227–242.

Kovats, R.S., M.J. Bouma, and A. Haines, 1999: *El Niño and Health*. WHO/SDE/PHE/99.4, World Health Organization, Geneva, Switzerland, 48 pp.

Kramer, L.D., J.L. Hardy, and S.B. Presser, 1983: Effect of temperature of extrinsic incubation on the vector competence of *Culex tarsalis* for western equine encephalomyelitis virus. *American Journal of Tropical Medicine and Hygiene,* **32,** 1130–1139.

Kríz, B., C. Benes, J. Cástková, and J. Helcl, 1998: Monitorování epidemiologické situace v zaplaven´ych oblastech v Ceské Republice v roce 1997. In: *Konference DDD '98; Kongresové Centrum Lázenská Kolonáda Podebrady, 11-13 Kvetna 1998* [Davidová, P. and V. Rupes (eds.)], pp.19–34 (in Czech).

Kuhn, K., 1997: *Climatic predictors of the abundance of sandfly vectors and the incidence of leishmaniasis in Italy*. Diss. London School of Hygiene and Tropical Medicine, University of London, London, United Kingdom, 95 pp.

Kunst, A.E., C.W.N. Looman, and J.P. Mackenbach, 1993: Outdoor air temperature and mortality in the Netherlands: a time-series analysis. *American Journal of Epidemiology*, **137,** 331–341.

Laake, K. and J.M. Sverre, 1996: Winter excess mortality: a comparison between Norway and England plus Wales. *Age and Ageing,* **25,** 343–348.

Langford, I.H. and G. Bentham, 1995: The potential effects of climate change on winter mortality in England and Wales. *International Journal of Biometeorology*, **38,** 141–147.

Lazzari, C.R., R.E. Gurtler, D. Canale, D.E. Mardo, and M.G. Lorenzo, 1998: Microclimatic properties of domestic and peridomestic Triatominae habitats in Northern Argentina. *Memorias do Instituto Oswaldo Cruz,* **93(II)**, 336.

Lederberg, J., S.C. Oates, and R.E. Shope, 1992: *Emerging Infections, Microbial Threats to Health in the United States*. Institute of Medicine, National Academy Press, Washington, DC, USA, 294 pp.

Lengeler, C., 1998: Insecticide treated bednets and curtains for malaria control, Cochrane Review 1998. In: *The Cochrane Library, Issue 3*. Oxford University Press, Oxford, United Kingdom.

Lindblade, K.A., E.D. Walker, A.W. Onapa, J. Katungu, and M.L. Wilson, 1999: Highland malaria in Uganda: prospective analysis of an epidemic associated with El Niño. *Transactions of the Royal Society of Tropical Medicine and Hygiene*, **93,** 480–487.

Lindgren, E., L. Tälleklint, and T. Polfeldt, 2000: Impact of climatic change on the northern latitude limit and population density of the disease-transmitting European tick, *Ixodes ricinus*. *Environmental Health Perspectives*, **108,** 119–123.

Lindsay, S.W. and M.H. Birley, 1996: Climate change and malaria transmission. *Annals of Tropical Medicine and Parasitology,* **90,** 573–588.

Lindsay, S.W. and W.J.M. Martens, 1998: Malaria in the African highlands: past, present and future. *Bulletin of the World Health Organization*, **76,** 33–45.

Lindsay, S.W., R. Bodker, R. Malima, H.A. Msangeni, and W. Kisinzia, 2000: Effect of 1997–98 El Niño on highland malaria in Tanzania. *Lancet*, **355,** 989–990.

Lipp, E.K. and J.B. Rose, 1997: The role of seafood in foodborne diseases in the United States of America. *Revue Scientific et Technicale, Office Internationale des Epizootics,* **16,** 620–640.

Lisle, J.T. and J.B. Rose, 1995: Cryptosporidium contamination of water in the USA and UK: a mini-review. *Aqua,* **44,** 103–117.

Lobitz, B., L. Beck, A. Huq, B. Wood, G. Fuchs, A.S.G. Faruque, and R. Colwell, 2000: Climate and infectious disease: use of remote sensing for detection of *Vibrio cholerae* by indirect measurement. *Proceedings of National Academy of Sciences*, **97,** 1438–1443.

Loevinsohn, M.E., 1994: Climate warming and increased malaria in Rwanda. *Lancet,* **343,** 714–748.

Lonergan, S., 1998: The role of environmental degradation in displacement. In: *Global Environmental Change and Human Security, Report 1*. University of Victoria, Victoria, BC, Canada, 2nd. ed.

Lonergan, S., 1999: *Global Environmental Change and Human Security: Science Plan*. International Human Dimensions Programme (IHDP) Report No. 11. IHDP, Bonn, Germany, 62 pp.

Loretti, A. and Y. Tegegn, 1996: Disasters in Africa: old and new hazards and growing vulnerability. *World Health Statistics Quarterly*, **49,** 179–184.

MacIver, D.C. and R.J.T. Klein (eds.), 1999: *Mitigation and Adaptation Strategies for Global Change (Special Issue on IPCC Workshop on Adaptation to Climate Variability and Change): Methodological Issues*, **4(3-4)**, 189–361.

Madico, G., J. McDonald, R.H. Gilman, L. Cabrera, and C. Sterling, 1997: Epidemiology and treatment of Cyclospora cayetanensis infection in Peruvian children. *Clinical Infectious Diseases*, **24,** 977–981.

Malakooti, M.A., K. Biomndo, and D.G. Shanks, 1998: Re-emergence of epidemic malaria in the highlands of western Kenya. *Emerging Infectious Diseases*, **4,** 671–676.

MARA, 1998: *Towards an Atlas of Malaria Risk. First Technical Report of the MARA/ARMA Collaboration*. Mapping Malaria Risk in Africa/Atlas du Risque de la Malaria en Afrique (MARA/ARMA), Durban, South Africa, 31 pp.

Marcondes, C.B., A.L. Lozovel, and J.H. Vilela, 1997: Geographical distribution of members of *Lutzomyia intermedia* (Lutz and Neiva, 1912) complex (Diptera: Psychodidae: Phlebotomidae). *Memorias do Instituto Oswaldo Cruz*, **92(I),** 317.

Martens, W.J.M., 1997: Climate change, thermal stress and mortality changes. *Social Science and Medicine*, **46,** 331–344.

Martens, W.J.M., T.H. Jetten, J. Rotmans, and L.W. Niessen, 1995: Climate change and vector-borne diseases: a global modelling perspective. *Global Environmental Change*, **5,** 195–209.

Martens, W.J.M., T.H. Jetten, and D.A. Focks, 1997: Sensitivity of malaria, schistosomiasis and dengue to global warming. *Climate Change,* **35,** 145–156.

Martens, W.J.M., R.S. Kovats, S. Nijhof, P. deVries, M.J.T. Livermore, A.J. McMichael, D. Bradley, and J. Cox, 1999: Climate change and future populations at risk of malaria. *Global Environmental Change*, **9,** S89–S107.

Martin, P. and M. Lefebvre, 1995: Malaria and climate: sensitivity of malaria potential transmission to climate. *Ambio,* **24,** 200–207.

McKenzie, R., B. Conner, and G. Bodeker, 1999: Increased summertime UV radiation in New Zealand in response to ozone loss. *Science*, **285,** 1709–1711.

McMichael, A.J. and R. Beaglehole, 2000: The changing global context of public health. *Lancet*, **356,** 495–499.

McMichael, A.J. and R.S. Kovats, 2000: Climate change and climate variability: adaptations to reduce adverse health impacts. *Environmental Monitoring and Assessment*, **61,** 49–64.

McMichael, A.J., M. Ando, R. Carcavallo, P. Epstein, A. Haines, G. Jendritzky, L. Kalkstein, R. Odongo, J. Patz, and W. Piver, 1996a: Human population health. In: *Climate Change 1995: Impacts, Adaptations, and Mitigation of Climate Change: Scientific-Technical Analyses. Contribution of Working Group II to the Second Assessment Report of the Intergovernmental Panel on Climate Change* [Watson, R.T., M.C. Zinyowera, and R.H. Moss (eds.)]. Cambridge University Press, Cambridge, United Kingdom and New York, NY, USA, pp. pp. 561–584.

McMichael, A.J., A. Haines, R. Slooff, and S. Kovats (eds.), 1996b: *Climate Change and Human Health*. WHO/EHG/96.7, an assessment prepared by a Task Group on behalf of the World Health Organization, the World Meteorological Organization, and the United Nations Environment Programme, Geneva, Switzerland, 297 pp.

McMichael, A.J., R.S. Kovats, P. Martens, S. Nijhof, M. Livermore, A. Cawthorne, and P. de Vries, 2000a: *Climate Change and Health*. Final Report for Department of Environment, Transport and the Regions, London School of Hygiene and Tropical Medicine, London, UK (unpublished report), 55 pp.

McMichael, A.J., U. Confalonieri, A. Githeko, A. Haines, R.S. Kovats, P. Martens, J. Patz, A. Sasaki, and A. Woodward, 2000b: Human health. In: *Special Report on Methodological and Technological Issues in Technology Transfer: A Special Report of IPCC Working Group III* [Metz, B., O.R. Davidson, J.M. Martens, S. van Rooijen, L. van Wie McGrory (eds.)]. Intergovernmental Panel on Climate Change, Cambridge University Press, Cambridge, United Kingdom and New York, NY, USA, pp. 329–347.

Meltzer, M.I., J.G. Rigau-Perez, G.G. Clark, P. Reiter, and D.J. Gubler, 1998: Using disability-adjusted life years to assess the economic impact of dengue in Puerto Rico: 1984–1994. *American Journal of Tropical Medicine and Hygiene*, **59,** 265–271.

Menne, B., K. Pond, E.K. Noji, and R. Bertollini, 1999: *Floods and Public Health Consequences, Prevention and Control Measures*. UNECE/MP.WAT/SEM.2/1999/22, discussion paper presented at the United Nations Economic Commission for Europe (UNCE) Seminar on Flood Prevention, Berlin, 7-8 October 1999. WHO European Centre for Environment and Health, Rome, Italy.

Michelozzi, P., F. Forastiere, D. Fusco, A. Tobias, and J. Anto, 1998: Air pollution and daily mortality in Rome, Italy. *Occupational and Environmental Medicine*, **55,** 605–611.

Mills, J.N. and J.E. Childs, 1998: Ecologic studies of rodent reservoirs: their relevance for human health. *Emerging Infectious Diseases*, **4,** 529–537.

Mills, J.N., T.L. Yates, T.G. Ksiazek, C.J. Peter, and J.E. Childs, 1999: Long-term studies of hantavirus reservoir populations in the southwestern United States. *Emerging Infectious Diseases*, **5,** 95–101.

Molineaux, D.H., 1988: The epidemiology of human malaria as an explanation of its distribution including some implications for its control. In: *Malaria: Principles and Practice of Malariology* [Wernsdorfer, W.H. and I. McGregor (eds.)]. Churchill Livingstone, Edinburgh, Scotland, United Kingdom, pp. 913–998.

Monath, T.P. (ed.), 1988: The arboviruses. *Epidemiology and Ecology, Vol. 1*. CRC Press, Boca Raton, FL, USA, pp. 87–126.

Moolgavkar, S.H., E.G. Luebeck, T.A. Hall, and E.L. Anderson, 1995: Air pollution and daily mortality in Philadelphia. *Epidemiology,* **6,** 476–484.

Morris, R., P. Guthrie, and C. Knopes, 1995: Photochemical modeling analysis under global warming conditions. In: *Proceedings of the American Waste Management Association Annual Meeting, San Antonio, Texas*. American Waste Management Association, USA.

Morris, R.E., M.W. Gery, M.K. Liu, G.E. Moore, C. Daly, and S.M. Greenfield, 1989: *Sensitivity of a Regional Oxidant Model to Variations in Climate Parameters*. U.S. Environmental Protection Agency, Washington, DC, USA.

Mouchet, J., O. Faye, F. Ousman, J. Julvez, and S. Manguin, 1996: Drought and malaria retreat in the Sahel, West Africa. *Lancet*, **348,** 1735–1736.

Mouchet, J., S. Manuin, S. Sircoulon, S. Laventure, O. Faye, A.W. Onapa, P. Carnavale, J. Julvez, and D. Fontenille, 1998: Evolution of malaria for the past 40 years: impact of climate and human factors. *Journal of the American Mosquito Control Association*, **14,** 121–130.

Munich Re, 1999: *Natural Catastrophes*. Munich Reinsurance Group, Munich, Germany.

Murray, C.J.L., 1994: Quantifying the burden of disease: the technical basis for disability-adjusted life years. World Health Organization, *Bulletin of WHO*, **72,** 429–445.

Murray, C.J.L. and A.D. Lopez (eds.), 1996: *The Global Burden of Disease: Global Burden of Disease and Injury Series, Vol. I*. Harvard School of Public Health, Harvard University, Boston, MA, USA 990 pp.

Najera, J.A., R.L. Kouznetzsov, and C. Delacollette, 1998: *Malaria epidemics: detection, control, forecasting and prevention*. WHO/MAL/98.184, World Health Organization, Geneva, Switzerland, 81 pp.

Nasci, R.S. and C.G. Moore, 1998: Vector-borne disease surveillance and natural disasters. *Emerging Infectious Diseases*, **4,** 333–334.

National Climate Centre, 1998: *Heavy Flood and Abnormal Climate in China in 1998*. Climate Publishing House, Beijing, People's Republic of China, pp. 2–4.

NRC, 1999: *From Monsoons to Microbes: Understanding the Ocean's Role in Human Health*. National Research Council, National Academy Press, Washington, DC, USA.

Newson, R., D. Strachan, E. Archibald, J. Emberlin, P. Hardaker, and C. Collier, 1998: Acute asthma epidemics, weather and pollen in England, 1987–1994. *European Respiratory Journal*, **11,** 694–701.

Noji, E. (ed.), 1997: *The Public Health Consequences of Disasters*. Oxford University Press, Oxford, United Kingdom and New York, NY, USA, 468 pp.

Norris, F.H., J.L. Perilla, J.K. Riad, K. Kaniasty, and E. Lavizzo, 1999: Stability and change in stress, resources, and psychological distress following natural disaster: findings from a longitudinal study of Hurricane Andrew. *Anxiety, Stress and Coping*, **12,** 363–396.

OECD, 2000: *Ancillary Benefits and Costs of Greenhouse Gas Mitigation: Proceedings from an IPCC Co-sponsored Workshop, 27–29 March, 2000, Washington, DC* [Davies, D.L., A.J. Krupnick, and G. McGlynn (eds.)]. Organisation for Economic Co-operation and Development, Paris, France, 592 pp.

Oke, T.R., 1987: *Boundary Layer Climates*. Cambridge University Press, Cambridge, United Kingdom and New York, NY, USA, 435 pp.

Ortega, Y.R., C.R. Sterling, R.H. Gilman, and F. Diaz, 1993: Cyclospora species—a new protozoan pathogen of humans. *New England Journal of Medicine*, **328,** 1308–1312.

Ortiz, P.L., M.E. Nieves Poveda, and A.V. Guevara Velasco, 1998: Models for setting up a biometeorological warning system over a populated area of Havana City. In: *Urban Ecology* [Breuste J., H. Feldmann, and O. Ohlmann (eds.)]. Springer-Verlag, Berlin, Germany, pp. 87–91.

PAHO, 1999: *Conclusions and Recommendations: Meeting on Evaluation of Preparedness and Response to Hurricanes George and Mitch*. Report of a meeting organized to evaluate the preparedness for and response to Hurricanes George and Mitch, Pan American Health Organization, Washington, DC, 39 pp.

PAHO, 1998: *Health in the Americas, Vol. 1*. Pan American Health Organization, Washington, DC, USA, 347 pp.

Parmenter, R.R., E.P. Yadav, C.A. Parmenter, P. Ettestad, and K.L. Gage, 1999: Incidence of plague associated with increased winter-spring precipitation in New Mexico. *American Journal of Tropical Medicine and Hygiene*, **6,** 814–821.

Parry, M.L., N. Arnell, M. Hulme, R. Nicholls, and M. Livermore, 1998: Adapting to the inevitable. *Nature*, **395,** 741.

Pascual, M., X. Rodo, S.P. Ellner, R.R. Colwell, and M.J. Bouma, 2000: Cholera dynamics and El Niño-Southern Oscillation. *Science,* **289,** 1766–1769.

Patz, J.A., M.A. McGeehin, S.M. Bernard, K.L. Ebi, P.R Epstein, A. Grambsch, D.J. Gubler, and P. Reiter, 2000: The potential health impacts of climate variability and change for the United States: executive summary of the report of the health sector of the U.S. National Assessment. *Environmental Health Perspectives,* **108,** 367–376.

Patz, J.A., W.J.M. Martens, D.A. Focks, and T.H. Jetten, 1998a: Dengue epidemic potential as projected by general circulation models of global climate change. *Environmental Health Persprectives,* **106,** 147–152.

Patz, J.A., K. Strzepec, S. Lele, M. Hedden, S. Green, B. Noden, S.I. Hay, L. Kalkstein, and J.C. Beier, 1998b: Predicting key malaria transmission factors, biting and entomological inoculation rates, using modelled soil moisture. *Tropical Medicine and International Health*, **3,** 818–827.

Penner, J.E., P.S. Connell, D.J. Wuebbles, and C.C. Covey, 1989: Climate change and its interactions with air chemistry: perspective and research needs. In: *The Potential Effects of Global Climate Change on the United States* [Smith, J.B. and D.A. Tirpak (eds.)]. U.S. Environmental Protection Agency, Office of Policy, Planning and Evaluation, Washington, DC, USA, pp. 1–78.

Peters, C.J., 1998: Hemorrhagic fevers: how they wax and wane. In: *Emerging Infections I* [Scheld, W.M., D. Armstrong, and J.M. Hughes (eds.)]. ASM Press, Washington, DC, USA, pp. 15–25.

Pfaff, A., K. Broad, and M. Glantz, 1999: Who benefits from climate forecasts? *Nature*, **397,** 645–646.

Pfluger, W., 1980: Experimental epidemiology of schistosomiasis, I: the prepatent period and cercarial production of *Schistosoma mansoni* in *Biomphalaria glabrata* at various constant temperatures. *Zeitschrift fur Parasitologie,* **63,** 159–69.

Philen, R.M., D.L. Combs, L. Miller, L.M. Sanderson, R.G. Parrish, and R. Ing, 1992: Hurricane Hugo-related deaths—South Carolina and Puerto Rico, 1989. *Disasters*, **16,** 53–59.

Piver, W.T., M. Ando, F. Ye, and C.J. Portier, 1999: Temperature and air pollution as risk factors for heat stroke in Tokyo, July and August 1980–1995. *Environmental Health Perspectives*, **107,** 911–916.

Poveda, G., N.E. Graham, P.R. Epstein, W. Rojas, M.L. Quiñonez, I.D. Vélez, and W.J.M. Martens, 2000: Climate and ENSO variability associated with vector-borne diseases in Colombia. *In: El Niño and the Southern Oscillation, Multiscale Variability and Global and Regional Impacts* [Diaz, H.F. and V. Markgraf (eds.)]. Cambridge University Press, Cambridge, United Kingdom and New York, NY, USA, pp. 183–204.

Prothero, R.M., 1994: Forced movements of population and health hazards in Tropical Africa. *International Journal of Epidemiology*, **23,** 657–663.

Purnell, R.E., 1966: Host-parasite relationships in schistosomiasis, I: the effect of temperature on the infection of *Biomphalaria [sudanica] tanganyicencis* with *Schistosoma mansoni* miracidia and of laboratory mice with *S. mansoni* cercariae. *Annals of Tropical Medicine and Parasitology*, **60,** 90–3.

Quattochi, D.A., J.C. Luvall, D.L. Rickman, M.G. Estes Jr., C.A. Laymon, and B.F. Howell, 2000: A decision support information system for urban landscape management using thermal infrared data. *Photogrammetric Engineering and Remote Sensing,* **66(10)**, 1195–1207.

Randolph, S.E. and D.J. Rogers, 2000: Fragile transmission cycles of tick-borne encephalitis virus may be distrupted by predicted climate change. *Proceedings of the Royal Society of London B*, **267,** 1741–1744.

Randolph, S.E., R.M. Green, M.F. Peacey, and D.J. Rogers, 2000: Seasonal synchrony: the key to tick-borne encephalitis foci identified by satellite. *Parasitology*, **121,** 15–23.

Rawlins, J.A., K.A. Hendricks, C.R. Burgess, R.M. Campman, G.G. Clark, L.J. Tabony, and M.A. Patterson, 1998: Dengue surveillance in Texas, 1995. *American Journal of Tropical Medicine and Hygiene*, **59,** 95–99.

Reeves, W.C., J.L. Hardy, W.K. Reisen, and M.M. Milby, 1994: Potential effect of global warming on mosquito-borne arboviruses. *Journal of Medical Entomology*, **31,** 323–332.

Reisen, W.K., 1995: Effect of temperature on Culex tarsalis (Diptera: Culicidae) from the Coachella and San Joaquin Valleys of California. *Journal of Medical Entomology*, **32(5),** 636–45.

Reisen, W.K., R.P. Meyer, S.B. Presser, and J.L. Hardy, 1993: Effect of temperature on the transmission of Western Equine encephalomyelitis and St. Louis encephalitis viruses by *Culex tarsalis* (Diptera: Culicidae). *Journal of Medical Entomology*, **30,** 151–160.

Reiter, P., 1998a: Global warming and vector-borne disease in temperate regions and at high altitude. *Lancet,* **351,** 839–840.

Reiter, P., 1998b: Global warming and vector-borne disease. *Lancet*, **351,** 1738.

Reiter, P., 1997: Could global warming bring "tropical" vectors and vector-borne diseases to temperate regions? In: *The Science and Culture Series: Nuclear Strategy and Peace Technology. International Seminar on Nuclear War and Planetary Emergencies, 22nd session* [Zichichi, A. (ed.)]. World Scientific, pp. 207–220.

Rodriguez, E., 1997: *Dengue Outbreak in Puerto Rico (1994–95): Hospitalization Cost Analysis*. World Health Organization, Geneva, Switzerland.

Rogers, D.J. and S.E. Randolph, 2000: The global spread of malaria in a future, warmer world. *Science*, **289,** 1763–1765.

Rooney, C., A.J. McMichael, R.S. Kovats, and M. Coleman, 1998: Excess mortality in England and Wales, and in Greater London, during the 1995 heatwave. *Journal of Epidemiology and Community Health*, **52,** 482–486.

Root, T.L. and S.H. Schneider, 1995: Ecology and climate: research strategies and implications. *Science*, **269,** 334.

Rose, J.B., S. Daeschner, D.R. Easterling, F.C. Curriero, S. Lele, and J.A. Patz, 2000: Climate and waterborne outbreaks in the U.S.: a preliminary descriptive analysis. *Journal of the American Water Works Association*, **92,** 77–86.

Rueda, L.M., K.J. Patel, R.C. Axtell, and R.E. Stinner, 1990: Temperature-dependent development and survival rates of *Culex* quinquefasciatus and *Aedes aegypti* (Diptera: Culicidae). *Journal of Medical Entomology*, **27,** 892–898.

Russell, R.C., 1998: Vector vs. humans in Australia: who is on top down under? An update on vector-borne disease and research on vectors in Australia. *Journal of Vector Ecology*, **23,** 1–46.

Sakamoto-Momiyama, M., 1977: *Seasonality in Human Mortality: A Medico-Geographical Study*. University of Tokyo Press, Tokyo, Japan, 181 pp.

Samet, J., S. Zeger, J. Kelsall, J. Xu, and L. Kalkstein, 1998: Does weather confound or modify the association of particulate air pollution with mortality? An analysis of the Philadelphia data, 1973–1980. *Environmental Research*, **77,** 9–19.

Sartor, F., R. Snacken, C. Demuth, and D. Walckiers, 1995: Temperature, ambient ozone levels, and mortality during summer 1994 in Belgium. *Environmental Research*, **70,** 105–113.

Schiff, C.J., W.C. Coutts, C. Yiannakis, and R.W. Holmes, 1979: Seasonal patterns in the transmission of *Schistosoma haematobium* in Rhodesia, and its control by winter application of molluscicide. *Transactions of the Royal Society of Tropical Medicine and Hygiene*, **73,** 375–380.

Schmaljohn, C. and B. Hjelle, 1997: Hantaviruses: a global disease problem. *Emerging Infectious Diseases*, **3,** 95–104.

Schneider, E., R.A. Hajjeh, R.A. Spiegel, R.W. Jibson, E.L. Harp, G.A. Marshall, R.A. Gunn, M.M. McNeil, R.W. Pinner, R.C. Baron, L.C. Hutwagner, C. Crump, L. Kaufman, S.E. Reef, G.M. Feldman, D. Pappagianis, and S.B. Werner, 1997: A coccidioidomycosis outbreak following the Northridge, California, earthquake. *Journal of the American Medical Association*, **277,** 904–908.

Schofield, C.J. and J.C. Dias, 1999: The Southern Cone Initiative against Chagas disease. *Advances in Parasitology*, **42,** 1–27.

Schofield, C.J., M.J. Lehane, P. McEwen, S.S. Catala, and D.E Gorla, 1992: Dispersive flight by *Triatoma infestans* under natural climatic conditions in Argentina. *Medical Veterinary Entomology*, **6,** 51–56.

Schorr, T.S., R.U. Carcavallo, D. Jenkins, and M.J. Jenkins (eds.), 1984: *Las Represas y sus Efectos Sobre la Salud* (Dams and Their Effects on Health). Pan American Health Organization, Mexico City, Mexico.

Seas, C., J. Miranda, A.I. Gil, R. Leon-Barua, J.A. Patz, A. Huq, R.R. Colwell, and R.B. Sack, 2000: New insights on the emergence of cholera in Latin America during 1991: the Peruvian experience. *American Journal of Tropical Medicine and Hygiene*, **62(4)**, (in press).

Semenza, J.C., J. McCullough, W.D. Flanders, M.A. McGeehin, C.H. Rubin, and J.R Lumpkin, 1999: Excess hospital admissions during the 1995 heat wave in Chicago. *American Journal of Preventive Medicine*, **16,** 269–277.

Semenza, J.C., C.H. Rubin, K.H. Falter, J.D. Selanikio, W.D. Flanders, H.L. Howe, and J.L. Wilhelm, 1996: Heat-related deaths during the July 1995 heat wave in Chicago. *New England Journal of Medicine*, **335,** 84–90.

Shindell, D.T., D. Rind, and P. Lonergan, 1998: Increased polar stratospheric ozone losses and delayed eventual recovery to increasing greenhouse gas concentrations. *Nature*, **392,** 589–592.

Shears, P., A.M. Berry, R. Murphy, and M.A. Nabil, 1985: Epidemiological assessment of the health and nutrition of Ethiopian refugees in emergency camps in Sudan. *British Medical Journal*, **295,** 214–318.

Siddique, A.K., A.H. Baqui, A. Eusof, and K. Zaman, 1991: 1988 floods in Bangladesh: pattern of illness and cause of death. *Journal of Diarrhoeal Disease Research*, **9,** 310–314.

Sillman, S. and P.J. Samson, 1995: Impact of temperature on oxidant photochemistry in urban, polluted rural, and remote environments. *Journal of Geophysical Research*, **100**, 11497–11508.

Smith, H.V., C.A. Paton, M.M. Mitambo, and R.W. Girdwood, 1997: Sporulation of *Cyclospora* sp. oocysts. *Applied Environmental Microbiology*, **63,** 1631–1632.

Smith, M.A., B. Jalaludin, J.E. Byles, L. Lim, and S.R. Leeder, 1996: Asthma presentations to emergency departments in Western Sydney during the January 1994 bushfires. *International Journal of Epidemiology*, **25,** 1227–1236.

Smithers, J. and B. Smit, 1997: Human adaptation to climatic variability and change. *Global Environmental Change*, **7,** 129–146.

Sornmani, S., K. Okanurak, and K. Indaratna, 1995: *Social and Economic Impact of Dengue Haemorrhagic Fever in Thailand*. Social and Economic Research Unit, Mahidol University, Bangkok, Thailand.

Sournia, A., 1995: Red tide and toxic marine phytoplankton of the world ocean: an inquiry into biodiversity. In: *Harmful Marine Algal Blooms, Technique et Documentation* [Lassus, P., G. Arzal, E. Erard-Le Den, P. Gentien, and C. Marcaillou-Le-Baut (eds.)]. Lavoisier, Paris, France.

Spieksma, F.T., J. Emberlin, M. Hjelmroos, S. Jager, and R.M. Leuschner, 1995: Atmospheric birch (*Betula*) pollen in Europe: trends and fluctuations in annual quantities and the starting dates of the seasons. *Grana*, **34,** 51–57.

Spitalnic, S.J., L. Jagminas, and J. Cox, 1996: An association between snowfall and ED presentation of cardiac arrest. *American Journal of Emergency Medicine*, **14,** 572–573.

Spix, C., H.R. Anderson, J. Schwartz, M.A. Vigotti, A. LeTertre, J.M. Vonk, G. Touloumi, F. Balducci, T. Piekarski, L. Bacharova, A. Tobias, A. Ponka, and K. Katsouyanni, 1998: Short-term effects of air pollution on hospital admissions of respiratory diseases in Europe: a quantitative summary of APHEA study results—air pollution and health: a European approach. *Archives of Environmental Health*, **53,** 54–64.

Stieb, D.M., R.T. Burnett, R.C. Beveridge, and J.R. Brook, 1996: Association between ozone and asthma emergency department visits in Saint John, New Brunswick, Canada. *Environmental Health Perspectives*, **104,** 1354–1360.

Stock, R., 1976: *Cholera in Africa*. African Environment Report No. 3, International African Institute, London, United Kingdom.

Sunyer, J., J. Castellsague, M. Saez, A. Tobias, and J.M. Anto, 1996: Air pollution and mortality in Barcelona. *Journal of Epidemiology and Community Health*, **50(1)**, 76–80.

Sutherst, R.W., 1998: Implications of global change and climate variability for vector-borne diseases: generic approaches to impact assessments. *International Journal for Parasitology*, **28,** 935–945.

Sutherst, R.W., J.S.I. Ingram, and H. Scherm, 1998: Global change and vector-borne diseases. *Parasitology Today*, **14,** 297–299.

Takahashi, Y., S. Kawashima, and S. Aikawa, 1996: Effects of global climate change on Japanese cedar pollen concentration in air—estimated results obtained from Yamagata City and its surrounding area. *Arerugi*, **45,** 1270–1276.

Talleklint, L. and T.G.T. Jaenson, 1998: Increasing geographical distribution and density of the common tick, Ixode ricinus, in central and northern Sweden? *Journal of Medical Entomology*, **35,** 521–526.

Tester, P.A., 1994: Harmful marine phytoplankton and shellfish toxicity: potential consequences of climate change. *Annals of New York Academy of Sciences*, **740,** 69–76.

Thiermann, A.B., 1980: Canine leptospirosis in Detroit. *American Journal of Veterinary Research*, **41,** 1659–1661.

Thomson, M.C., D.A. Elnaeim, R.W. Ashford, and S.J. Connor, 1999: Towards a Kala Azar risk map for Sudan: mapping the potential distribution of Phlebotomus orientalis using digital data of environmental variables. *Tropical Medicine and International Health*, **4,** 105–113.

Tikhoumirov, E., M. Santamaria, and K. Esteves, 1997: Meningococcal disease: public health burden and control. *World Health Statistics Quarterly*, **50,** 170–177.

Tong, S., P. Bi, K. Parton, K. Hobbs, and A.J. McMichael, 1998: Climate variability and transmission of epidemic polyarthritis. *Lancet*, **351,** 1100.

Touloumi, G., K. Katsouyanni, D. Zmirou, J. Schwartz, C. Spix, A.P. de Leon, A. Tobias, P. Quennel, D. Rabczenko, L. Bacharova, L. Bisanti, J.M. Vonk, and A. Ponka, 1997: Short-term effects of ambient oxidant exposure on mortality: a combined analysis with APHEA project. *American Journal of Epidemiology*, **146,** 177–185.

Trevejo, R.T., J.G. Rigau-Perez, D.A. Ashford, E.M. McClure, C. Jarquin-Gonzalez, J.J. Amador, J.O. de los Reyes, A. Gonzalez, S.R. Zaki, W.J. Shieh, R.G. McLean, R. S. Nasci, R.S. Weyant, C.A. Bolin, S.L. Bragg, B.A. Perkins, and R.A. Spiegel, 1998: Epidemic leptospirosis associated with pulmonary hemorrhage—Nicaragua, 1995. *Journal of Infectious Diseases*, **178,** 1457–1463.

UNDP, 1999: *World Development Report 1999*. United Nations Development Programme, New York, NY, USA, 300 pp.

UNEP, 1998: *Environmental Effects of Ozone Depletion: 1998 Assessment*. United Nations Environment Program, Nairobi, Kenya, 193 pp.

United Nations, 1982: *Sources and Biological Effects of Ionizing Radiation*. United Nations Scientific Committee on the Effects of Atomic Radiation, United Nations, New York, NY, USA.

Vargas, M.H., J.J. Sienra-Monge, G. Diaz-Mejia, and M. Deleon-Gonzalez, 1999: Asthma and geographical altitude: an inverse relationship in Mexico. *Journal of Asthma*, **36(6),** 511–517.

Walsh, J.F., D.H. Molineaux, and M.H. Birley, 1993: Deforestation: effects on vector-borne disease. *Parasitology*, **106,** S55–S75.

Walker, J., 1998: Malaria in a changing world: an Australian perspective. *International Journal of Parasitology*, **28,** 47–53.

Wang, X. and K.R. Smith, 1999: *Near-Term Health Benefits of Greenhouse Gas Reductions: A Proposed Assessment Method and Application in Two Energy Sectors in China*. WHO/SDE/PHE/99.1, World Health Organization, Geneva, Switzerland, 62 pp.

Waterlow, J., D.G. Armstrong, L. Fowden, and R. Riley (eds.), 1998: *Feeding a World Population of More than Eight Billion People*. Oxford University Press, Oxford, United Kingdom and New York, NY, USA, 280 pp.

Watson, R.T., J.A. Dixon, S.P. Hamburg, A.C. Janetos, and R.H. Moss, 1998: *Protecting Our Planet, Securing Our Future*. United Nations, NASA, and World Bank, Nairobi Kenya and Washington, DC, USA, 95 pp.

Watts, D.M., D.S. Burke, B.A. Harrison, R.E. Whitmire, and A. Nisalak, 1987: Effect of temperature on the vector efficiency of *Aedes aegypti* for Dengue 2 virus. *American Journal of Tropical Hygiene*, **36,** 143–152.

WFP, 1999: *World Food Programme*. Food Aid Organization of the United Nations, Rome, Italy. Available online at http://www.wfp.org/OP/countries/sudan.

Whitman, S., G. Good, E.R. Donoghue, N. Benbow, W. Shou, and S. Mou, 1997: Mortality in Chicago attributed to the July 1995 heat wave. American Journal of Public Health, **87,** 1515–1518.

WHO, 2000: *Climate Change and Stratospheric Ozone Depletion: Early Effects on Our Health in Europe* [Kovats, R.S., B. Menne, A.J. McMichael, R. Bertollini, and C. Soskolne (eds.)]. European Series No. 88, World Health Organization Regional Office for Europe, Copenhagen, Denmark, 200 pp.

WHO, 1999a: *The World Health Report, 1999: Making a Difference*. World Health Organization, Geneva, Switzerland, 122 pp.

WHO, 1999b: *Health Guidelines for Episodic Vegetation Fire Events*. WHO/EHG/99.7, World Health Organization, Geneva, Switzerland.

WHO, 1999c: Early human health effects of climate change and stratospheric ozone depletion in Europe. EUR/ICP/EHCO-02-02-05/15, World Health Organization, Copenagen, Denmark, 17 pp.

WHO, 1998a: *Malaria—WHO Fact Sheet No. 94*. World Health Organization, Geneva, Switzerland.

WHO, 1998b: Cholera in 1997. *Weekly Epidemiological Record*, **73,** 201–208.

WHO, 1997: *Plague—India 1994: Economic Loss*. World Health Organization, Geneva, Switzerland.

WHO, 1994: *Environmental Health Criteria 160: Ultraviolet Radiation*. Published under the joint sponsorship of the United Nations Environment Program, the World Health Organization, and the international Commission on Non-Ionizing Radiation Protection. World Health Organization, Geneva, Switzerland, 352 pp.

WHO, 1992: *Psychological Consequences of Disasters*. WHO/MNH/PSF 91.3.Rev 1, World Health Organization, Geneva, Switzerland.

WHO, 1984: *Environmental Health Criteria 37: Aquatic (Marine and Freshwater) Biotoxins*. World Health Organization, Geneva, Switzerland, 95 pp.

WHO-ECEH, 1998a: *Report of a WHO/EURO International Workshop on the Early Human Health Effects of Climate Change, 21–23 May, 1998*. World Health Organization European Centre for Environment and Health, Rome, Italy, 23 pp.

WHO-ECEH, 1998b: *Report of a WHO/EURO International Workshop on the Early Human Health Effects of Climate Change, 17–19 October, 1998*. WHO European Centre for Environment and Health, Rome, Italy.

WHO/UNAIDS, 1998: *Leishmania and HIV in Gridlock*. WHO/CTD/LEISH/ 98.9, World Health Organization/UNAIDS, Geneva, Switzerland.

Wijesunder, M.S., 1988: Malaria outbreaks in new foci in Sri Lanka. *Parasitology Today*, **4,** 147–150.

Williams, R.J., R.T. Bryan, J.N. Mills, R.E. Palma, I. Vera, F. De Velasquez, E. Baez, W.E. Schmidt, R.E. Figeroa, C.J. Peters, S.R. Zaki, A.S. Khan, and T.G. Ksiazek, 1997: An outbreak of hantavirus pulmonary syndrome in western Paraguay. *American Journal of Tropical Medicine and Hygiene*, **57,** 274–282.

Wilson, M.L., 1998: Distribution and abundance of *Ixodes scapularis* (Acari: Ixodidae) in North America: ecological processes and spatial analysis. *Journal of Medical Entomology*, **35,** 446–457.

WMO, 1997: *Reports to the 12th Session of the Commission for Climatology (Geneva, August 1997) and Report of the Meeting of Experts on Climate and Human Health (Freiburg, Germany, January 1997)*. WMO/TD-822, World Meteorological Organization, Geneva, Switzerland.

Woodruff, B.A., M.F. Toole, D.C. Rodrigue, E.W. Brink, E. Mahgoub, M.M. Ahmed, and A. Babikar, 1990: Disease surveillance and control after a flood: Khartoum, Sudan, 1988. *Disasters*, **14,** 151–163.

Woodward, A., S. Hales, and P. Weinstein, 1998: Climate change and human health in the Asia Pacific region: who will be the most vulnerable? *Climate Research*, **11,** 31–38.

Working Group on Public Health and Fossil-Fuel Combustion, 1997: Short-term improvements in public health from global-climate policies on fossil-fuel combustion: an interim report. *Lancet*, **350,** 1341–1349.

World Bank, 1997: *Clear water, blue skies*. World Bank, Washington, DC, USA.

Zeisberger. E., E. Schoenbaum, and P. Lomax, 1994: *Thermal Balance in Health and Disease: Recent Basic Research and Clinical Progress, Section III Adaptation*. Birkhaeuser Verlag, Basel/Berlin, Germany and Boston, MA, USA, 540 pp.

Zeledón, R. and J.E. Rabinovich, 1981: Chagas disease: an ecological approach with special emphasis on its insect vectors. *Annual Review of Entomology*, **26,** 101–133.

Ziska, L.H. and F.A. Caulfield, 2000a: The potential influence of rising atmospheric carbon dioxide (CO_2) on public health: pollen production of common ragweed as a test case. *World Resource Review*, **12(3),** 449–457.

Ziska, L.H. and F.A. Caulfield, 2000b: Rising CO_2 and pollen production of common ragweed (*Ambrosia artemisiifolia*), a known allergy-inducing species: implications for public health. *Australian Journal of Plant Physiology,* **27,** 893–898.

Zucker, J.R., 1996: Changing patterns of autochthonous malaria transmission in the United States: a review of recent outbreaks. *Emerging Infectious Diseases*, **2,** 37–43.

10

Africa

PAUL DESANKER (MALAWI) AND CHRISTOPHER MAGADZA (ZIMBABWE)

Lead Authors:
A. Allali (Morocco), C. Basalirwa (Uganda), M. Boko (Benin), G. Dieudonne (Niger), T.E. Downing (UK), P.O. Dube (Botswana), A. Githeko (Kenya), M. Githendu (Kenya), P. Gonzalez (USA), D. Gwary (Nigeria), B. Jallow (The Gambia), J. Nwafor (Nigeria), R. Scholes (South Africa)

Contributing Authors:
A. Amani (Niger), A. Bationo (Burkina Faso), R. Butterfield (UK), R. Chafil (Morocco), J. Feddema (The Netherlands), K. Hilmi (Morocco), G.M. Mailu (Kenya), G. Midgley (South Africa), T. Ngara (Zimbabwe), S. Nicholson (USA), D. Olago (Kenya), B. Orlando (USA), F. Semazzi (USA), L. Unganai (Zimbabwe), R. Washington (UK)

Review Editor:
I. Niang-Diop (Senegal)

CONTENTS

EXECUTIVE SUMMARY

Africa is highly vulnerable to the various manifestations of climate change. Six situations that are particularly important are:

- Water resources, especially in international shared basins where there is a potential for conflict and a need for regional coordination in water management
- Food security at risk from declines in agricultural production and uncertain climate
- Natural resources productivity at risk and biodiversity that might be irreversibly lost
- Vector- and water-borne diseases, especially in areas with inadequate health infrastructure
- Coastal zones vulnerable to sea-level rise, particularly roads, bridges, buildings, and other infrastructure that is exposed to flooding and other extreme events
- Exacerbation of desertification by changes in rainfall and intensified land use.

The historical climate record for Africa shows warming of approximately 0.7°C over most of the continent during the 20th century, a decrease in rainfall over large portions of the Sahel, and an increase in rainfall in east central Africa. Climate change scenarios for Africa, based on results from several general circulation models using data collated by the Intergovernmental Panel on Climate Change (IPCC) Data Distribution Center (DDC), indicate future warming across Africa ranging from 0.2°C per decade (low scenario) to more than 0.5°C per decade (high scenario). This warming is greatest over the interior of semi-arid margins of the Sahara and central southern Africa.

Projected future changes in mean seasonal rainfall in Africa are less well defined. Under the low-warming scenario, few areas show trends that significantly exceed natural 30-year variability. Under intermediate warming scenarios, most models project that by 2050 north Africa and the interior of southern Africa will experience decreases during the growing season that exceed one standard deviation of natural variability; in parts of equatorial east Africa, rainfall is predicted to increase in December–February and decrease in June–August. With a more rapid global warming scenario, large areas of Africa would experience changes in December–February or June–August rainfall that significantly exceed natural variability.

Water: Africa is the continent with the lowest conversion factor of precipitation to runoff, averaging 15%. Although the equatorial region and coastal areas of eastern and southern Africa are humid, the rest of the continent is dry subhumid to arid. The dominant impact of global warming is predicted to be a reduction in soil moisture in subhumid zones and a reduction in runoff. Current trends in major river basins indicate a decrease in runoff of about 17% over the past decade. Reservoir storage shows marked sensitivity to variations in runoff and periods of drought. Lake storage and major dams have reached critically low levels, threatening industrial activity. Model results indicate that global warming will increase the frequency of such low storage episodes.

Natural Resources Management and Biodiversity: Land-use changes as a result of population and development pressures will continue to be the major driver of land-cover change in Africa, with climate change becoming an increasingly important contributing factor by mid-century. Resultant changes in ecosystems will affect the distribution and productivity of plant and animal species, water supply, fuelwood, and other services. Losses of biodiversity are likely to be accelerated by climate change, such as in the Afromontane and Cape centers of plant endemism. Projected climate change is expected to lead to altered frequency, intensity, and extent of vegetation fires, with potential feedback effects on climate change.

Human Health: Human health is predicted to be adversely affected by projected climate change. Temperature rises will extend the habitats of vectors of diseases such as malaria. Droughts and flooding, where sanitary infrastucture is inadequate, will result in increased frequency of epidemics and enteric diseases. More frequent outbreaks of Rift Valley fever could result from increased rainfall. Increased temperatures of coastal waters could aggrevate cholera epidemics in coastal areas.

Food Security: There is wide consensus that climate change, through increased extremes, will worsen food security in Africa. The continent already experiences a major deficit in food production in many areas, and potential declines in soil moisture will be an added burden. Food-importing countries are at greater risk of adverse climate change, and impacts could have as much to do with changes in world markets as with changes in local and regional resources and national agricultural economy. As a result of water stress, inland fisheries will be rendered more vulnerable because of episodic drought and habitat destruction. Ocean warming also will modify ocean currents, with possible impacts on coastal marine fisheries.

Settlements and Infrastructure: The basic infrastructure for development—transport, housing, services—is inadequate now, yet it represents substantial investment by governments. An increase in damaging floods, dust storms, and other extremes would result in damage to settlements and infrastructure and affect human health.

Most of Africa's largest cities are along coasts. A large percentage of Africa's population is land-locked; thus, coastal facilities are economically significant. Sea-level rise, coastal erosion, saltwater intrusion, and flooding will have significant impacts on African communities and economies.

Desertification: Climate change and desertification remain inextricably linked through feedbacks between land degradation and precipitation. Climate change might exacerbate desertification through alteration of spatial and temporal patterns in temperature, rainfall, solar insolation, and winds. Conversely, desertification aggravates carbon dioxide (CO_2)-induced climate change through the release of CO_2 from cleared and dead vegetation and reduction of the carbon sequestration potential of desertified land. Although the relative importance of climatic and anthropogenic factors in causing desertification remains unresolved, evidence shows that certain arid, semi-arid, and dry subhumid areas have experienced declines in rainfall, resulting in decreases in soil fertility and agricultural, livestock, forest, and rangeland production. Ultimately, these adverse impacts lead to socioeconomic and political instability. Potential increases in the frequency and severity of drought are likely to exacerbate desertification.

Given the range and magnitude of the development constraints and challenges facing most African nations, the overall capacity for Africa to adapt to climate change is low. Although there is uncertainty in what the future holds, Africa must start planning now to adapt to climate change. National environmental action plans and implementation must incorporate long-term changes and pursue "no regret" strategies. Current technologies and approaches—especially in agriculture and water—are unlikely to be adequate to meet projected demands, and increased climate variability will be an additional stress. Seasonal forecasting—for example, linking sea-surface temperatures to outbreaks of major diseases—is a promising adaptive strategy that will help save lives. It is unlikely that African countries on their own will have sufficient resources to respond effectively.

Climate change also offers some opportunities. The process of adapting to global climate change, including technology transfer, offers new development pathways that could take advantage of Africa's resources and human potential. Examples would include competitive agricultural products, as a result of research in new crop varieties and increased international trade, and industrial developments such as solar energy. Regional cooperation in science, resource management, and development already are increasing.

This assessment of vulnerability to climate change is marked by uncertainty. The diversity of African climates, high rainfall variability, and a very sparse observational network make predictions of future climate change difficult at the subregional and local levels. Underlying exposure and vulnerability to climatic changes are well established. Sensitivity to climatic variations is established but incomplete. However, uncertainty over future conditions means that there is low confidence in projected costs of climate change.

Improvements in national and regional data and capacity to predict impacts is essential. Developing African capacity in environmental assessment will increase the effectiveness of aid. Regional assessments of vulnerability, impacts, and adaptation should be pursued to fill in the many gaps in information.

10.1. Introduction to African Region

10.1.1. Previous Syntheses of African Region

Previous assessments (Hulme, 1996; IPCC, 1998) concluded that the African continent is particularly vulnerable to the impacts of climate change because of factors such as widespread poverty, recurrent droughts, inequitable land distribution, and overdependence on rainfed agriculture. Timely response actions were considered to be beyond the economic means of some countries. Deterioration in terms of trade, inappropriate policies, high population growth rates, and lack of significant investment—coupled with a highly variable climate—have made it difficult for several countries to develop patterns of livelihood that would reduce pressure on the natural resource base. The reports fell short of assigning relative importance to these different factors in Africa's capacity to adapt to climate change. This still is not possible and presents a new challenge for future assessments.

10.1.2. What is Different about Africa?

The main background factors that need to be kept in mind in assessing the vulnerability of the African region to climate change—particularly the capacity of African governments to respond proactively to changes that are largely not of their making or under their control—are as follows:

- *Diversity:* The term "African region" is a geographical convenience only. There is as much diversity of climate, landform, biota, culture, and economic circumstance within the region as there is between it and, say, South America or Asia. Very few statements are valid for the entire continent. The generalities that follow must be read in that context.
- *Climate:* Africa is predominantly tropical, hot, and dry. There are small regions of temperate (cool) climates in the extreme south and north and at high altitudes in between. Parts of west Africa, as well as the western part of central Africa, are humid throughout the year. A large region north and south of this humid core is subhumid, with substantial rainfall during the wet season (or seasons, in the case of east Africa) but almost no rain during the extended dry season. Poleward from this zone is a large area of semi-arid climates, which permit marginal cropping during the wet season but are characterized by extreme unreliability of rainfall and few permanent surface-water sources. Most of the human population occurs in the subhumid and semi-arid zones. Corresponding to the tropics of Capricorn and Cancer are the vast desert regions of the Kalahari-Namib and the Sahara.
- *Development Status:* Measured by almost any index of human well-being, Africa contains the poorest and least-developed nations of the world. Per capita gross domestic product (GDP), life expectancy, infant mortality, and adult literacy are all in the bottom quartile globally when averaged across Africa, although individual nations may perform somewhat better on one or more of these indices. The general weakness of the science and technology infrastructure—in particular, the relatively small numbers of technically trained professionals—limits the rate at which adaptive research can be performed or implemented.
- *Food Supply:* More than half of the African population is rural and directly dependent on locally grown crops or foods harvested from the immediate environment. Per capita food production in Africa has been declining over the past 2 decades, contrary to the global trend. The result is widespread malnutrition, a recurrent need for emergency food aid, and increasing dependence on food grown outside the region.
- *Dependence on Natural Resources:* The formal and informal economies of most African countries are strongly based on natural resources: Agriculture, pastoralism, logging, ecotourism, and mining are dominant. Climatic variations that alter the viability of these activities, for better or for worse, have very high leverage on the economy.
- *Biodiversity:* About one-fifth of the world's plants, birds, and mammals originate or have major areas of present conservation in Africa. There are major "hot spots" of biodiversity within west, east, central, and southern Africa.
- *Low Capacity for State-Initiated Interventions:* Governance structures typically are underfunded and undercapacity. In many instances they have been undermined by military coups, despotism, tribalism, corruption, maladministration, and economic adjustment programs imposed by the international financial community. Communication from capitals to the remotest provinces—by road, rail, air, or telephone—often is unreliable and slow. State-centered political economies in their postcolonial sense are relatively recent over most of Africa, and their boundaries include wide ethnic diversity within single nations and cut across previous political territories.
- *Disease Burden:* Insect-vector diseases such as malaria and tryanosomiasis; water-borne diseases such as typhoid, cholera, and schistosomiasis; and poverty-related diseases such as tuberculosis are prevalent in Africa. Water and food security are closely linked with health. The HIV/AIDS pandemic is placing great strain on the health infrastructure. Heavy mortalities lead to great loss of productive potential.
- *Armed Conflict:* There has been chronic armed conflict in several regions of Africa almost continuously for the past 3 decades. This weakens the ability of the nations involved to respond to climate change and adds large refugee populations to the local population, which must be supported by the environment.
- *High External Trade and Aid Dependence:* Very little industrial beneficiation takes place in Africa. High volumes of relatively low-value goods dominate export economies. In general, there is no strong internal demand

(national or regional) to buffer the economies from changes in global trade. Trade linkages show the pattern established by the former colonial relationships. Many African countries have a negative trade balance, particularly as a result of heavy international debt-servicing burdens, and are chronically dependent on financial aid from the developed world.

10.1.3. *Past to Present*

10.1.3.1. Climatology

Africa is a vast continent, and it experiences a wide variety of climate regimes. The location, size, and shape of the African continent play key roles in determining climate. The poleward extremes of the continent experience winter rainfall associated with the passage of mid-latitude airmasses. Across the Kalahari and Sahara deserts, precipitation is inhibited by subsidence virtually throughout the year. In contrast, moderate to heavy precipitation associated with the Inter-Tropical Convergence Zone (ITCZ) characterizes equatorial and tropical areas. Because the movement of the ITCZ follows the position of maximum surface heating associated with meridional displacement of the overhead position of the sun, near-equatorial regions experience two rain seasons, whereas regions further poleward experience one distinct rainfall season. The mean climate of Africa is further modified by the presence of large contrasts in topography (Semazzi and Sun, 1995) and the existence of large lakes in some parts of the continent.

10.1.3.2. Interannual and Interdecadal Climate Variability

Humans have adapted to patterns of climate variability through land-use systems that minimize risk, with agricultural calendars that are closely tuned to typical conditions and choices of crops and animal husbandry that best reflect prevailing conditions. Rapid changes in this variability may severely disrupt production systems and livelihoods. Interannual variability of the African climate is determined by several factors. The El Niño-Southern Oscillation (ENSO) is the most dominant perturbation responsible for interannual climate variability over eastern and southern Africa (Nicholson and Entekhapi, 1986). The typical rainfall anomaly associated with ENSO is a dipole rainfall pattern: Eastern Africa is in phase with warm ENSO episodes, whereas southern Africa is negatively correlated with these events (Nicholson and Kim, 1997). The 1997–1998 ENSO event resulted in extreme wet conditions over eastern Africa (see Boxes 10-1 and 10-2), and the 1999–2000 La Niña may have caused devastating floods in Mozambique. Modeling exercises indicate that climate change may increase the frequency of ENSO warm phases by increasing the warm pool in the tropical western Pacific or by reducing the efficiency of heat loss (Trenberth and Hoar, 1997; Timmerman *et al.*, 1999).

In the Sahel and similar regions of west Africa, the problem is more complex. ENSO appears to influence year-to-year variations and reduces rainfall. Its influence appears to be greater within long dry intervals in the Sahel, but it is not the dominant factor controlling rainfall in this region (Ward, 1998).

Over northern Africa, the North Atlantic Oscillation (NAO) is a key factor that is responsible for interannual variability of the climate (Lamb, 1978). Across western Africa, year-to-year changes in seasonal climatic conditions are determined primarily by the Atlantic Ocean, although the rest of the world's oceans also play important roles. Low-lying islands and coastal regions receive significant amounts of rainfall from tropical cyclone activity, which is sensitive to interannual variability of SST conditions over adjacent ocean basins.

Box 10-1. The 1997–1998 ENSO Event

ENSO appears to play a major role in east Africa, but it masks the perhaps more important role of the other oceans, particularly the Indian Ocean. The 1961–1962 rains were spectacularly manifested as rapid rises in the levels of east African lakes. Lake Victoria rose 2 m in little more than a year (Flohn and Nicholson, 1980). This was not an ENSO year, but exceedingly high sea-surface temperatures (SSTs) occurred in the nearby Indian Ocean as well as the Atlantic. Such high SSTs are associated with most ENSO events, and it is probably SSTs in these regions, rather than the Pacific ENSO (Nicholson and Kim, 1997), that have the largest influence on east African rainfall. In another example, the dipole pattern anticipated to occur during ENSO events did not occur during the 1997–1998 event. There was a tremendous increase in rainfall in east Africa, but intense drought conditions did not occur throughout southern Africa. The reason appears to be an unusual pattern of SST in the Indian Ocean.

Box 10-2. Drought Conditions in the Sahel

One of the most significant climatic variations has been the persistent decline in rainfall in the Sahel since the late 1960s. The trend was abruptly interrupted by a return of adequate rainfall conditions in 1994. This was considered to be the wettest year of the past 30 and was thought to perhaps indicate the end of the drought. However, by the standard of the whole century, rainfall in 1994 barely exceeded the long-term mean. Also, the 1994 rainy season was unusual in that the anomalously wet conditions occurred toward the end of the rainy season and in the months following. Unfortunately, dry conditions returned after 1994. The persistent drying trend has caused concern among development planners regarding how to cope with losses of food production, episodes of food insecurity, displacements of populations, lack of water resources, and constraints on hydroelectricity.

The climate of Africa also exhibits high interdecadal variability. Rainfall variability in the Sahel derives from factors such as SST and atmospheric dynamics (Lamb, 1978; Folland *et al.*, 1986; Hulme and Kelly, 1997; Nicholson and Kim, 1997) and is modulated by land surface effects related to soil moisture, vegetation cover, dust, and so forth (Charney, 1975; Diedhiou and Mahfouf, 1996; Xue, 1997; Zeng *et al.*, 1999). Modeling evidence also suggests that orographic control plays a significant role in promoting climate teleconnections between global SST anomalies and west African interannual climate variability (Semazzi and Sun, 1997).

Besides ENSO, the NAO, and west African climate anomaly patterns, other continental-scale and subcontinental climate anomalies play significant roles in determining interannual and longer climate variability time scales (Nicholson *et al.*, 2000). For instance, the decade 1950–1959 was characterized by above-normal precipitation over most of Africa, although rainfall deficiencies prevailed over the near-equatorial region. Later, during the period 1960–1969, this rainfall anomaly pattern dramatically reversed in sign, with rainfall deficits observed for most of Africa while the equatorial region experienced widespread abundance of rainfall. These two time periods also coincide with a reversal in the sign of the Sahelian rainfall anomalies (Lamb and Peppler, 1992). More recently, the pattern has been one of increased aridity throughout most of the continent. Mean rainfall decreased by 20–49% in the Sahel between the periods 1931–1960 and 1968–1997 and generally 5–10% across the rest of the continent (see Figure 10-1).

In comparison with the period between 1950 and 1970, the average length of the rainy season has not changed significantly during the dry period 1970–1990. Instead, the decrease in rainfall in July and August explains most of the diminution of total annual rainfall over the Sahel since 1970. The average number of rainy events in August was reduced by about 30% (Le Barbé and Lebel, 1997).

There is emerging evidence that aerosols and dust also may be important factors in modulating the variability of the African climate (d'Almeida, 1986; Mohamed *et al.*, 1992; Pinker *et al.*, 1994). These studies provide overwhelming evidence of an extremely dense and deep (reaching up several kilometers) dust layer in the Sahel/Sudan during the main dust season from November to April.

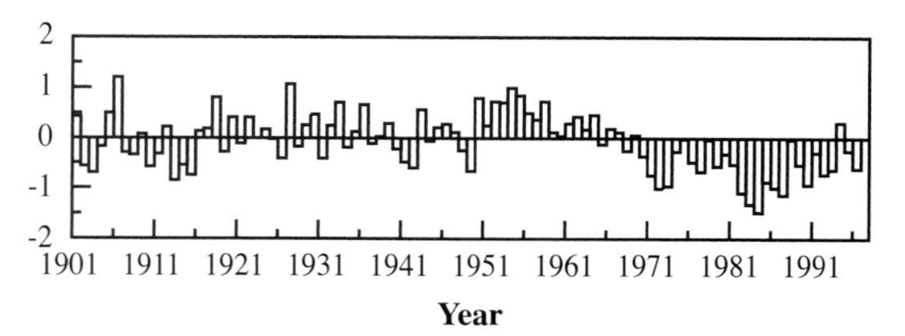

Figure 10-1: Rainfall fluctuations, 1901–1998, expressed as regionally averaged standard deviation (departure from long-term mean divided by standard deviation) for the Sahel.

10.1.3.3. Paleoclimate of Africa

Paleoclimatology and paleoenvironmental changes in Africa have been reconstructed from several lines of sedimentary evidence, such as fossil strand lines, diatom and pollen analyses, evidence of glaciation and fossil moraines on high mountains, sediment lithology, geochemistry and biogeochemistry, and so forth. Most records do not extend beyond 30,000 years before the present (BP), but they capture the climatic extremes of the last glacial-interglacial cycle—the last glacial maximum (22,000 to 14,000 years BP) through to the Holocene period (10,000 years BP to present) (Olago, 2001).

Temperatures during the last glacial maximum are estimated to have been 4–7°C lower than today, and they were coupled with intensive aridity and regression of lakes throughout the African continent, resulting from reduced precipitation as a consequence of weaker monsoons, stronger dry trade winds, and lowered SST (Coetzee, 1967; Flenley, 1979; COHMAP Members, 1988; Bonnefille *et al.*, 1990; Vincens *et al.*, 1993). Highland vegetation was depressed to significantly lower altitudes relative to today, and mountain glaciers were at their maximum extent. Grasslands were more widespread, lowland forests became fragmented, and subtropical desert margins advanced latitudinally by 300–700 km relative to their present positions (Flohn and Nicholson, 1980).

During the Holocene period of the past 10,000 years there was a "warm" climatic optimum roughly 5,000 years ago. At that time, more humid conditions generally were widespread, and deserts were markedly contracted. Lakes existed even in parts of the central Sahara. The current state of climate was reached roughly 3,000 years ago.

10.1.3.4. Recent Historical Record

Observational records show that the continent of Africa is warmer than it was 100 years ago (IPCC,1996). Warming through the 20th century has been at the rate of about 0.05°C per decade (see Figure 10-2), with slightly larger warming in the June, July, August (JJA) and September–November seasons than in December, January, February (DJF) and March–May (Hulme *et al.*, 2001). The 5 warmest years in Africa have all occurred since 1988, with 1988 and 1995 the two warmest years. This rate of warming is not dissimilar to that experienced globally, and the periods of most rapid warming—the 1910s to 1930s and the post-1970s—occur simultaneously in Africa and the world.

The climate of Africa has experienced wetter and drier intervals during the past 2 centuries. The most pronounced periods were during the 20th century. A very intense dry period, much like the current one, also prevailed for 2 to 3 decades during the first half of the 19th century. Humid conditions reminiscent of the 1950s prevailed around the 1870s or 1880s, but another milder arid interval of roughly 20 years commenced around 1895.

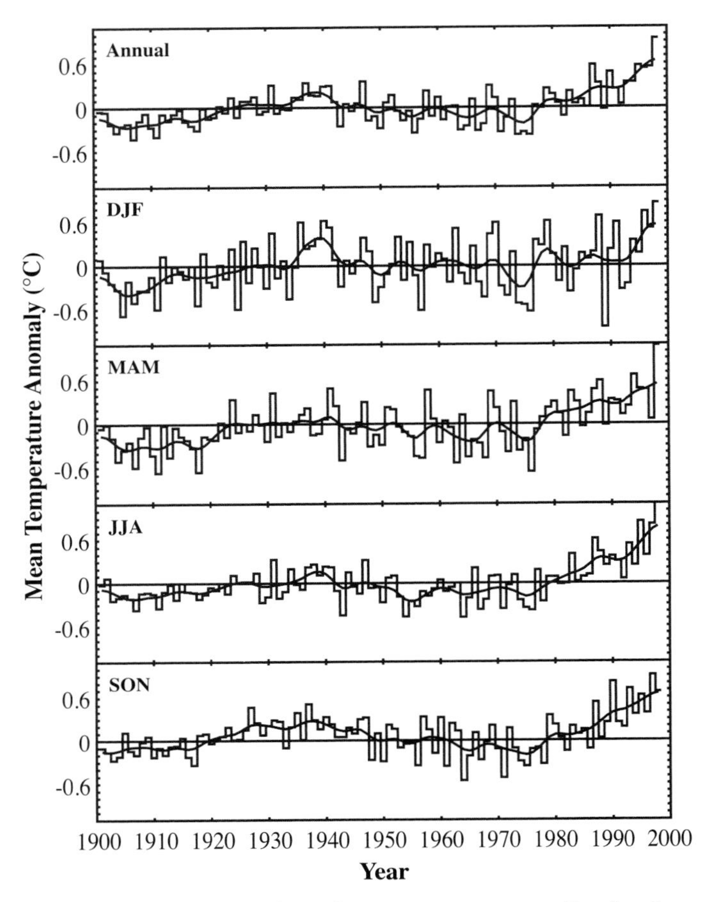

Figure 10-2: Mean surface air temperature anomalies for the African continent, 1901–1998, expressed with respect to 1961–1990 average, annual and four seasons (DJF, MAM, JJA, SON). Smooth curves result from applying a 10-year Gaussian filter.

10.1.4. Climate Change Scenarios

Carter *et al.* (2000) performed a comprehensive characterization of regional climate change projections for the 21st century. They assumed a range of atmospheric greenhouse gas (GHG) loadings according to the four draft marker scenarios developed for the IPCC *Special Report on Emissions Scenarios* (SRES), in combination with the IPCC range of climate sensitivities (1.5–4.5°C—IPCC, 1996). Their method of analysis involved combining simple climate model estimates of the global mean annual temperature response to four combinations of GHG forcing/climate sensitivity (see Table 3-9) with regional patterns of seasonal temperature and precipitation change obtained from 10 general circulation model (GCM) simulations for the end of the 21st century relative to 1961–1990. The GCM patterns of change were scaled up or down so that the global mean temperature change from the GCM coincided with that obtained from the simple climate model. Ten patterns of change were obtained for each emissions/climate sensitivity combination, and each was averaged over subcontinental regions, including five representing the African continent: southern Europe/north Africa, the Sahara, west Africa, east Africa, and southern Africa (Carter *et al.*, 2000; see Chapter 3). Ranges of projected rates of change in temperature and precipitation over these regions are depicted for each season, and projected changes over southern Africa by the 2050s are compared with modeled natural multi-tridecadal variability from the HadCM2 GCM 1,400-year control simulation (Tett *et al.*, 1997) for the summer (DJF) and winter (JJA) months.

An analysis using the similar methodology also has been conducted specifically for Africa (Hulme *et al.*, 2001). Future annual warming across Africa ranges from 0.2°C per decade (B1—low scenario) to more than 0.5°C per decade (A2—high scenario). This warming is greatest over the interior of semi-arid margins of the Sahara and central southern Africa. The intermodel range (an indicator of the extent of agreement between different GCMs) is smallest over north Africa and the equator and greatest over the interior of southern Africa.

Future changes in mean seasonal rainfall in Africa are less well defined. Under the lowest warming scenario, few areas experience changes in DJF or JJA that exceed two standard deviations of natural variability by 2050. The exceptions are parts of equatorial east Africa, where rainfall increases by 5–20% in DJF and decreases by 5–10% in JJA.

Under the two intermediate warming scenarios, significant decreases (10–20%) in rainfall during March to November, which includes the critical grain-filling period, are apparent in north Africa in almost all models by 2050, as are 5–15% decreases in growing-season (November to May) rainfall in southern Africa in most models.

Under the most rapid global warming scenario, increasing areas of Africa experience changes in summer or winter rainfall that exceed the one sigma level of natural variability. Large areas of equatorial Africa experience increases in DJF rainfall of 50–100% over parts of eastern Africa, with decreases in JJA over parts of the Horn of Africa. However, there are some JJA rainfall increases for the Sahel region.

Hulme *et al.* (2001) also analyzed future rainfall changes for three African regions—the Sahel, east Africa, and southeast Africa—to illustrate the extent of intermodel differences for these regions and to put future modeled changes in the context of past observed changes (see Figure 10-3). Although model results vary, there is a general consensus for wetting in East Africa, drying in southeast Africa, and a poorly specified outcome for the Sahel.

10.2. Key Regional Concerns

10.2.1. Water Resources

10.2.1.1 Overview of Regional Water Resources

Water resources are inextricably linked with climate, so the prospect of global climate change has serious implications for water resources and regional development (Riebsame *et al.*,

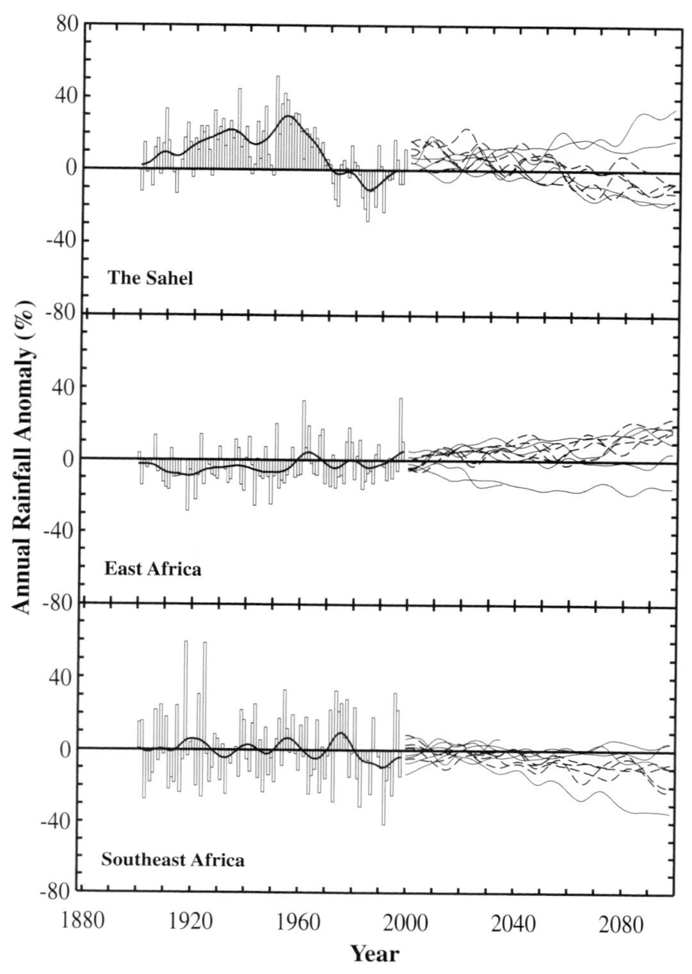

Figure 10-3: Observed annual rainfall anomalies for three African regions, 1900–1998, and model-simulated anomalies for 2000–2099. Model anomalies are for 10 model simulations derived from seven DDC GCM experiments; the four HadCM2 simulations are the dashed curves. All anomalies are expressed with respect to observed or model-simulated 1961–1990 average rainfall. Model curves are extracted directly from GCM experiments, and results are not scaled to SRES scenarios used in Chapter 3. Smooth curves result from applying a 20-year Gaussian filter (Hulme *et al.*, 2001).

1995). Efforts to provide adequate water resources for Africa will confront several challenges, including population pressure; problems associated with land use, such as erosion/siltation; and possible ecological consequences of land-use change on the hydrological cycle. Climate change—especially changes in climate variability through droughts and flooding—will make addressing these problems more complex. The greatest impact will continue to be felt by the poor, who have the most limited access to water resources.

Figure 10-4 shows that the hydrological performance of Africa results in much less runoff yield than in other regions. Apart from the Zambezi/Congo Rivers, the major African rivers (Nile, Niger, Senegal, Senqu/Orange, Rufiji) traverse semi-arid to arid lands on their way to the coast. Of the world's major rivers, the Nile has the lowest specific discharge (i.e., flow per unit catchment area), even if only the part of the catchment that receives precipitation is considered (Reibsame *et al.,* 1995). Furthermore, because these major rivers originate within the tropics, where temperatures are high, evaporative losses also are high in comparison to rivers in temperate regions. Elevated temperatures will enhance evaporative losses; unless they are compensated by increased precipitation, runoff is likely to be further reduced.

In Morocco, the northward displacement of the Azores high-pressure cell is a subject of study because of its asscociation with a drought cycle that is related to the dipole between the positioning of the Azores high-pressure cell and the Iceland low-pressure cell. These severe droughts seem to manifest themselves in Morocco in periodicities varying between 2 and 13 years (Stockton and Allali, 1992).

The Magreb region is characterized by erratic and variable rainfall, with a high rate of evapotranspiration (almost 80%). In addition, the Magreb region will have water scarcity by 2025, especially in Tunisia and Libya.

Associated with the poor hydrological performance of African river basins is the fact that most of the lakes in Africa have a delicate balance between precipitation and runoff; all of the

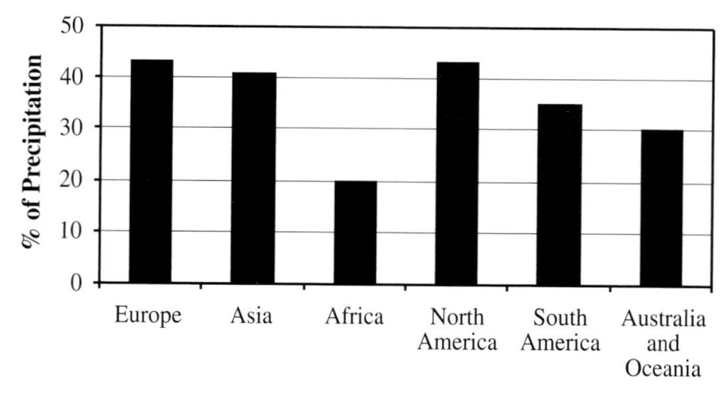

Figure 10-4: Comparative hydrology in world regions—total runoff as percentage of precipitation (GEMS, 1995).

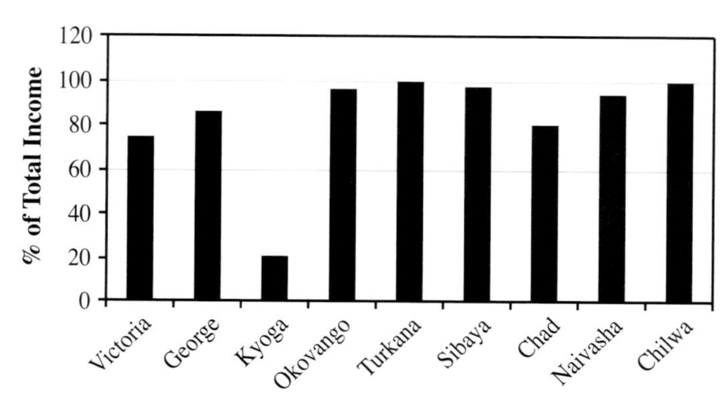

Figure 10-5: Evaporative losses as percentage of total hydrological income in selected African lakes (constructed from Talling and Lemoalle, 1998).

***Table 10-1**: Estimates of ranges of percentage changes in precipitation, potential evaporation, and runoff in African river basins, constructed from Figure 3 of Arnell (1999). In some basins, estimates given by HadCM3 simulation have been excluded where they appear to be outliers.*

Basin	Change in Precipitation (%)	Change in Potential Evaporation (%)	Change in Runoff (%)
Nile	10	10	0
Niger	10	10	10
Volta	0	4 to -5	0 to -15
Schebeli	-5 to 18	10 to 15	-10 to 40
Zaire	10	10 to 18	10 to 15
Ogooue	-2 to 20	10	-20 to 25
Rufiji	-10 to 10	20	-10 to 10
Zambezi	-10 to -20	10 to 25	-26 to -40
Ruvuma	-10 to 5	25	-30 to -40
Limpopo	-5 to -15	5 to 20	-25 to -35
Orange	-5 to 5	4 to 10	-10 to 10

large lakes show less than 10% runoff-to-precipitation ratio (Talling and Lemoalle, 1998), and important water basins like Lake Chad and the Okavango Delta have no outflow because evaporation and permiation balance runoff (see Figure 10-5).

In the savanna regions, the incidence of seasonal flow cessation may be on the increase, as shown by some streams in Zimbabwe (Magadza, 2000). Drought periods now translate into critical water shortages for industrial and urban domestic supplies (Magadza, 1996).

10.2.1.2. Major River Basin Systems

The heterogeneity of ground records in Africa imposes serious limitations in constructing future scenarios of water resources. Where consistent long-term climatic data are available, they indicate a trend toward reduced precipitation in current semi-arid to arid parts of Africa. Figure 10.3 of the Hulme *et al.* (2001) publication shows possible scenarios for different climatic regions of Africa. Although there is ambiguity in the Sahel, the simulations appear to indicate possible increases in precipitation in east Africa, whereas most simulations in southern Africa indicate reduced precipitation in the next 100 years.

Table 10-1 shows estimates of ranges of percentage changes in precipitation, potential evaporation, and runoff in African river basins as reconstructed from Arnell (1999, Figure 3). In some basins, estimates given by the HadCM3 simulation have been excluded where they appear to be outliers. A change in the hydrographs of large basins (Niger, Lake Chad, and Senegal) has been observed. Between the mean annual discharge of the humid and drought periods, the percentage of reduction varies from 40 to 60% (Olivry, 1993). Figure 10-6 shows the change in the hydrograph of the Niger River at the Niamey station. This illustrates a clear modification of the Niger River regime at Niamey. Similar situations are observed at the N'djamena station on the Chari at the entrance to Lake Chad. In the Nile basin, Sircoulon (1999) cites a reduction in runoff of 20% between 1972 and 1987, corresponding to a general decrease in precipitation in the tributary basins calculated by Conway and

Box 10-3. Impact of Drought in the Akasompo Dam (Graham, 1995)

Multiple droughts in recent decades have forced Ghana to reduce the generation of hydroelectricity, provoking a national debate about power supply.

The Akosombo and Kpong generating stations—commissioned in 1965 and 1982, respectively—provide the overwhelming bulk of Ghana's electricity. The two stations account for 1,072 MW of a total national power-generating capacity of 1,160 MW; Akosombo alone provides 833 MW.

Until recently, most of the minority of Ghanaians who use electricity tended to regard the Volta hydroelectric dams as sources of uninterruptible power. The unprecedented drought of 1982–1983, which compelled rationing of electricity until 1986, shattered that illusion. And if that drought's power cuts jolted the nation's complacency about hydroelectricity, the 1994 incident concentrated minds forcefully on the impermanence of power from Akosombo and Kpong and the need for alternative sources.

According to the Volta River Authority (VRA), the statutory power generating body, "cumulative inflow" into Volta Lake by the middle of August 1994 was "the worst...in the 50-year record of Volta river flows—worse than the same period in 1983." At its lowest, in early August, the level of Volta Lake was 73 m. This was well below the 75.6 m the VRA claims is the minimum level for generating power without risk of damaging the turbines.

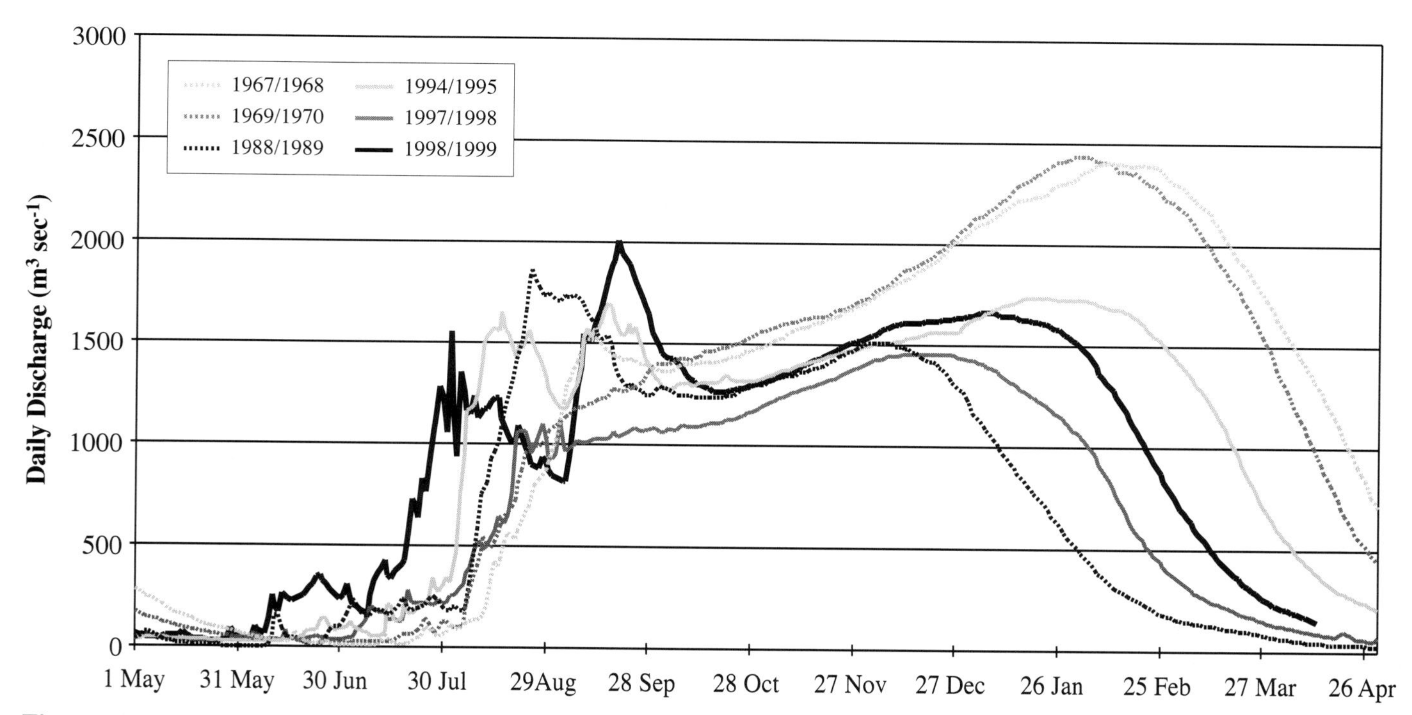

Figure 10-6: Decadal changes in hydrograph of the Niger River at Niamey Station between 1961 and 1999.

Hulme (1993). In recent years there have been significant interruptions in hydropower generation as a result of severe droughts (see Box 10-3).

Instrumental data and climate model simulations cited above indicate imminent water crisis in large parts of Africa. Several seminal works have appeared in the literature that analyze water for Africa, including Falkenmark (1989) and Gleick (1992, 1998).

Large basin-scale analyses often give the wrong impression that many areas of Africa are rich in water reserves, in which case local water problems could be solved easily by technology that would transfer water from the source to areas under stress, assuming that financial resources are available for such enterprises. Although in theory this may be a practical solution to many water problems in most of Africa, the very high costs associated with such projects make them impractical. Political goals such as self-sufficiency in food production and general socioeconomic development cannot be achieved under severe water scarcity (Falkenmark, 1989). Drought-prone zones of Africa already are water-limited, further increasing their vulnerability to water problems.

About 63% of the total land in Africa lies within transboundary river basins. Five major river basins—the Congo, Nile, Niger, Chad, and Zambezi—occupy about 42% of the geographical area and sustain more than 44% of the African population. Other shared basins in the continent are the Senegal, Gambia, Limpopo, Orange/Senqu, and Cunene Basins.

In west Africa, the dependency ratio—defined as the ratio between renewable water produced out of a country and the total renewable water of the same country—is more than 40% for seven of the nine countries comprising the Permanent Interstate Committee for Drought Control in the Sahel (CILSS). This ratio is nearly 90% for the Niger and the Mauritania. Similar transboundary dependencies are evident in southern Africa and on the Nile basin. The Congo basin is shared by the most countries (13), followed by the Niger and Nile basins (11 countries each) and the Zambezi and Chad basins (9 and 8 countries, respectively).

The impact of changes in precipitation and enhanced evaporation could have profound effects in some lakes and reservoirs. Conway and Hulme (1993) and Calder *et al.* (1995) have discussed the hydrology and paleohydrology of various African Lakes. Magadza (1996) has examined the impact of drought on reservoirs in Zimbabwe. Reports fron Ghana (Graham, 1995) indicate severe drought impacts on this large reservoir. Studies show that, in the paleoclimate of Africa and in the present climate, lakes and reservoirs respond to climate variability via pronounced changes in storage, leading to complete drying up in many cases. Furthermore, these studies also show that under the present climate regime several large lakes and wetlands show a delicate balance between inflow and outflow, such that evaporative increases of 40%, for example, could result in much reduced outflow. In the case of Lake Malawi, it has been reported that the lake had no outflow for more than a decade in the earlier part of this century (Calder *et al.*, 1995).

Predictions of response by the Nile to global warming are confounded by the fact that different simulations give conflicting results (Smith *et al.*, 1995), varying from 77% flow reduction in the Geophysical Fluid Dynamics Laboratory (GFDL) simulation to a 30% increase in the Goddard Institute for Space Studies

(GISS) model. Arnell's (1999) model results suggest increased precipitation in the Nile basin, but such gains are offset by evapotranspiration. Gleick (1992) projects that future climatic changes in the Nile basin would be significant and possibly severe. The response of the Nile basin to precipitation change is not linear, though it is symmetric for increased and decreased precipitation. Hulme (1992) shows a decline in total precipitation and overall warming of about 0.5°C over the last half on the 20th century. Conway and Hulme (1993) conclude that the effects of future climate change on Nile discharge would further increase uncertainties in Nile water planning and management, especially in Egypt. Nile precipitation responds more to changes in equatorial circulation, with little influence by the north African monsoon (Sestini, 1993).

Arnell (1999) shows that the greatest reduction in runoff by the year 2050 will be in the southern Africa region, also indicating that as the water use-to-resource ratio changes countries such as Zimbabwe and the Magreb region will shift into the high water-stress category. The Zambezi River has the worst scenario of decreased precipitation (about 15%), increased potential evaporative losses (about 15–25%), and diminished runoff (about 30–40%).

Lake Chad varies in extent between the rainy and dry seasons, from 50,000 to 20,000 km^2. Precise boundaries have been established between Chad, Nigeria, Cameroon, and Niger. Sectors of the boundaries that are located in the rivers that drain into Lake Chad have never been determined, and several complications are caused by flooding and the appearance or submergence of islands. A similar process on the Kovango River between Botswana and Namibia led to a military confrontation beteen the two states.

Vorosmarty and Moore (1991) have documented the potential impacts of impoundment, land-use change, and climatic change on the Zambezi and found that they can be substantial. Cambula (1999) has shown a decrease in surface and subsurface runoff of five streams in Mozambique, including the Zambezi, under various climate change scenarios. For the Zambezi basin, simulated runoff under climate change is projected to decrease by about 40% or more.

Growing water scarcity, increasing population, degradation of shared freshwater ecosystems, and competing demands for shrinking natural resources distributed over such a huge area involving so many countries have the potential for creating bilateral and multilateral conflicts (Gleick, 1992). Feddema (1998, 1999) has evaluated the impacts of soil degradation and global warming on water resources for Africa. All major watersheds are affected by global warming; although the trend is toward drying in most locations, there are significant differences in watershed-level responses, depending on timing and distribution of rainfall, as well as soil water-holding capacity. Soil water-holding capacity is modified by the degree of soil degradation.

10.2.1.3. Demography and Water Resources

Availability of water in sub-Saharan Africa (SSA) is highly variable. Only the humid tropical zones in central and west Africa have abundant water. Water availability varies considerably within countries as well, influenced by physical characteristics and seasonal patterns of rainfall. According to Sharma *et al.* (1996), eight countries were suffering from water stress or scarcity in 1990; this situation is getting worse as a consequence of rapid

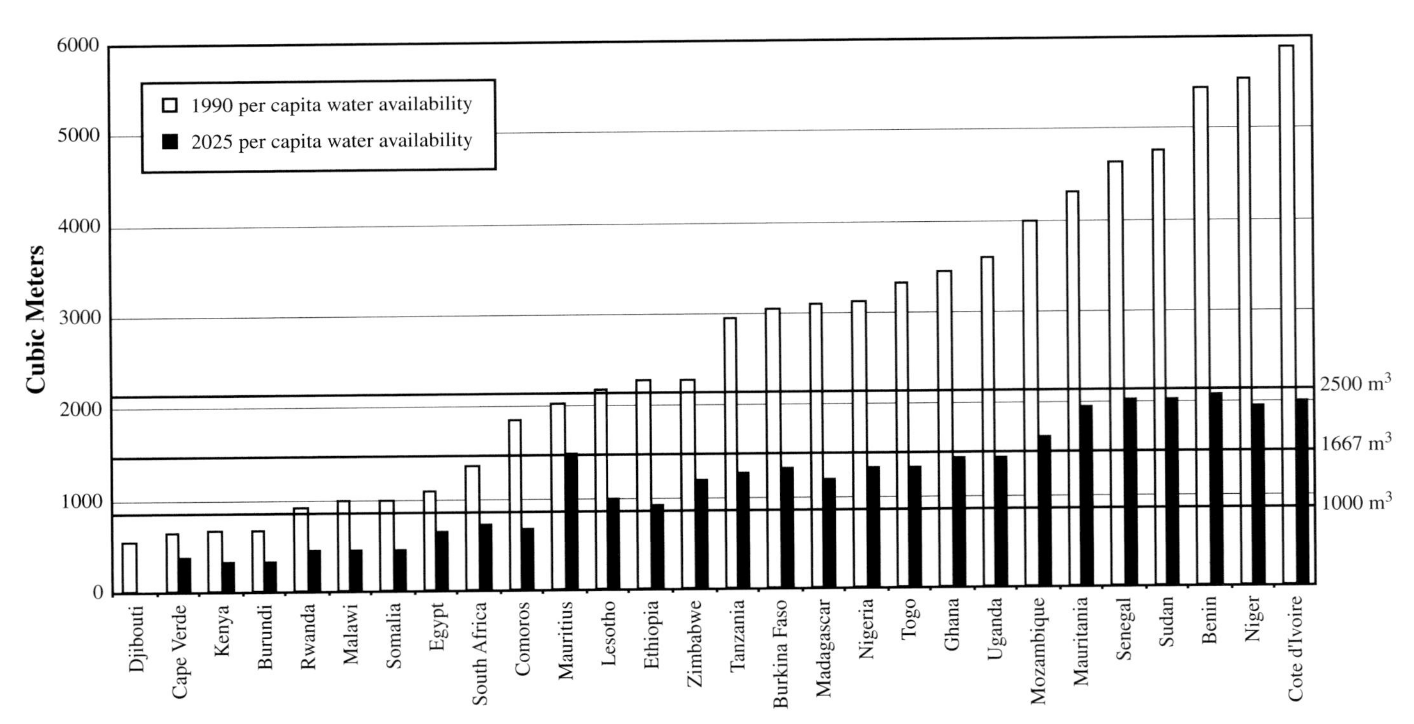

Figure 10-7: Water scarcity and people in Africa (Sharma *et al.*, 1996).

population growth, expanding urbanization, and increased economic development. By 2000, about 300 million Africans risk living in a water-scarce environment. Moreover, by 2025, the number of countries experiencing water stress will rise to 18—affecting 600 million people (World Bank, 1995). Figure 10-7 shows how countries will shift from water surplus to water scarcity as a result of population changes alone between 1990 and 2025, using a per capita water-scarcity limit of 1,000 m^3 yr^{-1}. Scarcity statistics also can be associated with challenges to international water resources: Many such basins face water stress or scarcity (Sharma *et al.*, 1996). Long-term precipitation records from the Sahara give a clear indication of declining precipitation in that region (UNEP, 1997). These declines in precipitation register as reduced hydrological discharges in major river basins in the subhumid zones.

Given the climate scenarios discussed in this report, it is apparent that several countries will face water availability restrictions by the middle of the 21st century, if current consumption trends persist. Pottinger (1997) discusses some concerns about future water availability in South Africa. These examples show that the combination of demographic trends and climate change is likely to cause economically significant connstraints in some parts of Africa.

10.2.1.4. Impacts and Vulnerability

Sensitivity analyses of major rivers of the continent indicate that these rivers are sensitive to climate change. Magadza (1996) examined changes in water storage in Zimbabwe's main water storage facilities during the 1991–1992 drought cycle. During this period, the mean temperature averaged 2°C above the long-term seasonal mean, and seasonal Penman evaporation exceeded the long-term seasonal mean by more than 30%—peaking at just less than 90% in February at the Kutsaga station. During this drought cycle, stored water resources dwindled to less than 10% of installed capacity.

Urbiztondo (1992) simulated the response of Lake Kariba power-generating capacity to various climate change scenarios. His results indicate that with no significant change in precipitation, the lake would regulary fail to meet installed generating capacity if the temperature rose by 4°C. During dry years, generating capacity would decrease by as much as 50%. Even during wet years, maximum generating capacity would barely exceed 50% of installed capacity.

Economic impacts from curtailment of hydropower generation from Lale Kariba, as a result of the 1991–1992 drought, were estimated to be US$102 million loss in GDP, US$36 million loss in export earnings, and loss of 3,000 jobs (Benson and Clay, 1998) The direct impacts on agriculture and the knockon impacts also were quite severe. These limited estimates provide a window on potential economic impacts of climate change-mediated water resources changes in the medium term (i.e., into the middle of the 21st century—a time span within the planning window of economic development strategies).

With an estimated error of about 8%, the Gambia River flow has been shown to be very sensive to climate change. Based on the results of river flow responses and vulnerability analysis, climate variables alone can cause a 50% change in runoff in the Gambia River catchment (Jallow *et al.*, 1999). In general, a 1% change in rainfall will result in a 3% change in runoff (Manneh, 1997). On the whole, this translates into saltwater intrusion into the Gambia River by 40 km at times of maximum intrusion. However, the flushing action of freshwater flow of the river will keep the saline/freshwater interface at an equilibrium, oscillating between 90 and 290 km. The rate of saline water intrusion during dry seasons would increase from about 20 km per month in January to about 40 km per month in April and May.

10.2.1.5. Adaptation

Likely changes in precipitation and discharge regimes call for a wide range of adaptations. Broadly, these adaptations would include:

- Refinement of early warning sytems to enable timely remedial measures
- Shared basin management, necessitating international agreements
- Water-use strategies—especially demand management—in industry, settlements, and agriculture
- Intensified monitoring to improve data reliability
- Intensive research into energy usage and alternate renewable energy at household and industrial levels
- Intensive research into design of infrastructure facilities, such as roads and telecommunications, to withstand extreme events
- Intensive research into flood control management technology
- Innovation in building designs (e.g., to minimize urban flooding)
- Research into and commencement of coastal defense facilities
- Research into adaptive agricultural startegies
- Reserarch into environmental flow requirements.

Although there are now subregional climate change scenarios for the African region, the quality of such scenarios varies with the intensity of historical data and the spatial distribution of monitoring stations. There is an urgent need to intensify the density of monitoring stations to improve climate change scenarios. The cost of rehabilitating stations that are in disrepair is not beyond the financial capability of African states. Appreciation of the strategic importance of these facilities and a sense of ownership of climate change concerns needs to be reawakened.

It has been noted that practically all of the major river basins of Africa include several states. In recognition of this fact, the past decade has seen the development of international river basin mangement protocols—such as the Southern Africa

Development Community (SADC) Protocol on Shared Waters, the Niger Basin Authority, and several others, including the more recent Lake Victoria Fisheries Authority. The United Nations Environment Programme International Environmental Technology Center (UNEP-ITEC) have emphasized the river basin as the fundamental unit of management (UNEP, 2000). We recommend that these international basin authorities be strengthened in terms of finance and human resources and that their perspectives should embrace near- and long-term climate variability and climate change issues in their work plans. Most important is that the legal framework for such river basin authorities should be robust to ensure equity in access to and accountability for water supply and water quality management. Failure to take these concepts on board could lead to water resources-related conflict.

In general, the African continent lacks technical strategies to optimize water resources. A few countries (e.g., South Africa and Zimbabwe) have begun to develop strategies for optimum use of water resources—using, for example, water pricing and demand management tools. Crop-watering technology is primitive and wasteful in most cases. Few industrial and household water-usage strategies incorporate water reuse. During drought periods, management authorities have resorted to supply restrictions, such as the 3-day supply per week in Mutare during the 1991–1992 drought. Water supply shortages conventionally are addressed through construction of more impoundments. Magadza (1996) remarks that in severe drought periods there would be a multiple failure syndrome of water storage facilities, especially where individual reservoirs are dedicated to defined communities. Although supply structures (banning of garden hoses) and construction of storage reservoirs are practical options, demand management—which reduces consumption per unit of product output—has proved increasingly to be a water-saving strategy that can allow communities to enter a drought cycle with adequate supplies.

Whatever strategies are adopted for optimizing water usage, successful development of such strategies is contingent on reliable meteorological and hydrological information. In many instances, application of hydrological models on a basin-wide scale is restricted by data density. Reliable impacts assessments and near-term predictions depend on robust databases. For example, flood propagation and thus flood warning capability, which are real-time processes, are a function of the density of measuring points. Similarly, crop yield forecasts could be made spatially more accurate by improving the intensity of climatic measurements.

Over the past half-century, Africa has invested heavily in hydroelectric power schemes. Recent drought episodes and demand escalation have highlighted the vulnerability of even the largest hydroelectric plants to climate variability (Magadza, 1996). This assessment has shown that future water resources, especially in the subhumid to semi-arid regions of the African savanna and subtropics, will be more restricted. Research into other forms of renewable energy and energy-use efficiency in industry and households is an essential inverstment in a more energy-secure future. At the regional level, energy resource sharing is a necessary strategy. Individual states must have trust and confidence to invest in neighboring countries for overall regional energy security.

Populations that live in flood-prone areas need to consider strategies for early warning procedures for flood events; strategic planning in the location of human habitations to minimize flood impacts; and strategies for robust protocols for alleviating impacts of drought events to minimize loss of human life, economic assets, and societal norms. On the other hand, early warning systems can be effective if impacted areas are accessible in the worst-case scenario, to enable either evacuation or relief supplies delivery.

Although there are major reservoirs on most large African river basins, these reservoirs were not designed for flood control. However, synchronization of operations of reservoirs that are located in the same basin can alleviate flood impacts. Nevertheless, there is a need to consider purposely building flood control facilities in some of the flood-prone areas of Africa, similar to those found on the Danube, which reduce flood crest intensity by sequestering floodwaters into temporary storage facilities along the river. This could be a subject of directed research for each African basin.

Research into coastal defense systems is an immediate need. Several African coastal areas already are experiencing sigificant coastal impacts. Coastal management infrastructures are likely to entail intergenerational investment programs in which each extent generation must make its contribution to minimize long-term costs. If our generation abrogates our responsibilities to future generations, we will impose immense costs on our posterity.

To minimize sensitivity to climate change, African economies should be more diversified, and agricultural technology should optimize water usage through efficient irrigation and crop development. Considerable advances have been made in agricultural industries of southern Africa, particularly in South Africa and Zimbabwe.

As water resource stresses become acute in future water-deficit areas of Africa as a result of a combination of climate impacts and escalating human demand, there will be intensifying conflict between human and environmental demands on water resources. Because maintanance of healthy ecosystems is an underpinning to economic sustainability, there is need in each water basin management unit to identify and factor into development projects the need for environmental flows.

10.2.2. Food Security

10.2.2.1. Context of Food Security

Present and future prospects for food security are significant determinants of the impacts of climate change. International

***Table 10-2**: Indicators of regional vulnerability in Africa (WRI, 1996).*

(a) Vulnerability Indicators

	Expenditure on Food (% of consumption)	Food Aid (Cereals) kg per capita	Refugees	Adult Female Literacy (%)	Infant Mortality (per 1000)
African Region[a]					
– Northern	42	18	221,450	45	59
– Sudano-Sahelian	42	13	974,800	17	119
– Gulf of Guinea	39	6	819,750	28	109
– Central	39	3	480,500	41	91
– Eastern	37	4	1,408,150	43	102
– Indian Ocean	57	12	0	73	66
– Southern	57	15	1,793,800	53	85
Total	57	10	5,698,450	35	97
Comparison Country					
– Bangladesh	59	12	245,300	22	108
– Thailand	30	2	255,000	90	26
– Mexico	35	3	47,300	85	35
– Greece	30	-1	1,900	89	8
– United Kingdom	12	-3	24,600		7

(b) Regional Agriculture in Africa

	Pop. Density (pop. km^{-2})	Pop. Growth (%)	Crop Land (% of total)	Irri-gated Land (% of total)	Avg. Yield of Cereals (kg ha^{-1})	Ferti-lizer Use (kg yr^{-1})	Food Prod. Index (1970= 100)	GNP per Capita (US$)	GNP in Agri-culture (%)	GNP Growth Rate (% yr^{-1})	Public Agri-cultural Invest-ment (US$)
Region[a]											
Northern	226	2.25	5	27	1,973	94	115	1,285	17	3.60	25
Sudano-Sahelian	106	2.72	4	7	727	5	90	860	34	2.36	7
Gulf of Guinea	891	2.83	21	2	892	6	100	760	39	1.87	15
Central	145	2.70	4	1	923	2	87	760	22	2.15	5
Eastern	451	2.88	10	2	1,363	12	92	593	47	3.05	13
Indian Ocean	262	1.96	5	23	1,988	140	98	280	22	3.85	6
Southern	208	2.56	6	7	929	27	76	333	21	3.38	7
Total	253	2.65	6	8	1,098	25	92	355	30	2.75	11
Comparison Country											
Bangladesh	9,853	2.18	72	31	2,572	101	96	205	37	4.20	68
Thailand	1141	0.92	45	19	2,052	39	109	1,697	13	7.80	78
Mexico	491	1.55	13	21	2,430	69	100	2,971	8	1.50	129
Greece	795	0.07	30	31	3,700	172	101	6,530	17	1.60	25
UK	2,404	0.19	28	2	6,332	350	112	33,850	2	2.80	347

[a] **Northern**: Algeria, Egypt, Libya, Morocco, Tunisia; **Sudano-Sahelian**: Burkina Faso, Cape Verde, Chad, Djibouti, Eritria, The Gambia, Mali, Mauritania, Niger, Senegal, Somalia, Sudan; **Gulf of Guinea**: Benin, Cote d'Ivoire, Ghana, Guinea, Guinea-Bissau, Liberia, Nigeria, Sierra Leone, Togo; **Central**: Angola, Cameroon, Central African Republic, Congo, Equatorial Guinea, Gabon, Sao Tome and Principe, Democratic Republic of Congo; **Eastern**: Burundi, Ethiopia, Kenya, Rwanda, Tanzania, Uganda; **Indian Ocean**: Comoros, Madagascar, Mautitius, Seychelles; **Southern**: Botswana, Lesotho, Malawi, Mozambique, Namibia, South Africa, Swaziland, Zambia, Zimbabwe.

agricultural systems and socioeconomic conditions at the household level are major elements of vulnerability. The consequences of present vulnerability for hunger and nutrition are marked. Regional indicators related to food security are shown in Table 10-2.

Food production in most of SSA has not kept pace with the population increase over the past 3 decades. In Africa as a whole, food consumption exceeded domestic production by 50% in the drought-prone mid-1980s and more than 30% in the mid-1990s (WRI, 1998). Food aid constitutes a major proportion of net food trade in Africa, and in many countries it constitutes more than half of net imports. In Kenya and Tanzania, for instance, food aid constituted two-thirds of food imports during the 1990s. Despite food imports, per capita dietary energy supply (DES) remains relatively low (Hulme, 1996); about one-third of the countries in Africa had per capita DES of less than 2,000 kcal day^{-1} in the 1990s—lower than the minimum recommended intake (data from WRI, 1998).

Agricultural and economic growth must rise—perhaps by 4% yr^{-1}—to realize basic development goals. Today, only a few countries achieve this rate of growth. One consequence of agricultural growth could be a doubling by the year 2050 of cultivated land area—at great cost to the natural environment—unless there is greater investment in agricultural management and technology on existing cropland (Anon, 1999). The scale of food imports fosters dependence on food production in the rest of the world. Africa faces the risk that supplies will fluctuate drastically with the rise and fall of grain reserves and prices on international markets. A major challenge facing Africa is to increase agricultural production and achieve sustainable economic growth; both are essential to improving food security.

Agriculture is not only a vital source of food in Africa; it also is the prevailing way of life. An average of 70% of the population lives by farming, and 40% of all exports are earned from agricultural products (WRI, 1996). One-third of the national income in Africa is generated by agriculture. Crop production and livestock husbandry account for about half of household income. The poorest members of society are those who are most dependent on agriculture for jobs and income. On average, the poor from developing countries of SSA spend 60–80% of their total income on food (see Odingo, 1990; WRI, 1998; FAO, 1999b). Although industry is significant in a few patches, it still is in its infancy. In many countries, the level of mechanization—including irrigation, processing, and storage facilities—is particularly low.

High-quality land resources per household have shrunk in Africa over the past 2 decades, often dramatically. Traditional, social, and legal status in the sub-Saharan region is responsible for unequal access to land. This, in turn, increases the risk of resource degradation. Lack of land tenure security reduces the motivation to invest in conservation of resources.

Agriculture and household incomes are characterized by large interannual and seasonal variations. The annual flow of income normally rises and peaks during the harvest season. Nonagricultural and migrant, off-farm wage incomes are substitutes during the dry season. The period preceding the harvest is critical: Farmers engage in unemployment-induced migration to urban centers as one of the strategies for coping with scarcity. Fluctuations in annual food production resulting from climate variability place a heavy reliance on food aid, at the national and household levels.

Reduced food supplies and high prices immediately affect landless laborers who have little or no savings. Poverty, population, and sometimes conflicts combine to affect education in many African countries. As a result of population pressure as well as rural and urban economic depression, population mobility sets in. Populations move from the savanna to the forest, from plateaus to drained valleys, from landlocked countries to coastal

***Table 10-3**: Undernourishment in Africa (FAO, 1999b).*

Region	Number of People (millions) 1996–1997	% of Population 1979–1981	% of Population 1995–1997
Africa			
– Central Africa	35.6	36	48
– East and southern Africa	112.9	33.5	43
– West Africa	31.1	40	16
– North Africa	5.4	8	4
Other Regions			
– Caribbean	9.3	19	31
– Central America	5.6	20	17
– South America	33.3	14	10
– Eastern Asia	176.8	29	14
– South Asia	283.9	38	23
– Southeast Asia	63.7	27	13
– Western Asia	27.5	10	12
– Indian Ocean	1.1	31	24

Table 10-4: Comparison of indices of human development and food security for regions in developing countries [Human Development Index (UNDP, 1998); data from WRI (1998) and UNDP (1998)].

Human Development Index (HDI)	Food Security			
	Low	Med-Low	Med-High	High
Low	West Africa, South Asia, East and South Africa, Central America, Central Africa			
Medium	South Pacific	Southeast Asia, North Africa		
High	East Asia, Caribbean	West Asia, South America, Central Asia	West Europe, Central and East Europe	Australasia, Indian Ocean, North America

areas and those with infrastructures, and from rural to urban centers within a country in search of better lands and opportunities (Davies, 1996). Internal mobility and its consequences vary from country to country: It is low in Ghana, Madagascar, Malawi, Burundi, and Rwanda but high in Burkina Faso and Kenya and very high in Cote d'Ivoire. Migrations lead to high and rising urban growth across the African region. This translates into increasing pressure on the environment, including social amenities.

The consequences of chronic and episodic food insecurity in Africa are evident in the prevalence of hunger. Nearly 200 million people in Africa are undernourished. In central, eastern, and southern Africa, more than 40% of the population is undernourished, and the number has risen over the past few decades. SSA is home to almost one-quarter of the developing world's food-deprived people, with variations across the continent (FAO, 1999b). Although west Africa has the largest total population of any of the African subregions, it has the fewest undernourished people. By contrast, east Africa has more than twice as many undernourished people. According to anthropometric surveys made between 1987 and 1998, 33% of African children are stunted, underweight, or wasted (FAO, 1999b) (see Table 10-3). Illiteracy and ignorance, along with poor housing and infrastructure, are predisposing conditions to ill health in many countries, which impacts food security.

African food security and potential to adapt to climate change can be portrayed by using national indicators. Of course, local conditions of vulnerability are critical; aggregate indicators are only one way of illustrating the relative risks and potential impacts of climate change. Two indices are shown in Table 10-4. The Human Development Index (HDI) is a composite of measures of life expectancy, literacy, education, and income (GDP per capita), as promoted by the United Nations Development Programme (UNDP). Except for north Africa, African regions score in the lowest group on the HDI. Among other world regions, only Central America is in the lowest group. A similar index of food security has been constructed, using indices of trends in food production, available food as a percentage of requirements, and arable land per capita (see Downing, 1991); north Africa is in the medium–low group. Relative to other regions, Africa clearly is among the regions with the lowest food security and the lowest ability to adapt to future changes (as indicated by the HDI).

The state of food security is not uniform, and there has been considerable progress in some countries (FAO, 1999b). For example, undernourishment in Ghana has decreased more rapidly than in any other country in the world, fueled by economic growth and consequent improvements in cropped area and yields. Plagued by population growth and conflict, Burundi is in stark contrast: Average daily food intake fell from 2,020 kcal in 1980 to 1,669 kcal in 1996.

The implications of this state of food insecurity in Africa for climate change are significant. The risks of adverse effects on agriculture, especially in semi-arid and subhumid regions and areas with more frequent and prolonged drought, become life-threatening risks. Internal coping mechanisms—through farm improvement, employment, and trade—are not likely to be adequate for many of the vulnerable populations. If food insecurity prevents private investment in agricultural economies (internal and from multinational corporations), resources for adapting to climate change may not keep pace with impacts. However, it also is clear that Africa has enormous resources—natural and human—that can be tapped to make rapid gains in food security and thus reduce the risk of adverse climate change.

10.2.2.2. Marine and Freshwater Fisheries

African nations possess a variety of lacustrine, riverine, and marine habitats with more than 800 freshwater and marine species (as noted in IPCC, 1998). GCMs do not provide direct information on water quality and other hydrological parameters that affect fisheries (Hlohowskyj *et al.*, 1996). As such, vulnerability assessments must translate projected atmospheric changes into changes in aquatic environments, making it possible for ecological and biological responses to climate change to be identified and evaluated. Most studies on the potential impacts of climate change on fisheries have been done for temperate-zone fisheries. In these studies, the emphasis has been to evaluate the impacts of changes in the availability of thermal habitat on fishery resources and evaluate the effects of temperature on physiological processes of fish.

Temperature increases may affect lake fisheries, although sensitivity across Africa is likely to vary. For example, Ntiba (1998) used empirically derived models to elucidate the relationship between long-term fishery yield data with climatic index value for Lakes Naivasha and Victoria in Kenya. The mean annual temperature is taken as the climatic index because of data availability and close correlation between air and water surface temperatures. The results indicate that in Lake Victoria, mean annual temperature has greater effect on fish yields than morphoedaphic index (MEI); the opposite is the case in Lake Naivasha. For Lake Naivasha, a rise of as much as 2°C above the current mean annual temperature may not even double the yield from the current 12 kg ha^{-1} yr^{-1} (yield regressed to temperature only). The additional effect of MEI will double and triple the yield with a rise of 1.5 and 2°C, respectively. For Lake Victoria, the maximum predicted yield under current climatic conditions is estimated to be 81.8 kg ha^{-1} yr^{-1} and is predicted to more than triple to 263.7 kg ha^{-1} yr^{-1} with a rise in temperature of 0.5°C. However, natural aquatic ecosystems have a finite carrying capacity at which fishery yields will reach a maximum sustainable yield (MSY) (Russell and Yonge, 1975); it is unlikely that the yield will exceed 84.2 kg ha^{-1} yr^{-1} in Lake Victoria, considering that the estimated MSY for the entire lake is 74.0 kg ha^{-1} yr^{-1} (Turner, 1996).

On Lake Kariba, Magadza (1996) found that drought years were accompanied by decreased fisheries catch. Hart and Rayner (1994) show that the distribution of copepods on the African continent is temperature-dependent, with species examined showing restricted temperature range preferences. In the laboratory, Magadza (1977) found that the optimum temperature for the reproduction of *Moinia dubia* is 24°C, with temperatures exceeding 28°C showing reproductive failure. Chifamba (2000) found negative relationships between catch per unit (CPU) effort and temperature in the pelargic *Limnothrissa miodon* fishery on Lake Kariba, whereas precipitation and river runoff were positively correlated with CPU. These observations indicate possible depression of planktivorous pelargic freshwater fisheries as a result of climate change impacts. Where such fisheries constitute a signifiant protein source, such impact will bear on food security.

A further consideration in the possible impacts of global warming on inland fisheries is the thermal behaviors of inland waters. In tropical areas, a unit change in temperature elicits a greater density change per unit temperature change. Thus, at the higher tropical temperatures indicated by climate models, thermal stratifications are likely to be more stable. In eutrophic lakes, anoxia and amonia intoxication leading to massive fish deaths—as repeatedly witnessed in Lake Chivero—are likely to be significant (Magadza, 1997; Moyo, 1997).

Aquaculture is a food production activity with one of the highest growth rates in the world. Risk management in aquaculture must take into account the level and frequencies of extreme events in assessing available technical options, based on the environment and climatic conditions.

Riverine fisheries will be affected. With a potential warming of 3–5°C projected for the next century, productivity of the Gambia River is estimated to increase by about 13–21% (Jallow *et al.*, 1999). There would be little or no effect on the suitability of the present habitat for some fish and shrimp species. In contrast, warming of more than 3°C will have negative impacts on habitats for catfish, and warming of more than 4°C will reduce the suitability of the present habitat for herring. Shrimp yield is estimated to increase by about 38–54%.

Upwelling of the Canary Current produced by the northeast and southwest tradewinds makes the fishery off the coast of Morocco one of the most productive in the world. Data from the Institut National de Recherche Haleutique du Maroc show that NAO weakens the upwelling, increases temperature, and reduces sardine stocks (Hilmi *et al.*, 1998).

The pelagic fishing industry in the southwest of Africa (based on the Benguela upwelling) contains several migratory species, such as the anchovy (Crawford *et al.*, 1987). These species are an important resource in their own right, but they also are a key element in the food chain of larger fish, seals, and birds. Recruitment in these species is influenced by water temperature, and growth is affected by the state of the upwelling, both of which are linked to global climatic conditions (Shannon *et al.*, 1990; Siegfried *et al.*, 1990). For instance, there is a phenomenon similar to the El Niño effect on the fisheries of Peru that leads to an intrusion of warm, nutrient-poor water from the north in some years, which severely depresses fish yields (Shannon *et al.*, 1990). The net impact of global climate change on southern African fisheries remains unclear because the potential impact on ocean circulation and wind shear in the coastal zone is uncertain. In the event of major reorganization of the circulation of the southern oceans (Hirst, 1999)—which is a possibility at high rates of warming—the impacts most likely would be severe.

10.2.2.3. Crop Production

Major impacts on food production will come from changes in temperature, moisture levels, ultraviolet (UV) radiation, CO_2 levels, and pests and diseases. CO_2 enrichment increases photosynthetic rates and water-use efficiency (WUE) (see Chapter 5). The direct effects are largest on crops with C_3 photosynthetic pathway (wheat, rice, and soybean) compared with C_4 crops (maize, sorghum, millet, and sugarcane). Increases in local temperatures may cause expansion of production into higher elevations. The grain filling period may be reduced as higher temperatures accelerate development, but high temperatures may have detrimental effects on sensitive development stages such as flowering, thereby reducing grain yield and quality. Crop water balances may be affected through changes in precipitation and other climatic elements, increased evapotranspiration, and increased WUE resulting from elevated CO_2.

Specific examples of impacts on crops are available. Pimentel (1993) notes that global warming is likely to alter production of rice, wheat, corn, beans, and potatoes—staples for millions of people and major food crops in Africa. Staple crops such as wheat and corn that are associated with subtropical latitudes may suffer a drop in yield as a result of increased temperature, and rice may disappear because of higher temperatures in the tropics (Odingo, 1990).

The possible impact of climate change on maize production in Zimbabwe was evaluated by simulating crop production under climate change scenarios generated by GCMs (Muchena and Iglesias, 1995). Temperature increases of 2 or 4°C reduced maize yields at all sites; yields also decreased under GCM climate change scenarios, even when the beneficial effects of CO_2 were included. It is suggested that major changes in farming systems can compensate for some yield decreases under climate change, but additional fertilizer, seed supplies, and irrigation will involve an extra cost. The semi-extensive farming zone was particularly sensitive to simulated changes in climate, and farmers in this zone would be further marginalized if risk increases as projected.

Analysis of potential impacts, using dynamic simulation and geographic databases, has been demonstrated for South Africa and the southern Africa region by Schulze *et al*. (1993) (see also Schulze *et al.,* 1995; Hulme, 1996; Schulze, 2000). Relatively homogenous climate and soil zones were used to run agrohydrological, primary productivity, and crop yield models. The results reaffirm the dependence of production and crop yield on intraseasonal and interannual variation of rainfall.

Impacts on crops need to be integrated with potential changes in the agricultural economy. Yates and Strzepek (1998) describe an integrated analysis for Egypt (see Figure 10-8). Their model is linked to a dynamic global food trade model, which is used to update the Egyptian sector model and includes socioeconomic trends and world market prices of agricultural goods. Impacts of climatic change on water resources, crop yields, and land resources are used as inputs into the economic model. The climate change scenarios generally had minor impacts on aggregated economic welfare (sum of consumer and producer surplus); the largest reduction was approximately 6%. In some climate change scenarios, economic welfare slightly improved or remained unchanged. Despite increased water availability and only moderate yield declines, several climate change scenarios showed producers being negatively affected by climate change. The analysis supports the hypothesis that smaller food-importing countries are at risk of adverse climate change, and impacts could have as much to do with changes in

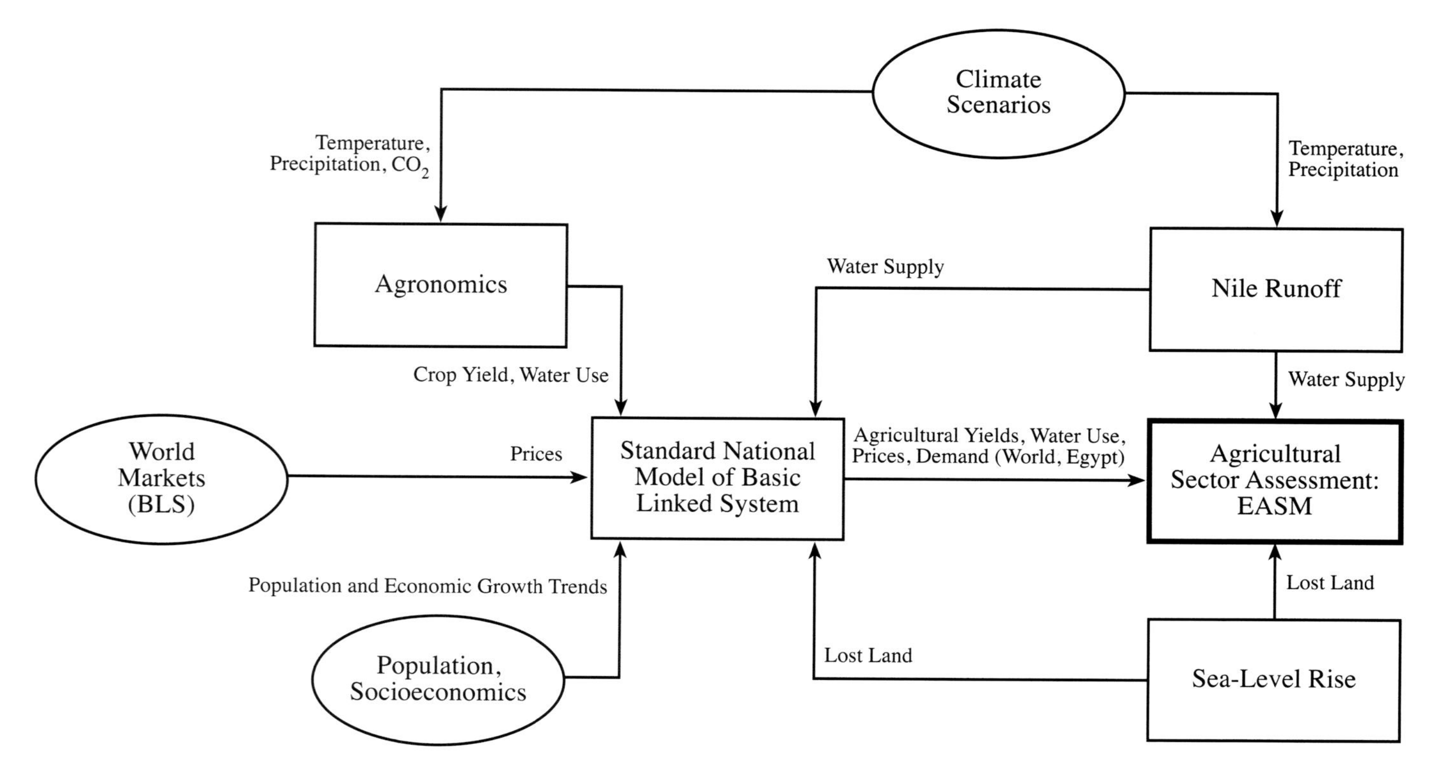

Figure 10-8: Schematic of forward-linkage approach to integrated assessment of climate change impacts on Egypt, using agricultural sector model (Yates and Strzepek, 1998).

world markets as with changes in local and regional biophysical systems and shifts in the national agricultural economy.

10.2.2.4. Livestock

Other than pigs, domestic livestock in Africa are concentrated in the arid and semi-arid zones. This is because the more humid areas were historically prone to livestock diseases such as *nagana* (a trypanosome carried by the tsetse fly—Ford and Katondo, 1977), typically support grasses of low digestibility (Scholes, 1990, 1993), and often are densely settled by crop agriculturalists. The overwhelming majority of these animals feed predominantly off natural grasslands and savannas, although crop residues are an important supplement during the dry season. Many urban and rural families also keep poultry.

Domestic livestock play a central role in many African cultures. Cattle and camels, in particular, have an importance that goes beyond the production of meat. Their value is based on the full set of services they supply (milk, meat, blood, hides, draft power), their asset value as a form of savings, and their cultural symbolism. It would be difficult and damaging for these cultures to abandon pastoralism in the event that it becomes climatically, environmentally, or economically unviable.

Although classical concepts of animal carrying capacity may not be very useful as local management tools in the context of African semi-arid systems with high interannual variability (Behnke *et al.*, 1993), they remain valid as indicators of animal production when they are applied over decadal periods and large areas. Many researchers have demonstrated a strong link between long-term, large-area herbivore biomass (LHB, kg km^{-2}) in African wildlife and pastoral systems and mean annual precipitation (MAP, mm) (Coe *et al.*, 1976; van den Berg, 1983). Subsequent studies have shown that this relationship is strongly influenced by soil type or its proxy, the underlying geology (East, 1984; Fritz and Duncan, 1993). These relationships have the form $\log(LHB) = A \times \log(MAP) + B$, where A and B are constants for a particular soil type. Because the values of A lie between 1.45 and 1.55, the relationships are exponentially increasing but approximately linear over the typical range of interannual variation at one site (see also Le Houerou, 1998, for a similar analysis in west Africa and the Mediterranean basin). Thus, in broad terms, changes in range-fed livestock numbers in any African region will be directly proportional to changes in annual precipitation. Given that several GCMs predict a decrease in MAP on the order of 10–20% in the main semi-arid zones of Africa, there is a real possibility that climate change will have a negative impact on pastoral livelihoods. The following additional factors must be considered.

The causal chain between rainfall and animal numbers passes through grass production, which also is approximately linearly related to rainfall (Breman, 1975; Le Houerou and Hoste, 1977; Rutherford, 1995). The slope of this relationship, which can be expressed as WUE, is a function of soil nutrient availability (De Ridder *et al.*, 1982; Scholes, 1993). WUE also is a function of CO_2 concentration in the atmosphere (Mooney *et al.*, 1999), especially in semi-arid regions. Because the CO_2 concentration will rise in the future, its positive impact on WUE (which is on the order of 20–30% for doubled CO_2, even in C_4-dominated grasslands such as these) will help to offset reduction in rainfall of the same magnitude. Simulations of grassland production in southern Africa indicate an almost exact balancing of these two effects for that region (Scholes *et al.*, 2000).

About 80% of the grazing lands of Africa are in savannas, a vegetation formation that consists of a mix of trees and grasses. Grass production in savannas is strongly depressed by tree cover (see the many references reviewed by Scholes and Archer, 1997). Because domestic livestock, with the exception of goats, predominantly eat the grass in these systems, future changes in tree cover are an important issue from the point of view of carbon sequestration and livestock production. Tree biomass ultimately is related to climate and soil type, but the mechanism appears to be via fire frequency and intensity. If fire frequency and intensity were to decrease—as a result of climate changes (for instance, an increase in dry-season rainfall) or, more likely, changes in land management—woody cover would increase. This has been demonstrated in numerous fire experiments in Africa (Trapnell *et al.*, 1976; Booysen and Tainton, 1984). Given the vast area of the savannas, the carbon sequestration potential is substantial (Scholes and van der Merwe, 1996). In addition, emissions of tropospheric ozone (O_3) precursors would decrease if savanna burning were reduced. The disadvantage would be a more than proportional decrease in livestock carrying capacity, as a result of the nonlinear suppressive effect of trees on grass production.

The bioclimatic limit of savannas in southern Africa is related to winter temperatures. An increase in temperature of 1–2°C—well within the range predicted for next century—would make the montane grasslands (highveld) of southern Africa susceptible to invasion by savanna trees (Ellery *et al.*, 1990).

In moister regions, animal productivity is limited not by the gross availability of fodder but by its protein (nitrogen) content (Ellery *et al.*, 1996). Increasing the CO_2 concentration or the rainfall will not increase the protein availability; thus, livestock in these regions are likely to be less responsive to the direct effects of atmospheric and climate change. Under elevated CO_2, the carbon-to-nitrogen ratio of forage will decrease, but this will not necessarily lead to decreased forage palatability despite the dilution of protein (Mooney *et al.*, 1999). This may be because in grasses, the bulk of the excess carbon is stored in the form of starch, which is readily digestible. Widespread use of protein and micronutrient feed supplements and new technology for the control of veterinary diseases will have a greater impact on livestock numbers and productivity, especially in the "miombo" region of south central Africa.

Domestic livestock, like other animals, have a climate envelope in which they perform optimally. The limits of the envelope are quite broad and can be extended by selecting for heat or cold tolerance, feed supplementation, or providing physical shelter for the animals. African cattle are mostly from the *Bos indicus* line, which is more heat-tolerant than the European line of *Bos taurus*. In extremely hot areas (mean daily warm-season temperatures greater than body temperature), even the *Bos indicus* breeds are beyond their thermal optimum (Robertshaw and Finch, 1976). Meat and milk production declines, largely because the animals remain in the shade instead of foraging. There is limited potential for extending this limit through breeding. Adaptation would require substitution by a species such as the oryx, which is physiologically equipped for high temperatures and low water supply.

In the higher altitude and higher latitude regions of Africa, livestock (typically sheep) currently are exposed to winter temperatures below their optimum. Mortality often results when cold periods coincide with wet periods, if the animals have not been herded to shelter. These episodes are likely to decrease in frequency and extent in the future.

Livestock distribution and productivity could be indirectly influenced via changes in the distribution of vector-borne livestock diseases, such as *nagana* (trypanosomiasis) and the tick-borne East Coast Fever and Corridor Disease (Hulme, 1996). Simulations of changes in the distribution of tsetse fly (*Glossina spp.*) indicate that with warming it could extend its range southward in Zimbabwe and Mozambique, westward in Angola, and northeast in Tanzania, although in all these simulations there were substantial reductions in the prevalence of tsetse in some current areas of distribution. The tick *Riphicephalus appendiculatus* was predicted to decrease its range in southern and eastern Africa and increase its range in the central and western part of southern Africa (Hulme, 1996).

One land-use model (IMAGE 2.0—Alcamo, 1994) projects that large parts of Africa will be transformed to pastoral systems during the 21st century. The model logic that leads to this conclusion is that increasing urbanization and a rising standard of living typically are associated with a change in dietary preference toward meat. The area that currently is used for meat production therefore would need to expand, assuming that meat demand was not met by import or by increased productivity of existing herds. These are reasonable but untested assumptions, and their consequences have major implications for biodiversity conservation and atmospheric composition. The areas indicated as being converted to pastures (largely the subhumid tropics) already support cattle to some degree. Increased cattle production would require widespread tree clearing (leading to conversion of a carbon sink into a carbon source), eradication of key cattle diseases, and the use of protein and micronutrient feed supplements. The quantity of fuel consumed by savanna fires would decrease (because it would be grazed), reducing the release of pyrogenic methane (CH_4) and O_3 precursors, but the production of methane from enteric fermentation would increase. Because methane production per unit of grass consumed is higher for enteric fermentation than for savanna fires (Scholes *et al.*, 1996), the result is likely to be a net increase in radiative forcing.

10.2.2.5. Impacts of Drought and Floods

Food security in Africa already is affected by extreme events, particularly droughts and floods (e.g., Kadomura, 1994; Scoones *et al.*, 1996). The ENSO floods in 1998 in east Africa resulted in human suffering and deaths, as well as extensive damage to infrastructure and crops in Kenya (Magadza, 2000). Floods in Mozambique in 2000 and in Kenya in 1997–1998 sparked major emergency relief as hundreds of people lost their lives and thousands were displaced from their homes (Brickett *et al.*, 1999; Ngecu and Mathu, 1999; see also <www.reliefweb.int>). The cost in Kenya alone was estimated at US$1 billion (Ngecu and Mathu, 1999). Droughts in 1991–1992 and 1997–1998 affected livelihoods and economies and heightened renewed interest in the impacts of climatic hazards (e.g., Kadomura, 1994; Campbell, 1999). For example, the impacts of the 1991–1992 drought in Zimbabwe are estimated to have been 9% of GDP (Benson and Clay, 1998).

Such climatic episodes can serve as an analog of climate change. Irrespective of whether climate change will cause more frequent or more intense extreme events, it is apparent that many aspects of African economies are still sensitive to climatic hazards. At the local level, some coping strategies are less reliable (Jallow, 1995)—for instance, Campbell (1999) notes that plants and trees used as food by pastoralists in southern Kenya declined between 1986 and 1996. National governments often struggle to provide food security during times of crisis (Ayalew, 1997; Gundry *et al.*, 1999). For national and international agencies, the cost of climatic hazards—impacts, recovery, and rehabilitation—may result in a shift in expenditure from reducing vulnerability to simply coping with immediate threats (e.g., Dilley and Heyman, 1995).

10.2.2.6. Adaptation Strategies

The nature and processes of human adaptation to long-range climate change are poorly understood, especially in Africa (but see Chemane *et al.*, 1997; Vogel, 1998). Often, human responses are assumed, or assumed to be rational with foresight and equity. Chapter 18, Smith and Lenhart (1996), and Smithers and Smit (1997) provide overviews; national assessments provide more detail.

A promising approach for much of Africa is to cope with current climate variability through the use of seasonal climate forecasting (e.g., Mason *et al.*, 1996; Mattes and Mason, 1998; Washington *et al.*, 1999; Dilley, 2000). If farmers can adapt to current year-to-year variability through the use of advance information on the future season's climate and institutional systems are in place to respond to short-term changes (such as early warning systems), communities will be in a position to adapt to longer term climate changes. For example, a seasonal maize water-stress forecast for the primary maize-growing regions of South Africa and Zimbabwe anticipates water stress 6 months prior to harvest, with hindcast correlation over 16 seasons of 0.92 for South Africa and 0.62 for Zimbabwe, based on correlations between water stress and historical global SST and sea-level pressure records (Martin *et al.*, 2000). Similar forecasts are possible in other regions and for other crops; however, seasonal forecasting by itself will not improve food security (e.g., Stack, 1998).

Better soil and water conservation practices, more tolerant crop varieties, improved pest and weed control, and more use of irrigation also are needed to adapt to changes in the weather. Omenda *et al.* (1998) recommend that in areas predicted to have a decline in precipitation, research into the development of maize varieties that are higher yielding, drought-resistant, early maturing, and disease- and pest-tolerant is desirable. They further suggest that methods of improving maize culture be studied, including use of inorganic fertilizers and manure and changes in planting dates. Better adaptation to climate change also will result from the use of improved technologies in agriculture—for example, in irrigation and crop husbandry. It has been suggested (Pinstrum-Anderson and Pandya-Lorch, 1999), for example, that by failing to capitalize on new opportunities that biotechnology offers, SSA may further add to its food insecurity and poverty problems.

Increased meat production can be achieved without massive expansion of area grazed or size of the herds, by application of modern herd and animal diet management. In many situations, this would require a cultural shift from regarding livestock principally as an asset and symbol to regarding them as a production system. Cow-calf systems with supplemental feeding can achieve offtakes of 40% yr^{-1} (compared to the 10–20% typical of most current herd management in Africa), permitting lower livestock numbers per hectare (Preston and Leng, 1987). This would achieve a reduction in CO_2 and CH_4 emissions. A limited amount of increased heat tolerance can be introduced through breeding programs.

The Senegal River basin (SRB) provides an illustration of sensitivity to climatic variations and opportunities for adaptation. The SRB is undergoing fundamental environmental, hydrologic, and socioeconomic transitions (Venema *et al.*, 1997). Senegal, Mauritania, and Mali—through the river basin development authority, the Organisation pour la Mise en Valeur du Fleuve Senegal (OMVS)—are promoting irrigated rice production for domestic consumption in the river basin to ease the severe foreign exchange shortfalls facing these riparian nations. With the recent completion of Manantali and Diama dams, year-round irrigated agriculture is now possible in the SRB. The full agricultural development potential of the SRB is constrained, however, by the basin's limited water resources, which are sensitive to climatic variations. An alternative approach to the SRB's scarce water resources is an agricultural development policy that is based on village-scale irrigation projects and intensive, irrigated agroforestry projects (Venema *et al.*, 1997). Village-scale irrigation is dedicated to low-water-consumption cereal grain crops and is managed by traditional sociopolitical structures. The proposed agroforestry production system has the dual objectives of using irrigation to reestablish a protective biomass cover in the desertifying river valley and reversing drought-induced migration from rural to urban areas. A comparative river system simulation was carried out to analyze the effects of the rice production development policy and the natural resources management policy on the SRB's full agricultural development potential. The simulation study compared three alternative hydrologic scenarios, using the pre-drought era, the 1970s-level drought, and the 1980s-level drought. Dynamic programming applied to water allocation in the Manantali reservoir showed that lower overall water demands for the natural resource policy scenarios had higher agricultural development potential than the proposed policy based on rice production.

A significant drawback in combating effects of climate variability is the failure of African governments to devolve power to people who are affected and to link environmental degradation to economic policy (Darkoh, 1998). Consequently, many programs lack local support or are undermined by conflicting trade and agricultural policies pursued by governments. It is contended that, for sustainable development strategies to work, policies should put the welfare of people at the center of the development agenda and give them the rights and power to determine their future. Policies should empower the people to develop adaptive strategies toward sustainable livelihoods. Moreover, threats posed to the environment and development by protection and overconsumption in the north and structural adjustment programs (SAPs) call for the removal of distortions created by the import barriers of developed countries, curbing of overconsumption, and a fundamental revision in the structure of SAPs to help alleviate poverty and protect the environment in these African countries. It is in the interest of the global community that the environment in Africa is protected.

Operational early warning systems in Africa—including the SADC Regional Early Warning Unit, the U.S. Agency for International Development (USAID) Famine Early Warning System Network, and the World Food Program (WFP) Vulnerability Assessment and Mapping unit—assess the vulnerability of rural households in many parts of Africa to food insecurity each year. Vulnerability assessments generally develop a picture of which geographic areas and which social groups will be unlikely to meet their subsistence needs before the next agricultural season, based on a convergence of available environmental and socioeconomic information (USAID, 1999). The USAID Sahel vulnerability assessment for 1999–2000 estimated that 3.8 million people were moderately food insecure in a high-rainfall year (USAID, 2000). This could signify chronic vulnerability resulting from structural weaknesses caused by desertification, climate change, and other long-term environmental and socioeconomic phenomena.

10.2.3. Natural Resource Management and Biodiversity

10.2.3.1. Forest and Woodland Resources

In Africa, forests—as defined and reported by FAO (1999a)—cover 5 million km², one-sixth of the continent's land area. The moist tropical forests of the Congo constitute the second most extensive rainforest in the world and a globally important reserve of carbon. Trees and shrubs constitute an important component of the more than 12 million km^2 of agricultural lands, pastures, shrublands, and savannas outside of closed-canopy forest areas.

Trees and shrubs provide ecosystem services of carbon sequestration, storing and transpiring water required for precipitation, maintaining soil fertility, and forming habitats for a diverse array of plant and animal species. Moreover, forest and woodland species also provide firewood, structural timber, traditional medicines, staple foods, and drought emergency foods. Because a large fraction of the population lives in rural areas, they depend on trees and shrubs for many of their subsistence needs. Indeed, firewood and charcoal provide approximately 70% of the energy used in Africa. Moreover, the export of timber, nuts, fruit, gum, and other forest products generates 6% of the economic product of African countries (FAO, 1999a). Thus, climate change renders vulnerable the large part of the African population that is dependent on forest species for subsistence needs and the nontrivial fraction of the economy that is based on forest products.

Because climate change alters the spatial and temporal patterns of temperature and precipitation, the two most fundamental factors determining the distribution and productivity of vegetation—geographical shifts in the ranges of individual species and changes in productivity—constitute the most likely impacts of CO_2-induced climate change on forest species. Research in Senegal (Gonzalez, 1997, 2001) has documented retraction of mesic species to areas of higher rainfall and lower temperature as a result of desertification in the last half of the 20th century. These changes have caused a 25–30 km southwest shift of Sahel, Sudan, and Guinean vegetation zones in half a century, proceeding at an average rate of 500–600 m yr^{-1}—foreshadowing the magnitude of projected vegetation shifts driven by CO_2-induced climate change (Davis and Zabinski, 1992). In northwest Senegal, the human population density is 45 people km^{-2}, whereas forest species can support only 13 people km^{-2} under altered conditions (Gonzalez, 1997).

Dry woodlands and savannas in semi-arid and subhumid areas will be increasingly subjected to drying in the next century, as well as increasing land-use intensity—including conversion to agriculture (Desanker *et al.*, 1997). Moreover, CO_2-induced climate change is very likely to alter the frequency, intensity, seasonality, and extent of vegetation fires that are critical to the maintenance of areas such as the Serengeti grasslands of east Africa, the miombo woodlands of southern Africa, and the fynbos of the Cape. Across the continent, farmers traditionally use fire to clear agricultural fields in forest areas and areas outside closed-canopy forest; pastoralists and hunter-gatherers use fire to improve the quality of plant resources available during the dry season. Satellite remote sensing reveals that more than half of the continent experiences a fire regime with a frequency greater than once per decade (Kendall *et al.*, 1997; Levine *et al.*, 1999).

Although the broad geographical pattern of fire-prone vegetation clearly is climatically related (van Wilgen and Scholes, 1997), the aspect of the fire regime that is most sensitive to the type and degree of climate change suggested for Africa is likely to be fire intensity, rather than its frequency or extent. Fire intensity is related largely to the available dry-season fuel load, which in turn is strongly and positively related to rainfall in the preceding wet season and nonlinearly related to woody plant cover. In the miombo woodlands, it is predicted that increased fire will expand savanna areas at the expense of wooded areas (Desanker *et al.*, 1997). Because emissions of CH_4, tropospheric O_3 precursors, and aerosols from vegetation fires in Africa constitute a significant contribution to the global budgets of these species (Crutzen and Andreae, 1990; Hao *et al.*, 1990; Scholes *et al.*, 1996), changes in the African fire regime could have consequences for global and regional climate.

Modeling of the distribution of forest species on the basis of the Holdridge (1967) life zone classification has projected changes from mesic vegetation to xeric vegetation in Tanzania and The Gambia (Jallow and Danso, 1997) but a shift from arid vegetation to moist vegetation in Mozambique (Bila, 1999). It is not suggested that vegetation formations and their associated fauna (biomes) will migrate as a unit. It is more likely that species will respond to changing climate and disturbance regimes individualistically, with substantial time lags and periods of reorganization. The broad pattern of productive potential of vegetation zones is likely to move with greater spatial integrity because there is a degree of functional redundancy in ecosystems.

The most promising adaptation strategies to declining tree resources include natural regeneration of local species, energy-efficient cookstoves, sustainable forest management, and community-based natural resource management. The most effective adaptation to the decline of trees and shrubs in semi-arid areas is natural regeneration of local species. In addition, the *ban ak suuk* cookstove in Senegal and the *jiko* ceramic stove in Kenya have both produced energy-efficient gains in semi-arid areas (Dutt and Ravindranath, 1993). These practices generally depend on the ability of local people to exercise power to inventory and manage local resources in systems of community-based natural resource management. Decentralization of decisionmaking and revenue allocation authority has promoted efficient forest management in small areas of Niger, Madagascar, and Zimbabwe (FAO, 1999a). All of these practices constitute "no regrets" strategies that society would want to undertake under any climate scenario for their intrinsic environmental and economic benefits.

10.2.3.2. Indigenous Biodiversity and Protected Areas

Africa occupies about one-fifth of the global land surface and contains about one-fifth of all known species of plants, mammals, and birds in the world, as well as one-sixth of amphibians and reptiles (Siegfried, 1989). This biodiversity is concentrated in several centers of endemism. The Cape Floral Kingdom (fynbos), which occupies only 37,000 km^2 at the southern tip of Africa, has 7,300 plant species—of which 68% occur nowhere else in the world (Gibbs, 1987). The adjacent Succulent Karoo biome contains an additional 4,000 species, of which 2,500 are endemic (Cowling *et al.,* 1998). These floristic biodiversity hotspots both occur in winter rainfall regions at the southern tip of the continent and are threatened particularly by a shift in rainfall seasonality (for instance, a reduction in winter rainfall amounts or an increase in summer rainfall, which would alter the fire regime that is critical to regeneration in the fynbos). Other major centers of plant endemism are Madagascar, the mountains of Cameroon, and the island-like Afromontane habitats that stretch from Ethiopia to South Africa at altitudes above about 2,000 m (Mace *et al.,* 1998). Montane centers of biodiversity are particularly threatened by increases in temperature because many represent isolated populations with no possibility of vertical or horizontal migration. Several thousand species of plants alone are potentially affected.

The broad patterns of African zoogeography also are climatically linked, but the location of concentrations of biodiversity and endemism, at least in the higher animals, is located in the savannas and tropical forests. World antelope and gazelle biodiversity (more than 90% of the global total of 80 species) is concentrated in Africa (Macdonald, 1987). Changes in climate of the magnitude predicted for the 21st century could alter the distribution range of antelope species (Hulme, 1996).

This biodiversity forms an important resource for African people. Uses are consumptive (food, fiber, fuel, shelter, medicinal, wildlife trade) and nonconsumptive (ecosystem services and the economically important tourism industry).

For a sample of 39 African countries, a median 4% of the continental land surface is in formally declared conservation areas in southern Africa (MacKinnon and MacKinnon, 1986). The fraction of landscape that is conserved varies greatly between countries (from 17% in Botswana to 0% in four countries), as does the degree of actual protection offered within nominally conserved areas (MacKinnon and MacKinnon, 1986). A very large fraction of African biodiversity occurs principally outside of formally conserved areas (especially in central and northern Africa), as a result of a relatively low rate of intensive agricultural transformation on the continent. This will no longer be true if massive extensification of agriculture and clearing of tropical forests occurs in the humid and subhumid zones, as is predicted to occur in the next century by some land-cover change models (Alcamo, 1994). Patterns of human pressure, including grazing by domestic stock, also will be altered and intensified by climate change. Land-use conversion effects on biodiversity in affected areas will overshadow climate change effects for some time to come.

In the medium term (~10–20 years), biodiversity of indigenous plants and animals in Africa is likely to be affected by all of the major environmental changes that constitute climate change. These include changes in ambient air temperature, rainfall and air vapor pressure deficit (which combine to cause altered water balance), rainfall variability, and atmospheric CO_2. Africa—like the other continents, though perhaps to a greater degree—is characterized by ecosystem control through disturbance, such as fire (Bond and van Wilgen, 1996) and grazing regimes. Changing disturbance regimes will interact with climate change in important ways to control biodiversity—for instance through rapid, discontinous ecosystem "switches." For example, changes in the grazing and fire regime during the past century are thought to have increased woody-plant density over large parts of southern Africa. Ecosystem switches are accompanied by drastic species shifts and even species extinction. Subtle changes in species composition of rich ecosystems such as forests will impact biodiversity resources. A significant reduction in rainfall or increase in evapotranspiration in Angola would threaten the Okavango delta wetland in Botswana. Much larger scale ecosystem switches (e.g., savanna to grassland, forest to savanna, shrubland to grassland) clearly occurred in the past (e.g., during the climatic amelioration dating from the last glacial maximum), but diversity losses were ameliorated by species and ecosystem geographical shifts. The geographical range shifts recquired to preserve biodiversity into the future will be strongly constrained by habitat fragmentation and cannot realistically be accommodated by a static nature reserve network with the low areal coverage evident in Africa.

Theory required to predict the extent and nature of future ecosystem switches and species geographical shifts in Africa is lacking, and case studies are few. The response of major vegetation types to changes such as rising atmospheric CO_2 are almost unstudied, although early evidence (Midgley *et al.,* 1999; Wand *et al.,* 1999) suggests, for example, that these responses may increase WUE in grass species significantly, which may increase grass fuel load or even increase water supply to deeper rooted trees. Recent analysis of tree/grass interactions in savannas suggests that rising atmospheric CO_2 may increase tree densities (Midgley *et al.,* 1999); this kind of ecosystem switch would have major implications for grazing and browsing animal guilds and their predators. For southern Africa, between one-quarter and one-third of current reserves were predicted to experience a biome shift (a major change in the dominant plant functional types) under the equilibrium climate resulting from a twice-preindustrial CO_2 concentration (Hulme, 1996). In South Africa, increased aridity in the interior Bushmanland plateau will introduce a desert-like environment to the country (Rutherford *et al.,* 1999). Analysis for South African conservation areas (Rutherford *et al.,* 1999) shows potentially large losses of plant species diversity in this semi-arid region with low landscape heterogeneity.

Thus, the vegetation and animal communities that many reserves aim to conserve will no longer be within their preferred bioclimatic region. Migration of animals to conserved areas with more suitable climate (if these exist) will be constrained by fragmentation of intervening ecosystems and potentially hostile landscapes. The required rate of migration may be too rapid for unassisted movement of most plant species, especially over relatively flat landscapes (Rutherford *et al.,* 1995). Without adaptive and mitigating strategies, the impact of climate change will be to reduce the effectiveness of the reserve network significantly, by altering ecosystem characteristics within it and causing species emigrations or extinctions.

At particular risk of major biodiversity loss are reserves on flat and extensive landscapes, those in areas where rainfall regime may change seasonality (e.g., the southern Cape), those where the tree/grass balance is sensitive to CO_2 conditions, and those where the fire regime may be altered. Species most at risk are those with limited distribution ranges and/or poor dispersal abilities, habitat specialists (soil specialists in the case of plants), and those that are responsive to specific disturbance regimes.

Mitigation and adaptation strategies will be greatly strengthened by a risk-sharing approach between countries, which could attempt to share the burden of conserving critical populations in a collaborative way. Part of this risk-sharing approach could include transboundary nature reserves, where this is appropriate for increasing connectivity in areas projected to change significantly. The corridor approach within and between countries would have the added benefit of increasing reserve resilience to current climate variability and would increase attractiveness to the tourism industry. Economic incentives, however, may differ across geographic scales. For the moist tropical forests of Masoala National Park in northeast Madagascar, economic incentives favor conservation at local and global scales, although logging provides more profit at a national scale (Kremen *et al.*, 2000).

A high degree of uncertainty is associated with predictions of the biodiversity effects of climate change. No systematic analysis of mechanisms of ecosystem switches, or areas exposed to them, has been carried out. Although fire and atmospheric CO_2 seem to be important determinants of ecosystem structure and function, little research is available to predict how these factors will interact with other environmental changes. The effects of CO_2 on grass water use may be an important mitigator of negative effects on productivity for grazer guilds in much of subtropical Africa. Effects of shifts of disease-prone areas on animal populations are unstudied.

10.2.3.3. Migratory Species

10.2.3.3.1. Large-mammal migratory systems

The vast herds of migratory ungulates in east and southern Africa remain a distinguishing ecological characteristic of the continent. A major migratory system is located in the Serengeti area of Tanzania and the Masai-Mara region of Kenya. Reduced large-mammal migratory systems persist in the Kalahari (Botswana, South Africa, and Namibia) and Etosha (Namibia) areas of southern Africa.

Migrations typically are regular, and between dry-season and wet-season grazing areas, and to that extent they are sensitive to climate change. There is currently no indication that the broad pattern of seasonality is likely to change in the Serengeti or the Kalahari, since they are controlled by gross features of the atmospheric circulation (the monsoon system and the position of the Hadley cells). The intensity of seasonality, and the absolute annual rainfall total could change, by about 15% in either direction (Hulme, 1996). This is well within the range of interannual variability. Thus the migratory systems are likely to persist if land-use pressures permit them to.

10.2.3.3.2. Bird migrations

About one-fifth of southern African bird species migrate on a seasonal basis within Africa, and a further one-tenth migrate annually between Africa and the rest of the world (Hockey, 2000). A similar proportion can be assumed for Africa as a whole. One of the main intra-Africa migratory patterns involves waterfowl, which spend the austral summer in southern Africa and winter in central Africa. Palearctic migrants spend the austral summer in locations such as Langebaan lagoon, near Cape Town, and the boreal summer in the wetlands of Siberia. If climatic conditions or very specific habitat conditions at either terminus of these migratory routes change beyond the tolerance of the species involved, significant losses of biodiversity could result. Although the species involved have some capacity to alter their destinations, in an increasingly intensively used world the probability of finding sufficient areas of suitable habitat in the new areas is small. The current system of protected habitats under the Ramsar Convention is based on the present distribution of climate.

10.2.3.3.3. Locust migrations

Aperiodic locust outbreaks characterize the desert/semi-arid fringe in southern Africa and the Sahelian region. The population biology of the outbreak phenomenon is strongly linked to climate, particularly the pattern of soil moisture and temperature (Hanrahan *et al.*, 1992). Outbreaks typically occur when a dry period is followed by good rains—for instance, in southern Africa following an El Niño episode. Changes in El Niño frequency would impact the timing, location, and extent of locust outbreaks in ways that presently are unpredictable.

10.2.3.3.4. Human migratory systems

Semi-arid areas of the Sahel, the Kalahari, and the Karoo historically have supported nomadic societies that respond to intra-annual rainfall seasonality and large interannual variability through migration. Nomadic pastoral systems are intrinsically quite robust to fluctuating and extreme climates (because that is what they evolved to cope with), provided they have sufficient scope for movement and other necessary elements in the system remain in place. The prolonged drying trend in the Sahel since the 1970s has demonstrated the vulnerability of such groups to climate change when they cannot simply move their axis of migration because the wetter end already is densely occupied and permanent water points fail at the drier end. The result has been widespread loss of human life and livestock and substantial changes to the social system.

10.2.4. Human Health

The IPCC *Special Report on Regional Impacts of Climate Change* (IPCC, 1998) acknowledges that climate will have an impact on vector-borne diseases. The assessment in that report is limited to a qualitative analysis of the impacts. The report identifies the scarcity of disease distribution maps and models as a handicap to establishing current baseline limits. In the case of malaria, however, a continental effort—Mapping Malaria Risk in Africa—is underway. No such parallel efforts, however, are underway for other diseases in the African continent that may be affected by climate change (e.g., arboviruses, trypanosomiasis, schistosomiasis). No specific references are made to water- and food-borne and epizootic/epidemic diseases in Africa.

In recent years it has become clear that climate change will have direct and indirect impacts on diseases that are endemic in Africa. Following the 1997–1998 El Niño event, malaria, Rift Valley fever, and cholera outbreaks were recorded in many countries in east Africa (see Table 10-5 for a summary of disease outbreaks for the 1997–1999 period). The meningitis belt in the drier parts of west and central Africa is expanding to the eastern region of the continent. These factors are superimposed upon existing weak infrastructure, land-use change, and drug resistance by pathogens such as *Plasmodium falciparum* and *Vibrio cholerae*.

10.2.4.1. Vector-Borne Diseases: Malaria

Although the principal causes of malaria epidemics in the African highlands still are a subject of debate in the literature (Mouchet *et al*., 1998), there is increasing evidence that climate change has a significant role (WHO, 1998). In a highland area of Rwanda, for example, malaria incidence increased by 337% in 1987, and 80% of this variation could be explained by rainfall and temperature (Loevinsohn, 1994). A similar association has been reported in Zimbabwe (Freeman and Bradley, 1996). Other epidemics in east Africa have been associated largely with El Niño. It can be expected that small changes in temperature and precipitation will support malaria epidemics at current altitudinal and latitudinal limits of transmission (Lindsay and Martens, 1998). Furthermore, flooding could facilitate breeding of malaria vectors and consequently malaria transmission in arid areas (Warsame *et al*., 1995). The Sahel region, which has suffered from drought in the past 30 years, has experienced a reduction in malaria transmission following the disappearance of suitable breeding habitats. Yet, there are risks of epidemics if flooding occurs (Faye *et al*., 1995).

***Table 10-5**: Summary of number of countries in Africa reporting diseases/outbreaks from 1997 to July 1999.[a] Note that outbreaks indicate above-normal disease prevalence.*

Disease	1997	1998	1999 (Jan–July)
Malaria	0	2	2
Rift Valley fever	0	4	1
Yellow fever	1	1	0
Meningits	3	2	
Plague	2	1	2
Cholera	8	10	7
Dengue	0	0	0

[a] (WHO: Outbreak, <http://www.who.int/emc/outbreak_news/n1997/feb>). No reports were available for schistosomiasis, trypanosomiasis, onchocerciasis, and filariasis.

10.2.4.2. Cholera

Cholera is a water- and food-borne disease and has a complex mode of transmission. Flood causes contamination of public water supplies, and drought encourages unhygienic practices because of water shortage. The seventh pandemic currently is active across Asia, Africa, and South America.

Colwell (1996) demonstrates the link between cholera and SST. Upwelling of the sea as a result of increased SST increases the abundance of phytoplankton, which in turn supports a large population of zooplankton—which serves as a reservoir of cholera bacteria. Besides other epidemiological factors, the effects of SST on the spread of cholera may be the most profound because they affect large areas of the tropical seas and lakes. During the 1997–1998 El Niño, a rise in SST and excessive flooding (WHO, 1998a) provided two conducive factors for cholera epidemics that were observed in Djibouti, Somalia, Kenya, Tanzania, and Mozambique—all lying along the Indian Ocean.

Cholera epidemics also have been observed in areas surrounding the Great Lakes in the Great Rift Valley region. Birmingham *et al*. (1997) found significant association between bathing, drinking water from Lake Tanganyika, and the risk of infection with cholera. Shapiro *et al*. (1999) have made a similar observation along the shores of Lake Victoria. It is likely that warming in these African lakes may cause conditions that increase the risk of cholera transmission. This is an area that urgently requires research. According to WHO (1998a), Africa accounted for 80% of the total reported number of cholera cases globally in 1997.

10.2.4.3. Meningitis

Major epidemics of meningococcal infections usually occur every 5–10 years within the African "meningitis belt;" they usually start in the middle of the dry season and end a few months later with the onset of the rains (Greenwood, 1984). Between February and April 1996, the disease affected thousands of people in parts of northern Nigeria, many of whom died (Angyo and Okpeh, 1997). This epidemic spread from the original meningitis belt to Kenya, Uganda, Rwanda, Zambia, and Tanzania (Hart and Cuevas, 1997). One of the environmental factors that predisposes to infection and epidemics is low humidity (Tikhomirov *et al.*, 1997). However, a climate-meningitis association was not clear in parts of the Gulf of Guinea (Besancenot *et al.*, 1997). That this disease has been limited to the semi-arid areas of Africa suggests that its transmission could be affected by warming and reduced precipitation.

10.2.4.4. Rift Valley Fever

From 1931 (when the disease was first described) until the end of the 1970s, Rift Valley fever (RVF) was considered to be a relatively benign zoonoses for humans that periodically developed in domestic animals (especially sheep) following heavy rains (Lefevre, 1997). Recent research indicates that although epizootics in east Africa are associated with an increase in rainfall, a similar association is unknown in west Africa (Zeller *et al.*, 1997). Recent data from west Africa indicate that the risk of a new epizootic is increasing in the region (Fontenille *et al.*, 1995), with significant exposure to the virus among livestock herders and wildlife rangers during the wet season (Olaleye *et al.*, 1996). Following the 1997–1998 El Niño event in east Africa, an RVF outbreak in Somalia and northern Kenya killed as much as 80% of the livestock and affected their owners (WHO, 1998b). Many cases also were reported in Tanzania. In Mauritania, the human epidemic was linked to the epizootic disease (Jouan *et al.*, 1989). Extensive research on mosquito vectors of RVF in Kenya (mainly *Aedes* and *Culex spp.*) has clearly linked the risk of outbreak with flooding (Linthicum *et al.*, 1990). It can be expected that increased precipitation as a consequence of climate change could increase the risk of infections in livestock and people. Such new risks could cause major economic and health problems for herding communities in Africa.

10.2.4.5. Plague

Plague is a flea-borne disease with rodents as reservoirs. The population of rodents can increase suddenly following heavy rains as a result of abundance of food (e.g., grain). During drought, rodents may migrate into human dwellings in search of food. Development of fleas and the pathogens they carry can be accelerated by increased temperature. Plague outbreaks recently have been reported in Mozambique, Namibia, Malawi, Zambia, and Uganda (see WHO Outbreak Web site: <www.who.int/disease-outbreak-news>).

10.2.4.6. Water-Associated Protozoal Diseases

Pollution of streams, wells, and other sources of rural water supplies by flooding could introduce parasites such as giardia, amoeba, and cryptosporidium into drinking water (Alterholf *et al.*, 1998). These parasites assume a new significance in HIV-infected individuals because of the latter's immunocompromised status (Mwachari *et al.*, 1998). Extreme weather events such as El Niño have been associated with increased episodes of diarrhea.

10.2.4.7. Other Major Parasitic Infections

Shifts in the epidemiology of schistosomiasis, onchocerciasis, and filariasis may take longer to become evident because these parasites are less sensitive to the effects of climate than diseases such as malaria. Changes in the impacts of climate on human trypnosomiasis may require substantial and permanent changes in tsetse fly ecology. Little or no data are available on this subject.

10.2.4.8. Air Pollution-Associated Diseases

Biomass burning and massive importation of badly maintained vehicles could result in increased air pollution—which, combined with increasing temperature, would exacerbate health risks such as respiratory problems and eye and skin infections (Boko, 1988).

10.2.4.9. Vulnerability

Evidence is emerging that many ecosystems on the African continent carry risks of climate-driven threats to human health. Predisposing factors include geographic location, socioeconomic status, and knowledge and attitude toward preventive measures. For example, populations living above 1,500 m in the east African highlands are at risk of epidemic malaria (Lindsay and Martens, 1998); those living along the shores of the Indian Ocean and the Great Lakes are at a risk of cholera infections when conditions for transmission are suitable (Birmingham *et al.,* 1997; Shapiro *et al.,* 1999).

Elsewhere in the Sahel and other arid areas where there are humidity deficits, populations are exposed to meningoccocal meningitis (Tikhomirov *et al.*, 1997) and, in flood-prone pastoral areas, RVF (Linthicum *et al.*, 1990). Vulnerability also can be increased by close habitation with animals that are reservoirs of zoonotic diseases such RVF and plague.

The socioeconomic status of communities may determine whether safe drinking water (piped water, rain-harvested water, and protected wells) is available (Sabwa and Githeko, 1985). The quality of housing is important because simple measures such as screening windows and doors will prevent the entry of disease vectors into human dwellings.

Human factors such as knowledge and attitude and practice will influence health care-seeking behavior of an individual (Karanja *et al.*, 1999). For example, individuals may choose to visit a local healer instead of a clinical facility, and this could affect the progression and outcome of an infection.

At the institutional level, the fragile infrastructure is unable to cope with the impacts of diseases. For example, flood areas often are inaccessible, and delivery of medical intervention is hampered considerably. Furthermore, pathogens such as those of malaria and cholera are resistant to commonly used medication. In the case of malaria, more than 60% of cases are treated at home (Reubush *et al.*, 1995) with drugs that may not be effective (Karanja *et al.*, 1999), particularly in nonimmune populations. Misdiagnoses of fevers, especially during epidemics of uncommon and unfamiliar diseases, leads to delayed treatment and consequently high morbidity and mortality (CDC, 1998). In many cases, foreign assistance is required, and this assistance may come too late. These factors increase the vulnerability of affected populations.

10.2.4.10. Adaptation

Understanding how climate affects the transmission of these diseases will lead to enhanced preparedness for early and effective interventions. Monitoring drug sensitivity to commonly used anti-malaria drugs and antibiotics will prevent the use of ineffective interventions. Communities that are exposed to water-borne diseases such as cholera could reduce the risk of infections by using safe drinking water technologies.

Several large-scale studies in Africa have demonstrated that insecticide-treated fabrics (e.g., bed-nets and curtains) can significantly reduce the risk of malaria infections (Lengeler, 1998). However, such interventions are not effective against day-biting mosquitoes that are vectors of RVF.

Remote sensing is increasingly becoming an important tool in forecasting the risks of transmission in malaria, RVF, and cholera. Hay *et al.* (1998) have shown that the normalized difference vegetation index (NDVI) correlated significantly with malaria presentation, with a lag period of 1 month. NDVI is a function of climatic factors that are similar to those that affect malaria transmission. Ability to use remote sensing to accurately detect parameters such as ground moisture that determine flooding could provide local officials with sufficient warning to allow for implementation of specific mosquito control measures before a disease (RVF) outbreak (Linthicum *et al.*, 1990, 1999). In the case of cholera, it is now possible to utilize remote sensing and computer processing to integrate ecological, epidemiological, and remotely sensed spatial data for the purpose of developing predictive models of cholera outbreaks (Colwell, 1996).

10.2.4.11. Technology for Safe Drinking Water

Flooding, which can be exacerbated by climate change, often results in increased contamination of drinking water. In other instances, drought and an increase in surface water temperatures have been associated with transmission of cholera. Although Cryptosporidium parvum is the more important water-borne pathogen in developed countries, Vibrio cholerae is more pervasive in developing countries. Giardia lumbria, a water-borne protozoa, has a universal distribution. These pathogens pose serious threat to individuals whose immune systems are compromised; furthermore, there are numerous records of resistance to antibiotics by cholera bacteria (e.g., Weber *et al.*, 1994). Therefore, it is essential that populations that are vulnerable to water-borne diseases should enhance safe drinking water technology.

Cryptosporidium parva oocytes are very resistant to chlorine and other drinking-water disinfectants (Venczel *et al.*, 1997). In addition, the cysts have a very low sedimentation rate (Medema *et al.*, 1998); consequently, boiling may be the most appropriate method of disinfecting water where risks of infection exist (Willocks *et al.*, 1998). However, the use of submicron point-of-use water filters may reduce the risk of water-borne cryptosporidiasis (Addis *et al.*, 1996).

Several simple and inexpensive techniques have been found to be effective in reducing the risk of infection with cholera from contaminated water. Huo *et al.* (1996) found that a simple filtration procedure involving the use of domestic sari material can reduce the number of cholera vibrio attached to plankton in raw water from ponds and rivers commonly used for drinking water. In Bolivia, the use of 5% calcium hypochlorite to disinfect water and subsequent storage of treated water in a narrow-mouthed jar produced drinking water from nonpotable sources that met WHO standards for microbiologic quality (Quick *et al.*, 1996). In many cases, boiling water is not possible because of scarcity of firewood and charcoal, particularly in flooded conditions. These examples of low-cost technologies should become widely available to populations that are likely to be impacted by contaminated water supplies, especially following extreme flooding events.

10.2.5. Settlements and Infrastructure

10.2.5.1. Overview of Issues

The main challenges that are likely to face African populations will emanate from the effects of extreme events such as tropical storms, floods, landslides, wind, cold waves, droughts, and abnormal sea-level rises that are expected as a result of climate change. These events are likely to exacerbate management problems relating to pollution, sanitation, waste disposal, water supply, public health, infrastructure, and technologies of production (IPCC, 1996).

The pattern of distribution of human settlements often reflects the uneven nature of resource endowments and availability between regions and within individual communities. In Africa, as elsewhere, there are heavy concentrations of human settlements within 100 km of coastal zones (Singh *et al.*, 1999), in areas of high economic potential, in river and lake basins, in close proximity to major transportation routes, and in places that enjoy hospitable climatic regimes. Changes in climate conditions would have severe impacts not only on the pattern of distribution of human settlements but also on the quality of life in particular areas.

The transport sector is based on long-term, immovable infrastructure such as roads, rails, and water. Road networks have tended to link industrial centers with major areas of agricultural activity; railways have been designed primarily with a sea-route orientation to facilitate international shipments of primary products. Climate change may lead to industrial relocation, resulting either from sea-level rise in coastal-zone areas or from transitions in agroecological zones. If sea-level rise occurs, the effect on the many harbors and ports around the continent will be quite devastating economically for many coastal-zone countries. Excessive precipitation, which may occur in some parts of Africa, is likely to have serious negative effects on road networks and air transport.

10.2.5.2. Coastal Settlements and Sea-Level Rise

More than one-quarter of the population of Africa resides within 100 km of a sea coast (Singh *et al.*, 1999), rendering a significant number of people vulnerable to rises in sea level as a result of climate change. Modeling the effects of a 38-cm mean global sea-level rise in 2080, Nicholls *et al.* (1999) estimate that the average annual number of people in Africa impacted by flooding could increase from 1 million in 1990 to a worst case of 70 million in 2080. Jallow *et al.* (1999) estimate that the capital of The Gambia, Banjul, could disappear in 50–60 years through coastal erosion and sea-level rise, putting more than 42,000 people at risk. El Raey *et al.* (1999) discuss threats to coastal areas of Egypt from sea-level changes. East Africa coastal settlements also are at risk from sea-level rise (Magadza, 2000).

There are three response strategies to rising sea level and its physical impacts: retreat, adapt, or defend. Retreat can involve chaotic abandonment of property and cultural investments, or it can be an ordered, planned program that minimizes losses from rising sea level and maximizes the cost-effectiveness of the operation. The operation also seeks to leave surrendered areas as aesthetic looking as possible and to avoid abandoned structures that are an operational hazard to other social and economic activities.

The capacity of individual states to undertake coastal defense work may be limited. However, if such works are planned on a long-term time scale, it is possible to develop such defenses well before the crisis occurs and thus to spread the total capital costs over many years.

Because the problem of coastal management is regional, such a process would require:

- Regional integration among coastal-zone states
- Recognition by all governments in the region of regional vulnerability to climate change impacts
- Political and institutional stability that allows inter-generational projects to be sustained without interruption from political upheavals.

10.2.5.3. Flooding

Because of their combination of several natural resources, such as fisheries and fertile alluvial soils, wetlands and floodplains often are sites of dense rural settlements as well as urban settlements, such as N'Djamena near Lake Chad and coastal areas of southern Mozambique. The east Africa floods of 1998 and the Mozambique floods in early 2000 caused considerable damage to property and infrastructure. The major infrastructure damage was road and rail network damage. Communications among human settlements in Kenya, Uganda, Rwanda, and Tanzania were seriously disrupted, impeding movement of goods and persons in the region (Magadza, 2000). Many refugees could not be reached by land in Somalia, resulting in significant depletion of their food and medical supplies and leading to mortalities. Road and rail links to seaports were disrupted. In both instances, relief operations were hampered by difficulty of access to affected communities. In Mozambique, the floods of early 2000 caused approximately 2 million people to be displaced or severely affected; about 600 died (<http://www.reliefweb.int>, accessed October 10, 2000). By October 2000, the estimated cost of the Mozambique floods stood at more than US$167 million in terms of emergency aid funds during the flooding and in immediate activities to rehabilitate the infrastructure and relocate displaced persons. The impact on the national economy is still being evaluated but is expected to be significant.

One identifiable adaptive measure against extreme events that are climate related is a state of preparedness to give adequate warning of imminent danger and deliver relief. Facilities to broadcast timely information of developing events such as storms to rural populations remain weak. National disaster plans are available in some countries, but financial resources to respond to emergencies are lacking. The ability to convey impacts quickly to the international media is a factor in the speed and amount of relief. Recent events in Mozambique and other countries of southern and east Africa will provide useful lessons in dealing with similar disasters; there is great value in studying these events not only from the physical point of view—as the Climate Variability and Predictability Program (CLIVAR) is proposing in a new Africa program; see <http://www.dkrz.de/clivar/hp.html>—but also from the social point of view.

There is a need for better understanding of the hydrology of river basins to identify vulnerable areas and plan coping mechanisms. Management of early warning systems depends on good understanding of the dynamics of flood systems in real time.

Because river basins sometimes involve more than one state, as in the Somalia floods, an effective flood management protocol will call for international cooperation. A regular bulletin of flood development in Ethiopia would have given the coastal inhabitants of Somalia time to prepare for damage minimization. In the Mozambique floods of 1996–1997, for example, the trigger was heavy rains in the Shire River basin. If the Shire and Zambezi Rivers were managed as one basin system, it would have been possible to alleviate flooding in the Zambezi delta by manipulating Zambezi river flow, using the flood control capacity of Lakes Kariba and Kabora Bassa.

A further need in formulating adaptive strategies is more refined regional climate change scenarios—especially a better understanding of extreme events. In southern Africa, for example, most of the regional climate change scenarios are rather ambivalent with regard to precipitation (Hulme, 1996). In the 1998–1999 season, the city of Harare suffered damage to its roads because sewer transport could not cope with entrained stormwater. If it were accepted that the frequency of such seasons would increase, future designs of infrastructure amenities would take cognizance of that prediction.

10.2.5.4. Energy

Threats to energy security from climate change impacts are outlined in Section 10.2.1. Disruption of energy supplies will have ripple effects in the social fabric through impacts in economic activity. Some adaptation options for the energy needs of settlements in the African region are in three broad areas:

- Regional cooperation in sharing hydroelectric potential of the continent, especially that of the Zaire River
- More intensive use of renewable energy, such as solar and wind energy and biogas
- Efficient use of biomass.

The countries of the southern and central African region (Democratic Republic of the Congo, Zambia, Zimbabwe, Botswana, South Africa, Mozambique, Namibia) already are networked on an electric power grid system. Climate change models to date indicate minimum changes in the hydrology of the Congo basin, whereas other basins have significant vulnerability to climate change. A regional project to develop the hydropower potential of the Congo basin could significantly increase the energy security of the region without resort to GHG-emitting, coal-driven thermal power plants.

Alternatives to biomass are wind-driven units, either as direct application of wind force—as in water-pumping windmills—or for generation of electricity, as in the windmill farms of Denmark. Biogas units, which utilize livestock dung, have been demonstrated successfully in rural areas of Zimbabwe.

10.2.5.5. Human Comfort

McMichael *et al.* (1996) found a relationship between ambient temperature and heat-related mortalities in Cairo, Egypt. This suggests a need to consider building technology and building materials' thermal properties to produce dwellings that are naturally climatically comfortable for tropical conditions. In Africa, there is a tendency to construct dwellings that do not take account of local climate because of inadequate natural ventilation and use of large decorative glass surfaces. Similarly, urban planners need to consider landscaping to avoid inner-city congestion that leads to unhealthy microclimate enclaves.

A combination of high temperatures and air pollution leads to increases in respiratory complaints. Clean air policies would not only alleviate health hazards but would be a contribution toward maintenance of the ozone layer in line with the Montreal Protocol. Sudden imposition of stringent air quality standards may cause undue strain to the economy, but graduated improvement of air quality standards, at a pace the economy can absorb, will be beneficial in terms of adaptation to climate change as well as general city health improvement.

10.2.5.6. Water Resources

Section 10.2.1 argues that future water resources for subhumid regions of Africa will be in jeopardy under global warming conditions. The impact of climate change to settlements, through the water resources pathway, will have multiple manifestations in all walks of life.

Adaptive measures include incentives for a water conservation culture, such as water pricing. However, many rural communities are not economically or culturally attuned to commercialization of water resources, which normally are administered in a common access mode. Commoditization of water resources as a strategy for efficient water use is contingent on comparable growth of economic activity and social well-being in all sectors of communities.

Land degradation has resulted in siltation thus disappearance of surface streamwater resources (Magadza, 1984). States are encouraged to consider measures that will rehabilitate streams, paying special attention to wetlands conservation, with an added bonus of biodiversity conservation.

Industrial water cycling in Africa is poorly developed. Processes that maximize water recycling should be encouraged.

At the regional level, there are beginnings of cooperation in interbasin transfers from water-surplus areas to water-deficit areas; the proposed diversion of Zambezi River waters toward the south is an example. Although this development will enhance the status of water and other natural resources as tradable commodities, the groundwork for legal regulation of water sharing between nations of the region must be developed sooner rather than later to avoid situations of water-related political tensions like those in the Middle East and north Africa (Caponera, 1996).

10.2.5.7. Sanitation

Section 10.2.4 draws attention to the possible health implications of climate change and climate variability through vector- and water-borne pathogens. In many African urban settlements, urban drift has outpaced the capacity of municipal authorities to provide civic works for sanitation and other health delivery services. The outbreak of cholera during recent floods in east Africa and Mozambique underscores the need for adequate sanitation. It should be noted that although the outbreaks were spread from as far north as Mombassa and Nairobi in the north to Beira in the south, incidences remained localized to the outbreak centers because of the isolated nature of the affected urban areas. If settlement conglomerations such as those envisaged for west Africa and the eastern seaboard of South Africa develop—as discussed by Nicholls *et al.* (1999)—vulnerable population and areas will tend to be regional, rather than local. Review of sanitary facilities now rather than later will not only be beneficial to communities now but in the long run will be cost saving for long-term health delivery services.

10.2.5.8. Food Security

Droughts in SSA often translate to famine, which leads to acceleration of urban drift to cities that are not equipped to absorb such migrations. Although maintenance of strategic food reserves is one coping mechanism, development policies increasingly must create other investment opportunities in rural areas besides agriculture, to diversify means of survival and, indeed, create rural wealth (De Lattre, 1988).

In semi-arid Africa, pastoralism is the main economic activity. Many pastoral communities include transnational migrants in search of new seasonal grazing. In drought situations, such pastoralists may come into conflict with settled agrarian systems (Anon, 1992; Lado, 1995; Cousins, 1996). Students of pastoralism note the lack of clear policies on pastoralists, who normally are marginalized in state agricultural policies.

10.2.6. Desertification

10.2.6.1. Context

The United Nations Convention to Combat Desertification (UNCCD) defines desertification as "land degradation in arid, semi-arid, and dry subhumid areas resulting from various factors, including climatic variations and human activities" (United Nations, 1994). Furthermore, UNCCD defines land degradation as a "reduction or loss, in arid, semi-arid, and dry subhumid areas, of the biological or economic productivity and complexity of rain-fed cropland, irrigated cropland, or range, pasture, forest, and woodlands resulting from land uses or from a process or combination of processes, including processes arising from human activities and habitation patterns, such as: (i) soil erosion caused by wind and/or water; (ii) deterioration of the physical, chemical, and biological or economic properties of soil; and (iii) long-term loss of natural vegetation."

Arid, semi-arid, and dry subhumid areas include those lands where the ratio of precipitation to potential evaporation (PET) ranges from 0.05 to 0.65. In Africa, these conditions cover 13 million km^2 (see Figure 10-9), or 43% of the continent's land area—on which 270 million people, or 40% of the continent's population, live (UNDP, 1997). Areas particularly at risk include the Sahel—a 3.5 million km^2 band of semi-arid lands stretching along the southern margin of the Sahara Desert—and some nations that consist entirely of drylands (e.g., Botswana and Eritrea). The death of as many as 250,000 people in the Sahel drought of 1968–1973 (UNCOD, 1977) demonstrated the tragic human toll of desertification.

Desertification in Africa has reduced by 25% the potential vegetative productivity of more than 7 million km^2, or one-quarter of the continent's land area (UNEP, 1997). Desertification consists more of degradation of the productive capacity of patches well outside open-sand deserts rather than the inexorable encroachment of open sand onto greenlands. Arid lands can respond quickly to seasonal fluctuations. Indeed, analysis of 1980–1990 NDVI data to track the limit of vegetative growth along the Sahara-Sahel margin revealed wide fluctuations: The 1990 limit of vegetative growth lay 130 km south of its 1980 position (Tucker *et al.*, 1991).

Unfortunately, the relative importance of climatic (see Section 10.2.6.3) and anthropogenic (see Section 10.2.6.2) factors in causing desertification remains unresolved. Some scientists have judged that anthropogenic factors outweigh climatic factors (Depierre and Gillet, 1971; Lamprey, 1975; Le Houérou, 1989; Westing, 1994), though others maintain that extended droughts remain the key factor (Mortimore, 1989; Hoffman and Cowling, 1990; Tucker *et al.*, 1991; Dodd, 1994). CO_2-induced climate change and desertification remain inextricably linked because of feedbacks between land degradation and precipitation (see Section 10.2.6.4).

10.2.6.2. Nonclimatic Driving Forces of Desertification

Unsustainable agricultural practices, overgrazing, and deforestation constitute the major anthropogenic factors among the forces that drive desertification. Unsustainable agricultural practices include short rotation of export crops, undisciplined use of fire, and removal of protective crop residues. Overgrazing consists of running livestock at higher densities or shorter rotations than an ecosystem sustainably can support. Finally, deforestation consists of permanent clearing of closed-canopy forests and cutting of single trees outside forests. Forest area in Africa decreased by approximately 37,000 km^2 yr^{-1} from 1990 to 1995 (FAO, 1999a). UNEP (1997) attributes two-thirds of the area already desertified in Africa to overgrazing and the remaining third to unsustainable agricultural and forestry practices.

Population growth ultimately can drive desertification if it intensifies agrosylvopastoral exploitation or if it increases the land area subjected to unsustainable agricultural practices, overgrazing, or deforestation. The total population of Africa grew from 220 million in 1950 to 750 million in 1998—a rate of 2.5% yr^{-1} (United Nations, 1999). Increasing food, wood, and forage needs accompanying this growth place an inordinate burden on the region's natural resources.

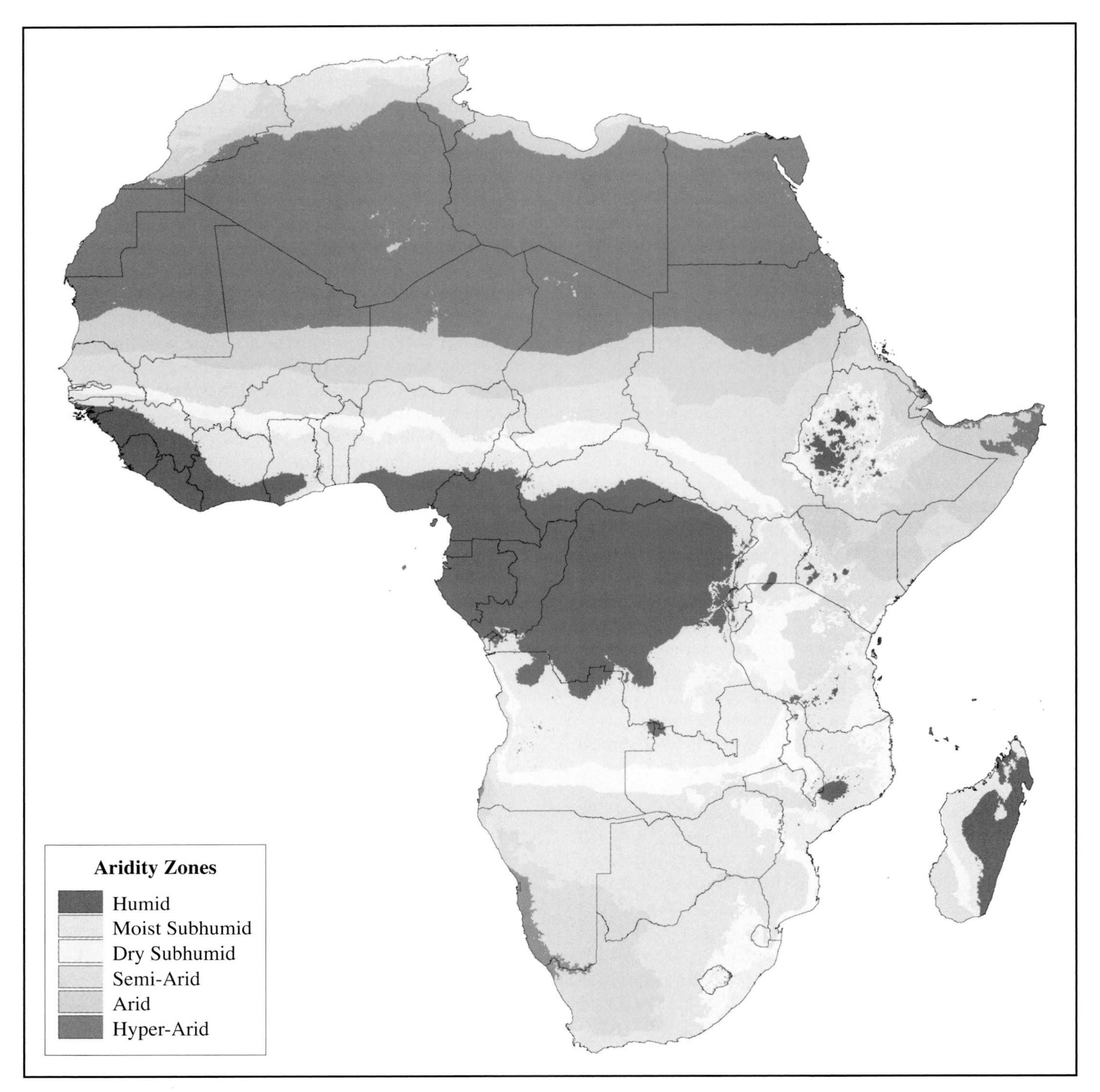

Figure 10-9: Aridity zones for Africa as derived from mean monthly precipitation and potential evapotranspiration surfaces included on *Spatial Characterization Tool for Africa* CD-ROM (UNDP, 1997).

10.2.6.3. Climatic Factors in Desertification

Precipitation and temperature determine the potential distribution of terrestrial vegetation and constitute principal factors in the genesis and evolution of soil. Extended droughts in certain arid lands have initiated or exacerbated desertification. In the past 25 years, the Sahel has experienced the most substantial and sustained decline in rainfall recorded anywhere in the world within the period of instrumental measurements (Hulme and Kelly, 1997). Linear regression of 1901–1990 rainfall data from 24 stations in the west African Sahel yields a negative slope amounting to a decline of 1.9 standard deviations in the period 1950–1985 (Nicholson and Palao, 1993). Since 1971, the average of all stations fell below the 89-year average and showed a persistent downward trend since 1951.

Because evapotranspiration constitutes the only local input to the hydrological cycle in areas without surface water, reduction in vegetative cover may lead to reduced precipitation, initiating a positive feedback cycle. Degradation of vegetation cover in moister areas south of the Sahel may have decreased continental evapotranspiration and reduced precipitation in the Sahel (Xue, 1997).

A positive feedback mechanism between vegetation cover and albedo may help to explain the Sahel drought (Charney, 1975). Some research supports an albedo-precipitation feedback mechanism (Otterman, 1974; Cunnington and Rowntree, 1986; Xue *et al.*, 1990; Diedhiou and Mahfouf, 1996; Zheng and Eltahir, 1997; Zeng *et al.*, 1999), although other research disputes the importance of albedo (Jackson and Idso, 1974; Ripley, 1976; Wendler and Eaton, 1983; Gornitz and NASA, 1985; Nicholson *et al.*, 1998; Nicholson, 2000).

Degraded land also may increase atmospheric dust and aerosols, which influence precipitation (see Section 10.1.3.2).

SST anomalies, often related to ENSO or NAO, also contribute to rainfall variability in the Sahel (Lamb, 1978; Folland *et al.*, 1986; Hulme and Kelly, 1997; Nicholson and Kim, 1997; Hulme *et al.*, 1999). Lamb (1978) observes that droughts in west Africa correlate with warm SST in the tropical south Atlantic. Examining oceanographic and meteorological data from the period 1901–1985, Folland *et al.* (1986) found that persistent wet and dry periods in the Sahel were related to contrasting patterns of SST anomalies on a near-global scale. When northern hemisphere oceans were cold, rainfall in the Sahel was low.

Street-Perrott and Perrott (1990) demonstrate that injections of freshwater into the north Atlantic (such as from glacial melt) decrease salinity—stabilizing the water column, inhibiting deep convection, and reducing northern transport of heat by the Atlantic thermohaline circulation, which is driven by a north-south SST gradient. This decreases evaporation from the ocean surface, causing drought in the Sahel and Mexico. From 1982 to 1990, Mynemi *et al.* (1996) found a correlation between ENSO-cycle SST anomalies and vegetative production in Africa. They found that warmer eastern equatorial Pacific waters during ENSO episodes correlated with rainfall of <1,000 mm yr^{-1} over certain African regions.

A combination of factors—including vegetation cover, soil moisture, and SST—best explains the reduction in rainfall in the Sahel. Diedhiou and Mahfouf (1996) modeled changes in albedo, soil moisture, land surface roughness, and SST anomalies and calculated a rainfall deficit over the Sahel similar to observed patterns. Eltahir and Gong (1996) suggest that a meridional distribution of boundary-layer entropy regulates the dynamics of monsoon circulation over west Africa, explaining observed correlations of SST to rainfall and the sensitivity of monsoon circulation to land-cover changes. A coupled surface-atmosphere model indicates that—whether anthropogenic factors or changes in SST initiated the Sahel drought of 1968–1973—permanent loss of Sahel savanna vegetation would permit drought conditions to persist (Wang and Eltahir, 2000). Zeng *et al.* (1999) compared actual rainfall data from the period 1950–1998 with the output of a coupled atmosphere-land-vegetation model incorporating SST, soil moisture, and vegetative cover. Their results indicate that actual rainfall anomalies are only weakly correlated to SST by itself. Only when the model includes variations in vegetative cover and soil moisture does it come close to matching actual rainfall data. Modeling the importance of SST, sea ice, and vegetative cover to the abrupt desertification of the Sahara 4,000–6,000 years ago, Claussen *et al.* (1999) show that changes in vegetative cover best explain changes in temperature and precipitation.

10.2.6.4. Linkages and Feedbacks between Desertification and Climate

CO_2-induced climate change might exacerbate desertification through alteration of spatial and temporal patterns in temperature, rainfall, solar insolation, and winds. Conversely, desertification aggravates CO_2-induced climate change through the release of CO_2 from cleared and dead vegetation and through the reduction of the carbon sequestration potential of desertified land.

Areas that experience reduced rainfall and increased temperature as a result of CO_2-induced climate change also could experience declines in agricultural yields, livestock yields, and tree cover, placing local people at risk of famine.

Lower soil moisture and sparser vegetative cover also would leave soil more susceptible to wind erosion. Reduction of organic matter inputs and increased oxidation of soil organic matter (SOM) could reduce the long-term water-retention capacity of soil, exacerbating desertification. Sample plots in Niger lost 46 t ha^{-1} in just four windstorms in 1993 (Sterk *et al.*, 1996), releasing 180 ± 80 kg ha^{-1} yr^{-1} of soil carbon (Buerkert *et al.*, 1996). Moreover, increased wind erosion increases wind-blown mineral dust, which may increase absorption of radiation in the atmosphere (Nicholson and Kim, 1997).

Desertification from anthropogenic and climatic factors in Senegal caused a fall in standing-wood biomass of 26 kg C ha^{-1} yr^{-1} in the period 1956–1993, releasing carbon at the rate of 60 kg C cap^{-1} yr^{-1} (Gonzalez, 1997).

Although altered surface albedo may increase surface air temperatures locally (Williams and Balling, 1996), the effect of desertification on global mean temperature is unlikely to have exceeded 0.05°C in the past century (Hulme and Kelly, 1997).

10.2.6.5. Impacts of Desertification

Desertification reduces soil fertility, particularly base cation content, organic matter content, pore space, and water-retention capacity. Desertification also reduces vegetative productivity, leading to long-term declines in agricultural yields, livestock yields, plant standing biomass, and plant biodiversity. These changes reduce the ability of the land to support people, often sparking an exodus of rural people to urban areas. Breaking the strong connection of people to the land produces profound changes in social structure, cultural identity, and political stability.

In Niger, on farmed land where organic carbon in the top 10 cm of the soil has fallen from 0.3 to 0.2% 4 years after coming out of fallow, millet yields fell from 280 to 75 kg ha^{-1} (Bationo *et al.*, 1993). Modeling of the 4,000 km^2 Mgeni River watershed in South Africa showed that conversion of more than one-quarter of the watershed from forest and rangeland to agriculture and exotic tree plantations since the area was colonized would double mean annual runoff in urban areas and other areas of reduced land cover (Schulze, 2000).

In the Senegal Sahel, the densities of trees with a height of >3 m declined from 10 trees ha^{-1} in 1954 to 7.8 trees ha^{-1} in 1989; the species richness of trees and shrubs fell from 16 species per 4 km^{-2} around 1945 to about 11 per 4 km^{-2} in 1993 (Gonzalez, 1997, 2001). These changes have caused a 25–30 km shift of the Sahel, Sudan, and Guinean vegetation zones in half a century, proceeding at an average rate of 500–600 m yr^{-1}. Arid Sahel species expanded in the northeast, tracking a concomitant retraction of mesic Sudan and Guinean species toward areas of higher rainfall and lower temperature to the southwest.

In the Senegal Sahel, human carrying capacity in 1993 stood at approximately 13 people km^{-2} at observed patterns of resource use, compared to an actual 1988 rural population density of 45 people km^{-2} (Gonzalez, 1997, 2001). This means that people with no other alternatives need to cut into their natural resource capital to survive. Such changes across Africa have pushed a rural exodus that may have displaced 3% of the population of Africa since the 1960s (Westing, 1994).

Desertification also will cause conversion of perennial grasslands to savannas dominated by annual grasses. Such changes have occurred in the Kalahari Gemsbok National Park in South Africa, where Landsat imagery showed increases in exposed soil surface (Palmer and van Rooyen, 1998). Such declines often are irreversible (Schlesinger *et al.*, 1990).

10.2.6.6. Vulnerability and Adaptation

The tragic death of as many as 250,000 people in the Sahel drought of 1968–1973 (UNCOD, 1977) demonstrates the vulnerability of humans to desertification. As desertification proceeds, agricultural and livestock yields decline, reducing people's options for survival. Furthermore, not only do local people lose the vital ecosystem services that dead trees and shrubs had provided; the loss of firewood, traditional medicine species, and emergency food species render them more vulnerable to future environmental change.

Adaptations by farmers and herders in Africa to climate change and desertification have involved diversification and intensification of resource use (Davies, 1996; Downing *et al.*, 1997). Resourceful diversification responses by women in Bambara and Fulbe households in Mali (Adams *et al.*, 1998) reflect the importance of women in guiding adaptation strategies across Africa. In southern Kenya, Maasai herders have adopted farming as a supplement to or replacement for livestock herding (Campbell, 1999). In Kano, Nigeria, peri-urban vegetable gardening has expanded (Adams and Mortimore, 1997), revealing a common diversification trend in small cities across west Africa. In northern Cameroon, Fulbe herders have increased the number of herd displacements between pasture areas and even resorted to long-distance migration, sometimes introducing significant changes to their way of life (Pamo, 1998).

In the future, seasonal climate forecasting (NOAA, 1999; Stern and Easterling, 1999) may assist farmers and herders to know times of higher probability of success of resource diversification or intensification. Seasonal forecasts for Africa currently exhibit moderate skill levels (Thiaw *et al.*, 1999) but skill levels and user communications are not yet high enough to permit users to confidently implement field applications (UNSO, 1999; Broad and Agrawala, 2000). Neither trade nor technology will likely avert the widespread nutritional and economic effects of desertification through the 2020s (Scherr, 1999).

Other adaptations to desertification involve more efficient management of resources. In Niger, farmers with access to credit will adopt low-cost, appropriate technologies for wind erosion control, including windbreaks, mulching, ridging, and rock bunds (Baidu-Forson and Napier, 1998). Across Africa, farmers traditionally have adapted to harsh environmental conditions by promoting natural regeneration of local trees and shrubs. Natural regeneration is a practice whereby farmers and herders seek to reconstitute vegetative cover by setting aside parcels of land or by selecting valued trees in their fields, pruning them, straightening them, and raising them to maturity. The Sereer in Senegal (Lericollais, 1973) and the Mossi in Burkina Faso (Kessler, 1992) have achieved doubling of tree densities in certain semi-arid areas with *Acacia albida* and *Butyrospermum parkii*, respectively.

10.3. Adaptation Potential and Vulnerability

The foregoing assessment highlights the high vulnerability of Africa to climate change as a result of limited adaptive capacity constrained by numerous factors at the national level. The floods of February 2000 in southern Africa (which affected Mozambique, South Africa, Botswana, and Zimbabwe) highlight huge differences in adaptive capacities between countries. Adaptive capacity was influenced largely by the ability to communicate potential risks to vulnerable communities and the ability to react as a result of perceived risks. The ability to mobilize emergency evacuation was critical in reducing adverse impacts. Although there may be high adaptive capacity locally or nationally, overall most countries in Africa have low capacity to adapt to abrupt and extreme events.

Scarce water resources are becoming increasingly critical for Africa; they determine food security as well as human and ecosystem health, and play a major role in political and socioeconomic development. Although parts of Africa have abundant water, shifting water to stressed areas is not an option in many cases. Groundwater resources are likely to be impacted by prolonged droughts and changes in land cover and land use, in a complex interaction of human activity and population growth rates, climate, and environmental responses. Adaptation will require small actions as well as major national approaches. At the management unit level (e.g., watershed), careful management of rainwater through damming will allow agricultural production. There is vast experience in arid regions of Africa such as Namibia, Botswana, and north Africa (such as Morocco), where brief periods of rain are utilized very efficiently for farming. The constraint will be in finding limits to water extraction that do not adversely impact communities downstream and result in conflicts. Regional bodies set up to negotiate international water rights will play an increasingly crucial role. At the national level, political goals such as self-sufficiency in food production will need to be reevaluated with reference to water resources available to the country and how they can be apportioned between food production, human needs, and ecosystem needs. Countries will need to be more open to fulfilling their food needs through imports and redistribution, using intensified production in areas where it is possible. Good communications within and between countries and major ports are critical to food security. These include roads, rail, and air transportation networks. For inland countries, large corridors being opened up or upgraded (such as the Maputo, Nacala, and Beira corridors between Mozambique and South Africa, Malawi, and Zimbabwe) will greatly enhance access to food and other imported goods. This places greater importance on international relations.

At the subregional scale, Africa is vulnerable to ENSO and related extreme events (drought, floods, changed patterns). As shown by Semazzi and Song (2000), deforestation is likely to alter circulations in distant places through teleconnected feedbacks, increasing the vulnerability of distant populations. Advances in seasonal forecasting, using climate models and satellite observations, has been shown to be a first-order response strategy to changing climate variability. Similar applications of satellite observations (such as for SST) also are useful in predicting disease outbreaks such as RVF. Effective communication of predicted extreme weather events and evaluation of potential risks is critical in minimizing human loss of life, where it is possible to react. Disaster management plans are required and need to be developed jointly with all members of a community.

There is great potential in investing in seasonal forecasting and development of tools (models) such as crop models that can be used to make adjustments in management. Although these models are still experimental, they offer a realistic response to changing climatic patterns. Data must be collected to calibrate and validate these models. In the longer term, governments will need to develop strategic plans that are based on solid foundations. This is an area that is underdeveloped in almost all of Africa.

External funding drives programs in many African countries, so agendas usually align closely with donor agency interests. This situation presents a dilemma for Africa. There is capacity in many countries now to evaluate effective strategies to adapt to adverse effects of climate change. However, these countries are at the mercy of donor agency representatives who often are less informed about issues of climate change. These representatives often regard immediate problems of poverty, erosion, health, and empowerment as the only priority issues for Africa. Longer term planning—for example, land-use planning in areas that are susceptible to flooding under infrequent cyclonic events—never receives the attention it deserves. Most African countries are unlikely to motivate internal funding for climate change; therefore, it is critical that funding agencies award high visibility to issues of climate change.

10.4. Synthesis

Select key impacts over the African continent are highlighted in Figure 10-10. Water resources are a key vulnerability in Africa for water supply for household use and for agricultural and industrial use. In shared river basins, regional cooperation protocols minimize adverse impacts and potential for conflicts. Land use in many of Africa's large and shared basins has long-lasting impacts through modifications of the water budget and through chemical and sediment input into waterways. Water issues in urban and rural areas are likely to become more critical, given increasing and competing demands, as well as rapid population growth. Infrastructure to store and serve water in major urban areas is mostly overstretched in capacity, and extreme events such as floods that cause physical damage add to the problem. Monitoring of water levels and water use is very poor and limits analysis of vulnerability.

There is wide consensus that climate change will worsen food security, mainly through increased extremes and temporal/spatial shifts. The continent already experiences a major deficit in food production in many areas, and potential declines in soil moisture will be an added burden. Food-insecure countries are at greater risk of adverse impacts of climate change.

Irreversible losses of biodiversity could be accelerated with climate change. Climate change is expected to lead to biome shifts, with drastic shifts of biodiversity-rich biomes such as the Succulent Karoo in South Africa, and many losses in species in other biomes. Analysis of potential biome shifts is possible in areas where good spatial databases of vegetation and biophysical variables and land use are available. It is expected that these analyses will improve as more data are available from new and better satellite sensing systems, as well as coordinated field studies in shared resource areas.

Changes in temperature and rainfall will have many negative impacts on human health. Changes in disease vector habitats will expose new populations to diseases such as malaria. Droughts and flooding, where sanitary infrastructure is inadequate, will result in increased frequency of water-borne diseases. Increased rainfall could lead to more frequent outbreaks of RVF. Many African nations do not have adequate financial resources for public health.

Sea-level rise, coastal erosion, saltwater intrusion, and flooding will have significant impacts for African communities and economies. Most of Africa's largest cities are along coasts and are highly vulnerable to extreme events, sea-level rise, and coastal erosion as a result of inadequate physical planning and escalating urban drift. Rapid unplanned expansion is likely to predispose large populations to infectious diseases from climate-related factors such as flooding.

Desertification is a critical threat to sustainable resource management in arid, semi-arid, and dry subhumid regions of Africa, undermining food and water security.

A diversity of constraints facing many nations limits overall adaptive capacity for Africa. Although there is uncertainty about what the future holds, Africa must start planning now to adapt to climate change. Current technologies and approaches—especially in agriculture and water—are unlikely to be adequate to meet projected demands, and increased climate variability will be an additional stress. It is unlikely that African countries on their own will have sufficient resources to respond effectively.

Climate change also offers some opportunities. The processes of adapting to global climate change, including technology transfer and carbon sequestration, offer new development pathways that could take advantage of Africa's resources and human potential. Regional cooperation in science, resource management, and development already are increasing, and access to international markets will diversify economies and increase food security.

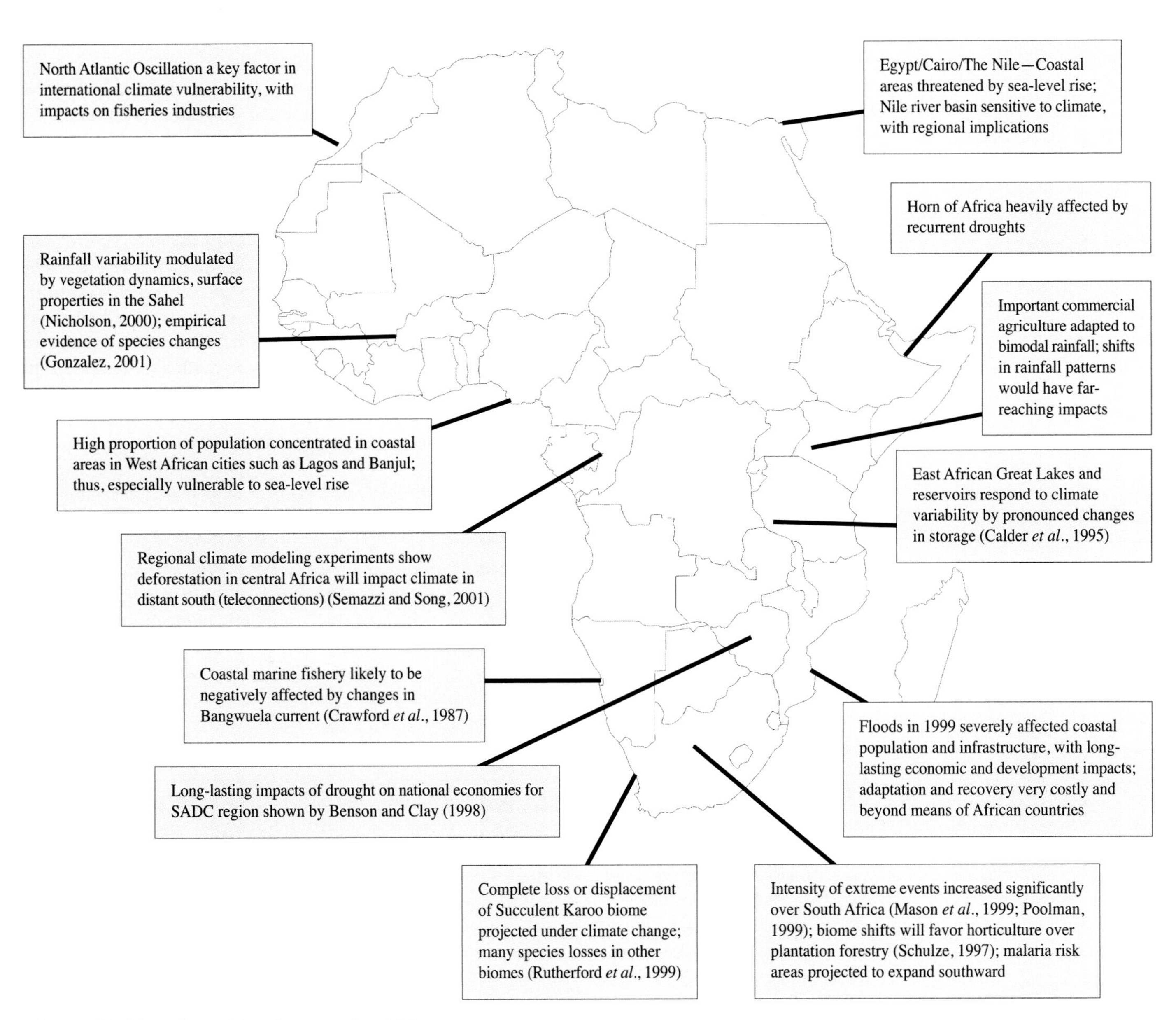

Figure 10-10: Selected key impacts for Africa.

10.4.1. Feedbacks and Interactions

Several integrated studies have looked at feedbacks and complex interactions in African regional systems. In the west African Sahel, land surface-atmosphere interactions have been examined in great detail to explore their role in interannual variability of rainfall since the long drought that started in the late 1960s. Reviews by Nicholson (2000) and Hunt (2000) summarize the state of knoweldge for the physical climate of the Sahel. In general, surface processes modulate rainfall variability, along with SST—but in complex ways. The Sahel is likely to remain a major study topic, and field-based observational studies are on the rise. A field campaign in 1992 called the Hydrologic Atmosphere Pilot Experiment (HAPEX)–Sahel was designed to find ways of improving modeling of land surface properties; a series of HAPEX-Sahel papers were published in a special issue of the *Journal of Hydrology* (Goutourbe *et al*., 1997). More recently, a regional land-atmosphere experiment is underway in southern Africa to study fires and emissions, with modeling studies planned to explore land-atmosphere linkages (see <http://safari.gecp.virginia.edu>, accessed October 2000). These studies—and integrated modeling in general—offer comprehensive tools for studying the integrated Earth system. More studies of this kind will assist in our understanding of linkages between land surface processes, regional and global linkages, and human activities.

Biomass burning plays an important role in global atmospheric chemistry (Andreae, 1991), particularly with respect to generation of trace gases that lead to the formation of tropospheric O_3, carbon monoxide (CO), nitrogen oxides, CH_4, volatile organic carbon (VOC, which also is a GHG), and smoke particles (aerosols), which have an anti-greenhouse effect. Africa is a significant location of biomass burning. In Africa, there are three main types of biomass fires: those associated with land clearing for agriculture, which are mostly located in humid tropical forests and the subhumid tropics; burning of wood for domestic energy (either directly or after first converting it to charcoal); and fires in natural and semi-natural vegetation, which are not associated with changes in land cover or use. Emissions from all three types are of broadly comparable magnitude (Scholes and Scholes, 1998), although the last category has received the most attention.

Fires in natural vegetation are not considered to be a net source or sink in the global carbon cycle because when integrated over large regions and over several years, CO_2 (as well as CO, CH_4, and VOC, which ultimately converts to CO_2) emitted by the fire is taken up again by vegetation regrowth. This is true if the overall fire frequency or intensity is not changing, but if fires become more frequent or consume more fuel over time, the result will be a net CO_2 source; conversely, if fire frequency is reduced or the fires burn less fuel, a carbon sink will result, manifested as an increase in woody biomass. There is no evidence that at a contental scale, fires in natural vegetation have increased or decreased in frequency or intensity in the historic period. For some subregions (such as parts of southern and east Africa), where there is clear thickening of the woody vegetation, it is likely that fire regimes have become less frequent and intense during the 20th century.

During years of regional drought, such as those in southern Africa associated with El Niño events, the area burned decreases by about half (Justice *et al*., 1996). It is believed that this is caused principally by a decrease in fuel availability.

The potential for teleconnections in impacts of land-use change on distant climates further increases the risk in communities that may be at low risk but will be impacted by actions taken in distant areas. For example, deforestation of the central African basin leads to climatic impacts in the savannas to the south in GCM modeling experiments (Semazzi and Song, 2001).

It is clear that rainfall (e.g., intensity) combined with land-use conversion in watershed areas leads to increased soil erosion. Enhanced siltation in rivers and increased use of chemicals also leaching into the runoff interferes with river chemistry (e.g., eutrophication), with major implications for water quality in lakes and coastal systems. Impacts on biodiversity—hence important economic fisheries, and consequent feedback on national economies—is an area of research that needs elucidation.

10.4.2. Uncertainties and Risks

There is great uncertainty about how climate might change at subregional scales in parts of Africa, especially how this might be influenced by human-driven factors such as deforestation and alternative land uses. Regional climate modeling will help reduce these uncertainties, and there are early results now of such modeling for parts of Africa. For these results to be useful, they will have to incorporate realistic disturbance regimes (e.g., realistic deforestation mechanisms and land-use characterizations).

Climate change will manifest itself through changes in climate variability and hence changes in extreme events. Given its socioeconomic status, Africa is unlikely to respond any better to extreme events than it has in the past. Flooding, droughts, and so forth are increasingly difficult for Africa to cope with given increasing pressures on resources from rapid population growth and dwindling resources. Most African countries remain largely unable to gather adequate data and information in a timely manner to address critical problems and surprises such as droughts and floods. Although progress is being made to design environmental information systems, models for analyzing impacts and policy options are largely nonexistent. Although adequate data probably exist, what is needed is the capacity to access large amounts of data and synthesize it into useful bits of information for decisionmaking. For example, satellite data have been collected over the past 2 to 3 decades, yet their use is largely restricted to mapping and short-term climate predictions. Effective information systems and monitoring have not been achieved.

Evaluation of impacts in monetary or other quantitative terms remains a major obstacle to comprehensive assessment of impacts of climate change for Africa. Regional integrated assessment modeling, such as in Egypt (Yates and Strzepek, 1998), offers a solution, and model development should be accelerated for subregions of Africa where building blocks exist (Desanker *et al.*, 2001).

ENSO-related impacts remain uncertain, given perceived changes in ENSO events in terms of frequency and duration. However, much progress is being reported in prediction of ENSO, and this information should be closely linked with case studies of how different regions and populations respond to specific climate-related events. These studies should document costs and benefits, as well as responses.

10.4.3. Cross-Cutting Issues

10.4.3.1. Costing

In some of the sectors affected by climate change—such as water resources, food security, and natural resources—it may be possible to attach financial values. At this point, there are few if any published data on the economic impacts of climate change on these sectors in Africa. However, there are indications of what it costs to support communities that have been affected by famine and floods. Because most of this is foreign aid, it is difficult to evaluate costs in relation to individual country economies. There are indications that water shortages have a negative impact on power generation and consequently economic activities. It is acknowledged that some elements such as health are difficult to cost. Nevertheless, the financial impacts of dealing with epidemics such as malaria, menengitis, and cholera are known by various governments. Governments may be able to translate the impacts described in this assessment to national costs and begin to plan how to develop adaptive financial measures. As methods for impact assessment involve more integrated modeling, it will be possible to quantify impacts of specific climate change scenarios, either in terms of goods and services or in monetary terms.

10.4.3.2. Development, Sustainability, and Equity

The great uncertainty is the political development of Africa. Africa has had its share of surprises in the political arena. What appeared to be a steady march to economic growth would be shattered by sudden political upheavals and festering corruption among national leadership, which tends to have a trickle-down effect until the process of governance is corrupted at its core. On the other hand, the sudden collapse of apartheid in South Africa brought a fresh wind of hope in southern Africa, with far-reaching impacts on the rest of the continent. The adaptive recommendations made in this assessment assume an underpinning of good governance and social responsibility. Achebe *et al.* (1990) tackle the issue of Africa's future political and economic demise. Using various approaches and assumptions, they conclude that Africa may not enjoy political and economic stability prior to 2020–2030 at the earliest. According to climate change impact scenarios, this is the period when climate impacts will have significant impacts on the economic and social fabric. Thus, the message is that Africa must get its house in order as a matter of urgency. We challenge the African political leadership to belie the Achebe *et al.* (1990) prognosis.

The impacts of climate change are expected to be severe, yet Africa's contribution to climate change through emissions is minimal. There are strong feelings among some people in the South who would like to see the North implement tangible emissions reductions and find ways of helping the South adapt to adverse impacts of climate change. From this synthesis, it is clear that Africa is highly vulnerable, with very low capacity to adapt. There is great potential to transfer technology to Africa that would help in developing sustainable agriculture, as well as other technologies that would assist in improving welfare and economic development.

Although the relative importance of climatic and anthropogenic factors in causing desertification remains unresolved, evidence shows that certain arid, semi-arid, and dry subhumid areas have experienced declines in rainfall, resulting in decreases in soil fertility and agricultural, livestock, forest, and rangeland production. Ultimately, these adverse impacts lead to political and socioeconomic instability.

Given the range and magnitude of development constraints and challenges facing most African nations, the overall capacity for Africa to adapt to climate change is low. Although there is uncertainty about what the future holds, Africa must start planning now to adapt to climate change. National environmental action plans and their implementation need to incorporate long-term changes and pursue "no regret" strategies. Current technologies and approaches—especially in agriculture and water—are unlikely to be adequate to meet projected demands, and increased climate variability will be an additional stress. Seasonal forecasting—for example, linking SST to outbreaks of diseases—is a promising adaptive strategy that will help to save lives.

Climate change offers some opportunities for development. The process of adapting to global climate change, including technology transfer, offers new development pathways that could take advantage of Africa's resources and human potential. Examples would include competitive agricultural products, resulting from research in new crop varieties and increased international trade, and industrial developments such as solar energy. Regional cooperation in science, resource management, and development already is increasing.

This assessment of vulnerability to climate change is marked by uncertainty. The diversity of African climates, high rainfall variability, and a very sparse observational network make predictions of future climate change difficult at the subregional and local levels. Underlying exposure and vulnerability to climatic changes are well established. Sensitivity to climatic variations is established but incomplete. However, uncertainty about future conditions means that there is low confidence in projected costs of climate change.

10.4.4. Future Needs

Issues related to Africa's capacity to understand projected impacts include:

- *Data Needs:* The potential exists to develop environmental information systems on the basis of satellite data products and geographic information systems at small management units such as river basins, with socioeconomic and biophysical attributes as required in analyses of impacts and management. Current impact assessment models are limited by input data, limiting their use to fairly general questions. Examples exist where detailed spatial databases have been built and are being used to run integrated agrohydrological models, such as in South Africa (Schulze, 1997; Schulze and Perks, 2000). Monitoring of environmental processes as well as increased weather observation are required. Coordinated collection of integrated data sets for subregions or in connection with an extreme event such as a drought or a flood is highly desirable and would contribute to understanding of adaptation and response strategies and regional integrated modeling.
- *Human Capacity:* There is great need for increased African capacity to study the more fundamental science issues of global change and its impacts. There is great capacity at the applied management level in Africa, and this must be strengthened by a strong science capacity. The increasing number of international environmental treaties and agreements will require an even greater capacity for analysis and delivery of timely reports. There also is great need to apply science findings in policy analysis and international negotiations.
- *Integrated Analysis:* It is becoming increasingly clear that most environmental problems such as climate change require integration of many disciplines and methods of analysis. There also is a shift in interest and focus from global scales to regional and local scales. Models that help to integrate science findings with management and policy issues are needed. These models, called integrated assessment models, are required at regional and subregional levels and should include all important linkages between the socioecological and economic sectors. Given the unique combinations of factors in subregions of Africa (climate, economics, infrastructure), it will be necessary in future assessments to develop and apply regional assessment models that reflect key factors for each subregion, and these models will need to be built around issues of sustainable development rather than emissions reduction. Linking climate change (and other environmental issues) to sustainable development is not going to be easy, but it should spawn a rich body of research to define methods and approaches that will work.
- *Literature Written in French:* It is recognized that there is a rich body of literature that is written in French, and although efforts were made to capture these studies, it simply was not possible to conduct an exhaustive synthesis of that body of work. This represents a major challenge for Africa-wide assessments.

References

Achebe, C., C. Magadza, G. Hyden, and A. Plala Okeyo (eds.), 1990: *Beyond Hunger in Africa*. Heinemann Kenya, James Currey, London, United Kingdom, 148 pp.

Adams, A.M., J. Cekan, and R. Sauerborn, 1998: A conceptual framework of household coping: reflections from rural West Africa. *Africa*, **68,** 263–283.

Adams, W.M. and M.J. Mortimore, 1997: Agricultural intensification and flexibility in the Nigerian Sahel. The Geographical Journal, 163, 150–160.

Addis, D.G., R.S. Pond, M. Remshak, D.D. Juranek, S. Stokes, and J.P. Davis, 1996: Reduction of risk of watery diarrhea with point-of-use water filters during a massive outbreak of water-borne Cryptosporidium infection, Milwaukee, Wisconsin, 1993: *American Journal of Tropical Medicine and Hygiene*, **54,** 549–553.

Alcamo, J. (ed.), 1994: *Image 2.0 Integrated Modelling System for Global Climate Change*. Kluwer Academic Publishers, Dordrecht, The Netherlands, 296 pp.

Alterholt, T.B., M.W. LeChevallier, W.D. Norton, and J.S. Rosen, 1998: Effects of rainfall on giardia and cryptosporidium. *Journal of American Water Works Association*, **90,** 66–80.

Andreae, M.O., 1991: Biomass burning: its history, use, distribution and impact on environmental quality and the global climate. In: *Global Biomass Burning* [Levine, J.S. (ed.)]. MIT Press, pp. 3–21.

Angyo, I.A. and Okpeh, 1997: Clinical predictor of epidemic outcome in meninggoccocal infections in Jos, Nigeria. *East Africa Medical Journal*, **74,** 423–426.

Anon, 1992: Pastoralists' progress: the future for the Sahel. *Spore*, **39,** 1–4.

Arnell, N.W., 1999: Climate change and global water resources. *Global Environmental Change*, **9,** S31–S50.

Ayalew, M., 1997: What is food security and famine and hunger? *Internet Journal for African Studies*, **2.** Available online at http://www.brad.ac.uk/research/ijas/ijasno2/ayalew.html.

Baidu-Forson, J. and T.L. Napier, 1998: Wind erosion control within Niger. *Journal of Soil and Water Conservation*, **55,** 120–125.

Bationo, A., B.C. Christianson, and M.C. Klaij, 1993: The effect of crop residue and fertilizer use on pearl millet yields in Niger. *Fertilizer Research*, **34,** 251–258.

Behnke, R.H., I. Scoones, and C. Kerven (eds.), 1993: *Range Ecology at Disequilibrium: New Models of Natural Variability and Pastoral Adaptation for African Savannas*. Overseas Development Institute, London, United Kingdom, 230 pp.

Benson, C. and E. Clay, 1998: The Impact of Drought on Sub-Saharan Economies. World Bank Technical Paper No. 401, World Bank, Washington, DC, USA, 91 pp.

Besancenot, J.P., M. Boko, and P.C. Oke, 1997: Weather conditions and cerebrospinal meningitis in Benin (Gulf of Guinea, West Africa). *European Journal of Epidemiology*, **13,** 807–815.

Bila, A., 1999: Impacts of climate change on forests and forestry sector of Mozambique. In: *Republic of Mozambique. Final Report of the Mozambique/U.S. Country Study Progam Project on Assessment of the Vulnerability of the Economy of Mozambique to Projected Climate Change*, Maputo, Mozambique, (unpublished).

Brickett, C., R. Murtugudde, and T. Allan, 1999: Indian Ocean climate event brings floods to East Africa's lakes and the Sudd marsh. *Geophysical Research Letters*, **26(8),** 1031–1034.

Birmingham, M.E., M.S. Deming, B.S. Hersh, L.A. Lee, N. Ndayimirije, S. Nkurikiye, and J.G. Wells, 1997: Epidemic cholera in Burundi: patterns in the Great Rift Valley Lake region. *Lancet*, **349(9057),** 981–985.

Boko, M., 1992: *Climats et Communautes Rurales du Benin*. Rythm Climatiques et Rythmes de Developpements Centre de Recherche de Climatologie, Universit, de Bourgogne, Dijon, France.

Bond, W.J. and B.W. van Wilgen, 1997: *Fire and Plants*. Population and Community Biology Series 14, Chapman & Hall, London, United Kingdom, 263 pp.

Bonnefille, R., J.C. Roeland, and J. Guiot, 1990: Temperature-rainfall estimates for the past 40,000 years in equatorial Africa. *Nature*, **346,** 347–349.

Booysen, P.deV. and N.M. Tainton (eds.), 1984: Ecological effects of fire in southern African ecosystems. In: *Ecological Studies 48*. Springer-Verlag, Berlin, Germany, 426 pp.

Breman, H., 1975: La capacite de charge maximale des pasturages maliens. In: *Inventaire et la Cartographie des Pasturages Tropicaux Africains, Actes du Colloque, Bamako, Mali, 3–8 March, 1975:* ILCA, Addis Ababa, Ethiopia, pp. 249–256.

Broad, K. and S. Agrawala, 2000: The Ethiopia food crisis—uses and limits of climate forecasts. *Science*, **289,** 1683–1684.

Buerkert, A., K. Michels, J.P.A. Lamers, H. Marshner, and A. Bationo, 1996: Anti-erosive, soil, physical, and nutritional effects of crop residues. In: *Wind Erosion in Niger: Implications and Control Measures in Amillet-Based Farming System* [Buerkert, B., B.E. Allison, and M. von Oppen (eds.)]. Developments in Plant and Soil Sciences 67, Kluwer Academic Publishers, Dordrecht, The Netherlands, 255 pp.

Calder, I.R., R.L. Hall, H.G. Bastable, H.R. Gunston, O. Shela, A. Chirwa, and R. Kafundu, 1995: The impact of land use change on water resources in sub-Saharan Africa: a modelling study of Lake Malawi. *Journal of Hydrology,* **170,** 123–135.

Cambula, P., 1999: Impacts of climate change on water resources of Mozambique. *Republic of Mozambique. Final Report of the Mozambique/U.S. Country Study Progam Project on Assessment of the Vulnerability of the Economy of Mozambique to Projected Climate Change*, Maputo, Mozambique, (unpublished).

Campbell, D.J., 1999: Response to drought among farmers and herders in southern Kajiado District, Kenya: a comparison of 1972–1976 and 1994–1995: *Human Ecology*, **27(3),** 377–416.

Caponera, D.A., 1996. Conflicts over international river basins in Africa, the Middle East, and Asia. *Review of European Community and International Environmnetal Law,* **5(2),** 97–106.

Carter, T.R., M. Hulme, J.F. Crossley, S. Malyshev, M.G. New, M.E. Schlesinger, and H. Tuomenvirta, 2000: *Climate Change in the 21st Century - Interim Characterizations based on the New IPCC Emissions Scenarios*. The Finnish Environment 433, Finnish Environment Institute, Helsinki, 148 pp.

CDC, 1998: Rift Valley Fever: east Africa, 1997–1998. Centers for Disease Control and Prevention, *Morbidity and Mortality Weekly Report*, **10(47),** 261–264.

Charney, J.G., 1975: Dynamics of deserts and drought in the Sahel. *Quarterly Journal of the Royal Meteorological Society*, **101,** 193–202.

Chemane, D., H. Mota, and M. Achimo, 1997: Vulnerability of coastal resources to climate changes in Mozambique: a call for integrated coastal zone management. *Ocean and Coastal Management*, **37(1),** 63–83.

Chifamba, F.C., 2000: The relationship of temperature and hydrological factors to catch per unit effort, condition and size of the freshwater sardine, *Limnothrissa miodon* (Boulenger), in Lake Kariba. *Fisheries Research*, **45,** 271–281.

Claussen, M., C. Kubatzki, V. Brovkin, A. Ganopolski, P. Hoelzmann, and H.-J. Pachur, 1999: Simulation of an abrupt change in Saharan vegetation in the mid-Holocene. *Geophysical Research Letters*, **26,** 2037–2040.

Coe, M.J., D.H. Cumming, and J. Phillipson, 1976: Biomass and production of large African herbivores in relation to rainfall and primary production. *Oecologia*, **22,** 341–354.

Coetzee, J.A., 1967: Pollen analytical studies in east and southern Africa. *Palaeoecology Africa*, **3,** 146 pp.

COHMAP Members, 1988: Climatic changes of the last 18,000 years: observations and model simulations. *Science,* **241,** 1043–1052.

Colwell, R.R., 1996: Global warming and infectious diseases. *Science*, **274,** 2025–2031.

Conway, D. and M. Hulme, 1993: Recent fluctuations in precipitation and runoff over the Nile sub-basins and their impact on main Nile discharge. *Climatic Change*, **25,** 127–152.

Cousins, B., 1996: Conflict management for multiple resource users in pastoralist and agro-pastoralist contexts. *IDS Bulletin,* **27(3),** 41–54.

Cowling, R.M., P.W. Rundel, P.G. Desmet, and K.J.E. Esler, 1998: Regional-scale plant diversity in southern African arid lands: subcontinental and global comparisons. *Biodiversity Research*, **4,** 27–36.

Crawford, R.J.M., L.V. Shannon, and D.E. Pollock, 1987: The Benguela ecosystem, part IV: the major fish and invertebrate resources. *Annual Review of Oceonography and Marine Biology,* **25,** 353–505.

Crutzen, P.J. and M.O. Andreae, 1990: Biomass burning in the Tropics: impact on atmospheric chemistry and biogeochemical cycles. *Science*, **250,** 1669–1678.

Cunnington, W.M. and P.R. Rowntree, 1986: Simulations of the Saharan atmosphere-dependence on moisture and albedo. *Quarterly Journal of the Royal Meteorological Society*, **112,** 971–999.

d'Almedia, G.A., 1986: A model for Saharan dust transport. *Journal of Climatology and Applied Meteorology*, **25,** 903–916.

Darkoh, M.B., 1998: The nature, causes and consequences of desertification in the dry lands of Africa. *Land Degradation and Development*, **9(1),** 1–20.

Davies, S., 1996: *Adaptable Livelihoods: Coping with Food Insecurity in the Malian Sahel*. St. Martin's Press, New York, NY, USA.

Davies, M.B. and C. Zabinski, 1992: Changes in geographical range resulting from greenhouse warming: Effects on biodiversity in forests. In: *Global Warming and Biological Diversity* [Peters, R.L. and T.E. Lovejoy (eds.)]. Yale University Press, New Haven, CT, USA, pp. 297–308.

De Lattre, A., 1988: What future for the Sahel? *OECD Observer*, **153,** 19–21.

De Ridder, N., L. Stroosnijder, A.M. Cisse, and H. van Keulen, 1982: *Productivity of Sahelian Rangelands, Vol. 1*. Department of Soil Science and Plant Nutrition, Wageningen Agricultural University, The Netherlands, 231 pp.

Depierre, D. and H. Gillet, 1971: Désertification de la zone sahélienne du Tchad. *Bois et Forêts des Tropiques*, **139,** 2–25.

Desanker, P.V., P.G.H. Frost, C.O. Justice, and R.J. Scholes (eds.), 1997: *The Miombo Network: Framework for a Terrestrial Transect Study of Land-Use and Land-Cover Change in the Miombo Ecosystems of Central Africa*. International Geosphere-Biosphere Programme (IGBP), Stockholm, Sweden, 109 pp.

Desanker, P.V., C.O. Justice, K. Masamvu, and G. Munthali, 2001: Requirements for integrated assessment modelling at the subregional and national levels in Africa to address climate change. In: *Climate Change for Africa: Science, Technology, Policy and Capacity Building* [Pak, S.L. (ed.)]. Kluwer Academic Publishers, Dordrecht, The Netherlands (in press).

Diedhiou, A. and J.-F. Mahfouf, 1996: Comparative influence of land and sea surfaces on the Sahelian drought: a numerical study. *Annales Geophysicae*, **14,** 115–130.

Dilley, M., 2000: Reducing vulnerability to climate variability in southern Africa: the growing role of climate information. *Climatic Change*, **45(1),** 63–73.

Dilley, M. and B.N. Heyman, 1995: ENSO and disaster—droughts, floods and El Niño Southern Oscillation warm events. *Disasters*, **19(3),** 181–193.

Dodd, J.L., 1994: Desertification and degradation in sub-Saharan Africa. *BioScience*, **44,** 28–34.

Downing, T.E., 1991: Vulnerability to hunger and coping with climate change in Africa. *Global Environmental Change*, **1(5),** 365–380.

Downing, T.E., L. Ringus, M. Hulme, and D. Waughray, 1997: Adapting to climate change in Africa: prospects and guidelines. *Mitigation and Adaptation Strategies for Global Change*, **2,** 19–44.

Dutt, G.S. and N.H. Ravindranath, 1993: Bioenergy: direct applications in cooking. In: *Renewable Energy* [Johansson, T.B., H. Kelly, A.K.N. Reddy, and R.H. Williams (eds.)]. Island Press, Washington, DC, USA, 1160 pp.

East, R., 1984: Rainfall, soil nutrient status and biomass of large African savanna animals. *African Journal of Ecology*, **22,** 245–270.

El Raey, M., K. Dewider, M. El Hattabb, 1999: Adaptation to the impacts of sea level rise in Egypt. In: *National Assessment Results of Climate Change: Impacts and Responses* [Mimura, N. (ed.)]. Oldendorf Luhe, Inter-Research, pp. 117–128.

Ellery, W.N., R.J. Scholes, and M.T. Mentis, 1991: An initial approach to predicting the sensitivity of the South African grassland biome to climate change. *South African Journal of Science*, **87,** 499–503.

Ellery, W., M.C. Scholes, and R.J. Scholes, 1996: The distribution of sweetveld and sourveld in South Africa's grassland biome in relation to environmental factors. *African Journal of Range and Forage Science*, **12,** 38–45.

Eltahir, E.A.B. and C. Gong, 1996: Dynamics of wet and dry years in West Africa. *Journal of Climate*, **9,** 1030–1042.

Falkenmark, M., 1989: The massive water scarcity now threatening Africa—why isn't it being addressed? *Ambio*, **18,** 112–118.

FAO, 1999a: *State of the World's Forests*. Food and Agricultural Organization of the United Nations, Rome, Italy, 154 pp. Available online at http://www.fao.org/forestry/FO/SOFO/SOFO99/sofo99-e.stm.

FAO, 1999b: *The State of Food Insecurity in the World*. Food and Agriculture Organization of the United Nations, Rome, Italy, 32 pp.

Faye, O., O. Gaye, D. Fontenille, L. Konate, J.P. Herve, Y. Toure, S. Diallo, J.F. Molez, and J. Mouchet, 1995: Drought and malaria decrease in the Niayes area of Senegal. *Sante*, **5,** 199–305.

Feddema, J.J., 1999: Future African water resources: interactions between soil degradation and global warming. *Climatic Change*, **42(3),** 561–596.

Feddema, J.J., 1998: Estimated impacts of soil degradation on the African water balance and climate. *Climate Research*, **10(2),** 127–141.

Flenley, J.R., 1979: The late Quaternary vegetational history of the equatorial mountains. *Progress in Physical Geography*, **3,** 488–509.

Flohn, H. and S. Nicholson, 1980: Climatic fluctuations in the arid belt of the "old world" since the last glacial maximum; possible causes and future implications. *Palaeoecology Africa*, **12,** 3–22.

Folland, C.K., T.N. Palmer, and D.E. Parker, 1986: Sahel rainfall and worldwide sea temperatures, 1901–1985. *Nature*, **320,** 602–607.

Fontenille, D., M. Traore-Lamizana, H. Zeller, M. Mondo, M. Diallo, and J.P. Gigoutte, 1995: Short report: Rift Valley fever in western Africa: isolations from Aedes mosquitoes during an interepizootic period. *American Journal of Tropical Medicine and Hygiene*, **52,** 403–404.

Ford, J. and K.M. Katondo, 1977: *The Distribution of Tsetse Flies in Africa*. Organisation of African Unity, Nairobi, Kenya, Cook, Hamond and Kell, London, United Kingdom, (plus maps).

Freeman, T. and M. Bradley, 1996: Temperature is predictive of severe malaria years in Ziambabwe. *Transactions of the Royal Society of Tropical Medicine and Hygiene*, **90,** 232.

Fritz, H. and P. Duncan, 1993: Large herbivores in rangelands. *Nature*, **364,** 292–293.

GEMS, 1989: *Environmental Data Report*. Global Environmental Monitoring and Assessment Research Center, Blackwell Publishers, 54 pp.

Gibbs, R.G.E., 1987: Preliminary floristic analysis of the major biomes of southern Africa. *Bothalia*, **17,** 213–227.

Gleick, P.H., 1998: *The World's Water: The Biennial Report on Freshwater Resources, 1998–1999*. Island Press, Washington, DC, USA, 307 pp.

Gleick, P.H. (ed.), 1993: *Water in Crisis: A Guide to the World's Water Resources*. Oxford University Press, Oxford, United Kingdom, 473 pp.

Gleick, P.H., 1992: *Water and Conflict: Occasional Papers Series on the Project on Environmental Change and Acute Conflict*. Security Studies Programme, American Academy of Arts and Sciences, University of Toronto, Toronto, ON, Canada, 62 pp.

Gonzalez, P., 2001: Desertification and a shift of forest species in the west African Sahel. *Climate Research*, (in press).

Gonzalez, P., 1997: Dynamics of biodiversity and human carrying capacity in the Senegal Sahel. Diss. University of California, Berkeley, CA, USA, 444 pp.

Gornitz, V. and NASA, 1985: A survey of anthropogenic vegetation changes in west Africa during the last century—climatic implications. *Climatic Change*, **7,** 285–325.

Goutorbe, J.P., P. Kabat, Y.H. Kerr, B. Monteny, S.D. Prince, J.N.M. Stricker, A. Tinga, J.S. Wallace, T. Lebel, A.J. Dolman, and J.H.C. Gash, 1997: An overview of HAPEX-Sahel: a study in climate and desertification. *Journal of Hydrology,* **188-189,** 4–17.

Graham, Y., 1995: Drought dims Ghana's hydroelectric power. *World Rivers Review,* **10,** 3. Available online at http://irn.org/pubs/wrr/9511.

Greenwood, B.M., 1984: Meningoccocal infections. In: *Weatherall: Oxford Textbook of Medicine* [Weatherall, D.J., J.G.G. Ledingham, and D.A. Warrel (eds.)]. Oxford University Press, Oxford, United Kingdom and New York, NY, USA, pp. 165–174.

Gundry, S., J. Wright, A. Ferro-Luzzi, G. Mudimu, and P. Vaze, 1999: *A Hierarchical Dimension to Food Security? The Multi-Level Structure of Spatial and Temporal Processes Influencing the Food and Nutrition System in a District of Zimbabwe*. Working Paper AAE 12/99, Department of Agricultural Economics and Extension, University of Zimbabwe, Harare, Zimbabwe.

Hanrahan, S.A., J. Lindesay, and P. Nailand, 1992: Swarming of the brown locust (Locustana pardalina) in relation to rainfall in southern Africa. In: *Proceedings of the 19th International Congress of Entomology, Beijing, China, 1992*.

Hao, W.M., M.H. Lui, and P.J. Crutzen, 1990: Estimates of annual releases of CO_2 and other trace gases to the atmosphere from fires in the tropics, based on FAO statistics for the period 1975–1980. In: *Fire in the Tropical Biota: Ecosystem Processes and Global Challenges* [Goldammer, J.G. (ed.)]. Springer-Verlag, Berlin, Germany, 497 pp.

Hart, C.A. and L.E. Cuevas, 1997: Meningoccocal disease in Africa. *Annals of Tropical Medicine and Parasitology*, **91,** 777–785.

Hart, R.C. and N.A. Rayner, 1994: Temperature-related distribution of *Metadiaptomus* and *Tropodiaptomus* (Copepoda; Calanoida), particularly in southern Africa. In: *Studies on the Ecology of Tropical Zooplankton—Developments in Hydrobiology 92, Hydrobiologia 272* [Dumont, H.J., J. Green, and H. Masundire (eds.)]. Kluwer Academic Publishers, Dordrecht, The Netherlands, pp. 77–86.

Hay, S.I., R.W. Snow, and D.J. Rogers, 1998: Predicting malaria seasons in Kenya using multi-temporal meteorological satellite sensor data. *Transactions of the Royal Society of Tropical Medicine and Hygiene*, **92,** 12–20.

Hilmi, K., J. Larissi, A. Makaoui, and S. Zizah, 1998: *Synthese Oceanographique de la Cote Atlantique Marocaine de 1994 a 1998*. Institut National de Recherche Halieutique, Royaume du Maroc, Rabat, Morrocco (unpublished project report).

Hirst, A., 1999: The Southern Ocean response to global warming in the CSIRO coupled ocean-atmosphere model. *Environmental Modeling and Software*, **14,** 227–241.

Hlohowskyj, I., M.S. Brody, and R.T. Lackey, 1996: Methods for assessing the vulnerability of African fisheries resources to climate change. *Climate Research*, **6(2),** 97–106.

Hockey, P.A.R., 2000: Patterns and correlates of bird migrations in sub-Saharan Africa. *Journal of Birds Australia,* **100,** 401–417.

Hoffman, M.T. and R.M. Cowling, 1990: Vegetation change in the semi-arid eastern Karoo over the last 200 years: an expanding Karoo, fact or fiction? *South African Journal of Science*, **86,** 462–463.

Holdridge, L.R., 1967: *Life Zone Ecology*. Tropical Science Center, San José, Costa Rica.

Hulme, M. (ed.), 1996: *Climate Change in Southern Africa: An Exploration of Some Potential Impacts and Implications in the SADC Region*. Climatic Research Unit, University of East Anglia, Norwich, United Kingdom, 96 pp.

Hulme, M., 1992: Rainfall changes in Africa (1931–1960 to 1961–1990). *International Journal of Climatology*, **12,** 658–690.

Hulme, M. and M. Kelly, 1997: Exploring the links between desertification and climate change. In: *Environmental Management: Readings and Case Studies* [Owen, L. and T.B.H. Unwin (eds.)]. Blackwell, Oxford, United Kingdom.

Hulme, M., R.M. Doherty, T. Ngara, M.G. New, and D. Lister, 2001: African climate change: 1900–2100. *Climate Research*, (in press).

Hulme, M., E.M. Barrow, N. Arnell, P.A. Harrison, T.E. Downing, and T.C. Johns, 1999: Relative impacts of human-induced climate change and natural climate variability. *Nature*, **397,** 688–691.

Hunt, B.G., 2000: Natural climatic variability and Sahelian rainfall trends. *Global and Planetary Change*, **24,** 107–131.

Huo, A., B. Xu, M.A. Chowdhury, M.S. Islam, R. Montilla, and R.R. Colwell, 1996: A simple filtration method to remove plankton-associated Vibrio cholerae in raw water supplies in developing countries. *Applied Environmental Microbiology*, **62,** 2508–2512.

IPCC, 1998: *The Regional Impacts of Climate Change: An Assessment of Vulnerability. Special Report of IPCC Working Group II* [Watson, R.T., M.C. Zinyowera, and R.H. Moss (eds.)]. Intergovernmental Panel on Climate Change, Cambridge University Press, Cambridge, United Kingdom and New York, NY, USA, 517 pp.

IPCC, 1996: *Climate Change 1995: Impacts, Adaptations, and Mitigation of Climate Change: Scientific-Technical Analyses. Contribution of Working Group II to the Second Assessment Report of the Intergovernmental Panel on Climate Change* [Watson, R.T., M.C. Zinyowera, and R.H. Moss (eds.)]. Cambridge University Press, Cambridge, United Kingdom and New York, NY, USA, 880 pp.

Jackson, R.D. and S.B. Idso, 1974: Surface albedo and desertification. *Science*, **189**, 1012–1013.

Jallow, B.P., 1995: Identification of and response to drought by local communities in Fulladu West District, The Gambia. *Singapore Journal of Tropical Geography*, **16(1)**, 22–41.

Jallow, B.P. and A.A. Danso, 1997: Assessment of the vulnerability of the forest resources of The Gambia to climate change. In: *Republic of The Gambia: Final Report of The Gambia/U.S. Country Study Program Project on Assessment of the Vulnerability of the Major Economic Sectors of The Gambia to the Projected Climate Change*. Banjul, The Gambia, (unpublished).

Jallow, B.P., S. Toure, M.M.K. Barrow, and A.A. Mathieu, 1999: Coastal zone of The Gambia and the Abidjan region in Cote d'Ivoire: sea level rise vulnerability, response strategies, and adaptation options. In: *National Assessment Results of Climate Change: Impacts and Responses* [Mimura, N. (ed.)]. Oldendorf Luhe, Inter-Research, Germany, pp. 129–136.

Jouan, A., I. Coulibaly, F. Adam, B. Philippe, B. Leguenno, R. Christie, N. Ould Merzoug, and J.P. Digoutte, 1989: Analytical study of a Rift Valley fever epidemic. *Research in Viroogyl.*, **140**, 175–186.

Justice, C.O., J.D. Kendall, P.R. Dowty, and R.J. Scholes, 1996: Satellite remote sensing of fires during the SAFARI campaign using NOAA-advanced very high resolution radiometer data. *Journal of Geophysical Research*, **101**, 23851–23863.

Kadomura, H., 1994: Climatic changes, droughts, desertification and land degradation in the Sudano-Sahelian Region—a historico-geographical perspective. In: *Savannization Processes in Tropical Africa II* [Kadomura, H. (ed.)]. Tokyo Metropolitan University, Tokyo, Japan, pp. 203–228.

Karanja, D.M.S., J. Alaii, K. Abok, N.I. Adungo, A.K. Githeko, I. Seroney, J.M. Vulule, P. Odada, and A.J. Oloo, 1999: Knowledge and attitude of malaria control and acceptability of permethrin impregnated sisal curtains. *East African Medical Journal*, **76**, 42–46.

Kendall, J.D., C.O. Justice, P.R. Dowty, C.D. Elvidge, and J.G. Goldammer, 1997: Remote sensing of fires during the SAFARI 92 Campaign. In: *Fire in Southern African Savannas* [van Wilgen, B.W., M.O. Andreae, J.G. Goldammer, and J.A. Lindesay (eds.)]. Witwatersrand University Press, Johannesburg, South Africa, pp. 89–134.

Kessler, J.J., 1992: The influence of karité (Vitellaria paradoxa) and néré (Parkia biglobosa) trees on sorghum production in Burkina Faso. *Agroforestry Systems*, **17**, 97–118.

Kremen, C., J.O. Niles, M.G. Dalton, G.C. Daily, P.R. Ehrlich, J.P. Fay, D. Grewal, and R.P. Guillery, 2000: Economic incentives for rain forest conservation across scales. *Science*, **288**, 1828–1832.

Lado, C. 1995: Some aspects of food security and social stress in the African arid and semi-arid environments. *Indonesian Journal of Geography*, **27(69)**, 31–49.

Lamb, P.J., 1978: Large-scale tropical Atlantic surface circulation patterns associated with sub-Saharan weather anomalies. *Tellus*, **30**, 240–251.

Lamb, P.J. and R.A. Peppler, 1992: Further case studies of tropical Atlantic surface atmospheric and oceanic patterns associated with sub-Saharan drought. *Journal of Climate*, **5**, 476–488.

Lamprey, H., 1975: *Report on the Desert Encroachment Reconnaissance in Northern Sudan*. Sudan National Council for Research and Sudan Ministry of Agriculture, Food and Natural Resources, Khartoum, Sudan.

Le Barbé, L. and T. Lebel, 1997 : Rainfall Climatology of the HAPEX-Sahel région during the years 1950–1990. *Journal of Hydrology*, **188–189**, 43–73.

Le Houérou, H.N., 1989: The grazing land ecosystems of the African Sahel. Springer-Verlag, Berlin, Germany, 282 pp.

Le Houérou, H.N., 1998: A probabilistic approach to assessing rangelands productivity, carrying capacity and stocking rate. In: *Drylands: Sustainable Use of Rangelands into the 21st Century* [Squires, V.E. and A.E. Sidahmed (eds.)]. IFAD, Rome, Italy, pp. 159–172.

Le Houérou, H.N. and C.N. Hoste, 1977: Rangeland production and annual rainfall relations in the Mediterranean basin and in the African Sahelo-Sudanian zone. *Journal of Range Management*, **30**, 181–189.

Lefevre, P.C., 1997: Current status of Rift Valley fever: what lessons to deduce from the epidemic of 1977 and 1987. *Medicine Tropicale*, **57(3)**, 61–64.

Lengeler, C., 1998: Insecticide treated bednets and curtains for malaria control, Cochrane Review 1998. In: *The Cochrane Library, Issue 3*. Oxford University Press, Oxford, United Kingdom, 54 pp.

Lericollais, A., 1973: *Sob. Étude Géographique d'un Terroir Serer (Sénégal)*. Office de la Recherche Scientifique et Technique Outre-Mer, Paris, France.

Levine, J.S., T. Bobbe, N. Ray, A. Singh, and R.G. Witt, 1999: *Wildland Fires and the Environment: A Global Synthesis*. U.N. Environment Program, Nairobi, Kenya.

Lindsay, S.W. and W.J.M. Martens, 1998: Malaria in the African highlands: past, present and future. *Bulletin of the World Health Organization*, **76**, 33–45.

Linthicum, K.J., A. Anyamba, C.J. Tucker, P.W. Kelley, M.F. Myers, and C.J. Peters, 1999: Climate and satellite indicators to forecast Rift Valley fever epidemics in Kenya. *Science*, **285**, 397–400.

Linthicum, K.J., C.L. Bailey, C.J. Tucker, K.D. Mitchell, T.M. Logan, F.G. Davis, C.W. Kamau, P.C. Thande, and J.N. Wagateh, 1990: Application of polar-orbiting, meteorological satellite data to detect flooding of Rift Valley Fever virus vector mosquito habitats in Kenya. *Medical Veterinary Entomology*, **4**, 433–438.

Loevinsohn, M.E., 1994: Climate warming and increased malaria in Rwanda. *Lancet*, **343**, 714–748.

Macdonald, D. (ed.), 1987: *The Encyclopaedia of Mammals*. Equinox, Oxford, United Kingdom, 895 pp.

Mace, G.M., A. Balmford, and J.R. Ginsberg (eds.), 1998: *Conservation in a Changing World*. Cambridge University Press, Cambridge, United Kingdom and New York, NY, USA, 308 pp.

MacKinnon, J. and K. MacKinnon, 1986: *Review of the Protected Areas System in the Afrotropical Realm*. International Union Conservation Networks, Gland, Switzerland.

Magadza, C.H.D., 2000: Climate change impacts and human settlements in Africa: prospects for adaptation. *Environmental Monitoring*, **61**, 193–205.

Magadza, C.H.D., 1997: Water pollution and catchment management in Lake Chivero. In: *Lake Chivero, A Polluted Lake* [Moyo, N.A.G. (ed.)]. University of Zimbabwe Publications, Zimbabwe, pp. 13–26.

Magadza, C.H.D., 1996: Climate change: some likely multiple impacts in southern Africa. In: *Climate Change and World Food Security* [Downing, T.E. (ed.)]. Springer-Verlag, Dordrecht, The Netherlands, pp. 449–483.

Magadza, C.H.D., 1977: Determination of development period at different temperatures in a tropical cladoceran, *Moina dubia* De Gurne and Richard. *Transactions of the Rhodesian Science Association*, **58(4)**, 24–27.

Magadza, C.H.D., 1984: An analysis of siltation rates in Zimbabwe. *Zimbabwe Science News*, **18 (6)**, 63–64.

Manneh, A., 1997: Vulnerability of the water resources sector of The Gambia to climate change. In: *Republic of The Gambia: Final Report of The Gambia/U.S. Country Study Program Project on Assessment of the Vulnerability of the Major Economic Sectors of The Gambia to the Projected Climate Change*. Banjul, The Gambia, (unpublished).

Martin, R., R. Washington, and T.E. Downing, 2000: Seasonal maize forecasting for South Africa and Zimbabwe derived from an agroclimatological model. *Journal of Applied Meteorology*, **39(9)**, 1473–1479.

Mason, S.J., A.M. Joubert, C. Dosign, and S.J. Crimp, 1996: Review of seasonal forecasting techniques and their applicability to southern Africa. *Water in South Africa*, **22(3)**, 203–209.

Mason, S.J., P.R. Waylen, G.M. Mimmack, B. Rajaratnam, and H.J. Harrison, 1999: Changes in extreme rainfall events in South Africa. *Climatic Change*, **41**, 249–257.

Mattes, M. and S.J. Mason, 1998: Evaluation of a seasonal forecasting procedure for Namibian rainfall. *South African Journal of Science*, **94(4)**, 183–185.

McMichael, A.J., A. Haines, R. Sloof, and S. Kovats (eds.), 1996: *Climate Change and Human Health: An Assessment Prepared by a Task Group on behalf of the World Health Organisation, the World Meteorological Organisation, and the United Nations Environment Programme*. World Health Organisation, Geneva, Switzerland, 297 pp.

Medema, G.J., D.D. Havelaar, F.M. Schets, and P.F.M. Teunis, 1998: Sedimentation of free and attached Cryptosporidium oocyst and Giardia cysts in water. *Applied Environmental Microbiology*, **64,** 4460–4466.

Midgley, G.F., W.J. Bond, S.J.E. Wand, and R. Roberts, 1999: Will Gullivers travel? Potential causes of changes in savanna tree success due to rising atmospheric CO_2. In: *Towards Sustainable Management in the Kalahari Region—Some Essential Background and Critical Issues* [Ringrose, S. and R. Chanda (eds.)]. University of Botswana, Gaberone, Botswana, 304 pp.

Mooney, H.A., J. Canadell, F.S. Chapin III, J.R. Ehleringer, C. Korner, R.E. McMurtrie, W.J. Parton, L.F. Pitelka, and E.-D. Schulze, 1999: Ecosystem physiology responses to global change. In: *The Terrestrial Biosphere and Global Change: Implications for Natural and Managed Ecosystems, Synthesis Volume* [Walker, B., W. Steffen J. Canadell, and J. Ingram (eds.)]. International Geosphere-Biosphere Programme Book Series 4, Cambridge University Press, Cambridge, United Kingdom and New York, NY, USA.

Mortimore, M., 1989: *Adapting to Drought: Farmers, Famines, and Desertification in West Africa*. Cambridge University Press, Cambridge, United Kingdom and New York, NY, USA, 289 pp.

Mouchet, J., S. Manuin, S. Sircoulon, S. Laventure, O. Faye, A.W. Onapa, P. Carnavale, J. Julvez, and D. Fontenille, 1998: Evolution of malaria for the past 40 years: impact of climate and human factors. *Journal of the American Mosquito Control Association*, **14,** 121–130.

Moyo, N.A.G., 1997: Causes of massive fish deaths in Lake Chivero. In: *Lake Chivero, A Polluted Lake* [Moyo, N.A.G. (ed.)]. University of Zimbabwe Publications, Harare, Zimbabwe, pp. 98–103.

Muchena, P. and A. Iglesias, 1995: Vulnerability of maize yields to climate change in different farming sectors in Zimbabwe: Plant Protection Research Institute, Causeway, Zimbabwe. In: *Climate Change and Agriculture—Analysis of Potential International Impacts: Proceedings of a Symposium Sponsored by the American Society of Agronomy in Minneapolis, MN, 4–5 November, 1992*. Agroclimatology and Agronomic Modeling and Division A-6 (International Agronomy), **59,** 229–239.

Mwachari, C., B.I. Batchelor, J. Paul, P.G. Wayaiki, and C.F. Gilks, 1998: Chronic diarrhoea among HIV-infected adult patients in Nairobi, Kenya. *Journal of Infections*, **37,** 48–53.

Myneni, R.B., S.O. Los, and C.J. Tucker, 1996: Sattelite-based identification of linked vegetation index and sea surface temperature anomaly areas from 1982–1990 for Africa, Australia, and South America. *Geophysical Research Letters*, **23,** 729–732.

Ngecu, W.M. and E.M. Mathu, 1999: The El Niño-triggered landslides and their socioeconomic impact on Kenya. *Environmental Geology*, **38(4),** 277–284.

Nicholls, R.J., F.M.J. Hoozemans, and M. Marchand, 1999: Increasing flood risk and wetland losses due to global sea-level rise: regional and global analyses. *Global Environmental Change*, **9,** S69–S87.

Nicholson, S.E., 2000: Land surface processes and Sahel climate. *Reviews of Geophysics*, **38,** 117–139.

Nicholson, S.E. and D. Entekhabi, 1986: The quasi-periodic behavior of rainfall variability in Africa and its relationship to the Southern Oscillation. *Journal of Climate and Applied Meteorology*, **34,** 331–348.

Nicholson, S.E. and J. Kim, 1997: The relationship of the El Niño Southern Oscillation to African rainfall. *International Journal of Climatology*, **17,** 117–135.

Nicholson, S.E. and I.M. Palao, 1993: A re-evaluation of rainfall variability in the Sahel, part I: characteristics of rainfall fluctuations. *International Journal of Climatology*, **13,** 371–389.

Nicholson, S.E., C.J. Tucker, and M.B. Ba (eds.), 1998: Desertification, drought and surface vegetation: an example from the West African Sahel. *Bulletin of the American Meteorological Society*, **79,** 815–829.

Nicholson, S.E., B. Some, and B. Kone, 2000: A note on recent rainfall conditions in West Africa, including the rainy season of the 1997 ENSO year. *Journal of Climate*, **13,** 2628–2640.

Ntiba, J.N., 1998: The potential impacts of climate change on fisheries in Lakes Naivasha and Victoria. In: *Vulnerability and Adaptation to Potential Impacts of Climate Change in Kenya* [Omenda, T.O., J.G. Kariuki, and P.N. Mbuthi (eds.)]. Kenya Country Study on Climate Change Project. Ministry of Research and Technology, P.O. Box 30568, Nairobi, Kenya.

NOAA, 1999: *An Experiment in the Application of Climate Forecasts: NOAA-OGP Activities Related to the 1997–98 El Niño Event*. Office of Global Programs, National Oceanic and Atmospheric Administration, Silver Spring, MD, USA.

Odingo, R.S., 1990: Implications for African agriculture of the greenhouse effect. In: *Soils on a Warmer Earth: Proceedings of an International Workshop on Effects of Expected Climate Change on Soil Processes in the Tropics and Subtropics, Nairobi, Kenya* [Scharpenseel, H.W., M. Schomaker, and A. Ayoub (eds.)]. Elsevier Press, New York, NY, USA, 274 pp.

Olago, D.O., 2001: Vegetation changes over palaeo-time scales in Africa. *Climate Research*, (in press).

Olaleye, O.D., O. Tomori, M.A. Ladipo, and H. Schmitz, 1996: Rift Valley Fever in Nigeria: infections in humans. *Review of Science Technologies*, **15,** 923–935.

Olivry J.C., J.P. Briquet, and G. Mahe, 1993. Vers un Apprauvrissment Durable des Resources en Eau de l'Afrique Humide? IAHS Publication No. 216, pp. 67–78.

Omenda, T.O., J.G. Kariuki, and P.N. Mbuthi, 1998: *Vulnerability and Adaptation to Potential Impacts of Climate Change in Kenya: Summary for Policy Makers*. Kenya Country Study Project, Ministry of Rresearch and Technology, Nairobi, Kenya.

Otterman, J., 1974: Baring high-albedo soils by overgrazing: a hypothesized desertification mechanism. *Science*, **186,** 531–533.

Palmer, A.R. and A.F. van Rooyen, 1998: Detecting vegetation change in the southern Kalahari using Landsat™ data. *Journal of Arid Environments*, **39,** 143–153.

Pamo, E.T., 1998: Herders and wildgame behavior as a strategy against desertification in northern Cameroon. *Journal of Arid Environments*, **39,** 170–190.

Pimentel, D., 1993: Climate changes and food supply. *Forum for Applied Research and Public Policy*, **8(4),** 54–60.

Pinker R.T., G. Idemudia, and T.O. Aro, 1994: Characteristic aerosol optical depths during the Harmattan season in sub-Saharan Africa. *Journal of Geophysical Research,* **21,** 685–692.

Pinstrum-Anderson, P. and R. Pandya-Lorch, 1999: The role of agriculture to alleviate poverty. In: *Agriculture and Rural Development* [Wilcke, A. (ed.)]. Technical Centre for Agricultural and Rural Cooperation (CTA), Frankfurt am Main, Germany, **6(2),** 53–56.

Poolman, E., 1999: *Heavy Rain Events Over South Africa*. Paper presented at the 15th Annual Conference of the South African Society of Atmospheric Sciences, Richards Bay, South Africa, 18–19 November 1999.

Pottinger, L., 1997: Coming up dry: South Africa's water crisis is trouble for rivers. *World Rivers Review*, **12(1)**.

Preston, T.R. and R.A. Lang, 1987: *Matching Ruminant Production Systems with Available Resources in the Tropics and Subtropics*. Penambul Books Ltd., Armidale, NSW, Australia.

Quick, R.E., L.V. Venzel, O. Gonzalez, E.D. Mintz, A.K. Highsmith, A. Espanda, N.H. Bean, E.H. Hannover, and R.V. Tauxe, 1996: Narrow-mouthed water storage vessels and in situ chlorination in a Bolivian community: a simple method to improve drinking water quality. *American Journal of Tropical Medicine and Hygiene*, **54,** 511–516.

Riebsame, W.E., K.M. Strzepek, J.L. Wescoat Jr., R. Perrit, G.L. Graile, J. Jacobs, R. Leichenko, C. Magadza, H. Phien, B.J. Urbiztondo, P. Restrepo, W.R. Rose, M. Saleh, L.H. Ti, C. Tucci, and D. Yates, 1995: Complex river basins. In: *As Climate Changes, International Impacts and Implications* [Strzepek, K.M. and J.B. Smith (eds.)]. Cambridge University Press, Cambridge, United Kingdom and New York, NY, USA, pp. 57–91.

Reubush, T.K., M.K. Kern, C.C. Campbell, and A.J. Oloo, 1995: Self-treatment of malaria in a rural area of western Kenya. *Bulletin of the World Health Organization*, **73,** 229–236.

Ripley, E.A., 1976: Drought in the Sahara: insufficient biogeophysical feedback? *Science*, **191,** 100.

Robertshaw, D. and V. Finch, 1976: The effects of climate on the productivity of beef cattle. In: *Cattle Production in Developing Countries* [Smith, A.J. (ed.)]. University of Edinburgh, Edinburgh, Scotland, United Kingdom, pp. 132–137.

Russell, S. and M. Yonge, 1975: *The Seas: An Introduction to Life in the Sea*. Butler and Tanner, London, United Kingdom, 283 pp.

Rutherford, M.C., G.F. Midgley, W.J. Bond, L.W. Powrie, C.F. Musil, R. Roberts, and J. Allsopp, 1999: *South African Country Study on Climate Change*. Terrestrial Plant Diversity Section, Vulnerability and Adaptation, Department of Environmental Affairs and Tourism, Pretoria, South Africa.

Rutherford, M.C., M. O'Callaghan, J.L. Hurford, L.W. Powrie, R.E. Schulze, R.P. Kunz, G.W. Davis, M.T. Hoffman, and F. Mack, 1995: Realized niche spaces and functional types: a framework for prediction of compositional change. *Journal of Biogeography*, **22,** 523–531.

Sabwa, D.M. and A.K. Githeko, 1985: Feacal contamination of urban community water supplies and its public health implications. *East African Medical Journal*, **62,** 794–801.

Scherr, S.J., 1999: *Soil Degradation: A Threat to Developing-Country Security by 2020?* International Food Policy Research Institute, Washington, DC, USA.

Schlesinger, W.H., J.F. Reynolds, G.I. Cunningham, L.F. Huennecke, W.M. Jarrell, R.A. Virginia, and W.G. Whitford, 1990: Biological feedbacks in global desertification. *Science*, **247,** 1043–1048.

Scholes, R.J., 1993: Nutrient cycling in semi-arid grasslands and savannas: its influence on pattern, productivity and stability. Proceedings of the XVII International Grassland Congress.Palmerston North: International Grasslands Society, 1331–1334.

Scholes, R.J., 1990: The effect of soil fertility on the ecology of African dry savannas. *Journal of Biogeography*, **17,** 415–419.

Scholes, R.J. and S. Archer, 1997: Interactions between woody plants and grasses in savannas. *Annual Review of Ecology and Systematics*, **28,** 517–544.

Scholes, R.J. and M.R. van der Merwe, 1996: Sequestration of carbon in savannas and woodlands. *The Environmental Professional*, **18,** 96–103.

Scholes, R.J. and M.C. Scholes, 1998: Natural and human-related sources of ozone-forming trace gases in southern Africa. *South African Journal of Science*, **94,** 422–425.

Scholes, R.J., C.O. Justice, and D. Ward, 1996: Emissions of trace gases and aerosol particles due to vegetation burning in southern-hemisphere Africa. *Journal of Geophysical Research*, **101,** 23677–23682.

Scholes, R.J., G. Midgeley, and S. Wand, 2000: The impacts of climate change on South African rangelands. In: *South African Country Studies on Climate Change*. Department of Environmental Affairs and Tourism, Pretoria, South Africa, (in press).

Schulze, R.E., 2000: Modelling hydrological responses to land use and climate change: a southern African perspective. *Ambio*, **29,** 12–22.

Schulze, R.E., 1997: Impacts of global climate change in a hydrologically vulnerable region: challenges to South African hydrologists. *Progress in Physical Geography*, **21,** 113–136.

Schulze, R.E. and L.A. Perks, 2000: *Assessment of the impact of Climate Change on Hydrology and Water Resources in South Africa*. ACRUcons Report 33, South African Country Studies for Climate Change Programme, School of Bioresources Engineering and Environmental Hydrology, University of Natal, Pietermaritzburg, South Africa, 118 pp.

Schulze, R.E., G.A. Kiker, and T.E. Downing, 1995: Global climate change and agricultural productivity in southern Africa: thought for food and food for thought. In: *Climate Change and World Food Security* [Downing, T.E. (ed.)]. Springer-Verlag, Dordrecht, The Netherlands, pp. 421–447.

Schulze, R.E., G.A. Kiker, and R.P. Kunz, 1993: Global climate change and agricultural productivity in southern Africa. *Global Environmental Change*, **3(4),** 330–349.

Scoones, I., C. Chibudu, S. Chikura, P. Jeranyama, D. Machaka, W. Machanja, B. Mavedzenge, B. Mombeshora, M. Mudhara, C. Mudziwo, F. Murimbarimba, and B. Zirereza, 1996: *Hazards and Opportunities—Farming Livelihoods in Dryland Africa: Lessons from Zimbabwe*. Zed Books, London, United Kingdom, 267 pp.

Semazzi, F.H.M. and L. Sun 1995. *On the Modulation of the Sahelian Summer Rainfall by Bottom Topography*. Paper presented at the Sixth Symposium on Global Change Studies, Amerian Meteorological Society, Dallas, Texas, USA, 1995.

Semazzi, F.H.M. and L. Song, 2000: Numerical simulation of regional climate variability over eastern Africa and implications for the rest of Africa. *Climate Research*, (in press).

Sestini, G., 1989: *The Implication of Climatic Changes for the Nile Delta*. Report WG 2/14, UNEP/OCA, Nairobi, Kenya.

Shannon, L.V., J.R.E. Lutjeharms, and G. Nelson, 1990: Causative mechanisms for intra-annual and interannual variability in the marine environment around southern Africa. *South African Journal of Science*, **86,** 356–373.

Shapiro, R.L., R.O. Muga, M.P. Adcock, A. Penelope, P. Howard, W.A. Hawley, L. Kumar, P. Waiyaki, B.L. Nahlen, and L. Slutsker, 1999: Transmission of epidemic Vibrio cholerae 01 in rural western Kenya associated with drinking water from Lake Victoria: an environmental reservior of cholera? *Amercian Journal of Tropical Medicine and Hygiene*, **60,** 271–276.

Sharma, N., T. Damhang, E. Gilgan-Hunt, D. Grey, V. Okaru, and D. Rothberg, 1996: *African Water Resources: Challenges and Opportunities for Sustainable Development*. World Bank Technical Paper No. 33, African Technical Department Series, The World Bank, Washington DC, USA, 115 pp.

Siegfried, W.R., 1989: Preservation of species in southern African nature reserves. In: *Biotic Diversity in Southern Africa: Concepts and Conservation* [Huntley, B.J. (ed.)]. Oxford University Press, Cape Town, South Africa, pp. 186–201.

Siegfried, W.R., R.J.M. Crawford, L.V. Shannon, D.E. Pollock, A.I.L. Payne, and R.G. Krohn, 1990: Scenarios for global warming induced change in the open ocean environment and selected fisheries of the west coast of Southern Africa. *South African Journal of Science*, **86,** 281–285.

Singh, A., A. Dieye, M. Finco, M.S.Chenoweth, E.A. Fosnight, and A. Allotey, 1999: *Early Warning of Selected Emerging Environmental Issues in Africa: Change and Correlation from a Geographic Perspective*. United Nations Environment Programme, Nairobi, Kenya.

Sircoulon, J., T. Lebel, and N.W. Arnell, 1999: Assessment of impacts of climate change variability and change on the hydrology of Africa. In: *Impacts of Climate Change and Climate Variability on Hydrological Regimes* [van Dam, J.C. (ed.)]. Cambridge University Press, Cambridge, United Kingdom and New York, NY, USA, 140 pp.

Smith, J.B. and S.S. Lenhart, 1996: Climate change adaptation policy options. *Climate Research*, **6(2),** 193–201.

Smithers, J. and B. Smit, 1997: Human adaptation to climatic variability and change. *Global Environmental Change*, **7,** 129–146.

Stack, J., 1998: *Drought Forecasts and Warnings in Zimbabwe: Actors, Linkages and Information Flows*. Working Paper AAE 2/98, Department of Agricultural Economics and Extension, University of Zimbabwe, Harare, Zimbabwe.

Sterk, G., A. Bationo, and L. Herrmann, 1996: Wind-blown nutrient transport and soil productivity changes in southwest Niger. *Land Degradation and Development*, **7,** 325–335.

Stern, P.C. and W.E. Easterling (eds.), 1999: *Making Climate Forecasts Matter*. National Academy Press, Washington, DC, USA, 175 pp.

Street-Perrott, F.A. and R.A. Perrott, 1990: Abrupt climate fluctuations in the tropics: the influence of Atlantic Ocean circulation. *Nature*, **343,** 607–612.

Stockton, R. and A. Allali, 1992: La secheresse au Maroc et sa relation avec le systeme Almoubarak. In: *Proceeding of the Symposium on Water Resources and Economy*. Taroudant, Morocco.

Talling, J.F. and J. Lemoalle, 1998: *Ecological Dynamics of Tropical Inland Waters*. Cambridge University Press, Cambridge, United Kingdom and New York, NY, USA, 441 pp.

Tett, S.F.B., T.C. Jones, and J.F.B. Mitchell, 1997: Global and regional variability in a coupled AOGCM. *Climate Dynamics*, **13,** 302–323.

Thiaw, W.M., A.G. Barnston, and V. Kumar, 1999: Predictions of African rainfall on the seasonal timescale. *Journal of Geophysical Research*, **104,** 31589–31597.

Tikhomirov, E., M. Santamaria, and K. Esteves, 1997: Meningoccocal disease: public health burden and control. *World Health Status Quarterly*, **50,** 170–177.

Timmermann, A., J. Oberhuber, A. Bacher, M. Esch, M. Latif, and E. Roeckner, 1999: Increased El Niño frequency in a climate model forced by future greenhouse warming. *Nature*, **398,** 694.

Trapnell, C.G., H.F. Birch, G.T. Chamberlain, and M.T. Friend, 1976: The effect of fire and termites of a Zambian woodland soil. *Journal of Ecology*, **64,** 577–588.

Trenberth, K.E. and T.J. Hoar, 1997: El Niño and climate change. *Geophysical Research Letters*, **24,** 3057–3060.

Tucker, C.J., H.E. Dregne, and W.W. Newcomb, 1991: Expansion and contraction of the Sahara Desert from 1980 to 1990. *Science*, **253,** 299–301.

Turner, F.F., 1996: Maximization of yield from African lakes. In: *Stock Assessment in Inland Fisheries* [Cowx, I.G. (ed.)]. Oxford University Press, Oxford and London, United Kingdomp, pp. 465–481.

United Nations, 1994: *United Nations Convention to Combat Desertification (UNCCD).* UN document number A/AC.241/27, United Nations, New York, NY, USA, 58 pp.

UNCOD, 1977: *Desertification: Its Causes and Consequences.* United Nations Conference on Desertification, Pergamon Press, Oxford, United Kingdom, 448 pp.

UNDP, 1998: *Human Development Report, 1998.* United Nations Development Programme, Oxford University Press, Oxford, United Kingdom and New York, NY, USA.

UNDP, 1997: *Aridity Zones and Dryland Populations: An Assessment of Population Levels in the World's Drylands.* United Nations Development Programme, Office to Combat Desertification and Drought, New York, NY, USA.

UNEP, 2000: *Planning and Mamnagement of Lakes and Reservoirs: An Intergrated Aproach to Uetrophication.* UNEP-IETC Technical Publications Series—Issue 11. UNEP International Environmental Technical Centre, Osaka/Shiga, Japan.

UNEP, 1997: *World Atlas of Desertification, 2nd. Ed.* United Nations Environment Program, Edward Arnold, London, United Kingdom, 69 pp.

UNSO, 1999: *Report from the International Workshop on Coping with Drought.* United nations, Office to Combat Desertification and Drought, New York, NY, USA.

Urbiztondo, R.J., 1992: *Modelling of Climate Change Impacts on the Upper Zambezi River Basin.* M.Sc. Thesis, University of Colorado, Boulder, CO, USA, 134 pp.

USAID, 2000: *FEWS Sahel 1999/2000 Current Vulnerability Assessment.* U.S. Agency for International Development, Washington, DC, USA. Available online at http://www.fews.org/va/vahome.html.

USAID, 1999: *FEWS Current vulnerability Assessment Guidance Manual.* U.S. Agency for International Development, Washington, DC, USA, Available online at http://www.fews.org/va/vahome.html.

van den Berg, J.A., 1983: *The Relationship Between the Long Term Average Rainfall and the Grazing Carrying Capacity of Natural Veld in the Dry Areas of South Africa.* Grassland Society of Southern Africa, **18,** 165–167.

van Wilgen, B.W. and R.J. Scholes, 1997: The vegetation and fire regimes of southern Africa. In: *Fire in Southern African Savannas* [van Wilgen, B.W., M.O. Andreae, J.G Goldammer, and J.A. Lindesay (eds.)]. Witwatersrand University Press, Johannesburg, South Africa, pp. 27–46.

Venczel, V.L., M. Arrowood, M. Hurd, and M.D. Sobsey, 1997: Inactivation of cryptosporidium parvum oocyst and clostridum perfrigens spores by a mixed-oxidant disinfectant and by free chlorine. *Applied Environmental Microbiology*, **63,** 1598–1601.

Venema, H.D., E.J. Schiller, K. Adamowski, and J.M. Thizy, 1997: A water resources planning response to the climate change in the Senegal River Basin. *Journal of Environmental Management*, **49(1),** 125–155.

Vincens, A., F. Chalié, R. Bonnefille, J. Guiot, and J.J. Tiercelin, 1993: Pollen-derived rainfall and temperature estimates for Lake Tanganyika and their implication for late Pleistocene water levels. *Quaternary Research*, **40,** 343–350.

Vogel, C., 1998: Disaster management in South Africa. *South African Journal of Science*, **94(3),** 98–100.

Vorosmarty, C.J. and B. Moore III, 1991: Modeling basin scale hydrology in support of physical climate and global biochemical studies: an example using the Zambezi River. *Surveys in Geophysics*, **12,** 271–311.

Wand, S.J.E., G.F. Midgley, M.H. Jones, and P.S. Curtis, 1999: Responses of wild C_4 and C_3 grass (Poaceae) species to elevated atmospheric CO_2 concentration: a test of current theories and perceptions. *Global Change Biology*, **5,** 723–741.

Wang, G. and E.A.B. Eltahir, 2000: Ecosystem dynamics and the Sahel drought. *Geophysical Research Letters*, **27,** 795–798.

Ward, M.N., 1998: Diagnosis and short-lead time prediction of summer rainfall in tropical northern Africa at interannual and multidecadal timescales. *Journal of Climate*, **11,** 3167–3191.

Warsame, M., W.H. Wernsdofer, G. Huldt, and A. Bjorkman, 1995: An epidemic of Plasmodium falciparum malaria in Balcad Somalia, and its causation. *Transactions of the Royal Society of Tropical Medicine and Hygiene*, **98,** 142–145.

Washington, R. and T.E. Downing, 1999: Seasonal forecasting of African rainfall: prediction, responses and household food security. *Geographical Journal,* **165(3),** 255–274.

Weber J.T., E.D. Mintz, R. Canizares, A. Semiglia, I. Gomez, R. Sempertegui, A. Davila, K.D. Greene, N.D. Puhr, D.N. Cameron, F.C. Tenover, T.J. Barrett, N.H. Bean, C. Ivey, R.V. Tauxe, and P.A. Blake, 1994: Epidemic cholera in Ecuador: multidrug-resistance and transmission by water and seafood. *Epidemiology Infections*, **112,** 1–11.

Wendler, G. and F. Eaton, 1983: On the desertification of the Sahel zone. *Climatic Change*, **5,** 365–380.

Westing, A.H., 1994: Population, desertification, and migration. *Environmental Conservation*, **21,** 109–114.

WHO, 1998a: *The State of World Health, 1997 Report.* World Health Organization, *World Health Forum*, **18,** 248–260.

WHO, 1998b: *Weekly Epidemiological Records,* **73 (20),** 145–152.

WHO, 1997: *Tropical Disease Research, 13th Progress Report.* World Health Organization, Washington, DC, USA.

Williams, M.A.J. and R.C. Balling Jr., 1996: *Interactions of Desertification and Climate.* Edward Arnold, London, United Kingdom, 270 pp.

Willocks, L., A. Crampin, L. Milne, C. Seng, M. Susman, R. Gair, M. Moulsdale, S. Shafi, R. Wall, R. Wiggins, and N. Lightfoot, 1998: A large outbreak of cryptosporidiosis within a public water supply from a deep chalk borehole. *Communicable Disease and Public Health*, **1,** 239–243.

WMO, 1995: *Global Climate System Review: Climate System Monitoring.* World Meteorological Organization, Geneva, Switzerland.

World Bank, 1995: *Towards Environmentally Sustainable Development in Sub-Saharan Africa: A Framework for Integrated Coastal Zone Management Building Blocks for Africa 2025.* Paper 4, Post UNCED Series, Environmentally Sustainable Development Division, Africa Technical Department, World Bank, Washington, DC, USA, 56 pp.

WRI, 1998: *1998–99 World Resources Database Diskette: A Guide to the Global Environment.* World Resources Institute, Washington, DC, USA, diskettes.

WRI, 1996: *World Resources: A Guide to Global Environment, 1996–1997.* World Resources Institute, United Nations Environment Program, World Bank, Oxford University Press, New York, NY, USA, 342 pp.

Xue, Y., 1997: Biosphere feedback on regional climate in tropical north Africa. *Quarterly Journal of the Royal Meteorological Society*, **123,** 1483–1515.

Xue, Y., K.-N. Liou, and A. Kasahara, 1990: Investigation of biogeophysical feedback on the African climate using a two-dimensional model. *Journal of Climate*, **3,** 337–352.

Yates, D.N. and K.M. Strzepek, 1998: An assessment of integrated climate change impacts on the agricultural economy of Egypt. *Climatic Change*, **38(3),** 261–287.

Zeller, H.G., D. Fontenille, M. Traore-Lamizana, Y. Thiongane, and J.P. Digoutte, 1997: Enzootic activity of Rift Valley Fever virus in Senegal. *American Journal of Tropical Medicine and Hygiene*, **56,** 265–272.

Zeng, N., J.D. Neelin, K.-M. Lau, and C.J. Tucker, 1999: Enhancement of interdecadal climate variability in the Sahel by vegetation interaction. *Science*, **286,** 1537–1540.

Zheng, X. and E.A.B. Eltahir, 1997: The response of deforestation and desertification in a model of West African Monsoons. *Geophysical Resource Letters*, **24(2),** 155–158.

11

Asia

MURARI LAL (INDIA), HIDEO HARASAWA (JAPAN), AND
DANIEL MURDIYARSO (INDONESIA)

Lead Authors:
W.N. Adger (UK), S. Adhikary (Nepal), M. Ando (Japan), Y. Anokhin (Russia), R.V. Cruz (Philippines), M. Ilyas (Malaysia), Z. Kopaliani (Russia), F. Lansigan (Philippines), Congxian Li (China), A. Patwardhan (India), U. Safriel (Israel), H. Suharyono (Indonesia), Xinshi Zhang (China)

Contributing Authors:
M. Badarch (Mongolia), Xiongwen Chen (China), S. Emori (Japan), Jingyun Fang (China), Qiong Gao (China), K. Hall (USA), T. Jarupongsakul (Thailand), R. Khanna-Chopra (India), R. Khosa (India), M.P. Kirpes (USA), A. Lelakin (Russia), N. Mimura (Japan), M.Q. Mirza (Bangladesh), S. Mizina (Kazakhstan), M. Nakagawa (Japan), M. Nakayama (Japan), Jian Ni (China), A. Nishat (Bangladesh), A. Novoplansky (Israel), T. Nozawa (Japan), W.T. Piver (USA), P.S. Ramakrishnan (India), E. Rankova (Russia), T.L. Root (USA), D. Saltz (Israel), K.P. Sharma (Nepal), M.L. Shrestha (Nepal), G. Srinivasan (India), T.S. Teh (Malaysia), Xiaoping Xin (China), M. Yoshino (Japan), A. Zangvil (Israel), Guangsheng Zhou (China)

Review Editors:
Su Jilan (China) and T. Ososkova (Uzbekistan)

CONTENTS

EXECUTIVE SUMMARY

Will the Climate in Asia Change?

Continuing emissions of greenhouse gases from human activities are likely to result in significant changes in mean climate and its intraseasonal and interannual variability in the Asian region. Given the current state of climate modeling, projections of future regional climate have only limited confidence. Currently available general circulation models (GCMs) suggest that the area-averaged annual mean warming would be about 3°C in the decade of the 2050s and about 5°C in the decade of the 2080s over the land regions of Asia as a result of future increases in atmospheric concentration of greenhouse gases. Under the combined influence of greenhouse gas and sulfate aerosols, surface warming would be restricted to about 2.5°C in the 2050s and about 4°C in the 2080s. In general, projected warming over Asia is higher during Northern Hemisphere (NH) winter than during summer for both time periods. The rise in surface air temperature is likely to be most pronounced over boreal Asia in all seasons. GCM simulations project relatively more pronounced increases in minimum temperature than in maximum temperature over Asia on an annual mean basis, as well as during winter, hence a decrease in diurnal temperature range (DTR). During summer, however, an increase in DTR is projected, suggesting that the maximum temperature would have more pronounced increases relative to the minimum temperature. The summertime increase in DTR over central Asia is likely to be significantly higher relative to that in other regions.

In general, all GCMs simulate an enhanced hydrological cycle and an increase in area-averaged annual mean rainfall over Asia. An annual mean increase in precipitation of approximately 7% in the 2050s and approximately 11% in the 2080s over the land regions of Asia is projected from future increases in atmospheric concentration of greenhouse gases. Under the combined influence of greenhouse gases and sulfate aerosols, the projected increase in precipitation is limited to about 3% and 7% in the 2050s and 2080s, respectively. The projected increase in precipitation is greatest during NH winter for both time periods. The increase in annual and winter mean precipitation is projected to be highest in boreal Asia; as a consequence, the annual runoff of major Siberian rivers is expected to increase significantly. Although area-averaged annual mean precipitation is projected to increase in temperate Asia, a decline in summer precipitation is likely over the central parts of arid and semi-arid Asia. Because the rainfall over this region is already low, severe water-stress conditions—leading to expansion of deserts—are quite possible, with rises in surface air temperature and depletion of soil moisture. GCMs show high uncertainty in future projections of winter and summer precipitation over south Asia (with or without aerosol forcings). Because much of tropical Asia is intrinsically linked with the annual monsoon cycle, research into a better understanding of the future behavior of the monsoon and its variability is warranted.

Is Asia Vulnerable to Projected Climate Change?

Climate change-induced vulnerabilities in Asia have to be understood against the backdrop of the physical, economic, and social environment of the countries in the region. They not only provide benchmarks against which vulnerabilities are to be assessed but also the potential for adaptation to them. The socioeconomic environment of many countries in Asia is characterized by high population density and relatively low rates of economic growth. Surface water and groundwater resources in Asian countries play vital roles in forestry, agriculture, fisheries, livestock production, and industrial activity. The water and agriculture sectors are likely to be most sensitive to climate change-induced impacts in Asia. Forest ecosystems in boreal Asia would suffer from floods and increased volume of runoff associated with melting of permafrost regions. The dangerous processes of permafrost degradation resulting from global warming strengthen the vulnerability of all relevant climate-dependent sectors affecting the economy in high-latitude Asia. Although the frequency and severity of floods eventually would increase in many countries of Asia, arid and semi-arid regions of Asia could experience severe water-stress conditions. The stresses of climate change are likely to disrupt the ecology of mountain and highland systems in Asia. Major changes in high-elevation ecosystems of Asia can be expected as a consequence of the impacts of climate change. Many species of mammals and birds and a large population of many other species in Asia could be exterminated as a result of the synergistic effects of climate change and habitat fragmentation. Glacial melt also is expected to increase under changed climate conditions, which would lead to increased summer flows in some river systems for a few decades, followed by a reduction in flow as the glaciers disappear.

Agricultural productivity in Asia is likely to suffer severe losses because of high temperature, severe drought, flood conditions, and soil degradation; food security of many developing countries in the region would be under tremendous threat. There are likely to be large-scale changes in productivity of warmwater and coolwater fish in many countries in Asia. Sea-level rise would cause large-scale inundation along the vast Asian coastline and recession of flat sandy beaches. The ecological security of mangroves and coral reefs around Asia would be put at risk.

The monsoons in tropical Asia could become more variable if El Niño-Southern Oscillation (ENSO) events become stronger and more frequent in a warmer atmosphere. Countries in temperate and tropical Asia are likely to have increased exposure to extreme events, including forest die-back and increased fire risk, typhoons and tropical storms, floods and landslide, and severe vector-borne diseases.

Major Risks in Asia from Climate Change

Based on present scientific research, the following risks linked to changes in climate and its variability for Asia are identified:[1]

- The dangerous processes of permafrost degradation resulting from global warming would increase the vulnerability of many climate-dependent sectors affecting the economy in boreal Asia. ***
- Surface runoff increases during spring and summer periods would be pronounced in boreal Asia. ***
- The frequency of forest fires is expected to increase in boreal Asia. ***
- The large deltas and coastal low-lying areas of Asia could be inundated by sea-level rise. ****
- The developing countries of temperate and tropical Asia already are quite vulnerable to extreme climate events such as droughts and floods; climate change and its variability could exacerbate these vulnerabilities. ****
- Increased precipitation intensity, particularly during the summer monsoon, could increase flood-prone areas in temperate and tropical Asia. There is a potential for drier conditions in arid and semi-arid Asia during summer, which could lead to more severe droughts. ***
- Freshwater availability is expected to be highly vulnerable to anticipated climate change. ****
- Tropical cyclones could become more intense. Combined with sea-level rise, this impact would result in enhanced risk of loss of life and properties in coastal low-lying areas of cyclone-prone countries of Asia. ***
- Crop production and aquaculture would be threatened by a combination of thermal and water stresses, sea-level rise, increased flooding, and strong winds associated with intense tropical cyclones. ****
- Warmer and wetter conditions would increase the potential for a higher incidence of heat-related and infectious diseases in tropical and temperate Asia. ***
- Climate change would exacerbate threats to biodiversity resulting from land-use/cover change and population pressure in Asia.***

[1]Uncertainties in observations, mechanisms, and scenarios are identified with a five-point scale, from "very low confidence" (*) to "very high confidence" (*****). The confidence scale has the following probability range: VL = Very Low (<5% probability); L = Low (5–33%); M = Medium (33–67%); H = High (67–95%); and VH = Very High (>95%).

Adaptation to Climate Change in Asia

The impacts of climate change are likely to be felt most severely in the majority of developing countries of Asia because of resource and infrastructure constraints (e.g., disparities in income level, technological gaps). These countries must develop and implement incremental adaptation strategies and policies to exploit "no regret" measures and "win-win" options. Detailed and reliable regional scenarios of climate change must be developed and used in rigorous vulnerability analysis (e.g., low-probability/high-consequence events vs. high-probability/ high-consequence events, risk perceptions). To understand which adaptation opportunities will be most cost-effective and have the greatest value, emphasis must be given to characteristics of system vulnerability such as resilience, critical thresholds, and coping ranges, which are highly dependent on regions and nations.

It is important to consider climate change in planning, designing, and implementing development activities in climate-sensitive resources for Asia, particularly where:

- Climate change may cause irreversible or catastrophic impacts
- Decisions with a long lifetime, such as building of infrastructure, are being made
- Development trends such as development of low-lying coastal areas increase vulnerability to climate change.

Two general strategies on adaptation can be used. The first is a macro strategy that involves rapid development. Sustainable and equitable development will increase income levels, education, and technical skills and improve public food distribution, disaster preparedness and management, and health care systems in developing countries of Asia. All of these changes could substantially enhance social capital and reduce the vulnerability of these countries to climate change.

The second strategy is a micro strategy that involves modifying the management of sectors that are most sensitive to climate change. This approach entails developing new institutions or modifying existing institutions related to these sectors that promote rather than discourage adaptation to climate change. It also involves modifying climate-sensitive infrastructures that are already planned or implemented or other long-term decisions that are sensitive to climate to incorporate the risks of climate change.

Based on the foregoing principles, specific adaptation strategies for countries in the Asian region have been identified in the relevant sectors (e.g., water resources, agriculture and food security, coastal resources, human health, and ecosystems and biodiversity). These strategies are summarized in Box 11-1.

Each subregion has its priority adaptation sector for its own situation. Food security, disaster preparedness and management, soil conservation, and human health sectors also appear to be crucial for countries with large populations (e.g., China, India,

Box 11-1. Summary of Potential Sector-Wide Adaptation Options for Subregions of Asia

Agriculture

- *Boreal Asia*: Adopt suitable crops and cultivars; make optimum use of fertilizers and adaptation of agro-technologies.
- *Arid and Semi-Arid Asia*: Shift from conventional crops to intensive greenhouse agriculture/aquaculture; protect against soil degradation.
- *Temperate Asia*: Adopt heat-resistant crops, water-efficient cultivars with resistance to pests and diseases, soil conservation.
- *Tropical Asia*: Adjust cropping calendar and crop rotation; develop and promote use of high-yielding varieties and sustainable technological applications.

Water Resources

- *Boreal Asia*: Develop flood-protection systems in north Asia (required because of permafrost melting and increased streamflow volume/surface runoff); enhance management of international rivers.
- *Arid and Semi-Arid Asia*: Enhance conservation of freshwater supply as option for extreme water-stress conditions.
- *Temperate Asia*: Flood and drought control measures required; improve flood warning and forecasting systems, including disaster management.
- *Tropical Asia*: Develop flood- and drought-control management systems; reduce future developments in floodplains; use appropriate measures for protection against soil erosion; conserve groundwater supply, water impoundments, and efficient water resource systems.

Ecosystems and Biodiversity

- Assess risks to endemic species and ecosystems.
- Introduce integrated ecosystem planning and management.
- Reduce habitat fragmentation and promote development of migration corridors and buffer zones.
- Encourage mixed-use strategies.
- Prevent deforestation and conserve natural habitats in climatic transition zones inhabited by genetic biodiversity of potential for ecosystem restoration.

Coastal Resources

- *Boreal Asia*: Modify infrastructures to accommodate sea-level rise.
- *Arid and Semi-Arid Asia*: Protect lakes and water reservoirs; develop aquaculture farming techniques.
- *Temperate Asia*: Follow setback examples for new coastal development; evaluate coastal subsidence rates in sensitive coastal regions; prepare contingency plans for migration in response to sea-level rise; improve emergency preparedness for weather extremes (e.g., typhoons and storm surges).
- *Tropical Asia*: Protect wetlands and allow for migration; prepare contingency plans for migration in response to sea-level rise; improve emergency preparedness for weather extremes (e.g., cyclones and storm surges); evaluate coastal subsidence rates in sensitive coastal regions.
- *Common Adaptation*: Implement coastal zone management; protect marine resources.

Human Health

- Build heat-resistant urban infrastructures and take additional measures to reduce air and water pollution.
- Adapt technological/engineering solutions to prevent vector-borne diseases/epidemics.
- Improve health care system, including surveillance, monitoring, and information dissemination.
- Improve public education and literacy rate in various communities.
- Increase infrastructure for waste disposal.
- Improve sanitation facilities in developing countries.

Cross-Cutting Issues

- Continue monitoring and analysis of variability and trends in key climatic elements.
- Improve weather forecasting systems in the region.
- Improve and implement reforms on land-use planning.
- Apply new techniques for confident projection of regional climate change and its variability, including extreme events.
- Improve coordination of climate change adaptation activities among countries in the region.
- Keep the nongovernmental organization (NGO) community and the public aware of developments on risks of climate change and involve them in planning, adaptation, and mitigation strategies.
- Take advantage of traditional knowledge in planning for the future.

Bangladesh). Adaptations proposed for human health, which essentially involve improving the health care system, are changes that are needed anyway to address the current human health situation in many Asian countries. Adaptations to deal with sea-level rise, potentially more intense cyclones, and threats to ecosystems and biodiversity should be considered high priority in temperate and tropical Asian countries.

The design of an optimal adaptation program in any country would have to be based on comparison of damages avoided with costs of adaptation. Other factors also should enter the decisionmaking process, such as the impacts of policies on society in terms of employment generation and opportunities, improved air and water quality, and the impacts of policies on broader concerns for equitable and sustainable development.

11.1. The Asian Region

11.1.1. Background

Following publication of its Second Assessment Report (SAR) and on recommendation of the Subsidiary Body for Scientific and Technological Advice of the Conference of the Parties, the Intergovernmental Panel on Climate Change (IPCC) published its *Special Report on the Regional Impacts of Climate Change* in early 1998, providing assessments of the vulnerability of natural ecosystems, socioeconomic sectors, and human health to climate change for 10 regions of the globe (IPCC, 1998). That Special Report served as guidance material, illustrating for the first time the potential character and magnitude of region-specific impacts—though often in qualitative sense only, based on a diverse range of methods and tools.

A key message of the regional assessments in that report was that "many systems and policies are not well adjusted even to today's climate and climate variability." Several examples cited, based on information gathered during country study and other projects, demonstrate current vulnerability in the Asian region as a result of increasing risks to human life and property from floods, storms, and droughts in recent decades. The report suggests that, as a consequence of climate change, there could be a large reduction in the area and productivity of forests in boreal Asia. In arid and semi-arid regions of Asia, water shortage—already a limiting factor for ecosystems, food and fiber production, human settlements, and human health—may be exacerbated by climate change. Limited water supplies and land degradation problems are likely to threaten the food security of some countries in this region. Major changes in the composition and distribution of vegetation types of semi-arid areas—for example, grasslands, rangelands, and woodlands—are anticipated. In temperate Asia, changes in temperature and precipitation may result in altered growing seasons and boundary shifts between grasslands, forests, and shrublands. An increase in temperature could lead to oxygen depletion in aquatic ecosystems, fish diseases, and introduction of unwanted species, as well as potential negative factors such as changes in established reproductive patterns, migration routes, and ecosystem relationships in temperate Asia. A rise in sea level will endanger sandy beaches in the coastal zones and add further to the problems of tectonically and anthropogenically induced land subsidence in deltaic regions of temperate Asia. Substantial elevational shifts of ecosystems in the mountains and uplands of tropical Asia are projected. Increases in temperature and seasonal variability in precipitation are expected to result in more rapid recession of Himalayan glaciers. Climate change impacts could result in significant changes in crop yields, production, storage, and distribution in this region. Densely settled and intensively used low-lying coastal plains, islands, and deltas in Asia are extremely vulnerable to coastal erosion and land loss, inundation and sea flooding, and upstream movement of saline water fronts as a result of sea-level rise. The incidence and extent of vector-borne diseases, which are significant causes of mortality and morbidity in tropical Asia, are likely to spread into new regions on the margins of present endemic areas as a result of climate change.

The 1998 Special Report underscores that, in many countries of Asia, economic policies and conditions (e.g., taxes, subsidies, and regulations) that shape decisionmaking, development strategies, and resource-use patterns (hence environmental conditions) hinder implementation of adaptation measures. For example, water is subsidized in most developing countries of Asia, encouraging overuse (which draws down existing sources) and discouraging conservation measures that may be elements of future adaptation strategies. Other examples are inappropriate land-use zoning and subsidized disaster insurance, which encourage infrastructure development in areas that are prone to flooding or other natural disasters—areas that could become even more vulnerable as a result of climate change. Adaptation and better incorporation of the long-term environmental consequences of resource use can be brought about through a range of approaches, including strengthening legal and institutional frameworks, removing pre-existing market distortions (e.g., subsidies), correcting market failures (e.g., failure to reflect environmental damage or resource depletion in prices or inadequate economic valuation of biodiversity), and promoting public participation and education. These types of actions would adjust resource-use patterns to current environmental conditions and better prepare systems for potential future changes.

The 1998 Special Report emphasizes that the challenge lies in identifying opportunities that would facilitate sustainable development by making use of existing technologies and developing policies that make climate-sensitive sectors resilient to today's climate variability. This strategy will require developing countries of Asia to have access to appropriate technologies, information, and adequate financing. In addition, adaptation will require anticipation and planning; failure to prepare systems for projected change in climate means, variability, and extremes could lead to capital-intensive development of infrastructure or technologies that are ill-suited to future conditions, as well as missed opportunities to lower the costs of adaptation.

Subsequent to publication of the 1998 Special Report, some advances in our ability to better understand the likely future state of social, economic, and environmental factors controlling the emission and concentration of greenhouse gases (GHGs) and aerosols that alter the radiative forcings of climate have been made. Details on future projections of climatic and environmental changes on finer scales also are now better understood. This chapter presents an update on the climate change projections for Asia and examines how projected changes in climate could affect social, environmental, and economic sectors in the region.

11.1.2. Physical and Ecological Features

11.1.2.1. Regional Zonation

The Asian continent is bounded on the north by the Arctic Ocean, on the east by the Pacific Ocean, and on the south by the Indian Ocean; the western boundary, with Europe, runs roughly north-south along the eastern Ural Mountains, the

Zhem River, the Caspian Sea, the Kuma-Manych Depression, the Black Sea, the Aegean Sea, the Mediterranean Sea, the Suez Canal, and the Red Sea. The world's largest plateau—the Tibetan Plateau, with an average elevation of more than 4,000 m—is located in Asia. Mt. Everest, the highest peak in the world (8,848 m), lies near the southern border of this plateau. The islands of Sri Lanka and Taiwan (China) and the archipelagoes of Indonesia, the Philippines, and Japan also are part of Asia. Asia is the most populous of the continents. Its total population in 1998 is estimated to be about 3,589 million, of which almost 65% is rural. The coastline of Asia is 252,770 km long.

Based on broad climatic and geographical features, the Asian region can be divided into four subregions: boreal Asia, arid and semi-arid Asia, temperate Asia, and tropical Asia (Figure 11-1). This chapter discusses key climatological, ecological, and socioeconomic features of each of these subregions and the countries falling within them, as well as various aspects of vulnerability, impacts, and adaptation in relation to climate change for each of these regions.

Boreal Asia is located on the northern margins of the Eurasian continent between 50°N and the Arctic Circle. Boreal forests cover most of this region. Siberia, which is a part of Russia (the main country in the region), has a mean monthly temperature of about -50°C in January and is the coldest region of the northern hemisphere in winter. Three Siberian rivers—the Ob, the Yenissey, and the Lena—contribute about 42% of total runoff from all rivers of the Arctic basin (CAFW, 1998). In the heart of this region is one of the world's largest and oldest lakes, Baikal, the age of which is estimated to be about 25–30 million years (Kuzmin *et al.*, 1997). Lake Baikal contains as much as 85% of Russian surface freshwaters (Izrael *et al.*, 1997a; Anokhin and Izrael, 2000). There is evidence of recession of permafrost in recent decades (ICRF, 1998).

Arid and semi-arid Asia extends from 22°N to 50°N and from 30°E to 105°E; it includes more than 20 countries of the Middle East and central Asia. Many of the countries in the region are landlocked. Many storms develop locally (*in situ*) over central Asia; some move into the region from the west. The maximum frequency of cyclones occurs in January and March. Despite this cyclonic activity, very little precipitation is recorded over most of the region because of the lack of moisture. Most of the region has a precipitation-to-potential evapotranspiration ratio of less than 0.45, which is typical of a semi-arid and arid

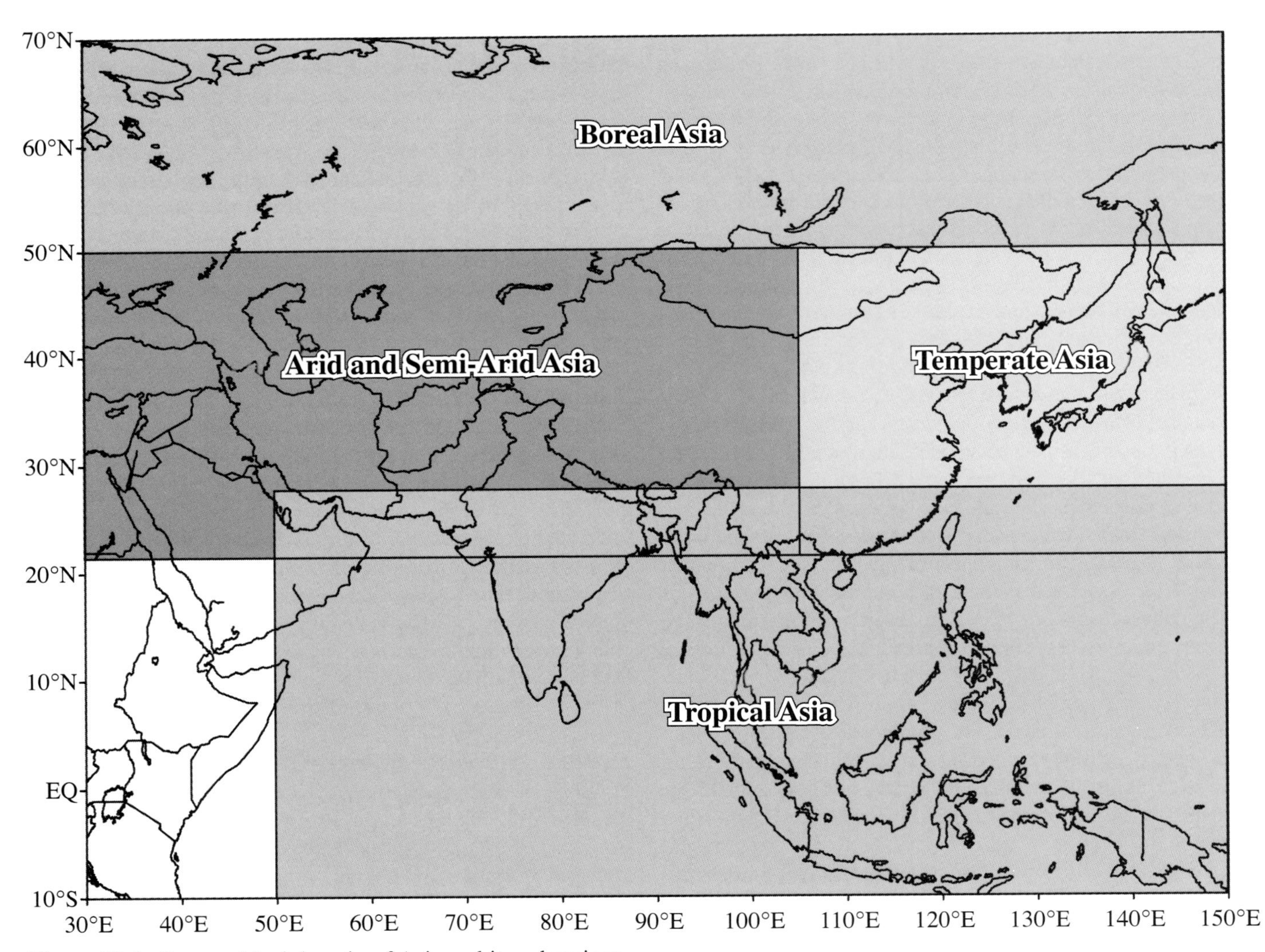

Figure 11-1: Geographical domain of Asia and its subregions.

climate. Grasslands, rangelands, and deserts dominate most of arid and semi-arid Asia.

Temperate Asia extends from 22°N to 50°N and from 105°E to 150°E; it includes eastern China, the Japanese islands, the Korean peninsula, Mongolia, and Taiwan (China). Geographically, the region is located on the eastern part of the Eurasian continent—the world's largest continent—and borders the Pacific, the world's largest ocean. The east-west distance of the area is about 5,000 km; its north-south extent is about 3,000 km. Much of the natural forest in the region has long been destroyed. Broad plains have been cultivated and irrigated, and natural grasslands have been used for animal husbandry for centuries.

Tropical Asia extends from 10°S to 28°N and from 50°E to 150°E; it includes several countries of south Asia, which are influenced predominantly by the monsoons. The region is physiographically diverse and ecologically rich in natural and crop-related biodiversity. Although the present population of the region is principally rural, the region includes seven of the 25 largest cities in the world. Agriculture is the main industry in several countries of this region. Exploitation of natural resources associated with rapid urbanization, industrialization, and economic development has led to increasing air and water pollution, land degradation, and other environmental problems in countries of this region. Climate change represents a further stress. Over the long period of human occupation in the region, human use systems have developed some resilience to a range of environmental stresses (IPCC, 1998).

Table 11-1 lists the total population, gross domestic product, total land area and land area classified under cropland, forest cover, annual internal water resources, and annual production of cereals, as well as per capita availability of fish and seafood in some of the countries of Asia.

11.1.2.2. Trends and Variability in Key Climate Variables

11.1.2.2.1. Surface air temperature

Climate differs widely within Asia. In boreal Asia, the climate generally is humid and cool; it is classified as continental type. Permafrost covers as much as 90% of boreal Asia. An average annual mean increase in surface air temperature of about 2.9°C in the past 100 years has been observed in boreal Asia. Warming was more pronounced during the period 1951–1995 in all of boreal Asia region except for the coast of the Arctic Ocean and Chukotka (Rankova, 1998). The mean surface air temperature increase is most pronounced in the winter, at a rate of approximately 4.4°C over the past 100 years (Gruza *et al.*, 1997). Summer temperatures in central Siberia have exhibited decreasing trends, however.

The climate of arid and semi-arid Asia is the warm temperate type, with hot and wet or dry summers. The highest value of DTR (on the order of 20°C) is experienced in this region. A mean maximum temperature of >45°C in July is not uncommon in some parts of arid Asia. In most of the Middle East, the long time series of surface air temperature shows a warming trend. An increasing tendency in spatially averaged seasonal and annual air temperatures has been observed in Kazakhstan over the past century. Maximal warming has occurred in spring. The mean annual surface temperature has risen by about 1.3°C during 1894–1997 (Pilifosova *et al.*, 1997); in the arid regions of China, air temperature has obviously increased since the 1970s (Chen, 1995). In Pakistan, annual mean surface temperature has a consistent rising trend since the beginning of 20th century (Chaudhari, 1994).

In temperate Asia, the average annual mean surface temperature in Mongolia has increased by approximately 0.7°C over the past 50 years; consequently, noticeable changes have taken place in the length of the cold and warm seasons (Khuldorj et al., 1998). Surface temperature in northeast China has increased in winter but decreased in summer since 1905 (Ren and Zhou, 1994; Ren, 1998); observations also reveal a 1–2°C decrease in temperature in some parts of southeastern China. In Japan, the surface air temperature has shown a warming trend during the past century (Yoshino, 1998a; Japan Meteorological Agency, 1999).

Tropical Asia has a unique climatological distinction because of the pervasive influence of the monsoon. In tropical Asia, climate uniformity is differentiated by three factors: latitude, relief, and continentality. The entire tropical Asia region stretches over 38° in latitude, so the differences resulting from this factor are pronounced. In spite of some differences, the climates of countries have one factor in common: The Asian monsoon modulates them all to a large extent. For countries near the Equator, only small seasonal variations occur, although most countries in this region experience clearly marked cold and warm seasons. The spatial range of temperature in this region is significantly large during winter. Extreme temperatures of over 45°C occur over the northwest part of the region during May-June. Several countries in this region have reported increasing surface temperature trends in recent decades. In Vietnam, annual mean surface temperature has increased over the period 1895–1996, with mean warming estimated at 0.32°C over the past 3 decades. Annual mean surface air temperature anomalies over Sri Lanka during the period 1869–1993 suggest a conspicuous and gradually increasing trend of about 0.30°C per 100 years (Rupakumar and Patil, 1996). The warming trend over India has been reported to be about 0.57°C per 100 years (Rupakumar *et al.*, 1994).

11.1.2.2.2. Precipitation

Rainfall in boreal Asia is highly variable on seasonal and interannual as well as spatial scales. The time series of annual mean precipitation in Russia suggests a decreasing trend; these tendencies have amplified during 1951–1995, especially in warm years (Rankova, 1998). In long-term mean precipitation, a decreasing trend of about -4.1 mm/month/100 years has been reported in boreal Asia. During the past 10–15 years, however,

***Table 11-1**: Key information on the socioeconomics of some of Asian countries (WRI, 1998; FAO, 1999a).*

Country/Region	Population, 1998 (1000s)	Gross Domestic Product, 1995 (per capita US$)	Land Area (10^3 ha)	Crop Land, 1992–1994 (10^3 ha)	Total Forest Cover, 1995 (10^3 ha)	Annual Internal RWR, 1998[a] (per capita m^3)	Cereal Food Prod., Avg. 1994–1996[a] (10^3 t)	Annual Sea Food, Avg. 1993–1995[a] (per capita kg)
Boreal Asia								
– Russia	147231	2333	1688850	133072	763500	29115	69524	15.4
Arid and Semi-Arid Asia								
– Afghanistan	23364	x	65209	8054	1398	2354	3019	x
– Armenia	3646	783	2820	582	334	2493	256	1.2
– Azerbaijan	7714	461	8660	1967	990	1069	981	4.8
– Georgia	5428	427	6970	1036	2988	10682	543	6.8
– Iran	73057	1756	162200	18500	1544	1755	16944	5.1
– Iraq	21795	2755	43737	5550	83	1615	2312	1.2
– Israel	5883	16645	2067	434	102	289	153	20.3
– Jordan	5956	1187	8893	405	45	114	93	3.4
– Kazakhstan	16854	1273	267073	35239	10504	4484	12340	3.5
– Kuwait	1809	15760	1782	5	5	11	2	10.4
– Kyrgyzstan	4497	685	19180	1387	730	10394	1132	0.1
– Lebanon	3194	3703	1023	306	52	1315	76	0.7
– Oman	2504	5483	21246	63	0	393	5	x
– Pakistan	147811	445	77088	21323	1748	1678	23818	2.4
– Saudi Arabia	20207	6875	214969	3777	222	119	3871	6.7
– Syrian Arab Republic	15335	1182	18378	5985	219	456	5816	0.6
– Tajikistan	6161	343	14060	846	410	11171	288	0.6
– Turkmenistan	4316	961	46993	1471	3754	232	988	8.4
– Turkey	63763	2709	76963	27611	8856	3074	28179	7.9
– United Arab Emirates	2354	17696	8360	75	60	64	7	21.5
– Uzbekistan	24105	947	41424	4618	9119	704	2718	1.0
Temperate Asia								
– China	1255091	572	929100	95145	133323	2231	416954	14.5
– Japan	125920	40846	37652	4467	25146	4344	14566	67.6
– Korea, DPR	23206	x	12041	2007	6170	2887	5022	46.3
– Korea, Republic	46115	10142	9873	2053	7626	1434	6877	51.2
– Mongolia	2624	349	156650	1357	9406	9375	264	0.9
Tropical Asia								
– Bangladesh	124043	246	13017	8849	1010	10940	27887	8.7
– Bhutan	1917	172	4700	136	2756	49557	111	x
– Cambodia	10754	276	17652	3832	9830	8195	3024	8.5
– India	975772	349	297319	169569	65005	1896	213326	4.0
– Indonesia	206522	1003	181157	31146	109791	12251	57197	15.7
– Laos	5358	361	23080	900	12435	50392	1485	6.6
– Malaysia	21450	4236	32855	7536	15471	21259	2158	28.3
– Myanmar	47625	x	65755	10067	27151	22719	20040	16.1
– Nepal	23168	197	14300	2556	4822	7338	5686	x
– Philippines	72164	1093	29817	9320	6766	4476	15119	33.5
– Singapore	3491	25156	61	1	4	172	x	x
– Sri Lanka	18450	720	6463	1889	1796	2341	2313	15.7
– Thailand	61400	2868	51311	21212	11630	1845	25759	26.3
– Vietnam	77896	279	32549	6738	9117	4827	26040	13.1

[a] Column 6 = annual internal renewable water resources (per capita, m^3), 1998; column 7 = cereal food production (thousands of metric tons), average 1994–96; and column 8 = per capita annual food supply from fish and seafood (kg), average 1993–95.

x = data not available.

precipitation has increased, mostly during the summer-autumn period (Izrael *et al*., 1997b). As a result of this increase in precipitation, water storage in a 1-m soil layer has grown by 10–30 mm (Robock *et al*., 2000). The large upward trends in soil moisture (of more than 1 cm/10 years) have created favorable conditions for infiltration into groundwater. The levels of major aquifers have risen by 50–100 cm; the growth of groundwater storage has resulted in increasing ground river recharge and considerable low-water runoff.

Annual mean rainfall is considerably low in most parts of the arid and semi-arid region of Asia. Moreover, temporal variability is quite high: Occasionally, as much as 90% of the annual total is recorded in just 2 months of the year at a few places in the region. Rainfall observations during the past 50 years in some countries in the northern parts of this region have shown an increasing trend on annual mean basis. A decreasing trend in annual precipitation for the period 1894–1997 has been observed in Kazakhstan. The precipitation in spring, summer, and autumn, however, has shown slight increasing trends. In Pakistan, seven of 10 stations have shown a tendency toward increasing rainfall during monsoon season (Chaudhari, 1994).

In temperate Asia, the East Asian monsoon greatly influences temporal and spatial variations in rainfall. Annual mean rainfall in Mongolia is 100–400 mm and is confined mainly to summer. Summer rainfall seems to have declined over the period 1970–1990 in Gobi; the number of days with relatively heavy rainfall events has dropped significantly (Rankova, 1998). In China, annual precipitation has been decreasing continuously since 1965; this decrease has become serious since the 1980s (Chen *et al*., 1992). The summer monsoon is reported to be stronger in northern China during globally warmer years (Ren *et al*., 2000). On the other hand, drier conditions have prevailed over most of the monsoon-affected area during globally colder years (Yu and Neil, 1991).

In tropical Asia, hills and mountain ranges cause striking spatial variations in rainfall. Approximately 70% of the total annual rainfall over the Indian subcontinent is confined to the southwest monsoon season (June-September). The western Himalayas get more snowfall than the eastern Himalayas during winter. There is more rainfall in the eastern Himalayas and Nepal than in the western Himalayas during the monsoon season (Kripalani *et al*., 1996). The annual mean rainfall in Sri Lanka is practically trendless; positive trends in February and negative trends in June have been reported, however (Chandrapala and Fernando, 1995). In India, long-term time series of summer monsoon rainfall have no discernible trends, but decadal departures are found above and below the long time averages alternatively for 3 consecutive decades (Kothyari and Singh, 1996). Recent decades have exhibited an increase in extreme rainfall events over northwest India during the summer monsoon (Singh and Sontakke, 2001). Moreover, the number of rainy days during the monsoon along east coastal stations has declined in the past decade. A long-term decreasing trend in rainfall in Thailand is reported (OEPP, 1996). In Bangladesh, decadal departures were below long-term averages until 1960; thereafter they have been much above normal (Mirza and Dixit, 1997).

11.1.2.3. Extreme Events and Severe Weather Systems

Apart from intraseasonal and interannual variability in climate, extreme weather events such as heat waves associated with extreme temperatures, extratropical and tropical cyclones, prolonged dry spells, intense rainfall, tornadoes, snow avalanches, thunderstorms, and dust storms are known to cause adverse effects in widely separated areas of Asia. There is some evidence of increases in the intensity or frequency of some of these extreme weather events on regional scales throughout the 20th century, although data analyses are relatively poor and not comprehensive (Balling and Idso, 1990; Bouchard, 1990; Agee, 1991; Yu and Neil, 1991; Chen *et al*., 1992; Ostby, 1993; Bardin, 1994; Born, 1996). For example, increases in climate extremes in the western Siberia-Baikal region and eastern parts of boreal Asia have been reported in recent decades (Gruza and Rankova, 1997; Gruza *et al*., 1999). There also are reports of an increase in thunderstorms over the land regions of tropical Asia (Karl *et al*., 1995). The frequency and severity of wildfires in grasslands and rangelands in arid and semi-arid Asia have increased in recent decades (Pilifosova *et al*., 1996).

Some mountains in Asia have permanent glaciers that have vacated large areas during the past few decades, resulting in increases in glacial runoff. As a consequence, an increased frequency of events such as mudflows and avalanches affecting human settlements has occurred (Rai, 1999). As mountain glaciers continue to disappear, the volume of summer runoff eventually will be reduced as a result of loss of ice resources. Consequences for downstream agriculture, which relies on this water for irrigation, will be unfavorable in some places. For example, low- and mid-lying parts of central Asia are likely to change gradually into more arid, interior deserts.

Countries in temperate Asia have been frequented by many droughts in the 20th century. In China, droughts in 1972, 1978, and 1997 have been recorded as the most serious and extensive. A large number of severe floods also have occurred in China, predominately over the middle and lower basins of the Yangtze (Changjiang), Huanghe, Huaihe, and Haihe Rivers (Ji *et al*., 1993). Severe flooding with daily rainfall exceeding 25 cm struck during July and August 1998 in Korea. In Japan, drought disasters are significantly more frequent during years following ENSO warm events than in normal years.

Floods, droughts, and cyclones are the key natural disasters in tropical Asia. The average annual flood covers vast areas throughout the region: In Bangladesh, floods cover 3.1 Mha; the total flood-prone area in India is about 40 Mha (Mirza and Ericksen, 1996). In India, chronically drought-affected areas cover the western parts of Rajasthan and the Kutch region of Gujarat. However, drought conditions also have been reported in Bihar and Orissa States in India. In Bangladesh, about 2.7

Mha are vulnerable to drought annually; there is about 10% probability that 41–50% of the country is experiencing drought in a given year (Mirza, 1998). Drought or near-drought conditions also occur in parts of Nepal, Papua New Guinea, and Indonesia, especially during El Niño years. In India, Laos, the Philippines, and Vietnam, drought disasters are more frequent during years following ENSO events. At least half of the severe failures of the Indian summer monsoon since 1871 have occurred during El Niño years (Webster *et al.* 1998). In the event of enhanced anomalous warming of the eastern equatorial Pacific Ocean, such as that observed during the 1998 El Niño, a higher frequency of intense extreme events all across Asia is possible.

11.1.3. Scenarios of Future Climate Change

Increases in atmospheric concentrations of GHGs from anthropogenic activities would warm the earth-atmosphere system. The radiative forcing inferred from likely future increases in GHGs and sulfate aerosols as prescribed under IS92a emission scenarios (Leggett *et al.*, 1992; IPCC, 1995) has been used in recent numerical experiments performed with coupled atmosphere-ocean global climate models (AOGCMs). Projections of future regional climate change and also most of the impact assessment studies for Asia cited herein are based on these numerical experiments (see Chapter 3 for further details).

To develop climate change scenarios on regional scales, it is first necessary to examine if the coupled AOGCMs are able to simulate the dynamics of present-day regional climate. The multi-century control integrations of AOGCMs unforced by anthropogenic changes in atmospheric composition offer an excellent opportunity to examine the skill of individual models in simulating the present-day climate and its variability on regional scales (Giorgi and Francisco, 2000; Lal *et al.*, 2000). A model validation exercise carried out for Asia and its subregions has indicated that each of these models shows large seasonal variations in surface air temperature over boreal Asia and only small seasonal variations over southeast tropical Asia (Figure 11-2). However, many of the AOGCMs have only limited ability for realistic portrayal of even large-scale precipitation distribution over Asia. Seasonal variations in observed precipitation over the South Asia region as a consequence of summer monsoon activity are poorly simulated by most of the models (Figure 11-3). Moreover, none of the seven models can reproduce the observed precipitation climatology over the Tibetan Plateau (Figure 11-4). Based on the pattern correlation coefficients and root mean square errors between the observed and model-simulated seasonal mean sea-level pressure, surface air temperature, and rainfall patterns over land regions of Asia and other relevant considerations in this validation exercise, the HadCM2, ECHAM4, CSIRO, and CCSR/NIES AOGCMs (developed at U.K. Hadley Climate Centre, German Climate Research Centre, Australian Commonwealth Scientific and Industrial Research Organisation, and Japanese Center for Climate System Research/National Institute for Environmental Studies, respectively) are rated to have some skill in simulating the broad features of present-day climate and its variability over Asia (Lal and Harasawa, 2000).

11.1.3.1. Surface Air Temperature

Climate change scenarios that are based on an ensemble of results as inferred from skilled AOGCMs for Asia and its subregions on annual and seasonal mean basis are presented in Table 11-2. Three future time periods centered around the 2020s (2010–2029), the 2050s (2040–2069), and the 2080s (2070–2099) have been considered here for developing scenarios of changes in surface air temperature and precipitation relative to the baseline period of 1961–1990. The projected area-averaged annual mean warming is 1.6±0.2°C in the 2020s, 3.1±0.3°C in the 2050s, and 4.6±0.4°C in the 2080s over land regions of

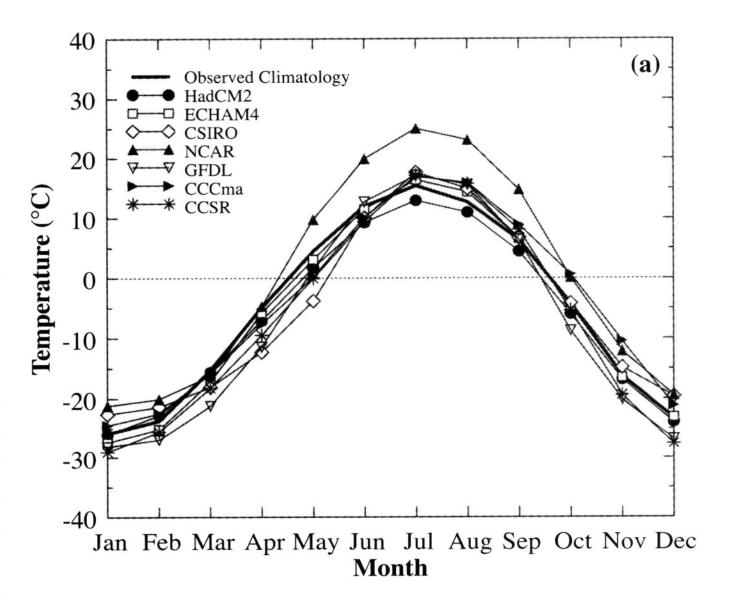

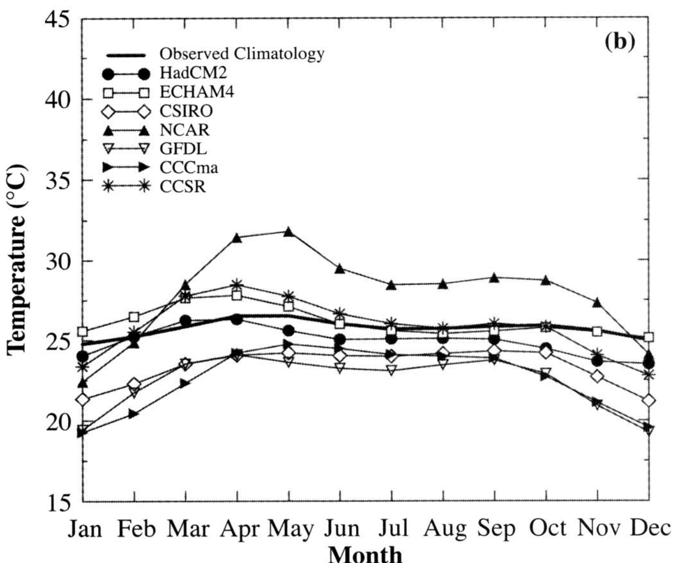

Figure 11-2: Validation of simulated and observed area-averaged annual cycles of surface air temperature over boreal (a) and southeast Asia (b) regions.

Asia as a result of increases in the atmospheric concentration of GHGs. Under the combined influence of GHGs and sulfate aerosols, surface warming will be restricted to 1.4±0.3°C in the 2020s, 2.5±0.4°C in the 2050s, and 3.8±0.5°C in the 2080s. In general, projected warming over Asia is higher during NH winter than during summer for all time periods (see also Giorgi and Francisco, 2000). The area-averaged increase in surface air temperature is likely to be most pronounced over boreal Asia and least in southeast Asia in all seasons (Lal and Harasawa, 2001). It is evident from Table 11-2 that even though aerosol forcing reduces surface warming, the magnitude of projected warming is still considerable and could substantially impact the Asian region.

In December 1998, the writing team for the IPCC *Special Report on Emission Scenarios* (SRES) released a preliminary set of four SRES "marker" scenarios: A1, A2, B1, and B2 (Nakicenovic *et al.*, 1998). Scenario B1 projects the most conservative future emissions of GHGs from Asia; scenario A2 is characteristic of scenarios with higher rates of GHG emissions in combination with higher sulfate and other aerosol emissions. The A1 scenario family has since been further divided into three groups that describe alternative directions of technological change in the energy system (IPCC, 2000). The three A1 groups are distinguished by their technological emphasis: fossil fuel-intensive (A1FI), nonfossil energy sources (A1T), or a balance across all sources (A1B). Projections of future aerosol loading as envisaged in SRES marker scenarios are significantly lower compared to the IS92a scenario.

Figure 11-5 illustrates the trends in area-averaged annual mean surface air temperature increase over land regions of Asia as

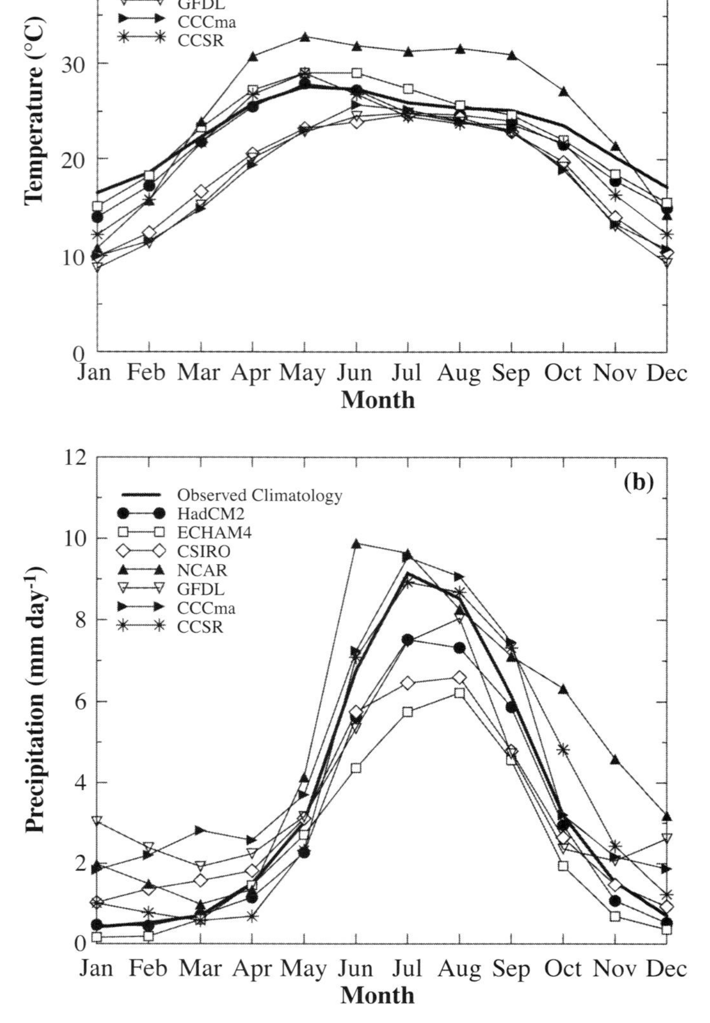

Figure 11-3: Validation of simulated and observed area-averaged annual cycles of surface air temperature (a) and precipitation (b) over south Asia.

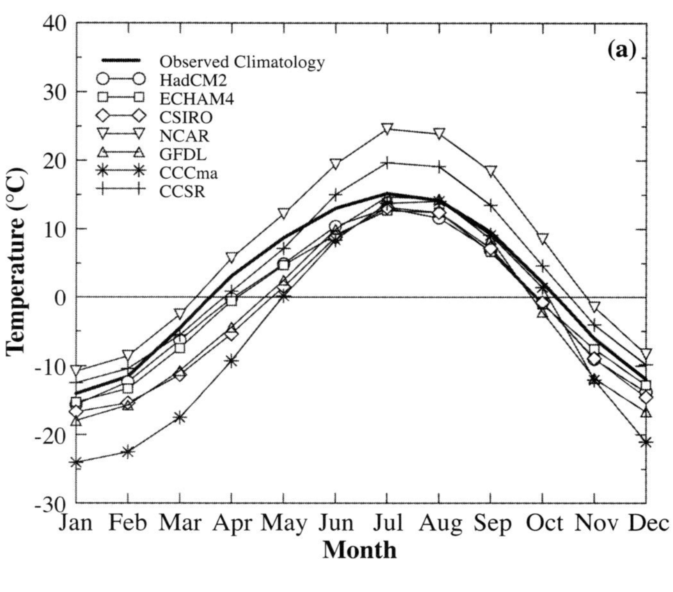

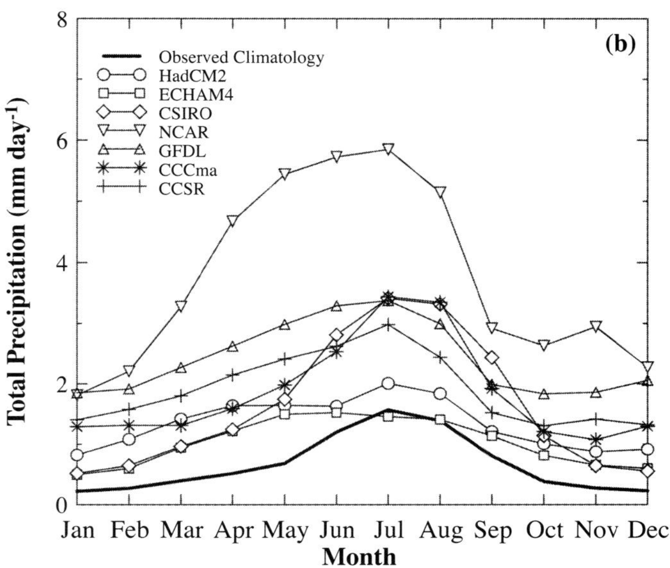

Figure 11-4: Validation of simulated and observed area-averaged annual cycles of surface air temperature (a) and precipitation (b) over Tibetan Plateau.

Table 11-2: Plausible changes in area-averaged surface air temperature (top) and precipitation (bottom) over Asia and its subregions as a result of future increases in greenhouse gases (under IS92a emission scenarios), as inferred from an ensemble of data generated in experiments with CCSR/NIES, CSIRO, ECHAM4, and HadCM2 AOGCMs. Numbers in parentheses are area-averaged changes when direct effects of sulfate aerosols are included.

	Temperature Change (°C)								
	2020s			**2050s**			**2080s**		
Regions	Annual	Winter	Summer	Annual	Winter	Summer	Annual	Winter	Summer
Asia	1.58 (1.36)	1.71 (1.52)	1.45 (1.23)	3.14 (2.49)	3.43 (2.77)	2.87 (2.23)	4.61 (3.78)	5.07 (4.05)	4.23 (3.49)
Boreal	2.17 (1.88)	2.66 (2.21)	1.71 (1.47)	4.32 (3.52)	5.52 (4.46)	3.29 (2.83)	6.24 (5.30)	8.04 (6.83)	4.82 (4.24)
Arid/Semi-Arid									
– Central Asia	1.61 (1.47)	1.56 (1.55)	1.77 (1.49)	3.18 (2.69)	2.81 (2.61)	3.55 (2.59)	4.83 (4.15)	4.41 (3.78)	5.34 (4.36)
– Tibet	1.77 (1.56)	1.90 (1.83)	1.62 (1.40)	3.38 (2.62)	3.55 (2.94)	3.19 (2.27)	5.04 (4.06)	5.39 (4.32)	4.69 (3.73)
Temperate	1.49 (1.19)	1.74 (1.50)	1.23 (0.99)	2.86 (2.10)	3.26 (2.40)	2.48 (1.72)	4.34 (3.31)	5.11 (3.83)	3.67 (2.77)
Tropical									
– South Asia	1.36 (1.06)	1.62 (1.19)	1.13 (0.97)	2.69 (1.92)	3.25 (2.08)	2.19 (1.81)	3.84 (2.98)	4.52 (3.25)	3.20 (2.67)
– SE Asia	1.05 (0.96)	1.12 (0.94)	1.01 (0.96)	2.15 (1.72)	2.28 (1.73)	2.01 (1.61)	3.03 (2.49)	3.23 (2.51)	2.82 (2.34)

	Precipitation Change (%)								
	2020s			**2050s**			**2080s**		
Regions	Annual	Winter	Summer	Annual	Winter	Summer	Annual	Winter	Summer
Asia	3.6 (2.3)	5.6 (4.3)	2.4 (1.8)	7.1 (2.9)	10.9 (6.5)	4.1 (1.5)	11.3 (7.0)	18.0 (12.1)	5.5 (3.5)
Boreal	6.1 (6.7)	11.1 (10.7)	2.6 (3.3)	12.8 (12.0)	23.8 (19.7)	5.1 (7.1)	20.7 (18.9)	39.5 (31.5)	7.7 (10.3)
Arid/Semi-Arid									
– Central Asia	1.3 (1.1)	3.0 (2.7)	-2.1 (5.9)	1.3 (0.6)	6.9 (1.4)	-2.3 (0.7)	-1.3 (-3.6)	6.9 (1.0)	-4.0 (-1.8)
– Tibet	5.9 (3.4)	8.9 (7.4)	4.4 (1.7)	9.0 (7.5)	19.2 (14.8)	4.7 (1.7)	12.8 (11.5)	25.6 (18.8)	5.7 (3.8)
Temperate	3.9 (0.9)	4.2 (0.4)	3.7 (1.2)	7.9 (1.3)	13.3 (4.3)	5.4 (0.7)	10.9 (4.8)	20.1 (7.1)	7.8 (3.1)
Tropical									
– South Asia	2.9 (1.0)	2.7 (-10.1)	2.5 (2.8)	6.8 (-2.4)	-2.1 (-14.8)	6.6 (0.1)	11.0 (-0.1)	5.3 (-11.2)	7.9 (2.5)
– SE Asia	2.4 (1.7)	1.4 (3.3)	2.1 (1.2)	4.6 (1.0)	3.5 (2.9)	3.4 (2.6)	8.5 (5.1)	7.3 (5.9)	6.1 (4.9)

simulated with CCSR/NIES AOGCMs for the IS92a scenario (GHG only and GHG + aerosols) and for the new SRES "marker" emission scenarios. Projections of future sulfate aerosol loading in SRES scenarios are significantly lower. Because the treatment of aerosol and GHG radiative effects in the old and new sets of scenario experiments is different, quantitative comparison of the results should be made with care. Nonetheless, the generally higher projected surface warming trend for SRES scenarios in the latter half of the 21st century is partially a result of intensive reduction of aerosol emissions. Projected surface warming trends for the IS92a and B1 emission pathways are close to each other. Maximum warming is simulated for the A1 emission scenario during the first half of the 21st century but is carried forward to the A2 emission scenario in the latter half of the 21st century. Projections of regional climate change that use these newer sets of emission scenarios for GHGs have not yet been thoroughly assessed for their applications in impact assessment studies.

11.1.3.2. Diurnal Temperature Range

One important aspect of the observed temperature change over the globe during the past century relates to its asymmetry during the day and night (Karl *et al.,* 1991). Observed warming in surface air temperatures over several regions of the globe has been reported to be associated with an increase in minimum temperatures (accompanied by increasing cloudiness) and a decrease in DTR (Hansen *et al.,* 1998).

AOGCM simulations with increasing concentrations of GHGs in the atmosphere suggest relatively more pronounced increases in minimum temperature than in maximum temperature over Asia on an annual mean basis, as well as during the winter, for the 2050s and the 2080s—hence a decrease in DTR (Table 11-3). During the summer, however, an increase in DTR is simulated—suggesting thereby that the maximum temperature would have a more pronounced increase relative to the minimum temperature (Lal and Harasawa, 2001). The summertime increase in DTR over central Asia is significantly higher relative to that in other regions. Most of the subregions follow the same pattern of change in DTR, except south and southeast Asia. A marginal increase in DTR during the winter and on an annual mean basis is simulated over southeast Asia. Over the south Asia region, a decrease in DTR on an annual mean basis and during the winter and a more pronounced decrease in DTR during the summer are projected. The significantly higher decrease in DTR over

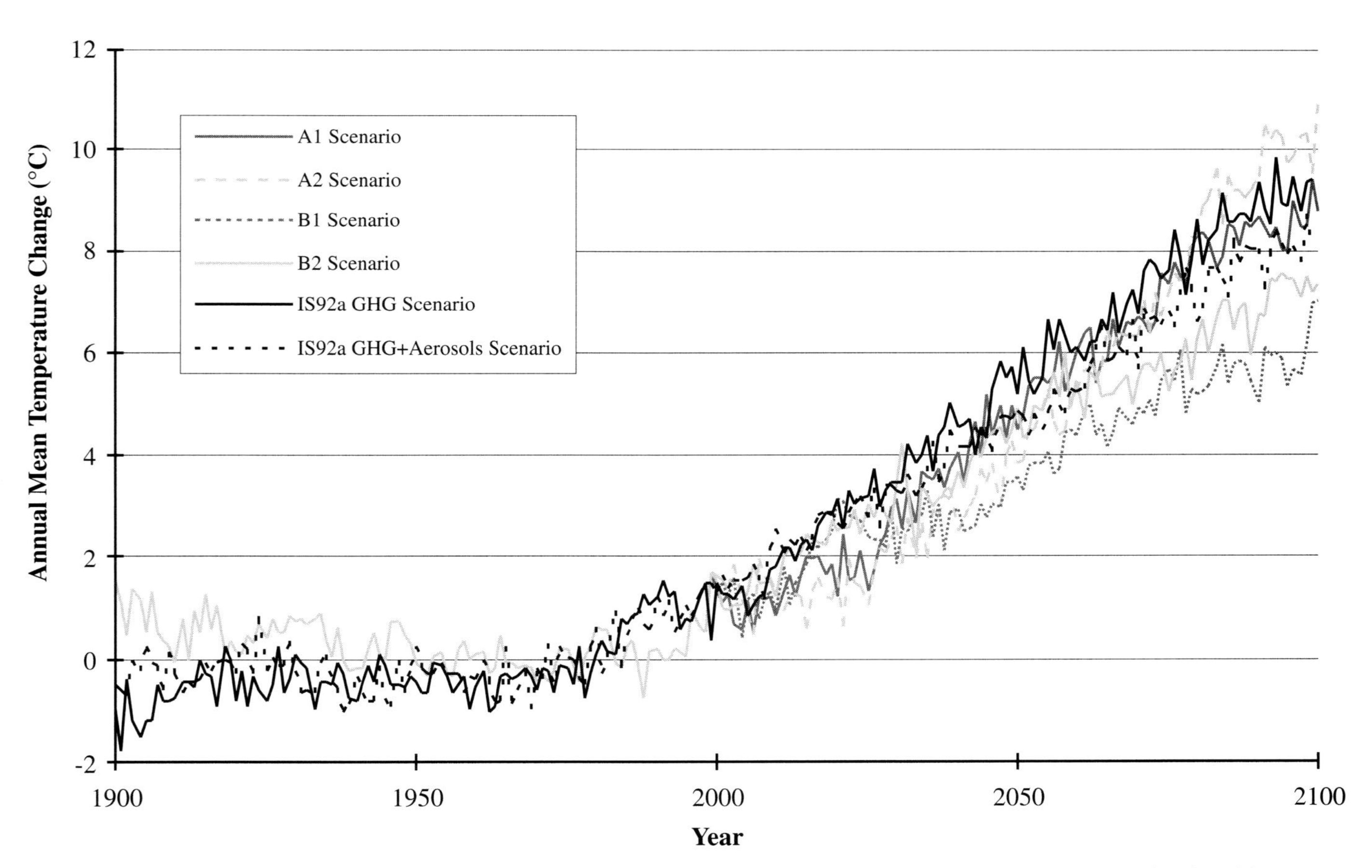

Figure 11-5: Future trends in area-averaged annual mean temperature increase over land regions of Asia as simulated by CCSR/NIES AOGCM for IS92a and SRES emission scenarios. Because the climate sensitivity of model versions used for the new (SRES) and old (IS92a) sets of simulations are different, temperature trends shown are scaled so they can be compared directly; trends for IS92a scenarios are scaled by a factor of 1.56, which is the ratio of climate sensitivity averaged over Asia of new model version to that of old version. Also note that only the direct effect of sulfate aerosols is considered in the IS92a GHG+Aerosols simulation, whereas the direct and indirect effects of sulfate and carbon aerosols are considered in SRES simulations.

***Table 11-3**: Projected changes in diurnal temperature range over Asia and its subregions under IS92a emission scenarios, as inferred from an ensemble of data generated in experiments with CCSR/NIES, CSIRO, ECHAM4, and HadCM2 AOGCMs.*

	Greenhouse Gases						***Greenhouse Gases + Sulfate Aerosols***					
	2050s			**2080s**			**2050s**			**2080s**		
Regions	Annual	Winter	Summer	Annual	Winter	Summer	Annual	Winter	Summer	Annual	Winter	Summer
Asia	-0.15	-0.26	1.42	-0.27	-0.45	1.36	-0.17	-0.13	-0.54	-0.27	-0.26	-0.54
Boreal	-0.38	-0.52	1.67	-0.53	-0.81	1.62	-0.45	-0.34	-0.49	-0.57	-0.53	-0.48
Arid/ Semi-Arid												
– Central	0.13	-0.07	4.64	0.17	-0.11	4.75	-0.02	-0.14	2.20	0.09	-0.04	2.36
– Tibet	-0.34	-0.60	2.01	-0.46	-0.81	1.91	-0.53	-0.63	0.18	-0.67	-0.80	0.13
Temperate	-0.18	-0.31	0.47	-0.23	-0.43	0.44	-0.19	-0.21	-0.83	-0.28	-0.30	-1.05
Tropical												
– South	-0.27	-0.27	-3.06	-0.45	-0.46	-2.89	-0.22	-0.14	-4.97	-0.31	-0.31	-4.95
– Southeast	0.15	0.24	-0.50	0.00	0.09	-0.66	0.35	0.42	-0.98	0.18	0.24	-1.09

south Asia during the summer is a result of the presence of monsoon clouds over the region (Lal *et al.*, 1996). In general, the decline in DTR is slightly moderated in the presence of sulfate aerosols (Lal and Harasawa, 2001). Changes in DTR over Asia in this case suggest a decrease in both seasons. Similar changes also are seen over boreal Asia and temperate Asia.

11.1.3.3. Precipitation

In general, all AOGCMs simulate an enhanced hydrological cycle and an increase in annual mean rainfall over most of Asia (Giorgi and Francisco, 2000). An area-averaged annual mean increase in precipitation of 3±1% in the 2020s, 7±2% in the 2050s, and 11±3% in the 2080s over the land regions of Asia is projected as a result of future increases in the atmospheric concentration of GHGs. Under the combined influence of GHGs and sulfate aerosols, the projected increase in precipitation is limited to 2±1% in the decade 2020s, 3±1% in the 2050s, and 7±3% in the 2080s. Figure 11-6 depicts projected changes in precipitation relative to changes in surface air temperature, averaged for land regions of Asia for each of the four skilled AOGCMs on an annual mean basis as well as during winter and summer for the 2050s and 2080s. The increase in precipitation is maximum during NH winter for both the time periods (Lal and Harasawa, 2000b). Clearly, intermodel differences in projections of precipitation are relatively large particularly during the winter even when they are averaged for the entire Asian continent—suggesting low confidence in projections of future precipitation in current AOGCMs.

The increase in annual mean precipitation is projected to be highest in boreal Asia. During the winter, boreal Asia and the Tibetan Plateau have the most pronounced increase in precipitation (Table 11-2). Over central Asia, an increase in winter precipitation and a decrease in summer precipitation are projected. Because the rainfall over this region is already low, severe water stress conditions—leading to expansion of deserts—are quite possible with a rise in surface air temperature here. The area-averaged annual mean and winter precipitation is projected to increase in temperate Asia. The models show high uncertainty in projections of future winter and summer precipitation over south Asia (with or without direct aerosol forcings). The effect of sulfate aerosols on Indian summer monsoon precipitation is to dampen the strength of the monsoon compared to that seen with GHGs only (Lal *et al.*, 1995a; Mitchell *et al.*, 1995; Cubasch *et al.*, 1996; Roeckner *et al.*, 1999). The overall effect of the combined forcing is at least partly dependent on the land/sea distribution of aerosol forcing and on whether the indirect effect is included along with the direct effect. To date, the effect of aerosol forcing (direct and indirect) on the variability of the monsoon has not been investigated.

Recent observations suggest that there is no appreciable long-term variation in the total number of tropical cyclones observed in the north Indian, southwest Indian, and southwest Pacific Oceans east of 160°E (Neumann, 1993; Lander and Guard, 1998). For the northwest subtropical Pacific basin, Chan and Shi (1996) found that the frequency of typhoons and the total number of tropical storms and typhoons has been more variable since about 1980. Several studies since the SAR have considered likely changes in tropical cyclones (Henderson-Sellers *et al.*, 1998; Knutson et al., 1998; Krishnamurti *et al.*, 1998; Royer *et al.*, 1998). Some of these studies suggest an increase in tropical storm intensities with carbon dioxide (CO_2)-induced warming.

Some of the most pronounced year-to-year variability in climate features in many parts of Asia has been linked to ENSO. Since the SAR, analysis of several new AOGCM results indicates that as global temperatures increase, the Pacific climate will

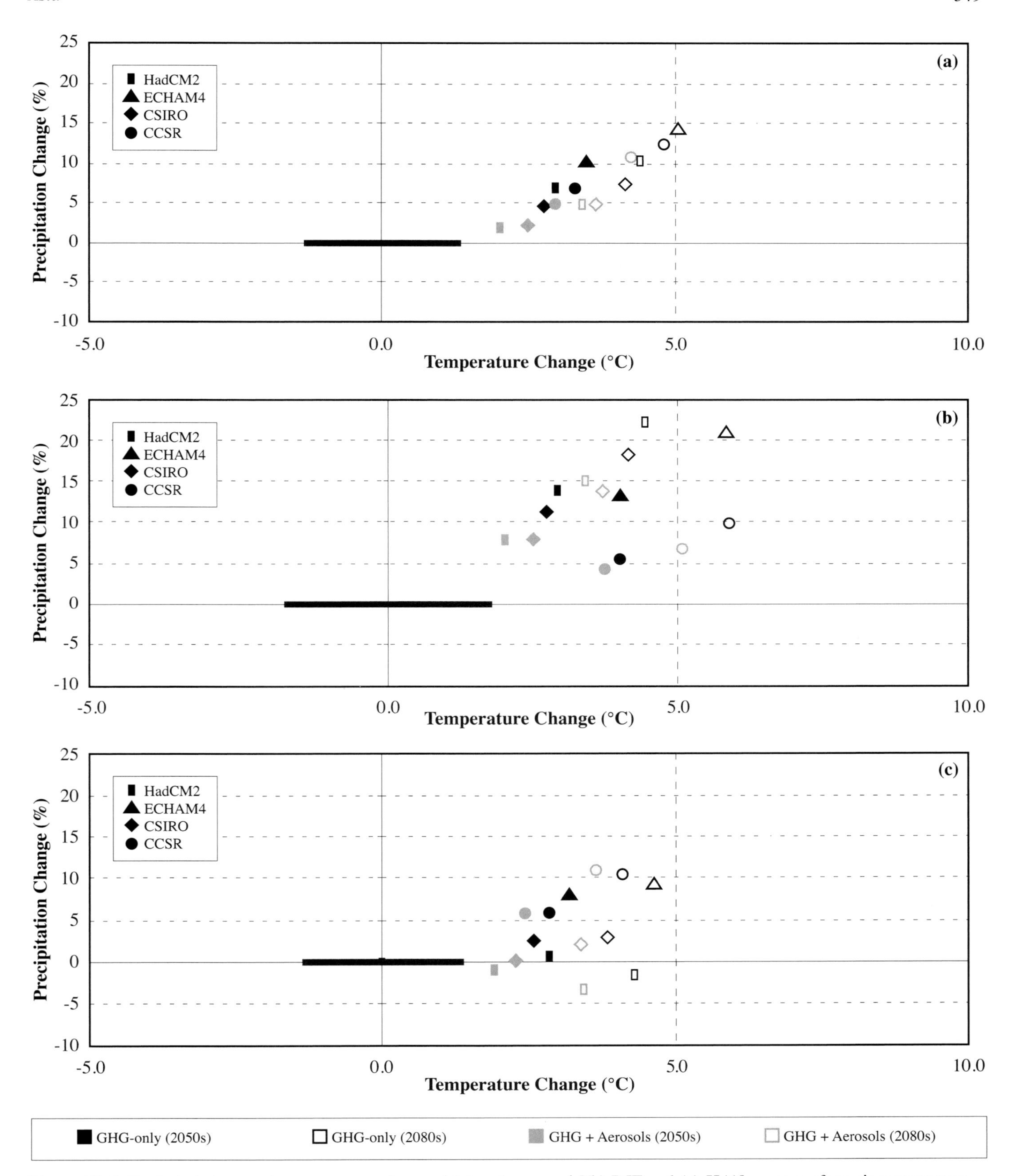

Figure 11-6: Projected changes in area-averaged annual (a) and seasonal [(b) DJF and (c) JJA)] mean surface air temperature and precipitation over land regions of Asia for the 2050s and 2080s as obtained in select AOGCMs.

tend to resemble a more El Niño-like state (Mitchell *et al*., 1995; Meehl and Washington, 1996; Knutson and Manabe, 1998; Boer *et al*., 1999; Timmermann *et al*., 1999). Collins (1999) finds an increased frequency of ENSO events and a shift in their seasonal cycle in a warmer atmosphere: The maximum occurs between August and October rather than around January as currently observed. Meehl and Washington (1996) indicate that future seasonal precipitation extremes associated with a given ENSO event are likely to be more intense in the tropical Indian Ocean region; anomalously wet areas could become wetter, and anomalously dry areas could become drier during future ENSO events.

Several recent studies (Kitoh *et al*., 1997; Lal *et al*., 2000) have confirmed earlier results (Kattenberg *et al*., 1996) indicating an increase in interannual variability of daily precipitation in the Asian summer monsoon with increased GHGs. Lal *et al*. (2000) also report an increase in intraseasonal precipitation variability and suggest that intraseasonal and interannual increases are associated with increased intraseasonal convective activity during the summer. The intensity of extreme rainfall events is projected to be higher in a warmer atmosphere, suggesting a decrease in return period for extreme precipitation events and the possibility of more frequent flash floods in parts of India, Nepal, and Bangladesh (Lal *et al*., 2000). However, Lal *et al*. (1995b) found no significant change in the number and intensity of monsoon depressions (which are largely responsible for the observed interannual variability of rainfall in the central plains of India) in the Bay of Bengal in a warmer climate. Because much of tropical Asia is intrinsically linked with the annual monsoon cycle, a better understanding of the future behavior of the monsoon and its variability is warranted for economic planning, disaster mitigation, and development of adaptation strategies to cope with climate variability and climate change.

11.1.3.4. High-Resolution Climate Change Experiments

Although these AOGCMs treat the complex interactions of atmospheric physics and planetary-scale dynamics fairly well, coarse horizontal resolution in the models restricts realistic simulation of climatic details on spatial variability. For example, tropical precipitation has high temporal and spatial variability, which cannot be resolved realistically in currently available AOGCMs. Many investigations on the ability of GCMs to simulate the Asian monsoon have been reported in the literature (e.g., Meehl and Washington, 1993; Chakraborty and Lal, 1994; Bhaskaran *et al*., 1995; Lal *et al*., 1995a, 1997, 1998a,b). These studies suggest that although most GCMs are able to simulate the large-scale monsoon circulation well, generally they are less successful with the summer monsoon rainfall. Since the SAR, nested modeling approaches have been followed to generate high-resolution regional climate scenarios (Jones *et al*., 1995; Lal *et al*., 1998c; Hassell and Jones, 1999) with more realistic mesoscale details and the response of GHG forcings to the surface climatology over the Asian monsoon region.

Large-scale patterns of temperature change simulated by GCMs and nested regional climate models (RCMs) are found to be generally similar under 2xCO_2 forcing, but the regional model results present some additional details associated with coastline and local topographical features (Hirakuchi and Giorgi, 1995). In addition, projected warming over eastern China region in the summer is less pronounced in the RCM

GCM GHG-Control, JJAS

-3 -1 -0.2 0 0.2 1 3 5 7

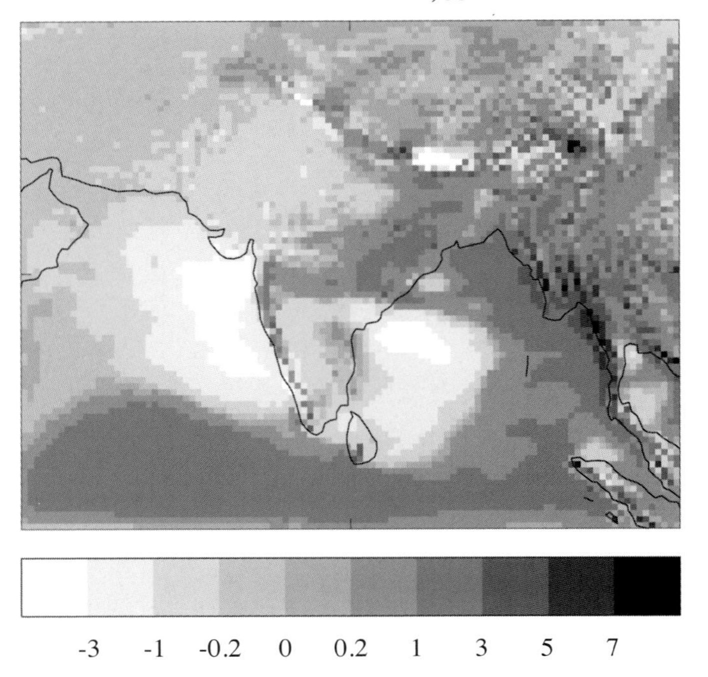

Figure 11-7: Spatial distribution of changes in monsoon rainfall over Indian subcontinent as simulated by Hadley Centre's global and regional climate models at the time of doubling of CO_2 in the atmosphere.

than in the GCM and is characterized by a different spatial pattern. This is related to an increase in monsoon precipitation simulated by the RCM there and associated surface cooling induced by evaporation and cloudiness. Nonetheless, the GCM and RCM simulations suggest a general increase in warming toward higher latitudes and greater warming in winter than in summer. The RCM produces a more pronounced increase in winter precipitation over southeastern China than does the GCM in response to orographic lifting of stronger low-level southerly onshore winds. The RCM also simulates increased precipitation in the monsoon rain belt over east China, Korea, and Japan during the summer.

The increase in surface air temperature simulated by the RCM over central and northern India is not as intense as in the GCM and does not extend as far south (Lal *et al.*, 1998c; Hassell and Jones, 1999). These anomalies are linked with changes in surface hydrological variables. Summer precipitation exhibits a more complex pattern of increases and decreases. Whereas an increase in rainfall is simulated over the eastern region of India, northwestern deserts see a small decrease in the absolute amount of rainfall in RCM simulation (Figure 11-7). Changes in soil moisture broadly follow those in precipitation except in eastern India, where they decrease as a result of enhanced drainage from the soil. The largest reductions (precipitation reduced to <1 mm day^{-1}; 60% decline in soil moisture) are simulated in the arid regions of northwest India and Pakistan. The projected increase in precipitation in flood-prone Bangladesh is approximately 20%. Nested RCM simulations have the potential to simulate the onset of the summer monsoon and its active/break cycle over India. The RCM captures the observed precipitation maximum over the southern tip of India during weak monsoon conditions, whereas the GCM does not.

Given the current state of climate modeling, projections of future regional climate have only limited confidence. The degree of confidence that could be attributed to RCM responses in terms of temporal and spatial changes resulting from GHG forcings would depend on more accurate simulation of the space and time evolution of large-scale monsoon circulation features in AOGCMs, as well as additional long-term RCM simulations with better skill. Current efforts on climate variability and climate change studies increasingly rely on diurnal, seasonal, latitudinal, and vertical patterns of temperature trends to provide evidence for anthropogenic signatures. Such approaches require increasingly detailed understanding of the spatial variability of all forcing mechanisms and their connections to global, hemispheric, and regional responses. Because the anthropogenic aerosol burden in the troposphere would have large spatial and temporal variations in the atmosphere, its future impact on regional scale would be in striking contrast to the impact from GHGs. It has also been suggested that aerosols produced by tropical biomass burning could lead to additional negative radiative forcing (Portmann *et al.*, 1997). Considerable uncertainty prevails about the indirect effect of aerosols on tropospheric clouds, which could strongly modulate the climate. The implications of localized radiative forcing on deep convection in tropical Asia and on Hadley circulation are still not understood (Lal *et al.*, 2000).

11.1.4. Sensitivity, Key Vulnerabilities, and Adaptability

11.1.4.1. Sensitivity and Key Vulnerabilities

An examination and analysis of the climate change-induced vulnerabilities in Asia has to be undertaken against the backdrop of the physical, economic, and social environment of the countries in the region. They provide not only the benchmark against which vulnerabilities are to be assessed but also the potential for adaptation to them. Surface water and groundwater resources in Asian countries play vital roles in forestry, agriculture, fisheries, livestock production, and industrial activity. The water and agriculture sectors are likely to be most sensitive to climate change-induced impacts in Asia.

As reported in IPCC (1998), climate change in boreal Asia could have serious effects on climate-dependent sectors such as agriculture, forestry, and water resources. Climate change and human activities, for example, may influence the levels of the Caspian and Aral Seas, with implications for the vulnerability of natural and social systems (Kelly *et al.*, 1983; Golubtsov *et al.,* 1996; Popov and Rice, 1997). The increase in surface temperature will have favorable effects on agriculture in the northernmost regions of Asia, and a general northward shift of crop zones is expected. However, as much as a 30% decrease in cereal production from the main agriculture regions of boreal Asia by 2050 has been projected (Budyko and Menzhulin, 1996; ICRF, 1998). A decrease in agriculture productivity of about 20% also is suggested in southwestern Siberia (ICRF, 1998). Forest ecosystems in boreal Asia could suffer from floods and increased volume of runoff, as well as melting of permafrost regions. Model-based assessments suggest that significant northward shifts (up to 400 km) in natural forest zones are likely in the next 50 years (Lelyakin *et al.*, 1997; Serebryanny and Khropov, 1997). There also is growing anxiety that significant increases in ultraviolet radiation, as observed in recent years, could have serious implications for ecosystems along the Arctic shore of Siberia (Makarov, 1999; Voskoboynikov, 1999; ACIA, 2000).

In arid and semi-arid Asia, the climate limits the portion of land that presently is available for agriculture and livestock production. Croplands in many of the countries in the region are irrigated because rainfall is low and highly variable. The agriculture sector here is potentially highly vulnerable to climate change because of degradation of the limited arable land. Almost two-thirds of domestic livestock are supported on rangelands, although in some countries a significant share of animal fodder also comes from crop residues (IPCC, 1998). The combination of elevated temperature and decreased precipitation in arid and semi-arid rangelands could cause a manifold increase in potential evapotranspiration, leading to severe water-stress conditions. Many desert organisms are near their limits of temperature tolerance. Because of the current marginality of soil-water and nutrient reserves, some ecosystems in semi-arid regions may be among the first to show the effects of climate change. Climate change has the potential to exacerbate the loss of biodiversity in this region.

The major impacts of global warming in temperate Asia will be large northward shifts of subtropical crop areas. Large increases in surface runoff—leading to soil erosion and degradation, frequent waterlogging in the south, and spring droughts in the north—ultimately will affect agriculture productivity (Arnell, 1999). As reported in IPCC (1998), the volume of runoff from glaciers in central Asia may increase three-fold by 2050. Permafrost in northeast China is expected to disappear if temperatures increase by 2°C or more. The northern part of China would be most vulnerable to hydrological impacts of climate change; future population growth and economic development here may exacerbate seriously the existing water shortage. Deltaic coasts in China would face severe problems from sea-level rise. Sea-level rise also will expand the flood-prone area and exacerbate beach erosion in Japan.

In monsoon Asia, the issue of sensitivity of physical and natural systems to the hydrological cycle is linked to major stresses caused by projected climate change on agricultural production and increased exposure of social and economic systems to impacts of extreme events, including forest die-back and increased fire risk; typhoons and tropical storms; floods and landslide; and human disease impacts. These stresses on physical systems translate into key social vulnerabilities, particularly in combination with unsustainable utilization of resources. For example, the drawing down of groundwater resources has increased the rate of relative sea-level change for many of the major cities of coastal Asia, such as Bangkok and Shanghai—thereby increasing the risk from climate change-induced sea-level rise (Jarupongsakul, 1999). Conversion of natural forests to palm oil plantations in many southeast Asian countries during recent decades (particularly Indonesia and Malaysia) increased the probability of uncontrolled forest fires and increased health and biodiversity impacts during the 1997 ENSO event (Yim, 1999; Barber and Schweithelm, 2000). The ecological security of mangroves and coral reefs may be put at risk by climate change. Sea-level rise could cause large-scale inundation along the coastline and recession of flat sandy beaches of south and southeast Asia. Monsoons in tropical Asia could become more variable if ENSO events become stronger and more frequent in a warmer atmosphere (Webster and Yang, 1992; Webster *et al.* 1998).

Vulnerability relates to social or natural systems and is delineated as such. The issues of social and physical vulnerability to climate change are directly related because the major sensitivities of ecological and natural systems across the regions of Asia translate into risks to socioeconomic systems. Social vulnerability is defined as the degree to which individuals or groups are susceptible to impacts; the determinants of social vulnerability are exposure to stress as a result of the impacts of climate change and the underlying social position (Adger, 1999a). On this basis, the key social vulnerabilities in the Asian context occur:

- Where resource-dependent communities have their resilience undermined by increased exposure to the physical stresses in water availability, agricultural impact, and other sensitivities mentioned above
- Where sustainability and equity (Munasinghe, 2000) are sacrificed for economic growth, exposing larger parts of the population to impacts.

11.1.4.2. Adaptability

Climate variability and change, both natural and anthropogenic, cause a wide range of direct and indirect impacts on natural and human systems. To understand which adaptation opportunities will be most cost-effective and have the greatest value, emphasis must be given to characteristics of system vulnerability, such as resilience, critical thresholds, and coping ranges, which are highly dependent on regions and nations. In this respect, lessons learned from past experiences regarding climate variability and change provide essential understanding of processes, actions, and successes.

The impacts of climate change and other drivers of environmental degradation (e.g., disparities in income level, technological gaps) are likely to be felt more severely in developing countries than in developed countries of Asia, irrespective of the magnitude of climate change, because of the poor resource and infrastructure bases. The developing countries need to scope the development of adaptation strategies incrementally to support development of existing policies that exploit "no regret" measures and "win-win" options (Smit *et al.*, 2000). Detailed and reliable regional scenarios of climate change need to be developed and used in rigorous vulnerability analysis (e.g., low-probability/high-consequence events versus high-probability/high-consequence events, risk perceptions). In developing countries, adaptation responses are closely linked to developmental activities. Consequently, there are likely to be large spillover effects between adaptation policies and developmental activities. Care in this regard must be taken in the evaluation of adaptation costs and benefits. In developing countries of Asia, options such as population growth control, poverty alleviation, and capacity building in food production, health care delivery, and water resource management hold great potential in creating more resilient social systems that are capable of withstanding the negative impacts of climate change.

11.2. Key Regional Concerns

Despite global technological and economic development, a large proportion of the nearly 1.5 billion people living in severe poverty at the dawn of the new millennium are located in Asia. Global per capita water supplies are declining and are now 30% lower than they were 25 years ago. By 2050, as much as 42% of the world's population may have to live in countries with insufficient freshwater stocks to meet the combined needs of agriculture, industry, and domestic use. The world's population will reach at least 8.9 billion by the middle of the 21st century (United Nations, 1998). India and China alone now account for 38% of the world's population. Most of the additional population would be in developing countries. These countries are likely to suffer adverse agricultural responses; significant changes in seasonal runoff; possibly severe vector-borne diseases; increased risks of severe tropical weather disturbances, including storms; vulnerability to sea-level rise; and other stresses. This section presents key regional concerns of various subregions of Asia related to climate change.

11.2.1. Ecosystems and Biodiversity

11.2.1.1. Mountain and Highland Systems

Relatively hospitable mountain regions in Asia are under pressure from human settlements and commercial cultivation, which have led to land degradation and adverse effects on water supply. Ongoing changes in different mountain systems within Asia include those associated with high crop production and those characterized by extensive animal husbandry and pastureland. Human encroachment in mountain regions has reduced vegetation cover, which has increased soil moisture evaporation, erosion, and siltation—with adverse effects on water quality and other resources. Changes in the snowfall pattern have been observed in mountain and highland systems, particularly in the Himalayas (Verghese and Iyer, 1993). These changes will have wider implications—from marked impact on the monsoon regime to seasonal runoff and vegetation cover, including agriculture. Changes in the hydrological regime also will trigger episodes of extreme events.

One-tenth of the world's known species of higher altitude plants and animals occur in the Himalayas. In addition, some countries in Asia are centers of origin for many crop and fruit-tree species; as such, they are important sources of genes for their wild relatives. Biodiversity is being lost in these regions because of human activities, especially land degradation and the overuse of resources. In 1995, approximately 10% of known species in the Himalayas were listed as threatened, and the number of species on the verge of extinction has increased since then. As a consequence of global warming, the present distribution of species in high-elevation ecosystems is projected to shift to higher elevations, although the rates of vegetation change are expected to be slow and colonization success would be constrained by increased erosion and overland flows in the highly dissected and steep terrains of the Himalayan mountain range. Weedy species with a wide ecological tolerance will have an advantage over others (Kitayama and Mueller-Dombois, 1995). High-elevation tree species—such as *Abies, Acer, and Betula*—prevail in cold climates because of their adaptations to chilling winters. In Japan, the area of suitable habitat at higher elevations has shrunk over the past 30 years, and the variety of alpine plants that grow there has been rapidly reduced (Masuzawa, 1997, 2000). Increases in temperature would result in competition between such species and new arrivals. The sensitivity of alpine flora to climatic factors and, in particular, water stress in the summit region of Mt. Kinabalu—the highest mountain in southeast Asia—already have been demonstrated (Kitayama, 1996; Aiba and Kitayama, 1999). The accumulated stresses of climate change are likely to disrupt the ecology of mountain and highland systems.

11.2.1.2. Lakes/Streams, Rivers, and Glaciers

Lakes, streams, glaciers, and other freshwater ecosystems in Asia are highly diversified in terms of plant and animal species. These freshwater ecosystems have been stressed by environmental burdens, exploitation of natural resources, transformation of lands, and recreational activities. There is growing concern that climate change may accelerate the damage to freshwater ecosystems such as lakes, marshes, and rivers. More than 50,000 ha of coastal territories, including 35,000 ha of delta in the Selenga River and 12,000 ha of delta in the Upper Angara River, have been damaged during the past few years by precipitation and riverflow increases (Anokhin and Izrael, 2000). With an increased amount of precipitation likely in the future, more incidences of flooding and other adverse impacts are possible. With a rise in temperature, a decrease in the amount of snowfall in the Lake Biwa catchment in Japan is projected—which might exacerbate the process of eutrophication (Fushimi, 2000a). Deterioration of lake water quality also is suggested in Kasumigaura Lake in eastern Japan (Fujimoto *et al.*, 1995; Fukushima *et al.*, 2000). The response of lakes and streams to climate change will involve complex interactions between the effects of climate on areal inputs, hydrology, and catchments and in-lake processes.

Many of the major rivers in Asia have long been targets for development projects related to the hydroelectric, water supply, agriculture, industry, and navigation sectors. As a consequence, there have been shifts between freshwater and estuarine conditions as a result of high freshwater flows during the rainy season and low to nonexistent freshwater flows in the dry season. Increasing literol vegetation is causing health risks for local habitats in many countries of south Asia. Changes in aquatic habitat also have affected fisheries in lower valleys and deltas; the absence of nutrient-rich sediments has detrimental effects on fish productivity. Reduced flows in lower valley catchments also have resulted in eutrophication and poor water quality.

Many rivers originate from the glaciers in the Tianshan mountain range, which create wide alluvial fans at the foot of the northern Tianshan. The Hindukush Himalayan ranges are the source of some major rivers. The total amount of water flowing from the Himalayas to the plains of the Indian subcontinent is estimated at about 8.6 x 10^6 m^3 per year. The Himalayas have nearly 1,500 glaciers; it is estimated that these glaciers cover an area of about 33,000 km^2 (Dyurgerov and Meier, 1997). These glaciers provide snow and the glacial meltwaters keep major rivers perennial throughout the year. In recent decades, the hydrological characteristics of watersheds in this region seem to have undergone substantial change as a result of extensive land-use change—leading to more frequent hydrological disasters, enhanced variability in rainfall and runoff, extensive reservoir sedimentation, and pollution of lakes (Ives and Messerli, 1989). Almost 67% of the glaciers in the Himalayan and Tienshan mountain ranges have retreated in the past decade (Ageta and Kadota, 1992; Yamada *et al.*, 1996; Fushimi, 2000b). The mean equilibrium-line altitude at which snow accumulation is equal to snow ablation for glaciers is estimated to be about 50–80 m higher than the altitude during the first half of the 19th century (Pender, 1995). Available records suggest that Gangotri glacier is retreating by about 30 m yr^{-1}. A warming is likely to increase melting far more rapidly than accumulation. As reported in IPCC (1998), glacial melt is expected to increase under changed climate conditions, which would lead to increased summer flows in some river systems for a few decades, followed by a reduction in flow as the glaciers disappear.

11.2.1.3. Forests, Grasslands, and Rangelands

Most of the frontier forests in Asia are endangered today by rapid population growth, ever-increasing demand for agricultural land, poverty, poor institutional capacity, and lack of effective community participation in forestry activities (Mackenzie *et al.*, 1998). Climate change is expected to affect the boundaries of forest types and areas, primary productivity, species populations and migration, the occurrence of pests and diseases, and forest regeneration. The increase in GHGs also affects species composition and the structure of ecosystems because the environment limits the types of organisms that can thrive and the amount of plant tissues that can be sustained (Melillo *et al.*, 1996). Compositional and structural changes, in turn, affect ecosystem function (Schulze, 1994). The interaction between elevated CO_2 and climate change plays an important role in the overall response of net primary productivity to climate change at elevated CO_2 (Xiao *et al.*, 1998).

Climate change will have a profound effect on the future distribution, productivity, and health of forests throughout Asia (see also Section 5.6). Because warming is expected to be particularly large at high latitudes, climate change could have substantial impact on boreal forests (Dixon *et al.*, 1996; IPCC, 1996; Krankina, 1997). Global warming will decrease permafrost areas, improve growing conditions, and decrease areas of disturbed stands and ecosystems in a general sense, although impacts would be significantly different at various locations within the boreal forests. Moreover, forest fire is expected to occur more frequently in boreal Asia as a result of increased mean temperature (Valendik, 1996; Lelyakin *et al.*, 1997). Pest activity also could increase with a rise in temperature, depending on the age composition of the boreal forests (Alfiorov *et al.*, 1998).

Asia's temperate forests are a globally important resource because of their high degree of endemism, biological diversity, ecological stability, and production potential. About 150 Mha of forests in central China have been cleared during the past several decades. Efforts are now underway to at least partially restore the area under forest cover in China through reforestation, soil recovery, and water conservation programs (Zhang *et al.*, 1997). Studies on projected impacts of climate change suggest that northeast China may be deprived of the conifer forests and its habitat, and broad-leaved forests in east China may shift northward by approximately 3° of latitude. These results are based on a 2°C increase in annual mean temperature and a 20% increase in annual precipitation (Omasa *et al.*, 1996; Tsunekawa *et al.*, 1996).

Tropical moist forests have trees with higher densities of wood and larger proportions of branch wood relative to those in temperate forests. As many as 16 countries of tropical Asia are located within the humid tropical forest region. These forests and woodlands are important resources that must be safeguarded, given the heavy use of wood as fuel in some countries. Past policies in the humid tropics have focused mainly on natural forest protection and conservation (Skole *et al.*, 1998). However, there is a need to shift emphasis from conservation alone to a strategy that involves sustained development, investment performance, and public accountability (see Section 5.6). Encouragingly, the current annual rate of reforestation is highest in tropical Asia as a result of relatively high investments in reforestation schemes, including social forestry.

Most semi-arid lands in Asia are classified as rangelands, with a cover of grassland or scrublands. Although the share of land area used for agricultural purposes is about 82% of the total area, it is mainly low-productive pastures. With an increase in temperature of 2–3°C combined with reduced precipitation as projected for the future in the semi-arid and arid regions of Asia, grassland productivity is expected to decrease by as much as 40–90% (Smith *et al.*, 1996). Approximately 70% of pastures are facing degradation, with dramatic decreases in fodder yield over recent decades in some parts of Mongolia (Khuldorj *et al.*, 1998). Rangelands in Nepal also have been subject to degradation in recent years (NBAP, 2000). Climate change is likely to represent an additional stress to rapid social change in many of Asia's rangelands.

11.2.1.4. Drylands

Precipitation is scarce and has a high annual variance in dryland areas. Very high daily temperature variance is recorded with frequent sand storms, dust ghost, and intense sunshine. Arid plants usually belong to drought escaper, drought evader, drought resister, or drought endurer categories. Evaporative losses and water limitations are the most prominent factors dictating animal life in arid environments. Low rainfall dictates the formation of shallow or extremely sallow soils that often are characterized by high content of airborne particles and small fractions of rock-erosion elements. Most of the soils are poor in or completely devoid of organic matter, and the nutrient pools of the soils are low. Apparently, humans not only utilize the ecosystem services of this region but are also influencing the evolution of some of its important biotic elements. In Mongolia, for example, while soil fertility has decreased by about 20% in the past 40 years, about one-third of the pasturage has been overgrazed and 5 Mha of arid land have constantly been threatened by moving sands (Khuldorj *et al.*, 1998). Soils exposed to degradation as a result of poor land management could become infertile as a result of climate change. Temperature increases would have negative impacts on natural vegetation in desert zones. Plants with surface root systems, which utilize mostly precipitation moisture, will be vulnerable. Climate change also would have negative impacts on sheep breeding and lamb wool productivity.

Just as shifts in vegetation belts are expected in non-drylands, in the drylands of Asia a shift in dryland types is expected as a result of climate change. Drylands are ranked along an aridity index, in relation to the ratio of precipitation to potential evapotranspiration (i.e., to a gradient in soil moisture available for driving production). Because soil moisture is likely to decline in this region, the least-dry land type (dry subhumid drylands) are expected to become semi-arid, and semi-arid land is expected to become arid. It is notable that population pressure on dryland resources is reduced with increasing aridity, but resistance to degradation and resilience following degradation also is reduced with increasing aridity. Therefore, semi-arid drylands, which are intermediate in aridity as compared to arid drylands and dry subhumid ones, are most susceptible to becoming further desertified (Safriel, 1995). Because semi-arid drylands are very common among Asian drylands, large areas will become not only dry but also desertified as a result of climate change,.

11.2.1.5. Cryosphere and Permafrost

Permafrost is highly responsive to climatic fluctuations at several temporal and spatial scales (Nelson and Anisimov, 1993). Evidence of spatially extensive episodes of permafrost thawing and poleward contraction has been documented (Halsey *et al.*, 1995; Anisimov and Nelson, 1996b; Anisimov and Nelson, 1997; WASI, 1997). Depending on regional climate and local biological, topographic, and edaphic parameters, pronounced warming in the high latitudes of Asia could lead to thinning or disappearance of permafrost in locations where it now exists (Anisimov and Nelson, 1996a). Poleward movement of the southern boundary of the sporadic permafrost area is likely in Mongolia and northeast China. Large-scale shrinkage of the permafrost region in boreal Asia also is likely. In northern regions of boreal Asia, the mean annual temperature of permafrost, hence the depth of seasonal thawing (active layer thickness), will increase (Izrael *et al.*, 1999). The perennially frozen rocks will completely degrade within the present southern regions (ICRF, 1998). The development of thermokarst and thermal erosion because of perennial thawing and increase in the depth of seasonal thawing of ice-rich grounds and monomineral ice accumulations is a critical process in permafrost regions of boreal Asia (Izrael *et al.*, 1999). The change in rock temperature will result in a change in the strength characteristics, bearing capacity, and compressibility of frozen rocks, generation of thermokarst, thermal erosion, and some other geocryological processes (Garagulia and Ershov, 2000). In response to projected climate change, four main economic sectors in permafrost regions—surface and underground construction, the mining industry, heating energy demand, and agricultural development—will be affected (ICRF, 1995, 1998; Anisimov, 1999). Because large quantities of carbon are sequestered in the permafrost of boreal peatlands and tundra regions (Botch *et al.*, 1995; Ping, 1996), changes in distribution of frozen ground and systematic increase in the thickness of seasonally thawed layer are likely to result in the release of large amounts of CO_2 and possibly methane (CH_4) into the atmosphere.

The permafrost area on the Tibetan Plateau has an average altitude of about 5,000 m and is one of the several regions not significantly affected by direct human activities. Because of this area's thermal and moisture conditions are on the edge of the ecological limitations of vegetation, it is believed to be highly sensitive to global warming (Zhang *et al.*, 1996). The boundary between continuous and discontinuous (intermittent or seasonal) permafrost areas on the Tibetan Plateau are likely to shift toward the center of the plateau along the eastern and western margins (Anisimov and Nelson, 1996b).

11.2.1.6. Protected Areas and Risks to Living Species

Protected areas usually are designated and managed to keep wild species that live within the area from becoming extinct. Even after an area has been set aside as protected habitat, extinction or population declines may still occur as a result of changes in environmental conditions related to climate change, land use in surrounding areas, or widespread pollution. Climate change is likely to induce vegetation change that will force wild plant and animal species to shift their distribution in response to the new conditions. For example, a variety of changes in butterflies, dragonflies, beetles, and other migratory insects have been recorded in green corridors of Japan in recent years (Ubukata, 2000), and shifts have been recorded in the ranges of many North American, European, Arctic, and Antarctic bird and insect species (see Section 5.4). If the protected area is not large enough to contain an area that will be suitable under the new climate conditions, a species may become locally extinct. In contrast, protected areas that are large enough to cover an elevation or a latitudinal gradient should allow species to make adjustments along the gradient as conditions change. In some cases, such as in coastal areas, habitat may simply be lost as a result of factors such as sea-level rise, with no potential area for species to migrate.

Frontier forests in Asia are home to more than 50% of the world's terrestrial plant and animal species (Rice, 1998). Risks to this rich array of living species are increasing. For instance, of the 436 species of mammals and 1,500 species of birds in Indonesia, more than 100 species each of mammals and birds have been declared threatened (UNEP, 1999). Similar trends also are seen in China, India, Malaysia, Myanmar, and Thailand. In India, as many as 1,256 higher plant species, of more than 15,000 species, are threatened (Sukumar *et al.*, 1995).

Coastal areas are likely to be at risk from climate change-induced sea-level rise. A rise in the water level of estuaries will reduce the size and connectedness of small islands and coastal and estuarine reserves and increase their isolation. The Yangtze (Changjiang) and Mekong deltas on mainland China and the Mai Po marshes in Hong Kong are refueling stops for migratory birds, especially ducks, geese, and shorebirds. The presence of these birds may be threatened by the disappearance of coastal marshes as a result of increases in sea level from global warming (Li *et al.*, 1991; Tang, 1995). Similarly, the Rann of Kutch in India supports one of the largest Greater Flamingo colonies in Asia (Ali, 1985; Bapat, 1992). With sea-level rise, these salt marshes and mudflats are likely to be submerged (Bandyopadhyay, 1993), which would result in decreased habitat for breeding flamingoes and lesser floricans (Sankaran *et al.*, 1992). In addition, about 2,000 Indian wild asses in the Rann of Kutch could lose their only habitat in India to rising sea level (Clark and Duncan, 1992).

The Sundarbans of Bangladesh, which support a diversity of wildlife, are at great risk from rising sea level. These coastal mangrove forests provide habitat for species such as Bengal tigers, Indian otters, spotted deer, wild boars, estuarine crocodiles, fiddler crabs, mud crabs, three marine lizard species, and five marine turtle species (Green, 1990). With a 1-m rise in sea level, the Sundarbans are likely to disappear, which may spell the demise of the tiger and other wildlife (Smith *et al.*, 1998).

Species that live in mountainous areas also are particularly at risk of losing habitat as a result of changes in climate. Extreme temperature conditions may cause these protected areas to undergo major changes, partly because of high rates of variation in habitat structure that naturally occur on mountain ranges as a result of changes in slope, steepness, and exposure. Protected areas also may be subjected to extreme surface runoffs because of rapid melting of winter snow. Because many Asian mountain ranges are east-west oriented, there will be little room for species to shift their ranges toward cooler mountainous habitat. An additional barrier to wildlife movements in mountainous habitat derives from the fact that many of the larger reserves in central Asia are located along international borders. Depending on topography, if these borders are heavily fenced, most of the larger terrestrial vertebrates will not be able to respond spatially to changes in their environment.

Besides loss of habitat, wild species are at risk from changes in environmental conditions that favor forest fires and drought. For example, forest fires under unseasonably high temperatures in Nepal may threaten local extinction for red pandas, leopards, monkeys, deer, bears, and other wild animals. If the frequency of these extreme events increases, the frequency of fire also may increase. Similarly, increases in the frequency of dry spells and local droughts may decrease populations. For example, drought-related decreases in the density and persistence of Green Leaf Warblers have been recorded on their wintering grounds in the Western Ghats of south India (Katti and Price, 1996). In desert ecosystems, protected areas often are located around oases, which are the basis for the existence of much of the local fauna. Protected oases often are far apart, so droughts that cause a decline in local forage often cause mass mortality because animals may not be able to move on to adjacent oases. The frequency of these droughts therefore is a key component in the viability of populations in such protected areas (Safriel, 1993).

Climate change is likely to act synergistically with many other stressors, such as land conversion and pollution, leading to major impacts on protected areas and species (see Chapter 5). Currently designated major protected areas in Asia need to be examined with respect to the ability of their species to shift in range in response to changing climate, as well as with respect to how much habitat could be lost. Many species—especially those in coastal areas and mountainous habitats—could experience large population declines; some may become extinct at least in part because of climate changes.

11.2.2. Agriculture and Food Security

11.2.2.1. Production Systems

Asia has the world's largest area under cereal cultivation and is the largest producer of staple foods (FAO, 1999a). Present crop yields in Asia are comparable to those in Europe and South America. Within Asia, India has the largest area under cereal cultivation. The total production of cereals in China is twice that of India, as a result of higher average productivity (FAO, 1999a). Most land that is suitable for cultivation is already in use; by 2010 per capita availability of land in developing countries of Asia will shrink from the present 0.8 ha to about 0.3 ha. Table 11-4 depicts the growth rate of rice cultivation area in select Asian countries. Current rates of land degradation suggest that a further 1.8 million km^2 of farmland could become unproductive in Asia by 2050, adding stress to a system that must ensure food security in the context of a rapidly growing population.

Rice is central to nutrition in Asia. In 1997, rice provided about 700 kilocalories per person per day or more for approximately 2.9 billion people, most of whom live in developing countries of Asia and Africa. During the 1990s, rice production and productivity in Asia grew at a much slower rate than did population. Yield deceleration of rice (the annual growth rate declined from 2.8% in the 1980s to 1.1% in the 1990s) in Asia has been attributed to water scarcity, indiscriminate addition and inefficient use of inputs such as inorganic fertilizers and pesticides, and policy issues and the reliance on a narrower genetic material base with impacts on variability (Hazell, 1985; Matson *et al.*, 1997; Naylor *et al.*, 1997). Several other factors also have contributed to productivity stagnation and the decline of rice (lower output/input ratio) in the intensive cropping system (two to three rice crops per year). Key factors currently contributing to the yield gap in different countries of Asia include biophysical, technical/management, socioeconomic, institutional/policy, technology transfer, and adoption/linkage problems.

Urbanization in Asia has accentuated increased demand for fresh vegetables; this demand is to be met by new production areas combined with more intensified horticulture crop management to raise the productivity per unit of land and water. In most cases, urban and peri-urban agriculture initiatives with uncontrolled use of agrochemicals are a high-risk activity.

***Table 11-4**: Changes in area under rice cultivation in select Asian countries, 1979–1999 (FAO, 1999a).*

Country	Period	Total Rice Cultivation Area (10^3 ha)	Change in Rice Cultivation Area (10^3 ha)	Rate of Change in Rice Cultivation Area (ha yr^{-1})
Bangladesh	1979	10,160	310	14,762
	1999	10,470		
Cambodia	1979	774	1,187	56,524
	1999	1,961		
China	1979	34,560	-2,840	-135,238
	1999	31,720		
India	1979	39,414	3,586	170,762
	1999	43,000		
Indonesia	1979	8,804	2,820	134,286
	1999	11,624		
Malaysia	1979	738	-93	-4,429
	1999	645		
Myanmar	1979	4,442	1,016	48,381
	1999	5,458		
Nepal	1979	1,254	260	12,381
	1999	1,514		
Sri Lanka	1979	790	39	1,857
	1999	829		
Pakistan	1979	2,035	365	17,381
	1999	2,400		
Philippines	1979	3,637	341	16,238
	1999	3,978		
Thailand	1979	8,654	1,346	64,095
	1999	10,000		
Vietnam	1979	5,485	2,163	103,000
	1999	7,648		

Adequate steps need to be taken at regional and local levels to safeguard specialized and diversified urban production systems (vegetables, fruits, and root crops) through sustainable intensification of natural resource use and strengthening of decision support systems. Increased productivity and sustained production of food grains and legumes, industrial crops (oil, gum and resins, beverage, fiber, medicines, aromatic plants), and horticultural crops through crop diversification is critical for food and nutritional security in Asia.

Even minor deviations outside the "normal" weather range seriously impair the efficiency of externally applied inputs and food production. Moisture stress from prolonged dry spells or thermal stress resulting from heat-wave conditions significantly affect the agricultural productivity when they occur in critical life stages of the crop (Rounsevell *et al*., 1999). As reported in IPCC (1998), stress on water availability in Asia is likely to be exacerbated by climate change. Several studies aimed at understanding the nature and magnitude of gains or losses in yield of particular crops at selected sites in Asia under elevated CO_2 conditions and associated climatic change have been reported in the literature (e.g., Lou and Lin, 1999). These studies suggest that, in general, areas in mid- and high latitudes will experience increases in crop yield, whereas yields in areas in the lower latitudes generally will decrease (see also Chapter 5). Climatic variability and change will seriously endanger sustained agricultural production in Asia in coming decades. The scheduling of the cropping season as well as the duration of the growing period of the crop also would be affected.

In general, increased CO_2 levels and a longer frost-free growing season are expected to enhance agricultural productivity in north Asia. The area under wheat cultivation is likely to expand in the north and west. The increase in surface temperature also may increase the growing season in temperate Asia, thereby prolonging the grain-filling period, which may result in higher yields (Rosenzweig and Hillel, 1998). In Japan, for example, simulation studies and field experiments indicate that enhanced CO_2 levels in a warmer atmosphere will substantially increase rice yields and yield stability in northern and north-central Japan (Horie *et al*, 1995a). In south central and southwestern Japan, however, rice yields are expected to decline by at least 30% because of spikelet sterility and shorter rice growing duration (Matsui and Horie, 1992). Climate change should be advantageous to wheat yield in northeast China. Because of an increase in respiration in a warmer atmosphere demanding more water availability, rice yield in China is expected to decline (Wang, 1996a). In central and north China, higher temperatures during teaseling and drawing stages and low soil moisture could result in reduced wheat yield. Increases in precipitation should be favorable for pests, diseases, and weeds in the south (Wang, 1996b; Dai, 1997). In tropical Asia, although wheat crops are likely to be sensitive to an increase in maximum temperature, rice crops would be vulnerable to an increase in minimum temperature. The adverse impacts of likely water shortage on wheat productivity in India could be minimized to a certain extent under elevated CO_2 levels; these impacts, however, would be largely maintained for rice crops, resulting in a net decline in rice yields (Aggarwal and Sinha, 1993; Rao and Sinha, 1994; Lal *et al*., 1998d). Acute water shortage conditions combined with thermal stress should adversely affect wheat and, more severely, rice productivity in India even under the positive effects of elevated CO_2 in the future.

Key findings on the impacts of an increase in surface temperature and elevated CO_2 on rice production in Asia—based on a study carried out for Bangladesh, China, India, Indonesia, Japan, Malaysia, Myanmar, the Philippines, South Korea, and Thailand under the Simulation and System Analysis for Rice Production Project at the International Rice Research Institute—are summarized in Table 11-5. Two process-based crop simulation models—the ORYZA1 model (Kropff *et al*., 1995) and the SIMRIW model (Horie *et al*., 1995b)—suggest that the positive effects of enhanced photosynthesis resulting from doubling of CO_2 are more than offset by increases in temperature greater than 2°C (Matthews *et al*., 1995a).

More than 10,000 different species of insect pest are found in the tropics, 90% of which are active in the humid tropics. The occurrence, development, and spread of crop diseases depend on integrated effects of pathogen, host, and environmental conditions. The survival rate of pathogens in winter or summer could vary with an increase in surface temperature (Patterson *et al*., 1999). Higher temperatures in winter will not only result in higher pathogen survival rates but also lead to extension of cropping area, which could provide more host plants for pathogens. Thus, the overall impact of climate change is likely to be an enlargement of the source, population, and size of pathogenic bacteria. Damage from diseases may be more serious because heat-stress conditions will weaken the disease-resistance of host plants and provide pathogenic bacteria with more favorable growth conditions. The growth, reproduction, and spread of disease bacteria also depend on air humidity; some diseases—such as wheat scab, rice blast, and sheath and culm blight of rice—will be more widespread in temperate and tropical regions of Asia if the climate becomes warmer and wetter.

***Table 11-5**: Model-simulated mean change (%) in potential yields of rice in Asia under fixed increments of air temperature and ambient CO_2 level (Matthews et al., 1995b).*

Model Used and Ambient CO_2 Levels	**Percent Change in Mean Potential Rice Yield in Asia resulting from Surface Air Temperature Increment of**			
	0°C	+1°C	+2°C	+4°C
ORYZA1 Model				
340 ppm	0.00	-7.25	-14.18	-31.00
1.5xCO_2	23.31	12.29	5.60	-15.66
2xCO_2	36.39	26.42	16.76	-6.99
SIMRIW Model				
340 ppm	0.00	-4.58	-9.81	-26.15
1.5xCO_2	12.99	7.81	1.89	-16.58
2xCO_2	23.92	18.23	11.74	-8.54

11.2.2.2. Grain Supply and Demand in Asia

Asia's total cereal production more than tripled between 1961 and 1999, from 329.5 to nearly 1,025.8 Mt; the increase in harvest area was only marginal (from 271.8 to 324.5 Mha) during this period. Rice paddy production in Asia jumped from 198.7 to 533.5 Mt while the harvest area increased from 106.9 Mha in 1961 to 136.5 Mha in 1999. Similarly, wheat production in Asia increased from 45.8 to 261.7 Mt while the harvest area increased from 61.2 Mha in 1961 to 97.2 Mha in 1999. A three-fold increase in production of sugarcane and almost five-fold increase in oil crops took place in Asia during 1961–1999. Asian production of starchy roots also doubled during this period. Even as the Asian population increased a little more than two-fold—from 1.70 billion to 3.58 billion people—during this time, per capita consumption of calories in Asia has increased since the 1960s (FAO, 1999a). This improvement in the food situation has been the consequence of an increase in production, resulting from technological advances, that has outpaced population growth in Asia. However, because there is a limit to the amount of arable land, while the Asian population continues to grow, the per capita area of harvest has consistently decreased.

The population growth rate has been a primary factor in the increase in demand for food grains in Asia. Since the mid-1980s, rapid changes in food supply and demand structures have been observed in most developing countries. Declines in self-sufficiency for grains have been particularly dramatic in countries with advanced economies, such as Japan, South Korea, Taiwan (China), and Malaysia, where the grain self-sufficiency rate has fallen below 30% in recent years. The rapid decline in grain self-sufficiency in these countries is unprecedented in any region of the world. Between 1970 and 1996, grain self-sufficiency fell from 45 to 27% in Japan, from 68 to 31% in Korea, from 61 to 19% in Taiwan (China), and from 60 to 25% in Malaysia (FAO, 1996). Declines in grain self-sufficiency also have been observed in the Philippines, Indonesia, and Sri Lanka. Populous countries such as China, India, Bangladesh, and Pakistan have long maintained grain self-sufficiency of more than 90%, but as these countries shift to industrialization it is possible that declines will be observed there as well. In contrast, Thailand and Vietnam, which encompass the great grain-producing regions of the Menam and Mekong River deltas, continue to maintain better than 100% self-sufficiency for grains.

The annual per capita consumption of meat in principal Asian countries also has shown an upward trend in recent years. Since the 1970s, South Korea and China have reported the most rapid increases in meat consumption. In South Korea, per capita meat consumption grew about seven-fold, from 5.2 to 40 kg between 1970 and 1997; in China, it increased nearly five-fold, from 8.8 to 42.5 kg. During the same period, per capita meat consumption more than doubled in Japan (42.2 kg in 1997) and Malaysia (51.9 kg in 1997). Meat consumption in Sri Lanka, Bangladesh, and India grew only marginally, remaining below or at 5 kg even in 1997—considerably lower than the level in east Asia. The increasing trend in meat consumption in temperate Asia means higher demand for livestock feed for the production of meat. Demand for this feed in Asia increased from 25.2 Mt in the 1970s to 61.2 Mt in 1980, 147.0 Mt in 1990, and 163.0 Mt in 1999. The demand for feed grain in Asia has grown at an annual rate of about 6.6% and has been a principal factor in the rapid increase in demand for world grain imports.

The world is moving toward a tighter grain supply and demand, and instability in the world grain market is rising. At the same time, Asia is becoming increasingly dependent on the world grain market. As shown in Figure 11-8, net grain import in Asia was 20.3 Mt in 1961; it increased to 80.9 Mt by 1998. The continuing increase in grain imports is a result of increasing

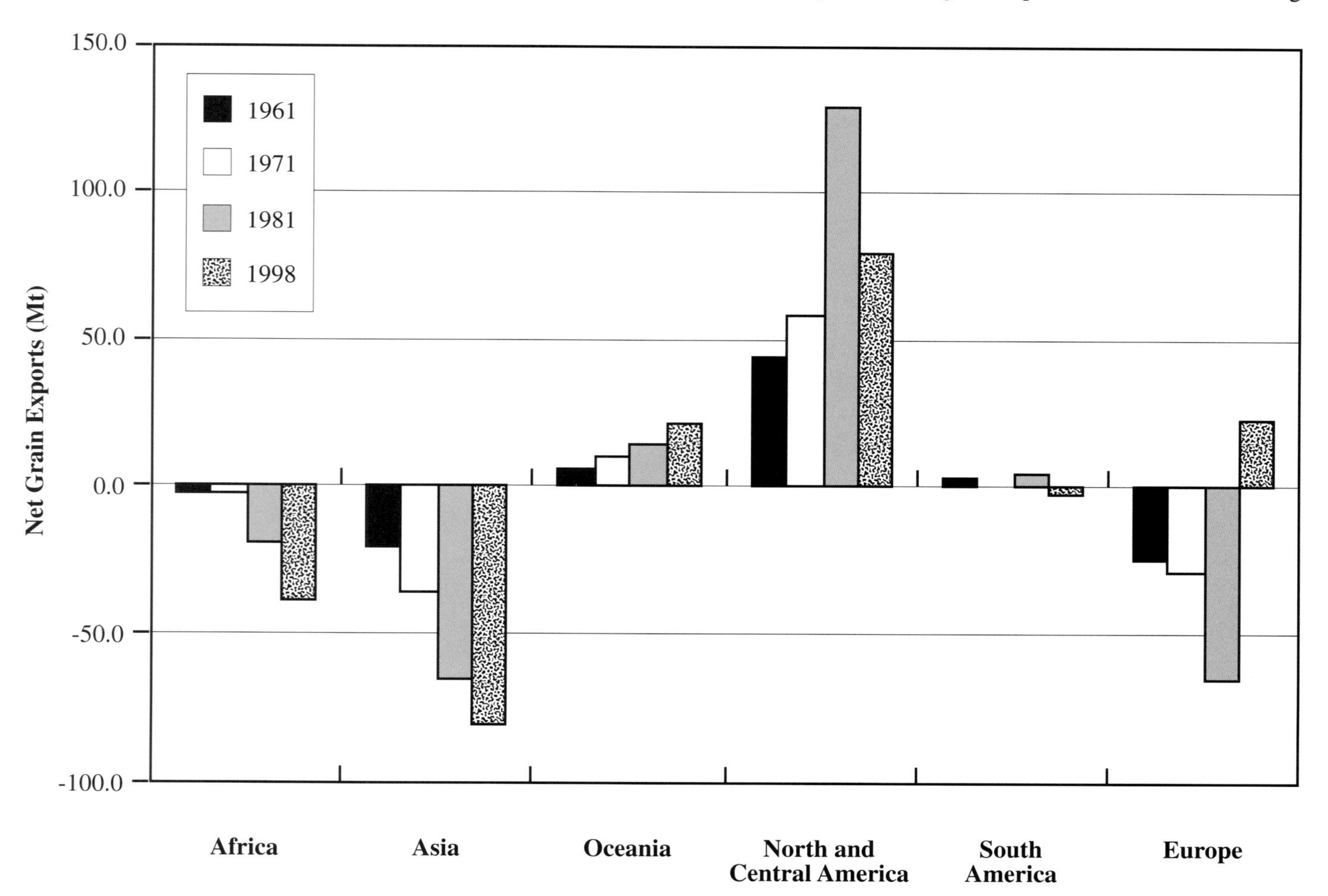

Figure 11-8: Trends in net grain exports in major world regions (USDA, 1999).

populations and increased demand for grain that accompanies economic growth, with which increases in grain production within the region cannot keep pace.

The impact of global warming on the international supply and demand of rice and wheat has been evaluated by Nakagawa *et al.* (1997) through the use of a partial equilibrium-type dynamic model for world supply and demand. Based on a world rice and wheat supply and demand model and a unit harvest scenario, the study projected that the serious impacts of global warming would be felt as early as the year 2020. The study suggests that the early impacts will not be so severe that humans are unable to control them. However, the gap between supply and demand may grow in some regions of Asia, resulting in considerably increased reliance on imports. Furthermore, the problem of short-term fluctuations in the market because of an increased frequency of droughts, floods, and other extreme weather events will be exacerbated, making it necessary to promote measures to combat climate change on a global scale.

11.2.2.3. Food Security

A major new challenge in future food production to meet the demand of the growing Asian population is in coping with the changing environment, which may alter the current optimum

***Table 11-6**: Prevalence of undernourishment in developing countries of Asia (FAO, 1999a; UNICEF, 1999).*

Country/Region	Population, 1996 (millions)	Main Cereal Consumed, 1995–1997	Dietary Energy Supply per Person, 1995–1997 (kcal day^{-1})	Access to Adequate Sanitation, 1990–1997 (%)	Under 5 Mortality Rate, 1995 (per 1000)	Number of Undernourished People (millions)	Fraction of Population Undernourished, 1979–1981 (%)	Fraction of Population Undernourished, 1995–1997 (%)
Arid and Semi-Arid Asia								
– Afghanistan	20.3	Wheat	1730	8	257	12.7	33	62
– Iran	63.5	Wheat	2830	81	40	3.7	9	6
– Iraq	20.6	Wheat	2370	75	71	3.2	4	15
– Jordan	4.4	Wheat	2910	77	25	0.1	6	3
– Kuwait	1.7	Wheat	3060	—	14	0.1	4	3
– Lebanon	3.1	Wheat	3270	63	40	0.1	8	2
– Pakistan	140.1	Wheat	2460	56	137	26.3	31	19
– Saudi Arabia	18.9	Wheat	2800	86	34	0.7	3	4
– Syrian Arab Republic	14.6	Wheat	3330	67	36	0.2	3	1
– Turkey	62.3	Wheat	3520	80	50	1.0	2	2
– United Arab Emirates	2.3	Rice/Wheat	3360	92	19	0.0	1	1
Temperate Asia								
– China	1238.8	Rice	2840	24	47	164.4	30	13
– Korea, DPR	22.6	Maize/Rice	1980	—	30	10.8	19	48
– Korea, Republic	45.3	Rice	3160	100	9	0.4	1	1
– Mongolia	2.5	Wheat	1920	86	74	1.2	27	48
South Asia								
– Bangladesh	120.6	Rice	2080	43	115	44.0	42	37
– India	950.0	Rice	2470	29	115	204.4	38	22
– Nepal	21.8	Rice	2320	16	114	4.6	46	21
– Sri Lanka	18.1	Rice	2290	63	19	4.6	22	25
Southeast Asia								
– Cambodia	10.2	Rice	2050	19	174	3.4	62	33
– Indonesia	200.4	Rice	2900	59	75	11.5	26	6
– Laos	4.9	Rice	2060	18	134	1.6	32	33
– Malaysia	20.5	Rice	2940	94	13	0.4	4	2
– Myanmar	43.4	Rice	2850	43	150	2.8	19	7
– Philippines	69.9	Rice	2360	75	53	15.6	27	22
– Thailand	59.2	Rice	2350	96	32	14.3	28	24
– Vietnam	75.1	Rice	2470	21	45	14.1	33	19

Box 11-2. Bangladesh: Food Insecurity in an Agrarian Nation

Malnutrition remains endemic in Bangladesh, an overwhelmingly agrarian country where most rural households do not own land and have few other opportunities to earn wage income. At barely 2,000 kilocalories per person per day, food availability falls short of meeting basic requirements. With extensive poverty, malnutrition, inadequate sanitation, and inadequate access to health care, the country is vulnerable to outbreaks of infectious, water-borne, or other types of diseases. Less than half the population of Bangladesh currently has access to adequate sanitation. Some areas of the country still face the risk of famine; others have frequent floods and often are devastated by cyclones and storm surges.

Overall, the rate of undernourishment is very high (37%), as is the prevalence of underweight, stunting, and wasting among children (Figure 11-9). Rates are high throughout the rural areas that are home to 80% of Bangladesh's population. More than 60% of rural households are functionally landless, and there are limited opportunities for income diversification (Mimura and Harasawa, 2000). The level of vulnerability is likely to increase as a result of severe land degradation, soil erosion, lack of appropriate technology, and the threat of sea-level rise from global warming. Climate change could result in a decreased supply of water and soil moisture during the dry season, increasing the demand for irrigation while supply drops. Improving irrigation efficiency and agricultural productivity will help make Bangladesh self-sufficient in crop production and reduce malnourishment. Higher yields may enable the country to store food supplies to carry it through low-harvest years (Azam, 1996). A switch to growing higher value crops and expansion of free market reforms in agriculture may enable Bangladesh to sell more crops for export. Diversification should help in providing robustness to withstand climate change and variability.

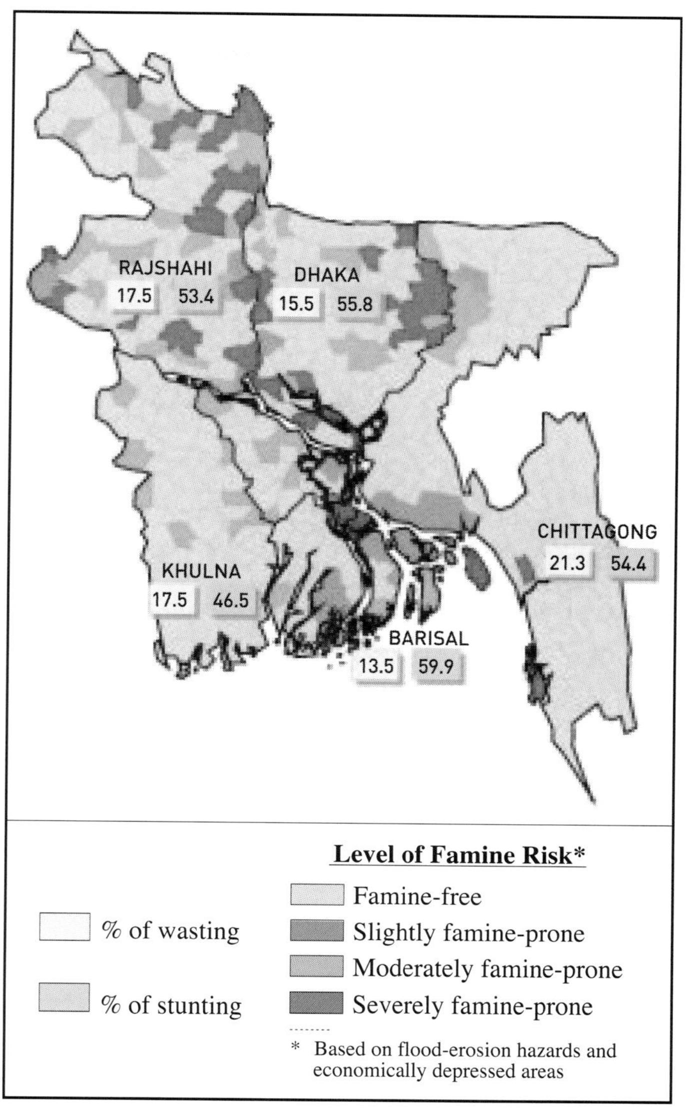

Figure 11-9: Food insecurity and malnutrition in Bangladesh (FAO, 1999b).

growing requirement of agricultural crops. The potential impacts of climate change on agriculture in Asia are crucial because of agriculture's ultimate role in providing food and fiber to Asia's human population. Countries in south and southeast Asia have shown strong reductions in undernourished population in the 1990s. Even so, it has been estimated that almost two-thirds (more than 500 million) of undernourished people live in developing countries of Asia and the Pacific. South Asia accounts for more than one-third of the world total (258 million). Another 64 million undernourished people live in southeast Asia; more than 160 million live in China (FAO, 1999b). The undernourished population almost doubled between 1995 and 1999—from 6 to 12%—as a result of the economic crisis in Indonesia. With the highest incidences of undernutrition and a very large population of children under the age of five, south Asia accounts for almost half of the world's underweight and stunted children (see Box 11-2). Table 11-6 lists the prevalence of undernourishment in Asian developing countries.

Ongoing studies on crop productivity in relation to global warming cover not only biophysical aspects but also socioeconomic drivers and consequences (Fischer *et al.*, 1995; Islam, 1995). The economic impacts of climate change on world agriculture are expected to be relatively minor because decreasing food production in some areas will be balanced by gains in others (e.g., Kane *et al.*, 1991; Tobey *et al.*, 1992; Rosenzweig and Parry, 1993). Such findings however, should be viewed as aggregate results that mask crucial differences in inter-country and intra-country production impacts and the distribution of food resources. In Asia, where rice is one of the main staple foods, production and distribution of rice-growing areas may be affected substantially by climate change. Disparity between rice-producing countries is already visible, and it is increasingly evident between developed and developing countries (Fischer *et al.*, 1996). The projected decline in potential yield and total production of rice in some Asian countries because of changes in climate and climate variability would have a significant effect on trade in agricultural commodities, hence on economic growth and stability (Matthews *et al.*, 1995b).

Increasing population growth and changing dietary patterns in Asia have resulted in more and more land moving from forests and grasslands into agricultural production. Regardless of the increased use of chemical fertilizers and pesticides, in addition to changes in irrigation practices and improved seed stock, yields for major cereal crops have stagnated in many Asian countries during recent years (Iglesias *et al.*, 1996; Sinha, 1997); further intensification of agriculture on area in cropland is certain, and conversion of more land to agricultural use is likely, especially in the developing countries of Asia. Both actions will have far-reaching implications with regard to increased soil erosion, loss of soil fertility, loss of genetic variability in crops, and depletion of water resources (Sinha *et al.*, 1998). Soil degradation is seemingly irreversible unless remedied through painstaking reconstruction of soil health.

A clear understanding of the relationship between climatic variability, crop management, and agricultural productivity is critical in assessing the impacts of climatic variability and change on crop production, the identification of adaptation strategies and appropriate management practices, and the formulation of mitigating measures to minimize the negative effects of climatic variability (including extreme events) on agricultural productivity. In the future, food security will be at the top of the agenda in Asian countries because of two emerging events: growing population, and many direct and indirect effects of climate change. Greatly enhanced efforts to understand the relationship between key climate elements and agriculture should provide a sound basis for meeting the challenges of optimizing the benefits of changing climatic resources.

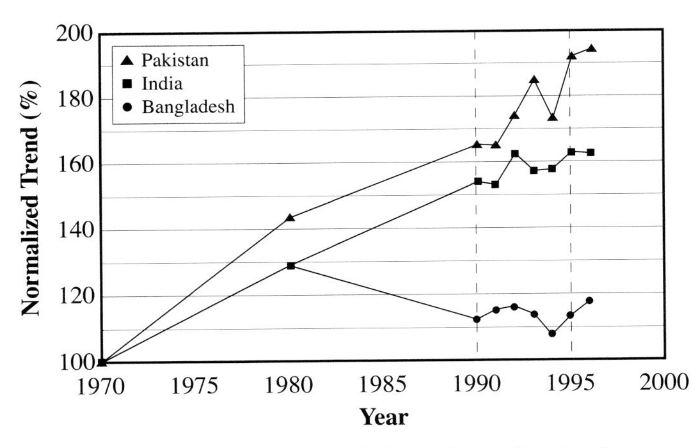

Figure 11-10: Normalized trends in grain production in Bangladesh, India, and Pakistan since 1970 (CIA, 1998).

In some Asian countries, the pace of food grain production has slowed in recent years as a result of depletion of soil nutrients and water resources, creation of salinity and waterlogging, resurgence of pests and diseases, and increased environmental pollution (Gadgil, 1995). Many natural as well as environmental factors—such as extremely dry or cold climates, erratic rainfall, storms and floods, topsoil erosion and severe land degradation, and poor investment and lack of appropriate technology—have played limiting roles in the agricultural potential of most developing countries of Asia (see also Section 5.3). For example, food grain production in Pakistan and India has continued to increase since the 1970s while it has stagnated in Bangladesh (Figure 11-10), largely because of increased losses to climate extremes and land degradation. In India, the estimated total requirement for food grains would be more than 250 Mt by 2010; the gross arable area is expected to increase from 191 to 215 Mha by 2010, which would require an increase of cropping intensity to approximately 150% (Sinha *et al.*, 1998). Because land is a fixed resource for agriculture, the need for more food in India could be met only through higher yield per units of land, water, energy, and time—such as through precision farming. To ensure food security in the developing countries of south and southeast Asia, it is necessary to expand agricultural production, develop the food distribution system, and promote nutrition education, as well as expand the economy and adjust the distribution of incomes.

11.2.3. Hydrology and Water Resources

11.2.3.1. Water Availability

One-third of the world's renewable water resources (13,500 km^3 yr^{-1} out of 42,700 km^3 yr^{-1}) are concentrated in Asia (Shiklomanov, 2001). Water availability varies widely across the different regions of Asia, however—from 77,000 m^3 yr^{-1} per capita to less than 1,000 m^3 yr^{-1} per capita. Table 11-7 provides available information on renewable local water resources, water availability, and water use dynamics by natural-economic region of Asia. The major share of Asian water resources (72%) is located in four countries: Russia (3,107 km^3 yr^{-1}), China (2,700 km^3 yr^{-1}), Indonesia (2,530 km^3 yr^{-1}), and India (1,456 km^3 yr^{-1}). Water resources of the two largest river systems of Asia—the Ganges (with Brahmaputra and Meghna) and the Yangtze (Changjiang)—attain 1,389 km^3 yr^{-1} (794 x 10^3 m^3 yr^{-1} km^{-2}) and 1,003 km^3 yr^{-1} (554 x 10^3 m^3 yr^{-1} km^{-2}). Figure 11-11 depicts the spatial distribution of currently estimated annual mean surface runoff over Asia. The runoff distribution within a year in Asia is most uniform in the rivers of southeast Asia. Over the rest of the continent, more than half of the annual runoff is discharged during the three summer months.

Water availability—in terms of temporal as well as spatial distribution—is expected to be highly vulnerable to anticipated climate change. Growing populations and concentration of population in urban areas will exert increasing pressures on water availability and water quality. As reported in IPCC (1998), runoff generally is expected to increase in the high latitudes and near the equator and decrease in the mid-latitudes under anticipated climate change scenarios.

An assessment of implications of climate change for global hydrological regimes and water resources, using climate change scenarios developed from Hadley Centre model simulations (Arnell, 1999), allows examination of the potential impacts on Asia. A macro-scale hydrological model was used to simulate river flows across the globe at a spatial resolution of 0.5°x0.5°, covering regions of 1,800–2,700 km². The study suggests that average annual runoff in the basins of the Tigris, Euphrates, Indus, and Brahmaputra rivers would decline by 22, 25, 27, and 14%, respectively by the year 2050. Runoff in the Yangtze (Changjiang) and Huang He Rivers have the potential to increase as much as 37 and 26%, respectively. Increases in annual runoff also are projected in the Siberian rivers: the Yenisey (15%), the Lena (27%), the Ob (12%), and the Amur (14%). Areas with particularly large percentage change in high flows include temperate Asia. Significant changes in monthly runoff regimes also are projected over most of Asia.

Some areas of the Asian continent are expected to experience increases in water availability; other areas will have reduced water resources available. Surface runoff is projected to decrease drastically in arid and semi-arid Asia under climate change scenarios and would significantly affect the volume of water available for irrigation and other purposes. Sensitivity

***Table 11-7**: Renewable local water resources, water availability, and water-use dynamics, by natural-economic regions of Asia (Shiklomanov, 2001).*

Region	Area (10^6 km²)	Water Resources (km³ yr⁻¹)	Potential Water Availability (10^3 m³ yr⁻¹)		Dynamics of Freshwater Use (km³ yr⁻¹)[a]						Water Use as a Percentage of Water Resources[a]	
			per km²	per capita	1900	1950	1995	2000 (Forecast)	2010 (Forecast)	2025 (Forecast)	1995	2025
Siberia and far east of Russia	12.76	3107	243	76.6	0.7 / 0.4	5.6 / 1.3	30.6 / 15	30 / 15	32 / 17	38 / 21	1.0 / 0.5	1.2 / 0.6
North China and Mongolia	8.29	1029	124	2.13	37 / 30	98 / 75	254 / 182	273 / 185	305 / 194	373 / 210	24.7 / 17.7	36.2 / 20.4
South Asia	4.49	1988	443	1.77	201 / 160	367 / 293	932 / 687	969 / 710	1060 / 767	1370 / 944	43.6 / 32.1	64.1 / 44.2
Southeast Asia	6.95	6646	956	4.77	99 / 77	230 / 170	525 / 388	551 / 393	617 / 413	781 / 425	7.8 / 5.8	11.6 / 6.3
Western Asia	6.82	490	71.8	2.11	43 / 34	91 / 71	238 / 174	248 / 181	283 / 201	346 / 229	48.5 / 35.5	70.6 / 46.7
Central Asia and Kazakhstan	3.99	181	45.4	3.78	29 / 19	57 / 37	154 / 102	151 / 102	160 / 110	169 / 122	75.5 / 50.0	82.8 / 59.8
Transcaucasia	0.19	67.9	358	4.63	4.2 / 2.1	11.4 / 7.1	23.7 / 17.5	23 / 17	26 / 19	27 / 20	32.0 / 23.7	36.5 / 27.0
Total	43.50	13510	311	3.92	414 / 322	860 / 654	2157 / 1565	2245 / 1603	2483 / 1721	3104 / 1971	16.0 / 11.6	23.0 / 14.6

[a]Nominator = total water withdrawal; denominator = water consumption.

analysis of water resources in Kazakhstan to projected climate change scenarios indicates that surface runoff would be substantially reduced as a result of an increase in surface air temperature of 2°C accompanied by a 5–10% decline in precipitation during summer (Gruza *et al.*, 1997). In temperate Asia, future changes in surface runoff would be highly spatially inhomogeneous. An increase in surface runoff seems likely in Mongolia and northern China. The hydrological characteristics

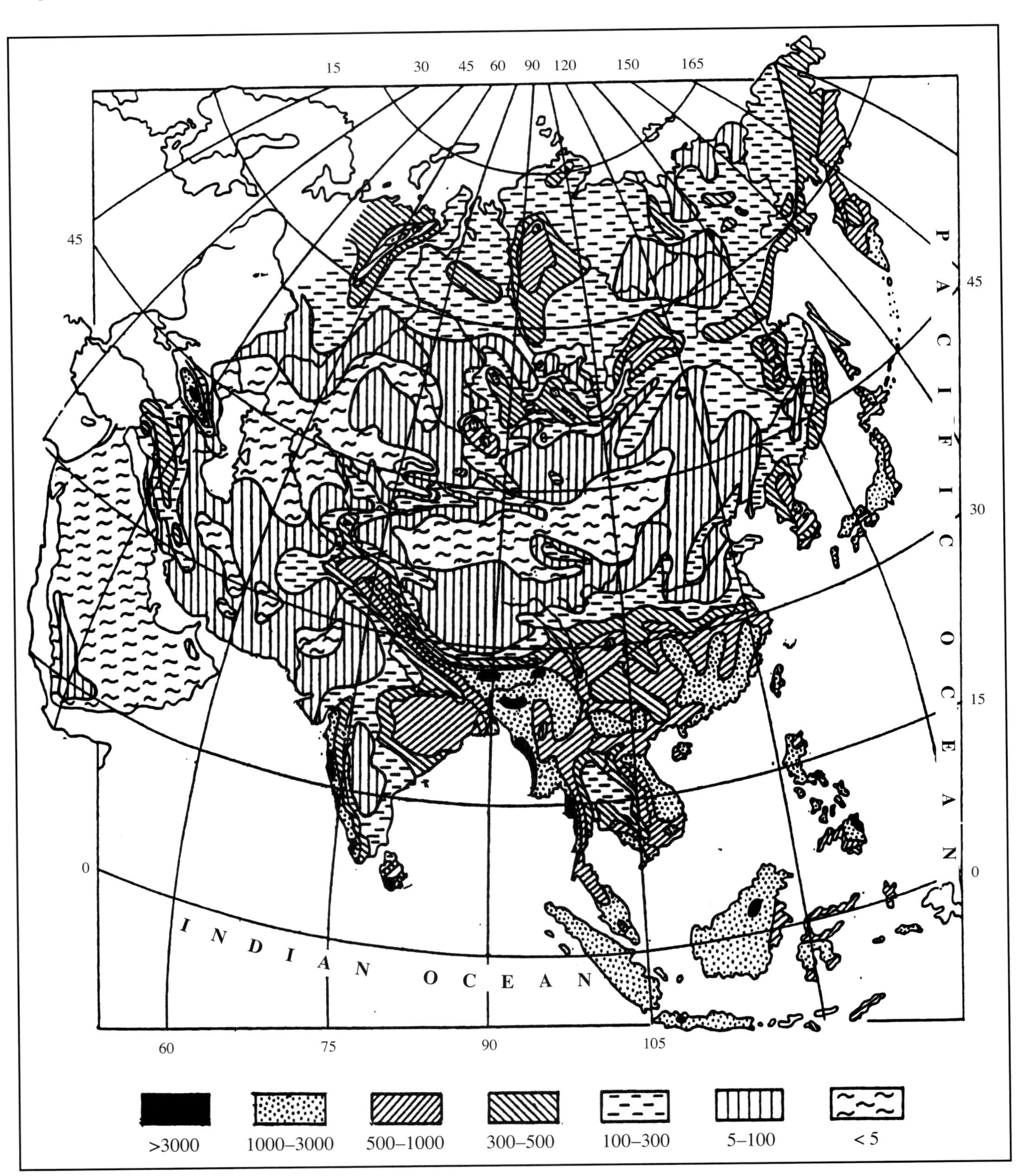

Figure 11-11: Spatial distribution of currently estimated surface runoff (mm) over Asian continent (Shiklomanov, 2001).

of Japanese rivers and lakes also are sensitive to climate change. Recent studies suggest that, on average, a 3°C increase in temperature coupled with a 10% increase in precipitation will increase river flows by approximately 15% in water-abundant areas. An increase in temperature also accelerates snow melting, which increases river flows from January through March but decreases flows from April through June (Hanaki *et al.*, 1998; Inoue and Yokoyama, 1998).

The perennial rivers originating in the high Himalayas receive water from snow and glaciers. Snow, ice, and glaciers in the region are approximately equivalent to about 1,400 km^3 of ice. The contribution of snow to the runoff of major rivers in the eastern Himalayas is about 10% (Sharma, 1993) but more than 60% in the western Himalayas (Vohra, 1981). Because the melting season of snow coincides with the summer monsoon season, any intensification of the monsoon is likely to contribute to flood disasters in Himalayan catchments. Such impacts will be observed more in the western Himalayas compared to the eastern Himalayas because of the higher contribution of snowmelt runoff in the west (Sharma, 1997). An increase in surface runoff during autumn and a decrease in springtime surface runoff are projected in highland regions of south Asia. The increase in surface temperature also will contribute to a rise in the snowline—which, in effect, reduces the capacity of the natural reservoir. This situation will increase the risk of flood in Nepal, Bangladesh, Pakistan, and north India during the wet season (Singh, 1998). No significant changes are projected for annual mean surface runoff in southeast Asia; an increase during winter and a decrease during summer season is likely, however.

Available data on the dynamics of freshwater use by natural-economic regions of Asia (Table 11-8) suggest that freshwater use—in terms of total water withdrawal and water consumption—have increased significantly in recent decades in all regions and is projected to increase further in the 21st century. Table 11-8 also suggests that water use in most regions of Asia (except Russia and southeast Asia) already has exceeded 20% of the available resources (Arnell, 1999) and will be increasing appreciably by 2025. It follows from this table that water is going to be a scarce commodity in Asia in the near future even without the threat of climate change.

11.2.3.2. Water Needs and Management Implications

At present, approximately 57% of total water withdrawal and 70% of water consumption in the world occurs within the countries of Asia. Table 11-8 presents the dynamics of freshwater use in Asia over the sectors of economic activities. As is evident from this table, agriculture (irrigation in particular) accounts for 81% of total water withdrawal and 91% of water consumption in Asia. The area of irrigated lands in Asia currently amounts to 175 Mha and may increase to 230 Mha by 2025. As Table 11-7 implies, the two most populated regions of south and southeast Asia account for about 68% of water withdrawal and about 69% of water consumption in Asia (see Chapter 4). More than

***Table 11-8**: Dynamics of freshwater use in Asia over sectors of economic activities, km^3 yr^{-1} (Shiklomanov, 2001).*

	Assessment				Forecast		
	1900	1950	1970	1995	2000	2010	2025
Population (million)		1464	2103	3498	3762	4291	4906
Irrigation area (Mha)	36.1	72.5	118	175	182	199	231
Water use[a]							
– Agriculture	408 320	816 643	1331 1066	1743 1434	1794 1457	1925 1553	2245 1762
– Industry	4 1	33 6	107 13	184 30	193 32	248 40	409 58
– Domestic	2 1	11 5	38 14	160 31	177 33	218 36	343 44
– Reservoirs (evaporation)	0	0.23	23	70	81	92	107
Total	414 322	860 650	1499 1116	2157 1565	2245 1603	2483 1721	3104 1971

[a]Nominator = total water withdrawal; denominator = water consumption.

75% of the total water available in India currently is used for irrigation. As much as 20% is required to meet domestic and municipal needs—leaving just 5% for industrial needs. The only river in north India that has surplus water to meet future needs of the country is the Brahmaputra. This river, however, is an international river; other countries such as Bangladesh may not approve of building a dam across some of its tributaries. In peninsular India, only the Mahanadi and Godavari have surplus water, but conveying it to drought-prone areas of the south is problematic. Many states in India need to adopt measures for restricting the use of groundwater to prevent a water famine in the future. China's rapid economic growth, industrialization, and urbanization—accompanied by inadequate infrastructure investment and management capacity—have contributed to widespread problems of water scarcity throughout the country. Of the 640 major cities in China, more than 300 face water shortages; 100 face severe scarcities (UNDP, 1997).

Taking into account projected dynamics of economic development in the temperate, tropical, and arid and semi-arid regions of Asia, combined with the climate change-imposed effect on hydrological regimes, agriculture and the public water supply would require priority attention in these regions to secure sustainable development and avoid potential intersectoral and international water conflicts. Radical changes in water management strategies and substantial investments will be required in Asia to cope with water problems in the 21st century. Adaptation measures will include legal, institutional, and technical initiatives such as modifying existing and constructing new infrastructure (reservoirs, interbasin water transfer schemes), introducing water-saving technologies, upgrading efficiency of irrigation systems, enhancing wastewater recycling systems, introducing low water-use crops, and implementing groundwater protection programs.

At least 14 major international river watersheds exist in Asia. An integrated and decentralized system of restoration and conservation of the water cycle in these drainage basins is vital to mitigate the negative consequences of natural and externally imposed perturbations. Watershed management is challenging in countries where the people-to-land ratio is high and policy and management are inadequate, prompting use of even the most fragile and unsuitable areas in the watersheds for residential, cultivation, and other intensive uses. This is particularly true for countries such as Bangladesh, Nepal, the Philippines, Indonesia, and Vietnam, where many watersheds suffer badly from deforestation, indiscriminate land conversion, excessive soil erosion, declining land productivity, erratic and unreliable surface and groundwater resources, and loss of biodiversity. Many watersheds in Asia already are stressed by intensive use of the land and other resources and by inhospitable climate (especially in arid and semi-arid Asia), beyond their ability to adequately supply water, prevent floods, and deliver other goods and services. In the absence of appropriate adaptation strategies, these watersheds are highly vulnerable to climate change. Global climate change also may have serious water management implications on the territory of boreal Asia. Recent assessments (Izrael *et al.*, 1997a; CAFW, 1998) for all major Siberian rivers (Ob, Yenisei, Lena) with 42% of the total freshwater inflow to the Arctic ocean show that the main water management problems by the year 2050 will be a consequence of significant annual runoff increases (up to 20%) and difficulties with seasonal inundation and flood control measures.

11.2.4. Oceanic and Coastal Ecosystems

11.2.4.1. Oceans and Coastal Zones

In line with global trends, more than half of the region's population—1.7 billion people—presently resides in the coastal zone of Asia (Middleton, 1999). Given the relentless and cumulative process of global environmental change driven by, among other factors, demographic changes, urbanization and industrial development, trade and transport demands, and lifestyle changes, the coastal zones of Asia are under increasing anthropogenic pressures (Turner *et al.*, 1996). The consequences of this process pose a significant threat to environmental and socioeconomic systems located in Asian coastal zones. All coastal areas are facing an increasing range of stresses and shocks, the scales of which now pose a threat to the resilience of human and environmental coastal systems and are likely to be exacerbated by climate change.

Climate impacts on coastal zones of Asia include accelerated sea-level rise and more frequent and severe storm events. Large populations in coastal areas of southeast and south Asia are vulnerable to pressure from unsustainable resource use and environmental degradation (e.g., Dow, 1999). These areas already are subject to numerous climate change-related stresses, including loss of coastal mangroves that act as major environmental determinants of coastal fisheries (Daily, 1997; Field *et al.*, 1998; Primavera, 1998). Major delta areas of Asia are likely to be subjected to stresses associated with sea-level rise, changes in water regimes, saltwater intrusion, siltation, and land loss (see Chapter 6). Low-lying coastal cities will be at the forefront of impacts; these cities include Shanghai, Tianjin, Guangzhou, Jakarta, Tokyo, Manila, Bangkok, Karachi, Mumbai, and Dhaka—all of which have witnessed significant environmental stresses in recent years. Jakarta, Bangkok, and Tianjin, for example, have experienced changes in relative sea level of as much as 5 cm yr^{-1} during the 1980s and 1990s as a result of subsidence associated with groundwater withdrawal (ESD-CAS, 1994; Nicholls, 1995). In addition, increases in temperature can lead to increased eutrophication in wetlands and freshwater supplies.

Tropical Asia experiences the impact of present-day climate variability associated with ENSO, therefore is more prone to changes in ENSO-related impacts with global climate change. The ENSO phenomenon is a major cause of year-to-year variability in the number of cyclones in the Asia Pacific region (Li, 1987; Nishimori and Yoshino, 1990; Lander, 1994). The numbers are higher during La Niña events, particularly late in the season over southeast Asia (Kelly and Adger, 2000). The 1982–1983 El Niño caused a decline in rainfall and associated impacts over large areas, including parts of Indonesia (Salafsky, 1994, 1998; Glantz, 1996). El Niño events are known to have significant impacts on coastal areas and water resources in southeast Asia through decreased precipitation in El Niño years and increased frequency of typhoons in subsequent La Niña years (Kelly and Adger, 2000).

Sea levels of interior seas in arid and semi-arid Asia have dramatically changed, driven either by climatic or anthropogenic factors. The level of the Caspian Sea lowered by about 4 m from 1930 to 1977; since 1978 it has begun rising (Mikhailov, 1998). The coastline also propagated landward by 20–40 km; 2 Mha of farming land and more than 100 oil wells have been inundated. Many cities in the coastal zone, such as Makhachkala and Kaspisk, are seriously damaged (Svitoch, 1997). The Volga delta propagated 17 km during the period of lowering sea level but has retreated 12–15 km during the period of the rise in sea level (Kasimov, 1997; Tian and Liu, 1999). Caspian Sea level fluctuations are attributed mainly to the Volga water discharge controlled by climate anomalies in its catchment area (Malinin, 1994). The Aral Sea area also reduced, from 60,000 to 28,500 km^2, in the period 1960–1989; the exposed seabed (particularly deltas) now have frequent dust and salt storms, and there are shortages of freshwater (Ellis, 1990; Tian and Liu, 1999).

11.2.4.2. Deltas, Estuarine, and Other Coastal Ecosystems

River-borne sediments have formed at least 10 deltas in the coastal zones of Asia with an area of more than 10,000 km^2 each (Coleman and Wright, 1975). Delta and estuarine ecosystems are sensitive to complex responses to agents associated with climate change (Sanchez-Arcilla and Jimenez, 1997). Low-lying deltas are especially vulnerable to sea-level rise and increasing shoreline wave action (Walker, 1998). A decrease in river water discharge, as projected under some climate change scenarios, could lead to hindrance of delta progradation and increase the risk of irreversible change for the ecosystem in estuarine-deltaic areas (Qian, *et al.*, 1993; Shi, 1995). Tidal rivers and estuaries will become more prone to saltwater intrusion as a result of projected sea-level rise (Huang, *et al.*, 1982; Li, 1984, 1985; Shi, 1995). Sea-level changes associated with global warming would be exacerbated by tectonic submergence, ground subsidence as a result of groundwater withdrawal, rise of water level created by delta progradation, and eustatic sea-level rise.

Low-lying muddy coastlines associated with large deltas form a significant resource and support large human populations. In China, for example, such low-lying deltas cover about 4,000 km—22% of the total coastline. These muddy coastal ecosystems are basically distributed along large deltas and partly in semi-closed bays (Ren, 1985). Erosion of muddy coastlines in Asia—as documented in China, for example—is triggered largely by sediment starvation resulting from human activities and delta evolution rather than sea-level rise (Ji *et al.*, 1993; Chen and Chen, 1998). Beach erosion is widespread in the coastal zone of Asia and has been reported in China, Japan, Indonesia, Sri Lanka, Thailand, Bangladesh, and Malaysia (Xia *et al.*, 1993; Sato and Mimura 1997; Teh 1997; Nishioka and Harasawa, 1998; Huq *et al.*, 1999; Middleton, 1999).

Coastal wetlands frequently are associated with deltas, estuaries, lagoons, and sheltered bays. Tidal flats of the muddy coast in Asia constitutes the main part of the coastal wetland (Bird, 1992). Large-scale wetland reclamation in the major deltas has taken place during the past few decades (Lang *et al.*, 1998; Liu *et al.*, 1998). Lagoons, which are important wetlands, are located across the coastal regions of India, Sri Lanka, Malaysia, Indonesia, China, and Russia. The transitional area between uplift and subsidence belts has favored the formation of lagoons along China's coastline (Li and Wang, 1991). These lagoons tend to be decreasing in area as a result of silting of sediments and plant growing (Zhu, 1991). Sea-level rise and reduction of river-borne sediments will decelerate delta progradation and wetland renewal. The rich biodiversity of wetlands in Asia is seriously threatened by loss of wetlands from sea-level rise (Nicholls *et al.*, 1999).

Mangroves are made up of salt-adapted evergreen trees; they are restricted to the intertidal zone along the vast coastlines of tropical countries in Asia and extend landward along tidal rivers. The Sundarbans in Bangladesh and adjacent areas in India, covering about 6,000 km^2, are the largest mangrove forests in the world (Allison, 1998). Depletion of mangrove forests by anthropogenic pressures has become a serious problem (Farnsworth and Ellison, 1997). Approxmately half of mangrove forests in Thailand were reduced by 56% during 1961–1996. In the Philippines, more than 75% of the mangrove forests have been lost in less than 70 years. Destruction of Indonesia's estimated 44,000 km^2 of mangroves has taken place mainly since 1975 (Middleton, 1999; UNEP, 1999). Mangrove forests are highly vulnerable to climate change-induced sea-level rise because it will change the salinity distribution and hence productivity. Large-scale changes in species composition and zonation in mangrove forests also are expected as a result of changes in sedimentation and organic accumulation, the nature of the coastal profile, and species interaction (Aksornkaoe and Paphavasit, 1993).

11.2.4.3. Coral Reefs

Coral reefs play a crucial role in fishery production and in protecting the coastline from wave action and erosion (Ruddle *et al.*, 1988; Middleton, 1999). Southeast Asia has almost one-third of the world's mapped coral reefs (Pennisi, 1997); these reefs extend to the northern extreme range in Japan (Nishioka and Harasawa, 1998). Coral reef productivity is a function of their structure, biological recycling, and high retention of nutrients. Reefs in Indonesia and the Philippines are noted for extraordinarily high levels of biodiversity: Each contains at least 2,500 species of fish. Severe coral bleaching can occur as a result of seawater warming and clear skies (resulting in higher incident solar radiation).

Major coral bleaching events have occurred in 1983 (Japan, Indonesia), 1987 (Maldives), 1991 (Thailand, Japan), 1995 (Thailand, Philippines), and 1998 (Maldives, Sri Lanka, India, Indonesia, Thailand, Japan, Malaysia, Philippines, Singapore, Vietnam, Cambodia). As a result of the major 1998 coral bleaching in the south Asia region, many reefs dominated by branching species have been severely damaged, with high mortality of these species. In coastal seas around the Maldives, Sri Lanka, the Andaman Islands of India, and Japan, reef community structure has switched from dominance by fast-growing branching species to monopolization by the more physically rigorous and slow-growing massive corals (Wilkinson, 1998). Deforestation in many island countries of Asia and quarrying of live corals for manufacture of calcium carbonate have led to significant coral decline or severe damage to the entire ecosystem.

Studies show that a moderate rise in sea level around the coast of Thailand would stimulate the growth of coral reef flats and extend corals shoreward. The enhanced growth potential is likely to be restricted by human infrastructure and development along the coast (Chansang, 1993). Asia's coral reefs are undergoing rapid destruction in terms of habitat richness (Cesar *et al.*, 1997; Nie *et al.*, 1997; Pennisi, 1998) as a result of several factors, including extreme temperatures and solar irradiance, subaerial exposure, sedimentation, freshwater dilution, contaminants, and diseases (Glynn, 1996). Virtually all of the Philippines' reefs and approximately 83% of Indonesia's reefs are at risk from destructive fishing techniques, reef mining, sedimentation, and marine pollution (Middleton, 1999; UNEP, 1999). The increase in atmospheric CO_2 concentration (resulting in higher $CaCO_3$ concentrations in seawater) and consequent rise in sea surface temperature (SST) is likely to have serious damaging effects on reef accretion and biodiversity.

11.2.4.4. Fisheries and Aquaculture

Asia dominates world aquaculture, producing four-fifths of all farmed fish, shrimp, and shellfish (FAO, 1997). Farming of fish, shrimp, shellfish, and seaweeds has become a vital source of food supply in Asia in recent decades. Fishery products are staples for the Asian population and are embedded in its culture. Fish, an important source of food protein, is critical to food security in many countries of Asia, particularly among poor communities in coastal areas. The annual fish catch and aquaculture production in Asia reached a peak at about 20.7 and 19.1 Mt, respectively, in the year 1998. Japan has the largest distant-waters fishery production. Inland fishery production is dominated by China and India, which have shown increases in recent years as a result of stock enhancement practices.

Fish farming requires land and water—two resources that already are in short supply in many countries in Asia. Nearly half of the land now used for shrimp ponds in Thailand was formerly used for rice paddies; water diversion for shrimp ponds has lowered groundwater levels noticeably in coastal areas of Thailand. In China, concern over the loss of arable land has led to restrictions on any further conversion of farmland to aquaculture ponds. Intensive production systems and large-scale facilities used to raise high-value shrimp, salmon, and other premium species has taken a heavy toll on coastal habitats, with mangrove swamps in southeast Asia being cleared at an alarming rate. Thailand lost more than 15% of its mangrove forests to shrimp ponds from 1987 to 1993 (World Bank, 1996). Destruction of mangroves has left these coastal areas exposed to erosion and flooding, altered natural drainage patterns, and increased salt intrusion.

The fishery resources of Japan, China, and many other countries of Asia are being depleted by overfishing, excessive use of pesticides, industrial pollution, red tide, and even construction of dikes and other coastal structures (Zou and Wu, 1993; Sato and Mimura, 1997). Loss of inshore fish nursery habitats to coastal development, as well as pollution from land-based activities, causes significant change to ecosystems supporting fisheries (see also Chapter 6). Marine productivity is greatly affected by temperature changes that control plankton shift, such as seasonal shifting of sardine in the Sea of Japan and induced during the cyclical occurrence of the ENSO in low latitudes (Chen and Shen, 1999; Piyakarnchana, 1999; Terazaki, 1999). The impact of global warming on fisheries will depend on the complicated food chain, which could be disturbed by sea-level rise, changes in ocean currents, and alteration of mixing layer thickness.

Anomalies in the water temperatures of major oceanic currents (e.g., declines in sardine catch in the Sea of Japan associated with changing patterns of the Kuroshio current in ENSO years) have resulted in low commercial fish catch in recent years (Yoshino, 1998b). The steady wintertime decrease in mean wind speed observed over the Sea of Japan between 1960 and 1990 has accelerated surface temperature increase and stagnated bottom water formation in recent years (Varlamov *et al.*, 1997). The rise in SST will shift the southern limit of salmon species further to the north (Seino *et al.*, 1998). It is also suggested that the Sea of Japan bottom water will become anoxic within a few hundred years; a decreased oxygen supply will lead to major losses in biological productivity in deep waters (Gamo, 1999). Fish production of certain species may decrease because of the decline of river volume (Zhou, 1991). Increases in marine culture products and declines in marine fishery output are current trends in most south Asian countries that are engaged in commercial fishery activity. It is likely that increased surface runoff and higher nutrient load might lead to potentially beneficial increases in plankton within the coastal zone of boreal Asia. However, increased frequency of El Niño events, which are likely in a warmer atmosphere, could lead to measurable declines in plankton biomass and fish larvae abundance in coastal waters of south and southeast Asia (see also Chapter 6). Such declines in lower levels of the food chain will have negative impacts on fisheries in Asia.

11.2.4.5. Tropical Cyclone and Storm Surges

Asia is close to the warm pool of the west equatorial Pacific Ocean, and tropical cyclones and associated storm surges strongly affect coastal zones of tropical and temperate Asia. Tropical cyclones and storm surges are one of most critical factors affecting loss of human lives in India and Bangladesh (Sato and Mimura, 1997). Approximately 76% of the total loss of human lives from cyclonic storms has occurred in India and Bangladesh (Ali, 1999). Several Asian countries are faced with cyclones and associated storm surges every year, which causes serious economic losses (Ali, 1999; Huang, 1999; Kelly and Adger, 2000).

There is concern that global warming may affect tropical cyclone characteristics, including intensity, because SST plays an important role in determining whether tropical disturbances form and intensify. Several researchers have used modeling techniques to examine the possible effects of global warming on tropical storms (Lighthill *et al.*, 1994; Sugi *et al.*, 1996; Henderson-Sellers and Zhang, 1997; Holland, 1997; Tonkin *et al.*, 1997; Henderson-Sellers *et al.*, 1998; Knutson *et al.*, 1998; Krishnamurti *et al.*, 1998; Royer *et al.*, 1998). Lighthill *et al.* (1994) conclude that there is no reason to expect any overall change in global tropical cyclone frequencies, although substantial regional changes may occur. Recent studies indicate that the maximum potential intensities of cyclones will possibly undergo a modest increase of as much as 10–20% in a warmer atmosphere (see Chapter 3 and TAR WGI Chapter 9). More recent analyses (Nakagawa *et al.*, 1998; Walsh and Pittock, 1998; Jones *et al.*, 1999) support the possibility of an increase in cyclone intensity. Coastal erosion in Asia should increase with sea-level rise, and storm surges could still exacerbate hazards, even if the number and intensities of tropical cyclones do not change (IPCC, 1998; Walsh and Pittock, 1998).

11.2.4.6. Potential Impacts and Coastal Zone Management

As outlined in IPCC (1998), climate-related stresses in coastal areas include loss and salinization of agricultural land resulting from changes in sea level, likely changes in the intensity of tropical cyclones, and the possibility of reduced productivity in coastal and oceanic fisheries. Table 11-9 lists estimates of potential land loss resulting from sea-level rise and the number of people exposed, assuming no adaptation (Mimura *et al.*, 1998; Nicholls and Mimura, 1998). These estimates of potential land loss and populations exposed demonstrate the scale of the issue for the major low-lying regions of coastal Asia. The results are most dramatic in Bangladesh and Vietnam, where 15 million and 17 million people, respectively, could be exposed given a relative change in sea level of 1 m (Brammer, 1993; Haque and Zaman, 1993)—though it should be recognized that a 1-m sea-level rise is at the extreme range of presently available scenarios. Nonetheless, these examples demonstrate the sensitivity of coastal areas to climate change impacts and unsustainable utilization of resources in these areas. The impacts could be exacerbated by continued population growth in low-lying agricultural and urban areas (Nicholls *et al.*, 1999). At the same time, adaptation strategies will alter the nature of the risk and change the socially differentiated nature of the vulnerability of populations living in hazardous regions. Response strategies that are based solely on tackling the physical parameters of risks from sea-level rise and tropical cyclones have been shown in some circumstances to enhance the vulnerability of certain parts of the population—usually those with least ability to influence decisionmaking (Blaikie *et al.*, 1994; Hewitt, 1997; Mustafa, 1998; Adger, 1999b).

Human activities, including protection facilities themselves, aggravate the vulnerability of the coastal regions to climate change and sea-level rise. There are complex interrelationships and feedbacks between human driving forces and impacts, on one hand, and climate- and sea level-induced changes and effects on the other (IPCC, 1996). At the interface between ocean and terrestrial resources, coastal ecosystems undergo stress from competing multi-usage demands, while having to retain their functional diversity and resilience in the face of global environmental change (Bower and Turner, 1998). To enhance coastal resilience and facilitate adaptation, integrated management of coastal zones must take into account the multiple resource demand and variety of stakeholders, as well as natural variability, recognizing the importance of the institutional, cultural, and historical context (Klein and Nicholls, 1999).

***Table 11-9**: Potential land loss and population exposed in Asian countries for selected magnitudes of sea-level rise and under no adaptation measures (modified from Nicholls and Mimura, 1998; Mimura et al., 1998).*

Country	Sea-Level Rise (cm)	Potential Land Loss (km^2)	Potential Land Loss (%)	Population Exposed (millions)	Population Exposed (%)
Bangladesh	45	15,668	10.9	5.5	5.0
	100	29,846	20.7	14.8	13.5
India	100	5,763	0.4	7.1	0.8
Indonesia	60	34,000	1.9	2.0	1.1
Japan	50	1,412	0.4	2.9	2.3
Malaysia	100	7,000	2.1	>0.05	>0.3
Pakistan	20	1,700	0.2	n.a.	n.a.
Vietnam	100	40,000	12.1	17.1	23.1

n.a. = not available.

Integrated coastal zone management (ICZM) is an iterative and evolutionary process for achieving sustainable development by developing and implementing a continuous management capability that can respond to changing conditions, including the effects of climate change (Bijlsma *et al.*, 1996). Essentially, ICZM is a cooperative effort on the part of coastal zone stakeholders that results in a "win-win" outcome. ICZM already has been developed and implemented in some Asian countries for the allocation of environmental, sociocultural, and institutional resources to achieve conservation and sustainable multiple use of the coastal zone (Sato and Mimura, 1997). Since the 1970s, the Philippines has formulated programs and projects on coastal management, covering fishery and mangrove reforestation (Perez *et al.*, 1999). Mangrove rehabilitation has been recommended to mitigate climate impacts in coastal zones of Vietnam (Tri *et al.*, 1998). Since the 1960s, a groundwater withdrawal/pumping-back system has been carried out to mitigate ground subsidence in Shanghai and Tianjing. Fishing in certain seasons has been banned and the annual quality of fish catch restricted in the coastal zone of China (ESD-CAS, 1994). Sri Lanka conserves coastal tourism resources by using ICZM principles (White *et al.*, 1997). Coastal natural conservation parks have been established in Bangladesh, Thailand, China, and other countries (Sato and Mimura, 1997; Allison, 1998). However, land ownership and management responsibility issues in Bangladesh have inhibited coastal zone management. Similarly, privatization and decentralization of storm protection systems in Vietnam have created a vacuum for strategic management and increased the potential impacts of climate variability (Adger, 1999b, 2000).

Given that many potential climate change impacts on coastal zones feature irreversible effects, surprise outcomes, and unpredictable changes, the appropriate policy response should be to maximize flexibility and enhance the resilience and adaptation potential of these areas (Pritchard *et al.*, 1998). By contrast, coastal management in Asia to date more often than not has been dominated by policies that have sought to buffer socioeconomic activities and assets from natural hazards and risks via hard engineering protection (Chua, 1998).

11.2.5. Human Health

11.2.5.1. Thermal Stress and Air Pollution-Related Diseases

A remarkable increase in the number of heatstroke patients and mortality was observed when maximum daily temperatures in Nanjing, China, exceeded 36°C for 17 days during July 1988 (Ando *et al.*, 1996). In Japan as well, intense heat in summer is now becoming more common in large cities. The numbers of heatstroke patients are reported to have increased exponentially with temperatures of more than 31°C in Tokyo, Japan, and 35°C in Nanjing, China. Beyond a maximum daily temperature of 36°C, a remarkable increase in the number of heatstroke patients and deaths caused by heatstroke has been observed (Honda *et al.*, 1995; Ando, 1998). Typical hyperthermia occurs during hot days and during physical exercise in the summer (Nylen *et al.*, 1997).

Climate change will have a wide range of health impacts all across Asia (see Chapter 9). Although a reduction in health stresses and wintertime deaths is anticipated as a result of less frequent occurrence of extreme cold temperatures in boreal and temperate Asia, an increase in the frequency and duration of severe heat waves and humid conditions during the summer will increase the risk of mortality and morbidity, principally in older age groups and urban poor populations of temperate and tropical Asia (Epstein *et al.*, 1995). Heat stress-related chronic health damages also are likely for physiological functions, metabolic processes, and immune systems (Bouchama *et al.*, 1991; Ando, 1998).

Adverse health impacts also result from the build-up of high concentrations of air pollutants such as nitrogen dioxide (NO_2), ozone, and air-borne particulates in large urban areas. Combined exposures to higher temperatures and air pollutants appear to be critical risk factors for cerebral infarction and cerebral ischemia during the summer months (Piver *et al.*, 1999). As summarized in Figure 11-12, the heat index (a combination of daily mean temperature and relative humidity) and the concentration of NO_2 are shown to be significant risk factors for heatstroke in males age 65 years and older residing in Tokyo, Japan (Piver *et al.*, 1999). The number of heatstroke emergency cases per million residents is found to be greater in males than in females in the same age groups. Global warming also will increase the incidences of some diseases, such as respiratory and cardiovascular diseases, in arid and semi-arid, temperate, and tropical Asia.

11.2.5.2. Vector-Borne Diseases

Health impacts secondary to the impacts of climate change on ecological and social systems should include changes in the occurrence of vector-borne infectious diseases in temperate

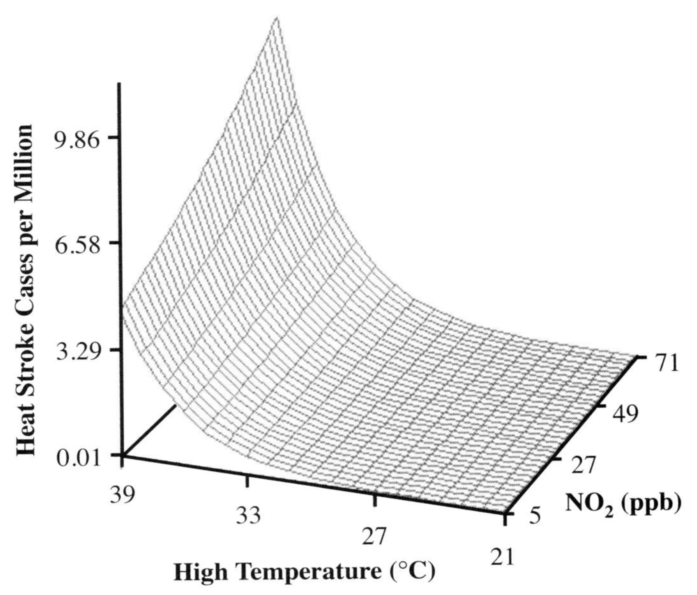

Figure 11-12: Heat stroke morbidity (cases per million; Tokyo, July-August, 1980–1995, males >65 years) (Piver *et al.*, 1999).

and tropical Asia (see Chapter 9). The distribution of diseases such as malaria is influenced by the spread of vectors and the climate dependence of infectious pathogens (Hales *et al.*, 1996; McMichael *et al.*, 1996; Epstein *et al.*, 1998). In recent years, the resistance of anopheline mosquito to pesticides and of malaria parasites to chloroquine has affected eradication activities (Trigg and Kondrachina, 1998). Malaria still is one of the most important diseases in countries in tropical Asia such as India (Bouma *et al.*, 1994; Akhtar and McMichael, 1996; Mukhopadhyay *et al.*, 1997), Bangladesh, Sri Lanka (Bouma and Van der Kaay, 1996; Gunawardena *et al.*, 1998), Myanmar, Thailand, Malaysia (Rahman *et al.*, 1997), Cambodia, Laos, Vietnam (Hien *et al.*, 1997), Indonesia (Fryauff *et al.*, 1997), Papua New Guinea (Genton *et al.*, 1998), and Yunnan, China (Jiao *et al.*, 1997; Xu and Liu, 1997) as a result of the presence of the mosquito vectors and the lack of effective control.

With a rise in surface temperature and changes in rainfall patterns, the distribution of vectors such as mosquito species may change (Patz and Martens, 1996; Reiter, 1998). Changes in environmental temperature and precipitation could expand vector-borne diseases into temperate and arid Asia. The spread of vector-borne diseases into more northern latitudes may pose a serious threat to human health. Climate change is likely to have principal impacts on epidemics of malaria, dengue, and other vector-borne diseases in Asia (Martens *et al.*, 1999). The epidemic areas of vector-borne diseases in Asia would depend on many demographic and societal factors, as well as environmental hygiene for vector control, available health infrastructure, and medical facilities (see also Chapter 9).

11.2.5.3. Diseases Resulting from Higher UV-B Exposures

Depletion of stratospheric ozone that normally filters out ultraviolet radiation in sunlight in the region from 280 to 320 nm (the UV-B region) has been linked to widespread use of volatile halogenated organic compounds, particularly chlorinated and brominated methanes and chlorofluorocarbons. Some of these compounds also are effective GHGs and therefore contribute to global warming as well. The quantitative relationship between UV-B dose and its physiological effect varies with the wavelength of UV-B exposure (Ilyas *et al.*, 1999). These effects include melanoma and non-melanoma skin cancers, cataracts and other ocular diseases, and dysfunction of the systemic and cutaneous immune systems (Kripke, 1994). The known effects of UV-B on the eye include inflammatory reactions from acute exposure, snow blindness (photo-kerato-conjunctivitis), and long-term damage to the cornea and lens (cataracts) from chronic exposure. It has been demonstrated that damage to melanocytes in human skin initiates the progression of changes leading to melanoma skin cancer (Kripke, 1994). Suppression of the immune response by UV-B radiation involves damage to Langerhans cells and subsequent activation of T-lymphocytes, thus increasing the severity of certain infectious diseases.

The impacts of greater exposure to shorter wavelength UV radiation on human health are cumulative and, for some effects, may have long latencies. Noticeable increases in UV-B radiation over high and mid-latitudes as a result of depletion of stratospheric ozone have occurred in recent decades (Mckenzie *et al.*, 1999; WMO, 1999; ACIA, 2000). Climate change could make conditions for the spread of diseases associated with higher UV-B doses more favorable.

11.2.5.4. Other Diseases

The distribution of water-borne infectious diseases is influenced mainly by the hygienic circumstance of water (Epstein, 1992; Echeverria *et al.*, 1995; Colwell, 1996; Esrey, 1996). Water-borne diseases—including cholera and the suite of diarrheal diseases caused by organisms such as giardia, salmonella, and cryptosporidium—are common with contamination of drinking water quality in many countries of south Asia (Echeverria *et al.*, 1995; Colwell, 1996; Esrey, 1996). Higher SSTs and rich nutrient load in major river deltas would support extended phytoplankton blooms in selected coastal areas of temperate and tropical Asia. These phytoplankton blooms are habitats for the survival and spread of infectious bacterial diseases. The cholera outbreak in Bangladesh during 1994 has been attributed to the presence of extended phytoplankton blooms (Colwell, 1996). The aforementioned water-borne diseases could become more common in many countries of south Asia in a warmer climate.

For preventive actions, impact assessments are necessary on various aspects such as the nutritional situation, drinking water supply, water salinity, and ecosystem damage (Kaye and Novell, 1994; Graczyk and Fried, 1998). Water-borne infectious diseases, natural disaster, environmental migration, nutritional deficiency, and environmental pollution should be major risk factors for human health (Thongkrajai *et al.*, 1990; Pazzaglia *et al.*, 1995, Colwell, 1996). The risk factor of diseases also will depend on infrastructure, economic conditions, the hygienic situation, and medical facilities. Risk could be reduced by awareness in the communities that are more vulnerable to instability in the future environment.

In Asia, economic and population growth will expand rapidly during the 21st century in many countries. The rapid increase in population will be accompanied by migration from rural communities to overcrowded large cities (Stephens, 1995). Disasters linked to climate extremes such as floods and droughts also would impact local and regional populations and enforce migration. The huge energy consumption by the expanding population in urban cities would result in degradation of air and water quality, whereas rapid expansion of the economy will bring about improvements in living standards, such as improved environmental sanitation, hygienic practice, and medical treatment facilities. Therefore, better understanding of the interaction among climate change and environmental and health status in communities at regional and local scales is crucial to forge physiological acclimatization and social adaptations in the future.

11.2.6. Human Dimensions

11.2.6.1. Climate Extremes and Migration

The impacts of climate change on Asia will place additional stress on socioeconomic and physical systems. These pressures may induce change in demographic processes. Demographic trends, including the stability and size of populations, will be influenced directly through the impacts of climate change on human health as described in Section 11.2.5 and indirectly through the impacts of climate change on food security and the viability of natural resource-based economic activity. A further demographic response will come about through the risk of extreme events on human settlements. If the incidence and magnitudes of events such as droughts and coastal floods increase, there could be large-scale demographic responses—for example, through migration. Migration in itself is not necessarily a signal of vulnerability to present-day extreme events. Motivations for migration are diverse; much rural-to-urban migration in Asia takes place as a result of increased economic opportunities in megacities (cities with at least 8 million inhabitants). Future increases in the frequency and intensity of severe weather systems as a consequence of climate change can trigger mass migration, however.

The annual rate of growth in migration on a global scale has been greatest in developing countries of south and southeast Asia. For instance, population growth and land scarcity has encouraged the migration of more than 10 million Bangladesh natives to neighboring Indian states during the past 2 decades. This migration has been exacerbated by a series of floods and droughts affecting the livelihoods of landless and poor farmers in this region. Land loss in coastal areas resulting from inundation from sea-level rise as a result of climate change is likely to lead to increased displacement of resident populations. Many south Asian countries increasingly expect the number of internally displaced persons to rise in future.

Immigrant labor often benefits both the donor and the host cities/countries (Connell and Conway, 2000). However, perceptions of regional/national identity, language/cultural differences, and fears of unemployment may contribute to increased hostilities between immigrants and nationals in years to come. Climate change will act in parallel with a complex array of social, cultural, and economic motivations for and impacts of migration (Pebley, 1998; Conway *et al.*, 2000; Kates, 2000). Irrespective of resource constraints in developing countries of Asia, they have to better equip themselves through appropriate public education and awareness programs with disaster preparedness measures, including infrastructures for effective resettlement of displaced people as a consequence of weather calamities.

11.2.6.2. Infrastructure Linkages

The urban population in Asia is growing at four to five times the rate of the rural population. At this rate, more than 60% of the people in Asia will be living in towns and cities by 2015. An estimated 80% of the increase will occur in developing countries. The number of megacities in Asia would grow to at least 23 of the world's 36 by the year 2015 (United Nations, 1998). Urbanization is rapid in fast-growing economies of south and southeast Asia, where the average annual urban growth rate is more than 4%. The current pace and scale of change often strains the capacity of local and national governments to provide even the most basic services to urban residents. An estimated 25–40% of urban inhabitants in developing countries today live in impoverished slums and squatter settlements, with little or no access to water, sanitation, or refuse collection (World Bank, 1997).

Basic infrastructure demand in urban corridors is likely to increase dramatically in the future. Already governments in several developing countries of Asia are introducing suites of acts and laws to ensure provision of adequate public services and minimize adverse effects on surrounding communities and ecosystems. For instance, Indonesia introduced the Spatial Use Management Act in 1992 for the identification of environmentally sensitive areas—where development activities would be restricted—and for improved planning for the location and support of activities such as industrial development (Djoekardi, 1995). Developing countries in Asia would soon need to develop new priorities and policies that try to address demands created by the increasing number of people in cities while capitalizing on the benefits of urbanization, such as economic growth and efficient delivery of services. Climate change has the potential to exacerbate basic infrastructure demands of urban inhabitants in many countries of Asia.

11.2.6.3. Industry, Energy, and Transportation

In many cities in the developing countries of Asia, movement of the labor force from dispersed agricultural centers to concentrated industrial sectors has increased urbanization and expansion of the suburban area where industries are located. This situation has caused serious traffic, housing, and sanitary problems. Moreover, a substantial share of industrial growth in developing countries revolves around the transformation of raw materials into industrial products such as steel, paper, and chemicals. Production of industrial chemicals has been shifting from developed countries to developing countries of Asia in recent years. Not only are these processes resource-intensive, in addition, industries such as electricity generation, chemicals and petroleum refining, mining, paper production, and leather tanning tend to produce a disproportionately large amount of hazardous and toxic wastes and already have caused serious air and water pollution. Moreover, because of poor regulatory capacity, developing countries have become targets for dumping of toxic wastes such as polychlorinated biphenyls (PCBs) by multinational companies.

In developing countries of Asia, an increase in paper consumption of more than 80% is expected to occur by 2010. However, a shortfall in the supply of all wood products and especially pulp and paper is likely in the 21st century as a result of declines in natural vegetation productivity. This trend will put immense pressure on unmanaged forests of Asia. Some countries in Asia have resorted to agroforestry and farm woodlots to meet their increasing demand for pulp and paper on a near-term basis. Over the long term, however, anticipated growth in demand for wood products of all types probably will necessitate changes in forest management practices such as greater reliance on industrial plantations. Heat stress to livestock herded in open areas would reduce animal weight gain, dairy and wool production, and feed conversion efficiency in arid and semi-arid Asia (IPCC, 1996). An increase in the frequency of drought and heavy rainfall could result in a decline in tea yield in Sri Lanka (IPCC, 1998). Climate change could have adverse impacts on agroindustries such as food, beverages, tobacco, natural fiber textiles, leather, wood, and rubber.

The manufacturing industry could be affected indirectly through the availability of water, energy supply, and transportation systems throughout Asia. Mining operations, which are required to produce raw materials such as coal (practiced intensively in China and India because of heavy demand), exposes sulfur- and iron-bearing rocks to weathering and erosion that would be aggravated by global warming. Mineral industries discharge large amounts of waste ore known as tailings. Manufacturing industries release effluent to the air and water during production processes. For instance, cement production and foundry operations release particulates; metal industries release SO_2, CO_2, HF, and organic solvents; and chemical plants release particulates, SO_2, and various hydrocarbons. Food processing, pulp and paper industries, brewing, and tanning industries release effluents in the form of contaminated water and sediments. This degradation of the environment may be exacerbated by climate change.

Climate change will adversely affect hydroelectric energy generation in Asia, exacerbating already depleted water resources in several major rivers in Asia. Increased humidity and hence more cloudiness in a warmer atmosphere could inhibit direct solar radiation. Photovoltaic technology, which is used extensively in Indonesia and a few other south Asian countries for remote area electrification programs, could be jeopardized. Changes in wind patterns could affect existing windmill installations. Availability of biomass, particularly fuel wood, depends on forest area and the quantity of rainfall. In areas with excess rainfall, CO_2 enrichment and higher temperatures could increase fuelwood production and supply. Geothermal resources also may be affected by changes in precipitation patterns. For instance, an increase in rainfall could increase recharge of groundwater for most geothermal fields (IPCC, 1996).

In general, no major impacts of climate change on the production and distribution facilities of fossil fuels are likely, other than policies aimed at reducing GHG emissions from burning of fossil fuels. However, offshore exploration, production, and distribution facilities of oil and natural gas would be influenced by sea-level rise and extreme weather events such as cyclonic storms and associated surges that are likely to be exacerbated by climate change. Most power plants and oil refineries in Asia also are located along the coastlines to facilitate transportation and easy access to cooling water supply. Construction of seawalls will be needed to protect these facilities against sea-level rise (Mimura *et al.*, 1998).

As reported in IPCC (1996), energy demand for heating, cooling, and agriculture activities can be influenced by climate change. Global warming would increase energy demand for cooling in the tropical Asia region and reduce energy demand for heating in boreal Asia. A similar pattern could be observed in temperate Asia, either in summer or winter. For agricultural activities, more energy would be needed for irrigation pumping during warmer weather as the soil becomes drier.

Tourism and outdoor recreational activities are likely to be disrupted in pattern by climate change. Global warming is expected to shorten the skiing season in many areas and affect the feasibility of some ski facilities (IPCC, 1998). Summer recreational activities in coastal recreational areas may be affected with the inundation of beaches. In Malaysia, for example, unprotected resorts located near the coastline would be lost through erosion; others that are not affected by erosion or inundation will cease to operate because of the loss of beaches (Teh, 1997). The increased frequency of forest fires because of drier conditions in Indonesia during the 1997 El Niño resulted in haze that affected the tourism industries of Indonesia, Singapore, and Malaysia (Schweithelm *et al.*, 1999). Some wildlife reservation areas that are famous as tourist spots may lose their attraction because of the disappearance of flora and fauna in the changed ecosystems. For example, it has been suggested that the flowering dates of cherry blossoms in Japan and Korea could move 3–4 days earlier if air temperatures during March increase by 1°C (Seino *et al.*, 1998). Extreme weather events in highland regions also would threaten rafting, mountain climbing, and other high-altitude tourism.

Transportation systems, including infrastructure and fuel, can be influenced directly and indirectly by climate change. A longer rainy season, melting of ice deposits, and sea-level rise could directly damage infrastructures such as roads, railways, runways, terminals, airports, and harbors, thereby disrupting transportation systems. Climate change would increase the inherent necessity for expansion of infrastructure and alter fuel consumption patterns. In Siberia, for instance, frozen rivers that currently are used as roads would require a shift to water transport or construction of permanent roads with shorter winters. As reported in IPCC (1996), similar impacts could be experienced from permafrost melting.

11.2.6.4. Financial Aspects

The financial services sector typically includes banking, insurance, stock exchanges, and brokerages, as well as financial services firms such as investment banks, advisory services, and asset management, among others. Banking crises, currency speculation, and devaluation have become common in recent years in Asia and now are key policy concerns. Climate change could have serious consequences for this sector because many adaptation and mitigation measures such as crop insurance against floods or droughts are mediated or implemented through this sector. The implications and role of risk management techniques such as derivatives, options, and swaps are some of the new developments in financial services to dampen the impact of disasters on national finances.

There is a growing body of literature on the economic impacts of global warming that takes adaptation into account in estimating the imposed costs of climate change, but these studies fall short of specifically estimating the costs and benefits of adaptation. The basis for estimating adaptation costs is the economic opportunity cost of a product or activity. Estimating this cost requires price and other data from market transactions such as sectoral coverage and assumptions about markets, behaviors, and policy instruments in addition to economic growth path as a function of socioeconomic conditions, resource endowment, and government policies. Multiple baseline choices are critical in the estimation and evaluation of the financial costs, based on accurate reporting of financial flows (see also Chapter 8).

Although global gross domestic product (GDP) has increased by a factor of three since 1960, the number of weather-related disasters has increased four-fold, real economic losses seven-fold, and insured losses 12-fold in the same period (see also Chapter 8). These losses have had some notable regional impacts—particularly in developing countries of Asia, where the impacts of climate change are expected to be greatest in terms of loss of life and effects on investment. Individual large events have shown visible short-term impacts on insurance profitability and pricing and public finance. There is only limited penetration of or access to insurance in developing countries. This situation makes them more vulnerable and will impair their ability to adapt. The property/casualty insurance segment and small specialized or undiversified companies have greater sensitivity. Coping mechanisms and adaptation strategies will depend largely on public or international support. Given finite financial resources and international aid, increased climate-related losses would compete with development efforts.

Adaptation to climate change presents complex challenges to the finance sector. Increasing risk could lead to a greater volume of traditional business, as well as development of new risk and financing products (e.g., catastrophe bonds). However, increased variability of loss events would result in greater actuarial uncertainty. The design of an optimal adaptation program in any country would have to be based on comparison of damages avoided with the costs of adaptation. Other factors, particularly in developing countries with incomplete markets, also enter the decisionmaking process—such as the impacts of policies on different social groups in society, particularly those that are vulnerable; employment generation and opportunities; improved air and water quality; and the impacts of policies on broader concerns such as sustainability.

11.3. Vulnerability and Adaptation Potential

11.3.1. Resilience of Resources, Populations, and Infrastructure

The adaptive capacity of a resource system or a human society depends on the resilience of these systems. Resilience in the face of climate change, as with resilience to present-day hazards such as floods and droughts, therefore depends on the scale, intensity, and rate of change of the climate system, as well as the inherent ability of ecosystems or communities to adjust to new circumstances (Riebsame *et al.*, 1995). Resilience is the ability of a system to return to a predisturbed state without incurring any lasting fundamental change. Resilient resource systems recover to some normal range of operation after a perturbation. The processes of short-term adjustment to changes in land productivity and food scarcity in traditional societies of Asian countries are resilient to perturbations. This resilience has been demonstrated in a range of resource systems throughout Asia, including highland agriculture, large-scale irrigated agriculture, and fishery-dependent communities (e.g., Bray, 1986; Bayliss-Smith, 1991; Tang 1992; Grove *et al.*, 1998; Ruddle, 1998; Adger, 1999a). Long-term adaptation to climate change requires anticipatory actions, which would require considerable investment of capital, labor, and time hence diversion from scarce available resources, existing services, and infrastructure. Constraints on such resources clearly are more acute in the developing countries of Asia. The three crucial sectors of land resources, water resources, and food productivity are of highest priority for planned adaptation, particularly for the poorer resource-dependent countries.

Adaptation to climate change in Asian countries depends on the real cost of adaptive measures, the existence and engagement of appropriate institutions, access to technology, and biophysical constraints such as land and water resource availability, soil characteristics, genetic diversity for crop breeding (e.g., development of heat-resistant rice cultivars), and topography. Demand for land and water already is increasing to support growing populations, increased agricultural activities, and expanding modern urban infrastructure. Most developing countries in Asia face significant impacts from present-day climatic hazards. Faced with impending floods, the economic insecurity of communities and rural households, lack of timely warnings, ignorance of the severity of danger from flood, and lack of efficient transport systems, some developing countries of Asia often choose not to evacuate homes to avoid climate-related disasters. Such circumstances also act as constraints for the alleviation of poverty and hence reinforce social vulnerability. For many developing countries in Asia, climate change is only one of a host of environmental problems; these countries have to individually and collectively evaluate the tradeoffs between climate change actions and nearer term needs (such as food security, air and water pollution, and energy demand). Adaptation measures designed to anticipate the potential effects of climate change can help to offset many of the negative effects (Burton, 1997). Adaptation measures that ameliorate the impacts of present-day climate variability include sea defenses, institutional adaptations, plant breeding, and adoption of new technologies in agriculture. Many countries in Asia already commit significant resources to ameliorating climate-related hazards (e.g., Golubtsov *et al*., 1996; Nishioka and Harasawa, 1998; Ali, 1999; Huq *et al*., 1999).

Development and broad application of integrated modeling efforts (those that consider interactions of biophysical and socioeconomic factors) and modeling approaches that are particularly applicable at the regional scale warrant increased attention. For example, mountain systems in Asia are vulnerable with respect to ecological and social systems, for reasons of high heterogeneity. Management of Asia's mountain landscape therefore demands diversified strategies that link ecological and social components for location-specific solutions (Ehrenfeld, 1991; Ramakrishnan, 1992). Inclusion of complex feedbacks between systems may change significantly the current "mean" estimate of impacts.

Sustainable development within Asia's agroecosystems is crucial to provide adequate food security for traditional farming communities in the lowlands and the uplands in developing countries and to ensure *in situ* conservation of crop biodiversity for sustaining high-input modern agriculture itself. However, conserving biodiversity with concerns for higher production from these complex agroecosystems is a challenging task, for which novel development alternatives are required (Ramakrishnan, 1992; Ramakrishnan *et al*., 1996; Swift *et al*., 1996). Traditional societies have always manipulated biodiversity to ensure ecosystem resilience and to cope with uncertainties in the environment, rather than to increase production on a short-term basis. There is increasing evidence now to suggest that we could learn from their traditional ecological knowledge base (Gadgil *et al*., 1993) for coping with uncertainties associated with global change.

The resilience of agricultural practices in the face of climate change depends on the nature and magnitude of region-specific climate change, regional sensitivity, or the threshold and social resilience and adaptive capacity of agricultural communities. Adjustment of planting dates to minimize the effect of temperature increase-induced spikelet sterility can be used to reduce yield instability, for example, by avoiding having the flowering period to coincide with the hottest period. Adaptation measures to reduce the negative effects of increased climatic variability may include changing the cropping calendar to take advantage of the wet period and to avoid extreme weather events (e.g., typhoons and storms) during the growing season. Crop varieties that are resistant to lodging (e.g., short rice cultivars) may withstand strong winds during the sensitive stage of crop growth. A combination of farm-level adaptations and economic adjustments such as increased investment in agriculture infrastructure and reallocation of existing land and water resources would be desired in the agriculture sector. Increasing demand for water by competing sectors may limit the viability of irrigation as a sustainable adaptation to climate change. Expansion of irrigation as a response to climate change will be difficult and costly in many of the countries in Asia even under favorable circumstances. Mounting societal pressures to reduce environmental degradation will likely foster an increase in protective regulatory policies, further complicating adaptations to climate change (Easterling, 1996).

A commonly prescribed adaptation to climate change in the water sector is to enhance characteristics that offer flexibility hence enhancing resilience. Flexibility issues are particularly important with regard to the development of water resources for industry or agriculture. Major projects such as dams actually may limit flexibility if they lose effectiveness as regional hydrological water balances undergo major changes. With likely changes in climate variability, dams and sea defenses built to withstand a 100-year extreme event may not be adequate thus leading to a risk of major catastrophe. If hydrological patterns change markedly and irrigated agriculture is required to relocate in response, prior investments may be lost as existing infrastructures become obsolete, and additional investments will be needed. This necessitates critical scrutiny of a range of available choices that incorporate economic and environmental concerns. The potential for adaptation should not lead to complacency (Rosenzweig and Hillel, 1995). Some adaptive measures may have detrimental impacts of their own.

The issue of natural resource management in Asia has a highly complex set of interconnections between natural and social systems. Natural resource management in largely rural tropical environments must reconcile ecological and social processes that operate at a range of scales, from species up to the landscape level. Studies have shown that ecologically important keystone species often are socially selected by many rural societies (Jodha, 1996). The possibility for species selection for rehabilitating a degraded ecosystem should be based on a value system that the local people understand and appreciate; therefore, their participation in the process of developmental activity is important. Community perceptions of soil and water management can be a powerful agent for sustainable management of natural resources (e.g., in the case of highly fragile and vulnerable Himalayan mountain systems) (Ramakrishnan *et al*., 1994). In other words, natural resource management in tropical Asia must be sensitive to social and even cultural perceptions (Ramakrishnan, 1998), as well as traditional resource management practices.

Major fishery-related environmental issues include the effects of trawling on sea-bottom habitats and the detrimental effects of catches of nontarget species on populations and ecosystems. Fishery resources also are threatened by activities other than commercial fishing. Loss of inshore fish nursery habitats from coastal development and pollution from land-based activities cause significant change to ecosystems that support fisheries. Effective conservation and sustainable management of marine and inland fisheries are needed at the regional level so that living aquatic resources can continue to meet regional and national nutritional needs. Asian economic growth has failed to alleviate poverty for a large share of Asian people to date. Achieving economic and industrial growth in Asia that is sustainable—both ecologically and economically viable over the long term—would require more than just cleaner, more efficient industrial processes; it demands a reorientation toward becoming less material-intensive and attempting to contribute toward protecting our environment and ecosystem.

11.3.2. Regional and Sectoral Strategies

Regions of the Asian continent differ widely in their biophysical characteristics hence in their physical vulnerability to climate change. Different regions also experience highly differentiated social vulnerability. Adaptation strategies therefore will be differentiated across regions and sectors, depending on their vulnerability profiles. In the following subsections, vulnerabilities and related adaptation strategies are discussed for four broad regions of Asia and for selected sectors.

11.3.2.1. Boreal Asia

At present, only areas of sporadic permafrost are used for agriculture. Global warming should play a positive role for agriculture in boreal Asia. The growing season is likely to expand by 1–1.5 months by 2100. The increase in mean monthly air temperatures during the summer will increase active soil temperatures. In addition, winter air temperatures will substantially increase (Sirotenko *et al.*, 1997). Shifts in the limit of the permafrost zone to the north, formation of vast areas of perennial ground thawing, and better soil climate will contribute to a northward shift of agriculture boundary. The key step for an agriculture adaptation strategy could be the choice of suitable crops and cultivars. Shifts in sowing date of spring crops will allow more effective use of the soil moisture content formed by snow melting. The dates of spring crop sowing could be moved forward in a crop rotation calendar in southern regions, and farmers could plant a second crop that could even be vegetable with a short growth period (Laverov, 1998). Optimum use of fertilizers and ecologically clean agrotechnologies would be beneficial for agriculture.

Climate change has the potential to exacerbate water resource stresses in some areas but ameliorate them in most parts of boreal Asia. The increase in surface temperatures will have considerable effect on the timing of snowmelt hence the timing of the flow regime (Arnell, 1999). Diversions of water systems would adversely impact fisheries and fishery habitat in the region (Rozengurt, 1991). A decrease in water flow during the dry summer season is likely in some parts of boreal Asia. Extraction of groundwaters from deep aquifers has been proposed as an option, keeping in view the likely surface water quality deterioration during dry periods (Laverov, 1998). It would be necessary to increase the capacity of recycled water supply systems and autonomous water-use systems. However, even with water-saving measures it may be necessary to cut water intake for industry needs during dry periods to meet increasing demand in the future. Bottom-deepening along navigation channels may be required to facilitate the transport of goods and material through rivers.

11.3.2.2. Arid and Semi-Arid Asia

The major impact of climate change in arid and semi-arid Asia is likely to be an acute shortage of water resources associated with significant increases in surface air temperature. Conservation of water used for irrigated agriculture therefore should be given priority attention. With increased evapotranspiration, any adaptation strategy in agriculture should be oriented toward a shift from conventional crops to types of agriculture that are not vulnerable to evapotranspiration (Safriel, 1995). These strategies entail either intensive agriculture in greenhouses—within which rates of evapotranspiration are much reduced—or developing alternatives such as aquaculture that will partly replace agriculture (e.g., fish for human/animal feed, crustaceans for human feed, and unicellular algae for fish/prawn feed, as well as for food additives and medicinal and cosmetic uses). All of these organisms are of high yield; they enjoy solar radiation and often heat, and they do not evaporate or transpire water. Expansion of commercial and artesian fisheries also could help reduce dependence on food productivity. Protection of soils from degradation should be given serious consideration.

Climate change would exacerbate threats to biodiversity resulting from land-use/cover change and population pressure in Asia. Ecosystem services can be impaired by loss of key species in arid and semi-arid Asia (Xiao *et al.*, 1998). Because intraspecies variation in response to environmental stress usually exists in populations subjected to year-to-year climate change, some genotypes in such populations are expected to be more resistant to climate change than others. Such genotypes are more common in peripheral populations than in core populations of species. Although the core population may become extinct because of global warming, resistant types in peripheral populations will survive and can be used to rehabilitate and restore affected ecosystems (Safriel *et al.*, 1994; Kark *et al.*, 1999). The geographic locations of the peripheral species population usually coincide with climatic transition zones, such as at the edges of drylands or along the transition between different types of drylands. Many countries in the region have more than one dryland type and hence should have peripheral populations—especially in desert and nondesert transitions, which often occur within semi-arid drylands. Identifying regions with concentrations of peripheral populations of species of interest and protecting their habitats from being lost to development therefore can play a role in enhancing planned adaptation for natural and semi-natural ecosystems.

11.3.2.3. Temperate Asia

Projected surface warming and shifts in rainfall in temperate Asia are significant and will induce increases in photorespiration, maintenance respiration, and saturation deficits—causing stomatal closure and decline in productivity (White *et al.*, 1999). An adaptive response in the agriculture sector should be an effort to breed heat-resistant crop varieties by utilizing genetic resources that may be better adapted to warmer and drier conditions. Improvements in farming systems, fertilizer management, and soil conservation form major adaptation strategies (Lou and Lin, 1999). Research is needed to define current limits to heat resistance and the feasibility of manipulating such attributes through modern genetic techniques. Crop architecture and physiology may be genetically altered to adapt to warmer environmental conditions. The genetic resources of seeds maintained in germplasm banks may be screened to find sources of resistance to changing diseases and insects, as well as tolerances to heat and water stress and better compatibility with new agriculture technologies. Genetic manipulation also may help to exploit the potentially beneficial effects of CO_2 enhancement on crop growth and water-use efficiency.

The process of rapid urbanization and industrialization in several Asian megacities has placed enormous stress on urban infrastructure, human well-being, cultural integrity, and socioeconomic arrangements. These urban cities are giant resource sinks and create a large "ecological footprint" on the surrounding countryside. The negative environmental impacts of expanding cities are already large; as they continue to grow and become more prosperous, these impacts are likely to increase. Rising levels of air and water pollution in many of the large cities are considerable. Production and consumption systems that sustain life in cities are largely responsible for many of these changes. As climate changes, the demand for basic infrastructure facilities such as housing, electricity, food supply and distribution, and drinking water supply will increase, and municipalities would have a difficult time managing waste recycling and waste disposal. Development policies that mitigate or avert some of these long-term problems would have to be country-specific and depend heavily on the availability of infrastructure resources, the size of the floating population, and sustainable behavioral changes in society.

Climate change will impinge on a diverse, complex, and dynamic form of climatic hazards such as floods, droughts, sea-level rise, and storm surges in the countries of temperate Asia. Preparation for changes in climate variability should include provision for the possibility of increased flooding, as well as incidences of drought. The present path of development in this region is placing more fixed infrastructures and economic activity within the coastal zone. This trend seems to offer limited scope for adjustments against flooding in the coastal zone resulting from sea-level rise. The likelihood of damage to infrastructure and loss of human life because of unexpected extreme events will rise. A wide range of precautionary measures at the regional and national levels—including awareness, perception, and the acceptability of risk factors among regional communities—are warranted to avert or reduce the impacts of such disasters on economic and social structures. Many current technical and socioeconomic barriers will need to be overcome to prevent risks to human health resulting from increases in disease incidences associated with climate change.

11.3.2.4. Tropical Asia

Agricultural adaptation to climatic variability is an evolving process. Planned interventions through research, extension, or pricing or marketing policies can have inadvertent detrimental impacts for poor farmers. Agricultural productivity in tropical Asia is sensitive not only to temperature increases but also to changes in the nature and characteristics of monsoons. An increase in leaf surface temperatures would have significant effects on crop metabolism and yields, and it may make crops more sensitive to moisture stress (Riha *et al*., 1996). Cropping systems may have to change to include growing suitable cultivars (to counteract compression of crop development), increasing crop intensities (i.e., the number of successive crops produced per unit area per year), or planting different types of crops (Sinha *et al*., 1998). Farmers will have to adapt to changing hydrological regimes by changing crops. For example, farmers in Pakistan may grow more sugarcane if additional water becomes available, and they may grow less rice if water supplies dwindle. The yield ceiling must be raised and the yield gap narrowed while maintaining sustainable production and a friendly environment. Development of new varieties with higher yield potential and stability is complementary to bridging the yield gap. Efficient production of a socially optimal level of agricultural output in this region ultimately may depend on biotechnological applications, but only if these applications prove to be environmentally sustainable.

Groundwater is the main source of freshwater in many parts of tropical Asia, particularly in semi-arid regions. Water resources already are limited in terms of supply and demand in this region. The aquifers in most countries have been depleted by high withdrawal and low recharge rates, and significant drawdown problems exist. Even with increases in precipitation, surface runoff may diminish in some river basins under projected climate change scenarios because of greater evaporation in a warmer atmosphere (higher hydrological elasticity). Increased runoff in some river basins can cause deleterious effects such as greater flooding, waterlogging, and salinity. More than 25% of the irrigated land in the Indus basin already is affected by waterlogging and salinization (Hillel, 1991). Freshwater availability in the coastal regions is likely to undergo substantial changes as a result of a series of chain effects. Improvements in runoff management and irrigation technology (e.g., river runoff control by reservoirs, water transfers, and land conservation practices) will be crucial. Increasing efforts should be directed toward rainwater harvesting and other water-conserving practices to slow the decline in water levels in aquifers. Recycling of wastewater should be encouraged in drought-prone countries in tropical Asia. However, major water development decisions to augment water supplies may have greater relative hydrological, environmental, and social impacts than climate change *per se* in the shorter term. Climate change will affect the benefits to be accrued by future water development projects and therefore should be taken into consideration in the context of water resource policies and planning.

With rapid development of the economy in several countries in tropical Asia during recent decades, the patterns of land use and land cover have been modified significantly; a sequence of transitions and conversions is discernable. As a result of different natural and socioeconomic conditions, the speed and scale of land-use change is very diverse in different parts of Asia. This process has contributed to significant losses in total forest cover; changes in standing biomass and the soil carbon budget; extinction of mammals, birds, and vascular plants; soil degradation; and threats to food security. Excessive human and livestock population pressure in association with inappropriate agricultural extension activities also lend an explanation for widespread land degradation in tropical Asia. Salinization and acidification of soil in low-lying coastal areas would adversely affect cropland in addition to land losses from permanent inundation of deltas from anticipated sea-level rise.

Many of the major rivers originating in mountains and highlands are charged with sediments, depending on the types of land uses in the watersheds (e.g., from forestry and agroforestry to open agriculture). Marginalization of production areas in highlands will continue to increase soil losses, land slips, and slides in the region. It has been suggested that upland micro-watersheds can be hydroecologically sustainable only if good forest cover and dense forests are maintained (Rai, 1999). Under changing climate conditions, the pressure toward small- and large-scale transfers of land for agricultural and urban uses may grow.

Sea-level rise poses the greatest threat and challenge for sustainable adaptation within south and southeast Asia. The sea level already is rising in many locations, primarily as a result of geological processes and anthropogenic manipulations. Projected sea-level rise along the Asian coastlines represents an increase of three to four times over present rates (Chansang, 1993; Midun and Lee, 1995; Mimura and Harasawa, 2000). The potential impacts of accelerated sea-level rise include inundation of low-lying deltas and estuaries, retreat of shorelines, and changes in the water table (Wong, 1992; Sivardhana, 1993). Episodic flooding from high storm surges would penetrate much further inland. Salinization and acidification of soil in low-lying coastal areas will adversely affect agricultural production, in addition to land losses from permanent inundation of deltas from anticipated sea-level rise. The impacts will vary from region to region because of local factors such as land subsidence, susceptibility to coastal erosion or sedimentation, varying tidal ranges, and cyclonicity (Bird, 1993). The Ganges-Brahmaputra in Bangladesh, the Irrawaddy in Myanmar, the Choo Phraya in Thailand, and the Mekong and Song Hong in Vietnam are among the key low-lying river deltas in tropical Asia that are most vulnerable to sea-level rise. Local-level social and institutional adaptations in sensitive regions and development and promotion of risk management can potentially prevent accelerated impacts to these deltas through protection of the ecology of the region from further human interventions so they can sustain themselves at least in a short time horizon.

11.3.3. Institutional and Financial Barriers

As reported in IPCC (1996), the global cost of weather-related disasters to insurers had risen rapidly since 1960 as a result of increased frequency and severity of extreme climate events. The property insurance industry is most likely to be directly affected by climate change because it is already vulnerable to variability in extreme weather events (Dlugolecki *et al.*, 1996). The trend of rapidly growing damage from weather-related disasters has continued. There have been several major climatic hazards in Asia in the late 1990s. There is, however, only very limited penetration of property insurance or agriculture crop insurance in many of the areas that are most affected by recent floods and cyclonic storms. The impacts of flooding are concentrated on the poorest sections of society and people living in marginal areas. There is a need, particularly in Asia, for increased recognition by the financial sector that climate change could affect its future. Climate change can be considered a threat as well as an opportunity for the insurance industry because an increase in risks and perceived risks implies more business opportunities for the sector.

Decisionmaking processes on the choice and capacity of adaptation to the impacts of climate change vary from country to country, depending on its social structure, culture, and economic capacity, as well as the level of environmental disruptions in the Asian region. Institutional inertia, a scarcity of technological adaptation options, and additional economic burdens in developing countries of Asia will be limiting factors for investment in environmental protection and would force people in these countries to face greater risks.

11.4. Synthesis

This chapter discusses the current status of our understanding of likely future changes in climate variables, such as surface temperature, precipitation, soil moisture, extreme events, their potential impacts on natural environment and human society, and possible adaptative options in the Asian region. These factors are delineated on a subregional, national, and sectoral basis. This section attempts to formulate a synthesis of climate change and its implications and consequences for Asia and identify commonality throughout the region, as well as subregional and national differences in terms of climate change and its consequences.

This synthesis also aims to identify the critical climatic threshold, feedbacks, interactions and nonlinearities involved in interactive systems of climate, natural environment, and human society. These analyses will provide a basis for further indepth understanding of the implications of climate change, setting policy-oriented goals to arrest human interventions to the climate system, and planning adaptive responses.

11.4.1. Key Observations and Uncertainties

Asia is characterized by an extreme diversity of natural environment, in longitudinal, latitudinal, and vertical directions. Each subregion and country is supported by natural resources such as water, forests, grasslands/rangelands, and fisheries, which have been utilized by diverse Asian societies in a sustainable manner over the centuries. Promotion of public awareness and participation in region-specific adaptation and mitigation strategies is essential, however, particularly in the developing countries of Asia, to effectively and collectively overcome problems associated with climate change. In general, climate change will impose significant stresses on available natural resources throughout Asia. At the same time, combinations of specific geographical settings and changes in climate will bring about different impacts on different subregions.

Asia has long been affected by natural hazards such as intense rainfall, flooding, droughts, extreme temperatures, snow avalanches, and other impacts of tropical cyclones, monsoons, and ENSO. The security and sustainability of the region is highly dependent on future trends of such extreme events and preparedness for them. However, there remains great uncertainty in projections of likely changes in tropical cyclones, monsoons, and El Niño. It should be noted that the key vulnerabilities related to such natural hazards are still largely qualitative, and there are possibilities of unforeseen surprises in the future.

Many systems are sensitive to natural climate variability in Asia as well as climate change and hence may not be resilient to climate change. These systems include mangroves, lakes, glaciers, deltas, rivers, and ecosystems within the permafrost region. Threshold levels of response are likely to be exceeded by projected climate change. Changes in these systems will occur at different climate thresholds in various regions of Asia. For example, a 0.5°C rise in mean temperature and 10-cm rise in sea level could lead to inundation of 15% (approximately 750 km^2) of the Bangladesh Sundarbans, the largest mangrove ecosystem in Asia. A 45-cm rise in sea level, corresponding to a 2°C rise in mean temperature, is required to introduce changes in the low saline zone of the Sundarbans (Smith *et al.*, 1998). Table 11-10 lists selected examples of regions that are sensitive to climate change, based on region- and country-specific studies.

Although many resource systems and populations exhibit high degrees of resilience and adaptive capacity, there is a significant risk that critical thresholds may be breached as a result of climate change, undermining this resilience. The environmental risks in a region could vary significantly for different sectors, depending on the degree of warming. To develop a risk profile for precautionary management of climate change, it is necessary to assess the critical climate threshold for a region for each of the relevant priority sectors, such as water and agriculture. For instance, a small degree of surface warming could be beneficial for agriculture in Russia but may be detrimental in India or Bangladesh. Increased glacier melt in the Himalayas may cause serious floods in Nepal, India, and Bangladesh. The potential environmental risks to water and marine resources in marginal seas of Asia such as the Sea of Japan, Bohai Bay, and the Yellow Sea also will depend on the degree of climate change—which could differ significantly in these regions. For many countries in Asia, there is sparse scientific research on the evaluation of sector-specific critical climatic thresholds. The response of driving forces such as economic growth, population increase, and technological progress in each country or region also should be included in the determination of thresholds for precautionary risk management as an adaptation strategy to climate change.

***Table 11-10**: Sensitivity of selected Asian regions to climate change (based on ICRF, 1998; Mirza, 1998; Smith et al., 1998; Mizina et al., 1999).*

Change in Climatic Elements and Sea-Level Rise	**Vulnerable Region**	**Primary Change**	**Impacts**	
			Primary	Secondary
0.5–2°C (10- to 45-cm sea-level rise)	Bangladesh Sundarbans	– Inundation of about 15% (~750 km^2) – Increase in salinity	– Loss of plant species – Loss of wildlife	– Economic loss – Exacerbated insecurity and loss of employment
4°C (+10% rainfall)	Siberian permafrosts	– Reduction in continuous permafrost – Shift in southern limit of Siberian permafrost by ~100–200 km northward	– Change in rock strength – Change in bearing capacity – Change in compressibility of frozen rocks – Thermal erosion	– Effects on construction industries – Effects on mining industry – Effects on agricultural development
>3°C (>+20% rainfall)	Water resources in Kazakhstan	– Change in runoff	– Increase in winter floods – Decrease in summer flows	– Risk to life and property – Summer water stress
~2°C (-5 to 10% rainfall; 45-cm sea-level rise)	Bangladesh lowlands	– About 23–29% increase in extent of inundation	– Change in flood depth category – Change in monsoon rice cropping pattern	– Risk to life and property – Increased health problems – Reduction in rice yield

Rapid demographic transition accompanying significant economic growth is likely in many Asian countries in the 21st century. These trends exacerbate pressures on resource use, the climate system, and the natural environment. Population growth and varying economic and technological conditions in Asia are likely to affect some societies and resources more than changes in climate *per se*. Moreover, socioeconomic and technological developments will interact with many vulnerable sectors. However, predicting population growth rates and future economic conditions is as uncertain an exercise as predicting the future climate. Institutional and legal structures may change and will co-evolve with climatic risks. GHG emission scenarios are especially important, but future emission trends will depend on population growth and prevailing economic and technological conditions. There have been limited studies that simultaneously take into account climate change and other human-induced stresses. For vulnerability and adaptation to be addressed comprehensively, they must be considered in a context of multiple stresses caused by climate change and other anthropogenic activities.

Table 11-11 summarizes the vulnerability of key sectors, from food and fiber through settlements, for subregions of Asia. This summary is derived from a synthesis of available scientific research reviewed in this chapter; it represents a consensus view of the authors. Levels of confidence assigned to vulnerability also are shown in Table 11-11, again based on iterative assessment of the available scientific evidence. Currently available research synthesized in this table reveals a wide range of vulnerability, as well as a wide range of uncertainties in the assessment of these vulnerabilities. Geographical resolution as well as integration and scaling of basic physical and biological responses are the key factors contributing to these uncertainties. The limited database also makes it difficult to aggregate various responses with only a few sample or spot assessments. Nevertheless, some key trends emerge from the synthesis. A consistent message is the vulnerability of south and southeast Asia across many sectors, as well as a high degree of confidence about this assertion.

Across virtually all subregions, water resources and natural ecosystems are appraised as highly vulnerable to climate change, although the level of confidence varies for some regions because of lack of data or greater complexity in the resource systems. There may be some minor benefits to boreal Asia, making it more resilient to potential impacts, but natural ecosystems clearly are under threat in this subregion.

11.4.2. Future Needs

The early signs of climate change already observed in some parts of Asia and elsewhere may become more prominent over the period of 1 or 2 decades. If this time is not used appropriately, it may be too late to avoid upheavals and significant human impacts for some nations. Climate change could lead either to cooperation or to conflict over the world's major resources. Integrated planning may be the greatest global challenge, now motivated by the potential for environmental and social transformation caused by climate change.

In the 21st century, Asian countries will have to produce more food and other agricultural commodities under conditions of diminishing per capita arable land and irrigation water resources and expanding biotic as well as abiotic stresses, including climatic constraints. The dual demands for food and ecological security would have to be based on appropriate use of biotechnology, information technology, and ecotechnology. Practical achievements in bringing about the desired paradigm shift in sustainable agriculture will depend on public policy support and political action. Critical areas for intervention would be:

- Improving the availability of seed/planting material of high-yielding varieties
- Developing and promoting the use of hybrids, especially for rainfed agro-ecosystems
- Expanding areas under different crops and commodities, through diversification of agriculture

***Table 11-11**: Vulnerability of key sectors to impacts of climate change for select subregions in Asia. Vulnerability scale is as follows: highly vulnerable (-2), moderately vulnerable (-1), slightly or not vulnerable (0), slightly resilient (+1), and most resilient (+2). Confidence levels abbreviated to VH (very high), H (high), M (medium), L (low), and VL (very low).*

Regions	Food and Fiber	Biodiversity	Water Resources	Coastal Ecosystems	Human Health	Settlements
Boreal Asia	+1 / H	-2 / M	+1 / M	+1 / L	-1 / L	0 / M
Arid and Semi-Arid Asia						
– Central Asia	-2 / H	-1 / L	-2 / H	-1 / L	-1 / M	-1 / M
– Tibetan Plateau	0 / L	-2 / M	-1 / L	Not applicable	No information	No information
Temperate Asia	-2 / H	-1 / M	-2 / H	-2 / H	-2 / M	-2 / H
Tropical Asia						
– South Asia	-2 / H	-2 / M	-2 / H	-2 / H	-1 / M	-2 / M
– Southeast Asia	-2 / H	-2 / M	-2 / H	-2 / H	-1 / M	-2 / M

- Improving the productivity of crops, existing plantations, and livestock
- Developing infrastructure for post-harvest management, marketing, and agribusiness
- Small farm mechanisms
- Transfer of technological inputs through assessments and refinements at regular time intervals in consonance with our understanding of climate variability and climate change.

Ensuring food security may remain an unaccomplished dream for many Asian countries unless appropriate strategies are put in place to ensure environmental and ecological protection and conservation of natural resources.

The food security issue is highly dependent on equitable guaranteed access to foods. Equitable access is highly differentiated across populations in the agrarian nations of Asia. This situation is further aggravated by natural disasters such as floods and droughts, which are known to have caused great famines in south Asian countries. Poverty in many south Asian countries seems to be the cause of not only hunger but even lack of shelter, access to clean drinking water, illiteracy, ill health, and other forms of human deprivation. Opportunities for assured and remunerative marketing at microenterprise levels should promote equitable use of available food resources.

In view of present uncertainties over the pace and magnitude of climate change, the most promising policy options are those for which benefits accrue even if no climate change takes place. Such policy actions include the following:

- Breeding of new crop varieties and species (heat- and salt-tolerant crops, low-water-use crops)
- Maintenance of seed banks, liberalization of trade of agricultural commodities, flexibility of commodity support programs, agricultural drought management
- Promotion of efficiency of irrigation and water use and dissemination of conservation management practices
- Trans-national cooperation to promote sustainable water resources management and flood risk management
- Rehabilitation of degraded forests and watersheds (such strategies can enhance biodiversity conservation and provide source of livelihood for many poor forest and upland watershed dwellers)
- Strengthening of biophysical and socioeconomic resources and resource use-related databases for natural and social systems and focused research to further our understanding of the climate-ecosystem-social system interaction. Data and information generated through these activities will be useful not only for designing appropriate mitigation and adaptation measures but also for management planning and decisionmaking.

The climate change issue has presented decisionmakers in Asian countries with a set of formidable complications: a considerable number of uncertainties (which are inherent in the complexity of the problem), the potential for irreversible damages to ecosystems, a very long planning horizon, long time lags between GHG emissions and effects, wide regional variation in causes and effects, the global scope of the problem, and the need to consider multiple GHGs and aerosols. The value of better information about climate change processes and impacts and responses to arrest these risks is likely to be great. A prudent strategy to deal with climate change would be to collectively reduce emission levels of GHGs through a portfolio of actions aimed at mitigation and adaptation measures. The agriculture and forestry sectors in several countries of Asia have a large GHG mitigation potential that should make a significant contribution to this strategy.

The principle of sustainable development must guide all future development strategies in developing and developed countries of Asia. Serious efforts toward promoting innovative research on efficient technology options and creative environmental literacy are needed while Asian countries adapt to new environmental policies and programs. The challenge lies in identifying opportunities that would facilitate sustainable development by making use of existing technologies and developing policies that make climate-sensitive sectors resilient to climate variability. This strategy will require developing countries in Asia to have more access to appropriate technologies, information, and adequate financing. In addition, adaptation will require anticipation and planning; failure to prepare systems for projected change in climate means, variability, and extremes could lead to capital-intensive development of infrastructures or technologies that are ill-suited to future conditions, as well as missed opportunities to lower the cost of adaptation.

References

ACIA, 2000: An assessment of consequences of climate variability and change and the effects of increased UV in the Arctic region. In: *State of the Arctic Environment Report*. Arctic Climate Impact Assessment, Oslo, Norway.

Adger, W.N., 1999a: Evolution of economy and environment: an application to land use in lowland Vietnam. *Ecological Economics*, **31,** 365–379.

Adger, W.N., 1999b: Social vulnerability to climate change and extremes in coastal Vietnam. *World Development*, **27,** 249–269.

Adger, W.N., 2000: Institutional adaptation to environmental risk under the transition in Vietnam. *Annals of the Association of American Geographers*, **90,** 738–758.

Agee, E.M., 1991: Trends in cyclone and anticyclone frequency and comparison with periods of warming and cooling over the Northern Hemisphere. *Journal of Climate*, **4,** 263–267.

Ageta, Y. and T. Kadota, 1992: Prediction of change of mass balance in the Nepal Himalayas and Tibetan Plateau: a case study of air temperature increase for three glaciers. *Annals of Glaciology*, **16,** 89–94.

Aggarwal, P.K. and S.K. Sinha, 1993: Effect of probable increase in carbon dioxide and temperature on productivity of wheat in India. *Journal of Agricultural Meteorology*, **48(5),** 811–814.

Aiba, S. and K. Kitayama, 1999: Species composition, structure and species diversity or rain forests in a matrix of altitude and substraites on Mt. Kinabalu, Borneo. *Plant Ecology*, **140,** 139–157.

Akhtar, R. and A.J. McMichael, 1996: Rainfall and malaria outbreaks in western Rajasthan. *Lancet*, **348,** 1457–1458.

Aksornkaoe, S. and N. Paphavasit, 1993: Effect of sea level rise on the mangrove ecosystem in Thailand. *Malaysian Journal of Tropical Geography*, **24(1–2),** 29–34.

Alfiorov, A.M., V.N. Busarov, G.V. Menzulin, S.A. Pegov, V.S. Savenko, V.A. Smirnova, V.A. Smolina, and P.M. Khomiakov, 1998: *Impact of Environmental and Climatic Global Changes on Russian Economy*. Ministry of Science and Technology of Russian Federation, Moscow, Russia, 102 pp. (in Russian).

Ali, A., 1999: Climate change impacts and adaptation assessment in Bangladesh. *Climate Research, CR Special 6*, **12(2/3),** 109–116.

Ali, S., 1985: *The Fall of a Sparrow*. Oxford University Press, New Delhi, India.

Allison, M.A., 1998: Geological framework and environmental status of the Ganges-Brahmaputra Delta. *Journal of Coastal Research*, **14(3),** 794–825.

Ando, M., 1998: Risk assessment of global warming on human health. *Global Environmental Research*, **2,** 69–78.

Ando, M., K. Tamura, S. Yamamoto, C.K. Liang, Y.P. Wu, J.P. Zhang, Z.C. Mao, M.M Yang, and A.L. Chen, 1996: Outline of health effects of global climate change. *Journal of Epidemiology*, **6,** 141–144.

Anisimov, O.A., 1999: Impact of climate change on heating and air-conditioning. *Meteorology and Hydrology*, **6,** 10–17 (in Russian).

Anisimov, O.A. and F.E. Nelson, 1997: Permafrost zonation and climate change in the Northern Hemisphere: results from transient general circulation models. *Climatic Change*, **35,** 241–258.

Anisimov, O.A. and F.E. Nelson, 1996a: Permafrost distribution in the Northern Hemisphere under scenarios of *Climatic Change*. *Global and Planetary Change*, **14,** 59–72.

Anisimov, O.A. and F.E. Nelson, 1996b: Permafrost and global warming: strategy of adaptation. In: *Adapting to Climate Change: Assessments and Issues* [Smith J., N. Bhatti, G. Menzhulin, R. Benioff, M.I. Budyko, M. Campos, B. Jallow, and F. Rijsberman (eds.)]. Springer-Verlag, New York, NY, USA, pp. 440–449.

Anokhin, Y.A. and Y.A. Izrael, 2000: Monitoring and assessment of the environment in the Lake Baikal region. *Aquatic Ecosystem Health and Management*, **3,** 199–201.

Arnell, N.W., 1999: Climate change and global water resources. *Global Environmental Change*, **9,** S51–S67.

Azam, J.P., 1996: The impact of floods on the adoption rate of high yielding rice varieties in Bangladesh. *Agricultural Economics*, **13,** 179–189.

Balling, R.C. and S.E. Idso, 1990: Effects of greenhouse warming on maximum summer temperatures. *Agricultural and Forest Meteorology*, **53,** 143–147.

Bandyopadhyay, M., 1993: Impact of rising sea levels along the Indian coastline. In: *Global Warming: Concern for Tomorrow* [Lal, M. (ed.)]. Tata McGraw-Hill Publishing Company Limited, New Delhi, India, pp. 153–66.

Bapat, N.N., 1992: A visit to the 'flamingo city' in the Great Rann of Kutch, Gujarat. *Journal of the Bombay Natural History Society*, **89,** 366–367.

Barber, C.V. and J. Schweithelm, 2000: *Trial by Fire: Forest Fire and Forestry Policy in Indonesia's Era of Crisis and Reform*. World Resource Institute, Forest Frontiers, World Wide Fund for Nature(WWF)-Indonesia, Telapak Indonesia Foundation, 448 pp.

Bardin, M.Y., 1994: Parameters of cyclonicity at 500 mb in the Northern Hemisphere extratropics. In: *Proceedings of XVIII Climate Diagnostic Workshop, Boulder, CO*. National Technical Information Service (NTIS), U.S. Department of Commerce, Springfield, VA, USA, 397 pp.

Bayliss-Smith, T., 1991: Food security and agricultural sustainability in the New Guinea Highlands: vulnerable people, vulnerable places. *IDS Bulletin*, **22(3),** 5–11.

Bhaskaran, B., J.F.B. Mitchell, M. Lal, and J. Lavery, 1995: Climatic response of Indian subcontinent to doubled CO_2 concentration. *International Journal of Climatology*, **15,** 873–892.

Bijlsma, L., C.N. Ehler, R.J.T. Klein, S.M. Kulshrestha, R.F. McLean, N. Mimura, R.J. Nicholls, L.A. Nurse, H. Pérez Nieto, E.Z. Stakhiv, R.K. Turner, and R.A. Warrick, 1996: Coastal zones and small islands. In: *Climate Change 1995: Impacts, Adaptations, and Mitigation of Climate Change: Scientific-Technical Analyses. Contribution of Working Group II to the Second Assessment Report of the Intergovernmental Panel on Climate Change* [Watson, R.T., M.C. Zinyowera, and R.H. Moss (eds.)]. Cambridge University Press, Cambridge, United Kingdom and New York, NY, USA, pp. 289–324.

Bird, E.C.F., 1993: Sea level rise impacts in Southeast Asia: an overview. *Malaysian Journal of Tropical Geography*, **24(1–2),** 107–110.

Bird, E.C.F., 1992: The impacts of sea level rise on coral reefs and reef islands. In: *Ocean Management in Global Change* [Fabbri, P. (ed.)]. Elsevier, London, United Kingdom and New York, NY, USA, pp. 90–107.

Blaikie, P., T. Cannon, I. Davis, and B. Wisner, 1994: *At Risk: Natural Hazards, People's Vulnerability and Disasters*. Routledge, London, United Kingdom, pp. 57–79.

Boer, G.J., G. Flato, C. Reader, and D. Ramsden, 1999: A transient climate change simulation with greenhouse gas and aerosol forcing: projected climate for the 21st century. *Climate Dynamics*, **16,** 405–425.

Born, K., 1996: Tropospheric warming and changes in weather variability over the Northern Hemisphere for the period 1967–91 using two data sets. *Meteorology and Atmospheric Physics*, **59,** 201–215.

Botch, M.S., K.I. Kobak, T.S. Vinson, and T.P. Kolchugina, 1995: Carbon pools and accumulation in peatlands of the former Soviet Union. *Global Biogeochemical Cycles*, **9,** 37–46.

Bouchama, A., R.S. Parhar, A. El-Yazigi, K. Sheth, and S. Al-Sedairy, 1991: Endotoxemia and release of tumor necrosis factor and interleukin 1 alpha in acute heatstroke. *Journal of Applied Physiology*, **70(6),** 2640–2644.

Bouchard, R.H., 1990: A climatology of very intense typhoons: or where have all the super typhoons gone? In: *1990 Annual Tropical Cyclone Report*. Joint Typhoon Warning Center, Guam, pp. 266–269.

Bouma, M.J. and H.J. Van der Kaay, 1996: The El Niño-Southern Oscillation and the historic malaria epidemics on the Indian subcontinent and Sri Lanka: an early warning system for future epidemics? *Tropical Medecine and International Health*, **1,** 86–96.

Bouma, M.J., H.E. Sondorp, and H.J. Van der Kaay, 1994: Climate change and periodic epidemic malaria. *Lancet*, **343,** 1440.

Bower, B.T. and R.K. Turner, 1998: Characterising and analysing benefits from integrated coastal management. *Ocean and Coastal Management*, **38,** 41–66.

Brammer, H., 1993: Geographical complexities of detailed impacts assessment for the Ganges-Brahmaputra-Meghna delta of Bangladesh. In: *Climate and Sea Level Change: Observations, Projections and Implications* [Warrick, R.A., E.M. Barrow, and T.M.L. Wigley (eds.)]. Cambridge University Press, Cambridge, United Kingdom and New York, NY, USA, 246–262.

Bray, F., 1986: *The Rice Economies: Technology and Development in Asian Societies*. Blackwell, Oxford, United Kingdom, 254 pp..

Budyko, M.I. and G.V. Menzhulin, 1996: Climate change impacts on agriculture and global food production: options for adaptive strategies. In: *Adapting to Climate Change: Assessment and Issues* [Smith J., R. Benioa, N. Bhatti, M.I. Budyko, M. Campos, B. Jallow, G. Menzhulin, and F. Rijsberman (eds.). Springer-Verlag, New York, NY, USA, pp. 188–203.

Burton, I., 1997: Vulnerability and adaptive response in the context of climate and climate change. *Climatic Change*, **36,** 185–196.

CAFW, 1998: Comprehensive assessment of the freshwater resources of the world. In: *Assessment of Water Resources and Water Availability in the World* [Shiklomanov, I.A. (ed.)]. World Meterological Organization, Geneva, Switzerland, 88 pp.

Cesar, H., C.G. Lundin, S. Bettencourt, and J. Dixon, 1997: Indonesia coral reefs—an economic analysis of a precious but threatened resource. *Ambio*, **26,** 345–350.

Chakraborty, B. and M. Lal, 1994: Monsoon climate and its change in a doubled CO_2 atmosphere as simulated by CSIRO9 model. *Terrestrial, Atmospheric and Oceanic Sciences*, **5(4),** 515–536.

Chan, J.C.L. and J. Shi, 1996: Long-term trends and interannual variability in tropical cyclone activity over the western North Pacific. *Geophysical Research Letters*, **23,** 2765–2767.

Chandrapala, L. and T.K. Fernando, 1995: Climate variability in Sri Lanka—a study of air temperature, rainfall and thunder activity. In: *Proceedings of the International Symposium on Climate and Life in the Asia-Pacific, University of Brunei, Darussalam, 10–13 April, 1995*.

Chansang, H., 1993: Effect of sea level rise on coral reefs in Thailand. *Malaysian Journal of Tropical Geography*, **24(1–2),** 21–28.

Chaudhari, Q.Z., 1994: Pakistan's summer monsoon rainfall associated with global and regional circulation features and its seasonal prediction. In: *Proceedings of the International Conference on Monsoon Variability and Prediction, Trieste, Italy, May 9–13, 1994*.

Chen, X.Q. and J.Y. Chen, 1998: A study of enclosure depth on the profiles of the Changjiang deltaic coast: on the fundamental problems associated with Bruun Rule and its application. *Acta Geographica Sinica*, **53(4),** 323–331 (in Chinese with English abstract).

Chen, Y.Q. and X.Q. Shen, 1999: Changes in the biomass of the East China Sea ecosystem. In: Large Marine Ecosystems of the Pacific Rim: Assessment, Sustainability, and Management [Sherman, K. and Q.S. Tang (eds.)]. Blackwell Science, Oxford, United Kingdom, pp. 221–239.

Chua, T.-E., 1998: Lessons learned from practicing integrated coastal management in southeast Asia. *Ambio*, **27,** 599–610.

CIA, 1998: *The World Factbook*. Central Intelligence Agency, Washington, DC, USA. Available online at http://www.odci.gov/cia/publications/factbook/index.html.

Clark, B. and P. Duncan, 1992: Asian wild asses—hemiones and kiangs. In: *Zebra, Asses and Horses: An Action Plan for the Conservation of Wild Equids*. The World Conservation Union (IUCN), Gland, Switzerland, pp. 17–22.

Coleman, J.M. and L.D. Wright, 1975: Modern river deltas: variability of processes and sand bodies. In: *Deltas: Models for Exploration* [Broussard, M.L. (ed.)]. Houston Geological Society, Houston, TX, USA, pp. 99–149.

Collins, M., 1999: The El-Niño Southern Oscillation in the second Hadley Centre coupled model and its response to greenhouse warming. *Journal of Climate*, **13,** 1299–1312.

Colwell, R., 1996: Global climate and infectious disease: the cholera paradigm. *Science*, **274,** 2025–2031.

Connell, J. and D. Conway, 2000: Migration and remittances in island microstates: a comparative perspective on the South Pacific and the Caribbean. *International Journal of Urban and Regional Research*, **24,** 52–78.

Conway, D., K. Bhattarai, and N.R. Shrestha, 2000: Population-environment relations at the forested frontier of Nepal. *Applied Geography*, **20,** 221–242.

Cubasch, U., G.C. Hegerl, and J. Waszkewitz, 1996: Prediction, detection and regional assessment of anthropogenic climate change. *Geophysica*, **32,** 77–96.

Dai, X.S., 1997: Potential effects of climatic variation on geographical distribution of wheat in China. *Quarterly Journal of Applied Meteorology*, **8(1),** 19–25.

Daily, G.C. (ed.) 1997: *Nature's Services: Societal Dependence on Natural Ecosystems*. Island Press, Washington, DC, USA, 412 pp.

Dixon, R.K, O.N. Krankina, and K.I. Kobak, 1996: Global climate change adaptation: examples from Russian boreal forests. In: *Adapting to Climate Change: Assessments and Issues* [Smith J., N. Bhatti, G. Menzhulin, R. Benioff, M.I. Budyko, M. Campos, B. Jallow, and F. Rijsberman (eds.)]. Springer-Verlag, New York, NY, USA, pp. 359–373.

Djoekardi, A.D., 1995: Urban land use planning policy in Indonesia. In: *Proceedings of the International Workshop on Policy Measures for Changing Consumption Patterns, Seoul, Republic of Korea, Aug 30–Sep 1, 1995*.

Dlugolecki, A.F., K.M. Clark, F. Knecht, D. McCauley, J.P. Palutikof, and W. Yambi, 1996: Financial services. In: *Climate Change 1995: Impacts, Adaptations, and Mitigation of Climate Change: Scientific-Technical Analyses. Contribution of Working Group II to the Second Assessment Report of the Intergovernmental Panel on Climate Change* [Watson, R.T., M.C. Zinyowera, and R.H. Moss (eds.)]. Cambridge University Press, Cambridge, United Kingdom and New York, NY, USA, pp. 536–560.

Dow, K., 1999: The extraordinary and the everyday in explanations on vulnerability to oil spill. *Geographical Review*, **89,** 74–93.

Dyurgerov, M. and M. Meier, 1997: Mass balance of mountain and subpolar glaciers: a new global assessment for 1961–1990. *Arctic and Alpine Research*, **29,** 379–391.

Easterling, W.E., 1996: Adapting North American agriculture to climate change: a review. *Agricultural and Forest Meteorology*, **80,** 1–53.

Echeverria, P., C.W. Hoge, L. Bodhidatta, O. Serichantalergs, A. Dalsgaard, B. Eampokalap, J. Perrault, G. Pazzaglia, P. O'Hanley, and C. English, 1995: Molecular characterization of Vibrio cholerae 0139 isolates from Asia. *American Journal of Tropical Medicine and Hygiene*, **52(2),** 124–127.

Ehrenfeld, D., 1991: The management of biodiversity—a conservation paradox. In: *Ecology, Economics, Ethics: The Broken Circle* [Bormann, F.H. and S.R. Kellert (eds.)]. Yale University Press, New Haven, CT, USA, pp. 26–39.

Ellis, W.S., 1990: The Aral—a Soviet sea lies dying. *National Geographic*, **177(2),** 73–92.

Epstein, P.R., 1992: Cholera and the environment. *Lancet*, **339,** 1167–1168.

Epstein, P.R., A. Haines, and P. Reiter, 1998: Global warming and vector-borne disease. *Lancet*, **351,** 1737–1738.

Epstein, Y., E. Sohar, and Y. Shapiro, 1995: Exceptional heatstroke: a preventable condition. *Israel Journal of Medical Science*, **31,** 454–462.

ESD-CAS, 1994: *Impact of Sea Level Rise on the Deltaic Regions of China and Its Mitigation*. Earth Science Division, Chinese Academy of Sciences, Science Press, Beijing, China, 355 pp. (in Chinese).

Esrey, S.A., 1996: Water, waste, and well being: a multi-country study. *American Journal of Epidemiology*, **143(6),** 608–623.

FAO, 1999a: *Production Yearbook 1999*. Food and Agriculture Organization of the United Nations, Rome, Italy.

FAO, 1999b: *The State of Food Insecurity in the World 1999*. Food and Agriculture Organization of the United Nations, Rome, Italy, 32 pp.

FAO, 1997: *The State of World Fisheries and Aquaculture, Yearbook Vol. 84*. Food and Agriculture Organization of the United Nations, Rome, Italy, 57 pp.

FAO, 1996: *World Food Survey—1996*. Food and Agriculture Organization of the United Nations, Rome, Italy.

Farnsworth, E.J. and A.M. Ellison, 1997: The global conservation status of mangroves. *Ambio*, **26(6),** 328–334.

Field, C.B., J.G. Osborn, L.L. Hoffman, J.F. Polsenberg, D.D. Ackerley, J.A. Berry, O. Bjorkman, A. Held, P.A. Matson, and H.A. Mooney, 1998: Mangrove biodiversity and ecosystem function. *Global Ecology and Biogeography Letters*, **7,** 3–14.

Fischer, G., K. Frohber, M. Parry, and C. Rosenzweig, 1996: Impacts of potential climate change on global and regional food production and vulnerability. In: *Climate Change and World Food Security* [Downing, T.E. (ed.)]. Springer-Verlag, Berlin, Germany, pp. 115–159.

Fischer, G., K. Frohber, M. Parry, and C. Rosenzweig, 1995: Climate change and world food supply, demand, and trade. In: *Climate Change and Agriculture: Analysis of Potential International Impacts* [Rosenzweig, C., J.T. Ritchie, J.W. Jones, G.Y. Tsuji, and P. Hildebrand (eds.)]. ASA Special Publication No. 59, American Society of Agronomy, Madison, WI, USA, pp. 341–382.

Fryauff, D.J., E. Gomez-Saladin, Purnomo, I. Sumawinata, M.A. Sutamihardja, S. Tuti, B. Subianto, and T.L. Richie, 1997: Comparative performance of the Parasite F test for detection of plasmodium falciparum in malaria-immune and non-immune populations in Irian Jaya, Indonesia. *Bulletin of the World Health Organization*, **75(6),** 547–552.

Fujimoto, N., T. Fukushima, Y. Inamori, and R. Sudo, 1995: Analytical evaluation of relationship between dominance of cyanobacteria and aquatic environmental factors in Japanese lakes. *Water Environment*, **18,** 901–908 (in Japanese with English abstract).

Fukushima, T., N. Ozaki, H. Kaminishi, H. Harasawa, and K. Matsushige, 2000: Forecasting the changes in lake water quality in response to climate changes, using past relationships between meteorological conditions and water quality. *Hydrological Processes,* **14,** 593–604.

Fushimi, H., 2000a: Influence of decreased snowfall due to warming climate trends on the water quality of Lake Biwa, Japan. In: A Threat to Life: The Impact of Climate Change on Japan's Biodiversity [Domoto, A., K. Iwatsuki, T. Kawamichi, and J. McNeely (eds.)]. Tsukiji-Shokan Publishing Company, Tokyo, Japan; The World Conservation Union (IUCN), Gland, Swizerland; and IUCN, Cambridge, United Kingdom, pp. 35–37.

Fushimi, H., 2000b: Recent changes in glacial phenomena in the Nepalese Himalayas. In: A Threat to Life: The Impact of Climate Change on Japan's Biodiversity [Domoto, A., K. Iwatsuki, T. Kawamichi, and J. McNeely (eds.)]. Tsukiji-Shokan Publishing Company, Tokyo, Japan; The World Conservation Union (IUCN), Gland, Swizerland; and IUCN, Cambridge, United Kingdom, pp. 42–45.

Gadgil, M., F. Berkes, and C. Folke, 1993: Indigineous knowledge for biodiversity conservation. *Ambio*, **22,** 151–156.

Gadgil, S., 1995: Climate change and agriculture—an Indian perspective. *Current Science*, **69(8),** 649–659.

Gamo, T., 1999: Global warming may have slowed down the deep conveyer belt of a marginal sea of the northwestern Pacific: Japan Sea. *Geophysical Research Letters*, **26(20)**, 3137–3140.

Garagulia, L.S. and E.D. Ershov (eds.), 2000: *Geocryological Hazards, Vol. 1 of the Series Natural Hazards of Russia*. Publishing House Kruk, Moscow, Russia, 315 pp. (in Russian).

Genton, B., F. Al-Yaman, M. Ginny, J. Taraika, and M.P. Alpers, 1998: Relation of anthropometry to malaria morbidity and immunity in Papua New Guinean children. *American Journal of Clinical Nutrition*, **68(3)**, 734–741.

Giorgi, F. and R. Francisco, 2000: Evaluating uncertainties in the prediction of regional climate. *Geophysical Research Letters*, **27(9)**, 1295–1298.

Glantz, M.H., 1996: *Currents of Change: El Niño's Impact on Climate and Society*. Cambridge University Press, Cambridge, United Kingdom, 194 pp.

Glynn, P.W., 1996: Coral reef bleaching: facts, hypotheses and implications. *Global Change Biology*, **2**, 495–509.

Golubtsov, V.V., V.I. Lee, and I.I. Scotselyas, 1996: Vulnerability assessment of the water resources of Kazakhstan to anthropogenic climate change and the structure of adaptation measures. *Water Resource Development*, **12**, 193–208.

Graczyk, T.K. and B. Fried, 1998: Echiostomiasis: a common but forgotten food-borne disease. *American Journal of Tropical Medicine and Hygiene*, **58(4)**, 501–504.

Green, J.B., 1990: *IUCN Directory of South Asian Protected Areas*. The World Conservation Union (IUCN), Gland, Switzerland.

Grove, R.H., V. Damodaran, and S. Sangwan, 1998: *Nature and the Orient: the Environmental History of South and Southeast Asia*. Oxford University Press, Oxford, United Kingdom.

Gruza, G. and E. Rankova, 1997: Indicators of Climate Change for the Russian Federation. In: *Proceedings of Workshop on Indices and Indicators for Climate Extremes, Session on Regional Extremes and Indicators Russia/China, Ashville, North Carolina, June 3–6, 1997*, pp. 1–20.

Gruza, G.V., L.K. Kletschenko, and L.N. Aristova, 1999: On relationships between climatic anomalies over the territory of Russia and ENSO event. *Meteorology and Hydrology*, **5**, 26–45 (in Russian).

Gruza, G.V., E. Rankova, M. Bardin, L. Korvkina, E. Rocheva, E. Semenjuk, and T. Platova, 1997: Modern state of the global climate system. In: *Global Changes of Environment and Climate: Collection of Selected Scientific Papers* [Laverov, N.P. (ed.)]. Russian Academy of Sciences, The Federal Research Program of Russia, Moscow, Russia, pp. 194–216.

Gunawardena, D.M., A.R. Wickremasinghe, L. Muthuwatta, S. Weerasingha, J. Rajakaruna, T. Senanayaka, P.K. Kotta, N. Attanayake, R. Carter, and K.N. Mendis, 1998: Malaria risk factor in an endemic region of Sri Lanka, and the impact and cost implications of risk factor-based interventions. *American Journal of Tropical Medicine and Hygiene*, **58(5)**, 533–542.

Hales, S., P. Weinstein, and A. Woodward, 1996: Dengue fever in the south Pacific: driven by El Niño-Southern Oscillation? *Lancet*, **348**, 1664–1665.

Halsey, L.A., D.H. Vitt, and S.C. Zoltai, 1995: Disequilibrium response of permafrost in boreal continental western Canada to climate change. *Climatic Change*, **30(1)**, 57–73.

Hanaki, K., K. Takara, T. Hanazato, H. Hirakuchi, and H. Kayanne, 1998: Impacts on hydrology/water resources and water environment. In: *Global Warming: The Potential Impact on Japan* [Nishioka, S. and H. Harasawa (eds.)]. Springer-Verlag, Tokyo, Japan, pp. 131–163.

Hansen, J.E., M. Sato, A. Lacis, R. Ruedy, I. Tegen, and E. Matthews, 1998: Climate forcings in the industrial era. *Proceedings of the National Academy of Sciences*, **95**, 12753–12758.

Haque, C.E. and M.Q. Zaman, 1993: Human responses to riverine hazards in Bangladesh: a proposal for sustainable floodplain development. *World Development*, **21**, 93–107.

Hassell, D. and R. Jones, 1999: *Simulating Climatic Change of the Southern Asian Monsoon Using a Nested Regional Climate Model*. Technical Note No. 8, Hadley Centre, Bracknell, Berks, United Kingdom, 16 pp.

Hazell, P.B.R., 1985: Sources of increased variability in world cereal production since the 1960s. *Journal of Agricultural Economics*, **36**, 145–159.

Henderson-Sellers, A. and H. Zhang, 1997: *Tropical Cyclones and global Climate Change. Report from the WMO/CAS/TMRP Committee on Climate Change Assessment*. Project TC-2, World Meteorological Organization, Geneva, Switzerland, 47 pp.

Henderson-Sellers, A., H. Zhang, G. Berz, K. Emanuel, W. Gray, C. Landsea, G. Holland, J. Lighthill, S.-L. Shieh, P. Webster, and K. McGuffie, 1998: Tropical cyclones and global climate change: a post-IPCC assessment. *Bulletin of the American Meteorological Society*, **79**, 19–38.

Hewitt, K., 1997: *Regions of Risk: A Geographical Introduction to Disasters*. Addison-Wesley Longman, Essex, United Kingdom, 389 pp.

Hien, T.T., N.V. Vinh Chau, N.N. Vinh, N.T. Hung, M.Q. Phung, L.M. Toan, P.P. Mai, N.T. Dung, D.T. Hoai Tam, and K. Arnold, 1997: Management of multiple drug-resistant malaria in Viet Nam. *Annals of the Academy of Medicine Singapore*, **26(5)**, 659–663.

Hillel, D., 1991: *Out of the Earth: Civilization and the Life of the Soil*. University of California Press, Berkeley, CA, USA, 321 pp.

Hirakuchi, H. and F. Giorgi, 1995: Multi-year present day and 2xCO_2 simulations of monsoon climate over eastern Asia and Japan with a regional climate model nested in a general circulation model. *Journal of Geophysical Research*, **100**, 21105–21126.

Holland, G.J., 1997: The maximum potential intensity of tropical cyclones. *Journal of Atmospheric Science*, **54**, 2519–2541.

Honda, Y., M. Ono, A. Sasaki, and I. Uchiyama, 1995: Relationship between daily maximum temperature and mortality in Kyusyu. *Japanese Journal of Public Health*, **42**, 260–268 (in Japanese with English abstract).

Horie, T., M.J. Kropff, H.G. Centeno, H. Nakagawa, J. Nakano, H.Y. Kim, and M. Ohnishi, 1995a: Effect of anticipated change in global environment on rice yields in Japan. In: Climate Change and Rice [Peng, S., K.T. Ingram, H. Neue, and L.H. Ziska (eds.)]. Springer-Verlag, Berlin, Germany, pp. 291–302.

Horie, T., H. Nakagawa, H.G. Centeno, and M.J. Kropff, 1995b: The rice crop simulation model SIMRIW and its testing. In: *Modeling the Impact of Climate Change on Rice Production in Asia* [Matthews, R.B., D. Bachelet, M.J. Kropff, and H.H. Van Laar (eds.)]. International Rice Research Institute (Philippines) and CAB International, Wallingford, Oxon, United Kingdom, pp. 51–66.

Huang, Z.G., 1999: *Sea Level Changes in Guangdong and Its Impacts and Strategies*. Guangdong Science and Technology Press, Guangzhou, China, 386 pp. (in Chinese).

Huang, Z.G., P.R. Li, and Z.Y. Zhang, 1982: *Zhujiang Delta: Formation, Development and Evolution*. Popular Science Press, Guangzhou, China, 274 pp. (in Chinese).

Huq, S., Z. Karim, M. Asaduzzaman, and F. Mahtab (eds.), 1999: *Vulnerability and Adaptation to Climate Change in Bangladesh*. J. Kluwer Academic Publishers, Dordrecht, The Netherlands, 147 pp.

ICRF, 1998: *Second National Communication*. Interagency Commission of the Russian Federation on Climate Change Problems, Roshydromet, Moscow, Russia, 67 pp.

ICRF, 1995: *First National Communication*. Interagency Commission of the Russian Federation on Climate Change Problems, Roshydromet, Moscow, Russia, 61 pp.

Iglesias, A., L. Erda, and C. Rosenzweig, 1996. Climate change in Asia: a review of the vulnerability and adaptation of crop production. *Water, Air, and Soil Pollution*, **92**, 13–27.

Ilyas, M., A. Pandy, and S.I.S. Hassan, 1999: UV-B radiation at Penang. *Atmospheric Research*, **51**, 141–152.

Inoue, S. and K. Yokoyama, 1998: Estimation of snowfall, maximum snow depth and snow cover condition in Japan under global climate change. *Snow and Ice*, **60**, 367–378 (in Japanese with English abstract).

IPCC, 2000: *Emissions Scenarios. A Special Report of Working Group III of the Intergovernmental Panel on Climate Change* [Nakicenovic, N., J. Alcamo, G. Davis, B. de Vries, J. Fenhann, S. Gaffin, K. Gregory, A. Grübler, T.Y. Jung, T. Kram, E.L. La Rovere, L. Michaelis, S. Mori, T. Morita, W. Pepper, H. Pitcher, L. Price, K. Raihi, A. Roehrl, H.-H. Rogner, A. Sankovski, M. Schlesinger, P. Shukla, S. Smith, R. Swart, S. van Rooijen, N. Victor, and Z. Dadi (eds.)]. Cambridge University Press, Cambridge, United Kingdom and New York, NY, USA, 599 pp.

IPCC, 1998: *The Regional Impacts of Climate Change: An Assessment of Vulnerability. Special Report of IPCC Working Group II* [Watson, R.T., M.C. Zinyowera, and R.H. Moss (eds.)]. Intergovernmental Panel on Climate Change, Cambridge University Press, Cambridge, United Kingdom and New York, NY, USA, 517 pp.

IPCC, 1996: *Climate Change 1995: Impacts, Adaptations, and Mitigation of Climate Change: Scientific-Technical Analyses. Contribution of Working Group II to the Second Assessment Report of the Intergovernmental Panel on Climate Change* [Watson, R.T., M.C. Zinyowera, and R.H. Moss (eds.)]. Cambridge University Press, Cambridge, United Kingdom and New York, NY, USA, 878 pp.

IPCC, 1995: *Climate Change 1994: Radiative Forcing of Climate Change and an Evaluation of the IPCC IS92 Emissions Scenarios* [Houghton, J.T., J. Bruce, B.A. Callander, E. Haites, N. Harris, H. Lee, K. Maskell, and L.G. Meira Filho (eds.)]. Cambridge University Press, Cambridge, United Kingdom and New York, NY, USA, 339 pp.

Islam, N. (ed.), 1995: *Population and Food in Early 21st Century: Meeting Future Food Demand of an Increasing Population.* International Food Policy Research Institute (IFPRI), Washington, DC, USA.

Ives, J.D. and B. Messerli, 1989: The Himalayan Dilemma: Reconciling Development and Conservation. Routledge, London, United Kingdom, 336 pp.

Izrael, Y., Y. Anokhin, and A.D. Eliseev, 1997a: Adaptation of water management to climate change. In: *Global Changes of Environment and Climate: Collection of Selected Scientific Papers* [Laverov, N.P. (eds.)]. The Federal Research Program of Russia, Russian Academy of Sciences, Moscow, Russia, pp. 373–392.

Izrael, Y., Y. Anokin, and A.D. Eliseev, 1997b: *Final Report of the Russian Country Study on Climate Problem, Russian Federal Service for Hydrometeorology and Environmental Monitoring, Vol. 3, Task 3: Vulnerability and Adaptation Assessments.* Roshydromet, Moscow, Russia, 105 pp.

Izrael, Y., A.V. Pavlov, and Y. Anokhin, 1999: Analysis of modern and predicted climate and permafrost changes in cold regions of Russia. *Meteorology and Hydrology*, **3,** 18–27 (in Russian).

Japan Meteorological Agency, 1999: *Extreme Weather Report '99.* Printing Bureau, Ministry of Finance, Tokyo, Japan, 341 pp. (in Japanese).

Jarupongsakul, T., 1999: Implications of sea level rise and coastal zone management in the upper Gulf of Thailand. In: *Proceedings of the IPCC Regional Experts Meeting, National Institute of Environmental Studies, Tsukuba, Japan, 21–23 June, 1999.*

Ji, Z.X., Z.X. Jiang, and J.W. Zhu, 1993: Impacts of sea level rise on coastal erosion in the Changjiang Delta Northern Jiangsu coastal plain. *Acta Geographica Sinica*, **48(6),** 516–526 (in Chinese with English abstract).

Jiao, X., G.Y. Liu, C.O. Shan, X. Zhao, X.W. Li, I. Gathmann, and C. Royce, 1997: Phase II trial in China of new, rapidly acting and effective oral antimalarial CGP 56697, for the treatment of plasmodium falciparum malaria. *Southeast Asian Journal of Tropical Medicine and Public Health*, **28(3),** 476–481.

Jodha, N.S., 1996: Property rights and development. In: *Rights to Nature* [Hanna, S.S., C. Folke, and K.-G. Maler (eds.)]. Island Press, Washington, DC, USA, pp. 205–220.

Jones, R.G., J.M. Murphy, and M. Noguer, 1995: Simulation of climate change over Europe using a nested regional climate model, part I: assessment of control climate, including sensitivity to location of lateral boundaries. *Quarterly Journal of Royal Meteorological Society,* **121,** 1413–1449.

Jones, R.N., K.J. Hennessy, C.M. Page, A.B. Pittock, R. Suppiah, K.J.E. Walsh, and P.H. Whetton, 1999: *An Analysis on the Effects of the Kyoto Protocol on Pacific Island Countries, Part Two: Regional Climate Change Scenarios and Risk Assessment Methods.* Commonwealth Scientific and Industrial Research Organisation (CSIRO), Atmospheric Research Report to the South Pacific Regional Environment Programme, Aspendale, Australia, 69 pp. (plus executive summary).

Kane, S., J. Reilly, and J. Tobey, 1991: *Climate Change: Economic Implication for World Agriculture.* AER-No. 647, U.S. Department of Agriculture, Economic Research Service, Washington, DC, USA, 21 pp.

Kark, S., U. Alkon, E. Randi, and U.N. Safriel, 1999: Conservation priorities for chukar partridge in Israel based on genetic diversity across an ecological gradient. *Conservation Biology*, **13,** 542–552.

Karl, T.R., R.W. Knight, and N. Plummer, 1995: Trends in high frequency climate variability in the twentieth century. *Nature*, **377,** 217–220.

Karl, T.R., G. Kukla, V.N. Razuvayev, M.J. Changery, K.G. Quayle, R.R. Heim, D.R. Easterling, and C.B. Fu, 1991: A new perspective on global warming: asymmetric increases of day and night temperatures. *Geophysical Research Letters*, **18,** 2253–2256.

Kasimov, N.S., 1997: *Geo-ecological Changes Under the Caspian Sea Level Fluctuation.* Moscow University Press, Moscow, Russia, 205 pp. (in Russian).

Kates, R.W., 2000: Cautionary tales: adaptation and the global poor. *Climatic Change*, **45,** 5–17.

Kattenberg, A., F. Giorgi, H. Grassl, G.A.Meehl, J.F.B.Mitchell, R.J. Stouffer, T. Tokioka, A.J.Weaver, and T.M.L.Wigley, 1996: Climate models—projections of future climate. In: *Climate Change 1995: The Science of Climate Change. Contribution of Working Group I to the Second Assessment Report of the Intergovernmental Panel on Climate Change* [Houghton, J.T., L.G. Meira Filho, B.A. Callander, N. Harris, A. Kattenberg, and K. Maskell (eds.)]. Cambridge University Press, Cambridge, United Kingdom and New York, NY, USA, pp. 285–357.

Katti, M. and T. Price, 1996: Effects of climate on palearctic warblers over wintering in India. *Journal of the Bombay Natural History Society*, **93,** 411–427.

Kaye, K. and M. Novell, 1994: Health practices and indices of a poor urban population in Indonesia, part E: immunization, nutrition, and incidence of diarrhea. *Asia Pacific Journal of Public Health*, **7(4),** 224–227.

Kelly, P.M. and W.N. Adger, 2000: Theory and practice in assessing vulnerability to climate change and facilitating adaptation. *Climatic Change*, **47,** 325–352.

Kelly, P.M., D.A. Campbell, P.P. Micklin, and J.R. Tarrant, 1983: Large scale water transfers in the USSR. *GeoJournal*, **7(3),** 201–214.

Khuldorj, M., M. Badarch, T. Tsetsegee, N. Oyun-erdene, and Chuluuntsetseg, 1998: *The Mongolian Action Programme for 21st Century.* Ulaanbaatar, Mongolia.

Kitayama, K., 1996: Climate of the summit region of Mount Kinabalu (Borneo) in 1992 and El Niño year. *Mountain Research and Development*, **16(1),** 65–75.

Kitayama, K. and D. Mueller-Dombois, 1995: Biological invasion on an oceanic island mountain: do alien plant species have wider ecological ranges than native species? *Journal of Vegetation Science*, **6,** 667–674.

Kitoh, A., S. Yukimoto, A. Noda, and T. Motoi, 1997: Simulated changes in the Asian summer monsoon at times of increased atmospheric CO_2. *Journal of the Meteorological Society of Japan*, **75,** 1019–1031.

Klein, R.J.T. and R.J. Nicholls, 1999: Assessment of coastal vulnerability to climate change. *Ambio*, **28(2),** 182–187.

Knutson, T.R. and S. Manabe, 1998: Model assessment of decadal variability and trends in the tropical Pacific Ocean. *Journal of Climate*, **11,** 2273–2296.

Knutson, T.R., R.E. Tuleya, and Y. Kurihara, 1998: Simulated increase of hurricane intensities in a CO_2-warmed climate. *Science*, **279,** 1018–1020.

Kothyari, U.C. and V.P. Singh, 1996: Rainfall and temperature trends in India. *Hydrological Processes*, **10,** 357–372.

Krankina, O.N., R.K. Dixon, A.P. Kirilenko, and K.I. Kobak, 1997: Global climate change adaptation: examples from Russian boreal forests. *Climatic Change*, **36(1–2),** 197–215.

Kripalani, R.H., S.R. Inamdar, and N.A. Sontakke, 1996: Rainfall variability over Bangladesh and Nepal: comparison and connection with features over India. *International Journal of Climatology*, **16,** 689–703.

Kripke, M.L., 1994: Ultraviolet radiation and immunology: something new under the sun. *Cancer Research*, **54,** 6102–6105.

Krishnamurti, T.N., R. Correa-Torres, M. Latif, and G. Daughenbaugh, 1998: The impact of current and possibly future SST anomalies on the frequency of Atlantic hurricanes. *Tellus*, **50A,** 186–210.

Kropff, M.J., R.B. Matthews, H.H. Van Laar, and H.F.M. Ten Berge, 1995: The rice model ORYZA1 and its testing. In: *Modeling the Impact of Climate Change on Rice Production in Asia* [Matthews, R.B., D. Bachelet, M.J. Kropff, and H.H. Van Laar (eds.)]. International Rice Research Institute (Philippines) and CAB International, Wallingford, Oxon, United Kingdom, pp. 27–50.

Kuzmin, M.I., D. Williams, and T. Kavai, 1997: Baikal drilling project: long-range climate reconstructions. In: *Global Changes of Environment and Climate: Collection of Selected Scientific Papers* [Laverov, N.P. (ed.)]. The Federal Research Program of Russia, Russian Academy of Sciences, Moscow, Russia, pp. 255–281.

Lal, M. and H. Harasawa, 2000: Comparison of the present-day climate simulation over Asia in selected coupled atmosphere-ocean global climate models. *Journal of the Meteorological Society of Japan*, **78(6),** 871–879.

Lal, M. and H. Harasawa, 2001: Future climate change scenarios for Asia as inferred from selected coupled atmosphere-ocean global climate models. *Journal of the Meteorological Society of Japan*, **79(1),** 219–227.

Lal, M., G.A. Meehl, and J.M. Arblaster, 2000: Simulation of Indian summer monsoon rainfall and its intraseasonal variability. *Regional Environmental Change*, **1(3/4),** 163–179.

Lal, M., G. Srinivasan, and U. Cubasch, 1996: Implications of global warming on the diurnal temperature cycle of the Indian subcontinent. *Current Science*, **71(10),** 746–752.

Lal, M., U. Cubasch, R. Voss, and J. Waszkewitz, 1995a: Effect of transient increases in greenhouse gases and sulphate aerosols on monsoon climate. *Current Science*, **69(9),** 752–763.

Lal, M., L. Bengtsson, U. Cubash, M. Esch, and U. Schlese, 1995b: Synoptic scale disturbances of Indian summer monsoon as simulated in a high resolution climate model. *Climate Research*, **5,** 243–258.

Lal, M., U. Cubasch, J. Perlwitz, and J. Waszkewitz, 1997: Simulation of the Indian monsoon climatology in ECHAM3 climate model: sensitivity to horizontal resolution. *International Journal of Climatology*, **17,** 1–12.

Lal, M., P.H. Whetton, A.B. Pittock, and B. Chakraborty, 1998a: Simulation of present-day climate over the Indian subcontinent by general circulation models. *Terrestrial, Atmospheric and Oceanic Sciences*, **9(1),** 69–96.

Lal, M., P.H. Whetton, A.B. Pittock, and B. Chakraborty, 1998b: The greenhouse gas-induced climate change over the Indian subcontinent as projected by general circulation model experiments. *Terrestrial, Atmospheric and Oceanic Sciences*, **9,** 673–690.

Lal, M., B. Bhaskaran, and S.K. Singh, 1998c: Indian summer monsoon variability as simulated by regional climate model nested in a global climate model. *Chinese Journal of Atmospheric Science*, **22(1),** 93–102.

Lal, M., K.K. Singh, L.S. Rathore, G. Srinivasan, and S.A. Saseendran, 1998d: Vulnerability of rice and wheat yields in north-west India to future changes in climate. *Agricultural and Forest Meteorology*, **89,** 101–114.

Lander, M., 1994: An exploratory analysis of the relationship between tropical storm formation in the western north Pacific and ENSO. *Monthly Weather Review*, **122,** 636–651.

Lander, M.A. and C.P. Guard, 1998: A look at global tropical cyclone activity during 1995: contrasting high Atlantic activity with low activity in other basins. *Monthly Weather Review*, **126,** 1163–1173.

Lang, H.Q., P. Lin, and J.J. Lu, 1998: *Conservation and Research of Wetlands in China*. East China Normal University Press, Shanghai, China, 420 pp. (in Chinese with English abstract).

Laverov, N.P. (ed.), 1998: *Global Changes of Environment and Climate: Collection of Selected Scientific Papers*. The Federal Research Program of Russia, Russian Academy of Sciences, Moscow, Russia, 433 pp.

Leggett, J., W.J. Pepper, and R.J. Swart, 1992: Emissions Scenarios for IPCC: an update. In: *Climate Change 1992. The Supplementary Report to the IPCC Scientific Assessment* [Houghton, J.T., B.A. Callander, and S.K. Varney (eds.)]. Cambridge University Press, Cambridge, United Kingdom and New York, NY, USA, pp. 69–95.

Lelyakin, A.L., A.O. Kokorin, and I.M. Nazarov, 1997: Vulnerability of Russian forests to climate changes, model estimation of CO_2, fluxes. *Climate Change*, **36(1–2),** 123–133.

Li, C.X., 1987: A study of the influence of El Niño upon typhoon action over the western Pacific. *Acta Meteorologica Sinica*, **45,** 229–236.

Li, C.X., 1985: *Luanhe Fan-Delta Depositional Systems*. Geological Publishing House, Beijing, China, 156 pp.

Li, C.X., 1984: Sedimentary processes in the Yangtze Delta since late Pleistocene. *Collected Oceanic Works*, **7(2),** 116–126.

Li, C.X. and P. Wang, 1991: Stratigraphy of the Late Quaternary barrier-lagoon depositional systems along the coast of China. *Sedimentary Geology*, **72,** 189–200.

Li, Z., R. Yang, D. Liu, Q. Fan, and L. Wang, 1991: Research on the relationship between migration of birds of prey and climate in the seaboard of eastern China. *Forest Research*, **4,** 10–14.

Lighthill, J., G.J. Holland, W.M. Gray, C. Landsea, K. Emanuel, G. Craig, J. Evans, Y. Kunihara, and C.P. Guard, 1994: Global climate change and tropical cyclones. *Bulletin of the American Meteorological Society*, **75,** 2147–2157.

Liu, Y.F., M.K. Han, L. Wu, and N. Mimura, 1998: Recent evolution of outlets in Zhujiang River Delta and the prospect for land reclamation. *Acta Geographica Sinica*, **53,** 492–499 (in Chinese with English abstract).

Lou, Q. and E. Lin, 1999: Agricultural vulnerability and adaptation in developing countries: the Asia Pacific region. *Climatic Change*, **43,** 729–743.

Mackenzie, F.T., L.M. Ver, and A. Lerman, 1998: Coupled biogeochemical cycles of carbon, nitrogen, phosphorous and sulfur in the land-ocean-atmosphere system. In: *Asian Change in the Context of Global Climate Change* [Galloway, J.N. and J.M. Melillo (eds.)]. Cambridge University Press, Cambridge, United Kingdom and New York, NY, USA, pp. 42–100.

Makarov, M.V., 1999: Influence of ultraviolet-radiation on growth of the dominant macroalgae of the Barents Sea. *Chemosphere Global Change Science: Climate Change effect on Northern Terrestrial and Freshwater Ecosystems*, **1(4),** 461–469.

Malinin, V.N., 1994: *Problems of the Prediction in Caspian Sea Level Changes*. SANKT Press, St. Petersburg, Russia, 159 pp. (in Russian).

Martens, P., R.S. Kovats, S. Nijhof, P. de Vries, M.T.J. Livermore, D.J. Bradley, J. Cox, and A.J. McMichael, 1999: Climate change and future populations at risk of malaria. *Global Environmental Change*, **9,** S89–S107.

Masuzawa, T., 2000: How will communities of alpine plants be affected by global warming? In: *A Threat to Life: The Impact of Climate Change on Japan's Biodiversity* [Domoto, A., K. Iwatsuki, T. Kawamichi, and J. McNeely (eds.)]. Tsukiji-Shokan Publishing Company, Tokyo, Japan; The World Conservation Union (IUCN), Gland, Swizerland; and IUCN, Cambridge, United Kingdom, pp. 46–51.

Masuzawa, T., 1997: *The Ecology of Alpine Plants*. Tokyo University Press, Tokyo, Japan, 220 pp. (in Japanese).

Matson, P.A., W.J. Parton, A.G. Power, and M.J. Swift, 1997: Agricultural intensification and ecosystem properties. *Science*, **277,** 504–509.

Matsui, T. and T. Horie, 1992: Effects of elevated CO_2 and high temperature on growth and yield of rice, part 2: sensitive period and pollen germination rate in high temperature sterility of rice spikelets at flowering. *Japan Journal of Crop Science*, **61,** 148–149.

Matthews, R.B., T. Horie, M.J. Kropff, D. Bachelet, H.G. Centeno, J.C. Shin, S. Mohandass, S. Singh, Z. Defeng, and H.L. Moon, 1995a: A regional evaluation of the effect of future climate change on rice production in Asia. In: *Modeling the Impact of Climate Change on Rice Production in Asia* [Matthews, R.B., M.J. Kropff, D. Bachelet, and H.H. Van Laar (eds.)]. International Rice Research Institute (Philippines) and CAB International, Wallingford, Oxon, United Kingdom, pp. 95–139.

Matthews, R.B., M.J. Kropff, D. Bachelet, and H.H. Van Laar, 1995b: *Modeling the Impact of Climate Change on Rice Production in Asia*. International Rice Research Institute (Philippines) and CAB International, Wallingford, Oxon, United Kingdom, 289 pp.

McKenzie, R.L., B. Connor, and G. Bodeker, 1999: Increased summertime UV radiation in New Zealand in response to ozone loss. *Science*, **285,** 1709–1711.

McMichael, A.J., A. Haines, R. Slooff, and S. Kovats (eds.), 1996: *Climate Change and Human Health*. WHO/EHG/96.7, an assessment prepared by a Task Group on behalf of the World Health Organization, the World Meteorological Organization, and the United Nations Environment Programme, Geneva, Switzerland, 297 pp.

Meehl, G.A. and W.M. Washington, 1996: El Niño-like climate change in a model with increased atmospheric CO_2 concentrations. *Nature*, **382,** 56–60.

Meehl, G.A. and W.M. Washington, 1993: South Asian summer monsoon variability in a model with doubled atmospheric carbon dioxide concentration. *Science*, **260,** 1101–1104.

Melillo, J.M., I.C. Printice, G.D. Farquhar, E.D. Schuze, and O.E. Sala, 1996: Terrestrial biotic responses to environmental change and feedbacks to climate. In: *Climate Change 1995: The Science of Climate Change. Contribution of Working Group I to the Second Assessment Report of the Intergovernmental Panel on Climate Change* [Houghton, J.T., L.G. Meira Filho, B.A. Callander, N. Harris, A. Kattenberg, and K. Maskell (eds.)]. Cambridge University Press, Cambridge, United Kingdom and New York, NY, USA, pp. 444–516.

Middleton, N., 1999: *The Global Casino: An Introduction to Environmental Issues*. 2nd. ed., Arnold, London, United Kingdom, 370 pp.

Midun, Z. and S.C. Lee, 1995: Implications of a greenhouse gas induced sea level rise: a national assessment for Malaysia. *Journal of Coastal Research*, **14,** 96–115.

Mikhailov, V.N., 1998: *The Volga River-Mouth area: Hydrological-Morphological Processes, Regime of Contaminants and Influence of the Caspian Sea Level Changes*. Geos Press, Moscow, Russia, 278 pp. (in Russian).

Mimura, N. and H. Harasawa, 2000: *Data Book of Sea-Level Rise 2000*. Centre for Global Environmetal Research, National Institute for Enviromental Studies, Environmetal Agency of Japan, Ibaraki, Japan, 280 pp.

Mimura, N., J. Tsutsui, T. Ichinose, H. Kato, and K. Sakaki, 1998: Impacts on infrastructure and socio-economic system. In: *Global Warming: The Potential Impact on Japan* [Nishioka, S. and H. Harasawa (eds.)]. Springer-Verlag, Tokyo, Japan, pp. 165–201.

Mirza, M.Q., 1998: *Modeling the Effects of Climate Change on Flooding in Bangladesh*. Diss. International Global Change Institute (IGCI), University of Waikato, Hamilton, New Zealand, 279 pp.

Mirza, M.Q. and A. Dixit, 1997: Climate change and water management in the GBM Basins. *Water Nepal*, **5,** 71–100.

Mirza, M.Q. and N.J. Ericksen, 1996: Impact of water control projects on fisheries resources in Bangladesh. *Enviornmental Management*, **20(4),** 527–539.

Mitchell, J.F.B., T.C. Johns, J.M. Gregory, and S.F.B. Tett, 1995: Climate response to increasing levels of greenhouse gases and sulphate aerosols. *Nature*, **376,** 501–504.

Mizina, S.V., J. Smith, E. Gossen, K. Spiecker, and S. Witowski, 1999: An evaluation of adaptation options for climate change impacts on agriculture in Kazakhstan. *Mitigation and Adaptation Strategies for Global Change*, **4,** 25–41.

Mukhopadhyay, A.K., P. Karmakar, A.K. Hati, and P. Dey, 1997: Recent epidemiological status of malaria in Calcutta Municipal Corporation area, West Bengal. *Indian Journal of Malariol*, **34(4),** 188–196.

Munasinghe, M., 2000: Development, equity and sustainabillity (DES) in the context of climate change. In: Climate Change and Its Linkages with Development, Equity and Sustainability: Proceedings of the IPCC Expert Meeting held in Colombo, Sri Lanka, 27–29 April, 1999 [Munasinghe, M. and R. Swart (eds.)]. LIFE, Colombo, Sri Lanka; RIVM, Bilthoven, The Netherlands; and World Bank, Washington, DC, USA, pp. 69–90.

Mustafa, D., 1998: Structural causes of vulnerability to flood hazard in Pakistan. *Economic Geography*, **74(3),** 289–305.

Nakagawa, A., J. Nagasawa, and S. Inoue, 1997: Projection of global warming impacts using world rice and wheat supply and demand model. *Quarterly Journal of National Research Institute of Agricultural Economy*, **36,** 37–50 (in Japanese).

Nakagawa, S., M. Sugi, T. Motoi, and S. Yukimoto, 1998: Climate change projections. In: *Global Warming: The Potential Impact on Japan* [Nishioka, S. and H. Harasawa (eds.)]. Springer-Verlag, Tokyo, Japan, pp. 1–33.

Nakicenovic, N., N. Victor, and T. Morita, 1998: Emissions scenarios database and review of scenarios. *Mitigations and Adaptation Strategies for Global Change*, **3,** 95–120.

Naylor, R, W. Falcon, and E. Zavaleta, 1997: Variability and growth in grain yields, 1950–94: does the record point to greater instability? *Population and Development Review*, **23(1),** 41–61.

NBAP, 2000: *National Biodiversity Action Plan*. Draft plan, His Majesty's Government of Nepal, Nepal.

Nelson, F.E. and O.A. Anisimov, 1993: Permafrost zonation in Russia under anthropogenic climatic change. *Permafrost Periglacial Processes*, **4(2),** 137–148.

Neumann, C.J., 1993: *Global Overview, Global Guide to Tropical Cyclone Forecasting*. WMO/TC No. 560, Report No. TCP-31, World Meteorological Organization, Geneva, Switzerland, Chapter 1, pp. 1–43.

Nicholls, R.J., 1995: Coastal megacities and climate change. *GeoJournal*, **37,** 369–379.

Nicholls, R.J. and N. Mimura, 1998: Regional issues raised by sea level rise and their policy implications. *Climate Research*, **11,** 5–18.

Nicholls, R.J., F.M.J. Hoozemans, and M. Marchand, 1999: Increasing flood risk and wetland losses due to global sea-level rise: regional and global analyses. *Global Environmental Change*, **9,** S69–S87.

Nie, B.F., T.G. Chen, and M.T. Lein, 1997: *The Relationship Between Reef Coral and Environmental Changes of Nasha Islands and Adjacent Regions*. China Science Press, Beijing, China, 101 pp. (in Chinese with English abstract).

Nishimori, M. and M. Yoshino, 1990: The relationship between ENSO events and the generation, development and movement of typhoons. *Geographic Review of Japan*, **63A-8,** 530–540 (in Japanese).

Nishioka, S. and H. Harasawa (eds.) 1998: *Global Warming: The Potential Impact on Japan*. Springer-Verlag, Tokyo, Japan, 244 pp.

Nylen, E.S., A. Al-Arifi, K.L. Becker, R.H. Snider Jr., and A. Alzeer, 1997: Effect of classic heatstroke on serum procalcitonin. *Critical Care Medicine*, **25(8),** 1362–1365.

OEPP, 1996: *Report on Environmental Conditions of the Year 1994*. Office of Environmental Policy and Planning, Ministry of Science, Technology and Energy, Bangkok, Thailand, 307 pp. (in Thai).

Omasa, K., K. Kai, H. Toda, Z. Uchijima, and M. Yoshino (eds.), 1996: *Climate Change and Plants in East Asia*. Springer-Verlag, Tokyo, Japan, 215 pp.

Ostby, F.P., 1993: The changing nature of tornado climatology. In: *Proceedings of the 17th Conference on Severe Local Storms, October 4–8, 1993, St. Louis, Missouri* , pp. 1–5.

Patterson, D.T., J.K. Westbrook, R.J.V. Joyce, P.D. Lingren, and J. Rogasik, 1999: Weeds, insects and diseases. *Climatic Change*, *43,* 711–727.

Patz, J.A. and W.J.M. Martens, 1996: Climate impacts on vector-borne disease transmission: global and site-specific. *Journal of Epidemiology*, **6,** S145–S148.

Pazzaglia, G., P. O'Hanley, and C. English, 1995: Molecular characterization of Vibro cholera 0139 isolates from Asia. *American Journal of Tropical Medicine and Hygiene*, **52(5),** 124–127.

Pebley, A.R., 1998: Demography and the environment. *Demography*, **35,** 377–389.

Pender, M., 1995: Recent retreat of the terminus of Rika Samba Glacier, Hidden Valley, Nepal. In: *Himalayan Climate Expedition: Final Report* [Wake, C.P. (ed.)]. Glacier Research Group, University of New Hampshire, pp. 32–39.

Pennisi, E., 1998: New threat seen from carbon dioxide. *Science*, **279,** 989–990.

Pennisi, E., 1997: Brighter prospects for the world's coral reefs? *Science*, **277,** 491–493.

Perez, R.T., L.A. Amodore, and R.B. Feir, 1999: Climate change impacts and responses in the Philippines coastal sector. *Climate Research, CR Special 6*, **12,** 97–107.

Pilifosova, O., I. Eserkepova, and S. Dolgih, 1997: Regional climate change scenarios under global warming in Kazakhstan. *Climatic Change*, **36,** 23–40.

Pilifosova, O., I. Eserkepova, and S. Mizina, 1996: Vulnerability and adaptation assessment for Kazakhstan. In: *Vulnerability and Adaptation to Climate Change—Interim Results from a U. S. Country Study Program* [Smith, J.B., S. Huq, S. Lenhart, L.J. Mata, I. Nemesova, and S. Toure (eds.)]. J. Kluwer Academic Publishers, Dordrecht, The Netherlands, pp. 161–181.

Ping, C.-L., 1996: Carbon storage in the Kuparuk River Basin. In: *Proceedings of the Arctic System Science/Land-Atmosphere-Ice Interactions Science Workshop, Seattle, Washington*.

Piver, W.T., M. Ando, F. Ye, and C.T. Portier, 1999: Temperature and air pollution as risk factors for heat stroke in Tokyo, July–August, 1980–1995. *Environmental Health Perspective*, **107(11),** 911–916.

Piyakarnchana, T., 1999: Change state and health of the Gulf of Thailand large marine ecosystem. In: *Large Marine Ecosystems of the Pacific Rim: Assessment, Sustainability, and Management* [Sherman K. and Tang Q.S. (eds.)]. Blackwell, Oxford, United Kingdom, pp. 240–250.

Popov, Y.M. and T.J. Rice, 1997: Economic and ecological problems of the Aral Sea region during transition to a market economy. *Ecology and Hydrometeorology*, **2,** 24–36.

Portmann, R.W., S. Solomon, J. Fishman, J.R. Olson, J.T. Kiehl, and B. Briegleb, 1997: Radiative forcing of the earth's climate system due to tropical ozone production. *Journal of Geophysical Research*, **102(D8),** 9409–9417.

Primavera, H., 1998: Mangroves as nurseries: shrimp populations in mangrove and non-mangrove habitats. *Estuarine, Coastal and Shelf Science*, **46,** 457–464.

Pritchard, L., J. Colding, F. Berkes, U. Svedin, and C. Folke, 1998: *The Problem of Fit Between Ecosystems and Institutions*. Working Paper No. 2, International Human Dimension Programme, Bonn, Germany. Available online at http://www.uni-bonne.de/ihdp/wp02mam.htm.

Qian, Y.Y., Q.C. Ye, and W.H. Zhou, 1993: *Water Discharge and Suspended Load Changes and Channel Evolution of Huanghe River*. China Building Material Industry Press, Beijing, China, 230 pp. (in Chinese).

Rahman, W.A., A. Che'Rus, and A.H. Ahmad, 1997: Malaria and Anopheles mosquitoes in Malaysia. *Southeast Asian Journal of Tropical Medicine and Public Health*, **28(3),** 599–605.

Rai, S.C., 1999: Land use change and hydrology of a Himalayan watershed. In: *Abstract Book of the 1999 IGES Open Meeting of the Human Dimensions of Global Environmental Change Research Community, Shonan Village, Japan, 24–26 June*, 200 pp.

Ramakrishnan, P.S., 1998: Sustainable development, climate change, and the tropical rain forest landscape. *Climatic Change*, **39,** 583–600.

Ramakrishnan, P.S., 1992: *Shifting Agriculture and Sustainable Development: An Interdisciplinary Study from North-Eastern India*. United nations Educational, Scientific, and Cultural Organization and the Man and the Biosphere Program (UNESCO-MAB) Series, Parthenon Publications, Carnforth, Lancaster, United Kingdom, 424 pp. (republished by Oxford University Press, New Delhi, India, 1993).

Ramakrishnan, P.S., A.N. Purohit, K.G. Saxena, and K.S. Rao, 1994: *Himalayan Environment and Sustainable Development*. Diamond Jubilee Publication, Indian National Science Academy, New Delhi, India, 84 pp.

Ramakrishnan, P.S., A.K. Das, and K.G. Saxena (eds.), 1996: *Conserving Biodiversity for Sustainable Development*. Indian National Science Academy, New Delhi, India, 246 pp.

Rankova, E., 1998: Climate change during the 20th century for the Russian Federation. In: *Abstract Book of the 7th International Meeting on Statistical Climatology, Whistler, British Columbia, Canada, May 25–29*. Abstract 102, p. 98.

Rao, G.D. and S.K. Sinha, 1994: Impact of climate change on simulated wheat production in India. In: *Implications of Climate Change for international agriculture: Crop Modelling Study* [Rosenzweig, C. and I. Iglesias (eds.)]. U.S. Environmental Protection Agency, EPA230-B-94-003, Washington, DC, USA, pp. 1–10.

Reiter, P., 1998: Global warming and vector-borne disease in temperate regions and at high altitude. *Lancet*, **351,** 839–840.

Ren, G., 1998: Temperature changes of the 20th century over Horqin region, Northeast China. *Scientia Meteorologica Sinica*, **18(4),** 373–380.

Ren, G. and W. Zhou, 1994: A preliminary study on temperature change since 1905 over Liaodong peninsula, Northeast China. *Acta Meteorologica Sinica*, **52(4),** 493–498.

Ren, G., H. Wu, and Z. Chen, 2000: Spatial pattern of precipitation change trend of the last 46 years over China. *Journal of Applied Meteorology*, **11(3),** 322–330.

Ren, M.E., 1985: *Modern Sedimentation in Coastal and Nearshore Zone of China*. China Ocean Press, Beijing, China and Springer-Verlag, Berlin and Heidelberg, Germany; New York, NY, USA; and Tokyo, Japan, 464 pp.

Rice, C.G., 1998: Habitat, population dynamics and conservation of Nilgiri tahr (Hemitragus hylocrius). *Biological Conservation*, **44,** 137–156.

Riebsame, W.E., K.M. Strzepek, J.L. Wescoat Jr., G.L. Gaile, J. Jacobs, R. Leichenko, C. Magadza, R. Perrit, H. Phien, B.J. Urbiztondo, P. Restrepo, W.R. Rose, M. Saleh, C. Tucci, L.H. Ti, and D. Yates, 1995: Complex river basins. In: *As Climate Changes, International Impacts and Implications* [Strzepek, K.M. and J.B. Smith (eds.)]. Cambridge University Press, Cambridge, United Kingdom and New York, NY, USA, pp. 57–91.

Riha, S.J., D.S. Wilks, and P. Simons, 1996: Impact of temperature and precipitation variability on crop model predictions. *Climatic Change*, **32,** 293–331.

Robock, A., K.Y. Vinnikov, G. Srinivasan, J.K. Entin, S.E. Hollinger, N.A. Speranskaya, S. Liu, and A. Namkhai, 2000: The global soil moisture data bank. *Bulletin of the American Meteorological Society*, **81,** 1281–1299.

Roeckner, E., L. Bengtsson, J. Feitcher, J. Lelieveld, and H. Rodhe, 1999: Transient climate change with a coupled atmosphere-ocean GCM, including the tropospheric sulphur cycle. *Journal of Climate*, **12,** 3004–3032.

Rosenzweig, C. and D. Hillel, 1998: *Climate Change and the Global Harvest: Potential Impacts of the Greenhouse Effect on Agriculture*. Oxford University Press, Oxford, United Kingdom, 324 pp.

Rosenzweig, C. and D. Hillel, 1995: Potential impacts of climate change on agriculture and food supply. *Consequences*, **1(2),** 23–32.

Rosenzweig, C. and M.L. Parry, 1993: Potential impacts of climate change on world food supply. In: *Agricultural Dimensions of Global Climate Change* [Keiser, H.M. and T.E. Drennen (eds.)]. St. Lucie Press, Delray Beach, FL, USA, pp. 87–116.

Rounsevell, M.D.A., S.P. Evans, and P. Bullick, 1999: Climate change and agricultural soils—impacts and adaptation. *Climatic Change*, **43,** 683–709.

Royer, J.-F., F. Chauvin, B. Timbal, P. Araspin, and D. Grimal, 1998: A GCM study of the impact of greenhouse gas increase on the frequency of occurrence of tropical cyclones. *Climatic Change*, **38,** 307–343.

Rozengurt, M.A., 1991: Strategy and ecological and societal results of extensive resources development in the south of the U.S.S.R. In: *Proceedings of the Soviet Union in the Year 2010*. USAIA and Georgetown University, Washington, DC, USA.

Ruddle, K., 1998: Traditional community-based coastal marine fisheries management in Vietnam. *Ocean and Coastal Management*, **40,** 1–22.

Ruddle, K., W.B. Morgan, and J.R. Pfafflin, 1988: *The Coastal Zone: Man's Response to Change*. Harwood Academic Publishers, Chur, Switzerland, 555 pp.

Rupakumar, K. and S.D. Patil, 1996: Long-term variations of rainfall and surface air temperature over Sri Lanka. In: *Climate Variability and Agriculture* [Abrol, Y.P., S. Gadgil, and G.B. Pant (eds.)]. Narosa Publishing House, New Delhi, India, pp. 135–152.

Rupakumar, K., K. Krishna Kumar and G.B. Pant, 1994: Diurnal asymmetry of surface temperature trends over India. *Geophysical Research Letters*, **21,** 677–680.

Safriel, U.N., 1995: The role of ecology in desert development. *Journal of Arid Land Studies*, **55,** 351–354.

Safriel, U.N., 1993: Climate change and the Israeli biota: little is known but much can be done. In: *Regional Implications of Future Climate Change* [Graber, M., A. Cohen, and M. Magaritz (eds.)]. The Israeli Academy of Sciences and Humanities, Jerusalem, Israel, pp. 252–260.

Safriel, U.N., S. Volis, and S. Kark, 1994: Core and peripheral populations and global climate change. *Israel Journal of Plant Sciences*, **42,** 331–345.

Salafsky, N., 1998: Drought in the rainforest, part II: an update based on the 1994 ENSO event. *Climatic Change*, **39,** 601–603.

Salafsky, N., 1994: Drought in the rainforest: effects of the 1991 El Niño-Southern Oscillation event on a rural economy in West Kalimantan, Indonesia. *Climatic Change*, **27,** 373–396.

Sanchez-Arcilla, A. and J.A. Jimenez, 1997: Physical impacts of climatic change on deltaic coastal systems: I: an approach. *Climatic Change*, **35,** 71–93.

Sankaran, R., A.R. Rahmani, and U. Ganguli-Lachungpa, 1992: The distribution and status of the lesse florican Sypheotides indica in the Indian subcontinent. *Journal of the Bombay Natural History Society*, **89,** 156–179.

Sato, Y. and N. Mimura, 1997: Environmental problems and current management issues in the coastal zones of south and southeast Asian developing countries. *Journal of Global Environmental Engineering*, **3,** 163–181.

Schulze, E.D., 1994: Flux control at the ecosystem level. *Trends in Ecology and Evolution*, **10,** 40–43.

Schweithelm, J., T. Jessup, and D. Glover, 1999: Conclusion and policy recommendation. In: *Indonesia's Fires and Haze: the Cost of Catastrophe* [Glover, D. and T. Jessup (eds.)]. Institute of Southeast Asia, Singapore, and International Development Research Centre, Ottawa, Ontario, Canada, pp. 130–153.

Seino, H., M. Amano, and K. Sasaki, 1998: Impacts on agriculture, forestry and fishery. In: *Global Warming: The Potential Impact on Japan* [Nishioka, S. and H. Harasawa (eds.)]. Springer-Verlag, Tokyo, Japan, pp. 101–124.

Serebryanny, L.R. and A.G. Khropov, 1997: Natural zones of north of Russia during a climatic holocene optimum. *Russian Academy of Science Report,* **357(6),** 826–827 (in Russian).

Sharma, K.P., 1997: *Impact of Land-Use and Climatic Changes on Hydrology of the Himalayan Basin: A Case Study of the Kosi Basin.* Diss. University of New Hampshire, Durham, NH, USA, 247 pp.

Sharma, K.P., 1993: Role of meltwater in major river systems of Nepal. In: *International Symposium on Snow and Glacier Hydrology* [Young, G.J. (ed.)]. International Association of Scientific Hydrology, Kathmandu, Nepal, pp. 113–122.

Shi, Y.F., 1995: Exacerbating coastal hazards and defensive countermeasures in China. In: *Geological Hazards and Environmental Studies of Chinese Offshore Areas* [Liu, S.Q. and M.S. Liang (eds.)]. Qindao Ocean University Press, Qindao, China, pp. 1–17.

Shiklomanov, I. (ed.), 2001: *World Water Resources at the Beginning of the 21st Century.* International Hydrological Series of the United Nations Educational, Scientific, and Cultural Organization (UNESCO), Cambridge University Press, Cambridge, United Kingdom and New York, NY, USA, 711 pp.

Singh, P., 1998: Effect of global warming on the streamflow of high-altitude Spiti River. In: *Ecohydrology of High Mountain Areas* [Chalise, S.R., A. Herrmann, N.R. Khanal, H. Lang, L. Molnar, and A.P. Pokhrel (ed.)]. International Centre for Integrated Mountain Development, Kathmandu, Nepal, pp. 103–114.

Singh, N. and N.A. Sontakke, 2001: Natural and anthropogenic environmental changes of the Indo-Gangetic Plains, India. *Climatic Change,* (communicated).

Sinha, S.K., 1997: Global change scenario—current and future with reference to land cover change and sustainable agriculture. *Current Science,* **72(11),** 846–854.

Sinha, S.K., M. Rai, and G.B. Singh, 1998: *Decline in Productivity in Punjab and Haryana: A Myth or Reality?* Indian Council of Agricultural Research (ICAR) Publication, New Delhi, India, 89 pp.

Sirotenko, O.D., H.V. Abashina, and V.N. Pavlova, 1997: Sensitivity of the Russian agriculture to changes in climate, CO_2, and tropospheric ozone concentrations and soil fertility. *Climatic Change,* **36(1–2),** 217–232.

Sivardhana, R., 1993: Methodology of assessing the economic impacts of sea level rise with reference to Ban Don Bay, Thailand. *Malaysian Journal of Tropical Geography,* **24(1–2),** 35–40.

Skole, D.L., W.A. Salas, and C. Silapathong, 1998: Interannual variation in the terrestrial carbon cycle: significance of Asian tropical forest conversions to imbalance in the global carbon budget. In: *Asian Change in the Context of Global Climate Change* [Galloway, J.N. and J.M. Melillo (eds.)]. Cambridge University Press, Cambridge, United Kingdom and New York, NY, USA, pp. 162–186.

Smit, B., I. Burton, R.J.T. Klein, and J. Wandel, 2000: An anatomy of adaptation to climate change and variability. *Climatic Change,* **45,** 223–251.

Smith, J.B., A. Rahman, S. Haq, M.Q. Mirza, 1998: *Considering Adaptation to Climate Change in the Sustainable Development of Bangladesh.* World Bank Report, World Bank, Washington, DC, USA, 103 pp.

Smith, J.B., N. Bhatti, G. Menzhulin, R. Benioff, M. Campos, B. Jallow, F. Rijsberman, M.I. Budyko, and R.K. Dixon (eds.), 1996: *Adapting to Climate Change: Assessment and Issues.* Springer-Verlag, New York, NY, USA, 475 pp.

Stephens, C., 1995: The urban environment, poverty and health in developing countries. *Health Policy and Planning,* **10(2),** 109.

Sugi, M., A. Noda, and N. Sato, 1996: Will the number of tropical cyclones be reduced by global warming? Implication from a numerical modeling experiment with the JMA global model. In: *Proceedings of the 1996 Spring Meeting of the Japanese Meteorological Society, 21-23 May 1996.* The Japan Meteorological Society, Ohomiya, Japan, p. 37 (in Japanese).

Sukumar, R., H.S. Suresh, and R. Ramesh, 1995: Climate change and its impact on tropical montane ecosystems in south India. *Journal of Biogeography,* **22,** 533–536.

Svitoch, A.A., 1997: *Extreme Rise of the Caspian Sea Catastrophe of Coasts of Dagestan Towns.* Moscow University Press, Moscow, Russia, 203 pp. (in Russian with English abstract).

Swift, M.J., J. Vandermeer, P.S. Ramakrishnan, J.M. Anderson, C.K. Ong, and B. Hawkins, 1996: Biodiversity and agroecosystem function. In: *Biodiversity and Ecosystem Properties: A Global Perspective* [Mooney, H.A., J.H. Cushman, E. Medina, O.E. Sala, and E.D. Schulze (eds.)]. John Wiley and Sons, Chichester, United Kingdom, pp. 261–298.

Tang, Q., 1995: The effects of climate change on resource populations in the Yellow Sea ecosystem. *Canadian Special Publication of Fisheries and Aquatic Sciences,* **121,** 97.

Tang, S.Y., 1992: *Institutions and Collective Action: Self-Governance in Irrigation.* Institute for Contemporary Studies Press, San Fransisco, CA, USA, 151 pp.

Teh, T.T., 1997: *Sea Level Rise Implications for Coastal and Island Resorts: Climate Change in Malaysia.* Universiti Putra, Selangor Darul Ehsan, Malaysia, pp. 83–102.

Terazaki, M., 1999: The Sea of Japan large marine ecosystem. In: *Large Marine Ecosystems of the Pacific Rim: Assessment, Sustainability, and Management* [Sherman, K. and Q.S. Tang (eds.)]. Blackwell, Oxford, United Kingdom, pp. 199–220.

Thongkrajai, E., P. Thongkrajai, J. Stoeckel, S. Na-nakhon, B. Karenjanabutr, and J. Sirivatanamethanont, 1990: Socioeconomic and heath programme effects upon the behavioral management of diarrhoeal disease in northeast Thailand. *Asia Pacific Journal of Public Health,* **4(1),** 45–52.

Tian, Y.J. and S. Liu, 1999: Humankind's desire and achievements in changing nature seen from three project cases. *Science and Technology Review,* **1,** 36–43 (in Chinese).

Timmermann, A., J. Oberhuber, A. Bacher, M. Esch, M. Latif, and E. Roeckner, 1999: Increased El Niño frequency in a climate model forced by future greenhouse warming. *Nature,* **398,** 694–696.

Tobey, J., J. Reilly, and S. Kane, 1992: Economic implication of global climate change for world agriculture. *Journal of Agriculture and Resource Economy,* **17,** 195–204.

Tonkin, H., C. Landsea, G.J. Holland, and S. Li, 1997: Tropical cyclones and climate change: a preliminary assessment. In: *Assessing Climate Change Results from the Model Evaluation Consortium for Climate Assessment* [Howe, W. and A. Henderson-Sellers (eds.)]. Gordon and Breach, London, United Kingdom, pp. 327–360.

Tri, N.H., N.W. Adger, and P.M. Kelly, 1998: Natural resource management in mitigating climate impacts: mangrove restoration in Vietnam. *Global Environmental Change,* **8(1),** 49–61.

Trigg, P.I. and A.V. Kondrachina, 1998: Commentary: malaria control in the 1990s. *Bulletin of the World Health Organization,* **76(1),** 11–16.

Tsunekawa, A., X. Zhang, G. Zhou, and K. Omasa, 1996: Climate change and its impacts on the vegetation distribution in China. In: *Climate Change and Plants in East Asia* [Omasa, K., K. Kai, H. Toda, Z. Uchijima, and H. Yoshino (eds.)]. Springer-Verlag, Tokyo, Japan, pp. 67–84.

Turner, R.K., S. Subak, and W.N. Adger, 1996: Pressures, trends and impacts in coastal zones: interactions between socio-economic and natural systems. *Environmental Management,* **20,** 159–173.

Ubukata, H., 2000: The impact of global warming on insects. In: *A Threat to Life: The Impact of Climate Change on Japan's Biodiversity* [Domoto, A., K. Iwatsuki, T. Kawamichi, and J. McNeely (eds.)]. Tsukiji-Shokan Publishing Company, Tokyo, Japan; The World Conservation Union (IUCN), Gland, Swizerland; and IUCN, Cambridge, United Kingdom, pp. 61–70.

UNDP, 1997: *Human Development Report 1997.* United Nations Development Programme, New York, NY, USA, 256 pp.

UNEP, 1999: *Global Environment Outlook 2000.* Earthscan, London, United Kingdom, 398 pp.

UNICEF, 1999: *UNICEF Statistical Data.* United Nations Children's Fund, New York, NY, USA. Available online at http://www.unicef.org/statis/indexr.htm.

United Nations, 1998: *UN World Population Prospects.* United Nations, Population Division, New York, NY, USA.

USDA, 1999: *Production, Supply, and Distribution Database.* Available online at http://usda.mannlib.cornell.edu/data-sets/international/.

Valendik, E.N., 1996: Ecological aspects of forest fires in Siberia. *Siberian Ecological Journal,* **1,** 1–8 (in Russian).

Varlamov, S.M., N.A. Dashko, and Y.S. Kim, 1997: Climate change in the far east and Japan Sea area for the last 50 years. In: *Proceedings of the CREAMS '97 International Symposium,* pp. 163–166.

Verghese, B.G. and R.R. Iyer, 1993: *Harnessing the Eastern Himalayan Rivers: Regional Cooperation in South Asia.* Konark Publishers, New Delhi, India.

Vohra, C.P., 1981: The climate of the Himalayas. In: *The Himalaya Aspect of Change* [Lall, J.S. (ed.)]. Oxford University Press, New Delhi, India, pp. 138–151.

Voskoboynikov, G.M., 1999: *The Change of UV Radiation and Its Impact on the Ecosystems of Russian Arctic Seas.* Report of the Russian Academy of Sciences, Moscow, Russia, Vol. 7, pp. 113–121 (in Russian).

Walker, H.J., 1998: Arctic deltas. *Journal of Coastal Research,* **14(3),** 718–738.

Walsh, K. and A.B. Pittock, 1998: Potential changes in tropical storms, hurricanes, and extreme rainfall events as a result of climate change. *Climatic Change,* **39,** 199–213.

Wang, F., 1996a: Climate change and crop production in China. *Economics of Chinese Country,* **11,** 19–23.

Wang, S., 1996b: Impacts of climate change on the water deficit status and growth of winter wheat in North China. *Quarterly Journal of Applied Meteorology,* **7(3),** 308–315.

WASI, 1997: *World Atlas of Snow and Ice Resources* [Kotliakov, B.M. (ed.)]. Institute of Geography, Russian Academy of Sciences, Moscow, Russia, 262 pp.

Webster, P.J. and S. Yang, 1992: Monsoon and ENSO—selective interactive systems. *Quarterly Journal of the Royal Meteorological Society,* **118,** 877–926.

Webster, P.J., V.O. Magana, T.N. Palmer, J. Shukla, R.A. Tomas, M. Yanagi, and T. Yasunari, 1998: Monsoons: processes, predictability and the prospects for prediction. *Journal of Geophysical Research,* **103(C7),** 14451–14510.

White, A., V. Barker, and G. Tantrigama, 1997: Using integrated coastal management and economics to conserve coastal tourism resources. *Ambio,* **26(6),** 335–343.

White, A., M.G.R. Cannell, and A.D. Friend, 1999: Climate change impacts on ecosystems and the terrestrial carbon sink: a new assessment. *Global Environmental Change,* **9,** S21–S30.

Wilkinson, C.R. (ed.), 1998: *Status of Coral Reefs of the World.* Australian Institute of Marine Science, Western Australia, Townsville, MC, Queensland, Australia, 184 pp.

WMO, 1999: *Scientific Assessment of Ozone Depletion 1998.* Global Ozone Research and Monitoring Project Report No. 44, World Meteorological Organization, Geneva, Switzerland, pp. 1.1–12.57.

Wong, P.P., 1992: Impacts of sea level rise on the coasts of Singapore: preliminary observations. *Journal of Southeast Asian Earth Sciences,* **7,** 65–70.

World Bank, 1997: *World Development Indicators.* World Bank, Washington, DC, USA, 231 pp.

World Bank, 1996: *Chao Phraya Flood Management Review, Final Report.* The Royal Thai Government, Thailand.

WRI, 1998: *World Resources, 1998–1999.* World Resources Institute, Oxford University Press, New York, NY, USA, 384 pp.

Xia, D.X., W.H. Wang, G.Q. Wu, J.R. Cui, and F.L. Li, 1993: Coastal erosion in China. *Acta Geographica Sinica,* **48(5),** 468–476 (in Chinese with English abstract).

Xiao, X.M., J.M. Melillo, D.W. Kicklighter, Y. Pan, A.D. McGuire, and J. Helirich, 1998: Net primary production of terrestrial ecosystems in China and its equilibrium responses to changes in climate and atmospheric CO_2 concentration. *Acta Phytoecologica Sinica,* **22(2),** 97–118.

Xu, J. and H. Liu, 1997: Border malaria in Yunnan, China. *Southeast Asian Journal of Tropical Medicine and Public Health,* **28(3),** 456–459.

Yamada, T., H. Fushimi, R. Aryal, T. Kadota, K. Fujita, K. Seko, and T. Yasunari, 1996: Report of avalanches accident at Pangka, Khumbu Region, Nepal in 1995. *Japanese Society of Snow and Ice,* **58(2),** 145–155 (in Japanese).

Yim, C.Y., 1999: The forest fires in Indonesia 1997–1998: possible causes and pervasive consequences. *Geography,* **84,** 251–260.

Yoshino, M., 1998a: Global warming and urban climates: report of the Research Center for Urban Safety and Security, Kobe University Special report no. 1. In: *Proceedings of the Second Japanese-German Meeting, Klima Analyse fur die Stadtplanung,* pp. 123–124.

Yoshino, M., 1998b: Deviation of catch in Japan's fishery in the ENSO years. In: *Climate and Environmental Change, International Geographical Union—Commission on Climatology, Evora, Portugal, 24–30 August, 1988,* 189 pp.

Yu, B. and D.T. Neil, 1991: Global warming and regional rainfall: the difference between average and high intensity rainfalls. *International Journal of Climatology,* **11,** 653–661.

Zhang, X.S., G.S. Zhou, Q. Gao, Z. Ni, and H.P. Tang, 1997: Study on global change and terrestrial ecosystems in China. *Earth Science Frontiers,* **4(1–2),** 137–144.

Zhang, X.S., D.A. Yang, G.S. Zhou, C.Y. Liu, and J. Zhang, 1996: Model expectation of impacts of global climate change on Biomes of the Tibetan Plateau. In: *Climate Change and Plants in East Asia* [Omasa, K., K. Kai, H. Taoda, Z. Uchijima, and M. Yoshino (eds.)]. Springer-Verlag, Tokyo, Japan, pp. 25–38.

Zhou, P., 1991: Likely impacts of climate warming on fresh water fishery. *Fresh Water Fishery,* **6,** 40–41.

Zhu, Z., 1991: Chinese fragile ecological ribbon and soil desertification. *Journal of Desert Research,* **11(4),** 11–21 (in Chinese).

Zou, J.Z. and Y. Wu, 1993: Marine environmental biology. In: *Marine Science Study and Its Prospect in China* [Tseng, C.K., H.Q. Zhou, and B.C. Li (eds.)]. Qingdao Publishing House, Qingdao, China, pp. 663–680.

12

Australia and New Zealand

BARRIE PITTOCK (AUSTRALIA) AND DAVID WRATT (NEW ZEALAND)

Lead Authors:
R. Basher (New Zealand), B. Bates (Australia), M. Finlayson (Australia), H. Gitay (Australia), A. Woodward (New Zealand)

Contributing Authors:
A. Arthington (Australia), P. Beets (New Zealand), B. Biggs (New Zealand), H. Clark (New Zealand), I. Cole (Australia), B. Collyer (Australia), S. Crimp (Australia), K. Day (Australia), J. Ford-Robertson (New Zealand), F. Ghassemi (Australia), J. Grieve (New Zealand), D. Griffin (Australia), A. Hall (New Zealand), W. Hall (Australia), G. Horgan (New Zealand), P.D. Jamieson (New Zealand), R. Jones (Australia), G. Kenny (New Zealand), S. Lake (Australia), R. Leigh (Australia), V. Lyne (Australia), M. McGlone (New Zealand), K. McInnes (Australia), G. McKeon (Australia), J. McKoy (New Zealand), B. Mullan (New Zealand), P. Newton (New Zealand), J. Renwick (New Zealand), D. Smith (Australia), B. Sutherst (Australia), K. Walsh (Australia), B. Watson (Australia), D. White (Australia), T. Yonow (Australia)

Review Editor:
M. Howden (Australia)

CONTENTS

EXECUTIVE SUMMARY

The Australia and New Zealand Region: This region spans the tropics to mid-latitudes and has varied climates and ecosystems, including deserts, rainforests, coral reefs, and alpine areas. The climate is strongly influenced by the surrounding oceans. The El Niño-Southern Oscillation (ENSO) phenomenon leads to floods and prolonged droughts, especially in eastern Australia and parts of New Zealand. The region therefore is sensitive to the possible changes toward a more El Niño-like mean state suggested by Working Group I. Extreme events are a major source of current climate impacts, and changes in extreme events are expected to dominate impacts of climate change. Return periods for heavy rains, floods, and storm surges of a given magnitude at particular locations would be modified by possible increases in intensity of tropical cyclones, mid-latitude storms, and heavy rain events (medium confidence) and changes in the location-specific frequency of tropical cyclones (low to medium confidence). Scenarios of climate change based on recent coupled atmosphere-ocean climate models suggest that large areas of mainland Australia will experience significant decreases in rainfall during the 21st century (low to medium confidence).

Before stabilization of greenhouse gas (GHG) concentrations, the north-south temperature gradient in mid-southern latitudes is expected to increase (medium to high confidence), strengthening the westerlies and the associated west-to-east gradient of rainfall across Tasmania and New Zealand. Following stabilization of GHG concentrations, these trends would be reversed (medium confidence).

Water Supply and Hydrology: In some areas, water resources already are stressed and are highly vulnerable, with intense competition for water supply. This is especially so with respect to salinization (parts of Australia) and competition for water between agriculture, power generation, urban areas, and environmental flows (high confidence). Increased evaporation and possible decreases of rainfall in many areas would adversely affect water supply, agriculture, and the survival and reproduction of key species in parts of Australia and New Zealand (medium confidence).

Ecosystems and Conservation: Warming of 1°C would threaten the survival of species currently growing near the upper limit of their temperature range, notably in some Australian alpine regions that already are near these limits, as well as in the southwest of western Australia. Other species that have restricted climatic niches and are unable to migrate because of fragmentation of the landscape, soil differences, or topography could become endangered or extinct. Other ecosystems that are particularly threatened by climate change include coral reefs (Australia) and freshwater wetlands in the coastal zone and inland.

Food and Fiber: Agricultural activities are particularly vulnerable to regional reductions in rainfall in southwest and inland Australia (medium confidence) and eastern New Zealand. Drought frequency and consequent stresses on agriculture are likely to increase in parts of Australia and New Zealand as a result of higher temperatures and possibly more frequent El Niños (medium confidence). Enhanced plant growth and water-use efficiency (WUE) resulting from carbon dioxide (CO_2) increases may provide initial benefits that offset any negative impacts from climate change (medium confidence), although the balance is expected to become negative with warmings in excess of 2–4°C and associated rainfall decreases (medium confidence). Reliance on exports of agricultural and forest products makes the region sensitive to changes in commodity prices induced by changes in climate elsewhere.

Australian and New Zealand fisheries are influenced by the extent and location of nutrient upwellings governed by prevailing winds and boundary currents. In addition, ENSO influences recruitment of some fish species and the incidence of toxic algal blooms. There is as yet insufficient knowledge about impacts of climate changes on regional ocean currents and about physical-biological linkages to enable confident predictions of changes in fisheries productivity.

Settlements, Industry, and Human Health: Marked trends to greater population and investment in exposed coastal regions are increasing vulnerability to tropical cyclones and storm surges. Thus, projected increases in tropical cyclone intensity and possible changes in their location-specific frequency, along with sea-level rise, would have major impacts—notably, increased storm-surge heights for a given return period (medium to high confidence). Increased frequency of high-intensity rainfall would increase flood damages to settlements and infrastructure (medium confidence). There is high confidence that projected climate changes will enhance the spread of some disease vectors, thereby increasing the potential for disease outbreaks, despite existing biosecurity and health services.

Vulnerability and Adaptation: Climate change will add to existing stresses on achievement of sustainable land use and conservation of terrestrial and aquatic biodiversity. These stresses include invasion by exotic animal and plant species, degradation and fragmentation of natural ecosystems through agricultural and urban development, dryland salinization

(Australia), removal of forest cover (Australia and New Zealand), and competition for scarce water resources. Within both countries there are economically and socially disadvantaged groups of people, especially indigenous peoples, that are particularly vulnerable to additional stresses on health and living conditions induced by climate change. Major exacerbating problems include rapid population and infrastructure growth in vulnerable coastal areas, inappropriate use of water resources, and complex institutional arrangements. Adaptation to climate change, as a means of maximizing gains and minimizing losses, is important for Australia and New Zealand but is relatively little explored. Options include improving water-use efficiency and effective trading mechanisms for water; more appropriate land-use policies; provision of climate information and seasonal forecasts to land users, to help them manage for climate variability and change; improved crop cultivars; revised engineering standards and zoning for infrastructure development; and improved biosecurity and health services. Such measures often will have other benefits, but they also may have costs and limits.

Integrated Assessments: Comprehensive cross-sectoral estimates of net climate change impact costs for various GHG emission scenarios, as well as for different societal scenarios, are not yet available. Confidence remains very low in the previously reported (Basher *et al.*, 1998) estimate for Australia and New Zealand of -1.2 to -3.8% of gross domestic product for an equivalent doubling of CO_2 concentrations. This out-of-date estimate did not account for many currently identified effects and adaptations.

Summary: Australia has significant vulnerability to changes in temperature and precipitation projected for the next 50–100 years (very high confidence) because it already has extensive arid and semi-arid areas and lies largely in the tropics and subtropics. New Zealand, a smaller and more mountainous country with a generally more temperate maritime climate, may be more resilient to climate changes than Australia, although considerable regional vulnerability remains (medium confidence).

12.1. The Australasian Region

12.1.1. Overview

Australasia is defined here as Australia, New Zealand, and their outlying tropical and mid-latitude islands. Australia is a large, relatively flat continent reaching from the tropics to mid-latitudes, with relatively nutrient-poor soils, a very arid interior, and rainfall that varies substantially on seasonal, annual, and decadal time scales, whereas New Zealand is much smaller, mountainous, and mostly well-watered. The ecosystems of both countries contain a large proportion of endemic species, reflecting their long evolutionary history and isolation from other land masses. They have been subject to significant human influences, before and after European settlement 200 years ago.

The region's climate is strongly influenced by the surrounding oceans. Key climatic features include tropical cyclones and monsoons in northern Australia; migratory mid-latitude storm systems in the south, including New Zealand; and the ENSO phenomenon, which causes floods and prolonged droughts, especially in eastern Australia.

The total land area is 8 million km^2, and the population is approximately 22 million. Much of the region is very sparsely populated; most people (85%) live in a relatively small number of coastal cities and towns. Both countries have significant populations of indigenous peoples who generally have lower economic and health status. The two countries have developed economies and are members of the Organisation for Economic Cooperation and Development (OECD); unlike other OECD countries, however, their export trade is dominated by commodity-based industries of agriculture and mining.

12.1.2. Previous Work

The Australasian chapter (Basher *et al.*, 1998) of the IPCC *Special Report on Regional Impacts of Climate Change* (RICC) (IPCC, 1998) provides an extensive assessment of likely climate change impacts and adaptation options for Australia and New Zealand, based on work published until early 1998. That report concludes that Australia's relatively low latitude makes it particularly vulnerable through impacts on its scarce water resources and crops that presently are growing near or above their optimum temperatures, whereas New Zealand—a cooler, wetter, mid-latitude country—may gain some benefit from the ready availability of suitable crops and a likely increase in agricultural production with regional warming. Nevertheless, a wide range of situations in which vulnerability was thought to be moderate to high were identified for both countries—particularly for ecosystems, hydrology, coastal zones, settlements and industry, and health. Indirect local impacts from possible climatically driven changes in international conditions—notably commodity prices and international trade—also was identified as a major issue in the 1998 report, as well as by the New Zealand Climate Change Programme (1990). Key points from Basher *et al.* (1998) follow.

Climate and Climate Trends: Climate trends were reported to be consistent with those in other parts of the world, with mean temperature increases of as much as 0.1°C per decade over the past century, a faster increase in nighttime than daytime temperatures, and sea level rising an average of about 20 mm per decade over the past 50 years. Increases in average rainfall and the frequency of heavy rainfalls were reported for large areas of Australia.

Climate Scenarios: Australian scenarios reported for 2030 exhibited temperature increases of 0.3–1.4°C, uncertain overall rainfall decrease of as much as 10%, and more high-intensity rainfall events. Projected changes for 2070 were about twice the 2030 changes. New Zealand projections included similar temperature increases, as well as stronger westerly air flow, with resulting precipitation increases in the west and decreases in the east.

Water Supply and Hydrology: Possible overall reduction in runoff, with changes in soil moisture and runoff varying considerably from place to place but reaching as much as ±20%, was suggested for parts of Australia by 2030. Sharpened competition was expected among water users, with the large Murray-Darling Basin river system facing strong constraints. Enhanced groundwater recharge and dam-filling events were expected from more frequent high-rainfall events, which also were expected to increase flooding, landslides, and erosion. A reduced snow season was expected to decrease the viability of the ski industry, although it would provide seasonally smoother hydroelectricity generation in New Zealand.

Ecosystems and Conservation: Significant potential impacts identified on Australasian land-based ecosystems included alteration in soil characteristics, water and nutrient cycling, plant productivity, species interactions, and ecosystem composition and function, exacerbated by any increases in fire occurrence and insect outbreaks. Aquatic systems would be affected by changes in runoff, river flow, and associated transport of nutrients, wastes, and sediments. These changes and sea-level rise would affect estuaries and mangroves. Australia's coral reefs were considered to be vulnerable to temperature-induced bleaching and possibly to sea-level rise and weather change.

Food and Fiber: Direct impacts on agriculture from CO_2 increases and climate changes were expected to vary widely in space and time, with perhaps beneficial effects early in the 21st century, followed by more detrimental effects in parts of Australia as warming increases. Any changes in global production and hence international food commodity prices would have major economic impacts. The net impact on production forestry from changes in tree productivity, forest operational conditions, weeds, disease, and wildfire incidence was not clear. The impact on fisheries could not be confidently predicted.

Settlements and Industry: Possible changes were noted in the frequency and magnitude of climatic "natural disaster" events affecting economically important infrastructure. Likely impacts of climate change were identified on water and air quality, water supply and drainage, waste disposal, energy production, transport operations, insurance, and tourism.

Human Health: Increases were expected in heat-stress mortality (particularly in Australia), the incidence of tropical vector-borne diseases such as dengue, and urban pollution-related respiratory problems.

Adaptation Potential and Vulnerability: Some of the region's ecosystems were identified as very vulnerable, with fragmentation and alteration of landscape by urban and agricultural development limiting natural adaptability. Land-use management was the primary adaptation option identified. Although coral reefs were identified as vulnerable, it was suggested that they might be able to keep pace with sea-level rise.

Techniques that already provide considerable adaptability of agriculture to existing climate variability may apply to climate change over the next few decades. However, at longer time horizons the climate was expected to become less favorable to agricultural production in Australia, leading to increased vulnerability. Scientifically based integrated fisheries and coastal zone management were regarded as principal adaptation options for fisheries.

Adaptation options identified for settlements and infrastructure included integrated catchment management, changes to water pricing systems, water efficiency initiatives, building or modifying engineering structures, relocation of buildings, and urban planning and management. Low-lying coastal settlements were regarded as highly vulnerable to high sea level and storm events. Adaptation options included integrated coastal zone management (ICZM); redesign, rebuilding, or relocation of capital assets; and protection of beaches and dunes. New Zealand is exposed to impacts on its Pacific island territories, including the eventual possibility of having to accept environmental refugees.

A moderate degree of vulnerability was identified for human health, with adaptation responses including strengthening existing public health infrastructure and meeting the needs of vulnerable groups such as isolated communities and the poor.

12.1.3. Socioeconomic Trends

The region's population is growing at a rate of about 1.2% yr^{-1}, approximately equally from natural increase and immigration. The population is progressively aging, in line with other OECD countries. Health status is improving, but indigenous peoples lag significantly. The main population centers are growing faster than rural areas. Australian lower latitude coastal zones are developing two to three times faster than the Australian average (see Section 12.6.4), for urban/suburban uses and for recreation and tourism. In New Zealand, there is a steady internal migration northward to Auckland.

Agricultural commodity prices have tended to fall, but yields per hectare have risen, and farm sizes and total volume of production have increased (ABARE, 1997; Wilson and Johnson, 1997). Average return on agricultural assets is low. Service industries are an increasing fraction of all industry, and there is a trend toward more intensive agriculture and forestry and to diversification of rural land use, including specialty crops and tourism. There is increasing competition for water in areas of low rainfall where irrigation is essential to intensive cropping; urban demands are rising quickly, and water is needed to maintain natural ecosystems (Hassall and Associates, *et al.*, 1998). Tourism is a major growth industry that is increasing the pressure on areas of attractions such as coastal zones and reefs.

Environmental concerns include air and water pollution from urban industries, land transport, and intensive farming and related processing and soil erosion, rising water tables, and salinization. Environmental management increasingly is based on the principle of sustainable management, as enshrined in New Zealand's Resource Management Act, and an integrated approach to environmental impacts in both countries. A major trend to a "user pays" principle and, especially in Australia, to market-driven water rights, with caps on irrigation supplies, is causing significant changes in rural industry. However, there still are many instances, particularly in coastal management, where these principles are not applied.

12.1.4. Climate Trends

Trends identified in the region continue to be generally consistent with those elsewhere in the world. Research on regional trends has been summarized in Salinger *et al.* (1996) and in specific studies by Plummer (1996), Torok and Nicholls (1996), Holbrook and Bindoff (1997, 2000), Lavery *et al.* (1997), Plummer *et al.* (1997), Zheng *et al.* (1997), McKeon *et al.* (1998), Collins and Della-Marta (1999), Hennessy *et al.* (1999), and Plummer *et al.* (1999).

Mean temperatures have risen by 0.05–0.1°C per decade over the past century, with a commensurate increase in the frequency of very warm days and a decrease in the frequency of frosts and very cold days (Plummer *et al.*, 1999; Collins *et al.*, 2000). Nighttime temperatures have risen faster than daytime temperatures; hence, the diurnal temperature range has decreased noticeably in most places. The past decade has seen the highest recorded mean annual temperatures.

Trends in rainfall are less clear. Australian annual mean rainfall has increased by a marginally significant amount over the past century (Collins and Della-Marta, 1999; Hennessy *et al.*, 1999). However, increases in the frequency of heavy rainfalls and average rainfall are significant in many parts of Australia. Average rainfall has increased most in the northwest and southeast quadrants (Collins and Della-Marta, 1999). The largest and most statistically significant change has been a decline in rainfall in the winter-rainfall-dominated region of the far southwest of western Australia, where in the period 1910–1995, winter (June-July-August, JJA) rainfall declined by 25%, mainly during the 1960s and 1970s. Previous studies (Wright, 1974;

Allan and Haylock, 1993; Yu and Neil, 1993), as well as a more recent one (Smith *et al*., 2000), have noted this decrease and attribute it to atmospheric circulation changes, predominantly resulting from natural variability.

There are marked interdecadal variations over northern and eastern Australia in summer half-year rainfall, which are dominated by ENSO-induced variations (Power *et al*., 1999a). There also are clear interannual and decadal variations in central and eastern Australian rainfall associated with Indian and Pacific Ocean sea surface temperatures (SSTs) (Power *et al*., 1999b). Some of the regional linear trends observed during the past century merely may reflect a particular pattern of decadal variation. Thus, the high degree of decadal variability may enhance or obscure a signal that is related to climatic change for several decades. A growing body of evidence is being obtained about past climate variability from coral cores (e.g., Lough and Barnes, 1997; Isdale *et al*., 1998; Quinn *et al*., 1998).

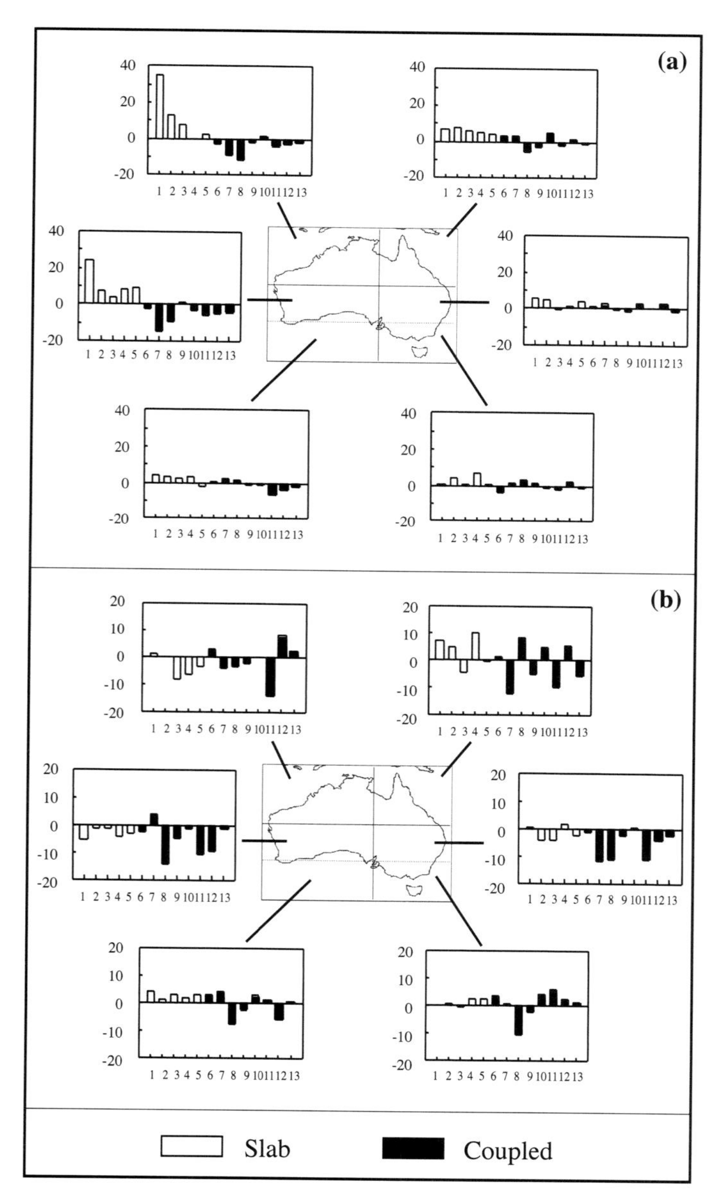

Figure 12-1: Enhanced greenhouse changes in rainfall (as % change per °C of global warming) for five slab-ocean GCM simulations (open bars) and eight coupled ocean-atmosphere GCM simulations (full bars), for six subregions of Australia in (a) summer (DJF) and (b) winter (JJA) Source: Whetton *et al*., 2001 (where individual models are identified). Spatial patterns and scatter plots are given for most of the same coupled models in Hulme and Sheard (1999) and Carter *et al*. (2000).

The strength of the relationship between eastern Australian climate and ENSO has been observed to vary over the past century. This seems to be linked to longer term climate oscillations such as the North Pacific Decadal Oscillation (NPDO) (e.g., Power *et al*., 1999a). Salinger and Mullan (1999) examined the 20-year periods before and after 1977 and showed increases after 1977 (some statistically significant) in mean rainfall for New Zealand's west coast, associated with strengthening westerly winds. These fluctuations in rainfall are partially explained by the increase in El Niño conditions over recent decades. There is some evidence of long-term variations in the Australasian region in storm frequency and tropical cyclones (Nicholls *et al*., 1996a; Radford *et al*., 1996; Hopkins and Holland, 1997; Leighton *et al*., 1997). Nicholls *et al*. (1998) show that although there has been a decrease in tropical cyclone numbers from 1969 to 1996 in the Australian region (105°E to 160°E), there has been an increase in the frequency of intense tropical cyclones with central pressures of less than 970 hPa.

The average rise in sea level in the Australia/New Zealand region over the past 50 years is about 20 mm per decade (Rintoul *et al*., 1996; Salinger *et al*., 1996), which is within the range of the current estimate of global sea-level rise (IPCC, 1996, WGI Section 7.2.1). However, the greater frequency and duration of El Niño episodes since the mid-1970s has reduced local New Zealand sea-level trends: The average sea-level change since 1975 at Auckland is close to zero (Bell *et al*., 1999). There has been a weak warming trend in ocean temperatures to 100-m depth in the southwest Pacific (39°S to 49°S, 141°E to 179°E) of about 0.13°C during the 34-year period 1955–1988 (Holbrook and Bindoff, 1997), and there have been shorter period SST fluctuations associated with ENSO.

12.1.5. Climate Scenarios Used in Regional Studies

12.1.5.1. Spatial Patterns of Temperature and Rainfall

Most recent impact studies for Australia and New Zealand have been based on scenarios released by the Commonwealth Scientific and Industrial Research Organisation (CSIRO, 1996a) or the National Institute of Water and Atmosphere (NIWA—Renwick *et al*., 1998b). The CSIRO (1996a) scenarios included two sets of rainfall scenarios that are based on results from equilibrium slab-ocean general circulation model (GCM) and transient coupled ocean-atmosphere GCM (AOGCM) simulations. Some impact and adaptation studies since RICC have used results from a 140-year simulation over the whole region that uses the CSIRO regional climate model (RCM) at

125-km resolution nested in the CSIRO Mark 2 coupled GCM run in transient mode from 1961 to 2100, as well as similar simulations at 60-km resolution over eastern Australia. Figure 12-1 (from Whetton *et al.*, 2001) summarizes results of slab-ocean and coupled AOGCM simulations of rainfall changes for various subregions of Australia in summer (December-January-February, DJF) and winter (JJA), respectively.

Figures 12-1a and 12-1b clearly show a variety of estimates for different model simulations; nevertheless, there is a tendency toward more decreases in winter rainfall in the southwest and the central eastern coast of Australia in the coupled model results than in the slab-ocean model results and a similar tendency for more agreement between coupled model results on less rainfall in the northwest and central west in summer.

In line with WGI conclusions, more recent scenarios, such as those of Hulme and Sheard (1999) and Carter *et al.* (2000) rely exclusively on coupled models; they also are based on the newer emissions scenarios from the IPCC's *Special Report on Emissions Scenarios* (SRES) rather than the IS92 scenarios described in IPCC (1996). This leads to some changes in the regional scenarios. According to Whetton (1999), the CSIRO (1996a) scenarios give warmings of 0.3–1.4°C in 2030 and 0.6–3.8°C in 2070 relative to 1990, compared to estimates by Hulme and Sheard (1999) of 0.8–3.9°C by the 2050s and 1.0–5.9°C in the 2080s, relative to the 1961–1990 averages. As discussed by Whetton (1999), both sets of scenarios use results from several coupled models, but the use of the SRES emissions scenarios leads to greater warmings in Hulme and Sheard (1999) than those that are based on the IS92 emission scenarios in CSIRO (1996). Nevertheless, the Hulme and Sheard (1999) results are preliminary in that they use scaled results from non-SRES simulations, rather than actual GCM simulations with SRES emissions.

Rainfall scenarios that are based on the coupled model results in CSIRO (1996a) show decreases of 0–8% in 2030 and 0–20% in 2070, except in southern Victoria and Tasmania in winter and eastern Australia in summer, where rainfall changes by –4 to +4% in 2030 and –10 to +10% in 2070. The Hulme and Sheard (1999) scenarios suggest that annual precipitation averaged over either northern or southern Australia is likely to change by –25 to +5%. Larger decreases are indicated for the 2080s and for some specific locations (e.g., a 50% decrease in parts of the southwest of western Australia in winter). This last finding suggests that even though the rainfall decrease in this region in the late 20th century was almost certainly dominated by natural variability (Smith *et al.*, 2000), a decreasing trend resulting from the enhanced greenhouse effect may dominate in this region by the mid- to late 21st century.

It is important to note that changing from joint use of equilibrium slab-ocean GCM and transient coupled AOGCM results to exclusive reliance on coupled model results in the Australian scenarios leads to a marked narrowing of the uncertainty range in rainfall changes predicted for southern Australia, with a tendency to more negative changes on the mainland (see Figure 12-1; Hulme and Sheard, 1999; Carter *et al.*, 2000). This means that the results of impact studies that used the wider range of rainfall scenarios, as in the CSIRO (1996a) scenarios (e.g., Schreider *et al.*, 1996), should be reinterpreted to focus on the drier end of the previous range. This is consistent with more recent studies by Kothavala (1999) and Arnell (1999).

To summarize the rainfall results, drier conditions are anticipated for most of Australia over the 21st century. However, consistent with conclusions in WGI, an increase in heavy rainfall also is projected, even in regions with small decreases in mean rainfall. This is a result of a shift in the frequency distribution of daily rainfall toward fewer light and moderate events and more heavy events. This could lead to more droughts and more floods.

Recent Australian impact studies have tended to use regional scenarios of temperature and rainfall changes per degree of global warming, based on the CSIRO RCM transient simulations, scaled to the range of uncertainty of the global warming derived from the IPCC Second Assessment Report (SAR) range of scenarios. For example, Hennessy *et al.* (1998) gives scenarios for six regions of New South Wales (NSW), based on a 60-km resolution simulation. Ranges are given for estimated changes in maximum and minimum temperatures, summer days over 35°C, winter days below 0°C, seasonal mean rainfall changes, and numbers of extremely wet or dry seasons per decade. Statistical downscaling (discussed in detail in Chapter 3) has not been used extensively in Australia apart from work by Charles *et al.* (1999) on the southwest of western Australia. It should be noted that in some cases, RCM results may change the sign of rainfall changes derived from coarser resolution GCMs, because of local topographic and other effects (Whetton *et al.*, 2001).

New Zealand scenarios reported in RICC (Basher *et al.*, 1998) are based on statistical downscaling from equilibrium GCM runs with CO_2 held constant at twice its present concentration. Renwick *et al.* (1998b, 2000) also have downscaled equilibrium GCM simulations over New Zealand, using a nested RCM.

A key factor in rainfall scenarios for New Zealand is the strength of the mid-latitude westerlies because of the strong orographic influence of the backbone mountain ranges lying across this flow (Wratt *et al.*, 1996). Earlier equilibrium slab-ocean GCM runs (applicable only long after stabilization of climate change) and the regional simulations obtained by downscaling them through statistical and nested modeling techniques predict a weakening of the westerlies across New Zealand, particularly in winter. As a result, downscaled equilibrium model runs predict that winter precipitation will remain constant or decrease slightly in the west but increase in Otago and Southland (Whetton *et al.*, 1996a; Renwick *et al.*, 2000). In contrast, transient coupled AOGCM runs suggest that over the next 100 years or more, the mean strength of the westerlies actually will increase in the New Zealand region, particularly in winter (Whetton *et al.*, 1996a; Russell and Rind, 1999; Mpelasoka *et al.*, 2001). As a result, downscaled transient model results predict that rainfalls will increase in the west of New Zealand but decrease in the east (Mullan *et al.*, 2001) as shown in Figure 12-2. Slightly larger temperature increases are predicted in the north of the country than in the south. Projected temporal changes in the westerlies before and after stabilization of climate also are likely to influence rainfall in southern Australia, especially in Tasmania.

GCM simulations extending beyond the stabilization of GHG concentrations indicate that global warming continues for centuries after stabilization of concentrations (Wigley, 1995; Whetton *et al.*, 1998; TAR WGI Chapters 9 and 11) but at a much reduced rate as the oceans gradually catch up with the stabilized radiative forcing. Importantly for the Australasian region, simulated patterns of warming and rainfall changes in the southern hemisphere change dramatically after stabilization of GHG concentrations. This is because of a reversal of the lag in warming of the Southern Ocean relative to the rest of the globe. This lag increases up to stabilization but decreases after stabilization (Whetton *et al.*, 1998), with consequent reversal of changes in the north-south temperature gradient, the mid-latitude westerlies, and associated rainfall patterns.

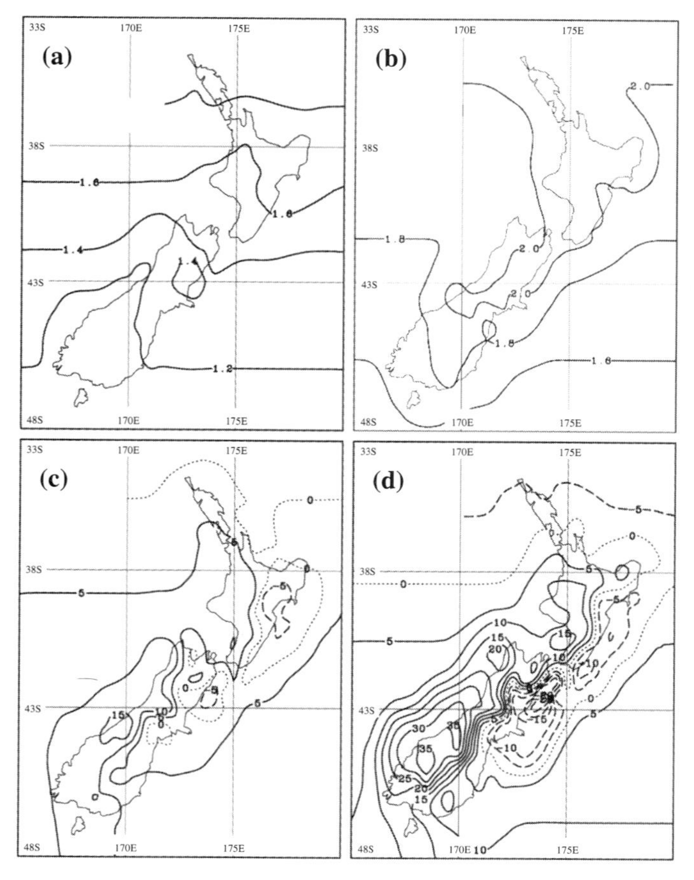

Figure 12-2: Scenarios for changes from the 1980s to the 2080s for New Zealand: (a) mean summer (DJF) temperature (°C), (b) mean winter (JJA) temperature (°C), (c) summer precipitation (%), and (d) winter precipitation (%). These plots were derived by averaging downscaled projections from four GCMs (CCC, CCSR, CSIRO9, and HadCM2) driven by CO_2 concentrations increasing at a compound rate of 1% yr^{-1}, plus specified sulfate aerosol concentrations (Mullan *et al.*, 2001).

12.1.5.2. Uncertainties and Probabilistic Scenarios

Some of the Australasian scenarios include uncertainty bands, based on the ranges of global warmings resulting from the IPCC IS92 emissions scenarios, the IPCC range of global sensitivity, or ranges of estimates of Australasian temperature or rainfall changes from different GCMs. Unquantified additional sources of uncertainty include changes in emission scenarios, such as to the new SRES scenarios; regional effects of biospheric feedback; and regional effects of global aerosol distributions. Quantification of potential changes in extreme events, tropical cyclones, and ENSO also are major uncertainties (see below), and uncertainty about the strength of the westerly circulation and hence rainfall regimes is a source of uncertainty for New Zealand. The modeled lag in warming in the Southern Ocean in the 20th century appears to be greater than that observed (Whetton *et al.*, 1996a), but this has not yet been thoroughly analyzed. Other conceivable lower probability, high-impact changes (see Chapter 3) such as changes in ocean circulation, ENSO behavior, or tropical cyclones could have important regional impacts.

Probabilistic scenarios for risk and adaptation analyses (see Section 12.8.4), based on the quantifiable range of uncertainties, have been explored by CSIRO (Pittock, 1999; Jones, 2000; Pittock and Jones, 2000).

12.1.5.3 Changes in Extreme Events and Sea Level

Pittock *et al.* (1999) have summarized the past importance of extreme events for Australia and prospects for the future. Major climatic hazards arise in Australia and New Zealand from tropical cyclones, floods, droughts, windstorm, snowstorm, wildfires, landslides, hail, lightning, heat waves, frost, and storm surges. Events that are directly related to temperature are more predictable (more heat waves, fewer frosts) than those associated with wind and rain; Chapter 3 discusses relevant projections and confidence levels (see Table 3-10). The incidence of wildfire in Australia is expected to increase with global warming (Beer and Williams, 1995; Pittock *et al.*, 1999; Williams *et al.*, 2001), as is that of landslides and storm surges (the latter because of both higher mean sea level and increased storm intensities). Changes in hail and lightning frequencies are uncertain, although there are some arguments for expected increases (Price and Rind, 1994; McMaster, 1999; Pittock *et al.*, 1999).

More intense tropical cyclones in the Australian region (see Table 3-10; Walsh and Ryan, 2000) would have serious implications for storm-surge heights, wind damages, and flooding. If they were to travel further poleward (Walsh and Katzfey, 2000), they would be more likely to impact on coastal regions in the southwest of western Australia, southern Queensland, and the northern NSW coastal region, as well as northern parts of New Zealand. The locations of tropical cyclone genesis in the region are correlated with ENSO (Evans and Allan, 1992; Basher and Zheng, 1995), so any change in the mean state of the tropical Pacific may affect the risk of tropical cyclone occurrence in particular locations.

Mid-latitude storms also may increase in intensity (see Table 3-10), and their frequency and location could change—for example, as a result of changes in the westerlies and ENSO. This would impact return periods for mid-latutude storm surges, high winds, and other phenomena.

Interannual variability in ENSO leads to major floods and droughts in Australia and New Zealand. Such variations are expected to continue under enhanced greenhouse conditions, though possibly with greater hydrological extremes as a result of more intense rainfall in La Niña years and more intense drought resulting from higher rates of evaporation during El Niño years (Walsh *et al.*, 1999). A more El Niño-like mean state of the tropical Pacific Ocean (see Table 3-10; Cai and Whetton, 2000) would imply greater drought frequency (Kothavala, 1999; Walsh *et al.*, 2000), as does the drying trend found over the Murray-Darling Basin in recent AOGCM simulations (Arnell, 1999).

Mean sea level is expected to increase, with local and regional variations as a result of land-sea movements and changes to ocean currents and climatic forcing (see Chapter 3). In addition, local and regional meteorological forcing leads to temporary fluctuations in sea level and extreme events that may cause coastal inundation. In New Zealand, storm surges of as much as about 1 m are possible at open-coast locations (Heath, 1979; Bell *et al.*, 1999). Storm surges in tropical Australia can be several meters as a result of tropical cyclonic forcing and shallow continental shelfs (Hubbert and McInnes, 1999a,b; McInnes *et al.*, 1999).

The actual height reached by a storm surge depends not only on the location and intensity of the storm but on its timing relative to the tides, coastal bathymetry and topography, and slower variations such as those from ENSO. The latter contribute to significant local sea-level variations around the coasts of Australia (Chiera *et al.*, 1997) and New Zealand (Bell *et al.*, 1999). In addition, any changes in storm intensities, frequencies, and locations will change the average time between surges of a given magnitude at particular locations.

12.1.5.4 New SRES Scenarios

Interim characterizations of regional climate changes to 2100 associated with the SRES emissions scenarios have been provided by Hulme and Sheard (1999) and Carter *et al.* (2000). However, they do not consider aerosol-induced spatial effects, and they use linear scaling of regional patterns of change from seven coupled GCM models, according to a range of global mean warmings generated using MAGICC (Wigley, 1995; Wigley *et al.*, 1997).

Over Australia, these studies show warmings in the 2080s higher than the IS92 scenarios, with similar spatial patterns. In New Zealand, warmings in the 2080s are estimated to be from 0.5 to >2.0°C. Projected precipitation changes are large (>1 standard deviation of the simulated 30-year variability) over much of southern Australia, with a decrease over the mainland in both summer and winter and an increase over Tasmania in winter. Over the South Island of New Zealand, an increase is predicted. For the 2080s, projected decreases in annual rainfall in the southwest of western Australia range from about zero (B1 low scenario) to between 30 and 50% (A2 high scenario). Projected rainfall increases over the South Island of New Zealand of 0–10% (B1) to 10–20% (A2) should be regarded with caution because the AOGCM simulations do not fully incorporate the important influence of the Southern Alps on South Island rainfall patterns.

The SRES scenarios have not yet been applied in any detailed studies of impacts in the region. Unlike parts of the northern hemisphere, high regional concentrations of sulfate aerosols are not expected in the Australasian region under any accepted scenario, so any increase in warming resulting from reduced sulfate aerosols will be less over Australia and New Zealand than in some regions of the northern hemisphere.

To date, impact and vulnerability studies in Australia and New Zealand in general have not taken account of specific socioeconomic scenarios for the future, such as those laid out in the SRES. Thus, vulnerabilities have been based on projected climate change impacts and adaptation, assuming the present socioeconomic situation, in some cases with a qualitative allowance for expected socioeconomic trends (e.g., increased competition for water supplies, increased population and investment in coastal zones).

12.2. Key Regional Concerns

This section summarizes some key regional concerns regarding vulnerability to climate change and impacts in Australia and New Zealand. They have not been prioritized. Supporting details and references are provided in Sections 12.3 through 12.8.

Drought, Flood, and Water Supply: Climate variability is a major factor in the economies of both countries, principally through the flow-on effects of ENSO-related major droughts on agriculture. Farmers in drought-sensitive parts of both countries will be increasingly vulnerable if interannual droughts occur more frequently or are more intense in the future. Less secure water supplies would accentuate competition between users and threaten allocations for environmental flows and future economic growth. Adelaide and Perth are the main cities with water supplies that are most vulnerable to climate change; increasing salinity in the Murray River is an increasing concern for Adelaide. Any increase in flood frequency would adversely affect the built environment. In New Zealand, floods and landslides are natural hazards that could increase in frequency and severity.

Ecosystem Uniqueness and Vulnerability: Australia and New Zealand have been isolated from the rest of the world for millions of years until relatively recent human settlement. Some species exhibit quite limited ranges of average climate. These two factors leave many of the region's ecosystems vulnerable to climatic change and to invasion by exotic animal and plant species introduced by human activity. This vulnerability has been exacerbated by fragmentation of ecosystems through land-use changes.

Coral Reefs: Australia has one of the greatest concentrations of coral reefs in the world. Rising sea level by itself may not be deleterious. However, the combination of sea-level rise with other induced stresses—notably, increasing atmospheric CO_2 (which leads to a decrease in calcification rates of corals); increasing sea temperatures, leading to coral bleaching; possibly increased riverine outflow events (low salinity and high pollution); and damage from tropical cyclones—may put much of this resource at risk.

Alpine Areas: In Australia, significant warming will raise snowlines, diminish the ski industry, and threaten alpine ecosystems. In New Zealand, snowline changes and the advance or retreat of glaciers also depend on changes in the strength and local orientation of mid-latitude westerlies. Options for relocation of the ski industry are limited by the relatively low altitude of Australia's alpine regions and by rugged terrain and conservation estate regulations in New Zealand.

Agricultural Commodities and Terms of Trade: A major fraction of exports from both Australia and New Zealand are agricultural and forestry products, production of which is sensitive to any changes in climate, water availability, CO_2 fertilization, and pests and diseases. Returns from these commodities could be affected by the projected increase in agricultural production in mid- to high-latitude northern hemisphere countries and resulting impacts on commodity prices and world trade.

Increasing Coastal and Tropical Exposure: Major population and economic growth in coastal areas, especially the tropical and subtropical east coast of Australia, are leading to greatly increased vulnerability to tropical cyclones and storm surges, as well as riverine and estuarine flooding. Rising sea level will accentuate these problems, as would any increase in storm intensities or a more poleward movement of tropical cyclones. Rising sea level also will increase the salinity of estuarine and coastal aquifer groundwater.

Indigenous People: In both countries, indigenous peoples (Aborigines and Torres Straits Islanders in Australia, Maori in New Zealand, as well as Pacific islanders) are among the most disadvantaged members of the population. They generally have lower incomes, and many live in isolated rural conditions or in the sometimes poorly serviced and low-lying margins of large towns and cities. They are more exposed to inadequate water supplies, climatic disasters, and thermal stress and are more vulnerable to an increase in the prevalence of pests and diseases.

12.3. Water Supply and Hydrology

12.3.1. Water Supply

Dry conditions in most parts of Australia tend to be associated with El Niño. The link between rainfall and streamflow and ENSO is statistically significant in most parts of eastern Australia (Chiew *et al.*, 1998; Power *et al.*, 1998). Relationships between river flows and ENSO also have been identified for some seasons in parts of New Zealand (McKerchar *et al.*, 1998). Because of the relatively high variability of Australian rainfall, the storage capacities of Australia's large dams are about six times larger than those of European dams for the same mean annual streamflow and probability of water shortfall. In contrast, New Zealand's hydroelectricity system has a total storage capacity of about 6 weeks of national demand because of its higher and more reliable rainfall (Basher *et al.*, 1998).

The Murray-Darling River basin is the largest in Australia and is heavily regulated by dams and weirs. About 40% of mean annual flow is used for human consumption, principally through irrigation; there is high interannual variability. Application of the CSIRO (1996a) scenarios, with their wide range of rainfall changes as a result of inclusion of both the older slab-ocean GCM and the more recent coupled AOGCM simulations, suggests a possible combination of small or larger decreases in mean annual rainfall, higher temperatures and evaporation, and a higher frequency of floods and droughts in northern Victorian rivers (Schreider *et al.*, 1996). A study of the Macquarie River basin in NSW indicates inflow reductions on the order of 10–30% for doubled CO_2 and reduced streamflows if irrigation demand remains constant or increases (Hassall and Associates *et al.*, 1998). Adelaide and Perth traditionally have been regarded as the most vulnerable metropolitan areas to future water supply problems, including increasing levels of salinity (Schofield *et al.*, 1988; Williams, 1992; PMSEIC, 1998; MDBC, 1999), although Perth recently has decided to spend AU$275 million for drought-proofing (Boer, 2000). Water supplies are adequate for many coastal regions in Australia. However, drier inland areas are vulnerable to water shortages during the annual dry season and drought.

Studies by Kothavala (1999) and Arnell (1999)—using results from the U.S. National Center for Atmospheric Research (NCAR) Community Climate Model (CCMO) GCM and the HadCM2 and HadCM3 AOGCMs, respectively—show increases in drought across eastern and southern Australia. Kothavala found that the Palmer Drought Index showed longer and more severe drought in northeastern and southeastern Australia. Arnell (1999) found marked decreases in runoff over most of mainland Australia but some increases over Tasmania. For the Murray-Darling basin, he found decreases in mean flow by the 2050s ranging from about 12 to 35%, with decreases in the magnitude of 10-year maximum and minimum monthly runoff.

The only recent water supply study for New Zealand is that by Fowler (1999), based on the RSNZ (1988) and Mullan and Renwick (1990) regional climate change scenarios and three equilibrium slab-ocean GCM simulations. These models give scenarios of rainfall increases in the Auckland region, leading to the conclusion that changes in water resources most likely would be positive. Scenarios based on recent AOGCM simulations have yet to be evaluated.

Atolls and low-lying islands (e.g., some in the Torres Strait and in association with New Zealand) rely on rainwater or limited groundwater resources for water supplies. These resources are sensitive to climate variations and in some cases already are stressed by increasingly unsustainable demand and pollution caused by human activity. Saltwater intrusion into aquifers might occur through sea-level rise, more frequent storm events, possible reductions in rainfall, and increased water demand as a result of higher temperatures (see Basher *et al.*, 1998; Chapter 17).

12.3.2. Water Allocation and Policy

Until recently, water planning in Australia was driven by demand and controlled by engineers, not by economics (Smith, 1998b). This situation has changed with growing population and demand (rural and urban/industrial), including rapid growth in irrigation of high-value crops such as cotton and vineyards. There also is an increasing awareness of stress on riverine ecosystems as a result of reduced mean flows, lower peak flows, and increasing salinity and algal blooms. Higher temperatures and changed precipitation as a result of climate change generally would exacerbate these problems and sharpen competition among water users (e.g., see Hassall and Associates *et al.*, 1998). In 1995, the Council of Australian Governments reviewed water resource policy in Australia and agreed to implement a strategic framework to achieve an efficient and sustainable water industry through processes to address water allocations, including provision of water for the environment and water-trading arrangements. The Agriculture and Resource Management Council of Australia and New Zealand subsequently commissioned a set of National Principles for the Provision of Water for Ecosystems, with the following stated goal: "To sustain and where necessary restore ecological processes and biodiversity of water-dependent ecosystems." Implementation of water reforms and national principles has resulted in the definition of conceptual frameworks and practical methods for assessing the water requirements of environmental systems.

In Australia, flow recommendations commonly are developed after water infrastructure projects and dams have been in place for some time and environmental flows implemented in river systems that already are experiencing a modified or regulated flow regime (Arthington, 1998; Arthington *et al.*, 1998). This situation is most applicable to adaptation to climate change in existing regulated flow regimes.

The Australian National Principles require that provision of water for ecosystems should use the best scientific information available on the hydrological regimes necessary to sustain aquatic ecosystems. Ideally, environmental flow recommendations are based on establishment of quantitative relationships between flow characteristics and desired geomorphological, ecological, or water-quality outcomes. Methods are available to estimate flow-related habitat requirements of aquatic invertebrates, fish, and aquatic and riparian plants (e.g., wetted perimeter, transect methods, instream flow incremental methodology (IFIM)—see Kinhill, 1988). However, there are no standard methods for assessing flows that are relevant to maintenance of key life history processes. In the absence of robust biological indicators of response to flow regulation, recent research has advocated the use of statistical descriptors of flow regimes. These methods include maintenance of critical flow characteristics within one or two standard deviations of mean parameters (Richter *et al.*, 1996).

In New Zealand, various pressures on riverine ecosystems have been recognized, including those from agriculture, urban usage and sewage, hydroelectricty and water supply dams, forestry and mining, and introduced pests and weeds. Management of water is covered by the Resource Management Act (RMA) of 1991. Under this Act, the intrinsic values of ecosystems, including their biodiversity and life-supporting capacity, must be considered. The emphasis has changed from multiple-use management to environmentally sustainable management (Taylor and Smith, 1997), and Maori values are explicitly recognized (Ministry for the Environment, 1999). Drought associated with ENSO has placed stress on water supplies in various parts of the country, and it is recognized that climate change could lead to further stresses, especially if there is an increase in the frequency of El Niño events (Taylor and Smith, 1997). The RMA provides a statutory basis for integrated catchment management in that regional Councils control land use, water use, and water quality. Regional plans under which water is allocated have a term of only 10 years, allowing for review to adapt to issues such as river flow changes caused by climate change.

In the context of climate change, it is relevant to ask: How much can critical features of the flow regime be changed before the system becomes seriously stressed? Finding answers to this question for a range of Australian rivers is central to the assessment and management of water allocations to sustain water-dependent systems. Climate change has yet to be systematically injected into this process, but at least the mechanisms are now in place to develop appropriate water allocations and price incentives to use water to the best advantage. There also is scope for increased application of seasonal climate forecasts in water resources management, as a tool to aid adaptation to climate variability.

12.3.3. Inland and Coastal Salinization

Natural salinity and high water tables have been present in Australia for centuries. However, because of changes in land management—notably land clearing and irrigation—salinity is now a major environmental issue in Australia (Ghassemi *et al.*, 1995; MDBC, 1999). About 2.5 Mha are affected in Australia, with the potential for this to increase to 12.5 Mha in the next 50 years (PMSEIC, 1999). Much of this area covers otherwise productive agricultural land. The area damaged by salinity to date represents about 4.5% of presently cultivated land, and known costs include US$130 million annually in lost agricultural production, US$100 million annually in damage to infrastructure (such as roads, fencing, and pipes); and at least US$40 million in lost environmental assets (Watson *et al.*, 1997; PMSEIC, 1998). The average salinity of the lower Murray River (from which Adelaide draws much of its water supply) is expected to exceed the 800 EC threshold for desirable drinking water about 50% of the time by 2020.

Although climate is a key factor affecting the rate of salinization and the severity of impacts, a comprehensive assessment of the effects of climate change on this problem has not yet been carried out. Revegetation policies and associated carbon credit motivational policies designed to increase carbon sinks are likely to have a significant impact on recharge. However, global warming and dryland salinity policies need to be coordinated to maximize synergistic impacts.

In many coastal areas and oceanic islands, development and management of fresh groundwater resources are seriously constrained by the presence of seawater intrusion. Seawater intrusion is a natural phenomenon that occurs as a consequence of the density contrast between fresh and saline groundwater. If conditions remain unperturbed, the saline water body will remain stationary unless it moves under tidal influences. However, when there is pumping of freshwater, sea-level change, or changing recharge conditions, the saline body will gradually move until a new equilibrium condition is achieved (Ghassemi *et al.*, 1996). If the sea level rises to its "best-guess" or extreme predicted value over the next century, this would significantly increase intrusion of seawater in coastal and island aquifers.

12.3.4. Water Quality

Water quality would be affected by changes in biota, particularly microfauna and flora; water temperature; CO_2 concentration; transport processes that place water, sediment, and chemicals in streams and aquifers; and the timing and volume of water flow. More intense rainfall events would increase fast runoff, soil erosion, and sediment loadings, and further deforestation and urbanization would tend to increase runoff amounts and flood wave speed. These effects would increase the risk of flash flooding, sediment load, and pollution (Basher *et al.*, 1998). On the other hand, increases in plantation and farm forestry—in part for carbon sequestration and greenhouse mitigation purposes—would tend to reduce soil erosion and sediment loads.

Eutrophication is a major water quality problem in Australia (State of the Environment, 1996). This is a natural process, but it has been greatly accelerated in Australia by human activities, including sewage effluent and runoff from animal farms, irrigation, and stormwater. Low flow, abundant light, clear water, and warmth all encourage algal growth, which affects the taste and odor of water and can be toxic to animals, fish, and humans. Thus, local climate warming and the potential for reduced streamflow may lead to increased risk of eutrophication.

12.4. Ecosystems and Conservation

12.4.1. Introduction

Until recent settlement, Australia and New Zealand were isolated for millions of years, and their ecosystems have evolved to cope with unique climate and biological circumstances (Kemp, 1981; Nix, 1981). Despite large year-to-year climatic variability, many Australian terrestrial species have quite limited ranges of long-term average climate, on the order or 1–2°C temperature and 20% in rainfall (Hughes *et al.*, 1996; Pouliquen-Young and Newman, 1999). Thus, these ecosystems are vulnerable to climatic change, as well as invasion by exotic animals and plants.

Rapid land clearance and subsequent land-use change have been occurring as a result of human activity over the past 500–1,000 years in New Zealand (McGlone, 1989; Wilmshurst, 1997) and, in Australia, subsequent to Aboriginal arrival tens of thousands of years ago—and especially since European settlers arrived 200 years ago. This has led to loss of biodiversity in many ecosystems as well as loss of some ecosystems as a whole. One of the major impacts has been an increase in weedy species in both countries. This is likely to continue and be exacerbated by climate change. Land-use change also has led to fragmentation of ecosystems and to salinization through rising water tables. These trends can inhibit natural adaptation to climate change via the dispersal/migration response. Systems therefore may be more vulnerable, and some might become extinct. For example, Mitchell and Williams (1996) have noted that habitat that is climatically suitable for the long-lived New Zealand kauri tree *Agathis australis* under a 4°C warming scenario would be at least 150 km from the nearest extant population. They suggest that survival of this species may require human intervention and relocation. Similar problems have been identified by Pouliquen-Young and Newman (1999) in relation to fragmented habitat for endangered species in the southwest of western Australia.

Many of the region's wetlands, riverine environments, and coastal and marine systems also are sensitive to climate variations and changes. A key issue is the effect on Australia's coral reefs of greenhouse-related stresses in addition to nonclimatic features such as overexploitation and increasing pollution and turbidity of coastal waters from sediment loading, fertilizers, pesticides, and herbicides (Larcombe *et al.*, 1996).

12.4.2. Forests and Woodlands

In Australia, some 50% of the forest cover in existence at the time of European settlement still exists, although about half of that has been logged (Graetz *et al.*, 1995; State of the Environment, 1996). Pressures on forests and woodlands as a whole are likely to decrease as a result of recent legislation relating to protection of forests in some Australian states, and as interest in carbon sequestration increases. In New Zealand (Taylor and Smith, 1987), 25% of the original forest cover remains, with 77% in the conservation estate, 21% in private hands, and 2% state owned. Legal constraints on native wood production mean that only about 4% currently is managed for production, and clear-felling without replacement has virtually ceased.

The present temperature range of 25% of Australian Eucalyptus trees is less than 1°C in mean annual temperature (Hughes *et al.*, 1996). Similarly, 23% have ranges of mean annual rainfall of less than 20% variation. The actual climate tolerances of many species are wider than the climate envelope they currently occupy and may be affected by increasing CO_2 concentrations, which change photosynthetic rates and water-use efficiency (WUE) and may affect the temperature response (Curtis, 1996). Such changes from increasing CO_2 would be moderated by nutrient stress and other stressors that are prevalent across Australian forests. Nevertheless, if present-day boundaries even approximately reflect actual thermal or rainfall tolerances, substantial changes in Australian native forests may be expected with climate change. Howden and Gorman (1999) suggest that adaptive responses would include monitoring of key indicators, flexibility in reserve allocation, increased reserve areas, and reduced fragmentation.

In a forested area in western Australia that is listed as one of 25 global "biodiversity hotspots" for conservation priority by Myers *et al.* (2000), Pouliquen-Young and Newman (1999) used the BIOCLIM program (Busby, 1991) to generate a climatic envelope from the present distribution of species. They assessed the effects of three incremental temperature and rainfall scenarios on three species of frogs, 15 species of endangered or threatened mammals, 92 varieties of the plant genus *Dryandra*, and 27 varieties of *Acacia* in the southwest of western Australia. The scenarios were based on the spatial pattern of change from the CSIRO RCM at 125-km resolution, scaled to the IS92 global scenarios. For plant species, suitability of soils also was considered. The results indicate that most species would suffer dramatic decreases in range with climate warming; all of the frog and mammal species studied would be restricted to small areas or would disappear with 0.5°C global-average warming above present annual averages, as would 28% of the *Dryandra* species and one *Acacia*. At 2°C global average warming, 66% of the *Dryandra* species, as well as all of the *Acacia,* would disappear. Adaptation opportunities were considered minimal, with some gain from linking present conservation reserves and reintroducing endangered species into a range of climatic zones.

Studies of the current distribution of New Zealand canopy trees in relation to climate suggest that major range changes can be expected with warming (Whitehead *et al.*, 1992; Leathwick *et al.* 1996; Mitchell and Williams, 1996). Trees in the highly diverse northern and lowland forests (e.g., *Beilschmiedia tawa*) are likely to expand their ranges southward and to higher altitudes. The extensive upland *Nothofagus* forests are likely to be invaded by broad-leafed species. Few tree species are confined to cool southern climates; those that are have a wide altitudinal range available for adjustment of their distribution, so no extinctions are expected. Most concern centers on the ability of tree species to achieve new distributions rapidly enough in a fragmented landscape, as well as invasion of natural intact forests by exotic tree, shrub, and liana species that are adapted to warm temperate or subtropical climates.

12.4.3. Rangelands

In Australia, rangelands are important for meat and wool production. In their natural state, rangelands are adapted to relatively large short-term variations in climatic conditions (mainly rainfall and temperature). However, they are under stress from human activity, mostly as a result of animal production, introduced animals such as rabbits, inappropriate management, and interactions between all of these factors (Abel *et al.*, 2000). These stresses, in combination with climatic factors, have led to problems of land degradation, salinization, and woody weed invasion and subsequent decreases in food production. In some cases, native dominant species (mostly plants) have been replaced by exotic species, leading to a decrease in population of many native animal species. Woody weed invasion also has changed the fire regime through formation of "thickets" that do not allow fires through, partly as a result of the fire resistance of some species (Noble *et al.*, 1996). Some Australian rangelands also are vulnerable to salinization resulting from rising water tables from irrigation and loss of native vegetation (see Section 12.3).

New Zealand rangelands are used predominantly for sheep grazing. Intensive use of indigenous grasslands and shrublands on land cleared of trees in the 19th century has increased vulnerability to invasion, especially by woody weeds (pine, broom, gorse, etc.) and herbaceous weeds (hawkweed, thistles, and subtropical grasses). Weed invasions are unlikely to further increase the susceptibility of the system to climatic disruptions but could themselves be accelerated by warming or increased climatic variability. Fire is now strictly regulated, although rangeland fires remain a serious problem, especially in ENSO drought years. Rangelands—in particular, in drier areas of eastern South Island—have many problems, including animal and plant pests and declining profitability of farming, leading to a decline in management and fertilizer inputs.

Increased CO_2 is likely to have beneficial effects for native pastures, with possible nitrogen limitation and increased subsoil drainage (Howden *et al.*, 1999d). Runoff and groundwater recharge also could increase (Krysanova *et al.*, 1999). This could lead to increased salinization problems in areas that are susceptible. However, decreases in rainfall in excess of about 10% at the time of CO_2 doubling would dominate over the CO_2 fertilization effect and lead to a decline in pasture productivity. This is more likely in the latest climate change scenarios that are based on coupled AOGCM results. Howden *et al.* (1999d) conclude that a doubling of CO_2 concentrations will result in only limited changes in the distribution of C_3 and C_4 grasses and that such changes will be moderated by warmer temperatures.

12.4.4. Alpine Systems

Basher *et al.* (1998) conclude that alpine systems are among the most vulnerable systems in the region. Despite the fact that they cover only a small area, they are important for many plant and animal species, many of which are listed as threatened. These systems also are under pressure from tourism activity. The Australian Alps are relatively low altitude (maximum about 2,000 m), and much of the Alpine ecosystem area and ski fields are marginal. Most year-to-year variability is related to large fluctuations in precipitation, but interannual temperature variations are small compared to warming anticipated in the 21st century. Studies by Hewitt (1994), Whetton *et al.* (1996b), and Whetton (1998) all point to a high degree of sensitivity of seasonal snow cover duration and depth. For Australia, Whetton (1998) estimates, for the full range of CSIRO (1996a) scenarios, an 18–66% reduction in the total area of snow cover by 2030 and a 39–96% reduction by 2070. This would seriously affect the range of certain alpine ecosystems and species (Bennett *et al.*, 1991). Decreases in precipitation and increased fire danger also would affect alpine ecosystems adversely.

There seems to be little opportunity for adaptation by alpine ecosystems in Australia, which cannot retreat upward very far because of the limited height of Australian hills and mountains. There are various options for the rapidly expanding mountain-based recreation industry, including increased summer recreation and artificial snowmaking. These adaptations would increase stress on alpine ecosystems and water resources.

The New Zealand Alps are of higher altitude (up to 3,700 m); about 9% of the New Zealand landmass is above the treeline. A large number of species (for example, 25% of vascular plants), which often are highly distinctive, grow there. Despite a 0.5°C rise in New Zealand's mean annual temperatures since the 1860s, there has been no significant rise in the treeline or shrubland expansion (Wardle and Coleman, 1992), and it seems unlikely that there will be any significant threat to alpine ecosystems from warming in the medium term.

12.4.5. Wetlands

The Australian State of the Environment Report (1996) states, "Wetlands continue to be under threat, and large numbers are already destroyed." For example, Johnson *et al.* (1999) estimate wetland loss of about 70% in the Herbert River catchment of Northern Queensland between 1943 and 1996. Wetland loss is caused by many processes, including water storage; hydroelectric and irrigation schemes; dams, weirs, and river management works; desnagging and channelization; changes to flow, water level, and thermal regimes; removal of instream cover; increased siltation; toxic pollution and destruction of nursery and spawning or breeding areas (Jackson, 1997); and use of wetlands for agriculture (Johnson *et al.*, 1999). Climate change will add to these factors through changes in inflow and increased water losses.

Specific threats to wetlands from climate change and sea-level rise have been studied as part of a national vulnerability assessment (Waterman, 1996). The best example is provided for Kakadu National Park in northern Australia. There are fears that World Heritage and Ramsar-recognized freshwater wetlands in this park could become saline, given current expectations of sea-level rise and climate change (Bayliss *et al.*, 1997; Eliot *et al.*, 1999). Although this analysis is supported by a large data resource, it is speculative, and efforts to develop more definite monitoring tools are needed. However, it does raise the possibility that many other Australian coastal wetlands could be similarly affected. Some of these wetlands may be unable to migrate upstream because of physical barriers in the landscape.

Many inland wetlands are subject to reduced frequency of filling as a result of water diversion for irrigation, and they also may be seriously affected by reductions in seasonal or annual rainfalls in the catchments as a result of climate change (Hassall and Associates *et al.*, 1998). This may threaten the reproduction of migratory birds (some species of which already are under threat), which rely on wetlands for their breeding cycle (Kingsford and Thomas, 1995; Kingsford and Johnson, 1998; Kingsford *et al.*, 1999). Large decreases in inflow predicted for the Macquarie River and several rivers in northern Victoria by Hassall and Associates *et al.* (1998) and Schreider *et al.* (1996, 1997) for scenarios that are consistent with the latest AOGCM simulations would have major impacts on wetland ecosystems.

Wetlands in New Zealand are the most threatened ecosystems; they have declined by 85% since European settlement (Stephenson, 1983). The vast majority have been drained or irretrievably modified by fire, grazing, flood control works, reclamation, or creation of reservoirs. Eutrophification, weed invasion, and pollution have greatly reduced their biodiversity (Taylor and Smith, 1997). More than 50% of the 73 significant wetlands that meet the Ramsar Convention standards for international wetlands are in coastal districts and will be impacted by rising sea levels. Most important wetlands are in highly urbanized or productive landscape settings and therefore have limited options for adaptation to decreased size or increased salinization.

12.4.6. Riverine Environments

Many Australian river systems, particularly in the southeast and southwest, have been degraded through diversion of water via dams, barrages, channels, and so forth, principally for irrigated agriculture. Many New Zealand rivers have been affected by hydroelectric generation; diversion of water for irrigation; agricultural, manufacturing, and urban pollution; and biotic invasion. Recent research has shown that river ecosystems are particularly sensitive to extremes in flow. Most research has been on the effects of flood flows. Droughts, as opposed to floods, have a slow onset and although recovery from floods by river flora and fauna is relatively rapid, recovery after droughts tends to be slow, may be incomplete, and may lag well behind the breaking of the drought (Lake, 2000). Floods and droughts interact with nutrient supply (Hildrew and Townsend, 1987; Biggs, 1996), so the effects of any possible changes in their frequency and magnitude need to be evaluated within the context of other human activities and climate-induced land-use change.

Current ranges of scenarios tend to suggest reductions in mean flow in many Australian rivers, similar to or greater than those in Schreider *et al.* (1997) and Hassall and Associates *et al.* (1998). In particular, any tendency toward more frequent or severe El Niño-like conditions beyond that already contained in the CSIRO (1996a) scenarios would further threaten many riverine and inland wetland systems in Australia and New Zealand. Findings of increased drought frequency and severity in eastern Australia under an NCAR CCMO transient simulation (Kothavala, 1999) and 12–35% reductions in mean flow by 2050 in the Murray-Darling basin by the 2050s in Arnell (1999), using results from the HadCM2 and HadCM3 GCM simulations, are cause for concern. Arnell (1999) found reductions in maximum and minimum flows. Walsh *et al.* (2000) found less severe increases in drought in Queensland, based on simulations with the CSIRO RCM nested in the CSIRO Mark 2 GCM.

Implications of these findings for riverine ecosystems and estuaries (Vance *et al.*, 1998; Loneragan and Bunn, 1999) and possible adaptations have yet to be investigated, although reduced diversions from rivers to increase environmental flows is one possibility. This could be achieved through increased WUE, imposition of caps on water diversions, or water pricing and trading, but the latter two measures are controversial and would have strong implications for rural industry (e.g., see ABARE, 1999). Increased efficiency in water delivery for irrigation currently is the favored option for restoring environmental flows in the heavily depleted Snowy River in southeastern Australia.

12.4.7. Coastal and Marine Systems

Australia has some of the finest examples of coral reefs in the world, stretching for thousands of kilometers along the northwest and northeast coasts (Ellison, 1996). Coral reefs in the Australian region are subject to greenhouse-related stresses (see Chapter 6 for a summary), including increasingly frequent bleaching episodes, changes in sea level, and probable decreases in calcification rates as a result of changes in ocean chemistry.

Mass bleaching has occurred on several occasions in Australia's Great Barrier Reef (GBR) and elsewhere since the 1970s (Glynn, 1993; Hoegh-Guldberg *et al.*, 1997; Jones *et al.*, 1997; Wilkinson, 1998). Particularly widespread bleaching, leading to death of some corals, occurred globally in 1997–1998 in association with a major El Niño event. Bleaching was severe on the inner GBR but less severe on the outer reef (Wilkinson, 1998; Berkelmans and Oliver, 1999). This episode was associated with generally record-high SSTs over most of the GBR region. This was a result of global warming trends resulting from the enhanced greenhouse effect and regional summer warming from the El Niño event, the combined effects of which caused SSTs to exceed bleaching thresholds (Lough, 1999). Three independent databases support the view that 1997–1998 SST anomalies were the most extreme in the past 95 years and that average SSTs off the northeast coast of Australia have significantly increased from 1903 to 1994. Lowered seawater salinity as a result of flooding of major rivers between Ayr and Cooktown early in 1998 also is believed to have been a major factor in exacerbating the effects in the inshore GBR (Berkelmans and Oliver, 1999). Solar radiation, which is affected by changes in cloud cover and thus by El Niño, also may have been a factor (Brown, 1997; Berkelmans and Oliver, 1999).

Although warming in Australia's coral reef regions on average is expected to be slightly less than the global average, according to the SRES global warming scenarios it may be in the range of 2–5°C by 2100. This suggests that unless Australian coral reefs can adapt quickly to these higher temperatures, they will experience temperatures above present bleaching thresholds (Berkelmans and Willis, 1999) almost every year, well before the end of the 21st century (Hoegh-Guldberg, 1999). Hoegh-Guldberg (1999) notes that apparent thresholds for coral bleaching are higher in the northern GBR than further south, suggesting that some very long-term adaptation has occurred. Coral reef biota may be able to adapt, at least initially, by selection for the more heat-tolerant host and symbiont species and genotypes that survived the 1997–1998 summer and by colonization of damaged sites by more heat-resistant genotypes from higher latitudes arriving as planktonic larvae. However, it is generally believed that the rate and extent of adaptation will be much slower than would be necessary for reef biota to resist the frequency and severity of high SST anomalies projected for the middle third of the 21st century (medium to high confidence). The most likely outlook is that mass bleaching, leading to death of corals, will become a more frequent event on Australian coral reefs in coming decades.

Increasing atmospheric CO_2 concentrations will decrease the carbonate concentration of the ocean, thereby reducing calcification rates of corals (Gattuso *et al.*, 1998, 1999; Kleypas *et al.*, 1999). This is complicated, however, by the effects of possible changes in light levels, freshwater discharge, current patterns, and temperature. For example, Lough and Barnes (2000) report a historic growth stimulus for the *Porites* coral that they correlate with increasing average SSTs. Thus, the net effect on Australian reefs up to 1980 appears to have been positive, but it is unclear whether decreased carbonate concentration resulting from rapidly increasing CO_2 concentration will outweigh the direct temperature effect later in the 21st century, especially if regional SSTs reach levels not experienced by the corals of the GBR during the Holocene.

As noted in Chapter 6, expected rates of sea-level rise to 2100 would not threaten healthy coral reefs (most Australian reefs) but could invigorate growth on reef flats. However, decreased calcification rates might reduce the potential ability of the reefs to keep up with rapid sea-level rise. Possible increases in tropical cyclone intensity with global warming also would impact coral reefs (high confidence), along with nonclimatic factors such as overexploitation and increasing pollution and turbidity of coastal waters by sediment loading, fertilizers, pesticides, and herbicides (Larcombe *et al.*, 1996). Climate change could affect riverine runoff and associated stresses of the reefs, including low-salinity episodes. Coupled with predicted rises in sea level and storminess, bleaching-induced coral death also could weaken the effectiveness of the reefs in protecting the Queensland coast and adversely affect the biodiversity of the reef complex.

On the whole, mangrove processes are less understood than those for coral reefs (Ellison, 1996). Mangroves occur on low-energy, sedimentary shorelines, generally between mean- and high-tide levels. Australian mangroves cover approximately 11,500 km^2 (Galloway, 1982). It is anticipated that they are highly vulnerable but also highly adaptable to climate change. Studies over glacial/interglacial cycles show that in the past mangroves have moved landward during periods of rising sea level (Woodroffe, 1993; Wolanski and Chappell, 1996; Mulrennan and Woodroffe, 1998). However, in many locations this will be inhibited now by coastal development. Coastal wetlands are thought to be nursery areas for many commercially important fish (e.g., barramundi), prawns, and mudcrabs.

In New Zealand, estuaries are the most heavily impacted of all coastal waters. Most are situated close to or within urban areas (Burns *et al.*, 1990). Most have been modified by reclamation or flood control works and have water-quality problems resulting from surrounding land use. Increasing coastal sedimentation is having a marked effect on many estuaries. This may increase with increased rainfall variability. In the South Island, increased coastal sedimentation has disrupted fish nursery grounds and destroyed weed beds, reef sponges, and kelp forests; in the North Island it has been linked to loss of seagrasses through worsening water clarity (RSNZ, 1993; Turner, 1995).

Over a long period, warming of the sea surface is expected (on average) to be associated with shoaling (thinning) of the mixing layer, lowering of phytoplankton growth-limiting dissolved inorganic nutrients in surface waters (Harris *et al.*, 1987; Hadfield and Sharples 1996), and biasing of the ecosystem toward microbial processes and lowered downward flux of organic carbon (Bradford-Grieve *et al.*, 1999). However, this would be modified regionally by any change in the Pacific Ocean to a more El Niño-like mean state. Warming also may lead to decreased storage of carbon in coastal ecosystems (Alongi *et al.*, 1998).

There is now palaeo-oceanographic evidence documenting environmental responses east of New Zealand to climatic warming, especially the Holocene "optimum" (~6–7 ka) and interglacial optimum (~120–125 ka), when SSTs were 1–2°C warmer than present. Immediately prior to and during those two periods, oceanic production appears to have increased, as manifested by greater amounts of calcareous nanoplankton and foraminifers (e.g., Lean and McCave, 1998; Weaver *et al.*, 1998). Other evidence suggests that storms in the New Zealand region may have been more frequent in warmer epochs (Eden and Page, 1998), affecting the influx of terrigenous material into the continental shelf (Foster and Carter, 1997). There also may be a relationship between strong El Niño events and the occurrence of toxic algal blooms in New Zealand waters (Chang *et al.*, 1998). Nevertheless, we do not know, over the longer term, how the oceanic biological system in the southwest Pacific will be influenced by the interaction of ENSO events with the overall warming trend.

South of the subtropical front, primary production is limited by iron availability (Boyd *et al.*, 1999), which has varied in the past. It is not known how or whether aeolian iron supply to the Southern Ocean in the southwest Pacific (Duce and Tindale, 1991) may be altered by climate change, although it could be affected by changes in aridity and thus vegetation cover over Australia as well as by strengthening of the westerlies. In any case, Harris *et al.* (1988) demonstrate that the strength of the zonal westerly winds is linked to recruitment of stocks of spiny lobsters over a wide area.

If reduction or cessation of North Atlantic or Antarctic bottomwater formation were to occur (Manabe and Stouffer, 1994; Hirst, 1999), this could lead to significant changes in deep ocean chemistry and dynamics, with wide ramifications for marine life. The common southern hemisphere copepod *Neocalanus tonsus* could be affected because it spends part of the year at depths between 500 and 1,300 m but migrates seasonally to surface waters, becoming the focus of feeding of animals such as sei whales and birds (Bradford-Grieve and Jillett, 1998).

The northern part of New Zealand is at the southern extension of the distribution of marine subtropical flora and fauna (Francis and Evans, 1993). With a warming climate, it is possible that many species would become a more permanent feature of the New Zealand flora and fauna and extend further south.

12.4.8. Landscape Management as a Goal for Conservation and Adaptation

Ecosystems that are used for food and fiber production form a mosaic in a landscape in which natural ecosystems also are represented. Aquatic systems, notably rivers and groundwater, often play a crucial role. Given the issues of fragmentation and salinization in many parts of the region, especially Australia, landscape management as an integrated approach (PMSEIC, 1999) may be one of the best ways of achieving conservation goals and human needs for food and fiber in the face of multiple stresses—of which climate change is only one.

This complex interconnection of issues in land management is evident in most parts of Australia and New Zealand—notably in the tropical coastal zone of Queensland, where rapid population and economic growth has to be managed alongside agricultural land use that impacts soil and riverine discharge into the waters of the GBR, a growing tourist industry, fisheries, indigenous people's rights, as well as the climatic hazards of tropical cyclones, floods, and droughts. Climate change and associated sea-level rise are just one of several major issues in this context that may be significant in adding stress to a complex system.

Similar complexities arise in managing other major areas such as the Murray-Darling basin, where control of land degradation through farm and plantation forestry is being considered as a major option, partly for its benefits in controlling salinization and waterlogging and possibly as a new economic option with the advent of incentives for carbon storage as a greenhouse mitigation measure. Similar problems and processes apply in New Zealand, where plantation forestry is regarded as a major option in land use and GHG mitigation.

12.5. Food and Fiber

12.5.1. *Introduction*

Some 60% of Australia is used for commercial agriculture (Pestana, 1993). Only 2% is used for broad acre and intensive crop production; 4% is sown to pastures. Soil and topography are major constraints on cropping, which produces about 50% of the gross value of farm production—the rest being divided equally between meat production and livestock products such as wool and milk.

Russell (1988) identifies climatic "frontiers" affected by climate change and interdecadal variability in rainfall—namely, the inland limit to agriculture (mean annual rainfall <300 mm in the south and <750 mm in the north), the southern limit of effective summer rainfall; and the lower rainfall limit of high value crops (on the order of 1,000 mm). Temperature also is limiting, with some temperate crops held south of their high-temperature (northern) limit, and other more tropical crops held at their low-temperature (southern or high-altitude) limit. Large areas of the interior and west are desert or very arid rangelands with low yields; much of this land is now returned to Aboriginal management.

There is great interannual variability, especially in the interior and more northern regions, associated mainly with ENSO, convective rainfall, and tropical cyclones. Australia is known as a land of droughts and flooding rains. Secondary factors such as wildfires also account for losses of fodder, animals, and farm infrastructure (sheds, fences, machinery), and hail causes significant crop losses. Accordingly, drought and disaster relief policies are matters of ongoing concern (O'Meagher *et al.*, 1998, 2000; Pittock *et al.*, 1999), as is sustainability in the face of economic pressures and global change (Abel *et al.*, 1997).

In New Zealand, pastoral agriculture provides more than 40% of the country's export earnings (Statistics New Zealand, 1998). Dairy farming is the major activity in wetter areas; sheep dominate hilly and drier areas, and beef cattle are widely distributed throughout the country. Pastures are highly productive—composed largely of introduced grass and nitrogen-fixing legume species—and support high stock numbers. Rainfall in New Zealand generally is not strongly seasonal, but high evapotranspiration rates in the summer make pastoral agriculture in the east vulnerable to largely ENSO-related variability in summer rainfall.

12.5.2. *Pastoral Farming*

Howden *et al.* (1999d) have summarized and updated work by Hall *et al.* (1998), McKeon *et al.* (1998), and Howden *et al.* (1999a,b). They find that although CO_2 increase alone is likely to increase pasture growth, particularly in water-limited environments, there also is strong sensitivity to rainfall, so that a 10% reduction in rainfall would counter the effect of a doubled CO_2 concentration. A 20% reduction in rainfall at doubled CO_2 is likely to reduce pasture productivity by about 15% and live-weight gain in cattle by 12% and substantially increase variability in stocking rates, reducing farm income. The latest scenarios, which have substantial reduction in rainfall in many parts of Australia, would tend to reduce productivity.

Howden *et al.* (1999d) also found that doubled CO_2 concentrations are likely to increase the deep drainage component under pastures, which may increase the risk and rates of salinization where the potential for this problem exists. Doubled CO_2 and increased temperature would result in only limited changes in C_3 and C_4 grass distributions (Howden *et al.*, 1999a).

In New Zealand, productivity of dairy farms might be adversely affected by a southward shift of undesirable subtropical grass species, such as *Paspalum dilatatum* (Campbell *et al.*, 1996). At present, *P. dilatatum* is recognized as a significant component of dairy pastures in Northland, Auckland, Waikato, and the Bay of Plenty. A "user-defined" management threshold for the probability of finding this grass in dairy pasture is predicted by a climate profile technique (Campbell and Mitchell, 1996). This can be considered the point at which adaptive management changes are regarded as necessary. This technique was applied with the IS92a (mid) and IS92e (high) climate change scenarios and the CSIRO4 GCM pattern, using the CLIMPACTS integrated assessment model (Kenny *et al.*, 1995, 2000). Results indicate a significant southward shift in the probability of occurrence of *P. dilatatum* with global warming; more southerly geographic thresholds are reached at later dates, but 25–30 years earlier with the higher emissions scenario.

Comprehensive assessment of the response of dairy cattle to heat stress in NSW and Queensland was carried out by Davison *et al.* (1996). Physiological effects of heat stress include reduced food intake, weight loss, decreased reproduction rates, reduction in milk yields, increased susceptibility to parasites, and, in extreme cases, collapse and death. Heat stress can be reduced by the use of shade and sprinklers, and thresholds for their use can be determined. Jones and Hennessy (2000) applied this adaptation to the Hunter Valley in NSW, using probabilistic estimates of temperature and dewpoint changes resulting from climate change for the IS92 range of scenarios to 2100. They then estimated the probabilities of given milk production losses as a function of time and calculated the economic benefits of provision of shade and sprinklers. They conclude that heat-stress management in the region would be cost-effective. However, such adaptation may not be as cost-effective in a hotter or more humid climate.

Howden and Turnpenny (1997) and Howden *et al.* (1999e) also have looked at heat stress in beef cattle. They find that heat stress already has increased significantly in subtropical Queensland over the past 40 years (where there has been a warming trend) and that it will increase further with greenhouse-induced global warming. They suggest a need for further selection for cattle lines with greater thermoregulatory control, but they point out that this may be difficult because it may not be consistent with high production potential (Finch *et al.*, 1982, 1984).

12.5.3. Cropping and Horticulture

Cropping in Australia recently has undergone great diversification, from predominantly wheat and barley to include a variety of other crops, including rice, cotton, pulses, and oilseeds. Cane sugar is grown extensively in coastal areas of Queensland. Many of these crops are subject to frost limitations on seasons and to water stress in dry spells; some are subject to direct heat stress or deterioration during heat waves. For example, wheat grain protein composition deteriorates after several days above 35°C (Burke *et al.*, 1988; Behl *et al.*, 1993), making it less suitable for high-value uses such as pasta and breadmaking. However, climate warming may allow earlier planting and faster phenological development, resulting in little change in heat shock risk up to a 4°C mean warming (Howden *et al.*, 1999c). Independently, increasing CO_2 can result in a decrease in wheat grain protein content—also leading to a decrease in breadmaking quality (Rogers *et al.*, 1998). A potential complication of these impacts is water stress that results in decreased yield and potentially increased protein.

Howden *et al.* (1999c) report a comprehensive study of global change impacts on Australian wheat cropping. Studies were conducted of changes in wheat yields, grain quality, and gross economic margins across 10 sites in the present Australian wheat belt. Results were scaled up to provide national estimates, with and without varying planting dates. Response surfaces were constructed across the full range of uncertainty in the CSIRO (1996a) scenarios and are shown in bar-graph form in Figure 12-3. The estimated increase in yield resulting from physiological effects of a doubling of actual atmospheric CO_2 is about 24%. The analysis assumes that the regional distribution of cropping is unaffected. (This is not completely accurate, but changes at the margins of present areas would not change the total yield much.) The best variety of wheat is used under each scenario, with current planting dates (a) and optimal planting dates (b) for each scenario. Note that yield reaches a maximum at about 1°C warming with current planting dates but about 2°C with optimal planting dates and that yield drops rapidly with decreases in rainfall. Under the SRES scenarios, warming in Australian wheat-growing areas would exceed 2°C and could be well in excess of 6°C by 2100; actual CO_2 concentrations could be between 540 and 970 ppm.

Doubling CO_2 alone produced national yield increases of 24% in currently cropped areas, but with a decline in grain nitrogen content of 9–15%, which would require increases in the use of nitrogen-based fertilizer of 40-220 kg ha^{-1} or increased rotations of nitrogen-fixing plants. Using the mid-range values from the CSIRO (1996a) scenario (which includes both slab-ocean and coupled GCMs and is now superseded—see below), climate change added to CO_2 increase led to national yield increases of 20% by 2100 under present planting practices or 26% with optimum planting dates (Howden *et al.*, 1999c). Regional changes varied widely.

Howden's response surfaces show that for doubled CO_2 but no change from historical rainfall, a 1°C increase in temperature would slightly increase national yield (see Figure 12-3a) when the best variety was used with the current planting window. However, the slope of the temperature curve turned negative

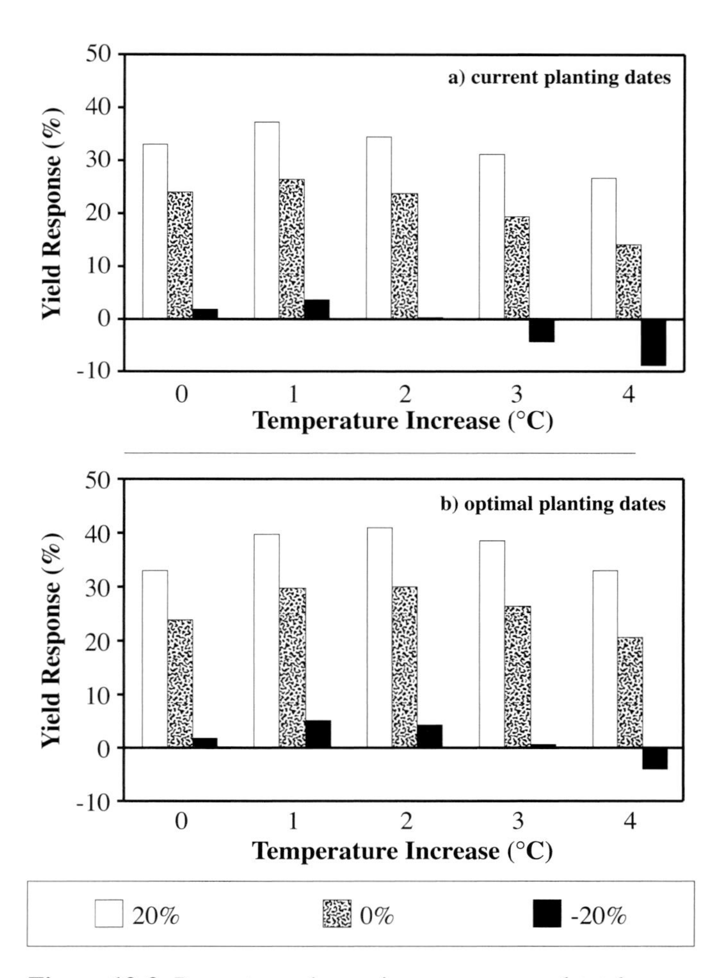

Figure 12-3: Percentage change in average annual total Australian wheat yield for doubling of actual CO_2 (to 700 ppm) and a range of changes in temperature and rainfall. Yield response is shown for rainfall changes of +20% (white), 0 (stippled), and -20% (black), for warmings of 0–4°C.

beyond 1°C, so the yield at 2°C was predicted to be similar to that at present temperatures, and the total yield declined below the current value for greater warmings. Adoption of earlier planting windows with climate change extended the yield plateau to 2°C warming before the slope of the temperature curve became negative (see Figure 12-3b). This was based on the present regional distribution of cropping, although cropping could expand into drier marginal areas with higher CO_2—but this may be countered by substantial reductions in rainfall or land degradation in currently cropped areas. Yield decreases rapidly with decreases in rainfall.

These response surfaces were used by Howden *et al.* (1999d) with the SRES scenarios (see Section 12.1.5.4), which use only recent coupled-ocean GCMs that show reductions in rainfall over most of mainland Australia in summer and winter. Results indicate that for mid-range scenarios (A1-mid and B2-mid) extended to 2100, national yields are reduced 3% without adaptation compared with current yields and increase only 3% with adaptation. Yields decline in western Australia and south Australia but increase in the eastern states. With the A2-high scenario, there are much larger negative impacts, with cropping becoming nonviable over entire regions, especially in western Australia. These results highlight the importance of the more negative rainfall scenarios found with the coupled-ocean GCMs.

In New Zealand, generally drier conditions and reductions in groundwater will have substantial impacts on cereal production in Canterbury (east coast South Island), the major wheat and barley production area of New Zealand. Other grain-producing areas (primarily Manawatu and Southland) are less likely to be affected. Grain phenological responses to warming and increased CO_2 are mostly positive, making grain filling slightly earlier and decreasing drought risk (Pyke *et al.*, 1998; Jamieson and Munro, 1999). Although grain-filling duration may be decreased by warmer temperatures, earlier flowering may compensate by shifting grain filling into an earlier, cooler period.

Maize production is mainly in Waikato (upper middle North Island) and Bay of Plenty and Poverty Bay (east coast North Island), with some production (more likely to be for silage than grain) further south in Manawatu and Canterbury. Rising temperatures make this crop less risky in the south, but water availability may become an issue in Canterbury.

Sweetcorn is grown mostly on the east coast of the North Island (Poverty Bay and Hawke's Bay) but increasingly in Canterbury. Climate warming is decreasing frost risk for late-sown crops, extending the season and moving the southern production margin further south. In the South Island, production is irrigated and is vulnerable to changes in river flows and underground water supply.

Horticulture in Australia includes cool,temperate fruit and vegetables in the south and at higher elevations, extensive areas of tropical fruits in the northeast and in irrigated areas in the northwest, and a rapidly expanding viticulture industry in cool and warm temperate zones. Many temperate fruits require winter chill or vernalization—which in some cases can be replaced by chemical treatments—and are strongly affected by disease and hail. Other more tropical fruit are subject to disease outbreaks and severe damage from hail, high winds, and heavy rain from tropical storms. These fruits are all likely to be affected by climate change, but few studies have been made (but see Hennessy and Clayton-Greene, 1995; Basher *et al.*, 1998).

In New Zealand, climate change may have mixed results on horticulture. Kiwifruit require some winter chill (Hall and McPherson, 1997a), and studies by Salinger and Kenny (1995) and Hall and McPherson (1997b) suggest that some varieties in some regions will become marginal; warmer summers and extended growing seasons may benefit others but may adversely affect timing for overseas markets.

Chilling requirements for most cultivars of pip-fruit are easily satisfied. However, some common cultivars have shown an adverse reaction to excessively warm conditions, with problems such as sunburn, water-core, and lack of color.

There has been a southern expansion of grapes in New Zealand over the past few decades, sometimes into more climatically marginal land. The New Zealand wine industry to date has shown a largely beneficial response to warm, dry conditions, which are expected to become more dominant in the east, but limitations on groundwater for irrigation may become a problem. Warmer conditions also are assisting expansion of the citrus industry in the north of New Zealand and are particularly beneficial for mandarins. However, this region would be susceptible to any increase in the location-specific frequency of subtropical storms reaching New Zealand.

12.5.4. Forestry

When Europeans arrived in Australia in 1788, there were approximately 70 Mha of forests. Since then, 40% has been cleared and a similar amount has been affected by logging; only about 25% remains relatively unaffected (Graetz *et al.*, 1995; State of the Environment, 1996). Nationally, land clearing still exceeds planting, although this varies greatly across the states, and is occurring mainly in areas defined as woodlands. Plantations have been expanding in Australia at an increasing rate since 1990, currently by more than 50,000 ha yr^{-1} (National Greenhouse Gas Inventory, 2000). Much of this planting is occurring on farmed land and receives federal government support (Race and Curtis, 1997). Additional plantings are occurring to ameliorate land degradation problems such as erosion, waterlogging, and salinization, and further plantings are associated with the establishment of carbon sinks (Howden *et al.*, 1999d).

Forests cover about 8.1 Mha (29%) of New Zealand's land area. Of this, about 6.4 Mha are in natural forest and 1.7 Mha in planted production forests. New forest establishment increased markedly during the 1990s. Almost all areas of harvested forest are replanted; during 1998, 52,000 ha of new forest plantings occurred (Statistics New Zealand, 1999).

Climatic factors are well known to influence species distributions (Hughes *et al.*, 1996; Austin *et al.*, 1997) and productivity (Landsberg and Waring, 1997). CO_2 concentrations also have a direct effect (Curtis and Wang, 1998). Kirschbaum (1999a,b) has used a forest growth model to assess response to climate change and CO_2 increases for a site near Canberra.

Howden and Gorman (1999) review this and other work on the impact of global change on Australian temperate forests. Productivity of exotic softwood and native hardwood plantations is likely to be increased by CO_2 fertilization effects, although the amount of increase is limited by various acclimation processes and environmental feedbacks through nutrient cycling. Where trees are not water-limited, warming may expand the growing season in southern Australia, but increased fire hazard and pests may negate some gains. Reduced rainfall in more recent scenarios would have adverse effects on productivity and increase fire risk. Increased rainfall intensity would exacerbate soil erosion problems and pollution of streams during forestry operations. In *Pinus radiata* and Eucalyptus plantations, fertile sites are more likely to have increased productivity for moderate warmings, whereas infertile sites could have decreased production. To date, large uncertainties have lowered the priority of climate change in management considerations.

Despite large year-to-year climatic variability, many Australian native species are confined in their natural climatic range to within 1 or 2°C average mean temperature (Hughes *et al.*, 1996; Pouliquen-Young and Newman, 1999), so without human intervention their survival will be threatened by warmings outside these ranges (see Section 12.4.1).

Of New Zealand's 13 Mha of land used for pastoral farming, at current prices 3–5 Mha of hill country would yield higher returns under forestry. Such land is being converted to plantation forestry at a rate of 40,000 ha yr^{-1}, from an initial rate averaging 60,000 ha yr^{-1} over the past decade (Statistics New Zealand, 1998); the recent decrease in planting rate reflects current lower wood prices. Carbon trading would facilitate increased planting rates, possibly up to 90,000 ha yr^{-1} (MAF, 1999). Steep land is particularly uneconomic to manage for pastoral farming and could be converted to forest or scrubland by planting or abandonment and regrowth. Control of possums (a pest introduced from Australia by early European settlers) to minimize transfer of diseases to farm animals has the added benefit of improving the health and regenerative capacity of some indigenous forests (Ministry for the Environment, 1997).

Biomass from forest residues and purpose-grown crops already provides 6% of New Zealand's primary energy supply (EECA, 1996) and significant energy resources for Australia. Biomass use is expected to increase substantially over the next decade (Sims, 1999), partly as a response to constraints on net carbon emissions. This also may encourage increased forestry planting rates.

The direct effects of elevated CO_2 on yield from *radiata* pine plantations are expected to be small in New Zealand. However, regional uncertainties remain with regard to possible increased growth loss under warmer, wetter conditions as a result of existing and new pests and diseases and losses from wind and fire associated with extreme weather events. Biosecurity mechanisms are being improved in New Zealand and abroad to better manage risk. Long-term trends in forest nutritional status are being examined. Indicators of sustainable forestry practices are receiving increasing attention. Risks from fire and wind also are being investigated in New Zealand. Systems to measure and predict effects on carbon sequestration in plantation forests are being improved.

12.5.5. Fisheries

Australia specializes in high-value, low-tonnage fisheries such as lobsters, pearl oysters, prawns, abalone, and tuna. Totaling about AU$2 billion yr^{-1} (ABARE, 2000), these fisheries are a significant local primary industry. Tonnage produced is very small by world standards because Australian surface waters generally are low in nutrients as a result of prevailing winds and boundary currents (Kailola *et al.,* 1993). New Zealand's Exclusive Economic Zone (EEZ) is one of the largest in the world (Statistics New Zealand, 1999), and its NZ$1.23 billion export revenues from fisheries in 1998 constituted 5.5% of total export revenue for that year (Seafood New Zealand, 2000).

For both countries, relationships have been established between recruitment of some fish species and climate variations, suggesting that fisheries in the region will be sensitive to climate change. However, it is uncertain how local winds and boundary currents that advect larvae and affect upwelling of nutrients might respond to GHG-induced climate changes, and downscaling from relevant global climate change model fields has not yet been done. Hence, this section concentrates on reporting studies of observed sensitivities of fisheries in the Australasian region to climate variability. There is insufficient information to date to project the impact of climate change on fisheries productivity.

Understanding of existing processes suggests that if El Niño were to become a more prevalent condition, the Indonesian throughflow and the Leeuwin current (Meyers, 1996) could weaken. If winds were favorable for upwelling, the west coast of Australia could undergo a dramatic shift from a low-production, high-biodiversity ecosystem to a more productive ecosystem typical of temperate shelves.

Australia's single largest fishery is western rock lobster (AU$260 million yr^{-1}—ABARE 2000). Presently, settlement of larval lobsters (and adult catch rates some years later) is much higher in La Niña years (high coastal sea level, high SST, strong Leeuwin current) than in El Niño years (Pearce and Phillips, 1994). Because the mechanism appears to be through larval advection processes, however, it is unclear whether the species' spawning strategy would adapt to a sustained shift to a weaker Leeuwin current. Many other western Australian fisheries also correlate (some positively, some negatively) with ENSO (Caputi *et al.,* 1996), through unknown mechanisms. Whether these mechanisms would continue to operate under the combined influence of a sustained weaker Leeuwin current (which tends to reduce temperatures) and a worldwide rise in SST is unknown. Southern bluefin tuna spawn where the Indonesian throughflow enters the Indian Ocean, but the impact of a possibly reduced throughflow also is unknown.

Conditions on the south coast of Australia also are influenced—but to a lesser degree—by the Leeuwin current, which tends to keep near-surface nutrient levels low. In addition, winds are favorable to downwelling, except during some summers, when Australia's only example of strong classical wind-driven coastal upwelling occurs off Portland, Victoria. Small meridional shifts of the subtropical high-pressure ridge modulate summer upwelling. Ecosystem impacts of this are poorly known.

On the east coast of Australia, the East Australian Current (EAC) is a dominant influence on coastal marine ecosystems. The EAC enhances upwelling and primary production (Hallegraeff and Jeffrey, 1993) and presumably fisheries, although this has yet to be demonstrated apart from its effect on the distribution of several tuna species (Lyne *et al.*, 1999). Farther north, Vance *et al.* (1985) report a correlation of catches of banana (but not tiger) prawns with rainfall, probably as a result of runoff-driven export of juveniles from estuary nursery beds.

Post hoc analyses (Smith, 1996) of a dramatic decline in the late 1980s in the Australian gemfish fishery suggest a combination of fishery pressure and poor recruitments as the cause. Recruitment appears to correlate with climatic cycles (Thresher, 1994). Smith (1996) developed a quantitative framework to evaluate management strategies for this and other fisheries.

For New Zealand, there is some evidence that fisheries recruitment may be enhanced by more frequent ENSO events (Harris *et al.*, 1988), although possible negative effects of increased incidence of toxic algal blooms also have been observed (Chang *et al.*, 1998). Changes in ENSO and ocean variability may combine with ocean warming in ways that are poorly understood. Recent New Zealand studies on snapper (Francis 1993, 1994a; Francis *et al.*, 1997), gemfish (Renwick *et al.*, 1998a), and hoki have shown that climatic variations may have a significant impact on spawning success or failure and subsequent recruitment into marine fish populations. Growth rates of juvenile and adult snapper appear to increase when SSTs are warmer (Francis, 1994b). This may have significant effects on the timing and scale of recruitment (e.g., Francis *et al.*, 1997). El Niño appears to have resulted in a westward shift of Chilean jack mackerel in the Pacific and subsequent invasion of this species into New Zealand waters in the mid-1980s (Elizarov *et al.,* 1993). This species now dominates the jack mackerel fishery in many areas. Variations in the abundance and distribution of pelagic large gamefish species in New Zealand may be closely correlated with variability in the ocean climate, with implications for recreational fishers as well as the tourist industry operating from charter boats.

Environmental temperature has a major influence on the population genetics of cold-blooded animals, selecting for temperature-sensitive alleles and genotypes. In New Zealand snapper, differences in allele frequencies at one enzyme marker have been found among year classes from warm and cold summers (Smith, 1979). Such differences could impact survival, growth rates, and reproductive success.

Finally, it should be noted that, if the wildcard of possible reduction or cessation of North Atlantic or Antarctic bottomwater formation were to occur (Manabe and Stouffer, 1994; Hirst, 1999), this could lead to significant changes in deep ocean chemistry, ocean dynamics, and nutrient levels on century time scales. This could have wide, but presently unknown, ramifications for fisheries in Australian and New Zealand waters.

12.5.6. Drought

In the Australia and New Zealand region, droughts are closely related to major drivers of year-to-year and decadal variability such as ENSO, Indian Ocean SSTs, the Antarctic Circumpolar Wave (White and Peterson, 1996; Cai *et al.*, 1999; White and Cherry, 1999), and the Interdecadal Pacific Oscillation (Mantua *et al.*, 1997; Power *et al.*, 1998; Salinger and Mullan, 1999), as well as more or less chaotic synoptic events. These are all likely to be affected by climate change (see Sections 12.1.5 and 12.2.3, and TAR WGI Chapters 9 and 10).

Using a transient simulation with the NCAR CCMO GCM at coarse resolution (R15) (Meehl and Washington, 1996), Kothavala (1999) found for northeastern and southeastern Australia that the Palmer Drought Severity Index indicated longer and more severe droughts in the transient simulation at about 2xCO_2 conditions than in the control simulation. This is consistent with a more El Niño-like average climate in the enhanced greenhouse simulation; it contrasts with a more ambivalent result by Whetton *et al.* (1993), who used results from several slab-ocean GCMs and a simple soil water balance model. Similar but less extreme results were found by Walsh *et al.* (2000) for estimates of meteorological drought in Queensland, based on simulations with the CSIRO RCM at 60-km resolution, nested in the CSIRO Mk2 GCM.

A global study by Arnell (1999), using results from an ensemble of four enhanced greenhouse simulations with the HadCM2 GCM and one with HadCM3, show marked decreases in runoff over most of mainland Australia, including a range of decreases in runoff in the Murray-Darling basin in the southeast by the 2050s of about 12–35%. HadCM3 results show large decreases in maximum and minimum monthly runoff. This implies large increases in drought frequency.

The decrease in rainfall predicted for the east of New Zealand by downscaling from coupled AOGCM runs for 2080 and the corresponding increase in temperature are likely to lead to more drought in eastern regions, from East Cape down to Southern Canterbury. Eastern droughts also could be favored by any move of the tropical Pacific into a more El Niño-like mean state (see Table 3-10). The sensitivity of New Zealand agriculture and the economy to drought events was illustrated by the 1997–1998 El Niño drought, which was estimated to result in a loss of NZ$618 million (0.9%) in GDP that year. A drought in north and central Otago and dry conditions in Southland associated with the 1998–1999 La Niña resulted in a loss of about NZ$539 million in GDP (MAF, 2000).

Recurring interest in Australia in policies on drought and disaster relief is evidence of a problem in managing existing climate variability and attempts to adapt (O'Meagher *et al.*, 1998). Present variability causes fluctuations in Australian GDP on the order of 1–2% (White, 2000). Drought and disaster relief helps immediate victims and their survival as producers (e.g., QDPI, 1996) but does not reduce costs to the whole community and in fact may prolong unsuitable or maladapted practices (Smith *et al.*, 1992; Daly, 1994), especially if there is climatic change. Farm productivity models are being used to simulate past and present farm production and to assess causes of and management options for coping with drought (Donnelly *et al.*, 1998). This is contributing to the fashioning of drought assistance and advisory policies.

The potential impact of drought on the Australian economy has declined, in relative economic terms, over time in parallel with the decline in the importance of agriculture to the economy (ABARE, 1997; Wilson and Johnson, 1997). In 1950–1951, the farm sector constituted 26.1% of GDP, whereas currently (1997–1998) it constitutes 2.5%. Similarly, the contribution of the farm sector to Australian exports has fallen from 85.3% (1950–1951) to 19.6% (1997–1998), with a reduction in the total farm sector labor force of about 6%. This despite the fact that farm production has increased over the same period. Thus, drought remains an important issue throughout Australia for social, political, geographical, and environmental reasons (Gibbs and Maher, 1967; West and Smith, 1996; Flood and Peacock, 1999).

Stehlik *et al.* (1999) studied the impact of the 1990 drought on more than 100 individuals from 56 properties in central Queensland and northern NSW to document the social experiences of dealing with drought. They conclude that there is strong evidence that the impact of the extended drought of the 1990s is such that rural Australia will never be the same again: "There is a decline in population: a closing down of small businesses, fewer and fewer opportunities for casual or itinerant work, more and more producers working 'off-farm' and a reduction in available services."

A change in climate toward drier conditions as a result of lower rainfall and higher evaporative demand would trigger more frequent or longer drought declarations under current Australian drought policy schemes, which rely on historical climate data and/or land-use practices on the basis of an expectation of historical climatic variability. A major issue for operational drought schemes is the choice of the most relevant historical period for the relative assessment of current conditions (Donnelly *et al.*, 1998).

Examples of Australian government involvement in rural industries that have been subject to decline in commodity prices over several decades (e.g., wool) suggest that the industries will be supported until the cost to the overall community is too high and the long duration or high frequency of drought declarations is perceived as evidence that the drought policy is no longer appropriate (Mercer, 1991; Daly, 1994). In the case of wool, the shift of government policy from that of support to facilitation of restructuring has involved a judgment about future demand and therefore prices (McLachlan *et al.*, 1999) and has only occurred after an extended period of low prices (Johnston *et al.*, 1999). With a change in climate toward drier conditions, drought policy probably would follow a similar path.

The New Zealand Government response to drought comes under Adverse Climatic Events and Natural Disasters Relief policy that was released in 1995. Government responds only when rare climatic or natural disasters occur on a scale that will seriously impact the national or regional economy and the scale of the response required is beyond the capacity of local resources. The policy is to encourage industry/community/individual response, rather than reliance on government support.

Science has a major role in assessing the probability that recent and current climatic conditions could be the result of natural variability or increased GHGs. At best, these assessments are presented in probabilistic terms (e.g., Trenberth and Hoar, 1997). The public and its representatives will have to judge what constitutes evidence of anthropogenic effects and to what extent future projections and their impacts should be acted on. Because of their impact, future droughts provide a very public focus for assessing the issues of climate change compared to natural variability. Appropriate land-use and management practices can be reassessed by using agricultural system models with CO_2 and climate projections from GCMs (Hall *et al.*, 1998; Howden *et al.*, 1999f; Johnston *et al.*, 1999). However, political judgments between the alternatives of supporting existing land use or facilitating reconstruction are likely to require greater certainty with regard to the accuracy of GCMs than is currently available (Henderson-Sellers, 1993).

One source of adaptation is seasonal and long-lead climate forecasting. This is one area in which climate science already is contributing to better agricultural management, profitability, and, to some extent, adaptation to climate change (Hammer *et al.*, 1991, 2000; Stone and McKeon, 1992; Stone *et al.*, 1996a; Johnston *et al.*, 1999). Indeed, empirical forecasting systems already are revealing the impact of global warming trends (Nicholls *et al.*, 1996b; Stone *et al.*, 1996b), and these systems already are adapting to climate change through regular revision and improvements in forecasting skill.

12.5.7. Pests and Diseases

Cropping, horticulture, and forestry in Australia and New Zealand are vulnerable to invasion by new pests and pathogens for which there are no local biological controls (Sutherst *et al.*,

1996; Ministry for the Environment, 1997). The likelihood that such pests and pathogens—particularly those of tropical or semi-tropical origin—will become established, once introduced to New Zealand, may increase with climate warming.

Indepth case studies are being conducted in Australia to test the performance of pest impact assessment methodologies for estimating the vulnerability of local rural industries to pests under climate change (Sutherst *et al.*, 1996). In New Zealand, pests that already are present may extend their ranges and cause more severe damage. For example, because of the reduced incidence of frosts in the north of New Zealand in recent years, the tropical grass webworm *(Herpetogramma licarisalis)* has increased in numbers and caused severe damage in some pastures in the far north.

The vulnerability of horticultural industries in Australia to the Queensland fruit fly *Bactrocera (Dacus) tryoni* under climate change was examined by Sutherst *et al.* (2000). Vulnerability was defined in terms of sensitivity and adaptation options. Regional estimates of fruit fly density, derived with the CLIMEX model, were fed into an economic model that took account of the costs of damage, management, regulation, and research. Sensitivity analyses were used to estimate potential future costs under climate change by recalculating costs with increases in temperature of 0.5, 1.0, and 2°C, assuming that the fruit fly will occur only in horticulture where there is sufficient rainfall or irrigation to allow the crop to grow. The most affected areas were the high-altitude apple-growing areas of southern Queensland and NSW and orange-growing areas in the Murrumbidgee Irrigation Area. Apples and pears in southern and central NSW also were affected. A belt from southern NSW across northern Victoria and into South Australia appeared to be the most vulnerable.

Adaptation options were investigated by considering, first, their sustainability under present conditions and, second, their robustness under climate variability and climate change. Bait spraying is ranked as the most sustainable, robust, and hence most promising adaptation option in boh the endemic and fruit fly exclusion zones, but it causes some public concern. The sterile insect technique is particularly safe, but there were concerns about costs, particularly with large infestations. Exclusion is a highly effective approach for minimizing the number of outbreaks of Queensland fruit fly in fly-free areas, although it is vulnerable to political pressure in relation to tourism. These three techniques have been given the highest priority.

12.5.8. Sustainability

Ecological and indeed economic sustainability has become a major issue in Australia and New Zealand (e.g., Moffatt, 1992). Australian government policy has been to integrate sustainability issues within a raft of polices and programs relating national heritage, land care, river care, wetlands, and carbon sequestration (Commonwealth of Australia, 1996). In both countries, land-use change and exotic pests and diseases, notably feral animals, are threatening many native species and ecosystems. In Australia, this is exacerbated by land degradation, notably soil erosion and increasing salinization brought about by loss of vegetative cover and rising water tables resulting from reduced evapotranspiration in catchments and irrigation with inadequate drainage.

These issues have been reviewed in several recent papers and reports, including a paper prepared for the Australian Prime Minister's Science, Engineering and Innovation Council (PMSEIC, 1999). This paper states that continuing degradation is costing Australia dearly in terms of lost production, increased costs of production and rehabilitation, possible damage to a market advantage as a producer of "clean and green" goods, increasing expenditures on building and repairing infrastructure, biodiversity losses, declining air and water quality, and declining aesthetic value of some landscapes.

PMSEIC (1999) states that the Australian community expects the use and management of resources to be economically, environmentally, and socially sustainable, which will require changes in management processes backed by science and engineering innovation. The report emphasizes that individual problems are linked and that an integrated approach is necessary to combat degradation and pursue remediation.

Climate change may exacerbate these problems by increasing opportunities for colonization by exotic species (e.g., woody weeds), by affecting the water balance and water tables, and by increasing erosion rates and flood flows through heavier rain events. Increased fire frequency also may threaten remnant forest and other ecosystems and impact soil degradation.

12.5.9. Global Markets

The impacts of climate change on food and fiber production in Australasia will be direct and indirect, the latter through changing global supply and demand influenced by climatic changes in other parts of the globe. Because a large proportion of food and fiber production in both countries is exported, the effect of commodity prices already is a major influence on the areas and mix of plantings and production, as well as profitability (Stafford Smith *et al.*, 1999). Adaptation in both countries has taken the form of changes in the mix of production between, for example, wool, lamb and beef, dairy products, horticulture and viticulture, and, most recently, farm and plantation forestry, as well as increasing exports of value-added and processed products. Increased variability of production resulting from climate change may restrict expansion of such added-value products. Response to markets has led to rapid changes in some sectors, but this is more difficult for commodities that require longer investment cycles, such as viticulture and forestry. Nevertheless, adaptation to a highly variable environment is a feature of Australian agriculture, and adaptations to climate change also may contribute to exports of agricultural technology. Improved forecasts of commodity prices and longer term trends in supply and demand, taking into account seasonal climate and ENSO forecasts, are a major means of adaptation. This will be especially important for climate change impacts and will require understanding of global effects.

In New Zealand, there are implications for future wood flows of scenarios for rates of increased areas of forest plantations. Currently, of the 18 million m^3 log volume harvested annually in New Zealand, one-third is used for domestic consumption and two-thirds is exported. Using a 50,000 ha yr^{-1} planting rate, by 2010 the ratio is expected to be 20:80, and by 2025 it is expected to be 5:95. Clearly, there is a need for an export focus. If sufficiently large export markets do not materialize for New Zealand, a major alternative use of wood within New Zealand could be for energy, especially if the economics change as a result of external considerations.

12.5.10. Indigenous Resource Management

Recognition of indigenous land rights in both countries recently has caused a much greater proportion of both countries to come back under the management control of Aboriginal and Maori peoples (Coombs *et al.*, 1990; Langton, 1997). In many situations, this has led to less intense economic exploitation, with more varied land use, (e.g., for low-intensity farming and pastoralism, combined with some horticulture, fishing, and ecotourism). Europeans in both countries have much to learn from traditional indigenous knowledge of land management, including the traditional custodianship ethic—particularly with regard to climatic fluctuations, extreme events, and sustainability. Indigenous knowledge may well lead to greater exploitation of indigenous species for nutritional and medicinal purposes. On the other hand, the indigenous people also have much to gain from greater economic and technical expertise related to markets and new technologies and products.

For example, Aboriginal traditional fire management regimes permitted reproduction of fire-dependent floral species and widespread savannas suitable for grazing. Through the creation of buffer zones, these regimes protected fire-intolerant communities such as monsoonal forests (Langton, 2000). Removal of Aboriginal groups into settlements led to areas where wildfires, fueled by accumulated biomass, cause extensive damage. Research in collaboration with traditional Aboriginal owners recently has played a key role in joint management of National Parks where customary Aboriginal burning is promoted to conserve biodiversity.

Andersen (1999) looks at the commonly accepted contrast between European ("scientific") and Aboriginal ("experiential") perspectives in fire management and concludes that in fact, European fire managers often lack clear land management goals and are no more "scientific" than Aboriginal fire managers. He argues that the task now is to introduce scientific goals into both European and Aboriginal fire management. This may be particularly applicable in adapting to changing vegetation patterns and increased fire danger in a changing climate.

12.6. Settlements and Industry

12.6.1 Infrastructure

Climate change will affect settlements and industry through changes in mean climate and changes in the frequency and intensity of extreme events. Obviously, changes in average climate affect design and performance, including variables such as heating and cooling demand, drainage, structural standards, and so forth. However, in many cases average climate is only a proxy for design standards that are developed to cope with extreme demands or stresses such as flooding rains, gale-force wind gusts, heat waves, and cold spells.

If the severity, frequency, or geographic spread of extreme events changes, the impact of such changes on infrastructure may be severe. For instance, movement of tropical cyclones further south into areas where infrastructure is not designed to cope with them would have significant consequences.

The rate and nature of degradation of infrastructure is directly related to climatic factors. Computer models to predict degradation as a function of location, materials, and design and construction factors have been developed (Cole *et al.*, 1999a,b,c,d). Because buildings and infrastructure that are being constructed now will have projected lives until 2050, placement, design, and construction changes to guarantee this life against climate change are needed. Moreover, the effects of degradation and severe meteorological events may have an unfortunate synergy. Increase in the rate of degradation as a result of climate change may promote additional failures when a severe event occurs. If the intensity or geographical spread of severe events changes, this effect may be compounded.

In New Zealand, a Climate Change Sustainability Index (CCSI) developed by Robinson (1999) rates the impact of climate change on a house and the contribution to climate change of GHG emissions from the house. The index includes GHG emissions from heating and cooling, comfort, tropical cyclone risk, and coastal and inland flooding. Basically, the closer the house is to sea level and/or a river or waterway, the lower the CCSI. Higher temperatures and rainfalls in general will shorten the life span of many buildings.

The impact of extreme climatic events already is very costly in both countries. This has been documented for Australia in Pittock *et al.* (1999), where it is shown that major causes of damage are hail, floods, tropical cyclones, and wildfire. In New Zealand, floods and landslides are the most costly climatically induced events, with strong winds and hail also important. The International Federation of Red Cross and Red Crescent Societies (1999) report estimates damages from a combination of drought, flood, and high wind (including cyclones, storms, and tornadoes) in Oceania (Australia, New Zealand, and the Pacific islands) to be about US$870 million yr^{-1} over the years 1988–1997. This figure apparently does not include hail damage, which is a major cost. Insured losses from a single severe hailstorm that struck Sydney in April 1999 were estimated at about AU$1.5 billion (roughly US$1 billion) (NHRC, 1999).

A scoping study for Queensland Transport (Queensland Transport *et al.*, 1999) has identified vulnerabilities for the Queensland transport infrastructure that will require adaptation. Infrastructure considered include coastal highways and railways, port installations and operations (as a result of high winds, sea-level rise, and storm surges), inland railways and roads (washouts and high temperatures), and some airports in low-lying areas. Key climate variables considered were extreme rainfall, winds, temperatures, storm surge, flood frequency and severity, sea waves, and sea level. Three weather systems combining extremes of several of these variables—tropical cyclones, east coast lows, and tropical depressions—also were assessed.

Regional projections for each of these variables were created, with levels of confidence, for four regions of Queensland for 2030, 2070, and 2100. Overall, the potential effects of climate change were assessed as noticeable by 2030 and likely to pose significant risks to transport infrastructure by 2070, if no adaptation were undertaken. Setting new standards, in the form of new design criteria and carrying out specific assessments in prioritized areas where infrastructure is vulnerable, was recommended for roads and rail under threat of flooding, bridges, and ports. Detailed risk assessments for airports in low-lying coastal locations were recommended.

Many ports and coastal communities already suffer from occasional storm-surge flooding and wave damage. A series of studies identifies particular vulnerabilities in some Queensland coastal cities (Smith and Greenaway, 1994; AGSO, 1999). Inland and coastal communities also are vulnerable to riverine flooding (Smith *et al.*, 1997; Smith, 1998a). A key feature of these studies is the nonlinear nature of damage response curves to increased magnitude and frequency of extreme events. This is partly because of exceedance of present design standards and the generally nonlinear nature of the damage/stress relationship, with the onset of building collapse and chain events from flying or floating debris (Smith, 1998a).

A study by McInnes *et al.* (2000; see also Walsh *et al.*, 2000) estimates the height of storm tides at the city of Cairns, in northern Queensland, for the present climate and for an enhanced greenhouse climate in which—based on the findings of Walsh and Ryan (2000)—the central pressure of tropical cyclones was lowered by about 10 hPa and the standard deviation of central pressure (a measure of variability) was increased by 5 hPa but the numbers were unchanged. Cairns is a low-lying city and tourist center, with a population of about 100,000 that is growing at about 3% yr^{-1}. Under present conditions, McInnes *et al.* (2000) found that the 1-in-100-year event is about 2.3 m in height; under the enhanced greenhouse conditions, it would increase to about 2.6 m, and with an additional 10- to 40-cm sea-level rise, the 1-in-100-year event would be about 2.7–3.0 m (see Figure 12-4). This would imply greatly increased inundation and wave damage in such an event, suggesting a possible need for changes in zoning, building regulations, and evacuation procedures.

Urban areas also are vulnerable to riverine flooding (Smith, 1998a) and flash floods exacerbated by fast runoff from paved and roofed areas (Abbs and Trinidad, 1996). Considerable effort has gone into methods to improve estimates of extreme

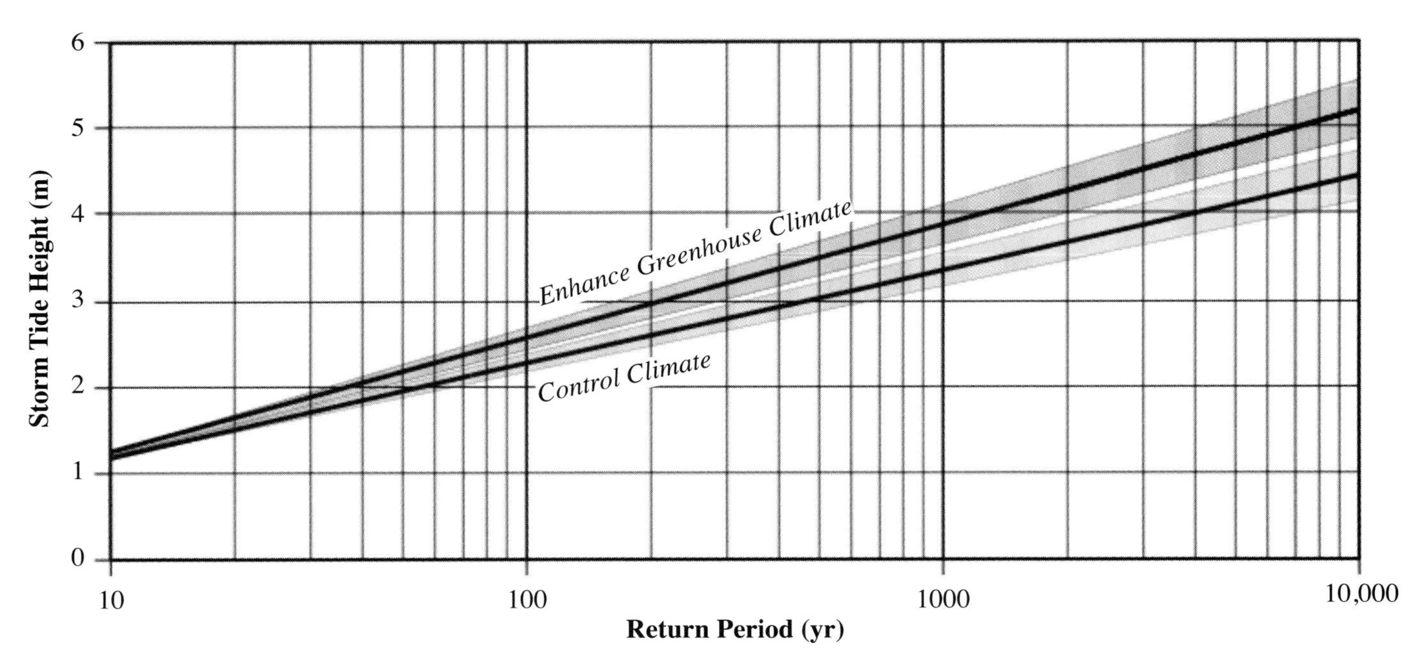

Figure 12-4: Simulated return periods (average time between events) of storm tides in Cairns, Queensland, for present climate (lower curve), and for enhanced greenhouse climate (upper curve), assuming 10 hPa lowering of central pressures and increased variability (additional 5 hPa standard deviation) of tropical cyclones. Anticipated mean sea-level rise should be added to these estimates. Uncertainty ranges of simulations are shown via grey shading (Walsh *et al.*, 2000).

precipitation under present conditions (Abbs and Ryan, 1997; Abbs, 1998). Schreider *et al.* (2000) applied a rainfall-runoff model to three different catchments upstream of Sydney and Canberra under doubled-CO_2 conditions. They found increases in the magnitude and frequency of flood events, but these effects differed widely between catchments because of the different physical characteristics of each catchment.

The safety of publicly owned and private dams also is a major issue (ANCOLD, 1986; Webster and Wark, 1987; Pisaniello and McKay, 1998) that is likely to be exacerbated by increases in rainfall intensity and probable maximum precipitation (Pearce and Kennedy, 1993; Fowler and Hennessy, 1995; Abbs and Ryan, 1997; Hennessy *et al.*, 1997).

Vulnerability depends not only on the severity of the potential impacts but on hazard mitigation measures put in place (including time- and location-specific hazard prediction), crisis management capability, and policies that avoid or minimize the hazard. These matters are discussed in Smith (1998a), Handmer (1997), and Kouzmin and Korac-Kakabadse (1999).

12.6.2. Investment and Insurance

According to Pittock *et al.* (1999), based on Insurance Council of Australia figures, major climatic catastrophe insurance losses from 1970 through 1996 averaged AU$208 million yr^{-1}. Of these losses, nearly half were from tropical cyclones; one-quarter were from hail. Other flooding and storm damage accounted for most of the rest; losses from fire were less than 10% of the total. Figures provided by the Insurance Council of New Zealand show that insurance industry payouts for New Zealand climatic catastrophes averaged NZ$23.5 million yr^{-1} (inflation-adjusted) between 1980 and 1998.

In an Australian study of insurance and climate change, Leigh *et al.* (1998a) examined four major climatic disasters: the Brisbane floods of 1974, the South Australian bushfires of 1983, the Nyngan floods of 1990, and the New South Wales bushfires of 1994. Total estimated damage from these four events was AU$178 million, $200–400 million, $47 million, and $168 million, respectively; however, the insurance industry bore only 39, 31, 9, and 33% of the cost, respectively. Government relief assistance was roughly equal to that from the insurance industry, and about 70–90% of that was provided by the federal government.

Leigh *et al.* (1998b) have reported on the potential for adaptation to climate change by the insurance industry in Australia by setting out an array of reactive and proactive options. Responses include reducing insurers' exposure or controlling claims through risk management to encourage disaster mitigation measures. The latter has the advantage of reducing overall losses to the community, rather than merely redistributing them among stakeholders. Natural disaster insurance can be more selective, so that good risks are rewarded and poor risks are penalized. Such rate-based incentives can motivate stakeholders to plan to more effectively minimize exposure to disasters. However, some individuals and businesses may have difficulties if some previously insurable properties become uninsurable against flood because of an increase in location-specific flood frequency. Government intervention and possible co-insurance between government and insurers also were canvassed. Cooperation between insurers and governments to ensure development and enforcement of more appropriate building codes and zoning regulations was regarded as desirable.

12.6.3. Energy and Minerals

Energy demand, essentially for air conditioning, is likely to increase in the summer and in more tropical parts of Australia and New Zealand (Lowe, 1988). However, winter demand for heating will similarly decrease in the winter and in cooler areas. Thus, increasing population in tropical and subtropical parts of Australia may combine with climate change to increase overall energy demand.

The other major uses of energy in Australia are transport and manufacturing. Transport demand generally will increase because of population growth but may be significantly affected by the changing distribution of growth across the continent, which in turn may be affected by climate change.

In general, warming will slightly reduce energy efficiency in most manufacturing, including electricity generation, but this is relatively minor compared with possible technological improvements in efficiency. Any decrease in water supply (see Section 12.3.1), such as is expected in the Murray-Darling basin in Australia, would impact adversely on hydroelectric generation and cooling of power stations, especially where there already is competition between water uses. Fitzharris and Garr (1996) predict benefits for hydroelectricity schemes in New Zealand's Southern Alps because they expect less water will be trapped as snow in the winter, which is the time of peak energy demand for heating.

12.6.4. Coastal Development and Management, Tourism

Economic development is proceeding rapidly in many coastal and tropical areas of Australia and New Zealand. This is fueled partly by general economic and population growth, but it is amplified in these regions by resource availability, shipping access for exports, attractive climates and landscapes, and the growth of the tourism industry. This selective growth in investment is leading to greater community risk and insurance exposure to present and future hazards, while many classes of hazard are expected to increase with global warming (see Table 3-10). Thus, present development trends are likely to make the impacts of climate change worse, especially for sea-level rise and increasing intensity of tropical cyclones. Particular attention should be paid to the implications for the risk to life and property of developments in coastal regions, as well as ways to reduce vulnerability to these hazards. Possible adaptations include improved design standards, zoning, early warning systems, evacuation plans, and emergency services.

Management of waste and pollution from settlements and industry will become more critical because of the potential for flood and waste discharge to impinge on water quality, including inland and coastal algal blooms, as well as adverse effects on ecotourism associated with damage to coral reefs (see Section 12.4.7). Sediment and pollution fluxes into the GBR lagoon already are a major concern (Larcombe *et al.*, 1996). This could be exacerbated by greater flood flows (see Sections 12.1.5.3 and 12.6.1) and increasing population and development. Higher temperatures will accentuate algal blooms.

The other major tourism and recreation sector that is likely to be seriously affected by climate change is the ski industry, which will be faced with significant reductions in natural snow cover (see Sections 12.2 and 12.4.4) and limited acceptance of artificial snow (Konig, 1998). Also, as the potential ranges of certain agricultural pests such as the fruit fly (see Section 12.5.7) and disease vectors such as mosquitos (see Section 12.7.1) increase, possible transfer of such pests and diseases through tourism may become an increasing issue.

12.6.5. Risk Management

As a result of the large uncertainties associated with possible future climate, as well as the stochastic nature of extreme events, there is great need for a risk management approach to development planning and engineering standards. In accordance with the precautionary principle, uncertainty should not be allowed to stand in the way of risk reduction measures, which in any case often will have other benefits such as protection of coastal and riverine environments.

Australia and New Zealand have jointly developed a risk management standard (Standards Australia and Standards New Zealand, 1999) that is designed to provide a consistent vocabulary and assist risk managers by delineating risk management as a four-step process that involves risk identification, risk analysis, risk evaluation, and risk treatment. Beer and Ziolkowski (1995) specifically examined environmental risk management and produced a risk management framework.

Examples of the application of a risk analysis approach are given in Sections 12.5.2 for pastures in New Zealand, 12.6.1 for storm surges, and 12.8.4 for irrigation water demand.

12.7. Human Health

12.7.1. Diseases and Injuries

Impacts of climate and climate change on health can be direct or indirect. Direct effects that are readily attributed to climate include heat stress and the consequences of natural disasters. However, the resulting burden of disease and injury may be less than that from indirect effects such as disrupted agriculture and reduced food security. Positive and negative effects can be anticipated, but there is insufficient evidence to state confidently what the balance will be (see Chapter 9). We have not attempted to estimate the overall economic costs of climate change impacts on health in the Australia-New Zealand region because there is considerable debate about the derivation and interpretation of monetary costs (see Chapter 19).

Guest *et al.* (1999) compared heat-related deaths in the five major Australian cities in the period 1977–1990 with those expected under different climate change scenarios (CSIRO, 1996a) for the year 2030. They estimate that greenhouse-induced climate change would increase climate-related deaths in the summer by a small amount, but this would be more than balanced by a reduction in climate-related deaths in the winter. Overall, this resulted in a decrease of 8–12% in climate-attributable mortality under the CSIRO "high" scenario compared to a scenario with no climate changes (but expected population changes).

The only study in New Zealand to date of elevated temperatures and mortality was conducted by Hales *et al.* (2000). Daily numbers of deaths in Christchurch were compared with measures of weather and ambient particulate pollution from June 1988 to October 1993. Above the third quartile (20.5°C) of summer maximum temperatures, an increase of 1°C was associated with a 1.3% increase (95% confidence interval, 0.4–2.3%) in all-cause mortality and a slightly greater increase in mortality from respiratory conditions. There was no evidence of interaction between the effects of temperature and particulate air pollution. Greater than expected numbers of deaths also occurred during winter days of low temperature, although this was not statistically significant. This suggests that cold-related deaths will be less common in a warmer climate. However, the mechanisms that explain excess winter mortality are not well understood.

The effects of solar ultraviolet (UV) radiation on skin cancer, skin aging, and cataracts of the eye are particularly important in New Zealand and Australia, which already have the highest skin cancer rates in the world (Marks *et al.*, 1989). The etiology of skin cancer is not fully understood, and factors other than sun exposure undoubtedly are involved. However, UV-B is a key factor. Present levels of UV radiation in this region are relatively high and have been increasing in the past 20 years (McKenzie *et al.*, 1999). It is expected with a high degree of confidence that if UV flux at ground level increases at a faster rate as a result of greenhouse-related cooling in the upper stratosphere and subsequent slowdown in the breakdown of ozone-depleting substances, the incidence of melanomas and other skin cancers will increase (Armstrong, 1994; Longstreth *et al*, 1998). Australians and New Zealanders of pale-skinned European descent will be particularly vulnerable to these effects. This topic is discussed in more detail in Chapter 9.

The numbers of notified cases of arbovirus infections (illnesses caused by insect-borne viruses) have increased in Australia in recent years (Russell, 1998), and exotic insect species such as *Aedes albopictus* and *Aedes camptorhynchus* that are competent vectors of viruses such as dengue and Ross River virus have been detected at New Zealand borders (Hearnden *et al.*, 1999).

There is good evidence that the frequency of mosquito-borne infections in this region is sensitive to short-term variations in climate. For example, outbreaks of Ross River fever and Murray Valley encephalitis in southeast Australia tend to follow heavy rainfall upstream in the Murray-Darling catchment (Nicholls, 1993; Maelzer *et al.*, 1999). In other parts of Australia where the predominant vector is the coastal mosquito *A. camptorhynchus*, variations in sea level also contribute to outbreaks of illness from Ross River virus (Mackenzie *et al.*, 1994). No quantitative estimates have been made of the possible impact of long-term climate change on rates of vector-borne infections. However, present climate change scenarios suggest that parts of Australia and New Zealand will experience conditions that are more favorable to breeding and development of mosquitoes (Bryan *et al*, 1996). In these areas, warmer conditions will tend to extend the range of reservoir hosts, decrease the extrinsic incubation period of arboviruses, and encourage outdoor exposure of humans (Weinstein, 1997). Therefore, it is expected with a high degree of confidence that the potential for insect-borne illness will increase. Whether this potential is translated into actual occurrence of disease will depend on many other factors, including border security, surveillance, vector eradication programs, and effectiveness of primary health care.

Endemic malaria was present in North Queensland and the Northern Territory until early in the 20th century (Ford, 1950; Black, 1972). Vectors to transmit the disease still are present in that part of Australia, and climate change will favor the spread of these mosquitoes southward (Bryan *et al.*, 1996). The disease-limiting factor at present is the effectiveness of local health services that ensure that parasitemic individuals are treated and removed from contact with mosquitoes. Therefore, climate change on its own is unlikely to cause the disease to return to Australia, unless services are overwhelmed. In New Zealand there currently are no mosquitoes that are capable of transmitting malaria; even under global warming scenarios, the possibility of an exotic vector becoming established is considered to be slight (Boyd and Weinstein, 1996).

Studies of the prevalence of asthma in New Zealand have shown an association with average temperature (Hales *et al.*, 1998). Electorates with lower mean temperatures tend to have lower levels of asthma, after adjusting for confounding factors. The reason is not clear but may be related to exposure to insect allergens. If this were so, warming may tend to increase the frequency of asthma, but too little is known about the causes of the disease to forecast the impact of climate change on asthma in Australia and New Zealand.

Climate change may influence the levels of several outdoor air pollutants. Ozone and other photochemical oxidants are a concern in several major Australian cities and in Auckland, New Zealand (Woodward *et al.*, 1995). In Brisbane, Australia, current levels of ozone and particulates have been associated with increased hospital admission rates (Petroeschevsky *et al.*, 1999) and daily mortality in persons ages 65 and over (Simpson *et al.*, 1999). Outdoor particulate pollution in the winter (largely generated by household fires) has been associated with increased daily mortality in Christchurch, New Zealand (Hales *et al.*, 2000). Formation of photochemical smog is promoted in warmer conditions, although there are many other climatic factors—such as windspeed and cloud cover—that are at least as important as temperature but more difficult to anticipate. A rise in overnight minimum temperatures may reduce the use of fires and hence emissions of particulates, but it is not known how this might affect pollution and population exposures.

Toxic algal blooms may affect humans as a result of direct contact and indirectly through consumption of contaminated fish and other seafood. At present this is not a major public health threat in Australia or New Zealand, but it is an economic issue (because of effects on livestock and shellfish). It could affect very large numbers of people (Oshima *et al.,* 1987; Sim and Wilson, 1997). No work has been carried out in Australia or New Zealand relating the health effects of algal blooms to climate. Elsewhere in the Pacific, it has been reported that the incidence of fish poisoning (resulting from ingestion of fish contaminated with ciguatoxins) is associated with ocean warming in some eastern islands, but not elsewhere (Hales *et al.*, 1999). It is uncertain whether these conditions will become more common in Australia and New Zealand with projected climate change.

Since 1800, deaths specifically ascribed to climatic hazards have averaged about 50 yr^{-1} in Australia (Pittock *et al.*, 1999), of which 40% are estimated to be caused by heat waves and 20% each from tropical cyclones and floods. Although this is not necessarily representative of present conditions because of changing population, statistical accounting, and technologies, it is an order of magnitude estimate. This suggests that if heat waves, floods, storm surges, and tropical cyclones do become more intense, some commensurate increase in deaths and injuries is possible. Whether this will occur will depend on the adequacy of hazard warnings and prevention. Statistics from other developed countries with larger populations indicate a recent trend toward increasing damages but decreasing death and injury from climatic hazards. Thus, hazard mitigation is possible, although it must more than outweigh increased exposure resulting from larger populations in hazardous areas.

12.7.2. Vulnerability

Climate impacts are determined not just by the magnitude of environmental change but also by the vulnerability of exposed populations. Examples of biophysical vulnerability are the susceptibility of pale-skinned populations to the effects of UV radiation and the vulnerability of isolated island ecosystems such as New Zealand's to invasion by exotic species (including disease vectors). Another example is infections spread to humans from animals, such as cryptosporidiosis. New Zealand has relatively high rates of notified cases of this infection (Russell *et al.*, 1998). This may be related to the conjunction of high densities of livestock and unprotected human drinking water supplies. In such a setting, increased rainfall intensity would promote transmission of the pathogen (by washing animal excreta containing the organism into the water supply).

Woodward *et al.* (1998) reviewed social determinants of vulnerability to health effects of climate change in the Pacific region. Australia and New Zealand are two of the wealthiest countries in the region, with relatively low population densities and well-developed social services. For these reasons, Australia and New Zealand are likely to be less vulnerable overall to many of the threats to health from climate change than neighboring countries. However, within both countries there are groups that are particularly susceptible to poor health. Sources of disadvantage include poverty, low housing standards, high-risk water supplies, lack of accessible health care, and lack of mobility. These factors tend to be concentrated in particular geographical locations and ethnic groups (Crampton and Davis, 1998) and carry with them increased vulnerability to most of the hazards that are associated with climate variability and climate change.

There have been no studies in New Zealand or Australia that have attempted to quantify vulnerability to disease and injury. Some work has been carried out on indices of environmental vulnerability (e.g., Kaly *et al.*, 1999). These studies have focused on measures of the resilience and integrity of ecosystems; with further development, they may assist in future forecasts of the impacts of climate change on human health.

12.7.3. Complexities of Forecasting Health Effects

There is no opportunity to study directly, in a conventional controlled fashion, the effects on health of climate change. Research to date in Australia and New Zealand has concentrated on the association of climate variability, at relatively restricted spatial and temporal scales, with the incidence of disease. It is not simple to extrapolate from these findings to long-term climate change. This problem is not particular to Australia and New Zealand, but there are informative local examples in the literature. These include analyses of roles of climate-related variables other than temperature (e.g., the critical effects of wind on air pollution, rainfall density on mosquito breeding, and humidity on heat stress); interactions with local ecosystems (e.g., the potential for amplification of certain arboviruses in New Zealand wildlife); and human behaviors that influence exposures (such as changing patterns of skin protection and the implications for UV exposures) (Hill *et al*, 1993; Russell, 1998b).

12.7.4. Public Health Infrastructure

A major challenge in Australia and New Zealand is how to protect and improve public health systems that deal with threats to health such as those that will potentially accompany climate change. Examples include border controls to prevent introduction of pathogens (including those from livestock and animal imports), measures required to ensure safe food and clean water, and primary health care services that reach the most disadvantaged and vulnerable members of the community. Threats to these systems include restrictions on government spending, increasing demands, and fragmented systems of purchase and provision of services.

With vector-borne diseases, the major challenge in Australia will be to control the expansion and spread of diseases that already are present in the country, such as Ross River virus and Murray Valley encephalitis—which are strongly influenced by climatic events. Introduction of new pathogens from close neighbors such as Papua New Guinea also is possible. Imported Japanese encephalitis and malaria remain serious threats, influenced principally by the numbers of people moving across the Torres Strait and the effectiveness of health services in the far north of Australia.

In New Zealand, the key issue for the health sector is how to prevent the introduction of vector-borne disease, particularly arboviruses carried by mosquitoes. Competent vectors for conditions such as Ross River virus and dengue have been detected frequently at entry points in recent years. This is likely to be a result of increasing trade and passenger traffic between New Zealand and other countries, as well as heightened awareness and better reporting. Eradication programs are expensive, and they are feasible only when the spread of exotic mosquitoes is confined. With repeated incursions and/or dispersion, the emphasis is likely to shift to control strategies.

12.7.5. Design of Human Environments

Several measures can be taken to better design human environments to cope with potential health stresses resulting from climate change. These measures include:

- Air conditioning and other measures to reduce exposure to heat
- Limiting exposure to disease vectors by measures such as use of screens on doors and windows and restriction of vector habitats (especially near waterways and urban wetlands)
- Land-use planning to minimize ecological factors that increase vulnerability to potential climate changes, such as deforestation, which increases runoff and the risk of flood-related injury and contamination of water supplies; animal stock pressures on water catchments; and settlement of marginal or hazardous areas such as semi-tropical coastal areas that are prone to storms and close to good vector breeding sites.

12.7.6. Vulnerable Populations, including Indigenous and Poor

Woodward *et al.* (1998) have argued that the effects of climate change on health will be most severe in populations that already are marginal. For these populations, climate change and sea-level rise impacts will be one more cause for "overload." In general, indigenous people in Australia and New Zealand are vulnerable to the effects of climate change because they tend to be excluded from mainstream economic activity and modern technological education and experience higher levels of poverty, lower rates of employment, and higher rates of incarceration than the overall population. These factors have widespread and long-term impacts on health (Braaf, 1999).

For example, Northern Territory health data for 1992–1994 show that the mortality rate for indigenous people was 3.5–4 times greater than that for nonindigenous people. Life expectation at birth was 14–20 years lower for indigenous Australians than for nonindigenous Australians (Anderson *et al.*, 1996). The indigenous population displays diseases and health problems that are typical of developed and developing nations. This includes high rates of circulatory diseases, obesity, and diabetes, as well as diarrheal diseases and meningococcal infections. High rates of chronic and infectious diseases affect individual and community well-being and reduce resilience to new health risks (Braaf, 1999).

A changing climate has implications for vector-borne and waterborne diseases in indigenous communities. In the "Top End" of the Northern Territory during the wet season, hot and humid conditions are conducive for vectors of infectious diseases endemic in the region. Vectors include flies, ticks, cockroaches, mites, and mosquitoes. Flies can spread scabies and other diseases. Mosquitoes are vectors for Australian encephalitis and endemic polyarthritis. Giardia and shigella are water-borne diseases that are common among indigenous children in the region. Both can be spread from infected people to others through consumption of infected food and untreated water. Existing and worsening overcrowded housing conditions, poor sanitation, and poor housing materials create breeding grounds for infection. Climate changes and sea-level rise—which create conditions that are suitable for new vectors (such as malaria) or expand distributions of existing vectors—may expose such vulnerable populations to increased risks.

In New Zealand, the gap between the health of indigenous people and the remainder of the population is less marked than in Australia but is substantial nevertheless. In 1996, life expectancy at birth was 8.1 years less for Maori females than for non-Maori females and 9.0 years less for Maori males than for non-Maori males (Statistics New Zealand, 1998a). As in Australia, this difference is associated with and partly caused by poorer economic circumstances and lack of appropriate, effective services (Durie, 1994). Consequently, Maori are at greater risk of health problems related to climate variability and climate change. An example is the lack of reticulated water supplies in the East Cape of the North Island, an area in which the population is predominantly Maori and in many cases cannot afford to truck in water in times of drought.

Impact assessments that consider only biophysical relationships between climate and health will be inadequate in evaluating indigenous health outcomes. The possibility—or, indeed, likelihood—that people may have very different views concerning what makes them vulnerable to climate change, which impacts may be significant, and what responses may be implemented also will need to be considered (Braaf, 1999).

The present social circumstances of indigenous peoples provide a poor basis on which to build adaptation responses to climate change threats. Thus, policies that aim to improve resilience to climate change impacts could encompass efforts to reduce relevant social liabilities—poverty, poor education, unemployment, and incarceration—and support mechanisms that maintain cultural integrity. Adaptive strategies could pursue economic development of these communities while sustaining the environments on which these populations are dependent (Howitt, 1993).

In other parts of the Pacific, there are many countries that are particularly susceptible to the effects of climate change—especially low-lying island states, which are likely to be severely affected by sea-level rise and increases in storm activity. Australia and New Zealand have close relations with many of the Pacific island states. For example, New Zealand has particular responsibilities for Niue, the Cook Islands, and Tokelau and contains substantial expatriate communities from most of the islands. Climate-related threats to these islands (see Chapter 17) would impact immediately on Australia and New Zealand.

12.8. Adaptation Potential and Vulnerability

12.8.1. Adaptation and Possible Benefits of Climate Change

It has not been assumed that all the impacts of climate change will be detrimental. Indeed, several studies have looked at possible benefits. Moreover, adaptation is a means of maximizing such gains as well as minimizing potential losses.

However, it must be said that potential gains have not been well documented, in part because of lack of stakeholder concern in such cases and consequent lack of special funding. Examples that have not been fully documented include the possible spread of tropical and subtropical horticulture further poleward (but see some New Zealand studies, on kiwi fruit, for example—Salinger and Kenny, 1995; Hall and McPherson, 1997b). In southern parts of Australia and New Zealand, notably Tasmania, there could be gains for the wine industry, increased comfort indices and thus tourism, and in some scenarios increased water for hydroelectric power generation.

Guest *et al.* (1999) have documented possible decreases in winter human mortality alongside possible increased summer mortality (see Section 12.7.1), and Howden *et al.* (1999d) have shown that Australian wheat yields may increase for 1 or 2°C warming, before showing declines at greater warmings (see Section 12.5.3 and Figure 12-3). A similar situation may apply to forestry (see Section 12.5.4). Such studies take account of gains from increased CO_2 concentrations. Changes in overseas production and thus in markets in some cases also could lead to greater demand and higher prices for Australian and New Zealand primary products (see Section 12.5.9), but only if such changes do not disrupt world trade in other ways (e.g., lower capacity to pay).

Vulnerability and adaptation to climate change must be considered in the context of the entire ecological and socioeconomic environment in which they will take place. Indeed, adaptations will be viable only if they have net social and economic benefits and are taken up by stakeholders. Adaptations should take account of any negative side effects, which would not only detract from their purpose but might lead to opposition to their implementation (PMSEIC, 1999).

Adaptation is the primary means for maximizing gains and minimizing losses. This is why it is important to include adaptation in impact and vulnerability studies, as well as in policy options. As discussed in Chapter 18, adaptation is necessary to help cope with inevitable climate change, but it has limits; therefore, it would be unwise to rely solely on adaptation to solve the climate change problem.

In some cases adaptation may have co-benefits. For example, reforestation to lower water tables and dryland salinization or to reduce storm runoff may provide additional income and help with mitigation (reduction of GHG emissions). However, other potential adaptations may be unattractive for other reasons (e.g., increased setbacks of development in coastal and riverine environments). These considerations have particular application in Australia and New Zealand. Studies of adaptation to climate change in Australia and New Zealand are still relatively few and far between. They are summarized in the remainder of this section.

12.8.2. *Integrated Assessments and Thresholds*

Over the past decade there have been several national and regional assessments of the possible impacts of climate change. A regional assessment for the Macquarie River basin was done by Hassall and Associates *et al.* (1998, reported in Basher *et al.*, 1998); Howden *et al.* (1999d) made a national assessment for terrestrial ecosystems (see Section 12.5). Two other preliminary regional assessments cover the Hunter Valley in NSW (Hennessy and Jones, 1999) and the Australian Capital Territory (Baker *et al.*, 2000). The former was based on a stakeholder assessment of climate change impacts that identified heat stress in dairy cattle as a subject for a demonstration risk assessment. Thresholds for heat stress and the probability of their being exceeded were evaluated, as were the economic value of adaptation through installation of shade and sprinklers (see Section 12.5.2; Jones and Hennessy, 2000). Baker *et al.* (2000) made a preliminary qualitative assessment of the impacts of scenarios on the basis of the CSIRO RCM at 60-km resolution (Hennessy *et al.*, 1998) on a wide range of sectors and activities.

However, most integrated studies in Australia and New Zealand have been "one-off" assessments, have lacked a time dimension, cannot readily be repeated to take account of advances in climate change science, and often have not placed the problem in its socioeconomic context. Several groups are collaborating on integrated modeling systems that overcome these drawbacks. In New Zealand this is called CLIMPACTS (Kenny *et al.*, 1995; Warrick *et al.*, 1996; Kenny *et al.*, 1999, 2000), and an Australian system called OZCLIM has been based on it. These integrated models contain a climate change scenario generator, climate and land surface data, and sectoral impact models. They provide a capacity for time-dependent analyses, a flexible scenario approach, a capability for rapid updating of scenarios; and inclusion of models for different sectors. One application is reported in Section 12.5.2.

OZCLIM contains regional climate patterns for monthly temperature and rainfall over Australia from several GCMs and the CSIRO RCM. They can be forced or scaled by the latest emission scenarios, and variables include potential evapotranspiration and relative humidity. It is being adapted to produce projected ranges of impact variables and to assess the risk of exceeding critical thresholds (CSIRO, 1996b; Jones, 2000; Pittock and Jones, 2000).

There are different levels and styles of integration in impact and adaptation assessment, and several of these have been attempted in Australia and New Zealand. Bottom-up integration was done for a range of climate change scenarios in the water supply, pasture, crop, and environmental flow sectors for the Macquarie River basin study by Hassall and Associates *et al.* (1998). It also has been done in a more probabilistic way to take account of uncertainty, with a focus on the probability of exceeding a user-defined threshold for performance and the need for adaptation (Jones, 2000).

Top-down integration has been attempted via the use of global impacts assessment models with some regional disaggregation—such as a regional analysis based on the Carnegie Mellon University ICAM model, which was used to examine adaptation strategies for the Australian agricultural sector (Graetz *et al.*, 1997). The principal conclusions were that climate matters and that the best strategy is to adapt better to climate variability.

Another top-down approach, based on an Australian regionalization of the DICE model of Nordhaus (1994), is that of Islam (1995). An initial application of this model to quantifying the economic impact of climate change damages on the Australian economy gave only a small estimate, but the authors expressed reservations about model assumptions and the need to better quantify climate impacts (Islam *et al.*, 1997). Others have examined the structure and behavior of the Integrated Model to Assess the Greenhouse Effect (IMAGE) but to date have not applied this to climate change impacts in Australia (Zapert *et al.*, 1998; Campolongo and Braddock, 1999).

A spatially explicit modeling system known as INSIGHT is being developed to evaluate a wide range of economic, social, environmental, and land-use impacts that could affect large areas (Walker *et al.*, 1996). It can map and summarize key social, economic, and environmental outcomes in annual steps to the year 2020. The need for such a system was identified through workshops involving potential stakeholders, and the system could factor in scenarios resulting from climate change.

As pointed out in PMSEIC (1999), much of Australia is subject to multiple environmental problems, of which climate change is only one. This leads to a logical emphasis on regional integrated assessments, which look for adaptations and policies that help to ameliorate more than one problem and have economic benefits.

12.8.3. Natural Systems

A large fraction of the region is composed of unmodified or nonintensively managed ecosystems where adaptation will depend mostly on natural processes. Vulnerability will occur when the magnitude or rate of climate variations lies outside the range of past variations. In some cases, adaptation processes may be very accommodating, whereas in others adaptation may be very limited. Ecosystems in the region handle a wide variety of climatic variability, in some cases with very large swings, but generally this variation occurs on short time scales—up to a few years. This does not necessarily confer adaptability to long-term changes of similar magnitude.

An important vulnerability identified by Basher *et al.* (1998) is the problem of temperatures in low to mid-latitudes that reach levels never before experienced and exceed the available tolerances of plants and animals, with no options for migration. The southwest of western Australia is a case in point (see Section 12.4.2). Another potential vulnerability arises from changes in the frequency of events. Examples include a climatic swing of duration exceeding a reproductive requirement (e.g., water birds in ephemeral lakes—Hassall and Associates *et al.*, 1998) and damaging events occurring too frequently to allow young organisms to mature and ecosystems to become reestablished.

Vulnerability also is expected to exist where species or ecosystems already are stressed or marginal, such as with threatened species; remnant vegetation; significantly modified systems; ecosystems already invaded by exotic organisms; and areas where physical characteristics set constraints, such as atolls, low-lying islands, and mountain tops. Coral reefs, for instance, may be able to survive short periods of rising sea level in clear water but are less likely to do so in turbid or polluted water or if their growth rates are reduced by acidification of the ocean (Pittock, 1999). Vulnerability of coastal freshwater wetlands in northern Australia to salinization resulting from increasing sea level and the inability of some of these wetlands to migrate upstream because of physical barriers in the landscape is described in Sections 12.4.5 and 12.4.7.

Unfortunately, there is relatively little specific information about the long-term capacity for and rates of adaptation of ecosystems in Australia and New Zealand that can be used to predict likely outcomes for the region. Therefore, a large degree of uncertainty inevitably exists about the future of the region's natural ecosystems under climate change.

12.8.4. Managed Systems

In the region's agriculture, many farming systems respond rapidly to external changes in markets and technology, through changes in cultivars, crops, or farm systems. Mid-latitude regions with adequate water supplies have many options available for adaptation to climate change, in terms of crop types and animal production systems drawn from other climatic zones. However, at low latitudes, where temperatures increasingly will lie outside past bounds, there will be no pool of new plant or animal options to draw from, and the productivity of available systems is likely to decline.

Adaptation options will be more limited where a climatic element is marginal, such as low rainfall, or where physical circumstances dictate, such as restricted soil types. Even where adaptations are possible, they may be feasible only in response to short-term or small variations. At some point, a need may arise for major and costly reconfiguration, such as a shift from or to irrigation or in farming activity. Indeed, one adaptation process already being implemented in Australia to cope with existing competition for water is water pricing and trading, which is likely to lead to considerable restructuring of rural industries (see Section 12.4.6).

Management of climate variability in Australia currently involves government subsidies in the form of drought or flood relief when a specific level of extreme that is classified as exceptional occurs (Stafford Smith and McKeon, 1998; see also Section 12.5.6). An adaptive measure being applied in Australia and New Zealand is to improve seasonal forecasting and to help farmers optimize their management strategies (Stone and McKeon, 1992; Stone *et al.*, 1996a; White, 2000), including reducing farm inputs in potentially poor years.

However, if a trend toward more frequent extremes were to occur, such measures might not allow farmers to make viable long-term incomes because there may be fewer good years. The question arises whether this is merely a string of coincidental extremes, for which assistance is appropriate, or whether it is part of an ongoing trend resulting from climate change. The alternative policy response to the latter possibility may well be to contribute to restructuring of the industry.

The PMSEIC (1999) report notes that there are opportunities for new sustainable production systems that simultaneously contribute to mitigation objectives through retention of vegetation and introduction of deeper rooted perennial pasture. Tree farming in the context of a carbon-trading scheme would provide additional opportunities, and this strategy could be linked to sustainability and alleviation of dryland salinity.

Nevertheless, the report recognizes that totally new production systems may be required for sustainability. These systems will need to capture water and nutrients that otherwise would pass the root zone and cause degradation problems. The design of such systems will entail research into rotating and mixing configurations of plants; manipulating phenology; modifying current crops and pastures through plant breeding, including molecular genetics; and possibly commercializing wildlife species and endemic biological resources. The report concedes, however, that there probably still will be agricultural areas where attempts to restore environmental and economic health will meet with little success.

Vulnerability and the potential for adaptation can be investigated quantitatively if a system and the climate change impacts on that system can be modeled. This was done by Pittock and Jones (2000) and Jones (2000) for climatically induced changes in irrigation water demand on a farm in northern Victoria. Here a window of opportunity is opened for adaptation via identification of a future time when adaptation would be necessary to reduce the risk of demand exceeding irrigation supply.

Jones (2000) used a model that was based on historical irrigation practice. Seasonal water use was used to estimate an annual farm cap of 12 Ml ha^{-1}, based on the annual allocated water right. Water demand in excess of this farm cap in 50% of the years was taken to represent a critical threshold beyond which the farmer cannot adapt. Conditional probabilities within projected ranges of regional rainfall and temperature change were utilized, combined with a sensitivity analysis, to construct risk response surfaces (see Figure 12-5). Monte Carlo sampling was used to calculate the probability of the annual farm cap being exceeded across ranges of temperature and rainfall change projected at intervals from 2000 to 2100. Some degree of adaptation was indicated as desirable by 2030, although the theoretical critical threshold was not approached until 2050, becoming probable by 2090 (see Figure 12-6).

In a full analysis, this example would need to be combined with an analysis of the likelihood of changes in water supply (e.g., Schreider *et al.*, 1997) affecting the allocated irrigation cap and an evaluation of possible adaptation measures.

12.8.5. Human Environments

The most vulnerable human environments in Australia and New Zealand are those that are subject to potential coastal or riverine flooding, landslides, or tropical cyclones and other intense storms. Adaptation to natural variability in these cases usually takes the form of planning zones, such as setbacks from coasts and flood levels of particular return periods or engineering standards for buildings and infrastructure. Many of these settlements and structures have long lifetimes—comparable to that of anthropogenic climate change. This means that many planning zones and design standards may become inappropriate in a changing climate.

Adaptation in these circumstances depends on costs and benefits, the lifetime of the structures, and the acceptability of redesigned measures or structures (e.g., seawalls). Thus, responses will depend in part on aesthetic and economic considerations; poorer communities, such as many indigenous settlements, will be particularly vulnerable. Conflicts will arise between investors with short time horizons and local government or other bodies who think on longer time scales and may bear responsibility for planning or emergency measures. Complex jurisdictional arrangements often will add to the difficulties of adopting rational adaptation measures (Waterman, 1996).

In an attempt to meet these problems for the Australian coastline, a guide has been developed for response to rising seas and climate change (May *et al.*, 1998), as well as good practice and coastal engineering guidelines (Institution of Engineers, 1998; RAPI, 1998).

Local governments in some parts of Australia and New Zealand are identifying measures they could implement to adapt to climate change. For example, the Wellington Regional Council is required by its Regional Policy Statement to periodically

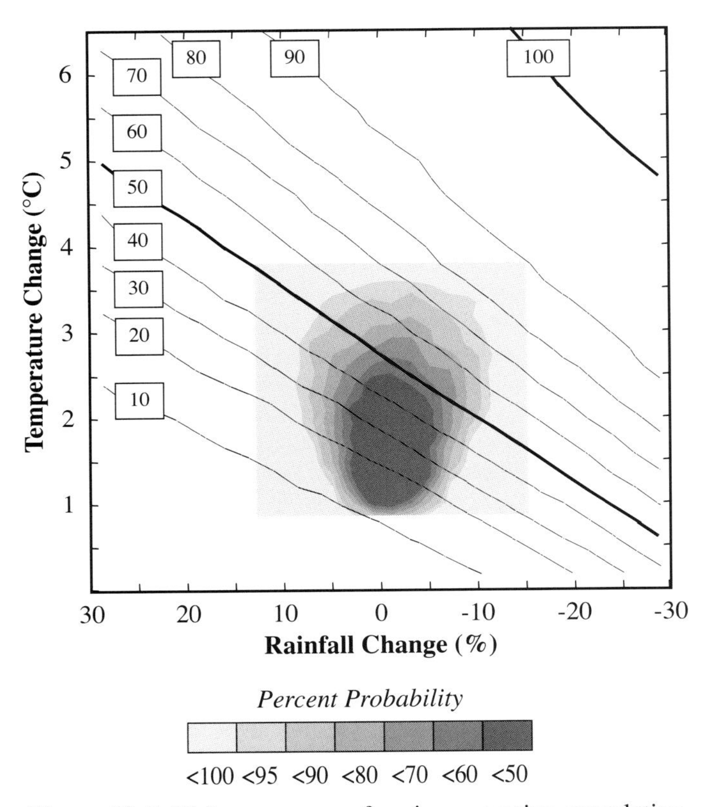

Figure 12-5: Risk response surface incorporating cumulative probability plots (in shaded box) for climate change magnitudes as indicated on x- and y-axes. Indicated percentage probabilities are probabilities of climate change in northern Victoria in 2070 lying within each shaded area (thus, there is a 100% probability of climate within the shaded square, and a 50% probability of climate within the innermost region). Probability (in percent) of irrigation water demand exceeding farm supply cap in any one year, for indicated climate change, is indicated by oblique lines. Critical threshold (heavy line) is set at a 50% chance of exceeding the cap (Jones, 2000).

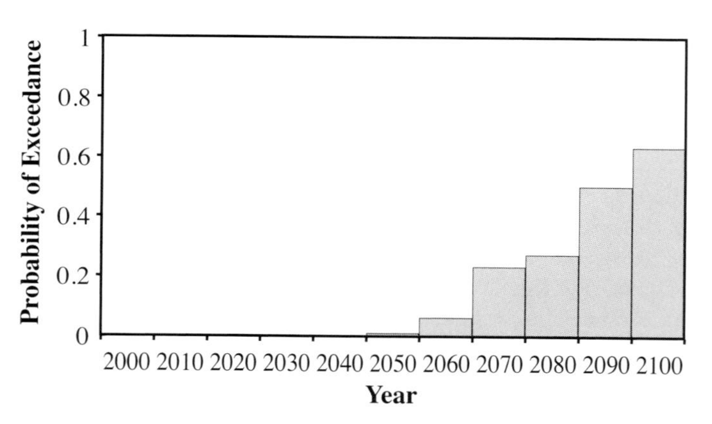

Figure 12-6: Change in probability of exceeding critical threshold (exceeding farm cap in 50% of all years) over time. Note that if farm cap were reduced through reductions in irrigation supply under climate change, the probability of the cap being exceeded would be increased (Jones, 2000).

review current knowledge on climate change and possible effects on natural hazards; the Council already allows for climate-induced variability in its flood protection activities (Green, 1999).

The region's health infrastructure is quite strong, and numerous existing adaptations, such as quarantine and eradication of disease vectors, are available to deal with the main changes expected. However, there is concern that already disadvantaged communities, especially indigenous people, may not have equitable access to adaptation measures. Another issue is the question of adaptations to deal with a climatic impact that may cause secondary effects. Examples include adaptations that require more energy production or higher water use, and vector controls that result in reduced population immunity to the disease carried.

12.8.6. Indigenous People

Traditional indigenous societies in the region have lived in a close and conscious relationship with their environment (Tunks, 1997; Skertchly and Skertchly, 2000). Australian Aborigines have modified and managed the landscape through the controlled use of low-intensity fire (Kohen, 1995). They have lived in Australia for at least 40,000 years. Thus, they have a long history of adaptation to sea-level rise, which rose by 130 m from the last glacial maximum 18,000 years ago until the present level was reached 6,000 years ago. Memories of these traumatic events are found in oral traditions recorded by early European settlers (Mulvaney and Kamminga, 1999).

With the recognition of indigenous land rights, indigenous people in both countries are now major land managers (Coombs *et al.*, 1990; Langton, 1997) and hence are impacted by and responsible for managing climatic impacts. The importance of their participation in the development of policy and response strategies to climate change is discussed in New Zealand Climate Change Programme (1990) and Tunks (1997), but to date their involvement has been minimal. This is a result partly of greater emphasis in Australia and elsewhere on mitigation of climate change rather than adaptation (Cohen, 1997), lack of indigenous community involvement in climate change research, and the array of other more pressing social issues for such communities (Braaf, 1999).

12.8.7. Extra-Regional Factors

Basher *et al.* (1998) draw attention to the vulnerability of the region to external influences arising from climate change—in particular, from likely changes in terms of trade (see also Stafford Smith *et al.*, 1999). Increased risk of invasion by exotic pests, weeds, and diseases and possible immigration from neighboring territories rendered uninhabitable by rising sea level are other factors. These issues exist independent of climate change but are likely to be exacerbated by climate change. A variety of adaptations to deal with each problem exist and can be strengthened, but the costs involved and remaining impacts could be considerable.

12.9. Synthesis

12.9.1 Introduction

Many activities and ecosystems in Australia and New Zealand are sensitive to climate change, with positive and negative effects. Much climate sensitivity information is still qualitative, and there are substantial uncertainties in predictions of regional- to local-scale climate changes—especially rainfall changes and changes in extreme events. Thus, comprehensive, quantitatively based cross-sectoral estimates of net Australasian costs of climate change impacts are not yet available. Confidence remains very low in the impact estimate for Australia and New Zealand of –1.2 to –3.8% of GDP for an equivalent doubling of CO_2 concentrations (Basher *et al.*, 1998). This estimate is based on climate change scenarios that are now outdated; does not include some potentially important impacts, including changes in weeds, pests and diseases, storm surges, and urban flooding; and does not account for possible adaptations to climate change.

Despite the uncertainties, there is a large body of knowledge and more agreement on changes than often is realized (e.g., between different coupled ocean-atmosphere climate models on the sign of the change in rainfall over large areas of Australia; see Section 12.1.5.1). There also is qualitative agreement on reduced water supplies in large agricultural regions of Australia (see Sections 12.3.1 and 12.5.6).

Potential net impacts on grazing, crops, and forests critically depend on the balance between the competing effects of warmings, positive or negative rainfall changes, direct physiological effects of higher CO_2 concentrations, and spatial and temporal variations in soil fertility. We expect with medium to high confidence that beneficial physiological effects of higher CO_2 concentration will become less dominant with time and effects of warming will become more damaging, especially given the expected tendency toward greater aridity over much of Australia. Impacts (and thus climate change benefits or damages) will not scale linearly with increasing GHG concentrations.

The following subsections draw together material from this chapter that is pertinent to the "policy-relevant questions" identified by the IPCC Bureau, as well as regional policy concerns.

12.9.2. Observed Consequences of Past and Current Climate Variability in the Region

Regional consequences documented in this chapter from floods, droughts, and temperature changes associated with recent ENSO events include losses in the pastoral agriculture sector from droughts (Australia, 1–2% of GDP; New Zealand, NZ$422 million estimated farm-gate costs from 1997 to 1999) and impacts on streamflow, water supply, stream ecology, horticulture, some commercial fisheries, and toxic algal blooms. Parts of the GBR have suffered mass coral bleaching from high SSTs (possibly related to ENSO and recent warming trends) and/or lowered salinity as a result of floods. Storms in the New Zealand region may have been more frequent in past warm epochs, leading to a greater influx of terrigenous material onto the continental shelf. Oceanic productivity east of New Zealand apparently increased immediately prior to and during warm periods 6,000–7,000 years ago and 120,000–125,000 years ago.

Extreme climatic events resulting from natural climatic variability in the past century have caused major damage and loss of life in Australia and New Zealand (see Sections 12.1.5.3 and 12.6.2). These extremes are expected to change in intensity, location-specific frequency, and sequence as a result of climate change (see Table 3-10 and Section 12.1.5.3), with major impacts on infrastructure and society unless strong adaptation measures are adopted.

12.9.3. Factors Influencing Vulnerability

12.9.3.1. Abrupt or Nonlinear Changes in Impacts

Several potential abrupt or nonlinear responses to climate change are listed in Table 12-1. In some cases, they involve a reversal of the sign of the effect with greater climate change; in others they result from the onset or acceleration of a biophysical or socioeconomic process that occurs or greatly accelerates beyond some threshold of climate change. Such thresholds can be quite location- and system-specific and need to be identified in collaboration between climatologists and stakeholders. The

Table 12-1: *Nonlinear or rapid climate change responses identified in this chapter.*

System	Description of Change	Certainty and Timing	Section
Great Barrier Reef	Reef death or damage from coral bleaching	Medium to high, next 20–50 years	12.4.7
Deep ocean	Chemical, dynamical, and biological changes from reduction in bottomwater formation	Low to medium, century time scale	12.4.7
Southwest of western Australia	Rapid loss of species with narrow annual mean temperature ranges	High, next 20–50 years	12.4.2
Australian alpine ecosystems	Loss of species as a result of warming and reduced snow cover	Medium to high, next 50 years	12.4.4
Insect-borne disease spread	Conditions more favorable to mosquitoes and other disease vectors	High potential, growing vulnerability	12.7.1
Agriculture	Shift from net profit to loss as a result of increased frequency of bad years	High potential in some places, next 20–50 years	12.8.4
Agriculture	Shift from positive to negative balance between benefits of increased CO_2 and losses from increasing aridity	High potential in parts of Australia, 50–100 years	12.5.2, 12.5.3
Built infrastructure	Change in magnitude and frequency of extremes to exceed design criteria, leading to rapid increases in potential damages to existing infrastructure	Medium to high in tropical coastal and riverine situations, next 30–50 years	12.6.1

frequency with which thresholds are exceeded often is used in engineering in the form of design criteria to withstand an event with a certain "return period," or average time between occurrences. It also can manifest itself as a change from an average profit to an average loss for a given enterprise such as a farm.

12.9.3.2. Interactions with Other Environmental and Social Factors

Some of the region's ecosystems are extremely vulnerable to invasion by exotic animal and plant species because of relative isolation before European settlement. Land-use changes have left some systems and areas more vulnerable to added stresses from climate changes, as a result of salinization or erosion, and because of ecosystem fragmentation, which lessens adaptation options for movement of species threatened by changing habitats (see Section 12.4.8).

Increasing human populations and development in coastal areas and on floodplains cause increasing vulnerability to tropical cyclones, storm surges, and riverine flooding episodes (see Sections 12.4.7 and 12.6.4), which may become more frequent with climatic change. Development has reduced the area and water quality of many estuaries, increasing the vulnerability of their ecosystems to sea-level rise and climate changes. Pressures on coral reefs including the GBR include increased coastal development, fisheries, tourism, and runoff of nutrients, chemicals, and sediment from land, as well as climate-related stresses such as rising sea level, rising temperatures, changes in tropical storm frequency, and acidification of the ocean from increasing CO_2 concentrations (see Section 12.4.7).

Poorer communities, including many indigenous settlements, are more vulnerable to climate-related natural hazards and stresses on health (see Section 12.7.6) because they often are in exposed areas and have less adequate housing, health, and other resources for adaptation.

Capacity to adapt is a function not only of the magnitude and critical nature of the climatic change but also of the demographics, economy, institutional capacity, and technology of a society. Thus, alternative socioeconomic futures will lead to different capacities to adapt. This has hardly been explored to date in Australia and New Zealand in relation to climate change.

12.9.3.3. Regional-Global Interactions

Reliance on exports of agricultural and forest products makes the region sensitive to changes in commodity prices produced by changes in climate elsewhere (see Section 12.5.9) and to increases in global forests as a result of carbon sink policies. Other extra-regional factors include increased risks of invasion by exotic pests, weeds, and diseases; pressures from immigration from neighboring low-lying Pacific island territories impacted by sea-level rise (see Section 12.7.6); and international agreements that constrain net emissions of GHGs. In an era of increasing globalization, these issues may assume more importance, although this report hardly touches on them.

12.9.4. Impacts for Differing Emissions Scenarios and Stabilization Pathways

Quantitative cross-sectoral impact assessments for differing scenarios are not yet available for Australia and New Zealand. Regional impacts will vary nonlinearly with time before and after stabilization of GHG concentrations. Warming will continue to increase after stabilization, but the beneficial effects of CO_2 on plants will no longer increase. Moreover, regional patterns of rainfall change, particularly in southern Australian and New Zealand areas, will tend to reverse after stabilization of GHGs (see Section 12.1.5.1). These complexities, together with the continuing post-stabilization rise in sea level, mean that estimated impacts at the time of stabilization may not be sufficient to determine whether the level of stabilization is a safe one.

12.9.5. Uncertainties and Risk Management

Given that some of the climate sensitivities listed in Section 12.9.6, and especially in Table 12-2, already have been observed for natural climate variations (such as El Niño), confidence is high that a range of impacts will occur in Australia and New Zealand as a result of climate change over the coming decades. This level of certainty, and the possibility that the early stages of greenhouse-related changes already may be occurring, justify prudent risk management through initiation of appropriate mitigation and adaptation strategies. Probabilistic assessments of risk, which account for the uncertainties, are regarded as a way forward. These assessments attempt to quantify the various sources of uncertainty to provide a conditional probability of climate change that would cause critical system performance thresholds to be exceeded and require adaptation or result in losses. Stakeholders may define their own subjective levels of acceptable risk and plan accordingly to adapt before or when the threshold is exceeded. Some examples for Australia and New Zealand are presented in this chapter (see Sections 12.5.2, 12.6.1, 12.8.2, and 12.8.4), but more are needed.

12.9.6. Vulnerability and Adaptability in Australia and New Zealand

The key regional concerns identified in this chapter regarding vulnerability to climate change impacts are ecosystem uniqueness, isolation, and vulnerability; agricultural commodities and terms of trade; droughts, floods, and water supply; increased coastal and tropical exposure to climate hazards; impacts on indigenous peoples and their involvement in adaptation planning; coral reefs; and Australian alpine areas.

Major expected impacts, vulnerability, and adaptability are summarized in Table 12-2. Note that although Australian and

Table 12-2: Main areas of vulnerability and adaptability to climate change impacts in Australia and New Zealand. Degree of confidence that tabulated impacts will occur is indicated by a letter in the second column (VH = very high, H = high, M = medium, L = low, VL = very low). These confidence levels, and the assessments of vulnerability and adaptability, are based on information reviewed in this chapter and assume continuation of present population and investment growth patterns.

Sector	Impact	Vulnerability	Adaptation	Adaptability	Section
Hydrology and water supply	– Irrigation and metropolitan supply constraints, increased salinization—**H**	High in some areas	– Planning, water allocation, and pricing	Medium	12.3.1, 12.3.2
	– Saltwater intrusion into some island and coastal aquifers—**H**	High in limited areas	– Alternative water supplies, retreat	Low	12.3.3
Terrestrial ecosystems	– Increased salinization of dryland farms and some streams (Australia)—**M**	High	– Changes in land-use practices	Low	12.3.3
	– Biodiversity loss, notably in fragmented regions, Australian alpine areas, and southwest of WA—**H**	Medium to high in some areas	– Landscape management; little possible in alpine areas	Medium to low	12.4.2, 12.4.4, 12.4.8
	– Increased risk of fires—**M**	Medium	– Land management, fire protection	Medium	12.1.5.3, 12.5.4, 12.5.10
	– Weed invasion—**M**	Medium	– Landscape management	Medium	12.4.3
Aquatic ecosystems	– Salinization of some coastal freshwater wetlands—**M**	High	– Physical intervention	Low	12.4.7
	– River and inland wetland ecosystem changes—**M**	Medium	– Change water allocations	Low	12.4.5, 12.4.6
	– Eutrophication—**M**	Medium in inland Aus. waters	– Change water allocations, reduce nutrient inflows	Medium to low	12.3.4
Coastal ecosystems	– Coral bleaching, especially Great Barrier Reef—**H**	High	– Seed coral?	Low	12.4.7
	—More toxic algal blooms?—**VL**	Unknown	—	—	12.4.7
Agriculture, grazing, and forestry	– Reduced productivity, increased stress on rural communities if droughts increase, increased forest fire risk—**M**	Location-dependent, worsens with time	– Management and policy changes, fire prevention, seasonal forecasts	Medium	12.5.2, 12.5.3, 12.5.4
	– Changes in global markets as a result of climate changes elsewhere—**H**, but sign uncertain	High, but sign uncertain	– Marketing, planning, niche and fuel crops, carbon trading	Medium	12.5.9
	– Increased spread of pests and diseases—**H**	Medium	– Exclusion, spraying	Medium	12.5.7
	– Increased CO_2 initially increases productivity, but offset by climate changes later—**L**	Changes with time	– Change farm practices, change industry		12.5.3, 12.5.4
Horticulture	– Mixed impacts (+ and -), depends on species and location—**H**	Low overall	– Relocate	High	12.5.3

Table 12-2 (continued)

Sector	Impact	Vulnerability	Adaptation	Adaptability	Section
Fish	– Recruitment changes (some species)—**L**	Unknown net effect	– Monitoring, management	—	12.5.5
Settlements and industry	– Increased impacts of flood, storm, storm surge, sea-level rise—**M**	High in some places	– Zoning, disaster planning	Moderate	12.6.1, 12.6.4
Human health	– Expansion and spread of vector-borne diseases—**H**	High	– Quarantine, eradication, or control	Moderate to high	12.7.1, 12.7.4
	– Increased photochemical air pollution—**H**	Moderate (some cities)	– Emission controls	High	12.7.1

(to a lesser extent) New Zealand farmers have adapted, at least in part, to existing El Niño-related droughts, they depend on good years for recovery. Thus, despite their adaptability, they are quite vulnerable to any increase in the frequency of drought or to a tendency for droughts to last for a longer period. This vulnerability flows through to the rural communities that service them (see Section 12.5.6).

Several of these vulnerabilities are likely to interact synergistically with each other and with other environmental stresses. Moreover, vulnerability is a result of exposure to hazard and capacity to adapt. Thus, vulnerability will be greatly affected by future changes in demography, economic and institutional capacity, technology, and the existence of other stresses.

There have been no rigorous studies for Australia or New Zealand that have taken all of these variables into account. Thus, Table 12-2 is based largely on studies that assume that the society that is being impacted is much like that of today. It should not be assumed, however, that socioeconomic changes in the future necessarily will reduce vulnerability in Australia and New Zealand. Indeed, many existing socioeconomic trends may exacerbate the problems. For instance, the bias toward population and economic growth in coastal areas, especially in the tropics and subtropics, by itself will increase exposure to sea-level rise and more intense tropical cyclones. If such trends are not to increase vulnerability, they will need to be accompanied by a conscious process of planning to reduce vulnerability by other means (e.g., changes in zoning and engineering design criteria). Thus, vulnerability estimates are based on present knowledge and assumptions and can be changed by new developments, including planned adaptation.

12.9.7. Knowledge Gaps

Significant knowledge gaps have become apparent during this assessment. These gaps mean that the net cost or benefit of unmitigated climate change in Australia and New Zealand is highly uncertain (see Section 12.9.1) and cannot be compared objectively with the cost of mitigation. Without this knowledge base, policymaking regarding adaptation and mitigation cannot be soundly based on economic considerations. Research priorities follow.

Indicators of climate impacts: Although long-term monitoring programs are in place for physical indicators (such as climate variables and sea level), work is still desirable on designing and implementing long-term monitoring programs that cover vulnerable animals, plants, and ecosystems and systematically examining them for the effects of climate changes. Flora and fauna with presently restricted or marginal climatic ranges would be most appropriate (see Section 12.4). (Candidate indicators for the UK are presented in Cannell *et al*., 1999.)

Underpinning physical knowledge and improved scenarios: Improved impact assessments will depend on better understanding of how climate change may influence factors such as the frequency and intensity of El Niños and related droughts, the intensity and location-specific frequency of tropical cyclones, and return periods for heavy rainfalls, floods, high winds, and hail. Oceanic issues include understanding differences between observed and modeled lag in warming of the Southern Ocean (see Section 12.1.5.2), possible impacts of cessation of bottomwater formation (see Section 12.4.7), and determining the influence of greenhouse warming on currents, upwelling, and nutrient supply. More knowledge is needed about influences of GHG-related cooling in the stratosphere on ozone depletion and regional UV radiation levels (see Section 12.7.1). All of this knowledge is needed to improve scenarios of regional climate (including ocean behavior).

Underpinning biological knowledge: The sensitivities of many plant and animal species and ecosystems in this region to climate changes, as well as the potential threats to biodiversity, are still unknown (see Section 12.4). Knowledge is required for assessment of potential impacts and for development of conservation strategies. This is important for marine and coastal environments, including coastal freshwater wetlands (see Section 12.4.7), as well as for terrestrial systems. Identification of relevant climatic thresholds for biological (and other) systems is needed (see Section 12.8.2). The effects of the time-varying balance between the beneficial physiological effects of increasing CO_2 concentrations and climate change on natural (indigenous) and managed ecosystems needs to be better understood, especially in light of recent regional scenarios.

Underpinning social knowledge: Better knowledge is required about the vulnerability of particular population groups (including indigenous people), about how people and organizations have adapted to past climate variability and changes, and about public attitudes to adaptation and mitigation options. Work is needed to understand how different socioeconomic futures (demography, economic capacity, and technological change) (see Section 12.1.5.4) would affect vulnerability (see Sections 12.8 and 12.9.2.2) and on socioeconomic thresholds for change, such as economic nonviability and unacceptable risk.

Fisheries: There is insufficient information to enable confident predictions of changes in fisheries productivity from climate change (see Section 12.5.5). This requires better knowledge of physical and biological processes in the ocean (as above) and improved information on the climate sensitivities of many species.

Health: Continuing work is needed on the potential for introduction (New Zealand) and spread (Australia) of significant disease vectors, including sensitivity to climate, population shifts, and effectiveness of health services and biosecurity procedures (see Section 12.7.4). Other health issues should be addressed, including potential effects of threats to water supply on remote communities (see Section 12.7.6).

Regional effects and integration: Better quantitative sectoral knowledge is required about, for example, influences of climate change on water supply and demand (see Section 12.3.1), salinization (see Section 12.3.3), and some crops and farming practices (see Sections 12.5.2 and 12.5.3). Because various sectors (e.g., agriculture, ecosystems, infrastructure, and hydrology) interact at the regional and national levels, continued work is needed on integrated assessment approaches and models that synthesize sectoral knowledge and draw on the social sciences (see Section 12.8.2), in rural and urban settings. Models of the physical economy that track fluxes and pools of materials, energy, land, and water are required for national analyses.

Global interactions: More understanding is needed of the interaction of global climate change impacts, and of mitigation policies, on Australian and New Zealand markets, sectoral change, and land use (see Sections 12.5.9 and 12.8.7).

Adaptation: Further objective studies are required, in close collaboration with stakeholders, on adaptation options and their acceptability, costs, co-benefits, side effects, and limits (see Section 12.8). Adaptation should be regarded as a means to maximize gains and minimize losses, with a greater exploration of opportunities (see Section 12.8.1).

Costing: More comprehensive and realistic costings are needed for impacts and adaptation options, taking account of human behavior and using up-to-date scenarios.

Communication of policy-relevant results: If climate change issues are to be addressed by decisionmakers, there will need to be better communication of results from research. This will come partly from consultation with decisionmakers and other stakeholders to ensure that the right policy-relevant questions are addressed (see Sections 12.8.2 and 12.9) and partly from effective communication of what is known, as well as the uncertainties. A risk-assessment approach geared to particular stakeholders seems likely to be most effective (see Section 12.9.5).

References

ABARE, 2000: *Australian Fisheries Statistics.* Australian Bureau of Agriculture and Resource Economics, Canberra, Australia, 13 pp. Available online at http://www.abareconomics.com/pdf/fishfacts.pdf.

ABARE, 1999: Irrigation water reforms: impact on horticulture farms in the southern Murray Darling basin. In: *Current Issues, No. 99.2.* Australian Bureau of Agriculture and Resource Economics, Canberra, Australia, 12 pp.

ABARE, 1997: Changing structure of farming in Australia. In: *Current Issues, No. 97.4.* Australian Bureau of Agriculture and Resource Economics, Canberra, Australia, 8 pp.

Abbs, D.J., 1998: A numerical modelling study to investigate the assumptions used in the calculation of probable maximum precipitation. *Water Resources Research,* **35,** 785–796.

Abbs, D.J. and B.F. Ryan, 1997: *Numerical Modelling of Extreme Precipitation Events.* Research Report No. 131, Urban Water Research Association of Australia, Melbourne, Australia, 71 pp.

Abbs, D.J. and G.S. Trinidad, 1996: Extreme weather and the built environment. In: *Proceedings of Workshop on Atmospheric Hazards: Process, Awareness and Response, The University of Queensland, Brisbane, 20–22 September, 1995.* The Workshop, Brisbane, Australia, 13 pp.

Abel, N., A. Langston, W. Tatnell, J. Ive, and M. Howden, 1997: Sustainable use of rangelands in the 21st century: a research and development project in western New South Wales. Unpublished paper to the *Australia New Zealand Ecological Economics Conference, Melbourne, 17–20 November, 1997.*

Abel, N., J. Ive, A. Langston, B. Tatnell, D. Tongway, B. Walker, and P. Walker, 2000: Resilience of NSW rangelands: a framework for analysing a complex adaptive system. In: *Management for Sustainable Ecosystems* [Hale, P., A. Petrie, D. Moloney, and P. Sattler (eds.)]. Centre for Conservation Biology, University of Queensland, Brisbane, Australia, pp. 58–70.

AGSO, 1999: *The Cities Project.* Australian Geological Survey Organisation, Australia. Available online at http://www.agso.gov.au/geohazards/grm/cities2.html.

Allan, R.J. and M.R. Haylock, 1993: Circulation features associated with the winter rainfall decrease in southwestern Australia. *Journal of Climate,* **7,** 1356–1367.

Alongi, D.M., T. Ayukai, G.J. Brunskill, B.F. Clough, and E. Wolanski, 1998: Sources, sinks, and export of organic carbon through a tropical, semi-enclosed delta (Hinchinbrook Channel, Australia). *Mangroves and Salt Marshes,* **2,** 237–242.

ANCOLD, 1986: *Guidelines on Design Floods for Dams.* Australian National Committee on Large Dams, Leederville, Australia, 42 pp.

Andersen, A., 1999: Cross-cultural conflicts in fire management in northern Australia: not so black and white. *Conservation Ecology,* **3(1),** 6. Available online at http://www.consecol.org./vol3/iss1/art6.

Anderson, P., K. Bhatia, and J. Cunningham, 1996: *Mortality of Indigenous Australians.* Occasional paper, Aboriginal and Torres Strait Islander Health and Welfare Information (a Joint Program of the Australian Bureau of Statistics and the Australian Institute of Health and Welfare), Australian Government Publishing Service, Canberra, Australia.

Armstrong, B.K., 1994: Stratospheric ozone and health. *International Journal of Epidemiology,* **23,** 873–85.

Arnell, N.W., 1999: Climate change and global water resources. *Global Environmental Change,* **9,** S31–S46.

Arthington, A.H., 1998: *Comparative Evaluation of Environmental Flow Assessment Techniques: Review of Holistic Methodologies*. Occasional Paper 25/98, Land and Water Resources Research and Development Corporation, Canberra, Australia, 46 pp.

Arthington, A.H., S.O. Brizga, and M.J. Kennard, 1998: *Comparative Evaluation of Environmental Flow Assessment Techniques: Best Practice Framework*. Occasional Paper 26/98, Land and Water Resources Research and Development Corporation, Canberra, Australia, 26 pp.

Austin, M.P., J.G. Pausas, and I.R. Noble, 1997: Modelling environmental and temporal niches of eucalypts. In: *Eucalypt Ecology: Individuals to Ecosystems* [Williams, J.E. and J.C. Woinarski (eds.)]. Cambridge University Press, Cambridge, United Kingdom and New York, NY, USA, pp. 129–150.

Baker, B., M. Austin, G. Barnett, A. Kearns, and H. Shugart, 2000: *Global Warming, Local Consequences: Can We Predict the Likely Impacts of Climate Change for the ACT Region?* A discussion paper prepared by CSIRO Wildlife and Ecology, Canberra, Australia, 36 pp.

Basher, R.E. and X. Zheng, 1995: Tropical cyclones in the southwest Pacific: spatial patterns and relationships to the Southern Oscillation and sea surface temperature. *Journal of Climate*, **8**, 1249–1260.

Basher, R.E., A.B. Pittock, B. Bates, T. Done, R.M. Gifford, S.M. Howden, R. Sutherst, R. Warrick, P. Whetton, D. Whitehead, J.E. Williams, and A. Woodward, 1998: Australasia. In: *The Regional Impacts of Climate Change. An Assessment of Vulnerability. A Special Report of IPCC Working Group II for the Intergovernmental Panel of Climate Change* [Watson, R.T, M.C. Zinyowera, R.H. Moss, and D.J. Dokken (eds.)]. Cambridge University Press, Cambridge, United Kingdom and New York, NY, USA, 878 pp.

Bayliss, B.L., K.G. Brennan, I. Eliot, C.M. Finlayson, R.N. Hall, T. House, R.W.J. Pidgeon, D. Walden, and P. Waterman, 1997: *Vulnerability Assessment of Predicted Climate Change and Sea Level Rise in the Alligator Rivers Region, Northern Territory, Australia*. Supervising Scientist Report 123, Supervising Scientist, Canberra, Australia, 134 pp.

Beer, T. and A. Williams, 1995: Estimating Australian forest fire danger under conditions of doubled carbon dioxide concentrations. *Climatic Change*, **29**, 169–188.

Beer, T. and F. Ziolkowski, 1995: *Risk Assessment: An Australian Perspective*. Report 102, Supervising Scientist, Barton, ACT, Australia. Available online at http://www.environment.gov.au/ssg/pubs/risk/risk_toc.html.

Behl, R.K., H.S. Nianawatee, and K.P. Singh, 1993: High temperature tolerance in wheat. In: *International Crop Science, Vol I*. Crop Science Society of America, Madison, WI, USA, pp. 349–355.

Bell, R.G., D.G. Goring, and W.P. de Lange, 1999: Rising and extreme sea levels around New Zealand. In: *Proceedings of the IPENZ Technical Conference, Auckland, 11-12 July 1999*. Institution of Professional Engineers, Wellington , New Zealand, pp.121-131.

Bennett, S., R. Brereton, I. Mansergh, S. Berwick, K. Sandiford, and C. Wellington, 1991: *The Potential Effect of Enhanced Greenhouse Climate Change on Selected Victorian Fauna*. Tech. Report Series No. 123, Arthur Rylah Institute, Heidelberg Victoria, Australia, 224 pp.

Berkelmans, R. and J.K. Oliver, 1999: Large scale bleaching of corals on the Great Barrier Reef. *Coral Reefs*, **18**, 55–60.

Berkelmans, R. and B.L. Willis, 1999: Seasonal and local spatial patterns in the upper thermal limits of corals on the inshore Central Great Barrier Reef. *Coral Reefs*, **18**, 219–228.

Biggs, B.J.F., 1996: Patterns in benthic algae of streams. In: *Algal Ecology: Freshwater Benthic Ecosystems* [Stevenson, R.J., M.L. Bothwell, and R.L. Lowe (eds.)]. Academic Press, San Diego, CA, USA, pp. 31–56.

Black, R.H., 1972: *Malaria in Australia*. Service Publication No. 9, School of Public Health and Tropical Medicine, University of Sydney, Australian Government Publishing Service, Canberra, Australia, 222 pp.

Boer, P., 2000: Huge water supply scheme for Perth. *Civil Engineers Australia*, **72(10)**, 58–98.

Boyd, A.M. and P. Weinstein, 1996: Anopheles annulipes: an under-rated temperate climate malaria vector. *NZ Entomologist*, **19**, 35–41.

Boyd, P., J. LaRoche, M. Gall, R. Frew, and R.M.L. McKay, 1999: The role of iron, light and silicate in controlling algal biomass in sub-Antarctic water SE of New Zealand. *Journal of Geophysical Research*, **104**, 13395–13408.

Braaf, R.R., 1999: Improving impact assessment methods: climate change and the health of indigenous Australians. *Global Environmental Change*, **9**, 95–104.

Bradford-Grieve, J.M. and J.B. Jillett, 1998: Ecological constraints on horizontal patterns, with special reference to the copepod Neocalanus tonsus. In: *Pelagic Biogeography ICoPB II. Proceedings of the Second International Conference* [Pierrot-Bults, A.C. and S. van der Spoel (eds.)]. Workshop Report No. 142, Intergovernmental Oceanographic Commission (IOC), UNESCO, pp. 65–77.

Bradford-Grieve, J.M., P.W. Boyd, F.H. Chang, S. Chiswell, M. Hadfield, J.A. Hall, M.R. James, S.D. Nodder, and E.A. Shushkina, 1999: Pelagic ecosystem structure and functioning in the Subtropical Front region east of New Zealand in austral winter and spring 1993. *Journal of Plankton Research*, **21**, 405–428.

Brown, B.E., 1997: Coral bleaching: causes and consequences. *Coral Reefs*, **16**, S129–S138.

Bryan, J.H., D.H. Foley, and R.W. Sutherst, 1996: Malaria transmission and climate change in Australia. *Medical Journal of Australia*, **164**, 345–347.

Burke, J.J., J.R. Mahan, and J.L. Hatfield, 1988: Crop-specific thermal kinetic windows in relation to wheat and cotton biomass production. *Agronomy Journal*, **80**, 553–556.

Burns, N.M., T.M. Hume, D.S. Roper, and R.K. Smith, 1990: Estuaries. In: *Climatic Change: Impacts on New Zealand*. Ministry for the Environment, Wellington, New Zealand, pp. 81–84.

Busby, J.R., 1991: BIOCLIM: a bioclimate analysis and prediction system. In: *Nature Conservation: Cost Effective* Biological *Surveys and Data Analysis* [Margules, C.R. and M.P. Austin (eds.)]. CSIRO, Melbourne, Australia, pp. 64–68.

Cai, W. and P.H. Whetton, 2000: Evidence for a time-varying pattern of greenhouse warming in the Pacific Ocean. *Geophysical Research Letters*, **27**, 2577–2580.

Cai, W., P.G. Baines, and H.B. Gordon, 1999: Southern mid-to-high latitude variability, a zonal wavenumber 3 pattern, and the Antarctic Circumpolar Wave in the CSIRO coupled model. *Journal of Climate*, **12**, 3087–3104.

Campbell, B.D. and N.D. Mitchell, 1996: *Report on the Development of Climate Profiles of Subtropical Grasses for CLIMPACTS*. Internal report, AgResearch, Palmerston North, New Zealand, 5 pp.

Campbell, B.D., G.M. McKeon, R.M. Gifford, H. Clark, M.S. Stafford Smith, P.C.D. Newton, and J.L. Lutze, 1996: Impacts of atmospheric composition and climate change on temperate and tropical pastoral agriculture. In: *Greenhouse: Coping With Climate Change* [Bouma, W.J., G.I. Pearman, and M.R. Manning (eds.)]. CSIRO Publishing, Collingwood, Victoria, Australia, pp. 171–189.

Campolongo, F. and R. Braddock, 1999: Sensitivity analysis of the IMAGE Greenhouse model. *Environmental Modelling and Software*, **14**, 275–282.

Cannell, M.G.R., J.P. Palutikof, and T.H. Sparks, 1999: *Indicators of Climate Change in the United Kingdom*. Department of Environment Transport Regions, Wetherby, United Kingdom, 87 pp.

Caputi, N., W.J. Fletcher, A. Pearce, and C.F. Chubb, 1996: Effect of the Leeuwin Current on the recruitment of fish and invertebrates along the west Australian coast. *Marine and Freshwater Research*, **47**, 147–55.

Carter, T.R., M. Hulme, J.F. Crossley, S. Malyshev, M.G. New, M.E. Schlesinger, and H. Tuomenvirta, 2000: *Climate Change in the 21st Century - Interim Characterizations based on the New IPCC Emissions Scenarios*. The Finnish Environment 433, Finnish Environment Institute, Helsinki, 148 pp.

Chang, F.H., J. Sharples, J.M. Grieve, M. Miles, and D.G. Till, 1998: Distribution of Gymnodinium cf. breve and shellfish toxicity from 1993 to 1995 in Hauralki Gulf, New Zealand. In: *Harmful Algae* [Reguera, B., J. Blanco, M.L. Fernandez, and T. Wyatt (eds.)]. Xunta de Galicia and Intergovernmental Oceanographic Commission of UNESCO, Paris, France, pp. 139–142.

Charles, S.P., B.C. Bates, P.H. Whetton, and J.P. Hughes, 1999: Validation of downscaling models for changed climate conditions: case study of southwestern Australia. *Climate Research*, **12**, 1–14.

Chiera, B.A., J.A. Filar, and T.S. Murty, 1997: The greenhouse effect and sea level changes around the Australian coastline. In: *Proceedings of the IASTED International Conference, Modelling and Simulation, 15-17 May 1997, Pittsburgh, PA, USA* [Hamza, M.H. (ed)]. ACTA Press, Anaheim, CA, USA, pp. 142-147.

Chiew, F.H.S., T.C. Piechota, J.A. Dracup, and T.A. McMahon, 1998: El Niño-Southern Oscillation and Australian rainfall, streamflow and drought—links and potential for forecasting. *Journal of Hydrology*, **204,** 138–149.

Cohen, S.J., 1997: Scientist-stakeholder collaboration in integrated assessment of climate change: lessons from a case study of northwest Canada. *Environmental Modeling and Assessment*, **2,** 281–293.

Cole, I.S., G.A. King,, G.S. Trinidad, W.Y Chan, and D.A. Paterson, 1999a: An Australia-Wide Map of Corrosivity: A GIS Approach. *Durability of Building Materials and Components 8* [Lacasse, M.A and D.J. Vasnier (eds.)]. NRC Research Press, Ottawa, ON, Canada. Vol. 2, pp. 901–911.

Cole, I.S., A. Neufeld, P. Kao, W.D. Ganther, L. Chotimongkol, C. Bhamornsut, N.V. Hue, and S. Bernado, 1999: Corrosion mechanisms and laboratory performance-based tests for tropical climates. In: *Proceedings of the 14th International Corrosion Congress, Cape Town, South Africa, 27 September - 1 October 1999*. Paper 265.2 (CD-ROM).

Cole, I.S., D.A. Paterson, S. Furman, A. Neufeld, and W. Ganther, 1999b: A holistic approach to modelling atmospheric corrosion. In: *Proceedings of the 14th International Corrosion Congress, Cape Town, South Africa, 27 September - 1 October 1999*. Paper 265.4 (CD-ROM).

Cole, I.S., G.S. Trinidad, and W.Y. Chan, 1999c: Prediction of the impact of the environment on timber components: a GIS-based approach. In: *Durability of Building Materials and Components 8* [Lacasse, M.A and D.J. Vasnier (eds.)]. NRC Research Press, Ottawa, ON, Canada. Vol. 1, pp. 693–703.

Collins, D.A. and P.M. Della-Marta, 1999d: Annual climate summary 1998: Australia's warmest year on record. *Australian Meteorological Magazine*, **48,** 273–383.

Collins, D.A., P.M. Della-Marta, N. Plummer, and B.C. Trewin, 2000: Trends in annual frequencies of extreme temperature events in Australia. *Australian Meteorological Magazine*, **49,** 277–292.

Commonwealth of Australia, 1996: *Australia: State of the Environment Executive Summary*. CSIRO Publishing, Collingwood, Victoria, Australia, 47 pp.

Coombs, H.C., J. Dargavel, J. Kesteven, H. Ross, D.I. Smith, and E. Young, 1990: *The Promise of the Land: Sustainable Use by Aboriginal Communities*. Working Paper 1990/1, Centre for Resource and Environmental Economics, Australian National University, Canberra, Australia, 19 pp.

Crampton, P. and P. Davis, 1998: Measuring deprivation and socioeconomic status: why and how? *New Zealand Public Health Report*, **5,** 81–83.

CSIRO, 1996a: *Climate Change Scenarios for the Australian Region*. Commonwealth Scientific and Industrial Research Organization, Atmospheric Research, Aspendale, Victoria, Australia, 8 pp.

CSIRO, 1996b: *OzClim: A climate Scenario Generator and Impacts Package for Australia*. CSIRO Atmospheric Research, Aspendale, Victoria, Australia, 5 pp. Available online at http://www.dar.csiro.au/publications/ozclim.htm, updated to 28 August 2000.

Curtis, P.S., 1996: A meta-analysis of leaf gas exchange and nitrogen in trees grown under elevated carbon dioxide. *Plant Cell and Environment*, **19,** 127–137.

Curtis, P.S. and X. Wang, 1998: A meta-analysis of elevated CO_2 effects on woody plant mass, form, and physiology. *Oecologia*, **113,** 299–313.

Daly, J.J., 1994: *Wet as a Shag, Dry as a Bone*. QI93028, ISSN 0727-6273, Queensland Department of Primary Industries, Brisbane, Australia, 150 pp.

Davison, T., M. McGowan, D. Mayer, B. Young, N. Jonsson, A. Hall, A. Matschoss, P. Goodwin, J. Goughan, and M. Lake, 1996: *Managing Hot Cows in Australia*. Queensland Department of Primary Industries, Brisbane, Australia, 58 pp.

Donnelly, J.R., M. Freer, and A.D. Moore, 1998: Using the GrassGro decision support tool to evaluate some objective criteria for the definition of exceptional drought. *Agricultural Systems*, **57(3),** 301–313.

Duce, R.A. and N.W. Tindale, 1991: Atmospheric transport of iron and its deposition in the ocean. *Limnology and Oceanography*, **36,** 1715–1726.

Durie, M., 1994: *Waiora: Maori Health Development*. Oxford University Press, Auckland, New Zealand, 238 pp.

Eden, D.N. and M.P. Page, 1998: Palaeoclimatic implications of a storm erosion record from late Holocene lake sediments, North Island, New Zealand. *Palaeogeography, Palaeoclimatology, Palaeoecology*, **139,** 37–58.

EECA, 1996: *New and Emerging Renewable Energy Opportunities in New Zealand*. Centre for Advanced Engineering and Energy Efficiency and Conservation Authority, Wellington, New Zealand, 266 pp.

Eliot, I., P. Waterman, and C.M. Finlayson, 1999: Predicted climate change, sea level rise and wetland management in the Australian wet-dry tropics. *Wetlands Ecology and Management*, **7,** 63–81.

Elizarov, A.A., A.S. Grechina, B.N. Kotenev, and A.N. Kuzetsov, 1993: Peruvian jack mackerel, Trachuru symmetricus murphyi, in the open waters of the South Pacific. *Journal of Ichthyology*, **33,** 86–104.

Ellison, J., 1996: The biodiversity of coastal zone wetlands in Oceania. In: *Oceania Day—Paradise Under Pressure: Conservation and Wise Use of Coastal Wetlands, Vol 11/12, Proceedings of the 6th Meeting of the Conference of Contracting Parties, Brisbane, Australia, 19–27 March, 1996*. Ramsar Convention Bureau, Gland, Switzerland, pp. 9–12.

Evans, J.L. and R.J. Allan, 1992: El Niño-Southern Oscillation modification to the structure of the monsoon and tropical cyclone activity in the Australasian region. *International Journal of Climatology*, **12,** 611–623.

Finch, V.A., I.L. Bennett, and C.R. Holmes, 1984: Coat colour in cattle: effect on thermal balance, behaviour and growth, and relationship with coat types. *Journal of Agricultural Science*, **102,** 141–147.

Finch, V.A., I.L. Bennett, and C.R. Holmes, 1982: Sweating response in cattle and its relation to rectal temperature, tolerance of sun and metabolic rate. *Journal of Agricultural Science*, **99,** 479–487.

Fitzharris, B.B. and C. Garr, 1996: Climate, water resources and electricity. In: *Greenhouse: Coping with Climate Change* [Bouma, W.J., G.I. Pearman, and M.R. Manning (eds.)]. CSIRO Publishing, Collingwood, Victoria, Australia, pp. 263–280.

Flood, N.R and A. Peacock, 1999: *Twelve Month Australian Rainfall (year from April to March) Relative to Historical Records*. Queensland Department of Natural Resources (QDNR), Queensland, Australia, wallposter.

Ford, E., 1950: The malaria problem in Australia and the Australian Pacific Territories. *Medical Journal of Australia*, **1,** 749–760.

Foster, G. and L. Carter, 1997: Mud sedimentation on the continental shelf at an accretionary margin—Poverty Bay, New Zealand. *New Zealand Journal of Geology and Geophysics*, **40,** 157–173.

Fowler, A., 1999: Potential climate impacts on water resources in the Auckland Region (New Zealand). *Climate Research*, **11,** 221–245.

Fowler, A.M. and K.J. Hennessy, 1995: Potential impacts of global warming on the frequency and magnitude of heavy precipitation. *Natural Hazards*, **11,** 283–303.

Francis, M.P., 1994a: Duration of larval and spawning periods in Pagrus auratus (Sparidae) determined from otolith daily increments. *Environmental Biology of Fishes*, **39(2),** 137–152.

Francis, M.P., 1994b: Growth of juvenile snapper, Pagrus auratus. *New Zealand Journal of Marine and Freshwater Research*, **28,** 201–218.

Francis, M.P., 1993: Does water temperature determine year class strength in New Zealand snapper (Pagrus auratus, Sparidae)? *Fisheries Oceanography*, **22,** 65–72.

Francis, M.P. and J. Evans, 1993: Immigration of subtropical and tropical animals into north-eastern New Zealand. In: *Proceedings of the Second International Temperate Reef Symposium, 7–10 January, 1992, Auckland, New Zealand* [Battershill, C.N. (ed.)]. NIWA, Wellington, New Zealand, pp. 131–136.

Francis, M.P., A.D. Langley, and D.J. Gilbert, 1997: Prediction of snapper (Pagrus auratus) recruitment from sea surface temperature. In: *Developing and Sustaining World Fisheries Resources: The State of Science and Management* [Hancock, D.A., D.C. Smith, A. Grant, and J.P. Beumer (eds.)]. Second World Fisheries Congress, CSIRO Publishing, Melbourne, Australia, pp. 67–71.

Galloway, R.W., 1982: Distribution and physiographic patterns of Australian mangroves. In: *Mangrove Ecosystems in Australia* [Clough, B. (ed.)]. Australian Institute for Marine Science and Australian National University Press, Canberra, Australia, 302 pp.

Gattuso, J.-P., D. Allemand, and M. Frankignoulle, 1999: Photosynthesis and calcification at cellular, organismal and community levels in coral reefs: a review of interactions and control by the carbonate chemistry. *American Zoologist*, **39,** 160–183.

Gattuso, J.-P., M. Frankignoulle, I. Bourge, S. Romaine, and R.W. Buddemeier, 1998: Effect of calcium carbonate saturation of seawater on coral calcification. *Global and Planetary Change*, **18,** 37–46.

Ghassemi, F., K.W.F. Howard, and A.J. Jakeman, 1996: Seawater intrusion in coastal aquifers and its numerical modelling. In: *Environmental Modelling* [Zannetti, P. (ed.)]. Computational Mechanics Publications, Southampton, United Kingdom, Vol. 3, pp. 299–328.

Ghassemi, F., A. Jakeman, and H. Nix, 1995: *Salinisation of Land and Water Resources: Humans Causes, Extent, Management and Case Studies.* University of New South Wales Press Ltd., Sydney, Australia, 562 pp.

Gibbs, W.J. and J.V. Maher, 1967: Rainfall deciles and drought indicators. *Bureau of Meteorology Bulletin*, **48,** 117.

Glynn, P.W., 1993: Coral bleaching ecological perspectives. *Coral Reefs*, **12,** 1–17.

Graetz, D., H. Dowlatabadi, J. Risbey, and M. Kandlikar, 1997: *Applying Frameworks for Assessing Agricultural Adaptation to Climate Change in Australia.* Report to Environment Australia, CSIRO Earth Observation Centre, Canberra, Australia, 136 pp.

Graetz, R.D., M.A. Wilson, and S.K. Campbell, 1995: *Landcover Disturbance over the Australian Continent: A Contemporary Assessment.* Department of Environment, Sport and Territories, Government of Australia, Canberra, Australia, 89 pp.

Green, M., 1999: *Climate Change—A Brief Review.* Publication No. WRC/RP-G-99/8, Wellington Regional Council, Wellington, New Zealand, 63 pp.

Guest, C.S., K. Willson, A. Woodward, K. Hennessy, L.S. Kalkstein, C. Skinner, and A.J. McMichael, 1999: Climate and mortality in Australia: retrospective study, 1979–1990, and predicted impacts in five major cities in 2030. *Climate Research*, **13,** 1–15.

Hadfield, M.G. and J. Sharples, 1996: Modelling mixed layer depth and plankton biomass off the west coast of South Island, New Zealand. *Journal of Marine Systems*, **8,** 1–29.

Hales, S., P. Weinstein, and A. Woodward, 1999: Ciguatera (fish poisoning), El Niño and Pacific sea surface temperatures. *Ecosystem Health*, **5,** 20–25.

Hales, S., S. Lewis, T. Slater, J. Crane, and N. Pearce, 1998: Prevalence of adult asthma symptoms in relation to climate in New Zealand. *Environmental Health Perspectives*, **106,** 607–610.

Hales, S., T. Kjellstrom, C. Salmond, G.I. Town, A. Woodward, 2000: Daily mortality in Christchurch, New Zealand in relation to weather and air pollution. *Australia and New Zealand Journal of Public Health*, **24,** 89–91.

Hall, A.J. and H.G. McPherson, 1997a: Modelling the influence of temperature on the timing of bud break in kiwifruit. *Acta Horticulturae*, **444,** 401–406.

Hall, A.J. and H.G. McPherson, 1997b: Predicting fruit maturation in kiwifruit (Actinidia deliciosa). *Journal of Horticultural Science*, **72,** 949–960.

Hall, W.B., G.M. McKeon, J.O. Carter, K.A. Day, S.M. Howden, J.C. Scanlan, P.W. Johnston, and W.H. Burrows, 1998: Climate change and Queensland's grazing lands, II: an assessment of impact on animal production from native pastures. *Rangeland Journal*, **20,** 177–205.

Hallegraeff, G.M. and S.W. Jeffrey, 1993: Annually recurrent diatom blooms in spring along along the New South Wales coast of Australia. *Australian Journal of Marine and Freshwater Research*, **44,** 325–334.

Hammer, G.L., N. Nicholls, and C. Mitchell (eds.), 2000: *Applications of Seasonal Climate Forecasting in Agricultural and Natural Ecosystems: The Australian Experience.* Atmospheric and Oceanographic Sciences Library, J. Kluwer Academic Publishers, Dordrecht, The Netherlands, Vol. 21, 469 pp.

Hammer, G.L., G.M. McKeon, J.F. Clewett, and D.R. Woodruff, 1991: Usefulness of seasonal climate forecasting in crop and pasture management. *Bulletin of the Australian Meteorological and Oceanographic Society*, **4,** 104–109.

Handmer, J. (ed.), 1997: *Flood Warning: The Issues and Practice in Total System Design.* Proceedings of Workshop held at Middlesex University, Sept. 1995. ISBN 1 85924 128 X. Flood Hazard Research Centre, Middlesex University, Enfield, United Kingdom.

Harris, G., P. Davies, M. Nunez, and G. Meyers, 1988: Interannual variability in climate and fisheries in Tasmania. *Nature*, **333,** 754–757.

Harris, G., C. Nilsson, L. Clementson, and D. Thomas, 1987: The water masses of the east coast of Tasmania: seasonal and interannual variability and the influence on phytoplankton biomass and productivity. *Australian Journal of Marine and Freshwater Research*, **38,** 569–590.

Hassall and Associates, 1998: *Climatic Change Scenarios and Managing the Scarce Water Resources of the Macquarie River. Australian Greenhouse Office.* Hassall and Associates, New South Wales Department of Land and Water Conservation, New South Wales National Parks and Wildlife Service, and CSIRO Atmospheric Research, Canberra, Australia, 113 pp.

Hearnden, M., C. Skelly, H. Dowler, and P. Weinstein, 1999: Improving the surveillance of mosquitoes with disease-vector potential in New Zealand. *New Zealand Public Health Report*, **6,** 25–27.

Heath, R.A., 1979: Significance of storm surges on the New Zealand coast. *New Zealand Journal of Geology and Geophysics*, **22,** 259–266.

Henderson-Sellers, A., 1993: An Antipodean climate of uncertainty. *Climatic Change*, **25,** 203–224.

Hennessy, K.J. and K. Clayton-Greene, 1995: Greenhouse warming and vernalisation of high-chill fruit in southern Australia. *Climatic Change*, **30,** 327–348.

Hennessy, K.J. and Jones, R.N., 1999: *Climate Change Impacts in the Hunter Valley—Stakeholder Workshop Report.* CSIRO Atmospheric Research, Aspendale, Victoria, Australia, 26 pp.

Hennessy, K.J., J.M. Gregory, and J.F.B. Mitchell, 1997: Changes in daily precipitation under enhanced greenhouse conditions. *Climate Dynamics*, **13,** 667–680.

Hennessy, K.J., R. Suppiah, and C.M. Page, 1999: Australian rainfall changes, 1910–1995. *Australian Meteorological Magazine*, **48,** 1–13.

Hennessy, K.J., P.H. Whetton, J.J. Katzfey, J.L. McGregor, R.N. Jones, C.M. Page, and K.C. Nguyen, 1998: *Fine Resolution Climate Change Scenarios for New South Wales: Annual Report 1997–98.* CSIRO Atmospheric Research, Aspendale, Victoria, Australia, 48 pp.

Hewitt, S.D., 1994: The impact of enhanced greenhouse conditions on the Australian snowpack. In: *Proceedings of Symposium on Snow and Climate, Geneva, 22–23 September, 1994.* Department of Geography, University of Geneva, Geneva, Switzerland.

Hildrew, A.G. and C.R. Townsend, 1987: Organization in freshwater benthic communities. In: *Organisation of Communities Past and Present* [Gee, J.H.R. and P.S. Giller (eds.)]. Blackwell Scientific Publications, Oxford, United Kingdom, pp. 347–371.

Hill, D., V. White, R. Marks, and R. Borland, 1993: Changes in sun-related attitudes and behaviours, and reduced sunburn prevalence in a population at high risk of melanoma. *European Journal Cancer Prevention*, **2,** 447–456.

Hirst, A., 1999: The Southern Ocean response to global warming in the CSIRO coupled ocean-atmosphere model. *Environmental Modelling and Software*, **14,** 227–241.

Hoegh-Guldberg, O., 1999: Climate change, coral bleaching and the future of the world's coral reefs. *Journal of Marine and Freshwater Research*, **50,** 839–866.

Hoegh-Guldberg, O., R. Berkelmans, and J. Oliver, 1997: Coral bleaching: implications for the Great Barrier Reef Marine Park. In: *Proceedings of The Great Barrier Reef Science, Use and Management Conference, 25–29 November, 1996, Townsville, Australia* [Turia, N. and C. Dalliston (eds.)]. Great Barrier Reef Marine Park Authority, Townsville, Australia, pp. 210–224.

Holbrook, N.J. and N.L. Bindoff, 2000: A digital upper ocean temperature atlas for the southwest Pacific: 1955–1988. *Australian Meteorological Magazine*, **49,** 37–49.

Holbrook, N.J. and N.L. Bindoff, 1997: Interannual and decadal variability in the southwest Pacific Ocean between 1955 and 1988. *Journal of Climate*, **10,** 1035–1048.

Hopkins, L.C. and G.J. Holland, 1997: Australian heavy-rain days and associated east coast cyclones: 1958–1992. *Journal of Climate*, **10,** 621–635.

Howden, S.M. and J.T. Gorman (eds.), 1999: *Impacts of Global Change on Australian Temperate Forests.* Working Paper Series 99/08, CSIRO Wildlife and Ecology, Canberra, Australia, 146 pp. Available online at http://www.dwe.csiro.au/research/futures/publications/WkgDocs99.htm.

Howden, S.M. and J. Turnpenny, 1997: Modelling heat stress and water loss of beef cattle in subtropical Queensland under current climates and climate change. In: *Modsim '97 International Congress on Modelling and Simulation, Proceedings, 8–11 December, University of Tasmania, Hobart* [McDonal D.A. and M. McAleer (eds.)]. Modelling and Simulation Society of Australia, Canberra, Australia, pp. 1103–1108.

Howden, S.M., G.M. McKeon, J.O. Carter, and A. Beswick, 1999a: Potential global change impacts on C_3-C_4 grass distributions in eastern Australian rangelands. In: *People and Rangelands: Building the Future. Proceedings of VI International Rangelands Congress, Townsville, July 1999* [Eldridge, D. and D. Freudenberger (eds.)]. 6th International Rangelands Congress Inc., Aitkenvale, Queensland, Australia, pp. 41–43.

Howden, S.M., G.M. McKeon, L. Walker, J.O. Carter, J.P. Conroy, K.A. Day, W.B. Hall, A.J. Ash, and O. Ghannoum, 1999b: Global change impacts on native pastures in south-east Queensland, Australia. *Environmental Modelling and Software*, **14,** 307–316.

Howden, S.M., P.J. Reyenga, and H. Meinke, 1999c: *Global Change Impacts on Australian Wheat Cropping*. Working Paper Series 99/04, CSIRO Wildlife and Ecology, Canberra, Australia, 121 pp. Available online at http://www.dwe.csiro.au/research/futures/publications/WkgDocs99.htm.

Howden, S.M., P.J. Reyenga, H. Meinke, and G.M. McKeon, 1999d: *Integrated Global Change Impact Assessment on Australian Terrestrial Ecosystems: Overview Report*. Working Paper Series 99/14, CSIRO Wildlife and Ecology, Canberra, Australia, 51 pp. Available online at http://www.dwe.csiro.au/research/futures/publications/WkgDocs99.htm.

Howden, S.M., W.B. Hall, and D. Bruget, 1999e: Heat stress and beef cattle in Australian rangelands: recent trends and climate change. In: *People and Rangelands: Building the Future. Proceedings of VI International Rangelands Congress, Townsville, July 1999* [Eldridge, D. and D. Freudenberger (eds.)]. 6th International Rangelands Congress Inc., Aitkenvale, Queensland, Australia, pp. 43–45.

Howden, S.M., P.J. Reyenga, and J.T. Gorman, 1999f: *Current Evidence of Global Change and its Impacts: Australian Forests and Other Ecosystems*. Working Paper Series 99/01, CSIRO Wildlife and Ecology, Canberra, Australia, 9 pp. Available online at http://www.dwe.csiro.au/research/futures/publications/WkgDocs99.htm.

Howitt, R., 1993: Social impact assessment as "applied peoples" geography. *Australian Geographical Studies*, **31,** 127–140.

Hubbert, G.D. and K.L. McInnes, 1999a: A storm surge inundation model for coastal planning and impact studies. *Journal of Coastal Research*, **15,** 168–185.

Hubbert, G.D. and K.L. McInnes, 1999b: Modelling storm surges and coastal ocean flooding. In: *Modelling Coastal Sea Processes* [Noye, J. (ed.)]. World Scientific Publishing, Singapore, pp. 159–188.

Hughes, L., E.M. Cawsey, and M. Westoby, 1996: Climatic range sizes of Eucalyptus species in relation to future climate change. *Global Ecology and Biogeography Letters*, **5,** 23–29.

Hulme, M. and N. Sheard, 1999: *Climate Change Scenarios for Australia*. Climatic Research Unit, University of East Anglia, Norwich, United Kingdom, 6 pp. Available online at http://www.cru.uea.ac.uk.

Institution of Engineers, 1998: *Coastal Engineering Guidelines for Working with the Australian Coast in an Ecologically Sustainable Way*. National Committee on Coastal and Ocean Engineering, Institution of Engineers, Barton, ACT, Australia, 82 pp.

International Federation of Red Cross and Red Crescent Societies, 1999: *World Disasters Report*. RCRCS, Geneva, Switzerland, 198 pp.

IPCC, 1998: *The Regional Impacts of Climate Change. An Assessment of Vulnerability. A Special Report of IPCC Working Group II for the Intergovernmental Panel of Climate Change* [Watson, R.T, M.C. Zinyowera, R.H. Moss, and D.J. Dokken (eds.)]. Cambridge University Press, Cambridge, United Kingdom and New York, NY, USA, 878 pp.

IPCC, 1996: *Climate Change 1995: The Science of Climate Change. Contribution of Working Group I to the Second Assessment Report of the Intergovernmental Panel on Climate Change* [Houghton, J.T., L.G. Meira Filho, B.A. Callander, N. Harris, A. Kattenberg, and K. Maskell (eds.)]. Cambridge University Press, Cambridge, United Kingdom and New York, NY, USA, 572 pp.

Isdale, P.J., B.J. Stewart, K.S. Tickle, and J.M. Lough, 1998: Palaeohydrological variation in a tropical river catchment: a reconstruction using fluorescent bands in corals of the Great Barrier Reef, Australia. *The Holocene*, **8,** 1–8.

Islam, S.M.N., P.J. Sheehan, and J. Sanderson, 1997: Costs of economic growth: estimates of climate change damages in Australia. In: *Proceedings of the 20th International Conference of the International Association for Energy Economics, New Delhi, India, 22–24 January 1997*. ICIAEE, New Delhi, India, Vol. 1, pp. 196–206.

Islam, S.N., 1995: *Australian Dynamic Integrated Climate Economy Model (ADICE): Model Specification, Numerical Implementation and Policy Implications*. Seminar Paper, Centre for Strategic Economic Studies, Victoria University, Melbourne, Australia, 43 pp.

Jackson, J., 1997: *State of Habitat Availability and Quality in Inland Waters*. Commonwealth of Australia, Canberra, Australia, 85 pp.

Jamieson, P.D. and C.A. Munro, 1999: A simple method for the phenological evaluation of new cereal cultivars. *Proceedings of the New Zealand Agronomy Society*, **29,** 63–68.

Johnson, A.K.L., S.P. Ebert, and A.E. Murray, 1999: Distribution of coastal freshwater wetlands and riparian forests in the Herbert River catchment and implications for management of catchments adjacent the Great Barrier Reef Marine Park. *Environmental Conservation*, **26,** 299–235.

Johnston, P.W., G.M. McKeon, R. Buxton, D.H. Cobon, K.A. Day, W.B. Hall, and J.C. Scanlan, 1999: Managing climatic variability in Queensland's grazing lands—new approaches. In: *Applications of Seasonal Climate Forecasting in Agricultural and Natural Ecosystems—The Australian Experience* [Hammer, G., N. Nicholls, and C. Mitchell (eds.)]. J. Kluwer Academic Publishers, The Netherlands, pp. 197–226.

Jones, R.J., R. Berkelmans, and J.K. Oliver, 1997: Recurrent bleaching of corals at Magnetic Island (Australia) relative to air and seawater temperature. *Marine Ecology Progress Series*, **158,** 289–292.

Jones, R.N., 2000: Analysing the risk of climate change using an irrigation demand model. *Climate Research*, **14,** 89–100.

Jones, R.N. and K.J. Hennessy, 2000: *Climate Change Impacts in the Hunter Valley: A Risk Assessment of Heat Stress Affecting Dairy Cattle*. CSIRO Atmospheric Research, Aspendale, Victoria, Australia, 22 pp.

Kailola, P.J., M.J. Williams, P.C. Stewart, R.E. Reichelt, A. McNee, and C. Grieve, 1993: *Australian Fisheries Resources*. Bureau of Resource Sciences and the Fisheries Research and Development Corporation, Canberra, ACT, Australia, 422 pp.

Kaly, U., L. Briguglio, H. McLeod, S. Schmall, C. Pratt, and R. Pal, 1999: *Environmental Vulnerability Index (EVI) to Summarise National Environmental Vulnerability Profiles*. SOPAC Technical Report 275, South Pacific Applied Geoscience Commission (SOPAC), Suva, Fiji.

Kemp, E.M., 1981: Tertiary palaeogeography and the evolution of Australian climate. In: *Ecological Biogeography of Australia* [Keast, A. (ed.)]. Monographiae Biologicae, Dr. W. Junk, The Hague, The Netherlands, Vol. 41, pp. 31–49.

Kenny, G.J., W. Ye, R.A. Warrick, and T. Flux, 1999: Climate variations and New Zealand agriculture: the CLIMPACTS system and issues of spatial and temporal scale. In: *Proceedings of MODSIM '99, International Congress on Modelling and Simulation, 6–9 December 1999, University of Waikato, New Zealand* [Oxley, L., F. Scrimgeour, and A. Jakeman (eds.)]. Modelling and Simulation Society of Australia and New Zealand Inc., Waikato, New Zealand, pp. 747–752.

Kenny, G.J., R.A. Warrick, N. Mitchell, A.B. Mullan, and M.J. Salinger, 1995: CLIMPACTS: an integrated model for assessment of the effects of climate change on the New Zealand environment. *Journal of Biogeography*, **22(4/5),** 883–895.

Kenny, G.J., R.A. Warrick, B.D. Campbell, G.C. Sims, M. Camilleri, P.D. Jamieson, N.D. Mitchell, H.G. McPherson, and M.J. Salinger, 2000: Investigating climate change impacts and thresholds: an application of the CLIMPACTS integrated assessment model for New Zealand agriculture. *Climatic Change*, **46,** 91–113.

Kingsford, R.T. and W. Johnson, 1998: Impact of water diversions on colonially-nesting waterbirds in the Macquarie Marshes of arid Australia. *Colonial Waterbirds*, **21,** 159–170.

Kingsford, R.T. and R.F. Thomas, 1995: The Macquarie Marshes in arid Australia and their waterbirds: a 50-year history of decline. *Environmental Management*, **19,** 867–878.

Kingsford, R.T., A.L. Curtin, and J. Porter, 1999: Water flows on Cooper Creek in arid Australia determine "boom" or "bust" periods for waterbirds. *Biological Conservation*, **88,** 231–248.

Kinhill, Pty. Ltd., 1988: *Techniques for Determining Environmental Water Requirements: A Review*. Report No. 40, Department of Water Resources, Victoria, Australia.

Kirschbaum, M.U.F., 1999a: Modelling forest growth and carbon storage with increasing CO_2 and temperature. *Tellus*, **51B,** 871–888.

Kirschbaum, M.U.F., 1999b: Forest growth and species distributions in a changing climate. *Tree Physiology*, **20,** 309–322.

Kleypas, J.A., R.W. Buddemeier, D. Archer, J.-P. Gattuso, C. Langdon, and B.N. Opdyke, 1999: Geochemical consequences of increased atmospheric carbon dioxide on coral reefs. *Science*, **284,** 118–120.

Kohen, J., 1995: *Aboriginal Environmental Impacts*. University of New South Wales Press, Sydney, Australia, 160 pp.

Konig, U., 1998: Climate change and the Australian ski industry. In: *Snow: A Natural History; An Uncertain Future* [Green, K. (ed.)]. Australian Alps Liaison Committee, Canberra, Australia, pp. 207–223.

Kothavala, Z., 1999: The duration and severity of drought over eastern Australia simulated by a coupled ocean-atmosphere GCM with a transient increase in CO_2. *Environmental Modelling and Software*, **14,** 243–252.

Kouzmin, A. and N. Korac-Kakabadse, 1999: From efficiency to risk sensitivity: reconstructing management capabilities after economic rationalism. *Australian Journal of Emergency Management*, **14(1),** 8–19.

Krysanova, V., F. Wechsung, A. Becker, W. Poschenrieder, and J. Gräfe, 1999: Mesoscale ecohydrological modelling to analyse regional effects of climate change. *Environmental Modelling and Assessment*, **4,** 259–271.

Lake, P.S., 2000: Disturbance, patchiness and diversity in streams. *Journal of the North American Benthological Society*, **19(4),** 573–592.

Landsberg, J.J. and R.H. Waring, 1997: A generalised model of forest productivity using simplified concepts of radiation-use efficiency, carbon balance and partitioning. *Forest Ecology and Management*, **95,** 209–228.

Langton, M., 1997: The future—strategies and action plans for economic development. In: *Proceedings of the Pathways to the Future Conference, Darwin, 12 September, 1997*. Northern Territory University, Centre for Indigenous Natural and Cultural Resource Management, Darwin, Australia. Available online at http://www.ntu.edu.au/cincrm/lecture2.html.

Langton, M., 2000: "The fire at the centre of each family": Aboriginal traditional fire regimes and the challenges for reproducing ancient fire management in the protected areas of northern Australia. In: *FIRE! The Australian Experience, Proceedings of the 1999 Seminar, National Academies Forum, Canberra, 30 September - 1 October, 1999*. National Academies Forum, Canberra, Australia, pp. 3–32.

Larcombe, P., K. Woolfe, and R. Purdon (eds.), 1996: *Great Barrier Reef: Terrigenous Sediment Flux and Human Impacts*. 2nd. ed., CRC Reef Research Centre, Townsville, Australia, 174 pp.

Lavery, B.M., G. Joung, and N. Nicholls, 1997: An extended high-quality historical rainfall dataset for Australia. *Australian Meteorological Magazine*, **46,** 27–38.

Lean, C.M.B. and I.N. McCave, 1998: Glacial to interglacial mineral magnetic and palaeoceanographic changes at Chatham Rise, southwest Pacific Ocean. *Earth and Planetary Science Letters*, **163,** 247–260.

Leathwick, J.R., D. Whitehead, and M. McLeod, 1996: Predicting changes in the composition of New Zealand's indigenous forests in response to global warming—a modelling approach. *Environmental Software*, **11,** 81–90.

Leigh, R., R. Taplin, and G. Walker, 1998a: Insurance and climate change: the implications for Australia with respect to natural hazards. *Australian Journal of Environmental Management*, **5,** 81–96.

Leigh, R., E. Cripps, R. Taplin, and G. Walker, 1998b: *Adaptation of the Insurance Industry to Climate Change and Consequent Implications*. Final Report for the Department of Environment, Sport and Territories, Government of Australia, Canberra, by Climatic Impacts Centre, Macquarie University, Sydney, Australia.

Leighton, R.M., K. Keay, and I. Simmonds, 1997: Variation in annual cyclonicity across the Australian region for the 29-year period 1965–1993 and the effect on annual all-Australia rainfall. In: *Proceedings Workshop on Climate Prediction for Agriculture and Resource Management*. Bureau of Resource Sciences, Canberra, Australia.

Loneragan, N.R. and S.E. Bunn, 1999: River flows and estuarine ecosystems: implications for coastal fisheries from a study of the Logan River, SE Queensland. *Australian Journal of Ecology*, **24,** 431–440.

Longstreth, J., F.R. de Gruijl, M.L. Kripke, S. Abseck, F. Arnold, H.I. Slaper, G. Velders, Y. Takizawa, and J.C. van der Leun, 1998: Health risks. *Journal of Photochemistry and Photobiology B: Biology*, **46,** 20–39.

Lough, J.M., 1999: *Sea Surface Temperatures Along the Great Barrier Reef; a Contribution to the Study of Coral Bleaching*. Research Publication No. 57, Great Barrier Reef Marine Park Authority, Townsville, Australia, 31 pp.

Lough, J.M. and D.J. Barnes, 1997: Several centuries of variation in skeletal extension, density, and calcification in massive *Porites* colonies from the Great Barrier Reef: a proxy for seawater temperature and a background of variability against which to identify unnatural change. *Journal of Experimental Marine Biology and Ecology*, **211,** 29–67.

Lough, J.M. and D.J. Barnes, 2000: Environmental controls on growth of massive coral Porites. *Journal of Experimental Marine Biology and Ecology*, **245,** 225–243.

Lowe, I., 1988: The energy policy implications of climate change. In: *Greenhouse: Planning for Climate Change* [Pearman, G.I. (ed.)]. CSIRO Publications, East Melbourne and E.J. Brill, Leiden, Australia, pp. 602–612.

Lyne, V.D., R.A. Scott, R.C. Gray, and R.W. Bradford, 1999: *Quantitative Interpretation of Find-Scale Catch-per-Unit-Effort for Southern Bluefin Tuna Off South Eastern Australia*. Project FRDC 93/077, ISBN 0-643-06160-6, Final report to the Fisheries Research and Development Corporation, Canberra, Australia, 88 pp.

Mackenzie, J., M. Lindsay, and R. Coelen, 1994: Arboviruses causing human disease in the Australasian zoogeographic region. *Archives of Virology*, **136,** 447–467.

Maelzer, D., S. Hales, P. Weinstein, M. Zalucki, and A. Woodward, 1999: El Niño and arboviral disease prediction. *Environmental Health Perspectives*, **107,** 817–818.

MAF, 2000: *Situation for New Zealand Agriculture and Forestry*. Ministry of Agriculture and Forestry, Wellington, New Zealand, 51 pp.

MAF, 1999: *A National Exotic Forest Description as of April 1998*. Ministry of Agriculture and Forestry, Wellington, New Zealand, 63 pp.

Manabe, S. and R.J. Stouffer, 1994: Multiple-century response of a coupled ocean-atmosphere model to an increase of atmospheric carbon dioxide. *Journal of Climate*, **4,** 5–23.

Mantua, N.J., S.R. Hare, Y. Zhang, J.M. Wallace, and R.C. Francis, 1997: A Pacific interdecadal climate oscillation with impacts on salmon production. *Bulletin of the American Meteorological Society*, **78,** 1069–1079.

Marks, R., D. Jolley, A.P. Dorevitch, and T.S. Selwood, 1989: The incidence of non-melanocytic skin cancers in an Australian population: results of a five-year prospective study. *Medical Journal Australia*, **150,** 475–478.

May, P., P. Waterman, and I. Eliot, 1998: *Responding to Rising Seas and Climate Change: A Guide for Coastal Areas*. Environment Australia, Canberra, Australia, 42 pp.

McGlone, M.S., 1989: The Polynesian settlement of New Zealand in relation to environmental and biotic changes. *New Zealand Journal of Ecology*, **12,** 115–129.

McInnes, K.L., K.J.E. Walsh, and A.B. Pittock, 2000: *Impact of Sea Level Rise and Storm Surges on Coastal Resorts*. CSIRO Tourism Research, Final Report. CSIRO Atmospheric Research, Aspendale, Victoria, Australia, 24 pp.

McInnes, K.L., K.J.E. Walsh, and A.B. Pittock, 1999: *Impact of Sea Level Rise and Storm Surges on Coastal Resorts*. CSIRO Tourism Research, Second Annual Report, CSIRO Atmospheric Research, Aspendale, Victoria, Australia, 29 pp.

McKenzie, R.L., B. Connor, and G. Bodeker, 1999: Increased summertime UV radiation in New Zealand in response to ozone loss. *Science*, **285,** 1709–1711.

McKeon, G.M., W.B. Hall, S.J. Crimp, S.M. Howden, R.C. Stone, and D.A. Jones, 1998: Climate change in Queensland's grazing lands, I: approaches and climatic trends. *Rangeland Journal*, **20,** 147–173.

McKerchar, A.I., B.B. Fitzharris, and C.P. Pearson, 1998: Dependency of summer lake inflows and precipitation on spring SOI. *Journal of Hydrology*, **205,** 66–80.

McLachlan, I., H. Clough, P. Gunner, M. Johnson, J. King, and D. Samson, 1999: *Diversity and Innovation for Australian Wool*. Report of the Wool Industry Future Directions Taskforce, Canberra, Australia.

McMaster, H.J., 1999: The potential impact of global warming on hail losses to winter cereal crops in New South Wales. *Climatic Change*, **43,** 455–476.

MDBC, 1999: *The Salinity Audit of the Murray-Darling Basin: A 100-Year Perspective*. Murray-Darling Basin Commission, Canberra, Australia, 48 pp.

Meehl, G.A. and W.M. Washington, 1996: El Niño-like climate change in a model with increased atmospheric CO_2 concentrations. *Nature*, **382,** 56–60.

Mercer, D., 1991: *A Question of Balance Natural Resources Conflict Issues in Australia*. The Federation Press, Leichhardt, Australia, 346 pp.

Meyers, G., 1996: Variation of Indonesian throughflow and the El Niño-Southern Oscillation. *Journal of Geophysical Research*, **101,** 12255–12263.

Ministry for the Environment, 1999: *Making Every Drop Count*. Ministry for the Environment, Wellington, New Zealand, 18 pp.

Ministry for the Environment, 1997: *Climate Change—The New Zealand Response II: New Zealand's Second National Communication Under the Framework Convention on Climate Change*. Ministry for the Environment, Wellington, New Zealand, 191 pp.

Mitchell, N.D. and J.E. Williams, 1996: The consequences for native biota of anthropogenic-induced climate change. In: *Greenhouse: Coping with Climate Change* [Bouma, W.J., G.I. Pearman, and M.R. Manning (eds.)]. CSIRO Publishing, Collingwood, Victoria, Australia, pp. 308–324.

Moffatt, I., 1992: The evolution of the sustainable development concept: a perspective from Australia. *Australian Geographical Studies*, **30,** 27–42.

Mpelasoka, F.S., A.B. Mullan, and R.G. Heerdegen, 2001: New Zealand climate change information derived by multivariate statistical and artificial neural networks approaches. *International Journal of Climatology*, (in press).

Mullan, A.B. and J.A. Renwick, 1990: *Climate Change in the New Zealand Region Inferred from General Circulation Models*. Report prepared for the New Zealand Ministry for the Environment, Wellington, 142 pp.

Mullan, A.B., D.S. Wratt, and J.A. Renwick, 2001: Transient model scenarios of climate changes for New Zealand. *Weather and Climate* (in press).

Mulrennan, M.E. and C.D. Woodroffe, 1998: Holocene development of the lower Mary River plains, Northern Territory, Australia. *The Holocene*, **8,** 565–579.

Mulvaney, J. and J. Kamminga, 1999: *Prehistory of Australia*. Allen and Unwin, Sydney, Australia, 480 pp.

Myers, N., R.A. Mittermeier, C.G. Mittermeier, G.A.B. da Fonseca, and J. Kent, 2000: Biodiversity hotspots for conservation priorities. *Nature*, **403,** 853–858.

National Greenhouse Gas Inventory, 2000: *Workbook for Carbon Dioxide from the Biosphere (Land Use Change and Forestry)*. Workbook 4.2, Inventory Supplement, National Greenhouse Gas Inventory Committee, Canberra, Australia.

New Zealand Climate Change Programme, 1990: *Climatic Change: A Review of Impacts on New Zealand*. Ministry for the Environment, Wellington, New Zealand, 32 pp.

NHRC, 1999: The April 1999 Sydney hailstorm. Natural Hazards Research Centre, *Natural Hazards Quarterly*, **5(2)**, 1–4.

Nicholls, N., 1993: El Niño-Southern Oscillation and vector-borne disease. *Lancet*, **342,** 1284–1285.

Nicholls, N., C. Landsea, and J. Gill, 1998: Recent trends in Australian region tropical cyclone activity. *Meteorological and Atmospheric Physics*, **65,** 197–205.

Nicholls, N., G.V. Gruza, J. Jouzel, T.R. Karl, L.A. Ogallo, and D.E. Parker, 1996a: Observed climate variability and change. In: *Climate Change 1995: The Science of Climate Change. Contribution of Working Group I to the Second Assessment Report of the Intergovernmental Panel on Climate Change* [Houghton, J.T., L.G. Meira Filho, B.A. Callander, N. Harris, A. Kattenberg, and K. Maskell (eds.)]. Cambridge University Press, Cambridge, United Kingdom and New York, NY, USA, pp. 137–192.

Nicholls, N., B. Lavery, C. Fredericksen, W. Drosdowsky, and S. Torok, 1996b: Recent apparent changes in relationships between the El Niño-Southern Oscillation and Australian rainfall and temperature. *Geophysical Research Letters*, **23,** 3357–3360.

Nix, H.A., 1981: The environment of Terra Australis. In: *Ecological Biogeography of Australia* [Keast, A. (ed.)]. Monographiae Biologicae, Dr. W. Junk, The Hague, The Netherlands, Vol. 41, pp. 103–133.

Noble, I.R., M. Barson, R. Dumsday, M. Friedel, R. Hacker, N. McKenzie, G. Smith, M. Young, M. Maliel, and C. Zammit, 1996: Land resources. In: *Australia: State of the Environment 1996*. CSIRO Publishing, Melbourne, Australia, pp. 6.1–6.55.

Nordhaus, W., 1994: *Managing the Global Commons: The Economics of Climate Change*. MIT Press, Cambridge, MA, USA, 213 pp.

O'Meagher, B., L. du Pisani, and D.H. White, 1998: Evolution of drought policy and related science in Australia and South Africa. *Agricultural Systems*, **57,** 231–258.

O'Meagher, B., M. Stafford Smith, and D.H. White, 2000: Approaches to integrated drought risk management: Australia's national drought policy. In: *Drought: A Global Assessment*. [Wilhite, D.A. (ed.)]. Routledge, London, United Kingdom, pp. 115–128.

Oshima, Y., M. Hasegawa, T. Yasumoto, G. Hallegraeff, and S. Blackburn, 1987: Dinoflagellate Gymnodinium catenatum as the source of paralytic shellfish toxins in Tasmanian shellfish. *Toxicon*, **25(10),** 1105–1111.

Pearce, A.F. and B.F. Phillips, 1994: Oceanic processes, puerulus settlement and recruitment of the Western Rock Lobster Panulirus cygnus. In: *The Bio-Physics of Marine Larval Dispersal* [Sammarco, P.W. and M.L. Heron (eds.)]. Coastal and Estuarine Studies 45, American Geophysical Union, Washington, DC, USA.

Pearce, H.J. and M.R. Kennedy, 1993: *Generalised Probable Maximum Precipitation Estimations Techniques for Australia*. Conf. Pub. No. 93/14, Hydrology and Water Resources Symposium, Institute of Engineers Australia, Newcastle, Australia.

Pestana, B., 1993: Agriculture in the Australian economy. In: *Australian Agriculture: The Complete Reference on Rural Industry, 4th ed*. National Farmers Federation, Morescope Publishing, Camberwell, Victoria, Australia, pp. 9–31.

Petroeschevsky, A., R.W. Simpson, L. Thalib, and S. Rutherford, 1999: Associations between outdoor air pollution and hospital admissions in Brisbane, Australia. *Epidemiology*, **10(4),** S96.

Pisaniello, J.D. and J. McKay, 1998: The need for private dam safety assurance policy—a demonstrative case study. *Australian Journal of Emergency Management*, **12,** 46–48.

Pittock, A.B., 1999: Coral reefs and environmental change: adaptation to what? *American Zoologist*, **39,** 10–29.

Pittock, A.B. and R.N. Jones, 2000: Adaptation to what and why? *Environmental Monitoring and Assessment*, **61,** 9–35.

Pittock, A.B., R.J. Allan, K.J. Hennessy, K.L. McInnes, R. Suppiah, K.J. Walsh, and P.H. Whetton, 1999: Climate change, climatic hazards and policy responses in Australia. In: *Climate, Change and Risk* [Downing, T.E., A.A. Oltshoorn, and R.S.L. Tol (eds.)]. Routledge, London, United Kingdom, pp. 19–59.

Plummer, N., 1996: Temperature variability and extremes over Australia, part 1: recent observed changes. *Australian Meteorological Magazine*, **45,** 233–250.

Plummer, N., B.C. Trewin, R. Hicks, N. Nicholls, S.J. Torok, B.M. Lavery, and R.M. Leighton, 1997: Australian data for documenting changes in climate extremes. In: *CLIVAR/GCOS/WMO Workshop on Indices and Indicators for Climate Extremes, National Climate Data Center, Ashville, NC, USA, 3–7 June 1997*.

Plummer, N., M.J. Salinger, N. Nicholls, R. Suppiah, K.J. Hennessy, R.M. Leighton, B. Trewin, C.M. Page, and J.M. Lough, 1999: Changes in climate extremes over the Australian region and New Zealand during the twentieth century. *Climatic Change*, **42,** 183–202.

PMSEIC, 1998: *Dryland Salinity and its Implications on Rural Industries and the Landscape*. Report prepared for the Prime Minister's Science, Engineering and Innovation Council, Canberra, Australia, 29 pp.

PMSEIC, 1999: *Moving Forward in Natural Resource Management: The Contribution that Science, Engineering and Innovation Can Make*. Report prepared for the Prime Minister's Science, Engineering and Innovation Council, Canberra, Australia, 33 pp.

Pouliquen-Young, O. and P. Newman, 1999: *The Implications of Climate Change for Land-Based Nature Conservation Strategies*. Final Report 96/1306, Australian Greenhouse Office, Environment Australia, Canberra, and Institute for Sustainability and Technology Policy, Murdoch University, Perth, Australia, 91 pp.

Power, S., T. Casey, C. Folland, A. Colman, and V. Mehta, 1999a: Decadal modulation of the impact of ENSO on Australia. *Climate Dynamics*, **15,** 319–324.

Power, S., F. Tseitkin, V. Mehta, B. Lavery, S. Torok, and N. Holbrook, 1999b: Decadal climate variability in Australia during the 20th century. *International Journal of Climatology*, **19,** 169–184.

Power, S., F. Tseitkin, S. Torok, B. Lavery, R. Dahni, and B. McAvaney, 1998: Australian temperature, Australian rainfall and the Southern Oscillation, 1910–1992: coherent variability and recent changes. *Australian Meteorology Magazine*, **47,** 85–101.

Price, C. and D. Rind, 1994: The impact of $2xCO_2$ climate on lightning-caused fires. *Journal of Climate*, **7,** 1484–1494.

Pyke, N.B., D.R. Wilson, P.J. Stone, and P.D. Jamieson, 1998: Climate change impacts on feed grain production and quality in New Zealand. *Proceedings of the New Zealand Agronomy Society*, **28,** 55–58.

QDPI, 1996: *Surviving the Drought, Returning to Profitability.* Attachment 4 to Submission No. 00585, Queensland Department of Primary Industries, Queensland, Australia, unpaginated.

Queensland Transport, CSIRO, and PPK, 1999: *The Effects of Climate Change on Transport Infrastructure in Regional Queensland: Synthesis Report.* Report prepared by Queensland Transport, CSIRO Atmospheric Research, and PPK Infrastructure and Environment Pty. Ltd., Queensland Transport, Brisbane, Australia, 18 pp.

Quinn, T.M., T.J. Crowley, F.W. Taylor, C. Henin, P. Joannot, and Y. Join, 1998: A multicentury stable isotope record from a New Caledonia coral: interannual and decadal sea surface temperature variability in the SW Pacific since 1657 AD. *Paleoceanography*, **13,** 412–426.

Race, D. and A. Curtis, 1997: Socio-economic considerations for regional farm forestry development. *Australian Forestry*, **60,** 233–239.

Radford, D., R. Blong, A.M. d'Aubert, I. Kuhnel, and P. Nunn, 1996: *Occurence of Tropical Cyclones in the Southwest Pacific Region 1920–1994.* Greenpeace International, Amsterdam, The Netherlands, 35 pp.

RAPI, 1998: *Good Practice Guidelines for Integrated Coastal Planning.* Royal Australian Planning Institute, Hawthorn, Victoria, Australia, 105 pp.

Renwick, J.A., J.J. Katzfey, J.L. McGregor, and K.C. Nguyen, 2000: On regional model simulations of climate change over New Zealand. *Weather and Climate*, **19,** 3–13.

Renwick, J.A., R.J. Hurst, and J.W. Kidson, 1998a: Climatic influences on the recruitment of southern gemfish (Rexea solandri, Gempylidae) in New Zealand waters. *International Journal of Climatology*, **18,** 1655–1667.

Renwick, J.A., J.J. Katzfey, K.C. Nguyen, and J.L. McGregor, 1998b: Regional model simulations of New Zealand climate. *Journal of Geophysical Research*, **103,** 5973–5982.

Richter, B.D., J.V. Baumgartner, J. Powell, and D.P. Braun, 1996: A method for assessing hydrologic alteration within ecosystems. *Conservation Biology*, **10,** 1–12.

Rintoul, S., G. Meyers, J. Church, S. Godfrey, M. Moore, and B. Stanton, 1996: Ocean processes, climate and sea level. In: *Greenhouse: Coping with Climate Change* [Bouma, W.J., G.I. Pearman, and M.R. Manning (eds.)]. CSIRO Publishing, Collingwood, Victoria, Australia, pp. 127–144.

Robinson, S., 1999: Mapping the effects of climate change. *Property Australia*, **July,** 18.

Rogers, G., P. Gras, L. Payne, P. Milham, and J. Conroy, 1998: The influence of carbon dioxide concentrations ranging from 280 to 900 uLL-1 on the protein starch and mixing properties of wheat flour (Triticum aestivum L. cv. Hartog and Rosella). *Australian Journal of Plant Physiology*, **25(3),** 387–393.

RSNZ, 1993: *Marine Resources, Their Management and Protection: A Review.* Miscellaneous Series 26, Royal Society of New Zealand, Wellington, New Zealand, 28 pp.

RSNZ, 1988: *Climate Change in New Zealand.* Miscellaneous Series 18, Royal Society of New Zealand, Wellington, New Zealand, 28 pp.

Russell, G.L. and D. Rind, 1999: Response to CO_2 transient increase in the GISS coupled model: regional coolings in a warming climate. *Journal of Climate*, **12,** 531–539.

Russell, J.S., 1988: The effect of climatic change on the productivity of Australian agroecosystems. In: *Greenhouse: Planning for Climate Change* [Pearman, G.I. (ed.)]. Brill, Leiden and CSIRO Publications, East Melbourne, Australia, pp. 491–505.

Russell, R.C., 1998: Mosquito-borne arboviruses in Australia: the current scene and implications of climate change for human health. *International Journal of Parasitology*, **28,** 955–969.

Russell, N., P. Weinstein, and A. Woodward, 1998: Cryptosporidiosis: an emerging microbial threat in the Pacific. *Pacific Health Dialog*, **5,** 137–141.

Salinger, M.J. and G.J. Kenny, 1995: Climate and kiwifruit, Hayward, 2: regions in New Zealand suited for production. *New Zealand Journal of Crop and Horticultural Science*, **23,** 173–184.

Salinger, M.J. and A.B. Mullan, 1999: New Zealand climate: temperature and precipitation variations and their link with atmospheric circulation 1930–1994. *International Journal of Climatology*, **19,** 1049–1071.

Salinger, M.J., R. Allan, N. Bindoff, J. Hannah, B.M. Lavery, Z. Lin, J. Lindesay, N. Nicholls, N. Plummer, and S.J. Torok, 1996: Observed variability and change in climate and sea-level in Australia, New Zealand and the South Pacific. In: *Greenhouse: Coping with Climate Change* [Bouma, W.J., G.I. Pearman, and M.R. Manning (eds.)]. CSIRO Publishing, Collingwood, Victoria, Australia, pp. 100–126.

Schofield, N.J., J.K. Ruprecht, and I.C. Loh, 1988: *The Impact of Agricultural Development on the Salinity of Surface Water Resources of South-West Western Australia.* Water Authority of Western Australia, John Tonkin Water Centre, Leederville, WA, Australia, 69 pp.

Schreider, S.Y., D.I. Smith, and A.J. Jakeman, 2000: Climate change impacts on urban flooding. *Climatic Change*, **47,** 91–115.

Schreider, S.Y., A.J. Jakeman, P.H. Whetton, and A.B. Pittock, 1997: Estimation of climate impact on water availability and extreme flow events for snow-free and snow-affected catchments of the Murray-Darling basin. *Australian Journal of Water Resources*, **2,** 35–46.

Schreider, S.Y., A.J. Jakeman, A.B. Pittock, and P.H. Whetton, 1996: Estimation of possible climate change impacts on water availability, extreme flow events and soil moisture in the Goulburn and Ovens basins, Victoria. *Climatic Change*, **34,** 513–546.

Seafood New Zealand, 2000: *New Zealand Seafood Species and Exports.* Available online at http://www.seafood.co.nz/exports.htm.

Sim, J. and N. Wilson, 1997: Surveillance of marine biotoxins, 1993–96. *New Zealand Public Health Report*, **4,** 9–11.

Sims, R.E.H., 1999: Bioenergy and carbon cycling—the New Zealand way. In: *Biomass: A Growth Opportunity on Green Energy and Value Added Products, Proceedings, Fourth Biomass Conference of the Americas, Oakland, September 1999* [Overend, R.P. and R. Chornet, eds.], Pergamon Press, Oxford, United Kingdom, pp. 401–406.

Simpson, R.W., G. Williams, A. Petroeschevsky, G. Morgan, and S. Rutherford, 1999: Associations between outdoor air pollution and daily mortality in Brisbane, Australia. *Epidemiology*, **10(4),** S96.

Skertchly, A. and K. Skertchly, 2000: Traditional Aboriginal knowledge and sustained human survival in the face of severe natural hazards in the Australian monsoon region: some lessons from the past for today and tomorrow. *Australian Journal of Emergency Management*, **14(4),** 42–50.

Smith, A.D.M., 1996: *Evaluation of Harvesting Strategies for Australian Fisheries at Different Levels of Risk from Economic Collapse.* Project No. T93/238, Final Report to Fisheries Research and Development Corporation, Canberra, Australia.

Smith, D.I., 1998a. Urban flood damage under greenhouse conditions: what does it mean for policy? *Australian Journal of Emergency Management*, **13(2),** 56–61.

Smith, D.I., 1998b: *Water in Australia: Resources and Management.* Oxford University Press, Melbourne, Australia, 384 pp.

Smith, D.I. and M.A. Greenaway, 1994: *Tropical Storm Surge, Damage Assessment and Emergency Planning: A Pilot Study for Mackay, Queensland.* Resource and Environmental Studies No. 8, Centre for Resource and Environmental Studies, Australian National University, Canberra, Australia, 59 pp.

Smith, D.I., M.F. Hutchinson, and R.J. McArthur, 1992: *Climatic and Agricultural Drought: Payments and Policy.* Australian National University, Report on Grant ANU-5A to Rural Industries Research and Development Corporation, Canberra, Australia, 103 pp.

Smith, D.I., S.Y. Schreider, A.J. Jakeman, A. Zerger, B.C. Bates, and S.P. Charles, 1997: *Urban Flooding: Greenhouse-Induced Impacts, Methodology and Case Studies.* Resource and Environmental Studies No. 17, Australian National University, Canberra, Australia, 68 pp.

Smith, I.N., P. McIntosh, T.J. Ansell, C.J.C. Reason, and K. McInnes, 2000: South-west Western Australia rainfall and its association with Indian Ocean climate variability. *International Journal of Climatology*, **20**, 1913–1930.

Smith, P.J., 1979: Esterase gene frequencies and temperature relationships in the New Zealand snapper Chrysophrys auratus. *Marine Biology*, **53**, 305–310.

Stafford Smith, M. and G.M. McKeon, 1998: Assessing the historical frequency of drought events on grazing properties in Australian rangelands. *Agricultural Systems*, **57**, 271–299.

Stafford Smith, D.M., R. Buxton, J. Breen, G.M. McKeon, A.J. Ash, S.M. Howden, and T.J. Hobbs, 1999: *Land Use Change in Northern Australia: The Impacts of Markets, Policy and Climate Change*. LUCNA Regional Report No.1, Charters Towers, CSIRO, Alice Springs, Australia, 135 pp.

Standards Australia and Standards New Zealand, 1999: *Risk Management*. AS/NZS 4360, Standards Australia, Strathfield, New South Wales, Australia.

State of the Environment, 1996: *Australia, State of the Environment 1996*. An independent report presented to the Commonwealth Minister for the Environment, State of the Environment Council, CSIRO Publishing, Collingwood, Victoria.

Statistics New Zealand, 1998: *New Zealand Official Yearbook 1998* [Zwartz , D. (ed.)]. G.P. Publications, Wellington, New Zealand, 607 pp.

Statistics New Zealand, 1999: *New Zealand Official Yearbook on the Web 1999*. Available online at http://www.stats.govt.nz/domino/external/PASfull/PASfull.nsf/Web/Yearbook+New+Zealand+Official+Yearbook+On+The+Web+1999?OpenDocument.

Stehlik, D., I. Gray, and G. Lawrence, 1999: *Drought in the 1990s: Australian Farm Families' Experiences*. Rural Industries Research and Development Corporation, Canberra, Australia, 120 pp.

Stephenson, G., 1983: *Wetlands, A Diminishing Resource*. Soil and Water Publication No. 58, Report to the Environmental Council, Ministry of Works and Development, Wellington, New Zealand, 62 pp.

Stone, R.C. and G.M. McKeon, 1992: Tropical pasture establishment: prospects for using weather predictions to reduce pasture establishment risk. *Tropical Grasslands*, **27**, 406–413.

Stone, R.C., G.L. Hammer, and T. Marcussen, 1996a: Prediction of global rainfall probabilities using phases of the Southern Oscillation Index. *Nature*, **384**, 252–255.

Stone, R., N. Nicholls, and G. Hammer, 1996b: Frost in NE Australia: trends and influence of phases of the Southern Oscillation. *Journal of Climate*, **9**, 1896–1909.

Sutherst, R.W., B.S. Collyer, and T. Yonow, 2000: The vulnerability of Australian horticulture to the Queensland fruit fly, Bactrocera (Dacus) tryoni, under climate change. *Australian Journal of Agricultural Research*, **51**, 467–480.

Sutherst, R.W., T. Yonow, S. Chakraborty, C. O'Donnell, and N. White, 1996: A generic approach to defining impacts of climate change on pests, weeds and diseases in Australasia. In: *Greenhouse: Coping with Climate Change* [Bouma, W.J., G.I. Pearman, and M.R. Manning (eds.)]. CSIRO Publishing, Victoria, Australia, pp. 190–204.

Taylor, R.T. and I. Smith, 1997: *The State of New Zealand's Environment 1997*. Ministry of the Environment, Wellington, New Zealand, 648 pp.

Thresher, R.E., 1994: Climatic cycles may help explain fish recruitment in south east Australia. *Australian Fisheries*, **53**, 20–22.

Torok, S. and N. Nicholls, 1996: A historical annual temperature data set for Australia. *Australian Meteorological Magazine*, **45**, 251–260.

Trenberth, K.E. and T.J. Hoar, 1997: El Niño and climate change. *Geophysical Research Letters*, **24**, 3057–3060.

Tunks, A., 1997: Tangata Whenua ethics and climate change. *New Zealand Journal of Environmental Law*, **1**, 67–123.

Turner, S., 1995: Restoring seagrass systems in New Zealand. *Water and Atmosphere*, **3**, 9–11.

Vance, D.J., M. Haywood, D. Heales, R. Kenyon, and N. Loneragan, 1998: Seasonal and annual variation in abundance of post-larval and juvenile banana prawns *Penaeus merguiensis* and environmental variation in two estuaries in tropical northeastern Queensland: a six year study. *Marine Ecology Progress Series*, 21–36.

Vance, D.J., D.J. Staples, and J. Kerr, 1985: Factors affecting year-to-year variation in the catch of banana prawns (Penaeus merguiensis) in the Gulf of Carpentaria. *Journal du Conseil International pour l'Exploration de la Mer*, **42**, 83–97.

Walker, P.A, M.D. Young, R.E. Smyth, and H. Lynch, 1996: *INSIGHT: A Future-Orientated and Spatially-Explicit Modelling System for Evaluation of Land Use Alternatives*. Report from the CSIRO Division of Wildlife & Ecology to the Rural Industries Research and Development Corporation, Canberra, Australia.

Walsh, K., R. Allan, R. Jones, B. Pittock, R. Suppiah, and P. Whetton, 1999: *Climate Change in Queensland Under Enhanced Greenhouse Conditions*. First Annual Report, 1997–1998, CSIRO Atmospheric Research, Aspendale, Victoria, Australia, 84 pp.

Walsh, K., K. Hennessy, R. Jones, B. Pittock, L. Rotstayn, R. Suppiah, and P. Whetton, 2000: *Climate Change in Queensland Under Enhanced Greenhouse Conditions*. Second Annual Report, 1998–1999, CSIRO Atmospheric Research, Aspendale, Victoria, Australia, 130 pp.

Walsh, K.J.E. and J.J. Katzfey, 2000: The impact of climate change on the poleward movement of tropical cyclone-like vortices in a regional climate model. *Journal of Climate*, **13**, 1116–1132.

Walsh, K.J.E. and B.F. Ryan, 2000: Tropical cyclone intensity increase near Australia as a result of climate change. *Journal of Climate*, **13**, 3029–3036.

Wardle, P. and M.C. Coleman, 1992: Evidence for rising upper limits of four native New Zealand forest trees. *New Zealand Journal of Botany*, **30**, 303–314.

Warrick, R.A., G.J. Kenny, G.C. Sims, N.J. Ericksen, Q.K. Ahmad, and M.Q. Mirza, 1996: Integrated model systems for national assessments of the effects of climate change: applications in New Zealand and Bangladesh. *Journal of Water, Air, and Soil Pollution*, **92**, 215–227.

Waterman, P., 1996: *Australian Coastal Vulnerability Project Report*. Department of the Environment, Sport and Territories, Canberra, Australia, 75 pp.

Watson, B., H. Morrisey, and N. Hall, 1997: Economic analysis of dryland salinity issues. In: *Proceedings of ABARE's National Agricultural Outlook Conference, Outlook 97, Canberra, 4–6 February*. Commodity Markets and Resource Management, Canberra, Australia, Vol. 1, pp. 157–166.

Weaver, P.P.E., L. Carter, and H.L. Neil, 1998: Response to surface water masses and circulation to late Quaternary climate change east of New Zealand. *Paleoceanography*, **13**, 70–83.

Webster, K.C. and R.J. Wark, 1987: Australian dam safety legislation. *ANCOLD Bulletin*, **78**, 63–78.

Weinstein, P., 1997: An ecological approach to public health intervention: Ross River virus in Australia. *Environmental Health Perspectives*, **105**, 364–366.

West, B. and P. Smith, 1996: Drought, discourse, and Durkheim: a research note. *The Australian and New Zealand Journal of Sociology*, **32**, 93–102.

Whetton, P.H., 1999: *Comment on the 1999 Climate Change Scenarios for Australia*. United Kingdom Climatic Research Unit and World Wildlife Fund, Climate Impact Team, CSIRO Atmospheric Research, Collingwood, Victoria, Australia, 3 pp. http://www.dar.csiro.au/publications/Hennessy_2000a.html.

Whetton, P.H., 1998: Climate change impacts on the spatial extent of snow-cover in the Australian Alps. In: *Snow: A Natural History; An Uncertain Future* [Green, K. (ed.)]. Australian Alps Liaison Committee, Canberra, Australia, pp. 195–206.

Whetton, P.H., Z. Long, and I.N. Smith, 1998: *Comparison of Simulated Climate Change Under Transient and Stabilised Increased CO_2 Conditions*. Research Report No. 69, Bureau of Meteorology Research Centre, Melbourne, Australia, pp. 93–96.

Whetton, P.H., M.H. England, S.P. O'Farrell, I.G. Watterson, and A.B. Pittock, 1996a: Global comparison of the regional rainfall results of enhanced greenhouse coupled and mixed layer ocean experiments: implications for climate change scenario development. *Climatic Change*, **33**, 497–519.

Whetton, P.H., M.R. Haylock, and R. Galloway, 1996b: Climate change and snow-cover duration in the Australian Alps. *Climatic Change*, **32**, 447–479.

Whetton, P.H., A.M. Fowler, M.R. Haylock, and A.B. Pittock, 1993: Implications for floods and droughts in Australia of climate change due to the enhanced greenhouse effect. *Climatic Change*, **25**, 289–317.

Whetton, P.H., J.J. Katzfey, K.J. Hennessy, X. Wu, J.L. McGregor, and K. Nguyen, 2001: Developing scenarios of climate change for southeastern Australia: an example using regional climate model output. *Climate Research*, **16**, 181–201.

White, B.J., 2000: The importance of climate variability and seasonal forecasting to the Australian economy. In: *Applications of Seasonal Climate Forecasting in Agricultural and Natural Ecosystems: The Australian Experience* [Hammer, G.L., N. Nicholls, and C. Mitchell (eds.)]. Atmospheric and Oceanographic Sciences Library, J. Kluwer Academic Publishers, Dordrecht, The Netherlands, Vol. 21, pp. 1–22.

White, B.W. and N.J. Cherry, 1999: Influence of the ACW upon winter temperature and precipitation over New Zealand. *Journal of Climate*, **12,** 960–976.

White, B.W. and R.G. Peterson, 1996: An Antarctic circumpolar wave in surface pressure, wind, temperature and sea-ice extent. *Nature*, **380,** 699–702.

Whitehead, D., J.R. Leathwick, and J.F.F. Hobbs, 1992: How will New Zealand's forests respond to climate change? Potential changes in response to increasing temperature. *New Zealand Journal of Forestry Science*, **22,** 39–53.

Wigley, T.M.L., 1995: Global-mean temperature and sea level consequences of greenhouse gas concentration stabilisation. *Geophysical Research Letters*, **22,** 45–48.

Wigley, T.M.L., S.C.B. Raper, M. Salmon and M. Hulme, 1997: *MAGICC: Model for the Assessment of Greenhouse-gas Induced Climate Change: Version 2.3*, Climatic Research Unit, Norwich, United Kingdom.

Wilkinson, C., 1998: *Status of Coral Reefs of the World: 1998*. Global Coral Reef Monitoring Network, Australian Institute of Marine Science, Townsville, Australia, 184 pp.

Williams, A.J., D.J. Karoly, and N. Tapper, 2001: The sensitivity of Australian fire danger to climate change. *Climatic Change*, (in press).

Williams, P., 1992: *The State of the Rivers of the South West: Water Resources Perspectives*. Publication No. WRC 2/92, Western Australian Water Resources Council, 70 pp.

Wilmshurst, J.M., 1997: The impact of human settlement on vegetation and soil stability in Hawke's Bay, New Zealand. *New Zealand Journal of Botany*, **35,** 97–111.

Wilson, L. and A. Johnson, 1997: Agriculture in the Australian economy. In: *Australian Agriculture: The Complete Reference on Rural Industry, 6th ed*. National Farmers Federation, Morescope Publishing, Hawthorn East, Victoria, Australia, pp. 7–18.

Woodroffe, C.D., 1993: Late Quaternary evolution of coastal and lowland riverine plains of southeast Asia and northern Australia: an overview. *Sedimentary Geology*, **83,** 163–175.

Woodward, A., S. Hales, and P. Weinstein, 1998: Climate change and human health in the Asia Pacific: who will be most vulnerable? *Climate Research*, **11,** 31–38.

Woodward, A., C. Guest, K. Steer, A. Harman, R. Scicchitano, D. Pisaniello, I. Calder, and A.J. McMichael, 1995: Tropospheric ozone: respiratory effects and Australian air quality goals. *Journal of Epidemiology and Community Health*, **49,** 401–407.

Wolanski, E. and J. Chappell, 1996: The response of tropical Australian estuaries to a sea level rise. *Journal of Marine Systems*, **7,** 267–279.

Wratt, D.S., R.N. Ridley, M.R. Sinclair, K. Larsen, S.M. Thompson, R. Henderson, G.L. Austin, S.G. Bradley, A. Auer, A.P. Sturman, I. Owens, B. Fitzharris, B.F. Ryan, and J.F. Gayet, 1996: The New Zealand Southern Alps Experiment. *Bulletin of the Amererican Meteoological Society*, **77,** 683–692.

Wright, P.B., 1974: Temporal variations in seasonal rainfall in southwestern Australia. *Monthly Weather Review*, **102,** 233–243.

Yu, B. and T.D. Neil, 1993: Long-term variations in regional rainfall in the south-west of Western Australia and the difference between average and high-intensity rainfall. *International Journal of Climatology*, **13,** 77–88.

Zapert, R., P.S. Gaertner, and J.A. Filar, 1998: Uncertainty propagation within an integrated model of climate change. *Energy Economics*, **20,** 571–598.

Zheng, X., R.E. Basher, and C.S. Thompson, 1997: Trend detection in regional-mean temperature: maximum, minimum, mean temperature, diurnal range and SST. *Journal of Climate*, **10,** 317–326.

13

Europe

ZBIGNIEW W. KUNDZEWICZ (POLAND) AND MARTIN L. PARRY (UK)

Lead Authors:
W. Cramer (Germany), J.I. Holten (Norway), Z. Kaczmarek (Poland), P. Martens (The Netherlands), R.J. Nicholls (UK), M. Öquist (Sweden), M.D.A. Rounsevell (Belgium), J. Szolgay (Slovakia)

Contributing Authors:
N.W. Arnell (UK), G. Balint (Hungary), M. Beniston (Switzerland), G. Berz (Germany), M. Bindi (Italy), T.R. Carter (Finland), S. des Clers (UK), H.Q.P. Crick (UK), A.F. Dlugolecki (UK), T. Dockerty (UK), M. Gottfried (Austria), G. Grabherr (Austria), A. Guisan (Switzerland), M. Hulme (UK), P. Imeson (Netherlands), J. Jenik (Czech Republic), A. Jordan (UK), A. Kedziora (Poland), S. Kovats (UK), S. Lavorel (France), M. Livermore (UK), J. Lowe (UK), J.P. Martinez Rica (Spain), U. Molau (Sweden), I. Nemesova (Czech Republic), J.E. Olesen (Denmark), J.P. Palutikof (UK), H. Pauli (Austria), A.H. Perry (UK), L. Ryszkowski (Poland), S.A. Shchuka (Russia), A.Z. Shvidenko (Austria), S. Tapsell (UK), J.-P. Theurillat (Switzerland), A. De la Vega-Leinert (UK), A.A. Velichko (Russia), L. Villar (Spain)

Review Editors:
M.J. de Seixas (Portugal) and S. Kellomäki (Finland)

CONTENTS

EXECUTIVE SUMMARY

- The adaptation potential of socioeconomic systems in Europe is relatively high because of economic conditions (high gross national product and stable growth); a stable population (with the capacity to move within the region); and well-developed political, institutional, and technological support systems. However, adaptation potential for natural systems generally is low. [very high confidence]

- Present-day weather conditions have effects on natural, social, and economic systems in Europe in ways that reveal sensitivities and vulnerabilities to climate change in these systems. Climate change may aggravate such effects. [very high confidence, well-established evidence]

- Vulnerability to climate change in Europe differs substantially between subregions; it is particularly high in the south and in the European Arctic. This has important equity implications. More marginal and less wealthy areas will be less able to adapt. [very high confidence, established but incomplete evidence]

- Water resources and their management in Europe are under pressure now, and these pressures are likely to be exacerbated by climate change [high confidence]. Flood hazard is likely to increase across much of Europe, except where snowmelt peak has been reduced, and the risk of water shortage is projected to increase particularly in southern Europe [medium to high confidence]. Climate change is likely to widen water resource differences between northern and southern Europe. [high confidence, well-established evidence]

- Soil properties will deteriorate under warmer and drier climate scenarios in southern Europe. The magnitude of this effect will vary markedly between geographic locations and may be modified by changes in precipitation. [medium confidence, established but incomplete evidence]

- Natural ecosystems will change as a result of increasing temperature and atmospheric concentration of carbon dioxide (CO_2). Permafrost will decline, trees and shrubs will encroach northern tundra, and broad-leaved trees may encroach coniferous forests. Net primary productivity in ecosystems is likely to increase (also as a result of nitrogen deposition). Diversity in nature reserves is under threat from rapid change. Loss of important habitats (wetlands, tundra, and isolated habitats) would threaten some species (including rare/endemic species and migratory birds). Faunal shifts as a result of ecosystem changes are expected in marine, aquatic, and terrestrial ecosystems. [high confidence, established but incomplete evidence]

- In mountain regions, higher temperatures will lead to an upward shift of biotic and cryospheric zones and perturb the hydrological cycle. There will be redistribution of species, with, in some instances, a threat of extinction. [high confidence]

- Timber harvest will increase in commercial forests in northern Europe [medium confidence, established but incomplete evidence], but reductions are likely in the Mediterranean, with increased drought and fire risk. [high confidence, well-established evidence]

- Agricultural yields will increase for most crops as a result of increasing atmospheric CO_2 concentration. This effect would be counteracted by the risk of water shortage in southern and eastern Europe and by shortening of growth duration in many grain crops as a result of increasing temperature. Northern Europe is likely to experience overall positive effects, whereas some agricultural production systems in southern Europe may be threatened. [medium confidence, established but incomplete evidence]

- Changes in fisheries and aquaculture production from climate change embrace faunal shifts affecting freshwater and marine fish and shellfish biodiversity. These changes will be aggravated by unsustainable exploitation levels and environmental change. [high confidence]

- The insurance industry faces potentially costly climate change impacts through the medium of property damage, but there is great scope for adaptive measures if initiatives are taken soon. [high confidence]

- Transport, energy, and other industries will face changing demand and market opportunities. Concentration of industry on the coast exposes it to sea-level rise and extreme events, necessitating protection or removal. [high confidence]

- Recreational preferences are likely to change with higher temperatures. Outdoor activities will be stimulated in northern Europe, but heat waves are likely to reduce the traditional peak summer demand at Mediterranean holiday destinations, and less reliable snow conditions could impact adversely on winter tourism. [medium confidence]

- A range of risks is posed for human health through increased exposure to heat episodes (exacerbated by air pollution in urban areas), extension of some vector-borne diseases, and coastal and riverine flooding. Based on current evidence, climate change would result in a reduction in wintertime deaths, at least in temperate countries. [medium confidence]

- In coastal areas, the risk of flooding, erosion, and wetland loss will increase substantially—with implications for human settlement, industry, tourism, agriculture, and coastal natural habitats. Southern Europe appears to be more vulnerable to these changes, although the North Sea coast already has high exposure to flooding. [high confidence]

The foregoing conclusions are broadly consistent with those expressed in the IPCC *Special Report on Regional Impacts of Climate Change* (1998) and the Second Assessment Report (1996). This survey incorporates much more information than previously reported, corroborating previous conclusions (with which it is broadly consistent) but extending knowledge into other sectors. It is more specific about subregional effects and includes new information concerning adaptive capacity.

13.1. The European Region

13.1.1. *Previous Work*

Western Europe was the subject of the first multi-country assessment of climate change impacts using general circulation model (GCM)-derived climate scenarios (Meinl and Bach, 1984). This included assessment of impacts on the agricultural, water, and energy sectors in the European Union (EU). Since that time, most assessments have been of single sectors or single countries. Most countries in Europe have now conducted climate impact studies, though these studies generally are based on expert reviews rather than new research. EU-wide assessments have been completed for water (Arnell *et al.,* 1999), agriculture (Harrison *et al.,* 1995a), forestry (Kellomäki, 1999), and coastal regions (European Commission, 1999).

The IPCC synthesis of regional impacts in its *Special Report on Regional Impacts of Climate Change* (RICC) captured some of the important likely effects—for example, on water resources, coastal regions, and agriculture—but drew no conclusions concerning effects on ecosystems, soils, forestry, insurance, and mountain regions (IPCC, 1998). The present survey refers to a much more extensive literature base (about three times as much as in RICC) and is able to cover additional fields and draw more specific conclusions. The main differences between the current assessment and the previous one are its coverage of the additional sectors noted above; the distinction it is able to draw between effects on different parts of Europe, particularly between northern and southern Europe; the greater degree of quantification achieved; and its evaluation of the adaptive capacity of different sectors to climate change impacts. This assessment draws substantially on the work of a 3-year review of impacts in Europe funded by the Commission of the European Communities, with extensive additional input of material for non-EU countries (Parry, 2000).

13.1.2. *What is Different about the European Region?*

13.1.2.1. Geography, Population, Environment

Europe, a continent with an area of 10.5 million km^2, extends from the Atlantic Ocean in the west to the Eastern Ural Mountains, the River Ural, and the Caspian Sea in the east and from the Arctic Sea in the north to the Caucasus Mountains, the Black Sea, and the Mediterranean Sea in the south.

Europe consists of large areas with low relief, including one of the world's largest uninterrupted plains: the European Plain. There are several mountain ranges; the highest peak is 5,642 m (Elbrus in the Caucasus Mountains). The continent is well-watered, with numerous permanent rivers, many of which flow outward from the central part of the continent.

There are five essential types of climate in Europe: maritime, transitional, continental, polar, and Mediterranean. The five major vegetation types are tundra, coniferous taiga (boreal forest), deciduous-mixed forest, steppe, and Mediterranean. A relatively large proportion of Europe is farmed, and about one-third of the area is arable; cereals are the predominant crop.

Europe has a total population of 720 million; it has a higher population density and lower birth rate than any other continent. In several countries of central and eastern Europe, population growth is negative at present, even as low as –1.2% (World Bank, 1999). High life expectancy and low infant mortality are results of advances in health care. Life expectancy at birth is among the highest in the world, in some countries reaching more than 75 years for men and more than 80 years for women (World Bank, 1999). European nations are aging faster than those of any other continent.

Key environmental pressures that are significant at the European scale are identified in the Dobris Report (Stanners and Bourdeau, 1995). They relate to areas such as biodiversity, landscape, soil and land degradation, forest degradation, natural hazards, water management, and recreational environment, among others. Most ecosystems in Europe are managed or semi-managed; they often are fragmented and under stress from pollution and other human impacts. Social concerns include issues such as competitiveness, employment, income, and social mobility (Parry, 2000). The relative importance of these issuses varies across Europe. Southern Europe, mountains, and coastal zones have their own sets of environmental concerns, some of which will be aggravated by climate change.

13.1.2.2. Economy

The pattern of wealth distribution in the European region is strongly nonhomogeneous. Values of gross national product (GNP) per capita range from US$540 to 44,320 in Moldova and Switzerland, respectively (World Bank, 1999).

The 15 states that belong to the European Union (EU) are developed countries with stable economies and high levels of productivity. Their industry is based on modern high technology. The EU has reached a high degree of integration and common economic policy.

Until 1990, several countries in central and eastern Europe (CEE) had centrally planned economies dominated by heavy industry. Since late 1989, the CEE has undergone dramatic economic and political changes toward market economy and democracy. CEE countries labeled as "economies in transition" have been overhauling outdated, ineffective, energy- and raw material-consuming and highly polluting industries. This has been a difficult and long-term process; as a result, in 1990–1992, a large drop in GNP was noted in all CEE countries. Subsequently, some countries have managed to achieve solid growth. Poland, for instance, has now experienced nine consecutive years of growth, and its mean annual GNP rise for 1990–1997 is 3.9% (World Bank, 1999). Yet for some other countries, mean annual GNP growth for 1990–1997 has been negative.

After the fall of the former political system in CEE, ties between these countries ceased to exist and new subregional links are being built, such as the Vysehrad Group created by Poland, the Czech Republic, Slovakia, and Hungary and the Central European Free Trade Agreement (CEFTA). Yet the tendency for many countries of CEE now is to seek access to Western institutions. Three countries—the Czech Republic, Hungary, and Poland—joined the North Atlantic Treaty Organization (NATO) in March 1999. Five countries (Czech Republic, Estonia, Hungary, Poland, and Slovenia) have started negotiations that are expected to lead to full access to the EU in a few years. Among the macro-level pressures in some countries are within-country ethnic tensions and difficulties of transition toward a democratic system with a market economy.

13.1.3. Recent Climate Variability in Europe, including Recent Warming

The climate of Europe exhibits large differences from west (maritime) to east (continental) and from north (Arctic) to south (Mediterranean). The climatic effects of the distribution of land and ocean are further complicated by numerous high mountain ranges, which act as physical barriers to atmospheric circulation and often introduce large precipitation gradients within small regions (e.g., Frei and Schär, 1998). There is no synoptic consistency to the behavior of the European atmosphere across such a heterogeneous climatic domain.

Europe possesses long instrumental data records. The central England temperature series, for example, commences in 1659 (Jones and Hulme, 1997). Reconstruction of regional climates in Europe with proxy data sets identifies the magnitudes of natural temperature variability that have occurred on even longer multi-century time scales: 1,400 years of summer temperatures from tree growth in Fennoscandia (Briffa *et al.*, 1990), 460 years of monthly temperature and precipitation patterns from hydrological and biological evidence in central Europe (Pfister, 1992), and 500 years of annual temperatures from ice cores in Greenland (Fischer *et al.*, 1998). Direct sea-level measurements are available from the 18th century. There have been several attempts at deciphering the remote past, based on proxy data (e.g., Velichko *et al.*, 1998; Zelikson *et al.*, 1998).

13.1.3.1. Temperature

Most of Europe has experienced increases in surface air temperature during the 20th century that, averaged across the continent, amount to about 0.8°C in annual temperature (see Figure 13-1) (ECSN, 1995; Beniston *et al.*, 1998; EEA, 1998). This warming has been largest over northwestern Russia and the Iberian peninsula (Nicholls *et al.*, 1996; Onate and Pou, 1996) and stronger in winter than in summer (Maugeri and Nanni, 1998; Brunetti *et al.*, 2000). An exception is Fennoscandia, which has recorded cooling in mean maximum and mean minimum temperature during 1910–1995 in winter but warming in summer (Tuomenvirta *et al.*, 1998). The past decade in Europe (1990–1999) has been the warmest in the instrumental record, annually and for winter. Increases in growing-season length also have been observed in Europe—for example, in western Russia (Jones and Briffa, 1995) and in Fennoscandia (Carter, 1998). Trends in the intensity of the growing season as measured by growing degree days, however, are more ambiguous in both of these studies. The evidence for longer growing seasons in Europe also is supported by phenological data collected in central Europe (Menzel and Fabian, 1999). These data point to increases of about 10 days in average growing-season length since the early 1960s. Other biological indicators of a changing growing season in Europe include poleward shifts of 35–240 km during this century in entire ranges of 34 different butterfly species (Parmesan *et al.*, 1999) and earlier breeding times for several species of amphibians and migratory birds during the past few decades (Forchhammer *et al.*, 1997).

Warming in annual mean temperature has occurred preferentially as a result of nighttime rather than daytime temperature increases (Brazdil *et al.*, 1996; Easterling *et al.*, 1997; Tuomenvirta *et al.*, 1998), reflecting similar tendencies to those in other world regions. There has been some evidence that this reduction in the diurnal temperature range (DTR) has been associated with increased cloudiness. This is especially true over parts of the former Soviet Union (FSU—Abakumova *et al.*, 1996; Groisman *et al.*, 1996), Fennoscandia (Kaas and Frich, 1995; Tuomenvirta *et al.*, 1998), and Switzerland, where Rebetez and Beniston (1998) found that a 20th-century decrease in DTR of about 1.5°C was strongly correlated with increased cloudiness, except at high elevations. In Italy (Brunetti *et al.*, 2000), daytime warming is higher than nighttime temperature rise, with a consequent increase in DTR.

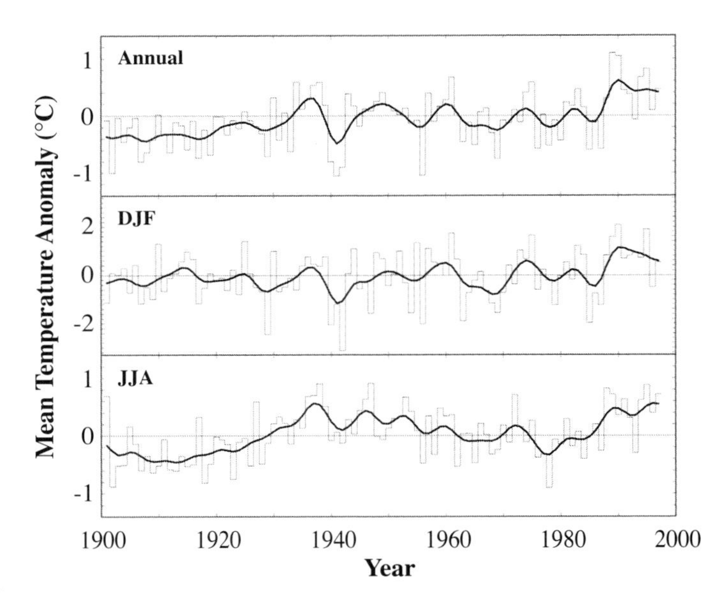

Figure 13-1: European annual mean temperature anomalies over land, 1901–1995, for the region 35°N to 75°N and 30°W to 50°E, with respect to 1961–90 mean. 10-pt filter applied. Data from IPCC Data Distribution Centre.

13.1.3.2. Precipitation

Trends in annual precipitation differ between northern Europe (wetting) and southern Europe (drying), reflecting a wider hemispheric pattern of contrasting zonal-mean precipitation trends between high and low latitudes (Dai *et al.*, 1997; Hulme *et al.*, 1998). Precipitation over northern Europe has increased by 10–40% in the 20th century, whereas some parts of southern Europe have dried by as much as 20% (see Figure 13-2). Romero *et al.* (1999) show that the numbers of days with precipitation over the Spanish southern coast and the Pyrenees region have decreased by 50 and 30%, respectively, in 1964–1993. In Italy, total precipitation in the 20th century has decreased by about 5% in the north and by about 15% in the south (Buffoni *et al.*, 1999; Brunetti *et al.*, 2001). Analysis of moisture extremes over Europe, using the Palmer Drought Severity Index (PDSI—Briffa *et al.*, 1994) showed strong decadal-scale variability in drought frequency; the 1940s and early 1950s experienced widespread and severe droughts—a pattern repeated in 1989 and 1990.

13.1.3.3. Extreme Events

Analyses of trends in extreme weather events in Europe generally have been limited to national studies, making it difficult to provide a continent-wide overview of changes in hot/cold day frequencies, precipitation intensities, or gale frequencies. Gruza *et al.* (1999) analyzed data over the whole of Russia, the western third of which falls into our definition of Europe, and found a slight increase over the 20th century in the Climate Extremes Index (CEI), which combines daily temperature, daily precipitation, and drought extremes. Analysis of 85 long-term maximum 1-day precipitation records in the Nordic countries indicates that there is a maximum in the 1930s and a tendency of increasing values during the 1980s and 1990s—decades with relatively high regional summer temperatures. In western Norway, the past 2 decades have been exceptional, with substantial increases in orographic precipitation during autumn, winter, and spring (Førland *et al.*, 1998). Elsewhere, daily precipitation intensities over the UK have increased in winter over recent decades (Osborn *et al.*, 1999), although not in other seasons. This increase in UK winter precipitation intensities has been paralleled by a marked decrease in the frequency of cold winter days in the UK (Jones *et al.*, 1999). Changes in storminess over the northeast Atlantic have been analyzed by Schmidt *et al.* (1998) and WASA (1998); they show that although storminess has increased in recent decades, storm intensities are no higher than they were early in the 20th century. Wave heights around the shores of northwest Europe also show large decadal variability, but no long-term trends emerge.

13.1.3.4. North Atlantic Oscillation

One important cause for interannual and perhaps interdecadal climate variability in Europe, particularly in winter, is the North Atlantic Oscillation (NAO). NAO is a measure of the strength of the westerly flow over the North Atlantic; records go back to the early 19th century (Jones *et al.*, 1997). Over the past 4 decades, the NAO pattern gradually has altered from the most extreme and persistent negative phase in the 1960s to the most extreme positive phase during the late 1980s and early 1990s—a trend that has been responsible for relatively mild and wet winters during the latter period over much of northwest Europe (Hurrell and van Loon, 1997) and can help explain the observed narrowing in DTR (Tuomenvirta *et al.*, 1998). Following its long period of amplification, the NAO index underwent a sharp decrease to a short-lived minimum in the winter of 1995–1996, with radical recognizable changes in the North Atlantic. It should be noted that the NAO appears to exert significant control on the export of ice and freshwater from the Arctic to the open Atlantic (WCRP, 1998). Since that temporary minimum, a recovery toward positive values of the NAO has been observed. NAO certainly is a more important influence on European climate than the El Niño-Southern Oscillation (ENSO), although several studies have explored the influence of ENSO on European precipitation variability, and links between NAO and ENSO are being sought. Evidence for such an influence is weak and shows phase differences through the continent (Fraedrich and Müller, 1992; Fraedrich *et al.*, 1992; Rodo *et al.*, 1997; Price *et al.*, 1998).

13.1.4. Key Sensitivities to Climate and Weather Now

The response of human activities and the natural environment to current weather perturbations provides a guide to where critical sensitivities to future climate change may lie. Even if adaptation may modify the response, analysis of present-day sensitivities helps us understand the likely impact of future climate changes on our environment and lifestyle.

The principal sensitivities in Europe to current climate and weather conditions are to:

- Extreme seasons
  - Exceptionally hot, dry summers
  - Mild winters
- Short-duration hazards
  - Windstorm (possibly associated with tidal surges)
  - Heavy rain leading to river-valley flooding and flash floods
- Slow, long-term change
  - Coastal squeeze.

Sensitivities vary across Europe as a result of differences in climate and topography, as well as the socioeconomic environment, but all of Europe is sensitive to the foregoing anomalous conditions.

13.1.4.1. Extreme Seasons

13.1.4.1.1. Exceptionally hot, dry summers

Extremely hot and dry summers were experienced in 1995 throughout much of western Europe (Palutikof *et al.*, 1997)

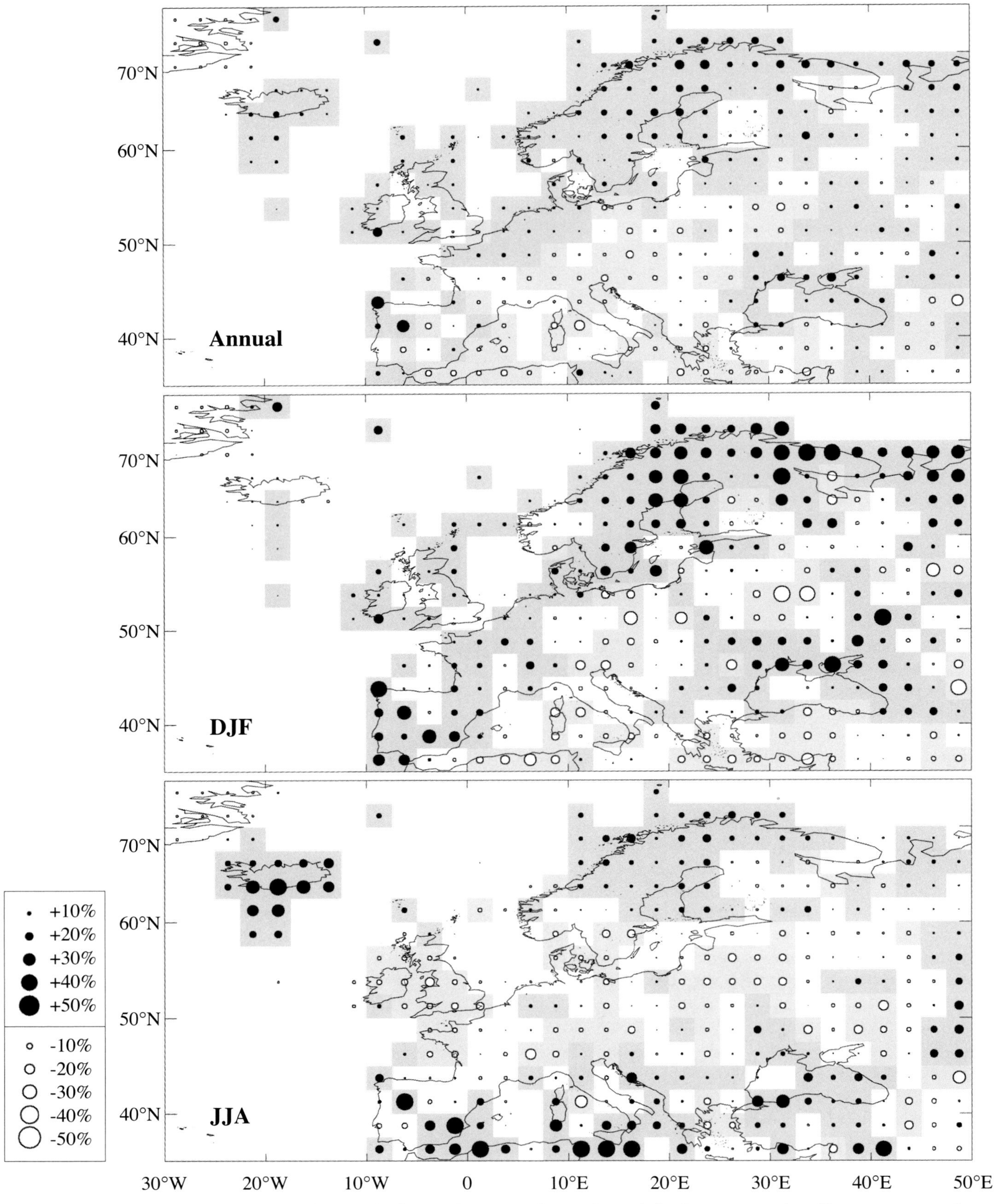

Figure 13-2: Trends in annual (top), winter (middle), and summer (bottom) precipitation expressed as % per century and calculated on 2.5° grid. Black circles denote wetting and white circles denote drying. Magnitude of trend is related to circle size. Shaded trends are significant at 90% (data from New *et al.*, 1999).

and in 1992 in CEE (Schellnhuber *et al.*, 1994). Summer drought conditions can have a devastating impact on the natural environment, particularly by reducing the availability of water for flora and fauna. Pollution levels may rise to high levels in the subsiding anticyclonic air, with add-on effects for human health in terms of increased numbers of asthma attacks and associated hospitalizations. High water temperatures often lead to algal blooms, which may make water bodies unusable for recreational purposes.

Agriculture and water supply are the two economic sectors that may be most severely affected by exceptional heat and drought in summer. Large proportions of the working populations in Mediterranean countries are engaged in agriculture, and even in "normal" years water shortages may be a limiting factor. These countries, therefore, appear to be more vulnerable to exceptional summer weather. An increased frequency of hot, dry summers is likely to reduce tree growth and affect timber yield and quality. Reduced tree vigor often favors outbreaks of pests and pathogens, and hot and dry conditions trigger wildfire.

13.1.4.1.2. Mild winters

Milder winters exert a major impact on the natural environment. Overwintering of species is more successful. However, failure to kill off pests and diseases that prey on wildlife, as well as failure to cull weaker members of the species—which then compete for food in the following springs and summers—in the end may be counterproductive. Cannell and Pitcairn (1993) show that during two mild winters in the UK, various life events occurred earlier—for example, insects and other animals moved about and fed more actively, causing widespread aphid damage on oak and mite damage on lime. Milder winters can affect fulfillment of chilling requirements in some plant species, which would have an impact on the formation of leaf and flower buds (e.g., Murray *et al.*, 1989). This could lead to regeneration failure; some susceptible species are common across much of Europe today (e.g., *Picea abies*) (Sykes and Prentice, 1996).

The skiing industry in Europe is vulnerable to variations in snowfall—that is, too little or too much snow. Many European ski resorts have suffered severe shortfalls in earnings when snow cover was not sufficiently deep during Christmas and winter school holidays. If temperatures remain below freezing, a more active hydrological cycle might lead to more episodes of heavy snowfall; the winter of 1998–1999 is a good proxy for this situation. Although the number of cold spells generally has decreased over the past few decades, cold spells still have a range of impacts on agriculture, settlements, the built environment, and transport.

13.1.4.2. Short-Duration Hazards

Severe windstorms affected Europe in the 1990s. The January 1990 storm (Daria) caused insured losses of about US$5.7 billion (at 1997 prices) and 95 deaths; in the following month, the storm of 26 February (Vivian) caused a further US$3.9 billion in insured losses (Swiss Re, 1998) and 64 deaths. The UK experienced a severe windstorm at Christmas 1997, causing 13 deaths and losses of US$500 million (Swiss Re, 1998).

The major impact of windstorm on the natural environment is on trees, woodland, and forest. It is estimated that 15 million trees (5 months' production of coniferous wood and 2 years' production of broadleaf timber production—see Quine, 1988) were blown down in the UK during the October 1987 storm.

Windstorm associated with tidal surge is a particularly deadly combination for low-lying coastal areas, as epitomized by the event of 1953. Severe northerly winds combined with an exceptionally high spring tide caused overtopping in The Netherlands and much of East Anglia, resulting in more than 2,000 deaths. Since then the policy has been to keep the sea out at all costs, and in fact several storm surges of similar size have occurred with no loss of life.

Several devastating river floods, of high severity, occurred in Europe in the 1990s (Kundzewicz and Takeuchi, 1999). Among the flood events to hit the headlines was the Odra flood in the summer of 1997, during which historic flow records were broken. The deluge caused more than 100 deaths and economic damage in excess of US$5 billion (Kundzewicz *et al.*, 1999).

13.1.4.3. Coastal Squeeze

Coastal squeeze occurs when coastal habitats are "squeezed" between rising sea level and fixed hard defenses (Bijlsma *et al.*, 1996). It is occurring in most northwest European countries, and it already is a coastal management issue (Rigg *et al.*, 1997). It is considered in more detail in Sections 13.2.1.3 and 13.3.5.

13.1.5. Climate Scenarios for the Future

Most of the impact studies evaluated in this chapter have attempted to characterize the future climate of a study region by using climate change scenarios. General reviews of the development and application of climate change scenarios are provided in Chapter 3 and in TAR WGI Chapter 13. It also should be noted that many recent impact studies in Europe have followed published guidelines concerning the use of scenarios (Carter *et al.*, 1994; USCSP, 1994; Smith and Hulme, 1998). This review offers a brief summary first of the types of scenario information provided in European impact assessments and second of research to improve this information.

13.1.5.1. Scenario Provision

One or more of three broad classes of climate change scenario generally have been adopted: synthetic scenarios, palaeoclimatic analogs, and scenarios that are based on outputs from GCMs.

The impacts of these scenario changes conventionally are assessed relative to conditions under a reference or "baseline" climate that represents present-day conditions (commonly 1961–1990).

Synthetic or incremental scenarios describe techniques whereby particular climatic (or related) elements are changed by a realistic but arbitrary amount, often according to an interpretation of climate model simulations for a region. They are simple to use and can offer a useful tool for exploring the sensitivity of an exposure unit to a plausible range of climatic variations. They commonly are applied prior to the adoption of more detailed GCM-based scenarios. For example, national assessments in CEE conducted as part of the U.S. Country Studies Program adopted adjustments of present-day temperatures by +1, +2, +3, and +4°C and baseline precipitation by ±5, +10, +15, and +20% (e.g., Smith and Pitts, 1997; Kalvová, 1995; Alexandrov, 1997).

Box 13-1. Some Key Features of Climate Scenarios for Europe

Temperature

- Annual temperatures over Europe warm at a rate of between 0.1 and 0.4°C per decade. This warming of future annual climate is greatest over southern Europe (Spain, Italy, Greece) and northeast Europe (Finland, western Russia) and least along the Atlantic coastline of the continent.
- In winter, the continential interior of eastern Europe and western Russia warms more rapidly (0.15–0.6°C per decade) than elsewhere. In summer, the pattern of warming displays a strong south-to-north gradient, with southern Europe warming at a rate of between 0.2 and 0.6°C per decade and northern Europe warming between 0.08 and 0.3°C per decade.
- Winters currently classified as cold (occurring 1 year in 10 during 1961–1990) become much rarer by the 2020s and disappear almost entirely by the 2080s. In contrast, hot summers become much more frequent. Under the 2080s scenario, nearly every summer is hotter than the 1-in-10 hot summer as defined under the present climate.
- The agreement between models about these future temperature changes is greatest over southern Europe in winter. In summer, however, this region shows the greatest level of disagreement between model simulations. All model simulations show warming in the future across the whole of Europe and in all seasons.

Precipitation

- The general pattern of future change in annual precipitation over Europe is for widespread increases in northern Europe (between +1 and +2% per decade), smaller decreases across southern Europe (maximum –1% per decade), and small or ambiguous changes in central Europe (France, Germany, Hungary, Belarus).
- There is a marked contrast between winter and summer patterns of precipitation change. Most of Europe gets wetter in the winter season (between +1 and +4% per decade); the exception is the Balkans and Turkey, where winters become drier. In summer, there is a strong gradient of change between northern Europe (wetting of as much as +2% per decade) and southern Europe (drying of as much as –5% per decade).
- The areas, however, where these changes are greater than the 2-standard deviation estimate of natural 30-year time scale climate variability are limited to the later periods (2050s and 2080s) and to the scenarios with the larger rates of global warming (B2, A1, and A2).
- Only for the A2-high scenario are there substantial areas in Europe (Fennoscandia and northwest Europe) where precipitation changes by the 2020s are larger than what might occur as a result of natural climate variability. Even for this scenario with rapid global warming, not all regions in Europe have well-defined precipitation signals from GHG-induced climate change by the 2080s.
- The intermodel range of seasonal precipitation changes generally is larger than the median change, implying that sign differences frequently exist between the precipitation changes simulated by different climate models. The largest intermodel differences tend to occur in southern and northern Europe. Intermodel differences are smallest across much of central Europe.

Weather Extremes

- The scenarios do not explicitly quantify changes in daily weather extremes. However, it is very likely that frequencies and intensities of summer heat waves will increase throughout Europe; likely that intense precipitation events will increase in frequency, especially in winter, and that summer drought risk will increase in central and southern Europe; and possible that gale frequencies will increase.

Sea Level

- Global-mean sea level rises by the 2050s by 13–68 cm. These estimates make no allowance for natural vertical land movements. Owing to tectonic adjustments following the last glaciation, there are regional differences across Europe in the natural rates of relative sea-level change.

Scenarios based on GCM simulations are the most widely adopted in impact studies reported from Europe. Some of these studies employ scenarios that are based on equilibrium 2xCO_2 model simulations that were conducted during the 1980s (e.g., Smith *et al.*, 1996; Tarand and Kallaste, 1998). The performance of some of these GCMs at simulating current climate over Europe was examined by Smith and Pitts (1997) and Kalvová and Nemesová (1997). Among the models considered, they

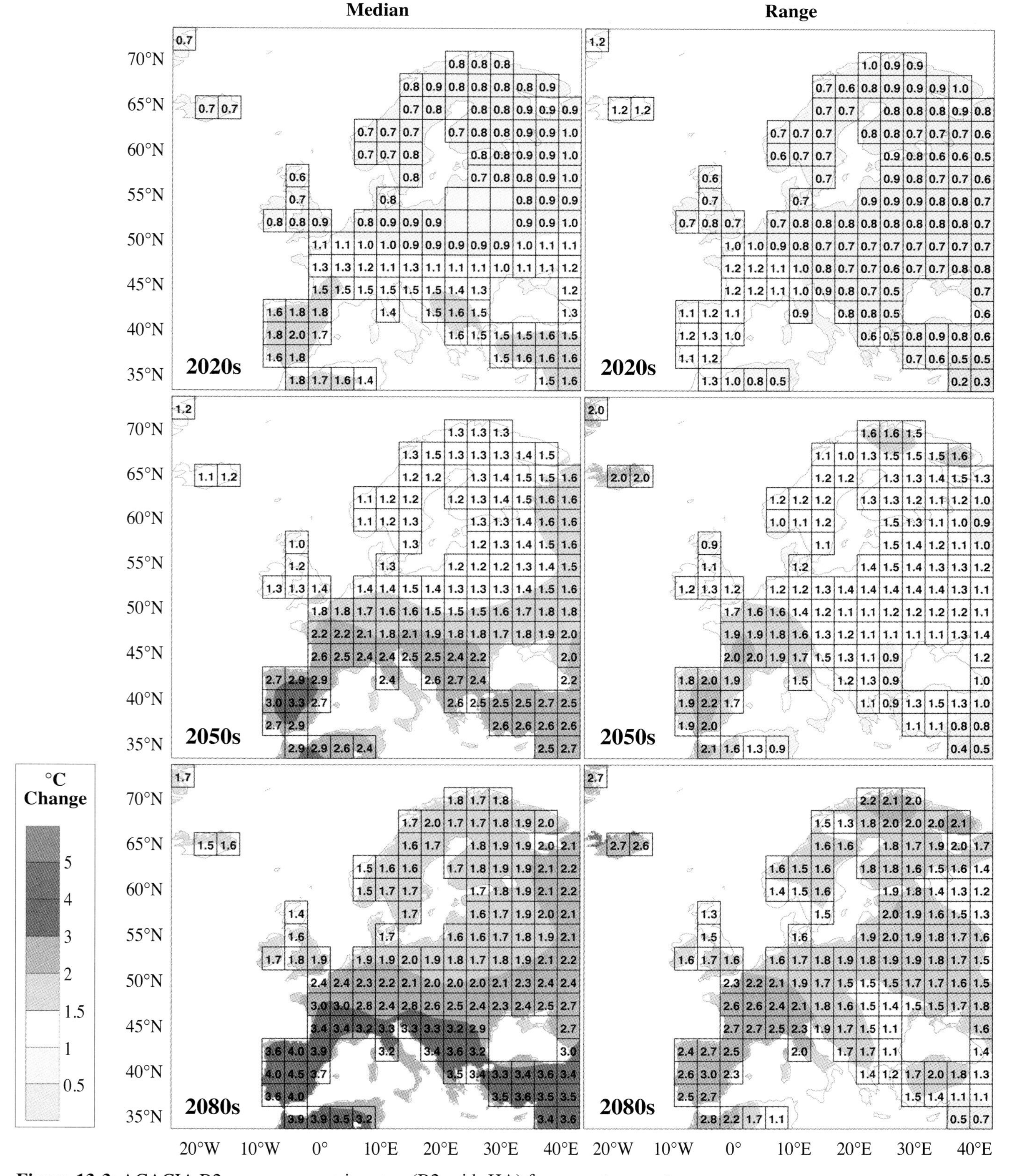

Figure 13-3: ACACIA B2 summer scenario maps (B2-mid, JJA) for mean temperature.

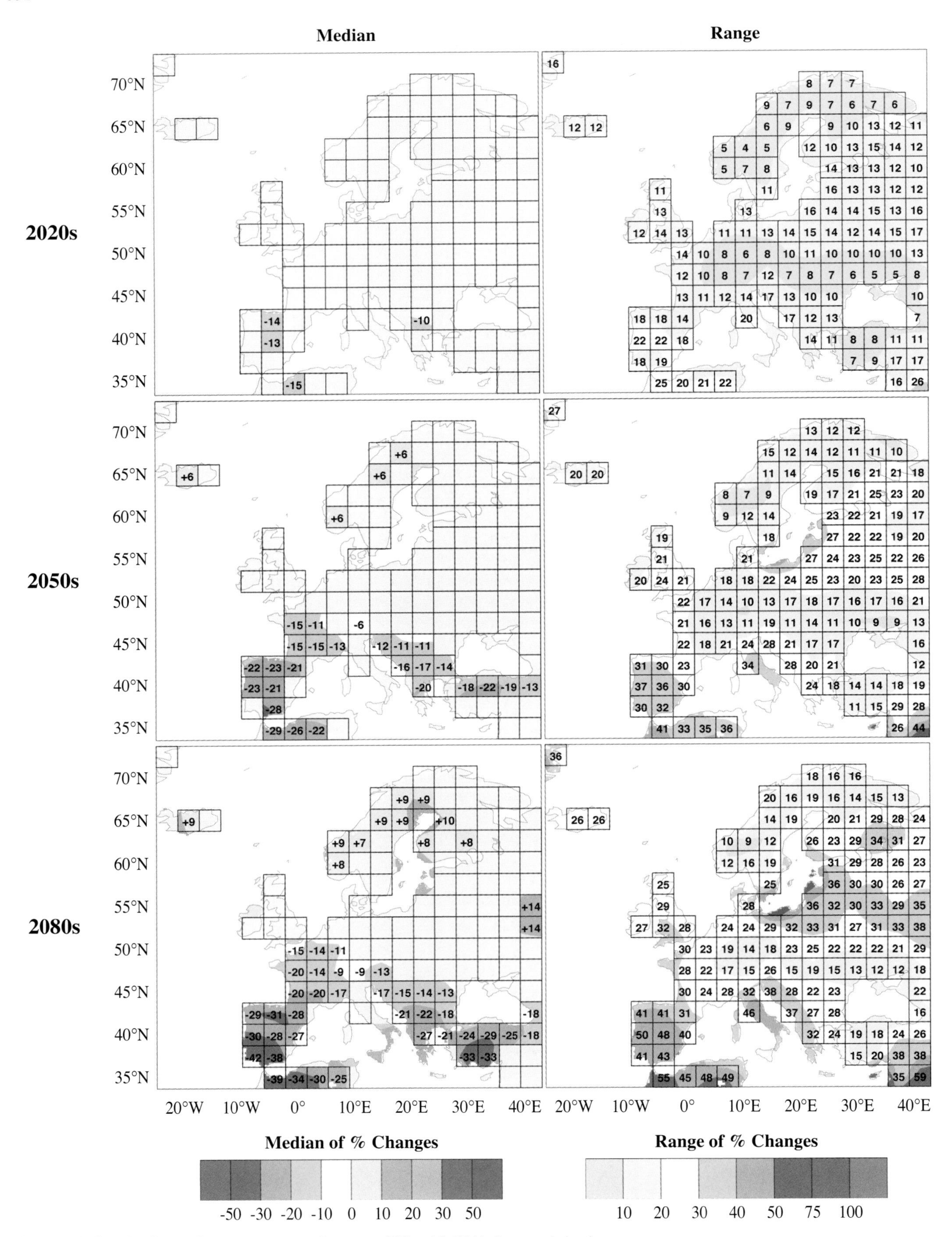

Figure 13-4: ACACIA B2 summer scenario maps (B2-mid, JJA) for precipitation.

found that the Goddard Institute for Space Studies (GISS) and Canadian Centre for Climate Modeling and Analysis (CCCM) models best simulated current temperature, whereas the GISS and UK89 models best simulated precipitation in northern regions and the CCCM model best simulated precipitation in southern regions.

Scenarios from the earliest transient-response experiments with coupled atmosphere-ocean GCMs (AOGCMs), which ignored historical greenhouse gas (GHG) forcing, were adopted in several studies that used direct model outputs (e.g., Harrison *et al.*, 1995b) or modified the outputs to account for historical forcing, using simple climate models (e.g., Carter *et al.*, 1996a; UK Department of the Environment, 1996). Scenarios from transient-response experiments that explicitly account for historical forcing have been adopted in the most recent impact studies (e.g., Arnell, 1999; Harrison and Butterfield, 1999; Hulme *et al.*, 1999). Some of these scenarios also incorporate aerosol effects, ensemble simulations, and multidecadal climatic variability. Many of these GCM results are lodged with the IPCC Data Distribution Centre (DDC) and were used in developing the European scenarios of A Concerted Action Towards A Comprehensive Climate Impacts and Adaptations Assessment for the European Union (ACACIA).

13.1.5.2. Scenarios for Europe

The climate change scenarios summarized here originally were prepared for the European ACACIA project and subsequently developed further for the IPCC (Hulme and Carter, 2000). The method by which they were developed is briefly described in Chapter 3 of this volume and more fully in Carter *et al*. (2000). These scenarios define a range of future European climates that embrace some of the major uncertainties in future climate prediction. The scenarios were placed in the context of model estimates of the natural variability of European climate. The baseline period selected was 1961–1990; changes in mean 30-year climates were calculated for the periods centered on the 2020s (2010–2039), the 2050s (2040–2069), and the 2080s (2070–2099). Each climate scenario is based on one of the four preliminary SRES98 marker emissions scenarios from the IPCC *Special Report on Emissions Scenarios* (SRES) (Nakicenovic, 2000).

For each scenario, season, variable, and time-slice, two maps were constructed (see Figures 13-3 and 13-4). One map shows the median change from the sample of eight standardized and scaled GCM responses; the other map shows the absolute range of these eight responses. The idea of signal-to-noise ratios in these regional responses was introduced by comparing median scaled-GCM changes against an estimate of natural multidecadal variability derived from the 1,400-year unforced climate simulation made with the HadCM2 model (Stott and Tett, 1998). In the maps showing median changes, only values *exceeding* the 2-standard deviation estimate of natural multidecadal variability are plotted. Figures 13-3 and 13-4 show an example of this information for the B2-mid scenario for summer temperature and precipitation. A complete set of illustrations appears in the report of the European ACACIA project (Hulme and Carter, 2000).

To condense this scenario information further, national-scale summary graphs for each European country or groups of countries have been calculated (see Figure 13-5 for an example covering Spain, Sweden, and Poland). Each country graph shows—for either winter or summer and for either the 2020s, the 2050s, or the 2080s—the distribution of mean changes in mean temperature and precipitation for each GCM simulation and for each scenario. As with the maps, these changes are compared with the natural multidecadal variability of temperature and precipitation extracted from the HadCM2 1,400-year unforced simulation. These graphs provide a quick assessment at a national scale of the likely range and significance of future climate change and the extent to which different GCMs agree with regard to their regional response to a given magnitude of global warming.

A contrasting future climate for Europe—the result of a rapid, nonlinear response of the climate system—has been suggested. This involves an abrupt collapse of the thermohaline circulation in the North Atlantic and consequent cooling in Europe at least for the first half of the 21st century (e.g., Alcamo *et al.*, 1994). Although this event has not been ruled out on theoretical grounds (see TAR WGI Chapter 11), it has not been simulated by any AOGCM (see TAR WGI Chapter 9) and therefore has not been included in this assessment.

13.1.6. Socioeconomic Scenarios for Europe

The four socioeconomic global futures (or SRES scenarios) described in Chapter 3 of this assessment have been characterized for Europe in the European ACACIA study (Jordan *et al.*, 2000), and are summarized below:

- Under a "World Market" (A1) scenario, the world becomes increasingly globalized, and materialist-consumerist social values predominate. Global societal values are primarily technocentric and short-termist. Nature therefore is assumed to be largely resilient to human stress. The emphasis is on pursuing economic growth in the narrow sense rather than sustainable development. Although rising income levels will make everyone in Europe richer, the poorest will gain relatively little. The EU functions as a single interconnected market, functionally integrated with other regional markets (e.g., in Asia and North America).
- Under a "Global Sustainability" (B1) scenario, Europe is run along more communitarian lines and within environmental resource limits. There is a strong emphasis on finding international solutions to globally interconnected problems. Thus, EU member states pool more and more of their sovereignty to address common environmental problems, the causes of which are considered to lie in the basic structure of

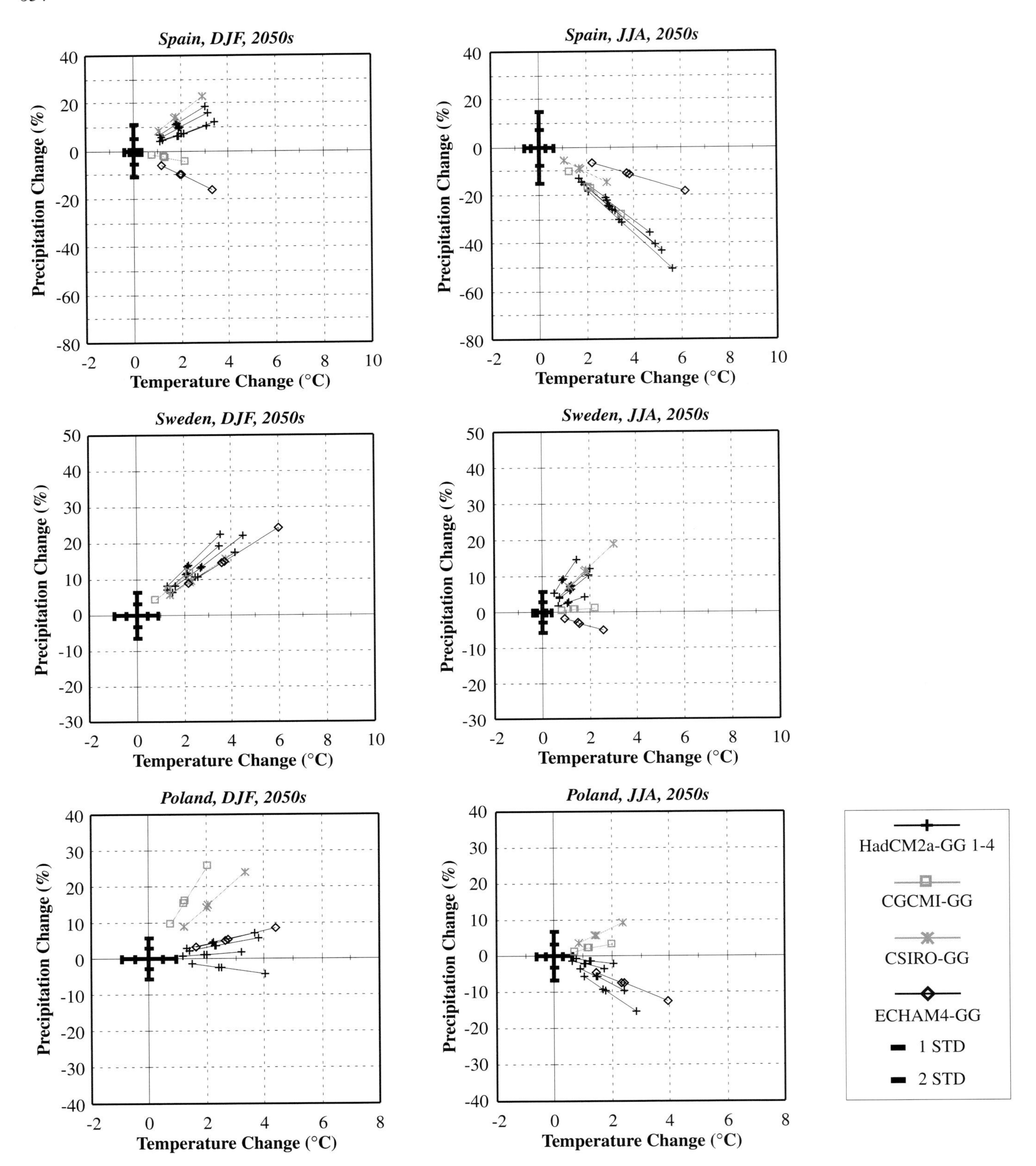

Figure 13-5: Scatter plot depicting scaled outputs of mean winter (left panels) and summer (right panels) temperature and precipitation change over land grid boxes representing Spain (top), Sweden (middle), and Poland (bottom) from each of seven GCM simulations (GFDL simulations are not shown because they extend only to 2025). Lines connect four points for each GCM simulation, each point representing standardized regional changes in climate from the GCM, linearly scaled according to global warming from each of the four ACACIA scenarios. The order of points along a line from the origin is the same for all plots: B1-low, B2-mid, A1-mid, A2-high. Also plotted are 1 and 2 standard deviation limits from 1,400-year HadCM2 unforced simulation, which is used to indicate natural multi-decadal variability.

society. The role of international institutions, including the EU, will extend from simply regulating environmental problems to tackling social inequality and social exclusion through adoption of social programs.

- In a "Provincial Enterprise" (A2) scenario, Europe is much more heterogeneous. Increasingly, the organization of society is dictated by short-term consumerist values. More policy decisions are taken at a national and subnational level. Europe adopts more protectionist economic and trade policies, which constrain innovation and stifle economic development, particularly in developing countries. Growth in global GDP is more modest than under the A1 scenario, and global inequality grows. Declining equity within and between member states of the EU produces tension and social exclusion. This typifies the uneasy tension in this scenario between the simultaneous desire for "free markets" (consumerist values) and protection of national state sovereignty. Politicians prioritize demands such as protecting the national economy and meeting short-term consumer demands for growth over environmental quality.
- In a "Local Sustainability" (B2) scenario, Europe is more committed to solving environmental problems by applying solutions that are attuned to local needs and circumstances. Thus, the dominant value system is more communitarian and ecocentric in nature, with greater commitment to longer term, strategic planning. In a sense, the shift in governance down to the local level accords with the principle of *subsidiarity*. National governments therefore are left to perform residual functions that cannot be undertaken at the subnational level. Small firms thrive under these conditions, whereas multinationals struggle to realign themselves to local needs. Overall, the world (including Europe) is more heterogeneous. Because of the lack of coordinated regional action, however, relative inequality may increase as local problems receive higher priority than those in other regions. Significantly, the environment benefits under this scenario, but not nearly as much as under B1 because of the limited extent of spatial coordination.

13.2. Key Regional Concerns

13.2.1. Water and Land Resources

13.2.1.1. Water Resources

Europe has a very diverse hydrological background, reflecting its varied climate and topography. In the south there is very significant variation in flow through the year, with long, dry summers. To the west there is less extreme variation, and in catchments underlain by absorbent aquifers flows remain reasonably constant through the year. In the north and east, much precipitation falls as snow, so much flow occurs during the spring snowmelt period. Major rivers such as the Rhine, Rhone, Po, and Danube distribute water from the "water tower" of the Alps. Superimposed on this varied hydrological base is a wide variety of water uses, pressures, and management approaches. A succession of floods and droughts has illustrated Europe's vulnerability to hydrological extremes. There are many other water-related pressures on Europe's environment, however (see the Dobris Assessment—Stanners and Bourdeau, 1995), such as increasing demand for water, particularly in the south, and subsequent increases in abstractions. River ecosystems and wetlands are increasingly at risk. The quality of Europe's rivers, lakes, and groundwater is being threatened by the discharge of sewage and industrial waste (often a legacy from past industrial development) and by excessive application of pesticides and fertilizers. At the same time, the institutional aspects of water resources management are changing. There is a shift in many countries from "supply-side" solutions to a "demand-side" approach, aimed at reducing demand for water or exposure to risk. Water managers in many European countries have adopted (at least in principle) a sustainable approach, and forthcoming EU directives are likely to encourage this further. Environmental demands are being taken increasingly seriously across most of Europe. Climate change adds another set of potential pressures on European water resources and their management.

13.2.1.1.1. Changes in hydrological cycle

Climate and land-use change influence the structure of the water balance (Ryszkowski and Kedziora, 1987; Ryszkowski *et al.*, 1990; Olejnik and Kedziora, 1991). Calculations at the European scale (Arnell, 1999) indicate that under most climate change scenarios, northern Europe would see an increase in annual average streamflow, but southern Europe would experience a reduction in streamflow. In much of mid-latitude Europe, annual runoff would decrease or increase by about 10% by the 2050s, but the change resulting from climate may be smaller than "natural" multidecadal variability in runoff. To the south and north, it may be substantially larger (Hulme *et al.*, 1999). Table 13-1 lists catchment-scale studies into potential hydrological changes in Europe that have been conducted since the IPCC's Second Assessment Report (SAR).

The consequences of climate change for the variation of flow through the year vary across Europe. In Mediterranean regions, climate change is likely to exaggerate considerably the range in flows between winter and summer. In maritime western Europe, the range also is likely to increase, but to a lesser extent. In more continental and upland areas, where snowfall makes up a large proportion of winter precipitation, a rise in temperature would mean that more precipitation falls as rain and therefore that winter runoff increases and spring snowmelt decreases. The timing of streamflow therefore alters significantly. Further east and at higher altitudes, most of the precipitation continues to fall as snow, so the distribution of flow through the year is altered little (Arnell, 1999). The substantial projected change in flow regime resulting from reduction in snowfall and snowmelt across large parts of CEE has been noted widely

Table 13-1: Catchment-scale studies into effects of climate change on runoff regimes in Europe (since SAR).

Country/Region	Reference
Albania	Bruci and Bicaj (1998)
Austria	Behr (1998)
Belgium	Gellens and Roulin (1998), Gellens *et al.* (1998)
Czech Republic	Dvorak *et al.* (1997), Hladny *et al.* (1997), Buchtele *et al.* (1998)
Danube	Starosolszky and Gauzer (1998)
Denmark	See Nordic region
Estonia	Bálint and Butina (1997), Jaagus (1998), Jarvet (1998), Roosaare (1998)
Finland	Vehviläinen and Huttunen (1997); see Nordic region
France	Mandelkern *et al.* (1998)
Germany	Daamen *et al.* (1998)
Greece	Panagoulia and Dimou (1996)
Hungary	Mika *et al.* (1997)
Latvia	Butina *et al.* (1998), Jansons and Butina (1998)
Nordic region	Sælthun *et al.* (1998)
Norway	See *Nordic region*
Poland	Kaczmarek *et al.* (1997)
Rhine basin	Grabs (1997)
Romania	Stanescu *et al.* (1998)
Russia	Kuchment (1998)
Slovakia	Pekárová (1996), Szolgay (1997), Hlavcová and Cunderlík (1998), Petrovic (1998), Hlavcová *et al.* (1999)
Spain	Avila *et al.* (1996)
Sweden	Xu (1998); see Nordic region
Switzerland	Seidel *et al.* (1998)
United Kingdom	Arnell (1996), Arnell and Reynard (1996), Reynard *et al.* (1998), Roberts (1998)

(e.g., Hladny *et al.*, 1996; Kasparék, 1998; Hlavcova and Cunderlik, 1998; Starosolszky and Gauzer, 1998).

Low-flow frequency generally will increase across most of Europe (Arnell, 1999), although in some areas where the minimum occurs during winter the absolute magnitude of low flows may increase because winter runoff increases: The season of lowest flow shifts toward summer. An implication of simulated changes in streamflow is that riverine flood risk generally would increase across much of Europe and that in some areas, the time of greatest risk would move from spring to winter. Effects on groundwater recharge (a major resource for many Europeans) are less clear because the general increase in winter rainfall may be offset by a reduction in the recharge season. Studies in the UK (Arnell and Reynard, 1999) and Estonia (Jarvet, 1998) indicate that groundwater recharge could be increased by climate change.

Implications of climate change for river water quality have been less well studied, but an increase in water temperature and widespread reductions in flow during summer are likely to lead to deterioration in many determinants of water quality (particularly dissolved oxygen concentrations). Jansons and Butina (1998) estimate nitrate and phosphate loads in a catchment in Latvia from streamflow; they use the relationships to infer an increase in nitrate and phosphate loads in winter (when flow increased) and a decrease in spring. Their model, however, did not account for possible temperature-related effects on nitrate and phosphate concentrations. Higher water temperatures are likely to increase the risk of blue-green algal blooms in rivers and lakes (e.g., Zalewski and Wagner, 1995).

13.2.1.1.2. Water management

The impacts of climate change on European water resources and their management depend not only on the change in hydrology but also (perhaps more particularly) on characteristics of the water management system. In general terms, the more stressed the system is under current conditions, the more sensitive it will be to climate change. Table 13-2 summarizes the key potential impacts on European water resources. These impacts should be considered against the background of other pressures and drivers on European water. Most of the potential impacts are self-explanatory.

Changes in the water resource base affect many sectors within Europe, including agriculture, industry, transport, power generation, the built environment, and ecosystems. Similarly, changes in many of these sectors will affect hydrology and the resource base. Changes in agricultural practices resulting from climate change, for example, will affect volumes and, more likely, quality of streamflow.

In many European countries, industrial water use has declined as a result of legislation, environmental protection, and economic change. The warmer climate may lead to significant increase in water demands in southern Europe—leading to possible overexploitation of groundwater resources, decrease of baseflow, and environmental degradation. Climate change is but one of the pressures facing European water managers. They will be played out against varying socioeconomic backgrounds, political demands for environmental improvements, and new trends in integrated water management. More detail on this and the foregoing conclusions appear in the water chapter of the European ACACIA report (Arnell, 2000).

13.2.1.2. Soils and Land Resources

Climate change will impact directly and, through land-use change, indirectly on a wide range of soil processes and properties that will determine the future ability of land to fulfill key functions that are important for all terrestrial ecosystems, as well as several socioeconomic activities, that underpin the well-being of society. The following subsections summarize key effects. Further information is summarized in the section on soils in the European ACACIA report (Rounsevell and Imeson, 2000).

Table 13-2*: Key impacts of climate change on European water resources.*

Sector	Potential Impacts	Sample Reference
Public water supply	– Reduction in reliability of direct river abstractions – Change in reservoir reliability (dependent on seasonal change in flows) – Reduction in reliability of water distribution network	Kaczmarek *et al.* (1996), Dvorak *et al.* (1997)
Demand for public water supplies	– Increasing domestic demand for washing and out-of-house use	Herrington (1996)
Water for irrigation	– Increasing demand – Reduced availability of summer water – Reduced reliability of reservoir systems	Kos (1993), Alexandrov (1998)
Power generation	– Change in hydropower potential through the year – Altered potential for run-of-river power – Reduced availability of cooling water in summer	Grabs (1997), Sælthun *et al.* (1998)
Navigation	– Change (reduction?) in navigation opportunities along major rivers	Grabs (1997)
Pollution risk and control	– Increased risk of pollution as a result of altered sensitivity of river system	Mänder and Kull (1998)
Flood risk	– Increased risk of loss and damage – Increased urban flooding from overflow of storm drains	Grabs (1997)
Environmental impacts	– Change in river and wetland habitats	

13.2.1.2.1. Soil physical properties

Soil water contents respond rapidly to variability in the amounts and distribution of precipitation or the addition of irrigation. Temperature changes affect soil water by influencing evapotranspiration, and plant water use is further influenced by elevated CO_2 concentrations, leading to lower stomatal conductance and increased leaf photosynthetic rates (Kirschbaum *et al.,* 1996). Soil water contents are highly variable in space (Rounsevell *et al*., 1999), so it is difficult to generalize about specific climate impacts.

Climate change can be expected to modify soil structure through the physical processes of shrink-swell (caused by wetting and drying) and freeze-thaw, as well as through changes in soil organic matter (SOM) contents (Carter and Stewart, 1996). Compaction of soils results from inappropriate timing of tillage operations during periods when the soil is too wet to be workable. Soil workability has a strong influence on the distribution and management of arable crops in temperate parts of Europe (Rounsevell, 1993; Rounsevell and Jones, 1993). Therefore, wet areas with heavy soils could benefit from climate change (MacDonald *et al*., 1994; Rounsevell and Brignall, 1994). In a similar way, grassland systems can suffer poaching by grazing livestock (i.e., damage caused by animal hooves) (Harrod, 1979). Thus, drier soil conditions for longer periods of the year would affect the distribution of intensive agricultural grassland in temperate Europe (Rounsevell *et al*., 1996a) and may result in intensification of currently wet upland grazing areas.

Soils with large clay contents shrink as they dry and swell when they become wet again, forming large cracks and fissures. Drier climatic conditions will increase the frequency and size of crack formation in soils, especially those in temperate regions of Europe, which currently do not reach their full shrinkage potential (Climate Change Impacts Review Group, 1991, 1996). Soils that shrink and swell cause damage to building foundations through subsidence, creating a problem for householders and the housing insurance industry (Building Research Establishment, 1990). Crack formation also results in more rapid and direct movement of water and solutes from surface soil to permeable substrata or drainage installations through bypass flow (Armstrong *et al.,* 1994; Flurry *et al.,* 1994). This will decrease the filtering function of soil and increase the possibility of nutrient losses and water pollution (Rounsevell *et al*., 1999).

13.2.1.2.2. Land degradation processes

Accumulation of salts in soils (salinization) results from capillary movement and dispersion of saline water because evapotranspiration is greater than precipitation and irrigation

(Vàrallyay, 1994). Such conditions, which are widespread throughout the warmer and drier regions of southern Europe, will be exacerbated by temperature rise coupled with reduced rainfall. Climate change also will increase flood incidence and salinity along coastal regions, through the influence of sea-level rise (Nicholls, 2000).

A decrease in precipitation and/or increase in temperature increases oxidation and loss of volume in lowland peat soils that are used for agriculture. It has been suggested that under climate change, the volume of peats in agricultural use will shrink by 40% (Kuntze, 1993). Some peat soils in western Europe are associated with acid sulfate conditions (Dent, 1986); strong acidity largely precludes agricultural use (Beek *et al.*, 1980). Soil acidification also can result from depletion of basic cations through leaching (Brinkman, 1990) where the soil is well drained and structurally stable and experiences high rainfall amounts and intensity—as in many upland areas of Europe. In wetter climate, soil acidification could increase if buffering pools become exhausted, although for most soils this will take a very long time.

Climate change is likely to increase wind and water erosion rates (Rosenberg and Tutwiler, 1988; Dregne, 1990; Botterweg, 1994), especially where the frequency and intensity of precipitation events grows (Phillips *et al.*, 1993). Erosion rates also will be affected by climate-induced changes in land use (Boardman *et al.*, 1990) and soil organic carbon contents (Bullock *et al.*, 1996). Relatively small changes in climate may push many Mediterranean areas into a more arid and eroded landscape (Lavee *et al.*, 1998) featuring decreases in organic matter content, aggregate size, and stability and increases in sodium adsorption ratio and runoff coefficient. However, increased erosion in response to climate change cannot be assumed for all parts of Europe. For example, in upland grazed areas, erosion rates will be reduced as a result of better soil surface cover and topsoil stability arising from higher temperatures that extend the duration of the growing season and reduce the number of frosts (Boardman *et al.*, 1990).

13.2.1.2.3. Biologically mediated soil properties

Climate change will impact directly on SOM through temperature and precipitation (Tinker and Ineson, 1990; Cole *et al.*, 1993; Pregitzer and Atkinson, 1993) and indirectly (and possibly more importantly) through changing land use (e.g., Hall and Scurlock, 1991). SOM contents increase with soil water content and decrease with temperature (Post *et al.*, 1982, 1985; Robinson *et al.*, 1995), although rates of decomposition vary widely between different soil carbon pools (van Veen and Paul, 1981; Parton *et al.*, 1987; Jenkinson, 1990). Changes in SOM contents depend on the balance between carbon inputs from vegetation and carbon losses through decomposition (Lloyd and Taylor, 1994); most SOM is respired by soil organisms within a few years. Net primary productivity (NPP) usually increases with increasing temperature and elevated atmospheric CO_2, leading to greater returns of carbon to soils (Loiseau *et al.*, 1994). However, increasing temperature strongly stimulates decomposition (Berg *et al.*, 1993; Lloyd and Taylor, 1994; Kirschbaum, 1995) at rates that are likely to outstrip NPP and lead to reduced SOM contents (Kirshbaum *et al.*, 1996). This effect will be strongest in cooler regions of Europe, where decomposition rates currently are slow (Jenny, 1980; Post *et al.*, 1982; Kirschbaum, 1995). Conversely, excess soil water resulting from increased precipitation will reduce decomposition rates (Kirschbaum, 1995) and thus increase SOM contents.

Plant growth and soil water use are strongly influenced by the availability of nutrients. Where climatic conditions are favorable for plant growth, the shortage of soil nutrients will have a more pronounced effect (Shaver *et al.*, 1992). Increased plant growth in a CO_2-enriched atmosphere may rapidly deplete soil nutrients; consequently, the positive effects of CO_2 increase may not persist as soil fertility decreases (Bhattacharya and Geyer, 1993). Increased SOM turnover rates over the long term are likely to cause a decline in soil organic nitrogen in temperate European arable systems (Bradbury and Powlson, 1994), although, in the short term, increased returns of carbon to soils would maintain soil organic nitrogen contents (Pregitzer and Atkinson, 1993; Bradbury and Powlson, 1994). Greater mineralization may cause an increase in nitrogen losses from the soil profile (e.g., Kolb and Rehfuess, 1997; Lukewille and Wright, 1997), although there is evidence to suggest that temperature-driven, increased nitrogen uptake by vegetation may reduce these losses (Ineson *et al.*, 1998).

There is great uncertainty surrounding the response of soil community function to global change and the potential effects of these responses at the ecosystem level (Smith *et al.*, 1998). Most soil biota have relatively large temperature optima and therefore are unlikely to be adversely affected by climate change (Tinker and Ineson, 1990), although some evidence exists to support changes in the balance between soil functional types (Swift *et al.*, 1998). Soil organisms will be affected by elevated atmospheric CO_2 concentrations where this changes litter supply to and fine roots in soils, as well as by changes in the soil moisture regime (Rounsevell *et al.*, 1996b). Furthermore, the distribution of individual species of soil biota will be affected by climate change where species are associated with specific vegetation and are unable to adapt at the rate of land-cover change (Kirschbaum *et al.*, 1996).

13.2.1.3. Coastal Zones

Coastal zones in Europe contain large human populations and significant socioeconomic activity. They also support diverse ecosystems that provide significant habitats and sources of food. Significant inhabited coastal areas in countries such as The Netherlands, England, Denmark, Germany, Italy, and Poland already are below normal high-tide levels, and more extensive areas are vulnerable to flooding from storm surges. Hard defenses to prevent such flooding, combined with the loss of the seaward edge of coastal habitats as a result of existing rates of sea-level rise, already are causing significant coastal squeeze in many locations (e.g., Pye and French, 1993; Rigg *et al.*, 1997; Lee, 1998). Deltaic areas often are particularly threatened because they naturally subside and may have been sediment-starved by dam construction (e.g., Sanchez-Arcilla *et al.*, 1998). Other nonclimate change factors such as pollution may condition the impacts of climate change. Information further to the summary given below appears in the chapter on coasts in the European ACACIA report (Nicholls, 2000).

Climate change could cause important impacts on coastal zones, particularly via sea-level rise and changes in the frequency and/or intensity of extreme events such as storms and associated surges. Under the SRES climate change scenarios, global sea level is expected to rise by 13–68 cm by the 2050s. Regional and local sea-level rise in Europe generally will differ from the global average because of vertical land movements (glacial isostatic rebound, tectonic activity, and subsidence). Deviations from the global mean sea level also will occur as a result of oceanic effects such as changes in oceanic circulation, water density, or wind and pressure patterns. Mediterranean sea levels have fallen by as much as 20 mm relative to the Atlantic since 1960, probably as a result of declining freshwater input and consequent seawater density increase (Tsimplis and Baker, 2000). Looking to the future, the net effect of these processes is likely to be as much as 10% of global mean change to the 2080s (Gregory and Lowe, 2000).

Sea-level rise can cause several direct impacts, including inundation and displacement of wetlands and lowlands, coastal erosion, increased storm flooding and damage, increased salinity in estuaries and coastal aquifers, and rising coastal water tables and impeded drainage (Bijlsma *et al.*, 1996). Potential indirect impacts are numerous; they include changes in the distribution of bottom sediments, changes in the functions of coastal ecosystems, and a wide range of socioeconomic impacts on human activities.

Other climate change factors also may be important. For example, rising air and sea temperatures may cause significant shifts in the timing and location of tourism (Perry, 1999) and recreational and commercial fisheries and decrease the incidence of sea ice during winter. These changes also may influence water quality through the occurrence of algal blooms, which would have adverse effects on tourism and human health (Kovats and Martens, 2000). Changes in the frequency and track of extratropical storms are less certain. It is worth noting that an analysis of the HadCM2 climate change simulations found a decrease in the number of northern hemisphere storms, but with a tendency for deeper low centers (Carnell and Senior, 1998). This would have important implications for coastal areas, including an additional increase in flood risk. Several studies suggest that storm surges in northwest Europe might change as a result of climate change (von Storch and Reichardt, 1997; Flather and Smith, 1998; Lowe and Gregory, 1998), but further investigation is required to produce definitive results. Storm occurrence has displayed significant interannual and interdecadal variability over the past 100 years (WASA, 1998); this could produce important and costly impacts without other changes (e.g., Peerbolte *et al.*, 1991) and might interact adversely with sea-level rise.

The impacts of sea-level rise would vary from place to place and would depend on the magnitude of relative sea-level rise, coastal morphology/topography, and human modifications. The most threatened coastal environments within Europe are deltas, low-lying coastal plains, islands and barrier islands, beaches, coastal wetlands, and estuaries (Beniston *et al.*, 1998). Tidal range is a key factor: In general, the smaller the tidal range, the greater the susceptibility to a given rise in sea level. The Mediterranean and Baltic coasts have a low tidal range (<1 m), which suggests that they will be more vulnerable to sea-level rise than the Atlantic Ocean and North Sea coasts (Nicholls and Mimura, 1998).

A regional/global model of flood and coastal wetland losses described by Nicholls *et al.* (1999) considers the interacting effects of sea-level rise, population growth, and improvements in protection standards. All other climate factors are assumed to be constant. This model allows the impacts of the SRES scenarios on Europe [excluding the former Soviet Union (FSU)] to be explored. Because increases in population and protection standards in Europe are minor, the major changes are caused by sea-level rise. In 1990, about 25 million people were estimated to live beneath the 1-in-1,000 year storm surge, with the largest exposure along the Atlantic/North Sea seaboard. However, these people generally are well protected from flooding now. The

***Table 13-3**: Estimates of flood exposure and incidence for Europe's coasts in 1990 and the 2080s (new runs using model described by Nicholls et al., 1999). Estimates of flood incidence are highly sensitive to assumed protection standard and should be interpreted in indicative terms only. Former Soviet Union is excluded.*

		Flood Incidence	
Region	***1990*** **Exposed Population (millions)**	***1990*** Average Number of People Experiencing Flooding (thousands yr^{-1})	***2080s*** Increase due to Sea-Level Rise, Assuming No Adaptation (%)
Atlantic coast	19.0	19	50 to 9,000
Baltic coast	1.4	1	0 to 3,000
Mediterranean coast	4.1	3	260 to 120,000

***Table 13-4**: Estimated coastal wetland losses by region in Europe by the 2080s (new runs using model described by Nicholls et al., 1999). Range of losses reflects range of SRES sea-level rise scenarios and uncertainty about wetland response to sea-level rise. Losses from other causes, such as direct human destruction, are likely. Former Soviet Union is excluded.*

Region	Minimum Wetland Stock in 1990 (km²)			Range of Losses by the 2080s (%)
	Saltmarsh	Unvegetated Intertidal Areas	Total	
Atlantic coast	2,306	6,272	8,578	0 to 17
Baltic coast	226	271	497	84 to 98
Mediterranean coast	347	136	483	31 to 100

changes in flooding, shown in Table 13-3, indicate a significant increase in the incidence of coastal flooding by the 2080s, assuming no adaptation, particularly around the Mediterranean.

Europe (excluding the FSU) is estimated to have at least 2,860 km² of saltmarshes and 6,690 km² of other unvegetated intertidal habitat, mainly composed of sites recognized in the Ramsar treaty. Based on coastal morphological type and the presence or absence of coastal flood defenses, Table 13-4 shows wetland losses resulting from sea-level rise. Wetland losses are most significant around the Mediterranean and Baltic. Under the A2-high scenario, wetlands in these regions could be eliminated. Any surviving wetlands may be substantially altered. Such losses could have serious consequences for biodiversity in Europe, particularly for wintering shorebird and marine fish populations.

Available national results emphasize the large human and ecological values that could be affected by sea-level rise. Table 13-5 shows results of national assessments in The Netherlands (Baarse *et al.*,1994; Bijlsma *et al.*, 1996), Poland (Zeidler, 1997), and Germany (Sterr and Simmering, 1996; Ebenhöh *et al.*, 1997) for existing development and all costs adjusted to 1990 US$. In Table 13-5, adaptation assumes protection except in areas with low population density. People at risk are the numbers of people flooded by storm surge in an average year. Adaptation/protection costs for Poland include capital and annual running costs; % GNP assumes that costs are all incurred in 1 year. Subnational and local studies from East Anglia, UK (Turner *et al.*, 1995); South Coast, UK (Ball *et al.*, 1991); Rochefort sur Mer, France (Auger, 1994); Estonia (Kont *et al.*, 1997); and Ukraine (Lenhart *et al.*, 1996), as well as regional reviews (Tooley and Jelgersma, 1992; Nicholls and Hoozemans, 1996) also support this conclusion. Many of Europe's largest cities—such as London, Hamburg, St. Petersburg, and Thessaloniki—are built on estuaries and lagoons (Frasetto, 1991). Such locations already are exposed to storm surges, and climate change is an important factor to consider for long-term planning and development.

Other values that may be affected include archaeological and cultural resources at the coast; these resources sometimes are being recognized only now (Fulford *et al.*, 1997; Pye and Allen, 2000). In Venice, a 30-cm relative rise in sea level in the 20th century has greatly increased the frequency of flooding and damage to this unique medieval city; solutions to this problem are the subject of a continuing debate and need to consider climate change (Consorzio Venezia Nuova, 1997; Penning-Rowsell *et al.*, 1998).

13.2.1.4. Mountains and Subarctic Environments

Mountain regions are characterized by sensitive ecosystems, enhanced occurrences of extreme weather events, and natural catastrophes. Often regarded as hostile and economically nonviable regions, mountains have attracted major economic investments for tourism, hydropower, and communication routes. The projected amplitude and rate of climatic change in coming decades is likely to lead to significant perturbations of natural systems as well as the social and economic structure of mountain societies, particularly where these are marginal. Because in many instances mountains and uplands are regions of conflicting interests between economic development and environmental conservation, shifts in climatic patterns probably will exarcerbate the potential for conflict (Beniston, 2000).

Nested GCM-regional climate model (RCM) techniques for a 2xCO_2 scenario have shown that the European Alps are likely to experience slightly milder winters, with more precipitation, than recently. Summer climate, however, may be much warmer and drier than today, as a result of northward shift of the Mediterranean climatic zones (Beniston *et al.*, 1995). These conditions are likely to have adverse effects on the alpine cryosphere and ecosystems.

Impacts of climatic change on physical systems will affect water, snow, and ice, and shifts in extremes will lead to changes in the frequency and intensity of natural hazards. Water availability in some regions may decline because of a reduction in precipitation amounts and because of reduced snow-pack and shorter snow season. Changes in snow amount will lead to significant shifts in the timing and amount of runoff in European river basins, most of which originate in mountains and uplands; this will have numerous consequences for the more populated lowland regions. Floods and droughts are likely to become more frequent. Indirect impacts include changes in erosion and sedimentation patterns, perhaps disrupting hydropower plants (Beniston *et al.*, 1996).

In most temperate mountain regions, the snowpack is close to its melting point, so it is very sensitive to changes in temperature. As warming progresses in the future, current regions of snow

***Table 13-5**: Impacts of sea-level rise in selected European countries, assuming no adaptation, plus adaptation costs (from Nicholls and de la Vega-Leinert, 2000).*

Country	Sea-Level Rise Scenario (m)	Coastal Floodplain Population		Population Flooded per Year		Capital Value Loss		Land Loss		Wetland Loss (km^2)	Adaptation Costs	
		# 10^3	% total	# 10^3	% total	US$ 10^9	% GNP	km^2	% total		US$ 10^9	% GNP
Netherlands	1.0	10,000	67	3,600	24	186	69	2,165	6.7	642	12.3	5.5
Germany	1.0	3,120	4	257	0.3	410	30	n.a.	n.a.	2,400	30	2.2
Poland	0.1	n.a.	n.a.	25	0.1	1.8	2	n.a.	n.a.	n.a.	0.7	2.1
Poland	0.3	n.a.	n.a.	58	0.1	4.7	5	845	0.25	n.a.	1.8	5.4
Poland	1.0	235	0.6	196	0.5	22.0	24	1,700	0.5	n.a.	4.8	14.5
Estonia	1.0	47	3	n.a.	n.a.	0.22	3	>580	>1.3	225	n.a.	n.a.
Turkey	1.0	2450	3.7	560	0.8	12	6	n.a.	n.a.	n.a.	20	10

precipitation increasingly will experience precipitation in the form of rain. For every 1°C increase in temperature, the snowline rises by about 150 m; as a result, less snow will accumulate at low elevations than today, although there could be greater snow accumulation above the freezing level because of increased precipitation in some regions. A warmer climate will lead to the upward shift of mountain glaciers, which will undergo further reductions. It is likely that in the 21st century, 30–50% of alpine glaciers will disappear (Haeberli, 1995). Permafrost also could be significantly perturbed by warming, leading to a reduction of slope stability and a consequent increase in the frequency and severity of rock and mudslides. These in turn would have adverse economic consequences for mountain communities.

According to Sætersdal and Birks (1997), mountain plants with narrow July and January temperature tolerances (typically centric species) are most vulnerable to climate warming. These species are characterized by small ranges and population sizes. Holten (1998) documents that these less common centric species have narrow altitudinal ranges—most between 400–600 m in southern Scandes (the Fennoscandian mountain range)—compared with widely distributed mountain plants that have vertical ranges of 800–1,800 m.

Temperature enhancement experiments in the northern Scandes (Henry and Molau, 1997; Molau and Alatalo, 1998) have shown disintegration of present plant communities; "arctic specialists," such as *Cassiope tetragona* and *Diapensia lapponica,* are least responsive to warming, but soon will suffer competitive exclusion from more competitive evergreen and deciduous dwarfshrubs such as *Empetrum hermaphroditum, Vaccinium vitis-idaea, Betula nana,* and *Salix spp*. The latter species are most favored by climatic amelioration (see also Jonasson *et al.*, 1996; Graglia *et al.*, 1997). The following vegetation types are regarded as most sensitive to climate change in the Scandes:

- *High-alpine fell-field vegetation:* Present plant cover is discontinuous. With available soil resources in combination with extensive seed rain (Molau and Larsson, 2000), rapid colonization by mid-alpine vegetation is expected.
- *Mid-alpine vegetation:* Because of a longer thaw season, spatial cover of snowbed communities—including species with high sensitivity to frost and drought—is anticipated to decrease rapidly.
- *Vegetation on cryosoils:* As a result of anticipated accelerated degradation of patchy permafrost, the discontinuous vegetation of wet cryosoils (patterned ground and tussock tundra) may be replaced rapidly by low-alpine heath scrub.

As a result of a longer growing season and higher temperatures, European alpine areas will shrink because of upward migration of tree species. After several centuries of invasion of forest into the alpine Scandes, the current alpine area might be reduced by as much as 40–60% (Holten, 1990; Holten and Carey, 1992). The speed and extent of upward migration will depend on species as well as physiographic conditions and climatic regimes. Fairly quick response is expected from pioneer species such as mountain birch *(Betula pubescens ssp. tortuosa)*. However, there seem to be competing explanations for the response time of migration of the timberline in European mountain ranges and how far timberlines will shift under specific climate scenarios (Woodward, 1992). Under optimal topographic/edaphic conditions, the mountain birch, Scots pine, and Norway spruce treelines might be elevated by as much as 300 m in the continental Scandes and less on the coastal slopes. This will probably take several hundreds of years, at least for the more slowly responding Scots pine and Norway spruce (Aas and Faarlund, 1995; Kullman, 1995). The interaction between climate change, acidification, and nitrogen deposition in alpine ecosystems in the Scandes certainly is very important; to date, however, it has been more or less overlooked (Keller *et al.*, 2000).

The predicted future climate suggests that the ranges of many species will extend to higher altitudes. In the topmost zone of the Middle Mountains and the lower external Alps, competition from closed-canopy forest of Norway spruce might restrict the area of small islands of arctic-alpine tundra, destroy patterned grounds and subarctic mires, and kill relict and endemic populations of arctic-alpine organisms (Jenik, 1997). Mountain ecosystems that are particularly vulnerable to climate change in Italy include, for example, shrub vegetation with *Pinus mugo* of the Apennines (Vaccinio-Piceetalia). This grows 1,500–2,300 m above sea level on glacial residue and is closely dependent on cold continental climate with long-lasting snow cover (Second National Communication on Climate, 1997).

Evidence from past climate changes indicates that species respond by migrating rather than by adapting genetically (Huntley, 1991). According to Scharfetter (1938), the warmest interglacial periods enabled forests to climb higher toward the summits of low mountains (1,800–2,300 m), thereby reducing high-elevation orophyte populations. This is relevant for many isolated endemics and orophytes that presently are living in refugia, such as tops of low mountains in the Alps. In such habitats, they will have no possibility to migrate upward, either because they cannot move rapidly enough or because the nival zone already is absent (Gottfried *et al.*, 1994; Grabherr *et al.*, 1994, 1995).

There is broad agreement that past climatic changes have had a strong impact on the distribution ranges of species, and the same can be expected in the future (Peters and Darling, 1985; Ozenda and Borel, 1991, 1995). However, some of these biotic changes are subject to considerable inertia, especially with long-lived plants such as trees. For treelines to expand upslope, a significantly warmer climate is required for at least 100 years (Holtmeier, 1994). Based principally on palynological and macro-fossil investigations, the forest limit did not extend upward more than 100–300 m during the warmest periods of the Boreal and Atlantic periods in the Holocene (e.g., Bortenschlager, 1993; Lang, 1993; Wick and Tinner, 1997). An increase in mean annual temperature of 1–2°C may not shift the present forest limit upward by much more than 100–200 m in the Alps.

An increase of 1–2°C is still likely to be in the range of tolerance of most alpine and nival species (Körner, 1995; Theurillat, 1995), whereas a greater increase (3–4°C) may not be (Theurillat, 1995; Lischke *et al.*, 1998; Theurillat *et al.*, 1998). This is particularly relevant for endemics and orophytes with widespread distributions throughout the Alps. Where ranges of species already are fragmented they may become even more fragmented, with regional disappearances if they cannot persist, adapt, or migrate. Some categories of vulnerable plants—for instance, isolated arctic, stenoicous, relict species that are pioneers in wet habitats—may disappear. Specialists in distinct relief situations can suffer from habitat loss through lack of suitable escape routes, as observed from modeling studies (Pauli *et al.*, 1999; Gottfried *et al.*, 2000). Such effects can be pronounced—for instance, in the northeastern Alps, where high numbers of endemics occur in narrow altitudinal ranges (Grabherr *et al.*, 1995).

Biogeographically, the Pyrenees are on the edge between the alpine, central European, and Mediterranean regions. In the eastern Pyrenees around Puigmal (2,910 m), some oro-Mediterranean communities occur, as well as some pasture types with the rude *Festuca supina* (Baudiere and Serve, 1974). On windier and drier areas, an ecological substitution is taking place: Alpine dense pastures are turning to discontinuous oro-Mediterranean communities. There is ongoing degradation of formerly stable soils and communities as a result of periglacial phenomena, including increased cryoturbation (Baudiere and Gauquelin, 1998). With regard to timberlines in the Pyrenees, Montserrat (1992) demonstrates that they have been moving upward in the postglacial period up to the present. Over 1940–1985, an increase in mean annual temperatures is suggested by changes in animal behavior. In Estangento Lake (eastern Pyrenees, 2,035 m), this happens especially in winter months, when the minimum temperatures reach 3°C. This phenomenon affects the hibernation time of a common bat *(Miniopterus shreibersi)*, so that populations enter the caves 1.5 months later than they did 20 years ago. Caves are regarded as stable environments that reflect only general climatic trends (White and Martinez Rica, 1996).

13.2.2. Semi-Natural Ecosystems and Forests

13.2.2.1. Forests

European forests belong to an important economic sector that is potentially affected by climate change and changes in atmospheric CO_2 concentrations. Forests also have important interactions with global change processes as a result of their sink potential.

Primarily, temperature and the availability of soil moisture limit the natural range of European tree species. Some forests (particularly in the north) also are nutrient-limited. The structure and composition of many forests is further influenced by the natural disturbance regime (e.g., fire, insects, windthrow). Most European forests are managed for one or several purposes, such as timber production, water resources, or recreation. This management has reduced forest area or strongly modified forest structure in most of Europe, and presently existing forests often consist of species that are different from those that would occur naturally.

In northern Europe, boreal forests are dominated by *Picea abies* and *Pinus sylvestris*, and these species grow well across most of their current distribution ranges. Under warmer conditions, these species are likely to invade tundra regions (Sykes and

Prentice, 1996). In the southern boreal forests, these species are expected to decline because of a concurrent increase of deciduous tree species (Kellomäki and Kolström, 1993). Most climate change scenarios suggest a possible overall displacement of the climatic zone that is suitable for boreal forests by 150–550 km over the next century (Kirschbaum *et al.*, 1996). This shift in climatic conditions would occur more rapidly than most species have ever migrated in the past (20–200 km per century—see Davis, 1981; Birks, 1989). Also questionable is whether soil structural development would be able to follow. Turnover of current tree populations may be enhanced, however, by changes in management practices and changing disturbance regimes, such as increased fire frequency or increased strong winds in late autumn and early spring (Solantie, 1986; Peltola *et al.*, 1999).

Under most recent climate change scenarios, winters are likely to be still cold enough to fulfill the chilling requirements of the main boreal tree species (Myking and Heide, 1995; Leinonen, 1996; Häkkinen *et al.*, 1998), but earlier budburst can be expected. If summer and winter precipitation increase (as indicated by some scenarios), boreal forests would become less susceptible to fire damages, which currently affect about 0.05% of forests yr^{-1} (Zackrisson and Östlund, 1991).

At present, cold winters in boreal and some temperate regions protect forests from many insects and fungi that are common further south (Straw, 1995). High summer temperatures and associated drought increase the growth of existing insect populations through enhanced physiological activity and turnover of insect populations. Throughout Europe, forests seem to be quite well buffered against new species coming from outside Europe, but the risk exists. A good example of new organisms with large potential to damage trees is *Bursaphelenchus xylophilus,* which originates in North America. This pine nematode is easily transported in fresh timber, but its success is related to temperature. Low summer temperatures and short growing seasons have effectively limited the success of this species in northern Europe (Tomminen and Nuorteva, 1987), although it frequently has occurred in imported timber. Reductions in these limitations may result in increased damage to trees.

In western and central Europe, the current forest structure and, in part, tree species composition are determined mainly by past land use and management rather than by natural factors (Ellenberg, 1986). Site-specific assessments of the future composition of near-natural forests suggests that conifers (e.g., *Picea abies*) may be replaced by deciduous species (e.g., *Fagus sylvatica*) at some sites (e.g., Kräuchi, 1995). Until recently, our capability to assess long-term forest dynamics at the regional scale was quite limited. Lindner *et al.* (1997) provided the first assessment of regional-scale patterns of forest composition under current and future climates. Their study suggests that future near-natural forests in the state of Brandenburg (east Germany) would be much more uniform, with little of the differentiation across different site types that shape today's landscape. A temperature increase of 1–3°C would advance budburst of many tree species by several weeks (Murray *et al.*, 1989). Introduction of phenologically suitable ecotypes or new species has been among the main tools to increase forest growth (Lines, 1987). Minimum winter temperatures seem to be critical for the survival of exotic species with insufficient winter frost hardiness—for example, *Nothofagus procera* in Britain (Cannell, 1985). Therefore, higher winter temperatures could broaden the potential distribution range of such species in Europe.

In southern Europe, most forests consist of sclerophyllous and some deciduous species that are adapted to summer soil water deficit. Climate scenarios indicate reduced water availability in the summer months and associated responses in forests (e.g., Gavilán and Fernández-González, 1997), although the interactions of this effect with enhanced CO_2 concentrations is uncertain. Temperature changes may allow expansion of some thermophilous tree species (e.g., *Quercus pyrenaica)* when water availability is sufficient. In the Pyrenees, a northward and upward movement of Mediterranean ecotypes is likely to occur with warming accompanied by drier conditions.

13.2.2.1.1. Growth trends

Forest growth has increased during the past several decades in northern forests (Lakida *et al.*, 1997; Lelyakin *et al.*, 1997; Myneni *et al.*, 1997) and elsewhere in Europe (Spiecker *et al.*, 1996). Climate warming, increasing CO_2, increased nitrogen deposition, and changes in management practices are factors that are assumed to be behind the increase. The impacts of temperature and CO_2 have been shown in experiments and are extrapolated by model calculations. For example, under an assumed increase of CO_2 by 3.5 μmol mol^{-1} yr^{-1} and temperature by 0.04°C yr^{-1} over 100 years, productivity of *Pinus sylvestris* increased by 5–15% as a result of the temperature elevation, 10–15% as a result of the increased CO_2, and 20–30% as a result of combined temperature and CO_2 (Kellomäki and Väisänen, 1997). In northern Europe, the effects of precipitation changes are likely to be much less important than the effects of temperature changes (Kellomäki and Väisänen, 1996; Talkkari and Hypén, 1996). Based on model computations that assume a seasonally uniform temperature increase, Proe *et al.* (1996) have suggested that growth of *Picea sitchensis* in Scotland could increase by 2.8 m^3 ha^{-1} yr^{-1} for each 1°C rise in temperature.

In Russian boreal forests, some studies predict large shifts in distribution (up to 19% area reduction) and productivity (e.g., Kondrashova and Kobak, 1996; Krankina *et al.*, 1997; Izrael, 1997; Raptsun, 1997). It could be concluded that climate change and CO_2 increase would be favorable for northern forests (e.g., as a result of increased regeneration capacity). Indeed, some studies suggest a significant increase in productivity of forests in higher latitudes. The largest changes are expected for forest tundra and the northern taiga, reaching 12–15% additional growth per 1°C warming (e.g., Karaban *et al.*, 1993; Shvidenko *et al.*, 1996; Lelyakin *et al.*, 1997). The same studies estimate the impact in more southern forest zones (southern taiga, mixed and deciduous forests) to be less: 3–8% per 1°C.

In central and southern Europe, limited moisture resulting from increasing temperature and (possibly) reduced summer rainfall may generate productivity declines regionally, but this cannot be predicted because of uncertain rainfall scenarios. In addition, CO_2 enrichment is likely to increase water-use efficiency (WUE), which makes growth less drought-sensitive. Forest growth conditions in the southern parts of eastern Europe (Russia, Ukraine, Moldova) are likely to decline as a result of increased drought, specifically in the steppe. Secondary problems could arise in the protective shelterbelts in the south of the forest-steppe zone that now covers about 3 Mha.

13.2.2.1.2. Disturbance regimes

In the Mediterranean region, elevation of summer temperature and reduction of precipitation may further increase fire risk. Colacino and Conte (1993a,b) examined the pattern of forest fires in the Mediterranean region in connection with the number of heat waves. An increase of 70% in the number of heat waves was recorded in the period 1980–1985 with respect to the period 1970–1975, and a similar increase was recorded in the extent of forest burned. In temperate eastern Europe, forest fire increase is less likely, but very dry and warm years could occur more frequently and promote pest and pathogen development. Large areas of pine forests in Ukraine, Belarus, and central regions of Russia might face some increased risk.

Increased forest fire risk is a crucial factor in the survival of boreal forests in Russia. Most dangerous are large forest fires, which occur during extremely dry and warm years. Such climatic conditions occur periodically (every 15–20 years) in parts of the Russian boreal zone. Currently these large fires account for about 1–2% of the total number of forest fires, but burned areas reach 70–80% and losses are as much as 90% of the total values. Most climate scenarios indicate that the probability of large fires will increase.

Estimates of the possible influence of climate change on insect infestation are uncertain because of complex interactions between forests, insects, and climate. The probability of outbreaks of pests such as *Dendrolimus sibirica* or *Limantria dispar* is expected to increase, especially in monocultures. Short-period warming also could promote infestation with new pest species that presently do not occur in the boreal zone. Increases of climate aridity would promote occurrence of some diseases (e.g., root and stem fungi decays).

13.2.2.2. Grasslands and Rangelands

Permanent grassland and heathland occupy a large proportion of the European agricultural area. The type of grassland varies greatly, from grass and shrub steppes in the Mediterranean region to moist heathland in western Europe. The annual cycle of many temperate grasses is limited by low temperature during the winter and spring and by water stress during the summer. Climate change can affect the productivity and composition of grasslands in two ways: directly through the effects of CO_2, or indirectly through changes in temperature and rainfall. Different species will differ in their responses to CO_2 and climate change, resulting in alterations in community structure (Jones and Jongen, 1996). Legumes, which are frequent in these communities, may benefit more from a CO_2 increase than nonfixing species (Schenk *et al.*, 1995).

Intensively managed and nutrient-rich grasslands will respond positively to the increase in CO_2 concentration and to rising temperature, as long as water resources are sufficient (Thornley and Cannell, 1997). The direct effect of doubling CO_2 concentration by itself may cause a 20–30% increase in productivity in nutrient-rich grasslands (Jones *et al.*, 1996; Cannell and Thornley, 1998). The importance of water management (including drainage) may be even more important, however, under changed climatic conditions in northern Europe (Armstrong and Castle, 1992). This positive effect of increased CO_2 on biomass production and WUE can be offset by climate change, depending on local climate and soil conditions (Topp and Doyle, 1996a; Riedo *et al.*, 1999). These effects also will determine the spatial distribution of agricultural grassland. An analysis by Rounsevell *et al.* (1996a) showed that grassland production in England and Wales is resilient to small perturbations in temperature and precipitation, but larger temperature increases may cause drought stress and reduced suitability for grassland production.

There is a greater controversy regarding the response of nitrogen-poor and species-rich grassland communities. Experimental studies in such grasslands have shown little response or even a reduction in production with CO_2 enrichment (Körner, 1996). On the other hand, simulation studies have shown that this could be just a transient response and that the long-term response of nitrogen-poor grassland ecosystems may be relatively larger than that of nitrogen-rich systems (Cannell and Thornley, 1998). This effect is caused by a reduction in nutrient losses and an increase in nitrogen fixation at elevated CO_2.

Because of its impacts on primary productivity and community structure, the long-term effect of elevated CO_2 on grasslands is an additional carbon sink. By contrast, increasing temperatures alone are likely to turn grasslands into a carbon source because soil respiration would be accelerated more than NPP. The net effect of current scenarios for CO_2 and temperature is likely to be a small carbon sink in European grasslands (Thornley and Cannell, 1997).

Arid and semi-arid environments (e.g., certain steppe-like habitats), which are well represented in the Mediterranean area, are crucial for the preservation of rich species diversity in this region. These regions seem to be the only places within Europe that certain insect species, such as *Lepidoptera*, can inhabit because of the abundant availability of their foodplants. Furthermore, the lack of winter climatic stress makes arid lands quite suitable as wintering grounds for birds. Overgrazing, fire, urbanization, and changes in land use can be considered the main threats to these regions. The potential distribution of these semi-arid environments may increase under drier and warmer climatic conditions, leading to landscape fragmentation at the local scale and consequent local extinctions (del Barrio and Moreno, 2000).

13.2.2.3. Freshwater Ecosystems: Inland Wetlands, Lakes, and Streams

European freshwater ecosystems encompass a varied assemblage of systems: lakes of various sizes and depth; streams with different hydrological characteristics; and wetlands, which by definition occupy a spatial continuum between aquatic and terrestrial environments. Wetlands are heterogeneous systems ranging from open-water surfaces to densely vegetated areas. Some wetlands are forested; shrubs, grasses, or mosses dominate others. European freshwater systems have been heavily subjected to and modified by damming, channeling, drainage, and other hydrological alterations; they also are influenced by humans through land and water use, pollution, erosion, and other factors.

The fundamental requirement for the existence of freshwater ecosystems is the spatial and temporal distribution of water in the landscape. The impacts of climate change on the future distribution and extent of these systems are analogous to those discussed in Section 13.2.1.1. However, apart from these impacts, changes in climatic parameters also will impact a range of chemical and biological functions, which combine with physical parameters to create the integrated ecological characteristics of future freshwater ecosystems in Europe. During the past 150 years, the winter ice cover of streams and lakes has declined, and in the northern hemisphere there has been a steady trend toward later freeze and earlier ice breakup (Magnuson *et al.*, 2000). Climate warming is likely to exaggerate this trend, and the timing and duration of freeze and breakup of ice in freshwater systems greatly affect inherent biological and ecological processes.

In the arctic and subarctic, freshwater systems are particularly sensitive to climate change—and most climate change scenarios indicate that the highest and most rapid temperature increases will occur in these regions. Increases in temperature may lead to changes in permafrost distribution (Anisimov and Nelson, 1996, 1997), with concomitant impacts on hydrology. In many wetland areas, permafrost acts as a drainage seal and promotes wetland development. However, Camill and Clark (1998) suggest that high-latitude systems might show lagged and complex dynamics in response to global warming, and local factors may exert more direct control over permafrost than regional ones. The estimated effect of climate change on average runoff in tundra regions is highly uncertain, but if water levels decrease, connections between tundra lakes could be severed. This would result in changes in community structure and possibly elimination of seasonal migrants to shallow, ice-covered winterkill lakes. Climate change impacts on ice breakup timing and intensity also will influence limnological characteristics by regulating the supply and flux of nutrients (Lesack *et al.*, 1991). Furthermore, such changes impact the influx of sunlight, which is a key factor in controlling primary productivity but also can have far-reaching effects on higher trophic levels. The populations of arctic char (and presumably other extreme coldwater fish species) are expected to decrease, especially in low-altitude, shallow lakes (Lehtonen, 1998).

In the boreal areas, scenarios typically display warmer winters (e.g., a shorter season of sub-zero temperatures). This would affect snow-cover conditions and lead to changes in the timing and intensity of snowmelt events and runoff characteristics, which would affect the ecological functions of freshwater systems. Wetland development might benefit if larger fractions of precipitation fell as rain, but the impact would depend on how temperature-induced higher evapotranspiration rates would counteract this effect, as well as topographical characteristics of the landscape. In response to higher temperatures, northern boreal populations of cyprinid and percid fish species are expected to increase at the expense of coldwater, salmonid species (Lehtonen, 1996). Shallow lakes would be most susceptible to these changes because of their lack of thermal stratification. Total freshwater fish production is expected to increase, but with the projected changes in the composition of fish fauna, the recreational and commercial value of catches will decrease (Lehtonen, 1996). A reduction in the spatial and temporal extent of lake and stream ice cover as a result of warmer winters can decrease light attenuation, which is a major limiting factor for production in boreal aquatic systems. Such a change could be expected to cause shifts in the biota of lakes and streams. It also can reduce winter anoxia that typically occurs in shallow lakes. Haapalea and Lepparänta (1997) modeled future ice-cover distribution in the Baltic Sea (which contains freshwater communities in the north). Simulations with a warming of 3.6°C to 2050 reduced the extent of ice cover from 38 to 10%, and a 6.6°C warming to 2100 resulted in no ice cover. The projected increase in biomass productivity in terrestrial systems also would affect lakes and streams because of alterations in the amount and quality of water and solid material inputs. Organic matter inputs are expected to increase when plant productivity increases and would be beneficial for heterotrophic organisms. Increases in organic matter concentrations also result in effects such as reduced light penetration (including damaging UV-B radiation—Schindler and Curtis, 1997) and changes in the vertical distribution of solar heating (Schindler *et al.*, 1996). In lakes, increased summer temperatures could lead to more pronounced thermal stratification, resulting in reduced secondary productivity as well as anoxic conditions in the hypolimnion. Warmer surface water can reduce the nutritional value of edible phytoplankton, but it also may shift primary production toward green algae and cyanobacteria, which are less favored by secondary consumers.

The dominating wetland types in the boreal regions are peatlands. Typically, the vegetation pattern and composition of boreal peatlands are governed by moisture regime rather than temperature and show high spatial variability in plant communities caused by variation in topography. It follows that a change in water balance could affect the function of boreal wetlands, including their carbon sequestering and carbon storage functions. A study in Finland suggests that very nutrient-poor peatlands can increase their long-term soil carbon accumulation after drainage (Minkkinen and Laine, 1998). In more nutrient-rich peatlands, however, soil carbon sequestering rates decrease and could shift to potential sources of atmospheric CO_2. Cao *et al.* (1998) have suggested that a temperature increase of less than 2°C could enhance methane (CH_4) emission rates from boreal wetlands, but greater warming might lead to reduction of fluxes because of decreasing soil moisture. Furthermore, field manipulation and laboratory experiments in Finland have shown that enhanced CO_2 concentrations (560 ppm) can lead to a 10–20% increase in CH_4 efflux from oligotrophic mire lawn communities (Saarnio and Silvola, 1999; Saarnio *et al.*, 2000). It is likely that boreal peatlands will expand further north into subarctic/arctic areas where the topography after permafrost disintegration still supports wetland formation.

In temperate Europe, the potential for precipitation decreases that result in lower flow rates could have major implications for lakes and streams. This could lead to changes in habitat and breeding locations of aquatic flora and fauna. These hydrological changes have the potential to be more significant for freshwater organisms than a temperature increase. The effect of warmer winters that lead to less extensive ice cover of lakes is expected to affect Europe's temperate lakes and streams as discussed above. Wetlands in the temperate regions of CEE are regarded as vulnerable to climate change (in combination with other anthropogenic threats—Best *et al.*, 1993; Hartig *et al.*, 1997). In the past, wetlands have been extensive in this area—for example, in the basins of the Pechora, Severnaya Dvina, and Upper Dnieper Rivers and in Karelia they have occupied 10–30% of the area. Now, many of them have been converted to agriculture, are affected by agricultural drainage, or are used in other ways, such as growing reed for thatch and livestock feed or collecting peat as a fuel for heating and cooking.

In the Mediterranean, the risk of acute water shortage in response to global warming would have severe impacts on freshwater ecosystems in the region. Hydrologically isolated systems, such as wetlands in topographical depressions, would be the most vulnerable, whereas those situated along larger rivers and lake shores might be less sensitive (Mortsch, 1998), although the extent of the latter may decrease as a result of lower flow rates. Increased competition for diminishing water resources also poses a potential threat to freshwater ecosystems. Although precipitation may increase during the winter—which is the main season for the seasonal wetlands in this area—this probably will be accompanied by comparably large increases in temperature, thus affecting net water availability. Summers are predicted to become warmer and drier, which would lead to deterioration of freshwater ecosystems (Haslam, 1997). Seasonal systems that presently can cope with occasional or periodic drought will experience additional stress that some species might not be able to survive (Brock and van Vierssen, 1992). The fact that wetlands in many parts of southern and central Europe are scattered in their location may prevent species migration to suitable climate conditions. In general, wetland plants with short life cycles are better adapted for geographical migration, indicating that this response is likely to occur faster in nonforested wetlands than in forested ones.

The risk of increased fire disturbance of terrestrial biota also will have consequences for lakes and streams in southern Europe. Freshwater systems adjacent to burned areas will receive an initial increase in solute input after fire; if the fire generates canopy gaps, the water bodies will be more influenced by wind mixing, inducing changes in thermal and chemical stratification characteristics.

13.2.2.4. Biodiversity and Nature Conservation

Europe is predominantly a region of fragmented natural or semi-natural habitats in a highly urbanized, agricultural landscape. A significant proportion of surviving semi-natural habitats of high conservation value is enclosed within protected sites, which are especially important as refuges for threatened species (Plowman, 1995). Nature reserves form a similarly important conservation investment across the whole of Europe. However, species distributions are projected to change in response to climate change (Huntley and Webb, 1989), and valued communities within reserves may disassociate, leaving species with nowhere to go (Peters and Darling, 1985; Peters and Lovejoy, 1992),

The impact of climate change on a particular reserve will depend on its location in relation to the climatic requirements of the species it accommodates. Sites that lie near the current maximum temperature limits of particular species could expect that if climate warms beyond those limits, species would become extinct at that site. Conversely, sites that lie close to the minimum temperature limits of species may assume greater importance for such species as the climate warms (Huntley, 1999). In Europe, nature reserves tend to form habitat "islands" for species in landscapes that are dominated by other land uses. The possibility of species colonizing other habitat islands could be limited. As a result of climate change, reserve communities may lose species at a faster rate than potential new species can colonize, leading to a long period of impoverishment for many reserves.

The requirements of a future conservation strategy in the advent of climate change have been considered by Huntley *et al.* (1997). They suggest that for Europe, where large-scale range changes are projected, a network of habitats and habitat corridors will be required to facilitate migration.

Questions that urgently must be asked are as follows: To what extent do rare and vulnerable species in Europe rely on protected areas for their survival in the present day? Do current policy measures being implemented throughout Europe under the Biodiversity Convention take into account the potential impacts of climate change? It will become increasingly important for conservation strategies to be developed on a pan-European scale to protect species in parts of their ranges that are least likely to be negatively impacted by climate change. Reevaluation of conservation priorities and the role of reserves is required for individual sites and in relation to national and international conservation strategies (Hendry and Grime, 1990; Parsons, 1991).

13.2.2.5. Migratory Animals

Insects: Parmesan *et al.* (1999) analyzed data for 35 nonmigratory butterflies with northern range limits in Great Britain, Sweden, Finland, or Estonia and southern boundaries in southeastern France, Catalonia (Spain), Algeria, Tunisia, or Morocco. More than 60% were found to have shifted north by 35–240 km in the 20th century, consistent with the 120-km northward shift of climatic isotherms reported in the SAR. This finding is contrary to the trend that might have been expected as a result of land-use change: Habitat loss has been greater in northern European countries over this period than southern ones. Scientific knowledge of butterfly biology supports the inference that this shift is in response to increased temperatures. Population eruptions of several species of forest lepidoptera in central Europe in the early 1990s, including the gypsy moth *Lymantria dispar,* have been linked to increased temperatures (Wulf and Graser, 1996), as have northward range expansions of several species of *Odonata* and *Orthoptera* (Kleukers *et al.,* 1996).

Insect pests: Most studies concur that insect pests are likely to become more abundant in Europe as temperature rises, as a result of increased rates of population development, growth, migration, and overwintering (Cannon, 1998). There has been little or no reported research at the level of pest population dynamics, however, about potential responses of insect pests to increased CO_2 (Cannon, 1998). Although changes in rainfall also could have a substantial effect (Lawton, 1995), this is difficult to quantify, particularly given uncertainties with regional precipitation scenarios. Migratory species may be able to extend their ranges as crop distributions change. For example, a 3°C rise in temperature would advance the limit for grain maize across much of Europe, which could be followed by a northward range expansion by the European corn borer *Ostrinia nubilalis* of as much as 1,220 km (Parry *et al.,* 1990; Porter *et al.,* 1991; Porter, 1995).

Birds: Climate change in Europe already has been demonstrated to be affecting migratory wild bird populations. In the UK, 20 of 65 species, including long-distance migrants, significantly advanced their egg-laying dates by 8 days, on average, between 1971 and 1995 (Crick *et al.,* 1997; Crick and Sparks, 1999). In general, species show advancement in average arrival and laying dates of about 3–5 days per 1°C. It is quite likely that birds will be able to adapt faster than most taxa to such changes, given their mobility and genetic variability (e.g., Berthold and Helbig, 1992). However, increased aridity in the Mediterranean region may be detrimental to the trans-Saharan migrants that use the area for foraging en route. Potentially great problems face the internationally important populations of waterfowl that use a relatively limited number of sites for wintering or while on passage in Europe. Where sea-level rise causes coastal squeeze on the availability of intertidal feeding areas (because sea defenses prohibit encroachment onto currently dry land), feeding resources available to wintering waterbirds may become limited and lead to population declines (Norris and Buisson, 1994). Arctic-breeding shorebirds are predicted to benefit in the short term as warmer temperatures increase the numbers of their insect food supplies, but in the long term they may suffer from the disappearance of their habitat as vegetation zones move northward toward the limit of any land (Lindström and Agrell, 1999).

13.2.3. Agriculture and Fisheries

13.2.3.1. Agriculture

Agriculture accounts for only a small part of GDP in Europe. Therefore, the vulnerability of the overall economy to changes that affect agriculture is low (Reilly, 1996). Locally, however, effects on society may be large. Europe as a whole is noted for its substantial output of arable crops and animal products (FAOSTAT, 1998).

Trends in European agriculture are dominated by the EU's Common Agricultural Policy (CAP). This occurs because EU member states account for a large proportion of agricultural production in Europe and because several countries currently are seeking membership in the EU and in this process are adjusting their policy to match the CAP. The EU seeks to integrate concerns for environmental protection and countryside livelihood into the CAP.

Many studies have assessed the effects of climate change on agricultural productivity (e.g., Ryszkowski and Kedziora, 1993; Harrison *et al.*, 1995b; Semenov and Porter, 1995; Alexandrov, 1999; Cuculeanu *et al.*, 1999; Harrison *et al.*, 1999). Relatively little work, however, has been done to link these results across sectors to identify vulnerable regions and farming systems. Such assessments are needed to identify appropriate policy responses to climate change. More extensive information than the summary presented in this section appears in the agriculture chapter of the European ACACIA report (Bindi and Olesen, 2000).

Cereals of different species and varieties are grown throughout Europe. Climatic warming will expand the area of cereals cultivation (e.g., wheat and maize) northward (Kenny *et al.*, 1993; Harrison and Butterfield, 1996; Carter *et al.*, 1996a). For wheat, a temperature rise will lead to a small yield reduction, whereas an increase in CO_2 will cause a large yield increase; the net effect of both for a moderate climate change is a large yield increase (Nonhebel, 1996; Harrison and Butterfield, 1999). Drier conditions and increasing temperatures in the Mediterranean region and parts of eastern Europe may lead to lower yields and the need for new varieties and cultivation methods. Yield reductions have been estimated for eastern

Europe, and yield variability may increase, especially in the steppe regions (Alexandrov, 1997; Sirotenko *et al.*, 1997). Figure 13-6a shows the response of wheat yields to change of climate and CO_2 concentration for a GCM scenario for 2050 that resembles the A1 scenario. The largest increases in yield occur in southern Europe—particularly northern Spain, southern France, Italy, and Greece. Relatively large yield increases (3–4 t ha^{-1}) also occur in Fenno-Scandinavia. In the rest of Europe, yields are 1–3 t ha^{-1} greater than at present. There are small areas where yields are predicted to decrease by as much as 3 t ha^{-1}, such as in southern Portugal, southern Spain, and Ukraine. For maize, future climate scenario analyses carried out for selected sites in Europe suggest mainly increases in yield for northern areas and decreases in southern areas (Wolf and van Diepen, 1995). This is a result of a small effect of increased CO_2 concentration on growth (maize is a C_4 plant that responds less positively to CO_2 increases than C_3 plants such as wheat and barley) and a negative effect of temperature on the duration of growing season. The latter effect, however, can be largely prevented by growing other maize varieties (Wolf and van Diepen, 1995).

Seed crops generally are determinate species, and the duration to maturity depends on temperature and day length. A temperature increase therefore will shorten the length of the growing period and possibly reduce yields (Peiris *et al.*, 1996). At the same time, the cropping areas of cooler season seed crops (e.g., pea, faba bean, and oilseed rape) probably will expand northward into Fenno-Scandinavia, leading to increased productivity of seed crops there. There also will be northward expansion of warmer season seed crops (e.g., soybean and sunflower). Harrison and Butterfield (1996) estimate this northward expansion for sunflower; they also found a general decrease in water-limited yield of sunflower in many regions, particularly in western Europe. Analysis of the effect of climatic change on soybean yield for selected sites in western Europe suggests mainly increases in yield (Wolf, 2000a). This is a result of a positive effect of CO_2 concentration on growth and only a small effect of temperature on crop duration.

Vegetables cover a wide range of species with a large variation in type of yield components, including leaves, stalks, inflorescence, bulbs, roots, and tubers. Most vegetables are high-value crops that are grown under ample water and nutrient supply. Their response to changes in temperature and CO_2 varies among species, mainly depending on the type of yield component and the response of phenological development to temperature change. For determinate crops such as onions, warming will reduce the duration of crop growth and hence yield (Harrison *et al.*, 1995b), whereas warming stimulates growth and yield in indeterminate crops such as carrots (Wheeler *et al.*, 1996). Onion yields are sensitive to the degree of warming (Harrison *et al.*, 1995b), with a yield decrease for warmer scenarios and a yield increase for cooler scenarios. There also is a spatial gradient, with yield increases in northwest Europe to decreases in southeast Europe. For lettuce, temperature has been found to have little influence on yield, whereas yield is stimulated by increasing CO_2 (Pearson *et al.*, 1997). For cool-season vegetable crops such as cauliflower, large temperature increases may decrease production during the summer period in southern Europe because of decreased yield quality (Olesen and Grevsen, 1993).

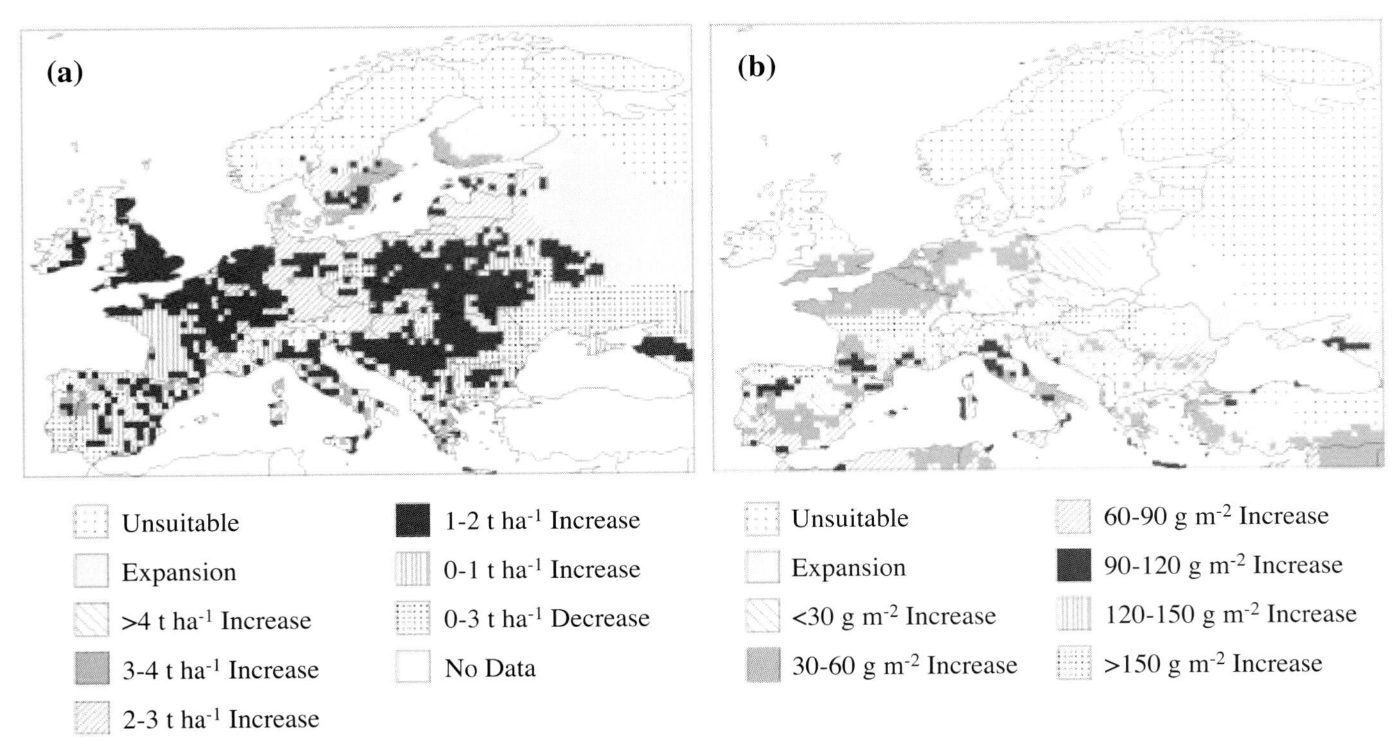

Figure 13-6: Change in water-limited yield for wheat (a) and potential yield for grapevine (b), using HadCM2 scenario for 2050 (Harrison and Butterfield, 1999).

Root and tuber crops are expected to show a large response to rising atmospheric CO_2 because of their large underground sinks for carbon and apoplastic mechanisms of phloem loading (Farrar, 1996; Komor *et al.*, 1996). On the other hand, warming may reduce the growing season and enhance water requirements, with consequences for yield. Climate change scenario studies performed with crop models show increases in potato yields in northern Europe and decreases or no change in the rest of Europe (Wolf, 2000b). Simulation results show an increase in potato yield variability for the whole of Europe, which enhances the risk for this crop. However, crop management strategies (e.g., advanced planting and cultivation of earlier varieties) seem to be effective in overcoming these changes (Wolf, 2000b). Indeterminate root crops such as sugar beet may be expected to benefit from warming and the increase in CO_2 concentrations. A study performed by Davies *et al.* (1996) for England and Wales indicates that the area of suitability for the growth of the crop moves westward.

Forage crops, including maize and whole crop cereals for silage, as well as root crops such as sugar beet and some Brassica species, also are described under cereals and root crops. When these crops are grown as forage crops, yield components and quality may change. Thus, there is a larger emphasis on total biomass yield and on digestability of biomass. The effects on production and quality of wheat whole crop silage will depend on the relative magnitudes of changes in CO_2 and temperature (Sinclair and Seligman, 1995). Yields of the different forage crop types will be affected differentially. Yields of indeterminate crops such as sugar beet and silage maize can be expected to show a larger increase than the yields of whole crop cereals, especially in northern Europe. This probably will lead to changes in the types of forage crops grown. Studies indicate an increase in suitability of the north and west UK to forage maize (Cooper and McGeechan, 1996; Davies *et al.*, 1996).

Perennial crops (e.g., grapevine, olive, and energy crops) have been relatively less studied in the context of climate change impacts. A study on the potential cultivation of grapevine in Europe under future climate scenarios has shown that there is potential for expansion of the wine-growing area in Europe and an increase in yield (see Figure 13-6b). Yet detailed predictions made for the main EU viticultural areas have shown an increase in yield variability (fruit production and quality). The quality of wine in good years is not guaranteed, and the demand for wine in poor years is not met, implying a higher economic risk for growers (Bindi *et al.*, 1996; Bindi and Fibbi, 2000). For olive, it was shown that in a $2xCO_2$ case, the suitable area for olive cultivation could be enlarged in France, Italy, Croatia, and Greece as a result of changes in temperature and precipitation patterns (Bindi *et al.*, 1992). For indeterminate energy crops that are favored by the longer growing season and by increased WUE resulting from higher CO_2 levels, higher temperatures and CO_2 concentrations generally would be favorable. A study of willow production in the UK found that warming generally would be beneficial for production, with increases in yield of as much as 40% for a temperature increase of 3°K (Evans *et al.*, 1995).

Livestock systems may be influenced by climate change directly by means of its effects on animal health, growth, and reproduction and indirectly through its impacts on productivity of pastures and forage crops. Heat stress has several negative effects on animal production, including reduced reproduction and milk production in dairy cows and reduced fertility in pigs (Furquay, 1989). This may negatively affect livestock production in summer in currently warm regions of Europe. Warming during the cold period for cooler regions is likely to be beneficial as a result of reduced feed requirements, increased survival, and lower energy costs. Impacts probably will be minor for intensive livestock systems where climate is controlled. Climate change may affect requirements for insulation and air-conditioning, however, and thus change housing expenses (Cooper *et al.*, 1998). The impact of climate change on grasslands will affect livestock living on these pastures. In Scotland, studies of the effect on grass-based milk production indicate that these effects vary by locality. For herds grazed on grass-clover swards, milk output may increase regardless of site, as a result of the CO_2 effect on nitrogen fixation (Topp and Doyle, 1996b).

Pest-disease-weed-host relationship can be affected by climate change in different ways. Pests, diseases, and weeds that currently are of minor significance may become key species, thereby causing serious losses. The distribution and intensity of current key pest, diseases, and weeds may be affected, leading to changed effects on yield and on control measures such as pesticides and integrated pest management. Competitive abilities in weed-plant interactions may be affected through changes in ecophysiology (i.e., CO_2 fertilization effects on C_3 and C_4 species). Pests and diseases generally will migrate as crops migrate (e.g., Lipa, 1997, 1999).

13.2.3.2. Fisheries

Detailed analyses of fish physiological response to water temperature have shown that the potential impact of climate change on freshwater and marine fish is large (Wood and McDonald, 1997). Unfortunately, current knowledge appears to be limited mostly to single key species, abstracted from the wider ecosystem context that supports fisheries production. It is likely that extrapolation from biological first principles will provide only limited foresight at a fisheries level in that context. However, it is likely that in the short term, fish will move to new habitats to find conditions to which they have adapted.

Two first-order effects—changes in biodiversity and changes in fisheries and aquaculture productivity—are examined. For each of these, key impacts in the European region are discussed with respect to temperature rise and other factors that are linked to climate change. For marine systems, these key factors are sea-level rise and changes in ice cover, salinity, and ocean currents; for inland waters, factors such as hydrological changes (e.g., dams, water abstraction), hydrochemical changes (including anoxia, water acidity, pollution and toxicity events), and eutrophication are key.

13.2.3.2.1. Freshwater and marine biodiversity

It has been possible, for the better-known species, to obtain a rough first picture of likely faunal movements and range shifts that would result from the temperature rise currently forecast for the European region.

This has been done, for example, for wild Atlantic salmon (*Salmo salar*), which sustains important recreational and commercial fisheries over most of northern Europe and has a high conservation indicator value over its entire range. Atlantic salmon spends its early and juvenile life in freshwater slowly moving away from the headwater spawning grounds in rivers out to sea, to grow and mature. Looking at the direct and indirect influence of temperature on protein synthesis on all life-cycle stages, McCarthy and Houlihan (1997) suggest that there will be a northward shift in the geographic distribution of Atlantic salmon in Europe, with likely local extinction at the southern edge of the current range and new habitats colonized in the north. The influence of temperature on overwintering survival of 1- and 2-year-old salmon and the distribution of post-smolts in the North Sea area already is apparent (Friedland *et al.*, 1998).

High sensitivity to water temperature of fish larval and juvenile stages, combined with the higher susceptibility of headwaters and smaller rivers to air temperature rise, implies important effects of climate change on cold and temperate anadromous species such as the sea trout, alewife (*Alosa alosa* Atlantic and Mediterranean), and sturgeon (Atlantic, Black Sea, Caspian Sea). Similar impacts are likely for all salmonid species in Europe, including those that do not migrate to sea. Fish species are likely to extend their range northward (e.g., Brander, 1997). To identify impacts on recreational fisheries and conservation efforts that can be attributed to climate change, changes in local species presence will need to be assessed at a pan-European level, beyond watershed and national levels.

Global environmental change and introduction of species make it difficult to identify impacts of climate change in freshwater systems. Direct evidence of species shift and long-range movement often is physically limited between watersheds; introductions of alien species and genotypes—either voluntary or accidentally—have been very extensive. The influence of global warming on marine fish species and migratory species that reproduce and spend their early life history in marine waters (catadromous fish—e.g., eel) is more complex to foresee because of the various spatial and long temporal scales involved, as well as the feedback loop between air and ocean temperatures.

Some insight may be gained by reviewing historical shifts in the geographical distribution of European fish species to gain a better estimation of how fast and how far species may change their distributions. For example, the period of warming between 1920 and 1940 resulted in widespread changes in the distribution of terrestrial and marine species from Greenland to the Barents Sea.

13.2.3.2.2. Fisheries and aquaculture productivity

Future impacts on aquatic systems productivity from climate change are mostly uncertain because of other, related or independent, pressures. Resource overexploitation appears to be the single most important factor directly threatening the sustainability of many commercial fisheries in Organisation for Economic Cooperation and Development (OECD) countries (OECD, 1997). Overexploitation increases the vulnerability of fisheries to climate variability because so few fish are left in the stock to grow and multiply in a year of poor recruitment. The North Sea cod fishing industry, for example, now relies on only 1- or 2-year classes (Cook *et al.*,1997) and therefore is vulnerable to a year or two of poor recruitment caused by adverse climatic conditions.

Productivity of some fish stocks may benefit from warming trends. For example, recruitment of cod in the Barents Sea is higher in warm years. This probably comes about as a result of indirect effects (on capelin and zooplankton) as much as through direct effects.

Cod in the North Sea are at the warm end of their thermal range, and their recruitment therefore seems to benefit during cold periods, such as the 1960s and 1970s (Dippner, 1997; Brander, 2000). The temperature and wind regime of the North Sea are strongly influenced by the NAO, records of which go back more than a century. Some of the variability in catches of North Sea cod may be explained by trends in the NAO, but changes in exploitation are a major factor.

Other factors are likely to combine with changes in temperature and decrease fish and shellfish productivity. Chronic levels of pollution are known to reduce marine and freshwater fish fecundity (Kime, 1995), decrease freshwater supply (which exacerbates low dissolved-oxygen concentrations), increase solid transport from erosion, and increase habitat fragmentation in inland waters. Development of marine aquaculture may be slowed by a decreasing availability of sites with cool enough surface water temperature and by increased susceptibility to disease.

The consequences of fisheries collapse are complex. For example, the North Sea herring fishery collapsed in 1977 and was closed for 5 years. Although the rapid recovery of the resource surprised fisheries biologists, the most dramatic effect on the industry resulted from a permanent change of consumer preferences (Bailey and Steele, 1992) away from kippers and fresh and pickled herring. The fishery recovered, but the market did not.

Finally, the dramatic effect of climate change on fisheries production is well documented (Durand, 1998) for the world's four main upwelling systems: the California current, the Peru current, the Canary current (off northwest Africa and southern Spain and Portugal), and the Benguela current (off the Atlantic coast of southern Africa). Although these currents are mostly outside European waters, the effects on Europe's distant water fishing fleet, fishing sector investments, and import prices for human consumption and aquaculture feed cannot be ignored.

13.2.4. Other Impact Areas

13.2.4.1. Energy

Increasing temperatures will have a direct impact on energy use, especially in the domestic sector, as Palutikof *et al*. (1997) note for the unusually warm year of 1995 in the UK. The main effects of climate variability are on markets for space-heating fuels and on the amount of electricity used for air conditioning and refrigeration. Air conditioning has become universal in new office buildings, and there is a strong likelihood that it may spread to new markets such as houses and flats in future years. The relationship between climate change and the "take up" of new air conditioning installations is largely unknown, especially in areas where the practice currently is not widely used.

Requirements for heating and cooling can be estimated from the change in the number of heating and cooling degree days per year. In general, we can expect increased temperatures in Europe to lead to an increase in space cooling and a decrease in space heating requirements (Climate Change Impacts Review Group, 1991, 1996). Very mild winters in the EU, such as those experienced between 1988 and 1990, resulted in a 2% drop of energy demand. Milbank (1989) estimates that for northern Europe, a 4.5°C temperature rise would more than double summer electricity consumption by air conditioning and refrigeration systems.

13.2.4.2. Insurance

The insurance industry in Europe has an annual turnover of 600 billion EUR, with assets of 4,000 billion EUR. The common view is that the sectors that are most germane to climate change impacts and adaptation are property insurance and reinsurance. There will be lesser effects on other branches, such as casualty, life, and pensions, and the industry also could be affected in its investment activities by shifts in the economics of other industries that are impacted by mitigation policies (Dlugolecki and Berz, 2000). Essentially, the insurance industry "recycles" other sectors' monetary risks, thereby focusing information on such impacts. However, there is a wide variety of insurance systems in place in Europe, so international comparisons or extrapolations are difficult. In addition, many economic risks are not handled through insurance currently—for example, standing crops, flood damage, and "pure" economic losses where no physical damage has occurred. Weather affects insurers through the medium of property damage caused by a variety of extreme events; storm, flood, freeze, drought, and hail are the prime ones (Dlugolecki and Berz, 2000).

In early 1990, a series of storms in northwest Europe resulted in insured damage of 10 billion EUR. Despite the use of reinsurance, much of this risk remains within the European insurance industry. The risk will continue to rise because of pressure from economic growth and economic wealth. The industry's assets are sufficiently large to cope with purely European climate change impacts. The industry's main vulnerability is likely to arise outside Europe, from events such as hurricanes, earthquakes, or a global stock-market collapse. Human health impacts will be minimal for insurers as employers or suppliers.

This analysis assumes that there will be no major change in the scope of insurance services to include impacts that currently are not insured, such as standing crops, flood (in many countries), and pure economic losses. Establishment of the common market will help to reduce national differences in insurance services, but only slowly.

13.2.4.3. Industry and Transport

A sophisticated transport system has evolved in Europe to move people and goods. Efficient, rapid, reliable, and dependable transport is an essential part of the continent's infrastructure, and disruptions and dislocations to transport systems have a rapid impact on most industrial and commercial activities. Air transport probably is the most sensitive sector to weather and climate change and rail transport the most tolerant. In manufacturing and retailing, "just-in-time" distribution systems are quickly disrupted by adverse weather conditions. Changes in consumer demand for many products, which influence indices of retail sales (Agnew and Thornes, 1995), are likely to accompany climate change.

Significant changes in the frequency of short-term climatic extremes such as windstorms impact transportation (Perry and Symons, 1994). The impact of wind and windstorms includes the effect on land-based terminals such as seaports and airports, as well as in-transit delays and damage to the means of transport itself. Instances of wind shear associated with summer thunderstorms, which pose a hazard to aircraft during takeoff and landing, could become more common in Europe. Increases in rainfall and increasing frequency of temperature oscillating around the freezing point can be expected to lead to higher levels of corrosion of transport infrastructures. Flooding in rivers and low water levels lead to interruptions of river navigation.

Coastal transport infrastructure can be damaged by a combination of sea-level rise and increased storminess. In many countries, a significant percentage of manufacturing industry is located along coastlines and estuaries and may require expensive coastal protection schemes. Transport infrastructure in river valleys also may be damaged or destroyed during floods. There are management implications for winter maintenance activities on roads and railways. Savings may be possible, especially in western Europe as winter temperatures rise, but more freeze-thaw activity is expected in CEE as winter minimum temperatures oscillate around the freezing point. Analysis of the levels of saving that can be expected in different areas are required, together with cost-benefit studies to examine whether further investment in comprehensive ice-detection systems can be justified.

Comprehensive studies of the likely impacts of climate change on transport are in their infancy in many countries. Broad-scale assessments are needed of likely climate-induced demands for transport as lifestyles and residential and migration patterns change. The evidence that currently is available (summarized in the Europe ACACIA report), suggests that fewer severe winters would be beneficial to the manufacturing industry, reducing disruption at all stages from the supply of raw materials through processing to marketing of finished goods. However, an increase in the frequency of hot summers could disrupt some industrial processes that use large quantities of water. There may be some absolute limiting factors in some countries, such as lack of water for power stations, that can be overcome only through massive capital investment or new technology (Perry, 2000).

13.2.4.4. Tourism

The tourist industry in Europe is expected to continue to grow, in part as a result of higher incomes and more leisure time. Predominant tourist flows presently are from north to south; at present these flows help to transfer capital. Changes in recreational habits and preferences will lead to opportunities for tourist investment in new areas, but existing major tourist flows to the Mediterranean might be weakened if summer heat waves increase in frequency or if prolonged droughts result in water supply problems and forest fires. Giles and Perry (1998) have shown that summers as good as that of 1995 in northern Europe can lead to a drop in the numbers of outbound tourists from countries such as the UK to traditional Mediterranean sun destinations.

Changing demographic patterns—particularly an aging, wealthy population—may lead to an increase in winter and shoulder season tourism to Mediterranean resorts and expansion of retirement to attractive areas, particularly coasts. "Health tourism" to spas and mountains is likely to increase in Europe. In northern Europe, short breaks are likely to be taken over a longer season as temperatures rise.

The coastal zone is the primary tourist resource of Europe, and associated tourist infrastructure is at risk from sea-level rise, including unique tourist attractions such as the city of Venice (Perry, 1999). Beaches, wetlands, and estuaries also are tourist resources that are at risk. Already, many tourist amenities, such as coastal golf courses and hotels, require protection. The need to maintain coastal amenity values, as well as protect infrastructure, is an important factor that is encouraging a shift from hard, rigid defenses to softer approaches, including sediment management and nourishment (Penning-Rowsell *et al.*, 1992; Hamm *et al.*, 1998).

Heat stress and poor urban air quality may render cities undesirable locations in summer, with more tourist traffic to the country and the coast. Outdoor recreational spending is likely to increase.

Mountainous zones are used extensively for recreation and are the main sites of the European winter sports industry, which is based on snow resources that are vulnerable to climate change. Mohnl (1996) has shown that there is a statistically significant trend in snow-cover reduction in the Alps over recent years. Abegg and Froesch (1994) have suggested that assuming a 3°C rise in mean temperatures, the snow line in winter will rise by 300 m in the central Alps, the first snowfall of the season will be delayed, and below an altitude of 1,200 m there will not be continuous winter snow cover. As the season contracts, there will be a need to bolster winter tourism with increased use of artificial snow and more alternatives to outdoor skiing. In Scotland, the skiing industry is likely to experience more frequent snow-deficient winters, with adverse impacts on the financial viability of the industry.

Jaagus (1997) analyzed the impact of climate change on the snow-cover pattern in Estonia, where snow cover is important for winter sports and tourism. Indications are that a considerable drop in snow-cover duration will take place on islands and in the coastal region of west Estonia.

13.2.4.5. Migrations

Migration caused by soil degradation is a very important issue in the Mediterranean region, the southern part of which is mostly arid and vulnerable to climate change. With perhaps 24% of total drylands in Africa in the process of desertification and 0.3% of the African population permanently displaced largely as a result of environmental degradation (LeHouerou, 1992), consequent in-migration pressures on neighboring regions such as southern Europe can be substantial (Goria, 1999).

13.2.5. Human Health

Climate change will influence human health in several ways. Impacts will reflect the conditions of the ecological and social environments in which humans live. The impacts of projected changes in climate will depend on current and future public health defenses. Difficult economic conditions during the past decade have had serious implications for the delivery of health care and the public health infrastructure in some countries in CEE. These countries are most at risk from potential health impacts of climate change.

13.2.5.1. Thermal Stress and Air Pollution

One can expect an increase in the frequency of heat waves, as well as warmer summers and milder winters. Analyses in European cities show that total mortality rises as summer temperatures increase (Katsouyanni *et al.*, 1993; Kunst *et al.*, 1993; Jendritzky *et al.*, 1997). Episodes of extreme high temperatures (heat waves) also have significant impacts on health. Heat waves in July 1976 and July–August 1995 were associated with a 15% increase in mortality in greater London (McMichael and Kovats, 1998; Rooney *et al.*, 1998). A major heat wave in July 1987 in Athens was associated with 2,000 excess deaths (Katsouyanni *et al.*, 1988, 1993). Much of the excess mortality attributable to heat waves is from cardiovascular, cerebrovascular, and respiratory disease, and the elderly are particularly vulnerable to heat-related illness and death (Faunt *et al.*, 1995; Sartor *et al.*, 1995).

In cold and temperate locations, daily deaths increase as daily wintertime temperature decreases (Khaw, 1995; Laake and Sverre, 1996). Based on current evidence, climate change (increase in mean winter temperatures) is likely to result in a reduction in wintertime deaths, at least in temperate countries. Langford and Bentham (1995) estimate that 9,000 wintertime deaths yr^{-1} could be avoided by the year 2025 in England and Wales under a 2.5°C increase in average winter temperature. A meta-analysis by Martens (1997) estimates that an increase in global temperature could result in a reduction in winter cardiovascular mortality in Europe, leading to a decrease in mortality rates in regions with cold/temperate climates. At this time, the literature does not enable quantitative comparison between changes in summer and winter mortality.

Climate change also is likely to affect air quality in urban areas. Formation (and destruction) of secondary air pollutants, such as ozone, increases at higher temperatures and increased levels of sunlight. Studies in the United States indicate that climate change would entail an increase in average ambient concentrations of ozone and an increase in the frequency of ozone pollution "episodes" (USEPA, 1989). Experiments with the Hadley Centre climate model indicate significant increases in baseline ozone concentrations in Europe. Ozone and other air pollutants have significant impacts on health; these pollutants are considered to be one of the most important environmental health problems in Europe. Finally, climate change is likely to change the seasonality of pollen-related disorders such as allergic rhinitis (hay fever) (Emberlin, 1994).

13.2.5.2. Vector-Borne Diseases

Climate plays a dominant role in determining the distribution and abundance of insects and tick species—directly, through its effects on vector and parasite development, and indirectly through its effects on host plants and animals and land-use changes (McMichael *et al.*, 1996). Therefore, it is anticipated that climate change will have an effect on the geographical range and seasonal activity of vector species and, potentially, disease transmission (Bradley, 1993; Martens *et al.*, 1997; Martens, 1998).

As a result of deterioration of health systems, the recent resurgence of malaria in eastern Europe is now a growing cause for concern. Climate change could exaggerate this increased risk. In regions of southern Europe, climate change would increase the current very small risk of local (autochthonous) transmission. A few such cases are reported in the Mediterranean area under current climate conditions (e.g., in Italy—Balderi *et al.*, 1998). Concomitant with increases in the volume of international travel, all countries in Europe have seen a steady increase in the number of imported cases of malaria, which provide the source of the pathogen (WHO, 1997). In the UK, however, local vectors are physiologically unable to transmit the most lethal form of malaria (falciparum) (Marchant *et al.*, 1998). Although localized outbreaks are more likely to occur under climate change, in northern and western Europe existing public health resources and a lack of breeding habitats necessary to maintain high densities of mosquitoes make re-emergent malaria unlikely.

Dengue currently is not present in Europe, although it has been present in the past (Gratz and Knudsen, 1996). However, one of the vectors (*Ae. albopictus*) currently is extending its range in Europe, and climate change could facilitate this expansion (Knudsen *et al.*, 1996). This vector has been reported in Italy since 1990 and in Albania since 1979.

In all countries bordering the Mediterranean, cutaneous and visceral leishmaniasis are transmitted to humans and dogs by phebotomine sandflies (Dedet *et al.*, 1995). Higher temperatures are likely to change the geographical distribution of the important sandfly vector species and accelerate maturation of the protozoal parasite, thereby increasing the risk of infection (Rioux *et al.*, 1985). Increased incidences of visceral leishmaniasis, unrelated to immune suppression, have been reported from regions of Italy and coastal Croatia (Gabutti *et al.,* 1998; Punda-Polic *et al.,* 1998). Several imported cases of canine leishmaniasis are reported in Germany, Switzerland, and Austria every year (Gothe *et al.*, 1997). Thus, imported canine cases are a potential source of the pathogen, if the vectors expand further north with climate change.

Ixodid ticks (e.g., *Ixodes ricinus* and *I. persulcatus*) are widely distributed in temperate regions and transmit tick-borne diseases in Europe, of which Lyme disease (Berglund *et al.*, 1995) and tick-borne encephalitis (TBE) are the most important (Tälleklint and Jaenson, 1998). In Sweden, TBE incidence has increased after milder winters, in combination with extended spring and autumn seasons during 2 successive years (Lindgren, 1998). There also is some evidence that the northern limit of the distribution of ticks in Sweden has moved northward between 1980 and 1994 as a result of the increased frequency of milder winters (Tälleklint and Jaenson, 1998; Lindgren *et al.*, 2000). Climate change may extend the length of the transmission season of tick-borne diseases and facilitate spread to higher latitude and altitudes in northern Europe. However, a model-based study indicates that overall the region suitable for TBE transmission may contract significantly (Randolph and Rogers, 2001).

13.2.5.3. Water-Related Diseases

Climate change could have a major impact on water resources and sanitation in situations where water supply is effectively reduced. Some populations in eastern Europe with restricted access to water in the home would be vulnerable to any climate-related decreases in freshwater availability. Decreased water availability also could lower the efficiency of local sewage systems and may necessitate use of poorer quality sources of freshwater, such as rivers. All of these factors could result in an increased incidence of water-borne diseases.

The most significant water-borne disease associated with the public water supply in western Europe is cryptosporidiosis. Increases in the frequency or intensity of extreme precipitation events can increase the risk of outbreaks of this disease. Cases of cercarial dermatitis (water-based parasitic disease) may increase if the climate becomes more favorable for the host—a water snail (de Gentile *et al.*, 1995).

13.2.5.4. Food-Borne Diseases

Warmer climate in combination with inappropriate food behavior may contribute to increased incidences of food-borne diseases. A study of food-borne illness in the UK found a strong relationship between incidence and temperature in the month preceding the illness (Bentham and Langford, 1995). The distribution and activity of flies, cockroaches, and rodents could change in response to climatic changes. These species are carriers of food-borne pathogens and are considered to be major hygienic pests in the domestic environment.

13.2.5.5. Health Implications of Floods

The risk of flooding (coastal and riverine) is likely to increase in Europe under climate change. With this risk comes additional risk to people's health as a consequence of flooding (Menne *et al.*, 1999). Some floods in Europe have been associated with an increased risk of leptospirosis—for example, outbreaks were reported following floods in Ukraine and the Czech Republic in 1997 (Kriz, 1998: Kriz *et al.*, 1998) and in Portugal in 1967 (Simoes, 1969). Many cases of post-flood food poisoning were noted during and after the Odra flood in 1997.

Some of the effects of flooding on mortality and ill health in developed countries is attributable to the distress and psychological effects of the event (Bennett, 1970). This was demonstrated in the Easter 1998 flooding in parts of England (the worst since 1947); the majority of flood victims interviewed cited stress associated with the event and post-flood recovery as the worst aspect of their experience (Tapsell and Tunstall, 2000). Cases of post-traumatic stress disorder, including 50 flood-linked suicides, were reported in the 2 months following the major floods in Poland in 1997 (IFRC, 1998). The health effects of flooding are complex and can result from a combination of factors. The impact of flooding on mental health therefore may be significant.

13.3. Adaptation Potential

In this section, our analysis is restricted to adaptation in the major primary sectors (water, soils, ecosystems, agriculture) and in coastal regions. These are the main sectors where adaptation will be needed.

13.3.1. Water

There are, broadly, two different approaches to adaptation in the water sector: "supply side" (change the water supply) and "demand side" (alter exposure to stress). Table 13-6 summarizes some supply-side and demand-side techniques that are available to cope with adverse impacts in the water sector and may be appropriate in the face of climate change.

***Table 13-6**: Adaptive techniques in the water sector: some examples.*

Impact	Supply Side	Demand Side
Reduced water-supply potential	– Change operating rules – Increased interconnections between sources – New sources – Improved seasonal forecasting	– Reduce demand (through pricing, publicity, statutory requirements, etc.)
Increased stress on irrigation water	– New sources (e.g., on-farm ponds storing winter runoff)	– Increase irrigation efficiency – Change cropping patterns
Increased flood risk	– Increase flood protection	– Accept higher risk of loss – Reduce exposure by relocation
Reduced navigation opportunities	– Enhance water-level management – Increase dredging	– Smaller ships
Reduced power generation opportunities	– Install/increase water storage	– Increase cooling water-use efficiency

"Supply-side" techniques to address water resources and risks are most widely known and used at present. In terms of water supply, these strategies include building new supply and distribution infrastructure and managing existing sources more efficiently. There is an increasing tendency toward conjunctive use of different sources within a region. Supply-side techniques in flood management are termed structural adjustments; they involve actions to lessen flood peaks and keep floodwaters away from at-risk property. Demand-side techniques historically have been less well used but are the focus of increasing attention. In terms of water supply, the most obvious demand-side approach is to reduce the demand (or slow the increase in demand) for the water resource through a range of measures, including differential pricing, public awareness campaigns, or statutory requirements for WUE (e.g., for domestic appliances).

Attention to date has been concentrated on supply-side options; demand-side techniques are less well-known and the subject of considerable technological advance. Although there is a good deal of awareness of broad options, there has been less research into applying these options in an uncertain future: How, for example, can a water distribution network be designed so that it can be incrementally updated as more information on climate change appears, rather than built to a single fixed standard?

An adaptation in one aspect of the water environment or in one area might have a severe effect on another aspect or area. For example, increased use of a river to maintain supplies may impact the instream environment. Increased storage of water in a reservoir over the winter might reduce its ability to prevent flooding. Finally, the time scales over which adaptation can occur vary. Some adaptive strategies can be implemented with little warning and can be easily amended. Others take longer to implement and, once completed, are harder to change. For example, it typically takes decades for a reservoir to move from the planning stage to completion. Urban storm drainage has a relatively short lead time but a very long design life. An ideal adaptive strategy is one that can be altered as more information becomes available. However, the adaptive response by a water management agency to climate change will depend on procedures for considering strategies, as well as adaptive capacity.

13.3.2. Soils

The use of land management techniques in adaptation to climate change will seek to maintain soil functions. A summary of appropriate techniques and their relationships to the soil functions and properties discussed earlier in this chapter is given in Table 13-7.

13.3.3. Ecosystems

Tundra areas have practically no adaptive options available. The only possible response is to better protect areas with particular value (e.g., for bird life) from other stresses such as land exploitation and tourism.

In boreal regions, many effects of climate change and CO_2 increase, as noted above, are positive in terms of productivity. Areas that are under threat (e.g., low-productivity mountain forests) have no adaptive options available. If changing water tables present a significant problem, boreal wetlands could receive technical measures for controlling water level, but such measures are unlikely to be technically and economically feasible in many areas. Other approaches to adaptation can include establishment of buffer areas around wetlands, promotion of sustainable uses of wetlands to minimize additional stresses brought on by climate change, and restoration of already destroyed wetland habitats (Hartig *et al.*, 1997).

Water shortage in parts of the temperate zone may be compensated to some extent by additional irrigation or river management. To achieve any large-scale effect, however, such measures would have to be applied in economically infeasible dimensions. Most other areas will have positive effects and therefore no adaptive problems.

In the Mediterranean, many landscapes face an acute management problem already without climate change because bush encroachment in earlier agricultural areas is affecting many ecosystems with respect to species richness and susceptibility to fire. Climate change may aggravate these developments in some areas. It is not likely that there are direct ways to adapt to these trends.

With regard to forests, the central problem is that any potential adaptation requires long-term planning. Essentially, adaptation to climate would require planting trees today that will be suitable for such a future climate. However, given our uncertainties in the prediction of future climate and the formulation of models that are used to assess its ecological impacts, it is unlikely that adaptation measures will be put into practice in a timely manner.

13.3.4. Agriculture

To avoid or at least reduce negative effects and exploit possible positive effects, several economic and agronomic adaptation strategies for agriculture have been suggested. Economic strategies are intended to render the agricultural costs of climate change small by comparison with overall expansion of agricultural products. Agronomic strategies intend to offset the loss of productivity caused by climate change, either partially or completely. Agronomic strategies include short-term adjustments and long-term adaptations.

13.3.4.1. Short-Term Adjustments

Short-term adjustments to climate change are efforts to optimize production with major system changes. They are autonomous in the sense that no other sectors (e.g., policy, research) are needed for their development and implementation. Thus, short-term adjustment can be considered the first defense tool against

Table 13-7: Land management options to mitigate impact of climate change on soils.

Function	Impact	Management/Adaptation Options
Land production	– Salinization	– Improved technology for application, better water quality, better water scheduling
	– Peat wastage	– No drainage of lowland peat soils
	– Acidification	– Soil pH management
	– Erosion	– Soil conservation techniques (expand)
	– Compaction	– Better timing of field operation, use of new tillage equipment
	– Soil biodiversity	– ???
Land regulation	– Soil water	– Irrigation (with improved technology and scheduling)
	– Soil organic matter	– Use of manures, reduced tillage, improved farming system methods, crop rotation management
	– Soil nutrients	– Sustainable use of fertilizers/manures, crop rotation management
	– Polluting chemicals	– Limits on use of polluting chemicals, clean-up of contaminated land
	– Soil temperature	– Mulching
	– Soil material resources	– No adaptation possible
Land carrier	– Water movement and soil structure	– Management of vertisols(?), timing of manure and sewage sludge applications
	– Nitrate leaching	– Change in fertilizer application rates, precision farming, crop selection (i.e., with different N requirements), breeding nitrogen-fixing crops, breeding crops to improve N-use efficiency (e.g., lower requirements, more efficient uptake), irrigation management, soil pH management, nitrification inhibitors, release rates (e.g., slow or timed release, coatings to limit or retard water solubility), improved fertilizer placement and timing (e.g., band placement, foliar applications), application placement (e.g., slurry injection), application timing, application amounts (e.g., controlled rate systems)
	– Volatilization	– See management options appropriate to reduce nitrate leaching
	– Carbon dioxide fluxes	– Land-use change for carbon sequestration
	– Methane fluxes	– Increased sink through fertilizer management
	– Nitrous oxide fluxes	– See management options appropriate to reduce nitrate leaching
Land information	– Historical record	– Conserve intact soil profiles and representative reference sites
Land consumption	– Shrink/swell damage	– Underpinning of building foundations, insure differently (e.g., location restrictions)

climate change. A large range of such adjustments have been reported, including:

- *Changes in planting dates and cultivars:* For spring crops, climate warming will allow earlier planting or sowing than at present. Crops that are planted earlier are more likely to have already matured when extreme high temperatures, such as in the middle of summer, can cause injury. Earlier planting in spring increases the length of the growing season; thus, earlier planting and use of long-season cultivars will increase yield potential, provided moisture is adequate and the risk of heat damage is low. Otherwise, earlier planting combined with a short-season cultivar would give the best assurance of avoiding heat and water stresses. Deeper planting of seeds also will contribute to making seed germination more likely. For winter crops (i.e., cereals), this approach may cause problems because their cycle length often is linked with cold temperature requirements (vernalization) that may not be completely fulfilled during warmer winters. Late cultivars also may not be able to escape heat and drought risks in the summer. Winter cereals must have a specific growth stage before the onset of winter to ensure winter survival, and they often are sown when temperatures approach the time when vernalization is most effective (Harrison and Butterfield, 1996). This may mean later sowings in northern Europe under climatic warming.
- *Changes in external inputs:* External inputs are used to optimize production of crops in terms of productivity and profitability. The use of fertilizers generally is adjusted to fit the removal of nutrients by the crop and any losses of nutrients that may occur during or between growing seasons. A change in yield level therefore, all other things being equal, will imply a corresponding change in fertilizer inputs. Projected

increases in atmospheric CO_2 concentration will cause a large nitrogen uptake by the crop and thus larger fertilizer applications. On the other hand, climatic constraints on yields may lead to less demand for fertilizers. Changes in climate also may cause larger (or smaller) losses of nitrogen through leaching or gaseous losses. This also may lead to changes in the demand for fertilizer. Global warming will lead to a higher incidence of weeds, pests, and diseases in many areas and thus to potentially increased use of chemical control measures (e.g., pesticides). These inputs can be kept low through adoption of integrated pest management systems, which adjust control measures to the observed problem and also take a range of influencing factors (including weather) into account. Current fertilizer and pesticide practices are based partly on models and partly on empirical functions obtained in field experiments. These models and functions are updated regularly with new experimental evidence. This process probably will capture the response of changes in the environment through CO_2 and climate. It is important, however, that agricultural researchers and advisors are aware of the possible impact of global change on the use of external inputs, so that older empirical data are used with proper caution.

- *Practices to conserve moisture:* Several water-conserving practices are commonly used to combat drought. These also may be used to reduce climate change impacts (Easterling, 1996). Such practices include conservation tillage and irrigation management. Conservation tillage (the practice of leaving some of the previous season's crop residues on the soil surface) may protect the soil from wind and water erosion and retain moisture by reducing evaporation and increasing infiltration of precipitation into the soil. Irrigation management can be used to improve considerably the utilization of applied water through proper timing of the amount of water distributed. For example, with irrigation scheduling practices, water is applied only when the crop needs it. This tunes the proper timing and amount of water to actual field conditions, allowing a reduction in water use as well as the cost of production.

13.3.4.2. Long-Term Adaptations

Long-term adaptations refer to major structural changes to overcome adversity caused by climate change. These may include:

- *Changes in land use* result from the differential response of crops to climate change. Studies reported by Parry *et al.* (1988) for central Europe show an "optimal land use" in which the area cultivated with winter wheat, maize, and vegetables increases while the allocation to spring wheat, barley, and potato, decreases. Changes in land allocation also may be used to stabilize production. In this case, crops with high interannual variability in production (e.g., wheat) may be replaced by crops with lower productivity but more stable yields (e.g., pasture).
- *Biotechnology* offers another possibility to adapt to stresses (heat, water, pests and disease, etc.) that are enhanced by climate change by allowing development of "designer cultivars" (Goodman *et al.*, 1987)—considering strictly the principles of biosafety to avoid possible negative impacts of this technique. Species that have not been used previously for agricultural purposes may be identified and others already identified may be more quickly brought into use.
- *Crop substitution* also may be useful for conservation of soil moisture. Some crops use a low amount of water and are more water- and heat-resistant, so they tolerate dry weather better than others do. For example, sorghum is more tolerant of hot and dry conditions than maize.
- *Microclimate modification* may be used to improve WUE in agriculture. Windbreaks, for example, reduce evaporative demand from the plants they shelter. Sheltered plants remain better hydrated and thus are better able to carry out photosynthesis (Rosenberg, 1979). A wide array of intercropping, multi-cropping, relay cropping, and other techniques provide greater production per unit area occupied and can be useful to improve WUE. Irrigation efficiency can be improved considerably with new land-field techniques (laser-leveling of fields, minimum tillage, chiseling compacted soils, stubble mulching, etc.) or new management strategies (irrigation scheduling, monitoring soil moisture status, etc.) (Kromm and White, 1990).
- *Changes in farming systems* may be necessary in some areas for farming to remain viable and competitive. Specialized arable farms with production of vegetables, cereals, seed crops, fruits, and other crops often have only a few species on the farm, depending on soil and climate conditions. These specialized farms, especially dairying and arable, may be more affected by climate change than mixed farms. Mixed farms with both livestock and arable production have more options for change and thus larger resilience to change in the environment.

Studies on adapting farming systems to climate change need to consider all of the agronomic decisions made at the farm level (Kaiser *et al.*, 1993). Economic considerations are very important in this context (Antle, 1996). Results of farm-level analyses on the impacts of climate change generally have shown a large reduction in adverse impacts when adaptation is fully implemented. However, this will result in land-use changes (Parry *et al.*, 1999).

13.3.5. Coastal Regions

Three broad response strategies are distinguished (Klein *et al.*, 2000):

- Reduce the risk of the event by decreasing its probability of occurrence

- Reduce the risk of the event by limiting its potential effects
- Increase society's ability to cope with the effects of the event.

These strategies have been termed "protect," "retreat," and "accommodate," respectively. Protection usually is associated with coastal squeeze and hence a decline in natural functions and values, although soft protection approaches may not raise this problem.

The actual strategy chosen will depend on local and national circumstances, including the economic and ecological importance of the coastline, technical and financial capabilities, and the legislative and political structure of the countries concerned. Although optimum response strategies have yet to be developed, it is likely that a range of responses will be the norm within any country (Bijlsma *et al.*, 1996). Turner *et al.* (1995) analyzed protection in East Anglia, England. At the aggregated scale of East Anglia, protection can be justified for the entire coast given a 50-cm rise in sea level by 2050. However, this is not the scale at which coastal management decisions are made. When the 113 individual flood compartments are evaluated independently, 20% optimally would be abandoned even for present rates of relative sea-level rise (10-cm rise in sea level by 2050). This analysis assumes that there is no interaction between flood compartments—which may not always be the case. However, it reinforces the conclusion that a range of responses will be appropriate.

Some national estimates of protection costs are given in Table 13-5, primarily for a 1-m rise in sea level. In terms of relative costs, Poland appears to be more vulnerable than Germany and The Netherlands. The absolute costs for Germany are larger than for The Netherlands, reflecting the much longer German coastline.

Adapting successfully to climate change requires more than a list of options. Klein *et al.* (1999) argue that successful adaptation must consider several issues, including recognizing the need for adaptation, planning, implementation, and evaluation. Although there is increasing recognition of the need for adaptation in some countries, this is not uniform across Europe (e.g., Nicholls and Hoozemans, 1996). It also is clear that coastal adaptation to climate change must happen in the broader context of coastal management.

Response options may be hindered by resource constraints. In Cyprus, the coast is no longer receiving new supplies of sand as a result of catchment regulation and management (Nicholls and Hoozemans, 1996). Erosion is expected in response to sea-level rise, but there are no ready sources of sand available for beach nourishment. Yet maintaining the beach is critical to the tourist industry. Although this problem has not been analyzed in detail, external (and hence costly) sources of sand may be required for beach nourishment. Many other Mediterranean islands appear to have similar problems, including those in Greece, Italy, France, and Spain.

Some adaptation that anticipates climate change already is being implemented, and this seems to be raising important questions about long-term coastal management (Klein *et al.*, 1999). The Netherlands is highly threatened by sea-level rise (see Table 13-5): About 60% of the country (23,600 km^2) is in the potential impact zone (Baarse *et al.*, 1994). A new national law "outlaws" erosion and mandates maintenance of the present shoreline position via ongoing beach nourishment (Koster and Hillen, 1995). However, a debate about the optimum response continues; some people advocate a more dynamic response, including a mixture of holding the line, allowing some retreat, and coastal advancement in areas where more land is required (de Ruig, 1998; Klein *et al.*, 1998). The debate includes explicit consideration of maintaining and enhancing coastal resilience.

In Britain, coastal cells have become the basis of shoreline management; about 40 shoreline management plans that cover the entire coastline of England and Wales are finished or nearing completion (UK Ministry of Agriculture, Fisheries, and Food, 1995; Leafe *et al.*, 1998). Managed realignment of sea defenses in estuaries also is attracting increasing attention, including some trial experiments (Klein *et al.*, 1999). Low-grade agricultural land is given up to the sea as flood defenses are abandoned or relocated inland. If managed realignment is practiced at a large scale, it will help to maintain natural values as well as the flood protection benefits of coastal wetlands under a rising sea level. However, there will be a corresponding net loss of freshwater ecosystems in the protected areas (Lee, 1998). About 100 ha yr^{-1} of land would need to be released just to counter present rates of salt marsh loss resulting from sea-level rise. Although the economic analysis of Turner *et al.* (1995) would suggest that there are many suitable sites available, this requires strategic planning to be put in place as part of shoreline management planning (UK Ministry of Agriculture, Fisheries, and Food, 1995; Leafe *et al.*, 1998). Political acceptance of managed retreat remains to be assessed.

Around the Mediterranean, models of deltaic response to sea-level rise and frameworks to analyze vulnerability and sustainability are being developed (e.g., Sanchez-Arcilla *et al.*, 1998). This raises the prospect of a dynamic management approach that harnesses natural inputs and processes within deltas to counter global and local (as a result of subsidence) sea-level rise, rather than a move to hard defenses.

One pertinent issue is cross-border transport of sediment, which has costs within the country of origin but reduces the vulnerability of the receiving country to sea-level rise. For instance, coastal erosion on the east coast of England provides mud that helps to sustain the Wadden Sea in The Netherlands and Germany under rising sea level (Dyer and Moffat, 1998; Nicholls and Mimura, 1998). This suggests a need for a regional perspective on the coastal impacts of sea-level rise.

In conclusion, a common problem across Europe is developing strategic management approaches that allow both continued human utilization of the coastal zone and preservation of coastal ecosystems, given sea-level rise (Nicholls, 2000). Slow but steady degradation of the coastal fringe in much of Europe has gone largely unnoticed until recently, and this will continue and accelerate with sea-level rise. Developing methods to balance protection of people and the economy against the costs of degradation of the coastal environment will require multidisciplinary research.

13.4. Synthesis

13.4.1. Key Impacts

Two important messages emerge from this review:

- Potential impacts on key resources (land, ecosystems, and particularly water) are significant.
- Even in prosperous Europe, adverse climate change impacts may aggravate equity issues.

Climate change in Europe will involve losses and gains to the natural resource base; in some cases, these changes have begun. These impacts will vary substantially from region to region (they will be particularly adverse in the south) and within regions, from sector to sector as well as within sectors. The significance of these effects, however, will depend to a considerable extent on nonclimatic drivers of environmental change, socioeconomic development, and policy evolution within Europe. The most significant impacts (and opportunities) that will require greatest attention with respect to policies of response may be summarized in general terms as follows (a fuller precis appears in the Executive Summary at the beginning of this chapter). Climate changes as characterized by the scenarios described in Chapter 3 of this report, if not adequately responded to through effective adaptation and policy development, would lead in Europe to:

- Increased pressures on water resources, particularly in the south
- Aggravated flood hazard
- Deterioration in soil quality
- Altered natural eosystems, with loss of some habitats and potential loss of species
- Increased productivity of northern commercial forests but reductions in the south
- Increased forest fire risk
- Positive effects on agriculture in the north but broadly negative effects in the south
- Altered fisheries potential
- Increased property damage
- Relatively minor effects on the transport, energy, and manufacturing sectors, some of which may be positive (though substantial effects may derive from policies of mitigation that are necessary to reduce impacts elsewhere)
- Changing tourist potential
- A range of human health implications
- Increased risk of flooding, erosion, wetland loss, and degradation in coastal zones
- Upward shift of biotic zones and snowlines in mountain regions.

In general, more adverse effects may be expected to occur in regions of Europe that are economically less developed because adaptive capacity will be less developed there. These areas would include more marginal areas of the EU and regions outside the EU such as the Balkans.

13.4.2. Uncertainty

The ability to adapt to adverse impacts and exploit opportunities as they emerge is significant. However, present uncertainties with regard to the magnitude and even the sign of the impacts, which characterize almost all of the conclusions reported in this chapter, often will hinder such efforts. These uncertainties stem from uncertainties concerning how climate may change in the future (a subject considered in Chapter 3) and how the

***Table 13-8**: Minimum, maximum, and median estimates of changes in water-limited wheat yield (t ha^{-1}) from baseline climate derived from running 24 climate change scenarios for 2050 through EuroWheat model. Estimates are summaries for selected regions in Europe (Harrison and Butterfield, 1999). Result for HadCM2 scenario also is shown.*

Region	Summary of 24 Scenarios: Minimum	Maximum	Median	HadCM2
Nordic countries	0.7	3.5	2.1	2.8
British Isles	0.9	2.6	1.7	1.8
Germany + Benelux	0.7	2.5	1.7	2.1
Alpine countries	1.5	3.2	2.1	2.8
France	-1.5	3.0	1.3	1.5
Portugal + Spain	-0.9	4.5	1.4	1.2
Italy + Greece	0.7	3.6	1.7	1.8
Poland	-0.5	2.2	1.3	1.7
Central Europe	-1.7	2.6	1.7	1.7
Bulgaria + Romania	-2.9	2.3	1.0	0.8

natural resource base may respond to such changes. In some cases, the combination of these uncertainties may mean that we are unsure about whether the total effect is broadly positive or negative. To illustrate, Table 13-8 shows the estimated effect of a wide range of possible climate futures on wheat yields in different regions of Europe. In half of the regions, yield responses range from negative to positive.

Low-probability/high-consequence events (sometimes termed "surprises") such as collapse of the thermohaline circulation of the North Atlantic (Hulme and Carter, 2000) also need to be considered (see Chapters 3 and 19). Therefore, caution must be exercised in interpreting currently available information, and there must be recognition that much more research is needed. This might include identifying flexible actions that are robust to a range of possible futures.

13.4.3. Research Needs

Further investigation is needed to understand how European climate is likely to change under various emission trends, as well as what the implications of these changes would be for Europe's human and ecological systems. There is a need to analyze outputs from climate models that relate to extreme events. The biophysical impacts of climate change on European water, soil, and land resources have been quite thoroughly researched, but we know less about their socioeconomic consequences (e.g., in agriculture, fisheries, nature conservation, transport systems, and tourism). Improved knowledge of European coastal systems is required for their sustainable management. A key research challenge is to evaluate the feasibility, costs, and benefits of potential adaptation options, measures, and technologies. In general, research programs could benefit from more integration between basic studies of the Earth system, climate change modeling, impact and adaptation assessments, and mitigation/policy analysis. Key research challenges include:

- Transboundary regional impact assessment of various natural and socioeconomic systems
- Transboundary monitoring of impacts on sensitive ecosystems
- Better understanding of likely changes in extreme weather events
- Quantification of natural and technological hazards resulting from climate change (especially flooding)
- Guidelines for water management decisions for different regions under climatic uncertainty
- Changes in European legislation for adaptation to and/or mitigation of climate change impacts.

13.4.4. Regional Issues

13.4.4.1. Extreme Events

One of the most important continent-wide issues is the effect of climate change on future water resources, particularly on extreme events—whether abundance or scarcity of water will increase in severity. A tendency toward increases in the frequency of extreme events is likely but largely unquantified. The implications of changes in the characteristics of extreme hydrological events extend beyond the water sector into agriculture, industry, settlements, coastal zones, transport, tourism, health, and insurance. Climate change is likely to lead to increased winter flows in much of Europe. In Mediterranean Europe, it is likely to increase variability in flow through the year, with summer flows reduced. In coastal areas, the risk of flooding and erosion will increase substantially. Instability of thermohaline circulation is a process whose probability of occurrence in the 21st century is low (see Chapters 3 and 19), yet potential consequences of this on the changing heat budget in Europe are very great.

13.4.4.2. Subregional Impacts

Climatic observations and the scenarios presented here suggest that southern Europe will be more adversely affected than northern Europe. A shift in climate-related resources from south to north may occur in sectors such as tourism, agriculture, and forestry. In particular, the Mediterranean region appears likely to be adversely affected. Among likely adverse effects are increased variability of river flow; increased flood risk; decreased summer runoff and recharge of aquifers; and reduced reliability of public water supply, power generation, and irrigation. Increased fire hazards affecting populated regions and forests and heat stress to humans, crops, and livestock may occur. Even a change of tourist destinations is possible if the present optimal summertime climate shifts northward, and heat waves and water shortages may jeopardize the attractiveness of the present southern summer destinations.

13.4.4.3. Sustainability and Equity

Sustainable development (understood as relaying natural capital to future generations in a nondepleted state) has been in jeopardy in Europe from several existing pressures, mostly nonclimatic (e.g., land-use change, environmental pollution, atmospheric deposition). Yet climate change adds an important element to the threat to the environment. Sea-level rise threatens coastal habitats with a squeeze between hard defenses and rising water levels. Most (50–90%) of the alpine glaciers could disappear by the end of the 21st century, and there may be local extinctions of species that require cold habitats for their survival (e.g., subarctic and montane species). Many ecosystems will respond to climate change via migration and change; a policy challenge is how to manage these changes.

Finally, climate change impacts will be differently distributed among different regions, generations, age classes, income groups, occupations, and genders. This has important equity implications, although these implications have not been investigated in detail. For example, elderly and sick people suffer more in heat waves. There is greater vulnerability, in general, in southern than in northern Europe. Mediterranean and mountain farmers are likely to be worse off in a warmer world. This presents a challenge to existing regional policies within the EU that are aimed at leveling up less-developed areas (peripheral Europe versus core Europe). In general, the more marginal and less wealthy areas will be less able to adapt, so climate change without appropriate policies of response may lead to greater inequity.

Possible climate change impacts on key resources are sufficient to warrant early consideration by European policymakers to ensure sustainable development. In general, the adaptation potential of socioeconomic systems in much of Europe is high because of economic conditions, a stable population with the capacity to move within the region, and well-developed political, institutional, and technological support systems.

References

Aas, B. and T. Faarlund, 1995: Forest limit development in Norway, with special regard to the 20th century. *AmS-Varia*, **24,** 89–100.

Abakumova, G.M., E.M. Feigelson, V. Russak, and V.V. Stadnik, 1996: Evaluation of long-term changes in radiation, cloudiness and surface temperature on the territory of the Former Soviet Union. *Journal of Climate*, **9,** 1319–1327.

Abegg, B. and R. Froesch, 1994: Climate change and winter tourism. In: *Mountain Environments in Changing Climates* [Beniston, M. (ed.)]. Routledge, London, United Kingdom, pp. 328–348.

Agnew, M.D. and J.E. Thornes, 1995: The weather sensitivity of the UK food retail and distribution industry. *Meteorological Applications*, **2,** 137–147.

Alcamo, J., G.J. van den Born, A.F. Bouwman, B.J. de Haan, K. Klein, Goldewijk, O. Klepper, J. Krabec, R. Leemans, J.G.J. Olivier, A.M.C. Toet, H.J.M. de Vries, and H.J. van der Woerd, 1994: Modeling the global society-biosphere-climate system, part 2: computed scenarios. *Water, Air, and Soil Pollution*, **76,** 37–78.

Alexandrov, V., 1997: Vulnerability of agronomic systems in Bulgaria. *Climatic Change*, **36**, 135–149.

Alexandrov, V., 1998: A strategy evaluation of irrigation management of maize crop under climate change in Bulgaria. In: *Proceedings of the Second International Conference on Climate and Water, Espoo, Finland, 17-20 August 1998* [Lemmelä, R. and N. Helenius (eds.)]. Vol. 3, pp. 1545–1555.

Alexandrov, V., 1999: Vulnerability and adaptation of agronomic systems in Bulgaria *Climate Research*, **12(2–3),** 161–173.

Anisimov, O.A. and F.E. Nelson, 1996: Permafrost distribution in the northern hemisphere under scenarios of climate change. *Global and Planetary Change*, **14,** 59–72.

Anisimov, O.A. and F.E. Nelson, 1997: Permafrost zonation and climatic change in the northern hemisphere: results from transient general circulation models. *Climatic Change*, **35,** 241–258.

Antle, J.M., 1996: Methodological issues in assessing potential impacts of climate change on agriculture. *Agricultural and Forest Meteorology*, **80,** 67–85.

Armstrong, A.C. and D.A. Castle, 1992: Potential impacts of climate change on patterns of production and the role of drainage in grassland. *Grass and Forage Science*, **47,** 50–61.

Armstrong, A.C., A.M. Matthews, A.M. Portwood, T.M. Addiscott, and P.B. Leeds-Harrison, 1994: Modelling the effects of climate change on the hydrology and water quality of structured soils. In: *Soil Responses to Climate Change. NATO ASI Series 23* [Rounsevell, M.D.A. and P.J. Loveland (eds.)]. Springer-Verlag, Heidelberg, Germany, pp. 113–136.

Arnell, N.W., 1996: *Global Warming, River Flows and Water Resources*. Wiley and Sons, Chichester, United Kingdom.

Arnell, N.W., 1999: The effect of climate change on hydrological regimes in Europe: a continental perspective. *Global Environmental Change*, **9,** 5–23.

Arnell, N., 2000: Water resources. In: *Assessment of Potential Effects and Adaptations for Climate Change in Europe: The Europe ACACIA Project* [Parry, M.L. (ed.)]. Jackson Environment Institute, University of East Anglia, Norwich, United Kingdom, 324 pp.

Arnell, N.W. and N.S. Reynard, 1996: The effects of climate change due to global warming on river flows in Great Britain. *Journal of Hydrology*, **183,** 397–424.

Arnell, N.W. and N.S. Reynard, 1999: Climate change and British hydrology. In: *The Hydrology of the UK* [Acreman, M.C. (ed.)]. Routledge, London, United Kingdom, pp. 3–29..

Arnell, N., W. Grabs, M. Mimikou, A. Shumann, N.S. Reynard, Z. Kaczmarek, V. Oancea, J. Buchtele, L. Kasparek, S. Blazkova, O. Starosolszky, and A. Kuzin,, 1999: *Effects of Climate Change on Hydrological Regimes and Water Resources in Europe*. Projects EV5V-CT93-0293 and EV5V-CT94-0114, summary brochure, European Commission, University of Southampton, Southampton, United Kingdom.

Auger, C., 1994: The impacts of sea-level rise on Rochefort sur Mer, France. In: *Global Climate Change and the Rising Challenge of the Sea: Proceedings of the third IPCC CZMS workshop, Margarita Island, March 1992* [O'Callahan, J. (ed.)]. National Oceanic and Atmospheric Administration, Silver Spring, MD, USA, pp. 317–328.

Baarse, G., E.B. Peerbolte, and L. Bijlsma, 1994: Assessment of the vulnerability of The Netherlands to sea-level rise. In: *Global Climate Change and the Rising Challenge of the Sea: Proceedings of the Third IPCC CZMS Workshop, Margarita Island, March 1992* [O'Callahan, J. (ed.)]. National Oceanic and Atmospheric Administration, Silver Spring, MD, USA, pp. 211–236.

Bailey, R.S. and J.H. Steele, 1992: North Sea herring fluctuations. In: *Climate Variability, Climate Change, and Fisheries* [Glantz, M.H. (ed.)]. Cambridge University Press, Cambridge, United Kingdom and New York, NY, USA.

Balderi, M., A. Tamburro, G. Sabatinelli, R. Romi, C. Severini, G. Cuccagna, G. Fiorilli, M.P. Allegri, C. Buriani, and M. Toti, 1998: Malaria in Maremma, Italy. *Lancet*, **351,** 1246–1247.

Bálint, G. and M. Butina, 1997: Application of the HBV model for the simulation of hydrological consequences of climate change in the Lielupe River basin in Latvia. In: *Proceedings of the Baltic States Hydrology Conference, Hydrology and Environment, May, 1997, Kaunas, Lithuania* [Gailiusis, D. (ed.)]. Lietuvos Energetikos Institutas, Kaunas, Lithuania, pp. 23–28.

Bálint, G. and B. Gauzer, 1994: A rainfall ruoff model as a tool to investigate the impact of climate change. In: *Proceedings, XVIIth Conference of Danube Countries, Budapest, 5-9 September 1994* [Hegedus, M. (ed.)]. Paper No. 3.3, Hungarian National Committee for IHP/UNESCO and OHP/WMO, Budapest, Hungary, pp. 471–479.

Ball, J.H., M.J. Clark, M.B. Collins, S. Gao, A. Ingham, and A. Ulph, 1991: *The Economic Consequences of Sea-Level Rise on the Central Coast of England*. Unpublished report to the UK Ministry of Agriculture, Fisheries, and Food, Geodata Unit, University of Southampton, 2 volumes.

Baudiere, A. and T. Gauquelin, 1998: Evolution récente des formations superficielles et de la végétation associée sur les hautes terres catalanes. In: *Procesos Biofisicos Actuales en Medios Frios* [Gomez, A., (ed.)]. Publicaciones de la Universidad de Barcelona, Spain, pp. 27–412 (in French).

Baudiere, A. and L. Serve, 1974: Les groupements oromediterraneenes des Pyrenees orientales et leurs relations avec les groupements similaires de la Sierra Nevada. In: *La Flore du Bassin Mediterranean: Essai de Systematique Synthetique*. Coll. Int. CNRS No. 235, Montpelier, France, pp. 457–468 (in French).

Beek, K.J., W.A. Blokhuis, P.M. Driessen, N. van Breemen, R. Brinkman, and L.J. Pons, 1980: Problem soils: their reclamation and management. In: *Land Reclamation and Water Management: Developments, Problems and Challenges*. International Institute for Land Reclamation and Improvement (ILRI), Publication 27, International Soil Reference and Information Centre, Wageningen, The Netherlands, pp. 43–72.

Behr, O., 1998: Possible climate impacts on the water resources of the Danube river basin. In: *Proceedings of the Second International Conference on Climate and Water, Espoo, Finland, 17-20 August 1998* [Lemmelä, R. and N. Helenius (eds.)]. Vol. 2, pp. 829-838.

Beniston, M., 2000: *Environmental Change in Mountains and Uplands*. Arnold/Hodder and Stoughton/Chapman and Hall Publishers, London, United Kingdom, 200 pp.

Beniston, M., D.G. Fox, S. Adhikary, R. Andressen, A. Guisan, J. Holten, J. Innes, J. Maitima, M. Price, and L. Tessier, 1996: The impacts of climate change on mountain regions. In: *Climate Change 1995: Impacts, Adaptions, and Mitigation of Climate Change: Scientific-Technical Analyses. Contribution of Working Group II to the Second Assessment Report for the Intergovernmental Panel on Climate Change* [Watson, R.T., M.C. Zinyowera, and R.H. Moss (eds.)]. Cambridge University Press, Cambridge, United Kingdom and New York, NY, USA, pp. 191–213.

Beniston, M., A. Ohmura, M. Rotach, P. Tschuck, M. Wild, and M.R. Marinucci, 1995: Simulation of climate trends over the Alpine Region. In: *Development of a Physically-Based Modeling System for Application to Regional Studies of Current and Future Climate*. Final Scientific Report No. 4031-33250, Swiss National Science Foundation, Bern, Switzerland.

Beniston, M., R.S.J. Tol, R. Delcolle, G. Hörmann, A. Iglesias, J. Innes, A.J. McMichael, A.J.M. Martens, I. Nemesova, R.J. Nicholls, and F.L. Toth, 1998: Europe. In: *The Regional Impacts of Climate Change: An Assessment of Vulnerability. Special Report of IPCC Working Group II* [Watson, R.T., M.C. Zinyowera, and R.H. Moss (eds.)]. Intergovernmental Panel on Climate Change, Cambridge University Press, Cambridge, United Kingdom and New York, NY, USA, pp. 149–185.

Bennett, G., 1970: Bristol floods 1968: controlled survey of effects on health of local community disaster. *British Medical Journal*, **3,** 454–458.

Bentham, G. and I.H. Langford, 1995: Climate change and the incidence of food poisoning in England and Wales. *International Journal of Biometeorology*, **39,** 81–86.

Berg, B., M.P. Berg, E. Box, P. Bottner, A. Breymeyer, R. Calvo de Anta, M.M. Couteaux, A. Gallardo, A. Escudero, W. Kratz, M. Medeira, C. McClaugherty, V. Meentemeyer, F. Muñoz, P. Piussi, J. Remacle, and A. Virzo de Santo, 1993: Litter mass loss in pine forests of Europe: relationship with climate and litter quality. In: *Geography of Organic Matter Production and Decay* [Breymeyer, A., A. Krawczyk, B. Kulikowski, R.J. Solon, M. Rosciszewski, and B. Jaworska (eds.)]. Polish Academy of Sciences, Warsaw, Poland, pp. 81–109.

Berglund, J., K. Eitrem, A. Ornstein, A. Lindberg, H. Rigner, M. Elmrud, A. Carlsson, C. Runehagen, A. Svanborg, and R. Norrby, 1995: An epidemiological study of Lyme disease in southern Sweden. *New England Journal of Medicine*, **333,** 1319-1327.

Berthold, P. and A.J. Helbig, 1992: The genetics of bird migration: stimulus, timing and direction. *Ibis*, **134(1),** 35–40.

Berthold, P., G. Mohr, and U. Querner, 1990: Control and evolutionary potential of obligate partial migration: results of a two-way selective breeding experiment with the blackcap (Sylvia atricapilla). *Journal für Ornithologie*, **131(1),** 33–46.

Best, E.P.H., J.T.A. Verhoeven, and W.J. Wolff, 1993: The ecology of the Netherlands wetlands: characteristics, threats, prospects and perspectives for ecological research. *Hydrobiologia*, **265,** 305–320.

Bhattacharya, N.C. and R.A. Geyer, 1993: Prospects of Agriculture in a Carbon Dioxide-Enriched Environment. In: *A Global Warming Forum: Scientific, Economic and Legal Overview*. CRC Press, Inc., Boca Raton, FL, USA, pp. 487–505.

Bijlsma, L., C.N. Ehler, R.J.T. Klein, S.M. Kulshrestha, R.F. McLean, N. Mimura, R.J. Nicholls, L.A. Nurse, H. Pérez Nieto, E.Z. Stakhiv, R.K. Turner, and R.A. Warrick, 1996: Coastal zones and small islands. In: *Climate Change 1995: Impacts, Adaptations and Mitigation of Climate Change: Scientific-Technical Analyses. Contribution of Working Group II to the Second Assessment Report of the Intergovernmental Panel on Climate Change* [Watson, R.T., M.C. Zinyowera, and R.H. Moss (eds.)]. Cambridge University Press, Cambridge, United Kingdom and New York, NY, USA, pp. 289–324.

Bindi, M., F. Ferrini, and F. Miglietta, 1992: Climatic change and the shift in the cultivated area of olive trees. *Journal of Agricultura Mediterranea*, **22,** 41–44.

Bindi, M. and L. Fibbi, 2000: Modelling climate change impacts at the site scale on grapevine. In: *Climate Change, Climate Variability, and Agriculture in Europe: An Integrated Assessment* [Downing, T.E., P.A. Harrison, R.E. Butterfield, and K.G. Lonsdale (eds.)]. Research Report 21, Environmental Change Unit, University of Oxford, Oxford, United Kingdom, pp. 117–134.

Bindi, M., L. Fibbi, B. Gozzini, S. Orlandini, and F. Miglietta, 1996: Modeling the impact of future climate scenarios on yield and yield variability of grapevine. *Climate Research*, **7,** 213–224.

Bindi, M. and J. Olesen, 2000: Agriculture. In: *Assessment of Potential Effects and Adaptations for Climate Change in Europe: The Europe ACACIA Project* [Parry, M.L. (ed.)]. Jackson Environment Institute, University of East Anglia, Norwich, United Kingdom, 324 pp.

Birks, H.J.B., 1989: Holocene isochrone maps and patterns of tree-spreading in the British Isles. *Journal of Biogeography*, **16,** 503–540.

Boardman, J., R. Evans, D.T. Favis-Mortlock, and T.M. Harris, 1990: Climate change and soil erosion on agricultural land in England and Wales. *Land Degradation and Rehabilitation*, **2,** 95–106.

Bortenschlager, S., 1993: Das hochst gelegene Moor der Ostalpen "Moor am Rofenberg" 2760 m. *Dissertationes Botanicae*, **196 (Festschrift Zoller),** 329–334 (in German).

Botterweg, P., 1994: Modelling the effects of climate change on runoff and erosion in central southern Norway. In: *Conserving Soil Resources: European Perspectives. Proceedings of the First International Conference, European Society for Soil Conservation, 1993* [Rickson, R.J. (ed.)]. European Society for Soil Conservation, pp. 273–285.

Bradbury, N.J. and D.S. Powlson, 1994: The potential impact of global environmental change on nitrogen dynamics in arable systems. In: *Soil Responses to Climate Change* [Rounsevell, M.D.A. and P.J. Loveland (eds.)]. NATO ASI Series 23, Springer-Verlag, Heidelberg, Germany, pp. 137–154.

Bradley, D.J., 1993: Human tropical diseases in a changing environment. In: *Environmental Change and Human Health*. Ciba Foundation Symposium 175, Wiley and Sons, Chichester, United Kingdom, pp. 147–170.

Brander, K.M., 1997: Effects of climate change on cod (Gadus morhua) stocks. In: *Global Warming: Implications for Freshwater and Marine Fish* [Wood, C.M. and D.G. McDonald (eds.)]. Cambridge University Press, Cambridge, United Kingdom and New York, NY, USA, pp. 255–278.

Brander, K.M., 2000: Detecting the effects of environmental change on growth and recruitment in cod (Gadus morhua) using a comparative approach. *Oceanologica Acta*, **23,** 485–496.

Brázdil, R., M. Pudiková, I. Auer, R. Bohm, T. Cegnar, P. Tasko, M. Lapin, M. Gajic-Capka, K. Zaninovic, E. Koleva, T. Niedzwiedz, S. Szalai, Z. Ustrnul, and R.O. Weber, 1996: Trends of maximum and minimum daily temperatures in central and southeastern Europe. *International Journal of Climatology*, **16,** 765–782.

Briffa, K.R., T.S. Bartholin, D. Eckstein, P.D. Jones, W. Karlén, F.H. Schweingruber, and P. Zetterberg, 1990: A 1,400-year tree-ring record of summer temperatures in Fennoscandia. *Nature*, **346,** 434–439.

Briffa, K.R., P.D. Jones, and M. Hulme, 1994: Summer moisture availability across Europe, 1892–1991: an analysis based on the Palmer Drought Severity Index. *International Journal of Climatology*, **14,** 475–506.

Brinkman, R., 1990: Resilience against climate change? Soil minerals, transformations and surface properties: EH and pH. In: *Soils on a Warmer Earth* [Scharpenseel, H.W., M. Schomaker, and A. Ayoub (eds.)]. Elsevier, London, United Kingdom, pp. 51–60.

Brock, T.C.M. W. van Vierssen, 1992: Climate change and hydrophyte-dominated communities in inland wetland ecosystems. *Wetlands Ecology and Management*, **2,** 37–49.

Bruci, E.D. and M. Bicaj, 1998: Implication of the expected climate change on the water resources in Albania. In: *Proceedings of the Second International Conference on Climate and Water, Espoo, Finland, 17-20 August 1998* [Lemmelä, R. and N. Helenius (eds.)]. Vol. 3, pp. 1471–1477.

Brunetti, M., L. Buffoni, M. Maugeri, and T. Nanni, 2001: Trend of minimum and maximum daily temperature in Italy from 1865 to 1996. *Theoretical and Applied Climatology*, (in press).

Brunetti, M., M. Maugeri, and T. Nanni, 2000: Variations of temperature and precipitation in Italy from 1866 to 1995. *Theoretical and Applied Climatology*, **65,** 165–174.

Buffoni, L., M. Maugeri, and T. Nanni, 1999: Precipitation in Italy from 1833 to 1996. *Theoretical and Applied Climatology*, **63,** 33–40.

Building Research Establishment, 1990: Assessment of damage in low rise buildings. Building Research Establishment, BRE Digest 251, London, United Kingdom, 110 pp.

Bullock, P., E. Le Houérou, M.T. Hoffman, M.D.A. Rounsevell, J. Sehgal, and G. Várallay, 1996: Land degradation and desertification. In: In: *Climate Change 1995: Impacts, Adaptations and Mitigation of Climate Change: Scientific-Technical Analyses. Contribution of Working Group II to the Second Assessment Report of the Intergovernmental Panel on Climate Change* [Watson, R.T., M.C. Zinyowera, and R.H. Moss (eds.)]. Cambridge University Press, Cambridge, United Kingdom and New York, NY, USA, pp. 170–190.

Butina, M., G. Melnikova, and I. Stikute, 1998: Potential impact of climate change on the hydrological regime in Latvia. In: *Proceedings of the Second International Conference on Climate and Water, Espoo, Finland, 17-20 August 1998* [Lemmelä, R. and N. Helenius (eds.)]. Vol. 3, pp. 1610–1617.

Camill, P. and J.S. Clark, 1998: Climate change disequilibrium of boreal permafrost peatlands caused by local processes. *The American Naturalist*, **151,** 207–222.

Cannell, M.G.J., 1985: Analysis of risks of frost damage to forest trees in Britain. In: *Crop Physiology of Forest Trees* [Tigerstedt, P.M.A., Puttonen, P. and Koski, V. (eds.)]. Helsinki University Press, Helsinki, Finland, pp. 153–166.

Cannell, M.G.R. and Pitcairn, C.E.R., 1993: *Impacts of the Mild Winters and Hot Summers in the United Kingdom in 1988–1990*. Department of the Environment, HMSO, London, 154.

Cannell, M.G.R. and J.H.M. Thornley, 1998: N-poor ecosystems may respond more to elevated CO_2 than N-rich ones in the long term: a model analysis of grassland. *Global Change Biology*, **4,** 431–442.

Cannon, R.J.C., 1998: The implications of predicted climate change for insect pests in the UK, with emphasis on non-indigenous species. *Global Change Biology*, **4,** 785–796.

Cao, M., K. Gregson, and S. Marshall, 1998: Global methane emission from wetlands and its sensitivity to climate change. *Atmospheric Environment*, **32,** 3293–3299.

Carnell, R.E. and C.A. Senior, 1998: Changes in mid-latitude variability due to increasing greenhouse gases and sulphate aerosols. *Climate Dynamics*, **14,** 369–383.

Carter, M.R. and B.A. Stewart, 1996: *Structure and Organic Matter Storage in Agricultural Soils*. CRC Press, London, United Kingdom, 125 pp.

Carter, T.R., 1998: Changes in the thermal growing season in Nordic countries during the past century and prospects for the future. *Agricultural and Food Science in Finland*, **7,** 161–179.

Carter, T.R., M.L. Parry, H. Harasawa, and S. Nishioka, 1994: *IPCC Technical Guidelines for Assessing Climate Change Impacts and Adaptations*. University College and Tsukuba University, London, United Kingdom and Tsukuba, Japan, 59 pp.

Carter, T.R., M. Posch, and H. Tuomenvirta, 1996a: The SILMU scenarios: specifying Finland's future climate for use in impact assessment. *Geophysica*, **32,** 235–260.

Climate Change Impacts Review Group, 1991: *The Potential Effects of Climate Change in the UK*. Department of the Environment, Her Majesty's Stationary Office, London, United Kingdom, 124 pp.

Climate Change Impacts Review Group, 1996: *The Potential Effects of Climate Change in the UK*. Department of the Environment, Her Majesty's Stationary Office, London, United Kingdom, 247 pp.

Colacino, M. and M. Conte, 1993a: Greenhouse effect and pressure patterns in the Mediterranean basin. *Nuovo Cimento C*, **16,** 67–76.

Colacino, M. and M. Conte, 1993b: *Clima Mediterraneo ed Incendi Boschivi*. Report 93-50, Consiglio Nazionale delle Ricerche, Istituto di Fisica dell'Atmosfera, Rome, Italy.

Cole, C.V., K. Paustian, E.T. Elliott, A.K. Metherell, D.S. Ojima, and W.J. Parton, 1993: Analysis of agroecosystem carbon pools. *Water, Air, and Soil Pollution*, **70,** 357–371.

Consorzio Venezia Nuova, 1997: Measures for the Protection of Venice and its Lagoon. Consorzio Venezia Nuova (concessionary of Ministry of Public Works, Water Authority of Venice), Venice, Italy.

Cook, R.M., A. Sinclair, and G. Stefansson, 1997: Potential collapse of North Sea cod stocks. *Nature*, **385,** 521–522.

Cooper, G. and M.B. McGechan, 1996: Implications of an altered climate for forage conservation. *Agricultural and Forest Meteorology*, **79,** 253–269.

Cooper, K., D.J. Parsons, and T. Demmers, 1998: A thermal balance model for livestock buildings for use in climate change studies. *Journal of Agricultural Engineering Research*, **69,** 43–52.

Crick, H.Q.P. and T.H. Sparks, 1999: Climate change related to egg-laying trends. *Nature*, **399,** 423–424.

Crick, H.Q.P., C. Dudley, and D.E. Glue, 1997: UK birds are laying eggs earlier. *Nature*, **399,** 526.

Cuculeanu, V, A. Marica, and C. Simota, 1999: Climate change impact on agricultural crops and adaptation options in Romania. *Climate Research*, **12 (2-3),** 153–160.

Daamen, K., P. Krahe, and K. Wilke, 1998: Impacts of possible climate change on the discharge in German low mountain range catchments. In: *Proceedings of the Second International Conference on Climate and Water, Espoo, Finland, 17-20 August 1998* [Lemmelä, R. and N. Helenius (eds.)]. Vol. 2, 805-819.

Dai, A., I.Y. Fung, and A.D. del Genio, 1997: Surface observed global land precipitation variations during 1900–1988. *Journal of Climate*, **10,** 2943–2962.

Davies, A., J. Shao, P. Brignall, R.D. Bardgett, M.L. Parry, and C.J. Pollock, 1996: Specification of climatic sensitivity of forage maize to climate change. *Grass and Forage Science*, **51,** 306–317.

Davis, M.B., 1981: Quaternary history and the stability of forest communities. In: *Forest Succession* [West, D.C., H.H. Shugart, and D.H. Botkin (eds.)]. Springer-Verlag, New York, NY, USA, pp. 132–153.

Dedet, J.P., M. Lambert, and F. Pratlong, 1995: Leishmaniasis and HIV infection. *Presse Medicale*, **24,** 1036–1040.

de Gentile, L., H. Picot, P. Bourdeau, R. Bardet, A. Kerjan, M. Piriou, A. le Guennic, C. Bayssade-Dufour, D. Chabasse, and K.E. Mott, 1995: La dermitate cercarienne en Europe: un probleme de santé publique nouveau (Cercarial dermatitis in Europe: a new public health problem). *Bulletin of the.World Health Organization*, **74(2),** 159–163 (in French).

del Barrio, G. and E. Moreno, 2000: Background data and information for the EEA report on biodiversity in Europe. In: *Species and Habitats in Arid Lands in the Mediterranean Parts of Europe*. Estacion Experimental de Zonas Aridas.

Dent, D.L., 1986: *Acid Sulphate Soils: A Baseline for Research and Development*. Publication 39, Institute for Land Reclamation and Improvement, Wageningen, The Netherlands, 45 pp.

de Ruig, J.H.M., 1998: Coastline management in The Netherlands: human use versus natural dynamics. *Journal of Coastal Conservation*, **4,** 127–134.

Dippner, J.W., 1997: Recruitment success of different fish stocks in the North Sea in relation to climate variability. *Deutsche Hydrographische Zeitschrift*, **49,** 277–293.

Dlugolecki, A. and G. Berz, 2000: Insurance. In: *Assessment of Potential Effects and Adaptations for Climate Change in Europe: The Europe ACACIA Project* [Parry, M.L. (ed.)]. Jackson Environment Institute, University of East Anglia, Norwich, United Kingdom, 324 pp.

Dregne, H.E., 1990: Impact of global warming on arid region soils. In: *Soils on a Warmer Earth* [Scharpenseel, H.W., M. Schomaker, and A. Ayoub (eds.)]. Elsevier, London, United Kingdom, pp. 177–184.

Durand, M.H. (ed.), 1998: *Global versus Local Changes in Upwelling Systems*. Collection Colloques et Séminaires. ORSTOM, Paris, France.

Dvorák, V., J. Hladny, and L. Kaspárek, 1997: Climate change hydrology and water resources impact and adaptation for selected river basins in the Czech Republic. *Climatic Change*, **36,** 93–106.

Dyer, K. and T.J. Moffat, 1998: Fluxes of suspended matter in the East Anglian Plume Southern North Sea. *Continental Shelf Research*, **18,** 1311–1331.

Easterling, D.R., B. Horton, P.D. Jones, T.C. Peterson, T.R. Karl, D.E. Parker, M.J. Salinger, V. Razuvayev, N. Plummer, P. Jameson, and C.K. Folland, 1997: Maximum and minimum temperature trends for the globe. *Science*, **277,** 364–367.

Easterling, W.E., 1996: Adapting North American agriculture to climate change. *Agricultural and Forest Meteorology*, **80,** 1–53.

Ebenhöh, W., H. Sterr, and F. Simmering, 1997: *Potentielle Gefährdung und Vulnerabilität der deutschen Nord- und Ostseeküste bei fortschreitenden Klimawandel (Case Study Based on the Common Methodology of the IPCC Subgroup B)*. Unpublished report, Hamburg, Germany, 138 pp. (in German).

ECSN, 1995: *Climate of Europe: Recent Variation, Present State and Future Prospects*. KNMI, de Bilt, The Netherlands, 72 pp.

EEA, 1998: *Europe's Environment: The Second Assessment*. European Environment Agency/Elsevier, Copenhagen, Denmark, 293 pp.

Ellenberg, H., 1986: *Vegetation Mitteleuropas mit den Alpen*. Ulmer, Stuttgart, Germany, 4th. ed., 989 pp. (in German).

Emberlin, J., 1994: The effects of patterns in climate and pollen abundance on allergy. *Allergy*, **94,** 15–20.

European Commission, 1999: *Towards a European Coastal Zone (ICZM) Strategy*. Office of Official Publications of the European Union, Luxembourg, 98 pp.

Evans, L.G., H. Eckersten, M.A. Semenov, and J.R. Porter, 1995: Modelling the effects of climate change and climatic variability on crops at site scale: effects on Willow. In: *Climate Change and Agriculture in Europe: Assessment of Impacts and Adaptations* [Harrison, P.A., R. Butterfield, and T.E. Downing (eds.)]. Resarch Report No. 9, Environmental Change Unit, University of Oxford, Oxford, United Kingdom, pp. 330–388.

FAOSTAT, 1998: *On-Line and Multilingual Database*. Food and Agriculture Organization, Rome, Italy, 120 pp.

Farrar, J.F., 1996: Sinks, integral parts of a whole plant. *Journal of Experimental Botany*, **47,** 1273–1280.

Faunt, J.D., T.J. Wilkinson, P. Aplin, P. Henschke, M. Webb, and R.K. Penhall, 1995: The effete in the heat: heat-related hospital presentations during a ten day heat wave. *Australia and New Zealand Journal of Medicine*, **25,** 117–120.

Fischer, H., M. Werner, D. Wagenbach, M. Schwager, T. Thorsteinnson, F. Wilhelms, and J. Kipfstuhl, 1998: Little ice age clearly recorded in northern Greenland ice cores. *Geophysical Research Letters*, **25,** 1749–1752.

Flather, R.A. and J.A. Smith, 1998: First estimates of changes in extreme storm surge elevations due to the doubling of CO_2. *The Global Atmosphere and Ocean System*, **6,** 193–208.

Flurry, M., J. Leuenberger, B. Studer, H. Flühler, W.A. Jury, and K. Roth, 1994: *Pesticide Transport through Unsaturated Field Soils: Preferential Flow*. CIBA Ltd., Basel, Switzerland, 293 pp.

Førland, E.J., H. Alexandersson, A. Drebs, I. Hanssen-Bauer, H. Vedin, and O.E. Tveito, 1998: *Trends in Maximum 1-Day Precipitation in the Nordic Region*. DNMI Report No. 14/98, Norwegian Meteorological Institute, Oslo, Norway, 55 pp.

Forchhammer, M.C., E. Post, and N.C. Stenseth, 1997: Breeding phenology and climate. *Nature*, **391,** 29–30.

Fraedrich, K. and K. Müller, 1992: Climate anomalies in Europe associated with ENSO extremes. *International Journal of Climatology,* **12,** 25–31.

Fraedrich, K., K. Müller, and R. Kuglin, 1992: Northern hemisphere circulation regimes during extremes of the El Niño/Southern Oscillation. *Tellus*, **44A,** 33–40.

Frasetto, R. (ed.), 1991: *Impacts of Sea-Level Rise on Cities and Regions*. Proceedings of the First International Meeting 'Cities on Water,' December 1989, Marsilio Editori, Venice, Italy.

Frei, C. and C. Schär, 1998: A precipitation climatology of the Alps from high-resolution rain-gauge observations. *International Journal of Climatology*, **18,** 873–900.

Friedland, K.D., D.G. Reddin, and D.A. Dunkley, 1998: Marine temperatures experienced by post-smolts and the survival of Atlantic salmon (Salmo Salar L.) in the North Sea area. *Fisheries Oceanography*, **7,** 22–34.

Fulford, M., T. Champion, and A. Long, 1997: *England's Coastal Heritage*. Archaeological Report 15, Royal Commission on the Historical Monuments of England and English Nature, London, United Kingdom, 268 pp.

Furquay, J.W., 1989: Heat stress as it affects animal production. *Journal of Animal Science*, **52,**164–174.

Gabutti, G., G. Balestra, G. Flego, and P. Crovari, 1998: Visceral leishmaniasis in Liguria, Italy. *Lancet*, **351,** 1136.

Gavilán, R. and E. Fernández-González, 1997: Climatic discrimination of Mediterranean broad-leaved sclerophyllous and deciduous forests in central Spain. *Journal of Vegetation Science*, **8,** 377–386.

Gellens, D., E. Roulin, and F. Gellens-Meulenberghs, 1998: Impact of climate change on the water balance in the river Meuse basin (Belgium). In: *Proceedings of the Second International Conference on Climate and Water, Espoo, Finland, 17-20 August 1998* [Lemmelä, R. and N. Helenius (eds.)]. Vol. 2, pp. 820-828.

Giles, A.R. and A.H. Perry, 1998: The use of a temporal analogue to investigate the possible impact of projected global warming on the UK tourist industry. *Tourism Management*, **19,** 75–80.

Goodman, R.M., H. Hauptli, A. Croosway, and V.C. Knauf, 1987: Gene transfer in crop improvement. *Science*, **236,** 48–54.

Goria, A., 1999: Impacts of climate change on migration in the Mediterranean basin: environmental dimension of South-North migration and directions for future policies. In: *Proceedings of the Conference on the Impacts of Climate Change on the Mediterranean Area: Regional Scenarios and Vulnerability Assessment, Venice, 9–10 December, 1999*. Venice, Italy, pp. 10–25.

Gothe, R., I. Nolte, and W. Kraft, 1997: Leishmaniasis of dogs in Germany: epidemiological case analysis and alternative to conventional causal therapy. *Tierarztiche Praxis*, **25,** 68–73.

Gottfried, M., H. Pauli, and G. Grabherr, 1994: Die Alpen im "Treibhaus": Nachweis füer das erwarmungsbedingte Hohersteigen der alpinen und nivalen Vegetation. *Jahrbuch des Vereins zum Schutz der Bergwelt*, **59,** 13–27 (in German).

Gottfried, M., H. Pauli, K. Reiter, and G. Grabherr, 2000: A fine-scaled predictive model for climate warming induced changes of high mountain plant species distribution patterns. *Diversity and Distributions*, **8,** 10–21.

Grabherr, G., M. Gottfried, and H. Pauli, 1994: Climate effects on mountain plants. *Nature*, **369,** 448.

Grabherr, G., M. Gottfried, and H. Pauli, 1995: Patterns and current changes in alpine plant diversity. In: *Arctic and Alpine Biodiversity: Patterns, Causes and Ecosystem Consequences* [Chapin, F.S. III and C. Korner (eds.)]. Springer-Verlag, Heidelberg, Germany, pp. 167–181.

Grabs, W., 1997: *Impact of Climate Change on Hydrological Regimes and Water Resource Management in the Rhine*. Cologne, Germany, 26 pp.

Graglia, E., S. Jonasson, A. Michelsen, and I.K. Schmidt, 1997: Effects of shading, nutrient application and warming on leaf growth and shoot densities of dwarf shrubs in two arctic/alpine plant communities. *Ecoscience*, **4,** 191–198.

Gratz, N.G. and A.B. Knudsen, 1996: *The Rise and Spread of Dengue and Dengue Hemorrhagic Fever and its Vectors: A Historical Review Up to 1995*. CTD/FIL(DEN)/96.7, World Health Organization, Geneva, Switzerland, 62 pp.

Gregory, J.M. and J.A. Lowe, 2000: Predictions of global and regional sea-level rise using AOGCMs with and without flux adjustment. *Geophysical Research Letters*, **27,** 3069–3072.

Groisman, P.Y., E.L. Genikhovich, and P.M. Zhai, 1996: "Overall" cloud and snow cover effects on internal climate variables: the use of clear sky climatology. *Bulletin of the American Meteorological Society*, **77,** 2055–2066.

Gruza, G., E. Rankova, V. Razuvaev, and O. Bulygina, 1999: Indicators of climate change for the Russian Federation. *Climatic Change*, **42(1),** 219–242.

Haapala, J. and M. Lepparänta, 1997: The Baltic Sea ice season in changing climate. *Boreal Environment Research*, **2(1),** 93–108.

Haeberli, W., 1995: Glacier fluctuations and climate change detection: operational elements of a worldwide monitoring strategy. *Bulletin of the World Meteorological Organization*, **44(1),** 23–31.

Häkkinen, R., T. Linkosalo, and P. Hari, 1998: Effects of dormancy and environmental factors on timing of bud burst in Betula pendula. *Tree Physiology*, **18,** 707–712.

Hall, D.O. and J.M.O. Scurlock, 1991: Climate change and the productivity of natural grasslands. *Annals of Botany*, **67,** 49–55.

Hamm, L., H. Hanson, M. Capobianco, H.H. Dette, A. Lecuga, and R. Spanhoff, 1998: *Beach Fills in Europe—Projects, Practices, and Objectives*. Proceedings of the 26th International Coastal Engineering Conference, Copenhagen, Denmark. American Society of Civil Engineers, New York, NY, USA, pp. 3060–3073.

Harrison, P.A. and R.E. Butterfield, 1996: Effects of climate change on Europe-wide winter wheat and sunflower productivity. *Climate Research*, **7,** 225–241.

Harrison, P.A. and R.E. Butterfield, 1999: Modelling climate change impacts on wheat, potato and grapevine in Europe. In: *Climate Change, Climate Variability and Agriculture in Europe: An Integrated Assessment* [Butterfield, R.E., P.E. Harrison, and T.E. Downing (eds.)]. Environmental Change Unit, Research Report No. 9, University of Oxford, Oxford, United Kingdom, 157 pp.

Harrison, P.A., R.E. Butterfield, and T.E. Downing, 1995a: *Climate Change and Agriculture in Europe: Assessment of Impacts and Adaptations*. Research Report No. 9, Environmental Change Unit, University of Oxford, Oxford, United Kingdom, 411 pp.

Harrison, P.A., R. Butterfield, and M.J. Gawith, 1995b: Modelling the effects of climate change on crops at the regional scale: effects on winter wheat, sunflower, onion and grassland in Europe. In: *Climate Change, Climate Variability and Agriculture in Europe: An Integrated Assessment* [Butterfield, R.E., P.E. Harrison, and T.E. Downing (eds.)]. Environmental Change Unit, Research Report No. 9, University of Oxford, Oxford, United Kingdom, pp. 330–388.

Harrod, T.R., 1979: Soil Suitability for Grassland. In: *Soil Survey Applications* [Jarvis, M.G. and D. Mackney (eds.)]. Soil Survey Technical Monograph No. 13, Harpenden, United Kingdom, pp. 51–70.

Hartig, E.K., O. Grozev, and C. Rosenzweig, 1997: Climate change, agriculture and wetlands in Eastern Europe: vulnerability, adaptation and policy. *Climatic Change*, **36,** 107–121.

Haslam, S.M., 1997: Deterioration and fragmentation of rivers in Malta. *Freshwater Biological Association*, **9,** 55–61.

Hendry, G.A.F. and J.P. Grime, 1990: Effects on plants: natural vegetation. In: *The Greenhouse Effect and Terrestrial Ecosystems of the UK* [Cannell, M.G.R. and M.D. Hooper (eds.)]. ITE Research Publication No. 4, Her Majesty's Stationary Office, London, United Kingdom, 82 pp.

Henry, G.H.R. and U. Molau, 1997: Tundra plants and climate change: the International Tundra Experiment (ITEX). *Global Change Biology,* **3(1),** 1–9.

Herrington, P., 1996: *Climate Change and the Demand for Water*. Her Majesty's Stationary Office, London, United Kingdom.

Hladny, J., J. Buchtele, M. Doubková, V. Dvorák, L. Kaspárek, O. Novicky, and E. Prenosilová, 1997: *Impact of a Potential Climate Change on Hydrology and Water Resources: Country Study of Climate Change for the Czech Republic*. Element 2, Publ. No. 20, National Climate Program of the Czech Republic, Prague, Czech Republic, 137 pp.

Hlavcová, K. and J. Cunderlík, 1998: Impact of climate change on the seasonal distribution of runoff in mountainous basins in Slovakia. In: *Hydrology, Water Resources, and Ecology in Headwaters*. IAHS Publication No. 248, pp. 39–46.

Hlavcová, K., J. Szolgay, J. Cunderlík, J. Parajka, and M. Lapin, 1999: *Impact of Climate Change on the Hydrological Regime of Rivers in Slovakia*. Slovak Committee for Hydrology, Bratislava, Slovakia.

Holten, J.I., 1990: Biological and ecological consequences of changes in climate in Norway. *NINA Utredning*, **11,** 1–59

Holten, J.I., 1998: Vertical distribution patterns of vascular plants in the Fennoscandian mountain range. *Ecologie*, **29(1–2),** 129–138.

Holten, J.I. and P.D. Carey, 1992: Responses of climate change on natural terrestrial ecosystems in Norway. *NINA Institute Research Report*, **29,** 1–59.

Holtmeier, F.K., 1994: Ecological aspects of climatically caused timberline fluctuations. In: *Mountain Environments in Changing Climates* [Beniston, M. (ed.)]. Routledge, London, United Kingdom, pp. 220–233.

Hulme, M. and T.R. Carter, 2000: The changing climate of Europe. In: *Assessment of Potential Effects and Adaptations for Climate Change in Europe: The Europe ACACIA Project* [Parry, M.L. (ed.)]. Jackson Environment Institute, University of East Anglia, Norwich, United Kingdom, 324 pp.

Hulme, M., E. Barrow, N.W. Arnell, P. Harrison, T. Downing, and R. Johns, 1999: Relative impacts of human-induced climate change and natural variability. *Nature*, **397,** 688–691.

Hulme, M., T.J. Osborn, and T.C. Johns, 1998: Precipitation sensitivity to global warming: comparison of observations with HadCM2 simulations. *Geophysical Research Letters*, **25,** 3379–3382.

Huntley, B., 1991: How plants respond to climate change: migration rates, individualism and the consequences for plant communities. *Annals of Botany*, **67(1),** 15–22.

Huntley, B., 1999: Species distribution and environmental change: considerations from the site to the landscape scale In: Ecosystem Management: Questions for Science and Society [Maltby, E., M. Holdgate, M. Acreman, and A. Weir (eds.)]. Royal Holloway Institute for Enviornmental Research, Virginia Water, United Kingdom, pp. 115–129.

Huntley, B., W. Cramer, A.V. Morgan, H.C. Prentice, and J.R.M. Allen, 1997: *Past and Future Rapid Environmental Changes: The Spatial and Evolutionary Responses of Terrestrial Biota*. NATO ASI Series, Series 1, Springer-Verlag, Berlin, Germany, 85 pp.

Huntley, B. and T. Webb III, 1989: Migration: species' response to climate variations caused by changes in the earth's orbit. *Journal of Biogeography*, **16,** 5–19.

Hurrell, J.W. and H. van Loon, 1997: Decadal variations in climate associated with the North Atlantic Oscillation. *Climatic Change*, **36,** 301–326.

IFRC, 1998: *World Disaster Report 1997*. Oxford University Press, New York, NY, USA.

Ineson, P., K. Taylor, A.F. Harrison, J. Poskitt, D.G. Benham, E. Tipping, and C. Woof, 1998: Effects of climate change on nitrogen dynamics in upland soils, part 1: a transplant approach. *Global Change Biology*, **4,** 143–152.

IPCC, 1996: *Climate Change 1995: Impacts, Adaptations, and Mitigation of Climate Change: Scientific-Technical Analyses. Contribution of Working Group II to the Second Assessment Report of the Intergovernmental Panel on Climate Change* [Watson, R.T., M.C. Zinyowera, and R.H. Moss (eds.)]. Cambridge University Press, Cambridge, United Kingdom and New York, NY, USA, 878 pp.

IPCC, 1998: *The Regional Impacts of Climate Change: An Assessment of Vulnerability. Special Report of IPCC Working Group II* [Watson, R.T., M.C. Zinyowera, and R.H. Moss (eds.)]. Intergovernmental Panel on Climate Change, Cambridge University Press, Cambridge, United Kingdom and New York, NY, USA, 517 pp.

Izrael, Y.A., 1997: *Russian Federation Climate Change Country Study. Final Report, Vol. 3: Vulnerability and Adaptation Assessments*. Russian Federal Service for Hydrometeorology and Environmental Monitoring, Moscow, Russia, 113 pp.

Jaagus, J., 1998: Modelled changes in river runoff using various climate scenarios for Estonia. In: *Proceedings of the Second International Conference on Climate and Water, Espoo, Finland, 17-20 August 1998* [Lemmelä, R. and N. Helenius (eds.)]. Vol. 1, pp. 94–103.

Jaagus, J., 1997: The impact of climate change on the snow cover pattern in Estonia. *Climatic Change*, **36,** 65–77.

Jansons, V. and M. Butina, 1998: Potential impacts of climate change on nutrient loads from small catchments. In: *Proceedings of the Second International Conference on Climate and Water, Espoo, Finland, 17-20 August 1998* [Lemmelä, R. and N. Helenius (eds.)]. Vol. 2, pp. 932–939.

Jarvet, A., 1998: Estimation on possible climate change impact on water management in Estonia. In: *Proceedings of the Second International Conference on Climate and Water, Espoo, Finland, 17-20 August 1998* [Lemmelä, R. and N. Helenius (eds.)].Vol. 3, pp. 1449–1458.

Jendritzky, G., K. Bucher, and F. Bendisch, 1997: Die Mortalitatsstudie des Deutschen Wetterdienstes. *Annalen der Meteorologie*, **33,** 46–51 (in German).

Jenik, J., 1997: Anemo-orographic systems in the Hercynian Mts. and their effects on biodiversity. *Acta Univ. Wratisl. No. 1959, Prace Instyt. Geogr., Seria C, Meteorologia i Klimatologia*, **4,** 9–21.

Jenkinson, D.S., 1990: The turnover of organic carbon and nitrogen in soil. *Proceedings of the Royal Society of London B*, **329,** 361–368.

Jenny, H., 1980: *The Soil Resource: Origin and Behaviour.* Ecological Studies No. 37, Springer-Verlag, New York, NY, USA, 377 pp.

Jonasson, S., J.A. Lee, T.V. Callaghan, M. Havström, and A. Parsons, 1996: Direct and indirect effects of increasing temperatures on subarctic ecosystems. *Ecological Bulletins*, **45,** 180–191.

Jones, M.B. M. Jongen, 1996: Sensitivity of temperate grassland species to elevated atmospheric CO_2 and the interaction with temperature and water stress. *Agricultural and Food Science in Finland*, **5,** 271–283.

Jones, M.B., M. Jongen, and T. Doyle, 1996: Effects of elevated carbon dioxide concentrations on agricultural grassland production. *Agricultural and Forest Meteorology*, **79,** 243–252.

Jones, P.D. and K.R. Briffa, 1995: Growing season temperatures over the former Soviet Union. *International Journal of Climatology*, **15,** 943–960.

Jones, P.D. and M. Hulme, 1997: The changing temperature of Central England. In: *Climates of the British Isles: Present, Past and Future* [Hulme, M. and E.M. Barrow (eds.)]. Routledge, London, United Kingdom, pp. 173–195.

Jones, P.D., T. Jonsson, and D. Wheeler, 1997: Extension of the North Atlantic Oscillation using early instrumental pressure observations from Gibralter and south-west Iceland. *International Journal of Climatology*, **17,** 1433–1450.

Jones, P.D., E.B. Horton, C.K. Folland, M. Hulme, D.E. Parker, and T.A. Basnett, 1999: The use of indices to identify changes in climatic extremes. *Climatic Change*, **42,** 131–149.

Jordan, A., T. O'Riordan, and I. Lorenzoni, 2000: Europe in the new millenium. In: *Assessment of Potential Effects and Adaptations for Climate Change in Europe: The Europe ACACIA Project* [Parry, M.L. (ed.)]. Jackson Environment Institute, University of East Anglia, Norwich, United Kingdom, 324 pp.

Kaas, E. and P. Frich, 1995: Diurnal temperature range and cloud cover in the Nordic countries: observed trends and estimates for the future. *Atmospheric Research*, **37,** 211–228.

Kaczmarek, Z., J.J. Napiórkowski, and K.M. Strzepek, 1996: Climate change impact on the water supply system in the Warta River catchment. *Water Resources Development*, **12(2),** 165–180.

Kaczmarek, Z., D. Jurak, and J.J. Napiórkowski, 1997: Impact of climate change on water resources in Poland. *Publications of the Institute of Geophysics*, **295,** 51.

Kaiser, H.M., S.J. Riha, D.S. Wilks, D.G. Rossiter, and R. Sampath, 1993: A farm-level analysis of economic and agronomic impacts of gradual climate warming. *American Journal of Agricultural Economics*, **75,** 387–398.

Kalvová, J., 1995: *Regional Study on Climate Change in the Czech Republic: Incremental Scenarios*. Country Study of Climate Change for the Czech Republic, Publ. No. 17 of the National Climate Program of the Czech Republic, Prague, Czech Republic, 101 pp.

Kalvová, J. and I. Nemesová, 1997: Projections of climate change for the Czech Republic. *Climatic Change*, **36,** 41–64.

Kasparék, L., 1998: Regional study on impacts of climate change on hydrological conditions in the Czech Republic. *Prace a Studie*, **193,** 70.

Katsouyanni, K., A. Pantazopoulu, G. Touloumi, I. Tselepidaki, K. Moustris, D. Asimakopoulos, G. Poulopoulou, and D. Trichopoulos, 1993: Evidence of interaction between air pollution and high temperatures in the causation of excess mortality. *Architecture and Environmental Health*, **48,** 235–242.

Katsouyanni, K., D. Trichopoulos, X. Zavitsanos, and G. Touloumi, 1988: The 1987 Athens heatwave. *Lancet*, **573,** ii.

Keller, F., F. Kienast, and M. Beniston, 2000: Evidence of the response of vegetation to environmental change at high elevation sites in the Swiss Alps. *Regional Environmental Change*, **2,** 70–77.

Kellomäki, S. and M. Kolström, 1993: Computations on the yield of timber by Scots pine when subjected to varying levels of thinning under a changing climate in southern Finland. *Forest Ecology and Management*, **59,** 237–255.

Kellomäki, S. and H. Väisänen, 1996: Model computations on the effect of rising temperature on soil moisture and water availability in forest ecosystems dominated by Scots pine in the boreal zone in Finland. *Climatic Change*, **32,** 423–445.

Kellomäki, S. and H. Väisänen, 1997: Modelling the dynamics of the boreal forest ecosystems for climate change studies in the boreal conditions. *Ecological Modelling*, **97(1–2),** 121–140.

Kenny, G.J., P.A. Harrison, J.E. Olesen, and M.L. Parry, 1993: The effects of climate change on land suitability of grain maize, winter wheat and cauliflower in Europe. *European Journal of Agronomy*, **2,** 325–338.

Khaw, K.T., 1995: Temperature and cardiovascular mortality. *Lancet*, **345,** 337–338.

Kime, D.E., 1995: The effects of pollution on reproduction in fish. *Reviews in Fish Biology and Fisheries*, **5,** 52–96.

Kirschbaum, M.U.F., 1995: The temperature dependence of soil organic matter decomposition and the effect of global warming on soil organic carbon storage. *Soil Biology and Biochemistry*, **27,** 753–760.

Kirschbaum, M.U.F., P. Bullock, J.R. Evans, K. Goulding, P.G. Jarvis, I.R. Noble, M.D.A. Rounsevell, and T.D. Sharkey, 1996a: Ecophysiological, ecological and soil processes in terrestrial ecosystems: a primer on general concepts and relationships. In: *Climate Change 1995: Impacts, Adaptations and Mitigation of Climate Change: Scientific-Technical Analyses. Contribution of Working Group II to the Second Assessment Report of the Intergovernmental Panel on Climate Change* [Watson, R.T., M.C. Zinyowera, and R.H. Moss (eds.)]. Cambridge University Press, Cambridge, United Kingdom and New York, NY, USA, pp. 57–76.

Kirschbaum, M.U.F., A. Fischlin, M.G.R. Cannell, R.V.O. Cruz, and W. Galinski, 1996b: Climate change impacts on forests. In: *Climate Change 1995: Impacts, Adaptations and Mitigation of Climate Change: Scientific-Technical Analyses. Contribution of Working Group II to the Second Assessment Report of the Intergovernmental Panel on Climate Change* [Watson, R.T., M.C. Zinyowera, and R.H. Moss (eds.)]. Cambridge University Press, Cambridge, United Kingdom and New York, NY, USA, pp. 95–129.

Klein, R.J.T., R.J. Nicholls, and N. Mimura, 1999: Coastal adaptation to climate change: can the IPCC Technical Guidelines be applied? *Mitigation and Adaptation Strategies for Global Change*, **4,** 51–64.

Klein, R.J.T., M.J. Smit, H. Goosen, and C.H. Hulsbergen, 1998: Resilience and vulnerability: coastal dynamics or Dutch dikes? *Geographic Journal*, **164(3),** 259–268.

Klein, R.J.T., J. Aston, E.N. Buckley, M. Capobianco, N. Mizutani, R.J. Nicholls, P.D. Nunn, and S. Ragoonaden, 2000: Coastal adaptation technologies. In: *IPCC Special Report on Methodological and Technological Issues in Technology Transfer* [Metz, B., O.R. Davidson, J.-W. Martens, S.N.M. van Rooijen, and L. Van Wie McGrory (eds.)]. Cambridge University Press, Cambridge, United Kingdom and New York, NY, USA, pp. 349–372.

Kleukers, R.M.J.C., K. Decleer, E.C.M. Haes, P. Kolshorn, and B. Thomas, 1996: The recent expansion of Conocephalus discolor (Thunberg) (Orthoptera: Tettigoniidae) in western Europe. *Entomologist's Gazette*, **47,** 37–49.

Knudsen, A.B., R. Romi, and G. Majori, 1996: Occurrence and spread in Italy of Aedes albopictus, with implications for its introduction into other parts of Europe. *Journal of the American Mosquito Control Association*, **12,** 177–183.

Kolb, E. and K.E. Rehfuess, 1997: Effects of a temperature increase in a field experiment on the nitrogen release from soil cores with different humus forms. *Zeitschrift fur Pflanzenernahrung und Bodenkunde*, **160,** 539–547.

Komor, E., G. Orlich, A. Weig, and W. Kockenberger, 1996: Phloem loading—not metaphysical, only complex: towards a unified model of phloem loading. *Journal of Experimental Botany*, **47,** 1155–1164.

Kondrashova, N.Y. and K.I. Kobak, 1996: Possible global warming caused changes of the nature zones boundaries in the Northern Hemisphere. In: *Problems of Ecological Monitoring and Ecosystem Modeling, Volume XVI* [Izrael, Y.A. (ed.)]. Institute of Global Climate and Ecology, RAS, St. Petersburg, Russia, pp. 90–99.

Kont, A., U. Ratas, and E. Puurmann, 1997: Sea-level rise impact on coastal areas of Estonia. *Climatic Change*, **36,** 175–184.

Körner, C., 1995: Impact of atmospheric changes on alpine vegetation: the ecophysiological perspective. In: *Potential Ecological Impacts of Climate Change in the Alps and Fennoscandian Mountains* [Guisan, A., J.I. Holten, R. Spichiger, and L. Tessier (eds.)]. Conserv. Jard. Bot., Geneva, Switzerland, pp. 113–120.

Körner, C., 1996: The response of complex multispecies systems to elevated CO_2. In: *Global Change and Terrestrial Ecosystems* [Walker, B. and W. Steffen (eds.)]. Cambridge University Press, Cambridge, United Kingdom and New York, NY, USA, pp. 20–42.

Kos, Z., 1993: Sensitivity of irrigation and water resources systems to climate change. *Journal of Water Management*, **41(4–5),** 247–269.

Koster, M.J. and R. Hillen, 1995: Combat erosion by law: coastal defence policy for The Netherlands. *Journal of Coastal Research*, **11,** 1221–1228.

Kovats, S. and P. Martens, 2000: Health. In: *Assessment of Potential Effects and Adaptations for Climate Change in Europe: The Europe ACACIA Project* [Parry, M.L. (ed.)]. Jackson Environment Institute, University of East Anglia, Norwich, United Kingdom, 324 pp.

Kozar, F., D.A.F. Sheble, and M.A. Fowjhan, 1995: Study on the further spread of Pseuda lacaspis pentagona in Central Europe, Israel. *Journal of Entomology*, **24,** 161–164.

Kräuchi, N., 1995: Application of the model FORSUM to the Solling spruce site. *Ecological Modelling*, **83,** 219–228.

Krankina, O.N., R.K. Dixon, A.P. Kirilenko, and K.I. Kobak, 1997: Global climate change adaptation: Examples from Russian boreal forests. *Climatic Change*, **36,** 197–215.

Kríz, B., 1998: *Infectious Disease Consequences of the Massive 1997 Summer Floods in the Czech Republic*. EHRO 020502/12, Working Group Paper.

Kríz, B., C. Benes, J. Cástková, and J. Helcl, 1998: *Monitoring of the epidemiological situation in flooded areas of the Czech Republic in year 1997*. Proceedings of the Conference DDD'98, 11–12th May 1998, Podebrady. Czech Republic.

Kromm, D.E. and S.E. White, 1990: Varability in adjustment preferences to groundwater depletion in the American High plains. *Water Resources Bulletin*, **22,** 791–801.

Kuchment, L.S., 1998: The estimation of the sensitivity of the river runoff to climate change for different physiogeographic regions of Russia and an experience of prediction of possible runoff change. In: *Proceedings of the Second International Conference on Climate and Water, Espoo, Finland, 17-20 August 1998* [Lemmelä, R. and N. Helenius (eds.)]. Vol. 1, pp. 171–177.

Kullman, L., 1995: Holocene tree limit and climate history from the Scandes mountains, Sweden. *Ecology*, **76(8),** 2490–2502.

Kundzewicz, Z.W. and K. Takeuchi, 1999: Flood protection and management—quo vadimus? *Hydrological Sciences Journal*, **44(3),** 417–432.

Kundzewicz, Z.W., K. Szamalek, and P. Kowalczak, 1999: The Great Flood of 1997 in Poland. *Hydrological Sciences Journal*, **44(5),** 855–870.

Kunst, A.E., C.W.N. Looman, and J.P. Mackenbach, 1993: Outdoor air temperature and mortality in The Netherlands: a time–series analysis. *American Journal of Epidemiology*, **137,** 331–341.

Kuntze, H., 1993: Bogs as sinks and sources of C and N. *Mitteilungen der Deutschen Bodenkundlichen Gesellschaft*, **69,** 277–280.

Laake, K. and J.M. Sverre, 1996: Winter excess mortality: a comparison between Norway and England plus Wales. *Age and Ageing*, **25,** 343–348.

Lakida, P., S. Nilsson, and A. Shvidenko, 1997: Forest phytomass and carbon in European Russia. *Biomass and Bioenergy*, **12(2),** 91–99.

Lang, G., 1993: Holozäne Veränderungen der Waldgrenze in der Schweizer Alpen-Methodische Ansatze und gegenwartiger Kenntnisstand. *Dissertationes Botanicae*, **196,** 147–148 (in German).

Langford, I.H. and G. Bentham, 1995: The potential effects of climate change on winter mortality in England and Wales. *International Journal of Biometeorology*, **38,** 141–147.

Lavee, H., A.C. Imeson, and P. Sarah, 1998: The impact of climate change on geomorphology and desertification along a Mediterranean arid transect. *Land Degradation and Development*, **9,** 407–422.

Lawton, J.H., 1995: The response of insects to environment change In: *Insects in a Changing Environment* [Harrington, R. and N.E. Stork (eds.)]. Academic Press, New York, NY, USA, 212 pp.

Leafe, R., J. Pethick, and I. Townsend, 1998: Realising the benefits of shoreline management. *Geographical Journal*, **164,** 282–290.

Lee, E.M., 1998: *The implication of Future Shoreline Management on Protected Habitats in England and Wales*. Technical Report W150, Environment Agency, Rand, Bristol, United Kingdom, 92 pp.

Lehtonen, H., 1996: Potential effects of global warming on northern European freshwater fish and fisheries. *Fisheries Management and Ecology*, **3,** 59–71.

Lehtonen, H., 1998: Does global warming threaten the existence of Arctic charrr, *Salvelnius alpinus* (Salmonidae), in northern Finland? *Italian Journal of Zoology*, **65,** 471–474.

Leinonen, I., 1996: Dependence of dormancy release on temperature in different origins of Pinus sylvestris and Betula pendula seedlings. *Scandinavian Journal of Forest Research*, **11,** 122–128.

Lelyakin, A.L., A.O. Kokorin, and I.M. Nazarov, 1997: Vulnerability of Russian forests to climate changes: model estimation of CO_2 fluxes. *Climatic Change*, **36(1–2),** 123–133.

Lenhart, S., S. Huq, L.J. Mata, I. Nemesova, S. Toure, and J.B. Smith (eds.), 1996: *Vulnerability and Adaptation to Climate Change – A Synthesis of Results from the U.S. Country Studies Program*. Interim Report. U.S. Country Studies Program, Washington, DC, USA, 366 pp.

Lesack, L.F.W., R.E. Hecky, and P. Marsh, 1991: The influence of frequency and duration of flooding on the nutrient chemistry of Mackenzie Delta lakes. In: *Mackenzie Delta, Environmental Interactions and Implications of Development*. National Hydrology Research Institute, Environment Canada, Saskatoon, Canada, NHRI Symposium No. 4, pp. 19–36.

Lindgren, E., 1998: Climate and tick–borne encephalitis in Sweden. *Conservation Ecology*, **2,** 5–7.

Lindgren, E., L. Tälleklint, and T. Polfeldt, 2000: Impact of climatic change on the northern latitude limit and population density of the disease-transmitting European tick, Ixodes ricinus. *Environmental Health Perspectives*, **108,** 119–123.

Lindner, M., H. Bugmann, P. Lasch, M. Flechsig, and W. Cramer, 1997: Regional impacts of climatic change on forests in the state of Brandenburg, Germany. *Agricultural and Forest Meteorology*, **84(1–2),** 123–135.

Lindström, Å. and J. Agrell, 1999: Global change and possible effects on the migration and reproduction of arctic-breeding waders. *Ecological Bulletins*, **47,** 145–159.

Lines, R., 1987: *Choice of Seed Origin for the Main Forest Species in Britain*. Forestry Commission Bulletin 66, Her Majesty's Stationary Office, London, United Kingdom, 42 pp.

Lipa, J.L., 1997: Zmiany klimatu Ziemi - konsekwencje dla rolnictwa i ochrony roslin. *Postepy w Ochronie Roslin*, **37(I),** 27–35 (in Polish).

Lipa, J.L., 1999: Do climate changes increase the threat to crops by pathogens, weeds, and pests? *Geographia Polonica*, **4,** 222–224.

Lischke, H., A. Guisan, A. Fischlin, J. Williams, and H. Bugmann, 1998: Vegetation responses to climate change in the Alps: modeling studies. In: *Views from the Alps: Regional Perspectives on Climate Change* [Cebon, P., U. Dahinden, H.C. Davies, D. Imboden, and C.C. Jaeger (eds.)]. MIT Press. Cambridge, MA, USA, pp. 309–350.

Lloyd, J. and J.A. Taylor, 1994: On the temperature dependence of soil respiration. *Functional Ecology*, **8,** 315–323.

Loiseau, P., J.F. Soussana, and E. Casella, 1994: Effects of climatic changes (CO_2, temperature) on grassland ecosystems: first five months, experimental results. In: *Soil Responses to Climate Change* [Rounsevell, M.D.A. and P.J. Loveland (eds.)]. Springer-Verlag, Heidelberg, Germany, pp. 223–228.

Lowe, J.A. and J.M. Gregory, 1998: *A Preliminary Report on Changes in the Occurrence of Storm Surges Around the United Kingdom Under a Future Climate Scenario*. Unpublished report, Hadley Centre for Climate Prediction and Research, Bracknell, United Kingdom, 65 pp.

Lukewille, A. and R.F. Wright, 1997: Experimentally increased soil temperature cause relese of nitrogen at a boreal forest catchment in southern Norway. *Global Change Biology*, **3,** 13–21.

MacDonald, A.M., K.B. Matthews, E. Paterson, and R.J. Aspinall, 1994: The impact of climate change on the soil-moisture regime of Scottish mineral soils. *Environmental Pollution*, **83,** 245–250.

Magnuson, J.J., D.M. Robertson, B.J. Benson, R.H. Wynne, D.M. Livingstone, T. Arai, R.A. Assel, R.G. Barry, V. Card, E. Kuusisto, N.G. Granin, T.D. Prowse, K.M. Stewart, and V.S. Vuglinski, 2000: Historical trends in lake and river ice cover in the Northern Hemisphere. *Science*, **289(5485),** 1743–1746.

Mandelkern, S., S. Parey, and N. Tauveron, 1998: Hydrological consequences of greenhouse effect, a case study of the river Doubs. In: *Proceedings of the Second International Conference on Climate and Water, Espoo, Finland, 17-20 August 1998* [Lemmelä, R. and N. Helenius (eds.)]. Vol. 3, pp. 1505-1514.

Mänder, Ü. and A. Kull, 1998: Impacts of climatic fluctuations and land use change on water budget and nutrient runoff: the Porijögi. In: *Proceedings of the Second International Conference on Climate and Water, Espoo, Finland, 17-20 August 1998* [Lemmelä, R. and N. Helenius (eds.)]. Vol. 2, 884–896.

Marchant, P., W. Eling, G-J. van Gemert, C.J. Leake, and C.F. Curtis, 1998: Could British mosquitoes transmit falciparum malaria? *Parasitology Today*, **14,** 344–345.

Martens, P., 1997: Climate change, thermal stress and mortality changes. *Social Science and Medicine*, **46,** 331–344.

Martens, P., 1998: *Health and Climate Change: Modelling the Impacts of Global Warming and Ozone Depletion*. Earthscan, London, United Kingdom, 176 pp.

Martens, P., T.H. Jetten, and D.A. Focks, 1997: Sensitivity of malaria, schistosomiasis and dengue to global warming. *Climatic Change*, **35(2),** 145–156.

Maugeri, M. and T. Nanni, 1998: Surface air temperature variations in Italy: recent trends and an update to 1998. *Theoretical and Applied Climatology*, **61,** 191–196.

McCarthy, I.D. and D.F.H. Houlihan, 1997: The effect of temperature on protein metabolism in fish: the possible consequences for wild Atlantic salmon (Salmo salar) stocks in Europe as a result of global warming. In: *Global Warming: Implications for Freshwater and Marine Fish* [Wood, C.M. and D.G. McDonald (eds.)]. Cambridge University Press, Cambridge, United Kingdom and New York, NY, USA, pp. 51–77.

McMichael, A.J., A. Haines, R. Slooff, and S. Kovats, 1996: *Climate Change and Human Health*. An assessment prepared by a Task Group on behalf of the World Health Organization, the World Meteorological Organization, and the United Nations Environment Program, Geneva, Switzerland

McMichael, A.J. and R.S. Kovats, 1998: *Assessment of the Impact on Mortality in England and Wales of the Heatwave and Associated Air Pollution Episode of 1976*. Report to the Department of Health, London School of Hygiene and Tropical Medicine, London, United Kingdom.

Meinl, H. and W. Bach (eds.), 1984: *Socio-Economic Impacts of Climatic Changes Due to a Doubling of Atmospheric CO_2 Content*. Report on CEC Contract No. CLI-063-D, Dornier System, 648 pp.

Menne, B., K. Pond, E.K. Noji, and R. Bertollini, 1999: *Floods and Public Health Consequences, Prevention and Control Measures: Discussion paper presented at the UNECE Seminar on Flood Prevention, Berlin, 7–8 October, 1999*. UNECE/MP.WAT/SEM.2/1999/22, UNECE, Berlin, Germany.

Menzel, A. and P. Fabian, 1999: Growing season extended in Europe. *Nature*, **397,** 659.

Mika, J., I. Dobi, and G. Bálint, 1997: Changes in means and variability of river flow in the Tisza Basin: scenarios and implications. *Annales Geophysicae*, **15(II),** C304.

Milbank, N., 1989: Building design and use: response to climate change. *Architects Journal*, **89,** 53–63.

Minkkinen, K. and J. Laine, 1998: Long-term effect of forest drainage on the peat carbon stores of pine mires in Finland. *Canadian Journal of Forest Research*, **28,** 1267–1275.

Mohnl, V., 1996: The fluctuation of winter-sport-related snow parameters of the last fifty years in the Austrian Alps. *Wetter und Leben*, **48,** 103–113.

Molau, U. and J.M. Alatalo, 1998: Responses of subarctic-alpine plant communities to simulated environmental change: biodiversity of bryophytes, lichens, and vascular plants. *Ambio*, **27,** 322–329.

Molau, U. and E.-L. Larsson, 2000: Seed rain and seed bank along an alpine altitudinal gradient in Swedish Lapland. *Canadian Journal of Botany*, **78,** 728–747.

Mortsch, L.D., 1998: Assessing the impact of climate change on the Great Lakes shoreline wetlands. *Climatic Change*, **40,** 391–416.

Murray, M.B., M.G.R. Cannell, and R.T. Smith, 1989: Date of budburst of fifteen tree species in Britain following climatic warming. *Journal of Applied Ecology*, **26,** 693–700.

Myking, T. and O.M. Heide, 1995: Dormancy release and chilling requirement of buds of latitudinal ecotypes of Betula pendula and B. pubescens. *Tree Physiology*, **15,** 697–704.

Myneni, R.B., C.D. Keeling, C.J. Tucker, G. Asrar, and R.R. Nemani, 1997: Increased plant growth in the northern high latitudes from 1981 to 1991. *Nature*, **386,** 698–702.

Nakicenovic, N., J. Alcamo, G. Davis, B. de Vries, J. Fenhann, S. Gaffin, K. Gregory, A. Grubler, T.Y. Jung, T. Kram, E.L. La Rovere, L. Michaelis, S. Mori, T. Morita, W. Pepper, H. Pitcher, L. Price, K. Raihi, A. Roehrl, H.-H. Rogner, A. Sankovski, M. Schlesinger, P. Shukla, S. Smith, R. Swart, S. van Rooijen, N. Victor, and Z. Dadi, 2000: An overview of the scenario literature. In: *Emissions Scenarios. A Special Report of Working Group III of the Intergovernmental Panel on Climate Change*. Cambridge University Press, Cambridge, United Kingdom and New York, NY, USA, 599 pp.

New, M., M. Hulme, and P.D. Jones, 1999: Representing twentieth century space-time climate variability, part 1: development of a 1961–90 mean monthly terrestrial climatology. *Journal of Climate*, **12,** 829–856.

Nicholls, R., 2000: Coastal zones. In: *Assessment of Potential Effects and Adaptations for Climate Change in Europe: The Europe ACACIA Project* [Parry, M.L. (ed.)]. Jackson Environment Institute, University of East Anglia, Norwich, United Kingdom, 324 pp.

Nicholls, R.J. and A. de la Vega-Leinert, 2000: Synthesis of sea-level rise impacts and adaptation costs for Europe. In: *Proceedings of the European Expert Workshop on European Vulnerability and Adaptation to the Impacts of ASLR, Hamburg, Germany, 19-21 June 2000* [de la Vega-Leinert, A., R.J. Nicholls, and R.S.J. Tol (eds.)]. Flood Hazard Research Centre, Enfield, United Kingdom. Available online at http://www.survas.mdx.ac.uk/.

Nicholls, R.J. and F.M.J. Hoozemans, 1996: The Mediterranean: vulnerability to coastal implications of climate change. *Ocean and Coastal Management*, **31,** 105–132.

Nicholls, R.J. and N. Mimura, 1998: Regional issues raised by sea-level rise and their policy implications. *Climate Research*, **11,** 5–18.

Nicholls, N., G.V. Gruza, J. Jouzel, T.R. Karl, L.A. Ogallo, and D.E. Parker, 1996: Observed climate variability and change. In: *Climate Change 1995: The Science of Climate Change. Contribution of Working Group I to the Second Assessment Report of the Intergovernmental Panel on Climate Change* [Houghton, J.T., L.G. Meira Filho, B.A. Callander, N. Harris, A. Kattenberg, and K. Maskell (eds.)]. Cambridge University Press, Cambridge, United Kingdom and New York, NY, USA, pp. 133–192.

Nicholls, R.J., F.M.J. Hoozemans, and M. Marchand, 1999: Increasing the flood risk and wetland losses due to global sea-level rise. *Global Environmental Change*, **9,** 569–587.

Nonhebel, S., 1996: Effects of temperature rise and increase in CO_2 concentration on simulated wheat yields in Europe. *Climatic Change*, **34,** 73–90.

Norris, K. and R. Buisson, 1994: Sea-level rise and its impact upon coastal birds in the UK. *RSPB Conservation Review*, **8,** 63–71.

OECD, 1997: *Towards Sustainable Fisheries: Economic Aspects of the Management of Living Marine Resources*. Main Report, Issue Papers, and Country Reports. Organisation for Economic Cooperation and Development, Paris, France.

Olesen, J.E. and K. Grevsen, 1993: Simulated effects of climate change on summer cauliflower production in Europe. *European Journal of Agronomy*, **2,** 313–323.

Olejnik, J., and A. Kedziora, 1991: A model for heat and water balance estimation and its application to land use and climate variation. *Earth Surface Processes and Landforms*, **16,** 601–617.

Onate, J.J. and A. Pou, 1996: Temperature variations in Spain since 1901: a preliminary analysis. *International Journal of Climatology*, **16,** 805–815.

Osborn, T.J., M. Hulme, P.D. Jones, and T. Basnett, 1999: Observed trends in the daily intensity of United Kingdom precipitation. *International Journal of Climatology*, **20,** 347–364.

Ozenda, P. and J.L. Borel, 1991: Les consequnces ecologiques possibles des changements climatiques dans l'Arc alpin. In: *Vol. Rapport FUTURALP 1*. Centre International poutr l'Environnement Alpin (ICALPE), Chambery, France (in French).

Ozenda, P. and J.L. Borel, 1995: Possible response of mountain vegetaion to a global climatic change: the case of the Western Alps. In: *Potential Ecological Impacts of Climate Change in the Alps and Fennoscandian Mountains* [Guisan, A., J.I. Holten, R. Spichiger, and L. Tessier (eds.).] Conserv. Jard. Bot., Geneva, Switzerland, pp. 137–144.

Palutikof, J.P., S. Subak, and M.D. Agnew, 1997: *Economic Impacts of the Hot Summer and Unusually Warm Year of 1995*. Department of the Environment, University of East Anglia, Norwich, United Kingdom, 196 pp.

Panagoulia, D. and G. Dimou, 1996: Sensitivities of groundwater-streamflow interaction to global climate change. *Hydrological Sciences Journal*, **41,** 781–796.

Parmesan, C., N. Ryrholm, C. Stephanescus, J.K. Hill, C.D. Thomas, H. Descimon, B. Huntley, L. Laili, K. Kullberg, T. Tammaru, W.J. Tennent, J.A. Thomas, and M. Warren, 1999: Poleward shifts in geographical ranges of butterfly species associated with regional warming. *Nature*, **399,** 579–583.

Parry, M.L. (ed.), 2000: *Assessment of Potential Effects and Adaptations for Climate Change in Europe: The Europe ACACIA Project*. Jackson Environment Institute, University of East Anglia, Norwich, United Kingdom, 324 pp.

Parry, M.L., I. Carson, T. Rehman, R. Tranter, P. Jones, D. Mortimer, M. Livermore, and J. Little, 1999: *Economic Implications of Climate Change on Agriculture in England and Wales*. Research Report No. 1, Jackson Environment Institute, University of East Anglia, Norwich, United Kingdom.

Parry, M.L., T.R. Carter, and N.T. Konijn (eds.), 1988: *The Impact of Climatic Variations on Agriculture*. Kluwer Academic Publishers. Dordrecht, The Netherlands.

Parsons, D.J., 1991: Planning for climate change in National Parks and other natural areas. *The Northwest Environmental Journal*, **7,** 255–269.

Parton, W.J., D.S. Schimel, C.V. Cole, and D.S. Ojima, 1987: Analysis of factors controlling soil organic matter levels in Great Plains grassland. *Soil Science Society of America Journal*, **51,** 1173–1179.

Pauli, H., M. Gottfried, and G. Grabherr, 1999: Vascular plant distribution patterns at the low-temperature limits of plant life - the alpine-nival ecotone of Mount Schrankogel (Tyrol, Austria). *Phytocoenologia*, **29(3),** 297–325.

Pearson, S., T.R. Wheeler, P. Hadley, and A.E. Wheldon, 1997: A validated model to predict the effects of environment on the growth of lettuce (Lactuca sativa L.): implications for climate change. *Journal of Horticultural Science*, **72,** 503–517.

Peerbolte, E.B., J.G. de Ronde, L.P.M. de Vrees, M. Mann, and G. Baarse, 1991: *Impact of Sea-Level Rise on Society: A Case Study for the Netherlands*. Delft Hydraulics and Rijkswaterstaat, Delft and The Hague, The Netherlands, 81 pp.

Peiris, D.R., J.W. Crawford, C. Grashoff, R.A. Jefferies, J.R. Porter, and B. Marshall, 1996: A simulation study of crop growth and development under climate change. *Agricultural and Forest Meteorology*, **79,** 271–287.

Pekárová, P., 1996: Simulation of runoff changes under changed climatic conditions in the Ondava catchment. *Journal of Hydrology and Hydromechanics*, **44(5),** 291–311 (in Slovak with English abstract).

Peltola, H., S. Kellomäki, and H. Väisänen, 1999: Model computations on the impacts of climatic change on soil frost with implications for windthrow risk of trees. *Climate Change*, **41,** 17–36.

Penning-Rowsell, E.C, C.H. Green, P.M. Thompson, A.M. Coker, S.M. Tunstall, C. Richards, and D.J. Perker, 1992: *The Economics of Coastal Management: A Manual of Benefit Assessment Techniques*. Belhaven Press, London, United Kingdom.

Penning-Rowsell, E., P. Winchester, and J.G. Gardiner, 1998: New approaches to sustainable hazard management for Venice. *Geographical Journal*, **164,** 1–18.

Perry, A., 1999: Impacts of climate change on tourism in the Mediterranean: adaptive responses. In: *Proceedings of the Conference on the Impacts of Climate Change on the Mediterranean Area: Regional Scenarios and Vulnerability Assessment, Venice, 9–10 December, 1999*. Venice, Italy.

Perry, A., 2000: Tourism and recreation. In: *Assessment of Potential Effects and Adaptations for Climate Change in Europe: The Europe ACACIA Project* [Parry, M.L. (ed.)]. Jackson Environment Institute, University of East Anglia, Norwich, United Kingdom, 324 pp.

Perry, A.H. and L. Symons, 1994: The wind hazard in the British Isles and its effects on transportation. *Journal of Transport Geography*, **2,** 122–130.

Peters, R.L. and J.D.S. Darling, 1985: The greenhouse effect and nature reserves: global warming could diminsih biological diversity by causing extinctions among reserve species. *Bioscience*, **35(11),** 707–717.

Peters, R.L. and T.E. Lovejoy (eds.), 1992: *Global Warming and Biological Diversity*. Yale University Press, New Haven, CT, USA, 211 pp.

Petrovic, P., 1998: Climate change impact on Hydrological Regime for two Profiles in the Nitra River Basin. In: *Proceedings, XIXth Conference of the Danube Countries on Hydrological Forecasting and Hydrological Bases of Water Management. Hrvatske Vode, Zagreb, 15-19 June 1998* [Bonacci, O. (ed.)]. Osijek, Croatia, pp. 117–122.

Pfister, C., 1992: Monthly temperature and precipitation patterns in Central Europe from 1525 to the present: a methodology for quantifying man-made evidence on weather and climate. In: *Climate since 1500* [Bradley, R.S. and P.D. Jones (eds.)]. Routledge, London, United Kingdom, pp. 118–143.

Phillips, D.L., D. White, and B. Johnson, 1993: Implications of climate change scenarios for soil erosion potential in the USA. *Land Degradation and Rehabilitation*, **4,** 61–72.

Plowman, J. (ed.), 1995: *Biodiversity: The UK Steering Group Report, Volume 1: Meeting the Rio Challenge*. Her Majesty's Stationary Office, London, United Kingdom, 81 pp.

Porter, J., 1995: The effects of climate change on the agricultural environment for crop insect pests with particular reference to the European corn borer and grain maize. In: *Insects in a Changing Environment* [Harrington, R. and N.E. Stork (eds.)]. Academic Press, San Diego, CA, USA, pp. 94–125.

Porter, J.H., M.L. Parry, and T.R. Carter, 1991: The potential effects of climatic change on agricultural insect pests. *Agricultural and Forest Meteorology*, **57(1–3),** 221–240.

Post, W.M., W.R. Emanuel, P.J. Zinke, and A.G. Stangenberger, 1982: Soil carbon pools and world life zones. *Nature*, **298,** 156–159.

Post, W.M., J. Pastor, P.J. Zinke, and A.G. Stangenberger, 1985: Global patterns of soil nitrogen storage. *Nature*, **317,** 613–616.

Pregitzer, K.S. and D. Atkinson, 1993: Impact of climate change on soil processes and soil biological activity. *BCPC Monograph*, **56,** 71–82.

Price, C., L. Stone, A. Huppert, B. Rajagopalan, and P. Alpert, 1998: A possible link between El Niño and precipitation in Israel. *Geophysical Research Letters*, **25,** 3963–3966.

Proe, M.F., S.M. Allison, and K.B. Matthews, 1996: Assessment of the impact of climate change on the growth of Sitka spruce in Scotland. *Canadian Journal of Forest Research*, **26,** 1914–1921.

Punda-Polic, V., S. Sardelic, and N. Bradaric, 1998: Visceral leishmaniasis in southern Croatia. *Lancet*, **351,** 188.

Pye, K. and J.R.L. Allen, 2000: *Coastal and Estuarine Environments: Sedimentology, Geomorphology and Geoarchaelogy*. Geological Society of London Special Publication, London, United Kingdom, 96 pp.

Pye, K. and P.W. French, 1993: *Targets for Coastal Habitat Recreation, Research and Survey in Nature Conservation*. No. 13, English Nature, Peterborough, United Kingdom, 101 pp.

Randolph, S.E. and D.J. Rogers, 2001: Fragile transmission cycles of tick-borne encephalitis virus may be distrupted by predicted climate change. *Proceedings of the Royal Society of London B*, (in press).

Raptsun, M., 1997: *Country Study on Climate Change in Ukraine*. US Country Study Initiative, Kiev, Russia, 100 pp.

Rebetez, M. and M. Beniston, 1998: Changes in sunshine duration are correlated with changes in daily temperature range this century: an analysis of Swiss climatological data. *Geophysical Research Letters*, **25,** 3611–3613.

Reilly, J., 1996: Agriculture in a changing climate: impacts and adaptation. In: *Climate Change 1995: Impacts, Adaptations, and Mitigation of Climate Change: Scientific-Technical Analyses. Contribution of Working Group II to the Second Assessment Report of the Intergovernmental Panel on Climate Change* [Watson, R.T., M.C. Zinyowera, and R.H. Moss (eds.)]. Cambridge University Press, Cambridge, United Kingdom and New York, NY, USA, pp. 429–467.

Reynard, N.S., C. Prudhomme, and S. Crooks, 1998: The potential effects of climate change on the flood characteristics of a large catchment in the UK. In: *Proceedings of the Second International Conference on Climate and Water, Espoo, Finland, 17-20 August 1998* [Lemmelä, R. and N. Helenius (eds.)]. Vol. 1, pp. 320–332.

Riedo, M., D. Gyalistras, A. Fischlin, and J. Fuhrer, 1999: Using an ecosystem model linked to GCM-derived local weather scenarios to analyse effects of climate change and elevated CO_2 on dry matter production and partitioning, and water use in temperate managed forests. *Climatic Change*, **23,** 213–223.

Rigg, K., A. Salman, D. Zanen, M. Taal, J. Kuperus, and J. Lourens, 1997: *Threats and Opportunities in the Coastal Areas of the European Union: a Scoping Study*. National Spatial Planning Agency, Ministry of Housing, Spatial Planning, and the Environment, Bilthoven, The Netherlands.

Rioux, J.A., J. Boulker, G. Lanotte, R. Killick-Hendrick, and A. Martini-Dumas, 1985: Ecologie des leishmanioses dans le sud de France, 21 influence de la temperature sur le development de Leishmania infantum Nicolle, 1908 chez Phlebotomus ariasi Tonnoir, 1921; etude experimentale. *Annales de Parasitologie*, **60(3),** 221–229.

Roberts, G., 1998: The effects of possible future climate change on evaporation losses from four contrasting UK water catchment areas. *Hydrological Processes*, **12,** 727–739.

Robinson, C.H., P.A. Wookey, A.N. Parsons, J.A. Potter, T.V. Callagnan, J.A. Lee, M.C. Press, and J.M. Welker, 1995: Responses of plant litter decomposition and nitrogen mineralisation to simulated environmental change in a high arctic polar semi-desert and a subarctic dwarf shrub heath. *Oikos*, **74,** 53–512.

Rodo, X., E. Baert, and F.A. Comin, 1997: Variations in seasonal rainfall in Southern Europe during the present century: relationships with the North Atlantic Oscillation and the El Niño-Southern Oscillation. *Climate Dynamics*, **13,** 275–284.

Romero, R., C. Ramis, and J.A. Guijarro, 1999: Daily rainfall patterns in the Spanish Mediterranean area: an objective classification. *International Journal of Climatology*, **19,** 95–112.

Rooney, C., A.J. McMichael, R.S. Kovats, and M. Coleman, 1998: Excess mortality in England and Wales, and in Greater London, during the 1995 heatwave. *Journal of Epidemiology and Community Health*, **52,** 482–486.

Roosaare, J., 1998: Local-scale spatial interpretation of climate change impact on river runoff in Estonia. In: *Proceedings of the Second International Conference on Climate and Water, Espoo, Finland, 17-20 August 1998* [Lemmelä, R. and N. Helenius (eds.)]. Vol. 1, pp. 86–93.

Rosenberg, N.J., 1979: Windbreaks for reducing moisture stress. In: *Modification of the Aerial Environment of Plants* [Barfield, B.J. and J.F. Gerber (eds.)]. American Society for Agricultural Engineering Monograph 2, pp. 394-408.

Rosenberg, N.J. and M.A. Tutwiler, 1988: Global climate change holds problems and uncertainties for agriculture. In: *US Agriculture in a Global Setting: An Agenda for the Future* [Tutwiler, M.A. (ed.)]. Resources for the Future, Washington, DC, USA, pp. 203–218.

Rounsevell, M. and A. Imeson, 2000: Soil and land resources. In: *Assessment of Potential Effects and Adaptations for Climate Change in Europe: The Europe ACACIA Project* [Parry, M.L. (ed.)]. Jackson Environment Institute, University of East Anglia, Norwich, United Kingdom, 324 pp.

Rounsevell, M.D.A., 1993: A review of soil workability models and their limitations in temperate regions. *Soil Use and Management*, **9(1),** 15–21.

Rounsevell, M.D.A. and A.P. Brignall, 1994: The potential effects of climate change on autumn soil tillage opportunities in England and Wales. *Soil and Tillage Research*, **32,** 275–289.

Rounsevell, M.D.A. and R.J.A. Jones, 1993: A soil and agroclimatic model for estimating machinery work days: the basic model and climatic sensitivity. *Soil and Tillage Research*, **26(3),** 179–191.

Rounsevell, M.D.A., A.P. Brignall, and P.A. Siddons, 1996a: Potential climate change effects on the distribution of agricultural grassland in England and Wales. *Soil Use and Management*, **12,** 44–51.

Rounsevell, M.D.A., P. Bullock, and J. Harris, 1996b: Climate change, soils and sustainability. In: *Soils and Sustainability and the Natural Heritage* [Taylor, A.G., J.E. Gordon, and M.B. Usher (eds.)]. Scottish Natural Heritage, Her Majesty's Stationary Office, London, United Kingdom, pp. 121–139.

Rounsevell, M.D.A., S.P. Evans, and P. Bullock, 1999: Climate change and agricultural soils: impacts and adaptation. *Climatic Change*, **43,** 683–709.

Ryszkowski, L. and A. Kedziora, 1987: Impact of agricultural landscape structure on energy flow and water cycling. *Landscape Ecology*, **1,** 85–94.

Ryszkowski, L. and A. Kedziora, 1993: Agriculture and greenhouse effect. *Kosmos*, **42,** 123–149.

Ryszkowski, L, A. Kedziora, and J. Olejnik, 1990: Potential effects of climate and land use changes on the water balance structure in Poland. In: *Processes of Change: Environmental Transformation and Future Patterns*. Kluwer Academic Publishers. Dordrecht, The Netherlands, pp. 253–274.

Saarnio, S., T. Saarinen, H. Vasander, and J. Silvola, 2000: A moderate increase in the annual CH_4 efflux by raised CO_2 or NH_4NO_3 supply in a boreal oligotrophic mire. *Global change Biology*, **6(2),** 137–144.

Saarnio, S. and J. Silvola, 1999: Effects of increased CO_2 and N on CH_4 efflux from a boreal mire: a growth chamber experiment. *Oecologia*, **119(3),** 349–356.

Sælthun, N.R., P. Aittoniemi, S. Bergström, K. Einarsson, T. Jóhannesson, G. Lindström, P.E. Ohlsson, T. Thomsen, B. Vehviläinen, and K.O. Aamodt, 1998: Climate change impacts on runoff and hydropower in the Nordic countries. *TemaNord*, **552,** 170.

Sætersdal, M. and H.J.B. Birks, 1997: A comparative ecological study of Norwegian mountain plants in relation to possible future climatic change. *Journal of Biogeography*, **24,** 127–149.

Sanchez-Arcilla, A., J. Jimenez, and H.I. Valdemoro, 1998: The Ebro delta: morphodynamics and vulnerability. *Journal of Coastal Research*, **14,** 754–772.

Schellnhuber, H.J., W. Enke, and M. Flechsig, 1994: *The Extreme Summer of 1992 in Northern Germany*. PIK Reports, Potsdam Institute for Climate Impact Research, Potsdam, The Netherlands, Vol. 2, 98 pp.

Schenk, U., R. Manderscheid, J. Hugen, and H.J. Weigel, 1995: Effects of CO_2 enrichment and intraspecific competition on biomass partitioning, nitrogen content and microbial biomass carbon in soil of perennial ryegrass and white clover. *Journal of Experimental Botany*, **46,** 987–993.

Schindler, D.W. and P.J. Curtis, 1997: The role of DOC in protecting freshwaters subjected to climatic warming and acidification from UV exposure. *Biogeochemistry*, **36(1),** 1–8.

Schindler, D.W., P.J. Curtis, B.R. Parker, and M.P. Stainton, 1996: Consequences of climate warming and lake acidification for UV-B penetration in North American boreal lakes. *Nature*, **379(6567),** 705–708.

Schmidt, T., E. Kaas, and T.S. Li, 1998: Northeast Atlantic winter storminess 1875–1995 re-analysed. *Climate Dynamics*, **14,** 529–536.

Second National Communication on Climate, 1997: *UN Framework Convention on Climate Change*. Report, Ministry of the Environment, Bratislava, Slovakia, 98 pp.

Seidel, K., C. Ehrler, and J. Martinec, 1998: Effects of climate change on water resources and runoff in an Alpine basin. *Hydrological Processes*, **12,** 1659–1669.

Semenov, M.A. and J.R. Porter, 1995: Climatic variability and the modelling of crop yields. *Agricultural and Forest Meteorology*, **73,** 265–283.

Shaver, G.R., W.D. Billings, F.S. Chapin III, A.E. Giblin, K.J. Nadelhoffer, W.C. Oechel, and E.B. Rastetter, 1992: Global change and the carbon balance of arctic ecosystems. *Bioscience*, **42,** 433–441.

Shvidenko, A., S. Venevsky, and S. Nilsson, 1996: *Increment and Mortality for Major Forest Species of Northern Eurasia with Variable Growing Stock*. IIASA, WP-96-98, Laxenburg,, Austria, 200 pp.

Sinclair, T.R. and N.G. Seligman, 1995: Global environment change and simulated forage quality of wheat, I: nonstressed conditions. *Field Crops Research*, **40,** 19–27.

Sirotenko, O.D., H.V. Abashina, and V.N. Pavlova, 1997: Sensitivity of the Russian agriculture to changes in climate, CO_2 and tropospheric ozone concentration and soil fertility. *Climatic Change*, **26,** 217–232.

Smith, J.B. and M. Hulme, 1998: Climate change scenarios. In: *Handbook on Methods of Climate Change Impacts Assessment and Adaptation Strategies* [Feenstra, J., I. Burton, J.B. Smith, and R.S.J. Tol (eds.)]. United Nations Environment Program, IES, Version 2.0, Amsterdam, The Netherlands, 310 pp.

Smith, J.B., S. Huq, S. Lenhart, L.J. Mata, I. Nemesová, and S. Toure, 1996: *Vulnerability and Adaptation to Climate Change: Interim Results from the U.S. Country Studies Program*. J. Kluwer Academic Publishers, Dordrecht, The Netherlands, 366 pp.

Smith, J.B. and G.J. Pitts, 1997: Regional climate change scenarios for vulnerability and adaptation assessments. *Climatic Change*, **36,** 3–21.

Smith, P., O. Andren, L. Brussard, M. Dangerfield, K. Ekscmidt, P. Lavelle, and K. Tate, 1998: Soil biota and global change at the ecosystem level: describing soil biota in mathematical models. *Global Change Biology*, **4,** 773–784.

Solantie, R., 1986: Hårda vindar och vindskador. *Skogsbruket*, **1,** 12–14 (in Norwegian).

Spiecker, H., K. Mielikäinen, M. Köhl, and J. Skovsgaard, 1996: *Growth Trends in European Forests: Studies from 12 Countries*. Springer-Verlag, Berlin, Germany, 372 pp.

Stanescu, A.V., C. Corbus, V. Ungureanu, and M. Simota, 1998: Quantification of the hydrological regime modification in the case of climatic changes. In: *Proceedings of the Second International Conference on Climate and Water, Espoo, Finland, 17-20 August 1998* [Lemmelä, R. and N. Helenius (eds.)]. Vol. 1, pp. 198–207.

Stanners, D. and P. Bourdeau, 1995: *Europe's Environment: The Dobris Assessment*. European Environment Agency, Copenhagen. Denmark, 385 pp.

Starosolszky, O. and B. Gauzer, 1998: Effect of precipitation and temperature changes on the flow regime of The Danube River. In: *Proceedings of the Second International Conference on Climate and Water, Espoo, Finland, 17-20 August 1998* [Lemmelä, R. and N. Helenius (eds.)]. Vol. 2, pp. 839–848.

Sterr, H. and F. Simmering, 1996: Die Küstenregionen im 21 Jahrhundert. In: *Beiträge zur Aktuellen Küstenforschung. Vechtaer Studien zur Angewandten Geographie und Regionalwissenschaft (VSAG)* [Sterr, H. and C. Preu (eds.)]. No. 18, Cologne, Germany, pp. 181–188 (in German).

Stott, P.A. and S.F.B. Tett, 1998: Scale-dependent detection of climate change. *Journal of Climate*, **11,** 3282–3294.

Straw, N.A., 1995: Climate change and the impact of the green spruce aphid, Elatobium abictinum (Walker), in the UK. *Scottish Forestry*, **49,** 134–145.

Swift, M.J., O. Andren, L. Bussard, M. Briones, M.M. Couteaux, K. Ekschmitt, A. Kjoller, P. Loiseau, and P. Smith, 1998: Global change, soil biodiversity and nitrogen cycling in terrestrial ecosystems: three case studies. *Global Change Biology*, **4,** 729–743.

Swiss Re, 1998: *Sigma, Vol. 3*. Economic Research, Swiss Reinsurance Company, Zurich, Switzerland, 38 pp.

Sykes, M.T. and I.C. Prentice, 1996: Climate change, tree species distributions and forest dynamics: a case study in the mixed conifer/northern hardwoods zone of Northern Europe. *Climatic Change*, **34,** 161–177.

Szolgay, J., 1997: Assessment of the potential impacts of climate change on river runoff in Slovakia. *Monographs of the Slovak National Climate Program*, **6,** 13–108 (in Slovak with English abstract).

Tälleklint, L. and T.G.T. Jaenson, 1998: Increasing geographical distribution and density of Ixodes ricinus (Acari: Ixodidae) in central and northern Sweden. *Journal of Medical Entomology*, **35,** 521–526.

Talkkari, A. and H. Hypén, 1996: Development and assessment of a gap-type model to predict the effects of climate change on forests based on spatial forest data. *Forest Ecology and Management*, **83,** 217–228.

Tapsell, S.M. and S.M. Tunstall, 2000: The health effects of floods: a case study of the Easter 1998 floods in England. In: *Floods* [Parker, D.J. (ed.)]. Routledge, London, United Kingdom, 181 pp.

Tarand, A. and T. Kallaste, 1998: *Country Case Study on Climate Change Impacts and Adaptation Assessments in the Republic of Estonia*. Stockholm Environment Institute, Tallinn Centre, and United Nations Environment Programme, Tallinn, Estonia, 146 pp.

Theurillat, J.P., 1995: Climate change and the alpine flora: some perspectives. In: *Potential Ecological Impacts of Climate Change in the Alps and Fennoscandian Mountains* [Guisan, A., J.I. Holten, R. Spichiger, and L. Tessier (eds.)]. Conserv. Jard. Bot., Geneva, Switzerland, pp. 121–127.

Theurillat, J.P., F. Felber, P. Geissler, J.M. Gobat, M. Fierz, A. Fischlin, P. Kuipfer, A. Schlussel, C. Velutti, and G.F. Zhao, 1998: Sensitivity of plant and soil ecosystems of the Alps to climate change. In: *Views from the Alps: Regional Perspectives on Climate Change* [Cebon, P., U. Dahinden, H.C. Davies, D. Imboden, and C.C. Jager, (eds.)]. MIT Press, Cambridge, MA, USA, 192 pp.

Thornley, J.H.M. and M.G.R. Cannell, 1997: Temperate grassland responses to climate change: an analysis using the Hurley pasture model. *Annals of Botany*, **80,** 205–221.

Tinker, P.B. and P. Ineson, 1990: Soil organic matter and biology in relation to climate change. In: *Soils on a Warmer Earth* [Scharpenseel, H.W., M. Schomaker, and A. Ayoub (eds.)]. Elsevier, London, United Kingdom, pp. 71–78.

Tomminen, J. and M. Nuorteva, 1987: *Pine Nematodes*. Report, Department of Agricultural and Forest Zoology, University of Helsinki, Vol. 11, pp. 1–18.

Tooley, M. and S. Jelgersma (eds.), 1992: *Impacts of Sea Level Rise on European Coastal Lowlands*. Blackwell Publishing, Oxford, United Kingdom.

Topp, C.F.E. and C.J. Doyle, 1996a: Simulating the impact of global warming on milk and forage production in Scotland, part 1: the effects on dry-matter yield of grass and grass-while clover swards. *Agricultural Systems*, **52,** 213–242.

Topp, C.F.E. and C.J. Doyle, 1996b: Simulating the impact of global warming on milk and forage production in Scotland, part 2: the effects on milk yields and grazing management of dairy herds. *Agricultural Systems*, **52,** 243–270.

Tsimplis, M.N. and T.F. Baker, 2000: Sea level drop in the Mediterranean Sea: an indicator of deep water salinity and temperature changes? *Geophysical Research Letters*, **27,** 1731–1734.

Tuomenvirta, H., H. Alexandersson, A. Drebs, P. Frich, and P.O. Nordli, 1998: *Trends in Nordic and Arctic Extreme Temperatures*. DNMI Report No. 13/98, Norwegian Meteorological Institute, Oslo, Norway, 37 pp.

Turner, R.K., P. Doktor, and W.N. Adger, 1995: Assessing the costs of sea-level rise. *Environment and Planning A*, **27,** 1777–1796.

UK Department of the Environment, 1996: *Review of the Potential Effects of Climate in the United Kingdom*. Her Majesty's Stationary Office, London, United Kingdom, 247 pp.

UK Ministry of Agriculture, Fisheries, and Food, 1995: *Shoreline Management Plans: A Guide for Coastal Defence Authorities*. Ministry of Agriculture, Fisheries and Food, The Welsh Office, Association of District Councils, English Nature and the National Rivers Authority, London, United Kingdom, 24 pp.

USCP, 1994: *Guidance for Vulnerability and Adaptation Assessment*. U.S. Country Studies Program, USCP Management Team, Washington DC, USA, 503 pp.

USEPA, 1989: *The Potential Effects of Global Climate Change on the United States: Appendix G, Health*. Report No. EPA 230-05-89-057, Office of Policy, Planning, and Evaluation, U.S. Environmental Protection Agency, Washington, DC, USA.

van Veen, J.A. and E.A. Paul, 1981: Organic carbon dynamics in grassland soils, part 1: background information and computer simulations. *Canadian Journal of Soil Science*, **61,** 185–201.

Várallyay, G., 1994: Climate change, soil salinity and alkalinity. In: *Soil Responses to Climate Change* [Rounsevell, M.D.A. and P.J. Loveland (eds.)]. NATO ASI Series 23, Springer-Verlag, Heidelberg, Germany, pp. 39–54.

Vehviläinen, B. and M. Huttunen, 1997: Climate change and water resources in Finland. *Boreal Environment Research*, **2,** 3–18.

Velichko, A.A., C.V. Krementski, O.K. Borisova, E.M. Zelikson, V.P. Nechaev, and H. Faure, 1998: Estimates of methane emission during the last 125, 000 years in Northern Eurasia. *Global and Planetary Change*, **16/17,** 159–180.

von Storch, H. and H. Reichardt, 1997: A scenario of storm surge statistics for the German Bight at the expected time of doubled atmospheric carbon dioxide concentration. *Journal of Climate*, **10,** 2653–2662.

WASA, 1998: Changing waves and storms in the northeast Atlantic. *Bulletin of the American Meteorological Society*, **79,** 741–760.

WCRP, 1998: *CLIVAR Initial Implementation Plan*. No. 103, World Climate Research Programme, Geneva, Switzerland, 81 pp.

Wheeler, T.R., R.H. Ellis, P. Hadley, J.I.L. Morison, G.R. Batts, and A.J. Daymond, 1996: Assessing the effects of climate change on field crop production. *Aspects of Applied Biology*, **45,** 49–54.

White, S. and J.P. Martinez Rica, 1996: Evidence and implications of climatic change in the Pyrenees. *Annals of Geophysical Research*, **2(14),** 336.

Wick, L. and W. Tinner, 1997: Vegtation changes and timberline fluctuations in the Central Alps as indicators of Holocene climate oscillations. *Arctic and Alpine Research*, **29,** 445–458.

Wolf, J., 1999: Modelling climate change impacts at the site scale on soybean. In: *Climate Change, Climate Variability and Agriculture in Europe: An Integrated Assessment* [Butterfield, R.E., P.E. Harrison, and T.E. Downing (eds.)]. Environmental Change Unit, Research Report No. 9, University of Oxford, Oxford, United Kingdom, 110 pp.

Wolf, J., 2000a: Modelling climate change impacts at the site scale on soybean. In: *Climate Change, Climate Variability, and Agriculture in Europe: An Integrated Assessment* [Downing, T.E., P.A. Harrison, R.E. Butterfield, and K.G. Lonsdale (eds.)]. Research Report 21, Environmental Change Unit, University of Oxford, Oxford, United Kingdom, pp. 103–116.

Wolf, J., 2000b: Modelling climate change impacts at the site scale on potato. In: *Climate Change, Climate Variability, and Agriculture in Europe: An Integrated Assessment* [Downing, T.E., P.A. Harrison, R.E. Butterfield, and K.G. Lonsdale (eds.)]. Research Report 21, Environmental Change Unit, University of Oxford, Oxford, United Kingdom, pp. 135–154.

Wolf, J. and C.A. van Diepen, 1995: Effects of climate change on grain maize yield potential in the European Community. *Climatic Change*, **29,** 299–331.

Wood, C.M. and D.G. McDonald (eds.), 1997: *Global Warming: Implications for Freshwater and Marine Fish*. Cambridge University Press, Cambridge, United Kingdom and New York, NY, USA.

Woodward, F.I., 1992: A review of the effects of the climate on vegetation: ranges, competition, and composition. In: *Global Warming and Biological Diversity* [Peters, R.L. and T.E. Lovejoy (eds.)]. Yale University Press, New Haven, CT, USA, pp. 105–123.

World Bank, 1999: *World Development Report 1998/1999*. Knowledge for Development, Oxford University Press, Oxford, United Kingdom, 113 pp.

Wulf, A. and E. Graser, 1996: Gypsy moth outbreaks in Germany and neighbouring countries. *Nachrichtenblatt Deutschen Pflanzenschutzdienstes Braunschweig*, **48,** 265–269.

Xu, C., 1998: Hydrological responses to climate change in central Sweden. In: *Proceedings of the Second International Conference on Climate and Water, Espoo, Finland, 17-20 August 1998* [Lemmelä, R. and N. Helenius (eds.)]. Vol. 1, pp. 188–197.

Zackrisson, O. and L. Östlund, 1991: Branden formade skoglandskapets mosaik. *Skog och Forskning*, **4,** 13–21 (in Norwegian).

Zalewski, M. and I. Wagner, 1995: Potential impact of global climate warming on biological processes in reservoirs. *Biblioteka Monitoringu Srodowiska*, 177–188 (in Polish).

Zeidler, R.B., 1997: Climate change variability and response strategies for the coastal zones of Poland. *Climatic Change*, **36,** 151–173.

Zelikson, E.M., O.K. Borisova, C.V. Kremenetsky, and A.A. Velichko, 1998: Phytomass and carbon storage during the Eemian optimum, late Weichselian maximum and Holocene optimum in Eastern Europe. *Global and Planetary Change*, **16/17,** 181–195.

14

Latin America

LUIS JOSE MATA (VENEZUELA) AND MAX CAMPOS (COSTA RICA)

Lead Authors:
E. Basso (Chile), R. Compagnucci (Argentina), P. Fearnside (Brazil), G. Magrin (Argentina), J. Marengo (Brazil), A.R. Moreno (Mexico), A. Suárez (Cuba), S. Solman (Argentina), A.Villamizar (Venezuela), L. Villers (Mexico)

Contributing Authors:
F. Argenal (Honduras), C. Artigas (Chile), M. Cabido (Argentina), J. Codignotto (Argentina), U. Confalonieri (Brazil), V. Magaña (Mexico), B. Morales-Arnao (Peru), O. Oropeza (Mexico), J.D. Pabón (Colombia), J. Paz (Peru), O. Paz (Bolivia), F. Picado (Nicaragua), G. Poveda (Colombia), J. Tarazona (Peru), W. Vargas (Argentina)

Review Editor:
C. Nobre (Brazil)

CONTENTS

EXECUTIVE SUMMARY

The Latin America region is remarkably heterogeneous in terms of climate, ecosystems, human population distribution, and cultural traditions. Land-use changes have become a major force driving ecosystem changes. Complex climatic patterns, which result in part from interactions of atmospheric flow with topography, intermingled with land-use and land-cover change, make it difficult to identify common patterns of vulnerability to climate change in the region. Water resources, ecosystems, agriculture and plantation forestry, sea-level rise, and human health may be considered the most important among the various sectors that may be impacted by climate change.

Climate, Extreme Events, and Water Resources

In most of Latin America, there are no clear long-term tendencies in mean surface temperature. Some changes in regional atmospheric circulation have been detected. For instance, the south Atlantic anticyclone has intensified, and the subtropical jet stream in South America has shifted south. These phenomena may be a sign of changes associated with climate change as they already are impacted by the El Niño-Southern Oscillation (ENSO) phenomena and extreme events.

The cryosphere in Latin America, which is composed of glaciers in the high Andes and three major ice areas in southern South America, may be severely affected by global warming. It has been well established that glaciers in Latin America, particularly along the tropical Andes, have receded in recent decades. Glaciers contribute to streamflow in rivers of semi-arid and arid areas of South America. Streamflows in Andean rivers—for example, in northwest Peru—exhibit tendencies that may be related to glacial extent changes before and after the mid-1970s. However, there also is evidence that rainfall and river flows in other regions of Latin America correspond only to interdecadal variability in the hydrological cycle. For instance, rainfall and streamflow in Amazonia and northeast Brazil exhibit interdecadal variability linked to Pacific and Atlantic Ocean influences. Warming in high mountains also could affect mountain sports and tourist activities, which represent an important source of income in the economies of some countries in the region.

At a subregional level, precipitation trends in Latin America vary. Precipitation trends depend on the location and the length of the time series under study. For instance, negative trends in some parts of Central America (e.g., Nicaragua) contrast with positive trends in northeastern Argentina, southern Brazil, and northwestern Mexico. Other regions, such as central-western Argentina, do not show well-defined trends. This suggests, for instance, an increase in precipitation in some regions of the mid-latitude Americas.

The Amazon River—by far the world's largest river in terms of streamflow—plays an important role in the water cycle and water balance of much of South America. Several model studies and field experiments show that about 50% of the rainfall in the region originates as water recycled in the forest. In the Amazon region, even small changes in evapotranspiration affect water vapor fluxes. Therefore, deforestation is likely to reduce precipitation because of a decrease in evapotranspiration, leading to important runoff losses in areas within and beyond this basin. Any reductions in rainfall would affect not only Amazonia but also Brazil's central-south region, where most of the country's agriculture and silviculture are located. However, with the current rate of deforestation of no more than 10% in Amazonia as a whole, discharge observations across the basin do not exhibit, to date, any significant trends.

Although it is expected that tropical storms (e.g., tropical cyclones) will increase their peak intensity under global warming (see Table 3-10), only some hints of such intensification in the tropical cyclones affecting the Americas have been detected. Some studies based on model experiments suggest that under climate change the hydrological cycle will be more intense, with changes in the distribution of extreme heavy rainfall, wet spells, and dry spells. Frequent severe droughts in Mexico during the past decade coincide with some of these model findings. Even though it is uncertain how global warming may affect the frequency and intensity of some extreme events, the infrequent overlapping of hydrological and weather/climate events historically have given rise to disasters, whose frequency may be enhanced by such warming.

El Niño

It also has been suggested that under climate change, more El Niño-like mean conditions will be experienced. El Niño differentially influences precipitation and temperature in different parts of the region; for example, it is related to dry conditions in northern northeast Brazil, northern Amazonia, and the Peruvian-Bolivian Altiplano. Southern Brazil and northwest Peru exhibit anomalously wet conditions during these periods. In Mexico and the Caribbean coast of Central America, there is compelling evidence of more winter precipitation and less summer precipitation during El Niño. Some of the most severe droughts in Mexico in recent decades have occurred during

El Niño years. During the 1997–1998 El Niño event, droughts occurred in Amazonia, Mexico, and Central America, favoring forest fires. Under climate change conditions, the number of forest fires may increase.

Natural Ecosystems

Latin America contains a large percentage of the world's biodiversity, and climate change could accelerate losses in biodiversity that already are occurring. Some adverse impacts on species that can be related to regional changes in climate have been observed. Studies show that climatic changes already are affecting frogs and small mammals in Central America. The tropical forests play an important role in the hydrological cycle of much of South America. Several model studies and field experiments show that a large part of the rainfall in the region originates as water recycled in the forest. Large-scale deforestation is likely to result in increased surface temperatures, decreased evapotranspiration, and reduced precipitation. Forest fragmentation and degradation has led to increased vulnerability of forests to fire. Global warming and regional climate change resulting from land-cover change may be acting synergistically to exacerbate stress over the region's tropical ecosystems.

Mountain ranges and plateaus play an important role in determining not only the Latin America climate and its hydrological cycle but also its large biodiversity. Mountains constitute source regions of major rivers such as the Amazon, the Parana, and the Orinoco. These river basins represent important habitats of biological diversity and endemism and are highly vulnerable to extreme climate conditions.

On decadal to centennial time scales, changes in precipitation and runoff may have significant impacts on mangrove forest communities. Sea-level rise would eliminate mangrove habitat at an approximate rate of 1% yr^{-1}. The rate is much faster in the Caribbean mainland (approximately 1.7% yr^{-1}). This problem is causing a decline in some of the region's fisheries at a similar rate because most commercial shellfish use mangroves as nurseries or refuges. Coastal inundation stemming from sea-level rise or flatland flooding resulting from climate change therefore may seriously affect mangrove ecology and associated human economy.

Agriculture

Agricultural lands (excluding pastures) represent approximately 19% of the total area of Latin America. For the past 40 years, the contribution of agriculture to the gross domestic product (GDP) of Latin American countries has been on the order of 10%. Therefore, this economic sector remains a key element of the regional economy. An important aspect of this sector is that a large percentage (30–40%) of the economically active population works in this sector. Agriculture also is an important element for the food security of the poorest sectors of the population. Under climate change conditions, subsistence farming could be severely threatened in some parts of Latin America, such as northeastern Brazil.

Studies in Argentina, Brazil, Chile, Mexico, and Uruguay based on general circulation models (GCMs) and crop models project decreased yields in several crops (e.g., maize, wheat, barley, grapes), even when the direct effect of carbon dioxide (CO_2) fertilization and the implementation of moderate adaptation measures at the farm level are considered. It is likely that increases in temperature will reduce crop yields in the region by shortening the crop cycle. However, the lack of consistency in the various GCM precipitation scenarios makes it difficult to have a precise scenario for crop production under climate change, even when the relationships between precipitation and crop yields are well known. Increased temperature, ultraviolet radiation, sea-level rise, and changes in pest ecology may threaten food production as well (e.g., in Argentina). Climate change may reduce silvicultural yields as a result of changes in water availability during the dry season.

Human Health

The magnitude of the impacts of climate change on health in Latin America primarily depends on the size, density, location, and wealth of the population. It has been established that exposure to heat or cold waves has an influence on mortality rates in risk groups in the region. The projected increase in temperature in polluted cities, such as Mexico City or Santiago, may have an influence on human health. There is evidence that the geographical distributions of vector-borne diseases (e.g., malaria, dengue) in Brazil, Colombia, Argentina, and Honduras and infectious diseases (e.g., cholera, meningitis) in Peru and Cuba change when temperature and precipitation increase. The exact distribution of these diseases, however, is not clear.

It is likely that extreme weather events will increase death and morbidity rates (injuries, infectious diseases, social problems, and damage to sanitary infrastructure), as during the heavy rains in Mexico in 1999 or in Venezuela in 1999. On longer time scales, El Niño and La Niña cause changes in disease vector populations and the incidence of water-borne diseases in Brazil, Peru, Bolivia, Argentina, and Venezuela. Extreme climate events appear to affect the incidence of allergies in Mexico. Some economic and health problems could be exacerbated in critical areas, fostering migrations from rural and small urban settlements into major cities and giving rise to additional stress at the national level and, at times, adversely affecting international relations between neighboring countries. Therefore, under climate change conditions the risks for human health in Latin America may increase.

Adaptation Potential and Vulnerability

The economy of Latin American countries can be severely affected by extremes of natural climate variability. There were more than 700 natural disasters in the region between 1980 and 1998. For instance, Hurricane Mitch in 1998 resulted in economic losses of approximately 40 and 70% of the gross national product (GNP) in Nicaragua and Honduras, respectively. Poverty and unequal distribution of wealth may increase as a result of the negative effects of climate change.

Trade agreements are becoming increasingly important for Latin American economies. In principle, they have been designed mainly to speed up socioeconomic development. However, various elements that relate to environmental issues gradually are being introduced into these agreements, including provisions related to climate change. Environmental issues are beginning to be reflected in the environmental legislation in the region. For instance, environmental impact assessments are now required in most project developments.

Although climate change may bring benefits for certain regions of Latin America, increasing environmental deterioration, combined with changes in water availability and agricultural lands, may reduce these benefits to a negligible level. The adaptive capacity of socioeconomic systems in Latin America is very low, particularly with respect to extreme climate events, and vulnerability is high.

14.1. The Latin America Region

14.1.1. *What is Unique about the Latin America Region?*

The Latin American population will increase to 838 million by the year 2050. Annual population growth rates will decrease from 1.68%, in the period 1995–2000, to an estimated 0.51% in the period 2040–2050, according to the medium prospect of the United Nations (Nawata, 1999). This signifies that the population explosion will continue, even if a decerease in population growth rates were possible. One of the critical difficulties caused by growing population is the problem of nutrition and availability of food. Global food supply is expected to meet the overall needs of the growing world population, but significant regional variation in crop yields as a result of climate change (Rosenzweig *et al.*, 1993) could lead to an increased risk of hunger for an additional 50 million people by the year 2050. Because most Latin American countries' economies depend on agricultural productivity, the issue of regional variation in crop yields is very relevant for the region.

The area of the Latin American region is approximately 19.93 million km^2—double that of Europe, but smaller than that of North America, Asia, or Africa. Latin America includes all of the continental countries of the Americas, from Mexico to Chile and Argentina, as well as adjacent seas (Canziani *et al.*, 1998). Even though the region has a predominantly southern location, Latin America also has a presence in the northern hemisphere, including Mexico, Central America, the Guyanas, Suriname, Venezuela, and parts of Colombia, Ecuador, and Brazil.

Latin America's orographical systems present a predominant north-south orientation, extending from the ranges of Mexico and Central America to the southern Andes. These features divide the region into two contrasting but interdependent geosystems, influencing the climatic and hydrological patterns and making primary productivity a direct dependent variable of the aforementioned environmental factors (Lieth, 1976).

These conditions initially led to a prevalence of agricultural activities near coastlines. From the end of the 19th century and the arrival of European migrations, these activities were extended to inner valleys and plateaus. Pre-Colombian cultures, however, had developed many of their community farming activities in the high plateaus, where the largest proportion of Latin America indigenous communities still are settled.

Mountain ranges and high plateaus play an important role in determining local climates that are conditioned by altitude and orientation, which in turn enhance biological diversity. Agricultural diversification also is coupled to habitat heterogeneity through varying crop species and agricultural time schedules. Morello (1976) has reported the way in which hunting, fishing, cattle and sheep grazing, and cropping activities are correlated to discontinued habitats along the altitudinal gradients in the humid tropical Andes in Colombia. Similar patterns have been reported for the Andes of Ecuador (Cornik and Kirby, 1981). Mountain ranges in Latin America also should be considered genetic/germoplasm banks for a wide variety of plants cultivated since the pre-Hispanic period, as well as for domesticated animals (llama, alpaca) and their wild relatives. The success of agriculture in the Andes is based on the genetic variability of plant populations and on the people themselves who have acquired the proper technology after centuries of agricultural practices. This genetic variability has resulted not only in a high number of cultivated species but also in a striking diversity of cultivars and genotypes adapted to the environmental heterogeneity of the mountain ecosystems (Blanco Galdós, 1981).

Each valley or mountain range has its own characteristics, especially in the tropical Andes, making the area one of the most diverse physical and biological mosaics in the world. At the same time, local populations in the Andes have developed appropriate technologies that are applicable in highlands agriculture; these technological reservoirs are comparable only to those existing in the high plateaus of Asia (Morello, 1984)

Earthquakes, volcanoes, and tectonic movements are common all over Latin America. Some of these events may be catastrophic because urban and rural settlements are likely to be devastated, but some may have positive effects. For example, floods in Argentina, which originate in the high Andean watersheds, have devastating effects on cultivated valleys. However, aquifers previously exhausted through alfalfa irrigation become recharged, and salinized soils are washed. Furthermore, volcanic eruption generates soil enrichment, which nourishes a wide range of crops, including coffee in Central America and Colombia.

Latin America contains a large variety of climates as a result of its geographical configuration. The climatic spectrum ranges from cold, icy high elevations, with some of the few glaciers still found in the tropics, to temperate and tropical climate. The region also has large arid and semi-arid areas. One of the most important characteristics of Latin America from the climatic point of view is its large sensitivity and vulnerability to ENSO events. From northern Mexico to Tierra del Fuego, every country in the continent exhibits anomalous conditions associated with ENSO.

The region also hosts the largest pluvial forest in the world: 7.5 million km^2 constitute Amazonia, of which 6.12 million km^2 are within the Amazon basin. The average rainfall in the Amazon basin is about 2,300 mm yr^{-1}, with real evaporation estimated at 1,146–1,260 mm yr^{-1}. The Amazon is undoubtedly the world's largest river in terms of its outflow, with an average annual flow rate of 209,000 m^3 sec^{-1}. The Amazon, the Parana-Plata, and the Orinoco carry into the Atlantic Ocean more than 30% of the freshwater of the world. However, these water resources are poorly distributed, and extensive zones have extremely limited water resources.

Latin America hosts one of the largest terrestrial and marine biological diversities in the world. South America has the largest fish catch on the eastern Pacific. There is an important flow of krill and other plankton species as a result of cold sea currents on both sides of the southern tip of South America. A combination of the prevailing atmospheric and oceanic circulation defines the climate and the land and sea productivity of the region. This explains the actual distribution of human settlements and the availability of basic services (e.g., water supply).

Overall, the health profile of the Latin American population can be classified as undergoing a slow epidemiological transition. At one extreme of the spectrum there is a high incidence of (and mortality from) chronic noninfectious diseases such as cardiovascular problems and cancer, which predominate in large metropolitan areas. On the other hand, infectious diseases still impose a heavy burden on the poverty-stricken parts of the population. The reasons for this dichotomy are two-fold: uneven socioeconomic development within countries and the extreme diversity of regional environments.

Latin America (and the Caribbean) has the greatest disparity in income distribution in the world. A mere 5% of the population receives 25% of all national income, and the top 10% receive 40%. Such proportions are comparable only to those found in some African countries (IDB, 1999).

Many problems that have affected Latin America adversely are now showing a wood-saw type of change, with some temporary improvements and downfalls. However, in some countries of the region, improvement in the macroeconomy is being observed, in spite of the negative impact from recent developments in Asia and some countries of the region (Mexico and Brazil). Improvement currently observed in the economies of Mexico, Brazil, Chile, and Argentina does not reflect the meso- and microeconomies net deterioration in the living standards of rural and peri-rural urban areas. The middle class, which had become a sign of progress in several countries, also is adversely affected. This situation, added to the effect of extreme events, has exacerbated migration toward richer cities and countries with relatively better economies. Shantytowns have grown steadily around big cities, and poverty belts have even tripled. Their location in flood-prone valleys and unstable hills results in a lack of potable water and sanitation services, which is posing a serious threat to these cities.

Cultural (language, traditions, religion), economic (degree of development, economic systems, wealth distribution), and social (demographic growth, political systems and practices, educational systems) similarities in Latin American countries indicate that they could address climate change with common (shared) methods.

14.1.2. Climate Variability and Change

There is ample evidence of climate variability at a wide range of time scales all over Latin America, from intraseasonal to long term. In many subregions of Latin America, this variability in climate normally is associated with phenomena that already produce impacts with important socioeconomic and environmental consequences that could be exacerbated by global warming and associated climate change. Signals that can be related to variability and/or change in climate conditions for Latin America have been identified in some of the analyses performed by researchers in the region, particularly for streamflow, precipitation, temperature, glacier oscillations, general circulation, and extreme events. Estimations of potential future climate conditions are based on climate change scenarios studies developed for some subregions of Latin America.

14.1.2.1. Past to Present

14.1.2.1.1. Glaciers, precipitation, and streamflow

Glaciers in Latin America have receded dramatically in the past decades, and many of them have disappeared completely (Williams and Ferrigno, 1998). In 18 glaciers in the Peruvian Andes, mass balances since 1968 and satellite images show a reduction of more than 20% of the glacial surface, corresponding to 11,300 million m^3 of ice (Morales-Arnao, 1969a,b; INAGGA-CONAM, 1999). Significant reductions also have occurred in southern Chile and Argentina (e.g., glacier Sarmiento) (Basso, 1997). Deglaciation may have contributed to observed negative trends in streamflows in that region (Morales-Arnao, 1999). For rivers in arid lands in northwest Peru and northeast and southeastern Brazil, significant negative trends also have been detected, but these variations seem to be related to human water management for irrigation purposes and increases in agricultural areas, rather than climate-induced changes (INRENA, 1994; Marengo, 1995; Marengo *et al.*, 1998).

Between 20°S and 40°S, precipitation around the Andes occurs mainly during the winter. Snow accumulates in the high parts of the cordillera and melts during the summer, becoming the main source of water for rivers in the region. Agricultural activities in central Chile and the Argentinean central western plains are maintained through irrigation. Therefore, it may be said with high confidence that fluctuations in winter precipitation have a strong socioeconomic impact in the region.

The precipitation record for Santiago, Chile, is highly correlated with snow depth in the cordillera. Recorded precipitation exhibited a decreasing trend from the late 19th century through the mid-1970s but has reverted since then. A similar trend has been detected in streamflow in the region (Minetti and Sierra, 1989; Carril *et al.*, 1997; Compagnucci and Vargas, 1998; Compagnucci *et al.*, 2000; Waylen *et al.*, 2000). In southern Chile and the Argentinean cordillera, a negative trend in precipitation and streamflow has been detected (Quintela *et al.*, 1993; Nuñez *et al.*, 1999).

In northwestern Mexico, there is a tendency for more winter precipitation, which has resulted in positive trends in river water levels. However, along with more intense winter precipitation, interannual climate variability has increased (Magaña and Conde, 2000). On the other hand, some parts of southern Mexico and Central America exhibit positive or negative rainfall trends, depending on the orientation of the catchment

(Aparicio, 1993; IPCC, 1996; Jáuregui, 1997; TAR WGI Chapter 3).

For Nicaragua, rainfall analysis for 1961–1995 showed negative trends in the north and northwest parts of the country. A systematic increment was detected on the Caribbean coast, and almost no variation was found along the central and the Pacific coastal regions (MARENA, 2000).

In Colombia, weak rainfall trends have been observed for the period 1955–1995, with no preferred sign at a regional level. For central Colombia, rainy seasons have been occurring earlier in recent years than 25 years ago (Mesa *et al.,* 1997). Trends in Colombian river streamflow are mixed, but the main river catchments such as the Cauca and Magdalena Rivers exhibit decreasing trends. Deforestation could account for such decreasing trends in river discharges (Poveda and Mesa, 1997).

For the Amazon region, Marengo *et al.* (2000) have identified multidecadal variations in rainfall in northern and southern portions of the basin, with opposite tendencies. Perhaps the most important finding is the presence of periods with relatively wetter or drier conditions that are more relevant than any unidirectional trends themselves. For instance, the period 1950–1976 was regionally wet in northern Amazonia, but since 1977 the region has been drier. This dryness does not seem to be related to regional deforestation (see Marengo *et al.*, 1998; Marengo and Nobre, 2000; TAR WGI Chapter 3). Similarly, streamflow series in Amazonian rivers also exhibit multidecadal variations; they do not display significant unidirectional trends (Richey *et al.*, 1989; Marengo, 1995).

In northeast Brazil, multidecadal variations in atmospheric circulation over the tropical Atlantic have been linked to similar time-scale variations in rainfall over the region (Hastenrath and Greischar, 1993; Nobre and Shukla, 1996; Wagner, 1996). On longer time scales, rainfall in northern northeast Brazil exhibits weak positive trends that are consistent with changes in decadal changes in circulation described in Wagner (1996).

Streamflow in the River Plate basin—particularly in the Negro, Paraguay, Paraná, and Uruguay Rivers—exhibits a negative trend from 1901 to 1970, which reverses after this period. Multidecadal variability also is observed in discharges (Garcia and Vargas, 1998; Genta *et al.*, 1998; Robertson and Mechoso, 1998). Moreover, there are written reports of alternating floods and droughts periods during the 16th–18th centuries, indicating high natural variability (Prieto and Herrera, 1992).

In subtropical Argentina, Paraguay, and Brazil, precipitation exhibits a long-term change, with a sharp increase in the period 1956–1990 after a dry period along 1921–1955 (Castañeda and Barros, 1996). In the Pampa region, there is a positive trend in precipitation during the period 1890–1984. This increase in annual rainfalls was accompanied by a relative increase in precipitation during the spring and summer (Penalba and Vargas, 1996; Hoffman *et al.*, 1997; Krepper and Sequeira, 1998).

At high elevations in northwest Argentina, paleoclimatic records suggest an increase in precipitation in the past 200 years (Villalba *et al.*, 1997). In the same region, as well as in Bolivia and southeast Peru, records show that the 17th-century climate was wetter and less variable (fewer floods and droughts), whereas the 18th century was highly unstable, with a large amplitude in the annual cycle and recurrent wet and dry periods (Prieto and Herrera, 1992).

Variations in precipitation in Latin America have a strong effect on runoff and streamflow, which also are affected by melting of glaciers and snow. Based on available information, there is evidence that these variations and their sign depend on the geographical subregion under consideration.

14.1.2.1.2. Temperature

Temperature in Latin America varies depending on the different subregions. For tropical Latin America, temperature depends on cloud cover and altitude; for other subregions, altitude, advection, and, at the southern cone, sea-surface temperature (SST) play more predominant roles. The following description is based on observed records; where possible, paleoclimatic information has been included to present a wider view.

Central America shows different signs for temperature trends, according to the specific area under analysis. For example, in Costa Rica, Alfaro (1993) identifies a positive trend in daily maximum temperature. Gómez and Fernández (1996) and OCCH (1999) have identified negative trends for large areas of Costa Rica and Honduras. MARENA's (2000) analyses of time series for Nicaragua find only a small increase in mean temperature for Managua, which might be associated with growth in urbanization.

For northwestern South America, monthly mean air temperature records show a warming of 0.5–0.8°C for the last decade of the 20th century (Pabón, 1995a; Pabón *et al.*, 1999; Quintana-Gomez, 1999). Colombia also presents increasing trends in the time series for the daily series of daily mean and minimum temperature for the past 30–40 years. Similar patterns have been observed in average monthly dew point and relative humidity (Mesa *et al.*, 1997; Pérez *et al.*, 1998). Coastal cities from northern Peru presented increases in air temperature since 1940, where 16 El Niño events were reported (Jaimes, 1997; SENAMHI, 1999).

In several cities in southern and southeastern Brazil, studies on long-term tendencies for air temperature, from the beginning of the 20th century, have indicated warming tendencies (Sansigolo *et al.*, 1992). This could be attributable to urbanization effects or to systematic warming observed in the south Atlantic since the beginning of the 1950 s (Venegas *et al.*, 1996, 1998). In the Amazon region, Victoria *et al.* (1998) have detected a significant warming trend of +0.63°C per 100 years.

Data since the beginning of the 20th century do not show a clear tendency in mean temperature in the southern cone, but there is a decrease in the thermal range. Moreover, south of 50°S there are indications of a positive tendency (Hoffman *et al.*, 1997). However, when a shorter record is used for the analysis, Argentina and Chile show a large warming rate of 1.2–3.0°C per 100 years (Rosenblüth *et al.*, 1997).

In south tropical Argentina, warming is observed only during the austral autumn season (Bejarán and Barros, 1998). The Argentina humid pampa, represented by Buenos Aires, presents a warming as a result of urban effects (Camilloni and Barros, 1997). Intensity and persistence of heat and cold waves present tendencies in which the sign depends on the region (Rusticucci and Vargas, 1998).

In extra-tropical west South America (Chile), surface air temperature has varied differently during the 20th century. South of approximately 45°S, temperatures have been increasing in stepwise fashion (Aceituno *et al.*, 1993). In the area spanning about 35°S to 45°S, the most significant feature is a well-defined cooling of 1–2°C from the 1950s to the mid-1970s.

At decadal scales, multiple climate records throughout Latin America consistently exhibit a shift in the mean during the mid-1970s. This could be a climatic consequence of sudden changes in the climatology of the Pacific Ocean (Trenberth, 1990).

Using tree-ring and glacial evidence, summer temperatures in northern Patagonia show distinct periods of higher and lower temperatures during the past 1,000 years (Villalba, 1994; Villalba *et al.*, 1997). For instance, there was a cold interval from AD 900 to 1070, followed by a warm period from AD 1080 to 1250 (coincident with the Medieval warm period). Warm climatic episodes similar to that observed during the 1980s may have occurred in the recent past under preindustrial CO_2 levels in northern Patagonia (Chile and Argentina) (Villalba *et al.*, 1997).

Latin America, in general, shows important variations in temperature, some of which might be connected to change in climate. At the same time, these variations might depend on the origin and quality of the data as well as the record periods used for the studies.

14.1.2.1.3. Large-scale atmospheric circulation

Latin America's climate is influenced mainly by the northern Atlantic anticyclone and the migration of the inter-tropical convergence zone, which also affects large areas of tropical South America. The southern part of the continent is more affected by Atlantic and Pacific anticyclones, the thermic low pressure of northwestern Argentina, and mid-latitudes westerlie. All of these circulation features interact strongly with the complex topography of Latin America.

Analysis of ice cores in west Antarctica indicate that meridional atmospheric circulation intensity between middle and high latitudes has experienced substantial strength variability, increasing in the Little Ice Age (Kreutz *et al.*, 1997; Leckenbush and Speth, 1999). At paleoclimatic time scales, analyses of fossilized pollen and lake sediments have shown more intense and frequent incursion of polar air from the Antarctic region 12,000–8,000 years before the present (BP) (Ledru *et al.*, 1994).

In southern Brazil, there has been a tendency over the past 20 years for fewer wintertime cold fronts and polar outbreaks (Marengo and Rogers, 2000), which is somewhat consistent with reported interdecadal variations in the mean position and intensity of the south Atlantic anticyclone (Venegas *et al.*, 1998).

For mid-latitude South America, important changes in zonal circulation have been observed between 1899 and 1986, with wintertime circulation weaker for the period 1939–1949 and strong during 1967–1977—suggesting interdecadal changes. Over Paraguay, southern Brazil, Uruguay, and northeast Argentina, northeasterly circulation associated with the subtropical Atlantic anticyclone increases after 1954 (Hoffman *et al.*, 1987; Minetti and Sierra, 1989; Cantañeda and Barros, 1993; Barros *et al.*, 1999).

Instrumental records, sounding information, and satellite data show changes, fluctuations, and "sudden jumps" in some features of atmospheric circulation over Latin America and its adjacent oceans, in connection with detected changes in the global climate system.

14.1.2.1.4. Variability and impacts from El Niño and the Southern Oscillation

The extremes of the Southern Oscillation are partly responsible for large portions of climate variability at interannual scales in Latin America. Therefore, some of the variations in the foregoing elements could be associated with manifestations of climate variability, such as the El Niño phenomenon, which represents the low phase of the Southern Oscillation; the positive phase is referred to as La Niña. Atmospheric circulation patterns are more perturbed during El Niño than during La Niña years (Salles and Compagnucci, 1995, 1997).

In Mexico and parts of the Caribbean, the ENSO signal corresponds to more winter precipitation and less summer precipitation (Magaña and Quintanar, 1997). Some of the most severe droughts in Mexico in recent decades have occurred during ENSO summers (Magaña *et al.*, 1998). The signal of La Niña is almost opposite to the ENSO signal. In Central America, orographic effects play an important role in understanding regional ENSO effects in precipitation. During El Niño years, the Pacific side of Central America suffers an important reduction in precipitation, whereas some parts of the Caribbean side experience more rain than usual.

Over Colombia, ENSO events are associated with reductions in precipitation, river streamflows, and soil moisture, whereas

La Niña is associated with heavier precipitation and floods (Poveda and Mesa, 1997). There also is a very high positive correlation between the Southern Oscillation Index (SOI) and river discharge in Colombia. This relationship is stronger during the December–January period and weaker during April–May. The influence of ENSO is stronger at river stations located in western Colombia and weaker for stations located in eastern Colombia. Over the eastern part of the Andes, Ecuador, and northern Peru, large positive anomalies in precipitation typically are observed during the warm episode.

Dry anomalous conditions affect the Amazon region of Brazil northward to the Caribbean through the latter half of the year (Ropelewski and Halpert, 1987, 1989, 1996; Díaz and Kiladis, 1992). In northern Amazonia and northeast Brazil, deficient rainy seasons have been observed during ENSO years (Aceituno, 1988; Marengo, 1992; Uvo, 1998). Droughts that led to forest fires were detected during the very strong ENSO events of 1911–1912, 1925–1926, 1982–1983, and 1997–1998. Extreme droughts also occurred during these years in northeast Brazil. In contrast, the ENSO signal in southern Brazil is opposite to that in northeast Brazil and northern Amazonia, with positive and sometimes extremely large anomalies of rainfall during the rainy season of ENSO years, whereas drought can occur during the positive Southern Oscillation phase (Ropelewski and Halpert, 1989; Grimm *et al.*, 1996, 2000).

Through northern and central Chile and at high altitudes of the Andes in Argentina, between 30°S and 40°S, most precipitation is recorded during the winter, with positive anomalies registered during early stages of the warm phase of ENSO. Because of the semi-arid conditions of this area, their economy is strongly affected (Quinn and Neal, 1982; Compagnucci, 1991; Ruttland and Fuenzalida, 1991; Canziani *et al.*, 1997; Compagnucci and Vargas, 1998). At the same time, strong rainfall events occur in low altitudes of Chile, triggering debris flows during the winter such as those in Santiago and its surrounding areas in 1991–1993 (Garreaud and Ruttland, 1996) and 1997.

At high altitudes of the Andes, large amounts of snow are recorded consistently. Melting of this accumulated snow is the main cause of river runoff during the summer. In Chile and central-western Argentina, north of 40°S, streamflows were normal or above normal during El Niño years (Waylen and Caviedes, 1990; Compagnucci and Vargas, 1998; Compagnucci, 2000). On the other hand, during cold events (La Niña), negative anomalies of rainfall and snowfall are present—with opposite consequences, including below-normal summer streamflow. For this region, the likelihood of dry conditions during La Niña is higher than that of wet conditions during El Niño (Compagnucci and Vargas, 1998).

14.1.2.1.5. Extreme events

Cold and warm fronts, tropical cyclones, and severe convergence are some of the most frequent phenomena that produce floods, droughts, mud and snow slides, heat waves, frosts, and climate-related fires throughout Latin America. These extreme events produce direct and indirect impacts on productivity and affect the quality of life for Latin Americans. A hazard (extreme climate phenomenon) becomes a disaster when it outstrips the ability of a country or region to cope.

There are subregions of Latin America where the occurrence of extreme events is very frequent. Central America and southern Mexico often experience the effect of tropical cyclones and associated heavy rain, flooding, and slides. For northwestern South America and northeastern Brazil, many of the extremes that occur are highly related to El Niño.

Sometimes these extreme events could be magnified to such a level (extreme of extremes) that the impact becomes a disaster. In Latin America, interaction with other complex phenomena, such as interannual or interdecadal oscillations, can contribute to create the appropriate conditions to produce a disastrous impact. Examples of these extraordinary extreme events include Hurricane Mitch in Central America, heavy rains in Venezuela, some of the most severe droughts in northeastern Brazil, and variations in ocean currents during El Niño for Peru and Ecuador.

Emanuel (1987, 1991) has suggested that warmer surface conditions and colder lower stratospheric temperatures would result in stronger hurricanes. Data for the eastern Pacific region indicate that the number of strong hurricanes in the region has been increasing since 1973 (Whitney and Hobgood, 1997). Such changes may represent a major environmental threat for countries such as Mexico (Jáuregui, 1995) and the Central American isthmus.

Some of the relatively weak cold surges may exhibit unusual intensity, causing frosts and low temperatures in coffee-growing areas of southeastern Brazil, resulting in heavy damage and losses in coffee production (Marengo *et al.*, 1997). In the Mexican Altiplano, dry atmospheric conditions result in radiative cooling and frosts even during the summer (Morales and Magaña, 1998).

Even though it is still uncertain how global warming may affect the frequency and intensity of extreme events, extraordinary combinations of hydrological and climatic conditions historically have produced disasters in Latin America. Thus, in assessing vulnerability and adaptation mechanisms, it is necessary to consider the potential influence that global warming might have on extreme events.

14.1.2.2. Future: Climate Scenarios

Climate modeling has proven to be extremely useful in building projections for climate change and scenarios of future climate under different forcings. General circulation models have demonstrated their ability to simulate realistically the large-scale features of observed climate; hence, they are widely used to assess the impact that increased loading of the atmosphere with greenhouse and other gases might have on the climate system. Although there are differences among models with regard to the way they represent the climate system processes, all of them yield comparable results on a global basis. However, they have difficulty in reproducing regional climate patterns, and large discrepancies exist among models. In several regions of the world, distributions of surface variables such as temperature and rainfall often are influenced by the local effects of topography and other thermal contrasts, and the coarse spatial resolution of GCMs cannot resolve these effects. Consequently, large-scale GCM scenarios should not be used directly for impact studies, especially at the regional and local levels (von Storch, 1994); downscaling techniques are required (see TAR WGI Chapters 10 and 13).

At the large scale, rates of mean annual temperature changes in the Latin American region for the next century are projected to be 0.2–2°C (Carter and Hulme, 2000) under the low-emissions scenario (B1) produced as part of the IPCC *Special Report on Emission Scenarios* (SRES). The warming rate could range between 2 and 6°C for the higher emissions case (A2). Most GCMs produce similar projections for temperature changes on a global basis; projected changes in precipitation remain highly uncertain.

For impact studies, it is crucial to have a projection of concurrent changes of temperature and precipitation at the regional scale. Various scenarios of climate change for Latin America have been put forward on the basis of GCM projections under the IS92a scenario. Most of these regional scenarios are based on GCM experiments that are downscaled through statistical techniques. Derived climate change scenarios for Mexico suggest that climate in Mexico will be drier and warmer (Perez, 1997). Several hydrological regions in Mexico are highly vulnerable to decreased precipitation and higher temperatures (Mendoza *et al.*, 1997). A regional climate change scenario for central Argentina in response to CO_2 doubling under the IS92 scenario for the year 2050, also obtained through a statistical downscaling approach, shows a smaller increase in minimum temperature as compared to the maximum and larger increases for summer than for winter months, which generates enhanced temperature amplitudes (Solman and Nuñez, 1999). In addition, a decrease in precipitation is projected over the region, which is larger for summer (12%) than for winter months (5%). This result highlights an important consequence in the rainfall regime over the region: A large decrease in rainfall projected for the rainy season will seriously affect soil moisture, hence agricultural production in the region.

Several climate change scenarios for other parts of Latin America rely on linear interpolation of GCM output to estimate increases in surface temperature and precipitation (Mata, 1996; Carril *et al.*, 1997; Hofstadter and Bidegain, 1997, Paz Rada *et al.*, 1997; Centella *et al.*, 1998; MARENA, 2000). In the case of Costa Rica (MINAE-IMN, 2000), under the IS92a scenario for the year 2100, the results show a small increase in precipitation for the southeastern Caribbean region and an important decrease—close to 25%—in the northwestern Pacific region. This latter region already experiences water problems as a result of El Niño and an increasing demand from infrastructure for tourism and irrigation. Under the same climate scenario, mean temperature in Costa Rica is expected to rise by more than 3°C by 2100, and tendencies in actual climate series (1957–1997) show already an increase of 0.4°C every 10 years for the more continental Central Valley areas. This last estimation may reflect signals other than the one related to climate change.

Results from climate scenarios for Nicaragua imply an additional pressure on productivity sectors and human activities. Under the IS92a emissions scenario, mean temperature for the Pacific watershed would be expected to rise, ranging from 0.9 for the year 2010 to 3.7°C for the year 2100, and precipitation would

***Table 14-1**: Estimated changes projected under IS92 scenario for some countries within Latin America region.*

Region	Temperature	Precipitation
Mexico	increase	decrease
Costa Rica		
– Pacific sector	+3°C	-25%
– Southeast Caribbean sector		small increase
Nicaragua		
– Pacific sector	+3.7°C	-36.6%
– Caribbean sector	+3.3°C	-35.7%
Brazil		
– Central and south central sector	+4°C	+10 to +15% for autumn reductions for summer
Central Argentina	summer: +1.57°C (+1.08–2.21°C) winter: +1.33°C (+1.12–1.57°C)	summer: -12% winter: -5%

decrease by 8.4% for the year 2010 and 36.6% for the year 2100. For the Caribbean watershed, mean temperature would increase, ranging from 0.8°C for the year 2010 to 3.3°C for the year 2100, and precipitation would decrease in a range between 8.2% for the year 2010 and 35.7% for the year 2100 (MARENA, 2000).

Potential effects of climate change in Brazil suggest changes of 4–4.5°C in surface temperature as a result of increased CO_2 concentrations (de Siqueira *et al.*, 1994, 1999). Central and south-central Brazil may experience increases of 10–15% in autumn rainfall; reductions could appear during December, with high risk of drought during summer, affecting crops (see Table 14-1).

Analysis of climate variations during the instrumental period and evidence suggested by paleoclimatic and other proxy climate information suggests that climate variations and change have been found in several regions in Latin America. Most climate records cover the past century; at this time scale, there have been indications of multidecadal and interannual variability, some linked to extremes of the Southern Oscillation. The lack of continuous and long-term records from the past does not allow one to identify climate patterns with a high degree of confidence to determine whether these climates were similar to or much different from that of present times—particularly with respect to the frequency and intensity of extreme events such as drought, floods, freezes, heat waves, and especially hurricanes and tropical storms. However, multidecadal variations have been identified in rainfall and streamflow records in the region, although no clear unidirectional trend indicators of climate change have been identified.

14.1.3. Socioeconomic and Trade Agreements Issues

14.1.3.1. Socioeconomic Issues

Latin America has one of the greatest disparities in income distribution in the world. From the social and environmental point of view, the region clearly is vulnerable to the effects of natural disasters. Inappropriate land use, for agriculture and human settlements, in watersheds causes serious damages during the occurrence of extraordinary climate extreme events, such as Hurricanes George and Mitch in Central America and intense rain events in Venezuela and Argentina. CEPAL (1999a,b,c,d,e) estimated losses from Hurricane Mitch (see Table 14-2). Losses in Honduras and Nicaragua were 70 and 45% of GNP, respectively, affecting the development and economic growth of both countries. There is strong evidence that the effects of natural disasters contribute to rising poverty and inequality in many regions of Latin America.

The year 1998 was one of the most problematic periods in recent times for Latin America and the Caribbean. The side effects of the international financial crisis that originated in Asia in the middle of 1997 limited the possibilities for the region to obtain external aid. In addition, many adverse climatic events

***Table 14-2**: Estimated losses from Hurricane Mitch.*

Country	Losses (US$ million)	Percentage of GNP
Costa Rica	92	1.0
El Salvador	388	6.1
Guatemala	748	1.5
Honduras	4,000	70.0
Nicaragua	988	45.0

worsened the socioeconomic conditions. The Latin American region hosts a myriad of socioeconomic conditions. Countries with a high level of development coexist with least-developed countries. Macro-economic figures indicate a moderately constant degree of growth in the region, even though more than 200 million people in Latin America are poor (CEPAL, 1998).

The Human Development Report for 1998 (UNDP, 1998), in its section on consumption patterns, notes that the overwhelming majority of people who die each year from pollution are poor people in developing countries. Large cities in Latin America such as Mexico City already have a serious problem with air pollution. The report also identifies several issues that affect most of Latin America which may be increased by global warming, including desertification, floods and storms, and harvest. Poor people are less resilient to these problems.

The World Bank (1997) analyzed the 1982–1983 El Niño event and, with a great degree of certainty, considered it to be the most intense in the 20th century. Losses from droughts, floods, and hurricanes were estimated at US$14 billion. Of these, US$2 billion were lost in the western coast of South America, half in Peru—mainly losses from fishing revenue and destruction of infrastructure. Social losses also were very high. Reconstruction and development of related activities in Peru depleted resources and resulted in losses of 6% in GNP. Damages in Peru in 1997–1998 were on the order of US$1 billion (of which 55% was transportation infrastructure, 15% agriculture, 14% energy, and 9% education), and more than 400 relief projects had to be implemented, requiring emergency attention in 14 of the 24 departments of Peru. Some positive effects resulted, such as increases in pastures (200,000 ha) and reforestation of 100,000 ha.

One sector that might be affected by climate change, based on previous experiences during the ENSO period, is fisheries. Fish capture was reduced by 53% during the 1998 ENSO event (IMARPE, 1998). However, adaptation measures such as changing the species captured and price increases reduced the losses to 40%. Some adaptation options, such as switching from anchovy to tuna, can significantly reduce losses from seawater warming (Arntz and Fahrbach, 1996; IMARPE,1998). Adaptation to flooding conditions may reduce damage from extreme floods. One way to address water-shortage areas is to store excess water; for example, a lake was created in north Peru in 1998 (IMARPE, 1998).

Insurance is an important financial aspect that may be impacted by climate change. A joint World Meteorological Organization/ Inter-American Development Bank (WMO/IDB) meeting—with participation by scientists, bankers, and insurers—recognized the importance of climate change in insurance policy and of the need for increased cooperation (WMO, 1994), but few concrete actions materialized, mainly because insurers feel that larger risks should be compensated with higher insurance premiums.

14.1.3.2. Trade Agreements

Several trade agreements exist in the region and have important effects on the economy and, indirectly, on the environment of the region. The most important of these agreements are listed in Table 14-3. Regional intergovernmental agreements could be an important mechanism for adaptation to climate change in Latin America. Agreements such as the one developed for conservation of water, the environment, and natural disaster management in the Central American isthmus (Central American Integration System—SICA) could be interpreted as a way to coordinate sectorial adaptation to different climatic conditions. During the South American presidential meeting in August 2000, leaders attempted to consolidate efforts for the region's political and economic integration, with emphasis on free trade. Because natural disasters cause a decline in productive capacity and increase demand for strong public and private investment in reconstruction, imports are likely to increase and exports are likely to fall, leading to a trade deficit. Disasters such as floods, droughts, and hurricanes also may lead to a significant increase in food product prices if production and distribution are disrupted (IDB, 2000).

There also have been other agreements on the technological and scientific agenda, such as initiatives to foster research on global and climate change and environmental issues. The Inter-American Institute for Global Change (IAI) supports research initiatives to document climate variability and change characteristics and their societal impacts in the Latin American and Caribbean region.

14.1.4. Environmental Legislation

Ratification of international and regional conventions and agreements brings the responsibilities and commitments associated with them into the text of national legislation. As a consequence, ratification of the recommendations of the United Nations Conference on Environment and Development (UNCED, 1992) by Latin American governments has initiated a path for preventive actions for natural hazards, as well as other recommendations included in Agenda 21 to be incorporated into national legislation. The same is true for the United Nations Framework Convention on Climate Change (UNFCCC), the United Nations Convention on Biodiversity, and the United Nations Convention to Combat Desertification and Drought.

***Table 14-3**: Important trade agreements in Latin America region.*

Agreement	Participating Countries	Notes
Asociación Latinoamericana de Libre Comercio (ALADI)	Spanish speaking countries of the region	Not a particularly active agreement.
Pacto Andino (Andean Pact)	Bolivia, Colombia, Ecuador, Peru, and Venezuela	Tends to eliminate duties in a gradual way. Has incorporated an environment unit and places increasing attention on environmental issues.
Mercosur	Argentina, Brazil, Paraguay, and Uruguay	Bolivia and Chile are associate members. Tends to eliminate duties in a gradual way. Future common currency under consideration.
Central American Common Market	Costa Rica, El Salvador, Guatemala, Honduras and Nicaragua	One of the oldest trade agreements. Plans to convert to a free-trade zone.
North America Free Trade Agreement (NAFTA)	Mexico, plus Canada and USA	Very successful for expanding trade. Strict environmental considerations. Plans call for expansion to all Latin American countries.
Bilateral agreements	Several pairs of countries or groups	Environmental considerations increasingly important, especially in agreements with the European Union.

This also has been the case with the adoption of the Vienna Convention on the Protection of the Stratospheric Ozone Layer and its Montreal Protocol. Similar steps will be needed should the Kyoto Protocol be ratified. The UNFCCC, particularly Articles 2 and 4, calls for action vis-à-vis climate variations, and the Kyoto Protocol opens avenues for mitigation of greenhouse gas (GHG) emissions; through Article 12, on Clean Development Mechanisms, the UNFCCC makes provision for assistance and transfer of technology from Annex I to non-Annex I Parties.

There also are a substantial number of regional agreements and a huge body of laws, rules, and regulations to ensure systematic and coordinated actions for protecting the environment, including flora and fauna and coastal and inland wetlands, as well as to promote sustainable development (PNUMA, 1991; Bertucci *et al.*, 1996; Solano, 1997). Most Latin American governments have a comprehensive environmental legal framework with relevant laws, rules, and procedures for specific resources and activities (e.g., water, forestry and mineral resources, marine resources and coastal areas, hunting and fishing, and tourism, as well as specific products and pesticides and pollution) (Sebastiani *et al.*, 1991, 1996a,b). Latin American governments also have developed national and regional environmental plans and strategies, as well as other sectoral or special programs. Although implementation in some countries is far from satisfactory, it also is very common to find legislation that regulates the use of natural resources and makes provision to punish noncompliance (Solano, 1997). In a large number of countries in the region, recent legal developments on environmental management include mandatory environmental impact assessment (EIA) studies. All Latin American countries are expected to adopt this policy. With regard to the climate component, the main shortcomings highlighted in the report to the IPCC Bureau on Systematic Observations are lack of sufficiently dense terrestrial and marine observing systems; lack of systematic observations on specific biological variables and socioeconomic impacts, as well as GHGs and aerosols; and consolidation of land surface observations (hydrology, ecosystems, and land use) (WMO, 1998). This means that, generally speaking, there are limitations that affect the value and reliability of EIAs in some areas of the region.

In line with United Nations Resolution A/52/629—which calls for cooperation to incorporate sustainable development programs at national, regional and global levels—countries in Latin America are engaged in fulfilling the objectives of such development programs. In this respect, policymakers and decisionmakers should be made aware of the role played by climate variation issues in such development strategies. The large majority of countries follow the recommendations made by the United Nations Commission for Sustainable Development (UNCSD), and the Economic Commission for Latin America and the Caribbean (ECLAC) is assisting them in integrating relevant disciplines and sectors, particularly for incorporating natural hazard response strategies in sustainable development policies. However, more work still is needed, as a result of the aforementioned shortcomings in the operation of observing and monitoring systems as well as the need for capacity-building on issues related to sustainable development and their well-known linkages with climate issues. Such action becomes relevant to efforts for mitigating environmental hazards and acting in line with the planning of the Pan-American Climate Information and Applications System (PACIS), as suggested by the Summit of the Americas in Santiago, Chile, in 1998.

As environmental concerns become more pressing, they are climbing higher on the international political agenda. Globalization in its many guises poses an enormous challenge for traditional governance structures. While nations—particularly developing nations—are losing ground in globalization, other actors are moving to the forefront, particularly international corporations and nongovernmental organizations (NGOs). Therefore, governments should not only take into account this new spectrum of participants in future environmental legislation but, above all, give them specific and urgently required roles in defense of the environment. New information and communication technologies are facilitating international networking, and activist groups, businesses, and international institutions are forging innovative partnerships.

14.1.5. Summary of Main Findings from the IPCC Special Report on Regional Impacts of Climate Change

Glaciers in the high Andes and three major ice fields in southern South America represent the cryosphere in Latin America. Warming in high mountain regions could lead to disappearance of significant snow and ice surfaces. In addition, changes in atmospheric circulation resulting from the ENSO phenomenon and climate change could modify snowfall rates—with a direct effect on the seasonal renewal of water supply—and surface and underground runoff in piedmont areas. This could affect mountain sports and tourist activities, which represent an important source of income in some economies. Glaciers are melting at an accelerated rate in the Venezuelan (Schubert, 1992; Hastenrath and Ames, 1995) and Peruvian Andes.

In the humid tropics, extreme precipitation events would increase the number of reservoirs silting up well before their design lives have been reached. Other areas affected by the impact of climate change on water resources could be those that rely on freshwater ecosystems (i.e., lakes and inland wetlands and their biota), including commercial and subsistence fisheries.

According to climate change projections, approximately 70% of the current temperate forest in Mexico could be affected by climate change (Villers, 1995). Other vulnerability studies (Gay-García and Ruiz Suarez, 1996)—carried out on the basis of Canadian Climate Centre (CCC)-J1 (Boer *et al.*, 1992; McFarlane *et al.*, 1992; Boer, 1993) and Geophysical Fluid Dynamics Laboratory (GFDL)-A3 (Wetherald and Manabe, 1988) GCMs—suggest that 10% of all vegetation types in northern Mexico's ecosystems—including forests and shrublands of southern Chihuahua, eastern Coahuila, northern Zacatecas, and San Luis Potosì—would be affected by drier and warmer conditions, resulting in expansion of dry and very dry tropical forests and xerophytic shrublands.

Studies of vulnerability to sea-level rise (Perdomo *et al.*, 1996) have suggested that countries such as Venezuela and Uruguay could suffer adverse impacts, leading to losses of coastal land and biodiversity, saltwater intrusion, and infrastructure damage. Impacts likely would be multiple and complex, with major economic implications. In Central America, impacts associated with sea-level rise would have their greatest effects on infrastructure, agriculture, and natural resources along the coastline, with immediate effects on socioeconomic conditions in the isthmus countries. Sea-level rise would exacerbate the processes of coastal erosion and salinization of aquifers and increase flooding risks and the impacts of severe storms along the coastline (Campos *et al.*, 1997). Flooding associated with sea-level rise is one of the main impacts in lowland areas such as the Amazon, Orinoco, and Parana River deltas and the mouth of other rivers, such as the Magdalena in Colombia. The report also identified the Rio de La Plata estuary as an area where saltwater intrusion could create problems in the freshwater supply.

In Latin America's extremely arid deserts (<100 mm annual precipitation)—the Chihuahuan, Sonoran, Peruvian, Atacama, Monte, and Patagonia—the impacts of climate change are not expected to be severe (Canziani *et al.*, 1998), because these systems already are adapted to wide fluctuations in rainfall. Therefore, hyper-arid lands are not as susceptible as drylands to climate change (Middleton and Thomas, 1997).

Studies developed in Costa Rica and Nicaragua (Halpin *et al.*, 1995) observed that shifts may occur in climatic zones that are associated with particular vegetation types in these countries. Global warming would have its greatest impacts on the cold temperate forest of southern Chile and Argentina, especially those neighboring xerophytic ecosystem types.

Projected changes in climate could increase the impacts of already serious chronic malnutrition and diseases affecting a large sector of the Latin American population. The geographical distribution of vector-borne diseases (e.g., malaria) would spread to higher elevations.

14.2. Key Regional Concerns

Studies developed after the *Special Report on Regional Impacts of Climate Change* (RICC) add new findings on the potential impact of climate change on different sectors in Latin America. Many of these new studies evolved from those mentioned in RICC, indicating that a process is in place in Latin America to reduce uncertainties associated with climate change.

14.2.1. Natural Ecosystems

Climate variations produce a variety of impacts on natural ecosystems (see Sections 5.4 to 5.9). Latin America possesses a large quantity of ecosystems, ranging from Amazonian tropical rainforest to cold Andean systems (Paramos). It also hosts remarkable rangelands, shrublands, deserts, savannas, grasslands, coastal wetlands (mainly along the Caribbean and Atlantic coastlines), and inland freshwater wetlands such as Pantanal and Iberá. Natural ecosystems in Latin America can be expected to suffer a variety of impacts from climate change. Latin American humid tropical forests represent an important group of ecosystems for which a great deal of information is offered (e.g., Canziani *et al.*, 1998).

14.2.1.1. Humid Tropical Forests

Conversion of large areas of tropical forest to pasture could reduce water cycling and precipitation in the region, in addition to its global role as a contribution to global warming. Pasture has much less leaf area than forest (McWilliam *et al.*, 1993). Because evapotranspiration is proportional to leaf area, water recycled through forest is much greater than that recycled through pasture—especially in the dry season, when pasture is dry but forest remains evergreen (Roberts *et al.*, 1996). This is aggravated by the much higher runoff under pasture, with measured increases of more than 1,000% in small (10 m^2) plots (Fearnside, 1989). Pasture grasses can partially compensate for reduced evapotranspiration by increasing their efficiency of water use when soil moisture is low, whereas forest trees maintain constant efficiency (McWilliam *et al.*, 1993). Soil under pasture quickly becomes highly compacted, inhibiting infiltration of rainwater into the soil (Schubart *et al.*, 1976). Rain falling on compacted soil runs off quickly, becoming unavailable for later release to the atmosphere through transpiration. The shallower root system of pasture, compared to that of forest, prevents pasture from transpiring during periods of drought (Nepstad *et al.*, 1994, 1999). Precipitation decreases therefore are greatest at the time of year when rain is most needed.

If the extent of deforestation were to expand to substantially larger areas, we have high confidence that reduced evapotranspiration will lead to less rainfall during dry periods in Amazonia and medium confidence that rainfall will be reduced in the center-west, center-south, and south regions of Brazil (Lean *et al.*, 1996). Although the annual rainfall total in Amazonia would decrease by only 7% from conversion to pasture, based on simulations with the Hadley Centre model, in August (dry season) the average rainfall would decrease from 2.2 mm day^{-1} with forest to 1.5 mm day^{-1} with pasture—a 32% decrease (Lean *et al.*, 1996). Simulations of conversions of Amazonian forest to pasture, using the Météo-France EMERAUDE GCM, indicate reduced volumetric soil moisture in the "arc of deforestation" where clearing activity is concentrated along the southern boundary of the Amazon forest. Rainfall reduction in southern Brazil is greatest for the January–March period (Manzi and Planton, 1996).

Greater dependence of Amazonian rainfall on water derived from evapotranspiration in the dry season means that conversion to pasture would cause this period to become longer and more severe—a change that could have harsh repercussions on the forest even if total annual precipitation were to remain unchanged (Fearnside, 1995). In patches of forest isolated by cattle pasture, trees on the edges of forest patches die at a much greater rate than do those in continuous forest (Laurance *et al.*, 1997, 1998; Laurance, 1998). Because many trees die while they are still standing, rather than being toppled by wind, dry conditions (particularly in the air) near reserve edges are a likely explanation for mortality. Soil water may partially counterbalance the effect of drier air; as trees die, soil water in the gaps they leave normally increases because the roots that would have removed water from the soil are gone. Increasing vines and decreasing forest biomass are strongly associated near forest edges, probably as a consequence of a positive feedback relationship between these factors (Laurance *et al.*, 2000). Drier microclimatic conditions have been found at forest edges (Kapos, 1989). Increased water stress—as indicated by altered ^{13}C in plant leaves—extends 60 m into the forest from an edge (Kapos *et al.*, 1993). Tree mortality increases significantly up to 100 m from the forest edge (Laurance *et al.*, 1998). Considering the length of forest edges measured by Skole and Tucker (1993) for the Brazilian Amazon in 1988, a 100-m disturbance buffer to these edges would represent a disturbed area in 1988 of 3.4 x 10^6 ha, or 15% of the area cleared by that year. Forest edges, which affect an increasingly large portion of the forest with the advance of deforestation, would be especially susceptible to the effects of reduced rainfall.

Greater severity of droughts reinforced by deforestation effects could lead to erosion of the remainder of the forest once a substantial portion of the region had been converted to pasture. The greatest effects are likely to occur during occasional complex phenomena such as El Niño (Tian *et al.*, 1998; Nepstad *et al.*, 1999). Precipitation in Amazonia is characterized by tremendous variability from 1 year to the next, even in the absence of massive deforestation (e.g., Fearnside, 1984; Walker *et al.*, 1995). If the forest's contribution to dry-season rainfall were to decrease, the result would be to increase the probability of droughts that are more severe than those experienced in the centuries or millennia over which the present forest became established. Occasional severe droughts would kill many trees of susceptible species. The result would be replacement of tropical moist forest with more drought-tolerant forms of scrubby, open vegetation resembling the *cerrado* (scrub savanna) of central Brazil (Shukla *et al.*, 1990).

Until recently, burning in Amazonia has been almost entirely restricted to areas where trees have been felled and allowed to dry before being set alight. Fire normally stops burning when it reaches the edge of a clearing rather than continuing into unfelled forest. Archaeological evidence suggests that catastrophic fires have occurred in Amazonia during major El Niño events four times over the past 2,000 years: 1,500, 1,000, 700, and 400 BP (Meggers, 1994). Human action could now turn less intensive El Niño events, which are much more frequent than major ones, into catastrophes. Increased fire initiation foci, together with increased forest flammability from logging, already have resulted in substantial incursions of fires into standing forest in eastern and southern Amazonia during dry years (Uhl and Buschbacher, 1985; Uhl and Kauffman, 1990; Cochrane and Schulze, 1999; Cochrane *et al.*, 1999; Nepstad *et al.*, 1999). The 1998–1999 fires in Roraima, in the far northern portion of Brazil, reflect the vulnerability of standing forests in Amazonia during El Niño events now that settlement areas in the forest provide permanent opportunities for fire initiation (Barbosa and Fearnside, 1999).

Increases in the amount of biomass burning could affect nutrient cycling in Amazonian forest ecosystems (medium confidence) (Fearnside, 1995). Droughts lead to increases in the area and completeness of burning in clearings in Amazonia, contributing to smoke and dust that function as sources of wind-borne nutrients to the surrounding forest (Talbot *et al.*, 1990). Climatic change also could increase nutrient supply via long-range transport of dust. African dust transported across the Atlantic Ocean by winds may be supplying significant amounts of phosphorus and calcium to Amazonia (Swap *et al.*, 1992). Amazonian soils are very poor in these elements. Soil nutrients are among the factors that limit growth, recruitment, and mortality of trees (Laurance *et al.*, 1998; Sollins, 1998). Smoke and ash particles from burning in savannas, possibly including those in Africa, also contribute nutrients (Talbot *et al.*, 1990). The extent to which these nutrient sources could increase the growth of Amazonian forests is not known. Increases undoubtedly differ by tree species, thereby altering forest composition. Burning is affected by climate, as well as by the size and behavior of the human population. Factors that influence the growth of intact forests in Amazonia are particularly important because of the large amounts of carbon that could be released to or removed from the atmosphere if the balance between forest growth and decay is altered.

CO_2 enrichment is believed to contribute to observed imbalances between CO_2 uptake and release by forest biomass in Amazonia; forest recovery from past disturbances also may contribute to these imbalances. Eddy correlation measurements (studies of gas movements in air flows inside and immediately above the forest) at one site in Rondônia indicated an uptake of 1.0 ± 0.2 t C ha^{-1} yr^{-1} (Grace *et al.*, 1995). A similar eddy correlation study near Manaus found uptake of 5.9 t C ha^{-1} yr^{-1} (Malhi *et al.*, 1999). An estimate based on reviewing existing measurements of tree growth and mortality in permanent plots found mean uptake of 0.62 ± 0.37 t C ha^{-1} yr^{-1} (Phillips *et al.*, 1998). On the other hand, at the Biological Dynamics of Forest Fragments site near Manaus (the largest and longest running study included in the forest growth measurement data set), no uptake or loss was found in 36 1-ha control plots located >100 m from a forest edge (Laurance *et al.*, 1997). The Rondônia and Manaus eddy correlation studies and the basin-wide review of tree growth indicate uptakes of 0.63, 3.66, and 0.38 x 10^9 t C, respectively, when extrapolated on the basis of consistent definitions of forest that indicate a total area of forest in the Amazon Basin of 620.5 x 10^6 ha (Fearnside, 2000). A process-based model of undisturbed ecosystems in the Amazon Basin, including savannas and forests (not necessarily defined as above), indicates wide interannual variations in net carbon flux from vegetation and soil, ranging from emissions of 0.2 x 10^9 t C in El Niño years to a sink of as much as 0.7 x 10^9 t C in other years; mean annual flux simulated over the 1980–1994 period gives an uptake of 0.2 x 10^9 t C (Tian *et al.*, 1998). If the frequency of El Niño events increases as a consequence of global warming (Timmermann *et al.*, 1999), these forests may release some of their large carbon stocks to the atmosphere. The future course of accumulation of CO_2 in the atmosphere—and consequently the time when concentrations would reach "dangerous" levels—depends heavily on continued uptake of carbon by the biosphere, including an important contribution from Amazonian forests. Climate effects contribute to making the sink in Amazonian forests unreliable as a brake on atmospheric carbon accumulation.

Although temperature changes from global warming are expected to be modest in the tropics as compared to temperate regions, it is important to realize that each degree of temperature alteration in a tropical environment may be "perceived" by forest species there as a greater change than would be the case for the same temperature shift in a temperate forest (Janzen, 1967). The direct effects of global warming on ecosystems at relatively low latitudes, if not at the equator itself, therefore may be greater than the small predicted temperature alterations at these sites might lead one to believe. In addition, direct effects of global warming through temperature change are likely to be less pronounced than effects that temperature can have through its influence on other climatic parameters, such as rainfall (Fearnside, 1995).

GCMs indicate a range of results for the effect of global warming on precipitation in Amazonia. Drying generally is expected; some models indicating greater drying than others. The Hadley Centre's HadCM2 model indicates especially dry climate over Amazonia. Process-based ecosystem models that use this simulated climate show large declines in net primary productivity (NPP) and release of carbon as a result of Amazonian forest dieback (Friend *et al.*, 1997). The varied GCM results suggest the need for a range of climate scenarios as inputs to ecosystem simulations (Bolin *et al.*, 2000). It should be noted that available scenarios (e.g., Nakicenovic *et al.*, 2000) represent the change in climate resulting from altered composition of the atmosphere only, not the additional impacts of regional land-use changes such as replacement of Amazonian forest by pasture.

Globally, models show that a doubling of GHGs may lead to a 10–15% expansion of the area that is suitable for tropical forests as equilibrium vegetation types (Solomon *et al.*, 1993). For tropical rainforest, the suitable area would expand 7–40%, depending on the GCM employed in estimating the future distribution of climatic zones. The GCM studies used by Solomon *et al.* (1993) assess the effects of doubled GHGs (i.e., CO_2-equivalence), through the direct effects of temperature and through temperature-driven alteration of precipitation regimes (but not rainfall changes provoked by deforestation). These results are indicators of *potential* for forest expansion and are not intended to reflect expected landscapes in the future; they do not include the influence of human populations in converting to other uses land that is climatically suitable for tropical forests.

One model that includes climate-induced and human changes to the year 2050 points to decreases in forest areas by about 5% in Latin America (Zuidema *et al.*, 1994). The deforestation estimates used in these calculations are based on the areas needed to satisfy expected demands for agricultural products. In the case of Brazil, deforestation is likely to exceed these forecasts because much of the forest clearing stems from motivations other than consumption of agricultural products (Hecht *et al.*, 1988; Reis and Margulis, 1991; Hecht, 1993; Fearnside, 1997). In any case, the combination of forces driving deforestation makes it unlikely that tropical forests will be permitted to expand to occupy the increased areas that are made climatically suitable for them by global warming. Land-use change interacts with climate through positive feedback processes that accelerate the loss of Brazil's Amazonian forests.

14.2.1.2. Dry Forests

Seasonally dry tropical forests have wide global distribution and coverage. Nearly 42% of tropical forests around the world are seasonally dry plant communities (Murphy and Lugo, 1986). Ancient Mesoamerican cultures developed in these regions. Domestication of animals and plants (e.g., maize, beans, sweet potato) has occurred mainly in dry forests (Challenger, 1998). Degradation of these seasonal forests is similar to or even greater than that of tropical rain forests, and only a small fraction remains intact (Janzen, 1988; Gentry, 1995; Murphy and Lugo, 1995). Janzen (1988) argues that because only a small proportion of the original distribution of dry forest remains intact in Mesoamerica, neotropical seasonally dry forests should be considered severely threatened. The estimated deforestation rate for Mexico, for the 1973–1989 period, is 1.4% yr^{-1}, which is equivalent to loss of 17.9 km^2 yr^{-1} (Trejo and Dirzo, 2000).

Costa Rican and Nicaraguan forests will be more severely affected by changes in precipitation than by changes in the annual mean temperature (see Section 14.1.5). In Venezuela, 40–50 Mha of moist forest will shift to dry or very dry forest under climate change scenarios (Mata, 1996). Between 44 and 51% of the total covered area of the Mexican deciduous tropical forest will be affected (Villers-Ruiz and Trejo-Vázquez, 1997).

Burning in the Cerrado shrubland that borders Amazonian forest to the south has increased in frequency in recent decades. This appears to create an unfavorable nutrient balance for the entire Amazonian ecosystem (Coutinho, 1990).

14.2.1.3. Savannas, Grasslands, and Deserts

Latin American dryland ecosystems are seriously threatened by desertification processes that have negative social, economic, ecological, cultural, and political consequences (Benedetti, 1997;

Table 14-4: *Estimated land use, drylands, desertification (modified from Dregne and Chou, 1992).*

Country	Irrigated Area (10^3 ha)	% Desertified	Rainfed Cropland Area (10^3 ha)	% Desertified	Rangeland Area (10^3 ha)	% Desertified	Hyperland Area (10^3 ha)	Total Drylands (10^3 ha)
Argentina	1,680	31	12,068	10	178,878	70	0	192,626
Bolivia	160	19	1,458	31	31,069	85	0	32,687
Brazil	2,300	11	3,904	69	74,558	90	0	80,762
Chile	1,257	8	1,281	47	20,976	80	11,740	35,254
Colombia	324	3	322	40	9,376	85	0	10,022
Cuba	390	1	35	14	10	90	0	435
Ecuador	540	7	400	62	7,986	90	0	8,926
El Salvador	110	5	10	10	15	93	0	135
Guatemala	75	8	88	11	719	89	0	882
Mexico	4,890	36	10,005	54	113,142	90	1,738	149,775
Paraguay	65	8	42	5	16,326	31	0	16,433
Peru	1,210	34	1,027	78	40,121	85	8,097	50,455
Venezuela	324	12	345	29	9,728	70	0	10,397

Anaya, 1998). Desertification is defined as land degradation in arid, semi-arid, and dry subhumid areas resulting from various factors, including climactic variations and human activities (conclusion from Earth Summit of Rio de Janeiro in 1992—UNCED, 1992). Evaluation of desertification around the world is complex because there is no unique measure of aridity. For example, using Thornthwaite's aridity index, 75% of Mexico is considered arid land (Thornthwaite, 1948). However, Garcia (1988) states that the arid region constitutes just more than 50% of Mexico. Large variability in the temporal and spatial distribution of precipitation complicates determination of arid and semi-arid region extension and consequently analysis of land degradation (Balling, 1994; Williams and Balling, 1996; Hernández and García, 1997).

At a global scale, the main desertification processes are degradation of vegetation, water and wind erosion, and salinization and waterlogging (Dregne and Chou, 1992). Major land-use activities in arid regions, such as irrigation and rainfed agriculture and livestock on rangelands, also are common factors in land degradation in various Latin American countries (see Table 14-4). In irrigated lands, salinization and waterlogging mainly cause the desertification. In rainfed cropland, the dominant processes for desertification are water and wind erosion; in this case, the percentage of the affected areas ranges from 10% for Argentina and El Salvador to 78% for Peru. Rangeland desertification is caused by overgrazing that results in vegetation degradation, as well as deforestation of woody species for fodder, fuel, charcoal production, and construction materials. Rangelands correspond to the major surface of drylands, where the percentage of desertified areas reach the highest levels (70–90%) compared to other land uses. On the other hand, according to the Global Assessment of Human-Induced Soil Degradation (GLASOD) survey (Middleton and Thomas, 1997), deforestation and removal of natural vegetation cover is the primary cause of soil degradation in South America, affecting 41.7% of the 79.1 Mha of drylands. The most affected regions are in northeast Brazil, along the Caribbean coasts of Venezuela and Colombia, and in northern Argentina (semi-arid Chacoan). Secondary causes of soil degradation are overgrazing (26.2%) and agricultural activities (11.6%) (<100 mm annual precipitation).

According Greco *et al.* (1994), precipitation changes projected under various climate change scenarios are unlikely to produce major ecosystem changes in this region. However, normal variations in rainfall patterns that are characteristic of this region may induce cyclic changes in vegetation physiognomy. These variations probably are more important than the total amount of precipitation. Very humid years may affect the vegetation of the region. For example, in San Luis Potosi, Mexico, in 1955, heavy rainfall from a large number of hurricanes caused the "mezquital" (*Prosopis*) shrub to disappear as a consequence of extremely wet soils, and the region became a grassland (Medellín-Leal and Gómez-González, 1979).

In the same way, the presence of El Niño in 1997–1998 in coastal arid zones of northern Peru generated drastic temporal changes in dry forest ecosystems (Torres Guevara, 1992). That area, where the historical average of annual precipitation is only 20–150 mm, received 1,000–3,000 mm of rainfall between December 1997 and May 1998. This precipitation had positive and negative effects in the region: NPP increased in all vegetation communities (Torres Guevara, 1992), reactivating rainfed agriculture activities. However, there was an outbreak of insect pests that reduced NPP.

14.2.1.4. Temperate Forests and Mountain and Polar Ecosystems

Studies carried out in Latin America on the potential impact of climate change in mountain ecosystems report an increase in mean temperature followed by a gradual reduction of glaciers in the high mountains (Flórez, 1992). As Flórez (1992) has reported, in Colombia there will be an ascent of the altitudinal limits of forest and agriculture, reducing the paramo life zone and possibly causing disappearance of current flora and fauna. The limits of the Andean and sub-Andean life zones also would ascend, as would the upper limit of the lowland tropical forest zone (Pabón, 1995b; van der Hammen, 1997). In the same way, studies in Costa Rica suggest the same effect on the tropical montane cloud forests, where biodiversity is very high. Halpin and Smith (1991) identify three types of changes in the areal arrangement of ecoclimatic zones in Costa Rica from four GCM models (UKMO, GISS, OSU, and GFDL). The first change is a strong trend toward displacement of montane and subalpine zones by warmer pre-montane climate types. The second change indicates potential heat stress in vegetation, and the third is a change in all altitudinal levels toward warmer climate types.

Villers-Ruiz and Trejo-Vázquez (1997) determined the vulnerability of Mexican forest ecosystems and forestry areas to climate change under two climate change scenarios, considering doubled-CO_2 concentrations (CCCM and GFDL-R30 models). They used Holdridge's life zones classification for their analysis. Their results showed that the most affected life zones would be temperate cold and warm forests. They conclude that increases in temperature and decreases in precipitation would reduce the extent of cool temperate and warm temperate life zones but would increase dry and very dry tropical forest zones.

The most affected natural protected areas in Mexico would be those located in the northern and western regions of the country (Villers-Ruiz and Trejo-Vázquez, 1998). Similarly, the most affected forest exploitation areas would be those located in the western part of Mexico. These changes suggest that life zones that sustain temperate desert, warm temperate desert, and cool temperate wet forest would disappear or would be severely reduced (Villers-Ruiz and Trejo-Vázquez, 1998). These changes would put national cellulose and paper production at risk because high and medium forestry production areas are located in the northern and western states of Mexico (Vargas-Pérez and Terrazas-Domínguez, 1991). Cool, temperate moist and wet forests (coniferous and oak forests) currently occupy these zones.

Natural protected areas would be affected by the change of their original vegetation, causing a reduction of animal populations. Sierra de Manantlán and the Monarch Butterfly reserves are examples of climate change impacts on natural protected areas (Villers-Ruiz and Trejo-Vázquez, 1998).

The severe drought during early 1998, associated with an El Niño event, resulted in an unusually large number of forest fires and severe economic losses (Palacio *et al.*, 1999). Cairns *et al.* (2000) estimates fuel consumption and coal emission in tropical Mexico for those events. The land-use/land-cover classes most extensively impacted were evergreen tropical forests and fragmented forests. They point out that similar fire events may be expected more frequently in the future if global change shifts toward a warmer and drier climate. The atmospheric consequences of those events were continuous emission of smoke to the United States for a large period of time (Wright, 1999). The costs from droughts and forest fires in Mexico and Central America during El Niño were approximately US$600 million.

The effects of a strong El Niño event on terrestrial ecosystems of Peru and Ecuador are as relevant (with high possibility) as in the ocean and shores (Arntz and Fahrbach, 1996). Increases in precipitation were recorded in Ecuador, Colombia, and northern Peru during the 1982–1983 El Niño event, with a simultaneous decrease in rainfall southward. Increases in precipitation also were recorded during the 1941 El Niño event. However, the increase in rainfall should not be assumed to be homogeneous. Whereas significant increases have been recorded in northern Peru, Ecuador, and even Colombia, severe droughts occurred around the Titicaca region and in northern Chile. In the highlands of the Andes as well, a north-south gradient in rainfall has been observed during El Niño events.

Increases in precipitation during El Niño events result in enhanced vegetation cover (from 5 to 89%) and primary productivity (from 0.005 to 3.5 t yr^{-1}) in coastal desert ecosystems. In northern Peru, strong El Niño events increase not only the ephemeral vegetation but also seed germination and seedling recruitment of woody species. Increases in precipitation from 20 mm (1996) to more than 1,000 mm (December 1997 to May 1998) have been recorded in Belizario in northern Peru. This increase in rainfall was correlated with a significant increase in annual NPP of the herb layer in a dry Prosopis pallida woodland, from almost zero in December 1997 to 0.51 g^{-2} yr^{-1} in February 1998. In the same region, the annual NPP of the shrub and tree layers also were significantly higher during the El Niño event, but fruit productivity decreased as a result of the mechanic effect of rain drops on buds, flowers, and immature fruits. Demographic explosions of two land snails were observed (Torres Guevara, 1992) during the increase in vegetation cover and annual NPP in northern Peru during 1998.

Besides the foregoing observations, little information has been reported about the explosive development of plant cover during strong El Niño events. Moreover, no data are available about the way in which this enhanced annual NPP impacts wildlife and range activities. According to Arntz and Fahrbach (1996), however, during this vegetation "explosion," insects, snails, and other invertebrates increase in number and diversity. Consequently, vertebrates such as rodents, birds, and foxes benefit from a diversified and enriched diet. Observational studies confirm a significant increase in the density of rodents (Muridae and Cricetidae), followed by an increase in the activity of foxes (*Dusicyon culpaeus*). The considerable increase in the density of insects and rodent populations has a direct impact on agriculture. Damage associated with the incidence of pests on crops is further aggravated by the occurrence of floods.

Extraordinarily strong precipitation during El Niño events is not restricted to the continent. The Galápagos Islands also were affected by increased rainfall during 1983. As in continental ecosystems, enhanced annual NPP had direct consequences on the densities of species in higher trophic levels (Tarazona and Valle, 1999).

As a consequence of the unusual combination of climatic and meteorological conditions attributed to the El Niño before and after the winter of 1997–1998, fires had a particularly strong effect on forests in Mexico. The number of fires was twice the average for the period 1992–1997 and 35% higher than the historical high mark recorded in 1988. The area affected was three times larger than the average for the same period (Barkin and García, 1999). Economic losses from those fires were estimated to be about US$230 million (Delgadillo *et al.*, 1999). Following the increase in rainfall associated with the strong El Niño event in 1982–1983, the next 2 years were exceptionally dry and cold in the Galápagos. Most of the plant biomass produced during the rainy years died back and accumulated as a result of the low decomposition rate. This increase in dry biomass is directly associated with the occurrence of fires affecting large areas. Recovery of plant cover after fires has been different in grasslands and woody vegetation. Whereas species diversity in grasslands has increased, fires had dramatic effects on woody vegetation because many trees and shrubs were severely affected by underground fires (Arntz and Fahrbach, 1996).

Scenarios of climate change for Latin America mountain areas are highly uncertain because available GCMs do not provide sufficiently accurate local predictions. Glacial retreat is underway in various parts of the Andes and in the ice fields at the southern tip of the continent (Canziani *et al.*, 1998). Shifting of ecosystems upslope is expected to result in loss of some vegetation types and increased vulnerability to genetic and environmental pressures.

14.2.1.5. Biodiversity

Latin America is known as home to some of the Earth's greatest concentrations of biodiversity (Heywood and Watson, 1995; Harcourt and Sayer, 1996). Seven of the world's most diverse and threatened areas are in Latin America and the Caribbean (Myers *et al.*, 2000). Of these, three rank among the world's five most critical hotspots. The tropical Andes qualifies as one of the world's two hyper-hot areas for its exceptional numbers of endemic plants and endemic vertebrates—the highest in the world. Maintenance of this diversity depends on the continued existence of representative areas of natural ecosystems (Fearnside and Ferraz, 1995; Fearnside, 1999). Dinerstein *et al.* (1995) have divided Latin America into 191 terrestrial "ecoregions" and collated information on the biodiversity importance and degree of risk of each in a systematic fashion to establish priorities for conservation. Many ecosystems already are at risk, without additional stresses expected from climatic change: 48% of all ecoregions are critical (18%) or endangered (30%); 32% are vulnerable, 16% are relatively stable, and 5% are relatively intact. Ecuador holds the distinction of being wholly covered by ecoregions with top priority at the regional level. The impacts of climate change can be expected to increase the risk of biodiversity loss in Latin America.

Central America has about 8% of the world's biodiversity concentrated in only 0.4% of the emerged surface of the planet. More than 15,000 species of plants and 1,800 species of vertebrates have been identified in the region. There are high quantities and variety of coastal wetlands in Central America. Central America's unique location between the Pacific Ocean and the Caribbean Sea—along with extreme climatic variations, tidal patterns, and geology—make these coastal wetlands among the most productive in the world (Tabilo-Valdivieso, 1997).

In addition to the loss of genetic resources, loss of productivity, and loss of ecosystem buffering against ecological perturbation, loss of biodiversity also may alter or impair the services that ecosystems provide (Naeem *et al.*, 1994). For a $2xCO_2$ scenario, surface relative humidity zones shift upward by hundreds of meters during the winter dry season, when these forests typically rely mostly on moisture from cloud contact (Leo, 1995). At the same time, an increase in the warmth index implies increased evapotranspiration; this combination of reduce cloud contact and increased evapotranspiration could have serious conservation implications, as indicated in studies in anurans (Donnelly and Crump, 1998). The results of Pounds *et al.* (1999) indicate the association in populations of birds, lizards, and anurans with the same climatic patterns, implying a broad response to regional climate change. Other studies inspired by the climate-linked epidemic hypothesis have found that dry weather in 1983 increased the vulnerability of harlequin frogs (*Atelopus varius*) to lethal parasites along one stream (Crump and Pounds, 1985, 1989).

Suárez *et al.* (1999) have found that the coastal biodiversity (flora and fauna) in Cuba will be most affected by sea-level rise. Adaptation options include establishing a national legal system and a national strategy for conserving biodiversity that includes terrestrial, marine, or coastal reserves.

14.2.2. Agriculture and Plantation Forestry

14.2.2.1. Arable Farming and Tree Crops

Agricultural lands (excluding pastures) represent approximately 19% of the land area of Latin America. Over the past 40 years, the contribution of agriculture to the GDP of Latin American countries has been on the order of 10%. Agriculture remains a key sector in the regional economy because it employs an important segment (30–40%) of the economically active population. It also is very important for the food security of the poorest sectors of the population.

Arable farming is based on annual crops of cereals (wheat, maize, barley, rice, oats), oil seeds (soybean, peanuts, sunflower), vegetables/tubercles (potatoes, cassava), and a variety of perennial grasses, including specialty crops such as cotton, tobacco, tea, coffee, cacao, sugarcane, and sugar beet. Major tree/shrub crops include a large variety of fruits, oil palm, and others. This farm production has given rise to associated activities—such as beekeeping and bee products—as well as important agro-industries that produce valuable incomes in countries that already have developed their own markets and exporting lines.

Although the more important commercial agriculture and agro-industry businesses are well developed in a few countries, many Latin American economies rely on small farming system production. In smaller and poorer countries, such as rural communities in Central America and the Andean valleys and plateaus, agriculture is the basis of subsistence lifestyles and the largest user of human capital. For these countries, agriculture is the main producing sector; it undoubtedly is severely affected by climate variations and would be seriously influenced by climate change (Rosenzweig and Hillel, 1998).

Extremes in climate variability (e.g., the Southern Oscillation) already severely affects agriculture in Latin America. In southeastern South America, maize and soybean yields tend to be higher than normal during the warm Southern Oscillation and lower during the cold phase (Berlato and Fontana, 1997; Grondona *et al.*, 1997; Magrin *et al.*, 1998; Baethgen and Romero, 2000). Contributions to variability as a result of global warming and/or reduction in evapotranspiration from forest loss would be added to this background variability, thereby aggravating losses caused by extreme events.

Land-use choices will be affected by climate change. For example, increasing precipitation in marginal areas could contribute to an increase in cropped lands (Viglizzo *et al.*, 1995). On the other hand, more favorable prices for grain crops relative to those for cattle are causing an increase in cultivated lands (Basualdo, 1995). The continued global trend to replace subsistence with market crops also creates an increasing threat to soil sustainability and enhances vulnerability to climate change.

Global warming and CO_2 fertilization effects on agricultural yields vary by region and by crop. Under certain conditions, the positive physiological effects of CO_2 enrichment could be countered by temperature increases—leading to shortening of the growth season and changes in precipitation, with consequent reductions in crop yields. Reduced availability of water is expected to have negative effects on agriculture in Mexico (Mundo and Martínez-Austria, 1993; Conde *et al.*, 1997b). However, increases in temperature would benefit maize yields at high altitudes and lower the risk of frost damage (Morales and Magaña, 1999). Several studies were carried out in the region to assess the impact of climate change on annual crop yields. Most of these studies use crop simulation models with GCMs and incremental (temperature and precipitation) scenarios as climatic inputs. Baethgen and Magrin (1995) have shown that winter crop yields in Uruguay and Argentina are more sensitive to expected variations in temperature than precipitation. Under nonlimiting water and nutrient conditions and doubled-CO_2, the results for Argentina have shown that maize, wheat, and sunflower yield variations are inversely related to temperature increments, whereas soybean would not be affected for temperature increments up to 3°C (Magrin *et al.*, 1997b, 1999a,b,c). Results obtained under rainfed conditions for different crops and management approaches in the region are summarized in Table 14-5; most of these results predict negative impacts, particularly for maize.

Adaptive measures to alleviate negative impacts have been assessed in the region. In Mexico, Conde *et al.* (1997a) found that increasing nitrogen fertilization would be the best option to increase maize yields, although it would not be economically feasible at all levels. In Argentina, the best option to improve wheat, maize, and sunflower yields would be to adjust planting dates to take advantage of the more favorable thermal conditions resulting from fewer late frosts (Travasso *et al.*, 1999). However, this adaptive measure would be insufficient for maintaining actual wheat and maize yield levels. Genetic improvement will be necessary to obtain cultivars that are better adapted to the new growing conditions. For wheat and barley crops in Uruguay and Argentina, a longer growth season could be achieved by increasing photoperiodical sensitivity (Hofstadter *et al.*, 1997; Travasso *et al.*, 1999).

Subsistence farming could be severely threatened in some parts of Latin America. The global agricultural model of Rosenzweig *et al.* (1993) identifies northeastern Brazil as suffering yield impacts that are among the most severe in the world (see Reilly *et al.*, 1996; Canziani *et al.*, 1998; Rosenzweig and Hillel, 1998). Because northeastern Brazil is home to more than 45 million people and is prone to periodic droughts and famines even in the absence of expected climate changes, any changes in this region would have major human consequences.

Climate changes can be expected to lead to changes in soil stocks of carbon and nitrogen. In the Argentinean pampas, chemical degradation of soils, based on climate changes predicted by the GISS GCM (Hansen *et al.*, 1988) at an atmospheric CO_2 concentration of 550 ppm, would reduce organic nitrogen by 6–10% and organic carbon by 7–20% in the topsoil as a result of lower dry-matter production and an increased mineralization rate (Díaz *et al.*, 1997).

Tree crops in locations where frost risk presents a limitation—such as coffee in Paraná, Brazil—benefit from higher minimum temperatures resulting from global warming (Marengo and Rogers, 2000).

14.2.2.2. Ranching

Ranching is a major land use in many parts of Latin America. In the three countries that dominate the region's agriculture and ranching sector (Brazil, Argentina, and Mexico), pastures occupy four to eight times more area than agriculture (Baethgen, 1997). In much of Latin America, livestock is almost exclusively raised on rangelands, with no storage of hay or other alternative feeds. Grass production in rangelands depends on rainfall, and limited grass availability during dry periods limits cattle stocking rates over most of the region. In areas that are subject to prolonged droughts, such as northeastern Brazil and many rangeland areas in Mexico, production would be negatively affected by increased variability of precipitation from climate change. In the case of cattle in central Amazonia (várzea), higher peak flood stages would cause losses to cattle kept on platforms (marombas) during the high-water period.

In Argentina, some cattle are fed on alfalfa and other forage crops. A 1°C rise in temperature would increase alfalfa yields by 4–8% on average for most varieties, but yields would be reduced by 16–25% in areas north of 36°S and increased by 50–100% south of this latitude (Magrin *et al.*, 1997b).

14.2.2.3. Plantation Silviculture

Plantation forestry is a major land use in Brazil and is expected to expand substantially over coming decades (Fearnside, 1998). Climatic change can be expected to reduce silvicultural yields to the extent that the climate becomes drier in major plantation states such as Minas Gerais, Espírito Santo, São Paulo, and Paraná as a result of global warming and/or reduced water vapor transport from Amazonia (e.g., Eagleson, 1986). Dry-season changes can be expected to have the greatest impact on silvicultural yields. Water often limits growth during this part of the year under present conditions, yet there may be water to spare during the rainiest part of the year. In areas outside of

***Table 14-5**: Assessments of climate change impacts on annual crops in Latin America.*

Study	Climate Scenario	Scope	Crop	Yield Impact (%)
Downing, 1992	+3°C -25% precipitation	Norte Chico, Chile	Wheat Maize Potato Grapes	decrease increase increase decrease
Baethgen, 1994	GISS, GFDL, UKMO	Uruguay	Wheat Barley	-30 -40 to -30
de Siqueira *et al.*, 1994	GISS, GFDL, UKMO	Brazil	Wheat Maize Soybeans	-50 to –15 -25 to –2 -10 to +40
Liverman and O' Brien, 1991	GFDL, GISS	Tlaltizapan, Mexico	Maize	-20 -24 -61
Liverman *et al.*, 1994	GISS, GFDL, UKMO	Mexico	Maize	-61 to –6
Sala and Paruelo, 1994	GISS, GFDL, UKMO	Argentina	Maize	-36 to -17
Baethgen and Magrin, 1995	UKMO	Argentina Uruguay (9 sites)	Wheat	-5 to -10
Conde *et al.*, 1997a	CCCM, GFDL	Mexico (7 sites)	Maize	increase-decrease
Magrin *et al.*, 1997a	GISS, UKMO, GFDL, MPI	Argentina (43 sites)	Maize Wheat Sunflower Soybean	-16 to +2 -8 to +7 -8 to +13 -22 to +21
Hofstadter *et al.*, 1997	Incremental	Uruguay	Barley Maize	-10[a] -8 to +5[b] -15[c] -13 to +10[b]

[a] For 1°C increase.
[b] Change of -20 to +20% in precipitation.
[c] For 2°C increase.

Brazil's extreme south, annual rings that are evident in the wood of plantation trees correspond to dry (as opposed to cold) seasons.

The effect of precipitation changes on plantation yields can be approximated by using a regression equation developed by Ferraz (1993) that relates biomass increment in Eucalyptus to precipitation at three sites in the state of São Paulo (Fearnside, 1999). UKMO model results (Gates *et al.*, 1992) indicate that annual rainfall changes for regions of Brazil would cause yields to decrease by 6% in Amazonia and 8% in southern Brazil and increase by 4% in the northeast. During the June-July-August (JJA) rainfall period, yields would decrease by 12% in Amazonia, 14% in southern Brazil, and 21% in the northeast (Fearnside, 1999).

The foregoing discussion of precipitation decreases considers only the effect of global warming. Brazil is likely to suffer additional losses of precipitation as a result of reductions in evapotranspiration caused by deforestation in Amazonia (see Section 14.5.1.1.1). Some of the water vapor originating in Amazonia is transported to southern Brazil (Salati and Vose, 1984; Eagleson, 1986). Decreased water vapor supply to southern Brazil, where most of the country's silviculture is located, would aggravate precipitation declines stemming from global warming.

The direct effects of rainfall reduction on yields are likely to underestimate the true effect of climate change. Synergistic effects with other factors could reduce yield substantially more—for example, through attack by pests (Cammell and Knight, 1992).

A drier climate in plantation areas also could be expected to lead to greater fire hazard. Fire is a problem in plantation silviculture even in the absence of climatic change, requiring a certain level of investment in fire control and a certain level of losses when burns occur. Pine plantations in Paraná require continuous vigilance (Soares, 1990). Eucalyptus also is fire-prone because of the high content of volatile oils in the leaves and bark.

Temperature changes can affect plantation yields. The models reviewed in the IPCC's Second Assessment Report (SAR) indicate a temperature increase of 2–3°C in Amazonia (Mitchell *et al.*, 1995; Kattenberg *et al.*, 1996). Considering a hypothetical increase of 1.5°C by the year 2050 in Espírito Santo and Minas Gerais, Reis *et al.* (1994) conclude that the present plantation area would have to be moved to a higher elevation (a shift that is considered impractical) or the genetic material would have to be completely replaced, following the global strategy proposed by Ledig and Kitzmiller (1992). In addition to direct effects of temperature considered by Reis *et al.* (1994), temperature increases have a synergistic effect with drought; the impact of dryness is worse at higher temperatures (lower elevations) as a result of higher water demands in plantations.

CO_2 enrichment would be beneficial for plantations. Higher atmospheric concentrations of CO_2 increase the water-use efficiency (WUE) of Eucalyptus. Photosynthetic rate increased in these experiments from 96% (*E. urophylla*) to 134% (*E. grandis*). Growth of different plant parts showed similar responses. Higher levels of CO_2 also stimulate nitrogen fixation, which could be expected to lower the fertilizer demands of plantations (Hall *et al.*, 1992).

Climatic change would require larger areas of plantations (and consequently greater expense) to meet the same levels of demand. The percentage increase in areas required can be greater than the percentage decline in per-hectare yields caused by climatic change because expansion of plantation area implies moving onto progressively poorer sites where productivity will be lower. Taking as examples rainfall reductions of 5, 10, 25, and 50%, plantation area requirements are calculated to increase as much as 38% over those without climatic change, which would bring the total plantation area by 2050 to 4.5 times the 1991 area (Fearnside, 1999).

14.2.3. Sea-Level Rise

14.2.3.1. General Impacts

Information on areas of land loss in several countries of Latin America as a result of sea-level rise is synthesized in Table 6-5 of the IPCC *Special Report on Regional Impacts of Climate Change* (IPCC, 1998). Fishing production is a sector that would suffer as a consequence of sea-level rise. Along the Central American coastline, sea-level rise will affect infrastructure, agriculture, and natural resources, as well as potentially exacerbate coastal erosion and salinization of aquifers and increase flood risks and the impact of severe storms (Campos *et al.*, 1997; MINAE-IMN, 2000).

Chapter 6 of the Special Report identified information on the economic cost of sea-level rise in Latin America as assessed by Saizar (1997) and Olivo (1997) for the Uruguayan and Venezuelan coastlines, respectively. Saizar (1997) assessed the potential impacts of a 0.5-m sea-level rise on the coast of Montevideo (Uruguay). Given no adaptive response, the cost of such a rise in sea level was estimated to be US$23 million, with a shoreline recession of 56 m and land loss of 6.8 ha. Olivo (1997) studied the potential economic impacts of a 0.5-m sea-level rise on the coast of Venezuela. At six study sites, she identified land and infrastructure at risk—such as oil infrastructure, urban areas, and tourist infrastructure. Evaluating four scenarios, Olivo (1997) suggests that Venezuela cannot afford the costs of sea-level rise, either in terms of land and infrastructure lost under a no-protection policy or in terms of the costs involved in any of three protection policies.

Coastal wetlands in the region endure the impact of population growth, expansion of the agricultural activity, and land-use changes.

Observed sea-level rise at the local or regional level in Latin America could be greater than the global average value (Field, 1995; Codignotto, 1997; Kjerve and Macintosh, 1997). Negative trends in river streamflow along the Patagonian coast may result in reduction of sediments toward deposition areas. Coastal erosion would be affected by this effect as well as increased sea level (Codignotto, 1997; Kokot, 1999).

14.2.3.1.1. Mangrove ecosystem

The response of mangrove forests to changes in sea level within 50–100 years under climate change conditions is complex and controversial; it depends on physiography as well as ecological and biological factors (Villamizar, 1994; Ellison and Farnsworth, 1996; Ewel and Twilley, 1998; Rull *et al.*, 1999).

The land-building function of mangrove vegetation has very important implications in coastal management because it works as a natural barrier to protect adjacent agricultural land by reducing erosion caused by wave action, tides, and river flow. This is important for shallow estuaries that are prone to flooding, especially where the land is below sea level (Twilley *et al.*, 1997; Villamizar and Fonseca, 1999).

In the tropical Americas, the loss of coastal forests, mainly mangroves, occurs at a rate of approximately 1% yr^{-1}. The rate is much faster in the Caribbean—approximately 1.7% yr^{-1} (Ellison and Farnsworth, 1997). Because most commercial shellfish and finfish use mangal for nurseries and refuge, fisheries in mangrove regions are declining at a similar rate as mangrove communities (Martínez *et al.*, 1995; Ewel and Twilley, 1998).

14.2.3.1.2. Coral reefs

The second largest coral reef system in the world dominates the offshore area of the western Caribbean (Milliman, 1993), and all but the northern Gulf coast have extensive reef systems. Growth of individual coral organisms is estimated to be 1–20 cm yr^{-1} (Vicent *et al.*, 1993), and reef growth rates as a whole are known to be up to 1.5 cm yr^{-1} (Hendry, 1993). Reefs that accumulate at these rates could keep pace with a sea-level rise of 20 cm by 2025 (UNEP, 1993) if other factors do not alter growth conditions.

Accurate predictions on the effect of sea-level rise may be possible in reefs that already have been physically and biologically monitored, such as in Panama, Jamaica, Puerto Rico, and Belize (UNEP, 1993; Gischler and Hudson, 1998).

14.2.3.1.3. Socioeconomic issues

Latin America coastal zones with economies that are based in fishing and tourism are particularly vulnerable to physical changes associated with sea-level rise.

Tourism is one of the most important industries in the region, especially in the Caribbean. Shoreline migration will create new areas of economic opportunity as new beaches are built, but protection, replenishment, and stabilization of existing beaches represents a principal socioeconomic impact. It is difficult to separate the impact of climate-induced sea-level rise from erosion associated with the persistent interaction of the sea on the coast. In addition, certain sand-mining practices (such as in Trinidad and Tobago) already have important effects on the ecosystem. Indirect socioeconomic effects on tourism from increasing pollution, coral reef mortality, and storm damage also are involved (UNEP, 1993).

Latin American economies could be severely affected by climate change. Coastal wetlands in Central America could generate US$750 million. Shrimp fisheries at the *Estero Real* in Nicaragua, which could provide US$60 million annually to the economy of the country, and the Gulf of Fonseca—which supplies important fishing, firewood, and transport to the rural communities of El Salvador, Honduras, and Nicaragua—could be affected by sea-level rise (Quesada and Jiménez, 1988).

Socioeconomic issues in the context of local response to global change—such as tourism, settlements and structures, and cultural heritage—and the influence of tropical storms are considered most important regarding levels of vulnerability (Mainardi, 1996; Tabilo-Baldivieso, 1997; Windovoxhel *et al.*, 1998). Approximately 1,600 km of coral reefs and 870 km of mangroves are located in the region of Central America (Tabilo-Valdivieso, 1997). More than 15,000 species of plants and 800 species of vertebrates identified in the region would be at risk from sea-level rise, along with resources for rural communities (about 450,000 people) that inhabit the coastal areas of Central America (Windovoxhel *et al.*, 1998).

14.2.3.2. Ecological and Local Community Values

Many local communities depend on coastal wetlands for survival; they use a wide range of natural products from the swamps and their surrounding waters (Field, 1997; RAMSAR, 1999). As Section 6.5.4 points out, in some coastal societies, cultural values are of equal—or even greater—significance than economic values. This is particularly true in Ecuador and Colombia on the Pacific coast, as well as the most northern part of Venezuela and Brazil on the Atlantic coast, where shrimp farming and timber exploitation represent the most common uses (Schaeffer-Novelli and Cintron, 1993; Sebastiani *et al.*, 1996; Trujillo, 1998). Girot (1991) identifies the landscape and the aesthetics and spirituality of local people as important social and cultural impacts in the Central American region. An increase in sea level could affect monuments and historic sites of Central America.

Patterns of human development and social organization in a community are important factors in determining the vulnerability of people and social institutions to sea-level rise and other coastal hazards. The most common problems in local subsistence economies in Latin America coastal zones relate to firewood, isolation from enforcement, shrimp ponds, cattle, clearing for village expansion, coconut plantation, and sewage and garbage disposal (Sebastiani *et al.*, 1996; Ellison and Fanrsworth, 1997).

14.2.4. Water Resources: Availability and Use

Water resources for domestic, industrial, and agricultural use, averaged per capita among Latin American countries, vary from 28,739 m^3 in Argentina (whose population is 34,587,000) to more than 472,813 m^3 in Suriname (whose population is 423,000) (see Appendix D of IPCC, 1998). These averages hide the enormous disparity in many areas, such as poor rural areas, which are ill-supplied. In some Latin American regions—especially in areas where it is possible that the combined effect of less rainfall and more evaporation could take place, leading to less runoff—global warming will substantially change the availability of freshwater. Watersheds in arid or semi-arid regions are especially sensitive because annual runoff already is highly variable (Medeiros, 1994). Watersheds in the southern hemisphere where snowmelt is an important source of runoff also can be severely affected (Basso, 1997). Vulnerability of oases between 29°S and 36°S to drier conditions in the high Andes can be observed (Canziani *et al.*, 1997)

Few specific water resources impact studies using climate change scenarios have been conducted in Latin America: Riebsame *et al.* (1995) studied the Uruguay River basin. In the study, all scenarios used indicated a shift of seasonality and a decrease in runoff during low-flow periods. For the Choqueyapu River (La Paz, Bolivia) under a UKMO-89 climate scenario, these studies projected an increase of discharge in the low-water period and in the months of December and January. For some other watersheds (Caire, Mamoré, Guadalquivir, and Miguillas), the magnitude and tendency of the results varies, depending on the scenario used (PNCC, 1997). For the Pirai River, located in a humid area and flowing through urban areas (Santa Cruz City), runoff increases under an increased precipitation scenario (PNCC, 2000). Estimates of water availability in Mexico and Central America (Izmailova and Moiseenko, 1998) indicate that about 70% of the population in those countries will live in regions with low water supply as soon as the first quarter of the 21st century. A study using climate scenarios from GFDL, GISS, and NCAR models, combined with incremental scenarios of 1–2°C temperature rise and 10% precipitation increase, found that decreasing precipitation in Mexico and El Salvador can cause a change in runoff by 5–7% but that in the winter runoff changes by only by 0.2–0.7%.

Potential changes in temperature and precipitation might have a dramatic impact on the pattern and magnitude of runoff, soil moisture, and evaporation, as well as the aridity level of some hydrological zones in Mexico (Mendoza *et al.*, 1997). A vulnerability study performed in conjunction with the National GHG Inventory in Argentina foresees a reduction in water availability as a result of changes in snowmelt in the high Andes. Similar studies in Peru show that warming has created several environmental hazards, such as avalanches in Peruvian Andean valleys, with a foreseen critical reduction of water resources for human and industrial consumption (Morales-Arnao, 1999).

Water use in Latin America is mainly for agricultural activities, averaging nearly 60% of total water use. It ranges from approximately 40% in Colombia and Venezuela to more than 75% in Argentina, Bolivia, Costa Rica, El Salvador, Ecuador, Guatemala, Honduras, Mexico, Panama, Paraguay, Suriname, and Uruguay (Canziani *et al.*, 1998). Even with a potential increase in water availability in some parts of northern Mexico (Mundo and Martínez-Austria, 1993; Magaña and Conde, 1998), demand for water from agricultural, urban, and industrial sectors has shown a much faster growth because of the rapid expansion of these sectors in recent decades.

Some hydrological scenarios for Cuba (Planos and Barros, 1999) show that a significant limitation of potential water resources will occur within the next century as a result of increments in evapotranspiration and changes in precipitation.

14.2.5. Human Health

Health impacts from climate change can arise via complex processes. The scale of these effects would depend primarily on the size, density, and wealth of human populations or communities (WHO, 1998). Extreme weather variability associated with climate change may add new stress to developing nations that already are vulnerable as a result of environmental degradation, resource depletion, overpopulation, or location (McMichael *et al.*, 1996).

Persistent poverty and population pressure accompanied by inadequate sanitation and inadequate public health infrastructure will limit many populations' capacity to adapt (Kovats *et al.*, 1998; Patz, 1998). Interaction between local environmental degradation and changes on a larger scale—climate change, population growth, and loss of biodiversity—may significantly influence effects on health (Haines and McMichael, 1997; WHO, 1998). The direct impacts of climate change depend mainly on exposure to heat or cold waves or extreme weather events, such as floods and droughts.

14.2.5.1. Effects of Changes in Climate Variables on Health

Kattenberg *et al.* (1996) made generalized tentative assessments concerning extreme weather and climate events. Studies in temperate and subtropical countries have shown increases in daily death rates associated with extreme outdoor temperatures (see Section 14.1.5; McMichael *et al.*, 1996). Climate change scenarios constructed from three models have been used to estimate human mortality with changes in baseline climate conditions for Buenos Aires, Caracas, San José, and Santiago. In Caracas and San José, where present temperatures are close to the comfort temperature for all months, mortality rates (total, cardiovascular, and respiratory) increase for most of the climate change scenarios employed. However, decreases in winter mortality may offset excess summer mortality in cities with relatively colder climates, such as Santiago (Martens, 1998). People who are more than 65 years old are more temperature sensitive than younger people (Martens, 1998).

There is evidence that people living in poor housing conditions (crowded and poorly ventilated) and urban populations in developing countries are particularly vulnerable to thermal stress enhanced by rapid urbanization (urban heat island) because of few social resources and low preexisting health status (Kilbourne, 1989; Martens, 1998). Furthermore, in rural areas, the relative importance of temperature on mortality may be different from its effect on urban populations (Martens, 1998).

Prolonged heat can enhance production of smog and dispersal of allergens. Both effects have been linked to respiratory symptoms (Epstein, 2000). High temperatures and air pollutants, especially particulates, act synergistically to influence human health. This effect is occurring in large cities, such as Mexico City, Santiago, and more recently Buenos Aires, where such conditions enhance the formation of secondary pollutants (e.g., ozone—Escudero, 1990; Katsouyanni *et al.*, 1993; Canziani, 1994). Saldivia (1994, 1995) has observed an increasing trend in the mortality of elderly people following peaks of air pollution in São Paulo. Daily mortality has been correlated mainly with temperature and ozone concentration, both measured the day before (Sartor *et al.*, 1995). Hyperthermic syndrome (heat stroke) affected children under 2 years old and people over 80 in Peru's coastal regions during the high temperatures of the El Niño phenomenon (Instituto Nacional de Salud, 1998a, 1999).

During droughts, the risk of wildfires increases, causing loss of green areas, property, livestock, and human life, as a result of increasing air pollution from suspended particles (OPS, 1998). The direct effects of wildfires on human health occur from burns and smoke inhalation (Kovats *et al.*, 1999). In *Alta Floresta*, Brazil, there was a 20-fold increase in outpatient visits for respiratory disease in 1997 during a biomass smoke episode (Brauer, 1998).

Increased ambient temperature may have significant effects on the distribution and overgrowth of allergenic plants. Higher temperatures and lower rainfall at the time of pollen dispersal are likely to result in higher concentrations of airborne pollen during the peak season (Emberlin, 1994; Rosas *et al.*, 1989). A relationship between concentrations of algae and weather parameters (temperature and vapor pressure) in Mexico has been correlated. Dispersion of algae has received much attention during recent years as a result of associated inhalant allergies and other respiratory disorders (Rosas *et al.*, 1989). The high degree of seasonality (dry-rainy season) in air-borne enteric bacterial, basidiomycete spore, particle, and protein concentrations also could be associated with temperature and vapor pressure in Mexico City (Rosas *et al.*, 1994, 1995; Calderón *et al.*, 1995). Thus, future changes in climatic variables might have an important effect in the distribution of air-borne bacteria, fungus, pollen, particles, and proteins whose allergenic properties have to be considered.

In 1998, there was an increase in cholera cases in Peru and Ecuador and other countries affected by weather disasters. Hurricane Mitch affected Guatemala (the number of cholera cases increased four-fold), Belize, and Nicaragua (the number of cholera cases was six times the usual number of reported cases) (OPS, 1999).

Global warming could increase the number and severity of extreme weather events, such as storms, floods, droughts, and hurricanes, along with related landslides and wildfires (IPCC, 1996). Such events tend to increase death and disease rates—directly through injuries or indirectly through infectious diseases brought about by damage to agriculture and sanitary infrastructure and potable water supplies (PAHO, 1998a). Floods and droughts could permanently or semi-permanently displace entire populations in developing countries, leading to overcrowding and diseases connected with it, such as tuberculosis and other air-borne and crowd diseases; adverse psychological effects; and other stresses (IPCC, 1996; McMichael and Kovats, 1998a,b; Epstein, 2000). Slums and shantytowns located on hills (e.g., Rio de Janeiro), as well as human settlements located in flood-prone areas, are particularly subject to periodic natural disasters that adversely affect human health and sanitary infrastructure (IPCC, 1996).

In developing countries, populations are becoming more rather than less vulnerable to disasters (McMichael and Kovats, 1998b). Natural disasters may be responsible for outbreaks of cholera, leptospirosis, malaria, and dengue (Moreira, 1986; PAHO, 1998a,b). Hurricane Mitch stalled over Central America in October 1998 for 3 days, claiming 11,000 lives. It was the most deadly hurricane to strike the western hemisphere in 2 centuries (Hellin *et al.*, 1999). After Mitch, Honduras reported thousands of cases of cholera, malaria, and dengue fever (PAHO, 1998a,b; OPS, 1999; Epstein, 2000). As the risk of flooding increases with climate change, so does the importance of the major drainage system, which will determine whether floodwaters drain in minutes, hours, or days (McMichael and Kovats, 1998b). Floodwaters can cause the release of dangerous chemicals from storage and waste disposal sites and precipitate outbreaks of vector- and water-borne diseases (Patz, 1998).

Indirect effects of disasters can damage the health care sector. After Hurricane Mitch, some countries in Central America regressed decades in health services and transport infrastructure, thereby making it more difficult to assist the affected population (OPS, 1999). Environmental refugees could present the most serious health consequences of climate change. Risks that stem from overcrowding include virtually absent sanitation; scarcity of shelter, food, and safe water; and heightened tensions—potentially leading to social conflicts (Patz, 1998). Weather disasters cause many deaths and have long-term impacts on communities, including psychological effects such as post-traumatic stress disorder (Kovats *et al.*, 1998). In 1999, heavy rains on the coast of Venezuela displaced 80,000–100,000 people and caused 20,000–50,000 deaths, as well as enormous damage in infrastructure (PNUD/CAF, 2000).

14.2.5.2. Vector-Borne Diseases

Arthropod vector organisms for vector-borne diseases (VBDs) are sensitive to climatic and hydrometeorological conditions (especially temperature and humidity, stagnant water pools, and ponds), as are life-cycle stages of the infecting parasite within the vector (Bradley, 1993; Haines *et al.*, 1993; Curto de Casas *et al.*, 1994; Ando *et al.*, 1998). Hence, the geographic range of potential transmission of VBDs may change under conditions of climate change (Leaf, 1989; Shope, 1991; Carcavallo and Curto de Casas, 1996; McMichael *et al.*, 1996; Patz *et al.*, 1996; WHO, 1998). Several studies have concluded that temperature affects the major components of vectorial capacity (Carcavallo *et al.*, 1998; Carcavallo, 1999; Moreno and Carcavallo, 1999).

Mosquitoes, in particular, are highly sensitive to climatic factors (Curto de Casas and Carcavallo, 1995). *Anopheline spp.* and *Aedes aegypti* mosquitoes have established temperature thresholds for survival, and there are temperature-dependent incubation periods for the parasites and viruses within them (the extrinsic incubation period) (Curto de Casas and Carcavallo, 1995; de Garín and Bejarán, 1998a,b; Epstein *et al.*, 1998; de Garín *et al.*, 2000). Climate change may influence the population dynamics of vectors for Chagas' disease (Burgos *et al.*, 1994), as well as the number of blood-feedings of mosquitoes and, therefore, the possibilities of infective direct contacts (Catalá, 1991; Catalá *et al.*, 1992).

Of relevance to infectious disease distribution, minimum temperatures are now increasing at a disproportionate rate compared to average and maximum temperature (Karl *et al.*, 1995). Such conditions may allow dengue and other climate-sensitive VBDs to extend into regions that previously have been free of disease or exacerbate transmission in endemic parts of the world. Temperature is one of the factors that can influence the seasonal transmission of malaria. Near the equator in Iquitos, Peru, seasonality in transmission is driven by small temperature fluctuation (1–2°C) (Patz *et al.*, 1998). The current and projected expansion of the range of vector species into the subtropics and to higher elevations warrant heightened entomological and epidemiological surveillance and control in highland areas and for populations living on the fringes of regions that now are affected (Epstein *et al.*, 1998).

Reemergence of dengue fever in Colombia followed reinvasion of the country by the principal mosquito vector (*Aedes aegypti*), and the disease hit with large upsurges following periods of heavy rain (Epstein *et al.*, 1995). The mosquito vector for dengue and yellow fever has been reported at an elevation of 2,200 m in Colombia (Suárez and Nelson, 1981).

Climate variability, environmental change, and lack of control of vector reproduction already have affected the distribution of VBDs. In Honduras, a sustained increase in ambient temperature makes the southern part too hot for anopheline mosquitoes, and reported cases of malaria have dropped off. Large areas of northeast tropical rainforest have been cleared, and migrants concentrated there tend not to be immune to malaria (Almendares *et al.*, 1993).

14.2.5.3. Water-Borne Diseases

Extremes of the hydrological cycle, such as water shortages and flooding, could worsen the diarrhea disease problem. In developing countries, water shortages cause diarrhea through poor hygiene. On the other extreme, flooding can contaminate drinking water from watershed runoff or sewage overflow (Patz, 1998). Depending on the disease agent and its transmission maintenance cycle, the effect may be an increase or a decrease in the incidence of infectious diseases (Gubler, 1998).

Between the first case of the current cholera outbreak—reported in Peru in 1991—and December 1996, cholera spread to more than 21 countries, resulting in almost 200,000 cases and more than 11,700 reported deaths (OPS, 1998). Colwell and Huq (1994) have collected data in Bangladesh and Peru suggesting that cholera has a complex route of transmission that is influenced by climate—in particular, SST and sea-level variations (Lobitz *et al.*, 2000). It has been suggested that the spread of *Vibrio cholerae* may be related to the development of various algae and zooplankton. Extensive studies during the past 25 years confirming the hypothesis that *V. cholerae* is autochthonous to the aquatic environment and is a commensal of zooplankton (i.e., copepods), combined with the findings of satellite data analyses, provide strong evidence that cholera epidemics are climate-linked (Lobitz *et al.*, 2000). Increased coastal algae blooms (which are sensitive to changes in climatic conditions) therefore may amplify *V. cholerae* and enhance transmission (Epstein, 2000). Furthermore, *V. cholerae* follows a salinity gradient, which might bring the disease to new shores if sea level rises (WHO, 1998).

In 1998, Ecuador's vulnerability to cholera increased as a result of climatic phenomena (OPS, 1999). In 1997–1998 in Peru, the same areas affected by climatic phenomena showed an increase in cholera cases, probably as a result of floods, problems with drainage, and food contamination in shelters (OPS, 1999). In Peru, persistence in transmission of diarrheal diseases such as *Salmonella typhi* and cholera was related to changes in environmental, climatic, and sanitary conditions (Carrillo, 1991a,b).

Floods foster fungal growth and provide new breeding sites for mosquitoes, whereas droughts concentrate microorganisms and encourage aphids, locusts, and whiteflies and—when interrupted by sudden rains—may spur explosions of rodent populations (Epstein and Chikwenhere, 1994).

The first recorded outbreak of Weil's disease (leptospirosis) in Colombia occurred mainly in children from poor neighborhoods. Symptoms of leptospirosis are similar to those of dengue, and the former can be fatal rapidly in patients not receiving proper treatment. The probable agents and disease seem to be linked with rodents escaping from floods (Epstein *et al.*, 1995).

In Cuba, acute diarrheal diseases occur more often during the warm and rainy period, when ecological conditions are favorable for reproduction of bacteria, viruses, and protozoa. Acute respiratory infection reports diminish after climatic conditions become warmer, more humid, and thermally less contrasting (Ortiz *et al.*, 1998). In Mexico, some rain in semi-arid zones has caused bubonic plague outbreaks (Parmenter *et al.*, 1999).

14.2.5.4. Effects of El Niño Phenomenon on Health

There is good evidence that the ENSO cycle is associated with increased risk of certain diseases (PAHO, 1998b). ENSO events may affect the distribution (reproduction and mortality) of disease vectors (Epstein *et al.*, 1998). El Niño events raise SST over the tropical Pacific, affecting some pathogenic agents. McMichael and Kovats (1998b), Patz (1998), and Instituto Nacional de Salud (1998a,b) have speculated that the last cholera outbreak in Peru, beginning in 1991, was linked with an El Niño phenomenon (1990–1995). The outbreak spread to most of the South American subcontinent, including places as far away as Buenos Aires (OPS, 1999). de Garín *et al.* (2000) have reported that El Niño has important impacts in andean population; they found large amounts of the insect, as well as eggs for the next period, in the northern part of Argentina.

Higher temperatures over coastal Peru associated with ENSO may have an impact on gastrointestinal infections. Salmonella infections increased after a flood in Bolivia, which resulted from the El Niño event of 1983 (Valencia Tellería, 1986). Salazar-Lindo *et al.* (1997) report that the number of patients with diarrhea and dehydration admitted to a rehydration unit in Lima, Peru, was 25% higher than usual during 1997, when temperatures where higher than normal as a result of the emerging El Niño. Increases in the incidence of acute diarrheas and acute respiratory diseases were recorded in Bolivia (Valencia Tellería, 1986) and Peru (Gueri *et al.*, 1986).

The effects of natural events vary by region, and the same weather condition may have the opposite effect in different areas for the same disease (e.g., a dry year may induce malaria epidemics in humid regions but cause malaria decreases in arid regions) (McMichael and Kovats, 1998b). For example, ENSO has been associated with severe drought in Iquitos, Peru, and the state of Roraima, Brazil, where malaria cases have drastically decreased (OPS, 1998; Confalonieri and Costa-Díaz, 2000). In Venezuela, malaria mortality has been shown to be more strongly related to the occurrence of drought in the year preceding outbreaks than to rainfall during epidemic years (Bouma and Dye, 1997). The El Niño phenomenon appears to be responsible in particular for serious epidemics in Peru, including one of malaria in 1983 (Moreira, 1986; Russac, 1986; Valencia Tellería, 1986), as well as cutaneous diseases, leptospirosis, and respiratory infections in 1998 (Instituto Nacional de Salud, 1998b). Compared with other years, malaria cases in Colombia increased 17.3% during an El Niño year and 35.1% in the post-El Niño year. Upsurges of malaria in Colombia during El Niño events are associated with its hydrometeorological variables—in particular, the increase in air temperature that enhances reproductive and biting rates and decreases the extrinsic incubation period, as well as changes in precipitation rates that favor formation of ponds and stagnant pools and thus create more mosquito breeding sites (Poveda and Rojas, 1997; Poveda *et al.*, 1999a,b).

Global analyses have shown no association between ENSO and the number of flood disasters (Dilley and Heyman, 1995) or between ENSO and the numbers of persons affected by floods and landslides (Bouma *et al.*, 1997). However, the number of persons affected by landslides, particularly in South America, increases in the year after the onset of El Niño (Bouma *et al.*, 1997). In 1983, the impacts of El Niño in Peru increased total mortality by nearly 40% and infant mortality by 103% (Toledo-Tito, 1997).

Predicting malaria risk associated with ENSO and related climate variables may serve as a short-term analog for predicting longer term effects posed by global climate change. The ability to predict years of high and low risk for malaria can be used to improve preventive measures (Bouma *et al.*, 1997).

14.2.5.5. Effects on Food Production and Safety

Climate change would affect human health indirectly by threatening food production, as a result of increased temperature, ultraviolet irradiation, sea-level rise, changes in pest ecology, ecological disruption in agricultural areas as a result of disasters, and socioeconomic shifts in land-use practices (Rosenzweig *et al.*, 1993; Siqueira *et al.*, 1994; Reilly *et al.*, 1996; Haines and McMichael, 1997; Magrin *et al.*, 1997c; Epstein *et al.*, 1998). A link between El Niño and variation of the inter-tropical convergence zone and drought in northeastern Brazil has been described for many years (Hastenrath and Heller, 1977). Periodic occurrences of severe droughts associated with El Niño in this agriculturally rich region have resulted in occasional famines (Kiladis and Díaz, 1986; Hastenrath, 1995). Severe food shortages have occurred in this region in 1988 and 1998 (Kovats *et al.*, 1999).

Developing countries already struggle with large and growing populations, and malnutrition rates would be particularly vulnerable to changes in food production (Patz, 1998). Changes in the distribution of plant pests have implications for food safety. Ocean warming could increase the number of temperature-sensitive toxins produced by phytoplankton, causing contamination of seafood more often and an increased frequency of poisoning. The rapid spread of cholera along the Peruvian coasts and the fact that the *V. cholerae* 01 isolates involved constitute a separate genetic variant that could be a result of environmental change (Wachsmuth *et al.*, 1991, 1993)—as well as the ability of *V. cholerae* to survive in seawater and freshwater—make cholera a persistent health hazard (Tamplin and Carrillo, 1991). Thus, climate-induced changes in the production of aquatic pathogens and biotoxins may jeopardize seafood safety (IPCC, 1996). Increased ambient temperature has been associated with food poisoning; multiplication of pathogenic microorganisms in food is strongly dependent on temperature (Colwell and Huq, 1994; Bentham and Langford, 1995; Patz, 1998). This indicates the importance of ambient conditions in the food production process, including animal husbandry and slaughtering, to avoid the adverse effects of a warmer climate.

In Argentina, the heavily populated Paraná Delta could be seriously affected by even small changes in sea level (Kovats *et al.*, 1998). The effects of sea-level rise may be counteracted by growing deltas as a result of the large amount of sediment coming down the Paraná and Uruguay Rivers from intense deforestation and consequential water erosion on the land of the upper basins.

Many glaciers and ice fields may soon disappear, potentially jeopardizing local water supplies that are critical for human consumption, regional agriculture, and hydroelectric power generation (Epstein *et al.*, 1998). There is high confidence in the effects of warming on glaciers, which already are disappearing in Peru and decreasing in the high Andes between 29°S and 36°S (Canziani *et al.*, 1997).

Climate changes are expected to have the greatest effect on health in developing nations in Latin America that already have poor and weak infrastructures. Linkages between local public health and issues of climate change must continue to be considered so that prevention and response mechanisms can be implemented against disease and other threats to human health (Kovats *et al.*, 1998).

14.3. Synthesis

There is ample evidence of climate variability at a wide range of time scales all over Latin America, from intraseasonal to long term. For instance, at decadal scales, multiple climate records throughout the region consistently exhibit a shift in the mean during the mid-1970s, which could be a consequence of

***Table 14-6**: Variability and impacts of El Niño and La Niña on several Latin American countries and subregions.*

Event	Climatic/Hydrological Variable	Subregion or Country	Reference(s)	Observation Period
El Niño[a]	Severe droughts[b] in recent decades	Mexico	Magaña and Conde, 2000	1958–1999[c]
	Severe droughts	Northeast Brazil	Silva Dias and Marengo, 1999	1901–1997
	Decrease in precipitation	Central America (Pacific)	Magaña and Conde, 2000	1958–1999
	Increase in precipitation	Central America (Atlantic)	Magaña and Conde, 2000	1958–1999
	Decrease in precipitation, soil moisture, river streamflow	Colombia	Poveda and Mesa, 1997 Carvajal *et al.*, 1999 Poveda *et al.*, 2001	1958–1995 1957–1997 1958–1998
	Increase in precipitation and floods	Northwest Peru	Marengo *et al.*, 1998	1930–1998
	Decrease in precipitation during rainy season	Northern Amazonia and northeast Brazil	Aceituno, 1988; Richey *et al.*, 1989; Marengo, 1992; Uvo, 1998	1931–1998
	Negative large anomalies of rainfall during rainy season	Northeast Brazil	Silva Dias and Marengo, 1999 Hastenrath and Greischar, 1993 Nobre and Shukla, 1996	1930–1998 1912–1989 1849–1984
	Increase in precipitation during November–January time frame	Argentina (Pampas region)	Barros *et al.*, 1996; Tanco and Berry, 1996; Vila and Berri, 1996; Vila and Grondona, 1996; Magrin *et al.*, 1998	1900–1996
	Intense snowfalls in high Andes mountains	Central western Argentina and central Chile	Canziani *et al.*, 1997	1900–1995
	Increase in runoff	Chile and central western Argentina	Compagnucci *et al.*, 2000 Compagnucci and Vargas, 1998	1906–1994 1909–1998

sudden changes in the climatology of the Pacific Ocean. These changes have important socioeconomic and environmental consequences that could be enhanced by global warming and its associated climate change.

Precipitation changes in Latin America do not follow a consistent trend. In northwestern Mexico there is a clear tendency for more winter precipitation, which has resulted in positive trends in river-water level. In north and northwestern Nicaragua, there is a negative trend in rainfall. In the Amazonian region, the most important finding is the presence of periods with relatively wetter or drier conditions that are more relevant than any unidirectional trend. Rainfall in north-northeast Brazil exhibits a weak positive trend. Precipitation in subtropical Argentina, Paraguay, and Brazil increased abruptly in the 1956–1990 period after a dry period from 1921 to 1995. In the Pampas, there was a positive trend in precipitation over the 1890–1984 period. At higher elevations in northwestern Argentina, records suggest an increase in precipitation over the past 200 years.

The Southern Oscillation is responsible for a large part of the climate variability at interannual scales in Latin America. The region is vulnerable to El Niño, with impacts varying across the continent. For example, El Niño is associated with dry conditions in northeast Brazil, northern Amazonia, the Peruvian-Bolivian Altiplano, and the Pacific coast of Central America, whereas southern Brazil and northwestern Peru exhibit anomalously wet conditions. Extensive studies of the Caribbean watersheds of Mexico and other countries show compelling evidence of more winter and less summer precipitation; in addition, the most severe droughts in Mexico in recent decades have occurred during El Niño years. In Colombia, La Niña is associated with heavy precipitation and flooding, whereas El Niño is associated with negative anomalies in precipitation and river streamflow. Drought also occurs in southern Brazil during the positive phase of the Southern Oscillation. If El Niño or La Niña were to increase, Latin America would be exposed to these conditions more often (see Table 14-6).

Warming in high mountain regions could lead to disappearance of significant snow and ice surfaces, which could affect mountain sport and tourist activities. Because these areas contribute to river streamflow, it also would affect water availability for irrigation, hydropower generation, and navigation, which represent important sources of income for some economies. It has been well-established that glaciers in Latin America have receded in recent decades.

It is well-established that Latin America accounts for one of the Earth's largest concentrations of biodiversity, and the impacts of climate change can be expected to increase the risk of biodiversity loss. Some adverse impacts on species that can be related to regional climate change have been observed, such as population declines in frogs and small mammals in Central America. Maintenance of remaining Amazonian forest is threatened by the combination of human disturbance and decreased precipitation from evapotranspiration loss, global warming, and El Niño. Fire in standing forest has increased in frequency and scale in Amazonia as a result of greater accumulation of deadwood in the forest from logging activity and from trees killed by past fires, more human settlement providing opportunities for fire initiation, and dry conditions

Table 14-6 (continued)

Event	Climatic/Hydrological Variable	Subregion or Country	Reference(s)	Observation Period
La Niña[d]	Heavier precipitation and floods	Colombia	Poveda and Mesa, 1997 Carvajal *et al.*, 1999	1972–1992 1957–1997
	Decrease in precipitation during October–December time frame	Argentina (Pampas region)	Barros *et al.*, 1996; Tanco and Berry, 1996; Vila and Berri, 1996; Vila and Grondona, 1996; Magrin *et al.*, 1998	1900–1996
	Increase in precipitation, higher runoff	Northern Amazonia Northeast Brazil	Marengo *et al.*, 1998 Meggers, 1994	1970–1997 Paleoclimate
	Severe droughts	Southern Brazil	Grimm *et al.*, 1996, 2000	1956–1992
	Negative anomalies of rainfall	Chile and central western Argentina	Compagnucci, 2000	Paleoclimate

[a] Extremes of the Southern Oscillation (SO) are responsible in part for a large portion of climate variability at interannual scales in Latin America. El Niño (or ENSO) events represent the negative (low) phase of the SO.
[b] Prolonged periods of reduced summer soil moisture.
[c] Six El Niño events occurred during this period.
[d] La Niña is the positive (high) phase of the SO.

during ENSO events. Neotropical seasonally dry forest should be considered severely threatened in Mesoamerica.

Mortality of trees has been observed to increase under dry conditions that prevail near newly formed edges in Amazonian forests. Edges, which affect an increasingly large portion of the forest with the advance of deforestation, would be especially susceptible to the effects of reduced rainfall. In Mexico, deciduous tropical forest would be affected in approximately 50% of the area presently covered by these forests. Heavy rain during the 1997–1998 ENSO event generated drastic changes in the dry ecosystems of northern Peru's coastal zone. Increases in biomass burning in Amazonia affected by human activity and by climate increase wind-borne smoke and dust that supply nutrients to the Amazon forests. Global warming would expand the area that is suitable for tropical forests as equilibrium vegetation types. However, the forces driving deforestation make it unlikely that tropical forests will be permitted to occupy these increased areas. Land-use change interacts with climate through positive feedback processes that accelerate loss of humid tropical forests.

Sea-level rise will eliminate the present habitats of mangroves and create new tidally inundated areas to which some mangrove species may shift. Coastal inundation stemming from sea-level rise and riverine and flatland flooding would affect water availability and agricultural land. These changes can exacerbate socioeconomic and health problems in critical areas.

On a scale of decades to centuries, it is well-established that changes in precipitation and catchment runoff may have significant effects on mangrove forest communities. This also would affect the region's fisheries because most commercial shellfish and finfish use mangroves as nurseries and for refuge.

Studies in Argentina, Brazil, Chile, Mexico, and Uruguay, based on GCMs and crop models, project decreased yield for several crops (e.g., maize, wheat, barley, grapes), even when the direct effects of CO_2 fertilization and implementation of moderate adaptation measures at the farm level are considered. Predicted increases in temperature will reduce crops yields in the region by shortening the crop cycle. Although relationships between the amount of precipitation and crop yields are well-established, the lack of consistency in the results of different GCMs means that confidence in the estimated impacts of future precipitation on crop production is necessarily limited.

Over the past 40 years, the contribution of agriculture to the GDP of Latin American countries has been on the order of 10%. Agriculture remains a key sector in the regional economy because it employs 30–40% of the economically active population. It also is very important for the food security of the poorest sectors of the population. Subsistence farming could be severely threatened in some parts of Latin America (e.g., northeastern Brazil).

Evidence is established but incomplete that climate change would reduce silvicultural yields because water often limits growth during the dry season, which is expected to become longer and more intense in many parts of Latin America.

The scale of health impacts from climate change in Latin America would depend primarily on the size, density, location, and wealth of populations. Evidence has been established but incomplete that exposure to heat or cold waves has impacts on mortality rates in risk groups in the region.

Increases in temperature would affect human health in polluted cities such as Mexico City and Santiago. Ample evidence provides high confidence that the geographical distribution of VBDs in Peru and Cuba (e.g., cholera, meningitis) would change if temperature and precipitation were to increase, although there is speculation about what the changes in patterns of diseases would be in different places. It is well-established that weather disasters (extreme events) tend to increase death and morbidity rates (injuries, infectious diseases, social problems, and damage to sanitary infrastructure)—as occurred in Central America with Hurricane Mitch at the end of 1998, heavy rains in Mexico and Venezuela at the end of 1999, and in Argentina in 2000. It is well-established that the ENSO phenomenon causes changes in disease-vector populations and in the incidence of water-borne diseases in Brazil, Peru, Bolivia, Argentina, and Venezuela. It has been speculated that weather factors and climate change may affect the incidence of diseases such as allergies in Mexico. Increased temperature, ultraviolet radiation, sea-level rise, and changes in pest ecology may threaten food production in Argentina.

In summary, climate change already has a diverse array of impacts in Latin America. Projected future changes in climate, together with future changes in the vulnerability of human and natural systems, could lead to impacts that are much larger than those experienced to date.

Acknowledgments

Both Coordinating Lead Authors (L.J. Mata and M. Campos) wish to thank their respective institutions—Zentrum für Entwicklungsforschung (ZEF) in Bonn and Comite Regional de los Recursos Hidraulicos de Centro America—for the time and support given to us during chapter preparation. We also wish to acknowledge the valuable comments and cooperation of O. Canziani (Argentina), J. Budhooram (Trinidad and Tobago), and the entire staff of the IPCC WGII Technical Support Unit.

References

Aceituno, P., 1988: On the functioning of the Southern Oscillation in the South American sector, part I: surface climate. *Monthly Weekly Review,* **116,** 505–524.

Aceituno, P., H. Fuenzalida, and B. Rosenbluth, 1993: Climate along the extratropical west coast of South America. In: *Part II: Climate Controls. Earth System Responses to Global Change: Contrasts Between North and South America* [Mooney, H., E. Fuentes, and B. Kronberg (eds).]. Kluwer Academic Publishers, Dortrecht, The Netherlands, pp. 61–72.

Alfaro, E., 1993: *Algunos Aspectos del Clima en Costa Rica en las últimas Décadas y su relación con Fenómenos de Escala Sinóptica y Planetaria.* Diss. (Tesis de Licenciatura), Universidad de Costa Rica, San José, Costa Rica, pp. 66–70 (in Spanish).

Almendares, J., M. Sierra, P.K. Anderson, and P.R. Epstein, 1993: Critical regions, a profile of Honduras. *Lancet*, **342,** 1400–1402.

Anaya, G.M., 1998: Cronología de la desertificación en México y lineamientos estratégicos para su prevención y control. In: *Ponencia, Reunión Anual sobre la "Conservación y Restauración de Suelos," Mexico, D.F., 26-28 de Octubre 1998* (in Spanish).

Ando, M., I. Uchiyama, and M. Masaji, 1998: Impacts on human health. In: *Global Warming: The Potential Impact on Japan* [Nishioka, S. and H. Harasawa (eds.)]. Springer-Verlag, Tokyo, Japan, pp. 203–213.

Aparicio, R., 1993: Meteorological and oceanographic conditions along the southern coastal boundary of the Caribbean Sea, 1951–1986. In: *Climate Change in the Intra-Americas Sea* [Maul, G.A. (ed.)]. United Nations Environment Program, IOC, Arnold, London, United Kingdom, pp. 100–114.

Arntz, W. and E. Fahrbach, 1996: *El Niño: Experimento Climático de la Naturaleza*. Primera Edición, Fondo de Cultura Económica, México, 312 pp. (in Spanish).

Baethgen, W.E., 1997: Vulnerability of the agricultural sector of Latin America to climate change. *Climate Research*, **9,** 1–7.

Baethgen, W.E., 1994: Impact of climate change on barley in Uruguay: yield changes and analysis of nitrogen management systems. In: *Implications of Climate Change for International Agriculture: Crop Modeling Study* [Rosenzweig, C. and A. Iglesias (eds.)]. U.S. Environmental Protection Agency, Washington, DC, pp. 1-13.

Baethgen, W.E. and G.O. Magrin, 1995: Assessing the impacts of climate change on winter crop production in Uruguay and Argentina using crop simulation models. In: *Implications of Climate Change for International Agriculture: Crop Modeling Study* [Rosenzweig, C. and A. Iglesias (eds.)]. American Society of Agronomy (ASA), Special Publication No. 59, Madison, WI, USA, pp. 207–228.

Baethgen, W.E. and R. Romero, 2000: Sea Surface Temperature in the El Niño region and crop yield in Uruguay. In: *Comisión Nacional sobre el Cambio Global (CNCG). Climate Variability and Agriculture in Argentina and Uruguay: Assessment of ENSO Effects and Perspectives for the Use of Climate Forecast: Final Report to the Inter-American Institute for Global Change Research.* Comisión Nacional sobre el Cambio Global, Montevideo, Uruguay.

Balling, R.C., 1994: *Analysis of Historical Temperature and Precipitation Trends in the Drylands of Mexico, the United States and Canada*. Report to United Nations IPED/INCD, Nairobi, Kenya.

Barbosa, R.I. and P.M. Fearnside, 1999: Incêndios na Amazônia brasileira: estimativa da emissão de gases do efeito estufa pela queima de diferentes ecossistemas de Roraima na passagem do evento "El Niño" (1997/98). *Acta Amazonica*, **29(4),** 513–534 (in Portuguese).

Barkin, D. and M.A. García, 1999: *The Social Construction of Deforestation in Mexico: A Case Study of the 1998 Fires in the Chimalapas Rain Forest*. Available online at http://www.wrm.org.uy/english/u_causes/regional/l_america/Chimalapas.html.

Barros, V., M.E. Castañeda, and M. Doyle, 1999: Recent precipitation trends in South America to the east of the Andes: an indication of climatic variability. In: *Southern Hemisphere Paleo and Neoclimates: Concepts and Problems* [Volheimer, W. and P. Smolka (eds.)]. Springer-Verlag, Berlin and Heidelberg, Germany.

Barros, V., M.E. Castañeda, and M. Doyle, 1996: Variabilidad interanual de la precipitación: señales del ENSO y del gradiente meridional hemisférico de temperatura. In: *Impacto de las Variaciones Climáticas en el Desarrollo Regional un Análisis Interdisciplinario*. VII Congreso Latinoamericano e Ibérico de Meteorología, pp. 321–322.

Basso, E., 1997: Southern Chile revisited. *World Meteorological Organization Bulletin*, **46(3),** 284–285.

Basualdo, E.M., 1995: El nuevo poder terrateniente: una respuesta. *Realidad Económica*, **132,** 126–149.

Bejarán, R.A. and V. Barros, 1998: Sobre el aumento de temperatura en los meses de otoño en la Argentina subtropical. *Meteorológica*, **23,** 15–26.

Benedetti, S., 1997: *Los Costos de la Desertificación, ¿Quien los Paga?*. Boletin informativo, Programa de Acción Nacional Contra la Desertificación, Ministerio de Agricultura, Corporacion Naciónal Forestal, (in Spanish).

Bentham, G. and I.A. Langford, 1995: Climate change and the incidence of food poisoning in England and Wales. *International Journal of Biometeorology*, **39,** 81–86.

Berlato, M.A. and D.C. Fontana, 1997: El Niño Oscilaçao Sul e a agricultura da regiao sul do Brasil. In: *Efectos de El Niño sobre la Variabilidad Climática, Agricultura y Recursos Hídricos en el Sudeste de Sudamérica (Impacts and Potential Applications of Climate Predictions in Southeastern South America), Workshop and Conference on the 1997–98 El Niño: 10–12 December, 1997, Montevideo, Uruguay* [Berry, G.J. (ed.)]. pp. 27–30 (in Spanish).

Bertucci, R., E. Cunha, L. Devia, M. Figueria, R. Ruiz, F. Diaz Labrano, and R. Vidal Perera, 1996: *Mercosur y Medio Ambienta, Edición Ciudad*. Argentina, (in Spanish).

Blanco Galdós, O.B., 1981: Recursos genéticos y tecnología de los Altos Andes. In: *Agricultura de ladera en América Tropical*. CATIE, Serie Técnica No. 11, (in Spanish).

Bolin, B., R. Sukumar, P. Ciais, W. Cramer, P. Jarvis, H. Kheshgi, C. Nobre, S. Semenov, and W. Steffen, 2000: Global perspective. In: *Land Use, Land-Use Change, and Forestry* [Watson, R.T., I.R. Noble, B. Bolin, N.H. Ravindranath, D.J. Verardo, and D.J. Dokken (eds.)]. Cambridge University Press, Cambridge, United Kingdom and New York, NY, USA, pp. 23–51.

Bouma, M.J. and C. Dye, 1997: Cycles of malaria associated with El Niño in Venezuela. *Journal of the American Medical Association*, **278,** 1772–1774 (in Spanish).

Bouma, M.J., G. Poveda, W. Rojas, D. Chavasse, M. Quiñones, J.S. Cox, and J.A. Patz, 1997: Predicting high-risk years for malaria in Colombia using parameters of El Niño Southern Oscillation. *Tropical Medicine and International Health*, **2(12),** 1122–1127.

Bradley, D.J., 1993: Human tropical diseases in a changing environment. In: *Environmental Change and Human Health* [Lake J., G. Bock, and K. Ackrill (eds.)]. Ciba Foundation Symposium, London, United Kingdom, pp. 146–162.

Brauer, M., 1998: Health impacts of biomass air pollution. In: *Report Prepared for the Bioregional Workshop on Health Impacts of Haze-Related Air Pollution, Kuala Lumpur, Malaysia, 1–4 June, 1998*. Malaysia.

Burgos, J.J., S.I. Curto de Casas, R.U. Carcavallo, and G.I. Galíndez-Girón, 1994: Global climate change influence in the distribution of some pathogenic complexes (malaria and chagas disease) in Argentina. *Entomología y Vectores*, **1(2),** 69–78.

Cairns, M.A., W.M. Hao, E. Alvarado, and P.K. Haggerty, 2000: Carbon emissions from spring 1998 fires in tropical Mexico. In: *Proceedings from the Joint Fire Science Conference and Workshop Vol I, University of Idaho, 15-17 June 1999* [Neuenschwander L.F., K.C. Ryan, G.E. Gollberg, and J.D. Greer (eds.)]. USA, pp. 242–248.

Calderón, C., J. Lacey, H.A. McCartney, and I. Rosas, 1995: Seasonal and diurnal variation of airborne basidiomycete spore concentrations in Mexico City. *Grana*, **34,** 260–268.

Camilloni, I. and V. Barros, 1997: On the urban heat island effect dependence on temperature trends. *Climatic Change*, **37,** 665–681.

Cammell, M.E. and J.D. Knight, 1992: Effects of climatic change on the population dynamics of crop pests. *Advances in Ecological Research*, **22,** 117–162.

Cantañeda, M.E. and V. Barros, 1993: Las tendencias de precipitación en el Cono Sur de América al este de los Andes. *Meteorológica*, **19,** (in Spanish).

Canziani, O.F., 1994: *La Problemática Ambiental Urbana*. Gestión Municipal de Residuos Urbanos, Instituto de Investigaciones sobre el Medio Ambiente, Buenos Aires, Argentina, pp. 19–50 (in Spanish).

Canziani, O.F., R.M. Quintela, and M. Prieto, 1997: *Estudio de Vulnerabilidad de los Oasis Comprendidos entre 29° y 36° S ante Condiciones más Secas en los Andes Altos*. Capítulo 10, Projecto UNDP ARG/95/G/31, Programa Naciones Unidos para el Desarrollo, Argentina, 116 pp. (in Spanish).

Canziani, O.F., S. Díaz, E. Calvo, M. Campos, R. Carcavallo, C.C. Cerri, C. Gay-García, L.J. Mata, A. Saizar, P. Aceituno, R. Andressen, V. Barros, M. Cabido, H. Fuenzalida-Ponce, G. Funes, C. Galvão, A.R. Moreno, W.M. Vargas, E.F. Viglizao, and M. de Zuviría, 1998: Latin America. In: *The Regional Impacts of Climate Change: An Assessment of Vulnerability. Special Report of IPCC Working Group II* [Watson, R.T., M.C. Zinyowera, and R.H. Moss (eds.)]. Intergovernmental Panel on Climate Change, Cambridge University Press, Cambridge, United Kingdom and New York, NY, USA, pp. 187–230.

Carcavallo, R.U., 1999: Climatic factors related to Chagas disease transmission. *Mem. Inst. Oswaldo Cruz,* **94(I),** 367–369.

Carcavallo, R.U. and S. Curto de Casas, 1996: Some health impacts of global warming in South America: vector-borne diseases. *Journal of Epidemiology,* **6,** S153–S157.

Carcavallo, R.U., C. Galvão, D.S. Rocha, J. Jurberg, and S.I. Curto de Casas, 1998: Predicted effects of warming on Chagas disease vector and epidemiology. *Entomología y Vectores,* **5,** 137.

Carril, A.F., M.E. Doyle, V.R. Barros, and M.N. Núñez, 1997: Impacts of climate change on the oases of the Argentinean cordillera. *Climate Research,* **9,** 121–129.

Carrillo, C., 1991a: El Instituto Nacional de Salud y la epidemia del cólera. *Revista Médica Herediana,* **2(2),** 89–93 (in Spanish).

Carrillo, C., 1991b: El Instituto Nacional de Salud y la epidemia del cólera: Simposio Internacional sobre cólera. B*iomédica,* **11,** 25-26 (in Spanish).

Carter, T.R., M. Hulme, J.F. Crossley, S. Malyshev, M.G. New, M.E. Schlesinger, and H. Tuomenvirta, 2000: *Climate Change in the 21st Century - Interim Characterizations based on the New IPCC Emissions Scenarios.* The Finnish Environment 433, Finnish Environment Institute, Helsinki, 148 pp.

Carvajal, Y., H. Jiménez, and H. Materón, 1999: *Incidencia del fenómeno El Niño en la Hidroclimatología del Valle del río Cauca en Colombia.* Available online at http://www.unesco.org.uy/phi/libros/enso/indice.html.

Castañeda, M.E. and V. Barros, 1996: *Sobre las Causas de las Tendencias de Precipitación en el Cono Sur al Este de los Andes.* Reporte 26, Center for Ocean and Atmospheric Studies, (in Spanish).

Catalá, S., 1991: The biting rate of *Triatoma infestans* in Argentina. *Medical Veterinary Entomology,* **5,** 325–333.

Catalá, S.S., D.E. Gorla, and M.A. Basombrio, 1992: Vectoral transmission of *Trypanosoma cruzi*—an experimental field study with susceptible and immunized hosts. *American Journal of Tropical Medicine and Hygiene,* **47,** 20–26.

Centella, A., L. Castillo, and A. Aguilar, 1998: *Escenarios de Cambio Climático para la Evaluación de los Impactos del Cambio Climático en El Salvador.* Proyecto GEF/ELS/97/G32, Primera Comunicación Nacional de Cambio Climático, Reporte Técnico, 49 pp. (in Spanish).

CEPAL, 1999a: *Guatemala: Evaluación de los Daños Ocasionados por El Huracán Mitch, 1998: Sus Implicaciones para el Desarrollo Económico y Social y el Medio Ambiente.* LC/MEX/L.370, Comisión Económica para América Latina y el Caribe, Mexico (in Spanish).

CEPAL, 1999b: *Honduras: Evaluación de los Daños Ocasionados por El Huracán Mitch, 1998: Sus Implicaciones para el Desarrollo Económico y Social y el Medio Ambiente.* LC/MEX/L.367, Comisión Económica para América Latina y el Caribe, Mexico, (in Spanish).

CEPAL, 1999c: *El Salvador: Evaluación de los Daños Ocasionados por El Huracán Mitch, 1998: Sus Implicaciones para el Desarrollo Económico y Social y el Medio Ambiente.* LC/MEX/L.371, Comisión Económica para América Latina y el Caribe, Mexico, (in Spanish).

CEPAL, 1999d: *Nicaragua: Evaluación de los Daños Ocasionados por El Huracán Mitch, 1998: Sus Implicaciones para el Desarrollo Económico y Social y el Medio Ambiente.* LC/MEX/L.372, Comisión Económica para América Latina y el Caribe, Mexico, (in Spanish).

CEPAL, 1999e: *Costa Rica: Evaluación de los Daños Ocasionados por El Huracán Mitch, 1998: Sus Implicaciones para el Desarrollo Económico y Social y el Medio Ambiente.* LC/MEX/L.373, Comisión Económica para América Latina y el Caribe, Mexico, (in Spanish).

CEPAL, 1998: Caída del crecimiento económico en la región. In: *Notas de la CEPAL, Comisión Económica para América Latina y el Caribe, 1 noviembre de 1998.* CEPAL, México, (in Spanish).

Challenger, A., 1998: *Utilización y Conservación de los Ecosistemas Terrestres de México: Pasado, Presente y Futuro.* CONABIO-IBUNAM-ASM-SC, Mexico, 847 pp. (in Spanish).

Cochrane, M.A. and M.D. Schulze, 1999: Fire as a recurrent event in tropical forests of the eastern Amazon: effects on forest structure, biomass, and species. *Biotropica,* **3,** 221–227.

Cochrane, M.A., A. Alencar, M.D. Schulze, C.M. Souza, D.C. Nepstad, P. Lefebvre, and E.A. Davidson, 1999: Positive feedbacks in the fire dynamic of closed canopy tropical forests. *Science,* **284,** 1832–1835.

Codignotto, J.O., 1997: *Capítulo Geomorfología y Dinámica Costera del Libro; El Mar Argentino y Sus Recursos Pesqueros.* Instituto Nacional de Investigación y Desarrollo Pesquero, I, Mar del Plata, Argentina, pp. 89–105 (in Spanish).

Colwell, R.R. and A. Huq, 1994: Environmental reservoir of *Vibrio cholerae. Annals of New York Academy of Sciences,* **15,** 44–54.

Compagnucci, R.H., 2000: Impact of ENSO events on the hydrological system of the Cordillera de los Andes during the last 450 years. In: *Southern Hemisphere Paleo and Neoclimates: Concepts, Methods and Problems* [Volheimer, W. and P. Smolka (eds.)]. Springer-Verlag, Berlin and Heidelberg, Germany.

Compagnucci, R.H., 1991: Influencia del ENSO en el desarrollo socio económico de Cuyo. In: *Anales CONGREMET VI, 23–27 March 1991, Buenos Aires, Argentina.* Centro Argentino de Meteorólogos, Argentina, pp. 95–96. (in Spanish).

Compagnucci, R.H. and W.M. Vargas, 1998: Interannual variability of the Cuyo River's streamflow in the Argentinian Andean mountains and ENSO events. *International Journal of Climatology,* **18,** 1593–1609.

Compagnucci, R.H., S.A. Blanco, M.A. Figliola, and P.M. Jacovkis, 2000: Variability in subtropical Andean Argentinean Atuel River: a wavelet approach. *Envirometrics,* **11,** 251–269.

Conde, C., D. Liverman, M. Flores, R. Ferrer, R. Araujo, E. Betancourt, G. Villareal, and C. Gay, 1997a: Vulnerability of rainfed maize crops in Mexico to climate change. *Climate Research,* **9,** 17–23.

Conde, C., D. Liverman, and V. Magaña, 1997b: *Climate Variability and Transboundary Freshwater Resources in North America: U.S.-Mexico Border Case Study.* Draft Final Report Prepared for the Commission on Environmental Cooperation, Montreal, QB, Canada, 44 pp.

Confalonieri, U.E.C. and R. Costa-Dias, 2000: Climate variability, land use/land cover change and malaria in the Amazon: preliminary results from the State of Roraima, Brazil. In: *Abstracts of the First LBA Scientific Conference, June 26–30, 2000, Belem, Para, Brazil.* CPTEC/INPE, 17 pp.

Cornik, T.R. and R.A. Kirby, 1981: Interacciones de cultivos y producción animal en la generación de tecnología en zonas de ladera. In: *Agricultura de Ladera en América Tropical.* CATIE, (in Spanish).

Coutinho, L.M., 1990: Fire in the ecology of the Brazilian Cerrado. In: *Fire in the Tropical Biota: Ecosystem Processes and Global Challenges* [Goldammer, J.G. (ed.)]. Springer-Verlag, Heidelberg, Germany, pp. 82–105.

Crump, M.L. and J.A. Pounds, 1989: Temporal variation in the dispersion of a tropical anuran. *Copeia,* **3,** 209–211.

Crump, M.L. and J.A. Pounds, 1985: Lethal parasitism of an aposematic anuran (*Atelopus varius*) by *Notochaeta bufonivora* (Diptera: Sarcophagidac). *L. Parasitol.,* **71,** 588–591.

Curto de Casas, S.I. and Carcavallo, R.U., 1995: Climate change and vector-borne diseases distribution. *Social Science and Medicine,* **40(11),** 1437–1440.

Curto de Casas, S.I., R.U. Carcavallo, C. Mena Segura, and I. Galíndez-Girón, 1994: Bioclimatic factors of Triatominae distribution: useful techniques for studies on climate change. *Entomología y Vectores,* **1(2),** 51–67.

de Garín, A., R. Bejarán, and N. Schweigmann, 2000: Eventos El Niño y La Niña y su relación con la abundancia potencial del vector del dengue en Argentina. In: *Actualizaciones de Artropodologiá Sanitaria Argentina* [Salomon, D. (ed.)]. Buenos Aires, Argentina, (in press), (in Spanish).

de Garín, A. and R. Bejarán, 1998a: *Variabilidad y Cambio Climático en Jujuy Aero y su Posible Impacto sobre la Dinámica Poblacional de Aedes aegypti.* Informe Técnico y Meteorológico, Ministerio de Salud Pública de la Nación, Buenos Aires, Argentina, 21 pp. (in Spanish).

de Garín, A. and R. Bejarán, 1998b: Estudio de la predictibilidad biometeorológica ante el riesgo de dengue en Posadas (Argentina). In: *Anales del VIII Congreso Latinoamericano e Ibérico de Meteorología* (CD-ROM). Brasil, 5 pp. (in Spanish).

de Siqueira, O., L. Salles, and J. Fernandes, 1999: Efeitos potenciais de mudancas climaticas na agricultura brasileira e estrategias adaptativas para algumas culturas. In: *Memorias do Workshop de Mudancas Climaticas Globais e a Agropecuaria Brasiloeira, 1–17 de Junho 1999, Campinas, SPO, Brasil*. Brasil, pp. 18–19 (in Portuguese).

de Siqueira, O.J.F., J.R.B. Farías, and L.M.A. Sans, 1994: Potential effects of global climate change for Brazilian agriculture: applied simulation studies for wheat, maize and soybeans. In: *Implications of Climate Change for International Agriculture: Crop Modeling Study* [Rosenzweig, C. and A. Iglesias (eds.)]. EPA 230-B-94-003, U.S. Environmental Protection Agency, Washington, DC, USA.

Delgadillo, M.J., O.T. Aguilar, and V.D. Rodríguez, 1999: Los aspectos económicos y sociales del El Niño. In: *Los Impactos de El Niño en México* [Magaña, V. (ed.)]. IAI, SEP-CONACYT, UNAM, México, (in Spanish).

Díaz, H.F. and G.N. Kiladis, 1992: Atmopheric teleconnections associated with the extreme phases of the Southern Oscillation. In: El Niño Historical and Paleoclimatic Aspects of the Southern Oscillation [Diaz, H.F. and V. Markgraf (eds.)]. Cambridge University Press, Cambridge, United Kingdom and New York, NY, USA, pp. 7–28.

Díaz, R.A., G.O. Magrin, M.I. Travasso, and R.O. Rodríguez, 1997: Climate change and its impact on the properties of agricultural soils in the Argentinean rolling pampas. *Climate Research*, **9,** 25–30.

Dilley, M. and B. Heyman, 1995: ENSO and disaster: droughts, floods, and El Niño/Southern Oscillation warm events. *Disasters*, **19,** 181–193.

Dinerstein, E., D.M. Olson, D.J. Graham, A.L. Webster, S.A. Primm, M.P. Bookbinder, and G. Ledec, 1995: *A Conservation Assessment of the Terrestrial Ecoregions of Latin America and the Caribbean*. The World Bank, Washington, DC, USA.

Donnelly, M.A. and M.L. Crump, 1998: Potential effects of climate change on two neotropical amphibian assemblages. *Climatic Change*, **39,** 541–561.

Downing, T.E., 1992: *Climate Change and Vulnerable Places: Global Food Security and Country Studies in Zimbabwe, Kenya, Senegal, and Chile*. Research Report No. 1, Environmental Change Unit, University of Oxford, Oxford, UK, 54 pp.

Dregne, H.E. and N.-T. Chou, 1992: Global desertification dimensions and costs. In*: Degradation and Restoration of Arid Lands* [Harold, E. and H. Dregne (eds.)]. International Center for Arid and Semiarid Land Studies, Texas Tech University, Lubbock, TX, USA, 298 pp.

Eagleson, P.S., 1986: The emergence of global-scale hydrology. *Water Resources Research*, **22(9),** 6–14.

Ellison, A.M. and E.J. Farnsworth, 1997: Simulated sea level change alters anatomy, physiology, growth, and reproduction of red mangrove (*Rhizophora mangle* L.). *Oecologia*, **112,** 435–446.

Ellison, A.M. and E.J. Farnsworth, 1996: Antropogenic disturbance of Caribbean mangrove ecosystem: Past impacts, present trends and future predictions. *Biotropica*, **28(4),** 549–565.

Emanuel, K.A., 1991: The theory of hurricanes. *Annals Rev Fluid Mech*, **23,** 179–196.

Emanuel, K.A., 1987: The dependence of hurricane intensity on climate. *Nature*, **326(2),** 483–485.

Emberlin, J., 1994: The effects of patterns in climate and pollen abundance on allergy. *Allergy*, **49,** 15–20.

Epstein, P.R., 2000: Is global warming harmful to health? *Scientific American*, **283(2),** 50–57.

Epstein, P.R. and G.P. Chikwenhere, 1994: Environmental factors in disease surveillance. *Lancet*, **343,** 1440–1441.

Epstein, P.R., H.F. Díaz, S. Elias, G. Grabherr, N.E. Graham, W.J.M. Martens, E. Mosley-Thompson, and J. Susskind, 1998: Biological and physical signs of climate change: focus on mosquito-borne diseases. *Bulletin of the American Meteorological Society*, **79(3),** 409–417.

Epstein, P.R., P.O. Calix, and R.J. Blanco, 1995: Climate and disease in Colombia. *Lancet*, **346,** 1243–1244.

Escudero, J., 1990: Control ambiental en grandes ciudades: caso de Santiago. In: *Seminario Latinoamericano sobre Medio Ambiente y Desarrollo, Bariloche, Octubre de 1990*. IEMA, Buenos Aires, Argentina, pp. 229–236 (in Spanish).

Ewell, K.C. and R.R. Twilley, 1998: Different kinds of mangrove forests provide different goods and services. *Global Ecology and Biogeography Letters*, **7(1),** 83–94.

Fearnside, P.M., 2000: Global warming and tropical land-use change: greenhouse gas emissions from biomass burning, decomposition and soils in forest conversion, shifting cultivation and secondary vegetation. *Climatic Change*, **46(1–2),** 115–158.

Fearnside, P.M., 1999: Plantation forestry in Brazil: the potential impacts of climatic change. *Biomass and Bioenergy*, **16(2),** 91–102.

Fearnside, P.M., 1998: Plantation forestry in Brazil: projections to 2050. *Biomass and Bioenergy*, **15(6),** 437–450.

Fearnside, P.M., 1997: Monitoring needs to transform Amazonian forest maintenance into a global warming mitigation option. *Mitigation and Adaptation Strategies for Global Change*, **2(2–3),** 285–302.

Fearnside, P.M., 1995: Potential impacts of climatic change on natural forests and forestry in Brazilian Amazonia. *Forest Ecology and Management*, **78,** 51–70.

Fearnside, P.M., 1989: *A Ocupação Humana de Rondônia: Impactos, Limites e Planejamento*. CNPq Relatórios de Pesquisa No. 5, Conselho Nacional de Desenvolvimento Científico e Tecnológico (CNPq), Brasilia, Brasil, 76 pp. (in Portuguese).

Fearnside, P.M., 1984: Simulation of meteorological parameters for estimating human carrying capacity in Brazil's Transamazon highway colonization area. *Tropical Ecology*, **25(1),** 134–142.

Fearnside, P.M., and J. Ferraz, 1995: A conservation gap analysis of Brazil's Amazonian vegetation. *Conservation Biology*, **9(5),** 1134–1147.

Ferraz, E.S.B., 1993: Influência da precipitação na produção de matéria seca de eucalipto. *IPEF Piracicaba*, **46,** 32–42 (in Portuguese).

Field, C., 1997: La restauración de ecosistemas de manglar. In: *Managua, Editora de Arte* [Field, C. (ed.)]. 280 pp. (in Spanish).

Field, C., 1995: Impact of expected climate change on mangroves. *Hydrobiologia*, **193,** 75–81.

Flórez, A., 1992: Los nevados de Colombia, glaciales y glaciaciones. *Análisis Geograficos*, **22,** 95 pp. (in Spanish).

Friend, A.D., A.K. Stevens, R.G. Knox, and M.G.R. Cannell, 1997: A process-based, terrestrial biosphere model of ecosystem dynamics (hybrid v. 3.0). *Ecological Modelling*, **95,** 249–287.

García, E., 1988: *Modificaciones al Sistema de Clasificación Climática de Köeppen (para adaptarlo a las codiciones de la República Mexicana)*. 4th. ed., Offset Larios, México, (in Spanish).

García, N.O. and W.M. Vargas, 1997: The temporal climatic variability in the "Rio de la Plata" basin displayed by river discharges. *Climate Change*, **102,** 929–945.

Garreaud, G. and J. Ruttland, 1996: Análisis meteorológico de los aluviones de Antofagasta y Santiago de Chile en el periódo 1991–1993. *Atmósfera*, **9,** 251–271 (in Spanish).

Gates, W.L., J.F.B. Mitchell, G.J. Boer, U. Cubasch, and V.P. Meleshko, 1992: Climate modelling, climate prediction and model validation. In: *Climate Change 1992: The Supplementary Report to the IPCC Scientific Assessment* [Houghton, J.T., B.A. Callander, and S.K. Varney (eds.)]. Cambridge University Press, Cambridge, United Kingdom and New York, NY, USA, pp. 97–134.

Gay-García, C. and L.G. Ruiz Suarez, 1996: *UNEP Preliminary Inventory of GHG Emissions: Mexico*. UNEP, Geneva, Switzerland.

Genta, J.L., G. Pérez-Iribarren, and C.R. Mechoso, 1998: A recent increasing trend in the streamflow of rivers in southern South America. *Journal of Climate*, **11,** 2858–2862.

Gentry, A.H., 1995: Diversity and floristic composition of neotropical dry forests. In: *Seasonally Dry Tropical Forests* [Bullock, S.H., H.A. Mooney, and E. Medina (eds.)]. Cambridge University Press, Cambridge, United Kingdom and New York, NY, USA, pp. 146–194.

Girot, P., 1991: *The Historical and Cultural Relations of the Maleku Indians with the Wetlands of Caño Negro*. Universidad de Costa Rica, San José, Costa Rica (reporte no publicado).

Gischler, E. and J.H. Hudson, 1998: Holocene development of three isolated carbonate platforms, Belize, Central America. *Marine Geology*, **144(4),** 333–347.

Gómez, I. and W. Fernández, 1996: Variación interanual de la temperatura en Costa Rica. *Tópicos Meteorológicos Oceanográficos*, **3(1),** 21–24 (in Spanish).

Grace, J., J. Lloyd, J. McIntyre, A. Miranda, P. Meir, H. Miranda, C. Nobre, J. Moncrieff, J. Massheder, M. Yadvinder, I. Wright, and J. Gash, 1995: Fluxes of carbon dioxide and water vapour over an undisturbed tropical rain forest in South-West Amazonia. *Global Change Biology,* **1,** 1–12.

Greco, S., R.H. Moss, D. Viner, and R. Jenne, 1994: *Climate Scenarios and Socio-Economic Projections for IPCC WGII Assessment.* Intergovernmental Panel on Climate Change, World Meteorological Organization, and United Nations Environment Program, Washington, DC, USA, 12 pp.

Grimm, A.M., V.R. Barros, and M.E. Doyle; 2000: Climate variability in southern South America associated with El Niño and La Niña events. *Journal of Climate,* **13,** 35–58.

Grimm, A., S.E. Teleginsky, and E.E.D. Freitas, 1996: Anomalias de precipitacao no dul do Brasil em eventos de El Nino. In: *Congreso Brasileiro de Meteorología, 1996, Campos de Jordao.* Anais, Sociedade Brasileira de Meteorología, Brasil, (in Portuguese).

Grondona, M.O., G.O. Magrin, M.I. Travasso, R.C. Moschini, G.R. Rodríguez, C. Messina, D.R. Boullón, G. Podestá, and J.W. Jones, 1997: Impacto del fenómeno El Niño sobre la producción de trigo y maíz en la región Pampeana Argentina. In: *Efectos de El Niño sobre la Variabilidad Climática, Agricultura y Recursos Hídricos en el Sudeste de Sudamérica (Impacts and Potential Applications of Climate Predictions in Southeastern South America), Workshop and Conference on the 1997–98 El Niño: 10–12 December, 1997, Montevideo, Uruguay* [Berry, G.J. (ed.)]. pp. 13–18 (in Spanish).

Gubler, J., 1998: Climate change: implications for human health. *Health and Environment Digest,* **12(7),** 54–56.

Gueri, M., C. González, and V. Morin, 1986: The effect of the floods caused by "El Niño" on health. *Disasters,* **10,** 118–124.

Haines, A. and A.J. McMichael, 1997: Climate change and health: implications for research, monitoring and policy. *British Medical Journal,* **315,** 870–874.

Haines, A., P.R. Epstein, and A. McMichael, 1993: Global health watch: monitoring impacts of environmental change. *Lancet,* **342,** 1464–1469.

Hall, D.O., R. Rosillo-Calle, R.H. Williams, and J. Woods, 1992: Biomass for energy: supply prospects. In: *Renewable Energy: Sources for Fuels and Electricity* [Johansson, T.B., H. Kelly, A.K.N. Reddy, and R.H. Williams (eds.)]. Island Press, Covelo, CA, USA, pp. 593–651.

Halpin, P.N. and T.M. Smith, 1991: *Potential Impacts of Climate Change on Forest Protection in the Humid Tropics: A Case Study of Costa Rica.* Department of Environmental Sciences, University of Virginia, Charlottesville, VA, USA, (unpublished report).

Halpin, P.N., P.M. Kelly, C.M. Secrett, and T.M. Schmidt, 1995: *Climate Change and Central America Forest System.* Background paper on the Nicaragua Pilot Project.

Hansen, J., I. Fung, A. Lascis, D. Rind, S. Lebedeff, R. Ruedez, and G. Russel, 1988: Global climate changes as forecast by Goddard Institute for Space Studies three-dimensional model. *Journal of Geophysical Research (Atmospheres),* **93,** 9341–9364.

Harcourt, C.S. and J.A. Sayer (eds.), 1996: *The Conservation Atlas of Tropical Forests: The Americas.* Simon and Schuster, New York, NY, USA, 335 pp.

Hastenrath, S., 1995: Recent advances in tropical climate prediction. *Journal of Climate,* **8,** 1519–1532.

Hastenrath, S. and A. Ames, 1995: Recession of Yanamarey Glacier in Cordillera Blanca, Peru, during the 20th century. *Journal of Glaciology,* **41(137),** 191–196.

Hastenrath, S. and L. Heller, 1977: Dynamics of climatic hazards in north-east Brazil. *Quarterly Journal of the Royal Meteorological Society,* **103,** 77–92.

Hastenrath, S. and L. Greischar, 1993: Further work on northeast Brazil rainfall anomalies. *Journal of Climate,* **6,** 743–758.

Hecht, S.B., 1993: The logic of livestock and deforestation in Amazonia. *BioScience,* **43(10),** 687–695.

Hecht, S.B., R.B. Norgaard, and C. Possio, 1988: The economics of cattle ranching in eastern Amazonia. *Interciencia,* **13(5),** 233–240.

Hellin, J., M. Haigh, and F. Marks, 1999: Rainfall characteristics of Hurricane Mitch. *Nature,* **399,** 21–22.

Hendry, M., 1993: Sea level movements and shoreline changes. In: *Climate Change in the Intra-Americas Sea* [Maul, G.A. (ed.)]. Arnold, London, United Kingdom, pp. 115–161.

Hernandez, C.M.E. and E. García, 1997: Condiciones climaticas de las zonas aridas de Mexico. *Geografia y Desarrollo,* **15,** 5–16 (in Spanish).

Heywood, V.H. and R.T. Watson (eds.), 1995: *Global Biodiversity Assessment.* Cambridge University Press, Cambridge, United Kingdom and New York, NY, USA, 1140 pp.

Hoffman, J.A.J., S.E. Nuñez, and W. Vargas, 1997: Temperature, humidity and precipitation variations in Agentina and the adjacent sub-Antarctic region during the present century. *Meteorol. Zeitschrift,* **6,** 3–11.

Hoffman, J.A.J., T.A. Gomez, and S.E. Nuñez, 1987: Los campos medios anuales de algunos fenómenos meteorológicos. In: *II Congreso de Meteorología.* CAM, AMS, SBM, OMMAC, and SOCOLMET, Buenos Aires, Argentina, (in Spanish).

Hofstadter, R. and M. Bidegain, 1997: Performance of General Circulation Models in southeastern South America. *Climate Research,* **9,** 101–105.

Hofstadter, R., M. Bidegain, W. Baetghen, C. Petraglia, J.H. Morfino, A. Califra, A. Hareau, R. Romero, J. Sauchik, R. Méndez, and A. Roel, 1997: Agriculture sector assessment. In: *Assessment of Climate Change Impacts in Uruguay.* Final report, Uruguay Climate Change Country Study, Montevideo, Uruguay.

IDB, 2000: *Social Protection for Equity and Growth.* Inter-American Development Bank, Washington, DC, USA, pp. 47–75.

IDB, 1999: *Facing Up to Inequality in Latin America, Economic and Social Progress in Latin America.* Inter-American Development Bank, Washington, DC, USA, 282 pp.

IMARPE, 1998: *El Fenómeno El Niño (El Evento 1997–1998).* Instituto del Mar del Perú, Perú, (in Spanish). Available online at www.imarpe.gob.pe.

INAGGA-CONAM, 1999: *Vulnerabilidad de Recursos Hídricos de Alta Montaña.*

INRENA, 1994: *Descargas de los Ríos y Almacenamiento de Reservorios y Represas de la Costa Peruana.* Ministerio de Agricultura, Instituto Nacional de Recursos Naturales, Lima, Peru, 80 pp. (in Spanish).

Instituto Nacional de Salud, 1999: *Memoria 1994–1998.* Instituto Nacional de Salud, Ministerio de Salud, Lima, Peru, 79 pp. (in Spanish).

Instituto Nacional de Salud, 1998a: Edición especial: Fenómeno El Niño. *Boletín,* **4(1),** 11–18 (in Spanish).

Instituto Nacional de Salud, 1998b: Edición especial: Fenómeno El Niño. *Boletín,* **4(3),** 11–18 (in Spanish).

IPCC, 1998: *The Regional Impacts of Climate Change: An Assessment of Vulnerability. Special Report of IPCC Working Group II* [Watson, R.T., M.C. Zinyowera, and R.H. Moss (eds.)]. Intergovernmental Panel on Climate Change, Cambridge University Press, Cambridge, United Kingdom and New York, NY, USA, pp. 201–204.

IPCC, 1996: *Climate Change 1995: Impacts, Adaptations and Mitigation of Climate Change: Scientific-Technical Analyses. Contribution of Working Group II to the Second Assessment Report of the Intergovernmental Panel on Climate Change* [Watson, R.T., M.C. Zinyowera, and R.H. Moss (eds.)]. Cambridge University Press, Cambridge, United Kingdom and New York, NY, USA, 880 pp.

Izmailova, A.V. and A.I. Moiseenkov, 1998: *Proceeding of the Second International Conference on Climate and Water, Espoo, Finland, 17–20 August 1998, Vol 3.*

Jaimes, E., 1997: Análisis de variaciones decadales de la temperatura del aire y su relación con el evento ENSO en la costa norte y centro del Peru 1940–94. In: *Gestión de Sistemas Oceanográficos del Pacífico Oriental. Comisión Oceanográfica Intergubernamental de la UNESCO* [Tarifeno, E. (ed.)]. pp. 259–266 (in Spanish).

Janzen, D.H., 1988: Tropical dry forests, the most endagered major tropical ecosystem. In: *Biodiversity* [Wilson, E.D. (ed.)]. National Academy Press, Washington, DC, USA, pp. 130–137.

Janzen, D.H., 1967: Why mountain passes are higher in the tropics. *American Naturalist,* **101,** 233–249.

Jáuregui, E., 1995: Rainfall fluctuations and tropical storm activity in Mexico, erkunde. *Archiv Für Wiseenschaftliche Geographie,* **49,** 39–48.

Jáuregui, E., 1997: Climate changes in Mexico during the historical and instrumented periods. *Quaternary International,* **43/44,** 7–17.

Kapos, V., 1989: Effects of isolation on the water status of forest patches in the Brazilian Amazon. *Journal of Tropical Ecology,* **5,** 173–185.

Kapos, V., G. Ganade, E. Matusi, and R.L. Victoria, 1993: Delta ^{13}C as an indicator of edge effects in tropical rainforest reserves. *Journal of Ecology,* **81,** 425–432.

Karl, T.R., R.W. Knight, and N. Plummer, 1995: Trends in high frequency climate variability in the XX Century. *Nature*, **377,** 217–220.

Katsouyanni, K., A. Pantazopoulo, G. Touloumi, I. Tselepidaki, K. Moustris, D. Asimakopoulos, G. Poulopoulou, and D. Trichopoulos, 1993: Evidence of interaction between air pollution and high temperatures in the causation of excess mortality. *Archives of Environmental Health*, **48,** 235–242.

Kattenberg, A., F. Giorgi, H. Grassl, G.A. Meehl, J.F.B. Mitchell, R.J. Stouffer, T. Tokioka, A.J. Weaver, and T.M.L. Wigley, 1996: Climate models—projections of future climate. In: *Climate Change 1995: The Science of Climate Change. Contribution of Working Group I to the Second Assessment Report of the Intergovernmental Panel on Climate Change* [Houghton, J.T., L.G. Meira Filho, B.A. Callander, N. Harris, A. Kattenberg, and K. Maskell (eds.)]. Cambridge University Press, Cambridge, United Kingdom and New York, NY, USA, pp. 289–357.

Kiladis, G.N. and H.F. Díaz, 1986: An analysis of the 1877–78 ENSO episode and comparison with 1982–83. *Monthly Weather Review*, **114,** 1035–1047.

Kilbourne, E.M., 1989: Heatwaves. In: *The Public Health Consequences of Disasters* [Gregg, M.B. (ed.)]. U.S. Department of Health and Human Services, Public Health Service, Centers for Disease Control and Prevention, Atlanta, GA, USA, pp. 51–61.

Kjerve, B. and D. Macintosh, 1997: Climate change on mangrove ecosystem. In: *Mangrove Ecosystem Studies in Latin America and Africa* [Kjerfve, B., L.D. de Lacerda, and E.H. Salif (eds.)]. UNESCO, New York, NY, USA, 349 pp.

Kokot, K.K., 1999: *Cambio Climático y Evolución Costera en Argentina*. Diss. Departamento de Ciencias Geológicas, Facultad Ciencias Exactas y Naturales, Universidad Buenos Aires, Buenos Aires, Argentina, 254 pp. (in Spanish).

Kovats, R.S., M.J. Bouma, and A. Haines, 1999: *El Niño and Health*. WHO/SDE/PHE/99.4, World Health Organization, Geneva, Switzerland, 48 pp.

Kovats, S., J.A. Patz, and D. Dobbins, 1998: Global climate change and environmental health. *International Journal of Occupational and Environmental Health*, **4(1),** 41–51.

Krepper, C.M. and M.E. Sequeira, 1998: Low frequency variability of rainfall in southeastern South America. *Theor. Appl. Climatol*., **61,** 19–28.

Kreutz, K.J., P.A. Mayewski, L.D. Meeker, M.S. Twickler, S.I. Whitlow, and I.I. Pittalwala, 1997: Bipolar changes in atmospheric circulation during the Little Ice Age. *Science*, **277,** 1294–1312.

Laurance, W.F., 1998: A crisis in the making: responses of Amazonian forests to land-use and climate change. *Trends in Ecology and Evolution*, **13,** 411–415.

Laurance, W.F., D. Perez, P. Delamonica, P.M. Fearnside, S. Agra, A. Jerozolinski, L. Pohl, and T.E. Lovejoy, 2000: Rain forest fragmentation and the structure of Amazonian liana communities. *Ecology*, (in press).

Laurance, W.F., S.G. Laurance, and P. Delamonica, 1998: Tropical forest fragmentation and greenhouse gas emissions. *Forest Ecology and Management*, **110(1–3),** 173–180.

Laurance, W.F., S.G. Laurance, L.V. Ferreira, J.M. Rankin-de-Merona, C. Gascon, and T.E. Lovejoy, 1997: Biomass collapse in Amazonian forest fragments. *Science*, **278,** 1117–1118.

Leaf, A., 1989: Potential health effects of global climate and environmental changes. *New England Journal of Medicine*, **321,** 577–583.

Lean, J., C.B. Bunton, C.A. Nobre, and P.R. Rowntree, 1996: The simulated impact of Amazonian deforestation on climate using measured ABRACOS vegetation characteristics. In: *Amazonian Deforestation and Climate* [Gash, J.H.C., C.A. Nobre, J.M. Roberts, and R.L. Victoria (eds.)]. John Wiley and Sons, Chichester, United Kingdom, pp. 549–576.

Leckenbush, G.C. and O. Speth, 1999: Meteorological interpretation of results from Antarctic ice core by using and AGCM under different paleoclimate boundary conditions. In: *Proceedings of the 10th Symposium on Global Change Studies, 10–15 January, 1999, Dallas, TX, Preprint Volume*. American Meteorological Society (AMS), Boston, MA, USA, pp. 294–295.

Ledig, F.T. and J.H. Kitzmiller, 1992: Genetic strategies for reforestation in the face of global climate change. *Forest Ecology and Management*, **50,** 153–169.

Ledru, M., H. Behling, M. Founier, L. Martin, and M. Servant, 1994: Localisation de la foret de Araucaria du Bresil au cours de l'Holocene: implications paleoclimatiques. *C.R. Acad. Sci. Paris, Sciences de la Vie/Life Sciences*, **317,** 521–527 (in French).

Leo, M., 1995: The importance of tropical montane cloud forest for preserving vertebrate emdemism in Peru: the Rio Abiseo National Park as a case study. In: *Tropical Montane Cloud Forest* [Hamilton, L.S. and J.O. Juvik (eds.)]. FN Scatena Editors, Springer-Verlag, New York, NY, USA, pp. 198–211.

Lieth, H., 1976: Present knowledge of productive patterns is best in temperate zones and poorest in the tropics.*Unasylvia*, **28,** 114.

Liverman, D., M. Dilley, K. O'Brien, and L. Menchaca, 1994: Possible impacts of climate change on maize yields in Mexico. In: *Implications of Climate Change for International Agriculture: Crop Modeling Study* [Rosenzweig, C. and A. Iglesias (eds.)]. U.S. Environmental Protection Agency, Mexico chapter, Washington, DC, pp. 1–14.

Liverman, D. and K. O'Brien, 1991: Global warming and climate change in Mexico. *Global Environmental Change*, **1(4),** 351–364.

Lobitz, B., L. Beck, A. Huq, B. Wood, G. Fuchs, A.S.G. Faruque, and R. Coldwell, 2000: Climate and infectious disease: use of remote sensing for detection of Vibrio cholerae by indirect measurement. *PNAS*, **97(4),** 1438–1443.

Magaña, V. and C. Conde, 2000: Climate and freshwater resources in northern Mexico, Sonora: a case study. *Environmental Monitoring and Assessment*, **61,** 167–185.

Magaña, V. and A. Quintanar, 1997: On the use of a general circulation model to study regional climate. In: *Proceedings of the Second UNAM-Cray Supercomputing Conference on Earth Sciences, Mexico City*. Cambridge University Press, Cambridge, United Kingdom and New York, NY, USA, pp. 39–48.

Magaña, V., J.L. Pérez, and C. Conde, 1998: *El Niño and its Impacts on Mexico*. Ciencias, School of Sciences, UNAM, Mexico.

Magrin, G.O., M.I. Travasso, G. Rodríguez, and D. Boullon, 1999a: Climate change assessment in Argentina, I: impacts upon crops production. In: *Proceedings of the Global Change and Terrestrial Ecosystems (GCTE) Focus 3 Conference on Food and Forestry: Global Change and Global Challenges. September, 20–23, 1999*. University of Reading, Reading, United Kingdom.

Magrin, G.O., M.I. Travasso, M. Grondona, C. Messina, G. Rodríguez, and D. Boullon, 1999b: *Climate Variability and Agriculture in Argentina and Uruguay: Assessment of ENSO Effects and Perspectives for the Use of Climate Forecast*. Final Report, Instituto de Clima y Agua (INTA), Castelar, Argentina, 23 pp.

Magrin, G.O., M. Grondona, M.I. Travasso, D. Boullon, G. Rodríguez, and C. Messina, 1999c: ENSO impacts on crop production in Argentina's Pampas Region. In: *Proceedings of the 10th Symposium on Global Change Studies, 10–15 January, 1999, Dallas, TX, Preprint Volume*. American Meteorological Society (AMS), Boston, MA, USA, pp. 65–66.

Magrin, G.O., M.O. Grondona, M.I. Travasso, D.R. Boullón, C.D. Rodriguez, and C.D. Messina, 1998: *Impacto del Fenómeno "El Niño" sobre la Producción de Cultivos en la Región Pampeana*. INTA-Boletín de divulgación, 16 pp. (in Spanish).

Magrin, G.O., M.I. Travasso, R.A. Días, and R.O. Rodríguez, 1997a: Vulnerability of the agricultural systems of Argentina to climate change. *Climate Research*, **9,** 31–36.

Magrin, G.O., R.A. Días, M.I. Travasso, R.O. Rodríguez, D. Boullon, M. Núñez, and S. Solman, 1997b: *Proyecto de Estudio sobre el Cambio Climático en Argentina*. Proyecto ARG/95/G/31-PNUD-SECYT, Informe Final del Sub-Proyecto Vulnerabilidad y Mitigación Relacionada con el Impacto del Cambio Global sobre la Producción Agrícola, Secretaría de Ciencia y Tecnología (SECYT), Buenos Aires, Argentina, 290 pp. (in Spanish).

Magrin, G.O., R.A. Díaz, M.I. Travasso, G. Rodriguez D. Boullón, M. Nuñez, and S. Solman, 1997c: Vulnerabilidad y mitigación relacionada con el impacto del cambio global sobre la producción agrícola. In: *Proyecto de Estudio sobre el Cambio Climático en Argentina* [Barros V., J.A. Hoffmann J.A., and W.M. Vargas (eds.)]. Proyecto ARG/95/G/31-PNUD-SECYT, Secretaría de Ciencia y Tecnología (SECYT), Buenos Aires, Argentina, 290 pp. (in Spanish).

Mainardi, V., 1996: *El Manglar de Térraba-Sierpe en Costa Rica*. CATIE, Proyecto Conservación para el Desarrollo Sostenible en América Central, 91 pp. (in Spanish).

Malhi, Y., A.D. Nobre, J. Grace, B. Cruijt, M.G.P. Pereira, A. Culf, and S. Scott, 1999: Carbon dioxide transfer over a Central Amazonian rain forest. *Journal of Geophysical Research*, **3,** 21–24.

Manzi, A.O. and S. Planton, 1996: A simulation of Amazonian deforestation using a GCM calibrated with ABRACOS and ARME data. In: *Amazonian Deforestation and Climate* [Gash, J.H.C., C.A. Nobre, J.M. Roberts, and R.L. Victoria (eds.)]. John Wiley and Sons, Chichester, United Kingdom, pp. 505–529.

MARENA, 2000: *Escenarios Climáticos y Socioeconómicos de Nicaragua para el Siglo XXI*. PNUD-NIC/98/G31, Ministerio del Ambiente y los Recursos Naturales de Nicaragua, Nicaragua, (in Spanish).

Marengo, J., 1995: Variations and change in South American streamflow. *Climate Change*, **31,** 99–117.

Marengo, J., 1992: Interannual variability of surface climate in the Amazon basin. *International Journal of Climatology*, **12,** 853–863.

Marengo, J. and C. Nobre, 2000: The hydroclimatological framework in Amazonia. In: *Biogeochemistry of Amazonia* [Richey, J., M. McClaine, and R. Victoria (eds.)]. Springer-Verlag, Berlin and Heidelberg, Germany, (in press).

Marengo, J. and J. Rogers, 2000: Cold front and polar air outbreaks in the Americas during modern climate assessments and impacts, and some past climate evidences. In: *Present and Past Inter-Hemispheric Climate Linkages in the Americas and Their Societal Effects* [Margraf, V. (ed.)]. Springer-Verlag, Berlin and Heidelberg, Germany, (in press).

Marengo, J., U. Bhatt, and C. Cunningham, 2000: Decadal and multidecadal variability of climate in the amazon basin. *International Journal of Climatolology*, (submitted).

Marengo, J., J. Tomasella, and C. Uvo, 1998: Long-term streamflow and rainfall fluctuations in tropical South America: Amazonia, eastern Brazil and northwest Peru. *Journal of Geophysical Research*, **103,**1775–1783.

Marengo, J., C.A. Nobre, and G. Sampaio, 1997: On the associations between hydrometeorological conditions in Amazonia and the extremes of the Southern Oscillation. In: *Consecuencias Climáticas e Hidrológicas del Evento El Niño a Escala Regional y Local. Extended Abstracts of "Memorias Técnicas," Seminario Internacional, 26–29 noviembre 1997, Quito, Ecuador*. pp. 257–266 (in Spanish).

Martens, W.J.M., 1998: Climate change, thermal stress and mortality changes. *Social Science and Medicine*, **46(3),** 331–344.

Martínez, J.O., J.L. González, O.H. Pilkey, and W.J. Neal, 1995: Tropical barrier islands of Colombia Pacific coast: sixty-two barrier islands. *Journal of Coastal Research*, **11(2),** 432–453.

Mata, L.J., 1996: A study of climate change impacts on the forest of Venezuela. In: *Adapting to Climate Change: An International Perspective* [Smith, J., N. Bhatti, G. Menzhulin, R. Benioff, M.I. Budyko, M. Campos, B. Jallow, and F. Rijsberman (eds.)]. Springer-Verlag, New York, NY, USA.

McMichael, A.J. and R.S. Kovats, 1998a: *Assessment of the Impact on Mortality in England and Wales of the Heatwave and Associated Air Pollution Episode of 1976*. Department of Epidemiology and Population Health, London School of Hygiene and Tropical Medicine, London, United Kingdom, 25 pp.

McMichael, A.J. and R.S. Kovats, 1998b: Climate change and climate variability: adaptation to reduce adverse health impacts. In: *Proceedings of IPCC Workshop, Adaptation to Climate Variability and Change, 29 March–1 April 1998, San José, Costa Rica*. Intergovernmental Panel on Climate Change, Washington, DC, USA.

McMichael, A.J., M. Ando, R. Carcavallo, P. Epstein, A. Haines, G. Jendritzky, L. Kalkstein, R. Odongo, J. Patz, and W. Piver, 1996: Human population health. In: *Climate Change 1995: Impacts, Adaptations, and Mitigation of Climate Change. Contribution of Working Group II to the Second Assessment Report of the Intergovernmental Panel on Climate Change* [Watson, R.T., MC. Zinyowera, and R.H. Moss (eds.)]. Cambridge University Press, Cambridge, United Kingdom and New York, NY, USA, pp. 561–584.

McWilliam, A., L.C. Roberts, J.M. Cabral, O.M.R. Leitão, M.V.B.R. de Costa, A.C.L. Maitelli, and C.A.G.P. Zamparoni, 1993: Leaf-area index and above-ground biomass of terra firme rain forest and adjacent clearings in Amazonia. *Functional Ecology*, **7(3),** 310–317.

Medeiros, Y., 1994: *Modeling the Hydrological Impacts of Climate Change on a Semi-Arid Region*. Diss. University of Newcastle, Newcastle, United Kingdom.

Medellín-Leal, F. and A. Gómez-González, 1979: Management of natural vegetation in the semi-arid ecosystems of Mexico. In: *Management of Semi-Arid Ecosystems* [Walker, B.H. (ed.)]. Elsevier Scientific Publication Company, pp. 351–376.

Meggers, B.J., 1994: Archeological evidence for the impact of mega-Niño events on Amazonia during the past two millennia. *Climatic Change*, **28(1–2),** 321–338.

Mendoza, M., E. Villanueva, and J. Adem, 1997: Vulnerability of basins and watersheds in Mexico to global climate change. *Climatic Research*, **9,** 139–145.

Mesa, O.J., G. Poveda, and L.F. Carvajal, 1997: *Indroducción al Clima de Colombia (Introduction to the Climate of Colombia)*. National University Press, Bogota, Colombia. (in Spanish).

Middleton, N. and D. Thomas, 1997: *World Atlas of Desertification*. Arnold, London, United Kingdom.

Milliman, J.D., 1993: Coral reefs and their response to global climate change. In: *Ecosystem and Socioeconomic Response to Future Climatic Conditions in the Marine and Coastal Regions of the Caribbean Sea, Gulf of Mexico, Bahamas, and the Northeast Coast of South America*. CEP Technical Report No. 22, United Nations Environment Program, Caribbean Environmental Programme, Kingston, Jamaica, pp. 306–321.

MINAE-IMN, 2000: *Estudios de Cambio Climático en Costa Rica*. Ministerio del Ambiente y Energía de Costa Rica—Instituto Meteorológico Nacional, (in Spanish).

Minetti, J.L. and E.M. Sierra, 1989: The influence of general circulation patterns on humid and dry years in the Cuyo Andean region of Argentina. *International Journal of Climatology*, **9,** 55–69.

Mitchell, J.F.B., R.A. Davis, W.J. Ingram, and C.A. Senior, 1995: On surface temperature, greenhouse gases and aerosols: models and observations. *Journal of Climate*, **10,** 2364–2386.

Morales, T. and V. Magaña, 1999: Unexpected frosts in central Mexico during summer. *Proceedings of the 10th Symposium on Global Change Studies, 10–15 January, 1999, Dallas, TX, Preprint Volume*. American Meteorological Society (AMS), Boston, MA, USA, pp. 262–263.

Morales, T. and V. Magaña, 1998: Climate variability and agriculture. *Claridades Agropecuarias*, **3,** 21–25 (in Spanish).

Morales-Arnao, B., 1999: Deglatiation in the Andes and its consequences. In: *Proceedings of the International Mining and Environment Meeting, Clean Technology: Third Millennium Challenges, July 12–16, 1999, Lima, Perú*, pp. 311–321.

Morales-Arnao, B., 1969a: Estudios de ablación en la Cordillera Blanca. *Boletín del Instituto Nacional de Glaciología del Peru*, **1,** 5 (in Spanish).

Morales-Arnao, B., 1969b: Estudio de la evolución de la lengua glaciar del Pucahiurca y de la laguna Safuna. *Boletín del Instituto Nacional de Glaciologiá del Perú*, **1,** 6 (in Spanish).

Moreira, C.J.E., 1986: Rainfall and flooding in the Guayas river basin and its effects on the incidence of malaria 1982–1985. *Disasters*, **10(2),** 107–111.

Morello, J., 1984: *Perfil Ecológico de América Latina*. Instituto de Cooperación Iberoamericana, Barcelona, Spain, 93 pp. (in Spanish).

Morello, J., 1976: Ecorregión del Macizo de Santa Marta, Colombia. In: *Proyecto Piloto de Ecodesarrollo en Colombia*. Programa Naciones para el Medio Ambiente, ORLA, (in Spanish).

Moreno, A.R. and R.U. Carcavallo, 1999: An ecological approach to Chagas Disease epidemiology. In: *Atlas of Chagas Disease Vectors in the Americas, Volume 3* [Carcavallo, R.U., I. Galíndez-Girón, J. Jurberg, and H. Lent (eds.)]. FIOCRUZ, Rio de Janeiro, Brazil, pp. 981–998.

Mundo, M.D. and P. Martínez-Austria, 1993: *Cambio Climático: Posibles Consecuencias y Algunas Sugerencias para Disminuir su Efecto en México, Enero–Abril 1993*. pp. 14–28 (in Spanish).

Murphy, P.G. and A.E. Lugo, 1995: Dry forest of Central America and the Caribbean. In: *Seasonally Dry Tropical Forests* [Bullock, S.H., H.A. Mooney, and E. Medina (eds.)]. Cambridge University Press, Cambridge, United Kingdom and New York, NY, USA, pp. 9–34.

Murphy, P.G. and A.E. Lugo, 1986: Ecology of tropical dry forest. *Annual Review of Ecology and Systematics*, **17,** 67–88.

Myers, N., R. Mittermeler, C. Mittermeler, G. da Fonseca, and J. Kent, 2000: Biodiversity hotspots for conservation priorities. *Nature*, **403,** 21–26.

Naeem, S., L.J. Thompson, S.P. Lawler, J.H. Lawton, and R.M. Woodfin, 1994: Declining biodiversity can alter the performance of ecosystems. *Nature*, **368,** 734–737.

Nakicenovic, N., J. Alcamo, G. Davis, B. de Vries, J. Fenhann, S. Gaffin, K. Gregory, A. Grubler, T.Y. Jung, T. Kram, E.L. La Rovere, L. Michaelis, S. Mori, T. Morita, W. Pepper, H. Pitcher, L. Price, K. Raihi, A. Roehrl, H.-H. Rogner, A. Sankovski, M. Schlesinger, P. Shukla, S. Smith, R. Swart, S. van Rooijen, N. Victor, and Z. Dadi, 2000: An overview of the scenario literature. In: *Emissions Scenarios. A Special Report of Working Group III of the Intergovernmental Panel on Climate Change*. Cambridge University Press, Cambridge, United Kingdom and New York, NY, USA, pp. 77-102.

Nawata, K. (ed.), 1999: *"Society" in 2050: Choice for Global Sustainability*. Global Industrial and Social Progress Research Institute, Asahi Glass Foundation, Japan, pp. 15–49.

Nepstad, D.C., A.A. Alencar, and A.G. Moreira, 1999: *Flames in the Rain Forest: Origins, Impacts and Alternatives to Amazonian Fires*. World Bank, Brasília, Brazil.

Nepstad, D., C.R. de Carvalho, E. Davidson, P. Jipp, P. Lefebvre, G.H. Negreiros, E.D. da Silva, T. Stone, S. Trumbore, and S. Vieira, 1994: The role of deep roots in the hydrological and carbon cycles of Amazonian forests and pastures. *Nature*, **372,** 666–669.

Nobre, P. and J. Shukla, 1996: Variations of sea surface temperature, wind stress, and rainfall over the tropical Atlantic and South America. *Journal of Climate*, **9,** 2464–2479.

Nuñez, R.H., T.S. Richards, and J.J. O'Briend, 1999: Statistical analysis of Chilean precipitation anomalies associated with El Niño-Southern Oscillation. *International Journal of Climatology*, **3,** 21–27.

OCCH, 1999: *Algunos Indicadores de Cambio Climático en Honduras*. Oficina de Cambio Climático de Honduras, Secretaría de Recursos Naturales y Ambiente, Tegucigalpa, Honduras, (in Spanish).

Olivo, M.L., 1997: Assessment of the vulnerability of Venezuela to sea level raise. *Climate Research*, **9,** 57–65.

OPS, 1999: *Situación del Cólera en las Américas*. OPA/HCP/HCT/AIEPI/99, Informe No. 19, Organización Panamericana de la Salud, Washington, DC, USA, 11 pp. (in Spanish).

OPS, 1998: *Repercusiones Sanitarias de la Oscilación del Sur (El Niño)*. CE122/10, Organización Panamericana de la Salud, Washington, DC, USA, 22 pp. (in Spanish).

Ortiz, P.L., M.E.N. Poveda, and A.V. Guevara-Velasco, 1998: Models for setting up a biometeorological warming system over a populated area of Havana City. In: *Urban Ecology*. Springer-Verlag, Berlin and Heidelberg, Germany, pp. 87–91.

Pabón, J.D., 1995a: Búsqueda de series de referencia para el seguimiento de la señal regional del calentamiento global. *Cuadernos de Geografía*, **2,** 164–173 (in Spanish).

Pabón, J.D., 1995b: *Perpectivas Globales Regionales del Cambio Climático y su Impacto en la Montaña Alta Colombiana en Memorias del Seminario Taller sobre Alta Monataña Colombiana* [Lozano, D.J.D. and J.D. Pabón (eds.)]. Academia Colombiana de Ciencias Exactas, Fisicas y Naturales, Vol. 3, pp. 19–32 (in Spanish).

Pabón, J.D., G.E. León, E.S. Rangel, J.E. Montealegre, G. Hurtado, and J.A. Zea, 1999: *El Cambio Climático en Colombia: Tendencias actuales y Proyecciones*. Nota Técnica del IDEAM, IDEAM/METEO/002-99, Santa Fe de Bogotá, Colombia, 20 pp. (in Spanish).

PAHO, 1998a: *Report on the Epidemiological Situation in Central America*. SC/XVII/40, Pan American Health Organization, Washington, DC, USA, 2 pp.

PAHO, 1998b: *Infectious Diseases Posing in the Greatest Epidemiological Risk Following Hurricane Mitch in Central America, 1998*. A Report of the Pan American Health Organization Emergency Task Force, Division of Disease Prevention and Control, Washington, DC, USA, 4 pp.

Palacio, P.J.L., G.L. Luna, and M.L. Macias, 1999: Detección de incendios en México utilizando imágenes AVHRR (temporada 1998). *Boletín del Instituto de Geografía, Investigaciones Geográficas*, **38,** 7–14 (in Spanish).

Parmenter, R.R., E.P. Yadav, C.A. Parmenter, P. Ettestad, and K.L. Gage, 1999: Incidence of plague associated with increased winter-spring precipitation in New Mexico. *American Journal of Tropical Medicine and Hygiene*, **6,** 814–821.

Patz, J.A., 1998: Climate change and health: new research challenges. *Health and Environment Digest*, **12(7),** 49–53.

Patz, J.A., P.R. Epstein, T.A. Burke, and J.M. Balbus, 1996: Global climate change and emerging infectious diseases. *Journal of the American Health Association*, **275(3),** 217–223.

Patz, J.A., W.J.M. Martens, D.A. Focks, and T.H. Jetten, 1998: Dengue fever epidemic potential as projected by general circulation models of global climate change. *Environmental Health Perspectives*, **106(3),** 147–153.

Paz Rada, O., S.R. Crespo, and F.T. Miranda, 1997: Analysis of climate scenarios for Bolivia. *Climate Research*, **9,** 115–120.

Penalba, O.C. and W.M. Vargas, 1996: Climatology of monthly annual rainfall in Buenos Aires, Argentina. *Meteorological Applications*, **3,** 275–282.

Pérez, C.A., G. Poveda, O.J. Mesa, LF. Carvajal, and P. Ochoa, 1998: *Evidence of Climate Change in Colombia: Trends, Phase and Amplitude Shifts of the Annual and Semi-Annual Cycles* [Cadier, E. (ed.)]. ORSTOM.

Pérez, J.L., 1997: *Variabilidad Climática Regional en México*. Diss. Facultad de Ciencias, UNAM, México, (in Spanish).

Phillips, O.L., Y. Malhi, N. Higuchi, W.F. Laurance, P.V. Núñez, R.M. Vásquez, S.G. Laurance, L.V. Ferreira, M. Stern, S. Brown, and J. Grace, 1998: Changes in the carbon balance of tropical forests: evidence from long-term plots. *Science*, **282,** 439–442.

Planos, E.O. and O. Barros, 1999: Impacto del cambio climático y medidas de adaptación en Cuba: sector recursos hídricos. In: *Impactos del Cambio Climático y Medidas de Adaptacion en Cuba* [Gutierrez, T., A. Centella, M. Limia, and M. López (eds.)]. Proyecto No. FP/CP/2200-97-12, UNEP INSMET, La Habana, Cuba, pp. 28–54.

PNCC, 2000: *Escenarios Climáticos, Estudios de Impactos y Opciones de Adaptación al Cambio Climático, Bolivia* [Paz, O. (ed.)]. PNCC, 253 pp. (in Spanish).

PNCC, 1997: *Vulnerabilidad y Adaptación al Cambio Climático, Bolivia* [Paz, O. (ed.)]. PNCC, 257 pp. (in Spanish).

PNUD/CAF, 2000: *Efectos de la Lluvias Caídas en Venezuela en Diciembre de 1999*. Programa de Naciones Unidas para el Desarrollo/Corporación Andina de Fomento, Caracas, Venezuela, 224 pp. (in Spanish).

PNUMA, 1991: Efectos de los cambios climáticos en los ecosistemas costeros y marinos del Pacífico Sudeste. In: *Informe de la Segunda Reunión del Grupo de Trabajo Regional sobre los Cambios Climáticos en el Pacífico Sudeste*. PNUMA/CPPS, Santiago, Chile, (in Spanish).

Pounds, J.A., M.P.L. Fogden, and J.H. Campbell, 1999: Biological response to climate change on a tropical mountain. *Nature*, **398,** 611–615.

Poveda, G. and O.J. Mesa, 1997: Feedbacks between hydrological processes in tropical South America and large scale oceanic-atmospheric phenomena. *Journal of Climate*, **10,** 2690–2702.

Poveda, G. and W. Rojas, 1997: Evidencias de la asociación entre brotes epidémicos de malaria en Colombia y el fenómeno El Niño-Oscilación del Sur. *Revista de la Academia Colombiana de Ciencias,* **21(81),** 421–429 (in Spanish).

Poveda, G., N.E. Graham, P. Epstein, W. Rojas, I.D. Vélez, M.L. Quiñonez, and P. Martens, 1999a: Climate and ENSO variability associated with vector-borne diseases in Colombia. In: *El Niño and the Southern Oscillation, Multiscale Variability and Global and Regional Impacts* [Diaz, H.F. and V. Markgraf (eds.)]. Cambridge, Cambridge University Press, United Kingdom and New York, NY, USA, pp. 177–198.

Poveda, G., N.E. Graham, P. Epstein, W. Rojas, I.D. Vélez, M.L. Quiñones, and P. Martens, 1999b: Climate and ENSO variability associated with malaria and Dengue Fever in Colombia. *Proceedings of the 10th Symposium on Global Change Studies, 10–15 January, 1999, Dallas, TX, Preprint Volume*. American Meteorological Society (AMS), Boston, MA, USA, pp. 173–176.

Poveda, G., A. Jaramillo, M.M. Gil, N. Quiceno, and R. Mantilla, 2001: Seasonality in ENSO-related precipitation, river discharge, soil moisture and vegetation index (NDVI) in Colombia. *Water Resources Research*, (in press).

Prieto, M.D.R. and R.G. Herrera, 1992: Las perturbaciones climáticas de fines del siglo XVIII Andina. In: *El Noreste Argentino como Región Histórica, Integración y Desintegración Regional: Estudio del País Interior*. Equipo de investigacion de la Universidad de Sevilla, Proyecto NOA, Sevilla, Spain, Vol. 1, pp. 7–35 (in Spanish).

Quesada, A. and L. Jiménez, 1988: Watershed management and a wetlands conservation strategy: the need for a cross-sectoral approach. In: *Ecology and Management of Wetlands* [Hook, D.D. (ed.)]. Timber Press, Portland, OR, USA, Vol. 2.

Quinn, W. and V. Neal, 1982: Long-term variation of the Southern Oscillation, El Niño and the Chilean subtropical rainfall. *Fishery Bulletin*, **81,** 363–374.

Quintana-Gomez, R.A., 1999: Trends of maximum and minimum temperatures in northern South America. *Journal of Climate*, **12(7),** 2104–2112.

Quintela, R.M., R.J. Broqua, and O.E. Scarpati, 1993: Posible impacto del cambio global en los recursos hídricos del Comahue (Argentina). In: *Proceedings of X Simpósio Brasileiro de Recursos Hídricos and I Simpósio de Recursos Hídricos do Cone Sul, 7–12 November, 1993, Brasil, Anais 3*. Brazil, pp. 320–329 (in Spanish).

RAMSAR, 1999: *Programa de Promoción, 1999–2002*. Propuesta No. 9, Proyecto de Resolución, RAMSAR, COP7 DOC.15.9, (in Spanish).

Reilly, J., W. Baethgen, F.E. Chege, S.C. van de Geijn, L. Erda, A. Iglesias, G. Kenny, D. Patterson, J. Rogasik, R. Rotter, C. Rosenzweig, W. Sombroek, J. Westbrook, D. Bachelet, M. Brklacich, U. Dammgen, M. Howden, R.J.V. Joyce, P.D. Lingren, D. Schimmelpfennig, U. Singh, O. Sirotenko, and E. Wheaton, 1996: Agriculture in a changing climate: impacts and adaptation. In: *Climate Change 1995: Impacts, Adaptations and Mitigation of Climate Change: Scientific-Technical Analyses. Contribution of Working Group II to the Second Assessment Report of the Intergovernmental Panel on Climate Change* [Watson, R.T., M.C. Zinyowera, and R.H. Moss (eds.)]. Cambridge University Press, Cambridge, United Kingdom and New York, NY, USA, pp. 427–467.

Reis, E.J. and S. Margulis, 1991: *Perspectivas Econômicas do Desflorestamento da Amazônia*. Textos para Discussão No. 215, Instituto de Pesquisa Econômica Aplicada (IPEA), Brasília, Brazil, 47 pp. (in Portuguese).

Reis, M.G.F., G.G. dos Reis, O.F. Valente, and H.A.C. Fernández, 1994: Sequestro e armazenamento de carbono em florestas nativas e plantadas dos Estados de Minas Gerais e Espírito Santo. In: *Emissão e Sequestro de CO_2: Uma Nova Oportunidade de Negócios para o Brasil* [Reis, M. and M. Borgonavi (eds.)]. Companhia Vale do Rio Doce (CVRD), Rio de Janeiro, Brazil, pp. 155–195 (in Portuguese).

Richey, J., C. Nobre, and C. Deser, 1989: Amazina River discharge and climate variability: 1903 to 1985. *Science*, **246,** 101–103.

Riebsame, W.E., K.M. Strzepek, J.L. Wescoat Jr., G.L. Graile, J. Jacobs, R. Leichenko, C. Magadza, R. Perrit, H. Phien, B.J. Urbiztondo, P. Restrepo, W.R. Rose, M. Saleh, C. Tucci, L.H. Ti, and D. Yates, 1995: Complex river basins. In: *As Climate Changes, International Impacts and Implications* [Strzepek, K.M. and J.B. Smith (eds.)]. Cambridge University Press, Cambridge, United Kingdom and New York, NY, USA, pp. 57–91.

Roberts, J.M., O.M.R. Cabral, J.P. da Costa, A.-L.C. McWilliam, and T.D. de A. Sá, 1996: An overview of the leaf area index and physiological measurements during ABRACOS. In: *Amazonian Deforestation and Climate* [Gash, J.H.C., C.A. Nobre, J.M. Roberts, and R.L. Victoria (eds.)]. John Wiley and Sons, Chichester, United Kingdom, pp. 565–285.

Robertson, A.W. and C.R. Mechoso, 1998: Interannual and decadal cycles in river flows of Southeastern South America. *Climate Change*, **11,** 2570–2581.

Ropelewski, C. and M. Halpert, 1989: Global and regional scale precipitation patterns associated with the high index phase of the Southern Oscillation. *Journal of Climate*, **2,** 268–284.

Ropelewski, C. and M. Halper, 1987: Global and regional scale precipitation patterns associated with the El Niño/Southern Oscillation. *Monthly Weather Review*, **115,** 1606–1626.

Ropelewski, C.F. and M. Halper, 1996: Quantifying Southern Oscillation-precipitation relationships. *Journal of Climate*, **9,** 1043–1059.

Rosas, I., G. Roy-Ocotla, and P. Mosiño, 1989: Meteorological effects on variation of airborne algae in Mexico. *International Journal of Biometeorology*, **33,** 173–179.

Rosas, I., A. Yela, and C. Santos-Burgoa, 1994: Occurrence of airborne enteric bacteria in Mexico City. *Aerobiología*, **10,** 39–45.

Rosas, I., A. Yela, E. Salinas, R. Arreguín, and A. Rodríguez-Romero, 1995: Preliminary assessment of protein associated with airborne particles in Mexico City. *Aerobiología*, **11,** 81–86.

Rosenblüth, B., H.A. Fuenzalida, and P. Aceituno, 1997: Recent temperature variations in southern South America. *International Journal of Climatology*, **17,** 67–85.

Rosenzweig, C. and D. Hillel, 1998: *Climate Change and the Global Harvest: Potential Impacts of the Greenhouse Effect on Agriculture*. Oxford University Press, Oxford, United Kingdom, 324 pp.

Rosenzweig, C., M.L. Parry, G. Fischer, and K. Frohberg, 1993: *Climate Change and World Food Supply*. Research Report No. 3, Environmental Change Unit, Oxford University, Oxford, United Kingdom, 28 pp.

Rull, V., T. Vegas-Vilarrubia, and N.E. de Pernia, 1999: Palynological record of an early-mid Holocene mangrove in eastern Venezuela: implications for sea-level rise and disturbance history. *Journal of Coastal Research*, **15(2),** 496–504.

Russac, P.A.A., 1986: Epidemiological surveillance: malaria epidemic following the El Niño phenomenon. *Disasters*, **10(2),** 112–117.

Rusticucci, M.M. and W.M. Vargas, 1998: *Variabilidad Interanual de las Olas de Calor y de Frío sobre la Argentina, su Relación con el ENSO*. Congreso Nacional del Agua, Santa Fé, Argentina, pp. 115–122 (in Spanish).

Ruttland, J. and H. Fuenzalida, 1991: Synoptic aspects of the central Chile rainfall variability associated with the Southern Oscillation. *International Journal of Climatology*, **11,** 63–76.

Saizar, A., 1997: Assessment of impacts of a potential sea level rise on the coast of Montevideo, Uruguay. *Climatic Research*, **9,** 73–79.

Sala, O.E. and J.M. Paruelo, 1994: Impacts of global climate change on maize production in Argentina. In: *Implications of Climate Change for International Agriculture: Crop Modeling Study* [Rosenzweig, C. and A. Iglesias (eds.)]. U.S. Environmental Protection Agency, Argentina chapter, Washington, DC, pp. 1-12.

Salati, E. and P.B. Vose, 1984: Amazon Basin: a system in equilibrium. *Science*, **225,** 129–138.

Salazar-Lindo, E., O. Pinell-Salles, and E. Chea Woo, 1997: El Niño and diarrhea and dehydratation in Lima, Peru. *Lancet*, **350,** 1597–1598.

Saldivia, P.H.N., 1995: Air pollution and mortality in elderly people: a time-series study in São Paulo, Brazil. *Archives of Environmental Health*, **50(2),** 159–164.

Saldivia, P.H.N., 1994: Association between air pollution and mortality due to respiratory diseases in children in São Paulo, Brazil: a preliminary report. *Environmental Research*, **65,** 218–225.

Salles, M.A. and R.H. Compagnucci, 1997: Características de la circulación de superficie durante el período diciembre de 1971–febrero de 1974 y sus relaciones con las anomalías ENSO en el sur de Sudamérica. *Meteorológica*, **22,** 35–48 (in Spanish).

Salles, M.A. and R.H. Compagnucci, 1995: Características de la circulación de superficie durante 1976–1977 y su relación con las anomalías en el sur de Sudamérica. *Meteorológica*, **20,** 7–16 (in Spanish).

Sansigolo, C., R. Rodriguez, and P. Etchichury, 1992: Tendencias nas temperaturas médias do Brasil. *Anais do VII Congreso Brasileiro de Meteorología*, **1,** 367–371 (in Portuguese).

Sartor, F., R. Snacken, C. Demuth, and D. Walckiers, 1995: Temperature, ambient ozone levels, and mortality during summer, 1994, in Belgium. *Environmental Research*, **70,** 105–113.

Schaeffer-Novelli, Y. and G. Cintron, 1993: Mangroves of arid environments of Latin America. In: *Towards the Rational Use of High Salinity Tolerant Plants* [Leith, H. and A. Al Masoom (eds.)]. J. Kluwer Academic Publishers, Dordrecht, The Netherlands, Vol. 1, pp. 107–116.

Schubart, H.O.R., W.J. Junk, and M. Petrere Jr., 1976: Sumário de ecología Amazónica. *Ciência e Cultura*, **28(5),** 507–509 (in Spanish).

Schubert, C., 1992: The glaciers of the Sierra Nevada de Merida (Venezuela): a photographic comparison of recent deglaciation. *Erdkunde*, **46,** 59–64.

Sebastiani, M., A. Villamizar, and M.L. Olivo, 1991: Coastal wetlands management in Venezuela. In: *Coastal Wetlands* [Bolton, S. and O.T. Magoon (eds.)]. ASCE Publishing, New York, NY, USA, pp. 503–512.

Sebastiani, M., A. Villamizar, H. Álvarez, and H. Fonseca, 1996a: Evolución de la fase inicial del procedimiento en Venezuela para el permiso de actividades a degradar el ambiente, con miras a un desarrollo sustentable. *Revista Geográfica de Venezuela*, **37(2),** 197–220 (in Spanish).

Sebastiani, M., A. Villamizar, and M.L. Olivo, 1996b: La protección del manglar en Venezuela y sus consecuencias para la gestión ambiental: retrospectiva. *Acta Científica Venezolana*, **47(2),** 77–84 (in Spanish).

SENAMHI, 1999: *Fenómeno El Niño*. Reporte Interno, SENAMHI, Lima, Perú, 83 pp. (in Spanish).

Shope, R., 1991: Global climate change and infectious diseases. *Environmental Health Perspectives*, **96,** 171–174.

Shukla, J., C. Nobre, and P. Sellers, 1990: Amazon deforestation and climate change. *Science*, **247,** 1322–1325.

Silva Dias, P. and J. Marengo, 1999: Aguas atmosféricas. In: *Aguas Doces no Brasil-capital Ecológico Usos Multiplos, Exploração Racional e Conservacão* [da Cunha Rebouças, A., B. Braga Jr., and J.G. Tundizi (eds.)]. IEA/USP, pp. 65–116.

Skole, D. and C. Tucker, 1993: Tropical deforestation and habitat fragmentation in the Amazon: satellite data from 1978 to 1988. *Science*, **260,** 1905–1910.

Soares, R.V., 1990: Fire in some tropical and subtropical South American vegetation types: An overview. In: *Fire in the Tropical Biota: Ecosystem Processes and Global Challenges* [Goldammer, J.G. (ed.)]. Springer-Verlag, Heidelberg, Germany, pp. 63–81.

Solano, P., 1997: *Legislación Ambiental Suramericana Aplicable a los Humedales*. UICN, Wetlands International SPDA, 204 pp. (in Spanish).

Sollins, P., 1998: Factors influencing species composition in tropical lowland rain forest: does soil matter? *Ecology*, **79(1),** 23–30.

Solman, S. and M. Núñez, 1999: *International Journal of Climatology*, **3,** 220–234.

Solomon, A.M., I.C. Prentice, R. Leemann, and W.P. Cramer, 1993: The interaction of climate and land use in future terrestrial carbon storage and release. *Water, Air, and Soil Pollution*, **70,** 595–614.

Suárez, M.F. and M.J. Nelson, 1981: Registro de altitud del Aedes aegypti en Colombia. *Biomédica*, **1,** 225–228 (in Spanish).

Suárez, A.G., A.J. López, A.R. Chamizo, H.Ferras, A. Martel, D. Vilamajo, and E. Mojena, 1999: Biodiversidad y vida silvestre. In: *Impactos del Cambio Climatico y Medidas de Adaptación en Cuba* [Gutierrez, T., A. Centella, M. Limia, and M. López (eds.)]. UNEP, La Habana, Cuba, pp. 164–178 (in Spanish).

Swap, R., M. Garstang, and S. Greco, 1992: Saharan dust in the Amazon Basin. *Tellus*, **44B,** 133–149.

Tabilo-Valdivieso, E., 1997: *El Beneficio de los Humedales en América Central: El Potencial de los Humedales para el Desarrollo*. 1st. ed., San José, Costa Rica, 48 pp. (in Spanish).

Talbot, R.W., M.O. Andreae, H. Berresheim, P. Artaxo, M. Garstang, R.C. Harriss, K.M. Beecher, and S.M. Li, 1990: Aerosol chemistry during the wet season in Central Amazonia: the influence of long-range transport. *Journal of Geophysical Research (Atmospheres)*, **95,** 16955–16969.

Tamplin, M. and C. Carrillo, 1991: Environmental spread of Vibrio cholerae in Peru. *Lancet*, **338,** 1216–1217.

Tanco, R. and G. Berri, 1996: Acerca del fenómeno El Niño sobre la precipitación en la Pampa Húmeda Argentina. In: *Impacto de las Variaciones Climáticas en el Desarrollo Regional un Análisis Interdisciplinario*. VII Congreso Latinoamericano e Ibérico de Meteorología, pp. 319–320.

Tarazona, J. and S. Valle, 1999: Impactos potenciales del cambio climatico global sobre el ecosistema peruano, en Perú vulnerabilidad sobre al cambio climatico: aproximaciones a la experiencia con el fenomeno El Niño. *CONAM*, **3,** 95–114 (in Spanish).

Thornthwaite, C.W., 1948: An approach towards a rational classification of climate. *Geographical Review*, **38(55),** 55–96.

Tian, H., J.M. Mellilo, D.W. Kicklighter, A.D. McGuire, J.V.K. Helfrich III, B. Moore III, and C. Vörösmarty, 1998: Effect of interanual climate variability on carbon storage in Amazonian ecosystems. *Nature*, **396,** 664–667.

Timmermann, A., 1999: Increased El Niño frequency in a climate model forced by future greenhouse warming. *Nature*, **398,** 694–696.

Toledo-Tito, J., 1997: Impacto en la Salud del Fenómeno de El Niño 1982–83 en el Perú. In: *Proceedings of The Health Impact of the El Niño Phenomenon, Central American workshop held in San José, Costa Rica, 3–5 November 1997*. Pan American Health Organization and the World Health Organization, Washington, DC, USA, (in Spanish).

Torres Guevara, J., 1992: *El Fenómeno El Niño 1997–98 y los Bosques Secos de La Costa Norte del Perú, caso Piura (Sechura y Tambogrande)*. (in Spanish).

Travasso, M.I., G.O. Magrin, G.R. Rodriguez, and D.R. Boullón, 1999: Climate change assessment in Argentina, II: adaptation strategies for agriculture. In: *Food and Forestry: Global Change and Global Challenges, GCTE Focus 3 Conference, Reading, U.K., September 1999*.

Trejo, I. and R. Dirzo, 2000: Deforestation of seasonally dry tropical forest: a national and local analysis in Mexico. *Biological Conservation*, **94,** 133–142.

Trenberth, K.Z., 1990: Recent observed interdecadal climate changes in the Northern Hemisphere. *Bulletin of the American Meteorological Society*, **71,** 988–993.

Trujillo, E., 1998: Descripción de la actividad pesquera que se desarrolla en el entorno de los islotes Los Lobos y Caribe, costa norte del Edo, Sucre, Venezuela. *Saber*, **10(1),** 21–29 (in Spanish).

Twilley, R.R., S.C. Snedaker, A. Yáñez-Arancibia, and E. Medina, 1997: Biodiversity and ecosystem processes in tropical estuaries: perspectives of mangrove ecosystems. In: *Functional Roles of Biodiversity: A Global Perspective* [Mooney, H.A., J.H. Cashman, E. Medina, O.E. Sala, and E.D. Schulze (eds.)]. John Wiley and Sons, Chichester, United Kingdom, pp. 327–368.

Uhl, C. and R. Buschbacher, 1985: A disturbing synergism between cattle-ranch burning practices and selective tree harvesting in the eastern Amazon. *Biotropica*, **17(4),** 265–268.

Uhl, C. and J.B. Kauffman, 1990: Deforestation, fire susceptibility, and potential tree responses to fire in the eastern Amazon. *Ecology*, **71(2),** 437–449.

UNCED, 1992: *Earth Summit, Rio de Janeiro*. United Nations Conference on Environment and Development, London Regency Press, London, United Kingdom.

UNDP, 1998: *Human Development Report 1998*. United Nations Development Programme, Oxford University Press, New York, NY, USA.

UNEP, 1993: *Ecosystem and Socioeconomic Response to Future Climatic Conditions in the Marine and Coastal Regions of the Caribbean Sea, Gulf of Mexico, Bahamas, and the Northeast Coast of South America*. CEP Technical Report No. 22, United Nations Environment Program, Caribbean Environmental Programme, Kingston, Jamaica.

Uvo, C., 1998: *Influence of Sea Surface Temperature on Rainfall and Runoff in Northeastern South America: Analysis and Modelling*. Diss. Department of Water Resources Engineering, Lund University, Lund, Sweden, 120 pp.

Valencia Tellería, A., 1986: Health consequences of floods in Bolivia in 1982. *Disasters*, **10(2),** 88–106.

van der Hammen, T., 1997: Cambios en el clima global: posibles efectos sobre la biodiversidad en Colombia. In: *Informe Nacional sobre el Estado de la Biodiversidad, Tomo II*. Instituto de Investigación de Recursos Biológicos "Alexander von Humbolt" and Ministerio del Medio Ambiente, Bogotá, Colombia, pp. 170–178 (in Spanish).

Vargas-Pérez, E. and S. Terrazas-Domínguez, 1991: Recursos y producción forestal. In: *Atlas Nacional de México, Vol. III, Hoja VI.4.1, Sección Economía, Escala: 1:8,000,000*. IGG, UNAM, México, (in Spanish).

Venegas, S., L. Mysak, and N. Straub, 1998: Atmospherre-ocean coupled variability in the south Atlantic. *Journal of Climate*, **10,** 2904–2920.

Venegas, S., L. Mysak, and N. Straub, 1996: Evidence for interannual and interdecadal climate variability in the south Atlantic. *Geophysical Research Letters*, **23,** 2673–2676.

Victoria, R., L. Matinelli, J. Moraes, M. Ballester, A. Krusche, G. Pellegrino, R. Almeida, and J. Richey, 1998: Surface air temperature variations in the Amazon region and its border during this century. *Journal of Climate*, **11,** 1105–1110.

Viglizzo, E.F., Z.E. Roberto, M.C. Filippin, and A.J. Pordomingo, 1995: Climate variability and agroecological change in the Central Pampas Region of Argentina. *Agriculture, Ecosystems and Environment*, **55,** 7–16.

Vila, D.A and G. Berri, 1996: Introducción al estudio de las relaciones entre los ciclos del ENSO y la temperatura media en la República Argentina. In: *Impacto de las Variaciones Climáticas en el Desarrollo Regional un Análisis Interdisciplinario*. VII Congreso Latinoamericano e Ibérico de Meteorología, pp. 311–312.

Vila, D.A. and M.O. Grondona, 1996: Estudio preliminar sobre las relaciones del ENSO y la frecuencia de días con lluvia en la Pampa Húmeda. In: *Impacto de las Variaciones Climáticas en el Desarrollo Regional un Análisis Interdisciplinario*. VII Congreso Latinoamericano e Ibérico de Meteorología, pp. 309–310.

Villalba, R., 1994: Tree-ring and glacial evidence for the Medieval Warm epoch and the Little Ice Age in southern South America. *Climatic Change*, **26,** 183–197.

Villalba, R., J.A. Boninsegna, T.T. Vebnlen, A. Schmelter, and S. Rubulis, 1997: Recent trends in tree ring records from high elevation sites in the Andes of Northern Patagonia. *Climatic Change*, **36,** 425–454.

Villamizar, A., 1994: *Desarrollo Estructural y Evolución Fisiográfica del Manglar del Río Hueque*. Diss. Universidad Simón Bolívar Caracas, Edo. Falcón, Caracas, Venezuela, 290 pp. (in Spanish).

Villamizar, A. and H. Fonseca, 1999: *Caracterización Ambiental de la Vegetación del Sector Costero Boca del Tocuyo-Chichiriviche*. Edo. Falcón, Venezuela, (in Spanish).

Villers, L., 1995: *Vulnerabilidad de los Ecosistemas Forestales*. Country Study Mexico Report 6.

Villers-Ruiz, L. and I. Trejo-Vázquez, 1997: Assessment of the vulnerability of forest ecosystems to climate change in Mexico. *Climate Research*, **9,** 87–93.

Villers-Ruiz, L. and I. Trejo-Vázquez, 1998: Climate change on Mexican forests and natural protected areas. *Global Environmental Change*, **8(2),** 141–157.

von Storch, H., 1994: *Inconsistencies at the Interface of Climate Impact Studies and Global Climate Research*. Technical Report 122, Max-Plank-Institut fuer Meteorologie, Hamburg, Germany, 25 pp.

Wachsmuth, I.K., C.A. Bopp, P.I. Fields, and C. Carrillo, 1991: Difference between toxigenic Vibrio cholerae 01 from South America and U.S. gulf coast. *Lancet*, **337,** 1097–1098.

Wachsmuth, I.K., G.M. Evins, P.I. Fields, Ø. Olsvik, T. Popovic, C.A. Bopp, J.G. Wells, C. Carrillo, and P.A. Blake, 1993: The molecular epidemiology of cholera in Latin America. *The Journal of Infectious Diseases*, **167,** 621–626.

Wagner, R., 1996: Decadal-scale trends in mechanisms controlling meridional sea surface temperature gradients in the tropical Atlantic. *Journal of Geophyical Research*, **101,** 16683–16694.

Walker, G.K., Y.C. Sud, and R. Atlas, 1995: Impact of the ongoing Amazonian deforestation on local precipitation: a GCM study. *Bulletin of the American Meteorological Society*, **26(3),** 346–361.

Waylen, P. and C. Caviedes, 1990: Annual and seasonal fluctuations of precipitation and streamflow in the Aconcagua river basin, Chile. *Journal of Hydrology*, **120,** 79–102.

Waylen, P., R. Compagnucci, and M. Caffera, 2000: Inter-annual and inter-decadal variability in streamflow from the Argentine Andes. *Physical Geography*, (in press).

Wetherald, R.T. and S. Manabe, 1988: Cloud feedback processes in a general circulation model. *Journal of the Atmospheric Sciences*, **45,** 1397–1415.

Whitney, L.D. and J.S. Hobgood, 1997: The relationship between sea surface temperature and maximum intensities of the tropical cyclones in the eastern north Pacific. *Journal of Climate*, **10,** 2921–2930.

WHO, 1998: Early human health effects on climate change. *Report of a WHO/EURO International Workshop, WHO/ECEH Rome Division, Rome, Italy, 21–23 May, 1998*. World Health Organization, 21 pp.

Williams, M.A.J. and R.C. Balling Jr., 1996: *Interactions of Desertification and Climate*. World Meteorological Organization and United Nations Environment Program.

Williams, R.S. and J.G. Ferrigno (eds.), 1998: *Satellite Image Atlas of Glaciers of the World: South America*. U.S. Geological Survey, Prof. Pap. 1386–I, U.S. Government Printing Office, Washington, DC, USA.

Windovoxhel, N., J. Rodríguez, and E. Lahmann, 1998: *Situación del Manejo Integrado en Zonas Costeras de América Central: Experiencias del Programa de Conservación de Humedales y Zonas Costeras de la UICN para la Región*. Serie Técnica Documento de Trabajo, No. 3, UICN/HORMA, 31 pp. (in Spanish).

WMO, 1998: *Report on the Status of Observing Networks*. Report submitted to CoP-4, World Meteorological Organization, Geneva, Switzerland.

WMO, 1994: *Conference on the Economic Benefits of Meteorological and Hydrological Services*. WMO/TD No. 630, World Meteorological Organization, Geneva, Switzerland, 307 pp.

World Bank, 1997: *The World Bank 1997 Annual Report*. World Bank, Washington, DC, USA.

Wright, G., 1999: *Análisis de Trayectorias de Parcelas en la Atmósfera de México*. Diss. UNAM, México, 80 pp. (in Spanish).

Zuidema, G., G.J. Van den Born, J. Alcamo, and G.J.J. Kreileman, 1994: Simulating changes in global land cover as affected by economic and climatic factors. *Water, Air, and Soil Pollution*, **76(1–2),** 163–198.

15

North America

STEWART COHEN (CANADA) AND KATHLEEN MILLER (USA)

Lead Authors:
K. Duncan (Canada), E. Gregorich (Canada), P. Groffman (USA), P. Kovacs (Canada), V. Magaña (Mexico), D. McKnight (USA), E. Mills (USA), D. Schimel (USA)

Contributing Authors:
G. Chichilnisky (USA), D. Etkin (Canada), R. Fleming (Canada), K. Hall (USA), S. Meyn (Canada), J. Patz (USA), R. Pulwarty (USA), D. Scott (Canada), G. Wall (Canada)

Review Editor:
E. Wheaton (Canada)

CONTENTS

EXECUTIVE SUMMARY

North America has experienced challenges posed by changing climates and changing patterns of regional development and will continue to do so. Varying impacts on ecosystems and human settlements will exacerbate subregional differences in climate-sensitive resource production and vulnerability to extreme events. Opportunities may arise from a warming climate, and some innovative adaptation strategies are being tested as a response to current challenges (e.g., water banks), but there are few studies on how these strategies could be implemented as regional climates continue to change. Recent experience demonstrates high capability in emergency response to extreme events, but long-term problems remain.

Climate Trends and Scenarios

North America has warmed by about 0.7°C during the past century and precipitation has increased, but both trends are heterogeneous (e.g., seasonal reductions in precipitation in some areas).

For a range of emission scenarios produced for this Third Assessment Report, model results suggest that North America could warm by 1–3°C over the next century for a low-emissions case (B1). Warming could be as much as 3.5–7.5°C for the higher emission A2 case. Published regional impact studies have used climate scenarios with global temperature changes that are similar to these new cases, but regional scenarios may not be directly comparable.

Key Regional Concerns

Water Resources

Changes in precipitation are highly uncertain. There is little agreement across climate scenarios regarding changes in total annual runoff across North America.

The modeled impact of increased temperatures on lake evaporation leads to consistent projections of reduced lake levels and outflows for the Great Lakes–St. Lawrence system under most climate change scenarios (medium confidence). The only exception is the HadCM2 transient scenario incorporating IS92a sulfate aerosol emissions, which suggests slight increases in lake levels and outflows.

Where snowmelt currently is an important part of the hydrological regime (e.g., Columbia basin), seasonal shifts in runoff are likely, with a larger proportion of runoff occurring in winter, together with possible reductions in summer flows (high confidence).

Adaptive responses to such seasonal runoff changes could include altered management of artificial storage capacity, increased reliance on conjunctive management of ground and surface water supplies, and voluntary water transfers between various water users. It may not be possible, however, to avoid adverse impacts on many aquatic ecosystems or to fully offset the impacts of reduced summer water availability for irrigation and other out-of-stream and instream water uses.

Where lower summer flows and higher water temperatures occur, there may be reduced water quality and increased stress on aquatic ecosystems (medium confidence).

Possible changes in the frequency/intensity/duration of heavy precipitation events may require changes in land-use planning and infrastructure design to avoid increased damages arising from flooding, landslides, sewerage overflows, and releases of contaminants to natural water bodies.

Responses to recurring and emerging water quality and quantity problems will provide opportunities to develop and test adaptive management options.

Natural Resources

Forests

Climate change is expected to increase the areal extent and productivity of forests over the next 50–100 years (medium confidence). Extreme and/or long-term climate change scenarios indicate the possibility of widespread decline (low confidence).

Climate change is likely to cause changes in the nature and extent of several "disturbance factors" (e.g., fire, insect outbreaks) (medium confidence). Of particular interest in North America are changes in fire regimes, including an earlier start to the fire season, and significant increases in the area experiencing high to extreme fire danger. The long-term effects of fire will depend heavily on changes in human fire management activities, which are uncertain, especially in remote boreal forests.

There is a strong need for a long-term comprehensive system to monitor forest "health" and disturbance regimes over regional scales that can function as an early warning system for climate change effects on forests.

Protected Areas

Climate change can lead to loss of specific ecosystem types, such as high alpine areas and specific coastal (e.g., salt marshes) and inland (e.g., prairie "potholes") wetland types (high confidence). There is moderate potential for adaptation to prevent these losses by planning conservation programs to identify and protect particularly threatened ecosystems.

Food and Fiber

Agriculture

Food production is projected to benefit from a warmer climate, but there probably will be strong regional effects, with some areas in North America suffering significant loss of comparative advantage to other regions (high confidence). There is potential for increased drought in the U.S. Great Plains/Canadian Prairies and opportunities for a limited northward shift in production areas in Canada (high confidence). Crop yield studies for the United States and Canada have indicated a wide range of impacts. Modeled yield results that include direct physiological effects of carbon dioxide (CO_2), with sufficient water and nutrients, are substantially different from those that do not account for such effects. Economic studies that include farm- and agricultural market-level adjustments (e.g., behavioral, economic, and institutional) indicate that the negative effects of climate change on agriculture probably have been overestimated by studies that do not account for these adjustments (medium confidence). However, the ability of farmers to adapt their input and output choices will depend on market and institutional signals, which may be partially influenced by climate change.

Production Forestry

Lands that are managed for timber production are likely to be less susceptible to climate change than unmanaged forests because of the potential for adaptive management. However, when the possibility of replanting with incorrect species is considered, economic impacts could become negative.

Carbon Sequestration—Adaptation Issues

Increased interest in agricultural sinks for carbon sequestration includes proposed use of reduced-tillage practices in North America. Negative consequences may include increased use of pesticides, reduced yields, and increased risk for farmers (medium confidence). Potential benefits include reduced input costs, increased soil moisture, and reduced soil erosion (high confidence).

Marine Fisheries

Climate-related variations in the marine environment—including changes in sea-surface temperatures, nutrient supply, and circulation dynamics—play an important role in determining the productivity of several North American fisheries (high confidence).

Projected climate changes have the potential to affect coastal and marine ecosystems, with impacts on the abundance and spatial distribution of species that are important to commercial and recreational fisheries. The degree of impact is likely to vary within a wide range, depending on species and community characteristics and region-specific conditions. These impacts are complex and difficult to observe, so climate variability constitutes a significant source of uncertainty for fishery managers. Recent experiences with Pacific salmon and Atlantic cod suggests that sustainable fisheries management will require timely and accurate scientific information on environmental conditions that affect fish stocks, as well as institutional flexibility to respond quickly to such information.

Human Health

Increased frequency and severity of heat waves may lead to an increase in illness and death, particularly among young, elderly, and frail people, especially in large urban centers. The net effect of reduced severity of extreme cold is likely to have a beneficial effect. Acclimatization may be slower than the rate of ambient temperature change.

Increased frequency of convective storms could lead to more cases of thunderstorm-associated asthma. More frequent flood events and other extreme events may result in an increase in deaths, injuries, infectious diseases, and stress-related disorders, as well as other adverse health effects associated with social disruption, environmentally forced migration, and settlement in urban slums.

Vector-borne diseases, including malaria and dengue fever, may expand their ranges in the United States and may develop in Canada. Tick-borne Lyme disease also may also expand its range in Canada. However, socioeconomic factors such as public health measures will play a large role in determining the existence or extent of such infections. Diseases associated with water may increase with warming of air and water temperatures, combined with heavy runoff events from agricultural and urban surfaces.

Respiratory disorders may be exacerbated by warming-induced increases in the frequency of smog (ground-level ozone) events, acidic deposition, and particulate air pollution.

Human Settlements and Infrastructure

Potential impacts of climate change on cities include fewer periods of extreme winter cold; increased frequency of extreme heat; rising sea levels and risk of storm surge; and changes in timing, frequency, and severity of flooding associated with storms and precipitation extremes.

Communities can reduce their vulnerability to potential adverse impacts from climate change through investments in adaptive infrastructure. These adaptations can be expensive. Rural, poor, and indigenous communities may not be able to make such investments. Furthermore, infrastructure investment decisions often are based on a variety of needs beyond climate change, including population growth and aging of existing systems.

Changes in the frequency, severity, and duration of extreme events may be among the most important risks associated with climate change. The rising cost of natural disasters in North America illustrates the vulnerability of current settlement practices. Human alterations of natural systems—such as drainage basins, barrier islands, and coastal margins—influence the impact of extreme weather hazards. Adaptations such as levees and dams often are successful in managing most variations in the weather, but they can increase vulnerability to the most extreme events.

Tourism

Shifts in temperature and precipitation patterns would lead to shifts in outdoor tourism and recreation opportunities (e.g., winter sports, fishing, parks, beaches). The extent to which ecological changes in parks will affect tourism is uncertain. Future shifts in water management, in response to development pressures as well as climate change, also could affect recreational opportunities and associated property values. Opportunities and challenges for recreational industries and destination areas need to be assessed in a systematic manner before net economic impacts can be reported with sufficient confidence.

Public and Private Insurance and Disaster Relief Systems

Inflation-corrected, weather-driven losses have been increasing in North America over the past 3 decades (high confidence). Both exposure and financial surplus of private insurers (especially property insurers) and reinsurers have been growing, and weather-related profit declines and insolvencies have been observed. Insured weather-related losses in North America (59% of the global total) are increasing with affluence and as populations continue to move into vulnerable areas. Insurer vulnerability to these changes varies considerably by region. Insurers are adversely affected by increased variability or actuarial uncertainty of weather-related events. Governments play a key role as insurers and/or providers of disaster relief, especially in cases in which risks are deemed to be uninsurable by the private sector.

Canada and the United States have different loss profiles (high confidence). In both countries, the nature of weather-related exposures and losses is diverse, ranging from property damages to business interruptions caused by electric power or communication system damage (as in the 1998 ice storm). U.S. government insurance programs for crops and floods have not been profitable and in some cases have encouraged more human activity in at-risk areas. In the absence of similar programs in Canada, government disaster relief programs have paid roughly 86% of flood losses over the past 2 decades. There remains an important tension between the allocation of such risks between private insurers and the public sector, and the effects of climate changes (e.g., coastal erosion) would increasingly stress government programs.

Recent extreme events have led to several responses by insurers, including increased attention to building codes and disaster preparedness, limiting insurance availability or increasing prices, and establishment of new risk-spreading mechanisms. Insurers can play an important role in climate adaptation and mitigation. However, because their actuarial outlook is based on past climatic experience and forward-looking modeling studies are just now beginning to be used, the potential for surprise is real.

Adaptation Potential and Vulnerability

Case studies for various North American subregions and border regions illustrate how the changing nature of climate-society relationships is influencing the nature of vulnerability, climate-related impacts, and adaptive responses. These cases include observed extreme events and potential responses to climate change scenarios. Increased levels of development may reduce vulnerability in some cases (e.g., agriculture) and increase or change vulnerability in others (e.g., Columbia basin water management).

The nature of observed damages reflects the increasing demands that society is placing on natural resources and systems that are sensitive to extreme events, as climate change is superimposed on complex environmental problems (e.g., coastal eutrophication in the Gulf of Mexico).

Climate-related consequences for water, health, food, energy, insurance, governments, and human settlements are likely to require substantial institutional and infrastructure changes in some cases. The short period of time required to make infrastructure adjustments is likely to require new institutions to cope with rapid and sweeping change. The example of new "water markets" in the western United States illustrates an adaptive measure that also may lead to concerns about accessibility of this essential commodity for lower income people, as well as conflicts about social priorities in its allocation (e.g., farms vs. residential use).

Determining responses to scenarios is a long process that requires dialogue with stakeholders. They understand the goals, objectives, and constraints driving the development, management, and operation of resource production and maintenance systems. This interdisciplinary and intercultural dialogue has the potential for improving sharing of information and enhancing its use in decisionmaking at various scales.

15.1. The North American Region

15.1.1. Previous Work

This chapter offers an assessment for Canada, the United States, and three border regions—the Arctic, U.S.–Mexican, and U.S.–Caribbean borders. More detailed assessments for the Arctic, Mexico, and the Caribbean appear in Chapters 16, 14, and 17, respectively.

The IPCC's *Special Report on Regional Impacts of Climate Change* (IPCC, 1998) provides a review of more than 450 research publications concerned with impacts in North America, generally covering work done during 1975–1997. That review focused on six key sectors:

- Hydrology and water resources
- Nonforest terrestrial and forested ecosystems
- Food and fiber, including agriculture, production forestry, and fisheries
- Coastal systems
- Human settlements and industry
- Human health.

Our update of key regional concerns (see Section 15.2) uses similar categories, with a couple of modifications. First, we consider coastal and marine ecosystems together with terrestrial ecosystems, within the review of natural resources (Section 15.2.2). Second, human settlements are reviewed as a separate category, with extended discussion of infrastructure (Section 15.2.5). Impacts on tourism and insurance are treated separately (Sections 15.2.6 and 15.2.7, respectively). We also include a series of subregional cases that highlight adaptation challenges and opportunities (see Section 15.3).

This categorization of key regional concerns in North America reflects growing interest in higher order impacts. Recent increases in financial losses from extreme weather events (hurricanes, winter storms, floods) have raised new questions about changing frequencies of atmospheric events superimposed on changing development and investment patterns. In the United States, there were 28 such events during the 1980–1997 period with losses exceeding US$1 billion. The most costly were the droughts and heat waves of 1988 and 1990, Hurricane Andrew in 1992, and the 1993 Mississippi River flood, with combined costs exceeding US$100 billion. In Canada, the 1996 Saguenay flood and 1998 ice storm in Ontario and Quebec each cost more than US$0.7 billion (CDN$ 1 billion) (Etkin, 1998). It is important to understand the extent to which these loss trends hinge on changes in climate or changes in vulnerability that may be independent of any changes in climate.

In recent years, the North American nations have undertaken intensive region-specific assessments of impacts and vulnerability (Canada Country Study, U.S. National Assessment, regional case studies). Although some of these assessments are still in progress at this writing, several themes have emerged. The involvement of stakeholders (commercial entities, local and regional governments, natural resource managers, and citizens) has brought new perspectives and refocused the definition of research questions. In nearly all regions, stakeholders perceive changes in variability to be far more threatening than decadal-scale gradual change. Many stakeholders see limited problems in adaptation and possible benefits to slow predictable change. However, there may be appreciable changes in ecosystem distributions and water resources, with significant economic impacts, leading to requirements for substantial changes in infrastructure (e.g., in coastal and permafrost regions, major urban centers). There clearly will be winners and losers.

Overall, available information suggests that if projected levels of climate changes occur, North America will become quite a different place.

15.1.2. What is Different about the North American Region?

North America includes economies and resources that might be comparable to those in developed countries and the more developed of the developing countries. Areas within highly urbanized and industrial zones or intensively managed agriculture, forests, and nonrenewable resource extraction all represent large-scale, highly managed resources and human-dominated ecosystems. Within this context, however, there continues to be a great deal of "extensive" land management. Many Canadian and U.S. forests are managed in relatively large tracts compared to other regions and are rather lightly managed aside from harvest schedules. Some are more intensively managed, including post-harvest site preparation for reforestation.

Table 15-1 shows a range of development indicators for Canada and the United States and compares these indicators to Mexico and the rest of the world. Higher levels of resource consumption, urbanization, life expectancy, and gross domestic product (GDP) are obvious, as are the higher rates of CO_2 emissions. At the same time, the two countries also are unusual in preserving extensive rangelands and other landscapes, so a unique feature is the juxtaposition of intensive management with "extensive" management of huge and sparsely populated areas. Some of these areas remain very much a part of the national economy, but other areas are used for subsistence. Areas that are undeveloped or are under aboriginal land claims that support traditional subsistence lifestyles include features that are found in developing country rural economies. These include low levels of built infrastructure, informal barter systems, close ties to the land, and strong historical and cultural attachments to subsistence-based communities, even in the face of opportunities in the modern wage economy.

In wage and nonwage circumstances, climate change scenarios become scenarios of changing opportunities and risks. Given the multi-objective and multi-stakeholder aspects of North American regions, a climate-related change in potential (e.g., longer growing season, higher fire risk, altered hydrological cycle) may not lead to an immediate, linear response by stakeholders (see Section 15.3). An important factor that must be considered

***Table 15-1**: Human development in North America.*[a]

Attribute	Canada	USA	Mexico	Rest of World
1995 land area (ha per capita)	31.4	3.4	2.2	2.1
1995 protected areas (ha per capita)	3	0.6	0.1	0.2
1998 internal renewable water (m^3 per capita yr^{-1})	94,373	8,983	4,508	6,332
1987/95 annual freshwater water withdrawal (m^3 per capita)	(Note b)	(Note b)	—	601
1995 electric consumption (kWhr per capita)	17,047	12,660	1,813	1,689
1994 commercial energy use (kg oil-equivalent per capita)	7,854	7,819	—	1,079
1995 CO_2 emission (t per capita)	14.8	20.5	3.5	3.2
1995 population over 65 (% of total)	12.0	12.6	4.5	6.2
1995 life expectancy at birth (yr)	79.1	76.4	73.0	62.9
1995 urban population (% of total)	77	76	61	43
1995 population in cities of 750,000 or more (% of total)	41	42	—	18
1990 labor force in agriculture (% of total)	3	3	22	52
1990 labor force in industry (% of total)	25	26	35	20
1990 labor force in services (% of total)	71	71	42	29
1995 GDP (US$ PPP per capita)	21,916	26,977	3,600	4,851

[a] Data from UNDP (1999), Organisation for Economic Cooperation and Development (<http://www.oecd.org/env/indicators/index.htm>); and Instituto Nacional de Estadistica Geografia e Informatica, Mexico (<http://www.inegi.gob.mx/poblacion/ingles/fipoblacion.html>).
[b] The United Nations has data for Canada and the United States, but they were not included in the report (UNDP, 1999). The report shows use of 1,069 m^3 per capita for industrial countries.
— = data not available.

is the extent to which individual rights are valued relative to community interests in the management of landscapes. Changing vulnerabilities may occur as a result of coupling of management to current climatic variability (e.g., watersheds developed for hydroelectric production and flood control must continue to adjust to seasonal and yearly variations in natural streamflow, as well as changing management objectives).

15.1.2.1. High Level of Intensive Water Management

The water resources of North America have been heavily modified and intensively managed to serve a variety of human purposes. Investments in water control and delivery infrastructure range from small, privately constructed impoundments, diversion works, and levees to major multi-purpose projects constructed by federal, state, or provincial governments. The usable human-made reservoir capacity in North America is equal to approximately 22% of average annual runoff, compared to a worldwide average of 10% (Dynesius and Nilsson, 1994). Development has been most intensive in the United States, where there are approximately 75,000 structures classified as dams, with a combined storage capacity approximating 70% of mean annual runoff, or about 1,300 km^3 (1 billion acre-feet) (Graf, 1999). Current water-management infrastructure has allowed the citizens of North America to make productive use of water and to reduce the adverse impacts of extreme high and low flows—but often at the cost of radically altering the natural functioning of aquatic ecosystems. The design of reservoirs and control structures and current operational protocols are based on the past hydrological record. Stationarity in the statistical characteristics of streamflow typically is assumed to apply for the future operational life of the facility. A change in mean flow or in variability could cause the physical infrastructure to be inadequate for the intended purposes or increase the risk of failure of the water resource system under extremes of drought or flood. In large water systems, such risks are buffered by robustness and resilience in the design of the system (Matalas, 1998); smaller systems may be more vulnerable under climate scenarios beyond those considered in their design.

In addition, North Americans have created laws and institutions that govern allocation of water among competing uses and define the rights and obligations of individuals, government entities, and other organizations with respect to particular water resources. These institutional aspects vary by region and have changed over time, reflecting differences in climatic and historical circumstances and changing societal values. In the western United States and western Canadian provinces, the economic importance of out-of-stream water uses drove the historical development of the prior appropriation system of water law, whereas U.S. states and Canadian provinces east of the 100th meridian generally adhere to riparian or riparian-based permit systems of water law (Chandler, 1913; Hutchins, 1971; Bates *et al.*, 1993; Scott and Coustalin, 1995). Recently, concerns about endangered species, water quality, and other public trust values have led to changes in permitted uses of some water rights in the United States, and environmental legislation is now a powerful force in determining the location, design, and feasibility of new water projects (California Supreme Court, 1983; Wilkinson, 1989; Butler, 1990; Miller *et al.*, 1996). A new development in the United States is that some dams are being removed or considered for removal to alleviate impairment of the aquatic ecosystem and to restore fisheries (WWPRAC, 1998).

Because North America's water management institutions and infrastructure evolved partly as adaptations to current climate variability, we expect these investments to be useful in fostering adaptability to some of the effects of long-term climate change. However, to the extent that water management facilities have impaired the health and diversity of aquatic ecosystems, they may exacerbate the adverse environmental consequences of global climate change by reducing the resilience of natural systems to further climatic stress.

15.1.2.2. Urbanization

Within developed regions of North America, one important feature is the high-value concentrated development center or corridor. Proliferation of these centers and corridors (e.g., U.S. Gulf coast, Boston-New York-Washington corridor) can lead to high damage costs from extreme events associated with weather phenomena (see Sections 15.2.5 and 15.2.7).

A second feature concerns extension of urban land use into previously undeveloped or less-developed areas, including agricultural lands, forested areas, wetlands, barrier islands, and other coastal margins. In the case of urban encroachment on forested areas, climate change effects on human use and value of forest ecosystems are likely to be significant but are very poorly understood (Binkley and Van Kooten, 1994). This is a challenge in considering traditional human uses of forests for wood products, recreational space, or environmental protection. A more recent phenomenon that must be considered, however, is the increase in human populations in many forested areas in the mid-latitude regions of North America. For example, in the Colorado Rocky Mountains in the United States, human population is expected to double in the next 20–40 years (Stohlgren, 1999). These population increases will amplify climate-induced stresses on forest habitat and species assemblages. Moreover, these increases will increase the exposure of human populations to climate-induced catastrophic events (e.g., fire, floods).

15.1.2.3. Continental Free Trade

The North American Free Trade Agreement (NAFTA) between Canada, the United States, and Mexico has contributed to several changes in economic activity. One indicator is the recent increase in the population of Mexican states along the U.S.-Mexican border (see Table 15-2), which is at a faster rate than for Mexico as a whole. Such shifts may alter the relationship

***Table 15-2**: Population of Mexican states along the U.S.–Mexican border (data from Instituto Nacional de Estadistica Geografia e Informatica, Mexico, <http://www.inegi.gob.mx/poblacion/ingles/fipoblacion.html>).*

Population	1990	1992	1997	% change
Baja California	1,660,855	1,908,434	2,241,029	+35
Sonora	1,823,606	1,866,757	2,183,108	+20
Chihuahua	2,441,873	2,503,515	2,895,672	+19
Coahuila	1,972,340	2,040,046	2,227,305	+13
Nuevo Leon	3,098,736	3,336,044	3,684,175	+19
Tamaulipas	2,249,581	2,351,663	2,628,839	+17
Total population in Mexico	81,249,645	85,627,971	93,716,332	+15

between regions and their climate, affecting the nature of future climate-related impacts and adaptation responses (e.g., see Section 15.3.2.9).

15.1.3. Past to Present

Society has long viewed climate as "constant"—that is, we expect year-to-year variations, but we expect that the chances of hurricanes, floods, or heat waves don't change with time. The probability of a 100-year event (defined by crossing some threshold—say, 5 cm of rain in an hour) should be a constant. As we have developed longer records of climate—from geologic records in ice cores and sediments to longer and longer instrumental records—we have learned that this assumption of constancy simply is not true. As we look back and see constant change, we should look to the future and expect additional changes to climate and to climate variability.

There have been significant changes (i.e., trends) in North America's climate over the past century. As a whole, North America has warmed by about 0.7°C (Carter *et al.*, 2000), although this warming has been quite heterogeneous. For example, the southeastern United States cooled slightly over that same period, although recent decades have seen some warming. There also have been trends in precipitation. Annual precipitation over North America decreased by about 50 mm in the early years of the century and since then has steadily increased by ~70 mm. These trends, like those of temperature, have been fairly heterogeneous. The largest increases have been in the northern Atlantic and Pacific coastal regions. Some regions have experienced seasonal decreases in precipitation. Accompanying these trends in temperature and precipitation, which are fairly well measured, have been changes in cloud cover and humidity, which affect plant growth as well as people's perception of temperature. Inferences from atmospheric CO_2 and satellite data suggest that the northern growing season (the period of active plant growth) has lengthened appreciably over the past decades (Myneni *et al.*, 1997), and data from Alaska show an unequivocal warming trend in the western Arctic (Bering Sea Impacts Study, 1999).

Over the course of the 20th century, there have been several extreme periods in the climate. The drought of the 1930s changed all of North American society. The damage to midcontinental agriculture and communities has consequences even today, through changes in land ownership, government policies, and farmer and merchant perceptions of risk. More recently, there have been major consequences of drought, flooding, tropical storms, and tornadoes. There is no clear evidence of trends in the meteorological risks from "heaviest rainfall category" events or that their frequency has changed over the period of instrumental measurements. However, evidence from the paleorecord documents changes over the past thousands of years that lie outside what our contemporary society can remember. For example, there have been several droughts in California, documented in tree-ring and lake-level data, of more than 80 years' duration (Meko *et al.*, 1991). Evidence from the Rocky Mountain and Great Plains regions suggests that there have been long periods in the Holocene (the time since the last glaciation) when these regions were much drier than today (Woodhouse and Overpeck, 1998). These historical events did not happen as a result of anthropogenic impacts but are within the range of natural climate variability; they provide a caution against assuming that tomorrow's climates will necessarily be no less benign than today's.

15.1.4. Scenarios for the Future

At the large scale, and for a range of emission scenarios produced for the IPCC's Third Assessment Report (see Section 3.8), climate model results suggest that North America could warm at a rate of 1–3°C over the next century for a low-emissions case (B1). Warming could be as rapid as 3.5–7.5°C for the higher emission A2 case. Even the B1 case suggests substantially more warming over the next century than we have observed over the past 100 years. In addition, we know that the past century saw changes in temperature, precipitation, and other variables. Climate model predictions of precipitation remain highly uncertain. Many models suggest higher rainfall over North America accompanying warming in simulations of the IS92a emission scenario (Schimel *et al.*, 1996). While some models suggest widespread and substantial increases in rainfall over most of North America, other models suggest a weaker increase in rainfall.

Understanding concurrent changes in regional temperature and precipitation is crucial. Warming with increasing precipitation is likely to increase plant growth, which may increase carbon storage (VEMAP Members, 1995) and may increase pest and pathogen invasion and expansion. Warming with a lesser increase or a decrease in precipitation could cause direct vegetation mortality and increase the risk of wildfire. Variability also plays a role. General increases in temperature and precipitation might increase plant growth, but occasional severe droughts would then maximize the chances of wildfire. Preliminary results from modeling of the United States suggest that projected changes in climate will cause very substantial changes to the distribution and productivity of ecosystems and to disturbance regimes (fire and drought probabilities). Subtropical conditions will extend further north into the United States, with accompanying changes to vegetation, hydrology, and the potential for disease. Changes at the Arctic border suggest changes to the forest-tundra transition region, losses of permafrost, and an altered growing season.

Concern about the spatial uncertainty of model-based climate scenarios has led to various attempts at downscaling global scenarios to regional scales. This can influence the results of impact studies, as illustrated by a recent case study of crop yields in the U.S. Great Plains (Mearns *et al.*, 1999). Furthermore, published regional impact studies have used older global climate scenarios similar to the temperatures resulting from the new emissions cases, but these may not be directly comparable.

15.2. Key Regional Characteristics

15.2.1. Water Resources

Available evidence suggests that global warming may lead to substantial changes in mean annual streamflows, seasonal distributions of flows, and the probabilities of extreme high- or low-flow conditions (Leavesley, 1994; Cubasch *et al.*, 1995; Mearns *et al.*, 1995; Trenberth and Shea, 1997). Runoff characteristics may change appreciably over the next several decades, but in the near term, the hydrological effects of global warming are likely to be masked by ongoing year-to-year climatic variability (Rogers, 1994; Miller, 1997; Matalas, 1998). There is some evidence that the intensity of rainfall events may increase under global warming, as a result of increases in the precipitable water content of the atmosphere (IPCC, 1996; Trenberth and Shea, 1997). This may increase flooding risks in some watersheds. Hydrological changes cannot yet be forecast reliably at the watershed scale, although numerous studies have addressed the potential effects of warming scenarios on water availability in North America (e.g., Mortsch and Quinn, 1996; Melack *et al.*, 1997; Moore *et al.*, 1997; Mulholland *et al.*, 1997; Woodhouse and Overpeck, 1998; Wilby *et al.*, 1999; Wolock and McCabe, 2000). Figure 15-1 summarizes some possible regional hydrological and ecological impacts of climate change identified in recent analyses.

In general, there is greater confidence in projections of seasonal shifts in runoff and related hydrological characteristics than there is in projections of changes in annual runoff. Regional patterns of precipitation change are highly uncertain. Runoff changes also will depend on changes in temperatures and other climatic variables. Warmer temperatures may cause runoff to decline even where precipitation increases (e.g., Nash and Gleick, 1993). Changes in vegetation characteristics will have further, complex impacts on streamflows (Callaway and Currie, 1985; Rosenberg *et al.*, 1990; Riley *et al.*, 1996). Wolock and McCabe (2000) computed annual runoff projections for the 18 major water-resource regions of the continental United States for two GCM scenarios used in the U.S. National Assessment. They found very little agreement between the models—the Canadian Centre for Climate Prediction and Analysis (CCC) model and the Hadley Centre for Climate Prediction and Research (HAD) model—regarding the direction of change in average annual runoff. In addition, most of the projected changes for the next century fell within the range of current variability.

Projections of shorter snow accumulation periods appear to be more robust. Many studies of snowmelt-dominated systems show similar seasonal shifts to greater winter runoff and reduced summer flow (e.g., Cooley, 1990; Lettenmaier and Gan, 1990; Rango and Van Katwijk, 1990; Duell, 1992, 1994; Lettenmaier *et al.*, 1992, 1996; Rango, 1995; Melack *et al.*, 1997; Fyfe and Flato, 1999; Wilby *et al.*, 1999). In mountainous areas of western North America, small high-elevation catchments may contribute the bulk of the flow of major river systems (Schaake, 1990; Redmond, 1998). Although some models predict decreases in snowpack, records from at least one long-term alpine site in the Rocky Mountains show an increase in annual precipitation since 1951 (Williams *et al.*, 1996). Earlier melt-off in combination with either lower or higher snowpack will tend to increase winter or spring flows and reduce summer flows. Warmer temperatures could increase the number of rain-on-snow events in some river basins, increasing the risk of winter and spring floods (Lettenmaier and Gan, 1990; Hughes *et al.*, 1993; Loukas and Quick, 1996). Lower summer flows, warmer summer water temperatures, and increased winter flows are results on which many of the regional ecological impacts identified in Figure 15-1 are based.

Studies based on climate change scenarios from older versions of GCMs that did not include aerosol effects suggest reductions in streamflow and lake levels in many Canadian watersheds, despite scenario increases in annual precipitation (Hofmann *et al.*, 1998). Bruce *et al.* (2000) examined a variety of newer evidence, including temperature and precipitation changes projected by transient runs of seven different atmosphere-ocean GCMs (AOGCMs) with business-as-usual greenhouse gas (GHG) and aerosol increases. They conclude that many areas of Canada, including southwestern Canada and the Great Lakes region, could experience "...reduced total flow, lower minimum flows and lower average annual peak flow" (Bruce *et al.*, 2000).

Open-water evaporation is an important part of the water balance of the North American Great Lakes. Increased evaporation as a result of warmer water temperatures therefore would likely affect future lake levels and outflow into the St. Lawrence River (Mortsch and Quinn, 1996; Mortsch *et al.*, 2000). Most analyses for the Great Lakes suggest declines in lake levels and outflows (Mortsch and Quinn, 1996; Chao, 1999). Chao (1999), for example, examined 10 different transient GCM scenarios (without aerosols) for IPCC decades 2 and 3 and concludes, "In general the decrease in inflows under all the GCM scenarios result in negative impacts to hydropower, navigation and coldwater habitat, and positive ones to shoreline damages." Mortsch *et al.* (2000) compared such early results to results based on transient runs of the CCC model and the HadCM2 model, both of which include aerosol impacts. Whereas the CCC model run suggests declines in lake levels and outflows comparable to the earlier doubled-CO_2 runs (e.g., a 1.01-m decline in the level of Lakes Michigan and Huron by 2050), the HadCM2 model indicates the possibility of a small rise in lake levels and outflows (0.03-m rise in Lakes Michigan and Huron by 2050). Caution is required in interpreting these results because there is substantial uncertainty regarding future sulfate emissions, and projections of aerosol concentrations have declined considerably since these runs were performed (Carter *et al.*, 2000).

Arid environments are characterized by highly nonlinear relationships between precipitation and runoff. Thus, streamflows in the arid and semi-arid western portions of North America will be particularly sensitive to any changes in temperature and precipitation (Schaake, 1990; Arnell *et al.*, 1996; Kaczmarek *et*

I. Alaska, Yukon, and Coastal British Columbia
Lightly settled/water-abundant region; potential ecological, hydropower, and flood impacts:
- Increased spring flood risks (1,2)
- Glacial retreat/disappearance in south, advance in north; impacts on flows, stream ecology (2,3,4)
- Increased stress on salmon, other fish species (2,5)
- Flooding of coastal wetlands (5)
- Changes in estuary salinity/ecology (2)

II. Pacific Coast States (USA)
Large and rapidly growing population; water abundance decreases north to south; intensive irrigated agriculture; massive water-control infrastructure; heavy reliance on hydropower; endangered species issues; increasing competition for water:
- More winter rainfall/less snowfall—earlier seasonal peak in runoff, increased fall/winter flooding, decreased summer water supply (6,7,8,9,10)
- Possible increases in annual runoff in Sierra Nevada and Cascades (9,11,12)
- Possible summer salinity increase in San Francisco Bay and Sacramento/San Joaquin Delta (6,13)
- Changes in lake and stream ecology—warmwater species benefitting; damage to coldwater species (e.g., trout and salmon) (6,8,14)

III. Rocky Mountains (USA and Canada)
Lightly populated in north, rapid population growth in south; irrigated agriculture, recreation, urban expansion increasingly competing for water; headwaters area for other regions:
- Rise in snowline in winter-spring, possible increases in snowfall, earlier snowmelt, more frequent rain-on-snow—changes in seasonal streamflow, possible reductions in summer streamflow, reduced summer soil moisture (4,15,16,17,18)
- Stream temperature changes affecting species composition; increased isolation of coldwater stream fish (19)

IV. Southwest
Rapid population growth, dependence on limited groundwater and surface water supplies, water quality concerns in border region, endangered species concerns, vulnerability to flash flooding:
- Possible changes in snowpacks and runoff (20)
- Possible declines in groundwater recharge—reduced water supplies (21)
- Increased water temperatures—further stress on aquatic species (22)
- Increased frequency of intense precipitation events—increased risk of flash floods (23)

V. Subarctic and Arctic
Sparse population (many dependent on natural systems); winter ice cover important feature of hydrological cycle:
- Thinner ice cover, 1- to 3-month increase in ice-free season, increased extent of open water (24,25)
- Increased lake-level variability, possible complete drying of some delta lakes (24,25)
- Changes in aquatic ecology and species distribution as a result of warmer temperatures and longer growing season (25,26,27)

VI. Midwest USA and Canadian Prairies
Agricultural heartland—mostly rainfed with some areas relying heavily on irrigation:
- Annual streamflow decreasing/increasing; possible large declines in summer streamflow (11,17,28,29,30,31,32)
- Increasing likelihood of severe droughts (4,33)
- Possible increasing aridity in semi-arid zones (34)
- Increases or decreases in irrigation demand and water availability—uncertain impacts on farm-sector income, groundwater levels, streamflows, and water quality (29,35,36)

VII. Great Lakes
Heavily populated and industrialized region; variations in lake levels/flows now affect hydropower, shipping, shoreline structures:
- Possible precipitation increases, coupled with reduced runoff and lake-level declines (37,38)
- Reduced hydropower production; reduced channel depths for shipping (4,38)
- Decreases in lake ice extent—some years w/out ice cover (39)
- Changes in phytoplankton/zooplankton biomass, northward migration of fish species, possible extirpations of coldwater species (39)

VIII. Northeast USA and Eastern Canada
Large, mostly urban population—generally adequate water supplies, large number of small dams but limited total reservoir capacity; heavily populated floodplains:
- Decreased snow-cover amount and duration (40)
- Possible large reductions in streamflow (40)
- Accelerated coastal erosion, saline intrusion into coastal aquifers (4,41,42)
- Changes in magnitude, timing of ice freeze-up/breakup, with impacts on spring flooding (41,43)
- Possible elimination of bog ecosystems (40)
- Shifts in fish species distributions, migration patterns (41)

IX. Southeast, Gulf, and Mid-Atlantic USA
Increasing population—especially in coastal areas, water quality/nonpoint-source pollution problems, stress on aquatic ecosystems:
- Heavily populated coastal floodplains at risk to flooding from extreme precipitation events, hurricanes (22,32)
- Possible lower base flows, larger peak flows, longer droughts (44)
- Possible precipitation increase—possible increases or decreases in runoff/river discharge, increased flow variability (32,44,45,46)
- Major expansion of northern Gulf of Mexico hypoxic zone possible—other impacts on coastal systems related to changes in precipitation/nonpoint-source pollutant loading (45,47)
- Changes in estuary systems and wetland extent, biotic processes, species distribution (44,48)

Figure 15-1: Potential water resources impacts.

al., 1996). Rivers that originate in mountainous regions will be particularly sensitive to winter precipitation in the headwaters, regardless of conditions in the downstream semi-arid zone (e.g., Cohen, 1991). In addition, "...severe flood events may be more damaging in drier climates where soils are more erodible..." (Arnell *et al*., 1996).

Little research attention has been given to the possible impacts of climate change on sediment transport and deposition, which could affect aquatic ecosystems, reservoir storage capacity, potential flood damages, and the need for dredging operations. However, projected increases in the intensity of precipitation events could contribute to increased erosion and sedimentation in some areas (Mount, 1995).

15.2.1.1. Impacts and Adaptation Options

The impacts of hydrological changes and options for effective adaptation will depend on the nature of the hydrological changes, the amount of buffering provided by natural and artificial storage capacity, the nature of demands on the resource, and the effectiveness with which institutions in place balance competing demands. Continued increases in demand are likely for the multiple services provided by North American water resources, encompassing out-of-stream and instream uses. Adaptations to the effects of climate change therefore will occur in conjunction with adjustments to changes in the level and characteristics of water demands.

A major comparative study of six U.S. water resource systems found that although climate change could have significant and often adverse impacts on system performance, adjustments to reservoir operations could be made, allowing those impacts to be smaller than the underlying hydrological changes (Lettenmaier *et al*., 1999). The authors also found that "The effects of anticipated demand growth and other plausible future operational considerations...would about equal or exceed the effects of climate change over system planning horizons" (Lettenmaier, *et al*., 1999).

The importance of future changes in demand is echoed by Frederick and Schwarz (1999). They found that even in the absence of climate change, the annual cost of supplying water to meet projected increases in U.S. water demands would be about US$13.8 billion higher in 2030 than in 1995, based on the current trend of increasingly using conservation to balance growing demands with supplies. This trend has been driven by "...the high costs of developing new supplies, environmental concerns, a growing appreciation for the values of instream flows and efforts to improve water quality" (Frederick and Schwarz, 1999). Using projected runoff changes calculated by Wolock and McCabe (2000) for 18 water resource regions, the study found that runoff increases projected by the HadCM2 model tended to reduce the cost of balancing future water supplies and demands, whereas the reduced flows projected by the CCC model resulted in large cost increases. With the CCC model projections, the study found that under an "efficient management" scenario, annual water costs could increase "...by nearly US$105 billion, about US$308 per person. Much higher cost increases would result from policies to maintain relatively high streamflow levels under such dry conditions" (Frederick and Schwarz, 1999).

This finding—that institutions can have significant impacts on the costs arising from reduced water availability—is supported by a study that assessed the possible consequences of a severe sustained drought on the Colorado River system (Lord *et al*., 1995). The study authors found that the "Law of the River," as currently interpreted and implemented, would leave sensitive biological resources, hydropower generation, recreational values, and Upper Colorado basin water users vulnerable to damages despite extraordinary engineering attempts to drought-proof the river. The study also found that certain proposed institutional and system operating changes could considerably alter the level and incidence of the damages (Booker, 1995; Kenney, 1995; Lord *et al*., 1995; Sangoyomi and Harding, 1995). For example, reallocation from low- to high-valued uses (e.g., through intrastate or interstate water marketing) and changes in reservoir storage policies to hold water in the Upper Colorado basin to reduce evaporative losses could reduce consumptive use damages by more than 90% (Booker, 1995).

Such "what if" analyses can foster creative thinking about adaptation options. Future adaptation strategies also are likely to build on current creative solutions to meeting water supply challenges. For example, when increases in agricultural and sewerage runoff in the Catskill/Delaware watershed region threatened the quality of New York City's water supplies, the city was faced with a choice of either building an artificial filtration plant or taking action to protect and restore the natural purification capacity of the watershed's ecosystem. The capital cost of the filtration plant would have been US$6–8 billion, plus annual operating costs of US$300 million. The city found that for a fraction of that cost it could enter into agreements with landowners in the watershed to adopt land-use practices

Figure 15-1 Notes

1. Loukas and Quick, 1999; **2.** Taylor and Taylor, 1997; **3.** Brugman *et al*., 1997; **4.** Hofmann *et al*., 1998; **5.** BESIS, 1999; **6.** Melack *et al*., 1997; **7.** Hamlet and Lettenmaier, 1999; **8.** Cohen *et al*., 2000; **9.** Wilby and Dettinger, 2000; **10.** Leung and Wigmosta, 1999; **11.** Wolock and McCabe, 2000; **12.** Felzer and Heard, 1999; **13.** Gleick and Chalecki, 1999; **14.** Thompson *et al*., 1998; **15.** Fyfe and Flato, 1999; **16.** McCabe and Wolock, 1999; **17.** Leith and Whitfield, 1998; **18.** Williams *et al*., 1996; **19.** Hauer *et al*., 1997; **20.** Wilby *et al*., 1999; **21.** USEPA, 1998b; **22.** Hurd *et al*., 1999; **23.** USEPA, 1998; **24.** Marsh and Lesack, 1996; **25.** Maxwell, 1997; **26.** Rouse *et al*., 1997; **27.** MacDonald *et al*., 1996; **28.** Herrington *et al*., 1997; **29.** Strzepek *et al*., 1999; **30.** Clair *et al*., 1998; **31.** Yulianti and Burn, 1998; **32.** Lettenmaier *et al*., 1999; **33.** Woodhouse and Overpeck, 1998; **34.** Evans and Prepas, 1996; **35.** Eheart *et al*., 1999; **36.** Hurd *et al*., 1998; **37.** Mortsch and Quinn, 1996; **38.** Chao, 1999; **39.** Magnuson *et al*., 1997; **40.** Moore *et al*., 1997; **41.** Abraham *et al*., 1997; **42.** Frederick and Gleick, 1999; **43.** Hare *et al*., 1997; **44.** Mulholland *et al*., 1997; **45.** Justic *et al*., 1996; **46.** Arnell, 1999; **47.** Cruise *et al*., 1999; **48.** Porter *et al*., 1996.

that would adequately protect water quality, making the filtration plant unnecessary (PCAST, 1998; Platt *et al.*, 2000).

Adjustments to the effects of climate change as well as to evolving demands for out-of-stream and instream water uses may entail impacts on the distribution of water-use benefits. The question of who will bear the costs of any reduction in water availability depends on the nature and ownership of existing water rights, how those rights are measured, and how they may be modified by other policies. In the United States, federal legislation—including the Endangered Species Act and the Clean Water Act—has regulated or constrained development of many new water projects and could be applied to modify existing water diversions (WWPRAC, 1998). Native American (aboriginal) communities often possess reserved water rights that, despite their high priority, have never been developed or clearly quantified. Substantial litigation and negotiation efforts have focused on clarifying these rights. Any significant change in water availability may heighten tensions between these communities and neighboring water users (Echohawk and Chambers, 1991). Under riparian water law and permit systems, governmental authorities may have substantial discretion in regulating water uses under drought conditions (Abrams, 1990; Flood, 1990; Sherk, 1990; Dellapenna, 1991; Scott and Coustalin, 1995). Under the prior appropriation system of water law, which is followed in western Canada and most states in the western portion of the United States, the risk of a shortfall generally is inversely proportional to the seniority of the right. Many junior rightholders, however, have access to water stored in reservoirs, which improves their security of supply.

Several studies have argued that improving the functioning of water markets could help to create the kind of flexibility needed to respond to uncertain changes in future water availability (e.g., Trelease, 1977; Tarlock, 1991; OTA, 1993; Miller *et al.*, 1997). If water supplies decline in particular locations or seasons, water markets could soften the impacts by moving water from lower to higher valued uses. In the western United States, where irrigation now accounts for more than 80% of consumptive water use, water market activity is likely to continue the current trend of movement of water out of irrigated agriculture to accommodate other water uses (WWPRAC, 1998). However, water markets are not likely to adequately protect instream flows and sensitive biological resources unless public agencies are given budgets to buy or rent water to protect those values (Wilkinson, 1989; NRC, 1992). In addition, water rights have to be clearly defined for water markets to function properly. Even in the western United States, where water markets have developed rapidly in recent years (Saliba and Bush, 1987; NRC, 1992; Colby, 1996; Yoskowitz, 1999), market development has been hampered by the fact that water rights often are not well documented (e.g., Costello and Kole, 1985; Gould, 1988; Colby, 1998). To avoid adverse impacts on other water users, water authorities typically review each transfer proposal. The entire process often entails substantial transaction costs, and these costs differ significantly across jurisdictions (Saliba and Bush, 1987; MacDonnell, 1990; NRC, 1992). This suggests that administrative practices and other institutional factors affect the efficiency of the water transfer process and that administrative reforms to reduce transaction costs could improve adaptability to the effects of climate change.

Finally, permanent water transfers are not particularly effective in promoting flexible adaptation to climatic variability or uncertain climate changes. Small, temporary transfers through water banks such as California's Emergency Drought Water Banks would provide greater flexibility (see Section 15.3.2.3).

Although well-functioning water markets may ameliorate the socioeconomic impacts of reduced water availability, they cannot completely eliminate the adverse impacts of a drying scenario. Using models that assume optimal allocation of water across all sectors (which implicitly assumes perfect, cost-free water markets), Hurd *et al.* (1998) estimated the water resources-related costs of climate change scenarios for four major river basins in the United States. They found that losses from moderate and severe drying scenarios tended to be greater in the semi-arid western United States than in the eastern states. From this analysis, they project that national water-related welfare losses arising from a temperature increase of 2.5°C coupled with a 7% increase in precipitation would be on the order of US$9.5 billion. A scenario with a temperature increase of 5°C coupled with no change in precipitation would result in losses of US$43 billion. Overall, they found that the economic costs probably would be dominated by impacts on water quality and instream nonconsumptive water uses, especially hydroelectricity generation. This result hinges on their assumption that irrigators—who currently dominate ownership of water in the western United States—would benefit by selling some of their water to other sectors. More realistic assumptions regarding the costs of transferring water through water markets would have produced larger estimated losses and a different distribution of losses across sectors.

Water rights are not always defined in terms that would permit the transfer of a specific amount of water from one use to another, particularly in eastern U.S. and Canadian jurisdictions that still follow a traditional riparian system of water law (Tarlock, 1989; Flood, 1990; Scott and Coustalin, 1995). In such areas, accommodation of competing demands and changing supplies will require local and regional planning and coordination. For example, if the levels and outflows of the Great Lakes–St. Lawrence system decline, as most available scenarios suggest, adaptation options are likely to raise sensitive interjurisdictional allocation issues and may require infrastructure investments that will increase the need for cooperative planning and management of the basin's water resources (Bruce *et al.*, 2000).

15.2.1.2. Water Quality

Water quality changes may be driven by changes in hydrological flowpaths in a watershed that are associated with changes in patterns of precipitation and evapotranspiration and changes in total flow in streams and rivers or in water level or duration of ice cover in temperate lakes. In regions such as the Precambrian shield, where watersheds are predicted to become drier in spring and summer, concentrations of dissolved organic material reaching lakes and streams from their catchments will decrease, increasing water clarity and changing physical and thermal regimes by increasing average thermocline depths in small, stratified lakes, for example (Snucins and Gunn, 1995; Schindler *et al.*, 1996; Perez-Fuentetaja *et al.*, 1999). In contrast, in the Great Lakes and other large lakes where dissolved organic carbon (DOC) concentrations are low, thermocline depths are determined by area or wind fetch and are not affected by DOC (Fee *et al.*, 1996). Models for the Great Lakes indicate that rapid spring warming may cause shallower and steeper thermoclines (reviewed by Magnuson *et al.*, 1997). For lakes and streams receiving flow from deep and shallow groundwater sources, drier watersheds could cause the major ion chemistry to be dominated more by the deep baseflow water sources (Webster *et al.*, 1996).

In several regions where warmer temperatures and longer growing seasons are expected, changes in water quality will be driven by increases in primary production, organic matter decomposition, and nutrient cycling within lake or stream ecosystems (e.g., Mulholland *et al.*, 1997). In the Great Lakes region, warming over the past 60 years already has moved forward by an average of 3 weeks the time of ice-cover breakup (ice-out), which has moved ahead the spring bloom of algal growth and changed the seasonal dynamics of nutrient utilization and production and decomposition of organic matter (Magnuson *et al.*, 1997).

Where streamflows and lake levels decline, water quality deterioration is likely as concentrations of nutrients and contaminants from wastewater treatment, agricultural and urban runoff, and direct industrial discharge increase in reduced volumes of carrying waters The extent of water quality deterioration will depend on adaptations in land use, population, and water use under changing climate. Warmer water temperatures may have further direct impacts on water quality—for example, by reducing dissolved oxygen concentrations. In the southeast, intensification of the summer temperature-dissolved oxygen squeeze (simultaneous high water temperatures and low dissolved oxygen concentrations) in many rivers and reservoirs is likely to cause a loss in habitat for coolwater fish species (Mulholland *et al.*, 1997).

Changes in the characteristics of precipitation events also may affect water quality, in complex ways. For example, increased incidence of heavy precipitation events as predicted for the southeast may result in increased leaching and sediment transport, causing greater sediment and nonpoint-source pollutant loadings to watercourses (Mulholland *et al.*, 1997). Because these high-flow events are episodic, the associated increase in nutrient and contaminant dilution during the event is not likely to offset the deleterious effects on water quality from low summer flows.

Warmer temperatures will increase the salinity of surface waters, especially lakes and reservoirs with high residence times, by increasing evaporative water losses. Higher initial salinities in reservoir water will then exacerbate salinity problems in irrigation return flow and degrade water quality in downstream habitats.

15.2.1.3. Flood Risks

Flooding poses risks to human life and property. Vulnerability to flood damages is highly location-specific, and there is considerable variability across watersheds in the value of developed property and population located within 500-year floodplains as defined by current understanding of current climate (Hurd *et al.*, 1999). Flood events also can have significant impacts on ecosystems. For example, heavy precipitation events may leach nitrogen and other nonpoint-source pollutants from agricultural lands, and the resulting nutrient pulse may severely stress coastal and estuarine ecosystems (Justic *et al.*, 1996; Rabalais *et al.*, 1996). In addition, the influx of freshwater during floods may affect estuary-dependent species. For example, oyster populations suffer severe declines when floods reduce the salinity of Galveston Bay (Hofmann and Powell, 1998).

Coupling of natural disasters, such as extreme storm events, with large-scale human disturbance of the landscape can cause extreme disturbance to freshwater and marine resources that would not be predicted by considering these effects independently. The consequences of Hurricanes Dennis and Floyd in eastern North Carolina in September 1999 provide a recent example of the importance of this coupling (see Section 15.3.2.7). Rainfall of almost 1 m generated highly polluted, organic-rich floodwaters as containment ponds for poultry and hog waste were breached and raw sewage, fertilizers, decaying vegetation, and other organic sources were entrained by the flood. One serious effect was contamination of shallow groundwater sources with fecal coliforms and organic pollutants, which may jeopardize local water supplies long after the floodwaters subsided. Of even greater economic impact for this region, the surge of floodwater caused the waters of the biologically rich estuary between the Carolina mainland and the Outer Banks to be the color of weak coffee and deposited large amounts of organic material in coastal sediments, especially in the estuaries and westernmost Pamlico Sound. The Albemarle-Pamlico Estuarine System provides fully half of the area used as nursery grounds for commercially important fish from Maine to Florida. These waters are a vitally important feeding area for small sport fish and menhaden and an important nursery for flounder, weakfish, shrimp, and crabs. At the time, there was considerable concern that the release of nutrients and consumption of oxygen as deposited organic material decomposed would cause physical stresses, disrupting the coastal food web and commercial fisheries for a significant period (Paerl *et al.*, 2000). As it turned out, the mesohaline estuaries west of the Pamlico Sound sustained the greatest damage from pollution that was washed in and deposited to the bottom muds (Burkholder *et al.*, 2000). Pamlico Sound was protected from high impacts because much of the pollution settled out in the estuaries and because of its high flushing exchange with the ocean relative to the estuaries. The high dilution provided by the extreme runoff associated with Hurricane Floyd was a "saving grace" that appeared to buffer the pollution effects, so no fish kills were reported throughout the system (Burkholder *et al.*, 2000). However, concerns remain about chronic, more long-term impacts from the pollution that remained behind in the estuaries.

Possible changes in runoff patterns, coupled with apparent recent trends in societal vulnerability to floods in parts of North America, suggest that flood risks may increase as a result of anthropogenic climate change (see Section 15.2.5). Changes in snowpack accumulation and the timing of melt-off are likely to affect the seasonal distribution and characteristics of flood events in some areas. For example, in mountainous western watersheds, winter and early spring flood events may become more frequent (Melack *et al.*, 1997; Lettenmaier *et al.*, 1999). In southeastern Canadian and northeastern U.S. watersheds, reductions in winter snowpacks and river ice will tend to reduce winter and spring flood risks (Bruce *et al.*, 2000), where at present "rain-on-snow and snowmelt floods can be the largest and most destructive stormflow events in the region" (Platt *et al.*, 2000). However, Canadian rivers in northern areas may begin to experience winter ice break-ups and associated flooding (Bruce *et al.*, 2000).

In inflation-adjusted terms, average annual flood damage has increased in the United States over the past few decades. This increasing trend in damages appears to be related to increases in population and the value of developed property in floodplains, as well as changes in precipitation characteristics, with perhaps as much as 80% of the trend attributable to population and wealth changes (Pielke and Downton, 2000). Measured as a proportion of real tangible wealth, average annual flood damages have been roughly constant over time (Pielke and Downton, 2000). This ongoing vulnerability comes despite the fact that various federal, state, and local governments and private entities have built approximately 40,000 km of levees along the rivers and streams of the United States—a combined total distance that is long enough to encircle the Earth at the equator (Pielke, 1999).

Recent severe flood events—particularly the 1993 Mississippi River floods, the 1996 Saguenay flood, the 1997 Red River flood, and winter flooding in California in 1997—have led to reexaminations of traditional approaches to flood management. For example, a U.S. federal interagency task force was formed in the wake of the 1993 floods, and its recommendations have contributed to altered federal practices (IFMRC, 1994). In an assessment of the 1993 floods, which caused on the order of US$18 billion in damages, Changnon (1996) notes that the extreme and prolonged flooding had significant and unexpected impacts that defied previous experience and design extremes. Changnon further concludes that many systems for monitoring and predicting flood conditions were inadequate; that incomplete or incorrect information was released during the flood; and that many previous approaches to mitigate flood losses failed. He also identifies benefits, including benefits to the natural ecosystem of the Mississippi floodplain.

It has been demonstrated that the efforts of one community to protect itself from floods (for example, through levee construction) may affect the likelihood of flood damages in other communities (Mount, 1995). Therefore, coordinated regional planning and management may allow more efficient adaptation to changing flood risks than uncoordinated efforts by individual communities. However, there are many things that individual communities can do to rationally adapt to flood risks and reduce the likelihood of serious damages (City of Tulsa, 1994). Many entities that are responsible for floodplain management are rethinking the design of levee systems and other flood management policies (City of Tulsa, 1994; IFMRC, 1994; Mount, 1995; Tobin, 1995; Pielke, 1996; Wright, 1996). These developments may improve resilience to future flooding events, but it is not yet clear if recent policy discussions will lead to substantial and effective changes in floodplain management or flood response practices.

15.2.2. Natural Resources

In this section, impacts studies on forests, grasslands, and protected areas are reviewed. Protected areas include mountains, wetlands, and coastal/marine areas.

15.2.2.1. Forests

We must consider two types of climate change effects on forests:

- Changes in the functions of existing forests relating to productivity, nutrient cycling, water quality, ecosystem carbon storage, trace gas fluxes, and biodiversity.
- Changes in composition as forests regenerate under altered conditions. Fundamental changes in forest ecosystem structure can lead to very dramatic changes in functions. Climate change effects on catastrophic events (e.g., fire, insect outbreaks, pathogens, storms) that have marked effects on ecosystem structure are particularly important to consider.

A general discussion of forest response to climate change appears in Chapter 5.

North America contains about 17% of the world's forests (Brooks, 1993), and these forests contain about 14–17% of the world's terrestrial biospheric carbon (Heath *et al.*, 1993). Key climate change issues related to forests in North America include:

- Changes in the geographic range of different forest types
- Increases in the frequency of fire and insect outbreaks
- Changes in the carbon storage function of forests (i.e., from sinks to sources)
- Evaluation of the importance of multiple stresses (ozone, nitrogen deposition, land-use change) that work in concert with climate change

- Changes in human interactions with forests (e.g., risk to settlements, recreational use)
- Concern for the boreal forests of Canada because of their large extent, carbon reserves, and commercial value, combined with the fact that climate change is expected to be most severe at high latitudes.

15.2.2.1.1. Changes in function of existing forests

There is strong evidence that there has been significant warming at high latitudes (Jacoby *et al.*, 1996) and that this warming has increased boreal forest productivity (Ciais *et al.*, 1995; Myneni *et al.*, 1997). However, carbon balance is not necessarily changed by increases in productivity. Net ecosystem carbon flux (or carbon storage) is a product of changes in ecosystem production and decomposition. Keyser *et al.* (2000) used long-term meteorological records to drive the BIOME-BGC model to evaluate changes in the carbon balance of North American high-latitude forests. They conclude that increases in net primary production and decomposition were roughly balanced and that net ecosystem production (i.e., total carbon storage) was not likely to shift significantly with climate change. In contrast, Goulden *et al.* (1998) and Lindroth *et al.* (1998) found that boreal forests could become net CO_2 sources. The key uncertainties in this area are the effects of permafrost melting on release of previously frozen carbon, the ability of more productive ecosystem types (aspen, white spruce) to expand in extent, and the importance of soil moisture. Evaluating changes in carbon balance in northern forests should be a priority topic for research.

There is consensus emerging that at mid-latitudes, site-specific conditions as well as history, human management, air pollution, and biotic effects (e.g., herbivory) are much stronger controllers of forest productivity, decomposition, and carbon balance than climate change or CO_2 enrichment (Eamus and Jarvis, 1989; Aber and Driscoll, 1997; Ollinger *et al.*, 1997; Goodale *et al.*, 1998; Stohlgren *et al.*, 1998).

There is general agreement that excess nitrogen deposition, which is most pronounced in the mid-latitudes, has increased carbon storage in mid-latitude forests by facilitating increases in production in response to elevated CO_2 (Townsend *et al.*, 1996). The ability of forests to continue to absorb excess nitrogen and CO_2 is not at all certain, however (Norby, 1998).

Evidence for climate change effects on forest ecosystem "services" (i.e., functions that are important to productivity, environmental quality, and other human concerns) are beginning to emerge in North America. Murdoch *et al.* (1998) suggest that climate warming increases soil acidification and stream nitrate (NO_3^-) concentrations, especially in forests with a history of high nitrogen deposition. Extreme climate events (e.g., soil freezing, which may increase as a result of warming-induced decreases in snow cover) also appear to lead to increases in soil and stream acidification and NO_3^- levels (Mitchell *et al.*, 1996; Groffman *et al.*, 1999). Evaluations of climate change effects on fluxes of trace gases other than CO_2 [methane (CH_4), nitrous oxide (N_2O)] have been inconclusive (Prather *et al.*, 1995).

Climate change effects on biogeochemical processes are likely to be small relative to site characteristics, land-use history, and atmospheric chemistry, especially in mid-latitudes (Aber and Driscoll, 1997).

15.2.2.1.2. Wholesale changes in forest structure and function

If climate change results in wholesale replacement of one forest community with another, effects on ecosystem functions ranging from carbon storage to wildlife habitat could be dramatic (Tilman, 1998). Possible wholesale changes in forest structure have been a great source of uncertainty and concern. The North America assessment in the IPCC's *Special Report on Regional Impacts of Climate Change* concluded that there were equal probabilities of considerable forest dieback and enhanced forest growth, given state-of-the-art models and GCM predictions (Shriner and Street, 1998).

Many types of models (biogeographic individual-based forest growth, gap, dynamic global vegetation, regression tree analysis, response surface, richness, and rare and endangered species) have been used with numerous climate scenarios to examine broad-scale climate change-induced changes in vegetation (VEMAP Members, 1995; Shugart and Smith, 1996; Aber *et al.*, 2001). There is great uncertainty about the precision and accuracy of each of these models and the assumptions that underlie them (Loehle and LeBlanc, 1996; Repo *et al.*, 1996; Beuker *et al.*, 1998). Many studies suggest that all major forest types in North America will expand northward and most will increase in extent in the next 50–100 years. The increase in forests is predicted to be driven by slight warming coupled with increases in water-use efficiency (WUE) associated with increased atmospheric CO_2 (Saxe *et al.*, 1998). However, with continued warming, increased water use associated with higher temperatures overwhelms the CO_2 effect, resulting in potentially important decreases in forest area (Aber *et al.*, 2001).

Wholesale changes in forest ecosystem structure over time are likely to be mediated by changes in disturbance regimes and/or catastrophic events that provide opportunities for forest regeneration over large areas (Suffling, 1995; Loehle and LeBlanc, 1996). Of particular interest in North America are changes in fire and insect outbreaks.

15.2.2.1.3. Fire

Stocks *et al.* (1998) used outputs from four GCMs to project forest fire danger levels in Canada under a warmer climate. Their analysis shows an earlier start to the fire season and significant increases in the area experiencing high to extreme fire danger. Increased lightning frequency associated with global warming also may increase fire frequency (Goldammer and Price, 1998). Changes in fire frequency have a wide range of effects, from production of tropospheric aerosols that influence climate (Clark *et al.*, 1996) to changes in ecosystem carbon storage and trace gas fluxes. The long-term effects of fire will depend heavily on changes in human fire management activities, which are uncertain—especially in remote boreal forests, where fire is a critical issue (e.g., see Section 15.3.2.8). In mid-latitudes, climate effects on fire frequency are much less important than human management factors (Veblen *et al.*, 2000).

15.2.2.1.4. Insects

Insects represent dominating disturbance factors (Hall and Moody, 1994) in North America's forests; during outbreaks, trees often are killed over vast areas (Hardy *et al.*, 1986; Candau *et al.*, 1998). Because the potential for wildfire often increases in stands after insect attack (Stocks, 1987; Wein, 1990), uncertainties in future insect damage patterns also lead to uncertainties in fire regimes. Insect outbreaks also lead to changes in ecosystem carbon and nutrient cycling, biomass decomposition, energy flow (Mattson and Addy, 1975; Schowalter *et al.*, 1986; Szujecki, 1987; Haukioja *et al.*, 1988; Chapin, 1993; Haack and Byler, 1993), and competitive relationships between plants (Morris, 1963; Holling, 1992)—hence successional pathways, species composition, and forest distribution.

Climate change already appears to be accelerating the seasonal development of some insects (Fleming and Tatchell, 1994). Forecasts based on historical relationships between outbreak patterns and climate in specific areas are likely to predict change as the climate in those areas changes. Such analyses suggest more frequent [mountain pine beetle (Thomson and Shrimpton, 1984), spruce budworm (Mattson and Haack, 1987), eastern hemlock looper (Carroll *et al.*, 1995), jack pine budworm (Volney and McCullough, 1994), western spruce budworm (Thomson *et al.*, 1984)] or longer [forest tent caterpillar (Roland *et al.*, 1998), spruce budworm (Cerezke and Volney, 1995)] outbreaks or range shifts northward and to higher elevations [spruce budworm (Williams and Liebhold, 1997)] as climate change progresses.

15.2.2.1.5. Vegetation In human settlements

Vegetation in human settlements plays two potentially important roles that are relevant to climate change: modification of local climate and sequestration of carbon.

Trees reduce demands for seasonal heating and cooling of the interiors of buildings and absorb air pollutants (Heisler, 1986; Nowak *et al.*, 1994). In the city of Chicago, calculations suggest that increasing tree cover by 10% could reduce building energy use by 5–10% (Nowak *et al.*, 1994). Thus, urban tree planting represents an adaptive strategy that can reduce energy use and associated CO_2 emissions and counteract temperature increases in urban areas, which are predicted to be extreme in some cases.

Carbon density in residential areas can be significant, amenable to management, and often overlooked in evaluation of landscape, regional, and national carbon budgets. Freedman *et al.* (1996) found that aboveground tree biomass of an old residential neighborhood in Halifax, Nova Scotia, was only slightly smaller than that of a natural forest in a nearby reserve. Nowak *et al.* (1994) estimated that carbon storage by urban forests in the United States was 440–990 Mt. There is a strong need for better estimates of "natural" carbon fluxes in human-dominated environments.

15.2.2.2. Protected Areas

15.2.2.2.1. Mountains

In the 20th century, there has been increasing human pressure on mountainous regions, initially through trapping, forestry, and reservoir construction and now through development of ski areas and other resorts and construction of residences, as well as forestry and continuing reservoir construction in northern Canada. At the same time, however, the national park systems in the United States and Canada and wilderness preserves have expanded to include many mountainous areas that are essentially pristine with respect to human development, especially in the Rocky Mountains. It is now recognized that these protected mountain ecosystems are still vulnerable to anthropogenic change through transport of atmospheric contaminants, such as nitrate and sulfate in acid rain, and through climate change. Warming of the climate eventually will cause two major changes—retreat of mountain glaciers and upward movement of treeline—and the response times for reaching new equilibrium conditions are on the order of 100 years or more, so these responses will lag continuing climate change.

Retreat of glaciers is driven by the rate of ablation, which includes melt, exceeding the rate of advance driven by snow accumulation over the glacier, and corresponds to a change in the shape of the glacier to a new equilibrium. Retreat of mountain glaciers already has begun in North America (Brugman *et al.*, 1997) and in other regions of the world, and this retreat will contribute to sea-level rise in an amount comparable in magnitude to expansion of ocean waters as a result of warming. On a regional scale, the retreat of glaciers will affect water resources by changing (probably decreasing) water supply from glacial melt during summer or changing the spatial location of the melt source [summer flows initially may increase but eventually will decline as glacier reservoir capacity declines (Pelto, 1993)]. Furthermore, glacial retreat will expose terrain that gradually will evolve with soil development and revegetation, and new lakes will form in exposed basins. These changes eventually will influence the water quality of drainage from these lakes. These sequences of glacial advance and retreat have occurred through the quaternary; the effect of climate change is to induce these changes. In terms of human vulnerability to climate change, retreat of mountain glaciers also is significant because it is an observable change that can be directly comprehended by the public as an indicator of warming—more so than warming of the open ocean or an increase in extreme hydrological events. For example, a recent article in a travel magazine (*Conde Nast*) outlined vacations to view retreating glaciers in North America, Europe, and Africa while they were still there.

From paleolimnological studies of alpine and subalpine lakes, the rise in treeline in response to past warming of climate is well-documented. The boundary between alpine tundra and subalpine forest is controlled by extremes of temperature, moisture, and wind. Vegetation in both ecosystems is long-lived, and changes will proceed slowly and in a manner that depends on whether total annual snowpack decreases or increases and whether melt occurs earlier; both factors control the growth of alpine and subalpine species. Movement of treeline could have a minor feedback on climate change by sequestering more carbon in subalpine forests. The eventual effect of upward movement of the treeline will be to shrink the extent of alpine tundra in North America, possibly causing species loss and ecosystem degradation through greater fragmentation (see Section 15.2.6. and Chapter 5).

15.2.2.2.2. Wetlands

Wetlands represent a variety of shallow water and upland water environments that are characterized by hydric soils and plant and animal species that are adapted to life in saturated conditions (NRC, 1995). These ecosystems are considered to be of great importance in a variety of functional contexts, including waterfowl habitat, carbon sequestration, CH_4 production, flood regulation, pollutant removal, and fish and shellfish propagation (Mitsch and Gosselink, 1993). About 14% of Canada's surface area is covered by wetlands, which is 24% of the global total (NWWG, 1988). Approximately 6% of the United States is wetland (Kusler *et al*., 1999).

Mid-latitude wetlands have been greatly affected by a variety of human activities over the past 200 years. More than 50% of the original wetlands in the United States have been destroyed for agriculture, impoundment, road building, and other activities (Dahl, 1990). Most of the remaining wetlands have been altered by harvest, grazing, pollution, hydrological changes, and invasion by exotic species (Kusler *et al*., 1999). High-latitude wetlands have experienced much lower levels of human disturbance (Schindler, 1998).

Climate change can have significant impacts on wetland structure and function, primarily through alterations in hydrology, especially water-table level (Clair *et al*., 1998; Clair and Ehrman, 1998). Wetland flora and fauna respond very dynamically to small changes in water-table levels (Poiani *et al*., 1996; Schindler, 1998). Moreover, climate change can exacerbate other stresses (e.g., pollution), especially in fragmented landscapes where wetlands have been cut off from other wetlands by a variety of landscape-level alterations (Mortsch, 1998; Kusler *et al*., 1999). With rising sea levels, shoreline development and efforts to protect private property from coastal erosion could lead to loss of public tidelands and coastal marshes, particularly along bayshores where preservation of natural shorelines has received less policy attention than is the case for most ocean beaches (Titus, 1998).

Specific changes predicted to occur in North American wetlands are wide ranging. Sea-level rise will result in loss of coastal wetlands in many areas, with potentially important effects on ocean fisheries (Michener *et al*., 1997; Turner, 1997). Increased drought conditions in the Prairie Pothole Region of the northern Great Plains, which are forecast to occur under nearly all GCM scenarios, will significantly reduce U.S. breeding duck populations (Sorenson *et al*., 1998). Tourism may benefit from extended seasons but will suffer if key processes (e.g., hunting, birding) are disrupted (Wall, 1998a). Alteration of water-table levels could affect the carbon sequestration function of the vast northern wetlands of Canada, but there is great uncertainty about the nature and extent of this effect (Moore *et al*., 1998; Waddington *et al*., 1998).

Wetlands have been the target of numerous protection and restoration efforts (NRC, 1995), which suggests that there is high potential for adaptive management in response to climate change, at least in mid-latitudes. Kusler *et al*. (1999) recommend a series of strategies for reducing the impacts of climate change on wetlands. These strategies include better control of filling and draining of wetlands, prevention of additional stresses, prevention of additional fragmentation, creation of upland buffers, control of exotic species, protection of low flows and residual water, enhanced efforts to restore and create wetlands, and aggressive efforts in stocking and captive breeding of critical wetland species. Five states in the United States have adopted rolling easement policies, which ensure that wetlands and/or beaches can migrate inland as sea level rises, instead of being squeezed between coastal development and the advancing sea (Titus, 1998). These efforts would be greatly enhanced by creation of regional inventories and management plans for wetlands at greatest risk from climate change.

15.2.2.2.3. Coastal/Marine

With the concentration of population on the coasts, various development constraints and environmental regulations have been enacted to protect areas of coast as wildlife preserves and for harvesting of shellfish. Coastal ecosystems clearly are vulnerable to change associated with eventual sea-level rise (see Section 15.2.2.3.4). Coastal and marine biota also are vulnerable to changes in upwelling, current dynamics, freshwater inflow, salinity, water temperatures, and other processes that affect food webs and nursery areas (Boesch *et al*., 2000). Moreover, long-term studies of estuaries such as San Francisco Bay have indicated that natural cycles of ecosystem processes, such as phytoplankton blooms, are being altered on a global scale by human activities. This includes manipulation of river flows, input of toxic contaminants and nutrients, and invasion of exotic species (Cloern, 1996). Thus, in the nearer term, estuaries and coastal ecosystems may be most vulnerable to hydrological changes in rivers and groundwater flows from shifts in inland precipitation, evapotranspiration, and river ecosystem dynamics.

The increased frequency and geographical range of incidents of hazardous blooms in estuaries on the Gulf, Atlantic, and Pacific coasts has caused serious economic impacts for numerous fisheries and poses a great public health challenge in closing beaches to shellfish. Consumption of toxic algae by shellfish causes the shellfish to contain concentrations of toxins that are high enough to cause paralysis and death of humans who consume the shellfish. This situation has raised concerns that increased nitrogen loading to watersheds has led not only to general coastal eutrophication but also to a greater probability of circumstances that are conducive to these blooms. Factors that appear to contribute to harmful algal bloom occurrence are warmer temperatures and high runoff from watersheds that feed the estuaries, although much remains to be learned and predictive capability has yet to be achieved. Thus, indirect effects of climate change, which exacerbate the hazardous algal bloom problem, could be significant and difficult to identify until better understanding is gained (Anderson, 1997; see Chapter 6).

15.2.2.2.4. Wildlife

The Endangered Species Act in the United States and other regulatory efforts preserve and manage wildlife populations in many regions of North America. These efforts have begun to involve habitat protection and ecosystem management rather than taking a strict population focus. Changes in habitats driven by climate change could further restrict wildlife populations. One process would be causing habitats to be less interconnected, restricting migration of individuals among different populations and causing loss of genetic diversity within more isolated populations. This process is a concern for aquatic and terrestrial wildlife.

Fish populations and other aquatic resources are likely to be affected by warmer water temperatures, changes in seasonal flow regimes, total flows, lake levels, and water quality. These changes will affect the health of aquatic ecosystems, with impacts on productivity, species diversity, and species distribution (Arnell *et al.*, 1996). For example, warming of lakes and deepening of thermoclines will cause a loss of habitat for coldwater fish in areas such as Wisconsin and Minnesota, and decreases in summer flow and increased temperatures will cause loss and fractionation of riverine habitat for coolwater fish species in the Rocky Mountain region (Rahel *et al.*, 1996; Cushing, 1997).

Wetlands and dependent wildlife resources may be adversely affected by general increases in evapotranspiration and reduced summer soil moisture, which may reduce the extent of semi-permanent and seasonal wetlands, particularly in the prairie regions of North America (Poiani *et al.*, 1995).

The state of terrestrial wildlife in North America varies geographically, by taxa, and by habitat association. In general, biodiversity increases from north to south, and species that are associated with rare habitats are most likely to be at high risk of extinction (Dobson *et al.*, 1997; Ricketts *et al.*, 1999). Many factors can cause a species to be at risk of extinction, but the most common causes include loss of habitat, pressures from introduced species or hunting, and reduced fitness as a result of chemical contaminants (Wilson, 1992; Meffe and Carroll, 1994). A minimum estimate of the number of species at risk comes from data for the United States, for which a recent summary suggests that 42 mammal species, 56 bird species, 28 reptile species, and 25 amphibian species are considered at least vulnerable to extinction (UNEP, 2000). Key additional pressures on wildlife associated with global climate change include changes in temperature and precipitation, changes in sea level, and changes in the frequency of extreme weather events (Peters, 1992; Parmesan *et al.*, 2000).

Climate-related pressures can act directly on wildlife through physiological effects (i.e., changes in growth rates, food demands, abilities to reproduce and survive) or indirectly through effects on other plant and animal species (e.g., Payette, 1987; Lewis, 1993; Post and Stenseth, 1999). These physiological effects can lead to changes in the range and abundance of North American species; recent studies suggest that we already are seeing climate-linked changes in butterflies (Parmesan, 1996) and desert-associated species (Brown *et al.*, 1997b; Smith *et al.*, 1998a). A key potential indirect effect of climate change on wildlife is loss of total habitat available as a result of changes in the distribution of a particular habitat type. An obvious example is loss of coastal habitat as a result of sea-level rise; in many places, coastal habitats will not be able to shift inland because adjacent lands already are developed (Harris and Cropper Jr., 1992; Daniels *et al.*, 1993). Similarly, potential shifts in the ranges of species in northern habitats are bounded by the Arctic Ocean (Kerr and Packer, 1998). A second key impact of climate-change related pressures on wildlife relate to how species interact with other species on which they depend. Because temperature and precipitation can be triggers for many wildlife behaviors, changes in these factors may differentially impact species in the same location. We already have evidence that the timing of bird migrations (Bradley *et al.*, 1999; Inouye *et al.*, 1999), bird breeding (Brown and Li, 1996; Brown *et al.*, 1999; Dunn and Winkler, 1999), and emergence of hibernating mammals (Inouye *et al.*, 1999) is becoming earlier. Depending on how food sources and other related species respond to changes, wildlife may become decoupled from the many ecological relationships of which they are a part.

15.2.3. Food and Fiber

This section includes a review of impacts on agriculture, production forestry, and marine fisheries.

15.2.3.1. Agriculture

Most global climate change scenarios indicate that higher latitudes in North America would undergo warming that would affect the growing season in this region. For example, estimates of increases in the frost-free season under climatic change range from a minimum of 1 week to a maximum of 9 weeks (Brklacich *et al.*, 1997a). For the Prairies, Ontario, and Quebec, most estimates suggest an extension of 3–5 weeks. Estimated temperature increases for the frost-free season in Ontario and Quebec are mostly between 1.5 and 5.0°C, and agricultural moisture regimes show an even broader range of estimates, indicating precipitation changes for the Prairies and Peace River regions ranging from decreases of 30% to increases of 80% (Brklacich *et al.*, 1997a).

Although warmer spring and summer temperatures might be beneficial to crop production in northern latitudes, they may adversely affect crop maturity in regions where summer temperature and water stress limit production (Rosenzweig and Tubiello, 1997). Predicted shifts in thermal regimes indicate a significant increase in potential evapotranspiration, implying increased seasonal moisture deficits. Modeling studies addressing the southeast United States have shown that changes in thermal regimes under conditions of doubled CO_2 would induce greater demand for irrigation water and lower energy efficiency of production (Peart *et al.*, 1995).

15.2.3.1.1. Change in land use

Drought may increase in the southern Prairies, and production areas may shift northward in Canada. In assessing the potential for expansion to areas in northern Canada (i.e., north of 55°N and west of 110°W) and Alaska, Mills (1994) identified 57 Mha of potentially arable land (class 1-5, based on Canada Land Inventory criteria) with agricultural potential for use in either annual cropping or perennial forage systems. This estimate drops to 39 Mha when climatic limitations are imposed but under a scenario of doubled atmospheric CO_2, increases to 55 Mha with an accompanying improvement of land class to class 3. Similar outcomes—expansion of agricultural land, especially expansion of the zone suitable for corn and soybean production—are expected for northern areas of eastern Canada (Brklacich *et al.*, 1997a). Other case studies conducted in the southern portion of the Mackenzie basin in northwestern Canada show that two different climate-change scenarios would relax the constraints imposed by a short and cool frost-free season but that drier conditions and accelerated crop development would offset the potential gains of a warmer climate (Brklacich *et al.*, 1996, 1997b).

Southern regions growing heat-tolerant crops such as citrus fruit and cotton might benefit from reduced incidence of killing frosts resulting from a change in climate (Miller and Downton, 1993; Mearns *et al.*, 2000). Results of simulations without CO_2-induced yield improvement indicate that production of citrus fruit would shift northward in the southern United States, but yields may decline in southern Florida and Texas because of excessive heat during the winter (Rosenzweig *et al.*, 1996).

Mexican agriculture appears to be particularly vulnerable to climate-induced changes in precipitation because most (about 85%) of its agricultural land is classified as arid or semi-arid. Recent national assessments of the impacts of climate change indicate that the northern and central regions of Mexico are most vulnerable in the agricultural sector (Conde, 1999) and that in these regions, the area of land that is unsuitable for rainfed maize production would expand under climate change (Conde *et al.*, 1997). On average, more than 90% of losses in Mexican agriculture are caused by drought (Appendini and Liverman, 1995). Using five GCM-based scenarios, it was estimated that potential evaporation may increase by 7–16% and the annual soil moisture deficit could increase by 18–45% in important maize-growing regions in eastern Mexico (Liverman and O'Brien, 1991). Rising levels of CO_2 can have the greatest relative beneficial impacts when water is limited. Therefore, rising CO_2 may be expected to have a significant positive impact because so much of Mexican crops are water-limited and rising CO_2 enhances water-use efficiency (see Chapter 14).

15.2.3.1.2. Crop yields and adaptation

Depending on existing conditions, global warming and CO_2 enrichment can have positive or negative effects on crop yields. It is believed that yield increases in mid- and high latitudes are caused by positive physiological effects of CO_2, longer growing season, and amelioration of the effects of cold temperature on growth. Decreases in yield could result from shortening of the growing period, reduced water availability, and/or poor vernalization.

Estimates of the impacts of climate change on crops across North America vary widely (see Table 15-3). In some studies, the impacts range from nearly total crop failure for wheat and soybeans at one U.S. site to wheat yield increases of 180–230% for other sites in the United States and Canada (Brklacich *et al.*, 1994; Rosenzweig *et al.*, 1994). Recent modeling efforts indicate that the impacts on yields for many crops grown under dryland conditions, even without adaptation, is positive (Reilly *et al.*, 2000). Threshold limits associated with temperature increases may be important. Rosenzweig *et al.* (1995) report generally positive crop yield responses to temperature increases of 2°C, but yield reductions occurred at increases of more than 4°C. Modeled yield results that include the direct physiological effects of CO_2 are substantially different from those that do not account for such effects (Fischer *et al.*, 1996).

Although it is known that the distribution and proliferation of weeds, crop diseases, and insects are determined to a large extent by climate, most crop modeling efforts have not thoroughly accounted for potential impacts of climate change and variability on pest populations and ranges. Interactions between crops and pests under changing climate conditions will be very complex and are difficult to predict because elevated CO_2, warmer temperatures, and increased climate variability would alter the relationships between crops, weeds, and insects significantly. Higher temperatures and warmer winters could reduce winterkill of insects as well as broaden the range of other temperature-sensitive pathogens (Rosenzweig *et al.*, 2000). Increases in the

***Table 15-3**: Range of climate change scenario impacts on agriculture.*

Crop Yield (% change from current)		Cropped Area	Change in Soil Carbon/Soil Quality	Pesticide Expenditures[6] (% change)	Irrigated Acreage	Livestock Production
Canada[1]		Increase[3] and decrease[4]	Increase[5]	Corn +10 to +20	Increase[7]	Decrease[8]
– Smallgrains	-24 to +14[a]					
	-35 to +66[b]			Wheat -15 to +15		
	-75 to +73[c]					
	-17 to 0[d]			Potato + 5 to +15		
	+21 to +124[e]					
US[2]				Soybean and Cotton +2 to +5		
– Spring Wheat	+17 to +23					
– Winter Wheat	-9 to +24					
– Corn	+11 to +20					
– Soybean	+7 to +49					
– Sorghum	+32 to +43					
– Potato	+7 to +8					
– Citrus (oranges/grapefruit)	+13 to +40					

[1] Data pertain to (a) Peace River/agricultural margin; (b) Alberta, Saskatchewan, Manitoba; (c) Ontario, Quebec; and (d) Atlantic region [adapted from Brklacich *et al.*, 1997a; based on scenarios from pre-1995 versions of four GCMs (CCC, GFDL, GISS, and UKMO) with different crop models (FAO and CERES), assuming no adaptation and no CO_2 fertilizer effects]. Data for note (e) represents yields of corn, spring and winter wheat, and canola [from McGinn *et al.*, 1999; based on CCC model (results also show growing degree days increase by 50%)].

[2] Weighted average yield impact for crops grown under dryland conditions with adaptation, percentage change from base conditions (Reilly *et al.*, 2000). Results based on simulations at 46 sites of current major production representing changes in climate predicted by the CCC, Hadley Centre, and Pacific Northwest National Laboratory models, and calculated using 20-yr averages centered around the year 2030, with an atmospheric CO_2 concentration of 445 ppm; crop yields were simulated by the DSSAT models (Tsuji *et al.*, 1994).

[3] For Alaska and northwestern Canada (Mills, 1994), and Peace River region, northern Ontario, and Quebec in northern Canada (Brklacich *et al.*, 1997b).

[4] For example, in citrus production in the southeastern United States, if risk of freeze damage increases with climate change (Miller and Downtown, 1993), area in cropland decreased 5–10% (Reilly *et al.*, 2000).

[5] If soil conservation practices (e.g., no tillage, increased forage production, higher cropping frequency) implemented as mitigation strategies (TAR WGIII).

[6] Reilly *et al.* (2000) results based on simulations at 45 sites of current major production representing changes in climate predicted by the CCCM and Hadley Centre models, and calculated using 20-yr averages centered around 2090, with an atmospheric CO_2 concentration of 660 ppm.

[7] Irrigated acreage estimated to increase by 0.8–7.3Mha in the United States (Adams *et al.*, 1990).

[8] Direct effects include warmer temperatures, which are estimated to suppress livestock appetite. If quality or supply of forage/feed grains is altered, production may be more affected by changes in pasture and grain prices (Adams *et al.*, 1999).

incidence of extreme weather events could reduce the efficacy of pesticide applications and result in more injury to nontarget organisms (Patterson *et al.*, 1999).

Modeling studies of changes in crop production show strong regional effects, with some areas suffering significant losses compared to other regions—suggesting that climate change may affect the comparative advantage of agricultural production regions within North America. For example, in scenarios investigated for the U.S. National Assessment (Reilly *et al.*, 2000), the lake states, mountain states, and Pacific region showed gains in production, whereas the southeast, delta, southern Plains, and Appalachia generally lost. The economic impact of these changes in crop production as a result of climate change is considered to be mostly beneficial to society as a whole. The effects are largely detrimental to producers because the overall positive effect on production leads to decreasing prices. Thus, climate change is beneficial for foreign trade surplus and for consumers. Analyses of the economic effects of various climate change scenarios on the welfare of consumers and producers in the United States show that agricultural welfare strictly increases with 1.5 °C warming, but further warming reduces the benefit at an increasing rate (Mendelsohn *et al.*, 1999). Additional precipitation is strictly beneficial (Adams *et al.*, 1999).

The costs and benefits of climate change must be evaluated concurrently with behavioral, economic, and institutional adjustments brought about by climate change. These adjustments occur at different levels. For example, farm-level adaptations can be made in plant and harvest dates, crop rotations, selections of crops and crop varieties for cultivation, water consumption for irrigation, use of fertilizers, and tillage practices. At the market level, prices are a strong signal to adapt as farmers make decisions about land use and which crops to grow.

Current economic studies of climate change that include farm- and/or market-level adjustments suggest that the negative effects of climate change on agriculture probably have been overestimated by studies that do not account for adjustments that will be made. This may be caused by the ability of the agricultural production community to respond with great flexibility to a gradually changing climate. Typically, extreme weather poses a significant challenge to individual farming operations that may lack the spatial diversity and financial resources of large, integrated, corporate enterprises with production capabilities in one or more areas.

Simulation modeling using four GCM-based scenarios showed that U.S. cereal production decreases by 21–38% when farmers continue to do what they are now doing (i.e., no adaptation) (see Table 15-4). When scenarios that involve adaptation by farmers are used, decreases in cereal production are not as large and the adaptations are shown to offset the initial climate-induced reduction by 35–60% (Schimmelpfennig *et al.*, 1996; Segerson and Dixon, 1999).

15.2.3.1.3. Response to climate variability and extreme events

The effects of changes in the variability of temperatures and precipitation on crop yields have been evaluated through simulation modeling. Changes in diurnal and interannual variability of temperature and moisture can result in substantial changes in the mean and variability of wheat yields. In Kansas, doubling of temperature variability resulted in greatly reduced average yield and increased variability of yield, primarily as a result of crop failure by winterkill (Mearns *et al.*, 1996). The main risk of climate change to some regions may be primarily from the potential for increased variability. Increased variability of temperature and precipitation results in substantially lower mean simulated yields, whereas decreased variability produces only small increases in yield that were insignificant (Reilly *et al.*, 2000). This asymmetric response to temperature variability underscores a major reason that the corn belt region of the United States is so productive: There generally is low variability in temperature across the region. It should be noted, therefore, that if minimum temperatures increase more than maximums, two outcomes could be suggested: Temperature variability may decline, and winterkill should be reduced.

These effects of diurnal and interannual climate variation may have important implications for farm values. Economic analysis has shown that greater interannual variation is harmful to farm values, and the marginal effect of temperature variation is relatively larger than the effect of variations in precipitation (Mendelsohn *et al.*, 1999).

15.2.3.1.4. Vulnerability of livestock

The effects of climate change on livestock can be direct (e.g., effects of higher temperature on livestock appetite) or indirect (e.g., effects of changes in quantity and quality of forage from grasslands and supplies of feed). In areas where livestock rely on surface water availability, water quality could have an impact on weight gain. This would be particularly important where fewer water sources become used by greater numbers of cattle.

***Table 15-4**: Percentage change in U.S. supply of cereals under various constraints, by climate change scenario (Schimmelpfennig et al., 1996).*

Scenario	No adaptation	With adaptation
GISS	-21.5	-8.7
GFDL	-37.8	-22.3
UKMO	-34.1	-19.4
OSU	-31.9	-20.9

Estimates of livestock production efficiency suggest that the negative effects of hotter weather in summer outweigh the positive effects of warmer winters (Adams *et al.*, 1999). The largest change occurred under a 5°C increase in temperature, when livestock yields fell by 10% in cow-calf and dairy operations in the Appalachia, southeast, Delta, and southern Plains regions of the United States. The smallest change was 1% under 1.5°C warming in the same regions. Livestock production also is affected by changes in temperature and extreme events. For example, an ice storm in eastern Canada and the northeast United States in the winter of 1998 had severe effects on livestock in the region (see Section 15.3.2.6).

15.2.3.1.5. Role of changing water resources

Although several studies have examined the potential implications of climate change for streamflows and water delivery reliability from reservoir systems in regions where irrigated agriculture is now important (see Section 15.2.1), there have been few direct analyses of the economic impacts on irrigated agriculture of changes in water availability. Some assessments of the impacts of climate change on agriculture in North America have relied on optimistic assumptions regarding the availability of irrigation water to offset precipitation deficiencies (Mendelsohn *et al.*, 1994). Other studies have attempted to estimate the impacts of projected climate change on the potential use of irrigation water. A study of potential climate change impacts on irrigation water use in the United States concluded, "The greatest impact of a warmer climate on the agricultural economy will be in the West where irrigators will be hard put to maintain even present levels of irrigation" (Peterson and Keller, 1990). That conclusion is based on first-order impacts of reduced water availability and does not consider possible earnings from sale or lease of water rights.

Studies of the impacts of drought events may provide useful insights into the impacts of substantial changes in seasonal streamflows that may result from climate warming—particularly in western North America, where mountain snowpacks now sustain streamflows into the summer months (see Section 15.2.1). However, the impacts of short-term droughts are an imperfect analog to long-term impacts of a drier climate because farmers are likely to adjust crop choices and farming practices as they acquire experience with any new climate regime.

Under some scenarios, demand for irrigation water declines (e.g., as a result of more rapid crop maturation and/or increased growing-season precipitation). Scenarios investigated for the U.S. National Assessment (Reilly *et al.*, 2000) suggest that demand for water resources by agriculture would decline nationwide on the order of 5–10% by 2030 and 30–40% by 2090. Land under irrigation showed similar magnitudes of decline. Crop yield studies generally favor rainfed over irrigated production and show declines in water demand on irrigated land. Such adaptations could help to relieve some of the stress on regional water resources by freeing water for other uses (Hurd *et al.*, 1999). However, the interplay between changes in irrigation demand and changes in water supplies has not been fully assessed.

15.2.3.1.6. Carbon sequestration

North American soils have lost large quantities of carbon since they first were converted to agricultural systems, leaving carbon levels in agricultural soils at about 75% of those in native soils (Bruce *et al.*, 1999c). Because carbon in agricultural soils is a manageable pool, it has been proposed that these soils be managed to sequester carbon from atmospheric CO_2.

15-1. Carbon Sequestration: Adaptation Issues

The Kyoto Protocol commits industrialized nations to take on binding targets for GHG emissions for the period 2008–2012. The Protocol mentions human-induced land-use changes and forestry activities (afforestation, reforestation, deforestation) as sinks of GHGs for which sequestration credits can be claimed; it also mentions that agricultural sinks may be considered in the future. As a result, a significant market is emerging in North America for ways to enhance carbon sequestration in these sectors. Although it is not within the purview of this section to deal with mitigation strategies, land management decisions impact a wide range of factors. There may be several consequent issues that result or are derived from implementation and adoption of these strategies. Negative consequences of reduced tillage implemented to enhance soil carbon sequestration may include (medium confidence):

- Increased use of pesticides for disease, insect, and weed management. This increased pesticide load may affect adjacent ecosystems and the quality of water within and outside agroecosystems.
- Capture of carbon in labile forms that are vulnerable to rapid oxidation if the system is changed. This may require that reduced-till systems be maintained for an extended period (which also would lengthen the beneficial aspects of reduced tillage).
- Reduced yields and cropping management options and increased risk for farmers (Would yield reductions and increased risk be compensated?).

Beneficial consequences of reduced tillage (especially no-till) may include (high confidence):

- Reduced input costs (e.g., fuel) for farmers, thereby increasing the economic profit margin
- Increased soil moisture and hence reductions in crop water stress in dry areas
- Reduction in soil erosion, which inhibits loss of carbon from erosional forces and preserves the natural land base
- The overall combination of these effects improves soil quality and the ability of soils to physically and chemically support plant growth, as well as conserving the continent's natural resources.

The extent to which carbon will be sequestered in agricultural and forest systems will be related to practical economics and land-use policies. For example:

- Expansion of agricultural lands for carbon sequestration may increase competition with use of agricultural lands for traditional food and fiber production. The effect may be decreased food and fiber production, with subsequent increases in prices and decreases in exports for agricultural commodities.
- Land prices may change (e.g., increase) as a consequence of competition between crops for food and crops for mitigation strategies.
- Reduction of agricultural lands by transferring these lands to forestry to enhance carbon sequestration also may increase competition for food and fiber production, but the influx of land into forestry subsequently may decrease forestry prices.

Thus, a focus on carbon sequestration in ecosystems may result in the transfer of large quantities of land between agriculture and forestry and change the management of existing agricultural and forest ecosystems. These changes may provide opportunities for landowners, but they also may have implications for food and fiber production and ecosystem functions.

The rate at which carbon is lost has subsided for most agricultural soils, and carbon levels in some soils have been maintained or even begun to increase as conservation farming practices have been adopted in the past 15–20 years. On cultivated land, these practices include conservation tillage (i.e., reduction or elimination of tillage) and residue management, use of winter cover crops, elimination of summer fallow, and methods to alleviate plant-nutrient and water deficiencies and increase primary production (Lal *et al.*, 1998). Revegetation of marginal lands and modified grazing practices on pastures can be used to increase soil carbon levels. On degraded soils, preventing and controlling erosion and reducing salinization help to maintain or increase soil carbon. Greater adoption of these measures in the United States and Canada could result in agricultural soils more effectively capturing carbon from atmospheric CO_2.

However, these agricultural practices that are effective in building soil carbon also may result in greater emissions of other GHGs (e.g., N_2O). Therefore, research is needed to weigh the positive and negative effects of building up soil carbon with respect to the overall goals of reducing GHG emissions. Moreover, implementation of such mitigation strategies and their effects on adaptation need further evaluation from the perspective of practical economics and land management decisions (see Box 15-1).

Some scenario studies suggest that interactions between soil and atmosphere will occur under a positive feedback system as temperatures increase: Higher temperatures will cause greater decomposition of soil carbon, in turn causing greater emissions from soil of CO_2, which will enhance the greenhouse effect and cause even higher temperatures. However, there is evidence that negative feedback mechanisms also exist. Some experiments indicate that more primary production is allocated to roots as atmospheric CO_2 rises (Schapendonk *et al.*, 1997), and these roots decompose more slowly than those grown at ambient CO_2 levels (Van Ginkel *et al.*, 1997). Recent comprehensive analyses of field data of forest soils suggests that increased temperature alone will not stimulate decomposition of forest-derived carbon in mineral soil (Giardina and Ryan, 2000).

Analysis of yield trends for 11 major crops over the period 1939–1994 indicates that the rate at which yield increased ranged from 1% on average to more than 3% yr^{-1} (Reilly and Fuglie, 1998). Conservative extrapolation of yields implies that the average annual increase in yield for the 11 crops between 1994 and 2020 would range from 0.7 to 1.3% yr^{-1}. More optimistic estimates of growth rates indicate that yield increases could be as high as 3% yr^{-1}. These yield increases could lead to substantial increases in soil carbon if crop residues are retained.

15.2.3.2. Production Forestry

Evaluation of effects of climate change on production forestry are constrained by uncertainties discussed in Section 15.2.2.1 (i.e., state-of-the-art forest models and GCM predictions produce equal probabilities of "considerable forest dieback" and "enhanced forest growth"). Moreover, in many cases, site-specific conditions as well as history, human management, air pollution, and biotic effects (e.g., herbivory) are much stronger controllers of forest productivity than climate change or CO_2 enrichment (Eamus and Jarvis, 1989; Aber and Driscoll, 1997; Ollinger *et al.*, 1997; Goodale *et al.*, 1998), especially in mid-latitudes. Finally, lands managed for timber production are likely to be less susceptible to climate change than unmanaged forests because of the potential for adaptive management (Binkley and Van Kooten, 1994).

In a broader assessment for the United States, Sohngen and Mendelsohn (1999) used several GCMs, a variety of ecological models, and a dynamic economic model (with adaptation) to assess climate change impacts on the U.S. timber market. Under a broad range of climate and ecosystem model predictions, economic changes were positive, as a result of generally positive impacts of climate change on U.S. forest production and the ability of producers to adapt. Disturbances from insects and fire were assumed to increase, but the study also assumed that there would be salvage logging followed by planting of the right species for a new climatic regime. In a sensitivity analysis, in which the possibility of replanting with incorrect species was considered, economic impacts became negative as a result of reductions in available stocks and increased regeneration costs.

The foregoing uncertainty raises questions about evaluating impacts and developing adaptation strategies. For example, Woodbury *et al.* (1998) used a climate change scenario derived from four GCMs, results from experimental studies, and a probabilistic regional modeling approach and estimate that there is a high likelihood that loblolly pine (a major timber production species) growth is likely to decrease slightly over a 12-state region of the southern United States. However, they also estimate that there is a substantial chance of either a large decrease or a large increase in growth. How can this information be used by the timber industry? Should managers assume that there would be no problems with loblolly pine plantations? Should they increase the area of these plantations? Should they convert plantations to a more mixed plantation community? Should plantations be converted to "natural succession"? Crippling uncertainty of this type may lead this production industry (as well as others) to disregard climate change as a factor in planning.

Similar questions have been raised in northwest Canada (see Section 15.3.2.8). Within the Mackenzie Basin Impact Study (MBIS), debate about forest management concerned the scenario of reduced spruce yield and increased risk of losses from fire and insect damage (Hartley and Marshall, 1997; Rothman and Herbert, 1997). In the short term, a large number of pressing issues divert attention from long-term climate change (e.g., land-use planning, British Columbia's Forest Practices Code, treaty negotiations with aboriginal people, trade with the United States, protected area strategy). Adaptation to climate change requires information that is relevant to the context of the industry, particularly if there are implications for harvesting (Barrett, 1997; Fletcher, 1997).

15.2.3.3. Marine Fisheries

Climate-related variations in marine/coastal environments are now recognized as playing an important role in determining the productivity of several North American fisheries. For example, large changes in species abundance and ecosystem dynamics off the coast of California have been associated with changes in sea-surface temperatures (SSTs), nutrient supply, and circulation dynamics (Ebbesmeyer *et al*., 1991; Roemmich and McGowan, 1995; Bakun, 1996). Similar relationships have been observed in the Bering Sea, the northeastern Pacific, and the Gulf of Alaska (Polovina *et al*., 1995; Ware, 1995; Shuntov *et al*., 1996; Beamish *et al*., 1997; Downton and Miller, 1998; Francis *et al*., 1998; Beamish *et al*., 1999a, 2000) and in the North Atlantic (Atkinson *et al*., 1997; Sinclair *et al*., 1997; Hofmann and Powell, 1998). In the Gulf of Mexico, variations in freshwater discharge affect harvests of some commercially important species (Hofmann and Powell, 1998). Projected climate changes have the potential to affect coastal and marine ecosystems through changes in coastal habitats, upwelling, temperature, salinity, and current regimes. Such changes may affect the abundance and spatial distribution of species that are important to commercial and recreational fisheries (Boesch *et al*., 2000).

Fishery management involves the difficult task of maintaining viable fish populations in the presence of difficult-to-predict shifts in resource availability, while regulating competition among harvesters for access to publicly managed, common-property fishery resources (McKay, 1995; Fujita *et al*., 1998; Myers and Mertz, 1998; Roughgarden, 1998). Attainment of management objectives may be confounded by the fact that some fish stocks tend to fluctuate widely from year to year. These fluctuations may arise from natural causes that are unrelated to fishing pressure or be exacerbated by harvesting. The exact cause of a sudden shift in abundance often is poorly understood. Climate variations often play a role in natural fluctuations, although their role may be complex and indirect. For example, a climatic variation may affect phytoplankton and zooplankton abundance in some part of the ocean, with cascading effects through a chain of predator-prey relationships (Bakun, 1996). These processes may result in multiple and lagged impacts on the abundance of a harvested species. Because it is difficult to identify and predict such effects, climate variability constitutes a significant source of uncertainty for fishery managers.

The potential impacts of climate change on fish populations are equally difficult to predict. Some work has focused on the direct impacts of warmer temperatures on marine species (e.g., Wood and McDonald, 1997; Welch *et al*., 1998a,b). However, Bakun (1996) notes that climate variables that are important on land (e.g., temperature and precipitation) may be relatively unimportant for organisms that live in the ocean. He identifies three basic processes (enrichment, concentration, and transport /retention) that influence the productivity and spatial distribution of marine fish populations but notes that very little is known regarding how these will change with global climate change.

Efforts to assess the impacts of climate change on the U.S. fishery sector are severely hampered by our current lack of understanding of possible changes in fish populations. Markowski *et al*. (1999) performed a sensitivity analysis that examined the potential economic impacts of hypothetical changes in the abundance of selected fish populations, but the analysis is too hypothetical for use here.

Uncertainty regarding the magnitude and sources of variations in fish stocks also creates political stumbling blocks to effective fisheries management. Within single jurisdictions, competing harvesters and gear groups vie for shares of a "pie" whose dimensions are imperfectly known. In the case of international fisheries, cooperative harvesting agreements often have degenerated into mutually destructive fish wars when expectations have been upset by unforeseen changes in abundance or the spatial pattern of availability (McKelvey, 1997). For example, the Pacific Salmon Treaty foundered for several years because declining runs of southern coho and chinook salmon and increasing salmon abundance in Alaskan waters frustrated efforts to achieve a mutually acceptable balance of U.S. and Canadian interceptions of one another's salmon stocks (Munro *et al*., 1998; Miller, 2000a).

Accounts of the collapse of cod stocks off Newfoundland on Canada's east coast have cited the inability of governments to effectively control fishing pressure and a natural shift to less favorable environmental conditions (Hutchings and Myers, 1994; Sinclair *et al*., 1997; Hofmann and Powell, 1998). This case suggests that sustainable fisheries management will require timely and accurate scientific information on the environmental conditions that affect fish stocks and institutional flexibility to respond quickly to such information.

The western U.S.-Mexican border region is located between subtropical and mid-latitude ocean regions. Variations in temperature in this transition zone result in major fluctuations in fisheries productivity (Lluch *et al*., 1991). In recent decades, this region of the Pacific has shown a trend toward warming and changes in regional productivity, independent of overexploitation. In a global warming scenario, the sardine population may decrease along the U.S.-Mexican Pacific Ocean border region, whereas the shrimp population may increase. Interdecadal natural climate variability, however, appears to be the most important sardine population modulator (Lluch-Cota *et al*., 1997).

Available evidence suggests that there are likely to be impacts on fisheries arising, for example, from changes in current dynamics, temperature-dependent distribution, and food web dynamics. These impacts will be variable across species and locations and are difficult to forecast with any precision. Because the effects of exploitation and environmental change can be synergistic, it will be increasingly important to consider changing environmental conditions in future fisheries management (Boesch *et al*., 2000; see Chapter 6 for further discussion).

15.2.4. Human Health

Global climate change would disturb the Earth's physical systems (e.g., weather patterns) and ecosystems (e.g., disease vector habitats); these disturbances, in turn, would pose direct and indirect risks to human health. Direct risks involve climatic factors that impinge directly on human biology. Indirect risks do not entail direct causal connections between climatic factors and human biology (McMichael, 1996; McMichael *et al*., 1996). Health care will significantly help people to adapt to climate change. Unfortunately, not everyone has adequate health care; for example, in 1996, nearly 18% of Americans did not have access to a doctor's office, clinic, health center, or other source of health advice or treatment (Miller *et al*., 2000).

15.2.4.1. Potential Direct Health Impacts of Climate Change

15.2.4.1.1. Health impacts of thermal extremes

In a warmer world, heat waves are expected to become more frequent and severe, with cold waves becoming less frequent (Kattenberg *et al*., 1996). Increased frequency and severity of heat waves may lead to an increase in illness and death, particularly among the young (CDC, 1993), the elderly (Ramlow and Kuller, 1990; CDC, 1993; Semenza, 1999; Patz *et al*., 2000), the poor (Schuman, 1972; Applegate *et al*., 1981), the frail and the ill, and those who live in the top floors of apartment buildings and lack access to air conditioning (Patz *et al*., 2000), especially in large urban areas (CDC, 1989; Grant, 1991; Canadian Public Health Association, 1992; Kalkstein, 1993, 1995; Kalkstein and Smoyer, 1993a,b; Canadian Global Change Program, 1995; Environment Canada, 1995; Guidotti, 1996; Kalkstein *et al*., 1996a,b; Tavares, 1996; Last *et al*., 1998). Other vulnerable people are those who take medications that affect the body's thermoregulatory ability (Marzuk *et al*., 1998; Patz *et al*., 2000).

Heat waves affect existing medical problems, not just those related to problems of the respiratory or cardiovascular systems (Canadian Global Change Program, 1995). Morbidity—such as heat exhaustion, heat cramps, heat syncope or fainting, and heat rash—also results from heat waves (Shriner and Street, 1998; Patz *et al*., 2000).

In the United States, populations in northeastern and midwestern cities may experience the greatest number of heat-related illnesses and deaths in response to increased summer temperatures (Patz *et al*., 2000). Recent episodes include the heat-related deaths of 118 persons in Philadelphia in 1993 (CDC, 1993), 91 persons in Milwaukee in 1995, and 726 persons in Chicago in 1995 (CDC, 1995; Phelps, 1996; Semenza *et al*., 1996, 1999). This follows several episodes in the 1980s, particularly in 1980, 1983, and 1988 (CDC, 1995).

In Canada, urbanized areas in southeastern Ontario and southern Quebec could be "impacted very negatively" by warmer temperatures. An "average" summer in 2050 could result in 240–1,140 additional heat-related deaths yr^{-1} in Montreal, 230–1,220 in Toronto, and 80–500 in Ottawa, assuming no acclimatization (Kalkstein and Smoyer, 1993b). The significance of these estimates is demonstrated by the fact that a total of only 183 Canadians died as a result of excessive heat for the years 1965–1992 (Duncan *et al*., 1998). Heat-related illness and death are largely preventable through behavioral adaptations, such as use of air conditioners and increased intake of fluids. In the United States, use of air conditioning is expected to become nearly universal by the year 2050 (U.S. Census Bureau, 1997a,b). Other adaptive measures include development of community-wide heat emergency plans, improved heat warning systems, and better heat-related illness management plans (Patz *et al*., 2000).

Finally, it is important to note that in a warmer world, cold waves are expected to become less frequent. For example, in Saskatoon, Canada, the number of January days with temperature below –35°C could decrease from the current average of 3 days yr^{-1} to 1 day every 4 years (Hengeveld, 1995). Currently, more people die of cold exposure than heat waves. Therefore, an expected decrease in cold waves is likely to have a beneficial effect—a decrease in weather-related mortality.

15.2.4.1.2. Health impacts of extreme weather events

It has been postulated that there will be increases in the frequency and severity of extreme events, which may result in an increase in deaths, injuries, toxic contamination or ingestion, infectious diseases, and stress-related disorders, as well as other adverse health effects associated with social disruption, environmentally forced migration, and settlement in poorer urban areas (McMichael *et al*., 1996). Adaptive measures to counter the health impacts of extreme events include improved building codes, disaster policies, warning systems, evacuation plans, and disaster relief (Noji, 1997).

15.2.4.1.2.1. Convective storms

There is some evidence of increases in the intensity or frequency of some extreme events at regional scales throughout the 20th century. Frequencies of heavy precipitation events have been increasing in the United States and southern Canada (Easterling *et al*., 2000). Unfortunately, it is difficult to predict where these storms will occur and to identify vulnerable populations. In 1997, severe storms caused 600 deaths and 3,799 reported injuries in the United States.

Patients with specific allergies to grass pollen are at risk of thunderstorm-related asthma (Venables *et al*., 1994; Celenza *et al*., 1996; Hajat *et al*., 1997; Knox *et al*., 1997; Suphioglu, 1998). Thunderstorm-associated asthma epidemics in Melbourne, Australia (1987/1989), and London, England (1994), placed considerable demands on the health system. Several London health departments ran out of drugs, equipment, and doctors (Davidson *et al*., 1996). It is unclear if this situation could arise in North America.

15.2.4.1.2.2. Floods

Floods also may become more frequent (see Section 15.2.1.3). All rivers are susceptible to flooding, and nearby populations are potentially vulnerable. In the United States, floods are the most frequent natural disaster, as well as the leading cause of death from natural disasters. The mean annual loss of life is estimated to be 146 deaths yr^{-1} (National Weather Service, 1992; Patz *et al.*, 2000). In 1997, the Canadian Red River flood displaced more than 25,000 people (Francis and Hengeveld, 1998; Manitoba Water Commission, 1998).

During a flood, disaster relief workers may be at risk of injury. For example, 119 injuries were identified from medical claims of people engaged in sandbagging activities in the 1993 Midwest floods. Heat-related injury or illness (HRI), which occurs when the body can no longer maintain a healthy core temperature, was the most frequently reported injury diagnosis; a total of 23 HRI (19.3% of the 119 total injuries) were reported (Dellinger *et al.*, 1996). HRI therefore is a potential problem in disaster relief situations, particularly if high ambient temperature and high humidity exist. Following a flood, flood victims may be at risk of post-traumatic stress disorder (PTSD) and depression, which are risk factors for suicide. Krug *et al.* (1998) showed that suicide rates increased from 12.1 to 13.8 per 100,000 population in the 4 years after floods. Inundations of sites that contain toxic wastes, sewage, animal wastes, or agrochemical products may result in immediate human exposure to wastes from floodwaters, contamination of edible fish, and long-term contamination of flooded living structures (see Sections 15.2.1.3 and 15.2.4.2.2.2).

15.2.4.1.2.3. Hurricanes

Climate models currently are unable to project accurately how hurricanes will change in the future. Today, an average of two hurricanes make landfall each year along the coastline of the continental United States (Hebert *et al.*, 1993). There has been considerable interdecadal variability in the number of landfalling hurricanes in the United States (Pielke and Pielke, 1997). The Federal Emergency Management Agency (FEMA) declared fewer than 20 natural disasters annually in the 1950s and 1960s but more than 40 yr^{-1} in the 1990s (Miller *et al.*, 2000). Hurricanes' strong winds and heavy rains cause injury, death, and psychological disorders (Logue *et al.*, 1979; Patz *et al.*, 2000). A total of 20–30% of adults who lived through Hurricane Andrew showed evidence of PTSD at 6 months and 2 years after the event (Norris *et al.*, 1999).

15.2.4.1.2.4. Ice Storms

Milder winter temperatures will decrease heavy snowstorms but could cause an increase in freezing rain if average daily temperatures fluctuate about the freezing point. It is difficult to predict where ice storms will occur and identify vulnerable populations. The ice storm of January 1998 (see Section 15.3.2.6) left 45 people dead and nearly 5 million people without heat or electricity in Ontario, Quebec, and New York (CDC, 1998; Francis and Hengeveld, 1998; Kerry *et al.*, 1999). The storm had a huge impact on medical services and human health. Doctors' offices were forced to close, and a large number of surgeries were cancelled (Blair, 1998; Hamilton, 1998). One urban emergency department reported 327 injuries resulting from falls in a group of 257 patients (Smith *et al.*, 1998b).

15.2.4.1.2.5. Tornadoes

Although some evidence is available regarding increases in the intensity and frequency of some extreme weather events, it is not yet clear how tornadoes will be affected. The tornado of July 31, 1987, in Edmonton, Alberta, killed 27 people and injured 253 (Etkin *et al.*, 1998). Trends in the United States appear to show a decreasing number of deaths since the 1950s, although data on the number of events causing deaths do not show a trend (Kunkel *et al.*, 1999). Godleski (1997) reports that persons who endure tornadoes often experience a variety of stress responses, including depression, acute and post-traumatic stress disorders, substance abuse, anxiety, and somatization.

15.2.4.2. Potential Indirect Health Impacts of Climate Change

15.2.4.2.1. Vector-borne diseases

15.2.4.2.1.1. Encephalitis

In the midwestern United States, outbreaks of St. Louis encephalitis (SLE) appear to be associated with a sequence of warm, wet winters; cold springs; and hot, dry summers (Monath, 1980). In the western United States, a 3–5°C increase in mean temperature may cause a northern shift in the distribution of western equine encephalitis (WEE) and SLE outbreaks and a decrease in the range of WEE in southern California (Reeves *et al.*, 1994). In Canada, the ranges of eastern equine encephalitis (EEE), snowshoe hare virus (SHV), and WEE probably would expand with global warming. All three already have been reported in Canada or adjacent U.S. states, albeit sporadically (McLean *et al.*, 1985; Artsob, 1986; Tourangeau *et al.*, 1986; Keane and Little, 1987; Heath *et al.*, 1989; Carman *et al.*, 1995; Duncan *et al.*, 1998).

15.2.4.2.1.2. Malaria

In the United States, sporadic autochthonous malaria transmission was observed in New York and New Jersey during the 1990s (Layton *et al.*, 1995; Zucker, 1996). Malaria is imported into Canada, however.

According to studies by Martens *et al.* (1995), Martin and Lefebvre (1995), and Duncan (1996), with global warming, malaria may extend northward into temperate countries. These studies note, however, that many of these countries had regular epidemics of malaria in the 19th century and the first half of the 20th century and that continued and increased application of control measures—such as water management, disease surveillance and prompt treatment of cases—probably would counteract any increase in vectorial capacity.

Malaria once prevailed throughout the United States and southern Canada (Bruce-Chwatt, 1988). As recently as 1890, the census recorded more than 7,000 malaria deaths per 100,000 people across the American South and more than 1,000 malaria deaths per 100,000 people in states such as Michigan and Illinois. It is important to note that diagnoses and reporting did not meet today's standards. By 1930, malaria had been controlled in the northern and western United States and generally caused fewer than 25 deaths per 100,000 people in the South. In 1970, the World Health Organization (WHO) Expert Advisory Panel on Malaria recommended that the United States be included in the WHO official register of areas where malaria had been eradicated.

In Canada, vivax malaria became widespread at the end of the 18th century, when refugees from the southern United States settled in large numbers as far north as "the Huron" in the aftermath of the American War of Independence. Malaria was further spread with the building of the Rideau Canal (1826–1832) (Duncan, 1996). By the middle of the 19th century, malaria extended as far north as 50°N. In 1873, the great malarious district of western Ontario was only a fraction of a large endemic area, extending between Ontario and the state of Michigan. In Canada, the disease disappeared at the end of the 19th century (Bruce-Chwatt, 1988; Haworth, 1988; Duncan, 1996).

The history of malaria in North America underscores the fact that increased temperatures may lead to conditions that are suitable for the reintroduction of malaria to North America. Socioeconomic factors such as public health measures will continue to play a large role in determining the existence or extent of such infections (Shriner and Street, 1998).

15.2.4.2.1.3. Dengue and Yellow Fever

Although the *Aedes aegypti* mosquito already is found in the southern United States, socioeconomic factors play a large role in determining the actual risk of climate-sensitive diseases (see Chapter 9). For example, socioeconomic differences between Texas and bordering Mexico determine disease incidence: A total of 43 cases of dengue were recorded in Texas during 1980–1996, compared to 50,333 in the three contiguous border states in Mexico (Reiter, 1999).

15.2.4.2.1.4. Lyme Disease and Rocky Mountain Spotted Fever

Lyme disease—the most common vector-borne disease in the United States—currently circulates among white-footed mice in woodland areas of the Mid-Atlantic, Northeast, upper Midwest, and West Coast of the United States (Gubler, 1998). In 1994, more than 10,000 cases of the disease were reported (Shriner and Street, 1998). Although possible tick vectors have been reported in various parts of Canada, self-reproducing populations of infected ticks are believed to occur only in Long Point, Ontario (Barker *et al*., 1992; Duncan *et al*., 1998). Lyme disease has been predicted to spread within Canada with increased temperatures (Grant, 1991; Canadian Global Change Program, 1995; Environment Canada, 1995; Guidotti, 1996; Hancock and Davies, 1997). However, in assessing climate-induced risks for Lyme disease, the ecology of two mammalian species along with projections for land use make predictions very difficult (see Chapter 9). Finally, Grant (1991) has suggested that, with warming, Rocky Mountain Spotted Fever might increase in some localities in Canada.

15.2.4.2.2. Rodent-borne diseases

15.2.4.2.2.1. Hantavirus

In 1993, Sin Nombre virus, which causes hantavirus pulmonary syndrome (HPS), emerged in the Four Corners region of the southwestern United States. Unusually prolonged rainfall associated with the 1991–1992 El Niño was implicated as a causal factor in the outbreak (Engelthaler *et al*., 1999; Glass *et al*., 2000). As of 1999, 231 cases had been confirmed in the United States, with a mortality rate of 42% (Patz *et al*., 2000). A total of 16 cases of HPS have been identified in Canada. These cases occurred in British Columbia, Alberta, and Saskatchewan (Stephen *et al*., 1994; Werker *et al*., 1995; Duncan *et al*., 1998).

HPS could undergo changes in occurrence related to increased contacts between rodents and people. Because the virus is present in rodents in the United States and Canada and changes in climate and ecology are known to affect rodent behavior, it is assumed that changes in the incidence of HPS would result with global warming (Duncan *et al*., 1998), but they will be difficult to predict because of local rainfall variability.

Adaptive measures to reduce the risks of contracting vector- and rodent-borne diseases include providing information, vaccination, and drug prophylaxis for travellers, as well as use of repellants, surveillance, and monitoring (Patz *et al*., 2000).

15.2.4.2.2.2. Diseases associated with water

More than 200 million people in the United States have direct access to treated public water supply systems. Nevertheless, 9 million cases of water-borne diseases are estimated to occur each year (Bennett *et al*., 1987). Although most of the water-borne disease involves mild gastrointestinal illness, some disease causes severe outcomes, such as myocarditis (Patz *et al*., 2000).

Giardia occurs in American and Canadian watersheds, resulting in widespread human exposure (Schantz, 1991; Chow, 1993; Moore *et al*., 1993; Wallis *et al*., 1996; Olson *et al*., 1997;

Duncan *et al.*, 1998). A study of 1,760 water samples from 72 municipalities across Canada showed that giardia cysts were found in 73% of the raw sewage samples, 21% of the raw water samples, and 18.2% of the treated water samples (Wallis *et al.*, 1996). Cryptosporidium—considered to be one of the most common enteric pathogens worldwide (Meinhardt *et al.*, 1996)—is less common in Canada than giardia cysts, however. Cryptosporidium was found in only 6.1% of raw sewage samples, 4.5% of raw water samples, and 3.5% of treated water samples (Wallis *et al.*, 1996).

Increases in ambient temperatures, a prolonged summer season, increased heavy rainfall and/or runoff events, and many watersheds with mixes of intensive agriculture and urbanization led to recent large outbreaks of cryptosporidium in the United States (MacKenzie *et al.*, 1994; Goldstein *et al.*, 1996; Osewe *et al.*, 1996) and the UK (Bridgeman *et al.*, 1995) and may be indicative of the future. In 1993, more than 400,000 cases (including 54 deaths) resulted from a cryptosporidium outbreak in the Milwaukee, Wisconsin, water supply (MacKenzie *et al.*, 1994). A positive correlation between rainfall, cyrptosporidium oocyst and giardia cyst concentrations in river water, and human disease outbreaks has been noted (Weniger *et al.*, 1983).

The largest ever reported outbreak of toxoplasmosis was traced to the municipal water supply of the greater Victoria area of British Columbia (British Columbia Toxoplasmosis Team, 1995; Den Hollander and Noteboom, 1996; Mullens, 1996; Bowie *et al.*, 1997; Duncan *et al.*, 1998). There already is evidence that exposure to the causative parasite in Canada is widespread (Tizard *et al.*, 1977, 1978). Given the increasing number of feral cats in Canada and the persistence of sporulated oocysts in a variety of environments (Dubey and Beattie, 1988), areas in Canada that are hospitable to oocyst survival are likely to expand as a result of climate change.

Warm marine water may favor growth of toxic organisms such as red tides, which cause three varieties of shellfish poisoning: paralytic, diarrheic, and amnesic. Domoic acid—a toxin produced by the *Nitzchia pungens* diatom that causes amnesic shellfish poisoning—appeared on Prince Edward Island for the first time in 1987. A total of 107 patients were identified, of whom 19 were hospitalized. Of those requiring hospitalization, 12 people required intensive care because of coma, profuse respiratory secretions, or unstable blood pressure. A total of four people died as a result of eating contaminated mussels (Perl *et al.*, 1990). The outbreak coincided with an El Niño year, when warm eddies of the Gulf Stream neared the shore and heavy rains increased nutrient-rich runoff (Glavin *et al.*, 1990; Perl *et al.*, 1990; Teitelbaum *et al.*, 1990; Hatfield *et al.*, 1994; Shriner and Street, 1998). In the United States, marine-related illness increased during El Niño events over the past 25 years. During the strong El Niño event of 1997–1998, precipitation and runoff greatly increased counts of fecal bacteria and viruses in local coastal waters in Florida (Harvell *et al.*, 1999). Contamination of water bodies by animal and human wastes can stimulate harmful algae such as *Pfiesteria* that have been demonstrated to cause illness in humans and death in some species of fish (Burkholder *et al.*, 2000).

Potential adaptive measures to reduce water-borne disease include improved water safety criteria, monitoring, treatment of surface water, and sewage/sanitation systems (Patz *et al.*, 2000). Land-use management should include consideration of water supply and quality.

15.2.4.2.3. Respiratory disorders

In 1997, approximately 107 million people in the United States lived in counties that did not meet air quality standards for at least one regulated pollutant (Patz *et al.*, 2000). Climate change increases smog (NRC, 1991; Sillman and Samson, 1995; USEPA, 1998a; Patz *et al.*, 2000) and acidic deposition. Climate change is likely to have a positive (worsening) effect on suspended particulates (Maarouf and Smith, 1997). These changes would have an impact on human health. However, at this time there are too few studies on the effect climate change will have on all pollutants to project human health impacts. Studies (Bates and Sitzo, 1987, 1989; Tseng *et al.*, 1992; Burnett *et al.*, 1994; Delfino *et al.*, 1994, 1997; Schwartz, 1994; Thurston *et al.*, 1994) have demonstrated that hospital admissions for respiratory illnesses are increased during contemporary air pollution episodes, when levels of ozone, acid aerosols, or particulates are elevated (Campbell *et al.*, 1995).

Adaptive measures to changing pollution levels include federal legislation and warnings for the general population and susceptible individuals (Patz *et al.*, 2000).

15.2.4.2.3.1. Smog

More than half of all Canadians live in areas in which ground-level ozone may reach unacceptable levels during the summer months (Duncan *et al.*, 1998). Peak 1-hour concentrations during typical pollution episodes in the Windsor-Quebec City corridor often reach 150 ppb. Windsor exceeds standards for ozone air quality (82 ppb) 30 days yr^{-1} on average. In the Lower Fraser Valley, ozone concentrations typically are in the 90–110 ppb range during pollution episodes. In the Southern Atlantic region, peak hourly ozone concentrations are in the 90–150 ppb range (Duncan *et al.*, 1998).

Two expert panels from the Canadian Smog Advisory Program have listed a wide range of health effects of ground-level ozone at levels that plausibly may occur in Canada. These effects include pulmonary inflammation, pulmonary function decrements, airway hyper-reactivity, respiratory symptoms, possible increased medication use and physician/emergency room visits among individuals with heart or lung disease, reduced exercise capacity, increased hospital admissions, and possible increased mortality (Stieb *et al.*, 1995). The panels conceptualized potential health effects of air pollution as occurring in a logical "cascade" or "pyramid," ranging from severe, uncommon events (e.g., death) to mild, common effects (e.g., eye, nose, and throat irritation) and asymptomatic changes of unclear clinical significance (e.g., small pulmonary function decrements and pulmonary inflammation) (American Thoracic Society, 1985; Bates, 1992).

Healthy persons can demonstrate effects from ozone exposure when they have an increased respiratory rate (e.g., when they are involved in strenuous activities outdoors) (Brauer *et al.*, 1996). Ozone may pose a particular health threat, however, to those who already suffer from respiratory problems such as asthma, emphysema, or chronic bronchitis (Stieb *et al.*, 1996). These three conditions affect about 7.5% of the Canadian population (Ontario Lung Association, 1991). Ozone also may pose a health threat to young, elderly, and cardiovascular patients (Duncan *et al.*, 1998). An increase in smog also would pose a greater risk to African Americans, who have consistently higher rates of deaths and emergency room visits than caucasians (Mannino *et al.*, 1998).

15.2.4.2.3.2. Acidic deposition

Acidic aerosols—such as sulfur dioxide (SO_2), sulfates, and nitrogen dioxide (NO_2)—have a colloidal affinity to fine particulates, which provide the vector needed to penetrate deeply into the distal lung and airspaces. In general, NO_2 and SO_2 have acute negative impacts on the respiratory system (Campbell *et al.*, 1995).

Several studies (Bates and Sitzo, 1987, 1989; Tseng *et al.*, 1992; Burnett *et al.*, 1994; Delfino *et al.*, 1994; Schwartz, 1994; Thurston *et al.*, 1994) have demonstrated that hospital admissions for respiratory illnesses increase during contemporary air pollution episodes when levels of ozone, acid aerosols, or particulates are elevated (Campbell *et al.*, 1995). A study by Raizenne *et al.* (1996) that examined the health effects of acid aerosols on children living in 24 communities in the United States and Canada found that long-term exposure had a deleterious effect on lung growth, development, and function. Dockery *et al.* (1996) found that children living in communities with the highest levels of strong particle acidity were significantly more likely to report at least one episode of bronchitis in the past year compared to children living in the least polluted communities.

15.2.4.2.3.3. Suspended particulates

Fine particulates are associated with respiratory symptoms, airway hyperreactivity, impaired lung function, reduced exercise capacity, pulmonary inflammation, pulmonary function decrements, increased number of emergency room visits for asthma, increased hospitalizations, increased absence from school or work, and increased mortality from cardiopulmonary disease and lung cancer. Children, the elderly, smokers, asthmatics, and others with respiratory disorders are especially vulnerable to particulate air pollution (Stieb *et al.*, 1995; Seaton, 1996; Choudry *et al.*, 1997; Duncan *et al.*, 1998; USEPA, 1998a).

15.2.4.2.4. Nutritional health

In the United States, food-borne diseases are estimated to cause 76 million cases of illness annually, with 325,000 hospitalizations and 5,000 deaths (Mead *et al.*, 1999). Future food importations are likely to be associated with increases in outbreaks of some viral, parasitic, and bacterial diseases, such as hepatitis A (Duncan *et al.*, 1998).

Many aboriginal communities undertake hunting, fishing, and other resource-based activities for subsistence. Climate change is likely to dramatically alter the abundance and distribution of wildlife, fish, and vegetation. As a result, food supplies and economic livelihoods of many First Nations peoples would be in jeopardy (Last *et al.*, 1998; Weller and Lange, 1999). Disappearance of traditional medicinal plants from areas populated by Native American and other indigenous peoples may likewise affect physical, mental, and spiritual well-being.

15.2.5. Human Settlements and Infrastructure

Large metropolitan centers and industrial areas are particularly vulnerable to global environmental change (Schmandt and Clarkson, 1992). Large cities are considered to be areas of high risk because warming could lead to problems such as heat stress, water scarcity, and intense rainfall. Other potential impacts vary with location. In Canada and the northern United States, for example, people in larger cities are expected to experience fewer periods of extreme winter cold (Born, 1996). Many coastal communities will be affected by rising sea levels and increased risk of storm surge, but the impacts will differ because of variations in local and regional factors (Nicholls and Mimura, 1998). Most people in North America live in land that now is considered coastal and subject to coastal weather extremes. This sector of the urban population is growing faster than the population as a whole—a trend that is expected to continue (Boesch *et al.*, 2000). Indeed, "25 percent of the buildings within 500 feet of U.S. coastlines are predicted to fall victim to erosion in the next six decades" (Associated Press, 2000). Cities that are vulnerable to regular flooding may experience changes in the timing, frequency, and severity of this hazard (Weijers and Vellinga, 1995). In addition, the risk of increased periods of drought will be a challenge, particularly for communities that already are struggling to cope with water management issues. Overall, climate change should reduce vulnerability to some hazards such as cold waves but increase it with respect to others such as sustained periods of extreme heat.

15.2.5.1. Demographic Pressures

The number of people in North America increased from 83 million in 1901 to 301 million in 1998 (Statistics Canada, 1999a; U.S. Census Bureau, 1999b). The number living in large urban communities of more than 750,000 people increased over this period from 6 million to more than 140 million. A higher share of the population in North America lives in large urban centers

(45% in 1995) than in any other region of the world (UNDP, 1999). These large population centers may be vulnerable to climate change.

Aging of the population is another important demographic trend in North America. Europe and Japan currently are adapting to an aging of the population that is just emerging in North America. These demographic trends also appear in native communities with individuals living to older ages: Elderly people who have lived away from reservations may return home for retirement. An older population typically is more vulnerable to climate extremes (McMichael, 1997).

15.2.5.2. *Infrastructure Investments in Adaptation*

Many systems have the potential to be impacted negatively by climate change, including drainage and water systems, roads and bridges, and mass transit (Miller, 1989). Communities can reduce their vulnerability and increase their resilience to adverse impacts from climate change through investments in adaptive infrastructure (Bruce *et al.*, 1999a).

Across North America, however, government spending on public infrastructure has been declining for some time, measured as a share of economic activity. Almost 3.5% of GDP was spent on a broad range of infrastructure projects in the early 1960s, compared to less than 2% in the 1990s (Statistics Canada, 1999b; USBEA, 1999).

Sustained lower spending has increased society's vulnerability to some hazards. The American Society of Civil Engineers, for example, has warned that many dams in the United States have exceeded their intended lifespan (Plate and Duckstein, 1998). More than 9,000 regulated dams have been identified as being at high risk of failing, and there may be significant loss of life and property from future failures.

Highways, bridges, culverts, residences, commercial structures, schools, hospitals, airports, coastal ports, drainage systems, communications cables, transmission lines, and other infrastructure have been built on the basis of historical climate experience (Bruce *et al.*, 1999a). Similarly, land-use practices and building codes have been developed to provide effective protection from the existing climate. Emerging knowledge about future climate pressures was not available when most investment decisions were made. This includes, for example, recent research into actions that can be taken to protect underground transit systems from the increased risk of intense rainfall and subsequent flooding (Liebig, 1997). Furthermore, building codes have been modified frequently over the years to reflect emerging information about safer construction techniques, but these changes have not been applied to existing structures.

Coastal communities have developed a variety of systems to manage exposure to erosion, flooding, and other hazards affected by rising sea levels, but some of these systems have not been maintained—increasing the difficulty of putting in place enhancements to address future risk. Similarly, consideration of investments in larger diameter storm sewers in communities that expect more periods of intense rainfall may be affected by the age of existing systems (Fowler and Hennessey, 1995; Trenberth, 1998). Urban expansion and population growth further complicate decisionmaking with respect to infrastructure investments.

15.2.5.3. *Coastal Regions Particularly Vulnerable*

The prospect of rising sea level is one of the most widely recognized potential impacts of climate change. Some parts of North America have been experiencing sea-level rise for thousands of years (Hendry, 1993; Lavoie and Asselin, 1998). Most climate models, however, project that the pace will accelerate in many regions. This acceleration would increase the difficulty of adaptation for human settlements and natural systems. The greatest vulnerability is expected in areas that recently have become much more developed, such as Florida and much of the U.S. Gulf and Atlantic coasts. Insured property value in Florida alone exceeds US$1 trillion (Nutter, 1999).

Titus and Richman (2001) have developed a data set of coastal land elevations by using digital-elevation models and printed topographic maps to determine areas that are vulnerable to sea-level rise along the U.S. Atlantic and Gulf coasts. Louisiana, Florida, Texas, and North Carolina account for more than 80% of the 58,000 km^2 that are vulnerable to sea-level rise.

Rising sea levels, in turn, can cause increased erosion to shores and habitat and may contaminate some freshwater bodies with salt (Mason, 1999). Climate extremes, such as hurricanes, can add to the adverse impact (Michener *et al.*, 1997). Sea-level rise and climate and weather extremes cause problems associated with beach erosion, siltation of waterways, and flood risk in coastal communities (Hanson and Lindh, 1996; Leatherman, 1996).

More than 65% of people in North America live in coastal communities (Changnon, 1992; see also Figure 15-2). This includes those who live near the Atlantic or Pacific oceans as well as those near the Great Lakes, where the impact of climate change is very different (see Sections 15.2.1 and 15.3.2.5). Accordingly, there is vulnerability across most of the region. Particular concern arises with regard to the combination of sea-level rise with other risks, such as storm surge. Salt intrusion and frequent flooding may adversely impact farming and manufacturing activity in low-lying areas (Gough and Grace, 1998).

Tourism frequently is the major industry in many coastal communities. The risk of beach erosion, siltation, and flooding may become an important challenge for some existing tourism sites, yet these same changes may open new opportunities for some other communities. Tourism has been and will continue to be affected by beach closures resulting from coliform from septic systems and sewage outflows during and following

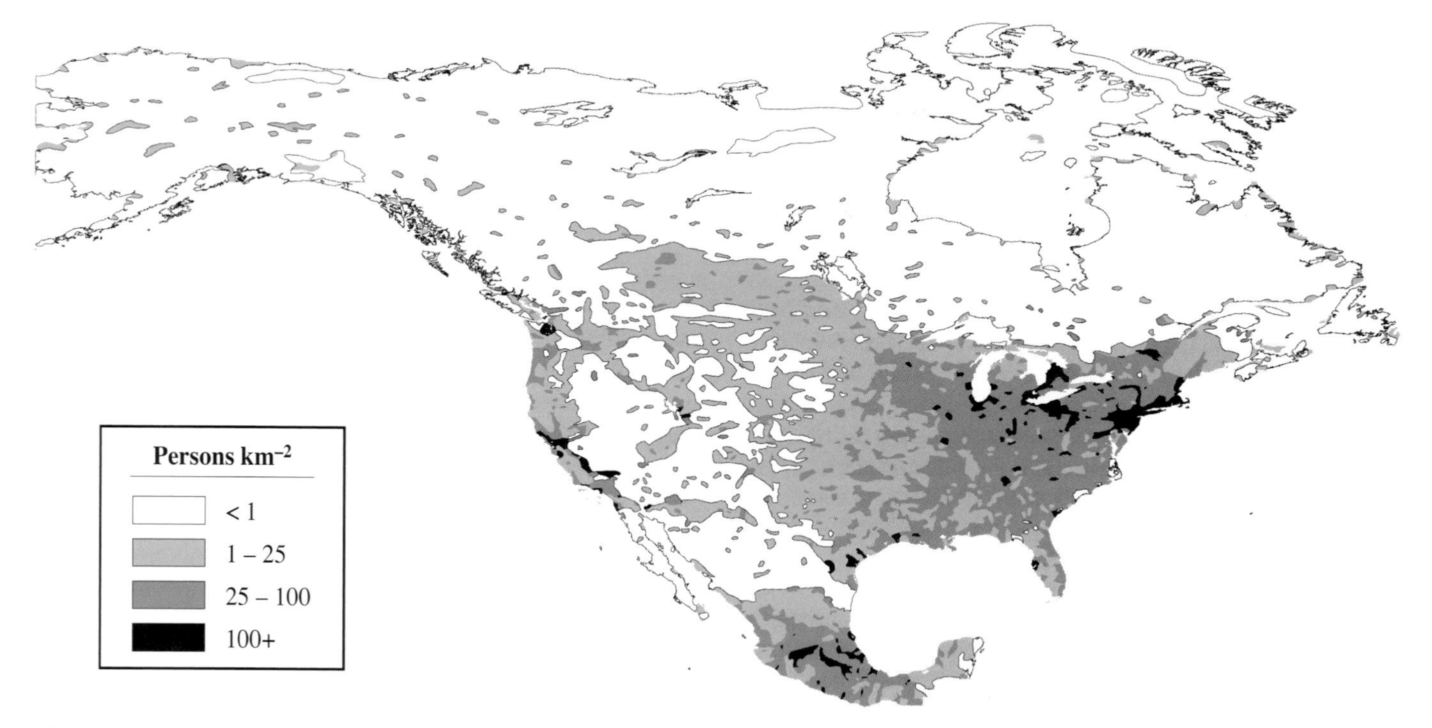

Figure 15-2: North American population density (ESRI, 1998).

storms and extreme events (see also Section 15.2.6). Sea-level rise also will be an important challenge for major ports in many parts of North America.

A study of the impact on the United States of the increase of sea levels through 2065 found losses of US$370 million for dryland, US$893 million for wetlands, and US$57–524 million in transient cost (Yohe *et al.*, 1999). This estimate is much lower than those of earlier U.S. studies because of assumed adaptation responses, including decisions not to protect certain areas. No comparable information is available for Canada, although several vulnerable areas have been identified—particularly the Fraser River delta, Nova Scotia, and the Beaufort Sea region (Shaw *et al.*, 1998).

15.2.5.4. Vulnerability to System Failure

Investments in summer cooling, winter heating, and shelter from the elements are common for most people and businesses in North America. Similar investments are evident around the world, but the scope and scale often is greater in North America. For example, the percentage of homes with air conditioning units in the United States increased from 35% in 1970 (U.S. Census Bureau, 1975) to 76% in 1997 (U.S. Census Bureau, 1999a). Similarly, more vehicles, schools, hospitals, and businesses now have climate control mechanisms. Even traditional tasks have changed, such as air-conditioned cabs on farm vehicles.

The trend of warming across North America should reduce the cost of heating and the cost of investing in heating systems. At the same time, demand for summer cooling is expected to rise. In turn, changes in the weather affect demand for power: Peak demand for electricity is strongly correlated to swings in summer temperature (Colombo *et al.*, 1999).

Investments to manage normal fluctuations in the weather presumably have been effective in increasing the comfort of people in North America. These investments, however, also have increased vulnerability to systems failure. This could become most evident for most people in North America during a summer heat wave or a winter storm, when a failure in major support systems could place many people at risk. For example, in January 1998, a severe ice storm caused a power failure in Quebec and eastern Ontario (see Section 15.3.2.6).

A similar investment is in human alteration of hydrological drainage systems, including building of dikes, which may be an important factor in determining the severity of flooding (Changnon and Demissie, 1996). These interventions appear to be successful in managing most variations in the weather, but they can increase vulnerability to extreme events as new investments are made in regions that were thought to be sheltered from severe weather.

15.2.5.5. Development and Vulnerability to Extreme Events

The cost of natural disasters in North America has increased dramatically since the mid-1980s, in spite of the goal of the International Decade for Natural Disaster Reduction (the 1990s) to reduce the costs of natural disasters by 50% by the year 2000. Although an increase in the number of storms may play a role in these disasters, there is no question that the increasing costs are largely the result of increased vulnerability (Pielke and Downton, 2000).

Natural disasters occur when social vulnerability is triggered by a natural event (see Figure 15-3). Society responds to a disaster through three overlapping activities: response and recovery, mitigation, and preparedness. These activities alter future vulnerability (and therefore the construct of future disasters)—reducing risk if they are done wisely, or not if they are done otherwise. Often, however, mitigation that is designed to protect against natural hazards increases long-term vulnerability (Burton *et al.*, 1993). An example is the levee system planned for the Grand Forks region, as part of the post-1997 Red River flood mitigation scheme. NHC/DRI (1999) notes, "If the levee system is constructed as planned...[it will] have the potential to greatly increase damage and losses if the levee is breached or overtopped in a future flood."

Superimposed on the changing development landscape is the possibility that climate warming will reduce the risk of some extreme events and increase the risk of others (Karl *et al.*, 1996; White and Etkin, 1997). Changes in the frequency, severity, and duration of extreme events may be among the most important risks associated with climate change. In some parts of North America, this includes fewer periods of extreme cold, fewer snowstorms (although there may be an increase in the number of intense storms), increased spring flooding, more frequent summer droughts, and more wildfires. Studies also suggest that there will be a more thermodynamically unstable atmosphere in the future, which probably will result in more frequent heavy rainfalls and possibly increased hail risk and more tornadoes and downbursts (Etkin, 1996).

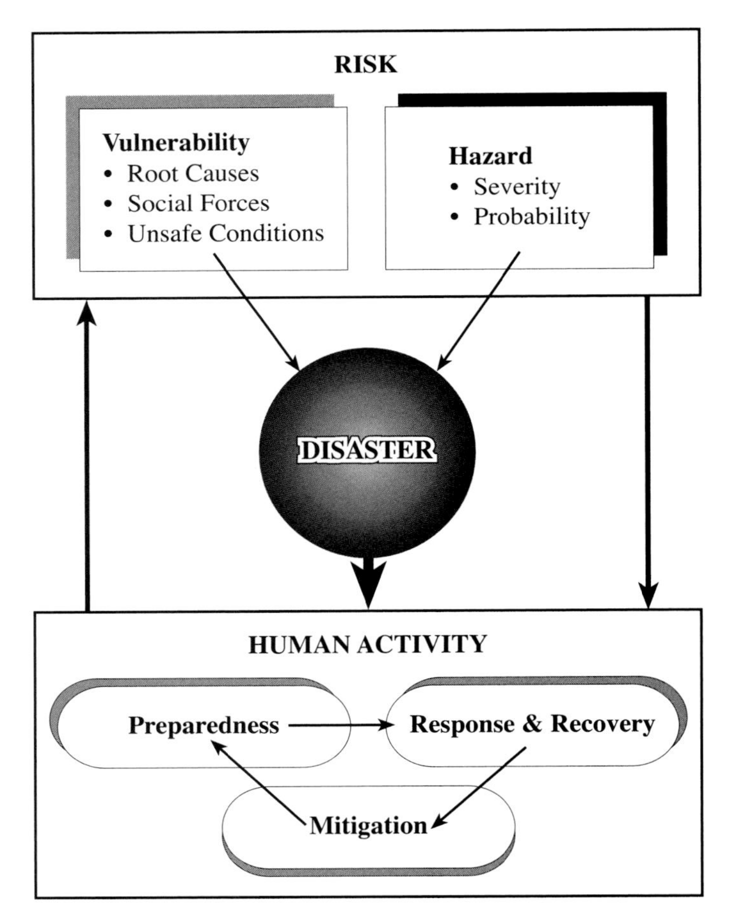

Figure 15-3: Cycle of human response to natural hazards, including response and recovery, mitigation, and preparedness. This response can alter vulnerability and thereby influence future disasters (Etkin, 1999).

Studies suggest that El Niño and La Niña events may become more frequent in the future, although there is still much uncertainty regarding predictions of tropical phenomena (Henderson-Sellers *et al.*, 1998; Timmermann and Oberhuber, 1999). Any changes would affect the number of extreme events, including the occurrence of Atlantic hurricanes because during El Niños (La Niña), Atlantic hurricanes are less (more) frequent. Landsea *et al.* (1999) point out that recent intensification of urban development along the U.S. Atlantic coast parallels a reduction of hurricane activity from 1966 to the early 1990s; they ask whether the United States is prepared for a recurrence of higher frequencies.

North American measures of property damage from severe storms are consistently higher and the loss of life lower, compared to less-developed countries. Investments in warning systems and disaster preparedness save lives. When these systems are overwhelmed by extreme events, however, the damage can be catastrophic. It is interesting that housing losses from extreme events generally are on the same scale in North America as those in developing countries, particularly when disaster strikes large urban centers (Comerio, 1997).

Older buildings often experience the greatest damage during extreme events. New buildings, however, also are at risk if insufficient effort is made to ensure compliance with building codes. Studies show that compliance with building codes may have eliminated up to 25% of insured losses resulting from major hurricanes (IBHS, 1999).

Damage from flooding in North America has increased in the past few decades (Mileti, 1999). Land-use planning remains a powerful tool to help reduce the loss of life and property. One study that indirectly illustrates this concept is Brown *et al.* (1997a). They found that nonagricultural flood damage from a set of storms moving through Michigan exceeded that of southwestern Ontario by a factor of about 900, even though the flood yields in Ontario were greater than those in Michigan. Their analysis ascribes the cause as greater development in flood-prone areas. In Michigan, the storms generally exceeded land-use design thresholds, whereas in Ontario they did not.

Munich Re (1997) found that "the extent of loss has not infrequently been increased through a false assessment of the risk circumstances or in the pursuit of profit." If the amount of development arising from the perception of reduced short-term risk tends to be disproportionate to the real increased risk from rare events, long-term vulnerability is increased. Mileti (1999) argues that "...our national [U.S.] path also is leading us toward natural and related technological catastrophes in the next millennium that are larger than any we have ever experienced ...that is the future we have designed for ourselves.... [We] have done more to postpone losses into the future than to eliminate them." Social and economic trends suggest overall increases in vulnerability; if natural triggers of disasters are exacerbated as a result of climate change, it seems likely that more frequent and severe natural disasters are going to occur in the future (see Section 15.2.7).

15.2.6. Tourism and Recreation

Tourism is a major sector of the global economy, with global receipts from international tourism of US$439 billion in 1998. With a projected annual growth rate of 6.7%, annual international tourism expenditures are expected to surpass US$2 trillion by 2020 (WTO, 1998). The United States and Canada were among the top 10 tourism destinations in terms of international tourist arrivals, with related 1998 receipts of US$71 billion and US$9 billion, respectively (WTO, 1999). Domestic tourism is many times more important in terms of participation and economic activity. The magnitude of the implications of climate change for the tourism and recreation sector will depend on the distribution and importance of tourism and recreation phenomena and the characteristics of climate change and variability. In the United States, the Pacific and south Atlantic regions are major providers of tourism and recreation opportunities. However, if population distribution is taken into consideration, less populated areas often have high economic dependence on tourism and recreation.

One critically important dimension of the tourism and recreation sector that will be sensitive to climate change is the length of the operating season. Any changes in season length would have considerable implications for the short- and long-term viability of tourism and recreation enterprises. In Canada, where 43% of domestic and 62% of international tourism expenditures take place in July through September, an extended warm-weather recreation season is likely to be economically beneficial (Wilton and Wirjanto, 1998). The limited regional studies of golfing, camping, and boating that are available reinforce this conclusion (Wall, 1998a,b). This positive outlook must be tempered with the possibility that economic benefits may occur at the expense of increased environmental deterioration, as destinations host more visitors for longer periods of time. These concerns will be particularly relevant for national parks and other natural areas, which may need to establish "no-go" zones or visitation limits.

Conversely, winter recreations—such as downhill skiing, cross-country skiing, snowmobiling, ice fishing, and other activities that are dependent on snow or ice—as well as the businesses and destination areas associated with them are likely to be impacted negatively. Case studies from the Great Lakes (McBoyle and Wall, 1992) and New England (Bloomfield and Hamburg, 1997) regions indicate that the vulnerability of ski resorts will differ considerably, depending on location (latitude and altitude) and adaptations (snowmaking) to offset or compensate for the effects of less reliable snow conditions. Another potential adaptive strategy is diversification of activities to ensure that investments in property and infrastructure generate income and employment for much of the year. Without analysis of the impacts of altered snow conditions for major ski areas in the mountain ranges of western Canada and the United States and the large snowmobile industry in both nations, the economic impact to the winter recreation sector in North America remains uncertain.

Perhaps as important as changes to season length will be the impact of climate change on the availability and quality of the resource base on which recreational activities depend. Below-average Great Lakes water levels in 1999 again revealed the sensitivity of marinas and the substantial recreational boating industry to climate variability. Similarly, the impact of the 1988 drought on Prairie wetlands and waterfowl breeding success is illustrative of the potential impact of climate change for this sport-hunting resource. Global warming is anticipated to modify many other ecosystems on which outdoor recreations depend. Parks and other natural areas are important tourism and recreation resources whose attractions are based to a considerable extent on the species they conserve and the ecological processes they sustain. A climate change assessment of Canada's National Park system (Scott and Suffling, 2000) indicates that 75–80% of the parks would experience a shift in dominant vegetation under $2xCO_2$ scenarios. Analysis of vegetation response in the Yellowstone National Park region in the United States revealed regional extinctions and the emergence of communities with no current analog (Bartlein *et al.*, 1997). Moreover, a global analysis of habitat change resulting from climate change found that more than 50% of the territory of seven Canadian provinces and greater than 33% of the territory in 11 U.S. states are at risk (Malcolm and Markham, 2000). Although this will pose an unprecedented challenge to the conservation mandate of protected areas, the impact of ecological changes on tourism remains uncertain.

Changes in the magnitude and frequency of extreme events such as hurricanes, avalanches, fires, and floods have considerable implications for tourism and recreation. One likely consequence of global climate change is sea-level rise. This may have considerable consequences for the provision of recreational opportunities in coastal communities, particularly if it is associated with increased storm frequency.

Coastal zones are among the most highly valued recreational areas and are primary tourist destinations. Houston (1996) estimates that 85% of all tourist revenues in the United States are earned by coastal states, and there are as many as 180 million recreational visitors to U.S. coasts every year (Boesch *et al.*, 2000). Sea-level rise in beach areas backed by seawalls or other development that precludes landward migration would lead to loss of beach area through inundation or erosion and pose an increased threat to the recreation infrastructure concentrated along the coast (sea-front resorts, marinas, piers, etc.). Beach nourishment is widely used to protect highly valued recreational beaches. One study estimates that this adaptation strategy would cost US$14–21 billion to preserve major U.S. recreational beaches from a 50-cm sea-level rise (Wall, 1998c). Furthermore, impacts to ecologically important wetlands and coral reefs also could have major implications for sport fishing and diving-related tourism activities in coastal regions. The risk to coastal recreation is most prominent in warm-weather destinations in the southern United States and small island nations in the Caribbean (see Chapter 17 and Section 15.3.2.10), where tourism is a leading sector of the economy.

Intersectoral resource competition also may become more pronounced, particularly with respect to water resources. Climate change scenarios for the Trent-Severn Waterway (Walker, 1996) indicate that regulation of flows for adequate downstream municipal demand would diminish the recreational boating industry, with attendant impacts for riparian recreational home property values. Increased municipal and agricultural water demand in arid regions may outweigh development of new golf courses and, in severe cases, diminish the capacity to irrigate existing facilities economically. Like declining resource availability and quality, increased resource competition may constrain the opportunities afforded by a longer recreational season.

Two main groups can be considered with respect to the potential to adapt to climate change: participants themselves and businesses and communities that cater to them. The former may be able to adapt to climate change much more readily than the latter, which are more likely to have large amounts of capital invested in fixed locations. The effect of and potential for substitution as an adaptive strategy, including locations (beach, ski resort), preferred species (coldwater vs. warmwater fish), and recreational activities broadly (skiing to mountain hiking, snowmobile to all-terrain vehicle use, golf to sailing) require more detailed investigation.

Tourism and recreation are not regarded as major net generators of greenhouse gases (except, perhaps, in the travel phase), but GHG reduction policies (e.g., carbon taxes) may increase the cost of travel, with substantial implications for destination areas (particularly isolated destinations with little domestic tourism demand).

Improvement in climate change projections, although helpful, will be insufficient to improve understanding of the implications of climate change for tourism and recreation. Even if climate change could be reliably forecast now, it is doubtful if the industry has sufficient understanding of its sensitivity to climatic variability to plan rationally for future conditions. Furthermore, the salience of climate change versus other long-term influencing variables in this sector (globalization and economic fluctuations, fuel prices, aging populations in industrialized countries, increasing travel safety and health concerns, increased environmental and cultural awareness, advances in information and transportation technology, environmental limitations—water supply and pollution) remains a critical source of uncertainty.

Global climate change will present challenges and opportunities for recreational industries and destination areas. The net economic impact of altered competitive relationships within the tourism and recreation sector is highly uncertain. Studies by Mendelsohn and Markowski (1999) and Loomis and Crespi (1999) attempt to put an economic value on climate change impacts in the United States. Although these were pioneering efforts, the assumptions and methods employed limit the confidence that can be placed in the findings. Until systematic national-level analyses of economically important recreation industries and integrated sectoral assessments for major tourism regions have been completed, there will be insufficient confidence in the magnitude of potential economic impacts to report a range (based on disparate climate, social, technical, and economic assumptions) of possible implications for this sector.

15.2.7 Public and Private Insurance Systems

The North American economy is widely affected by weather conditions. The Chicago Mercantile Exchange stated recently that weather affects US$2 trillion of the US$9 trillion gross national product (GNP). Insurance is a critical part of the vulnerability/adaptation equation because many of the economic risks and impacts of weather-related events are diversified and ultimately paid through insurance.

The discussion in the following subsections embraces public and private insurance systems, as well as disaster relief. Within the private insurance sector are many actors, including property/casualty (P/C) insurers, life/health insurers, reinsurers, self-insurers, and various trade allies (risk managers, brokers, agents, etc.). Within the public sector are direct insurance programs, as well as disaster preparedness and recovery activities. Other segments of the financial services sector appear to be less vulnerable and are treated in Chapter 8.

15.2.7.1 Private-Sector Insurance Systems

Private insurance is among the largest economic sectors in North America, with about 40% (US$780 billion) of global premium revenues in 1998 (Swiss Re, 1999). North American premiums represent 9% of GDP, or about US$2,600 per capita. Despite its size, the industry is hardly a monolith; there are numerous types of insurance companies and market segments (Mills *et al*., 2001).

Weather-related loss data presented here are based on diverse sources, and the particular costs included can vary somewhat among countries and over time. In some cases, definitions set minimum thresholds for inclusion; for example, because of the minimum cost threshold of US$25 million in the U.S. (formerly US$5 million), no winter storms were included in the statistics from 1949–1974, and few were included thereafter (Kunkel *et al*., 1999). Although large in aggregate, highly diffuse losses from structural damages as a result of land subsidence also would be captured rarely in these statistics. Data-gathering conventions can result in omission of certain types of costs (e.g., weather-related vehicle losses). Thus, the totals presented here are inherently underestimates of actual losses.

Although North America experienced 59% of global weather-related insurance losses and 36% of total economic losses during the 1985–1999 period, it experienced only 20% of the events and 1% of associated mortalities (see Figure 15-4). Total economic losses (insured and uninsured) from weather-related events represented US$253 billion (current dollars) during this period. Of that total, 38% (US$96 billion) were privately insured, with considerable year-to-year fluctuations in the ratio.

During this period, weather-related natural disasters represented 82% of total natural disaster losses in the United States and virtually the entire total in Canada (where significant earthquake losses have not occurred). Although considerable attention is given to catastrophic losses, half of all insured weather-related losses are from relatively small events (see Chapter 8).

Inflation-adjusted catastrophe losses have been growing in North America over the past 3 decades (see Figure 15-5). Corresponding exposures, measured as the inverse of the ratio of premium income to losses, have been growing (i.e., if the ratio goes down, exposure goes up); the ratio has varied by a factor of six in the United States between 1974 and 1999, and by a factor of four in Canada between 1987 and 1999 (see Figure 15-6).

Although many of the upward trends in weather-related losses are consistent with what would be expected under climate change (see Chapter 8), efforts to disentangle socioeconomic and demographic effects from climatic factors have had mixed results in the United States (Changnon *et al.*, 1997, 2000; Karl and Knight, 1998; Pielke and Landsea, 1998; Changnon, 1999). In Canada, there is a stronger sense that both factors are at work (White and Etkin, 1997; Hengeveld, 1999).

Irrespective of climate changes, it is clear that human exposure is increasing with affluence and as populations continue to move into harm's way. The estimated value of insured coastal property exposure for the first tier of counties along the Atlantic and Gulf Coasts as of 1993 was US$3.15 trillion (IIPLR and IRC, 1995).

Most types of weather-related events—rain, hail, ice storms, tidal surges, mudslides, avalanche, windstorm, drought, land subsidence, lightning, and wildfire—are of concern to insurers (Ross, 2000). Corresponding losses can range from property damage to business interruptions and temporary housing costs as a result of loss of electric power. Coastal erosion is an important consequence of sea-level rise and already is responsible for a considerable and growing rate of losses (Heinz Center, 2000). Insurance in North America originally focused on fire peril; only since the late 1930s have insurers provided broad-based coverage for weather-related events (Mills *et al.*, 2001).

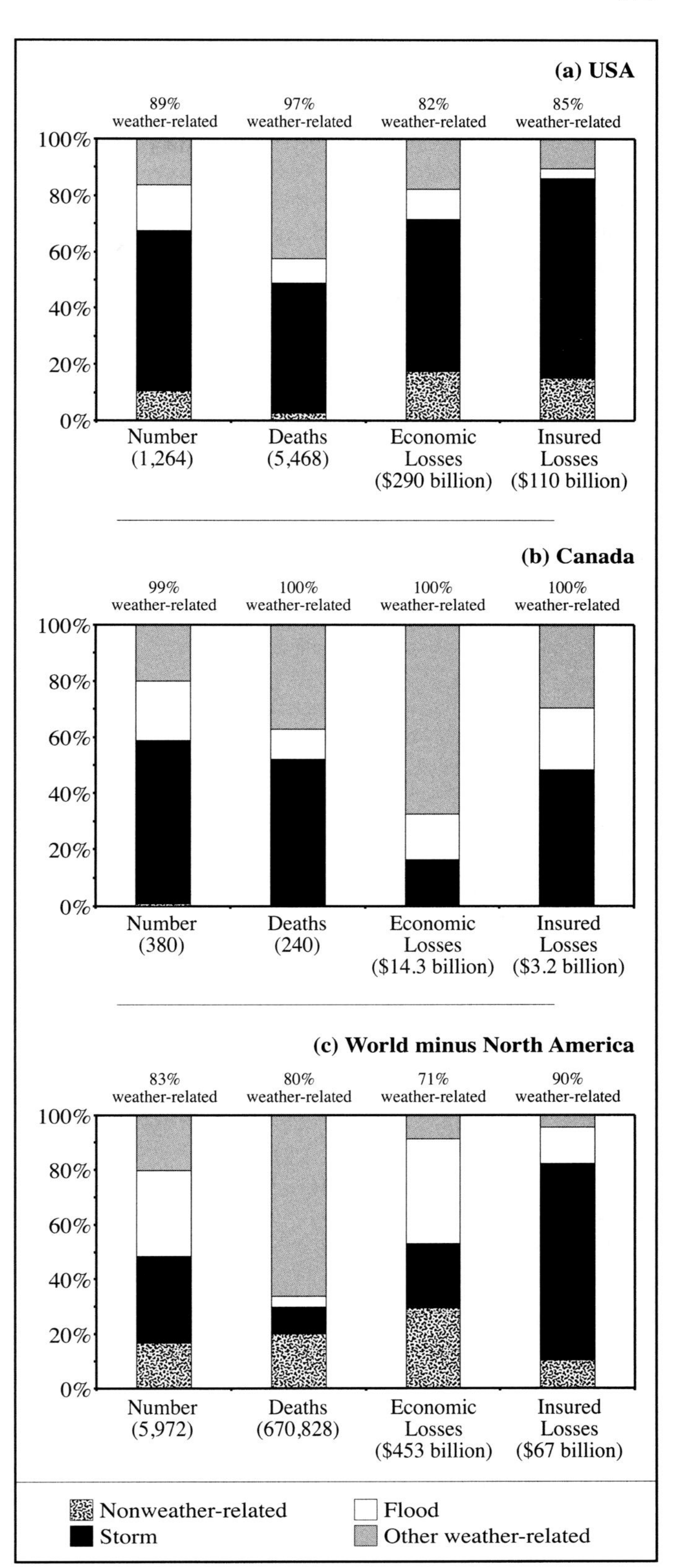

Figure 15-4: Distribution of natural-disaster losses in (a) Canada, (b) the United States, and (c) globally: 1985–1999 (Munich Re, 2000). "Storm" includes hurricanes, tornadoes, and high winds. "Other" includes weather-related events such as wildfire, landslides, avalanches, extreme temperature events, droughts, lightning, frost, and ice/snow damages.

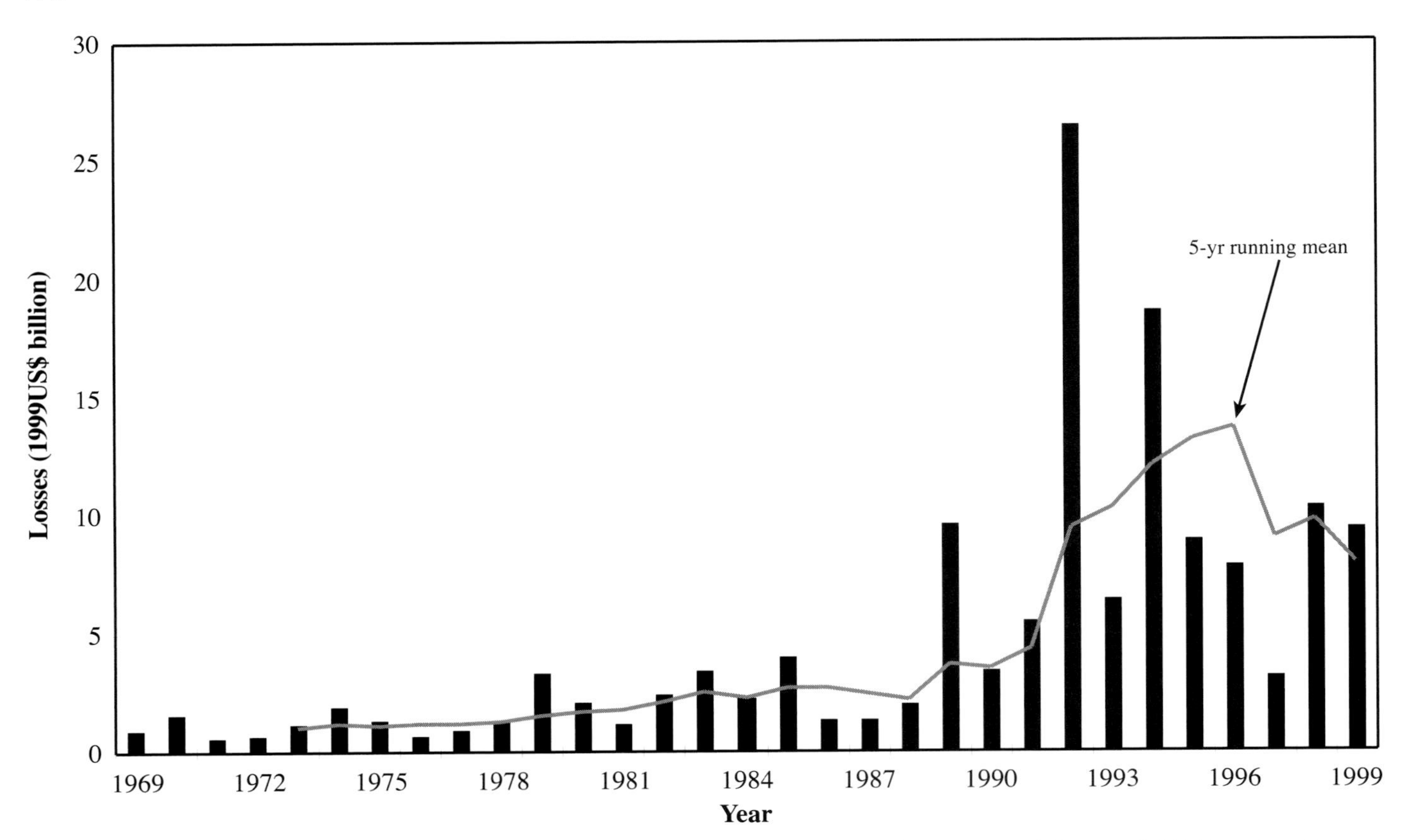

Figure 15-5: Insured natural disaster losses have increased 10-fold in North America between 1969 and 1999 (based on 5-year running mean). Data include nonweather-related losses (~15% of total). Sources: Emergency Preparedness Canada, 2000; Kunreuther and Roth, 1998 (United States).

The types of weather-related losses vary considerably between the United States and Canada. As shown in Figure 15-4, storm-related losses, including hurricanes, represent a larger share of losses in the United States, whereas flood and other weather-related events represent a far higher share in Canada. Large hurricane losses are substantial in the United States but have been virtually absent in Canada since Hurricane Hazel in 1954. Flood and storm events (other than hurricanes) have accounted for 95% of Canada's recent weather-related losses.

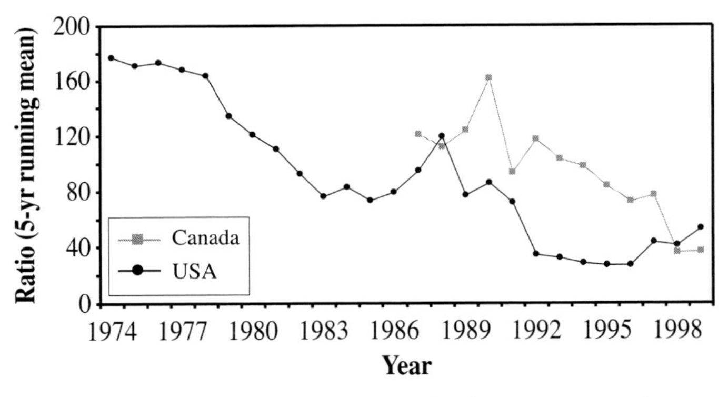

Figure 15-6: Ratio of property/casualty insurance premiums to catastrophe losses. This measure of exposure ranges from 180:1 in 1974 to 52:1 in 1999, with a minimum value of 27:1 in 1992 (the year of Hurricane Andrew). For Canada, the high value is 161:1 in 1990, and the low value is 37:1 in 1999. U.S. premiums are from AM Best and insured natural catastrophe loss figures are from Property Claims Service (Kunreuther and Roth, 1998). Canadian premiums are from Insurance Bureau of Canada (2000) and losses are from Emergency Preparedness Canada (2000). Note that premiums include revenues from nonweather-related business segments.

North American insurers have demonstrated sensitivity to the extremes and uncertainties of weather-related events. The trend in recent decades is toward increasing adverse impacts such as rising losses, upward pressure on prices, availability problems, company insolvencies, depressed stock prices, and increased reliance on government-provided insurance and disaster preparedness/recovery resources (see Chapter 8). P/C industries in the United States (see Chapter 8) and Canada (Emergency Preparedness Canada, 2000) have observed reduced and even negative operating results in years with large natural disaster losses.

A U.S.-based insurance trade association estimates that in 1997, 17% of U.S. insurance P/C premiums were associated with "significant" exposure to weather-related loss, 2% with "moderate" exposure, 66% with "minor" exposure, 10% with "minor to no" exposure, and 4% with "no" exposure (American Insurance Association, 1999). In these estimates, "exposure" is measured in terms of insurer premiums as opposed to the value of exposed property. Most flood and crop risks are not included because government insurance programs assume them. The "minor exposure" segment is mostly vehicle insurance. However, in the United States, 16% of automobile accidents are attributed to adverse weather conditions (NHTSA, 1998), as are one-third of the accidents in Canada (White and Etkin, 1997). Vehicles also sustain privately insured losses from floods and hailstorms.

In practice, insurer surplus is not poolable. Even at an industry-wide level, insurance pricing can be inadequate to cover future losses. This was evident in the case of the Northridge earthquake in California, where the US$3.4 billion in earthquake premiums collected during 25 years prior to the event fell far short of the US$15.3 billion loss (Gastel, 1999). Individual firms can come under stress long before the industry does as a whole. U.S. P/C insurers experienced approximately 650 insolvencies (bankruptcies) between 1969 and 1998, 8% of which were caused primarily by natural disasters (Matthews *et al.*, 1999). These disaster-related insolvencies include small as well as very large firms (see Chapter 8). Meanwhile, no recorded Canadian insurance insolvencies have been attributed to extreme events.

Insurers' ability to withstand weather-related losses is based on a combination of event magnitude and resources (often referred to as assets or surplus) with which they can pay claims.

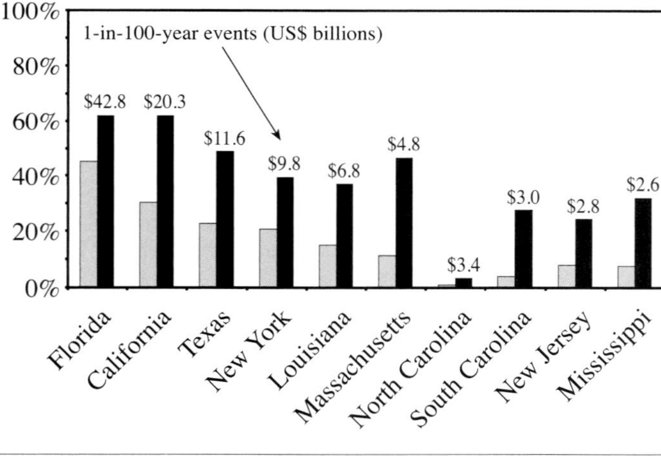

Figure 15-7: Vulnerability of U.S. insurers to 100-year events, represented as combined effect of loss magnitude and insurance company capacity (GAO, 2000). This analysis assumes that all insurers place and price policies identically. It excludes reinsurance, as well as local government-supported insurance or reinsurance programs in California and Florida. It also excludes effects of catastrophes striking more than one state (e.g., estimated 1-in-100-year loss for the entire United States is $155 billion). Capacity implied may include some surplus amounts that are not available for paying natural catastrophe claims. Losses that result in claims of more than 20% of surplus trigger initial stage of formal solvency review by National Association of Insurance Commissioners. Puerto Rico (not shown) has a 1-in-100-year loss of US$27.1 billion.

Probable maximum losses (PMLs) and vulnerability vary by locality and have been on the rise (Davidson, 1996; Cummins, 1999; GAO, 2000). As illustrated for the United States, natural disasters that are limited to the boundaries of a given state have widely varying potential impacts on insurers, depending on the capacity for paying losses (see Figure 15-7). Insurers have not performed quantitative analyses of the potential effects of climate change on PMLs.

Surplus—a measure of adaptive capacity—tends to be unstable over time, sometimes changing abruptly in response to perturbations in stock and bond market valuations or interest rates (Mooney, 1999; GAO, 2000; Swiss Re, 2000). Although U.S. insurer capacity increased considerably through the 1990s, as illustrated in Figure 15-8, large events still can result in a substantial volume of unpaid claims. Adaptive capacity can be depleted by multiple sequential events (AIRAC, 1986), as well as by past nonweather-related losses [e.g., from environmental liability (Superfund)]. There also are potential losses stemming from issues such as tobacco liability or the combination of increasing reliance on information technology and emerging questions regarding the reliability of the electricity grid. The finding that U.S. electric cooling demand would increase under global warming (Morrison and Mendelsohn, 1999) would be a compounding factor in the latter example, and adaptation (e.g., increased use of air conditioning) would result in increased GHG emissions. The trend toward increased competition is an additional stress on insurers, although consolidation also can—under the proper circumstances—increase resilience of firms.

An insurer's vulnerability often extends beyond the borders of the country in which it resides. For example, U.S. insurers collected US$35 billion in premiums for overseas insurance sales in 1997, and such insurance has been growing faster than overall premiums in recent years (III, 1999). Canadian insurers also write policies outside the country, including vulnerable parts of the United States (White and Etkin, 1997). Reinsurers (many which are based in North America) have a high degree of vulnerability given the geographical diversity of their risk portfolios.

15.2.7.2 Government-Based Insurance and Disaster-Relief Systems

Government entities are vulnerable to climate change as providers of insurance and/or disaster relief; providers of domestic and international disaster preparedness/services; and managers of property and weather-sensitive activities.

Private insurers find certain weather-related risks to be technically uninsurable as a result of their spatial concentration, actuarial uncertainty, and associated difficulties in pricing. In the United States, this problem arises mostly with respect to crop and flood risks, although government flood insurance is limited to residential and very small commercial customers. There is ongoing tension between government and private-sector players

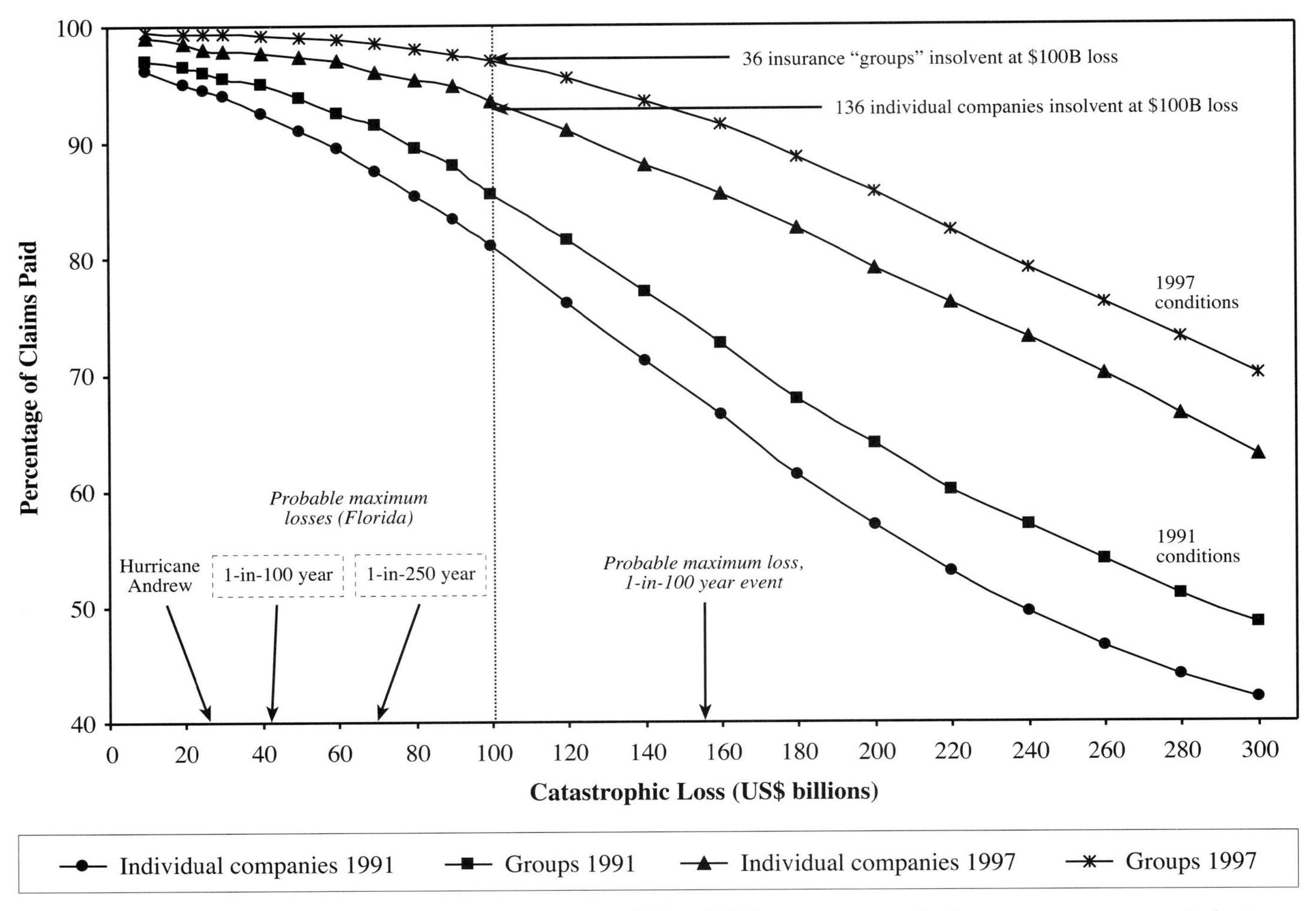

Figure 15-8: Reductions in claims payable, by size of loss. Ability of U.S. property-casualty insurance sector, as a whole, to pay claims over a wide range of losses. The chart shows four views, encompassing changes in capacity between 1991 and 1997 and whether companies have access to resources of groups that own them. Groups are not obligated to pay losses experienced by individual member firms but retain the option to do so. Together, these four scenarios represent a range of ability to pay losses. For example, for a US$155 billion loss year—a recent estimate of probable maximum loss (PML) (GAO, 2000) could be several events combined, including weather- and nonweather-related ones—65 to 90% of claims would be paid (adapted from Cummins *et al*., 1999). PML benchmarks are from GAO (2000).

in the natural disaster arena, with each looking to the other to clarify the somewhat ambiguous division of responsibility.

In Canada, government (federal, provincial, and municipal) has paid most of the natural disaster payments, amounting to US$15 billion (CDN$22 billion) in disaster relief between 1982 and 1999, including 86% of flood-related losses (Emergency Preparedness Canada, 2000). Consolidated data for U.S. government payments for natural disasters are not readily available. One source reports US$119 billion (1993US$) in government outlays between 1977 and 1993 (Anderson, 2000).

Government insurance programs can have difficulty achieving solvency, as exemplified by the US$810 million deficit in the U.S. flood insurance program in the mid-1990s (Anderson, 2000). U.S. crop and flood insurance programs have never been profitable, and growing coastal erosion risks will require doubling of rates in the United States (GAO, 2000; Heinz Center, 2000). Government insurance programs also have resulted in cases of maladaptation (e.g., inducing people to settle in vulnerable areas) (Heinz Center, 2000).

15.2.7.3 Creating and Maintaining Adaptive Capacity

Private and public insurers alike have at their disposal a variety of tools to increase adaptive capacity. These tools can be divided into the broad categories of risk spreading and risk reduction (see Chapter 8). Some of these strategies involve shifting of risk between the public and private spheres, or even back to the consumer, and in that regard have considerable public policy ramifications that are beyond the scope of this chapter.

Given their sensitivity and potential vulnerability to weather-related losses, North American insurers and their regulators have designed a variety of adaptation mechanisms. These strategies include economic measures—raising premiums or deductibles, withdrawing coverage, creating systems for pooling risks among multiple insurers (e.g., U.S. government-mandated FAIR Plans or beach/windstorm plans and state-mandated "guaranty funds" based on mandatory contributions of insurers, to pay the claims of insolvent insurers)—and the use of capital market alternatives to finance risk (see Chapter 8). Responses also include more engineering-oriented risk-reduction strategies, such as land-use planning, flood control, cloud seeding (see Section 15.3.2.4), and early warning systems.

Hurricane Andrew in 1992 and the 1998 ice storm (see Section 15.3.2.6) were wake-up calls for the North American insurance industry. Among the numerous responses have been increased utilization of catastrophe modeling, a more proactive stance toward disaster preparedness/recovery, new efforts in land-use planning, increased attention to building codes, and increased use of incentives to implement loss-prevention measures (Bruce *et al.*, 1999b).

One embedded vulnerability facing insurers and government risk managers alike is that, by design, their actuarial outlook and disaster reserving conventions are based on past experience. There is an emerging and encouraging trend to forward-looking catastrophe modeling, although it has yet to integrate knowledge from climate change modeling. Thus, under climate change, the potential for surprise is real. Vulnerability is compounded by the potential for changes in the frequency, variability, and/or spatial distribution of extreme weather events. Increased uncertainties of this nature can complicate and confound insurer methods for preparing for future events (Chichilnisky and Heal, 1998).

15.2.7.4 Equity and Sustainability Issues in Relation to Insurance

Although North America is a wealthy region, significant equity considerations arise from its cultural and economic diversity (Hooke, 2000). This raises important issues concerning the design and deployment of effective natural disaster preparedness programs (Solis *et al.*, 1997). Lower income groups tend to live in more vulnerable housing and are least able to afford potential increases in insurance costs by insurers or cross-subsidies utilized to spread risks (Miller *et al.*, 2000). Poorer individuals also tend to carry little if any life and health insurance and thus may be uninsured against public health risks such as urban heat catastrophes or urban air quality risks posed by climate change (see Chapter 9).

In some cases, sustainable development and environmental protection complement insurance loss reduction objectives. Examples include protection or restoration of wetland areas that increases tidal flooding defense, sustainable forest practices that reduce risk of inland flooding and mudslides, or agricultural practices that increase drought resistance or reduce soil erosion (see Chapters 5 and 8; Scott and Coustalin, 1995; Hamilton, 2000). Similarly on the end-user side, a variety of energy conservation and renewable energy measures have been found to offer insurance loss-control co-benefits (Mills, 1996, 1999; Nutter, 1996; Mills and Knoepfel, 1997; American Insurance Association, 1999; Vine *et al.*, 1999). The insurance sector itself has begun to examine its activities within the broader framework of sustainability (Mileti, 1997; Kunreuther and Roth, 1998).

15.3. Adaptation Potential and Vulnerability

15.3.1. Generic Issues

Subregions within North America can be affected by global climate change through direct first-order effects on land and water resources and through indirect impacts that are related to reactive and proactive responses by actors within and outside the subregion. Direct effects include incremental processes and extreme weather events with long-term impact on the region's environmental, social, or economic systems. Indirect effects would include societal changes that could be set in motion, primarily by policies generated outside the region, to mitigate greenhouse warming or adapt to perceived threats.

Calculations of climate-related impacts within subregions must consider both direct and indirect effects. Actual impacts will depend on the effectiveness of adaptation. For example, changes in growing season, natural streamflow, or climate-related demand for heating and cooling energy can be calculated directly from changes in climatic parameters. However, changes in potential may not necessarily lead to a response by stakeholders. Other factors may intervene to prevent a subregion from adapting to a new climate-related opportunity or risk, including changes in market conditions, institutional arrangements, and management objectives.

Because North America includes areas of intensive urban and landscape management, as well as areas of "extensive" management, adaptation capabilities and vulnerabilities are likely to vary between subregions. There is potential for maladaptation, in part as a result of the availability of insurance and disaster relief measures (MacIver, 1998). Vulnerabilities also exist in highly developed regions with well-maintained infrastructure (e.g., dams designed for flood control) because of the growing need to manage resources to achieve multiple objectives. Would climate change ameliorate or exacerbate these risks? Analysis of case studies can provide guidance.

15.3.2. Subregional and Extra-Regional Cases

15.3.2.1. Introduction

The discussion that follows offers a sample of subregional and extra-regional cases that reflect the unique and changing nature of the relationship between climate and places in the North American context. Six subregions have been defined: Pacific, Rocky Mountains–Southwest United States, Prairies–Great Plains,

Great Lakes–St. Lawrence, North Atlantic, and Southeast United States (see Figure 15-9). These subregions represent unique landscapes that have experienced development paths that reflect natural potential for resource development as well as region-specific histories of evolving economies and communities. Some of these cases describe responses to recent extreme events. Others focus on scenarios of future climatic changes. Canada and the United States have a long history of cooperating on resource management concerns in transboundary areas, including the Boundary Waters Treaty of 1909 and more recent agreements on transboundary air pollution, watershed development, and the environmental agreement associated with NAFTA. There also have been binational and subregional conflicts over a wide range of issues that are climate-sensitive, including water resources, wetland protection, trading of forest and agricultural commodities, and harvesting of transboundary fish stocks. Various cooperative management approaches are being considered. Could climate change make a difference in any of these conflicts? Extra-regional issues are considered for the Arctic, U.S.-Mexican, and U.S.-Caribbean borders.

15.3.2.2. *Pacific Subregion*

This subregion includes the Columbia River basin—the fourth-largest in North America, and one of the most valuable and heavily developed watersheds. The Columbia system produces 18,500 MW of hydroelectricity; it also is managed for flood control, agriculture, navigation, sport and commercial fisheries, log transportation, recreation, wildlife habitat, and urban, industrial, and aboriginal uses. Climate variability affects the total volume and temporal pattern of natural runoff, although binational management agreements have substantially altered this pattern. Any hydrological changes caused by long-term climatic change will have impacts that may affect the management of the system and the nature of conflicts among various users.

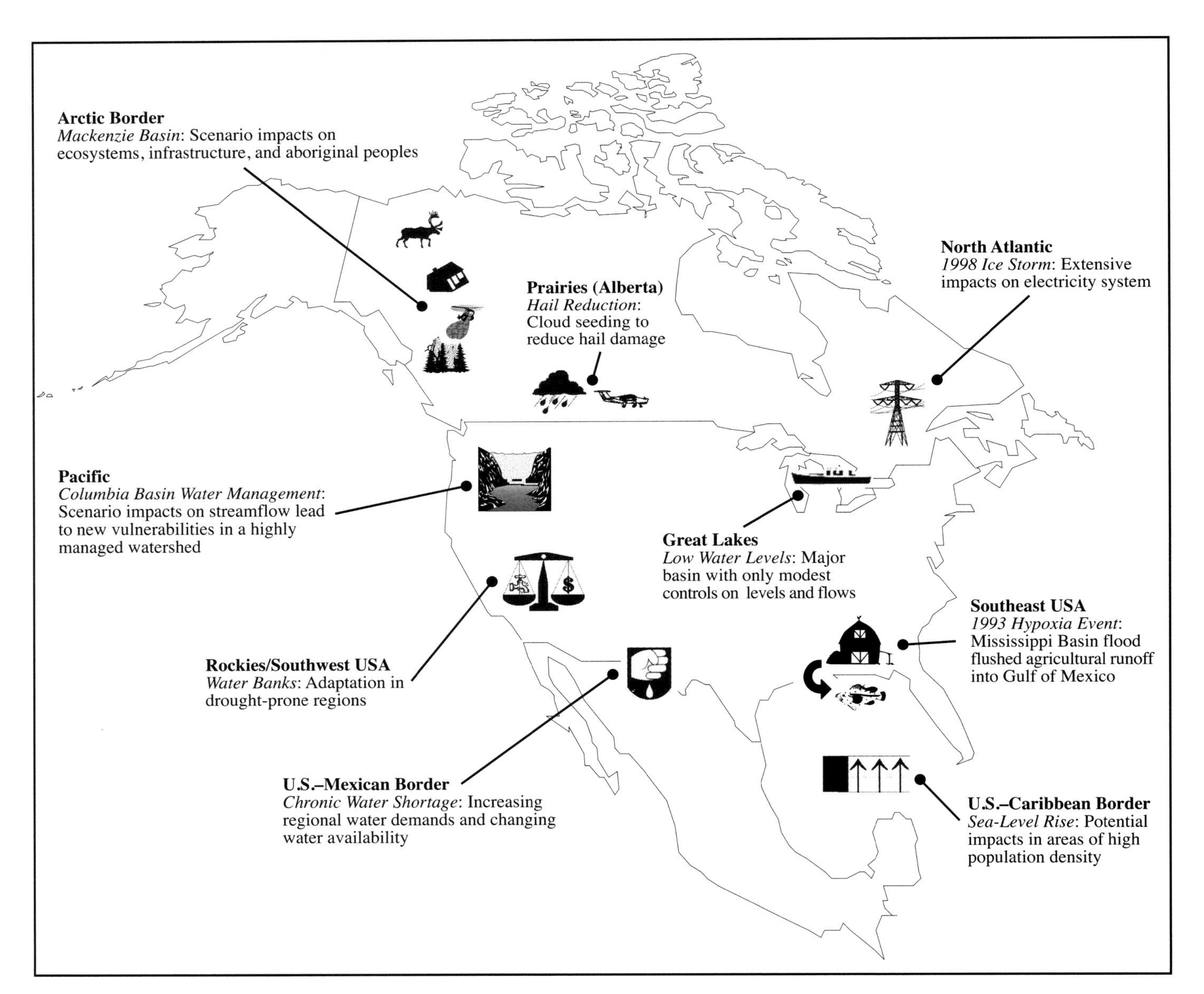

Figure 15-9: Subregional cases from North America (described in Section 15.3.2).

Recent management and policy changes have led to changes in system operations; fish habitat protection now is regarded as a high priority—on par with hydroelectricity production and flood control (Cohen *et al.*, 2000).

Climatic change is expected to lead to reductions in annual streamflow. Estimates for three warming scenarios range from a small increase to a decrease of 16% by 2050, with an earlier annual peak, higher winter flows, and lower summer flows (Lettenmaier *et al.*, 1996, 1999; Hamlet and Lettenmaier, 1999; Miles *et al.*, 2000). This would affect system reliability. Results from a reservoir system model for the driest of the three scenarios [based on the Max Planck Institute (MPI) GCM, ECHAM-4] indicate reductions of 15% in reliability for Snake River irrigation, 11% for McNary Dam "biological flow" (to support fish habitat), 14% for firm energy production, and 19% for nonfirm energy production. Reliability for flood control would remain relatively unchanged in this case but would decrease in the wetter scenarios (Hamlet and Lettenmaier, 1999; Miles *et al.*, 2000).

Interviews were held in 1997 with water resources managers and other stakeholders in the upstream portion of the Columbia, on the Canadian side. Their responses to the MPI scenario were similar to those obtained from the reservoir system model. For example, they noted that flood risk would not necessarily be reduced because upstream storage facilities may be releasing high discharges for fish habitat protection or other reasons. Similarly, irrigation supply and hydroelectricity production would be reduced because of lower annual flow combined with fish protection requirements. In these circumstances, regional utilities might have to purchase electricity from fossil fuel sources, thereby affecting the region's ability to meet GHG emission targets (Cohen *et al.*, 2000).

This case study suggests that despite the high level of development and management in the Columbia basin, vulnerabilities would still exist and impacts could still occur in scenarios of natural streamflow changes caused by global climate change. It also suggests that expanded development of hydroelectric facilities, as a means of reducing GHG emissions, will create considerable challenges to management, given increases in the sensitivity of regional energy supplies and other water-related resources to climate.

15.3.2.3. Rocky Mountains–Southwest U.S. Subregion

Rapid population growth coupled with water supplies that are limited, heavily utilized, and highly variable present significant challenges to governments and the private sector in this region. The effects of climate change may add further stresses (Frederick and Schwarz, 1999). One strategy that has been developed to cope with these issues is use of water banks. Water banks can be used to mitigate the economic impacts of drought periods by increasing the reliability of water supply or by facilitating short-term reallocation of water among users. Water banks can be used for environmental purposes—for example, Idaho's water banks have provided significant quantities of water to assist anadromous fish passage (Miller, 2000b). However, to the extent that water banking entails increased water withdrawals from surface sources, there could be further adverse impacts on aquatic ecosystems.

There are two types of arrangements to which the term "water bank" has been applied. "Groundwater storage banks" include active conjunctive-use programs whereby surface water is used to recharge an aquifer, which is then used as a source of water supply during periods when surface water is less abundant. Another use of the term "water bank" refers to a formal mechanism created to facilitate voluntary changes in the use of water under existing rights. We can distinguish these as "water transfer banks." The defining feature of this type of water bank is that it provides an established process or procedure for accomplishing such transfers. California has pioneered the development of both types of water banks. Various groundwater recharge programs have been developed in California, some of which incorporate features of groundwater banks and water transfer banks. One such example, the Bakersfield Recharge Facility in Kern County, is discussed below. The Emergency Drought Water Banks set up by the state of California in 1991, 1992, and 1994 are examples of water transfer banks.

In 1987, at the beginning of a severe multi-year drought, California had relatively little experience with privately arranged water transfers outside the confines of individual irrigation districts and major project service areas. The lack of market development may have been related, in part, to the predominance of large state and federal water projects in California and, in part, to the difficulty of obtaining approval for such transfers from the California State Water Resources Control Board (SWRCB) (MacDonnell, 1990; NRC, 1992). As the drought persisted, it became clear that neither within-project water transfers nor those requiring approval of the SWRCB were likely to occur in sufficient volume to alleviate the very uneven impacts of the drought on major urban centers and irrigated agriculture. In 1991, the state responded to the growing crisis by creating the first Emergency Drought Water Bank. California's Department of Water Resources (DWR) acted as the manager of these water banks because most of the anticipated transfers required conveyance of the water through State Water Project (SWP) facilities (primarily the state's pumping plant in the Sacramento/San Joaquin delta and the California aqueduct). DWR personnel negotiated the purchase contracts, monitored compliance with those contracts, obtained SWRCB approval where needed, and coordinated deliveries of water to purchasers. Despite initial difficulties in estimating the quantity of water needed and the appropriate price, as well as some concerns about uncompensated third-party impacts, the program generally is considered to have been a major success—with net economic benefits far exceeding any negative impacts (Howitt *et al.*, 1992; Dixon *et al.*, 1993; MacDonnell *et al.*, 1994).

Another successful California water bank is the 1,135-ha recharge facility operated by the city of Bakersfield. This groundwater-banking program also facilitates water transfers.

The city owns Kern River water rights, which yield a highly variable quantity of water from 1 year to the next. A portion of these rights is leased on long-term contracts to neighboring irrigation districts. The city has constructed a recharge facility into which it spreads much of its remaining Kern River water. This activity has increased the amount of water that potentially is available for extraction in the underlying aquifer. Three other water districts also are allowed to bank their surplus water by spreading it in the recharge facility. Each bank participant can withdraw the water in its own account, as needed (e.g., during drought periods), or transfer the water to another party:

> "The banked water can be extracted and transported for direct use, sold or exchanged, with the stipulation that extracted water must be used in the San Joaquin Valley portion of Kern County for irrigation, light commercial and industrial, and municipal or domestic purposes, unless otherwise specified. Located upstream of the facility, the city of Bakersfield often sells its banked water to users downstream, or exchanges it for water from other sources" (Wong, 1999).

Operation of this recharge facility provides county water users with drought insurance and flood control benefits. "Banking recharge operations during flood release periods serve to minimize downstream flooding problems while maximizing recharge of water on the Kern River for local benefit" (Wong, 1999). For example, the bank participants successfully used the recharge facility in 1995 to manage heavy runoff.

15.3.2.4. Prairies–Great Plains Subregion

An attempt at proactive "mitigation" of an atmospheric threat is the cloud seeding program underway in Alberta since 1996 to reduce the severity of hailstorms. Central North America is one of the most hail-prone regions in the world. A 1995 hailstorm did more than US$1.1 billion damage in Dallas, a 1990 storm did US$600 million damage in Denver, and a 1991 hailstorm did about US$240 million (CDN$340 million) damage in Calgary. Hailstorms account for more than half of the catastrophic events in this region. In the Alberta program, storms with the potential to generate hail damage are intensively treated through cloud seeding. Since the program began, there has been an increase in the number of potential hailstorms, yet a pronounced reduction in actual hail damage. Indeed, during the first 4 years of the program, only one storm has done damage in the area under supervision (IBC, 2000).

Throughout the long history of hail suppression in North America and elsewhere, success has been mixed (e.g., Changnon *et al.*, 1978), so it is not yet clear whether this approach will work over the long term. The experimental program in Alberta is expected to continue indefinitely because the modest US$1 million (CDN$1.5 million) annual operating cost appears to be more than offset by the reduction in observed hail damage (IBC, 2000).

15.3.2.5. Great Lakes–St. Lawrence Subregion

Scenario-based studies in the Great Lakes–St. Lawrence basin conducted over the past 15 years (see Section 15.2.1) have indicated consistently that a warmer climate would lead to reductions in water supply and lake levels (Cohen, 1986; Croley, 1990; Hartmann, 1990; Mortsch and Quinn, 1996; Mortsch, 1998). Observed low-level events have been rare during the past several decades, but when they have occurred (e.g., 1963–1965, 1988), conflicts related to existing water diversions and competing stakeholders have led to legal challenges and political stress (Changnon and Glantz, 1996).

In 1998–1999, dry and record warm weather combined to reduce ice extent and lower water levels (Assel *et al.*, 2000). In November 1998 and again in spring 1999, cargo limits were placed on ships travelling through the Great Lakes and the St. Lawrence Seaway when water levels fell by more than 25 cm. Some commercial ships ran aground. Lower water levels significantly increased the distance between docks and waters at some tourism facilities and marinas. Recreational boaters had access to fewer waterways. In addition, hydroelectric facilities are very dependent on water flows and have expressed concern about maintaining production (Mittelstaedt, 1999). These observed impacts are similar to the impact scenarios described earlier. The reduction in ice cover did provide some offsetting benefits by reducing the need for icebreakers during the winter and spring (Assel *et al.*, 2000).

Water resources management, ecosystem and land management, and health concerns are highlighted in the Great Lakes–St. Lawrence Basin Project (GLSLB) and the Toronto–Niagara Region Study (TNR). The former is a binational exercise that used analogs and model-based scenarios to assess impacts and adaptation responses. Reduced lake levels continue to be projected, leading to potential conflicts over water regulation and diversions, water quality, and rural water use (Mortsch *et al.*, 1998). The TNR exercise is in its initial stages. Within the GLSLB region, the TNR represents a highly urbanized area that has a significant impact on surrounding agricultural and forest landscapes. Urban influences on land use, air quality, and human health will be a major focus. In addition, concerns about vulnerabilities of the built environment (transportation and energy infrastructure, etc.) have been identified as high-priority research questions (Mills and Craig, 1999).

There also have been studies of alternative scenarios for lake-level management. Decision support systems can facilitate this process (Chao *et al.*, 1999).

15.3.2.6. North Atlantic Subregion

Severe winter storms combined with a loss of power can have devastating consequences, even in a highly developed region such as Ontario-Quebec and the northeast United States. In January 1998, a severe winter storm struck; instead of snow, some areas accumulated more than 80 mm of freezing rain—double the

***Table 15-5**: Estimated damages from 1998 ice storm.*[a]

Type of Loss	Canada (CDN$)	United States (US$)	Total (US$)
Insured losses	$1.44 billion	$0.2 billion	$1.2 billion
Insurance claims	696,590	139,650	835,240
Deaths	28	17	45
People (customers) without power	4,700,000 (1,673,000)	546,000	5,246,000
Electricity transmission towers/distribution poles toppled	130/30,000	unknown	unknown
Electric transmission system damage	$1 billion	unknown	unknown
Manufacturing, transportation, communications, and retail business losses	$1.6 billion	unknown	unknown
Forests damaged	unknown	17.5 million acres	unknown
Loss of worker income	$1 billion	unknown	unknown
Dairy producers experiencing business disruption	5,500	unknown	unknown
Loss of milk	$7.3 million	$12.7 million	$18 million
Agricultural sector (poultry, livestock, maple syrup)	$25 million	$10.5 million	$28 million
Quebec and Ontario governments	$1.1 billion		

[a] Based on analysis conducted by the Canadian Institute for Catastrophic Loss Reduction and U.S.-based Institute for Business and Home Safety, both of which are insurance industry organizations (Lecomte *et al.*, 1998). Losses as of 1 October 1998 (1 CDN$ = 0.7 US$).

amount of precipitation experienced in any prior ice storm. The result was a catastrophe that produced the largest estimated insured loss in the history of Canada. The same storm ran across northern New York and parts of Vermont, New Hampshire, and Maine in the United States (see Table 15-5). The basic electric power infrastructure, with its lengthy transmission lines, was severely damaged, stranding some residents and farmers without power for as many as 4 weeks. Almost 5 million people were without power at some point during the storm, and consideration was given to evacuation of Montreal. In Canada, there were 28 deaths, and damages were US$2–3 billion (CDN$3–4 billion) (Kovacs, 1998; Kerry *et al.*, 1999). The same storm caused 17 deaths in the United States, as well as damages exceeding US$1 billion in New York and the New England states—one-third of which was from losses to electric utilities and communications (DeGaetano, 2000). Combined Canadian and U.S. insured losses stood in excess of US$1.2 billion as of October 1, 1998. Total Canadian insured and uninsured economic losses were approximately US$4 billion (CDN$6.4 billion).

The ice storm produced more than 835,000 insurance claims from policyholders in Canada and the United States. This was 20% more claims than were created by Hurricane Andrew, the costliest natural disaster in the history of the United States.

The event served as a grim learning laboratory for the insurance and disaster recovery communities. It revealed the wide spectrum of insured and noninsured losses that can materialize from a single natural catastrophe, including:

- Property losses (e.g., roof damages and destruction of perishable goods as a result of loss of electric power)
- Business interruption losses (19% of the employed Canadian workforce was unable to get to work)
- Health/life losses (including losses incurred during recovery operations)
- Additional living expense costs for people relocated to temporary housing
- A host of agricultural losses, ranging from livestock deaths, to interrupted maple syrup production, to milk production
- Disruption and damage to recreation and tourism infrastructure

- Disaster recovery costs, including personnel and overtime expenses, provision of backup electric generators and fuel, debris clearing, temporary shelter for displaced citizens, and disaster assistance payments to victims.

Ice storms occur regularly in North America, although severe and prolonged damage is rare. Several urban centers are vulnerable to major storms, including Minneapolis, Winnipeg, Chicago, Detroit, Toronto, Buffalo, Montreal, Boston, and even New York. Many communities are not prepared for an extreme winter storm, particularly combined with the loss of electric power. In the 1998 storm, total losses exceeded insured losses by a substantial margin. The event also raised questions about the connection between such events, the El Niño phenomenon, and global climate change.

15.3.2.7. Southeast United States

The Gulf of Mexico has a problem with hypoxia (low oxygen) during summer over an area of approximately 15,000 km^2 (Turner and Rabalais, 1991; Rabalais *et al*., 1996). The hypoxia has been shown to be a result of excess nutrients—primarily nitrogen—transported to the Gulf from the Mississippi River basin. This basin is heavily fertilized for crop production. The causes of, and proposed solutions for, this hypoxia problem are an excellent example of a complex interaction between climate and other human stresses on natural ecosystems. Specific "lessons" from this problem include the following:

- The hypoxia problem was greatly exacerbated by the 1993 Mississippi basin flood that appeared to "flush" large amounts of residual nitrogen from agriculture to the Gulf (Rabalais *et al*., 1996). Increases in extreme precipitation events, which are likely to occur with climate change, are likely to exacerbate coastal eutrophication problems in many locations.
- Recommended solutions to the hypoxia problem include creation of approximately 60,000 km^2 of wetlands to remove ~1,000 t yr^{-1} of nitrogen by denitrification (Mitsch *et al*., 2001). If such "eco-technologies" become a common solution to eutrophication problems on a global basis, they could result in a significant flux of N_2O to the atmosphere.

Management practices designed to reduce nitrogen losses from agricultural watersheds—from improved fertilizer management to the construction of wetlands—will be strongly affected by climate (NRC, 1993). Climate change will decrease the reliability of these practices. Increases in climate extremes almost certainly will decrease their long-term performance.

15.3.2.8. Arctic Border

A case study of the regional impacts of climate change scenarios has been completed in the Mackenzie basin, a watershed that extends from the mid-latitudes to the subarctic in northwest Canada. A lengthy description of this case study, known as the Mackenzie Basin Impact Study, is available in Cohen (1994, 1996, 1997a,b,c). A sketch of the MBIS integrating framework is shown in Figure 15-10. Within this process, several types of integration exercises were used, including models, stakeholder consultation, and thematic discussions.

As a high-latitude watershed, the Mackenzie basin has been regarded as an area that might benefit in certain ways from a warmer climate. Taken individually, economic impacts could be quantified, and these impacts might show substantial benefits for the region. Other factors must be considered, however, and some of these factors may constrain the potential benefits:

- The current system of land transportation, much of which is based on a stable ice and snow cover for winter roads
- Current ranges and habitats of wildlife, which underpin conservation plans and native land claims
- Scientific uncertainty, which hampers anticipatory responses to projected beneficial conditions.

Potential negative impacts of climate warming also must be considered because they may offset possible benefits. An example is the implication of hydrological and landscape changes on water management agreements. Initial projections of runoff and lake levels are for declines below observed minima (Soulis *et al*., 1994; Kerr, 1997). Peace River ice cover will be affected by temperature changes and changes in outflow from the Bennett Dam in northeast British Columbia (Andres, 1994). There continue to be uncertainties in projections of hydrological impacts; the Global Energy and Water Cycle Experiment (GEWEX) is addressing these uncertainties (see Chapter 16). There has been a strong warming trend in the region during the past 40 years, and Great Slave Lake experienced new record minimum lake levels in 1995.

It would appear that the other main threats to the Mackenzie landscape are accelerated erosion and landslides caused by permafrost thaw and extreme events (fire, storm surges), especially in sloping terrain and the Beaufort Sea coastal zone (Aylsworth and Egginton, 1994; Solomon, 1994; Aylsworth and Duk-Rodkin, 1997; Dyke *et al*., 1997); increased fire hazard (Hartley and Marshall, 1997; Kadonaga, 1997); changes in climate conditions that influence the development of peatlands (Nicholson *et al*., 1996, 1997; Gignac *et al*., 1998); and invasion of new pests and diseases from warmer regions (Sieben *et al*., 1997).

Impacts on fisheries and wildlife are difficult to project, as a result of lack of long-term data, complexity of life cycles, and incomplete information on responses to previous environmental changes (Brotton and Wall, 1997; Gratto-Trevor, 1997; Latour and MacLean, 1997; Maarouf and Boyd, 1997; Melville, 1997). Outside of MBIS, there have been few impact studies on North American boreal and Arctic freshwater fisheries (Weatherhead and Morseth, 1998). Some information is available on terrestrial wildlife and Arctic marine fisheries (see Chapter 16). Others have outlined the potential for freshwater ecosystem

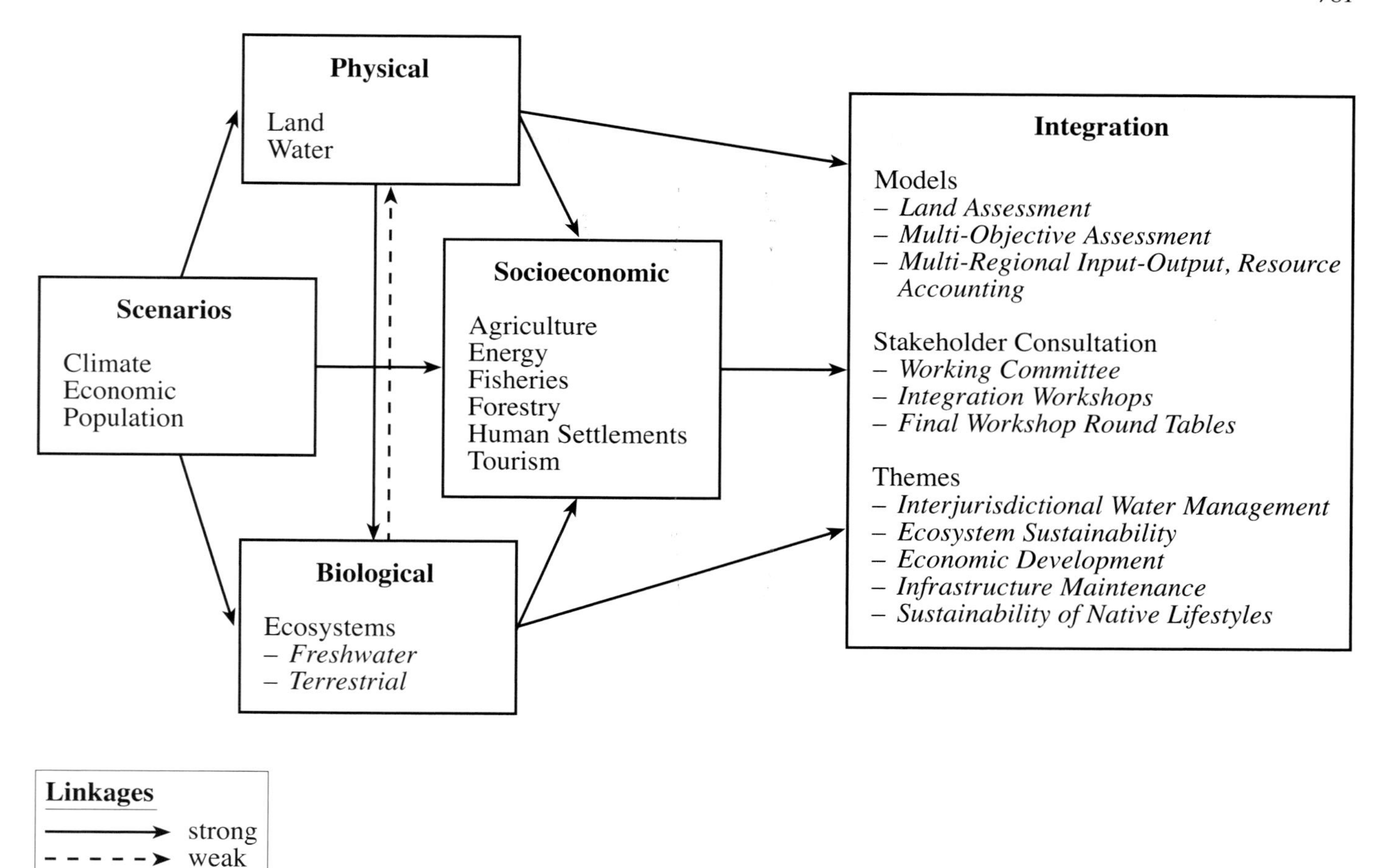

Figure 15-10: Research framework for MBIS (Cohen, 1997b). Scenarios were generated for use by physical and biological system studies. Socioeconomic component used results from these assessments, as well as climate scenario information where appropriate. Integration exercises were based on outputs from sectoral studies.

impacts, including loss or reduction of deltaic lakes, increased pondwater temperatures, side effects of permafrost thaw (including sedimentation of rivers), and changes in primary productivity depending on nutrient levels (Rouse *et al.,* 1997; Schindler *et al.,* 1997; Meyer *et al.,* 1999).

First-order and second-order impacts eventually lead to others that are considerably more difficult to address. Will land claims or water resources agreements be affected? Would it be appropriate to artificially maintain historic water levels in the Peace-Athabasca delta within this scenario of climate change (see Chapter 16)? Could there be new conflicts over land use, especially if agriculture expands northward to take advantage of improved soil capability to support crop production (Brklacich *et al.*, 1996, 1997b)? What might be the effects on parks and other protected areas (Pollard and Benton, 1994) and tourism (Staple and Wall, 1996; Brotton and Wall, 1997; Brotton *et al.*, 1997; Wall, 1998a,b)? Could climate change affect the economics of commercial forestry (Rothman and Herbert, 1997) or oil and gas production in the Beaufort Sea (Anderson and DiFrancesco, 1997)?

Two exercises dealt with individual perceptions of aboriginal responses to future scenarios. The first asked for a listing of physical and biological impacts, as well as how aboriginal people would be affected if they continued to pursue a traditional aboriginal lifestyle or if they became more active in the formal wage economy (similar to the dual-economy situation described by Shukla, 1997). Results showed that perceptions of impact and vulnerability were influenced by visions of future lifestyles (Aharonian, 1994). The second exercise used input-output modeling and a community survey to look at responses to a potential benefit of warming—an opportunity for expanded activity in the formal wage economy (because of the longer summer) that would force people to relocate from their traditional community if they wanted the employment. Results showed a willingness to accept the opportunity, but there were concerns about social impacts on the community of relocation or creation of commuter workers who would be absent for extended periods (Lonergan and Kavanagh, 1997).

Potential land-use impacts and adaptation responses also were considered. One issue was whether the region could take advantage of a longer growing season to expand wheat production. Results showed that this was possible (given the constraints of the model), but only if expanded irrigation services would be made available (Brklacich *et al.*, 1997a). The side effects of this are unknown. The second case was a forestry study that showed that commercial yields of softwood lumber would decline because of changing growth potential for these species, combined with increased fire frequency. This would offset improvements for hardwood species (Hartley and Marshall, 1997; Rothman and Herbert, 1997). Integrated modeling exercises, using these sectoral outputs, indicated that agriculture production would be able to achieve its goals, but there would be shortfalls in softwood (spruce) production and a substantial increase in erosion (Yin and Cohen, 1994; Huang *et al*., 1998; Yin *et al*., 2000).

Stakeholders were asked to react to the various research results through thematic roundtable discussions. One of these themes was infrastructure maintenance. Engineers suggested that physical infrastructure—such as roads, airstrips, and pipelines—could be repaired or designed differently to meet changing climatic conditions, but there certainly would be increased costs in doing so. It also was suggested that the notion of infrastructure should include education and insurance mechanisms, emergency response, and social and cultural institutions that pool risk and support people in times of stress (Zdan, 1997). Changes in community development already have influenced emergency response to floods and could affect adaptation to future climate changes (Newton, 1995, 1997).

In another thematic context—ecosystem sustainability—it is unclear whether current institutions in the region would be capable of coping with a scenario of simultaneous changes in renewable resources and dependent communities, superimposed on other nonclimate stresses affecting the region. During the past few years in northern Canada, co-management processes have been developing for aboriginal and non-aboriginal (and government) stakeholders concerned with wildlife and water resources. This has been part of a larger trend of devolution of responsibility from central to regional authorities in Canada—a trend that is taking place for reasons other than climate change. Regional stakeholders regard these processes as their best approach to coping with climate change, although it was noted that climate change would be a new experience for everyone (Irlbacher, 1997).

Climate change joins a long list of factors that affect the lifestyles and livelihoods of people who live in the north. With or without climate change, however, native lifestyles are in flux, and any potential impacts should be considered in that context. Communities already are being affected by changes in wildlife harvesting opportunities and wage-based employment. Education and transportation have improved, and some native people are opting for jobs in the wage economy. As the "two economies" continue to evolve, communities will be trying to provide opportunities in traditional and wage-based activities, but if climate change affects wildlife numbers or habitats, traditional patterns of wildlife harvesting would have to be modified.

Native people have adapted to change, but the predictability of the extent, duration, and speed made such adaptation possible. There is real concern that if future changes were to be fast, dramatic, and surprising, native communities would be left in a very vulnerable position. Traditional lifestyles would be at risk of disappearing.

Stakeholders have proposed that monitoring and adaptation could be undertaken through more effective partnerships between governments and native communities. Traditional knowledge, along with western science, should be used in modern management practices as well as in traditional activities. Training in renewable resource management for aboriginal youth would be an important component of preparing for the future (Pinter, 1997).

15.3.2.9. U.S.–Mexican Border

The U.S.-Mexican border is characterized by a semi-arid climate, with higher precipitation toward both coasts. Evaporation exceeds precipitation for many months of the year, causing a soil moisture deficit. Many studies relate climate and hydrology along the U.S. border states (Green and Sellers, 1964; Hastings and Turner, 1965; Norwine, 1978; Cayan and Peterson, 1989) and along the Mexican border states (Mosiño and García, 1973; Schmidt, 1975; Acosta, 1988; Cavazos, 1998). The western border has an annual rainfall of about 250 mm yr^{-1}, with temperatures ranging from 10 to 28°C. The Arizona-Sonora border has a much more extreme and desert climate, with only 80 mm yr^{-1} of precipitation and maximum temperatures reaching 45°C. Around the Ciudad Juarez (Mexico)–El Paso (Texas) border, annual precipitation is around 220 mm, with maximum temperatures reaching 40°C. Finally, in the easternmost part of the border (Brownsville, Texas–Matamoros, Mexico), annual precipitation is close to 675 mm, with temperatures ranging from 11 to 38°C.

There is large climate variability along the U.S.–Mexican border states. Drought and floods occur frequently (Powell Consortium, 1995), some influenced by the occurrence of El Niño (Cavazos and Hastenrath, 1990; Cayan and Webb, 1992). The Mexican states along the border are highly vulnerable to drought, particularly along the Rio Grande region, areas with low precipitation rates (less then a 100 mm yr^{-1}) (Mundo and Martínez Austria, 1993; Hernández, 1995; Mendoza *et al*., 1997). Strong El Niño events during the summer result in severe water deficit in northern Mexico (Magaña and Quintanar, 1997), and serious negative impacts in agricultural activities. Such vulnerable conditions during drought periods have resulted in legal disputes between Mexico and the state of Texas with regard to rights to Rio Grande water.

It is unclear what the signs of future changes in precipitation and water availability will be along the border (Mearns *et al*., 1995; Magaña *et al*., 1997; Mendoza *et al*., 1997). Some scenarios suggest increased winter precipitation and decreased summer precipitation (Magaña *et al*., 1998). Without reliable predictions of precipitation changes across drainage basins, little confidence can be placed in hypothesized effects of global warming on annual runoff (Karl and Riebsame, 1989).

Therefore, to examine potential impacts of climate change on water availability, most studies make use of long instrumental records of precipitation and streamflow. Some analyses indicate that in recent decades there is a positive trend in streamflow and even precipitation, corresponding to more water availability along the U.S.–Mexican border (Magaña *et al.*, 1998; Magaña and Conde, 2000). However, demand for water in agriculture, industries, and cities is increasing steadily, surpassing recent increases in water availability (Mundo and Martínez Austria, 1994).

The signal of climate change along the U.S.–Mexican border may be exacerbated along the Mexican side by differences in land use. Along the U.S. side, the presence of gardens and green areas lead to a cooler environment relative to the Mexican side. Currently, maximum temperatures in contiguous border towns may differ by as much as 3°C as a result of different characteristics in vegetation (Balling, 1988). This cross-border contrast is possible because water consumption in U.S. border cities is four times larger than in Mexican border cities. Surface water distribution along the border (Rio Grande and Colorado watersheds) is regulated by the Binational Treaty of 1944. Yet, there are no regulations for subsurface water (Sanchez, 1994).

15.3.2.10. The U.S.–Caribbean Border

Some of the U.S.–Caribbean border problems that may increase as a result of climate change are related to:

- Effects of sea-level changes on coastal ecosystems
- Effects of temperature increases on terrestrial and aquatic ecosystems, including possible effects on economically important species
- Effects of climate change on socioeconomic structures and activities.

Considering the high population density in part of the Caribbean islands and coastal areas, human settlements will be highly affected by sea-level rise and saline intrusion (Vincente *et al.*, 1993). Coral reefs in the region may be severely affected by coral bleaching induced by warmer water temperatures (Milliman, 1993).

Tourism is a major economic activity in most of the region. A change in rainfall patterns, tropical storms, and warming of the climate in temperate countries may affect the comparative advantages of this sector in the region (Alm *et al.*, 1993). Narrow beaches combined with projected sea-level rise contribute to the vulnerability of the tourism sector to changes in climate (Gable, 1997).

Potential consequences of changes in extreme events (e.g., hurricanes) are not well-defined. Other trends, such as changing demographic patterns, may exacerbate impacts. Recent large losses of life from rainfall-induced floods, mudflows, and landslides reflect increasing concentrations of residents in high risk areas (Rodriguez, 1997; Pulwarty, 1998).

15.4. Synthesis

Our impacts and adaptation assessment indicates that changing climate and changing patterns of regional development have challenged North America and will continue to do so. Damage costs from recent climatic events have increased, and conflicts over climate-sensitive resources (e.g., fish) continue to occur. Future climatic changes may exacerbate these problems and indeed create new ones. Opportunities may arise from a warming climate, and some innovative adaptation mechanisms (e.g., water banks) are being tested to address current problems. However, the literature provides few cases on how adaptive strategies could be implemented in the future as regional climates continue to change.

Recent studies in agriculture point to greater crop diversity and possibilities for adaptation as factors in reducing estimates of future damage. Production forestry is regarded in much the same way, except in high-latitude regions where vulnerabilities to changing climate-related risks (pests, fire, permafrost thaw) still are considered significant challenges for management. Estimates of future damage to U.S. coastal zones have declined despite the recent series of high-cost extreme events. However, studies also have documented changing vulnerabilities of built (urban) environments to atmospheric events, risks to indigenous lifestyles, and the challenge of managing resources within a changing climatic regime to meet multiple objectives. Climate change along the U.S.–Mexican border may result in social and economic problems related to transboundary water resources. Unique natural ecosystems such as prairie wetlands, alpine tundra, and coldwater ecosystems are at risk, and effective mitigation is unlikely.

As impact studies of North America attempt to include adaptation in a more significant way, damage estimates in some sectors decline or become benefits. Mendelsohn and Neumann (1999) illustrate this trend for agriculture, production forestry, and coastal zone impacts in the United States. What also emerges, however, is that high damage costs are still calculated for water-related activities and summer energy demand. Most important, there are several outstanding methodological issues related to the evolution of economic activities, adaptation assumptions, and the value of nonmarket components, resulting in continued difficulties in estimating the cost of impacts and the costs and benefits of adaptation responses. If optimistic views about adaptation were realistic, North America should have been better able to plan for extreme events. Recent experience demonstrates high capability in emergency response, but long-term problems remain.

During recent extreme weather events, synergies between development pressures and environmental changes have resulted in substantial damages. Examples documented in Sections 15.2.1, 15.2.3, 15.2.7, and 15.3.2 include water-quality problems resulting from North Carolina hurricanes, damage from the 1998 ice storm, reductions in fish stocks as a result of climatic shifts combined with fishing pressure, and various impacts

Table 15-6: *Climate change adaptation issues in North American subregions. Unique issues for certain locations also are indicated.*

North American Subregions	Development Context	Climate Change Adaptation Options and Challenges
Most or all subregions	– Changing commodity markets – Intensive water resources development over large areas—domestic and transboundary – Lengthy entitlement/land claim/treaty agreements—domestic and transboundary – Urban expansion – Transportation expansion	– Role of water/environmental markets – Changing design and operations of water and energy systems – New technology/practices in agriculture and forestry – Protection of threatened ecosystems or adaptation to new landscapes – Increased role for summer (warm weather) tourism – Risks to water quality from extreme events – Managing community health for changing risk factors – Changing roles of public emergency assistance and private insurance
Arctic border	– Winter transport system – Indigenous lifestyles	– Design for changing permafrost and ice conditions – Role of two economies and co-management bodies
Coastal regions	– Declines in some commercial marine resources (cod, salmon) – Intensive coastal zone development	– Aquaculture, habitat protection, fleet reductions – Coastal zone planning in high-demand areas
Great Lakes	– Sensitivity to lake-level fluctuations	– Managing for reduction in mean levels without increased shoreline encroachment

from recent water-level fluctuations in the Great Lakes. Unfortunately, only a few climate change scenario studies explore these kinds of synergies. Besides the challenge of projecting future economies, scenario studies of impacts and adaptation also have to consider the varied demands of regional stakeholders from the same resource and/or location. A summary of adaptation issues for North American subregions is provided in Table 15-6.

A wide range of factors—including changing political and institutional arrangements, technologies, and perceptions—may influence regional development futures. Concerns about development, equity, and sustainability are emerging in international negotiations of the United Nations Framework Convention on Climate Change and the Kyoto Protocol, and regional impacts/adaptation concerns can play an important role in this debate (Munasinghe and Swart, 2000). How could development paths alter a region's vulnerability to climate change? Could incentives related to emission reduction also enhance adaptation? Or, could management for certain environmental objectives lead to difficulties in meeting others? The example from the Columbia basin (Section 15.3.2), in which managing for fish protection could lead to increases in GHG emissions (Cohen *et al*., 2000), illustrates the dilemma that many subregions could face as they consider options for responding to climate change scenarios. How can these interactions be addressed in scenario studies? How can the various economic and social dimensions be accounted for so that there could be more confidence in estimates of impacts damages and adaptation costs/benefits?

Additional information on alternative climate scenarios and regional development futures will be needed in order to improve estimates of the costs of subregional impacts of climate change and the costs and benefits of adaptation responses. As measures associated with the Kyoto Protocol attract more attention from governments and industries, consideration also should be directed at studies on the interplay of mitigation and adaptation at the subregional scale.

References

Aber, J., R. Nielson, S. McNulty, J. Lenihan, D. Bachelet, and R. Drapek, 2001: Forest processes and global environmental change: predicting the effects of individual and multiple stressors. *BioScience*, (in press).

Aber, J.D. and C.T. Driscoll, 1997: Effects of land use, climate variation and N deposition on N cycling and C storage in northern hardwood forests. *Global Biogeochemical Cycles*, **11,** 639–648.

Abraham, J., T. Canavan, and R. Shaw (eds.), 1997: *Climate Change and Climate Variability in Atlantic Canada. Volume IV of the Canada Country Study: Climate Impacts and Adaptation*. Atmospheric Science Division, Atlantic Region, Environment Canada, Toronto, ON, Canada, 130 pp.

Abrams, R.H., 1990: Water allocation by comprehensive permit systems in the east: considering a move away from orthodoxy. *Virginia Environmental Law Journal*, **9(2)**, 255–285.

Acosta, A., 1988: El Niño: sus efectos sobre el Norte de Mexico. *Ingeniería Hidráulica en México*, **3(1)**, 13–29 (in Spanish).

Adams, R.M., B.A. McCarl, K. Segerson, C. Rosenzweig, K.J. Bryant, B.L. Dixon, R. Connor, R.E. Evenson, and D. Ojima, 1999: The economic effects of climate change on U.S. agriculture. In: *The Economics of Climate Change* [Mendelsohn, R. and J. Neumann (eds.)]. Cambridge University Press, Cambridge, United Kingdom and New York, NY, USA, pp. 19–54.

Adams, R.M., C. Rosenzweig, R.M. Peart, J.T. Ritchie, B.A. McCarl, J.D. Glyer, R.B. Curry, J.W. Jones, K.J. Boote, and L.H. Allen Jr., 1990: Global change and U.S. agriculture. *Nature*, **345**, 219–224.

Aharonian, D., 1994: Land use and climate change: an assessment of climate-society interactions in Aklavik, NT. In: *Mackenzie Basin Impact Study Interim Report No. 2* [Cohen, S.J. (ed.)]. Environment Canada, Downsview, ON, Canada, pp. 410–420.

AIRAC, 1986: *Catastrophic Losses: How the Insurance System Would Handle Two $7 Billion Hurricanes*. All-Industry Research Advisory Council (now the Insurance Research Council), Malvern, PA, USA, 73 pp.

Alm, A., E. Blommestein, and J.M. Broadus, 1993: Climate change and socioeconomic impacts. In: *Climate Change in the Intra Americas Sea* [Maul, G.A. (ed.)]. Arnold, London, United Kingdom, pp. 333–349.

American Insurance Association, 1999: *Property-Casualty Insurance and the Climate Change Debate: A Risk Assessment*. American Insurance Association, Washington, DC, USA, 13 pp.

American Thoracic Society, 1985: Guidelines as to what constitutes an adverse respiratory effect, with special reference to epidemiologic studies of air pollution. *American Review of Respiratory Diseases*, **131**, 666–668.

Anderson, D.M., 1997: Bloom dynamics of toxic Alexandrium species in the northeastern U.S. *Limnology and Oceanography*, **42(5)**, 1009–1022.

Anderson, D.R., 2000: Catastrophe insurance and compensation: remembering basic principles. *CPCU Journal*, **53(2)**, 76–89.

Anderson, W.P. and R.J. DiFrancesco, 1997: Potential impacts of climate warming on hydrocarbon production in the northern Mackenzie Basin. In: *Mackenzie Basin Impact Study (MBIS) Final Report* [Cohen, S.J. (ed.)]. Environment Canada, Toronto, ON, Canada, pp. 247–252.

Andres, D., 1994: Peace River ice regime: an interim report. In: *Mackenzie Basin Impact Study Interim Report No. 2* [Cohen, S.J. (ed.)]. Environment Canada, Toronto, ON, Canada, pp. 237–245.

Appendini, K. and D. Liverman, 1995: Agricultural policy and climate change in Mexico. *Climate Change and World Food Security* [Downing, T.E. (ed.)]. Springer-Verlag, Berlin, Germany, 662 pp.

Applegate, W., J. Rynyan, L. Brasfield, M. Williams, C. Konigsberg, and C. Fouche, 1981: Analysis of the 1980 heat wave in Memphis. *Journal of the American Geriatric Society*, **29**, 337–420.

Arnell, N., B. Bates, H. Lang, J.J. Magnuson, P. Mulholland, S. Fisher, C. Liu, D. McKnight, O. Starosolszky, and M. Taylor, 1996: Hydrology and freshwater ecology. In: *Climate Change 1995: Impacts, Adaptions, and Mitigation of Climate Change: Scientific- Technical Analyses. Contribution of Working Group II to the Second Assessment Report for the Intergovernmental Panel on Climate Change* [Watson, R.T., M.C. Zinyowera, and R.H. Moss (eds.)]. Cambridge University Press, Cambridge, United Kingdom and New York, NY, USA, pp. 325–363.

Arnell, N.W., 1999: Climate change and global water resources. *Global Environmental Change*, **9**, S31–S49.

Artsob, H., 1986: Arbovirus activity in Canada 1985. *Canada Diseases Weekly Report*, **12**, 109–110.

Assel, R.A., J.E. Janowiak, D. Boyce, C. O'Connors, F.H. Quinn, and D.C. Norton, 2000: Laurentian Great Lakes ice and weather conditions for the 1998 El Niño Winter. *Bulletin of the American Meteorological Society*, **81(4)**, 703–717.

Associated Press, 2000: *FEMA: Erosion Threatens 87,000 Coastal Buildings in U.S.* Associated Press, June 27, 1 p. Available online at http://www.cnn.com/2000/NATURE/06/27/fema.erosion.02/index.html.

Atkinson, D.B., G.A. Rose, E.F. Murphy, and C.A. Bishop, 1997: Distribution changes and abundance of northern cod (Gadus morhua), 1981–1993. *Canadian Journal of Fisheries and Aquatic Science*, **54(1)**, 132–138.

Aylsworth, J.M. and P.A. Egginton, 1994: Sensitivity of slopes to climate change. In: *Mackenzie Basin Impact Study Interim Report No. 2* [Cohen, S.J. (ed.)]. Environment Canada, Toronto, ON, Canada, pp. 278–283.

Aylsworth, J.M. and A. Duk-Rodkin, 1997: Landslides and permafrost in the Mackenzie Valley. In: *Mackenzie Basin Impact Study (MBIS) Final Report* [Cohen, S.J. (ed.)]. Environment Canada, Toronto, ON, Canada, pp. 278–283.

Bakun, A., 1996: *Patterns in the Ocean: Ocean Process and Marine Population Dynamics*. California Sea Grant/CIB, La Paz, Mexico, 323 pp.

Balling, R.C., 1988: The climate impact of Sonoran vegetation discontinuity. *Climate Change*, **13**, 99–109.

Barker, I., G. Surgeoner, and H. Artsob, 1992: Distribution of the Lyme Disease vector, Ixodes dammini (Acari: Ixodae) and isolation of Borrelia burgdorferi in Ontario, Canada. *Journal of Medical Entomology*, **78**, 171–176.

Barrett, R., 1997: Round table #3: economic development. In: *Mackenzie Basin Impact Study (MBIS) Final Report* [Cohen, S.J. (ed.)]. Environment Canada, Toronto, ON, Canada, pp. 55–58.

Bartlein, P.J., C. Whitlock, and S. Shafer, 1997: Future climate in the Yellowstone Nation Park region and its potential impact on vegetation. *Conservation Biology*, **11(3)**, 782–792.

Bates, D., 1992: Health indices of the adverse effects of air pollution: the question of coherence. *Environmental Research*, **59**, 336–349.

Bates, D. and R. Sitzo, 1989: The Ontario Air Pollution Study: identification of the causative agent. *Environmental Health Perspectives*, **79**, 69–72.

Bates, D. and R. Sitzo, 1987: Air pollution and hospital admissions in southern Ontario: the acid summer haze effect. *Environmental Research*, **43**, 317–331.

Bates, S.F., D.H. Getches, L.J. MacDonnell, and C.F. Wilkinson, 1993: *Searching Out the Headwaters: Change and Rediscovery in Western Water Policy*. Island Press, Covelo, CA, USA, 242 pp.

Beamish, R.J., C. Mahnken, and C.M. Neville, 1997: Hatchery and wild production of Pacific salmon in relation to large-scale, natural shifts in the productivity of the marine environment. *ICES Journal of Marine Science*, **54**, 1200–1215.

Beamish, R.J., G.A. McFarlane, and J.R. King, 1999a: Fisheries climatology: understanding decadal scale processes that naturally regulate British Columbia fish populations. In: *Fisheries Oceanography: An Integrative Approach to Fisheries Ecology and Management* [Parsons, T. and P. Harrison (eds.)]. Blackwell Science Ltd., Osney, Oxford, United Kingdom, 94–145.

Beamish, R.J., D.J. Noakes, G.A. McFarlane, L. Klyashtorin, V.V. Ivanov, and V. Kurashov, 1999b: The regime concept and natural trends in the production of Pacific salmon. *Canadian Journal of Fisheries and Aquatic Sciences*, **56(3)**, 516–526.

Beamish, R.J., D.J. Noakes, G.A. McFarlane, W. Pinnix, R. Sweeting, and J. King, 2000: Trends in coho marine survival in relation to regime concept. *Fisheries Oceanography*, **9(1)**, 114–119.

Bennett, J., S. Homberg, M. Rogers, and S. Soloman, 1987: Infectious and parasitic diseases. *American Journal of Preventative Medicine*, **55**, 102–114.

BESIS, 1999: Assessing the consequences of climate change in Alaska and the Bering Sea region. In: *Proceedings of a workshop at the University of Alaska Fairbanks, 29–30 October, 1998* [Weller, G. and P.A. Anderson (eds.)]. Bering Sea Impacts Study, Center for Global Change and Arctic System Research, Fairbanks, AK, USA, 94 pp. Available online at http://www.besis.uaf.edu/overview.html.

Beuker, E., E. Valtonen, and T. Repo, 1998: Seasonal variation in the forest hardiness of Scots pine and Norway spruce in old provenance experiments in Finland. *Forest Ecology and Management*, **197**, 87–98.

Binkley, C.S. and G.C. Van Kooten, 1994: Integrated climatic change and forests: economic and ecologic assessments. *Climatic Change*, **28**, 91–110.

Blair, L., 1998: Never the same again: family doctors' priorities challenged by ice storm. *Canadian Family Physician*, **44,** 721–728.

Bloomfield, J. and S. Hamburg, 1997: *Global Warming and New England's White Mountains*. Environmental Defense Fund, Washington, DC, USA, 33 pp.

Boesch, D.F., J.C. Field, and D. Scavia (eds.), 2000: *The Potential Consequences of Climate Variability and Change on Coastal Areas and Marine Resources*. Report of the Coastal Areas and Marine Resources Sector Team, U.S. National Assessment of the Potential Consequences of Climate Variability and Change, U.S. Global Change Research Program. NOAA Coastal Ocean Program Decision Analysis Series Number 21, NOAA Coastal Ocean Program, Silver Spring, MD, USA, 163 pp.

Booker, J.F., 1995: Hydrologic and economic impacts of drought under alternative policy responses. *Water Resources Bulletin*, **31(5),** 889–906.

Born, K., 1996: Tropospheric warming and changes in weather variability over the Northern Hemisphere during the period 1967–1991. *Meteorology and Atmospheric Physics*, **56,** 201–215.

Bowie, W., A. King, and D. Werker, 1997: Outbreak of toxoplasmosis associated with municipal drinking water: the BC Toxoplasma Investigation Team. *Lancet*, **350(9072),** 173–177.

Bradley, N.L., A.C. Leopold, J. Ross, and W. Huffaker, 1999: Phenological changes reflect climate change in Wisconsin. *Proceedings of the National Academy of Sciences of the United States of America*, **96,** 9701–9704.

Brauer, M., J. Blair, and S. Vedal, 1996: Effect of ambient ozone exposure on lung function in farm workers. *American Journal of Respiratory and Critical Care Medicine*, **154(4/1),** 981–987.

Bridgeman, S., R. Robertson, and Q. Syed, 1995: Outbreak of cryptosporidiosis associated with a disinfected groundwater supply. *Epidemiology and Infection*, **115,** 555–566.

British Columbia Toxoplasmosis Team, 1995: Outbreak of toxoplasmosis associated with municipal drinking water—British Columbia. *Canada Communicable Disease Report*, **21,** 161–164.

Brklacich, M., P. Curran, and D. Brunt, 1996: The application of agricultural land rating and crop models to CO_2 and climate change issues in northern regions: the Mackenzie Basin case study. *Agricultural and Food Science in Finland*, **5,** 351–365.

Brklacich, M., C. Bryant, B. Veenhof, and A. Beauchesne, 1997a: Implications of global climatic change for Canadian agriculture: a review and appraisal of research from 1984–1997. In: *Chapter 4, Canada Country Study: Climate Impacts and Adaptation*. Environment Canada, Toronto, ON, Canada, pp. 220–256.

Brklacich, M., P. Curran, D. Brunt, and D. McNabb, 1997b: CO_2 induced climatic change and agricultural potential in the Mackenzie Basin case. In: *Mackenzie Basin Impact Study (MBIS) Final Report* [Cohen, S.J. (ed.)]. Environment Canada, Toronto, ON, Canada, pp. 242–246.

Brklacich, M., R. Stewart, V. Kirkwood, and R. Muma, 1994: Effects of global climate change on wheat yields in the Canadian prairie. In: *Implications of Climate Change for International Agriculture: Crop Modelling Study* [Rosenzweig, C. and A. Iglesias (eds.)]. U.S. Environmental Protection Agency, Washington, DC, USA, pp. 1–23.

Brooks, D.J., 1993: *U.S. Forests in a Global Context*. General Technical Report RM-228, USDA Forest Service, Rocky Mountain Research Station, Fort Collins, CO, USA, 24 pp.

Brotton, J., T. Staple, and G. Wall, 1997: Climate change and tourism in the Mackenzie Basin. In: *Mackenzie Basin Impact Study (MBIS) Final Report* [Cohen, S.J. (ed.)]. Environment Canada, Toronto, ON, Canada, pp. 253–264.

Brotton, J. and G. Wall, 1997: Climate change and the Bathurst caribou herd in the Northwest Territories. *Climatic Change*, **35(1),** 35–52.

Brown, D., S. Moin, and M. Nicholson, 1997a: A comparison of flooding in Michigan and Ontario: soft data to support soft water management approaches. *Canadian Water Resources Journal*, **22(2),** 125–139.

Brown, J.H., T.J. Valone, and C.G. Curtin, 1997b: Reorganization of an arid ecosystem in response to recent climate change. *Proceedings of the National Academy of Sciences of the United States of America*, **94,** 9729–9733.

Brown, J.L. and S.-H. Li, 1996: Delayed effect of monsoon rains influences laying date of passerine bird living in an arid environment. *Condor*, **98,** 879–884.

Brown, J.L., S.-H. Li, and N. Bhagabati, 1999: Long-term trend toward earlier breeding in an American bird: a response to global warming? *Proceedings of the National Academy of Sciences of the United States of America*, **96,** 5565–5569.

Bruce, J., I. Burton, H. Martin, B. Mills, and L. Mortsch, 2000: *Water Sector: Vulnerability and Adaptation to Climate Change, Final Report*. Global Change Strategies International Inc. and Atmospheric Environment Service, Environment Canada, Ottawa, ON, Canada, 141 pp.

Bruce, J.P., I. Burton, I.D.M. Egener, and J. Thelen, 1999a: *Municipal Risks Assessment: Investigation of the Potential Municipal Impacts and Adaptation Measures Envisioned as a Result of Climate Change*. The National Secretariate on Climate Change, Ottawa, ON, Canada, 36 pp.

Bruce, J.P., I. Burton and M. Egener, 1999b: *Disaster Mitigation and Preparedness in a Changing Climate*. Institute for Catastrophic Loss Reduction, Research Paper No. 2, Ottawa, ON, Canada, 37 pp. Available online at http://www.epc-pcc.gc.ca/research/down/DisMit e.pdf.

Bruce, J.P., M. Frome, E. Haites, H.H. Janzen, R. Lal, and K. Paustian, 1999c: Carbon sequestration in soils. *Journal of Soil and Water Conservation*, **54,** 382–389.

Bruce-Chwatt, L., 1988: History of malaria from prehistory to eradication. In: *Malaria: Principles and Practice of Malariology* [Wersdorfer, W. (ed.)]. Churchill Livingstone, Edinburgh, Scotland, pp. 913–999.

Brugman, M., P. Raistrick, and A. Pietroniro, 1997: Glacier related impacts of doubling carbon dioxide concentrations on British Columbia and Yukon. *Responding to Global Climate Change in British Columbia and Yukon. Volume I of the Canada Country Study: Climate Impacts and Adaptation* [Taylor, E. and B. Taylor (eds.)]. Environment Canada, Vancouver and British Columbia Ministry of Environment, Land and Parks, Victoria, BC, Canada, Chapter 6, pp. 1–9.

Burkholder, J.M., H.B. Glasgow, and R. Reed, 2000: Impacts from Hurricane Floyd on water quality in the Neuse River and estuary, and the Pamlico Sound. In: *North Carolina Water Resources—The Year of the Hurricanes: Abstracts and Presentations/Posters at the Annual North Carolina Water Resources Research Conference, 30 March 2000*. North Carolina State University and the Water Resources Research Institute of the University of North Carolina, Raleigh, NC, USA, p. 26.

Burnett, R., R. Dales, and M. Raizenne, 1994: Effects of low ambient levels of ozone and sulfates on the frequency of respiratory admissions to Ontario hospitals. *Environmental Research*, **65(2),** 172–194.

Burton, I., R. Kates, and G.F. White, 1993: *The Environment as Hazard*. The Guilford Press, New York, NY, USA, 290 pp.

Butler, L.L., 1990: Environmental water rights: an evolving concept of public property. *Virginia Environmental Law Journal*, **9(2),** 323–380.

California Supreme Court, 1983: *National Audubon Society v. Superior Court of Alpine County*. 658 P.652d, California Supreme Court, Sacramento, CA, USA, 709 pp.

Callaway, J.M. and J.W. Currie, 1985: Water resource systems and changes in climate and vegetation. In: *Characterization of Information Requirements for Studies of CO_2 Effects: Water Resources, Agriculture, Fisheries, Forests and Human Health* [White, M. (ed.)]. DOE/ER-0236, U.S. Department of Energy, Washington, DC, USA, pp. 23–68.

Campbell, M., B. Benson, and M. Muir, 1995: Urban air quality and human health: a Toronto perspective. *Canadian Journal of Public Health*, **86(5),** 351–357.

Canadian Global Change Program, 1995: *Implications of Global Change for Human Health*. Final Report of the Health Issues Panel, The Royal Society of Canada, Ottawa, ON, Canada, 30 pp.

Canadian Public Health Association, 1992: *Human and Ecosystem Health: Canadian Perspectives, Canadian Action*. Canadian Public Health Association, Ottawa, ON, Canada, 21 pp.

Candau, J.-N., R.A., Fleming and A.A. Hopkin, 1998: Spatio-temporal patterns of large-scale defoliation caused by the spruce budworm in Ontario since 1941. *Canadian Journal of Forest Research*, **28,** 1–9.

Carman, S., H. Artsob, and S. Emery, 1995: Eastern equine encephalitis in a horse from Southwestern Ontario. *Canadian Veterinary Journal*, **36,** 170–171.

Carroll, A.L., J. Hudak, J.P. Meades, J.M. Power, T. Gillis, P.J. McNamee, C.H.R. Wedeles, and G. D. Sutherland, 1995: EHLDSS: a decision support system for management of the eastern hemlock looper. In: *Proceedings of Decision Support 2001, Toronto, Canada, 13-15 September 1994*. American Society for Photogrammetry and Remote Sensing, Bethesda, MD, USA, pp. 807–824.

Carter, T.R., M. Hulme, J.F. Crossley, S. Malyshev, M.G. New, M.E. Schlesinger, and H. Tuomenvirta, 2000: *Climate Change in the 21st Century - Interim Characterizations based on the New IPCC Emissions Scenarios*. The Finnish Environment 433, Finnish Environment Institute, Helsinki, 148 pp.

Cavazos, T., 1998: Large-scale circulation anomalies conducive to extreme precipitation events and derivation of daily rainfall in northeastern Mexico and southeastern Texas. *Journal of Climate*, **12(5),** 1506–1523.

Cavazos, T. and S. Hastenrath, 1990: Convection and rainfall over Mexico and their modulation by the Southern Oscillation. *International Journal of Climatology*, **10(4),** 377–386.

Cayan, D.R. and D.H. Peterson, 1989: The influence of north Pacific atmospheric circulation on streamflow in the West. In: *Aspects of Climate Variability in the Pacific and the Western Americas, Geophysical Monograph* [Peterson, D.H. (ed.)]. American Geophysical Union, Washington, DC, USA, Vol. 55, pp. 375–397.

Cayan, D.R. and R.H. Webb, 1992: El Niño/Southern Oscillation and streamflow in the western United States. In: *El Niño: Historical and Paleoclimatic Aspects of the Southern Oscillation* [Diaz, H.F. and V. Markgraf (eds.)]. Cambridge University Press, Cambridge, United Kingdom and New York, NY, USA, pp. 29–68.

Celenza, A., J. Forthergill, and E. Kupek, 1996: Thunderstorm associated asthma: a detailed analysis of environmental factors. *Bristish Medical Journal*, **312(7031),** 604–607.

CDC, 1998: Community needs assessment and morbidity surveillance following an ice storm—Maine, January 1998. *Morbidity and Mortality Weekly Report*, **47(17),** 351–354.

CDC, 1995: Heat-related mortality: Chicago, July 1995. Centers for Disease Control and Prevention, *Morbidity and Mortality Weekly Report*, **44,** 577–579.

CDC, 1993: Heat-related deaths: United States, 1993. Centers for Disease Control and Prevention, *Morbidity and Mortality Weekly Report*, **42,** 558–560.

CDC, 1989: Heat-related deaths: Missouri, 1979–1989. Centers for Disease Control and Prevention, *Morbidity and Mortality Weekly Report*, **38,** 437–439.

Cerezke, H.F. and W.J.A. Volney, 1995: Forest insect pests in the northwest region. In: *Forest Insect Pests in Canada* [Armstrong, J.A. and W.G.H. Ives (eds.)]. Canadian Forest Service, Ottawa, ON, Canada, pp. 59–72.

Chandler, A.E., 1913: *Elements of Western Water Law*. Technical Publishing Co., San Francisco, CA, USA, 150 pp.

Changnon, S.A., 1999: Factors affecting temporal fluctuations in damaging storm activity in the United States based on insurance loss data. *Meteorological Applications*, **6,** 1–10.

Changnon, S.A. (ed.), 1996: *The Great Flood of 1993: Causes, Impacts and Responses*. Westview Press, Boulder, CO, USA, 321 pp.

Changnon, S.A., 1992: Inadvertent weather modification in urban areas: lessons for global climate change. *Bulletin of the American Meteorological Society*, **73(5),** 619–627.

Changnon, S.A. and M. Demissie, 1996: Detection of changes in streamflow and floods resulting from climate fluctuations and land-use drainage changes. *Climate Change*, **32,** 411–421.

Changnon, S.A. and M.H. Glantz, 1996: The Great Lakes diversion at Chicago and its implications for climate change. *Climatic Change*, **32,** 199–214.

Changnon, S.A., B.C. Farhar, and E.R. Swanson, 1978: Hail suppression and society. *Science*, **200,** 387–394.

Changnon, S.A., R.A. Pielke, D. Changnon, R.T. Sylves, and R. Pulwarthy, 2000: Human factors explain the increased losses from weather and climate extremes. *Bulletin of the American Meteorological Society*, **81(3),** 437–442.

Changnon, S.A., D. Changnon, E.R. Fosse, D.C. Hoganson, R.J. Roth Sr., and J.M. Totsch, 1997: Effects of recent weather extremes on the insurance industry: major implications for the atmospheric sciences. *Bulletin of the American Meteorological Society*, **78(3),** 425–431.

Chao, P., 1999: Great Lakes water resources: Climate change impact analysis with transient GCM scenarios. *Journal of the American Water Resources Association*, **35(6),** 1499–1508.

Chao, P.T., B.F. Hobbs, and B.N. Venkatesh, 1999: How climate uncertainty should be included in Great Lakes management: Modeling workshop results. *Journal of the American Water Resources Association*, **35(6),** 1485–1497.

Chapin, F.S.I., 1993: Functional role of growth forms in ecosystem and global processes. In: *Scaling Physiological Processes* [Ehleringer, J.R. and C.B. Field (eds.)]. Academic Press, San Diego, CA, USA, pp. 287–312.

Chichilnisky, G. and G. Heal, 1998: Managing unknown risks: the future of global reinsurance. *Journal of Portfolio Management*, **Summer,** 85–91.

Choudry, A., M. Gordian, and S. Morris, 1997: Associations between respiratory illness and PM10 air pollution. *Archives of Environmental Health*, **52,** 113–117.

Chow, M., 1993: *A Case-Control Study of Endemic Giardiasis in Waterloo Region*. MSc. Thesis. Department of Population Medicine, University of Guelph, Guelph, ON, Canada, (unpublished).

Ciais, P., P.P. Tans, M. Trolier, J.W.C. White, and R.J. Francey, 1995: A large northern hemisphere terrestrial CO_2 sink indicated by the $^{13}C/^{12}C$ ratio for atmospheric CO_2. *Science*, **269,** 1098–1102.

City of Tulsa, 1994: *From Rooftop to River: Tulsa's Approach to Floodplain and Stormwater Management*. City of Tulsa Stormwater Drainage Advisory Board and Public Works Department, Tulsa, OK, USA, 26 pp.

Clair, T.A. and J.M. Ehrman, 1998: Using neural networks to assess the influence of changing seasonal climates in modifying discharge, dissolved organic carbon, and nitrogen export in eastern Canadian rivers. *Water Resources Research*, **34(3),** 447–455.

Clair, T.A., J. Ehrman, and K. Higuchi, 1998: Changes to the runoff of Canadian ecozones under a doubled CO_2 atmosphere. *Canadian Journal of Fisheries and Aquatic Science*, **55,** 2464–2477.

Clark, J.S., B.J. Stocks, and P.J.H. Richard, 1996: Climate implications of biomass burning since the 19th century in eastern North America. *Global Change Biology*, **2,** 443–452.

Cloern, J.E., 1996: Phytoplankton bloom dynamics in coastal ecosystems: a review with some general lessons from sustained investigation of San Francisco Bay. *Reviews of Geophysics*, **34(2),** 127–168.

Cohen, S.J., 1997a: Scientist-stakeholder collaboration in integrated assessment of climate change: lessons from a case study of Northwest Canada. *Environmental Modeling and Assessment*, **2,** 281–293.

Cohen, S.J., 1997b: What if and so what in northwest Canada: could climate change make a difference to the future of the Mackenzie Basin? *Arctic*, **50(4),** 293–307.

Cohen, S.J. (ed.), 1997c: *Mackenzie Basin Impact Study (MBIS) Final Report*. Environment Canada, Toronto, ON, Canada, 372 pp.

Cohen, S.J., 1996: Integrated regional assessment of global climatic change: lessons from the Mackenzie Basin Impact Study (MBIS). *Global and Planetary Change*, **11,** 179–185.

Cohen, S.J. (ed.), 1994: *Mackenzie Basin Impact Study Interim Report No. 2*. Environment Canada, Downsview, ON, Canada, 485 pp.

Cohen, S.J., 1991: Possible impacts of climatic warming scenarios on water resources in the Saskatchewan River Sub-basin, Canada. *Climatic Change*, **19,** 291–317.

Cohen, S.J., 1986: Impacts of CO_2-induced climatic change on water resources of the Great Lakes basin. *Climatic Change*, **8,** 135–153.

Cohen, S.J., K.A. Miller, A.F. Hamlet, and W. Avis, 2000: Climate change and resource management in the Columbia River Basin. *Water International*, **25(2),** 253–272.

Colby, B.G., 1998: Book review: water markets: priming the invisible pump. *Land Economics*, **74(4),** 575–579.

Colby, B.G., 1996: Markets as a response to water scarcity: policy changes and economic implications. *Advances in the Economics of Environmental Resources, Vol. 1, Marginal Cost Rate Design and Wholesale Water Markets* [Hall, D.C. (ed.)]. JAI Press, Greenwich, CT, USA, pp. 211–224.

Colombo, A.F., D. Etkin, and B.W. Karney, 1999: Climate variability and the frequency of extreme temperature events for nine sites across Canada: implications for power usage. *Journal of Climate*, **12,** 2490–2502.

Comerio, M.C., 1997: Housing issues after disasters. *Journal of Contigencies and Crisis Management*, **5(3),** 166–178.

Conde, C., 1999: *Impacts of Climate Change and Climate Variability in Mexico*. U.S. National Assessment, U.S. Global Change Research Program, Washington, DC, USA. Available online at http://www.nacc.usgcrp.gov/newsletter/1999.10/Mexico.html.

Conde, C., D. Liverman, M. Flores, R. Ferrer, R. Araujo, E. Betancourt, G. Villareal, and C. Gay, 1997: Vulnerability of rainfed maize crops in Mexico to climate change. *Climate Research*, **9,** 17–23.

Cooley, K.R., 1990: Effects of CO_2-induced climatic changes on snowpack and streamflow. *Hydrological Sciences Journal*, **35(5),** 511–522.

Costello, P.D. and P.J. Kole, 1985: Commentary on the Swan Falls Resolution. *Western Natural Resource Litigation Digest*, **Summer,** 11–18.

Croley, T.E., 1990: Laurentian Great Lakes double-CO_2 climate change hydrological impacts. *Climatic Change*, **17,** 27–47.

Cruise, J.F., A.S. Limaye, and N. Al-Abed, 1999: Assessment of impacts of climate change on water quality in the Southeastern United States. *Journal of the American Water Resources Association*, **35(6),** 1539–1550.

Cubasch, U., G. Waszkewitz, G. Hegerl, and J. Perlwitz, 1995: Regional climate changes as simulated in time-slice experiments. *Climatic Change*, **31(2–4),** 273–304.

Cummins, J.D., N.A. Doherty, and A. Lo, 1999: *Can Insurers Pay for the Big One?* Wharton School, University of Pennsylvania, Philadelphia, PA, USA, 46 pp. Available online at http://fic.wharton.upenn.edu/fic/wfic/papers/98/pcat01.html.

Cushing, C.E., 1997: Freshwater ecosystems and climate change: a regional assessment. *Hydrologic Processes*, **11,** 819–1067.

Dahl, T.E., 1990: *Wetlands: Losses in the United States 1780's to 1980's*. U.S. Department of the Interior, Fish and Wildlife Service, Washington, DC, USA.

Daniels, R.C., T.W. White, and K.K. Chapman, 1993: Sea-level rise: destruction of threatened and endangered species habitat in South Carolina. *Environmental Management*, **17,** 373–385.

Davidson, A., J. Emberlin, and A. Cook, 1996: A major outbreak of asthma associated with a thunderstorm: experience of accident and emergency departments and patients' characteristics, Thames Regions Accident and Emergency Trainees Association. *British Medical Journal*, **312(7031),** 601–604.

Davidson, R.J. Jr., 1996: Tax-deductible, pre-event catastrophe reserves. *Journal of Insurance Regulation*, **15(2),** 175–190.

DeGaetano, A.T., 2000: Climatic perspective and impacts of the 1998 northern New York and new England ice storm. *Bulletin of the American Meterological Society*, **81(2),** 237–254.

Delfino, R., M. Becklake, and J. Hanley, 1994: Estimation of unmeasured particulate air pollution data for epidemiological study of daily respiratory morbidity. *Environmental Research*, **67(1),** 20–38.

Delfino, R., A. Murphy-Moulton, and R. Burnett, 1997: Effects of air pollution on emergency room visits for respiratory illnesses in Montreal, Quebec. *American Journal of Respiratory and Critical Care Medicine*, **155(2),** 568–576.

Dellapenna, J.W., 1991: Regulated riparianism. In: *Waters and Water Rights* [Beck, R.E. (ed.)]. Michie, Charlottsville, VA, USA, pp. 413–579.

Dellinger, A., S. Kachur, and E. Sternberg, 1996: Risk of heat-related injury to disaster relief workers in a slow-onset flood disaster. *Journal of Occupational and Environmental Medicine*, **38(7),** 689–692.

Den Hollander, N. and R. Noteboom, 1996: Toxoplasma gondii in Ontario and water-borne toxoplasmosis in Victoria, B.C. *Public Health and Epidemiology Report Ontario*, **7,** 28–31.

Dixon, L.S., N.Y. Moore, and S.W. Schechter, 1993: *California's 1991 Drought Water Bank: Economic Impacts in the Selling Regions*. Rand Corporation, Santa Monica, CA, USA, 88 pp.

Dobson, A.P., J.P. Rodriguez, W.M. Roberts, and D.S. Wilcove, 1997: Geographic distribution of endangered species in the United States. *Science*, **275,** 550–553.

Dockery, D., J. Cunningham, and A. Damokosh, 1996: Health effects of acid aerosols on North American children: respiratory symptoms. *Environmental Health Perspectives*, **104(5),** 500–505.

Downton, M.W. and K.A. Miller, 1998: Relationships between Alaskan salmon catch and north Pacific climate on interannual and decadal time scales. *Canadian Journal of Fisheries and Aquatic Sciences*, **55,** 2255–2265.

Dubey, J. and C. Beattie, 1988: *Toxoplasmosis of Animals and Man*. CRC Press, Boca Raton, FL, USA, 220 pp.

Duell, L.F.W.J., 1994: The sensitivity of northern Sierra Nevada streamflow to climate change. *Water Resources Bulletin*, **30(5),** 841–859.

Duell, L.F.W.J., 1992: Use of regression models to estimate effects of climate change on seasonal streamflow in the American and Carson River Basins, California-Nevada. In: *Managing Water Resources During Global Change, American Water Resources Association 28th Annual Conference Proceedings, Reno, NV, USA, 1-5 November 1991* [Herrmann, R. (ed.)]. American Water Resources Association, Bethesda, MD, USA, pp. 731–740.

Duncan, K., 1996: Anthropogenic greenhouse gas-induced warming: suitability of temperatures for the development of vivax and faliparum malaria in the Toronto Region of Ontario. In: *Great Lakes-St. Lawrence Basin Project Progress Report #1: Adapting to the Impacts of Climate Change and Variability* [Mortsch, L. and B. Mills (eds.)]. Environment Canada, Burlington, ON, Canada, pp. 112–118.

Duncan, K., T. Guidotti, W. Cheng, K. Naidoo, G. Gibson, L. Kalkstein, S. Sheridan, D. Waltner-Toews, S. MacEachern, and J. Last, 1998: Health sector. In: *The Canada Country Study: Climate Impacts and Adaptation* [Koshida, G. and W. Avis (eds.)]. Environment Canada, Toronto, ON, Canada, pp. 501–590.

Dunn, P.O. and D.W. Winkler, 1999: Climate change has affected the breeding date of tree swallow throughout North America. *Proceedings of the Royal Society of London B*, **266,** 2487–2490.

Dyke, L.D., J.M. Aylsworth, M.M. Burgess, F.M. Nixon, and F. Wright, 1997: Permafrost in the Mackenzie Basin, its influence on land-altering processes, and its relationship to climate change. In: *Mackenzie Basin Impact Study (MBIS) Final Report* [Cohen, S.J. (ed.)]. Environment Canada, Toronto, ON, Canada, pp. 112–117.

Dynesius, M.L. and C. Nilsson, 1994: Fragmentation and flow regulation of river systems in the northern third of the world. *Science*, **266(4),** 753–762.

Eamus, D. and P.G. Jarvis, 1989: The direct effects of increase in the global atmospheric CO_2 concentration on natural and commercial temperate trees and forests. In: *Advances in Ecological Research* [Cragg, J.B. (ed.)]. Academic Press, London, United Kingdom, 2–57 pp.

Easterling, D.R., H.F. Diaz, A.V. Douglas, W.D. Hogg, K.E. Kunkel, J.C. Rogers and J.F. Wilkinson, 2000: Long-term observations for monitoring extremes in the Americas. *Climatic Change*, **42(1),** 285–308.

Ebbesmeyer, C.C., D.R. Cayan, D.R. McLain, F.H. Nichols, D.H. Peterson, and K.T. Redmond, 1991: 1976 step in the Pacific climate: forty environmental changes between 1968–1975 and 1977–1984. In: *Proceedings of the Seventh Annual Pacific Climate (PACLIM) Workshop, April 1990, Pacific Grove, CA* [Betancourt, J.L. and V.L. Tharp (eds.)]. California Department of Water Resources, Interagency Ecological Studies Program, Sacramento, CA, USA, pp. 115–126.

Echohawk, J.E. and R.P. Chambers, 1991: *Implementing Winters Doctrine Indian Reserved Water Rights*. Western Water Policy Discussion Series Papers, DP10, Natural Resources Law Center, University of Colorado, Boulder, CO, USA, 19 pp.

Eheart, J.W., A.J. Wildermuth, and E.E. Herricks, 1999: The effects of climate change and irrigation on criterion low streamflows used for determining total maximum daily loads. *Journal of the American Water Resources Association*, **35(6),** 1365–1372.

Emergency Preparedness Canada, 2000: *Disaster Database, Version 3.0*. Emergency Preparedness Canada, Ottawa, ON, Canada. Available online at http://www.epc-pcc.gc.ca/research/epcdatab.html.

Engelthaler, D., D. Mosely, J. Cheek, C. Levy, K. Komatsu, P. Ettestad, T. Davis, T. Tanda, L. Miller, and J. Frampton, 1999: Climatic and environmental patterns associated with hantavirus pulmonary syndrome, Four Corners region, United States. *Emerging Infectious Diseases*, **5,** 87–94.

Environment Canada, 1995: *Climate Change Impacts: An Ontario Perspective*. Smith and Lavendar Environmental Consultants, Sustainable Futures, and Environment Canada, Toronto, ON, Canada, 76 pp.

ESRI, 1998: *ESRI World Thematic Data*. Environmental Systems Research Institute Inc., Redlands, CA, USA. Available online at http://www.esri.com/data/online/esri.

Etkin, D.A., 1999: Risk transferance and related trends: driving forces towards more mega-disasters. *Environmental Hazards*, **1,** 69–75.

Etkin, D., 1998: Climate change and extreme events: Canada. In: *Canada Country Study: Climate Impacts and Adaptation* [Mayer, N. and W. Avis (eds.)]. Environment Canada, Toronto, ON, Canada, Vol. VIII, pp. 31–80.

Etkin, D., 1996: Beyond the year 2000: more tornadoes in western Canada? Implications from the historic record. *Natural Hazards*, **12,** 19–27.

Etkin, D., M. Vazquez, and I. Kelman, 1998: *Natural Disasters and Human Activity*. North American Commission on Environmental Cooperation, The North American Institute, Santa Fe, NM, USA, 8 pp.

Evans, J.C. and E.E. Prepas, 1996: Potential effects of climate change on ion chemistry and phytoplankton communities in prairie saline lakes. *Limnology and Oceanography*, **41(5),** 1063–1076.

Fee, E.J., R.E. Hecky, S.E.M. Kasian, and D.R. Cruikshank, 1996: Effects of lake size, water clarity, and climatic variability on mixing depths in Canadian Shield lakes. *Limnology and Oceanography*, **41,** 912–920.

Felzer, B. and P. Heard, 1999: Precipitation differences amongst GCMs used for the U.S. National Assessment. *Journal of the American Water Resources Association*, **35(6),** 1327–1340.

Fischer, G., K. Frohberg, M.L. Parry, and C. Rosenzweig, 1996: The potential effects of climate change on world food production and security. In: *Global Climate Change and Agricultural Production* [Bazzaz, F. and W. Sombroek (eds.)]. John Wiley and Sons, Chichester, United Kingdom, 345 pp.

Fleming, R.A. and G.M. Tatchell, 1994: Long term trends in aphid flight phenology consistent with global warming: methods and preliminary results. In: *Individuals, Populations and Patterns in Ecology* [Leather, S.R., A.D. Watt, N.J. Mills, and K.A.F. Walters (eds.)]. Intercept, Andover, Hants, United Kingdom, pp. 63–71.

Fletcher, C., 1997: The Mackenzie Basin Impact Study: a British Columbia Ministry of Forests perspective. In: *Mackenzie Basin Impact Study (MBIS) Final Report* [Cohen, S.J. (ed.)]. Environment Canada, Toronto, ON, Canada, pp. 340–343.

Flood, P.K., 1990: Water rights of the fifty states and territories. In: *Water Rights of the Fifty States and Territories* [Wright, K.R. (ed.)]. American Water Works Association, Denver, CO, USA, pp. 31–73.

Fowler, A.M. and K.J. Hennessey, 1995: Potential impacts of global warming on the frequency and magnitude of heavy precipitations. *Natural Hazards*, **12,** 283–303.

Francis, D. and H. Hengeveld, 1998: *Extreme Weather and Climate Change*. Environment Canada, Toronto, ON, Canada, 31 pp.

Francis, R.C., S.R. Hare, A.B. Hollowed, and W.S. Wooster, 1998: Effects of interdecadal climate variability on the oceanic ecosystems of the northeast Pacific. *Fisheries Oceanography*, **7,** 1–21.

Frederick, K.D. and P.H. Gleick, 1999: *Water and Global Climate Change: Potential Impacts on U.S. Water Resources*. The Pew Center on Global Climate Change, Arlington, VA, USA, 55 pp.

Frederick, K.D. and G.E. Schwarz, 1999: Socioeconomic impacts of climate change on U.S. water supplies. *Journal of American Water Resources Association*, **35(6),** 1563–1584.

Freedman, B., S. Love, and B. O'Neil, 1996: Tree species composition, structure and carbon storage in stands of urban forest of varying character in Halifax, Nova Scotia. *Canadian Field Naturalist*, **110,** 675–682.

Fujita, R.M., T. Foran, and I. Zevos, 1998: Innovative approaches for fostering conservation in marine fisheries. *Ecological Applications*, **8(1),** S139–S150.

Fyfe, J.C. and G.M. Flato, 1999: Enhanced climate change and its detection over the Rocky Mountains. *Journal of Climate*, **12(1),** 230–243.

Gable, F., 1997: Climate change impacts on Caribbean coastal areas and tourism. *Journal of Coastal Research*, **24,** 49–70.

GAO, 2000: *Insurers' Ability to Pay Catastrophe Claims*. U.S. General Accounting Office, Washington, DC, USA, 32 pp. Available online at http://www.gao.gov/corresp/gg00057r.pdf.

Gastel, R., 1999: Catastrophes: insurance issues. *Insurance Issues Update*.

Giardina, C.P. and M.G. Ryan, 2000: Evidence that decomposition rates of organic carbon in mineral soil do not vary with temperature. *Nature*, **404,** 858–861.

Gignac, L.D., B.J. Nicholson, and S.E. Bayley, 1998: The utilization of bryophytes in bioclimatic modeling: predicted northward migration of peatlands in the Mackenzie River Basin, Canada, as a result of global warming. *The Bryologist*, **101(4),** 572–587.

Glass, G., J. Cheek, J.A. Patz, T.M. Shields, T.J. Doyle, D.A. Thoroughman, D.K. Hunt, R.E. Ensore, K.L. Gage, C. Ireland, C.J. Peters, and R. Bryan, 2000: Predicting high risk areas for Hantavirus Pulmonary Syndrome with remotely sensed data: the Four Courners outbreak, 1993. *Journal of Emerging Infectious Diseases*, **6(3),** 239–246.

Glavin, G., C. Pinsky, and R. Bose, 1990: Gastrointestinal effects of contaminated mussels and putative antidotes thereof. *Canada Diseases Weekly Report*, **16(1E),** 111–115.

Gleick, P.H. and E.L. Chalecki, 1999: The impacts of climatic changes for water resources of the Colorado and Sacramento-San Joaquin river basins. *Journal of the American Water Resources Association*, **35(6),** 1429–1442.

Godleski, L., 1997: Tornado disasters and stress responses. *Journal of the Kentucky Medical Association*, **95(4),** 145–148.

Goldammer, J.G. and C. Price, 1998: Potential impacts of climate change on fire regimes in the tropics based on MAGICC and a GISS GCM-derived lightning model. *Climatic Change*, **39,** 273–296.

Goldstein, S., D. Juranek, and O. Raveholt, 1996: Cryptosporidiosis: an outbreak associated with drinking water despite state-of-the-art water treatment. *Annals of Internal Medicine*, **124,** 459–468.

Goodale, C.L., J.D. Aber, and E.P. Farrell, 1998: Applying an uncalibrated, physiologically based model of forest productivity to Ireland. *Climate Research*, **10,** 51–67.

Gough, L. and J.B. Grace, 1998: Effects of flooding, salinity and herbivory on coastal plant communities, Louisiana, United States. *Oecologia*, **117(4),** 527–535.

Gould, G., 1988: Water rights transfers and third-party effects. *Land and Water Law Review*, **23(1),** 1–41.

Goulden, M.L., S.C. Wofsy, and J.W. Harden, 1998: Sensitivity of boreal forest carbon balance to soil thaw. *Science*, **279,** 214–216.

Graf, W.L., 1999: Dam nation: a geographic census of American dams and their large-scale hydrologic impacts. *Water Resources Research*, **35(4),** 1305–1311.

Grant, L., 1991: Human health effects of climate change and stratospheric ozone depletion. In: *Global Climate Change: Health Issues and Priorities*. Health and Welfare Canada, Ottawa, ON, Canada, pp. 72–77.

Gratto-Trevor, C., 1997: Mackenzie Delta shorebirds. In: *Mackenzie Basin Impact Study (MBIS) Final Report* [Cohen, S.J. (ed.)]. Environment Canada, Toronto, ON, Canada, pp. 205–210.

Green, C.R. and W.D. Sellers, 1964: *Arizona Climate*. University of Arizona Press, Institute of Atmospheric Physics,Tucson, AZ, USA.

Groffman, P.M., J.P. Hardy, S. Nolan, C.T. Driscoll, R.D. Fitzhugh, and T.J. Fahey, 1999: Snow depth, soil frost and nutrient loss in a northern hardwood forest. *Hydrological Processes*, **13(14–15),** 2275–2286.

Gubler, D.J., 1998: Resurgent vector-borne diseases as a global health problem. *Emerging Infectious Diseases*, **4(3),** 442–450.

Guidotti, T., 1996: *Implications for Human Health of Global Ecological Change*. Canadian Association of Physicians for the Environment, Edmonton, AB, Canada.

Haack, R.A. and J.W. Byler, 1993: Insects and pathogens: regulators of forest ecosystems. *Journal of Forestry*, **91,** 32–37.

Hajat, S., S. Goubet, and A. Haines, 1997: Thunderstorm-associated asthma: the effect on GP consultations. *British Journal of General Practice*, **47(423),** 39–41.

Hall, J.P. and B.H. Moody, 1994: *Forest Depletions Caused by Insects and Diseases in Canada, 1982–1987*. Information Report ST-X-8, Canadian Forest Service, Ottawa, ON, Canada, 14 pp.

Hamilton, J., 1998: Quebec's ice storm '98: "all cards wild, all rules broken" in Quebec's shell-shocked hospitals. *Canadian Medical Association Journal*, **158(4),** 520–525.

Hamilton, R.M., 2000: Science and technology for natural disaster reduction. *Natural Hazards Review*, **1(1),** 56–60.

Hamlet, A.F. and D.P. Lettenmaier, 1999: Effects of climate change on hydrology and water resources in the Columbia River Basin. *Journal of the American Water Resources Association*, **35(6),** 1597–1624.

Hancock, T. and K. Davies, 1997: *An Overview of the Health Implications of Global Environmental Change: A Canadian Perspective*. Canadian Global Change Program, Royal Society of Canada, Ottawa, ON, Canada, 27 pp.

Hanson, H. and G. Lindh, 1996: The rising risk of rising tides. *Forum for Applied Research and Public Policy*, **11(2),** 86–88.

Hardy, Y., M. Mainville, and D.M. Schmitt, 1986: *An Atlas of Spruce Budworm Defoliation in Eastern North America, 1938–1980*. Miscellaneous Publication No. 1449, U.S. Department of Agriculture, Forest Service, Cooperative State Research Service, Washington, DC, USA, 51 pp.

Hare, F.K., R.B.B. Dickinson, and S. Ismail, 1997: *Climatic Variation over the Saint John Basin: An Examination of Regional Behaviour*. Climate Change Digest 97-02, Atmospheric Environment Service, Environment Canada, Ottowa, ON, Canada.

Harris, L.D. and W.P. Cropper Jr., 1992: Between the devil and the deep blue sea: implications of climate change for Florida's fauna. In: *Global Warming and Biological Diversity* [Peters, R.L. and T.E. Lovejoy (eds.)]. Yale University Press, New Haven, CT, USA, pp. 309–324.

Hartley, I. and P. Marshall, 1997: Modelling forest dynamics in the Mackenzie Basin under a changing climate. In: *Mackenzie Basin Impact Study (MBIS) Final Report* [Cohen, S.J. (ed.)]. Environment Canada, Toronto, ON, Canada, pp. 146–156.

Hartmann, H.C., 1990: Climate change impacts on Laurentian Great Lakes levels. *Climatic Change*, **17,** 49–67.

Harvell, C., K. Kim, J. Burkholder, R. Colwell, P. Epstein, D. Brimes, E. Hofmann, E. Lipp, A. Osterhaus, and R. Overstreet, 1999: Emerging marine diseases: climate links and anthropogenic factors. *Science*, **285,** 1505–1510.

Hastings, J.R. and R.M. Turner, 1965: Seasonal precipitation regimes in Baja California, Mexico. *Geografiska Annaler*, **47(4),** 204–222.

Hatfield, C., J. Wekell, and E. Gauglitz, 1994: Salt clean-up procedure for the determination of domoic acid by HPLC. *Natural Toxins*, **2(4),** 206–211.

Hauer, F.R., J.S. Baron, D.H. Campbell, K.D. Fausch, S.W. Hostetler, G.H. Leavesley, P.R. Leavitt, D.M. McKnight, and J.A. Stanford, 1997: Assessment of climate change and freshwater ecosystems of the Rocky Mountains, USA and Canada. *Hydrological Processes*, **11,** 903–924.

Haukioja, E., S. Neuvonen, S. Hanhimaki, and P. Niemela, 1988: The autumnal moth in Fennoscandia. In: *Dynamics of Forest Insect Populations: Patterns, Causes and Management Strategies* [Berryman, A.A. (ed.)]. Plenum Press, New York, NY, USA, pp. 163–178.

Haworth, J., 1988: The global distribution of malaria and the present control effort. *Malaria: Principles and Practice of Malariology* [Wernsdorfer, W. (ed.)]. Churchill Livingstone, Edinburgh, Scotland, pp. 1379–1420.

Heath, L.S., P.E. Kauppi, P. Burschel, H.D. Gregor, R. Guderian, G.H. Kohlmaier, S. Lorenz, D. Overdieck, F. Scholz, H. Thomasius, and M. Weber, 1993: Contribution of temperate forests to the world's carbon budget. *Water, Air, and Soil Pollution*, **70,** 55–69.

Heath, S., H. Artsob, and R. Bell, 1989: Equine encephalitis caused by snowshoe hare (California serogroup) virus. *Canadian Veterinary Journal*, **30,** 669–672.

Hebert, P.J., J.D. Jarrell, and M. Mayfield, 1993: *The Deadliest, Costliest, and Most Damaging United States Hurricanes of this Century (and Other Frequently Requested Hurricane Facts)*. NOAA Technical Memorandum NWS NHC-31, February 1993, National Hurricane Center, Coral Gables, FL, USA, 41 pp.

Heinz Center, 2000: *Evaluation of Erosion Hazards*. Federal Emergency Management Agency, Washington, DC, USA, 252 pp. Available online at http://www.fema.gov/nwz00/erosion.htm.

Heisler, G.M., 1986: Energy savings with trees. *Journal of Arboriculture*, **12,** 113–125.

Henderson-Sellers, A., H. Zhang, G. Berz, K. Emmanuel, W. Gray, C. Landsea, G. Holland, J. Lighthill, S.-L. Shieh, P. Webster, and K. McGuffie, 1998: Tropical cyclones and global climate change: a post-IPCC assessment. *Bulletin of the American Meteorological Society*, **79,** 19–38.

Hendry, M., 1993: Sea-level movements and shoreline changes. In: *Climate Change in the Intra-Americas Sea* [Maul, G. (ed.)]. Arnold Publishers, London, United Kingdom, pp. 115–161.

Hengeveld, H., 1995: *Understanding Atmospheric Change: A Survey of the Background Science and Implications of Climate Change and Ozone Depletion*. Environment Canada, Toronto, ON, Canada, 68 pp.

Hengeveld, H., 1999: *Climate Change and Extreme Weather*. Atmospheric Environment Service, Environment Canada, Ottawa, ON, Canada, 27 pp.

Hernández, M.E., 1995: La sequía en México, Mexico ante el cambio climático: primer taller de estudios de país. *Memorias do Instituto Oswaldo Cruz*, **1,** 141–149 (in Spanish).

Herrington, R., B. Johnson, and F. Hunter, 1997: *Responding to Global Climate Change in the Prairies. The Canada Country Study: Climate Impacts and Adaptation*. Environment Canada, Toronto, ON, Canada, 75 pp.

Hofmann, E.E. and T.M. Powell, 1998: Environmental variability effects on marine fisheries: four case histories. *Ecological Applications*, **8(1),** S23–S32.

Hofmann, N., L. Mortsch, S. Donner, K. Duncan, R. Kreutzwiser, A. Kulshreshtha, A. Piggott, S. Schellenberg, B. Schertzer, and M. Slivitzky, 1998: Climate change and variability: impacts on Canadian water. In: *Canada Country Study: Climate Impacts and Adaptation. Volume VII, National Sectoral Issues* [Koshida, G. and W. Avis (eds.)]. Environment Canada, Toronto, ON, Canada, pp. 1–120.

Holling, C.S., 1992: The role of forest insects in structuring the boreal landscape. In: *Analysis of the Global Boreal Forest* [Shugart, H.H., R. Leemans, and G.B. Bonan (eds.)]. Cambridge University Press, Cambridge, United Kingdom and New York, NY, USA, 545 pp.

Hooke, W.H., 2000: U.S. participation in international decade for natural disaster reduction. *Natural Hazards Review*, **1(1),** 2–9.

Houston, J., 1996: International tourism and United States beaches. *Shore and Beach*, **64(2),** 3–4.

Howitt, R., N. Moore, and R.T. Smith, 1992: *A Retrospective on California's 1991 Emergency Drought Water Bank*. California Department of Water Resources, Sacramento, CA, USA, 76 pp.

Huang, G.H., S.J. Cohen, Y.Y. Yin, and B. Bass, 1998: Land resources adaptation planning under changing climate: a study for the Mackenzie Basin. *Resources, Conservation and Recycling*, **24,** 95–119.

Hughes, J.P., D.P. Lettenmaier, and E.F. Wood, 1993: An approach for assessing the sensitivity of floods to regional climate change. In: *The World at Risk: Natural Hazards and Climate Change, American Institute of Physics Conference Proceedings 277, Cambridge, MA, USA, 14-16 January 1992* [Bras, R. (ed.)]. American Institute of Physics, New York, NY, USA, pp. 112–124.

Hurd, B.H., J.M. Callaway, J.B. Smith, and P. Kirshen, 1998: Economic effects of climate change on U.S. water resources. In: *The Economic Impacts of Climate Change on the U.S. Economy* [Mendelsohn, R. and J.E. Neumann (eds.)]. Cambridge University Press, Cambridge, United Kingdom and New York, NY, USA, pp. 133–177.

Hurd, B.H., J. Smith, R. Jones, and K. Spiecker, 1999: *Water and Climate Change: A National Assessment of Regional Vulnerability*. Report to U.S. Environmental Protection Agency, Washington, DC, USA.

Hutchings, J.H. and R.H. Myers, 1994: What can be learned from the collapse of a renewable resource: Atlantic Cod (Gadus morhua), off Newfoundland and Labrador. *Canadian Journal of Fisheries and Aquatic Science*, **51,** 2126–2146.

Hutchins, W.A., 1971: *Water Rights Laws in the Nineteen Western States, Volume 1*. U.S. Department of Agriculture, Washington, DC, USA, 650 pp.

Inouye, D.W., B. Barr, K.B. Armitage, and B.D. Inouye, 1999: Climate change is affecting altitudinal migrants and hibernating species. *Proceedings of the National Academy of Sciences of the United States of America*, **97,** 1630–1633.

IBHS, 1999: *Coastal Exposure and Community Protection: Hurricane Andrew's Legacy*. Institute of Business and Home Safety, Tampa, FL, USA, 48 pp.

IBC, 2000: *Facts of the General Insurance Industry of Canada*. Insurance Bureau of Canada, Toronto, ON, Canada, 44 pp. Available online at http://www.ibc.ca.

IFMRC, 1994: *A Blueprint for Change—Sharing the Challenge: Floodplain Management into the 21st Century*. Report of the Interagency Floodplain Management Review Committee to the Administration Floodplain Management Task Force, U.S. Government Printing Office, Washington, DC, USA.

III, 1999: *The Insurance Fact Book: 2000*. Insurance Information Institute, New York, NY, USA, 146 pp.

IIPLR and IRC, 1995: *Coastal Exposure and Community Protection: Hurricane Andrew's Legacy*. Insurance Institute for Property Loss Reduction and Insurance Research Council, Boston, MA, USA.

IPCC, 1998: *The Regional Impacts of Climate Change: An Assessment of Vulnerability. Special Report of IPCC Working Group II* [Watson, R.T., M.C. Zinyowera, and R.H. Moss (eds.)]. Intergovernmental Panel on Climate Change, Cambridge University Press, Cambridge, United Kingdom and New York, NY, USA, 517 pp.

IPCC, 1996: *Climate Change 1995: The Science of Climate Change. Contribution of Working Group I to the Second Assessment Report of the Intergovernmental Panel on Climate Change* [Houghton, J.T., L.G. Meira Filho, B.A. Callander, N. Harris, A. Kattenberg, and K. Maskell (eds.)]. Cambridge University Press, Cambridge, United Kingdom and New York, NY, USA, 572 pp.

Irlbacher, S., 1997: Sustainability of ecosystems. In: *Mackenzie Basin Impact Study (MBIS) Final Report* [Cohen, S.J. (ed.)]. Environment Canada, Toronto, ON, Canada, pp. 50–54.

Jacoby, G.C., R.D. D'Arrigo, and T. Davaajamts, 1996: Mongolian tree rings and 20th-century warming. *Science*, **273,** 771–773.

Justic, D., N.N. Rabalais, and R.E. Turner, 1996: Effects of climate change on hypoxia in coastal waters: a doubled CO_2 scenario for the northern Gulf of Mexico. *Limnology and Oceanography*, **41(5),** 992–1003.

Kaczmarek, Z., N.W. Arnell, E.Z. Stakhiv, K. Hanaki, G.M. Mailu, L. Somlyody, and K. Strzepek, 1996: Water resources management. In: *Climate Change 1995: Impacts, Adaptions, and Mitigation of Climate Change: Scientific- Technical Analyses. Contribution of Working Group II to the Second Assessment Report for the Intergovernmental Panel on Climate Change* [Watson, R.T., M.C. Zinyowera, and R.H. Moss (eds.)]. Cambridge University Press, Cambridge, United Kingdom and New York, NY, USA, pp. 471–486.

Kadonaga, L., 1997: Forecasting future fire susceptibility in the Mackenzie Basin. In: *Mackenzie Basin Impact Study (MBIS) Final Report* [Cohen, S.J. (ed.)]. Environment Canada, Toronto, ON, Canada, pp. 157–165.

Kalkstein, L., 1995: Lessons from a very hot summer. *Lancet*, **346,** 857–859.

Kalkstein, L., 1993: Health and climate change: direct impacts in cities. *Lancet*, **342,** 1397–1399.

Kalkstein, L. and K. Smoyer, 1993a: The impact of climate change on human health: some international implications. *Experientia*, **49,** 969–979.

Kalkstein, L. and K. Smoyer, 1993b: *The Impact of Climate on Canadian Mortality: Present Relationships and Future Scenarios*. Canadian climate Centre Report No. 93-7, Atmospheric Environment Service, Environment Canada, Downsview, ON, Canada, 50 pp.

Kalkstein, L., P. Jamason, and J. Greene, 1996a: The Philadelphia hot weather-health watch/warning system: development and application, summer 1995. *Bulletin of the American Meteorological Society*, **77,** 1519–1528.

Kalkstein, L., W. Maunder, and G. Jendritzky, 1996b: *Climate and Human Health*. World Meteorological Organization, World Health Organization, and United Nations Environment Program, New York, NY, USA, 24 pp.

Karl, T. and W. Riebsame, 1989: The impact of decadal fluctuations in mean precipitation and temperature on runoff: a sensitivity study over the United States. *Climatic Change*, **15,** 423–447.

Karl, T.R. and R.W. Knight, 1998: Secular trends of precipitation amount, frequency, and intensity in the United States. *Bulletin of the American Meteorological Society*, **79(2),** 231–241.

Karl, T.R., R.W. Knight, D.R. Easterling, and R.G. Quayle, 1996: Indices of climate change for the United States. *Bulletin of the American Meteorological Society*, **77,** 279–292.

Kattenberg, A., F. Giorgi, H. Grassl, G.A. Meehl, J.F.B. Mitchell, R.J. Stouffer, T. Kokioka, A.J. Weaver, and T.M.L. Wigley, 1996: Climate models: projections of future climate. In: *Climate Change 1995: The Science of Climate Change. Contribution of Working Group I to the Second Assessment Report of the Intergovernmental Panel on Climate Change* [Houghton, J.T., L.G. Meira Filho, B.A. Callander, N. Harris, A. Kattenberg, and K. Maskell (eds.)]. Cambridge University Press, Cambridge, United Kingdom and New York, NY, USA, pp. 285–357.

Keane, D. and P. Little, 1987: Equine viral encephalomyelitis in Canada: a review of known and potential causes. *Canadian Veterinary Journal*, **28,** 497–503.

Kenney, D.S., 1995: Institutional options for the Colorado River. *Water Resources Bulletin*, **31(5),** 837–850.

Kerr, J.A., 1997: Future water levels and flows for Great Slave and Great Bear Lakes, Mackenzie River and Mackenzie Delta. In: *Mackenzie Basin Impact Study (MBIS) Final Report* [Cohen, S.J. (ed.)]. Environment Canada, Toronto, ON, Canada, pp. 73–91.

Kerr, J. and L. Packer, 1998: The impact of climate change on mammal diversity in Canada. *Environmental Monitoring and Assessment*, **49,** 263–270.

Kerry, L., G. Kelk, D. Etkin, I. Burton, and S. Kalhok, 1999: Glazed over: Canada copes with the ice storm of 1998. *Environment*, **41(1),** 6–11 and 28–33.

Keyser, A.R., J.S. Kimball, R.R. Nemani, and R.S.W., 2000: Simulating the effects of climate change on the carbon balance of North American high latitude forests. *Global Change Biology*, **6(1),** 185–195.

Knox, R., C. Suphioglu, and P. Taylor, 1997: Major grass pollen allergen Loi p i binds to diesel exhaust particles: implications for asthma and air pollution. *Clinical and Experimental Allergy*, **27(3),** 246–251.

Kovacs, P., 1998: *Ice Storm '98*. Institute for Catastrophic Loss Reduction, Ottawa, ON, Canada.

Krug, E., M. Kresnow, and J. Peddicord, 1998: Suicide after natural disasters. *New England Journal of Medicine*, **338(6),** 373–378.

Kunkel, K.E., R.A. Pielke, and S.A. Changnon, 1999: Temporal fluctuations in weather and climate extremes that cause economic and human health impacts: a review. *Bulletin of the American Meteorological Society*, **80(6),** 1077–1098.

Kunreuther, H. and R. Roth (eds.), 1998: *Paying the Price: The Status and Role of Insurance Against Natural Disasters in the United States*. Joseph Henry Press, Washington, DC, USA, 320 pp.

Kusler, J., M. Brinson, W. Niering, J. Patterson, V. Burkett, and D. Willard, 1999: *Wetlands and Climate Change: Scientific Knowledge and Management Options*. Institute for Wetland Science and Public Policy, Association of Wetland Managers, Berne, NY, USA.

Lal, R., J.M. Kimble, C.V. Cole, and R.F. Follett, 1998: *The Potential of U.S. Cropland to Sequester Carbon and Mitigate the Greenhouse Effect*. Sleeping Bear Press, Ann Arbor Press, Chelsea, MI, USA, 128 pp.

Landsea, C.W., R.A. Pielke Jr., A.M. Mestas-Nunez, and J.A. Knaff, 1999: Atlantic Basin hurricanes: indices of climatic changes. *Climatic Change*, **42(1),** 89–129.

Last, J., 1993: Global change: ozone depletions, greenhouse warming and public health. *Annual Review of Public Health*, **14,** 115–136.

Last, J., K. Trouton, and D. Pengelly, 1998: *Taking Our Breath Away: The Health Effects of Air Pollution and Climate Change*. David Suzuki Foundation, Vancouver, BC, Canada, 55 pp.

Latour, P. and N. MacLean, 1997: Climate warming and marten, lynx and red fox. In: *Mackenzie Basin Impact Study (MBIS) Final Report* [Cohen, S.J. (ed.)]. Environment Canada, Toronto, ON, Canada, pp. 179–188.

Lavoie, D. and E. Asselin, 1998: Upper Ordovician facies in the Lac Saint-Jean outlier, Quebec (eastern Canada): paleoenvironmental significance for Late Ordovician oceanography. *Sedimentology*, **45(5),** 817–832.

Layton, M., M. Parise, C. Campbell, R. Advani, J. Sexton, E. Bosler, and J. Zucker, 1995: Mosquito transmitted malaria in New York. *Lancet*, **346(8977),** 729–731.

Leatherman, S.P., 1996: Shoreline stabilization approaches in response to sea level rise: U.S. experience and implications for Pacific island and Asian nations. *Water, Air, and Soil Pollution*, **92(1–2),** 42–46.

Leatherman, S., 1989: Beach response strategies to accelerated sea-level rise. In: *Coping with Climate Change* [Topping, J. (ed.)]. Climate Institute, Washington, DC, USA.

Leavesley, G.H., 1994: Modeling the effects of climate change on water resources: a review. *Climatic Change*, **28,** 159–177.

Lecomte, E., A.W. Pang, and J.W. Russell, 1998: *Ice Storm '98*. Institute for Catastrophic Loss Reduction, Ottawa, ON, Canada and Institute for Business and Home Safety, Boston, MA, USA, 39 pp.

Leith, R.M.M. and P.H. Whitfield, 1998: Evidence of climate change effects on the hydrology of streams in south-central B.C. *Canadian Water Resources Journal*, **23(3),** 219–230.

Lettenmaier, D.P. and T.Y. Gan, 1990: Hydrologic sensitivities of the Sacramento-San Joaquin River Basin, California, to global warming. *Water Resources Research*, **26(1),** 69–86.

Lettenmaier, D.P., K.L. Brettmann, L.W. Vail, S.B. Yabusaki, and M.J. Scott, 1992: Sensitivity of Pacific Northwest water resources to global warming. *The Northwest Environmental Journal*, **8,** 265–283.

Lettenmaier, D.P., D. Ford, S.M. Fisher, J.P. Hughes, and B. Nijssen, 1996: *Water Management Implications of Global Warming, 4: The Columbia River Basin*. Report to Interstate Commission on the Potomac River Basin and Institute for Water Resources, U.S. Army Corps of Engineers, University of Washington, Seattle, WA, USA.

Lettenmaier, D.P., A.W. Wood, R.N. Palmer, E.F. Wood, and E.Z. Stakhiv, 1999: Water resources implications of global warming: a U.S. regional perspective. *Climatic Change*, **43(3),** 537–579.

Leung, L.R. and M.S. Wigmosta, 1999: Potential climate change impacts on mountain watersheds in the Pacific Northwest. *Journal of the American Water Resources Association*, **35(6),** 1463–1472.

Lewis, S.E., 1993: Effect of climatic variation on reproduction by pallid bats (Antrozous pallidus). *Canadian Journal of Zoology*, **71(7),** 1429–1433.

Liebig, A., 1997: Overhead contact lines at underground flood barrier gates. *Elektrische Bahnan*, **95(1–20),** 42–46.

Lindroth, A., A. Grelle, and A.-S. Moren, 1998: Long-term measurements of boreal forest carbon balance reveal large temperature sensitivity. *Global Change Biology*, **4,** 443–450.

Liverman, D. and K. O'Brien, 1991: Global warming and climate change in Mexico. *Global Environmental Management*, **1,** 351–364.

Lluch, B., D.S. Hernández, D. Lluch-Cota, C.A. Salinas, F. Magallon, and F. Lachina, 1991: Variación climática y oceanográfica global: sus efectos en el noroeste mexicano. *Ciencia y Desarrollo*, **98,** 79–88 (in Spanish).

Lluch-Cota, D.B., S. Hernández, and S. Lluch-Cota, 1997: *Empirical Investigation on the Relationship Between Climate and Small Pelagic Global Regimes and El Niño-Southern Oscillation (ENSO)*. FIRM/C9334, FAO Fisheries Circular No. 934, Food and Agriculture Organization, Rome, Italy, 48 pp.

Loehle, C. and D. LeBlanc, 1996: Model-based assessments of climate change effects on forests: a critical review. *Ecological Modelling*, **90,** 1–31.

Logue, J., H. Hansen, and E. Struening, 1979: Emotional and physical distress following Hurricane Agnes in Wyoming Valley of Pennsylvania. *Public Health Report*, **94,** 495–502.

Lonergan, S. and B. Kavanagh, 1997: Global environmental change and the dual economy of the north. In: *Mackenzie Basin Impact Study (MBIS) Final Report* [Cohen, S.J. (ed.)]. Environment Canada, Toronto, ON, Canada, pp. 298–306.

Loomis, J. and J. Crespi, 1999: Estimated effects of climate change on selected outdoor recreation activities in the United States. In: *The Impact of Climate Change on the United States Economy* [Mendelsohn, R. and J.E. Neumann (eds.)]. Cambridge University Press, Cambridge, United Kingdom and New York, NY, USA, pp. 289–314.

Lord, W.B., J.F. Booker, D.H. Getches, B.L. Harding, D.S. Kenney, and R.A. Young, 1995: Managing the Colorado River in a severe sustained drought: an evaluation of institutional options. *Water Resources Bulletin*, **31(5),** 939–944.

Loukas, A. and M.C. Quick, 1999: The effect of climate change on floods in British Columbia. *Nordic Hydrology*, **30,** 231–256.

Loukas, A. and M.C. Quick, 1996: Effect of climate change on hydrologic regime of two climatically different watersheds. *Journal of Hydrologic Engineering*, **1(2),** 77–87.

Maarouf, A. and H. Boyd, 1997: Influences of climatic conditions in the Mackenzie Basin on the success of northern-nesting geese. In: *Mackenzie Basin Impact Study (MBIS) Final Report* [Cohen, S.J. (ed.)]. Environment Canada, Toronto, ON, Canada, pp. 211–216.

Maarouf, A. and J. Smith, 1997: Interactions amongst policies designed to resolve individual air issues. *Environmental Monitoring and Assessment*, **46,** 5–21.

MacDonald, M.E., A.E. Hershey, and M.C. Miller, 1996: Global warming impacts on lake trout in arctic lakes. *Limnology and Oceanography*, **41(5),** 1102–1108.

MacDonnell, L.J., 1990: *The Water Transfer Process as a Management Option for Meeting Changing Water Demands*. U.S. Geological Survey Grant #14-08-0001-G1538, Natural Resources Law Center, University of Colorado, Boulder, CO, USA, Vol. 1, 70 pp. and Vol. 72, 358 pp.

MacDonnell, L.J., C.W. Howe, K.A. Miller, T. Rice, and S.F. Bates, 1994: *Water Banking in the West*. U.S. Geological Survey Grant Award #1434-92-2253, Natural Resources Law Center, University of Colorado, Boulder, CO, USA, 270 pp.

MacIver, D.C. (ed.), 1998: Workshop summary. In: *IPCC Workshop on Adaptation to Climate Variability and Change, San Jose, Costa Rica, 29 March–1 April, 1998*. Atmospheric Environment Service, Environment Canada, Ottowa, ON, Canada, 55 pp.

MacKenzie, W., N. Hoxie, and M. Proctor, 1994: A massive outbreak in Milwaukee of Cryptosporidium infection transmitted through the public water supply. *New England Journal of Medicine*, **331,** 161–167.

Magaña, V. and C. Conde, 2000: Climate and freshwater resources in Mexico: Sonora, a case study. *Environmental Monitoring and Assessment*, **61,** 167–185.

Magaña, V. and A. Quintanar, 1997: On the use of a general circulation model to study regional climate. In: *Numerical Simulations in the Environmental and Earth Sciences, Proceedings of the Second UNAM-Cray Supercomputing Conference, Mexico City, 21-24 June 1995* [García, F., G. Cisneros, A. Fernández-Equiarte, and R. Alvarez (eds.)]. Cambridge University Press, Cambridge, United Kingdom and New York, NY, USA, pp. 39–48.

Magaña, V., C. Conde, O. Sánchez, and C. Gay, 1998: Evaluación de escenarios regionales de clima actual y de cambio climático futuro para México. In: *México, Una Visión en el Siglo XXI. El Cambio Climático en México*. Programa Universitario de Medio Ambiente (PUMA), Mexico, (in Spanish).

Magaña, V., C. Conde, O. Sánchez, and C. Gay, 1997: Assessment of current and future regional climate scenarios for México. *Climate Research*, **9(1),** 107–114.

Magnuson, J.J., R.A. Assel, C.J. Bouser, P.J. Dillon, J.G. Eaton, H.E. Evans, E.J. Fee, R.I. Hall, L.R. Mortsch, D.W. Schindler, F.H. Quinn, and K.H. Webster, 1997. Potential effects of climate change on aquatic systems: Laurentian Great Lakes and Precambrian Shield region. *Hydrological Processes*, **11,** 825–872.

Malcolm, J. and A. Markham, 2000: *Global Warming and Terrestrial Biodiversity Decline*. World Wildlife Fund for Nature, Gland, Switzerland. 34 pp.

Manitoba Water Commission, 1998: *An Independent Review of Actions Taken During the 1997 Red River Flood*. Minister of Natural Resources, Winnipeg, MB, Canada.

Mannino, D.M., D.M. Homa, C.A. Pertowski, A. Ashizawa, L.L. Nixon, C.A. Johnson, L.B. Ball, E. Jack, and D.S. Kang, 1998: Surveillance for asthma: United States, 1960–1995. *MMWR Surveillance Summaries*, **47(SS-1),** 1–28. Available online at http://www.cdc.gov/epo/mmwr/preview/mmwrhtml/00052262.htm.

Markowski, M., A. Knapp, J.E. Neumann, and J. Gates, 1999: The economic impact of climate change on the U.S. commercial fishing industry. In: *The Impact of Climate Change on the United States Economy* [Mendelsohn, R. and J.E. Neumann (eds.)]. Cambridge University Press, Cambridge, United Kingdom and New York, NY, USA, pp. 237–264.

Marsh, P. and L.F.W. Lesack, 1996: The hydrologic response of perched lakes in the Mackenzie delta: potential responses to climate change. *Limnology and Oceanography*, **41(5),** 849–856.

Martens, W., T. Jetten, and J. Rotman, 1995: Climate change and vector-borne diseases: a global modelling perspective. *Global Environmental Change*, **5,** 195–209.

Martin, P. and M. Lefebvre, 1995: Malaria and climate: sensitivity of malaria potential transmission to climate. *Ambio*, **24,** 200–207.

Marzuk, P., K. Tardiff, A. Leon, C. Hirsch, L. Portera, M. Iqbal, M. Nock, and N. Hartwell, 1998: Ambient temperature and mortality from unintentional cocaine overdose. *Journal of the American Medical Association*, **279,** 1795–1800.

Mason, C., 1999: The ocean's role in climate variability and change and the resulting impact on coasts. *Natural Resources Forum*, **23(2),** 123–134.

Matalas, N.C., 1998: Note on the assumption of hydrologic stationarity. *Water Resources Update*, **112,** 64–72.

Matthews, P.B., M.P. Sheffield, J.E. Andre, J.H. Lafayette, J.M. Roethen, and E. Dobkin, 1999: Insolvency: will historic trends return? *Best's Review: Property/Casualty Edition*, **March,** 59–67. Available online at http://www.bestreview.com/pc/1999-03/trends.html.

Mattson, W.J. and N.D. Addy, 1975: Phytophagous insects as regulators of forest primary production. *Science*, **190,** 515–522.

Mattson, W.J. and R.A. Haack, 1987: The role of drought in outbreaks of plant-eating insects. *BioScience*, **37,** 110–118.

Maxwell, B. (ed.), 1997: *Responding to Global Climate Change in Canada's Arctic. Volume II of the Canada Country Study: Climate Impacts and Adaptation*. Environment Canada, Toronto, ON, Canada, 82 pp.

McBoyle, G. and G. Wall, 1992: Great Lakes skiing and climate change. In: *Mountain Resort Development: Proceedings of the Vail Conference, 1991, Vail, Colorado* [Gill, A. and R. Hartman (eds.)]. Centre for Tourism Policy and Research, Simon Fraser University, Vancouver, BC, Canada, pp. 82–92.

McCabe, G.J. and D.M. Wolock, 1999: General circulation model simulations of future snowpack in the western United States. *Journal of the American Water Resources Association*, **35(6)**, 1473–1484.

McGinn, S.M., A. Toure, O.O. Akinremi, D.J. Major, and A. G. Barr, 1999: Agroclimate and crop response to climate change in Alberta, Canada. *Outlook on Agriculture*, **28(1)**, 19–28.

McKay, B.J., 1995: Property rights and fisheries management. *Ocean and Coastal Management (Special Issue)*, **28,** 189 pp.

McKelvey, R., 1997: Game theoretic insights into the international management of fisheries. *Natural Resource Modeling*, **10(2),** 129–171.

McLean, R., G. Frier, and G. Parham, 1985: Investigations of the vertebrate hosts of Eastern equine encephalitis during an epizootic in Michigan, 1980. *American Journal of Tropical Medicine and Hygiene*, **34,** 1190–1202.

McMichael, A., 1996: Human population health. In: *Climate Change 1995: Impacts, Adaptions and Mitigation of Climate Change: Scientific-Technical Analyses. Contribution of Working Group II to the Second Assessment Report for the Intergovernmental Panel on Climate Change* [Watson, R.T., M.C. Zinyowera, and R.H. Moss (eds.)]. Cambridge University Press, Cambridge, United Kingdom and New York, NY, USA, pp. 561–584.

McMichael, A., A. Haines, and R. Sloof, 1996: *Climate Change and Human Health*. World Health Organization, Geneva, Switzerland, 297 pp.

McMichael, A.J., 1997: Global environmental change and human health: impact assessment, population vulnerability, and research priorities. *Ecosystem Health*, **2(4),** 200–210.

Mead, P., L. Slutsker, V. Dietz, L. McCaig, J. Bresee, C. Shapiro, P. Griffen, and R. Tauxe, 1999: Food-related illness and death in the United States. *Emerging Infectious Diseases*, **5,** 607–625.

Mearns, L.O., C. Rosenzweig, and R. Goldberg, 1996: The effects of changes in daily and interannual climate variability in CERES—wheat: a sensitivity study. *Climatic Change*, **32(1)**, 257–292.

Mearns, L.O., F. Giorgi, L. McDaniel, and C. Shields, 1995: Analysis of daily variability of precipitation in a nested regional climate model: comparison with observations and doubled CO_2 results. *Global and Planetary Change*, **10,** 55–78.

Mearns, L.O., T. Mavromatis, E. Tsvetsinskaya, C. Hays, and W. Easterling, 1999: Comparative responses of EPIC and CERES crop models to high and low spatial resolution climate change scenarios. *Journal of Geophysical Research*, **104(D6),** 6623–6646.

Mearns, L.O., G. Carbone, W. Gao, L. McDaniel, E. Tsvetinskaya, B. McCarl, and R. Adams, 2000: The issue of spatial scale in integrated assessments: an example of agriculture in the southeastern U.S. In: *Preprints of the Annual Meeting of the American Meteorological Society, Jan. 2000, Long Beach, CA*. American Meteorological Society, Boston, MA, USA, 408 pp.

Meffe, G.K. and C.R. Carroll, 1994: *Principles of Conservation Biology*. Sinauer Associates, Sunderland, MA, USA, 600 pp.

Meinhardt, P., D. Casemore, and K. Miller, 1996: Epidemiological aspects of human cryptosporidiosis and the role of waterborne transmission. *Epidemiological Review*, **18,** 118–136.

Meko, D., M. Hughes, and C. Stockton, 1991: Climate change and climate vulnerability: the paleo record. In: *Managing Water Resources in the West under Conditions of Climate Uncertainty, 14-16 November 1990, Scottsdale, AZ*. National Research Council, National Academy Press, Washington, DC, USA, pp. 71-100

Melack, J.M., J. Dozier, C.R. Goldman, D. Greenland, A.M. Milner, and R.J. Naiman, 1997: Effects of climate change on inland waters of the Pacific coastal mountains and western Great basin of North America. *Hydrological Processes*, **11,** 971–992.

Melville, G.E., 1997: Climate change and yield considerations for cold-water fish based on measures of thermal habitat: lake trout in the Mackenzie Great Lakes. In: *Mackenzie Basin Impact Study (MBIS) Final Report* [Cohen, S.J. (ed.)]. Environment Canada, Toronto, ON, Canada, pp. 189–204.

Mendelsohn, R. and M. Markowski, 1999: The impact of climate change on outdoor recreation. In: *The Impact of Climate Change on the United States Economy* [Mendelsohn, R. and J.E. Neumann (eds.)]. Cambridge University Press, Cambridge, United Kingdom and New York, NY, USA, pp. 267–288.

Mendelsohn, R. and J. Neumann (eds.), 1999: *The Impact of Climate Change on the United States Economy*. Cambridge University Press, Cambridge, United Kingdom and New York, NY, USA, 331 pp.

Mendelsohn, R., W. Nordhaus, and D. Shaw, 1999: The impact of climate variation on U.S. agriculture. In: *The Impact of Climate Change on the United States Economy* [Mendelsohn, R. and J.E. Neumann (eds.)]. Cambridge University Press, Cambridge, United Kingdom and New York, NY, USA, pp. 55–74.

Mendelsohn, R., W.D. Nordhaus, and D. Shaw, 1994: The impact of global warming on agriculture: a Ricardian analysis. *The American Economic Review*, **84(4),** 753–771.

Mendoza, V.M., E.E. Villanueva, and J. Adem, 1997: Vulnerability of basins and watersheds in Mexico to global climate change. *Climate Research*, **9(1),** 139–145.

Meyer, J.L., M.J. Sale, P.J. Mulholland, and N.L. Poff, 1999: Impacts of climate change on aquatic ecosystem functioning and health. *Journal of the American Water Resources Association*, **35,** 1373–1386.

Michener, W.K., E.R. Blood, and K.L. Bilstein, 1997: Climate change, hurricanes, and tropical storms, and rising sea level in coastal wetlands. *Ecological Applications*, **7,** 770–801.

Miles, E.L., A.F. Hamlet, A.K. Snover, B. Callahan, and D. Fluharty, 2000: Pacific Northwest regional assessment: the impacts of climate variability and climate change on the water resources of the Columbia River Basin. *Journal of the American Water Resources Association*, **36(2),** 399–420.

Mileti, D.S., 1999: *Disasters by Design: A Reassessment of Natural Hazards in the United States*. Joseph Henry Press, Washington, DC, USA, 351 pp.

Mileti, D.S., 1997: Managing hazards into the next century. In: *Proceedings of the 1997 Association of State Flood Plain Managers 21st Annual Conference, 28 April - 2 May 1997, Little Rock, AR*. Association of State Flood Plain Managers, Madison, WI, USA, pp. 17–18.

Miller, A., G. Sethi, and G.H. Wolff, 2000: *What's Fair? Consumers and Climate Change*. Redefining Progress, San Francisco, CA, USA, 67 pp.

Miller, K.A., 2000a: Pacific salmon fisheries: climate, information and adaptation in a conflict-ridden context. *Climatic Change*, **45(1),** 37–61.

Miller, K.A., 2000b: Managing supply variability: the use of water banks in the western United States. *Drought: A Global Assessment, Volume II* [Wilhite, D.A. (ed.)]. Routledge, London, United Kingdom, pp. 70–86.

Miller, K.A., 1997: *Climate Variability, Climate Change and Western Water*. Western Water Policy Review Advisory Commission, National Technical Information Service, Springfield, VA, USA, 54 pp.

Miller, K.A. and M.W. Downton, 1993: The freeze risk to Florida citrus, part I: investment decisions. *Journal of Climate*, **6,** 354–363.

Miller, K.A., S.L. Rhodes, and L.J. MacDonnell, 1997: Water allocation in a changing climate: institutions and adaptations. *Climatic Change*, **35,** 157–177.

Miller, K.A., S.L. Rhodes, and L.J. MacDonnell, 1996: Global change in microcosm: the case of U.S. water institutions. *Climatic Change*, **29(4),** 271–290.

Miller, T.R., 1989: Urban infrastructure. In: *The Potential Effects of Global Climate Change on the United States* [Smith, J.B. and D.A. Tirpak (eds.)]. U.S. Environmental Protection Agency, Office of Policy, Planning and Evaluation, Washington, DC, USA, 689 pp.

Milliman, J.D., 1993: Coral reefs and their response to global climate change. *Climate Change in the Intra Americas Sea* [Maul, G.A. (ed.)]. United Nations Environment Program, Arnold, London, United Kingdom, pp. 306–321.

Mills, B. and L. Craig (eds.), 1999: Atmospheric change in the Toronto-Niagara region: towards an understanding of science, impacts and responses. In: *Proceedings of a workshop held May 27–28, 1998, Toronto, ON, Canada*. Environment Canada, Toronto, ON, Canada, 226 pp.

Mills, E., 1999: The insurance and risk management industries: new players in the delivery of energy-efficient products and services. In: *Proceedings of the ECEEE 1999 Summer Study, European Council for an Energy-Efficient Economy, 31 May - 4 June 1999, Mandelieu, France*. United Nations Environment Programme's 4th International Conference of the Insurance Industry Initiative, Natural Capital at Risk: Sharing Practical Experiences from the Insurance and Investment Industries, 10-11 July 1999, Oslo, Norway.

Mills, E., 1996: Energy efficiency: no-regrets climate change insurance for the insurance industry. *Journal of the Society of Insurance Research*, **Fall**, 21–58. Available online at http://eande.lbl.gov/CBS/Insurance/ClimateInsurance.html.

Mills, E. and I. Knoepfel, 1997: Energy-efficiency options for insurance loss-prevention. In: *Proceedings of the 1997 ECEEE Summer Study, European Council for an Energy-Efficient Economy, Copenhagen, Denmark (Refereed)*. Lawrence Berkeley National Laboratory Report No. 40426, Berkeley, CA, USA. Available online at http://eande.lbl.gov/CBS/PUBS/no-regrets.html.

Mills, E., E. Lecomte, and A. Peara, 2001: *U.S. Insurance Industry Perspectives on Climate Change*. Lawrence Berkeley National Laboratory Technical Report No. LBNL-45185, Berkeley, CA, USA, 185 pp.

Mills, P.F., 1994: The agricultural potential of northwestern Canada and Alaska and the impact of climatic change. *Arctic*, **47(2)**, 115–123.

Mitchell, M.J., C.T. Driscoll, J.S. Kahl, G.E. Likens, P.S. Murdoch, and L.H. Pardo, 1996: Climatic control of nitrate loss from forested watersheds in the northeastern United States. *Environmental Science and Technology*, **30**, 2609–2612.

Mitsch, W.J. and J.G. Gosselink, 1993: *Wetlands*. Van Nostrand Reinhold, New York, NY, USA.

Mitsch, W.J., J.W.J. Day, J.W. Gilliam, P.M. Groffman, D.L. Hey, G.W. Randall, and N. Wang, 2001: Reducing nutrient loads, especially nitrate-nitrogen, to surface water, groundwater, and the Gulf of Mexico. *BioScience*, (in press).

Mittelstaedt, M., 1999: Great Lakes water levels fall to lowest point in 30 years. *The Globe and Mail*, **May 15 (Metro edition)**, A9.

Monath, T., 1980: Epidemiology. In: *St. Louis Encephalitis* [Monath, T. (ed.)]. American Public Health Association, Washington, DC, USA, pp. 239–312.

Mooney, S., 1999: Should we worry about loss reserve drop? *National Underwriter*, **103(13)**, 15.

Moore, A., B. Helwaldt, and G. Craun, 1993: Surveillance for waterborne disease outbreaks: United States, 1991–1992. *Morbidity and Mortality Weekly Report*, **42**, 1–22.

Moore, M.V., P.M.L., J.R. Mather, P.S. Murdoch, R.W. Howarth, C.L. Folt, C.Y. Chen, H.F. Hemond, P.A. Flebbe, and C.T. Driscoll, 1997: Potential effects of climate change on freshwater ecosystems of the New England/Mid-Atlantic region. *Hydrologic Processes*, **11**, 925–947.

Moore, T.R., N.T. Roulet, and J.M. Waddington, 1998: Uncertainty in predicting the effect of climatic change on the carbon cycling of Canadian peatlands. *Climatic Change*, **40**, 229–245.

Morris, R.F. (ed.), 1963: The dynamics of epidemic spruce budworm populations. *Memoirs of the Entomological Society of Canada*, **31**, 332 pp.

Morrison, W.M. and R. Mendelsohn, 1999: The impact of global warming on U.S. energy expenditures. In: *The Impact of Climate Change on the United States Economy* [Mendelsohn, R. and J. Neumann (eds.)]. Cambridge University Press, Cambridge, United Kingdom and New York, NY, USA, pp. 209–236.

Mortsch, L. and F. Quinn, 1996: Climate change scenarios for Great Lakes Basin ecosystem studies. *Limnology and Oceanography*, **41(5)**, 903–911.

Mortsch, L., H. Hengeveld, M. Lister, B. Lofgren, F. Quinn, M. Slivitzky, and L. Wenger, 2000: Climate change impacts on the hydrology of the Great Lakes-St. Lawrence System. *Canadian Water Resources Journal*, **25(2)**, 153–179.

Mortsch, L.D., 1998: Assessing the impact of climate change on the Great Lakes shoreline wetlands. *Climatic Change*, **40**, 391–406.

Mortsch, L.D., S. Quon, L. Craig, B. Mills, and B. Wrenn (eds.), 1998: Adapting to climate change and variability in the Great Lakes - St. Lawrence Basin. In: *Proceedings of a binational symposium, 13-15 May 1997,Toronto, ON, Canada*. Environment Canada, Toronto, ON, Canada, 193 pp.

Mosiño, P. and E. García, 1973: *The Climate of Mexico*.World Survey of Climatology, Elsevier, London, United Kingdom, pp. 345–404.

Mount, J.F., 1995: *California Rivers and Streams: Conflict Between Fluvial Process and Land Use*. University of California Press, Berkeley, CA, USA, 359 pp.

Mulholland, P.J., G.R. Best, C.C. Coutant, G.M. Hornberger, J.L. Meyer, P.J. Robinson, J.R. Stenberg, R.E. Turner, F. Verra-Herrara, and R.G. Wetzel, 1997: Effects of climate change on freshwater ecosystems of the south-eastern United States and Gulf of Mexico. *Hydrological Processes*, **11**, 131–152.

Mullens, A., 1996: "I think we have a problem in Victoria." MDs respond quickly to toxoplasmosis outbreak in BC. *Canadian Medical Association Journal*, **154**, 1721–1724.

Munasinghe, M. and R. Swart (eds.), 2000: *Climate Change and Its Linkages with Development, Equity, and Sustainability. Proceedings of the IPCC Expert Meeting held in Columbo, Sri Lanka, April 1999*. LIFE, Colombo, Sri Lanka; RIVM, Bilthoven, The Netherlands; and World Bank, Washington, DC, USA, 319 pp.

Mundo, M.D. and P. Martínez Austria, 1994: El cambio climático y sus efectos potenciales en los recursos hídricos y la agricultura del Valle del Yaui, Sonora (estudio preliminar indicativo). *Ingeniería Hidráulica en México*, **IX(1)**, 13–33 (in Spanish).

Mundo, M.D. and P. Martínez Austria, 1993: Cambio climático: posibles consecuencias y algunas sugerencias para disminuir su efecto en México. *Ingeniería Hidráulica en México*, **18(1)**, 14–28 (in Spanish).

Munich Re, 2000: *Annual Review of Natural Disasters: 2000*. Munich Reinsurance Group, Report 2946-M-e. Supplementary data and analyses provided by Munich Re, MRNatCatSERVICE, Munich, Germany.

Munich Re, 1997: *Flooding and Insurance*. Munich Reinsurance Company, Munich, Germany, 77 pp.

Munro, G.R., T. McDorman, and R. McKelvey, 1998: *Transboundary Fishery Resources and the Canada-United Sates Pacific Salmon Treaty*. Occasional Papers, Canadian-American Public Policy, Canadian-American Center, University of Maine, Orono, ME, USA, 43 pp.

Murdoch, P.S., D.A. Burns, and G.B. Lawrence, 1998: Relation of climate change to the acidification of surface waters by nitrogen deposition. *Environmental Science and Technology*, **32**, 1642–1647.

Myers, R.A. and G. Mertz, 1998: The limits of exploitation: a precautionary approach. *Ecological Applications*, **8(1)**, S165–S169.

Myneni, R.B., C.D. Keeling, C.J. Tucker, G. Asrar, and R.R. Nemani, 1997: Increased plant growth in the northern high latitudes from 1981 to 1991. *Nature*, **386**, 698–702.

Nash, L.L. and P.H. Gleick, 1993: *The Colorado River Basin and Climatic Change: The Sensitivity of Streamflow and Water Supply to Variations in Temperature and Precipitation*. EPA 230-R-93-009, report prepared for the U.S. Environmental Protection Agency, Office of Policy, Planning and Evaluation, Climate Change Division, by Pacific Institute for Studies in Development, Environment, and Security, Oakland, CA, USA.

National Weather Service, 1993: *Hurricane Andrew: South Florida and Louisiana, August 23–26, 1992. Natural Disaster Survey Report*. National Weather Service, Rockville, MD, USA.

National Weather Service, 1992: *Summary of Natural Hazard Deaths for 1991*. National Weather Service, Rockville, MD, USA.

Newton, J., 1997: Coping with floods: An analogue for dealing with the transition to a modified climate in the northern sector of the Mackenzie Basin. In: *Mackenzie Basin Impact Study (MBIS) Final Report* [Cohen, S.J. (ed.)]. Environment Canada, Toronto, ON, Canada, pp. 219–224.

Newton, J., 1995: An assessment of coping with environmental hazards in Northern aboriginal communities. *Canadian Geographer*, **39(2)**, 112–120.

NHC/DRI, 1999: *An Assessment of Recovery Assistance Provided in Canada and the United States After the 1997 Floods in the Red River Basin*. Report submitted to the International Joint Commission, Natural Hazards Center, University of Colorado and the Disaster Research Institute, University of Manitoba, Winnipeg, Manitoba, Canada, 64 pp.

NHTSA, 1998: *Traffic Safety Facts*. U.S. Department of Transportation, National Highway Traffic Safety Administration, Washington, DC, USA, 226 pp. Available online at http://www.nhtsa.dot.gov/people/ncsa/tsf-1998.pdf.

Nicholls, R.J. and N. Mimura, 1998: Regional issues raised by sea-level rise and their policy implications. *Climate Research*, **11(1)**, 5–18.

Nicholson, B.J., L.D. Gignac, and S.E. Bayley, 1996: Peatland distribution along a north-south transect in the Mackenzie River Basin in relation to climatic and environmental gradients. *Vegetation*, **126**, 119–133.

Nicholson, B.J., L.D. Gignac, S.E. Bayley, and D.H. Vitt, 1997: Vegetation response to global warming: interactions between boreal forest, wetlands and regional hydrology. In: *Mackenzie Basin Impact Study (MBIS) Final Report* [Cohen, S.J. (ed.)]. Environment Canada, Toronto, ON, Canada, pp. 125–145.

Noji, E., 1997: The nature of disaster: general characteristics and public health effects. In: *The Public Health Consequences of Disasters* [Noji, E. (ed.)]. Oxford University Press, Oxford, United Kingdom, pp. 3–20.

Norby, R.J., 1998: Nitrogen deposition: a component of global change analyses. *New Phytologist*, **139,** 189–200.

Norris, F., J. Perilla, J. Riad, K. Kaniasty, and E. Lavizzo, 1999: Stability and change in stress, resources, and psychological distress following natural disaster: findings from a longitudinal study of Hurricane Andrew. *Anxiety Stress Coping*, **12,** 363–396.

Norwine, J., 1978: Twentieth-century semi-arid climates and climatic fluctuations in Texas and northeastern Mexico. *Journal of Arid Environments*, **1,** 313–325.

Nowak, D.J., E.G. McPherson, and R.A. Rowntree, 1994: Executive summary. In: *Chicago's Urban Forest Ecosystem. Results of the Chicago Urban Forest Climate Project*. U.S. Department of Agriculture, Forest Service, Northeastern Forest Experimentation Station, Radnor, PA, USA, pp. iii-vi.

NRC, 1995: *Wetlands: Characteristics and Boundaries*. National Research Council, National Academy Press, Washington, DC, USA.

NRC, 1993: *Soil and Water Quality: An Agenda for Agriculture*. National Research Council, National Academy Press, Washington, DC, USA.

NRC, 1992: *Water Transfers in the West: Efficiency, Equity and the Environment*. National Research Council, National Academy Press, Washington, DC, USA, 300 pp.

NRC, 1991: *Rethinking the Ozone Problem in Urban and Regional Air Pollution*. National Research Council, National Academy Press, Washington, DC, USA, 500 pp.

Nutter, F.W., 1999: Global climate change: why U.S. insurers care. *Climatic Change*, **42(1),** 45–49.

Nutter, F.W., 1996: Insurance and natural sciences: partners in the public interest. *Journal of Society of Insurance Research*, **Fall,** 15–19.

NWWG, 1988: *Wetlands of Canada*. Ecological Land Classification Series #24, National Wetlands Working Group, Environment Canada, Ottawa, ON, Canada, 452 pp.

Ollinger, S.V., J.D. Aber, and P.B. Reich, 1997: Simulating ozone effects on forest productivity: interactions between leaf-, canopy- and stand-level processes. *Ecological Applications*, **7,** 1237–1251.

Olson, M., P. Roach, and M. Stabler, 1997: Giardiasis in ringed seals from the western arctic. *Journal of Wildlife Diseases*, **33(3),** 646–648.

Ontario Lung Association, 1991: *Lung Facts*. Ontario Lung Association, Toronto, ON, Canada.

Osewe, P., D. Addiss, and K. Blair, 1996: Cryptosporidiosis in Wisconsin: a case-control study of post-outbreak transmission. *Epidemiology and Infection*, **117,** 297–304.

OTA, 1993: *Preparing for an Uncertain Climate*. Office of Technology Assessment, U.S. Congress, U.S. Government Printing Office, Washington, DC, USA, Vol. 1, 359 pp.

Paerl, H.W., J.D. Bales, L.W. Ausley, C.P. Buzzelli, L.B. Crowder, L.A. Eby, M. Go, B.L. Peierts, T.L. Richardson, and J.S. Ramus, 2000: Hurricanes' hydrological, ecological effects linger in major U.S. estuary. *EOS, Transactions, American Geophysical Union*, **81(40),** 457–462.

Parmesan, C., 1996: Climate and species' range. *Nature*, **382,** 765–766.

Parmesan, C., T.L. Root, and M.R. Willig, 2000: Impacts of extreme weather and climate on terrestrial biota. *Bulletin of the American Meteorological Society*, **81,** 443–450.

Patterson, D.T., J.K. Westbrook, R.J.V. Joyce, and P.D. Lindgren, 1999: Weeds, insects, and diseases. *Climatic Change*, **43,** 711–727.

Patz, J., M. McGeehin, S. Bernard, K. Ebi, P. Epstein, A. Grambsch, D. Gubler, P. Reiter, I. Romieu, J. Rose, J. Samet, and J. Trtanj, 2000: The potential health impacts of climate variability and change for the United States: executive summary of the report of the health sector of the U.S. National Assessment. *Environmental Health Perspectives*, **108(4),** 367–376.

Payette, S., 1987: Recent porcupine expansion at tree line: a dendroecological analysis. *Canadian Journal of Zoology*, **65,** 551–557.

PCAST, 1998: *Teaming with Life: Investing in Science to Understand and Use America's Living Capital*. President's Committee of Advisors on Science and Technology and Ecosystems, Executive Secretariat, Washington, DC, USA, 86 pp.

Peart, R.M., R.B. Curry, C. Rozenzweig, J.W. Jones, K.J. Boote, and L.H.J. Allen, 1995: Energy and irrigation in southeastern U.S. agriculture under climate change. *Journal of Biogeography*, **22,** 635–642.

Pelto, M.S., 1993: Changes in water supply in alpine regions due to glacier retreat. In: *The World at Risk: Natural Hazards and Climate Change, New York, NY, American Institue of Physics Conference Proceedings 277* [Bras, R. (ed.)]. American Institute of Physics, New York, NY, USA, pp. 61–67.

Pérez-Fuentetaja, A., P.J. Dillon, N.D. Yan, and D.J. McQueen, 1999. Significance of dissolved organic carbon in the prediction of thermocline depth in small Canadian Shield lakes. *Aquatic Ecology*, **33,** 127–133.

Perl, T., L. Bedard, and T. Kosatsky, 1990: An outbreak of toxic encephalopathy caused by eating mussels contaminated with domoic acid. *New England Journal of Medicine*, **322(25),** 1775–1780.

Peters, R.L., 1992: Conservation of biological diversity in the face of climate change. *Global Warming and Biological Diversity* [Peters, R.L. and T.E. Lovejoy (eds.)]. Yale University Press, New Haven, CT, USA, pp. 15–30.

Peterson, D.F. and A.A. Keller, 1990: Irrigation. *Climate Change and U.S. Water Resources* [Waggoner, P.E. (ed.)]. John Wiley and Sons, New York, NY, USA, pp. 269–306.

Phelps, P., 1996: *Conference on Human Health and Global Climate Change: Summary of the Proceedings*. National Academy Press, Washington, DC, USA, 64 pp.

Pielke, R.A., 1999: Nine fallacies of floods. *Climatic Change*, **42,** 413–438.

Pielke, R.A., 1996: *Midwest Flood of 1993: Weather Climate and Societal Impacts*. Environmental and Societal Impacts Group, National Center for Atmospheric Research, Boulder, CO, USA, 159 pp.

Pielke, R.A. and M.W. Downton, 2000: Precipitation and damaging floods: trends in the United States, 1932–1997. *Journal of Climate*, **13(20),** 3625–3637.

Pielke, R.J. and C.W. Landsea, 1998: Normalized hurricane damages in the United States: 1925–95. *Weather and Forecasting*, **13,** 621–631.

Pielke, R.A. Jr. and R.A. Pielke Sr., 1997: *Hurricanes: Their Nature and Impacts on Society*. Wiley and Sons, Chichester, United Kingdom, 279 pp.

Pinter, L., 1997: Sustainability of native lifestyles. In: *Mackenzie Basin Impact Study (MBIS) Final Report* [Cohen, S.J. (ed.)]. Environment Canada, Toronto, ON, Canada, pp. 62–65.

Plate, E.J. and L. Duckstein, 1998: Reliability-based design concepts in hydraulic engineering. *Water Resources Bulletin*, **24(2),** 235–245.

Platt, R.H., P.K. Barten, and M.J. Pfeffer, 2000: A full clean glass? Managing New York City's watersheds. *Environment*, **42(5),** 8–20.

Poiani, K.A., W.C. Johnson, and T.G. Kittle, 1995: Sensitivity of a prairie wetland to increased temperature and seasonal precipitation changes. *Water Resources Bulletin*, **31(2),** 283–294.

Poiani, K.A., W.C. Johnson, G.A. Swanson, and T.C. Winter, 1996: Climate change and northern prairie wetlands: simulations of long-term dynamics. *Limnology and Oceanography*, **41,** 871–881.

Pollard, D.F.W. and R.A. Benton, 1994: The status of protected areas in the Mackenzie Basin. In: *Mackenzie Basin Impact Study Interim Report No. 2* [Cohen, S.J. (ed.)]. Environment Canada, Downsview, ON, Canada, pp. 23–27.

Polovina, J.J., G.T. Mitchum, and G.T. Evans, 1995: Decadal and basin-scale variation in mixed layer depth and the impact on biological production in the central and north Pacific, 1960–1988. *Deep-Sea Research*, **42,** 1201–1716.

Porter, K.G., P.A. Saunders, K.A. Haberyan, A.E. Macubbin, T.R. Jacobsen, and R.E. Hodson, 1996: Annual cycle of autotrophic and heterotrophic production in a small, monomictic Piedmont lake (Lake Oglethorpe): analog for the effects of climatic warming on dimictic lakes. *Limnology and Oceanography*, **41(5),** 1041–1051.

Post, E. and N.C. Stenseth, 1999: Climatic variability, plant phenology, and northern ungulates. *Ecology*, **80,** 1322–1339.

Powell Consortium, 1995: *Severe Sustained Drought: Managing the Colorado River System in Times of Shortage*. Arizona Water Resources Center, Tucson, AZ, USA, Vol. 3, p. 5.

Pulwarty, R., 1999: Hurricane impacts in the context of climate variability, climate change and coastal management policy on the eastern U.S. seaboard. In: *Climate Change and Risk* [Downing, T., A. Olsthoorn, and R. Tol (eds.)]. Routledge, London, United Kingdom, pp. 173–204.

Prather, M., R. Derwent, D. Ehhalt, P. Fraser, E. Sanhueza, and X. Zhou, 1995: Other trace gases and atmospheric chemistry. *Climate Change 1995: Radiative Forcing of Climate Changes and an Evaluation of IPCC IS92 Emission Scenarios* [Houghton, J., L.G. Meira, E. Haites, N. Harris, and K. Maskell (eds.)]. Cambridge University Press, Cambridge, United Kingdom and New York, NY, USA, pp. 86–102.

Rabalais, N.N., R.E. Turner, D. Justic, Q. Dortch, W.J. Wiseman, and B.K.S. Gupta, 1996: Nutrient changes in the Mississippi River system responses on the adjacent continental shelf. *Estuaries*, **19,** 386–407.

Rahel, F.J., C.J. Keheler, and J.L. Anderson, 1996: Potential habitat loss and population fragmentation for cold water fish in the North Platte River drainage of the Rocky Mountains: response to climate warming. *Limnology and Oceanography*, **41,** 1116–1123.

Raizenne, M., L.M. Neas, and F.E. Speizer, 1996: Health effects of acid aerosols on North American children: pulmonary function. *Environmental Health Perspectives*, **104(5),** 506–514.

Ramlow, J. and L. Kuller, 1990: Effects of the summer heat wave of 1988 on daily mortality in Alleghany County, PA. *Public Health Report*, **105,** 283–289.

Rango, A., 1995: Effects of climate change on water supplies in mountainous snowmelt regions. *World Resources Review*, **7(3),** 315–325.

Rango, A. and V. Van Katwijk, 1990: Water supply implications of climate change in western North American basins. In: *International and Transboundary Water Resources Issues, American Water Resources Association 27th Annual Conference Proceedings, Toronto, ON, 1-4 April 1990* [Fitzgibbon, J.E. (ed.)]. American Water Resources Association, Bethesda, MD, USA, pp. 577–586.

Redmond, K.T., 1998: Climate change issues in the mountainous and intermountain west. In: *Proceedings of the Rocky Mountain/Great Basin Regional Climate Change Workshop, Feb. 16–18, 1998, Salt Lake City, Utah* [Wagner, F. and J. Baron (eds.)]. U.S. National Assessment of the Consequences of Climate Change, Washington, DC, USA, 166 pp.

Reeves, W., J. Hardy, W. Reisen, and M. Milby, 1994: Potential effects of global warming on mosquito-borne arboviruses. *Journal of Medical Entomology*, **31,** 323–332.

Reilly, J., F. Tubiello, B. McCarl, and J. Melillo, 2000: Climate change and agriculture in the United States. In: *Climate Change Impacts on the United States: The Potential Consequences of Climate Variability and Change*. Report for the U.S. Global Change Research Program. Cambridge University Press, Cambridge, United Kingdom and New York, NY, USA, pp. 379–403.

Reilly, J.M. and K.O. Fuglie, 1998: Future yield growth in field crops: what evidence exists? *Soil and Tillage Research*, **47,** 275–290.

Reiter, P., 1999: Global climate change and mosquito-borne diseases. *Encyclopedia of Human Ecology* [Watt, K. (ed.)]. Academic Press, London, United Kingdom, pp. 245–255.

Repo, T., H. Hänninen, and S. Kellomäki, 1996: The effects of long-term elevation of air temperature and CO_2 on the frost hardiness of Scots pine. *Plant Cell and Environment*, **19,** 209–216.

Ricketts, T.H., E. Dinerstein, D.M. Olson, C.J. Loucks, W. Eichbaum, D. DellaSala, K. Kavanagh, P. Hedao, P.T. Hurley, K.M. Carney, R. Abell, and S. Walters, 1999: *Terrestrial Ecoregions of North America: A Conservation Assessment*. Island Press, Washington, DC, USA, 485 pp.

Riley, J.P., A.K. Sikka, A.S. Limaye, R.W. Sunderson, G.E. Bingham, and R.D. Hansen, 1996: *Water Yield in Semiarid Environment Under Projected Climate Change*. U.S. Bureau of Reclamation, Provo, UT, USA, 67 pp.

Rodriguez, H., 1997: A socio-economic analysis of hurricanes in Puerto Rico: an overview of disaster mitigation and preparedness. In: *Hurricanes: Climate and Socio-Economic Impact* [Diaz, H. and R. Pulwarty (eds.)]. Springer-Verlag, Heidelberg, Germany, pp. 121–146.

Roemmich, D. and J. McGowan, 1995: Climate warming and the decline of zooplankton in the California current. *Science*, **267,** 1324–1326.

Rogers, P., 1994: Assessing the socioeconomic consequences of climate change on water resources. *Climatic Change*, **28,** 179–208.

Roland, J., B.G. Mackey, and B. Cooke, 1998: Effects of climate and forest structure on duration of forest tent caterpillar outbreaks across central Ontario, Canada. *Canadian Entomologist*, **130,** 703–714.

Rosenberg, N.J., A. Kimball, P. Martin, and C.F. Cooper, 1990: From climate and CO_2 enrichment to evapotranspiration. In: *Climate Change and U.S. Water Resources* [Waggoner, P.E. (ed.)]. John Wiley and Sons, New York, NY, USA, pp. 151–176.

Rosenzweig, C. and F.N. Tubiello, 1997: Impacts of future climate change on Mediterranean agriculture: current methodologies and future directions. *Mitigating Adaptive Strategies in Climate Change*, **1,** 219–232.

Rosenzweig, C., M.L. Parry, and G. Fischer, 1995: World food supply. In: *As Climate Changes: International Impacts and Implications* [Strzepek, K.M. and J.B. Smith (eds.)]. Cambridge University Press, Cambridge, United Kingdom and New York, NY, USA, pp. 27–56.

Rosenzweig, C., B. Curry, J.T. Richie, J.W. Jones, T.Y. Chou, R. Goldberg, and A. Iglesias, 1994: The effects of potential climate change on simulated grain crops in the United States. In: *Implications of Climate Change for International Agriculture: Crop Modelling Study* [Rosenzweig, C. and A. Iglesias (eds.)]. U.S. Environmental Protection Agency, Washington, DC, USA, pp. 100–124.

Rosenzweig, C., J. Phillips, R. Goldberg, R. Carroll, and T. Hodges, 1996: Potential impacts of climate change on citrus and potato production in the U.S. *Agricultural Systems*, **52,** 455–479.

Rosenzweig, C., A. Iglesias, X.B. Yang, P.R. Epstein, and E. Chivian, 2000: *Climate Change and U.S. Agriculture: The Impacts of Warming and Extreme Weather Events on Productivity, Plant Diseases, and Pests.* Center for Health and the Global Environment, Harvard University, Cambridge, MA, USA, 46 pp.

Ross, A., 2000: *Reflections on the Future: Climate Change and its Impacts on the Insurance Industry*. Paper No. 8, University of Western Ontario and Institute for Catastrophic Loss Reduction, Toronto, ON, Canada, 9 pp.

Rothman, D.S. and D. Herbert, 1997: The socio-economic implications of climate change in the forest sector of the Mackenzie Basin. In: *Mackenzie Basin Impact Study (MBIS) Final Report* [Cohen, S.J. (ed.)]. Environment Canada, Toronto, ON, Canada, pp. 225–241.

Roughgarden, J., 1998: How to manage fisheries. *Ecological Applications*, **8(1),** S160–S164.

Rouse, W.R., M.S.V. Douglas, R.E. Hecky, A.E. Hershey, G.W. Kling, L. Lesack, P. Marsh, M. McDonald, B.J. Nicholson, N.T. Roulet, and J.P. Smol, 1997: Effects of climate change on freshwaters of arctic and subarctic North America. *Hydrological Processes*, **11,** 873–902.

Saliba, B.C. and D.B. Bush, 1987: *Water Markets in Theory and Practice: Market Transfers, Water Values, and Public Policy*. Westview Press, Boulder, CO, USA, 273 pp.

Sanchez, R., 1994: *Cambio Climatico y Sus Posibles Consecuencias en Ciudades de México*. Taller Estudio de Pais, Cuernavaca, Morelos, Mexico, pp. 213–220 (in Spanish).

Sangoyomi, T.B. and B.L. Harding, 1995: Mitigating impacts of a severe sustained drought on Colorado River water resources. *Water Resources Bulletin*, **31(5),** 925–938.

Saxe, H., D.S. Ellsworth, and J. Heath, 1998: Tree and forest functioning in an enriched CO_2 atmosphere. *New Phytologist*, **139,** 395–436.

Schaake, J.C., 1990: From climate to flow. In: *Climate Change and U.S. Water Resources* [Waggoner, P.E. (ed.)]. John Wiley and Sons, New York, NY, USA, pp. 177–206.

Schantz, P., 1991: Parasitic zoonoses in perspective. *International Journal for Parasitology*, **21,** 161–170.

Schapendonk, A.H.C.M., P. Kijkstra, J. Groenwold, C.S. Pot, and S.C. Van de Geijn, 1997: Carbon balance and water use efficiency of frequently cut Lolium perenne L. swards at elevated carbon dioxide. *Global Change Biology*, **3,** 207–216.

Schimel, D., D. Alves, I. Enting, M. Heimann, F. Joos, D. Raynaud, T. Wigley, M. Prather, R. Derwent, D. Ehhalt, P. Fraser, E. Sanhueza, X. Zhou, P. Jonas, R. Charlson, H. Rodhe, S. Sadasivan, K. P. Shine, Y. Fouquart, V. Ramaswamy, S. Solomon, J. Srinivasan, D. Albritton, I. Isaksen, M. Lal, and D. Wuebbles, 1996: Radiative forcing of climate change. In: *Climate Change 1995: The Science of Climate Change. Contribution of Working Group I to the Second Assessment Report of the Intergovernmental Panel on Climate Change* [Houghton, J.T., L.G. Meira Filho, B.A. Callander, N. Harris, A. Kattenberg, and K. Maskell (eds.)]. Cambridge University Press, Cambridge, United Kingdom and New York, NY, USA, pp. 65–131.

Schimmelpfennig, D.E., J. Lewandrowski, J.M. Reilly, M. Tsigas, and I. Parry, 1996: Agricultural adaptation to climate change: issues of longrun sustainability. In: *Agriculture Economic Report 740*. Natural Resources and Environment Division, Economic Research Service, U.S. Department of Agriculture, Washington, DC, USA, 57 pp.

Schindler, D.W., 1998: A dim future for boreal watershed landscapes. *BioScience*, **48,** 157–164.

Schindler, D.W., S.E. Bayley, B.R. Parker, K.G. Beaty, D.R. Cruikshank, E.J. Fee, E.U. Schindler, and M.P. Stainton, 1996: The effects of climatic warming on the properties of boreal lakes and streams at the Experimental Lakes Area, Northwestern Ontario. *Limnology and Oceanography*, **41,** 1004–1017.

Schindler, D.W., P.J. Curtis, S.E. Bayley, B.R. Parker, K.G. Beaty, and M.P. Stainton, 1997: Climate induced changes in the dissolved organic carbon budgets of boreal lakes. *Biogeochemistry*, **36,** 9–28.

Schmandt, J. and J. Clarkson, 1992: Introduction: global warming as a regional issue. In: *The Regions and Global Warming* [Schmandt, J. and J. Clarkson (eds.)]. Oxford University Press, Oxford, United Kingdom, pp. 3–11.

Schmidt, R.H., 1975: *The Climate of Chihuahua, Mexico*. Technical Reports on the Meteorology and Climatology of Arid Regions No. 23, Institute of Atmospheric Physics, University of Arizona, Tucson, AZ, USA.

Schowalter, T.D., W.W. Hargrove, and D.A.J. Crossley, 1986: Herbivory in forested ecosystems. *Annual Review of Entomology*, **31,** 177–196.

Schuman, S., 1972: Patterns of urban heat-wave deaths and implications for prevention: data from New York and St. Louis during 1966. *Environmental Research*, **5,** 59–75.

Schwartz, J., 1994: Air pollution and daily mortality: a review and meta-analysis. *Environmental Research*, **64,** 36–52.

Scott, A. and G. Coustalin, 1995: The evolution of water rights. *Natural Resources Journal*, **35(4),** 831–979.

Scott, D. and R. Suffling, 2000: *Climate Change and Canada's National Park System: A Screening Level Assessment*. Cat. No. En56-155/2000E, Adaptation and Impacts Research Group, Environment Canada, Toronto, ON, Canada, 183 pp.

Seaton, A., 1996: Particles in the air: the enigma of urban air pollution. *Journal of the Royal Society of Medicine*, **89,** 604–607.

Segerson, K. and B.L. Dixon, 1999: Climate change and agriculture: the role of farmer adaptation. In: *The Impact of Climate Change on the United States Economy* [Mendelsohn, R. and J. Neumann (eds.)]. Cambridge University Press, Cambridge, United Kingdom and New York, NY, USA, pp. 75–93.

Semenza, J., 1999: Are electronic emergency department data predictive of heat-related mortality? *Journal of Medical Systems*, **23,** 419–424.

Semenza, J., C. Rubin, and K. Falter, 1996: Heat-related deaths during the July 1995 heat wave in Chicago. *New England Journal of Medicine*, **335,** 84–90.

Semenza, J., J. McCullough, D. Flanders, M. McGeehin, and J. Lumpkin, 1999: Excess hospital admissions during the 1995 heat wave in Chicago. *American Journal of Preventive Medicine*, **16,** 269–277.

Shaw, J., R.B. Taylor, S. Solomon, H.A. Christian, and D.L. Forbes, 1998: Potential impacts of global sea level rise on Canadian coasts. *The Canadian Geographer*, **42(4),** 365–379.

Sherk, G.W., 1990: Eastern water law: trends in state legislation. *Virginia Environmental Law Journal*, **9(2),** 287–321.

Shriner, D. and R. Street, 1998: North America. In: *The Regional Impacts of Climate Change: An Assessment of Vulnerability. Special Report of IPCC Working Group II* [Watson, R.T., M.C. Zinyowera, and R.H. Moss (eds.)]. Intergovernmental Panel on Climate Change, Cambridge University Press, Cambridge, United Kingdom and New York, NY, USA, pp. 253–330.

Shugart, H.H. and T.M. Smith, 1996: A review of forest patch models and their application to global change research. *Climatic Change*, **34,** 131–153.

Shukla, P.R., 1997: Socio-economic dynamics of developing countries: some ignored dimensions in integrated assessment. In: *Climate Change and Integrated Assessment Models (IAMs): Bridging the Gaps* [Cameron, O.K., K. Fukuwatari, and T. Morita (eds.)]. Center for Global Environmental Research, Tsukuba, Japan, pp. 165–182.

Shuntov, V.P., E.P. Dulepova, V.I. Radchenko, and V.V. Lapko, 1996: New data about communities of plankton and nekton of the far-eastern seas in connection with climate-oceanological reorganization. *Fisheries Oceanography*, **5(1),** 38–44.

Sieben, B.G., D.L. Spittlehouse, R.A. Benton, and J.A. McLean, 1997: A first approximation of the effect of climate warming on the white pine weevil hazard in the Mackenzie River Drainage Basin. In: *Mackenzie Basin Impact Study (MBIS) Final Report* [Cohen, S.J. (ed.)]. Environment Canada, North York, ON, Canada, pp. 166–177.

Sillman, S. and P. Samson, 1995: Impact of temperature on oxidant photochemistry in urban, polluted rural, and remote environments. *Journal of Geophysical Research*, **100,** 11497–11508.

Sinclair, M., R.O. Boyle, D.L. Burke, and G. Peacock, 1997: Why do some fisheries survive and others collapse? In: *Developing and Sustaining World Fisheries Resources: The State of Science and Management, Second World Fisheries Congress* [Hancock, D.A., D.C. Smith, A. Grant, and J.P. Beumer (eds.)]. CSIRO Publishing, Brisbaine, Australia, pp. 283–290.

Smith, F.A., H. Browning, and U.L. Shepherd, 1998a: The influence of climate change on the body mass of woodrats Neotoma in an arid region of New Mexico. *Ecography*, **21,** 140–148.

Smith, J., B. Lavender, H. Auld, D. Broadhurst, and T. Bullock, 1998b: Adapting to climate variability and change in Ontario. In: *The Canada Country Study: Climate Impacts and Adaptation*. Environment Canada, Toronto, ON, Canada, Vol. IV, 117 pp.

Snucins, E.J. and J.M. Gunn, 1995: Coping with a warm environment: behavioral thermoregulation by lake trout. *Transactions of the American Fisheries Society*, **124,** 118–123.

Sohngen, B.L. and R. Mendelsohn, 1999: The impacts of climate change on the U.S. timber market. In: *The Impact of Climate Change on the United States Economy* [Mendelsohn, R. and J.E. Neumann (eds.)]. Cambridge University Press, Cambridge, United Kingdom and New York, NY, USA, pp. 94–132.

Solis, G., H.C. Hightower, and J. Kawaguchi, 1997: *Guidelines on Cultural Diversity and Disaster Management*. Prepared by The Disaster Preparedness Resources Centre, University of British Columbia, for Emergency Preparedness Canada within the Canadian Framework for the International Decade for Natural Disaster Reduction, Ottawa, ON, Canada. Available online at http://www.epc-pcc.gc.ca/research/scie tech/guid cult.html.

Solomon, S., 1994: Storminess and coastal erosion at Tuktoyaktuk. In: *Mackenzie Basin Impact Study Interim Report No. 2* [Cohen, S.J. (ed.)]. Environment Canada, Downsview, ON, Canada, pp. 286–292.

Sorenson, L.G., R. Goldberg, T.L. Root, and M.G. Anderson, 1998: Potential effects of global warming on waterfowl populations breeding in the northern Great Plains. *Climatic Change*, **40,** 343–369.

Soulis, E.D., S.I. Solomon, M. Lee, and N. Kouwen, 1994: Changes to the distribution of monthly and annual runoff in the Mackenzie Basin using a modified square grid approach. In: *Mackenzie Basin Impact Study (MBIS) Interim Report No. 2* [Cohen, S.J. (ed.)]. Environment Canada, Downsview, ON, Canada, pp. 197–209.

Staple, T.L. and G. Wall, 1996: Climate change and recreation in Nahanni National Park. *Canadian Geographer*, **40(2),** 109–120.

Statistics Canada, 1999a: *Census of Canada*. Ottawa, ON, Canada.

Statistics Canada, 1999b: *Canadian Economic Observer: Historical Statistical Supplement*. Statistics Canada, Ottawa, ON, Canada.

Stephen, C., M. Johnson, and A. Bell, 1994: First reported cases of Hantavirus Pulmonary Syndrome in Canada. *Canada Communicable Disease Report*, **20,** 121–125.

Stieb, D., R. Burnett, and R. Beveridoe, 1996: Association between ozone and asthma emergency department visits in Saint John, New Brunswick, Canada. *Environmental and Health Perspectives*, **104(12),** 1354–1360.

Stieb, D., L. Pengelly, and N. Arron, 1995: Health effects of air pollution in Canada: expert panel findings for the Canadian Smog Advisory Program. *Canadian Respiratory Journal*, **2,** 155–160.

Stocks, B.J., 1987: Fire potential in the spruce budworm-damaged forests of Ontario. *Forestry Chronicle*, **63,** 8–14.

Stocks, B.J., M.A. Fosberg, T.J. Lynham, L. Mearns, B.M. Wotton, Q. Yang, J.Z. Jin, K. Lawrence, G.R. Hartley, J.A. Mason, and D.W. McKenney, 1998: Climate change and forest fire potential in Russian and Canadian boreal forests. *Climatic Change*, **38,** 1–13.

Stohlgren, T.J., 1999: The Rocky Mountains. In: *Status and Trends of the Nation's Biological Resources* [Mac, M.J., P.A. Opler, C.E. Puckett Haecker, and P.D. Doran (eds.)]. Biological Resources Division, U.S. Geological Survey, Reston, VA, USA, pp. 1–32.

Stohlgren, T.J., T.N. Chase, R.A. Pielke, Sr., T.G.F. Kittel, and J.S. Baron, 1998: Evidence that local land use practices influence regional climate, vegetation, and streamflow patterns in adjacent natural areas. *Global Change Biology*, **4,** 495–504.

Strzepek, K.M., D.C. Major, C. Rosenzweig, A. Iglesias, D.N. Yates, A. Holt, and D. Hillel, 1999: New methods of modeling water availability for agriculture under climate change: the U.S. cornbelt. *Journal of the American Water Resources Association*, **35(6),** 1639–1656.

Suffling, R., 1995: Can disturbance determine vegetation distribution during climate warming? A boreal test. *Journal of Biogeography*, **22,** 501–508.

Suphioglu, C., 1998: Thunderstorm asthma due to grass pollen. *International Archives of Allergy and Immunology*, **116(4),** 253–260.

Swiss Re, 2000: *Solvency of Non-Life Insurers: Balancing Security and Profitability Expectations*. Sigma Report No. 1/2000, Swiss Reinsurance Company, Zurich, Switzerland, 39 pp. Available online at http://www.swissre.com/e/publications/publications/sigma1/sigma1/sigma010300.html.

Swiss Re, 1999: *World Insurance in 1998*. Sigma Report No. 7, Swiss Reinsurance Company, Zurich, Switzerland, 31 pp. Available online at http://www.swissre.com/e/publications/publications/sigma1/sigma9907.html.

Szujecki, A., 1987: *Ecology of Forest Insects*. Junk, Boston, MA, USA, 600 pp.

Tarlock, A.D., 1991: Western water law, global climate change and risk allocation. In: *Managing Water Resources in the West Under Conditions of Climate Uncertainty* [Council, N.R. (ed.)]. National Academy Press, Washington, DC, USA, pp. 239–251.

Tarlock, A.D., 1989: *Law of Water Rights and Resources*. Clark Boardman, New York, NY, USA.

Tavares, D., 1996: Weather and heat-related morbidity relationships in Toronto (1979–1989). In: *Great Lakes-St. Lawrence Basin Project Progress Report #1: Adapting to the Impacts of Climate Change and Variability* [Mortsch, L. and B. Mills (eds.)]. Environment Canada, Toronto, ON, Canada, pp. 110–112.

Taylor, E. and B. Taylor (eds.), 1997: *Responding to Global Climate Change in the British Columbia and Yukon Region. Volume I of the Canada Country Study: Climate Impacts and Adaptation*. Environment Canada, Toronto, ON, Canada, pp. 1-1 to A-38.

Teitelbaum, J., R. Zatorre, and S. Carpenter, 1990: Neurologic sequelae of domoic acid intoxication due to the ingestion of contaminated mussels. *New England Journal of Medicine*, **322(25),** 1781–1787.

Thomson, A.J. and D.M. Shrimpton, 1984: Weather associated with the start of mountain pine beetle outbreaks. *Canadian Journal of Forest Research*, **14,** 255–258.

Thomson, A.J., R.F. Shepherd, J.W.E. Harris, and R.H. Silversides, 1984: Relating weather to outbreaks of western spruce budworm, Choristoneura occidentalis (Lepidoptera: Tortricidae), in British Columbia. *Canadian Entomologist*, **116,** 375–381.

Thompson, R.S., S.W. Hostetler, P.J. Bartlein, and K.H. Anderson, 1998: *A Strategy for Assessing Potential Future Change in Climate, Hydrology, and Vegetation in the Western United States*. Circular 1153, U.S. Geological Survey, Washington, DC, USA, 20 pp.

Thurston, G., K. Ito, and C. Hayes, 1994: Respiratory hospital admissions and summertime haze air pollution in Toronto, Ontario: consideration of the role of acid aerosols. *Environmental Research*, **65,** 271–290.

Tilman, D., 1998: Species composition, species diversity and ecosystem processes: understanding the impacts of global change. In: *Successes, Limitations and Frontiers in Ecosystem Science* [Pace, M.L. and P.M. Groffman (eds.)]. Springer-Verlag, New York, NY, USA, pp. 452–472.

Timmermann, A. and J. Oberhuber, 1999: Increased El Niño frequency in a climate model forced by future greenhoue warming. *Nature*, **398,** 634–696.

Titus, J.G., 1998: Rising seas, coastal erosion, and Takings Clause: how to save wetlands and beaches without hurting property owners. *Maryland Law Review*, **57(4),** 1279–1399.

Titus, J.G. and C. Richman, 2001: Maps of lands vulnerable to sea level rise: modeled elevations along the U.S. Atlantic and Gulf coasts. *Climate Research*, (in press).

Tizard, I., S. Chauhan, and C. Lai, 1977: The prevalence and epidemiology of toxoplasmosis in Ontario. *Journal of Hygiene Cambridge*, **78,** 275–289.

Tizard, I., J. Harmeson, and C. Lai, 1978: The prevalence of serum antibodies to Toxoplasma gondii in Ontario mammals. *Canadian Journal of Community Medicine*, **42,** 177–183.

Tobin, G.A., 1995: The levee love affair: a stormy relationship. *Water Resources Bulletin*, **31,** 359–367.

Tourangeau, F., G. Delage, and M. Gauthier-Chouisnard, 1986: California Encephalities—Quebec. *Canada Diseases Weekly Report*, **12–25,** 110–111.

Townsend, A.R., B.H. Braswell, E.A. Holland, and J.E. Penner, 1996: Spatial and temporal patterns in terrestrial carbon storage due to deposition of fossil fuel nitrogen. *Ecological Applications*, **6,** 806–814.

Trelease, F.J., 1977: Climatic change and water law. In: *Climate, Climatic Change, and Water Supply* [Hart, T. and O. Shapiro (eds.)]. National Academy of Sciences, Washington, DC, USA, pp. 70–84.

Trenberth, K.E., 1998: Atmospheric moisture residence times and cycling: Implications for rainfall rates and climate change. *Climate Change*, **39(4),** 667–694.

Trenberth, K.E. and D.J. Shea, 1997: Atmospheric circulation changes and links to changes in rainfall and drought. In: *Conference Preprints, AMS Thirteenth Conference on Hydrology, 2–7 February, 1997, Long Beach, CA*. American Meteorological Society, Boston, MA, USA, pp. J14–J17

Tseng, R., C. Li, and J. Spinks, 1992: Particulate air pollution and hospitalization for asthma. *Annals of Allergy*, **68,** 425–432.

Tsuji, G.Y., G. Uehara, and S. Balas (eds.), 1994: *DSSAT v3*. University of Hawaii, Honolulu, Hawaii.

Turner, R.E., 1997: Wetland loss in the northern Gulf of Mexico: multiple working hypotheses. *Estuaries*, **20,** 1–13.

Turner, R.E. and N.N. Rabalais, 1991: Changes in Mississippi River water quality this century. *BioScience*, **41,** 140–147.

UNDP, 1999: *Human Development Index 1999*. United Nations Development Programme, Oxford University Press, Oxford, United Kingdom.

UNEP, 2000: *Global Environment Outlook 2000*. United Nations Environment Program, Nairobi, Kenya, 398 pp.

USBEA, 1999: *National Income and Product Accounts, Survey of Current Business*. United States Bureau of Economic Analysis, Washington, DC, USA.

U.S. Census Bureau, 1999a: *American Housing Survey for the United States: 1997*. Series H1 50/97, U.S. Census Bureau, Washington, DC, USA.

U.S. Census Bureau, 1999b: *Population Division, Current Population Reports*. Series P-25, U.S. Census Bureau, Washington, DC, USA.

U.S. Census Bureau, 1997a: *American Housing Survey for the United States in 1995*. H-150-95RV, U.S. Census Bureau, Washington, DC, USA.

U.S. Census Bureau, 1997b: *Statistical Abstracts of the United States: 1997*. U.S. Census Bureau, Washington, DC, USA.

U.S. Census Bureau, 1975: *American Housing Survey for the United States: 1973, Current Housing Reports*. U.S. Census Bureau, Washington, DC, USA.

USEPA, 1998a: *National Air Quality and Emission Trends Report, 1997. EPA Trends 1998*. Office of Air Quality Planning and Standards, U.S. Environmental Protection Agency, Washington, DC, USA.

USEPA, 1998b: *Climate Change and Arizona*. Publication EPA 236-F-98-007c, U.S. Environmental Protection Agency, Washington, DC, USA, 5 pp.

USEPA, 1998c: *Climate Change and New Mexico*. Publication EPA 236-F-98-007p, U.S. Environmental Protection Agency, Washington, DC, USA.

Van Ginkel, J.H., A. Gorissen, and J.A. Van Veen, 1997: Carbon and nitrogen allocation in Lolium perenne in resonse to elevated atmospheric CO_2 with emphasis on soil carbon dynamics. *Plant Soil*, **188,** 299–308.

Veblen, T.T., T. Kitzberger, and J. Donnegan, 2000: Climatic and human influences on fire regimes in ponderosa pine forests in the Colorado Front Range. *Ecological Applications*, **10,** 1178–1195.

VEMAP, 1995: Vegetation/ecosystem modeling and analysis project: comparing biogeography and biogeochemistry models in a continental-scale study of terrestrial ecosystem responses to climate change and a CO_2 doubling. *Global Biogeochemical Cycles*, **9,** 407–437.

Venables, K., U. Allitt, and C. Collier, 1994: Thunderstorm-related asthma: the epidemic of 25/25 June 1994. *Clinical and Experimental Allergy*, **27(7),** 725–726.

Vincente, V.P., N.C. Singh, and A.V. Botello, 1993: Ecological implications of potential climate change and sea-level rise. In: *Climate Change in the Intra Americas Sea* [Maul, G.A. (ed.)]. United Nations Environment Program, Arnold, London, United Kingdom, pp. 262–281.

Vine, E., E. Mills, and A. Chen, 1999: Tapping into energy: new technologies and procedures that use energy more efficiently or supply renewable energy offer a largely untapped path to achieving risk management objectives. *Best's Review: Property/Casualty Edition*, **May,** 83–85. Available online at http://eetd.lbl.gov/cbs/insurance/lbnl-41432.html.

Volney, W.J.A. and D.G. McCullough, 1994: Jack pine budworm population behaviour in northwestern Wisconsin. *Canadian Journal of Forest Research*, **24,** 502–510.

Waddington, J.M., T.J. Griffis, and W.R. Rouse, 1998: Northern Canadian wetlands: net ecosystem CO_2 exchange and climatic change. *Climatic Change*, **40,** 267–275.

Walker, R., 1996: *Assessment of Climate Change Impacts in the Bay of Quinte, Ontario*. Prepared by Beak Consultants, Ltd. for Environment Canada, Burlington, ON, Canada, 58 pp.

Wall, G., 1998a: Implications of global climate change for tourism and recreation in wetland areas. *Climatic Change*, **40,** 371–389.

Wall, G., 1998b: Impacts of climate change on recreation and tourism. In: *Responding to Global Climate Change: National Sectoral Issues. Volume XII of the Canada Country Study: Climate Impacts and Adaptation* [Koshida, G. and W. Avis (eds.)]. Environment Canada, Toronto, ON, Canada, pp. 591–620.

Wall, G., 1998c: Climate change, tourism, and the IPCC. *Tourism Recreation Review*, **23(2),** 65–68.

Wallis, P., S. Erlandsen, and J. Isaac-Renton, 1996: Prevalence of Giardia cysts and Cryptosporidium oocysts and characterization of Giardia spp. isolated from drinking water in Canada. *Applied and Environmental Microbiology*, **62(8),** 2789–2797.

Ware, D.M., 1995: A century and a half of change in the climate of the NE Pacific. *Fisheries Oceanography*, **4,** 267–277.

Weatherhead, E.C. and C.M. Morseth (eds.), 1998: Climate change, ozone and ultraviolet radiation. In: *AMAP Assessment Report: Arctic Pollution Issues*. Arcitic Monitoring and Assessment Programme (AMAP), Oslo, Norway, pp. 717–774.

Webster, K.E., T.K. Kratz, C.J. Bowser, and J.J. Magnuson, 1996: The influence of landscape position on lake chemical responses to drought in northern Wisconsin. *Limnology and Oceanography*, **41,** 977–984.

Weijers, E. and P. Vellinga, 1995: *Climate Change and River Flooding: Changes in Rainfall Processes and Flooding Regimes Due to an Enhanced Greenhouse Effect*. Institute for Environmental Studies, Vrije Universiteit, Amsterdam, The Netherlands, 42 pp.

Wein, R.W., 1990: The importance of wildfire to climate change: hypotheses for the taiga. In: *Fire in Ecosystem Dynamics* [Goldammer, J.G. and M.J. Jenkins (eds.)]. Academic Press, The Hague, The Netherlands, pp. 185–190.

Welch, D.W., Y. Ishida, and K. Kagasawa, 1998a: Thermal limits and ocean migrations of sockeye salmon (Oncorchynuchus nerka): long-term consequences of global warming. *Canadian Journal of Fisheries and Aquatic Sciences*, **55,** 937–948.

Welch, D.W., Y. Ishida, K. Nagasawa and J.P. Eveson, 1998b: Thermal limits on the ocean distribution of steelhead trout (Onchorhynchus mykiss). *North Pacific Anadromous Fisheries Commission Bulletin*, **1,** 396–404.

Weller, G. and M. Lange (eds.), 1999: *Impacts of Global Climate Change in the Arctic Regions: Report from a Workshop on the Impacts of Global Change, 25–26 April 1999*. Published for the International Arctic Science Commiteee (IASC) by Center for Global Change and Arctic System Research, University of Fairbanks, Alaska, Tromso, Norway, 59 pp.

Weniger, B., M. Blaser, J. Gedrose, E. Lippy, and D. Juranek, 1983: An outbreak of waterborne giardiasis associated with heavy water runoff due to warm weather and volcanic ashfall. *American Journal of Public Health*, **73,** 868–872.

Werker, D., A. Singh, and H. Artsob, 1995: Hantavirus Pulmonary Syndrome in North America. *Canada Communicable Disease Report*, **21,** 77–79.

White, R. and D. Etkin, 1997: Climate change, extreme events and the Canadian insurance industry. *Natural Hazards*, **16,** 135–163.

Wilby, R.L. and M.D. Dettinger, 2000: Streamflow changes in the Sierra Nevada, California, simulated using a statistically downscaled general circulation model scenario of climate change. In: *Linking Climate Change to Land Surface Change* [McLaren, S. and D. Kniveton (eds.)]. J. Kluwer Academic Publishers, Dordrecht, The Netherlands, 276 pp.

Wilby, R.L., L.E. Hay, and G.H. Leavesley, 1999: A comparison of downscaled and raw GCM output: implications for climate change scenarios in the San Juan River basin, Colorado. *Journal of Hydrology*, **225(1–2),** 67–91.

Wilkinson, C.F., 1989: Aldo Leopold and western water law: thinking perpendicular to the prior appropriation doctrine. *Land and Water Law Review*, **24(1),** 1–38.

Williams, D.W. and A.M. Liebhold, 1997: Latitudinal shifts in spruce budworm (Lipidoptera: Tortricidae) outbreaks and spruce-fir forest distributions with climate change. *Acta Phytopathologica et Entomologica Hungarica*, **32,** 205–215.

Williams, M.W., M. Losleban, N. Caine, and D. Greenland, 1996: Changes in climate and hydrochemical responses in a high-elevation catchment in the Rocky Mountains, U.S.A. *Limnology and Oceanography*, **41,** 939–946.

Wilson, E.O., 1992: *The Diversity of Life*. Harvard University Press, Cambridge, MA, USA, 424 pp.

Wilton, D. and T. Wirjanto, 1998: *An Analysis of the Seasonal Variation in the National Tourism Indicators*. Canadian Tourism Commission, Ottawa, ON, Canada, 49 pp.

Wolock, D.M. and G.J. McCabe, 2000: Estimates of runoff using water-balance and atmospheric general circulation models. *Journal of the American Water Resources Association*, **35(6),** 1341–1442.

Wong, A.K., 1999: Improving water management through groundwater banking: Bakersfield 2800 acre recharge facility and semitropic groundwater banking program. In: *The Sustainable Use of Water: California Success Stories* [Owens-Viani, L., A.K. Wong, and P.H. Gleick (eds.)]. Pacific Institute for Studies in Development, Environment and Security, Oakland, CA, USA, pp. 317–334.

Wood, C.M. and D.G. McDonald (eds.), 1997: *Global Warming: Implications for Freshwater and Marine Fish*. Cambridge University Press, Cambridge, United Kingdom and New York, NY, USA, 441.

Woodbury, P.B., J.E. Smith, D.A. Weinstein, and J.A. Laurence, 1998: Assessing potential climate change effects on loblolly pine growth: a probabilistic regional modeling approach. *Forest Ecology and Management*, **107,** 99–116.

Woodhouse, C.A. and J.T. Overpeck, 1998: 2000 years of drought variability in the central United States. *American Meteorological Society Bulletin*, **79,** 2693–2714.

Wright, J.M., 1996: Effects of the flood on national policy: some achievements, major challenges remain. In: *The Great Flood of 1993: Causes, Impacts and Responses* [Changnon, S.A. (ed.)]. Westview Press, Boulder, CO, USA, pp. 245–275.

WTO, 1999: *Tourism Highlights: 1999*. World Tourism Organization, Madrid, Spain, 17 pp.

WTO, 1998: *Tourism 2020 Vision*. World Tourism Organization, Madrid, Spain. 2nd. ed., 48 pp.

WWPRAC, 1998: *Water in the West: Challenge for the Next Century*. National Technical Information Service, Western Water Policy Review Advisory Commission, Springfield, VA, USA, 420 pp.

Yin, Y. and S.J. Cohen, 1994: Identifying regional goals and policy concerns associated with global climate change. *Global Environmental Change*, **4(3),** 246–260.

Yin, Y., S. Cohen, and G.H. Huang, 2000: Global climate change and regional sustainable development: the case of Mackenzie Basin in Canada. *Integrated Assessment*, **1,** 21–36.

Yohe, G., J.E. Neumann, and P. Marshall, 1999: The economic damage induced by sea level rise in the United States. In: *The Impact of Climate Change on the United States Economy* [Mendelsohn, R. and J.E. Neumann (eds.)]. Cambridge University Press, Cambridge, United Kingdom and New York, NY, USA, pp. 178–208.

Yoskowitz, D.W., 1999: Spot market for water along the Texas Rio Grande: opportunities for water management. *Natural Resources Journal*, **39(2),** 345–355.

Yulianti, J.S. and D.H. Burn, 1998: Investigating links between climatic warming and low streamflow in the Prairies region of Canada. *Canadian Water Resources Journal*, **23(1),** 45–60.

Zdan, T., 1997: Maintenance of infrastructure. In: *Mackenzie Basin Impact Study (MBIS) Final Report* [Cohen, S.J. (ed.)]. Environment Canada, Toronto, ON, Canada, pp. 59–61.

Zucker, J., 1996: Changing patterns of autochthonous malaria transmission in the United States: a review of recent outbreaks. *Emerging Infectious Diseases*, **2(1),** 37–43.

16

Polar Regions (Arctic and Antarctic)

OLEG ANISIMOV (RUSSIA) AND BLAIR FITZHARRIS (NEW ZEALAND)

Lead Authors:
J.O. Hagen (Norway), R. Jefferies (Canada), H. Marchant (Australia), F. Nelson (USA), T. Prowse (Canada), D.G. Vaughan (UK)

Contributing Authors:
I. Borzenkova (Russia), D. Forbes (Canada), K.M. Hinkel (USA), K. Kobak (Russia), H. Loeng (Norway), T. Root (USA), N. Shiklomanov (Russia), B. Sinclair (New Zealand), P. Skvarca (Argentina)

Review Editors:
Qin Dahe (China) and B. Maxwell (Canada)

CONTENTS

EXECUTIVE SUMMARY

In this summary we indicate our uncertainty in observations, mechanisms, and scenarios by using a five-point scale, from "very high confidence" (*****) to "very low confidence" (*).

Climate Changes in the 20th Century

Although there are some regional anomalies, there is strong evidence that climate change has had an impact in the Arctic and the Antarctic. Many documented changes already parallel those forecast to result from climate change:

- In the Arctic, extensive land areas show a 20th-century warming trend in air temperature of as much as 5°C. Over sea ice, there has been slight warming in the 1961–1990 period.***** Precipitation has increased.**
- Arctic sea-ice extent has decreased by 2.9% per decade over the 1978–1996 period; sea ice has thinned, and there are now more melt days per summer. Sea-ice extent in the Nordic seas has decreased by 30% over the past 130 years.***** It is not yet clear whether changes in sea ice of the past few decades are linked to a natural cycle in climate variability or have resulted explicitly from global warming.
- Atlantic water flowing into the Arctic Ocean has warmed, and the surface layer has become thinner. The mixed layer in the Beaufort Sea has become less saline.****
- Regions underlain by permafrost have been reduced in extent, and a general warming of ground temperatures has been observed in many areas.*****
- There has been a statistically significant decrease in spring snow extent over Eurasia since 1915.****
- In summary, many observations of environmental change in the Arctic show a trend that is consistent with warming and similar to that predicted by general circulation models (GCMs).
- In the Antarctic, over the past half-century there has been a marked warming trend in the Antarctic Peninsula.**** Elsewhere there is a general but not unambiguous warming trend.**
- Precipitation in the Antarctic has increased.*
- Satellite observations show no significant change in Antarctic sea-ice extent over the 1973–1996 period.***** Analysis of whaling records and modeling studies indicate that Antarctic sea ice retreated south by 2.8 degrees of latitude between the mid-1950s and the early 1970s.***
- Surface waters of the Southern Ocean have warmed and become less saline.***

Impacts

Substantial warming and increases in precipitation are projected for polar regions over the 21st century by almost all climate models. There are eight key concerns related to the impact of this climate change in the Arctic and Antarctic. Associated with these concerns will be changes to the atmosphere and the oceans that will propagate to other regions of the world:

1) *Changes in ice sheets and polar glaciers:* Increased melting is expected on Arctic glaciers and the Greenland ice sheet, and they will retreat and thin close to their margins. Most of the Antarctic ice sheet is likely to thicken as a result of increased precipitation. There is a small risk, however, that the West Antarctic and Greenland ice sheets will retreat in coming centuries. Together, these cryospheric changes may make a significant contribution to sea-level rise.****
2) *Changes around the Antarctic Peninsula:* This region has experienced spectacular retreat and collapse of ice shelves, which has been related to a southerly migration of the January 0°C isotherm resulting from regional warming. The loss of these ice shelves has few direct impacts. Projected warming is likely, however, to break up ice shelves further south on the Antarctic Peninsula, expose more bare ground, and cause changes in terrestrial biology, such as introduction of exotic plants and animals.****
3) *Changes in the Southern Ocean and impacts on its life:* Climate change is likely to produce long-term—perhaps irreversible—changes in the physical oceanography and ecology of the Southern Ocean. Projected reductions in sea-ice extent will alter under-ice biota and spring bloom in the sea-ice marginal zone and will cause profound impacts at all levels in the food chain, from algae to krill to the great whales. Marine mammals and birds, which have life histories that tie them to specific breeding sites, will be severely affected by shifts in their foraging habitats and migration of prey species. Warmer water will potentially intensify biological activity and growth rates of fish. Ultimately, this should lead to an increase in the catch of marketable fish, and retreat of sea ice will provide easier access to southern fisheries.***
4) *Changes in sea ice:* There will be substantial loss of sea ice in the Arctic Ocean. Predictions for summer ice indicate that its extent could shrink by 60% for a doubling of carbon dioxide (CO_2), opening new sea routes. This will have major trading and strategic implications. With more open water, there will be a moderation of temperatures

and an increase in precipitation in Arctic lands. Antarctic sea-ice volume is predicted to decrease by 25% or more for a doubling of CO_2, with sea ice retreating about 2 degrees of latitude.****

5) *Changes in permafrost:* Thickening of the seasonally thawed layer above permafrost (active layer) is expected. Modeling studies indicate that large areas of permafrost terrain will begin to thaw, leading to changes in drainage, increased mass movements, thermal erosion, and altered landscapes in much of the Arctic and subarctic. Warming of permafrost, thawing of ground ice, and development of thermokarst terrain have been documented over the past several decades. In developed areas of the Arctic, continuation of such changes may lead to costly damage to human infrastructure.****
6) *Changes in Arctic hydrology:* The hydrology of the Arctic is particularly susceptible to warming because small rises in temperature will result in increased melting of snow and ice, with consequent impacts on the water cycle. There will be a shift to a runoff regime that is driven increasingly by rainfall, with less seasonal variation in runoff. There will be more ponding of water in some areas, but peatlands may dry out because of increased evaporation and transpiration from plants. In some areas, thawing of permafrost will improve infiltration. An expected reduction in ice-jam flooding will have serious impacts on riverbank ecosystems and aquatic ecology, particularly in the highly productive Arctic river deltas. Changes in Arctic runoff will affect sea-ice production, deepwater formation in the North Atlantic, and regional climate. A major impact would result from a weakening of the global thermohaline circulation as a result of a net increase in river flow and the resulting increased flux of freshwater from the Arctic Ocean.***
7) *Changes in Arctic biota:* Warming should increase biological production; however, the effects of increased precipitation on biological production are unclear. As warming occurs, there will be changes in species compositions on land and in the sea, with a tendency for poleward shifts in species assemblages and loss of some polar species. Changes in sea ice will alter the seasonal distributions, geographic ranges, patterns of migration, nutritional status, reproductive success, and ultimately the abundance and balance of species. Animals that are dependent on sea ice—such as seals, walrus, and polar bears—will be disadvantaged. High-arctic plants will show a strong growth response to summer warming. It is unlikely that elevated CO_2 levels will increase carbon accumulation in plants, but they may be damaged by higher ultraviolet-B radiation. Biological production in lakes and ponds will increase.***
8) *Impacts on human communities:* Climate change, in combination with other stresses, will affect human communities in the Arctic. The impacts may be particularly disruptive for communities of indigenous peoples following traditional lifestyles. Changes in sea ice, seasonality of snow, and habitat and diversity of food species will affect hunting and gathering practices and could threaten longstanding traditions and ways of life. On the other hand, communities that practice these lifestyles may be sufficiently resilient to cope with these changes. Increased economic costs are expected to affect infrastructure, in response to thawing of permafrost and reduced transportation capabilities across frozen ground and water. However, there will be economic benefits—including new opportunities for trade and shipping across the Arctic Ocean, lower operational costs for the oil and gas industry, lower heating costs, and easier access for ship-based tourism.*****

Feedbacks and Interactions

Climate change and global warming will affect key polar drivers of further climate change. These effects will have impacts that affect other regions of the world. Models indicate that once triggered, these impacts will continue for centuries and lead to further change elsewhere in the world:

- Warming will reduce sea-ice and snow extent, particularly in the Arctic, causing additional heating of the surface—which, in turn, will further reduce ice/snow cover.*****
- Deep ocean water around the Antarctic and in the north Atlantic is a crucial part of the ocean's thermohaline circulation. Its rate of production is likely to decrease because of freshening of waters from increased Arctic runoff from glacial icemelt, from increases in precipitation over evaporation, and from reduced sea-ice formation. Models indicate that the impact will be a prolonged, major slowing of the thermohaline circulation and ocean ventilation, even with stabilization of greenhouse gases (GHGs).***
- Polar regions have oceans, wetlands, and permafrost that act as major sources and sinks for atmospheric CO_2 and methane (CH_4) over vast areas. Projected climate change will alter these features and increase their contributions to GHGs. The Southern Ocean's uptake is projected to decline; CO_2 emissions from Arctic tundra may rise initially as a result of changes in water content, peat decomposition, and thawing of permafrost.***

Vulnerability and Adaptation

- Localities within the Antarctic and Arctic where water is close to its melting point are highly sensitive to climate change; this sensitivity renders their biota and socioeconomic life particularly vulnerable to climate change. In the Antarctic Peninsula, as ice melts, changes are likely to be rapid, but overall the Antarctic and the Southern Ocean are likely to respond relatively slowly to climate change, so there will be less impact in this region compared with elsewhere by 2100. Nevertheless, climate change in the Antarctic will initiate processes that could last for millennia—long after greenhouse emissions have stabilized—and these changes will cause irreversible impacts on ice sheets, oceanic circulation of water, and sea-level rise.

- The Arctic is extremely vulnerable to climate change, and major ecological, sociological, and economic impacts are expected. A variety of positive feedback mechanisms induced by climate change are likely to operate in the Arctic; these mechanisms will cause rapid and amplified responses, with consequential impacts on the thickness and extent of sea ice, thawing of permafrost, runoff into the Arctic Ocean, and coastal erosion.
- Biota are particularly vulnerable to climate change in the polar regions. Less sea ice will reduce ice edges, which are prime habitats for marine organisms. Habitat loss for some species of seal, walrus, and polar bear results from ice melt, and apex consumers—with their low-reproductive outputs—are vulnerable to changes in the long polar marine food chains.
- Adaptation to climate change in natural polar ecosystems is likely to occur through migration and changing species assemblages, but the details of these effects are unknown. Some animals may be threatened (e.g., walrus, polar bear, and some species of seal), whereas others may flourish (e.g., some species of fish and penguins).
- Loss of sea ice in the Arctic will provide increased opportunities for new sea routes, fishing, and new settlements, but also for wider dispersal of pollutants. Collectively, these changes emphasize the need for an adequate infrastructure to be in place before they occur. Disputes over jurisdiction in Arctic waters, sustainable development of fisheries and other marine resources, and construction of navigational aids and harbor facilities, as well as problems arising from oil and gas development, including pollution and environmental monitoring, will all have to be resolved by polar and associated nations as climate-induced change becomes widespread. Just as important is the need for new building codes for roads, railways, runways, and buildings to cope with the effects of permafrost thawing.
- Although most indigenous peoples are highly resilient, the combined impacts of climate change and globalization create new and unexpected challenges. Because their livelihood and economy increasingly are tied to distant markets, they will be affected not only by climate change in the Arctic but also by other changes elsewhere. Local adjustments in harvest strategies and in allocation of labor and capital will be necessary. Perhaps the greatest threat of all is to maintenance of self-esteem, social cohesion, and cultural identity of communities.

16.1. Polar Regions

For the purposes of this assessment, the Arctic is defined as the area within the Arctic Circle. It covers the Arctic Ocean and the islands and northern continental land areas (see Figure 16-1). Thus, it extends far enough south to include parts of the boreal forest and discontinuous permafrost zone. Note that the Arctic, thus defined, overlaps with other regions covered in this report—namely, North America, Asia, and Europe—and that important physical and biological processes that are typical of the Arctic also occur south of the Arctic Circle. The Antarctic is defined here as the Antarctic continent, together with the surrounding Southern Ocean south of the Antarctic Convergence (polar front), an oceanographic barrier that shifts with time and longitude but generally is close to 58°S. Also included in the polar regions are sub-Antarctic islands such as Campbell Island, Heard Island, and South Georgia, some of which are north of the Antarctic Convergence (see Figure 16-2).

The two polar regions are dominated by cold conditions and the presence of ice, snow, and water. They are different in that the Arctic is a frozen ocean surrounded by continental landmasses and open oceans, whereas Antarctica is a frozen continent surrounded solely by oceans. Antarctica tends to be thermally isolated from the rest of the planet by the surrounding Southern Ocean and the atmospheric polar vortex, whereas the Arctic is influenced strongly by seasonal atmospheric transport and river flows from surrounding continents. Both regions have major influences on the global ocean.

The Arctic and Antarctic influence climate over a significant part of the globe. Many unique climatic processes operate in these regions. Some involve complex interactions and feedback loops (Simmonds, 1998) that may lead ultimately to glacial-interglacial climate transitions (Petit *et al.*, 1999). Processes in polar regions greatly influence sea level. The Arctic and Antarctic have food webs and natural ecosystems with remarkable productivity. The Arctic is on the periphery of human settlement, where people must adapt to harsh, cold regimes; the Antarctic is uninhabited apart from research bases.

16.1.1. Previous Work—Summary of Special Report on Regional Impacts of Climate Change

The IPCC, in its *Special Report on Regional Impacts of Climate Change* (RICC), produced an assessment of the impacts of climate change on the Arctic and the Antarctic (Everett and Fitzharris, 1998). In addition, the impact of climate change on the cryosphere is discussed in the IPCC Second Assessment Report (SAR) (Fitzharris, 1996). The main points arising from the regional assessment were that the Arctic is extremely vulnerable to projected climate change—major physical, ecological, sociological, and economic impacts are expected. Because of a variety of positive feedback mechanisms, the Arctic is likely to respond rapidly and more severely than any other area on Earth, with consequent effects on sea ice, permafrost, and hydrology. On the other hand, the Antarctic would respond relatively slowly to climate change, with much smaller impacts expected by 2100, except in the Antarctic Peninsula.

RICC noted that substantial loss of sea ice in the Arctic Ocean would have major implications for trade and defense (Everett and Fitzharris, 1998) . With more open water, there would be moderation of temperatures and increased precipitation. Considerable thawing of permafrost would lead to changes in drainage, increased slumping, and altered landscapes over vast areas of northern parts of North America and Eurasia. RICC also purported that polar warming probably should increase biological production, but different species compositions are likely on land and in the sea, with a tendency for poleward shifts in major biomes and associated animals. Animals that are dependent on ice would be disadvantaged. Human communities in the Arctic—especially indigenous peoples following traditional lifestyles—would be affected by these changes.

RICC also pointed out that changes in polar climate are likely to affect other parts of the world through changes in sea level, decreased oceanic heat transport, and increased emissions of GHGs from thawing permafrost. However, there would be economic benefits as well as costs. Potential benefits include new opportunities for shipping across the Arctic Ocean, lower operational costs for the oil and gas industry, lower heating costs, and easier access for tourism. Increased costs could be expected from changes such as disruptions to land and infrastructure caused by thawing of permafrost and reduced transportation capabilities across frozen ground and water.

Since these IPCC reports, there have been important advances in knowledge about climate change in polar regions. These advances include more information about decreases in Arctic and Antarctic sea-ice extent, verification of substantial thinning of Arctic sea ice, documentation of important changes in polar oceans, and more analyses of continental snow-cover trends. Many observations of environmental change in the Arctic show a trend that is consistent with GHG warming and similar to that predicted by climate models.

This report is different from earlier IPCC reports in that it focuses on eight key concerns of climate impact. The risk of collapse of the West Antarctic ice sheet is now considered to be lower than first thought. Changes around the Antarctic Peninsula are given prominence, and the implications for the whole continent are discussed in more detail. Previous reports said little about the impacts on polar oceans and consequences for marine life. New information, especially for the Southern Ocean and its role in the global thermohaline circulation, is presented here. There also are improved predictions of sea ice over the 21st century, and the role of Arctic hydrology and the implications of altered river flows into the Arctic Ocean for the Atlantic driver of the thermohaline circulation are addressed. More information is now available regarding the impacts of climate change on Arctic biota. More detail is supplied about impacts on human communities, especially indigenous peoples following traditional lifestyles. This research has highlighted the role of climate change in the Arctic and Antarctic and its impact on polar drivers of the global system. New modeling has identified the large role of polar regions in affecting the global thermohaline circulation, sea-level rise, and greenhouse exchanges between the atmosphere, cold oceans, and tundra. It is now clearer that climate change could initiate processes that could last for millennia and persist long after greenhouse emissions have stabilized.

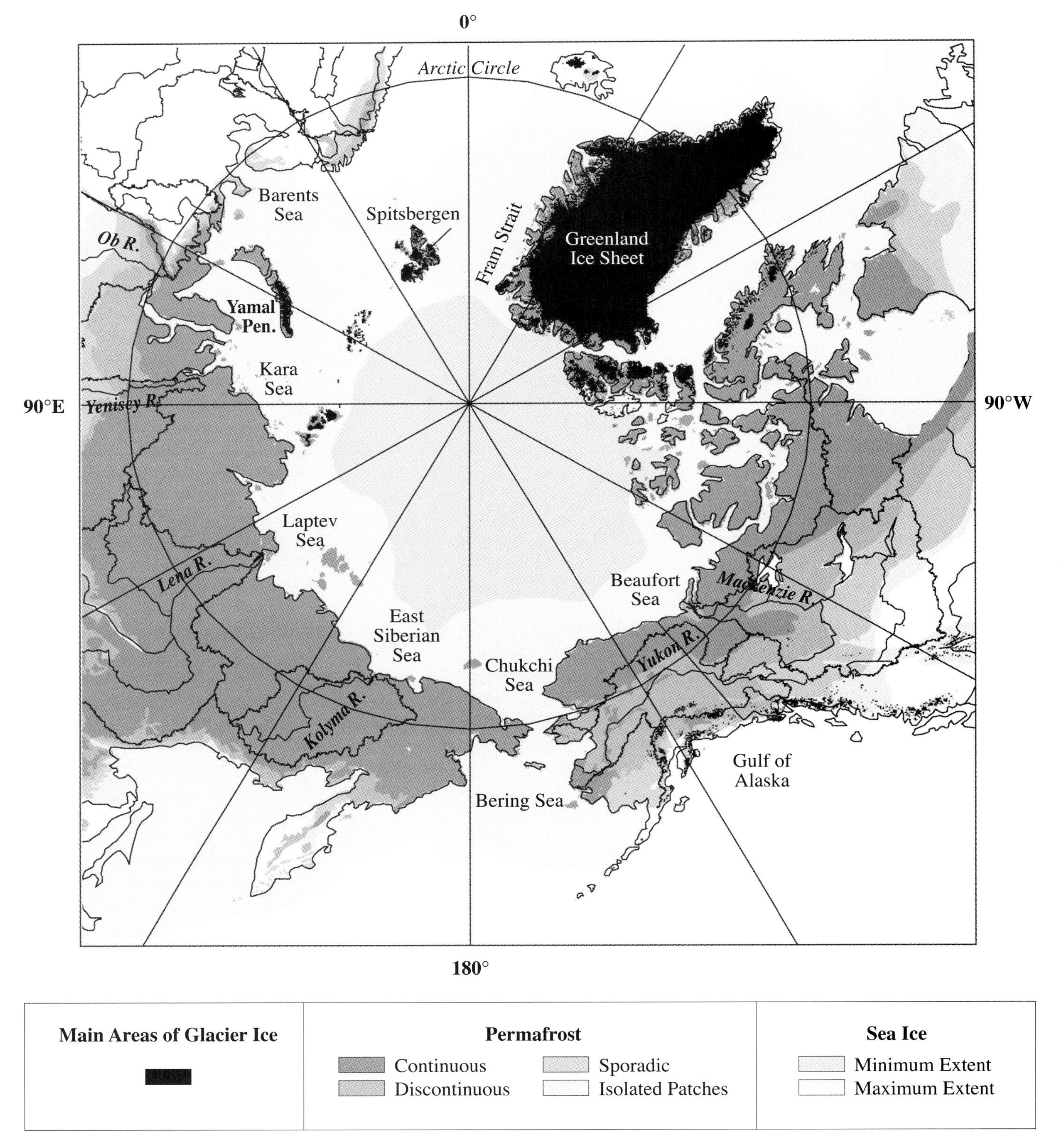

Figure 16-1: Location map for the Arctic. Permafrost zonation, drainage basins, sea ice, and main areas of glacier ice are shown. Drainage basins are delimited by solid black lines. Permafrost zonation is based on a digital version of the map by Brown *et al*. (1997).

16.1.2. Distinctive Characteristics of Polar Regions

The most distinctive feature of polar regions is the large seasonal variation in incoming solar radiation—from very little in winter to 24 hours of continuous sunlight in summer. Although the poles receive less solar radiation annually than equatorial locations, near the time of the solstices they receive more per day. The high albedo of the snow- and ice-covered polar regions, together with the large loss of long-wave radiation through the very clear and dry atmosphere, ensures a net loss of radiation in most months of the year. These losses of radiation are particularly large during the long polar night and help

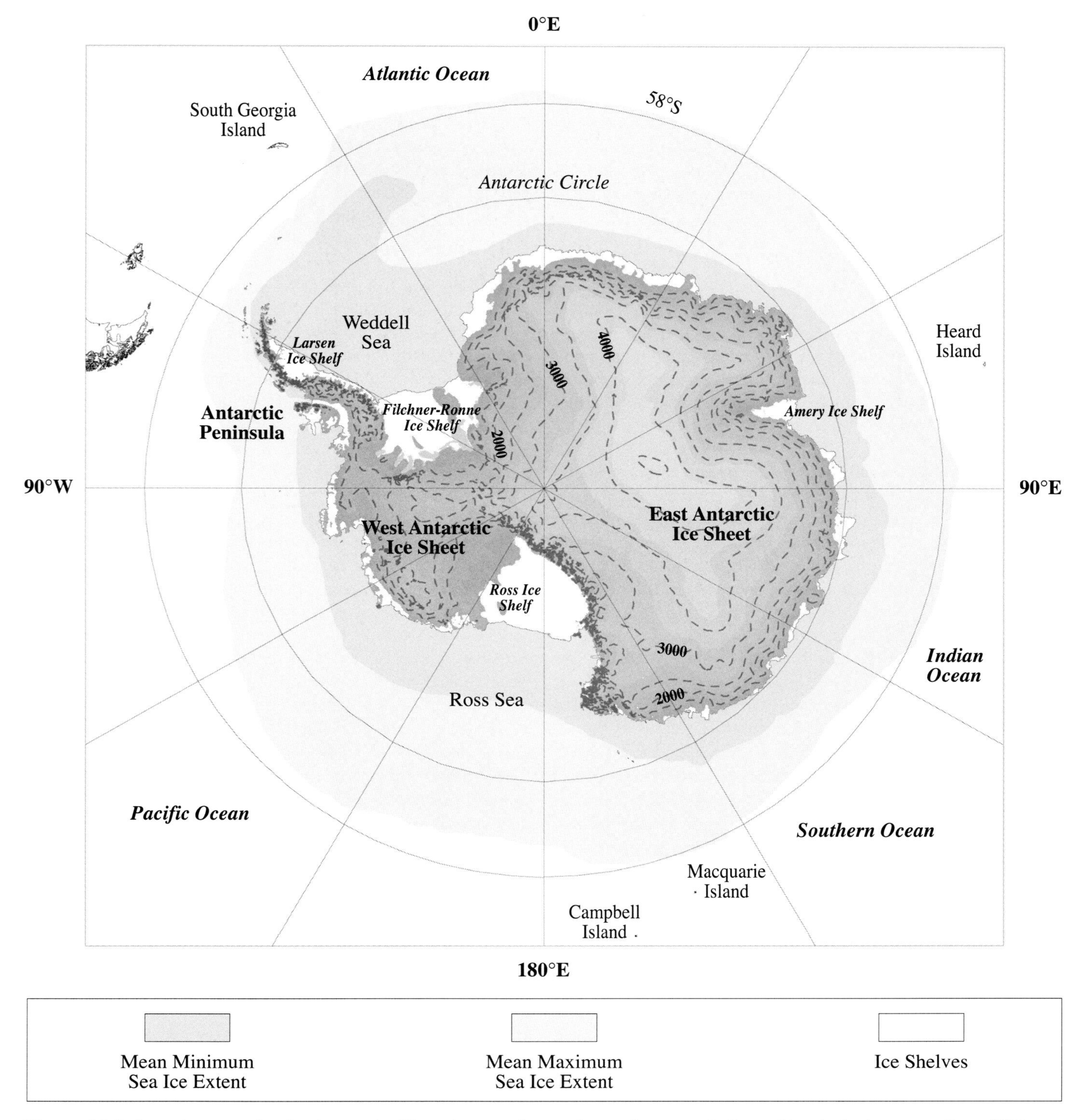

Figure 16-2: Location map for the Antarctic. Elevations on Antarctic continent are indicated by dashed contour lines (500-m interval). Topography and ice shelf outlines are from U.S. Geological Survey and British Antarctic Survey data. Sea-ice extent data are from Couper (1989).

create a deep, stable boundary layer of very cold, dry air at the surface.

These processes ensure sustained, very low temperatures in winter. Even in the summer, many areas—especially in the Antarctic—remain below freezing. Thus, much water remains frozen throughout the year. In Antarctica, snow and ice have continued to accumulate on the continent for millions of years. On the other hand, there are large volumes of liquid water in polar regions in the Arctic and Southern Oceans, although their surface is partially covered with sea ice, whose extent varies seasonally over vast areas. Sea ice insulates the underlying water from heat loss and, because of its high albedo, reflects much of the incoming solar radiation.

Formation of sea ice has important oceanographic consequences in that much latent heat is released, and highly saline, dense water is created. The dense water sinks in the north Atlantic and Southern Oceans, helping maintain the ocean conveyor belt and carrying nutrients and carbon. Production of deep ocean water is a consequence of processes that are operating in both polar regions. There is little evaporation from the vast tracts of sea ice and glacier ice. Extensive areas of the polar regions have very low precipitation; most comes from cyclonic storms that penetrate from the surrounding belt of subpolar depressions.

The polar regions incorporate important environmental thresholds, many of which are associated with water-phase changes. Together with the large seasonal changes in solar radiation, polar regions stimulate important geophysical and biological processes with high sensitivity to impacts. Sustained warming or cooling across the freezing threshold brings dramatic physical changes over land and sea. As a result, the physical environment, biota, and socioeconomic factors are all vulnerable to climate change. The climate and vast areas of ice make the polar regions very inhospitable and marginal for many species, including humans. However, specially adapted species thrive in some terrestrial and marine polar ecosystems. Consequently, the Arctic and Antarctic are characterized by the presence of highly distinctive biomes and are important places for many migratory species. Although the Antarctic continent essentially is a pristine wilderness, the great whales and some species of seals in the Southern Ocean were commercially exploited to virtual commercial extinction in the 19th and 20th centuries. Human activities on many sub-Antarctic islands have altered their biota dramatically.

The Antarctic has limited resource use, apart from growing fishing and tourism industries. There is a multinational approach to natural resources and environmental management, with mineral exploration and exploitation banned by international agreement. The area is managed by the Consultative Parties to the Antarctic Treaty to the dedication of science and peace (UNEP, 1997). By contrast, the Arctic has been populated for thousands of years. There is considerable economic activity, based on fishing, herding, and shipping. Recent decades have seen the establishment of urban areas and resource developments, based on the petroleum, gas, and mining industries. The extreme environment requires unique cold-region engineering and infrastructure solutions. There is a distinct contrast—and sometimes conflict—between the developments of modern society and indigenous peoples. The Arctic lies within the political boundaries of some of the world's richest and most powerful nations. During the Cold War, it was a critical strategic area, and substantial defense establishments remain in the region.

16.1.3. Climate Change in the 20th Century

There has been substantial climate change during the past 2 million years in the Arctic and the Antarctic. These changes are well documented in several natural archives, such as ice cores and marine sediments. The quasi-periodic sequence of glacial-interglacial periods and corresponding changes in GHGs is shown clearly in the record of atmospheric composition and climate derived from the ice cores for the past 420,000 years (e.g.,White and Steig, 1998; Petit *et al.*, 1999). Longer term climate changes are not discussed in this chapter, which focuses on observed 20th-century changes as well as those predicted for the 21st century. The following subsections provide a brief discussion of the main climatic changes in the 20th century. More details are provided in TAR WGI Chapter 2.

16.1.3.1. The Arctic

Instrumental observations of climatic parameters over the 20th century are available from standard climate stations on land and measurements taken on drifting ice floes in the Arctic Ocean. The land stations show that warming in the Arctic, as indicated by daily maximum and minimum temperatures, has been as great as in any other part of the world. Although not geographically uniform, the magnitude of the warming is about 5°C per century, with areas of cooling in eastern Canada, the north Atlantic, and Greenland (Koerner and Lundgaard, 1996; Borzenkova, 1999a,b; Jones *et al.*, 1999; Serreze *et al.*, 2000). Data from ice floe measurements show a slight warming on an annual basis, with statistically significant warming in May and June over the 1961–1990 period. Air temperature anomalies in the Arctic basin have been strongly positive since 1993. In the period 1987–1997, air temperature in the Arctic increased by 0.9°C (Alexandrov and Maistrova, 1998).

Significant warming from the beginning of the 20th century has been confirmed by many different proxy measurements (Maslanik *et al.*, 1996; Bjorgo *et al.*, 1997; Smith, 1998). Magnuson *et al.* (2000) found consistent evidence of later freeze-up (5.8 days per 100 years) and earlier breakup (6.5 days per 100 years) of ice on lakes and rivers around the northern hemisphere from 1846 to 1995 and an increase since 1950 in interannual variability of both dates. Glaciers and ice caps in the Arctic also have shown retreat in low-lying areas since about 1920. Numerous small, low-altitude glaciers and perennial snow patches have disappeared. However, no increasing melting trend has been observed during the past 40 years (Jania and Hagen, 1996; Koerner and Lundegaard, 1996; Dowdeswell *et al.*, 1997). Glaciers in Alaska have receded, with typical ice-thickness decreases of 10 m over the past 40 years, but some glaciers have thickened in their upper regions (BESIS, 1997). Greenland's ice sheet has thinned dramatically around its southern and eastern margins. Above 2,000-m elevation, the ice sheet is in balance, on average. The net effect is a loss of about 51 $km^3 yr^{-1}$. (Krabill *et al.*, 1999, 2000).

Snow-cover extent in the northern hemisphere has been reduced since 1972 by about 10%, largely as a result of spring and summer deficits since the mid-1980s (Brown, 2000; Serreze *et al.*, 2000). Most Arctic regions have experienced increases in precipitation since at least the 1950s (Groisman *et al.*, 1991; Groisman and Easterling, 1994; Georgiyevskii, 1998). Measurements from Spitsbergen show a statistically significant increase in precipitation during all seasons, except winter (Hanssen-Bauer and Forland, 1998).

Groisman *et al.* (1994) analyzed seasonal snow extent in the northern hemisphere and demonstrated an inverse relationship with near-surface air temperature. Recent findings have provided evidence of a significant decrease in spring snow extent since 1915 over Eurasia (Brown, 1997) and southern Canada (Brown and Goodison, 1996). Such trends may be related to low-frequency fluctuations in hemispheric atmospheric circulation patterns (Serreze *et al.*, 2000).

Arctic sea-ice extent decreased by approximately 3% per decade between 1978 and 1996 (Cavalieri *et al.*, 1997; Parkinson *et al.*, 1999; Johannessen *et al.*, 1999; Serreze *et al.*, 2000). The most significant contractions were detected in 1990, 1993, and 1995 (Maslanik *et al.*, 1996). Ice composition also has changed, with a reduction in the area of multi-year ice in winter. Summer sea-ice extent has shrunk by 20% (880,000 km^2) over the past 30 years in the Atlantic part of the Arctic Ocean (Johannessen *et al.*, 1999), but the shrinkage has only been 5% in the Canadian Arctic Sea. Sea-ice extent in the Bering Sea experienced a dramatic reduction when a regime shift occurred in 1976 and has continued to decrease (BESIS, 1997). New compilations of Arctic sea-ice extent, using historical data from the past 135 years (Vinje, 2001), show that overall sea-ice extent in April has been reduced in the Nordic seas by 0.79 x 10^6 km^2 (33%) (see Figure 16-3). Nearly half of this reduction took place between 1860 and 1900. Although there are large interannual and seasonal variations in sea-ice extent, the reduction is greatest in spring. This is consistent with the 1912–1996 temperature record from Svalbard, which shows significant warming (3°C) in spring (Hanssen-Bauer and Forland, 1998). There is an approximate 10-year climate signal in the Arctic and subarctic, with a clockwise propagating signal in sea-ice concentration anomalies and a standing oscillation in sea-level pressure anomalies—the latter linked to the two phases of the North Atlantic Oscillation (NAO) (Mysak and Venegas, 1998). Comparison of these trends with outputs from climate models (forced by observed GHGs and tropospheric sulfate aerosols) reveals that the observed decrease in northern hemisphere sea-ice extent agrees with transient simulations (see Figure 16-6) and that both trends are much larger than would be expected from natural climate variations (Vinnikov *et al.*, 1999).

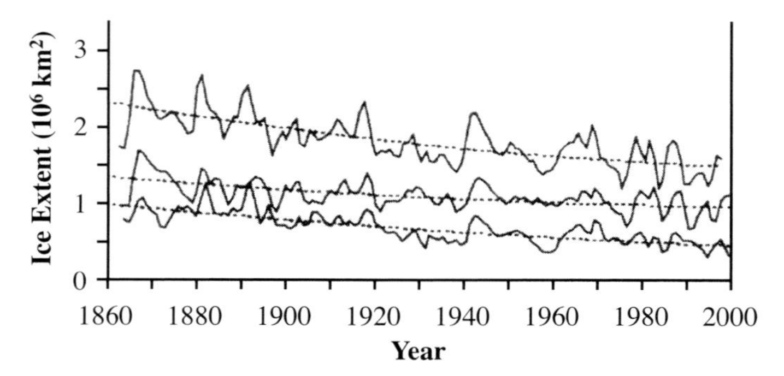

Figure 16-3: Time series of April sea-ice extent in Nordic Sea (1864–1998) given by 2-year running mean and second-order polynomial curves. Top: Nordic Sea; middle: eastern area; bottom: western area (after Vinje, 2000).

The most effective method for determining the thickness of sea ice is to use sonar, directed upward at the floating ice from submarines or moorings (Melling *et al.*, 1995). The results indicate great variability. Average thinning of sea ice has been observed since 1976 on some transects of the Atlantic and Arctic Oceans (Wadhams, 1997). A thinning by about 0.13 m over the period 1970–1992,with the maximum decrease detected in the eastern Siberian Sea, was found by Russian scientists (Nagurduy, 1995). Rothrock *et al.* (1999) found that ice draft at the end of the melt season has decreased by about 1.3 m in most of the deepwater portion of the Arctic Ocean, from 3.1 m in 1958–1976 to 1.8 m in the 1990s (~15% per decade). The decrease is greater in the central and eastern Arctic than in the Beaufort and Chukchi seas (Rothrock *et al.*, 1999). Vinje *et al.* (1998), however, claim that no significant change in ice thickness can be observed in Fram Strait. Their conclusion is in agreement with submarine observations from 1960–1982. Although there is large variability, ice cover has continued to become thinner and has decreased by 40% over the past 3 decades (Rothrock *et al.*, 1999). Analysis of the duration of summer melt over a large fraction of the perennial Arctic sea ice from 1979 to 1996 reveals an increase of 5.3 days (8%) per decade in the number of melt days each summer (Smith, 1998).

Changes to below 1,000 m in the Arctic Ocean have been observed, with less sea ice and freshening of the Beaufort Sea mixed layer (Maslanik *et al.*, 1996; and McPhee *et al.*, 1998). Atlantic water inflow to the Arctic Ocean has warmed (Carmack *et al.*, 1995). The halocline, which isolates the surface from this warmer Atlantic water, has grown thinner (Steele and Boyd, 1998). Data gathered from submarines indicate that sea-surface temperature (SST) in the Arctic basin increased by 1°C over the past 20 years, and the area of warm Atlantic water in the polar basin increased by almost 500,000 km^2 (Kotlyakov, 1997). Field measurements in 1994 and 1995 showed consistent Arctic seawater warming of 0.5–1.0°C, with a maximum detected in the Kara Sea (Alekseev *et al.*, 1997). It is not yet clear whether these changes are part of low-frequency natural variability or whether they represent the early impacts of long-term climate change.

Thickening of the active layer and changes in the distribution and temperature of permafrost are further important indicators of warming in the Arctic. Permafrost thawing has led to major erosion problems and landslides in the Mackenzie basin (Cohen, 1997a) and has caused major landscape changes from forest to bogs, grasslands, and wetland ecosystems in Alaska, affecting land use. In central and western Canada and in Alaska, contraction in the extent of permafrost has occurred during the 20th century (Halsey *et al.*, 1995; Weller, 1998; Weller and Lange, 1999; Osterkamp *et al.*, 2000). However, a cooling trend has been noted for permafrost in eastern Canada (Allard *et al.*, 1995; Allard and Kasper, 1998). Permafrost degradation during the 20th century in subarctic and boreal peatlands of Quebec is discussed by Laberge and Payette (1995).

Temperatures at the top of permafrost along a several hundred kilometer north-south transect in central Alaska warmed by 0.5–1.5°C between the late 1980s and 1996. Thawing rates of about 0.1 m per year were calculated for two sites (Osterkamp and Romanovsky, 1999). On the north slope of Alaska and in northwestern Canada, the temperature-depth profile indicates a temperature rise of 2–4°C over the past century (Lachenbruch and Marshall 1986). In the western Yamal Peninsula, temperatures at a depth of 10 m in ice-rich permafrost increased 0.1–1.0°C between 1980 and 1998; the largest increases in Siberian permafrost temperatures have occurred in relatively cold permafrost in the continuous zone.

Palaeoclimate reconstructions based on dendrochronology also indicate a steady increase in temperature during the past 150 years (Bradley and Jones 1993; Szeicz and MacDonald 1995; see TAR WGI Chapter 2). Growth rings in Larix Sibirica from northern Siberia show that mean temperatures during the 20th century were the highest in 1,000 years, exceeding even those during the Medieval Warm Period (Briffa *et al.*, 1995). Changes in diatom assemblages in lake sediments in the High Arctic over the same period also are thought to be a consequence of climatic warming (Douglas *et al.*, 1994).

Other changes in terrestrial ecosystems include northward movement of the treeline, reduced nutritional value of browsing for caribou and moose, decreased water availability, and increased forest fire tendencies (Weller and Lange, 1999). There are altered plant species composition, especially forbs and lichen, on the tundra. In the marine ecosystem, northern fur seal pups on the major Bering Sea breeding grounds declined by half between the 1950s and the 1980s. In parts of the Gulf of Alaska, harbor seal numbers are as much as 90% below what they were in the 1970s. There have been significant declines in the populations of some seabird species, including common murres, thick-billed murres, and kittiwakes. Comprehensive interviews by Gibson and Scullinger (1998) have revealed notable impacts on food sources and natural environments of native Alaskan communities. Warmer climate has increased growing degree days by 20% for agriculture and forestry in Alaska, and boreal forests are expanding northward at a rate of about 100 km per °C (Weller and Lange, 1999).

16.1.3.2. The Antarctic

Instrumental records analyzed by Jacka and Budd (1998) and summarized in Figure 16-4 have shown overall warming at

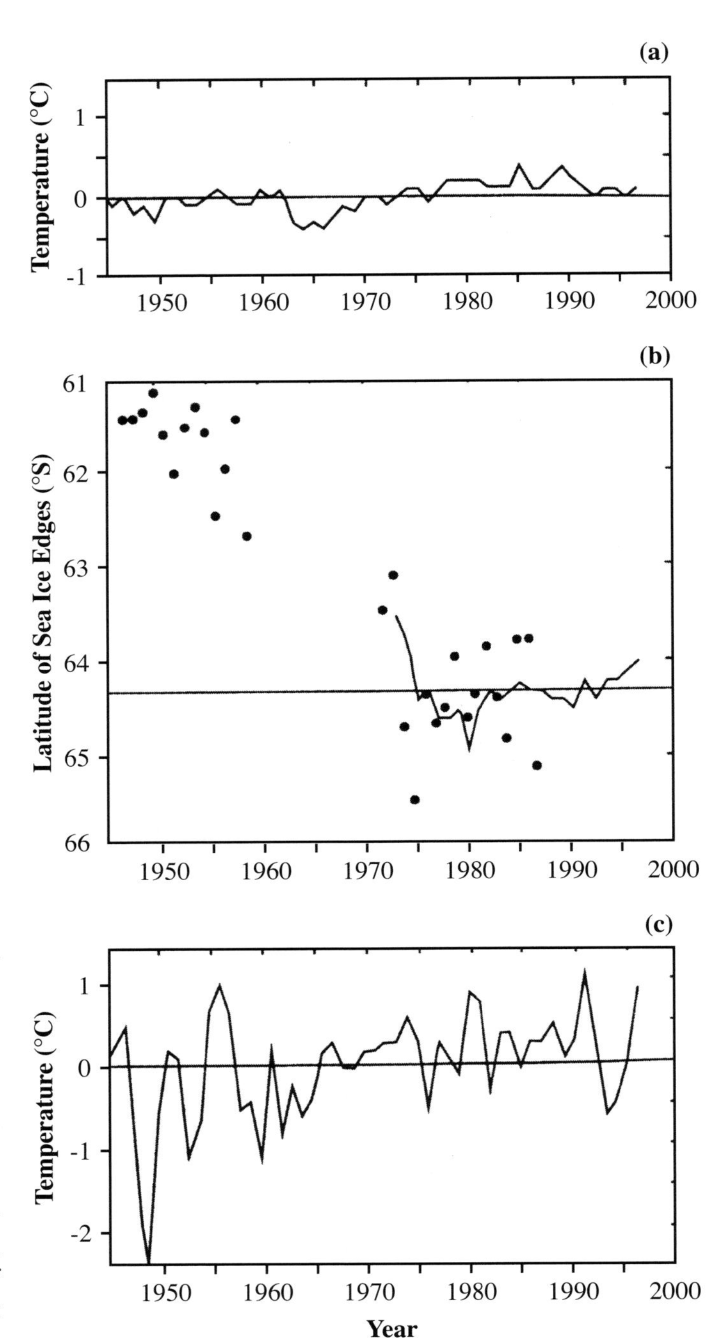

Figure 16-4: (a) Mean temperature anomalies for Southern Ocean climate stations; (b) summer sea-ice extent [dots are from estimates based on southernmost positions of whaling-factory ships (de la Mare, 1997)]; (c) mean temperature anomalies of coastal Antarctic climate stations (after Jacka and Budd, 1998).

permanently occupied stations on the Antarctic continent (1959–1996) and Southern Ocean island stations (1949–1996). Using two different statistical techniques, they found that the 16 Antarctic stations have warmed at a mean rate of 0.9–1.2°C per century, and the 22 Southern Ocean stations have warmed at 0.7–1.0°C per century. Antarctic Peninsula stations show a consistent regional rate of warming that is more than twice the average for other Antarctic stations. King and Harangozo (1998) suggest that this warming is associated with an increase in the northerly component of the atmospheric circulation over the Peninsula and perhaps changes in sea-ice extent. For Antarctic stations, warming trends are largest in winter and smallest in autumn. For the Southern Ocean stations, warming trends are largest in autumn and smallest in spring and summer. However, three GCMs were unable to reproduce these trends (Connolley and O'Farrell, 1998).

Another analysis of a 21-station data set from Antarctica by Comiso (1999) found a warming trend equivalent to 1.25°C per century for a 45-year record beginning in the 1950s but a slight cooling trend from 1979 to 1998. The slight cooling trend for this later 20-year period also was confirmed via analysis of surface temperatures over the whole continent, as inferred from satellite data. These changes can be placed in a long-term context by comparison with results from the high-accumulation ice coring sites on Law Dome, East Antarctica, which show clear climate signals with sufficient resolution to identify seasonal variations (van Ommen and Morgan, 1996, 1997; Curran *et al.*, 1998). There was a cooling period in the late 1700s and 1800s and a warming over the 19th century, with greater variability and change in winter than in summer.

Changes in precipitation in the Antarctic are more poorly understood. Model estimates from Smith *et al.* (1998a) indicate that the accumulation rate for the East Antarctic ice sheet surface has increased by a rate of 1.9 mm yr^{-1} (water equivalent) over the period 1950–1991. Their estimate of sensitivity is 12.5 mm yr^{-1} per degree of warming. Examination of water-mass properties of oceans shows that significant changes have occurred over the past 30 years. Bindoff and McDougall (2000) and Wong *et al.* (1999) point out that sub-Antarctic mode water (SAMW) and Antarctic intermediate water (AAIW) have become less saline and cooler, and both water masses are now deeper. These changes indicate surface warming in the source region of SAMW and increased precipitation in the source region of AAIW.

Jacka and Budd (1998) found no significant trends in Antarctic sea-ice data over the satellite era (1973–1996). Although the mean trend is zero, the sector from 0° to 40°E has a clear trend toward increased sea ice. This is matched by a larger sector of decreasing extent near the Bellingshausen and Amundsen Seas, from about 65°W to 160°W. Elsewhere, sea-ice extent trends are relatively small (see Figure 16-4)—a finding that also is supported by Cavalieri *et al.* (1997). Analysis of whaling records by de la Mare (1997) suggests that the Antarctic summer sea-ice edge has moved southward by 2.8 degrees of latitude between the mid-1950s and the early-1970s. This suggests a decline in the area covered by sea ice of 25%. It should be noted, however, that the data used in this analysis span two distinct periods, during which differing whale species were harvested. Using atmosphere-ocean sea-ice models, the computations of Wu and Budd (1998) indicate that sea ice was more extensive over the past century, on the annual average by 0.7–1.2 degrees of latitude. It also was thicker by 7–13 cm than at present. Wu *et al.* (1999) conclude that the sea-ice extent reduced by 0.4–1.8 degrees of latitude over the 20th century.

16.1.4. Scenarios of Future Change

The IPCC commissioned a *Special Report on Emissions Scenarios* (SRES). Four "marker scenarios" representing different world storylines are used to estimate emissions and climate change to 2100 (IPCC, 2000). Table 16-1 summarizes these climate projections for the polar regions. In almost all cases, predicted climates are well beyond the range of variability of current climate. However, these estimates cover a very large range of precipitation and temperatures, so future climate remains uncertain except that it will be wetter and warmer. Some of the projected increases in precipitation and temperatures are larger than for any other part of the globe.

***Table 16-1**: IPCC SRES climate scenarios for 2080. Values are changes from present climate summarized from Carter et al. (2000), and are scaled output of nine AOGCMs.*

	Summer		Winter	
Region	Precipitation (%)	Temperature (°C)	Precipitation (%)	Temperature (°C)
Arctic				
– Land	+10–20[a]	+4.0–7.5	+5–80	+2.5–14.0
– Arctic Ocean	+2–25[b]	+0.5–4.5	+2–45[c]	+3.0–16.0
Antarctica Land	+1–28	+1.0–4.8	+4–32	+1.0–5.0
Southern Ocean	+2–17	+0.0–2.8	+5–20	+0.5–5.0

[a] CSIRO-mk2 model predicts +38%.
[b] CSIRO-mk2 model predicts +42%.
[c] ECHAM4 model predicts +70%.

Models predict that land areas in the Arctic will receive substantially increased snowfall in winter and that the climate will be markedly warmer. Summer could be much warmer and wetter than present. The climate over the Arctic Ocean does not change as dramatically, but it will become warmer and wetter by 2080. For the Antarctic continent, the models tend to predict more snow in winter and summer. Although temperatures are forecast to increase by 0.5°C, there will be little impact on melt because they will remain well below freezing, except in limited coastal localities. The Southern Ocean warms least, especially in summer. Precipitation increases by as much as 20%, so there will be more freshwater input to the ocean surface. This chapter also refers to other climate models. Some are equilibrium models for the atmosphere only; others are transient, coupled atmosphere-ocean models. Some deal with aerosols and other do not. In polar regions there can be large differences in predictions, depending on the model chosen, although most predict large changes in climate over the next 100 years. Assessments of impacts will vary, depending on the climate model chosen. This should be kept in mind in assessing the impacts described in this chapter.

Discrepancies among climate models and problems of downscaling (Shackley *et al.*, 1998) mean that alternative methods of prediction that are based on analysis of empirical climate data (e.g., palaeoclimatic analogs and extrapolation of recent instrumental records) still have value. Anisimov and Poljakov (1999) analyzed modern temperature trends over the northern hemisphere; they suggest that warming in the Arctic will be most pronounced in the continental parts of North America and Eurasia. The potential impacts of continued deepening of the winter polar vortex would include weakening of the wind-driven Beaufort Gyre. This would further reduce the extent and thickness of the Arctic pack ice (McPhee *et al.*, 1998) and change ocean temperatures and sea-ice boundaries (Dickson *et al.*, 1999).

16.2. Key Regional Concerns

This chapter follows the agreed TAR template for WGII regional assessments. Rather than assessing all possible impacts of climate change, emphasis is placed on eight key regional concerns. These concerns are chosen for the Arctic and the Antarctic on the basis of earlier findings by the IPCC (Everett and Fitzharris, 1998) and on discussions at various workshops of the TAR. They represent what are considered to be the most important impacts of future climate change in the Arctic and Antarctic in the wider global perspective. Although some other impacts (e.g., Arctic glaciers and terrestrial Antarctic biota) may be very important locally, space does not permit a comprehensive review.

16.2.1. Changes in Ice Sheets and Glaciers

Changes in the polar climate will have a direct impact on the great ice sheets, ice caps, and glaciers of the polar regions. How each responds will depend on several climatological parameters; some will grow, whereas others shrink. We have high confidence that their overall contribution to rising sea level will be positive, with glaciers and the Greenland ice sheet shrinking. The contribution from Antarctica, however, is uncertain at present. There is a high likelihood that increasing temperatures over the continent and changing storm tracks will cause increased precipitation and thickening of the ice sheet, but there still exists at low confidence the possibility that the West Antarctic ice sheet will retreat dramatically in coming centuries (Vaughan and Spouge, 2000). Such a change would not result from recent and future climate change (Bentley, 1998) but more probably from continuing readjustment to the end of the last glacial period (Bindschadler, 1998), as a result of internal dynamics of the ice sheet (MacAyeal, 1992), or as a result of ice shelf-ocean interaction. This subject and the general issue of sea-level rise are reviewed more comprehensively in TAR WGI Chapter 11; we include only a summary of the main points here.

The Greenland ice sheet suffers melting in summer at its margin. There is a trend toward an increase in the area and duration of this melt (Abdalati and Steffen, 1997). This trend is likely to continue. Airborne altimetric monitoring has shown that over the period 1993–1998, the Greenland ice sheet was slowly thickening at higher elevations; at lower elevations, thinning (about 1 m yr^{-1}) was underway (Krabill *et al.*, 1999, 2000). If warming continues, the Greenland ice sheet will shrink considerably, as occurred in previous interglacial periods (Cuffey and Marshall, 2000), and if the warming is sustained, the ice sheet will melt completely (see TAR WGI Chapter 11).

Over the Antarctic ice sheet, where only a few limited areas show summer melting (Zwally and Fiegles, 1994), a slight thickening is likely as precipitation rates increase (e.g., Ohmura *et al.*, 1996; Smith *et al.*, 1998a; Vaughan *et al.*, 1999). In the past decade there has been some change in the ice cover in local areas (e.g., on the Antarctic Peninsula; see Section 16.2.2), but the majority of the Antarctic ice sheet appears, from satellite altimetry, to be close to a state of balance (Wingham *et al.*, 1998). Only the Thwaites and Pine Island glacier basins show any spatially coherent trend, but it is not yet known if their thinning is related to a decrease in precipitation or some dynamic change in the ice sheet. Chinn (1998) reports that recession is the dominant change trend of recent decades for glaciers of the Dry Valleys area of Antarctica. The future of glaciers in the Arctic will be primarily one of shrinkage, although it is possible that in a few cases they will grow as a result of increased precipitation.

This report is concerned primarily with the impacts of climate change on particular regions. An important question is: How will changes in glacial ice in the polar regions impact local human populations and ecological systems, and what will be the socioeconomic consequences? The short answer is that impacts on ice systems will be substantial, but because the populations of humans and other biota within polar region are low, impacts may be relatively minor. In Antarctica, the continental human population is only a few thousand. A few localities may undergo changes such as that at Stonington Island on the Antarctic Peninsula (Splettoesser, 1992), where retreat of the ice sheet has left the station stranded on an island. In general, however, changes to ice sheets will directly cause few significant life-threatening problems.

16.2.2. Changes around the Antarctic Peninsula

At least five meteorological records from scientific stations on the Antarctic Peninsula show marked decadal warming trends (King, 1994; Harangozo *et al.*, 1997; King and Harangozo, 1998; Marshall and King, 1998; Skvarca *et al.*, 1998). Although the periods and seasons of observations have been different, the records show consistent warming trends—as much as 0.07°C yr^{-1}, considerably higher than the global mean. This atmospheric warming has caused several notable changes in the ice cover of the Antarctic Peninsula, including changes in snow elevation (Morris and Mulvaney, 1995; Smith *et al.*, 1999a) and the extent of surface snow cover (Fox and Cooper, 1998). The most important change has been the retreat of ice shelves. Seven ice shelves on the Antarctic Peninsula have shown significant, progressive, and continued retreat. On the West Coast, the Wordie Ice Shelf, Müller Ice Shelf, George VI Ice Shelf, and Wilkins Ice Shelf have retreated (Ward, 1995; Vaughan and Doake, 1996; Luchitta and Rosanova, 1998). On the east coast, the ice shelves that occupied Prince Gustav Channel and Larsen Inlet and Larsen Ice Shelf A have retreated (Rott *et al.*, 1996; Vaughan and Doake, 1996; Skvarca *et al.*, 1998). Following Mercer (1978), Vaughan and Doake (1996) show that the pattern of retreat could be explained by a southerly movement of the 0°C January isotherm, which appears to define a limit of viability for ice shelves. To date, about 10,000 km^2 of ice shelf have been lost.

Retreat of the ice shelves on the Antarctic Peninsula has attracted considerable media coverage, and environmental campaigns of some nongovernmental organizations (NGOs) have expressed concern that these events presage a more important collapse of the West Antarctic ice sheet. However, few direct impacts result from the loss of these ice shelves. The ice shelves were floating, so their melting does not directly add to sea level, and they usually are replaced by sea-ice cover, so overall albedo changes very little. Because the Antarctic Peninsula is steep and rugged, there is no evidence that removal of ice shelves will cause melting of the glaciers that fed them to accelerate and add to sea-level rise (Vaughan, 1993). Terrestrial ecosystems generally will be unaffected by ice-shelf retreat. Most polar benthic organisms, especially in the Antarctic, grow extremely slowly, so colonization of exposed seabed will be slow.

Because warming on the Antarctic Peninsula exceeds that over much of the rest of the continent (Jacka and Budd, 1998), migration of the limit of viability for ice shelves probably will not affect the Ronne-Filchner or Ross ice shelves in the next 100 years (Vaughan and Doake, 1996). A substantial increase in Antarctic summer temperatures, however, would threaten other large ice shelves beyond the Antarctic Peninsula. The real implications of ice-shelf retreat are that it highlights issues of risk perception and public understanding of climate change, rather than real physical impacts. These issues include questions such as: Does the Antarctic Peninsula warming result from a global effect, or simply from natural regional climate variations? Such problems of attribution of climate change will recur, especially if the costs of adaptation are to be spread across nations. The inherent nonlinearity of natural systems sometimes causes exaggerated local responses to small climate changes. In the present generation of climate models, the Antarctic Peninsula is not well resolved, so local effects on these scales cannot yet be reproduced. Confidence in predicting such changes in the natural ecosystems of the Antarctic Peninsula comes from observed changes on the terrestrial biota of southern ocean islands (Bergstrom and Chown, 1999).

Rapidly increasing temperatures have been directly implicated in colonization by introduced species and displacement of indigenous biota. Thus, the extent of higher terrestrial vegetation on the Antarctic Peninsula currently is increasing (Fowber and Lewis Smith, 1994). In the marine environment, the warming trend on the Antarctic Peninsula has been linked statistically to an associated change in sea ice for this region (Stammerjohn and Smith, 1996). For example, Smith *et al.* (1999b) show that rapid climate warming and associated reduction in sea ice are concurrent with a shift in the population size and distribution of penguin species in the Antarctic Peninsula region.

16.2.3. Changes in the Southern Ocean and Impacts on its Life

16.2.3.1. Overview

The Southern Ocean is an active component of the climate system that causes natural climate variability on time scales from years to centuries. It is a major thermal regulator and has the potential to play an active role in climate change by providing important feedbacks. If climate change is sustained, it is likely to produce long-term and perhaps irreversible changes in the Southern Ocean. These changes will alter the nature and amount of life in the ocean and on surrounding islands, shores, and ice. The impact of climate change is likely to be manifest in the large-scale physical environment and in biological dynamics. There even is the possibility that changes in the Southern Ocean could trigger abrupt and very long-term changes that affect the climate of the entire globe.

16.2.3.2. Role of Ocean Changes

The Southern Ocean is of special significance to climate change because of the variety of water masses produced there and the ability of these waters to spread throughout the global ocean. In particular, shelf waters with temperatures near the freezing point (about -1.9°C) are produced along the margin of the Antarctic continent. They contribute to the formation of Antarctic bottomwater and continuing ventilation of this water in the world's oceans (Whitworth *et al.*, 1998). There already is evidence that climate change has altered SAMW and AAIW (Johnson and Orsi, 1997; Wong *et al.*, 1999; Bindoff and McDougall, 2000). Wong *et al.* (1999) compared trans-Pacific sections 20 years apart and found temperature and salinity changes that are consistent with surface warming and freshening in formation regions of the water masses in the Southern Ocean and their subsequent subduction into the ocean.

The main contributions to the production of Antarctic bottomwater come from cooling and transformation of seawater under the floating Filchner-Ronne and Ross ice shelves, as well as from dense water formed as a result of freezing of sea ice. Polynyas also are important. These areas of combined open water and thin sea ice (surrounded by sea ice or land ice) play an important role in ocean-atmosphere heat transfer, ice production, formation of dense shelf water, spring disintegration of sea ice, and sustenance of primary and secondary productivity in polar regions. Their formation is linked to the strength and persistence of cold outflow winds off the Antarctic continent and the position of the polar jet stream and Southern Ocean atmospheric circulation patterns (Bromwich *et al.*, 1998).

Research by Wu *et al.*, (1997), Budd and Wu (1998), and Hirst (1999)—using the CSIRO coupled, transient model and IPCC IS92a emissions scenario—suggests that certain aspects of Southern Ocean circulation may be very sensitive to climate change. With an increase in atmospheric equivalent CO_2, there is a substantial increase in the strength of the near-surface halocline and a potential reduction in the formation of Antarctic bottomwater and convection in the Southern Ocean. The decrease in downwelling of cold, dense water, as shown in Figure 16-5, leads to a slowdown in the thermohaline circulation

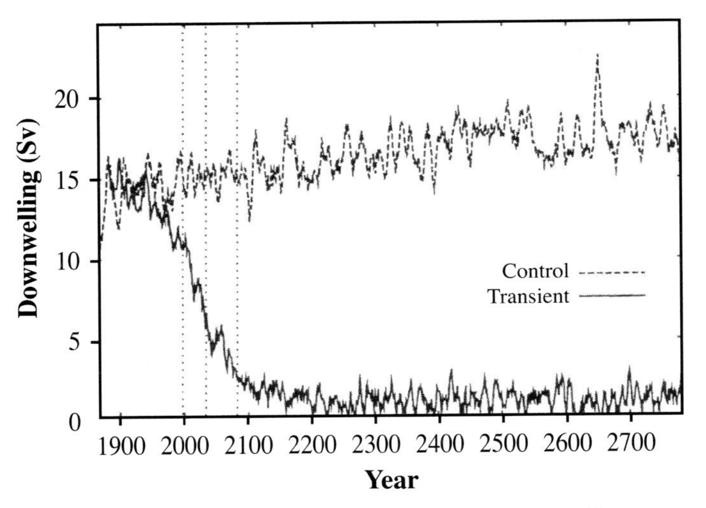

Figure 16-5: Volume transport of Antarctic downwelling past 1,250-m depth level for control and transient climate integrations. Values are in Sverdrups (Sv), filtered using a 7-year running mean, and are for the area of the Southern Ocean between Antarctic coast and 45°S (from Hirst, 1999, Figure 11).

(Broecker *et al.*, 1985). A tripling of GHGs is sufficient to induce a shutdown of the thermohaline circulation. Furthermore, this shutdown continues for at least several centuries, to the end of the model integration. The main processes causing a decline in the thermohaline circulation are an increase in precipitation minus evaporation, an increase in glacial melt from the Antarctic, and reduced sea-ice formation, so there is less brine injection into the ocean.

According to this simulation, if concentrations of GHGs continue to grow to the end of the 21st century, there will be significant weakening of the ocean's thermohaline circulation and even potential for its shutdown. These changes might become irreversible in subsequent centuries, despite possible stabilization of GHG concentrations. Modeling by Cai and Baines (1996) indicates that the Antarctic Circumpolar Current is driven at least partly by the thermohaline circulation associated with the formation of Antarctic bottomwater, so this major current of the Southern Ocean also can be expected to slow down. This result is confirmed by Cai and Gordon (1998), who found a clear decreasing trend in the magnitude of the Antarctic Circumpolar Current as CO_2 increases. The foregoing results indicate an ocean circulation that is significantly more sensitive to climate change than that found by Manabe and Stouffer (1994). The possibility remains that certain factors and parameterizations in the Hirst (1999) model could make it overly sensitive and that future changes may not be as dramatic as indicated. For example, in other models the ocean component is much less diffusive, allowing for prolonged maintenance of an enhanced halocline. None of the present generation of climate models can resolve polynyas, where the majority of brine rejection takes place and dense water masses form. These processes need to be better represented in coupled ocean-climate models before the timing of any substantial slowing of Antarctic bottomwater production or of any shutdown of the ocean thermohaline circulation can be predicted confidently. Changes in this circulation also may be driven by variable fluxes at lower latitudes, and results from different ocean models indicate uncertainty about the magnitude and timing of thermohaline slowdown (see TAR WGI Chapter 7 for more detailed discussion of the global thermohaline circulation).

16.2.3.3. Role of Ice Shelves

The sensitivity of the Southern Ocean to changes in ice shelves remains uncertain (Williams *et al.*, 1998a,b); it depends on whether there is free access of oceanic water to their underside (Nicholls, 1997). Circulation of meltwater under the ice shelves is complex. Any increase in melting as a result of global warming initially will introduce a fresher layer of water (see Jacobs *et al.*, 1992; Jenkins *et al.*, 1997; O'Farrell *et al.*, 1997). This could restrict further melting, unless the current or tides are strong enough to flush the fresher layer away or entrain it into other water masses. Using a three-dimensional model adapted to the underside of the Amery Ice Shelf, Williams *et al.* (1998b) conclude that when the adjacent seas were warmed by 1°C, net melt increased more than three-fold.

Grosfeld and Gerdes (1998) also show increased net melt with warming, but the situation stabilizes if there is less sea-ice formation. In considering the long-term response of the Antarctic ice sheet to global warming of 3°C, Warner and Budd (1998) show that it takes at least several centuries for the ice shelves to disappear in their model.

16.2.3.4. Impacts on Biology of Southern Ocean

No single factor controls overall primary production in the Southern Ocean. The organisms in Antarctic marine communities are similar to the inhabitants of marine systems at lower latitudes, although there is substantial endemism in Antarctica (Knox, 1994). Ice cover and vertical mixing influence algae growth rates by modulating the flux of solar radiation (Priddle *et al.*, 1992). Micronutrients, especially iron, are likely to limit phytoplankton growth in some areas. Experiments involving addition of iron to the ocean show dramatic increases in the biological activity of phytoplankton (de Baar *et al.*, 1995; Coale *et al.*, 1996; Sedwick *et al.*, 1999). Findings by Boyd *et al.* (2000) demonstrate that iron supply controls phytoplankton growth and community composition during the summer, but the fate of algal carbon remains unknown and depends on the interplay between processes that control export, remineralization, and water-mass subduction. Grazing by zooplankton also may be important.

Several of the physical controls on phytoplankton production are sensitive to climate change. Although it is presently impossible to make numerical predictions, these controls have been outlined in a qualitative way by Priddle *et al.* (1992). They consider that projected changes in water temperature and wind-induced mixing of the Southern Ocean will be too small to exert much effect but that changes in sea ice are likely to be more important. Release of low-salinity water from sea ice in spring and summer is responsible for developing the shallow mixed layer in the sea-ice marginal zone—an area of the Southern Ocean that is nearly as productive as the coastal zone (Arrigo *et al.*, 1998)—and plays a major role in supporting other marine life. Projected reductions in the amount of sea ice (Section 16.2.4.2) may limit the development of the sea-ice marginal zone, with a consequential impact on biota there. On the other hand, greater freshening of the mixed ocean layer from increased precipitation, ice-sheet runoff, and ice-shelf melting might have a compensating effect. It seems that the sea-ice marginal zone, under-ice biota, and subsequent spring bloom will continue, but shift to more southern latitudes, as a consequence of the retreat of the ice edge.

Research also demonstrates that the biological production of the Antarctic food web is linked closely to physical aspects of the ocean and ice ecosystem. Matear and Hirst (1999) point out that changes in ocean circulation will impact ocean biological production. They project a reduction in biological export from the upper ocean and an expansion of the ocean's oligotrophic regions. This will alter the structure and composition of the marine ecosystem. For example, interdecadal variations in sponge/predator population and in anchor/platelet ice at depths shallower than 30 m appear to be related to alterations in regional currents and ocean climate shifts (Jacobs and Giulivi, 1998). Changes in ocean currents could bathe new areas of the sea floor in near-freezing water, so that anchor ice and ice crystals will rise through the water column. This will be a liability for some benthic species. On the other hand, there could be a fresher and more stable layer, which could lead to changes in phytoplankton community structure (Arrigo *et al.*, 1999), and stronger ocean fronts. Both of these physical changes would be beneficial to many parts of the marine ecosystem. A 20% decline in winter and summer sea ice since 1973 west of the Antarctic Peninsula region (Jacobs and Comiso, 1997) has led to a decline in Adelie penguins, which are obligate inhabitants of pack ice. By contrast, Chinstrap penguins in open water have increased in numbers (Fraser *et al.*, 1992; Ainley *et al.*, 1994). Krill recruitment around the Antarctic Peninsula seems to be dependent on the strength of the westerlies and sea-ice cover, with a 1-year lag (Naganobu *et al.*, 2000). Both will decrease in the future, so there will be less krill.

The direct effect of a change in temperature is known for very few Antarctic organisms. Much of the investigation on ecophysiology has concentrated on adaptations to living at low temperature, with relatively little attention devoted to their response to increasing temperatures. Few data are available to assess quantitatively the direct and indirect impacts of climate change. Perhaps the best-studied example of temperature affecting the abundance of marine microorganisms is the increased rate of production of the cyanobacterium *Synechococcus* with increasing temperature, which approximately doubles for an increase in temperature of 2.5°C (Marchant *et al.*, 1987).

The virtual absence of cyanobacteria represents a fundamental difference between the microbial loop in Antarctic waters compared to that in temperate and tropical waters. As discussed by Azam *et al.* (1991), metazoan herbivores apparently cannot directly graze *Synechococcus*; their production must be channelled through heterotrophic protists able to consume this procaryote. Adding another trophic step reduces the energy available to higher trophic levels. Coupled with the direct utilization of nanoplankton by grazers, this may account in part for high levels of tertiary production in the Southern Ocean, despite relatively low levels of primary production (but see Arrigo *et al.*, 1998). Any increase in water temperature will increase the concentration of cyanobacteria and the heterographs that graze them. It is possible that the prey for krill and other grazers also will change, but the ultimate effects are unknown. Changes in the microbial loop may lessen carbon drawdown because of increased respiration by heterotrophs. Furthermore, there is an apparent uncoupling of bacterioplankton and phytoplankton assemblages that contrasts with temperate aquatic ecosystems (Bird and Kalff, 1984; Cole *et al.*, 1988; Karl *et al.*, 1996). The structure and efficiency of the Antarctic marine food web is temporally variable, and Karl (1993) has suggested that it is reasonable to expect several independent (possibly overlapping in space and time) food webs. There is no consensus with respect to the importance of bacteria and their consumers within this food web or their degree of interaction with photoautotrops. Bird and Karl (1999) have demonstrated, however, that uncoupling of the microbial loop in coastal waters during the spring bloom period was the direct result of protistan grazing. Although underlying mechanisms remain unclear, the distinct difference of the microbial loop in Antarctic waters, compared to more temperate waters, suggests that climate change will have profound effects on the structure and efficiency of the Southern Ocean food web.

An increase in temperature is likely to lead to shifts in species assemblages. Organisms that are unable to tolerate the present low-temperature regime will invade the Southern Ocean. Some that already are there will exhibit increased rates of growth. Predicted reductions in the extent and thickness of sea ice will have ramifications not only for the organisms directly associated with sea ice but also for those that rely on oceanographic processes that are driven by sea-ice production. In the open ocean, there is a correlation between the standing crop and productivity of phytoplankton with wind speed (Dickson *et al.*, 1999). Diatoms—the dominant phytoplanktonic organisms in the Southern Ocean—have high sinking rates and require a turbulent mixed zone to remain in the photic zone. If climate change results in diminution of wind forcing of surface mixing, a reduced biomass of diatoms can be expected. This would lead to less available food for the higher trophic levels and diminution in the vertical flux of carbon and silicon. Together, these effects are likely to have a profound impact on Antarctic organisms at all trophic levels, from algae to the great whales.

Any reduction in sea ice clearly represents a change in habitat for organisms that are dependent on sea ice, such as Crabeater seals and Emperor penguins. Some species of penguins and seals are dependent on krill production. Increased ultraviolet irradiance from ozone depletion is likely to favor the growth of organisms with UV-protecting pigments and/or repair mechanisms (Marchant, 1997; Davidson, 1998). This will lead to a change in species composition and impact trophodynamics and vertical carbon flux. Naganobu *et al.* (2000) show evidence that ozone depletion impacts directly and indirectly on krill density. The growth, survival, and hatching rates of penguin chicks and seal pups are directly influenced by krill abundance in the sea. Animals that migrate great distances, such as the great whales and seabirds, are subject to possible disruptions in the timing and distribution of their food sources. Contraction of sea ice may alter migration patterns but would not be expected to present a major problem to such mobile animals. Other marine mammals (seals, sea lions) have life histories that tie them to specific geographic features such as pupping beaches, ice fields, or sub-Antarctic islands; they may be more severely affected by changes in the availability of necessary habitats and prey species that result from climate change.

Box 16-1. Climate Change and Fisheries

The Southern Ocean has large and productive fisheries that constitute part of the global food reserve. Currently, there are concerns about sustainability, especially with regard to species such as Patagonian toothfish. There are likely to be considerable changes in such fisheries under the combined pressures of exploitation and climate change. Spawning grounds of coldwater fish species are very sensitive to temperature change. Warming and infusion of more freshwater is likely to intensify biological activity and growth rates of fish (Everett and Fitzharris, 1998). Ultimately, this is expected to lead to an increase in the catch of marketable fish and the food reserve. This could be offset in the long term by nutrient loss resulting from reduced deepwater exchange. Fisheries on the margin of profitability could prosper because the retreat of sea ice will provide easier access to southern waters. Everett and Fitzharris (1998) discuss catch-per-unit-effort (CPUE) statistics from the commercial krill fishery operating around South Georgia and demonstrate that there is correlation with ice-edge position. The further south the ice, the lower the CPUE in the following fishing season. Fedoulov *et al.* (1996) report that CPUE also is related to water temperature and atmospheric circulation patterns, and Loeb *et al.* (1997) document the close relationships between seasonal sea-ice cover and dominance of either krill or salps (pelagic tunicates). Ross *et al.* (2000) identify that maximum krill growth rates are possible only during diatom blooms and that production in Antarctic krill is limited by food quantity and quality. Consequently, differences in the composition of the phytoplankton community caused by changes in environmental conditions, including climate change, will be reflected at higher trophic levels in the grazer community and their levels of productivity.

Arctic fisheries are among the most productive in the world. Changes in the velocity and direction of ocean currents affect the availability of nutrients and disposition of larval and juvenile organisms, thereby influencing recruitment, growth, and mortality. Many groundfish stocks also have shown a positive response to recent climate change (NRC, 1996), but Greenland turbot—a species that is more adapted to colder climates—and King crab stocks in the eastern Bering Sea and Kodiak have declined (Weller and Lange, 1999). Projected climate change could halve or double average harvests of any given species; some fisheries may disappear, and other new ones may develop. More warmer water species will migrate poleward and compete for existing niches, and some existing populations may take on a new dominance. These factors may change the population distribution and value of the catch. This could increase or decrease local economies by hundreds of millions of dollars annually.

16.2.4. Changes in Sea Ice

Sea ice is a predominant feature of the polar oceans. Its extent expands and contracts markedly from winter to summer. Sea ice has a dramatic effect on the physical characteristics of the ocean surface. The normal exchange of heat and mass between the atmosphere and ocean is strongly modulated by sea ice, which isolates the sea surface from the usual atmospheric forcing (Williams *et al.*, 1998a). In addition, sea ice affects albedo, the exchange of heat and moisture with the atmosphere, and the habitats of marine life. Finally, sea ice plays a significant role in the thermohaline circulation of the ocean. Changes in air temperature expected with projected climate change are likely to alter the sea-ice regime and hence have impacts on the foregoing mechanisms.

Warming is expected to cause a reduction in the area covered by sea ice, which in turn will allow increased absorption of solar radiation and a further increase in temperature. In some climate models, this sea ice-albedo feedback has led to amplification of projected warming at higher latitudes, more in the Arctic Ocean than in the Southern Ocean. At some point, with prolonged warming a transition to an Arctic Ocean that is ice-free in summer—and perhaps even in the winter—could take place. The possibility of a transition to an ice-free Arctic Ocean that is irreversible also must be considered. Climate models predict a wide range of changes for polar sea ice by the year 2100 (see TAR WGI Chapter 1). Many of the earlier models treated sea ice very simplistically and were not coupled with the ocean, so their results are unlikely to be reliable. More recent climate models include sophisticated sea-ice routines that take into account the dynamics and thermodynamics that control sea-ice formation, transport, and melt (Everett and Fitzharris, 1998). These models, however, still are limited in their ability to reproduce detailed aspects of sea-ice distribution and timing.

Snow on sea ice controls most of the radiative exchange between the ocean and atmosphere, but its exact role in energy transfer is uncertain (Iacozza and Barber, 1999) and difficult to model because of a lack of adequate data (Hanesiak *et al.*, 1999; Wu *et al.*, 1999). A further complication is the timing of snowfall in relation to formation of sea ice (Barber and Nghiem, 1999). Accumulation of snow on sea ice also plays a significant role in sub-ice primary production and for habitats of marine mammals (e.g., polar bear and seals).

16.2.4.1. Sea Ice in the Arctic Ocean

Whether the sea ice in the Arctic Ocean will shrink depends on changes in the overall ice and salinity budget, the rate of sea-ice production, the rate of melt, and advection of sea ice into and out of the Arctic Basin. The most important exit route is through Fram Strait (Vinje *et al.*, 1998). The mean annual export of sea ice through Fram Strait was ~2,850 km^3 for the period 1990–1996, but there is high interannual variability caused by atmospheric forcing and, to a lesser degree, ice thickness variations. Other important passages are the northern Barents Sea and through the Canadian Arctic Archipelago (Rothrock *et al.*, 1999). Analyses by Gordon and O'Farrell (1997), using a dynamic ice routine with a transient coupled atmosphere-ocean climate model (Commonwealth Scientific and Industrial Research Organisation—CSIRO), predict a 60% loss in summer sea ice in the Arctic for a doubling of CO_2. The summer season, during which ice retreats far offshore, increases from 60 to 150 days. The likely distance between northern coasts and Arctic pack ice will increase from the current 150–200 to 500–800 km.

In a more recent study, there is good agreement between Arctic sea-ice trends and those simulated by control and transient integrations from the Geophysical Fluid Dynamics Laboratory (GFDL) and the Hadley Centre (see Figure 16-6). Although the Hadley Centre climate model underestimates sea-ice extent and thickness, the trends of the two models are similar. Both models predict continued decreases in sea-ice thickness and extent (Vinnikov *et al.*, 1999), so that by 2050, sea-ice extent is reduced to about 80% of area it covered at the mid-20th century.

GCM simulations for Arctic sea ice predict that warming will cause a decrease in maximum ice thickness of about 0.06 m per °C and an increase in open water duration of about 7.5 days per °C (Flato and Brown, 1996). These projections are somewhat lower than changes observed during the latter part of the 20th century (see Section 16.1.3). Increased snowfall initially causes a decrease in maximum thickness (and corresponding increase in open-water duration), but beyond 4 mm per day (1.33 mm per day water equivalent), formation of "slush ice" by surface

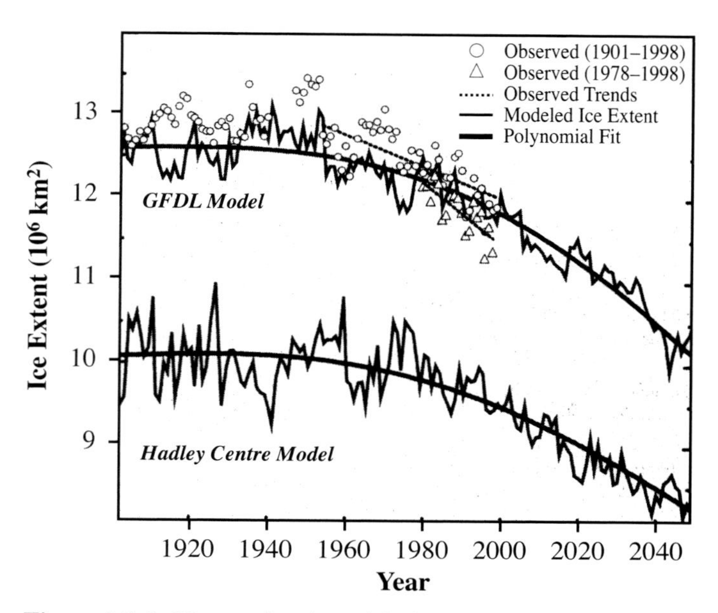

Figure 16-6: Observed and modeled variation of annual averages of Arctic sea-ice extent, based on Vinnikov *et al.* (1999). Observed data are from Chapman and Walsh (1993) and Parkinson *et al.* (1999). Sea-ice curves are produced by GFDL low-resolution R15 climate model and by HADCM2 climate model, both forced by CO_2 and aerosols.

flooding offsets the insulating effect of snow and causes an increase in maximum thickness and a decrease in the duration of open water.

Without sea ice, wave heights will increase, and the Arctic coast will be more exposed to severe weather events such as storm surges that cause increased coastal erosion, inundation, and threat to structures. Areas of ice-rich permafrost are particularly vulnerable to coastal erosion (Nairn *et al.*, 1998; Wolfe *et al.*, 1998). A portent of the future is severe coastal erosion—as much as 40 m yr^{-1}—along the Siberian coast. Deposited organic material changes the entire biogeochemistry of the nearshore waters (Weller and Lange, 1999). More open water may lead to increased precipitation and further amelioration of temperatures. Projected losses in sea ice are likely to have considerable impacts on Arctic biology through the entire food chain, from algae to higher predators (e.g., polar bears and seals). Loss of sea ice also will affect indigenous peoples and their traditional ways of life. A more open ocean will favor increased shipping along high-latitude routes and could lead to faster and cheaper ship transport between eastern Asia, Europe, and eastern North Amercia.

16.2.4.2. Sea Ice in the Southern Ocean

Antarctic sea ice is not confined by land margins but is open to the Southern Ocean. Sea-ice extent contracts and expands on an annual cycle in a roughly concentric zone around Antarctica. The ultimate extent is controlled by a balance of air temperature, leads, wind direction, upper ocean structure, and pycnocline depth. Some of these parameters are controlled in the atmosphere by the relative position of the subpolar trough with respect to the sea ice. In the ocean, variations in the Antarctic Circumpolar Current are important. The extent and thickness of Antarctic sea ice are sensitive to the depth and thermal properties of overlying snow, about which relatively little is known.

A reduction in Antarctic sea ice volume of about 25–45% is predicted for a doubling of CO_2, with sea ice retreating fairly evenly around the continent (Gordon and O'Farrell, 1997). This CSIRO model assumes a 1% yr^{-1} compounding increase of CO_2, corresponding to global warming of 2.1°C. Using a similar but modified model that has a higher albedo feedback and predicted global warming of 2.8°C, Wu *et al.* (1999) calculate a reduction in mean sea-ice extent of nearly two degrees of latitude, corresponding to 45% of sea-ice volume. These estimates do not represent the equilibrium state, and sea ice can be expected to shrink further, even if GHGs are stabilized. Changes in Antarctic sea ice will have little impact on human activity except where they allow shipping (mostly research, fishing, and tourist vessels) to get closer to the Antarctic continent. However, there are important biological and oceanographic impacts derived from reductions in sea ice, as well as significant ecological consequences attributable to changes in the magnitude and timing of seasonal sea-ice advance and retreat (Smith *et al.*, 1999b).

16.2.5. Permafrost

The term *permafrost* refers to layers of earth materials at relatively shallow depth that remain below 0°C throughout 2 or more consecutive years, independent of material composition or ice content. The role of permafrost in the context of global change is three-fold (Nelson *et al.*, 1993): It records temperature changes (Lachenbruch and Marshall, 1986), translates changes to other components of the environment (Osterkamp *et al.*, 2000), and facilitates further climatic changes through release of trace gases (Rivkin, 1998; Robinson and Moore, 1999). Permafrost is highly susceptible to long-term warming. Although it is an important factor in ice-free parts of Antarctica (Bockheim, 1995), permafrost is far more extensive in the continental areas of the subarctic and Arctic (Brown *et al.*, 1997; Zhang *et al.*, 1999), and changes there have great potential to affect human activities adversely. Accordingly, frozen ground activity has been designated as a "geoindicator" for monitoring and assessing environmental change (Berger and Iams, 1996).

Most biological and hydrological processes in permafrost terrain are confined to the active (seasonally thawed) layer, which forms a boundary across which exchanges of heat, moisture, and gases occur between the atmospheric and terrestrial systems. Its thickness is a response to a complex interplay among several factors, including aboveground climate, vegetation type and density, snow-cover properties, thermal properties of the substrate, and moisture content. If other conditions remain constant, the thickness of the active layer could be expected to increase in response to climate warming. Long-term records of active-layer thickness are rare; those that do exist, however, indicate that variations occur at multiple temporal scales and are determined by climatic trends and local conditions at the surface and in the substrate (Nelson *et al.*, 1998b).

16.2.5.1. Temperature Archive

Because heat transfer within thick permafrost occurs primarily by conduction, the shallow earth acts as a low-pass filter and "remembers" past temperatures. Temperature trends are recorded in the temperature-depth profile over time scales of a century or more (e.g., Lachenbruch and Marshall, 1986; Clow *et al.*, 1998; Osterkamp *et al.*, 1998; Taylor and Burgess, 1998). Permafrost is affected primarily by long-term temperature changes and thus contains a valuable archive of climate change (Lachenbruch and Marshall, 1986), although changes in snow cover must be taken into account. In contrast, the temperature regime of the overlying seasonally thawed active layer is highly complex, owing to nonconductive heat-transfer processes that operate much of the year (Hinkel *et al.*, 1997).

Multi-decadal increases in permafrost temperature have been reported from many locations in the Arctic, including northern and central Alaska (Lachenbruch and Marshall, 1986; Osterkamp and Romanovsky, 1999), northwestern Canada (Majoriwicz and Skinner, 1997), and Siberia (Pavlov, 1996). Temperature increases are not uniform, however, and recent cooling of permafrost has been reported in northern Quebec (Wang and Allard, 1995). To obtain a more comprehensive picture of the spatial structure and variability of long-term changes in permafrost, the Global Terrestrial Network for Permafrost or (GTNet-P) (Brown *et al.*, 2000; Burgess *et al.*, 2000b) was developed in the 1990s. The program has two components: long-term monitoring of the thermal state of permafrost in a network of boreholes, and monitoring of active-layer thickness and processes at representative locations. The latter network—the Circumpolar Active Layer Monitoring (CALM) program (Nelson and Brown, 1997)—has been in operation since the mid-1990s and incorporates more than 80 stations in the Arctic and Antarctic. The CALM network includes components in which regional mapping (Nelson *et al.*, 1999) and spatial time series analysis of active-layer thickness (Nelson *et al.*, 1998a) are performed.

16.2.5.2. Predicted Changes in Permafrost

Figure 16-1 illustrates the geographic distribution of permafrost and ground ice in the northern hemisphere. Computations based on a digital version of the map by Brown *et al.* (1997) indicate that permafrost currently underlies 24.5% of the exposed land area of the northern hemisphere (Zhang *et al.*, 1999). Under climatic warming, much of this terrain would be vulnerable to subsidence, particularly in areas of relatively warm, discontinuous permafrost (Osterkamp *et al.,* 2000). Anisimov and Nelson (1997) conducted experiments with a simple model of permafrost distribution driven by three transient GCMs (as reported by Greco *et al.*, 1994). Figure 16-7 shows the mid-range result from this study; the area of the northern hemisphere occupied by permafrost eventually could be reduced by 12–22% of its current extent. Experiments by Smith and Burgess (1999) with a 2xCO_2 GCM indicated that permafrost eventually could disappear from half of the present-day Canadian permafrost region. Thawing of ice-rich permafrost is subject, however, to considerable lag resulting from the large latent heat of fusion of ice. Simulations by Riseborough and Smith (1993), for example, indicate that areas of 5 m thick ice-rich permafrost near the southern limit of the discontinuous zone in subarctic Canada could thaw in less than 70 years. Where permafrost currently is thick, it could persist in relict form for centuries or millennia. Figure 16-7 therefore is best regarded as indicating areas in which considerable potential exists for changes in permafrost distribution. Nonetheless, pronounced changes in permafrost temperature, surface morphology, and distribution are expected (Smith and Burgess, 1998, 1999; Osterkamp and Romanovsky, 1999).

16.2.5.3. Environmental Impacts

The geological record of the Arctic shows abundant evidence for regional-scale deterioration of permafrost in response to climatic change. Thawing of ice-rich permafrost can result in subsidence of the ground surface and development of uneven and frequently unstable topography known as thermokarst terrain. In unglaciated parts of Siberia, alases (coalescing thaw depressions) formed during Holocene warm intervals occupy areas as large as 25 km^2. Cryostratigraphic studies in northwestern Canada have documented thaw unconformities in the form of truncated ice wedges associated with climatic warming during the early Holocene (Burn, 1997, 1998). Warming of permafrost and thaw subsidence has been reported widely from subarctic regions in recent years (Fedorov, 1996; Osterkamp and Romanovsky, 1999; Osterkamp *et al.*, 2000; Wolfe *et al.*, 2000; Zuidhoff and Kolstrup, 2000).

Changes in active-layer thickness could have marked impacts on local environments (Anisimov *et al.*, 1997; Burgess *et al.*, 2000a; Dyke, 2000). Thawing of perennially frozen subsurface materials will create few problems where permafrost is "dry" (Bockheim and Tarnocai, 1998). However, thawing of ice-rich permafrost can be accompanied by mass movements and subsidence of the surface, possibly increasing delivery of sediment to watercourses and causing damage to infrastructure in developed regions. Many recent examples of such phenomena already have been reported, and a poll conducted for the Canadian Climate Centre showed that most of the engineers interviewed think that climate warming is a very important factor for the stability of structures in permafrost regions (Williams, 1995; Vyalov *et al.*, 1998; Weller and Anderson, 1999). Weller and Lange (1999) note that the bearing capacity of permafrost has decreased with warming, resulting in failure of pilings for buildings and pipelines, as well as roadbeds. Accelerated permafrost thawing has led to costly increases in road damage and maintenance. In Alaska, for example, it costs as much as US$1.5 million to replace 1 km of road system. However, eventual disappearance of permafrost reduces construction problems.

Degradation of permafrost, which depends on the precipitation regime and drainage conditions, poses a serious threat to Arctic biota through oversaturation or desiccation of the surface (Callaghan and Jonasson, 1995). Changes in permafrost and active-layer thickness will produce a complicated response. Development of thermokarst in relatively warm, discontinuous permafrost in central Alaska has transformed some upland forests into extensive wetlands (Osterkamp *et al.*, 2000). In contrast, climatic warming in the northern foothills of the Brooks Range could lead to increasing plant density, resulting in evolutionary transitions from nonacidic to acidic tundra or shrub tundra. Despite increasing air temperature, such changes could prevent increases in active-layer thickness (Walker *et al.*, 1998).

16.2.6. Arctic Hydrology

The hydrological regime of the Arctic is particularly susceptible to predicted climate change because of the dominance of the thermally sensitive cryosphere and its controlling influence on the water cycle. Virtually all major hydrological processes and related aquatic ecosystems are affected by snow and ice processes, including the major Arctic rivers. Although these rivers originate in more temperate southern latitudes, they pass

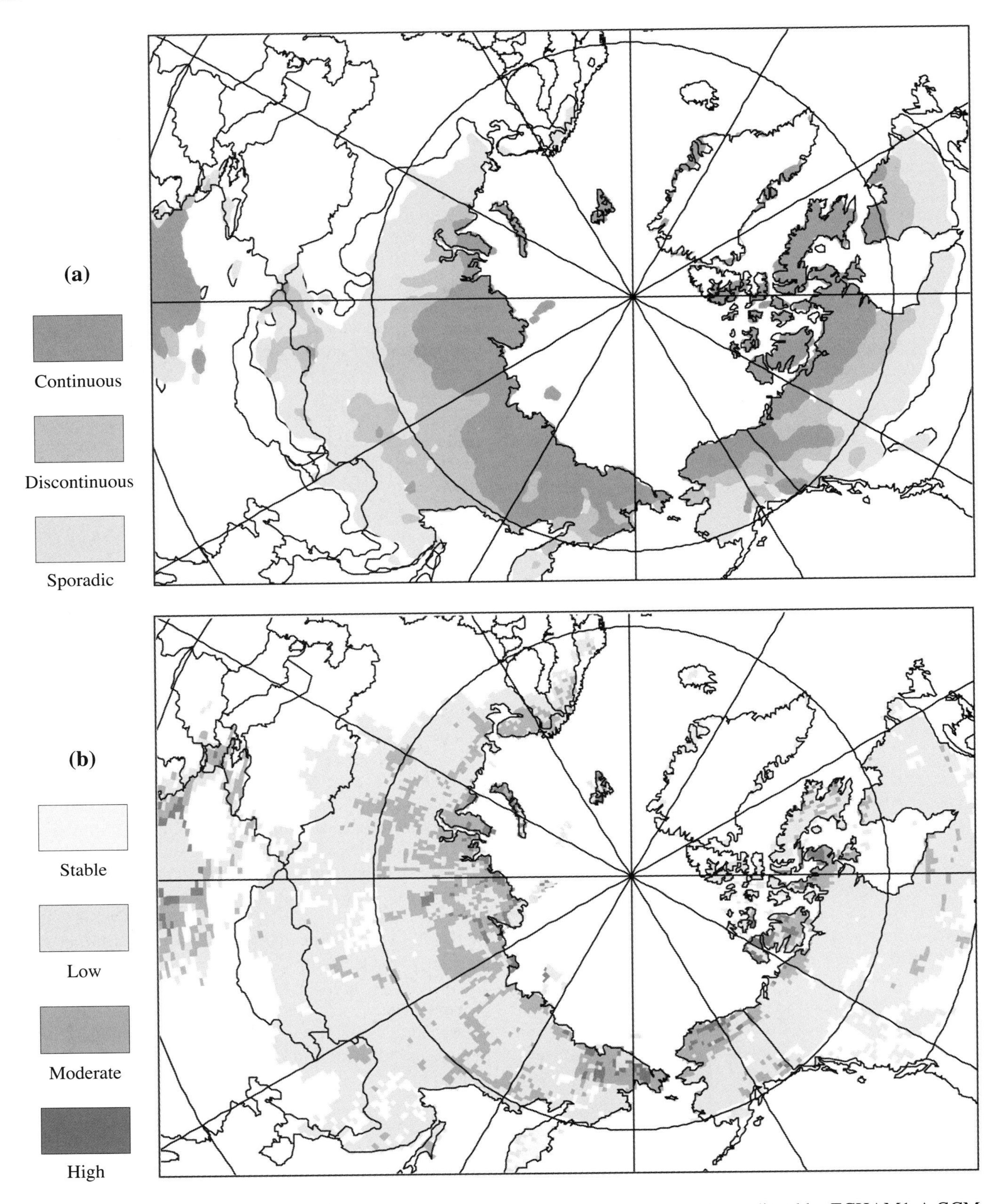

Figure 16-7: (a) Zonation of permafrost in the northern hemisphere under climate scenario predicted by ECHAM1-A GCM (Cubasch *et al*., 1992). (b) Hazard potential associated with degradation of permafrost under ECHAM1-A climate change scenario. Map represents areas of stable permafrost and low, moderate, and high susceptibility to subsidence. Classification is based on a "thaw-settlement index" calculated as the product of existing ground-ice content (Brown *et al*., 1997) and predicted increases in depth of thaw (Anisimov *et al*., 1997). Hazard-zone intervals were derived through division of resulting frequency distribution, using a nested-means procedure (Scripter, 1970).

through cold regions before reaching the Arctic Ocean. Changes in precipitation also have great potential for modifying the hydrology and sediment load in rivers, which could affect aquatic life (Michel and van Everdingen, 1994).

16.2.6.1. Changes in Precipitation, Snow Accumulation, and Spring Melt

Almost all climate models forecast increases in precipitation; but estimates vary widely and are complicated by many other factors, such as clouds (Stewart, 2000). Combined changes in temperature and precipitation will produce changes in the pattern of snow storage and, ultimately, Arctic hydrology. Warmer conditions generally will reduce the length of winter. Seasonal snow accumulation could increase in higher elevation zones, especially above the freezing level, where there would be little summer melt. Increased summer storminess will reduce melt at intermediate elevations as a result of increased cloud and summer snowfall, so many existing semi-permanent snowpacks will continue to exist (Woo, 1996). At lower elevations, particularly in the more temperate maritime zones, rainfall and rain-on-snow melt events probably will increase.

An earlier transition from winter to spring will mean that snowmelt will be more protracted, with less intense runoff. This will be accentuated where increased active-layer thickness and loss of permafrost increases the water storage capacity of the ground, leading to a decrease in runoff. Summer base flow will increase. Overall, this means that there will be less seasonal fluctuation in runoff through the year (Rouse *et al.*, 1997).

16.2.6.2. Surface Water Budgets and Wetlands

Predicted climate change will lead to greater evaporation and transpiration. Transpiration should increase as nontranspiring lichens and mosses that currently dominate the tundra regions are replaced by a denser cover of vascular plants (Rouse *et al.*, 1997). These changes are likely to reduce the amount of ponded water and runoff. The unglaciated lowlands of many Arctic islands, where special ecological niches occur, are likely to be especially sensitive. For coastal areas, changes to local micro-climates may occur because longer open-water seasons in the adjacent sea will lead to more frequent fog and low clouds, as well as reduced incident solar radiation. These changes may limit the expected increase in evaporation and transpiration (Rouse *et al.*, 1997).

The ability of Arctic wetlands to act as a source or sink of organic carbon and CH_4 depends on the position of the water. Analysis of subarctic sedge fens to increases in temperature of 4°C suggests reduced water storage of 10–20 cm over the summer in northern peatlands, even with a small increase in precipitation (Rouse, 1998). This is significant, given that Moore and Roulet (1993) suggest that a reduction of 0.1 m in water storage in northern forested peatland is sufficient to convert these areas from a source to a sink of atmospheric CH_4. Wider questions concerning the carbon budget, including the level of uncertainty, are discussed in TAR WGI Chapter 3. Further predictions based on $2xCO_2$ changes indicate a 200–300 km retreat of the southern boundary of peatlands in Canada toward the Arctic coast and significant changes in their structure (Gignac and Vitt, 1994). Degradation of permafrost, which currently forms an impermeable layer, will couple many ponds with the groundwater system and lead to their eventual drainage. Areas of special sensitivity include patchy Arctic wetlands on continuous permafrost and those along the southern limit of permafrost (Woo and Young, 1998). In contrast, warming of surface permafrost also could lead to formation of new wetlands, ponds, and drainage networks through the process of thermokarst development, especially in areas with high concentrations of ground ice.

16.2.6.3. Ecological Impact of Changing Runoff Regimes

A warmer climate will create a more pluvial runoff regime as a greater proportion of the annual precipitation is delivered by rain rather than snow and a flattening of the seasonal runoff cycle occurs. Enhancement of winter flow will mean streams that currently freeze to their beds will retain a layer of water beneath the ice. This will be beneficial to invertebrates and fish populations. However, such rivers will then be prone to ice jamming and hence larger annual flood peaks. Warming will lead to a shortened ice season and thinner ice cover. For large northward-flowing rivers, this could reduce the severity of ice jamming in spring, especially if the magnitude of the peak snowmelt that drives breakup also is reduced (Beltaos and Prowse, 2000). Decreased ice-jam flooding will benefit many northern communities located near river floodplains. In contrast, reductions in the frequency and severity of ice-jam flooding would have a serious impact on northern riparian ecosystems, particularly the highly productive river deltas, where periodic flooding has been shown to be critical to the survival of adjacent lakes and ponds (Marsh and Hey, 1989; Prowse and Conly, 1998).

Decreases in winter snowpack and subsequent spring runoff from upstream tributaries has led to reduced frequency and severity of ice jams affecting the very large Peace-Athabasca Delta (Prowse *et al.*, 1996). These changes are analogous to those expected with predicted climate warming. Ice breakup is a major control over aquatic ecology, affecting numerous physical and biochemical processes and the biodiversity and productivity of such northern rivers (Prowse, 1994; Scrimgeour *et al.*, 1994). Major adjustments in their ecology are expected in the future. A similar situation exists for northern ponds and lakes, where ice cover will be thinner, form later, and break up earlier—with consequent limnological changes. Total primary production should increase in all Arctic lakes and ponds with an extended and warmer ice-free season (Douglas *et al.*, 1994). Primary productivity of Arctic aquatic systems also should be boosted by a greater supply of organic matter and nutrients draining from a more biologically productive terrestrial landscape (Schindler, 1997; Hobbie *et al.*, 1999). Thinner ice cover will increase the solar radiation penetrating to the underlying water, thereby increasing photosythnetic production of oxygen and reducing the potential for winter fish kills. However, a longer ice-free season will increase the depth of mixing and lead to lower oxygen concentrations and increased stress on coldwater organisms (Rouse *et al.*, 1997).

16.2.6.4. Sensitivity of Arctic Ocean to River Flow

Increased future melting from the Greenland ice sheet and Arctic glaciers and ice caps has potential to increase significantly freshwater runoff to the northern circumpolar seas on time scales as short as decades. Combined with changes in Arctic runoff, there will be an effect on Arctic sea ice. Freshwater input to the Arctic Ocean is important for growth, duration, and melt of sea ice, particularly on the large, shallow continental shelf areas that constitute 30% of the area of the Arctic basin. Spreading over the ocean of lower density freshwater from summer runoff, together with sinking of denser brine rejected during autumn and winter sea-ice production, maintain characteristic salinity and temperature layering. This leads to the Arctic halocline—a cold, salt-stratified layer that separates the upper mixed layer from underlying warm saline waters. The strength and position of surface and deep currents in the slope water south of Newfoundland are thought to vary as a coupled system in relation to the dipole in atmospheric sea-level pressure known as the NAO (Keigwin and Pickart, 1999).

Overall, runoff to the Arctic Ocean is approximately 3–4 x 10^3 km^3 yr^{-1} (Prowse and Flegg, 2000)—twice that produced by precipitation minus evaporation for the Arctic Ocean or the influx of low-salinity ocean water (Barry and Serreze, 2000). Export of freshwater (as low-salinity water and sea ice) from the Arctic Ocean into the North Atlantic couples northern latitudes to the world thermohaline circulation.

Almost 70% of the total runoff to the Arctic Ocean is provided by four rivers: the Ob, Yenisey, and Lena in Russia and the Mackenzie in Canada. The three Russian rivers have flow measurements dating back to the 1930s, whereas the Mackenzie record begins only in 1972. Although there is no significant long-term trend in flow for these rivers, there has been a slight increase in runoff beginning in the late-1970s for other rivers in European Russia and western Siberia, with some dependence on season (Georgievskii *et al.*, 1996; Georgievskii, 1998; Grabs *et al.*, 2000; Shiklomanov *et al.*, 2000). Shiklomanov *et al.* (2000) found that historical temperature shifts of -0.5 to +1.0°C had little effect on river inflows to the Arctic Ocean, which remained within 3–5% of their long-term average.

For a range of future climate scenarios, results from hydrological routines embedded in climate models and from independent hydrological models that are driven by climate model output indicate that discharge from the major Arctic rivers will increase significantly (Miller and Russell, 1992; Shiklomanov, 1994; van Blarcum *et al.*, 1995; Shiklomanov, 1997; Hagemann and Dümenil, 1998; Lewis *et al.*, 2000). Mackenzie basin studies, however, have projected some reduction in streamflow as a result of expected climate warming (MBIS, 1997). Summarizing the general results, Shiklomanov *et al.* (2000) note that with an atmospheric CO_2 doubling, total annual inflow to the Arctic Ocean will increase by 10–20%, with a 1.5- to 2.0-fold increase during winter, although most of the flow still will occur in summer. Such predictions must be treated cautiously, however, because most climate models have been found to overestimate Arctic precipitation and simulate soil moisture and evaporation poorly (Robock *et al.*, 1998; Walsh *et al.*, 1998). Uncertainty in hydrological modeling of high-latitude river systems must be resolved before the risk of changes to the Arctic Ocean freshwater budget and thermohaline circulation can be quantified accurately (Carmack, 2000; Prowse and Flegg, 2000).

16.2.7. Changes in Arctic Biota

16.2.7.1. Impacts of Climate Change on Arctic Terrestrial Environments

Northern environments are thought to be particularly sensitive to climate warming, owing to changes in surface albedo degradation of permafrost, increased active-layer thickness, and earlier snowmelt. Recent warming reflects partial replacement of dry Arctic air masses by wetter Atlantic, Pacific, and southerly air masses, particularly in the subarctic and southern Arctic (Rouse *et al.*, 1997). Most of the increase is related to higher late-winter and early-spring air temperatures (+4.5°C), whereas summer temperatures in the region have increased by only about 2°C (Chapman and Walsh, 1993). In parts of Alaska, growing-season length has increased over the past 65 years as a result of reductions in snow and ice cover (Sharratt, 1992; Groisman *et al.*, 1994). The annual amplitude of the seasonal CO_2 cycle also has increased recently by as much as 40% (Keeling *et al.*, 1996). There has been an accompanying advance in the timing of spring by about 7 days and a delay in autumn by the same period, leading to an expected lengthening of the growing season (Keeling *et al.*, 1996). Such trends are likely to continue with projected climate change.

Precipitation will increase during the summer months, but evaporation and transpiration rates also are predicted to increase; this, together with earlier snowmelt, will lead to soil moisture deficits later in the summer (Oechel *et al.*, 1997). These changes in moisture and the thermal regimes of arctic soils ultimately will determine the fate of soil carbon stored in permafrost and the active layer (Miller *et al.*, 1983, Billings, 1987; Gorham, 1991; Schlesinger, 1991). Any change in physical properties in the active layer will have an impact on organisms and abiotic variables (Gates *et al.*, 1992; Kane *et al.*, 1992; Waelbroeck, 1993; Groisman *et al.*, 1994). The effects of changed drainage patterns and active-layer detachments (Dyke, 2000)—increasing sediment-nutrient loads in lakes and rivers—will alter biological productivity in aquatic ecosystems considerably (McDonald *et al.*, 1996).

Changes in the relative abundance of soil biota are predicted to affect soil processes, but only in a modest way. The rate of decomposition of organic matter is expected to rise with an increase in soil temperature and a longer growing season. This is likely to lead to enhanced rates of gross mineralization. However, the high carbon-to-nitrogen ratio of organic soils may still limit the availability of nitrogen to plants. If the thickened active layer extends into mineral layers, the release of mineral nutrients may be increased greatly, causing the upper organic layer to become more alkaline (Heal, 1998).

In summary, climate change probably will release more soil nutrients to biota, but nutrient limitations—especially of nitrogen and phosphorus—still are likely to occur. Studies have identified that the combination of nutrient and temperature increases results in the increasing importance of dominant plant species and suppression of subordinate ones. Species richness increased in an Arctic site, as a result of invasion of nitrophilous species, but decreased in a subarctic site. The impact of higher nutrient supplies and temperatures are likely to exceed those of elevated CO_2 and UV-B as measured in a related experiment (Press *et al.*, 1998). Because little is known about interactions between species within the soil and hence overall community response, the effect of climate change on rates of tundra soil decomposition and carbon loss is hard to predict (Smith *et al.*, 1998b).

16.2.7.2. Response of Arctic Plant Communities

There are a wide range of physiological responses of Arctic plants to climate change. Evidence indicates that changes will be at the level of individual species rather than groups of species (Chapin *et al.*, 1997; Callaghan *et al.*, 1998). Some responses to increased nutrient supply are specific to particular growth forms, such as the positive response of forbs, graminoids, and deciduous shrubs and the negative responses of mosses, lichens, and evergreen shrubs (Jonasson, 1992; Chapin *et al.*, 1995, 1996; Callaghan *et al.*, 1998; Shaver and Jonasson, 1999). The direct growth responses of evergreen dwarf shrubs to increased temperatures are small (Havström *et al.*, 1993; Wookey *et al.*, 1993; Chapin *et al.*, 1995, 1996). This is in contrast to the growth rates of graminoids, forbs, and deciduous dwarf shrubs (Wookey *et al.*, 1994; Henry and Molau, 1997; Arft *et al.*, 1999). In the Arctic, where these growth forms are abundant, there is a significant increase in plant biomass in response to increased temperatures in summer (Henry and Molau, 1997). The response of reproductive and vegetative structures to warming will vary, depending on abiotic constraints (Wookey *et al.*, 1994; Arft *et al.*, 1999). Warming in the winter and spring may encourage premature growth, so subsequent frost can lead to damage in plants (e.g., as discussed for *Vaccinium myrtillus* by Laine *et al.*, 1995, and for *Diapensia lapponica* by Molau, 1996). In northern Europe, processes that control the transition from boreal forest to tundra are more complicated than in North America because of infrequent large-scale defoliation of birch forest by geometrid moths. The eggs of the moths, which are killed at low temperatures, are likely to survive if winter temperatures increase (Neuvonen *et al.*, 1999).

Increased CO_2 concentrations in the atmosphere also could directly affect photosynthesis and growth rates of Arctic plants, but the effects are not easily predictable. For example, a rise in CO_2 concentration in the air leads to an initial increase in photosynthesis in individual tussocks of *Eriophorum vaginatum*. However, this effect can disappear in as little as 3 weeks as a result of homeostatic adjustment, indicating a loss of photosynthetic potential has occurred (Oechel *et al.*, 1997). Such short-lived responses are thought to be caused by limitations in nutrient (especially nitrogen) availability (Chapin and Shaver, 1985; Oechel *et al.*, 1997). At the ecosystem level in tussock tundra plots, the homeostatic adjustment took 3 years—by which time enhanced rates of net CO_2 uptake had disappeared entirely, in spite of substantial photosynthetic carbon gain during the period (Oechel *et al.*, 1994). Adjustment of the photosynthetic rate was thought to be caused by nutrient limitation. Körner *et al.* (1996) also have reported an increase in net CO_2 accumulation of 41% in *Carex curvula* after 2 years of exposure to elevated CO_2, but plant growth changes were minimal. Exposure of plants to increased temperatures and levels of CO_2 indicated that the initial stimulation in net CO_2 flux persisted, for reasons that are not entirely clear, although it may be linked to increased nutrient availability (Oechel and Vourlitis, 1994; Oechel *et al.*, 1994, 1997).

Results from plants of *Nardus stricta* growing adjacent to springs in Iceland indicate that they have been exposed to elevated levels of CO_2 for at least 150 years. There was a reduction in photosynthetic capacity of high-CO_2 grown plants of this grass from the vicinity of the spring, compared with plants grown at ambient CO_2 concentations—again linked to reduction in Rubisco content and the availability of nutrients (Oechel *et al.*, 1997). Collectively, these results indicate that it is unlikely that carbon accumulation will increase markedly over the long term as a result of the direct effects of CO_2 alone (Oechel *et al.*, 1997). A return to summer sink activity has occurred during the warmest and driest period in the past 4 decades in Alaskan arctic ecosystems, thereby eliminating a net summer CO_2 flux to the atmosphere that was characteristic of the early 1980s. The mechanisms are likely to include nutrient cycling, physiological acclimatization, and reorganization of populations and communities (Oechel *et al.*, 2000), but these systems are still net sources of CO_2.

A compounding consideration for Arctic plants is the impact of increased UV-B radiation. In Arctic regions, UV-B radiation is low, but the relative increase from ozone depletion is large, although the ancestors of present-day Arctic plants were growing at lower latitudes with higher UV-B exposure. Over the past 20 years, there has been a trend of decreasing stratospheric ozone of approximately 10–15% in northern polar regions (Thompson and Wallace, 2000). As a first approximation, a 1% decrease in ozone results in a 1.5–2% increase in UV-B radiation. Damage processes to organisms are temperature-independent, whereas repair processes are slowed at low temperatures. Hence, it is predicted that Arctic plants may be sensitive to increased UV-B radiation, especially because many individuals are long-lived and the effects are cumulative. In a study of responses by *Ericaceous* plants to UV-B radiation, responses varied from species to species and were more evident in the second year of exposure (Bjorn *et al.*, 1997; Callaghan *et al.*, 1998). For unknown reasons, however, the growth of the moss *Hylocomium splendens* is strongly stimulated by increased UV-B, provided adequate moisture is available (Gehrke *et al.*, 1996). Increased UV-B radiation also may alter plant chemistry that could reduce decomposition rates and nutrient availability (Bjorn *et al.*, 1997, 1999). Soil fungi differ with regard to their sensitivity to UV-B radiation, and their response also will affect the processes of decomposition (Gehrke *et al.*, 1995).

Climate change is likely to result in alterations to major biomes in the Arctic. Ecosystem models suggest that the tundra will decrease by as much as two-thirds of its present size (Everett and Fitzharris, 1998). On a broad scale—and subject to suitable edaphic conditions—there will be northward expansion of boreal forest into the tundra region, such that it may eventually cover more than 1.6 million km^2 of the Arctic. In northern Europe, vegetation change is likely to be more complicated. This is because of the influence of the geometrid moths, *Epirrita autumnata* and *Operophtera spp.*, which can cause large-scale defoliation of boreal forests when winter temperatures are above 3.6°C (Neuvonen *et al.*, 1999). Boreal forests are protected from geometrid moths only during cold winters. Empirical models estimate that by 2050, only one-third of the boreal forests of northern Europe will be protected by low winter temperatures in comparsion to the proportion protected during the period 1961–1991 (Virtanen *et al.*, 1998). However, the northward movement of forest may lag changes in temperature by decades to centuries (Starfield and Chapin, 1996; Chapin and Starfield, 1997), as occurred for migration of different tree species after the last glaciation (Delcourt and Delcourt, 1987). The species composition of forests is likely to change, entire forest types may disappear, and new assemblages of species may be established. Significant land areas in the Arctic could have entirely different ecosystems with predicted climate changes (Everett and Fitzharris, 1998). However, note that locally, climate change may affect boreal forest through decreases in effective soil moisture (Weller and Lange, 1999), tree mortality from insect outbreaks (Fleming and Volney, 1995; Juday, 1996), probability of an increase of large fires, and changes caused by thawing of permafrost.

16.2.7.3. Changes in Arctic Animals

In the immediate future, the greatest environmental change for some parts of the Arctic is likely to result from increased herding of reindeer rather than climate change (Crete and Huot, 1993; Manseau *et al.*, 1996; Callaghan *et al.*, 1998). Winter lichen pastures are particularly susceptible to grazing and trampling, and recovery is slow, although summer pastures in tundra meadows and shrub-forb assemblages are less vulnerable (Vilchek, 1997). The overall impact of climatic warming on the population dynamics of reindeer and caribou ungulates is controversial. One view is that there will a decline in caribou and muskoxen, particularly if the climate becomes more variable (Gunn, 1995; Gunn and Skogland, 1997). An alternative view is that because caribou are generalist feeders and appear to be highly resilient, they should be able to tolerate climate change (Callaghan *et al.*, 1998). Arctic island caribou migrate seasonally across the sea ice between Arctic islands in late spring and autumn. Less sea ice could disrupt these migrations, with unforeseen consequences for species survival and gene flow.

The decrease in the extent and thickness of Arctic sea ice in recent decades may lead to changes in the distribution, age structure, and size of populations of marine mammals. Seal species that use ice for resting, pup-rearing, and molting, as well as polar bears that feed on seals, are particularly at risk (Tynan and DeMaster, 1997). If break-up of annual ice occurs too early, seal pups are less accessible to polar bears (Stirling and Lunn, 1997; Stirling *et al.*, 1999). According to observational data, recent decreases in sea ice are more extensive in the Siberian Arctic than in the Beaufort Sea, and marine mammal populations there may be the first to experience climate-induced geographic shifts or altered reproductive capacity (Tynan and DeMaster, 1997).

Ice edges are biologically productive systems, with diatoms and other algae forming a dense layer on the surface that sustains secondary production. Of concern as ice melts is the loss of prey species of marine mammals, such as Arctic cod (*Boreogadus saida)* and amphipods, that are associated with ice edges (Tynan and DeMaster, 1997). The degree of plasticity within and between species to adapt to these possible long-term changes in ice conditions and prey availability is poorly known and requires study. Regime shifts in the ocean will impact the distribution of commercially important fish stocks. Recruitment seems to be significantly better in warm years than in cold years, and the same is valid for growth (Loeng, 1989). The distribution of fish stocks and their migration routes also could vary considerably (Buch *et al.*, 1994; Vilhjalmsson, 1997).

For other species, such as the lesser snow goose, reproductive success seems to be dependent on early-season climatic variables, especially early snowmelt (Skinner *et al.*, 1998). Insects will benefit from temperature increases in the Arctic (Danks, 1992; Ring, 1994). Many insects are constrained from expanding northward by cold temperatures, and they may quickly take advantage of a temperature increase by expanding their range (Parmesan, 1998).

16.2.8. Impact on Human Communities in the Arctic

16.2.8.1. Impacts on Indigenous Peoples

Historically, most indigenous groups have shown resilience and ability to survive changes in resource availability (e.g., the transition from Dorset to Thule cultures), but they may be less well equipped to cope with the combined impacts of climate change and globalization (Peterson and Johnson, 1995). Indigenous peoples, who number 1.5 million of a total Arctic population of 10 million, have a mixture of formal economies (e.g., commercial harvesting of fish, oil and mineral extraction, forestry, and tourism) and informal economies (e.g., harvesting of natural renewable resources). Increasingly, the overall economy is tied to distant markets. For example, in Alaska, gross income from tourism is US$1.4 billion, and in Russia 92% of exported oil is extracted from wells north of the Arctic Circle (Nuttall, 1998). The distinction between formal and informal economies becomes blurred by transfer payments and income derived from commercial ventures. The value of native harvests of renewable resources has been estimated to be 33–57% of the total economy of some northern communities (Quigley and McBridge, 1987; Brody, 1991). However, harvesting of renewable resources also must be considered in terms of maintaining cultural activities (Duerden, 1992). Harvesting contributes to community cohesion and self-esteem, and knowledge of wildlife and the environment strengthens social relationships (Warren *et al.*, 1995; Berkes, 1998). For example, hunting of wildlife is an essential part of Inuit tradition (Wenzel, 1995).

Climate change and economic development associated with oil extraction, mining, and fish farming will result in changes in diet and nutritional health and exposure to air-, water-, and food-borne contaminants in northern people (Bernes, 1996; Rees and Williams, 1997; Vilchek and Tishkov, 1997; AMAP, 1998; Weller and Lange, 1999; Freese, 2000). People who rely on marine systems for food resources are particularly at risk because Arctic marine food chains are long (Welch, 1995; AMAP, 1997). Low-lying Arctic coasts of western Canada, Alaska, and the Russian Far East are particularly sensitive to sea-level rise. Coastal erosion and retreat as a result of thawing of ice-rich permafrost already are threatening communities, heritage sites, and oil and gas facilities (Forbes and Taylor, 1994; Dallimore *et al.*, 1996; Cohen, 1997a,b; Shaw *et al.*, 1998; Wolfe *et al.*, 1998; Are, 1999).

Along the Bering and Chukchi Sea coasts, indigenous peoples report thinning and retreating sea ice, drying tundra, increased storms, reduced summer rainfall, warmer winters, and changes in the distribution, migration patterns, and numbers of some wildlife species. These peoples testify that they already are feeling some of the impacts of a changing, warming climate (Mulvaney, 1998). For example, when sea ice is late in forming, certain forms of hunting are delayed or may not take place at all. When sea ice in the spring melts or deteriorates too rapidly, it greatly decreases the length of the hunting season. Many traditional foods are dried (e.g., walrus, whale, seal, fish, and birds) in the spring and summer to preserve them for consumption over the long winter months. When the air is too damp and wet during the "drying" seasons, food becomes moldy and sour. The length of the wet season also affects the ability to gather greens such as willow leaves, beach greens, dock, and wild celery. These testimonies reflect the kinds of changes that could be expected as global warming affects the Arctic (Mulvaney, 1998). As climate continues to change, there will be significant impacts on the availability of key subsistence marine and terrestrial species. At a minimum, salmon, herring, walrus, seals, whales, caribou, moose, and various species of waterfowl are likely to undergo shifts in range and abundance. This will entail local adjustments in harvest strategies as well as in allocations of labor and resources (e.g., boats, snowmobiles, weapons). As the climate changes, community involvement in decisionmaking has the potential to promote sustainable harvesting of renewable resources, thereby avoiding deterioration of common property. However, factors that are beyond the control of the local community may frustrate this ideal. For example, many migratory animals are beyond the hunters' geographical range for much of the year—and thus beyond the management of small, isolated communities. Traditional subsistence activities are being progressively marginalized by increasing populations and by transnational commercial activities (Sklair, 1991; Nuttall, 1998).

In the past, when population densities of indigenous people were lower and economic and social structures were linked only weakly to those in the south, northern peoples showed significant flexibility in coping with climate variability (Sabo, 1991; Odner, 1992). Now, commercial, local, and conservation interests have reduced their options. Predicted climate change is likely to have impacts on marine and terrestrial animal populations; changes in population size, structure, and migration routes also are probable (Beamish, 1995; Gunn, 1995; Ono, 1995). Careful management of these resources will be required within a properly consultative framework, similar to recent agreements that are wide-ranging and endeavor to underpin the culture and economy of indigenous peoples (Nuttall, 1998). Langdon (1995) claims that "the combination of alternative cultural lifestyles and altered subsistence opportunities resulting from a warmer climate may pose the greatest threat of all to the continuity of indigenous cultures in northern North America." An alternative view is that northern people live with uncertainty and learn to cope with it; this view suggests that "for indigenous people, climate change is often not a top priority, but a luxury, and Western scientists may well be indoctrinating Natives with their own terminology and agenda on climate change" (BESIS, 1999).

16.2.8.2. Impacts on Economic Activity

The following subsections summarize the impacts of climate change and adaptive responses for different economic sectors that are relevant to the Arctic. Within different regions of the Arctic, important economic sectors differ substantially; some sectors are underrepresented, and others are just developing in certain regions (e.g., tourism). This latter topic (like agriculture and forestry) receives little or no comment in this section because there is insufficient literature to address the effects of climate change from a polar perspective.

16.2.8.2.1. Oil and gas extraction

Exploration, production, transportation, and associated construction of processing facilities are likely to be affected by climatic change (Maxwell, 1997). Changes in a large number of climate and related variables will affect on- and offshore oil and gas operations. Use of oil drilling structures or ice-strengthened drillships designed to resist ice, use of the ice itself as a drilling platform, and construction of artificial islands are likely to give way to more conventional drilling techniques employed in ice-free waters (Maxwell, 1997). These likely changes are not without concerns. Although the use of regular drillships may reduce operating costs by as much as 50% (Croasdale, 1993), increased wave action, storm surges, and coastal erosion may necessitate design changes to conventional offshore and coastal facilities (McGillivray *et al.*, 1993; Anderson and DiFrancesco, 1997). This may increase the costs of pipeline construction because extensive trenching may be needed to combat the effects of coastal instability and erosion, especially that caused by permafrost melting (Croasdale, 1993; Maxwell, 1997). Design needs for onshore oil and gas facilities and winter roads are strongly linked to accelerated permafrost instability and flooding. The impact of climate change is likely to lead to increased costs in the industry associated with design and operational changes (Maxwell, 1997).

16.2.8.2.2. Buildings and industrial facilities

The capacity of permafrost to support buildings, pipelines, and roads has decreased with atmospheric warming, so pilings fail to support even insulated structures (Weller and Lange, 1999). The problem is particularly severe in the Russian Federation, where a large number of five-story buildings constructed in the permanent permafrost zone between 1950 and 1990 already are weakened or damaged, probably as a result of climate change (Khroustalev, 1999). For example, a 2°C rise in soil temperature in the Yakutsk region has led to a decrease of 50% in the bearing capacity of frozen ground under buildings. Khroustalev (1999) has predicted that by 2030, most buildings in cities such as Tiksi and Yakutsk will be lost, unless protective measures are taken. The impact of warming is likely to lead to increased building costs, at least in the short term, as new designs are produced that cope with permafrost instability. Snow loads and wind strengths may increase, which also would require modifications to existing building codes (Maxwell, 1997). There will be reduced demand for heating energy with warmer climate (Anisimov, 1999).

16.2.8.2.3. Transportation and communications

The impact of climate warming on transportation and communications in Arctic regions is likely to be considerable. Within and between most polar countries, air transport by major commercial carriers is widely used to move people and freight. Irrespective of climate warming, the number of scheduled flights in polar regions is likely to increase. This will require an adequate infrastructure over designated routes, including establishment of suitable runways, roads, buildings, and weather stations. These installations will require improved engineering designs to cope with permafrost instability. Because paved and snow-plowed roads and airfield runways tend to absorb heat, the mean annual surface temperature may rise by 1–6°C, and this warming may exacerbate climate-driven permafrost instability (Maxwell, 1997). Cloud cover, wind speeds and direction, and patterns of precipitation may be expected to change at the regional level in response to global warming. At present, the density of weather stations is relatively low in Arctic regions. Increased air (and shipping) travel under a changing climate will require a more extensive weather recording network and navigational aids than now exists.

The impact of climate warming on marine systems is predicted to lead to loss of sea ice and opening of sea routes such as the Northeast and Northwest passages. Ships will be able to use these routes without strengthened hulls. There will be new opportunities for shipping associated with movement of resources (oil, gas, minerals, timber), freight, and people (tourists). However, improved navigational aids will be needed, and harbor facilities probably will have to be developed. The increase in shipping raises questions of maritime law that will need to be resolved quickly. These issues include accident and collision insurance, which authority is responsible for removal of oil or toxic material in the event of a spill, and which authority or agency pays expenses incurred in an environmental cleanup. These questions are important because sovereignty over Arctic waters is disputed among polar nations, and increased ship access could raise many destabilizing international issues. Increased storm surges are predicted that will affect transport schedules.

16.2.8.2.4. Pollution associated with increased economic activity

There already are a large number of case studies in the Arctic that indicate the effects of different pollutants on terrestrial, freshwater, and marine ecosystems (Crawford, 1997). In the event of increased industrial activity (e.g., mining, oil and gas extraction) under climate warming, new codes will be needed for retention of toxic wastes and to limit emissions of pollutants from processing plants. In the oil industry, considerable progress has been made in revegetating disturbed and polluted sites (McKendrick, 1997), often with plant species that can survive at northern sites under climatic warming. Changes in hydrology, possible increases in catchment rates, and melting of ice may result in wider dispersion of pollutants from accidents. Current ice cover and the low productivity of Arctic lakes restrict sequestration of contaminants (Barrie *et al.*, 1997; Gregor *et al.*, 1998). Projected changes to ice cover and the hydrology of these lakes may cause them to become greater sinks for river-borne contaminants, similar to those in more temperate regions. In the Arctic Ocean, many persistent organic pollutants (e.g., hexachlorocyclohexane) are trapped under ice as "ghosts of the past" (de March *et al.*, 1998). Reductions in sea-ice cover may speed their introduction to the Arctic food chain and pose risks for the human population. Long-lived apex consumers with high lipid content have a high potential for long-term accumulation of contaminants (Alexander, 1995;

Tynan and DeMaster, 1997), and not all of these pollutants are derived within the Arctic. Development of Arctic haze is thought to result from aerosol loading (primarily sulfate particles) of the atmosphere in mid-latitude regions. These particles are then carried northward to the Arctic. The haze is most pronounced in the winter because the particles have a longer residence time in the stable Arctic air masses at that time of year. The increased presence of sulfate particles in the atmosphere is of concern because of their ability to reduce the flow of energy through the atmosphere to the Earth's surface (Shaw, 1987).

16.2.8.2.5. Fisheries

High-latitude marine fisheries are very productive. For example, those in the northern Pacific Ocean and the Bering Sea contribute more than 28% to total world landings of fish, mollusks, and crustaceans. In some northern countries, fisheries and fish products account for a large proportion of gross domestic product. In Greenland, the shrimp industry contributed more than 30% to GDP in 1986. Landings of fish in the Northwest Territories and in Nunavut, in the commercial and the subsistence sectors of the economy, are valued at CDN$12 million.

Shifts in oceanic circulation associated with global warming are likely to affect the distribution of commercially important fish and their migration routes (Buch *et al.*, 1994; Vilhjalmsson, 1997). For example, the first catches of two species of Pacific salmon *(Oncorhnchus nerka, O. gorbuscha)* recently have been made in estuaries on Banks Island, Canada. These locations are well outside the known range for these Pacific species (Babaluk *et al.*, 2000). Changes in ocean currents, nutrient availability, salinity, and the temperature of ocean waters can be expected to influence the disposition of larval and juvenile organisms, the growth rates of individuals, and the population structure of different fish species (Otterson and Loeng, 2000). For example, recruitment appears to be significantly better in warm years (Loeng, 1989), an example of which is increased landings of cod (*Gadus morhua*) associated with warmer air and surface water temperatures (Brander, 1996). During a warm phase between the 1920s and the 1960s, Norwegian herring fed in Icelandic waters but disappeared when the water temperature cooled by 1°C (Vilhjalmsson, 1997). Quinn and Marshall (1989) report positive correlations between salmon returns and reduced sea ice. However, species that are adapted to cold water, such as the Greenland turbot and the Alaskan King crab, declined in numbers during these warm phases, although other factors also contributed to the decline of crab stocks (Weller and Lange, 1999). However, the underlying mechanisms that account for changes in population sizes are poorly understood. This topic is a high research priority, particularly because plankton production and trophic interactions may be significantly altered by changes in climate. Research and management advice for fish stocks is provided by the International Council for the Exploration of the Seas (ICES). This authority and others face formidable challenges if the distributions of fish populations change in response to global warming (Hønneland *et al.*, 1999; Freese, 2000).

16.2.8.2.6. Reindeer husbandry

Husbandry of different subspecies of *Rangifer tarandus* is widely practiced in different regions of the Arctic, particularly in Eurasia. Between 1991 and 1997, Russia's domestic reindeer stock declined from 2.3 million to 1.6 million animals. Whether climate change contributed to this decline is uncertain (Weller and Lange, 1999), but climate warming is likely to alter husdandry practices. Concerns include the presence of deep snow with an ice surface that stops animals from obtaining forage, lichens and graminoids that are ice-covered, destruction of vegetation as a result of "overgrazing," exposure of soil that encourages establishment of southerly weedy species under a warmer climate (Vilchek, 1997), and an increased likelihood of damage from more frequent tundra fires.

16.3. Synthesis

16.3.1. Feedbacks and Interactions—Polar Drivers

Climate change will affect some key polar drivers, creating impacts in the wider global arena. Many of these impacts will be self-amplifying and, once triggered, will affect other regions of the world for centuries to come. These impacts relate to probable changes in the cryosphere, sea level, thermohaline circulation and ocean ventilation, exchange of GHGs, and cloudiness:

- *Snow/ice–albedo feedback:* The amount of absorbed solar radiation, and thus surface heating, depends on surface albedo, which is very high for snow or ice surfaces and much lower in the absence of snow and ice. Thus, absorbed shortwave radiation over the vast areas of snow and ice in polar regions is about three times lower than over non-snow-covered surfaces. Warming will shrink the cryosphere, particularly in the Arctic, causing additional heating of the surface, which in turn will further reduce ice and snow cover. Thus, any significant alteration in albedo over large areas will have the potential to produce a nonlinear, accelerated change.
- *Sea-level rise:* Projected climate change in polar regions will have a critical impact on global sea levels. Expected sea-level rise is in the range 0.09–0.88 m by 2100, using the SRES scenarios (TAR WGI Chapter 11). Increased melting of the Greenland ice sheet and Arctic glaciers, as well as possible thinning of the West Antarctic ice sheet, is expected to make important contributions. However, increased snow accumulation on the East and West Antarctic ice sheets is a major process that can offset sea-level rise. Sea level will continue to rise long after atmospheric GHGs are stabilized, primarily because of the large heat capacity of the ocean and the slow response of glaciers and ice sheets. The rate of downwelling of cold, dense waters in polar regions is a major control on thermal expansion of the ocean and hence the rate of sea-level rise over the next centuries.

Feedbacks that link sea-level rise to the size and health the West Antarctic ice sheet remain the subject of research and debate and are discussed in more depth in TAR WGI Chapter 11.

- *Ocean circulation:* Ocean-climate models predict increased stability of the surface mixed layer, reduction in salt flux, less ocean convection, and less deepwater formation. This could lead to a prolonged, major reduction in thermohaline circulation and ocean ventilation (O'Farrell *et al.*, 1997; Budd and Wu, 1998; Hirst, 1999). Such changes will affect surface ocean currents and climates of Europe and mid-latitude landmasses in the southern hemisphere, where it could slow warming in some regions (Murphy and Mitchell, 1995; Whetton *et al.*, 1996) but amplify it in others. Changes in runoff from large Arctic rivers (especially in Siberia) and increased melt from the Greenland ice sheet and glaciers will to lead to more freshwater in the Arctic Ocean. This may further weaken the thermohaline circulation in the North Atlantic. With less overturning in the ocean, there will be a reduction of upwelling in temperate and subtropical latitudes. Wood *et al.* (1999) present simulations of present-day thermohaline circulation, using a coupled ocean-atmosphere climate model without flux adjustments. The model responds to forcing with increasing atmospheric concentrations of GHGs with a collapse of circulation and convection in the Labrador Sea. These changes are similar in two separate simulations with different rates of increase of CO_2 concentrations. Although various models give differing results, any changes in the thermohaline circulation will have profound consequences for marine biology and fisheries because of inevitable changes in habitat and nutrient supply. Perturbations caused by projected climate change, such as a marked increase in freshwater inputs in polar regions, may cause reorganization of the global ocean thermohaline circulation, leading to abrupt climate change (e.g., Manabe and Stouffer, 1993; Wright and Stocker, 1993; Stocker and Schmittner, 1997). Palaeoclimatic effects of past large freshwater inputs are widely discussed for the Atlantic (e.g., Broecker *et al.*, 1990; Rasmussen *et al.*, 1996; Bianchi and McCave, 1999) and for extra melt from the Antarctic ice sheet (Mikolajewicz, 1998). These studies show that with past climate changes, shifts from one circulation mode to another have caused large, and sometimes abrupt, regional climate changes. Although there is low confidence that such events will occur, the associated impacts would be substantial.
- *Greenhouse gases—reduced uptake by the Southern Ocean:* Projected climate change will alter vast areas of oceans, wetlands, and permafrost in the polar regions that act as major sources and sinks for atmospheric CO_2 and CH_4. Projected climate change will alter these features, thereby altering the exchange of these gases. Model results (Sarmiento and Le Quere, 1996) show that of all oceans, the Southern Ocean is likely to experience the greatest slowing in CO_2 uptake with climate change. Reduced downwelling also will limit the ability of the ocean to sequester anthropogenic CO_2 (Sarmiento *et al.*, 1998). Changing marine biology also must be considered. Using coupled climate model output under the IPCC IS92a GHG scenario, Matear and Hirst (1999) calculate that by the year 2100, there could be a reduction in cumulative oceanic uptake of carbon of 56 Gt. This reduced uptake is equivalent to a 4% yr^{-1} increase in CO_2 emissions for 1995–2100.
- *Greenhouse gases—emission by Arctic landscapes:* Whether the Arctic will be a net sink or a net source of CO_2 will depend largely on the magnitude and direction of hydrological changes and the rate of decomposition of exposed peat in response to temperature rise (Oechel *et al.*, 1993; McKane *et al.*, 1997a,b). Tundra ecosystems have large stores of nutrients and carbon bound in permafrost, soil, and microbial biomass and have low rates of CO_2 uptake as a result of low net primary production (Callaghan and Jonasson, 1995). The 25-year pattern of net CO_2 flux indicates that tundra in Alaska was a net sink during the cool, wet years of the 1970s; a net source of CO_2 during the warm, dry 1980s; and a net sink during the warm but less dry 1990s (Vourlitis and Oechel, 1999). These responses also may reflect a decrease in the rate of decomposition of soil organic matter because decay potential decreases with depth, as older, more recalcitrant carbon is exposed in soil profiles (Christensen *et al.*, 1998, 1999a,b). Sink activity will be altered by changes in soil water content, temperature, or longer term adjustment of biotic processes (Oechel and Billings, 1992; Shaver *et al.*, 1992; Oechel and Vourlitis, 1994; Chapin *et al.*, 1995; Jonasson, 1996; Nadelhoffer *et al.*, 1996; Rastetter *et al.*, 1996; Waelbroeck *et al.*, 1997). Increased frequency of disturbances such as fire in boreal forest also could contribute to increased seasonal amplitude of atmospheric CO_2 (Zimov *et al.*, 1999). CH_4 production also is related to the position of the water table in the active layer (Torn and Chapin, 1993; Vourlitis *et al.*, 1993; Johnson *et al.*, 1996). The gas is oxidized in unsaturated soils and in the uppermost layer of the soil water column (Gilmanov and Oechel, 1995; Rouse *et al.*, 1995; Tenhunen, 1996). Hence, the future magnitude of soil emissions of CH_4 and CO_2 will reflect the net outcome of anaerobic and aerobic processes. Thawing of permafrost has the potential to release considerable quantities of CH_4 and CO_2 (Fukuda, 1994; Michaelson *et al.*, 1996; Anisimov *et al.*, 1997; Goulden *et al.*, 1998; Bockheim *et al.*, 1999). Considering all of these effects, future warming is likely to further increase natural GHG emissions. Fluxes of these gases eventually may revert, however, to current levels after an initial pulse (Waelbroeck *et al.*, 1997).
- *Hydrates of greenhouse gases:* Substantial amounts of natural gas may be released to the atmosphere as a result of climate-induced destabilization of gas hydrates beneath the surface of the Earth. On the continents, stable gas hydrates can be found only at depths of several hundreds of meters, making it unlikely that they will be

released by climate change in the coming centuries. In the northern seas, gas hydrates may be deposited in the near-bottom zone, and their decomposition is likely to occur if deepwater temperature rises by even a few degrees. Because there are large uncertanties in the estimated amounts of the near-bottom gas hydrates, their role in providing positive feedback to the climate system cannot be evaluated with reasonable accuracy. There is evidence of methane hydrate destabilization and release with warming of coastal ocean bottomwater from other parts of the world (Kennett *et al.*, 2000).

16.3.2 Adaptation Potential and Vulnerability

Parts of the Arctic and Antarctic where water is close to its melting point are highly sensitive to climate change, rendering their biota and socioeconomic life particularly vulnerable. Adaptation to climate change will occur in natural polar ecosystems mainly through migration and changing mixes of species. This may cause some species to become threatened (e.g., walrus, seals, polar bears), whereas others may flourish (e.g., fish, penguins). Although such changes may be disruptive to many local ecological systems and particular species, the possibility remains that predicted climate change eventually will increase the overall productivity of natural systems in polar regions.

For people, successful future adaptation to change depends on technological advances, institutional arrangements, availability of financing, and information exchange. Stakeholders must be involved in studies from the beginning as well as in discussions of any adaptive and mitigative measures (Weller and Lange, 1999). For indigenous communities following traditional lifestyles, opportunities for adaptation to climate change appear to be limited. Long-term climate change, combined with other stresses, may cause the decline and eventual disappearance of communities. Technologically developed communities are likely to adapt quite readily to climate change by adopting altered modes of transport and by increased investment to take advantage of new commercial and trade opportunities.

Except in the Antarctic Peninsula, the Antarctic and the Southern Ocean probably will respond slowly to climate change; consequently, there will be less obvious impact in this region by 2100. Nevertheless, these areas are vulnerable because climate change could initiate millennial-scale processes with the potential to cause irreversible impacts on ice sheets, global ocean circulation, and sea-level rise. Antarctic drivers of sea-level rise, slowdown of the ocean thermohaline circulation, and changes in marine ecological habitats will continue for several centuries, long after GHG emissions are stabilized.

16.3.3. Development, Sustainability, and Equity

Distinctive patterns of development in the Arctic arise from the special nature of northern communities. The region is marked by decentralized administration and the presence of relict military establishments. The main forms of resource use are oil, gas, and mineral mining (e.g., lead, zinc, gold, diamonds), ecotourism, fishing, and traditional hunting and gathering by indigenous peoples. Further development of these resources is likely. Maintenance of existing infrastructure is likely to be more costly. Transportation may be affected as permafrost thaws and ice disappears. Waste disposal strategies also will have to change. Reduced sea ice will change strategic defense situations, especially for navies of the large powers flanking the Arctic. Sovereignty issues are of concern because of confusion over northern boundaries, the increased likelihood of territorial disputes as ice gives way to open water, and new northern sea routes create new trade patterns. Changes in sea ice and easier navigation may bring new policy initiatives, and improved sea access will greatly increase ecotourism. Overall, there will be increased human activity in the Arctic.

There are large regional differences across the Arctic in development, infrastructure, and ability of people to cope with climate change. Increasingly, Arctic communities are sustainable only with support from the south. Indigenous peoples are more sensitive to climate change than nonindigenous peoples. Their homelands and hunting habitats will be directly affected, and they cannot easily retreat to less affected areas. Some native peoples may be able to adapt, but probably at the expense of traditional lifestyles. Nonindigenous peoples also are vulnerable where links with the south are broken by changes in the physical environment and altered political circumstances. Their lifestyles require high capital investment, which will have to be maintained or even increased for them to be adaptable to climate change. With climate change, economies that rely on support from the south may become more expensive because of disrupted land-based transport, and this may not be sustainable. However, new transport opportunities, growing communities, and easier mining will create new wealth—but only for those who move away from traditional lifestyles.

In Antarctica, future use of the continent is governed by the Antarctic Treaty, and there are no permanent residents. With regard to policy issues, changes in the climate may mean less sea ice, easier access for ecotourism, and increased pressure on the environment. Sustaining the Antarctic's pristine nature may become more difficult.

16.3.4. Uncertainties and Risks

The polar regions will play a substantial role in driving global climate change through positive feedbacks to global warming. They also provide us with unparalleled records of that change. The most important uncertainties, risks, and thus targets for research are as follows:

- *Ocean thermohaline circulation:* Will downwelling in the North Atlantic and the Southern Ocean cease, causing a shutdown of the circulation of the global ocean? Or will this downwelling simply reduce and eventually recover with stabilization of GHGs? What is the role of changing input of freshwater from Arctic rivers?

- *Antarctic ice sheet:* What will be the contribution of the Antarctic ice sheets to global sea-level rise over the coming centuries? Although it is likely that the Antarctic ice sheet will provide some degree of mitigation of predicted sea-level rise, there is a small risk that the West Antarctic ice sheet (or portions of it) may retreat rapidly, causing a greater than predicted rise.
- *Marine biology:* What will be the response to predicted climate change of the structure of marine communities and the overall productivity of polar oceans? Uncertainty arises because existing biological models are not yet sufficiently developed to provide authoritative and quantitative estimates. There is some risk of unforeseen collapse of parts of the marine biological system, with consequent global effects—particularly on fisheries.
- *Arctic ice sheets, glaciers, and ice caps:* What is the current and future magnitude of freshwater input to the oceans from ice masses? What is the impact on global sea level and the thermohaline circulation on decadal to century time scales, and how can uncertainties in estimates be reduced?
- *Permafrost:* As ice-rich permafrost degrades, what will be the magnitude, spatial extent, and variability of its impacts? Will increases in the thickness of the active layer in currently cold, continuous permafrost be sufficient to cause widespread damage to human infrastructure? How important is soil organic carbon sequestered in the upper layer of permafrost in the context of the world carbon balance?
- *Arctic hydrology:* Will the balance between increased precipitation and evapotranspiration lead to a drier or wetter Arctic landscape? What will be the water balances of the large river basins that generate freshwater inflow to the Arctic Ocean?
- *Arctic sea ice:* Is it possible that with increased open water in the Arctic Ocean, summer sea ice in the Arctic eventually could disappear completely? The risk is that substantially more open water could generate large changes in regional climate for countries on the Arctic rim.
- *Fluxes of greenhouse gases:* What are the current and future fluxes of GHGs from polar oceans and landscapes? In particular, what is the likely future role of gas hydrates?
- *Stresses on human communities in the Arctic:* Can Arctic communities survive the combined stresses of globalization and marked changes in their local environments that may result from climate change? Traditional lifestyles will be threatened, but the communities that practice these lifestyles may be sufficiently resilient to cope with these changes, as they have in the recent and distant past.

References

Abdalati, W. and K. Steffen, 1997: Snowmelt on the Greenland ice sheet as derived from passive microwave satellite data. *Journal of Climate*, **10,** 165–175.

Ainley, D.G., C.R. Ribic, and W.R. Fraser, 1994: Ecological structure among migrant and resident seabirds of the Scotia-Weddell confluence region. *Journal of Animal Ecology*, **63,** 347–364.

Alekseev, G.V., Z.V. Bulatov, V.F. Zakharov, and V.P. Ivanov, 1997: Influx of very warm water in the Arctic basin. *Doklady Akademii Nauk SSSR*, **356,** 401–403 (in Russian).

Alexander, V., 1995: The influence of the structure and function of the marine food web on the dynamics of contaminants in Arctic Ocean ecosystems. *The Science of the Total Environment,* **160/161,** 593–603.

Alexandrov, Ye.I. and V.V. Maistrova, 1998: The comparison of the air temperature measurements for the polar regions. *The Antarctica,* **34,** 60–72.

Allard, M. and J.N. Kasper, 1998: Ice wedge activity and climate changes during the recent Holocene near Salluit, Northern Quebec, Canada. *Transactions of the American Geophyiscal Union*, **79,** F834.

Allard, M., B. Wang, and J.A. Pilon, 1995: Recent cooling along the southern shore of Hudson Strait Quebec, Canada, documented from permafrost temperature measurements. *Arctic and Alpine Research*, **27,** 157–166.

AMAP, 1998: *Arctic Pollution Issues* [Wilson, S.J., J.L. Murray, and H.P. Huntington (eds.)]. Arctic Monitoring and Assessment Programme, Oslo, Norway, 859 pp.

AMAP, 1997: *Arctic Pollution Issues: A State of the Arctic Environment Report*. Arctic Monitoring and Assessment Program, Oslo, Norway, 188 pp.

Anderson, W.P. and R.J. DiFrancesco, 1997: Potential impacts of climate warming on hydrocarbon production in the northern Mackenzie Basin. In: *Mackenzie Basin Impact Study Final Report* [Cohen, S.J. (ed.)]. Atmospheric Environment Service, Environment Canada, Downsview, Ontario, Canada, pp. 247–252.

Anisimov, O., 1999: Impacts of anthropogenic climate change on heating and air conditioning of buildings. *Meteorology and Hydrology*, **6,** 10–17 (in Russian).

Anisimov, O.A. and F.E. Nelson, 1997: Permafrost zonation and climate change: results from transient general circulation models. *Climatic Change*, **35,** 241–258.

Anisimov, O.A. and V.Y. Poljakov, 1999: Predicting changes of the air temperature in the first quarter of the 21st century. *Meteorology and Hydrology,* **2,** 25–31 (in Russian).

Anisimov, O.A., N.I. Shiklomanov, and F.E. Nelson, 1997: Effects of global warming on permafrost and active layer thickness: results from transient general circulation models. *Global and Planetary Change*, **61,** 61–77.

Are, F.E., 1999: The role of coastal retreat for sedimentation in the Laptev Sea. In: *Land-Ocean Systems in the Siberian Arctic: Dynamics and History* [Kassens, H., H.A. Bauch, I. Dmitrenko, H. Eicken, H.-W. Hubberten, M. Melles, J. Thiede, and L. Timokhov (eds.)]. Springer-Verlag, Berlin, Germany, pp. 287–295.

Arft, A.M., M.D. Walker, J. Gurevitch, J.M. Alatalo, M.S. Bret-Harte, M. Dale, M. Diemer, F. Gugerli, G.H.R. Henry, M.H. Jones, R.D. Hollister, I.S. Jonsdottir, K. Laine, E. Levesque, G.M. Marion, U. Molau, P. Molgaard, U. Nordenhall, V. Raszhivin, C.H. Robinson, G. Starr, A. Stenstrom, M. Stenstrom, O. Totland, P.L. Turner, L.J. Walker, P.J. Webber, J.M. Welker, and P.A. Wookey, 1999: Responses of tundra plants to experimental warming: meta-analysis of the international tundra experiment. *Ecological Monographs*, **69,** 491–511.

Arrigo, K.R., D. Worthen, A. Schnell, and M.P. Lizotte, 1998: Primary production in Southern Ocean waters. *Journal of Geophysical Research*, **103,** 15587–15600.

Arrigo, K.R., D.H. Robinson, D.L. Worthen, R.B. Dunbar, G.R. DiTullio, M. van Woert, and M.P. Lizotte, 1999: Phytoplankton community structure and the drawdown of nutrients and CO_2 in the Southern Ocean. *Science*, **283,** 365–367.

Azam, F., D.C. Smith, and J.T. Hollibaugh, 1991: The role of the microbial loop in Antarctic pelagic ecosystems. *Polar Research*, **10,** 239–243.

Babaluk, J.A., J.D. Reist, J.D. Johnson, and L. Johnson, 2000: First records of sockeye (*Oncorhynchus nerka*) and pink salmon (*O. gorbuscha*) from Banks Island and other records of Pacific salmon in Northwest Territories, Canada. *Arctic*, **53,** 161–164.

Barber, D.G. and S.V. Nghiem, 1999. The role of snow on the thermal dependence of backscatter over sea ice. *Journal of Geophysical Research,* **104(C11),** 25789–25803.

Barrie, L., R. Macdonald, T. Bidleman, M. Diamond, D. Gregor, R. Semkin, W. Strachan, M. Alaee, S. Backus, M. Bewers, C. Gobeil, C. Halsall, J. Hoff, A. Li, L. Lockhart, D. Mackay, D. Muir, J. Pudykiewicz, K. Reimer, J. Smith, G. Stern, W. Schroeder, R. Wagemann, F. Wania, and M. Yunker, 1997: Sources, occurrences and pathways. In: *Canadian Arctic Contaminants Assessment Report* [Jensen, J., K. Adare, and R. Shearer (eds.)]. Department of Indian and Northern Affairs, Government of Canada, Ottawa, Ontario, Canada, pp. 25–182.

Barry, R.G. and M.C. Serreze, 2000: Atmospheric component of the Arctic Ocean freshwater balance and their interannual variability. In: *The Freshwater Budget of the Arctic Ocean* [Lewis, E.L., E.P. Jones, P. Lemke, T.D. Prowse, and P. Wadhams (eds.)]. NATO Science Series, No. 2, Environmental Security, J. Kluwer Academic Publishers, Dordrecht, The Netherlands, Vol. 70, pp. 45–56.

Beamish, R.J., 1995: Response of anadromous fish to climate change in the North Pacific. In: *Human Ecology and Climate Change* [Peterson, D.L. and D.R. Johnson (eds.)]. Taylor & Francis, Washington, DC, USA, pp. 123–136.

Beltaos, S. and T.D. Prowse, 2000: Climate impacts on extreme ice jam events in Canadian rivers. In: *Contributions to IHP-V by Canadian Experts*. International Hydrological Programme, Technical Documents in Hydrology, United Nations Educational, Scientific, and Cultural Organization (UNESCO), Paris, France, Vol. 33, pp. 22–46.

Bentley, C.R., 1998: Rapid sea-level rise from a west Antarctic ice-sheet collapse: a short-term perspective. *Journal of Glaciology*, **44,** 157–163.

Berger, A.R. and W.J. Iams, 1996: *Geoindicators: Assessing Rapid Environmental Change in EarthSystems*. A.A. Balkema, Rotterdam, The Netherlands, 466 pp. Available online at http//www.gcrio.org/geo/title.html.

Bergstrom, D.M. and S.L. Chown, 1999: Life at the front: history of ecology and change on Southern Ocean Islands. *Trends in Ecology and Evolution*, **14 (12),** 472–477.

Berkes, F., 1998: Indigenous knowledge and resource management systems in the Canadian sub-Arctic. In: *Linking Social and Ecological Systems* [Berkes, F., C. Folke, and J. Colding (eds.)]. Cambridge University Press, Cambridge, United Kingdom and New York, NY, USA, pp. 98–128.

Bernes, C., 1996: *The Nordic Arctic Environment: Unspoilt, Exploited, Polluted*? The Nordic Council of Ministers, Copenhagen, Denmark, 140 pp.

BESIS, 1999: *The Impacts of Global Climate Change in the Arctic Regions: An Initial Assessment*. Bering Sea Impact Study, International Arctic Science Committee, Oslo, Norway, 41 pp.

BESIS, 1997: *The Impacts of Global Climate Change in the Bering Sea Region: An Assessment Conducted by the International Arctic Science Committee Under Its Bering Sea Impact Study (BESIS)*. International Arctic Science Committee, Oslo, Norway, 41 pp.

Bianchi, G.G. and I N. McCave, 1999: Holocene periodicity in north Atlantic climate and deep ocean flow south of Iceland. *Nature*, **397,** 515–517.

Billings, W.D., 1987: Carbon balance of Alaskan tundra and taiga ecosystems: past, present, and future. *Quaternary Science Reviews*, **6,** 165–177.

Bindoff, N.L. and T.J. McDougall, 2000: Decadal changes along an Indian Ocean section at 32°S and their interpretation. *Journal of Physical Oceanography*, 30(6), 1207–1222.

Bindschadler, R.A., 1998: Future of the west Antarctic ice sheet. *Science,* **282,** 428–429.

Bird, D.F. and J. Kalff, 1984: Empirical relationships between bacterial abundance and chlorophyll concentration in fresh and marine waters. *Canadian Journal of Fisheries and Aquatic Science,* **41,** 1015–1023.

Bird, D.F. and D. Karl, 1999: Uncoupling of bacteria and phytoplankton during the austral spring bloom in Gerlache Strait, Antarctic Peninsula. *Aquatic Microbial Ecology,* **19,** 13–27.

Bjorgo, E., O.M. Johannessen, and M.W. Miles, 1997: Analysis of merged SMMR SSMI time series of Arctic and Antarctic sea ice parameters, 1978–1995. *Geophysical Research Letters*, **24,** 413–416.

Björn, L.O., T.V. Callagan, C. Gehrke, D. Gwynn-Jones, B. Holmgren, U. Johanson, and M. Sonesson, 1997: Effects of enhanced UV-B radiation on sub-Arctic vegetation. In: *Ecology of Arctic Environments* [Woodin, S.J. and M. Marquiss (eds.)]. Blackwell Science, Oxford, United Kingdom, pp. 241–253.

Björn, L.O., T.V. Callagan, C. Gehrke, D. Gwynn-Jones, J.A. Lee, U. Johanson, M. Sonesson, and N.D. Buck, 1999: Effects of ozone depletion and increased ultraviolet-B radiation on northern vegetation. *Polar Research*, **18,** 331–337.

Bockheim, J.G., 1995: Permafrost distribution in the southern circumpolar region and its relation to the environment: a review and recommendations for further research. *Permafrost and Periglacial Processes*, **6,** 27–45.

Bockheim, J.G, and C. Tarnocai, 1998: Recognition of cryoturbation for classifying permafrost-affected soils. *Geoderma*, **81,** 281–293.

Bockheim, J.G, L.R. Everett, K.M. Hinkel, F.E. Nelson, and J. Brown, 1999: Soil organic carbon storage and distribution in Arctic tundra, Barrow, Alaska. *Soil Science Society of America Journal*, **63,** 934–940.

Borzenkova, I.I., 1999a: Environmental indicators of recent global warming. In: *Environmental Indices* [Pykh, Y.A., A. Yuri, D.E. Hyatt, and R.J.M. Lenz (eds.)]. EOLSS Publishing Company, Baldwin House, London, United Kingdom, pp. 455–465.

Borzenkova, I.I., 1999b: About natural indicators of the present global warming. *Meteorology and Hydrology*, **6,** 98–116 (in Russian).

Boyd, P.W., A.J. Watson, C.S. Law, E.R. Abraham, T. Trull, R. Murdoch, D.C.E. Bakker, A.R. Bowie, K.O. Buesseler, H. Chang, M. Charette, P. Croot, K. Downing, R. Frew, M. Gall, M. Hadfield, J. Hall, M. Harvey, G. Jameson, J. LaRoche, M. Liddicoat, R. Ling, M.T. Maldonado, R.M. McKay, and S. Nodder, 2000: A mesoscale phytoplankton bloom in the polar Southern Ocean stimulated by iron fertilization. *Nature,* **407,** 695–702.

Bradley, R.S. and P.D. Jones, 1993: Little Ice Age summer temperature variations, their nature and relevance to recent global warming trends. *The Holocene*, **3,** 367–376.

Brander, K.M., 1996: Effects of climate change on cod (*Gadus morhua*) stocks. In: *Global Warming: Implications for Freshwater and Marine Fish* [Wood, C.M. and D.G. McDonald (eds.)]. Cambridge University Press, Cambridge, United Kingdom and New York, NY, USA, pp. 255–278.

Briffa, K.P., P.D. Jons, F.N. Schweigruber, and S.G. Shiyatov, 1995: Unusual twentieth-century summer warmth in a 1000-year temperature record from Siberia. *Nature*, **376,** 156–159.

Brody, H., 1991: *Maps and Dreams: Indians and the British Columbia Frontier*. Douglas and McIntyre Press, Toronto, Ontario, Canada, 297 pp.

Broecker, W.S., G. Bond, and M.A. Klas, 1990: A salt oscillator in the glacial Atlantic? 1: the concept. *Paleoceanography,* **5,** 469–477.

Broecker, W.S., D.M. Peteet, and D. Rind, 1985: Does the ocean-atmosphere system have more than one stable mode of operation. *Nature*, **315,** 21–26.

Bromwich, D., L. Zhong, and A.N. Rogers, 1998: Winter atmospheric forcing of the Ross Sea polynya. In: *Ocean, Ice and Atmosphere, Interactions at the Antarctic Continental Margin* [Jacobs, S.S. and R.F. Weiss (eds.)]. *American Geophysical Union, Antarctic Research Series*, **75,** 101–133.

Brown, J., 1997: Circumpolar active layer monitoring program. *Frozen Ground*, **21,** 22–23.

Brown, J., O.J.J. Ferrians, J.A. Heginbottom, and E.S. Melnikov, 1997: *International Permafrost Association Circum Arctic Map of Permafrost and Ground Ice Conditions*. U.S. Geological Survey CircumPacific Map Series, Map CP 45, Scale 1:10,000,000, Washington, DC, USA.

Brown, J., M.M. Burgess, A. Pavlov, V. Romanovsky, and S. Smith, 2000: The Global Terrestrial Network for Permafrost (GTN-P): a progress report. In: *Rhythms of Natural Processes in the Earth Cryosphere*. Russian Academy of Sciences, Pushchino, Russia, pp. 203–204.

Brown, R.D., 2000: Northern Hemisphere snow cover variability and change, 1915–1997. *Journal of Climate*, **13,** 2339–2355.

Brown, R.D. and B.E. Goodison, 1996: Interannual variability in reconstructed Canadian Snow Cover, 1915–1992. *Journal of Climate*, **9,** 1299–1318.

Buch, E., S.A. Horsted, and H. Hovgard, 1994: Fluctuations in the occurrence of cod in Greenland waters and their possible causes. *International Council for the Exploration of the Sea (ICES) Marine Science Symposia*, **198,** 158–174.

Budd, W.F. and X.Wu, 1998: Modeling long term global and Antarctic changes resulting from increased greenhouse gases. In: *Coupled Climate Modelling* [Meighen, P.J. (ed.)]. Bureau of Meteorology, Canberra, Australia, Research Report, **69,** 71–74.

Burgess, M.M., D.T. Desrochers, and R. Saunders, 2000a: Potential changes in thaw depth and thaw settlement for three locations in the Mackenzie Valley. In: *The Physical Environment of the Mackenzie Valley: a Baseline for the Assessment of Environmental Change* [Dyke, L.D. and G.R. Brooks, (eds.)]. *Geological Survey of Canada Bulletin,* **547,** 187–195.

Burgess, M.M., S.L. Smith, J. Brown, V. Romanovsky, and K. Hinkel, 2000b: The Global Terrestrial Network for Permafrost (GTNet-P): permafrost monitoring contributing to global climate observations. In: *Current Research 2000-E14*. Geological Survey of Canada, Ottawa, Ontario, Canada, 8 pp. Available online at http://www.nrcan.gc.ca/gsc/bookstore.

Burn, C.R., 1998: Field investigations of permafrost and climatic change in northwest North America. In: *Proceedings of the Seventh International Conference on Permafrost, Yellowknife, NWT, 23-27 June 1998* [Lewkowicz, A.G. and M. Allard (eds.)]. Collection Nordiana, No. 57, Centre d'Etudes Nordiques, Université Laval, Quebec, Canada, pp. 107–120.

Burn, C.R., 1997: Cryostratigraphy, paleogeography, and climate change during the early Holocene warm interval, western Arctic coast, Canada. *Canadian Journal of Earth Sciences*, **34,** 912–935.

Cai, W. and P.G. Baines, 1996: Interactions between thermohaline- and wind-driven circulations and their relevance to the dynamics of the Antarctic Circumpolar Current, in a coarse-resolution global ocean general circulation model. *Journal of Geophysical Research*, **101,** 14073–14093.

Cai, W. and H.B. Gordon, 1998: Transient responses of the CSIRO climate model to two different rates of CO_2 increase. *Climate Dynamics*, **14,** 503–506.

Callaghan, T.V. and S. Jonasson, 1995: Implications for changes in arctic plant biodiversity from environmental manipulation experiments. In: *Arctic and Alpine Biodiversity: Patterns, Causes, and Ecosystem Consequences* [Chapin, F.S. III and C.H. Körner (eds.)]. Springer-Verlag, Heidelberg, Germany, pp. 151–164.

Callaghan, T.V., C. Körner, S.E. Lee, and J.H.C. Cornelison, 1998: Part 1: scenarios for ecosystem responses to global change. In: *Global Change in Europe's Cold Regions* [Heal, O.W., T.V. Callaghan, J.H.C. Cornelissen, C. Körner, and S.E. Lee (eds.)]. European Commission Ecosystems Research Report, 27, L-2985, Luxembourg, pp. 11–63.

Carmack, E.C., 2000: Review of the Arctic Oceans freshwater budget: sources, storage and export. In: *The Freshwater Budget of the Arctic Ocean* [Lewis, E.L, E.P. Jones, P. Lemke, T.D. Prowse, and P. Wadhams (eds.)]. Kluwer Academic Publishers, Dordrecht, The Netherlands, pp. 91–126.

Carmack, E.C., R.W. Macdonald, R.G. Perkin, F.A. McLaughlin, and R.J. Pearson, 1995: Evidence for warming of Atlantic water in the southern Canadian basin of the Arctic Ocean: results from the Larsen-93 exhibition. *Geophysical Research Letters*, **22,** 1061–1064.

Carter, T.R., M. Hulme, J.F. Crossley, S. Malyshev, M.G. New, M.E. Schlesinger, and H. Tuomenvirta, 2000: *Climate Change in the 21st Century - Interim Characterizations based on the New IPCC Emissions Scenarios*. The Finnish Environment 433, Finnish Environment Institute, Helsinki, 148 pp.

Cavalieri, D.J., P. Gloersen, C.L. Parkinson, J.C. Comiso, and H.J. Zwally, 1997: Observed hemispheric asymmetry in global sea ice changes. *Science*, **278,** 1104–1106.

Chapin, F.S. III and G.R. Shaver, 1985: Individualistic growth response of tundra plant species to environmental manipulations in the field. *Ecology*, **66,** 564–576.

Chapin, F.S. III and A.M. Starfield, 1997: Time lags and novel ecosystems in response to transient climatic change in arctic Alaska. *Climatic Change*, **35,** 449–461.

Chapin, F.S. III, J.P. McFadden, and S.E. Hobbie, 1997: The role of arctic vegetation in ecosystem and global processes. In: *Ecology of Arctic Environments* [Woodin, S.J. and M. Marquiss (eds.)]. Blackwell Science, Oxford, United Kingdom, pp. 97–112.

Chapin, F.S. III, M.S. Bret-Harte, S.E. Hobbie, and H. Zhong, 1996: Plant functional types as predictors of the transient response of arctic vegetation to global change. *Journal of Vegetation Science*, **7,** 347–358.

Chapin, F.S. III, G.R. Shaver, A.E. Giblin, K.G. Nadelhoffer, and J.A. Laundre, 1995: Response of arctic tundra to experimental and observed changes in climate. *Ecology*, **66,** 564–576.

Chapman, W.L. and J.E. Walsh, 1993: Recent variations of sea ice and air temperature in high latitudes. *Bulletin of the American Meteorological Society*, **74,** 33–47.

Chinn, T.J., 1998: Recent fluctuations of the Dry Valleys glaciers, McMurdo Sound, Antarctica. *Annals of Glaciology*, **27,** 119–125.

Christensen, T.R., S. Jonasson, T.V. Callaghan, and M. Havstrom, 1999a: On the poterntial CO_2 releases from tundra soils in a changing climate. *Applied Soil Ecology*, **11,** 127–134.

Christensen, T.R., S. Jonasson, T.V. Callaghan, M. Harström, and F.R. Livens, 1999b: Carbon cycling and methane exchange in Eurasian tundra ecosystems. *Ambio*, **28,** 239–244.

Christensen, T.R., S. Jonasson, A. Michelsen, T.V. Callaghan, and M. Harström, 1998: Environmental controls on soil respiration in the Eurasian and Greenlandic Arctic. *Journal of Geophysical Research*, **103(D22),** 29015–29021.

Clow, G.D., R.W. Saltus, A.H. Lachenbruch, and M.C. Brewer, 1998: Arctic Alaska climate change estimated from borehole temperature: past, present, future. *EOS, Transactions, American Geophysical Union*, **79,** F883.

Coale, K.H., K.S. Johnson, S.E. Fitzwater, R.M. Gordon, S. Tanner, F.P. Chavez, L. Ferioli, C. Sakamoto, P. Rogers, F. Millero, P. Steinberg, P. Nightingale, D. Cooper, W.P. Cochlan, M.R. Landry, J. Constantinou, G. Rollwagen, A. Trasvina, and R. Kudela, 1996: A massive phytoplankton bloom induced by an ecosystem scale iron fertilisation experiment in the equatorial Pacific Ocean. *Nature*, **383,** 495–501.

Cohen, S.J., 1997a: What if and so what in northwest Canada: could climate change make a difference to the future of the Mackenzie Basin? *Arctic*, **50,** 293–307.

Cohen, S.J., 1997b: Scientist-stake holder collaboration in integrated assessment of climate change: lessons from a case study of northwest Canada. *Environmental Modelling and Assessment*, **2,** 281–293.

Cole, J.J., S. Findlay, and M.L. Pace, 1988: Bacterial production in fresh and saltwater ecosystems: a cross-system overview. *Marine Ecology Progress Series*, **43,** 1–10.

Comiso, J.C., 1999: Variability and trends in Antarctic surface temperatures from in situ and satellite infrared measurements. *Journal of Climate*, **13(10),** 1674–1696.

Connolley, W.M. and S.P. O'Farrell, 1998: Comparison of warming trends over the last century around Antarctica from three couple models. *Annals of Glaciology*, **27,** 565–570.

Couper, A., 1989: *The Times Atlas and Encyclopaedia of the Sea*. Harper & Row, New York, NY, USA, 272 pp.

Crawford, R.M.M. (ed.), 1997: *Disturbance and Recovery in Arctic Lands*. NATO ASI Series 2: Environment, Vol. 25, J. Kluwer Academic Publishers, Dordrecht, The Netherlands, 621 pp.

Crete, M. and J. Huot, 1993: Regulation of a large herd of migratory caribou: summer nutrition affects calf growth and body reserves of dams. *Canadian Journal of Zoology*, **71,** 2291–2296.

Croasdale, K.R., 1993: Climate change impacts on northern offshore petroleum operation. In: *Impacts of Climate Change on Resource Management in the North* [Wall, G. (ed.)]. Occasional Paper No. 16, Department of Geography, University of Waterloo, pp. 175–184.

Cubasch, U.K. Hasselmann, H. Hock, E. Maierreimer, U. Mikolajewicz, B.D. Santer, and R. Sausen, 1992: Time-dependent greenhouse warming computations with a coupled ocean-atmosphere model. *Climate Dynamics*, **8,** 55–69.

Cuffey, K.M. and S.J. Marshall, 2000: Substantial contribution to sea-level rise during the last interglacial from the Greenland ice-sheet. *Nature*, **404,** 591–594.

Curran, M.A.J., T.D. van Ommen, and V. Morgan, 1998: Seasonal characteristics of the major ions in the high accumulation DSS ice core Law Dome, Antarctica. *Annals of Glaciology*, **27,** 385–390.

Dallimore, S.R., S. Wolfe, and S.M. Solomon, 1996: Influence of ground ice and permafrost on coastal evolution, Richards Island, Beaufort Sea Coast, NWT, Canada. *Canadian Journal of Earth Sciences*, **33,** 664–675.

Danks, H.V., 1992: Arctic insects as indicators of environmental change. *Arctic*, **45,** 159–166.

Davidson, A.T, 1998: The impact of UVB radiation on marine plankton. *Mutation Research—Fundamental and Molecular Mechanisms of Mutagenesis*, **422,** 119–129.

de Baar, H.J., J.T.M. de Jong, D.C.E. Bakker, B.M. Loscher, C. Veth, U. Bathmann, and V. Smetaceck, 1995: Importance of iron for phytoplankton blooms and carbon dioxide drawdown in the Southern Ocean. *Nature*, **373,** 412–415.

de la Mare, W.K., 1997: Abrupt mid-twentieth century decline in Antarctic sea ice extent from whaling records. *Nature*, **389,** 57–61.

de March, B.G.D., C.A. de Wit, and D.C.G. Muir, 1998: Persistent organic pollutants. In: *Arctic Monitoring and Assessment Report: Arctic Pollution Issues* [Wilson, S.J., J.L. Murray, and H.P. Huntington (eds.)]. Arctic Monitoring and Assessment Programme (AMAP), Oslo, Norway, pp. 183–371.

Delcourt, P.A. and H.R. Delcourt, 1987: Late-Quaternary dynamics of temperate forests: applications of paleocology to issues of global environment change. *Quaternary Science Reviews*, **6,** 129–146.

Dickson, B., L. Anderson, M. Bergmann, R. Colony, and P. Malkki, 1999: Arctic-subarctic linkages to form a new focus of AOSB activity. *News from the Arctic Ocean Science Board*, **3,** 1–5.

Douglas, M.S.V., J.P. Smol, and W. Blake Jr., 1994: Marked post-18th century environmental change in high Arctic ecosystems. *Science*, **266,** 416–419.

Dowdeswell, J., J.O. Hagen, H. Bjornsson, A. Glazovsky, P. Holmlund, J. Jania, E. Josberger, R. Koerner, S. Ommanney, and B. Thomas, 1997: The mass balance of circum Arctic glaciers and recent climate change. *Quaternary Research*, **48,** 1–14.

Duerden, F., 1992: A critical look at sustainable development in the Canadian North. *Arctic*, **45,** 219–255.

Dyke, L.D., 2000: Stability of permafrost slopes in the Mackenzie Valley. In: *The Physical Environment of the Mackenzie Valley: A Baseline for the Assessment of Environmental Change* [Dyke, L.D. and G.R. Brooks (eds.)]. *Geological Survey of Canada Bulletin*, **547,** 161–169.

Everett, J.T and B.B. Fitzharris, 1998: The Arctic and the Antarctic. In: *The Regional Impacts of Climate Change. An Assessment of Vulnerability. A Special Report of IPCC Working Group II for the Intergovernmental Panel of Climate Change* [Watson, R.T, M.C. Zinyowera, R.H. Moss, and D.J. Dokken (eds.)]. Cambridge University Press, Cambridge, United Kingdom and New York, NY, USA, pp. 85–103.

Fedorov, A.N., 1996: Effects of recent climate change on permafrost landscapes in central Sakha. *Polar Geography*, **20,** 99–108.

Fedoulov, P.P., E. Murphy, and K.E. Shulgovsky, 1996: Environment krill relations in the South Georgia marine system. *CCAMLR Science*, **3,** 13–30.

Fitzharris, B.B., 1996: The cryosphere: Changes and their impacts. In: *Climate Change 1995: Impacts, Adaptions, and Mitigation of Climate Change: Scientific- Technical Analyses. Contribution of Working Group II to the Second Assessment Report for the Intergovernmental Panel on Climate Change* [Watson, R.T., M.C. Zinyowera, and R.H. Moss (eds.)]. Cambridge University Press, Cambridge, United Kingdom and New York, NY, USA, 880 pp.

Flato, G.M. and R.D. Brown, 1996: Variability and climate sensitivity of landfast Arctic sea ice. *Journal of Geophysical Research*, **101,** 25767–25777.

Fleming, R.A. and J.A. Volney, 1995: Effects of climate change on insect defoliator population processes in Canada's boreal forest: some plausible scenarios. *Water, Air, and Soil Pollution*, **82,** 445–454.

Forbes, D.L. and R.B. Taylor, 1994: Ice in the shore zone and the geomorphology of cold coasts. *Progress in Physical Geography*, **18,** 59–89.

Fowber, J.A. and R.I. Lewis Smith, 1994: Rapid population increases in native vascular plants in the Argentine Islands, Antarctic Peninsula. *Arctic and Alpine Research*, **26,** 290–296.

Fox, A.J. and A.P.R. Cooper, 1998: Climate-change indicators from archival aerial photography of the Antarctic Peninsula. *Annals of Glaciology,* **27,** 636–642.

Fraser, W.R.,W.Z. Trivelpiece, D.G. Ainley, and S.G. Trivelpiece, 1992: Increases in Antarctic penguin populations: reduced competition with whales or a loss of sea ice due to environmental warming? *Polar Biology*, **11,** 525–531.

Freese, C.H., 2000: *The Consumptive Use of Wild Species in the Arctic: Challenges and Opportunities for Ecological Sustainability.* Report submitted to World Wildlife Fund, Arctic Programme, Toronto, Ontario, Canada, 145 pp.

Fukuda, M., 1994: Methane flux from thawing Siberian permafrost (ice complexes) results from field observations. *EOS, Transactions, American Geophysical Union*, **75,** 86.

Gates, W.L., J.F.B. Mitchell, G.J. Boer, U.K. Cusasch, and V.P. Meleshko, 1992: Climate modelling, climate prediction, and model validation. In: *Climate Change 1992: The Supplementary Report to the IPCC Scientific Assessment* [Houghton, J.T., B.A. Callander, and S.K. Varney (eds.)]. Intergovernmental Panel on Climate Change, Cambridge University Press, Cambridge, United Kingdom and New York, NY, USA, pp. 97–135.

Gehrke, C., U. Johanson, T.V. Callaghan, D. Chadwick, and C.H. Robinson, 1995: The impact of enhanced ultraviolet B radiation on litter quality and decomposition processes in *Vaccinium* leaves from the Subarctic. *Oikos*, **72,** 213–222.

Gehrke, C., U. Johanson, D. Gwynn Jones, L.O. Bjorn, T.V. Callaghan, and J.A. Lee, 1996: Effects of enhanced ultraviolet B radiation on terrestrial subarctic ecosystems and implications for interactions with increased atmospheric CO_2. *Ecological Bulletin*, **45,** 192–203.

Georgievskii, V.Y., 1998: On global climate warming effects on water resources. In: *Water: a looming crisis*? Technical Documents in Hydrology, No. 18, UNESCO, Paris, pp. 37–46.

Georgievskii, V.Y., A.V. Ezhov, A.L. Shalygin, I.A. Shiklomanov, and A.I. Shiklomanov, 1996: Assessment of the effect of possible climate changes on hydrological regime and water resources of rivers in the former USSR. *Meteorology and Hydrology*, **11,** 66–74 (in Russian).

Gibson, M.A. and S.A. Schullinger, 1998: *Answers from the Ice Edge: The Consequences of Climate Change on Life in the Bering and Chukchi Seas*. Greenpeace Arctic Network, Anchorage, AK, USA, pp. 32.

Gignac, L.D. and D.H. Vitt, 1994: Responses of northern peatlands to climatic change, effects on bryophytes. *Journal of the Hattori Botanical Laboratory*, **75,** 119–132.

Gilmanov, T.G. and W.C. Oechel, 1995: New estimates of organic matter reserves and net primary productivity of the North American tundra ecosystems. *Journal of Biogeography*, **22,** 723–741.

Gordon, H.B. and S.P. O'Farrell, 1997: Transient climate change in the CSIRO coupled model with dynamic sea ice. *Monthly Weather Review*, **25,** 875–907.

Gorham, E., 1991: Northern peatlands: role in the carbon cycle and probable responses to climate warming. *Ecological Applications*, **1,** 182–195.

Goulden, M.L., S.C. Wofsy, J.W. Harden, S.E. Trumbore, P.M. Crill, S.T. Gower, T. Fries, B.C. Daube, S M. Fan, D.J. Sutton, A. Bazzaz, and J.W. Munger, 1998: Sensitivity of boreal forest carbon balance to soil thaw. *Science*, **279,** 214–217.

Grabs, W.E., F. Portmann, and T. De Couet, 2000: Discharge observation networks in Arctic regions: computation of the river runoff into the Arctic Ocean, its seasonality and variability. In: *The Freshwater Budget of the Arctic Ocean* [Lewis, E.L., E.P. Jones, P. Lemke, T.D. Prowse, and P. Wadhams (eds.)]. NATO Science Series, No. 2, Environmental Security, J. Kluwer Academic Publishers, Dordrecht, The Netherlands, Vol. 70, pp. 249–267.

Greco, S., R.H. Moss, D.Viner, and R. Jenne, 1994: *Climate Scenarios and Socioeconomic Projections for IPCC WGII Assessment*. Consortium for International Earth Science Informations Network, Washington, DC, USA, 290 pp.

Gregor, D.J., H. Loeng, and L. Barrie, 1998: The influence of physical and chemical processes on contaminant transport into and within the Arctic. In: *Arctic Monitoring and Assessment Report: Arctic Pollution Issues* [Wilson, S.J., J.L. Murray, and H.P. Huntington (eds.)]. Arctic Monitoring and Assessment Programme, Oslo, Norway, pp. 25–116.

Groisman, P.Y., V.V. Koknaeva, T.A. Belokrylova, and T.R. Karl, 1991: Overcoming biases of precipitation measurments: a history of the USSR experience. *Bulletin of the American Meterological Society*, **72,** 1725–1733.

Groisman, P.Y. and D.R. Easterling, 1994: Variability and trends of precipitation and snowfall over the United States and Canada. *Journal of Climate*, **7,** 184–205.

Groisman, P.Y., T.R. Karl, and R.W. Knight, 1994: Observed impact of snow cover on the heat balance and the rise of continental spring temperatures. *Science*, **263,** 198–200.

Grosfeld, K. and R. Gerdes, 1998: Circulation beneath the Filchner Ice Shelf, Antarctica, and its sensitivity to changes in the oceanic environment: a case study. *Annals of Glaciology*, **27,** 99–104.

Gunn, A., 1995: Responses of Arctic ungulates to climate change. In: *Human Ecology and Climate Change* [Peterson, D.L. and D.R. Johnson (eds.)]. Taylor & Francis, Washington, DC, USA, pp. 89–104.

Gunn, A. and T. Skogland, 1997: Responses of caribou and reindeer to global warming. *Ecological Studies*, **124,** 189–200.

Hagemann, S. and L. Dümenil, 1998: A parameterization of the lateral waterflow for the global scale. *Climate Dynamics*, **14,** 17–31.

Halsey, L.A., D.H. Vitt, and S.C. Zoltai, 1995: Disequilibrium response of permafrost in boreal continental western Canada to climate change. *Climate Change*, **30,** 57–73.

Hanesiak, J.M., D.G. Barber, and G.M. Flato, 1999: The role of diurnal processes in the seasonal evolution of sea ice and its snow cover. *Journal of Geophysical Research (Oceans),* **104(C6),** 13593–13604.

Hanssen-Bauer, I. and E.J. Forland, 1998: Long term trends in precipitation and temperature in the Norwegian Arctic: can they be explained by changes in atmospheric circulation patterns? *Climate Research*, **10,** 143–153.

Harangozo, S.A., S.R. Colwell, and J.C. King, 1997: An analysis of a 34-year air temperature record from Fossil Bluff (71°S, 68°W), Antarctica. *Antarctic Science*, **9,** 355–363.

Havström, M., T.V. Callaghan, and S. Jonasson, 1993: Differential growth responses of *Cassiope tetragona*, an arctic dwarf shrub, to environmental perturbations among three contrasting high and sub-arctic sites. *Oikos*, **66,** 389–402.

Heal, O.W., 1998: Executive summary: introduction, general summary and research priorities. In: *Global Change in Europe's Cold Regions* [Heal, O.W., T.V. Callaghan, J.H.C. Cornelissen, C.H. Korner, and S.E. Lee (eds.)]. European Commission, Ecosystems Research Report 27, Luxembourg, pp. 66–78.

Henry, G.H.R. and U. Molau, 1997: Tundra plants and climate change: the International Tundra Experiment (ITEX). *Global Change Biology*, **3,** 1–9.

Hinkel, K.M., S.I. Outcalt, and A.E. Taylor, 1997: Seasonal patterns of coupled flow in the active layer at three sites in northwest North America. *Canadian Journal of Earth Sciences*, **34,** 667–678.

Hirst, A.C., 1999: The Southern Ocean response to global warming in the CSIRO coupled ocean atmosphere model: special issue on global change. *Environmental Modelling and Software*, **14,** 227–242.

Hobbie, J.E., B.J. Petersen, N. Bettez, L. Deegan, W.J. O'Brian, G.W. Kling, G.W. Kipphut, W.B. Bowden, and A.E. Hershey, 1999: Impact of global change on the biogeochemistry and ecology of an Arctic freshwater system. *Polar Research*, **18,** 207–214.

Hønneland, G., A.K. Jørgensen, and K. Kovacs, 1999: *Barents Sea Ecoregion: Reconnaissance Report*. Report to the World Wildlife Fund for Nature, Oslo, Norway, 32 pp.

Iacozza, J. and D.G. Barber, 1999: Modelling the distribution of snow on sea ice using variograms. *Atmosphere-Oceans,* **37,** 21–51.

IPCC, 2000: *Emissions Scenarios. A Special Report of Working Group III of the Intergovernmental Panel on Climate Change* [Nakicenovic, N., J. Alcamo, G. Davis, B. de Vries, J. Fenham, S. Gaffin, K. Gregory, A. Grubler, T.Y. Jung, T. Kram, E.L. La Rovere, L. Michaelis, S. Mori, T. Morita, W. Pepper, H. Pitcher, L. Price, K. Raihi, A. Roehrl, H.-H. Rogner, A. Sankovski, M. Schlesinger, P. Shukla, S. Smith, R. Swart, S. van Rooijen, N. Victor, and Z. Dadi (eds.)]. Cambridge University Press, Cambridge, United Kingdom and New York, NY, USA, 599 pp.

Jacka, T.H. and W.F. Budd, 1998: Detection of temperature and sea ice extent changes in the Antarctic and Southern Ocean, 1949–96. *Annals of Glaciology*, **27,** 553–559.

Jacobs, S.S., H.H. Helmer, C.S.M. Doake, A. Jenkins, and R.M. Frolich, 1992: Melting of ice shelves and the mass balance of Antarctica. *Journal of Glaciology*, **38,** 375–387.

Jacobs, S.S. and J.C. Comiso, 1997: Climate variability in the Amundsen and Bellingshausen Seas. *Journal of Climate*, **10,** 697–709.

Jacobs, S.S. and C.F. Giulivi, 1998: Interannual ocean and ice variability in the Ross Sea. *American Geophysical Union, Antarctic Research Series*, **75,** 135–150.

Jania, J. and J.O. Hagen (eds.), 1996: *Mass Balance Review of Arctic Glaciers*. Report No. 5, International Arctic Science Committee (IASC), Oslo, Norway, 95 pp.

Jenkins, A., D.G. Vaughan, S.S. Jacobs, H.H. Hellmer, and J.R. Keys, 1997: Glaciological and oceanographic evidence of high melt rates beneath Pine Island Glacier, West Antarctica. *Journal of Glaciology,* **43,** 114–121.

Johannessen, O.M., E.V. Shalina, and M.W. Miles, 1999: Satellite evidence for an arctic sea ice cover in transformation. *Science*, **286,** 1937–1939.

Johnson, G.C. and A. Orsi, 1997: Southwest Pacific Ocean water-mass changes between 1968/69 and 1990/91. *Journal of Climate*, **10,** 301–316.

Johnson, L.C., G.R. Shaver, A.E. Giblin, K.L. Nadelhoffer, E.R. Rastetter, J.A. Laundre, and G.L. Murray, 1996: Effects of drainage and temperature on carbon balance of tussock tundra microcosms. *Oecologia*, **108,** 737–748.

Jonasson, S., 1996: Buffering of Arctic plant responses to a changing climate. In: *Global Change and Arctic Terrestrial Ecosystems* [Oechel, W.C., T. Callaghan, T. Gilmanov, J.I. Holten, B. Maxwell, U. Molau, and B. Svein Bjornsson (eds.)]. Ecological Studies, Vol. 124, Springer-Verlag, Berlin, Germany, pp. 365–380.

Jonasson, S., 1992: Plant responses to fertilization and species removal in tundra related to community structure and clonality. *Oikos*, **63,** 420–429.

Jones, P.D., M. New, D.E. Parker, S. Martin, and I.G. Rigor, 1999: Surface air temperature and its changes over the past 150 years. *Reviews of Geophysics*, **37,** 173–199.

Juday, G.P., 1996: Boreal forests (Taiga). In: *The Biosphere and Concepts of Ecology. Encyclopedia Britannica, Volume 14*. Encyclopedia Britannica, Inc., Chicago, IL, USA, 15th ed., pp. 1210–1216.

Kane, D.L., L.D. Hinzman, M.K. Woo, and K.R. Everett, 1992: Arctic hydrology and climate change physiological ecology of Arctic plants. In: *Arctic Ecosystems in a Changing Climate: An Ecophysiological Perspective* [Chapin, F.S. III, R.L. Jefferies, J. Reynolds, G. Shaver, and J. Svoboda (eds.)]. Academic Press, New York, NY, USA, pp. 35–57.

Karl, D.M., 1993: Microbial processes in the southern oceans. In: *Antarctic Microbiology* [Friedmann, E.I. (ed.)]. Wiley-Liss, New York, NY, USA, pp. 1–63.

Karl, D.M., J.R. Christian, J.E. Dore, and R.M. Letelier, 1996: Microbiological oceanography in the region west of the Anarctic Peninsula: microbial dynamics, nitrogen cycle and carbon flux. *American Geophysical Union, Antarctic Research Series*, **70,** 303–322.

Keeling, C.D., J.F.S. Chin, and T.P. Whorf, 1996: Increased activity on northern vegetation inferred from atmospheric CO_2 measurements. *Nature*, **382,** 146–149.

Keigwin, L.D. and R.S. Pickart, 1999: Slope water current over the Laurentian Fan on interannual to millenial timescales. *Science*, **286(5444),** 1479.

Kennett, J.P., K.G. Cannariato, I.L. Hendy, and R.J. Behl, 2000: Carbon isotopic evidence for methane hydrate instability during Quaternary interstadials. *Science*, **288,** 128–133.

Khroustalev, L.N., 1999: Personal communication. In: *Impacts of Global Climate Change in the Arctic Regions* [Weller, G. and M. Lange (eds.)]. International Arctic Science Committee, Center for Global Change and Arctic System Research, University of Alaska, Fairbanks, AK, USA, pp. 35–36.

King, J.C., 1994: Recent climate variability in the vicinity of the Antarctic Peninsula. *International Journal of Climatology*, **14,** 357–369.

King, J.C. and S.A. Harangozo, 1998: Climate change in the western Antarctic Peninsula since 1945: observations and possible causes. *Annals of Glaciology*, **27,** 571–576.

Knox, G.A, 1994: *The Biology of the Southern Ocean*. Cambridge University Press, Cambridge, United Kingdom and New York, NY, USA, 444 pp.

Koerner, R.M. and L. Lundgaard, 1996: Glaciers and global warming. *Geographie Physique et Qquaternaire,* **49,** 429–434.

Körner, C., M. Diemer, B. Schäppi, and L. Zimmerman, 1996: Responses of alpine vegetation to elevated CO_2. In: *Terrestrial Ecosystem Response to Elevated Carbon Dioxide* [Koch, G. (ed.)]. Academic Press, San Diego, CA, USA, pp. 177–196.

Kotlyakov, B.M., 1997: Quickly warming Arctic basin. *The Earth and the World*, **4,** 107.

Krabill, W., E. Frederick, S. Manizade, C. Martin, J. Sonntag, R. Swift, R. Thomas, W. Wright, and J. Jungel, 1999: Rapid thinning of parts of the southern Greenland ice sheet. *Science*, **283,** 1522–1524.

Krabill, W., W. Abdalati, E. Frederick, S. Manizade, C. Martin, J. Sonntag, R. Swift, R. Thomas, W.Wright, and J. Jungel, 2000: Greenland ice sheet: high elevation-balance and peripheral thinning. *Science*, **289,** 428–430.

Lachenbruch, A.H. and B.V. Marshall, 1986: Changing climate: geothermal evidence from permafrost in the Alaskan Arctic. *Science*, **234,** 689–696.

Laine, K., P. Lahdesmaki, E. Menpaa, T. Pakonen, E. Saari, P. Havas, and O. Juntilla, 1995: Effect of changing snow cover on some ecophysiological parameters in *Vaccinium myrtillus* under nitrogen overdose conditions. In: *Global Change and Arctic Terrestrial Ecosystems* [Callaghan, T.V., U. Molau, M.J. Tyson, J.I. Holten, W.C. Oechel, T. Gilmanov, B. Maxwell, and B. Sveinbjorrnsson (eds.)]. Ecosystems Research Report 10, European Commission, Luxembourg, pp. 123–124.

Langdon, S.J., 1995: Increments, ranges and thresholds: human population responses to climate change in northern Alaska. In: *Human Ecology and Climate Change* [Peterson, D.L. and D.R. Johnson (eds.)]. Taylor & Francis, Washington, DC, USA, pp. 139–154.

Loeb, V.V.S., O. Holm Hansen, R. Hewitt, W. Fraser, W. Trivelpiece, and S. Trivelpiece, 1997: Effects of sea ice extent and krill or salp dominance on the Antarctic food web. *Nature*, **387,** 897–900.

Loeng, H., 1989: The influence of temperature on some fish population parameters in the Barents Sea. *Journal of the Northwest Atlantic Fisheries Science*, **9,** 103–113.

Luchitta, B.K. and C.E. Rosanova, 1998: Retreat of northern margins of George VI and Wilkins ice shelves, Antarctic Peninsula. *Annuals of Glaciology*, **27,** 41–46.

MacAyeal, D.R., 1992: Irregular oscillations of the West Anarctic ice sheet. *Nature*, **359,** 29–32.

Magnuson, J.J., D.M. Robinson, R.H. Wynne, B.J. Benson, D.M. Livingstone, T. Arai, R.A. Assel, R.D. Barry, V. Card, E. Kuusisto, N.G. Granin, T.D. Prowse, K.M. Stewart, and V.S. Vuglinski, 2000: Ice cover phenologies of lakes and rivers in the Northern Hemisphere and climate warming. *Science,* **289,** 1743–1746.

Majorowicz, J.A. and W.R. Skinner, 1997: Potential causes of differences between ground and surface air temperature warming across different ecozones in Alberta, Canada. *Global and Planetary Change,* **15(3-4),** 79–91.

Manabe, S. and R.J. Stouffer, 1994: Multiple century response of a coupled ocean atmosphere model to an increase of atmospheric carbon dioxide. *Journal of Climate*, **4,** 5–23.

Manabe, S. and R.J. Stouffer, 1993: Century-scale effects of increased atmospheric CO_2 on the atmosphere ocean system. *Nature*, **364,** 215–218.

Manseau, M., J. Huot, and M. Crete, 1996: Effects of summer grazing by caribou on composition and productivity of vegetation: community and landscape level. *Journal of Ecology*, **84,** 503–513.

Marchant, H.J, 1997: The effects of enhanced UV-B irradiance on Antarctic organisms. In: *Antarctic Communities: Species, Structure and Survival* [Battaglia, B., J. Valencia, and D.W.H. Walton (eds.)]. Cambridge University Press, Cambridge, United Kingdom and New York, NY, USA, pp. 367–374.

Marchant, H.J., A.T. Davidson, and S.W. Wright, 1987: The distribution and abundance of chroococoid cyanobacteria in the Southern Ocean. In: *Proceedings of the NIPR Symposium on Polar Biology, Tokyo, Japan* [Matsuda, T., T. Hoshiai, and M. Fukuchi (eds.)]. National Institute of Polar Research, Vol. 1, pp. 1–9.

Marsh, P. and M. Hey, 1989: The flooding hydrology of Mackenzie Delta lakes near Inuvik, NWT, Canada. *Arctic*, **42,** 41–49.

Marshall, G.J. and J.C. King, 1998: Southern Hemisphere circulation anomalies associated with extreme Antarctic Peninsula winter temperatures. *Geophysical Research Letters*, **25,** 2437–2440.

Maslanik, J.A., M.C. Serreze, and R.G. Barry, 1996: Recent decreases in Arctic summer ice cover and linkages to atmospheric circulation anomalies. *Geophysical Research Letters*, **23,** 1677–1680.

Matear, R.J and A.C. Hirst, 1999: Climate change feedback on the future oceanic CO_2 uptake. *Tellus*, **51B,** 722–733.

Maxwell, B., 1997: *Responding to Global Climate Change in Canada's Arctic. Volume II of the Canada Country Study: Climatic Impacts and Adaptation*. Environment Canada, Downsview, Ontario, Canada, 82 pp.

MBIS, 1997: *Mackenzie Basin Impact Study (MBIS), Final Report* [Cohen, S. (ed.)]. Environment Canada, Downsview, Ontario, Canada, 372 pp.

McDonald, M.E., A.E. Hershey, and M.C. Miller, 1996: Global warming impacts on trophic structure in Arctic lakes. *Limnology and Oceanography*, **41,** 1102–1108.

McGillivray, D.G., T.A Agnew, G.A. McKay, G.R. Pilkington, and M.C. Hill, 1993: Impacts of climatic change on the Beaufort sea-ice regime: implications for the Arctic petroleum industry. In: *Climate Change Digest CCD 93–01*. Environment Canada, Downsview, Ontario, Canada, 36 pp.

McKane, R.B., E.B. Rasetter, G.R. Shaver, K.J. Nadelhoffer, A.E. Giblin, J.A. Laundre, and F.S. Chapin III, 1997a: Climate effects of tundra carbon inferred from experimental data and a model. *Ecology*, **78,** 1170–1187.

McKane, R.B., E.B. Rasetter, G.R. Shaver, K.J. Nadelhoffer, A.E. Giblin, J.A. Laundre, and F.S. Chapin III, 1997b: Reconstruction and analysis of historical changes in carbon storage in Arctic tundra. *Ecology*, **78,** 1188–1198.

McKendrick, J.D., 1997: Long-term tundra recovery in northern Alaska. In: *Disturbance and Recovery in Arctic Lands* [Crawford, R.M.M. (ed.)]. NATO ASI Series 2: Environment, Vol. 25, J. Kluwer Academic Publishers, Dordrecht, The Netherlands, The Netherlands, pp. 503–518.

McPhee, M.G., T.P. Stanton, J.H. Morison, and D.G. Martinson, 1998: Freshening of the upper ocean in the Central Arctic: is perennial sea ice disappearing? *Geophysical Research Letters*, **25,** 1729–1732.

Melling, H., P.H. Johnston, and D.A. Riedel, 1995: Measurement of the draft and underside topography of sea ice by moored subsea sonar. *Journal of Atmospheric and Oceanic Technology*, **12,** 591–602.

Mercer, J.H., 1978: West Antarctic ice sheet and CO_2 greenhouse effect: a threat of disaster? *Nature*, **27,** 321–325.

Michaelson, G.J., C.L. Ping, and J.M. Kimble, 1996: Carbon storage and distribution in tundra soils of Arctic Alaska, USA. *Arctic and Alpine Research*, **28,** 414–424.

Michel, F.A. and R.O. van Everdingen, 1994: Changes in hydrogeologic regimes in permafrost regions due to climate change. *Permafrost and Periglacial Processes,* **5,** 191–195.

Mikolajewicz, U., 1998: Effect of meltwater input from the Antarctic ice sheet on the thermohaline circulation. *Annals of Glaciology*, **27,** 311–315.

Miller, J.R. and G.L. Russell, 1992: The impact of global warming on river runoff. *Journal of Geophysical Research*, **97,** 2757–2764.

Miller, P.C., R. Kendall, and W.C. Oechel, 1983: Simulating carbon accumulation in northern ecosystems. *Simulation*, **40,** 119–131.

Molau, U., 1996: Climatic impacts on flowering, growth, and vigour in an arctic-alpine cushion plant, *Diapensia lapponica*, under different snow cover regimes. *Ecological Bulletins*, **45,** 210–219.

Moore, T.R. and N.T. Roulet, 1993: Methane flux, water table relations in northern wetlands. *Geophysical Research Letters*, **20,** 587–590.

Morris, E.M. and R. Mulvaney, 1995: Recent changes in surface elevation of the Antarctic Peninsula ice sheet. *Zeitschrift für Gletscherkunde und Glazialgeologie*, **31,** 7–15.

Mulvaney, K., 1998: Arctic voices: global warming is changing the traditional Eskimo environment. *New Scientist*, **160(2160),** 55.

Murphy, J.M. and J.F.B. Mitchell, 1995: Transient response of the Hadley Centre coupled ocean atmosphere model to increasing carbon dioxide, part II: spatial and temporal structure of response. *Journal of Climate*, **8,** 57–80.

Mysak, L.A. and S.S. Venegas, 1998: Decadal climate oscillations in the Arctic: a new feedback loop for atmosphere-ice-ocean interactions. *Geophysical Research Letters*, **25,** 3609–3610.

Nadelhoffer, K.J., G.R. Shaver, A. Giblin, and E.B. Rastetter, 1996: Potential impacts of climate change on nutrient cycling, decomposition, and productivity in Arctic ecosystems. In: *Global Change and Arctic Terrestrial Ecosystems* [Oechel, W.C., T. Callaghan, T. Gilmanov, J.I. Holten, B. Maxwell, U. Molau, and B. Sveinbjornsson (eds.)]. Ecological Studies, **124,** Springer-Verlag, Berlin, Germany, pp. 349–364.

Naganobu, M., K. Kutsuwada, Y. Sasui, S. Taguchiand, and V. Siegel, 2000: Relationships between Antarctic krill (*Euphasauia superba*) variability and westerly fluctuations and ozone depletion in the Antarctic Peninsula area. *Journal of Geophysical Research*, **104(C9),** 20651–20665.

Nagurduy, A.P., 1995: Long standing changes of the ice thickness the Arctic basin. *Meteorology and Hydrology*, **6,** 80–83 (in Russian).

Nairn, R.B., S. Solomon, N. Kobayashi, J. Virdrine, 1998: Development and testing of a thermal-mechanical numerical model for predicting Arctic shore erosion processes. In: *Proceedings of the Seventh International Conference on Permafrost, Yellowknife, NWT, 23-27 June 1998* [Lewkowicz, A.G. and M. Allard (eds.)]. Collection Nordiana, No. 57, Centre d'Etudes Nordiques, Université Laval, Quebec, Canada, pp. 789–795.

Nelson, F.E. and J. Brown, 1997: Global change and permafrost. *Frozen Ground*, **21,** 21–24.

Nelson, F.E., N.I. Shiklomanov, and G.R. Mueller, 1999: Variability of active-layer thickness at multiple spatial scales, north-central Alaska, USA. *Arctic and Alpine Research*, **31,** 158–165.

Nelson, F.E., K.M. Hinkel, N.I. Shiklomanov, G.R. Mueller, L.L. Miller, and D.A. Walker, 1998a: Active-layer thickness in north-central Alaska: systematic sampling, scale, and spatial autocorrelation. *Journal of Geophysical Research*, **103**, 28963–28973.

Nelson, F.E., S.I. Outcalt, J. Brown, N.I. Shiklomanov, and K.M. Hinkel, 1998b: Spatial and temporal attributes of the active layer thickness record, Barrow, Alaska, USA. In: *Proceedings of the Seventh International Conference on Permafrost, Yellowknife, NWT, 23-27 June 1998* [Lewkowicz, A.G. and M. Allard (eds.)]. Collection Nordiana, No. 57, Centre d'Etudes Nordiques, Université Laval, Quebec, Canada, pp. 797–802.

Nelson, F.E., A.H. Lachenbruch, M.-K. Woo, E.A. Koster, T.E. Osterkamp, M.K. Gavrilova, and G.D. Cheng, 1993: Permafrost and changing climate. In: *Proceedings of the Sixth International Conference on Permafrost, Beijing, PRC, July 1993*. South China University of Technology Press, Wushan, Guangzhou, China, pp. 987–1005.

Neuvonen, S., P. Niemelä, and T. Virtanen, 1999: Climate change and insect outbreaks in boreal forests: the role of winter temperatures. *Ecological Bulletins*, **47,** 63–67.

Nicholls, K.W., 1997: Predicted reduction in basal melt rates of an Antarctic ice shelf in a warmer climate. *Nature*, **388,** 460–461.

NRC, 1996: *The Bering Sea Ecosystem*. National Academy Press, Washington, DC, USA, 307 pp.

Nuttall, M., 1998: *Protecting the Arctic: Indigenous Peoples and Cultural Survival*. Harwood Academic Publishers, Amsterdam, The Netherlands, 195 pp.

Odner, K., 1992: *The Varanger Saami: Habitation and Economy AD 1200–1900*. Scandinavian University Press, Oslo, Norway, 320 pp.

Oechel, W.C. and W.D. Billings, 1992: Anticipated effects of global change on carbon balance of arctic plants and ecosystems. In: *Arctic Ecosystems in a Changing Climate: An Ecophysiological Perspective* [Chapin, F.S. III, R.L. Jefferies, J. Reynolds, G. Shaver, and J. Svoboda (eds.)]. Academic Press, San Diego, CA, USA, pp. 139–168.

Oechel, W.C. and G.L.Vourlitis, 1994: The effects of climate change on land-atmosphere feedbacks in Arctic tundra regions. *Trends in Ecology and Evolution*, **9,** 324–329.

Oechel, W.C., A.C. Cook, S.J. Hastings, and G.L. Vourlitis, 1997: Effects of CO_2 and climate change on arctic ecosystems. In: *Ecology of Arctic Environments* [Woodin, S.J. and M. Marquiss (eds.)]. Blackwell Science, Oxford, United Kingdom, pp. 255–273.

Oechel, W.C., S.J. Hastings, G. Vourlitis, M. Jenkins, G. Riechers, and N. Grulke, 1993: Recent change of Arctic tundra ecosystems from a net carbon sink to a source. *Nature*, **361,** 520–523.

Oechel, W.C., S. Cowles, N. Grulke, S.J. Hastings, B. Lawrence, T. Prudhomme, G. Riechers, B.Strain, D. Tissue, and G. Vourlitis, 1994: Transient nature of CO_2 fertilization in Arctic tundra. *Nature*, **371,** 500–503.

O'Farrell, S.P., J.L. McGregor, L.D. Rotstayn, W.F. Budd, C. Zweck, and R. Warner, 1997: Impact of transient increases in atmospheric CO_2 on the accumulation and mass balance of the Antarctic ice sheet. *Annals of Glaciology*, **25,** 137–144.

Ohmura, A., M. Wild, and L. Bengtsson, 1996: A possible change in mass balance of Greenland and Antarctic ice sheets in the coming century. *Journal of Climate*, **9,** 2124–2135.

Ono, K.A., 1995: Effects of climate change on marine mammals in the far north. In: *Human Ecology and Climate Change* [Peterson, D.L. and D.R. Johnson (eds.)]. Taylor & Francis, Washington, DC, USA, pp. 105–122.

Osterkamp, T.E. and V.E. Romanovsky, 1999: Evidence for warming and thawing of discontinuous permafrost in Alaska. *Permafrost and Periglacial Processes*, **10,** 17–37.

Osterkamp, T.E., V.E. Romanovsky, T. Zhang, V. Gruol, J.K. Peterson, T. Matava, and G.C. Baker, 1998: A history of continuous permafrost conditions in northern Alaska. *EOS, Transactions, American Geophysical Union*, **79,** 833.

Osterkamp, T.E., L. Viereck, Y. Shur, M.T. Jorgenson, C. Racine, A. Doyle, and R.D. Boone, 2000: Observations of thermokarst and its impact on boreal forests in Alaska, U.S.A. *Arctic, Antarctic, and Alpine Research*, **32,** 303–315.

Otterson, G. and H. Loeng, 2000: Covariability in early growth and year-class strength of Barents Sea cod, haddock and herring: the environmental link. *ICES Journal of Marine Science*, **57(2),** 339–348.

Parkinson, C., D. Cavalieri, P. Gloersen, H. Zwally, and J. Comiso, 1999: Arctic sea ice extents, areas, and trends, 1978–1996. *Journal of Geophysical Research (Oceans),* **C9,** 20837–20856.

Parmesan, C., 1998: Climate and species' range. *Nature*, **382,** 765–766.

Pavlov, A.V., 1996: Permafrost-climatic monitoring of Russia: analysis of field data and forecast. *Polar Geography*, **20,** 44–64.

Peterson, D.L. and D.R. Johnson (eds.), 1995: *Human Ecology and Climate Change*. Taylor & Francis, Washington, DC, USA.

Petit, J.R., J. Jouzel, D. Raynaud, N.I. Barkov, J.M. Barnola, I. Basile, M. Bender, J. Chappellaz, M. Davis, G. Delaygue, M. Delmotte, V.M. Kotlyakov, M. Legrand, V.Y. Lipenkov, C. Lorius, L. Pepin, C. Ritz, E. Saltzman, and M. Stievenard, 1999: Climate and atmospheric history of the past 420,000 years from the Vostok ice core, Antarctica. *Nature*, **399,** 429–437.

Press, M.C., T.V. Callaghan, and J.A. Lee, 1998: How will European Arctic ecosystems respond to projected global environmental change? *Ambio*, **4,** 77–90.

Priddle, J., V. Smetacek, and U.V. Bathmann, 1992: Antarctic marine primary production, biogeochemical carbon cycles and climatic change. *Philosophical Transactions of the Royal Society of London*, **338B,** 289–297.

Prowse, T.D., 1994: The environmental significance of ice to cold regions streamflow. *Freshwater Biology*, **32,** 241–260.

Prowse, T.D. and M. Conly, 1998: Impacts of climatic variability and flow regulation on ice jam flooding of a northern delta. *Hydrological Processes*, **12,** 1589–1610.

Prowse, T.D. and P.O. Flegg, 2000: Arctic river flow: a review of contributing areas. In: *The Freshwater Budget of the Arctic Ocean* [Lewis, E.L., E.P. Jones, P. Lemke, T.D. Prowse, and P. Wadhams (eds.)]. NATO Science Series, No. 2, Environmental Security, J. Kluwer Academic Publishers, Dordrecht, The Netherlands, Vol. 70, pp. 269–280.

Prowse, T.D., B. Aitken, M.N. Demuth, and M. Peterson, 1996: Strategies for restoring spring flooding to a drying northern delta. *Regulated Rivers*, **12,** 237–250.

Quigley, N.C. and N.J. Bridge, 1987: The structure of an arctic microeconomy: the traditional sector in community economic development. *Arctic*, **40,** 204–210.

Quinn, T.J. and R.P. Marshall, 1989: Time series analysis: quantifying variability and correlation in Alaska salmon catches and environmental data. *Canadian Special Publications in Fisheries and Aquatic Science*, **108,** 67–80.

Rasmussen, T.L., E. Thomsen, T.C.E. van Weering, and L. Labeyrie, 1996: Rapid changes in surface and deep water conditions at the Faroe Margin during the last 58,000 years. *Paleoceanography*, **11,** 757–771.

Rastetter, E.B., R.B. McKane, G.R. Shaver, K.J. Nadelhoffer, and A.Giblin, 1996: Analysis of CO_2, temperature, and moisture effects on carbon storage in Alaskan Arctic tundra using a general ecosystem model. In: *Global Change and Arctic Terrestrial Ecosystems* [Oechel, W.C., T. Callaghan, T. Gilmanov, J.I. Holten, B. Maxwell, U. Molau, and B. Sveinbjornsson (eds.). Ecological Studies, Vol. 124, Springer-Verlag, Berlin, Germany, pp. 437–451.

Rees, W.G. and M. Williams, 1997: Satellite remote sensing of the impact of industrial pollution on tundra biodiversity. In: *Disturbance and Recovery in Arctic Lands, An Ecological Perspective* [Crawford, R.M.M. (ed.)]. Springer Verlag, Berlin, Germany, pp. 253–282.

Ring, R., 1994: Arctic insects and global change. In: *Biological Implications of Global Change: Northern Perspective* [Riewe, R. and J. Oakes (eds.)]. Canadian Global Change Program and Association of Canadian Universities for Northern Studies, Canadian Circumpolar Institute, and the Royal Society of Canada, Ottawa, Ontario, Canada, pp. 61–66.

Riseborough, D.W. and M.W. Smith, 1993: Modelling permafrost response to climate change and climate variability. In: *Proceedings, Fourth International Symposium on Thermal Engineering & Science for Cold Regions* [Lunardinai, V.J. and S.L. Bowen (eds.)]. U.S. Army Cold Regions Research and Engineering Laboratory, Special Report 93–22, Hanover, NH, USA, pp. 179–187.

Rivkin, F.M., 1998: Release of methane from permafrost as a result of global warming and other disturbances. *Polar Geography*, **22,** 105–118.

Robinson, S.D. and T.R. Moore, 1999: Carbon and peat accumulation over the past 1200 years in a landscape with discontinuous permafrost, northwestern Canada. *Global Biogeochemical Cycles*, **13,** 591–601.

Robock, A., C.A. Schlosser, K.Y. Vinnikov, N.A. Speranskaya, J.K. Entin, and S. Qiu, 1998: Evaluation of the AMIP soil moisture simulations. *Global and Planetary Change*, **19,** 181–208.

Ross, R.M., L.B. Quetin, K.S. Baker, M. Vernet, and R.C. Smith, 2000: Growth limitation in young *Euphasusia superba* under field conditions. *Limnology and Oceanography*, **45,** 31–43.

Rothrock, D.A., Y. Yu, and G.A. Maykut, 1999: Thinning of the Arctic sea-ice cover. *Geophysical Research Letters*, **26,** 3469–3472.

Rott, H., P. Skvarca, and T. Nagler, 1996: Rapid collapse of Northern Larsen Ice Shelf, Antarctica. *Science*, **271,** 788–792.

Rouse, W.R., 1998: A water balance model for a subarctic sedge fen and its application to climate change. *Climatic Change*, **38,** 207–234.

Rouse, W.R., S. Holland, and T.R. Moore, 1995: Variability in methane emissions from wetlands at northern treeline near Churchill, Manitoba, Canada. *Arctic and Alpine Research*, **27,** 146–156.

Rouse, W.R., M.S.V. Douglas, R.E. Hecky, A.E. Hershey, G.W. Kling, L. Lesack, P. Marsh, M. McDonald, B.J. Nicholson, N.T. Roulet, and J.P. Smol, 1997: Effects of climate change on the fresh waters of Arctic and sub-Arctic North America. *Hydrological Processes*, **11,** 873–902.

Sabo, G. III., 1991: *Long-Term Adaptations Among Arctic Hunter-Gatherers*. Garland Publishing, London, United Kingdom, 403 pp.

Sarmiento, J.L. and C. Le Quere, 1996: Oceanic carbon dioxide uptake in a model of century scale global warming. *Science*, **274,** 1346–1350.

Sarmiento, J.L., T.M.C. Hughes, R.J. Stouffer, and S. Manabe, 1998: Simulated response of the ocean carbon cycle to anthropogenic climate warming. *Nature*, **393,** 245–249.

Schindler, D.W., 1997: Widespread effects of climatic warming on freshwater ecosystems in North America. *Hydrological Processes*, **11,** 1043–1067.

Schlesinger, W.H., 1991: *Biogeochemistry: An Analysis of Global Change*. Academic Press, San Diego, CA, USA, 588 pp.

Scrimgeour, G.A., T.D. Prowse, J.M. Culp, and P.A. Chambers, 1994: Ecological effects of river ice break up: a review and perspective. *Freshwater Biology*, **32,** 261–276.

Scripter, M.W., 1970: Nested-means map classes for statistical maps. *Annals of the Association of American Geographers*, **60,** 385–393.

Sedwick, P.N., G.R. DiTullio, D.A. Hutchins, P.W. Boyd, F.B. Griffiths, A.C. Crossley, T.W.Trull, and B. Queguiner, 1999: Limitation of algal growth by iron deficiency in the Australian Subantarctic region. *Geophysical Research Letters*, **26,** 2865–2868.

Serreze, M.C., J.E. Walsh, F.S. Chapin III, T. Osterkamp, M. Dyurgerov, V. Romanovsky, W.C. Oechel, J. Morison, T. Zhang, and R.G. Barry, 2000: Observational evidence of recent change in the northern high latitude environment. *Climatic Change*, **46,** 159–207.

Shackley, S., P. Young, S. Parkinson, and B. Wynne, 1998: Uncertainty, complexity and concepts of good science in climate change modelling: are GCMs the best tools? *Climatic Change*, **38,** 159–205.

Sharratt, B.S., 1992: Growing season trends in the Alaskan climate record. *Arctic*, **45,** 124–127.

Shaver, G.R. and S. Jonasson, 1999: Response of Arctic ecosystems to climate change: results of long-term field experiments in Sweden and Alaska. *Polar Research*, **18,** 245–252.

Shaver, G.R., W.D. Billings, F.S. Chapin III, A.E. Gibbin, K.J. Nadelhoffer, W.C. Oechel, and E.B. Rastetter, 1992: Global changes and the carbon balance of Arctic ecosystems. *BioScience*, **61,** 415–435.

Shaw, J., R.B. Taylor, S. Solomon, H.A. Christian, and D.L. Forbes, 1998: Potential impacts of global sea-level rise on Canadian coasts. *The Canadian Geographer*, **42,** 365–379.

Shaw, R.W., 1987: Air pollution by particles. *Scientific American,* **257,** 96–103.

Shiklomanov, I.A., 1997: On the effect of anthropogenic change in the global climate on river runoff in the Yenisei basin. In: *Runoff Computations for Water Projects* [Rozhdestvensky, A.V. (ed.)]. Technical Documents in Hydrology, No. 9, International Hydrological Programme (IHP-V), United Nations Educational, Scientific, and Cultural Organization (UNESCO), Paris, France, pp. 113–199.

Shiklomanov, I.A., 1994: Influence of anthropogenic changes in global climate on the Yenisey River runoff. *Meteorology and Hydrology*, **2,** 68–75 (in Russian).

Shiklomanov, I.A., A.I. Shiklomanov, R.B. Lammers, B.J. Peterson, A.I. Shiklomanov, and C.J. Vorosmarty, 2000: The dynamics of river water inflow to the Arctic Ocean. In: *The Freshwater Budget of the Arctic Ocean* [Lewis, E.L., E.P. Jones, P. Lemke, T.D. Prowse, and P. Wadhams (eds.)]. NATO Science Series, No. 2, Environmental Security, J. Kluwer Academic Publishers, Dordrecht, The Netherlands, Vol. 70, pp. 281–296.

Simmonds, I., 1998: The climate of the Antarctic region. In: *Climates of the Southern Continents* [Hobbs, J.E., J.A.Lindesay, and H.A. Bridgemann (eds.)]. John Wiley and Sons, New York, NY, USA, pp. 137–159.

Skinner, W.R., R.L. Jefferies, T.J. Carleton, R.F. Rockwell, and K.F. Abraham, 1998: Prediction of reproductive success and failure in lesser snow geese based on early season climatic variables. *Global Change Biology*, **4,** 3–16.

Sklair, L., 1991: *Sociology of the Global System*. Harvester Wheatsheaf, London, United Kingdom, 269 pp.

Skvarca, P., W. Rack, H. Rott, and T. Ibarzabaly Donangelo, 1998: Evidence of recent climatic warming on the eastern Antarctic Peninsula. *Annals of Glaciology*, **27,** 628–632.

Smith, D.M., 1998: Recent increase in the length of the melt season of perennial Arctic sea ice. *Geophysical Research Letters*, **25,** 655–658.

Smith, A.M., D.G. Vaughan, C.S.M. Doake, and A.C. Johnson, 1999a: Surface lowering of the ice ramp at Rothera Point, Antarctic Peninsula, in response to regional climate change. *Annals of Glaciology*, **27,** 113–118.

Smith, R.C., D. Ainley, K. Baker, E. Domack, S. Emslie, B. Fraser, J. Kennett, A. Leventer, E. Mosley-Thompson, S. Stammerjohn, and M. Vernet, 1999b: Marine ecosystem sensitivity to climate change. *Bioscience,* **49,** 393–404.

Smith, I.N., W.F. Budd, and P. Reid, 1998a: Model estimates of Antarctic accumulation rates and relationship to temperature changes. *Annals of Glaciology*, **27,** 246–250.

Smith, P., O. Andren, L. Brussard, M. Dangerfield, K. Ekschmitt, P. Lavelle, and K. Tate, 1998b: Soil biota and global change at the ecosystem level: describing soil biota in mathematical models. *Global Change Biology*, **4,** 773–784.

Smith, S.L. and M.M. Burgess, 1999: Mapping the sensitivity of Canadian permafrost to climate warming. In: *Interactions Between the Cryosphere, Climate and Greenhouse Gases. Proceedings of International Union of Geodesy and Geophysics (IUGG) 99 Symposium HS2, Birmingham, July 1999* [Tranter, M., R. Armstrong, E. Brun, G. Jones, M. Sharp, and M. Williams (eds.)]. International Association of Hydrological Sciences (IAHS), Birmingham, United Kingdom, Vol. 256, pp. 71–80.

Smith, S.L. and M.M. Burgess, 1998: Mapping the response of permafrost in Canada to climate warming. In: *Current Research 1998-E*. Geological Survey of Canada, Ottawa, Ontario, Canada, pp. 163–171.

Splettoesser, J., 1992: Antarctic Global Warming? *Nature*, **355,** 503.

Stammerjohn, S.E. and R.C. Smith, 1996: Spatial and temporal variability of western Anarctic Peninsula sea ice coverage. In: *Foundations for Ecological Research West of the Antarctic Peninsula, Antarctic Research Series* 70 [Ross, R.M., E.E. Hofmann, and L.B. Quetin (eds.)]. American Geophysical Union, Washington, DC, USA, pp. 81–104.

Starfield, A.M. and F.S. Chapin, III, 1996: A dynamic model of arctic and boreal vegetation change in response to global changes in climate and land-use. *Ecological Applications*, **6,** 842–864.

Steele, M. and T. Boyd, 1998: Retreat of the cold halocline layer in the Arctic Ocean. *Journal of Geophysical Research*, **103,** 10419–10435.

Stewart, R.E., 2000: The variable climate of the Mackenzie River Basin: its water cycle, feedbacks and fresh water discharge. In: *The Freshwater Budget of the Arctic Ocean* [Lewis, E.L., E.P. Jones, P. Lemke, T.D. Prowse, and P. Wadhams (eds.)]. NATO Science Series, No. 2, Environmental Security, J. Kluwer Academic Publishers, Dordrecht, The Netherlands, Vol. 70, pp. 367–381.

Stirling, I. and N.J. Lunn, 1997: Environmental fluctuations in arctic marine ecosystems reflected by variability in reproduction of polar bears and ringed seals. In: *Ecology of Arctic Environments* [Woodin, S.J. and M. Marquiss (eds.)]. Blackwell Science, Oxford, United Kingdom, pp. 167–181.

Stirling, I., N.J. Lunn, and J. Iacozza, 1999: Long-term trends in the population ecology of polar bears in western Hudson Bay in relation to climate change. *Arctic*, **52,** 294–306.

Stocker, T.F. and A. Schmittner, 1997: Influence of CO_2 emission rates on the stability of the thermohaline circulation. *Nature*, **388,** 862–865.

Szeicz, J.M. and G.M. MacDonald, 1995: Dendroclimatic reconstruction of summer temperatures in northwestern Canada since A.D. 1638 based on age-dependent modeling. *Quaternary Research*, **44,** 257–266.

Taylor, A. and M. Burgess, 1998: Canada's deep permafrost temperatures and the research inspiration of Arthur H. Lachenbruch. *EOS, Transactions, American Geophysical Union*, **79,** 833.

Tenhunen, J.D., 1996: Diurnal and seasonal patterns of ecosystem CO_2 efflux from upland tundra in the foothills of the Brooks Range, Alaska, USA. *Arctic and Alpine Research*, **28,** 328–338.

Thompson, D.W.J. and J.M. Wallace, 2000: Annual modes in the extratropical circulation, part II: trends. *Journal of Climate*, **13,** 1018–1036.

Torn, M.S. and F.S. Chapin III, 1993: Environmental and biotic controls over methane flux from arctic tundra. *Chemosphere*, **26,** 357–368.

Tynan, C.T. and D.P. DeMaster, 1997: Observations and predictions of Arctic climatic change: potential effects on marine mammals. *Arctic*, **50,** 308–322.

UNEP, 1997: *Global Environment Outlook: An Overview*. United Nations Environment Program, in collaboration with the Stockholm Environment Institute, Oxford University Press, Oxford, United Kingdom, 264 pp.

van Blarcum, S.C., J.R. Miller, and G.L. Russell, 1995: High latitude river runoff in a doubled CO_2 climate. *Climatic Change*, **30,** 7–26.

van Ommen, T.D. and V. Morgan, 1996: Peroxide concentrations in the DSS ice core, Law Dome, Antarctica. *Journal of Geophysical Research*, **101,** 15147–15152.

van Ommen, T.D. and V. Morgan, 1997: Calibrating the ice core paleothermometer using seasonality. *Journal of Geophysical Research*, **102,** 9351–9352.

Vaughan, D.G., 1993: Implications of the break up of Wordie Ice Shelf, Antarctica for sea level. *Antarctic Science*, **5,** 403–408.

Vaughan, D.G. and C.S.M. Doake, 1996: Recent atmospheric warming and retreat of ice shelves on the Antarctic Peninsula. *Nature*, **379,** 328–331.

Vaughan, D.G. and J.R. Spouge, 2000: Risk estimation of the collapse of the West Antarctic ice sheet. *Climatic Change* (in press).

Vaughan, D.G., J.L. Bamber, M. Giovinetto, J. Russell, and A.P.R. Cooper, 1999: Reassessment of net surface mass balance in Antarctica. *Journal of Climate*, **12,** 933–946.

Vilchek, G.E., 1997: Arctic ecosystem stability and disturbance. In: *Disturbance and Recovery in Arctic Lands, An Ecological Perspective* [Crawford, R.M.M. (ed.)]. NATO-ASI Series, No. 2, Environment, J. Kluwer Academic Publishers, Dordrecht, The Netherlands, Vol. 25, pp. 179–189.

Vilcheck, G.E. and A.A. Tishkov, 1997: Usinsk oil spill: Environmental catastrophe or routine event? In: *Disturbance and Recovery in Arctic Lands, An Ecological Perspective* [Crawford, R.M.M. (ed.)]. NATO-ASI Series, No. 2, Environment, J. Kluwer Academic Publishers, Dordrecht, The Netherlands, Vol. 25, pp. 411–420.

Vilhjalmsson, H., 1997: Climatic variations and some examples of their effects on the marine ecology of Icelandic and Greenland waters, in particular during the present century. *Rit Fiskideildar Journal of the Marine Research Institude, Rykjavik*, **XV,** 9–29.

Vinje, T., 2001: Anomalies and trends of sea ice extent and atmosphere circulation in the Nordic Seas during the period 1864–1998. *Journal of Climate*, **14,** 255–267.

Vinje, T., N. Nordlund, and Å. Kvambekk, 1998: Monitoring ice thickness in Fram Strait. *Journal of Geophysical Research*, **103,** 10437–10449.

Vinnikov, K.Y., A. Robock, R.J. Stouffer, J.E. Walsh, C.L. Parkinson, D.J. Cavalieri, J.F.B. Mitchell, D. Garrett, and V.F. Zakharov, 1999: Global warming and Northern Hemisphere sea ice extent. *Science*, **286,** 1934–1937.

Virtanen, T., S. Neuvon, and A. Nikula, 1998: Modelling topoclimatic patterns of egg mortality of Epirrita Autumnata (Lepidoptera, Geometridae) with a geographical information system-predictions for current climate and warmer climate scenarios. *Journal of Applied Ecology*, **35(2),** 311–322.

Vourlitis, G.L. and W.C. Oechel, 1999: Eddy covariance measurements of CO_2 and energy fluxes of an Alaskan tussock tundra ecosystem. *Ecology*, **80,** 686–701.

Vourlitis, G.L., W.C. Oechel, S.J. Hastings, and M.A. Jenkins, 1993: The effect of soil moisture and thaw depth on CH_4 flux from wet coastal tundra ecosystems on the north slope of Alaska. *Chemosphere*, **26,** 329–337.

Vyalov, S.S., A.S. Gerasimov, and S.M. Fotiev, 1998: Influence of global warming on the state and geotechnical properties of permafrost. In: *Proceedings of the Seventh International Conference on Permafrost, Yellowknife, NWT, 23-27 June 1998* [Lewkowicz, A.G. and M. Allard (eds.)]. Collection Nordiana, No. 57, Centre d'Etudes Nordiques, Université Laval, Quebec, Canada, pp. 1097–1102.

Wadhams, P., 1997: Ice thickness in the Arctic Ocean: the statistical reliability of experimental data. *Journal of Geophysical Research*, **102,** 27951–27959.

Waelbroeck, C., 1993: Climate soil processes in the presence of permafrost: a systems modeling approach. *Ecological Modeling*, **69,** 185–225.

Waelbroeck, C., P. Monfray, W.C. Oechel, S. Hastings, and G. Vourlitis, 1997: The impact of permafrost thawing on the carbon dynamics of tundra. *Geophysical Research Letters*, **24,** 229–232.

Walker, D.A., J.G. Bockheim, F.S. Chapin III, W. Eugster, J.Y. King, J.P. McFadden, G.J. Michaelson, F.E. Nelson, W.C. Oechel, C.L.Ping, W.S. Reeburgh, S. Regli, N.I. Shiklomanov, and G.L. Vourlitis, 1998: A major arctic soil pH boundary: implications for energy and trace gas fluxes. *Nature*, **394,** 469–472.

Walsh, J.E., V. Kattsov, D. Portis, and V. Meleshko, 1998: Arctic precipitation and evaporation: model results and observational estimates. *Journal of Climate*, **11,** 72–87.

Wang, B. and M. Allard, 1995: Recent climatic trend and thermal response of permafrost in Salluit, northern Quebec, Canada. *Permafrost and Periglacial Processes*, **6,** 221–233.

Ward, C.G., 1995: The mapping of ice front changes on Müller Ice Shelf, Antarctic Peninsula. *Antarctic Science*, **7,** 197–198.

Warner, R.C. and W.F. Budd, 1998: Modelling the long-term response of the Antarctic ice sheet to global warming. *Annals of Glaciology*, **27,** 161–168.

Warren, D.M., L.J. Slikkerveer, and D. Brokensha (eds.), 1995: *The Cultural Dimension of Development: Indigenous Knowledge Systems*. Intermediate Technology Publications, London, United Kingdom, 582 pp.

Welch, H.E., 1995: Marine conservation in the Canadian Arctic: a regional overview. *Northern Perspectives*, **23,** 1–19.

Weller, G., 1998: Regional impacts of climate change in the Arctic and Antarctic. *Annals of Glaciology*, **27,** 543–552.

Weller, G. and P.A. Anderson (eds.), 1999: *Assessing the Consequences of Climate Change for Alaska and the Bering Sea Region. Proceedings of a Workshop at the University of Alaska, Fairbanks, 29-30 October 1998*. Center for Global Change and Arctic System Research, University of Alaska, Fairbanks, AK, USA, 94 pp.

Weller, G. and M. Lange, 1999: *Impacts of Global Climate Change in the Arctic Regions*. International Arctic Science Committee, Center for Global Change and Arctic System Research, University of Alaska, Fairbanks, AK, USA, pp. 1–59.

Wenzel, G.W., 1995: Warming the Arctic: environmentalism and the Canadian Inuit. In: *Human Ecology and Climate Change* [Peterson, D.L. and D.R. Johnson (eds.)]. Taylor & Francis, Washington, DC, USA, pp. 169–184.

Whetton, P.H., M.H. England, S.P. O'Farrell, I.G. Watterson, and A.B. Pittock, 1996: Global comparisons of the regional rainfall results of enhanced greenhouse coupled and mixed layer experiments: implications for climate change scenario development for Australia. *Climatic Change*, **33,** 497–519.

White, J.W.C. and E.J. Steig, 1998: Timing of everything in a game of two hemispheres. *Nature*, **394,** 717–718.

Whitworth, T., A.H. Orsi, S.J. Kim, and W.D. Nowlin, 1998: Water masses and mixing near the Antarctic slope front. In: *Ocean, Ice and Atmosphere, Interactions at the Antarctic Continental Margin, Antarctic Research Series 75* [Jacobs, S.S. and R.F. Weiss (eds.)]. American Geophysical Union, Washington, DC, USA, pp. 1–27.

Williams, M.J.M., A. Jenkins, and J. Determann, 1998a: Physical controls on ocean circulation beneath ice shelves revealed by numerical models. In: *Ocean, Ice and Atmosphere, Interactions at the Antarctic Continental Margin, Antarctic Research Series 75* [Jacobs, S.S. and R.F. Weiss, (eds.)]. American Geophysical Union, Washington, DC, USA, pp. 285–299.

Williams, M.J.M., R.C. Warner, and W.F. Budd, 1998b: The effects of ocean warming on melting and ocean circulation under the Amery Ice Shelf, east Antarctica. *Annals of Glaciology*, **27,** 75–80.

Williams, P.J., 1995: Permafrost and climate change: geotechnical implications. *Philosophical Transactions of the Royal Society of London*, **352A,** 347–358.

Wingham, D.J., A.J. Ridout, R. Scharroo, R.J. Arthern, and C.K. Schum, 1998: Antarctic elevation change from 1992 to 1996. *Science*, **282,** 456–458.

Wolfe, S.A., S.R., Dallimore, and S.M. Solomon, 1998: Coastal permafrost investigations along a rapidly eroding shoreline, Tuktoyaktuk, NWT, Canada. In: *Proceedings of the Seventh International Conference on Permafrost, Yellowknife, NWT, 23-27 June 1998* [Lewkowicz, A.G. and M. Allard (eds.)]. Collection Nordicana No. 57, Centre d'Etudes Nordiques, Université Laval, Québec, Canada, pp. 1125–1131.

Wolfe, S.A., E. Kotler, and F.M. Nixon, 2000: Recent warming impacts in the Mackenzie Delta, Northwest Territories and northern Yukon Territory coastal areas. In: *Current Research 2000-B1*. Geological Survey of Canada, Ottawa, Ontario, Canada, 9 pp. Available online at http://www.nrcan.gc.ca/gsc/bookstore.

Wong, A.P.S., N.L. Bindoff, and J.A. Church, 1999: Large-scale freshening of intermediate waters in the Pacific and Indian Oceans. *Nature*, **400,** 440–443.

Woo, M.K., 1996: Hydrology of northern North America under global warming. In: *Regional Hydrological Responses to Climate Change* [Jones, J.A.A., C. Liu, M.-K. Woo, and H.-T. Kung (eds.)]. J. Kluwer Academic Publishers, Dortrecht, The Netherlands, pp. 73–86.

Woo, M.K. and K.L. Young, 1998: Characteristics of patchy wetlands in a polar desert environment, Arctic Canada. In: *Proceedings of the Seventh International Conference on Permafrost, Yellowknife, NWT, 23-27 June 1998* [Lewkowicz, A.G. and M. Allard (eds.)]. Collection Nordicana No. 57, Centre d'Etudes Nordiques, Université Laval, Québec, Canada, pp. 1141–1146.

Wood, R.A., A.B. Keen, J.F.B. Mitchell, and J.M. Gregory, 1999: Changing spatial structure of the thermohaline circulation in response to atmospheric CO_2 forcing in a climate model. *Nature*, **399,** 572–575.

Wookey, P.A., J.M. Welker, A.N. Parsons, M.C. Press, T.V. Callaghan, and J.A. Lee, 1994: Differential growth, allocation and photosynthetic responses of *Polygoruim viviparum* to similated environmental change at a high arctic polar semi-desert. *Oikos*, **70,** 131–139.

Wookey, P.A., A.N. Parsons, J.M. Welker, J.A. Potter, T.V. Callaghan, J.A. Lee, and M.C. Press, 1993: Comparitive responses of phenology and reproductive development to simulated environmental change in sub-Arctic and high Arctic plants. *Oikos*, **67,** 490–502.

Wright, J.F. and T.F. Stocker, 1993: Younger Dryas experiments. In: *Ice in the Climate System* [Peltier, W.R. (ed.)]. NATO-ASI Series, Springer-Verlag, Berlin, Germany, Vol. 112, pp. 395–416.

Wu, X. and W.F. Budd, 1998: Modelling global warming and Antarctic sea ice changes over the past century. *Annals of Glaciology*, **27,** 413–419.

Wu, X., I. Simmonds, and W.F. Budd, 1997: Modelling of Antarctic sea ice in a general circulation model. *Journal of Climate*, **10,** 593–609.

Wu, X., W.F. Budd, V.I. Lytle, and R.A Massom, 1999: The effect of snow on Antarctic sea ice simulations in a coupled atmosphere sea ice model. *Climate Dynamics*, **15(2),** 127–143.

Zhang, T., R.G. Barry, K. Knowles, J.A. Heginbottom, and J. Brown, 1999: Statistics and characteristics of permafrost and ground-ice distribution in the Northern Hemisphere. *Polar Geography*, **23,** 132–154.

Zimov, S.A., S.P. Davidov, G.M. Zimova, A.I. Davidova, F.S. Chapin III, M.C. Chapin, and J.F. Reynolds, 1999: Contribution of disturbance to high-latitude amplification of atmospheric CO_2. *Science*, **284,** 1973–1976.

Zuidhoff, F.S. and E. Kolstrup, 2000: Changes in palsa distribution in relation to climate change in Laivadalen, northern Sweden, especially 1960–1997. *Permafrost and Periglacial Processes*, **11,** 55–69.

Zwally, H.J. and S. Fiegles, 1994: Extent and duration of Antarctic surface melting. *Journal of Glaciology*, **40,** 463–476.

17

Small Island States

LEONARD A. NURSE (BARBADOS) AND GRAHAM SEM (PAPUA NEW GUINEA)

Lead Authors:
J.E. Hay (New Zealand), A.G. Suarez (Cuba), Poh Poh Wong (Singapore), L. Briguglio (Malta), S. Ragoonaden (Mauritius)

Contributing Authors:
A. Githeko (Kenya), J. Gregory (UK), V. Ittekkot (Germany), U. Kaly (Australia), R. Klein (Germany/The Netherlands), M. Lal (India), A. McKenzie (Jamaica), H. McLeod (UK), N. Mimura (Japan), J. Price (USA), Dahe Qin (China), B. Singh (Trinidad and Tobago), P. Weech (The Bahamas)

Review Editors:
R. Payet (Seychelles) and S. Saeed (Maldives)

CONTENTS

EXECUTIVE SUMMARY

The small island states considered in this chapter are located mainly in the tropics and the subtropics. These island states span the ocean regions of the Pacific, Indian, and Atlantic, as well as the Caribbean and Mediterranean Seas. Because of the very nature of these states, the ocean exerts a major influence on their physical, natural, and socioeconomic infrastructure and activities.

Although small island states are not a homogeneous group, they share many common features that serve to increase their vulnerability to projected impacts of climate change. These characteristics include their small physical size and the fact that they are surrounded by large expanses of ocean; limited natural resources; proneness to natural disasters and extreme events; relative isolation; extreme openness of their economies, which are highly sensitive to external shocks; large populations with high growth rates and densities; poorly developed infrastructure; and limited funds, human resources, and skills. These characteristics limit the capacity of small island states to mitigate and adapt to future climate and sea-level change.

The most significant and immediate consequences for small island states are likely to be related to changes in sea levels, rainfall regimes, soil moisture budgets, and prevailing winds (speed and direction) and short-term variations in regional and local patterns of wave action. Owing to their coastal location, the majority of socioeconomic activities and infrastructure and the population are likely to be highly vulnerable to the impacts of climate change and sea-level rise.

Review of past and present trends of climate and climate variability indicates that temperatures have been increasing by as much as 0.1°C per decade, and sea level has risen by 2 mm yr^{-1} in regions in which small island states are located. Analysis of observational data for these regions suggests that increases in surface air temperatures have been greater than global rates of warming (e.g., in the Pacific Ocean and the Caribbean Sea regions). Observational evidence also suggests that much of the variability in the rainfall record of Caribbean and Pacific islands appears to be closely related to the onset of El Niño-Southern Oscillation (ENSO). However, part of the variability in these areas also may be attributable to the influence of the Inter-Tropical Convergence Zone (ITCZ) and the South Pacific Convergence Zone (SPCZ). It is acknowledged however, that for some small islands it is difficult to establish clear trends of sea-level change because of limitations of observational records, especially geodetic-controlled tide gauge records.

The use of the state-of-the-art coupled atmosphere-ocean general circulation models (AOGCMs) to estimate future response of climate to anthropogenic radiative forcing suggests an enhanced climate change in the future. Several AOGCMs have been analyzed for the Atlantic, Pacific, and Indian Ocean regions and the Caribbean and Mediterranean Seas. The outputs from these models indicate general increases in surface air temperature for the 2050s and 2080s and an increase in rainfall of about 0.3% for the 2050s and 0.7% for the 2080s for the Pacific region. However, a marginal decline in rainfall is projected for the other regions, with a possible reduction of water availability. The diurnal temperature range is projected to decrease marginally for the regions of the small island states for both time horizons.

With respect to extreme events, AOGCM (CSIRO and ECHAM) transient experiments project that by the 2050s and 2080s, there will be increased thermal stress during summer, as well as more frequent droughts and floods in all four tropical ocean regions in which small island states are located.This projection implies that in the future these regions are likely to experience floods during wet seasons and droughts during dry seasons. Furthermore, warming in some regions (e.g., the Pacific Ocean) is likened to an El Niño pattern, suggesting that climate variability associated with the ENSO phenomenon will continue on a seasonal and decadal time scale. It is probable that such an association may dominate over any effects attributable to global warming. Given their high vulnerability and low adaptive capacity to climate change, communities in the small island states have legitimate concerns about their future on the basis of the past observational record and climate model projections. In this Third Assessment Report, analysis of the scientific-technical literature identifies the following key issues among the priority concerns of small island states.

Development, sustainability, and equity issues. The small island states account for less than 1% of global greenhouse gas (GHG) emissions but are among the most vulnerable of all locations to the potential adverse effects of climate change and sea-level rise. Economic development and alleviation of poverty constitute the single most critical concern of many small island states. Thus, with limited resources and low adaptive capacity, these islands face the considerable challenge of meeting the social and economic needs of their populations in a manner that is sustainable. At the same time, they are forced to implement appropriate strategies to adapt to increasing threats resulting from greenhouse gas forcing of the climate system, to which they contribute little.

Sea-level rise. Although there will be regional variation in the signal, it is projected that sea level will rise by as much as 5 mm yr^{-1} over the next 100 years as a result of GHG-induced global warming. This change in sea level will have serious consequences for the social and economic development of many small island states. For some islands, the most serious consideration will be whether they will have adequate potential to adapt to sea-level rise within their own national boundaries.

Beach and coastal changes. Most coastal changes currently experienced in the small island states are attributable to human activity. With the projected increase in sea level over the next 50–100 years superimposed on further shoreline development, however, the coastal assets of these states will be further stressed. This added stress, in turn, will increase the vulnerability of coastal environments by reducing natural resilience, while increasing the economic and social "costs" of adaptation.

Biological systems. Coral reefs, mangroves, and seagrass beds, which provide the economic foundation for many small islands, often rely on "stable" coastal environments to sustain themselves. Although it is acknowledged that human-induced stresses are contributing to their degradation, these systems will be adversely affected by rising air temperature and sea levels. In most small islands, coral reefs already are undergoing great stress from episodic warming of the sea surface, causing widespread bleaching. Mangroves—which are common on low-energy, nutrient/sediment-rich coasts and embayments in the tropics—have been altered by human activities. Changes in sea levels are likely to affect landward and longshore migration of remnants of mangrove forests, which provide some protection for the coasts and backshore infrastructure. It is projected that changes in the availability of sediment supply, coupled with increases in temperature and water depth as a consequence of sea-level rise, will adversely impact the productivity and physiological functions of seagrasses. Consequently, this would have a negative downstream effect on fish populations that feed on these communities.

Biodiversity. It is estimated that 33% of known threatened plants are island endemics, and 23% of bird species found on islands also are threatened. Although there is still some uncertainty about precisely how and to what extent biodiversity and wildlife in small islands will be affected, available projections suggest that climate change and sea-level rise will cause unfavorable shifts in biotic composition and adversely affect competition among some species.

Water resources, agriculture, and fisheries. The availability of water resources and food remain critical concerns in island communities. In many countries, water already is in short supply because islands (many of which are drought-prone) rely heavily on rainwater from small catchments or limited freshwater lenses. Arable land for crop agriculture often is in short supply; thus, the likely prospect of land loss and soil salinization as a consequence of climate change and sea-level rise will threaten the sustainability of both subsistence and commercial agriculture in these islands. Because water resources and agriculture are so climate sensitive, it is expected that these sectors also will be adversely affected by future climate and sea-level change. Although climate change is not expected to have a significant impact on world fisheries output, it is projected to have a severe impact on the abundance and distribution of reef fish population on the islands.

Human health, settlement and infrastructure, and tourism. Several human systems are likely to be affected by projected changes in climate and sea levels in many small island states. Human health is a major concern in many tropical islands, which currently are experiencing a high incidence of vector- and water-borne diseases. This is attributable partly to changes in temperature and rainfall, which may be linked to the ENSO phenomenon, and partly to changes in the patterns of droughts and floods. Climate extremes also place a huge burden on human welfare; this burden is likely to increase in the future. Almost all settlements, socioeconomic infrastructure, and activities such as tourism in many island states are located at or near coastal areas. Their location alone renders them highly vulnerable to future climate change and sea-level rise. Tourism is a major revenue earner and generates significant employment in many small islands. Changes in temperature and rainfall regimes, as well as loss of beaches, could be devastating for the economies that rely on this sector. Because climate change and sea-level rise are inevitable in the future, it is vital that beach and coastal assets in the small island states are managed wisely. Integrated coastal management has been identified and proposed as an effective framework for accomplishing this goal.

Sociocultural and traditional assets. Other island assets, such as know-how and traditional skills (technologies), are under threat from climate change and sea-level rise. In some societies, community structures and assets such as important traditional sites of worship, ritual, and ceremony—particularly those at or near the coasts—could be adversely affected by future climate change and sea-level rise.

It is significant to note that although many vulnerability assessment methodologies have been applied to different regions of the world with varying degrees of success, global assessments have consistently identified the small island states as one of the most high-risk areas, irrespective of methodology employed. This evidently robust finding must be of considerable concern to these states. It is further established that climate change is inevitable as a result of past GHG emissions and that small islands are likely to suffer disproportionately from the enhanced effects of climate change and sea-level rise. Hence, identification and implementation of effective adaptation measures and avoidance of maladaptation (i.e., measures that increase exposure rather than decrease vulnerability) are critical for small islands, even if there is swift implementation of any global agreement to reduce future emissions.

For most small islands, the reality of climate change is just one of many serious challenges with which they are confronted. Such pressing socioeconomic concerns as poverty alleviation; high unemployment; and the improvement of housing, education, and health care facilities all compete for the slender resources available to these countries. In these circumstances, progress in adaptation to climate change almost certainly will require integration of appropriate risk reduction strategies with other sectoral policy initiatives in areas such as sustainable development planning, disaster prevention and management, integrated coastal management, and health care planning.

17.1. Regional Characteristics

17.1.1. Review of Previous Work

Almost without exception, small island states have been shown to be at great risk from projected impacts of climate change, particularly sea-level rise. The projected global rate of rise of 5 mm yr^{-1} (±2–9 mm yr^{-1}) is two to four times greater than the rate experienced in the previous 100 years (IPCC, 1998). Many of these islands rarely exceed 3–4 m above present mean sea level; even on the higher islands, most of the settlements, economic activity, infrastructure, and services are located at or near the coast.

Reliable instrumental records indicate that on average, Caribbean islands have experienced an increase in temperature exceeding 0.5°C since 1900. During the same period, there has been a significant increase in rainfall variability, with mean annual total rainfall declining by approximately 250 mm. For Pacific islands, the post-1900 temperature increase has been slightly lower than in the Caribbean: less than 0.5°C. In the case of rainfall, no clear trend emerges from the record, which shows decadal fluctuations of ±200 mm and 50–100 mm for mean annual and mean seasonal totals, respectively (IPCC, 1998).

Given the strong influence of the ocean on the climate of islands, and based on projected warming of the oceans (1–2°C for the Caribbean Sea and Atlantic, Pacific, and Indian Oceans) with a doubling of carbon dioxide (CO_2), small islands in these regions are expected to continue to experience moderate warming in the future. Under a similar scenario, mean rainfall intensity also is projected to increase by 20–30% over the tropical oceans, where most small island states are located. However, a decrease in mean summer precipitation over the Mediterranean Sea, where the small islands of Malta and Cyprus are located, is projected (IPCC, 1998).

Based on these projections, certain marine and coastal ecosystems are most likely to be adversely affected. Corals—many species of which currently exist near the upper limits of their tolerance to temperature—are projected to experience more frequent bleaching episodes as a result of elevated sea surface temperatures (SSTs). Bleaching therefore will pose a distinct threat to the productivity and survival of these valuable ecosystems (Wilkinson, 1996; Brown, 1997a,b; CARICOMP, 1997; Woodley *et al.*, 1997). On some islands, mangroves also are expected to be threatened by the impacts of climate change. Where the rate of sedimentation is slower than the projected rate of sea-level rise and where mangroves cannot naturally adapt and migrate landward, the vulnerability of these ecosystems will increase (IPCC, 1996, 1998). In some cases, the natural resilience of these ecosystems already has been impaired by anthropogenic stresses; thus, their capacity to cope with an additional stressor such as climate change will be further compromised.

Sea-level rise poses by far the greatest threat to small island states relative to other countries. Although the severity of the threat will vary from island to island, it is projected that beach erosion and coastal land loss, inundation, flooding, and salinization of coastal aquifers and soils will be widespread. Moreover, protection costs for settlement, critical infrastructure, and economic activities that are at risk from sea-level rise will be burdensome for many small island states. Similarly, tourism—the leading revenue earner in many states—is projected to suffer severe disruption as a consequence of adverse impacts expected to accompany sea-level rise (Teh, 1997; IPCC, 1998).

17.1.2. Special Circumstances of Small Islands

Small island states are by no means a homogeneous group of countries. They vary by geography; physical, climatic, social, political, cultural, and ethnic character; and stage of economic development. Yet they tend to share several common characteristics that not only identify them as a distinct group but underscore their overall vulnerability in the context of sustainable development (Maul, 1996; Leatherman, 1997). These common characterists include the following:

- Limited physical size, which effectively reduces some adaptation options to climate change and sea-level rise (e.g., retreat; in some cases entire islands could be eliminated, so abandonment would be the only option)
- Generally limited natural resources, many of which already are heavily stressed from unsustainable human activities
- High susceptibility to natural hazards such as tropical cyclones (hurricanes) and associated storm surge, droughts, tsunamis, and volcanic eruptions
- Relatively thin water lenses that are highly sensitive to sea-level changes
- In some cases, relative isolation and great distance to major markets
- Extreme openness of small economies and high sensitivity to external market shocks, over which they exert little or no control (low economic resilience)
- Generally high population densities and in some cases high population growth rates
- Frequently poorly developed infrastructure (except for major foreign exchange-earning sectors such as tourism)
- Limited funds and human resource skills, which may severely limit the capacity of small islands to mitigate and adapt to the effects of climate change.

17.1.3. Past and Present Trends

The Intergovernmental Panel on Climate Change (IPCC) in its Second Assessment Report (SAR) projected a global average temperature increase of 1.0–3.5°C, and a consequential rise in global mean sea level of 15–95 cm by the year 2100. Results from observational data show that temperatures have been increasing by as much as 0.1°C per decade and sea levels by approximately 2 mm yr^{-1} in the regions where most of the small island states are located—namely, the Pacific, Indian, and Atlantic Oceans and the Caribbean Sea (see Figure 17-1). There also is evidence that the ENSO phenomenon will continue

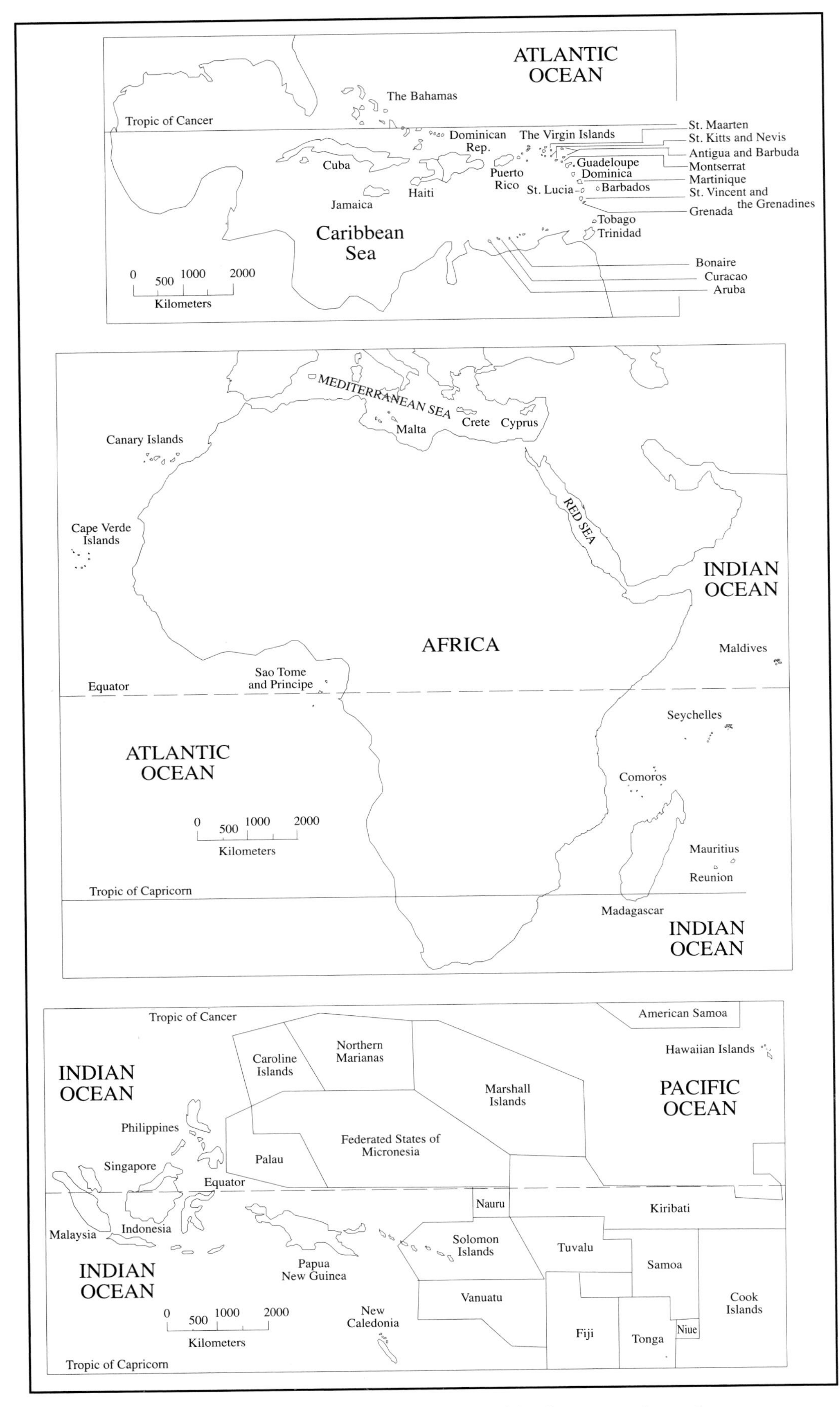

Figure 17-1: Main regions of the world in which small island states are located.

to have a major influence on climate variability in these regions.

Analyses conducted by the New Zealand Meteorological Services reveal that, since 1920, temperature has risen by 0.6–0.7°C in Noumea (New Caledonia) and Rarotonga (Cook Islands), which is greater than the mean global increase. Based on data from 34 stations in the Pacific from about 160°E and mostly south of the equator, surface air temperatures have increased by 0.3–0.8°C during the 20th century, with the greatest increase in the zone southwest of the SPCZ. This is well in excess of global rates of warming. Further recent work undertaken by the New Zealand Institute of Water and Atmospheric Research (NIWA) shows a noticeable change in aspects of the South Pacific climate since the mid-1970s. For instance, western Kiribati, the northern Cook Islands, Tokelau, and northern French Polynesia have become wetter, whereas New Caledonia, Fiji, and Tonga have become drier. Meanwhile, Samoa, eastern Kiribati, Tokelau, and northeast French Polynesia have become cloudier, with warmer nighttime temperatures; New Caledonia, Fiji, Tonga, the southern Cook islands, and southwest French Polynesia and Tuvalu have become warmer and sunnier (Salinger *et al*., 1995). It also might be noted that in the Pacific generally, observed changes in temperature and cloudiness appear to be closely influenced by the pattern of ENSO events (Salinger, 1999).

The records also indicate that rainfall has increased in the northeast Pacific but has decreased in the southwest Pacific region. Interannual variations in temperature and rainfall are strongly associated with ENSO, resulting in water shortages and drought in Papua New Guinea, the Marshall Islands, the Federated States of Micronesia, American Samoa, Samoa, and Fiji. Although a causal link has yet to be established, all of the foregoing changes have coincided with an eastward shift of the SPCZ since 1970. Research now suggests that some of the foregoing changes (including the shift in the SPCZ) may be closely correlated with interdecadal patterns of variability—for example, the Pacific Decadal Oscillation (PDO) (Salinger and Mullen, 1999). It should be noted, nevertheless, that the changes observed in the 20th century are considered to be consistent with patterns related to anthropogenic GHG-induced climate change (Salinger *et al.*, 1995; Hay, 2000).

The most significant and more immediate consequences for small island states are likely to be related to changes in sea levels, rainfall regimes, soil moisture budgets, and prevailing winds (speed and direction), as well as short-term variations in regional and local sea levels and patterns of wave action (Sem *et al.*, 1996). The short-term (including interannual) variations are likely to be strengthened by the ENSO phenomenon. Vulnerability assessment studies undertaken in some small islands suggest that climate change will impose diverse and significant impacts on small island states (Leatherman, 1997). In most small islands (including the high islands), the majority of the population, socioeconomic activities, and infrastructure are located within a few hundred meters of the coast; therefore, they are highly vulnerable to the impacts of climate change and sea-level rise (Hay *et al.*, 1995; Bijlsma, 1996; Nurse *et al.*, 1998; Burns, 2000). In this regard, an increase in the frequency and magnitude of tropical cyclones would be a major concern for small island states. This would increase the risk of flooding, accelerate existing rates of beach erosion, and cause displacement of settlements and infrastructure.

The key questions, therefore, are how will the impacts manifest themselves, and what are the most appropriate responses for avoiding, minimizing, or adapting to these impacts? Because small island states traditionally experience some of the greatest interannual variations in climatic and oceanic conditions, many of their natural systems are well adapted to the stresses that result. Thus, many strategies that small island states might employ to adapt to climate change usually are the same as those that constitute sound environmental management, wise use of resources, and appropriate responses to present-day climate variability.

Although the full extent of climate change impacts in the small island states is far from certain, mostly adverse consequences are projected for several systems. The combined effect of GHG-induced climate change and sea-level rise can contribute to coastal erosion and land loss, flooding, soil salinization, and intrusion of saltwater into groundwater aquifers. The quantity and quality of available water supplies can affect agricultural production and human health. Similarly, changes in SST, ocean circulation, and upwelling could affect marine organisms such as corals, seagrasses, and fish stocks. Tourism—which is a very important economic activity in many island states—could be affected through beach erosion, loss of land, and degraded reef ecosystems, as well as changes in seasonal patterns of rainfall.

17.1.4. Scenarios of Future Climate Change and Variability

17.1.4.1. The Models Used

Projections of future climate change discussed in this chapter essentially are based on data sets available at IPCC Data Distribution Centers (DDCs) at Hamburg and Norwich and currently available numerical experiments with state-of-the-art global climate models that consider a near-identical GHG

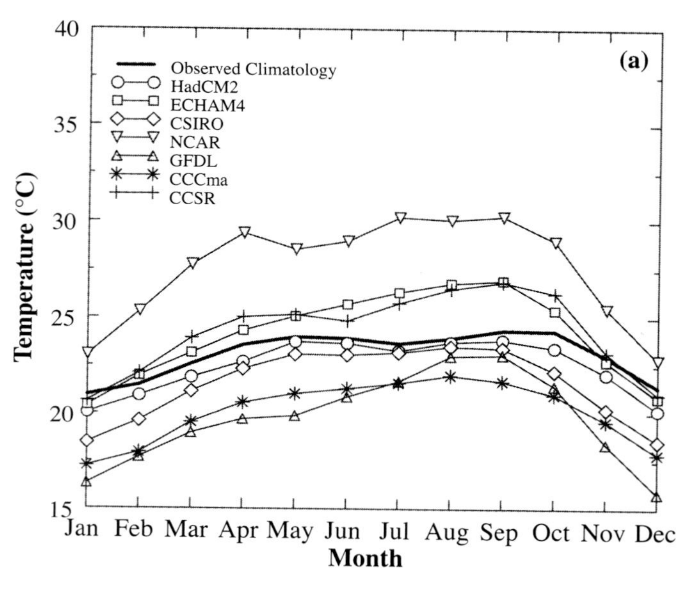

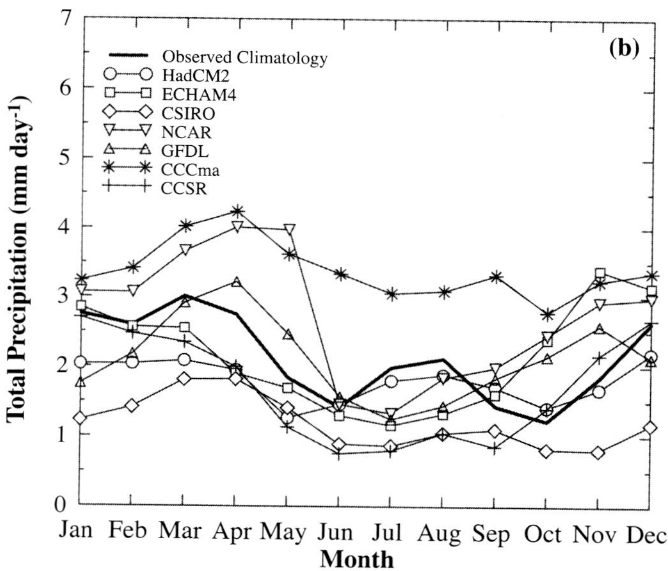

Figure 17-2: Validation of seven AOGCM-simulated and observed climatology [(a) temperature and (b) rainfall] for Mediterranean Sea islands.

forcing: 1% yr^{-1} growth in GHG concentrations (IS92a) after 1990. Coupled AOGCMs offer the most credible tools for estimating the future response of climate to anthropogenic radiative forcings. The DDCs have compiled outputs generated in transient experiments with a set of seven recent AOGCMs that reflect the state-of-the-art of model experiments and provide a representative range of results from different AOGCMs.

17.1.4.2. Region-Specific Model Validation

A model validation exercise was undertaken for the four main regions (the Atlantic Ocean and Caribbean Sea, the Pacific Ocean, the Indian Ocean, and the Mediterranean Sea region) in which the majority of small island states are located. The results indicate that five of the seven AOGCMs [HadCM2 (UK), ECHAM4 (Germany), CSIRO (Australia), CCSR/NIES (Japan), and CCCma (Canada)] have reasonable capability in simulating the broad features of present-day climate and its variability over these regions. Comparison of the monthly mean observed and model-simulated climatology of surface air temperature and rainfall over the Mediterranean Sea and Indian Ocean regions is shown in Figures 17-2 and 17-3. The time period of 30 years (1961–1990) in baseline climatology has been used for the purpose of these model validation exercises and to generate climate change scenarios for these island regions. Two future time periods centered around the 2050s (2040–2069) and the 2080s (2070–2099) have been considered here in developing scenarios of changes in surface air temperature and precipitation.

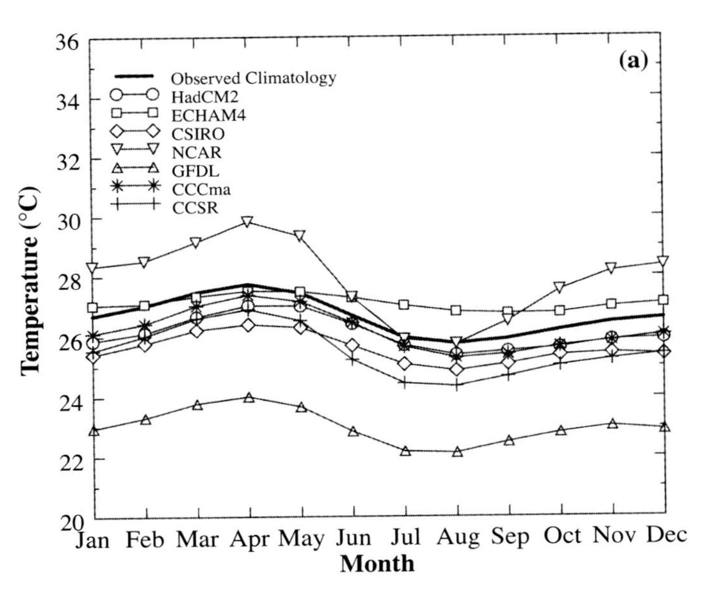

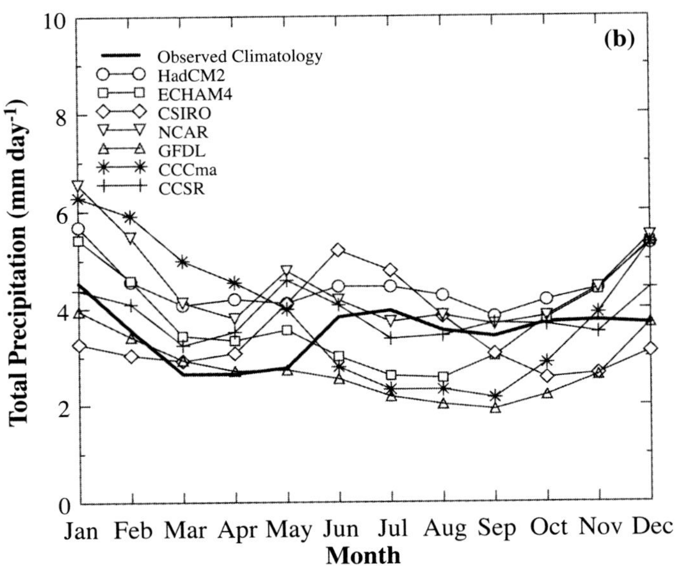

Figure 17-3: Validation of seven AOGCM-simulated and observed climatology [(a) temperature and (b) rainfall] for Indian Ocean islands.

For the four regions identified above, the HadCM2, ECHAM4, CSIRO, CCSR/NIES, and CCCma models have demonstrated good capability in simulating the present-day area-averaged monthly mean climatology in terms of surface temperature, diurnal temperature range, and rainfall (Lal *et al.*, 2001). The model projections discussed here are the scenarios arising from GHG-induced positive radiative forcings and those that take into account the negative radiative forcing of sulfate aerosols (direct effects).

17.1.4.3. Climate Change Projections for Small Island States

17.1.4.3.1. Surface air temperature

Tables 17-1, 17-2, and 17-3 provide an ensemble of climate change scenario results inferred from the five AOGCMs referred to above for four regions. These tables summarize projected seasonal mean changes in surface air temperature and precipitation during the 2050s and 2080s. The projected area-averaged annual mean warming as a consequence of increases in atmospheric concentrations of GHGs over the Atlantic Ocean and Caribbean Sea, the Pacific Ocean , the Indian Ocean, and the Mediterranean Sea is approximately 2.0, 2.0, 2.1, and 2.8°C, respectively, for the 2050s, and approximately 3.1, 3.0, 3.2, and 4.3°C, respectively, for the 2080s (see Table 17-1). Projected warming over the Mediterranean Sea area is marginally higher during Northern Hemisphere (NH) summer than during NH winter for both time periods. The increase in surface air temperature is more or less uniform in both seasons in the other three regions (Tables 17-2 and 17-3). The scatter of projected annual mean surface temperature increase over each of the regions, as simulated by each of the five GCMs, is depicted in Figures 17-4 to 17-7. The area-averaged annual increase in mean surface temperature resulting from increases in GHGs is projected to be smallest over the Pacific Ocean region and highest over the Mediterranean Sea. It is evident that even though aerosol forcing marginally reduces surface warming, the magnitude of projected warming is still considerable and could substantially impact the small island states.

With respect to diurnal changes, GCM simulations with increasing concentrations of GHGs in the atmosphere suggest relatively more

***Table 17-1**: Ensemble of annual mean climate change scenarios for small islands for the 2050s and 2080s as inferred from AOGCMs (numbers in brackets show standard deviation between model projections).*

Regions	Annual Mean Temperature Change (°C)				Annual Mean Precipitation Change (%)			
	2050s		*2080s*		*2050s*		*2080s*	
	GHG	GHG+A	GHG	GHG+A	GHG	GHG+A	GHG	GHG+A
Atlantic Ocean and Caribbean	2.03 (±0.43)	1.71 (±0.25)	3.06 (±0.84)	2.64 (±0.61)	-5.2 (±11.9)	-1.3 (±7.8)	-6.8 (±15.8)	-0.7 (±12.3)
Pacific Ocean	1.98 (±0.41)	1.63 (±0.23)	2.99 (±0.87)	2.54 (±0.63)	5.5 (±2.5)	4.9 (±0.8)	7.6 (±3.3)	7.0 (±1.9)
Indian Ocean	2.10 (±0.43)	1.64 (±0.23)	3.16 (±0.89)	2.61 (±0.65)	3.1 (±4.5)	1.6 (±3.9)	5.1 (±4.3)	4.3 (±4.9)
Mediterranean	2.83 (±0.62)	2.31 (±0.29)	4.27 (±1.26)	3.57 (±0.83)	1.0 (±11.0)	-2.4 (±8.6)	4.3 (±14.9)	-0.1 (±12.9)

***Table 17-2**: Ensemble of seasonal mean climate change scenarios for small islands for the 2050s as inferred from AOGCMs (numbers in brackets show standard deviation between model projections).*

Regions	Temperature Change (°C)				Precipitation Change (%)			
	December–February		*June–August*		*December–February*		*June–August*	
	GHG	GHG+A	GHG	GHG+A	GHG	GHG+A	GHG	GHG+A
Atlantic Ocean and Caribbean	2.00 (±0.46)	1.68 (±0.32)	2.01 (±0.44)	1.71 (±0.21)	3.4 (±14.3)	5.9 (±7.4)	-14.4 (±12.2)	-6.9 (±11.5)
Pacific Ocean	1.98 (±0.39)	1.65 (±0.20)	1.98 (±0.43)	1.61 (±0.27)	4.3 (±1.9)	3.7 (±1.2)	7.2 (±4.8)	6.8 (±3.3)
Indian Ocean	2.11 (±0.43)	1.67 (±0.15)	2.09 (±0.44)	1.63 (±0.30)	3.5 (±6.0)	2.0 (±7.5)	-1.8 (±10.0)	-4.7 (±4.5)
Mediterranean	2.64 (±0.72)	2.27 (±0.44)	2.93 (±0.53)	2.27 (±0.17)	8.1 (±14.7)	2.6 (±15.7)	-4.8 (±10.3)	-8.9 (±6.0)

***Table 17-3**: Ensemble of seasonal mean climate change scenarios for small islands for the 2080s as inferred from AOGCMs (numbers in brackets show standard deviation between model projections).*

Regions	Temperature Change (°C)				Precipitation Change (%)			
	December–February		*June–August*		*December–February*		*June–August*	
	GHG	GHG+A	GHG	GHG+A	GHG	GHG+A	GHG	GHG+A
Atlantic Ocean and Caribbean	3.01 (±0.87)	2.61 (±0.66)	3.07 (±0.86)	2.64 (±0.61)	4.8 (±14.6)	8.5 (±12.9)	-19.2 (±18.8)	-8.2 (±17.1)
Pacific Ocean	2.97 (±0.82)	2.56 (±0.57)	2.98 (±0.91)	2.52 (±0.67)	6.0 (±1.5)	5.6 (±1.6)	10.2 (±5.9)	8.9 (±4.9)
Indian Ocean	3.18 (±0.88)	2.61 (±0.60)	3.16 (±0.91)	2.62 (±0.69)	5.9 (±10.3)	6.2 (±10.5)	-2.6 (±12.6)	-5.9 (±7.4)
Mediterranean	3.94 (±1.34)	3.31 (±1.01)	4.52 (±1.16)	3.70 (±0.72)	16.1 (±21.1)	9.9 (±21.6)	-7.4 (±16.2)	-11.6 (±10.7)

pronounced increases in minimum temperature than in maximum temperature over the regions where the small island states are located, on an annual mean basis as well as during winter, for both the 2050s and the 2080s. Hence, a marginal decrease in diurnal temperature range (between 0.3 and 0.7°C) is projected.

17.1.4.3.2. Precipitation

In general, all AOGCMs simulate only a marginal increase or decrease in annual rainfall. An area-averaged annual mean increase in precipitation of approximately 0.3% for the 2050s and 0.7% for the 2080s over the Pacific Ocean area is projected, either as a consequence of increases in atmospheric concentrations of GHGs or because of the combined influence of GHGs and sulfate aerosols. The projected increase in precipitation is at a maximum during NH summer (June-July-August) for both time periods. A marginal decline in precipitation is projected for the other three regions, particularly during NH summer—suggesting the possibility of reduced water availability (Lal *et al.*, 2001).

17.1.4.3.3. Extreme high temperature and precipitation events

An analysis of model-simulated daily temperature and precipitation data (from CSIRO and ECHAM model experiments) for the present-day atmosphere and for the two future time slices (2050s and 2080s) projects that the frequency of extreme temperatures during the summer is likely to be higher in all four regions. This implies an increased likelihood of thermal stress conditions during the 2050s and more so during the 2080s. Similarly, although the models project a lesser number of annual rainy days, an increase in the daily intensity of precipitation also is projected (Lal *et al.*, 2001). This suggests an increase in the probability of occurrence of more frequent droughts, as well as floods, in the Atlantic Ocean and Caribbean Sea, the Mediterranean Sea, and the Indian and Pacific Oceans.

17.1.4.3.4. ENSO and precipitation variability

Although it is difficult to obtain reliable regional projections of climate change from GCMs, some consistent patterns are beginning to emerge for the Pacific with regard to ENSO and precipitation variability. Although ocean temperatures over most of the western and southern Pacific warm more slowly than the global average, the eastern central Pacific warms faster than the global average in outputs from several AOGCMs (Meehl and Washington, 1996; Timmerman *et al.*, 1997; Knutson and Manabe, 1998; Jones *et al.*, 1999). These results have been broadly interpreted as an El Niño-like pattern (e.g., Cai and Whetton, 2000) because some simulations appear

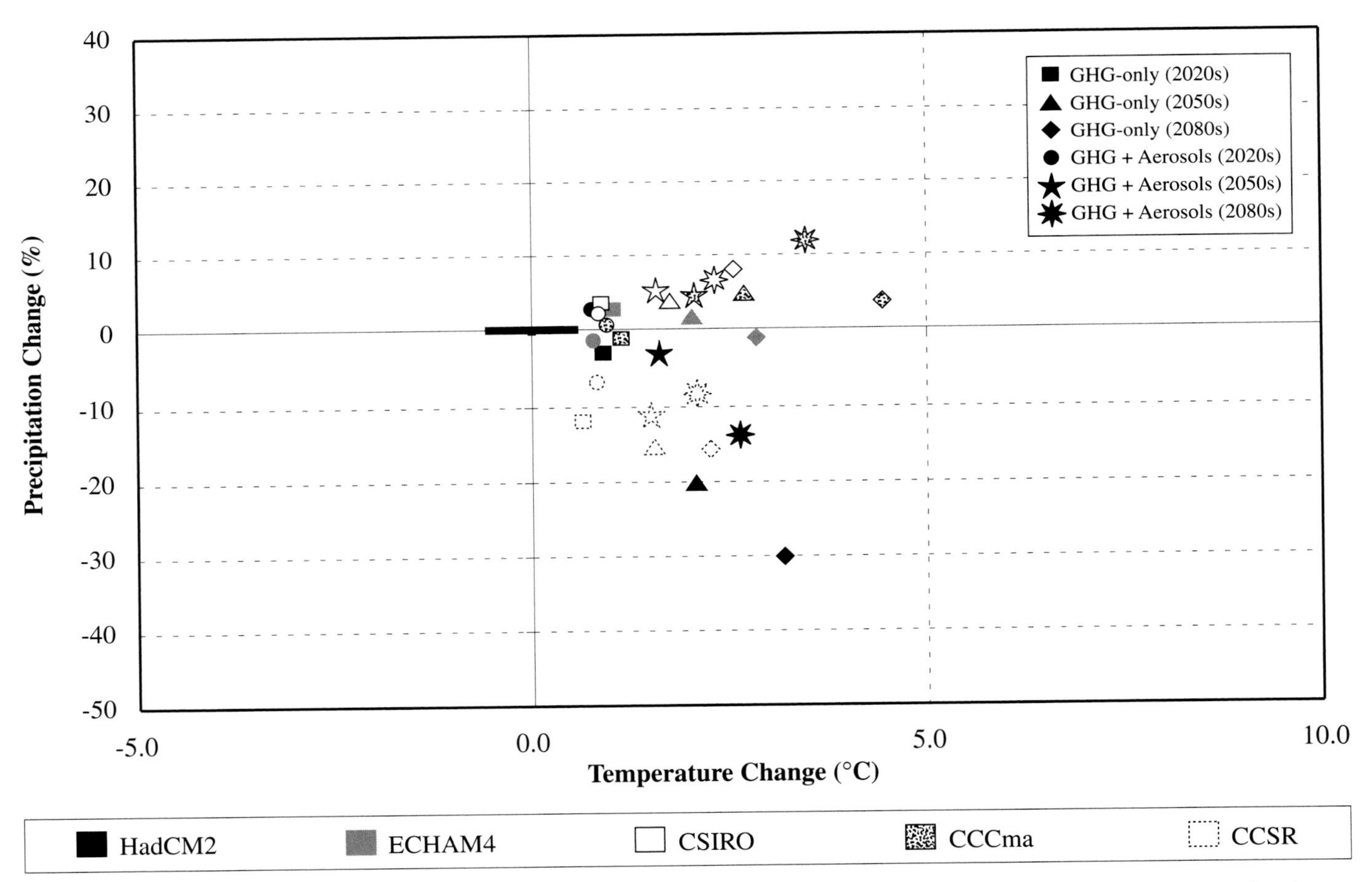

Figure 17-4: Climate change scenarios for Atlantic Ocean and Caribbean Sea islands as simulated by five AOGCMs for the 2020s, 2050s, and 2080s.

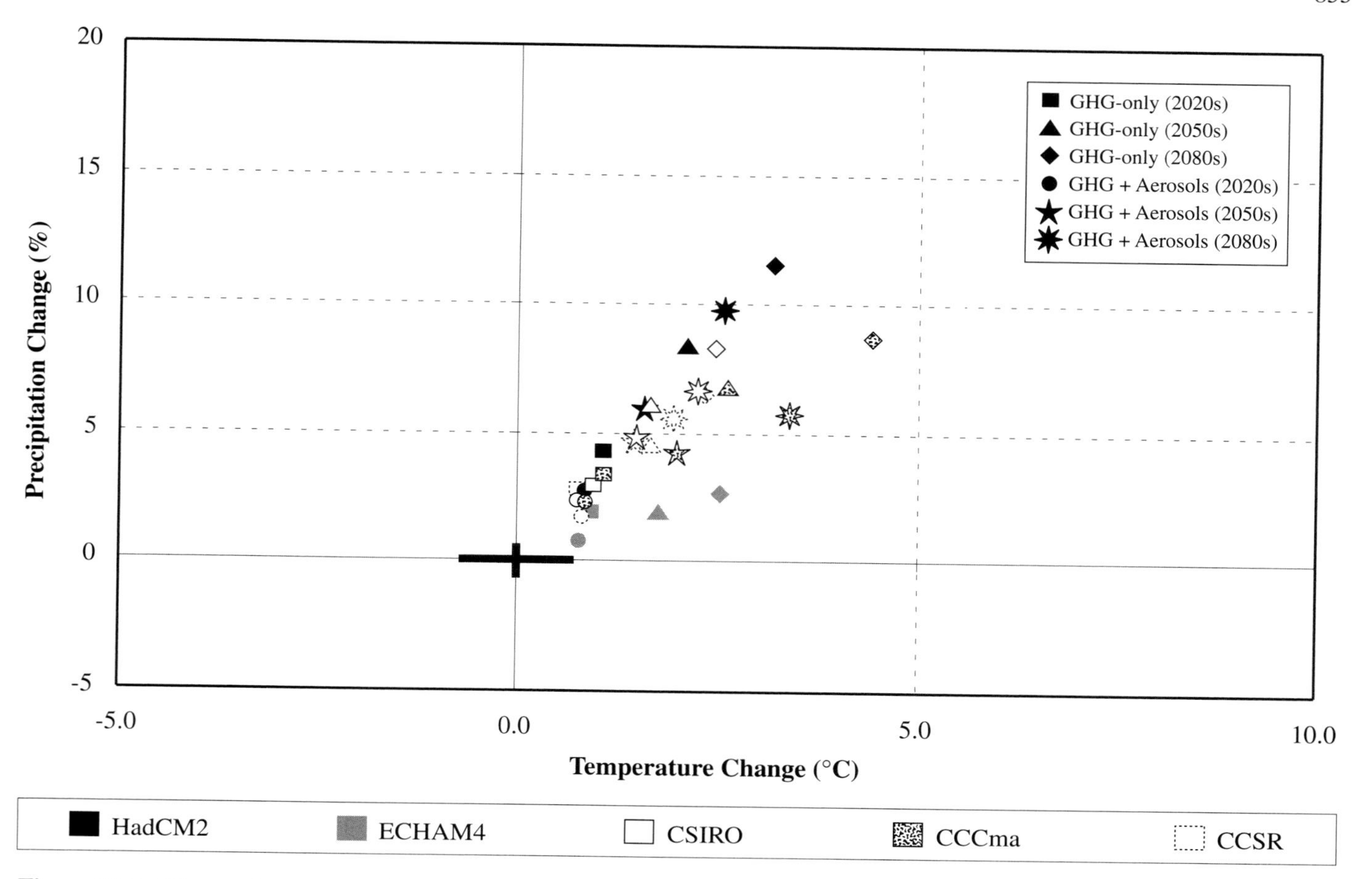

Figure 17-5: Climate change scenarios for Pacific Ocean islands as simulated by five AOGCMs for the 2020s, 2050s, and 2080s.

to project more frequent ENSO-like patterns (Timmerman *et al.*, 1997). In the Pacific, climate variability associated with the ENSO phenomenon manifests itself on a seasonal to decadal time scale. This, in turn, affects factors such as temperature, rainfall, wind speed and direction, sea level, and tropical cyclone climatology that are directly associated with numerous climate-related impacts in the region.

Jones *et al.* (1999) have produced projected ranges of change in average temperature and rainfall for the western Pacific, based on scaled patterns from three independent AOGCMs and a suite of two AOGCMs and a regional circulation model (RCM). These scaled patterns were produced by the regression method described in Hennessy *et al.* (1998) and Giorgi and Mearns (1999), in which the influences of decadal variability are significantly reduced. Jones *et al.* (1999) showed that warming in the Pacific region is projected to increase by less than the global average in most cases. Projections of rainfall are constrained by the models' ability to simulate the SPCZ and ITCZ. All models were able to produce aspects of both features, although the eastern part of the ITCZ was not well captured by any of the models; the higher resolution models were the most realistic. The models also did not produce consistent changes to these features, execept for a large increase in rainfall over the central and east-central Pacific. Most of the projected changes across the western Pacific were increases, with significant increases along the equatorial belt from North Polynesia to further east. Possible decreases were noted in some models for Melanesia and South Polynesia in both halves of the year (April–October and November–March).

Climate variability in the Pacific is a combination of seasonal, multi-annual variability associated with the ENSO phenomenon and decadal variability, the latter influencing the ENSO phenomenon itself. The major concern for impacts in the region is not with the mean climate changes described above but with the extremes that are superimposed on those mean changes. Numerous studies describe the likely intensification of rainfall when the mean change ranges from a slight decrease to an increase. For instance, mean decreases of 3.5% over South Polynesia from the mixed-layer GCM (UKHI) produced little change in intensity, whereas an increase of 7.5% over Micronesia halved the return periods of heavy rainfall events (Jones *et al.*, 1999).

GCMs currently project an increase in SSTs of approximately 1°C by the 2050s and increased rainfall intensity in the central equatorial Pacific, which would impact many small island states in that region. Recent variations over the tropical Pacific Ocean and surrounding land areas are related to the fact that since the mid-1970s, warm episodes (El Niño) have been relatively more frequent or persistent than the opposite phase (La Niña). There are indications that the ENSO phenomenon may be the primary mode of climate variability on the 2- to 5-year

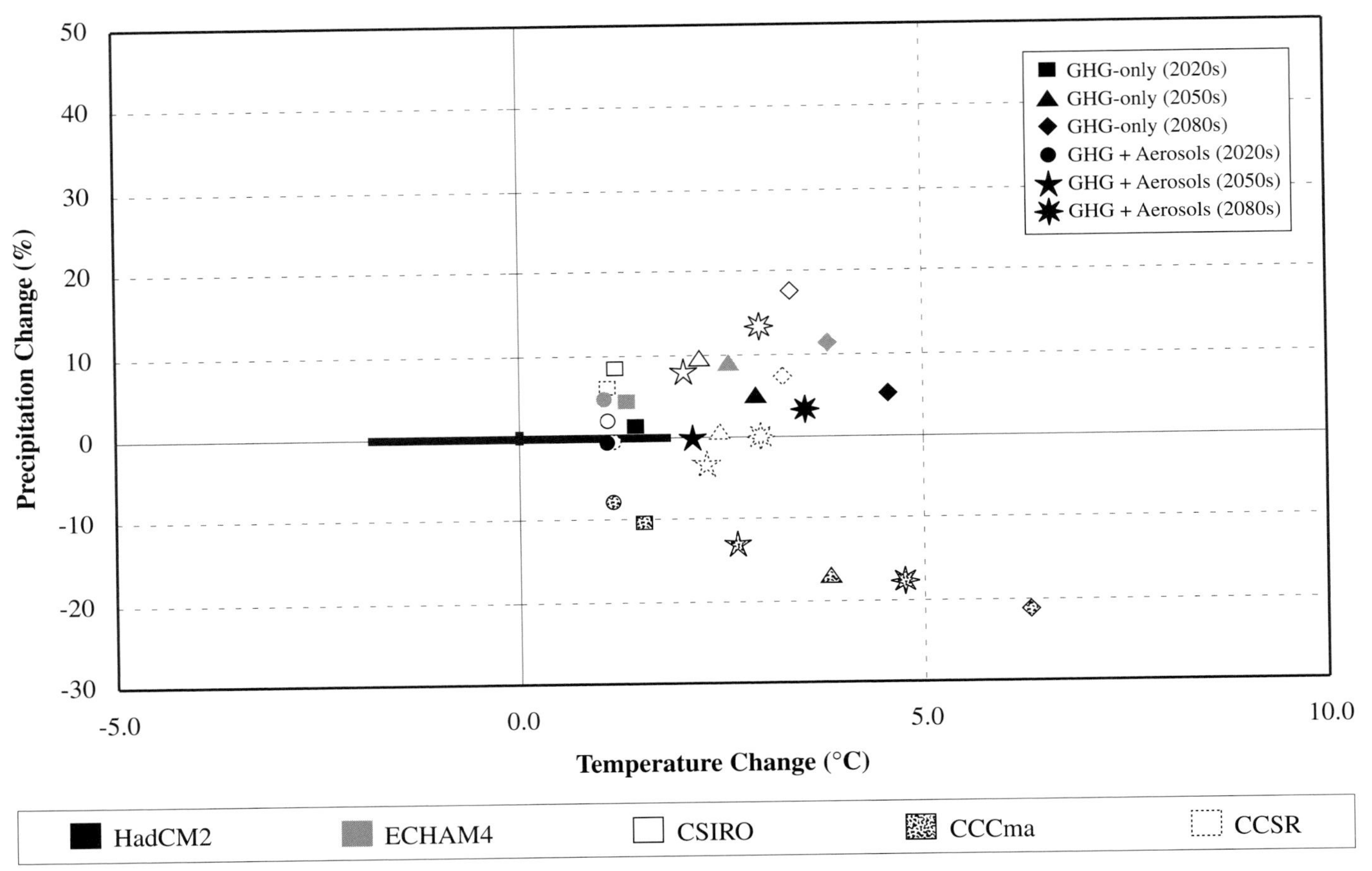

Figure 17-6: Climate change scenarios for Mediterranean Sea islands as simulated by five AOGCMs for the 2020s, 2050s, and 2080s.

time scale and that the current large interannual variability in the rainfall associated with ENSO is likely to dominate over any mean effects attributable to global warming (Jones *et al*., 1999).

Comparisons between observations and model simulations for the Pacific region further indicate that regional warming would be mostly less than the global average because of the large expanse of ocean. However, there has been a strong indication from several model simulations that the least warming would occur in the southern ocean; the greatest warming could be expected in the far west, central, and eastern equatorial Pacific. In the case of rainfall, increases are likely to be greater where warming over the ocean is greatest, although one GCM output showed no increase in rainfall variability between 1960 and 2100. Model variability is likely to be much lower than historical variability because of the great influence of ENSO, particularly in the Pacific (Jones *et al*., 1999).

17.1.4.3.5. Tropical cyclones

There is no consensus regarding the conclusions of studies related to the behavior of tropical cyclones in a warmer world. Working Group I concludes that current information is insufficient to assess recent trends, and confidence in understanding and models is inadequate to make firm projections (see TAR WGI). Royer *et al*. (1998), using a downscaled AOGCM coupled with Gray's method for hurricane forecasting, found no significant change in hurricane frequency or geographical extent for the north Atlantic Ocean, the Pacific Ocean, or the Indian Ocean.

Notwithstanding the foregoing conclusions, individual studies have reported the likelihood of a possible increase of approximately 10–20% in *intensity* of tropical cyclones under enhanced CO_2 conditions (Holland, 1997; Tonkin *et al*. 1997). This finding is supported by Jones *et al*. (1999), who conducted an analysis of tropical cyclones from a 140-year simulation of an RCM nested in a coupled AOGCM for the Pacific region (see Box 17-1). Although the preliminary analysis implies that there might be a small decrease in cyclone formation, an increase in system intensity is projected. The pattern of cyclones during phases of ENSO was unchanged, suggesting that the relationship between cyclone distribution and ENSO may continue. The study by Jones *et al*. (1999) considers that increases in cyclone intensity (10–20%) estimated by Tonkin *et al*. (1997) and Holland (1997) are highly likely.

17.2. Key Regional Concerns

17.2.1. Development, Sustainability, and Equity Issues

The small island states account for less than 1% of global GHG emissions but are among the most vulnerable of all areas to the

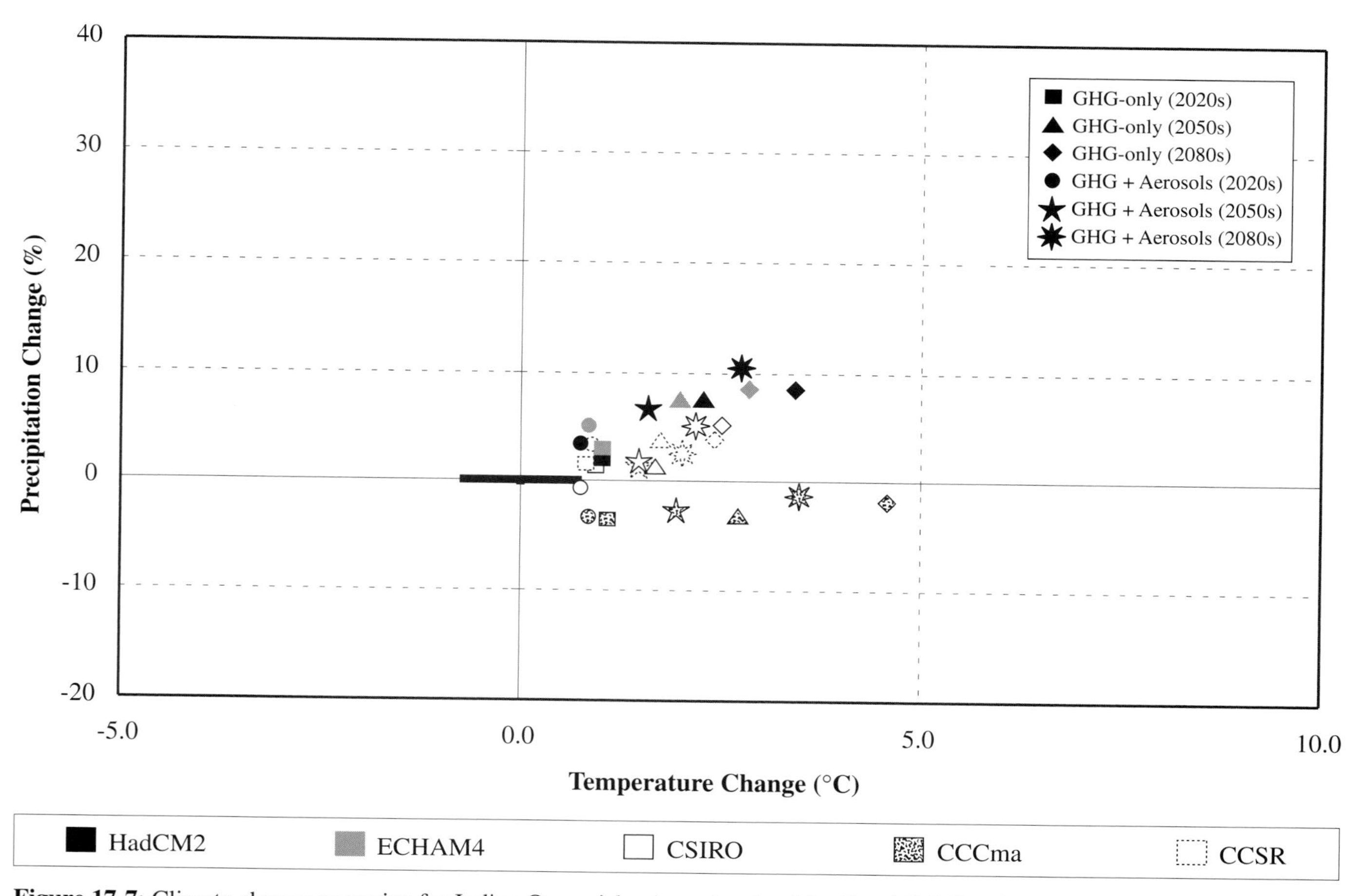

Figure 17-7: Climate change scenarios for Indian Ocean islands as simulated by five AOGCMs for the 2020s, 2050s, and 2080s.

potential adverse effects of climate change and sea-level rise (Jones, 1998; Nurse *et al.*, 1998). It has been established that there already is a global commitment to climate change and sea-level rise as a result of greenhouse forcing arising from historic emissions (Warwick *et al.*, 1996; Jones, 1998; Nicholls *et al.*, 1999; Parry *et al.*, 1999). Moreover, analysis has shown that even with a fully implemented Kyoto Protocol, by 2050 warming would be only about 1/20th of a degree less than what is projected by the IPCC (Parry *et al.*, 1999). Therefore, climate change impacts are inevitable.

Thus, owing to their high vulnerability and low adaptive capacity to climate change, communities in small island states have legitimate concerns about their future on the basis of the past observational record and present climate model projections. Economic development, quality of life, and alleviation of poverty presently constitute the most pressing concerns of many small island states. Thus, with limited resources and low adaptive capacity, these islands face the considerable challenge of charting development paths that are sustainable and controlling GHG emissions, without jeopardizing prospects for economic development and improvements in human welfare (Munasinghe, 2000; Toth, 2000). At the same time, given the inevitability of climate change and sea-level rise, they are forced to find resources to implement strategies to adapt to increasing threats resulting from GHG forcing of the climate system, to which they contribute little (Hay and Sem, 1999; Sachs, 2000). Consequently, the already meager resources of these island states will be placed under further pressure.

17.2.2 Water-Level Changes

17.2.2.1. Sea-Level Rise

Although the severity of the threat will vary regionally, sea-level rise of the magnitude currently projected (i.e., 5 mm yr^{-1}, with a range of 2–9 mm yr^{-1}), is expected to have disproportionately great effects on the economic and social development of many small island states (Granger, 1997; IPCC, 1998). Coastal land loss already is projected to have widespread adverse consequences. Indeed, it is argued that land loss from sea-level rise, especially on atolls (e.g., those in the Pacific and Indian Oceans) and low limestone islands (e.g., those in the Caribbean), is likely to be of a magnitude that would disrupt virtually all economic and social sectors in these countries (Leatherman, 1997). Recent estimates indicate that with a 1-m rise in sea level, 10.3 km^2 of land in Tongatapu island, Tonga, would be lost (Mimura and Pelesikoti, 1997). This figure would increase to 37.3 km^2 (14%) with storm surge superimposed on a 1-m sea-level rise scenario. For some main Yap Island (Federated States of Micronesia) sites, a retreat of 9 to 96 m is projected with a 1-m rise in sea level (Richmond *et al.*, 1997). On Majuro Atoll, Marshall Islands, land loss from one area based on the Bruun rule is estimated

Box 17-1. Climate Change Scenarios for the South Pacific Region

As part of the Pacific Islands Climate Change Assistance Programme, the Commonwealth Scientific and Industrial Research Organisation (CSIRO) has prepared regional climate change scenarios for the four South Pacific regions of Micronesia, Melanesia, and north and south Polynesia (Jones *et al.*, 1999). Six coupled atmosphere-ocean climate simulations were included in the analysis of regional climate change scenarios: CSIRO Mark 2 GCM with and without sulfates, CSIRO DARLAM 125 km, DKRZ ECHAM4/OPYC3 GCM, Hadley Centre HADCM2, and the Canadian CGCM1. The regional scenarios derived can be considered as projections that represent a range of possible future climates.

Generally, the models project a temperature increase that is less than the global mean. Results show the least warming in the South Pacific; regional maximum warming is projected in the far west, central, and eastern equatorial Pacific. Four of the models show an increase in rainfall over the central and eastern Pacific over both half-years (i.e., May to October and November to April). Movements of both the ITCZ and the SPCZ were not consistent between models, but rainfall consistently increased. Increases in daily rainfall intensity are expected in regions where rainfall increases, remains the same, or decreases slightly, as derived from several models and studies. Thus, high confidence is attached to this result.

Historical sea-level rise over the Pacific from tide gauge records adjusted for postglacial rebound is consistent with global estimates of 1–2 mm yr^{-1}. ENSO is the dominant influence on climate variability in the Pacific, and model outputs show that the ENSO phenomenon is likely to continue to 2100. The results also suggest that under climate change, there is likely to be a more El Niño-like mean state over the Pacific. There is no evidence that tropical cyclone numbers may change, but a general increase in *tropical cyclone intensity*, expressed as possible increases in wind speed and central pressure of 10–20% with $2xCO_2$ equivalent, now appears likely. Moderate confidence is attached to this result. No significant change in regions of formation was noted in the DARLAM 125 km resolution simulation, although this may alter in response to long-term changes to ENSO. Although there appears to be no major change in regions of origin, tropical cyclones showed a tendency to track further poleward. Low confidence is attached to this result.

to be nearly 65 ha of dry land from a 1-m rise in sea level (Holthus *et al.*, 1992).

One of the most serious considerations for some small islands is whether they will have adequate potential to adapt to sea-level rise within their own national boundaries (Nurse, 1992; IPCC, 1998). In tiny islands where physical space already is very scarce, adaptation measures such as retreat to higher ground and use of building set-backs appear to have little practical utility. In extreme circumstances, sea-level rise and its associated consequences could trigger abandonment and significant "off-island migration," at great economic and social costs (Leatherman, 1997; Nicholls and Mimura, 1998).

17.2.2.2. Storm Surge and Flood Risks

Changes in the highest sea levels at a given locality will result from the change in mean sea level at that location and changes in storm-surge heights. If mean sea level rises, present extreme levels will be attained more frequently, all else being equal. The increase in maximum heights will be equal to the change in the mean, which implies a significant increase in the area threatened with inundation. This will be especially true in areas with a small surge envelope, which is typical in most small islands. Under such circumstances, even incrementally small elevations in sea level would have severely negative effects on atolls and low islands (Forbes and Solomon, 1997; Nicholls *et al.*, 1999).

Changes in storm-surge heights also would result from alterations in the occurrence of strong winds and low pressures, as would occur during the passage of tropical storms and cyclones. It is already known that tropical cyclones are the major cause of storm surges that impact small islands in the Atlantic, Pacific, and Indian Oceans. Changes in the frequency and intensity of tropical cyclones could result from alterations to SST, large-scale atmospheric circulation, and the characteristics of ENSO (Pittock *et al.*, 1996). Although there is no consensus yet with regard to whether there will be changes in the behavior of these systems (Royer *et al.*, 1998; Jones *et al.*, 1999), the prospect of extreme sea levels (related to storm surges and higher wave amplitudes) is a concern that small island states cannot easily ignore.

Based on global sea-level rise scenarios produced by the Hadley Centre (HADCM2 and HADCM3), Nicholls *et al.* (1999) estimate that global sea levels are expected to rise by about 38 cm between 1990 and the 2080s. They project that many coastal areas are likely to experience annual or more frequent flooding, with the islands of the Caribbean and the Indian and Pacific Oceans facing the largest relative increase in flood risk. Projected out to the 2080s, the number of people facing high flood risk from sea-level rise in these regions would be 200 times higher than in the case of no climate change (Nicholls *et al.*, 1999). Recent studies for Cuba (based on HADCM2 and IS92a scenarios) also project that 98 coastal settlements with a combined population exceeding 50,000 persons would be inundated by a 1-m rise in sea level (Perez *et al.*, 1999).

17.2.3. Beach and Coastal Changes

The morphology, characteristics, and classification of beaches are influenced by a variety of factors, including island origin, geologic structure and composition (e.g., volcanic, coral atoll, raised atoll, reef island, or emergent limestone), age, elevation, and size (Gillie, 1997; Solomon and Forbes, 1999). Thus, given their varied origins, a wide range of beach types and characteristics is represented in the small island states of the Caribbean and Mediterranean Seas and the Atlantic, Pacific, and Indian Oceans.

Coastal erosion—partly the result of anthropogenic factors such as sand mining (Gillie, 1997; Ragoonaden, 1997)—already is a problem on many islands and evidently may be exacerbated by sea-level rise (Mimura and Nunn, 1998). On many atolls (as in the Pacific) and low reef islands (as in the Caribbean), carbonate beaches are maintained by sand produced from productive reefs whose degradation already is causing accelerated beach erosion. Similarly, in the Mediterranean Sea, where the islands are periodically susceptible to flooding and scour from storm surges, an increase in storminess would further stress natural and human systems located at the coast (Nicholls and Hoozemans, 1996). The impact of the equivalent of a 1-m rise in sea level for a cumulative few hours on the coast was observed in Singapore on 7-14 February 1974. The event indicates clearly the vulnerability of the island's low-lying area to flooding and coastal erosion if sea-level rise takes place (Wong, 1992).

Beach erosion rates of approximately 2–4 m yr^{-1}, on average, have been reported for several beaches in Trinidad and Tobago, where mean relative sea-level rise of 8–10 mm yr^{-1} has been recorded by two gauges during the past 15 years. Although beach erosion results from multiple forces, sea-level rise is considered a contributory factor (Singh, 1997a,b). Ragoonaden (1997) measured shoreline retreat of 2.7 m yr^{-1} at Flic-en-Flac (Mauritius), and Nunn and Mimura (1997) report that the coasts of some islands in Fiji have retreated by more than 30 m in the past 70 years. In the specific case of Viti Levu and Taveuni, Fiji, beach erosion has been attributed to a combination of human-induced causes (including loss of the mangrove fringe and other natural protection) and elevated sea level, which has been rising at a rate of approximately 1–1.5 mm yr^{-1} since 1960 (Forbes and Solomon, 1997; Nunn and Mimura, 1997).

17.2.3.1. Response, Adaptation, and Management

It is now widely accepted that strategies for adaptation to sea-level rise tend to fall into three main categories: retreat, accommodate, and protect (Bijlsma, 1996; IPCC, 1996, 1998). Hard engineering—involving the construction of groynes, seawalls, breakwaters, and bulkheads—has long been the traditional response to coastal erosion and flooding in many small island states. Unfortunately, this approach has not always been efficiently implemented and has even helped to increase coastal vulnerability in some cases (Mimura and Nunn, 1998; Solomon and Forbes, 1999). In these specific circumstances, the term "maladaptation" (which refers to a response that does not succeed in reducing vulnerability but increases it instead; see Chapter 18) may be applied. Realistically, however, for some islands the application of hard solutions may be the only practical option along well-developed coasts, where vital infrastructure is at immediate risk.

There are other potential options available to small island states, including enhancement and preservation of natural protection (e.g., replanting of mangroves and protection of coral reefs), use of softer options such as artificial nourishment, and raising the height of the ground of coastal villages (Nunn and Mimura, 1997). Raising the height of the ground requires additional aggregate such as sand and stone and a lot of pumping, in which many small islands are seriously deficient. Removal of materials from "unimportant" islands to build up important islands via sand transfer by pipes and barges has been suggested by the IPCC (1990). Some island states may be faced with few practical options. Thus, it might be necessary for them to lose some islands so that the entire nation is not completely inundated (Nicholls and Mimura, 1998).

Similarly, beach nourishment may not be a practical or economical option for many island nations because sand often is a scarce resource (Leatherman, 1997; IPCC, 1998). Moreover, beach nourishment requires maintenance in the form of periodic sand replenishment, sometimes every 5–10 years or less (Amadore *et al.*, 1996). Such a requirement could prove to be unsustainable in small economies. In contrast, on some islands such as Singapore, where the technology and resources are more readily available, beach fill projects (used in combination with offshore breakwaters to form artificial headlands) is a feasible option (Wong, 1985). As a general strategy to respond to sea-level rise, it is likely that Singapore will focus on three main types of responses: coastal protection for developed or heavily populated areas and reclaimed land, anti-salt-intrusion measures for coastal reservoirs, and flood prevention measures (such as tidal gates) for major canals (Wong, 1992).

In some islands, such as those in the Caribbean, more emphasis is being placed on the application of "precautionary" approaches, such as enforcement of building set-backs, land-use regulations, building codes, and insurance coverage. In addition, application of traditional, appropriate responses (e.g., building on stilts and use of expendable, readily available indigenous building materials), which have proven to be effective responses in many islands in the past, ought to be more widely considered (Forbes and Solomon, 1997; Mimura and Nunn, 1998).

Given the vulnerability of many small island states to various aspects of global change, integrated coastal management (ICM) is rapidly becoming an attractive paradigm for planning adaptation (Bijlsma, 1996; Cicin-Sain, 1998; Nicholls and Mimura, 1998). Furthermore, ICM can be regarded as both an anticipatory and a predictive tool, with the capability to plan for and respond to medium- and long-term concerns such as sea-level rise as well as short-term, present-day needs (Nurse, 1999; Solomon and Forbes, 1999). In addition, ICM can provide an effective framework for resolving potential conflict among competing stakeholder interests, in a manner that is equitable to all groups. In this context, it is noteworthy that all of the small island states that recently participated in the U.S. Country Studies Program (i.e., Federated States of Micronesia, Samoa, Fiji, Kiribati, Marshall Islands, Sri Lanka, and Mauritius) concluded that ICM was the most appropriate adaptation strategy and should form an essential part of their climate change national action plans (Huang, 1997).

Enhancing the resilience of coastal systems has been suggested (e.g., Bijlsma, 1996) as an appropriate proactive adaptive response to reduce vulnerability. Klein and Nicholls (1998) agree that this could be a more cost-effective way to prepare for uncertain changes such as sea-level rise, rather than relying entirely on building traditional, more costly coastal defenses. Helmer *et al.* (1996) strongly support the notion of enhancement of coastal resilience whereby dynamic systems (e.g., dunes, lagoons, and estuaries) should be allowed to utilize their natural capacity to grow in response to rising sea levels. This philosophy could be applied through pilot studies in small islands. One of the ways in which a dynamic and resilient coast can be created is by managed retreat, based on an enforced building set-back that allows the coastline to recede to a new line of defense, thus restoring natural coastal processes and systems. An orderly plan to retreat could be a feasible option on larger islands that cannot commit the resources necessary to prevent coastal land loss in the face of rising sea levels (Leatherman, 1997).

One recommended approach to planning adaptation to sea-level rise and coastal change involves an estimation of the costs of protecting or abandoning developed properties. For developed coasts, West and Dowlatabadi (1999) propose that the real economic evaluation of sea-level rise should be regarded as the difference in utility (not just damages) with and without a sea-level rise scenario, in which it is assumed that physical conditions (e.g., erosion and storms) and human behavior remain constant. This is an improvement on past approaches, which provided the estimated costs of sea-level rise on the basis of market value of inundated land and property or the cost of structural protection.

17.2.4. Coral Reefs, Mangroves, and Seagrasses

17.2.4.1. Coral Reefs

Coral reefs represent one of the most important natural resources of many tropical islands. They are a source of food, beach sand, and building materials and function as natural breakwaters along the coasts of many tropical islands. They also provide habitats for many marine animals and reef fish and generate significant revenues for many small island economies through avenues such as tourism (e.g., snorkeling and scuba diving). On many islands, coral reefs are facing severe threats from climate- and non-climate-related stressors. The total areal extent of living coral reefs has been estimated at about 255,000–1,500,000 km^2 (Spalding and Grenfell, 1997), of which 58% are considered at risk from human activities, according to a global assessment (Bryant *et al.*, 1997).

Owing to their narrow temperature tolerances, some species of corals currently live at or near their thermal limits (Goreau, 1992; IPCC, 1998). SST projections (based on three variants of the Max Planck Institute ECHAM and CSIRO GCMs) suggest that the thermal tolerance of reef-building corals will be exceeded within the next few decades. Moreover, the incidence of bleaching will rise rapidly, with the rate of increase highest in the Caribbean and slowest in the central Pacific region (Hoegh-Guldberg, 1999).

There is now substantial evidence that indicates that "episodic" warming of the ocean surface, as occurs in El Niño years, leads to significant coral bleaching (Brown and Ogden, 1993; Glynn, 1993; Goreau and Hayes, 1994; Wilkinson and Buddemeier, 1994; Brown, 1997a,b; CARICOMP, 1997; Goreau *et al.*, 1997). The major coral bleaching episodes in the past 20 years were found to be associated with periods when ocean temperature were about 1°C higher than the summer maximum. It also has been suggested that bleaching events could occur annually in most tropical oceans in the next 30–50 years (Hoegh-Guldberg, 1999). Bleaching was particularly severe and widespread during the period of the most recent El Niño episode of 1997–1998, which was considered to be the most intense such event on record. On some islands, more than 90% of all live reefs have been affected (Goreau and Hayes, 1994), with branching species generally most severely impacted (Wilkinson 1998, 1999). An assessment of the literature on coral bleaching is provided in Chapter 6.

The impact of increasing CO_2 concentrations in the oceans on coral reefs is now the focus of an emerging though as yet unresolved debate. Since publication of the SAR and the *Special Report on Regional Impacts of Climate Change* (IPCC, 1998), it has been suggested that the ability of reef plants and animals to make the limestone skeletons that build the reefs is being reduced by rising atmospheric CO_2 concentrations. Indeed, some authors suggest that based on projected CO_2 concentration in the atmosphere, the calcification rate of corals would decline by approximately 14–30% by 2050 (Gattuso *et al.*, 1999; Kleypas *et al.*, 1999). Again, see Chapter 6 for a full evaluation of the main issues in this debate.

Chapter 6 points out that earlier IPCC assessments have concluded that the threat of sea-level rise to reefs (as opposed to reef islands) is negligible. This conclusion was based on projected rates of global sea-level rise from Warrick *et al.* (1996) on the order of 2–9 mm yr^{-1} over the next 100 years. It has been suggested that healthy reef flats will be able to keep pace with projected sea-level rise, given an approximate upper limit of vertical reef growth during the Holocene of 10 mm yr^{-1} (Schlager, 1999). However, the prognosis is far less positive in many small island states (e.g., in the Caribbean Sea and the Indian Ocean), where reef structures have been weakened by a variety of anthropogenic stresses. This concern also is applicable to many island countries, where reefs in close proximity to major settlements have been severely stressed. The ability of reefs to keep pace with sea-level rise also will be adversely affected by more frequent coral bleaching episodes and by reduced calcification rates resulting from higher CO_2 concentrations.

17.2.4.2. Mangroves

Mangroves provide important functions as protection against storms, tides, cyclones, and storm surges and are used as "filters" against the introduction of pests and exotic insects (Menendez and Priego, 1994; Suman, 1994). Mangroves have important ecological and socioeconomic functions as well, particularly in relation to animal and plant productivity, as nutrient sinks, for substrate stabilization, and as a source of wood products. These functions sometimes may be in conflict and differ in importance between riverine, basin, and coastal fringe mangroves; the latter are important primarily for shoreline protection.

Many mangrove forests are under stress from excessive exploitation, reducing resilience in the face of sea-level rise. The importance of sediment flux in determining mangrove response to rising sea levels is well established in the literature. Ellison and Stoddart (1991), Ellison (1993), and Parkinson and Delaune (1994) have suggested that mangrove accretion in low- and high-island settings with low sediment supply may not be able to keep up with future rates of sea-level rise, but Snedaker and Meeder (1994) have suggested that low-island mangroves may be able to accommodate much higher rates. These observations may not necessarily represent conflicting views because the resilience of mangroves to sea-level rise also is conditioned by the composition and status of the stands and other factors such as tidal range and sediment supply (Woodroffe, 1995; Ewel *et al.*, 1998; Farnsworth, 1998). In some protected coastal settings, inundation of low-lying coastal land actually may promote progressive expansion of mangrove forest with rising sea level (Richmond *et al.*, 1997), provided vertical accretion keeps pace.

Notwithstanding the foregoing, studies have shown that mangrove forests in some small islands will be lost as a result of elevated sea levels. For example, it is projected that with a 1-m sea-level rise in Cuba, more than 300 ha of mangroves, representing approximately 3% of that country's forests, would be at risk (Perez *et al.*, 1999). Under similar conditions, Alleng (1998) projects a complete collapse of the Port Royal mangrove wetland in Jamaica, which has shown little capacity to migrate in the past 300 years; Suman (1994) envisages that accelerated sea-level rise would adversely affect mangroves in Puerto Rico, where 62% already has been eliminated by direct human activity.

17.2.4.3. Seagrasses

Seagrass communities provide useful habitat for many marine fish, particularly in the shallow, intertidal environments of many islands. It is postulated that an increase in SST will adversely affect seagrass communities because these ecosystems already are sensitive to land-based pollution and runoff in coastal environments (Edwards, 1995). It is argued further that the distribution of seagrasses will shift as a result of temperature stress, which in turn can cause changes in sexual reproduction patterns (Short and Neckles, 1999). In addition, an increase in temperature will alter seagrasses' growth rates and other physiological functions. Sea-level rise would mean increasing water depth and reduction of the amount of light reaching the seagrass beds, which would reduce plant productivity.

The effect of increased CO_2 in the water column will vary according to species and environmental circumstances but will likely alter the competition between species, as well as between seagrasses and algal populations (Beer and Koch, 1996). Laboratory experiments suggest that some seagrasses, such as *Zostera marina,* are able to respond positively to increased CO_2 levels by increasing their rate of photosynthesis (Zimmerman *et al.*, 1997), although earlier research on field-collected samples of *Thallassia testudinum* suggests that maximum photosynthesis is lower with elevated CO_2 (Durako, 1993).

As with other submerged aquatic plants, seagrasses are sensitive to ultraviolet-B (UV-B) radiation because such radiation can penetrate depths of up to 10 m (Larkum and Wood, 1993). Laboratory experiments have demonstrated that the response can vary from strong photosynthetic tolerance, in the case of *Halophila wrightii*; to moderate tolerance, as with *Syringodium filiforme*; to little photosynthetic tolerance, as exhibited by *Halophila engelmanni* (Hader, 1993; Short and Neckles, 1999).

17.2.5. Biodiversity of Islands

Small islands are highly variable with respect to their biological diversity. Some states, such as the low reef islands, have low biodiversity and low endemism. Coral reefs exhibit the highest known diversity among marine ecosystems, with 91,000 described species of reef taxa. Table 17-4 gives the diversity of mammals, birds, plants, and endemism for some select small island states. In general, small islands in absolute figures tend to have high terrestrial diversity and endemism. In Cuba, for instance, 50% of the flora and 41% of the fauna are endemic (Vales *et al.*, 1998). In the Canary Islands, 45% of all bird species are endemic. In the Hawaiian islands—the most isolated of all floristic regions—more than 90% of plant species are endemic (Biagini, 1999). When relative biodiversity and endemism are calculated in relation to area, the figures for many small island states tend to be higher than those for most other regions of the world.

Although there generally is high diversity associated with the ecosystems (marine and terrestrial) of islands, their long-term survival is threatened by anthropogenic stresses including pollution, overexploitation, and generally poor management. As in other regions, it is expected that climate change will affect the biodiversity of small islands directly and indirectly. Rising atmospheric CO_2 concentrations are projected to increase the productivity of some communities and alter competition

***Table 17-4**: Biodiversity status for selected small island states, 1990 (extracted from IPCC, 1998).*

Country	Known Mammal Species	Known Endemic Mammal Species	Known Bird Species	Known Endemic Bird Species	Known Plant Species	Known Endemic Plant Species
Cuba	31	12	342	22	6,004	3,229
Dominican Republic	20	—	254	0	5,000	1,800
Fiji	4	1	109	26	1,307	760
Haiti	3	0	220	0	4,685	1,623
Jamaica	24	3	262	25	2,746	923
Mauritius	4	2	81	9	700	325
Palau	214	57	708	80	10,000	—
Solomon Islands	53	19	223	44	2,780	30
Trinidad and Tobago	100	1	433	1	1,982	236

among others by eliminating some species and introducing new species to take their place (McIver, 1998). In marine ecosystems such as coral reefs, incremental increases in atmospheric CO_2 would be expected to threaten the diversity of these systems by the process described in Section 17.2.4.1 (Hatcher, 1997).

The impacts of climate change and sea-level rise on biota in island states are much greater than the impacts on continental areas. For example, sea-level rise could impact the habitats of the endangered Tuamotu sandpiper (*Prosobonia cancellata*) and bristle-thighed curlew (*Numenius tahitiensis*), as well as the seabird colony of 18 species on Laysan Island (Hawaii). Other potentially vulnerable areas (not inclusive) for breeding seabirds include the Kerguelen and Crozet Islands (seabirds), the Galapagos Islands (Galapagos penguin, *Spheniscus mendiculus*), and the nesting habitat for the Bermuda petrel (*Pterodroma cahow*) (Sattersfield *et al.*, 1998). Based on outputs from HADCM2 and scenarios IS92a and Kyoto 1, Suarez *et al.* (1999) also have found that in the eastern region of Cuba, certain endemic species of flora would face extinction.

Inundation and flooding of low-lying forested islets with species such as the Manus fantail (*Rhipidura semirubra*) also might be lost. The majority of threatened bird species on islands are found in forested habitats (Sattersfield *et al.*, 1998). Impacts of climate change on these species likely would be from physiological stress and changes and loss in habitat, especially from fires and cyclones. For example, 30% of the forested area on the Santa Cruz islands was lost in one cyclone in 1993. Some vulnerable species and areas include the endangered New Caledonian lorikeet (*Charmosyna diadema*) and critically endangered New Caledonian rail *(Gallirallus lafresnayanus)* on New Caledonia, the Samoan white-eye (*Zosterops samoensis*) and critically endangered Samoan moorhen (*Gallinula pacifica*) on Savai'i (Samoa), and the Santo Mountain starling (*Aplonis santovestris*) on Espiritu Santo (Sattersfield *et al.*, 1998). In Samoa, most species of flowering plants are pollinated by a few species of animals; nearly 100% of seed dispersal in the dry season is mediated by flying foxes (*Pteropus spp.*) (Cox *et al.*, 1991). Likewise, flying foxes are the key pollinators and seed dispersers on many islands in the South Pacific. If their habitat is threatened by climate change, the result would be the loss of many dependent plant species.

Generally on islands, one of every three known threatened plants is endemic. Among birds, approximately 23% of the species found on islands are threatened, compared with only 11% of the global bird population (McNeely *et al.*, 1993). Establishment of terrestrial, marine, or coastal reserves such as the Morne Trois Pitons Park and Forestry Reserve (Dominica, 1986), Bonaire Marine Park (Netherlands Antilles, 1979), Tobago Cays Marine Park (St. Vincent and the Grenadines), Soufriere Marine Management Area (St. Lucia, 1997), and other similar management units ("biosphere reserves") established elsewhere is a useful management option. It has been demonstrated that the creation of such reserves helps to preserve endangered habitats and ecosystems of small islands and will contribute to maintenance of biological diversity, while increasing the resilience of these systems to cope with climate change.

17.2.6. Water Resources

Availability of water resources is a limiting factor for economic and social development in small island states. Many such countries rely almost entirely on a single source of supply, such as groundwater (Barbados, Antigua, The Bahamas, Kiribati), rainwater (Tuvalu, northern atolls of Cook Islands, and the Maldives), surface reservoirs and imports (Singapore), or rivers and other surface flows (Seychelles, Dominica).

Water supply is most vulnerable in the atoll states of the Pacific, where water supply is sensitive to precipitation patterns and changes in storm tracks (Salinger *et al.*, 1995). Captured rainwater is stored in cisterns; the only backup reserve for these islands is a thin wedge of fresh groundwater that sits on top of the saltwater lens. A reduction in precipitation coupled with sea-level rise would not only cause a diminution of the volume of potable water but would reduce the size of the narrow freshwater lens (Amadore *et al.*, 1996).

The situation is equally critical in the low limestone islands of the eastern Caribbean, where seasonality of rainfall (a marked dry and wet season regime) is pronounced. On islands such as Anguilla, Antigua and Barbuda, Grenada, and Barbados, more than 65% of total annual rainfall may be recorded in the wet season, which spans the 6-month period of June to December. Moreover, most of the rainfall is strongly associated with the genesis and passage of easterly waves, tropical depressions, and storms (Gray, 1993; Nurse *et al.*, 1998). Thus, changes in the occurrence of these heavy rainfall events will certainly impact the water supply of many Caribbean islands. The situation is further exacerbated in Barbados, where recent research has shown that groundwater recharge is restricted to the three wettest months of the year, and only 15–30 % of annual rainfall reaches the aquifer (Jones *et al.*, 1998).

Reduced availability of adequate water supply in a changing climate also poses a potential threat to the Mediterranean islands of Cyprus and Malta. Because these two countries already experience water shortages (Nicholls and Hoozemans, 1996), and given a projected decrease in mean summer precipitation over the Mediterranean Sea region (IPCC, 1998), the water resources of these states could be placed under considerable pressure in the future (Tables 17-2 and 17-3). The threat is equally strong in some parts of the eastern Caribbean. Within the past few decades in Dominica, for instance, an apparent tendency toward more extended periods of drought is well correlated with reduced flows in the Castle Comfort, Roseau, Layou, and Geneva Rivers (Government of Commonwealth of Dominica, 2000). Because rivers are the main source of potable and irrigation water on the island and are also harnessed for power generation, declining flows have become a matter of serious national concern.

Climate change can present additional water management and related challenges. Such challenges may arise from a variety of sources, including increased flood risks and impeded drainage and the presence of elevated water tables—which may pose special engineering problems. It is projected that on Andros island, the Bahamas, where the water table presently is only 30 cm below the surface, high evaporation rates and increasing brackishness will eventuate with continued sea-level rise (Martin and Bruce, 1999). Similar projections also have been made for Cuba, where underground water supplies already are stressed (Planos and Barros, 1999). For many small island states, the prospect of salinity intrusion into the freshwater lens would be a matter of great concern. In many of these islands where salinization from overpumping of aquifers is already occurring (e.g., The Bahamas and Barbados), sea-level rise would compound the risk. In some cases, higher salinity would be experienced not only in coastal aquifers but also inland at freshwater pumping plants as the salty groundwater rises. Singh (1997a,b) has reported a recent increase in salinity levels for several coastal aquifers in Trinidad and Tobago in the southern Caribbean, attributable mainly to rapid drawdown exacerbated by sea-level rise.

There now is substantial evidence to support the view that precipitation variability in various parts of the world is linked to ENSO events. During the 1982–1983 ENSO event, rainfall in many parts of the western Pacific was a mere 10–30% of the long-term mean average (Falkland, 1992). More specifically, it is well established that ENSO has a strong influence on rainfall patterns in the tropics and low-latitude regions of the Southern Hemisphere (Shea, 1994; Whetton *et al.*, 1996). In the Caribbean islands, droughts appear to be more frequent in El Niño years, whereas conditions tend to be wetter in La Niña years. The devastating drought in the region in 1998 coincided with what is believed to be the strongest El Niño signal on record. ENSO-related droughts also are known to occur in the low-lying atolls of the tropical Pacific. Thus, in countries such as the Federated States of Micronesia and the Marshall Islands where rainwater is the main source of supply, more frequent and intense ENSO events will impose further stress on already meager water resources (Meehl and Washington, 1996); other islands in the central and eastern tropical Pacific will experience heavy rains (Jones *et al.*, 1999).

Realistically, the options available to many small islands for reducing the adverse effects of climate change on water availability are limited. This implies that greater urgency and emphasis will have to be placed on improving water resource management efforts, including inventorying of resources and rational and equitable allocation. Implementation of more efficient rainwater harvesting methods, efficient leak detection and repair, use of water-saving devices, and aggressive recycling efforts are strategies worth considering. Desalination also is becoming an increasingly attractive option, especially where the necessary technical and financial capacity is available and in cases in which more traditional strategies are inadequate or not feasible (e.g., in Singapore, Malta, Cyprus, Barbados, Antigua and Barbuda, St. Kitts and Nevis, and Grenada). As part of their long-term adaptation strategy, these water-scarce islands might wish to include "no-regrets" measures, which will promote sustainability with or without climate change. Some countries may even wish to consider application of market-based systems to allocate water supplies, which could result in less wasteful practices under current conditions and thus enable water users to more efficiently adapt to climate change (Amadore *et al.*, 1996).

17.2.7 Tourism

Tourism is a major economic sector in many small island states. Table 17-5, which expresses tourist arrivals in terms of the populations of small island states for which data were available, shows that in most small island states, the numbers of visitors substantially exceed the number of inhabitants. Table 17-5 also expresses tourism receipts in terms of gross national product (GNP) and in terms of foreign exchange inflows from exports of goods and services. Again, the ratios tend to be relatively high in most small island states; the Maldives, Antigua and Barbuda, and the Bahamas exhibit the highest ratios. In many countries, tourism also makes a significant contribution to employment. For example, the

***Table 17-5**: Tourist inflows and receipts, various years, for select small island states (Waters, 1998).*

Country	Number of Tourists (000s)[a]	Tourists as % of Population[a]	Tourist Receipts[b] as % of GNP	as % of Exports
Antigua and Barbuda	232	364.2	63.4	73.5
Bahamas	1618	586.4	42.0	75.6
Barbados	472	182.4	39.2	56.2
Cape Verde	45	11.4	11.5	37.3
Comoros	26	4.9	10.6	47.8
Cuba	1153	10.5	8.8	n/a
Cyprus	2088	280.7	24.0	49.1
Dominica	65	97.6	15.9	32.5
Dominican Republic	2211	28.1	13.6	30.2
Fiji	359	45.3	19.2	29.1
Grenada	111	116.2	27.0	60.6
Haiti	149	2.2	3.9	50.5
Jamaica	1192	45.6	31.6	39.8
Maldives	366	130.7	95.0	68.4
Malta	1111	294.7	22.9	28.7
Mauritius	536	46.4	15.7	26.8
Papua New Guinea	66	1.5	2.1	3.0
St. Kitts and Nevis	88	210.5	30.6	63.6
St. Lucia	248	164.7	41.1	66.6
St. Vincent	65	54.6	23.8	45.9
Samoa	68	31.1	19.6	48.8
Seychelles	130	166.7	34.6	52.2
Singapore	7198	209.2	6.2	4.1
Solomon Islands	16	3.7	2.8	4.2
Trinidad and Tobago	324	28.7	4.2	8.3
Vanuatu	49	27.1	19.3	40.9

[a]Data on tourist inflows and ratio to population pertain to 1997.
[b]Data for tourist receipts pertain to 1997 for the Bahamas, Cape Verde, Jamaica, the Maldives, Malta, Mauritius, Samoa, Seychelles, Singapore, and Solomon Islands; to 1996 for Antigua and Barbuda, Cuba, Dominica, Dominican Republic, Fiji, Grenada, Haiti, Papua New Guinea, St. Lucia, and St. Vincent; to 1995 for Barbados, Comoros, Cyprus, Trinidad and Tobago, and Vanuatu; and to 1994 for St. Kitts and Nevis.

industry provides jobs for 70% of the labor force in the Bahamas, 40% in Malta, and 20% in Seychelles (Waters, 1998).

It is widely acknowledged that the effects of climate change on tourism will be both direct and indirect. For instance, sea-level rise would disrupt the sector through loss of beaches, inundation, degradation of coastal ecosystems, saline intrusion, and damage to critical infrastructure (Nicholls and Hoozemans, 1996; Teh, 1997; Perez *et al.*, 1999). Because many small islands are so heavily dependent on the tourism sector for their economic survival, adverse impacts on the industry, from climate change or other causes, would be of great concern in these countries. In many small islands, the industry also would be sensitive to other climate-related impacts, such as loss of attractiveness of coral reefs as a result of bleaching.

A high proportion of tourism in small island states is motivated by the desire of visitors from developed countries of the north (their largest market) to escape cold winters. Small island states are becoming increasingly concerned that projected milder winters in these markets could reduce the appeal of these islands as tourist destinations (Martin and Bruce, 1999). It is projected that tourism could be further harmed by increased airline fares if GHG mitigation measures (e.g., levies and emission charges) were to result in higher costs to airlines servicing routes between the main markets and small island states (Wall, 1996).

To ensure the sustainability of the tourist industry in Cyprus, it has been recommended that a strategy of protection of infrastructure combined with planned retreat would be effective and appropriate to local circumstances. The overall goal would be to maintain the limited beach area to sustain the vital tourist industry, specifically by erecting hard structures, enforcing building set-backs, and use of artificial nourishment, although the latter measure may require external sources of sand (Nicholls and Hoozemans, 1996). Although not all these strategies may be applicable to the atoll states, many other island nations—such as Barbados, Jamaica, Grenada, St. Lucia, and Singapore—already have begun to implement similar approaches as part of the ICM process.

17.2.8. Food Security

17.2.8.1. Crop Agriculture

Subsistence agricultural production is vital to the economies, nutritional status, and social well-being of small islands—particularly the small, low-lying, atoll states where food security is a major concern. The main subsistence crops include taro, sweet potato, yam, breadfruit, bananas, coconut, and a variety of vegetables. Production of cash crops such as sugarcane, copra, coffee, cocoa, rubber, and tea (grown at higher elevations on high islands) also is important because export of these products earns valuable foreign exchange. Climate change could precipitate heat stress, changes in soil moisture and temperature, evapotranspiration, and rainfall that might affect the growth of some subsistence root crops and vegetables. The consequences of such changes for agriculture are likely to be more severe in areas that already are under stress—for example, water-scarce islands. Crop agriculture also can be affected by tropical cyclones and other extreme events, such as floods and droughts. To the extent that many small islands are susceptible to these phenomena, it is highly likely that crop production in these states would be impacted by alterations in the patterns of these events as a consequence of climate change.

On low islands and atolls in the Pacific, practically all crop agriculture is concentrated at or near the coast. Thus, changes in the height of the water table and salinization as a result of sea-level rise would be stressful for most varieties of taro and other crops, which have low tolerance for salt. It has been suggested that in general, C_3 crops, which include many tropical crops, will benefit more from the effect of CO_2 fertilization than C_4 plants. However, recent findings indicate that the impact on sugarcane and maize yields would be adverse (Jones *et al.*, 1999).

Singh and El Maayar (1998), using GCM (CCC 11) outputs and high, medium, and low CO_2 emission scenarios coupled with a crop model (FAO) to simulate crop yields, found that sugarcane yields may decrease by 20–40% under a $2xCO_2$ climate change scenario in Trinidad and Tobago in the southern Caribbean. The decrease in yields is attributed to increased moisture stress caused by the warmer climate. These reductions in sugarcane yields deriving from climate change are similar to those found for maize—another C_4 crop—in nearby Venezuela (Maytin *et al.*, 1995). These results are supported by similar findings in Mauritius, which are derived from the Agricultural Production Systems Simulator Model (APSIM-Sugarcane) developed by the Agricultural Production Systems Research Unit, Australia. The study projects a decline in sucrose yield by more than 50% with a doubling of CO_2 (Cheeroo-Nayamuth and Nayamuth, 1999).

17.2.8.2. Fisheries

Although fishing is largely artisanal or small-scale commercial, it is an important activity on most small islands and makes a significant contribution to the protein intake of island inhabitants (Blommestein *et al.*, 1996; Mahon, 1996). The impacts of climate change on fisheries are complex and in some cases are indirect. As with other renewable resources, an assessment of climate change impacts on fisheries is complicated by the presence of anthropogenic and other non-climate-related stresses, such as habitat loss and overexploitation (Challenger, 1997).

Many breeding grounds for commercially important fish and shellfish are located in shallow waters near coasts. These areas include mangroves, coral reefs, seagrass beds, and salt ponds—all of which are likely to be affected by climate change. Generally, fisheries in the small island states are not expected to be adversely affected by sea-level rise *per se*. Higher sea level would be a critical factor for fisheries only if the rate of rise were far more rapid than the current succession of coastal ecosystems (e.g., mangroves, seagrasses, corals) on which some fish species depend (Everett, 1996). In tropical islands, these ecosystems function as nurseries and forage sites for a variety of important commercial and subsistence species. In this context, the unfavorable effects of higher CO_2 concentrations on coral reef development, coupled with widespread coral bleaching, must be considered a significant threat in many small island states (see Section 17.2.4.1). Fish production obviously would suffer if these habitats were endangered or lost (Costa *et al.*, 1994).

On a global scale, it is not expected that climate change and climate variability will lead to any significant reduction in fisheries production. However, important changes in the abundance and distribution of local stocks (which may be of direct concern to some small islands) are likely to occur (IPCC, 1996). For example, Lehodey *et al.* (1997) have shown that spatial shifts in the abundance of skipjack tuna in the Pacific are linked to the ENSO cycle. They note that catches are highest in the western equatorial Pacific warm pool, which can be displaced by as much as 50° of longitude eastward during El Niño episodes and westward in La Niña years (Lehodey *et al.*, 1997). This must be a concern to Pacific islanders whose access to the skipjack stocks now appears to be largely controlled by the periodicity of ENSO events.

Several management strategies for minimizing the adverse effects of climate change on fish stocks have been proposed. These measures—many of which already are being implemented in some island states—include conservation, restoration, and enhancement of vital habitats such as mangroves, coral reefs, and seagrass beds; establishment and management of marine reserves and protected areas for identified critical species; and implementation of bilateral and multilateral agreements and protocols for exploitation and management of shared fisheries (migratory and straddling stocks) (IPCC, 1998; Berkes *et al.*, 2001). Aquaculture also may be considered by island states as another means of reducing stress on wild stocks. However, great precaution must be taken to ensure that this measure does not exacerbate existing problems of habitat loss and competition for nutrients (Carvalho and Clarke, 1998; see also Section 6.6.4).

17.2.9. Human Welfare

17.2.9.1. Settlement and Infrastructure

In most small island states (including high islands such as Seychelles, Reunion, and Fiji), narrow coastal plains provide attractive locations for settlement and a variety of infrastructure to support economic and social needs. Most of the population, settlements, and economic activities are concentrated in areas where competition for space is acute and where fragile ecosystems, aquatic and terrestrial, coexist. In most Caribbean islands, for example, more than 50% of the population live within 2 km of the coast. On atolls, most of the important infrastructure and population clusters often are less than 100 m from the shoreline. As the shortage of coastal space becomes increasingly acute in many small islands, land reclamation often is practiced as a solution to this need. Ironically, this practice exposes these islands to greater risk by attracting more settlement and infrastructure to already highly vulnerable locations.

As elsewhere, coastal development in most small islands has been undertaken in the past without taking climate change and sea-level projections into consideration. With currently projected rates of sea-level rise and flooding, coupled with the possibility of more intense and frequent extreme events such as cyclones (hurricanes) and associated storm surge, critical infrastructure such as social services, airports, port facilities, roads, coastal protection structures, tourism facilities, and vital utilities will be at severe risk. Furthermore, the capacity of most small island states to respond effectively to these threats is limited by their low adaptive capacity, which results from a combination of factors—including physical size (little opportunity to retreat), limited access to capital and technology, and a shortage of human resource skills.

In some countries, particularly the low islands and micro-atolls, resettlement within national boundaries may have to be considered as the only viable option. However, implementation of this strategy could become extremely complicated, especially for densely populated coastal lowlands such as in the Federated States of Micronesia (950 persons km^{-2}), Majuro, Marshall Islands (2,188 persons km^{-2}) and Male, Republic of Maldives (35,000 persons km^{-2}). In extreme circumstances, it may even become necessary to abandon some atolls altogether (Nurse *et al.*, 1998). Such an option would be socially and culturally disruptive and would require access to substantial resources—which most of these countries may be unable to afford.

17.2.9.2. Human Health

Increased instability of weather patterns and large interannual variability in climate enhanced by ENSO forcing and GHG-induced climate change have catalyzed a new focus on possible health consequences in a changing climate (Epstein, 1997; Epstein *et al.*, 1997; Hales *et al.*, 1997; Woodward *et al.*, 1998). Many tropical islands are now experiencing high incidences of vector- and water-borne diseases that are attributed to changes in temperature and rainfall regimes, which may be linked to events such as ENSO, droughts, and floods. In the Pacific, there is growing evidence that outbreaks of dengue are becoming more frequent and appear to be strongly correlated with the ENSO phenomenon (Hales *et al.*, 1997, 1999a). Many of the small island states lie in the tropical zone, where the climate is suitable for the transmission of tropical diseases such as malaria, dengue, filariasis, and schistosomiasis.

Some of the small island states, such as the Bahamas, Kiribati, the Marshall Islands, and the Maldives, are a mere 3–4 m above mean sea level, which predisposes them to inundation with seawater and, as a consequence, salinization of freshwater supplies and flooding from sea-level rise. Furthermore, low-lying islands are particularly vulnerable to storms and cyclones; these also can adversely affect public water supplies. Vector-borne diseases such as malaria and dengue are particularly sensitive to warming and flooding. Filariasis and schistosomiasis are less sensitive to short-term seasonal climatic changes, but the epidemiology of these diseases could change with long-term effects of climate change. Water-borne diseases such as shigella, cryptosporidium, giardia, and amoebiasis could increase as a result of disruption of sewage and water systems by flooding (see Chapter 9).

It is also projected that, with temperature and rainfall changes, some vectors could extend their range, so there is likely to be wider transmission of some diseases (McMichael, 1996). For example, malaria—which previously tended to be confined largely to the western and central Pacific region—now appears to be extending east as far as Fiji. It also is worth noting that the interior uplands of many islands, which now are virtually free of vectors (e.g., *Aedes aegypti* mosquito) that transmit malaria, dengue, and other tropical illnesses—could become favorable breeding sites in a changing climate (disease-specific details appear in Chapter 9).

In some regions—for example, the Pacific—it has been noted that extreme weather events appear to be occurring at a higher frequency than elsewhere (Timmerman *et al.*, 1997). As a consequence, physical injuries arising from these events can be expected to increase.

It is well-established that vulnerability to such health risks will vary according to factors such as availability of quality health care, the present health status of the population, and availability of technical and other resources (McMichael, 1993; WHO, 1996). Unfortunately, health care facilities and related infrastructure in many small island states and other developing countries often are inadequate. Hence, the resilience of such states and their capacity to respond effectively to (or mitigate) increasing health threats posed by climate change is likely to be low.

A range of adaptation strategies for reducing the severity of possible climate change-related health threats has been proposed, and many of these measures may be successfully pursued in the small island states. Such measures include, *inter alia*, implementation of effective health education programs, preventive maintenance and improvement of health care facilities, cost-effective sewerage and solid waste management practices, and disaster preparedness plans (McMichael, 1996).

Adoption of efficient early warning systems also would be beneficial in vulnerable small islands (Stern and Easterling, 1999). This process would involve monitoring of health-risk indicators by improving the forecasting of conditions that are favorable to the outbreak of climate-sensitive diseases such as dengue, cholera, and malaria (Patz *et al.,* 1996; IPCC, 1998; Epstein, 1999). Where large populations exist, as in urban areas, simple, low-cost measures could be implemented to control the vectors of dengue and other diseases, where a risk of transmission exists. At the individual level, insecticide-treated bed nets could provide protection against vectors of malaria and filariasis. Simple technology, such as the use of sari cloth to filter drinking water, reduces the risk of cholera transmission at the household level. Although some measures for adapting to climate change can be deferred, early implementation of preventive strategies could reduce current and future health "costs."

17.2.10. Other Economic and Sociocultural Impacts

Climate change could have direct and indirect impacts on other economic and social sectors in some small islands. The insurance industry is one sector that is highly sensitive to the magnitude and frequency of various hazards, including climate-related phenomena such as tropical storms and floods. Because insurance premiums are based on assessment of risk of occurrence of a particular event, any indication of an increase in the frequency or intensity of phenomena such as tropical cyclones and floods is likely to trigger an increase in the cost of insurance. Within the past decade, insurance costs in the Caribbean have increased significantly, following the passage of a series of severe hurricanes that caused widespread socioeconomic dislocation, injury, and loss of life. Claims were so high that some reinsurance companies withdrew from the market; others imposed higher deductibles, separate conditions for windstorms, and a premium structure to minimize the risk of underinsurance (Murray, 1993; Saunders, 1993). Even in cases in which these systems did not make landfall in the insular Caribbean itself—as with Hurricane Andrew, which devastated southern Florida in the United States—an increase in insurance premiums in the islands subsequently occurred.

Certain traditional island assets (goods and services) also will be at risk from climate change and sea-level rise. These assets include subsistence and traditional technologies (skills and knowledge), community structure, and coastal villages and settlements. Sea-level rise and climate changes, coupled with environmental changes, have destroyed some very important and unique cultural and spiritual sites, coastal protected areas, and traditional heritage sites in the Federated States of Micronesia, Tuvalu, the Marshall Islands, Niue, and Kiribati and continue to threaten others (Kaluwin and Smith, 1997).

Although some of these assets fall into the category of nonmarket goods and services, they are still considered to be of vital importance in small island states. In Tuvalu, for instance (as in other Pacific atoll states), strong traditional ties to land and sea constitute a vital component of local cosmology (Sem *et al*., 1996). Some of these values and traditions are compatible with modern conservation and environmental practices; therefore, priority action is needed in the following areas: research into traditional knowledge and practices of conservation and environment control—which have sustained these societies for generations, even in the face of hazards, risk, and uncertainty; inventorying of traditional, heritage, and other cultural sites; encouragement of practices that marry use of modern science and technology with traditional wisdom; and more effective transmission of traditional knowledge to younger generations.

17.3. Vulnerability and Adaptation Potential

17.3.1. Setting the Context

The potential impacts of climate change on small island states discussed in Section 17.2 have given rise to considerable concern. Relevant chapters in the First Assessment Report (FAR; Tsyban *et al*., 1990) and SAR (Bijlsma, 1996) and in the *Special Report on Regional Impacts of Climate Change* (Nurse *et al*., 1998) already concluded that low-lying small islands are among the most vulnerable countries in the world. A similar conclusion is reached in this Third Assessment Report (see Section 19.3.7). Their overall vulnerability is shown to be a function of the *degree of exposure* of these states to climate change and their *limited capacity to adapt* to projected impacts.

This section assesses the relevant literature on the vulnerability of small island states that has become available since the SAR, including an overview of available country studies. In addition, this section pays particular attention to adaptation. Whereas Bijlsma (1996) and Nurse *et al*. (1998) focuse on available adaptation options, this section takes a process-oriented approach to adaptation and recognizes adaptive capacity as an important determinant of vulnerability.

17.3.2. Generic Issues

Many definitions of vulnerability and adaptation exist in the literature (see Chapter 18). However, this report defines vulnerability to climate change as "the degree to which a system is sensitive to and unable to cope with adverse impacts of climatic stimuli. Vulnerability is a function of a system's exposure and its adaptive capacity." Adaptation is the "adjustment in natural or human systems in response to actual or expected climatic stimuli, or their effects" (see Chapter 2). Vulnerability therefore is a function of potential impacts and adaptive capacity, and adaptation refers to both natural and human system responses. Whereas the previous section focused on impacts and responses of natural systems, this section discusses the relevance of adaptation for human systems in a small islands context.

17.3.3. Vulnerability

Despite their heterogeneity, small island states share some common characteristics that help to define their high vulnerability and low adaptation potential to climate change effects (Nurse *et al.*, 1998; see also Section 17.1.2). Vulnerability assessment typically seeks to achieve three main goals: to identify the degree of future risks induced by climate change and sea-level rise; to identify the key vulnerable sectors and areas within a country; and to provide a sound basis for designing adaptation strategies and their implementation.

The IPCC Common Methodology was the first method to be widely applied to assess the vulnerability of countries to sea-level rise (IPCC, 1992). However, the methodology lacks the flexibility to consider factors of critical significance for small islands (e.g., so-called nonmarket goods and services) and requires certain quantitative data that often are not easily available in many small island states. An index-based method was developed for use in the south Pacific (Yamada *et al.*, 1995, based on Kay and Hay, 1993). In addition, alternative assessment methodologies were developed in conjunction with the various country study programs (e.g., Leatherman, 1996; Klein and Nicholls, 1998). What is significant, however, is that all available assessments confirm the high vulnerability of small island states to climate change, independent of the methodology applied. As already noted, global assessments come to the same conclusions (Nicholls *et al.*, 1999). This is therefore a robust finding, which must be of considerable concern to these countries.

Climate change is expected to be one factor among many that affect ecological systems and economic development. Other factors that interact with climate change include overexploitation of resources, pollution, increasing nutrient fluxes, decreasing freshwater availability, sediment starvation, and urbanization (Goldberg, 1994; Viles and Spencer, 1995). Particularly relevant for small island states are rapid population growth, intra- and inter-island migration, rapid changes in social structure, and effects of economic globalization. These nonclimate stresses can decrease the resilience of natural and human systems, increasing their vulnerability to climate variability and anticipated climate change (Nicholls and Branson, 1998; Klein and Nicholls, 1999).

Most vulnerability indices developed to date have focused on economic and social systems, although studies by Ehrlich and Ehrlich (1991) and Atkins *et al.* (1998) focus on environmental vulnerability. The economic vulnerability indices include those developed by Briguglio (1995, 1997), the Commonwealth Secretariat (Wells, 1996, 1997; Atkins *et al.*, 1998), Pantin (1997), and the Caribbean Development Bank (Crowards, 1999). Another index, the Environmental Vulnerability Index, has been developed recently for small island states, particularly countries for which data availability is limited. It incorporates climate, nonclimate, and human stresses on the environment and seeks to reflect relative vulnerability as a function of these combined factors (Kaly e*t al.*, 1999). All vulnerability indices consistently identify small states—sometimes more specifically small island states—as being more economically vulnerable than larger states.

Box 17-2. Tools for Vulnerability Assessment and Adaptation Policy Development

The numerous and well-developed interactions between the natural and human systems of island countries underscore the relevance of integrated assessment as a meaningful analytical tool for designing adaptation strategies. One such tool that has proven particularly beneficial is VANDACLIM, an integrated assessment model developed by the International Global Change Institute (University of Waikato, New Zealand), in collaboration with the South Pacific Regional Environment Programme (SPREP) and United Nations Institute for Training and Research (UNITAR) (Warrick *et al.*, 1999). Enhanced and country-specific versions of VANDACLIM currently are being developed.

Development of VANDACLIM involved linking a regional scenario generator with selected impact models for four key sectors: agriculture, coastal zones, human health, and water resources. The user has considerable flexibility in generating scenarios; the user can choose among a large range of projections from GHG emission scenarios; the low, mid, or high cases from each projection (which encompasses the range of uncertainty in model parameter values); several GCM patterns; and the year of interest (in 5-year increments from 1990 to 2100).

VANDACLIM integrates a variant of the "Bruun rule" with a simple inundation model that is suitable for flat, low-lying deltaic coastal plains. Health impacts projections are derived from a biophysical index that estimates potential incidence of malaria and a simple threshold index for estimating change in the risk of cholera outbreaks related to extreme flooding events. For water resources, three models are included: an atmospheric water balance model for assessing the overall water resource situation for the country; a water balance-river discharge model for estimating monthly mean discharge for estimating wet and dry season river flow; and a discharge-flood area model for defining the areal extent of flooding. For agriculture, various crop models and indices are integrated, including degree-day models, rainfall (soil moisture), a land suitability index, and temperature for a variety of tropical crops.

Most existing studies have fulfilled the first two of the three aforementioned goals of vulnerability assessment, although information on socioeconomic impacts of climate change often is limited. The third goal—to provide a basis to guide possible adaptation—usually is met only in general terms. Effective planning and design of adaptation strategies requires more detailed information on crucial vulnerable sectors and areas. Such information may be partly derived from analysis at an integrated level (see Box 17-2)—as suggested, for instance, by Klein and Nicholls (1999).

17.3.4. Adaptation and Adaptive Capacity

As Campbell (1996) notes, a key misconception is that adaptation is a task carried out by governments. Insofar as governments have property and are responsible for carrying out a variety of activities, they will be required to take adaptive action. Most adaptation, however, will be carried out by individual stakeholders and communities, urban or rural, that inhabit island countries. Therefore, the government's primary role is to facilitate and steer this process—ideally in a manner that benefits the wider community.

Small island states often are susceptible to the impacts of a wide range of natural hazards, including climatic extremes. In the south Pacific region alone, island states suffered a total of 79 tropical cyclones, 95 storm surges, 12 floods, 31 droughts, four earthquakes, five landslides, two tsunamis, and four volcanic eruptions during the 1990s (Burns, 2000; Gillespie and Burns, 2000; Hay, 2000). The World Conference on Natural Disaster Reduction and the Global Conference on Sustainable Development of Small Island States noted several issues that influence adaptation to such impacts. These issues include the limited capacity of developing small island states to respond to and recover from natural and environmental disasters, owing to their narrow resource base and small size. Another issue is the decline in traditional coping mechanisms employed by island states, such as food preservation and storage techniques and disaster-resistant housing designs.

Given their high vulnerability, it is generally accepted that a proactive approach to adaptation planning would be especially beneficial to small islands, to minimize the adverse effects of climate change and sea-level rise (Campbell and de Wet, 2000). One essential prerequisite for implementing adaptive measures is support from policymakers and the general public. Thus, raising public awareness and understanding about the threats of climate change and sea-level rise and the need for appropriate adaptation require urgent and consistent attention. Because strong social and kinship ties exist in many small island states—for example, in the Pacific—a community-based approach to adaptation could be vital if adaptation policies and options are to be successfully pursued.

It also should be noted that small island states have faced many hazards in the past; as a consequence, their inhabitants have developed some capacity to cope by resorting to a combination of strategies, including application of traditional knowledge, locally appropriate technology (e.g., construction on stilts in flood-prone areas), use of indigenous materials, and other customary practices. Thus, for these states, it would be mandatory for any climate change adaptation policy and implementation plan to incorporate these traditional coping skills.

One of the obstacles to implementation of adaptation strategies stems from the uncertainties associated with the projection of future climate change and its impacts, at scales appropriate to small islands. Therefore, better guidance is needed for policy development in the face of uncertainties, together with more reliable climate projections at a scale that is relevant to the small island states (Edwards, 2000).

Many island states confront a range of pressing socioeconomic concerns (e.g., poverty alleviation, unemployment, health, and education), and climate change tends to be assigned a low priority on most national agendas. Thus, given the long lead time for implementing and assesing adaptation (as much as 50–100 years), progress in realizing its goals almost certainly will require integration of adaptation strategies with other sectoral and national policies, such as economic development, disaster prevention and management, integrated coastal management, and sustainable development frameworks.

17.3.5. Regional and External Factors

Small island states account for a small percentage of world energy consumption and extremely low levels of global GHG emissions and on balance are likely to be severely impacted by the effects of climate change (Yu *et al.*, 1997; see Box 17-3). In most states, the bulk of the energy requirements are met

Box 17-3. Greenhouse Gas Emissions from Small Island States in the Pacific

Based on application of the *IPCC Guidelines for National Greenhouse Gas Inventories*, Pacific island countries are responsible for a per capita equivalent emission of approximately 0.96 t of CO_2 yr^{-1}. Hence, the total Pacific island population of 7.1 million in 22 countries produces 6.816 Mt of CO_2 yr^{-1}. In contrast, based on International Energy Agency data for 1996, global CO_2 emissions arising from fossil fuel combustion alone are 22,620.46 Mt of CO_2 yr^{-1}, or 4.02 t of CO_2 yr^{-1} per capita. Thus, on average, Pacific islanders produce approximately one-quarter of the CO_2 emissions attributable to the average person worldwide. Expressed another way, the Pacific islands region as a whole accounts for 0.03% of the global emissions of CO_2 from fuel combustion despite having approximately 0.12% of the world's population.

Source: Hay and Sem, 1999.

Box 17-4. Renewable Energy Use in Small Island States: A "Win-Win" Strategy

Most small islands are heavily dependent on imported fossil fuels for the majority of their energy requirements, particularly transport and electricity production. This is clearly demonstrated in the case of the Caribbean and Pacific islands, where petroleum imports are responsible for more than 75 and 88%, respectively, of primary energy demand. The cost of fossil fuel imports also places a considerable economic burden on small island states, accounting on average for almost 15% of all imports in these countries. In addition, the cost of electricity production (US$0.10–0.15 and 0.20 kWh^{-1} for the Caribbean and the Pacific, respectively) can be as much as three to four times higher than in developed countries.

In many islands, the high unit cost of conventional power production versus the increasingly competitive cost of renewable energy technologies (especially solar and wind), make the latter economically viable and environmentally friendly options. For these reasons, several small island states are making a significant contribution to global utilization of renewable energy resources. These include, *inter alia*, the following countries:

- Barbados, where approximately 33% of all households use solar water heaters
- La Desirade, Guadeloupe, where more than 75% of all electricity is generated from wind power
- Fiji and Dominica, where hydropower accounts for more than 30% of electricity production
- Tuvalu, where photovoltaics supply 45% of the electricity
- Reunion, where almost 20% of the electricity is biomass-generated (from bagasse, a by-product of sugarcane); bagasse also is becoming increasingly important as an energy source in Jamaica and Fiji.

Sources: Jensen, 1999; Ellis and Fifita, 1999.

from imported fossil fuels, which places a heavy burden on island economies (Yu *et al*., 1997). Adaptation and mitigation strategies in these countries, as elsewhere, will necessitate more economic and efficient energy use and greater emphasis on development of renewable energy sources (see Box 17-4).

To implement these strategies, many small islands, will require external technical, financial, and other assistance (Rijsberman, 1996). Given these states' size and limited individual capacities, pooling of resources through regional cooperation has been proposed as an effective means of designing and implementing some adaptation measures (Nicholls and Mimura, 1998). Some island groupings already have begun to implement regional projects aimed at building capacity to respond to climate change. Two projects—Caribbean Planning for Adaptation to Climate Change (CPACC), which is being implemented by 12 Caribbean states, and Pacific Islands Climate Change Assistance Program (PICCAP), which is being executed by SPREP for 10 Pacific island countries—are outstanding models of regional cooperation.

17.4. Synthesis

17.4.1. Feedbacks, Interactions, and Resilience: The Relevance of these Concepts to Small Islands

Feedback mechanisms in changing systems are poorly understood. These mechanisms could produce a series of *downstream effects* as conditions within a system deviate from "normal" and can include mechanisms that bring conditions back toward "normal" (correcting) or push conditions further away from "normal" (compounding). Feedback of interactions between humans and the environment in the face of climate change are likely to be negative and are expected to increase. As humans have to take adaptive actions to preserve their systems in the face of climate change and sea-level rise, there is a risk that impacts on the environment will increase, despite better awareness of the issues. The short-term needs of humans are likely to take precedence over longer-term needs, which are intimately tied up with the environment. This could mean greater-than-expected and unpredictable indirect impacts resulting from climate change and sea-level rise.

Small island states urgently need sensible predictive information and tools for minimizing the likely outcomes of climate change and sea-level rise because it is now generally accepted that small islands are intrinsically more vulnerable than larger countries (Briguglio, 1992, 1993, 1995, 1997; Wells, 1996, 1997; Atkins *et al*., 1998; UNDP, 1998). The vulnerability of natural systems is still under investigation, although work is underway (Yamada *et al*., 1995; Sem *et al*., 1996; Kaly *et al*., 1999). This makes the requirement for information and tools for generating such information a top priority. The outcome of climate change on human and natural systems in small island states will depend on changes induced by climate itself, feedback mechanisms, human adaptive capacity, and the resilience of biophysical and human systems.

Indicators of high resilience might include the presence of healthy, intact ecosystems; the ability of species to acclimatize to new temperature regimes; the presence of land higher than the maximum predicted transgression plus storm surges; high productivity; reproduction and recruitment of species; and high rates of natural recovery (Kaly *et al*., 1999).

Resilience refers to the innate ability of biophysical and human systems to maintain their integrity when subjected to disturbance (Holling, 1973; Ludwig *et al.*, 1997). For most natural systems, knowledge of resilience to climate change and sea-level rise is inadequate. For example, there are insufficient data to describe the ability of a reef to withstand sea-level rise of 20–40 cm over the next 50 years (but see Hoegh-Guldberg, 1999). Predicting which ecological variables (e.g., species, processes) might be affected and what effect this would have on ecosystem diversity, function, and future resilience may be difficult or impossible (Lubchenco *et al.*, 1993). Proxy estimates have been made, most notably using recent El Niño events in the eastern Pacific; these estimates have many of the expected effects of climate change (Castilla *et al.*, 1993; Burns, 2000). For oceans, these and other proxy estimates predict the extinction of species, mass mortality and bleaching of corals, changes in the geographic range of species, increases in disease, unexpected predation, decreases in productivity, and increases in harmful algal blooms (Glynn, 1984, 1988, 1991; Bak *et al.*, 1984; Hallegraeff, 1993; Lessios, 1998; Hoegh-Guldberg, 1999).

Although earlier work suggested that systems exposed to perturbations tend to have greater capacity to recover from shock (e.g., Holling, 1973), research in a small island context challenges this conventional wisdom. Kaly *et al.* (1999) demonstrate that for many systems, the greater the number and intensity of hazards (human-induced as well as natural) that have impacted them in the past, the greater is their level of vulnerability to future stresses. Furthermore, because neither the natural resilience nor the altered resilience of any ecosystem is known—let alone the resilience that might arise as a result of summed or interactive effects—it is impossible to directly estimate overall resilience. This finding is very disturbing for most small island states, where ecosystems already are severely stressed from natural and anthropogenic forces.

17.4.2. Uncertainties and Risks

17.4.2.1. Inconsistencies and Limitations in Projected Changes

Present ability to accurately assess the effects of climate change in small island states, as in other regions, is further impeded by conflicting predictions of how natural systems may respond. For example, some studies cite the responses of ecosystems to El Niño events in the 1980s and 1990s as indicative of the likely effects of climate change, with conflicting outcomes (Salinger, 1999; Gillespie and Burns, 2000; Hay, 2000). On one hand, it is expected that primary productivity in oceans will *decrease* as a result of global warming when upwelled waters become shoaled (Roemmich and McGowan, 1995; Barber *et al.*, 1996). On the other hand, it has been predicted that harmful algal blooms will *increase* with increasing temperatures (Colwell, 1996; Nurse *et al.*, 1998; see also Hales *et al.*, 1999b). Clearly, forecasting of likely outcomes will be more complex than may have been expected initially. Thus, planning of appropriate responses in regions of low adaptive capacity, such as small island states, presents an even greater challenge.

17.4.2.2. Ecosystem Collapse

One of the likely outcomes of climate change and sea-level rise in natural systems is their collapse. Although ecosystem collapse is not presently well-defined or understood, it is a concern of island countries, which depend heavily on resources provided by these systems to sustain their economic and social well-being (Burns, 2000; Gillespie and Burns, 2000). If mangrove systems are to migrate inland, for example, what is involved is the collapse of the original mangrove system on the existing shore and its eventual replacement with another ecosystem. It is unclear whether the old mangrove area would be replaced by a viable alternative marine ecosystem or whether the new mangroves produced inland would be a viable ecosystem, with all of the same functional characteristics of the original. These and similar questions are central to understanding the ability of human and natural systems to adapt to climate change and sea-level rise.

17.5. Future Requirements, Information, and Research Needs

Although good progress has been made in understanding the vulnerability and adaptation potential of small island states to climate change, the foregoing discussion highlights critical information gaps and uncertainties that still exist. It has been established that small island states constitute a very high-risk group of countries as a consequence of their high vulnerability and low adaptive capacity. Climate change is inevitable, even if any global agreement to limit GHG emissions were swiftly implemented. Thus, the need to focus on adaptation options and requirements already is critical for small islands, given that these countries are projected to suffer disproportionately from the effects of climate change (Bijlsma, 1996; Nurse *et al.*, 1998; Nicholls *et al.*, 1999; Gillespie and Burns, 2000). The agenda set out below therefore is designed not only to fill existing knowledge gaps but also to help identify opportunities for minimizing the adverse effects of climate change (including avoidance of maladaptation), as an important component of adaptation planning in these islands:

- For most small islands, the lack of geographical detail is a critical shortcoming. Outputs from GCMs currently used to assess climate change impacts in small islands are coarse and do not provide adequate information for countries at the scale of small islands. Hence, there is an urgent need for downscaling the outputs of the GCMs to better define and understand island-scale processes and impacts.
- Research into the sensitivity of small islands to climate change, employing an integrated approach, should continue. Studies on the vulnerability of human and biophysical systems to climate change and their interaction with and response to natural and human stresses (including extreme events) need to be integrated because the more common "reductionist" approaches tend to be deficient in their treatment of interactive effects.

- Some small island states have initiated efforts to reduce the impact of natural disasters and to use seasonal to interannual climate analysis and climate forecasts to reduce the impacts of natural hazards. Because long-term climate change may result in a more El Niño-like state, with more frequent and severe extreme climate events, support to build on these past efforts could help to reduce the vulnerability of these islands to climate change.
- Although small islands have many characteristics in common, the heterogeneity factor should not be overlooked. Local conditions on widely varying island types (e.g., tectonic changes, shorelines with large sediment availability versus those with sediment deficits, highly fragmented versus single-island states) may increase or decrease climate change impacts, so the outcomes could be dramatically different in each small island setting. Climate change assessments under such varying circumstances would improve present understanding of vulnerability and adaptation requirements in small islands.
- Given their wide geographical dispersion, there is a need for a coordinated monitoring program for small islands that evaluates the long-term response of ecosystems to climate variability and change. The focus of such an effort should be on the complex interactions that may occur within human and natural systems to modify the frequency and magnitude of impacts expected and to identify ecosystems that may be in danger of collapsing, so that timely adaptive action might be taken. Such work would help to improve our understanding of the concepts of homeostasis, resilience, and feedback mechanisms, which—though frequently alluded to in the literature—are poorly understood and are critical for adaptation planning at the local and regional scale.
- Although the susceptibility of small islands to climate change impacts is high and adaptive capacity is low, the overall level of vulnerability varies within and among states. Thus, vulnerability indices being developed and refined specifically for small islands [Crowards, 1999 (Caribbean Development Bank); Kaly *et al.*, 1999 (SOPAC)] have the potential to make a significant contribution to adaptation planning and implementation in this constituency. Countries therefore might wish to consider continuation of this work among their research priorities.

Finally, there is some uneasiness in the small island states about perceived overreliance on the use of outputs from climate models as a basis for planning risk reduction and adaptation to climate change. There is a perception that insufficient resources are being allocated to relevant empirical research and observation in small islands. Climate models are simplifications of very complex natural systems; they are severely limited in their ability to project changes at small spatial scales, although they are becoming increasingly reliable for identifying general trends. In the face of these concerns, therefore, it would seem that the needs of small island states can best be accommodated by a balanced approach that combines the outputs of downscaled models with analyses from empirical research and observation undertaken in these countries.

References

Alleng, G.P., 1998: Historical development of the Port Royal Mangrove Wetland, Jamaica. *Journal of Tropical Research*, **14(3),** 951–959.

Amadore, L., W.C. Bolhofer, R.V. Cruz, R.B. Feir, C.A. Freysinger, S. Guill S, K.F. Jalal, A. Iglesias, A. Jose, S. Leatherman, S. Lenhart, S. Mukherjee, J.B. Smith, and J. Wisniewski, 1996: Climate change vulnerability and adaptation in Asia and the Pacific: workshop summary. *Water, Air, and Soil Pollution*, **92,** 1–12.

Atkins, J., S. Mazzi, and C. Ramlogan, 1998: *A Composite Index of Vulnerability*. Commonwealth Secretariat, London, United Kingdom, 64 pp.

Bak, R.P.M., M.J.E. Carpay, and E.D. de Ruyter van Steveninck, 1984: Densities of the sea urchin *Diadema antillarum* before and after mass mortalities on the coral reefs of Curacao. *Marine Ecology Progress Series*, **17,** 105–108.

Barber, R.T., M.P. Sanderson, S.T. Lindley, F. Chai, T. Newton, C.C. Trees, D.G. Foley, and F.P. Chavez, 1996: Primary productivity and its regulation in the equatorial Pacific during and following the 1991–1992 El Niño. *Deep-Sea Research Part II*, **43(4–6),** 933–969.

Beer, S. and E. Koch, 1996: Photosynthesis of marine macroalgae and seagrass in globally changing CO_2 environments. *Marine Ecology Progress Series*, **141,** 199–204.

Berkes, F., R. Mahon, P. McConney, R. Pollnac, and R. Pomeroy, 2001: *Managing Small-Scale Fisheries—Alternative Directions and Methods*. International Development Research Centre, Ottawa, Ontario, Canada, 250 pp.

Biagini, E., 1999: Island environments. In: *Insularity and Development: International Perspectives on Islands* [Biagini, E. and B. Hoyle (eds.)]. Pinter, London, United Kingdom, pp. 17–41.

Bijlsma, L., 1996: Coastal zones and small islands. In: *Climate Change 1995: Impacts, Adaptations, and Mitigation of Climate Change: Scientific-Technical Analyses. Contribution of Working Group II to the Second Assessment Report of the Intergovernmental Panel on Climate Change* [Watson, R.T., M.C. Zinyowera, and R.H. Moss (eds.)]. Cambridge University Press, Cambridge, United Kingdom and New York, NY, USA, pp. 289–324.

Blommestein, E., B. Boland, T. Harker, S. Lestrade, and J. Towle, 1996: Sustainable development and small island states of the Caribbean. In: *Small Islands: Marine Science and Sustainable Development* [Maul, G.A. (ed.)]. American Geophysical Union, Washington, DC, USA, pp. 385–419.

Briguglio, L., 1997: Alternative economic vulnerability indicators for developing countries with special reference to SIDS. In: *Report Prepared for the Expert Group on Vulnerability Indices, UN-DESA, 17–19 December 1997*. United Nations, New York, NY, USA, 49 pp.

Briguglio, L., 1995: Small island states and their economic vulnerabilities. *World Development*, **23,** 1615–1632.

Briguglio, L., 1993: The economic vulnerabilities of small island developing states. In: *Study Commissioned by CARICOM for the Regional Technical Meeting of the Global Conference on the Sustainable Development of Small Island Developing States, Port of Spain, Trinidad and Tobago, July 1993*.

Briguglio, L., 1992: P*reliminary Study on the Construction of an Index for Ranking Countries According to their Economic Vulnerability*. UNCTAD/LDC/Misc.4.

Brown, B.E., 1997a: Coral bleaching: causes and consequences. *Coral Reefs*, **16,** 129–138.

Brown, B.E., 1997b: Adaptations of reef corals to physical environmental stress. *Advances in Marine Biology*, **31,** 221–299.

Brown, B. and J. Ogden, 1993: Coral bleaching. *Scientific American*, **2,** 64–70.

Bryant, D., L. Burke, J. McManus, and M. Spalding, 1997: *Reefs at Risk: A Map-Based Indicator of Potential Threats to the World's Coral Reefs*. World Resources Institute, Washington, DC, USA, 56 pp.

Burns, W.C.G., 2000: The impact of climate change on Pacific island developing countries in the 21st century. In: *Climate Change in the South Pacific: Impacts and Responses in Australia, New Zealand and Small Island States* [Gillespie, A. and W.C.G. Burns (eds.)]. Kluwer Academic Publishers, Dordrecht, The Netherlands, pp. 233–250.

Cai, W.J. and P.H. Whetton, 2000: Evidence for a time-varying pattern of greenhouse warming in the Pacific Ocean. *Geological Research Letters*, **27(16),** 2577–2580.

Campbell, J.R., 1996: Contexualizing the effects of climate change in Pacific island countries. In: *Climate Change: Developing Southern Hemisphere Perspectives* [Giambelluca, T.W. and A. Henderson-Sellers (eds.)]. John Wiley and Sons, Brisbane, Australia, pp. 349–374.

Campbell, J.R. and N. de Wet, 2000: *Adapting to Climate Change: Incorporating Climate Change Adaptation into Development Activities in Pacific Island Countries: A Set of Guidelines for Policy Makers and Development Planners*. South Pacific Regional Environment Programme (SPREP), Apia, Samoa, 35 pp.

CARICOMP, 1997: Studies in Caribbean bleaching. In: *Proceedings of the 8th International Coral Reef Symposium*. pp. 673–678.

Carvalho, P. and B. Clarke, 1998: Ecological sustainability of the South Australian coastal aquaculture management policies. *Coastal Management*, **26,** 281-290.

Castilla, J.C., S.A. Navarrete, and J. Lubchenco, 1993. Southeastern Pacific coastal environments: main features, large-scale perturbations, and global climate change. In: *Earth Systems Responses to Global Change: Contrasts Between North and South America* [Mooney, H.H., B. Kronberg, and E.R. Fuentes (eds.)]. Academic Press, San Diego, CA, USA, 167 pp.

Challenger, B., 1997: *Adaptation to Climate Change in Antigua and Barbuda*. Report prepared for the Government of Antigua and Barbuda, United Nations Environment Programme, Global Environment Facility Country Case Study on Climate Change Impacts and Adaptation Assessments in Antigua and Barbuda, 43 pp.

Cheeroo-Nayamuth, B.F. and A.R. Nayamuth, 1999: *Vulnerability and Adaptation Assessment of the Sugar Cane Crop to Climate Change in Mauritius*. Mauritius Sugar Industry Research Institute and the National Climate Committee, Mauritius, 40 pp.

Cicin-Sain, B. 1998. Integrated *Coastal and Ocean Management: Concepts and Practices* [Cicin-Sain, B. and R.W. Knecht (eds.)]. University of Delaware, Center for the Study of Marine Policy, and Island Press, Washington, DC, USA, 517 pp.

Colwell, R.R., 1996: Global climate and infectious disease: the cholera paradigm. *Science*, **274,** 2025–2031.

Costa, M.J., J.L. Costa, P.R. Almeida, and C.A. Assis, 1994: Do eel grass beds and salt marsh borders act as preferential nurseries and spawning grounds for fish? An example of the Mira estuary in Portugal. *Ecological Engineering*, **3,** 187–195.

Cox, P.A., T. Elmqvist, E.D. Pierson, and W.E. Rainey, 1991: Flying foxes as strong interactors in south-pacific island ecosystems—a conservation hypothesis. *Conservation Biology*, **5(4),** 448–454.

Crowards, T., 1999: *An Economic Vulnerability Index for Developing Countries, with Special Reference to the Caribbean*. Caribbean Development Bank, Bridgetown, Barbados, 26 pp. (plus appendices).

Durako, M.J., 1993: Photosynthetic utilization of CO_2 and HCO_3 in Thallassia testudinum (Hydrocharitaceae). *Marine Biology*, **115,** 373–380.

Edwards, A.J., 1995: Impact of climate change on coral reefs, mangroves and tropical seagrass ecosystems. In: *Climate Change: Impact on Coastal Habitation* [Eisma, D. (ed.)]. Lewis Publishers, Boca Raton, FL, USA, pp. 209–234.

Edwards, M., 2000: Parochialism and empowerment. In: *Climate Change in the South Pacific: Impacts and Responses in Australia, New Zealand and Small Island States* [Gillespie, A. and W.C.G. Burns (eds.)]. Kluwer Academic Publishers, Dordrecht, The Netherlands, pp. 251–268.

Ehrlich, P.R and A.H. Ehrlich, 1991: *Healing the Planet*. Addisson-Wesley Publication Co. Inc., Menlo Park, CA, USA, 212 pp.

Ellis, M. and S. Fifita, 1999: *Greenhouse Gas Mitigation: A Regional Analysis for Pacific Island Countries*. South Pacific Regional Environment Programme, Apia, Samoa, 64 pp.

Ellison, J.C., 1993: Mangrove retreat with rising sea level, Bermuda. *Estuarine Coastal and Shelf Science*, **37(1),** 75–87.

Ellison, J.C. and D.R. Stoddart, 1991: Mangrove ecosystem collapse during predicted sea-level rise: holocene analogs and implications. *Journal of Coastal Research*, **7(1),** 151–165.

Epstein, P.R., 1999: Climate and health. *Science*, **285,** 347.

Epstein, P.R, 1997: Environmental changes and human health. *Consequences*, **3,** 2.

Epstein, P.R., H.F. Diaz, S.A. Elias, G. Grabherr, N.E. Graham, W.J.M. Martens, E. Moseley-Thompson, and J. Susskind, 1997: Biological and physical signs of climate change: focus on mosquito-borne diseases. *Bulletin of the American Meteorological Society*, **78,** 409–417.

Everett, J.T., 1996: Fisheries. In: *Climate Change 1995: Impacts, Adaptations, and Mitigation of Climate Change: Scientific-Technical Analyses. Contribution of Working Group II to the Second Assessment Report of the Intergovernmental Panel on Climate Change* [Watson, R.T., M.C. Zinyowera, and R.H. Moss (eds.)]. Cambridge University Press, Cambridge, United Kingdom and New York, NY, USA, pp. pp. 511–537.

Ewel, K.C., J.A. Bourgeois, T.G. Cole, and S.F. Zheng, 1998: Variation in environmental characteristics and vegetation in high-rainfall mangrove forests, Kosrae, Micronesia. *Global Ecology and Biogeography Letters*, **7(1),** 49–56.

Falkland, A.C., 1992: *Small Tropical Islands*. International Hydrological Programme Humid Tropics Programme Series No. 2, United Nations Educational, Scientific and Cultural Organisation (UNESCO) Paris, France.

Farnsworth, E.J., 1998: Issues of spatial, taxonomic, and temporal scale in delineating links between mangrove diversity and ecosystem function. *Global Ecology and Biogeography Letters*, **7(1),** 15–25.

Forbes, D.L. and S.M. Solomon, 1997: *Approaches to Vulnerability Assessment on Pacific Island Coasts: Examples from Southeast Viti Levu (Fiji) and Tarawa (Kiribati)*. Miscellaneous Report 277, SOPAC, Suva, 21 pp.

Gattuso, J.-P., M. Frankignouille, and I. Bourge, 1999: Effects of calcium carbonate saturation of seawater on coral calcification. *Global Planetary Change*, **18,** 37–47.

Gillespie, A. and W.C.G. Burns, 2000: *Climate Change in the South Pacific: Impacts and Responses in Australia, New Zealand and Small Island States*. Kluwer Academic Publishers, Dordrecht, The Netherlands, 385 pp.

Gillie, R.D., 1997: Causes of coastal erosion in Pacific island nations. *Journal of Coastal Research*, **24,** 174–204.

Giorgi, F and L.O. Mearns, 1999: Introduction to special section: regional climate modeling revisited. *Journal of Geophysical Research*, **104(D6),** 6335–6352.

Glynn, P.W., 1993: Coral reef bleaching: ecological perspectives. *Coral Reefs*, **12,** 1–17.

Glynn, P.W., 1991: Coral reef bleaching in the 1980s and possible connections with global warming. *Trends in Ecology and Evolution*, **6,** 175–179.

Glynn, P.W., 1988: El Niño-Southern Oscillation 1982–1983: nearshore population, community, and ecosystem responses. *Annual Review of Ecology and Systematics*, **19,** 309–345.

Glynn, P.W., 1984: Widespread coral mortality and the 1982/83 El Niño warming event. *Environmental Conservation*, **11,** 133–146.

Goldberg, E.D., 1994: *Coastal Zone Space—Prelude to Conflict?* IOC Ocean Forum I, UNESCO Publishing, Paris, France, 138 pp.

Goreau, T.J., 1992: Bleaching and reef community change in Jamaica: 1951–1991. *American Zoology*, **32,** 683–695.

Goreau, T.J. and R.M. Hayes, 1994: Coral bleaching and ocean "hot spots." *Ambio*, **23,** 176–180.

Goreau, T.J., R.M. Hayes, and A.E. Strong, 1997: Tracking south Pacific coral reef bleaching by satellite and field observations. In: *Proceedings of the 8th International Coral Reef Symposium, Panama City, Panama*. pp. 1491–1494.

Government of Commonwealth of Dominica, 2000: *The Commonwealth of Dominica's First National Report on the Implementation of the United Nations Convention to Combat Desertification (UNCCD)*. Environmental Coordinating Unit, Ministry of Agriculture, Planning and Environment, Roseau, Commonwealth of Dominica, 20 pp.

Granger, O.E., 1997: Caribbean island states: perils and prospects in a changing global environment. *Journal of Coastal Research*, **24,** 71–94.

Gray, C.R., 1993: Regional meteorology and hurricanes. In: *Climatic Change in the Intra-Americas Sea* [Maul, G.A. (ed.)]. Edward Arnold, London, United Kingdom, pp. 87–99.

Hader, D.P., 1993: Effects of enhanced solar ultraviolet radiation on aquatic ecosystems. In: *UV-B Radiation and Ozone: Effects on Humans, Animals, Plants, Microorganisms and Materials* [Tevini, M. (ed.)]. Lewis Publishers, pp. 155–192.

Hales, S., Y. Souares, P. Weinstein, and A. Woodward, 1999a: El Niño and the dynamics of vector-borne disease transmission. *Environmental Health Perspectives,* **107(2),** 3–6.

Hales, S., P. Weinstein, and A. Woodward, 1999b: Ciguatera fish poisoning, El Niño and Pacific sea surface temperatures. *Ecosystem Health*, **5,** 20–25.

Hales, S., P. Weinstein, and A. Woodward, 1997: Public health impacts of global climate change. *Reviews on Environmental Health,* **12(3),** 191–199.

Hallegraeff, G.M., 1993: A review of harmful algal blooms and their apparent global increase. *Phycologia*, **32(2),** 79–99.

Hatcher, B.G., 1997: Coral reef ecosystems: how much greater is the whole than the sum of the parts? *Coral Reefs*, **16(5),** 577–591.

Hatcher, B G., 1996: Coral reef ecosystems: how much greater is the whole than the sum of the parts? In: *Proceedings of the 8th International Coral Reef Symposium, Panama City, Panama*. pp. 43–56.

Hay, J.E., 2000: Climate change in the Pacific: Science-based information and understanding. In: *Climate Change in the South Pacific: Impacts and Responses in Australia, New Zealand and Small Island States* [Gillespie, A. and W.C.G. Burns (eds.)]. Kluwer Academic Publishers, Dordrecht, The Netherlands, pp. 269–287.

Hay, J.E. and G. Sem, 1999: *A Regional Synthesis of National Greenhouse Gas Inventories*. South Pacific Regional Environment Programme, Apia, Samoa, 29 pp.

Hay, J.E., C. Kaluwin, and N. Koop, 1995: Implications of climate change and sea level rise for small island nations of the South Pacific: a regional synthesis. *Weather and Climate*, **15,** 5–20.

Helmer, W., P. Vellinga, G. Litjens, E.C.M. Ruijgrok, H. Goosen, and W. Overmars, 1996: *Meegroeien met de Zee*. Wereld Natuur Fonds, Zeist, 35 pp.

Hennessy, K.J., P.H. Whetton, J.J. Katzfey, J.L. McGregor, R.N. Jones, C.M. Page, and K.C. Nguyen, 1998: *Fine Resolution Climate Change Scenarios for New South Wales, Annual Report 1997–1998*. Commonwealth Scientific and Industrial Research Organisation, Atmospheric Research, Aspendale, Victoria, Australia, 48 pp.

Hoegh-Guldberg, O., 1999: *Climate Change, Coral Bleaching and the Future of the World's Coral Reefs*. Greenpeace International, Sydney, New South Wales, Australia, 27 pp.

Holland, G.T., 1997: The maximum potential intensity of tropical cyclones. *Journal of Atmospheric Science,* **54,** 2519–2541.

Holling, C.S., 1973: Resilience and stability of ecological systems. *Annual Review of Ecology and Systematics*, **4,** 1–23.

Holthus, P., M. Crawford, C. Makroro, and S. Sullivan, 1992: *Vulnerability Assessment of Accelerated Sea-Level Rise—Case Study: Majuro Atoll, Marshall Islands*. South Pacific Regional Environment Programme (SPREP) Reports and Studies Series No. 60. South Pacific Regional Environment Programme, Apia, Western Samoa, 107 pp.

Huang, J.C.K., 1997: Climate change and integrated coastal management: a challenge for small island nations. *Ocean and Coastal Management*, **37,** 95–107.

IPCC, 1998: *The Regional Impacts of Climate Change: An Assessment of Vulnerability. A Special Report of IPCC Working Group II* [Watson, R.T., M.C. Zinyowera, and R.H. Moss (eds.)]. Cambridge University Press, Cambridge, United Kingdom and New York, NY, USA, 517 pp.

IPCC, 1996: *Climate Change 1995: Impacts, Adaptations, and Mitigation of Climate Change: Scientific-Technical Analyses. Contribution of Working Group II to the Second Assessment Report* [Watson, R.T., M.C. Zinyowera, and R.H. Moss (eds.)]. Cambridge University Press, Cambridge, United Kingdom and New York, NY, USA, 880 pp.

IPCC, 1992: A common methodology for assessing vulnerability to sea-level rise—second revision. In: *Global Climate Change and the Rising Challenge of the Sea*. Report of the Coastal Zone Management Subgroup, Response Strategies Working Group of the Intergovernmental Panel on Climate Change, Ministry of Transport, Public Works and Water Management, The Hague, The Netherlands, 27 pp.

IPCC, 1990: *Strategies for Adaptation to Sea Level Rise*. IPCC Response Strategies Working Group, Ministry of Transport and Public Works, Ministry of Housing,, The Hague, The Netherlands, 122 pp.

Jensen, T.J., 1999: Renewable energy on small islands. *Tiempo*, **32,** 11–14.

Jones, I.C., J.L. Banner, and B.J. Mwansa, 1998: Geochemical constraints on recharge and groundwater evolution: the Pleistocene Aquifer of Barbados. In: *Proceedings of the Third Annual Symposium, American Water Resources Association, and Fifth Caribbean Islands Water Resources Congress, San Juan, Puerto Rico, July 12–16, 1998*. 6 pp.

Jones, R.N., 1998: *An Analysis of the Impacts of the Kyoto Protocol on Pacific Island Countries, Part 1: Identification of Latent Sea-Level Rise Within the Climate System at 1995 and 2020*. South Pacific Regional Environment Programme, Apia, Samoa and Commonwealth Scientific and Industrial Research Organistaion, Canberra, Australia, 9 pp.

Jones, R.N., K.L. Hennessy, C.M. Page, K.J.E. Walsh, and P.H. Whetton, 1999: *An Analysis of the Effects of Kyoto Protocol on Pacific Island Countries, Part 2: Regional Climate Change Scenarios and Risk Assessment Methods*. A research report prepared for the South Pacific Regional Environment Programme, Apia, Samoa and CSIRO Atmospheric Research, Canberra, Australia, 61 pp.

Kaluwin, C. and J.E. Hay (eds.), 1999: *Climate Change and Sea Level Rise in the South Pacific Region. Proceedings of the Third SPREP Meeting, New Caledonia, August, 1997*. South Pacific Regional Environment Programme, Apia, Western Samoa, 347 pp.

Kaly, U., L. Briguglio, H. McLeod, S. Schmall, C. Pratt, and R. Pal, 1999: *Environmental Vulnerability Index (EVI) to Summarise National Environmental Vulnerability Profiles*. Technical Report 275, South Pacific Applied Geoscience Commission, Suva, Fiji, 67 pp.

Kay, R.C. and J.E. Hay, 1993: A decision support approach to coastal vulnerability and resilience assessment: a tool for integrated coastal zone management. In: *Vulnerability Assessment to Sea-Level Rise and Coastal Zone Management. Proceedings of the IPCC/WCC'93 Eastern Hemisphere Workshop, Tsukuba, Japan, 3–6 August, 1993* [McLean, R.F. and N. Mimura (eds.)]. Department of Environment, Sport and Territories, Canberra, Australia, pp. 213–225.

Klein, R.J.T. and R.J. Nicholls, 1999: Assessment of coastal vulnerability to climate change. *Ambio*, **28(2),** 182–187.

Klein, R.J.T. and R.J. Nicholls, 1998: Coastal zones. In: *Handbook on Climate Change Impact Assessment and Adaptation Strategies* [Feenstra, J.F., I. Burton, J.B. Smith, and R.S.J. Tol (eds.)]. Version 2.0, United Nations Environment Programme and Institute for Environmental Studies, Vrije Universiteit, Nairobi, Kenya and Amsterdam, The Netherlands, Chapter 7, pp. 1–35. Available online at http://www.vu.nl/english/o_o/instituten/IVM/research/climatechange/Handbook.htm.

Kleypas, J.A., R.W. Buddemeier, D. Archer, J.P. Gattuso, C. Langdon, and B.N. Opdyke, 1999: Geochemical consequences of increased atmospheric carbon dioxide in coral reefs. *Science*, **284,** 118–120.

Knutson, T.R. and S. Manabe, 1998: Model assessment of decadal variability and trends in the tropical Pacific Ocean. *Journal of Climatology*, **11,** 503–519.

Lal, M., H. Harasawa, and K. Takahashi, 2001: Future climate change and its impacts over small island states. *Climate Research*, (in press).

Larkham, A.W.D. and W.F. Wood, 1993: The effect of UV-B radiation on photosynthesis and respiration of phytoplankton, benthic macroalgae and seagrasses. *Photosynthesis Research*, **36,** 17–23.

Leatherman, S.P., 1997. Beach ratings: a methodological approach. *Journal of Coastal Research*, **13,** 1050–1063.

Leatherman, S.P., 1996: Shoreline stabilization approaches in response to sea-level rise: U.S. experience and implications for Pacific island and Asian nations. *Water, Air, and Soil Pollution*, **92,** 149–157.

Lehodey, P., M. Bertignac, J. Hampton, A. Lewis, and J. Picaut, 1997: El Niño Southern Oscillation and tuna in the western Pacific. *Nature*, **389,** 715–718.

Lessios, H.A., 1998: Mass mortality of *Diadema antillarum* in the Caribbean: what have we learned? *Annual Review of Ecological Systems*, **19,** 371–393.

Lubchenco, J., S.A. Navarrete, B.N. Tissot, and J.C. Castilla, 1993: Possible ecological responses to global climate change: near shore benthic biota of Northeastern Pacific coastal ecosystems. In: *Earth System Responses to Global Change: Contrasts Between North and South America* [Mooney, H.H., B. Kronberg, and E.R. Fuentes (eds.)]. Academic Press, San Diego, CA, USA, pp. 147–166.

Ludwig, D., B. Walker, and C.S. Holling, 1997: Sustainability, stability and resilience. *Conservation Ecology* **1(1),** 7. Available online at http://www.consecol.org/vol1/iss1/art7.

Mahon, R., 1996: Fisheries of small island states and their oceanographic research and information needs. In: *Small Islands: Marine Science and Sustainable Development* [Maul, G.A. (ed.)]. American Geophysical Union, Washington, DC, USA, pp. 298–322.

Martin, H. and J.P. Bruce, 1999: *Effects of Climate Change: Hydrometeorological and Land-Based Effects in The Bahamas*. Global Change Strategies International Inc., Ottawa, ON, Canada, 34 pp.

Maul, G.A., 1996: *Marine Science and Sustainable Development*. Coastal and Estuarine Studies, America Geophysical Union, Washington, DC, USA, 467 pp.

Maytin, C.E., M.F. Acevedo, R. Jaimez, R. Anderson, M.A. Harwell, A. Robock, and A. Azocar, 1995: Potential effects of global climate change on the phenology and yield of maize in Venezuela. *Climatic Change*, **29,** 189–211.

McIver, D.C. (ed.), 1998: *Adaptation to Climate Variability and Change*. IPCC Workshop Summary, San José, Costa Rica, 29 March–1 April, 1998. Atmospheric Environment Service, Environment Canada, Montreal, Canada, p. 55.

McMichael, A.J., 1996: Human population health. In: *Climate Change 1995: Impacts, Adaptations, and Mitigation of Climate Change: Scientific-Technical Analyses. Contribution of Working Group II to the Second Assessment Report of the Intergovernmental Panel on Climate Change* [Watson, R.T., M.C. Zinyowera, and R.H. Moss (eds.)]. Cambridge University Press, Cambridge, United Kingdom and New York, NY, USA, pp. 561–584.

McMichael, A.J. 1993: *Planetary Overload: Global Environmental Change and the Health of the Human Species*. Cambridge University Press, Cambridge, United Kingdom and New York, NY, USA, 212 pp.

McNeely, J.A. (ed.), 1993: *Parks for Life*. Report of the IVth World Congress on National Parks and Protected Areas, Caracas, Venezuela, 10-12 February 1992. World Conservation Union, Gland, Switzerland.

McNeely, J.A., M. Gadgil, C. Leveque, C. Padoch, and K. Redford, 1993: Human influences on biodiversity. In: *Global Biodiversity Assessment* [Heyword, V.H. and R.T. Watson (eds.)]. United Nations Environmental Programme, Cambridge University Press, Cambridge, United Kingdom, pp. 711–821.

McNeely, J.A., K.R. Miller, W.V. Reid, R.A. Mittermeier, and T.B. Werner (eds.), 1990: *Conserving the World's Biological Diversity*. International Union for Conservation of Nature and Natural Resources, Gland, Switzerland; and World Resources Institute, Conservation International, World Wildlife Fund–U.S., and the World Bank, Washington, DC, USA.

Meehl, G.A. and W.M. Washington, 1996: Vulnerability of freshwater resources to climate change in the tropical Pacific region. *Water, Air, and Soil Pollution*, **92,** 203–213.

Menendez, L. and A. Priego, 1994: Los manglares de Cuba: ecologia. In: *El Ecosistema de Manglar en America Latina y la Cuenta del Caribe: Su Manejo y Conservacion* [Suman, D.O. (ed.)]. Rosentiel School of Marine and Atmospheric Science, University of Miami, Miami, FL, USA, and the Tinker Foundation, New York, NY, USA, pp. 64–75.

Mieremet, B., B.M. Bruce, and T.E. Reiss, 1997. Yap Islands natural coastal systems and vulnerability to potential accelerated sea-level rise. *Journal of Coastal Research*, **27,** 151–170.

Mimura, N. and P.D. Nunn, 1998: Trends of beach erosion and shoreline protection in rural Fiji. *Journal of Coastal Research*, **14(1),** 37–46.

Mimura, N. and N. Pelesikoti, 1997: Vulnerability of Tonga and future sea-level rise. *Journal of Coastal Research*, **24,** 117–132.

Munasinghe, M., 2000: Development, equity and sustainability (DES) in the context of climate change. In: *Climate Change and Its Linkages with Development, Equity and Sustainability: Proceedings of the IPCC Expert Meeting held in Colombo, Sri Lanka, 27–29 April 1999* [Munasinghe, M. and R. Swart (eds.)]. LIFE, Colombo, Sri Lanka; RIVM, Bilthoven, The Netherlands; and World Bank, Washington, DC, USA, pp. 13–66.

Murray, C., 1993: Catastrophe reinsurance crisis in the Caribbean. *Catastrophe Reinsurance Newsletter*, **6,** 14–18.

Nicholls, R.J. and J. Branson (eds.), 1998: Enhancing coastal resilience—planning for an uncertain future. *The Geographical Journal*, **164(3),** 255–343.

Nicholls, R.J. and M.J. Hoozemans, 1996: The Mediterranean: vulnerability to coastal implications of climate change. *Ocean and Coastal Management*, **31,** 105–132.

Nicholls, R.J. and M. Mimura, 1998: Regional issues raised by sea-level rise and their policy implications. *Climate Change*, **11,** 5–18.

Nicholls, R.J., F.M.J. Hoozemans, and M. Marchand, 1999: Increasing flood risk and wetland losses due to global sea-level rise: regional and global analyses. *Global Environmental Change*, **9,** 69–87.

Nunn, P.D. and N. Mimura, 1997: Vulnerability of South Pacific island nations to sea-level rise. *Journal of Coastal Research*, **24,** 133–151.

Nurse, L.A., 1999: Integrated coastal zone management in the Caribbean: lessons borrowed from empirical experience. *Natural Resources Management in the Caribbean: Discussion Papers*. Organization of Eastern Caribbean States–Natural Resources Management Unit/ Caribbean Centre for Administration Development/Organization of American States, Castries, St. Lucia, pp. 1–5.

Nurse, L.A., R.F. McLean, and A.G. Suarez, 1998: Small island states. In: *The Regional Impacts of Climate Change: An Assessment of Vulnerability. A Special Report of IPCC Working Group II* [Watson, R.T., M.C. Zinyowera, and R.H. Moss (eds.)]. Cambridge University Press, Cambridge, United Kingdom and New York, NY, USA, pp. 331–354.

Nurse, L.A., 1992: Predicted sea-level rise in the wider Caribbean: likely consequences and response options. In: *Semi-Enclosed Seas* [Fabbri, P. and G. Fierro (eds.)]. Elsevier Science, Essex, United Kingdom, pp. 52–78.

Nurse, L.A., 1985: A review of selected aspects of the geography of the Caribbean and their implications for environmental impact assessment. In: *Proceedings of the Caribbean Seminar on Environmental Impact Assessment, May 27–June 7, 1985* [Geoghegan, T. (ed.)]. Centre for Resource Management and Environmental Studies, University of the West Indies; Caribbean Conservation Association; Institute for Resource and Environmental Studies-Dalhousie University; and the Canadian International Development Agency, Bridgetown, Barbados, pp. 10–18.

Pantin, D.A., 1997: *Alternative Ecological Vulnerability Indices for Developing Countries with Special Reference to Small Island Developing States (SIDS)*. Report for the United Nations Department of Economic and Social Affairs, Sustainable Economic Development Unit, University of West Indies, Trinidad, 59pp.

Parkinson, R.W. and R.D. Delaune, 1994: Holocene sea-level rise and the fate of mangrove forests within the wider Caribbean region. *Journal of Coastal Research*, **10(4),** 1077–1086.

Parry, M., N. Arnell, M. Hulme, R. Nicholls, and M. Livermore, 1999: Buenos Aires and Kyoto targets do little to reduce climate change impacts. *Global Environment Change*, **8(4),** 1–5.

Patz, J.A., P.R. Epstein, T.A. Burke, and J.M. Balbus, 1996: Global climate change and emerging infectious diseases. *Journal of American Medical Association*, **275,** 217–223.

Perez, A.L., C. Rodriguez, C.A. Alvarez, and A.D. Boquet, 1999: Asentamientos humanaos y uso de la tierra. In: *Impactos del Cambio Climatico y Medidas de Adaptacion en Cuba* [Gutierrez, T., A. Centella, M. Limia, and M. Lopez (eds.)]. Proyecto No. FP/CP/2200-97-12, United Nations Environment Programme/INSMET, La Habana, Cuba, pp. 130–163.

Pittock, A.B., K.J. Walsh, and K.L. McInnes, 1996: Tropical cyclones and coastal inundation under enhanced greenhouse conditions. *Water, Air, and Soil Pollution*, **92,** 159–169.

Planos, E.O. and O. Barros, 1999: Impacto del cambio climatico y medidas de adaptacion en Cuba: sector recursos hidricos. In: *Impactos del Cambio Climatico y Medidas de Adaptacion en Cuba* [Gutierrez, T., A. Centella, M. Limia, and M. Lopez (eds.)]. Proyecto No. FP/CP/2200-97-12, United Nations Environment Programme/Institute of Meteorology, La Habana, Cuba, pp. 28–54.

Ragoonaden, S., 1997: Impact of sea-level rise on Mauritius. *Journal of Coastal Research Special Issue*, **24,** 206–223.

Richmond, B.M., B. Mieremet, and T. Reiss, 1997: Yap Islands natural coastal systems and vulnerability to potential accelerated sea-level rise. *Journal of Coastal Research*, **24,** 153–172.

Rijsberman, F., 1996: Rapporteur's statement. In: *Adapting to Climate Change: An International Perspective* [Smith, J.B., N. Bhatti, G. Menzhulin, R. Beniof, M.I. Budyko, M. Campos, B. Jallow, and F. Rijsberman (eds.)]. Springer-Verlag, New York, NY, USA, pp. 279–282.

Roemmich, D. and J. McGowan, 1995: Climate warming and the decline of zooplankton in the California current. *Science*, **267(5202),** 1324–1326.

Royer, F., B. Chauvin, P. Timbal, P. Araspin, and D. Grimal, 1998: A GCM study of the impact of greenhouse gas increases on the frequency of occurrence of tropical cyclones. *Climatic Change*, **38,** 307–343.

Sachs, W., 2000: Development patterns in the north and their implications for climate change. In: *Climate Change and Its Linkages with Development, Equity and Sustainability: Proceedings of the IPCC Expert Meeting held in Colombo, Sri Lanka, 27–29 April, 1999* [Munasinghe, M. and R. Swart (eds.)]. LIFE, Colombo, Sri Lanka; RIVM, Bilthoven, The Netherlands; and World Bank, Washington, DC, USA, pp. 163–176.

Salinger, M.J., 1999: Variability in the southwest Pacific. In: *Climate Change and Sea Level Rise in the South Pacific Region. Proceedings of the Third SPREP Meeting, New Caledonia, August, 1997* [Kaluwin, C. and J.E. Hay (eds.)]. South Pacific Regional Environment Programme, Apia, Western Samoa, pp. 47–65.

Salinger, M.J. and A.B. Mullen, 1999: New Zealand climate: temperature and precipitation variations and their links to atmospheric circulation, 1930–1994. *International Journal of Climatology*, **19,** 1049–1071.

Salinger, M.J., R. Basher, B. Fitzharris, J. Hay, P.D. Jones, I.P. Macveigh, and I. Schmideley-Lelu, 1995: Climate trends in the south-west Pacific. *International Journal of Climatology*, **15,** 285–302.

Sattersfield, A.J., M.J. Crosby, A.J. Long, and D.C. Wege, 1998: *Endemic Bird Areas of the World: Priorities for Biodiversity Conservation*. BirdLife Conservation Series No. 7, BirdLife International, Cambridge, United Kingdom.

Saunders, A., 1993: *Underwriting Guidelines*. Paper presented at the 13th Caribbean Insurance Conference, Association of British Insurers, London, United Kingdom, 7 pp.

Schlager, W., 1999: Scaling of sedimentation rates and drowning of reefs and carbonate platforms. *Geology*, **27,** 183–186.

Sem, G., J.R. Campbell, J.E. Hay, N. Mimura, S. Nishioka, K. Yamada, E. Ohno, and M. Serizawa, 1996: Coastal vulnerability and resilience in Tuvalu. In: *Assessment of Climate Change Impacts and Adaptation*. South Pacific Regional Environment Programme, Apia, Western Samoa, and Environment Agency of Japan, Overseas Environmental Cooperation Centre, Tokyo, Japan, 130 pp.

Shea, E., 1994: Climate change reality: current trends and implications. In: *Proceedings of the Climate Change Implications and Adaptation Strategies for the Indo-Pacific Island Nations Workshop, Honolulu, Hawaii, 26–30 September 1994* [Rappa, P., A. Tomlinson, and S. Ziegler (eds.)]. U.S. Country Studies Program, Washington, DC, USA, pp. 26–28.

Short, F.T. and H.A. Neckles, 1999: The effects of global climate change on seagrasses. *Aquatic Botany*, **63,** 169–196.

Singh, B., 1997a: Climate-related global changes in the southern Caribbean: Trinidad and Tobago. *Global and Planetary Change*, **15,** 93–111.

Singh, B., 1997b: Climate changes in the greater and southern Caribbean. *International Journal of Climatology*, **17,** 1093–1114.

Singh, B. and M. El Maayar, 1998: Potential impacts of a greenhose gas climate change scenarios on sugar cane yields in Trinidad, southern Caribbean. *Tropical Agriculture*, **75(3),** 348–355.

Snedaker, S.C. and J.F. Meeder, 1994: Mangrove ecosystem collapse during predicted sea-level rise: holocene analogs and implications discussion. *Journal of Coastal Research*, **10(2),** 497–498.

Solomon, S.M. and D.L. Forbes, 1999: Coastal hazards and associated management issues on the South Pacific islands. *Oceans and Coastal Management*, **42,** 523–554.

Spalding, M.D. and A.M. Grenfell, 1997: New estimates of global and regional coral reef areas. *Coral Reefs*, **16,** 225–230.

Stern, P.C. and D. Easterling, 1999: *Making Climate Forecasts Matter*. National Academy Press, Washington, DC, USA.

Suarez, A.G., A. Lopez, H. Ferras, A. Chamizo, D. Vilamajo, A. Martel, and E. Mojena, 1999: Biodiversidad y vida Silvestre. In: *Impactos del Cambio Climatico y Medidas de Adaptacion en Cuba* [Gutierrez, T., A. Centella, M. Limia, and M. Lopez (eds.)]. Proyecto No. FP/CP/2200-97-12, United Nations Environment Programme, La Habana, Cuba, pp. 164–178.

Suman, D.O., 1994: Status of mangroves in Latin America and the Caribbean basin. In: *El Ecosistema de Manglar en America Latina y la Cuenta del Caribe: Su Manejo y Conservacion* [Suman, D.O. (ed.)]. Rosentiel School of Marine and Atmospheric Science, Universidad de Miami, FL and the Tinker Foundation, New York, NY, USA, pp. 11–20.

Teh, T.S., 1997: Sea-level rise implications for coastal and island resorts. In: *Climate Change in Malaysia, Proceedings of the National Conference on Climate Change, 12–13 August 1996*. Universiti Putra Malaysia, Penang, Malaysia, pp. 83–102.

Timmerman, A., M. Latif, and A. Bacher, 1997: Increased El Niño frequency in a climate model forced by future greenhouse warming. *Nature*, **398,** 694–696.

Tonkin, H., C. Landsea, G.J. Holland, and S. Li, 1997: Tropical cyclones and climate change: a preliminary assessment. In: *Assessing Climate Change Results from the Model Evaluation Consortium for Climate Assessment* [Howe, W. and A. Henderson-Sellers (eds.)]. Gordon and Breach, Sydney, Australia. pp. 327–360.

Toth, F., 2000: Development, equity and sustainability concerns in climate change decisions. In: *Climate Change and Its Linkages with Development, Equity and Sustainability: Proceedings of the IPCC Expert Meeting held in Colombo, Sri Lanka, 27–29 April 1999* [Munasinghe, M. and R. Swart (eds.)]. LIFE, Colombo, Sri Lanka; RIVM, Bilthoven, The Netherlands; and World Bank, Washington, DC, USA, pp. 263–288.

Tsyban, A., J.T. Everett, and J.G. Titus, 1990: World oceans and coastal zones. In: *Climate Change: The IPCC Impacts Assessment. Contribution of Working Group II to the First Assessment Report of the Intergovernmental Panel on Climate Change* [Tegart, W.J.McG., D.C. Griffiths, and G.W. Sheldon (eds.)]. Australian Government Publishing Service, Canberra, Australia, pp. 1–28.

UNDP, 1998: *Human Development Report*. United Nations Development Programme, New York, NY, USA, 228 pp.

Vales, M., L. Alvarez, L. Montes, and A. Avila, 1998: *Estudio Nacional Sobre La Diversidad Biologica de La Republica de Cuba*. CENBIO-IES, AMA, United Nations Environment Programme, La Habana, Cuba, 480 pp.

Viles, H. and T. Spencer, 1995: *Coastal Problems—Geomorphology, Ecology and Society at the Coast*. Edward Arnold, London, United Kingdom, 350 pp.

Wall, G., 1996: The implications of climate change for tourism in small island states. In: *Sustainable Tourism in Islands and Small States: Issues and Policies* [Briguglio, L., B. Archer, J. Jafari, and G. Wall (eds.)]. Pinter, London, United Kingdom, pp. 205–216.

Warrick, R.A., G.J. Kenny, G.C. Sims, W. Ye, and G. Sem, 1999: VANDACLIM: a training tool for climate change vulnerability and adaptation assessment. In: *Climate Change and Sea Level Rise in the South Pacific Region. Proceedings of the Third SPREP Meeting, New Caledonia, August, 1997* [Kaluwin, C. and J.E. Hay (eds.)]. South Pacific Regional Environment Programme, Apia, Western Samoa, pp. 147–156.

Warrick, R.A., C. Le Provost, M.F. Meier, J. Oerlemans, and P.L. Woodworth, 1996: Changes in sea level. In: *Climate Change 1995: The Science of Climate Change. Contribution of Working Group I to the Second Assessment Report of the Intergovernmental Panel on Climate Change* [Houghton, J.T., L.G. Meira Filho, B.A. Callander, N. Harris, A. Kattenberg, and K. Maskell (eds.)]. Cambridge University Press, Cambridge, United Kingdom and New York, NY, USA, pp. 359–405.

Waters, S.R.,1998: *Travel Industry World Yearbook: The Big Picture 1997-98*. Vol. 41, Child and Waters, New York, NY, USA.

Wells, J., 1997: *Composite Vulnerability Index: A Revised Report*. Commonwealth Secretariat, London, United Kingdom, 51 pp.

Wells, J., 1996: *Composite Vulnerability Index: A Preliminary Report*. Commonwealth Secretariat, London, United Kingdom, 29 pp.

West, J.J. and H. Dowlatabadi, 1999: On assessing the economic impacts of sea level rise on developing coasts. In: *Climate, Change and Risk* [Downing, T.E., A.A. Olsthoorn, and R.S.J. Tol (ed.)]. Routledge, London, United Kingdom, pp. 205–220.

Whetton, P.H., M.H. England, S.P. O'Farrell, I.G. Watterson, and A.B. Pittock, 1996: Global comparison of the regional rainfall results of enhanced greenhouse coupled and mixed layer ocean experiments: implications for climate scenario development. *Climate Change*, **33,** 497–519.

WHO, 1996: *Climate Change and Human Health* [McMichael, A.J., A. Haines, R. Slooff, and S. Kovats (eds.)]. WHO/EHG/96.7, an assessment prepared by a Task Group on behalf of the World Health Organization, the World Meteorological Organization, and the United Nations Environment Programme, Geneva, Switzerland, 297 pp.

Wilkinson, C.R., 1999: Global and local threats to coral reef functioning and existence: review and predictions. *Marine and Freshwater Research*, **50,** 867–878.

Wilkinson, C.R. (ed.), 1998: *Status of Coral Reefs of the World*. Australian Institute of Marine Science, Western Australia, Australia, 184 pp.

Wilkinson, C.R. 1996. Global change and coral reefs: impacts on reefs, economies and human cultures. *Global Change Biology*, **2(6),** 547–558.

Wilkinson, C.R. and R.W. Buddemeier, 1994: *Global Climate Change and Coral Reefs: Implications for People and Reefs*. Report of the United Nations Environment Programme-Intergovernmental Oceanographic Commission-Association of South Pacific Environmental Institutions, Global Task Team on the Implications of Climate Change on Coral Reefs, International Union for Conservation of Nature, Gland, Switzerland, 124 pp.

Wong, P.P., 1992: Impact of a sea level rise on the coast of Singapore: preliminary observations. *Journal of Southeast Asian Earth Sciences*, **7,** 65–70.

Wong, P.P., 1985: Artificial coastlines: the example of Singapore. *Zeitschrift fur Geomorphologie N.F.*, **57,** 175–219.

Woodley, J.D., D. Bone, K. Buchan, P. Bush, K. de Meyer, J. Garzon-Ferreira, P. Gayle, G.T. Gerace, L. Grober, E. Klein, K. Koltes, F. Losada, M.D. McField, T. McGrath, J.M. Mendes, I. Nagelkerken, G. Ostrander, L.P.J.J. Pora, A. Rodriguez, R. Rodriguez, F. Ruiz-Renteria, G. Smith, J. Tschirky, P. Alcolado, K. Bonair, J.R. Garcia, F. Geraldes, H. Guzman, C. Parker, and S.R. Smith, 1997: *Studies on Caribbean Coral Bleaching, 1995–1996. Proceedings of the 8th International Coral Reef Symposium, Panama City, Panama*. pp 673–678.

Woodroffe, C.D., 1995: Response of tide-dominated mangrove shorelines in northern Australia to anticipated sea-level rise. *Earth Surface Processes and Landforms*, **20(1),** 65–85.

Woodward, A., S. Hales, and P. Weinstein, 1998: Climate change and human health in the Asia Pacific region: who will be most vulnerable? *Climate Research*, **11,** 31–38.

Yamada, K., P.D. Nunn, N. Mimura, S. Machida, and M. Yamamoto, 1995: Methodology for the assessment of vulnerability of South Pacific island countries to sea-level rise and climate change. *Journal of Global Environment Engineering*, **1(1),** 101–125.

Yu, X., R. Taplin, and A.J. Gilmour, 1997: Climate change response and renewable energy systems in Pacific Islands region. *Environmental Management*.

Zimmerman, R.C., D.G. Kohrs, D.L. Steller, and R.S. Alberte, 1997: Impacts of CO_2 enrichment on productivity and light requirements of eelgrass. *Plant Physiology*, **115(2),** 599–607.

18

Adaptation to Climate Change in the Context of Sustainable Development and Equity

BARRY SMIT (CANADA) AND OLGA PILIFOSOVA (KAZAKHSTAN)

Lead Authors:

I. Burton (Canada), B. Challenger (Antigua and Barbuda), S. Huq (Bangladesh), R.J.T. Klein (Germany/The Netherlands), G. Yohe (USA)

Contributing Authors:

N. Adger (UK), T. Downing (UK), E. Harvey (Canada), S. Kane (USA), M. Parry (UK), M. Skinner (Canada), J. Smith (USA), J. Wandel (Canada)

Review Editors:

A. Patwardhan (India) and J.-F. Soussana (France)

CONTENTS

EXECUTIVE SUMMARY

Adaptation refers to adjustments in ecological, social, or economic systems in response to actual or expected climatic stimuli and their effects or impacts. It refers to changes in processes, practices, and structures to moderate potential damages or to benefit from opportunities associated with climate change.

Estimates of likely future adaptations are an essential ingredient in *impact and vulnerability assessments*. The extent to which ecosystems, food supplies, and sustainable development are vulnerable or "in danger" depends both on exposure to changes in climate and on the ability of the impacted system to adapt. In addition, adaptation is an important policy *response* option, along with mitigation. There is a need for the development and assessment of planned adaptation initiatives to help manage the risks of climate change.

Adaptations vary according to the system in which they occur, who undertakes them, the climatic stimuli that prompts them, and their timing, functions, forms, and effects. In unmanaged natural systems, adaptation is autonomous and reactive; it is the process by which species and ecosystems respond to changed conditions. This chapter focuses on adaptations consciously undertaken by humans, including those in economic sectors, managed ecosystems, resource use systems, settlements, communities, and regions. In human systems, adaptation is undertaken by private decisionmakers and by public agencies or governments.

Adaptation depends greatly on the *adaptive capacity* or adaptability of an affected system, region, or community to cope with the impacts and risks of climate change. The adaptive capacity of communities is determined by their socioeconomic characteristics. Enhancement of adaptive capacity represents a practical means of coping with changes and uncertainties in climate, including variability and extremes. In this way, enhancement of adaptive capacity reduces vulnerabilities and promotes sustainable development.

Adaptation to climate change has the potential to substantially reduce many of the adverse impacts of climate change and enhance beneficial impacts—though neither without cost nor without leaving residual damage.

The key features of climate change for vulnerability and adaptation are those related to variability and extremes, not simply changed average conditions. Most sectors and regions and communities are reasonably adaptable to changes in average conditions, particularly if they are gradual. However, these communities are more vulnerable and less adaptable to changes in the frequency and/or magnitude of conditions other than average, especially extremes.

Sectors and regions will tend to adapt autonomously to changes in climate conditions. Human systems have evolved a wide range of strategies to cope with climatic risks; these strategies have potential applications to climate change vulnerabilities. However, losses from climatic variations and extremes are substantial and, in some sectors, increasing. These losses indicate that autonomous adaptation has not been sufficient to offset damages associated with temporal variations in climatic conditions. The ecological, social, and economic costs of relying on reactive, autonomous adaptation to the cumulative effects of climate change are substantial.

Planned anticipatory adaptation has the potential to reduce vulnerability and realize opportunities associated with climate change, regardless of autonomous adaptation. Implementation of adaptation policies, programs, and measures usually will have immediate benefits, as well as future benefits. Adaptation measures are likely to be implemented only if they are consistent with or integrated with decisions or programs that address nonclimatic stresses. The costs of adaptation often are marginal to other management or development costs.

The capacity to adapt varies considerably among regions, countries, and socioeconomic groups and will vary over time. The most vulnerable regions and communities are those that are highly exposed to hazardous climate change effects and have limited adaptive capacity. Countries with limited economic resources, low levels of technology, poor information and skills, poor infrastructure, unstable or weak institutions, and inequitable empowerment and access to resources have little capacity to adapt and are highly vulnerable.

Enhancement of adaptive capacity is a necessary condition for reducing vulnerability, particularly for the most vulnerable regions, nations, and socioeconomic groups. Activities required for the enhancement of adaptive capacity are essentially equivalent to those promoting sustainable development. Climate adaptation and equity goals can be jointly pursued by initiatives that promote the welfare of the poorest members of society—for example, by improving food security, facilitating access to safe water and health care, and providing shelter and access to other resources. Development decisions, activities, and programs play important roles in modifying the adaptive capacity of communities and regions, yet they tend not to take into account risks associated with climate variability and change. Inclusion of climatic risks in the design and implementation of development initiatives is necessary to reduce vulnerability and enhance sustainability.

Current knowledge of adaptation and adaptive capacity is insufficient for reliable prediction of adaptations; it also is insufficient for rigorous evaluation of planned adaptation options, measures, and policies of governments. Climate change vulnerability studies now usually consider adaptation, but they rarely go beyond identifying adaptation options that might be possible; there is little research on the dynamics of adaptation in human systems, the processes of adaptation decisionmaking, conditions that stimulate or constrain adaptation, and the role of nonclimatic factors. There are serious limitations in existing evaluations of adaptation options: Economic benefits and costs are important criteria but are not sufficient to adequately determine the appropriateness of adaptation measures; there also has been little research to date on the roles and responsibilities in adaptation of individuals, communities, corporations, private and public institutions, governments, and international organizations. Given the scope and variety of specific adaptation options across sectors, individuals, communities, and locations, as well as the variety of participants—private and public—involved in most adaptation initiatives, it is probably infeasible to systematically evaluate lists of particular adaptation measures; improving and applying knowledge on the constraints and opportunities for enhancing adaptive capacity is necessary to reduce vulnerabilities associated with climate change.

18.1. Introduction: Adaptation and Adaptive Capacity

Adaptation is adjustment in ecological, social, or economic systems in response to actual or expected climatic stimuli and their effects or impacts. This term refers to changes in processes, practices, or structures to moderate or offset potential damages or to take advantage of opportunities associated with changes in climate. It involves adjustments to reduce the vulnerability of communities, regions, or activities to climatic change and variability. Adaptation is important in the climate change issue in two ways—one relating to the assessment of impacts and vulnerabilities, the other to the development and evaluation of response options.

Understanding expected adaptations is essential to *impact and vulnerability assessment* and hence is fundamental to estimating the costs or risks of climate change (Fankhauser, 1996; Yohe *et al.*, 1996; Tol *et al.*, 1998; UNEP, 1998; Smit *et al.*, 1999; Pittock and Jones, 2000). Article 2 of the United Nations Framework Convention on Climate Change (UNFCCC) refers to "dangerous" human influences on climate in terms of whether they would "allow ecosystems to adapt, ensure food production is not threatened, and enable economic development to proceed in a sustainable manner." The extent to which ecosystems, food supplies, and sustainable development are vulnerable or "in danger" depends on their exposure to climate change effects and on the ability of impacted systems to adapt. Thus, to assess the dangerousness of climate change, impact and vulnerability assessments must address the likelihood of autonomous adaptations (see Figure 18-1).

Adaptation also is considered an important *response option or strategy*, along with mitigation (Fankhauser, 1996; Smith, 1996; Pielke, 1998; Kane and Shogren, 2000). Even with reductions in greenhouse gas (GHG) emissions, global temperatures are expected to increase, other changes in climate—including extremes—are likely, and sea level will continue to rise (Raper *et al.*, 1996; White and Etkin, 1997; Wigley, 1999). Hence, development of planned adaptation strategies to deal with these risks is regarded as a necessary complement to mitigation actions (Burton, 1996; Smith *et al.*, 1996; Parry *et al.*, 1998; Smit *et al.*, 1999) (see Figure 18-1). Article 4.1 of the UNFCCC commits parties to formulating, cooperating on, and implementing "measures to facilitate adequate adaptation to climate change." The Kyoto Protocol (Article 10) also commits parties to promote and facilitate adaptation and deploy adaptation technologies to address climate change.

Adaptive capacity is the potential or ability of a system, region, or community to adapt to the effects or impacts of climate change. Enhancement of adaptive capacity represents a practical means of coping with changes and uncertainties in climate, including variability and extremes. In this way, enhancement of adaptive capacity reduces vulnerabilities and promotes

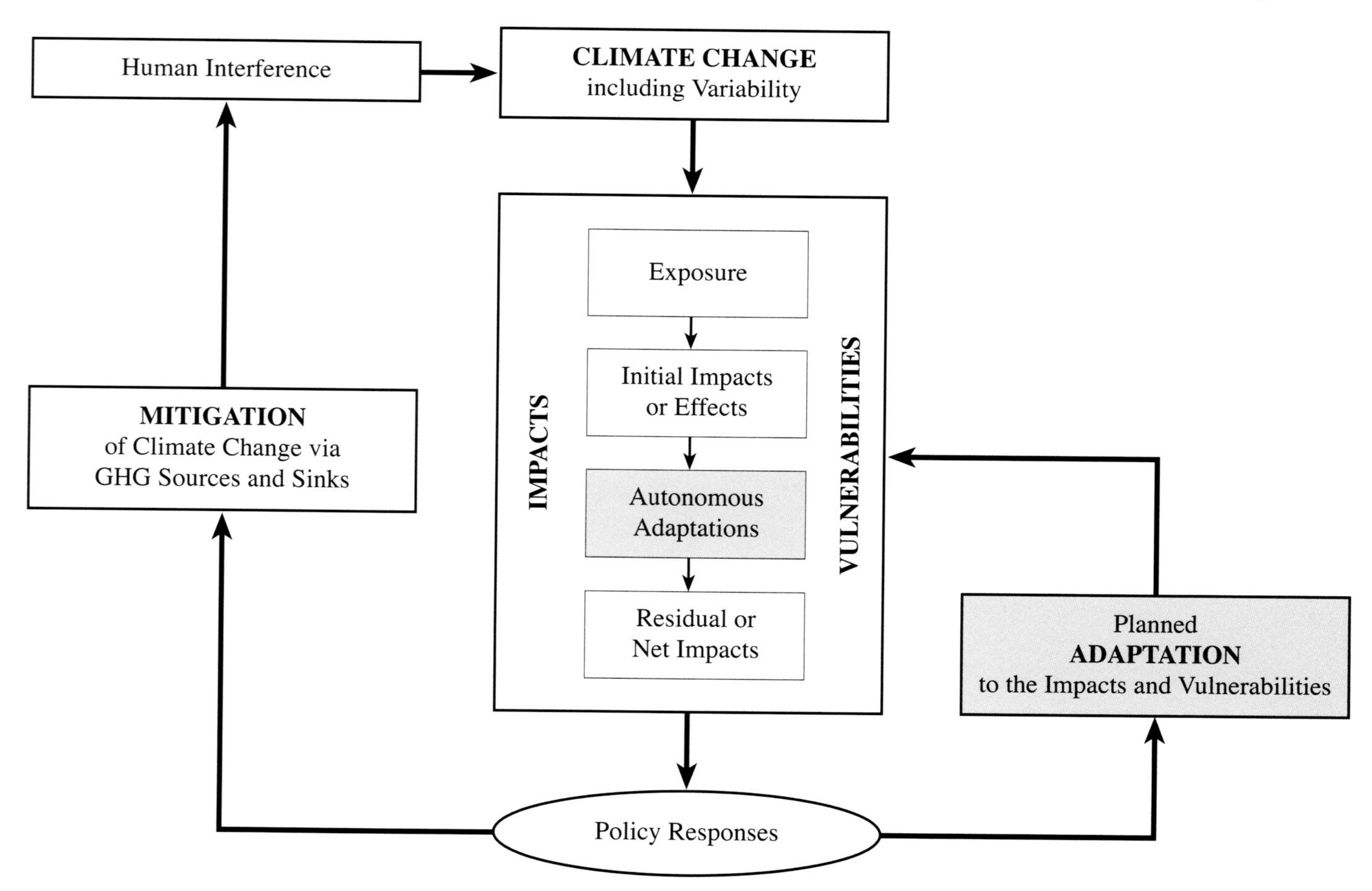

Figure 18-1: Places of adapatation in the climate change issue (Smit *et al.*, 1999).

sustainable development (Goklany, 1995; Burton, 1997; Cohen *et al.*, 1998; Klein, 1998; Rayner and Malone, 1998; Munasinghe, 2000; Smit *et al.*, 2000).

Considerable attention has been devoted to the characteristics of communities, countries, and regions that influence their propensity or ability to adapt and hence their vulnerability to risks associated with climate change. These *determinants of adaptive capacity* relate to the economic, social, institutional, and technological conditions that facilitate or constrain the development and deployment of adaptive measures (e.g., Bohle *et al.*, 1994; Rayner and Malone, 1998; Kelly and Adger, 1999).

18.2. Adaptation Characteristics and Processes

Adaptation refers both to the *process* of adapting and to the *condition* of being adapted. The term has specific interpretations in particular disciplines. In ecology, for example, adaptation refers to changes by which an organism or species becomes fitted to its environment (Lawrence, 1995; Abercrombie *et al.*, 1997); whereas in the social sciences, adaptation refers to adjustments by individuals and the collective behavior of socioeconomic systems (Denevan, 1983; Hardesty, 1983). This chapter follows Carter *et al.* (1994), IPCC (1996), UNEP (1998), and Smit *et al.* (2000) in a broad interpretation of adaptation to include adjustment in natural or human systems in response to experienced or future climatic conditions or their effects or impacts—which may be beneficial or adverse.

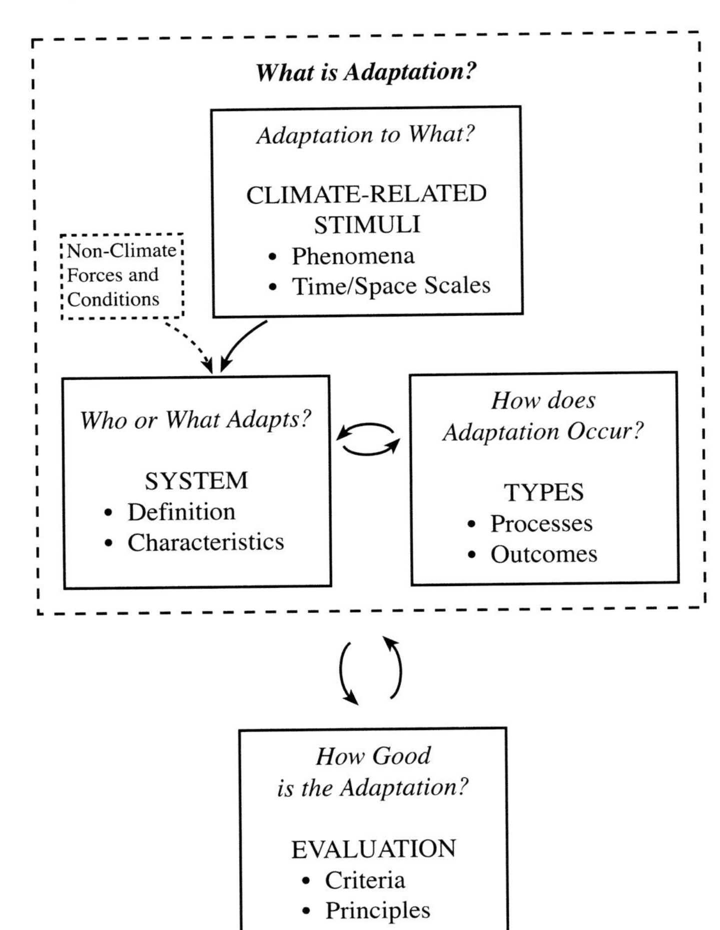

Figure 18-2: Adaptation to climate change and variability (from Smit *et al.*, 2000).

18.2.1. Components and Forms of Adaptation

As both a process and a condition, adaptation is a relative term: It involves an alteration in something (the system of interest, activity, sector, community, or region) to something (the climate-related stress or stimulus). Description of an adaptation requires specification of who or what adapts, the stimulus for which the adaptation is undertaken, and the process and form it takes (Downing *et al.*, 1997; Krankina *et al.*, 1997; UNEP, 1998; Pittock *et al.*, 1999; Risbey *et al.*, 1999; Reilly and Schimmelpfennig, 2000). These elements are summarized in Figure 18-2 and addressed in turn in subsequent subsections.

18.2.2. Climate Stimuli for Adaptation

Most impact and adaptation studies to date have been based on climate change scenarios that provide a limited set of possible future climates—invariably specified as average annual conditions, such as temperature and moisture. Yet the climate change-related stimuli for which adaptations are undertaken (i.e., adaptation to what?) are not limited to changes in average annual conditions; they include variability and associated extremes. Climatic conditions are inherently variable, from year to year and decade to decade. Variability goes along with, and is an integral part of, climate change (Mearns *et al.*, 1997; Karl and Knight, 1998; Berz, 1999; Hulme *et al.*, 1999): A change in mean conditions actually is experienced through changes in the nature and frequency of particular yearly conditions, including extremes (see Figure 18-3). Thus, adaptation to climate change necessarily includes adaptation to variability (Hewitt and Burton, 1971; Parry, 1986; Kane *et al.*, 1992b; Katz and Brown, 1992; Downing, 1996; Yohe *et al.*, 1996; Smithers and Smit, 1997; Smit *et al.*, 1999). Downing *et al.* (1996), Etkin (1998), Mileti (1999), and others use the term "climate hazards" to capture those climate stimuli, in addition to changes in annual averages, to which the system of interest is vulnerable. Climate change stimuli are described in terms of "changes in mean climate and climatic hazards," and adaptation may be warranted when either of these changes has significant consequences (Downing *et al.*, 1997). In water resource management, changes in the recurrence interval of extreme conditions, which are associated with changes in means, are the key stimuli (Beran and Arnell, 1995; Kundzewicz and Takeuchi, 1999).

Furthermore, for most systems and communities, changes in the mean condition commonly fall within the coping range (see Figure 18-3), whereas many systems are particularly vulnerable to changes in the frequency and magnitude of extreme events or conditions outside the coping range (Baethgen, 1997; Schneider, 1997; Rayner and Malone, 1998; Kelly and Adger,

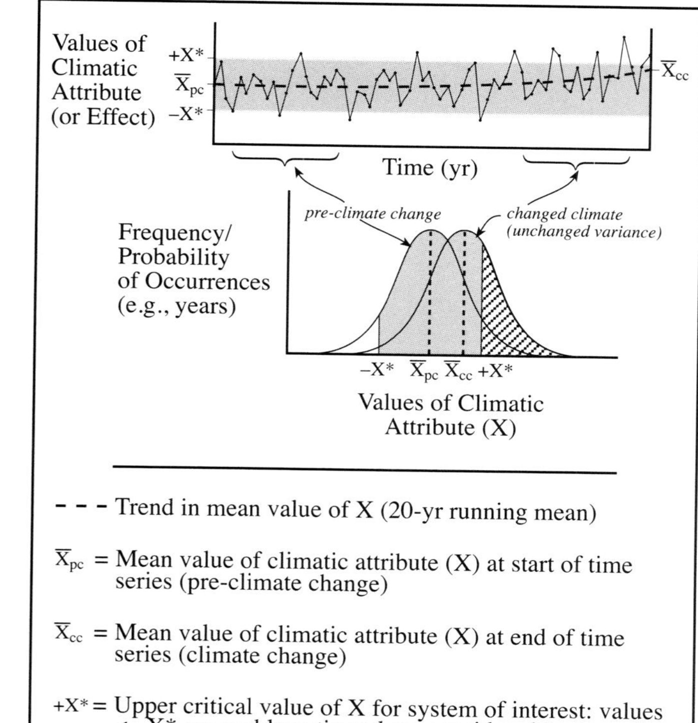

Figure 18-3: Climate change, variability, extremes, and coping range (after Hewitt and Burton, 1971; Fukui, 1979; Smit *et al.*, 1999; and others).

1999). Interannual variations are key stimuli in many sectors (Rosenzweig, 1994; Adams *et al.*, 1995; Mearns *et al.*, 1997; Bryant *et al.*, 2000).

Natural and human systems have adapted to spatial differences in climate. There also are examples of adaptation (with varying degrees of success) to temporal variations—notably, deviations from the annual average conditions on which climate change scenarios focus. Many social and economic systems—including agriculture, forestry, settlements, industry, transportation, human health, and water resource management—have evolved to accommodate some deviations from "normal" conditions, but rarely the extremes. This capacity of systems to accommodate variations in climatic conditions from year to year is captured in Figure 18-3 in the shaded "coping range." This capacity also is referred to as the vulnerability or damage threshold (Pittock and Jones, 2000). The coping range, which varies among systems and regions, need not remain static, as depicted in Figure 18-3. The coping range itself may change (move up or down, expand or contract), reflecting new adaptations in the system (De Vries, 1985; de Freitas, 1989; Smit *et al.*, 2000). The coping range indicated in Figure 18-3 can be regarded as the adaptive capacity of a system to deal with current variability. Adaptive capacity to climate change would refer to both the ability inherent in the coping range and the ability to move or expand the coping range with new or modified adaptations. Initiatives to enhance adaptive capacity (Section 18.6) would expand the coping range.

18.2.3. Adaptation Types and Forms

Adaptations come in a huge variety of forms. Adaptation *types* (i.e., how adaptation occurs) have been differentiated according to numerous attributes (Carter *et al.*, 1994; Stakhiv, 1994; Bijlsma *et al.*, 1996; Smithers and Smit, 1997; UNEP, 1998; Leary, 1999; Bryant *et al.*, 2000; Reilly and Schimmelpfennig, 2000). Commonly used distinctions are purposefulness and timing. Autonomous or spontaneous adaptations are considered to be those that take place—invariably in reactive response (after initial impacts are manifest) to climatic stimuli—as a matter of course, without the directed intervention of a public agency. Estimates of these autonomous adaptations are now used in impact and vulnerability assessment. Planned adaptations can be either reactive or anticipatory (undertaken before impacts are apparent). In addition, adaptations can be short or long term, localized or widespread, and they can serve various functions and take numerous forms (see Table 18-1).

Adaptations have been distinguished according to individuals' choice options as well, including "bear losses," "share losses," "modify threats," "prevent effects," "change use," and "change location" (Burton *et al.*, 1993; Rayner and Malone, 1998). The choice typology has been extended to include the role of community structures, institutional arrangements, and public policies (Downing *et al.*, 1997; UNEP, 1998; see Figure 18-4).

18.2.4. Systems, Scales, and Actors

Adaptations occur in something (i.e., who or what adapts?), which is called the "system of interest," "unit of analysis," "exposure unit," "activity of interest," or "sensitive system" (Carter *et al.*, 1994; Smithers and Smit, 1997; UNEP, 1998; Reilly and Schimmelpfennig, 2000). In *unmanaged natural systems,* adaptation is autonomous and reactive and is the means by which species and communities respond to changed conditions. In these situations, adaptation assessment is essentially equivalent to natural system impact assessment (addressed in other WGII chapters). This chapter focuses on adaptations consciously undertaken by *humans*, including those in economic sectors, settlements, communities, regions, and managed ecosystems.

Human system adaptation can be motivated by private or public interest (i.e., who adapts?). *Private* decisionmakers include individuals, households, businesses, and corporations; *public* interests are served by governments at all levels. The roles of

Table 18-1: Bases for characterizing and differentiating adaptation to climate change (Smit et al., 1999).

General Differentiating Concept or Attribute	Examples of Terms Used		
Purposefulness	Autonomous	←→	Planned
	Spontaneous	←→	Purposeful
	Automatic	←→	Intentional
	Natural	←→	Policy
	Passive	←→	Active
			Strategic
Timing	Anticipatory	←→	Responsive
	Proactive	←→	Reactive
	Ex ante	←→	*Ex post*
Temporal Scope	Short term	←→	Long term
	Tactical	←→	Strategic
	Instantaneous	←→	Cumulative
	Contingency		
	Routine		
Spatial Scope	Localized	←→	Widespread
Function/Effects	Retreat - Accommodate - Protect Prevent - Tolerate - Spread - Change - Restore		
Form	Structural - Legal - Institutional - Regulatory - Financial - Technological		
Performance	Cost - Effectiveness - Efficiency - Implementability - Equity		

public and private participants are distinct but not unrelated. Figure 18-5 shows examples of types of adaptation differentiated according to timing, natural or human systems, and public or private decisionmakers.

Planned adaptation often is interpreted as the result of a deliberate policy decision on the part of a public agency, based on an awareness that conditions are about to change or have changed and that action is required to minimize losses or benefit from opportunities (Pittock and Jones, 2000). Autonomous adaptations are widely interpreted as initiatives by private actors rather than by governments, usually triggered by market or welfare changes induced by actual or anticipated climate change (Leary, 1999). Smith *et al*. (1996) describe autonomous adaptations as those that occur "naturally," without interventions by public agencies, whereas planned adaptations are called "intervention strategies." Thus defined, autonomous and planned adaptation largely correspond with private and public adaptation, respectively (see Figure 18-5).

The extent to which society can rely on autonomous, private, or market adaptation to reduce the costs of climate change impacts to an acceptable or nondangerous level is an issue of great interest. Autonomous adaptation forms a baseline against which the need for planned anticipatory adaptation can be evaluated.

Distinguishing among the various decisionmakers involved in adaptation is important. The case of African agriculture and water resources illustrates that stakeholders and potential adapters range from vulnerable consumers to international organizations charged with relief and research (Eele, 1996; Magadza, 1996; Downing *et al*., 1997). Poor and landless households have limited resources, yet failure to adapt can lead to significant deprivation, displacement, morbidity, and mortality. Subsistence farmers do not have the same adaptation options as commercial producers. Water supply adaptations may involve landowners, private traders, local authorities, water-dependent businesses, national governments, and international organizations. Each stakeholder has distinct interests, information, risks, and resources and hence would consider distinct types of adaptive responses (Downing *et al*., 1997).

18.2.5. Processes and Evaluation of Adaptations

In order to predict autonomous adaptations and provide input to adaptation policies, there is a need for improved knowledge about processes involved in adaptation decisions. This knowledge includes information on steps in the process, decision rationales, handling of uncertainties, choices of adaptation types and timing, conditions that stimulate or dampen adaptation, and the consequences or performance of adaptation strategies or

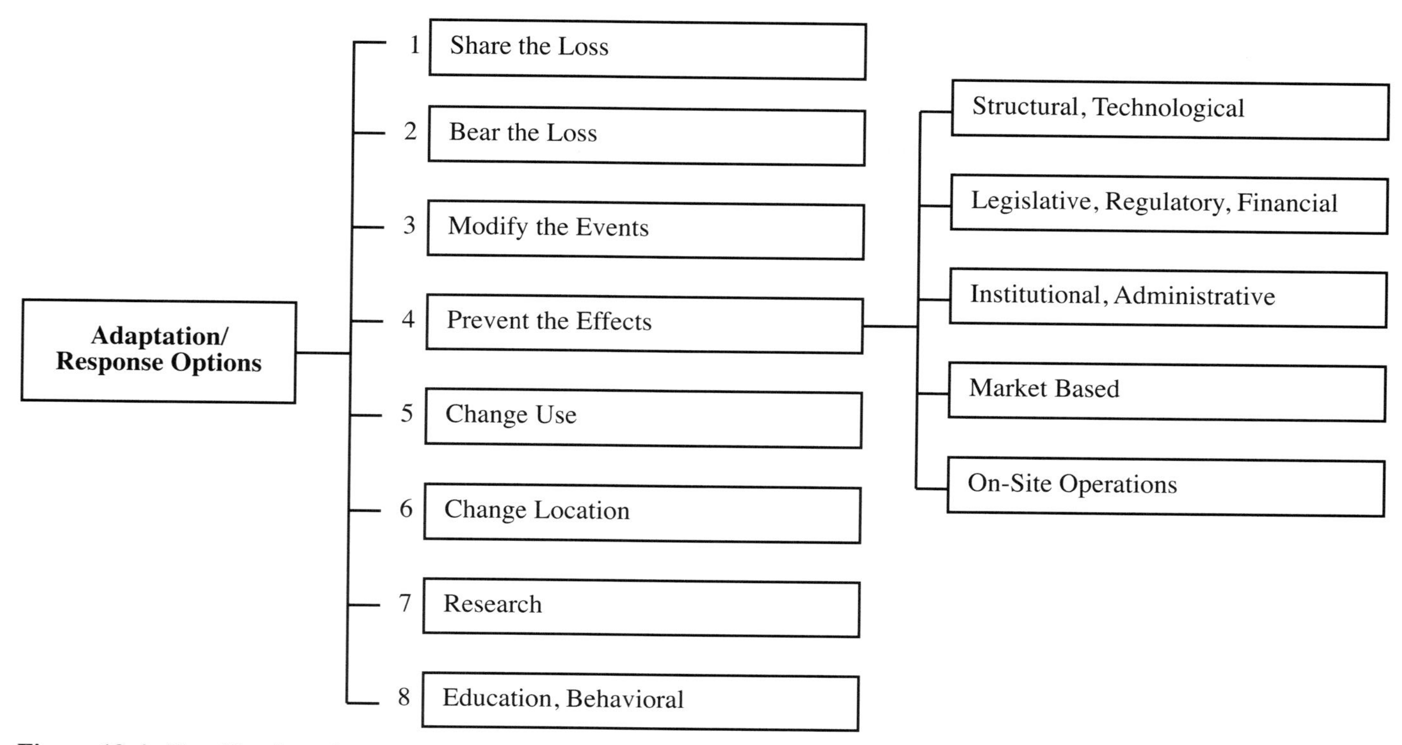

Figure 18-4: Classification of adaptation options (Burton, 1996).

measures (Burton, 1997; Rayner and Malone, 1998; Tol *et al.*, 1998; Basher, 1999; Klein *et al.*, 1999; Pittock, 1999; Smit *et al.*, 1999).

Decisions regarding adaptations can be undertaken at any of several scales, by private individuals, local communities or institutions, national governments, and international organizations. Where these adaptations are consciously planned activities, whether by public agencies or individuals, there is an interest in assessing the performance or relative merits of alternative measures and strategies (see Figure 18-4). This *evaluation* (i.e., how good is the adaptation?) can be based on criteria such as costs, benefits, equity, efficiency, and implementability (see Sections 18.3.5 and 18.4.3).

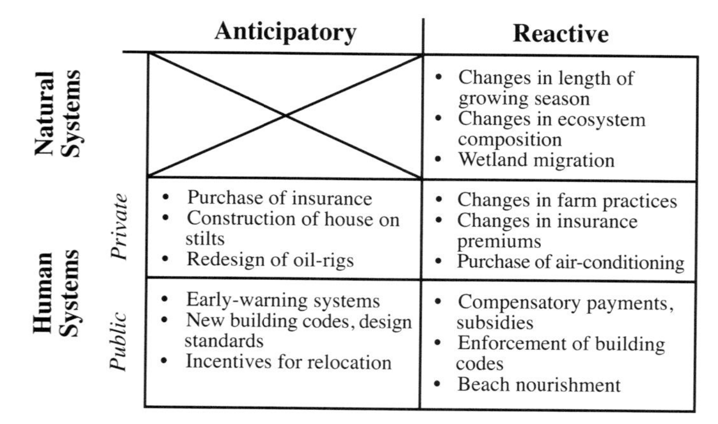

Figure 18-5: Types of adaptation to climate change, including examples (from Klein, 1999).

18.3. Future Adaptations

Predictions or estimates of likely future adaptations are an essential element of climate change impact and vulnerability assessment. The degree to which a future climate change risk is dangerous depends greatly on the likelihood and effectiveness of adaptations in that system. Studies that ignore or assume no adaptation are likely to overestimate residual or net impacts and vulnerabilities, whereas those that assume full and effective adaptation are likely to underestimate residual impacts and vulnerabilities (Reilly, 1999; Reilly and Schimmelpfennig, 1999; Risbey *et al.*, 1999; Smit *et al.*, 2000). Hence, it is important to have an improved understanding of the process of adaptation and better information on the conditions under which adaptations of various types are expected to occur. Such scholarship on the "how, when, and why" of adaptation is necessary to make informed judgments on the vulnerabilities of sectors, regions, and communities (Ausubel, 1991a; Kane *et al.*, 1992a; Reilly, 1995; Burton, 1997; Smithers and Smit, 1997; Tol *et al.*, 1998; Klein *et al.*, 1999). Insights into processes of adaptation have been gained from several types of analysis, including listing of possible adaptation measures, impact assessment models, adaptation process models, historical and spatial analogs, and empirical analysis of contemporary adaptation processes.

18.3.1. Possible Adaptation Measures

There are many arbitrary lists of possible adaptation measures, initiatives, or strategies that have a potential to moderate impacts,

***Table 18-2**: Examples of multilevel adaptive measures for some anticipated health outcomes of global climate change (Patz, 1996).*

Adaptive Measure	Heat-Related Illness	Vector-Borne Diseases	Health and Extreme Weather Events
Administrative/legal	– Implement weather watch/warning systems – Plant trees in urban areas – Implement education campaigns	– Implement vaccination programs – Enforce vaccination laws – Implement education campaigns to eliminate breeding sites	– Create disaster preparedness programs – Employ land-use planning to reduce flash floods – Ban precarious residential placements
Engineering	– Insulate buildings – Install high-albedo materials for roads	– Install window screens – Release sterile male vectors	– Construct strong seawalls – Fortify sanitation systems
Personal behavior	– Maintain hydration – Schedule work breaks during peak daytime temperatures	– Use topical insect repellents – Use pyrethroid-impregnated bed nets	– Heed weather advisories

if they were implemented (e.g., Benioff *et al.*, 1996; Smith *et al.*, 1996; Mimura, 1999b). Such *possible* adaptations are based on experience, observation, and speculation about alternatives that might be created (Carter, 1996); they cover a wide range of types and take numerous forms (UNEP, 1998). For example, possible adaptive measures for health risks associated with climate change listed by Patz (1996) appear in Table 18-2.

Similarly, in coastal zone studies, comprehensive lists of potential adaptation measures are presented; these adaptations include a wide array of engineering measures, improvements, or changes, including agricultural practices that are more flood-resistant; negotiating regional water-sharing agreements; providing efficient mechanisms for disaster management; developing desalination techniques; planting mangrove belts to provide flood protection; planting salt-tolerant varieties of vegetation; improving drainage facilities; establishing setback policies for new developments; developing food insurance schemes; devising flood early warning systems; and so forth (Al-Farouq and Huq, 1996; Jallow, 1996; Rijsberman and van Velzen, 1996; Teves *et al.*, 1996; Mimura and Harasawa, 2000). In many other sectors and regions, arbitrary lists of possible adaptations are common (Erda, 1996; Iglesias *et al.*, 1996). In the Canadian agricultural sector alone, 96 different adaptation measures have been identified, as summarized in Table 18-3.

Such lists indicate the range of strategies and measures that represent possible adaptations to climate change risks in particular sectors and regions. They show that there is a large variety and number of possible adaptations, including many with the potential to reduce adverse climatic change impacts. Many of these adaptations—especially in agriculture, water resources, and coastal zone applications—essentially represent improved resource management, and many would have benefits in dealing with current climatic hazards as well as with future climatic risks (El Shaer *et al.*, 1996; Harrington, 1996; Huang, 1996; Stakhiv, 1996; Frederick, 1997; Hartig *et al.*, 1997; Mendelsohn and Bennett, 1997; Major, 1998). In only a few cases are such lists of possible adaptations considered according to who might undertake them, under what conditions might they be implemented, and how effective might they be (Easterling, 1996; Harrington, 1996; Frederick, 1997; Major, 1998; Moss, 1998).

18.3.2. Impact Assessment Models

Estimates of likely future adaptations are essential parts of climate change impact models. Integrated assessment models also include assumptions about adaptations in the impact

***Table 18-3**: Adaptation strategies for the agricultural sector (adapted from Smit, 1993; Carter, 1996).*

Adaptation Strategy	Number of Measures
Change topography of land	11
Use artificial systems to improve water use/ availability and protect against soil erosion	29
Change farming practices	21
Change timing of farm operations	2
Use different crop varieties	7
Governmental and institutional policies and programs	16
Research into new technologies	10

components (Leemans, 1992; Rotmans *et al.*, 1994; Dowlatabadi, 1995; Hulme and Raper, 1995; West and Dowlatabadi, 1999). Some early studies of impacts assumed no adaptation (Tol *et al.*, 1998), invoking the so-called "naive" or "dumb farmer" assumption. The "dumb farmer" assumption—which is not unique to agriculture—is a metaphor for any impacted agent that is assumed not to anticipate or respond to changed climate conditions but continues to act as if nothing has changed (Rosenberg, 1992; Easterling *et al.*, 1993; Smit *et al.*, 1996). By ignoring autonomous and planned adaptations, such studies do not distinguish between potential and residual net impacts and are of limited utility in assessing vulnerability.

An alternative approach that is common in more recent impact modeling has been to assume levels of adaptation. Applications include Nicholls and Leatherman (1995) for coastal zones, Mendelsohn *et al.* (1994) and Rosenzweig and Parry (1994) for agriculture, Sohngen and Mendelsohn (1998) for timber, and Rosenthal *et al.* (1995) for space conditioning in buildings. These studies demonstrate that adaptive measures have the potential to significantly alleviate adverse impacts of climate change and to benefit from opportunities associated with changed climatic conditions (Helms *et al.*, 1996; Schimmelpfennig, 1996; Mendelsohn and Neumann, 1999). The models of Rosenzweig and Parry (1994) show that, with adaptations assumed, food production could be increased under climate change in many regions of the world. Stuczyinski *et al.* (2000) conclude that climate change would reduce Polish agriculture production by 5–25% without adaptation; with adaptation assumed, production is estimated to change by –5 to +5% of current levels. Downing (1991) demonstrates the potential of adaptations to reduce food deficits in Africa from 50 to 20%. Mendelsohn and Dinar (1999) estimate that private adaptation could reduce potential climate damages in India's agriculture from 25 to 15–23%. Reilly *et al.* (1994) estimate global "welfare" losses in the agri-food sector of between US$0.1 billion and 61.2 billion without adaptation, compared to +US$70 to –37 billion with adaptation assumed. These studies indicate *potential* rather than the *likelihood* of adaptation to alleviate damages (or benefit from opportunities) associated with changes in climatic mean conditions (rather than changing conditions that include variability and extremes of climate).

Impact models invariably are based on climate scenarios that focus on adaptation to changed average conditions, with little attention given to interannual variations and extremes. Limited research suggests that the potential of adaptation to cope with changes in average conditions is greater than its potential to cope with climate change-related variability. For example, Mendelsohn *et al.* (1999) show that, assuming adaptation, increases in average temperature would be beneficial for U.S. agriculture, but increases in interannual variation would be harmful. West and Dowlatabadi (1999) demonstrate that considering variability and extremes can lead to estimates of "optimal" adaptation and damages that differ considerably from those based on gradual changes in mean climatic conditions. The importance of considering variability, not just mean climate, when estimating adaptation is widely recognized (Robock *et al.*, 1993; Mearns *et al.*, 1997; Alderwish and Al-Eryani, 1999; Alexandrov, 1999; Luo and Lin, 1999; Murdiyarso, 2000).

In numerical impact models, assumptions about perception and adaptation are more commonly arbitrary or based on principles of efficiency and rationality and assume full information (Yohe *et al.*, 1996; Hurd *et al.*, 1997; Mendelsohn *et al.*, 1999). As Tol *et al.* (1998), Schneider *et al.* (2000), and others have noted, however, actual and assumed behavior do not necessarily match. In an analysis of global food production, Parry *et al.* (1999) assume farm-level and economic system adaptations but recognize that the "adoption of efficient adaptation techniques is far from certain." In addition to questions relating to rationality principles, adaptation behavior is known to vary according to the amount and type of information available, as well as the ability to act. Hence, rational behavior that is based on assumed perfect information differs from rational behavior under uncertainty (Yohe *et al.*, 1996; Yohe and Neumann, 1997; West and Dowlatabadi, 1999). Replacing the "no adaptation" model with one that assumes rational, unconstrained actors with full information replaces the "dumb farmer" assumption with the "clairvoyant farmer" assumption (Smit *et al.*, 1996; Risbey *et al.*, 1999). Reilly (1998) questions the ability and hence the likelihood of agents to detect and respond efficiently to the manifestations of climate change. Tol (1998b) also questions whether perfect foresight and rational behavior are realistic assumptions for predictive models. Schneider (1997) explores further the assumptions that underlie equilibrium approaches (ergodic economics), including the equivalence of temporal and spatial variations.

Numerical impact assessment models tend to *use*, rather than *generate*, information on adaptations to estimate future impacts of climate stimuli, after the effects of adaptation have been factored in. They indicate the potential of human systems to adapt autonomously and thus to moderate climate change damages.

18.3.3. Models, Analogs, and Empirical Analysis of Autonomous Adaptation

Adaptation to rapid anthropogenic climate change may be a new challenge, but individuals, societies, and economies have adapted—in various ways and with various degrees of success—to changed and variable environmental conditions throughout history. These experiences in adaptive behavior provide information on the processes, constraints, and consequences of adaptations.

Knowledge of the processes by which individuals or communities actually adapt to changes in conditions over time comes largely from analog and other empirical analyses (Wigley *et al.*, 1981; Glantz, 1996; Meyer *et al.*, 1998; Tol *et al.*, 1998; Smit *et al.*, 1999; Yohe and Dowlatabadi, 1999; Bryant *et al.*, 2000). These studies indicate that autonomous adaptations tend to be incremental and ad hoc, to take multiple forms, to be in response to multiple stimuli (usually involving a particular catalyst and rarely climate alone), and to be constrained by economic, social, technological, institutional, and political conditions.

Conceptual models of adaptation processes describe sequential relationships and feedback involving climatic and nonclimatic stimuli, system sensitivities and impacts, tactical and strategic adaptations, and net or residual impacts. They also indicate conditions that constrain or facilitate various kinds of adaptation (e.g., Carter, 1996; Smit *et al.*, 1996; UNEP, 1998; Schneider *et al.*, 2000). Spatial analogs have been used to gain insight into adaptation, by transferring experience from existing climatic regions to places where such climate may be found in the future. The contributions and limits of spatial analogs are known (Schneider, 1997; Rayner and Malone, 1998). Some ecological and paleoecological studies reconstruct species or community dynamics over hundreds and thousands of years (e.g., MacDonald *et al.*, 1993).

Temporal analog or case studies document adaptive responses to climatic stimuli in resource-based economic sectors and communities over periods of several decades (e.g., Glantz, 1988; Olsthoorn *et al.*, 1996; Changnon *et al.*, 1997). Other empirical analyses have examined adaptive behavior in key sectors such as agriculture in light of climatic variability and extremes over even shorter time periods (e.g., Appendi and Liverman, 1996; Smit *et al.*, 1997; Bryant *et al.*, 2000).

These direct empirical analyses of adaptation processes tend to start with the system of interest, then assess its sensitivity and adaptability to climate and other stimuli. This analytical strategy is consistent with vulnerability assessment (Downing *et al.*, 1996; Adger, 1999; Handmer *et al.*, 1999; Kelly and Adger, 1999), the "adjoint approach" (Parry, 1986), and "shift-in-risk" perspectives (Warrick *et al.*, 1986). These studies have yielded some important insights about adaptation.

For systems such as agriculture, forestry, water resources, and coastal zone settlements, the key climatic stimuli are not average conditions but variability and extremes. A direct climatic condition prompts adaptation less often than the economic and social effects or implications of the climatic stimuli that are fundamental in triggering adaptive responses. Nonclimatic conditions are important in moderating and sometimes overwhelming the influence of climate stimuli in the decisionmaking of resource users. Decisions on adaptation are rarely made in response to climate stimuli alone. These findings are important for predicting autonomous adaptations and for improving adaptation assumptions in impact models.

In estimating future adaptations and developing adaptation policies (see Section 18.4), it is helpful to understand factors and circumstances that hinder or promote adaptation. As Rayner and Malone (1998) conclude, the consequences of a climate event are not direct functions of its physical characteristics; they also are functions of "the ways in which a society has organized its relation to its resource base, its relations with other societies, and the relations among its members." To understand vulnerability in archeological, historical, and contemporary contexts, Rayner and Malone (1998) identify the most promising research strategy:

> "...explicitly to focus attention on the process of adaptation—or, on the other hand, of failure to adapt—that partly condition the impact of the climatic stress in particular societies...cases in which societies appear to have been seriously damaged by, or even totally succumbed to, climatic stress should not be taken to demonstrate the determining influence of climate. It is essential to consider ways in which these societies might have coped better, and to focus on the political, cultural, and socioeconomic factors which inhibited them from doing so" (Ingram *et al.*, 1981).

Following this approach, McGovern's (1991) reexamination of the extinction of Greenland settlements found that the stress imposed by climate shifts was indeed severe but was within the theoretical ability of the colonies to have coped, by means that were available to them. Why they failed to employ those adaptive means emerges as the key question, still incompletely answered, in explaining the collapse: "It did get cold and they did die out, but why?" (McGovern, 1991). Intervening between the physical events and the social consequences is the adaptive capacity and hence vulnerability of the society and its different groups and individual members.

18.3.4. Costs of Autonomous Adaptation

As assessments of climate impacts (commonly measured as "costs" that include damages and benefits) increasingly have incorporated expected adaptations, and particularly as impact models and "integrated assessment" models have shown the potential of adaptation to offset initial impact costs, interest has grown in calculating the costs of autonomous adaptations. Whether climate change or another climate stimulus is expected to have problematic or "dangerous" impacts depends on the adaptations and their costs (Leary, 1999). Climate change impact cost studies that assume adaptation also should include the "adjustment of costs" of these adaptations (Reilly, 1998).

Tol and Fankhauser (1997) provide a comprehensive summary of analyses of the costs of autonomous, mainly (but not exclusively) reactive adaptations, undertaken privately (i.e., not adaptation policies of government). A common basis for evaluating impact costs is to sum adaptation costs and residual damage costs (Fankhauser, 1996; Rothman *et al.*, 1998). Procedures for defining and calculating such adaptation costs are subject to ongoing debate. Tol and Fankhauser (1997) note that most approaches consider equilibrium adaptation costs but ignore transition costs. Hurd *et al.* (1997) include market and nonmarket adaptation in their assessment of impact costs. Most research to date on adaptation "costs" is limited to particular economic measures of well-being (Brown, 1998). Any comprehensive assessment of adaptation costs (including benefits) would consider not only economic criteria but also social welfare and equity.

Cost estimation for autonomous adaptations is not only important for impact assessment; it also is a necessary ingredient in the "base case," "reference scenario," or "do-nothing option" for evaluations of policy initiatives, with respect to both adaptation and mitigation (Rayner and Malone, 1998; Leary, 1999; Smit *et al*., 2000).

18.3.5. Lessons from Adaptation Experiences

Research in many sectors and regions indicates an impressive human capacity to adapt to long-term mean climate conditions but less success in adapting to extremes and to year-to-year variations in climatic conditions. Climate change will be experienced via conditions that vary from year to year, as well as for ecosystems (Sprengers *et al*., 1994) and human systems (Downing *et al*., 1996); these variations are important for adaptation. Thus, although human settlements and agricultural systems, for example, have adapted to be viable in a huge variety of climatic zones around the world, those settlements and systems often are vulnerable (with limited adaptive capacity) to temporal deviations from normal conditions (particularly extremes). As a result, adaptations designed to address changed mean conditions may or may not be helpful in coping with the variability that is inherent in climate change.

All socioeconomic systems (especially climate-dependent systems such as agriculture, pastoralism, forestry, water resources, and human health) are continually in a state of flux in response to changing circumstances, including climatic conditions. The evidence shows that there is considerable potential for adaptation to reduce the impacts of climate change and to realize new opportunities. In China's Yantze Valley, 18th-century regional expansions and contractions on the double-cropping system for rice represented adaptive responses to the frequency of production successes and failures associated with climatic variations (Smit and Cai, 1996). Adaptation options occur generally in socioeconomic sectors and systems in which the turnover of capital investment and operating costs is shorter and less often where long-term investment is required (Yohe *et al*., 1996; Sohngen and Mendelsohn, 1998).

Although an impressive variety of adaptation initiatives have been undertaken across sectors and regions, the responses are not universally or equally available (Rayner and Malone, 1998). For example, the viability of crop insurance depends heavily on the degree of information, organization, and subsidy available to support it. Similarly, the option of changing location in the face of hazard depends on the resources and mobility of the affected part and on the availability and conditions in potential destination areas (McGregor, 1993). Many response strategies have become less available; many others have become more available. Individual cultivator response to climate risk in India has long relied on a diverse mix of strategies, from land use to outside employment (sometimes requiring temporary migration) to reciprocal obligations for support; many of these strategies have been undermined by changes such as population pressure and government policy, without being fully replaced by others—illustrating the oft-remarked vulnerability of regions and populations in transition (Gadgil *et al*., 1988; Johda, 1989). In areas of China, many historical adaptations in agriculture (e.g., relocating production or employing irrigation) are no longer available as population pressures increase on limited land and water resources (Fang and Liu, 1992; Cai and Smit, 1996). In Kenya, effective smallholder response to drought has shifted from traditional planting strategies to employment diversification (Downing *et al*., 1989).

Not only is there rarely only one adaptation option available to decisionmakers (Burton and Cohen, 1993) but also "rarely do people choose the best responses—the ones among those available that would most effectively reduce losses—often because of an established preference for, or aversion to, certain options" (Rayner and Malone, 1998). In some cases there is limited knowledge of risks or alternative adaptation strategies. In other cases, adoption of adaptive measures is constrained by other priorities, limited resources, or economic or institutional barriers (Eele, 1996; Bryant *et al*., 2000; de Loë and Kreutzwiser, 2000). Recurrent vulnerabilities, in many cases with increasing damages, illustrate less-than-perfect adaptation of systems to climatic variations and risks. There is some evidence that the costs of adaptations to climate conditions are growing (Burton, 1997; Etkin, 1998). There is strong evidence of a sharp increase in damage costs of extreme climatic or weather events (Berz, 1999; Bruce, 1999). Growing adaptation costs reflect, at least in part, increases in populations and/or improvements in standards of living, with more disposable income being used to improve levels of comfort, health, and safety in the short run. It is not clear whether the expansion in adaptations is likely to be effective and sustainable in the long run. In any event, although adaptations to changed and variable climatic conditions are undertaken, they are not necessarily effective or without costs.

Many adaptations to reduce vulnerability to climate change risks also reduce vulnerability to current climate variability, extremes, and hazards (El Shaer *et al*., 1996; Rayner and Malone, 1998). Measures that are likely to reduce current sensitivity of climate variations in Africa also are likely to reduce the threat of adverse impacts of climate change (Ominde and Juna, 1991):

> "Most analysts in the less-developed countries believe that the urgent need, in the face of both climate variation and prospective climate change, is to identify policies which reduce recurrent vulnerability and increase resilience. Prescriptions for reducing vulnerability span drought-proofing the economy, stimulating economic diversification, adjusting land and water uses, providing social support for dependent populations, and providing financial instruments that spread the risk of adverse consequences form individual to society and over longer periods. For the near term, development strategies should ensure that livelihoods are resilient to a wide range of perturbations." (Rayner and Malone, 1998)

Examples of current adaptation strategies in agriculture with clear applications to climate change are given by Easterling (1996) and Smit *et al*. (1997), including moisture-conserving practices, hybrid selection, and crop substitution. In the water resources sector, Stakhiv (1996) shows how current management practices represent useful adaptive strategies for climate change. Some analysts go further to point out that certain adaptations to climate change not only address current hazards but may be additionally beneficial for other reasons (e.g., "no regrets" or "win-win" strategies) (Carter, 1996).

Societal responses to large environmental challenges tend to be incremental and ad hoc rather than fundamental (Rayner and Malone, 1998). In all of the climate analog cases examined by Glantz (1988), "Ad hoc responses were favored over long term planned responses. As a result, there has been a tendency to 'muddle through.' This has not necessarily been an inappropriate response, but it is probably more costly in the long term than putting a long-term strategy together in order to cope with climate-related environmental change." In each case, moreover, action was not taken without a catalyst or trigger that dramatically indicated the seriousness of a threat (Glantz, 1988). Other studies also indicate the ad hoc nature of adaptations and the importance of a catalyst (Wilhite *et al*., 1986; Glantz, 1992; Kasperson *et al*., 1995). These findings suggest that problems that demand early or long-term attention often fail to receive it, and the most efficient responses are not taken. That the earlier action would have been more efficacious, however, presupposes that the best strategy was evident to the decisionmakers and that premature responses closing off useful options would not have been taken instead (Rayner and Malone, 1998). There is little evidence that efficient and effective adaptations to climate change risks will be undertaken autonomously.

A consistent lesson from adaptation research is that climate is not the singular driving force of human affairs that is sometimes assumed—but neither is it a trivial factor. Climate is an important resource for human activities and an important hazard. Climate change is a source of significant stresses (and perhaps significant opportunities) for societies, yet it has always been only one factor among many. The consequences of a shift in climate are not calculable from the physical dimensions of the shift alone; they require attention to human dimensions through which they are experienced (Rayner and Malone, 1998; Bryant *et al*., 2000). The significance of climate change for regions depends fundamentally on the ability and likelihood of those regions to adapt.

To what degree are societies likely to adapt autonomously to avoid climate change damages? Some studies show faith in market mechanisms and suggest considerable capacity of human systems to adapt autonomously (Ausabel, 1991b; Mendelsohn *et al*., 1996; Yohe *et al*., 1996; Mendelsohn, 1998; Mendelsohn and Neumann, 1999). Other studies highlight the constraints on "optimal" autonomous adaptation, such as limited information and access to resources, adaptation costs, and residual damages; these studies emphasize the need for planned, especially anticipatory, adaptations undertaken or facilitated by public agencies (Smith *et al*., 1996; Reilly, 1998; Tol, 1998a; Fankhauser *et al*., 1999; Bryant *et al*., 2000; Schneider *et al*., 2000)

18.4. Planned Adaptations and Evaluation of Policy Options

This section considers *planned*, mainly (but not exclusively) *anticipatory* adaptations, undertaken or directly influenced by *governments* or collectives as a public policy initiative. These adaptations represent conscious policy options or response strategies to concerns about climate change (Benioff *et al*., 1996; Fankhauser, 1996; Smith, 1997; Pielke, 1998; UNEP, 1998). Public adaptation initiatives may be direct or indirect, such as when they encourage or facilitate private actions (Leary, 1999). Planned adaptation by public agencies represents an alternative or complementary response strategy to mitigation (of net GHG emissions). Analyses of such planned adaptations are essentially normative exercises involving identification of possible policy strategies and evaluation of the relative merit of alternatives, as an aid to policy development.

18.4.1. Rationale and Objectives for Planned Adaptations

Numerous reasons have been given for pursuing planned adaptations at this time (see Table 18-4). Public adaptation initiatives are regarded not as a substitute for reducing GHG emissions but as a necessary strategy to manage the impacts of climate change (Burton, 1996; Pielke, 1998). Adaptation can yield benefits regardless of the uncertainty and nature of climate change (Ali, 1999). Fankhauser *et al*. (1998) and Leary (1999) outline rationales for public adaptation policies or projects relative to relying on private actions. Leary concludes that "we

Table 18-4*: Six reasons to adapt to climate change now (Burton, 1996).*

1) Climate change cannot be totally avoided.
2) Anticipatory and precautionary adaptation is more effective and less costly than forced, last-minute, emergency adaptation or retrofitting.
3) Climate change may be more rapid and more pronounced than current estimates suggest. Unexpected events are possible.
4) Immediate benefits can be gained from better adaptation to climate variability and extreme atmospheric events.
5) Immediate benefits also can be gained by removing maladaptive policies and practices.
6) Climate change brings opportunities as well as threats. Future benefits can result from climate change.

cannot rely solely or heavily on autonomous adjustments of private agents to protect public goods and should examine public policy responses to do so." Planned anticipatory adaptation, as recognized in the UNFCCC (Article 3.3), is aimed at reducing a system's vulnerability by diminishing risk or improving adaptive capacity.

There has been work on the process by which public agencies might or should undertake planned adaptation strategies, particularly noting the steps to be followed, relationships with other policy and management objectives, and the criteria with which options might be evaluated (Louisse and Van der Meulen, 1991; Carter *et al.*, 1994; Smith and Lenhart, 1996; Stakhiv, 1996; Major and Frederick, 1997; Smith, 1997). Klein and Tol (1997) identify five generic objectives of adaptation:

1) Increasing robustness of infrastructural designs and long-term investments—for example, by extending the range of temperature or precipitation a system can withstand without failure and changing the tolerance of loss or failure (e.g., by increasing economic reserves or by insurance)
2) Increasing the flexibility of vulnerable managed systems—for example, by allowing mid-term adjustments (including change of activities or location) and reducing economic lifetimes (including increasing depreciation)
3) Enhancing the adaptability of vulnerable natural systems—for example, by reducing other (nonclimatic) stresses and removing barriers to migration (including establishing eco-corridors)
4) Reversing trends that increase vulnerability (also termed "maladaptation")—for example, by introducing setbacks for development in vulnerable areas such as floodplains and coastal zones
5) Improving societal awareness and preparedness—for example, by informing the public of the risks and possible consequences of climate change and setting up early-warning systems.

18.4.2. Identification of Adaptation Policy Options

Research addressing future adaptations to climate change tends to be normative, suggesting anticipatory adaptive strategies to be implemented through public policy. Generally, such adaptation recommendations are based on forecasts of expected (though still largely unpredictable) climate change. Recommended adaptations:

- Tend to be in response to changes in *long-term mean climate*, though more specific elements of climate change (e.g., sea-level change) gain focus when sector-specific adaptations are proposed (e.g., integrated coastal zone management) (Al-Farouq and Huq, 1996; Smith *et al.*, 1996), and some studies specifically examine potential adaptations to variability and extreme events (e.g., Appendi and Liverman, 1996; Yang, 1996; Yim, 1996).
- *Range in scope* from very broad strategies for adaptation (e.g., enhancing decisionmakers' awareness of climatic change and variability) to recommendations of sector-specific policy. Sectors receiving particular attention include water resources, coastal resources, agriculture, and forest resources (Smith and Lenhart, 1996; Smith *et al.*, 1996; Hartig et al., 1997; Mendelsohn and Bennett, 1997).
- Tend to be *regionally focused* (Smith and Lenhart, 1996), in recognition of the fact that vulnerability to the impacts of climate change is highly spatially variable. There is interest in developing countries and nations with economies in transition, given their greater reliance on natural systems-based economic activity (such as agriculture) (e.g., Magalhães, 1996; Smith *et al.*, 1996; Kelly and Adger, 1999).

Because no single set of adaptive policy recommendations can be universally appropriate, several studies suggest means by which proposed adaptations may be selected and evaluated. At a very basic level, the success of potential adaptations is seen to depend on the *flexibility* or effectiveness of the measures, such as their ability to meet stated objectives given a range of future climate scenarios (through either robustness or resilience), and their potential to produce *benefits that outweigh costs* (financial, physical, human, or otherwise) (Smith and Lenhart, 1996). Clearly, these are difficult criteria to assess, given the complexity of adaptation measures, the variable sensitivities and capacities of regions, and uncertainties associated with climate change and variability. Some research (e.g., Carter, 1996; Smith and Lenhart, 1996; Smith *et al.*, 1996; de Loë and Kreutzwiser, 2000) offers supplementary characteristics of, or criteria for, the identification of adaptations:

- The measure generates benefits to the economy, environment, or society under current conditions (i.e., independent of climate change).
- The measure addresses high-priority adaptation issues such as irreversible or catastrophic impacts of climate change (e.g., species extinction), long-term planning for adaptation (e.g., infrastructure), and unfavorable trends (e.g., deforestation, which may inhibit future adaptive flexibility).
- The measure targets current areas of opportunity (e.g., land purchases, revision of national environmental action or development plans, research and development).
- The measure is feasible—that is, its adoption is not significantly constrained by institutional, social/cultural, financial, or technological barriers.
- The measure is consistent with, or even complementary to, adaptation or mitigation efforts in other sectors.

18.4.3. Evaluation of Adaptation Options and Adaptation Costs

Some very general steps for identifying and evaluating planned adaptations are given in Carter *et al.* (1994) and UNEP (1998).

Somewhat more detailed procedures for evaluating anticipatory adaptation policies in the climate change context are outlined in Smith and Lenhart (1996) and Smith (1997). This approach addresses management of institutional processes and players and proposes net benefits and implementability as central evaluative criteria. Numerous other considerations are noted, including flexibility, benefits independent of climate change ("no regrets"), local priorities, levels of risk, and time frames of decisions. From a disaster management perspective, Tol *et al.* (1996) argue that policies must be evaluated with respect to economic viability, environmental sustainability, public acceptability, and behavioral flexibility. Tol *et al.* (1999) apply these observations in an examination of adaptation to increased risk of river floods in The Netherlands. They note several possible adaptations, but none could be accomplished without creating significant distributional and/or ecological impacts. None, therefore, would be feasible without enormous political will and institutional reform. Klein and Tol (1997) and UNEP (1998) describe methodologies for evaluation, including cost-benefit, cost-effectiveness, risk-benefit, and multi-criteria methods. Multi-criteria methods to evaluate possible adaptation options have been demonstrated for coastal zones (El-Raey *et al.*, 1999) and agriculture (Mizina *et al.*, 1999).

Fankhauser (1996) provides an economic efficiency framework in which adaptation actions are considered justified as long as the additional costs of adaptation are lower than the additional benefits from the associated reduced damages. Optimal levels of adaptation (in an economic efficiency sense) are based on minimizing the sum of adaptation costs and residual damage costs. Such studies require the definition of a base case that involves estimation of autonomous adaptations. These and other normative studies (e.g., Titus, 1990; Goklany, 1995) illustrate the range of principles and methods that have been proposed for identifying, evaluating, and recommending (planned) adaptation measures.

There are, however, few comprehensive estimates of the costs of adaptation. Mimura and Harasawa (2000) report estimates of 11.5–20 trillion Yen as the cost of maintaining the functions of Japanese infrastructure against a 1-m rise in sea level. Yohe and Schlesinger (1998) applied a cost-benefit rule to adaptation decisions across a sample of the developed coastline of the United States. With a 3% discount rate, their national estimates of the expected discounted cost of protecting or abandoning developed coastal property in response to sea-level rise that is based on a mean greenhouse emissions scenario is US$1.3 billion with foresight and US$1.8 billion without. Their estimates climb to more than US$4 billion and 5 billion, respectively, along the 1-m sea-level rise scenario that matches the Mimura and Harasawa study. Between 55 and 70% of these costs were attributed to planned adaptation. The remainder reflect estimates of residual damage associated with abandoning property with and without completely efficient autonomous adaptation. Indeed, the differences between the foresight and non-foresight estimates can be regarded as estimates of the incremental cost of incomplete autonomous adaptation in advance of planned responses.

On a more local scale, Smith *et al.* (1998) report cost estimates that are clearly sensitive to design and evaluation criteria. For example, none of the five flood protection strategies for the southernmost part of the Dutch Meuse (assuming 10% more winter precipitation and a warming of 2°C) would achieve economic benefits that exceed their costs of DGL 243–1,505 million, given a 5% discount rate. Moreover, only building quays would meet the benefit-cost standard with a 5% discount rate. Nonetheless, the government chose a wildlife renovation strategy on the basis of additional benefits for nature and recreation. Smith *et al.* (1998) also report that the cost of raising the Northumberland Bridge between Prince Edward Island and New Brunswick to accommodate a 1-m sea-level rise would be US$1 billion or 250,000, depending on whether the entire bridge or only the portion that spanned the shipping lanes were raised.

Klein *et al.* (1999) develop a conceptual framework of the process of planned adaptations, aimed at changing existing

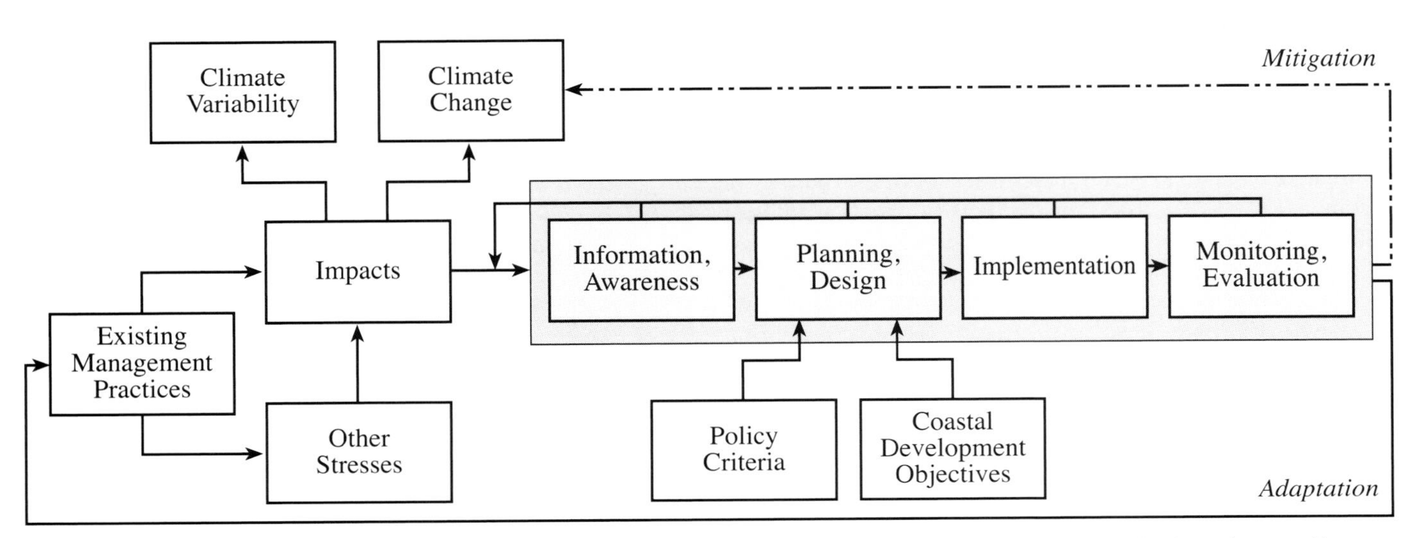

Figure 18-6: Conceptual framework showing, in shaded area, iterative steps involved in planned coastal adaptation to climate variability and change (Klein *et al.*, 1999).

management practices in coastal zones. In this model, adaptation is a continuous and iterative cycle, involving several steps: information collection and awareness raising, planning and design (incorporating policy criteria and development objectives), implementation, monitoring, and evaluation (see Figure 18-6).

18.4.4. Public Adaptation Decisions, Uncertainty, and Risk Management

Research increasingly addresses how adaptation is considered in actual policy decisionmaking. Stakhiv (1996) and Frederick (1997), dealing with the U.S. water resources sector, conclude that existing institutions and planning processes can deal with climate change; such processes essentially represent adaptive management. As in many other sectors and circumstances, adaptation to climate change hazards in the coastal zone is part of ongoing coastal zone management. Adaptation to sea-level rise and extreme climate events is being included in Japanese coastal policies (Mimura and Kawaguchi, 1997), British shoreline management (Leafe *et al.*, 1998), and Dutch law and coastal zone management (Koster and Hillen, 1995; Helmer *et al.*, 1996; Klein *et al.*, 1998).

Planning of adaptation invariably is complicated by multiple policy criteria and interests that may be in conflict (Hareau *et al.*, 1999). For example, the economically most efficient path to implement an adaptation option might not be the most effective or equitable one. Moreover, decisions have to be made in the face of uncertainty (Lempert *et al.*, 2000). Uncertainties that are pertinent to adaptation are associated with climate change itself, its associated extremes, their effects, the vulnerability of systems and regions, conditions that influence vulnerability, and many attributes of adaptations, including their costs, implementability, consequences, and effectiveness (Campos *et al.*, 1996; Lansigan *et al.*, 1997; Handmer *et al.*, 1999; Murdiyarso, 2000).

Given these uncertainties, it is not surprising that adaptation strategies frequently are described as forms of risk management. For example, adaptations to deal with climate change impacts or risks to human health can be biological (acquired immunity), individual (risk-aversion options), or social (McMichael *et al.*, 1996). Most social adaptation strategies are measures to reduce health risks via public health programs (Patz, 1996; McMichael and Kovats, 2000). Similarly, public adaptations via "disaster loss mitigation" (Bruce, 1999) are mainly risk management initiatives such as improved warning and preparedness systems, less vulnerable buildings and infrastructure, risk-averse land-use planning, and more resilient water supply systems. Nguyen *et al.* (1998), Hisschemöller and Olsthoorn (1999), and Perez *et al.* (1999) also describe adaptations to climate change and extremes as modifications to existing risk management programs. As de Loë and Kreutzwiser (2000) and others point out, it remains unclear whether practices designed for historical climatic variability will be able to cope with future variability.

To recognize these uncertainties, decision tools to help evaluate adaptation options include risk-benefit and multi-criteria analyses (Klein and Tol, 1997). Such evaluations are further complicated by the existence of secondary impacts related to the adaptation itself. For example, water development projects (adaptations to water supply risks) can have significant effects on local transmission of parasitic diseases, including malaria, lymphatic filasiasis, and schistosomiasis (Hunter *et al.*, 1993; McMichael and Kovats, 2000). Improved water supply in some rural areas of Asia has resulted in a dramatic increase in Aedes mosquito breeding sites and, consequently, outbreaks of dengue (WHO, 1997). Langen and Tol (1999) provide examples of technical response options to climate hazards that are counterproductive in the longer term. Existing resource management programs do not necessarily consider changed risks or recognize local interests and inequities (Primo, 1996). Wilhite's (1996) analysis of programs in the United States, Australia, and Brazil shows the ineffectiveness of reactive crisis management approaches and the need for proactive and cooperative planning.

Nonetheless, it is widely accepted that planned adaptations to climate risks are most likely to be implemented when they are developed as components of (or as modifications to) existing resource management programs or as part of national or regional strategies for sustainable development (Campos *et al.*, 1996; Magalhães, 1996; Theu *et al.*, 1996; Mimura, 1999a; Apuuli *et al.*, 2000; Munasinghe, 2000; Osokova *et al.*, 2000).

18.5. Adaptive Capacity and its Determinants

18.5.1. Vulnerability and Adaptive Capacity

Considerable attention has been devoted to the characteristics of systems (communities or regions) that influence their propensity or ability to adapt (as part of impact and vulnerability assessment) and/or their priority for adaptation measures (as a basis for policy development). These characteristics have been called *determinants of adaptation*. Generic concepts such as sensitivity, vulnerability, susceptibility, coping range, critical levels, adaptive capacity, stability, robustness, resilience, and flexibility have been used to differentiate systems according to their likelihood, need, or ability for adaptation (Sprengers *et al.*, 1994; De Ruig, 1997; Klein and Tol, 1997; Smithers and Smit, 1997; Adger and Kelly, 1999; Kelly and Adger, 1999). These characteristics influence (promote, inhibit, stimulate, dampen, or exaggerate) the occurrence and nature of adaptations and thereby circumscribe the vulnerability of systems and their residual impacts. In the hazards literature, these characteristics are reflected in socially constructed or endogenous risks (Blaikie *et al.*, 1994; Hewitt, 1997). Together (in whole or part), they represent the adaptive capacity of a system.

Table 18-5 lists terms that are commonly used to characterize the adaptive propensity of systems to climate stimuli. There is considerable overlap in the basic concepts captured in these terms. Particular terms have been employed to distinguish natural from socioeconomic systems or to differentiate the condition of a system before adaptation from its condition after adaptation (Klein and Nicholls, 1998). These distinctions are important and can be captured without narrowing the meaning of widely used terms. Thus, ecosystem vulnerability is different from socioeconomic vulnerability.

Adaptive capacity refers to the potential, capability, or ability of a system to adapt to climate change stimuli or their effects or impacts. Adaptive capacity greatly influences the vulnerability of communities and regions to climate change effects and hazards (Bohle *et al.*, 1994; Downing *et al.*, 1999; Kelly and Adger, 1999; Mileti, 1999; Kates, 2000). Vulnerability has been described as the "capacity to be wounded" (Kates *et al.*, 1985). Human activities and groups are considered sensitive to climate to the degree that they can be affected by it and vulnerable to the degree that they can be harmed (Rayner and Malone, 1998). Because vulnerability and its causes play essential roles in determining impacts, understanding the dynamics of vulnerability is as important as understanding climate itself (Liverman, 1990; Handmer *et al.*, 1999).

With regard to climate change, the vulnerability of a given system or society is a function of its physical exposure to climate change effects and its ability to adapt to these conditions. Chambers (1989) distinguishes between these two aspects of differential vulnerability: physical exposure to the hazardous agent and the ability to cope with its impacts. Thus, vulnerability recognizes the role of socioeconomic systems in amplifying or moderating the impacts of climate change and "emphasizes the degree to which the risks of climate catastrophe can be cushioned or ameliorated by adaptive actions that or can be brought within the reach of populations at risk" (Downing, 1991).

The significance of climate variation or change depends on the change itself and the characteristics of the society exposed to it

***Table 18-5**: Terms to describe characteristics of systems that are pertinent to adaptation[a] (from Smit et al., 1999).*

Term	Description
Sensitivity	Degree to which a system is affected by or responsive to climate stimuli (note that sensitivity includes responsiveness to both problematic stimuli and beneficial stimuli)
Susceptibility	Degree to which a system is open, liable, or sensitive to climate stimuli (similar to sensitivity, with some connotations toward damage)
Vulnerability	Degree to which a system is susceptible to injury, damage, or harm (one part—the problematic or detrimental part—of sensitivity)
Impact Potential	Degree to which a system is sensitive or susceptible to climate stimuli (essentially synonymous with *sensitivity*)
Stability	Degree to which a system is not easily moved or modified
Robustness	Strength; degree to which a system is not given to influence
Resilience	Degree to which a system rebounds, recoups, or recovers from a stimulus
Resistance	Degree to which a system opposes or prevents an effect of a stimulus
Flexibility	Degree to which a system is pliable or compliant (similar to *adaptability*, but more absolute than relative)
Coping Ability	Degree to which a system can successfully grapple with a stimulus (similar to *adaptability*, but includes more than adaptive means of "grappling")
Responsiveness	Degree to which a system reacts to stimuli (broader than *coping ability* and *adaptability* because responses need not be "successful")
Adaptive Capacity	The potential or capability of a system to adapt to (to alter to better suit) climatic stimuli or their effects or impacts
Adaptability	The ability, competency, or capacity of a system to adapt to (to alter to better suit) climatic stimuli (essentially synonymous with *adaptive capacity*)

[a] These definitions of systems characteristics are based on widely (but not unanimously) held conventions. They focus on distinguishing generic properties and do not include factors that might influence the state of a property or the forms it might take. The terms "climate stimulus" and "system" are used as established earlier.

(Ausubel, 1991a; Rayner and Malone, 1998; Munasinghe, 2000). These characteristics of society determine its adaptive capacity and its adaptability. Adaptive capacity refers to the ability to prepare for hazards and opportunities in advance (as in anticipatory adaptation) and to respond or cope with the effects (as in reactive adaptation).

Studies of similar hazardous events recurring at different times in a given region show vastly different consequences because of societal transformations that occurred between the events. For example, rainfall and temperature fluctuations in western Europe have far milder effects on human well-being today (society generally is less vulnerable) than they did in the medieval and early modern periods, essentially as a result of enhanced adaptive capacity that reflects changes in practices, economics, and government programs (Abel, 1976; De Vries, 1977; Rayner and Malone, 1998). Similarly, particular climate events or hazards can have "vastly different consequences for those on whom they infringe because of differences in coping ability" (Rayner and Malone, 1998). An extreme climatic event will result in higher losses of life in a developing country than in a developed country because of differential adaptive capacity (Burton, *et al.,* 1993; Blaikie *et al*., 1994; Kundzewicz and Takeuchi, 1999). Martens *et al.* (1999) describe potential adaptations to deal with increases in disease incidence associated with climate change but note that in most poor developing countries, socioeconomic, technical, and political barriers will mean that the changed health risks will not be addressed.

> "In developing countries overall social, environmental, and economic vulnerability enhances the effects of droughts and other climatic events. Overpopulation (relative to current productivity, income, and natural resources), poverty, and land degradation translate into a poor capacity to face any kind of crisis. Poor people have no insurance against loss of income. Weak economic structures mean difficulties in maintaining jobs during an economic failure. Degraded marginal lands become totally unproductive when precipitation decreases. As a result, these regions have difficulty in facing climatic crises, although such crises are recurrent. Any extreme climatic event can become a social catastrophe when combined with the social-political characteristics of the region. For example, the droughts and internecine wars in Ethiopia interact to increase the adverse effects of both. Although developing regions are more vulnerable to climate changes than are developed countries, the degree of vulnerability varies in each specific region." (Magalhães, 1996)

Research on comparative adaptive capacity and vulnerability is evolving, and its difficulties are well recognized (Bohle *et al*., 1994; Downing, 1996; Handmer *et al*., 1999; Kelly and Adger, 1999). Estimates of adaptive capacity tend to be based on premises such as the position that highly managed systems (such as agriculture), given sufficient resources, are likely to be more adaptable (and at a lower cost) than less managed ecosystems (Strzepek and Smith, 1995; Burton, 1996; Toman and Bierbaum, 1996). It is also widely accepted that systems with high levels of capacity to cope with historical and/or existing stresses can be expected to have high adaptive capacity for stresses associated with climatic change (Ausubel, 1991a). Such premises have formed the basis for broad assessments of sensitivity and adaptability (USNAS, 1992). Of course, sensitivity and adaptive capacity vary according to the climate change-related stress being considered. Thus, adaptive capacity to gradual changes in mean temperature may be high (or not much needed), but adaptive capacity to changes in the magnitude or frequency of extreme climatic conditions may not be so high (Appendi and Liverman, 1996).

18.5.2. Determinants of Adaptive Capacity

Adaptation to climate change and risks takes place in a dynamic social, economic, technological, biophysical, and political context that varies over time, location, and sector. This complex mix of conditions determines the capacity of systems to adapt. Although scholarship on adaptive capacity is extremely limited in the climate change field, there is considerable understanding of the conditions that influence the adaptability of societies to climate stimuli in the fields of hazards, resource management, and sustainable development. From this literature, it is possible to identify the main features of communities or regions that seem to determine their adaptive capacity: economic wealth, technology, information and skills, infrastructure, institutions, and equity.

18.5.2.1. Economic Resources

Whether it is expressed as the economic assets, capital resources, financial means, wealth, or poverty, the economic condition of nations and groups clearly is a determinant of adaptive capacity (Burton *et al*., 1998; Kates, 2000). It is widely accepted that wealthy nations are better prepared to bear the costs of adaptation to climate change impacts and risks than poorer nations (Goklany, 1995; Burton, 1996). It is also recognized that poverty is directly related to vulnerability (Chan and Parker, 1996; Fankhauser and Tol, 1997; Rayner and Malone, 1998). Although poverty should not be considered synonymous with vulnerability, it is "a rough indicator of the ability to cope" (Dow, 1992). Holmes (1996) recognizes that Hong Kong's financial strength has contributed in the past to its ability to better manage environmental hazards through conservation and pollution control. Bohle *et al*. (1994) state that, by definition, it usually is the poor who are among the most vulnerable to famine, malnutrition, and hunger. Deschingkar (1998) describes a situation in India in which pastoralist communities are "locked into" a vulnerable situation in part because of a lack of financial power that would allow them to diversify and engage in other sources of income. At a local level, Pelling (1998) concludes that the highest levels of household vulnerability in coastal Guyana also are characterized by low household incomes in conjunction with poor housing quality and little community organization. Neighborhoods with higher levels of household income are better able to manage vulnerability through the transfer of flood impacts from health to economic investment and loss. Kelly and Adger (1999) demonstrate the influence of poverty on a region's coping capacity; poor regions tend to have less diverse and more restricted entitlements and a lack of empowerment to adapt. There is ample evidence that poorer nations and disadvantaged groups within nations are especially vulnerable to disasters (Banuri, 1998; Munasinghe, 2000).

18.5.2.2. Technology

Lack of technology has the potential to seriously impede a nation's ability to implement adaptation options by limiting the range of possible responses (Scheraga and Grambsch, 1998). Adaptive capacity is likely to vary, depending on availability and access to technology at various levels (i.e., from local to national) and in all sectors (Burton, 1996). Many of the adaptive strategies identified as possible in the management of climate change directly or indirectly involve technology (e.g., warning systems, protective structures, crop breeding and irrigation, settlement and relocation or redesign, flood control measures). Hence, a community's current level of technology and its ability to develop technologies are important determinants of adaptive capacity. Moreover, openness to the development and utilization of new technologies for sustainable extraction, use, and development of natural resources is key to strengthening adaptive capacity (Goklany, 1995). For example, in the context of Asian agriculture and the impact of future climate change, Iglesias *et al.* (1996) note that the development of heat-resistant rice cultivars will be especially crucial. Regions with the ability to develop technology have enhanced adaptive capacity.

18.5.2.3. Information and Skills

"Successful adaptation requires a recognition of the necessity to adapt, knowledge about available options, the capacity to assess them, and the ability to implement the most suitable ones" (Fankhauser and Tol, 1997). In the context of climate variability and change, this idea may be better understood through the example of the insurance industry: As information on weather hazards becomes more available and understood, it is possible to study, discuss, and implement adaptation measures (Downing, 1996). Building adaptive capacity requires a strong, unifying vision; scientific understanding of the problems; an openness to face challenges; pragmatism in developing solutions; community involvement; and commitment at the highest political level (Holmes, 1996). Lack of trained and skilled personnel can limit a nation's ability to implement adaptation options (Scheraga and Grambsch, 1998). In general, countries with higher levels of stores of human knowledge are considered to have greater adaptive capacity than developing nations and those in transition (Smith and Lenhart, 1996). Magalhães (1996) includes illiteracy along with poverty as a key determinant of low adaptive capacity in northeast Brazil. Such findings have prompted Gupta and Hisschemöller (1997) to conclude that it is important, therefore, to ensure that systems are in place for the dissemination of climate change and adaptation information nationally and regionally and that there are forums for discussion and innovation of adaptation strategies at various levels.

18.5.2.4. Infrastructure

Adaptive capacity is likely to vary with social infrastructure (Toman and Bierbaum, 1996). Some researchers regard the adaptive capacity of a system as a function of *availability of* and *access to* resources by decisionmakers, as well as vulnerable subsectors of a population (Kelly and Adger, 1999). For example, the Philippine island of Mindanao uses hydroelectric power to generate more than 90% of its electricity, which in turn supports local development and industry. During El Niño, drought conditions resulted in suspension of production by the hydroelectric plant and severely increased the economic vulnerability of the region (Tiglao, 1992). In the coastal area of Hong Kong, the capacity to adapt to the risk of typhoons differs for existing urban areas and for new coastal land reclamation. For existing urban areas, there is no possibility of retreat or accommodation, although during urban renewal the formation level of the ground could be raised, thereby decreasing the vulnerability of settlements (Yim, 1996). At the community level, Pelling (1997) notes that the lack of flexibility "in formal housing areas where dwelling form and drainage infrastructure were more fixed" reduced the capacity to respond to contemporary environmental conditions.

18.5.2.5. Institutions

O'Riordan and Jordan (1999) describe the role of institutions "as a means for holding society together, giving it sense and purpose and enabling it to adapt." In general, countries with well-developed social institutions are considered to have greater adaptive capacity than those with less effective institutional arrangements—commonly, developing nations and those in transition (Smith and Lenhart, 1996). The role of inadequate institutional support is frequently cited in the literature as a hindrance to adaptation. Kelly and Adger (1999) show how institutional constraints limit entitlements and access to resources for communities in coastal Vietnam and thereby increase vulnerability. Huq *et al.* (1999) demonstrate that Bangladesh is particularly vulnerable to climate change—especially in the areas of food production, settlements, and human life—reflecting serious constraints on adaptive capacity in the "existing institutional arrangements (which) is not conducive to ease the hardship of the people. Due to inherent institutional deficiencies and weaknesses in managerial capacities to cope with the anticipated natural event, it would be extremely difficult for the country to reduce vulnerability to climate change" (Ahmed *et al.*, 1999). Baethgen (1997) discusses an example in which the presence of inconsistent and unstable agricultural policies has increased the vulnerability of the food production sector in Latin America. Drastic changes in economic and policy conditions are expected to make agricultural systems more vulnerable to changes in climate. Parrish and Funnell (1999) note that although the local agro-ecosystem in the Moroccon High Atlas region may prove resilient to climate change initially, it is possible that the need to change tenure conditions and other arrangements may create conflicts that are beyond the capacity of local institutions to resolve. Magadza (2000) shows how adaptation options in southern Africa are precluded by political and institutional inefficiencies and resulting resource inequities.

It is generally held that established institutions in developed countries not only facilitate management of contemporary climate-related risks but also provide an institutional capacity to help deal with risks associated with future climate change. Stakhiv (1994) states that in the water resource sector, present-day strategies, demand management tools, and measures (i.e., institutions) have evolved over the past 25 years and are capable of serving as a basis for adaptive response strategies to climate change: "[T]he accumulation of numerous small changes in the present range of water resources management practices and procedures increases the flexibility for adaptation to current climate uncertainty and serves as a precursor to future possible responses with an ill-defined, changing climatic regime." However, some analyses are less sanguine about the ability of existing institutions to efficiently deal with climate change hazards. For example, Miller *et al.* (1997) note that "the time has come for innovative thinking on the question of how our water allocation institutions should function to improve our capacity to adapt to the uncertain but potentially large impacts of global climate change on regional water supplies. Given the climatic uncertainties and the very different institutional settings that have developed in this country, there is no simple prescription for adaptation."

18.5.2.6. Equity

It is frequently argued that adaptive capacity will be greater if social institutions and arrangements governing the allocation of power and access to resources within a community, nation, or the globe assure that access to resources is equitably distributed (Ribot *et al.*, 1996; Mustafa, 1998; Adger, 1999; Handmer *et al.*, 1999; Kelly and Adger, 1999; Rayner and Malone, 1999; Toth, 1999). The extent to which nations or communities are "entitled" to draw on resources greatly influences their adaptive capacity and their ability to cope (Adger and Kelly, 1999). Some people regard the adaptive capacity of a system as a function not only of the *availability* of resources but of *access* to those resources by decisionmakers and vulnerable subsectors of a population (Kelly and Adger, 1999). In the case of technological innovation, Cyert and Kumar (1996) show that differential distribution of information within an organization can impose constraints on adaptation strategies. Differentiation in demographic variables such as age, gender, ethnicity, educational attainment, and health often are cited in the literature as being related to the ability to cope with risk (Chan and Parker, 1996; Burton *et al.*, 1998; Scheraga and Grambsch, 1998). Wisner's (1998) study of homeless people in Tokyo provides an example of a situation in which inequality in access to resources results in a diminished capacity to adapt to environmental risk. Homeless people generally occupy marginal areas that are more vulnerable to environmental hazards. An associated lack of financial resources and infrastructure restricts the availability of adaptation options. A study by Bolin and Stanford (1991) draws parallel conclusions about the marginalization of minority groups.

These determinants of adaptive capacity are not independent of each other, nor are they mutually exclusive. Adaptive capacity is the outcome of a combination of determinants and varies widely between countries and groups, as well as over time. "Vulnerability varies spatially because national environments, housing and social structure vary spatially. It varies temporally because people move through different life stages with varying mixes of resources and liabilities" (Uitto, 1998). Bohle *et al.* (1994) document variable vulnerability to climatic variations of groups in Zimbabwe and its association with poverty, the macro-political economy, and inequitable land distribution. Not only are conditions for adaptive capacity diverse, they also behave differently in different countries and regions, particularly depending on the level of development. These determinants represent conditions that constrain or enhance the adaptive capacity and hence the vulnerability of regions, nations, and communities.

18.5.3. Adaptive Capacity of Regions

At the global scale, there is considerable variation among countries with regard to their capacity to adapt to climate change. Given their economic affluence and stability; their institutions and infrastructures; and their access to capital, information, and technology, developed nations are broadly considered to have greater capacity to adapt than developing regions or countries in economic transition (Goklany, 1995; Burton, 1996; Magalhães, 1996; Toman and Bierbaum, 1996). In general, countries with well-developed social institutions supported by higher levels of capital and stores of human knowledge are considered to have greater adaptive capacity (Smith and Lenhart, 1996). Adaptation options—including traditional coping strategies—often are available in developing countries and countries in transition; in practice, however, those countries' capacity to effect timely response actions may be beyond their infrastructure and economic means (IPCC, 1997). For those countries, the main barriers are (Smith, 1996; IPCC, 1998; Mizina *et al.*, 1999):

- Financial/market (uncertain pricing, availability of capital, lack of credit)
- Institutional/legal (weak institutional structure, institutional instability)
- Social/cultural (rigidity in land-use practices, social conflicts)
- Technological (existence, access)
- Informational/educational (lack of information, trained personnel).

A study by Rosenzweig and Parry (1994) found considerable disparity between developed and developing countries in terms of potential adverse effects of climate change on agricultural systems; developing countries suffer the greatest losses. In addition, poorer, developing regions presumably will face stricter constraints on technology and institutions (Fankhauser and Tol, 1997) and that measures taken in response to climate change may be very demanding financially (Dvorak *et al.*, 1997; Deschingkar, 1998). Researchers also believe that compared to industrialized countries, developing countries possess a lower adaptive capacity as a result of greater reliance on climatic resources (Schelling, 1992; Fankhauser and Tol, 1997).

There is some suggestion, however, that the complex, multi-species, low- to middle-intensity farming systems that characterize agricultural endeavors in the developing world may have greater adaptive capacity under conditions of global climate change than western monocultures (Ramakrishnan, 1998). An example is found in the village of Maatisar (India), where local institutions in the past have operated on the principle of "moral economy," or guaranteed subsistence to all households in the village. These institutions have eroded over time, however, giving way to competitive market relations that do not guarantee subsistence during times of drought. As a result, the capacity of individual households to withstand seasonal fluctuations has decreased over time (Chen, 1991). Magalhães (1996) describes how northeast Brazil has become more vulnerable to droughts as inappropriate land use overstresses natural land and water resources and as the capacity to cope is limited by poverty.

Acceptance of western economic ideals coupled with increasing and rapid development may reduce the capacity of traditional societies to adapt (Watts, 1983; Chan and Parker, 1996). In the case of traditional or indigenous societies, the pursuit of western/European-style development trajectories may modify the nature of adaptive capacity (some improved, some diminished) by introducing greater technology dependence and higher density settlement and by devaluing traditional ecological knowledge and cultural values (Newton, 1995). For example, notwithstanding remarkable adaptations to a harsh climate, the North American Inuit continue to be vulnerable to climate change as a result of their dependence on wildlife (which are climate-sensitive). This vulnerability has been reduced by technological enhancement of adaptive capacity through the acquisition of snowmobiles, motorized boats, and even sonar. Such technological advances have allowed Inuit communities to become far more "fixed" than before. Many of the most densely populated areas lie at least partially within a few meters of sea level. This lack of "semi-permanence" may actually increase the Inuits' vulnerability to potential climate-induced sea-level rise by decreasing their capacity to adapt through retreat or migration (Rayner and Malone, 1998).

Although there is considerable literature on the determinants of adaptive capacity and examples of how they influence the adaptability of particular communities, there is little scholarship (and even less agreement) on criteria or variables by which adaptive capacity can be measured and by which the adaptive capacity of global regions can be quantitatively compared. Various studies have attempted to identify overall trends that cause increased or decreased vulnerability to environmental hazards (Torry, 1979; Lamb, 1982; Warrick and Reibsame, 1983; Ausubel, 1991b; Blaikie *et al.,* 1994); unfortunately, however, the concept of vulnerability "does not rest well on a developed theory, nor is it associated with widely accepted indicators or methods of measurement" (Bohle *et al.*, 1994).

Even less progress has been made in measuring adaptive capacity. In the context of African agriculture, Downing *et al.* (1997) attempt to quantitatively measure relative adaptive capacity of regions by using crude surrogates such as gross national product (GNP). Empirical local-level studies of vulnerability are so complex, however, that attempts to describe patterns or estimate trends at global or regional scales are extremely difficult (Liverman, 1990; Downing, 1991; Dow, 1992). These "difficulties in generalizing about levels of vulnerability even in a relatively small community" are demonstrated by Adger and Kelly's (1999) study of vulnerability to climate change in Vietnam. Because vulnerability is a composite concept, social change has the potential to make individuals or activities more vulnerable in some ways and less vulnerable in others (Rayner and Malone, 1998). The influence of changes in the determinants of adaptive capacity are not necessarily direct or clear, rendering the attempt to develop systematic indices for measurement and comparison a difficult task.

18.6. Enhancing Adaptive Capacity

The adaptive capacity of a system or nation is likely to be greater when the following requirements are met:

1) The nation has a stable and prosperous economy. Regardless of biophysical vulnerability to the impacts of climate change, developed and wealthy nations are better prepared to bear the costs of adaptation than developing countries (Goklany, 1995; Burton, 1996).
2) There is a high degree of access to technology at various levels (i.e., from local to national) and in all sectors (Burton, 1996). Moreover, openness to development and utilization of new technologies for sustainable extraction, use, and development of natural resources is key to strengthening adaptive capacity (Goklany, 1995).
3) The roles and responsibilities for implementation of adaptation strategies are well delineated by central governments and are clearly understood at national, regional, and local levels (Burton, 1996).
4) Systems are in place for the dissemination of climate change and adaptation information, nationally and regionally, and there are forums for the discussion and innovation of adaptation strategies at various levels (Gupta and Hisschemöller, 1997).
5) Social institutions and arrangements governing the allocation of power and access to resources within a nation, region, or community assure that access to

resources is equitably distributed because the presence of power differentials can contribute to reduced adaptive capacity (Mustafa, 1998; Handmer *et al.*, 1999; Kelly and Adger, 1999).

6) Existing systems with high adaptive capacity are not compromised. For example, in the case of traditional or indigenous societies, pursuit of western/European-style development trajectories may reduce adaptive capacity by introducing greater technology dependence and higher density settlement and by devaluing traditional ecological knowledge and cultural values.

18.6.1. Adaptive Capacity and Sustainable Development

Ability to adapt clearly depends on the state of development (Berke, 1995; Munasinghe, 1998). As Ribot *et al.* (1996) illustrate, underdevelopment fundamentally constrains adaptive capacity, especially because of a lack of resources to hedge against extreme but expected events. The events are not surprises: "It is not that the risk is unknown, not that the methods for coping do not exist...rather inability to cope is due to lack of—or systematic alienation from—resources needed to guard against these events" (Ribot *et al.*, 1996).

The process of enhancing adaptive capacity is not simple; it involves "spurts of growth inter-dispersed with periods of consolidation, refocusing and redirection" (Holmes, 1996). Enhancement of adaptive capacity involves similar requirements as promotion of sustainable development, including:

- Improved access to resources (Ribot *et al.*, 1996; Kelly and Adger, 1999; Kates, 2000)
- Reduction of poverty (Berke, 1995; Eele, 1996; Karim, 1996; Kates, 2000)
- Lowering of inequities in resources and wealth among groups (Berke, 1995; Torvanger, 1998)
- Improved education and information (Zhao, 1996)
- Improved infrastructure (Magalhães and Glantz, 1992; Ribot *et al.*, 1996)
- Diminished intergenerational inequities (Berke, 1995; Munasinghe, 2000)
- Respect for accumulated local experience (Primo, 1996)
- Moderate long-standing structural inequities (Magadza, 2000)
- Assurance that responses are comprehensive and integrative, not just technical (Ribot *et al.*, 1996; Cohen *et al.*, 1998; Rayner and Malone, 1998; Munasinghe and Swart, 2000)
- Active participation by concerned parties, especially to ensure that actions match local needs and resources (Berke, 1995; Ribot *et al.*, 1996; Rayner and Malone, 1998; Ramakrishnan, 1999)
- Improved institutional capacity and efficiency (Handmer *et al.*, 1999; Magadza, 2000).

Because actions taken without reference to climate have the potential to affect vulnerability to it, enhancement of adaptive capacity to climate change can be regarded as one component of broader sustainable development initiatives (Ahmad and Ahmed, 2000; Munasinghe, 2000; Robinson and Herbert, 2000). Hazards associated with climate change have the potential to undermine progress with sustainable development (Berke, 1995; Wang'ati, 1996). Therefore, it is important for sustainable development initiatives to explicitly consider hazards and risks associated with climate change (Apuuli *et al.*, 2000).

Clearly, adaptive capacity to deal with climate risks is closely related to sustainable development and equity. Enhancement of adaptive capacity is fundamental to sustainable development. For example, in the drought-stricken region of northeastern Brazil, an assessment of past successes and failures has indicated that a comprehensive sustainable development strategy is needed to increase regional and societal capacity to face present and future climate variability (Magalhães, 1996). By assessing differences in vulnerability among regions and groups and by working to improve the adaptive capacity of those regions and groups, planned adaptation can contribute to equity considerations of sustainable development. In the context of African agriculture, Downing *et al.* (1997) conclude that enhancement of present resource management activities is necessary to prepare for potential impacts of climate change. In Malawi, as in many other places, the UNFCCC's objectives to "ensure food production is not threatened, and to enable economic development to proceed in a sustainable manner" also are central to the nation's development policies (Theu *et al.*, 1996). Thus, progress to reducing vulnerability to climate risks is consistent with Malawi's planning and development initiatives.

Notwithstanding the considerable literature on the impacts of climate change as described throughout this volume, very little attention has been devoted to the interaction of adaptation to climate change with ongoing development projects and programs. Because vulnerability to climate depends on the adaptive capacity of a wide range of attributes, it may be unrealistic to focus on development programs that deal with adaptation to climate alone (Cohen, *et al.*, 1998; Rayner and Malone, 1998). Yet there is surprisingly little recognition of climate hazards and risks associated with climate change in established development projects and programs (Berke, 1995; Burton and Van Aalst, 1999). O'Brian and Liverman (1996) show how climate change can have serious implications for development projects planned or underway in Mexico, including hydroelectric and irrigation initiatives. Torvanger (1998) shows how climate flexibility considerations that can be built into development investments at modest incremental costs are applicable regardless of the uncertainties of climate change and with immediate value because of existing risks.

18.6.2. Capacity Enhancement by Scale

The vulnerabilities and anticipated impacts of climate change will be observed at different scales and levels of society—and enhancement of adaptive capacity can be initiated at different

***Table 18-6**: Adaptation and adaptive capacity in sectors (key findings from Chapters 4 through 9).*

Sector	Key Findings
Water Resources	– Water managers have experience adapting to change. Many techniques exist to assess and implement adaptive options. However, the pervasiveness of climate change may preclude some traditional adaptive strategies, and available adaptations often are not used. – Adaptation can involve management on the supply side (e.g., altering infrastructure or institutional arrangements) and on the demand side (changing demand or risk reduction). Numerous no-regret policies exist that will generate net social benefits regardless of climate change. – Climate change is just one of numerous pressures facing water managers. Nowhere are water management decisions taken solely to cope with climate change, although it is increasingly considered for future resource management. Some vulnerabilities are outside the conventional responsibility of water managers. – Estimates of the economic costs of climate change impacts on water resources depend strongly on assumptions made about adaptation. Economically optimum adaptation may be prevented by constraints associated with uncertainty, institutions, and equity. – Extreme events often are catalysts for changes in water management, by exposing vulnerabilities and raising awareness of climate risks. Climate change modifies indicators of extremes and variability, complicating adaptation decisions. – Ability to adapt is affected by institutional capacity, wealth, management philosophy, planning time scale, organizational and legal framework, technology, and population mobility. – Water managers need research and management tools aimed at adapting to uncertainty and change, rather than improving climate scenarios.
Ecosystems and Their Services	– Adaptation to loss of some ecosystem services may be possible, especially in managed ecosystems. However, adaptation to losses in wild ecosystems and biodiversity may be difficult or impossible. – There is considerable capacity for adaptation in agriculture, including crop changes and resource substitutions, but adaptation to evolving climate change and interannual variability is uncertain. – Adaptations in agriculture are possible, but they will not happen without considerable transition costs and equilibrium (or residual) costs. – Greater adverse impacts are expected in areas where resource endowments are poorest and the ability of farmers to adapt is most limited. – In many countries where rangelands are important, lack of infrastructure and investment in resource management limit options for adaptation. – Commercial forestry is adaptable, reflecting a history of long-term management decisions under uncertainty. Adaptations are expected in both land-use management (species-selection silviculture) and product management (processing-marketing). – Adaptation in developed countries will fare better, while developing countries and countries in transition, especially in the tropics and subtropics, will fare worse.
Coastal Zones	– Without adaptations, the consequences of global warming and sea-level rise would be disastrous. – Coastal adaptation entails more than just selecting one of the technical options to respond to sea-level rise (strategies can aim to protect, accommodate, or retreat). It is a complex and iterative process rather than a simple choice. – Adaptation options are more acceptable and effective when they are incorporated into coastal zone management, disaster mitigation programs, land-use planning, and sustainable development strategies. – Adaptation choices will be conditioned by existing policies and development objectives, requiring researchers and policymakers to work toward a commonly acceptable framework for adaptation. – The adaptive capacity of coastal systems to perturbations is related to coastal resilience, which has morphological, ecological, and socioeconomic components. Enhancing resilience—including the technical, institutional, economic, and cultural capability to cope with impacts—is a particularly appropriate adaptive strategy given future uncertainties and the desire to maintain development opportunities. – Coastal communities and marine-based economic sectors with either low exposure or high adaptive capacity will be least affected. Communities with less economic resources, poorer infrastructure, less developed communications and transportation systems, and weak social support systems have less access to adaptation options and are more vulnerable.

Table 18-6 (continued)

Sector	Key Findings
Human Settlements, Energy, and Industry	– The larger and more costly impacts of climate change occur through changed probability of extreme weather events that overwhelm the design resiliency of human systems. – There are many adaptation options available to reduce the vulnerability of settlements. However, urban managers, especially in developing countries, have so little capacity to deal with current problems (housing, sanitation, water, and power) that dealing with climate change risks is beyond their means. – Lack of financial resources, weak institutions, and inadequate or inappropriate planning are major barriers to adaptation in human settlements. – Successful environmental adaptation cannot occur without locally based, technically competent, and politically supported leadership. – Uncertainty with respect to capacity and the will to respond hinder the assessment of adaptations and vulnerability.
Insurance and Other Financial Services	– Adaptation in financial and insurance services in the short term is likely to be to changing frequencies and intensities of extreme weather events. – Increasing risk could lead to a greater volume of traditional business and the development of new financial risk management products, but increased variability of loss events would heighten actuarial uncertainty. – Financial services firms have adaptability to external shocks, but there is little evidence that climate change has been incorporated into investment decisions. – The adaptive capacity of the financial sector is influenced by regulatory involvement, the ability of firms to withdraw from at-risk markets, and fiscal policy regarding catastrophe reserves. – Adaptation will involve changes in the roles of private and public insurance. Changes in the timing, intensity, frequency, and/or spatial distribution of climate-related losses will generate increased demand on already overburdened government insurance and disaster assistance programs. – Developing countries seeking to adapt in a timely manner face particular difficulties, including limited availability of capital, poor access to technology, and absence of government programs. – Insurers' adaptations include raising prices, nonrenewal of policies, cessation of new policies, limiting maximum claims, and raising deductibles—actions that can seriously affect investment in developing countries. – Developed countries generally have greater adaptive capacity, including technology and economic means to bear the costs.
Human Health	– Adaptation involves changes in society, institutions, technology, or behavior to reduce potential negative impacts or to increase positive ones. There are numerous adaptation options, which may occur at the population, community, or personal levels. – The most important and cost-effective adaptation measure is to rebuild public health infrastructure—which, in much of the world, has declined in recent years. Many diseases and health problems that may be exacerbated by climate change can be effectively prevented with adequate financial and human public health resources, including training, surveillance and emergency response, and prevention and control programs. – Adaptation effectiveness will depend on timing. "Primary" prevention aims to reduce risks before cases occur, whereas "secondary" interventions are designed to prevent further cases. – Determinants of adaptive capacity to climate-related threats to health include the level of material resources, the effectiveness of governance and civil institutions, the quality of public health infrastructure, and the preexisting burden of disease. – Capacity to adapt also will depend on research to understand associations between climate, weather, extreme events, and vector-borne diseases.

social scales (Ribot *et al.*, 1996; Handmer *et al.*, 1999). In Bangladesh, Ahmed *et al.* (1999) distinguish between four scales: mega, macro, meso, and micro. Using the example of sea-level rise as a climate change impact, the authors describe adaptation options at each scale. The process of sea-level rise occurs at the mega-scale and is global in its effect. At the macro-scale, an associated increase in surface water and groundwater has the potential to similarly effect neighboring rivers and flood plains in China, Nepal, India, Bhutan, and Pakistan. Adaptive capacity at this scale is a function of international economic and political structures, with implications for the nations' capital and technological resources and institutions. At the meso-scale, different communities within Bangladesh are differentially vulnerable, depending on adaptive capacity and physiographic characteristics. At this scale, location-specific adaptation options would need to be considered. Finally, at a micro-scale, family units and individuals would experience vulnerabilities irrespective of the origin of the processes and would employ adaptations within their particular economic and sociocultural constraints.

Because the vulnerabilities of climate change occur at various scales, successful adaptation will depend on actions taken at a number of levels. Examples of initiatives to enhance adaptive capacity at various scales follow:

- At a *global* scale
  - Greater cooperation between industrialized and developing countries to align global and local priorities by improving policy/science interactions and working toward greater public awareness of climate change and adaptation issues (Wang'ati, 1996; Gupta and Hisschemöller, 1997)
  - Inclusion of global institutions for global-level adaptation, which would include research and facilitation of policy, funding, and monitoring at all levels (Ahmed *et al.*, 1999)
  - Removal of barriers to international trade; it is argued that improving market conditions, reducing the exploitation of marginal land, accelerating the transfer of technology, and contributing to overall economic growth will promote both sustainability and adaptive capacity (Goklany, 1995)
  - Effective global economic participation. Benefits go beyond direct financial gain and include technology transfers, technical and managerial skills transfers, and other skills transfers associated with the "learning and doing" process (Ebohon *et al.*, 1997)
- At a *national* level
  - Development of climate change policy that is specifically geared toward more vulnerable sectors in the country (Mustafa, 1998), with an emphasis on poverty reduction (Kelly and Adger, 1999)
  - Establishment of broadly based monitoring and communication systems (e.g., integrated drought monitoring and information system, as suggested in Wilhite, 1997)
  - Establishment of public policy that encourages and supports adaptation at local or community levels and in the private sector (Burton, 1996)
  - Pursuit of sustainable economic growth—which, in turn, allows for greater dedication of resources to development of adaptive technologies and innovations (Goklany, 1995)
- Via *local* means
  - Establishment of social institutions and arrangements that discourage concentration of power in a few hands and prevent marginalization of sections of the local population (Mustafa, 1998); arrangements need to consider representativeness of decision-making bodies and maintenance of flexibility in the functioning of local institutions (Ramakrishnan, 1998)
  - Encouragement of diversification of income sources (and therefore risk-spreading), particularly for poorer sectors of society (Wang'ati, 1996; Adger and Kelly, 1999)
  - Encouragement of formal or informal arrangements for collective security (Kelly and Adger, 1999)
  - Identification and prioritization of local adaptation measures and provision of feedback to higher levels of government. These efforts would have to be reinforced by the adequate provision of knowledge, technology, policy, and financial support (Ahmed *et al.*, 1999).

18.7. Sectoral and Regional Findings

Insights gained about adaptation and adaptive capacity from the sector chapters and the regional chapters are summarized in Tables 18-6 and 18-7, respectively.

Increasingly, adaptation and adaptive capacity are explicitly considered in impact and vulnerability assessments, and there are some consistent findings across sectors and regions (see Section 18.8). However, there is insufficient basis to rank systematically countries according to their adaptive capacity or to list the "most vulnerable" overall. Analyses to date indicate that adaptive capacity and vulnerability are multidimensional, so that one country (or, more often, a group within a country) may be extremely vulnerable economically whereas another country (or community) is extremely vulnerable in terms of life and livelihood. These different types of vulnerability reflect different types of exposures and adaptive capacities.

18.8. Conclusions

Adaptation can significantly reduce adverse impacts of climate change. Adaptation is an important part of societal response to global climate change. Planned, anticipatory adaptation has the potential to reduce vulnerability and realize opportunities associated with climate change effects and hazards. There are numerous examples of successful adaptations that would apply to climate change risks and opportunities. Substantial reductions in climate change damages can be achieved, especially in the most vulnerable regions, through timely deployment of adaptation measures.

In the absence of planned adaptation, communities will adapt autonomously to changing climatic conditions, but not without costs and residual damages. Societies and economies have been making adaptations to climate for centuries. However, losses from climate-related extreme events are substantial and, in some sectors, increasing—indicating patterns of development that remain vulnerable to temporal variations in climatic

***Table 18-7**: Adaptation and capacity in regions (key findings from Chapters 10 through 17).*

Sector	Key Findings
Africa	– Adaptive measures would enhance flexibility and have net benefits in water resources (irrigation and water reuse, aquifer and groundwater management, desalinization), agriculture (crop changes, technology, irrigation, husbandry), and forestry (regeneration of local species, energy-efficient cook stoves, sustainable community management). – Without adaptation, climate change will reduce the wildlife reserve network significantly by altering ecosystems and causing species emigration and extinctions. This represents an important ecological and economic vulnerability in Africa. – A risk-sharing approach between countries will strengthen adaptation strategies, including disaster management, risk communication, emergency evacuation, and cooperative water resource management. – Most countries in Africa are particularly vulnerable to climate change because of limited adaptive capacity, as a result of widespread poverty, recurrent droughts, inequitable land distribution, and dependence on rainfed agriculture. – Enhancement of adaptive capacity requires local empowerment in decisionmaking and incorporation of climate adaptation within broader sustainable development strategies.
Asia	– Priority areas for adaptation are land and water resources, food productivity, and disaster preparedness and planning—particularly for poorer, resource-dependent countries. – Adaptations already are required to deal with vulnerabilities associated with climate variability, in human health, coastal settlements, infrastructure, and food security. The resilience of most sectors in Asia to climate change is very poor. Expansion of irrigation will be difficult and costly in many countries. – For many developing countries in Asia, climate change is only one of a host of problems to deal with, including nearer term needs such as hunger, water supply and pollution, and energy. Resources available for adaptation to climate are limited. Adaptation responses are closely linked to development activities, which should be considered in evaluating adaptation options. – Early signs of climate change already are observed and may become more prominent over 1 or 2 decades. If this time is not used to design and implement adaptations, it may be too late to avoid upheavals. Long-term adaptation requires anticipatory actions. – A wide range of precautionary measures are available at the regional and national level to reduce economic and social impacts of disasters. These measures include awareness building and expansion of the insurance industry. – Development of effective adaptation strategies requires local involvement, inclusion of community perceptions, and recognition of multiple stresses on sustainable management of resources. – Adaptive capacities vary between countries, depending on social structure, culture, economic capacity, and level of environmental disruptions. Limiting factors include poor resource and infrastructure bases, poverty and disparities in income, weak institutions, and limited technology. – The challenge in Asia lies in identifying opportunities to facilitate sustainable development with strategies that make climate-sensitive sectors resilient to climate variability. – Adaptation strategies would benefit from taking a more systems-oriented approach, emphasizing multiple interactive stresses, with less dependence on climate scenarios.
Australia and New Zealand	– Adaptations are needed to manage risks from climatic variability and extremes. Pastoral economies and communities have considerable adaptability but are vulnerable to any increase in the frequency or duration of droughts. – Adaptation options include water management, land-use practices and policies, engineering standards for infrastructure, and health services. – Adaptations will be viable only if they are compatible with the broader ecological and socioeconomic environment, have net social and economic benefits, and are taken up by stakeholders. – Adaptation responses may be constrained by conflicting short- and long-term planning horizons. – Poorer communities, including many indigenous settlements, are particularly vulnerable to climate-related hazards and stresses on health because they often are in exposed areas and have less adequate housing, health care, and other resources for adaptation.

Table 18-7 (continued)

Sector	Key Findings
Europe	– Adaptation potential in socioeconomic systems is relatively high as a result of strong economic conditions; stable population (with capacity to migrate); and well-developed political, institutional, and technological support systems. – The response of human activities and the natural environment to current weather perturbations provides a guide to critical sensitivities under future climate change. – Adaptation in forests requires long-term planning; it is unlikely that adaptation measures will be put in place in a timely manner. – Farm-level analyses show that if adaptation is fully implemented large reductions in adverse impacts are possible. – Adaptation for natural systems generally is low. – More marginal and less wealthy areas will be less able to adapt, so without appropriate policies of response climate change may lead to greater inequities.
Latin America	– Adaptation measures have potential to reduce climate-related losses in agriculture and forestry. – There are opportunities for adapting to water shortages and flooding through water resource management. – Adaptation measures in the fishery sector include changing species captured and increasing prices to reduce losses.
North America	– Strain on social and economic systems from rapid climate and sea-level changes will increase the need for explicit adaptation strategies. In some cases, adaptation may yield net benefits, especially if climate change is slow. – Stakeholders in most sectors believe that technology is available to adapt, although at some social and economic cost. – Adaptation is expected to be more successful in agriculture and forestry. However, adaptations for the water, health, food, and energy sectors and the cities are likely to require substantial institutional and infrastructure changes. – In the water sector, adaptations to seasonal runoff changes include storage, conjunctive supply management, and transfer. It may not be possible to continue current high levels of reliability of water supply, especially with transfers to high-valued uses. Adaptive measures such as "water markets" may lead to concerns about accessibility and conflicts over allocation priorities. – Adaptations such as levees and dams often are successful in managing most variations in the weather but can increase vulnerability to the most extreme events. – There is moderate potential for adaptation through conservation programs that protect particularly threatened ecosystems, such as high alpines and wetlands. It may be difficult or impossible to offset adverse impacts on aquatic systems.

conditions and to climate change. The ecological, social, and economic costs of relying on reactive, autonomous adaptation to the cumulative effects of climate change are substantial and largely avoidable through planned, anticipatory adaptation.

The key features of climate change for vulnerability and adaptation are those that relate to variability and extremes, not simply changed average conditions. In addition, the speed of changes in event frequency is important. Most communities, sectors, and regions are reasonably adaptable to changes in average conditions, unless those changes are particularly sudden or not smooth. However, these communities are more vulnerable and less adaptable to changes in the frequency and/or magnitude of conditions other than average, especially extremes. Changes in the frequency and magnitude of extremes underlie changes in mean conditions and thus are inherent in climate change; adaptation initiatives to these hazards are of particular need.

Implementation of adaptation policies, programs, and measures usually will have immediate as well as future benefits. Adaptations to current climate and climate-related risks (recurring droughts, storms, floods, and other extremes) generally are consistent with adaptation to changing and changed climatic conditions.

Adaptations to changing climatic conditions are more likely to be implemented if they are consistent with or integrated with decisions or programs that address nonclimatic stresses. Vulnerabilities associated with climate change rarely are experienced independent of nonclimatic conditions. Impacts of climatic stimuli are felt via economic or social stresses, and adaptations to climate (by individuals, communities, and governments) are evaluated and undertaken in light of these conditions. The costs of adaptation often are marginal to other management or development costs. To be effective, climate change adaptation must consider nonclimatic stresses and be

Table 18-7 (continued)

Sector	Key Findings
Polar Regions	– Adaptation will occur in natural polar ecosystems through migration and changing mixes of species. Species such as walrus, seals, and polar bears will be threatened, although others (such as fish) may flourish. – Potential for adaptation is limited in indigenous communities that follow traditional lifestyles. – Technologically developed communities are likely to adapt quite readily, although the high capital investment required may result in costs in maintaining lifestyles. – Adaptation depends on technological advances, institutional arrangements, availability of financing, and information exchange.
Small Island States	– The need for adaptation has become increasingly urgent, even if swift implementation of global agreements to reduce future emissions occurs. – Most adaptation will be carried out by people and communities who inhabit island countries; support from governments is essential for implementing adaptive measures. – Progress will require integration of appropriate risk-reduction strategies with other sectoral policy initiatives in areas such as sustainable development planning, disaster prevention and management, integrated coastal zone management, and health care planning. – Strategies for adaptation to sea-level rise are retreat, accommodate, and protect. Measures such as retreat to higher ground, raising of the land, and use of building set-backs appear to have little practical utility, especially when hindered by limited physical size. – Measures for reducing the severity of health threats include health education programs, improved health care facilities, sewerage and solid waste management, and disaster preparedness plans. – Islanders have developed some capacity to adapt by application of traditional knowledge, locally appropriate technology, and customary practice. Overall, however, adaptive capacity is low because of the physical size of nations, limited access to capital and technology, shortage of human resource skills, lack of tenure security, overcrowding, and limited access to resources for construction. – Many small islands require external financial, technical, and other assistance to adapt. Adaptive capacity may be enhanced by regional cooperation and pooling of limited resources.

consistent with existing policy criteria, development objectives, and management structures.

Adaptive capacity varies considerably among regions, countries, and socioeconomic groups. The ability to adapt and cope with climate change impacts is a function of wealth, technology, information, skills, infrastructure, institutions, and equity. Groups and regions with limited adaptive capacity are more vulnerable to climate change damages.

Development decisions, activities, and programs play important roles in modifying the adaptive capacity of communities and regions, yet they tend not to take into account risks associated with climate variability and change. This omission in the design and implementation of many recent and current development initiatives results in unnecessary additional losses to life, well-being, and investments in the short and longer terms.

Enhancement of adaptive capacity is necessary to reduce vulnerability, particularly for the most vulnerable regions, nations, and socioeconomic groups. Activities required for the enhancement of adaptive capacity are essentially equivalent to those that promote sustainable development and equity.

Current knowledge about adaptation and adaptive capacity is insufficient for reliable prediction of adaptations and for rigorous evaluation of planned adaptation options, measures, and policies of governments:

- Although climate change vulnerability studies now usually consider adaptation, they rarely go beyond identifying adaptation options that might be possible. There is little research on the dynamics of adaptation in human systems, the processes of adaptation decisionmaking, the conditions that stimulate or constrain adaptation, and the role of nonclimatic factors.
- There are serious limitations in existing evaluations of adaptation options. Economic benefits and costs are key criteria, but they are not sufficient to adequately determine the appropriateness of adaptation measures. There also has been little research to date on the roles and responsibilities of individuals, communities, corporations, private and public institutions, governments, and international organizations in adaptation.
- Given the scope and variety of specific adaptation options across sectors, individuals, communities, and locations and the variety of participants—private and public—involved in most adaptation initiatives, it is probably infeasible to systematically evaluate lists of particular adaptation measures. Improving and applying knowledge on the constraints and opportunities for enhancing adaptive capacity is necessary to reduce vulnerabilities.

References

Abel, W., 1976: *Massenarmut und Hungerkrisen im Vorindustriellen Europa*. Rev. ed., Paul Parey, Hamburg, Germany, 233 pp.

Abercrombie, M., C.J. Hickman, and M.L. Johnson, 1997: *A Dictionary of Biology*. Penguin Books, Harmondsworth, United Kingdom, 309 pp.

Adams, R.M., K.J. Bryant, D.M. Legler, B.A. McCarl, J. O'Brian, A. Solow, and R. Weiher, 1995: Value of improved long-range weather information. *Contemporary Economic Policy*, **13,** 10–19.

Adger, W.N., 1999: Exploring income inequality in rural, coastal Vietnam. *Journal of Development Studies*, **35,** 96–119.

Adger, W.N. and P.M. Kelly, 1999: Social vulnerability to climate change and the architecture of entitlements. *Mitigation and Adaptation Strategies for Global Change*, **4(3–4),** 253–266.

Ahmad, Q.K. and A.U. Ahmed, 2000: Social sustainability, indicators and climate change. In: *Climate Change and Its Linkages with Development, Equity and Sustainability: Proceedings of the IPCC Expert Meeting held in Colombo, Sri Lanka, 27–29 April, 1999* [Munasinghe, M. and R. Swart (eds.)]. LIFE, Colombo, Sri Lanka; RIVM, Bilthoven, The Netherlands; and World Bank, Washington, DC, USA, pp. 95–108.

Ahmed, A.U., M. Alam, and A.A. Rahman, 1999: Adaptation to climate change in Bangladesh: future outlook. In: *Vulnerability and Adaptation to Climate Change for Bangladesh* [Huq, S., M. Asaduzzaman, Z. Karim, and F. Mahtab (eds.)]. J. Kluwer Academic Publishers, Dordrecht, The Netherlands, pp. 125–143.

Alderwish, A. and M. Al-Eryani, 1999: An approach for assessing the vulnerability of the water resources of Yemen to climate change. In: *National Assessment Results of Climate Change: Impacts and Responses: CR Special 6* [Mimura, N. (ed.)]. Inter-research, Oldendorf, Germany, pp. 85–89.

Alexandrov, V., 1999: Vulnerability and adaptation of agronomic systems in Bulgaria. In: *National Assessment Results of Climate Change: Impacts and Responses: CR Special 6* [Mimura, N. (ed.)]. Inter-research, Oldendorf, Germany, pp. 161–173.

Al-Farouq, A. and S. Huq, 1996: Adaptation to climate change in the coastal resources sector of Bangladesh: some issues and problems. In: *Adapting to Climate Change: An International Perspective* [Smith, J., N. Bhatti, G. Menzhulin, R. Benioff, M.I. Budyko, M. Campos, B. Jallow, and F. Rijsberman (eds.)]. Springer-Verlag, New York, NY, USA, pp. 335–344.

Ali, A., 1999: Climate change impacts and adaptation assessment in Bangladesh. In: *National Assessment Results of Climate Change: Impacts and Responses: CR Special 6* [Mimura, N. (ed.)]. Inter-research, Oldendorf, Germany, pp. 109–116.

Appendi, K. and D. Liverman, 1996: Agricultural policy and climate change in Mexico. In: *Climate Change and World Food Security* [Downing, T.E. (ed.)]. Springer-Verlag, Berlin, Germany, pp. 525–550.

Apuuli, B., J. Wright, C. Elias, and I. Burton, 2000: Reconciling national and global priorities in adaptation to climate change: with an illustration from Uganda. *Environmental Monitoring and Assessment*, **61(1),** pp. 145–159.

Ausubel, J., 1991a: A second look at the impacts of climate change. *American Scientist*, **79,** 211–221.

Ausubel, J.H., 1991b: Does climate still matter? *Nature*, **350,** 649–652.

Baethgen, W., 1997: Vulnerability of the agricultural sector of Latin America to climate change. *Climate Research*, **9,** 1–7.

Banuri, T., 1998: Human and environmental security. *Policy Matters*, **3,** 196–205.

Basher, R.E., 1999: Data requirements for developing adaptations to climate variability and change. *Mitigation and Adaptation Strategies for Global Change*, **4(3–4),** 227–225.

Benioff, F., S. Guill, and J. Lee, 1996: *Vulnerability and Adaptation Assessments: An International Handbook*. J. Kluwer Academic Publishers, Boston, MA, USA, 177 pp.

Beran, M. and N. Arnell, 1995: Climate change and hydrological disasters. In: *Hydrology of Disasters* [Singh, V.P. (ed.)]. J. Kluwer Academic Publishers, Dordrecht, The Netherlands, 53–77.

Berke, P.R., 1995: Natural-hazard reduction and sustainable development: a global assessment. *Journal of Planning Literature*, **9(4),** 370–382.

Berke, P.R., J. Kartez, and D. Wenger, 1993: Recovery after disaster: achieving sustainable development, mitigation and equity. *Disasters: The Journal of Disaster Studies and Management*, **17(2),** 93–109.

Berz, G.A., 1999: Catastrophes and climate change: concerns and possible countermeasures of the insurance industry. *Mitigation and Adaptation Strategies for Global Change*, **4(3–4),** 283–293.

Bijlsma, L., C.N. Ehler, R.J.T. Klein, S.M. Kulshrestha, R.F. McLean, N. Mimura, R.J. Nicholls, L.A. Nurse, H. Pérez Nieto, E.Z. Stakhiv, R.K. Turner, and R.A. Warrick, 1996: Coastal zones and small islands. In: *Climate Change 1995: Impacts, Adaptations, and Mitigation of Climate Change: Scientific-Technical Analyses. Contribution of Working Group II to the Second Assessment Report of the Intergovernmental Panel on Climate Change* [Watson, R.T., M.C. Zinyowera, and R.H. Moss (eds.)]. Cambridge University Press, Cambridge, United Kingdom and New York, NY, USA, pp. 289–324.

Blaikie, P., T. Cannon, I. Davies, and B. Wisner, 1994: *At Risk: Natural Hazards, People's Vulnerability, and Disasters*. Routledge, New York, NY, USA, 284 pp.

Bohle, H.G., T.E. Downing, and M.J. Watts, 1994: Climate change and social vulnerability: toward a sociology and geography of food insecurity. *Global Environmental Change*, **4(1),** 37–48.

Bolin, R. and L. Stanford, 1991: Shelter, housing and recovery: a comparison of U.S. disasters. *Disasters*, **15(1),** 24–34.

Brown, P.G., 1998: Toward and economics of stewardship: the case of climate. *Ecological Economics*, **26,** 11–21.

Bruce, J.P., 1999: Disaster loss mitigation as an adaptation to climate variability and change. *Mitigation and Adaptation Strategies for Global Change*, **4(3–4),** 295–306.

Bryant, C.R., B. Smit, M. Brklacich, T.R. Johnston, J. Smithers, Q. Chiotti, and B. Singh, 2000: Adaptation in Canadian agriculture to climatic variability and change. *Climatic Change*, **45(1),** 181–201.

Burton, I., 1997: Vulnerability and adaptive response in the context of climate and climate change. *Climatic Change*, **36,** 185–196.

Burton, I., 1996: The growth of adaptation capacity: practice and policy. In: *Adapting to Climate Change: An International Perspective* [Smith, J., N. Bhatti, G. Menzhulin, R. Benioff, M.I. Budyko, M. Campos, B. Jallow, and F. Rijsberman (eds.)]. Springer-Verlag, New York, NY, USA, pp. 55–67.

Burton, I. and M. van Aalst, 1999: *Come Hell or High Water: Integrating Climate Change Vulnerability and Adaptation into Bank Work*. World Bank Environment Department Paper No. 72, Climate Change Series, Washington D.C., USA, 60 pp.

Burton, I. and S.J. Cohen, 1993: Adapting to global warming: regional options. In: *Proceedings of the International Conference on the Impacts of Climatic Variations and Sustainable Development in the Semi-Arid Regions, A Contribution to UNCED* [Magelhães, A.R. and A.F. Bezerra (eds.)]. Esquel Brazil Foundation, Brasilia, Brazil, pp. 871–886.

Burton, I., R.W. Kates, and G.F. White, 1993: *The Environment As Hazard*. The Guilford Press, New York, NY, USA, 290 pp.

Burton, I., J.B. Smith, and S. Lenhart, 1998: Adaptation to climate change: theory and assessment. In: *Handbook on Methods for Climate Change Impact Assessment and Adaptation Strategies* [Feenstra, J.F., I. Burton, J.B. Smith, and R.S.J. Tol (eds.)]. United Nations Environment Programme and Institute for Environmental Studies, Free University of Amsterdam, Amsterdam, pp. 5.1–5.20.

Cai, Y. and B. Smit, 1996: Sensitivity and adaptation of Chinese agriculture under global climate change. *Acta Geographica Sinica*, **51(3),** 202–212.

Campos, M., A. Sánchez, and D. Espinosa, 1996: Adaptation of hydropower generation in Costa Rica and Panama to climate change. In: *Adapting to Climate Change: An International Perspective* [Smith, J., N. Bhatti, G. Menzhulin, R. Benioff, M.I. Budyko, M. Campos, B. Jallow, and F. Rijsberman (eds.)]. Springer-Verlag, New York, NY, USA, pp. 232–242.

Carter, T.R., 1996: Assessing climate change adaptations: the IPCC guidelines. In: *Adapting to Climate Change: An International Perspective* [Smith, J., N. Bhatti, G. Menzhulin, R. Benioff, M.I. Budyko, M. Campos, B. Jallow, and F. Rijsberman (eds.)]. Springer-Verlag, New York, NY, USA, pp. 27–43.

Carter, T.R., M.L. Parry, H. Harasawa, and S. Nishioka, 1994: *IPCC Technical Guidelines for Assessing Climate Change Impacts and Adaptations*. University College, London, United Kingdom, and Centre for Global Environmental Research, Tsukuba, Japan, 59 pp.

Chambers, R., 1989: Editorial introduction: vulnerability, coping and policy. *IDS Bulletin*, **21,** 1–7.

Chan, N. and D. Parker, 1996: Response to dynamic flood hazard factors in peninsular Malaysia. *The Geographic Journal*, **162(3),** 313–325.

Changnon, S.A., D. Changnon, E.R. Fosse, D.C. Hoganson, R.J. Roth, and J.M. Totsch, 1997: Effects of recent weather extremes on the insurance industry: major implications for the atmospheric science. *Bulletin of the American Meteorological Society,* **78(3),** 425–435.

Chen, M., 1991: *Coping with Seasonality and Drought*. Sage Publications, New Delhi, India, 254 pp.

Cohen, S., D. Demeritt, J. Robinson, and D. Rothman, 1998: Climate change and sustainable development: towards dialogue. *Global Environmental Change*, **8(4),** 341–371.

Cyert, R. and P. Kumar, 1996: Strategies for technological innovation with learning and adaptation costs. *Journal of Economics and Management Strategy*, **5(1),** 25–67.

De Loë, R.C. and R. Kreutzwiser, 2000: Climate variability, climate change and water resource management in the Great Lakes. *Climatic Change*, **45(1),** 163–179.

De Freitas, C.R., 1989: The hazard potential of drought for the population of the Sahel. In: *Population and Disaster* [Clarke, J.I., P. Curson, S.L. Kayastha, and P. Nag (eds.)]. Blackwell, Oxford, United Kingdom, pp. 98–113.

De Ruig, J.H.M., 1997: Resilience in ditch coastline management. *Coastline*, **6(2),** 4–8.

Denevan, W.M., 1983: Adaptation, variation, and cultural geography. *Professional Geographer*, **35,** 399–407.

Deschingkar, P., 1998: Climate change adaptation in India: a case study of forest systems in Himachal Pradesh. *International Journal of Environment and Pollutio*n, **9(2/3),** 186–197.

De Vries, J., 1985: Analysis of historical climate-society interaction. In: *Climate Impact Assessment* [Kates, R.W., J.H. Ausubel, and M. Berberian (eds.)]. John Wiley and Sons, New York, NY, USA, pp. 273–291.

De Vries, J., 1977: Histoire du climat et économie: des faits nouveaux, une interprétation différente. *Annales: Economies, Sociétés, Civilisations*, **32,** 198–227.

Dow, K., 1992: Exploring the differences in our common future(s): the meaning of vulnerability to global environmental change. *Geoforum*, **23(3),** 417–436.

Dowlatabadi, H., 1995: Integrated assessment models of climate change. *Energy Policy*, **23(4/5),** 289–296.

Downing, T.E. (ed.), 1996: *Climate Change and World Food Security*. Springer-Verlag, Berlin, Germany, pp. 662.

Downing, T.E., 1991: Vulnerability to hunger in Africa. *Global Environmental Change*, **1,** 365–380.

Downing, T.E., K.W. Gitu, and C.M. Kaman (eds.), 1989: *Coping with Drought in Kenya: National and Local Strategies*. Lynne Rienner, Boulder, CO, USA, 411 pp.

Downing, T.E., A.A. Olsthoorn, and R.S.J. Tol, 1996: *Climate Change and Extreme Events: Altered Risks, Socio-Economic Impacts and Policy Responses*. Vrije Universiteit, Amsterdam, The Netherlands, 411 pp.

Downing, T.E., L. Ringius, M. Hulme, and D. Waughray, 1997: Adapting to climate change in Africa. *Mitigation and Adaptation Strategies for Global Change*, **2(1),** 19–44.

Downing, T.E., M.J. Gawaith, A.A. Olsthoorn, R.S.J. Tol, and P. Vellinga, 1999: Introduction. In: *Climate Change and Risk* [Downing, T.E., A.A. Olsthoorn, and R.S.J. Tol (eds.)]. Routledge, London, United Kingdom, pp. 1–19.

Dvorak, V., J. Hladmy, and L. Kasparek, 1997: Climate change, hydrology and water resources impact and adaptation for selected river basins in the Czeck Republic. *Climatic Change*, **36,** 93–106.

Easterling, W.E., 1996: Adapting North American agriculture to climate change in review. *Agricultural and Forest Meteorology*, **80(1),** 1–54.

Easterling, W.E., P.R. Crosson, N.J. Rosenberg, M.S. McKenney, L.A. Katz, and K.M. Lemon, 1993: Agricultural impacts of and responses to climate change in the Missouri-Iowa-Nebraska-Kansas (MINK) Region. *Climatic Change*, **24,** 23–61.

Ebohon, O.J., B.G. Field, and R. Ford, 1997: Institutional deficiencies and capacity building constraints: the dilemma for environmentally sustainable development in Africa. *International Journal of Sustainable Development and World Ecology*, **4(3),** 204–213.

Eele, G., 1996: Policy lessons from communities under pressure. In: *Climate Change and World Food Security* [Downing, T.E. (ed.)]. Springer-Verlag, Berlin, Germany, pp. 611–624.

El-Raey, M., K. Dewidar, and M. El-Hattab, 1999: Adaptation to the impacts of sea level rise in Egypt. *Mitigation and Adaptation Strategies for Global Change*, **4(3–4),** 343–361.

El-Shaer, M.H., H.M. Eid, C. Rosenzweig, A. Iglesias, and D. Hillel, 1996: Agricultural adaptation to climate change in Egypt. In: *Adapting to Climate Change: An International Perspective* [Smith, J., N. Bhatti, G. Menzhulin, R. Benioff, M.I. Budyko, M. Campos, B. Jallow, and F. Rijsberman (eds.)]. Springer-Verlag, New York, NY, USA, pp. 109–127.

Erda, L., 1996: Agricultural vulnerability and adaptation to global warming in China. *Water, Air and Soil Pollution*, **92,** 63–75.

Etkin, D., 1998: Climate change and extreme events: Canada. *Canada Country Study: Climate Impacts and Adaptation*, **8,** 31–80.

Fang, J-Q. and G. Liu, 1992: Relationship between climatic change and the nomadic southward migration in eastern Asia during historical times. *Climatic Change*, **22,** 151–169.

Fankhauser, S., 1996: The potential costs of climate change adaptation. In: *Adapting to Climate Change: An International Perspective* [Smith, J., N. Bhatti, G. Menzhulin, R. Benioff, M.I. Budyko, M. Campos, B. Jallow, and F. Rijsberman (eds.)]. Springer-Verlag, New York, NY, USA, pp. 80–96.

Fankhauser, S. and R.S.J. Tol, 1997: The social costs of climate change: the IPCC second assessment report and beyond. *Mitigation and Adaptation Strategies for Global Change*, **1,** 385–403.

Fankhauser, S., J.B. Smith, and R.S.J. Tol, 1999: Weathering climate change: some simple rules to guide adaptation decisions. *Ecological Economics*, **30(1),** 67–78.

Fankhauser, S., J.B. Smith, and R.S.J. Tol, 1998: *Weathering Climate Change: Some Simple Rules to Guide Adaptation Decisions*. Institute for Environmental Studies, Amsterdam, The Netherlands, 12 pp.

Frederick, K.D., 1997: Adapting to climate impacts in the supply and demand for water. *Climatic Change*, **37,** 141–156.

Fukui, H., 1979: Climate variability and agriculture in tropical moist regions. In: *Proceedings of the World Climate Conference*. World Meteorological Association Report No. 537, Geneva, Switzerland, pp 407–426.

Gadgil, S., A.K.S. Huda, N.S. Johda, R.P. Singh, and S.M. Virmani, 1988: The effects of climatic variations on agriculture in dry tropical regions of India. In: *The Impact of Climatic Variations on Agriculture: Assessments in Semi-Arid Regions* [Parry, M.L., T.R. Carter, and N.T. Konjin (eds.)]. J. Kluwer Academic Publishers, Dordrecht, The Netherlands, 764 pp.

Glantz, M.H., 1996: Forecasting by analogy: local responses to global climate change. In: *Adapting to Climate Change: An International Perspective* [Smith, J., N. Bhatti, G. Menzhulin, R. Benioff, M.I. Budyko, M. Campos, B. Jallow, and F. Rijsberman (eds.)]. Springer-Verlag, New York, NY, USA, pp. 407–426.

Glantz, M.H. (ed.), 1992: *Climate variability, climate change, and fisheries*. Cambridge University Press, Cambridge, United Kingdom and New York, NY, USA, 450 pp.

Glantz, M.H., 1988: *Societal Responses to Climate Change: Forecasting by Analogy*. Westview Press, Boulder, CO, USA, 428 pp.

Goklany, I.M., 1995: Strategies to enhance adaptability: technological change, sustainable growth and free trade. *Climatic Change*, **30,** 427–449.

Gupta, J. and M. Hisschemöller, 1997: Issue linkage as a global strategy toward sustainable development: a comparative case study of climate change. *International Environmental Affairs*, **9(4),** 289–308.

Handmer, J., S. Dovers, and T.E. Downing, 1999: Societal vulnerability to climate change and variability. *Mitigation and Adaptation Strategies for Global Change*, **4(3–4),** 267–281.

Hardesty, D.L., 1983: Rethinking cultural adaptation. *Professional Geographer*, **35,** 399–406.

Hareau, A., R. Hofstadter, and A. Saizar, 1999: Vulnerability to climate change in Uruguay: potential impacts on the agricultural and coastal resource sectors and response capabilities. In: *National Assessment Results of Climate Change: Impacts and Responses: CR Special 6* [Mimura, N. (ed.)]. Inter-research, Oldendorf, Germany, pp. 185–193.

Harrington, K., 1996: The risk management dilemma of climate change. *Water Resources Update*, **103,** 35–45.

Hartig, E.K., O. Grosev, and C. Rosenzweig, 1997: Climate change, agriculture and wetlands in eastern europe: vulnerability adaptation and policy. *Climatic Change*, **36,** 101–121.

Helmer, W.P., P. Vellinga, G. Litjens, H. Goosen, E. Ruijgrok, and W. Overmars, 1996: *Growing with the Sea—Creating a Resilient Coastline*. World Wildlife Fund for Nature, Zeist, The Netherlands, 39 pp.

Helms, S., R. Mendelsohn, and J. Neumann, 1996: The impact of climate change on agriculture: editorial essay. *Climatic Change*, **33,** 1–6.

Hewitt, K., 1997: *Regions of Risk: A Geographical Introduction to Disasters*. Addison-Wesley Longman, Essex, United Kingdom, 389 pp.

Hewitt, K.and I. Burton, 1971: *The Hazardousness of a Place: A Regional Ecology of Damaging Events*. University of Toronto, Toronto, ON, Canada, 154 pp.

Hisschemöller, M. and A.A. Olsthoorn, 1999: Identifying barriers and opportunities for policy responses to changing climatic risks. In: *Climate, Change and Risk* [Downing, T.E., A.A. Olsthoorn, and R.S.J. Tol (eds.)]. Routledge, London, United Kingdom, pp. 365–390.

Holmes, P., 1996: Building capacity for environmental management in Hong Kong. *Water Resources Development*, **12(4),** 461–472.

Huang, J.C.K., 1996: Climate change adaptation strategies: a challenge for small island nations. In: *Adapting to Climate Change: An International Perspective* [Smith, J., N. Bhatti, G. Menzhulin, R. Benioff, M.I. Budyko, M. Campos, B. Jallow, and F. Rijsberman (eds.)]. Springer-Verlag, New York, NY, USA, pp. 311–321.

Hulme, M. and S. Raper, 1995: An integrated framework to address climate change (ESCAPE) and further developments of the global and regional climate modules (MAGICC). *Energy Policy*, **23(4/5),** 347–355.

Hulme, M., E.M. Barrow, N.W. Arnell, P.A. Harrison, T.C. Johns, and T.E. Downing, 1999: Relative impacts of human-induced climate change and natural climate variability. *Nature*, **397,** 688–691.

Hunter, J.M., L. Rey, K.Y. Chu, E.O. Adekolu-John, and K.E. Mott, 1993: *Parasitic Diseases in Water Resource Development: The Need for Intersectoral Negotiation*. World Health organization, Geneva, Switzerland, 152 pp.

Huq, S., Z. Karim, M. Asaduzzaman, and F. Mahtab (eds.), 1999: *Vulnerability and Adaptation to Climate Change in Bangladesh*. J. Kluwer Academic Publishers, Dordrecht, The Netherlands, 147 pp.

Hurd, B., J. Callaway, P. Kirshen, and J. Smith, 1997: Economic effects of climate change on U.S. water resources. In: *The Impacts of Climate Change on the U.S. Economy* [Mendelsohn, R. and J. Newmann (eds.)]. Cambridge University Press, Cambridge, United Kingdom and New York, NY, USA, 331 pp.

Iglesias, A., L. Erda, and C. Rosenzweig, 1996: Climate change in Asia: a review of the vulnerability and adaptation of crop production. *Water, Air, and Soil Pollution*, **92,** 13–27.

Ingram, M.J., G. Farmer, and T.M.L. Wigley, 1981: Past climates and their impacts on man: a review. In: *Climate and History* [Wigley, T.M.L., M.J. Ingram, and G. Farmer (eds.)]. Cambridge University Press, Cambridge, United Kingdom and New York, NY, USA, 530 pp.

IPCC, 1998: *The Regional Impacts of Climate Change: An Assessment of Vulnerability. Special Report of IPCC Working Group II* [Watson, R.T., M.C. Zinyowera, and R.H. Moss (eds.)]. Intergovernmental Panel on Climate Change, Cambridge University Press, Cambridge, United Kingdom and New York, NY, USA, 517 pp.

IPCC, 1997: *Technologies, Policies, and Measures for Mitigating Climate Change: IPCC Technical Paper I. Intergovernmental Panel on Climate Change, Working Group II* [Watson, R.T., M.C. Zinyowera, and R.H. Moss (eds.)]. World Meteorological Organization, Geneva, Switzerland, 84 pp.

IPCC, 1996: *Climate Change 1995: Impacts, Adaptations, and Mitigation of Climate Change: Scientific-Technical Analyses. Contribution of Working Group II to the Second Assessment Report of the Intergovernmental Panel on Climate Change* [Watson, R.T., M.C. Zinyowera, and R.H. Moss (eds.)]. Cambridge University Press, Cambridge, United Kingdom and New York, NY, USA, pp. 1–18.

Jallow, B.P., 1996: Response strategies and adaptive measures to potential sea-level rise in the Gambia. In: *Adapting to Climate Change: An International Perspective* [Smith, J., N. Bhatti, G. Menzhulin, R. Benioff, M.I. Budyko, M. Campos, B. Jallow, and F. Rijsberman (eds.)]. Springer-Verlag, New York, NY, USA, pp. 299–310.

Johda, N.S., 1989: Potential strategies for adapting to greenhouse warming: perspectives from the developing world. In: *Greenhouse Warming: Abatement and Adaptation, Resources for the Future* [Rosenberg, N.J., W.E. Easterling, P.R. Crosson, and J. Darmstadter (eds.)]. Resources for the Future, Washington, DC, USA, pp. 147–158.

Kane, S.M., J. Reilly, and J. Tobey, 1992a: An empirical study of the economic effects of climate change on world agriculture. *Climatic Change*, **21,** 17–35.

Kane, S.M., J. Reilly, and J. Tobey, 1992b: A sensitivity analysis of the implications of climate change for world agriculture. In: *Economic Issues in Global Climate Change* [Reilly, J.M. and M. Anderson (eds.)]. Westview Press, Boulder, CO, USA, pp. 117–131.

Kane, S.M. and J.F. Shogren, 2000: Linking adaptation and mitigation in climate change policy. *Climatic Change*, **45(1),** 75–102.

Karim, Z., 1996: Agricultural vulnerability and poverty alleviation in Bangladesh. In: *Climate Change and World Food Security* [Downing, T.E. (ed.)]. Springer-Verlag, Berlin, Germany, pp. 307–346.

Karl, T.R. and R.W. Knight, 1998: Secular trends of precipitation amount, frequency, and intensity in the United States. *Bulletin of the American Meteorological Society*, **79(2),** 231–241.

Kasperson, J.X., R.E. Kasperson, and B.L. Turner II (eds.), 1995: *Regions at Risk: Comparisons of Threatened Environments*. United Nations University Press, Tokyo, Japan, 588 pp.

Kates, R.W., 2000: Cautionary tales: adaptation and the global poor. *Climatic Change*, **45(1),** 5–17.

Kates, R.W., J.H. Ausubel, and M. Berberian (eds.), 1985: *Climate Impact Assessment: Studies of the Impact of Climate and Society*. John Wiley and Sons, Chichester, United Kingdom, 625 pp.

Katz, R. and B. Brown, 1992: Extreme events in a changing climate: variability is more important than averages. *Climatic Change*, **21,** 289–302.

Kelly, P. and W.N. Adger, 1999: *Assessing Vulnerability to Climate Change and Facilitating Adaptation*. Working Paper GEC 99–07, Centre for Social and Economic Research on the Global Environment, University of East Anglia, Norwich, United Kingdom, 32 pp.

Klein, R.J.T., 1998: Towards better understanding, assessment and funding of climate adaptation. *Change*, **44,** 15–19.

Klein, R.J.T. and R.J. Nicholls, 1998: Coastal zones. In: *UNEP Handbook on Methods for Climate Change Impact Assessment and Adaptation Studies* [Burton, I., J.F. Feenstra, J.B. Smith, and R.S.J. Tol (eds.)]. Version 2.0, United Nations Environment Programme and Institute for Environmental Studies, Vrije Universiteit, Nairobi, Kenya and Amsterdam, The Netherlands, Chapter 7, pp. 1–36. Available online at http://www.vu.nl/english/o_o/instituten/IVM/research/climatechange/Handbook.htm.

Klein, R.J.T. and R.S.J. Tol, 1997: *Adaptation to Climate Change: Options and Technologies, An Overview Paper*. Technical Paper FCCC/TP/1997/3, United Nations Framework Convention on Climate Change Secretariat, Bonn, Germany, 33 pp. Available online at http://www.unfccc.int/resource/docs/tp/tp3.pdf.

Klein, R.J.T., R.J. Nicholls, and N. Mimura, 1999: Coastal adaptation to climate change: Can the IPCC Technical Guidelines be applied? *Mitigation and Adaptation Strategies for Global Change*, **4(3-4),** 239–252.

Klein, R.J.T., M.J. Smit, H. Goosen, and C.H. Hulsbergen, 1998: Resilience and vulnerability: coastal dynamics or Dutch dikes? *The Geographical Journal*, **164(3),** 259–268.

Koster, M.J. and R. Hillen, 1995: Combat erosion by law coastal defence policy for the Netherlands. *Journal of Coastal Research*, **11(4),** 1221–1228.

Krankina, O.N., R.K. Dixon, A.P. Kirilenko, and K.I. Kobak, 1997: Global climate change adaptation: examples from Russian boreal forests. *Climatic Change*, **36,** 197–215.

Kundzewicz, Z. and K. Takeuchi, 1999: Flood protection and management: quo vadimus? *Hydrological Sciences*, **44(3),** 417–432.

Lamb, H.H., 1982: *Climate, History and the Modern World*. Routledge, London, United Kingdom, 387 pp.

Langen, A. and R.S.J. Tol, 1999: A concise history of riverine floods and flood management in the Dutch Rhine delta. In: *Climate, Change and Risk* [Downing, T.E., A.A. Olsthoorn, and R.S.J. Tol (eds.)]. Routledge, London, United Kingdom, pp. 162–172.

Lansigan, F.P., S. Pandey, and M.A.M. Bouman, 1997: Combining crop modeling with economic risk-analysis for the evaluation of crop management. *Journal of Field Crop Research*, **51,** 133–145.

Lawrence, E., 1995: *Henderson's Dictionary of Biological Terms.* Longman Scientific and Technical, Harlow, United Kingdom, 693 pp.

Leafe, R., J. Pethick, and I. Townsend, 1998: Realizing the benefits of shoreline management. *The Geographical Journal,* **164(3),** 282–290.

Leary, N.A., 1999: A framework for benefit-cost analysis of adaptation to climate change and climate variability. *Mitigation and Adaptation Strategies for Global Change,* **4(3–4),** 307–318.

Leemans, R., 1992: Modelling ecological and agricultural impacts of global change on a global scale. *Journal of Scientific and Industrial Research,* **51,** 709–724.

Lempert, R.J., M.E. Schesinger, S.C. Bankes, and N.G. Andronova, 2000: The impacts of climate variability on near-term policy choices and the value of information. *Climatic Change,* **45(1),** 129–161.

Liverman, D.M., 1990: Vulnerability to global environmental change. In: *Environmental Risks and Hazards* [Cutter, S. (ed.)]. Prentice-Hall, Englewood Cliffs, NJ, USA, pp. 326–342.

Louisse, C.J. and F. Van der Meulen, 1991: Future coastal defence in The Netherlands: strategies for protection and sustainable development. *Journal of Coastal Research,* **7(4),** 1027–1041.

Luo, Q. and E. Lin, 1999: Agricultural vulnerability and adaptation in developing countires: the Asia-Pacific region. *Climatic Change,* **43,** 729–743.

MacDonald, G.M., T.W.D. Edwards, K.A. Moser, R. Pienitz, and J.P. Smol, 1993: Rapid response of treeline vegetation and lakes to past climate warming. *Nature,* **361,** 243–246.

Magadza, C.H.D., 2000: Climate change impacts and human settlements in Africa: prospects for adaptation. *Environmental Monitoring and Assesment,* **61(1),** 193–205.

Magadza, C.H.D, 1996: Climate change: some likely multiple impacts in southern Africa. In: *Climate Change and World Food Security* [Downing, T.E. (ed.)]. Springer-Verlag, Berlin, Germany, pp. 449–484.

Magalhães, A.R., 1996: Adapting to climate variations in developing regions: a planning framework. In: *Adapting to Climate Change: An International Perspective* [Smith, J., N. Bhatti, G. Menzhulin, R. Benioff, M.I. Budyko, M. Campos, B. Jallow, and F. Rijsberman (eds.)]. Springer-Verlag, New York, NY, USA, pp. 44–54.

Magalhães, A.R. and M.H. Glantz, 1992: *Socio-Economic Impacts of Climate Variations and Policy Responses in Brazil.* United Nations Environment Program, Esquel, Brasilia, Brazil, 155 pp.

Major, D.D. and K.D. Frederick, 1997: Water resources planning and climate change assessment methods. *Climatic Change,* **37(1),** 25–40.

Major, D.C., 1998: Climate change and water resources: the role of risk management methods. *Water Resources Update,* **112,** 47–50.

Martens, P., R.S. Kovats, S. Nijhof, P. de Vries, M.T.J. Livermore, D.J. Bradley, J. Cox, and A.J. McMichael, 1999: Climate change and future populations at risk of malaria. *Global Environmental Change,* **9,** S89–S107.

McGovern, T.H., 1991: Climate, correlation, and causation in Norse Greenland. *Arctic Anthroplogy,* **28(2),** 77–100.

McGregor, J., 1993: Refugees and the environment. In: *Geography and Refugees: Patterns and Processes of Change* [Black, R. and V. Robinson (eds.)]. Belhaven, Chichester, United Kingdom, 220 pp.

McMichael, A.J. and R.S. Kovats, 2000: Climate change and climate variability: adaptations to reduce adverse health impacts. *Environmental Monitoring and Assessment,* **61,** 49–64.

McMichael, A.J., A. Haines, R. Slooff, and S. Kovats (eds.), 1996: *Climate Change and Human Health.* WHO/EHG/96.7, an assessment prepared by a Task Group on behalf of the World Health Organization, the World Meteorological Organization, and the United Nations Environment Programme, Geneva, Switzerland, 270 pp.

Mearns, L.O., C. Rosenzweig, and R. Goldberg, 1997: Mean and variance change in climate scenarios: methods, agricultural applications, and measures of uncertainty. *Climatic Change,* **34(4),** 367–396.

Mendelsohn, R., 1998: Climate-change damages. In: *Economics and Policy Issues in Climate Change* [Nordhaus, W.D. (ed.)]. Resources for the Future, Washington, DC, USA, pp. 219–236.

Mendelsohn, R. and L.L. Bennett, 1997: Global warming and water management: water allocation and project evaluation. *Climatic Change,* **37(1),** 271–290.

Mendelsohn, R. and A. Dinar, 1999: Climate change, agriculture, and developing countries: does adaptation matter? *The World Bank Research Observer,* **14(2),** 277–293.

Mendelsohn, R. and J. Neumann, 1999: *The Impact of Climate Change on the United States Economy.* Cambridge University Press, Cambridge, United Kingdom and New York, NY, USA, 344 pp.

Mendelsohn, R., W. Nordhaus, and D. Shaw, 1999: The impact of climate variation on U.S. agriculture. In: *The Impact of Climate Change on the United States Economy* [Mendelsohn, R. and J. Neumann (eds.)]. Cambridge University Press, Cambridge, United Kingdom and New York, NY, pp. 55–74.

Mendelsohn, R., W. Nordhaus, and D. Shaw, 1996: Climate impacts on aggregate farm value: accounting for adaptation. *Agricultural and Forest Meteorology,* **80(1),** 55–66.

Mendelsohn, R., W. Nordhaus, and D. Shaw, 1994: The impact of global warming on agriculture: a Ricardian analysis. *The American Economic Review,* **84(4),** 753–771.

Meyer, W.B., K.W. Butzer, T.E. Downing, B.L. Turner II, G.W. Wenzel, and J.L. Wescoat, 1998: Reasoning by analogy. In: *Human Choice and Climate Change, Volume 3: Tools for Analysis* [Rayner, S. and E.L. Malone (eds.)]. Battelle Press, Columbus, OH, USA, pp. 217–289.

Mileti, D.S., 1999: *Disasters by Design: A Reassessment of Natural Hazards in the United States.* Joseph Henry Press, Washington, DC, USA, 351 pp.

Miller, K.A., S.L. Rhodes, and L.J. MacDonnell, 1997: Water allocation in a changing climate: institutions and adaptation. *Climatic Change,* **35,** 157–177.

Mimura, N., 1999a: Vulnerability of island countries in the South Pacific to sea level rise in climate change. In: *National Assessment Results of Climate Change: Impacts and Responses: CR Special 6* [Mimura, N. (ed.)]. Inter-research, Oldendorf, Germany, pp. 137–143.

Mimura, N. (ed.), 1999b: *National Assessment Results of Climate Change: Impacts and Responses: CR Special 6.* Inter-research, Oldendorf, Germany, 230 pp.

Mimura, N. and H. Harasawa, 2000: *Data Book of Sea-Level Rise 2000.* Centre for Global Environmetal Research, National Institute for Enviromental Studies, Environmetal Agency of Japan, Ibaraki, Japan, 280 pp.

Mimura, N. and E. Kawaguchi, 1997: Responses of coastal topography to sea-level rise. In: *Coastal Engineering. 1996 Proceedings of the Twenty-Fifth International Conference, Orlando, FL, USA, September 1996* [Edge, B.L. (ed.)]. American Society of Civil Engineers, New York, NY, USA, pp. 1349–1360.

Mizina, S.V., J. Smith, E. Gossen, K. Spiecker, and S. Witkowski, 1999: An evaluation of adaptation options for climate change impacts on agriculture in Kazakhstan. *Mitigation and Adaptation Strategies for Global Change,* **4,** 25–41.

Moss, R.H., 1998: Water and the challenge of linked environmental changes. *Water Resources Update,* **112,** 6–9.

Munasinghe, M., 2000: Development, equity and sustainabillity (DES) in the context of climate change. In: *Climate Change and Its Linkages with Development, Equity and Sustainability: Proceedings of the IPCC Expert Meeting held in Colombo, Sri Lanka, 27–29 April, 1999* [Munasinghe, M. and R. Swart (eds.)]. LIFE, Colombo, Sri Lanka; RIVM, Bilthoven, The Netherlands; and World Bank, Washington, DC, USA, pp. 13–66.

Munasinghe, M., 1998: Climate change decision-making: science, policy and economics. *International Journal of Environment and Pollution,* **10(2),** 188–239.

Munasinghe, M. and R. Swart (eds.), 2000: *Climate Change and Its Linkages with Development, Equity, and Sustainability. Proceedings of the IPCC Expert Meeting held in Columbo, Sri Lanka, April 1999.* LIFE, Colombo, Sri Lanka; RIVM, Bilthoven, The Netherlands; and World Bank, Washington, DC, USA, 319 pp.

Murdiyarso, D., 2000: Adaptation to climatic variability and change: Asian perspectives on agriculture and food security. *Environmental Monitoring and Assessment,* **61(1),** 123–131.

Mustafa, D., 1998: Structural causes of vulnerability to flood hazard in Pakistan. *Economic Geography,* **74(3),** 289–305.

Newton, J., 1995: An assessment of coping with environmental hazards in northern aboriginal communities. *The Canadian Geographer,* **39(2),** 112–120.

Nguyen, H.T., W.N. Adger, and P.M. Kelly, 1998: Natural resource management in mitigating impacts: the example of mangrove restoration in Vietnam. *Global Environmental Change,* **8(1),** 49–61.

Nicholls, N. and S.P. Leatherman, 1995: The implications of accelerated sea-level rise for developing countries: a discussion. *Journal of Coastal Research*, **14,** 303–323.

O'Brian, K. and D. Liverman, 1996: Climate change and variability in Mexico. In: *Climate Variability, Climate Change and Social Vulnerability in the Semi-Arid Tropics* [Ribot, J.C., A.R. Magalhães, and S.S. Panagides (eds.)]. Cambridge University Press, Cambridge, United Kingdom and New York, NY, USA, pp. 55–70.

Ominde, S.H. and C. Juma, 1991: Stemming the tide: an action agenda. In: *A Change in the Weather: African Perpectives on Climate Change* [Ominde, S.H. and C. Juma (eds.)]. ACTS Press, Nairobi, Kenya, pp. 125–153.

O'Riordan, T. and A. Jordan, 1999: Institutions, climate change and cultural theory: towards a common analytical framework. *Global Environmental Change*, **9,** 81–93.

Olsthoorn, A.A., W.J. Maunder, and R.S.J. Tol, 1996: Tropical cyclones in the southwest Pacific: Impacts on Pacific island countries with particular reference to Fiji. In: *Climate Change and Extreme Events: Altered Risks, Socio-Economic Impacts and Policy Responses* [Downing, T.E., A.A. Olsthoorn, and R.S.J. Tol (eds.)]. Institute for Environmental Management, Vrije Universiteit, Amsterdam, The Netherlands, pp. 185–208.

Osokova, T., N. Gorelkin, and V. Chub, 2000: Water resources of central Asia and adaptation measures for climate change. *Environmental Monitoring and Assessment*, **61(1),** 161–166.

Parrish, R. and D.C. Funnell, 1999: Climate change in mountain regions: some possible consequences in the Moroccan High Atlas. *Global Environmental Change*, **9,** 45–58.

Parry, M.L., 1986: Some implications of climatic change for human development. In: *Sustainable Development of the Biosphere* [Clark, W.C. and R.E. Munn (eds.)]. International Institute for Applied Systems Analysis, Laxenburg, Austria, pp. 378–406.

Parry, M., C. Rosenzweig, A. Iglesias, G. Fischer, and M. Livermore, 1999: Climate change and world food security: a new assessment. *Global Environmental Change*, **9,** S51–S68.

Parry, M., N. Arnell, M. Hulme, R. Nicholls, and M. Livermore, 1998: Adapting to the inevitable. *Nature*, **395,** 741–742.

Patz, J.A., 1996: Health adaptation to climate change: need for far-sighted, integrated approaches. In: *Adapting to Climate Change: An International Perspective* [Smith, J., N. Bhatti, G. Menzhulin, R. Benioff, M.I. Budyko, M. Campos, B. Jallow, and F. Rijsberman (eds.)]. Springer-Verlag, New York, NY, USA, pp. 450–464.

Pelling, M., 1998: Participation, social capital and vulnerability to urban flooding in Guyana. *Journal of International Development*, **10,** 469–486.

Pelling, M., 1997: What determines vulnerability to floods; a case study in Georgetown, Guyana. *Environment and Urbanization*, **9(1),** 203–226.

Perez, R.T., L.A. Amadore, and R.B. Feir, 1999: Climate change impacts and responses in the Philippines coastal sector. In: *National Assessment Results of Climate Change: Impacts and Responses: CR Special 6* [Mimura, N. (ed.)]. Inter-research, Oldendorf, Germany, pp. 97–107.

Pielke, R.A., 1998: Rethinking the role of adaptation in climate policy. *Global Environmental Change*, **8(2),** 159–170.

Pittock, B., 1999: The question of significance. *Nature*, **397,** 657–658.

Pittock, B. and R.N. Jones, 2000: Adaptation to what and why? *Environmental Monitoring and Assessment*, **61(1),** 9–35.

Pittock, B., R.J. Allan, K.J. Hennessy, K.L. McInnes, R. Suppia, K.J. Walsh, P.H. Whetton, H. McMaster, and R. Taplin, 1999: Climate change, climatic hazards and policy responses in Australia. In: *Climate, Change and Risk* [Downing, T.E., A.A. Olsthoorn, and R.S.J. Tol (eds.)]. Routledge, London, United Kingdom, pp. 19–59.

Primo, L.H., 1996: Anticipated effects of climate change on commercial pelagic and artisanal coastal fisheries in the Federated States of Micronesia. In: *Adapting to Climate Change: An International Perspective* [Smith, J.B. (ed.).]. Springer-Verlag, New York, NY, USA, pp. 427–436.

Ramakrishnan, P.S., 1999: Lessons from the Earth Summit: protecting and managing biodiversity in the Tropics. In: *Global Environmental Economics* [Dore, M.H.I. and T.D. Mount (eds.)]. Blackwell Publishers, Oxford, United Kingdom, pp. 240–264.

Ramakrishnan, P.S., 1998: Sustainable development, climate change, and the tropical rain forest landscape. *Climatic Change*, **39,** 583–600.

Raper, S.C.B., T.M.L. Wigley, and R.A. Warrick, 1996: Global sea level rise: past and future. In: *Sea Level Rise and Coastal Subsidence: Causes, Consequences and Strategies* [Milliman, J.D. and B.U. Haq (eds.)]. J. Kluwer Academic Publishers, Dordrecht, The Netherlands, pp. 11–45.

Rayner, S. and E.L. Malone, 1999: Climate change, poverty and intragenerational equity: the national level. In: *Climate Change and Its Linkages with Development, Equity and Sustainability: Proceedings of the IPCC Expert Meeting held in Colombo, Sri Lanka, 27–29 April, 1999* [Munasinghe, M. and R. Swart (eds.)]. LIFE, Colombo, Sri Lanka; RIVM, Bilthoven, The Netherlands; and World Bank, Washington, DC, USA, pp. 215–242.

Rayner, S. and E.L. Malone (eds.), 1998: *Human Choice and Climate Change Volume 3: The Tools for Policy Analysis*. Battelle Press, Columbus, OH, USA, 429 pp.

Reilly, J., 1999: What does climate change mean for agriculture in developing countries? A comment on Mendelsohn and Dinar. *The World Bank Research Observer*, **14(2),** 295–305.

Reilly, J., 1998: Comments: climate-change damages. In: *Economics and Policy Issues in Climate Change* [Nordhaus, W.D. (ed.)]. Resources for the Future, Washington, DC, USA, pp. 243–256.

Reilly, J., 1995: Climate change and global agriculture: recent findings and issues. *American Journal of Agricultural Economics*, **77,** 727–733.

Reilly, J. and D. Schimmelpfennig, 2000: Irreversibility, uncertainty, and learning: portraits of adaptation to long-term climate change. *Climatic Change*, **45(1),** 253–278.

Reilly, J. and D. Schimmelpfennig, 1999: Agricultural impact assessment, vulnerability, and the scope for adaptation. *Climatic Change*, **43,** 745–788.

Reilly, J., N. Hohmann, and S. Kane, 1994: Climate change and agricultural trade: who benefits, who loses? *Global Environmental Change*, **4(1),** 24–36.

Ribot, J.C., A. Najam, and G. Watson, 1996: Climate variation, vulnerability and sustainable development in the semi-arid tropics. In: *Climate Variability, Climate Change and Social Vulnerability in the Semi-Arid Tropics* [Ribot, J.C., A.R. Magalhães, and S.S. Panagides (eds.)]. Cambridge University Press, Cambridge, United Kingdom and New York, NY, USA, pp. 13–54.

Rijsberman, F.R. and A. van Velzen, 1996: Vulnerability and adaptation assessment of climate change and sea-level rise in the coastal zone: perspective from the Netherlands and Bangladesh. In: *Adapting to Climate Change: An International Perspective* [Smith, J., N. Bhatti, G. Menzhulin, R. Benioff, M.I. Budyko, M. Campos, B. Jallow, and F. Rijsberman (eds.)]. Springer-Verlag, New York, NY, USA, pp. 322–334.

Risbey, J., M. Kandlikar, H. Dowlatabadi, and D. Graetz, 1999: Scale, context, and decision making in agricultural adaptation to climate variability and change. *Mitigation and Adaptation Strategies for Global Change*, **4,** 137–165.

Robinson, J.B. and D. Herbert, 2000: Integrating climate change and sustainable development. In: *Climate Change and Its Linkages with Development, Equity and Sustainability: Proceedings of the IPCC Expert Meeting held in Colombo, Sri Lanka, 27–29 April, 1999* [Munasinghe, M. and R. Swart (eds.)]. LIFE, Colombo, Sri Lanka; RIVM, Bilthoven, The Netherlands; and World Bank, Washington, DC, USA, pp. 143–162.

Robock, A., R.P. Turco, M.A. Harwell, T.P. Ackerman, R. Andressen, H.-S. Chang, and M.V.K. Sivakumar, 1993: Use of general circulation model output in the creation of climate change scenarios for impact analysis. *Climatic Change*, **23,** 293–355.

Rosenberg, N.J., 1992: Adaptation of agriculture to climate change. *Climatic Change*, **21,** 385–405.

Rosenthal, D.H., H.G. Gruenspecht, and E.A. Moran, 1995: Effects of global warming on energy use for space heating and cooling in the United States. *Energy Journal*, **16(2),** 77–96.

Rosenzweig, C., 1994: Maize suffers a sea change. *Nature*, **370,** 175–176.

Rosenzweig, C. and M.L. Parry, 1994: Potential impacts of climate change on world food supply. *Nature*, **367,** 133–139.

Rothman, D.S., D. Demeritt, Q. Chiotti, and I. Burton, 1998: Costing climate change: the economics of adaptations and residual impacts for Canada. *Canada Country Study: Climate Impacts and Adaptation*, **8,** 1–29.

Rotmans, J., M. Hulme, and T.E. Downing, 1994: Climate change implications for Europe: an application of the ESCAPE model. *Global Environmental Change*, **4(2),** 97–124.

Schelling, T.C., 1992: Some economics of global warming. *The American Economic Review*, **82(1),** 1–14.

Scheraga, J. and A. Grambsch, 1998: Risks, opportunities, and adaptation to climate change. *Climate Research*, **10**, 85–95.

Schimmelpfennig, D., J. Lewandrowski, J.Reilly, M. Stigas, and I. Parry, 1996: *Agricultural Adaptation to Climate Change: Issues of Long Run Sustainability*. Agricultural Economic Report Number 740, U.S. Department of Agriculture, Washington, DC, USA, 53 pp.

Schneider, S.H., 1997: Integrated assessment modeling of global climate change: transparent rational tool for policymaking or opaque screen hiding value-laden assumptions? *Environmental Modeling and Assessment*, **2(4),** 229–248.

Schneider, S.H., W.E. Easterling, and L.O. Mearns, 2000: Adaptation: sensitivity to natural variability, agent assumptions and dynamic climate changes. *Climatic Change*, **45(1),** 203–221.

Smit, B. (ed.), 1993: *Adaptation to Climatic Variability and Change: Report of the Task Force on Climate Adaptation*. Guelph, Environment Canada, 53 pp.

Smit, B. and Y. Cai, 1996: Climate change and agriculture in China. *Global Environmental Change*, **6(3),** 205–214.

Smit, B., R. Blain, and P. Keddie, 1997: Corn hybrid selection and climatic variability: gambling with nature? *The Canadian Geographer*, **42(1),** 429–438.

Smit, B., D. McNabb, and J. Smithers, 1996: Agricultural adaptation to climate change. *Climatic Change*, **33,** 7–29.

Smit, B., I. Burton, R.J.T. Klein, and R. Street, 1999: The science of adaptation: a framework for assessment. *Mitigation and Adaptation Strategies for Global Change*, **4,** 199–213.

Smit, B., I. Burton, R.J.T. Klein, and J. Wandel, 2000: An anatomy of adaptation to climate change and variability. *Climatic Change*, **45,** 223–251.

Smith, J.B., 1997: Setting priorities for adaptation to climate change. *Global Environmental Change*, **7(3),** 251–264.

Smith, J.B. and S.S. Lenhart, 1996: Climate change adaptation policy options. *Climate Research*, **6(2),** 193–201.

Smith, J.B., R.S.J. Tol, S. Ragland, and S. Fankhauser, 1998: *Proactive Adaptations to Climate Change: Three Case Studies on Infrastructure Investments*. IVM Discussion Paper D98/03, Institute for Environmental Studies, Vrije Universiteit, Amsterdam, The Netherlands, 14 pp.

Smith, J.B., N. Bhatti, G. Menzhulin, R. Benioff, M.I. Budyko, M. Campos, B. Jallow, and F. Rijsberman (eds.), 1996: *Adapting to Climate Change: An International Perspective*. Springer-Verlag, New York, NY, USA, 475 pp.

Smith, K., 1996: *Environmental Hazards: Assessing Risk and Reducing Disaster*. Routledge, London, United Kingdom, 389 pp.

Smith, K.R., 1990: Risk transition and global warming. *Journal of Energy Engineering*, **116(3),** 178–188.

Smithers, J. and B. Smit, 1997: Human adaptation to climatic variability and change. *Global Environmental Change*, **7(2),** 129–146.

Sohngen, B. and R. Mendelsohn, 1998: Valuing the impact of large-scale ecological change in a market: the effect of climate change on U.S. timber. *The American Economic Review*, **88(4),** 686–710.

Sprengers, S.A., L.K. Slager, and H. Aiking, 1994: *Biodiversity and Climate Change Part 1: Establishment of Ecological Goals for the Climate Convention*. Vrije Universiteit, Amsterdam, The Netherlands, 172 pp.

Stakhiv, E.Z., 1996: Managing water resources for climate change adaptation. In: *Adapting to Climate Change: An International Perspective* [Smith, J., N. Bhatti, G. Menzhulin, R. Benioff, M.I. Budyko, M. Campos, B. Jallow, and F. Rijsberman (eds.)]. Springer-Verlag, New York, NY, USA, pp. 243–264.

Stakhiv, E.Z., 1994: Managing water resources for adaptation to climate change. *Engineering Risk in Natural Resources Management*, **275,** 379–393.

Strzepek, K.M. and J.B. Smith (eds.), 1995: *As Climate Changes, International Impacts and Implications*. Cambridge University Press, Cambridge, United Kingdom and New York, NY, USA, 213 pp.

Stuczyinski, T., G. Demidowicz, T. Deputat, T. Górski, S. Krazowicz, and J. Kus, 2000: Adaptation scenarios of agriculture in Poland to future climate changes. *Environmental Monitoring and Assessment*, **61(1),** 133–144.

Teves, N., G. Laos, S. Carrasco, C. San Roman, L. Pizarro, G. Cardenas, and A. Romero, 1996: Sea-level rise along the Lima Coastal Zone, Peru, as a result of global warming: environmental impacts and mitigation measures. In: *Adapting to Climate Change: An International Perspective* [Smith, J., N. Bhatti, G. Menzhulin, R. Benioff, M.I. Budyko, M. Campos, B. Jallow, and F. Rijsberman (eds.)]. Springer-Verlag, New York, NY, USA, pp. 283–298.

Theu, J., G. Chavula, and C. Elias Malawi, 1996: How climate change adaptation options fit within the UNFCCC national communication and national development plans. In: *Adapting to Climate Change: An International Perspective* [Smith, J., N. Bhatti, G. Menzhulin, R. Benioff, M.I. Budyko, M. Campos, B. Jallow, and F. Rijsberman (eds.)]. Springer-Verlag, New York, NY, USA, pp. 97–104.

Tiglao, R., 1992: Grinding to a halt. *Far East Economic Review*, **16,** 50.

Titus, J.G., 1990: Strategies for adapting to the greenhouse effect. *APA Journal*, **56(3),** 311–323.

Tol, R.S.J., 1998a: Comment: climate-change damages. In: *Economics and Policy Issues in Climate Change* [Nordhaus, W.D. (ed.)]. Resources for the Future, Washington, DC, USA, pp. 237–242.

Tol, R.S.J., 1998b: Short-term decisions under long-term uncertainty. *Energy Economics*, **20,** 557–569.

Tol, R.S.J. and S. Fankhauser, 1997: On the representation of impact in integrated assessment models of climate change. *Environmental Modeling and Assessment*, **3,** 63–74.

Tol, R.S.J., S. Fankhauser, and J.B. Smith, 1998: The scope for adaptation to climate change: what can we learn from the impact literature? *Global Environmental Change*, **8(2),** 109–123.

Tol, R.S.J., H.M.A. Jansen, R.J.T. Klein, and H. Verbruggen, 1996: Some economic considerations on the importance of proactive integrated coastal zone management. *Ocean and Coastal Management*, **32(1),** 39–55.

Tol, R.S.J., N.M. van der Grijp, A.A. Olsthoorn, and P.E. van der Werff, 1999: *Adapting to Climate Change: A Case Study on Riverine Flood Risks in the Netherlands*. Societal and Institutional Response to Climate Change and Climate Hazards (SIRCH) Working Paper 5, Institute for Environmental Studies, Vrije Universiteit, Amsterdam, The Netherlands, 9 pp.

Toman, M. and R. Bierbaum, 1996: An overview of adaptation to climate change. In: *Adapting to Climate Change: An International Perspective* [Smith, J., N. Bhatti, G. Menzhulin, R. Benioff, M.I. Budyko, M. Campos, B. Jallow, and F. Rijsberman (eds.)]. Springer-Verlag, New York, NY, USA, pp. 5–15.

Torry, W.I., 1979: Anthropological studies in hazardous environments: past trends and new horizons. *Current Anthropology*, **20,** 517–540.

Torvanger, A., 1998: Burden sharing and adaptation beyond Kyoto: a more systematic approach essential for global climate success. *Environment and Development Economics*, **3(406),** 406–409.

Toth, F., 1999: Development, equity and sustainability concerns in climate change decisions. In: *Climate Change and Its Linkages with Development, Equity and Sustainability: Proceedings of the IPCC Expert Meeting held in Colombo, Sri Lanka, 27–29 April, 1999* [Munasinghe, M. and R. Swart (eds.)]. LIFE, Colombo, Sri Lanka; RIVM, Bilthoven, The Netherlands; and World Bank, Washington, DC, USA, pp. 263–288.

Uitto, J., 1998: The geography of disaster vulnerability in megacities: a theoretical framework. *Applied Geography*, **18,** 7–16.

UNEP, 1998: *Handbook on Methods for Climate Impact Assessment and Adaptation Strategies, 2* [Feenstra, J., I. Burton, J. Smith, and R. Tol (eds.)]. United Nations Environment Program, Institute for Environmental Studies, Amsterdam, The Netherlands, 359 pp.

USNAS, 1992: *Policy Implications of Greenhouse Warming: Mitigation, Adaptation, and the Science Base*. U.S. National Academies of Science and Engineering, Institute of Medicine, National Academy Press, Washington, DC, 944 pp.

Wang'ati, F.J., 1996: The impact of climate variation and sustainable development in the Sudano-Sahelian region. In: *Climate Variability, Climate Change and Social Vulnerability in the Semi-Arid Tropics* [Ribot, J.C., A.R. Magalhães, and S.S. Panagides (eds.)]. Cambridge University Press, Cambridge, United Kingdom and New York, NY, USA, pp. 71–91.

Warrick, R.A. and W.E. Riebsame, 1983: Societal response to CO_2-induced climate change: opportunities. In: *Societal Research and Climate Change: An Interdisciplinary Appraisal* [Chen, R.S., E. Boulding, and S.H. Schneider (eds.)]. D. Reidel, Dordrecht, The Netherlands, pp. 20–60.

Warrick, R.A, R.M. Gifford, and M.L. Parry, 1986: CO_2 , climatic change and agriculture. In: *The Greenhouse Effect, Climatic Change and Ecosystems* [Bolin, B., B.R. Döös, J. Jager, and R.A. Warrick (eds.)]. John Wiley and Sons, New York, NY, USA, pp. 393–473.

Watts, M., 1983: On the poverty theory: natural hazards research in context. In: *Interpretations of Calamity from the Viewpoint of Human Ecology* [Hewitt, K. (ed.)]. Allen & Unwin, Boston, MA, USA, pp. 231–260.

West, J.J. and H. Dowlatabadi, 1999: On assessing the economic impacts of sea-level rise on developed coasts. In: *Climate, Change and Risk* [Downing, T.E., A.A. Olsthoorn, and R.S.J. Tol (eds.)]. Routledge, London, United Kingdom, pp. 205–220.

White, R. and D. Etkin, 1997: Climate change, extreme events and the Canadian insurance industry. *Natural Hazards*, **16,** 135–163.

WHO, 1997: *Division of Control of Tropical Diseases (CTD) Progress Report 1996*. CTD/PR/97.1, World Health Organization, Geneva, Switzerland, 147 pp.

Wigley, T.M.L., 1999: *The Science of Climate Change: Global and U.S. Perspectives*. Pew Center for Climate Change, Washington, DC, USA, 48 pp.

Wigley, T., M. Ingram, and G. Farmer, 1981: Past climates and their impact on man: a review. In: *Climate and History* [Ingram, M., G. Farmer, and T. Wigley (eds.)]. Cambridge University Press, Cambridge, United Kingdom and New York, NY, USA, pp. 3–50.

Wilhite, D.A., 1997: Responding to drought: common threads from the past, visions for the future. *Journal of the American Water Resources Association*, **33(5),** 951–959.

Wilhite, D.A., 1996: A methodology for drought preparedness. *Natural Hazards*, **13,** 229–252.

Wilhite, D.A., N.J. Rosenberg, and M.H. Glantz, 1986: Improving federal response to drought. *Journal of Climate and Applied Meteorology*, **25,** 332–342.

Wisner, B., 1998: Marginality and vulnerability: why the homeless in Tokyo don't count in disaster preparations. *Applied Geography*, **18(1),** 25–33.

Yang, H., 1996: Potential effects of sea-level rise in the Pearl River delta area: preliminary study results and a comprehensive adaptation strategy. In: *Adapting to Climate Change: An International Perspective* [Smith, J., N. Bhatti, G. Menzhulin, R. Benioff, M.I. Budyko, M. Campos, B. Jallow, and F. Rijsberman (eds.)]. Springer-Verlag, New York, NY, USA, 475 pp.

Yim, W.W.S., 1996: Vulnerability and adaptation of Hong Kong to hazards under climatic change conditions. *Water, Air, and Soil Pollution*, **92,** 181–190.

Yohe, G. and H. Dowlatabadi, 1999. Risk and uncertainties, analysis and evaluation: lessons for adaptation and integration. *Mitigation and Adaptation Strategies for Global Change*, **4(3-4),** 319–330.

Yohe, G.W. and J. Neumann, 1997: Planning for sea-level rise and shore protection under climate uncertainty. *Climatic Change*, **37(1),** 243–270.

Yohe, G.W. and M.E. Schlesinger, 1998: Sea level change: the expected economic cost of protection or abandonment in the United States. *Climatic Change*, **38,** 447–472.

Yohe, G., J. Neumann, P. Marshall, and A. Ameden, 1996: The economic costs of sea-level rise on developed property in the United States. *Climate Change*, **32,** 387–410.

Zhao, Z.C., 1996: Climate change and sustainable development in China's semi-arid regions. In: *Climate Variability, Climate Change and Social Vulnerability in the Semi-arid Tropics* [Ribot, J.C., A.R. Magalhães, and S.S. Panagides (eds.)]. Cambridge University Press, Cambridge, United Kingdom and New York, NY, USA, pp. 92–108.

19

Vulnerability to Climate Change and Reasons for Concern: A Synthesis

JOEL B. SMITH (USA), HANS-JOACHIM SCHELLNHUBER (GERMANY), AND M. MONIRUL QADER MIRZA (BANGLADESH)

Lead Authors:

S. Fankhauser (Switzerland), R. Leemans (The Netherlands), Lin Erda (China), L. Ogallo (Kenya), B. Pittock (Australia), R. Richels (USA), C. Rosenzweig (USA), U. Safriel (Israel), R.S.J. Tol (The Netherlands), J. Weyant (USA), G. Yohe (USA)

Contributing Authors:

W. Bond (South Africa), T. Bruckner (Germany), A. Iglesias (Spain), A.J. McMichael (UK), C. Parmesan (USA), J. Price (USA), S. Rahmstorf (Germany), T. Root (USA), T. Wigley (USA), K. Zickfeld (Germany)

Review Editors:

C. Hope (United Kingdom) and S.K. Sinha (India)

CONTENTS

EXECUTIVE SUMMARY

This chapter synthesizes the results of Work Group II of the Third Assessment Report (TAR) and assesses the state of knowledge concerning Article 2 of the United Nations Framework Convention on Climate Change (UNFCCC). The TAR's task is to define what is known about the effects of climate change: how sensitive systems are, what adaptive capacity they have, and what their vulnerability is. It is not the goal of this assessment to determine whether these effects are tolerable or are considered dangerous.

The goal of this chapter is to synthesize information on climate change impacts in a manner that will enable readers to evaluate the relationship between increases in global mean temperature and impacts. The chapter focuses on certain "reasons for concern" that may aid readers in making their own determination about what is a "dangerous" climate change. Each reason for concern is consistent with a paradigm that can be used by itself or in combination with other paradigms to help determine what level of climate change is dangerous. The reasons for concern are:

1) The relationship between global mean temperature increase and damage to or irreparable loss of unique and threatened systems
2) The relationship between global mean temperature increase and the distribution of impacts
3) The relationship between global mean temperature increase and global aggregate damages
4) The relationship between global mean temperature increase and the probability of extreme weather events
5) The relationship between global mean temperature increase and the probability of large-scale singular events such as the breakup of the West Antarctic Ice Sheet or the collapse of the North Atlantic thermohaline circulation.

In addition, we examine what observed effects of climate change tell us with regard to Article 2 of the UNFCCC. Increase in global mean temperature since 1900 (i.e., mean global warming) is used as the common metric against which impacts are measured. This metric is closely related to greenhouse gas (GHG) concentrations but is more relevant for impact assessments.

Some general caveats apply to all of the reasons for concern:

- In spite of many studies on climate change impacts, there still is substantial uncertainty about how effective adaptation will be (and could be) in ameliorating negative effects of climate change and taking advantage of positive effects.
- The effect of changes in baseline conditions, such as population and economic growth and development of new technologies that could change vulnerability, has not been adequately considered in most impact studies.
- Most impact studies assess the effects of a stable climate, so our understanding of what rates of change may be dangerous is limited.

It does not appear to be possible to combine the different reasons for concern into a unified reason for concern that has meaning and is credible. However, we can review the relationship between impacts and temperature for each reason for concern and draw some preliminary conclusions about the potential severity and risk of impacts for the individual reasons for concern. Note that the following findings do not incorporate the costs of limiting GHG emissions to levels that are sufficient to avoid changes that may be considered dangerous. Also note that there is substantial uncertainty regarding the impacts of climate change at the temperatures mentioned. These temperatures should be taken as approximate indications of impacts, not as absolute thresholds. In addition, change in global mean temperature does not describe all relevant aspects of climate change impacts, such as rate and pattern of change and changes in precipitation, extreme climate events, or lagged (or latent) effects such as rising sea levels. For simplification, we group different levels of temperature increase into "small," "medium," and "large." "Small" denotes a global mean temperature increase of as much as approximately 2°C; "medium" denotes a global mean temperature increase of approximately 2–3°C; and "large" denotes a global mean temperature increase of more than approximately 3°C.

Based on a review of the literature of observations of climate change impacts, as reflected in other chapters in the TAR, we conclude the following:

- *Observations:* Statistically significant associations between trends in regional climate and impacts have been documented in ~100 physical processes and ~450 biological species or communities in terrestrial and polar environments. Although the presence of multiple factors (e.g., land-use change, pollution, biotic invasion) makes attribution of observed impacts to regional climate change difficult, more than 90% (~99% physical, ~80% biophysical) of the changes documented worldwide are consistent with how physical and biological processes are known to respond to climate. Based on expert judgment, we have high confidence that the overall patterns and processes of observations reveal a widespread and coherent impact of 20th-century climate changes on many physical and biological systems. Signals of regional climate change impacts may be clearer in physical and biological systems than in socioeconomic systems, which also are simultaneously undergoing many complex changes that are not related to climate change, such as population growth and urbanization. Socioeconomic systems have complex and varying mechanisms for adapting to climate change. There are preliminary indications that some social and economic systems have been affected in part by 20th-century regional climate changes (e.g., increased damages from flooding and droughts in some locations). It generally is difficult to separate climate change effects from coincident or alternative explanations for such observed regional impacts.

- *Unique and Threatened Systems:* Tropical glaciers, coral reefs, mangroves, ecotones, and biodiversity "hot spots" are examples of unique and threatened entities that are confined to narrow geographical ranges and are very sensitive to climate change. However, their degradation or loss could affect regions outside their range. There is medium confidence that several of these systems will be affected by a small temperature increase; for example, coral reefs will bleach and glaciers will recede. At higher magnitudes of temperature increase, other and more numerous unique and threatened systems would be adversely affected.
- *Distribution of Impacts:* The impacts of climate change will not be evenly distributed among the peoples of the world. There is high confidence that developing countries will be more vulnerable to climate change than developed countries, and there is medium confidence that climate change would exacerbate income inequalities between and within countries. There also is medium confidence that a small temperature increase would have net negative impacts on market sectors in many developing countries and net positive impacts on market sectors in many developed countries. However, there is high confidence that with medium to high increases in temperature, net positive impacts would start to decline and eventually would turn negative, and negative impacts would be exacerbated. Estimates of distributional effects are uncertain because of aggregation and comparison methods, assumptions about climate variability, adaptation, levels of development, and other factors.
- *Aggregate Impacts:* With a small temperature increase, there is medium confidence that aggregate market sector impacts would amount to plus or minus a few percent of world gross domestic product (GDP), and there is low confidence that aggregate nonmarket impacts would be negative. Most people in the world would be negatively affected by a small to medium temperature increase. Most studies of aggregate impacts find that there are net damages at the global scale beyond a medium temperature increase and that damages increase from there with further temperature increases. The important qualifications raised with regard to distributional analysis (previous bullet item) also apply to aggregate analysis. By its nature, aggregate analysis masks potentially serious equity differences. Estimates of aggregate impacts are controversial because they treat gains for some as canceling out losses for others and because the weights that are used to aggregate over individuals are necessarily subjective.
- *Extreme Climate Effects:* The frequency and magnitude of many extreme climate events increase even with a small temperature increase and will become greater at higher temperatures (high confidence). Extreme events include, for example, floods, soil moisture deficits, tropical and other storms, anomalous temperatures, and fires. The impacts of extreme events often are large locally and could strongly affect specific sectors and regions. Increases in extreme events can cause critical design or natural thresholds to be exceeded, beyond which the magnitude of impacts increases rapidly (high confidence).
- *Large-Scale Singularities:* Large-scale singularities in the response of the climate system to external forcing, such as shutdown of the North Atlantic thermohaline circulation or collapse of the West Antarctic ice sheet, have occurred in the past as a result of complex forcings. Similar events in the future could have substantial impacts on natural and socioeconomic systems, but the implications have not been well studied. Determining the timing and probability of occurrence of large-scale singularities is difficult because these events are triggered by complex interactions between components of the climate system. The actual impact could lag the climate change cause (involving the magnitude and the rate of climate change) by decades to millenia. There is low to medium confidence that rapid and large temperature increases would exceed thresholds that would lead to large-scale singularities in the climate system.

19.1. Introduction

This chapter draws on the results of the entire TAR to assess the state of knowledge concerning Article 2 of the UNFCCC. Article 2 of the UNFCCC states that:

> "...the ultimate objective of this Convention...is to achieve...stabilization of greenhouse gas concentrations in the atmosphere at a level that would prevent dangerous anthropogenic interference with the climate system. Such a level should be achieved with a time frame sufficient to allow ecosystems to adapt naturally to climate change, to ensure that food production is not threatened and to enable economic development to proceed in a sustainable manner." (UNEP/WMO, 1992).

The ultimate goal for stabilizing GHG concentrations is to avoid "dangerous anthropogenic interference with the climate system." The question of what is dangerous is one that the authors of this chapter cannot answer. Danger is a function of the degree to which effects are negative and the degree to which those effects are unacceptable. The latter is a value judgment. The TAR's task is to define what is known about the effects of climate change—to identify their character and their implications and whether they are negative or positive. It is not about determining whether these effects are acceptable.

The preceding chapters review the literature about vulnerability to climate change in regions and sectors. The goal of this chapter is to draw on very disparate reasons for concern regarding climate change impacts in a manner that will enable readers to evaluate the relationship between increases in global mean temperature and impacts (for an explanation of why change in global mean temperature is used as an indicator, see Section 19.1.2). It attempts to enable readers to understand the risks of higher magnitudes of increased global mean temperature.

19.1.1. Reasons for Concern

To provide information to readers in a manner that will enable them to make judgments about what level of climate change may be dangerous, this chapter addresses "reasons for concern," which represent a way for readers to think about the seriousness of climate change impacts. These reasons for concern are taken from debates and literature about the risks of climate change. The authors of this chapter make no judgment regarding whether one or several reasons for concern are more important than others. Nor do we attempt to combine the reasons for concern to produce a single "bottom line."

The reasons for concern are as follows:

1) *The relationship between global mean temperature increase and damage to or irreparable loss of unique and threatened systems:* Some unique and threatened systems may be irreparably harmed by changes in climate beyond certain thresholds.
2) *The relationship between global mean temperature increase and the distribution of impacts:* Some regions, countries, islands, and cultures may be adversely affected by climate change, whereas others could benefit, at least up to a point. For example, in some sectors, adverse effects may be experienced in some parts of the world while other parts may have net gains. Within countries, some regions or groups of people could be harmed while others benefit or experience less harm.
3) *The relationship between global mean temperature increase and global aggregated impacts:* Using a consistent method of measurement and aggregation of climate change impacts, we address how aggregate impacts change as global mean temperature increases, whether aggregate impacts are positive at some levels of temperature increase and negative at others, whether change will occur smoothly or in a more complex dynamic pattern, and whether aggregate impacts mask unequal distribution of impacts.
4) *The relationship between global mean temperature increase and the probability of extreme weather events:* As mean climate changes, so too will the probability of extreme weather events such as days with very high or very low temperatures, extreme floods, droughts, tropical cyclones, and storms. This chapter addresses how the probability and consequences of such events may change as global mean temperature increases.
5) *The relationship between global mean temperature increase and the probability of large-scale singular events, such as collapse of the West Antarctic ice sheet (WAIS) or shutdown of the North Atlantic thermohaline circulation (THC):* This chapter addresses what is known about how the probabilities of such events change as the magnitude of climate change increases.

In addition, this chapter addresses whether changes in climate during the 20th century have resulted in observed impacts. The IPCC has documented these changes, and an important question is whether these changes have resulted in measurable impacts on nature or society. Important questions include the following:

- Are the observed effects of climate change consistent with model predictions, particularly those that estimate more serious impacts at larger GHG concentrations?
- Even if it is not clear whether observed effects are caused primarily by climate change, do these effects give us information about the potential vulnerability of systems to climate change?

Observations are not a reason for concern. Instead, they help us determine whether impacts that are relevant to any of the five reasons for concern have occurred.

19.1.2. Choice of Indicator

A critical issue is the indicator of climate change against which we measure impacts. A common measure allows consistent discussion about the relationship between climate change and impacts. Several indicators could be used:

1) GHG emission levels
2) Atmospheric GHG concentration levels
3) Changes in global mean temperature and sea-level rise
4) Changes in regional climate variables
5) Changes in the intensity or frequency of extreme events.

Several considerations must be taken into account in selecting an indicator. Using GHG emission levels (1) or even concentration levels (2) implies examining impacts beyond the 21st century. Published estimates of time frames for stabilizing GHG atmospheric concentration levels tend to assume such levels will not be stabilized until after the end of the 21st century (Enting *et al*., 1994; Wigley *et al*., 1996; Schimel *et al*., 1997).

The problem with using such levels as an indicator is that most of the impact literature examines potential impacts only as far as 2100. In addition, most studies are based on scenarios of specific changes in global mean or, more typically, regional climate variables such as temperature or precipitation.[1] It is difficult to relate a specific level of GHG concentration to a specific change in global average climate or regional climate. For each GHG concentration level, there is a range of potential changes in global mean temperature (see Box 19-1). And for each change in global mean temperature, there is a range of potential changes in average regional temperature, precipitation, and extreme events.

The problem with indicators 3, 4, and 5 is the inverse of the foregoing problem. For each change in global or regional climate or extreme events, there is a range of levels of GHG concentrations that could cause such a change in climate. Thus, using these indicators makes it more difficult to work back to defining atmospheric concentrations of GHGs, as required by Article 2 of the UNFCCC. In addition, as one gets to finer levels of spatial and temporal resolution, such as changes in regional climates and extreme events, it becomes more difficult to attribute such changes to changes in GHG concentrations.

Thus, whatever the indicator selected, there will be problems in using it to relate impacts to the level of GHG concentrations. The choice of indicator depends on two factors:

1) What does the literature on climate change impacts allow us to consider?
2) What indicator can be most directly related to GHG concentrations?

We selected change in global mean temperature as our indicator for two reasons. The first is that the impact literature can be directly related to a change in global mean temperature. Many studies are based on specific results from general circulation models (GCMs), which estimate a change in global mean temperature. Other studies can be related to a change in global mean temperature by inversely using the scaling method from Chapter 4. The second reason is that, as discussed in Box 19-1, it is most feasible to relate changes in global mean temperature to GHG concentrations. It is harder to relate the other indicators directly to GHG concentrations. Thus, global mean temperature increase is the indicator that can be used most readily to relate GHG emissions (and emissions control) to changes in climate and impacts.

For any change in global mean temperature, there are many possible changes in regional climate and climate variability, which could have quite different results. Thus, a 2°C increase in global mean temperature may result in a particular region being much wetter or drier or having more or fewer extreme climate events. Whether the region gets wetter or drier or has more severe climate is likely to have much greater bearing on impacts than a change in mean temperature. Hence, although the use of global mean temperature as an indicator is preferable to the other options because it has fewer problems in implementation, it has its own limitations.

This chapter does not address the effect of different rates of change in climate on vulnerability. There is no doubt that a 3°C increase in global mean temperature realized in 50 years could be far worse than the same amount of warming realized in 100 or 200 years. In addition, changes in extreme events such as more intense El Niño-Southern Oscillation (ENSO) events (see, e.g., Timmermann *et al*., 1999) could lead to more adverse impacts than a monotonic and gradual change in climate. Thus, rate of change is an important factor affecting what climate change is considered to be dangerous. Unfortunately, most of the impact literature has addressed only static or equilibrium changes in climate. These studies have not examined what rates of change various sensitive systems can adapt to. Future research should address this matter.

19.1.3. Role of Adaptation

Successful adaptation reduces vulnerability to an extent that depends greatly on adaptive capacity—the ability of an affected system, region, or community to cope with the impacts and risks of climate change (see Chapter 18). Enhancement of adaptive capacity can reduce vulnerability and promote sustainable development across many dimensions.

Adaptive capacity in human systems varies considerably among regions, countries, and socioeconomic groups. The ability to adapt to and cope with climate change impacts is a function of wealth, technology, information, skills, infrastructure, institutions, equity, empowerment, and ability to spread risk. Groups and regions with adaptive capacity that is limited along any of these dimensions are more vulnerable to climate change damages, just as they are more vulnerable to other stresses. Enhancement of adaptive capacity is a necessary condition for reducing vulnerability, particularly for the most vulnerable regions, nations,

[1]One recent exception is DETR (1999), which examines changes in global impacts at different CO_2 stabilization levels.

Box 19-1. Uncertainties in Future Warming

Does a given atmospheric concentration of GHGs cause a specific change in global mean temperature (or other climate variables, for that matter)? To answer this question, we quantify uncertainties in the change in global mean temperature for a given CO_2 concentration level. This is accomplished by using the same simple models that are used in the TAR Working Group I report (TAR WGI Chapter 9). These models are updated versions of models used previously by the IPCC in the Second Assessment Report (SAR) (Kattenberg *et al.*, 1996; see also Raper *et al.*, 1996). We consider the effects of uncertainties in future emissions of *all* radiatively important gases (particularly the relative importance of CO_2 to other forcing factors) and climate sensitivity, but not uncertainties in translating emissions to concentrations.

These uncertainty issues are addressed by comparing CO_2 concentrations (not other GHGs) and the corresponding temperature projections for 5-year time steps from 1990 to 2100 (i.e., using results for 1995, 2000, 2005, etc.) for the six illustrative emissions scenarios from the IPCC *Special Report on Emissions Scenarios* (SRES) (Nakicenovic *et al.*, 2000) under a range of climate sensitivity assumptions. The six emissions scenarios provide a sampling of the space of the relative effects of CO_2 compared with other GHGs and sulfur dioxide (SO_2)-derived sulfate aerosols. Climate sensitivity (ΔT_{2x}) values of 1.5, 2.5, and 4.5°C are used.

The results are plotted as a simple scatter diagram of temperature change against CO_2 concentration (see Figure 19-1). The scatter plot has 22 5-year values (1990 values are zero in each case) by six scenarios by three sensitivities (396 points). The diagram is meant only to illustrate a range of possibilities. One cannot associate any specific confidence intervals with the ranges shown; however, simultaneous use of realistic values in several input parameters with the judgment that the climate sensitivity range of 1.5–4.5°C represents approximately the 90% confidence interval (see, e.g., Morgan and Keith, 1995) suggests that the probability of a result *outside* the ranges shown, during the interval 1990–2100, is less than 10%.

The results are shown in Figure 19-1. For example, for a future CO_2 level of 550 ppmv, the global mean warming range is 1–3°C relative to 1990. Thus, a specific CO_2 concentration could lead to a range of increases in global mean temperature. Note that this is a transient result; in other words, *if CO_2 concentrations were stabilized at 550 ppmv, substantial additional warming would occur beyond this range* as the climate system slowly relaxed toward a new equilibrium state. The levels of increase in global mean temperature displayed in the diagram are less than what would eventually happen if CO_2 concentrations were stabilized at a particular level. Note also that there is no time (or date) associated with any particular concentration level. For, example, in the SRES scenarios, 550 ppmv is reached at a range of dates from about 2050 onward.

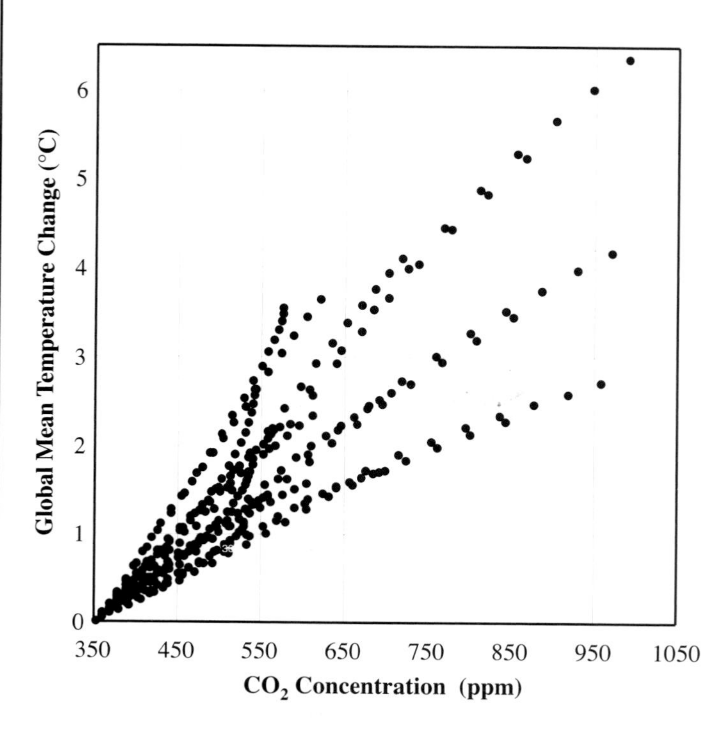

< **Figure 19-1**: Global mean temperature change (from 1990) as a function of CO_2 concentration for SRES scenarios. For any given CO_2 level, uncertainties in temperature arise through several factors. The three most important are accounted for here: First, different temperatures for a given future CO_2 level may arise because each emissions scenario has different levels of other GHGs and different levels of SO_2 emissions—factors that lead to a range of possible non-CO_2 forcings (results here consider all six SRES illustrative scenarios); second, different temperatures arise because of uncertainties in climate sensitivity (three values—1.5, 2.5, and 4.5°C equilibrium warming for a CO_2 doubling—are used here); and third, different temperatures arise because different rates of radiative forcing change and different climate sensitivities lead to different levels of damping of the instantaneous equilibrium response.

and socioeconomic groups. To be sure, some development paths can increase some types of vulnerabilities, whereas others can reduce those vulnerabilities.

Adaptive capacity in natural systems tends to be more limited than adaptive capacity in human systems. Many species have limited ability to migrate or change behavior in response to climate change. What may be of greater concern is the harm that already has been done to natural systems by societal development. Habitat fragmentation and destruction, as well as creation of barriers to migration, will make it much more difficult for species to cope with climate change than if natural systems were undisturbed.

We do not address adaptation explicitly in this chapter, except to the extent that the literature cited here considers adaptation. Adaptation may have the potential to reduce vulnerability and, in many cases, shift the threshold for negative impacts to higher magnitudes of climate change. The degree to which adaptation can do so is not addressed here; it should be the subject of future research.

19.1.4. Chapter Organization

The chapter is organized into the following sections:

- Section 19.2 addresses the insights we can gain by examining observed effects of climate change. Are we seeing impacts of climate change on nature and society?
- Section 19.3 addresses what changes in global mean temperature may cause harm to threatened and unique systems. For example, are threatened and unique systems at risk from even low levels of increase in global mean temperature? Are some societies at particular risk at low levels of temperature increase?
- Section 19.4 addresses the evidence regarding the relationship between change in global mean temperature and distribution of impacts. Are adverse or positive impacts from climate change distributed equally around the world and within countries? Are some regions harmed at certain levels of climate change while others benefit? Are some subgroups or cultures at greater risk than the population as a whole?
- Section 19.5 addresses what insights we gain from aggregate or comprehensive approaches to measuring impacts. What do approaches such as monetization or looking at the number of people who are harmed or benefited tell us about the relationship between aggregate impacts and higher temperatures? This section also addresses insights gained from integrated assessment models (IAMs).
- Section 19.6 addresses the potential for increases in extreme climate events and large-scale singular effects. As temperatures increase or the rate of temperature rise increases, does the potential for extreme climate events and singular effects such as a change in ocean circulation patterns or the collapse of ice sheets increase? Can thresholds of change in terms of magnitudes or rates of change be identified?
- Section 19.7 addresses the limitations of the information used in this chapter to address observations and the reasons for concern. It also addresses future research that is needed to narrow these uncertainties.
- Section 19.8 summarizes the findings on observations and the reasons for concern.

Sections 19.2 and 19.3 draw most heavily on the TAR. Examples can be found in the region and sector chapters of this report; the sections in this chapter do not introduce new information. Instead, they synthesize that information in ways that the other chapters are unable to because they do not examine all regions and sectors. Sections 19.4, 19.5, and 19.6 draw on information that is not found in the regional and sectoral chapters. They do so because they address issues that those chapters cannot:

- Comparison of impacts across regions (Section 19.4). The sectoral chapters do this for each sector, but this can be done comprehensively only in this chapter.
- Aggregation of impacts (Section 19.5). This requires use of common metrics to aggregate impacts across sectors and regions. None of the other chapters can do this.
- Examination of changes in extreme events and large-scale discontinuities (Section 19.6). This generally is not addressed in the region and sector chapters because the climate change scenarios that are used most commonly in impacts studies examine only changes in average conditions, not changes in extreme events or large-scale discontinuities (see Chapter 4).

Thus, this chapter contains much new information in a framework that can help readers judge what may considered to be a dangerous level of climate change.

19.2. Observations of Climate Change Impacts

It is well established from physical, ecological, and physiological studies that climate strongly influences physical and biological systems. This section addresses whether changes in regional climate during the 20th century, documented by WGI, have resulted in measurable impacts on physical and biological systems. We also consider the potential for detecting observed impacts of regional climate change in socioeconomic systems. The objective here is to evaluate the accumulating body of evidence with regard to the following questions:

1) Is there a coherent signal in patterns of observed impacts?
2) Are observed effects of regional climate changes consistent with functional understanding and modeled predictions of impacts?
3) Do observed effects provide information about the potential vulnerability of systems to climate change?
4) How do impacts observed over the past century relate to the five reasons for concern brought forward by this chapter?

In relation to the five reasons for concern, the accumulating body of studies documenting observed impacts of regional climate changes may contribute to understanding of:

- Actual and potential climate change effects on *unique and threatened systems*
- Relationships of impacts to changes in *extreme events*
- Functional and geographical *distribution* of current and future climate change effects
- *Aggregation of impacts*
- Potential effects of *large-scale singularities*.

In this section, we focus on observed impacts that have been associated with regional climate changes over the past 100 years. We examine evidence in physical and biological systems in terrestrial, coastal and marine, and freshwater environments, as well as in socioeconomic systems, including agriculture, commercial fisheries, human settlements, insurance and financial services, and human health (see also other chapters in this report).

The studies reviewed document an observed impact in a physical, biological, or socioeconomic system associated with changes in one or more regional climate variables (most often temperature rise). The effects are examined with regard to the range and geographical extent of processes and species involved, their consistency with functional understanding of mechanisms or processes involved in climate-impact relationships, and the possibility of alternative explanations and confounding factors. Expected directions of change relating to regional climate warming for physical systems include shrinkage of glaciers, decrease in snow cover, shortening of duration of lake and river ice cover, declines in sea-ice extent and thickness, lengthening of frost-free seasons, and intensification of the hydrological cycle. Expected directions of change relating to regional climate warming for biological systems include poleward and elevational shifts in distribution and earlier phenology (i.e., earlier breeding, emergence, flowering) in plant and animal species.

We follow the WGI definition of climate change as a statistically significant variation in the mean state of the climate or its variability, persisting for an extended period (typically decades or longer). Climate change, as defined here, may be caused by natural internal processes or external forcings or by persistent anthropogenic changes in the composition of the atmosphere or land use.

Since 1860, the global mean temperature has warmed 0.6 ± 0.2°C; regional temperature changes have varied, ranging from greater than 0.6°C to cooling in some regions (TAR WGI Chapter 2). Annual land-surface precipitation has increased (0.5–1% per decade) in most middle and high latitudes of the northern hemisphere, except over eastern Asia. In contrast, over much of the subtropical land areas, rainfall has decreased during the 20th century (0.3% per decade), although it has been recovering in recent years (TAR WGI Chapter 2). The recent warming period began in 1976, with pronounced warming observed in northwestern North America, central northern Asia, and the southern Pacific Ocean. Detection of climate change and attribution of causes are discussed in TAR WGI Chapter 12.

19.2.1. Methods of Analysis

Accumulation of evidence over time and space, based on numerous individual studies, is needed to detect and characterize patterns and processes of observed climate change impacts on a global basis (see Chapter 2). In many studies, changes in impact systems are compared with trends in climate variables over the same period and location. Many studies establish statistically significant trends in the observed impact and the climate variable, as well as a statistically significant association between the two (e.g., Beebee, 1995; Brown *et al.*, 1999; Barber *et al.*, 2000). Others refer to trends in climate documented elsewhere (e.g., Menzel and Fabian, 1999; Thomas and Lennon, 1999). When multiple species or locations are examined, cases are reported that exhibit no change, change that is consistent with understanding of climate-impact relationships, and change that is inconsistent with understanding of climate-impact relationships. This allows for assessment of whether observed changes are significantly different from random chance and are consistent with functional understanding of climate responses (e.g., Ellis, 1997; Ellis *et al.*, 1997; Bradley *et al.*, 1999; Pounds *et al.*, 1999).

Individual studies that link observed impacts to regional climate change may be hampered by methodological problems such as length of time-series data of observed impacts; number of replications of populations, census sites, or species; availability of climate data to which to compare observed changes; and uncertainty about whether observed impacts and regional climate variables are measured at appropriate spatial scales (Chapter 2). In some regions, several individual studies have focused on differing aspects of a common ecosystem, providing evidence for associations between climate change and multiple responses in a given geographical area (e.g., Smith *et al.*, 1999); in other regions, however, studies examine more isolated responses.

Because changing climate and ecological responses are linked over a range of temporal scales, long periods of study allow more accurate conclusions regarding the significance of observed ecosystem changes. Large-amplitude temporal changes usually involve large spatial dimensions, so broad-scale spatial/temporal ecosystem studies tend to be more robust. The majority of studies document trends for periods of more than 20 years (e.g., Post *et al.*, 1997; Winkel and Hudde, 1997; Post and Stenseth, 1999); a few studies document trends for 10–19 years (e.g., Jarvinen, 1994; Forchhammer *et al.*, 1998); and several studies analyze data from two periods with a gap between them (Bradley *et al.*, 1999; Sagarin *et al.*, 1999).

Climate Trends: The various studies of observed impacts of recorded regional temperature change over the past century, which include the recent warm decades of the 1980s and 1990s, often differentiate responses to mean, minimum, and maximum temperatures. Regional precipitation changes and periods of droughts and floods are much more variable in observed records and more uncertain with regard to future predictions and are not the primary focus here. Studies also have considered possible observed responses to the rising atmospheric concentrations of CO_2 over the past century, but these studies are not included in this review.

To the extent that periodicities or trends are found in the climate record, nonzero autocorrelations are to be expected on the interannual time scale. Their importance depends on the percentage of variance associated with the periodicities and the magnitude of the trend relative to interannual noise. Often the periodicities represent only a small proportion of the total variance; this is especially true on a local level, where the noise is likely to be higher than at broader spatial scales. A nonzero autocorrelation does not automatically mean the year-to-year ecological impact is not meaningful because if year-to-year climate variability is associated with a periodic or steadily increasing climate forcing, so too would be the ecological response.

Processes and Mechanisms: Beyond statistical association, an important aspect of many studies is comparison of documented changes to known relationships between climate and impact systems. For example, under regional warming, retreat of glaciers is expected because of shifts in the energy balance of glaciers, as is poleward expansion of species' ranges when temperatures exceed physiological thresholds. If documented changes are consistent with known processes that link climate and the impact system, confidence in the associations between changes in regional climate and observed changes is enhanced.

Multiple Causal Factors: The presence of multiple causal factors (e.g., land-use change, pollution, biotic invasion) makes attribution of many observed impacts to regional climate change a complex challenge at the individual study and meta-analysis levels (e.g., Prop *et al.*, 1998; Körner, 1999). Some of the competing explanations for observed impacts themselves could have a common driver that would make them strongly correlated; identifying these drivers is a methodological challenge. Studies seek to document observed climate change impacts by ruling out other possible contributing causative factors, ecological or anthropogenic, through study design and sampling techniques (e.g., Parmesan, 1996; Menzel and Fabian, 1999; Parmesan *et al.*, 1999), statistical analyses (e.g., Prop *et al.*, 1998; Reading, 1998), or expert judgment (De Jong and Brakefield, 1998; Brown *et al.*, 1999). Sometimes, different studies offer alternative explanations for observed impacts (e.g., Körner, 1999).

Signals of regional climate change impacts may be clearer in physical systems than in biological systems, which are simultaneously undergoing many complex changes that are not related to climate, including land-use change and pollution processes such as eutrophication and acidification. Observed impacts in high-latitude and high-altitude physical systems, such as melting of glaciers, may be more straightforward to detect, whereas biological responses to climate tend to be more complex and may be masked by the presence of the aforementioned multiple causal factors. To deal with these ecological complexities, confounding factors often are minimized by conducting studies away from large urban or agricultural areas, in large natural areas (e.g., northern Canada, Australia), or in preserved areas.

Signals of regional climate change impacts probably are most difficult to detect in socioeconomic systems because such systems are strongly affected by simultaneous trends in population and income growth and urbanization and because of the presence of adaptive capacity (see Chapter 18). Observed climate change impacts in socioecosystems may be adaptations in many cases, such as farmers sowing crops earlier in response to warmer spring temperatures.

An example of these methodological complexities in climate change impact detection may be drawn from the human health sector. Although climate is known to influence many disease vectors (such as the range of anopheline mosquitos that carry malaria), the presence or absence of sanitation systems, vaccination programs, adequate nutritional conditions, animal husbandry, irrigation, and land-use management also influences whether the presence of a disease in wild vectors leads to disease outbreaks in human populations (see Chapter 9).

Evaluating Patterns of Change: Grouping individual studies to evaluate patterns and processes of change on larger spatial scales reduces the influences of study-specific biases and local nonclimatic factors. Comparing expected geographical patterns of responses to regional climate changes and to changes that are not related to climate helps distinguish among multiple possible causations. For example, regional warming would be expected to skew the distribution of insect extinctions to be greater at the southern boundaries rather than at the northern boundaries; land-use change, in contrast, would be expected to cause approximately equal extinctions at both range boundaries (Parmesan, 1996; Parmesan *et al.*, 1999). Care must be taken to ensure that the sample of studies is representative across time and space, is not biased in reporting, and uses appropriate statistical tests. Spottiness of evidence in other regions may indicate that observed impacts of regional climate change are not occurring, have not yet been detected, or are being masked by other changes, such as urbanization.

Some studies of observed impacts have used a "fingerprint" approach, based on the definition of expected biological changes arising from regional climate change (e.g., Epstein *et al.*, 1998). This approach is similar to that used in detection of climate changes (see TAR WGI Chapter 12) but differs in that fingerprint studies of ecosystem impacts use selected data and that long-term monitoring of changes in ecosystems generally is lacking at regional or global scales.

19.2.2. Synthesis of Observed Impacts

There is an accumulating body of evidence of observed impacts relating to regional climate changes—primarily rising temperature across a broad range of affected physical processes and biological taxa—and widespread geographical distribution of reported effects (see Figure 19-2 and accompanying notes). In many cases, reported changes are consistent with functional understanding of the climate-impact processes involved. Cases of no change or change in unexpected direction are noted, as are possible alternative explanations and confounding factors, where available.

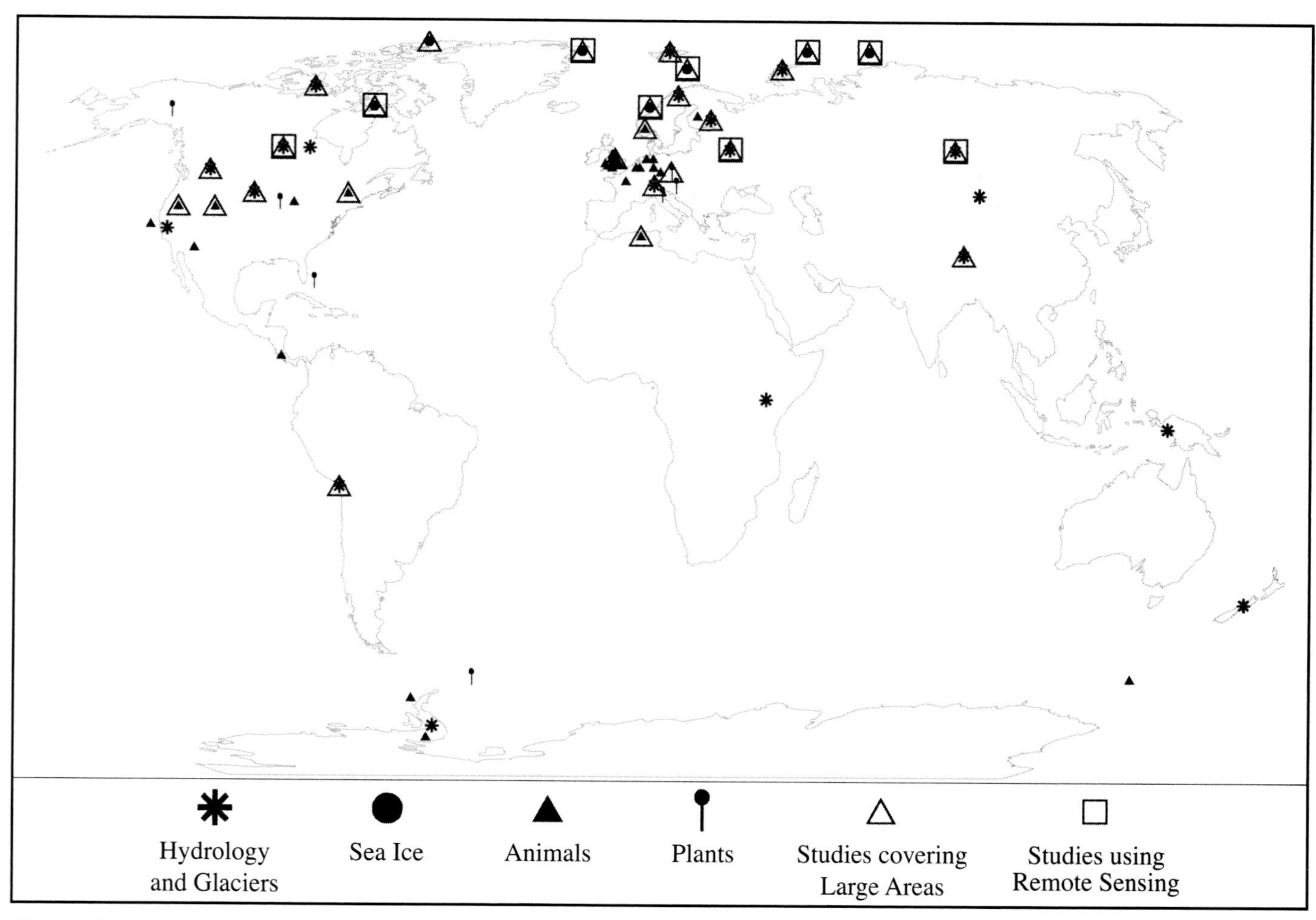

Figure 19-2: Observed impacts of temperature-related regional climate change in the 20th century:

- *Hydrology and Glaciers*—Glacier retreat, decrease in snow-cover extent/earlier snowmelt, reduction in annual duration of lake and river ice
- *Sea Ice*—Decline in sea-ice extent and thickness
- *Animals*—Poleward and elevational shifts in range, alteration in species abundance, changes in phenology (including earlier reproduction and migration), physiological and morphological adaptation
- *Plants*—Change in abundance and diversity, change in phenology (including earlier flowering), change in growth.

Studies that cover large areas and use remote-sensing methods are indicated. About 50 studies were selected, according to the following criteria: (1) hydrology/sea-ice studies that report long-term trends in observed variables (time periods of studies range from ~20 to 150 years), and (2) terrestrial and marine ecosystem studies that associate trends in observed change(s) with trends in regional climate data for ≥20 years (time periods of studies range from ~20 to 50 years). Of the ~100 physical processes and ~450 biophysical species that exhibited change, more than 90% (~99% physical, ~80% biophysical) are consistent with well-known mechanisms of system responses to climate.

Sources: *Hydrology and Glaciers, and Sea Ice*—Ames and Hastenrath (1996), Cavalieri *et al.* (1997), Dettinger and Cayan (1995), Dowdeswell *et al.* (1997), Dyurgerov and Meier (1997), Greene *et al.* (1999), Groisman *et al.* (1994), Haeberli and Beniston (1998), Hastenrath (1995), Johannessen *et al.* (1999), Kaser (1999), Kratz *et al.* (2001), Magnuson *et al.* (2000), Maslanik *et al.* (1996), Rothrock *et al.* (1999), Schindler *et al.* (1990), and Vinnikov *et al.* (1999); *Animals and Plants*—Barber *et al.* (2000), Bergmann (1999), Bezzel and Jetz (1995), Bradley *et al.* (1999), Brown *et al.* (1999), Crick and Sparks (1999), Crick *et al.* (1997), Cunningham and Moors (1994), Dunn and Winkler (1999), Ellis (1997), Ellis *et al.* (1997), Fleming and Tatchell (1995), Forchhammer *et al.* (1998), Fraser *et al.* (1992), Gatter (1992), Grabherr *et al.* (1994), Hasenauer *et al.* (1999), Jarvinen (1994), Loeb *et al.* (1997), Ludwichowski (1997), Mason (1995), McCleery and Perrins (1998), Menzel and Fabian (1999), Pauli *et al.* (1996), Parmesan (1996, 2001), Parmesan *et al.* (1999), Post and Stenseth (1999), Post *et al.* (1997), Pounds *et al.* (1999), Ross *et al.* (1994), Sagarin *et al.* (1999), Slater (1999), Smith (1994), Smith *et al.* (1999), Sparks (1999), Thomas and Lennon (1999), Visser *et al.* (1998), Winkel and Hudde (1996, 1997), Zhou *et al.* (1995).

19.2.2.1. Hydrology

The hydrological cycle is expected to respond to regional climate warming through changes in the energy balance of glaciers and the depth and extent of snow cover, earlier snowmelt runoff, seasonal changes in freezing and thawing of lakes and rivers, and intensification of precipitation and evaporative processes. For the most part, evidence of regional climate change impacts on elements of the hydrological cycle is consistent with expected responses to warming temperatures and intensification of hydrological regimes (see Chapters 4 and 5, and TAR WGI).

Evidence for such changes in the 20th century includes recession of glaciers on all continents (e.g., Hastenrath, 1995; Ames and Hastenrath, 1996; Dowdeswell *et al.*, 1997; Dyurgerov and Meier, 1997; Haeberli and Beniston, 1998; Greene *et al.*, 1999; Kaser, 1999; Krabill *et al.*, 1999; Serreze *et al.*, 2001). There have been decreases in the extent of snow cover (10% since the late 1960s and 1970s) in the northern hemisphere (e.g., Groisman *et al.*, 1994; Serreze *et al.*, 2001). Since the late 1940s, snowmelt and runoff have occurred increasingly earlier in northern and central California (Dettinger and Cayan, 1995). Annual duration of lake- and river-ice cover in the middle and high latitudes of the northern hemisphere has been reduced by about 2 weeks and is more variable (Schindler *et al.*, 1990; Magnuson *et al.*, 2000; Kratz *et al.*, 2001).

Also reported is increased frequency of extreme rainfall in the middle and high latitudes of the northern hemisphere, including the United States (Karl and Knight, 1998), the UK (Osborn *et al.*, 2000), and most extratropical land areas except China (Groisman *et al.*, 1999).

19.2.2.2. Terrestrial Ecosystems

Ecological theory predicts several types of species and community responses to changing regional climate in plants and animals: changes in ecosystem structure and dynamics, including shifts in ranges and distributions; altered phenology; effects on physiology; and genetic evolutionary responses (see Chapters 2 and 5). Changes in disturbance (e.g., fires, wind damage) also may be occurring but are not included in this review (see Chapters 5 and 6). Evidence from plants and animals documents all of these types of ecological responses to regional warming, especially poleward and elevational shifts in species ranges and earlier timing of reproduction. Reviews of recent changes in biological systems also have documented examples of these different types of responses, consistent with process-level understanding (Hughes, 2000).

19.2.2.2.1. Vegetation

Much of the evidence of vegetation change relating to regional climate change comes from responses to warming at high-latitude and high-altitude environments, where confounding factors such as land-use change may be minimized and where climate signals may be strongest (see TAR WGI Chapter 12). Increases in species richness were found at 21 of 30 high summits in the Alps; remaining summits exhibited stagnation or a slight decrease (Grabherr *et al.*, 1994; Pauli *et al.*, 1996). However, Körner (1999) suggests that grazing, tourism, and nitrogen deposition may be contributing to such observed migrations. Hasenauer *et al.* (1999) found significant increases in diameter increments of Norway spruce across Austria related to increased temperatures from 1961 to 1990. In North America, Barber *et al.* (2000) linked reduced growth of Alaskan white spruce to temperature-induced drought stress, and Hamburg and Cogbill (1988) propose that historical declines in red spruce in the northeastern United States are related to climatic warming, possibly aggravated by pollution and pathogen factors.

In more temperate ecosystems, Bradley *et al.* (1999) documented phenological advances in flowering date in 10 herbaceous and tree species and no change in 26 such species related to local warming in southern Wisconsin over the periods 1936–1947 and 1976–1998. Menzel and Fabian (1999) document extension of the growing season for 12 tree and shrub species at a network of sites throughout Europe, which they attribute to warming temperature. Alward and Detling (1999) found reorganization of a shortgrass steppe ecosystem in a semi-arid site in Colorado related to increased spring minimum temperatures, although the responses of C_3 and C_4 species did not occur as expected.

Regarding regional changes in precipitation—which are much more uncertain with regard to future climate—reorganization of a semi-arid ecosystem in Arizona, including increases in woody shrubs, has been associated with increases in winter precipitation (Brown *et al.*, 1997); retraction of mesic species to areas of higher rainfall and lower temperature has been attributed to a long-term decline in rainfall in the West African Sahel (Gonzalez, 2001).

19.2.2.2.2. Animals

Temperature change-related effects in animals have been documented within all major taxonomic groups (amphibians, birds, insects, mammals, reptiles, and invertebrates) and on all continents (see Chapter 5). Terrestrial evidence in animals that follows process-level understanding of responses to warming includes poleward and elevational changes in spatial distribution, alterations in species abundance and diversity, earlier phenology (including advances in timing of reproduction), and physiological and genetic adaptations.

Poleward and elevational shifts associated with regional warming have been documented in the ranges of North American, British, and European butterfly species (Parmesan, 1996; Ellis, 1997; Ellis *et al.*, 1997; Parmesan *et al.*, 1999), birds (Thomas and Lennon, 1999), and insects (Fleming and Tatchell, 1995). Prop *et al.* (1998) found that increasing spring temperatures and changes in agricultural practices in Norway have allowed barnacle geese (*Branta leucopsis*) to move northward and invade active agricultural areas. Changes in species distribution and abundance of amphibians, birds, and reptiles in Costa Rica have been associated with changing patterns of dry-season mist frequency and Pacific sea-surface temperatures (SST) (Pounds *et al.*, 1999; Still *et al.*, 1999).

Earlier timing of reproduction has been found for many bird species (Mason, 1995; Crick *et al.*, 1997; McCleery and Perrins, 1998; Crick and Sparks, 1999; Slater, 1999) and amphibians (Beebee, 1995; Reading, 1998) in the UK and Europe (Winkel and Hudde, 1996, 1997; Ludwichowski, 1997; Forchhammer *et al.*, 1998; Visser *et al.*, 1998; Bergmann, 1999). Zhou *et al.* (1995) found a warming trend in the spring to be associated with earlier aphid flights in the UK. Also in the UK, Sparks (1999) has associated arrival times of bird migration to warmer spring temperature. Bezzel and Jetz (1995) and Gatter (1992) document delays in the autumn migratory period in the Alps and Germany, respectively.

In North America, Brown *et al.* (1999) document earlier egg-laying in Mexican jays (*Aphelocoma ultramarina*) associated with significant trends toward increased monthly minimum temperatures in Arizona. Dunn and Winkler (1999) found that the egg-laying date of North American tree swallows advanced by as much as 9 days, associated with increasing air temperatures at the time of breeding. Bradley *et al.* (1999) document phenological advances in arrival dates for migratory birds in southern Wisconsin, associated with earlier icemelt of a local lake and higher spring temperature.

Post *et al.* (1997) and Post and Stenseth (1999) document differential selection of body size in red deer throughout Norway from 1965 to 1995. Male red deer have been getting larger and females smaller, correlated with warming trends and variations in the North Atlantic Oscillation (NAO). Post and Stenseth (1999) also report on the interactions of plant phenology, northern ungulates (red deer, reindeer, moose, white-tailed deer, musk oxen, caribou, and Soay sheep), and the NAO. Jarvinen (1994) found that increased mean spring temperatures in Finnish Lapland are associated with mean egg volume of the pied flycatcher. De Jong and Brakefield (1998) found shifts in color patterns (black with red spots versus red with black spots), most likely related to thermal budgets of ladybird beetles (*Adalia bipunctata*) in The Netherlands, coinciding with an increase in local ambient spring temperatures. The potential for rapid adaptive responses and their genetic costs to populations has been studied by Rodriguez-Trelles and Rodriguez (1998), who found microevolution and loss of chromosomal diversity in *Drosophila* in northwestern Spain as the local climate warmed.

19.2.2.3. Coastal Zones and Marine Ecosystems

In coastal zones and marine ecosystems, there is evidence of changes in physical and biological systems associated with regional trends in climate, especially warming of air temperatures and SST (see Chapters 4, 5, and 6). However, separating out responses of marine ecosystems to variability caused by large-scale ocean-atmosphere phenomena, such as ENSO and NAO, from regional climate changes is a challenge (e.g., Southward *et al.*, 1995; McGowan *et al.*, 1998, 1999; Sagarin *et al.*, 1999). Variations caused by ENSO and NAO *per se* are not considered climate change, but multi-decadal trends of change in ENSO or NAO frequency and intensity are climate changes, according to the IPCC definition.

19.2.2.3.1. Physical processes

Changes in the physical systems of coastal zones related to regional warming trends include trends in sea ice and coastal erosion. Since the 1950s, Arctic sea-ice extent has declined by about 10–15%; in recent decades, there has been about a 40% decline in Arctic sea-ice thickness during late summer to early autumn and a considerably slower decline in winter (e.g., Maslanik *et al.*, 1996; Cavalieri *et al.*, 1997; Johannessen *et al.*, 1999; Rothrock *et al.*, 1999; Vinnikov *et al.*, 1999; Serreze *et al.*, 2001). No significant trends in Antarctic sea-ice extent are apparent (see TAR WGI).

19.2.2.3.2. Marine ecosystems

Evidence from marine ecosystems documents changes in species abundance and diversity and spatial distributions associated with air and ocean temperature rises (Chapters 5 and 6). Several studies document changes from the Antarctic region: Increases in chinstrap (*Pygoscelis antarctica*) penguins, stability or slow declines in Adelie (*Pygoscelis adeliae*) penguins, and declines in rockhopper penguins in recent decades are attributed in part to differential responses to warming climate conditions that are altering bird habitats (Fraser *et al.*, 1992; Cunningham and Moors, 1994; Smith *et al.*, 1999). Loeb *et al.* (1997) report effects on the Antarctic food web resulting from decreased frequency of winters with extensive sea-ice development; krill abundance is positively correlated with sea-ice extent, and salp abundance is negatively correlated. Smith (1994) reports a significant and relatively rapid increase in the numbers of individuals and populations of the only two native Antarctic vascular plant species at two widely separated localities in the maritime Antarctic.

Increases in abundance of southern macroinvertebrate species and declines in northern species in a rocky intertidal community on the California coast are consistent with recent climate warming (Sagarin *et al.*, 1999). Warming annual temperature has been suggested as a possible cause of increases in abundance of plankton in the German Bight, but numerous factors, including regional eutrophication, also have been noted (Greve *et al.*, 1996). Lehman (2000) found that the distribution of phytoplankton biomass in northern San Francisco Bay Estuary was influenced by environmental conditions resulting from an interdecadal climate regime shift between 1975 and 1993; precipitation regimes were primarily implicated, with water temperatures also playing an important role. Ross *et al.* (1994) document the loss of low-elevation pine forests in the Florida Keys because of rising sea level.

19.2.2.4. Socioeconomic Systems

Evidence of observed impacts of regional climate changes from socioeconomic systems is much sparser than from physical and biological systems, and methodologically it is much more difficult to separate climate effects from other factors such as technological change and economic development, given the complexities of these systems. Vulnerability to climate change and climate variability is a function of exposure and adaptive capacity (see Chapter 18). Exposure varies from region to region, sector to sector, and community to community, and adaptive capacity may be even more variable. The adaptive capacity of socioeconomic systems also contributes to the difficulty of documenting effects of regional climate changes; observable effects may be adaptations to a climate change rather than direct impacts. Evidence of observed adaptation of many of these systems to multiple stresses, including climate variability, suggests that complexities inherent in socioeconomic systems could be a source of resilience, with potential for beneficial adaptations in some cases. Studies that have explored some of these complex relationships are briefly reviewed in the following subsections, but they are not included in the summary tabulation or figure.

19.2.2.4.1. Agriculture and commercial fisheries

It has been proposed that observed impacts of changes in regional climate warming that are relevant to agriculture are related to increasing yield trends in Australia, lengthening growing seasons at high latitudes, improved wine quality in California, and expansion and advanced phenologies of agricultural pests. However, links between changes in regional climate variables and such changes are hard to prove because agriculture is a multifactored biophysical and socioeconomic system (see Chapter 5).

Nicholls (1997) analyzed Australian wheat yields from 1952 to 1992 and concluded that climate trends appear to be responsible for 30–50% of observed increases, with increases in minimum temperatures (decreases in frosts) the dominant influence (Nicholls, 1997); this conclusion has been questioned, however, by Godden *et al.* (1998) and Gifford *et al.* (1998). Possible confounding socioeconomic factors in identifying the effects of climate change on crop yields are responses of farmers to growing conditions (e.g., farmers may increase fertilizer application in good years, thereby exaggerating the impact of climate variables on yield), technological progress, changes in market structure, and changes in agricultural subsidies. Crop responses to increasing atmospheric CO_2 concentrations also may affect yield trends.

Carter (1998) found that the growing season in the Nordic region (Iceland, Denmark, Norway, Sweden, and Finland) lengthened between 1890 and 1995 at all sites except Iceland, with likely but undocumented impacts on crop phenologies and timing of farm operations.

Nemani *et al.* (2001) relate warming at night and during spring in California over the period 1951–1997 (especially since 1976) to improved vintage quantity and quality.

Recent expansion and advances in insect phenologies may be associated with regional increases in mean or minimum temperatures (e.g., advances in flight phenology of aphid species in Britain) (Fleming and Tatchell, 1995; Zhou *et al.*, 1995). Such increases in insect pests may be contributing to agricultural losses at least partially related to recent climate trends, but these effects have not been examined analytically.

Some changes in marine and coastal ecosystems have links to commercial fisheries, but it is difficult to separate regional climate effects from human use of fish stocks (see Chapter 6). Recent warming trends and coincident overfishing and eutrophication have been noted in the English Channel and North Sea, with potential future consequences for fish of high mass-market value (e.g., haddock, cod, plaice, lemon-sole cod—Southward *et al.*, 1995; O'Brien *et al.*, 2000). Diminished krill supplies in the Antarctic associated with decreases in annual sea-ice cover and warmer air temperature documented by Loeb *et al.* (1997) between 1976 and 1996 may have long-term negative effects on upper tropic levels, affecting commercial harvests. These observations, in part, have prompted the Commission for the Conservation of Antarctic Marine Living Resources (CCAMLR) to request updated krill data currently used in krill management. CCAMLR manages and sets limits on the international harvest of Antarctic krill (Loeb *et al.*, 1997).

19.2.2.4.2. Energy, industry, human settlements, and financial and insurance services

Associations between regional climate trends and impacts related to energy, industry, and human settlements are sparse. One documented example is rapid coastal retreat along ice-rich coasts of the Beaufort Sea in northwestern Canada (Dallimore *et al.*, 1996). Where communities are located in ice-rich terrain along the shore, warmer temperatures combined with increased shoreline erosion can have a very severe impact (see Chapter 6).

Determining the relationship between regional climate trends and impacts relating to financial and insurance services is difficult because of concurrent changes in population growth, economic development, and urbanization. Trends have been analyzed regarding increased damages by flooding and droughts in some locations. Global direct losses resulting from large weather-related disasters have increased in recent decades (see Chapter 8). Socioeconomic factors such as increased coverage against losses account for part of these trends; in some regions, increases in floods, hailstorms, droughts, subsidence, and wind-related events also may be partly responsible (see Chapter 8). Attribution is still unclear, however, and there are regional differences in the balance of these two causes. Hurricane and flood damages in the United States have been studied by Changnon *et al.* (1997), Changnon (1998), and Pielke and Downton (2001). Pielke and Downton (2001) found that increases in recent decades in total flooding damage in the United States are related to climate factors and societal factors: increased precipitation and increasing population and wealth. Hurricane damages, on the other hand, are unaffected by observed climate change (Changnon *et al.*, 1997; Changnon, 1998).

19.2.2.4.3. Human health

There is little evidence that recent trends in regional climates have affected health outcomes in human populations (see Chapter 9). This could reflect a lack of such effects to date or difficulty in detecting them against a noisy background containing other more potent influences on health (Kovats *et al.*, 1999). The causation of most human health disorders is multifactorial, and the socioeconomic, demographic, and environmental context varies constantly. With respect to infectious diseases, for example, no single epidemiological study has clearly related recent climate trends to a particular disease.

Various studies of the correlation between interannual fluctuations in climatic conditions and the occurrence of malaria, dengue, cholera, and several other infectious diseases have been reported. Pascual *et al.* (2000) report a relationship between cholera and El Niño events. Such studies confirm the climate sensitivity of many infectious diseases, but they do not provide quantitative information about the impact of decadal-level climate change. Fingerprint studies examine the patterns of collocated change in infectious diseases and their vectors (if applicable) in simpler physical and ecological systems. This is an exercise in pattern recognition across qualitatively different systems.

One example is the set of competing explanations for recent increases in malaria in the highlands (see Chapter 9). A fingerprint study has hypothesized possible connections of plant and insect data, glacier observations, and temperature records to global climate change in high-altitude locations, with implications for patterns of mosquito-borne diseases (Epstein *et al.*, 1998). Loevinsohn (1994) notes a connection between climate warming and increased rainfall with increased malaria incidence in Rwanda, whereas Mouchet *et al.* (1998) emphasizes the importance of nonwarming factors (e.g., land-use change in response to population growth, climate variability related to ENSO) in explaining variations in malaria in Africa.

Changes in disease vectors (e.g., mosquitoes, ticks) are likely to be detected before changes in human disease outcomes. Furthermore, a change in vector does not necessarily entail an increase in health impacts because of simultaneous processes related to the disease itself and the human population at risk. For example, the presence or absence of sanitation systems, vaccination programs, adequate nutritional conditions, animal husbandry, irrigation, and land-use management influences whether the presence of a disease in wild vectors leads to disease outbreaks in human populations. The effects of changes in frequency of extreme events may entail changes in health impacts, but these have not been documented to date.

19.2.3. Conclusions

Statistically significant associations between trends in regional climate and impacts have been documented in ~100 physical processes and ~450 biological species or communities in terrestrial and polar environments. More than 90% of the changes (~99% physical, ~80% biophysical) documented worldwide are consistent with how physical and biological processes are known to respond to climate. There are systematic trends of ecological change across major taxonomic groups (amphibians, birds, insects, mammals, reptiles, and invertebrates) inhabiting diverse climatic zones and habitats. The overall processes and patterns of observations reveal a widespread and coherent impact of 20th-century climate changes on many physical and biological systems (see Figure 19-2).

Expected directions of change relating to regional climate warming for physical systems have been reported in studies documenting shrinkage of glaciers, decreases in snow cover, shortening of duration of lake- and river-ice cover, declines in sea-ice extent and thickness, lengthening of frost-free seasons, and intensification of the hydrological cycle. Expected directions of change relating to regional climate warming for biological systems have been reported in studies documenting poleward and elevational shifts in distribution and earlier phenology (i.e., earlier breeding, emergence, flowering) in plant and animal species.

In general, geographic patterns of responses also conform to expectations relating to regional climate change, as opposed to alternative explanations. Reported cases of observed impacts are concentrated in high-latitude and high-altitude physical and biological systems and tend to be in regions where observed regional warming has been greatest and confounding factors often are at least partially minimized. Although land-use change, pollution, and biotic invasions are widespread anthropogenic influences, they are unlikely to cause the spatial patterns (e.g., skewed poleward and elevational range shifts) and temporal patterns (e.g., earlier breeding and flowering) that are documented over the set of reported studies.

The sample of studies shown in Figure 19-2 was drawn from a literature survey with keywords relating to climate trends and observed trends in impacts. The time period of most of the studies includes the recent warm period beginning in the late 1970s. The geographical distribution of studies to date is biased toward Europe and North America but does include evidence of observed impacts of regional climate change relating to physical processes from all continents. The spottiness of biological evidence in other regions may indicate that observed impacts of regional climate change are not occurring, have not yet been detected, or are being masked by other changes, such as urbanization. Many studies include multiple species and report on the number of species that responded to regional climate changes as expected, not as expected, or exhibited no change. Most of the biophysical studies included in Figure 19-2 report on statistical tests of trends in climate variable, trends in observed impacts, and relationships between the two (see Chapter 5).

In Figure 19-2, ~16 studies examining glaciers, sea ice, snow-cover extent/snowmelt, or ice on lakes or streams at more than 150 sites were selected. Of these ~150 sites, 67% (~100) show change in one or more variable(s) over time. Of these ~100 sites, about 99% exhibit trends in a direction that is expected, given scientific understanding of known mechanisms that relate temperatures to physical processes that affect change in that variable. The probability that this proportion of sites would show directional changes by chance is much less than 0.00001.

There are preliminary indications that some social and economic systems have been affected in part by 20th-century regional climate changes (e.g., increased damages from flooding and droughts in some locations). It generally is difficult to separate climate change effects from coincident or alternative explanations for such observed regional impacts. Evidence from studies relating regional climate change impacts on socioeconomic systems has been reviewed but is not included in the summary figure because of the complexities inherent in those systems.

The effects of regional climate change observed to date provide information about the potential vulnerability of physical, biological, and socioeconomic systems to climate change in terms of exposure, sensitivity, and adaptive capacity. Some of the observed effects are adaptations. In some cases, observed impacts are large relative to the levels of regional climate changes (e.g., large changes in ecosystem dynamics with small changes in regional climate). In general, observations of impacts agree with predictions that estimate more serious impacts at higher GHG concentrations because the greater regional climate changes are associated with stronger impacts.

Relating the observed impacts summarized here to the reasons for concern analyzed in this chapter, we find the following:

1) There is preliminary evidence that *unique and threatened systems* are beginning to be affected by regional climate change (e.g., glaciers, polar environments, rare species).
2) With regard to the *distributional effects* of observed impacts relating to regional climate changes, most evidence to date comes from high-latitude and high-altitude environments, where regional warming has been and is expected to be more pronounced.
3) *Aggregate impacts* of regional climate changes at the global level are difficult to define, except in sectors in which there is a common metric, as in market sectors. The many simultaneous factors and varying adaptive capacities make extracting aggregate effects attributable to observed climate change difficult. What can be stated in summary regarding the diverse set of impacts reported to date is that there are cases of observed impacts in many diverse environments; that they occur in a wide array of processes, species, and ecosystems; and that the overall patterns and processes reveal a coherent impact of 20th-century climate changes on many physical and biological systems.
4) Impacts of *extreme events* have been implicated in many of the observations summarized in this section, including increases in extreme precipitation events in some locations.
5) There is no current evidence in observed impacts that *large-scale abrupt changes* already are occurring. Yet, paleoclimate evidence (see TAR WGI Chapter 2) shows that such changes have occurred in physical and biological systems in the past and therefore may occur with a continuation of the current warming trend.

19.3. Impacts on Unique and Threatened Systems

19.3.1. What are Unique and Threatened Systems?

Unique systems are restricted to a relatively narrow geographical range but can affect other entities beyond their range. Indeed, many unique systems have global significance. The fact that these unique entities are restricted geographically points to their sensitivity to environmental variables, including climate, and therefore attests to their potential vulnerability to climate change.

Identification of these unique entities provides the first reason for concern regarding vulnerability to climate change. In this section, we provide examples of unique entities that are likely to be threatened by future changes in climate. From those treated by WGII, we address physical, biological, and human systems. We offer a few examples in each system: tropical glaciers, coral reefs, mangrove ecosystems, biodiversity "hot spots," ecotones, and indigenous communities. These are meant only as illustrative examples; there are many unique and threatened entities. Table 19-1 lists some unique and potentially threatened systems in relation to climate change thresholds that may cause adverse effects. Table 19-2 lists some of the unique and threatened systems that are discussed elsewhere in the TAR.

19.3.2. Physical Systems

A number of physical systems are threatened by climate change. Among the most prominent are those in regions dominated by cold temperatures, such as glaciers. Many glaciers already are receding, and many are threatened by climate change. Other physical systems, such as small lakes in areas that will become drier (see Chapter 4), also are threatened by climate change. Changes in unique physical systems can have serious consequences for unique biological and human systems.

19.3.2.1. Tropical Glaciers

Tropical glaciers are present on several mountains in Asia, Africa, and Latin America. These glaciers are valuable because, among other reasons, they are a major source of water for people living below them. For example, through a network of mountain streams, meltwater of the Himalayan glaciers contributes a sizeable portion of river flows to the Ganges, Brahmaputra, Indus, and other river systems in south Asia.

***Table 19-1**: Vulnerability of wildlife to climate change (compiled from Chapter 5).*

Geographic Area	Impact	Vulnerable to
Most continents, marine, polar regions	– Poleward/elevational shifts in ranges	– Already observed in many species in response to regional climate change
Most continents, marine, polar regions	– Shifts in phenology (e.g., breeding, arrival dates, flowering)	– Already observed in response to regional climate change
Sunderbans, Bangladesh	– Loss of only remaining habitat of Royal Bengal tiger	– Sea-level rise
Caribbean, South Pacific Islands	– Habitat loss, direct mortality of birds	– Hurricanes
Marine	– Reproductive failure in seabirds	– Increased sea-surface temperature (ENSO)
Galapagos, Ecuador, Latin America	– Reduced survival of iguanas	– ENSO
Africa	– Reduced overwinter survival of palearctic migratory birds	– Extreme drought in the Sahel
Monteverde Reserve, Costa Rica	– Extirpation of some cloud forest reptiles and amphibians (already has occurred), elevational shift in some birds	– ENSO, warming, increased frequency of dry season mist
Norway	– Poleward shift of spring range of barnacles geese	– Increase in number of April and May days with temperatures above 6°C
Australia	– Susceptibility of quokka to salmonella infections	– Environmental conditions
United Kingdom	– Earlier hatching of spittlebugs	– Winter-warmed (3°C) grassland plots (experimental)
Scotland	– Faster growth in juvenile red deer, leading to increased body size	– Warmer springs
Isle Royale National Park, United States	– Increased wolf pack size, increased moose mortality, greater growth of understory balsam fir	– Reduction in winter snow cover
Western Antarctic Peninsula	– Reductions in Adelie penguin populations, increases in chinstrap penguin populations	– Increased midwinter surface air temperature, reductions in pack ice, increased snowfall
Northern Hemisphere	– Increased winter survival of some boreal insect pests	– Increased nighttime winter temperatures
Great Plains, USA, and Canada	– Reductions in waterfowl breeding populations as a result of wetland loss	– Increased drought
Africa and Australia	– Wetland loss	– Increased drought
Africa and Australia	– Reduced populations of some mammals	– Increased drought

Table 19-1 (continued)

Geographic Area	Impact	Vulnerable to
Canada	– Loss of 60% of available habitat (habitat migration blocked by Arctic Ocean)	– Climate change
USA and Canada	– Reductions in populations of caribou	– Increased temperatures, snowfall, shifts in precipitation timing
Mexico	– Loss of wintering habitat for eastern population of monarch butterfly	– Climate change leading to habitat change
USA	– Loss of migratory shorebird habitat	– Sea-level rise tied to 2°C temperature increase
Arctic	– Reduced habitat availability and accessibility hampering migration and survival of polar bears, muskox, caribou, and some birds	– Increased temperatures, changing sea-ice regimes
United Kingdom	– Loss of habitat in 10% of designated nature reserves within 30–40 years	– Climate change

Similarly, snow accumulates in winter in the high parts of the cordillera in Peru and melts during summer, becoming the main source of water for many rivers in Latin America (see Section 14.1.3.1.1). In addition, glaciers act as buffers that regulate runoff water supply from mountains to plains during dry and wet spells. Thus, tropical glaciers are instrumental in securing agricultural productivity and livelihoods and provide cultural inspiration for millions of people who live remote from their sources.

Because of the narrow range of ambient temperatures in the tropics, tropical glaciers are more sensitive to climate change than glaciers elsewhere (see Section 4.3.11). Indeed, records spanning several decades show accelerated retreat of several Himalayan and other tropical glaciers (see Section 11.2.1.2).

In the transient phase of melting, increasing discharge will generate floods in the mountains and immediate vicinity, increased siltation of rivers, and larger sediment load in dams and reservoirs. Riparian mountain ecosystems will be impacted during their dry seasons—in the transient phase by a significant increment of downstream flow, as well as following the transient phase—by significant reduction of this flow. These changes will have tangible economic and cultural implications (see Section 11.2.1.2). This example of a tropical unique entity provides an "early warning" for nontropical glaciers and their potential impacts.

19.3.3. Biological Systems

As discussed in Section 19.2, change in climate already appears to be affecting many biological systems. Continued climate changes can threaten a large number of unique biological systems. This section identifies specific characteristics of some of the most unique and threatened systems, which explain why many are at risk from climate change. In addition, some specific examples of unique and threatened biological systems are presented. Many others also are threatened by climate change; these are discussed in detail in other chapters of this report. Examples of natural systems that may be threatened include montane ecosystems that are restricted to upper 200–300 m of mountainous areas, prairie wetlands, remnant native grasslands, coldwater and some coolwater fish habitat, ecosystems that overlie permafrost, and ice-edge ecosystems that provide habitat for polar bears and penguins. Examples of species that may be threatened by changes in climate include forest birds in Tanzania, the resplendent quetzal in Central America, the mountain gorilla in Africa, amphibians that are endemic to the cloud forests of the neotropics, and the spectacled bear of the Andes.

19.3.3.1. Risks to Species and Ecosystems

Laboratory and field studies have demonstrated that climate plays a strong role in limiting the ranges of species and ecosystems. Species already are responding to changes in regional climate, with altered population sizes and breeding times or flowering dates that occur earlier in the season (see Chapter 5). These responses suggest that many unique species will undergo complex changes with a few degrees of warming, which could lead to extinction in many locations. Such species can be found across various regions (see Table 19-1). Other chapters in this report list many examples (see Table 19-2). However, projecting possible responses of wild animal and plant species is extremely difficult for most species because there are many possible biological interactions and confounding factors, such as habitat destruction and invasive species.

***Table 19-2**: Threatened and unique entities identified in WGII TAR.*

Chapter	Entity
4. Water Resources	– Endorheic lakes: Caspian and Aral Seas, Lake Balkash, Lake Chad, Lake Titicaca, Great Salt Lake – Glaciers (in general, no particular reference)
5. Ecosystems and Their Services	– Some butterfly species in United States and Europe – Leadbetters's possum in Australia – Cape Floral Kingdom, South Africa
6. Coastal Zones and Marine Ecosystems	– Coral reefs
7. Human Settlements	– Coastal settlements along North Sea coast in northwest Europe, the Seychelles, parts of Micronesia, Gulf Coast of United States and Mexico, Nile delta, and Bay of Bengal
10. Africa	– Cape Floral Kingdom and Succulent Karoo
11. Asia	– Biodiversity of Lake Baikal – Glaciers in the Tianshan, Hindukush Himalayas; permafrost in Tibet – Mangroves
12. Australasia	– Alpine ecosystems, snow and glaciers in New Zealand, wetlands in Kakadu National Park, Queensland fruit fly – Indigenous communities
13. Europe	– Snowpack and permafrost in the mountains
14. Latin America	– Mountain glaciers
15. North America	– Mountain glaciers – Sardine population – Indigenous communities
16. Polar Regions (Arctic and Antarctic)	– Indigenous communities
17. Small Island States	– Mangroves and seagrass beds – Coral reefs

Species that make up a natural community, however, most likely will not shift together (Davis, 1986; Overpeck *et al*., 1994; Root, 2000). This could break apart established natural communities and create newly evolving assemblages. Depending on the magnitude and duration of the environmental disturbance, some or all individuals of a given species may shift out of an area. This, in turn, can cause a local (or even the overall) population size to decline. Superimposed on these potential changes are those caused by land-use change, which frequently fragments populations into patches throughout their ranges.

Species with wide nonpatchy ranges, rapid dispersal mechanisms, and a large population normally are not in danger of extinction [e.g., European house sparrow (*Passer domesticus)* and many weedy plant species]. Those with narrow patchy ranges and small populations frequently are endangered and may require management for survival [e.g., most crane species (*Gruidea spp*.)]. In summary, species tend to become rarer when ranges shift from wide to narrow, available habitat becomes patchier, and population size declines (Huntley *et al*., 1997). Indeed, a species is likely to become extinct if it is forced into a narrow patchy range and its population declines—a probability that is enhanced when environmental disturbances such as climate change, along with companion transient changes, occur.

Even when conservation management of rare species is effective, survival still may be problematic because in a small population, genetically similar individuals may breed, which decreases genetic variability. This, in turn, may reduce adaptability to stresses, thereby further lowering population size and decreasing the types of habitat within which the species could survive. Environmental catastrophes such as hurricanes, oil spills, extreme temperatures, and drought can trigger the extinction of even well-managed rare species. The only way to reduce the risk of extinction brought about by catastrophes is to increase population sizes and maintain corridors between isolated populations.

19.3.3.2. Biodiversity Hot Spots

Biodiversity "hot spots" are areas that feature exceptional concentrations of species, including many endemic species. Unfortunately, many such hot spots also experience large habitat losses. In addition to a hot spot's economic, social, and cultural significance to local people, the uniqueness of its biodiversity and its high share of global biodiversity give the hot spot a global value. Thus, biodiversity hot spots qualify as unique and threatened entities.

Myers *et al.* (2000) define a hot spot as an area featuring a biogeographic unit that contains at least 0.5% of the world's 300,000 vascular plant species as endemics and has lost 70% or more of its primary vegetation. Table 19-3 shows that two-thirds of the hot spots listed in Myers *et al.* (2000) are in the tropics, some of which have the highest percentage of global plants (6.7%) and as much as 28% of area of habitat with primary vegetation. Arctic and boreal biomes, however—which are devoid of hot spots—will have the greatest changes in temperature and precipitation by 2100, whereas the exposure of nearly all hot spots to a global change of 4°C and/or 30% of precipitation is ranked only 3 (on a 1 to 5 scale proposed by Sala *et al.*, 2000). With respect to biome-specific exposure, climate is expected to warm most dramatically at high latitudes, change least in the tropics, and show intermediate changes in other biomes. Indeed, Table 19-3 shows that the tropical hot spots are least vulnerable to climate change and elevated CO_2 (0.12 and 0.10, respectively, on a scale of 0 to 1), whereas the eight Mediterranean and savanna hot spots are at least twice as vulnerable (0.24 and 0.30 for climate change and elevated CO_2, respectively—Sala *et al.*, 2000).

The Cape Floral Kingdom (also called the Cape Floristic Province) and the adjacent succulent Karoo in South Africa are examples of Mediterranean and savanna biodiversity hot spots that very much qualify as unique and threatened entities. The Cape Floral Kingdom is sixth in the world in plant richness of species (5,682 endemic species—Cowling and Hilton-Taylor, 1997). These hotspots are vulnerable for the following reasons:

- Their mountains have no permanent snow cover to which high montane species can retreat as climate warms.
- Montane endemic plants already are concentrated near the peaks, with little or no possibility for altitudinal expansion.
- Endemics are concentrated in the southwestern corner of Africa, with no possibility for latitudinal shifts farther south (except for the extreme southern tip of the continent, which is intensively farmed).
- Increased frequency of fires and drought will affect many short-lived and fire-sensitive species; seedlings that germinate after fires will be exposed to successively more extreme climate conditions.

The succulent Karoo flora may be effectively lost with a mean annual temperature increase of 3–4°C (Rutherford *et al.*, 1999), owing to changing fire regimes, loss of specialist pollinators, and increased frequency of drought. Tropic hot spots that are not as sensitive as the Cape Floral Kingdom also will be seriously affected if other anthropogenic drivers act synergistically (Sala *et al.*, 2000). Thus, although the hot spot analysis (Myers *et al.*, 2000) indicates that much of the problem of current and projected mass extinction could be countered by protection of the 25 hot spots, the ability of these hot spots to be sources of biodiversity is threatened by climate change.

19.3.3.3. Ecotones

Ecotones are transition areas between adjacent but different environments: habitats, ecosystems, landscapes, biomes, or ecoclimatic regions (Risser, 1993). Ecotones that are unique

***Table 19-3**: Sensitivity of biodiversity hot spots (Myers et al., 2000; Sala et al., 2000).*

Biome	Number of Hotspots Biome	% of Global Plants (range)	% of Remaining Habitats with Primary Vegetation (range)	Impact by 2100 (of a large change in driver, scale 1–5)		Effect by 2100 (expected change in driver x impact, scale 0–1)	
				of climate change	of elevated CO_2	of climate change	of elevated CO_2
Tropical forests	15	0.5–6.7	3–28	3	1	0.12	0.10
Mediterranean	5	0.7–4.3	5–30	3	2	0.24	0.20
Savanna, grassland	3	0.6–1.5	20–27	3	3	0.23	0.30
North temperate forest	2	0.5–1.2	8–10	2	1.5	0.17	0.15

entities in the context of climate change are transition zones between ecoclimatic regions. Ecotones have narrow spatial extent, a steep ecological gradient and hence high species richness (Risser, 1993), a unique species combination, genetically unique populations (Lesica and Allendorf, 1994), and high intra-species genetic diversity (Safriel *et al.*, 1994).

Ecotones affect distant and larger areas: They regulate interactions between biomes by modifying flows between them (Johnston, 1993; Risser, 1993); they generate evolutionary diversity (Lesica and Allendorf, 1994); and they serve as repositories of genetic diversity to be used for rehabilitation of ecosystems in adjacent ecoclimatic regions if and when these ecosystems lose species because of climate change (see Section 11.3.2.2.2; Volis *et al.*, 1998; Kark *et al.*, 1999). Conservation of ecotone biodiversity therefore is an adaptation. Finally, although ecological changes in response to climate change will occur everywhere, the signals will be detectable first in ecotones (Neilson, 1993). This sensitivity makes them indicators that provide early warning for other regions (Risser, 1993).

Although ecotones are unique in provision of climate change-related services, they are threatened. Conservation traditionally is aimed at "prime" core areas of biomes rather than ecotones. Even conservation efforts that are directed at ecotones may not suffice, however: 47–77% of the areas of biosphere reserves are predicted to experience change in ecosystem types, compared to only 39–55% of the total global terrestrial area that will undergo such changes (Leemans and Halpin, 1992; Halpin, 1997).

An example of a threatened ecotone is the desert/nondesert ecoclimatic transition zone—the semi-arid drylands sandwiched between arid and the dry subhumid drylands (Middleton and Thomas, 1997). Semi-arid drylands are prone to desertification, expressed as irreversible loss of soil productivity because of topsoil erosion (see Section 11.2.1.4). Already affected by extreme soil degradation are 67 Mha of semi-arid drylands (2.9% of global semi-arid area)—nearly as much as affected dry-subhumid drylands (28 Mha, 2.2%) and arid drylands (43 Mha, 2.7%—Middleton and Thomas, 1997). This degradation is destroying the habitats of the biodiversity assets of these ecotones, including those to be conserved as an adaptation to climate change (Safriel, 1999a,b).

Climate change is expected to exacerbate desertification (see Section 11.2.1.4; Schlesinger *et al.*, 1990; Middleton and Thomas, 1997). Reduced precipitation and increased evapotranspiration will change ecotones' spatial features (e.g., coalescence of patches at one side and increased fragmentation at the other—Neilson, 1993). Furthermore, overexploitation of vegetation that is typical in semi-arid drylands (UNDP, 1998; ICCD, 1999), in synergy with climate change, will further increase habitat loss and hence loss of biodiversity, ecosystem services, and the potential for adaptation. Similar synergies between climate change effects and other anthropogenic impacts are projected for alpine ecotones (Rusek, 1993). To conclude, ecotones between biomes and within climatic transition areas are unique entities; they are important for monitoring climate change and for adapting to climate change, yet they are highly threatened by climate change interacting with other anthropogenic stresses.

19.3.3.4. Coral Reefs

Coral reefs are restricted to narrow latitudinal, horizontal, and vertical ranges along the tropical continental shelves. Their contribution to global coastal biodiversity is disproportionate to their spatial extent: Although they cover less than 1% of the world's oceans, they are inhabited by one-third of globally known marine species (Reaka-Kudla, 1996). Coral reefs have far-reaching effects; they are nurseries for many ocean fish species, and they protect coastlines from wave impact and erosion (see Section 11.2.4.3). Thus, fisheries, tourism, infrastructures, societies, and cultures depend on the well-being of this unique entity that is impacted by increased temperature, atmospheric CO_2, and sea level, synergistically combined with anthropogenic stresses that are independent of climate change.

Many reef-building coral species already live close to their upper thermal limit (see Section 6.4.5). If they are exposed to moderate increases (1–2°C) in water temperature, they become stressed and experience bleaching (Goreau *et al.*, 1998; Hoegh-Guldberg, 1999). The increasing frequency of coral bleaching events during the past decade provides a reason for concern for this warming-induced impact (see Section 12.4.2.3). For example, 50–90% bleaching-induced mortality in the Indian Ocean reefs was associated with a 2–6°C above-normal sea-surface maximum triggered by El Niño during 1997–1998; several other severe bleaching events occurred in the 1982–1983 and 1987 El Niño years (Glynn, 1991; Wilkinson *et al.*, 1999; see Figure 19-3).

Defining the upper thermal thresholds of corals and using global warming scenarios, Hoegh-Guldberg (1999) found that the frequency of bleaching is expected to rise until they become annual events in most oceans (at about a 1°C warming). In some areas, bleaching events would happen more frequently as early as 2020 (with less than 0.5°C warming); within the next 30–50 years, bleaching could be triggered by normal seasonal changes in seawater temperature, and most regions may experience severe bleaching conditions every year. This trend exceeds the frequency at which corals can effectively recover from bleaching-related mortality (Hoegh-Guldberg, 1999).

Besides the detrimental effect of temperature rise, increased atmospheric CO_2 concentrations reduce coral calcification rates (Gattuso *et al.*, 1999; Kleypas *et al.*, 1999; see Sections 6.4.5 and 12.4.1.6), which already might have decreased and could decrease an additional 10–30% by 2100 (see Chapter 6). A 10–20% decrease in calcium carbonate production may impair expansion of coral reefs into higher latitudes as a response to predicted increasing SST (Kleypas *et al.*, 1999). Healthy reef flats may benefit from projected increased sea level because they would be able to keep up with the projected rise in sea level. However, any increase in the frequency of El Niño and

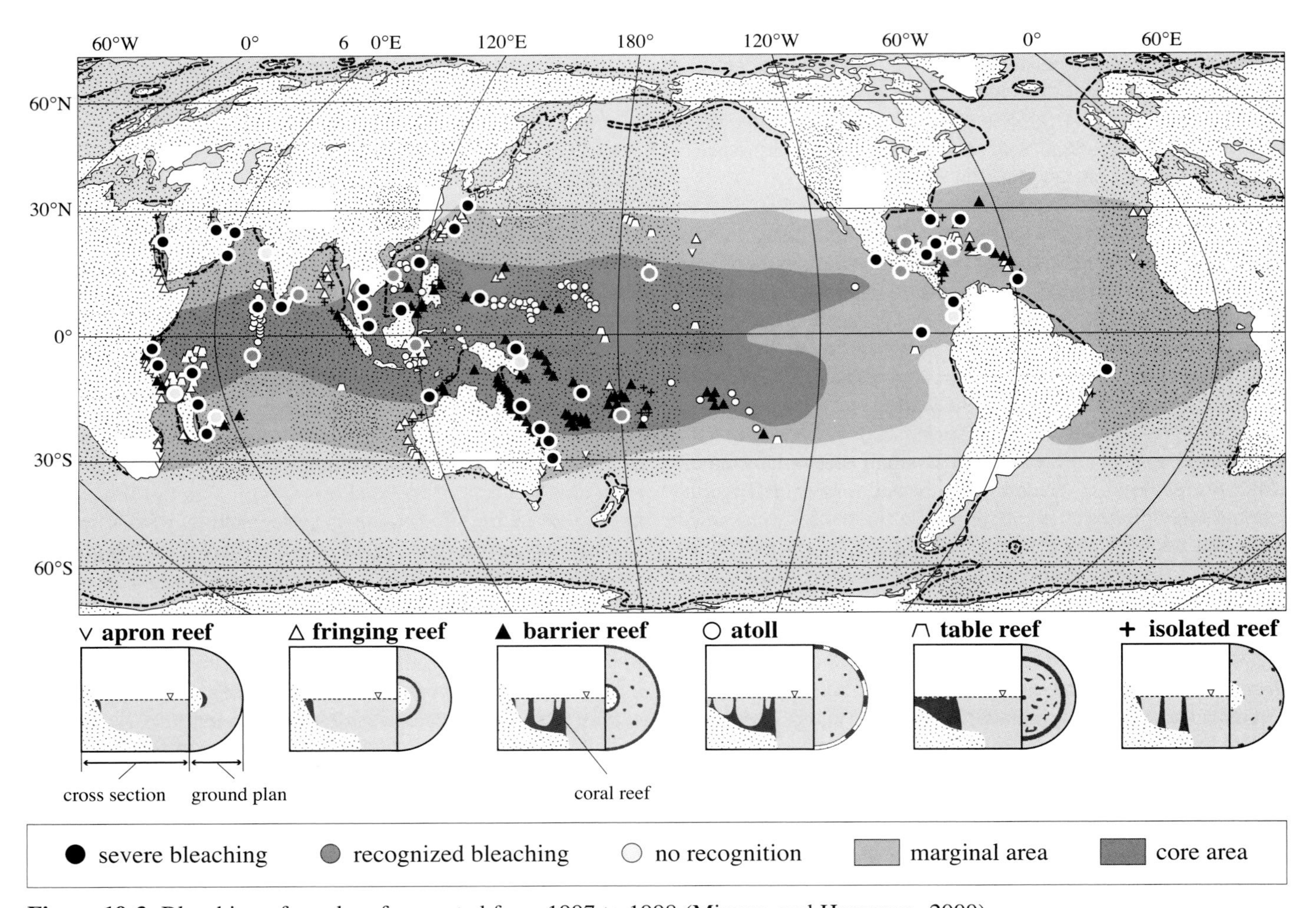

Figure 19-3: Bleaching of coral reefs reported from 1997 to 1998 (Mimura and Harasawa, 2000).

other ocean-atmosphere interactions, such as Indian Ocean dipole events, will lead to regular and prolonged sea-level depressions (10–40 cm) in the western Pacific and eastern Indian Ocean, with adverse effects on shallow reefs in their regions (see Chapter 6).

19.3.3.5. Mangrove Ecosystems

Mangrove ecosystems resemble coral reefs in that they have a narrow global distribution: They cover 11,500 km^2 in all of Australia, and the largest mangrove forest in the world—the Sundarbans of Bangladesh and India—covers 6,000 km^2 (see Section 11.2.4.1). Yet they have a rich biodiversity and a significant effect on adjacent and distant systems. Mangroves are made up of salt-adapted, intertidal evergreen trees on low-energy, sedimentary tropical shorelines, extending landward in lagoons, estuary margins, and tidal rivers (see Sections 6.4.4 and 11.2.4.1).

Mangrove ecosystems are highly vulnerable to sea-level rise induced by climate change, which will change the salinity distribution and inundate mangroves. For example, a 45-cm rise could inundate 75% of the Sundarbans (see Section 11.4.1); a 1-m rise will completely inundate the Sundarbans (see Section 11.2.1.6). In addition, redistribution of species whose habitats will be affected by inundation may be impaired because migration, especially to the north, will be blocked by human settlements. Loss of productivity, species, and ecosystem goods and services therefore is expected. Climate change effects will be further exacerbated, and vulnerability to climate change will increase human-induced damage. For example, between 56 and 75% of different Asian mangrove forests have been lost during the 20th century because of overexploitation and replacement by aquaculture installations (see Section 11.1.3.1). Like the Sundarbans, the Port Royal mangrove wetland in Jamaica may completely collapse with a 1-m sea-level rise (see Section 17.2.4.2).

On the other hand, although mangroves are vulnerable, some may be adaptable to climate change (see Section 12.4.1.6) because they could survive in areas where vertical accretion equals sea-level rise. Because sediment flux determines mangrove response to sea-level rise and fluxes vary between regions and locations, the fate of the world's mangrove ecosystems will not be uniform (see Section 6.4.4). Yet even where accretion will offset sea-level rise, any infrastructure will limit the potential for landward migration of coastal mangrove species and habitats (see Section 6.2). Thus, at least some if not many of the world's mangrove ecosystems are unique and threatened entities.

19.3.4. Human Systems

Some human systems also are unique and threatened by climate change. These tend to be poor and isolated communities that are tied to specific locations or ecosystems. Among the unique and threatened human systems are some small island states and indigenous communities.

19.3.4.1. Threatened Small Island States

Because of their low elevation and small size, many small island states are threatened with partial or virtually total inundation by future rises in sea level. In addition, increased intensity or frequency of cyclones could harm many of these islands. The existence or well-being of many small island states is threatened by climate change and sea-level rise over the next century and beyond.

Many small island states—especially the atoll nations of the Pacific and Indian Oceans—are among the most vulnerable to climate change, seasonal-to-interannual climate variability, and sea-level rise. Much of their critical infrastructure and many socioeconomic activities tend to be located along the coastline—in many cases at or close to present sea level (Nurse, 1992; Pernetta, 1992; Hay and Kaluwin, 1993). Coastal erosion, saline intrusion, sea flooding, and land-based pollution already are serious problems in many of these islands. Among these factors, sea-level rise will pose a serious threat to the ecosystems, economy, and, in some cases, existence of many small island states. It is estimated that 30% of known threatened plant species are endemic to such islands, and 23% of bird species found on these islands are threatened (Nurse *et al.*, 1998). Projected future climate change and sea-level rise will lead to shifts in species composition (see Chapter 17).

Many small island nations are only a few meters above present sea level. These states may face serious threat of permanent inundation from sea-level rise. Among the most vulnerable of these island states are the Marshall Islands, Kiribati, Tuvalu, Tonga, the Federated States of Micronesia, and the Cook Islands (in the Pacific Ocean); Antigua and Nevis (in the Caribbean Sea); and the Maldives (in the Indian Ocean). Small island states may face the following types of impacts from sea-level rise and climate change (Gaffin, 1997; Nurse *et al.*, 1998):

- Increased coastal erosion
- Changes in aquifer volume and water quality with increased saline intrusion
- Coral reef deterioration resulting from sea-level rise and thermal stress
- Outmigration caused by permanent inundation
- Social instability related to inter-island migration
- Loss of income resulting from negative effects on tourist industry
- Increased vulnerability of human settlement due to decrease in land area
- Loss of agriculture and vegetation.

Gaffin (1997) concludes that without planned adaptation, the vulnerabilities of small island states are as follows:

- An 80-cm sea-level rise could inundate two-thirds of the Marshall Islands and Kiribati.[2]
- A 90-cm sea-level rise could cause 85% of Male, the capital of the Maldives, to be inundated (Pernetta, 1989).

19.3.4.2. Indigenous Communities

Indigenous people often live in harsh climatic environments to which their culture and traditions are well adapted. Indigenous people generally have low incomes and inhabit isolated rural environments and low-lying margins of large towns and cities. Therefore, they are more exposed to social problems of economic insecurity, inadequate water supplies, and lower health standards (see Sections 12.2.5 and 15.3.2.8). These inadequacies in social safety nets indeed put them at greater risk of climate-related disasters and their effects (see Section 12.7.2.4).

For many reasons, indigenous communities are unique and threatened by climate change. First, they are more vulnerable to climate-related disasters such as storms, floods, and droughts because of inadequate structural protection measures and services, as well as to any increase in the prevalence of pests and diseases—especially vector-borne, respiratory, or otherwise infectious diseases (Woodward *et al.*, 1998; Braaf, 1999). Second, their lifestyles are tied to current climate and vegetation and wildlife. Third, changes in current climate could threaten these lifestyles and would present these peoples with difficult choices concerning their future.

Native peoples in the Mackenzie basin in Canada are an example of an indigenous community that is threatened by climate change (Cohen 1994, 1996, 1997a,b,c). The Mackenzie basin is a watershed that extends from the mid-latitudes to the subarctic in northwest Canada. Over the past 35 years, the area has been experiencing a rapid temperature increase of about 1°C per decade. The changes in temperature also are changing the landscape of the basin as permafrost melts, landslides and forest fires increase, and water levels are lowered.

For the native people in the basin, wildlife is the important natural resource; it is harvested by hunting, fishing, and trapping. It is critically important in economic terms—primarily as a source of food, income, and traditional clothing—but inseparable from the cultural importance for maintaining traditional systems

[2]The estimate of land loss based on a 1-m sea-level rise, resulting in 80% island losses (IPCC, 1996).

of knowledge and identity (Pinter, 1997). As noted, changes in the climate in the basin would have substantial impacts on water resources and vegetation. Changes in forest fire frequencies would lead to cumulative impacts on wildlife, including terrestrial, aquatic, and bird species. For example, because of a decrease in water availability, muskrats already have disappeared from the Peace Athabasca delta (Pinter, 1997). In this area, trapping once was a major industry, but this economic activity has now disappeared. Thus, changes in ecosystem resource bases will have direct impacts on native lifestyles in the Mackenzie basin (Cohen *et al.*, 1997a).

Some important changes are expected in native lifestyles in the Mackenzie Basin regardless of climate change. For example, an increasing number of people will seek their livelihoods in the wage economy, and migration to other areas will intensify. These changes could result in a decline in cultural values and heritage that are thousands of years old. If climate change adversely affects the lifestyle of the indigenous community, this decline could be accelerated.

19.3.5. Conclusions

There are many unique and threatened systems distributed over various regions of the world. Although they are restricted to relatively narrow geographical ranges, they can affect other entities beyond their range. The existence or functioning of some of these systems is threatened by a small temperature change; the existence or functioning of many others will be threatened by a medium to large temperature change. These effects include impacts such as loss of many species and ecosystems, disappearance of tropical glaciers, damage to coral reefs, inundation of some low-lying islands, loss of coastal wetlands, and potential harm to aboriginal societies and their cultures.

Many of these systems already are stressed by development, including pollution, habitat destruction, encroachment for expansion of human habitation, and overextraction of natural resources. The combination of climate change and societal development puts these systems at greater risk. In some cases, climate change hastens the destruction of these systems; in other cases it may result in the destruction of systems that could survive societal stresses alone (e.g., small island states and some mangrove ecosystems such as the Sundarbans).

Removing societal stresses and managing resources in a sustainable manner may help some unique and threatened systems cope better with climate change.

19.4. Distribution of Impacts

A second reason for concern is the distribution of impacts among people and across regions. The impacts of climate change will not be distributed equally. Some individuals, sectors, systems, and regions will be less affected—or may even benefit; other individuals, sectors, systems, and regions may suffer significant losses. This pattern of relative benefits or losses is not likely to remain constant over time. It will be different with different magnitudes of climate change. Some regions may have gains only for certain changes in temperature and precipitation and not for others. As a result, some regions that may first see net benefits eventually may face losses as well as the climate continues to warm.

19.4.1. Analysis of Distributional Incidence: State of the Art

Research into the distribution of impacts of climate change is in its infancy, in large measure because this research poses several methodological challenges.

A first difficulty is synthesis—the need to reduce the complex pattern of individual impacts to a more tractable set of regional or sectoral indicators. The challenge is to identify a set of indicators that can summarize and make comparable the impacts in different regions, sectors, or systems in a meaningful way. A range of indicators and methods have been put forward. Many models use physical measures such as the number of people affected (e.g., Hoozemans *et al.*, 1993), change in net primary productivity (NPP) (White *et al.*, 1999), or the number of systems undergoing change (e.g., Alcamo *et al.*, 1995).

The most widespread numeraire, however, is economic cost (Nordhaus, 1991, 1994a; Cline, 1992; Hohmeyer and Gaertner, 1992; Titus, 1992; Downing *et al.*, 1995, 1996; Fankhauser, 1995; Tol, 1995; Mendelsohn and Neumann, 1999). This numeraire is particularly well-suited to measure market impacts—that is, impacts that are linked to market transactions and directly affect GDP (i.e., a country's national accounts). The costs of sea-level rise, for example, can be expressed as the capital cost of protection plus the economic value of land and structures at loss or at risk; agricultural impacts can be expressed as costs or benefits to producers and consumers, including the incremental costs of adaptation. Using a monetary numeraire to express nonmarket impacts such as effects on ecosystems or human health is more difficult. It is possible in principle, however. There is a broad and established literature on valuation theory and its application, including studies (mostly in a nonclimate change context) on the monetary value of lower mortality risk, ecosystems, quality of life, and so forth. However, economic valuation can be controversial and requires sophisticated analysis, which still is mostly lacking in a climate change context.

Physical metrics—such as NPP or percentage of systems affected—on the other hand, are best suited for natural systems. When they are applied to systems under human management, they suffer from being poorly linked to human welfare, the ultimate indicator of concern. Some researchers therefore recommend different numeraires for market impacts, mortality, ecosystems, quality of life, and equity (Schneider *et al.*, 2000b). They recognize, however, that final comparisons across different numeraires nonetheless are required; they regard this as the job of policymakers, however.

Persistent knowledge gaps is a second source of difficulty. Distributional analysis depends heavily on the geographical details of climate change, but these details are one of the major uncertainties in the outputs of climate change models. This is particularly true for estimates of precipitation; for example, estimates of water-sector impact can vary widely depending on the choice of GCM.[3] Uncertainties continue at the level of impact analysis. Despite a growing number of country-level case studies, our knowledge of local impacts is still too uneven and incomplete for a careful, detailed comparison across regions. Furthermore, differences in assumptions often make it difficult to compare case studies across countries. Only a few studies try to provide a coherent global picture on the basis of a uniform set of assumptions. The basis of most such global impact assessments tends to be studies undertaken in developed countries—often the United States—which are then extrapolated to other regions. Such extrapolation is difficult and will be successful only if regional circumstances are carefully taken into account, including differences in geography, level of development, value systems, and adaptive capacity. Not all analyses are equally careful in undertaking this task.

There are other shortcomings that affect the quality of analysis. Although our understanding of the vulnerability of developed countries is improving—at least with respect to market impacts—information about developing countries is quite limited. Nonmarket damages, indirect effects (e.g., the effect of changed agricultural output on the food-processing industry), the link between market and nonmarket effects (e.g., how the loss of ecosystem functions will affect GDP), and the sociopolitical implications of change also are still poorly understood. Uncertainty, transient effects (the impact of a changing rather than a changed and static climate), and the influence of climate variability are other factors that deserve more attention. Because of these knowledge gaps, distributional analysis has to rely on (difficult) expert judgment and extrapolation if it is to provide a comprehensive picture.

A third problem is adaptation. There has been substantial progress in the treatment of adaptation since the SAR, but adaptation is difficult to capture adequately in an impact assessment. Adaptation will entail complex behavioral, technological, and institutional adjustments at all levels of society, and the capacity to undertake them will vary considerably (see Chapter 18). Various approaches are used to model adaptation (e.g., spatial analogs, microeconomic modeling), but they are prone to systematic errors about its effectiveness. The standard approach used in coastal impact assessment and in many agricultural models is to include in the analysis a limited number of "prominent" but ultimately arbitrary adaptations. This underestimates adaptive capacity because many potentially effective adaptations are excluded (Tol *et al*., 1998). On the other hand, approaches that are based on analogs—such as the Ricardian approach used by, for example, Mendelsohn *et al*. (1994), Mendelsohn and Dinar (1998), and Darwin (1999)—probably overestimate adaptive capacity because they neglect the cost of transition and learning. This is especially true for cases in which adaptation in developed countries today is used as a proxy for worldwide adaptation to an uncertain future climate. Only a very few studies model adaptation as an optimization process in which agents trade off the costs and benefits of different adaptation options (Fankhauser, 1995; Yohe *et al*., 1995, 1996).

The analysis is further complicated by the strong link between adaptation and other socioeconomic trends. The world will change substantially in the future, and this will affect vulnerability to climate change. For example, a successful effort to roll back malaria (as promoted by the development community) could reduce the negative health effects of climate change. On the other hand, growing pressure on natural resources from unsustainable economic development is likely to exacerbate the impacts of

Box 19-2. The Impact of Climate Change on Coastal Zones

The impact of sea-level rise has been widely studied for many parts of the world. Although uncertainties remain, several generic conclusions can be drawn. First, impacts will not be distributed evenly. Islands and deltas are particularly vulnerable. Second, forward-looking and sustainable economic development, coupled with efficient adaptation (mostly protection of vulnerable shores), can significantly reduce the economic impact of sea-level rise. Some analysts have even found that coastal vulnerability may decrease if the rate of economic development is sufficiently high and climate change sufficiently slow. However, not all countries will be able to undertake the necessary adaptation investments without outside financial assistance, and uncertainty about sea levels (e.g., as a result of storm surges) may make it difficult to identify efficient policies. Third, coastal wetlands can cope with a relatively modest rate of sea-level rise, but not with a fast one. Additional wetlands could be lost if their migration is blocked by hard structures built to protect developed coastal areas. Fourth, most of the impact will not be through gradual sea-level rise but through extreme events such as floods and storms. This makes people without insurance or a strong social network especially vulnerable. Thus, as a whole, sea-level rise is likely to have strong negative effects on some people, even if the aggregate impact is limited. Fifth, the aggregate impact of sea-level rise could be roughly proportional to the observed rise. At a local scale, however, sea-level rise is more likely to be felt through successive crossings of thresholds.

[3]For example, Frederick and Schwarz (1999) found that climate changes estimated in the southeastern United States in the 2030s under the Canadian Climate Centre scenario result in an estimated US$100 billion yr^{-1} in damages. This estimate may be the result of internal model variability and does not fully account for adaptive responses or lower damages from reduced flood risks. Nonetheless, it demonstrates the high sensitivity of water resources to extreme changes in climate.

climate change on natural systems. Even without explicit adaptation, impact assessments therefore depend on the "type" of socioeconomic development expected in the future. The sensitivity of estimates to such baseline trends can be strong enough in some cases to reverse the sign (i.e., a potentially negative impact can become positive under a suitable development path, or vice versa) (Mendelsohn and Neumann, 1999).

Despite the limits in knowledge, a few general patterns emerge with regard to the distribution of climate change impacts. These patterns are derived from general principles, observations of past vulnerabilities, and limited modeling studies.

19.4.2. Distribution of Impacts by Sector

Susceptibility to climate change differs across sectors and regions. A clear example is sea-level rise, which mostly affects coastal zones (see Box 19-2). People living in the coastal zone generally will be negatively affected by sea-level rise, but the numbers of people differ by region. For example, Nicholls *et al.* (1999) found that under a sea-level rise of about 40 cm by the 2080s, assuming increased coastal protection, 55 million people would be flooded annually in south Asia; 21 million in southeast Asia, the Philippines, Indonesia, and New Guinea; 14 million in Africa; and 3 million in the rest of the world. The relative impacts in small island states also are significant (see Section 19.3). In addition, the Atlantic coast of North and Central America, the Mediterranean, and the Baltic are projected to have the greatest loss of wetlands. Inland areas face only secondary effects—which, unlike the negative primary effects, may be either negative or positive (Yohe *et al.*, 1996; Darwin and Tol, 2001).

Agriculture, to turn to another example, is a major economic sector in some countries and a small one in others. Agriculture is one of the sectors that is most susceptible to climate change, so countries with a large portion of the economy in agriculture face a larger exposure to climate change than countries with a lower share, and these shares vary widely. Whereas countries of the Organisation for Economic Cooperation and Development (OECD) generate about 2–3% of their GDP from agriculture, African countries generate 5–58% (WRI, 1998).

Activities at the margin of climatic suitability have the most to lose from climate change, if local conditions worsen, and the most to win if conditions improve. One example is subsistence farming under severe water stress—for instance, in semi-arid regions of Africa or south Asia. A decrease of precipitation, an increase in evapotranspiration, or higher interannual variability (particularly longer droughts) could tip the balance from a meager livelihood to no livelihood at all, and the unique cultures often found in marginal areas could be lost. An increase in precipitation, on the other hand, could reduce pressure on marginal areas. Numerous modeling studies of shifts in production of global agriculture—including Kane *et al.* (1992), Rosenzweig and Parry (1994), Darwin *et al.* (1995), Leemans (1997), Parry *et al.* (1999), and Darwin (1999)—have estimated that production in high-latitude countries is likely to increase and production in low-latitude countries is likely to decrease, even though changes in total global output of agriculture could be small. Results in the temperate zone are mixed. Low-latitude countries tend to be least developed and depend heavily on subsistence farming. Under current development trends they will continue to have a relatively high share of GDP in agriculture. Thus, the impacts of declines in agricultural output on low-latitude countries are likely to be proportionately greater than any gains in high-latitude countries (see Box 19-3).

Box 19-3. The Impact of Climate Change on Agriculture

The pressures of climate change on the world's food system are better understood than most other impacts. Research has focused on crop yields; on the basis of those insights, many studies also look at farm productivity, and a smaller number look at national and international agricultural markets.

Climate change is expected to increase yields at higher latitudes and decrease yields at lower latitudes. Changes in precipitation, however, also can affect yields and alter this general pattern locally and regionally. Studies of the economic impact of this change (in all cases, climate change associated with $2xCO_2$) conclude that the aggregated global impact on the agricultural sector may be slightly negative to moderately positive, depending on underlying assumptions (e.g., Rosenzweig and Parry, 1994; Darwin, 1999; Parry *et al.*, 1999; Mendelsohn *et al.*, 2000). Most studies on which these findings are based include the positive effect of carbon fertilization but exclude the negative impact of pests, diseases, and other disturbances related to climate change (e.g., droughts, water availability). The aggregate also hides substantial regional differences. Beneficial effects are expected predominantly in the developed world; strongly negative effects are expected for populations that are poorly connected to regional and global trading systems. Regions that will get drier or already are quite hot for agriculture also will suffer, as will countries that are less well prepared to adapt (e.g., because of lack of infrastructure, capital, or education). Losses may occur even if adaptive capacity is only comparatively weak because trade patterns will shift in favor of those adapting best. Overall, climate change is likely to tip agriculture production in favor of well-to-do and well-fed regions—which either benefit, under moderate warming, or suffer less severe losses—at the expense of less-well-to-do and less well-fed regions. Some studies indicate that the number of hungry and malnourished people in the world may increase, because of climate change, by about 10% relative to the baseline (i.e., an additional 80–90 million people) later in the 21st century (e.g., Parry *et al.*, 1999).

Vulnerability to the health effects of climate change also differs across regions and within countries, and differences in adaptive capacity again are important. Box 19-4 notes that wealthier countries will be better able to cope with risks to human health than less wealthy countries. Risks also vary within countries, however. In a country such as the United States, the very young and the very old are most sensitive to heat waves and cold spells, so regions with a rapidly growing or rapidly aging population would have relatively large exposure to potential health impacts. In addition, poor people in wealthy countries may be more vulnerable to health impacts than those with average incomes in the same countries. For example, Kalkstein and Greene (1997) found that in the United States, residents of inner cities, which have a higher proportion of low-income people, are at greater risk of heat-stress mortality than others. Differences among income groups may be more pronounced in developing and transition countries because of the absence of the elaborate safety nets that developed countries have constructed in response to other, nonclimate stresses.

These observations underscore one of the critical insights in Chapter 18: Adaptive capacity differs considerably between sectors and systems. The ability to adapt to and cope with climate change impacts is a function of wealth, technology, information, skills, infrastructure, institutions, equity, empowerment, and ability to spread risk. The poorest segments of societies are most vulnerable to climate change. Poverty determines vulnerability via several mechanisms, principally in access to resources to allow coping with extreme weather events and through marginalization from decisionmaking and social security (Kelly and Adger, 2000). Vulnerability is likely to be differentiated by gender—for example, through the "feminization of poverty" brought about by differential gender roles in natural resource management (Agarwal, 1991). If climate change increases water scarcity, women are likely to bear the labor and nutritional impacts.

The suggested distribution of vulnerability to climate change can be observed clearly in the pattern of vulnerability to natural disasters (e.g., Burton *et al.*, 1993). The poor are more vulnerable to natural disasters than the rich because they live in more hazardous places, have less protection, and have less reserves, insurance, and alternatives. Adger (1999), for instance, shows that marginalized populations within coastal communities in northern Vietnam are more susceptible to the impacts of present-day weather hazards and that, importantly, the wider policy context can exacerbate this vulnerability. In the Vietnamese case, the transition to market-based agriculture has decreased

Box 19-4. The Health Impacts of Climate Change

Global climate change will have diverse impacts on human health—some positive, most negative. Changes in the frequency and intensity of extreme heat and cold, floods and droughts, and the profile of local air pollution and aeroallergens will directly affect population health. Other effects on population health will result from the impacts of climate change on ecological and social systems. These impacts include changes in infectious disease occurrence, local food production and nutritional adequacy, and the various health consequences of population displacement and economic disruption. Health impacts will occur very unevenly around the world. In general, rich populations will be better protected against physical damage, changes in patterns of heat and cold, introduction or spread of infectious diseases, and any adverse changes in world food supplies.

The geographic range and seasonality of various vector-borne infectious diseases (spread via organisms such as mosquitoes and ticks) will change, affecting some populations that currently are at the margins of disease distribution. The proportion of the world's population living in regions of potential transmission of malaria and dengue fever, for example, will increase. In areas where the disease currently is present, the seasonal duration of transmission will increase. Decreases in transmission may occur where precipitation decreases reduce vector survival, for example.

An increased frequency of heat waves will increase the risk of death and serious illness, principally in older age groups and the urban poor. The greatest increases in thermal stress are forecast for mid- to high-latitude (temperate) cities, especially in populations with limited air conditioning. Warmer winters and fewer cold spells, because of climate change, will decrease cold-related mortality in many temperate countries. Basic research to estimate the aggregate impact of these changes has yet been limited largely to the United States and parts of Europe. Recent modeling of heat-wave impacts in 44 U.S. urban populations, allowing for acclimatization, suggests that large U.S. cities may experience, on average, several hundred extra deaths per summer. Although the impact of climate change on thermal stress-related mortality in developing country cities may be significant, there has been little research in such populations.

For each anticipated adverse health impact, there is a range of social, institutional, technological, and behavioral adaptation options that could lessen that impact. The extent to which health care systems will have the capacity to adopt them is unclear, however, particularly in developing countries. There is a basic and general need for public health infrastructure (programs, services, surveillance systems) to be strengthened and maintained. The ability of affected communities to adapt to risks to health also depends on social, political, and economic circumstances.

the access of the poor to social safety nets and facilitated the ability of rich households to overexploit mangroves, which previously provided protection from storms. Similarly, Mustafa (1998) demonstrates differentiation of flood hazards in lowland Pakistan by social group: Insecure tenure leads to greater impacts on poorer communities. See Chapter 18 for further examples. The natural disaster literature also concludes that organization, information, and preparation can help mitigate large damages at a moderate cost (e.g., Burton *et al.*, 1993). This underscores the need for adaptation, particularly in poor countries.

19.4.3. Distribution of Total Impacts

Several studies have estimated the total impact (aggregated across sectors) in different regions of the world. Table 19-4 shows aggregate, monetized impact estimates for a doubling of atmospheric CO_2 on the current economy and population from four studies. Clearly, in all of these studies there are substantial uncertainties about the total impacts to regions and whether some regions will have net benefits or net damages at certain changes in global average temperature. Most studies, however, show the following:

- Developing countries, on the whole, are more vulnerable to climate change than developed countries.
- At low magnitudes of temperature change, damages are more likely to be mixed across regions, but at higher magnitudes virtually all regions have net damages.
- The distribution of risk may change at different changes in temperature.

Developing countries tend to be more vulnerable to climate change because their economies rely more heavily on climate-sensitive activities (particularly agriculture), and many already

***Table 19-4**: Indicative world impacts, by region (% of current GDP). Estimates are incomplete, and confidence in individual numbers is very low. See list of caveats in Section 19.4.1. There is a considerable range of uncertainty around estimates. Tol's (1999a) estimated standard deviations are lower bounds to real uncertainty. Figures are expressed as impacts on a society with today's economic structure, population, laws, etc. Mendelsohn et al. (2000) estimates denote impact on a future economy. Positive numbers denote benefits; negative numbers denote costs (Pearce et al., 1996; Tol, 1999a; Mendelsohn et al., 2000; Nordhaus and Boyer, 2000).*

	IPCC SAR	**Mendelsohn *et al.***		**Nordhaus and Boyer**	**Tol**
	2.5°C Warming	1.5°C Warming	2.5°C Warming	2.5°C Warming	1°C Warming[a]
North America					3.4 (1.2)
– United States			0.3	-0.5	
OECD Europe					3.7 (2.2)
– EU				-2.8	
OECD Pacific					1.0 (1.1)
– Japan			-0.1	-0.5	
Eastern Europe/FSU					2.0 (3.8)
– Eastern Europe				-0.7	
– Russia			11.1	0.7	
Middle East				-2.0[b]	1.1 (2.2)
Latin America					-0.1 (0.6)
– Brazil			-1.4		
South, Southeast Asia					-1.7 (1.1)
– India			-2.0	-4.9	
China			1.8	-0.2	2.1 (5.0)[c]
Africa				-3.9	-4.1 (2.2)
Developed countries	-1.0 to -1.5	0.12	0.03		
Developing countries	-2.0 to -9.0	0.05	-0.17		
World					
– Output weighted	-1.5 to -2.0	0.09	0.1	-1.5	2.3 (1.0)
– Population weighted				-1.9	
– At world average prices					-2.7 (0.8)
– Equity weighted					0.2 (1.3)

[a] Figures in brackets denote standard deviations.
[b] High-income countries in Organization of Petroleum Exporting countries (OPEC).
[c] China, Laos, North Korea, Vietnam.

operate close to environmental and climatic tolerance levels (e.g., with respect to coastal and water resources). If current development trends continue, few developing countries will have the financial, technical, and institutional capacity and knowledge base for efficient adaptation (a key reason for higher health impacts). For temperature increases of less than 2–3°C, some regions may have net benefits and some may have net damages. If temperature increases more than 2–3°C, most regions have net damages, and damages for all regions increase at higher changes in global average temperature.

19.5. Aggregate Impacts

The third reason for concern relates to the overall (i.e., worldwide or aggregate) economic and ecological implications of climate change. Numerous studies have addressed aggregate impacts, particularly in the context of integrated assessment.

19.5.1. Aggregate Analysis: An Assessment

Estimating the aggregate impact of climate change is an intricate task that requires careful professional judgment and skills. Aggregate analysis is based on the same tools as most distributional analysis and uses regional data as inputs. Consequently, it shares with distributional analysis the methodological difficulties and shortcomings discussed more fully in Section 19.4:

- Choice of an appropriate (set of) numeraire(s) in which to express impacts
- Need to overcome knowledge gaps and scientific uncertainties to provide a comprehensive picture
- Difficulties in modeling the effects of adaptation
- Difficulties in forecasting baseline developments (such as economic and population growth, technical progress).

In addition, analysts have to grapple with some issues that are generic to aggregate analysis. The most important issue is spatial and temporal comparison of impacts. Aggregating impacts requires an understanding of (or assumptions about) the relative importance of impacts in different sectors, in different regions, and at different times. Developing this understanding implicitly involves value judgments. The task is simplified if impacts can be expressed in a common metric, but even then aggregation is not possible without value judgments. The value judgments that underlie regional aggregation are discussed and made explicit in Azar and Sterner (1996), Fankhauser *et al*. (1997, 1998), and Azar (1999). Aggregation across time and the issue of discounting are discussed in more detail in TAR WGIII Chapter 7. Aggregate impact estimates can be very sensitive to the aggregation method and the choice of numeraire (see Chapter 1).

All of these factors make aggregate analysis difficult to carry out and reduce our overall confidence in aggregate results. Nevertheless, aggregate studies provide important and policy-relevant information.

19.5.2. Insights and Lessons: The Static Picture

Most impact studies assess the consequences of climate change at a particular concentration level or a particular point in time, thereby providing a static "snapshot" of an evolving, dynamic process. The SAR suggested that the aggregate impact of $2xCO_2$—expressed in monetary terms—might be equivalent to 1.5–2.0% of world GDP. Estimated damages are slightly lower (relative to GDP) in developed countries but significantly higher in developing countries—particularly in small island states and other highly vulnerable countries, where impacts could be catastrophic (Pearce *et al*., 1996). The SAR was careful, however, to point out the low quality of these numbers and the many shortcomings of the underlying studies.

Since publication of the SAR, our understanding of aggregate impacts has improved, but it remains limited. Some sectors and impacts have received more analytical attention than others and as a result are better understood. Agricultural and coastal impacts in particular are now well studied (see Boxes 19-2 and 19-3). Knowledge about the health impacts of climate change also is growing (see Box 19-4). Several attempts have been made to identify other nonmarket impacts, such as changes in aquatic and terrestrial ecological systems and ecosystem services, but a clear and consistent quantification has not yet emerged.

Table 19-4 contains a summary of results from aggregate studies that use money as their numeraire. The numerical results as such remain speculative, but they can provide insights on signs, orders of magnitude, and patterns of vulnerability. Results are difficult to compare because different studies assume different climate scenarios, make different assumptions about adaptation, use different regional disaggregation, and include different impacts. The estimates by Nordhaus and Boyer (2000), for example, are more negative than others because they factor in the possibility of catastrophic impact. The estimates by Mendelsohn *et al*. (2000), on the other hand, are driven by optimistic assumptions about adaptive capacity and baseline development trends, which result in mostly beneficial impacts.

Standard deviations rarely are reported, but they are likely to be several times larger than the "best guess." They are larger for developing countries, where results generally are derived through extrapolation rather than direct estimation. This is illustrated by the standard deviations estimated by Tol (2001b), also reproduced in Table 19-4. These estimates probably still underestimate the true uncertainty—for example, because they exclude omitted impacts and severe climate change scenarios. Note that the aggregation can mask large standard deviations in estimates of damages to individual sectors (Rothman, 2000).

An alternative indicator of climate change impact (excluding ecosystems) is the number of people affected. Few studies directly calculate this figure, but it is possible to compare the population of regions experiencing negative impacts with that of positively affected regions. Such calculations suggest that a majority of people may be negatively affected already at average global warming of 1–2°C. This may be true even if the net aggregate monetary impact is positive because developed economies, many of which could have positive impacts, contribute the majority of global production but account for a smaller fraction of world population. The quality of estimates of affected population is still poor, however. They are essentially "back-of-the envelope" extensions of monetary models, and the qualifications outlined in that context also apply here. In addition, they do not consider the distribution of positive and negative effects within countries.

On the whole, our confidence in the numerical results of aggregate studies remains low. Nevertheless, a few generic patterns and trends are emerging in which we have more confidence:

- Market impacts are estimated to be lower than initially thought and in some cases are estimated to be positive, at least in developed countries. The downward adjustment is largely a result of the effect of adaptation, which is more fully (although far from perfectly) captured in the latest estimates. Efficient adaptation reduces the net costs of climate change because the cost of such measures is lower than the concomitant reduction in impacts. However, impact uncertainty and lack of capacity may make efficient and error-free adaptation difficult.
- Nonmarket impacts are likely to be pronounced, and many (but not all) of the effects that have not yet been quantified could be negative. In particular, there is concern about the impact on human health and mortality. Although few studies have taken adequate account of adaptation, the literature suggests substantial negative health impacts in developing countries, mainly because of insufficient basic health care (e.g., Martens *et al.*, 1997). There also is concern about the impact on water resources (e.g., Arnell, 1999; Frederick and Schwarz, 1999) and ecosystems (e.g., Markham, 1996; White *et al.*, 1999).
- "Horizontal" interlinkages such as the interplay between different impact categories (e.g., water supply and agriculture), the effect of stress factors that are not related to climate, adaptation, and exogenous development trends are crucial determinants of impact but have not been fully considered in many studies.
- Estimates of global impact are sensitive to the way numbers are aggregated. Because the most severe impacts are expected in developing countries, aggregate impacts are more severe and thus more weight is assigned to developing countries. Using a simple summing of impacts, some studies estimate small net positive impacts at a few degrees of warming; others estimate small net negative impacts. Net aggregate benefits do not preclude the possibility that a majority of people will be negatively affected—some of them severely so.

Overall, the current generation of aggregate estimates may understate the true cost of climate change because they tend to ignore extreme weather events, underestimate the compounding effect of multiple stresses, and ignore the costs of transition and learning. However, studies also may have overlooked positive impacts of climate change. Our current understanding of (future) adaptive capacity, particularly in developing countries, is too limited, and the treatment of adaptation in current studies is too varied, to allow a firm conclusion about the direction of the estimation bias.

19.5.3. Insights and Lessons: Vulnerability over Time

One of the main challenges of impact assessments is to move from the static analysis of certain benchmarks to a dynamic representation of impacts as a function of shifting climatic parameters, adaptation measures, and exogenous trends such as economic and population growth. Little progress has been made in this respect, and our understanding of the time path that aggregate impacts will follow under different warming and development scenarios still is extremely limited. Among the few explicitly dynamic analyses are Sohngen and Mendelsohn (1999) and Yohe *et al.* (1996).

Some information about impacts over time is available for individual sectors. Scenarios derived from IAMs can provide comprehensive emissions, concentrations, and climate change estimates that can be linked to impact models. Table 19-5 summarizes estimates of global ecosystem impacts that were derived from such a model (IMAGE 2.1—Leemans *et al.*, 1998; Swart *et al.*, 1998). The metric used is percentage change. The example illustrates the clearly nonlinear dynamics of nonmarket impacts with different pathways for positive (escalating) and negative (saturating) impacts. The impact levels in this model evolve gradually, and there are impacts even at low levels of climate change. Although this finding is consistent with observed change (see Section 19.2), it is sensitive to the choice of metric. White *et al.* (1999), for example, found that carbon storage in terrestrial vegetation would expand under moderate warming because increases in productivity are enough to offset reductions elsewhere. They show that as higher GHG concentrations and magnitudes of climate change are reached, carbon storage eventually will decline.

Little is known about the shape of the aggregate impact function. Dynamic functions remain highly speculative at this point because the underlying models provide only a very rough reflection of real-world complexities. Figure 19-4 provides examples from three studies. Although some analysts still work with relatively smooth impact functions (e.g., Nordhaus and Boyer, 2000), there is growing recognition (e.g., Mendelsohn and Schlesinger, 1997; Tol, 2001c) that climate change dynamics in fact might be more complex and may not follow a monotonic path. Generic patterns that are emerging include the following:

- Moderate climate change may have positive and negative effects, with most positive effects occurring in the market sector of developed countries. For higher levels of warming, impacts are likely to become predominantly negative. However, the overall pattern is complex, estimates remain uncertain, and the possibility of highly deleterious outcomes cannot be excluded (medium confidence).

***Table 19-5**: Aggregate impact of climate change on ecosystems (Swart et al., 1998). See also list of caveats in Section 19.4.1.*

Impact Indicator	Scenario					
	0.5°C	1.0°C	1.5°C	2.0°C	2.5°C	3.0°C
Temperate cereals, area experiencing						
– Yield decrease[a]	12	16	18	20	20	22
– Yield increase[a]	2	3	4	8	12	15
Maize, land area experiencing						
– Yield decrease[a]	13	18	22	26	29	33
– Yield increase[a]	2	4	6	9	13	17
Change in natural vegetation[b]	11	19	26	32	37	43
Endangered nature reserves[c]	9	17	24	32	37	42

[a] Yield decrease and increase are percentage area with at least 10% change in potential rainfed yield. Reference area is current crop area.
[b] Change in natural vegetation is percentage of land area that shifts from one vegetation type to another. Reference area is global land area.
[c] Endangered nature reserves are percentage of reserves, where original vegetation disappears, so that conservation objectives cannot be met. Reference is total reserve number.

- Impacts in different sectors may unfold along different paths. Coastal impacts, for example, are expected to grow continuously over time, more or less in proportion to the rise in sea level. The prospects for agriculture, by contrast, are more complex. Whereas some models predict aggregate damages already for moderate warming, many studies suggest that under some (but not all) scenarios the impact curve might be hump-shaped, with short-term (aggregate) benefits under modest climate change turning into losses under more substantial change (e.g., Mendelsohn and Schlesinger, 1997) (low confidence).

Aggregating intertemporal impacts into a single indicator is extremely difficult, perhaps elusive. The marginal damages caused by 1 t of CO_2 emissions in the near future were estimated in the SAR at US$5–125 t^{-1} C. Most estimates are in the lower part of that range; higher estimates occur only through the combination of high vulnerability with a low discount rate (see Pearce *et al.*, 1996). Plambeck and Hope (1996), Eyre *et al.* (1997), and Tol (1999a) have since reassessed the marginal costs of GHG emissions. Performing extensive sensitivity and uncertainty analyses, they arrive at essentially the same range of numbers as Pearce *et al.* (1996). In the complex dynamics that determine marginal damage costs, the more optimistic estimates of market damages used in recent studies are balanced out by other factors such as higher nonmarket impacts and a better capture of uncertainties. Overall, the SAR assessment still is a good reflection of our understanding of marginal damage costs; our confidence in marginal damage numbers remains very low.

19.5.4. Sensitivity of Aggregate Estimates

At a time when the quality of numerical results still is low, a key benefit of aggregate impact analysis lies in the insights it provides regarding the sensitivity of impacts. Sensitivity analysis offers critical information about attributes of the damage function that are likely to be most influential for the

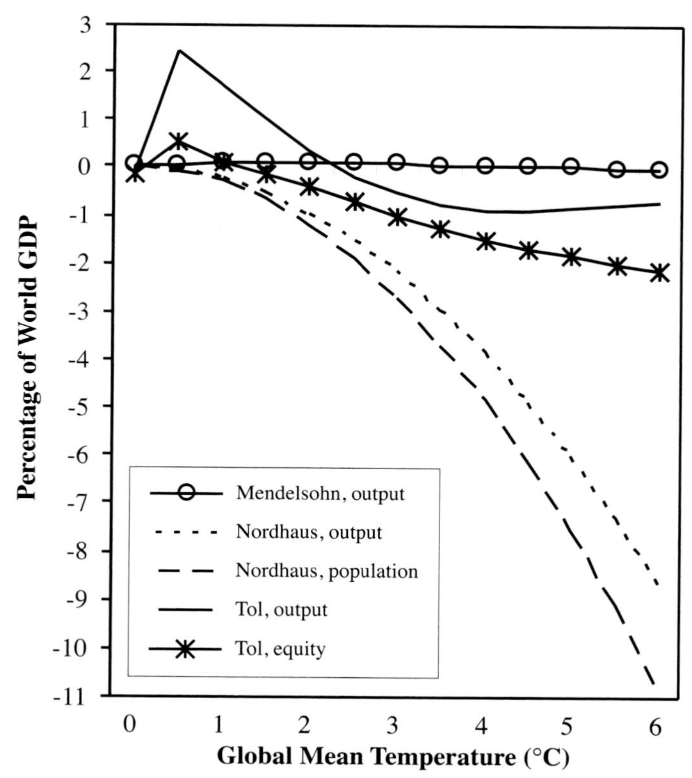

Figure 19-4: Monetary impacts as a function of level of climate change (measured as percentage of global GDP). Although there is confidence that higher magnitudes and rates of increase in global mean temperature will lead to increasing damages, there is uncertainty about whether aggregate damages are positive or negative at relatively low increases in global mean temperature.

choice of policy and, by implication, where additional climate change impacts research is most needed.

19.5.4.1. Composition of Impact Function

Most aggregate analysis is based on IAMs (see Chapter 2). Impact functions used in IAMs vary greatly with respect to the level of modeling sophistication, the degree of regional aggregation, the choice of numeraire, and other characteristics (see Tol and Fankhauser, 1998). Many models have used monetary terms (e.g., U.S. dollars) to measure impacts. Spatially detailed models (e.g., Alcamo *et al.*, 1998) pay some attention to unique ecosystems. Disruptive climate changes have received little attention, except for a survey of expert opinions (Nordhaus, 1994b) and analytical work (e.g., Gjerde *et al.*, 1999). Some climate change impact studies restrict themselves to sectors and countries that are relatively well studied (e.g., Mendelsohn and Neumann, 1999). Others try to be comprehensive, despite the additional uncertainties (e.g., Hohmeyer and Gaertner, 1992). Some studies rely on an aggregate description of all climate change impacts for the world as a whole (e.g., Nordhaus, 1994a). Other studies disaggregate impacts with substantial spatial and regional detail (e.g., Alcamo *et al.*, 1998). The aggregate approaches tend to point out implications for efficiency and in practice often ignore equity (see Tol, 2001a, for an exception). The detailed approaches tend to identify distributional issues, but working out the equity implications typically is left to the reader.

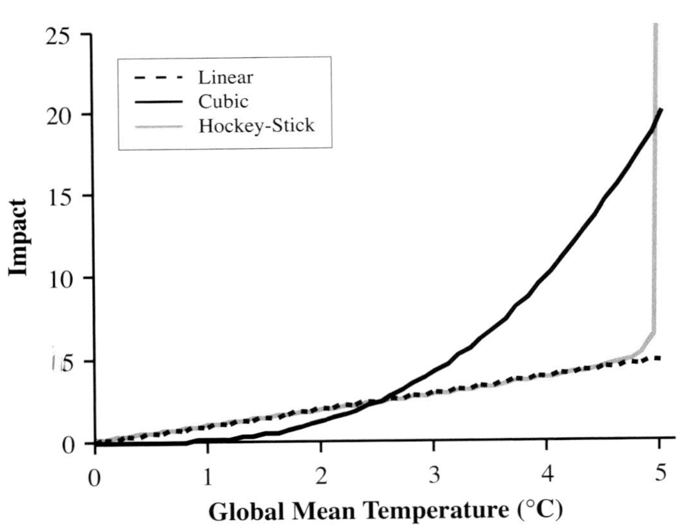

Figure 19-5: Aggregate impact of climate change as a function of global mean temperature. Displayed are hypothetical examples of a linear function, which assumes that impacts are proportional to temperature change since preindustrial times; a cubic function, which assumes that impacts are proportional to temperature change to the power of three; and a hockey-stick function, which assumes that impacts are approximately proportional to temperature change until a critical threshold is approached. Aggregate damage functions used in integrated assessments are mostly illustrative. They should be regarded as "placeholders" that will be replaced by more accurate functional forms as our knowledge of impact dynamics improves.

19.5.4.2. Shape of Damage Function

Most impact studies still look at the equilibrium effect of one particular level of GHG concentration, usually $2xCO_2$. Full analysis, however, requires impacts to be expressed as a function of change in GHG concentrations. With so little information to estimate this function, studies have to rely on sensitivity analyses. Different damage functions can lead to profoundly different policy recommendations. Compare, for example, the profile of impacts under a linear and a cubic damage function (see Figure 19-5). Relative to the linear specification, a cubic function implies low near-term impacts but rapidly increasing impacts further in the future. Using conventional discounting, this means that early emissions under a cubic damage function will cause less damage over their atmospheric lifetime, compared to a scenario with linear damages. The marginal damage caused by emissions further in the future, on the other hand, is much higher if we assume a cubic damage function (Peck and Teisberg, 1994).

Some studies explore the implications of more nonlinear impact functions. For instance, Manne and Richels (1995) use a "hockey-stick" function that suddenly turns upward at arbitrarily chosen thresholds. Such studies are designed to reflect relatively small impacts before $2xCO_2$ and rapidly worsening impacts beyond $2xCO_2$. In this analysis, it is economically efficient to stabilize CO_2 concentrations, but the desired level of stabilization depends on the shape of the hockey stick and the location of its kink. Other analyses, which rely on more linear impact functions, have difficulty justifying concentration stabilization at any level.

19.5.4.3. Rate of Change

Although most impact studies focus on the level of climate change, the rate of climate change generally is believed to be an important determinant, in many instances because it affects the time that is available for adaptation. Again, the paucity of underlying impact studies forces integrated assessors to use exploratory modeling. Under most "business-as-usual" scenarios, the rate of climate change is greater in the short run than in the long run because emissions increase faster in the short run; this is even more pronounced in emission reduction policy scenarios. Indeed, in considering the rate of change, tolerable window and safe-landing analyses (Alcamo and Kreileman, 1996; Toth *et al.*, 1997; Petschel-Held *et al.*, 1999) often find the rate of change to be the binding constraint in the first half of the 21st century.

19.5.4.4. Discount Rate and Time Horizon

Aggregate models suggest that the most severe impacts of climate change will occur further in the future. The chance of large-scale discontinuities (thermohaline circulation, West Antarctic ice sheet) also is higher in the future. The outcome of policy analysis therefore is sensitive to the weight afforded to events occurring in the remote future. In other words, estimates are sensitive to the choice of time horizon (Cline, 1992; Azar and Sterner, 1996; Hasselmann *et al.*, 1997) and the discount rate (i.e., the value of future consumption relative to today's value). The literature on discounting is reviewed in Portney and Weyant (1999) and in TAR WGIII Chapter 7. Numerical analysis (e.g., Tol, 1999a) has shown that estimates of marginal damage (i.e., the additional damage caused by an extra ton of emissions) can vary by as much as a factor of 10 for different (and reasonable) assumptions about the discount rate. This makes the discount rate the second-most important parameter for marginal damage. The most important parameter is the degree of cooperation in reducing emissions (Nordhaus and Yang, 1996; Tol, 1999b).

19.5.4.5. Welfare Criteria

Comparison of impacts (i.e., the relative weight assigned to impacts in different regions and at different times) is one of the most sensitive aspects of aggregate analysis. With the exception of the discount rate, little explicit attention is paid to this aspect of climate change impacts, although studies differ considerably in their implicit assumptions. Fankhauser *et al.* (1997) and Azar (1999) are among the few studies that make their aggregation assumptions explicit. They find that, in general, the higher the concerns about the distribution of the impacts of climate change, the more severe the aggregate impacts. Fankhauser's (1995) estimate of the annual global damage of $2xCO_2$, for instance, is based on the implicit assumption that people are neutral with respect to distribution (that is, losses to the poor can be compensated by equal gains to the rich) and risk (that is, a 1:1,000,000 chance of losing $1 million is equivalent to losing $1 with certainty). Replacing these assumptions with standard risk aversion or mild inequity aversion, the global damage estimate increases by about one-third (Fankhauser *et al.*, 1997). Marginal impacts are more sensitive. For the same changes in assumptions, Tol (1999a) finds a three-fold increase in the marginal damage estimate. The sensitivity of aggregate impact estimates is further illustrated in Figure 19-4.

19.5.4.6. The Treatment of Uncertainty

Sensitivity analysis is the standard approach to deal with impact uncertainty. Some studies, however, have gone one step further and explicitly model uncertainty as a hedging problem. The premise underlying these models is that today's policymakers are not required to make once-and-for-all decisions binding their successors over the next century. There will be opportunities for mid-course adjustments. Climate negotiations are best viewed as an ongoing process of "act, then learn." Today's decisionmakers, in this view, must aim at evolving an acceptable hedging strategy that balances the risks of premature actions against those of waiting too long.

The first step, then, is to determine the sensitivity of *today's* decisions to major uncertainties in the greenhouse debate. How important is it to be able to predict impacts for the second half of this century? Or to know what energy demands will be in 30 years and identify the technologies that will be in place to meet those demands? An exhaustive analysis of these questions has yet to be undertaken, but considerable insight can be gleaned from an Energy Modeling Forum study conducted several years ago (EMF, 1997). In the study, seven modeling teams addressed a key consideration in climate policymaking: concerns about events with low probability but high consequences.

The study assumed uncertainty would not be resolved until 2020. Two parameters were varied: the mean temperature sensitivity factor and the cost of damages associated with climate change and variability. The unfavorable high-consequence scenario was defined as the top 5% of each of these two distributions. Two surveys of expert opinion were used for choosing the distribution of these variables (for climate sensitivity, see Morgan and Keith, 1995; for damages, see Nordhaus, 1994a).

The analysis showed that the degree of hedging depends on the stakes, the odds, and nonimpact parameters such as society's attitude toward risk and the cost of greenhouse insurance. Also critical is the timing of the resolution of key uncertainties. This underscores the importance of scientific research.

19.6. Extreme and Irreversible Effects

19.6.1. The Irregular Face of Climate Change

Natura non facit saltus—nature does not take jumps. Modern science has thoroughly shattered this tenet of the Aristotelian school of thought. Long-term observations and experimental insights have demonstrated convincingly that smooth, or *regular,* behavior is an exception rather than a rule. Available records of climate variability, for example, reveal sudden fluctuations of key variables at all time scales. Large, abrupt climate changes evident in Greenland ice-core records (known as Dansgaard-Oeschger oscillations—Dansgaard *et al.,* 1993) and episodic, massive discharges of icebergs into the North Atlantic (known as Heinrich events—Bond *et al.*, 1992) are obvious examples of *irregular* behavior as a result of weak external forcing. Ecosystems also display discontinuous responses to changing ambient conditions, such as changes in disturbance regimes (Holling, 1992a; Peterson *et al.*, 1998) and species extinctions (Pounds *et al.*, 1999). Irreversible changes in ecosystems are triggered by disturbances (e.g., Gill, 1998), pests (e.g., Holling, 1992b), and shifts in species distributions (Huntley *et al.*, 1997). Irregular behavior is accepted as a major aspect of the dynamics of complex systems (Berry, 1978; Schuster, 1988; Wiggins, 1996; Badii and Politi, 1997).

A quantitative entity behaves "irregularly" when its dynamics are discontinuous, nondifferentiable, unbounded, wildly varying, or otherwise ill-defined. Such behavior often is termed *singular*, particularly in catastrophe theory (Saunders, 1982), and illustrates how smooth variations of driving forces can cause abrupt and drastic system responses. The occurrence, magnitude, and timing of singularities are relatively difficult to predict, which is why they often are called "surprises" in the literature.

It is important to emphasize that singular behavior is not restricted to natural systems. There has been speculation, for example, about possible destabilization of food markets, public health systems, and multilateral political agreements on resource use, but solid evidence rarely has been provided (e.g., Döös, 1994; Hsu, 1998). Rigorous scientific analysis of certain classes of singular socioeconomic phenomena is emerging (Bunde and Schellnhuber, 2000), but huge cognitive gaps remain in this field.

Singularities have large consequences for climate change vulnerability assessments. Unfortunately, most of the vulnerability assessment literature still is focusing on a smooth transition from what is assumed to be an equilibrium climate toward another equilibrium climate (often $1xCO_2$ to $2xCO_2$). This means that most impact assessments still implicitly assume that climate change basically is a "well-behaved" process. Until recently, only a few authors have emphasized the importance of discontinuous, irreversible, and extreme events to the climate problem (e.g., Lempert *et al*., 1994; Nordhaus, 1994a; Schellnhuber, 1997); concerns about the impacts of these events and their consequences for society now are becoming much more common. Singularities could lead to rapid, large, and unexpected impacts on local, regional, and global scales. Anticipating and adapting to such events and their impacts would be much more difficult than responding to smooth change, even if these responses must be made in the face of uncertainty. Furthermore, singularities considerably complicate the search for optimal emissions reduction strategies that are based on, for example, benefit-cost analysis or tolerable emissions strategies that are based on, as another example, the precautionary principle.

This section reviews and synthesizes relevant available information on the impacts of singular behavior of (components of) the climate system or singular impacts of climate change and draws conclusions about the consequences for vulnerability assessments. Because no generally accepted framework to assess singularities of climate change exists, an illustrative typology of singularities is discussed first. The different characteristics of each class in this typology justify why insights from this section contribute to two separate reasons for concern: extreme weather events and large-scale singularities.

19.6.2. Characteristics of Singularities

The causes of singularities are diverse, but most can be grouped in the categories of *nonlinearity*, *complexity*, and *stochasticity*. Choices about how to assess singular climate impacts depend strongly on the factors generating such behavior. The first two categories arise in a largely *deterministic* context, so their incidence can be assessed with proper models. The latter is probabilistic, however, rendering its incidence basically unpredictable. Only statistical properties can be analyzed. Predictability (and consequently adaptability) is directly related to the *stochastic* nature of the underlying dynamics.

The first, and most obvious, class of singularities is caused by strongly nonlinear or discontinuous functional relationships. A conspicuous case is the critical threshold, where responses to a continuous change in a driving variable bring about sudden and severe impacts, such as extinction events. Changes in mean climate can increase the likelihood of crossing these thresholds. Even one of the simplest physical thresholds in the climate system—the melting point of ice—could induce singular impacts. For example, thawing of permafrost regions would be induced by only a few degrees of warming (Pavlov, 1997) and would severely affect soil and slope stability, with disastrous effects on Arctic infrastructure such as oil pipelines (see Section 16.2.5 and SAR WGII Section 11.5.3). Section 19.3 extensively illustrates the occurrence of critical thresholds that are relevant for bleaching of coral reefs (a temperature threshold) and coastal mangroves (a sea-level rise threshold).

Complexity itself is a second potential cause for singular behavior in many systems. Complex systems, of course, are composed of many elements that interact in many different ways. Anomalies in driving forces of these systems generally distort interactions between constituents of the system. Positive feedback loops then can easily push the systems into a singular response. (Note that complexity is by no means synonymous with nonlinearity!)

Complex interactions and feedbacks gradually have become a focal point of global and climate change investigations: Several illustrative studies, for example, deal with the interplay between atmosphere, oceans, cryosphere, and vegetation cover that brought about the rapid transition in the mid-Holocene from a "green" Sahara to a desert (Brovkin *et al*., 1998; Ganopolski *et al*., 1998; Claussen *et al*., 1999), with the mutual amplification of regional climate modification and unsustainable use of tropical forests as mediated by fire (Cochrane, 1999; Goldammer, 1999; Nepstad *et al*., 1999) and with the dramatic disruptions possibly inflicted on Southern Ocean food webs and ecological services by krill depletion resulting from dwindling sea-ice cover (see Brierly and Reid, 1999; see also Section 16.2.3).

The third category, stochasticity, captures a class of singularities that are triggered by exceptional events. In the climate context, these are, by definition, extreme weather events such as cyclones and heavy rains (see Table 3-10). Their occurrence is governed by a generally well-behaved statistical distribution. The irregular character of extreme events stems mainly from the fact that, although they reside in the far tails of this distribution, they nonetheless occur from time to time. Therefore, they could affect downstream systems by surprise and trigger effects that are vastly disproportional to their strength. Climate change also could lead, however, to changes in probability distributions for extreme events. Such changes actually could cause serious problems because the risk and consequences of these transitions are difficult to quantify and identify in advance. The impacts caused by these events have not yet been explored, although they should constitute an essential aspect of any impact and adaptation assessment.

The impacts of extreme event consequences of stochastic climate variability, however, have begun to attract researchers' attention in a related context. As noted in Chapter 18, changes in mean climate can increase the likelihood that distributed weather will cross thresholds where the consequences and impacts are severe and extreme. This variant of stochastic singularity therefore can change in frequency even if the probability of extreme weather events, measured against the mean, is unaffected by long-term trends.

There also is a fourth type that generally arises from a combination of all other singularity categories. This type—sometimes referred to as "imaginable surprises" (Schneider *et al.*, 1998; see also Chapter 1)—represents conceivable global or regional disruptions of the operational mode of the Earth system. Such *macro-discontinuities* may cause damages to natural and human systems that exceed the negative impacts of "ordinary" disasters by several orders of magnitudes.

Responses to climate change can alter their character from singular to regular—and vice versa—as they cascade down the causal chain: *geophysical perturbations*, *environmental impacts*, *sectoral and socioeconomic impacts*, and *societal responses*. Only the last three are climate change effects in the proper sense, but the first is important because it translates highly averaged indicators of climate change into the actual trigger acting at the relevant scale. Most singular geophysical perturbations create singular impacts—which may, in turn, activate singular responses. One therefore might assume that singularities tend to be preserved down such a cascade. Singular events also can arise further down the causal chain. Purely regular geophysical forcing, for example, can cause singular impacts, and singular socioeconomic responses may result from regular impacts.

Harmful impacts of climate change generally can be alleviated by adaptation or exacerbated by mismanagement (see, e.g., West and Dowlatabadi, 1999; Schneider *et al.*, 2000a; see also Chapter 18). Climate-triggered singular phenomena can generate substantial impacts because their predictability and manageability are low. Such impacts would be considerably reduced if they could be "regularized" by appropriate measures. For example, an ingenious array of seawalls and dikes could transform an extreme storm surge into a mundane inundation that could be controlled by routine contingency procedures. So too could a long-term policy of retreat from the sea. However, inappropriate flood control structures could wreak havoc, particularly because they foster a false sense of security and actually inspire further coastal development.

In summary, singularities tend to produce singularities, as a rule; regularities may turn into singularities under specific conditions, and singularities can be regularized by autonomous ecological processes or judicious societal measures. Defining the propagation of singular events in the causal cascade or opportunities to convert them into regular events remains a major research challenge.

19.6.3. Impacts of Climate Change Singularities

This subsection sketches the most evident singularities discussed in the context of climate change and reviews the pertinent literature on their potential impacts.

19.6.3.1. Extreme Weather Events

That the occurrence of weather events is essentially stochastic is a well-established fact (e.g., Lorenz, 1982; Somerville, 1987). Most climatic impacts arise from extreme weather events or from climatic variables exceeding some critical level and thereby affecting the performance or behavior of a biological or physical system (e.g., Downing *et al.*, 1999). The same holds for the impacts of climate change (see Chapters 1, 2, and 3, especially Table 3-9; Pittock and Jones, 2000).

For many important climate impacts, we are interested in the effects of specific extreme events or threshold magnitudes that have design or performance implications. To help in zoning and locating developments or in developing design criteria for the capacities of spillways and drainage structures, the heights of levee banks, and/or the strengths of buildings, for example, planners and engineers routinely use estimated "return periods" (the average time between events) at particular locations for events of particular magnitudes. Such event magnitudes include flood levels (Hansen, 1987; Handmer *et al.*, 1999) and storm-surge heights (Middleton and Thompson, 1986; Hubbert and McInnes, 1999). Return period estimates normally are based on recent instrumental records, sometimes augmented by estimates from other locations, or statistical or physically based modeling (Middleton and Thompson, 1986; Hansen, 1987; Beer *et al.*, 1993; National Research Council, 1994; Pearce and Kennedy, 1994; Zhao *et al.*, 1997; Abbs, 1999). The assumption usually is made that these statistics, based on past events, are applicable to the future—but climate change means that this often will not be the case.

Thus, a central problem in planning for or adapting to climate change and estimating the impacts of climate change is how these statistics of extreme events are likely to change. Similar problems arise in nonengineering applications such as assessing the economic performance or viability of particular enterprises that are affected by weather—for example, farming (Hall *et al.*, 1998; Kenny *et al.*, 1999; Jones, 2000)—or health effects (Patz *et al.*, 1998; McMichael and Kovats, 2000; see also Chapter 9).

Relatively rapid changes in the magnitude and frequency of specified extreme events arise because extremes lie in the low-frequency tails of frequency distributions, which change rapidly with shifts in the means. Moreover, there also can be changes in the shape of frequency distributions, which may add to or subtract from the rate of change of extremes in particular circumstances (Mearns *et al.*, 1984; Wigley, 1985, 1988; Hennessy and Pittock, 1995; Schreider *et al.*, 1997). Such changes in the shape of frequency distributions require special attention. Evidence suggests that they are particularly important for changes in extreme rainfall (Fowler and Hennessy, 1995; Gregory and Mitchell, 1995; Walsh and Pittock, 1998), possibly in the intensities of tropical cyclones (Knutson *et al.*, 1998; Walsh and Ryan, 2000), and in ENSO behavior (Dilley and Heyman, 1995; Bouma *et al.*, 1997; Bouma, 1999; Timmermann *et al.*, 1999; Fedorov and Philander, 2000). Return periods can shorten, however, even if none of these higher moment effects emerge; simply moving mean precipitation higher, for example, could make the 100-year flood a 25-year flood.

It is noteworthy that the central role in impact assessments of the occurrence of extreme weather events gives rise to multiple sources of uncertainty in relation to climate change. The stochastic nature of the occurrence of extremes and the limited historical record on which to base the frequency distribution for such events give rise, even in a stationary climate, to a major uncertainty. Beyond that, any estimate of a change in the frequency distribution under a changing climate introduces new uncertainties. Additional uncertainties relate to our limited understanding of the impacted systems and their relevant thresholds, as well as the possible effects of adaptation, or societal change, in changing these thresholds. If this were not complicated enough, many impacts of weather extremes arise from sequences of extremes of the same or opposite sign—such as sequences of droughts and floods affecting agriculture, settlements, pests, and pathogens (e.g., Epstein, 2000) or multiple droughts affecting the economic viability of farmers (e.g., Voortman, 1998).

Planned adaptation to climate change therefore faces particular difficulty in this environment because projections of changes in the frequency of extreme events and threshold exceedence require a multi-decadal to century-long projected (or "recent" observed) data series, or multiple ensemble predictions (which is one way of generating improved statistics). Thus, it is difficult to base planned adaptation on the record of the recent past, even if there is evidence of a climate change trend in the average data. Planned adaptation therefore must rely on model predictions of changes in the occurrence of extreme and threshold events (e.g., see Pittock *et al.*, 1999), with all their attendant uncertainties. Real-life adaptation therefore will most likely be less optimal (more costly or less effective) than if more precise information on future changes in such thresholds and extremes were available.

Nonetheless, planned adaptation will most likely proceed in response to changes in the perceived relative frequency of extreme events. Properly done, it can have immediate benefit by reducing vulnerability to current climate as well as future benefit in reducing exposure to future climate change. As suggested above, however, there are many ways to respond inappropriately if care is not taken. In short, changes in extremes and in the frequency of exceeding impacts thresholds are a vital feature of vulnerability to climate change that is likely to increase rapidly in importance because the frequency and magnitude of such events will increase as global mean temperature rises.

19.6.3.2. Large-Scale Singularities

Singularities that occur in complex systems with multiple thresholds can be assessed with appropriate models. In real systems, however, there always are stochastic elements that influence the behavior of these systems, which are difficult to model. The runaway greenhouse effect, for example, consists of a series of positive feedback loops that result from systemic interactions or can be triggered by stochastic events (Woodwell *et al.*, 1998). Table 19-6 lists examples of such singularities that are triggered by different causes. All of these examples have regional or global consequences. The systemic insights in their behavior generally are based on different simulation approaches. Although local examples (e.g., species extinction) also are abundant in the scientific literature, they are ignored here because climate change does not (yet) seem to be the sole cause, and the processes involved generally are not modeled systematically.

19.6.3.2.1. Nonlinear response of North Atlantic thermohaline circulation

Many model studies (reviewed in Weaver *et al.*, 1993; Rahmstorf *et al.*, 1996) have analyzed the nonlinear response of the worldwide ocean circulation—the so-called conveyor belt. This system transports heat and influences regional climate patterns. One component of this system is the current in the Atlantic Ocean. Warm surface currents flow northward. Heat release and evaporation from the ocean surface lowers the temperature and increases the density and salinity of the water. In the North Atlantic, this denser water sinks at the Labrador and Greenland convection sites and flows back south as deepwater. This so-called North Atlantic THC could slow down or even shut down under climate change (see TAR WGI Chapters 7 and 9).

The paleoclimatic record shows clear evidence of rapid climatic fluctuations in the North Atlantic region (with possible connections to other regions) during the last glaciation and in the early Holocene (see TAR WGI Section 2.4.3). At least some of these events—notably the Younger Dryas event, when postglacial warming was interrupted by a sudden return to colder conditions within a few decades about 11,000 years ago—are thought to be caused by changes in the stability of North Atlantic waters. These changes, which are recorded in the central Greenland ice cores and elsewhere, were accompanied by large changes in pollen and other records of flora and fauna in western Europe, indicating that they had widespread effects on European regional climate and ecosystems (Ammann, 2000; Ammann *et al.*, 2000). The likely cause for these fluctuations is changes in the stability of the THC brought about by an influx of freshwater from melting icebergs and/or ice caps (see TAR WGI Section 7.3.7). As discussed in WGI, enhanced greenhouse

***Table 19-6**: Examples of different singular events and their impacts.*

Singularity	Causal Process	Impacts	Reference
Nonlinear response of thermohaline circulation (THC)	– Changes in thermal and freshwater forcing could result in complete shutdown of North Atlantic THC or regional shutdown in the Labrador and Greenland Seas. In the Southern Ocean, formation of Antarctic bottomwater could shut down. Such events are found in the paleoclimatic record, so they are plausible.	– Consequences for marine ecosystems and fisheries could be severe. Complete shutdown would lead to a stagnant deep ocean, with reducing deepwater oxygen levels and carbon uptake, affecting marine ecosystems. Such a shutdown also would represent a major change in heat budget and climate of northwestern Europe.	WGI TAR Chapters 2.4, 7, and 9; see Section 19.6.4.2.1
Disintegration of West Antarctic Ice Sheet (WAIS), with subsequent large sea-level rise	– WAIS may be vulnerable to climate change because it is grounded below sea level. Its disintegration could raise global sea level by 4–6 m. Disintegration could be initiated irreversibly in the 21st century, although it may take much longer to complete.	– Considerable and historically rapid sea-level rise would widely exceed adaptive capacity of most coastal structures and ecosystems.	WGI TAR Chapters 7 and 11; see Section 19.6.4.2.2; Oppenheimer, 1998
Runaway carbon dynamics	– Climate change could reduce the efficiency of current oceanic and biospheric carbon sinks. Under some conditions, the biosphere could even become a source. – Gas hydrate reservoirs also may be destabilized, releasing large amounts of methane to the atmosphere. – These processes would generate a positive feedback, accelerating buildup of atmospheric GHG concentrations.	– Rapid, largely uncontrollable increases in atmospheric carbon concentrations and subsequent climate change would increase all impact levels and strongly limit adaptation possibilities.	WGI TAR Chapter 3; Smith and Shugart, 1993; Sarmiento *et al.*, 1998; Woodwell *et al.*, 1998; Bains *et al.*, 1999; Joos *et al.*, 1999; Katz *et al.*, 1999; Norris and Rohl, 1999; Walker *et al.* 1999; White *et al.*, 1999
Transformation of continental monsoons	– Increased GHGs could intensify Asian summer monsoon. Sulfate aerosols partially compensate this effect, although dampening is dependent on regional patterns of aerosol forcing. Some studies find intensification of the monsoon to be accompanied by increase in interseasonal precipitation variability.	– Major changes in intensity and spatial and temporal variability would have severe impacts on food production and flood and drought occurrences in Asia.	TAR WGI Sections 9.3.6.2 and 9.3.5.2.2; TAR WGII Section 11.5.1; Lal *et al.*, 1995; Mudur, 1995; Meehl and Washington, 1996; Bhaskaran and Mitchell, 1998
Qualitative modification of crucial climate-system patterns such as ENSO, NAO, AAO, and AO	– ENSO could shift toward a more El Niño-like mean state under increased GHGs, with eastward shift of precipitation. Also, ENSO's variability could increase. – There is a growing attempt to investigate changes in other major atmospheric regimes [NAO, Arctic Oscillation (AO), and Antarctic Oscillation (AAO)]. Several studies show positive trend in NAO and AO indices with increasing GHGs.	– Changing ENSO-related precipitation patterns could lead to changed drought and flood patterns and changed distribution of tropical cyclones. – A positive NAO/AO phase is thought to be correlated with increased storminess over western Europe.	TAR WGI Sections 7.7.3 and 9.3.5.2; Corti *et al.*, 1999; Fyfe *et al.*, 1999; Shindell *et al.*, 1999; Timmermann *et al.*, 1999

Table 19-6 (continued)

Singularity	Causal Process	Impacts	Reference
Rearrangement of biome distribution as a result of rising CO_2 concentrations and climate change	– Many studies show large redistribution of vegetation patterns. Some simulate rapid dieback of tropical forests and other biomes; others depict more gradual shifts. More frequent fire could accelerate ecosystem changes.	– All models initially simulate an increase in biospheric carbon uptake, which levels out later. Only a few models simulate carbon release.	White *et al.*, 1999; Cramer *et al.*, 2000
Destabilization of international order by environmental refugees and emergence of conflicts as a result of multiple climate change impacts	– Climate change—alone or in combination with other environmental pressures—may exacerbate resource scarcities in developing countries. These effects are thought to be highly nonlinear, with potential to exceed critical thresholds along each branch of the causal chain.	– This could have severe social effects, which, in turn, may cause several types of conflict, including scarcity disputes between countries, clashes between ethnic groups, and civil strife and insurgency, each with potentially serious repercussions for the security interests of the developed world.	Homer-Dixon, 1991; Myers, 1993; Schellnhuber and Sprinz, 1995; Biermann *et al.*, 1998; Homer-Dixon and Blitt, 1998

warming could produce similar changes in stability in the North Atlantic because of warming and freshening of North Atlantic surface waters.

The current operation of THC is self-sustaining within limits that are defined by specific thresholds. If these thresholds were exceeded, two responses would be possible: shutdown of a regional component of the system or complete shutdown of the THC. Both responses have been simulated. A complete shutdown was simulated by Manabe and Stouffer (1993) for a quadrupling of atmospheric CO_2 and by Rahmstorf and Ganopolski (1999) for a transient peak in CO_2 content. These studies suggest that the threat of such complete shutdown increases beyond a global mean annual warming of 4–5°C, but this is still speculative. It took several centuries until the circulation was shut down completely in both studies. A regional shutdown in the Labrador Sea (while the second major Atlantic convection site in the Greenland Sea continued to operate) was simulated by Wood *et al.* (1999). Simulated regional shutdown can occur early in the 21st century and can happen rapidly—in less than a decade. Simulations by Manabe and Stouffer (1993) and Hirst (1999) show further the possibility of a shutdown of the formation of Antarctic bottomwater, which is the second major deepwater source of the world ocean.

These simulations clearly identify possible instability for the THC. Determining appropriate threshold values, however, requires analysis of many scenarios with different forcings and sensitivity studies of important model parameters. Stocker and Schmittner (1997), for example, have shown that the THC is sensitive not only to the final level of atmospheric CO_2 concentration but also to the rate of change. Rahmstorf and Ganopolski (1999) show that uncertainties in the hydrological cycle are a prime reason for uncertainty in forecasting, whether a threshold is crossed or not (see Figure 19-6). Further parameters are climate sensitivity (high values increase the likelihood of a circulation change) and the preindustrial rate of Atlantic overturning (an already weak circulation is more liable to break down) (e.g., Schneider and Thompson, 2000). These simulations suggest that global warming over the next 100 years could lead to a sudden breakdown of the THC decades to centuries later, which would lead irrevocably to major effects on future generations.

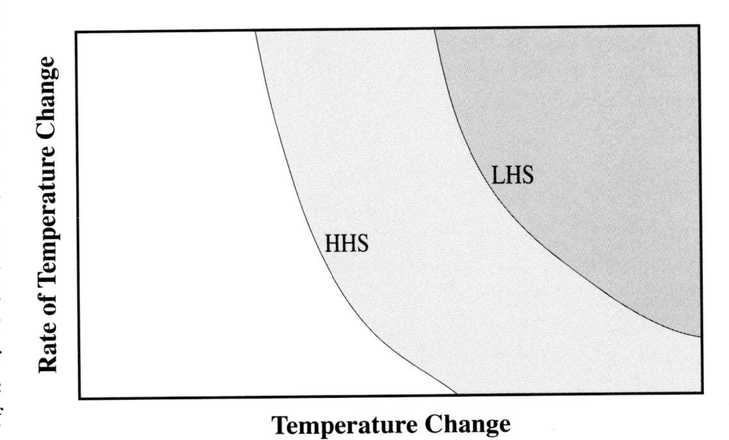

Figure 19-6: Stability of North Atlantic thermohaline circulation (THC) computed with the CLIMBER model (Petoukhov *et al.*, 2000). Degree of shading indicates probability of THC collapse. Light shading denotes low probability; dark shading denotes high probability. The higher the hydrological sensitivity (HHS = high hydrological sensitivity, LHS = low hydrological sensitivity), the faster the rate of temperature increase, or the greater the magnitude of temperature increase, the more likely that the North Atlantic THC becomes unstable.

The possible impacts of these circulation changes have not yet been studied systematically. Complete shutdown of the THC would represent a major change in the heat budget of the northern hemisphere because this circulation currently warms northwestern Europe by 5–10°C (Manabe and Stouffer, 1988; Rahmstorf and Ganopolski, 1999). Consequently, shutdown would lead to sudden reversal of the warming trend in this region. The impacts of a regional shutdown would be much smaller but probably still serious. For the European climate, loss of the Greenland Sea branch probably would have a much stronger effect than loss of the Labrador Sea branch because the northward extent of the warm North Atlantic current depends mainly on the former. In either case, the consequences of circulation changes for marine ecosystems and fisheries could be severe (see Section 6.3). Shutdown of the major deepwater sources in the North Atlantic and Southern Ocean would lead to an almost stagnant deep ocean, with as-yet unexplored consequences (e.g., for deepwater oxygen levels, carbon uptake, and marine ecosystems).

Neither the probability and timing of a major ocean circulation change nor its impacts can be predicted with confidence yet, but such an event presents a plausible, non-negligible risk. The change would be largely irreversible on a time scale of centuries, the onset could be relatively sudden, and the damage potential could be very high.

19.6.3.2.2. Disintegration of West Antarctic ice sheet

The WAIS contains 3.8 million km^3 of ice, which, if released to the ocean, would raise global sea level by 4–6 m. The WAIS has been the subject of attention since analysis of paleodata (Hughes, 1973) and ice sheet models (Weertman, 1974) predicted that such a marine-grounded ice sheet is inherently unstable.

Analysis of ice sediments indicates that in the past 1.3 million years, the WAIS has collapsed at least once (Scherer *et al.*, 1998). It was inferred from marine sediments that the WAIS is still dynamic. Since the last glacial maximum, the grounding line (i.e., the boundary between the floating ice shelves and the grounded ice) has retreated considerably (Hughes, 1998), and this process continues. It probably reflects dynamics that were set in motion in the early Holocene (Conway *et al.*, 1999). This has important implications because it points toward the long equilibration time scales involved in WAIS dynamics.

Fast-flowing ice streams, which feed the shelves from the interior, dominate the discharge of the WAIS (see TAR WGI Section 11.5). These ice flows are constrained at various boundaries. Whereas early studies emphasized the role of ice-shelf boundaries in such ice flow, more recent work points to the importance of different boundaries (i.e., the ice-stream bed, the lateral margins, and the inland end—Anandakrishnan *et al.*, 1998; Bell *et al.*, 1998; Joughin *et al.*, 1999; Payne, 1999). With respect to the time scales of an eventual WAIS disintegration, this distinction is crucial because the ice shelves respond to changes in climate within centuries, whereas the conditions at the ice-stream margins and beds have response times on the order of millennia (e.g., McAyeal, 1992). Whether proper incorporation of ice-stream dynamics into ice-sheet models generally eliminates the presumed instability cannot be conclusively resolved. McAyeal (1992), for example, incorporated ice-stream dynamics and deformable bed conditions explicitly into his ice-sheet model and showed that under periodic climate and sea-level forcing (100,000-year cycles), the WAIS collapsed and regrew sporadically throughout a period of 1 million years.

Even if accelerated loss of grounded ice were unlikely to occur over the 21st century, changes in ice dynamics could result in increased outflow of ice into the ice shelves and trigger a grounding-line retreat. An in-water temperature of a few degrees Celsius could cause the demise of the WAIS ice shelves in a few centuries and float its marine-based parts over a period of 1,000–2,000 years (Warner and Budd, 1998). This would produce a sea-level rise of 2–3 m. Huybrechts and de Wolde (1999) evaluate a climate change scenario that stabilizes GHG concentrations at four times the present value in 2150. They show that melting of the WAIS would contribute to 1-m sea-level rise by 2500—a rate of rise that would be sustained at 2 mm yr^{-1} for centuries thereafter. The response of the Greenland ice sheet contributed to several meters of sea-level rise by 3000. Even under this stabilization scenario, melting of the Greenland ice sheet would be irreversible. Both studies, however, simply assume no change in ocean circulation and an immediate warming of water in the sub-ice-shelf cavity with a warming climate. Both assumptions still await full validation.

Global warming projected for the 21st century could set in motion an irreversible melting of the West Antarctic and Greenland ice sheets, implying sustained sea-level rise and irreversible losses. The impacts of complete disintegration of the WAIS and subsequent sea-level rise by 4–6 m, however, have not been fully explored. As summarized by Oppenheimer (1998), the disintegration time scales predicted by models vary widely, between 400–500 years (Thomas *et al.*, 1979) and 1,600–2,400 years (McAyeal, 1992). These time scales correspond to a mean contribution to sea level of 10–15 and 2.5 mm yr^{-1}, respectively. Whereas an estimate in the lower range is approximately equal to the present-day rate of sea-level rise, a value in the middle to high range lies outside human experience and would widely exceed the adaptive capacity of most coastal structures and ecosystems (see Sections 19.3 and 6.5).

19.6.4. Climate Protection in an Irregular World

The predictability and manageability of singular phenomena is low. Their impacts can be sudden, large, and irreversible on a time scale of centuries. Regularizing such impacts would be an appropriate response, but this would require much better understanding of the statistics and characteristics of the complex processes involved. The presence of singularities therefore makes analytic and political treatment of the climate change problem particularly difficult.

Little is known, in quantitative terms, about the potential damages that could be inflicted by singularities on ecosystems and market sectors across the globe. This deficit has two main reasons (see also Moss and Schneider, 2000). First, extensive research on the causes, mechanisms, and impacts of singular events in the context of climate change is just getting started. Second, mechanistic and probabilistic analysis of complex nonlinear systems is more demanding—by orders of magnitude—than investigation of simple linear ones.

The knowledge base for assessing consequences of singularities will probably be broadened considerably over the next 5–10 years. Further advances in simulation modeling soon will allow better projections of future climate variability down to modified extreme events statistics (CLIVAR, 1998), as well as better translations of those projections into impacts on natural and societal systems (e.g., Weyant *et al.*, 1996; Alcamo *et al.*, 1998; Rotmans and Dowlatabadi, 1998). Earth system analysis, as supported by the big international research programs—World Climate Research Programme (WCRP), International Geosphere-Biosphere Programme (IGBP), and International Human Dimensions Programme (IHDP)—will bring about more complete understanding of macro-singularities within the responses of the Earth system under pertinent forcing (Schellnhuber, 1999). A major source of information and comprehension, in this context, will be evidence provided by paleorecords (IGBP, 1998). These scientific efforts should assist the decisionmaking process by creating a clearer picture of the future. Unfortunately, creating plausible projections is always tricky in practice (Sarewitz *et al.*, 2000).

A major challenge is to make responsible use of available information regarding the likelihood and the consequences of conceivable singular events. Responsibility here means the obligation of decisionmakers to make the "right" decision, taking into account the diverse societal values and wide ranges of individual interests that are at stake and that may be mutually contradictory. Thus, the standard challenge is to develop proper policies under uncertainty (i.e., neither ignorance nor omniscience) to achieve the objectives of the UNFCCC and to satisfy affected stakeholders as well as possible.

A broad and intensive discourse on the ethical aspects of singular responses to climate change (e.g., Markandya and Halsnaes, 2000; Munasinghe, 2000; Toth, 2000) is rediscovering many of the arguments put forward in traditional moral philosophy and risk policy. Ethical and procedural aspects of this type have been examined in various other contexts before, where certain concepts (such as human rights) act as a constraint on economic activity (emphasizing utilitarian goals), even when the cost-benefit ratio is unfavorable (e.g., the review of the agricultural situation by Aiken, 1986).

One of the crucial questions is how to deal with high-consequence impacts that *may* wipe out entire systems or cultures. Such non-implausible "nightmare" or "doomsday" scenarios could result from the speculative but consistent concatenation of individually possible causal relationships (e.g., Schellnhuber and Yohe, 1997). A vexing question is whether the lack of credible scientific evidence for such a scenario provides justification to ignore its possibility completely. Some argue that such effects have to be avoided by all means, irrespective of the economic burdens involved. Others argue that the uncertainties involved do not provide enough support for extensive measures and their economic costs. Within the climate-change framework, however, many incalculable risks could be reduced considerably by more sensible measures. The debate on the "legitimacy" of the different perspectives is impossible to resolve, however (Jasanoff, 1990).

The vague evidence provided by the present state of research supports the notion that even relatively small changes in mean climate could lead to large changes in the occurrence of stochastic extreme events. Furthermore, it suggests that large-scale discontinuities are unlikely below a 2°C warming but relatively plausible for a sustained warming of 8–10°C. The relatively small set of investigations discussed above lead to the conclusion that a warming range of 4–5°C seems to represent a critical disturbance regime where macro-discontinuities may start to emerge. This temperature threshold appears to be sensitive to the rate of change at which this level is reached.

19.7. Limitations of Methods and Directions for Future Research

This section discusses the strengths and limitations of the analytic approaches used to address the reasons for concern, mainly with regard to whether they can, with the confidence levels given, indicate the severity of impact or risk as a function of increase in global mean temperature. This discussion identifies key uncertainties inherent in each method and offers directions for future research that could improve our confidence in the results produced with each approach.

The organization of this section parallels that of the previous sections of this chapter. The strengths, limitations, uncertainties, and directions for each approach are discussed in the same order in which they were discussed in the preceding sections. However, integrated assessment frameworks are considered separately from aggregate approaches. Last is a discussion of integration across methods and reasons for concern.

19.7.1. Observations

Advantages: Because observations are based on observed effects rather than models, they can be used to indicate whether climate change is causing impacts and whether impacts lead to positive, negative, or indeterminate outcomes. They also can be used to validate hypotheses and models that formalize hypotheses on cause and effect.

Disadvantages: The problem with relying on observations to determine the severity of impacts or risk from climate change is that there has been only 0.7°C of mean global warming over the past century (although some regions have experienced much more warming). Because many impact thresholds may not be crossed until greater magnitudes or rates of warming are reached, it is not clear how to interpret an observed effect of warming or a group of such observations. Such observed impacts to date often will be of only minor consequence, even though they may tend to confirm our understanding of impact processes. Moreover, lack of observed impacts may be simply because climate change has not yet reached critical thresholds for such effects. Finally, attribution of causality is very difficult with observed effects or groups of effects. One must be able to demonstrate that a regional change in climate is a significant cause of an observed effect and that the regional change in climate is linked to global climate change.

Uncertainties: Uncertainties include the magnitude of climate change that has occurred, the extent to which impacts can be attributed to climate change that has occurred, and whether the relationship between climate change and possible impacts is linear or nonlinear and continuous or discontinuous.

Research Needs: For climate change impact detection to advance, there is a need for continued, improved, and augmented data collection and further development of analytical techniques. Geographical diversity is needed to balance the current bias of study locations in North America and Europe; more observation studies are needed in developing countries, with emphasis on those where physical, biological, and socioeconomic systems have higher vulnerability to climate change (see Chapter 18).

Because climate and impact systems are linked over a range of temporal scales, longer time series of data allow better understanding of the relative magnitudes of short- and long-term responses (Duarte *et al.*, 1992; McGowan *et al.*, 1999). Large-amplitude temporal changes usually involve large spatial dimensions, so broad-scale spatial/temporal studies are necessary as well. Satellite measurements of the Earth's surface provide a very useful monitoring capability for ocean, ecosystem, and land-cover changes. For example, satellite measurements of the Earth's surface offer the potential for aggregation of observed impacts with regard to broad-scale ecological responses such as vegetative responses to increasing lengths of growing seasons (e.g., Myneni *et al.*, 1997), complemented by meteorological and vegetation data (e.g., Schwartz, 1998).

For ecosystem impacts, continuing observations are needed at sites where studies already have been conducted, at long-term ecological research sites (e.g., Chapin *et al.*, 1995), and in protected areas. Programs that provide continued long-term monitoring of marine and terrestrial environments also are important (Duarte *et al.*, 1992; Southward *et al.*, 1995). Large-scale spatial/temporal ecosystem studies are necessary because effects from local changes cannot be extrapolated to large areas without evidence (McGowan *et al.*, 1999; Parmesan *et al.*, 1999).

Definition of indicator species or systems is a useful element of detection studies (e.g., Beebee, 1995; Nehring, 1998; Cannell *et al.*, 1999). Coupled with monitoring programs, such data may then provide a consistent set of evidence with which to study past, present, and future impacts of climate changes.

A further critical research need is to strengthen analytical tools for understanding and evaluating observed climate change impacts. Robust meta-analyses of studies that present good quality, multivariate data from a diversity of settings around the world will help to define further the global coherence among impacts now observed. Care also must be taken to ensure that the sample of studies is representative across time and space, is not biased in its reporting, and uses appropriate statistical tests. Also needed is development of methods to analyze differential effects of climate across a range or sector. Individual and grouped studies need to address possible correlations with competing explanations in a methodologically rigorous manner.

Also needed are refinements in the fingerprint approach (e.g., Epstein *et al.*, 1998), including more precise definition of expected changes and quantitative measurement techniques, similar to that used in detection of climate changes (see TAR WGI Chapter 12). For climate, fingerprint elements include warming in the mid-troposphere in the southern hemisphere, a disproportionate rise in nighttime and winter temperatures, and statistical increases in extreme weather events in many locations. These aspects of climate change and climate variability have implications for ecological, hydrological, and human systems that may be used to define a clear and robust multidimensional "expected impact signal" to be tested in a range of observations. A more refined and robust fingerprint approach may aid in the study of difficult-to-detect, partially causal climate effects on socioeconomic systems such as agriculture and health.

19.7.2. Studies of Unique and Threatened Systems

Assessments of unique and threatened systems tend to be based on studies of particular exposure units such as coral reefs, small islands, and individual species.

Advantages: These studies contain richness of detail and involve many researchers, often from developing and transition countries. In contrast to aggregate studies, studies of exposure units can be used, at least in principle, to analyze distributional effects by focusing on impacts on particular systems, species, regions, or demographic groups.

Disadvantages: One of the main disadvantages is that exposure-unit studies often are not carried out in a consistent manner. Exposure-unit studies often examine related sectors in isolation and do not examine linkages or integration among sectors and regions; for example, studies of the effects of climate change on ecosystems or individual species often are conducted without examining the potential effects of societal development on such systems. Local processes and forces (e.g., urbanization, local air pollution) often can be more important than global ones at the local scale, complicating the task of measuring the influence of global climate change at the local scale.

Another key disadvantage is incompleteness of coverage. For example, in spite of many and extensive country studies, there still are many gaps in coverage in terms of countries, regions within countries, and unique and important potential impacts that have not been assessed. The choice of exposure units may not necessarily cover the most vulnerable systems. Topics such as impacts on biodiversity or unique ecosystems often are not covered. There also has been little attention to impacts on poor and disadvantaged members of society. Even where particular critical exposure units have been covered, there may be just a single study. Drawing conclusions with high confidence on the basis of one study may be inappropriate.

Uncertainties: Uncertainties include the likely magnitude of climate change at the spatial resolution required by the study of the particular unique and threatened system, masking of global change effects by nonclimate factors, the degree of linearity/nonlinearity in the relationship between stimulus and response, and the degree to which results from individual studies can be extrapolated or aggregated.

Research Needs: It would be desirable to have more studies of individual systems, according to some set of priorities concerning the likely immediacy of the impacts. Additional work on standardizing methods and reporting of results also would be extremely useful. It also would be useful to devote more effort to integration of results from existing studies. Again, it would be especially useful to increase monitoring of changes in organisms, species, and systems that have limited range now or are near their limits and to try to separate out or consider other causal mechanisms such as local air pollution, loss of habitat, and competition from invading pests and weeds.

19.7.3. Distributional Impacts

Advantages: Distributional impact studies draw attention to likely heterogeneity in impacts among different regions and social and economic groups. They also help to identify and assess the situation of the "most vulnerable" people and systems. Thus, such studies bring equity considerations to center stage.

Disadvantages: Distributional impact studies require regional climate change projections and impact projections at the regional to local scale, where GCMs may not be very accurate. They also require projections of demographics and socioeconomic structure over a long time horizon.

Uncertainties: Research into the distribution of impacts of climate change is recent (see Section 19.4). There are some findings on which there is virtual unanimity. Some findings are broad conclusions—such as that more resource-constrained regions are likely to suffer more negative impacts, as are people whose geographic location exposes them to the greatest hazards from climate change. (Such people often live in regions with marginal climate for food growing or in highly exposed coastal zones.) Others are more specific but to date have been more conclusive with regard to the direction of different impacts among regions, rather than the magnitudes. For example, we know that impacts in developing low-latitude countries are more negative—in part because those countries tend to be operating at or above optimum temperatures already—and, in some cases, in regions where rainfall will decrease, leading to water stress. There also is limited capacity for adaptation in these areas. In some mid-latitude developed countries, agriculture would benefit initially from warmer conditions and longer growing seasons. Beyond such sweeping statements, uncertainties are vast. Resource constraints and (climatic) marginality are multidimensional and complex phenomena. Currently, it is not known which components of resource constraints or climatic marginality are more important or which components may compensate for others or may have synergistic effects. There are suggestions in other literature, but these have not been systematically applied to the impacts of climate change, conceptually or empirically.

In sum, there is virtual consensus about the broad patterns. There is much less knowledge about the details, although that situation is slowly improving.

Research Needs: Development of appropriate indicators of differences in regional impacts and ways of comparing them across regions and socioeconomic groups would be extremely useful. Improved methods for characterizing baseline demographics and socioeconomic conditions in the absence of climate change or climate change-motivated policies also would be useful. There is a need to quantify regional differences and to develop estimates of the cost of inequity in monetary or other terms (e.g., effect on poverty rates and trade, social and political instability, and conflict). More accurate projections of regional climate change would increase confidence in predictions of regional climate change impacts.

19.7.4. Aggregate Approaches

Advantages: Aggregate analyses synthesize climate change impacts in an internally consistent manner, using relatively comprehensive global indicators or metrics. These often are expressed in U.S. dollars (e.g., Tol, 2001b) or other common metrics such as changes in vegetation cover (Alcamo *et al.*, 1998). This enables direct comparisons of impacts among sector systems and regions and with other environmental problems and emission control costs. Some aggregate analyses have assessed differences in relative impacts in developed and developing regions of the world and have shown that regional differences in impacts may be substantial.

Disadvantages: Aggregate analyses lack richness of detail. Partly this is inherent because aggregation explicitly seeks to synthesize complex information. Partly this is because aggregate analyses tend to rely on reduced-form models. Condensing the diverse pattern of impacts into a small number of damage indicators is difficult. Some metrics may not accurately capture the value of certain impacts; for example, nonmarket impacts such as mortality and loss of species diversity or cultural heritage often are not well captured in monetization approaches, and change in vegetation cover may not clearly indicate threats to biodiversity. Other complicating issues concern comparison of impacts across time (impact today and several generations from now) and between regions (e.g., impact in developing and developed countries), as well as how much importance to assign to different effects. In addition, many aggregate studies examine a static world rather than a dynamic one and do not consider the effects of changes in extreme events or large-scale discontinuities. The aggregation process is not possible without value judgments, and different ethical views imply different aggregate measures across socioeconomic groupings and generations (see Azar and Sterner, 1996; Fankhauser *et al.*, 1997). Choice of discount rates can affect valuation of damages. In addition, general shortcomings that affect all reasons for concern are particularly prominent in aggregate analysis (e.g., accounting for baseline development, changes in variability and extreme events, and costs and benefits of adaptation).

Uncertainties: Uncertainties include whether all climate change impacts (positive and negative) are included, the implications of various aggregation and valuation methods, and implicit or explicit assumptions of methods, including possible mis-specifications of nonlinearities and interaction effects.

Research Needs: The next generation of aggregate estimates will have to account better for baseline developments, transient effects, climate variations, and multiple stresses. Further progress also is still needed in the treatment of adaptation. A broader set of primary studies on impacts in developing countries and nonmarket sectors would reduce the need for difficult extrapolation. More work also is needed on the ethical underpinnings of aggregation and on alternative aggregation schemes. Work on reflecting information from the other reasons for concern into the aggregate approach is underway, but proceeding slowly.

19.7.5. Integrated Assessment Frameworks

Advantages: Integrated assessment frameworks or models provide a means of structuring the enormous amount of and often conflicting data available from disaggregated studies. They offer internally consistent and globally comprehensive analysis of impacts; provide "vertical integration" (i.e., cover the entire "causal chain" from socioeconomic activities giving rise to GHG emissions to concentration, climate, impacts, and adaptations); provide "horizontal integration" (i.e., account for interlinkages between different impact categories, adaptations, and exogenous factors such as economic development and population growth); and allow for consistent treatment of uncertainties. IAMs have been used primarily for benefit-cost and inverse (or threshold) analyses. The latter have the advantage of being directly related to Article 2 because they define impacts that may be considered "dangerous" (through specification of thresholds related to, e.g., harm to unique and threatened systems or the probability of large-scale discontinuities).

Disadvantages: The main disadvantages with most IAMs are those associated with aggregate approaches: reliance on a single or a limited number of universal measures of impacts. These may not adequately measure impacts in meaningful ways. This is partly because IAMs rely on reduced-form equations to represent the complexities of more detailed models. Their usefulness is highly dependent on how well they are able to capture the complexities of more disaggregated approaches. Some of the IAMs used for benefit-cost analyses have considered large-scale irregularities (e.g., Gjerde *et al.*, 1999), but inclusion of such outcomes is preliminary. Few have accounted for loss of or substantial harm to unique and threatened systems. Although inverse (or threshold) approaches allow researchers to overcome these problems, the disadvantages of this kind of analysis include the difficulty of explicitly specifying thresholds and combining them within and across sectors and regions.

Uncertainties: Uncertainties are the same as those for the aggregate approach or for unique and threatened systems, depending on the structure and objectives of the model. This also would include the effects of different assumptions, methods, and value choices.

Research Needs: Among the biggest challenges facing integrated assessment modelers (see Weyant *et al.*, 1996) are developing a credible way to represent and value the impacts of climate change; a credible way to handle low-probability but potentially catastrophic events; a credible way to incorporate changes in extreme weather events; and realistic representations of changes in socioeconomic and institutional conditions, particularly in developing countries. In addition, they must decide how to allow explicitly for effects of different value choices, systems, and assumptions; how to quantify uncertainties; and how to credibly incorporate planned adaptation, including costs and limitations.

19.7.6. Extreme Events

Advantages: Extreme events are recognized as major contributors to the impacts of climate variability now and to potential impacts of climate change in the future. Thus, realistic climate change impact assessments must take them into account even though they may change in complex ways—such as in frequency, magnitude, location, and sequences (e.g., increased variability may lead to more frequent floods and droughts). Better understanding of changes in extreme events and adaptation measures for coping with them also will help in coping with present variability.

Disadvantages: Extreme events are more difficult to model and characterize than average climates. Changes in extreme events will be complex and uncertain, in part because extremes occur in a chaotic manner even in the present climate. Large data series are needed to characterize their occurrence because, by definition, they are rare events. This means that long time scale model simulations are needed to develop relevant statistics from long time slices or multiple realizations. Extreme events need to be considered in terms of probabilities or risks of occurrence rather than predictions. This chaotic element adds to other sources of uncertainty. It means that engineering or other design standards based on climatology that normally use long data series of observations will need a synthetic data set that simulates potential changes in future climate. It also makes adaptation to changes in extremes more difficult because planned adaptation must rely on necessarily uncertain projections into the future from theory and thus requires greater faith in the science before the information will be acted on.

Uncertainties and Research Needs: Better knowledge of the behavior of extremes will require long or multiple simulations at finer spatial and temporal scales, to capture the scale, intensity, and frequency of the events. Some types of extreme events (e.g., hail and extreme wind bursts) are poorly simulated at present; others, such as ENSO and tropical cyclones, are extremely complex and only now are beginning to be better simulated. Arguments for changes in their behavior are still often largely theoretical, qualitative, or circumstantial, rather than well based in verified models. Moreover, much more work is needed on how they will affect natural and human systems and how much of the recent trend to greater damages from extreme events is related to changes in exposure (e.g., greater populations, larger investments, more insurance cover, or greater reporting) rather than changes in the number and intensity of those extremes. More work is needed on how best to adapt to changes in extreme events, especially on how planners and decisionmakers can best take information on projected changes in extremes into consideration. This may be done best by focusing on projected change in the risk of exceeding prescribed natural, engineering, or socioeconomic impacts thresholds.

19.7.7. Large-Scale Singular Events

Advantages: Consideration of strongly nonlinear or even disruptive effects accompanying climate change is a critical component of the "dangerous interference" debate. The basic idea is to corroborate any non-negligible probability for high-consequence impacts that may be triggered by human climate perturbations. The political process to avoid high-consequence impacts may be facilitated by the global scope of such effects (e.g., disintegration of the WAIS generating a planetary sea-level rise of approximately 5 m). Inclusion of extreme events in the analysis helps, in general, to pursue all other reasons for concern in a realistic way because irregular impacts may dominate impacts on unique and threatened systems, distributional impacts, and aggregate impacts.

Disadvantages: This is an emerging area of research, facing several serious challenges because of the complexity of nonlinear interactions to be considered. The prevailing lack of knowledge is reflected in use of the term "surprises" for disruptive events. The potentials for climate change-induced transformations of extreme events regimes and for large-scale discontinuities in the Earth system are still highly uncertain. The search for irregularities might turn out to be futile and distract scientific resources from other important topics, such as the distributional aspects of regular climate change impacts.

Uncertainties and Research Needs: By definition, uncertainties are most severe in this realm of impact research. At present, there is no way of estimating the probabilities of certain disruptive events or assigning confidence levels to those probabilities. As a consequence, a strong research program should be launched that combines the best paleoclimate observations with the strongest simulation models representing full and intermediate complexity.

19.7.8. Looking across Analytic Approaches

Looking across the different analytic approaches (implicitly, the different reasons for concern), it is clear that to a great extent they complement and in many respects do not overlap each other. Combining these approaches into an integrated framework is the ambition of IAMs, at least in principle. However, this process is just starting. Because observed evidence has not been incorporated in the other analytic approaches, impacts to unique and threatened systems have not been accounted for in aggregate and IAM approaches, they are difficult to sum, and large-scale irregular impacts have only begun to be addressed, it does not appear to be feasible yet to combine these approaches into a comprehensive analytic approach. Thus, those who are seeking to implement climate policies must currently do their own integration of information from the alternative lines of inquiry.

19.8. Conclusions

This chapter focuses on certain reasons for concern with regard to what might be considered a "dangerous" climate change (reported as increases in global mean temperature; see Section 19.1.2). Each reason for concern can be used by itself or in combination with other reasons for concern to examine different aspects of vulnerability to climate change. We offer no judgment about how to use some or all of these reasons for concern to determine what is a dangerous level of climate change. The reasons for concern are as follows:

1) The relationship between global mean temperature increase and damage to or irreparable loss of unique and threatened systems
2) The relationship between global mean temperature increase and the distribution of impacts
3) The relationship between global mean temperature increase and globally aggregated impacts
4) The relationship between global mean temperature increase and the probability of extreme weather events
5) The relationship between greenhouse concentrations and the probability of large-scale singular events.

In addition, we address what observed effects of climate change tell us with regard to Article 2 of the UNFCCC. We review the state of knowledge with regard to what observations and each reason for concern tell us about climate change impacts.

19.8.1. Observations

Based on a review of the literature of observations of climate change impacts, as reflected in other TAR chapters, we conclude:

- Statistically significant associations between trends in regional climate and impacts have been documented in ~10 physical processes and ~450 biological species, in terrestrial and marine environments on all continents. Although the presence of multiple factors (e.g., land-use change, pollution, biotic invasion) makes attribution of observed impacts to regional climate change difficult, more than 90% (~99% physical, ~80% biophysical) of the changes documented worldwide are consistent with how physical and biological processes are known to respond to climate. Based on expert judgment, we have high confidence that the overall patterns and processes of observations reveal a widespread and coherent impact of 20th-century climate changes on many physical and biological systems.
- Signals of regional climate change impacts may be clearer in physical and biological systems than in socioeconomic systems, which also are simultaneously undergoing many complex changes that are not related to climate, such as population growth and urbanization. There are preliminary indications that some social and economic systems have been affected in part by 20th-century regional climate changes (e.g., increased damages from flooding and droughts in some locations). It generally is difficult to separate climate change effects from coincident or alternative explanations for such observed regional impacts.

There is preliminary evidence that unique and threatened systems are beginning to be affected by regional climate change and that some systems have been affected by recent increases in extreme climate events in some areas. Many high-latitude and high-altitude systems are displaying the effects of regional climate change. It is difficult to define observed impacts at aggregate levels, and evidence of large-scale singular events occurring as a result of recent climate change is lacking.

19.8.2. What does Each Reason for Concern Indicate?

Looking across these different reasons for concern, what can we conclude about what change in global average temperature is "dangerous"? A few general caveats apply:

- In spite of many studies on climate change impacts, there is still substantial uncertainty about how effective adaptation will be (and could be) in ameliorating negative effects of climate change and taking advantage of positive effects.
- The effect of changes in baseline conditions, such as economic growth and development of new technologies, that could reduce vulnerability has not been adequately considered in most impact studies.
- Most impact studies assess the effects of a stable climate, so our understanding of what rates of change may be dangerous is limited.

It does not appear to be possible—or perhaps even appropriate—to combine the different reasons for concern into a unified reason for concern that has meaning and is credible. However, we can review the relationship between impacts and temperature over the 21st century for each reason for concern and draw some preliminary conclusions about what change may be dangerous for each reason for concern. *Note that the following findings do not incorporate the costs of limiting climate change to these levels. Also note that there is substantial uncertainty regarding the temperatures mentioned below. These magnitudes of change in global mean temperature should be taken as an approximate indicator of when various categories of impacts might happen; they are not intended to define absolute thresholds.*

For simplification, we group different levels of global mean temperature increase into "small," "medium," and "large." "Small" denotes a global mean temperature increase of up to approximately 2°C;[4] "medium" denotes a global mean temperature increase of approximately 2–3°C; and "large" denotes a global mean temperature increase of more than approximately 3°C. In addition, changes in global mean temperature do not describe all relevant aspects of climate-change impacts, such as rates and patterns of change and changes in precipitation, extreme climate events, or lagged (or latent) effects such as rising sea levels.

19.8.2.1. Unique and Threatened Systems

Tropical glaciers, coral reefs, mangroves, biodiversity "hot spots," and ecotones are examples of unique and threatened entities that are confined to narrow geographical ranges and are very sensitive to climate change. However, their degradation or loss could affect regions outside their range. There is medium confidence that many of these unique and threatened systems will be affected by a small temperature increase. For example, coral reefs will bleach and glaciers will recede; at higher magnitudes of temperature increase, other and more numerous unique and threatened systems would become adversely affected.

19.8.2.2. Distributional Impacts

The impact of climate change will not be evenly distributed among the peoples of the world. There is high confidence that

[4]A 2°C warming from 1990 to 2100 would be a magnitude of warming greater than any that human civilization has ever experienced. Thus, "small" does not necessarily mean negligible.

developing countries tend to be more vulnerable to climate change than developed countries, and there is medium confidence that climate change would exacerbate income inequalities between and within countries. There also is medium confidence that a small temperature increase would have net negative impacts on market sectors in many developing countries and net positive impacts on market sectors in many developed countries. However, there is high confidence that with medium to high increases in temperature, net positive impacts would start to decline and eventually turn negative, and negative impacts would be exacerbated. Estimates of distributional effects are uncertain because of aggregation and comparison methods, assumptions about climate variability, adaptation, levels of development, and other factors. In addition, impacts are likely to vary between and within countries. Thus, not all developing or developed countries will necessarily have benefits or damages in unison.

19.8.2.3. Aggregate Impacts

With a small temperature increase, there is medium confidence that aggregate market sector impacts would amount to plus or minus a few percent of world GDP; there is low confidence that aggregate nonmarket impacts would be negative. Some studies find a potential for small net positive market impacts under a small to medium temperature increase. However, given the uncertainties about aggregate estimates, the possibility of negative effects cannot be excluded. In addition, most people in the world would be negatively affected by a small to medium temperature increase. Most studies of aggregate impacts find that there are net damages at the global scale beyond a medium temperature increase and that damages increase from there with further temperature increases. The important qualifications raised regarding distributional analysis also apply to aggregate analysis. By its nature, aggregate analysis masks potentially serious equity differences. Estimates of aggregate impacts are controversial because they treat gains for some as cancelling out losses for others and because weights that are used to aggregate over individuals are necessarily subjective.

19.8.2.4. Extreme Climate Effects

The frequency and magnitude of many extreme climate events increase even with a small temperature increase and will become greater at higher temperatures (high confidence). Extreme events include, for example, floods, soil moisture deficits, tropical and other storms, anomalous temperatures, and fires. The impacts of extreme events often are large locally and

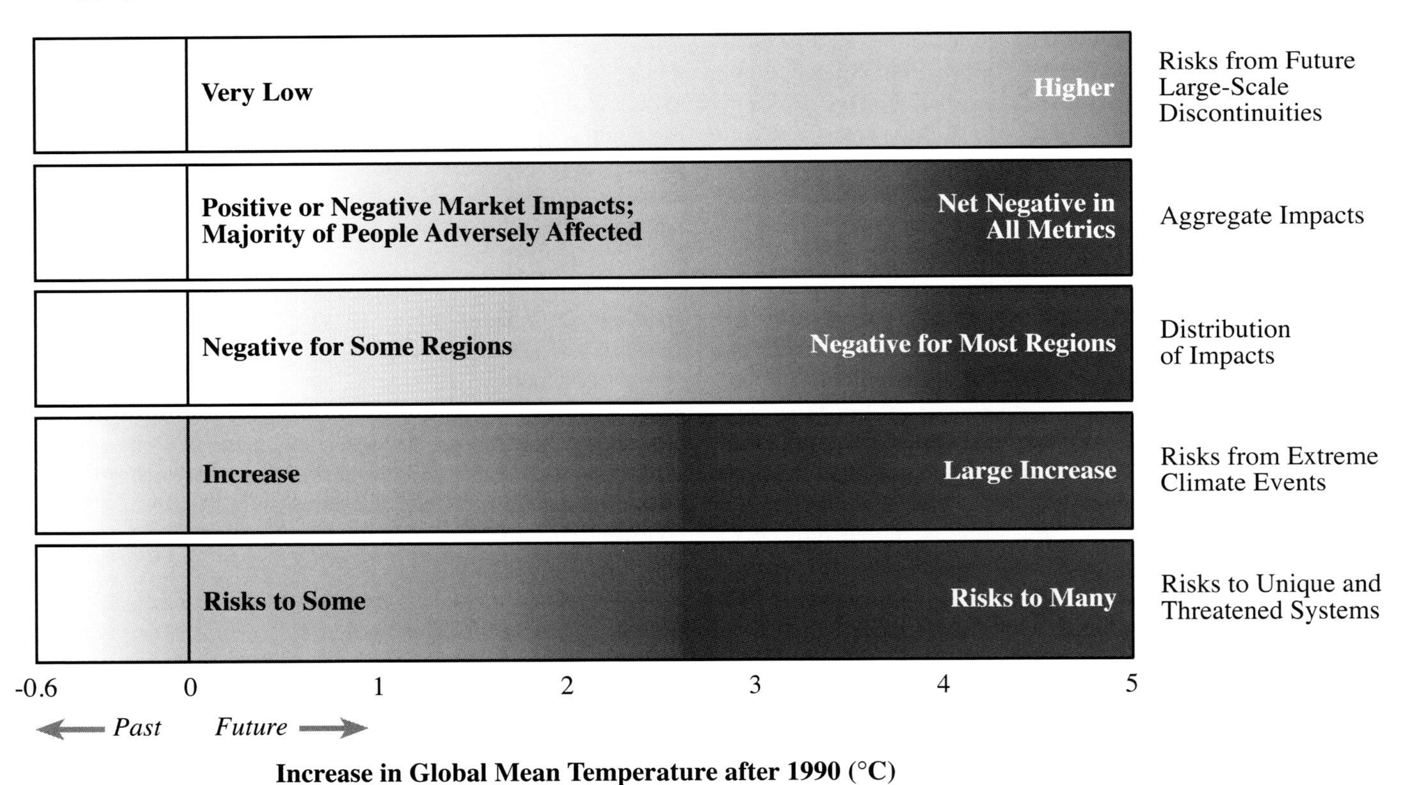

Figure 19-7: Impacts of or risks from climate change, by reason for concern. Each row corresponds to a reason for concern; shades correspond to severity of impact or risk. White means no or virtually neutral impact or risk, light gray means somewhat negative impacts or low risks, and dark gray means more negative impacts or higher risks. Global average temperatures in the 20th century increased by 0.6°C and led to some impacts. Impacts are plotted against increases in global mean temperature after 1990. This figure addresses only how impacts or risks change as thresholds of increase in global mean temperature are crossed, not how impacts or risks change at different rates of change in climate. Temperatures should be taken as approximate indications of impacts, not as absolute thresholds.

could strongly affect specific sectors and regions. Increases in extreme events can cause critical design or natural thresholds to be exceeded, beyond which the magnitude of impacts increases rapidly (high confidence).

19.8.2.5. Large-Scale Singularities

Large-scale singularities in the response of the climate system to external forcing, such as shutdown of the North Atlantic THC or collapse of the WAIS, have occurred in the past as a result of complex forcings. Similar events in the future could have substantial impacts on natural and socioeconomic systems, but the implications have not been well studied. Determining the timing and probability of occurrence of large-scale singularities is difficult because these events are triggered by complex interactions between components of the climate system. The actual impact could lag the climate change cause (involving the magnitude and the rate of climate change) by decades to millenia. There is low to medium confidence that rapid and large temperature increases would exceed thresholds that would lead to large-scale singularities in the climate system.

Figure 19-7 sums up the reasons for concern regarding impacts relative to change in temperature. Each row corresponds to a reason for concern, and the shades correspond to the severity of impact or risk. White means no or virtually neutral impact or risk, light gray means somewhat negative impacts or low risks, and dark gray means more negative impacts or higher risks. The period 1850–1990 warmed by 0.6°C and led to some impacts. Unique and threatened systems were affected, and the magnitude and frequency of some extreme events have changed. Future impacts are plotted against increases in global mean temperature after 1990.

Adverse impacts are estimated to occur in three reasons for concern even at a small increase in temperature: unique and threatened systems, extreme weather events, and distributional impacts. For the other two reasons for concern—adverse impacts and large-scale discontinuities—adverse impacts begin at the medium level of temperature increase for the former and a large temperature increase for the latter.

References

Abbs, D.J., 1999: A numerical modeling study to investigate the assumptions used in the calculation of probable maximum precipitation. *Water Resources Research*, **35,** 785–796.

Adger, W.N., 1999: Social vulnerability to climate change and extremes in coastal Vietnam. *World Development*, **27,** 249–269.

Agarwal, B., 1991: *A Field of Her Own*. Oxford University Press, Oxford, United Kingdom.

Aiken, W.H., 1986: On evaluating agricultural research. In: *New Directions for Agriculture and Agricultural Research: Neglected Dimensions and Emerging Alternatives*. Roman and Allanheld, Totowa, NJ, USA, pp. 31–41.

Alcamo, J. and G.J.J. Kreileman, 1996: Emission scenarios and global climate protection. *Global Environmental Change*, **6(4),** 305–334.

Alcamo, J., M. Krol, and R. Leemans, 1995: *Stabilizing greenhouse gases: global and regional consequences*. National Institute for Public Health and the Environment, Bilthoven, The Netherlands. Backgroup Report for FCCC-CoP, Berlin, Germany, 12 pp.

Alcamo, J., R. Leemans, and E. Kreileman, 1998: *Global Change Scenarios of the 21st Century. Results from the IMAGE 2.1 Model*. Pergamon and Elsevier Science, London, United Kingdom, 296pp.

Alward, R.D. and J.K. Detling, 1999: Grassland vegetation changes and nocturnal global warming. *Science*, **283,** 229–231.

Ames, A. and S. Hastenrath, 1996: Diagnosing the imbalance of Glacier Santa Rosa, Cordillera Raura, Peru. *Journal of Glaciology*, **42,** 212–218.

Ammann, B., 2000: Biotic responses to rapid climatic changes: introduction to a multidisciplinary study of the Younger Dryas and minor oscillations on an altitudinal transect in the Swiss Alps. *Palaeogeography, Palaeoclimatology, Palaeoecology*, **159,** 191–201.

Ammann, B., H.J.B. Birks, S.J. Brooks, U. Eicher, U. von Grafenstein, W. Hofmann, G. Lemdahl, J. Schwander, K. Tobolski, and L. Wick, 2000: Quantification of biotic responses to rapid climatic changes around the Younger Dryas—a synthesis. *Palaeogeography, Palaeoclimatology, Palaeoecology*, **159,** 313–347.

Anandakrishnan, S., D.D. Blankenship, R.B. Alley, and P.L. Stoffa, 1998: Influence of subglacial geology on the position of a West Antarctic ice stream from seismic observations. *Nature*, **394,** 62–65.

Arnell, N.W., 1999: Climate change and global water resources. *Global Environmental Change*, **9,** S31–S50.

Azar, C., 1999: Weight factors in cost-benefit analysis of climate change. *Environmental and Resource Economics* **13,** 249–268.

Azar, C. and T. Sterner, 1996: Discounting and distributional considerations in the context of global warming. *Ecological Economics,* **19,** 169–184.

Badii, R. and A. Politi, 1997: *Complexity—Hierarchical Structures and Scaling in Physics*. Cambridge University Press, Cambridge, United Kingdom, 316 pp.

Bains, S., R.M. Corfield, and R.D. Norris, 1999: Mechanisms of climate warming at the end of the Paleocene. *Science*, **285,** 724–727.

Barber, V.A., G.P. Juday, and B.P. Finney, 2000: Reduced growth of Alaskan white spruce in the twentieth century from temperature-induced drought stress. *Nature*, **405,** 668–673.

Beebee, T.J.C., 1995: Amphibian breeding and climate. *Nature*, **374,** 219–220.

Beer, T., R. Allan, and B. Ryan, 1993: Modelling climatic hazards. In: *Modelling Change in Environmental Systems* [Jakeman, A.J., M.B. Beck, and M.J. McAleer (eds.)]. John Wiley and Sons, New York, NY, USA, pp. 453–477.

Bell, R.E., D.D. Blankenship, C.A. Finn, D.L. Morse, T.A. Scambos, J.M. Brozena, and S.M. Hodge, 1998: Influence of subglacial geology on the onset of a West Antarctic ice stream from aerogeophysical observations. *Nature*, **394,** 58–62.

Bergmann, F., 1999: Long-term increase in numbers of early-fledged reed warblers (*Acrocephalus scirpaceus*) at Lake Constance (Southern Germany). *Journal Fuer Ornithologie*, **140,** 81–86 (in German).

Berry, M.V., 1978: Regular and irregular motion. In: *Topics in Nonlinear Dynamics, A Tribute to Sir Edward Bullard, American Institute of Physics Conference Proceedings* [Jorna, S. (ed.)]. American Institue of Physics, La Jolla, CA, USA, No. 46, 16–120.

Bezzel, E. and W. Jetz, 1995: Delay of the autumn migratory period in the blackcap (*Sylvia atricappila*) 1966–1993: a reaction to global warming? *Journal Fuer Ornithologie*, **136,** 83–87.

Biermann, F., G. Petschel-Held, and C. Roloff, 1998: Umweltzerstoerung als Konfliktursache? *Zeitschrift fuer Internationale Beziehungen*, **2,** 273–308.

Bond, G., H. Heinrich, W. Broecker, L. Labeyrie, J. McManus, J. Andrews, S. Huon, R. Jantschik, S. Clasen, S. Sirnet, K. Tedesco, M. Klas, G. Bonani, and S. Ivy, 1992: Evidence for massive discharges of icebergs into the North Atlantic ocean during the last glacial period. *Nature*, **360,** 245–249.

Bouma, M.J., 1999: Extreme events and subtle effects of El Niño: can short-term climate variability help us assess the health impact of global change? In: *Climate Change and Human Health, A Discussion Meeting, 20 October, 1998, London* [Haines, A. and A.J. McMichael (eds.)]. Royal Society, London, United Kingdom, pp. 71–78.

Bouma, M.J., S. Kovats, S.A. Goubet, J. Cox, and A. Haines, 1997: Global assessment of El Nino's disaster burden. *Lancet*, **350(9089),** 1435–1438.

Braaf, R.R., 1999: Improving impact assessment methods: climate change and the health of indigenous Australians. *Global Environmental Change*, **9,** 95–104.

Bradley, N.L., A.C. Leopold, J. Ross, and W. Huffaker, 1999: Phenological changes reflect climate change in Wisconsin. *Proceedings of the National Academy of Sciences of the United States of America*, **96,** 9701–9704.

Brierly, A. and K. Reid, 1999: Kingdom of the krill. *New Scientist*, **162(2182),** 36–41.

Brovkin, V., M. Claussen, V. Petoukhov, and A. Ganopolski, 1998: On the stability of the atmosphere–vegetation system in the Sahara/Sahel region. *Journal of Geophysical Research*, **103(D24),** 31613–31624.

Brown, J.H., T.J. Valone, and C.G. Curtin, 1997: Reorganization of an arid ecosystem in response to recent climate change. *Proceedings of the National Academy of Sciences of the United States of America*, **94,** 9729–9733.

Brown, J.L., S.-H. Li, and N. Bhagabati, 1999: Long-term trend toward earlier breeding in an American bird: a response to global warming? *Proceedings of the National Academy of Sciences of the United States of America*, **96,** 5565–5569.

Bunde, A., J. Kropp, and H.-J. Schellnhuber, 2001: *Science of Disaster. Climate disruptions heart attacks and market crashes,* Springer-Verlag, Berlin and Heidelberg, Germany, (in press).

Burton, I., R.W. Kates, and G.F. White, 1993: *The Environment as Hazard.* Guilford Press, New York, NY, USA.

Cannell, M.G.R., J.P. Palutikof, and T.H. Sparks (eds.), 1999: *Indicators of Climate Change in the UK.* National Environmental Research Council (DETR) Wetherby, United Kingdom, 87 pp.

Carter, T.R., 1998: Changes in the thermal growing season in Nordic countries during the past century and prospects for the future. *Agricultural and Food Science in Finland*, **7,** 161–179.

Cavalieri, D.J., P. Gloersen, C.L. Parkinson, J.C. Comiso, and H.J. Zwally, 1997: Observed hemispheric asymmetry in global sea ice changes. *Science*, **278,** 1104–1106.

Changnon, S.A., 1998: Comments on "secular trends of precipitation amount, frequency, and intensity in the United States" by Karl and Knight. *Bulletin of the American Meteorological Society*, **79,** 2550–2552.

Changnon, S.A., D. Changnon, E.R. Fosse, D.C. Hoganson, R.J. Roth, and J.M. Totsch, 1997: Effects of recent weather extremes on the insurance industry; major implications for the atmospheric sciences. *Bulletin of the American Meteorological Society*, **78,** 425–435.

Chapin, F.S. III, G.R. Shaver, A.E. Giblin, K.J. Nadelhoffer, and J.A. Laundre, 1995: Responses of arctic tundra to experimental and observed changes in climate. *Ecology*, **76(3),** 694–711.

Claussen, M., C. Kubatzki, V. Brovkin, A. Ganopolski, P. Hoelzmann, and H.-J. Pachur, 1999: Simulation of an abrupt change in Saharan vegetation in the mid-Holocene. *Geophysical Research Letters*, **26(14),** 2037–2040.

Cline, W.R., 1992: *The Economics of Global Warming.* Institute for International Economics, Washington, DC, USA, 399 pp.

CLIVAR, 1997: *A Research Programme on Climate Variability and Predictability for the 21st Century.* WCRP Report No. 101, World Climate Research Programme, Geneva, Switzerland, 48 pp.

Cochrane, M.A., 1999: Positive feedbacks in the fire dynamic of closed canopy tropical forests. *Science*, **284,** 1832–1835.

Cohen, S.J. (ed.), 1997a: *Mackenzie Basin Impact Study (MBIS) Final Report: Summary Results.* Environment Canada, Toronto, ON, Canada, 20 pp.

Cohen, S.J., 1997b: Scientist-stakeholder collaboration in integrated assessment of climate change: lessons from a case study of Northwest Canada. *Environmental Modelling and Assessment*, **2,** 281–293.

Cohen, S.J., 1997c: What if and what in northwest Canada: could climate change make a difference to the future of the Mackenzie Basin? *Arctic* **50(4),** 293–307.

Cohen, S.J., 1996: Integrated regional assessment of global climatic change: lessons from the Mackenzie Basin Impact Study (MBIS). *Global and Planetary Change*, **11,** 179–185.

Cohen, S.J. (ed.), 1994: *Mackenzie Basin Impact Study (MBIS) Interim Report No. 2.* Environment Canada, Toronto, ON, Canada, 485 pp.

Cohen, S.J., R. Barret, S. Irlbacher, P. Kertland, L. Mortch, L. Pinter, and T. Zdan, 1997: Executive summary. In: The *Mackenzie Basin Impact Study (MBIS) Final Report* [Cohen, S.J. (ed.)]. Environment Canada, Toronto, ON, Canada, pp. 1–18.

Conway, H., B.L. Hall, G.H. Denton, A.M. Gades, and E.D. Waddington, 1999: Past and future grounding-line retreat of the West Antarctic ice sheet. *Science*, **286,** 280–283.

Corti, S., F. Molteni, and T.N. Palmer, 1999: Signature of recent climate change in frequencies of natural atmospheric circulation regimes. *Nature*, **398,** 799–802.

Cowling, R.M. and C. Hilton-Taylor, 1997: Phytogeography, flora and endemism. In: *Vegetation of Southern Africa* [Cowling, R.M., D.M. Richardson, and S.M. Pierce (eds.)]. Cambridge University Press, Cambridge, United Kingdom and New York, NY, USA, pp. 43–61.

Cramer, W., A. Bondeau, F. I. Woodward, I. C. Prentice, R. A. Betts, V. Brovkin, P. M. Cox, V. Fisher, J. Foley, A.D. Friend, C. Kucharik, M.R. Lomas, N. Ramankutty, S. Sitch, B. Smith, A. White, and C. Young-Molling, 2001: Global response of terrestrial ecosystem structure and function to CO_2 and climate change: results from six dynamic global vegatation models. *Global Change Biology*, (in press).

Crick, H.Q.P. and T.H. Sparks, 1999: Climate change related to egg-laying trends. *Nature*, **399,** 423–424.

Crick, H.Q.P., C. Dudley, D.E. Glue, and D.L. Thomson, 1997: UK birds are laying eggs earlier. *Nature*, **388** (6442): p.526.

Cunningham, D.M. and P.J. Moors, 1994: The decline of rockhopper penguins (*Eudyptes chrysocome*) at Campbell Island, Southern Ocean and the influence of rising sea temperatures. *Emu*, **94,** 27–36.

Dallimore, S.R., S. Wolfe, and S.M. Solomon, 1996: Influence of ground ice and permafrost on coastal evolution, Richards Island, Beaufort Sea Coast, N.W.T., Canada. *Canadian Journal of Earth Sciences*, **33,** 664–675.

Dansgaard, W., S.J. Johnsen, H.B. Clausen, D. Dahl-Jensen, N.S. Gundestrup, C.U. Hammer, C.S. Hvidberg, J.P. Steffensen, A.E. Sveinbjörnsdottir, J. Jouzel, and G. Bond, 1993: Evidence for general instability of past climate from a 250–kyr ice-core record. *Nature*, **364,** 218–220.

Darwin, R., 1999: A FARMer's view of the Ricardian approach to measuring agricultural effects of climatic change. *Climatic Change*, **41,** 371–411.

Darwin, R.F. and R.S.J. Tol, 2001: Estimates of the economic effects of sea level rise. *Environmental and Resource Economics*, (in press).

Darwin, R., M. Tsigas, J. Lewandrowski, and A. Raneses, 1995: *World Agriculture and Climate Change: Economic Adaptations.* Agricultural Economic Report No. 703, U.S. Department of Agriculture, Washington, DC, USA, 86 pp.

Davis, M.B., 1986: Climatic instability, time lags, and community disequilibrium. In: *Community Ecology* [Diamond, J. and T.J. Case (eds.)]. Harper and Row, New York, NY, USA, pp. 269–284.

De Jong, P.W. and P.M. Brakefield, 1998: Climate and change inclines for melanism in the two-spot ladybird, *Adalia bipunctata* (Coleoptera: Coccinellidae). *Proceedings of the Royal Society of London B*, **265,** 39–43.

DETR, 1999: *Climate Change and its Impacts: Stabilisation of CO_2 in the Atmosphere.* Department of Environment, Transport, and Regions, Hadley Centre for Climate Prediction and Research, Bracknell, United Kingdom, Report 99/812, 28 pp.

Dettinger, M.D. and D.R. Cayan, 1995: Large-scale atmospheric forcing of recent trends toward early snowmelt runoff in California. *Journal of Climate*, **8,** 606–623.

Dilley, M. and B. Heyman, 1995: ENSO and disaster: droughts, floods, and El Niño/Southern Oscillation warm events. *Disasters*, **19,** 181–193.

Döös, B.R., 1994: Environmental degradation, global food production, and risk for large-scale migrations. *Ambio*, **23,** 124–130.

Dowdeswell, J., J.O. Hagen, H. Bjornsson, A. Glazovsky, P. Holmlund, J. Jania, E. Josberger, R. Koerner, S. Ommanney, and B. Thomas, 1997: The mass balance of circum Arctic glaciers and recent climate change. *Quaternary Research*, **48,** 1–14.

Downing, T.E., R.A. Greener, and N. Eyre, 1995: *The Economic Impacts of Climate Change: Assessment of Fossil Fuel Cycles for the ExternE Project.* Environmental Change Unit and Eyre Energy Environment, Oxford and Lonsdale, United Kingdom, 48 pp.

Downing, T.E., A.A. Olsthoorn, and R.S.J. Tol, 1999: *Climate, Change and Risk*. Routledge, London, United Kingdom and New York, NY, USA, 407 pp.

Downing, T.E., A.A. Olsthoorn, and R.S.J. Tol (eds.), 1996: *Climate Change and Extreme Events: Altered Risk, Socioeconomic Impacts and Policy Responses*. Environmental Change Unit, Oxford, Institute for Environmental Studies, Amsterdam, The Netherlands, 309 pp.

Duarte, C.M., J. Cebrián, and N. Marba, 1992: Uncertainty of detecting sea change. *Nature*, **356,** 190.

Dunn, P.O. and D.W. Winkler, 1999: Climate change has affected the breeding date of tree swallow throughout North America. *Proceedings of the Royal Society of London B*, **266,** 2487–2490.

Dyurgerov, M. and M. Meier, 1997: Mass balance of mountain and subpolar glaciers: a new global assessment for 1961–1990. *Arctic and Alpine Research*, **29,** 379–391.

Ellis, W.N., 1997: Recent shifts in phenology of Microlepidoptera, related to climatic change (Lepidoptera). *Entomologische Berichten (Amsterdam)*, **57,** 66–72.

Ellis, W.N., J.H. Donner, and J.H. Kuchlein, 1997: Recent shifts in distribution of microlepidoptera in The Netherlands. *Entomologische Berichten (Amsterdam)*, **57,** 119–125.

EMF, 1997: *Preliminary Results from EMF 14 on Integrated Assessment of Climate Change*. EMF Occasional Paper 48, Stanford Energy Modeling Forum, Stanford University, Palo Alto, CA, USA, 16 pp.

Enting, I.G., T.M.L. Wigley, and M. Heimann, 1994: *Future Emissions and Concentrations of Carbon Dioxide: Key Ocean/Atmosphere/Land Analyses*. Technical Paper No. 31, Commonwealth Scientific and Industrial Research Organisation, Division of Atmospheric Research, Aspendale, Australia, 118 pp.

Epstein, P.R., 2000: Is global warming harmful to health? *Scientific American*, **283**(2)50–57.

Epstein, P.R., H.F. Diaz, S. Elias, G. Grabherr, N.E. Graham, W.J.M. Martens, E. Mosley-Thompson, and J. Susskind, 1998: Biological and physical signs of climate change: focus on mosquito-borne diseases. *Bulletin of the American Meteorological Society*, **79,** 409–417.

Eyre, N., T.E. Downing, R. Hoekstra, and K. Rennings, 1997: *Global Warming Damages*. ExternE, Brussels, Belgium.

Fankhauser, S., 1995: Protection vs. retreat: the economic costs of sea level rise. *Environment and Planning A*, **27,** 299–319.

Fankhauser, S., R.S.J. Tol, and D.W. Pearce, 1998: Extensions and alternatives to climate change impact valuation: on the critique of IPCC Working Group III's impact estimates. *Environment and Development Economics*, **3,** 59–81.

Fankhauser, S., R.S.J. Tol, and D.W. Pearce, 1997: The aggregation of climate change damages: a welfare theoretic approach. *Environmental and Resource Economics*, **10,** 249–266.

Fedorov, A.V. and S.G. Philander, 2000: Is El Niño changing? *Science*, **288,** 1997–2002.

Fleming, R.A. and G.M. Tatchell, 1995: Shifts in the flight periods of British aphids: a response to climate warming? In: *Insects in a Changing Environment* [Harrington, R. and N.E. Stork (eds.)]. Academic Press, San Diego, CA, USA, pp. 505–508.

Forchhammer, M.C., E. Post, and N.C. Stenseth, 1998: Breeding phenology and climate. *Nature*, **391,** 29–30.

Fowler, A.M. and K.J. Hennessy, 1995: Potential impacts of global warming on the frequency and magnitude of heavy precipitation. *Natural Hazards*, **11,** 283–303.

Fraser, W.R., W.Z. Trivelpiece, D.G. Ainley, and S.G. Trivelpiece, 1992: Increase in Antarctic penguin populations: reduced competition with whales or a loss of sea ice due to environmental warming? *Polar Biology*, **11,** 525–531.

Frederick, K.D. and G.E. Schwarz, 1999: Socioeconomic impacts of climate change on U.S. water supplies. *Journal of the American Water Resources Association*, **35(6),** 1563–1583.

Fyfe, J.C., G.J. Boer, and G.M. Flato, 1999: The Arctic and Antarctic oscillations and their projected changes under global warming. *Geophysical Research Letters*, **26(11),** 1601–1604.

Gaffin, S.R., 1997: *Impacts of Sea Level Rise on Selected Coasts and Islands*. The Environmental Defense Fund, New York, NY, USA, 34 pp.

Ganopolski, A., C. Kubatzki, M. Claussen, V. Brovkin, and V. Petoukhov, 1998: The influence of vegetation-atmosphere-ocean interaction on climate during the mid-Holocene. *Science*, **280,** 1916–1919.

Gatter, W., 1992: Timing and patterns of visible autumn migration: can effects of global warming be detected? *Journal Fuer Ornithologie*, **133,** 427–436.

Gattuso, J.P., D. Allemand, and M. Frankignoulle, 1999: Photosynthesis and calcification at cellular, organismal and community levels in coral reefs: a review on interactions and control by carbonate chemistry. *American Zoologist*, **39(1),** 160–183.

Gifford, R., J. Angus, D. Barrett, J. Passioura, H. Rawson, R. Tichards, and M. Stapper, 1998: Climate and Australian wheat yield. *Nature*, **391,** 448–449.

Gill, A.M., 1998: Big versus small fires: the bushfires of Greater Sydney, January 1994. In: *Large Forest Fires* [Moreno, J.M. (ed.)]. Backhuys Publishers, Leiden, Germany, pp. 49–68.

Gjerde, J., S. Grepperud, and S. Kverndok, 1999: Optimal climate policy under the possibility of a catastrophe. *Resource and Energy Economics*, **21,** 289–317.

Glynn, P.W., 1991: Coral reef bleaching in the 1980s and possible connections with global warming. *Trends in Ecology and Evolution*, **6,** 175–179.

Godden, D., R. Batterham, and R. Drynan, 1998: Climate change and Australian wheat yield. *Nature*, **391,** 447–448.

Goldammer, J.G., 1999: Forests on fire. *Science*, **284,** 1782–1783.

Gonzalez, P., 2001: Desertification and a shift of forest species in the West African Sahel. *Climate Research*, (in press)

Goreau, T.J., R. Hayes, A. Strong, E. Williams, G. Smith, J. Cervino, and M. Goreau, 1998: Coral reefs and global change: impacts of temperature, bleaching, and emerging diseases. *Sea Wind*, **12(3),** 2–6.

Grabherr, G., M. Gottfried, and H. Pauli, 1994: Climate effects on mountain plants. *Nature*, **369,** 448.

Greene, A.M., W.S. Broecker, and D. Rind, 1999: Swiss glacier recession since the Little Ice Age: reconciliation with climate records. *Geophysical Research Letters*, **26,** 1909–1912.

Gregory, J.M. and J.F.B. Mitchell, 1995: Simulation of daily variability of surface temperature and precipitation over Europe in the current and $2xCO_2$ climates using the UKMO model. *Quarterly Journal of the Royal Meteorology Society*, **121,** 1451–1476.

Greve, W., F. Reiners, and J. Nast, 1996: Biocoenotic changes of the zooplankton in the German Bight: the possible effects of eutrophication and climate. *Journal of Marine Science*, **53,** 951–956.

Groisman, P.Y., T.R. Karl, and R.W. Knight, 1994: Observed impact of snow cover on the heat balance and the rise of continental spring temperatures. *Science*, **263,** 198–200.

Groisman, P.Y., T.R. Karl, D.R. Easterling, R.W. Knight, P.F. Jamason, K.J. Hennessy, R. Suppiah, C.M. Page, J. Wibig, K. Fortuniak, V.N. Razuvaev, A. Douglas, E. Forland, and P.M. Zhai, 1999: Changes in the probability of heavy precipitation: important indicators of climatic change. *Climatic Change*, **42,** 243–283.

Haerberli, W. and M. Beniston, 1998: Climate change and its impacts on glaciers and permafrost in the Alps. *Ambio*, **27,** 258–265.

Hall, W.B, G.M. McKeon, J.O. Carter, K.A. Day, S.M. Howden, J.C. Scanlan, P.W. Johnston, and W.H. Burrows, 1998: Climate change and Queensland's grazing lands, II: an assessment of impact on animal production from native pastures. *Rangeland Journal*, **20(2),** 177–205.

Halpin, P.N., 1997: Global climate change and natural-area protection: management responses and research directions. *Ecological Applications*, **7,** 828–843.

Hamburg, S.P. and C.V. Cogbill, 1988: Historical decline of red spruce populations and climatic warming. *Nature*, **331,** 428–430.

Handmer, J., E. Penning-Rowsell, and S. Tapsell, 1999: Flooding in a warmer world: the view from Europe. In: *Climate, Change and Risk* [Downing, T.E., A.A. Olsthoorn, and R.S.J. Tol (eds.)]. Routledge, London, United Kingdom and New York, NY, USA, 125-161.

Hansen, E.M., 1987: Probable maximum precipitation for design floods in the United States. *Journal of Hydrology*, **96,** 267–278.

Hasenauer, H., R.R. Nemani, K. Schadauer, and S.W. Running, 1999: Forest growth response to changing climate between 1961 and 1990 in Austria. *Forest Ecology and Management*, **122,** 209–219.

Hasselmann, K., S. Hasselmann, R. Giering, V. Ocana, and H. Von Storch, 1997: Sensitivity study of optimal CO_2 emission paths using a simplified structural integrated assessment model (SIAM). *Climatic Change*, **37,** 345–386.

Hastenrath, S., 1995: Glacier recession on Mount Kenya in the context of the global tropics. *Bulletin Institut Francais d'Etudes Andines*, (Bulletin of the French Institue for Andean Studies) **24(3),** 633–638.

Hay, J.E. and C. Kaluwin (eds.), 1993: *Climate Change and Sea Level Rise in the South Pacific Region, Proceedings of the Second SPREP Meeting, Nomea* New Caledonia , *6–10 April, 1993*. South Pacific Regional Environment Programme, Apria, West Samoa, 238 pp.

Hennessy, K.J. and A.B. Pittock, 1995: Greenhouse warming and threshold temperature events in Victoria, Australia. *International Journal of Climatology*, **15,** 591–612.

Hirst, A.C., 1999: The Southern Ocean response to global warming in the CSIRO coupled ocean-atmosphere model. *Environmental Modelling and Software*, **14,** 227–241.

Hoegh-Guldberg, O., 1999: Climate change, coral bleaching and the future of the world's coral reefs. *CSIRO Marine and Freshwater Research*, **50(8),** 839–866.

Hohmeyer, O. and M. Gaertner, 1992: *The Costs of Climate Change—A Rough Estimate of Orders of Magnitude*. Fraunhofer-Institut fur Systemtechnik und Innovationsforschung, Karlsruhe, Germany.

Holling, C.S., 1992a: Cross-scale morphology, geometry and dynamics of ecosystems. *Ecological Monographs*, **62,** 447–502.

Holling, C.S., 1992b: The role of boreal forest insects in structuring the boreal landscape. In: *A Systems Analysis of the Global Boreal Forest* [Shugart, H.H., R. Leemans, and G.B. Bonan (eds.)]. Cambridge University Press, Cambridge, United Kingdom and New York, NY, USA, pp. 170–191.

Homer-Dixon, T.F., 1991: On the threshold. *International Security*, **16(2),** 76–117.

Homer-Dixon, T.F. and J. Blitt, 1998: *Ecoviolence: Links Among Environment, Population and Security*. Rowman and Littlefield Publishers, Lanham, United Kingdom, 238 pp.

Hoozemans, F.M.J., M. Marchand, and H.A. Pennekamp, 1993: *A Global Vulnerability Analysis: Vulnerability Assessment for Population, Coastal Wetlands and Rice Production and a Global Scale (Second, Revised Edition)*. Delft Hydraulics, Delft, The Netherlands, 184 pp.

Hsu, K.J., 1998: Sun, climate, hunger, and mass migration. *Science in China Series D—Earth Sciences*, **41(5),** 449–472.

Hubbert, G.D. and K.L. McInnes, 1999: A storm surge inundation model for coastal planning and impact studies. *Journal of Coastal Research*, **15,** 168–185.

Hughes, L., 2000: Biological consequences of global warming: is the signal already apparent? *Trends in Ecology and Evolution*, **15(2),** 56–61.

Hughes, T.J., 1998: *Ice Sheets*. Oxford University Press, Oxford, United Kingdom, 343 pp.

Hughes, T., 1973: Is the West Antarctic ice sheet disintegrating? *Journal of Geophysical Research*, **78(33),** 7884–7910.

Huntley, B., W.P. Cramer, A.V. Morgan, H.C. Prentice, and J.R.M. Allen (eds.), 1997: *Past and Future Rapid Environmental Changes: The Spatial and Evolutionary Responses of Terrestrial Biota*. Springer-Verlag, Berlin, Germany, 523 pp.

Huybrechts, P. and J. deWolde, 1999: The dynamic response of the Greenland and Antarctic ice sheets to multiple-century climatic warming. *Journal of Climate*, **12(8),** 2169–2188.

ICCD, 1999: *Collaboration and Synergies Among Rio Conventions for the Implementation of the UNCCD*. International Convention to Combat Desertification (ICCD)/Conference of Parties (COP)3/9, Item 11 of the Provisional Agenda, United Nations International Convention to Combat Desertification (UNCCD), New York, NY, USA, pp.1-50.

IGBP, 1998: PAGES status report and implementation plan. International Geosphere-Biosphere Programme (IGBP). Oldfield, F. ed.Rep. 45, IGBP Report Series, Stockholm, Sweden, 236 pp.

IPCC, 1996: *Climate Change 1995: Impacts, Adaptation and Mitigation of Climate Change: Scientific-Technical Analysis. Contribution of Working Group II to the Second Assessment Report of the Intergovernmental Panel on Climate Change* [Watson, R.T., M.C. Zinyowera, and R.H. Moss (eds.)]. Cambridge University Press, Cambridge, United Kingdom and New York, NY, USA, 880 pp.

Jarvinen, A., 1994: Global warming and egg size of birds. *Ecography*, **17(1),** 108–110.

Jasanoff, S.S., 1990: *The Fifth Branch: Science Advisors as Policymakers*. Harvard University Press, Cambridge, MA, USA, 302pp.

Johannessen, O.M., E.V. Shalina, and M.W. Miles, 1999: Satellite evidence for an Arctic sea ice cover in transformation. *Science*, **286,** 1937-1939.

Johnston, C.A., 1993: Material fluxes across wetland ecotones in northern landscapes. *Ecological Applications*, **3,** 424–440.

Jones, R.N., 2000: Analysing the risk of climate change using an irrigation demand model. *Climate Research*, **14,** 89–100.

Joos, F., G.-K. Plattner, T.F. Stocker, O. Marchal, and A. Schmittner, 1999: Global warming and marine carbon cycle feedbacks on future atmospheric CO_2. *Science*, **284,** 464–467.

Joughin, I., L. Gray, R. Bindschadler, S. Price, D. Morse, C. Hulbe, K. Mattar, and C. Werner, 1999: Tributaries of West Antarctic ice streams revealed by RADARSAT interferometry. *Science*, **286,** 283–286.

Kalkstein, L.S. and J.S. Greene, 1997: An evaluation of climate/mortality relationships in large U.S. cities and the possible impacts of a climate change. *Environmental Health Perspectives*, **105(1),** 2–11.

Kane, S., J. Reilly, and J. Tobey, 1992: An empirical study of the economic effects of climate change on world agriculture. *Climatic Change*, **21,** 17–35.

Kark, S., U. Alkon, U.N. Safriel, and E. Randi, 1999: Conservation priorities for chukar partridge in Israel based on genetic diversity across an ecological gradient. *Conservation Biology*, **13,** 542–552.

Karl, T.R. and R.W. Knight, 1998: Secular trends of precipitation amount, frequency and intensity in the United States. *Bulletin of the American Meteorological Society*, **79,** 231–241.

Kaser, G., 1999: A review of the modem fluctuations of tropical glaciers. *Global and Planetary Change*, **22,** 93–103.

Kattenberg, A., F. Giorgi, H. Grassl, G.A. Meehl, J.F.B. Mitchell, R.J. Stouffer, T. Tokioka, A.J. Weaver, and T.M.L. Wigley, 1996: Climate models—projections of future climate. In: *Climate Change 1995: The Science of Climate Change. Contribution of Working Group I to the Second Assessment Report of the Intergovernmental Panel on Climate Change* [Houghton, J.T., L.G. Meira Filho, B.A. Callander, N. Harris, A. Kattenberg, and K. Maskell (eds.)]. Cambridge University Press, Cambridge, United Kingdom and New York, NY, USA, pp. 289–357.

Katz, M.E., D.K. Pak, G.R. Dickens, and K.G. Miller, 1999: The source and fate of massive carbon input during the Latest Paleocene Thermal Maximum. *Science*, **286,** 1531–1533.

Kelly, P.M. and W.N. Adger, 2000: Theory and practice in assessing vulnerability to climate change and facilitating adaptation. *Climatic Change*, **47(4),** 325-352.

Kenny, G.J., R.A. Warrick, B.D. Campbell, G.C. Sims, M. Camilleri, P.D. Jamieson, H.G. McPherson, and M.J. Salinger, 1999: Investigating climate change impacts and thresholds: an application of the CLIMPACTS integrated assessment model for New Zealand agriculture. *Climatic Change*, **46,** 91–113.

Kleypas, J.A., R.W. Buddemeier, D. Archer, J.P. Gattuso, C. Langdon, and B.N. Opdyke, 1999: Geochemical consequences of increased atmospheric carbon dioxide on coral reefs. *Science*, **284,** 118–120.

Knutson, T.R., R.E. Tuleya, and Y. Kurihara, 1998: Simulated increase of hurricane intensities in a CO_2–warmed climate. *Science*, **279,** 1018–1020.

Körner, C.H., 1999: Alpine plant diversity: a global survey and functional interpretations. In: *Arctic and Alpine Biodiversity: Patterns, Causes and Ecosystem Consequences* [Chapin, F.S. III and C.H. Körner (eds.)]. Ecological Studies 113, Springer-Verlag, Heidelberg, Germany, pp. 45–62.

Kovats, R.S., M.J. Bouma, and A. Haines, 1999: *El Niño and Health*. WHO/SDE/PHE/99.4, World Health Organization, Geneva, Switzerland, 48 pp.

Krabill, W., E. Frederick, S. Manizade, C. Martin, J. Sonntag, R. Swift, R. Thomas, W. Wright, and J. Jungel, 1999: Rapid thinning of parts of the Southern Greenland ice sheet. *Science*, **283,** 1522–1524.

Kratz, T.K., B.P. Hayden, B.J. Benson, and W.Y.B. Chang, 2001: Patterns in the interannual variability of lake freeze and thaw dates. *Verhandlungen Internationale Vereinigung Für Theoretische Und Angewandte Limnology*, (in press).

Lal, M., U. Cubasch, R. Voss, and J. Waszkewitz, 1995: Effect of transient increase in greenhouse gases and sulphate aerosols on monsoon climate. *Current Science*, **69(9),** 752–763.

Leemans, R., 1997: Effects of global change on agricultural land use: scaling up from physiological processes to ecosystem dynamics. In: *Ecology in Agriculture* [Jackson, L.E. (ed.)]. Academic Press, San Diego, CA, USA, pp. 415–452.

Leemans, R. and P. Halpin, 1992: Global change and biodiversity. In *Biodiversity 1992: Status of the Earth's Living Resources* [Groombridge, B. (ed.)]. Chapman and Hall, London, United Kingdom, pp. 254–255.

Leemans, R., E. Kreileman, G. Zuidema, J. Alcamo, M. Berk, G.J. van den Born, M. den Elzen, R. Hootsmans, M. Janssen, M. Schaeffer, A.M.C. Toet, and H.J.M. de Vries, 1998: *The IMAGE User Support System: Global Change Scenarios from IMAGE 2.1*. RIVM Publication (CD-ROM) 4815006, National Institute of Public Health and the Environment (RIVM), Bilthoven, The Netherlands.

Lehman, P.W., 2000: The influence of climate on phytoplankton community biomass in San Franscisco Bay Estuary. *Limnology and Ocean*ography, **45**(3), 580–590.

Lempert, R.J., M.E. Schlesinger, and J.K. Hammitt, 1994: The impact of potential abrupt climate changes on near-term policy choices. *Climatic Change*, **26,** 351–376.

Lesica, P. and F.W. Allendorf, 1994: When are peripheral populations valuable for conservation? *Conservation Biology*, **9,** 753–760.

Loeb, V., V. Siegel, O. Holm-Hansen, R, Hewitt, W. Fraser, W. Trivelpiece, and S. Trivelpiece, 1997: Effects of sea-ice extent and krill or salp dominance on the Antarctic food web. *Nature*, **387,** 897–900.

Loevinsohn, M.E., 1994: Climate warming and increased malaria in Rwanda. *Lancet*, **343,** 714–748.

Lorenz, E.D., 1982: Atmospheric predictability experiments with a large numerical model. *Tellus*, **34,** 505–513.

Ludwichowski, I., 1997: Long-term changes of wing-length, body mass and breeding parameters in first-time breeding females of goldeneyes (*Bucephala clangula clangula*) in northern Germany. *Vogelwarte*, **39,** 103–116.

Magnuson, J.J., D.M. Robertson, B.J. Benson, R.H. Wynne, D.M. Livingston, R. Arai, R.A. Assel, R.G. Barry, V. Card, E. Kuusisto, N.G. Granin, T.D. Prowse, K.M. Stewart, and V.S. Vuglinski, 2000: Historical trends in land and river ice cover in the Northern Hemisphere. *Science*, **289,** 1743–1746.

Manabe, S. and R.J. Stouffer, 1993: Century-scale effects of increased atmospheric CO_2 on the ocean-atmosphere system. *Nature*, **364,** 215–218.

Manabe, S. and R.J. Stouffer, 1988: Two stable equilibria of a coupled ocean-atmosphere model. *Journal of Climate*, **1,** 841–866.

Manne, A. and R. Richels, 1995: The greenhouse debate: economic efficiency, burden sharing and hedging strategies. *The Energy Journal*, **16(4),** 1–37.

Markandya, A. and K. Halsnaes, 2000: Costing methodologies. In: *Guidance Papers on the Cross Cutting Issues of the Third Assessment Report of the IPCC* [Pachauri, R., T. Taniguchi, and K. Tanaka (eds.)]. Intergovernmental Panel on Climate Change, Geneva, Switzerland, pp. 15–31.

Markham, A., 1996: Potential impacts of climate change on ecosystems: a review of implications for policymakers and conservation biologists. *Climate Research*, **6(2),** 179–191.

Martens, W.J.M., T.H. Jetten, and D.A. Focks, 1997: Sensitivity of malaria, schistosomiasis and dengue to global warming. *Climatic Change*, **35,** 145–156.

Maslanik, J.A., M.C. Serreze, and R.G. Barry, 1996: Recent decreases in Arctic summer ice cover and linkages to atmospheric circulation anomalies. *Geophysical Research Letters*, **23,** 1677–1680.

Mason, C.F., 1995: Long-term trends in the arrival dates of spring migrants. *Bird Study*, **42,** 182–189.

McAyeal, D.R., 1992: Irregular oscillations of the West Antarctic ice sheet. *Nature*, **359,** 29–32.

McCleery, R.H. and C.M. Perrins, 1998: Temperature and egg-laying trends. *Nature*, **391,** 30–31.

McGowan, J.A., D.R. Cayan, and L.M. Dorman, 1998: Climate-ocean variability and ecosystem response in the Northeast pacific. *Science*, **281,** 210–217.

McGowan, J.A., D.B. Chelton, and A. Conversi, 1999: Plankton patterns, climate, and change in the California current. In: *Large Marine Ecosystems of the Pacific Rim: Assessment, Sustainability, and Management* [Sherman, K. and Q. Tang (eds.)]. Blackwell Science, Malden, MA, USA.

McMichael, A.J. and R.S. Kovats, 2000: Climate change and climate variability: adaptations to reduce adverse health impacts. *Environmental Monitoring and Assessment*, **61,** 49–64.

Mearns, L.O., R.W. Katz, and S.H. Schneider, 1984: Extreme high-temperature events: changes in their probabilities with changes in mean temperature. *Journal of Applied Meteorology*, **23,** 1601–1613.

Meehl, G.A. and W.M. Washington, 1996: El Niño-like climate change in a model with increased atmospheric CO_2 concentrations. *Nature*, **382,** 56–60.

Mendelsohn, R. and A. Dinar, 1998: *The Impact of Climate Change on Agriculture in Developing Countries: Case Studies of India and Brazil*. The World Bank, Washington, DC, USA, 266 pp.

Mendelsohn, R. and J.E. Neumann (eds.), 1999: *The Impact of Climate Change on the United States Economy*. Cambridge University Press, Cambridge, United Kingdom and New York, NY, USA, 331 pp.

Mendelsohn, R. and M.E. Schlesinger, 1997: *Climate Response Functions*. Yale University, New Haven, CT and University of Urbana-Champaign, Urbana Champain, IL, USA.

Mendelsohn, R.O., W.D. Nordhaus, and D. Shaw, 1994: The impact of climate on agriculture: a Ricardian analysis. *American Economic Review*, **84(4),** 753–771.

Mendelsohn, R., W. Morrison, M.E. Schlesinger, and N.G. Andronova, 2000: Country-specific market impacts of climate change. *Climatic Change*, **45,** 553–569.

Menzel, A. and P. Fabian, 1999: Growing season extended in Europe. *Nature*, **397,** 659.

Middleton, J.F. and K.R. Thompson, 1986: Return periods of extreme sea level events from short records. *Journal of Geophysical Research*, **91(C10),** 11707–11716.

Middleton, N. and D. Thomas, 1997: *World Atlas of Desertification*. Arnold, London, United Kingdom, pp. 1-69.

Mimura, N. and H. Harasawa, 2000: *Data Book of Seal-Level Rise*. Center for Global Environmental Research, National Institute for Environmental Studies, Environment Agency of Japan, Tokyo, Japan.

Morgan, M.G. and D.W. Keith, 1995: Subjective judgements by climate experts. *Environmental Science and Technology*, **29,** 468A–476A.

Moss, R., and S.H. Schneider, 2000: Uncertainties in the IPCC TAR: recommendations to lead authors for a more consistent assessment and reporting. In: *Guidance Papers on the Cross Cutting Iissues of the Third Assessment Report of the IPCC* [Pachauri, R., T. Taniguchi, and K. Tanaka (eds.)]. Intergovernmental Panel on Climate Change, Geneva, Switzerland, pp. 33–51.

Mouchet, J., S. Manguin, J. Sircoulon, S. Laventure, O. Faye, A.W. Onapa, P. Carnevale, J. Julvez, and D. Fontenille, 1998: Evolution of malaria in Africa for the past 40 years: impact of climatic and human factors. *Journal of the American Mosquito Control Association*, **14(2),** 121–130.

Mudur, G., 1995: Monsoon shrinks with aerosol model. *Science*, **270,** 1922.

Munasinghe, M., 2000: Development, equity and sustainability (DES) and climate change. In: *Guidance Papers on the Cross Cutting Issues of the Third Assessment Report of the IPCC* [Pachauri, R., T. Taniguchi, and K. Tanaka (eds.)]. IPCC, Geneva, Switzerland, pp. 69–110.

Mustafa, D., 1998: Structural causes of vulnerability to flood hazards in Pakistan. *Economic Geography*, **74,** 289–305.

Myers, N., 1993: *Ultimate Security*. W.W. Norton, New York, NY, USA, 308 pp.

Myers, N., R.A. Mittermeier, C.G. Mittermeier, G.A.B. da Fonseca, and J. Kent, 2000: Boidiversity hotspots for conservation priorities. *Nature*, **403,** 853–858.

Myneni, R.B., C.D. Keeling, C.J. Tucker, G. Asrar, and R.R. Nemani, 1997: Increased plant growth in the northern high latitudes from 1981–1991. *Nature*, **386,** 698–702.

Nakicenovic, N., J. Alcamo, G. Davis, J. Fenhann, S. Gaffin, K. Gregory, A. Grübler, T.Y. Jung, T. Kram, E.L. La Rovere, L. Michaelis, S. Mori, T. Morita, W. Pepper, H. Pitcher, L. Price, K. Raihi, A. Roehrl, H.-H. Rogner, A. Sankovski, M. Schlesinger, P. Shukla, S. Smith, R. Swart, S. van Rooijen, N. Victor, B. de Vries, and Z. Dadi, 2000: *Emissions Scenarios. A Special Report of Working Group III of the Intergovernmental Panel on Climate Change*. Cambridge University Press, Cambridge, United Kingdom and New York, NY, USA, 599 pp.

National Research Council, 1994: *Estimating Bounds on Extreme Precipitation Events: A Brief Assessment.* National Academy Press, Washington, DC, USA, 30 pp.

Nehring, S., 1998: Establishment of thermophilic phytoplankton species on the North Sea: biological indicators of climatic changes? *ICES Journal of Marine Science*, **55,** 818–823.

Neilson, R.P., 1993: Transient ecotone response to climate change: some conceptual and modelling approaches. *Ecological Applications*, **3,** 385–395.

Nemani, R.R., M.A. White, D.R. Cayan, G.V. Jones, S.W. Running, and J.C. Coughlan, 2001: asymmetric climatic warming improves California vintages. *Climate Research*, (in press).

Nepstad, D.C., A. Verissimo, A. Alencar, C. Nobre, E. Lima, P. Lefebvre, P. Schlesinger, C. Potter, P. Moutinho, E. Mendoza, M. Cochrane, and V. Brooks, 1999: Large-scale impoverishment of Amazonian forests by logging and fire. *Nature*, **398,** 505–508.

Nicholls, N., 1997: Increased Australian wheat yield due to recent climate trends. *Nature*, **387,** 484–485.

Nicholls, R.J., F.M.J. Hoozemans, and M. Marchand, 1999: Increasing flood risk and wetland losses due to global sea-level rise: regional and global analyses. *Global Environmental Change*, **9,** S69–S87.

Nordhaus, W.D., 1991: To slow or not to slow: the economics of the greenhouse effect. *Economic Journal*, **101,** 920–937.

Nordhaus, W.D., 1994a: *Managing the Global Commons: The Economics of Climate Change.* MIT Press, Cambridge, MA, USA, 213 pp.

Nordhaus, W.D., 1994b: Expert opinion on climatic change. *American Scientist*, **82(1),** 45–51.

Nordhaus, W. and J. Boyer, 2000: *Warming the World: Economic Models of Climate Change.* MIT Press, Cambridge, MA, USA.

Nordhaus, W.D. and Z. Yang, 1996: RICE: a regional dynamic general equilibrium model of optimal climate-change policy. *American Economic Review*, **86(4),** 741–765.

Norris, R.D. and U. Rohl, 1999: Carbon cycling and chronology of climate warming during the Paleocene/Eocene transition. *Nature*, **401,** 775–778.

Nurse, L., 1992: Predicted sea-level rise in the wider Carribean: likely consequences and response options. In: *Semi Enclosed Seas* [Fabbri, P. and G. Fierro (eds.)]. Elsevier Applied Science, Essex, United Kingdom, pp. 52–78.

Nurse, L., R.F. McLean, and A.G. Suarez, 1998: Small island states. In: *The Regional Impacts of Climate Change: An Assessment of Vulnerability. Special Report of IPCC Working Group II* [Watson, R.T., M.C. Zinyowera, and R.H. Moss (eds.)]. Intergovernmental Panel on Climate Change, Cambridge University Press, Cambridge, United Kingdom and New York, NY, USA, pp. 331–354.

O'Brien, C.M., C.J. Fox, B. Planque, and J. Casey, 2000: Climate variability and North Sea cod. *Nature*, **404,** 142.

Oppenheimer, M., 1998: Global warming and the stability of the West Antarctic ice sheet. *Nature*, **393,** 325–332.

Osborn, T.J., M. Hulme, P.D. Jones, and T.A. Basnet, 2000: Observed trends in the daily intensity of United Kingdom precipitation. *International Journal of Climatology*, **20,** 347–364.

Overpeck, J. T., R.S. Webb, and T. Webb III, 1994: Mapping eastern North American vegetation change of the past 18,000 years: no-analogs and the future. *Geology*, **20,** 1071–1074.

Parmesan, C., 2001: Butterflies as bio-indicators for climate change impacts. In: *Evolution and Ecology Taking Flight: Butterflies as Model Systems* [Boggs, C.L., W.B. Watt, and P.R. Erlich (eds.)]. University of Chicago Press, Chicago, IL, USA, (in press).

Parmesan, C., 1996: Climate and species' range. *Nature*, **382,** 765–766.

Parmesan, C., N. Ryrholm, C. Stefanescu, J.L. Hill, C.D. Thomas, H. Descimon, B. Huntley, L. Kaila, J. Kullberg, T. Tammaru, W.J. Tennent, J.A. Thomas, and M. Warren, 1999: Poleward shifts in geographical ranges of butterfly species associated with regional warming. *Nature*, **399,** 579–583.

Parry, M., C. Rosenzweig, A. Iglesias, and G. Fischer, 1999: Climate change and world food security: a new assessment. *Global Environmental Change*, **9,** S51–S67.

Pascual, M., X. Rodo, S.P. Ellner, R. Colwell, and M.J. Bouma, 2000: Cholera dynamics and El Niño-Southern Oscillation. *Science*, **289,** 1766–1769.

Patz, J.A., W.J.M. Martens, D.A. Focks, and T.H. Jetten, 1998: Dengue fever epidemic potential as projected by general circulation models of global climate change. *Environmental Health Perspectives*, **106,** 147–153.

Pauli, H., M. Gottfried, and G. Grabherr, 1996: Effects of climate change on mountain ecosystems—upward shifting of alpine plants. *World Resource Review*, **8(3),** 382–390.

Pavlov, A.V., 1997: Patterns of frozen ground formation accompanying recent climate changes. *Polar Geography*, **21(2),** 137–153.

Payne, A.J., 1999: A thermomechanical model of ice flow in West Antarctica. *Climate Dynamics*, **15,** 115–125.

Pearce, D.W, W.R. Cline, A.N. Achanta, S. Fankhauser, R.K. Pachauri, R.S.J. Tol, and P. Vellinga, 1996: The social costs of climate change: greenhouse damage and the benefits of control. In: *Economic and Social Dimensions of Climate Change, Equity and Social Considerations. Contribution of Working Group III to the Second Assessment Report of the Intergovernmental Panel on Climate Change* [Bruce, J.P., H. Lee, and E.F. Haites (eds.)]. Cambridge University Press, Cambridge, United Kingdom and New York, NY, USA, pp. 179–224.

Pearce, H.J., and M.R. Kennedy, 1994: Generalised probable maximum precipitation estimation methods for Australia. *Australian Civil Engineering Transactions*, **CE34,** 97–104.

Peck, S.C. and T.J. Teisberg, 1994: Optimal carbon emissions trajectories when damages depend on the rate or level of global warming. *Climatic Change*, **28,** 289–314.

Pernetta, J.C., 1992: Impacts of climate change and sea-level rise on small island states: national and international responses. *Global Environmental Change*, **2,** 19–31.

Pernetta, J.C., 1989: Cities on oceanic islands: a case study of Male, capital of the Republic of Maldives. In: *Impacts of Sea Level Rise on Cities and Regions, Proceedings of the First Annual International Meeting, "Cities on Water," 11-13 December,1989,Vanice* [Frassetto, R. (ed.)]. Marcilio Editors, Venice, Italy, pp. 169–182.

Peterson, G., C.R. Allen, and C.S. Holling, 1998: Ecological resilience, biodiversity, and scale. *Ecosystems*, **1,** 6–18.

Petoukhov, V., A. Ganopolski, V. Brovkin, M. Claussen, A. Eliseev, C. Kubatzki, and S. Rahmstorf, 2000: CLIMBER-2: a climate system model of intermediate complexity, part 1: model description and performance for present climate. *Climate Dynamics*, **16,** 1–17.

Petschel-Held, G., H.-J. Schellnhuber, T. Bruckner, F.L. Toth, and K. Hasselmann, 1999: The tolerable windows approach: theoretical and methodological foundations. *Climatic Change*, **41(3–4),** 303–331.

Pielke, R.A. and M.W. Downton, 2001: Precipitation and damaging floods: trends in the United States, 1932–1997. *Journal of Climate*, **13(20),** 3625-3637.

Pinter, L., 1997: Impact of climate change on sustainability of native lifestyles in the Mackenzie River Basin. In: *The Mackenzie Basin Impact Study (MBIS) Final Report* [Cohen, S.J. (ed.)]. Environment Canada, Toronto, ON, Canada, pp. 62–65.

Pittock, A.B. and R.N. Jones, 2000: Adaptation to what? *Environmental Monitoring and Assessment*, **61,** 9–35.

Pittock, A.B., R.J. Allan, K.J. Hennessy, K.L. McInnes, R. Suppiah, K.J. Walsh, P.H. Whetton, H. McMaster, and R. Taplin, 1999: Climate change, climatic hazards and policy responses in Australia. In: *Climate, Change and Risk* [Downing, T.E., A.A. Olsthoorn, and R.S.J. Tol (eds.)]. Routledge, London, United Kingdom and New York, NY, USA, pp. 19–59.

Plambeck, E.L. and C. Hope, 1996: An updated valuation of the impacts of global warming. *Energy Policy*, **24(9),** 783–793.

Portney, P. and J. Weyant (eds.), 1999: *Discounting and Intergenerational Equity*. Johns Hopkins University Press, Baltimore, MD, USA.

Post, E. and N.C. Stenseth, 1999: Climatic variability, plant phenology, and northern ungulates. *Ecology*, **80,** 1322–1339.

Post, E., N.C. Stenseth, R. Langvatn, and J.M. Fromentin, 1997: Global climate change and phenotypic variation among red deer cohorts. *Proceedings of the Royal Society of London B*, **264,** 1317–1324.

Pounds, J.A., M.P.L. Fogden, and J.H. Campbell, 1999: Biological response to climate change on a tropical mountain. *Nature*, **398,** 611–615.

Prop, J., J.M. Black, P. Shimmings, and M. Owen, 1998: The spring range of barnacle geese *Branta leucopsis* in relation to changes in land management and climate. *Biological Conservation*, **86,** 339–346.

Rahmstorf, S. and A. Ganopolski, 1999: Long-term global warming scenarios computed with an efficient coupled climate model. *Climatic Change*, **43(2),** 353–367.

Rahmstorf, S., J. Marotzke, and J. Willebrand, 1996: Stability of the thermohaline circulation. In: *The Warm Water Sphere of the North Atlantic Ocean* [Krauss, W. (ed.)]. Borntraeger, Stuttgart, Germany, pp. 129–158.

Raper, S.C.B., T.M.L. Wigley, and R.A. Warrick, 1996: Global sea level rise: past and future. In: *Sea-Level Rise and Coastal Subsidence: Causes, Consequences and Strategies* [Milliman, J. and B.U. Huq (eds.)]. J. Kluwer Academic Publishers, Dordrecht, The Netherlands, pp. 11–45.

Reading, C.J., 1998: The effect of winter temperatures on the timing of breeding activity in the common toad *Bufo bufo*. *Oecologia*, **117,** 469–475.

Reaka-Kudla, M.L., 1996: The global biodiversity of coral reefs: a comparison with rainforests. In: *Biodiversity II: Understanding and Protecting our Natural Resources* [Reaka-Kudla, M.L., D.E. Wilson, and E.O. Wilson (eds.)]. Joseph Henry and National Academy Press, Washington, DC, USA, pp. 83–108.

Risser, P.G., 1993: Ecotones. *Ecological Applications*, **3,** 367–368.

Rodriguez-Trelles, F. and M.A. Rodriguez, 1998: Rapid micro-evolution and loss of chromosomal diversity in *Drosophila* in response to climate warming. *Evolutionary Ecology*, **12,** 829–838.

Root, T.L., 2000: Ecology: Possible consequences of rapid global change. In: *Earth Systems: Processes and Issues* [Ernst, G. (ed.)]. Cambridge University Press, Cambridge, United Kingdom and New York, NY, USA, pp. 315–324.

Rosenzweig, C. and M.L. Parry, 1994: Potential impact of climate change on world food supply. *Nature*, **367,** 133–138.

Ross, M.S., J.J. O'Brien, and L. da Silveira Lobo Sternberg, 1994: Sea-level rise and the reduction in pine forests in the Florida Keys. *Ecological Applications*, **4,** 144–156.

Rothman, D.S., 2000: Measuring environmental values and environmental impacts: going from the local to the global. *Climatic Change*, **44,** 351–376.

Rothrock, D.A., Y. Yu, and G.A. Maykut, 1999: Thinning of the Arctic sea-ice cover. *Geophysical Research Letters*, **26,** 3469-3472.

Rotmans, J. and H. Dowlatabadi, 1998: Integrated assessment modelling. In: *Human Choice and Climate Change. Volume 3: The Tools for Policy Analysis* [Rayner, S. and E.L. Malone (eds.)]. Batelle Press, Columbus, OH, USA, pp. 291–377.

Rusek, J., 1993: Air-pollution-mediated changes in alpine ecosystems and ecotones. *Ecological Applications*, **3,** 409–416.

Rutherford, M.C., G.F. Midgley, W.J. Bond, L.W. Powrie, R. Roberts, and J. Allsopp, 1999: *Plant Biodiversity: Vulnerability and Adaptation Assessment: South African Country Study on Climate Change*. National Botanical Institute, P/Bag X7, Claremont 7735, South Africa.

Safriel, U.N., 1999a: Science and strategies to combat desertification. In: *Drylands, Poverty, and Development, Proceedings of the June 15 and 16, 1999, World Bank Round Table, Washington, DC*. Esikuri, Enos (ed.).The World Bank, Washington, DC, USA, pp. 99–107.

Safriel, U.N., 1999b: The concept of sustainability in dryland ecosystems. In: *Arid Lands Management—Towards Ecological Sustainability* [Hoekstra, T.W. and M. Shachak (eds.)]. University of Illinois Press, Urbana, IL, USA, pp. 117–140.

Safriel, U.N., S. Volis, and S. Kark, 1994: Core and peripheral populations and global climate change. *Israel Journal of Plant Sciences*, **42,** 331–345.

Sagarin, R.D., J.P. Barry, S.E. Gilman, and C.H. Baxter, 1999: Climate-related change in an intertidal community over short and long time scales. *Ecological Monographs*, **69(4),** 465–490.

Sala ,O.E., Chapin, F.S. III, Armesto, J.J., Berlow, E., Bloomfield, J., Dirzo, R., Huber-Sanwald, E., Huenneke, L.F., Jackson, R., Kinzig, A., Leemans, R., Lodge, D., Mooney, H.A., Oesterheld, M., Poff, L., Sykes, M.T., Walker, B.H., Walker, M., Wall, D. 2000. Global biodiversity scenarios for the year 2100. *Science* 287: 1770-1774

Sarewitz, D., R.A. Pielke, and R. Byerly, 2000: *Science, Decision Making, and the Future of Nature*. Island Press, Washington, DC, USA, 400 pp.

Sarmiento, J.L., T. Hughes, R. Stouffer, and S. Manabe, 1998: Simulated response of the ocean carbon cycle to anthropogenic climate warming. *Nature*, **393,** 245–248.

Saunders, P.T., 1982: *An Introduction to Catastrophe Theory*. Cambridge University Press, Cambridge, United Kingdom, 144 pp.

Schellnhuber, H.-J., 1997: Integrated assessment of climate change: regularity and singularity. In: *Proceedings of "Climate Impact Research: Why, How and When?" symposium, Berlin-Brandenburg Academy of Sciences and German Academy Leopoldina, Berlin, Germany, 28–29 October, 1997*.

Schellnhuber, H.-J., 1999: Earth system analysis and the second Copernican revolution. *Nature*, **402,** C19–C23.

Schellnhuber, H.-J. and D.F. Sprinz, 1995: Umweltkrisen und internationale Sicherheit (Environmental crises and international security). In: *Deutschlands neue Außenpolitik (Germany's new foreign policy), Vol. 2* [Kaiser, K. and H.W. Maull (eds.)]. R. Oldenbourg Verlag, Munich, Germany, pp. 239–260.

Schellnhuber, H.-J. and G.W. Yohe, 1998: Comprehending the economic and social dimensions of climate change by integrated assessment. In: Proceedings of the WCRP Conference: achievements, benefits and challenges, 26-28 August 1997, Geneva, Switzerland WMO, 179.

Scherer, R.P., A. Aldahan, S. Tulaczyk, G. Possnert, H. Engelhardt, and B. Kamb, 1998: Pleistocene collapse of the West Antarctic ice sheet. *Science*, **281,** 82–85.

Schimel, D., M. Grubb, F. Joos, R. Kaufmann, R. Moss, W. Ogana, R. Richels, T. Wigley, R. Cannon, J. Edmonds, E. Haites, D. Harvey, A. Jain, R. Leemans, K. Miller, R. Parkin, E. Sulzman, R. Tol, J. de Wolde, and M. Bruno, 1997: *Stabilization of Atmospheric Greenhouse Gases: Physical, Biological and Socio-Economic Implications*. IPCC Technical Paper, Intergovernmental Panel on Climate Change, Geneva, Switzerland, 50 pp.

Schindler, D.W., K.G. Beaty, E.J., Fee, D.R., Cruikshank, E.R Debruyn, D.L. Findlay, G.A. Linsey, J.A. Shearer, M.P Stainton, and M.A. Turner, 1990: Effects of climatic warming on lakes of the central boreal forest. *Science*, **258,** 967–970.

Schlesinger, W.H., J.F. Reynolds, and J.L. Cunningham, L.F. Huenneke, W.M. Jarrell, R.A. Virginia, and W.G. Whitford, 1990: Biological feedbacks in global desertification. *Science*, **247,** 1043–1048.

Schneider, S.H. and S.L. Thompson, 2000: A simple climate model used in economic studies of global change. In: *New Directions in the Economics and Integrated Assessment of Global Climate Change* [DeCanio, S.J., R.B. Howarth, A.H. Sanstad, S.H. Schneider, and S.L. Thompson (eds.)]. The Pew Center on Global Climate Change, Washington, DC, USA, pp. 59–80.

Schneider, S.H., W. Easterling, and L. Mearns, 2000a: Adaptation: Sensitivity to natural variability, agent assumptions and dynamic climate changes. *Climatic Change*, **45,** 203–221.

Schneider, S.H., K. Kuntz-Duriseti, and C. Azar, 2000b: Costing non-linearities, surprises, and irreversible events. *Pacific and Asian Journal of Energy*, **10(1),** 81–106.

Schneider, S.H., B.L. Turner II, and H. Morehouse Garriga, 1998: Imaginable surprise in global change science. *Journal of Risk Research*, **1(2),** 165–185.

Schreider, S.Y., A.J. Jakeman, P.H. Whetton, and A.B. Pittock, 1997: Estimation of climate impact on water availability and extreme events for snow-free and snow-affected catchments of the Murray-Darling Basin. *Australian Journal of Water Resources*, **2,** 35–46.

Schuster, H.G., 1988: *Deterministic Chaos: An Introduction*. VCH, Weinheim, Federal Republic of Germany, 270 pp.

Serreze, M.C., J.E. Walsh, F.S. Chapin III, T. Osterkamp, M. Dyurgerov, V. Romanovsky, W.C. Oechel, J. Morison, T. Zhang, and R.G. Barry, 2000: Observational evidence of recent change in the northern high-latitude environment. *Climatic Change*, **46**, 159-207.

Shindell, D.T., R.L. Miller, G.A. Schmidt, and L. Pandolfo, 1999: Simulation of recent northern winter climate trends by greenhouse-gas forcing. *Nature*, **399,** 452–455.

Slater, F.M., 1999: First-egg date fluctuations for the pied flycatcher *Ficedula hypoleuca* in the woodlands of mid-Wales in the twentieth century. *Ibis*, **141,** 497–499.

Smith, R.I.L., 1994: Vascular plants as bioindicators of regional warming in Antarctic. *Oecologia*, **99,** 322–328.

Smith, R.C., D. Ainley, K. Baker, E. Domack, S. Emslie, B. Fraser, J. Kennett, A. Leventer, E. Mosley-Thompson, S. Stammerjohn, and M. Vernet, 1999: Marine ecosystem sensitivity to climate change. *BioScience*, **49(5),** 393–404.

Smith, T.M., and H.H. Shugart, 1993: The transient response of terrestrial carbon storage to a perturbed climate. *Nature*, **361,** 523–526.

Sohngen, B.L. and R. Mendelsohn, 1999: The impacts of climate change on the U.S. timber market. In: *The Impact of Climate Change on the United States Economy* [Mendelsohn, R. and J.E. Neumann (eds.)]. Cambridge University Press, Cambridge, United Kingdom and New York, NY, USA, pp. 94–132.

Somerville, R., 1987: The predictability of weather and climate. *Climatic Change*, **11,** 239–246.

Southward, A.J., S.J. Hawkins, and M.T. Burrows, 1995: Seventy years' observations of changes in distribution and abundance of zooplankton and intertidal organism in the western English Channel in relation to rising sea temperature. *Journal of Thermal Biology*, **20(1/2),** 127–155.

Sparks, T.H., 1999: Phenology and the changing pattern of bird migration in Britain. *International Journal of Biometeorology*, **42,** 134–138.

Still, C.J., P.N. Foster, and S.H. Schneider, 1999: Simulating the effects of climate change on tropical montane cloud forests. *Nature*, **398,** 608–610.

Stocker, T.F. and A. Schmittner, 1997: Influence of CO_2 emission rates on the stability of the thermohaline circulation. *Nature*, **388,** 862–865.

Swart, R., M.M. Berk, M. Janssen, E. Kreileman, and R. Leemans, 1998: The safe landing analysis: risks and trade-offs in climate change. In: *Global Change Scenarios of the 21st Century. Results from the IMAGE 2.1 Model* [Alcamo, J., R. Leemans, and E. Kreileman (eds.)]. Pergamon and Elsevier Science, London, United Kingdom, pp. 193–218.

Thomas, C.D. and J.J. Lennon, 1999: Birds extend their ranges northwards. *Nature*, **399,** 213.

Thomas, R.H., T.J.O. Sanderson, and K.E. Rose, 1979: Effect of climatic warming on the West Antarctic ice sheet. *Nature*, **277,** 355–358.

Timmermann, A., J. Oberhuber, A. Bacher, M. Esch, M. Latif, and E. Roeckner, 1999: Increased El Niño frequency in a climate model forced by future greenhouse warming. *Nature*, **398,** 694–696.

Titus, J.G., 1992: The costs of climate change to the United States. In: *Global Climate Change: Implications, Challenges and Mitigation Measures* [Majumdar, S.K., L.S. Kalkstein, B.M. Yarnal, E.W. Miller, and L.M. Rosenfeld (eds.)]. Pennsylvania Academy of Science, Easton, PA, USA, pp. 384–409.

Tol, R.S.J., 2001a: Equitable cost-benefit analysis of climate change. *Ecological Economics*, (in press).

Tol, R.S.J., 2001b: *Estimates of the Damage Costs of Climate Change, Part I: Benchmark Estimates, Environmental and Resource Economics*. Working Paper D99–01, Institute for Environmental Studies, Vrije Universiteit, Amsterdam, The Netherlands, (in press).

Tol, R.S.J., 2001c: *Estimates of the Damage Costs of Climate Change, Part II: Dynamic Estimates, Environmental and Resource Economics*. Working Paper D99–02, Institute for Environmental Studies, Vrije Universiteit Amsterdam, The Netherlands, (in press).

Tol, R.S.J., 1999a: The marginal costs of greenhouse gas emissions. *The Energy Journal*, **20(1),** 61–81.

Tol, R.S.J., 1999b: Time discounting and optimal control of climate change—an application of FUND. *Climatic Change*, **41(3–4),** 351–362.

Tol, R.S.J., 1995: The damage costs of climate change toward more comprehensive calculations. *Environmental and Resource Economics*, **5,** 353–374.

Tol, R.S.J. and S. Fankhauser, 1998: On the representation of impact in integrated assessment models of climate change. *Environmental Modeling and Assessment*, **3,** 63–74.

Tol, R.S.J., S. Fankhauser, and J.B. Smith, 1998: The scope for adaptation to climate change: what can we learn from the impact literature? *Global Environmental Change*, **8,** 109–123.

Toth, F., 2000: Decision analysis frameworks in TAR: a guidance paper for IPCC. In: *Guidance Papers on the Cross Cutting Issues of the Third Assessment Report of the IPCC* [Pachauri, R., T. Taniguchi, and K. Tanaka (eds.)]. IPCC, Geneva, Switzerland, pp. 53–68.

Toth, F.L., T. Bruckner, H.-M. Füssel, M. Leimbach, G. Petschel-Held and H.-J. Schellnhuber, 1997: The tolerable windows approach to integrated assessments. In: *Climate Change and Integrated Assessment Models: Bridging the Gaps. Proceedings of the IPCC Asia-Pacific Workshop on Integrated Assessment Models, United Nations University, Tokyo, Japan, March 10–12, 1997* [Cameron, O.K., K. Fukuwatari, and T. Morita (eds.)]. Center for Global Environmental Research, National Institute for Environmental Studies, Tsukuba, Japan, pp. 403–430.

UNDP, 1998: *Synergies in National Implementation: the Rio Agreements*. Sustainable Energy and Environment Division (SEED), United Nations Development Programme, New York, NY, USA, pp. 1-69.

UNEP/WMO, 1992: *United Nations Framework Convention on Climate Change*. United Nations Environment Programme and World Meteorological Organization (UNEP/WMO), Information Unit on Climate Change, Geneva, Switzerland, 29pp.

Vinnikov, K.Y., A. Robock, R.J. Stouffer, J.E. Walsh, C.L. Parkinson, D.J. Cavalieri, J.F.B. Mitchell, D. Garrett, and V.F. Zakharov, 1999: Global warming and Northern Hemisphere sea ice extent. *Science*, **286,** 1934–1937.

Visser, M.E., A.J. Vannoordwijk, J.M. Tinbergen, and C.M. Lessells, 1998: Warmer springs lead to mistimed reproduction in great tits (*Parus major*). *Proceedings of the Royal Society of London B*, **265,** 1867–1870.

Volis, S., S. Mendlinger, U.N. Safriel, and N. Orlovsky, 1998: Phenotypic variation and stress resistance in core and peripheral populations of *Hordeum spontaneum*. *Biodiversity and Conservation*, **7,** 799–813.

Voortman, R.L., 1998: Recent historical climate change and its effect on land use in the eastern part of West Africa. *Physics and Chemistry of the Earth*, **23,** 385–391.

Walker, B., W. Steffen, J. Canadell, and J. Ingram (eds.), 1999: *The Terrestrial Biosphere and Global Change: Implications for Natural and Managed Ecosystems*. Cambridge University Press, Cambridge, United Kingdom and New York, NY, USA, 439 pp.

Walsh, J.E. and B.F. Ryan, 2000: Tropical cyclone intensity increase near Australia as a result of climate change. *Journal of Climate*, **13,** 3029–2036.

Walsh, K. and A.B. Pittock, 1998: Potential changes in tropical storms, hurricanes, and extreme rainfall events as a result of climate change. *Climatic Change*, **39,** 199–213.

Warner, R.C., and W.F. Budd, 1998: Modelling the long-term response of the Antarctic ice sheet to global warming. *Annals of Glaciology*, **27,** 161–168.

Weaver, A.J., J. Marotzke, P.F. Cummins, and E.S. Sarachik, 1993: Stability and variability of the thermohaline circulation. *Journal of Physical Oceanography*, **23,** 39–60.

Weertman, J., 1974: Stability of the junction of an ice sheet and an ice shelf. *Journal of Glaciology*, **13,** 3–11.

West, J.J., and H. Dowlatabadi 1999: On assessing the economic impacts of sea-level rise on developed coasts. In: *Climate Change and Risk* [Downing, T.E., A.A. Olsthoorn, and R.S.J. Tol (eds.)]. Routledge, London, United Kingdom, pp. 205–220.

Weyant, J., O. Davidson, H. Dowlabathi, J. Edmonds, M. Grubb, E.A. Parson, R. Richels, J. Rotmans, P.R. Shukla, R.S.J. Tol, W. Cline, and S. Fankhauser, 1996: Integrated assessment of climate change: an overview and comparison of approaches and results. In: *Economic and Social Dimensions of Climate Change, Equity and Social Considerations. Contribution of Working Group III to the Second Assessment Report of the Intergovernmental Panel on Climate Change* [Bruce, J.P., H. Lee, and E.F. Haites (eds.)]. Cambridge University Press, Cambridge, United Kingdom and New York, NY, USA, pp. 367–396.

White, A., G.R. Melvin, and A.D. Friend, 1999: Climate change impacts on ecosystems and the terrestrial carbon sink: a new assessment. *Global Environmental Change*, **9,** S21–S30.

Wiggins, S., 1996: *Introduction to Applied Nonlinear Dynamical Systems and Chaos*. Springer-Verlag, Berlin, Germany, 672 pp.

Wigley, T.M.L., 1988: The effect of changing climate on the frequency of absolute extreme events. *Climate Monitor*, **17(2),** 44–55.

Wigley, T.M.L., 1985: Impact of extreme events. *Nature*, **316,** 106–107.

Wigley, T., R. Richels, and J. Edmonds, 1996: Economic and environmental choices in the stabilization of atmospheric CO_2 concentration. *Nature*, **379,** 242–245.

Wilkinson, C., R. Linden, H. Cesar, G. Hodgson, J. Rubens, and A.E. Strong, 1999: Ecological and socioeconomic impacts of 1998 coral mortality in the Indian Ocean: an ENSO impact and a warning of future change? *Ambio*, **28,** 188–196.

Winkel, W. and H. Hudde, 1997: Long-term trends in reproductive traits of tits (*Parus major, P. caeruleus*) and pied flycatchers *Ficedula hypoleuca*. *Journal of Avian Biology*, **282,** 187–190.

Winkel, W. and H. Hudde, 1996: Long-term changes of breeding parameters of nuthatches *Sitta europaea* in two study areas of northern Germany. *Journal Fuer Ornithologie*, **137,** 193–202.

Wood, R.A., A.B. Keen, J.F.B. Mitchell, and J.M. Gregory, 1999: Changing spatial structure of the thermohaline circulation in response to atmospheric CO_2 forcing in a climate model. *Nature*, **399,** 572–575.

Woodward, A., S. Hales, and P. Weinstein, 1998: Climate change and human health in the Asia Pacific: who will be most vulnerable? *Climate Research*, **11,** 11–39.

Woodwell, G., F. Mackenzie, R. Houghton, M. Apps, E. Gorham, and E. Davidson, 1998: Biotic feedbacks in the warming of the earth. *Climatic Change*, **40,** 495–518.

WRI, 1998: *World Resources 1998–99*. World Resources Institute, Oxford University Press, Oxford, United Kingdom.

Yohe, G., J.E. Neumann, and H. Ameden, 1995: Assessing the economic cost of greenhouse-induced sea-level rise: methods and application in support of a national survey. *Journal of Environmental Economics and Management*, **29,** S78–S97.

Yohe, G., J.E. Neumann, P. Marshall, and H. Amaden 1996: The economic cost of greenhouse induced sea level rise for developed property in the United States. *Climatic Change*, **32(4),** 387–410.

Zhao, W., J.A. Smith, and A.A. Bradley, 1997: Numerical simulation of a heavy rainfall event during the PRE-STORM experiment. *Water Resources Research*, **33,** 783–799.

Zhou, X., R. Harrington, I.P. Woiwod, J.N. Perry, J.S. Bale, and S.J. Clark, 1995: Effects of temperature on aphid phenology. *Global Change Biology*, **1,** 303–313.

Climate Change 2001:
Impacts, Adaptation, and Vulnerability — Annexes

Prepared by IPCC Working Group II

A

Authors and Expert Reviewers

American Samoa

Lelei Peau	NOAA

Antigua

Brian Challenger	APUA

Argentina

Osvaldo F. Canziani	Co-Chair, WGII
Rodolfo Carcavallo	Department of Entomology
Jorge O. Codignotto	Laboratorio Geologia y Dinamica Costera
Rosa Hilda Compagnucci	Dpto de Ciencia de la Atmosfera
Sandra Myrna Diaz	Instituto Multidisciplinario de Biologia Vegetal
Jorge Frangi	Universidad Nacional de la Plata
Graciela Magrin	Instituto Nacional de Tecnologia Agropecuaria
Gabriel Soler	Fundación Instituto Latinoamericano de Políticas Sociales
Silvina Solman	Ciudad Universitaria
Eduardo Usunoff	Instituto de Hidrologie de Llanuras
Walter Vargas	University of Buenos Aires
Ernesto F. Viglizzo	PROCISUR/INTO/CONICET

Australia

Kay Abel	Australian Greenhouse Office
Bryson Bates	CSIRO
Ian Carruthers	Department of the Environment, Sport and Territories
Steven Crimp	Queensland Centre for Climate Applications
Max Finlayson	Environmental Research Institute
Alistair Gilmour	Center for Environmental and Urban Studies
Habiba Gitay	Australian National University
Ann Henderson-Sellers	Australian Nuclear Science and Technology Office
David Hopley	James Cook University of North Queensland
Mark Howden	Bureau of Rural Sciences, Agriculture, Fisheries, and Forestry
Richard Hoy	Electricity Supply Association of Australia
Roger Jones	CSIRO Atmospheric Research
Lawrence Leung	BHP Research-Newcastle Laboratories
Janice Lough	Australian Institute of Marine Science
Harvey J. Marchant	Australian Antarctic Division
Roger McLean	University of New South Wales
Ian Noble	Australian National University
Barrie Pittock	CSIRO Climate Impact Group
Andy Reisinger	Ministry for the Environment
David Shearman	University of Adelaide
Brian H. Walker	CSIRO Division of Wildlife and Ecology

Austria

Paul Freeman	International Institute for Applied Systems Analysis
Helmut Hojesky	Federal Ministry for Environment
Nebojsa Nakicenovic	International Institute for Applied Systems Analysis
Klaus Radunsky	Federal Environment Agency

Bangladesh

Q.K. Ahmad	Bangladesh Unnayan Parishad
Saleemul Huq	Imperial College London
M. Monirul Qader Mirza	The Institute for Environmental Studies, University of Toronto

Barbados

Leonard Nurse	Coastal Zone Management Unit

Belgium

Alain Dassargues	University of Liege
Mark Rounsevell	Universite Catholique de Louvain
Jean-Pascal van Ypersele	Institut d'Astronomie et de Geophysique G. Lemaitre

Benin

Epiphane Dotou Ahlonsou	Service Météorologique National

Botswana

Michel Boko	Universite de Bourgogne
Pauline O. Dube	University of Botswana

Brazil

Heraldo C.N.S. Campos	University of Vale do Rio dos Sinos
Ulisses Confalonieri	National School of Public Health
Philip M. Fearnside	Instituto Nacional de Pesquisas da Amazonia
Emilio Lebre La Rovere	Federal University of Rio de Janeiro
J.A. Marengo Orsini	CPTEC/INPE
Y.D.P. Medeiros	Universidade Federal da Bahia
Carlos A. Nobre	CPTEC/INPE

Bulgaria

Vesselin Alexandrov	National Institute of Meteorology and Hydrology

Canada

Michael Apps	Northern Forest Centre
David Barber	Environment Canada
Richard J. Beamish	Pacific Biological Station
James P. Bruce	Canadian Climate Program Board
Ian Burton	University of Toronto
Ian Campbell	Canadian Forest Service
Wenjun Chen	Canada Centre for Remote Sensing
Josef Cihlar	Canada Centre for Remote Sensing

Stewart J. Cohen	Environmental Adaptation Research Group, University of British Columbia
Jean Cooper	Meteorological Service of Canada
Alexandre Desbarats	Environment Canada
Kirsty Duncan	University of Windsor
Patti Edwards	Meteorological Service of Canada
David Etkin	Environment Canada
Donald Forbes	Bedford Institute of Oceanography
Edward G. Gregorich	Agriculture Canada
Marc Hinton	Environment Canada
Robert L. Jefferies	University of Toronto
Mark Johnston	Saskatchewan Environment and Resource Management
Paul Kay	University of Waterloo
Roy M. Koerner	Geological Survey of Canada
Paul Kovacs	Institute for Catastrophic Loss Reduction
Joan Masterton	Environment Canada
Barrie Maxwell	Environment Canada
Linda Mortsch	Environment Canada
Terry Prowse	National Water Research Institute
Alfonso Rivera	Environment Canada
David W. Schindler	University of Alberta
John Shaw	Bedford Institute of Oceanography
Barry Smit	University of Guelph
Sharon Smith	Environment Canada
David L. Spittlehouse	B.C. Ministry of Forests
John M.R. Stone	Policy, Program, and International Affairs Directorate
J.A. Trofymow	Environment Canada
Elaine Wheaton	Saskatchewan Research Council
J.M. White	Environment Canada
G. Daniel Williams	Environment Canada (retired)
Louise Wilson	B.C. Ministry of Employment and Investment
Raymond Wong	Alberta Environment

Chile

Eduardo Basso	Independent Consultant

China

Liping Bai	Agrometeorology Institute
Youmin Chen	Agrometeorology Institute
Jiaqi Chen	Institute of Water Resources and Hydropower Research
Liu Chunzhen	Hydrological Forecasting and Water Control Center
Li Congxian	Tongji University
Gao Deming	Chinese Academy of Agricultural Sciences
Bilan Du	China Institute for Marine Development Strategy
Lin Erda	Agrometeorology Institute
Su Jilan	Second Institute of Oceanography
Hui Ju	Chinese Academy of Agricultural Science
Wen Kegang	China Meteorological Administration
Bai Keyu	Chinese Academy of Agricultural Sciences
Yue Li	Agrometeorology Instutite
Dahe Qin	Chinese Academy of Sciences
Guoyu Ren	China Meteorological Administration
Chengguo Shen	Chinese Academy of Agricultural Sciences
Fulu Tao	Chinese Academy of Sciences
Changrong Yan	Agrometeorology Instutite
Xiu Yang	Agrometeorology Instutite
Ding Yihui	China Meteorological Administration
Aiwen Ying	Ministry of Water Resources
Lv Yingyun	Institute of Global Climate Change
Xinshi Zhang	Chinese Academy of Science
Jian-yun Zhang	Ministry of Water Resources
Guangsheng Zhou	Chinese Academy of Sciences

Costa Rica

Max Campos	National Meteorological Institute
Jorge Cortes	Universidad de Costa Rica

Cuba

Avelino G. Suarez	Institute of Ecology and Systematics

Denmark

Jens H. Christensen	Danish Meteorological Institute

El Salvador

J. Roberto Jovel	Independent Consultant

Finland

Timothy Carter	Finnish Environment Institute
Paula Kankaala	University of Helsinki
Seppo Kellomaki	University of Joensuu
Pirkko Kortelainen	Finnish Environment Institute
Jukka Laine	University of Helsinki
Seppo Neuvonen	University of Turku

France

Marc Gillet	Mission Interministerielle de l'Effet de Serre
Arnaud Hequette	Universite du Littoral
Benoit Lesaffre	Ministere de l'Amenagement du Territoire et de l'Environnement
Martin Pecheux	Nice University
Michel Petit	Vice-Chair, WGII
Francois Rodhain	Institut Pasteur
Jean-Francois Soussana	INRA, Unite d'Agronomie

Gambia

Bubu Pateh Jallow	Department of Water Resources

Germany

Alfred Becker	Potsdam Institute for Climate Impact Research
Rosemarie Benndorf	Federal Environmental Agency
Cornelia Berns	Federal Ministry for Food, Agriculture, and Forestry
Gerhard Berz	Munich Reinsurance Company
Markus Breuer	German Agency for Technical Cooperation
Harald Bugmann	Swiss Federal Institute of Technology
Wolfgang P. Cramer	Potsdam Institute for Climate Impact Research
Thomas Frisch	Federal Environmental Agency
Birgit Georgi	Federal Environmental Agency
Anke Herold	Oko-Institut
Venugopalan Ittekkot	Centre for Tropical Marine Ecology
Richard J.T. Klein	Potsdam Institute for Climate Impact Research
Harald Kohl	Federal Ministry of the Environment, Nature Conservation, and Nuclear Safety
Marcus Linder	Potsdam Institute for Climate Impact Research
Petra Mahrenholz	Federal Environmental Agency
I. Colin Prentice	Max-Planck Institute for Biogeochemistry
Rolf Sartorius	Federal Environmental Agency
Bernd Schanzenbacher	Credit Suisse
H.-J. Schellnhuber	Potsdam Institute for Climate Impact Research
Ernst-Detlef Schulze	University of Bayreuth
Ferenc Toth	Potsdam Institute for Climate Impact Research
Thomas Voigt	Federal Environmental Agency
Horst Wingrich	Dresden University of Technology

Guatemala

Joao S. de Queiroz	USAID/G-CAP
Roberto Morales	USAID/G-CAP

Guinea

C. Ibrahima Sory	University of Conakry
Joseph Sylla	Direction Nationale de l'Environnement

India

Rais Akhtar	University of Kashmir
Sujata Gupta	Tata Energy Research Institute
Shreekant Gupta	University of Delhi
Murari Lal	Indian Institute of Technology
Kirit Parikh	Indira Gandhi Institute of Development Research
Anand Patwardhan	Indian Institute of Technology
S.S. Prihar	Independent Consultant
P.S. Ramakrishnan	School of Environmental Sciences
S.K. Sinha	Indian Agricultural Research Institute

Indonesia

Daniel Murdiyarso	State Ministry for Environment
Hari Suharyono	BPPT

Israel

Z. Alperson	Israel Meteorological Service
Tartakovsky Leonid	Israel Institute of Technology
Uriel N. Safriel	The Blaustein Institute for Desert Research
Leonid Tartakovsky	Israel Institute of Technology

Italy

Gaetano Borrelli	National Agency for New Technologies, Energy, and the Environment
Mauro Centritto	Instituto di Biochimica ed Ecofisiologia Vegetali
Domenico Gaudioso	National Environmental Protection Agency
Filippo Giorgi	Abdus Salam International Centre for Theoretical Physics
Wulf Killmann	Food and Agriculture Organization
Bettina Menne	WHO-European Centre for Environment and Health
Teresa Nanni	Institute of Atmospheric and Oceanic Sciences of the National Research Council
Astrid Raudner	National Environmental Protection Agency

Jamaica

A. Anthony Chen	University of the West Indies

Japan

Mitsuru Ando	National Institute for Environmental Studies
Yoshitaka Fukuoka	Risshou University
Keisuke Hanaki	University of Tokyo
Hideo Harasawa	Social and Environmental Systems Division
Takeshi Horie	Kyoto University
Masami Iriki	Yamanashi Institute of Environmental Sciences
Hiroshi Kadomura	Rissho University
Kanehiro Kitayama	Japanese Forestry and Forest Products Research Institute
Nobuo Mimura	Ibaraki University
Toyohiko Miyagi	Tohoku Gakuin University
Hisayoshi Morisugi	Tohoku University
Shinichi Nagata	Environment Agency
Masahisa Nakamura	Lake Biwa Research Institute

Shuzo Nishioka	Keio University
Eiji Ohno	Meijo University
Kenji Omasa	The University of Tokyo
Akihiko Sasaki	National Institute of Public Health
Tsuguyoshi Suzuki	Japan Science and Technology Cooperation
Nobuyuki Tanaka	Regeneration Process Laboratory Forestry and Forest Products Research Institute
Satoshi Tanaka	Embassy of Japan
Masatomo Umitsu	Nagoya University
Masatoshi Yoshino	Tsukuba University

Kazakhstan

Olga Pilifosova	UNFCCC Secretariat

Kenya

Andrew Githeko	Kenya Medical Research Institute
Mukiri Githendu	Ministry of Research and Technology
Wilson Kimani	Kenya Meteorological Department
Gabriel M. Mailu	Ministry of Research and Technology
F.M. Mutua	University of Nairobi
John Nganga	University of Nairobi
Joseph Njihia	Kenya Meteorological Department
William Nyakwada	Kenya Meteorological Department
Laban Ogallo	University of Nairobi
R.E. Okoola	University of Nairobi
Christopher Oludhe	University of Nairobi
Joyce Onyango	National Environment Secretariat
Helida Oyieke	Kenya Museum
Mwakio P. Tole	Moi University
Joshua Wairoto	Kenya Meteorological Department

Kiribati

Nakibae Teuatabo	Ministry of Environment and Social Development

Malawi

Paul Desanker	University of Virginia

Malaysia

Chan Ah Kee	Malaysian Meteorological Service
Mohammad Ilyas	Universiti Sains Malaysia

Maldives

Simad Saeed	Ministry of Home Affairs, Housing, and Environment

Malta

Lino Briguglio	Foundation for International Studies

Mauritius

Sachooda Ragoonaden	Mauritius Meteorological Services

Mexico

A. de la Vega Navarro	National Autonomous University of Mexico
M. de Lourdes Villers-Ruiz	College of Forest Resources
Ernesto Jauregui	National University
Victor Magana	Centro de Ciencias de la Atmosfera
Ana Rosa Moreno	North American Center for Environmental Information and Communication
Jose Sarukhan	Ciudad Universitaria

Morocco

Abdelkader Allali	Ministry of Agriculture, Rural Development, and Fishing

Nepal

Sharad P. Adhikary	Water and Energy Commission Secretariat

New Zealand

Jon Barnett	Macmillan Brown Centre for Pacific Studies, University of Canterbury
Reid Basher	National Institute of Water and Atmospheric Research
John R. Campbell	University of Waikato
B. Blair Fitzharris	University of Otago
John Hay	The University of Waikato
Richard Ibbitt	National Institute of Water and Atmospheric Research
Neil Mitchell	The University of Auckland
Talbot Murray	National Institute of Water and Atmospheric Research
Gerald Rys	Ministry of Research, Science, and Technology
Richard Warrick	University of Waikato
Alastair Woodward	Wellington School of Medicine
David Wratt	National Institute of Water and Atmospheric Research

Niger

Abdelkrim Ben Mohamed	Universite de Niamey
Garba Goudou Dieudonne	Office of the Prime Minister

Nigeria

James O. Adejuwon	Obafemi Awolowo University
Daniel M. Gwary	University of Maiduguri
James C. Nwafor	University of Nigeria

Norway

Torgrim Asphjell	Norwegian Pollution Control Authority
Torstein Bye	Statistics Norway
Oyvind Christophersen	Ministry of Environment
Jon Ove Hagen	University of Oslo
Jarle Inge Holten	Terrestrial Ecology Research Institute
Bjorn Fossli Johansen	Norwegian Polar Institute

Else Lobersli	Directorate for Nature Management
Harald Loeng	Institute of Marine Research
Sophia Mylona	Norwegian Pollution Control Authority
Petter Nilsen	Norwegian Forest Research Institute
Karen O'Brien	Center for International Climate and Environmental Research
Marit Viktoria Pettersen	Norwegian Pollution Control Authority
Kare Venn	Norwegian Forest Research Institute

Peru

Humberto Guerra	Universidad Peruana Cayetano Heredia

Philippines

Maximo W. Baradas	Philippine Rice Research Institute
Rex Victor Cruz	University of the Phillipines at Los Banos
Felino Lansigan	University of the Philippines at Los Banos

Poland

Jan Dobrowolski	Department of Environmental Management and Protection
Zdzislaw Kaczmarek	Polish Academy of Sciences
Zbigniew Kundzewicz	Polish Academy of Sciences
Wojciech Suchorzewski	Warsaw University of Technology

Portugal

Julia de Seixas	Universidade Nova de Lisboa
Maria Rosa Paiva	Universidade Nova de Lisboa
Luis Veiga da Cunha	Universidade Nova de Lisboa

Romania

Constanta Boroneant	National Institute of Meteorology and Hydrology
Aristita Busuioc	National Institute of Meteorology and Hydrology
Ciprian Stelian Corbus	National Institute of Meteorology and Hydrology
Vasile Cuculeanu	National Institute of Meteorology and Hydrology
Anton Geicu	National Institute of Meteorology and Hydrology
Adriana Marica	National Institute of Meteorology and Hydrology
Marinela Simota	National Institute of Meteorology and Hydrology
Viorel A. Stanescu	National Institute of Meteorology and Hydrology

Russian Federation

Oleg Anisimov	State Hydrological Institute
Yurij Anokhin	Institute of Global Climate and Ecology
Irena Borzenkova	State Hydrological Institute
Zurab Kopaliani	State Hydrological Institute
Serguei Semenov	Institute of Global Climate and Ecology
Alla Tsyban	Institute of Global Climate and Ecology
Igor Shiklomanov	State Hydrological Institute
Oleg Sirotenko	All Russian Research Institute of Agricultural Meteorology

Samoa

Graham Sem	UNFCCC Secretariat

Saudi Arabia

Mohammad Al Sabban	Government of Saudi Arabia

Scotland

Peter S. Maitland	RAMSAR Convention

Senegal

Isabelle Niang-Diop	University of Dakar

Seychelles

Rolph Payet	Ministry of Industries and International Business

Singapore

Poh Poh Wong	National University of Singapore

Slovakia

Jan Szolgay	Slovak Technical University

South Africa

Gerrie Coetzee	Department of Environmental Affairs and Tourism
Robert J. Scholes	CSIR
Clive R. Turner	Eskom Resources and Strategy Group

Spain

Sergio Alonso	University of the Balearic Islands
Francisco Ayala-Carcedo	Geomining Technological Institute of Spain
Ferran Ballester	Institut Valencia d'Estudis en Salut Publica
Lucila Candela	Tecnical University of Catalonia
E. Crespo de Nogueira	Organismo Autonomo Parques Nacionales
Francisco Diaz-Fierros	University of Santiago
Francesc Gallart	Institute of Earth Sciences Jaume Almera
M.-C. Llasat Botija	University of Barcelona
Pilar Llorens	Institute of Earth Sciences Jaume Almera
Josep Penuelas	Center for Ecological Research and Forestry Applications
Javier Zapata	Organismo Autonomo Parques Nacionales

Sri Lanka

Y.A.D.S. Wanasinghe — University of Sri Jayew Ardenepura

Sweden

Christian Azar — Chalmers University of Technology
Sten Bergström — Swedish Meteorological and Hydrological Institute
Torben R. Christensen — Lund University
Ulf Molau — University of Gothenburg
Mats Olsson — Swedish University of Agricultural Sciences
Mats Oquist — Department of Water and Environmental Studies

Switzerland

Aiko Bode — UNEP Insurance Industry Initiative
Michael J. Coughlan — World Meteorological Organization
Andreas Fischlin — Terrestrial Systems Ecology
Frank Oldfield — IGBP Past Global Changes Office
Jose Romero — Office Federal de l'Environnement des Forets et du Paysage

Thailand

Somsri Arumin — Land Development Department
Asdaporn Krairapanond — Office of Environmental Policy and Planning
Bundit Limmeechokchai — Sirindhorn International Institute of Technology
Ed Sarobol — Kasetsart University
Duangkae Vilainerun — Ministry of Public Health

The Netherlands

Luitzen Bijlsma — National Institute for Coastal and Marine Management
P.S. Bindraban — Plant Research International
Laurens Bouwer — Institute for Environmental Studies
A.J. Dietz — University of Amsterdam
H. Dolman — Wageningen University and Research Centre
W.L. Hare — Greenpeace International
Jean-Paul Hettelingh — RIVM
Michiel A. Keyzer — Vrije Universiteit
Rik Leemans — National Institute of Public Health and Environmental Protection
Marcel Marchand — WllDelft Hydraulics
Pim Martens — Maastricht University
Hans Nieuwenhuis — Ministry of Housing, Spatial Planning and the Environment
Maresa Oosterman — Ministerie van Buitenlandse Zaken
J.H.J. Spiertz — Wageningen University and Research Centre
H.F.M. Ten Berge — Plant Research International
Richard S.J. Tol — Vrije Universiteit
S.C. van de Geijn — Landbouw Universiteit Wageningen
A. van Hoorn — Ministerie van Landbouw
G. van Tol — Expertisecentrum LNV
A. Veldkamp — Wageningen University
Pier Vellinga — Institute for Environmental Studies
J. Verbeek — Ministry of Transport, Public Works, and Water Management
Jan A. Verhagen — Plant Research International

Togo

Ayite-Lo Ajavon — Atmospheric Chemistry Laboratory

Uganda

Charles Basalirwa — Makerere University

United Kingdom

Neil Adger — University of East Anglia
Nigel Arnell — University of Southampton
Charlotte Benson — Overseas Development Institute
Barnaby Briggs — Environmental Resources Management
Barbara Brown — University of New Castle upon Tyne
Humphrey Crick — British Trust for Ornithology
Andrew F. Dlugolecki — Independent Consultant
Thomas E. Downing — Environmental Change Institute, University of Oxford
Samuel Fankhauser — European Bank of Reconstruction and Development
Jonathan Grant — IPIECA
Andrew Haines — London Medical School
Christopher Hope — University of Cambridge
Mike Hulme — University of East Anglia
P.M. Kelly — University of East Anglia
Sari Kovats — London School of Hygiene and Tropical Medicine
David Mansell-Moullin — IPIECA
Greg Masters — Climate Change Research Initiative
M. McKenzie Hedger — University of Oxford
Anthony J. McMichael — London School of Hygiene and Tropical Medicine
Robert J. Nicholls — Middlesex University
Jean P. Palutikof — University of East Anglia
Martin Parry — Jackson Environment Institute
Allen Perry — University of Wales Swansea
Sarah Randolph — University of Oxford
David Satterthwaite — International Institute for Environment and Development
Jim F. Skea — University of Sussex
David Vaughan — British Antarctic Survey
Richard Vincent — Department of the Environment, Transport, and the Regions
David A. Warrilow — Department of the Environment

Uruguay

Walter Baethgen	IFDC

United States

Shardul Agrawala	International Research Institute
L.H. Allen, Jr.	U.S Department of Agriculture
Richard B. Alley	Pennsylvania State University
Jeffrey S. Amthor	Oak Ridge National Laboratory
Mark Anderson	Office of Science and Technology Policy
John Antle	Montana State University
Phillip L. Antweiler	U.S. Department of State
Assaf Anyamba	National Aeronautics and Space Administration
Phillip A. Arkin	International Research Institute for Climate Prediction
Joan L. Aron	Science Communication Studies
Kevin R. Arrigo	Stanford University
Mitchell Baer	Department of Energy
Sam Baldwin	Office of Science and Technology Policy
Rosina Bierbaum	Office of Science and Technology Policy
Terence Jack Blasing	Oak Ridge National Laboratory
Suzanne Bolton	National Marine Fisheries Service
Mark Bove	Florida State University
Jeff Brokaw	U.S. Agency for International Development
David H. Bromwich	Ohio State University
Sandra Brown	Winrock International
Ronald Brunner	University of Colorado
Earle N. Buckley	National Oceanic and Atmospheric Administration
James L. Buizer	Office of Global Programs
Virginia Burkett	U.S. Geological Survey
Antonio J. Busalacchi	NASA Goddard Space Flight Center
David Campbell	Michigan State University
Terry Chapin III	University of Alaska
William C. Clark	Harvard University
Louis A. Codispoti	Horn Point Laboratory
Roy Darwin	U.S. Department of Agriculture
Margaret Davidson	NOAA Coastal Services Center
Jonathan Davis	The National Academies, Institute of Medicine
Benjamin DeAngelo	U.S. Environmental Protection Agency
Stephen Decanio	University of California at Santa Barbara
Dennis Devlin	Exxon Mobil Biomedical Sciences, Inc.
David J. Dokken	University Corporation for Atmospheric Research
James J. Dooley	Pacific Northwest National Laboratory
Robert B. Dunbar	Stanford University
William Easterling	The Pennsylvania State University
Richard S. Eckaus	Massachusetts Institute of Technology
Sylvia Edgerton	Pacific Northwest National Laboratory
Paul N Edwards	University of Michigan
Gary S. Eilerts	Famine Early Warning System Project
Hugh Ellis	The Johns Hopkins University
Kerry Emanuel	Massachusetts Institute of Technology
Paul R. Epstein	Harvard Medical School
John Everett	National Marine Fisheries Service
James Fahn	NOAA, Office of Global Programs
Lisa Farrow	NOAA, Office of Global Programs
Howard Feldman	American Petroleum Institute
Lauren Flejzor	U.S. Department of State
Joshua Foster	NOAA, Office of Global Programs
Franco Furger	George Mason University
John Furlow	U.S. Environmental Protection Agency
Janet Line Gamble	U.S. Environmental Protection Agency
Mary Gant	U.S. Environmental Protection Agency
Luis E. Garcia	Inter-American Development Bank
Laurie Geller	National Research Council
Lee C. Gerhard	Kansas Geological Survey, University of Kansas
Suzanne Giannini-Spohn	U.S. Environmental Protection Agency
Michael H. Glantz	National Center for Atmospheric Research
Peter Gleick	Pacific Institute
Per Gloersen	NASA Goddard Space Flight Center
Patrick Gonzalez	U.S. Geological Survey
William Gore	Boehringer Ingelheim Pharmaceuticals. Inc.1
Vivien Gornitz	Columbia University
Anne Grambsch	U.S. Environmental Protection Agency
Kenneth Green	Reason Public Policy Institute
Peter M. Groffman	Institute of Ecosystem Studies
Duane J. Gubler	Centers for Disease Control and Prevention
Stephen Guptill	U.S. Geological Survey
Peter M. Haas	University of Massachusetts
Kimberly Hall	The University of Michigan
Michael P. Hamnett	Social Science Research Institute
P.J. Hanson	Oak Ridge National Laboratory
Arthur Hawkins	RAMSAR Convention
Wanda Haxton	U.S. Environmental Protection Agency
A.S. Heagle	U.S. Department of Agriculture
Geoffrey Heal	Columbia Business School
Daniel Hellerstein	U.S. Department of Agriculture
Rachelle D. Hollander	National Science Foundation

Paul Houser	National Aeronautics and Space Administration
Schuyler Houser	Sinte Gleska University
Richard B. Howarth	Dartmouth College
Charles Howe	University of Colorado
Joseph Huang	U.S. Country Studies Program
M.A. Huston	Oak Ridge National Laboratory
Charles F. Hutchinson	Arizona Remote Sensing Center
Mare Imhoff	National Aeronautics and Space Administration
Kevin Ingram	U.S. Department of Agriculture
Dale Jamieson	Carleton College
Sheila Jasanoff	Harvard University
Julie D. Jastrow	Argonne National Laboratory
Carol Jones	U.S. Department of Agriculture
Susan Herrod Julius	NCEA, ORD, EPA
Daniel Kammen	University of California, Berkeley
Sally Kane	National Oceanic and Atmospheric Administration
Bob Kates	Independent Consultant
Richard W. Katz	National Center for Atmospheric Research
Glen Kelly	Global Climate Coalition
Victor Kennedy	University of Maryland
Mojdeh Keykhah	J.F. Kennedy School of Government
Bruce Kimball	U.S. Department of Agriculture
Jane Kinsel	National Institutes of Health
Lee Ann Kozak	Southern Company Services
Alan Krupnick	Resources for the Future
Kristin Kuntz-Duriseti	Stanford University
Rattan Lal	Ohio State Universtiy
Chris Landsea	NOAA Hurricane Research Division
William K.M. Lau	NASA Goddard Space Flight Center
Jim Lazorchak	U.S. Environmental Protection Agency
Neil Leary	University Corporation for Atmospheric Research
Michael Ledbetter	National Science Foundation
Robert Lempert	Rand Corporation
Clement D. Lewsey	National Oceanic and Atmospheric Administration
Frances C. Li	National Science Foundation
Sven B. Lundstedt	The Ohio State University
Michael C. MacCracken	National Assessment Coordination Office
John J. Magnuson	University of Wisconsin
David C. Major	Columbia University
Elizabeth Malone	Pacific Northwest National Laboratory
Karen Marsh	Federal Emergency Management Agency
Paul Mayewski	Institute for Quaternary
James J. McCarthy	Co-Chair, WGII
John McCarty	U.S. Environmental Protection Agency
Robert McFadden	GHG Associates
Michael A. McGeehin	National Center for Environmental Health
Diane McKnight	INSTAAR
S.B. McLaughlin	Oak Ridge National Laboratory
Linda Mearns	National Center for Atmospheric Research
Robert Mendelsohn	Yale University
William Meyer	Clark University and Harvard University
Elizabeth Middleton	NASA Goddard Space Flight Center
Richard B. Mieremet	National Oceanic and Atmospheric Administration
Edward L. Miles	College of Ocean and Fisheries Sciences, University of Washington
Kathleen Miller	National Center for Atmospheric Research
Ansje Miller	Redefining Progress
N.L. Miller	Lawrence Berkeley National Laboratory
R.M. Miller	Argonne National Laboratory
Evan Mills	Lawrence Berkeley National Laboratory
Jeff Miotke	U.S. Department of State
James W. Mjelde	Texas A&M University
Jack A. Morgan	U.S Department of Agriculture
James Morison	University of Washington
Susanne C. Moser	Union of Concerned Scientists
Richard H. Moss	Battelle Pacific Northwest National Laboratory
Thomas Muir	Office of Science and Technology Policy
Daniel Mullarkey	U.S. Department of Agriculture
Frederick E. Nelson	University of Delaware
Claudia Nierenberg	NOAA, Office of Global Programs
Richard Norgaard	University of California, Berkeley
Jim O'Brien	Center for Ocean-Atmospheric Prediction
Walter Oechel	San Diego State University
Martin Offutt	Office of Science and Technology Policy
Camille Parmesan	University of Texas
Edward Parson	J.F. Kennedy School of Government
Kim Partington	National Aeronautics and Space Administration
Anthony Patt	Harvard University
Jonathan Patz	Johns Hopkins School of Public Health
Andrew Peara	Independent Consultant
Roger A. Pielke	Colorado State University
Mark Pineda	National Institutes of Health
Stephen R. Piotrowicz	NOAA, Office of Oceanic and Atmospheric Research
Warren T. Piver	National Institute of Environmental Health Sciences
Wayne Polley	Agricultural Research Service

Robert S. Pomeroy	World Resources Institute
Jeff Price	American Bird Conservancy
Roger Pulwarty	NOAA, Office of Global Programs
Jonathan Pundsack	National Aeronautics and Space Administration
Robert L. Randall	The RainForest ReGeneration Institute
John M. Reilly	Massachusetts Institute of Technology
Charles Revelle	Johns Hopkins University
Jim Reynolds	Duke University
Charles W. Rice	Kansas State University
Richard Richels	Electric Power Research Institute
Alan Robock	Rutgers University
Michael Rodemeyer	U.S. House of Representatives
Catriona Rogers	Global Change Research Program
Terry Root	University of Michigan
C.P. Ropelewski	International Research Institute for Climate Prediction
Cynthia Rosenzweig	Goddard Institute for Space Studies
Dale Rothman	Columbia University
R. Bradley Sack	Johns Hopkins University
Ted Scambos	University of Colorado
Tom Schelling	University of Maryland
David Schimel	National Center for Atmospheric Research
David Schimmelpfennig	U.S. Department of Agriculture
Stephen Schneider	Stanford University
Russ Schnell	National Oceanic and Atmospheric Administration
Michael J. Scott	Battelle Pacific Northwest National Laboratory
Mark C. Serreze	University of Colorado
Steven R. Shafer	U.S. Department of Agriculture
Caitlin Simpson	NOAA, Office of Global Programs
Joel B. Smith	Stratus Consulting Inc.
Raymond C. Smith	University of California, Santa Barbara
Amy Snover	JISAO/SMA Climate Impacts Group
Brent Sohngen	Ohio State University
Allen M. Solomon	Office of Science and Technology Policy
Lisa Sorenson	Boston University
Eugene Z. Stakhiv	U.S. Army Institute for Water Resources
Macol Stewart	NOAA, Office of Global Programs
Peter Stone	Goddard Institute for Space Studies
A.B. Sullivan	Oak Ridge National Laboratory
Tonna-Marie Surgeon	NOAA, Office of Global Programs
George Teslioudisand	National Aeronautics and Space Administration
James Titus	U.S. Environmental Protection Agency
Michael Toman	Resources for the Future
Kevin Trenberth	National Center for Atmospheric Research
Juli M. Trtanj	NOAA, Office of Global Programs
George Tselioudisand	Goddard Institute for Space Studies
Jan C. Vermeiren	Organization of American States
John E. Walsh	University of Illinois
Larry Weber	Office of Science and Technology Policy
J. Jason West	Massachusetts Institute of Technology
T.O. West	Oak Ridge National Laboratory
John P. Weyant	Energy Modeling Forum, Stanford University
Kasey S. White	University Corporation for Atmospheric Research
Tom Wilbanks	Oak Ridge National Laboratory
Stan Wilson	NOAA, Office of Chief Scientist
Elizabeth Wilson	U.S. Environmental Protection Agency
Darrell Winner	U.S. Environmental Protection Agency
Robert C. Worrest	Consortium for International Earth Science Information Network
Christina Wright	U.S. Environmental Protection Agency
Stan Wullschleger	Oak Ridge National Laboratory
Gary Yohe	Wesleyan University
Richard Zepp	U.S. Environmental Protection Agency
Dorothy Zukor	NASA Goddard Space Flight Center

Uzbekistan

Tatyana Ososkova	KazNIIMOSK

Venezuela

Luis Jose Mata	Nord-Sued Zentrum fru Enmtwicklungsforschung
Alicia Villamizar	Universidad Simon Bolivar

Zimbabwe

Chris H.D. Magadza	Universitry of Zimbabwe

B

Glossary of Terms

Ablation
All processes by which snow and ice are lost from a glacier, floating ice, or snow cover.

Acclimatization
The physiological adaptation to climatic variations.

Active Layer
The top layer of soil in *permafrost* that is subjected to seasonal freezing and thawing.

Adaptability
See *adaptive capacity*.

Adaptation
Adjustment in natural or human systems in response to actual or expected climatic *stimuli* or their effects, which moderates harm or exploits beneficial opportunities. Various types of adaptation can be distinguished, including anticipatory and reactive adaptation, private and public adaptation, and autonomous and planned adaptation:

- ***Anticipatory Adaptation***—Adaptation that takes place before impacts of climate change are observed. Also referred to as proactive adaptation.
- ***Autonomous Adaptation***—Adaptation that does not constitute a conscious response to climatic stimuli but is triggered by ecological changes in natural systems and by market or welfare changes in human systems. Also referred to as spontaneous adaptation.
- ***Planned Adaptation***—Adaptation that is the result of a deliberate policy decision, based on an awareness that conditions have changed or are about to change and that action is required to return to, maintain, or achieve a desired state.
- ***Private Adaptation***—Adaptation that is initiated and implemented by individuals, households or private companies. Private adaptation is usually in the actor's rational self-interest.
- ***Public Adaptation***—Adaptation that is initiated and implemented by governments at all levels. Public adaptation is usually directed at collective needs.
- ***Reactive Adaptation***—Adaptation that takes place after impacts of climate change have been observed.

See also *adaptation assessment*, *adaptation benefits*, *adaptation costs*, *adaptive capacity*, and *maladaptation*.

Adaptation Assessment
The practice of identifying options to adapt to climate change and evaluating them in terms of criteria such as availability, benefits, costs, effectiveness, efficiency, and feasibility.

Adaptation Benefits
The avoided damage costs or the accrued benefits following the adoption and implementation of *adaptation* measures.

Adaptation Costs
Costs of planning, preparing for, facilitating, and implementing *adaptation* measures, including transition costs.

Adaptive Capacity
The ability of a system to adjust to *climate change* (including *climate variability* and extremes) to moderate potential damages, to take advantage of opportunities, or to cope with the consequences.

Aero-Allergens
Allergens present in the air.

Aerosols
A collection of airborne solid or liquid particles, with a typical size between 0.01 and 10 mm that reside in the atmosphere for at least several hours. Aerosols may be of either natural or anthropogenic origin. Aerosols may influence climate in two ways: directly through scattering and absorbing radiation, and indirectly through acting as condensation nuclei for cloud formation or modifying the optical properties and lifetime of clouds.

Afforestation
Planting of new forests on lands that historically have not contained forests. For a discussion of the term *forest* and related terms such as *afforestation*, *reforestation*, and *deforestation*, see the IPCC *Special Report on Land Use, Land-Use Change, and Forestry* (IPCC, 2000).

Aggregate Impacts
Total impacts summed up across sectors and/or regions. The aggregation of impacts requires knowledge of (or assumptions about) the relative importance of impacts in different sectors and regions. Measures of aggregate impacts include, for example, the total number of people affected, change in net primary productivity, number of systems undergoing change, or total economic costs.

Agronomy
The branch of agriculture that deals with the theory and practice of field-crop production and the scientific management of soil.

Alases
Coalescing thaw depressions.

Albedo
The fraction of solar radiation reflected by a surface or object, often expressed as a percentage. Snow-covered surfaces have a high albedo; the albedo of soils ranges from high to low; vegetation-covered surfaces and oceans have a low albedo. The Earth's albedo varies mainly through varying cloudiness, snow, ice, leaf area, and land-cover changes.

Algal Blooms
A reproductive explosion of algae in a lake, river, or ocean.

Alkalinity
A measure of the capacity of water to neutralize acids.

Allergens
Antigenic substances capable of producing immediate-type hypersensitivity.

Alpine
The *biogeographic* zone made up of slopes above timberline and characterized by the presence of rosette-forming herbaceous plants and low shrubby slow-growing woody plants.

Alternative Risk Transfer
Capital-market alternatives to traditional insurance (e.g., catastrophe bonds).

Anadromous Species
A species of fish, such as salmon, that spawn in freshwater then migrate into the ocean to grow to maturity.

Anaerobic
Living, active, or occurring in the absence of free oxygen.

Anoxia
A deficiency of oxygen, especially of such severity as to result in permanent damage.

Antarctic Bottomwater
A type of water in the seas surrounding Antarctica with temperatures ranging from 0 to -0.8°C, salinities from 34.6 to 34.7 PSU, and a density near 27.88. This is the densest water in the free ocean.

Antarctic Circumpolar Current
A Southern Ocean current that flows around the entire globe driven by the circumpolar westerlies.

Antarctic Intermediate Water
Created through large-scale cooling and Ekman convergence in the Southern Ocean.

Anthropogenic
Resulting from or produced by human beings.

AOGCM
See *climate model*.

Apex Consumers
Organisms at the top of food chains; top predators.

Aquaculture
Breeding and rearing fish, shellfish, etc., or growing plants for food in special ponds.

Aquifer
A stratum of permeable rock that bears water. An unconfined aquifer is recharged directly by local rainfall, rivers, and lakes, and the rate of recharge will be influenced by the permeability of the overlying rocks and soils. A confined aquifer is characterized by an overlying bed that is impermeable and the local rainfall does not influence the aquifer.

Arbovirus
Any of various viruses transmitted by arthropods and including the causative agents of dengue fever, yellow fever, and some types of encephalitis.

Arid Regions
Ecosystems with <250 mm precipitation per year.

Autotrophic
Organisms independent of external sources of organic carbon (compounds) for provision of their own organic constituents, which they can manufacture entirely from inorganic material. Plants are autotrophic (photoautotrophs) using the energy of sunlight to produce organic carbon compounds from inorganic carbon and water in the process of *photosynthesis*.

Baseflow
Sustained flow in a river or stream that is primarily produced by groundwater runoff, delayed subsurface runoff, and/or lake outflow.

Baseline/Reference
The baseline (or reference) is any datum against which change is measured. It might be a "current baseline," in which case it represents observable, present-day conditions. It might also be a "future baseline," which is a projected future set of conditions excluding the driving factor of interest. Alternative interpretations of the reference conditions can give rise to multiple baselines.

Basin
The drainage area of a stream, river, or lake.

Benthic Organisms
The *biota* living on, or very near, the bottom of the sea, river, or lake.

Biodiversity
The numbers and relative abundances of different genes (genetic diversity), species, and ecosystems (communities) in a particular area. See also *functional diversity*.

Biodiversity Hot Spots
Areas with high concentrations of *endemic* species facing extraordinary habitat destruction.

Biofuels
A fuel produced from dry organic matter or combustible oils produced by plants. Examples of biofuel include alcohol (from fermented sugar), black liquor from the paper manufacturing process, wood, and soybean oil.

Biomass
The total mass of living organisms in a given area or volume; recently dead plant material is often included as dead biomass.

Biome
A grouping of similar plant and animal communities into broad landscape units that occur under similar environmental conditions.

Biosphere
The part of the Earth system comprising all ecosystems and living organisms in the atmosphere, on land (terrestrial biosphere), or in the oceans (marine biosphere), including derived dead organic matter, such as litter, soil organic matter, and oceanic detritus.

Biota
All living organisms of an area; the flora and fauna considered as a unit.

Bog
A poorly drained area rich in accumulated plant material, frequently surrounding a body of open water and having a characteristic flora (such as sedges, heaths, and sphagnum).

Boreal Forest
Forests of pine, spruce, fir, and larch stretching from the east coast of Canada westward to Alaska and continuing from Siberia westward across the entire extent of Russia to the European Plain.

Breakwater
An offshore structure (such as a wall or jetty) that, by breaking the force of the wave, protects a harbor, anchorage, beach, or shore area.

C_3 Plants
Plants that produce a three-carbon compound during photosynthesis, including most trees and agricultural crops such as rice, wheat, soybeans, potatoes, and vegetables.

C_4 Plants
Plants that produce a four-carbon compound during photosynthesis (mainly of tropical origin), including grasses and the agriculturally important crops maize, sugar cane, millet, and sorghum.

Carbon Cycle
The term used to describe the flow of carbon (in various forms, e.g., as in carbon dioxide) through the atmosphere, ocean, terrestrial biosphere, and lithosphere.

Carbon Dioxide (CO_2)
A naturally occurring gas, also a by-product of burning fossil fuels and *biomass*, as well as from land-use changes and other industrial processes. It is the principal *anthropogenic greenhouse gas* that affects the Earth's radiative balance. It is the reference gas against which other greenhouse gases are measured and therefore has a Global Warming Potential of 1.

Carbon Dioxide Fertilization
The enhancement of the growth of plants as a result of increased atmospheric *carbon dioxide* concentration. Depending on their mechanism of *photosynthesis*, certain types of plants are more sensitive to changes in atmospheric CO_2 concentration. In particular, *C_3 plants* generally show a larger response to CO_2 than *C_4 plants*.

Carbon Flux
Transfer of carbon from one carbon pool to another in units of measurement of mass per unit area and time (e.g., t C).

Carrying Capacity
The number of individuals in a population that the resources of a *habitat* can support.

Catchment
An area that collects and drains rainwater.

Chagas' Disease
A parasitic disease caused by the *Trypanosoma cruzi* and transmitted by triatomine bugs in the Americas, with two clinical periods: acute (fever, swelling of the spleen, edemas) and chronic (digestive syndrome, potentially fatal heart condition).

Cholera
An intestinal infection that results in frequent watery stools, cramping abdominal pain, and eventual collapse from dehydration.

Climate
Climate in a narrow sense is usually defined as the "average weather," or more rigorously, as the statistical description in terms of the mean and variability of relevant quantities over a period of time ranging from months to thousands of years. The classical period is 3 decades, as defined by the World Meteorological Organization (WMO). These quantities are most often surface variables such as temperature, precipitation, and wind. Climate in a wider sense is the state, including a statistical description, of the climate system.

Climate Change
Climate change refers to any change in climate over time, whether due to natural variability or as a result of human activity. This usage differs from that in the *United Nations Framework Convention on Climate Change (UNFCCC)*, which defines "climate change" as: "a change of climate which is attributed directly or indirectly to human activity that alters the composition of the global atmosphere and which is in addition to natural climate variability observed over comparable time periods." See also *climate variability*.

Climate Model (Hierarchy)
A numerical representation of the climate system based on the physical, chemical, and biological properties of its components, their interactions and feedback processes, and accounting for all or some of its known properties. The climate system can be represented by models of varying

complexity (i.e., for any one component or combination of components a hierarchy of models can be identified, differing in such aspects as the number of spatial dimensions; the extent to which physical, chemical, or biological processes are explicitly represented; or the level at which empirical parameterizations are involved. Coupled atmosphere/ocean/sea-ice General Circulation Models (AOGCMs) provide a comprehensive representation of the climate system. There is an evolution towards more complex models with active chemistry and biology. Climate models are applied, as a research tool, to study and simulate the climate, but also for operational purposes, including monthly, seasonal, and interannual climate predictions.

Climate Prediction
A climate prediction or climate forecast is the result of an attempt to produce a most likely description or estimate of the actual evolution of the climate in the future (e.g., at seasonal, interannual, or long-term time scales. See also *climate projection* and *climate scenario*.

Climate Projection
A projection of the response of the climate system to emission or concentration scenarios of *greenhouse gases* and *aerosols*, or *radiative forcing* scenarios, often based upon simulations by climate models. Climate projections are distinguished from *climate predictions* in order to emphasize that climate projections depend upon the emission/concentration/radiative forcing scenario used, which are based on assumptions, concerning, for example, future socioeconomic and technological developments that may or may not be realized and are therefore subject to substantial uncertainty.

Climate Scenario
A plausible and often simplified representation of the future *climate*, based on an internally consistent set of climatological relationships, that has been constructed for explicit use in investigating the potential consequences of anthropogenic climate change, often serving as input to impact models. Climate projections often serve as the raw material for constructing climate scenarios, but climate scenarios usually require additional information such as about the observed current climate. A "climate change scenario" is the difference between a climate scenario and the current climate.

Climate System
The climate system is the highly complex system consisting of five major components: the atmosphere, the hydrosphere, the *cryosphere*, the land surface, and the *biosphere*, and the interactions between them. The climate system evolves in time under the influence of its own internal dynamics and because of external forcings such as volcanic eruptions, solar variations and human-induced forcings such as the changing composition of the atmosphere and *land use*.

Climate Variability
Climate variability refers to variations in the mean state and other statistics (such as standard deviations, the occurrence of extremes, etc.) of the climate on all temporal and spatial scales beyond that of individual weather events. Variability may be due to natural internal processes within the climate system (internal variability), or to variations in natural or anthropogenic external forcing (external variability). See also *climate change*.

CO_2 Fertilization
See *carbon dioxide fertilization*.

Communicable Disease
An infectious disease caused by transmission of an infective biological agent (virus, bacterium, protozoan, or multicellular macroparasite).

Coping Range
The variation in climatic *stimuli* that a system can absorb without producing significant impacts.

Coral Bleaching
The paling in color of corals resulting from a loss of symbiotic algae. Bleaching occurs in response to physiological shock in response to abrupt changes in temperature, salinity, and turbidity.

Cordillera
An individual mountain chain with closely connected, distinct summits. In South America, "cordillera" refers to an individual mountain range.

Cryosphere
The component of the climate system consisting of all snow, ice, and *permafrost* on and beneath the surface of the earth and ocean.

Cryptosporidiosis
An opportunistic infection caused by an intestinal parasite common in animals. Transmission occurs through ingestion of food or water contaminated with animal feces. The parasite causes severe chronic diarrhea, especially in people with HIV.

Deepwater Formation
Occurs when seawater freezes to form sea ice. The local release of salt and consequent increase in water density leads to the formation of saline coldwater that sinks to the ocean floor. See *Antarctic bottomwater*.

Deforestation
Conversion of forest to non-forest. For a discussion of the term *forest* and related terms such as *afforestation*, *reforestation*, and *deforestation*, see the IPCC *Special Report on Land Use, Land-Use Change, and Forestry* (IPCC, 2000).

Dengue Fever
An infectious viral disease spread by mosquitoes, often called breakbone fever because it is characterized by severe pain in joints and back. Subsequent infections of the virus may lead to dengue haemorrhagic fever (DHF) and dengue shock syndrome (DSS), which may be fatal.

Desert
An ecosystem with <100 mm precipitation per year.

Desertification
Land degradation in arid, semi-arid, and dry sub-humid areas resulting from various factors, including climatic variations and human activities. Further, the United Nations Convention to Combat Desertification (UNCCD) defines land degradation as a reduction or loss in arid, semi-arid, and dry sub-humid areas of the biological or economic productivity and complexity of rain-fed cropland, irrigated cropland, or range, pasture, forest, and woodlands resulting from land uses or from a process or combination of processes, including those arising from human activities and habitation patterns, such as: (i) soil erosion caused by wind and/or water; (ii) deterioration of the physical, chemical, and biological or economic properties of soil; and (iii) long-term loss of natural vegetation.

Diatom
A class of unicellular algae (Bacillariophyceae) that are widespread on soil surfaces and in freshwater and marine systems, especially cold waters of relatively low salinity. These have cell sizes ranging from 5 to 2000 μm.

Disturbance Regime
Frequency, intensity, and types of disturbances, such as fires, inspect or pest outbreaks, floods, and *droughts*.

Diurnal Temperature Range
The difference between the maximum and minimum temperature during a day.

Downscaling
Reducing the scale of a model from a global to regional level.

Drought
The phenomenon that exists when precipitation has been significantly below normal recorded levels, causing serious hydrological imbalances that adversely affect land resource production systems.

Ecosystem
A distinct system of interacting living organisms, together with their physical environment. The boundaries of what could be called an ecosystem are somewhat arbitrary, depending on the focus of interest or study. Thus the extent of an ecosystem may range from very small spatial scales to, ultimately, the entire Earth.

Ecosystem Services
Ecological processes or functions which have value to individuals or society

Ecotone
Transition area between adjacent ecological communities (e.g., between forests and grasslands), usually involving competition between organisms common to both.

Edaphic
Of or relating to the soil; factors inherent in the soil.

Effective Rainfall
The portion of the total rainfall that becomes available for plant growth.

El Niño-Southern Oscillation (ENSO)
El Niño, in its original sense, is a warmwater current that periodically flows along the coast of Ecuador and Peru, disrupting the local fishery. This oceanic event is associated with a fluctuation of the intertropical surface pressure pattern and circulation in the Indian and Pacific Oceans, called the *Southern Oscillation*. This coupled atmosphere-ocean phenomenon is collectively known as El Niño-Southern Oscillation. During an El Niño event, the prevailing trade winds weaken and the equatorial countercurrent strengthens, causing warm surface waters in the Indonesian area to flow eastward to overlie the cold waters of the Peru current. This event has great impact on the wind, sea surface temperature, and precipitation patterns in the tropical Pacific. It has climatic effects throughout the Pacific region and in many other parts of the world. The opposite of an El Niño event is called La Niña.

Emission Scenario
A plausible representation of the future development of emissions of substances that are potentially radiatively active (e.g., *greenhouse gases*, *aerosols*), based on a coherent and internally consistent set of assumptions about driving forces (such as demographic and socioeconomic development, technological change) and their key relationships. In 1992, the IPCC presented a set of emission scenarios that were used as a basis for the climate projections in the Second Assessment Report (IPCC, 1996). These emission scenarios are referred to as the IS92 scenarios. In the IPCC *Special Report on Emission Scenarios* (Nakicenovic *et al.*, 2000), new emission scenarios—the so-called SRES scenarios—were published.

Endemic
Restricted or peculiar to a locality or region. With regard to human health, endemic can refer to a disease or agent present or usually prevalent in a population or geographical area at all times.

Endorheic Lake
A lake with no outflow; also known as a closed lake.

Enzootic
A disease affecting the animals in an area. It corresponds to an *endemic* disease among humans.

Epidemic
Occurring suddenly in numbers clearly in excess of normal expectancy, said especially of infectious diseases but applied also to any disease, injury, or other health-related event occurring in such outbreaks.

Erosion
The process of removal and transport of soil and rock by weathering, mass wasting, and the action of streams, glaciers, waves, winds, and underground water.

Eustatic Sea-Level Rise
See *sea-level rise*.

Eutrophication
The process by which a body of water (often shallow) becomes (either naturally or by pollution) rich in dissolved nutrients with a seasonal deficiency in dissolved oxygen.

Evaporation
The process by which a liquid becomes a gas.

Evapotranspiration
The combined process of *evaporation* from the Earth's surface and *transpiration* from vegetation.

Exoheic Lake
A lake drained by outflowing rivers.

Exotic Species
See *introduced species*.

Exposure
The nature and degree to which a system is exposed to significant climatic variations.

Exposure Unit
An activity, group, region, or resource that is subjected to climatic *stimuli*.

Externalities
By-products of activities that affect the well-being of people or the environment, where those impacts are not reflected in market prices. The costs (or benefits) associated with externalities do not enter cost-accounting schemes.

Extinction
The complete disappearance of an entire species.

Extirpation
The disappearance of a species from part of its range; local extinction.

Extreme Weather Event
An event that is rare within its statistical reference distribution at a particular place. Definitions of "rare" vary, but an extreme weather event would normally be as rare as or rarer than the 10th or 90th percentile. By definition, the characteristics of what is called "extreme weather" may vary from place to place. An "extreme climate event" is an average of a number of weather events over a certain period of time, an average which is itself extreme (e.g., rainfall over a season).

Extrinsic Incubation Period
In blood-feeding anthropod vectors, the time between acquisition of the infectious blood meal and the time when the anthropod becomes capable of transmitting the agent. In the case of malaria, the life stages of the plasmodium parasite spent within the female mosquito vector (i.e., outside the human host).

Feedback
A process that triggers changes in a second process that in turn influences the original one; a positive feedback intensifies the original process, and a negative feedback reduces it.

Fen
Low land covered wholly or partly with water unless artificially drained.

Fiber
Wood, fuelwood (either woody or non-woody).

Food Insecurity
A situation that exists when people lack secure access to sufficient amounts of safe and nutritious food for normal growth and development and an active and healthy life. It may be caused by the unavailability of food, insufficient purchasing power, inappropriate distribution, or inadequate use of food at the household level. Food insecurity may be chronic, seasonal, or transitory.

Forecast
See *climate prediction* and *climate projection*.

Forest
A vegetation type dominated by trees. Many definitions of the term *forest* are in use throughout the world, reflecting wide differences in biogeophysical conditions, social structure, and economics. For a discussion of the term *forest* and related terms such as *afforestation*, *reforestation*, and *deforestation*, see the IPCC *Special Report on Land Use, Land-Use Change, and Forestry* (IPCC, 2000).

Freshwater Lens
A lenticular fresh groundwater body that underlies an oceanic island. It is underlain by saline water.

Functional Diversity
The number of functionally different organisms in an ecosystem (also referred to as "functional types" and "functional groups").

General Circulation Model (GCM)
See *climate model*.

General Equilibrium Analysis
An approach that considers simultaneously all the markets in an economy, allowing for feedback effects between individual markets.

Geomorphic
Pertaining to the form of the Earth or its surface features.

Glacier
A mass of land ice flowing downhill (by internal deformation and sliding at the base) and constrained by the surrounding topography (e.g., the sides of a valley or surrounding peaks); the bedrock topography is the major influence on the dynamics and surface slope of a glacier. A glacier is maintained by accumulation of snow at high altitudes, balanced by melting at low altitudes or discharge into the sea.

Greenhouse Effect
Greenhouse gases effectively absorb infrared radiation emitted by the Earth's surface, by the atmosphere itself due to the same gases, and by clouds. Atmospheric radiation is emitted to all sides, including downward to the Earth's surface. Thus greenhouse gases trap heat within the surface-troposphere system. This is called the "natural greenhouse effect." Atmospheric radiation is strongly coupled to the temperature of the level at which it is emitted. In the troposphere, the temperature generally decreases with height. Effectively, infrared radiation emitted to space originates from an altitude with a temperature of on average -19°C, in balance with the net incoming solar radiation, whereas the Earth's surface is kept at a much higher temperature of on average 14°C. An increase in the concentration of greenhouse gases leads to an increased infrared opacity of the atmosphere, and therefore to an effective radiation into space from a higher altitude at a lower temperature. This causes a radiative forcing, an imbalance that can only be compensated for by an increase of the temperature of the surface-troposphere system. This is called the "enhanced greenhouse effect."

Greenhouse Gas
Greenhouse gases are those gaseous constituents of the atmosphere, both natural and anthropogenic, that absorb and emit radiation at specific wavelengths within the spectrum of infrared radiation emitted by the Earth's surface, the atmosphere, and clouds. This property causes the *greenhouse effect*. Water vapor (H_2O), carbon dioxide (CO_2), nitrous oxide (N_2O), methane (CH_4), and ozone (O_3) are the primary greenhouse gases in the Earth's atmosphere. Moreover, there are a number of entirely human-made greenhouse gases in the atmosphere, such as the halocarbons and other chlorine- and bromine-containing substances which are dealt with under the Montreal Protocol. Beside CO_2, N_2O, and CH_4, the *Kyoto Protocol* deals with the greenhouse gases sulfur hexaflouride (SF_6), hydrofluorocarbons (HFCs) and perfluorocarbons (PFCs).

Groin
A low, narrow jetty, usually extending roughly perpendicular to the shoreline, designed to protect the shore from erosion by currents, tides, or waves, or to trap sand for the purpose of building up or making a beach.

Gross Primary Production
The amount of carbon fixed from the atmosphere through *photosynthesis*.

Groundwater Recharge
The process by which external water is added to the zone of saturation of an *aquifer*, either directly into a formation or indirectly by way of another formation.

Habitat
The particular environment or place where an organism or species tends to live; a more locally circumscribed portion of the total environment.

Halocline
A layer in the ocean in which the rate of salinity variation with depth is much larger than layers immediately above or below it.

Hantavirus
A virus in the family Bunyaviridae that causes a type of haemorrhagic fever. It is thought that humans catch the disease mainly from infected rodents, either through direct contact with the animals or by inhaling or ingesting dust that contains their dried urine.

Heath
Any of the various low-growing shrubby plants of open wastelands, usually growing on acidic, poorly drained soils.

Heat Island
An area within an urban area characterized by ambient temperatures higher than those of the surrounding area because of the absorption of solar energy by materials like asphalt.

Herbaceous
Flowering, non-woody plants.

Heterotrophic Respiration
The release of CO_2 from decomposition of organic matter.

Highland Malaria
Malaria that occurs around the altitudinal limits of its distribution.

Human Settlement
A place or area occupied by settlers.

Human System
Any system in which human organizations play a major role. Often, but not always, the term is synonymous with "society" or "social system" (e.g., agricultural system, political system, technological system, economic system); all are human systems in the sense applied in the TAR.

Hypolimnion
The part of a lake below the *thermocline* made up of water that is stagnant and of essentially uniform temperature except during the period of overturn.

Ice Cap
A dome-shaped ice mass covering a highland area that is considerably smaller in extent than *ice sheets*.

Ice Jam
An accumulation of broken river or sea ice caught in a narrow channel.

Ice Sheet
A mass of land ice which is sufficiently deep to cover most of the underlying bedrock topography, so that its shape is mainly determined by its internal dynamics (the flow of the ice as it deforms internally and slides at its base). An ice sheet flows outwards from a high central plateau with a small average surface slope. The margins slope steeply, and the ice is discharged through fast-flowing ice streams or outlet glaciers, in some cases into the sea or into ice shelves floating on the sea. There are only two large ice sheets in the modern world—on Greenland and Antarctica, the Antarctic ice sheet being divided into east and west by the Transantarctic Mountains; during glacial periods there were others.

Ice Shelf
A floating *ice sheet* of considerable thickness attached to a coast (usually of great horizontal extent with a level or gently undulating surface); often a seaward extension of ice sheets.

Immunosuppression
Reduced functioning of an individual's immune system.

(Climate) Impact Assessment
The practice of identifying and evaluating the detrimental and beneficial consequences of climate change on natural and human systems.

(Climate) Impacts
Consequences of climate change on natural and human systems. Depending on the consideration of adaptation, one can distinguish between potential impacts and residual impacts.

- ***Potential Impacts***—All impacts that may occur given a projected change in climate, without considering adaptation.
- ***Residual Impacts***—The impacts of climate change that would occur after adaptation.

See also *aggregate impacts*, *market impacts*, and *non-market impacts*.

Indigenous Peoples
People whose ancestors inhabited a place or a country when persons from another culture or ethnic background arrived on the scene and dominated them through conquest, settlement, or other means and who today live more in conformity with their own social, economic, and cultural customs and traditions than those of the country of which they now form a part (also referred to as "native," "aboriginal," or "tribal" peoples)

Industrial Revolution
A period of rapid industrial growth with far-reaching social and economic consequences, beginning in England during the second half of the 18th century and spreading to Europe and later to other countries including the United States. The industrial revolution marks the beginning of a strong increase in the use of fossil fuels and emission of in particular fossil carbon dioxide. In the TAR, the terms "pre-industrial" and "industrial" refer, somewhat arbitrarily, to the periods before and after 1750, respectively.

Infectious Diseases
Any disease that can be transmitted from one person to another. This may occur by direct physical contact, by common handling of an object that has picked up infective organisms, through a disease carrier, or by spread of infected droplets coughed or exhaled into the air.

Infrastructure
The basic equipment, utilities, productive enterprises, installations, and services essential for the development, operation, and growth of an organization, city, or nation.

Insolvency
Inability to meet financial obligations; bankruptcy.

Integrated Assessment
A method of analysis that combines results and models from the physical, biological, economic, and social sciences, and the interactions between these components, in a consistent framework to evaluate the status and the consequences of environmental change and the policy responses to it.

Introduced Species
A species occurring in an area outside its historically known natural range as a result of accidental dispersal by humans (also referred to as "exotic species" or "alien species").

Invasive Species
An introduced species that invades natural habitats.

Keystone Species
A species that has a central servicing role affecting many other organisms and whose demise is likely to result in the loss of a number of species and lead to major changes in ecosystem function.

Kyoto Protocol
The Kyoto Protocol was adopted at the Third Session of the Conference of the Parties (COP) to the *UN Framework Convention on Climate Change* (UNFCCC) in 1997 in Kyoto, Japan. It contains legally binding commitments, in addition to those included in the UNFCCC. Countries included in Annex B of the Protocol (most OECD countries and EITs) agreed to reduce their anthropogenic GHG emissions (CO_2, CH_4, N_2O, HFCs, PFCs, and SF_6) by at least 5% below 1990 levels in the commitment period 2008 to 2012. The Kyoto Protocol has not yet entered into force (as of June 2001).

La Niña
See *El Niño-Southern Oscillation (ENSO)*.

Land Use
The total of arrangements, activities, and inputs undertaken in a certain land-cover type (a set of human actions). The social and economic purposes for which land is managed (e.g., grazing, timber extraction, conservation).

Landslide
A mass of material that has slipped downhill by gravity, often assisted by water when the material is saturated; rapid movement of a mass of soil, rock, or debris down a slope.

Large-Scale Singularities
Abrupt and dramatic changes in systems in response to smooth changes in driving forces. For example, a gradual increase in atmospheric greenhouse gas concentrations may lead to such large-scale singularities as slowdown or collapse of the thermohaline circulation or collapse of the West Antarctic Ice Sheet. The occurrence, magnitude, and timing of large-scale singularities are difficult to predict.

Leaching
The removal of soil elements or applied chemicals through percolation.

Legume
Plants that are able to fix nitrogen from the air through a symbiotic relationship with soil bacteria (e.g., peas, beans, alfalfa, clovers).

Limnology
Study of lakes and their *biota*.

Littoral Zone
A coastal region; the shore zone between high and low watermarks.

Local Agenda 21
Local Agenda 21s are the local plans for environment and development that each local authority is meant to develop through a consultative process with their populations, with particular attention paid to involving women and youth. Many local authorities have developed Local Agenda 21s through consultative processes as a means of reorienting their policies, plans, and operations towards the achievement of *sustainable development* goals. The term comes from Chapter 28 of Agenda 21—the document formally endorsed by all government representatives attending the UN Conference on Environment and Development (also known as the Earth Summit) in Rio de Janeiro in 1992.

Maladaptation
Any changes in natural or human systems that inadvertently increase vulnerability to climatic stimuli; an adaptation that does not succeed in reducing vulnerability but increases it instead.

Malaria
Endemic or epidemic parasitic disease caused by species of the genus Plasmodium (protozoa) and transmitted by mosquitoes of the genus Anopheles; produces high fever attacks and systemic disorders, and kills approximately 2 million people every year.

Market Impacts
Impacts that are linked to market transactions and directly affect gross domestic product (GDP, a country's national accounts)—for example, changes in the supply and price of agricultural goods. See also *non-market impacts*.

Mass Movement
Applies to all unit movements of land material propelled and controlled by gravity.

Meningitis
Inflammation of the meninges (part of the covering of the brain).

Metazoan
An animal whose body consists of many cells. See also *protozoan*

Microbial Loop
Complex food web involving bacteria, single-celled animals and plants, viruses, and dissolved and particulate organic material. Dissolved and particulate material, released from organisms, is utilized by bacteria, which are grazed by protozoa which in turn are grazed by metazoa. Around 50% (often more) of primary production passes through the microbial loop rather than along the classical food chain of phytoplankton to herbivore.

Microclimate
Local climate at or near the Earth's surface. See also *climate*.

Mitigation
An anthropogenic intervention to reduce the *sources* or enhance the *sinks* of *greenhouse gases*.

Mixed Layer
The upper region of the ocean well-mixed by interaction with the overlying atmosphere.

Monsoon
Wind in the general atmospheric circulation typified by a seasonal persistent wind direction and by a pronounced change in direction from one season to the next.

Montane
The biogeographic zone made up of relatively moist, cool upland slopes below timberline and characterized by the presence of large evergreen trees as a dominant life form.

Montreal Protocol
The Montreal Protocol on Substances that Deplete the Ozone Layer was adopted in Montreal in 1987, and subsequently adjusted and amended in London (1990), Copenhagen (1992), Vienna (1995), Montreal (1997), and Beijing (1999). It controls the consumption and production of chlorine- and bromine-containing chemicals that destroy stratospheric *ozone*, such as CFCs, methyl chloroform, carbon tetrachloride, and many others.

Morbidity
Rate of occurrence of disease or other health disorder within a population, taking account of the age-specific morbidity rates. Health outcomes include chronic disease incidence/prevalence, rates of hospitalization, primary care consultations, disability-days (i.e., days when absent from work), and prevalence of symptoms.

Morphology
The form and structure of an organism or any of its parts.

Mortality
Rate of occurrence of death within a population within a specified time period; calculation of mortality takes account of age-specific death rates, and can thus yield measures of life expectancy and the extent of premature death.

Nanoplankton
Phytoplankton whose lengths range from 10-50 μm.

Net Biome Production (NBP)
Net gain or loss of carbon from a region. NBP is equal to *Net Ecosystem Production* minus the carbon lost due to a disturbance (e.g., a forest fire or a forest harvest).

Net Ecosystem Production (NEP)
Net gain or loss of carbon from an ecosystem. NEP is equal to *Net Primary Production* minus the carbon lost through heterotrophic respiration.

Net Primary Production (NPP)
The increase in plant biomass or carbon of a unit of a landscape. NPP is equal to *Gross Primary Production* minus carbon lost through autotrophic respiration.

Nitrogen Oxides (NO_x)
Any of several oxides of nitrogen.

Non-Linearity
A process is called "non-linear" when there is no simple proportional relation between cause and effect.

Non-Market Impacts
Impacts that affect ecosystems or human welfare, but that are not directly linked to market transactions—for example, an increased risk of premature death. See also *market impacts*.

Non-Point-Source Pollution
Pollution from sources that cannot be defined as discrete points, such as areas of crop production, timber, surface mining, disposal of refuse, and construction. See also *point-source pollution*.

No Regrets Policy
One that would generate net social benefits whether or not there is *anthropogenic* climate change.

North Atlantic Oscillation (NAO)
The North Atlantic Oscillation consists of opposing variations of barometric pressure near Iceland and near the Azores. It is the dominant mode of winter climate variability in the North Atlantic region ranging from central North America to Europe.

Obligate Species
Species restricted to one particularly characteristic mode of life.

Ocean Conveyor Belt
The theoretical route by which water circulates around the entire global ocean, driven by wind and the *thermohaline circulation*.

Ocean Ventilation
Downwelling of water from near the surface to the deep ocean. See also *deepwater formation*.

Oligotrophic
Relatively unproductive areas of the sea, lakes, and rivers with low nutrient content. See also *eutrophic*.

Opportunity Costs
The cost of an economic activity forgone by the choice of another activity.

Orography
The study of the physical geography of mountains and mountain systems.

Ozone
Ozone, the triatomic form of oxygen (O_3), is a gaseous atmospheric constituent. In the troposphere, it is created both naturally and by photochemical reactions involving gases resulting from human activities *(photochemical smog)*. In high concentrations, tropospheric ozone can be harmful to a wide-range of living organisms. Tropospheric ozone acts as a *greenhouse gas*. In the stratosphere, ozone is created by the interaction between solar ultraviolet radiation and molecular oxygen (O_2). Stratospheric ozone plays a decisive role in the stratospheric radiative balance. Depletion of stratospheric ozone, due to chemical reactions that may be enhanced by *climate change*, results in an increased ground-level flux of *ultraviolet (UV-) B radiation*. See also *Montreal Protocol*.

Particulates
Very small solid exhaust particles emitted during the combustion of fossil and biomass fuels. Particulates may consist of a wide variety of substances. Of greatest concern for health are particulates of less than or equal to 10 nm in diameter, usually designated as PM_{10}.

Peat
Unconsolidated soil material consisting largely of partially decomposed organic matter accumulated under conditions of excess moisture or other conditions that decrease decomposition rates.

Pelagic
Of, relating to, or living or occurring in the open sea.

Permafrost
Perennially frozen ground that occurs wherever the temperature remains below 0°C for several years.

Phenology
The study of natural phenomena that recur periodically (e.g., blooming, migrating) and their relation to climate and seasonal changes.

Photic Zone
The upper waters of lakes, rivers, and sea sufficiently illuminated for *photosynthesis* to occur.

Photochemical Smog
A mix of photochemical oxidant air pollutants produced by the reaction of sunlight with primary air pollutants, especially hydrocarbons.

Photosynthate
The product of *photosynthesis*.

Photosynthesis
The process by which plants take carbon dioxide from the air (or bicarbonate in water) to build carbohydrates, releasing oxygen in the process. There are several pathways of photosynthesis with different responses to atmospheric CO_2 concentrations. See also *CO_2 fertilization*, *C_3 plants*, and *C_4 plants*.

Physiographic
Of, relating to, or employing a description of nature or natural phenomena.

Phytophagous Insects
Insects that feed on plants.

Phytoplankton
The plant forms of *plankton* (e.g., *diatoms*). Phytoplankton are the dominant plants in the sea, and are the bast of the entire marine food web. These single-celled organisms are the principal agents for photosynthetic carbon fixation in the ocean. See also *zooplankton*.

Plankton
Aquatic organisms that drift or swim weakly. See also *phytoplankton* and *zooplankton*.

Point-Source Pollution
Pollution resulting from any confined, discrete source, such as a pipe, ditch, tunnel, well, container, concentrated animal-feeding operation, or floating craft. See also *non-point-source pollution*.

Polynyas
Areas of open water in pack ice or sea ice.

Pool
See *reservoir*.

Potential Production
Estimated production of a crop under conditions when nutrients and water are available at optimum levels for plant growth and development; other conditions such as day length, temperature, soil characteristics, etc., determined by site characteristics.

Pre-Industrial
See *Industrial Revolution*.

Primary Energy
Energy embodied in natural resources (e.g., coal, crude oil, sunlight, uranium) that has not undergone any *anthropogenic* conversion or transformation.

Producer Surplus
Returns beyond the cost of production that provide compensation for owners of skills or assets that are scarce (e.g., agriculturally productive land).

Projection (Generic)
A projection is a potential future evolution of a quality or set of quantities, often computed with the aid of a model. Projections are distinguished from predictions in order to emphasize that projections involve assumptions—concerning, for example, future socioeconomic and technological developments that may or may not be realized—and are therefore subject to substantial uncertainty. See also *climate projection* and *climate prediction*.

Protozoan
A single-celled animal.

Radiative Forcing
Radiative forcing is the change in the net vertical irradiance [expressed in Watts per square meter (Wm^{-2})] at the tropopause due to an internal change or a change in the external forcing of the climate system, such as a change in the concentration of CO_2 or the output of the Sun. Usually radiative forcing is computed after allowing for stratospheric temperatures to readjust to radiative equilibrium, but with all tropospheric properties held fixed at their unperturbed values.

Rangeland
Unimproved grasslands, shrublands, savannas, and tundra.

Reference Scenario
See *baseline/reference*.

Reforestation
Planting of *forests* on lands that have previously contained forests but that have been converted to some other use. For a discussion of the term *forest* and related terms such as *afforestation*, *reforestation*, and *deforestation*, see the IPCC *Special Report on Land Use, Land-Use Change, and Forestry* (IPCC, 2000).

Regeneration
The renewal of a stand of trees through either natural means (seeded onsite or adjacent stands or deposited by wind, birds, or animals) or artificial means (by planting seedlings or direct seeding).

Reinsurance
The transfer of a portion of primary insurance risks to a secondary tier of insurers (reinsurers); essentially "insurance for insurers."

Reservoir
A component of the climate system, other than the atmosphere, that has the capacity to store, accumulate, or release a substance of concern (e.g., carbon, a *greenhouse gas*, or precursor). Oceans, soils, and *forests* are examples of reservoirs of carbon. "Pool" is an equivalent term (note that the definition of pool often includes the atmosphere). The absolute quantity of substances of concern held within a reservoir at a specified time is called the "stock." The term also means an artificial or natural storage place for water, such as a lake, pond, or *aquifer*, from which the water may be withdrawn for such purposes as irrigation, water supply, or irrigation.

Reservoir Host
Any animal, plant, soil, or inanimate matter in which a pathogen normally lives and multiplies, and on which it depends primarily for survival (e.g., foxes are a reservoir for rabies). Reservoir hosts may be asymptomatic.

Resilience
Amount of change a system can undergo without changing state.

Respiration
The process whereby living organisms convert organic matter to carbon dioxide, releasing energy and consuming oxygen.

Riparian
Relating to or living or located on the bank of a natural watercourse (as a river) or sometimes of a lake or a tidewater.

Runoff
That part of precipitation that does not evaporate. In some countries, runoff implies *surface runoff* only.

Salinization
The accumulation of salts in soils.

Saltwater Intrusion/Encroachment
Displacement of fresh surface water or groundwater by the advance of saltwater due to its greater density, usually in coastal and estuarine areas.

Scenario (Generic)
A plausible and often simplified description of how the future may develop, based on a coherent and internally consistent set of assumptions about driving forces and key relationships. Scenarios may be derived from projections, but are often based on additional information from other sources, sometimes combined with a "narrative storyline." See also *climate scenario* and *emissions scenario*.

Sea-Level Rise
An increase in the mean level of the ocean. Eustatic sea-level rise is a change in global average sea level brought about by an alteration to the volume of the world ocean. Relative sea-level rise occurs where there is a net increase in the level of the ocean relative to local land movements. Climate modelers largely concentrate on estimating eustatic sea-level change. Impact researchers focus on relative sea-level change.

Seawall
A human-made wall or embankment along a shore to prevent wave erosion.

Semi-Arid Regions
Ecosystems that have >250 mm precipitation per year, but are not highly productive; usually classified as rangelands.

Sensitivity
Sensitivity is the degree to which a system is affected, either adversely or beneficially, by climate-related *stimuli*. The effect may be direct (e.g., a change in crop yield in response to a change in the mean, range, or variability of temperature) or indirect (e.g., damages caused by an increase in the frequency of coastal flooding due to *sea level rise*).

Sequestration
The process of increasing the carbon content of a carbon pool other than the atmosphere.

Silt
Unconsolidated or loose sedimentary material whose constituent rock particles are finer than grains of sand and larger than clay particles.

Silviculture
Development and care of forests.

Sink
Any process, activity, or mechanism that removes a *greenhouse gas*, an *aerosol*, or a precursor of a greenhouse gas or aerosol from the atmosphere.

Snowpacks
A seasonal accumulation of slow-melting snow.

Soil Carbon Pool
Refers to the relevant carbon in the soil. It includes various forms of soil organic carbon (humus) and inorganic soil carbon and charcoat. It excludes soil biomass (e.g., roots, bulbs, etc.) as well as the soil fauna (animals).

Source
Any process, activity, or mechanism that releases a *greenhouse gas*, an *aerosol*, or a precursor of a greenhouse gas or aerosol into the atmosphere.

Southern Oscillation
A large-scale atmospheric and hydrospheric fluctuation centered in the equatorial Pacific Ocean, exhibiting a pressure anomaly, alternatively high over the Indian Ocean and high over the South Pacific. Its period is slightly variable, averaging 2.33 years. The variation in pressure is accompanied by variations in wind strengths, ocean currents, sea-surface temperatures, and precipitation in the surrounding areas.

Stakeholders
Person or entity holding grants, concessions, or any other type of value that would be affected by a particular action or policy.

Stimuli (Climate-Related)
All the elements of climate change, including mean climate characteristics, climate variability, and the frequency and magnitude of extremes.

Stochastic Events
Events involving a random variable, chance, or probability.

Stock
See *reservoir*.

Stratosphere
Highly stratified region of atmosphere above the *troposphere* extending from about 10 km (ranging from 9 km in high latitudes to 16 km in the tropics on average) to about 50 km.

Streamflow
Water within a river channel, usually expressed in m^3 sec^{-1}.

Sub-Antarctic Mode Water (SAMW)
A type of water in the Sub-Antarctic Zone of the Southern Ocean. The SAMW is the deep surface layer of water with uniform temperature and salinity created by convective processes in the winter. It can be identified by a temperature of around -1.8°C and a salinity of around 34.4 PSU, and is separated from the overlying surface water by a halocline at around 50 m in the summer. Although it is not considered to be a water mass, it contributes to the Central Water of the Southern Hemisphere, and is additionally responsible for the formation of Antarctic Intermediate Water in the eastern part of the South Pacific Ocean. It is also known as Winter Water.

Submergence
A rise in the water level in relation to the land, so that areas of formerly dry land become inundated; it results either from a sinking of the land or from a rise of the water level.

Subsidence
The sudden sinking or gradual downward settling of the Earth's surface with little or no horizontal motion.

Succession
Transition in the composition of plant communities following disturbance.

Surface Runoff
The water that travels over the soil surface to the nearest surface stream; *runoff* of a drainage basin that has not passed beneath the surface since precipitation.

Sustainable Development
Development that meets the needs of the present without compromising the ability of future generations to meet their own needs.

Synoptic
Relating to or displaying atmospheric and weather conditions as they exist simultaneously over a broad area.

Taiga
Coniferous forests of northern North America and Eurasia.

Thermal Erosion
The erosion of ice-rich permafrost by the combined thermal and mechanical action of moving water.

Thermal Expansion
In connection with *sea-level rise*, this refers to the increase in volume (and decrease in density) that results from warming water. A warming of the ocean leads to an expansion of the ocean volume and hence an increase in sea level.

Thermocline
The region in the world's ocean, typically at a depth of 1 km, where temperature decreases rapidly with depth and which marks the boundary between the surface and the ocean.

Thermohaline Circulation
Large-scale density-driven circulation in the ocean, caused by differences in temperature and salinity. In the North Atlantic, the thermohaline circulation consists of warm surface water flowing northward and cold deepwater flowing southward, resulting in a net poleward transport of heat. The surface water sinks in highly restricted sinking regions located in high latitudes.

Thermokarst
Irregular, hummocky topography in frozen ground caused by melting of ice.

Timberline
The upper limit of tree growth in mountains or high latitudes.

Transpiration
The emission of water vapor from the surfaces of leaves or other plant parts.

Troposphere
The lowest part of the atmosphere from the surface to about 10 km in altitude in mid-latitudes (ranging from 9 km in high latitudes to 16 km in the tropics on average) where clouds and "weather" phenomena occur. In the troposphere, temperatures generally decrease with height.

Tsunami
A large tidal wave produced by a submarine earthquake, landslide, or volcanic eruption.

Tundra
A treeless, level, or gently undulating plain characteristic of arctic and subarctic regions.

Ultraviolet (UV)-B Radiation
Solar radiation within a wavelength range of 280–320 nm, the greater part of which is absorbed by *stratospheric ozone*. Enhanced UV-B radiation suppresses the immune system and can have other adverse effects on living organisms.

Uncertainty
An expression of the degree to which a value (e.g., the future state of the climate system) is unknown. Uncertainty can result from lack of information or from disagreement about what is known or even knowable. It may have many types of sources, from quantifiable errors in the data to ambiguously defined concepts or terminology, or uncertain projections of human behavior. Uncertainty can therefore be represented by quantitative measures (e.g., a range of values calculated by various models) or by qualitative statements (e.g., reflecting the judgment of a team of experts).

Undernutrition
The result of food intake that is insufficient to meet dietary energy requirements continuously, poor absorption, and/or poor biological use of nutrients consumed.

Unique and Threatened Systems
Entities that are confined to a relatively narrow geographical range but can affect other, often larger entities beyond their range; narrow geographical range points to sensitivity to environmental variables, including climate, and therefore attests to potential vulnerability to *climate change*.

United Nations Framework Convention on Climate Change (UNFCCC)
The Convention was adopted on 9 May 1992, in New York, and signed at the 1992 Earth Summit in Rio de Janeiro by more than 150 countries and the European Community. Its ultimate objective is the "stabilization of greenhouse gas concentrations in the atmosphere at a level that would prevent dangerous anthropogenic interference with the climate system." It contains commitments for all Parties. Under the Convention, Parties included in Annex I aim to return greenhouse gas emissions not controlled by the Montreal Protocol to 1990 levels by the year 2000. The Convention entered in force in March 1994. See also *Kyoto Protocol*.

Ungulate
A hoofed, typically herbivorous, quadruped mammal (such as a ruminant, swine, camel, hippopotamus, horse, rhinoceros, or elephant).

Upwelling
Transport of deeper water to the surface, usually caused by horizontal movements of surface water.

Urbanization
The conversion of land from a natural state or managed natural state (such as agriculture) to cities; a process driven by net rural-to-urban migration through which an increasing percentage of the population in any nation or region come to live in settlements that are defined as "urban centers."

Vector
An organism, such as an insect, that transmits a pathogen from one host to another. See also *vector-borne diseases* and *vectorial capacity*.

Vector-Borne Diseases
Disease that is transmitted between hosts by a *vector* organism (such as a mosquito or tick— for example, malaria, dengue fever, and leishmaniasis.

Vectorial Capacity
Quantitative term used in the study of the transmission dynamics of malaria to express the average number of potentially infective bites of all vectors feeding upon one host in one day, or the number of new inoculations with a vector-borne disease transmitted by one vector species from one infective host in one day.

Vernalization
The act or process of hastening the flowering and fruiting of plants by treating seeds, bulbs, or seedlings so as to induce a shortening of the vegetative period.

Vulnerability
The degree to which a system is susceptible to, or unable to cope with, adverse effects of climate change, including climate variability and extremes. Vulnerability is a function of the character, magnitude, and rate of climate variation to which a system is exposed, its sensitivity, and its adaptive capacity.

Water Consumption
Amount of extracted water irretrievably lost at a given territory during it's use (evaporation and goods production). Water consumption is equal to water withdrawal minus return flow.

Water Stress
A country is water stressed if the available freshwater supply relative to *water withdrawals* acts as an important constraint on development. Withdrawals exceeding 20% of renewable water supply has been used as an indicator of water stress.

Water Withdrawal
Amount of water extracted from water bodies.

Water Use Efficiency
Carbon gain in photosynthesis per unit water lost in evapotranspiration. It can be expressed on a short-term basis as the ratio of photosynthetic carbon gain per unit transpirational water loss, or on a seasonal basis as the ratio of *net primary production* or agricultural yield to the amount of available water.

Xeric
Requiring only a small amount of moisture.

Zoonosis
The transmission of a disease from an animal or nonhuman species to humans. The natural reservoir is a nonhuman animal.

Zooplankton
The animal forms of *plankton*. They consume *phytoplankton* or other zooplankton. See also *phytoplankton*.

Sources

IPCC, 1996: *Climate Change 1995: The Science of Climate Change. Contribution of Working Group I to the Second Assessment Report of the Intergovernmental Panel on Climate Change* [Houghton, J.T., L.G. Meira Filho, B.A. Callander, N. Harris, A. Kattenberg, and K. Maskell (eds.)]. Cambridge University Press, Cambridge, United Kingdom and New York, NY, USA, 572 pp.

IPCC, 1998: *The Regional Impacts of Climate Change: An Assessment of Vulnerability. A Special Report of IPCC Working Group II* [Watson, R.T., M.C. Zinyowera, and R.H. Moss (eds.)]. Cambridge University Press, Cambridge, United Kingdom and New York, NY, USA, 517 pp.

IPCC, 2000: *Land Use, Land-Use Change, and Forestry. A Special Report of the IPCC* [Watson, R.T., I.R. Noble, B. Bolin, N.H. Ravindranath, D.J. Verardo, and D.J. Dokken (eds.)]. Cambridge University Press, Cambridge, United Kingdom and New York, NY, USA, 377 pp.

Jackson, J. (ed.), 1997: *Glossary of Geology*. American Geological Institute, Alexandria, Virginia.

Moss, R.H. and S.H. Schneider, 2000: Uncertainties in the IPCC TAR: recommendations to lead authors for more consistent assessment and reporting. In: *Guidance Papers on the Cross Cutting Issues of the Third Assessment Report of the IPCC* [Pachauri, R., K. Tanaka, and T. Taniguchi (eds.)]. Intergovernmental Panel on Climate Change, Geneva, Switzerland, pp. 33–51. Available online at http://www.gispri.or.jp.

Nakicenovic, N., J. Alcamo, G. Davis, B. de Vries, J. Fenhann, S. Gaffin, K. Gregory, A. Grübler, T.Y. Jung, T. Kram, E.L. La Rovere, L. Michaelis, S. Mori, T. Morita, W. Pepper, H. Pitcher, L. Price, K. Raihi, A. Roehrl, H.-H. Rogner, A. Sankovski, M. Schlesinger, P. Shukla, S. Smith, R. Swart, S. van Rooijen, N. Victor, and Z. Dadi, 2000: *Emissions Scenarios. A Special Report of Working Group III of the Intergovernmental Panel on Climate Change*. Cambridge University Press, Cambridge, United Kingdom and New York, NY, USA, 599 pp.

United Nations Environment Programme, 1995: Global Biodiversity Assessment [Heywood, V.H. and R.T. Watson (eds.)]. Cambridge University Press, Cambridge, United Kingdom and New York, NY, USA, 1140 pp.

C

Acronyms, Abbreviations, and Units

ACRONYMS AND ABBREVIATIONS

AAIW Antarctic Intermediate Water
ACACIA A Concerted Action Towards A Comprehensive Climate Impacts and Adaptations Assessment for the European Union
ADB Asian Development Bank
AGBM Ad Hoc Group on the Berline Mandate
A-O atmosphere-ocean
APSIM Agricultural Production Systems Simulator Model
ARMCANZ Agriculture and Resource Management Council of Australia and New Zealand
ART alternative risk transfer
BAPMoN Background Atmospheric Pollution Monitoring Network
BP before the present
CALM Circumpolar Active Layer Monitoring
CAP Common Agricultural Policy
CBA cost-benefit analysis
CBAA cost-benefit absent adaptation
CBWAF cost-benefit with adaptation foresight
CCAMLR Commission for the Conservation of Antarctic Marine Living Resources
CCC Canadian Centre for Climate Prediction and Analysis
CCSI Climate Change Sustainability Index
CCSR Center for Climate System Research
CDM Clean Development Mechanism
CDS Conventional Development Scenario
CEA cost-effectiveness analysis
CEE central and eastern Europe
CEI Climate Extremes Index
CFC chlorofluorocarbon
CIS Commonwealth of Independent States
COAG Council of Australian Governments
CPACC Caribbean Planning for Adaptation to Global Climate Change
CPU catch per unit
CPUE catch per unit effort
CSIRO Commonwealth Scientific and Industrial Research Organisation (Australia)
CV compensated variation
CVM contingent valuation method
DA decision analysis
DAF decision analytic framework
DALY disability-adjusted life year
DDC Data Distribution Centre
DES dietary energy supply
DOC dissolved organic carbon
DSE development, sustainability, and equity
DTR diurnal temperature range
DWR Department of Water Resources
EAC East Australian Current
ECLAC Economic Commission for Latin America and the Caribbean
EEE eastern equine encephalitis
EEZ Exclusive Economic Zone
EIA environmental impact assessment
EIP extrinsic incubation period
ELA equilibrium line altitude
ENSO El Niño-Southern Oscillation
EPIC Erosion Productivity Impact Calculator
EU European Union
EV equivalent variation
FACE Free-Air CO_2 Enrichment
FAIR Fair Access to Insurance Requirements
FAO Food and Agriculture Organization
FAR First Assessment Report
FEMA Federal Emergency Management Agency
FEWS NET Famine Early Warning System Network
GARP Global Atmospheric Research Program
GBR Great Barrier Reef
GCM general circulation model
GDP gross domestic product
GEF Global Environment Facility
GEWEX Global Energy and Water Cycle Experiment
GFDL Geophysical Fluid Dynamics Laboratory
GHG greenhouse gas
GIS Geographic Information System
GISS Goddard Institute for Space Studies
GLASOD Global Assessment of Human-Induced Soil Degradation
GLSLB Great Lakes-St. Lawrence Basin Project
GNP gross national product
GPP gross primary productivity
GTNet-P Global Terrestrial Network for Permafrost
HAPEX Hydrologic Atmosphere Pilot Experiment
HDI human development index
HIV human immunodeficiency virus
HMGP Hazard Mitigation Grant Program
HPS hantavirus pulmonary syndrome
HRI heat-related injury or illness
IAI Inter-American Institute for Global Change
IAM integrated assessment model
ICES International Council for the Exploration of the Seas
ICLEI International Council for Local Environmental Initiatives
ICZM integrated coastal zone management
ICM integrated coastal management
IDNDR International Decade for Natural Disaster Reduction
IFRC-RCS International Federation of Red Cross and Red Crescent Societies
IGBP International Geosphere-Biosphere Programme
IHDP International Human Dimensions Programme
IMAGE Integrated Model to Assess the Greenhouse Effect
IPCC Intergovernmental Panel on Climate Change

ISEW Index of Sustainable Economic Welfare
ITCZ Inter-Tropical Convergence Zone
IWRM integrated water resources management
JE Japanese encephalitis
JI joint implementation
JJA June, July, August
JUA Joint Underwriting Association
LAC La Crosse encephalitis
LCC land-cover change
LGM last glacial maximum
LHB large-area herbivore biomass
LUC land-use change
MAP mean annual precipitation
MBIS Mackenzie Basin Impact Study
MEI morphoedaphic index
MPI Max-Planck Institute
MSF Médecins sans Frontières
MSY maximum sustainable yield
MSX multinucleated spore unknown
MVE Murray Valley encephalitis
NAFTA North American Free Trade Agreement
NAO North Atlantic Oscillation
NAST National Assessment Synthesis Team
NATO North Atlantic Treaty Organization
NBP net biome productivity
NCAR National Center for Atmospheric Research
NDVI Normalized Difference Vegetation Index
NEP net ecosystem productivity
NGO nongovernmental organization
NH Northern Hemisphere
NIES National Institute for Environmental Studies (Japan)
NIWA New Zealand Institute of Water and Atmospheric Research
NPDO North Pacific Decadal Oscillation
NPP net primary productivity
NWFP non-wood forest products
NSW New South Wales
OECD Organisation for Economic Cooperation and Development
OMVS Organisation pour la Mise en Valeur du Fleuve Senegal
PACIS Pan-American Climate Information and Applications System
PAN peroxyacetyl nitrate
P/C property-casualty
PCB polychlorinated biphenyl
PDO Pacific Decadal Oscillation
PDSI Palmer Drought Severity Index
PE policy exercise
PET potential evapotranspiration
PG protection guaranteed
PML probable maximum loss
PPP purchasing power parity
PQLI physical quality of life index
P-S-I-R pressure-state-impact-responses
PSMSL Permanent Service for Mean Sea Level
PTSD post-traumatic stress disorder
RA risk assessment
RCM regional climate model
R&D research and development
REWU Regional Early Warning Unit
RICC Regional Impacts of Climate Change
RMA Resource Management Act (New Zealand)
RVF Rift Valley fever
SADC Southern Africa Development Community
SAMW sub-Antarctic mode water
SAP structural adjustment program
SAR Second Assessment Report
SARP Simulation and System Analysis for Rice Production
SCS strategic cyclical scaling
SHV snowshoe hare virus
SICA Central American Integration System
SLE St. Louis encephalitis
SO Southern Oscillation
SOC soil organic carbon
SOI Southern Oscillation Index
SOM soil organic matter
SPC Secretariat of the Pacific Community
SPCZ Southern Pacific Convergence Zone
SPREP South Pacific Regional Environment Programme
SRB Senegal River basin
SRES Special Report on Emissions Scenarios
SSA sub-Saharan Africa
SST sea-surface temperature
SWRCB State Water Resource Control Board
TAR Third Assessment Report
TBE tick-borne encephalitis
TEM Terrestrial Ecosystem Model
THC thermohaline circulation
TNR Toronto-Niagara Region Study
UKCIP United Kingdom Climate Impacts Programme
UKMO United Kingdom Meteorological Office
UN United Nations
UNCCD UN Convention to Combat Desertification
UNCED UN Conference on Environment and Development
UNCHS UN Center for Human Settlements
UNCSD UN Commission for Sustainable Development
UNDP UN Development Programme
UNEP UN Environmental Programme
UNFCCC UN Framework Convention on Climate Change
UN-OCHA UN Office for the Coordination of Humanitarian Affairs
USAID U.S. Agency for International Development
UV ultraviolet
V&A vulnerability and adaptation
VAM Vulnerability Assessment and Mapping
VBD vector-borne disease
VEE Venezuelan equine encephalitis
VC vectorial capacity

VOC volatile organic carbon
VRA Volta River Authority
WAIS West Antarctic ice sheet
WAP western Antarctic peninsula
WCED World Commission on Environment and Development
WCRP World Climate Research Programme
WEE western equine encephalitis
WFP World Food Program
WGI Working Group I
WGII Working Group II
WGIII Working Group III
WHO World Health Organization
WMO World Meteorological Organization
WN West Nile
WTA willingness to accept payment
WTP willingness to pay
WUE water-use efficiency

UNITS

SI (Systéme Internationale) Units

Physical Quantity	Name of Unit	Symbol
length	meter	m
mass	kilogram	kg
time	second	s
thermodynamic temperature	kelvin	K
amount of substance	mole	mol

Special Names and Symbols for Certain SI-Derived Units

Physical Quantity	Name of SI Unit	Symbol for SI Unit	Definition of Unit
force	newton	N	kg m s^{-2}
pressure	pascal	Pa	kg m^{-1} s^{-2} (= Nm^{-2})
energy	joule	J	kg m^2 s^{-2}
power	watt	W	kg m^2 s^{-3} (= Js^{-1})
frequency	hertz	Hz	s^{-1} (cycle per second)

Decimal Fractions and Multiples of SI Units Having Special Names

Physical Quantity	Name of Unit	Symbol for Unit	Definition of Unit
length	ångstrom	Å	10^{-10} m = 10^{-8}cm
length	micrometer	μm	10^{-6}m = μm
area	hectare	ha	10^4 m^2
force	dyne	dyn	10^{-5} N
pressure	bar	bar	10^5 N m^{-2}
pressure	millibar	mb	1hPa
weight	ton	t	10^3 kg

D

List of Major IPCC Reports

Climate Change—The IPCC Scientific Assessment
The 1990 Report of the IPCC Scientific Assessment Working Group (also in Chinese, French, Russian, and Spanish)

Climate Change—The IPCC Impacts Assessment
The 1990 Report of the IPCC Impacts Assessment Working Group (also in Chinese, French, Russian, and Spanish)

Climate Change—The IPCC Response Strategies
The 1990 Report of the IPCC Response Strategies Working Group (also in Chinese, French, Russian, and Spanish)

Emissions Scenarios
Prepared for the IPCC Response Strategies Working Group, 1990

Assessment of the Vulnerability of Coastal Areas to Sea Level Rise–A Common Methodology
1991 (also in Arabic and French)

Climate Change 1992—The Supplementary Report to the IPCC Scientific Assessment
The 1992 Report of the IPCC Scientific Assessment Working Group

Climate Change 1992—The Supplementary Report to the IPCC Impacts Assessment
The 1992 Report of the IPCC Impacts Assessment Working Group

Climate Change: The IPCC 1990 and 1992 Assessments
IPCC First Assessment Report Overview and Policymaker Summaries, and 1992 IPCC Supplement

Global Climate Change and the Rising Challenge of the Sea
Coastal Zone Management Subgroup of the IPCC Response Strategies Working Group, 1992

Report of the IPCC Country Studies Workshop
1992

Preliminary Guidelines for Assessing Impacts of Climate Change
1992

IPCC Guidelines for National Greenhouse Gas Inventories
Three volumes, 1994 (also in French, Russian, and Spanish)

IPCC Technical Guidelines for Assessing Climate Change Impacts and Adaptations
1995 (also in Arabic, Chinese, French, Russian, and Spanish)

Climate Change 1994—Radiative Forcing of Climate Change and an Evaluation of the IPCC IS92 Emission Scenarios
1995

Climate Change 1995—The Science of Climate Change – Contribution of Working Group I to the IPCC Second Assessment Report
1996

Climate Change 1995—Impacts, Adaptations, and Mitigation of Climate Change: Scientific-Technical Analyses – Contribution of Working Group II to the IPCC Second Assessment Report
1996

Climate Change 1995—Economic and Social Dimensions of Climate Change – Contribution of Working Group III to the IPCC Second Assessment Report
1996

Climate Change 1995—IPCC Second Assessment Synthesis of Scientific-Technical Information Relevant to Interpreting Article 2 of the UN Framework Convention on Climate Change
1996 (also in Arabic, Chinese, French, Russian, and Spanish)

Technologies, Policies, and Measures for Mitigating Climate Change – IPCC Technical Paper I
1996 (also in French and Spanish)

An Introduction to Simple Climate Models used in the IPCC Second Assessment Report – IPCC Technical Paper II
1997 (also in French and Spanish)

Stabilization of Atmospheric Greenhouse Gases: Physical, Biological and Socio-economic Implications – IPCC Technical Paper III
1997 (also in French and Spanish)

Implications of Proposed CO_2 Emissions Limitations – IPCC Technical Paper IV
1997 (also in French and Spanish)

The Regional Impacts of Climate Change: An Assessment of Vulnerability – IPCC Special Report
1998

Aviation and the Global Atmosphere – IPCC Special Report
1999

Methodological and Technological Issues in Technology Transfer – IPCC Special Report
2000

Land Use, Land-Use Change, and Forestry – IPCC Special Report
2000

Emission Scenarios – IPCC Special Report
2000

Good Practice Guidance and Uncertainty Management in National Greenhouse Gas Inventories
2000

Climate Change 2001: The Scientific Basis – Contribution of Working Group I to the IPCC Third Assessment Report
2001

Climate Change 2001: Impacts, Adaptation, and Vulnerability – Contribution of Working Group II to the IPCC Third Assessment Report
2001

Climate Change 2001: Mitigation – Contribution of Working Group III to the IPCC Third Assessment Report
2001

Climate Change 2001: IPCC Third Assessment Synthesis Report
2001

ENQUIRIES: IPCC Secretariat, c/o World Meteorological Organization, 7 bis, Avenue de la Paix, Case Postale 2300, 1211 Geneva 2, Switzerland

E

Index

Note: Numbers in *italics* indicate a reference to a table or diagram.

A

B

C

D

E

G

I

J

K

L

M

N

O

Q

R

T

U

Y

Z